Contents

Metals Handbook Ninth Edition

Volume 3 Properties and Selection:

Stainless Steels, Tool Materials and Special-Purpose Metals

Prepared under the direction of the ASM Handbook Committee

William H. Cubberly, Director of Reference Publications
Paul M. Unterweiser, Assistant Director of Reference Publications
David Benjamin, Senior Editor
Craig W. Kirkpatrick, Associate Editor
Vicki Knoll, Production Manager
Kathy Nieman, Editorial Assistant

AMERICAN SOCIETY FOR METALS
METALS PARK, OHIO 44073

First printing, December 1980
Second printing, June 1981
Third printing, April 1985
Fourth printing, October 1986
Fifth printing, April 1988

Metals Handbook is a collective effort involving thousands of technical specialists. It brings together in one book a wealth of information from world-wide sources to help scientists, engineers and technicians solve current and long-range problems.

Great care is taken in the compilation and production of this volume, but it should be made clear that no warranties, express or implied, are given in connection with the accuracy or completeness of this publication, and no responsibility can be taken for any claims that may arise.

Nothing contained in the Metals Handbook shall be construed as a grant of any right of manufacture, sale, use or reproduction, in connection with any method, process, apparatus, product, composition or system, whether or not covered by letters patent, copyright or trademark, and nothing contained in the Metals Handbook shall be construed as a defense against any alleged infringement of letters patent, copyright or trademark, or as a defense against any liability for such infringement.

Comments, criticisms and suggestions are invited, and should be forwarded to the American Society for Metals.

Library of Congress Cataloging in Publication Data

American Society for Metals
Properties and selection: stainless steels, tool materials, and special-purpose metals.

(Metals handbook; 9th ed., v. 3)
Includes bibliographical references and index.

1. Steel, Stainless. 2. Steel, Heat resistant.
3. Machine-tools—Materials. 4. Metals. 5. Alloys.
I. Benjamin, David, 1936- II. Kirkpatrick, Craig W.
III. Title. IV. Series: American Society for Metals.
Metals handbook; 9th ed., v. 3.
TA459.A5 9th ed., vol. 3 [TA479.S7] 669s [620.1'6]
ISBN 0-87170-009-3 80-26336

SAN 204-7586

Printed in the United States of America

Foreword

Publication of Volume 3 of the 9th edition of Metals Handbook marks the completion of a fundamental trilogy devoted to properties and selection of metals. In common with its single-volume counterpart of the 8th edition, the trilogy can be viewed as the cornerstone of an evolving series which, in the case of the 8th edition, totalled eleven volumes in all and which, as always, shall continue to be guided by and to reflect the technical needs of the metallurgical, metalworking and engineering communities. In the long journey that the 9th edition is committed to travel toward completion, no first step could be more welcome or appropriate than the publication of this exhaustive review of properties and selection of metals.

Although we are a technical society whose dedication to metals is explicit in our name, our interest in specific metals and alloys tends for the most part to vary directly with their current industrial importance. Like the Franklin stove and its uncanny inventor, ours is a practical interest for a practical benefit more often than not. We confess a certain preference for the principal subjects of Volume 3—stainless steels, tool materials and special-purpose metals, materials that were not available to the technical community of Dr. Franklin's time but that, we suspect, would have stirred his boundless ingenuity and fired his inventive genius. Herein are many of our "high technology" materials that, coupled with the fruits of human ingenuity, comprise the bases of major industries and thousands of products that were unknown to our readers in 1923, the year the very first Metals Handbook appeared.

That historic volume and every succeeding volume over the intervening years has reflected the combined efforts and talents of technically expert and singularly devoted people. This third volume of the 9th edition has been similarly favored by those who continue the tradition—the ASM Handbook Committee, author committees, individual contributors and editorial staff. For their valuable contribution from which we all so richly benefit, we extend to all of them our sincere thanks.

Raymond L. Smith
President

Allan Ray Putnam
Managing Director

The Ninth Edition of Metals Handbook
is dedicated to the memory of
TAYLOR LYMAN, A. B. (Eng.), S. M., Ph. D.
(1917-1973)
Editor, Metals Handbook 1945-1973

Preface

This is the third and last volume within the Ninth Edition of Metals Handbook that supersedes Volume 1 of the Eighth Edition. This third volume marks the culmination of a project begun almost six years ago. In bringing the project to completion, we have been continually aware that man's quest for new ideas does not invalidate his previous body of knowledge, but, rather, expands it. Indeed, no one should be surprised that it has taken three volumes—more than 2300 pages—to replace the 1236 pages in Volume 1 of the Eighth Edition.

The current volume contains exhaustive data on the properties, metallurgical characteristics, uses and selection of corrosion-resistant metals, tool materials, heat-resistant metals and a wide variety of metals and material systems for special purposes. Among the last group are electrical resistance alloys, electric-contact materials, bearing alloys, magnetic materials and wear-resistant materials. All of the articles on these subjects were substantially rewritten for this edition. Coverage of the special-purpose materials has been extended to encompass articles on depleted uranium, machinable tungsten alloys, titanium alloy castings and corrosion-resistant fasteners. None of these four subjects had been addressed in previous editions. In other sections—most notably superhard tool materials, titanium and wrought superalloys—the information in this volume is vastly improved over the information available in the Eighth Edition.

By far the greatest amount of space in this volume is devoted to stainless steels, heat-resistant alloys and tool materials. In terms of articles, there are nine on stainless steels and other corrosion-resistant metals; four on heat-resistant metals; and twenty-three on tool materials, nineteen of those on the selection of materials for specific types of metalworking tools.

The information in this volume is again given in two distinct formats: descriptive articles and data compilations. The descriptive articles discuss alloy families or material systems in a logical and orderly fashion, relating composition and structure to properties, and correlating properties with general uses and specific applications. In many of the articles, unique metallurgical behavior of an alloy class is described and put in perspective. For instance, exposure to elevated temperature results in precipitation of carbides and topologically close-packed phases in certain classes of alloys. These precipitation phenomena are mentioned in the articles on stainless steels and heat-resistant alloys, and are discussed at length in an appendix to the article on wrought superalloys.

In the data compilations are listed specific property values for individual alloys. Not all groups of metals are covered in the data compilations because it seemed more appropriate—in stainless steels, for example—to present the property data in tabular form within the basic article rather than to use formatted data compilations. In most instances, the data compilations include not only mechanical and physical properties most often required for the solution of ordinary engineering problems, but also properties less often of interest, especially where the latter could be of vital importance in certain unique applications. Besides specific values of mechanical and physical properties, the data compilations include listings of trade names, specifications, composition limits and in many instances a summary of fabrication characteristics. These data compilations are not intended to provide extensive design information, but merely enough of the pertinent data to allow intelligent material selection.

Volume 3 is divided into seven sections: Corrosion-Resistant Materials, Heat-Resistant Materials, Titanium and Titanium Alloys, Tool Materials, Wear-Resistant Materials, Materials for Special Applications, and Special Topics in Materials Engineering. Special note is due four articles in the last two sections, all of which were written especially for this edition of the Handbook. "Thermocouples for Industrial Applications" and "Alloys for Structural Applications at Subzero Temperatures" are major new contributions to the technical literature. Each of these articles represents the accumulated knowledge of individuals who have spent a lifetime in that field. "Concepts and Criteria in Materials Engineering" and "Guidelines for Selection of Material" present two distinct but compatible viewpoints on the need for a rational approach to selecting the right material for the job.

Numerical values presented in this volume are given largely in dual engineering units. Those of Système Internationale d'Unités (SI) are the primary units of measure throughout the Ninth Edition of Metals Handbook. Equivalent U.S. customary units are given where appropriate. (The Metals Handbook policy on units of measure is set forth more fully immediately following this preface.)

Because much of the information that forms the body of knowledge about metals was gathered together for Volume 1 of the Eighth Edition, we hereby acknowledge the efforts of past committees and individuals whose work would otherwise go unheralded. We are even more deeply indebted to the many contributors who have prepared and revised the articles and data compilations that comprise Volume 3 of the Ninth Edition. They and many of their unnamed colleagues have given freely of their time and expertise that members and friends of ASM may be better able to solve nagging technical problems. Our heartfelt gratitude also is extended to Howard E. Boyer, former Editor of Metals Handbook, and to Dr. John A. Fellows, Consultant and former President of ASM, who are chiefly responsible for acquiring information on material properties for these first three volumes of the Ninth Edition. Lastly, we are most grateful for the efforts of our consulting editors—Dr. Matthew J. Donachie, Jr. (Cast Heat-Resistant Alloys), T. C. Du Mond (Wear-Resistant Materials), Prof. A. E. Focke (Tool Materials), Russell B. Gunia (Wrought Stainless Steels), C. H. Junge (Sleeve-Bearing Materials), Dr. F. R. Morral (Wrought Superalloys) and Dr. H. P. Munger (Materials Selection)—who contributed both editorial skills and technical acumen. Without the skill, knowledge and dedication of our contributors and consulting editors, Metals Handbook would not exist. Upon their collective competence and experience rest the accuracy and authority of the information contained herein.

David Benjamin
Senior Editor
Metals Handbook

Policy on Units of Measure

By a resolution of its Board of Trustees, the American Society for Metals has adopted the practice of publishing data in both metric and customary U.S. units of measure. In preparing this Handbook, the editors have attempted to present data primarily in metric units based on Système Internationale d'Unités (SI), with secondary mention of the corresponding values in customary U.S. units. The decision to use SI as the primary system of units was based on the aforementioned resolution of the Board of Trustees, the widespread use of metric units throughout the world, and the expectation that the use of metric units in the United States will increase substantially during the anticipated lifetime of this Handbook.

For the most part, numerical engineering data in the text and in tables are presented in SI-based units with the customary U.S. equivalents in parentheses (text) or adjoining columns (tables). For example, pressure, stress and strength are shown in both SI units, which are pascals (Pa) with a suitable prefix (see the description of SI at the back of the volume), and in customary U.S. units, which are pounds per square inch (psi). To save space, large values of psi have been changed to kips per square inch (ksi), where one kip equals 1000 pounds. Some strictly scientific data are presented in SI units only.

To clarify some illustrations that depict machine parts described in the text, only one set of dimensions is presented on artwork. References in the accompanying text to dimensions in the illustrations are presented in both SI-based and customary U.S. units.

On graphs and charts, grids correspond to SI-based units, which appear along the left and bottom edges; where appropriate, corresponding customary U.S. units appear along the top and right edges. Some previously published charts, particularly histograms depicting statistical distribution of values of mechanical properties, could not be redrawn because of the absence of the original data points; these have been reproduced in their original forms, with SI equivalents on the top and right edges.

Data pertaining to a specification published by a specification-writing group may be given in only the units used in that specification or in dual units, depending on the nature of the data. For example, the typical yield strength of aluminum sheet made to a specification written in customary U.S. units would be presented in dual units, but the thickness ranges listed in that specification might be presented only in inches.

Data obtained according to specified test methods for which the specification implies a particular system of units are presented in the units of that system. Wherever feasible, equivalent units are also presented.

Conversions and rounding have been done in accordance with ASTM Standard E-380, with careful attention to the number of significant digits in the original data. For example, an annealing temperature of 1575 °F contains three significant digits (and possibly only two), because few commercial heat treatment systems can control the temperature of an entire load of parts within a spread of 10 °F. In this instance, the equivalent temperature would be given as 860 °C, or perhaps 850 °C depending on the degree of accuracy meant to be conveyed in the conversion; the exact conversion to 857.22 °C would not be appropriate. In many instances (especially in tables and data compilations), temperature values in °C and °F are alternatives rather than conversions.

The policy on units of measure in this Handbook contains several exceptions to strict conformance to ASTM E380; in each instance, the exception has been made to improve the clarity of the Handbook. Three examples of such exceptions are the use of "L", rather than "l" as the abbreviation for litre, reporting temperature in °C rather than K and reporting stress intensity in $MPa\sqrt{m}$ rather than $MNm^{-3/2}$.

SI practice requires that only one virgule (diagonal) appear in units formed by combination of several basic units. Therefore, all of the units preceding the virgule are in the numerator and all units following the virgule are in the denominator of the expression (and no parentheses are required to prevent ambiguity).

Handbook Committee, Officers and Trustees

Members of the ASM Handbook Committee (1975-1980)

Officers and Trustees of the American Society for Metals

Previous Chairmen of ASM Handbook Committee

R. S. Archer
(1940-1942) (Member, 1937-1942)

L. B. Case
(1931-1933) (Member, 1927-1933)

E. O. Dixon
(1952-1954) (Member, 1947-1955)

R. L. Dowdell
(1938-1939) (Member, 1935-1939)

J. P. Gill
(1937) (Member, 1934-1937)

J. D. Graham
(1965-1968) (Member, 1961-1970)

J. F. Harper
(1923-1926) (Member, 1923-1926)

C. H. Herty, Jr.
(1934-1936) (Member, 1930-1936)

J. B. Johnson
(1948-1951) (Member, 1944-1951)

R. W. E. Leiter
(1962-1963)
(Member, 1955-1958, 1960-1964)

G. V. Luerssen
(1943-1947) (Member, 1942-1947)

Gunvant N. Maniar
(1978-1980) (Member, 1974-1980)

W. J. Merten
(1927-1930) (Member, 1923-1933)

N. E. Promisel
(1955-1961) (Member, 1954-1963)

G. J. Shubat
(1972-1975) (Member, 1966-1975)

W. A. Stadtler
(1968-1972) (Member, 1962-1972)

Raymond Ward
(1976-1978) (Member, 1972-1978)

D. J. Wright
(1964-1965) (Member, 1959-1967)

Author and Review Committees

ASM Committee on Wrought Stainless Steels

Russell B. Gunia,
Chairman
Engineering Consultant

Alvin G. Cook
Consultant

John L. Giove
Manager, Plate Structural and Railroad Product Metallurgy
U. S. Steel Corporation

John B. Guernsey
Vice President—Technical Services
Jessop Steel Company

Joseph Jasper
Senior Staff Engineer
Stainless Steel Research
Research & Technology
Armco Inc.

Mark J. Johnson
Research Specialist
Allegheny Ludlum Steel Corp.

Lewis P. Myers
Specialist, Stainless Alloys
Carpenter Steel Division
Carpenter Technology Corporation

Robert N. Peterson
Manager, Metallurgical Services and Quality Control
Enduro Division
Republic Steel Corporation

James D. Redmond
Manager, Stainless Steel Development
Climax Molybdenum Company

Laurence C. Shaheen
Director, Mechanical Development
Superior Tube Company

Martin H. Webster
Coordinator, Market Planning
Falconbridge Canada

W. I. Weed
Metallurgical Engineer
Jones & Laughlin Steel Corp.

Gordon R. Woodrow
Manager, Quality Control
G. O. Carlson, Inc.

David Benjamin, *Secretary*
Senior Editor, Metals Handbook
American Society for Metals

ASM Committee on Cast Corrosion-Resisting and Heat-Resisting Alloys

Charles R. Bird
Chief Metallurgist
Stainless Foundry and Engineering Inc.

Matthew J. Donachie, Jr.
Senior Materials Engineer
Pratt & Whitney Aircraft Group
United Technologies Corporation

A. M. Hall
Executive Director
Materials Technology Institute

C. S. Kortovich
Equipment Division
TRW, Inc.

E. A. Schoefer
Consultant

Harold C. Templeton
Consultant

ASM Committee on Heat-Resisting Alloys and Refractory Metals

Thomas H. Bassford
Chief Mechanical Testing Engineer
Huntington Alloys, Inc.

Elihu F. Bradley
(retired) formerly Chief Materials Engineer
Pratt & Whitney Aircraft Group
United Technologies Corporation

Alfons J. DeRidder
Chief Metallurgist
Ladish Co.

Matthew J. Donachie, Jr.
Senior Materials Engineer
Pratt & Whitney Aircraft Group
United Technologies Corporation

R. W. Fraser
Manager—Physical Metallurgy Research
Sherritt Gordon Research Center

John C. Freche
(retired) formerly Acting Chief
Materials and Structures & Mechanical Technologies Div.
National Aeronautics & Space Administration
Lewis Research Center

Ronald C. Gebeau
Carpenter Steel Division
Carpenter Technology Corporation

R. B. H. Herchenroeder
Senior Member, Technical Staff
Cabot Corporation

George W. King
Fellow Research Engineer
Lamp Divisions
Westinghouse Electric Corporation

Ray A. Lula
Research Center
Allegheny Ludlum Steel Corporation

Frank J. Rizzo
Research Engineering Supervisor
Crucible Incorporated

Mortimer Schussler
Senior Scientist and Corporate Director of Patents
Fansteel Inc.

M. Semchyshen
Vice President
Metallurgical Development
Climax Molybdenum Company of Michigan

ASM Committee on Titanium and Titanium Alloys

Donald R. Betner
Development Engineer
Materials Research and Engineering
Detroit Diesel Allison Division
General Motors Corporation

Roger G. Broadwell
Manager, Quality Control
Timet Division
Titanium Metals Corp. of America

R. Frank Malone
Consulting Engineer
Crucible Research Center
Colt Industries/Crucible Incorporated

Robert E. Newcomer
Technical Specialist
Engineering Technology
McDonnell Aircraft

Duane L. Ruckle
Assistant Materials Project Engineer
Materials Engineering and Research Laboratory
Pratt & Whitney Aircraft Group
United Technologies Corporation

T. K. Redden
Manager, Material Development Projects
Aircraft Engine Group
General Electric Company

John J. Shaw
Technical Director
Frankel Company

Albert Tobin
Senior Research Scientist
Research Department
Grumman Aerospace Corporation

Douglas Wilson
Manager, Customer Technical Services
RMI Company

Richard A. Wood
Research Metallurgist
Applied Metallurgy Section
Battelle Columbus Laboratories

Robert Zanoni
Supervisor, Aerospace Metallurgy
Metallurgy and Research
Ladish Co.

Howard W. Zoeller
Materials Engineer
U.S. Air Force Materials Laboratory
Wright-Patterson Air Force Base

David Benjamin, *Secretary*
Senior Editor, Metals Handbook
American Society for Metals

ASM Committee on Tooling Materials

Arthur E. Focke, *Chairman*
Professor Emeritus of Metallurgical Engineering
University of Cincinnati

Robert A. Cary
Vice President—Technical Director
Teledyne VASCO

James E. Denton
Metallurgical Engineer
Materials Technology
Cincinnati Milacron, Inc.

Walter T. Haswell
Vice President, Technology
Crucible Specialty Metals Division
Colt Industries

Lawrence A. Hauser
Marketing Manager, Toolmaker Sales
Universal-Cyclops Specialty Steel Division
Cyclops Corporation

Bernard S. Lement
Materials Engineering Consultant

James J. McCarthy
Metallurgical Engineer
Bethlehem Steel Corporation

John F. McGraw
Associate Metallurgist, Cold Finish Bar & Wire
Carpenter Steel Division
Carpenter Technology Corporation

William A. Reich
Manager—Technical Resources
Carboloy Systems Department
General Electric Company

Emlyn N. Smith
Metallurgical Consultant
Kennametal Inc.

Richard A. Weaver
Superintendent of Engineering
Cleveland Twist Drill Co.

Bruce E. Wright
General Supervisor of Specifications
Buick Motor Division
General Motors Corporation

Hugh Baker, *Secretary*
American Society for Metals

ASM Committee on Austenitic Manganese Steels

Howard S. Avery
Consulting Engineer
Research Center
Abex Corporation

Douglas V. Doane
Associate Director of Research
Climax Molybdenum Company of Michigan

Herman A. Fabert, Jr.
Technical Director
Amsco Division
Abex Corporation

Leroy Finch
Senior Metallurgist
Esco Corporation

Telfer E. Norman
Consultant

Franklin A. Sabatini
Process Metallurgist
The Frog Switch & Manufacturing Co.

ASM Committee on Magnetically Soft Materials

Gilbert Y. Chin
Head, Physical Metallurgy and Ceramics Research and Development Department
Bell Laboratories

Leslie L. Harner
Carpenter Steel Division
Carpenter Technology Corporation

Martin F. Littmann
Principal Research Engineer
Research & Technology
Armco Inc.

Jack W. Shilling
Manager, Physical Metallurgy Research
Allegheny Ludlum Steel Corp.

ASM Committee on Permanent Magnets

Melvin A. Bohlmann
Manager of Research
Indiana General Division
Electronic Memories & Magnetics Corp.

Fred G. Jones
Senior Engineer
Hitachi Magnetics Corporation

Fred E. Luborsky
Corporate Research & Development
General Electric Company

ASM Committee on Electrical Resistance

Roger R. Giler
Vice President, Engineering
The Kanthal Corporation

John H. Lang, Jr.
Materials and Processes Engineering
Space Systems Division
Lockheed Missiles & Space Co., Inc.

Richard K. Pitler
Vice President
Technical Director
Allegheny Ludlum Steel Corporation

Robert H. Pry
Executive Vice President—Research & Development
Gould Inc.

C. Dean Starr
Vice President
Engineering & Research
AMAX Specialty Metals Corporation

Thomas E. Wells
(retired) formerly Physicist, Electricity & Magnetism
Electrical Reference Standards Section
National Bureau of Standards

ASM Committee on Electric-Contact Materials

David Edwin Brown
Research Supervisor, Metallurgy
Eaton Corporation

Frank S. Brugner
Manager, Research & Development Laboratories
Square D Company

Yuan-Shou Shen
Engelhard Industries Division
Engelhard Minerals & Chemical Corporation

Hendrick J. Slaats
Supervisor of Materials Process Engineering
Texas Instruments, Inc.

ASM Committee on Sliding Bearings

George R. Kingsbury, *Chairman*
Manager—International Operations
Engine Parts Division
Gould Inc.

Rudolph A. Braun
Senior Process & Materials Development Engineer
Engine Parts Division
Gould Inc.

James H. Doubrava
Manager of Materials & Process Development
Engine Parts Division
Gould Inc.

Henry M. T. Harris
Chief Engineer, Bearings
Castings Group
Abex Corporation

Charles H. Junge
Consultant

S. Frank Murray
Senior Research Engineer
Mechanical Engineering Dept.
Aeronautical Engineering & Mechanics
Rensselaer Polytechnic Institute

ASM Committee on Selection for Economy in Manufacture

G. F. Bush
Manager, Materials Engineering
Ford Motor Co.

Robert C. Barch
Metallurgical Engineer
The Hoover Co.

Andrew M. Belavic
Manager of Project Planning
Engineered Fasteners Division
Eaton Corporation

Robert E. Cook
Vice-President and General Manager
Mills Alloy Steel Co.

Subi Dinda
Senior Materials Development Engineer
Chrysler Corporation

H. E. Fairman
Bristol Corporation

Daniel J. Hayes
Metallurgical Engineer
United States Steel Corporation

Victor A. Kortesoja
Chief Metallurgist
Ford Motor Co.

Roger G. Mack
Director, Research and Development
Abar Corporation

Bruce L. McMillan
Plant Metallurgist
International Harvester Co.

Eugene Ross
Plant Metallurgist
General Motors Corporation

R. E. Seward
Research Associate
General Electric Co.

Kenneth E. Spray
Section Head, Metallurgy
Clark Equipment Co.

J. A. Sweet
Senior Development Engineer
Great Lakes Steel

Victor J. Turk
Supervisor
Terex Division
General Motors Corporation

Francis R. Varrese
Principal Materials Engineer
Honeywell Corporation

George Wood
Advisory Engineer
IBM Corporation

Daniel S. Zamborsky
Corporate Metallurgist
Warner & Swasey Corporation

Contents

Corrosion-Resistant Materials

Wrought Stainless Steels

By the ASM Committee on Wrought Stainless Steels*

STAINLESS STEELS are more resistant to rusting and staining than plain carbon and low-alloy steels. They have superior corrosion resistance because they contain relatively large amounts of chromium. Although other elements such as copper, aluminum, silicon, nickel and molybdenum also increase the corrosion resistance of steel, their usefulness in this respect is limited.

Stainless steels may be defined as alloy steels containing at least 10% chromium—with or without other elements. It has been customary in the United States to include with stainless steels those alloys that contain as little as 4% chromium. Together, these steels form a family known as stainless and heat-resisting steels. Few of these alloys contain more than 30% chromium or less than 50% iron. This section includes some information on alloys that contain from 4 to 10% chromium as well as on those with chromium contents of 30% or more. Steels in both categories are used mainly for elevated-temperature applications; additional information concerning them appears in the article on heat-resisting alloys in this volume.

Original discoveries and developments in stainless steel technology began in England and Germany about 1910. Commercial production and use of stainless steels in the United States began in the 1920's, with Allegheny, Armco, Carpenter, Crucible, Firth-Sterling, Jessop, Ludlum, Republic, Rustless and U. S. Steel being among the early producers.

Only modest tonnages of stainless steel were produced in the United States in the middle 1920's, but annual production has risen steadily since that time. Even so, tonnage has never exceeded about 1.5% of total production for the steel industry. Table 1 shows shipments of stainless steel over a recent 10-year period. Figure 1 shows how stainless steel production rose from 1929 (when recordkeeping began) through 1978. Production tonnages are shown only for U. S. domestic production. France, Italy, Japan, Sweden, the United Kingdom and West Germany produce substantial tonnages of steel, and data on production in these countries also are available. However, other free-world countries do not make their figures public, and production statistics are not available from the USSR or other Communist nations, which makes it impossible to accurately estimate total world production of stainless steel.

The development of precipitation-hardenable stainless steels was spearheaded by successful production of Stainless W by U. S. Steel in 1945. Since then, Armco, Allegheny-Ludlum and Carpenter Technology have succeeded in developing a series of precipitation-hardenable alloys.

Table 1 Total U.S. shipments of stainless steel over the period 1970 to 1979

Year	Shipments kt	Shipments 1000 tons
1970	643	709
1971	651	718
1972	776	855
1973	1029	1134
1974	1220	1345
1975	687	757
1976	924	1019
1977	1014	1118
1978	1080	1191
1979	1234	1361

Problems in supplying raw materials have been encountered from time to time in stainless steel production. One of the most serious of these was the nickel shortage associated with war conditions in the 1950's. Earlier work had indicated that manganese and nitrogen might be substituted for nickel. Some commercial steels in the AISI 200 series were first produced during that time. Production of these stainless steels continues.

Improvements in melting technology have led to production of alloys with very low carbon contents and remarkable freedom from nonmetallic in-

*See page X for committee list.

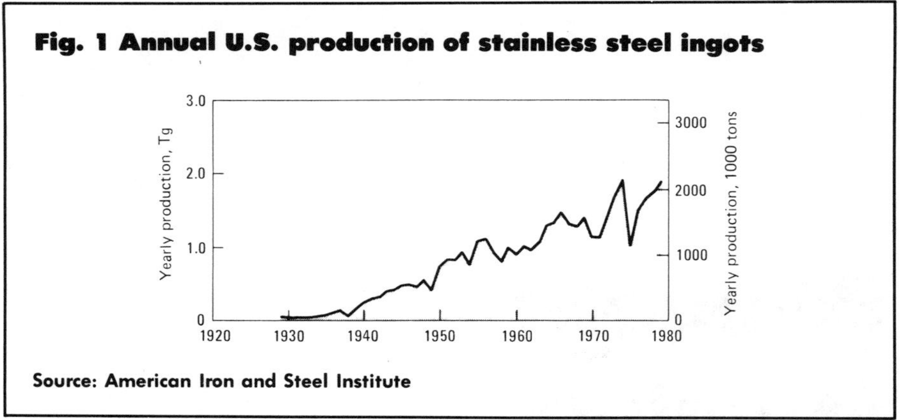

Fig. 1 Annual U.S. production of stainless steel ingots

Source: American Iron and Steel Institute

clusions. Inert-atmosphere melting techniques and vacuum remelting processes have made it possible to produce stainless steels with carbon levels below 0.010%.

Over the years, stainless steels have become firmly established as materials for cooking utensils, fasteners, cutlery, flatware, decorative architectural hardware, and equipment for use in chemical plants, dairy and food-processing plants, health and sanitation applications, petroleum and petrochemical plants, textile plants, and the pharmaceutical and transportation industries. Some of these applications involve exposure to either elevated or cryogenic temperatures; stainless steels are well suited to either type of service. Properties of stainless steels at elevated temperatures are discussed in this article; properties at cryogenic temperatures are discussed in the article in this volume entitled "Alloys for Structural Applications at Subzero Temperatures."

Modifications in composition sometimes are made to facilitate production: for instance, basic compositions are altered to make it easier to produce stainless steel tubing and castings. Similar modifications are made for manufacture of stainless steel welding electrodes; here, combinations of electrode coating and wire composition are used to produce desired compositions in deposited weld metal.

Classification of Stainless Steels

Stainless steels are commonly divided into four groups: (*a*) martensitic stainless steels, (*b*) ferritic stainless steels, (*c*) austenitic stainless steels and (*d*) precipitation-hardening stainless steels.

Standard Types. Although the American Iron and Steel Institute does not write specifications, it has published a list of "standard types" of stainless steel. The compositions of AISI standard types are given in Table 2. The criteria used by AISI to establish which types of stainless steel are standard types are rather loosely defined but include tonnage produced during a specific period, availability (number of producers), and composition limits. Specification-writing organizations such as ASTM and SAE include AISI standard types in their specifications. In referring to specific compositions, the term "type" is preferred over the term "grade". Some specifications establish a series of grades within a given type, which makes it possible to specify properties more precisely for a given nominal composition.

In each of the three original groups of stainless steels—austenitic, ferritic and martensitic—there is one composition that represents the basic, general-purpose alloy. All other compositions derive from this basic alloy, with specific variations in composition being made to impart very specific properties. The so-called family relationships for these three groups are summarized in Fig. 2 to 4.

Nonstandard Types. Besides AISI standard types, many proprietary stainless steels are used for specific applications. Compositions of the more popular "nonstandard" stainless steels are given in Table 3; some of the nonstandard grades are identified by AISI type numbers.

A recent cooperative study by ASTM and SAE has resulted in the Unified Numbering System (UNS) for designation and identification of metals and alloys in commercial use in the United States. In UNS listings, stainless steels are identified by the letter "S" followed by five digits. A few stainless alloys are classified as nickel alloys in the UNS system (identification letter "N") because of their very high nickel contents.

Use of UNS numbers and AISI standard-type numbers ensures that a consumer can obtain suitable material time after time even from different producers or suppliers. Nevertheless, some variation in fabrication and service characteristics can be expected, even with material obtained from a single producer.

Factors in Selection

The first and most important step toward successful use of a stainless or heat-resisting steel is selection of a type that is appropriate for the application. There are a large number of standard types that differ from one another in composition, corrosion resistance, physical properties and mechanical properties; selection of the optimum type for a specific application is the key to satisfactory performance at minimum total cost.

The characteristics and properties of individual types discussed in this article and elsewhere in this volume provide some of the information useful in steel selection. For a more detailed discussion, the reader is referred to *Design Guidelines for the Selection and Use of Stainless Steel,* published by the Committee of Stainless Steel Producers and available through AISI.

Below is a checklist of characteristics to be considered in selection of the proper type of stainless steel for a specific application.

- Corrosion resistance
- Resistance to oxidation and sulfidation
- Strength and ductility at ambient and service temperatures
- Suitability for intended fabrication techniques
- Suitability for intended cleaning procedures
- Stability of properties in service
- Toughness
- Resistance to abrasion and erosion
- Resistance to galling and seizing
- Surface finish and/or reflectivity
- Magnetic properties
- Thermal conductivity
- Electrical resistivity
- Sharpness (retention of cutting edge)
- Rigidity
- Dimensional stability.

Table 2 Compositions of standard stainless steels

Type	UNS number	C	Mn	Si	Cr	Ni(b)	P	S	Others
		Composition, %(a)							
Austenitic types									
201	S20100	0.15	5.5-7.5	1.00	16.0-18.0	3.5-5.5	0.06	0.03	0.25 N
202	S20200	0.15	7.5-10.0	1.00	17.0-19.0	4.0-6.0	0.06	0.03	0.25 N
205	S20500	0.12-0.25	14.0-15.5	1.00	16.5-18.0	1.0-1.75	0.06	0.03	0.32-0.40 N
301	S30100	0.15	2.00	1.00	16.0-18.0	6.0-8.0	0.045	0.03	....
302	S30200	0.15	2.00	1.00	17.0-19.0	8.0-10.0	0.045	0.03	....
302B	S30215	0.15	2.00	2.0-3.0	17.0-19.0	8.0-10.0	0.045	0.03	....
303	S30300	0.15	2.00	1.00	17.0-19.0	8.0-10.0	0.20	0.15 min	0.6 Mo(c)
303Se	S30323	0.15	2.00	1.00	17.0-19.0	8.0-10.0	0.20	0.06	0.15 min Se
304	S30400	0.08	2.00	1.00	18.0-20.0	8.0-10.5	0.045	0.03	....
304H	S30409	0.04-0.10	2.00	1.00	18.0-20.0	8.0-10.5	0.045	0.03	....
304L	S30403	0.03	2.00	1.00	18.0-20.0	8.0-12.0	0.045	0.03	....
304LN	....	0.03	2.00	1.00	18.0-20.0	8.0-10.5	0.045	0.03	0.10-0.15 N
S30430	S30430	0.08	2.00	1.00	17.0-19.0	8.0-10.0	0.045	0.03	3.0-4.0 Cu
304N	S30451	0.08	2.00	1.00	18.0-20.0	8.0-10.5	0.045	0.03	0.10-0.16 N
305	S30500	0.12	2.00	1.00	17.0-19.0	10.5-13.0	0.045	0.03	....
308	S30800	0.08	2.00	1.00	19.0-21.0	10.0-12.0	0.045	0.03	....
309	S30900	0.20	2.00	1.00	22.0-24.0	12.0-15.0	0.045	0.03	....
309S	S30908	0.08	2.00	1.00	22.0-24.0	12.0-15.0	0.045	0.03	....
310	S31000	0.25	2.00	1.50	24.0-26.0	19.0-22.0	0.045	0.03	....
310S	S31008	0.08	2.00	1.50	24.0-26.0	19.0-22.0	0.045	0.03	....
314	S31400	0.25	2.00	1.5-3.0	23.0-26.0	19.0-22.0	0.045	0.03	....
316	S31600	0.08	2.00	1.00	16.0-18.0	10.0-14.0	0.045	0.03	2.0-3.0 Mo
316F	S31620	0.08	2.00	1.00	16.0-18.0	10.0-14.0	0.20	0.10 min	1.75-2.5 Mo
316H	S31609	0.04-0.10	2.00	1.00	16.0-18.0	10.0-14.0	0.045	0.03	2.0-3.0 Mo
316L	S31603	0.03	2.00	1.00	16.0-18.0	10.0-14.0	0.045	0.03	2.0-3.0 Mo
316LN	....	0.03	2.00	1.00	16.0-18.0	10.0-14.0	0.045	0.03	2.0-3.0 Mo; 0.10-0.30N
316N	S31651	0.08	2.00	1.00	16.0-18.0	10.0-14.0	0.045	0.03	2.0-3.0 Mo; 0.10-0.16 N
317	S31700	0.08	2.00	1.00	18.0-20.0	11.0-15.0	0.045	0.03	3.0-4.0 Mo
317L	S31703	0.03	2.00	1.00	18.0-20.0	11.0-15.0	0.045	0.03	3.0-4.0 Mo
321	S32100	0.08	2.00	1.00	17.0-19.0	9.0-12.0	0.045	0.03	5×%C min Ti
321H	S32109	0.04-0.10	2.00	1.00	17.0-19.0	9.0-12.0	0.045	0.03	5×%C min Ti
329	S32900	0.10	2.00	1.00	25.0-30.0	3.0-6.0	0.045	0.03	1.0-2.0 Mo
330	N08330	0.08	2.00	0.75-1.5	17.0-20.0	34.0-37.0	0.04	0.03	....
347	S34700	0.08	2.00	1.00	17.0-19.0	9.0-13.0	0.045	0.03	10×%C min Nb + Ta(c)
347H	S34709	0.04-0.10	2.00	1.00	17.0-19.0	9.0-13.0	0.045	0.03	10×%C min Nb + Ta
348	S34800	0.08	2.00	1.00	17.0-19.0	9.0-13.0	0.045	0.03	0.2 Cu; 10×%C min Nb + Ta(c)
348H	S34809	0.04-0.10	2.00	1.00	17.0-19.0	9.0-13.0	0.045	0.03	0.2 Cu; 10×%C min Nb + Ta(c)
384	S38400	0.08	2.00	1.00	15.0-17.0	17.0-19.0	0.045	0.03	....
Ferritic types									
405	S40500	0.08	1.00	1.00	11.5-14.5	....	0.04	0.03	0.10-0.30 Al
409	S40900	0.08	1.00	1.00	10.5-11.75	....	0.045	0.045	6×% C min Ti(e)
429	S42900	0.12	1.00	1.00	14.0-16.0	....	0.04	0.03	....
430	S43000	0.12	1.00	1.00	16.0-18.0	....	0.04	0.03	....
430F	S43020	0.12	1.25	1.00	16.0-18.0	....	0.06	0.15 min	0.6 Mo(c)
430FSe	S43023	0.12	1.25	1.00	16.0-18.0	....	0.06	0.06	0.15 min Se
434	S43400	0.12	1.00	1.00	16.0-18.0	....	0.04	0.03	0.75-1.25 Mo
436	S43600	0.12	1.00	1.00	16.0-18.0	....	0.04	0.03	0.75-1.25 Mo; 5×%C min Nb + Ta(f)
442	S44200	0.20	1.00	1.00	18.0-23.0	....	0.04	0.03	....
446	S44600	0.20	1.50	1.00	23.0-27.0	....	0.04	0.03	0.25 N

(continued)

(a) Single values are maximum values unless otherwise indicated. (b) For some tubemaking processes, the nickel content of certain austenitic types must be slightly higher than shown. (c) Optional. (d) 0.10% max Ta. (e) 0.75% maximum. (f) 0.70% maximum.

Table 2 (continued)

Type	UNS number	C	Mn	Si	Cr	Ni(b)	P	S	Others
					Composition, %(a)				
Martensitic types									
403	S40300	0.15	1.00	0.50	11.5-13.0	....	0.04	0.03	....
410	S41000	0.15	1.00	1.00	11.5-13.0	....	0.04	0.03	....
414	S41400	0.15	1.00	1.00	11.5-13.5	1.25-2.50	0.04	0.03	....
416	S41600	0.15	1.25	1.00	12.0-14.0	....	0.04	0.03	0.6 Mo(c)
416Se	S41623	0.15	1.25	1.00	12.0-14.0	....	0.06	0.06	0.15 min Se
420	S42000	0.15 min	1.00	1.00	12.0-14.0	....	0.04	0.03	....
420F	S42020	0.15 min	1.25	1.00	12.0-14.0	....	0.06	0.15 min	0.6 Mo(c)
422	S42200	0.20-0.25	1.00	0.75	11.0-13.0	0.5-1.0	0.025	0.025	0.75-1.25 Mo; 0.75-1.25 W; 0.15-0.3 V
431	S43100	0.20	1.00	1.00	15.0-17.0	1.25-2.50	0.04	0.03	....
440A	S44002	0.60-0.75	1.00	1.00	16.0-18.0	....	0.04	0.03	0.75 Mo
440B	S44003	0.75-0.95	1.00	1.00	16.0-18.0	....	0.04	0.03	0.75 Mo
440C	S44004	0.95-1.20	1.00	1.00	16.0-18.0	....	0.04	0.03	0.75 Mo
501	S50100	0.10 min	1.00	1.00	4.0-6.0	....	0.04	0.03	0.40-0.65 Mo
501A	S50300	0.15	0.30-0.60	0.50-1.00	6.0-8.0	....	0.03	0.03	0.45-0.65 Mo
501B	S50400	0.15	0.30-0.60	0.50-1.00	8.0-10.0	....	0.03	0.03	0.9-1.1 Mo
502	S50200	0.10	1.00	1.00	4.0-6.0	....	0.04	0.03	0.40-0.65 Mo
503	S50300	0.15	1.00	1.00	6.0-8.0	....	0.04	0.04	0.45-0.65 Mo
504	S50400	0.15	1.00	1.00	8.0-10.0	....	0.04	0.04	0.9-1.1 Mo
Precipitation-hardening types									
PH 13-8 Mo	S13800	0.05	0.10	0.10	12.25-13.25	7.5-8.5	0.01	0.008	2.0-2.5 Mo; 0.90-1.35 Al; 0.01 N
15-5 PH	S15500	0.07	1.00	1.00	14.0-15.5	3.5-5.5	0.04	0.03	2.5-4.5 Cu; 0.15-0.45 Nb + Ta
17-4 PH	S17400	0.07	1.00	1.00	15.5-17.5	3.0-5.0	0.04	0.03	3.0-5.0 Cu; 0.15-0.45 Nb + Ta
17-7 PH	S17700	0.09	1.00	1.00	16.0-18.0	6.5-7.75	0.04	0.03	0.75-1.5 Al

(a) Single values are maximum values unless otherwise indicated. (b) For some tubemaking processes, the nickel content of certain austenitic types must be slightly higher than shown. (c) Optional. (d) 0.10% max Ta. (e) 0.75% maximum. (f) 0.70% maximum.

Corrosion resistance frequently is the most important characteristic of a stainless steel, but often is also the most difficult to assess for a specific application. General corrosion resistance to natural conditions and to pure chemical solutions is comparatively easy to determine, and Table 4 shows resistance of standard types to various common media. (For detailed information about the corrosion of stainless steels, see the article in this volume entitled "Stainless Steels in Corrosion Service".)

General corrosion is often much less serious than localized forms such as stress-corrosion cracking, crevice corrosion in tight spaces or under deposits, pitting attack, and intergranular attack in sensitized material such as weld heat-affected zones. Such localized corrosion can cause unexpected and sometimes catastrophic failure while most of the structure remains unaffected, and therefore must be considered carefully in design and in selection of the proper grade of stainless steel. Corrosive attack can also be increased dramatically by seemingly minor impurities in the medium that may be difficult to anticipate but that can have major effects, even when present in only parts-per-million concentrations; by heat transfer through the steel to or from the corrosive medium; by contact with dissimilar metallic materials; by stray electrical currents; and by many other subtle factors. At elevated temperatures, attack can be accelerated significantly by seemingly minor changes in atmosphere that affect scaling, sulfidation or carburization.

Despite these complications, a suitable steel can be selected for most applications on the basis of experience, perhaps with assistance from the steel producer. Laboratory corrosion data can be misleading in predicting service performance. Even actual service data have limitations, because "similar" corrosive media may differ substantially due to slight variations in some of the corrosion factors listed above. For difficult applications, extensive study of comparative data may be necessary, sometimes followed by pilot-plant or in-service testing.

Mechanical properties at service temperature obviously are important, but satisfactory performance at other temperatures must be considered also. Thus, a product for arctic service must have suitable properties at subzero temperatures even though steady-state operating temperature may be much higher; room-temperature properties after extended service at elevated temperature can be important for applications such as boilers and jet engines, which are intermittently shut down.

Fabrication and Cleaning. Frequently a particular stainless steel is chosen for a fabrication characteristic such as formability or weldability. Even a required or preferred cleaning procedure may dictate selection of a specific type. For instance, a weldment that is to be cleaned in a medium such

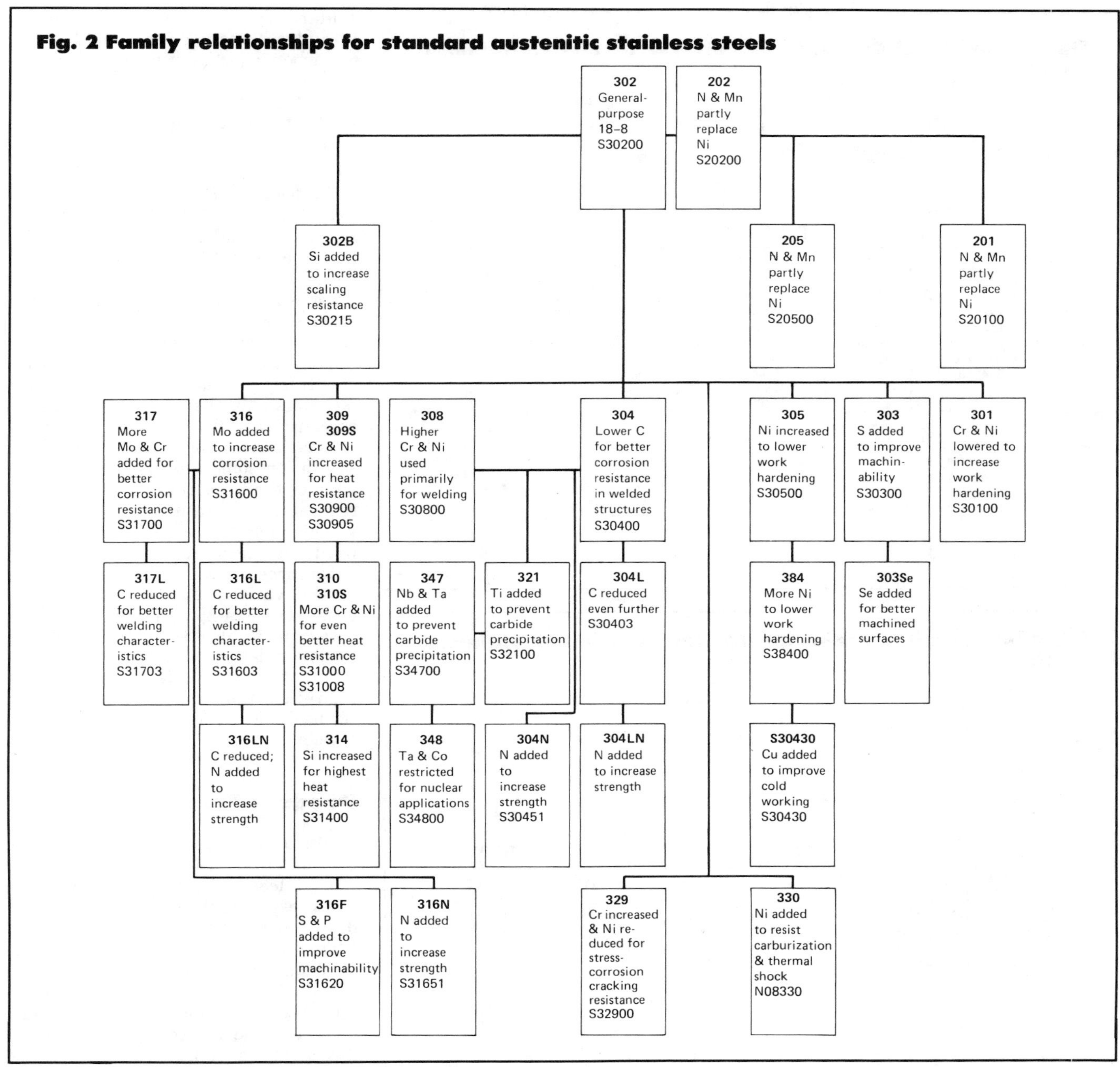

Fig. 2 Family relationships for standard austenitic stainless steels

as nitric-hydrofluoric acid, which attacks sensitized stainless steel, should be produced from stabilized or low-carbon stainless steel even though sensitization may not affect performance under service conditions.

Experience in the use of stainless steels indicates that many factors can affect their corrosion resistance. Some of the more prominent factors are:

- Chemical composition of the corrosive medium, including impurities
- Physical state of the medium—liquid, gaseous, solid, or combinations thereof
- Temperature variations
- Aeration of the medium
- Oxygen content of the medium
- Bacteria content of the medium
- Ionization of the medium
- Repeated formation and collapse of bubbles in the medium
- Relative motion of the medium with respect to the steel
- Chemical composition of the metal
- Nature and distribution of microstructural constituents
- Continuity of exposure of the metal to the medium
- Surface condition of the metal
- Stresses in the metal during exposure to the medium
- Contact of the metal with one or more dissimilar metallic materials
- Stray electric currents
- Differences in electric potential

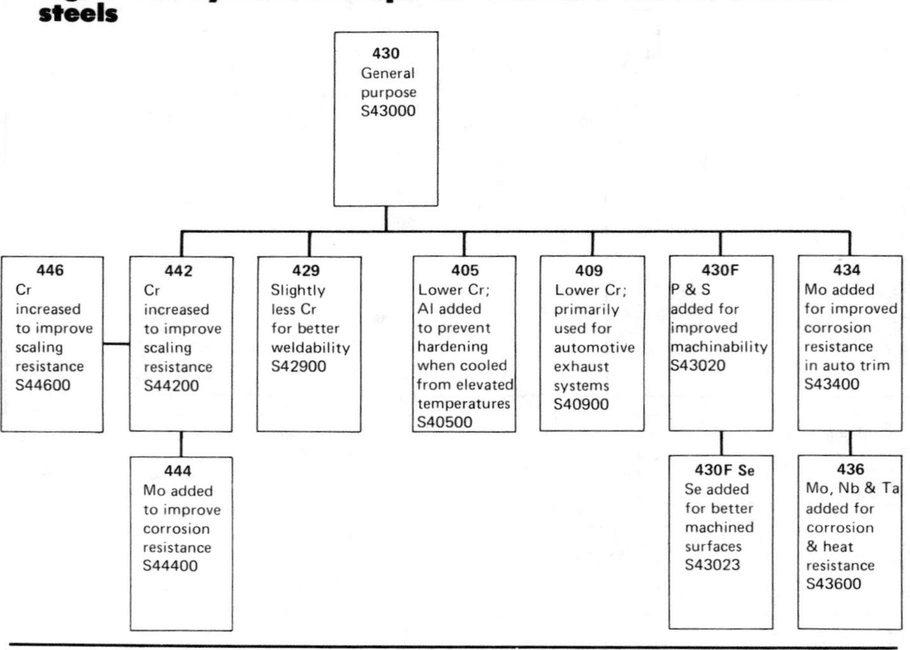

Fig. 3 Family relationships for standard ferritic stainless steels

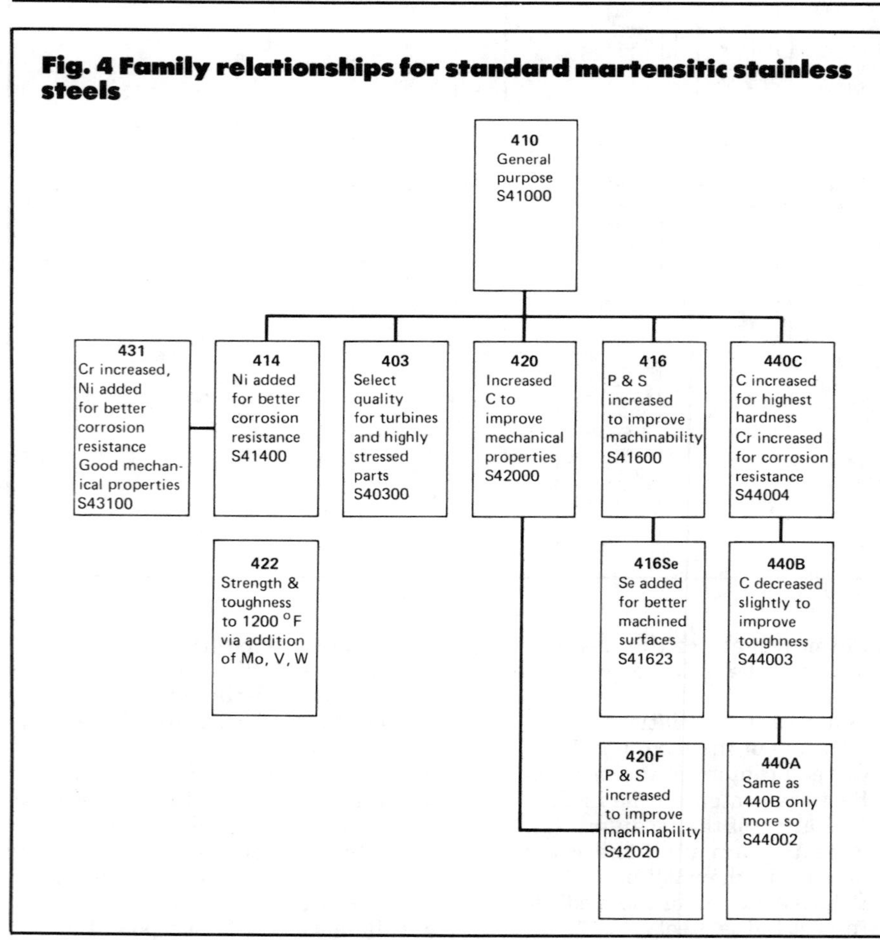

Fig. 4 Family relationships for standard martensitic stainless steels

- Marine growths such as barnacles
- Sludge deposits on the metal
- Carbon deposits from heated organic compounds
- Dust on exposed surfaces
- Effects of welding, brazing and soldering.

Surface Finish. Other characteristics in the checklist on page 4 are vital for some specialized applications but of little concern for many applications. Among these characteristics, surface finish is important more often than any other except corrosion resistance, and stainless steels sometimes are selected because they are available in a variety of attractive finishes. Selection of surface finish may be made on the basis of appearance, frictional characteristics or sanitation. The effect of finish on sanitation sometimes is thought to be simpler than it actually is, and tests of several candidate finishes may be advisable. Selection of finish may in turn influence selection of alloy because of differences in availability or durability of the various finishes for different types. For example, a more corrosion-resistant stainless steel will maintain a bright finish in a corrosive environment that would dull a lower-alloy type. Selection among finishes is described in more detail elsewhere in this article.

Product Forms

Plate is a flat-rolled or forged product more than 254 mm (10 in.) in width and at least 4.76 mm (0.1875 in.) in thickness. Stainless steel plate is produced in all types shown in Table 2. Plate usually is produced by hot rolling from slabs that have been directly cast or rolled from ingots and that usually have been conditioned to improve plate surface. Some plate may be produced by direct rolling from ingot. It is referred to as sheared plate or sheared mill plate when rolled between horizontal rolls and trimmed on all edges, and as universal plate or universal mill plate when rolled between horizontal and vertical rolls and trimmed only on the ends. Universal plate sometimes is rolled between grooved rolls.

Stainless steel plate generally is produced in the annealed condition and is either blast cleaned or pickled. Blast cleaning generally is followed by further cleaning in appropriate acids to remove surface contaminants such as

Table 3 Compositions of nonstandard stainless steels

Designation(a)	UNS number	Composition, %(b) C	Mn	Si	Cr	Ni	P	S	Others
Austenitic stainless steels									
Type 216 (XM-17)	S21600	0.08	7.5-9.0	1.00	17.5-22.0	5.0-7.0	0.045	0.03	2.0-3.0 Mo; 0.25-0.50 N
Type 304HN	S30452	0.04-0.10	2.00	1.00	18.0-20.0	8.0-10.5	0.045	0.03	0.10-0.16 N
Type 308	S30800	0.08	2.00	1.00	19.0-21.0	10.0-12.0	0.045	0.03	....
Type 308L	...	0.03	2.00	1.00	19.0-21.0	10.0-12.0	0.045	0.03	....
Type 309S	S30908	0.08	2.00	1.00	22.0-24.0	12.0-15.0	0.045	0.03	....
Type 309S Cb	S30940	0.08	2.00	1.00	22.0-24.0	12.0-15.0	0.045	0.03	8×%C min Nb
Type 309 Cb + Ta	...	0.08	2.00	1.00	22.0-24.0	12.0-15.0	0.045	0.03	8×%C min Nb + Ta
Type 312	...	0.15	2.00	1.00	30.0 nom	9.0 nom	0.045	0.03	....
Type 317LM	...	0.03	2.00	1.00	18.0-20.0	12.0-16.0	0.045	0.03	4.0-5.0 Mo
Type 330HC	...	0.40	1.50	1.25	19.0 nom	35.0 nom	....	....	....
Type 332	...	0.04	1.00	0.50	21.5 nom	32.0 nom	0.045	0.03	....
Type 385	...	0.08	2.00	1.00	11.5-13.5	14.0-16.0	0.045	0.03	....
904L	N08904	0.02	2.00	1.00	19.0-23.0	23.0-28.0	0.045	0.035	4.0-5.0 Mo; 1.0-2.0 Cu
18-18-2 (XM-15)	S38100	0.08	2.00	1.5-2.5	17.0-19.0	17.5-18.5	0.03	0.03	0.08-0.18 N
18-18 Plus	S28200	0.15	17.0-19.0	1.00	17.5-19.5	....	0.045	0.03	0.5-1.5 Mo; 0.5-1.5 Cu; 0.4-0.6 N
20Cb-3	N08020	0.07	2.00	1.00	19.0-21.0	32.0-38.0	0.045	0.035	2.0-3.0 Mo; 3.0-4.0 Cu; 8×%C min Nb(c)
AL-6X	N08366	0.03	2.00	0.75	20.0-22.0	23.5-25.5	0.030	0.003	6.0-7.0 Mo
303 Plus X (XM-5)	...	0.15	2.5-4.5	1.00	17.0-19.0	7.0-10.0	0.20	0.25 min	0.6 Mo
HNM (d)	...	0.30	3.5	0.5	18.5	9.5	0.25	....	....
Crutemp 25 (d)	...	0.05	1.5	0.4	25.0	25.0	....	....	....
JS-700	N08700	0.04	2.00	1.00	19.0-23.0	24.0-26.0	0.04	0.03	4.3-5.0 Mo; 0.5 Cu; 8×%C min Nb(e); 0.005 Pb; 0.035 Sn
JS-777	...	0.04	2.00	1.00	19.0-23.0	24.0-26.0	0.045	0.035	4.0-5.0 Mo; 1.9-2.5 Cu
Nitronic 32 (d)	S24100	0.10	12.0	0.5	18.0	1.6	....	....	0.35 N
Nitronic 33 (d)	S24000	0.06	13.0	0.5	18.0	3.0	....	....	0.30 N
Nitronic 40 (21-6-9) (XM-10)	S21900	0.08	8.0-10.0	1.00	18.0-20.0	5.0-7.0	0.06	0.03	0.15-0.40 N
Nitronic 50 (22-13-5) (XM-19)	S20910	0.06	4.0-6.0	1.00	20.5-23.5	11.5-13.5	0.04	0.03	1.5-3.0 Mo; 0.2-0.4 N; 0.1-0.3 Nb; 0.1-0.3 V
Nitronic 60	S21800	0.10	7.0-9.0	3.5-4.5	16.0-18.0	8.0-9.0	0.04	0.03	....
Tenelon (XM-31)	S21400	0.12	14.5-16.0	0.3-1.0	17.0-18.5	0.75	0.045	0.03	0.35 N
Cryogenic Tenelon (XM-14)	S21460	0.12	14.0-16.0	1.00	17.0-19.0	5.0-6.0	0.06	0.03	0.35-0.50 N

(continued)

(a) XM designations in this column are ASTM designations for the listed alloy. Type numbers in parentheses are obsolete AISI designations. (b) Single values are maximum values unless otherwise indicated. (c) 1.00% maximum. (d) Nominal composition; composition limits not available. (e) 0.50% maximum. (f) 0.75% maximum. (g) 0.80 maximum. (h) 7 N minimum.

Table 3 (continued)

Designation(a)	UNS number	C	Mn	Si	Composition, %(b) Cr	Ni	P	S	Others
Ferritic stainless steels									
Type 404	...	0.05	1.00	0.50	11.0-12.5	1.25-2.00	0.03	0.03	....
Type 430Ti	S43036	0.10	1.00	1.00	16.0-19.5	0.75	0.04	0.03	5×%C min Ti(f)
Type 444 (18-2)	S44400	0.025	1.00	1.00	17.5-19.5	1.00	0.04	0.03	1.75-2.5 Mo; 0.035 max N; 0.2 + 4(%C + %N) min (Ti + Nb)
18SR (d)	...	0.04	0.3	1.00	18.0	....	....	....	2.0 Al; 0.4 Ti
18-2 FM	S18200	0.08	2.50	1.00	17.5-19.5	....	0.04	0.15 min	....
E Brite 26-1 (XM-27)	S44625	0.01	0.40	0.40	25.0-27.5	0.50	0.02	0.02	0.75-1.5 Mo; 0.015 N; 0.2 Cu; 0.5 Ni + Cu
26-1 Ti (XM-33)	S44626	0.06	0.75	0.75	25.0-27.0	0.50	0.04	0.02	0.75-1.5 Mo; 0.04 N; 0.2 Cu; 0.2-1.0 Ti(g)
29-4	S44700	0.010	0.30	0.20	28.0-30.0	0.15	0.025	0.02	3.5-4.2 Mo
29-4-2	S44800	0.010	0.30	0.20	28.0-30.0	2.0-2.5	0.025	0.02	3.5-4.2 Mo
Monit	S44635	0.25	1.00	0.75	24.5-26.0	3.5-4.5	0.04	0.03	3.5-4.5 Mo; 0.3-0.6 (Ti + Nb)
Sea-cure/Sc-1	S44660	0.025	1.00	0.75	25.0-27.0	1.5-3.5	0.04	0.03	2.5-3.5 Mo; 0.2 + 4(%C + %N) min (Ti + Nb)
Martensitic stainless steels									
Type 410Cb (XM-30)	S41040	0.18	1.00	1.00	11.5-13.5	....	0.04	0.03	0.05-0.30 Nb
Type 410S	S41008	0.08	1.00	1.00	11.5-13.5	0.60	0.04	0.03	....
Type 414L	...	0.06	0.50	0.15	12.5-13.0	2.5-3.0	0.04	0.03	0.5 Mo; 0.03 Al
416 Plus X (XM-6)	S41610	0.15	1.5-2.5	1.00	12.0-14.0	....	0.06	0.15 min	0.6 Mo
Precipitation-hardening stainless steels									
AM-350 (Type 633)	S35000	0.07-0.11	0.5-1.25	0.50	16.0-17.0	4.0-5.0	0.04	0.03	2.5-3.25 Mo; 0.07-0.13 N
AM-355 (Type 634)	S35500	0.10-0.15	0.5-1.25	0.50	15.0-16.0	4.0-5.0	0.04	0.03	2.5-3.25 Mo
AM-363 (d)	...	0.04	0.15	0.05	11.0	4.0	....	....	0.25 Ti
Custom 450 (XM-25)	S45000	0.05	1.00	1.00	14.0-16.0	5.0-7.0	0.03	0.03	1.25-1.75 Cu; 0.5-1.0 Mo; 8×%C min Nb
Custom 455 (XM-16)	S45500	0.05	0.50	0.50	11.0-12.5	7.5-9.5	0.04	0.03	0.5 Mo; 1.5-2.5 Cu; 0.8-1.4 Ti; 0.1-0.5 Nb
PH 15-7 Mo (Type 632)	S15700	0.09	1.00	1.00	14.0-16.0	6.5-7.75	0.04	0.03	2.0-3.0 Mo; 0.75-1.5 Al
Stainless W (Type 635)	S17600	0.08	1.00	1.00	16.0-17.5	6.0-7.5	0.04	0.03	0.4 Al; 0.4-1.2 Ti
17-10 P (d)	...	0.07	0.75	0.5	17.0	10.5	0.28	....	....

(a) XM designations in this column are ASTM designations for the listed alloy. Type numbers in parentheses are obsolete AISI designations. (b) Single values are maximum values unless otherwise indicated. (c) 1.00% maximum. (d) Nominal composition; composition limits not available. (e) 0.50% maximum. (f) 0.75% maximum. (g) 0.80 maximum. (h) 7 N minimum.

particles of steel picked up from the mill rolls.

Plate can be produced with "mill edge" and uncropped ends.

Sheet is a flat-rolled product in coils or cut lengths at least 610 mm (24 in.) wide and less than 4.76 mm (0.1875 in.) thick.

Stainless steel sheet is produced in all types shown in Table 2 except 303, 303Se, 416, 416Se, 420F, 422, 430F, 430FSe, 431, 440A, 440B and 440C. It is produced on continuous mills or on hand mills. The steel is cast in ingots, and the ingots are rolled on a slabbing mill or a blooming mill into slabs or sheet bars. The slabs or sheet bars then are conditioned prior to being hot rolled on a finishing mill. Alternatively, the steel may be continuous cast directly into slabs that are ready for hot rolling on a finishing mill. The current trend worldwide is toward greater production from continuous cast slabs. It has been estimated that more continuous cast than ingot cast stainless steel will be produced in the U. S. from 1980 on.

Sheet produced from slabs on continuous rolling mills is coiled directly off the mill. After they are annealed and descaled, these "hot bands" are cold rolled to the required thickness, and coils off the cold mill are either annealed and descaled or bright annealed. Full coils or lengths cut from coils may then be lightly cold rolled on either dull or bright rolls to produce the required finish. Sheet may be shipped in coils, or cut sheets may be produced by shearing lengths from a coil and flattening them by roller leveling or stretcher leveling.

Sheet produced on hand mills from sheet bars is rolled in lengths and then annealed and descaled. It may be subject to additional operations, including cold reduction, annealing, descaling, light cold rolling for finish, and flattening.

A specified minimum tensile strength, minimum yield strength or hardness level higher than that normally obtained on sheet in the annealed condition, or a combination thereof, can be attained by controlled cold rolling.

Sheet made of chromium-nickel stainless steel (often type 301) or of chromium-nickel-manganese stainless steel (often type 201) is produced in the following cold rolled tempers:

Table 4 Resistance of standard types of stainless steel to various classes of environments

Type	Mild atmospheric and fresh water	Atmospheric		Salt water	Chemical		
		Industrial	Marine		Mild	Oxidizing	Reducing
Austenitic stainless steels							
201	x	x	x		x	x	
202	x	x	x		x	x	
205	x	x	x		x	x	
301	x	x	x		x	x	
302	x	x	x		x	x	
302B	x	x	x		x	x	
303	x	x			x		
303Se	x	x			x		
304	x	x	x		x	x	
304H	x	x	x		x	x	
304L	x	x	x		x	x	
304N	x	x	x		x	x	
S30430	x	x	x		x	x	
305	x	x	x		x	x	
308	x	x	x		x	x	
309	x	x	x		x	x	
309S	x	x	x		x	x	
310	x	x	x		x	x	
310S	x	x	x		x	x	
314	x	x	x		x	x	
316	x	x	x	x	x	x	x
316F	x	x	x	x	x	x	x
316H	x	x	x	x	x	x	x
316L	x	x	x	x	x	x	x
316N	x	x	x	x	x	x	x
317	x	x	x	x	x	x	x
317L	x	x	x	x	x	x	x
321	x	x	x		x	x	
321H	x	x	x		x	x	
329	x	x	x	x	x	x	x
330	x	x	x	x	x	x	x
347	x	x	x		x	x	
347H	x	x	x		x	x	
348	x	x	x		x	x	
348H	x	x	x		x	x	
384	x	x	x		x	x	
Ferritic stainless steels							
405	x				x		
409	x				x		
429	x	x			x	x	
430	x	x			x	x	
430F	x	x			x		
430FSe	x	x			x		
434	x	x	x		x	x	
436	x	x	x		x	x	
442	x	x			x	x	
446	x	x	x		x	x	
Martensitic stainless steels							
403	x				x		
410	x				x		
414	x				x		
416	x						
416Se	x						
420	x						

(continued)

An "x" notation above indicates that the specific type is considered resistant to the corrosive environment.

Table 4 (continued)

Type	Mild atmospheric and fresh water	Atmospheric: Industrial	Atmospheric: Marine	Salt water	Chemical: Mild	Chemical: Oxidizing	Chemical: Reducing
Martensitic stainless steels (continued)							
420F	x						
422	x						
431	x	x	x		x		
440A	x				x		
440B	x						
440C	x						
501							
502							
503							
504							
Precipitation-hardening stainless steels							
PH 13-8 Mo	x	x			x	x	
15-5 PH	x	x	x		x	x	
17-4 PH	x	x	x		x	x	
17-7 PH	x	x	x		x	x	

An "x" notation above indicates that the specific type is considered resistant to the corrosive environment.

Temper	Minimum tensile strength MPa	Minimum tensile strength ksi	Minimum yield strength MPa	Minimum yield strength ksi
1/4 hard	860	125	515	75
1/2 hard	1035	150	760	110
3/4 hard	1205	175	930	135
Full hard	1275	185	965	140

Strip is a flat-rolled product, in coils or cut lengths, less than 610 mm (24 in.) wide and 0.13 to 4.76 mm (0.005 to 0.1875 in.) thick. Cold finished material 0.13 mm (0.005 in.) thick and less than 610 mm (24 in.) wide fits the definitions of both strip and foil, and may be referred to using either term.

Cold rolled stainless steel strip is manufactured from hot rolled, annealed and pickled strip (or from slit sheet) by rolling between polished rolls. Depending on desired thickness, various numbers of cold rolling passes through the mill are required for effecting the necessary reduction and for securing the desired surface characteristics and mechanical properties.

Hot rolled stainless steel strip is a semifinished product obtained by hot rolling slabs or billets, and is produced for conversion to finished strip by cold rolling.

Strip of all types of stainless steel commonly is either annealed or annealed and skin passed, depending on requirements. When severe forming, bending and drawing operations are involved it is recommended that such requirements be indicated so that the producer will have all the information necessary to ensure that he supplies the proper type and condition. When stretcher strains are objectionable in ferritic stainless steels such as type 430, they can be minimized by specifying a No. 2 finish. Cold rolled strip in types 410, 414, 416, 420, 431, 440A, 440B, 440C, 501 and 502 can be produced in the hardened and tempered condition.

Strip made of chromium-nickel stainless steel (often type 301) or of chromium-nickel-manganese stainless steel (often type 201) is produced in the same cold rolled tempers in which sheet is produced.

For strip, edge condition often is important—more important than it usually is for sheet. Strip can be furnished with a mill edge (as produced, condition unspecified), a No. 1 edge (edge rolled, rounded or square), a No. 3 edge (as slit) or a No. 5 edge (square edge produced by rolling or filing after slitting). Mill edge is the least expensive edge condition, and is adequate for many purposes. No. 1 edge provides improved width tolerance over mill edge plus a cold rolled edge condition; rounded edges are preferred for applications requiring the lowest degree of stress concentration at corners. No. 3 and No. 5 edges give progressively better width tolerance and squareness over No. 1 edge.

Foil is a flat-rolled product, in coil form, up to 0.13 mm (0.005 in.) thick and less than 610 mm (24 in.) wide. Foil is produced in slit widths with edge conditions corresponding to No. 3 and No. 5 edge conditions for strip.

Foil is made from types 201, 202, 301, 302, 304, 304L, 305, 316, 316L, 321, 347, 430 and 442, as well as from certain proprietary alloys.

The finishes, tolerances and mechanical properties of foil differ from those of strip because of limitations associated with the way in which foil is manufactured. Nomenclature for finishes, and for width and thickness tolerances, vary among producers.

Finishes for foil are described by the finishing operations employed in their manufacture. However, each finish in itself is a category of finishes, with variations in appearance and smoothness that depend on composition, thickness and method of manufacture. Chromium-nickel and chromium-nickel-manganese stainless steels have a characteristic appearance different from that of straight chromium types for corresponding finish designations.

In general, mechanical properties of foil vary with thickness. Tensile strength is increased somewhat, and ductility is lowered, by a decrease in thickness.

Bar is a product supplied in straight lengths; it is either hot or cold finished, and is available in various shapes, sizes and surface finishes. This category includes (*a*) small shapes whose dimensions do not exceed 76 mm (3 in.) and (*b*) hot rolled flat stock at least 3.2 mm (0.125 in.) thick and up to 254 mm (10 in.) wide.

Hot finished bar commonly is produced by hot rolling, forging or pressing ingots to blooms or billets of intermediate size, which are subsequently hot rolled, forged or extruded to final dimensions. Whether rolling, forging or extrusion is selected as the finishing method depends on several factors, including composition and final size.

Following hot rolling or forging, hot finished bar may be subject to various operations, including (*a*) annealing or other heat treatment, (*b*) descaling by pickling, blast cleaning or other methods, (*c*) surface conditioning by grinding or rough turning and (*d*) machine straightening.

Cold finished bar is produced from hot finished bar or rod by additional operations such as cold rolling or cold drawing, which result in close control of dimensions, smooth surface finish, and higher tensile and yield strengths. Sizes and shapes of cold reduced stock classified as bar are essentially the same as for hot finished bar, except that all cold reduced flat stock less than 4.76 mm (0.1875 in.) thick and over 9.5 mm (0.375 in.) wide is classified as strip.

Cold finished round bar commonly is machine straightened; afterward, it can be centerless ground, or centerless ground and polished. Centerless grinding and polishing do not alter the mechanical properties of cold finished bar, and are used only to improve surface finish or provide closer tolerances. Some increase in hardness, more marked at the surface and particularly in 2*xx* and 3*xx* stainless steels, results from machine straightening. The amount of increase varies chiefly with composition, size and amount of cold work necessary to straighten the bar.

Cold finished bars that are square, flat, hexagonal, octagonal or of certain special shapes are produced from hot finished bars by cold drawing or cold rolling.

When cold finished bar is required to have high strength and hardness, it is cold drawn or heat treated depending on composition, section size and required properties. Round sections can be subsequently centerless ground or centerless ground and polished.

"Free-machining wire" is a bar commodity used for making parts in automatic screw machines or other types of machining equipment. The principal types used are 303, 303Se, 416, 416Se, 420F, 430F, and 430FSe. "Free-machining wire" is commonly produced with a cold drawn or centerless ground finish and with selected hardness depending on the machining operation involved.

Hot rolled, bar-size structural shapes are produced in angles, channels, tees and zees. They can be purchased in the following conditions:

- Hot rolled
- Hot rolled and annealed
- Hot rolled, annealed and blast cleaned
- Hot rolled, annealed and chemically cleaned
- Hot rolled, annealed, blast cleaned and chemically cleaned.

Wire is a coiled product derived by cold finishing hot rolled and annealed rod. Cold finishing imparts excellent dimensional accuracy, good surface smoothness, fine finish and specific mechanical properties. Wire is produced in several tempers and finishes.

Wire is customarily referred to as round wire when the contour is completely cylindrical and as shape wire when the contour is other than cylindrical: for example, wires that are half round, half oval, oval, square, rectangular, hexagonal, octagonal or triangular in cross section are all referred to as shape wire. Shape wire is cold finished either by drawing or by a combination of drawing and rolling.

In production of wire, rod (which is a coiled hot rolled product approximately round in cross section) is drawn through the tapered hole of a die or a series of dies. The smallest size of hot rolled rod commonly made is 6.4 mm (0.25 in.). Rod smaller than this is produced by cold work, the number of dies employed depending on the finished diameter required.

Round stainless steel wire is commonly produced within the approximate size range 0.08 to 15.9 mm (0.003 to 0.625 in.).

Shape wire, except cold finished flat wire, is commonly produced within the approximate size range of 1.12 to 12.7 mm (0.044 to 0.500 in.), although the particular shape governs the specific sizes that can be produced.

Tempers of Wire. Annealed temper describes soft wire that has undergone no further cold drawing after the last annealing treatment. Wire in this temper is made by annealing in open-fired furnaces or molten salt, and annealing ordinarily is followed by pickling that produces a clean, gray, matte finish. It is also made with a bright finish by annealing in a protective atmosphere, and sometimes is described as bright annealed wire.

Soft-temper wire is given a single light draft following the final annealing operation and generally is produced to a defined upper limit of tensile strength or hardness. Wire in this temper is produced with various dry-drawn finishes, including lime soap, lead, copper and oxide. It also may be given a bright finish produced by oil or grease drawing.

Intermediate-temper wire is drawn one or more drafts after annealing as required to produce a specific minimum strength or hardness. The properties of such wire can be varied between the properties of soft-temper wire and properties approaching those of spring-temper wire. Intermediate-temper wire usually is produced with one of the dry-drawn finishes.

Spring-temper wire is drawn several drafts as required to produce high tensile strengths.

Special Wire Commodities. There are many classes of stainless steel wire that have been developed for specific components or for particular applications. The unique properties of each of these individual wire commodities are developed by employing a particular combination of composition, steel quality, process heat treatment and cold drawing practice. The details of manufacture may vary slightly from one wire manufacturer to another, but the finished wire will fulfill the specified requirements.

Cold heading wire is produced in any of the various types of stainless steel. In all instances, cold heading wire is subjected to special testing and inspection to ensure satisfactory performance in cold heading and cold forging operations.

Of the chromium-nickel group, type 305 is the all-purpose type used for cold heading wire and generally is necessary for severe upsetting. Other grades commonly cold formed include 303Se, 304, 316, 321, 347 and 384.

Of the 4*xx* series, types 410, 420, 430 and 431 are used for a variety of cold headed products. Types 430 and 410 are commonly used for severe upsetting and for recessed-head screws and bolts. Types 416, 416Se, 430F and 430FSe are intended primarily for free cutting and are not recommended for cold heading.

Cold heading wire is manufactured using a closely controlled annealing treatment that produces optimum softness and still permits a very light finishing draft after pickling. The purposes of the finishing draft are (*a*) to provide a lubricating coating that will aid the cold heading operation and (*b*) to produce a kink-free coil of wire having more uniform dimensions.

Cold heading wire is produced with a variety of finishes, all of which have the function of providing proper lubrication in the header dies. The finish or coating should be suitably adherent to prevent galling and excessively rapid die wear. A copper coating, which is applied after the annealing treatment and just prior to the finishing draft, is available; the copper-coated wire is then lime coated and drawn using soap as the drawing lubricant. Coatings of lime and soap or of oxide and soap are also employed.

Spring wire is drawn from annealed rod and is subjected to mill tests and inspection that ensure the quality required for extension and compression springs. The types of stainless steel in which spring wire is commonly produced include 302, 304 and, for additional corrosion resistance, 316.

Spring wire in large sizes can be furnished in a variety of finishes such as dry-drawn lead, copper, lime and soap, or oxide and soap. Fine sizes usually are wet drawn, although they can be dry drawn.

Tensile-strength ranges or minimums for type 302, 304, 305 and 316 spring wire in various sizes are given in Table 5.

The torsional modulus for stainless steel spring wire may range from 59 to 76 GPa (8.5 to 11 × 10^6 psi), depending on alloy and wire size. Magnetic permeability is extremely low compared to that of carbon steel wire. Springs made from stainless steel wire retain their physical and mechanical properties at temperatures up to about 315 °C (600 °F).

Rope wire is used to make rope, cable and cord for a variety of uses such as aircraft control cables, marine ropes, elevator cables, slings and anchor cables. Because of special requirements for fatigue strength, rope wire is produced from specially selected and processed material.

Rope wire is made of type 302 or type 304 unless a higher level of corrosion resistance is required, in which case type 316 is generally selected. Special nonmagnetic characteristics may be required, which necessitates selection of heats having little or no ferrite or martensite in the microstructure, and use of special drawing practices to limit or avoid deformation-induced transformation to martensite.

Tensile properties of regular rope wire are slightly lower than those of stainless steel spring wire. Finishes for rope wire vary from a gray matte finish to a bright finish, and include a series of bright to dark soap finishes. Soap finishes afford some lubrication that facilitates laying up of rope and also lubricates the strands of rope to some extent in service.

Weaving wire is used in weaving of screens for many different applications in coal mines, sand-and-gravel pits, paper mills, chemical plants, dairy plants, oil refineries and food-processing plants. Annealing and final drawing must be carefully controlled to maintain uniform temper and finish throughout each coil or spool. Because weaving wire must be ductile, it usually is furnished in the annealed temper with a bright annealed finish, or in the soft temper with either a lime-soap finish or an oil- or grease-drawn finish.

Most types of stainless steel are available in weaving wire; the types most generally used are 302, 304, 309, 310, 316, 410 and 430. Annealed wire in the 3*xx* series commonly has a tensile strength of 655 to 860 MPa (95 to 125 ksi) and an elongation (in 50 mm or 2 in.) of 35 to 60%. Soft-temper wire, which is commonly specified for sizes over 0.75 mm (0.030 in.), averages 860

Table 5 Room temperature tensile strength of stainless steel spring wire

Diameter		Tensile strength	
mm	in.	MPa	ksi
Types 302 and 304			
Up to 0.23	Up to 0.009	2241 to 2448	325 to 355
Over 0.23 to 0.25	Over 0.009 to 0.010	2206 to 2413	320 to 350
Over 0.25 to 0.28	Over 0.010 to 0.011	2192 to 2399	318 to 348
Over 0.28 to 0.30	Over 0.011 to 0.012	2179 to 2385	316 to 346
Over 0.30 to 0.33	Over 0.012 to 0.013	2165 to 2372	314 to 344
Over 0.33 to 0.36	Over 0.013 to 0.014	2151 to 2358	312 to 342
Over 0.36 to 0.38	Over 0.014 to 0.015	2137 to 2344	310 to 340
Over 0.38 to 0.41	Over 0.015 to 0.016	2124 to 2330	308 to 338
Over 0.41 to 0.43	Over 0.016 to 0.017	2110 to 2317	306 to 336
Over 0.43 to 0.46	Over 0.017 to 0.018	2096 to 2303	304 to 334
Over 0.46 to 0.51	Over 0.018 to 0.020	2068 to 2275	300 to 330
Over 0.51 to 0.56	Over 0.020 to 0.022	2041 to 2248	296 to 326
Over 0.56 to 0.61	Over 0.022 to 0.024	2013 to 2220	292 to 322
Over 0.61 to 0.66	Over 0.024 to 0.026	2006 to 2206	291 to 320
Over 0.66 to 0.71	Over 0.026 to 0.028	1993 to 2192	289 to 318
Over 0.71 to 0.79	Over 0.028 to 0.031	1965 to 2172	285 to 315
Over 0.79 to 0.86	Over 0.031 to 0.034	1944 to 2137	282 to 310
Over 0.86 to 0.94	Over 0.034 to 0.037	1930 to 2124	280 to 308
Over 0.94 to 1.04	Over 0.037 to 0.041	1896 to 2096	275 to 304
Over 1.04 to 1.14	Over 0.041 to 0.045	1875 to 2068	272 to 300
Over 1.14 to 1.27	Over 0.045 to 0.050	1841 to 2034	267 to 295
Over 1.27 to 1.37	Over 0.050 to 0.054	1827 to 2020	265 to 293
Over 1.37 to 1.47	Over 0.054 to 0.058	1800 to 1993	261 to 289
Over 1.47 to 1.60	Over 0.058 to 0.063	1779 to 1965	258 to 285
Over 1.60 to 1.78	Over 0.063 to 0.070	1737 to 1937	252 to 281
Over 1.78 to 1.90	Over 0.070 to 0.075	1724 to 1917	250 to 278
Over 1.90 to 2.03	Over 0.075 to 0.080	1696 to 1896	246 to 275
Over 2.03 to 2.21	Over 0.080 to 0.087	1668 to 1868	242 to 271
Over 2.21 to 2.41	Over 0.087 to 0.095	1641 to 1848	238 to 268
Over 2.41 to 2.67	Over 0.095 to 0.105	1600 to 1806	232 to 262
Over 2.67 to 2.92	Over 0.105 to 0.115	1565 to 1772	227 to 257
Over 2.92 to 3.18	Over 0.115 to 0.125	1531 to 1744	222 to 253
Over 3.18 to 3.43	Over 0.125 to 0.135	1496 to 1710	217 to 248
Over 3.43 to 3.76	Over 0.135 to 0.148	1448 to 1662	210 to 241
Over 3.76 to 4.12	Over 0.148 to 0.162	1413 to 1620	205 to 235
Over 4.12 to 4.50	Over 0.162 to 0.177	1365 to 1572	198 to 228
Over 4.50 to 4.88	Over 0.177 to 0.192	1338 to 1551	194 to 225
Over 4.88 to 5.26	Over 0.192 to 0.207	1296 to 1517	188 to 220
Over 5.26 to 5.72	Over 0.207 to 0.225	1255 to 1475	182 to 214
Over 5.72 to 6.35	Over 0.225 to 0.250	1207 to 1413	175 to 205
Over 6.35 to 7.06	Over 0.250 to 0.278	1158 to 1365	168 to 198
Over 7.06 to 7.77	Over 0.278 to 0.306	1110 to 1324	161 to 192
Over 7.77 to 8.41	Over 0.306 to 0.331	1069 to 1282	155 to 186
Over 8.41 to 9.20	Over 0.331 to 0.362	1020 to 1241	148 to 180
Over 9.20 to 10.01	Over 0.362 to 0.394	979 to 1193	142 to 173
Over 10.01 to 11.12	Over 0.394 to 0.438	931 to 1138	135 to 165
Over 11.12 to 12.70	Over 0.438 to 0.500	862 to 1069	125 to 155

(continued)

Table 5 (continued)

Diameter mm	Diameter in.	Tensile strength MPa	Tensile strength ksi
Types 305 and 316			
Up to 0.25	Up to 0.010	1689 to 1896	245 to 275
Over 0.25 to 0.38	Over 0.010 to 0.015	1655 to 1862	240 to 270
Over 0.38 to 1.04	Over 0.015 to 0.041	1620 to 1827	235 to 265
Over 1.04 to 1.19	Over 0.041 to 0.047	1586 to 1723	230 to 260
Over 1.19 to 1.37	Over 0.047 to 0.054	1551 to 1758	225 to 255
Over 1.37 to 1.58	Over 0.054 to 0.062	1517 to 1724	220 to 250
Over 1.58 to 1.85	Over 0.062 to 0.072	1482 to 1689	215 to 245
Over 1.85 to 2.03	Over 0.072 to 0.080	1448 to 1655	210 to 240
Over 2.03 to 2.34	Over 0.080 to 0.092	1413 to 1620	205 to 235
Over 2.34 to 2.67	Over 0.092 to 0.105	1379 to 1586	200 to 230
Over 2.67 to 3.05	Over 0.105 to 0.120	1344 to 1551	195 to 225
Over 3.05 to 3.76	Over 0.120 to 0.148	1276 to 1482	185 to 215
Over 3.76 to 4.22	Over 0.148 to 0.166	1241 to 1448	180 to 210
Over 4.22 to 4.50	Over 0.166 to 0.177	1172 to 1379	170 to 200
Over 4.50 to 5.26	Over 0.177 to 0.207	1103 to 1310	160 to 190
Over 5.26 to 5.72	Over 0.207 to 0.225	1069 to 1276	155 to 185
Over 5.72 to 6.35	Over 0.225 to 0.250	1034 to 1241	150 to 180
Over 6.35 to 7.92	Over 0.250 to 0.312	931 to 1138	135 to 165
Over 7.92 to 12.68	Over 0.312 to 0.499	793 to 1000	115 to 145
Over 12.68	Over 0.499	Consult producer	

to 1035 MPa (125 to 150 ksi) in tensile strength and exhibits 15 to 40% elongation. For annealed wire in types 410 and 430, tensile strength averages 495 to 585 MPa (72 to 85 ksi) and elongation averages 17 to 23%.

Armature binding wire is produced in type 302 or 304 stainless steel of a composition that is balanced to produce high tensile and yield strengths and low magnetic permeability. Minimum tensile strength of 1515 MPa (220 ksi), minimum yield strength (0.2% offset) of 1170 MPa (170 ksi) and maximum permeability of 4.0 at 16 kA/m (200 oersteds) usually are specified. The wire must be strong enough to withstand the centrifugal forces encountered in use, yet ductile enough to withstand being bent sharply back on itself without cracking when a hook is formed to hold the armature wire during the binding operation. Armature binding wire is furnished on spools, and has a smooth, tightly adherent tinned coating that facilitates soldering.

Slide forming wire is produced in all standard types, particularly in types 302, 304, 316, 410 and 430. It can be produced in any temper suitable for forming any of the numerous shapes made on slide-type wireforming machines.

Wool wire is designed for production of wool by shredding. It is commonly furnished in an intermediate temper and produced to rigid standards so that it will perform satisfactorily in the wool-cutting operation. Wool wire usually is made from type 430 and has a lime-soap finish.

Reed wire is high-quality wire produced for manufacture of dents for reeds that, once assembled, are used in weaving textiles and other products. Dents are made by rolling the round reed wire into a flat section, and then machining and polishing the edges to a very smooth and accurate contour before cutting the wire into individual dents. Accuracy in size and shape are necessary because of the various processes that the wire must undergo.

Reed wire usually is made from type 430 in an intermediate temper, which must be uniform in properties throughout each coil and each shipment. The finish also must be uniform and bright.

Lashing wire is designed for lashing electric power transmission lines to support cables. Lashing wire commonly is made from type 430; it is furnished in the annealed temper with a bright finish, and it has a maximum tensile strength of 655 MPa (95 ksi) and minimum elongation of 17% in 250 mm or 10 in. It is normally furnished on coreless spools.

Cotter-pin wire is approximately half-round wire designed for fabrication of cotter pins. It is generally produced by rolling round wire between power-driven rolls, by drawing it between power-driven rolls or by drawing it through a die or turkshead. To facilitate spreading of the ends of the cotter pins, it is desirable that the flat side of the wire have a small radius rather than sharp corners at the edges.

Cotter-pin wire is commonly furnished in vibrated or hank-wound coils with the flat side of the wire facing inward. Ordinarily it is produced in the soft temper, to prevent undesirable spring-back in the legs of formed cotter pins. Usually it is furnished with a bright finish, but it is also available with a metallic coating.

Semifinished Products. Blooms, billets and slabs are hot rolled, hot forged or hot pressed to approximate cross-sectional dimensions, and generally have rounded corners. These semifinished products, as well as tube rounds, are produced in random lengths or are cut to specified lengths or to specified weights. There are no invariable criteria for distinguishing between the terms "bloom" and "billet", and they often are used interchangeably.

The nominal cross-sectional dimensions of blooms, billets and slabs are designated in inches and fractions of an inch. The size ranges commonly listed as hot rolled stainless steel blooms, billets and slabs include square sections 100 by 100 mm (4 by 4 in.) and larger, and rectangular sections at least 10 300 mm^2 (16 $in.^2$) in cross-sectional area.

Blooms, billets and slabs made of 4*xx* stainless steels that are highly hardenable (types 414, 420, 420F, 422, 431, 440A, 440B and 440C) are annealed before shipment, to prevent cracking. Other hardenable types, such as 403, 410, 416 and 416Se, also may be furnished in the annealed condition, depending on composition and size.

In general practice, blooms, billets and slabs are cut to length by hot shearing. Hot sawing and flame cutting also are used. When the end distortion or burrs normally encountered in regular mill cutting are not acceptable, ends can be prepared for subsequent operations by any method that does not leave distortion or burrs: usually, by grinding. Blooms, billets, tube rounds and slabs are surface conditioned by grinding or turning prior to being processed by hot rolling, hot forging, hot extruding or hot piercing. After being conditioned, they can be tested for internal soundness by macroetch testing or ultrasonic inspection. At the time an order is placed, producer and customer should come to an agreement regarding the manner in which testing or inspection is to be conducted and results interpreted.

Pipe, tubes and tubing are hollow products made either by piercing rounds or by rolling and welding strip. They are used for conveying gases, liquids and solids, and for divers mechanical and structural purposes. (Cylindrical forms intended for use as containers for storage and shipping purposes, and products cast to tubular shape, are not included in this category.) The number of terms used in describing sizes and other characteristics of stainless steel tubular products has grown with the industry, and in some cases terms may be difficult to define or to distinguish from one another. For example, the terms "pipe", "tubes" and "tubing" are distinguished from one another only by general usage, and not by clear-cut rules. Pipe is distinguished from tubes chiefly by the fact that it is commonly produced in relatively few standard sizes. Tubing is generally made to more exacting specifications regarding dimensions, finish, chemical composition and mechanical properties than either pipe or tubes.

Stainless steel tubular products are classified according to intended service, as described in the following paragraphs and tabular matter.

Stainless Steel Tubing for General Corrosion-Resisting Service. Straight chromium (ferritic or martensitic) types are produced in the annealed or heat treated condition, and chromium-nickel (austenitic) types are produced in the annealed or cold worked condition. Austenitic types are inherently tougher and more ductile than ferritic types for similar material conditions or tempers.

ASTM specifications A268 and A269 apply to stainless steel tubing for general service: A268 to ferritic grades and A269 to austenitic grades. Most ferritic grades also are covered by ASME SA268, which sets forth the same material requirements as ASTM A268.

Stainless steel pressure pipe is made from straight chromium and chromium-nickel types, and is governed by the following specifications:

Specifications ASTM	ASME	Description
A312	SA312	Seamless and welded pipe
A358	SA358	Electric-fusion-welded pipe for high-temperature service
A376	SA376	Seamless pipe for high-temperature central-station service
A409	...	Large-diameter welded pipe for corrosion or high-temperature service

Seamless steel pressure tubes include boiler, superheater, condenser and heat-exchanger tubes, which commonly are manufactured from chromium-nickel types; requirements are set forth in the following specifications:

Specifications ASTM	ASME	Description
A213	SA213	Ferritic and austenitic alloy seamless tubes for boilers, superheaters and heat exchangers
A249	SA249	Austenitic alloy welded tubes for boilers, superheaters, heat exchangers and condensers
A271	SA271	Austenitic alloy seamless still tubes for refinery service
A498	...	Ferritic and austenitic alloy seamless and welded tubes with integral fins

Stainless steel sanitary tubing is used extensively in the dairy and food industries, where cleanness and exceptional corrosion resistance are important surface characteristics. In many instances, even the slight amounts of corrosion that result in tarnishing or in release of a few ppm of metallic ions into the process stream are objectionable. Sanitary tubing may be polished on the outside or the inside, or both, to provide smooth, easily cleaned surfaces. Special finishes, and close dimensional tolerances to accommodate special fittings, sometimes are required. ASTM A270 is in common use for this tubing.

Stainless steel mechanical tubing is produced in round, square, rectangular and special-shape cross sections. It is used for many different applications, most of which do not require the tubing to be pressurized. Mechanical tubing is used for bushings, small cylinders, bearing parts, fittings, various types of hollow cylindrical or ringlike formed parts, and structural members such as furniture frames, machinery frames and architectural members. ASTM A511 and A554 apply to seamless and welded mechanical tubing, respectively.

Stainless steel aircraft tubing, produced from various chromium-nickel types, has many structural and hydraulic applications in aircraft construction because of its high resistance to both heat and corrosion. Work-hardened tubing can be used in high-strength applications, but it is not recommended for parts that may be exposed to certain corrosive substances or to certain combinations of corrodent and static or fluctuating stress. Low-carbon types or compositions stabilized by titanium or by niobium with or without tantalum are commonly used when welding is to be done without subsequent heat treatment.

Aircraft tubing is made to close tolerances, and with special surface finishes, special mechanical properties and stringent requirements for testing and inspection. It is used for structural components of aircraft fuselages, engine mounts, engine oil lines, landing gear and engine parts, and is finding increasing application in parts for hydraulic, fuel-injection, exhaust and heating systems.

Aircraft structural tubing is seamless and welded stainless steel tubing in sizes larger than those referred to as aircraft tubing. It is commonly used in exhaust systems (including stacks), cross headers, collector rings, various engine parts, heaters and pressurizers. Sometimes, stainless steel aircraft structural tubing is produced especially for parts that are to be machined. Stabilized types are used for welded and brazed structures.

Seamless and welded stainless steel aircraft structural tubing is made in sizes ranging from 1.6 to 125 mm (1/16 to 5 in.) in outside diameter and from 0.25 to 6.35 mm (0.010 to 0.250 in.) in wall thickness. It is ordinarily produced to the government and AMS specifications listed below. However, the U. S. government has embarked on a program of replacing MIL specifications with AMS and ASTM specifications, so the MIL specifications listed may no longer apply.

Specification	UNS number and composition and condition
Seamless tubing	
AMS 5560	S30400; 19Cr-9Ni, annealed
AMS 5561	S21900; 21Cr-6Ni-9Mn, annealed
AMS 5570	S32100; 18Cr-11Ni (Ti stabilized), annealed
AMS 5571	S34700; 18Cr-11Ni (Nb + Ta stabilized), annealed
AMS 5572	S31008; 25Cr-20Ni, annealed
AMS 5573	S31600; 17Cr-12.5Ni-2.5Mo, annealed
AMS 5574	S30908; 23Cr-13.5Ni, annealed
AMS 5578	S45500; 12.5Cr-8.5Ni-0.03 (Nb + Ta)-1.1Ti-2.0Cu, annealed
Welded tubing	
MIL-T-6737	18-8 (stabilized), annealed
AMS 5565	S30400; 19Cr-9Ni, annealed
AMS 5575	S34700; 18Cr-11Ni (Nb + Ta stabilized), annealed
AMS 5576	S32100; 18Cr-10Ni (Ti stabilized), annealed
AMS 5577	S31008; 25Cr-20Ni, annealed
Seamless and welded tubing	
MIL-T-5695	18-8, hardened (cold worked)
MIL-T-8506	S30400; 18-8, annealed
MIL-T-8686	18-8 (stabilized), annealed

Aircraft Hydraulic-Line Tubing. Stainless steel tubing is used widely in aircraft and aerospace vehicles for fuel-injection lines and hydraulic systems. Most of the tubing used for such applications is relatively small; types 304 304L, 321, 347 and 21-6-9 are most often specified. Aircraft hydraulic-line tubing must have high strength, high ductility, high fatigue resistance, high corrosion resistance and good cold working qualities. Ability to be flared for use with standard flare fittings ability to be bent without excessive distortion or fracture, and cleanness of inside surface are important requirements.

Stainless steel aircraft hydraulic. line tubing is produced in either the annealed or the cold worked (1/8 hard) condition. The 1/8 hard temper is used wherever possible to save weight. Spec.-ifications for stainless steel aircraft hydraulic-line tubing, either seamless or welded, are listed below:

Specification	UNS number, or type, and condition
MIL-T-6845	S30400, 1/8 hard
MIL-T-8504	S30400, annealed
MIL-T-8808	321 or 347, annealed
AMS 5556	S34700, annealed
AMS 5557	S32100, annealed
AMS 5560	S30400, annealed
AMS 5566	S30400, 1/8 hard

Tensile Properties

Mechanical properties of most stainless steels, especially ductility and toughness, are higher than the same properties of carbon steels. Strength and hardness can be raised by cold work for ferritic and austenitic types, and by heat treatment for precipitation-hardening and martensitic types. Certain ferritic stainless steels also can be hardened slightly by heat treatment.

Austenitic Types. Basic room-temperature properties of standard austenitic stainless steels and of several nonstandard austenitic stainless steels are given in Tables 6 and 7. Certain austenitic stainless steels—the so-called "metastable" types—can develop higher strengths and hardnesses than other "stable" types for a given amount of cold work. In metastable austenitic stainless steels, deformation triggers transformation of austenite to martensite. The effect of this transformation on strength is illustrated in Fig. 5, which compares the stress-strain curve for stable type 304 with that for metastable type 301. The parabolic shape of the curve for type 304 indicates that strain hardening occurs throughout the duration of the application of stress, but that the amount of strain hardening for a given increment of stress decreases as stress increases. On the other hand, the stress-strain relationship for type 301 indicates an accelerated rate of strain hardening after an initial increment of 10 to 15% plastic strain. This accelerated strain hardening is a direct result of deformation-induced transformation to martensite.

Fig. 5 Typical stress-strain curves for types 301 and 304 stainless steel

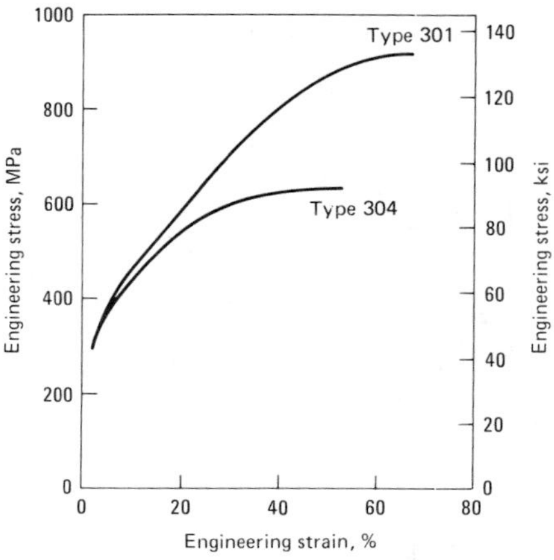

Ferritic types of stainless steel are defined as those that contain at least 10% chromium and that have microstructures of ferrite plus carbides. These steels are less ductile than the austenitic types. Basic room-temperature mechanical properties of ferritic stainless steels are given in Table 8. Strength is enhanced only moderately by cold working.

Martensitic types are iron-chromium steels with or without small additions of other alloying elements. They are ferritic in the annealed condition, but are martensitic after rapid cooling in air or a liquid medium from above the critical temperature. Steels in this group usually contain no more than 14% Cr—except types 440A, 440B and 440C, which contain 16 to 18% Cr—and an amount of carbon sufficient to promote hardening. They may or may not contain other elements; if they do, the total concentration usually is no more than 2 to 3%. Martensitic stainless steels may be hardened and tempered in the same manner as alloy steels. They have excellent strength and are magnetic. Basic room-temperature properties of the martensitic types are given in Table 9 and Fig. 6.

Martensitic stainless steels harden when cooled off the mill after hot pro-

Table 6 Minimum room-temperature mechanical properties of austenitic stainless steels

Product form(a)	Condition	Tensile strength MPa	Tensile strength ksi	0.2% yield strength MPa	0.2% yield strength ksi	Elongation, %	Reduction in area, %	Hardness, HRB	ASTM specification
Type 301 (UNS S30100)									
B, W, P, Sh, St	Annealed	515	75	205	30	40	...	88 max	A167
Sh, St	1/4 hard	860	125	515	75	25	...	...	A177
Sh, St	1/2 hard	1030	150	760	110	18	...	...	A177
Sh, St	3/4 hard	1210	175	930	135	12	...	...	A177
Sh, St	Full hard	1280	185	965	140	9	...	...	A177
Type 302 (UNS S30200)									
B	Hot finished and annealed	515	75	205	30	40	50	...	A276
B	Cold finished(b) and annealed	620	90	310	45	30	40	...	A276
B	Cold finished(c) and annealed	515	75	205	30	30	40	...	A276
W	Annealed	515	75	205	30	...	...	...	A580
W	Cold finished	620	90	310	45	...	...	...	A580
P, Sh, St	Annealed	515	75	205	30	40	...	88 max	A167
P, Sh, St	High tensile, grade B	585	85	310	45	40	...	...	A666
P, Sh, St	High tensile, grade C	860	125	515	75	...	...	...	A666
P, Sh, St	High tensile, grade D	1030	150	760	110	...	...	...	A666
B, W	High tensile(d)	2240	325	...	...	...	...	...	A313
Type 302B (UNS S30215)									
B	Hot finished and annealed	515	75	205	30	40	50	...	A276
B	Cold finished(b) and annealed	620	90	310	45	30	40	...	A276
B	Cold finished(c) and annealed	515	75	205	30	30	40	...	A276
W	Annealed	515	75	205	30	...	...	...	A580
W	Cold finished	620	90	310	45	...	...	...	A580
P, Sh, St	Annealed	515	75	205	30	...	...	...	A167
Type 302Cu (UNS S30430)									
B	Annealed	450 to 585	65 to 85	...	...	...	...	...	A493
B	Lightly drafted	485 to 620	70 to 90	...	...	...	...	...	A493
W(e)	Annealed	485 to 620	70 to 90	...	...	...	...	...	A493
W(e)	Lightly drafted	485 to 620	70 to 90	...	...	...	...	...	A493
W(f)	Annealed	485 to 690	70 to 100	...	...	...	...	...	A493
W(f)	Lightly drafted	520 to 725	75 to 105	...	...	...	...	...	A493

(continued)

(a) B, bar; W, wire; P, plate; Sh, sheet; St, strip. (b) Up to 13 mm (0.5 in.) thick. (c) Over 13 mm (0.5 in.) thick. (d) Depending on size and amount of cold reduction. (e) 4 mm (0.156 in.) in diameter and over. (f) Under 4 mm (0.156 in.) in diameter. (g) Values given are typical. (h) Not a basis for acceptance or rejection.

Table 6 (continued)

Product form(a)	Condition	Tensile strength MPa	Tensile strength ksi	0.2% yield strength MPa	0.2% yield strength ksi	Elongation, %	Reduction in area, %	Hardness, HRB	ASTM specification
Types 303 (UNS S30300) and 303Se (UNS S30323)									
B	Annealed	585(g)	85(g)	240(g)	35(g)	50(g)	55(g)	...	A581
W	Annealed	585 to 860	85 to 125	...	...	...	...	...	A581
W	Cold worked	790 to 1000	115 to 145	...	...	...	...	...	A581
Type 304 (UNS S30400)									
B	Hot finished and annealed	515	75	205	30	40	50	...	A276
B	Cold finished(b) and annealed	620	90	310	45	30	40	...	A276
B	Cold finished(c) and annealed	515	75	205	30	30	40	...	A276
W	Annealed	515	75	205	30	...	...	...	A580
W	Cold finished	620	90	310	45	...	...	...	A580
P, Sh, St	Annealed	515	75	205	30	40	...	88 max	A167
Sh, St	High tensile, grade B	550	80	310	45	...	...	...	A666
Sh, St	High tensile, grade C	860	125	515	75	...	...	...	A666
Sh, St	High tensile, grade D	1030	150	690	110	...	...	...	A666
B, W	High tensile(d)	2240	325	...	...	...	...	...	A313
Type 304L (UNS S30403)									
B	Hot finished and annealed	480	70	170	25	40	50	...	A276
B	Cold finished(b) and annealed	620	90	310	45	30	40	...	A276
B	Cold finished(c) and annealed	480	70	170	25	30	40	...	A276
W	Annealed	480	70	170	25	...	...	...	A580
W	Cold finished	620	90	310	45	...	...	...	A580
P, Sh, St	Annealed	480	70	170	25	40	...	88 max	A167
Types 304N (UNS S30451) and 316N (UNS S31651)									
B	Annealed	550	80	240	35	30	...	...	A276
Type 304LN									
B	Annealed	515	75	205	30	...	...	...	...

(continued)

(a) B, bar; W, wire; P, plate; Sh, sheet; St, strip. (b) Up to 13 mm (0.5 in.) thick. (c) Over 13 mm (0.5 in.) thick. (d) Depending on size and amount of cold reduction. (e) 4 mm (0.156 in.) in diameter and over. (f) Under 4 mm (0.156 in.) in diameter. (g) Values given are typical. (h) Not a basis for acceptance or rejection.

Table 6 (continued)

Product form(a)	Condition	Tensile strength MPa	Tensile strength ksi	0.2% yield strength MPa	0.2% yield strength ksi	Elongation, %	Reduction in area, %	Hardness, HRB	ASTM specification
Type 305 (UNS S30500)									
B	Hot finished and annealed	515	75	205	30	40	50	...	A276
B	Cold finished(b) and annealed	260	90	310	45	30	40	...	A276
B	Cold finished(c) and annealed	515	75	205	30	30	40	...	A276
W	Annealed	515	75	205	30	...	...	...	A580
W	Cold finished	620	90	310	45	...	...	...	A580
P, Sh, St	Annealed	480	70	170	25	40	...	88 max	A167
B, W	High tensile(d)	1690	245	...	...	...	...	...	...
Types 308 (UNS S30800), 321 (UNS S32100), 347 (UNS S34700) and 348 (UNS S34800)									
B	Hot finished and annealed	515	75	205	30	40	50	...	A276
B	Cold finished(b) and annealed	620	90	310	45	30	40	...	A276
B	Cold finished(c) and annealed	515	75	205	30	30	40	...	A276
W	Annealed	515	75	205	30	...	...	...	A580
W	Cold finished	620	90	310	45	...	...	...	A580
P, Sh, St	Annealed	515	75	205	30	40	...	88 max	A167
Type 308L									
B	Annealed	550(g)	80(g)	207(g)	30(g)	60(g)	70(g)	...	...
Types 309 (UNS S30900), 309S (UNS S30908), 310 (UNS S31000) and 310S (UNS S31008)									
B	Hot finished and annealed	515	75	205	30	40	50	...	A276
B	Cold finished(b) and annealed	620	90	310	45	30	40	...	A276
B	Cold finished(c) and annealed	515	75	205	30	30	40	...	A276
W	Annealed	515	75	205	30	...	...	...	A580
W	Cold finished	620	90	310	45	...	...	...	A580
P, Sh, St	Annealed	515	75	205	30	40	...	95 max	A167
Type 312									
Weld metal		655	95			20			MIL-E-19933
Type 314 (UNS S31400)									
B	Hot finished and annealed	515	75	205	30	40	50	...	A276
B	Cold finished(b) and annealed	620	90	310	45	30	40	...	A276
B	Cold finished(c) and annealed	515	75	205	30	30	40	...	A276
W	Annealed	515	75	205	30	...	...	...	A580
W	Cold finished	620	90	310	45	...	...	...	A580

(continued)

(a) B, bar; W, wire; P, plate; Sh, sheet; St, strip. (b) Up to 13 mm (0.5 in.) thick. (c) Over 13 mm (0.5 in.) thick. (d) Depending on size and amount of cold reduction. (e) 4 mm (0.156 in.) in diameter and over. (f) Under 4 mm (0.156 in.) in diameter. (g) Values given are typical. (h) Not a basis for acceptance or rejection.

Table 6 (continued)

Product form(a)	Condition	Tensile strength MPa	Tensile strength ksi	0.2% yield strength MPa	0.2% yield strength ksi	Elongation, %	Reduction in area, %	Hardness, HRB	ASTM specification
Type 316 (UNS S31600)									
B	Hot finished and annealed	515	75	205	30	40	50	...	A276
B	Cold finished(b) and annealed	620	90	310	45	30	40	...	A276
B	Cold finished(c) and annealed	515	75	205	30	30	40	...	A276
W	Annealed	515	75	205	30	...	...	...	
W	Cold finished	620	90	310	45	...	...	...	A580
P, Sh, St	Annealed	515	75	205	30	40	...	95 max	A580
B, W	High tensile(d)	1690	245	...	...	...	...	...	A167
Type 316F (UNS S31620)									
B	Annealed	585(g)	85(g)	240(g)	35(g)	40(g)	55(g)	...	
Type 316L (UNS S31603)									
B	Hot finished and annealed	480	70	170	25	40	50	...	A276
B	Cold finished(b) and annealed	620	90	310	45	30	40	...	A276
B	Cold finished(c) and annealed	480	70	170	25	30	40	...	A276
W	Annealed	480	70	170	25	...	...	...	A580
W	Cold finished	620	90	310	45	...	...	...	A580
Type 316LN									
B	Annealed	515(g)	75(g)	205(g)	30(g)	60(g)	70(g)	...	...
Type 317 (UNS S31700)									
B	Hot finished and annealed	515	75	205	30	40	50	...	A276
B	Cold finished(b) and annealed	620	90	310	45	30	40	...	A276
B	Cold finished(c) and annealed	515	75	205	30	30	40	...	A276
W	Annealed	515	75	205	30	...	...	...	A580
W	Cold finished	620	90	310	45	...	...	...	A580
P, Sh, St	Annealed	515	75	205	30	35	...	95 max	A167
Type 317L (UNS S31703)									
B	Annealed	585(g)	85(g)	240(g)	35(g)	55(g)	65(g)	85 max(g)	
P, Sh, St	Annealed	515	75	205	30	35	...	95 max	A167
Type 317LM									
B, P, Sh, St	Annealed	515	75	205	30	35	50	95 max	...

(continued)

(a) B, bar; W, wire; P, plate; Sh, sheet; St, strip. (b) Up to 13 mm (0.5 in.) thick. (c) Over 13 mm (0.5 in.) thick. (d) Depending on size and amount of cold reduction. (e) 4 mm (0.156 in.) in diameter and over. (f) Under 4 mm (0.156 in.) in diameter. (g) Values given are typical. (h) Not a basis for acceptance or rejection.

Table 6 (continued)

Product form(a)	Condition	Tensile strength MPa	Tensile strength ksi	0.2% yield strength MPa	0.2% yield strength ksi	Elongation, %	Reduction in area, %	Hardness, HRB	ASTM specification
Type 329 (UNS S32900)									
B	Annealed	724(g)	105(g)	550(g)	80(g)	25(g)	50(g)	...	
Type 330 (UNS N08330)									
B	Annealed	480	70	210	30	30	...	...	B511
P, Sh, St	Annealed	480	70	210	30	30	...	75 to 85(h)	B536
Type 330HC									
B, W, St	Annealed	585(g)	85(g)	290(g)	42(g)	45(g)	65(g)	...	...
Type 332									
B, W, Sh, St	Annealed	550(g)	80(g)	240(g)	35(g)	45(g)	70(g)	...	...
Types 384 (UNS S38400) and 385 (UNS S38500)									
B	Annealed	415 to 550	60 to 80	...	...	...	...	...	A493
B	Lightly drafted	450 to 585	65 to 85	...	...	...	...	...	A493
W(e)	Annealed	450 to 585	65 to 85	...	...	...	...	...	A493
W(e)	Lightly drafted	485 to 620	70 to 90	...	...	...	...	...	A493
W(f)	Annealed	450 to 655	65 to 95	...	...	...	...	...	A493
W(f)	Lightly drafted	485 to 690	70 to 100	...	...	...	...	...	A493
904L (UNS N08904)									
B, P, Sh, St	Annealed	490	71	220	31	35	...	95 max	B625
AL-6X (UNS N08366)									
Sh, St	Annealed	515	75	205	30	30	...	...	B676
18-18-2 (UNS S38100)									
P, Sh, St	Annealed	515	75	205	30	40	...	96 max	A167
Crutemp 25									
P, Sh, St	Annealed	615(g)	89(g)	275(g)	40(g)	40(g)	...	...	
JS-700 (UNS N08700)									
P, Sh, St	Annealed	550	80	205	30	30	40	...	B599
JS-777									
B, P, Sh, St	Annealed	550	80	240	35	30	40	95 max	
20Cb-3 (UNS N08020)									
B	Annealed	585	85	240	35	30	50	...	B473
Shapes	Cold finished and annealed	585	85	240	35	15	50	...	B473
W	Annealed	620 to 825	90 to 120	...	...	...	...	...	B473
P, Sh, St	Annealed	585	85	275	40	30	...	95 max	B463

(a) B, bar; W, wire; P, plate; Sh, sheet; St, strip. (b) Up to 13 mm (0.5 in.) thick. (c) Over 13 mm (0.5 in.) thick. (d) Depending on size and amount of cold reduction. (e) 4 mm (0.156 in.) in diameter and over. (f) Under 4 mm (0.156 in.) in diameter. (g) Values given are typical. (h) Not a basis for acceptance or rejection.

Table 7 Minimum mechanical properties of high-nitrogen austenitic stainless steels

Product form(a)	Condition	Tensile strength MPa	Tensile strength ksi	0.2% yield strength MPa	0.2% yield strength ksi	Elongation, %	Reduction in area, %	Hardness, HRB	ASTM specification
Type 201 (UNS S20100)									
B	Annealed	515	75	275	40	40	45	...	A276
W, P, Sh, St	Annealed	655	95	310	45	40	...	...	A276, A412
Sh, St	1/4 hard	860	125	515	75	20	...	...	A412
Sh, St	1/2 hard	1030	150	760	110	10	...	...	A412
Sh, St	3/4 hard	1210	175	930	135	7	...	...	A412
Sh, St	Full hard	1280	185	965	140	5	...	...	A412
Type 202 (UNS S20200)									
B	Annealed	515	75	275	40	40	...	...	A276
W, P, Sh, St	Annealed	655	95	310	45	40	...	...	A412
Sh, St	1/4 hard	860	125	515	75	12	...	...	A412
Sh, St	1/2 hard	1030	150	760	110	10	...	...	A666
Type 205 (UNS S20500)									
P	Annealed(b)	830(b)	120(b)	475(b)	69(b)	58(b)	62(b)	98 max(b)	
Type 216 (UNS S21600)									
Sh, St	Annealed	690	100	415	60	40	...	100 max	A240
P	Annealed	620	90	345	50	40	...	100 max	A240
Type 304N (UNS S30451)									
B	Annealed	550	80	240	35	30	...	...	A276
P, Sh, St	Annealed	550	80	240	35	30	...	88 max	A240
Type 304HN (UNS S30452)									
B	Annealed	620	90	345	50	30	50		
Sh, St	Annealed	620	90	345	50	30	...	100 max	A240
P	Annealed	585	85	275	40	30	...	100 max	A240
Type 316N (UNS S31651)									
B	Annealed	550	80	240	35	30	...	...	A276
P, Sh, St	Annealed	550	80	240	35	30	...	95 max	A240
Nitronic 32 (UNS S24100)									
B	Annealed	690	100	380	55	30	50	...	A276
W	Annealed	690	100	380	55	30	50	...	A580
Nitronic 33 (UNS S24000)									
B	Annealed	690	100	380	55	30	50	...	A276
W	Annealed	690	100	380	55	30	50	...	A580
Sh, St	Annealed	690	100	415	60	40	...	...	A412
P	Annealed	690	100	380	55	40	...	...	A412
Nitronic 40 (UNS S21900)									
B	Annealed	550	80	345	50	45	...	...	A276
W	Annealed	620	90	345	50	45	60	...	A580
Sh, St	Annealed	690	100	415	60	40	...	...	A412
Sh, St	10% cold rolled	895	130	795	115	15	...	...	A412
P	Annealed	620	90	345	50	45	...	...	A412
Nitronic 50 (UNS S20910)									
B	Annealed	690	100	380	55	35	55	...	A276
W	Annealed	690	100	380	55	35	55	...	A580
Sh, St	Annealed	825	120	515	75	30	...	...	A412
P	Annealed	690	100	380	55	35	...	...	A412

(continued)

(a) B, bar; W, wire; P, plate; Sh, sheet; St, strip. (b) Typical values.

Table 7 (continued)

Product form(a)	Condition	Tensile strength MPa	Tensile strength ksi	0.2% yield strength MPa	0.2% yield strength ksi	Elongation, %	Reduction in area, %	Hardness, HRB	ASTM specification
Nitronic 60 (UNS S21800)									
B	Annealed	655	95	345	50	35	55	...	A276
W	Annealed	655	95	345	50	35	55	...	A580
18-18 Plus (UNS S28200)									
B	Annealed	825(b)	120(b)	450(b)	65(b)	60(b)	70(b)	95 min(b)	
B	Annealed	760	110	415	60	35	55	...	A276
W	Annealed	760 to 930	110 to 135	...	...	...	...	...	A493
Tenelon (UNS S21400)									
Sh	Annealed	860	125	485	70	40	...	...	A240
St	Annealed	725	105	380	55	40	...	...	A240

(a) B, bar; W, wire; P, plate; Sh, sheet; St, strip. (b) Typical values.

Table 8 Minimum mechanical properties of ferritic stainless steels

Product form(a)	Condition	Tensile strength MPa	Tensile strength ksi	0.2% yield strength MPa	0.2% yield strength ksi	Elongation, %	Reduction in area, %	Hardness, HRB	ASTM specification
Type 405 (UNS S40500)									
W	Annealed	480	70	275	40	20	45	...	A580
W	Annealed, cold finished	480	70	275	40	16	45	...	A580
P, Sh, St	Annealed	415	60	170	25	20	...	88 max	A176
Type 409 (UNS S40900)									
B	Annealed	450(b)	65(b)	240(b)	35(b)	25(b)	...	75 max(b)	
P, Sh, St	Annealed	415	60	205	30	22(c)	...	80 max	A176
Type 429 (UNS S42900)									
B	Annealed	490(b)	71(b)	310(b)	45(b)	30(b)	65(b)	...	
P, Sh, St	Annealed	450	65	205	30	22(c)	...	88 max	A176
Type 430 (UNS S43000)									
B	Annealed, hot finished	480	70	275	40	20	45	...	A276
B	Annealed, cold finished	480	70	275	40	16	45	...	A276
W	Annealed	480	70	275	40	20	45	...	A580
W	Annealed, cold finished	480	70	275	40	16	45	...	A580
P, Sh, St	Annealed	450	65	205	30	22(c)	...	88 max	A176
Type 430F (UNS S43020)									
W	Annealed	585 to 860	85 to 125	...	...	...	...	...	A581
Type 430Ti (UNS S43036)									
B	Annealed	515(b)	75(b)	310(b)	45(b)	30(b)	65(b)	...	...
Type 434 (UNS S43400)									
W	Annealed	545(b)	79(b)	415(b)	60(b)	33(b)	78(b)	90 max(b)	...
Sh	Annealed	530(b)	77(b)	365(b)	53(b)	23(b)	...	83 max(b)	...
Type 436 (UNS S43600)									
Sh, St	Annealed	530(b)	77(b)	365(b)	53(b)	23(b)	...	83 max(b)	...
Type 442 (UNS S44200)									
B	Annealed	550(b)	80(b)	310(b)	45(b)	20(b)	40(b)	90 max(b)	...
P, Sh, St	Annealed	515	75	275	40	20	...	95 max	A176

(continued)

(a) B, bar; W, wire; P, plate; Sh, sheet; St, strip. (b) Typical values. (c) 20% reduction for 1.3 mm (0.050 in.) in thickness and under.

Table 8 (continued)

Product form(a)	Condition	Tensile strength MPa	ksi	0.2% yield strength MPa	ksi	Elongation, %	Reduction in area, %	Hardness, HRB	ASTM specification
Type 444 (UNS S44400)									
P, Sh, St	Annealed	415	60	275	40	20	...	95 max	A176
Type 446 (UNS S44600)									
B	Annealed, hot finished	480	70	275	40	20	45	...	A276
B	Annealed, cold finished	480	70	275	40	16	45	...	A276
W	Annealed	480	70	275	40	20	45	...	A580
W	Annealed, cold finished	480	70	275	40	16	45	...	A580
P, Sh, St	Annealed	515	75	275	40	20	...	95 max	A176
18 SR									
Sh, St	Annealed	620(b)	90(b)	450(b)	65(b)	25(b)	...	90 min(b)	
E-Brite 26-1 (UNS S44625)									
B	Annealed, hot finished	450	65	275	40	20	45	...	A276
B	Annealed, cold finished	450	65	275	40	16	45	...	A276
P, Sh, St	Annealed	450	65	275	40	22(c)	...	90 max	A176
26-1 Ti (UNS S44626)									
P, Sh, St	Annealed	470	68	310	45	20	...	95 max	A176
Monit (UNS S44635)									
B, P, Sh, St	Annealed	650	94	550	80	20	...	100 max	A176
Sea-cure/SC-1 (UNS S44600)									
B, P, Sh, St	Annealed	550	80	380	55	20	...	100 max	A176
29-4 (UNS S44700)									
B, P, Sh, St	Annealed	550	80	415	60	20	...	98 max	A176
29-4-2 (UNS S44800)									
B, P, Sh, St	Annealed	550	80	415	60	20	...	98 max	A176

(a) B, bar; W, wire; P, plate; Sh, sheet; St, strip. (b) Typical values. (c) 20% reduction for 1.3 mm (0.050 in.) in thickness and under.

Fig. 6 Typical hardness of selected martensitic stainless steels

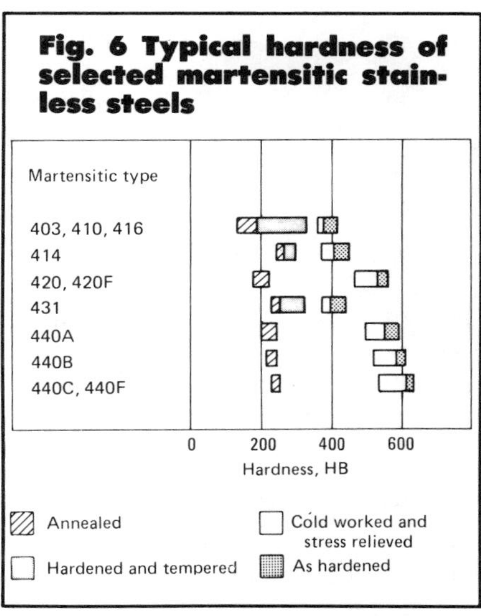

cessing, so they often are given a process anneal at 650 to 760 °C (1200 to 1400 °F) for about four hours. Process annealing differs from full annealing, which is done by heating at 815 to 870 °C (1500 to 1600 °F), cooling in the furnace at a rate of 40 to 55 °C/h (75 to 100 °F/h) to about 540 °C (1000 °F) and then cooling in air to room temperature. Occasionally, martensitic types are purchased in the tempered condition; this condition is achieved by cooling directly off the mill to harden the steel and then reheating to a tempering temperature of 540 to 650 °C (1000 to 1200 °F), or by reheating the steel to a hardening temperature of 1010 to 1065 °C (1850 to 1950 °F), cooling it, and then tempering it. The influence of tempering temperature on properties of hardened martensitic stainless steels is shown in Fig. 7. In heat treating of martensitic stainless steels, temperatures up to about 480 °C (900 °F) are referred to as stress-relieving temperatures because little change in tensile properties occurs upon heating hardened material to these temperatures. Temperatures of 540 to 650 °C are referred to as tempering temperatures, and temperatures of 650 to 760 °C are called annealing temperatures.

Precipitation-hardening types generally are heat treated to final properties by the fabricator. Table 10 summarizes the minimum properties that can be expected in both material as received from the mill and material that has been properly heat treated.

The precipitation-hardening stainless steels are of two general classes:

Table 9 Minimum mechanical properties of martensitic stainless steels

Product form(a)	Condition	Tensile strength MPa	Tensile strength ksi	0.2% yield strength MPa	0.2% yield strength ksi	Elongation, %	Reduction in area, %	Rockwell hardness	ASTM specification
Type 403 (UNS S40300)									
B	Annealed, hot finished	485	70	275	40	20	45	...	A276
B	Annealed, cold finished	485	70	275	40	16	45	...	A276
B	Intermediate temper, hot finished	690	100	550	80	15	45	...	A276
B	Intermediate temper, cold finished	690	100	550	80	12	40	...	A276
B	Hard temper, hot finished	825	120	620	90	12	40	...	A276
B	Hard temper, cold finished	825	120	620	90	12	40	...	A276
W	Annealed	485	70	275	40	20	45	...	A580
W	Annealed, cold finished	485	70	275	40	16	45	...	A580
W	Intermediate temper, cold finished	690	100	550	80	12	40	...	A580
W	Hard temper, cold finished	825	120	620	90	12	40	...	A580
P, Sh, St	Annealed	485	70	205	30	25(b)	...	88 HRB max	A176
Type 404 (UNS S40400)									
B	Tempered 260 °C (500 °F)	1120(c)	162(c)	910(c)	132(c)	15(c)	50(c)	35 HRC(c)	...
B	Tempered 593 °C (1100 °F)	745(c)	108(c)	655(c)	95(c)	23(c)	70(c)	20 HRC(c)	...
Type 410 (UNS S41000)									
B	Annealed, hot finished	485	70	275	40	20	45	...	A276
B	Annealed, cold finished	485	70	275	40	16	45	...	A276
B	Intermediate temper, hot finished	690	100	550	80	15	45	...	A276
B	Intermediate temper, cold finished	690	100	550	80	12	40	...	A276
B	Hard temper, hot finished	825	120	620	90	12	40	...	A276
B	Hard temper, cold finished	825	120	620	90	12	40	...	A276
W	Annealed	485	70	275	40	20	45	...	A580
W	Annealed, cold finished	485	70	275	40	16	45	...	A580
W	Intermediate temper, cold finished	690	100	550	80	12	40	...	A580
W	Hard temper, cold finished	825	120	620	90	12	40	...	A580
P, Sh, St	Annealed	450	65	205	30	22(b)	...	95 HRB max	A176
Type 410S (UNS S41008)									
P, Sh, St	Annealed	415	60	205	30	22	...	95 HRB max	A176

(continued)

(a) B, bar; W, wire; P, plate; Sh, sheet; St, strip. (b) 20% elongation for 1.3 mm (0.050 in.) and under in thickness. (c) Typical values. (d) Heat treated for high-temperature service.

Table 9 (continued)

Product form(a)	Condition	Tensile strength MPa	Tensile strength ksi	0.2% yield strength MPa	0.2% yield strength ksi	Elongation, %	Reduction in area, %	Rockwell hardness	ASTM specification
Type 410Cb (UNS S41040)									
B	Annealed, hot finished	485	70	275	40	13	45	...	A276
B	Annealed, cold finished	485	70	275	40	12	35	...	A276
B	Intermediate temper, hot finished	860	125	690	100	13	45	...	A276
B	Intermediate temper, cold finished	860	125	690	100	12	35	...	A276
Type 414 (UNS S41400)									
B	Intermediate temper, hot finished	795	115	620	90	15	45	...	A276
B	Intermediate temper, cold finished	795	115	620	90	15	45	...	A276
W	Annealed, cold finished	1030 max	150 max	...	...	...	...	...	A580
Type 414L									
B	Annealed	795(c)	115(c)	550(c)	80(c)	20(c)	60(c)	...	...
Types 416 (UNS S41600) and 416Se (UNS S41623)									
W	Annealed	585 to 860	85 to 125	...	...	...	...	...	A581
W	Intermediate temper	795 to 1000	115 to 145	...	...	...	...	...	A581
W	Hard temper	965 to 1210	140 to 175	...	...	...	...	...	A581
Type 416 Plus X									
B	Annealed	515	75	275	40	30	60	...	...
Type 418 (UNS S41800)									
B	Tempered 260 °C (500 °F)	1450(c)	210(c)	1210(c)	175(c)	18(c)	52(c)	...	...
B	Tempered 649 °C (1200 °F)	930(c)	135(c)	725(c)	105(c)	20(c)	60(c)	...	...
Type 420 (UNS S42000)									
B	Tempered 204 °C (400 °F)	1720	250	1480(c)	215(c)	8(c)	25(c)	52 HRC(c)	...
W	Annealed, cold finished	860 max	125 max	...	...	...	...	...	A580
Type 422 (UNS S42200)									
B	Intermediate and hard tempers(d)	965	140	760	110	13	30	...	A565
Type 431 (UNS S43100)									
B	Tempered 260 °C (500 °F)	1370(c)	198(c)	1030(c)	149(c)	16(c)	55(c)	...	...
B	Tempered 593 °C (1100 °F)	965(c)	140(c)	795(c)	115(c)	19(c)	57(c)	...	...
W	Annealed, cold finished	965 max	140 max	...	...	...	...	...	A580
Type 440A (UNS S44002)									
B	Annealed	725(c)	105(c)	415(c)	60(c)	20(c)	...	95 HRB(c)	...
B	Tempered 316 °C (600 °F)	1790(c)	260(c)	1650(c)	240(c)	5(c)	20(c)	51 HRC(c)	...
W	Annealed, cold finished	965 max	140 max	...	...	...	...	...	A580

(continued)

(a) B, bar; W, wire; P, plate; Sh, sheet; St, strip. (b) 20% elongation for 1.3 mm (0.050 in.) and under in thickness. (c) Typical values. (d) Heat treated for high-temperature service.

Table 9 (continued)

Product form(a)	Condition	Tensile strength MPa	Tensile strength ksi	0.2% yield strength MPa	0.2% yield strength ksi	Elongation, %	Reduction in area, %	Rockwell hardness	ASTM specification
Type 440B (UNS S44003)									
B	Annealed	740(c)	107(c)	425(c)	62(c)	18(c)	...	96 HRB(c)	...
B	Tempered 316 °C (600 °F)	1930(c)	280(c)	1860(c)	270(c)	3(c)	15(c)	55 HRC(c)	...
W	Annealed, cold finished	965 max	140 max	...	...	...	...	...	A580
Type 440C (UNS S44004)									
B	Annealed	760(c)	110(c)	450(c)	65(c)	14(c)	...	97 HRB(c)	...
B	Tempered 316 °C (600 °F)	1970(c)	285(c)	1900(c)	275(c)	2(c)	10(c)	57 HRC(c)	...
W	Annealed, cold finished	965 max	140 max	...	...	...	...	...	A580
Type 501 (UNS S50100)									
B, P	Annealed	485(c)	70(c)	205(c)	30(c)	28(c)	65(c)	...	...
B, P	Tempered (1000 °F)	1210(c)	175(c)	965(c)	140(c)	15(c)	50(c)	...	...
Type 502 (UNS S50200)									
B, P	Annealed	485(c)	70(c)	205(c)	30(c)	30(c)	70(c)	...	...

(a) B, bar; W, wire; P, plate; Sh, sheet; St, strip. (b) 20% elongation for 1.3 mm (0.050 in.) and under in thickness. (c) Typical values. (d) Heat treated for high-temperature service.

single-treatment alloys and double-treatment alloys. Single-treatment alloys, such as Custom 450, 17-4 PH and 15-5 PH, are solution annealed at about 1040 °C (1900 °F) to dissolve the hardening agent. On cooling to room temperature, the structure transforms to martensite that is supersaturated with respect to the hardening agent. A single tempering treatment at about 480 to 620 °C (900 to 1150 °F) is all that is required to precipitate a secondary phase to strengthen the alloy. As shown in Table 10, different tempering temperatures within this range produce different properties.

Double-treatment alloys such as 17-7 PH are solution treated at about 1040 °C and then water quenched to retain the hardening agent in solution in an austenitic structure. The austenite is "conditioned" by heating to 760 °C (1400 °F) to precipitate carbides and thereby unbalance the austenite so that it transforms to martensite on cooling to a temperature below 15 °C (60 °F); this treatment produces condition T. Alternatively, the austenite may be conditioned at a higher temperature, 925 °C (1700 °F), where fewer carbides precipitate, and then transformed to martensite by cooling to room temperature followed by refrigerating to −75 °C (−100 °F); this treatment produces condition R. Transformation also can be effected by severe cold work (about 60 to 70% reduction), and such treatment produces condition C. Once the structure has been transformed to martensite by one of these three processes, tempering at 480 to 620 °C (900 to 1150 °F) will induce precipitation of a secondary metallic phase, which will strengthen the alloy. Properties developed by typical TH, RH and CH treatments are given in Table 10 for 17-7 PH.

Notch Toughness and Transition Temperature

Notched bar impact testing of stainless steels is likely to show a wide scatter in test results, regardless of type or test conditions. Because of this wide scatter, only general behavior of the different classes can be described.

Austenitic types have good notched bar impact resistance. Charpy impact energies of 135 J (100 ft·lb) or greater are typical of all types at room temperature. Cryogenic temperatures have little or no effect on notch toughness; ordinarily, austenitic stainless steels maintain values exceeding 135 J even at very low temperatures. On the other hand, cold work lowers the resistance to impact at all temperatures.

Martensitic stainless steels exhibit a decreasing resistance to impact with decreasing temperature, and the fracture appearance changes from a ductile mode at mildly elevated temperatures to a brittle mode at subzero temperatures. This fracture transition is characteristic of martensitic materials, although for martensitic stainless steels the upper-shelf energy is greater than for most other martensitic steels. Both the upper-shelf energy and the lower-shelf energy are not greatly influenced by heat treatment in martensitic stainless steels. However, the temperature range over which transition occurs is affected by heat treatment, minor variations in composition, and prior cold work. Heat treatments that result in high hardness move the transition range to higher temperatures, and those that result in low hardness move the transition range to lower temperatures. As indicated in Fig. 8, transition generally occurs in the range −75 to +95 °C (−100 to +200 °F), which is the temperature range in which martensitic stainless steels are ordinarily used. Consequently, it may be necessary to thoroughly investigate fracture behavior before specifying a martensitic stainless steel for a particular application.

Fracture-toughness data are not

Fig. 7 Variation of tensile properties and hardness with tempering temperature for three martensitic stainless steels

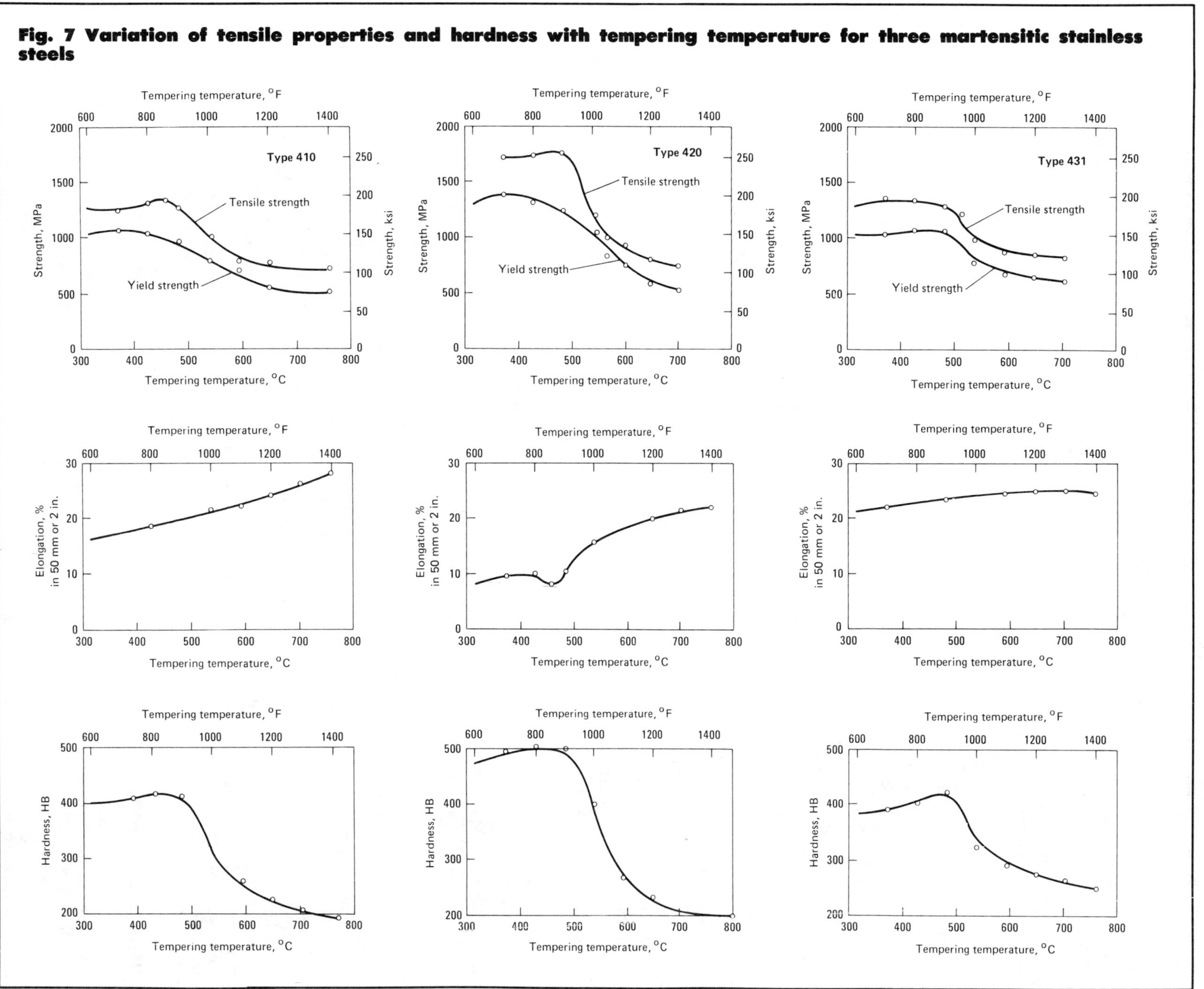

Table 10 Minimum mechanical properties of precipitation-hardening stainless steels

Product form(a)	Condition	Tensile strength MPa	Tensile strength ksi	Yield strength MPa	Yield strength ksi	Elongation, %	Reduction in area(b), %	Hardness(c), HRC Min	Hardness(c), HRC Max
PH 13-8 Mo (UNS S13800)									
B, P, Sh, St	H950	1520	220	1410	205	6-10(c)	45	45	...
B, P, Sh, St	H1000	1380	200	1310	190	6-10(c)	45	43	...
15-5 PH (UNS S15500) and 17-4 PH (UNS S17400)									
B, P, Sh, St	H900	1310	190	1170	170	10(d)	35(e)	40	48
B, P, Sh, St	H925	1170	170	1070	155	10(d)	38(e)	38	47(d)
B, P, Sh, St	H1025	1070	155	1000	145	12(d)	45(e)	35(d)	42(d)
B, P, Sh, St	H1075	1000	145	860	125	13(d)	45(e)	32(d)	38(d)
B, P, Sh, St	H1100	965	140	795	115	14(d)	45(e)	31(d)	38(d)
B, P, Sh, St	H1150	930	135	725	105	16(d)	50(e)	28(d)	36(d)
B, P, Sh, St	H1150M	795	115	515	75	18(d)	55(e)	24(d)	34(d)
17-7 PH (UNS S17700)									
B	RH950	1275	185	1030	150	6	10	41	...
B	TH1050	1170	170	965	140	6	25	38	...
P, Sh, St	RH950	1450	210	1310	190	1-6(d)	...	41(d)	44(d)
P, Sh, St	TH1050	1240	180	1030	150	3-7(d)	...	38	
P, Sh, St	Cold rolled (condition C)	1380	200	1210	175	1	...	41	...
P, Sh, St	CH900	1650	240	1590	230	1	...	46	...
Custom 450 (UNS S45000)									
B	Annealed	860	125	655	95	10	40	...	33
P, Sh, St	Annealed	895	130	620	90	4	...	25	...
B, P, Sh, St	H900	1240	180	1170	170	10	40	40	...
B, P, Sh, St	H1000	1100	160	1030	150	12	45	36	...
B, P, Sh, St	H1150	860	125	515	75	15	50	26	...
Custom 455 (UNS S45500)									
B	H900	1620	235	1520	220	8	30	47	...
B	H950	1520	220	1410	205	10	40	44	...
P, Sh, St	H950	1530	222	1410	205	Up to 4	...	44	...

(continued)

(a) B, bar; P, plate; Sh, sheet; St, strip; W, wire; F, forgings. (b) Values are for bar products. (c) Where minimum value is also given, maximum value applies only to flat-rolled products. Both max and min values may vary with thickness for flat-rolled products. (d) Value varies with thickness for flat-rolled products. (e) Value generally lower for flat-rolled products and varies with thickness. (f) Values are typical. (g) Rockwell B hardness.

available for many of the standard types of stainless steel. Most testing has been concentrated on the high-strength precipitation-hardening stainless steels, because these materials have been used in critical applications where fracture-toughness testing has been found most useful for evaluating materials. Table 11 lists typical fracture toughness for several of the high-strength stainless steels for which this property has been determined.

Fatigue Strength

Three types of fatigue tests are used to develop data on the fatigue behavior of stainless steels. The most common of these tests is the rotating-beam test, which most closely approximates the kind of loading to which shafts and axles are subjected. The flexural fatigue test is used to evaluate the behavior of sheet, and most closely simulates the action of leaf springs, which are expected to flex without deforming or breaking. The axial-load fatigue test subjects a fatigue specimen to unidirectional loading that can range from full reversal (tension-compression) to tension-tension loading, and can have virtually any conceivable ratio of maximum stress to minimum stress.

Fatigue data can be given in the form of *S-N* curves (Fig. 9) or constant-life diagrams (Fig. 10). Data from any of the three types of tests can be presented as *S-N* curves, but only data from flexural fatigue and axial fatigue tests can be presented in the form of a constant-life diagram. In analyzing fatigue data, and particularly in selecting materials on the basis of fatigue life, it is important to understand the influence of stress ratio on fatigue life. In general, fatigue conditions involving tension-compression loading (stress ratio, R, between 0 and -1) lead to shorter fatigue lives than conditions involving tension-tension loading (stress ratio, R, between 0 and $+1$) at the same value of maximum stress.

Elevated-Temperature Properties

Many stainless steels—particularly the austenitic types 304, 309, 310, 316, 321 and 347 and certain precipitation-hardening types such as PH 15-7 Mo, 15-5 PH, 17-4 PH, 17-7 PH, AM-350 and AM-355—are used extensively for elevated-temperature applications such as chemical processing equipment, high-temperature heat exchang-

Table 10 (continued)

Product form(a)	Condition	Tensile strength MPa	ksi	Yield strength MPa	ksi	Elongation, %	Reduction in area(b), %	Hardness(c), HRC Min	Max
AM-350 (UNS S35000)									
P, Sh, St	H850	1275	185	1030	150	2-8	...	42	...
P, Sh, St	H1000	1140	165	1000	145	2-8	...	36	...
AM-355 (UNS S35500)									
B	Equalize plus overtemper 537 °C (1000 °F)	1170	170	1070	155	12	25	39	...
P, Sh, St	H850	1310	190	1140	165	10	...	37	...
P, Sh, St	H1000	1170	170	1030	150	12	...	28(f)	...
AM-363									
St	Annealed	850(f)	123(f)	730(f)	106(f)	12(f)	...	...	...
Stainless W (UNS S17600)									
B, P, Sh, St	H950	1310	190	1170	170	8	25	39	...
B, P, Sh, St	H1000	1240	180	1100	160	8	30	37	...
B, P, Sh, St	H1050	1170	170	1070	150	10	40	35	...
PH 15-7 Mo (UNS S15700)									
B	RH950	1380	200	1210	175	7	25	...	...
B	TH1050	1240	180	1100	160	8	25	...	...
P, Sh, St	RH950	1550	225	1380	200	1-5(d)	...	43(d)	46(d)
P, Sh, St	TH1050	1310	190	1170	170	2-5(d)	...	38(d)	40(d)
P, Sh, St	Cold rolled	1380	200	1210	175	1	...	41	...
P, Sh, St	Cold rolled and aged	1650	240	1590	230	1	...	46	...
17-10 P									
B	Annealed	615(f)	89(f)	255(f)	37(f)	70(f)	76(f)	82(f)(g)	...
B	Aged	945(f)	137(f)	605(f)	88(f)	25(f)	39(f)	30(f)	...
HNM									
B, W, F	Aged at 704 °C (1300 °F)	825	120	550	80	18	30	...	...

(a) B, bar; P, plate; Sh, sheet; St, strip; W, wire; F, forgings. (b) Values are for bar products. (c) Where minimum value is also given, maximum value applies only to flat-rolled products. Both max and min values may vary with thickness for flat-rolled products. (d) Value varies with thickness for flat-rolled products. (e) Value generally lower for flat-rolled products and varies with thickness. (f) Values are typical. (g) Rockwell B hardness.

ers and superheater tubes for power boilers. For more detail on elevated-temperature properties of selected types, see "Iron-Base Heat-Resistant Alloys", in this volume.

Extended service at elevated temperature can result in embrittlement of austenitic stainless steels or in "sensitization" which degrades the ability of the material to withstand corrosion, particularly in acid media. Most often, such degradation is caused by the precipitation of secondary phases such as carbides and sigma phase. Precipitation depends on both time and temperature—longer times at temperature and higher temperatures both promote more extensive precipitation. The problems arising from embrittlement and sensitization, and the remedies that can help combat them are discussed in more detail in the article following this one, entitled "Fabrication of Wrought Stainless Steels".

Influence of Product Form on Properties

The mechanical properties of cast or wrought stainless steels vary widely from group to group, vary less widely from type to type within groups and may vary with product form for a given type. Because of the wide variation from group to group, one must first decide whether a martensitic, ferritic, austenitic or precipitation-hardening stainless steel is most suitable for a given application. Once the appropriate group is selected, method of fabrication or service conditions may then dictate which specific type is required.

Before typical properties of the various product forms are discussed, it is important that two key points about stainless steels be recognized. First, many stainless steels are manufactured and/or used in a heat treated condition—that is, in some thermally treated condition other than "process annealed" or typically mill processed. When this is the case, a tabulation of typical properties may not give all the required information. Second, in many products strain hardening during fabrication is a very important consideration. All stainless steels strain harden to some degree depending on structure, alloy content and amount of cold working. Consequently, in applications where service performance of the finished product depends on enhancement of properties during fabrication, it is

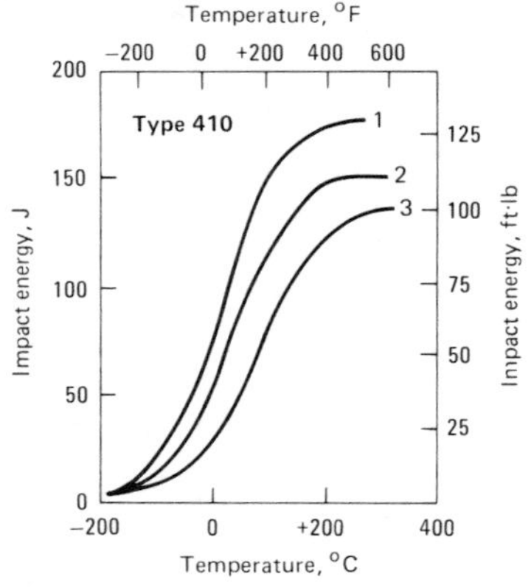

Fig. 8 Typical transition behavior of type 410 martensitic stainless steel

All data from Charpy V-notch tests. Curve 1 represents material tempered at 790 °C (1450 °F); final hardness, 95 HRB. Curve 2 represents material tempered at 663 °C (1225 °F); final hardness, 24 HRC. Curve 3 represents material tempered at 595 °C (1100 °F); final hardness, 30 HRC.

Table 11 Longitudinal fracture toughness of PH stainless steels

Alloy designation	Condition	Hardness, HRC	Fracture toughness MPa$\sqrt{m}$	Fracture toughness ksi$\sqrt{in.}$
17-4 PH	H900	44	53	48
17-7 PH	RH950	44	76	69
Custom 450	Aged at 480 °C (900 °F)	43	81	74
Custom 455	Aged at 480 °C (900 °F)	50	47	43
	Aged at 510 °C (950 °F)	48	80	73
	Aged at 540 °C (1000 °F)	44	110	100
PH 13-8 Mo	H950	47	99	90
	H1000	46	121	110
PH 15-7 Mo	TH1080	42	55	50

Fig. 9 Typical rotating-beam fatigue behavior of types 304 and 310 stainless steel

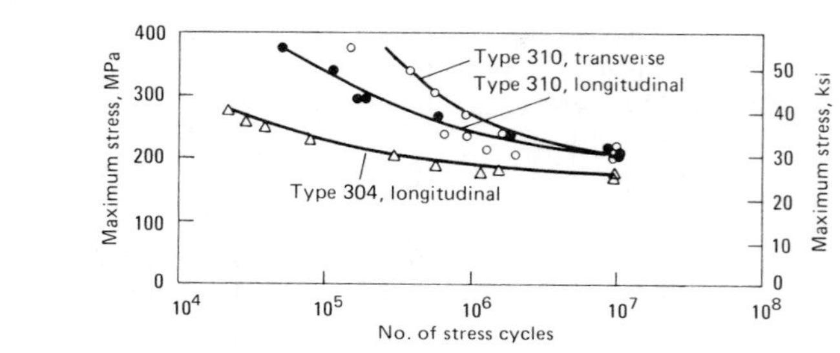

Fig. 10 Constant-life fatigue diagram for PH 13-8 Mo stainless steel, condition H1000

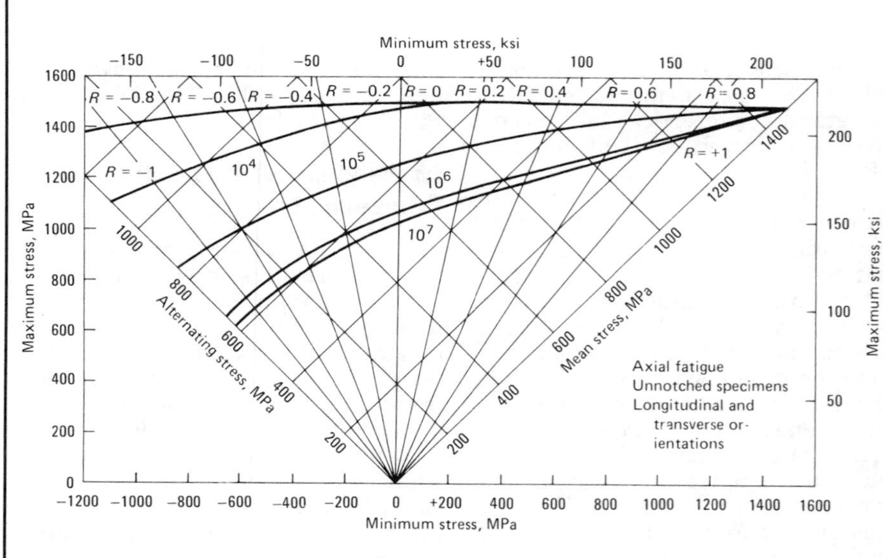

essential that the manufacturer determine this effect independently for each individual product. Here, techniques such as statistical-reliability testing are invaluable.

Cast Structures. Whether produced as ingot, slab or billet in a mill or as shape castings in a foundry, cast structures can exhibit wide variations in properties. Because of the possible existence of large dendritic grains, intergranular phases and alloy segregation, typical mechanical properties cannot be stated precisely and generally are inferior to those of any wrought structure.

Hot Processing. The initial purpose of hot rolling or forging an ingot, slab or billet is to refine cast structure and improve mechanical properties. Because recrystallization occurs during deformation at high temperatures, hot reduced products and hot reduced and annealed products exhibit coarser grain structures and lower strengths than cold processed products. Grain size and orientation depend chiefly on start and finish temperatures and on method of hot reduction. For instance, cross-rolled hand-mill plate will exhibit a more equiaxed grain structure than continuous hot rolled strip.

Hot reduction may be a final sizing

operation, as in the case of hot rolled bar, billet, plate, sheet or bar flats, or it may be an intermediate processing step for products such as cold finished bar, rod and wire, and cold rolled sheet and strip.

Typical properties of hot processed products and of hot processed and annealed products are different from those of either cast or cold reduced products. Both hot processing and annealing tend to produce coarser grain sizes.

Cold Reduced Products. When strained at ambient temperatures, all stainless steels tend to work harden, as shown in Fig. 11. Some grades have specifically balanced compositions so that high strengths can be attained with reasonably low cold reductions. Large amounts of cold reduction, which are typical in production of sheet and strip, impart a hard metastable structure that results in a variety of structures and properties following subsequent thermal processing. Because recrystallization does not occur during cold working, final properties of thermally treated products depend on (*a*) the amount of cold reduction (which helps determine the number of potential recrystallization sites), (*b*) the type of mill thermal treatment (subcritical annealing, normalizing or solution treatment) and (*c*) time at any given temperature. Wrought products that have been cold reduced and annealed generally have higher strengths than hot processed products. Cold reduced products sometimes exhibit greater differences between transverse and longitudinal properties than hot processed products.

Cold finishing generally is done to improve dimensional tolerances or surface finish, or to raise mechanical strength. Cold finished products—whether they have been previously hot worked and annealed or have been hot worked, cold worked and annealed—have higher mechanical strength and slightly lower ductility than their process-annealed counterparts.

Fig. 11 Typical effect of cold rolling on tensile strength of selected stainless steels

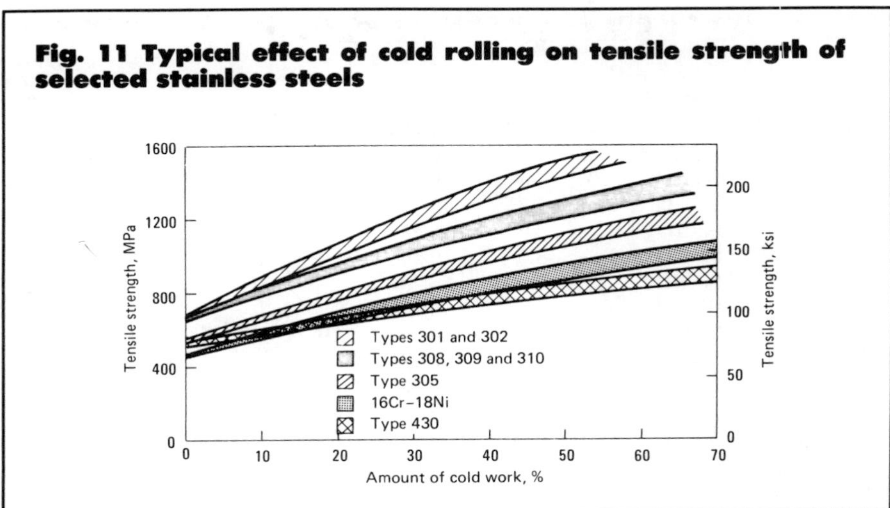

Physical Properties

There are relatively few applications for stainless steels where physical properties are the determining factors in selection. However, there are many applications where physical properties are important in product design. For instance, stainless steels are used for many elevated-temperature applications, often in conjunction with steels of lesser alloy content. Because many stainless steels have higher coefficients of thermal expansion and lower thermal conductivities than carbon and alloy steels, these characteristics must be taken into account in design of stainless steel — carbon steel or stainless steel — alloy steel products such as heat exchangers. In such products, differential thermal expansion will impose stresses on the unit that would not be present were the unit made entirely of carbon or alloy steel; also, if the heat-transfer surface is made of stainless steel, it must be larger than if it were made of carbon or alloy steel.

Typical physical properties of selected grades of annealed wrought stainless steels are given in Table 12. Physical properties may vary slightly with product form and size, but such variations usually are not of critical importance to the application.

Surface Finishing of Stainless Steel

Stainless steels are unique in their ability to maintain bright finished surfaces. Lacquers and other protective coatings are seldom recommended, because unprotected stainless steel surfaces are more serviceable and easier to clean than coated surfaces. Any of several surface textures and degrees of finish can be specified, depending on the particular requirements of the application.

Although not used for protection, paint may be applied to stainless steels when contrasting colors are required, such as for nameplates or signs. Surfaces should be chemically cleaned and/or etched for good paint adherence.

After fabrication, stainless steels may be bright finished by any of the following methods: mechanical grinding and polishing, Tampico brushing, buffing, electrolytic polishing, barrel finishing, or wet or dry blasting. Sometimes, two or more of these methods are used in combination. In addition, any of several processes that involve combinations of heating and chemical treatment can be used to produce dull black surfaces having good abrasion resistance; this is done most often on small parts. The most widely used finishing process, however, is grinding and/or polishing.

When fabrication does not involve extensive machining, severe deformation or general surface marring, sheet and round bar often are polished before fabrication. Welds in prepolished parts may be finished by grinding the welds flush and then polishing contrasting stripes over the weld seams. Alternatively, the entire surface may be given a final polish in the direction of previous polishing to render weld areas virtually invisible.

Grinding, Polishing and Buffing. The term grinding applies to rough, often dry, finishing operations in which significant amounts of metal are removed using loose abrasives or abrasive belts coarser than 100 grit. Polishing refers to finishing operations using abrasives finer than 100-grit suspended in lubricants. Buffing is a special type of polishing that is done with high-speed cloth wheels charged with extremely fine abrasives.

Although equipment used for grind-

Table 12 Typical physical properties of wrought stainless steels, annealed condition

Type	UNS designation	Density Mg/m³	Density lb/in.³	Elastic modulus GPa	Elastic modulus 10⁶ psi	0 °C to: / 32 °F to:	Mean coefficient of thermal expansion µm/m·°C 100 / 212	µm/m·°C 315 / 600	µm/m·°C 538 / 1000	µin./in.·°F 100 / 212	µin./in.·°F 315 / 600	µin./in.·°F 538 / 1000
201	S20100	7.8	0.28	197	28.6		15.7	17.5	18.4	8.7	9.7	10.2
202	S20200	7.8	0.28	...	...		17.5	18.4	19.2	9.7	10.2	10.7
205	S20500	7.8	0.28	197	28.6		...	17.9	19.1	...	9.9	10.6
301	S30100	8.0	0.29	193	28.0		17.0	17.2	18.2	9.4	9.6	10.1
302	S30200	8.0	0.29	193	28.0		17.2	17.8	18.4	9.6	9.9	10.2
302B	S30215	8.0	0.29	193	28.0		16.2	18.0	19.4	9.0	10.0	10.8
303	S30300	8.0	0.29	193	28.0		17.2	17.8	18.4	9.6	9.9	10.2
304	S30400	8.0	0.29	193	28.0		17.2	17.8	18.4	9.6	9.9	10.2
304L	S30403	8.0	0.29	...	...		...	...	...	...	...	...
S30430	S30430	8.0	0.29	193	28.0		17.2	17.8	...	9.6	9.9	...
304N	S30451	8.0	0.29	196	28.5		...	...	...	...	...	...
305	S30500	8.0	0.29	193	28.0		17.2	17.8	18.4	9.6	9.9	10.2
308	S30800	8.0	0.29	193	28.0		17.2	17.8	18.4	9.6	9.9	10.2
309	S30900	8.0	0.29	200	29.0		15.0	16.6	17.2	8.3	9.2	9.6
310	S31000	8.0	0.29	200	29.0		15.9	16.2	17.0	8.8	9.0	9.4
314	S31400	7.8	0.28	200	29.0		...	15.1	...	...	8.4	...
316	S31600	8.0	0.29	193	28.0		15.9	16.2	17.5	8.8	9.0	9.7
316L	S31603	8.0	0.29	...	...		...	...	...	...	...	...
316N	S31651	8.0	0.29	196	28.5		...	...	...	...	...	...
317	S31700	8.0	0.29	193	28.0		15.9	16.2	17.5	8.8	9.0	9.7
317L	S31703	8.0	0.29	200	29.0		16.5	...	18.1	9.2	...	10.1
321	S32100	8.0	0.29	193	28.0		16.6	17.2	18.6	9.2	9.6	10.3
329	S32900	7.8	0.28	...	...		...	...	...	...	...	...
330	N08330	8.0	0.29	196	28.5		14.4	16.0	16.7	8.0	8.9	9.3
347	S34700	8.0	0.29	193	28.0		16.6	17.2	18.6	9.2	9.6	10.3
384	S38400	8.0	0.29	193	28.0		17.2	17.8	18.4	9.6	9.9	10.2
405	S40500	7.8	0.28	200	29.0		10.8	11.6	12.1	6.0	6.4	6.7
409	S40900	7.8	0.28	...	...		11.7	...	...	6.5	...	...
410	S41000	7.8	0.28	200	29.0		9.9	11.4	11.6	5.5	6.3	6.4
414	S41400	7.8	0.28	200	29.0		10.4	11.0	12.1	5.8	6.1	6.7
416	S41600	7.8	0.28	200	29.0		9.9	11.0	11.6	5.5	6.1	6.4
420	S42000	7.8	0.28	200	29.0		10.3	10.8	11.7	5.7	6.0	6.5
422	S42200	7.8	0.28	...	...		11.2	11.4	11.9	6.2	6.3	6.6
429	S42900	7.8	0.28	200	29.0		10.3	...	...	5.7	...	...
430	S43000	7.8	0.28	200	29.0		10.4	11.0	11.4	5.8	6.1	6.3
430F	S43020	7.8	0.28	200	29.0		10.4	11.0	11.4	5.8	6.1	6.3
431	S43100	7.8	0.28	200	29.0		10.2	12.1	...	5.7	6.7	...
434	S43400	7.8	0.28	200	29.0		10.4	11.0	11.4	5.8	6.1	6.3
436	S43600	7.8	0.28	200	29.0		9.3	...	...	5.2	...	...
440A	S44002	7.8	0.28	200	29.0		10.2	...	...	5.7	...	...
440C	S44004	7.8	0.28	200	29.0		10.2	...	...	5.7	...	...
444	S44400	7.8	0.28	200	29.0		10.0	10.6	11.4	5.6	5.9	6.3
446	S44600	7.5	0.27	200	29.0		10.4	10.8	11.2	5.8	6.0	6.2
PH 13-8 Mo	S13800	7.8	0.28	203	29.4		10.6	11.2	11.9	5.9	6.2	6.6
15-5 PH	S15500	7.8	0.28	196	28.5		10.8	11.4	...	6.0	6.3	...
17-4 PH	S17400	7.8	0.28	196	28.5		10.8	11.6	...	6.0	6.4	...
17-7 PH	S17700	7.8	0.28	204	29.5		11.0	11.6	...	6.1	6.4	...

(a) At 0 to 100 °C (32 to 212 °F). (b) Approximate values.

Thermal conductivity, W/m·K, °C: 100 / °F: 212	W/m·K, 500 / 932	Btu/h·ft·°F, 100 / 212	Btu/h·ft·°F, 500 / 932	Specific heat(a), J/kg·K	Specific heat(a), Btu/lb·°F	Electrical resistivity, nΩ·m	Magnetic permea-bility(b)	Melting range, °C	Melting range, °F	Type
16.2	21.5	9.4	12.4	500	0.12	690	1.02	1400-1450	2550-2650	201
16.2	21.6	9.4	12.5	500	0.12	690	1.02	1400-1450	2550-2650	202
...	...	...	...	500	0.12	...	...	...	...	205
16.2	21.5	9.4	12.4	500	0.12	720	1.02	1400-1420	2550-2590	301
16.2	21.5	9.4	12.4	500	0.12	720	1.02	1400-1420	2550-2590	302
15.9	21.6	9.2	12.5	500	0.12	720	1.02	1375-1400	2500-2550	302B
16.2	21.5	9.4	12.4	500	0.12	720	1.02	1400-1420	2550-2590	303
16.2	21.5	9.4	12.4	500	0.12	720	1.02	1400-1450	2550-2650	304
...	...	...	...	...	...	...	1.02	1400-1450	2550-2650	304L
11.2	21.5	6.5	12.4	500	0.12	720	1.02	1400-1450	2550-2650	S30430
...	...	...	...	500	0.12	720	1.02	1400-1450	2550-2650	304N
16.2	21.5	9.4	12.4	500	0.12	720	1.02	1400-1450	2550-2650	305
15.2	21.6	8.8	12.5	500	0.12	720	...	1400-1420	2550-2590	308
15.6	18.7	9.0	10.8	500	0.12	780	1.02	1400-1450	2550-2650	309
14.2	18.7	8.2	10.8	500	0.12	780	1.02	1400-1450	2550-2650	310
17.5	20.9	10.1	12.1	500	0.12	770	1.02	...	...	314
16.2	21.5	9.4	12.4	500	0.12	740	1.02	1375-1400	2500-2550	316
...	...	...	...	...	...	...	1.02	1375-1400	2500-2550	316L
...	...	...	...	500	0.12	740	1.02	1375-1400	2500-2550	316N
16.2	21.5	9.4	12.4	500	0.12	740	1.02	1375-1400	2500-2550	317
14.4	...	8.3	...	500	0.12	790	...	1375-1400	2500-2550	317L
16.1	22.2	9.3	12.8	500	0.12	720	1.02	1400-1425	2550-2600	321
...	...	...	...	460	0.11	750	...	...	...	329
...	...	...	...	460	0.11	1020	1.02	1400-1425	2550-2600	330
16.1	22.2	9.3	12.8	500	0.12	730	1.02	1400-1425	2550-2600	347
16.2	21.5	9.4	12.4	500	0.12	790	1.02	1400-1450	2550-2650	384
27.0	...	15.6	...	460	0.11	600	...	1480-1530	2700-2790	405
...	...	...	...	...	...	...	...	1480-1530	2700-2790	409
24.9	28.7	14.4	16.6	460	0.11	570	700-1000	1480-1530	2700-2790	410
24.9	28.7	14.4	16.6	460	0.11	700	...	1425-1480	2600-2700	414
24.9	28.7	14.4	16.6	460	0.11	570	700-1000	1480-1530	2700-2790	416
24.9	...	14.4	...	460	0.11	550	...	1450-1510	2650-2750	420
23.9	27.3	13.8	15.8	460	0.11	...	...	1470-1480	2675-2700	422
25.6	...	14.8	...	460	0.11	590	...	1450-1510	2650-2750	429
26.1	26.3	15.1	15.2	460	0.11	600	600-1100	1425-1510	2600-2750	430
26.1	26.3	15.1	15.2	460	0.11	600	...	1425-1510	2600-2750	430F
20.2	...	11.7	...	460	0.11	720	...	...	...	431
...	26.3	...	15.2	460	0.11	600	600-1100	1425-1510	2600-2750	434
23.9	26.0	13.8	15.0	460	0.11	600	600-1100	1425-1510	2600-2750	436
24.2	...	14.0	...	460	0.11	600	...	1370-1480	2500-2700	440A
24.2	...	14.0	...	460	0.11	600	...	1370-1480	2500-2700	440C
26.8	...	15.5	...	420	0.10	620	...	...	...	444
20.9	24.4	12.1	14.1	500	0.12	670	400-700	1425-1510	2600-2750	446
14.0	22.0	8.1	12.7	460	0.11	1020	...	1400-1440	2560-2625	PH 13-8 Mo
17.8	23.0	10.3	13.1	420	0.10	770	95	1400-1440	2560-2625	15-5 PH
18.3	23.0	10.6	13.1	460	0.11	800	95	1400-1440	2560-2625	17-4 PH
16.4	21.8	9.5	12.6	460	0.11	830	...	1400-1440	2560-2625	17-7 PH

ing and polishing stainless steel is the same as standard equipment used for grinding and polishing other metals, the characteristics of stainless steel necessitate some modifications in technique. The suggestions given below for austenitic Cr-Ni types apply to straight chromium types as well, even though the latter have certain characteristics similar to those of carbon steels. Variations in procedure may be necessary because of differing characteristics of the various types. The chromium-nickel types have about half the thermal conductivity and nearly twice the coefficient of expansion of carbon steels. This combination of characteristics means that austenitic stainless steels tend to heat up locally when ground or polished. When this occurs, they expand excessively in the heated area. It is very important, therefore, that polishing heads should not be allowed to dwell on one spot, and excessive polishing pressure should be avoided. Attempts to hasten polishing will aggravate overheating and may cause the part to warp or buckle.

Straight chromium stainless steels tend to "load" the abrasive, and this can result in scoring or galling of the polished surface. Frequently, chromium stainless steel appears slightly rougher than chromium-nickel stainless steel when the same polishing grits have been used. A slight color difference exists between polished finishes on chromium and chromium-nickel types, although this difference usually is not noticeable unless pieces are adjacent and in the same general viewing plane.

Polishing in different directions in different areas of the same surface produces contrasts in appearance. A pleasing contrast can be obtained by polishing one area in a direction perpendicular to the polishing direction for another area. However, there sometimes is little or no contrast when different areas are polished in directions opposite (180°) to each other. If uniformity is desired, grit lines should be parallel over the entire surface; long, even strokes are required when hand tools are used.

Buffing, when used to produce a fine, scratch-free, mirror finish, usually consists of two operations: the "cutting" operation, in which a high wheel speed and a "cutting" compound are used; and the "coloring" operation, in which a slightly lower wheel speed and a "color" compound are employed. A satin finish, which is really the result of fine polishing rather than a true buffed finish, is produced by final buffing with "satin finish" compounds. Usually, grinding and polishing are necessary to obtain a surface suitable for buffing.

Grinding of welds using solid grinding wheels of suitable grit will remove excess weld metal and prepare weld surfaces for subsequent polishing operations.

Welds should be ground in the direction of the weld bead without allowing the wheel to dwell in one spot. Grinding across the weld is objectionable, because cross grinding tends to cut into parent-metal surfaces. Small-diameter wheels are controlled more easily than large-diameter wheels. It is advisable to stop rough grinding before the weld reinforcement is entirely removed, thus leaving some stock for finishing. When only one surface of a weld must be finished, the side opposite that from which the weld was made usually has less surplus metal and requires less grinding. Light-gage sheet or strip should be rigidly backed up to provide a firm base for grinding.

Subsequent polishing and buffing can then be done in the usual manner. Where feasible, welds in mill-polished sheet should run in the direction of the polishing lines. This will make it easier to polish the weld to match the existing finish.

Electropolishing provides, at low cost, the combination of an attractive high-luster finish plus deburring. In electropolishing, which is the reverse of electroplating, a small amount of surface metal is removed. Wire and bar products, stampings and small forgings have all been electropolished with good results. Recesses in intricately shaped pieces may be readily brightened, which makes electropolishing particularly suitable for springs, wire racks, and the inside surfaces of deep drawn bowls and pans.

Because it removes only a small amount of metal, electropolishing does not eliminate heavy scratches, deep die scoring or embedded nonmetallic particles. The process is not recommended for descaling, although it removes discoloration such as that resulting from spot welding. An electropolished finish does not simulate a mechanically polished finish.

Mill Finishes

Sheet, strip, plate, bar and wire made of stainless steel all have different designations of mill finish, each representing a standardized appearance that is characteristic of the process used to impart final mechanical properties. Although the various mill finishes are standardized, there is sufficient variability in mill processing that exact matching of color and reflectivity cannot be expected from lot to lot. Even wider differences can be expected between mill products from different producers.

Sheet finishes are designated by a system of numbers: No. 1, 2D, 2B and 2BA for unpolished finishes; and No. 3, 4, 6, 7 and 8 for polished finishes.

Each of the unpolished finishes is in itself a category of finishes, with variations in appearance and smoothness depending on composition, sheet thickness and method of manufacture. Generally, the thinner the sheet, the smoother the surface. Chromium-nickel and chromium-nickel-manganese stainless steels are characteristically different in appearance from straight chromium types having the same finish. Furthermore, sheet produced continuously (in coil form) generally differs in appearance from sheet produced as individual pieces on hand mills.

The appearance or "color" of polished finishes may differ slightly among 2*xx*, 3*xx* and 4*xx* series stainless steels. Sheet can be produced with one or both sides polished; when only one side is polished, the other side often is rough ground to obtain better flatness.

No. 1 finish is produced by hot rolling followed by annealing and descaling. It is generally used in industrial applications where smoothness of finish is not particularly important, such as equipment for elevated-temperature or corrosion service.

No. 2D finish is a dull cold rolled finish produced by cold rolling, annealing and descaling. The dull finish may result from descaling or pickling or may be developed by a subsequent final light cold rolling pass using dull rolls. No. 2D finish is favorable for retention of lubricants in deep drawing, and is generally preferred for deep drawn articles that will be polished after fabrication.

No. 2B finish is a bright cold rolled finish commonly produced in the same manner as No. 2D, except that the final light cold rolling pass is done using polished rolls. No. 2B is a general-purpose cold rolled finish commonly used for all but exceptionally difficult deep drawing applications. It is more readily polished to high luster than a No. 1 or No. 2D finish.

No. 2BA finish is a mirrorlike appearance produced by cold rolling, then bright annealing or double bright annealing. The final appearance is developed by a single light skin pass through a cold mill over highly polished rolls, but is also dependent on additional millwork, such as grinding the surface at an intermediate gage. A No. 2BA finish is often specified for architectural applications and for other uses where a highly reflective surface is desired on the as-fabricated part.

No. 3 finish is a polished finish obtained with abrasives approximately 100 mesh in particle size. It is used for articles that may or may not receive additional polishing during fabrication.

No. 4 finish is a general-purpose polished finish widely used for restaurant equipment, kitchen equipment, storefronts and dairy equipment. Following initial grinding with coarser abrasives, final finishing generally is done with abrasives having a particle size of approximately 120 to 150 mesh.

No. 6 finish is a dull satin finish having lower reflectivity than No. 4 finish. It is produced by Tampico brushing No. 4 finish sheet in a medium of abrasive and oil. It is used in architectural and ornamental applications where high luster is undesirable; it is also used effectively for contrast with brighter finishes.

No. 7 finish has a high degree of reflectivity. It is produced by buffing finely ground surfaces, but not to the extent that existing grit lines are removed. It is used chiefly for architectural and ornamental parts.

No. 8 finish, the most reflective finish that is commonly produced on sheet, is obtained by polishing with successively finer abrasives and buffing extensively with very fine buffing rouges. The surface is essentially free of grit lines from preliminary grinding operations. No. 8 finish is most widely used for press plates, small mirrors and reflectors.

Strip Finishes. Only three unpolished finishes (No. 1, No. 2 and bright annealed) and one polished finish (mill buffed) are commonly supplied on stainless steel strip. As with finishes on stainless steel sheet, each unpolished strip finish comprises a category of finishes that vary in appearance and smoothness depending on composition, thickness and method of manufacture. Generally, the thinner the strip, the smoother the surface. Chromium-nickel and chromium-nickel-manganese stainless steels are characteristically different in appearance from straight-chromium types having the same finish.

No. 1 finish is produced by cold rolling, annealing and pickling. Appearance varies from dull gray matte to fairly reflective, depending largely on stainless steel type. This finish is used for severely drawn or formed parts, as well as for applications where the brighter No. 2 finish is not required, such as parts to be used at high temperatures. No. 1 finish for strip approximates No. 2D finish for sheet in corresponding chromium-nickel or chromium-nickel-manganese types.

No. 2 finish is produced by the same treatment used for No. 1 finish followed by a final light cold rolling pass, which generally is done using highly polished rolls. This final pass produces a smoother and more reflective surface, the appearance of which varies with stainless steel type. No. 2 finish for strip is a general-purpose finish widely used for household appliances, automotive trim, tableware and utensils. No. 2 finish for strip approximates No. 2B finish for sheet in corresponding chromium-nickel or chromium-nickel-manganese stainless steels.

Bright annealed finish is a bright, cold rolled, highly reflective finish retained by final annealing in a controlled-atmosphere furnace. The purpose of atmosphere control is to prevent scaling or oxidation during annealing. The atmosphere usually consists of either dry hydrogen or dissociated ammonia. Bright annealed strip is used most extensively for automotive trim.

Mill-buffed finish is a highly reflective finish obtained by subjecting either No. 2 or bright annealed coiled strip to a continuous buffing pass. The purpose of mill buffing is to provide a finish uniform in color and reflectivity. It also can provide a surface receptive to chromium plating. This type of finish is used chiefly for automotive trim, household trim, tableware, utensils, fire extinguishers and plumbing fixtures.

Plate Finishes. Stainless steel plate can be produced in the following conditions and surface finishes:

Condition and finish	Description and remarks
Hot rolled	Scale not removed; not heat treated; plate not recommended for final use in this condition(a)
Hot or cold rolled, annealed or heat treated	Scale not removed; use of plate in this condition generally confined to heat-resisting applications; scale impairs corrosion resistance(a)
Hot or cold rolled, annealed or heat treated, blast cleaned or pickled	Condition and finish commonly preferred for corrosion-resisting and most heat-resisting applications
Hot or cold rolled, annealed, descaled and temper passed	Smoother finish for specialized applications
Hot rolled, annealed, descaled, cold rolled, annealed, descaled, optionally temper passed	Smooth finish with greater freedom from surface imperfections than any of the above
Hot or cold rolled, annealed or heat treated, surface cleaned and polished	Polished finishes similar to the polished finishes on sheet.

(a) Surface inspection is not practicable for plate that has not been pickled or otherwise descaled.

Plate commonly is conditioned by localized grinding to remove surface imperfections on either or both surfaces; ground areas are well flared and the thickness is not reduced below the allowable tolerance in any of these areas.

Bar Finishes. Stainless steel bar is produced in the conditions and surface finishes given in the table below. It is important that both condition and finish be specified, because each finish is applicable only to certain conditions.

Condition	Surface finish
Hot worked only	Scale not removed (except for spot conditioning)
	Rough turned(a)(b)
	Blast cleaned
Annealed or otherwise heat treated	Scale not removed (except for spot conditioning)
	Rough turned(a)
	Pickled or blast cleaned and pickled
	Cold drawn or cold rolled
	Centerless ground(a)
	Polished(a)
Annealed and cold worked to high tensile strength(c)	Cold drawn or cold rolled
	Centerless ground(a)
	Polished(a)

(a) Applicable to round bar only
(b) Bar of 4*xx* series stainless steels that are highly hardenable, such as type 414, 420, 420F, 431, 440A, 440B and 440C, are annealed before rough turning. Other hardenable types, such as types 403, 410, 416 and 416Se, also may require annealing depending on composition and size.
(c) Produced only in mill orders; made predominantly in types 301, 302, 303Se, 304, 304N, 316 and 316N.

Wire Finishes. *Oil- or grease-drawn finish* is a special bright finish for wire intended for uses, such as racks and handles, where the finish supplied is to be the final finish of the end product. In producing this finish, lower drawing speeds are necessary and additional care in processing is needed to provide a surface with few scratches and with only a very light residue of lubricant.

Diamond drawn finish is a very bright finish generally limited to wet drawn stainless steel wire in fine sizes. Drafting speeds are necessarily reduced to obtain the desired brightness.

Copper-coated wire is supplied when a special finish is required for lubrication in an operation such as spring coiling or cold heading. Generally, copper coated wire is drawn after coating, the amount depending on the desired cold worked temper of the wire.

Tinned wire is coated by passing single strands through a bath of molten tin. Tinned wire is used in soldering applications. The temper of the finished wire is controlled by processing prior to tinning.

Lead-coated wire is coated by passing single strands through, or immersing bundles of wire in, a bath of molten lead. The wire is then drawn to final size, with the lead forming a thin coating over the entire surface. This coating is useful on wire for coil springs, where it serves as a lubricant during coiling operations.

Interim Surface Protection

Finishes on stainless steel often require protection during shipment of mill products to fabrication plants. Otherwise, it is inevitable that scratches, dings and other evidence of material handling operations will mar the appearance of some end products. Furthermore, if a mill finish is intended to give the end product its appearance, the finish must be protected from incidental damage during fabrication of parts. Although finishes on completed parts and assemblies can be protected by proper packaging, such packaging can be made much simpler in design if the need to protect the finish from certain kinds of incidental damage during shipment can be reduced.

All of these considerations have led to the development of various masking materials for protection of finishes on stainless steel surfaces prior to end use. Masking materials used in mills are most often obtained in the form of rolls of adhesive-backed protective film that can be applied by hand or machine to stainless steel mill products. In addition, water-base and solvent-base strippable coatings are available; these are applied by dip-dry methods. Sometimes, special adhesive-backed protective coverings of the customer's choosing can be applied at the mill prior to shipment.

Most often, adhesives are based on latex or proprietary organic compounds. Masking materials include paper, rubber and plastics. Plastics such as polyvinyl chloride, polyolefin, polyethylene and polypropylene are the most popular. Coverings range in thickness from about 0.06 to 0.13 mm (2.5 to 5 mils) and are available in several colors (generally white, clear, black and light blue). Sometimes, paper coverings are printed with a company logo or a proprietary message such as instructions for in-service care of the finish.

Adhesion to stainless steel generally is measured as the force per unit width required to remove the covering from sheet having a No. 4 finish. Other data available from the masking-material manufacturer include tensile strength, elongation and unwinding force.

Plastic and rubber coverings have sufficient elongation and adhesion to enable them to survive all but the most severe fabrication operations, including drawing, bending and roll forming.

Of interest to the end user, of course, is the ease with which the masking material can be removed. Often, a compromise must be reached between degree of protection afforded the finish and ease of removal. Usually, adhesive-backed coverings are easiest to remove when the coated steel is first received. They often become more difficult to remove if they are exposed to temperatures significantly above or below normal room ambient, to ultraviolet light or to outdoor (weathering) environments. Excessively prolonged indoor storage also can make it quite difficult to remove adhesive-backed coverings.

In-Service Care

Despite the fact that stainless steel surfaces are generally considered nontarnishing, they still need a certain amount of care to maintain a given surface appearance under normal conditions of service. Table 13 summarizes methods of cleansing stainless steel for a wide variety of applications.

Architectural Applications. For exposed exterior surfaces in inland, light industrial areas, minimum maintenance is needed. Ordinarily, normal rainfall is adequate to maintain the desired appearance, and only sheltered areas such as entryways need occasional washing with a scrub brush or a pressurized stream of water. In marine atmospheres and heavy industrial areas, periodic cleaning with detergents and water to remove salt and dirt deposits is advisable. Heavy or stubborn deposits may have to be removed with strong industrial cleaners.

For interior surfaces, only occasional cleaning with detergent and water is required for maintenance of finish. Where fingerprints are a problem, a commercial glass cleaner or wax is suggested. Often, a No. 4 sheet finish is specified to minimize the effect of fingerprints on appearance.

Food-Handling Applications. Stainless steel is widely specified for food-handling equipment because of its excellent "bacterial cleanability". In many instances, strong sanitizing or

Table 13 Typical methods of cleansing stainless steel surfaces

Cleansing problem	Cleansing agent(a)	Method of application(b)
Routine cleansing	Soap, ammonia, detergent and water	Sponge with cloth, then rinse with clear water and wipe dry. Satisfactory for use on all finishes
Smears and fingerprints	Arcal 20, Lac-O-Nu, Lumin Wash, O'Cedar Cream Polish, Stainless Shine, Wind-O-Shine	Rub with cloth as directed on the package. Satisfactory for use on all finishes. Provides barrier film to minimize prints
Stubborn spots, stains and other light discolorations	Allchem Concentrated Cleaner, Samae, Cameo Copper Cleaner, Cooper's Stainless Steel Cleaner, Revere Stainless Steel Cleaner, Paste NuSteel, DuBois Temp, Aerogroom Household cleansers, such as Old Dutch, Bab-O, Sapolio, Bon Ami, Ajax, Comet; grade F Italian pumice, Steel Bright, Lumin Cleaner, Zud, Restoro, Sta-Clean, Highlite, Penny-Brite, Copper-Brite, DuBois Stainless Steel Polish	Apply with sponge or cloth. Satisfactory for use on all finishes. Use in direction of polish lines on No. 4 finish. Use light pressure on No. 2, 7 and 8 finishes. May scratch No. 2, 7 and 8 finishes
Burnt-on foods and grease, fatty acids, milkstone (where swabbing or rubbing is not practical)	Easy-Off, De-Grease-It, 4 to 6% hot solution of such agents as trisodium phosphate or sodium tripolyphosphate, 5 to 15% caustic soda solution	Apply generous coating, allow to stand for 10-15 minutes, rinse. Repeated application may be necessary. Satisfactory for use on all finishes
Tenacious deposits, rusty discolorations, industrial atmospheric stains	Oakite No. 33, Dilac, Flash-Flenz, Caddy Cleaner, Turco Scale 4368, Permag 57	Swab and soak with clean cloth, let stand 15 minutes or more according to directions on package, then rinse and dry. Satisfactory for use on all finishes
Hard water spots and scale	Vinegar, 5% oxalic acid, 5% sulfamic acid, 5 to 10% phosphoric acid, Dilac, Oakite No. 33, Texo 12	Swab or soak with cloth, let stand 10-15 minutes. Always follow with neutralizer rinse and dry. Satisfactory for use on all finishes
Grease and oil	Organic solvents, detergents, caustic cleaners	Rub with cloth. (Organic solvents may be flammable and/or toxic.) Satisfactory for use on all finishes

(a) Use of proprietary names is intended to indicate type of cleanser, and does not constitute an endorsement. Omission of any proprietary cleanser does not imply inadequacy. All products should be used in strict accordance with instructions on package. (b) In all applications, stainless steel wool, a sponge, or a fibrous brush or pad is recommended for scouring stainless steel. Use of ordinary steel wool or steel brushes will leave a residue and result in corrosion and/or rust staining.

sterilizing solutions are used for cleaning the equipment to prevent bacterial contamination of the food products being processed. Where this is done, it should be standard practice to monitor exposure time and thoroughly flush the cleaned surfaces with water. Burnt-on foods and grease spots can be removed by soaking in hot water and detergent. Stubborn spots can be removed by scrubbing with a nonabrasive cleanser and a fiber brush, a sponge, or a pad of stainless steel wool or nickel-silver wool.

Chemical, textile and drug applications often require high purity in the product being processed. Stainless steel is used in equipment for these industries not only because it is chemically inert to the products, which effectively eliminates corrosion as a possible source of low-level contamination, but also because the surfaces of stainless steel equipment are easy to clean and sterilize, which effectively eliminates bacteria and residues as sources of contamination. Equipment usually is cleaned with strong chemical cleaners, then repeatedly rinsed with water. To facilitate cleaning, equipment usually has rounded corners and fillets, welded construction is used instead of mechanical seams, and all welds and other protrusions are ground flush and polished.

Selected references for latest specification limits and standards

- ASTM Book of Standards
- AISI Steel Products Manual
- Bulletins of the Welded Steel Tube Institute
- AWS Standards
- Producers' literature, especially data sheets and product bulletins

For General Reading

- *Stainless Iron and Steel* (2 Volumes), by J. H. G. Monypenny: Chapman and Hall, London, 1951
- *The Alloys of Iron and Chromium* (2 Volumes), by A. B. Kinzel *et al:* McGraw-Hill, New York, 1937
- *The Book of Stainless Steels,* Edited by E. E. Thum: American Society for Metals, Cleveland, 1935
- *Stainless Steel for Architectural Use:* STP454, American Society for Testing and Materials, Philadelphia, 1969

- *The Making, Shaping and Treating of Steel,* 9th Ed., Edited by H. E. McGannon, United States Steel Corporation, Pittsburgh, 1971
- *The Metallurgical Evolution of Stainless Steels,* Edited by F. B. Pickering: American Society for Metals, Metals Park, OH, 1979
- *Source Book on Stainless Steels:* American Society for Metals, Metals Park, OH, 1976
- *Stainless Steels,* by C. A. Zapffe: American Society for Metals, Cleveland, 1949
- *Handbook of Stainless Steels,* by D. Peckner and I. M. Bernstein: McGraw-Hill, New York, 1977
- *Corrosion of Stainless Steels,* by A. J. Sedriks: Wiley-Interscience, New York, 1979
- *Elevated Temperature Properties in Austenitic Stainless Steels,* Edited by A. O. Schaefer: American Society of Mechanical Engineers, New York, 1974
- *An Introduction to Stainless Steel,* by J. G. Parr and A. Hanson: American Society for Metals, Metals Park, OH, 1965
- *Stainless Steel '77,* Edited by R. Q. Barr: Climax Molybdenum Company, Ann Arbor, MI, 1978

Fabrication of Wrought Stainless Steels

By the ASM Committee on Wrought Stainless Steels*

FABRICATION of wrought stainless steels differs from fabrication of carbon and low-alloy steels primarily because stainless steels (*a*) are stronger, harder and more ductile, (*b*) work harden more readily and (*c*) generally must present a corrosion-resistant surface in the finished product. These characteristics dictate use of greater power, more frequent repair or replacement of processing equipment, and application of procedures to minimize or correct surface contamination. The specific process used for fabrication of stainless steel depends on the type of stainless steel being processed.

Forming

The method chosen for forming stainless steel should be based on the characteristics of the type to be used and the thickness of the part to be formed. As indicated above, power requirements are higher for forming stainless steels than for forming carbon steels—particularly austenitic types, which work harden more rapidly than ferritic types. Warm or hot forming may be necessary for thicknesses that can be formed cold in carbon steel. More detailed information on forming may be found in a separate volume of this Handbook. Because of their high ductility, austenitic types are the most readily formed stainless steels: austenitic stainless steel sheet can be severely drawn. Austenitic types 201 and 301 can be formed with biaxial stretching in excess of 35% because partial transformation to martensite during deformation helps the metal resist necking and deform more uniformly. (For unusually severe forming, composition may have to be adjusted to suit the particular job, and slow forming may be necessary to prevent buildup of heat and loss of the martensite effect.) Ferritic and lower-alloy martensitic types also can be cold formed extensively. However, they are less ductile than austenitic types, and thus forming of these alloys is more limited and intermediate annealing is more likely to be needed. The higher-carbon martensitic types such as 440A, 440B and 440C have only limited cold formability. Stainless steels generally gall more readily than other steels and thus require more attention to lubrication during forming. Straight mineral oils rarely provide adequate lubrication where sliding contact occurs in forming, and lubricants with extreme-pressure additives are often used.

Surface contamination during forming and handling should be kept to a minimum by thoroughly cleaning equipment and providing proper lubrication. Contamination arises from pickup of residues of other metals formed on the same equipment or from pickup of die material—particularly by heavy sections such as plate and bar, which require high contact pressures. Carbon or low-alloy steel embedded in the surface of a formed stainless steel part quickly produces superficial rusting, which impairs appearance and can initiate corrosion of the stainless steel itself under some conditions. If it is not practical to keep the equipment clean and the stock well lubricated, formed parts may be acid cleaned to remove surface contamination, as discussed later in this article.

Cold formed stainless steel parts usually are used in the as-formed condition. However, for applications in which stress-corrosion cracking may occur, austenitic types susceptible to

*See page X for committee list.

this failure process should be solution annealed after forming to remove residual stresses. Under all but the mildest service conditions, hot formed parts require postforming annealing to restore corrosion resistance and/or ductility.

Forging

All standard types of stainless steel can be forged. However, as alloy content increases within a given group, forging becomes more difficult. Forging difficulties are most common in initial breakdown of high-alloy ingots, and precautions may be needed to avoid surface ruptures.

Working Temperatures. Typical forging-temperature ranges for most standard stainless steels are indicated in Fig. 1. A wide range of forging temperatures may be used for most of the common austenitic types because of the natural workability of austenite and the absence of allotropic transformation. The conventional 18-8 types often are forged at temperatures up to 1260 °C (2300 °F). However, the upper temperature limit is lower for the higher-alloy grades due to metallurgical changes at higher temperatures that can cause surface ruptures. Maximum temperature is lowest for types 309, 310 and 330. Adherence to maximum temperature limits is particularly important for ingot breakdown, where severe tearing along grain boundaries in the cast metal may occur if temperature is too high.

Small amounts of delta ferrite can impair the forgeability of some austenitic types—particularly in upset forging, where some of the tangential tensile forces are perpendicular to ferrite stringers. Types 304, 309, 316, 317 and 321 are especially likely to contain significant amounts of ferrite, and it may be advisable to limit ferrite content of these grades for severe forging applications. Ferrite can be particularly troublesome in initial ingot breakdown; a homogenization treatment at about 1150 °C (2100 °F), to transform some of the delta ferrite to austenite before heating to forging temperature, can be helpful.

Austenitic stainless steel forgings should be solution annealed to restore corrosion resistance and maximum ductility. For some applications, annealing may be omitted if working is finished above 870 °C (1600 °F) and the forging is rapidly cooled below 425 °C (800 °F) in order to prevent carbide precipitation. This approach should be used with caution, and should incorporate metallurgical checks to determine that detrimental carbide precipitation has actually been avoided. The stabilized types (321, 347 and 348) and extra-low-carbon types (304L and 316L) will not precipitate carbides as readily, and may be used in the as-forged condition with fewer precautions if maximum corrosion resistance and ductility are not essential.

Fig. 1 Typical temperature ranges for forging of stainless steels

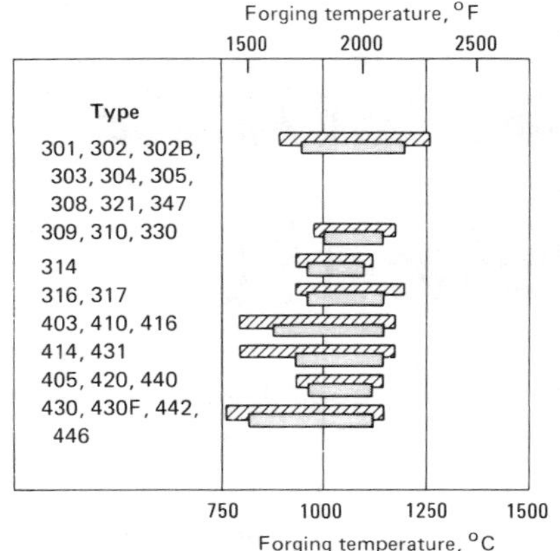

Crosshatched bars show temperature ranges that have been recommended by some but on which there is not general agreement. Solid bars are generally accepted.

The lower-carbon martensitic stainless steels can be hot worked over a wide temperature range. Finishing temperature is limited only by the allotropic transformation, which begins near 800 °C (1500 °F). The higher-carbon types (420 and 440) have a more limited forging-temperature range because of lower hot ductility. Because all martensitic types are highly hardenable, they require some type of heat treatment (generally annealing) after forging. Except for type 410, which may be used in the annealed condition, martensitic types should be further heat treated, if not for higher hardness then for best corrosion resistance, before being put into service.

A broad range of forging temperatures is shown in Fig. 1 for the ferritic steels. Forging is done at slightly lower temperatures than for austenitic types because the ferritic types tend to exhibit grain growth and structural weakness. Finishing temperatures are closely restricted only for types 405, 430 and 442. These alloys require special consideration because of grain-boundary weakness due to development of a small amount of austenite. The other ferritic types are commonly finished at temperatures as low as 700 °C (1300 °F). For fully ferritic types such as type 446, at least the final 10% reduction (and preferably more) should be performed below 870 °C (1600 °F) to achieve grain refinement and to develop optimum room-temperature ductility. Annealing after forging is recommended for ferritic types, because most

Fig. 2 Hypothetical closed-die forgings illustrating three degrees of severity

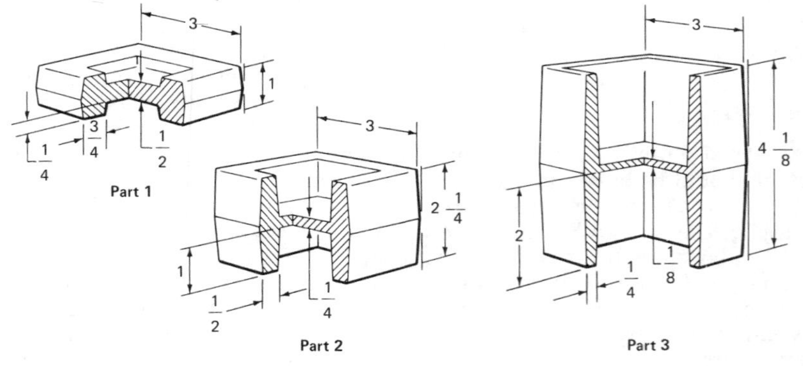

Dimensions are in inches; for equivalent dimensions in millimetres, multiply by 25.

of them contain substantial amounts of brittle martensite in the as-forged condition even though this may not be apparent from hardness measurements.

Forging Design. All standard 300 and 400 series stainless steels can be forged into any of the hypothetical parts shown in Fig. 2. For any given shape, die life will be shorter in forging a stainless steel than in forging a carbon or low-alloy steel. Die life and cost generally vary with the hot strength of the grade, the austenitic types giving shorter die life than ferritic or martensitic types, and the highly alloyed austenitic types giving the shortest life.

Shapes with mild contours, such as that of part 1 in Fig. 2, can be forged economically from any type of stainless steel with a single heating and about five hammer blows. A single die made from conventional prehardened die steel (hardness, 41 to 45 HRC) is recommended for quantities up to 10 000 parts. Forms approximating the severity of part 2 can be forged from any type of stainless steel with a single heating and about 10 hammer blows. Dies can be made from prehardened steels such as those recommended for part 1, but inserts of high-alloy hot work steel such as H12 (hardness, 46 to 48 HRC) can extend die life significantly. Part 3 represents maximum severity for parts forged from any stainless steel—especially from one with high strength at forging temperature, such as type 309, 310, 314 or 330. Straight chromium types are the easiest to forge into such a severe shape, with type 440 being the least practical of the 400 series because of its high carbon content. The austenitic types other than 309, 310, 314 and 330 are intermediate in forgeability for this shape. For any stainless steel, at least one reheating would be required to forge part 3, and a thicker web and more generous radii might be necessary for the higher-alloy grades. Regardless of the alloy being forged, a softer die-block material (36 to 40 HRC) should be used to lessen the probability of the die splitting. Die inserts made of H12 are recommended if more than a very few parts are to be forged.

Die life generally is determined more by severity of shape than by the alloy being forged. Life of dies used in forging parts similar to part 2 normally is no more than half that of dies used in forging shapes similar to part 1. In one extreme instance, average die life in forging a mildly contoured part from type 431 was 8000 pieces. In forging the same steel with the same die material to a shape about halfway in severity between part 2 and part 3, die life was only 1500 pieces per die.

Fig. 3 Hypothetical cold headed parts illustrating three degrees of severity

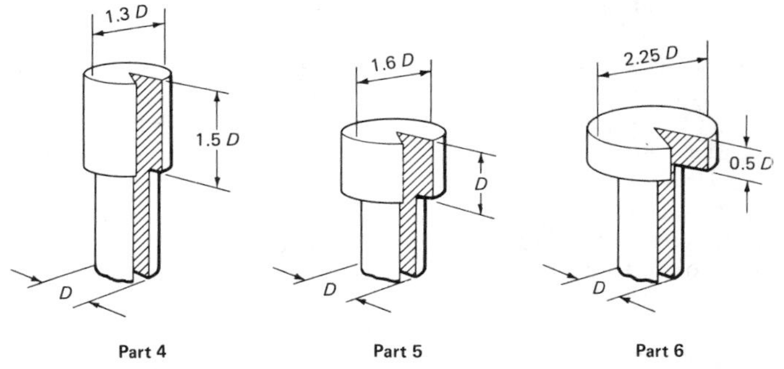

1.3 D
1.5 D
D
Part 4
1.6 D
D
D
Part 5
2.25 D
0.5 D
D
Part 6

Hot Heading. Hypothetical parts 4, 5 and 6 in Fig. 3 can be hot upset in one blow from any type of stainless steel. Conditions are similar to those encountered in hot die forging. First, upsetting part 6 from stainless steel will result in several times as much die wear as will upsetting part 4. Second, die wear from forming any shape will increase as the elevated-temperature strength of the alloy increases. For this reason, type 410, which has the lowest strength at high temperature, would perform best in hot upsetting of any of the three hot upset parts, especially part 6; type 330 would be the least practical for hot upsetting.

Both gripper dies and heading tools sustain more wear in upsetting stainless steels than in upsetting carbon or low-alloy steels. For example, 8000 pieces of 9310 steel similar in shape to part 6 were obtained using a single set of upsetting tools made of H12 steel. In hot upsetting the identical shape from type 304, only 1500 pieces were produced using a similar set of tools.

Cold Working

The cold working operations that can be successfully performed on most stainless steels include cold heading, cold drawing, cold extrusion and cold riveting. Cold working of stainless steel is more difficult than cold working of carbon steel because of differences in strength and work hardening, and power requirements are proportionately greater.

Cold Heading. As a result of the wide differences in work-hardening capabilities among stainless steels, some types are much more adaptable than others to cold heading. More detailed information can be found in Volume 4 of the 8th Edition of this Handbook. With the exception of free-machining types such as 303 and 416, which are almost never recommended for cold heading, all stainless steels can be cold headed into parts similar in shape to part 4 in Fig. 3 with two blows using machine-oil lubricant. The surface finish of the part depends largely on the finish of the wire, which suggests the use of lightly drawn rather than annealed wire.

With sufficiently powerful equipment, parts with shapes similar to that of part 5 in Fig. 3 can be cold headed in three blows from any stainless steel except the free-machining and higher-carbon martensitic types. However, forming pressure increases with severity of shape, and lubrication of dies is difficult when upsetting is severe. Ferritic and lower-carbon martensitic types behave like carbon or low-alloy steels in cold heading, and can be headed to shapes like that of part 5 using machine-oil lubricant. Austenitic types require at least extreme-pressure lubricants for such heading operations, and copper plating or lime coating with machine-oil lubricant is frequently more satisfactory.

Type 430 is the only grade of the 400 series recommended for cold heading to shapes as severe as that of part 6. Machine-oil lubrication is generally satisfactory. Of the austenitic types, only the low-alloy, low-work-hardening grades 305 and 304Cu, coated with copper or lime, are recommended for such severe cold heading. The other austenitic types can be cold headed to this degree, but short tool life, and high scrap rates due to edge cracking, normally make them too costly for such parts. Often, however, they can be warm headed at about 200 °C (400 °F) to shapes like that of part 6. At this temperature, work hardening is drastically reduced because no martensite forms during deformation, minimizing tool loads and the likelihood of cracking.

Cold headed parts made of austenitic and ferritic stainless steels are most often used in the as-headed condition, although annealing may be needed if service conditions are likely to cause stress-corrosion cracking. Martensitic types are usually heat treated after heading.

Cold Drawing. Stainless steel bar and wire can be cold drawn in the conventional manner except for the greater power requirements and the need for greater attention to lubrication. Round, square, hexagonal and other sections up to 75 mm (3 in.) and more in diameter and up to 12.2 m (40 ft) long can be cold drawn with proper lubrication (usually over a lime coating) and slow drawbench speed. In wiredrawing, a set of carbide dies will produce about 180 000 kg (400 000 lb) of wire 1.25 mm (0.050 in.) in diameter. Diamond dies often are used for sizes under 0.8 mm (0.032 in.). Sizes as small as 0.075 mm (0.003 in.) are readily drawn, although intermediate annealing is usually necessary for such fine wire. Any of the stainless steels can be butt welded to join one length to another, but welding of austenitic types is the most trouble-free.

Type 304 is the type most commonly drawn into wire, but other types are drawn when special characteristics are needed. Ferritic and martensitic types may require special care in drawing. The free-machining types require small reductions per pass because of their high content of nonmetallic inclusions. The high-carbon martensitic types are susceptible to breakage even after thorough annealing. Martensitic types in general, and high-carbon martensitic types in particular, are also likely to be damaged by pickup of hydrogen during annealing, pickling or electroplating. To a somewhat lesser extent, hydrogen can also be troublesome in metastable austenitic types such as 301. Unless the material is baked at about 175 °C (350 °F), breakage due to hydrogen pickup can occur in drawing and even in coiling, sometimes to a severe degree.

Cold Extrusion. Despite its high work-hardening rate and tendency to gall, stainless steel can be cold extruded into parts such as tubular rivets. Depth of extrusion is limited to about 85% of that obtainable in plain carbon steels, and drafts lighter than those used for other materials are required for reasonable punch life. If these limitations can be accepted, substantial savings can be realized by making such parts by cold extrusion rather than by machining.

Riveting. Rivets of most stainless steels can be driven cold. Exceptions are those made of the free-machining and high-carbon types, which tend to crack or split unless driven hot. If rivets are to be driven hot, certain precautions must be observed. Martensitic and ferritic types must never be heated to a temperature that exceeds 785 °C (1450 °F). Heating above this temperature results in undesirable hardening and embrittlement. Unstabilized austenitic stainless steels are susceptible to harmful carbide precipitation at temperatures from about 425 to about 870 °C (800 to 1600 °F), and therefore rivets made of these types should be heated to about 1100 °C (2000 °F) for hot driving. Rivets of stabilized or extra-low-carbon types can be driven at any convenient temperature, but heating time should be kept short. The high thermal expansion of austenitic types ensures a tight axial set on cooling, and thus these types are usually preferred for hot riveting.

It is preferable that ferritic and martensitic steel rivets be designed with generous fillets and be driven into chamfered holes. For rivets made of austenitic types, which are more ductile, these precautions are not necessary.

Machining

Stainless steels as a class are more difficult to machine than carbon and low-alloy steels because of their higher strength and higher work-hardening rates ("gummy" nature). These characteristics require greater power and lower machining speed, shorten tool life, and sometimes lead to difficulty in obtaining a fine finish on the machined surface. Wide variations exist in these characteristics among the different stainless steels. More detailed information about machining of stainless steels can be found in the article entitled "Machining Stainless Steels" in the Metals Handbook volume on machining.

Procedures. In machining of stainless steels, special attention must be paid to equipment in order to control the effects of strength and work hardening. Rigid equipment and tooling are necessary to prevent chatter. Chip-curler tools are generally recommended because of the tough, stringy chips produced—particularly in machining austenitic and high-alloy ferritic types. Carbide cutting tools are preferred because they provide acceptable tool life at production machining speeds.

The following precautions should be instituted to avoid work hardening. Tools should never be permitted to ride or glaze without cutting, because the surface can work harden to the extent that cutting tools will become burned before they penetrate the surface. Care should be taken to ensure that hardening from one operation does not interfere with subsequent machining. For example, a tripod punch rather than a conventional center punch is preferred for hole location to prevent work hardening at the spot that will be touched first by the drill.

Heavier feeds and lower speeds than those used in machining low-alloy steels are used to minimize work hardening. Table 1 shows typical machining speeds for most standard stainless steels. The speeds listed represent generalizations based on years of field experience, and this table should be used only as a guide in setting up particular jobs. Optimum speed can vary widely depending on specific conditions. In general, the tabulated values are based on one tool regrind per eight-hour shift with the tool cutting 60% of the time.

Type Variations. Low-alloy martensitic and ferritic stainless steels have machining characteristics much like those of low-alloy steels, whereas the higher-carbon martensitic stainless steels are among the most difficult metals to machine. Austenitic and precipitation-hardening stainless steels vary more widely in machining characteristics within each class than do the ferritic and martensitic grades. Most easily machined are the free-machining types. The unique characteristics of

Table 1 Machinability of wrought stainless steels

Operation or type of cutting machine	Machining speeds, ft/min(a), for type: 403 (b), 405, 410 (b): 180 to 240 HB	416 (b): 180 to 240 HB	420, 420F (c): 180 to 230 HB	430: 170 to 230 HB	430F: 170 to 230 HB	414 (b), 431: 230 to 280 HB	440A, 440B, 440C, 440F (c): 200 to 265 HB	446: 170 to 230 HB
Automatic screw machine(d)	90 to 100	120 to 150	80 to 100	90 to 100	120 to 150	80 to 100	60 to 80	80 to 100
Heavy-duty single or multiple spindle(d)	80 to 100	110 to 130	60 to 80	80 to 100	110 to 130	70 to 90	50 to 70	60 to 80
Turret lathe(d)	80 to 100	100 to 130	60 to 80	80 to 100	110 to 130	70 to 90	50 to 70	60 to 80
Automatic screw machine(e)	110 to 140	120 to 150	90 to 120	110 to 140	120 to 150	100 to 140	60 to 100	100 to 140
Milling(f)	40 to 60	50 to 80	30 to 50	40 to 60	50 to 80	40 to 60	30 to 50	40 to 60
Reaming(f)								
Smooth finish	15 to 40	15 to 40	15 to 40	15 to 40	15 to 40	15 to 40	15 to 40	15 to 40
Work sizing	40 to 120	40 to 120	40 to 120	40 to 120	40 to 120	40 to 120	40 to 120	40 to 120
Threading(g)	10 to 25	10 to 25	10 to 25	10 to 25	10 to 25	10 to 25	10 to 25	10 to 25
Tapping(g)	10 to 25	10 to 25	10 to 25	10 to 25	10 to 25	10 to 25	10 to 25	10 to 25
Drill press(g)	40 to 80	60 to 90	30 to 50	40 to 80	60 to 90	40 to 60	30 to 50	40 to 60
Single-point turning:								
Carbide tooling:								
Roughing	150 to 200	150 to 200	100 to 150	150 to 200	150 to 200	140 to 180	100 to 150	140 to 180
Finishing	200 to 400	200 to 400	150 to 250	200 to 400	200 to 400	150 to 350	150 to 200	150 to 350
High-cobalt or cast alloy tooling:								
Roughing	100 to 130	100 to 150	80 to 100	100 to 130	100 to 150	90 to 120	60 to 80	100 to 130
Finishing	100 to 150	150 to 200	100 to 150	100 to 150	150 to 200	90 to 140	80 to 100	100 to 150
High speed steel tooling:								
Roughing	80 to 100	80 to 100	60 to 80	80 to 100	80 to 100	60 to 80	40 to 60	60 to 90
Finishing	80 to 130	100 to 150	80 to 120	80 to 130	100 to 150	80 to 100	60 to 80	90 to 120

Operation or type of cutting machine	Machining speeds, ft/min(a), for type: 301, 302, 304, 304L: 150 to 250 HB	303: 150 to 240 HB	309, 309S, 310, 310S, 316, 316L: 150 to 240 HB	321, 347: 150 to 240 HB	347F: 150 to 240 HB	17-4 PH: 300 to 360 HB	17-7 PH: 150 to 240 HB
Automatic screw machine(d)	70 to 90	100 to 130	60 to 80	70 to 90	90 to 110	60 to 80	60 to 80
Heavy-duty single or multiple spindle(d)	60 to 80	90 to 120	60 to 80	60 to 80	80 to 100	50 to 70	50 to 70
Turret lathe(d)	60 to 80	90 to 120	60 to 80	60 to 80	80 to 100	50 to 70	50 to 70
Automatic screw machine(e)	80 to 120	110 to 130	80 to 120	80 to 120	100 to 120	80 to 120	80 to 120
Milling(f)	40 to 60	40 to 60	30 to 50	40 to 60	40 to 60	40 to 60	40 to 60
Reaming(f)							
Smooth finish	15 to 40	15 to 40	15 to 40	15 to 40	15 to 40	15 to 40	15 to 40
Work sizing	40 to 80	40 to 120	40 to 80	40 to 80	40 to 80	40 to 80	40 to 80
Threading(g)	10 to 25	10 to 25	10 to 25	10 to 25	10 to 25	10 to 25	10 to 25
Tapping(g)	10 to 25	10 to 25	10 to 25	10 to 25	10 to 25	10 to 25	10 to 25
Drill press(g)	30 to 50	50 to 80	30 to 50	30 to 50	30 to 50	40 to 60	40 to 60
Single-point turning:							
Carbide tooling:							
Roughing	130 to 180	150 to 250	130 to 180	130 to 180	150 to 250	130 to 180	130 to 180
Finishing	150 to 300	200 to 400	150 to 300	150 to 300	200 to 400	150 to 300	150 to 300
High-cobalt or cast alloy tooling:							
Roughing	100 to 130	100 to 150	100 to 130	100 to 130	100 to 140	100 to 130	100 to 130
Finishing	100 to 150	150 to 200	100 to 150	100 to 150	140 to 190	100 to 150	100 to 150
High speed steel tooling:							
Roughing	60 to 90	70 to 90	60 to 90	60 to 90	60 to 90	60 to 90	60 to 90
Finishing	100 to 120	100 to 140	100 to 120	100 to 120	100 to 130	100 to 120	100 to 120

(a) To obtain equivalent values in m/min, multiply listed values by 0.3. (b) Harder stock in the 260 to 320 HB range may be machined by reducing these speeds approximately 20%. (c) When using an automatic screw machine, cutting speeds may be increased about 10% over those shown. (d) Based on tungsten or molybdenum high speed steel tooling. Rates may be increased 15 to 30% with high-cobalt or cast alloys. (e) Based on the use of tools made of cemented carbide or cast cobalt-chromium-tungsten alloy. (f) Based on tungsten or molybdenum high speed steel tooling. Greatly increased speeds can be used with carbide tooling. (g) Based on tungsten or molybdenum high speed steel tooling.

free-machining stainless steels should be considered, particularly as part of an effort to control strength and work hardening.

The structures of low-alloy martensitic and ferritic stainless steels make these types somewhat brittle, resulting in reasonably good chip breakage. However, hardness levels generally are higher than those of annealed low-alloy steels. The low-alloy martensitic stainless steels often must be machined in the hardened-and-tempered condition (up to 38 HRC), which produces excellent dimensional accuracy and surface finish.

Higher-carbon martensitic stainless steels such as types 420 and 440, and particularly type 440C, are progressively more difficult to machine because of their high annealed hardnesses (up to 240 HB) and the presence of hard, abrasive chromium carbides in their microstructures. The high-chromium ferritic stainless steels such as type 446 are difficult to machine because, like austenitic types, they are "gummy" and produce stringy chips.

Austenitic steels such as types 304 and 316 have tensile strengths of 550 to 620 MPa (80 to 90 ksi) in the annealed condition—the same range of strength as that of annealed 1050 carbon steel. However, austenitic stainless steels exhibit much greater spreads between yield and ultimate strengths and much higher work-hardening rates—particularly the leaner alloys such as types 302 and 304.

Precipitation-hardening types vary considerably in machining characteristics because of differences in structure. They may be ferritic, martensitic, austenitic or two-phase, so machining characteristics will be characteristic of the structure that exists at the time of machining. Like martensitic types, precipitation-hardening stainless steels sometimes are machined after being heat treated to high strength in order to produce parts with closer tolerances than those obtainable by machining before heat treatment.

Free-machining stainless steels such as types 416, 430F and 303 are significantly more machinable than their non-free-machining counterparts because they contain small amounts of various free-machining additives. The most common additive is sulfur, which minimizes buildup of metal on cutting edges and promotes chip breakage, thereby permitting higher machining speeds and lower power consumption, and promoting longer tool life. The sulfur is present in nonmetallic inclusions, usually complex manganese sulfides. Selenium has beneficial effects similar to those of sulfur, but generally gives a better surface finish. Selenium also imparts greater ductility to free-machining stainless steels than does sulfur, and thus selenium-bearing types are better adapted to applications that require both good ductility, such as for cold upsetting, and good machinability. Other free-machining additives are phosphorus, which reduces matrix "gumminess", and lead, which lubricates tools and facilitates chip breakage.

Martensitic type 416 and ferritic type 430F can be machined at speeds up to 85% of those used for B1112 screw stock—a significant gain over equivalent types that are not free-machining. Chips produced from free-machining types are short, brittle and easily disposable. Of the austenitic stainless steels, type 303 has about 30% better machinability than type 304. Free-machining variants of other types are also available on a limited basis. Increases in machinability may be somewhat less for the higher alloys—particularly for the high-carbon martensitic stainless steels such as types 440A, 440B and 440C, for which abrasion from chromium carbides is a primary limiting factor.

Because of the high costs of labor and capital, the economic benefits of free-machining stainless steels can be substantial. The free-machining types are slightly more expensive, but this is more than offset by savings in machining costs when extensive machining is required. As a rule of thumb, a free-machining type may be cost-effective in any application where more than 10% of the material must be machined away.

The characteristics of free-machining types must be carefully considered to ensure that parts will perform satisfactorily in service. These steels have somewhat lower corrosion resistance than their unmodified counterparts, particularly under conditions where nonmetallic inclusions may initiate pitting. However, the corrosion resistance of free-machining types is probably adequate for the majority of applications where stainless steel is specified. Free-machining stainless steels have a limited ability to be cold headed, but selenium-bearing types are more readily cold headed than other free-machining types. They all have limited weldability, and sound weldments have been produced only under closely controlled conditions.

Passivation. Machined stainless steel parts often are given a cleaning/passivation treatment by immersing them for 10 minutes to one hour in an oxidizing acid bath. (The subject of acid cleaning is discussed in greater detail in the section of this article entitled "Cleaning and Finishing".) Such treatment actually does not enhance the natural surface passivity of stainless steel but rather restores corrosion resistance by removing surface contamination such as embedded tool material.

Heat Treating

Stainless steels are subjected to various heat treatments depending on the type and on the requirements of the application. These treatments, which include annealing, hardening and stress relieving, restore desirable properties such as corrosion resistance and ductility to metal altered by prior fabrication operations. Heat treatment is often performed in controlled atmospheres to prevent detrimental surface effects.

Annealing. All types of stainless steel can be annealed; specific characteristics determine the process used. The various types differ in amount of heating required, length of time at temperature, cooling process used, and required cooling rate.

Annealing of austenitic types not only recrystallizes the grains and softens the metal, but also takes chromium carbides into solution in the austenite. Because of the latter effect, the process is sometimes referred to as solution annealing. Temperatures must exceed an intermediate range to avoid sensitization due to carbide precipitation along grain boundaries. Annealing temperatures usually are above 1040 °C (1900 °F), although some types may be annealed at closely controlled temperatures as low as 1010 °C (1850 °F) when fine grain size is important. Time at temperature is kept short to hold surface scaling to a minimum and to control grain growth, which can lead to "orange peel" in forming.

Annealing of austenitic stainless steel is occasionally called quench annealing because the metal must be cooled rapidly, usually by water quenching, to prevent sensitization (except for stabilized and extra-low-carbon types). Precipitation of chromium carbides can severely impair corrosion resistance because chromium is depleted in the matrix immediately adjacent to the carbides and/or because the carbides themselves may induce galvanic corrosion. Therefore, if water quenching is not used, thorough investigation is needed to ensure that sensitization does not occur. The investigation must take actual composition into account, because the rate of carbide precipitation varies markedly with composition: a heat of type 304 containing 0.05% carbon may be free of precipitation under cooling conditions that would produce heavy sensitization in the same alloy containing 0.08% carbon. Austenitic stainless steels are softened by recrystallization at the annealing temperature and, unlike most other steels, are not hardened by quenching.

A stabilizing anneal is sometimes performed after conventional annealing for types 321, 347 and 348. Most of the carbon content is combined with titanium in type 321 or with niobium in types 347 and 348 when these types are annealed in the usual manner. A further anneal at 870 to 900 °C (1600 to 1650 °F) for 2 to 4 h followed by rapid cooling precipitates all possible carbon as a titanium or niobium carbide and prevents subsequent precipitation of chromium carbide. It is believed that this special protective treatment is sometimes useful when service conditions are rigorously corrosive—especially when service also involves temperatures from about 400 to 870 °C (750 to 1600 °F).

Before annealing or other heat treating operations are performed on austenitic stainless steels, the steel should be cleaned to remove oil, grease and other carbonaceous residues. Such residues lead to carburization during heat treating, which degrades corrosion resistance.

All martensitic and most ferritic stainless steels can be subcritical annealed (process annealed) by heating into the upper part of the ferrite temperature range, or full annealed by heating above the critical temperature into the austenite range, followed by slow cooling. Usual temperatures are 760 to 830 °C (1400 to 1525 °F) for subcritical annealing, and 845 to 900 °C (1550 to 1650 °F) for full annealing. When material has been previously heated above the critical temperature, such as in hot working, at least some martensite is present even in ferritic stainless steels such as type 430. Relatively slow cooling at about 25 °C/h (50 °F/h) from full annealing temperature, or holding for one hour or more at subcritical annealing temperature, is required to produce the desired soft structure of ferrite and spheroidized carbides. However, parts that have undergone only cold working after full annealing can be subcritically annealed satisfactorily in less than 30 minutes.

The ferritic types that retain predominantly single-phase structures throughout the working temperature range (types 409, 442, 446 and 26Cr-1Mo) require only short recrystallization annealing in the range 760 to 955 °C (1400 to 1750 °F). The higher-chromium types such as 446 and 26Cr-1Mo require rapid cooling through the range from 540 to 370 °C (1000 to 700 °F) to avoid "885 °F" embrittlement and consequent loss of ductility.

Hardening. Martensitic stainless steels are hardened by austenitizing, quenching and tempering much like lower-alloy steels. Austenitizing temperatures normally are 980 to 1010 °C (1800 to 1850 °F)—well above the critical temperature. As-quenched hardness increases with austenitizing temperature to about 980 °C (1800 °F) and then decreases due to retention of austenite, as shown in Fig. 4. For some types, the optimum austenitizing temperature may depend on the subsequent tempering temperature. For type 431 that is to be tempered at 315 °C (600 °F), best toughness is obtained by austenitizing at 1065 °C (1950 °F); but for 595 °C (1100 °F) tempering, austenitizing at 980 °C (1800 °F) gives maximum toughness.

Preheating before austenitizing is recommended to prevent cracking in high-carbon types and in intricate sections of low-carbon types. Preheating at 790 °C (1450 °F), and then heating to the austenitizing temperature is the most common practice, but very large or extremely intricate parts sometimes are successively preheated at 540 °C (1000 °F) and then at 790 °C (1450 °F) before austenitizing.

Martensitic stainless steels have high hardenability because of their high alloy content. Air cooling from the austenitizing temperature is usually adequate to produce full hardness. Oil quenching is sometimes used, particularly for larger sections. As discussed earlier, tempering temperature must be chosen for the optimum combination of hardness, toughness and corrosion resistance. Parts should be tempered as soon as they have cooled to room temperature—particularly if oil quenching has been used—to avoid delayed cracking. Parts sometimes are refrigerated to −75 °C (−100 °F) before tempering to transform retained austenite—particularly where dimensional stability is important, such as in gage blocks made of type 440C. Tempering at temperatures above 510 °C (950 °F) should be followed by relatively rapid cooling to below 400 °C (750 °F) to avoid "885 °F" embrittlement.

Fig. 4 Typical as-quenched hardness for type 403, 410 and 416 martensitic stainless steels

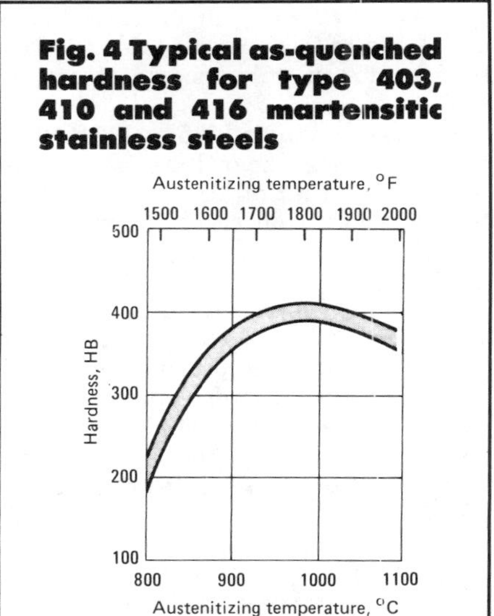

Some precipitation-hardening stainless steels require more complicated heat treatments than standard martensitic types. For instance, a semiaustenitic precipitation-hardening type may require annealing, trigger annealing (to condition austenite for transformation on cooling to room temperature), subzero cooling (to complete the transformation of austenite) and aging (to fully harden the alloy). On the other hand, martensitic precipitation-hardening types often require nothing more than a simple aging treatment.

Stress Relieving. Stainless steel weldments generally are heated to temperatures below the usual annealing temperature to decrease high residual stresses when full annealing after welding is impossible. Most often, stress relieving is performed on weldments that are too large or intricate for full annealing or on dissimilar-metal weldments consisting of austenitic stainless steel welded to alloy steel.

Stress relieving at temperatures below 400 °C (750 °F) is an acceptable practice but results in only modest stress relief.

Stress relieving at 425 to 925 °C (800 to 1700 °F) significantly reduces residual stresses that otherwise might lead to stress-corrosion cracking or dimensional instability in service. One hour at 870 °C (1600 °F) typically relieves about 85% of the residual stresses. However, stress relieving in this temperature range also precipitates grain-boundary carbides, resulting in sensitization that severely impairs corrosion resistance in many media. Sensitized austenitic stainless steels are susceptible to intergranular corrosion or stress-assisted intergranular corrosion even in some media considered mild. To avoid these effects, it is strongly recommended that a stabilized stainless steel (type 321, 347 or 348) or a low-carbon type (304L or 316L) be used, particularly when lengthy stress relieving is required.

When austenitic stainless steels have been cold worked to develop high strength, low-temperature stress relieving will increase the proportional limit and yield strength (particularly compressive yield strength). A two-hour treatment at 345 to 370 °C (650 to 700 °F) is normally used; temperatures up to 425 °C (800 °F) may be used if resistance to intercrystalline corrosion is not required for the application. Higher temperatures will reduce strength and sensitize the metal, and generally are not used for stress relieving cold worked products.

Stress relieving of martensitic or ferritic stainless steel weldments will simultaneously temper weld and heat-affected zones, and for most types will restore corrosion resistance to some degree. However, annealing temperatures are relatively low for these grades, and normal subcritical annealing is the heat treatment usually selected if the weldment is to be heat treated at all.

Atmospheres. Most heat treatment of stainless steel is carried out in conventional air furnaces. Heat treatment in air results in an oxidized and decarburized surface, but this is of little consequence for the many parts that are machined and/or cleaned following heat treatment. However, in some instances, stringent controls must be adopted to ensure that all of the decarburized surface is machined away. In others, acid pickling may be considered too expensive, may create waste-disposal problems and may degrade surface appearance. Because of all these difficulties associated with heat treating in an air furnace, much heat treating of stainless steels is now carried out in controlled atmospheres or in vacuum.

It is most common to use hydrogen, dissociated ammonia or vacuum, but increasing use is being made of "inert" high-nitrogen atmospheres or, for martensitic types only, endothermic atmospheres. Controlled-atmosphere heat treatment is sound practice, particularly when a bright annealed finish is desired. However, the atmosphere must be selected carefully for the application. Hydrogen pickup from a hydrogen or dissociated-ammonia atmosphere can affect toughness, particularly of martensitic types hardened to high strength, and has been known to increase breakage in drawing wire of martensitic or metastable austenitic stainless steels. Inadvertent nitriding due to poor control of high-nitrogen or dissociated-ammonia atmospheres can reduce the ductility of austenitic stainless steel foil, and can lower the M_s temperature in the surface layers of martensitic stainless steel parts.

Joining Processes

Stainless steels are commonly joined by welding, brazing and soldering. Arc welding is the overwhelming choice for joining stainless steel to stainless steel because it gives a relatively crevice-free joint of high joint efficiency. Resistance welding also is often used for austenitic types, because it produces a mechanically strong joint quickly and inexpensively. Procedures and precautions appropriate for the various types are important if optimum corrosion resistance and mechanical properties are to be attained in the completed assembly. Brazing usually is preferred for joining stainless steels to other metals.

Welding. All stainless steels can be readily arc welded by any common welding process (see the article on arc welding of stainless steel in Volume 6 of the 8th Edition of this Handbook for details). Table 2 lists filler metals commonly used for welding various types. Major factors that must be considered are (*a*) corrosion resistance in the weld and heat-affected zones, (*b*) residual stress, which can lead to distortion, weld cracking or fissuring, and (*c*) for martensitic and ferritic types, mechanical properties in the weld and heat-affected zones.

Resistance welding of stainless steels also is quite common. In fact, austenitic stainless steels are resistance welded more often than any other metal except carbon steel. The welding currents used in welding austenitic stainless steels are significantly lower than those required for welding carbon steels because of the lower thermal conductivity, higher electrical resistivity and nonmagnetic character of the austenitic types. Distortion is somewhat more troublesome than in resistance welding of low-alloy steels because austenitic stainless steels have greater coefficients of thermal expansion. Corrosion resistance may not be seriously impaired by carbide precipitation, because welding times are too short to produce extensive sensitization. Crevice corrosion, a form of localized corrosion resulting from limited access of oxygen, can be a problem in spot welded sheets when they are exposed in certain media. Martensitic and ferritic types also can be joined by resistance welding, although these types are used for resistance welded components much less commonly than austenitic types. For martensitic types and those ferritic types that form some martensite on cooling after welding, a second heating pulse can be programmed into the welding cycle to temper the weld area.

Stainless steels are rarely joined by gas welding. In oxyacetylene welding of stainless steels, great skill in controlling the welding atmosphere is required to prevent either oxidation or carburization of the weld puddle. Gas welding with an atomic hydrogen torch is less demanding, but such equipment is no longer widely used.

Austenitic types are the most weldable stainless steels, but are also the ones most different in welding behavior from carbon and low-alloy steels. Probably the most important metallurgical factor to consider in planning for austenitic stainless weldments is susceptibility to grain-boundary carbide precipitation (sensitization) at moderately elevated temperatures. Material immediately adjacent to the weld will be heated to or above the annealing temperature and will be free of precipitation. At some distance away—perhaps 3 mm (1/8 in.) or more, depending on welding parameters—the base metal is heated to 650 to 870 °C (1200 to

Table 2 Filler metals for arc welding of stainless steels

Base metal	Condition of weldment for service(a)	Electrode or welding rod(b)	Notes
Austenitic stainless steels			
301, 302, 304, 305, 308	1 or 2	308	(c)
302B	1	309	(d)
304L	1 or 4	347, 308L	. . .
303, 303(Se)	1 or 2	312	(e)
309, 309S	1	309	. . .
310, 310S	1	310	. . .
316	1 or 2	316	(f)
316L	1 or 4	318, 316L	(f)
317	1 or 2	317	(f)
317L	1 or 4	317(Cb)	(f)
318 [316(Cb)]	1 or 5	318	(f)
321	1 or 5	347	(g)
347	1 or 5	347	(h)
348	1 or 5	347	(j)
Martensitic stainless steels			
403, 410, 416, 416(Se)	2 or 3	410	(k)
403, 410	1	308, 309, 310	(m)
416, 416(Se)	1	308, 309, 312	(m)
420	2 or 3	420	(n)
431	2 or 3	410	(n)
431	1	308, 309, 310	(p)
Ferritic stainless steels			
405	2	405(Cb), 430	(q)
405, 430	1	308, 309, 310	(m)
430F, 430F(Se)	1	308, 309, 312	(m)
430, 430F, 430F(Se)	2	430	(r)
446	2	446	. . .
446	1	308, 309, 310	(s)

Source—Welding Characteristics of Stainless Steels, by George E. Linnert: *Metals Engineering Quarterly,* Nov 1967.
(a) 1, as welded; 2, annealed; 3, hardened and stress relieved; 4, stress relieved; 5, stabilized and stress relieved. (b) Prefix E or ER omitted. (c) Type 308 weld metal also referred to as 18-8 and 19-9 compositions. Actual weld requirements are 0.08% max C, 19.0% min Cr and 9.0% min Ni. (d) Type 310 (1.50% max Si) may be used as filler metal, but pickup of silicon from the base metal may result in hot cracking of the weld metal. (e) Free-machining base metal increases the probability of hot cracking of the weld metal. Type 312 filler metal provides weld deposits that contain large amounts of ferrite to prevent hot cracking. (f) Welds made with types 316, 316L, 317 and 317(Cb) electrodes or welding rods may occasionally display poor corrosion resistance in the as-welded condition. In such instances, corrosion resistance of the weld metal may be restored by the following heat treatments: for types 316 and 317 base metal, full anneal at 1065 to 1120 °C (1950 to 2050 °F); for types 316L and 317L base metal, 870 °C (1600 °F) stress relief; for type 318 base metal, 870 to 900 °C (1600 to 1650 °F) stabilizing treatment. Where postweld heat treatment is not possible, other filler metals may be specially selected to meet the requirements of the application for corrosion resistance. (g) Type 321 covered electrodes are not regularly manufactured, because titanium is not readily recovered during deposition. (h) Caution is needed in welding thick sections, because of cracking problems in heat-affected zones. (j) In base metal and weld metal, for nuclear service, tantalum is restricted to 0.10% max, and cobalt to 0.20% max. (k) Annealing softens and imparts ductility to heat-affected zones and weld metal. Weld metal responds to heat treatment in a manner similar to the base metal. (m) These austenitic weld metals are soft and ductile in the as-welded condition, but heat-affected zones have limited ductility. (n) Requires careful preheating and postweld heat treatment to avoid cracking. (p) Requires careful preheating. Service in as-welded condition requires consideration of hardened heat-affected zones. (q) Annealing increases ductility of heat-affected zones and weld metal. Type 405(Cb) weld metal contains niobium (columbium) rather than aluminum, to reduce hardening. (r) Annealing is employed to increase ductility of the welded joint. (s) Type 308 filler metal will not display scaling resistance equal to that of the base metal. Consideration must be given to differences in the coefficients of thermal expansion of the base metal and the weld metal.

1600 °F) and grain-boundary carbides can precipitate despite the short time at temperature. Carbide precipitation severely impairs corrosion resistance in many media, including the acids most often used for pickling to remove oxide. Nevertheless, because sensitization occurs in such a narrow region of the heat-affected zone, many austenitic stainless steel weldments are used in the as-welded condition without concern.

Free-machining stainless steels are quite difficult to weld, with porosity and segregation being common problems. However, these types can be welded successfully with special consumable electrodes (type 312 electrodes, for instance) if precautions are taken to exclude all traces of hydrogen.

Most precipitation-hardening stainless steels can be arc welded by techniques similar to those used for welding austenitic types. Normally, filler-metal composition is selected to match or nearly match base-metal composition so that response to heat treatment is similar. Welding usually is followed by full heat treatment, although aging alone may be adequate for single-pass welds in some types (such as 17-4 PH). In heat treated weldments, weld strength closely approaches base-metal strength, but toughness and ductility of the weld may be somewhat inferior, especially for material heat treated to high strength. Weldability of the fully austenitic precipitation-hardening stainless steels HNM and 17-10P is limited because these steels are hot short above 1180 °C (2150 °F) and therefore are susceptible to underbead cracking. HNM and 17-10P have been flash butt welded successfully.

Brazing. All stainless steels can be brazed, and this process is frequently used for joining stainless steels to other metals. All brazing techniques can be used, but furnace brazing is employed more often than any other technique because it permits brazing to be done in a protective environment (usually hydrogen or vacuum) that prevents oxidation of the stainless steel. Most brazing of austenitic types is performed at temperatures in the sensitization range, and solution annealing after brazing is impossible. Therefore, stabilized or extra-low-carbon types must be used if service conditions might lead to intergranular corrosion of sensitized material. In brazing of martensitic and ferritic types, a brazing filler metal that melts below the critical temperature of 830 °C (1525 °F) is normally used to avoid martensitic hardening. The heat of brazing will temper, and possibly soften, hardened martensitic stainless steels.

Virtually all types of brazing filler metals are used, including silver, nickel, gold and copper alloys. High-phosphorus copper-base filler metals should be avoided, however, because they have harmful effects on stainless steels. Certain austenitic types (21Cr-6Ni-9Mn, for instance) should not be brazed with any copper-base filler metal because molten copper attacks the stainless steel during brazing.

Destructive penetration of stainless steel is possible during brazing. A form of stress-corrosion cracking develops in some types if they are brazed while in a highly stressed condition. Penetration during brazing may be prevented by annealing before brazing, by heating parts slowly enough to relieve stresses before the brazing temperature is reached, or by selecting a brazing filler metal that inhibits such penetration. Grain-boundary penetration may also occur if high-temperature brazing alloys that melt above 980 °C (1800 °F) are used. This problem can be minimized by selecting a favorable brazing alloy, using as little brazing alloy as possible and keeping time at temperature short. Welding through or near a brazed joint can also cause penetration of stainless steel and should be avoided.

Use of a silver brazing alloy to which nickel has been added helps minimize crevice corrosion of brazed joints exposed to certain corrosive media such as seawater. Complete elimination of this problem in a type 430 brazement can be attained only by using a special filler metal that contains tin and nickel.

Soldering with tin-lead solders does not reduce mechanical properties of stainless steel components. However, strength of soldered joints obviously is inferior to base-metal strength, and soldering is used only where high joining temperatures must be avoided. The relatively poor wetting characteristics of stainless steels cause problems in soldering; the thin oxide film that accounts for the passivity and good corrosion resistance of stainless steel interferes with wetting by solder. Removal of this tenacious film before soldering generally requires use of special fluxes. These fluxes are highly corrosive, and it is vital that the flux-contaminated surface be thoroughly washed and neutralized as soon as possible after soldering. Molten solder can be destructive to stainless steel at high temperatures, and welding should not be performed through or near a soldered joint.

Precautions in Welding

The most obvious problem associated with welding of stainless steel is maintenance of uniform resistance to corrosion across the weld zone and the adjacent base-metal zones. This problem is commonly overcome by close control of composition and welding conditions. Sometimes, postweld heat treatment is required to restore corrosion properties altered during welding—particularly in material subject to sensitization. This and other problems associated with stainless steel weldments are discussed in the following paragraphs.

Sensitization (harmful carbide precipitation) can be avoided, where necessary, by any of three methods:

- *Solution annealing after welding* relieves any residual stresses and improves the structure and corrosion resistance of the weld metal itself. However, solution annealing and the postanneal pickling that it necessitates are costly and inconvenient. Distortion may be a serious problem, particularly if water quenching is used to ensure that general reprecipitation of carbides does not occur during cooling. Postweld annealing is impossible for large weldments such as tanks and pressure vessels.
- *Limitation of carbon content.* Carbides will not precipitate in a weld heat-affected zone if carbon content is low enough. Extra-low-carbon stainless steels such as types 304L and 316L (0.030% max carbon) are commonly used in the United States, and such low carbon contents are necessary for weldments that are to be stress relieved. Even the restriction of carbon content to 0.030% max may not completely protect against sensitization and loss of corrosion resistance in the higher-nickel alloys or in any austenitic stainless steel that has been mildly cold worked. Limitation of carbon to 0.05% max will prevent sensitization in the heat-affected zone under most welding conditions, particularly in light sections. This is common practice in Europe for applications requiring good corrosion resistance in the as-welded condition. When exposed to strongly oxidizing corrosive media such as hot nitric acid, molybdenum-bearing types 316 and 317 are susceptible to intergranular corrosion in the heat-affected zone due to formation of grain-boundary sigma phase, even when carbides are not precipitated. When weldments of molybdenum-bearing grades are to be used in oxidizing media, it is common practice to subject sensitized samples from each heat to corrosion testing for assessment of susceptibility to intergranular corrosion.
- *Stabilization of carbon.* Sensitization in the heat-affected zone can be prevented by tying up the carbon in titanium carbides (as in type 321) or in niobium carbides (as in types 347 and 348) so that detrimental chromium carbides cannot form. The stabilized grades are also recommended for applications involving long-time exposure to temperatures in the sensitization range, such as in prolonged stress relief or service at high temperatures. However, weldments of stabilized grades heated into the sensitization range and later exposed to certain corrosive media can suffer from severe localized corrosion called knifeline attack. The stable carbides are partly dissolved in a very narrow zone immediately adjacent to the weld, and during later exposure in the sensitization range chromium carbides can precipitate. This unusual type of sensitization can be prevented by annealing or stabilization annealing, even locally, after welding.

Distortion and Residual Stress. Austenitic stainless steels have a coefficient of thermal expansion at least 50% greater than that of low-alloy steels or straight chromium stainless steels. Thermal conductivity of the austenitic types also is very low—about one-third that of carbon steel. These differences in physical properties significantly affect distortion due to welding, and fixturing to control warping is more often needed for austenitic stainless steels than for other steels.

An austenitic stainless steel weldment typically contains residual stresses in or near the weld that are approximately equal to the room-temperature yield strength. Control of welding technique can reduce the extent of stressed areas, but generally cannot reduce peak stresses. Residual stresses may combine with service stresses to cause distortion or cracking in service. More importantly, residual tensile stresses promote stress-corrosion cracking in some media, particularly when service tensile stresses also are imposed. Stress-relief annealing often is used to reduce residual stresses. However, as discussed earlier, this is a dangerous practice because it can cause sensitization. Full annealing after welding, or use of low-carbon or stabilized types for weldments that are to be stress relieved, is preferred practice if intergranular attack at precipitated carbides is even remotely possible. Shot peening to produce residual compressive stresses in the surface has also been used to minimize stress-corrosion cracking in stainless steel weldments. This practice also can help minimize intergranular attack in sensitized austenitic stainless steel.

Distortion and residual stress are less troublesome in ferritic types than in austenitic types because the ferritic types are lower in coefficient of thermal expansion and higher in thermal conductivity. Ferritic types generally are not susceptible to stress-corrosion cracking, so residual tensile stresses are of less concern. Where necessary, stress relieving can be performed with fewer complications than for austenitic types, except that precautions must be taken to avoid "885 °F" embrittlement in the high-chromium alloys.

Fig. 5 Modified constitution diagram for stainless steel weld metal

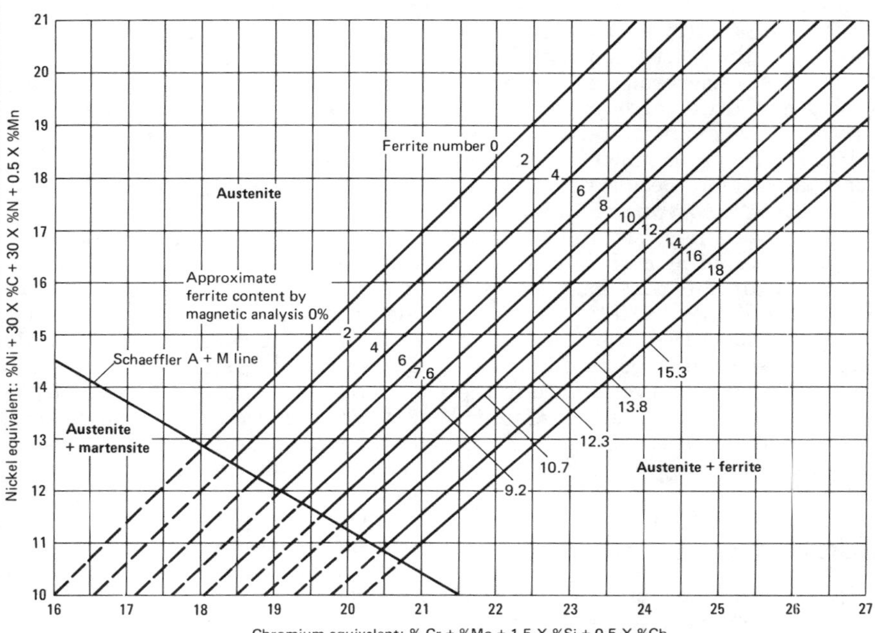

Constitution diagram for austenitic stainless steel weld metal that includes an allowance for nitrogen content in the nickel equivalent. Adapted from a diagram by W. T. DeLong published in the 1974 Metal Progress Data Book (mid-June 1974, p 226).

Weld Cracking. Austenitic stainless steel welds are extremely tough and ductile, and thus cold weld cracking is almost never a problem. However, austenitic stainless steels are susceptible to hot cracking or microfissuring as they cool from the solidus to about 980 °C (1800 °F). Microfissuring can be prevented or kept to a minimum by eliminating or reducing tensile stress imposed on the weld during cooling through this range. To some degree, microfissuring can be controlled by controlling concentrations of residual elements such as phosphorus. However, the most common control measure is to ensure the presence of at least 3 to 4% ferrite in the as-deposited weld. Small amounts of this phase seem to prevent the cracking that often occurs in fully austenitic weld metal. Ferrite content is usually estimated on the basis of composition by use of the DeLong diagram shown in Fig. 5, which is a modification of the long-used Schaeffler diagram. DeLong's modification takes into account the potent austenitic stabilization effect of nitrogen. Because ferrite contents calculated in this manner are not completely precise, it is recommended that for critical applications actual ferrite content be determined by magnetic analysis of as-deposited weld metal. For production welds, measurement is especially preferred to calculation in the common instance where a high-ferrite welding electrode is used to weld lower-ferrite base metal. Weld composition then varies with the degree of dilution.

Control of ferrite content is not always an acceptable solution to microfissuring. Ferrite is a magnetic phase, reduces corrosion resistance in some media and may lead to embrittlement in long-time elevated-temperature service exposure due to precipitation of sigma phase. Ferrite content in the weld can be reduced significantly (typically by 2 to 4%) by annealing after welding; but where postweld annealing is not possible, fully austenitic welds may be required. Some steels such as type 310 are fully austenitic through the entire specified composition range. Weld cracking can be minimized in fully austenitic stainless steels by welding with low heat input, minimizing restraint, designing for low constraint and keeping residual elements at low concentrations.

Ferritic types are less ductile than austenitic types and therefore are more susceptible to weld cracking. Certain ferritic stainless steels (type 430, for instance) form significant amounts of martensite on cooling after welding, which increases susceptibility to cold cracking. Preheating at 150 to 230 °C (300 to 450 °F) is recommended to minimize weld cracking in all ferritic types.

In the fully ferritic types such as 409, 446 and 26Cr-1Mo, welding causes grain coarsening in the base metal immediately adjacent to the weld. Toughness therefore is reduced, particularly in heavy sections, and cannot be restored by postweld heat treatment. Ferritic stainless steels that form austenite at elevated temperatures are not coarsened significantly, but postweld annealing is recommended to transform the resulting martensite and enhance ductility in the heat-affected zone.

Martensitic stainless steels are even more susceptible to weld cracking than ferritic types. Preheating at 200 to 300 °C (400 to 600 °F) generally is required. Postweld annealing is standard practice, particularly for steels with carbon contents greater than 0.20%.

Underbead cracking can occur in the heat-affected zones adjacent to welds in austenitic stainless steel, particularly when the weld zone is heavily restrained or the section thickness is greater than 19 mm (3/4 in.). Such cracking is most common in type 347 because of strain-induced precipitation of niobium carbides, but it has been reported in other types as well. Weld restraint is the most important factor in control of underbead cracking. Preheating is of little value in preventing cracking of austenitic stainless steel and can cause other problems, such as increased carbide precipitation.

Corrosion resistance of weld heat-affected zones in ferritic stainless steels may be impaired by carbide precipitation in material that has reached temperatures above 980 °C (1800 °F). Corrosion resistance can be restored to near-normal levels by postweld annealing. Loss of corrosion resistance in the heat-affected zone also can be minimized by specifying types that are stabilized or that are very low in carbon content.

Cleaning and Finishing

Proper cleaning and finishing of stainless steel parts are essential for maintenance of the corrosion resistance and appearance for which stainless steels are specified. The degree of care required depends on the nature of the application; the most stringent precautions (such as clean-room assembly and sophisticated postassembly cleaning) are used for critical applications such as nuclear-reactor cores, pharmaceutical and food-handling equipment, and some aerospace applications. Cleaning and finishing practices for stainless steels are discussed in more detail in another volume of this handbook (Volume 2 of the 8th Edition, or Volume 5 of the 9th Edition) and in ASTM A380.

Mechanical Cleaning. Even heavy or tightly adhering scale can be removed quickly from stainless steels by various mechanical means. Sand blasting is perhaps most widely used, but shot blasting is also a common practice. Barrel finishing (tumbling in abrasive media) is also advantageous for smaller parts. Mechanical cleaning—particularly shot blasting—may cold work the surface sufficiently to reduce ductility and must be used with care if cold forming is to follow.

If the stainless steel is to be used in the mechanically cleaned condition (often true of large welded assemblies for which mechanical cleaning is used to remove relatively light oxide from joint areas), great care must be taken to prevent the stainless steel surface from becoming contaminated with a residue of iron or steel particles. For sand blasting, only clean silica sand free of iron contamination and scale particles can be used. Shot blasting must be done using stainless steel shot equal in corrosion resistance to the metal being cleaned, and scale particles must be continuously separated from the shot. Wire brushes, which can be used to remove heat tint, must be made of stainless steel equal in corrosion resistance to the type being cleaned, and brushes must not be used on materials other than stainless steels. Even with these precautions, mechanically cleaned stainless steel will not be quite as corrosion resistant as acid-pickled material because mechanical cleaning leaves some residue of scale as well as some contamination from the cleaning operation. Final pickling is mandatory for applications that require maximum corrosion resistance, particularly when hot working or annealing scale must be removed.

Mechanical cleaning is used most often as a preparation for acid pickling, and can reduce total cleaning cost by decreasing pickling time. Mechanical precleaning also permits use of milder pickling solutions, which minimizes safety hazards and pollution problems. When pickling is to follow mechanical cleaning, the precautions listed above regarding mechanical cleaning techniques are not as vital to subsequent performance.

Soil Removal. Greases, oils and soils frequently must be removed from stainless steels before annealing (to prevent attack of the metal or development of refractory scales), before pickling (to permit full access of pickling acid) or before service. Cleaning can be done with either alkaline solutions or organic solvents. Mechanical action, such as brushing during cleaning or spraying with steam or high-pressure water jets, can greatly accelerate the cleaning action.

Alkaline cleaning may be done with proprietary mixtures or with such common cleaning chemicals as trisodium phosphate. Care must be taken to avoid chloride contamination. Stress-corrosion cracking can develop during the cleaning operation itself when hot, high-chloride detergent solution is used, and any residual chloride can cause corrosion in service. For maximum effectiveness, alkaline cleaning solutions generally are applied hot—at about 70 °C (160 °F)—whether circulated through an assembly, sprayed or swabbed. Parts should be thoroughly rinsed after cleaning, because alkaline residues can cause corrosion under some processing or service conditions.

Organic solvents can be applied by spraying (sometimes with steam jets), swabbing or vapor degreasing, or by washing with solvent-water emulsions stabilized with soaps. Chlorinated solvents are often used, but care is required to prevent contamination with water, which can react with some solvents to form hydrochloric acid. Chlorinated solvents are not recommended for closed systems or for components with internal crevices or voids that can trap the solvent.

Acid descaling (pickling) is generally used, alone or as a final operation, to remove the relatively heavy oxides left after hot working or annealing. Treatment A, B or C in Table 3 is required. Any grease, oil or heavy soil must be removed by solvent or alkaline cleaning before acid pickling, because they interfere with the pickling action. In addition, solvent or alkaline cleaning often is required before annealing to ensure that a relatively thin uniform scale will form. Even with such precautions, obtaining a uniform white finish by acid pickling can be a difficult and frustrating task.

In addition to mechanical cleaning before acid pickling, stainless steels also may be subjected to pretreatment in various proprietary molten salts based on sodium or potassium hydroxides. These salt baths, most of which are operated at about 425 °C (800 °F), remove scale and/or convert it to an acid-soluble form. Except for some low-temperature variants, molten salt bath treatments may embrittle high-chromium ferritic stainless steels such as type 446 and may reduce the strength of high-strength martensitic stainless steels. After mechanical cleaning or salt bath pretreatment, stainless steel parts can be acid cleaned in relatively short times using treatment D or E in Table 3.

Pickling in nitric-hydrofluoric acid should never be used on sensitized austenitic stainless steel. This acid rapidly attacks grain boundaries containing precipitated carbides. Effective quenching from the annealing temperature is therefore important for all austenitic stainless steels except the stabilized and extra-low-carbon types.

Table 3 Chemical descaling, cleaning and passivation treatments for stainless steels
Adapted from ASTM A380

Treatment code	Applicable alloy classes	Material condition	Chemical and concentration in vol %	Temperature °C	Temperature °F	Time, min
Acid descaling (pickling) treatments						
A(a)	All except free-machining alloys	Fully annealed only	8-11% H_2SO_4	65-80	150-180	5-45
B	200 and 300 series, 400 series containing 16% Cr or more, precipitation-hardening grades, but not free-machining grades	Fully annealed only	15-25% HNO_3 plus 1-8% HF	20-60	70-140	5-30(b)
C	All free-machining grades and 400 series containing less than 16% Cr	Fully annealed only	10-15% HNO_3 plus ½-1½% HF	Room(c)	Room(c)	5-30(b)
Acid cleaning treatments						
D(d)	200 and 300 series, 400 series containing 16% Cr or more, precipitation-hardening grades, but not free-machining grades	Fully annealed only	6-25% HNO_3 plus ½-8% HF	20-60	70-140	As necessary
E(d)	Free-machining grades, maraging grades and 400 series containing less than 16% Cr	Fully annealed only	10% HNO_3 plus ½-1½% HF	Room(c)	Room(c)	1-2
Cleaning/passivation treatments						
F	All alloys containing 16% Cr or more, but not free-machining grades	Annealed, cold rolled or work hardened, with dull or nonreflective finishes	20-50% HNO_3	50-70 or 20-40	120-160 or 70-100	10-30 or 30-60(b)
G	All alloys containing 16% Cr or more, but not free-machining grades	Annealed, cold rolled or work hardened, with bright machined or polished finishes	20-40% HNO_3 plus 2-6 wt% $Na_2Cr_2O_7 \cdot 2H_2O$	50-70 or 20-40	120-155 or 70-100	10-30 or 30-60(b)
H	400 series, maraging grades and precipitation-hardening grades containing less than 16% Cr; all high-carbon straight Cr grades; but not free-machining grades	Annealed or hardened, with dull or nonreflective finishes	20-50% HNO_3	45-55 or 20-40	110-130 or 70-100	20-30 or 60
I	400 series, maraging grades and precipitation-hardening grades containing less than 16% Cr; all high-carbon straight Cr grades; but not free-machining grades	Annealed or hardened, with bright machined or polished finishes	20-25% HNO_3 plus 2-6 wt% $Na_2Cr_2O_7 \cdot 2H_2O$	50-55	120-130	15-30
J	200, 300 and 400 series free-machining grades	Annealed or hardened, with bright machined or polished finishes	20-50% HNO_3 plus 2-6 wt% $Na_2Cr_2O_7 \cdot 2H_2O$	20-50	70-120	25-40
K	200, 300 and 400 series free-machining grades	Annealed or hardened, with bright machined or polished finishes	1-2% HNO_3 plus 1-5 wt% $Na_2Cr_2O_7 \cdot 2H_2O$	50-60	120-140	10
L	200, 300 and 400 series free-machining grades	Annealed or hardened, with bright machined or polished finishes	12% HNO_3 plus 4 wt% $CuSO_45H_2O$	50-60	120-140	10
M	Special 400 series free-machining grades containing more than 1.25% Mn or more than 0.40% S	Annealed or hardened, with bright machined or polished finishes	40-60% HNO_3 plus 2-6 wt% $Na_2Cr_2O_7 \cdot 2H_2O$	50-70	120-160	20-30

(continued)

Table 3 (continued)

Treatment code	Applicable alloy classes	Material condition	Chemical and concentration in vol %	Temperature °C	Temperature °F	Time, min
General-purpose cleaning treatments						
N..........	All alloys except free-machining grades	Fully annealed only	1 wt% citric acid plus 1 wt% sodium nitrate	20	70	60
O..........	All alloys except free-machining grades	Fully annealed only	5-10 wt% ammonium citrate	50-70	120-160	10-60
P..........	Assemblies of stainless steel and carbon steel	Sensitized	Inhibited solution of 2 wt% hydroxyacetic acid plus 1 wt% formic acid	95	200	6 h
Q..........	Assemblies of stainless steel and carbon steel	Sensitized	Inhibited ammonia-neutralized solution of EDTA(e), followed by hot-water rinse and dip in solution of 10 ppm NH_4OH plus 100 ppm hydrazine	Up to 120	Up to 250	6 h

(a) For removal of tight scale, parts should be dipped in this solution for a few minutes, rinsed in water and then cleaned using treatment D or E. (b) Smut remaining after treatment may be removed by immersing the parts in caustic permanganate solution at 70 to 80 °C (160 to 180 °F) for 5 to 60 min, followed by thorough rinsing in water and drying. (c) Temperatures up to 60 °C (140 °F) may be used with caution. (d) For removal of tight scale, parts may be subjected to treatment A for a few minutes and rinsed in water before being acid cleaned using this treatment. (e) Ethylene diamine tetra-acetic acid.

Austenitic stainless steel weldments that must be pickled in the as-welded condition should be made from stabilized or extra-low-carbon types. Alternatively, weldments may be mechanically cleaned and then (if necessary) passivated in a plain nitric acid solution, or in acid P or Q in Table 3, for improved corrosion resistance.

Acid pickling can introduce hydrogen into the metal, which causes severe embrittlement of martensitic stainless steel parts. The degree of embrittlement increases with carbon content and hardness level; therefore, type 440C at high hardness is affected to the greatest degree. Addition of inhibitors to the acid to reduce attack actually increases hydrogen pickup in the metal. Probably the best method of controlling hydrogen embrittlement is to use mechanical cleaning as a pretreatment and to keep pickling time short. Baking treatments at temperatures up to the tempering temperature will partly restore toughness and ductility.

A mixture of sulfuric and hydrochloric acids, or a sulfuric acid solution to which sodium chloride has been added, occasionally is used instead of the nitric-hydrofluoric acid mixtures listed in Table 3. Use of sulfuric-hydrochloric acid baths avoids potential dangers to personnel resulting from handling of hydrofluoric acid or exposure to nitrogen tetroxide fumes. However, pickling in acid solutions that also contain chloride ions is very risky and generally is not recommended. (Acid solutions containing chloride ions usually are forbidden for critical applications such as nuclear work.) Chloride pitting can develop during pickling if conditions are not optimum. Furthermore, stress-corrosion cracking is likely during pickling of stressed austenitic parts such as as-welded structures. Even if pitting and cracking are avoided in the pickling operation itself, failure to remove all traces of the chloride-containing pickling bath can cause such problems in service.

Acid cleaning (passivation cleaning) is recommended where the surface must be free of iron and other metallic contamination and where parts have not been acid pickled. The cleaning/passivation baths recommended in Table 3 (treatments F through M) effectively remove particles of iron or low-alloy steel embedded during manufacturing operations such as forming and machining, and therefore improve corrosion resistance in service. Because it was once thought that treatment in an oxidizing acid bath helped form the passive film, such treatment is often called "passivation". However, it is now known that mere exposure to air develops a fully passive film, which is enhanced only slightly by chemical treatment. Removal of surface contaminants, on the other hand, does make a substantial contribution to corrosion resistance.

Parts contaminated with any greases, oils or waxes—such as lubricants from machining or forming operations—must be cleaned to remove these contaminants before the metal is acid cleaned. Acid solutions cannot penetrate organic layers and therefore cannot remove metallic contamination covered by an organic contaminant.

Best surface quality is obtained by minimizing the duration of acid cleaning. Periodic inspection during removal of metallic contamination is helpful in determining the time required for acid cleaning. Such inspection usually is done by exposing specimens in a humidity cabinet, or by dipping representative parts in a solution of copper sulfate so that copper will plate out spontaneously on areas where iron contamination is present.

After acid cleaning, parts must be thoroughly rinsed in clean water to remove all traces of acid. Under some conditions, a neutralizing treatment followed by several water rinses may be preferred.

Solutions N, O, P and Q in Table 3 sometimes are used for general-purpose cleaning. Solutions P and Q are intended specifically for removal of oxide from welded joints without the grain-boundary attack on sensitized austenitic stainless steels that sometimes results when nitric-hydrofluoric acid solutions are used.

Hydrochloric acid, inhibited to prevent pitting attack, sometimes is recommended for cleaning stainless steel assemblies. This solution has been used successfully under carefully controlled conditions for assemblies that can be flushed thoroughly after cleaning. However, residual chloride ions from such cleaning can cause severe subsequent corrosion. Hydrochloric acid cleaning of stainless steel should be done only with extreme caution, and should be avoided where crevices, blind holes or pockets may prevent the rinsing step from flushing out all traces of chloride. Hydrochloric acid cleaning is sometimes forbidden by specifications—for instance, those that apply to nuclear assemblies.

Grinding and Polishing. Stainless steels can be readily ground, polished and buffed, but certain characteristics of these materials require some modification of standard techniques for best results. Most notably, the high strength, tendency to "load up" abrasive grains, and low thermal conductivity of stainless steels all lead to buildup of surface heat. This in turn can produce heat tinting (surface oxidation) or surface smearing, and in extreme cases even sensitization of austenitic stainless steels or "burning" (rehardening) of heat treated martensitic grades. Techniques that help prevent buildup of surface heat include (*a*) use of lower speeds and feeds, and (*b*) careful selection of lubricants, and of proper grit size and type, so as to minimize loading of the abrasive.

Corrosion resistance of stainless steels may be adversely affected by polishing with coarse abrasives. Corrosion resistance is normally restored to the level of 2B sheet by polishing to a #4 (180-grit) finish. Polishing with fine alumina or chromium oxide to obtain still higher finishes—such as buffed finishes #7 and #8—removes fine pits and surface imperfections and generally improves corrosion resistance.

Iron contamination must be avoided or removed if polished stainless steel surfaces are to have good corrosion resistance. Abrasives and polishing compounds must be essentially iron-free (less than 0.01% for best results), and equipment used for processing stainless steels must not be used for other metals. If these conditions cannot be met, a cleaning/passivation treatment (after precleaning to remove polishing compounds and lubricants) will be required to restore good corrosion resistance.

Plating

Stainless steels may be electroplated for various reasons, such as color matching, lubrication and enhancement of corrosion resistance. Plating is not particularly difficult, except that plating schedules must include activation (removal of the passive film) just prior to plating.

Hydrogen embrittlement during electroplating can cause serious problems in ferritic and martensitic stainless steels—particularly in martensitic types heat treated to high hardness and in cold worked metastable austenitic types (such as 301) that have partly transformed to martensite during cold working. Hydrogen pickup and consequent embrittlement is particularly likely when metals with low cathode efficiency (such as chromium) are plated onto stainless steel. When cathode efficiency is low, substantial amounts of hydrogen are deposited on the surface of the part along with the metal being plated. When such metals must be plated on hardened stainless steels, hydrogen damage can be reduced by baking after plating at temperatures up to the tempering temperature. Such baking is least effective for the highest hardness levels (lowest tempering temperatures), and electroplating of such parts should be avoided if the same effect can be achieved by other means. For cold worked metastable austenitic stainless steels, baking at about 175 °C (350 °F) generally is recommended.

Stainless Steels in Corrosion Service

By the ASM Committee on Wrought Stainless Steels*

THE CORROSION RESISTANCE of stainless steels is believed, although not by all investigators, to result from the presence of a thin hydrous oxide film on the surface of the metal. Thor N. Rhodin, Jr. (Ref 1) has stated that the passive film on type 304 stainless steel is $4M_3O_4 \cdot SiO_2 \cdot nH_2O$, where the 12 metal atoms represented by $4M_3$ consist of about 7 Fe plus 2 Ni plus 3 Cr, and n is an integer whose average value is about 9. The film varies in composition from alloy to alloy and with different treatments, such as rolling, pickling and heat treating. For any stainless steel, this film, stabilized by chromium, is considered to be continuous, nonporous, insoluble and self-healing. If broken, the film will repair itself when re-exposed to air or a suitable oxidizing agent.

Passivity is the corrosion-resistant behavior produced by the presence of a "passive" oxide film. It is not a constant state; it exists only in certain environments or under certain conditions. The range of conditions over which a stainless steel exhibits passivity may be broad or narrow, and passivity may be destroyed by slight changes in conditions. Under circumstances favorable to passivity, stainless steels have solution potentials approaching those of noble metals. When passivity is destroyed, the potential approximates that of ordinary iron.

Stainless steels are normally passive, but when exposed to mildly oxidizing corrosive solutions, they become active. (Anything that will cause stainless steel to corrode is oxidizing in the chemical sense that atoms have lost electrons; throughout this article, however, oxidizing means furnishing oxygen to produce oxygenated compounds.) Hence, oxygenating agents must be present and must be replenished constantly to maintain passivity. Otherwise, localized corrosion frequently results—for example, the crevice corrosion that occurs beneath barnacles in seawater. The cell formed between a well-aerated area and the oxygen-depleted area beneath a barnacle is known as a differential aeration cell.

An increase in the velocity of a corrosive solution increases the rate at which oxygen dissolved in the solution is brought in contact with the steel. Because of this, the rate at which a stainless steel undergoes electrochemical corrosion tends to decrease as velocity increases. Increased velocity, however, tends to increase mechanical actions such as erosion and cavitation, which can prevent formation of a passive film or remove a film originally present. Thus for most stainless steels, corrosion rates tend to decrease with increasing velocity up to some limiting velocity, then increase again as mechanical effects begin to compete with purely electrochemical corrosion. The limiting velocity varies not only with steel composition, but also with temperature, amount of suspended solids, type and concentration of corrodent, and other environmental factors.

Stainless steels would have a high solution tendency wherever the metal is in direct contact with a corrodent were it not for the presence and maintenance of an inert, passive film. The metal itself, with no covering, does not have good corrosion resistance. Thus, by nature, the resistance of stainless steel is usually either quite good or very bad. In corrosive service, it is seldom intermediate in performance. Two examples may be cited to illustrate the extremes. A 12%-Cr grade, 2.8 mm

*See page X for committee list.

(0.11 in.) thick, exposed in sour crude oil, developed holes as great as 13 mm (1/2 in.) in diameter in less than one year. Ordinary steel had not become perforated in the same period, even though the weight loss was higher than that of the steel containing chromium. Neither material is suitable for tanks to hold sour crude oil. On the other hand, railroad passenger cars made of type 304 stainless steel hardly corrode at all and when properly designed require only slight upkeep.

The high solution tendency is also evident when the passive film is destroyed locally. If that happens, stainless steels can fail by localized mechanisms such as pitting, crevice corrosion, intergranular corrosion or stress-corrosion cracking. Localized corrosion can be catastrophic. Usually a very small portion of the stainless steel area is involved, and the damage may be difficult to detect before failure occurs.

Effect of Composition

Stainless steels derive their resistance to corrosion from the presence of chromium. Increasing the chromium in steel progressively enhances resistance to rusting in the atmosphere. The rate at which steel develops a passive film in the atmosphere depends on its chromium content. However, only the grades containing about 10% Cr or more develop passivity. Some grades, even in this passive range, may show evidence of superficial attack in severe marine and industrial atmospheres (Fig. 1).

The presence of nickel in high-chromium stainless steels greatly improves their resistance to certain nonoxygenating media. Nickel may not be needed for corrosion resistance in some atmospheric environments, but will impart other desired properties.

Manganese is an effective austenite stabilizer that does not significantly alter the resistance to corrosion of high-chromium steels. In the 200-series stainless steels, manganese is used as a substitute for part of the nickel that would otherwise be required to make the steels austenitic at room temperature.

The presence of molybdenum in stainless steels greatly improves their resistance to solutions of halogen salts and to pitting in seawater (Fig. 2). Additions of molybdenum strengthen the passive film in some media where it is otherwise likely to fail. Addition of molybdenum greatly reduces the incidence of pitting. However, when pits do occur—in type 316 for example—they become just as deep as those in molybdenum-free grades (such as type 304). As already mentioned, pitting corrosion will also develop under foreign deposits that prevent access of oxygen to the surface of the steel. Pitting, however, may occur in environments that are free from foreign deposits, if oxygenating conditions are borderline or halogen salts are present in sufficient concentration.

Effect of Heat Treatment

The changes in microstructure produced by different heat treatments have considerable influence on the corrosion resistance of stainless steels. These steels normally exhibit greater resistance when all the carbon is in solution, producing a homogeneous single-phase structure.

Fig. 1 Atmospheric corrosion resistance as a function of chromium content

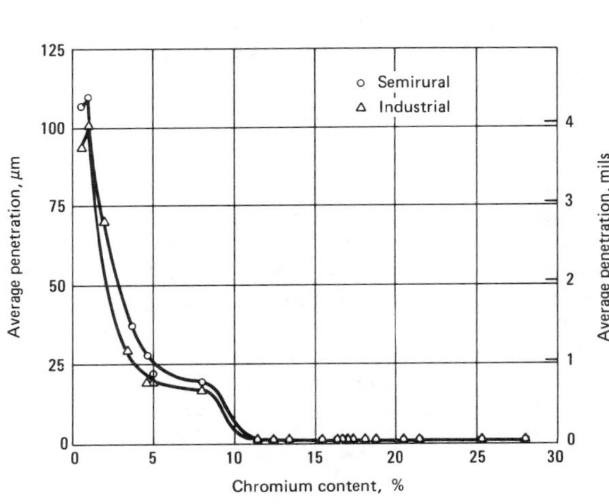

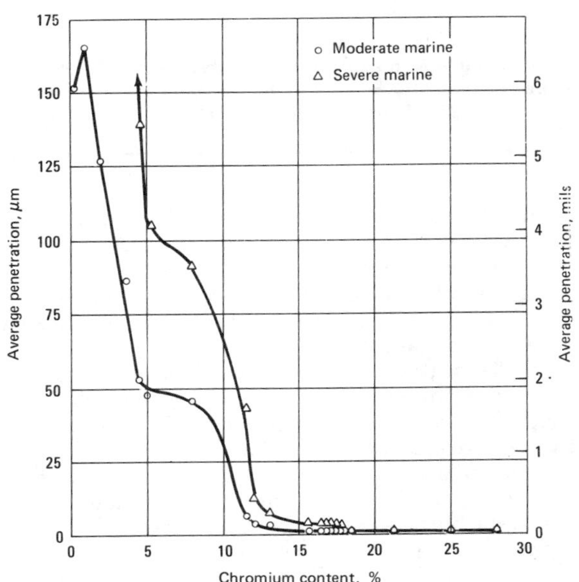

Straight chromium steels were exposed for 8 yr in semirural and industrial atmospheres (left), and in moderate and severe marine atmospheres (right). (Ref 2)

Fig. 2 Relationship between critical pitting potential and resistance to crevice corrosion for Fe-Cr-Mo alloys

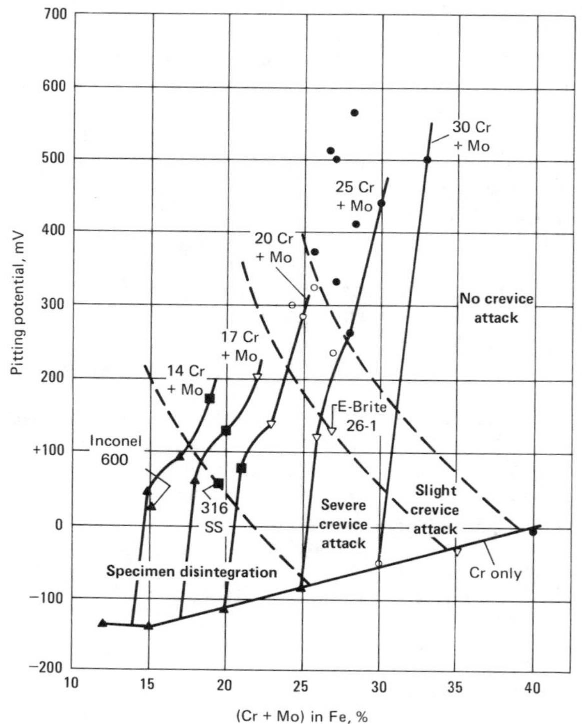

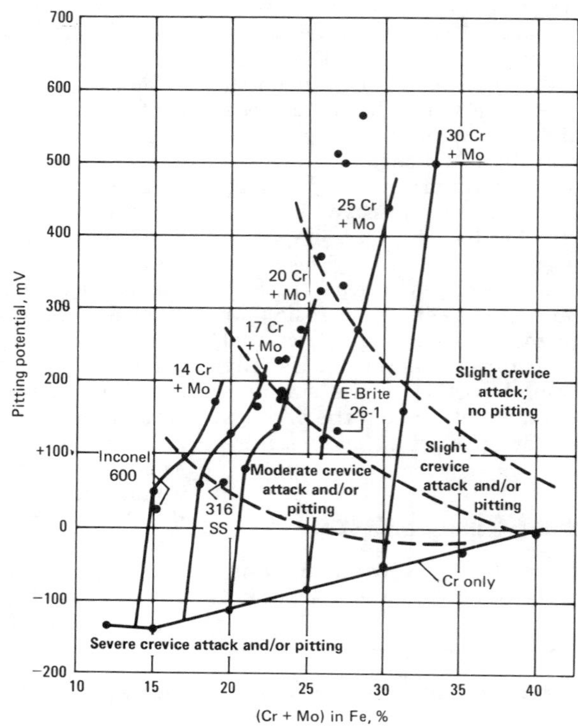

Pitting potentials were measured in deaerated synthetic seawater (pH of 7.2 ± 0.2) at 90 °C (194 °F) vs a saturated calomel electrode. Left: resistance to crevice attack after 6 days in oxygenated 10% ferric chloride at 21 ± 1 °C (70 ± 2 °F). Right: resistance to crevice attack after 14 days at 121 °C (250 °F) in synthetic seawater containing about 60 ppm oxygen. (Ref 3)

Unstabilized austenitic stainless steels become subject to severe attack along grain boundaries at room temperature in a number of corrosive media if the metal is heated to temperatures in the range 550 to 850 °C (1000 to 1550 °F). This attack is known as intergranular corrosion and results from precipitation of chromium carbide and consequent depletion of chromium in the areas adjacent to the grain boundaries. Decreasing carbon content reduces susceptibility of the steel to carbide precipitation and subsequent intergranular attack.

Besides carbon content, time at the sensitizing temperature is important. Precipitation of carbides at the grain boundaries may take place very rapidly. For example, after welding, the metal adjacent to the weld may become subject to intergranular attack, whereas the weld metal itself and the base metal outside the heat-affected zone may remain virtually free from attack. Intergranular susceptibility may be removed by annealing, or may be avoided by using stabilized compositions (types 321 and 347) or extra-low-carbon grades. (Types 321 and 347 are subject to "knife-line attack" under certain conditions, as discussed in a later section of this article.)

Mechanical properties of stainless steels are not severely affected by sensitization. However, subsequent intergranular corrosion has a deleterious effect on the ability of a stainless steel structure to sustain a load.

Martensitic steels must be properly heat treated to give optimum resistance to corrosion, and normally they show maximum resistance to atmospheric corrosion in the fully hardened condition. Tempering at 375 °C (700 °F) or less relieves quenching stresses and greatly improves ductility and toughness without severely reducing resistance to atmospheric corrosion. However, tempering between 375 and 560 °C (700 and 1050 °F) should be avoided, because temperatures in this range impart low toughness as well as low resistance to corrosion.

Corrosion resistance of the ferritic grades may be adversely affected by some thermal treatments. Therefore, it is advisable to anneal the nonhardenable 10 to 29% Cr grades after welding, in order to obtain maximum resistance to corrosion and to obtain ductility in the heat-affected zones adjacent to welds. As with austenitic steels, these problems are minimized by using stabilized compositions such as types 321 and 347 or extra-low-carbon grades such as types 304L and 316L.

If properly solution heat treated, austenitic chromium-nickel steels exhibit passivity in a wide variety of corrosive environments. Austenitic steels resist corrosion most effectively when they are heated to temperatures from 1040 to 1150 °C (1900 to 2100 °F) and cooled rapidly; this heat treatment produces a homogeneous austenitic structure. The lower side of the above temperature range is usually preferred to minimize warpage and to avoid difficulty in scale removal. Rapid cooling from the heat treating temperature is essential in order to keep the carbides in solution. Very small parts may be air cooled, but larger parts must be water quenched to prevent sensitization. Specific guidelines for avoiding sensitization during heat treatment are difficult to establish because sensitization in a given part is affected by factors such as thickness and mass of the part, as well as by composition (especially carbon content and amounts of carbide formers other than chromium). With regard to cooling rate, it is not cooling rate *per se* that is important, but rather the amount of time that a part spends within the sensitization range. Here, a few minutes at a temperature near the middle of the range is equivalent to several hours at temperatures near the extremes of the range.

Cold work sometimes reduces the corrosion resistance of stainless steels. This effect is rather specific and depends on the composition of the steel, the extent and uniformity of cold work, and particularly the nature of the environment. For example, local cold work, such as that produced by imprinting identification numbers with metal stamps, seems to have a deleterious effect on the corrosion resistance of stainless steel.

Effect of Welding

The degree of sensitization induced by welding varies chiefly with heat input per unit length of weld. Low heat input, which is characteristic of arc welding processes and high travel speeds, produces lower amounts of sensitization than high heat input, which is characteristic of oxyfuel-gas welding and low travel speeds. Oxyfuel-gas welding is seldom used to weld stainless steels; not only does it exhibit high heat input, but also it has a high potential for carburizing the stainless steel and thereby increasing the sensitization effect in the weld and heat-affected zones.

Effect of Surface Condition

To ensure satisfactory service life, the surface condition of stainless steel must be given careful attention. Smooth surfaces, plus freedom from surface imperfections, blemishes and all traces of scale and other foreign material, reduce the probability of corrosion. Generally, a smooth, highly polished reflective surface has greater resistance to corrosion. Rough surfaces are more likely to catch dust, salts and moisture, which tend to localize corrosive attack.

Oil and grease can be removed by either hydrocarbon solvents or alkaline cleaners, but these cleaners must be removed before heat treatment. Hydrochloric acid formed from residual amounts of trichloroethylene, used for degreasing, has caused severe attack of stainless steels. Surface contamination may be caused by machining, shearing and drawing operations. Small particles of metal from tools become embedded in the steel surface and, unless removed, may cause localized galvanic corrosion. These particles are removed best by immersing the surface in a solution containing approximately 20% nitric acid, heated to a temperature between 50 and 60 °C (120 and 140 °F). Some believe this treatment promotes development of passivity. Additional information on cleaning and descaling of stainless steel may be found in ASTM A380 and in the Metals Handbook volume on cleaning and finishing.

Shot blasting or sand blasting should be avoided unless iron-free silica sand is used; metal shot in particular will contaminate the stainless steel surface. If shot blasting or shot peening with metal grit is unavoidable, it is essential to clean the parts after blasting or peening by immersing them in a solution of about 20% nitric acid.

Effect of Design and Fabrication

Failure due to corrosion often can be eliminated by suitable changes in design without changing the type of steel. Factors to be considered include joint design, surface continuity and concentration of stress. Welds should be well spaced, and any applicable codes or standards should be followed for location and manner of making attachments. Weldments should be designed for economical cutting of plates and for good fit-up at joints. Butt welds are preferred over lap joints. If lap welds are necessary, they should be sealed completely against penetration by corrosive solutions; otherwise, crevice or concentration-cell corrosion may result. Use of certain attachments, such as doubling plates or reinforcing plates encircled by fillet welds, should be minimized; such attachments induce biaxial residual stresses that generally are difficult to relieve by heat treatment. In order to avoid diffusion of carbon into stainless steel tanks from low-carbon steel supporting legs, which can occur during service at elevated temperatures, the legs should be welded to a stainless steel saddle, and the saddle welded to the tank.

Wherever possible, welding should be done in the downhand position; this generally produces the most consistently sound welds. Stringer beads are preferred for manual welding because the heat input and residual stress are lower than for other types of beads. Welds less than 5 mm (3/16 in.) thick may be water cooled to lessen the effects of welding heat and to reduce distortion. Control of heat effects to avoid sensitization is generally unnecessary with stabilized stainless steels (types 321 and 347) or extra-low-carbon grades (types 304L and 316L), but freedom from residual fabricating stresses is important, particularly where there is a possibility that stress-corrosion cracking may occur in service. It is especially important to provide for good fit-up at weld joints; poor fit-up causes residual stresses, because either the parts have to be forced into position and then welded or welds are uneven and nonuniform due to gaps and unevenness at the joint. Residual stresses can also be kept to a minimum by using weld sequences that do not impose excessive constraint on the joint.

Stainless steel weldments should be

thoroughly cleaned, by blast cleaning with iron-free grit or by pickling, to remove all contaminants such as oxides and iron dust, weld spatter welding flux, dirt and organic matter. This should be followed by final cleaning with a solution of 10% nitric – 1% hydrofluoric acid. The final acid cleaning step should be done with care because severe attack along the heat-affected zone can occur if the cleaning time is prolonged or the amount of hydrofluoric acid in the solution is above 1%. Alternatively, one of the less aggressive cleaning solutions described in ASTM A380 can be used; more details on cleaning can be found in the article in this volume on fabrication of stainless steels.

Elimination of notches, threads, grinding or abrasive scratches, corners, grooves and similar stress raisers is essential. Generous fillets at corners, smooth contour welds without undercuts, ground and polished welds, rounded corners, ground and polished edges, and flat surfaces will eliminate causes of stress concentration.

Cold forming operations, such as rolling of tubes into tube sheets, should be held to a minimum. Severe forming operations, such as rolling of dished heads for tanks should be done by hot working above 870 °C (1600 °F). Castings, bolts, tank ends and other hot formed components should be annealed at 1100 °C (2000 °F) and quenched in water or in an air blast.

If heat treatment is required, consideration should be given to structural stability at the heat treating temperature. Surfaces must be clean and free from carbonaceous materials such as oil, grease and paint, which may be absorbed at high temperature and may thereby increase the carbon content. Availability of heat treating facilities in the locality where the stainless steel is to be fabricated will influence cost. Sometimes the higher-priced stabilized or extra-low-carbon grades may reduce total cost by eliminating heat treatment.

Designs that tend to localize concentrated solutions should be avoided. For example, the inlet of a tank should be designed in such a way that concentrated solutions are mixed and diluted as they are introduced (Fig. 3). Otherwise, localized pockets of concentrated solutions can cause excessive corrosion.

Similar problems can be created by poor design of heaters, such as those that cause hot spots and thereby accelerate corrosion. Heaters should be centrally located (Fig. 4); if a tank is to be heated externally, heaters should be distributed over as large a surface area as possible.

Fig. 3 Design for mixing of concentrated and dilute solutions

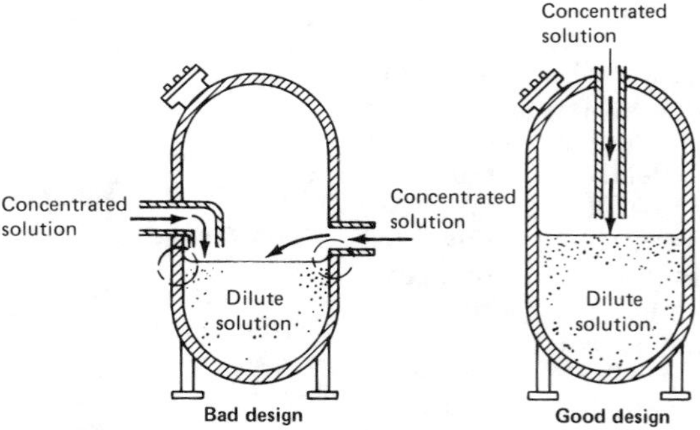

Bad design (left) causes concentration and uneven mixing of incoming chemicals along vessel wall (circled areas). Good design (right) allows concentrated solution to mix into solution in vessel away from vessel walls.

Fig. 4 Design for heating of solutions

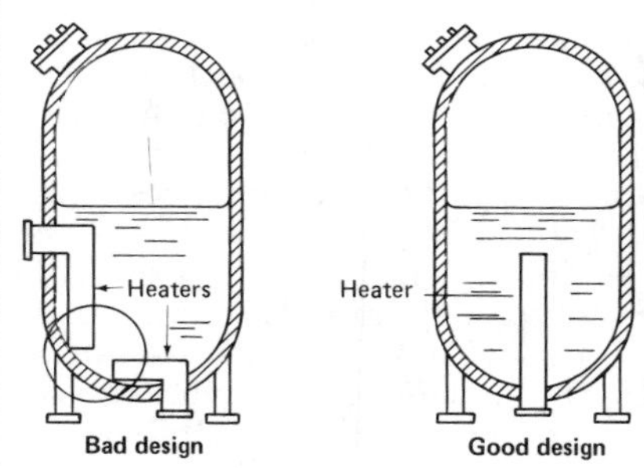

Bad design (left) creates hot spots (in circled area), may induce boiling underneath heater in bottom of vessel or may result in formation of deposits between heaters and vessel walls. Good design (right) avoids hot spots and pockets where a small volume of liquid can become trapped between heater and vessel wall.

Hot gases that are innocuous to the stainless steel may form corrosive condensates on cold portions of a poorly insulated unit. Proper design or insulation can prevent such localized cooling (Fig. 5). Conversely, vapors from noncorrosive liquids may cause attack, and exhausts and overflows should be designed to prevent hot vapor pockets. Generally, open ends of inlets, outlets and tubes in heat exchangers should be flush with tank walls or tube sheets to avoid build-up of harmful corrodents, sludges and deposits.

Tanks and tank supports should be designed to prevent or minimize corrosion due to spills and overflows (Fig. 6). A tank support structure may not be as corrosion resistant as the tank itself, but the support structure is a very important part of the unit and should not be made vulnerable to spilled corrodents.

Intergranular Corrosion

Improper heat treatment of stainless steels can lead to early failure under severely corrosive conditions and can greatly reduce service life in many relatively mild environments. Unstabilized austenitic stainless steels that contain more than 0.03% C become susceptible to intergranular corrosion because complex chromium carbides

Fig. 5 Design to prevent localized cooling

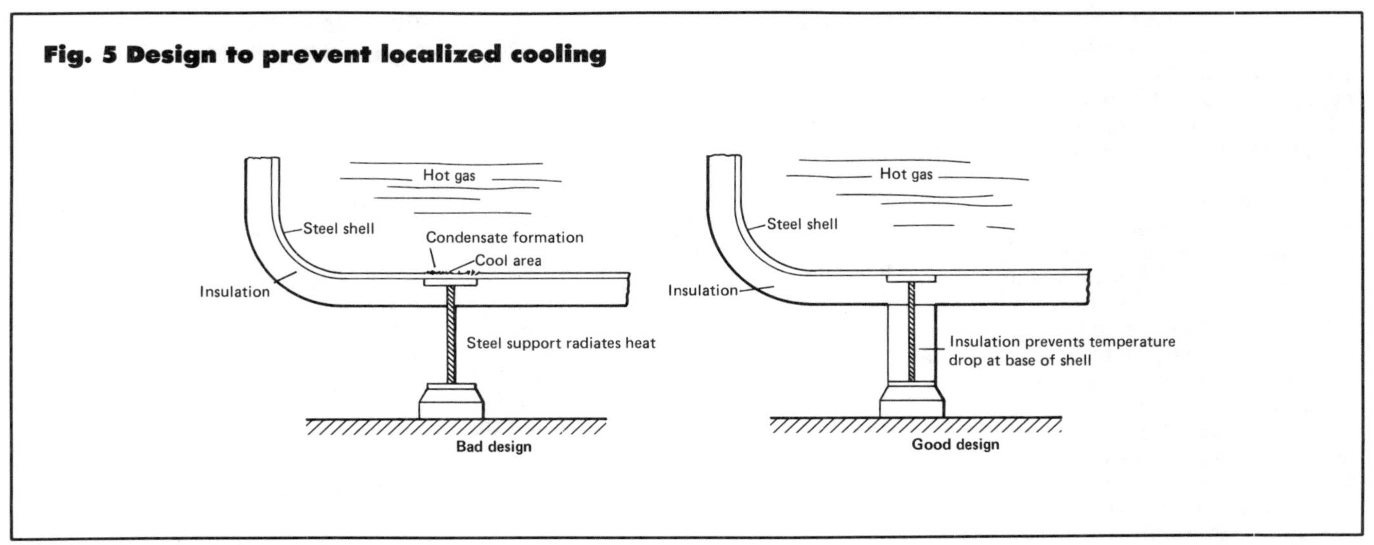

Fig. 6 Design to prevent external corrosion from spills and overflows

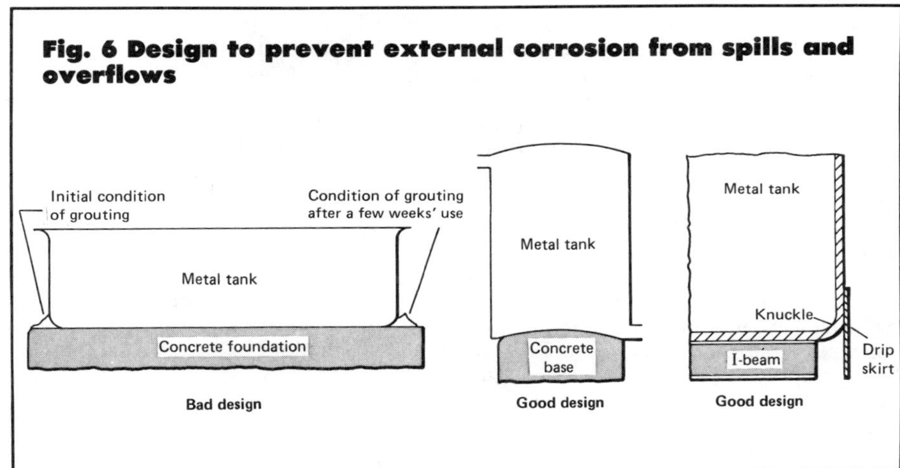

precipitate along grain boundaries when the steel is exposed to temperatures from about 550 to 850 °C (1000 to 1550 °F). Normal welding procedures also induce susceptibility. The fact that a stainless steel is susceptible to intergranular corrosion does not necessarily mean that it will be so attacked. However, many corrosive media, including some considered only mildly corrosive, will corrode susceptible steels intergranularly. The possibility of intergranular attack must be recognized unless ruled out by previous experience. Chromium carbide precipitation in austenitic stainless steels can be eliminated by:

1 Heating the steel, after final fabrication or processing, to a temperature high enough to dissolve the carbides (usually, 1040 to 1150 °C, or 1900 to 2100 °F), and cooling rapidly enough to avoid reprecipitation. Localized heat treatment of the area immediately adjacent to a weld is not satisfactory for prevention of chromium carbide precipitation. For effective heat treatment, the entire unit must be heated and quenched.
2 Using a stainless steel that has been stabilized with niobium or titanium, which combine with the carbon and thereby prevent harmful precipitation of chromium carbides.
3 Reducing carbon content to such a low level that difficulty is avoided. The availability of extra-low-carbon stainless steels (types 304L and 316L) has made this method of great practical importance.

Among the methods available for detecting susceptibility to intergranular corrosion are the boiling 65% nitric acid (Huey) test, the copper – copper sulfate – sulfuric acid test, the nitric acid – hydrofluoric acid test, the oxalic acid etch test and the ferric sulfate – sulfuric acid test. Details for performing all of these tests are given in ASTM A262. In tests on unstabilized stainless steels, specimens must represent the condition in which the steel is to be used. In tests performed on stabilized or extra-low-carbon grades to determine whether the steel as supplied is resistant to sensitization, specimens are sensitized by heat treatment in the temperature-sensitive range (for example, 1 h at 675 °C, or 1250 °F) to reveal any latent susceptibility.

The nitric acid test is not ordinarily used for routine evaluation of types 316L and 321 because phenomena other than chromium carbide precipitation can cause these steels to corrode intergranularly in boiling 65% nitric acid. Intermetallic compounds of iron and chromium known as sigma and chi phases (or perhaps transition phases intermediate in the formation of sigma and chi) may be responsible for this effect.

The danger of chromium carbide precipitation is now so widely recognized and guarded against that relatively few failures occur from this cause. Nevertheless, costly intergranular corrosion still is sometimes encountered.

In atmospheric exposure or mildly corrosive conditions where freedom from contamination is the primary objective, precautions against intergranular corrosion usually are not required.

"Knife-line attack" is a special form of intergranular corrosion sometimes encountered in type 321 and to a lesser degree in type 347. It usually occurs when weldments are subjected for a considerable period of time to temperatures within the sensitizing range after fabrication—as, for instance, in a stress-relieving treatment. After such a treatment, metal immediately adjacent to welds may be attacked by corrosive media. The explanation is that the affected area has been heated to a temperature high enough to decompose the titanium or niobium carbides. Consequently, part of the carbon combines with chromium during subsequent exposure within the temperature range for chromium carbide precipitation. The metal is therefore made susceptible to intergranular corrosion. A "stabilizing" treatment at 900 °C (1650 °F) after welding may be beneficial if such problems are encountered.

Pitting Corrosion

Because stainless steels are passive under almost all conditions in which they are normally used, any localized corrosion, under circumstances that prevent restoration of passivity, may cause rapid penetration at the point of initiation. This occurs because an active-passive electrolytic cell is formed between the large cathodic (passive) area and the small anodic area under attack; the surrounding oxygen serves as a depolarizer, and pitting proceeds.

Solutions containing chlorides are especially detrimental because they promote formation of active-passive electrolytic cells. Acid chlorides in their higher-valence state (such as cupric chloride and ferric chloride) are particularly severe, but any chloride in appreciable concentration is a possible source of trouble. Solutions of other halide salts and of some sulfates may cause pitting.

As shown in Fig. 2, molybdenum in stainless steels dramatically increases resistance to pitting.

Crevice Corrosion

Crevices formed by joints and connections, or at points of contact between metals and nonmetals, are most frequently subject to attack. Similarly, deposits of foreign matter may promote local attack. When a stainless steel is immersed in seawater, fouling by growth of barnacles promotes local corrosion.

In a crevice, the oxygen supply is limited and cannot repair the passive oxide film, and a so-called "differential concentration cell" appears. Porous substances in the crevice may trap corrosive solutions such as seawater or moisture condensed from the atmosphere. A crevice stays damp longer than a fully exposed surface. Salts are likely to accumulate in crevices and under deposits, particularly if the area around the crevice is alternately wet and dry. Gasket materials containing sulfur or graphite contribute to this type of corrosion. On the other hand, corrosion will seldom occur if the crevice is sealed completely to prevent the intrusion of moisture.

If the oxygen concentration decreases below a level necessary to maintain passivity in the anodic area, there is a double electrolytic effect. The difference in oxygen concentration alone will tend to accentuate attack on the anodic area; then the electrolytic potential between the active anode and the passive cathode is quite high, in itself, leading to a further degree of sustained continuous corrosion. There is a tendency for the cathode to become polarized with "plated hydrogen", but because oxygen is immediately available, the hydrogen film is destroyed, permitting attack to proceed, often at an unacceptably high rate. Chlorides and other nonoxygenating salts, being electrolytes that will not contribute to passivity, assist in this type of corrosion, often leading to destructive pitting. Furthermore, the solution within the crevice becomes very acidic (pH of 1.2 has been reported), which adds to the accelerated attack.

In contact with seawater, stainless steels perform well only if design prevents fouling growths or deposits of silt, or if the stainless steels contain molybdenum and high chromium. For example, clean, continuously flowing seawater has been carried successfully in stainless steel piping if velocity is 1.5 m/s (5 ft/s) or more. Reducing design stresses and residual stresses helps prevent stress-corrosion cracking. Stainless steel ship propellers are also being used successfully in seawater and brackish water. Sacrificial anodes are used in such instances to protect the stern of the ship.

Galvanic and Concentration-Cell Corrosion

The factors that influence galvanic corrosion include conductivity of the circuit, potential between anode and cathode, polarization, relative areas of cathode and anode, geometrical relationships between dissimilar-metal surfaces and contact between metals. Of these, relative areas of anodic and cathodic surfaces have the most pronounced effect on the extent of damage produced by galvanic action, because a small anode and a large cathode result in an increase in current density at the anode with a great consequent increase in the rate of corrosion. Thus, small differences in potential under these conditions may produce extensive corrosion because of increased current density in the anodic areas.

A carbon steel bolt in a stainless steel plate will corrode at a rapid rate because the stainless acts as a large cathode. A stainless bolt in a large carbon steel plate will, in general, cause the carbon steel to corrode at only a slightly increased rate. It is generally considered poor practice to couple an anodic material such as carbon steel to a large cathode of stainless steel for exposure in a nonpassivating electrolyte; in a few instances, it is poor practice to couple carbon steel to stainless steel even if the area of exposed stainless steel is small.

If the solution promotes an active state, a galvanically coupled metal such as copper or bronze will accelerate corrosion of stainless steel, particularly if the area of activated stainless steel is small in comparison with the area of copper or bronze. On the other hand, passivated stainless steels are extremely stable in contact with these metals if the solution sustains the passive state.

In humid atmospheres, stainless steel usually will not corrode extensively; metals such as carbon steel, low-alloy steel, cast iron and zinc may corrode more rapidly than usual if they are coupled to stainless steel. There are reports that bronze, brass, copper, graphite and cast iron have been coupled to an austenitic grade of stainless steel (type 304) without significantly affecting their corrosion behavior. In some instances, stainless steel and aluminum alloys may also be coupled together with no serious effect on corrosion behavior. This is demonstrated by the fact that they are used together for building trim in high-humidity locations. In seawater, the rate of attack on aluminum, lead and certain lead alloys is increased by coupling to stainless steel. Despite its nobility, however, stainless steel, when coupled to nonstainless steel, zinc, cadmium, tin and other less-noble metals in the presence of seawater or 3.5% NaCl, causes less rapid corrosion of these metals than do other metals, such as copper, titanium and nickel.

Stress-Corrosion Cracking

Stress-corrosion cracking in austenitic stainless steels has been reported in chloride solutions when high stresses (residual or applied) are present in the metal. Solutions of sodium, magnesium, calcium, zinc or lithium chlorides, and most ethyl chlorides, are among the most aggressive. Of these, a boiling concentrated solution of magnesium chloride is very aggressive and will cause austenitic steels to fail by stress-corrosion cracking in short periods of time (½ to 2 h) when the applied stress is near the yield strength of the steel. A boiling 42% magnesium chloride solution is especially severe in this respect (Fig. 7) and has been used for laboratory comparison testing.

Fig. 7 Stress-corrosion cracking in type 316 stainless steel

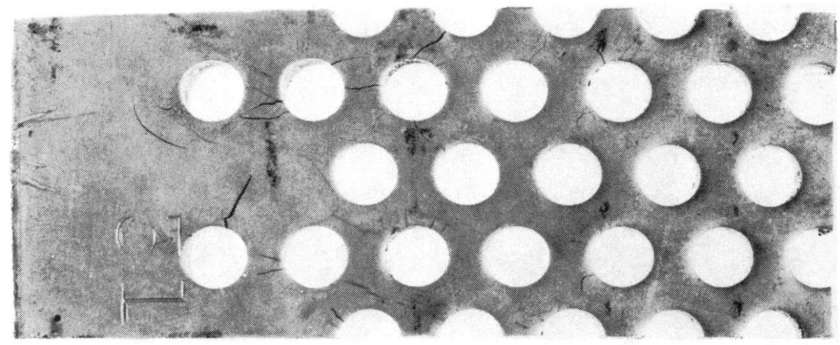

Tube-sheet specimen that cracked after 18 h in boiling 42% magnesium chloride

Stress-corrosion cracking in annealed and quenched alloys is characterized by branching transgranular cracks, and this type of cracking is often characteristic of chloride-induced stress-corrosion cracking in sensitized material. However, sensitized material may display stress-assisted intergranular corrosion having a cracklike appearance under conditions where intergranular attack in unstressed material would be minor. This latter failure mode is often referred to as intergranular stress-corrosion cracking.

The corrosive environments that can cause stress-corrosion cracking are rather specific and are usually environments that cause little or no general attack. Conventional transgranular stress-corrosion cracking usually occurs only in chloride or caustic solutions. Intergranular stress-corrosion cracking occurs most often in polythionic acid solutions or in high-temperature oxygenated water. Because failure is caused by the combined action of stress, temperature and the specific corrodent, it can be controlled or avoided by reducing or eliminating any one or more of the contributing factors.

Fig. 8 Cracked overflow float from a deaerating water heater

Type 304 stainless steel float failed from stress-corrosion cracking in vapor above the water level in the tank.

A 250-mm (10-in.) overflow float of 304 steel cracked badly after six months of exposure to vapor in a deaerating water heater (Fig. 8). The water was treated city water, and no evidence of chemical attack was present. It was presumed that cracking had resulted from extreme cold work in manufacture and from the changing temperature within the water heater.

Figure 9 shows a perforated backing plate of a salt (NaCl) drum filter that failed by stress corrosion cracking. The majority of the cracks radiate from the cold pierced holes. Many slip lines were present in the vicinity of the holes, indicating extreme cold work without heat treatment. Cracking was chiefly transgranular.

Cracking has been observed even in hot water of relatively low chloride content, especially where cold worked parts were involved or where chloride was concentrated in crevices or pockets.

As-drawn or severely cold worked stainless steels crack readily in contact with aqueous hydrogen sulfide, but annealed stainless steels are not susceptible unless stressed well above their yield strength. Several media, including strong hot caustic solutions under pressure, have been reported to cause cracking, although most failures in hot caustic solutions may have been caused by unreported chloride impurities. Chloride compounds cause most stress-corrosion cracking in stainless steel.

Ferritic stainless steels are not considered susceptible to stress-corrosion cracking in chloride solutions but will pit badly. They also appear to resist cracking in caustic solutions. Martensitic grades are known to crack under stress when exposed to chlorides.

Fig. 9 Stress-corrosion cracking of a type 347 stainless steel backing plate from a salt (NaCl) drum filter

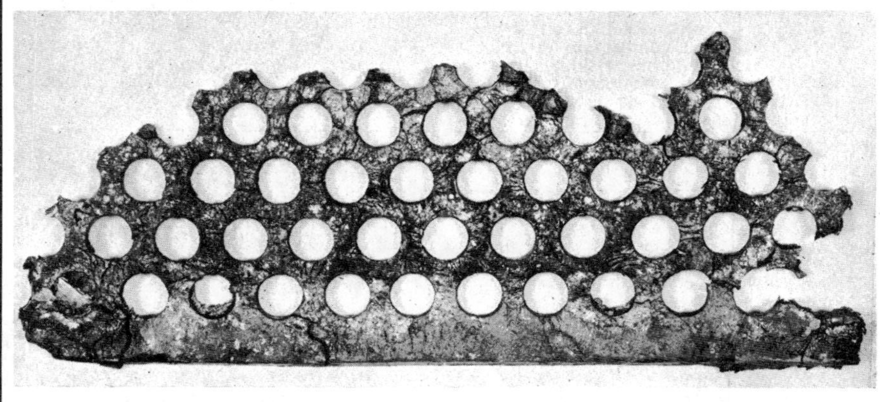

Table 1 Residual stresses in austenitic stainless steel after various treatments (Ref 4)

Stress-relieving temperature, °C	Stress-relieving temperature, °F	Time, h	Residual stress, MPa	Residual stress, ksi
After welding 9¼-in.-OD, 6½-in.-ID pipe				
As welded		...	206-177	30.0-25.7
600	1100	16	138	20.0
600	1100	48	138	20.0
600	1100	72	159	23.0
650	1200	4	148-165	21.5-24.0
After welding 5-in.-OD, 4-in.-ID pipe				
As welded		...	127-101	18.5-14.7
650	1200	4	94.5-105	13.7-15.3
650	1200	12	110	16.0
650	1200	36	108	15.6
900	1650	2	nil	nil
1000	1850	1	nil	nil

To avoid stress-corrosion cracking, some equipment fabricated from types 304L and 316L has been stress relieved at temperatures ordinarily used for carbon steel (540 to 650 °C, or 1000 to 1200 °F). Although this temperature range would be harmful to ordinary 18-8 because of carbide precipitation, it has no harmful effect on extra-low-carbon types. Table 1 shows the residual stresses in solid austenitic stainless steels after various stress-relieving treatments. Integrally bonded or attached stainless linings on carbon steel cannot be stress relieved effectively by heating because of the large difference in coefficient of expansion.

Specimens of type 304L and 316L heated to 650 °C (1200 °F) for one to three days and then tested in the copper – copper sulfate – sulfuric acid test as prescribed in ASTM A262 have shown no intergranular corrosion after three 72-h immersions. Exposure of the low-carbon grades to temperatures in the carbide-precipitation range may create harmful effects in certain corrosive media. For example, type 316L may fail in the Huey test (boiling 65% nitric acid) after heating for only 1 h at 650 °C (1200 °F); this may be caused by formation of sigma phase at grain boundaries. The mechanism of sigma-phase formation is similar to carbide precipitation in that there is a depletion of chromium in the adjacent areas. Depletion caused by sigma-phase formation may be less severe than that caused by carbide precipitation. This would account for intergranular attack occurring in boiling nitric acid but not in acidified copper sulfate.

Although low-temperature stress relief may be adequate to avoid stress-corrosion cracking in certain corrosives, stresses are not completely removed. For complete relief of residual stress, a temperature of 900 °C (1650 °F) is required. If the threshold stress for stress-corrosion cracking is known to be higher than the residual stress remaining after stress relieving at a low temperature, the low-temperature stress relief should be practical. If the stress levels are not known, high-temperature stress relief should be considered.

Weld deposits exposed to a sensitizing temperature sometimes are subject to stress-corrosion cracking; the use of niobium-stabilized welding electrodes may prevent this. Use of extra-low-carbon weld metal (less than 0.03% C) is not as effective in preventing cracking. In many instances, residual stresses resulting from welding are sufficient to sustain stress-corrosion cracking, even in the absence of appreciable service stresses. Consequently, stress relieving of weldments is generally considered good practice for applications where the environment is known or suspected to contain a substance that induces stress-corrosion cracking.

Corrosion Fatigue

Stainless steels have higher corrosion-fatigue limits than other steels, chromium being more effective than nickel for increasing resistance to corrosion fatigue. However, prior thermal treatment may have a marked influence. For example, a plain chromium steel containing 0.38% C and 14.5% Cr showed a damage ratio (corrosion fatigue limit divided by endurance limit) of 0.69 in both well water and salt water when the steel was annealed. When the same steel was quenched and tempered, the damage ratio remained nearly the same when testing was done in well water, but was 0.51 in salt water. An austenitic grade (essentially type 302) showed a damage ratio of 1.00 (no reduction) when tested in well water. However, the damage ratio was

0.50 when this steel was tested in salt water (Ref 5).

Corrosion Testing

Because actual service conditions are difficult or impossible to reproduce in standard laboratory tests, results of such tests usually can serve only as a guide. Chemical conditions, temperature, velocity and aeration should parallel those in the proposed process; therefore, field tests in existing equipment in a comparable process should be used wherever possible, in order to duplicate anticipated conditions. If the installation will be expensive to fabricate and install, simulated service tests may be made in a pilot plant.

Sometimes, stainless steel may be so adversely affected during fabrication that it fails prematurely in service; tests to evaluate the effect of proposed fabrication methods should be included in the corrosion-test program. It is advisable to include welded specimens typical of the job, and also sensitized specimens, in order to evaluate weld deposits, heat effects in the weld zone, and the possibility of stress-corrosion cracking from residual stresses. Stress-corrosion specimens loaded to various levels of stress should be tested to evaluate stress-corrosion cracking susceptibility and to indicate whether fabricated equipment should be stress relieved. Specimens heat treated to represent job conditions should be tested. It is also important that surface finish be typical of the job and the same on all specimens. Frayed surfaces or cold worked edges produced in preparation of test specimens should be removed before testing.

Careful evaluation of test results is of utmost importance. Microscopic examination of the surface of the specimen for attack at grain boundaries is helpful when corrosion rates are low or available testing time is short. Grain-boundary attack generally indicates poor serviceability.

Recently developed electrochemical tests also may be used for screening materials; as with most laboratory corrosion tests, these should be considered only for initial screening of candidate materials.

Atmospheric Environments

Geographic location markedly affects the rate and type of corrosion of stainless steel and must be considered when selecting a specific grade. Atmospheric conditions can vary from dry, noncorrosive rural atmospheres to contaminated, highly corrosive industrial-marine environments. Even within the same geographic area, local conditions such as dusts, winds and insects may further influence choice of grade.

Rural Atmospheres

All grades of stainless steel are suitable for exposure to rural or other atmospheres that are essentially free from contamination, even when relative humidity may approach 100%. Selection for such applications depends on cost, availability, mechanical properties, fabricability and appearance.

Selection for exposure in a dry rural area often involves merely choosing the grade that results in the lowest total cost. Even straight low-chromium grades (10 to 13.5% Cr) will serve indefinitely without significant change in appearance. An almost identical statement can be made for exposure in rural areas where relative humidity may approach 100%. The lower-alloy grades may rust slightly and reveal a sensitivity to variations in original finish, but for most practical purposes all grades will serve satisfactorily in uncontaminated atmospheres, regardless of humidity. On the other hand, choice of grade, fabrication procedures and finishing techniques are quite important for exposure in contaminated industrial atmospheres or salt-laden, humid marine atmospheres, where corrosion is much more severe.

Industrial Atmospheres

Most grades of stainless steel are suitable for use in industrial atmospheres, although lower-chromium grades may be unsuitable for more severely contaminated atmospheres. Application often depends on the appearance required. Lower-chromium grades may fulfill service requirements but will tarnish severely. If appearance is important, type 430 is the lowest-alloy grade that can be used, and a higher-alloy grade usually is required. Type 302 has given excellent service in most applications.

In atmospheres free from chloride contamination, stainless steels have excellent corrosion resistance. Types 430, 302, 304 and 316 normally do not show even superficial rust. Some rusting may occur in marine atmospheres or in industrial exposures where surfaces become contaminated with chloride salts. Rusting is most likely to be severe on sheltered surfaces that are not well washed by rain.

In 1953, Subcommittee IV of ASTM Committee A-10 started a series of test exposures of 0.6-metre-square (2-foot-square) panels of chromium, chromium-nickel and chromium-manganese-nickel stainless steels at Pittsburgh and on the roof of the New York Port Authority Building. In 1956, the latter panels were transferred to the ASTM test site at Newark, New Jersey. Inspection of panels in 1957 showed the following results:

Type	Result
At Newark, New Jersey	
410	A few rust specks on skyward surface and many pits visible at a magnification of 40×; heavy rust on groundward surface
430	No rust on skyward surface; slight rusting on groundward surface
442, 446, 301, 304, 201	No rusting on skyward or groundward surface
At Pittsburgh	
410	Many minute pits and a few scattered rust specks on skyward surface; scattered rust areas on groundward surface, but less than at Newark
430, 442, 446, 301, 304, 201	No rust on skyward or groundward surface

Except for type 410, these stainless steels retained their original appearance after long exposure to atmospheres without chloride contamination. However, the panels at the Pittsburgh site were dirtier than those at the Newark site.

In choosing a grade of stainless steel for a specific industrial atmospheric application, all possible abnormal local conditions should be considered. Experiences of other users of stainless steels in the same region should be evaluated.

Austenitic or Ferritic Grades. Table 2 shows the advantage of choosing a grade with higher chromium content when a ferritic stainless steel is to be used; types 410, 430 and 446, in that order, show progressively better resistance to corrosion.

Whereas stainless steel specimens

Table 2 Atmospheric corrosion of stainless steels at two industrial sites

	New York City (industrial)		Niagara Falls (industrial-chemical)	
Type	Exposure time, yr	Specimen-surface evaluation	Exposure time, yr	Specimen-surface evaluation
Austenitic stainless steels(a)				
302	5	Free from rust stains	<2/3	Rust stains
302	26	Free from rust stains	...	...
304	26	Free from rust stains	<1	Rust stains
304	...	...	6	Covered with rust spots and pitted
347	26	Free from rust stains	...	...
316	23	Free from rust stains	<2/3	Slight stains
316	...	...	6	Slight rust spots, slightly pitted
317	...	...	<2/3	Slight stains
317	...	...	6	Slight stains
310	...	...	<1	Rust stains
310	...	...	6	Rust spots; pitted
Ferritic stainless steels (b)				
410	15	Dark, discolored on front; rusty on back	1	Some rust
410	...	...	15½	Severely stained and rusted; pitted
430	15	Dull on front; slight stains on back	<1	Rust spots
430	...	...	15½	Uniform light rust; pitted
430	7	No rust or stains on front; slight rust on back	<1	Rust spots
446	7	No rust stains	<1	Rust stains

(a) Solution-annealed sheet, 1.6 mm (1/16 in.) thick. (b) Annealed sheet, 1.6 mm (1/16 in.) thick.

exposed to the atmosphere in midtown New York City remained practically unaffected for more than 25 years (Table 2), specimens tested at Niagara Falls in the vicinity of plants producing chemicals and alloys showed marked staining and rusting. The conditions prevailing at this location were as follows: The racks were about 550 m (1800 ft) directly north of a plant manufacturing chlorine and hydrochloric acid, 120 m (400 ft) directly south of another chlorine plant, and 150 m (500 ft) east of a plant producing alloys and calcium carbide. A strong odor of chlorine and other chemicals was noticed frequently in this area. There are probably no more than a few sites in the United States where the atmosphere is as polluted with chlorine and hydrochloric acid as it is at this particular location.

As shown in Table 2, stainless steel specimens were susceptible to staining and rusting in this corrosive environment after a few months of exposure. The molybdenum-bearing steels were more resistant than the other types, but they were not completely immune to attack by this highly contaminated atmosphere. The conditions prevailing at this test site were highly localized even within the city itself; specimens exposed for several years at a site two miles north remained practically unaffected.

At the most severely corrosive site in Niagara Falls, the stainless steels appeared more likely to stain and rust in areas that were partly protected or recessed. To follow up this observation, specimens were exposed under a canopy, where they were more likely to accumulate particles of dust than when freely exposed to the washing action of rainfall. Both fully exposed and covered specimens were affected, but the attack was more severe on those under the canopy.

A building marquee constructed of type 430 stainless steel, near an industrial section in Chicago, rusted at protected or sheltered areas on the underside after only a few months of exposure, and cleaning did not prevent the rapid recurrence of rust. The marquee was replaced by one made of type 302, which is occasionally cleaned to remove dirt deposits but which has not rusted.

Atmospheric Contaminants. Chlorides and metallic iron dust are the contaminants most often responsible for rusting of structural stainless steels. Surface contamination by iron particles during fabrication or erection, or upon subsequent atmospheric exposure, will almost invariably result in rust stains. Chloride contamination has been traced to the calcium chloride used in making concrete. In one instance, discoloration of the stainless steel used in an architectural application in Columbus, Ohio was traced to a chloride residue left after the protective adhesive paper used in shipping the steel was removed. Removing the adhesive residue and then washing the stainless steel with a nitric acid solution eliminated the discoloration.

Results of atmospheric exposure of several stainless steels for 12 years at Bayonne, New Jersey showed an approximate inverse relationship between chromium content and extent of atmospheric corrosion. The experimental steels used in this study had higher carbon contents than are normal today for atmospheric exposure. Best resistance was exhibited by stainless steels with 19% Cr or more. Steels with less than 12% Cr were rusted over their entire surface after 12 years. Test results for 11 austenitic grades and one ferritic grade are shown in Fig. 10. Effectiveness of the molybdenum contents of types 316 and 317 is evident. In recent years, newly developed molybdenum-bearing austenitic and ferritic alloys have also shown similarly outstanding corrosion resistance.

Retention of appearance may be the decisive requirement in choosing grades for exposure to contaminated atmospheres. Often a specific grade will, for all practical purposes, meet service requirements and yet not retain its original appearance. For example, a carbon steel rooftop in an industrial area where the atmosphere was contaminated with chlorine and flue dust became perforated after five years and was replaced with one made of type 410 stainless steel. The stainless roof was as badly blackened after its first five years of service as the carbon steel had been, but was not significantly damaged as far as desired service was concerned. After 25 years, the stainless roof was still intact and looked as

Fig. 10 Corrosion of stainless steels in an industrial atmosphere (Bayonne, New Jersey)

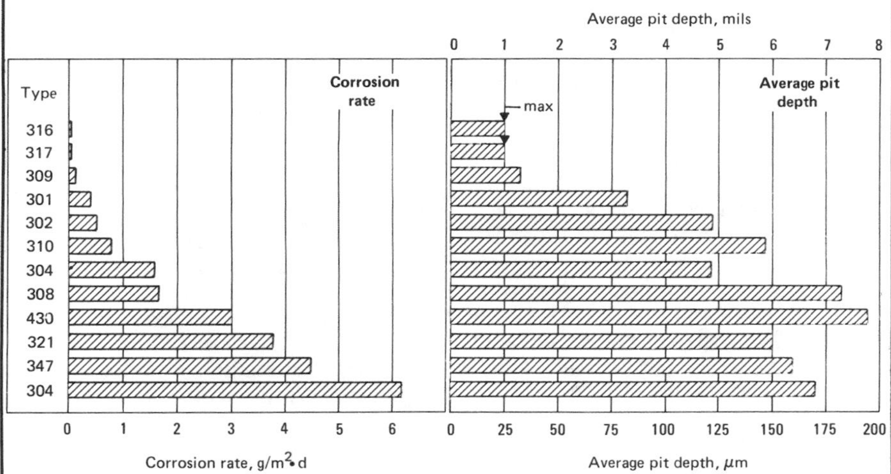

Data represent tests on panels about 0.019 m² in area (both sides of panels 3 × 4⅞ in.) exposed for 11.88 yr at sheltered sites (roofed, but open-sided) in a vertical position. Maximum pit depth is plotted at right for types 316 and 317 because these two steels were essentially unattacked; for all other steels, the chart represents average pit depth. All steels except types 316 and 317 developed rust spots that spread in from the edges until 50% or more of the specimen area appeared rusty. (Specimens of the same steels openly exposed nearby remained essentially unattacked.)

though it might be serviceable for another 25 years.

Architectural Applications. Type 302 stainless steel has been exposed for up to 50 years as architectural paneling on well-known buildings such as the Chrysler Building, the Empire State Building and others. The surfaces become very dirty but, after cleaning, are virtually free from corrosion. Inspection reports on buildings in New York, Philadelphia and Pittsburgh demonstrate that type 302 gives long and dependable service in industrial atmospheres.

Stainless steel curtain-wall construction is becoming an increasingly popular use of stainless steel in building construction. Trim, frames and molding for store fronts, banks and other buildings have been made of stainless steels, as have certain art objects. Periodic cleaning to maintain architectural appeal is normally recommended.

Marine Atmospheres

Stainless steels for service in marine atmospheres have been evaluated by the American Society for Testing and Materials, agencies of the United States Government and several private companies. In choosing grades for service in marine atmospheres, environmental factors such as wind direction, distance from the ocean (in both horizontal and vertical directions), rainfall, sunshine, dew, air pollution, sand and dust must be considered. Because of the above variables, atmospheric test data can be used to rank materials, but not to predict the life of an actual installation.

To obtain meaningful and comparable rankings, ASTM and other cooperating societies have standardized the manner in which test panels are to be exposed. Panels of specific composition exposed at a given site may be compared with panels of another composition only when exposed for the identical calendar period.

Type 430 is satisfactory for hardware on boats used in fresh water, but seldom for use on seagoing vessels. Here, the conventional 18-8 grades are commonly used and usually give long and satisfactory service. The high-manganese grades, such as 201 and 202, have been reported satisfactory, although some doubt remains as to whether they can always be substituted for types 301, 302 and 304. In close proximity to seawater, type 316 is superior to all of the above grades, and this molybdenum-containing grade or one of the higher-chromium grades may be required for satisfactory service.

For corrosion resistance of stainless steels in marine atmospheres, as in other environments, chromium content is very important (see Fig. 1). A gradual decrease in rust staining and pitting can be observed as chromium is increased from 10 to 29%. Nickel confers some improvement in resistance to marine atmospheres. Molybdenum, however, has a more striking effect. At a test site 24 m (80 ft) from the ocean at Kure Beach, North Carolina, a type 316 (2 to 3% Mo) panel was in excellent condition after 11 years of exposure, whereas some areas of a type 302 panel were stained with a superficial film of rust. Types 301, 304 and 321 showed spotty rust staining after 11 years of exposure at the 24-m test site. Ferritic stainless steels and austenitic stainless steels with more than 3% Mo, resisted any rust staining for up to five years of exposure at a site 244 m (800 ft) from the ocean.

It is important that austenitic stainless steels not be sensitized when exposed to marine environments. As previously mentioned, these steels are susceptible to stress-corrosion cracking and have been known to crack at ambient temperatures when exposed to marine environments in the sensitized condition.

Studies by J. J. Halbig and O. B. Ellis at Kure Beach demonstrated that precipitation-hardening stainless steels show greater resistance to marine atmospheres than quench-hardened straight chromium grades. However, there is considerable variation in the performance of precipitation-hardening grades with alterations in heat treatment, as judged by visual estimates of rust staining. No numerical data are available.

Nitriding reduces the corrosion resistance of type 302 stainless steel. This conclusion was based on 18-month exposure 244 m (800 ft) from the ocean at Kure Beach. Therefore, nitriding is not recommended for stainless steels exposed to marine atmospheres.

Experimental shipboard installations in washrooms incorporated types 304 and 430 for partitions, supports,

Table 3 Atmospheric corrosion of stainless steel boat hardware in a severe marine environment

Fitting	Material	Condition(a)
Plain block	Sides, strap: type 304; socket, rivets: type 303(b)	Sheave turned freely. Rust stains darkest on burred ends of rivets. Surfaces 25 to 60% covered with rust stains. Entire surface dulled by light tarnish. Surfaces remained in good condition; easily cleaned with mild abrasive.
Swivel block	Sides, strap: type 304 socket, rivets: type 303(c)	General appearance same as above. Rust stains generally lighter except for swivel rivet and washers. Sheave frozen but easily freed.
Barrel	Type 303 tubing(d)	Metal surface in good condition. Rust stains covered 75% of surface; heaviest at ends. Remaining 25% of surface had a dull finish. Stains easily removed with mild abrasive.
Gooseneck	Strap: type 304; clevis pins: type 303(b)	All surfaces 50 to 60% covered with rust stain. Stains heaviest at clevis pins and adjacent areas. Areas not rust stained were tarnished. Components functioned freely. Rust and stains removable with mild abrasive.

(a) After 3 years at a test site 24 m (80 ft) from the ocean at Kure Beach, North Carolina. (b) Components were electropolished. (c) Electropolished and buffed. (d) Pickled, annealed, electropolished and buffed.

coamings, deck covering and bulkhead sheathing. Evaluation was based on appearance, rust stains or discoloration, degree of corrosion (if any), ease of cleaning, and required maintenance. The stainless steel equipment was normally cleaned with mild detergents, although abrasive cleaners were sometimes used. After one year of service, it was concluded that stainless steel was suitable for these applications. Type 304 was selected because it retained better appearance and required less maintenance than type 430.

Under conditions of exposure to marine atmospheres and occasional wetting with seawater, items such as rub rails and deck fittings show only superficial rust that can be removed with most household cleaners. In installations such as rub rails that permit salt-water splash to concentrate in crevices between the hull or deck and the rail, stainless steel will sometimes "rust bleed". This can be eliminated by coating the back of the rail or fitting with a corrosion barrier such as zinc chromate primer. A rub rail installed without a corrosion barrier showed rust bleeding after limited use. A similar rub rail installed with primer showed no bleeding after seven years of service.

Types 430, 304 and 316 have given satisfactory performance for small-boat hardware; types 304 and 316 are better for boats in salt-water service. All three types are satisfactory for use around fresh water. Other stainless steels of the 200 series and 300 series also have been found satisfactory for fresh-water service. In six-month exposure tests at Kure Beach, boat hardware of alloy CF-20 (cast counterpart of type 302) showed many small rust spots, stains and incipient pits. All rust was readily removed with a mild abrasive. White corrosion was noted at the silver-brazed joints after three weeks, but had not progressed further after six months. Other results observed on boat hardware at Kure Beach after exposure for three years are given in Table 3.

Transportation Equipment

It is often difficult to choose grades of stainless steel for transportation equipment, such as trucks and trains, because the equipment may be exposed to the entire range of atmospheric conditions during service life, and different components on the same vehicle may have to meet different service or appearance requirements. The difficulty has become progressively more acute since the early 1950's, when use of deicing salt and salt mixtures in the United States increased dramatically (Fig. 11). The extensive use of salts, coupled with industrial pollution, has made the northeast region of the United States a very corrosive region for transportation equipment.

Several types of stainless steels, chiefly types 409, 430, 434, 201, 301 and 304, are used extensively in transportation service, in both exterior and interior applications.

Railroad passenger cars constructed of ferritic type 430 and austenitic types 201 and 301 provide a shiny appearance and give good performance in service. More than 100 railroad cars, built from austenitic stainless as long ago as 1937, are still free from pitting and other severe corrosion on roofs and panels. These passenger cars have been exposed not only to a wide range of atmospheric environments, but also to a variety of chemical cleaning agents. Acidic salt solutions are indispensable for removing brake-shoe dust, but in spite of the potentially corrosive nature of such solutions on stainless steel surfaces, no serious attack has been observed.

From December 1941 through July 1942, several all-stainless-steel passenger cars were constructed. In many of these cars, type 18-4-4 chromium-nickel-manganese stainless steel (a forerunner of type 201) and type 301 chromium-nickel stainless steel were used in both exposed and unexposed locations. In some, type 430 ferritic stainless steel was used for fluted side panels. Inspection in 1954 showed that the two types of austenitic stainless steels (18-4-4 and 301) had equivalent durability in the same environment; etching and pitting were negligible at contacting surfaces and almost nonexistent on exposed surfaces. Inspection of three of the cars with the type 430 side panels showed marked pitting of exposed surfaces, but this was reported to be "arrested and inactive". In crevices, pits were smaller, shallower and farther apart than those on exposed surfaces.

Six all-stainless-steel diesel railroad cars constructed of types 301 and 201 steels proved the good corrosion resistance of both grades. Before examination, these cars had been in use about 1½ years in the northeastern United States, with some exposure to coastal atmospheres. Exposure to either ma-

Fig. 11 Increase in road-salt usage in the United States (Ref 6)

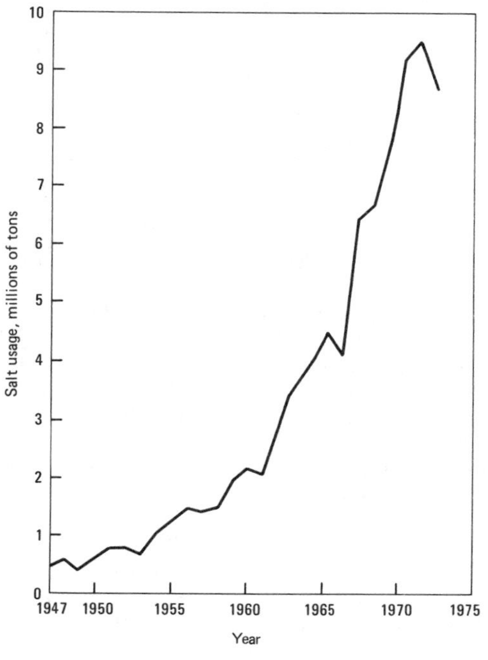

rine environments or corrosive washing solutions may produce rust stains or pits in a few months, and for this reason the good condition of the steel in this instance is significant, even though the period of service was comparatively short.

Buses. Stainless steel has been used for its appearance and corrosion resistance in several external and internal parts of buses. One bus manufacturer made engine-closure doors of 18-9 expanded stainless steel, welded to 18-8 frames. Weld scale was removed by pickling and electropolishing. Iron contamination from steel wire brushing (contrary to specification) led to discoloration by rust.

Among other bus applications, type 430 stainless steel piano hinges have been used in contact with anodized aluminum, and no corrosion has resulted. Carbon steel tubing, completely wrapped with type 430 stainless steel with a 2B finish, has served satisfactorily for 10 to 15 years as grab rails in interstate coach service.

Truck Trailers. Experience with commercial highway trailers made of stainless steel demonstrates that the austenitic types withstand most atmospheric environments to which such equipment is exposed. Field examinations have disclosed bright, unetched, pit-free surfaces where a reasonable cleaning schedule has been followed. Generally, the sides, front end, and rear doors are more carefully maintained than the roof. Nevertheless, the condition of most roof panels has been excellent. Scattered shallow pits and light etching have sometimes appeared on underframe beams where road salt and dirt accumulate on flanges, but such attack has been mild. Cleaning is invariably less thorough under the trailer, but spray and splash thrown up from wet roads during rainstorms help to dilute and wash away dirt and salt deposits.

Even on neglected equipment, a bright, as-rolled sheen is revealed when soil is wiped off. Some shallow-to-moderate pitting and light etching may be found in those areas of the trailer exposed to salty road spray and dirt thrown up by the wheels. The pits and etched areas are usually so small as to be clearly discernible only at a magnification of about 10×.

Wherever severe corrosive attack has occurred, (*a*) the soil on the panel has been caked, and the trailer has not been washed recently; (*b*) chlorides have been present in the soil; and (*c*) solutions collected by leaching soluble material from the soil have had pH values of 2 to 5. Of the limited amount of outside-surface pitting attack, that on roofs invariably has been the most severe.

This type of transportation equipment is exposed not only to highway conditions but also to industrial environments that vary in corrosivity. Moreover, cargoes range from harmless ones to highly corrosive chemicals such as aluminum chloride. All the above observations on truck trailers apply to both chromium-nickel and chromium-nickel-manganese types of austenitic stainless steel.

During the period when nickel was in short supply, type 430 sheet with a 2D finish was substituted for types 301 and 201 in the fabrication of rear door panels for trailers. The rear doors were subject to road spray in winter and to salt, dirt and water in all seasons; they were seldom washed. After the third winter of service, complaints of pitting and rusting became so numerous that door panels of type 430 were discontinued, and the austenitic types 301 and 201 were again used for this application.

Automobiles. Stainless steel types 409, 430, 434, 202, 301 and 304 are used for automotive parts. Type 434 usually is preferred for trim, partly because it more closely resembles chromium plate in color and partly because its corrosion resistance is better than that of type 430.

Due to the increased use of de-icing salts and general pollutants, use of type 430 for automotive trim has diminished considerably in recent years, having been replaced by type 434. Nonropping alloys such as type 436 also have replaced type 430 in some applications.

Type 301 is generally preferred for wheel covers because of its excellent corrosion resistance and because of its mechanical properties. During forming the alloy develops a work hardened condition that makes it an excellent spring retainer as well as a decorative, corrosion-resistant cover.

Since 1960, type 409 has found extensive use in automotive exhaust systems. Initially, mufflers made entirely of type 409 were used in some automobiles, with excellent service performance. In more recent years, noncriti-

Table 4 Corrosion of type 304 stainless steel in flowing seawater

Specimen(a)	Weight loss, g	Corrosion rate			Depth of pitting			
					Maximum		Average	
		g/m²·d	μm/yr	mils/yr	mm	in.	mm	in.
A	1.11	0.3	13	0.5	1.27(b)	0.050(b)	0.99	0.039
B	0.46	0.1	5	0.2	1.17	0.046	0.76	0.030
C	0.70	0.2	10	0.4	1.19	0.047	0.58	0.023
D	0.76	0.2	10	0.4	...	...	...	...

(a) Test panels 102 by 102 by 2.1 mm (4 by 4 by 0.083 in.) were exposed for 6 months to seawater flowing at 0.3 to 0.6 m/s (1 to 2 ft/s). (b) One pit on top side, underside pitted in the form of grooves 13 to 19 mm (1/2 to 3/4 in.) long.

cal structural parts or muffler parts that exhibit very little corrosion have been redesigned in less corrosion-resistant materials. All catalytic converters, used as pollution-control equipment in automobiles, are constructed of type 409.

Seawater Environments

Exposure to seawater introduces a number of factors not present to any great extent in atmospheric exposure. Selection of a stainless steel grade for seawater immersion is far more complex than selection for atmospheric service. Types 304 and 316 (especially 316) usually give the best service. However, in stagnant seawater (particularly badly contaminated harbor waters), all types are likely to pit severely from biofouling. Even type 316 may be completely unsatisfactory if water velocity is less than 1.5 m/s (5 ft/s). Recently developed highly alloyed ferritic and austenitic types such as 26Cr-1Mo 29Cr-4Mo, JS700, AL-6X and 22Cr-13Ni-5Mn are extending the use of stainless steels in such highly corrosive applications.

Applications of stainless steels in seawater include structures, heat exchangers, piping and special items such as propellers. If stainless steel is used for a structural or a heat-exchanger application in seawater, the design should avoid crevices at lapped joints or tube-to-tube-sheet joints. To maintain good corrosion resistance, stainless steels must be kept in contact with constantly flowing seawater and must be free from deposits and biofouling. The latter factors interfere with the maintenance of a passive film. When stainless steels fail in condensers, it is from pitting attack rather than the corrosion-erosion common to many condenser alloys. Weldments in unstabilized grades are subject to localized attack in seawater because welding upsets the condition of heat treatment necessary for good corrosion resistance.

Most types of stainless steel are likely to pit or groove in seawater. For example, type 304 showed little loss of weight in six months of service, and the average penetration of 102-by-102-mm (4-by-4-in.) panels was only 5 to 13 μm/yr, or 0.2 to 0.5 mil/yr (Table 4). However, maximum pitting was 1270 μm (50 mils) in six months. A container, pipe, or heat-exchanger tube, if not properly exposed to seawater, would probably fail from local perforation rather than from loss of wall thickness. A constant seawater velocity of 1.5 m/s (5 ft/s) or more tends to prevent local attack.

Composition Effects. Results of a series of tests on stainless steels exposed to flowing seawater are given in Table 5. Not all the data are comparable, because the time of exposure varied from 106 to 1255 days. Nevertheless, some comparisons can be made. For example, type 405 showed more weight loss than type 347, but the maximum pit depth was almost 50% greater for 347 in one test. Types 304, 309 and 310, in that order, exhibited decreasing weight loss and decreasing maximum depth of pitting, although type 304 was the least affected by contact with the washer used in attaching the specimen to the test rack.

Test results on the crevice-corrosion behavior of some of the newly developed ferritic stainless steels in low-velocity seawater are given in Table 6. The data in this table, and those in Tables 7 and 8, illustrate the effects of chromium, molybdenum and nickel content on performance of stainless steel in seawater. In general, both ferritic and austenitic alloys show improved resistance to pitting and crevice corrosion with increased amounts of chromium and/or molybdenum. AL-6X (20Cr-24Ni-6Mo) has been used for seawater condenser tubes with no reported problems during six years of service.

Intergranular attack of austenitic stainless steels increases as carbon content is raised unless the steel is fully annealed. Fully quench-annealed structures, free from carbide precipitation at grain boundaries, are preferred for seawater use, as well as for highly corrosive service in chemical equipment. However, experiments suggest that carbides in cast alloys do not play a major role in seawater service. Seven cast 19Cr-9Ni steels were tested in a marine environment; their compositions were essentially the same, except for carbon content, which varied from 0.06 to 0.45%. All specimens were welded without postweld heat treatment and were exposed to seacoast atmospheres. None showed any intergranular corrosion after five months; after one year, however, some intergranular corrosion had occurred in specimens containing 0.28% C or more.

Hardened surfaces sometimes are desirable on stainless steel in corrosion service. Hardening by carburizing results in severe intergranular corrosion in seawater and is not recommended. If hard surfaces are needed, they are best obtained by plating with hard nickel or chromium.

Addition of sulfur or selenium to either type 410 or type 430 to promote machinability usually makes the steel less corrosion resistant in severely corrosive environments. In one experiment, test specimens measuring 30 by 73 mm (1 3/16 by 2 7/8 in.) were finished with 180-grit cloth and cleaned with a wet bristle brush and powdered pumice until free from a water break. After drying, each specimen was coated on the back, on the edges and on a 3.2-mm (1/8-in.) border on the exposed face to prevent corrosion at the interface between the specimen and the supporting member of the test unit. Tests were carried out in seawater flowing at 4 m/s (13 ft/s) for 15-day periods, at water temperatures between 20 and 30 °C (65 and 85 °F). Treatments of the steel tested are given in the caption of Fig. 12.

Electrode potentials were measured against a saturated calomel half cell, using a portable potentiometer with an accuracy of ±0.5 mV for values higher than 10 mV. The calomel half cell was located 89 mm (3 1/2 in.) downstream from the nearest point on the test specimen. The location of the reference elec-

Table 5 Corrosion of stainless steels in flowing seawater

Type(a)	Original surface condition	Test period, days	Weight loss, g	Depth of pitting: Surface, Maximum, mm	Surface, Maximum, in.	Surface, Average(b), mm	Surface, Average(b), in.	Under Bakelite washers, Maximum, mm	Under Bakelite washers, Maximum, in.	Under Bakelite washers, Average(b), mm	Under Bakelite washers, Average(b), in.	Condition of edge
405	Hot rolled, No. 4 finish	755(c)	135	2.5	0.100	1.5	0.059	4.75(d)	0.187(d)	2.6	0.103	Badly pitted
403	Hot rolled, No. 4 finish	388(c)	118	1.9	0.075	1.0	0.038	1.2(e)	0.046(e)	1.0	0.040	Badly pitted
430	Hot rolled, pickled	568(c)	109	3.4	0.135	1.45	0.057	1.85(e)	0.072(e)	1.55	0.061	Badly pitted
308	Hot rolled, No. 4 finish	755(c)	28	5.2	0.205	2.1	0.083	3.6	0.141	1.7	0.066	Badly pitted
347	Hot rolled, No. 4 finish	755(c)	26	3.8	0.148	1.5	0.59	1.55	0.061	1.2	0.046	Badly pitted
	Hot rolled, pickled	755(c)	30	2.0	0.079	1.4	0.054	1.2	0.047	1.0	0.038	Few pits
321	Hot rolled, welded, 1/2 pickled, 1/2 No. 4 finish	944	...	1.35(f)	0.053(f)	0.38	0.015	1.4	0.056	0.6	0.024	No pitting
316	Hot rolled, welded, 1/2 pickled, 1/2 No. 4 finish	944	...	0.18	0.007	1 pit	1 pit	0.25	0.010	0.08	0.003	No pitting
	Hot rolled, pickled	1255	3	1.3	0.050	0.56	0.022	4.3	0.170	1.2	0.046	No pitting
317	Hot rolled, No. 4 finish	1075	5	0.58	0.023	0.30(g)	0.012(g)	1.1	0.045	0.7	0.027	Pitted
304	Pickled in HNO_3-HF	320	12	0.89	0.035	0.56	0.022	0.9	0.036	0.6	0.024	No pitting
309	Pickled in HNO_3-HF	320	11	0.51	0.020	0.30	0.012	1.4(d)	0.055(d)	1.4(d)	0.055(d)	No pitting
310	Pickled in HNO_2-HF	320	4	0.15	0.006	0.08	0.003	0.6(d)	0.024(d)	0.6(d)	0.024(d)	No pitting
325	Hot rolled, pickled, No. 4 finish	106	15	0.81	0.032	0.18	0.007	None	None	None	None	Grooved to 1.3 mm (0.05 in.)
	Hot rolled, pickled, No. 1 finish	106	60	1.4	0.056	1 pit	1 pit	None	None	None	None	Grooved to 2.5 mm (0.10 in.)
329	Hot rolled, pickled, No. 4 finish	106	0	0.0	0.000	0.0	0.000	0.25	0.010	2 pits	2 pits	No pitting
	Hot rolled, pickled, No. 1 finish	106	7	0.0	0.000	0.0	0.000	0.9	0.037	2 pits	2 pits	No pitting

(a) Panels 305 by 305 mm (12 by 12 in.) were completely immersed at Kure Beach, North Carolina, in seawater flowing at 0.3 to 0.6 m/s (1 to 2 ft/s). (b) These values are averages of the ten deepest pits. (c) Specimens withdrawn from test due to failure in period indicated. (d) Specimen became perforated. (e) Local attack directly under washer; holes were greatly enlarged. (f) One pit in weld, 3.2 mm (0.125 in.) deep. (g) Five pits.

trode was chosen to give an indication of the over-all potential of the entire surface of the test specimen rather than that of any small anodic or cathodic area. (Locating the tip of the half cell close to a particular spot might have given results that were not representative of the entire surface.)

Type 416 alloy, which contains sulfur, was found, unexpectedly, to be more noble after a tempering heat treatment than type 410 (Fig. 12, left and center). Also, the sulfur-free type 430 was more noble, as might be expected, than the sulfur-containing type 430F (Fig. 12, right). Type 416 was less noble than type 410 after hardening and stress relieving.

Surface Condition. It is usually undesirable to paint stainless steel, especially with antifouling paints, because of the danger that pitting will start at pores in the paint.

Five 18-8 stainless steel panels, with finishes ranging from rough pickled No. 1 to a buffed mirror No. 8, were tested to show the effect of finish on corrosion resistance in relatively quiet seawater (Table 9). It was hoped that barnacles would be unable to adhere to a mirror-smooth surface. However, the same number of barnacles were attached to the mirror-finished specimen as to the specimens with the rougher finishes. The weight-loss data given in Table 9 indicate that external factors, such as barnacle attachment, are more important than surface finish in determining the extent and distribution of corrosive attack.

Included in the data are laboratory counts of the numbers of pits that were

Table 6 Crevice corrosion of ferritic stainless steels in low-velocity seawater (Ref 7)

Nominal composition, %			Probability of initiation(a), %		Maximum depth			
					61 days		272 days	
Cr	Mo	Ni	61 days	272 days	mm	in.	mm	in.
18	2	...	13	...	0.64(b)	0.025(b)	...	...
26	1	...	0	...	0	0	...	...
26	1	...	3	...	0.43	0.017	...	...
28	2	...	0	0.8	0	0	0.14	0.006
28	2	4	0	7.5	0	0	0.2	0.008
25	3.5	...	0	0	0	0	0(c)	0(c)
25	3.5	2	...	0	...	...	0	0
25	3.5	4	0	0	0	0	0	0
29	4	...	0	0	0	0	0	0
29	4	2	1.4	0	<0.02	<0.0008	0	0

(a) At ambient temperature. (b) Perforated. (c) 186 days.

Table 7 Effect of composition on corrosion resistance in relatively quiescent seawater

Type(a)	Exposure time(b), days	Weight loss, g	Corrosion rate		Depth of pitting			
					Maximum		Average	
			μm/yr	mils/yr	mm	in.	mm	in.
13% Cr	388	118	79	3.1	1.91	0.075	0.97	0.038
15% Cr	605	109	48	1.9	3.43	0.135	1.45	0.057
17% Cr	605	102	46	1.8	2.72	0.107	1.78	0.070
18-8	643	24	5	0.2	(c)	(c)	0.79	0.031
18-8(Mo)	2773	8	<2.5	<0.1	0.86	0.034	0.41	0.016
18-8(Mo)	3164	3	<2.5	<0.1	0.64	0.025	0.28	0.011

Composition, %:	C	Mn	P	S	Si	Cr	Ni	Mo
13% Cr	0.08	0.42	0.02	0.018	0.30	13.75	0.22	...
15% Cr	0.10	0.41	0.02	0.010	0.30	14.69	...	...
17% Cr	0.10	0.49	0.02	0.013	0.44	17.28	...	...
18-8	0.06	0.46	0.012	0.009	0.34	18.34	9.13	...
18-8(Mo)	0.07	1.48	...	...	...	19.06	10.83	2.4
18-8(Mo)	0.07	0.35	0.01	0.013	0.43	18.19	9.7	3.18

(a) All specimens were 305 mm (12 in.) square. Panel of 18-8 was 1.27 mm (0.050 in.) thick; 18-8(Mo) panels were 6.4 mm (0.25 in.) thick. (b) All specimens except the 13% Cr stainless steel were removed from test, cleaned, reweighed and replaced several times during the period of exposure. (c) Test specimen became perforated.

Table 8 Corrosion of wrought stainless steels with No. 2B finish in flowing seawater

Type	Exposure time(a), days	Weight loss, g	Depth of pitting(b)			
			Maximum		Average(c)	
			mm	in.	mm	in.
304	195	12	1.63	0.064	1.2	0.049
316	195	2	0.38	0.015	0.1	0.004

(a) Test was conducted at Kure Beach, North Carolina, on panels 305 by 305 mm (12 by 12 in.) totally immersed in seawater flowing at 0.3 to 0.6 m/s (1 to 2 ft/s). (b) The values shown are for the exposed surfaces of the panels. The metal under the Bakelite washers was perforated. The edges of the panels were pitted. (c) These values are averages of the ten deepest pits.

formed in a circulating solution of ferric chloride. Although this accelerated test indicated that the smoother finishes should have resisted pitting much better than the rougher finishes, no such result was obtained under the natural conditions of exposure to seawater. Once again, external factors (such as marine organisms) were of greater importance than the factors covered by the laboratory procedure.

Galvanic couples with stainless steels may result in severe corrosion of the other metal when the couple is immersed in seawater. Although this is a good general rule, stainless steels generally cause less galvanic corrosion of less-noble metals than do other noble metals (Table 10). Still, it is preferable to couple only those metals that are closely related in electrode potential. This principle also applies to the coupling of different grades of stainless steel to each other. One report indicates that coupling of titanium to types 302 and 316 stainless appeared to have a negligible effect because the potentials of the metals involved were nearly the same.

Both composition and condition have marked effects on the corrosion potentials of stainless steels. Data given in Fig. 13 show that, in the study reported, average potential varied from 150 to 350 mV for the various combinations of composition and condition investigated. It is noteworthy that an equally large variation was encountered among supposedly identical specimens of the same steel in several of the tests. Test procedure and specimen size were the same as those described above in the discussion of the influence of sulfur on potential.

Crevice Corrosion. Stainless steel corrodes in crevices when exposed to seawater. Crevices were formed in each of eight test specimens, by overlapping two pieces of type 430 stainless steel, to determine the influence of the area of freely exposed metal outside the crevice on extent of corrosion. To confine the action to surfaces of known area and location, all other surfaces were coated with two layers of a baked phenolic varnish, each layer being 0.05 mm (0.002 in.) thick. The assemblies were exposed for 87 days in seawater flowing at a rate of 0.3 to 0.6 m/s (1 to 2 ft/s). The results of this study (Fig. 14) illustrate that the greatest danger exists when a crevice is associated with a large exposed metal surface and that, when the area outside the crevice is small, pitting within the crevice approaches a minimum value.

To prevent crevice attack, sealing compounds have been tried. The corrosion in crevices is highly localized and occurs at joints between similar or dissimilar metals. Test results of efforts to find suitable sealing materials are shown in Fig. 15. Tests were also conducted on the following materials without success:

- Teflon adhesive
- Synthetic rubber
- Anaerobic Permafil
- Polyester resin

Fig. 12 Effect of sulfur on electrode potential of ferritic stainless steel exposed in flowing seawater

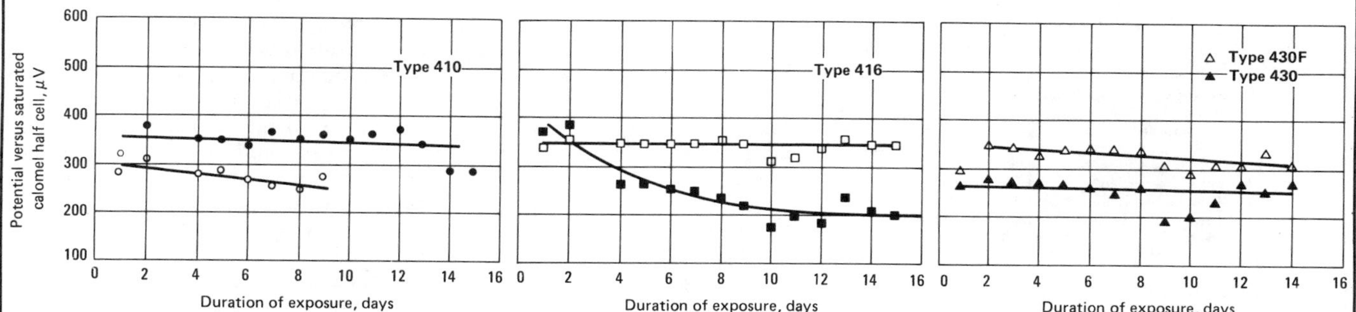

Type 410 (sulfur free) and type 416 (sulfur bearing) quenched from 980 °C (1800 °F); open symbols are for hardened and stress relieved material, solid symbols for hardened and tempered material. Type 430 (sulfur free) and type 430F (sulfur bearing) annealed 1 h at 730 °C (1350 °F). (Ref 8)

Table 9 Effect of surface finish on corrosion of 18-8 stainless steel in seawater

			Depth of pitting						
		Weight	Surface(b),		Under washer				Number
	Surface	loss,	average		Maximum		Average		of
Specimen(a)	finish	g	mm	in.	mm	in.	mm	in.	pits(c)
A	No. 1	28	1.02	0.040	1.73	0.068	1.42	0.056	80
B	No. 2B	35	1.67	0.046	2.16	0.085	0.97	0.038	6
C	No. 4	35	0.94	0.037	1.96	0.077	1.35	0.053	7
D	No. 6	42	1.02	0.040	(b)	(b)	1.65	0.065	8
E	No. 8	31	1.65	0.065	2.79	0.110	1.52	0.060	7

Composition, %:	C	Mn	Si	Cr	Ni
A	0.11	0.92	0.42	17.86	8.52
B	0.11	0.92	0.42	17.86	8.52
C	0.05	0.47	0.47	18.54	9.01
D	0.05	0.47	0.47	18.54	9.01
E	0.05	0.47	0.47	18.54	9.01

(a) Specimens 305 by 305 by 3.2 mm (12 by 12 by 1/8 in.) were immersed for 685 days in relatively quiescent seawater. (b) Specimen became perforated at one or more locations. (c) Number of pits developed in specimen during pit test using 0.7*M* $FeCl_2$ solution.

Table 10 Galvanic corrosion of medium-carbon steel in contact with type 304 stainless steel, titanium or copper in seawater (Ref 9)

	Corrosion rate of steel		Increase caused by galvanic effect	
Couple(a)	µm/yr	mils/yr	µm/yr	mils/yr
Steel (uncoupled)	790	31	...	...
Steel–type 304	910	36	130	5
Steel–titanium	1170	42	280	11
Steel–copper	2540	100	1750	69

(a) Area of steel equaled area of other metal in couple. Seawater flow rate, 2.4 m/s (7.8 ft/s); temperature, 10 °C (50 °F).

- Liquid neoprene. (In one test, the neoprene took a light set before it was used, and crevice corrosion was severe; in another it was used immediately after mixing, and no attack occurred.)
- Copper washers.

Copper plating and 50Pb-50Sn solder showed some promise as effective sealing compounds. Of these, the lead-tin solder performed best in eliminating crevice corrosion between abutting surfaces of type 304 stainless steel after exposure to seawater for 335 days.

Data relating to under-washer corrosion of stainless steel test panels in seawater are given in Table 11.

Temperature sometimes affects corrosion rate of stainless steels in seawater, especially where crevices are present. Comparative corrosion rates for several types of stainless steel at different temperatures are given in Table 12. The corrosion rates of both 410 and 430 increased markedly when temperature was increased from 15 °C (60 °F) to 25 °C (77 °F). Figure 16 compares the corrosion rates of four stainless steels at three different temperatures.

Cathodic protection is useful in preventing crevice corrosion and pitting of types 302 and 316 stainless steel immersed in quiescent seawater. Types 410 and 430 stainless steel cannot be completely protected by cathodic currents because hydrogen blisters devel-

Fig. 13 Corrosion potentials of stainless steels in flowing seawater

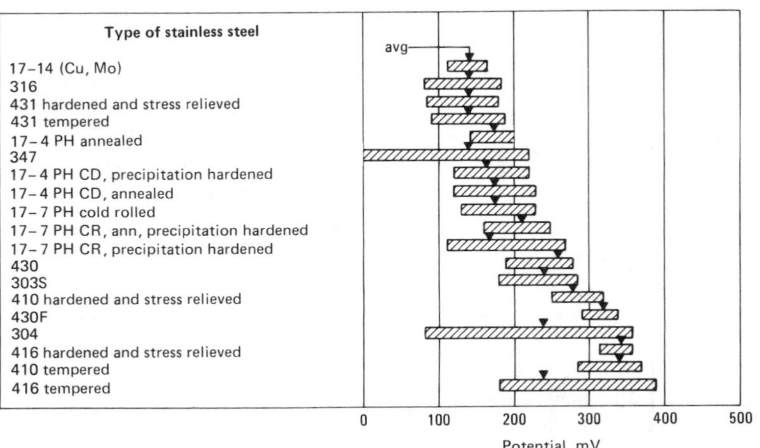

Corrosion potentials vs saturated calomel electrode in seawater flowing at 4 m/s (13 ft/s) vary with composition and condition of the stainless steels. Temperature for these tests was 18 to 29 °C (64 to 84 °F).

Fig. 14 Variation of pitting depth with exposed area outside a crevice for type 430 stainless steel in flowing seawater

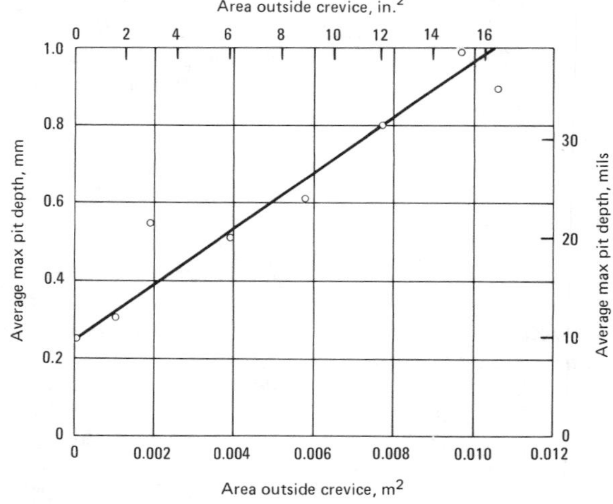

Seawater flow rate was 0.3 to 0.6 m/s (1 to 2 ft/s). (Ref 10)

op at current densities below those required for complete protection. Polarization curves for four stainless steels are given in Fig. 17 for the test conditions indicated.

Cathodic protection is credited with preserving a beaded chain made of type 302 stainless steel, such as those used in fishing tackle, for one year in tidal seawater. All parts (beads, links and clips) were bright, with no visible rust or rust stain, and all operated in a satisfactory manner. The good performance was believed to have been made possible by cathodic protection due to sacrificial corrosion of the lead sinker attached to the chain.

At low velocities, where pitting of stainless steel is almost certain to occur in natural seawater, cathodic protection has been shown to be effective in reducing the intensity of pitting and crevice corrosion. The data in Table 13 indicate the degree to which stainless steel was cathodically protected by magnesium anodes in seawater flowing at a velocity of less than 0.3 m/s (1 ft/s). Several specimens of types 410, 430, 302 and 316 were immersed in the basin at Kure Beach. The weight losses given are attributable to both pitting of exposed surfaces and crevice corrosion beneath the insulating washers. In all instances except for type 316, total weight loss decreased as current density was increased. However, both 410 and 430 suffered hydrogen blistering when the current density was increased to 215 mA/m^2 (20 mA/ft^2), which resulted in polarization that was nearly complete for type 430, but that was still insufficient to arrest pitting and weight loss for type 410.

Considerable reduction in corrosion occurred without distinct polarization of the materials in the test. At current densities below the requirements for complete polarization of the specimens, the potentials varied in an erratic manner during the exposure time, as shown by the variation in pitting. The data in Table 13 correspond reasonably well with the curves in Fig. 17, which indicate the effect of current density on weight loss—especially for type 316, which shows nearly complete polarization for a current density higher than 108 mA/m^2 (10 mA/ft^2).

Velocity Effects. At flow rates above 3 m/s (10 ft/s), type 316 generally remains passive. Below 1.5 m/s (5 ft/s), type 316 may become active, and some other material, such as Monel, is usually preferred. At velocities between 1.5

Fig. 15 Effectiveness of sealing compounds in preventing crevice attack on type 304 stainless steel in seawater

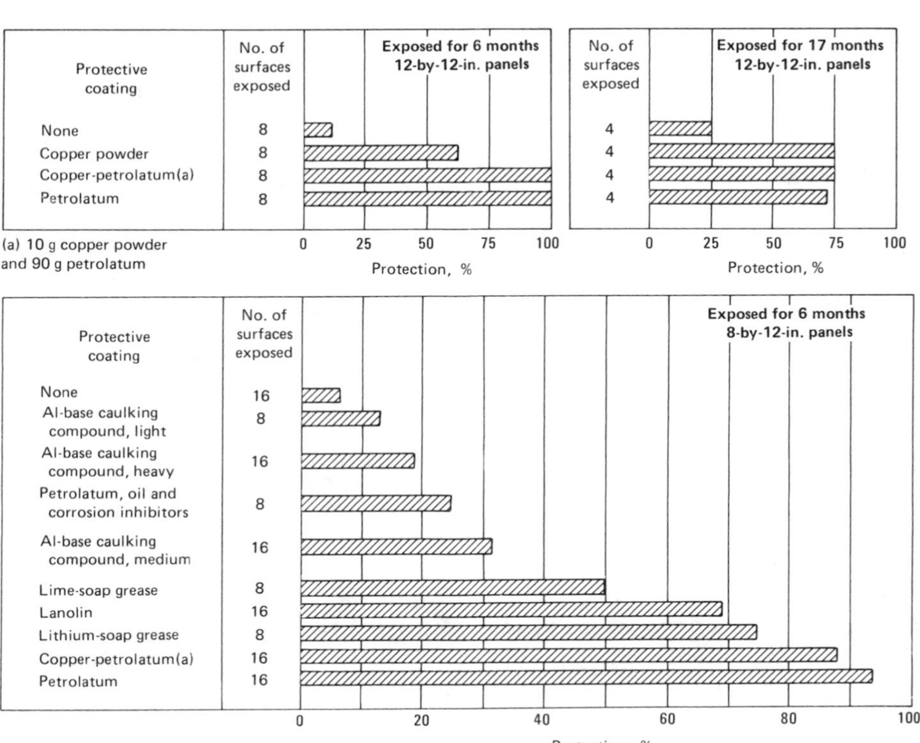

Effectiveness evaluated as a percentage based on the number of unattacked surfaces. See text under the heading "Crevice Corrosion" for a list of six materials that were tested and found not protective. Exposure was done in seawater flowing at 0.3 to 0.6 m/s (1 to 2 ft/s).

and 3 m/s (5 and 10 ft/s), the choice of type 316 is questionable if based on velocity alone. Sometimes use of this grade is mandatory, such as for the process side of a heat exchanger.

The importance of using stainless steel in seawater service only where sufficient velocity is maintained is illustrated by the following examples:

Example 1. Immersion Test Panels. A type 302 stainless steel panel measuring 152 by 38 by 1.6 mm (6 by 1½ by 1/16 in.), immersed in quiescent seawater under tidal flow conditions, lost 1.25 g in 371 days of exposure. The panel was perforated in four locations. Similar panels immersed in seawater flowing at 0.6 m/s (2 ft/s) lost 1.03 g in 343 days of exposure. There were perforations near the edges of the panels and crevice corrosion to a depth of 0.6 mm (0.025 in.) under fouling organisms.

Example 2. Test Panels Immersed in a Harbor Basin and in a Flume (Table 14). The steels in this study were divided into two groups. One group was exposed in relatively quiescent water (velocity less than 0.3 m/s, or 1 ft/s) in the basin at Freeport, Texas; the other was exposed in a flume where seawater velocity was about 1.2 m/s (4 ft/s). The panels exposed in the basin were subject to fouling by marine organisms.

Because of the presence of chlorine and the higher velocity of the water, those exposed in the flume did not become fouled.

Major differences in corrosion were observed for the two exposure conditions (see Table 14). Based on previous experience, these differences were attributed to the difference in velocity of water flow rather than to any direct corrosive effect of the small amount of chlorine added to the water in the flume. The chlorine would be expected to prevent fouling, which, in turn, would prevent local corrosion.

The stainless steels corroded less in the flowing water; the considerable local attack in the basin exposure was evidently promoted by the organisms that became attached to the panels. This was particularly true for the austenitic stainless steels, which remained free from attack in the flume. The ferritic 13%-Cr steel underwent considerable corrosion in both locations, but

Table 11 Evaluation of under-washer corrosion of stainless steel test panels in seawater

Type(a)	Surface finish	Heat treatment	Attack under Bakelite washers
304	No. 2	Annealed	Attack under 2 washers: depth, 0.89 mm (0.035 in.)
316	No. 2	Annealed	Attack under 1 washer: depth, 1.3 mm (0.051 in.)
430	No. 2	Annealed	Perforation under 2 washers: depth, 6.4 mm (0.25 in.)
431	No. 2	Hardened, stress relieved	Free from attack
431	No. 2	Hardened, tempered	Attack under 3 washers: depth, 0.48 mm (0.019 in.)
304	No. 4	Annealed	Attack under 3 washers: depth, 1.3 mm (0.053 in.)
430	No. 4	Annealed	Attack under 4 washers: depth, 1.5 mm (0.060 in.); perforation under 1 washer: depth, 6.4 mm (0.25 in.)
431	No. 4	Hardened, stress relieved	Attack under 2 washers: depth, 0.23 mm (0.009 in.)
431	No. 4	Hardened, tempered	Attack under 4 washers: depth, 5.0 mm (0.199 in.); perforation under 1 washer: depth, 6.4 mm (0.25 in.)

(a) Test panels 305 by 305 by 6.4 mm (12 by 12 by 1/4 in.) were exposed for 178 days to seawater flowing at 0.3 to 0.6 m/s (1 to 2 ft/s). Eight washers were used to attach each panel to a seawater-immersion test rack. Silicone grease was used under the washers.

Table 12 Corrosion resistance of stainless steels exposed to flowing seawater

Type(a)	Weight loss(b), mg	Maximum pitting, mm	Maximum pitting, in.
At 13 to 16 °C (55 to 61 °F)			
303S (c)	136.3	None	None
316(c)	1.8	None	None
410(d)	77.3	0.013	0.0005(e)
416(d)	93.3	None	None
430(f)	72.1	0.013 to 0.005	0.0005 to 0.0002(g)
430F (f)	89.4	None	None
431(d)	35.5	None	None
At 25 °C (77 °F)			
410	771.5	0.94	0.037
430	629.9	1.45	0.057

(a) Specimens 29 by 73 mm (1.125 by 2.875 in.) were exposed to seawater flowing at 2.4 m/s (7.8 ft/s) for 15 days. All specimens were passivated in HNO_3 before testing. (b) Virtually all weight loss caused by crevice corrosion. (c) Annealed. (d) Hardened and stress relieved. (e) Only one pit present. (f) Process annealed. Major weight loss caused by crevice corrosion. (g) Three pits present.

attack was more severe in the quiescent water.

Example 3. Austenitic Stainless Steel Panels Exposed in Flowing Seawater (Table 15). When specimens of types 302 and 316 stainless steel were exposed to seawater flowing at about 1 m/s (3 ft/s), pits 0.5 to 0.8 mm (0.02 to 0.03 in.) deep developed in the faces of the specimens after 483 days. These pits reached depths of 1.3 to 1.5 mm (0.05 to 0.06 in.) after an exposure time of 4½ years. Pitting at the edges was greater than at the faces of the specimens in all instances (see Table 15). When corrosion was evaluated in terms of depth of penetration per year for the 483-day exposure, type 316 showed a higher rate than type 302, but type 316 exhibited a lower rate for the longer exposure of 4½ years. This may be attributed to the difference in initial surface condition—bright cold rolled for type 302 and hot rolled and pickled for type 316.

Aeration is usually associated with velocity. Under stagnant conditions, oxygen is depleted rapidly, the hydrous oxide film cannot repair itself and the steel becomes active and starts to pit. Stainless steels give best service when exposed to aerated, rapidly flowing seawater. Lack of aeration at a specific site often leads to severe attack.

Biofouling. Stainless steel is highly susceptible to attack under foreign deposits where oxygen dissolved in the water cannot reach the surface of the metal. Fouling organisms settle readily on stainless steel, regardless of surface finish. In one instance, the rate of pitting under a barnacle was about 13 mm (0.5 in.) per year.

Pollution. Impurities in water can result in conditions that definitely require stainless steel. Stainless steels may be especially suitable if conditions are oxidizing. Other types of pollution may make use of stainless inadvisable.

For land-based installations using polluted seawater, type 316 is satisfactory. This type has also been used successfully in harbor installations at Los Angeles, San Francisco and New York. Type 316 is used for pump parts where erosion-corrosion is involved. By contrast, a ladder made of type 304 lasted only two years in stagnant seawater.

Impingement Attack. Stainless steels generally show excellent resistance to impingement attack in both clean and polluted seawater. In the example that follows, the severity of crevice corrosion is not indicated, although it should be kept in mind that stainless steels are particularly susceptible to this type of attack.

Example 4. Impingement Attack in a Seawater Jet (Table 16). Type 304 steel was exposed to clean seawater and to seawater containing 0.5 and 1.0 ppm of added hydrogen sulfide. Specimens 89 mm (3½ in.) long, 13 mm (½ in.) wide and 0.8 to 3.0 mm (0.03 to 0.12 in.) thick were exposed to a jet stream at 4.6 m/s (15 ft/s) for 28 days. Air in amounts of 2 to 4% by volume was added. Average temperature of the seawater was 10 °C (50 °F). Results were identical for specimens tested in clean water and in polluted water. Weight loss was 1 mg and attack did not progress to a measurable depth either at the jet or elsewhere.

Further testing of types 304, 316 and 430 demonstrated the excellent resistance of stainless steels to impingement attack, as given in Table 16. The

Fig. 16 Corrosion rate versus seawater temperature for four stainless steels

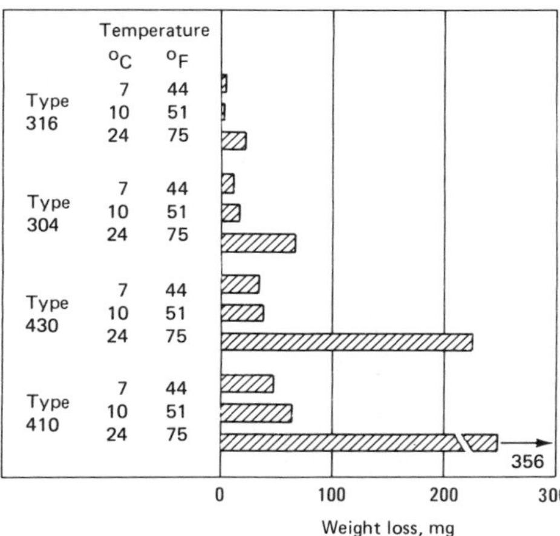

Specimens were exposed 17 days at 7 °C, and 15 days at 10 and 24 °C; the seawater flow rate was about 4 m/s (13 ft/s). All specimens suffered crevice corrosion, which accounted for most of the weight loss. Stainless steels of the 400 series were more susceptible to crevice corrosion than the 300-series stainless steels.

Table 13 Cathodic protection of stainless steels in seawater by coupling to sacrificial magnesium anodes (Ref 11)

Specimen No.(a)	Induced current density mA/m²	mA/ft²	Weight loss, g	Maximum depth of pitting mm	in.	Maximum penetration under washer mm	in.
Type 410 (13% Cr)							
1	0	0	37.8	0	0	0	0
2	54	5	19.4	0(b)	0(b)	Perforated	
3	108	10	6.6	0	0	Perforated	
4	215	20	3.7	0	0	0.97	0.038
Type 430 (17% Cr)							
5	0	0	29.6	0(b)	0(b)	Perforated	
6	54	5	13.3	0(b)	0(b)	Perforated	
7	108	10	14.6	0(b)	0(b)	Perforated	
8	215	20	0.9	0	0	0	0
Type 302 (18Cr-8Ni)							
9	0	0	12.0	0(b)	0(b)	Perforated	
10	32	3	9.3	0	0	2.29	0.090
11	75	7	1.2	0.25	0.010	0.99	0.039
12	161	15	0.7	0	0	1.17	0.046
Type 316 (18Cr-12Ni-3Mo)							
13	0	0	2.1	0.18	0.007	1.40	0.055
14	32	3	2.8	0	0	0.94	0.037
15	54	5	1.6	0.05	0.002	0.81	0.032
16	161	15	0.7	0	0	0	0

(a) Test panels were immersed in seawater flowing at a velocity of less than 0.3 m/s (1 ft/s) at Kure Beach, North Carolina, for 180 days, except for specimens 1 and 4, which were exposed for only 110 days. Specimens were 152 by 305 mm (6 by 12 in.), except for specimens 1 and 4, which were 305 by 305 mm (12 by 12 in.). (b) Severe attack along panel edges.

area that was attacked by crevice corrosion was under supporting pins and rubber bands used to hold the specimens in place.

Cavitation Erosion. Stainless steel has excellent resistance to cavitation erosion in seawater. Austenitic grades are recommended for uses such as pump impellers and ship propellers.

Because of good resistance to cavitation erosion, two proprietary stainless steels (type 384 and Carpenter Cb-3) have been found especially satisfactory for use in pumps that handle seawater.

Type 304 stainless steel has been used extensively for sonar domes because of its resistance to cavitation erosion. These domes are carried beneath the hulls of ships. They are painted on the outside and cathodically painted on the inside, but paint usually comes off the outside after one day to three months of operation. They are frequently scraped to remove biofouling that interferes with sonar reception and nucleates pitting. Deterioration is not significant provided the thin (1 to 3 mm, or 0.04 to 0.125 in.) outer skin is kept clean. Some rusting may occur in welded areas, but since 1954 hundreds of these units made from type 304 stainless have been in service with satisfactory results.

Ship Propellers and Shafts. In addition to resistance to general corrosion, cavitation erosion and impingement corrosion, a ship propeller should have high tensile and yield strength to permit minimum blade thickness for greater efficiency. A propeller blade should bend, not break, if it strikes an object, and impact strength should be high to reduce damage. A high modulus of elasticity is desirable to reduce deflection and vibration in service. Propellers damaged in service should be capable of being repaired by welding.

Ship propellers cast in CF-4 have given satisfactory service in both river water and seawater. These propellers were stress relieved after casting. (CF-4 is a little-used alloy roughly equivalent to wrought type 304L and only slightly different in composition from CF-3.)

Ships normally are fitted with protective zinc or magnesium anodes to prevent corrosion of steel rudders, rudder posts and stern sections. Some cathodic current is consumed by the propeller. The data in Table 17 indicate the practical value of cathodic protec-

Fig. 17 Polarization curves for four stainless steels in seawater

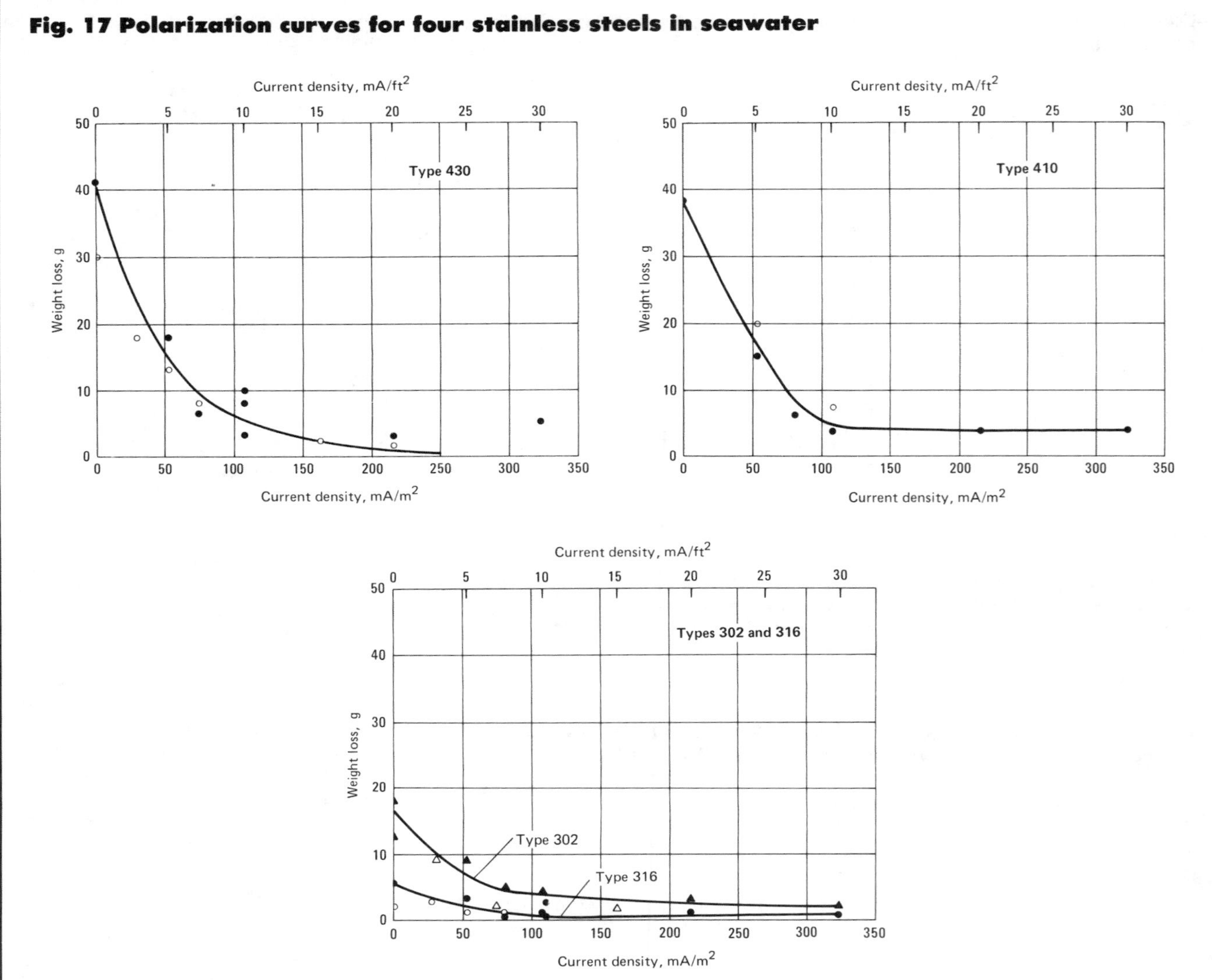

Weight loss is plotted as a function of cathodic polarization current during exposure in quiescent seawater. Open symbols represent flat panels with surface area of about 0.093 m^2 (6 by 12 in.) exposed for 186 days; solid symbols represent flat panels with surface area of about 0.186 m^2 (12 by 12 in.) exposed for 109 days. (Ref 11)

tion of a stainless steel propeller and the cutless bearings during layup.

Chemical Environments

The performance of stainless steels in the chemical process industries can be affected adversely by (*a*) general corrosion, (*b*) intergranular corrosion, (*c*) stress-corrosion cracking, (*d*) pitting, (*e*) crevice corrosion and (*f*) galvanic corrosion. Frequently, minor variations in environment have marked effects on service life. Design and fabrication practices also exert an influence.

The applications discussed in the remainder of this article are limited to chemical solutions and other substances that carbon steel and other metals cheaper than stainless steel do not resist satisfactorily. One of the more useful summaries of information on corrosion in chemical environments is "Corrosion Data Survey" (1967 Edition), by G. A. Nelson, published by the National Association of Corrosion Engineers. These data are compiled from various sources, and encompass laboratory and field tests conducted under a variety of conditions. Any data reported in the literature should be used only as a guide for selecting materials to consider for an application. Field tests are of the greatest value because the effects of heat transfer, minor contaminants, operational cycles and the like can be evaluated only under actual plant conditions.

Acetic Acid. Austenitic stainless

steels are fully resistant to attack by all concentrations of acetic acid at ambient temperatures. At higher temperatures, differences in resistance among austenitic types become apparent (see Table 18). Types 304 and 347 act similarly in acetic acid. Both show rates of 38 μm/yr (1.5 mils/yr) or less in all concentrations of refined acid up to 99% at temperatures below 50 °C (120 °F). Both can be used in refined acid at concentrations as high as about 50% at temperatures up to the boiling point of the solution. Tests should be made to determine the suitability of these two steels at concentrations above 50% and temperatures above 50 °C (120 °F) (Table 18).

Types 309 and 310 may be used safely in acid concentrations up to about 50% at temperatures up to the boiling point of the solution, or in all concentrations of refined acid up to 99% at temperatures below 75 °C (170 °F). Tests should be made to determine their suitability in acid concentrations above 50% and at temperatures greater than 75 °C.

Because they contain molybdenum, types 316 (Table 18) and 317, as well as the cast stainless steel CN-7M, have very good corrosion resistance. They may be used in all concentrations of acid up to 99% and at temperatures up to the boiling point. These data apply to pure, refined acid; contaminants—even in small amounts—limit the use of 18-8 stainless steels to temperatures below 50 °C (120 °F). The effect of ionized halogens on the rate of corrosion is shown in Tables 19 and 20.

Type 317 is more resistant than 316 to corrosive attack in hot acetic acid and is used considerably in distillation equipment, in spite of being somewhat difficult to fabricate. Cast stainless steels of the CN-7M type are used to make pumps and valves for handling hot acetic acid.

Molybdenum-containing ferritic stainless steels also show good corrosion resistance to acetic acid solutions. The data in Table 21 indicate that 18Cr-2Mo and 26Cr-1Mo have corrosion resistance at least as good as that of type 316.

As the concentration approaches that of glacial acetic acid at boiling temperatures and at superheated vapor temperatures, none of the stainless steels are sufficiently resistant. For these concentrations and conditions, nickel-base alloys must be used.

Types 304 and 347 are used for a wide variety of acetic acid equipment, including stills, base heaters, holding tanks, heat exchangers, pipelines, valves and pumps, under conditions ranging from dilute solutions (below 50%) at room temperature to concentrated solutions (up to 99%) at 135 °C (275 °F). Type 316, heat treated after fabrication, is suitable for fractionating equipment for acetic acid in concentrations of 30 to 99%; type 304 is not suitable. Halogen ions must be eliminated from the solutions because they have an adverse effect on corrosion rates. Type 316 is used for storage vessels, pumps and process piping to handle glacial acetic acid; type 304 is not satisfactory because it discolors the acid.

Table 14 Corrosion rates of welded stainless steels in seawater flowing at low and moderate rates

Type(a)	Exposed in flume(b) Weight loss, g	Corrosion rate μm/yr	mils/yr	Exposed in basin(c) Weight loss, g	Corrosion rate μm/yr	mils/yr
410	2(d)	<2.5	<0.1	112(d)	30.5	1.2
316	0	0	0	20(e)	5.1	0.2
310	0	0	0	35(d)	10.2	0.4
Composition, %:	**C**	**Mn**	**Si**	**Cr**	**Ni**	**Mo**
410 (plate)	0.10	0.30	0.37	12.95	0.20	...
410 (weld)	0.10	0.60	0.33	12.95	0.19	...
316 (plate)	0.07	1.67	0.32	17.33	12	2.36
316 (weld)	0.07	1.67	0.44	20.52	12	2.84
310 (plate)	0.12	1.48	0.39	25.54	20.92	...
310 (weld)	0.09	1.67	0.38	27.13	20.60	...

(a) Sand blasted, welded test panels 102 by 305 mm (4 by 12 in.) were exposed for 1257 days at Freeport, Texas. (b) Seawater velocity in a flume was approximately 1.2 m/s (4 ft/s). (c) Seawater velocity in the basin was less than 0.3 m/s (1 ft/s). (d) Specimens perforated at weld. (e) Specimen corroded at weld.

Table 15 Corrosion resistance of stainless steels in flowing seawater (Example 3) (Ref 12)

Type(a)	Surface condition	Corrosion rate μm/yr	mils/yr	Max face pitting mm	in.	Max edge pitting mm	in.
Exposed 483 days							
302	Bright, cold rolled	2.0	0.08	0.75	0.03	1.5	0.06
316	Hot rolled, pickled	4.1	0.16	0.5	0.02	2.5	0.10
Exposed 4½ years							
302	Bright, cold rolled	2.3	0.09	1.5	0.06	5.0	0.20
316	Hot rolled, pickled	1.5	0.06	1.3	0.05	2.5	0.10

(a) Test specimens were panels, 38 by 152 mm (1½ by 6 in.). Seawater flow rate was 1 m/s (3 ft/s).

Table 16 Resistance of stainless steels to impingement attack in seawater jet tests

Type	Composition, % C	Mn	P	S	Si	Cr	Ni	Cu	Mo	Weight loss(a), mg	Max depth of crevice corrosion mm	in.
304	0.058	0.63	0.023	0.016	0.62	18.37	9.06	0.14	0.20	265.3	0.86	0.034
316	0.054	1.51	0.022	0.016	0.54	17.68	12.32	...	2.19	67.4	0.74	0.029
430	0.072	0.41	0.015	0.019	0.56	16.88	0.33	...	...	839.5	(b)	(b)

(a) Test specimens were 1.6 by 13 by 89 mm (1/16 by ½ by 3½ in.). Jet velocity was 4.6 m/s (15 ft/s). Stream was 2 mm in diameter, with 2 to 4 vol % air. Duration of test was 28 days. Average temperature of water was 24 °C (75 °F). Depth of attack at jet in all specimens was nil. (b) Specimen became perforated in test.

Considerable difficulty has been encountered with castings such as pump volutes and impellers because of contamination by iron picked up from the sand or from abrasive cleaners. Castings should be pickled in a nitric-hydrofluoric acid solution before being installed.

For temperatures above 50 °C (120 °F) and applications involving acetic acid that is mixed or contaminated with other substances, types 316 and 317 are generally satisfactory in equipment such as stills, heat exchangers and holding tanks. The superiority of the molybdenum-bearing types is illustrated by the results of plant tests in boiling 99% acetic acid. The corrosion rates for types 304, 321 and 347 varied from 510 to 1525 μm/yr (20 to 60 mils/yr), compared with 38 μm/yr (1.5 mils/yr) for type 316. Results of other tests in 99% acid vapors at 110 °C (230 °F) gave 1600 μm/yr (63 mils/yr) for type 304 and 50 μm/yr (2.0 mils/yr) for type 317. Type 317 has somewhat greater resistance than type 316 to severely corrosive conditions.

Castings of CN-7M are used extensively as pumps, valves, and related equipment for all concentrations of acetic acid at temperatures up to the boiling point. These alloys have good resistance to mixtures of acetic acid with small amounts of sulfuric or formic acid.

The presence of reducing agents, either as impurities or as necessary constituents in the process, destroys the passivity of stainless steels. Mixtures of acetic acid with other acids—especially sulfuric, hydrochloric and formic—may be more corrosive than acetic acid itself, particularly at high temperatures. The data in Table 22 emphasize that slight changes in solution concentrations can have significant effects on corrosion rates and that suitable corrosion tests must be made whenever any change in operating conditions is contemplated.

Stainless steels are not always satisfactory for contact with hot solutions of acetic acid in concentrations greater than about 25% and containing 2% or more reducing agent (such as formic acid). If oxidizing agents such as sodium dichromate may be added to the acetic acid, the useful lives of stainless steel components can be appreciably extended.

In a still handling boiling 99.5% acetic acid liquors and vapors, the service lives of heating coils made of type 316 varied from ten months to five years. Corrosion tests have also shown that in

Table 17 Effect of cathodic protection on corrosion of stainless steel propeller shafts

Type	Cathodic protection	Current density(a), mA/m²	Current density(a), mA/ft²	Shaft potential(a), mV	Depth of attack beneath nonmetallic bearings(b): Maximum, mm	Maximum, mils	Average(c), mm	Average(c), mils
304 (d)	No	0	0	−185	0.7–1.1	27–45	0.5–1.0	20–40
304 (e)	Yes	56	5.2	−823	0	0	0	0
303 (f)	No	0	0	−260	0.5–3.1	20–123	0.35–0.75	14–30
303 (e)	Yes	58	5.4	−822	0	0	0	0

(a) Values averaged over entire test period. All potentials determined versus a saturated calomel half cell. (b) Cutless bearings, 127 mm (5 in.) long, covering 0.013 m^2 (0.14 ft^2) of shaft surface. Shafts were exposed for 2 yr without rotation in quiescent seawater, during which time they became completely fouled with marine organisms. (c) Average values of five deepest areas under four bearings on two shafts (two bearings per shaft). (d) Extensive corrosion beneath bearings and slight crevice corrosion beneath fouling elsewhere on shaft surface. (e) No evidence of corrosion. (f) Extensive corrosion beneath bearings and slight crevice corrosion beneath fouling elsewhere on shaft surface; one end of shaft was severely corroded.

Table 18 Corrosion in acetic acid solutions

Acetic acid, %	Temperature, °C	Temperature, °F	Type 304, μm/yr	Type 304, mils/yr	Type 316, μm/yr	Type 316, mils/yr	Type 347, μm/yr	Type 347, mils/yr
10	Boiling	Boiling	<38	<1.5	...	...	...	...
25	104	219	<130	<5.0	<130	<5.0	<130	<5.0
50	Boiling	Boiling	12.7	0.5	2.5	0.1	...	...
70	Boiling	Boiling	840	33.0	...	...	...	...
75	30	86	2.5	0.1	...	...	...	...
	145	293	400	15.8	104	4.1	...	...
99.5	140	284	...	...	5	0.2	287	11.3
	Boiling	Boiling	686	27.0	23	0.9	175	6.9

Table 19 Effect of ionized halogens in acetic acid on corrosion of type 316 stainless steel

Ionized halogens, ppm	Corrosion rate(a), μm/yr	Corrosion rate(a), mils/yr
3.5	180	7
22.8	760	30
25.0	940	37
200	3120	123

(a) At 90% acid concentration and 110 °C (230 °F).

Table 20 Corrosion in impure acetic acid

Acetic acid(a), %	Temperature, °C	Temperature, °F	Type 304, μm/yr	Type 304, mils/yr	Type 316, μm/yr	Type 316, mils/yr	Type 316(b), μm/yr	Type 316(b), mils/yr	Type 317, μm/yr	Type 317, mils/yr	No. 20, μm/yr	No. 20, mils/yr
10	106	223	...	...	50	2.0	102	4.0	7.6	0.3	102	4.0
24	110	230	...	...	69	2.7	...	...	69	2.7	102	4.0
83	116	241	1680	66	230	9.0	560	22	102	4.0	127	5.0
87	122	252	1070	42	406	16.0	...	...	25	1.0	203	8.0
98	128	262	180	7.0	50	2.0	102	4.0	25	1.0	50	2.0
99.5	130	266	...	...	7.6	0.3	...	...	10	0.4	13	0.5

(a) Data were obtained during processing of acetic acid containing ionized halogens; duration of tests, 51 days. Concentration of halide ions varied during period of observation; estimated range of concentration, 5 to 10 ppm. (b) All specimens sensitized at 650 °C (1200 °F); attacked intergranularly.

Table 21 Corrosion resistance of stainless steels in boiling acetic acid

Acetic acid concentration	Type 304 µm/yr	Type 304 mils/yr	Type 316 µm/yr	Type 316 mils/yr	18Cr-2Mo µm/yr	18Cr-2Mo mils/yr	26Cr-1Mo µm/yr	26Cr-1Mo mils/yr	29Cr-4Mo µm/yr	29Cr-4Mo mils/yr
	Corrosion rate									
20%	760	30	7.6	0.3	5	0.2	0.0	0.0	...	0.0
80%	...	...	...	...	5	0.2	...	...	...	...
Concentrated	81	3.2	13	0.5	...	...	13	0.5	...	...
Concentrated + 220 ppm Cl^-	6860	270	4570	180	...	...	500	20	...	...

Table 22 Corrosion in formic acid–acetic acid mixtures

Type	Test solution(a) Acetic acid	Formic acid	Temperature °C	°F	Corrosion rate µm/yr	mils/yr
304, 347	30 to 50%	2 to 10%	106	223	(b)	(b)
316	30 to 50%	2 to 10%	106	223	76 to 500	3 to 20
317	30 to 50%	2 to 10%	106	223	50 to 280	2 to 11
316	Glacial	None	93 to 110	200 to 230	<38	<1.5
	Glacial	4%	93 to 110	200 to 230	84	3.3
	25%	1.25%	104	220	38	1.5
317	25%	1.25%	104	220	<25	<1.0
316	25%	4%	104	220	76	3.0
317	25%	4%	104	220	50	2.0
321, 347	25%	4%	104	220	(b)	(b)

(a) Solution consisting of indicated concentration of acetic acid to which indicated amount of formic acid was added. (b) Completely corroded.

Table 23 Corrosion of type 304 in wet chlorinated solvents

Solvent	Corrosion rate(a) µm/yr	mils/yr
Methylene chloride	2.5	0.1
Carbon tetrachloride	130	5.0
Methyl chloroform	250	10.0
Trichloroethylene	20	0.8
Ethylene dichloride	8	0.3
Perchloroethylene	2.5	0.1
Propylene dichloride	430	17.0

(a) In 12-day tests at refluxing temperatures.

boiling 75% acetic acid vapors, both 304 and 316 were susceptible to severe pitting corrosion. In some boiling liquors where severely corrosive conditions exist (for example, where acetic acid is contaminated with various chlorides), nickel-base alloys such as Hastelloy C have proved more resistant than stainless steels, and their use is justified economically by the longer life of equipment.

In equipment constructed of both wrought 18-8 stainless steels and stainless steel castings, CF-8M castings are generally used instead of CF-8 castings because CF-8M has greater resistance to pitting attack. To obtain service comparable to that of molybdenum-bearing wrought 18-8 stainless steels, the chromium content of cast alloys should be on the high side of the composition range. Castings should be fully annealed for best service.

Amino Acids. Generally, amino acids have the same corrosive characteristics as acetic acid. Austenitic stainless steels have sufficient corrosion resistance to handle aqueous amino acid solutions at temperatures up to the boiling point.

Ammonia. Stainless steels show good resistance to all concentrations of ammonia up to the boiling point and find numerous applications in production of ammonia. For instance, in desulfurizers, types 304, 316 and 20 Cb-3 are used for wire mesh screens and support grating. These are critical parts that must resist any corrosion.

Tubes for primary reformer furnaces usually are HK-40 castings (comparable to wrought type 310). For shrouds in secondary reformers, for crossover lines, for piping and for other equipment operating below 900 °C (1650 °F), types 304 and 321 have been used. In shift converters, types 304 and 410 are used for wire mesh screens and grating, while type 304 is used for piping, including tees and ells. Wire mesh screens and grating in methanators are mostly of type 304.

Because of the possibility of stress-corrosion cracking, especially from chlorinated cooling water, type 430 tubes are used in water-cooled exchangers in the synthesis step. Pitting may be encountered if the chloride content of the cooling water is too high. Under such conditions, the more pit-resistant ferritics such as 18Cr-2Mo, 26Cr-1Mo and 29Cr-4Mo can be used. Catalyst cartridges and integral heat exchangers in synthesis converters are constructed of type 304.

Ammonium Sulfate. Types 316 and CN-7M are used to contain ammonium sulfate plus free sulfuric acid. Before these stainless steels were available, construction was almost entirely of lead, which was subject to fatigue cracking, and occasionally of silicon iron, which is brittle and thus requires continuous maintenance. Straight 18Cr and 18Cr-8Ni steels became pitted and underwent severe corrosion.

Type 316 must be annealed after welding; otherwise, heavy intergranular corrosion occurs adjacent to the welds. When the extra-low-carbon grade (type 316L) is used, heat treatment is unnecessary and large tanks can be fabricated. Such tanks have been in service for several years with no evidence of either general corrosion or intergranular attack. CF-8M and CN-7M castings also must be annealed, but if the casting skin is damaged, corrosion may occur anyway. Castings repaired by welding must be heat treated again after welding; otherwise they will corrode intergranularly adjacent to the welds. Passivation pretreatment is not required.

Bromoform. Type 304 generally is satisfactory for handling bromoform, either wet or dry, at ambient temperatures; however, wet bromoform will become slightly discolored. If a water-white product is required, stainless steel is not suitable.

Chlorinated Solvents. The halogen derivatives of methane, ethane, ethylene, propane and benzene are widely used in dry cleaning, metal cleaning, vapor degreasing and solvent extraction processes, and as chemical intermediates. The compounds of primary interest are methylene chloride, chloroform, carbon tetrachloride, ethylene dichloride, trichloroethylene, perchloroethylene, methyl chloroform, propylene dichloride, dichloroethyl eth-

Table 24 Corrosion of stainless steels in chromic acid solutions

Type	Test duration, days	Corrosion rate µm/yr	mils/yr
1% solution at 100 °C (212 °F)			
12Cr	2	10	0.4
27Cr	2	10	0.4
18-8Ti	2	13	0.5
18-8Mo	2	8	0.3
5% solution at 82 °C (180 °F)			
304	3	18	0.7
316	3	74	2.9
10% solution at 82 °C (180 °F)			
304	3	150	5.9
316	3	305	12
15% solution at 82 °C (180 °F)			
304	3	1 420	56
316	3	460	18
15% solution at 100 °C (212 °F)			
304	2.5	2 490	98
316	2.5	9 960	392
12Cr	2	860	34
27Cr	2	970	38
18-8Ti	2	5 840	230
18-8Mo	2	2 570	101
25% solution at 24 °C (75 °F)			
304	3	nil	nil
316	3	18	0.7
25% solution at 82° C (180 °F)			
304	3	18 500	730
316	3	27 400	1080

Table 25 Corrosion of stainless steels in citric acid solutions

Type	Temperature °C	°F	Corrosion rate µm/yr	mils/yr
6% acid concentration				
409	Room		0	0
	Boiling		2.5	0.1
10% acid concentration				
409	Room		0	0
	Boiling		2.5	0.1
410	99-102	210-215	260	10.3
430	99-102	210-215	20	0.8
304	99-102	210-215	210	8.3
316	99-102	210-215	13	0.5

er, monochlorbenzene and orthodichlorobenzene. They are used individually as chemically pure or commercial grades, as mixtures, or with other compounds to control boiling point, freezing point, solvency and flammability of mixtures.

Stainless steels are not corroded by chlorinated solvents when water is absent; but when a water phase is present, the compounds hydrolyze to form hydrochloric acid and sometimes organic acids. Although the presence of metallic materials usually increases the rate of decomposition, contact with stainless steels does not, and their use in equipment for handling chlorinated solvents is generally satisfactory. The corrosion rate of type 304 after 12 days in wet chlorinated solvents at refluxing temperatures is shown in Table 23. These data are from laboratory tests with mixtures of solvent and water, in which one-third of the specimen was in the solvent layer, one-third in the water layer and one-third in the vapor phase.

Consideration should be given to the use of type 316, type 317 or Carpenter 20 Cb-3 for applications where pitting is encountered. Intergranular attack sometimes occurs at welded joints; tests should be made to evaluate this possibility. Stress-corrosion cracking also may be encountered in equipment handling chlorinated solvents.

Chlorosulfonic Acid. Although carbon steels are satisfactory for exposure to liquid chlorosulfonic acid, type 317 or higher-alloy stainless steels are recommended for withstanding vapors at and above the liquid level. The products of sulfation, which may retain chlorides, cause stress corrosion and sometimes severe pitting of stainless steels. Pipe welds and vessels of the austenitic steels should be fully annealed whenever this is feasible.

Chromic Acid. Although chromic acid is a highly oxidizing acid, it will corrode stainless steels. As indicated by the corrosion rates in Table 24, stainless steels can be used in low concentrations and/or at low temperatures.

Fig. 18 Stress-corrosion cracking in type 304 stainless steel caused by citric acid and salt at 100 °C (212 °F)

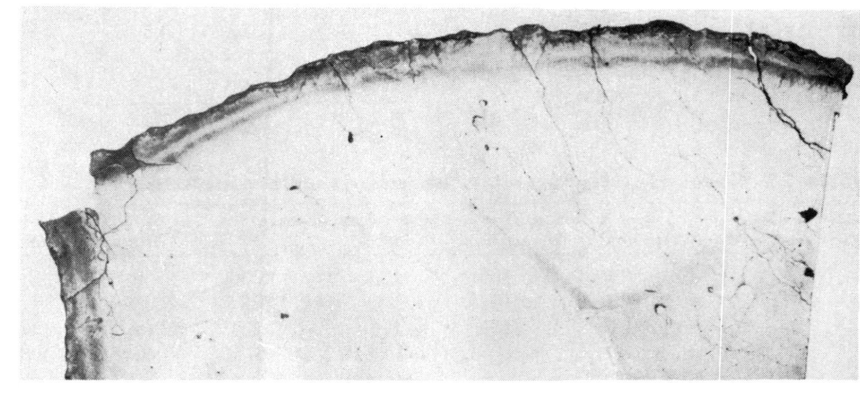

Citric acid is slightly less corrosive than acetic acid; however, it also is a nonoxidizing acid. All stainless steels will resist attack by citric acid at low concentrations or temperatures (see Table 25). On the other hand, at elevated temperatures, at high concentrations or in the presence of contaminants such as chlorides, the more highly alloyed stainless steels are preferred. Chlorides can cause pitting, crevice corrosion and possibly stress-corrosion cracking of austenitic stainless steels (Fig. 18).

Dyes. Stainless steels do not suffer general corrosion when exposed to dye solutions. However, certain dyes containing chlorides or hydrochloric acid can cause pitting and crevice corrosion.

Epichlorohydrin. The fact that processing equipment for use with epichlorohydrin failed by stress-corrosion cracking was confirmed by laboratory tests. Horseshoe specimens of types 304, 316 and 347 cracked in seven days or less at 60 °C (140 °F).

Esters. Pure esters, such as methyl acetate, ethyl acetate, propyl acetate and vinyl acetate, are not corrosive toward stainless steels. However, in the esterification step during the manufacture of esters, catalysts such as sulfuric acid, which can cause extensive

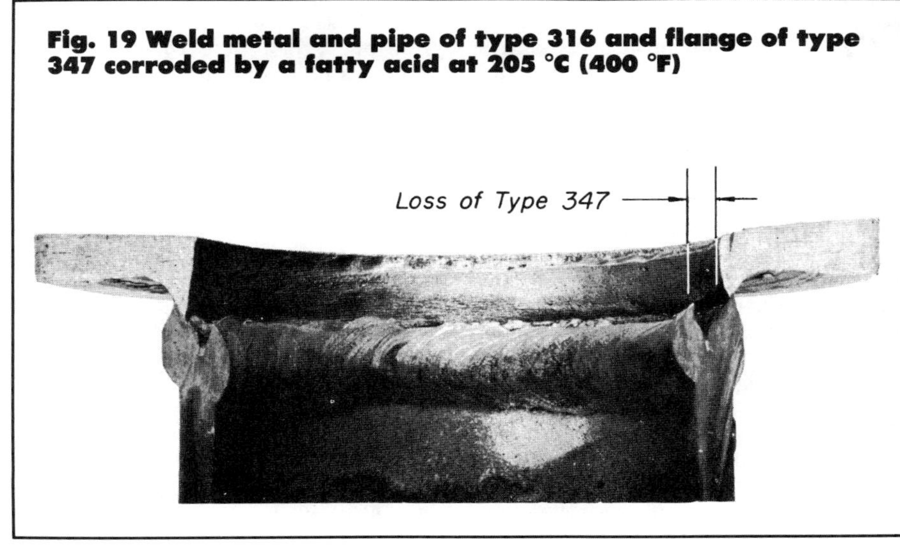

Fig. 19 Weld metal and pipe of type 316 and flange of type 347 corroded by a fatty acid at 205 °C (400 °F)

Table 26 Corrosion of stainless steels in refined tall oil

Type	Corrosion rate μm/yr	mils/yr
At 285 °C (545 °F)		
302	4 570	180
316	100	4
317	25	1
At 300 °C (575 °F)		
302	12 700	500
316	100	4
317	25	1
At 315 °C (600 °F)		
302	20 300	800
316	1 340	53
317	530	21
At 330 °C (625 °F)		
316	12 700	500

corrosion, are used. Beyond this step, the process streams become less corrosive and stainless steels may be used.

Fatty Acids. Fatty acids of lower molecular weight, such as acetic and formic acids, require use of 18-8 stainless steels (see section on acetic acid). The following discussion is concerned with the acids of higher molecular weight, such as lauric, myristic, palmitic and stearic acids, which are less corrosive. At temperatures up to 65 °C (150 °F), cheaper metals such as carbon steel and aluminum are moderately corroded, but if color and absence of contamination of the acid are important, 18-8 stainless steels should be used.

At temperatures below 175 °C (350 °F), all standard 18-8 types are satisfactory; above 175 °C, type 316 is needed to avoid pitting and general corrosion, as is indicated for tall oil in Table 26.

Corrosion in fatty-acid vapors is no greater than in liquid, except at high vapor velocities. Under these conditions, erosion-corrosion rates are lower in type 316 than in types 304, 321 and 347.

Pitting and loss of surface metal are caused by high-temperature plant processes (Fig. 19). There are no reports of straight fatty acids having caused intergranular failures in 18-8 stainless steels.

Cast alloys, including type CN-7M, have been satisfactory. Wrought molybdenum-bearing stainless steels and the newer precipitation-hardening stainless steels have been used for pump and valve parts to prevent galling or to provide increased hardness. High-nickel cast iron has given satisfactory service in fatty acids at 260 °C (500 °F).

Table 27 Corrosion in fatty acids acidulated with sulfuric acid and agitated by steam

Sulfuric acid, %	Temperature °C	°F	Corrosion rate μm/yr	mils/yr
Type 304 stainless steel				
0.01	96	205	390	15.4
0.1	107	225	190	7.55
5.0	93	200	1220	48.0
10	102	215	...	...
25	116	240	...	...
Type 316 stainless steel				
0.01	96	205	150	6.0
0.1	107	225	97	3.8
5.0	93	200	410	19.2
10	102	215	1970	77.5
25	116	240	1190	467

Fatty acids mixed with chlorides cause failure by stress-corrosion cracking. Acidulation of fatty acids by sulfuric acid produces wide variation in the corrosion rates of stainless steels. Factors that contribute to such variation include (*a*) unknown dilution of the concentrated sulfuric acid, (*b*) moisture inherent in the fatty acids, (*c*) temperature and (*d*) methods of agitation. Corrosion rates encountered with acidulated fatty acids of higher molecular weight with steam agitation are reported in Table 27.

Fertilizers. Stainless steels are used extensively for handling fertilizers. For dry fertilizers, type 409 has found wide acceptance. Type 304 is preferred for liquid types. The major corrosion problems are encountered with potash. Depending on the source, potash sometimes contains substantial amounts of potassium chloride, which can cause extensive pitting of stainless steels.

Formic Acid. The behavior of stainless steels in contact with formic acid is similar to its reaction to acetic acid. In many instances, a slightly higher corrosion rate can be expected. Impurities such as formaldehyde in formic acid can cause pitting of stainless steels.

Any of the austenitic stainless steels can safely handle solutions of formic acid at room temperature. High-chromium ferritic stainless steels containing molybdenum show some promise in hot formic acid solutions, as the laboratory test data in Table 28 indicate.

Hydrochloric Acid. Although types 316, 317 and 329; Carpenter 20 Cb-3;

Table 28 Corrosion of stainless steels in boiling 45% formic acid

Type	Corrosion rate µm/yr	mils/yr
304	1220	48
316	280	11
26Cr-1Mo	76	3
29Cr-4Mo	50	2

and the cast alloys CN-7M and CF-8M find some use in very dilute aerated hydrochloric acid environments, stainless steels are not usually recommended for this service.

Solutions containing chloride salts at pH values below 7.0 are essentially hydrochloric acid environments. Pitting and stress-corrosion cracking are encountered at acid concentrations less than 1%, depending on temperature, aeration and agitation.

Bimetallic couples between stainless steels and other alloys should be avoided, because corrosion may be accelerated at the junction. In such couples, the stainless steel may become the anode in dilute hydrochloric acid, resulting in loss of passivity and rapid corrosion.

Hydrochloric acid at pH values of 2.0 to 4.0 and temperatures of 50 to 80 °C (120 to 180 °F) has caused pitting and subsequent failure of stainless steel heat-exchanger tubing and heating coils. Calcareous scale in hydrochloric acid vessels has induced pitting failures. Activated carbon that settled out of HCl solutions has caused pitting of heating coils and tank bottoms made of type 316 stainless steel.

Stress-corrosion cracking of heat-exchanger tubes has occurred at 70 °C (160 °F) and a pH of 4.0. Excessive stresses were induced in the tubes when a floating head on the heat exchanger became "fixed". Bending of tubes between baffle supports has induced stress-corrosion cracking in tubes made of type 316 stainless steel. Excessive rolling of tubes into tube sheets has induced stress-corrosion cracking of the tubes just adjacent to the sheets.

Hydrochloric acid solutions can corrode weld deposits of type 316 on sheet of similar grade, weakening the joint. Use of electrodes of type 317 or 310 Mo improves the corrosion resistance of these welds. Weld-zone attack has been observed in type 316 linings for steel tanks handling acidified starch slurry at a pH of 2.0 and a temperature of 50 °C (120 °F).

Covers and vents for acidified-starch slurry tanks usually corrode rapidly. Condensed vapors of dilute hydrochloric acid usually are more corrosive than the liquid phase. Stainless steels are usually unsatisfactory for tank covers or vent piping for such tanks.

Hydrocyanic Acid. Pure hydrocyanic acid is not corrosive to most constructional materials, but when stabilized against polymerization at elevated temperatures by the addition of acidic materials, it becomes corrosive to steel, copper and aluminum. Straight-chromium stainless steels are not recommended for use with stabilized hydrocyanic acid solutions.

Austenitic stainless steels resist corrosion by hydrocyanic acid that contains small amounts of sulfur dioxide as a polymerization inhibitor, at all concentrations and at temperatures up to the boiling point. Types 316 and 317, as well as CN-7M and CF-8M, have greater corrosion resistance than stainless steels without molybdenum. Unstabilized stainless steels should be fully annealed to prevent intergranular attack in these solutions.

Hydrofluoric Acid. At room temperature, stainless steels are extensively corroded by hydrofluoric acid except at very low and very high concentrations.

Lactic Acid. The use of types 304, 316 and 317; Carpenter 20 Cb-3; and the cast alloys CN-7M and CF-8M is limited in lactic acid solutions. Molybdenum-containing varieties generally have greater corrosion resistance than type 304.

Purity, concentration, temperature, aeration and agitation are environmental factors that influence selection of a particular type of stainless steel for use in process equipment. The presence of chlorides or sulfates in lactic acid solutions increases the severity of corrosion. Impure solutions from which lactic acid is ultimately separated and concentrated are usually more corrosive than purified solutions. Stainless steels are not suitable for use with lactic acid above 95 °C (200 °F).

Heating coils or heat exchangers for lactic acid should be designed for use with hot water or low-pressure steam. Decomposition of lactic acid and subsequent formation of a carbonaceous deposit on heating coils can result in pitting and perforation under the deposit.

Temperatures above 95 °C (200 °F), concentrations of lactic acid ranging from 30 to 70%, and the presence of chlorides or inorganic impurities usually increase severity of corrosion. Use of type 304 should be limited to vessels for storing pure solutions at temperatures below 40 °C (100 °F). Distillation of lactic acid causes corrosion by the vapor phase, and if lactic acid esters and volatile acid impurities are present, pitting attack will result.

Pitting failures in heat exchangers for lactic acid solutions have been reported. One failure of this type was limited to surfaces covered by the liquid, particularly in the parts of the tubing where solids had settled out.

Weld-zone attack and corrosion failures have been reported for type 304, but rarely for type 316. The weld-zone failures were in stainless steels that were not of the extra-low-carbon variety and had not been annealed.

Monoethanolamine. Stainless steels have excellent resistance to corrosion by monoethanolamine, and by monoethanolamine saturated with carbon dioxide plus oxygen, at temperatures up to 95 °C (200 °F). Stainless steel is used in preference to carbon steel in process steps where carbon dioxide is stripped from monoethanolamine—for example, in reboilers, heat exchangers and parts of fractionating columns. For heat exchangers, a common practice is to specify stainless steel only for tube bundles rated for 1.0-MPa (150-psi) steam inside and monoethanolamine that is rich in carbon dioxide outside. Type 304 is adequate.

Experience has been variable with stainless steel in monoethanolamine solutions in processes for removing hydrogen sulfide or carbon dioxide from natural and refinery gases. Probably about one-fourth of the monoethanolamine gas treating plants make some use of stainless steel piping and vessels. Remedial process changes often can be devised to avoid the use of stainless steel.

Nitric Acid. Stainless steels, first used commercially on a large scale in service involving nitric acid, continue to be used in such installations. These first applications involved an alloy containing 15 to 18% chromium (now type 430) and, soon thereafter, 18Cr-8Ni steel (now type 304). The necessity for full annealing to prevent accelerated corrosion and intergranular attack in nitric acid was soon demonstrated by service failures of improperly heat

Table 29 Corrosion of stainless steels commonly used with nitric acid

Nitric acid, %	Corrosion rate(a) Liquid µm/yr	mils/yr	Vapor µm/yr	mils/yr
Type 347 stainless steel				
98.5	1500	61.0	1200	47.0
96.3	400	16.0	970	38.0
94.2	58	2.3	760	30.0
92.6	2.5	0.1	710	28.0
90.1	2.5	0.1	180	7.0
85.1	2.5	0.1	2.5	0.1
Type 430 stainless steel				
98.5	1700	67.0	1300	51.0
96.3	500	20.0	1070	42.0
94.2	120	4.6	740	29.0
92.6	28	1.1	690	27.0
90.1	15	0.6	690	27.0
85.1	10	0.4	2.5	0.1

(a) At 22 °C (77 °F).

Table 30 Corrosion of annealed stainless steels in boiling 65% nitric acid

Type	Corrosion rate µm/yr	mils/yr
304	180	7.0
304L	180	7.0
309	100	4.0
310	100	4.0
316	200	8.0
347	180	7.0

treated and as-welded equipment. These difficulties were eliminated by postfabrication heat treatments involving slow cooling from about 790 °C (1450 °F) for type 430 and rapid cooling from about 1095 °C (2000 °F) for type 304. Subsequently, for the austenitic grades, the use of stabilizing elements (particularly niobium in type 347) and, more recently, reduction of carbon content to 0.03% max (type 304L) have been effective in controlling this problem without the necessity for quenching fabricated equipment from a high-temperature heat treatment. In the as-welded condition, types 304L and 347 show satisfactory resistance to corrosion by nitric acid and are therefore suitable for field-erected equipment.

Where corrosion rates must be held to less than 125 µm/yr (5 mils/yr), types 304L and 347 can be used with nitric acid in concentrations up to about 40% at the atmospheric boiling point, from 40 to 70% up to about 80 °C (175 °F) and from 70 to 90% at temperatures up to about 50 °C (120 °F). For a corrosion rate of 1270 µm/yr (50 mils/yr) max, the corresponding concentration limits are approximately 40 to 70% at the boiling point, 70 to 90% at 70 °C (160 °F) and 90% at 30 °C (85 °F). If the acid is recirculated so that corrosion products accumulate, attack in hot solutions at the higher concentrations is accelerated when the chromium in the acid exceeds a certain level. For boiling 65% nitric acid, the limiting chromium content of the solution is about 0.004%, above which corrosion increases rapidly. Under these conditions, corrosion is intergranular, even with stabilized or extra-low-carbon grades.

Corrosion by nitric acid in storage is slight for concentrations up to about 94%, but the acid condensate is of higher concentration, and attack becomes appreciable on parts of the tank exposed to the condensate (Table 29). (Aluminum is commonly used for storing 95 and 98% nitric acid, but its resistance decreases rapidly with decreasing concentration; consequently, exposure to the dilute acid must be avoided.) Corrosion data for specimens of 347 and 430 stainless steels in various concentrations of nitric acid at 22 °C (72 °F) are compared in Table 29. In hot concentrated solutions where attack is too severe to be tolerated, high-silicon iron can be used if its mechanical properties are suitable.

In reactions under pressure and at temperatures considerably above the atmospheric boiling point, corrosion rates of all stainless steels increase rapidly with both temperature and concentration, and only very dilute nitric acid solutions can be handled suitably in equipment made of stainless steel.

Type 304 in the annealed-and-water-quenched condition has essentially the same resistance to corrosion by nitric acid as types 304L and 347, but type 304 should be heat treated after fabrication to prevent intergranular corrosion. Types 316 and 316L in the annealed condition have similar resistance to nitric acid, but unless they are required for reasons other than resistance to corrosion, their use usually is not justified because of higher cost.

Type 309S-Cb is somewhat more resistant under the most severe conditions and is occasionally used where lower-alloy stainless steels are not satisfactory. If properly annealed and water quenched, types 309 and 310 have about the same resistance as type 309S-Cb. However, unless their carbon content is less than about 0.10%, these alloys cannot be cooled fast enough in commercial heat treatments to avoid susceptibility to intergranular corrosion.

Typical corrosion rates of annealed austenitic stainless steels in boiling 65% nitric acid are given in Table 30.

Type 430 is still widely used for various kinds of equipment in the ammonia oxidation process for manufacture of nitric acid, and for tank cars, storage tanks, forged valves and other components. Type 430 costs less than the austenitic grades and, although temperature ranges are somewhat more limited at various nitric acid concentrations, this alloy is adequate for many applications. Its principal limitation is that it requires heat treatment after fabrication and is therefore not suitable for equipment erected in the field or for equipment repair. For this steel, limits of nitric acid concentration for a maximum corrosion rate of 125 µm/yr (5 mils/yr) are as high as about 20% at the atmospheric boiling point, 20 to 40% up to about 70 °C (160 °F), 40 to 70% up to about 60 °C (140 °F) and 70 to 90% at temperatures as high as 30 °C (85 °F). Corresponding limits for a maximum rate of 1270 µm/yr (50 mils/yr) are approximately 20 to 40% at the boiling point, 40 to 70% up to 90 °C (195 °F), 70 to 90% up to 50 °C (120 °F) and more than 90% up to 30 °C (85 °F).

The resistance of chromium steels to nitric acid is related directly to chromium content, as indicated by the data in Table 31 for tests in boiling 65% nitric acid.

Table 31 Corrosion of chromium steels in boiling 65% nitric acid

Chromium, %	Corrosion rate µm/yr	mils/yr
10	10 700	420
12	3 810	150
16	1 020	40
18	660	26
20	460	18
24	305	12
26	180	7

Table 32 Corrosion of stainless steels in boiling 10% oxalic acid

Type	Corrosion rate µm/yr	mils/yr
409	47 800	1 880
304	1 220	48
316	915	36
26Cr-1Mo	60	2.4
29Cr-4Mo	60	2.4

Type 446 is comparable to type 304L in resistance to nitric acid, but because it is more difficult to fabricate it is employed in only a few special applications. With the exception of type 430, none of the other chromium steels are used to any appreciable extent in contact with nitric acid.

Stainless steels are relatively insensitive to factors such as aeration, velocity and agitation, because nitric acid is oxidizing and tends to favor passivity. Neither pitting nor stress-corrosion cracking occurs under these circumstances. However, nitric acid causes intergranular attack in unstabilized stainless steels that contain more than 0.03% C, unless they have been properly heat treated. The presence of hydrofluoric acid in nitric acid, as in certain pickling solutions, increases such attack. Hydrofluoric acid also increases the rate of general corrosion, as do appreciable amounts of other halides.

In hot dilute mixtures of nitric acid and sulfuric acids, no appreciable attack of stainless steels occurs when the ratio of nitric acid to sulfuric acid is about 2 to 1, or higher. This is one of several examples where sufficient nitric acid will prevent attack that would otherwise occur. With very dilute hot mixtures of sulfuric acid and nitric acid (about 1 to 1.5% total acid), where the proportion of nitric acid will not maintain passivity for the austenitic grades, type 443 (20Cr-1Cu) has greater resistance. Cast grades such as CN-7M containing 3 to 4% copper, CF-8 and CF-8M are widely used for valves, pumps and other castings in nitric acid service. Addition of stabilizing elements (usually niobium) or restriction of carbon content to 0.03% is not ordinarily justified, because most stainless castings can be quenched readily in water, and they are seldom welded in place in the field. Types CF-8M and CN-7M, containing molybdenum or molybdenum and copper, are no more resistant to nitric acid than the CF-8 (18-8) grade but will handle a wider variety of process solutions. High molybdenum (and

Fig. 20 Corrosion of austenitic stainless steels in phosphoric acid

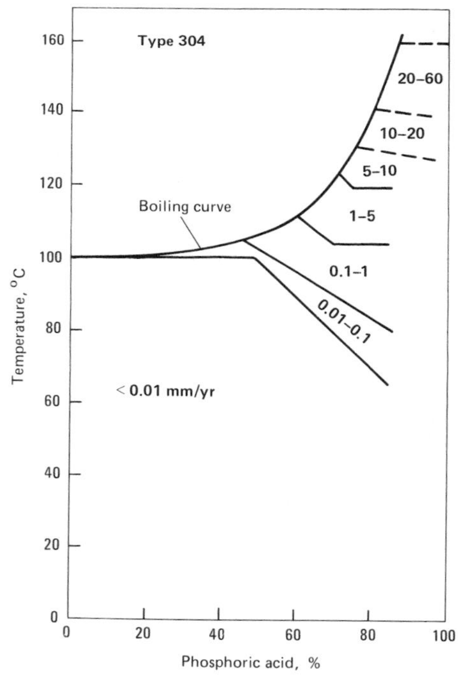

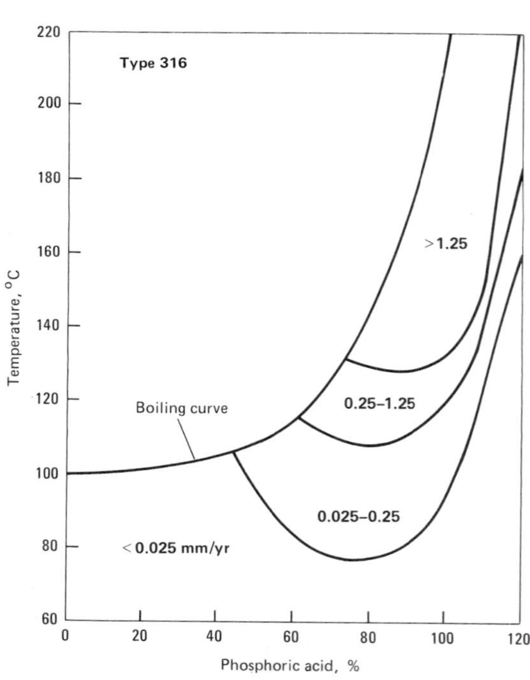

Left: isocorrosion chart for type 304 stainless steel. Right: isocorrosion chart for type 316 stainless steel in mildly agitated furnace-grade phosphoric acid.

Fig. 21 Isocorrosion chart for austenitic Cr-Ni stainless steels in sodium hydroxide

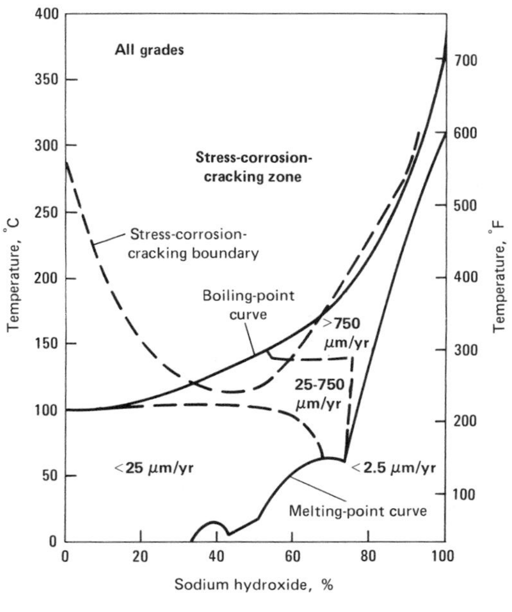

Table 33 Corrosion of stainless steels in phenol

Type	Phenol concentration, %	Temperature °C	Temperature °F	Corrosion rate μm/yr	Corrosion rate mils/yr
430, 316	10	105	220	0	0
302, 316	88	53	125	5	0.2
302, 316	88	185	365	5	0.2
304, 310, 316	100	315	600	25	1

Table 34 Corrosion of stainless steels in boiling phosphoric acid

Type	Corrosion rate μm/yr	Corrosion rate mils/yr
20% acid concentration		
316	180	7.2
29Cr-4Mo	0	0
29Cr-4Mo-2Ni	0	0
54% acid concentration		
316	550	21.6
29Cr-4Mo	30	1.2
29Cr-4Mo-2Ni	30	1.2
60% acid concentration		
316	550	21.6
29Cr-4Mo	120	4.8
29Cr-4Mo-2Ni	120	4.8

silicon) contents are somewhat detrimental to resistance to hot nitric acid in intermediate and high concentrations. Straight chromium cast stainless steels are seldom used in nitric acid service.

Oxalic acid is more corrosive to stainless steel than acetic acid. Although types 304 and 316 may be used at room temperature, they have very limited use at elevated temperatures. As with formic acid, the high-chromium ferritic stainless steels containing molybdenum appear to offer improved corrosion resistance (see Table 32).

Phenol is a rather weak organic acid that is not corrosive to stainless steels. All stainless steels have excellent resistance to this acid (Table 33), and material selection depends on factors other than corrosion.

Phosphoric Acid. Resistance of stainless steels to corrosion by phosphoric acid depends on acid concentration, temperature, contamination and alloy composition. The isocorrosion charts shown in Fig. 20 provide a general idea regarding the corrosion resistance of types 304 and 316.

The higher-molybdenum special stainless steels, such as JS700 and Uddeholm 904L, and the high-chromium-molybdenum ferritics in Table 34 appear to extend the usefulness of stainless steels into higher phosphoric acid concentrations.

Silver Nitrate. In production of silver nitrate, where contamination must be held to a minimum, corrosion cannot exceed 25 μm/yr (1 mil/yr). Type 310, and a modification of type 309 in which chromium content is 24% min, meet this requirement. Types 304 and 316 are suitable for aqueous solutions of silver nitrate at room temperature but are unsatisfactory at elevated temperatures or for acidified solutions. This is illustrated by an installation in which a CF-8M casting was used for valve bodies, mixer shafts and blades in equipment that handles, at room temperature, a 70% silver nitrate solution containing traces of nitric acid. After 15 years of service, the equipment was still in excellent condition; however, when valve bodies of this composition were used in 85% silver nitrate solutions at 95 °C (200 °F), corrosion failure resulted in three years.

Type 310 or CK-20 (or a modification of type 309 or CH-10, both of which contain 24% Cr min) should be used at elevated temperatures and in acidified solutions such as those encountered in production of silver nitrate from nitric acid and silver.

Cast alloys corrode more readily than wrought alloys. This is atrributed to the lower corrosion resistance of the ferrite in castings, especially in CH-10. Service is satisfactory with CK-20.

Correct annealing and pickling are necessary to hold corrosion rate to 25 μm/yr (1 mil/yr). In a unit for production of silver nitrate crystals at a solution temperature of 95 °C (200 °F), dissolving kettles fabricated from a high-chromium modification of type 309 stainless steel (properly annealed and pickled) were still in excellent condition after 20 years. Similar kettles installed at the same time, having less than 24% Cr, lasted six years. These kettles handled solutions in which silver nitrate concentration varied from 0 to 85% and nitric acid concentration varied from traces to 50%.

Corrosion tests are essential in evaluating stainless steels for service in silver nitrate solutions. Because very low corrosion rates are required, perfor-

Table 35 Corrosion of stainless steels in solutions of inorganic compounds

Compound and concentration(a)	Type(b)	Temperature °C	Temperature °F	Corrosion rate(c) µm/yr	Corrosion rate(c) mils/yr
Ammonium arsenate, 30%	304	Boiling	Boiling	125	5
Ammonium chloride, 10%	316	Boiling	Boiling	<25	<1
Hydrochloric acid, 0.4%	304	27	80	250	10
	316	49	120	125	5
Manganese chloride, 30%	304	90	194	18	0.7
	347	90	194	18	0.7
Nickel sulfate, 28%	304	50	122	1.0	0.04
Sodium bisulfide, 40%	304	Boiling	Boiling	40	1.6
	316	Boiling	Boiling	28	1.1
Sodium sulfide, 30%	304	54	130	230	9.1
	316	54	130	530	21
Sodium sulfide, 40%	304	Boiling	Boiling	38	1.5
	316	Boiling	Boiling	64	2.5
Sulfurous acid, 6%	304	40	104	2.5	0.1
	304	90	194	460	18
Sulfur chloride, 100%	304	Boiling	Boiling	33	1.3
	410	...	...	79	3.1

(a) All solutions from CP chemicals. (b) All steels in the annealed condition. (c) Tests made in the laboratory.

Table 36 Corrosion of stainless steels in stannous fluoride solutions

Solution strength, %	Type 304 µm/yr	Type 304 mils/yr	Type 316 µm/yr	Type 316 mils/yr	Type 347 µm/yr	Type 347 mils/yr
2.0	100	3.92	...	...	...	...
15	23	0.9	None	None	9.9	0.39
30	14	0.56	None	None	20	0.79
50	46	1.81(a)	...	...	...	...

(a) Stained black.

mance cannot be predicted solely on the basis of chemical composition.

Sodium Hydroxide. All stainless steels resist corrosion by all concentrations of sodium hydroxide up to about 65 °C (150 °F). At higher temperatures, various stainless steels undergo varying degrees of corrosion. The isocorrosion chart in Fig. 21 provides a general idea of the corrosion performance of Cr-Ni stainless steels. It also outlines the possible stress-corrosion-cracking boundary for such steels.

In recent years, the high-chromium-molybdenum ferritics, particularly E-Brite 26-1, have found extensive use in production of caustics. E-Brite 26-1 tubing is used extensively for the first-, second- and third-effect evaporators handling 16% NaOH – 14% NaCl, 26% NaOH – 9% NaCl and 45% NaOH – 5% NaCl at temperatures as high as 150 °C (300 °F) with no reported problems.

Sodium Sulfide. Type 304 can be used satisfactorily in contact with sodium sulfide solutions at concentrations up to 50% and at temperatures up to the boiling point (Table 35). Other inorganic compounds are also shown in Table 35 for comparison.

In repair of equipment that has been in sulfide service, surfaces should be cleaned thoroughly by abrasive blasting before welding to avoid weld cracking.

Stannic Chloride. Types 304 and 316 have satisfactory resistance to aqueous solutions of stannic chloride at temperatures up to 95 °C (200 °F) for concentrations not exceeding 1%. Type 316 is more resistant than 304 and has fair resistance to solutions of 10 to 15% at 20 °C (70 °F), but is unsatisfactory at higher temperatures and concentrations.

Stannous Fluoride. Laboratory tests made at 95 °C (200 °F) with aqueous solutions ranging from 2 to 50% by weight indicate that stannous fluoride solutions can be handled in equipment made of 300-series stainless steels. A maximum rate of 100 µm/yr (3.9 mils/yr) was obtained in a 2% solution. With allowance for experimental error, all rates either decreased or remained the same with increasing concentration. No evidence of pitting could be found, and stressed horseshoe-type specimens of 304 and 316 tested for stress-corrosion cracking did not fail. Table 36 summarizes the results of these tests.

Table 37 Corrosion by 78% sulfuric acid mixed with sulfonation products

Alloy	Corrosion rate µm/yr	Corrosion rate mils/yr
At 27 °C (80 °F)		
Type 316	5.1	0.2
Carpenter 20 Cb-3	None	None
Inconel	None	None
Mild steel	510	20
At 60 °C (140 °F)		
Type 316	510	20
Hastelloy B	15	0.6
Hastelloy C	38	1.5
Carpenter 20 Cb-3	76	3.0
Inconel	205	8.0
Mild steel	3400	134

Sulfation and Sulfonation Products. Austenitic stainless steels and carbon steels have low corrosion rates in oleum (fuming sulfuric acid) and sulfuric acid in concentrations higher than 80% at room temperature. At 100 to 103% there is a distinct rise in the corrosion rate for carbon steel. Above 103%, both stainless and carbon steels have satisfactory corrosion rates. Steels of the 300 series are satisfactory for sulfonation practice at room temperature with 78% sulfuric acid mixed with sulfonation products. At 60 °C (140 °F), corrosion of 300-series steels is excessive. Corrosion rates at these temperatures are reported in Tables 37 and 38.

If accuracy of parts is essential, as in valves and control instruments, or if velocity of liquid is high, as in pumps or mixing equipment, corrosion rates are excessive, and steels such as CN-7M and nickel-base alloys are needed. The neutralized products of sulfonation may separate if the solution becomes stagnant, and 300-series steels can be severely pitted by the resulting diluted acid.

Sulfuric Acid. The 18-8 varieties of stainless steel are resistant to corrosive attack by sulfuric acid within rather narrow ranges of concentration and temperature. Although stainless steels

Fig. 22 Weld-zone failures in type 316 stainless steel kettle that contained dilute sulfuric acid

Failures occurred after 1½ yr of service in 4% H_2SO_4 in methanol. Kettle was steam jacketed.

Fig. 23 Pitted flange from Van Stone end of a continuous converter containing sulfuric acid and sugar solution

Solution pH was 1.8; temperature was 135 to 165 °C (275 to 325 °F).

Table 38 Corrosion by spent sulfuric acid after separation of sulfonation products

Alloy	Corrosion rate µm/yr	Corrosion rate mils/yr
At 27 °C (80 °F)		
Type 316	5.3	0.21
Carpenter 20 Cb-3	1.5	0.06
Inconel	2.3	0.09
Mild steel	480	18.9
At 60 °C (140 °F)		
Type 316	385	15.1
Carpenter 20 Cb-3	150	5.9
Inconel	490	19.3
Mild steel	1615	63.6

may be used safely in contact with 80 to 100% sulfuric acid at ambient temperature (carbon steel is ordinarily used in this range), they are attacked at slightly higher temperatures. Sulfuric acid in concentrations of 1 to 5% at ambient temperature should not be stored in vessels of molybdenum-free stainless steels. Type 316 may be used for this purpose; 317, with a higher molybdenum content, may be used safely in this range of acid concentration at temperatures as high as 65 °C (150 °F).

Alloys such as Carpenter 20 Cb-3 and CN-7M resist all concentrations of sulfuric acid at temperatures up to 60 °C (140 °F), and concentrations as high as 10% up to the boiling point, but do not resist all concentrations over a wide range of temperatures.

The preceding data pertain to pure sulfuric acid. Addition of oxidizing agents (such as nitric acid, air and copper salts) will widen the range of applicability for all stainless steels; reducing agents (such as hydrogen) will narrow the range of usefulness. If other than pure sulfuric acid is used with stainless steel, corrosion tests must be made under service conditions in order to evaluate the usefulness of specific alloys. Applications should be restricted to only those concentrations of sulfuric acid and temperatures that have proved satisfactory in corrosion tests. Tests should include annealed, stressed and crevice-type specimens, as well as specimens sensitized by heating 1 h or longer at 650 °C (1200 °F).

Agitation and aeration in stainless steel equipment and the velocities of sulfuric acid solutions in piping should be adequate to keep all solids suspended; 1.5 to 4.5 m/s (5 to 15 ft/s) is usually sufficient. Deposits such as charred organic matter or calcium sulfate scale may induce pitting and perforation. Surfaces should be kept clean during shutdown periods.

Organic acids and traces of inorganic salts contributed to corrosion failure of welded zones in a steam-jacketed kettle of type 316 used to heat 4% sulfuric acid in methanol. The interior surfaces of the kettle are shown in Fig. 22. This failure could have been avoided or delayed by the use of 316L or by fully annealing the kettle after fabrication.

The corrosion failure of Van Stone ends (type 316) on a continuous converter for sugar solutions is shown in Fig. 23. Crevice-type corrosion in the flanged ends was enhanced by a carbonaceous deposit at these points. The solution contained sulfuric acid at a pH of 1.8 and was held between 135 °C (275 °F) and 165 °C (325 °F). Because the ends had not been annealed after forming, corrosion was increased by the severe stresses developed in the flanged areas. The corrosion could have been curtailed by using ends with thicker walls and annealing them, and by cleaning to remove the deposit.

Oxidizing agents such as nitric acid, chromic acid and sodium dichromate have dramatic effects on corrosion of stainless steels in sulfuric acid. Figure 24 shows the substantial reductions in corrosion rates for types 304 and 316 that occur when such agents are added even in relatively small amounts. Aeration of sulfuric acid solutions has a similar inhibiting effect (compare Fig. 25 and 26). Advantage can be taken of these inhibiting effects when designing with stainless steels, as long as substantial safeguards are taken to maintain the proper inhibiting effect.

Sulfurous Acid and Sulfur Dioxide. Carpenter 20 Cb-3 and types 316, 317, CF-8M and CN-7M have been used in equipment for sulfur dioxide (wet) and sulfurous acid environments. The molybdenum in these alloys is responsible for their good resistance to the reducing environments of sulfurous acid. Wrought type 316 and cast CF-8M are the most widely used.

Complete suspension of any solids present is necessary to avoid crevice-type pitting. Figure 27 shows pitting and perforation of a Van Stone flanged end of type 304 stainless steel welded to a tube of type 316; the latter did not corrode. Crevice pockets, lapped joints, 90° corner intersections and similar obstructions should be avoided, and the surfaces should be clean and smooth.

Fig. 24 Corrosion rates of types 304 and 316 stainless steels in 30% sulfuric acid containing oxidizing agents

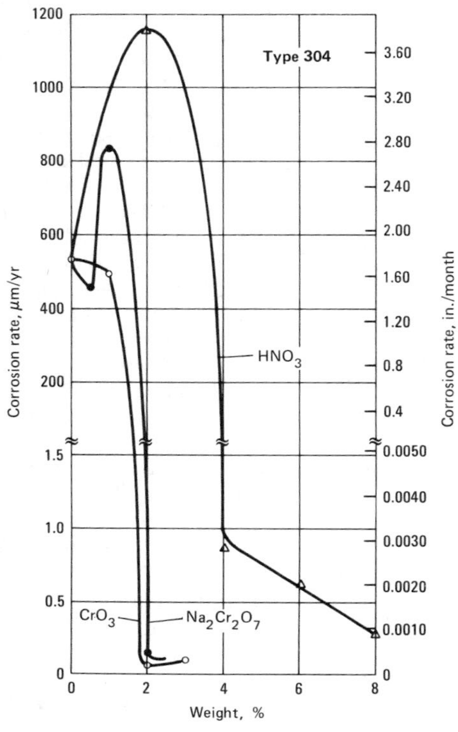

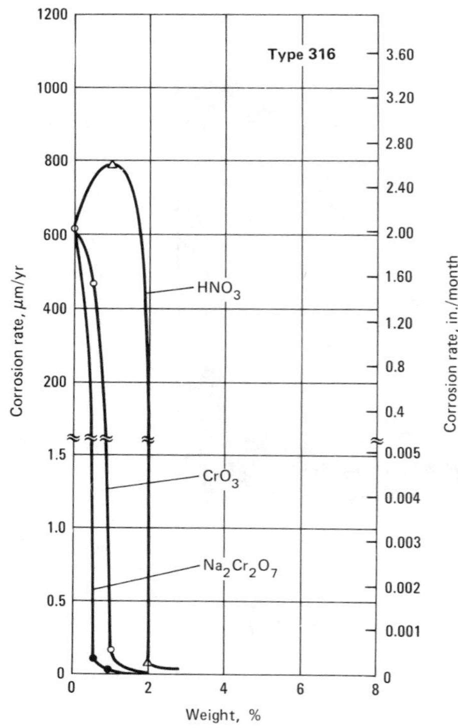

Temperature: 93 °C (200 °F).

Fig. 25 Isocorrosion chart for type 316 stainless steel in air-free sulfuric acid

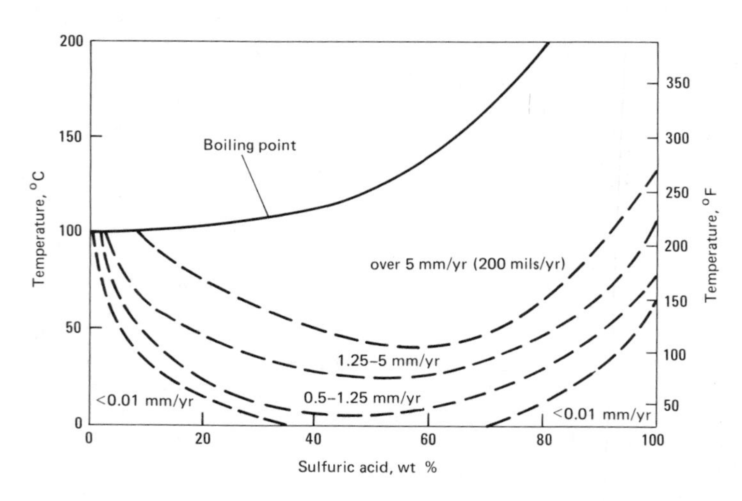

Cold and hot working should be limited to minor forming operations that keep the hardness of the steel below 96 HRB. Stress-corrosion cracking can occur in steel exposed to sulfurous acid solutions containing 100 ppm or more of metal chlorides.

To ensure against failure of weld zones in severely corrosive environments (more than 1.0% sulfur dioxide), extra-low-carbon alloys should be used for equipment that requires significant amounts of welding.

Crevice-type pitting has caused failure of tank bottoms and perforation of heating coils. Solids such as filtering aids, activated carbon and bentonite, which settle on the bottoms of tanks and the tops of heating coils, cause pitting and perforation of these surfaces.

Tubing of type 316, handling sulfur dioxide and sulfur trioxide vapors between a sulfur burner and a sulfurous acid absorber, underwent accelerated

Fig. 26 Active and passive corrosion regions for stainless steels in aerated sulfuric acid solutions

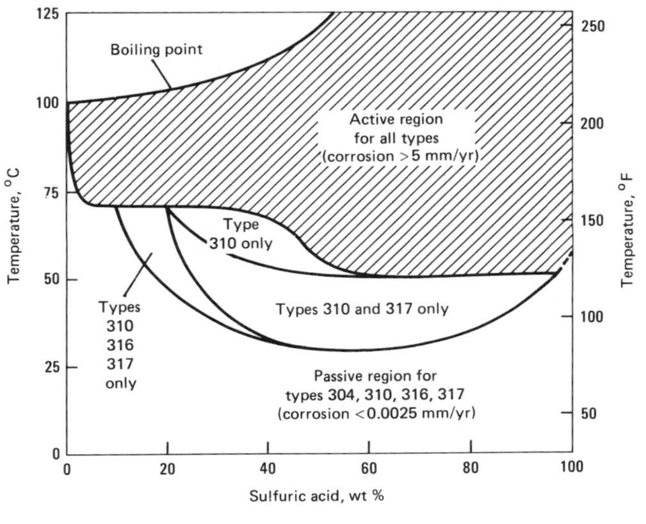

Fig. 27 Van Stone flange of type 304 stainless steel that was pitted and perforated by sulfurous acid

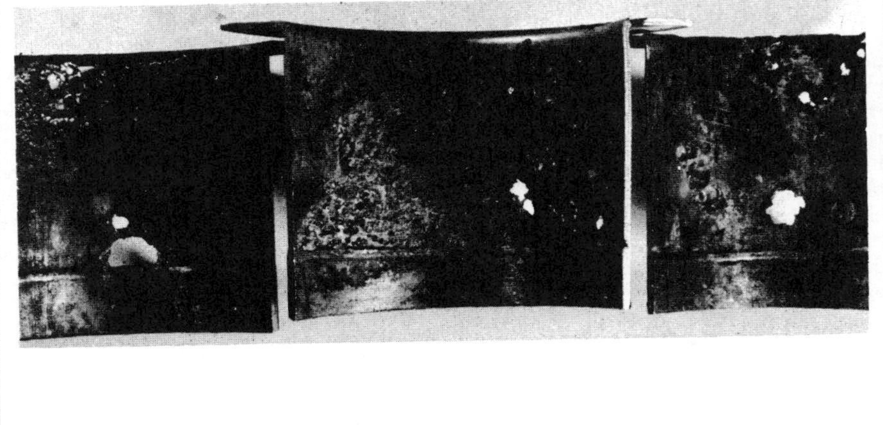

corrosion caused by condensation of moist sulfurous and sulfuric acids at 65 to 120 °C (150 to 250 °F). A field corrosion test had indicated that stainless steels were unsatisfactory for this environment.

Severe erosion-corrosion of stainless steel pump impellers and valve bodies has been caused by sulfurous acid slurries that contain suspended solids. Pump impellers may fail from erosion-corrosion in a few weeks. Alloys such as 316, CF-8M and CN-7M generally give the longest service in erosion-corrosion environments. The useful service life of stainless steels can be predicted from data obtained in laboratory corrosion testing of circular specimens centered on a stainless steel shaft revolving at high speed (10 000 rpm) in the process slurry.

Pharmaceuticals and Fine Chemicals

Stainless steels, principally the 18-8 types, are used by the fine chemical and pharmaceutical industries for maintaining sanitary conditions in corrosive and noncorrosive environments, as well as for their basic resistance to corrosion. Sanitation requirements dictate selection of stainless steels for most installations. Although the effect of corrosion on the life of equipment that handles and processes the product is important, it is usually less important than the potential for corrosion products or residues to contaminate the chemical or biological products being processed.

Purity, color and stability of pharmaceutical products may be greatly affected by the presence of trace quantities of metallic ions. For example, traces of iron affect vitamin B_6 by forming a highly colored complex; therefore, stainless steel is unsuitable for handling vitamin B_6 hydrochloride, even though corrosion rates may be very low. Stainless steel has proved satisfactory for processing of vitamin C; however, all traces of copper must be eliminated, because copper in aqueous solutions accelerates the rate of decomposition of vitamin C by a factor of about 3000.

Although carbon steel is satisfactory for handling dry chloroform at room temperature, it will not serve in processing operations involving chloroform that contains acid. Where a carbon steel pipeline was used with a chloroform processing system made of type 316 stainless steel, it was found necessary to install type 316 pipe. A carbon steel pipe that carried the solution from the condensation reaction unit, which contained a free base neutralized with hydrochloric acid at a pH of 5.8, gave satisfactory service until a process at pH 5.0 was initiated. Then

Fig. 28 Performance of ten stainless alloys in paper-mill bleach plant service

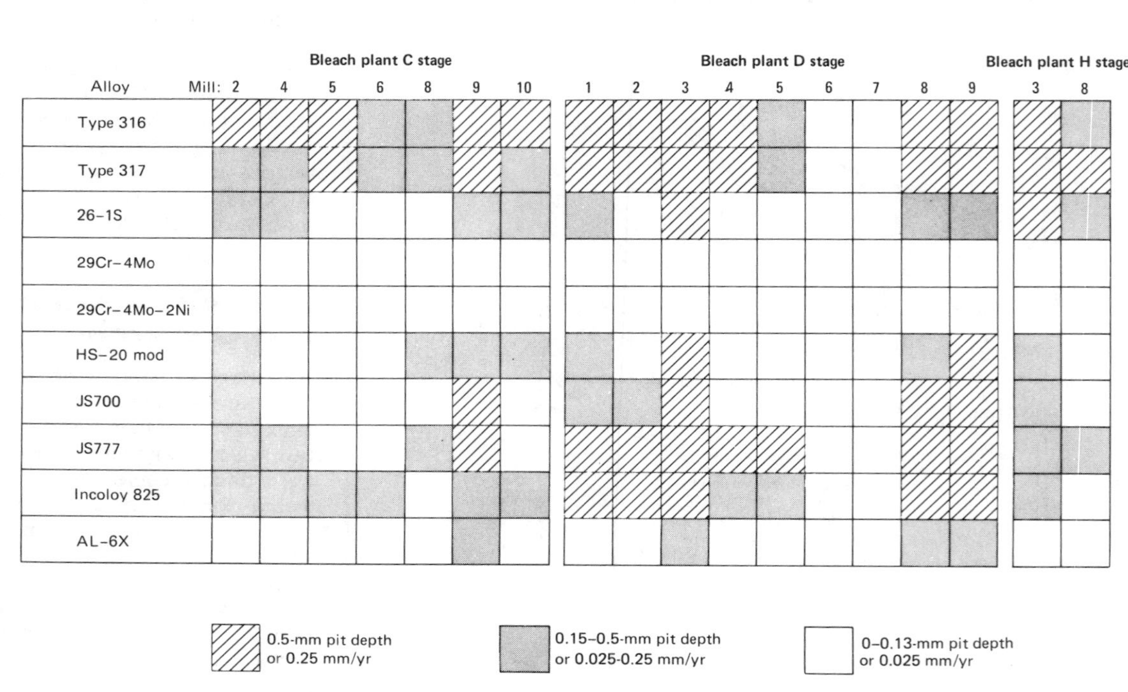

Charts represent selected results from corrosion tests in comparable stages of the bleach plants in ten different paper mills. (Ref 13)

corrosion immediately became noticeable.

Stainless steel pipe (Carpenter 20 Cb-3), installed as a replacement, gave satisfactory service although some corrosion still occurred. Generally, Carpenter 20 Cb-3 can be used with solutions acidified with hydrochloric acid even if the solution is more acidic than pH 5.0.

A complex of metaphosphates called "hyphos" has been used as a chelating agent. In an emergency, the solution was transferred in a pipe of carbon steel. After a short time the pipe had corroded wafer thin. A replacement pipe made of type 316 showed no corrosion.

Among the various fine chemicals and pharmaceuticals handled in stainless steel equipment are: 35% ammonium sulfate (pH 4 to 7.5), 5% butanol at 20 °C (70 °F), 25% caustic at 20 °C, 20% phenol at 20 °C, 20% sodium sulfite at 25 °C (80 °F), 50% sodium hydroxide, organic acids from protein extraction (pH not less than 4.5), and biological media.

Pulp and Paper

The pulp and paper industry, like many other industries, is plagued with corrosion problems. One investigator indicated that corrosion of equipment added $8.00 to the cost of each ton of product. Corrosion severity has been increasing because pollution-control requirements and the more widespread use of recycled paper instead of wood pulp as raw material have led to changes in traditional papermaking processes. Historically, types 304 and 316 have been used for washers, evaporators, green- and white-liquor equipment, head boxes, save-alls, pipe and tubing. Types 316 and 317 have been used in pulp washers and rotary-drum filter equipment. Closed systems require materials more resistant to pitting and crevice corrosion. Recent corrosion tests conducted by the Metals Task Force of the TAPPI Corrosion and Materials Engineering Committee in bleach plant stages C, D and H of 10 different paper mills (Ref 13) indicate the need for such alloys (see Fig. 28).

Of course, use of stainless steels in the pulp and paper industry is not confined to solving corrosion problems. Stainless steels often are selected because they resist scaling and sliming, prevent contamination and have good mechanical and physical properties.

Foods

Stainless steels are used in the food industry for their corrosion resistance wherever metal comes in contact with food products during processing, and because they can be quickly and easily cleaned and made sanitary. Applications include pumps, tubing, tanks, kettles, filling machines, heat exchangers and vacuum tanks. Process temperatures may be as high as 150 °C (300 °F), and processing may be done in vessels operating under pressure or vacuum.

Stainless steels also are used where food-processing equipment is subjected to water spray, continual flow of water, or severe clean-up procedures; equipment includes bottle washers, continuous coolers, continuous cookers and conveyors. In such applications, the product does not contact the metal, but use of carbon steel may (*a*) cause rust to appear on the surface of the product container or (*b*) result in severe corrosion of equipment and lead to costly replacement or repair.

In the primary stages of processing, certain concentrated ingredient solutions that contain salt and vinegar, such as pickle liquors and sauces, are highly corrosive. Food products such as vegetables contain little acid, and their effects on stainless steels are negligible. Handling of sauces or pickle liquors has caused much difficulty due to pitting of tubing (type 304) at the level of the liquid. Type 316 is more resistant, but tubing of this alloy can become pitted in 4 to 12 months. Proprietary alloys such as JS700, AL-6X and 29-4, which are more resistant to pitting and crevice corrosion, should be considered for applications where type 316 becomes pitted.

Pumps of type CF-8 stainless steel are used for less-corrosive solutions, such as vegetables; for sauces, pickle liquors and items containing vinegar, type CF-8M is necessary. Because pumps and impellers may need to be interchanged between different processes, type CF-8M is preferred for such units. Type 304 is used for exhaust hoods over open kettles and tanks because of its ability to withstand corrosive fumes and vapors from sauces and liquors containing vinegar. Impellers for handling fumes from these tanks and kettles are made of type 316. Type 304 is also used for agitators in corrosive solutions and is recommended for tanks for storing all products except pickle liquors and sauces, which should be stored in tanks made of 316L, or of 316 that has been heat treated, sand blasted and passivated.

Drums for holding dry ingredients such as vegetables usually are made of type 304. Tanks used as filler bowls should be made of extra-low-carbon stainless steels to prevent pitting attack adjacent to welded areas. For filler bowls that handle food products less corrosive than sauces, type 304 is recommended.

Food-handling equipment must be designed so that the surfaces are kept clean by the product as it is continually flushed across the surface; there must be no crevices or corners where the product can become lodged. Welds in corners, which cannot be ground smooth, should be avoided.

Many corrosion difficulties can be avoided by frequent cleaning of stainless steel equipment. At the conclusion of an operation, the equipment is flushed thoroughly with fresh water, then scrubbed with a nylon brush, using a detergent and hot water. After this, the vessel is washed with hot water and rinsed with cold water.

In equipment for the more corrosive food products, extra-low-carbon stainless steels should be used wherever possible. Generally, it costs less to use these than to use stainless steels that require annealing to combat sensitization.

For food-processing equipment, experience has shown a 2B finish to be superior to a No. 4 finish. The bright, cold rolled 2B finish has fewer grooves where tiny food particles can cling.

Data indicating the corrosion resistance of stainless steels to food-process solutions can be obtained by using one or more of the following test solutions:

- **Solution A**
  - Acetic acid 2.1%
  - Salt 6.25%
  - Garlic 0.33%
- **Solution B**
  - Salt 7.5%
- **Solution C**
  - Salt 2.0%
  - Acetic acid 3.0%
  - Sugar 22.0%

All three solutions should be used at 60 °C (140 °F). Test duration may vary from a few days to as long as several months. Test specimens should be totally submerged beneath the liquid level, partly submerged in the liquid, and suspended immediately above the liquid. Welded specimens should be included in the tests. Although laboratory tests are important, test results do not always correlate with actual experience, and it may be desirable to conduct tests under actual service conditions. As in laboratory testing, specimens should be totally submerged in, partly submerged in, and suspended above the process solutions.

REFERENCES

1. Oxide Films on Stainless Steels, by T. N. Rhodin, Jr.: *Annals of the New York Academy of Sciences,* Vol 58, 1954, p 855
2. Influence of Chromium on the Atmospheric-Corrosion Behavior of Steel, by R. J. Schmitt and C. X. Mullen: *Stainless Steel for Architectural Use,* STP 454, American Society for Testing and Materials, Philadelphia, 1969, p 124-136
3. Development of Ferritic Steels for Use in Desalination Plants, by N. Pessall and J. I. Nurminen: *Corrosion,* Vol 30, No. 11, Nov 1974
4. "Heat Treatments of Welded Structures for the Relief of Residual Stresses with Particular Reference to Type 347 Stainless Steel Weldments", by W. L. Fleischmann: Knolls Atomic Power Laboratory, 17 Aug 1953
5. *Corrosion Handbook,* by H. H. Uhlig: John Wiley & Sons, New York, 1948, p 584
6. Corrosion, De-Icing Salts and the Environment, by R. L. Chance: *Materials Performance,* Vol 13, No. 10, Oct 1974, p 16-22
7. The New Ferritic Stainless Steels, by R. M. Davison and R. F. Steigerwald: *Metal Progress,* Vol 115, No. 6, June 1979, p 42
8. Some Observations of the Potentials of Stainless Steels in Flowing Sea Water, by K. M. Huston and R. B. Teel: *Corrosion,* Vol 8, July 1952, p 251-256
9. "Corrosion Resistance of the Austenitic Chromium-Nickel Stainless Steels in Marine Environments": The International Nickel Company, Inc., New York, 1963
10. Area Effects in Crevice Corrosion, by O. B. Ellis and F. L. LaQue: *Corrosion,* Vol 7, Nov 1951, p 362-364
11. Effectiveness of Cathodic Currents in Reducing Crevice Corrosion and Pitting of Several Materials in Sea Water, by T. P. May and H. A. Humble: *Corrosion,* Vol 8, Feb 1952, p 50-56
12. Corrosion Properties of Titanium in Marine Environments, by H. B. Bomberger, P. J. Cambourelis and G. E. Hutchinson: *Journal of the Electrochemical Society,* Vol 101, Sept 1954, p 442
13. Corrosion Resistance of Alloys to Bleach Plant Environments, by A. H. Tuthill, J. D. Rushton, J. J. Geisler, R. H. Heasley and L. L. Edwards: *TAPPI,* Vol 62, No. 11, Nov 1979

Corrosion-Resistant Steel Castings

Edited by the ASM
Committee on
Cast Corrosion-Resisting
and Heat-Resisting Alloys*

CORROSION-RESISTANT STEEL CASTINGS are used where carbon and low-alloy steels would be destroyed by the corrosive action of the environment. Cast corrosion-resistant steels resist attack by aqueous solutions at or near room temperature, and by hot gases and high-boiling-point liquids at temperatures up to 650 °C (1200 °F).

All cast corrosion-resistant steels contain more than 11% chromium, and most contain from 1 to 30% nickel (a few have less than 1% Ni). Carbon content, especially important in its influence on both corrosion resistance and strength, is usually less than 0.20% and sometimes lower than 0.03%.

Chemical compositions of these steels are given in Table 1. Physical and mechanical properties, melting temperatures, microstructures, heat treatments and other important fabricating considerations are presented in the data compilations immediately following this article, for alloys standardized by the Alloy Casting Institute Division of the Steel Founders' Society of America.

Although sixteen of the ACI casting grades (Table 1) have counterparts among AISI wrought stainless steels, the composition ranges for the cast and wrought alloys differ. Therefore, the casting alloys should be referred to by their ACI designations, a system of nomenclature that is used in ASTM specifications and by many individual producers.

About two-thirds of the corrosion-resistant steel castings produced in the United States are of grades that contain 18 to 22% Cr and 8 to 12% Ni; the straight chromium compositions are also produced in considerable quantity, particularly the steel with 11.5 to 14.0% Cr.

Chromium imparts passivity to ferrous alloys when present in amounts of more than about 11%, particularly if conditions are strongly oxidizing. Corrosion resistance improves as chromium content is increased.

In general, addition of nickel to iron-chromium alloys improves ductility and impact strength. An increase in nickel content increases resistance to corrosion by neutral chloride solutions and weakly oxidizing acids.

Addition of molybdenum increases resistance to pitting attack by chloride solutions. It also extends the range of passivity in solutions of low oxidizing characteristics.

Addition of copper to duplex-structure Ni-Cr alloys produces alloys that can be precipitation hardened to higher strength and hardness. Addition of copper to single-phase austenitic alloys greatly improves their resistance to corrosion by sulfuric acid. In all Fe-Cr-Ni stainless alloys, resistance to corrosion by environments that cause intergranular attack can be improved by lowering the carbon content.

The type CA-15 alloy (11.5 to 14% Cr) can be hardened by quenching and tempering for wear applications. It is widely used for trim in carbon steel valves, for pumps handling acid mine water containing abrasives, and in paper-mill applications where wear is as important as corrosion resistance. CA-6NM alloy has corrosion resistance similar to that of CA-15, plus a much higher impact strength and resistance to cavitation and the ability to be welded with little or no preheating. Type CB-30 alloy (18 to 22% Cr) is ferritic at all temperatures, and thus, as

*See page X for committee list.

Table 1 Standard designations and composition ranges for corrosion-resistant steel castings

ACI type(a)	Wrought alloy type(b)	C (max)	Mn (max)	Si (max)	Cr	Ni	Others(c)
		Composition, %					
CA-6NM	...	0.06	1.00	1.00	11.5 to 14.0	3.5 to 4.5	0.40 to 1.0 Mo
CA-15	410	0.15	1.00	1.50	11.5 to 14.0	1.0 max	0.5 max Mo(d)
CA-40	420	0.40	1.00	1.50	11.5 to 14.0	1.0 max	0.5 max Mo(d)
CB-7Cu-1	...	0.07	0.70	1.00	15.5 to 17.7	3.6 to 4.6	2.5 to 3.2 Cu; 0.20 to 0.35 Nb; 0.05 max N
CB-7Cu-2	...	0.07	0.70	1.00	14.0 to 15.5	4.5 to 5.5	2.5 to 3.2 Cu; 0.20 to 0.35 Nb; 0.05 max N
CB-30	431	0.30	1.00	1.50	18.0 to 22.0	2.0 max	...
CC-50	446	0.50	1.00	1.50	26.0 to 30.0	4.0 max	...
CD-4MCu	...	0.04	1.00	1.00	25.0 to 26.5	4.75 to 6.0	1.75 to 2.25 Mo; 2.75 to 3.25 Cu
CE-30	312	0.30	1.50	2.00	26.0 to 30.0	8.0 to 11.0	...
CF-3 (e)	304L	0.03	1.50	2.00	17.0 to 21.0	8.0 to 12.0	...
CF-3M (e)	316L	0.03	1.50	2.00	17.0 to 21.0	8.0 to 12.0	2.0 to 3.0 Mo
CF-8 (e)	304	0.08	1.50	2.00	18.0 to 21.0	8.0 to 11.0	...
CF-8C	347	0.08	1.50	2.00	18.0 to 21.0	9.0 to 12.0	Nb(f)
CF-8M	316	0.08	1.50	2.00	18.0 to 21.0	9.0 to 12.0	2.0 to 3.0 Mo
CF-12M	316	0.12	1.50	2.00	18.0 to 21.0	9.0 to 12.0	2.0 to 3.0 Mo
CF-16F	303	0.16	1.50	2.00	18.0 to 21.0	9.0 to 12.0	1.50 max Mo; 0.20 to 0.35 Se
CF-20	302	0.20	1.50	2.00	18.0 to 21.0	8.0 to 11.0	...
CG-8M	317	0.08	1.50	1.50	18.0 to 21.0	9.0 to 13.0	3.0 to 4.0 Mo
CH-20	309	0.20	1.50	2.00	22.0 to 26.0	12.0 to 15.0	...
CK-20	310	0.20	2.00	2.00	23.0 to 27.0	19.0 to 22.0	...
CN-7M	...	0.07	1.50	1.50	19.0 to 22.0	27.5 to 30.5	2.0 to 3.0 Mo; 3.0 to 4.0 Cu
CN-7MS	...	0.07	1.50	3.50(g)	18.0 to 20.0	22.0 to 25.0	2.5 to 3.0 Mo; 1.5 to 2.0 Cu

(a) Most of these standard grades are covered by ASTM A743 and A744. (b) Type numbers of wrought alloys are listed only for nominal identification of corresponding wrought and cast grades. Composition ranges of cast alloys are not the same as for corresponding wrought alloys; cast alloy designations should be used for castings only. (c) Phosphorus content is 0.04% max except in CF-16F, which has 0.17% max P; sulfur content is 0.04% max in all grades. (d) Molybdenum not intentionally added. (e) CF-3A, CF-3MA and CF-8A have the same composition ranges as CF-3, CF-3M and CF-8, respectively, but have balanced compositions so that ferrite contents are at levels that permit higher mechanical-property specifications than those for related grades. They are covered by ASTM A351. (f) Nb, 8 × %C min (1.0% max); or Nb + Ta, 9 × %C (1.1% max). (g) For CN-7MS, silicon ranges from 2.50 to 3.50.

normally made, cannot be hardened by heat treatment. However, because of its higher chromium content, it resists acids in a wider range of concentrations and at a wider range and temperatures than do the 12% Cr alloys.

Iron-Chromium-Nickel Alloys

Iron-chromium-nickel alloys have found wide acceptance and constitute about 60% of total production of high-alloy castings. They generally are both austenitic and slightly magnetic. However, they may vary from nonmagnetic to rather strongly magnetic depending on the composition of the heat or on effects of metal thickness (solidification rate) and heat treatment. The most popular alloys of this type are CF-8 and CF-8M. These alloys are nominally 18-8 stainless steels, and are the cast counterparts of wrought types 304 and 316, respectively. Their carbon contents are maintained at 0.08% max.

Alloys CF-3M and CF-8M are modifications of CF-3 and CF-8 containing 2 to 3% molybdenum to enhance general corrosion resistance. Their passivity under weakly oxidizing conditions is more stable than that of CF-3 and CF-8. They have good resistance to such corrosive media as sulfurous and acetic acids, and they are more resistant to pitting by mild chlorides. These alloys are suitable for use in flowing seawater, but will pit under stagnant conditions. Alloy CG-8M, a modification of CF-8M that contains 3 to 4% Mo, has increased resistance to sulfurous and sulfuric acid solutions and to halogen compounds. It is not suitable for use in nitric acid or other strongly oxidizing media.

Precipitation-Hardening Alloys

Corrosion-resistant alloys capable of being hardened by low-temperature treatment to obtain improved mechanical properties usually are duplex-structure alloys with much more chromium than nickel. Addition of copper enables these alloys to be strengthened by precipitation hardening. These alloys are significantly higher in strength than the other corrosion-resistant alloys even without hardening.

CB-7Cu-1 and CB-7Cu-2 alloys have

Fig. 1 Distribution of room-temperature tensile data for four corrosion-resistant casting alloys

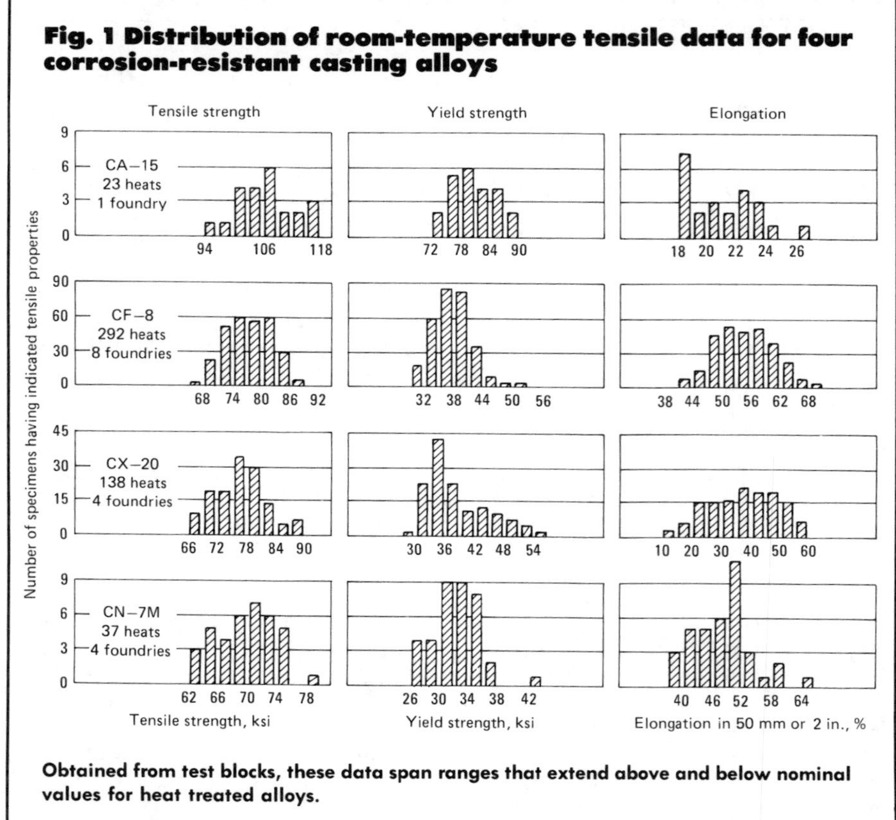

Obtained from test blocks, these data span ranges that extend above and below nominal values for heat treated alloys.

corrosion resistance between those of CA-15 and CF-8. They are widely used for structural components requiring moderate corrosion resistance, as well as for components requiring resistance to erosion and wear.

CD-4MCu alloy is widely used in many applications where its good corrosion resistance (which often equals or even exceeds that of CF-8M) and excellent resistance to erosion make it the most desirable alloy. CD-4MCu has outstanding resistance to nitric acid and mixtures of nitric acid and organic acids as well as excellent resistance to a wide range of corrosive chemical process conditions. This alloy normally is used in the solution-annealed condition, but it can be precipitation hardened for carefully selected applications where lower corrosion resistance can be tolerated and where there is no potential for stress-corrosion cracking.

Iron-Nickel-Chromium Alloys

For some types of service, extensive use is made of iron-nickel-chromium alloys that contain more nickel than chromium. Most important among this group is alloy CN-7M, which has a nominal composition of 28 Ni, 20 Cr, 3.5 Cu, 2.5 Mo and 0.07 max C. In effect, this alloy is made by adding 20% Ni and 3.5% Cu to alloy CF-8M, which greatly improves resistance to hot, concentrated, weakly oxidizing solutions such as sulfuric acid and also improves resistance to severely oxidizing media. Alloys of this type can withstand all concentrations of sulfuric acid at temperatures up to 65 °C (150 °F) and many concentrations up to 79 °C (175 °F). They are widely used in nitric-hydrofluoric pickling solutions, phosphoric acid, cold dilute hydrochloric acid, hot acetic acid, strong hot caustic solutions, brines, and many complex plating solutions and rayon spin baths.

CN-7MS alloy has corrosion resistance much the same as that of CN-7M. CN-7MS has outstanding resistance to corrosion by high-strength nitric acid (over 90%).

Influence of contaminants is one of the most important considerations in selecting an alloy for a particular process application. Ferric chloride in relatively small amounts, for example, will cause concentration-cell corrosion and pitting. Buildup of corrosion products in a chloride solution may increase the iron concentration to a level high enough to be destructive. Thus, chlorine salts, wet chlorine gas and unstable chlorinated organic compounds cannot be handled by any of the iron-base alloys, which creates a need for nickel-base alloys.

Mechanical Properties

Figure 1 gives tensile data for 490 heats of four alloys, determined in tests of specimens from separately cast test blocks representing metal poured into production castings. Tensile properties of metal in thicker and thinner sections of intricate castings are affected by local cooling rates in the mold and may differ from those determined for test blocks.

Corrosion-resistant steel castings are somewhat lower in modulus of elasticity in tension than carbon or low-alloy steels, varying from about 165 GPa (24×10^6 psi) for high-nickel grades to 200 GPa (29×10^6 psi) for high-chromium, low-nickel alloys. This compares with 207 GPa (30×10^6 psi) for carbon and low-alloy steels.

A comparison of mechanical properties, after heat treatment, of test bars taken from different locations in CB-7Cu-1 and CA-15 castings is presented in Fig. 2.

Heat Treatment. Stainless steel castings are almost always heat treated. For the hardenable ferritic-martensitic straight chromium compositions, heat treating is done primarily to obtain desired mechanical properties. Castings such as CA-15, a 12% Cr alloy, are air cooled or oil quenched from about 980 °C (1800 °F) and tempered between 540 and 760 °C (1000 and 1400 °F), depending on specified properties. Tempering between 370 and 540 °C (700 and 1000 °F) causes a marked loss in impact resistance and should be avoided.

Ferrite in Austenitic Grades

Ferrite content is controlled by proper balance of composition. With other elements unchanged, ferrite content decreases as carbon content increases.

Chromium, molybdenum and silicon promote formation of ferrite (magnetic); carbon, nickel, nitrogen and man-

Fig. 2 Mechanical properties at different locations in two heat treated corrosion-resistant steel castings

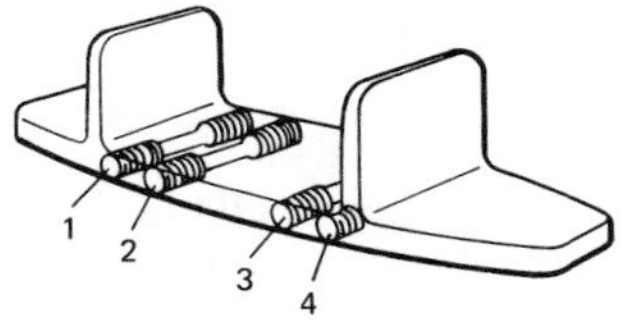

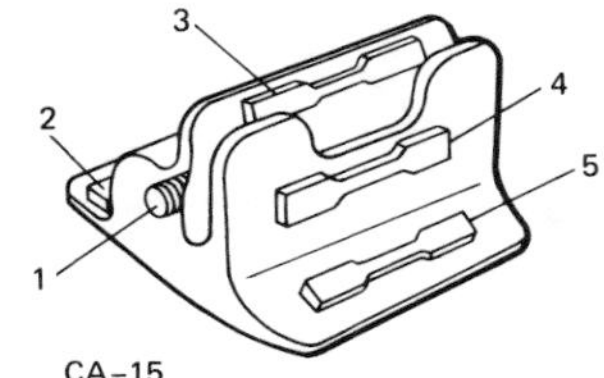

Test bar	Tensile strength MPa	Tensile strength ksi	Yield strength MPa	Yield strength ksi	Elongation, %	Reduction in area, %
CB-7Cu-1 (a)						
1	1335	193.6	1113	161.5	17	48.1
2	1342	194.7	1136	164.8	16	50.0
3	1325	192.2	1095	158.8	15	41.6
4	1343	194.8	1128	163.6	12	34.9
CA-15 (b)						
1	1017	147.5	869	126.1	2	...
2	1300	188.5	1019	147.8	6	...
3	1293	187.5	1025	148.7	8	...
4	1342	194.7	1059	153.6	6	...
5	1300	188.6	1006	145.9	9	...

(a) Solution treated and aged to a hardness of 43 HRC. (b) Hardened and tempered to a hardness of 42 HRC.

Table 2 Effect of ferrite content on tensile properties of 19Cr-9Ni alloys

Ferrite content, %	Tensile strength MPa	Tensile strength ksi	Yield strength at 0.2% offset MPa	Yield strength at 0.2% offset ksi	Elongation in 50 mm or 2 in., %	Reduction in area, %
Tested at room temperature						
3	465	67.4	216	31.3	60.5	64.2
10	498	72.2	234	34.0	61.0	73.0
20	584	84.7	296	43.0	53.5	58.5
41	634	91.9	331	48.0	45.5	47.9
Tested at 355 °C (670 °F)						
3	339	49.1	104	15.1	45.5	63.2
10	350	50.8	109	15.8	43.0	69.7
20	457	66.3	183	26.5	36.5	47.5
41	487	70.8	188	27.3	33.8	49.4

ganese favor formation of austenite (nonmagnetic). For example, a cast extra-low-carbon grade (0.03 max C) cannot be completely nonmagnetic unless it contains 12 to 15% Ni. The wrought grades of these alloys normally contain about 13% Ni. They are made fully austenitic to improve rolling and forging characteristics.

Cast austenitic alloys usually have from 5 to 20% ferrite distributed in discontinuous pools throughout the matrix. In ordinary service, where these steels may be heated in the range from 425 to 650 °C (800 to 1200 °F), carbide precipitation occurs at the edges of the ferrite pools in preference to the austenite grain boundaries. When the steel is heated above 650 °C, the ferrite pools transform to chi or sigma phase. If these pools are distributed in such a way that a continuous network is formed, embrittlement or a network of corrosion penetration may result. Also, if the amount of ferrite is too great, it may form continuous stringers where corrosion can take place, producing a condition similar to grain-boundary attack.

Some solutions attack the austenite phase in heat treated alloys, whereas others attack the ferrite. For instance, calcium chloride solutions attack the austenite. On the other hand, a 10° Baumé cornstarch solution, acidified to a pH of 1.8 with sulfuric acid and heated to a temperature of 135 °C (275 °F), attacks the ferrite.

The Schoefer diagram, shown in Fig. 3, is used to estimate the ferrite contents of castings from their chemical compositions. Ferrite content also can be estimated by metallographic procedures and by measurement of magnetic response. All methods are subject to error, and differences of plus-or-minus 6% between values estimated by different techniques are not uncommon.

The tensile properties of 19Cr-9Ni alloys are affected by ferrite content. An increase in ferrite content increases strength (see Table 2) and decreases ductility and impact strength.

At temperatures above 540 °C (1000 °F), austenite has better creep resistance than ferrite. The weaker ferrite phase may lend better plasticity to the alloy, but after long exposure at temperatures in the 540 to 760 °C (1000 to 1400 °F) range it may transform to sigma or chi phase, which reduces resistance to impact. In some instances, the alloy is deliberately aged to form the sigma or chi phase and thus increase strength. Austenite can transform directly to sigma or chi without going through the ferrite phase.

Magnetic properties of high-alloy castings depend on microstructure. The straight chromium types are ferritic and ferromagnetic. All other grades are mainly austenitic, with or without minor amounts of ferrite, and are either weakly magnetic or wholly nonmagnetic.

Cast nonmagnetic parts for applications in radar and in mine sweepers require close control of ferrite content. Thicker sections have higher permeability than thinner sections. Therefore, to ensure low magnetic permeability in all areas of a casting, magnetic

Fig. 3 Constitution diagram for corrosion-resistant steel castings

Courtesy of Ernest A. Schoefer, Consultant.

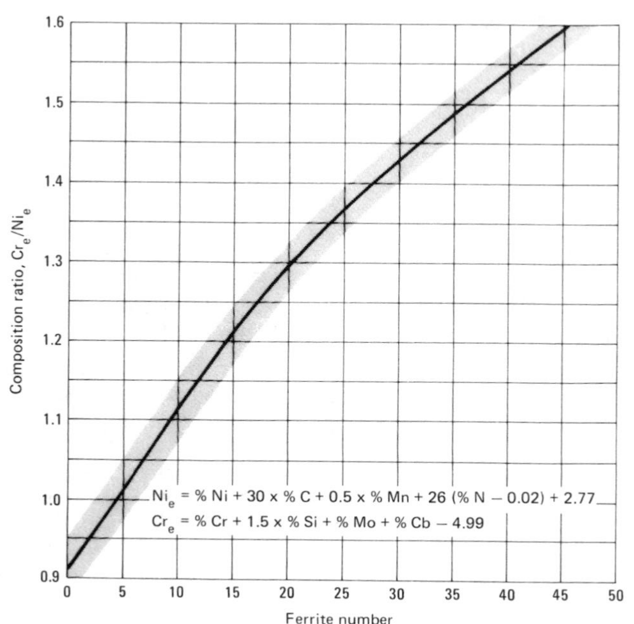

To estimate ferrite content, calculate Cr_e/Ni_e ratio from the equations given on the face of the graph. Next, estimate nominal ferrite content by finding the value of ferrite content corresponding to the calculated ratio of chromium and nickel equivalents. To find the limits of accuracy for this estimate, find the values of ferrite number corresponding to the values of Cr_e/Ni_e ratio at which a vertical line through the nominal value intersects the limits of the scatter band (shaded region). The scatter band on this graph exists because of the inherent uncertainty in determining exact concentrations for certain elements.

permeability checks should be made on the thicker sections.

Corrosion

The electrolytic oxalic acid etch test is widely used for rapid evaluation of the corrosion resistance of stainless steels. This recommended practice, and other methods for determining susceptibility of stainless steels to intergranular attack, are covered in ASTM A262.

In alloys of the CF type, the effects of composition on rates of general corrosion attack have been studied, and certain definite relationships have been established. Through use of the Huey test (five 48-h periods of exposure to boiling 65% nitric acid, as described in Practice C of ASTM A262), it has been shown that, in this standardized environment, carbide-free quench-annealed alloys of various nickel, chromium, silicon, carbon and manganese contents have corrosion rates directly related to these contents.

Figure 4 shows the influences on corrosion rate exerted by various elements in a 19Cr-9Ni casting alloy. Variations in nickel, manganese and nitrogen contents for the ranges shown have relatively slight influences, but variations in chromium, carbon and silicon have marked effects. The relationship between composition and corrosion rate

Fig. 4 Effects of various elements in a 19Cr-9Ni casting alloy on corrosion rate in boiling 65% nitric acid

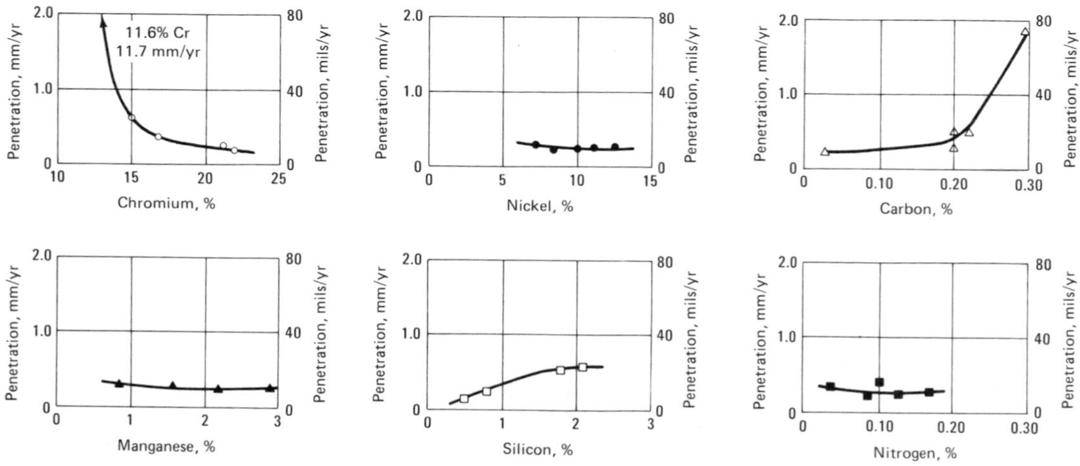

Data were determined for solution annealed and quenched specimens. Composition of base alloy: 19 Cr, 9 Ni, 0.09 C, 0.8 Mn, 1.0 Si, 0.04 max P, 0.03 max S, 0.06 N.

Fig. 5 Nomograph for determining corrosion rate in boiling 65% nitric acid for solution annealed and quenched type CF casting alloys

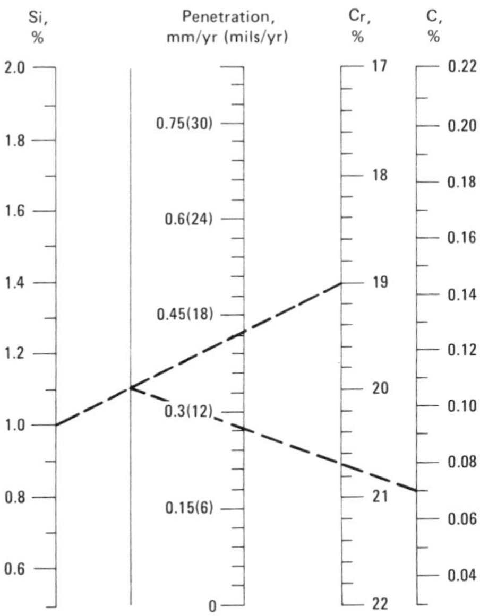

for properly heat treated CF alloys in boiling 65% nitric acid is summarized in the nomograph presented in Fig. 5.

Intergranular attack can be avoided in CF alloys (*a*) by addition of the stabilizing element niobium (columbium), (*b*) by use of an extra-low-carbon grade such as CF-3 or CF-3M, or (*c*) possibly by formation of small amounts of ferrite, which can be induced either by reducing some of the austenite-stabilizing elements such as nickel and carbon or by increasing such ferrite-stabilizing elements as molybdenum, silicon and chromium. Addition of niobium to molybdenum-containing type CF alloys has been found unsatisfactory for castings. When both niobium and molybdenum are present, the ferrite phase tends to form as an interconnected network and is especially likely to transform into the brittle sigma phase. As a result, castings in the as-cast condition become embrittled and have a tendency to crack.

In general, intergranular corrosion is of less concern in the straight chromium alloys, especially those containing 25% chromium or more.

In-Plant Service Tests. Results of in-plant corrosion testing of CF-8, CF-8M and CN-7M alloys are shown in Table 3. These tests illustrate the specific effect of molybdenum on 19Cr-9Ni alloys in reducing selective attack and pitting, and the over-all corrosion rate computed from loss in weight. The higher nickel plus copper and molybdenum in the CN-7M alloy reduces the rate of corrosion compared with that of the CF-8M alloy.

The austenitic steel used in equipment handling sulfite pulp requires a controlled molybdenum content for optimum serviceability. It varies somewhat with the composition of the acid cooking solution, but a molybdenum content from 2.25 to 2.5% is desirable. In one application where the chloride content of the cooking solution was higher than normal because the logs had been floated on salt water, pitting was encountered in valves made of CF-8M alloy containing 2.15% molybdenum. For the same installation, it was found that the wrought evaporator heater tubes of type 316 required a minimum molybdenum content of 2.75%.

Molybdenum may produce a detrimental catalytic reaction. For example, the residual molybdenum in CF-8 alloy must be held below 0.5% in the presence of hydrazine.

Influence of Heat Treatment. The optimum corrosion resistance of 18-8 grades of stainless steel is obtained by heating in the range from 1065 to 1135 °C (1950 to 2075 °F) and then quenching in water, oil or air to ensure total solution of carbides. Water is generally the accepted quenching medium; air is used only for relatively thin sections. Holding time at temperature varies with the thickness of the casting section, but should be long enough to heat all sections uniformly throughout. This process is known as solution quenching, solution annealing or quench annealing. If this process is omitted or is not done properly, or if the alloy is exposed to temperatures from 425 to 870 °C (800 to 1600 °F) after solution quenching, complex chromium carbides may form at grain boundaries. These carbides are attacked selectively in oxidizing solutions and will in time lead to failure by intergranular corrosion. These alloys are vulnerable to formation of both sigma and chi phases at about 650 °C (1200 °F).

Austenitic alloys such as 19Cr-9Ni are solution heat treated for corrosion resistance, especially the intergranular type. However, welding may cause carbides to reprecipitate with resulting impairment of corrosion resistance in the region of the weld. This difficulty can be avoided by using the 0.03% max C or niobium-stabilized grades.

The relationships among rates of corrosion in 65% boiling nitric acid, quenching temperatures, and carbon, silicon and molybdenum contents of a series of 19Cr-9Ni alloys are shown in Fig. 6. The solid curves on these charts form the boundaries between alloys with constant rates of corrosion and those with rates that increase with time. All alloys to the left and above the broken curves have average rates of corrosion below 0.75 mm/yr (30 mils/yr), as measured by five 48-h test periods in boiling nitric acid. It is advisable to select a quenching temperature at least 28 °C (50 °F) above the solid line for the composition under consideration. In no instance should the heat treating temperature be below 1065 °C (1950 °F) for maximum resistance to intergranular corrosion.

Where the usual quenching treatment is difficult or impossible, holding for 24 to 48 h at 870 to 980 °C (1600 to 1800 °F) and air cooling is helpful in improving the resistance of castings to intergranular corrosion. However, except for alloys of very low carbon content and castings with thin sections, this treatment fails to produce material with as good resistance to intergranu-

Table 3 Results of in-plant corrosion testing of CF-8, CF-8M and CN-7M alloys

Type and composition of corroding solution	Temperature of solution, °C	°F	Alloy	Metal loss on surface µm/yr	mils/yr	Surface condition by visual examination	Remarks
Neutralizer after formation of ammonium sulfate: ammonium sulfate plus small excess of sulfuric acid, ammonia vapor and steam	100	212	CF-8 CF-8M CN-7M	665 28 18	26.2 1.1 0.7	Very heavy etch(a) Light tarnish(b) Bright	CF-8M was installed for low-corrosion-tolerance equipment in this service and performed satisfactorily.
Settling tank after neutralizer: ammonium sulfate plus excess of sulfuric acid	50	122	CF-8 CF-8M CN-7M	385 10 2.5	15.2 0.4 0.1	Very heavy etch(a) Slight tarnish Bright(b)	CF-8 in service showed excessive corrosion rate plus heavy concentration-cell attack.
Ammonium sulfate processing solution: ammonium sulfate at pH of 8.0	50	122	CF-8 CF-8M CN-7M	685 175 5	27.0 6.8 2.0	Heavy etch Moderate etch Light etch	CF-8M had too high a corrosion rate in service for good valve life, although suitable for equipment of greater corrosion tolerance. CN-7M was installed in this service.
99 to 100% fuming nitric acid	20	68	CF-8 CN-7M CF-8M	245 79 345	9.6 3.1 13.5	Moderate etch Light etch Moderate etch	CF-8 was satisfactory except for low-tolerance equipment such as valves. CN-7M valves performed satisfactorily in service.
Saturated solution of sodium chloride plus 15% sodium sulfate; pH, 4.5	60	140	CF-8M CF-8	2.5 240	0.1 9.5	Bright Concentration-cell corrosion at various small areas of specimen	CF-8M was installed for valves in service.

(a) Concentration-cell attack under insulating washer. (b) Slight concentration-cell attack under insulating washer.

Table 4 Speeds and feeds for machining corrosion-resistant steel castings

Operation	Approximate feed, in./rev(a)	Machining speed, ft/min(b): CF-20 CF-8	CF-16F	CE-30 CF-8M CH-20 CK-20	CF-8C	CN-7M	CA-6NM CA-15	CA-40	CB-30	CC-50
Broaching	0.001 to 0.005	8 to 15	10 to 20	8 to 15	8 to 15	8 to 15	10 to 20	8 to 15	10 to 20	10 to 20
Tapping	0.003 to 0.007	10 to 20	15 to 30	10 to 25	10 to 25	12 to 20	10 to 25	10 to 20	10 to 25	10 to 25
Threading	0.003 to 0.008	10 to 20	10 to 25	10 to 25	10 to 25	10 to 20	10 to 20	10 to 20	10 to 25	10 to 25
Reaming	0.003 to 0.008	20 to 60	30 to 100	40 to 80	40 to 80	20 to 60	20 to 60	20 to 60	40 to 120	40 to 120
Drilling	0.003 to 0.007	15 to 40	35 to 85	30 to 50	30 to 50	30 to 60	35 to 75	30 to 60	40 to 60	40 to 60
Turret lathe	0.003 to 0.008	60 to 90	90 to 130	60 to 80	60 to 90	60 to 80	80 to 110	60 to 100	70 to 100	60 to 100
Milling	0.003 to 0.008	35 to 65	75 to 110	30 to 50	40 to 60	35 to 70	70 to 105	35 to 70	40 to 60	40 to 60
Turning, boring	0.003 to 0.008	40 to 85	85 to 120	60 to 120	60 to 120	60 to 80	80 to 115	40 to 80	60 to 100	60 to 120
Screw machine	0.003 to 0.008	60 to 90	90 to 130	60 to 80	60 to 90	60 to 80	80 to 110	60 to 100	70 to 100	60 to 100
Hack sawing	Use a coarse-tooth blade (not over 10 teeth per in.) at about 50 strokes per minute with positive pressure.									

(a) For feeds in mm/rev, multiply listed values by 25. (b) For speeds in m/min, multiply listed values by 0.3.

Table 5 Welding conditions for corrosion-resistant steel castings(a)

ACI designation	Type of electrodes used(b)	Preheat °C	Preheat °F	Postweld heat treatment
CA-6NM	Same composition	100 to 150	212 to 300	590 to 620 °C (1100 to 1150 °F)
CA-15	410	200 to 315	400 to 600	610 to 760 °C (1125 to 1400 °F), air cool
CA-40	410 or 420	200 to 315	400 to 600	610 to 760 °C (1125 to 1400 °F), air cool
CB-7Cu	Same composition or 308	Not required		480 to 590 °C (900 to 1100 °F), air cool
CB-30	442	315 to 425	600 to 800	780 °C (1450 °F) min, air cool
CC-50	446	200 to 700	400 to 1300	900 °C (1650 °F), air cool
CD-4MCu	Same composition	Not required		Heat to 1120 °C (2050 °F), cool to 1040 °C (1900 °F), quench
CE-30	312	Not required		Quench from 1090 to 1120 °C (2000 to 2050 °F)
CF-3	308L	Not required		Usually unnecessary
CF-8	308	Not required		Quench from 1040 to 1120 °C (1900 to 2050 °F)
CF-8C	347	Not required		Usually unnecessary
CF-3M	316L	Not required		Usually unnecessary
CF-8M	316	Not required		Quench from 1070 to 1150 °C (1950 to 2100 °F)
CF-12M	316	Not required		Quench from 1070 to 1150 °C (1950 to 2100 °F)
CF-16F	308 or 308L	Not required		Quench from 1090 to 1150 °C (2000 to 2100 °F)
CF-20	308	Not required		Quench from 1090 to 1150 °C (2000 to 2100 °F)
CG-8M	317	Not required		Quench from 1040 to 1120 °C (1900 to 2050 °F)
CH-20	309	Not required		Quench from 1090 to 1150 °C (2000 to 2100 °F)
CK-20	310	Not required		Quench from 1090 to 1180 °C (2000 to 2150 °F)
CN-7M	320	200	400	Quench from 1120 °C (2050 °F)

(a) Metal arc, inert-gas arc and electroslag welding methods can be used. The following table lists suggested electrical settings and electrode sizes for various section thicknesses:

Section thickness, in.	Electrode diameter, in.	Current, A	Maximum arc voltage, V
1/8 to 1/4	3/32	45 to 70	24
1/8 to 1/4	1/8	70 to 105	25
1/8 to 1/4	5/32	100 to 140	25
1/4 to 1/2	3/16	130 to 180	26
1/2 and over	1/4	210 to 290	27

(b) Lime-coated electrodes are recommended.

lar corrosion as properly quench-annealed material.

The extra-low-carbon grades, CF-3 and CF-3M, which contain 0.03% max C, are made in quantity for applications where heat treatment is impractical or fabrication by welding is required after machining. Because of the low carbon content, the amount of chromium carbide present in the as-cast condition is not sufficient for selective corrosive attack to occur. Therefore, these grades are relatively immune to intergranular corrosion failure.

The niobium-modified grade of 18-8, known as CF-8C, is produced for similar applications where heat treatment is impractical. When this alloy is in the as-cast condition, most of its carbon is in the form of niobium carbide, thus precluding chromium carbide precipitation in the critical temperature range from 425 to 870 °C (800 to 1600 °F) and particularly from 565 to 650 °C (1050 to 1200 °F). CF-8C is solution treated at 1120 °C (2050 °F), quenched to room temperature and then reheated to 870 to 925 °C (1600 to 1700 °F), at which temperature precipitation of niobium carbide occurs. An alternative method is solution treating at 1120 °C (2050 °F), cooling to the 870 to 925 °C (1600 to 1700 °F) range, and then holding at this temperature before cooling to room temperature. For maximum corrosion resistance, it is recommended that this alloy be solution treated before being stabilized.

Niobium-containing alloys that have been heated to sensitizing temperatures around 650 °C (1200 °F) are not susceptible to intergranular corrosion. They are more susceptible to over-all corrosion when tested in nitric acid, compared with the niobium-free, quench-annealed alloys of the same nickel, chromium and carbon contents.

Weld crack sensitivity of CF alloys containing niobium (CF-8C) is more pronounced in the fully austenitic grade, cracking may be alleviated through the introduction into the weld deposit of a small amount of ferrite, usually between 4 and 10%. However, appreciable amounts of ferrite in niobium-bearing corrosion-resistant steels will transform, at least partly, to the sigma or chi phase on heating between 540 and 925 °C (1000 and 1700 °F).

In weld deposits, the presence of sigma or chi phase is extremely detrimental to ductility. When welding for service at room temperature or up to 540 °C (1000 °F), 4 to 10% ferrite may be present and will greatly reduce the tendency toward weld cracking. However, for service at temperatures between 540 and 815 °C (1000 and 1500 °F), the amount of ferrite in the weld must be reduced to less than 5% to avoid em-

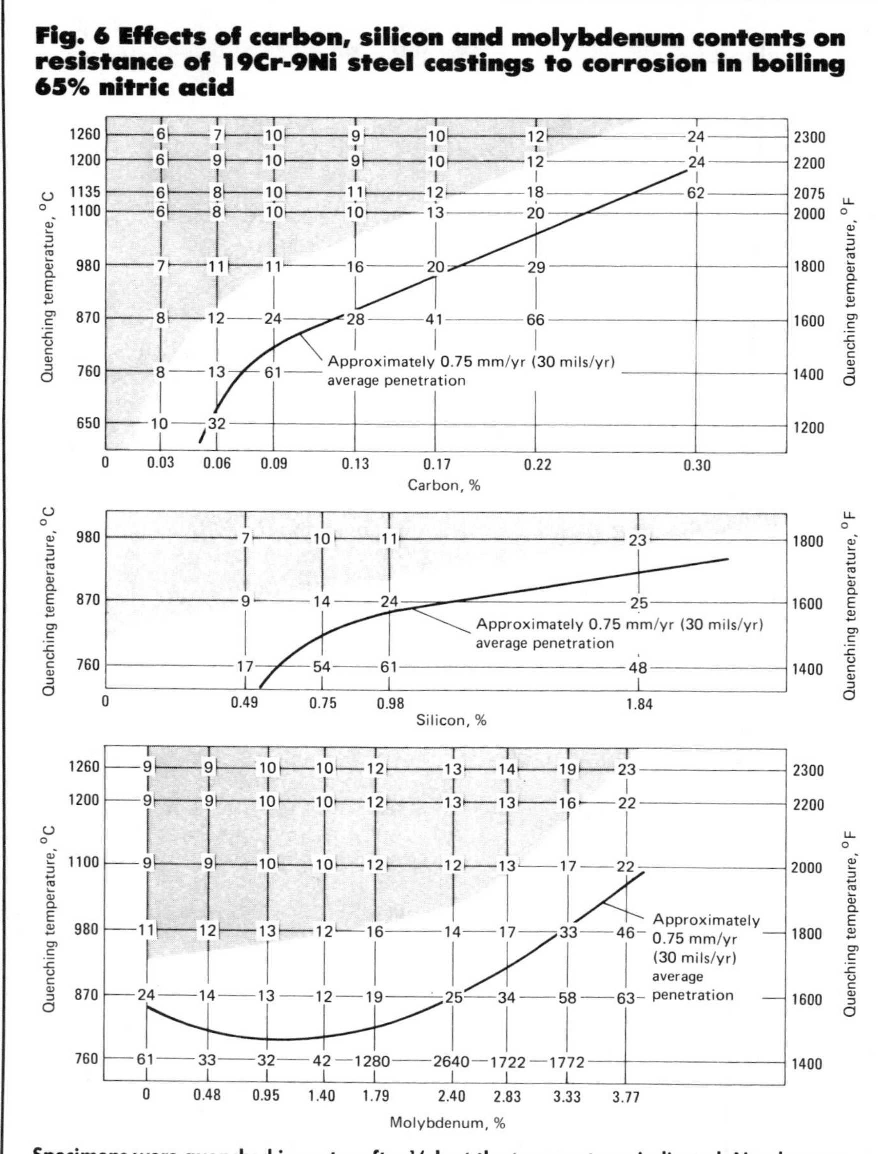

Fig. 6 Effects of carbon, silicon and molybdenum contents on resistance of 19Cr-9Ni steel castings to corrosion in boiling 65% nitric acid

Specimens were quenched in water after ½ h at the temperatures indicated. Numbers on graphs indicate penetration in ten-thousandths of an inch per month, calculated from the average loss in weight during five 48-h periods in boiling nitric acid. To convert these values to mm/yr, multiply by 0.03. In the shaded areas, rate of corrosion is constant with time; elsewhere, rate of corrosion increases with time. Base composition of alloy: 19 Cr, 9 Ni, 0.8 Mn, 1.0 Si, 0.02 S, 0.02 P, 0.06 N.

brittlement from excessive sigma or chi phase.

Galling

Corrosion-resistant steel castings are susceptible to galling and seizing when dry surfaces slide or chafe against each other. However, the surfaces of the castings can be nitrided so that they are hard and wear resistant. Tensile properties are not impaired. After being nitrided, the 19Cr-9Ni and straight chromium alloys are resistant to superheated steam (10 MPa at 500 °C, or 1.5 ksi at 930 °F), saturated steam, boiler feedwater and petroleum-base fuels; they are not resistant to halogen acids or salts, nor to any corrosive medium that will attack the untreated alloy. Nitriding reduces resistance to corrosion by concentrated nitric or mixed acids.

Parts such as gate disks for gate valves usually are furnished in the solution-treated condition but may be nitrided to reduce susceptibility to seizure in service. Similar results are obtained by hard facing with Co-Cr-W alloys.

Machining

The machinability of straight chromium alloys is as good as or better than that of annealed 19Cr-9Ni alloys. The machining characteristics of all iron-chromium-nickel alloys (chromium in excess of nickel) are about on a par with quench-annealed CF alloys. CE alloys, and alloys that contain niobium, are somewhat easier to machine; CH is slightly less machinable than CF. More detailed machining information for several alloys and operations is given in Table 4.

Steels are available that contain selenium or sulfur (about 0.30%) to improve machinability; however, these steels are not widely used. Techniques for machining the conventional CF-8 alloy have been developed to the point where the slight assistance obtained by adding selenium or sulfur to the composition generally is not needed.

Welding

Corrosion-resistant steel castings can be welded by shielded metal-arc welding (SMAW), gas tungsten-arc welding (GTAW), gas metal-arc welding (GMAW) and electroslag (submerged-arc) welding. Austenitic castings normally are welded without preheat, and are solution annealed after welding. Martensitic castings require preheating to avoid cracking during welding, and are given an appropriate postweld heat treatment. Specific conditions for welding specific alloys are listed in Table 5.

When welds are properly made, tensile and yield strengths of the welded joint are similar to those of the unwelded castings. Elongation generally is lower for specimens taken perpendicular to the weld bead (Table 6).

Table 6 Short-time tensile properties of peripheral-welded cylinders of CF-8 alloy

Cylinders were 38 mm (1½ in.) thick; specimens were machined with longitudinal axes perpendicular to welded seam and with seam at middle of gage length.

Testing temperature		Tensile strength		Yield strength at 0.2% offset		Proportional limit(a)		Reduction in area, %	Elongation in 50 mm or 2 in., %	Modulus of elasticity(a),		Location of final rupture
°C	°F	MPa	ksi	MPa	ksi	MPa	ksi			GPa	10^6 psi	
Base metal												
Keel block(b)		500	72.5	238	34.5	...	...	59.0	49	...	...	...
Room		500	72.5	261	37.8	179	26	62.1	58	186	27	...
315	600	330	47.8	169	24.5	90	13	54.9	33.5	152	22	...
425	800	339	49.2	167	24.2	59	8.5	58.6	37.5	134	19.5	...
540	1000	291	42.2	140	20.3	55	8	60.8	32.5	117	17	...
595	1100	279	40.4	130	18.8	45	6.5	59.1	38	110	16	...
Welded joint												
Room		490	71.0	247	35.8	148	21.5	70.8	42	186	27	Base metal
315	600	341	49.5	199	28.8	72	10.5	58.3	15.5	152	22	Base metal
425	800	355	51.5	171	24.8	69	10	46.3	24.5	131	19	Base metal
540	1000	326	47.3	188	27.3	62	9	62.8	23.5	114	16.5	Base metal
595	1100	272	39.4	134	19.5	55	8	70.4	31	107	15.5	Base metal

(a) Values of proportional limit and modulus of elasticity at elevated temperatures are apparent values because creep occurs. (b) Separately cast from same heat as cylinders.

Properties of Cast Stainless Steels

Adapted from data compiled for the Alloy Casting Institute Division of the Steel Founders' Society of America by Ernest A. Schoefer

CA-6NM
12Cr-4Ni-0.7Mo

Commercial Name

UNS number. J91540

Specifications

ASTM. A487 and A351

Chemical Composition

Composition limits. 0.06 max C; 1.00 max Mn; 1.00 max Si; 0.04 max P; 0.04 max S; 11.5 to 14.0 Cr; 3.5 to 4.5 Ni; 0.40 to 1.0 Mo; rem Fe

Applications

Typical uses. Major use in large hydraulic turbine runners for power generation; other uses include casings, compressor impellers, diaphragms, diffusers, discharge spacers, impulse wheels, packing housings, propellers, pump impellers, suction spacers, valve bodies and parts in chemical marine, petroleum, pollution control, and power plant industries.
Precautions in use. Tempering in the vicinity of 480 °C (900 °F) should be avoided because lowered impact strength will result.

Mechanical Properties

Tensile properties. Typical (a). Tensile strength, 825 MPa (120 ksi); yield strength at 0.2% offset, 690 MPa (100 ksi); elongation, 24% in 50 mm or 2 in; reduction in area, 60%. See also Fig. 1.
Hardness. Typical (a). 269 HB
Elastic modulus. Tension, 200 GPa (29×10^6 psi)
Impact strength. Charpy, V-notch, see Fig. 2.

(a) Typical values when air cooled from above 955 °C (1750 °F) and tempered at 590 to 620 °C (1100 to 1150 °F).

Structure

Microstructure. Essentially 100% martensite in the normalized and tempered condition

Mass Characteristics

Density. 7.7 Mg/m³ (0.278 lb/in.³) at 20 °C (70 °F)
Patternmaker's shrinkage. 21 mm/m (¹/₄ in./ft)

Thermal Properties

Melting temperature. 1510 °C (2750 °F)
Coefficient of thermal expansion. Linear, 10.8 µm/m·K (6.0 µin./in.·°F)

Fig. 1 Short-time elevated-temperature tensile properties of CA-6NM

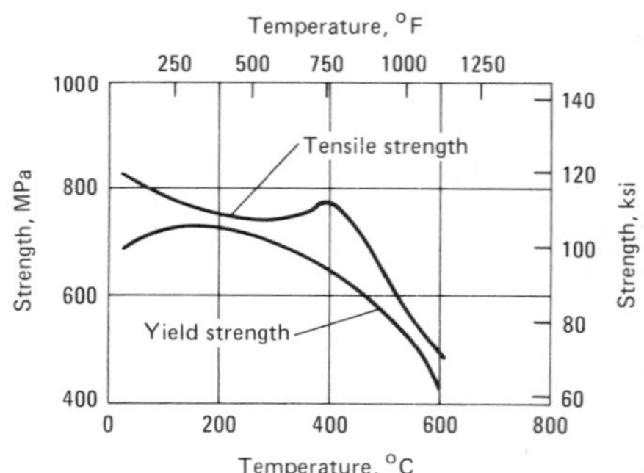

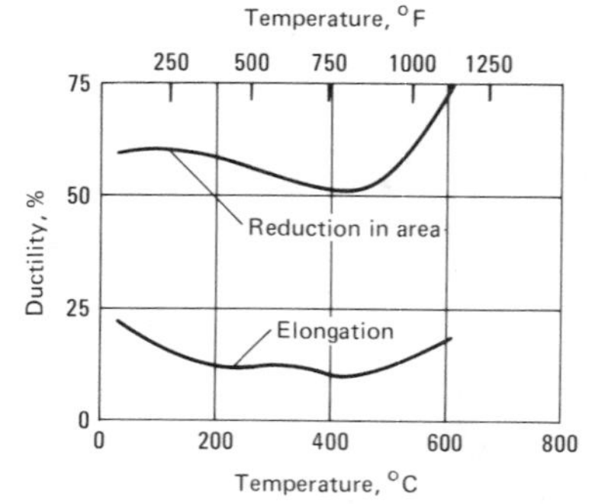

at 20 to 100 °C (68 to 212 °F), 12.6 μm/m·K (7.0 μin./in.· °F) at 20 to 540 °C (68 to 1000 °F)
Specific heat. 460 J/kg·K (0.11 Btu/lb·°F) at 20 °C (70 °F)
Thermal conductivity. 25.1 W/m·K (14.5 Btu/ft·h·°F) at 100 °C (212 °F); 28.9 W/m·K (16.7 Btu/ft·h·°F) at 540 °C (1000 °F)

Electrical Properties

Electrical resistivity. 780 nΩ·m at 20 °C (70 °F)

Magnetic Properties

Magnetic permeability. Ferromagnetic

Chemical Properties

General corrosion behavior. Similar in corrosion resistance to CA-15, but the addition of nickel and molybdenum improves its resistance to attack by seawater

Fabrication Characteristics

Machinability. Good. Carbide or high speed steel cutters may be used, and it is advisable that the tool be kept continually entering into the metal. Slow feeds, deep cuts and powerful, rigid machines are necessary for best results. Typical conditions using high speed steel cutters: roughing speed, 13 to 15 m/min (40 to 50 ft/min) with a feed of 0.3 to 0.8 mm/rev (0.010 to 0.030 in./rev); finishing speed, 24 to 30 m/min (80 to 100 ft/min) with a feed of 0.1 to 0.3 mm/rev (0.005 to 0.010 in./rev)
Weldability. Shielded metal-arc, inert-gas arc and electroslag methods: good
Heat treatment. To obtain maximum softness, castings may be annealed at 790 to 815 °C (1450 to 1500 °F) and slowly furnace cooled. The alloy is hardened by heating to 1040 to 1060 °C (1900 to 1950 °F) and cooling in oil or air. After hardening, castings should be tempered as soon as possible at 315 °C (600 °F) or in the range 595 to 620 °C (1100 to 1150 °F). Tempering in the vicinity of 480 °C (900 °F) should be avoided because lowered impact strength will result. Some retransformation to austenite may occur if tempering temperatures above 650 °C (1200 °F) are used and, upon cooling, the microstructure may then contain untempered martensite. Highest strength and hardness is obtained by tempering at 315 °C (600 °F).

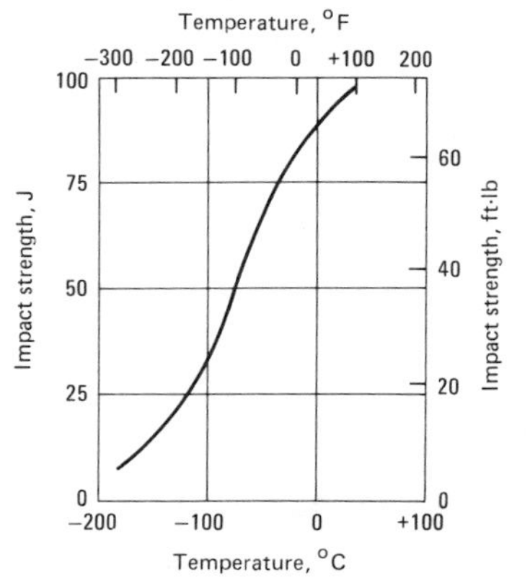

Fig. 2 Charpy V-notch impact strength of CA-6NM

CA-15, CA-15M
12Cr-0.1C

Commercial Name

UNS number. J91150

Specifications

AMS. 5351
ASTM. A217, grade CA-15; A743 grades CA-15 and CA-15M
SAE. 60410
Government. MIL-S-16993A, class I and II

Chemical Composition

Composition limits. CA-15: 0.15 max C; 1.00 max Mn; 1.50 max Si; 0.04 max P; 0.04 max S; 11.5 to 14.0 Cr; 1.0 max Ni; 0.5 max Mo; rem Fe. CA-15M: 0.15 max C; 1.00 max Mn; 0.65 max Si; 0.04 max P, 0.04 max S; 11.5 to 14.0 Cr; 1.0 max Ni; 0.15 to 1.0 Mo; rem Fe

Applications

Typical uses. Burning torch gas distributor heads, bushings and liners, catalyst trays, fittings, furnace burner tips and pilot cones, gears, hydrafiner parts, impellers, jet engine components, letters, placques, pump casings, railings, shafts, ship propellers, skimmer ladles, stuffing boxes, turbine blades, valve bodies, valve trim

Precautions in use. Tempering in the vicinity of 480 °C (900 °F) will result in lowered impact strength.

Mechanical Properties

Tensile properties. See Table 1.
Hardness. See Table 1.
Elastic modulus. Tension, 200 GPa (29 × 10^6 psi)
Impact strength. Charpy keyhole, see Table 1.

Structure

Microstructure. Ferritic structure can be varied by heat treatment so that a wide range of hardness (144 to about 400 HB) and other mechanical properties may be obtained. In the annealed condition the ferrite matrix contains agglomerated carbide particles. Depending on the temperature of heat treatment, the hardened alloy exhibits a pearlitic to martensitic structure that results in a tough, erosion-resisting material.

Mass Characteristics

Density. 7.61 Mg/m³ (0.275 lb/in.³) at 20 °C (70 °F)
Patternmaker's shrinkage. 21 mm/m (1/4 in./ft)

Thermal Properties

Coefficient of thermal expansion:

Temperature °C	°F	Mean coefficient μm/m·K	μin./in.·°F
20–100	70– 212	9.9	5.5
20–540	70–1000	11.5	6.4
20–705	70–1300	12.1	6.7

Specific heat. 460 J/kg·K (0.11 Btu/lb·°F) at 20 °C (70 °F)
Thermal conductivity. 25.1 W/m·K (14.5 Btu/ft·h·°F) at 100 °C (212 ° F); 28.9 W/m·K (16.7 Btu/ft·h·°F) at 540 °C (1000 °F)
Melting temperature. 1510 °C (2750 °F)

Electrical Properties

Electrical resistivity. 780 nΩ·m at 20 °C (70 °F)

Magnetic Properties

Magnetic permeability. Ferromagnetic

Chemical Properties

General corrosion behavior. Excellent corrosion resistance

Table 1 Typical mechanical properties of CA-15 at room temperature

Tempered at (a): °C	°F	Tensile strength MPa	ksi	Yield strength, 0.2% offset MPa	ksi	Elongation in 50 mm or 2 in., %	Reduction in area,%	Hardness, HB	Impact strength, Charpy keyhole J	ft·lb
315	600	1380	200	1035	150	7	25	390	20	15
595	1100	930	135	795	115	17	55	260	14	10
650	1200	795	115	690	100	22	55	225	27	20
790	1450	690	100	515	75	30	60	185	47	35

(a) Air cooled from 980 °C (1800 °F).

Resistance to specific corroding agents. Good resistance to atmospheric corrosion and excellent resistance to corrosion or staining by organic media in relatively mild service

Fabrication Characteristics

Machinability. Fair to good. Selenium may be added to composition to improve machinability. For typical conditions, see CA-6NM.

Weldability. Shielded metal-arc, inert-gas arc and electroslag methods: fair to good. Shielded metal-arc is most frequently used.

Annealing temperature. 845 to 900 °C (1550 to 1650 °F)

Heat treatment. For maximum softness, castings may be annealed at 790 °C (1450 °F) min, usually 845 to 900 °C (1550 to 1650 °F) and slowly cooled. The alloy is hardened by heating to 980 to 1010 °C (1800 to 1850 °F) and cooling in oil or air. After hardening, castings should be tempered as soon as possible at 315 °C (600 °F) max or in the range 595 to 815 °C (1100 to 1500 °F). Tempering in the vicinity of 480 °C (900 °F) should be avoided because low impact strength will result.

Highest strength and hardness are obtained by tempering at 315 °C (600 °F) or below, and the alloy has best corrosion resistance in this fully hardened condition. When tempered above 595 °C (1100 °F) castings have improved ductility and impact strength but corrosion resistance is somewhat decreased. Poorest corrosion resistance results from tempering around 595 °C (1100 °F).

CA-40
12Cr-0.3C

Commercial Name

UNS number. J91153

Specifications

ASTM. A743
SAE. 60420

Chemical Composition

Composition limits. 0.20 to 0.40 C; 1.00 max Mn; 1.50 max Si; 0.04 max P; 0.04 max S; 11.5 to 14.0 Cr; 1.0 max Ni; 0.5 max Mo; rem Fe

Applications

Typical uses. Various castings used in food processing, glass, oil refining, power plants, pulp and papermaking
Precautions in use. Tempering in the vicinity of 480 °C (900 °F) will lower impact strength.

Mechanical Properties

Tensile properties. See Table 2.
Hardness. See Table 2.
Elastic modulus. Tension, 200 GPa (29 × 10^6 psi)
Impact strength. Charpy keyhole, see Table 2.

Structure

Microstructure. Ferritic structure that can be varied by heat treatment so that a wide range of hardness (up to about 500 HB) and other mechanical properties may be obtained. In the annealed condition the ferrite matrix contains agglomerated carbide particles. Depending on the temperature of heat treatment, the hardened alloy exhibits a pearlitic to martensitic structure that results in a tough, erosion-resisting material.

Mass Characteristics

Density. 7.61 Mg/m³ (0.275 lb/in.³) at 20 °C (70 °F)
Patternmaker's shrinkage. 21 mm/m (1/4 in./ft)

Thermal Properties

Melting temperature. 1495 °C (2725 °F)
Coefficient of thermal expansion:

Temperature °C	°F	Mean coefficient µm/m·K	µin./in.·°F
20–100	70–212	9.9	5.5
20–540	70–1000	11.5	6.4
20–705	70–1300	12.1	6.7

Specific heat. 460 J/kg·K (0.11 Btu/lb·°F) at 20 °C (70 °F)
Thermal conductivity. 25.1 W/m·K (14.5 Btu/ft·h·°F) at 100 °C (212 °F); 28.9 W/m·K (16.7 Btu/ft·h·°F) at 540 °C (1000 °F)

Electrical Properties

Electrical resistivity. 760 nΩ·m at 20 °C (70 °F)

Magnetic Properties

Magnetic permeability. Ferromagnetic

Chemical Properties

General corrosion behavior. This alloy has excellent corrosion resistance.
Resistance to specific corroding agents. Same as CA-15.

Fabrication Characteristics

Machinability. Carbide or high speed steel cutters may be used, and it is advisable that the tool be kept continually entering into the metal. Slow feeds, deep cuts, and powerful,

Table 2 Typical mechanical properties of CA-40 at room temperature

Tempered at (a): °C	°F	Tensile strength MPa	ksi	Yield strength, 0.2% offset MPa	ksi	Elongation, in 50 mm or 2 in.,%	Hardness, HB	Impact strength, Charpy keyhole J	ft·lb
315	600	1517	220	1138	165	1	470	1.4	1
595	1100	1034	150	862	125	10	310	2.7	2
650	1200	965	140	780	113	14	267	5.4	4
760	1400	758	110	462	67	18	212	4.1	3

(a) After air cooling from 980 °C (1800 °F).

rigid machines are necessary for best results. Typical conditions using high speed steel cutters: roughing speed, 7.5 to 10 m/min (25 to 35 ft/min) with a feed of 0.8 to 1.0 mm/rev (0.030 to 0.040 in./rev); finishing speed, 15 to 21 m/min (50 to 70 ft/min) with a feed of 0.4 to 0.5 mm/rev (0.015 to 0.020 in./rev)

Weldability. Can be welded by shielded metal-arc and by inert-gas arc methods. Shielded metal-arc welding is most frequently used.

Annealing temperature. 845 to 900 °C (1550 to 1650 °F)

Heat treatment. To obtain maximum softness, castings may be annealed to 790 °C (1450 °F) min, usually 845 to 900 °C (1550 to 1650 °F) and furnace cooled. The alloy is hardened by heating to 980 to 1010 °C (1800 to 1850 °F) and cooling in oil or air. After hardening, castings should be tempered as soon as possible at 315 °C (600 °F) max or at 595 to 815 °C (1100 to 1500 °F). Tempering in the vicinity of 480 °C (900 °F) should be avoided. Highest strength and hardness is obtained by tempering at 315 °C (600 °F) or below, and the alloy has best corrosion resistance in this fully hardened condition. When tempered above 595 °C (1100 °F) castings have improved ductility and impact strength, but corrosion resistance is somewhat decreased. Poorest corrosion resistance results from tempering around 595 °C (1100 °F).

CB-7Cu
16Cr-4Ni-3Cu

Commercial Name

UNS number.

Specifications

AMS. 5398

ASTM. A747, grades CB-7Cu-1 and CB-7Cu-2

Chemical Composition

Composition limits. CB-7Cu-1: 0.07 max C; 0.70 max Mn; 1.00 max Si; 0.035 max P; 0.03 max S; 15.50 to 17.70 Cr; 3.60 to 4.60 Ni; 2.50 to 3.20 Cu; 0.20 to 0.35 Nb (a); 0.05 max N; rem Fe. CB-7Cu-2: 0.07 max C; 0.70 max Mn; 1.00 max Si; 0.035 max P; 0.03 max S; 14.00 to 15.50 Cr; 4.50 to 5.50 Ni; 2.50 to 3.20 Cu; 0.20 to 0.35 Nb (a); 0.05 max N; rem Fe

(a) Niobium not added when alloy is to be hardened by 480 °C (900 °F) aging treatment.

Applications

Typical uses. Various castings where machining is involved; in aerospace, aircraft, chemical, food processing, gas turbine, marine, petrochemical, pulp and paper industries

Precautions in use. Castings intended to be welded should have the copper content below 3.00% to avoid underbead cracking

Mechanical Properties

Tensile properties. See Table 3 and Fig. 3.

Hardness. See Table 3.

Elastic modulus. Tension, 196.5 GPa (28.5 × 10^6 psi); see Fig. 4 for effect of temperature on modulus.

Impact strength. Charpy V-notch, see Table 3.

Structure

Microstructure. In the solution annealed state, the microstructure consists of a soft martensite formed upon cooling the casting from the solution temperature at which the original as-cast structure was transformed to austenite containing dissolved copper. This copper remains in the martensite in super-saturated solution but, if the alloy is later reheated to the range 480 to 620 °C (900 to 1150 °F), it precipitates submicroscopically and substantially increases the strength and hardness of the casting.

Fig. 3 Short-time elevated-temperature tensile properties of CB-7Cu aged at 480 °C (900 °F)

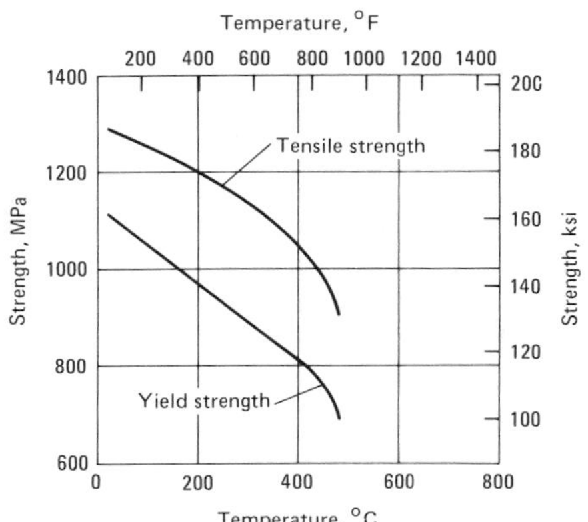

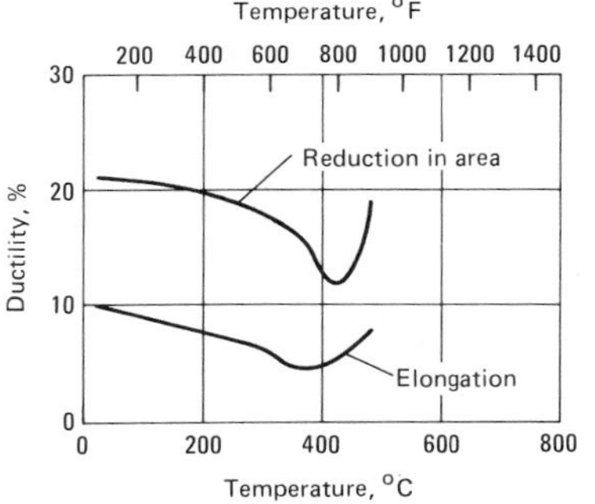

Mass Characteristics

Density. 7.75 Mg/m³ (0.28 lb/in.³) at 20 °C (70 °F)

Patternmaker's shrinkage. 21 mm/m (¼ in./ft)

Thermal Properties

Melting temperature. 1510 °C (2750 °F)

Coefficient of thermal expansion:

Table 3 Typical mechanical properties of CB-7 Cu at room temperature

Aged at (a): °C	°F	Tensile strength MPa	ksi	Yield strength, 0.2% offset MPa	ksi	Elongation, in 50 mm or 2 in.,%	Reduction in area,%	Hardness, HB	Impact strength, Charpy V-notch J	ft·lb
480	900	1290	187	1110	161	10	21	412	9.5	7
495	925	1303	189	1138	165	11	26	412	16.3	12
550	1025	1138	165	1089	158	14	35	350	29.8	22
580	1075	1069	155	972	141	14	35	319	36.6	27
595	1100	1000	145	910	132	15	39	315	40.7	30
620	1150	965	140	827	120	16	42	307	50.2	37

(a) After air cooling from 1050 °C (1925 °F). For elevated temperatures, see Fig.4.

Fig. 4 Variation of modulus of elasticity with temperature for aged CB-7Cu

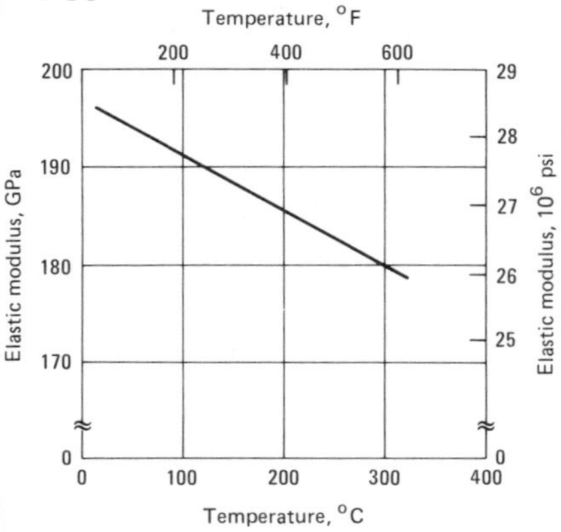

Temperature range °C	°F	Mean coefficient µm/m·K	µin./in.·°F
Aged at 480 °C (900 °F)			
20–95	70–200	10.8	6.0
20–205	70–400	11.0	6.1
20–315	70–600	11.3	6.3
20–425	70–800	11.7	6.5
Aged at 595 °C (1100 °F)			
20–95	70–200	11.9	6.6
20–205	70–400	12.4	6.9
20–315	70–600	12.8	7.1
20–425	70–800	13.0	7.2

Specific heat. 460 J/kg·K (0.11 Btu/lb·°F) at 20 °C (70 °F)

Thermal conductivity:

Temperature °C	°F	Conductivity W/m·K	Btu/ft·h·°F
100	212	17.1	9.9
260	500	19.6	11.3
460	860	22.5	13.0
480	900	22.7	13.1

Electrical Properties

Electrical resistivity. 770 nΩ·m at 20 °C (70 °F), for material aged at 480 °C (900 °F)

Magnetic Properties

Magnetic permeability. Ferromagnetic

Chemical Properties

General corrosion behavior. Good resistance to atmospheric corrosion and many aqueous corrodents including seawater, food products, and paper-mill liquors

Fabrication Characteristics

Machinability. Carbide and high speed steel cutters may be used, and it is advisable that the tool be kept continuously entering into the metal to avoid work hardening the surface. Slow feeds, deep cuts and powerful, rigid machines are necessary for best results. Typical conditions using high speed steel cutters: roughing speed, 7.5 to 21 m/min (25 to 70 ft/min) with a feed of 0.5 to 0.6 mm/rev (0.020 to 0.025 in./rev); finishing speed, 15 to 42 m/min (50 to 140 ft/min) with a feed of 0.1 to 0.3 mm/rev (0.005 to 0.010 in./rev).

Weldability. Can be welded by shielded metal-arc and inert-gas arc methods. Castings having a copper content above 3.00% may suffer underbead cracking when welded. Lime coated electrodes of similar composition should be used for welding by shielded metal-arc process.

Heat treatment. Castings are supplied in either the solution annealed or hardened condition. Solution annealing consists of heating the castings to 1050 °C (1925 °F) ± 30 °C (50 °F), holding them for 30 minutes per 25 mm (1 in.) of thickness in the heaviest section (30 minutes minimum), and then cooling them to below 30 °C (90 °F). Castings are intended to be used only in the precipitation hardened condition, but may be supplied in the solution annealed condition if machining is to be done prior to hardening. Precipitation hardening involves heating the solution annealed castings (*a*) at 480 °C (900 °F) for 1 h, (*b*) at 495 °C (925 °F) for 1.5 h or (*c*) at 550 °C (1025 °F), 580 °C (1075 °F), 595 °C (1100 °F) or 620 °C (1150 °F) for 4 h. After the required time at temperature the castings are air cooled.

CB-30
20Cr

Commercial Name

UNS number. J91803

Specifications

ASTM. A743
SAE. 60442

Chemical Composition

Composition limits. 0.30 max C; 1.00 max Mn; 1.50 max Si; 0.04 max P; 0.04 max S; 18.0 to 21.0 Cr; 2.0 max Ni; rem Fe

Applications

Typical uses. Various castings used in chemical processing, food processing, heat treating, oil refining, ore roasting and power plant equipment

Mechanical Properties

Tensile properties. (a) Tensile strength, 655 MPa (95 ksi), yield strength, 0.2% offset, 415 MPa (60 ksi); elongation 15% in 50 mm or 2 in.

Hardness. (a) 195 HB
Elastic modulus. Tension, 200 GPa (29×10^6 psi)
Impact strength. (a) Charpy keyhole, 2.7 J (2 ft·lb)

(a) Annealed at 790 °C (1450 °F), furnace cooled to 540 °C (1000 °F), then air cooled. Values may vary considerably depending on composition and heat treatment.

Structure

Microstructure. Ferritic, and even at the higher carbon levels only a small amount of ferrite transforms at elevated temperature to austenite for subsequent change to martensite upon rapid cooling. In contrast to the hardenable CA-15 grade, the CB-30 type is virtually nonhardenable.

Mass Characteristics

Density. 7.53 Mg/m³ (0.272 lb/in.³) at 20 °C (70 °F)
Patternmaker's shrinkage. 21 mm/m (1/4 in./ft)

Thermal Properties

Melting temperature. 1495 °C (2725 °F)
Coefficient of thermal conductivity. Linear:

Temperature °C	Temperature °F	Mean coefficient µm/m·K	Mean coefficient µin./in.·°F
20–100	70– 212	10.3	5.7
20–540	70–1000	11.7	6.5
20–705	70–1300	12.1	6.7

Specific heat. 460 J/kg·K (0.11 Btu/lb·°F) at 20 °C (70 °F)
Thermal conductivity. 22.2 W/m·K (12.8 Btu/ft·h·°F) at 100 °C (212 °F); 25.1 W/m·K (12.8 Btu/ft ·h·°F) at 540 °C (1000 °F)

Electrical Properties

Electrical resistivity. 760 nΩ·m at 20 °C (70 °F)

Magnetic Properties

Magnetic permeability. Ferromagnetic

Chemical Properties

Resistance to specific corroding agents. Excellent resistance to corrosion by nitric acid, alkaline solutions and many organic chemicals (see Fig. 5).

Fabrication Characteristics

Machinability. High speed steel or carbide cutters may be used, and it is advisable that the tool be kept continuously entering into the metal. Slow feeds, deep cuts and powerful, rigid machines are necessary for best results. Typical conditions using high speed steel cutters: roughing speed, 12 to 15 m/min (40 to 50 ft/min) with a feed of 0.5 to 0.8 mm/rev (0.020 to 0.030 in./rev); finishing speed, 24 to 30 m/min (80 to 100 ft/min) with a feed of 0.3 to 0.4 mm/rev (0.010 to 0.015 in./rev)
Weldability. Can be welded by shielded metal-arc, inert-gas arc and electroslag methods. Shielded metal-arc is most frequently used.
Annealing temperature. 790 °C (1450 °F) min
Heat treatment. Castings are normally used in the annealed condition. Annealing consists of heating castings to 790 °C (1450 °F) min, furnace cooling to about 540 °C (1000 °F), then air cooling.

Fig. 5 Isocorrosion chart for CB-30 in nitric acid

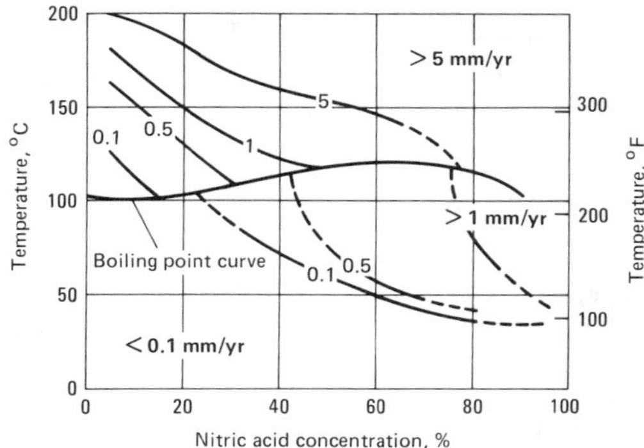

Castings were annealed at 790 °C (1450 °F), furnace cooled to 540 °C (1000 °F), and then air cooled to room temperature.

CC-50
28Cr

Commercial Name

UNS number. J92616

Specifications

ASTM. A743
SAE. 60446

Chemical Composition

Composition limits. 0.50 max C; 1.00 max Mn; 1.50 max Si; 0.04 max P; 0.04 max S; 26.0 to 30.0 Cr; 4.0 max Ni; rem Fe

Applications

Typical uses. Various castings used in chemical manufacturing, mining, pulp and papermaking, and synthetic fiber manufacturing equipment

Mechanical Properties

Tensile properties. See Table 4.
Hardness. See Table 4.
Elastic modulus. Tension, 200 GPa (29×10^6 psi)
Impact strength. See Table 4.

Structure

Microstructure. Ferritic at all temperatures; hence, it cannot be hardened by heat treatment

Mass Characteristics

Density. 7.53 Mg/m³ (0.272 lb/in.³) at 20 °C (70 °F)
Patternmaker's shrinkage. 18 mm/m (7/32 in./ft)

Thermal Properties

Melting temperature. 1495 °C (2725 °F)
Coefficient of thermal expansion. Linear: 10.6 µm/m·K (5.9 µin./in.· °F) at 20 to 100 °C (68 to 212 °F); 11.5 µm/m·K (6.4 µin./in.·°F) at 20 to 540 °C (68 to 1000 °F)
Specific heat. 500 J/kg·K (0.12 Btu/lb· °F) at 20 °C (70 °F)
Thermal conductivity. 21.8 W/m·K (12.6 Btu/ft·h·°F) at 100 °C (212 °F); 40.0 W/m·K (17.9 Btu/ft·h·°F) at 540 °C (1000 °F)

Table 4 Typical mechanical properties of CC-50 at room temperature

Condition	Tensile strength MPa	Tensile strength ksi	Yield strength, 0.2% offset MPa	Yield strength, 0.2% offset ksi	Elongation, in 50 mm or 2 in.,%	Hardness, HB	Impact strength, Izod V-notch J	Impact strength, Izod V-notch ft·lb
As cast (a)	483	70	448	65	2	212	2.7	2
As cast (b)	655	95	414	60	15	193	61	45
Air cooled from 1040 °C (1900 °F) (b)	669	97	448	65	18	210	...	...

(a) Under 1.0% Ni. (b) Over 2.0% Ni with 0.15% min N.

Electrical Properties

Electrical resistivity. 770 nΩ·m at 20 °C (70 °F)

Magnetic Properties

Magnetic permeability. Ferromagnetic

Chemical Properties

Resistance to specific corroding agents. Excellent resistance to dilute sulfuric acid in mine waters, mixed nitric and sulfuric acids, and all oxidizing acids

Fabrication Characteristics

Machinability. Readily machinable using high speed steel or carbide cutters, and it is advisable that the tool be kept continuously entering the metal. Slow feeds, deep cuts and powerful, rigid machines are necessary for best results. Typical conditions using high speed steel cutters: roughing speed, 12 to 15 m/min (40 to 50 ft/min) with a feed of 0.6 to 0.9 mm/rev (0.025 to 0.035 in./rev); finishing speed, 24 to 30 m/min (80 to 100 ft/min) with a feed of 0.3 to 0.4 mm/rev (0.010 to 0.015 in./rev)

Weldability. Can be welded by shielded metal-arc, inert-gas arc and electroslag methods

Annealing temperature. 790 °C (1450 °F) min

Heat treatment. Castings are normally used in the annealed condition, achieved by heating to 790 °C (1450 °F) min, followed by air or furnace cooling.

Fig. 6 Tensile properties of CD-4MCu at 315 °C (600 °F) after prolonged exposure at test temperature

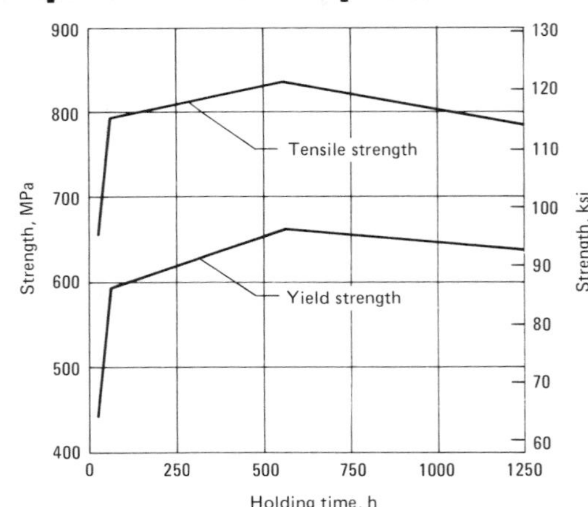

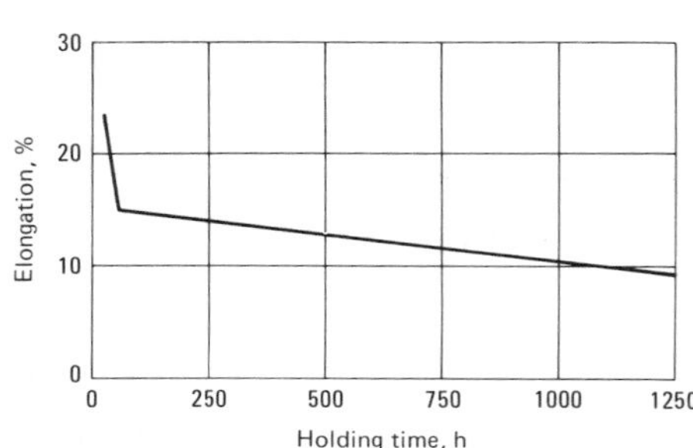

Properties are shown for a single heat whose ductility at room temperature was on the high side of the normal range for CD-4MCu.

CD-4MCu 26Cr-5Ni-2Mo-3Cu

Commercial Name

UNS number.

Specifications

ASTM. A351, A743 and A744

Chemical Composition

Composition limits. 0.04 max C; 1.00 max Mn; 1.00 max Si; 0.04 max P; 0.04 max S; 25.0 to 26.5 Cr; 4.75 to 6.00 Ni; 1.75 to 2.25 Mo; 2.75 to 3.25 Cu; rem Fe

Applications

Typical uses. Various castings used in the chemical processing, marine, municipal water supply, paint, petroleum refining, power plant, pulp and paper, soap manufacturing, textile and transportation industries

Precautions in use. Recommended for use only in the solution annealed condition

Mechanical Properties

Tensile properties. (a) At room temperature: tensile strength, 745 MPa (108 ksi); yield strength, 0.2% offset, 562 MPa (81.5 ksi); elongation, 35% in 50 mm or 2 in. At 315 °C (600 °F): tensile strength, 635 MPa (92 ksi); yield strength, 420 MPa (61 ksi); elongation, 25% in 50 mm or 2 in. See also Fig. 6.

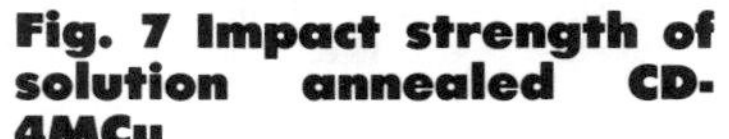

Fig. 7 Impact strength of solution annealed CD-4MCu

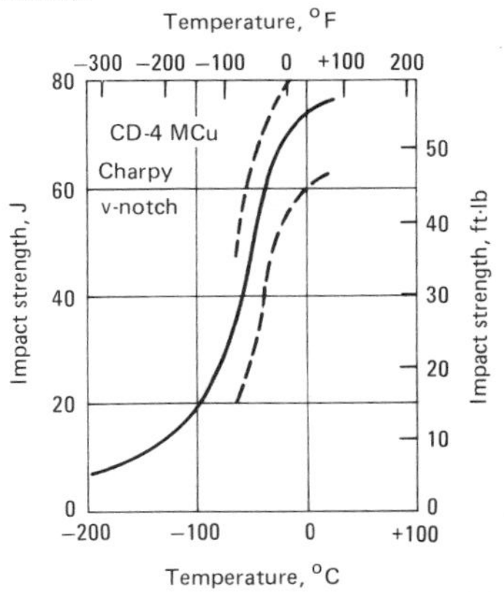

Castings were annealed at 1120 °C (2050 °F), furnace cooled to 1035 °C (1900 °F) and water quenched. Dashed lines indicate statistical spread of individual values.

Fig. 8 Isocorrosion chart for CD-4MCu in nitric acid

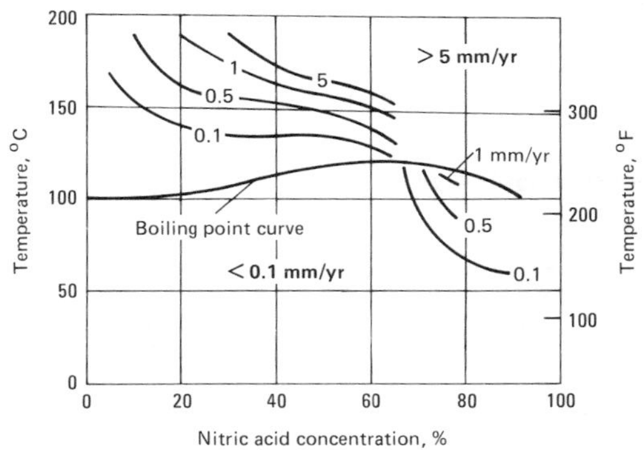

Material was solution annealed at 1120 °C (2050 °F) and water quenched.

Fig. 9 Isocorrosion chart for CD-4MCu in sulfuric acid

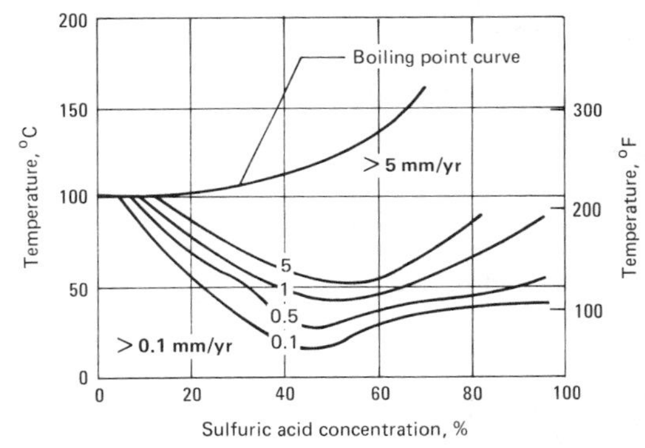

Material was solution annealed at 1120 °C (2050 °F) and water quenched.

Hardness. (a) 253 HB
Elastic modulus. Tension, 200 GPa (29×10^6 psi)
Impact strength. See Fig. 7.

(a) Typical values when solution annealed at 1120 °C (2050 °F) min and water quenched from 1040 °C (1900 °F).

Structure

Microstructure. As cast: two-phase, ferrite plus austenite. Low carbon content results in only small amounts of chromium carbide, distributed throughout the matrix, which must be dissolved by suitable heat treatment for maximum corrosion resistance. Increasing amounts of austenite form as the alloy cools from the carbide solution temperature of 1120 °C (2050 °F), and rapid cooling tends to cause cracking of heavy sections. Hence, it is recommended that castings be slowly cooled to 1040 °C (1900 °F) before quenching. If reheated, precipitates are formed that increase tensile strength and hardness. The temperature and holding time for precipitation hardening and the temperature from which the alloy is quenched have a complex relationship that has a critical bearing on the resultant properties of the metal.

Mass Characteristics

Density. 7.75 Mg/m^3 (0.280 $lb/in.^3$) at 20 °C (70 °F)

Patternmaker's shrinkage. 21 mm/m (1/4 in./ft)

Thermal Properties

Melting temperature. 1480 °C (2700 °F)
Coefficient of thermal expansion. Linear:

Temperature °C	°F	Mean coefficient µm/m·K	µin./in.·°F
20–100	70–212	11.3	6.3
20–315	70–600	11.8	6.6
20–540	70–1000	12.4	6.9
20–650	70–1200	12.6	7.0

Specific heat. 460 J/kg·K (0.11 Btu/lb·°F) at 20 °C (70 °F)

Thermal conductivity. 15.2 W/m·K (8.8 Btu/ft·h·°F) at 100 °C (212 °F); 23.2 W/m·K (13.4 Btu/ft·h·°F) at 540 °C (1000 °F)

Electrical Properties

Electrical resistivity. 750 nΩ·m at 20 °C (70 °F)

Magnetic Properties

Magnetic permeability. Ferromagnetic

Chemical Properties

General corrosion behavior. Superior to types CF-8 and CF-8M in many media.
Resistance to specific corroding agents. Excellent resistance to environments involving abrasion or

erosion-corrosion. Exceptional resistance to stress-corrosion cracking in chloride-containing solutions or vapors and is usefully employed in handling both oxidizing and reducing corrodents. See Fig. 8 and 9 for corrosion rates in nitric and sulfuric acids.

Fabrication Characteristics

Machinability. Good, using high speed steel or carbide cutters. It is important to keep the tool continuously entering into the metal, and slow speeds, deep cuts and powerful, rigid machines are necessary for best results. Typical conditions using high speed steel cutters: roughing speed, 12 to 15 m/min (40 to 50 ft/min) with a feed of 0.5 to 0.6 mm/rev (0.020 to 0.025 in./rev); finishing speed, 24 to 30 m/min (80 to 100 ft/min) with a feed of 0.1 to 0.25 mm/rev (0.005 to 0.010 in./rev).
Weldability. Shielded metal-arc and inert-gas arc methods: good
Heat treatment. For complete solution of carbides and maximum corrosion resistance, castings should be heated until sections are uniformly at 1120 °C (2050 °F) min, held for three hours and then slowly cooled to the range 1040 to 955 °C (1900 to 1750 °F), held there for ½ h followed by quenching in water, oil or air.

CE-30
27Cr-9Ni

Commercial Name

UNS number. J93423

Specifications

ASTM. A743
SAE. 60312

Chemical Composition

Composition limits. 0.30 max C; 1.50 max Mn; 2.0 max Si; 0.04 max P; 0.04 max S; 26.0 to 30.0 Cr; 8.0 to 11.0 Ni; rem Fe

Applications

Typical uses. Various castings used in chemical processing, mining, oil refining, pulp and papermaking, and synthetic fiber manufacturing

Mechanical Properties

Tensile properties. As cast: tensile strength, 655 MPa (95 ksi) yield strength, 0.2% offset, 310 MPa (45 ksi); elongation 15% in 50 mm or 2 in. Water quenched from 1095 to 1120 °C (2000 to 2050 °F): tensile strength, 670 MPa (97 ksi); yield strength 0.2% offset, 434 MPa (63 ksi); elongation, 18% in 50 mm or 2 in.
Hardness. 190 HB
Elastic modulus. Tension, 172 GPa (25 × 10^6 psi)
Impact strength. Charpy keyhole: as cast, 13.6 J (10 ft·lb); water quenched, 9.5 J (7 ft·lb)

Structure

Microstructure. As cast: two-phase austenite plus ferrite containing carbides. The high chromium content and duplex structure permit a fairly high carbon content without serious loss of corrosion resistance when the alloy is exposed to temperatures in the precipitation range 425 to 870 °C (800 to 1600 °F). A modification having the composition balanced to obtain ferrite contents of 5 to 20% is being used in oil refinery applications at temperatures around 440 °C (825 °F). Designated CE-30A, it successfully resists stress corrosion cracking in an environment containing polythionic acid and some chlorides.

Mass Characteristics

Density. 7.67 Mg/m³ (0.277 lb/in.³) at 20 °C (70 °F)
Patternmaker's shrinkage. 26 mm/m (5/16 in./ft)

Thermal Properties

Melting temperature. 1454 °C (2650 °F)
Coefficient of thermal expansion. Linear:

Temperature °C	°F	Mean coefficient µm/m·K	µin./in.·°F
20–540	70–1000	17.3	9.6
20–650	70–1200	17.8	9.9
20–760	70–1400	18.4	10.2
20–870	70–1600	18.9	10.5

Specific heat. 590 J/kg·K (0.14 Btu/lb·°F) at 20 °C (70 °F)
Thermal conductivity:

Temperature °C	°F	Conductivity W/m·K	Btu/ft·h·°F
100	212	14.7	8.5
315	600	18.2	10.5
540	1000	21.5	12.4
650	1200	23.4	13.5

Electrical Properties

Electrical resistivity. 850 nΩ·m at 20 °C (70 °F)

Magnetic Properties

Magnetic permeability. Over 1.5

Chemical Properties

Resistance to specific corroding agents. Resistant to sulfurous acid and sulfites in the paper industry, dilute sulfuric acid with sulfurous acid, and sulfuric with nitric acid.

Fabrication Characteristics

Machinability. Good, using high speed steel or carbide cutters. It is important that the tool be kept continuously entering the metal; slow feeds, deep cuts and powerful, rigid machines are necessary for best results. Typical conditions using high speed steel cutters: roughing speed, 9.1 to 12.2 m/min (30 to 40 ft/min) with a feed of 0.5 to 0.6 mm/rev (0.020 to 0.025 in./rev); finishing speed, 18.3 to 24.4 m/min (60 to 80 ft/min) with a feed of 0.1 to 0.3 mm/rev (0.005 to 0.010 in./rev).
Weldability. Shielded metal-arc, inert-gas arc and electroslag methods: good. Shielded metal-arc is most frequently used.
Heat treatment. Castings often used as-cast. For maximum corrosion resistance and improved ductility, castings should be heated to 1065 to 1120 °C (1950 to 2050 °F) and quenched in water, oil or air.

CF-3, CF-3A
19Cr-10Ni

Commercial Name

UNS number. J92700

Specifications

ASTM. A743 and A744, grade CF-3; A351, grades CF-3 and CF-3A

Chemical Composition

Composition limits. 0.03 max C; 1.50 max Mn; 2.00 max Si; 0.04 max P; 0.04 max S; 17.0 to 21.0 Cr; 8.0 to 12.0 Ni; rem Fe

Applications

Typical uses. Widely used in riverboat service; also used in the beverage, brewery, distillery, food, heavy water manufacturing, marine, nuclear power, petroleum, pipe line, soap and detergent industries. CF-3A is used extensively in nuclear power plant construction.

Precautions in use. Because of the thermal instability of CF-3A, it is not recommended for service at temperatures over 425 °C (800 °F)

Mechanical Properties (a)

Tensile properties. CF-3: tensile strength, 530 MPa (77 ksi); yield strength, 0.2% offset, 248 MPa (36 ksi); elongation, 60% in 50 mm or 2 in. CF-3A: tensile strength, 600 MPa (87 ksi); yield strength, 0.2% offset, 290 MPa (42 ksi); elongation, 50% in 50 mm or 2 in.

Impact strength. Charpy V-notch, CF-3: 149 J (110 ft·lb); CF-3A: 136 J (100 ft·lb)

(a) Water quenched from above 1040 °C (1900 °F).

Structure

Microstructure. As cast: austenitic structure containing 10 to 20% ferrite in the form of discontinuous pools but with virtually no chromium carbides

Mass Characteristics

Density. 7.75 Mg/m^3 (0.280 lb/in.^{3}) at 20 °C (70 °F)

Patternmaker's shrinkage. 26 mm/m (5/16 in./ft)

Thermal Properties

Melting temperature. 1455 °C (2650 °F)

Coefficient of thermal expansion. Linear, 16.2 μm/m·K (9.0 μin./in.·°F) at 20 to 540 °C (68 to 1000 °F)

Specific heat. 500 J/kg·K (0.12 Btu/lb·°F) at 20 °C (70 °F)

Thermal conductivity. 15.9 W/m·K (9.2 Btu/ft·h·°F) at 100 °C (212 °F) 20.9 W/m·K (12.1 Btu/ft·h·°F) at 540 °C (1000 °F)

Electrical Properties

Electrical resistivity. 762 nΩ·m at 20 °C (70 °F)

Fig. 10 Isocorrosion chart for solution annealed, quenched and sensitized CF-3 in nitric acid

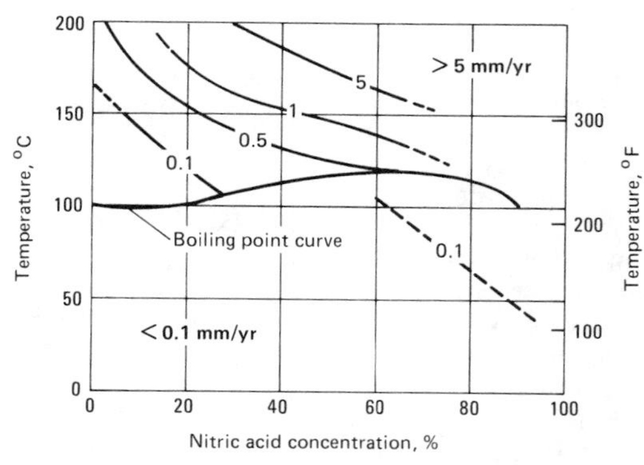

Magnetic Properties

Magnetic permeability. 1.20 to 3.00

Chemical Properties

General corrosion behavior. Equal to or better than type CF-8, it is particularly useful where postweld heat treatment is inconvenient or impossible with no impairment of corrosion resistance.

Resistance to specific corroding agents. For corrosion rate in nitric acid, see Fig. 10.

Fabrication Characteristics

Machinability. Good, using high speed steel and carbide cutters. It is important to keep tool continuously entering the metal to avoid work-hardening the surface. Slow feeds, deep cuts, and powerful, rigid machines are necessary for best results. Typical conditions using high speed steel cutters: roughing speed, 7.5 to 10 m/min (25 to 35 ft/min) with a feed of 0.5 to 0.6 mm/rev (0.020 to 0.025 in./rev); finishing speed, 15 to 21 m/min (50 to 70 ft/min) with a feed of 0.1 to 0.3 mm/rev (0.005 to 0.010 in./rev)

Weldability. Shielded metal-arc, inert-gas arc and electroslag methods: good. Shielded metal-arc is most frequently used.

Heat treatment. Castings should be heated in the range 1040 to 1120 °C (1900 to 2050 °F) and then quenched in water, oil or air to ensure complete solution of carbides.

CF-3M, CF-3MA 19Cr-11Ni-2.5Mo

Commercial Names

UNS number. J92800

Specifications

ASTM. A743 and A744, grade CF-3M; A351, grades CF-3M and CF-3MA

Chemical Composition

Composition limits. 0.03 C; 1.50 Mn; 1.50 Si; 0.04 P; 0.04 S; 17.0 to 21.0 Cr; 9.0 to 13.0 Ni; 2.0 to 3.0 Mo; rem Fe

Applications

Typical uses. Mixer parts, pump casings and impellers, tubes, valve bodies and parts requiring field welding where postweld heat treatment is inconvenient or impossible

Precautions in use. Because of the thermal instability of CF-3MA, it is not recommended for service at temperatures over 425 °C (800 °F)

Mechanical Properties

Tensile properties. (a) CF-3M; tensile strength, 552 MPa (80 ksi); yield strength, 0.2% offset, 262 MPa (38 ksi); elongation, 55% in 50 mm or 2 in. CF-3MA: tensile strength, 620 MPa (90 ksi); yield strength, 0.2% offset, 310 MPa (45 ksi); elongation, 45% in 50 mm or 2 in.

Hardness. (a) CF-3M: 150 HB; CF-3MA: 170 HB

Elastic modulus. Tension, 193 GPa (28 x 10^6 psi)

Impact strength. (a) Charpy V-notch, CF-3M: 163 J (120 ft·lb). CF-3MA: 136 J (100 ft·lb)

(a) Water quenched from above 1040 °C (1900 °F).

Structure

Microstructure. Austenitic, containing discrete ferrite pools or stringers about 20% by volume. These ferrite pools provide a preferred location for precipitation of carbides that may form when exposed to welding temperatures, thus reducing the sensitivity to intergranular corrosion caused by grain boundary precipitates. Controlled ferrite grade CF-3MA is made to obtain a 25% increase in yield strength over CF-3M, but thermal instability of the microstructure makes the alloy unsuitable for operation at temperatures above 425 °C (800 °F).

Mass Characteristics

Density. 7.75 Mg/m^3 (0.280 lb/in.^{3}) at 20 °C (70 °F)
Patternmaker's shrinkage. 26 mm/m (5/16 in./ft)

Thermal Properties

Melting temperature. 1425 °C (2600 °F)
Coefficient of thermal expansion. Linear, 16.0 μm/m·K (8.9 μin./in.·°F) at 20 to 100 °C (68 to 212 °F); 17.5 μm/m·K (9.7 μin./in.·°F) at 20 to 540 °C (68 to 1000 °F)
Specific heat. 500 J/kg·K (0.12 Btu/lb·°F) at 20 °C (70 °F)
Thermal conductivity. 16.3 W/m·K (9.4 Btu/ft·h·°F) at 100 °C (212 °F)

Electrical Properties

Electrical resistivity. 820 nΩ·m at 20 °C (70 °F)

Magnetic Properties

Magnetic permeability. 1.50 to 3.00

Chemical Properties

General corrosion behavior. Essentially the same as CF-8M

Fabrication Characteristics

Machinability. Good; for typical conditions, see CF-3.
Weldability. Shielded metal-arc, inert-gas arc and electroslag methods: good. Shielded metal-arc is most frequently used.
Heat treatment. For maximum corrosion resistance, castings should be heated in the range 1040 to 1120 °C (1900 to 2050 °F) and then quenched in water, oil or air to ensure complete solution of carbides.

CF-8, CF-8A 19Cr-9Ni

Commercial Name

UNS number. J92600

Specifications

ASTM. A743 and A744, grade CF-8; A351, grades CF-8 and CF-8A
SAE. 60304
Government. MIL-S-867 (ships), class I

Chemical Composition

Composition limits. 0.08 max C; 1.50 max Mn; 2.00 max Si; 0.04 max P; 0.04 max S; 18.0 to 21.0 Cr; 8.0 to 11.0 Ni; rem Fe

Applications

Typical uses. Various castings used in the aircraft, aerospace, architectural, beverage and brewing, brass mill, chemical processing, electronic food processing, marine, military and naval, nuclear power, oil refining, oxygen manufacturing, pharmaceutical, photographic, plastics, power plant, pulp and paper, sewage, soap manufacturing, steel mill, synthetic fiber, and textile industries
Precautions in use. Because of the thermal instability of CF-8A, it is not recommended for service at temperatures over 425 °C (800 °F)

Mechanical Properties

Tensile properties. (a) CF-8: tensile strength, 530 MPa (77 ksi); yield strength, 0.2% offset, 255 MPa (37 ksi); elongation, 55% in 50 mm or 2 in; CF-8A: 586 MPa (85 ksi); yield strength, 0.2% offset, 310 MPa (45 ksi); elongation, 50% in 50 mm or 2 in.
Hardness. (a) CF-8: 140 HB; CF-8A: 156 HB
Elastic modulus. Tension, 193 GPa (28 × 10^6 psi)
Impact strength. (a) See Fig. 11.

(a) Water quenched from 1040 to 1120 °C (1900 to 2050 °F).

Structure

Microstructure. As cast: austenitic with chromium carbides and varying amounts of ferrite distributed throughout the matrix. The carbides must be put into solution by heat treatment to provide maximum corrosion resistance. If the heat treated material is later exposed to temperatures in the range 425 to 870 °C (800 to 1600 °F) carbides will be reprecipitated; this takes place rapidly around 650 °C (1200 °F). Castings thus sensitized, as in welding, must be solution treated again to restore full corrosion resistance. As normally produced: contains about 10% ferrite that takes the form of discrete pools in a matrix of austenite. This ferrite is helpful in avoiding intergranular corrosion in castings exposed to temperatures in the sensitizing range and reduces the tendency for cracking or microfissuring of welds experienced with wholly austenitic alloys. "Controlled ferrite" grade CF-8A has considerably higher tensile properties, but is not suitable for service temperatures above 425 °C (800 °F).

Mass Characteristics

Density. 7.75 Mg/m^3 (0.280 lb/in.^{3}) at 20 °C (70 °F)
Patternmaker's shrinkage. 26 mm/m (5/16 in./ft)

Thermal Properties

Melting temperature. 1425 °C (2600 °F)
Coefficient of thermal expansion. Linear:

Temperature °C	Temperature °F	Mean coefficient μm/m·K	Mean coefficient μin./in.·°F
−200–+20	−325–+70	14.6	8.1
−160–+20	−260–+70	14.8	8.2
−100–+20	−150–+70	15.5	8.6
20–100	70–212	16.2	9.0
20–540	70–1000	18.0	10.0
20–650	70–1200	18.4	10.2

Specific heat. 500 J/kg·K (0.12 Btu/lb·°F) at 20 °C (70 °F)
Thermal conductivity. 15.9 W/m·K (9.2 Btu/ft·h·°F) at 100 °C (212 °F); 20.9 W/m·K (12.1 Btu/ft·h·°F) at 540 °C (1000 °F)

Electrical Properties

Electrical resistivity. 762 nΩ·m at 20 °C (70 °F)

Fig. 11 Charpy keyhole notch impact strength of CF-8

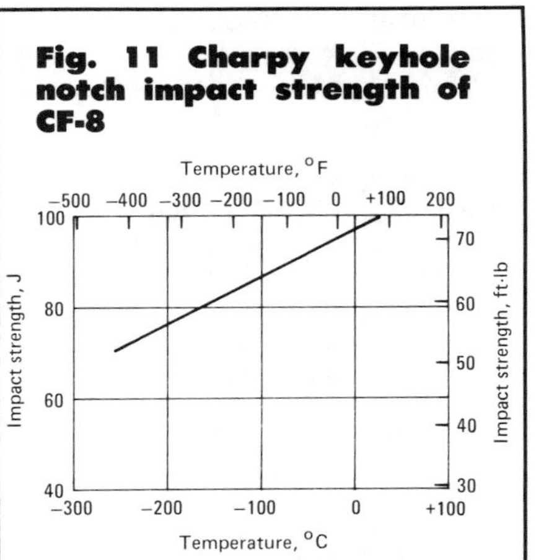

Fig. 12 Isocorrosion charts for CF-8 in nitric acid, phosphoric acid and sodium hydroxide solutions

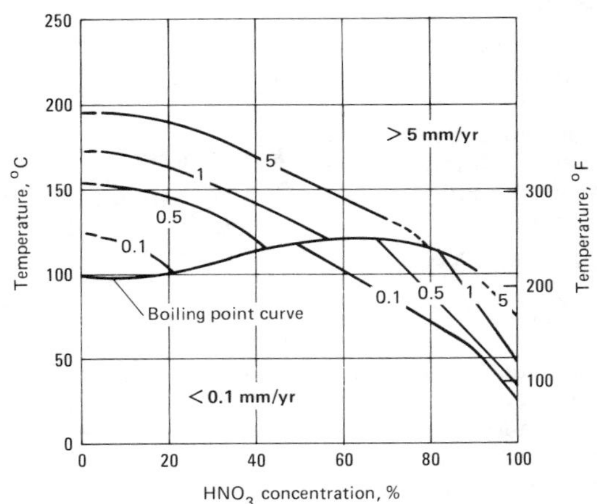

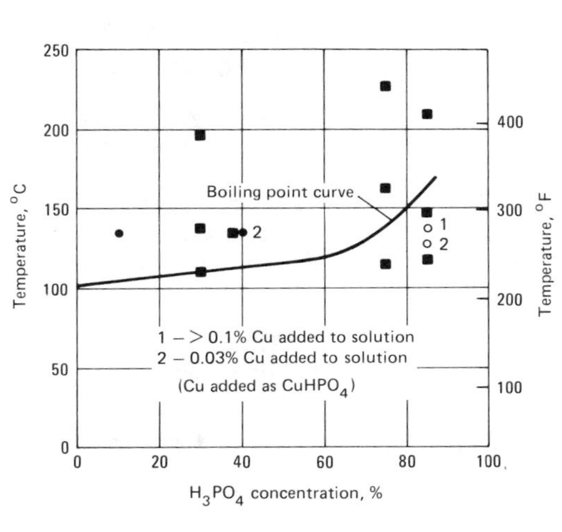

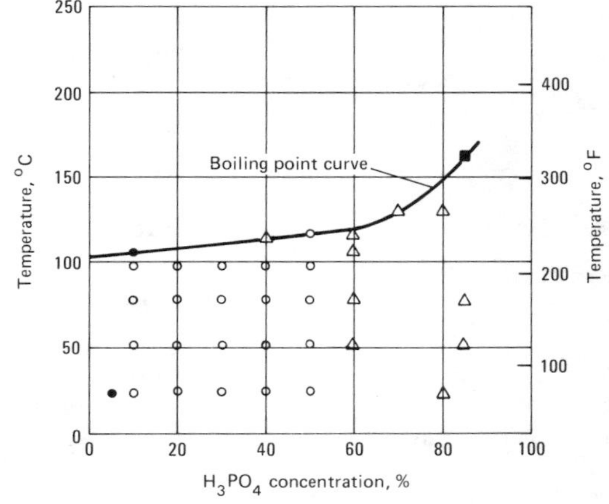

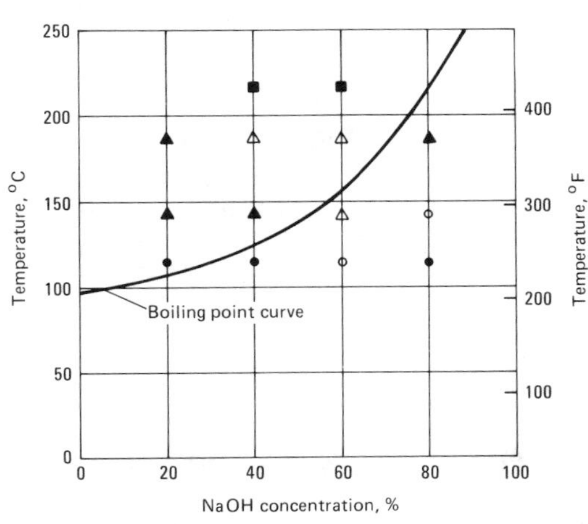

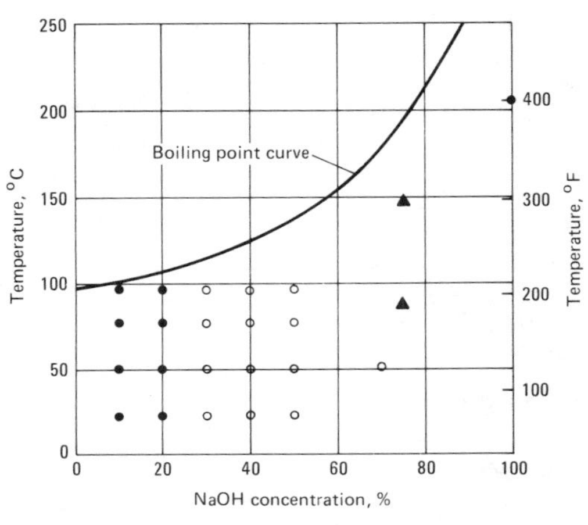

Magnetic Properties

Magnetic permeability. After heat treatment, 1.00 to 1.30

Chemical Properties

General corrosion behavior. Excellent resistance to a wide variety of corrodents (see Fig. 12).
Resistance to specific corroding agents. Especially useful in resisting attack by strongly oxidizing media such as boiling nitric acid

Fabrication Characteristics

Machinability. Good. For typical conditions, see CF-3.
Weldability. Shielded metal-arc, inert-gas arc and electroslag methods: good. Shielded metal-arc is most frequently used.
Heat treatment. For maximum corrosion resistance, castings should be heated in the range 1040 to 1120 °C (1900 to 2050 °F) and then quenched in water, oil or air to ensure complete solution of carbides.

CF-8C
19Cr-10Ni-1Nb

Commercial Name

UNS number. J92710

Specifications

AMS. 5363B
ASTM. A743, A744 and A351
SAE. 60347
Government. MIL-S-867 (ships), class II

Chemical Composition

Composition limits. 0.08 max C; 1.50 max Mn; 2.00 max Si; 0.04 max P; 0.04 max S; 18.0 to 21.0 Cr; 9.0 to 12.0 Ni; 8 × C min to 1.0 max Nb (or 9 × C min to 1.1 max Nb-Ta); rem Fe

Applications

Typical uses. Various castings used in the aircraft, nuclear, chemical processing, marine, oil refining and plastics industries

Mechanical Properties

Tensile properties. (a) Tensile strength, 530 MPa (77 ksi); yield strength, 0.2% offset, 262 MPa (38 ksi), elongation, 39% in 50 mm or 2 in.
Hardness. (a) 149 HB
Elastic modulus. Tension, 193 GPa (28×10^6 psi)
Impact strength. (a) Charpy keyhole, 41 J (30 ft·lb)

(a) Water quenched from 1065 to 1120 °C (1950 to 2050 °F).

Structure

Microstructure. In the heat treated condition, austenitic with some ferrite (5 to 20%) distributed throughout the matrix in the form of discontinuous pools. When stabilized at temperatures in the range 870 to 900 °C (1600 to 1650 °F), niobium carbides are precipitated instead of chromium carbides, thus preventing depletion of chromium along the grain boundary network during exposures to temperatures in the range 425 to 815 °C (800 to 1500 °F) for short times (as in welding) or for long times (as in elevated temperature service), thereby protecting the alloy against intergranular corrosion attack.

Mass Characteristics

Density. 7.75 Mg/m^3 (0.280 $lb/in.^3$) at 20 °C (70 °F)
Patternmaker's shrinkage. 29 mm/m (11/32 in./ft)

Thermal Properties

Melting temperature. 1425 °C (2600 °F)
Coefficient of thermal expansion. Linear, 16.7 μm/m·K (9.3 μin./in.·°F at 20 to 540 °C (68 to 1000 °F)
Specific heat. 500 J/kg·K (0.12 Btu/lb·°F) at 20 °C (70 °F)
Thermal conductivity. 16.1 W/m·K (9.3 Btu/ft·h·°F) at 100 °C (212 °F); 22.2 W/m·K (12.8 Btu/ft·h·°F) at 540 °C (1000 °F)

Electrical Properties

Electrical resistivity. 710 nΩ·m at 20 °C (70 °F)

Magnetic Properties

Magnetic permeability. 1.20 to 1.80

Chemical Properties

General corrosion behavior. Approximately equivalent to CF-8

Fabrication Characteristics

Machinability. Somewhat easier to machine than CF-8. For typical conditions, see CE-30.
Weldability. Shielded metal-arc, inert-gas arc and electroslag methods: excellent. Shielded metal-arc is most frequently used.
Heat treatment. Castings can be used as cast but are normally supplied in the heat treated condition, which consists of heating at 1065 to 1120 °C (1950 to 2050 °F) followed by quenching in water, oil or air to ensure complete solution of any chromium carbides that may have formed in the casting process. A stabilizing treatment at 870 to 900 °C (1600 to 1650 °F) following the solution treatment will cause the preferential precipitation of niobium carbides, and is desirable if castings are for service in the temperature range 425 to 815 °C (800 to 1500 °F).

CF-8M, CF-12M
19Cr-10Ni-2.5Mo

Commercial Name

UNS number. CF-8M: J92900

Specifications

AMS. 5361
ASTM. A743 and A744, grades CF-8M and CF-12M; A351, grade CF-8M
SAE. 60316
Government. MIL-S-867 (ships), class III

Chemical Composition

Composition limits. 0.08 max C in CF-8M, 0.12 max in CF-12M; 1.50 max Mn; 2.00 max Si(a); 0.04 max P; 0.04 max S; 18.0 to 21.0 Cr; 9.0 to 12.0 Ni; 2.0 to 3.0 Mo; rem Fe

(a) Silicon limited to 1.50 max for CF-8M in ASTM A351.

Applications

Typical uses. Various castings for use in the aircraft, chemical processing, electronic, nuclear, fertilizer, food processing, guided missile, marine, mining, oil refining, pharmaceutical, photographic, plastics, power plant, soap, synthetic fiber, synthetic rubber and textile industries

Mechanical Properties

Tensile properties. (a) Tensile strength, 552 MPa (80 ksi); yield strength, 0.2% offset, 290 MPa (42

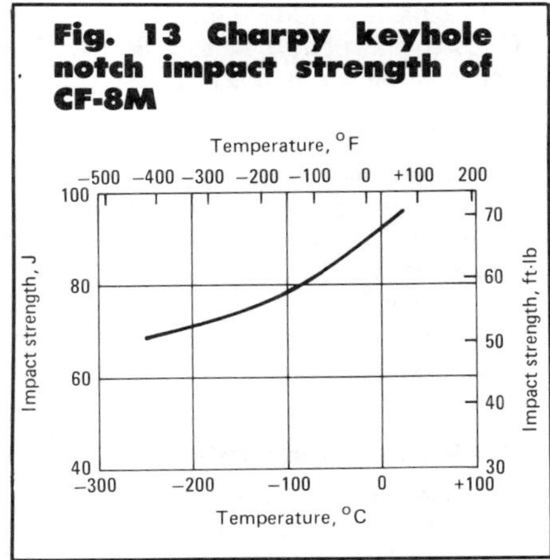

Fig. 13 Charpy keyhole notch impact strength of CF-8M

ksi); elongation, 50% in 50 mm or 2 in.

Hardness. (a) 156 to 210 HB

Elastic modulus. Tension, 193 GPa (28×10^6 psi)

Impact strength. (a) Charpy keyhole, see Fig. 13.

(a) CF-8M: water quenched from 1065 to 1150 °C (1950 to 2100 °F). CF-12M: quenched from above 1095 °C (2000 °F).

Structure

Microstructure. In the heat treated condition, austenite with some ferrite (15 to 25%) distributed throughout the matrix in the form of discontinuous pools. When heated at 425 to 870 °C (800 to 1600 °F), as in welding, these pools provide a preferred location for carbides to precipitate, thus tending to reduce susceptibility of the alloy to intergranular corrosion caused by precipitation of carbides at austenite grain boundaries. The amount of ferrite present decreases as carbon content of the alloy is increased. Thus, the low-carbon CF-8M grade is more strongly magnetic than the higher-carbon CF-12M type. By balancing the compositions, both types can be made wholly austenitic and nonmagnetic, but with a concurrent decrease in strength. At operating temperatures of 650 °C (1200 °F) or higher, ferrite may transform to the brittle sigma phase, and under such conditions the balanced CF-12M type is generally preferred. Below 540 °C (1000 °F), CF-8M is usually employed.

Mass Characteristics

Density. 7.75 Mg/m³ (0.280 lb/in.³) at 20 °C (70 °F)

Patternmaker's shrinkage. 26 mm/m (5/16 in./ft)

Fig. 14 Isocorrosion charts for CF-8M and CF-12M in phosphoric acid and sodium hydroxide solutions

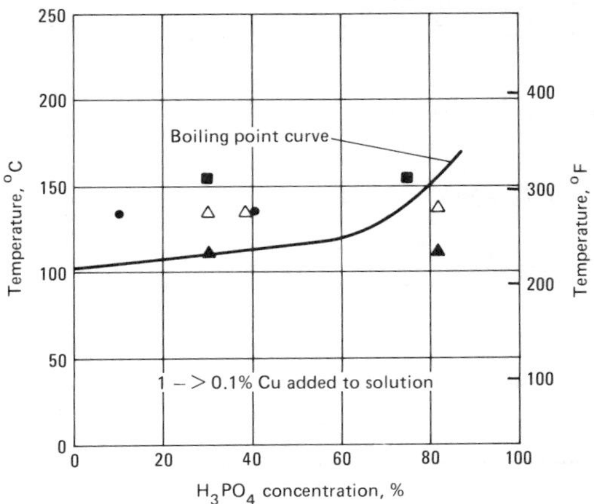

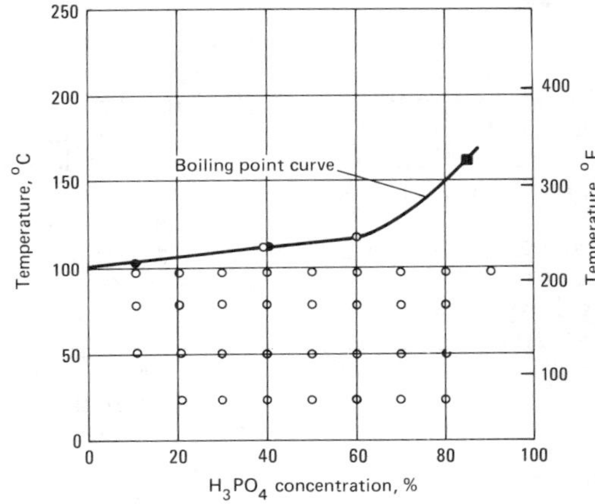

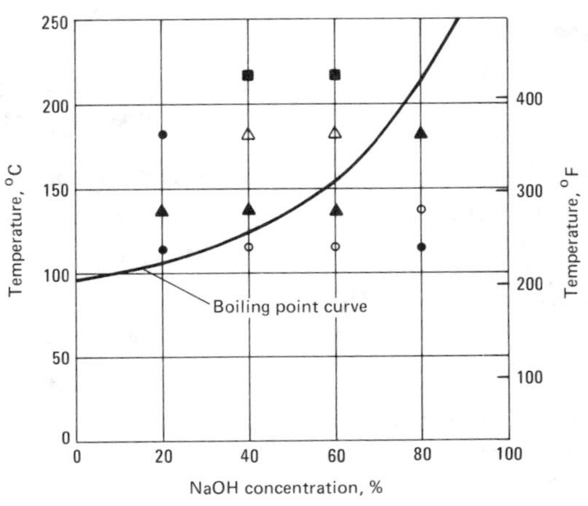

Fig. 14 (continued)

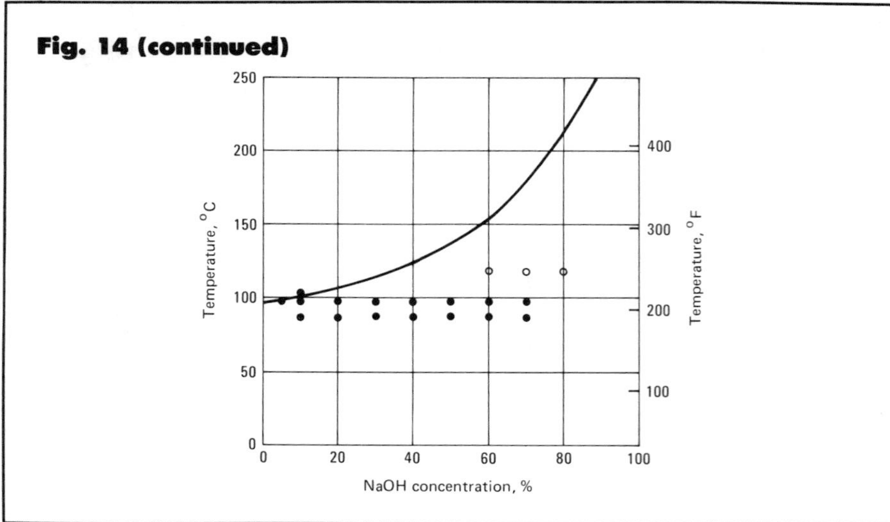

Thermal Properties

Melting temperature. 1400 °C (2550 °F)
Coefficient of thermal expansion. Linear, 16.0 μm/m·K (8.9 μin./in.·°F) at 20 to 100 °C (68 to 212 °F); 17.5 μm/m·K (9.7 μin./in.·°F) at 20 to 540 °C (68 to 1000 °F)
Specific heat. 500 J/kg·K (0.12 Btu/lb·°F) at 20 °C (70 °F)
Thermal conductivity. 16.3 W/m·K (9.4 Btu/ft·h·°F) at 100 °C (212 °F); 21.3 W/m·K (12.3 Btu/ft·h·°F) at 540 °C (1000 °F)

Electrical Properties

Electrical resistivity. 820 nΩ·m at 20 °C (70 °F)

Magnetic Properties

Magnetic permeability. 1.50 to 2.50

Chemical Properties

General corrosion behavior. Good resistance to reducing corrosive media
Resistance to specific corroding agents. Substantially more resistant to pitting corrosion than CF-8 when exposed to chlorides, as in seawater. Although not quite so resistant to strongly oxidizing corrodents such as boiling nitric acid, molybdenum-containing alloys are more stably passive than CF-8 under weakly oxidizing conditions. See Fig. 14.

Fabrication Characteristics

Machinability. Good. For typical conditions, see CF-3.
Weldability. Shielded metal-arc, inert-gas arc and electroslag methods: good. Shielded metal-arc is most frequently used.
Heat treatment. For maximum corrosion resistance, castings should be heated in the range of 1065 to 1150 °C (1950 to 2100 °F) and quenched in water, oil or air to ensure complete solution of carbides. The low side of the solution treating range may be used for CF-8M castings, but CF-12M alloy should be quenched from above 1095 °C (2000 °F).

CF-16F, CF-16Fa 19Cr-10Ni-0.3Se-0.12C

Commercial Name

UNS number. J92701

Specifications

ASTM. A743, grade CF-16F
SAE. 60303 and 60303a

Chemical Composition

Composition limits. CF-16F: 0.16 max C; 1.50 max Mn; 2.00 max Si; 0.17 max P; 0.04 max S; 18.0 to 21.0 Cr; 9.0 to 12.0 Ni; 0.20 to 0.35 Se; 1.50 max Mo; rem Fe. CF-16Fa: 0.16 max C; 1.50 max Mn; 2.00 max Si; 0.04 max P; 0.20 to 0.40 S; 18 to 21.0 Cr; 9 to 12.0 Ni; 0.40 to 0.80 Mo; Se not specified; rem Fe.

Applications

Typical uses. Bearings, bushings, fitting, flanges, machinery parts, pump casings and valves
Precautions in use. The presence of complex selenides makes an acceptable polished surface, as for sanitary applications, very difficult to obtain.

Mechanical Properties

Tensile properties. (a) Tensile strength, 530 MPa (77 ksi); yield strength, 0.2% offset, 275 MPa (40 ksi); elongation, 52% in 50 mm or 2 in.
Hardness. (a) 150 HB
Elastic modulus. Tension, 193 GPa (28×10^6 psi)
Impact strength. (a) Charpy keyhole, 102 J (75 ft·lb)

(a) Water quenched from above 1095 °C (2000 °F).

Structure

Microstructure. As cast: austenite containing chromium carbides and some ferrite (up to 15%) distributed throughout the matrix. The carbides must be put into solution by heat treatment to provide maximum corrosion resistance. Complex selenides present in the as-cast and heat treated material contribute a free-machining quality to these alloys by serving as chip breakers. If heat treated material is later exposed to temperatures in the range of 425 to 870 °C (800 to 1600 °F) carbides will be reprecipitated; this takes place rapidly around 650 °C (1200 °F). Castings thus "sensitized", as in welding, must be solution treated again to restore full corrosion resistance. Type CF-16F cannot be hardened by heat treatment but ductility is improved.

Mass Characteristics

Density. 7.75 Mg/m³ (0.280 lb/in.³) at 20 °C (70 °F)
Patternmaker's shrinkage. 26 mm/m (5/16 in./ft)

Thermal Properties

Melting temperature. 1400 °C (2550 °F)
Coefficient of thermal expansion. Linear, 16.2 μm/m·K (9.0 μin./in.·°F) at 20 to 100 °C (68 to 212 °F); 17.8

μm/m·K (9.9 μin./in.·°F) at 20 to 540 °C (68 to 1000 °F)
Specific heat. 500 J/kg·K (0.12 Btu/lb·°F) at 20 °C (70 °F)
Thermal conductivity. 16.3 W/m·K (9.4 Btu/ft·h·°F) at 100 °C (212 °F); 21.3 W/m·K (12.3 Btu/ft·h·°F) at 540 °C (1000 °F)

Electrical Properties

Electrical resistivity. 720 nΩ·m at 20 °C (70 °F)

Magnetic Properties

Magnetic permeability. 1.00 to 2.00

Chemical Properties

General corrosion behavior. Adequate for many purposes, but inferior to CF-20

Fabrication Characteristics

Machinability. Excellent, using high speed steel or carbide cutters. It is important that the tool be kept continuously entering the metal to avoid work-hardening; slow feeds, deep cuts and powerful, rigid machines are necessary for best results. Typical conditions using high speed steel cutters: roughing speed, 14 to 17 m/min (45 to 55 ft/min) with a feed of 0.5 to 0.6 mm/rev (0.020 to 0.025 in./rev); finishing speed, 27 to 34 m/min (90 to 110 ft/min) with a feed of 0.1 to 0.3 mm/rev (0.005 to 0.010 in./rev)
Weldability. Shielded metal-arc, inert-gas arc and electroslag methods: good. Shielded metal-arc is most frequently used
Heat treatment. For maximum corrosion resistance, castings should be heated at 1095 to 1150 °C (2000 to 2100 °F) and quenched in water, oil or air to ensure complete solution of carbides.

CF-20
19Cr-9Ni-0.20C

Commercial Names

UNS number. J92602
Trade name. CF-20

Specifications

AMS. 5358
ASTM. A743
SAE. 60302

Chemical Composition

Composition limits. 0.20 max C; 1.50 max Mn; 2.00 max Si; 0.04 max P; 0.04 max S; 18.0 to 21.0 Cr; 8.0 to 11.0 Ni; rem Fe

Applications

Typical uses. Various castings used in the architectural, chemical processing, explosives manufacturing, food and dairy, marine, oil refinery, pharmaceutical, power plant, pulp and paper, and textile industries

Mechanical Properties

Tensile properties. (a) Tensile strength, 530 MPa (77 ksi); yield strength, 0.2% offset, 250 MPa (36 ksi); elongation, 50% in 50 mm or 2 in.
Hardness. (a) 163 HB
Elastic modulus. 195 GPa (28×10^6 psi)
Impact strength. (a) Charpy keyhole, 81 J (60 ft·lb)

(a) Water quenched from above 1095 °C (2000 °F).

Structure

Microstructure. As cast: austenite with chromium carbides. The carbides must be put into solution by heat treatment to provide maximum corrosion resistance. If the heat treated material is later exposed to temperatures of 425 to 870 °C (800 to 1600 °F) carbides will be reprecipitated, which takes place rapidly around 650 °C (1200 °F). Castings thus "sensitized", as in welding, must be solution heat treated again to restore full corrosion resistance. Type CF-20 alloy cannot be hardened by heat treatment.

Mass Characteristics

Density. 7.75 Mg/m³ (0.280 lb/in.³) at 20 °C (70 °F)
Patternmaker's shrinkage. 29 mm/m (11/32 in./ft)

Thermal Properties

Melting temperature. 1415 °C (2575 °F)
Coefficient of thermal expansion. Linear, 17.3 μm/m·K (9.6 μin./in.·°F) at 20 to 100 °C (68 to 212 °F); 18.7 μm/m·K (10.4 μin./in.·°F) at 20 to 540 °C (68 to 1000 °F)
Specific heat. 500 J/kg·K (0.12 Btu/lb·°F) at 20 °C (70 °F)
Thermal conductivity. 15.9 W/m·K (9.2 Btu/ft·h·°F) at 100 °C (212 °F); 20.9 W/m·K (12.1 Btu/ft·h·°F) at 540 °C (1000 °F)

Electrical Properties

Electrical resistivity. 779 nΩ·m at 20 °C (70 °F)

Magnetic Properties

Magnetic permeability. 1.01

Chemical Properties

General corrosion behavior. Satisfactory in many types of oxidizing corrosion service.

Fabrication Characteristics

Machinability. Good. For typical conditions, see CF-3.
Weldability. Shielded metal-arc, inert-gas arc and electroslag methods: good. Shielded metal-arc is the most frequently used.
Heat treatment. For maximum corrosion resistance, castings should be heated at 1095 to 1150 °C (2000 to 2100 °F) and quenched in water, oil or air to ensure completed solution of carbides.

CG-8M
19Cr-11Ni-3.5Mo

Commercial Name

UNS number.

Specifications

ASTM. A743 and A744

Chemical Composition

Composition limits. 0.08 max C; 1.50 max Mn; 1.50 max Si; 0.04 max P; 0.04 max S; 18 to 21.0 Cr; 9.0 to 13.0 Ni; 3.0 to 4.0 Mo; rem Fe

Applications

Typical uses. Dyeing equipment, flow meter components, propellers, pump parts, valve bodies and parts used in heavy water manufacturing, nuclear, petroleum, pipe line, power, pulp and paper, printing and textile industries

Mechanical Properties

Tensile properties. (a) Tensile strength, 565 MPa (82 ksi); yield

strength, 0.2% offset, 305 MPa (44 ksi); elongation, 45% in 50 mm or 2 in.
Hardness. (a) 176 HB
Elastic modulus. Tension, 193 GPa (28 × 10^6 psi)
Impact strength. (a) Charpy V-notch, 108 J (80 ft·lb)

(a) Water quenched from above 1040 °C (1900 °F).

Structure

Microstructure. After heat treatment, austenitic matrix in which 15 to 35% ferrite is distributed in the form of discontinuous pools, which gives considerable resistance to stress corrosion cracking and high strength at room and elevated temperatures. Long exposure to temperatures above 650 °C (1200 °F) may cause embrittlement due to transformation of some of the ferrite to sigma phase.

Mass Characteristics

Density. 7.75 Mg/m³ (0.280 lb/in.³) at 20 °C (70 °F)
Patternmaker's shrinkage. 26 mm/m (5/16 in./ft)

Thermal Properties

Melting temperature. 1400 °C (2550 °F)
Coefficient of thermal expansion. Linear, 16.0 μm/m·K (8.9 μin./in.·°F) at 20 to 70 °C (68 to 212 °F); 17.5 μm/m·K (17.5 μin./in.·°F) at 20 to 540 °C (68 to 1000 °F)
Specific heat. 500 J/kg·K (0.12 Btu/lb·°F) at 20 °C (70 °F)
Thermal conductivity. 16.3 W/m·K (9.4 Btu/ft·h·°F) at 1000 °C (212 °F); 21.3 W/m·K (12.3 Btu/ft·h·°F) at 540 °C (1000 °F)

Electrical Properties

Electrical resistivity. 820 nΩ·m at 20 °C (70 °F)

Magnetic Properties

Magnetic permeability. 1.50 to 3.00

Chemical Properties

General corrosion behavior. Excellent resistance to corrosion by reducing media
Resistance to specific corroding agents. Increased resistance to sulfurous and sulfuric acid solutions and to the pitting action of halogen compounds over CF-8M. Not suitable for use in nitric acid or other strongly oxidizing environments.

Fabrication Characteristics

Machinability. Good. For typical conditions, see CF-3.
Weldability. Shielded metal-arc, inert-gas arc and electroslag methods: good. Shielded metal-arc is most frequently used.
Heat treatment. For maximum corrosion resistance, castings should be heated in the range 1040 to 1120 °C

CH-10, CH-20
24Cr-13Ni

Commercial Name

UNS number. CH-10: J93401 CH-20: J93402

Specifications

ASTM. A743 and A351
SAE. 60309

Chemical Composition

Composition limits. CH-10: 0.10 max C. CH-20: 0.20 max C; 1.50 max Mn, 2.00 max Si; 0.04 max P; 0.04 max S; 22.0 to 26.0 Cr; 12.0 to 15.0 Ni; rem Fe

Applications

Typical uses. Digester fittings, pumps and parts, roasting equipment, valves and water strainers for use in chemical processing, power plants and pump and papermaking

Mechanical Properties

Tensile properties. (a) Tensile strength, 607 MPa (88 ksi); yield strength, 0.2% offset, 348 MPa (50 ksi); elongation, 38% in 50 mm or 2 in.
Hardness. (a) 190 HB
Elastic modulus. Tension, 193 GPa (28 × 10^6 psi)
Impact strength. (a) Charpy keyhole, 41 J (30 ft·lb)

(a) Water quenched from above 1095 °C (2000 °F).

Structure

Microstructure. As cast: austenitic with chromium carbides and some ferrite distributed throughout the matrix. The carbides must be put into solution by heat treatment to provide maximum corrosion resistance. Heat treated material exposed in the range of 425 to 870 °C (800 to 1600 °F) will precipitate carbides. Type CH alloys are not hardened by heat treatment.

Mass Characteristics

Density. 7.72 Mg/m³ (0.279 lb/in.³) at 20 °C (70 °C)
Patternmaker's shrinkage. 26 mm/m (5/16 in./ft)

Thermal Properties

Melting temperature. 1427 °C (2600 °F)
Coefficient of thermal expansion. Linear:

Temperature °C	°F	Mean coefficient μm/m·K	μin./in.·°F
20–100	70– 212	15.5	8.6
20–315	70– 600	15.7	8.7
20–540	70–1000	17.1	9.5

Specific heat. 500 J/kg·K (0.12 Btu/lb·°F) at 20 °C (70 °F)
Thermal conductivity:

Temperature °C	°F	Conductivity W/m·K	Btu/ft·h·°F
100	212	14.2	8.2
315	600	17.5	10.1
540	1000	20.8	12.0

Electrical Properties

Electrical resistivity. 840 nΩ·m at 20 °C (70 °F)

Magnetic Properties

Magnetic permeability. After heat treatment, 1.71

Chemical Properties

General corrosion behavior. The high nickel and chromium contents impart excellent resistance to corrosives and make this alloy less susceptible to intergranular corrosion after exposure to carbide-precipitation temperatures.
Resistance to specific corroding agents. Resistant to hot dilute sulfuric acid

Fabrication Characteristics

Machinability. Fair. For typical conditions, see CF-3.
Weldability. Shielded metal-arc, inert-gas arc and electroslag meth-

ods: good. Shielded metal-arc is most frequently used.
Heat treatment. For maximum corrosion resistance, castings should be heated at 1095 to 1150 °C (2000 to 2100 °F) and then quenched in water, oil or air to ensure complete solution of carbides.

CK-20
25 Cr-20 Ni

Commercial Name

UNS number. J94202

Specifications

AMS. 5365A
ASTM. A743 and A351
SAE. 60310
Government. MIL 20150 (ships)

Chemical Composition

Composition limits. 0.20 max C; 2.00 max Mn (a); 2.00 max Si (a); 0.04 max P; 0.04 max S; 23.0 to 27.0 Cr; 19.0 to 22.0 Ni; rem Fe

(a) 1.50 max Mn and 1.75 max Si in ASTM A351.

Applications

Typical uses. Various castings used for special service conditions at high temperatures where specific requirements warrant the cost, in aircraft, chemical processing, oil refining, and pulp and papermaking

Mechanical Properties

Tensile properties. (a) Tensile strength, 524 MPa (76 ksi); yield strength, 0.2% offset, 262 MPa (38 ksi); elongation, 37% in 50 mm or 2 in.
Hardness. (a) 144 HB
Elastic modulus. Tension, 200 GPa (29×10^6 psi)
Impact strength. Izod V-notch, see Fig. 15.

(a) Water quenched from 1150 °C (2100 °F).

Structure

Microstructure. As cast: austenitic with chromium carbides distributed throughout the matrix. The carbides must be put into solution by heat treatment for maximum corrosion resistance. If the heat treated material is later exposed to temperatures in

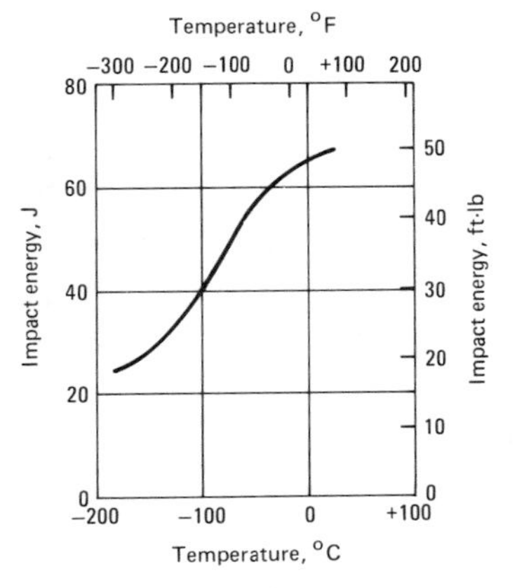

Fig. 15 Izod impact strength of CK-20

the range 425 to 870 °C (800 to 1600 °F) carbides will be reprecipitated. Type CK-20 alloy suffers little from intergranular attack, if the elevated-temperature exposure is brief. This alloy is not hardened by heat treatment but ductility and strength are improved.

Mass Characteristics

Density. 7.75 Mg/m^3 (0.280 lb/in.^{3}) at 20 °C (70 °F)
Patternmaker's shrinkage. 26 mm/m (5/16 in./ft)

Thermal Properties

Melting temperature. 1425 °C (2600 °F)
Coefficient of thermal expansion. Linear:

Temperature °C	°F	Mean coefficient µm/m·K	µin./in.·°F
20–100	70– 212	14.9	8.3
20–315	70– 600	16.0	8.9
20–540	70–1000	16.9	9.4

Specific heat. 50 J/kg·K (0.12 Btu/lb·°F) at 20 °C (70 °F)
Thermal conductivity:

Temperature °C	°F	Conductivity W/m·K	Btu/ft·h·°F
100	212	13.7	7.9
315	600	17.0	9.8
540	1000	20.4	11.8

Electrical Properties

Electrical resistivity. 900 nΩ·m at 20 °C (70 °F)

Magnetic Properties

Magnetic permeability. 1.02

Chemical Properties

General corrosion behavior. Good resistance to many corrodents, including dilute sulfuric acid

Fabrication Characteristics

Machinability. Good. For typical conditions, see CF-3.
Weldability. Shielded metal-arc, inert-gas arc and electroslag methods: good. Shielded metal-arc is most frequently used.
Heat treatment. For maximum corrosion resistance, castings should be heat treated at 1095 to 1175 °C (2000 to 2150 °F) and quenched in water, oil or air to ensure complete solution of carbides.

CN-7M, CN-7MS
20Cr-29Ni-2.5Mo-3.5Cu

Commercial Name

UNS number. J95150

Specifications

ASTM. A743, A744 and A351

Chemical Composition

Composition limits. CN-7M: 0.07 max C; 1.50 max Mn; 1.50 max Si; 0.04 max P; 0.04 max S; 19.0 to 22.0 Cr; 27.5 to 30.5 Ni; 2.0 to 3.0 Mo; 3.0 to 4.0 Cu; rem Fe. CN-7MS: 0.07 max C; 1.0 max Mn; 2.5 to 3.5 Si; 0.04 max P; 0.03 max S; 18.0 to 20.0 Cr; 22.0 to 25.0 Ni; 2.5 to 3.0 Mo; 1.5 to 2.0 Cu.

Applications

Typical uses. Various castings used in chemical processing, food processing, metal cleaning and plating, mining, munitions manufacturing, oil refining, paint and pigment, pharmaceutical, plastics, pulp and paper, soap and detergent, steel mill, synthetic rubber, textile and dye industries

Mechanical Properties

Tensile properties. (a) Tensile strength, 475 MPa (69 ksi); yield strength, 0.2% offset, 217 MPa (31.5 ksi); elongation, 48% in 50 mm or 2 in.
Hardness. (a) 130 HB

Fig. 16 Isocorrosion charts for solution annealed and quenched CN-7M in sulfuric acid, nitric acid, phosphoric acid and sodium hydroxide

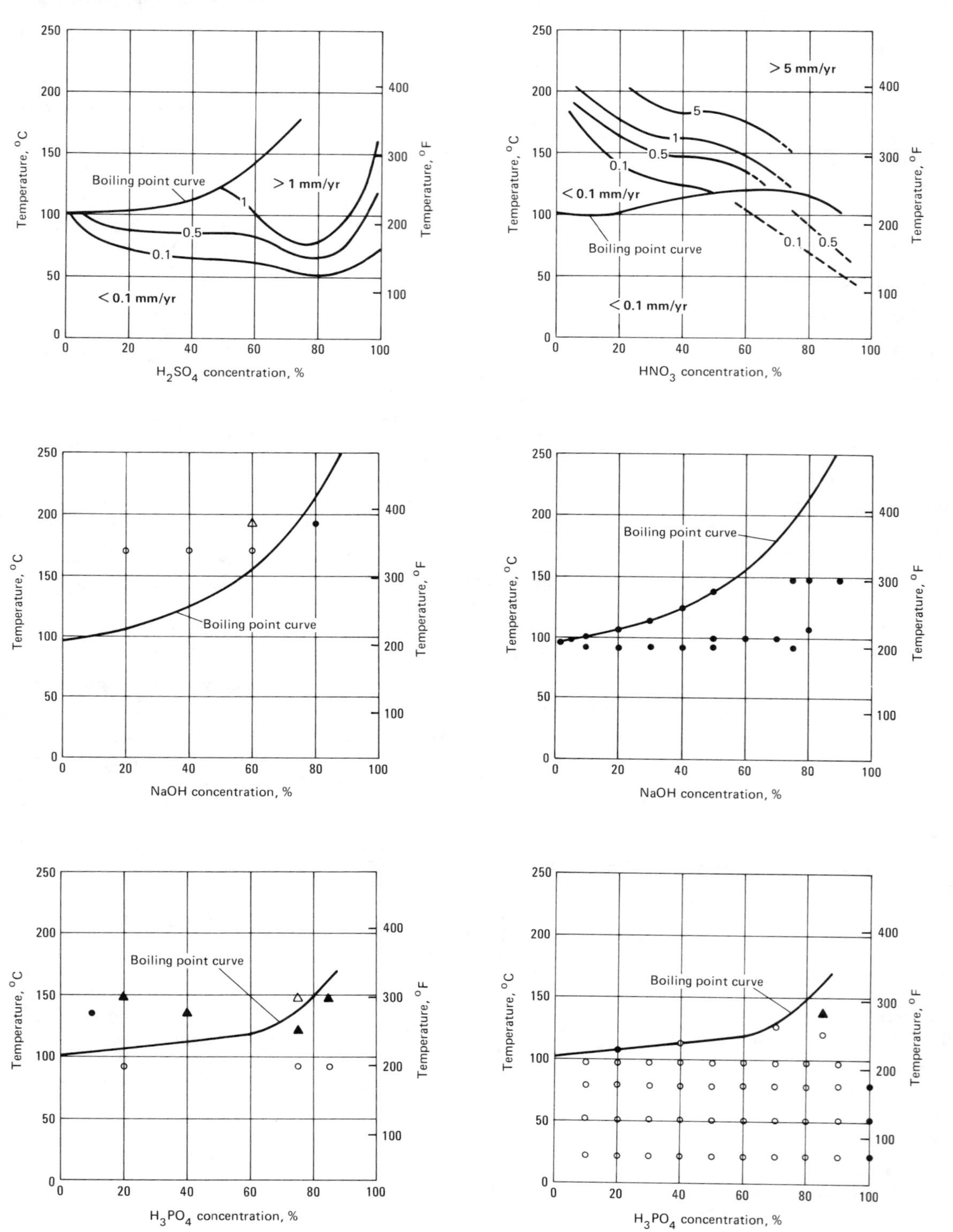

Elastic modulus. Tension, 165 GPa (24×10^6 psi)
Impact strength. (a) Charpy keyhole, 95 J (70 ft·lb)

(a) Water quenched from 1065 to 1120 °C (1950 to 2050 °F).

Structure

Microstructure. Austenitic in the heat treated condition. Carbides must be put into solution by heat treatment to provide maximum corrosion resistance and to eliminate susceptibility to intergranular attack. Castings later exposed to temperatures in the range 425 to 870 °C (800 to 1600 °F) must be heat treated again to restore full corrosion resistance. Type CN-7M cannot be hardened by heat treatment.

Mass Characteristics

Density. 8.00 Mg/m³ (0.289 lb/in.³) at 20 °C (70 °F)
Patternmaker's shrinkage. 26 mm/m (5/16 in./ft)

Thermal Properties

Melting temperature. 1455 °C (2650 °F)
Coefficient of thermal expansion. Linear, 15.5 μm/m·K (8.6 μin./in.· °F) at 20 to 100 °C (68 to 212 °F); 17.5 μm/m·K (9.7 μin./in.· °F) at 20 to 540 °C (68 to 1000 °F)
Specific heat. 460 J/kg·K (0.11 Btu/lb·°F) at 20 °C (70 °F)
Thermal conductivity. 20.9 W/m·K (12.1 Btu/ft·h· °F) at 100 °C (212 °F)

Electrical Properties

Electrical resistivity. 896 nΩ·m at 20 °C (70 °F)

Magnetic Properties

Magnetic permeability. 1.01 to 1.10

Chemical Properties

Resistance to specific corroding agents. Good resistance to sulfuric acid, dilute and hot chloride salt solutions, nitric acid, and many reducing chemicals. See Fig. 16 for data on corrosion in sulfuric, nitric and phosphoric acids, and NaOH.

Fabrication Characteristics

Machinability. Excellent. For typical conditions, see CF-16F.
Weldability. Shielded metal-arc, inert-gas arc and electroslag methods: fair. Shielded metal-arc is most frequently used.
Heat treatment. For maximum corrosion resistance, castings should be heated to 1120 °C (2050 °F) min and quenched in water, oil or air to ensure complete solution of carbides.

Illium P
28Cr-8Ni-2Mo-3Cu

Compiled by C. R. Bird
Chief Metallurgist
Stainless Foundry & Engineering, Inc.

Commercial Name

Trade name. Illium P

Chemical Composition

Nominal composition. 0.2 C; 28.0 Cr; 8.0 Ni; 3.0 Cu; 2.0 Mo; rem Fe

Applications

Typical uses. Castings used in phosphoric acid and other chemicals

Mechanical Properties

Tensile properties. (a) Tensile strength, 670 to 760 MPa (97 to 110 ksi); yield strength at 0.2% offset, 365 to 545 MPa (53 to 79 ksi); elongation, 10 to 18 in 50 mm or 2 in.; reduction in area, 11 to 25%
Hardness. (a) 217 to 255 HB

(a) After water quench from 1080 °C (1975 °F).

Structure

Microstructure. Austenite matrix with ferrite distributed throughout. Carbides are dissolved by heat treatment for maximum corrosion resistance. Exposure to temperatures in the range of 600 to 925 °C (1100 to 1700 °F) precipitates sigma phase which reduces ductility. Castings may be restored to original condition by solution heat treatment.

Mass Characteristics

Density. 7.61 Mg/m³ (0.275 lb/in.³) at 20 °C (70 °F)
Patternmaker's shrinkage. 18 mm/m (7/32 in./ft)

Thermal Properties

Melting temperature. 1425 to 1450 °C (2600 to 2650 °F)

Magnetic Properties

Magnetic permeability. Ferromagnetic

Chemical Properties

General corrosion behavior. Resistant to attack from mineral acids, especially those containing abrasive solids
Resistance to specific corroding agents. Excellent resistance to corrosion/erosion from processes where phosphate rock is reacted with phosphoric and/or sulfuric acid. Also highly resistant to corrosive attack from high temperature phosphoric acid present in evaporation concentrators.

Fabrication Characteristics

Machinability. Better than the austenitic grades because of reduced tendency to gall and smear

Illium PD
27Cr-5Ni-2.5Mo-7Co

Compiled by C. R. Bird
Chief Metallurgist
Stainless Foundry & Engineering, Inc.

Commercial Name

Trade name. Illium PD

Chemical Composition

Nominal composition. 0.06 C; 27.0 Cr; 7.0 Co; 5.0 Ni; 2.5 Mo; rem Fe

Applications

Typical uses. Castings used in food processing, chemical, alumina refining, pulp and papermaking industries

Mechanical Properties

Tensile properties. See Table 5.
Hardness. See Table 5.

Structure

Microstructure. Austenite matrix with ferrite distributed throughout. Carbides are dissolved by heat treatment for maximum corrosion resistance. Exposure to temperatures in the range of 600 to 925 °C (1100 to 1700 °F) precipitates sigma phase which reduces ductility. Castings may be restored to original condition by solution heat treatment.

Mass Characteristics

Density. 7.95 Mg/m³ (0.287 lb/in.³) at 20 °C (70 °F)

Patternmaker's shrinkage. 18 mm/m (7/32 in./ft)

Thermal Properties

Coefficient of thermal expansion. Linear:

Temperature °C	°F	Average coefficient µm/m·K	µin./in.·°F
20–200	68– 400	13.07	7.26
20–315	68– 600	13.70	7.61
20–425	68– 800	14.39	7.99
20–540	68–1000	14.68	8.15

Thermal Properties

Melting temperature. 1425 to 1450 °C (2600 to 2650 °F)

Magnetic Properties

Magnetic permeability. Ferromagnetic

Table 5 Typical mechanical properties of Illium PD

Condition	Tensile strength MPa	ksi	Yield strength, 0.2% MPa	ksi	Elongation in 50 mm or 2 in.,%	Hardness, HB
Water quenched (a)	655–724	95–105	448–552	65–80	35–41	201–217
Water quenched (a) and aged (b)	793–862	115–125	552–655	80–95	28–36	255–269

(a) After water quench from 1080 °C (1975 °F). (b) Aged at 510 °C (950 °F).

Chemical Properties

General corrosion behavior. Resistant to corrosion in atmospheric and marine applications as well as acids and alkalies with suspended abrasive solids; has up to ten times the corrosion resistance of the CF austenitic cast stainless alloys

Resistance to specific corroding agents. Excellent resistance to erosion-corrosion in hot caustic and acid digesting liquors used in paper pulping applications, and the caustic extraction of alumina from bauxite. Resists attack from nitric acid and food acids; resists stress corrosion cracking in seawater.

Fabrication Characteristics

Machinability. Better than the austenitic grades because of reduced tendency to gall and smear.

Heat treatment. For maximum corrosion resistance, castings should be heated in the range of 1065 to 1150 °C (1950 to 2100 °F) and quenched in water to ensure solution of carbides and sigma phase.

Nickel and Its Uses

By Donald L. Pasquine
The International Nickel Company, Inc.

NICKEL ALLOYS were known and used for thousands of years before nickel was isolated and identified as an element. Use of iron-nickel alloys of meteoric origin by primitive man has been well documented. The legendary swords of Damascus, often described as being "heaven-sent", probably were made from iron-nickel meteorites. Alloys of copper and nickel were used by the ancient Chinese to make such objects as coins, swords and arrowheads.

Nickel is a hard, tough and malleable silver-white metal that has good resistance to oxidation and corrosion. It is used principally in association with other elements to impart strength, toughness and corrosion resistance to a wide range of ferrous and nonferrous alloys. Nickel, like iron and cobalt, is magnetic, and thus nickel alloys often are used in magnetic applications.

Mining and Refining

Although nickel ores are widely distributed over the face of the earth, workable deposits of nickel-containing minerals exist in relatively few localities. Even in these, the concentration of nickel is so low that profitable recovery usually depends on the ability to also recover valuable by-products.

Pure nickel in native form is unknown. Small quantities of nickel (combined with iron) have been found in meteorites. Sulfide ores and laterites (oxide ores) are the main sources of nickel. Other important nickel minerals are arsenides and silicates (including nickeliferous iron ores). Approximately 60% of the nickel currently produced comes from sulfide ores, and about 40% from laterite deposits. It is estimated that 80% of the world's known land-based nickel resources are in laterite ores and only about 20% in sulfide ores. Nickel is present in significant quantities in the ocean, particularly in deep-sea nodules that contain nickel, copper, cobalt and manganese. If undersea mining becomes economically feasible, the ocean floor may provide a future source of nickel to supplement land-based reserves.

Sulfide ores, mined principally in Canada, Finland, USSR, several Comecon countries, Botswana, South Africa and Western Australia, account for about 60% of current nickel production. In sulfide ores, pentlandite is the principal nickel mineral. It is almost invariably associated with pyrrhotite, an iron sulfide, and usually with chalcopyrite ($CuFeS_2$). In addition to nickel, iron and copper, these ores contain important amounts of cobalt and precious metals (including platinum-group metals, gold and silver). Typically, sulfide ores contain about 1 to 3% nickel and a similar amount of copper.

Laterites (oxide ores), amounting to about 40% of current world nickel ore production, come from the United States (Oregon), Cuba, the Dominican Republic, Guatemala, Brazil, USSR (where they account for only a minor share of production), Comecon countries, Greece, the Philippines, Indonesia, New Caledonia and Australia (Queensland). Oxide ores are principally of two types: a silicate ore in which the nickel occurs as hydrous nickel-magnesium silicate (of which garnierite is the most common mineral); and nickeliferous limonitic ore, predominantly a hydrated ferric oxide goethite in which nickel is dispersed. In addition to nickel, magnesium and iron, oxide ores contain significant amounts of cobalt and chromium. Typically, oxide ores contain about 1 to 3% nickel.

World production of nickel ore for recent years is shown in Table 1. In 1950, the vast majority of nickel ore came from three countries—Canada, USSR and New Caledonia. According to 1977 production statistics, nickel is mined in at least 20 countries; the principal countries are shown in Table 1. In addition, nickel is refined solely from imported raw materials in the United States (Louisiana), Norway, the United Kingdom, France and Japan. Estimated total world nickel-refining capacity for 1978 was approximately 1.5-

Table 1 World mine production of nickel

Country	Nickel content of ores, short tons							
	1960	1965	1970	1973	1974	1975	1976	1977(a)
Australia	...	...	32 870	44 308	44 500	60 000	91 652	94 500
Botswana	...	...	...	600	13 042	18 314	13 866	13 000
Canada	214 506	267 308	305 296	268 908	296 600	269 826	289 348	259 500
Dominican Republic	...	...	...	33 200	34 400	34 400	26 896	27 000
Finland	2495	3295	5545	6371	6575	6299	7309	6400
Greece	...	...	...	28 980	31 440	31 014	30 380	10 600
Indonesia	...	...	...	22 946	23 250	23 000	19 600	17 700
New Caledonia	43 325	53 054	116 120	118 363	148 333	146 767	117 506	127 300
Philippines	...	...	...	440(b)	359(b)	13 200(b)	29 723(b)	40 600
South Africa	3200(b)	3300	12 739(b)	21 413	24 361	22 877	24 660	24 300
Southern Rhodesia	...	...	...	13 000(b)	12 700(b)	12 100(b)	16 500(b)	14 300
United States	12 530	13 510	15 319	18 272	16 618	16 987	16 469	14 300
Cuba	12 547	20 200(b)	38 800(b)	40 200(b)	40 200(b)	40 300(b)	40 800(b)	40 800
USSR	58 000(b)	90 000(b)	121 000(b)	125 000	135 000	168 000	154 000	148 000(b)
Total(c)	346 600	468 270	647 689	757 905	842 690	878 201	886 337	860 200

(a) From World Bureau of Metal Statistics (March 1979). (b) Estimated. (c) Includes ores mined by other small producers.
Source: U.S. Bureau of Mines.

billion pounds. By the end of 1980, world nickel refining capacity is expected to reach 1.7-billion pounds with the addition of new capacity, primarily in Indonesia, Guatemala and Greece.

Uses of Nickel

Estimated world consumption of nickel in 1977 is shown in Table 2. The bulk of this consumption was in the United States, Western Europe and Japan. Of the total amount of nickel consumed, about 55% ordinarily is used as alloying additions to stainless and alloy steels, with the major portion going into austenitic chromium-nickel stainless steels. Another important market for nickel, comprising about 20% of total consumption, is in nonferrous alloys, where nickel is the major component in a large number of nickel-base, cobalt-base and copper-base alloys as well as other families of nonferrous alloys such as those containing substantial amounts of molybdenum. The major nonalloying use of nickel is in electroplating, which traditionally comprises about 10 to 15% of total demand. The balance of the nickel consumption is in the foundry industry and in a wide range of minor applications.

Stainless steels, principally the 18Cr-8Ni varieties that comprise the 300 series, are by far the largest consumers of nickel, accounting for 43% of consumption. Nickel is used in about 70% of all stainless steel production. Nickel is added to stainless steels primarily to increase corrosion resistance, improve fabricability and improve strength at elevated temperature. Austenitic stainless steels find widespread application in consumer products and in equipment for the process, chemical, energy and transportation industries.

Table 2 Estimated world consumption of nickel in 1977

Use	Amount consumed, %(a)
Stainless steels	43
Nonferrous alloys	20
Electroplating	12
Alloy steels	10
Foundry products	8
All others	7

(a) Based on 1.1-billion pounds non-Communist world consumption.
Source: J. E. Carter, Inco Ltd., presentation to the Canadian Institute of Mining and Metallurgy and the Winnipeg Society of Financial Analysts, Winnipeg, Canada, November 9, 1978.

Nonferrous alloys, comprising about 20% of total nickel consumption, are used primarily for resistance to corrosion and heat. Included in this group are nickel-base, copper-base, cobalt-base and molybdenum-base alloys. Nickel-base superalloys constitute one important class of nonferrous alloys and are used chiefly for jet-engine parts and other advanced propulsion systems. High nickel alloys also are important materials of construction for chemical and petrochemical process equipment. Nonferrous alloys are used in a broad variety of applications in the transportation, power, process, chemical and electronics industries.

Electroplating, the major nonalloying use of nickel, accounts for 12% of consumption. It is used primarily for decorative coatings in which nickel is overplated with chromium, although industrial coatings and electroforming applications are significant. The principal objectives of nickel plating are to improve appearance, surface finish and corrosion resistance, but nickel also is used to build up worn or mismachined parts and to improve wear resistance. Nickel/chromium coatings are applied to such basis materials as steel, zinc-alloy die castings, copper alloys, plastics, magnesium alloys and aluminum alloys. Nickel is electroplated on a wide variety of products such as automobile parts, consumer appliances and equipment for the food-processing and chemical industries.

Alloy steels other than austenitic stainless steels account for 10% of nickel consumption. Nickel in amounts up to 18% is an important alloying addition to steel. It normally is used in conjunction with one or more other alloying elements (such as chromium and molybdenum) to permit development of optimum hardness, strength, ductility and processing characteristics. Alloy steels are used in automotive, construction and machinery industries for parts requiring high strength, toughness and wear resistance. Nickel improves the atmospheric corrosion resistance of steels, particularly when in combination with other elements such as copper and chromium. Nickel lowers the ductile-to-brittle transition temperature of steel, making nickel-containing steels useful for transporting and handling liquefied gases and for machinery and structures used at subzero temperatures.

Foundry products, both ferrous and nonferrous, account for 8% of nickel consumption. Nickel, in combination with chromium, is an important constituent in cast heat-resistant and corrosion-resistant stainless steels. In the cast heat-resistant alloys, its principal function is to strengthen and toughen the matrix. In cast corrosion-resistant alloys, nickel supplements the passivating effects of chromium under oxidizing conditions, and increases basic corrosion resistance under reducing conditions. The addition of molybdenum to the latter alloys reduces the tendency toward pitting in seawater and other chloride solutions. Nickel, in amounts up to 3.5%, is also used in engineering grades of steel castings, where it improves strength and ductility in much the same manner as it improves these properties in wrought products. Nickel in amounts up to about 35% is added to improve the properties of a wide range of cast irons, including gray iron, ductile iron, white iron and the various austenitic cast irons. Nickel is added to gray iron in amounts up to 3.5%, with or without other alloying elements, to improve strength, wear resistance and machinability. In several abrasion-resistant chilled and white cast irons, additions of up to 5% nickel, alone or in combination with chromium and molybdenum, improve wear resistance. Several important series of alloy cast irons can be either martensitic or austenitic depending on nickel content. Martensitic white irons, which contain up to 7% nickel in combination with chromium, are widely used in applications requiring resistance to wear and abrasion. The high-nickel alloy cast irons that contain sufficient nickel to produce austenitic structures are used for their unique ability to resist both wear and corrosion at elevated temperatures. The austenitic cast irons contain up to 36% nickel and varying amounts of chromium and copper; these alloys are used for gas turbines, exhaust manifolds, turbocharger components, and corrosion-resisting equipment such as pumps, valves and compressors. In production of ductile iron, magnesium, frequently in the form of a nickel-magnesium alloy, is the nodulizing agent added to molten cast iron that causes eutectic graphite to grow as spheres rather than flakes when the iron solidifies.

Miscellaneous uses, including chemicals, salts, catalysts, ceramics, coinage and permanent magnets, account for about 7% of total consumption of nickel.

SELECTED REFERENCES

K. Sproule, Nickel and Its Uses, *Metals Handbook* (8th Ed.), Vol 1, American Society for Metals, Metals Park, OH, 1961, p 1113-1114

Nickel and Its Alloys, Monograph 106, U.S. National Bureau of Standards, 1968

Minerals Yearbook, U.S. Bureau of Mines, Washington, 1977

Properties of Nickels and Nickel Alloys

Wrought Nickel Alloys

By John Gadbut and Donald E. Wenschhof
Huntington Alloys, Inc.
and
Robert B. Herchenroeder
High Technology Materials Division
Cabot Corporation

Nickel 200
99.5Ni-0.08C

Commercial Names

Trade name. Nickel 200
Common name. Commercially pure nickel
UNS number. N02200

Specifications

ANSI. H34.1, H34.7, H34.8, H34.15, H34.42
ASME. SB160, SB161, SB162, SB163
ASTM. B160, B161, B162, B163, B366

Chemical Composition

Composition limits. 99.0 min Ni + Co, 0.15 max C, 0.25 max Cu, 0.40 max Fe, 0.35 max Mn, 0.35 max Si, 0.01 max S

Applications

Typical uses. Chemical and food processing, electronic parts, aerospace equipment
Precaution in use. Should not be used at service temperatures above 315 °C (600 °F)

Mechanical Properties

Tensile properties. See Tables 1 and 2.
Compressive properties. See Table 3.
Hardness. See Table 3.
Poisson's ratio. 0.264 at 20 °C (68 °F)
Elastic modulus. Tension, 204 GPa (29.6×10^6 psi) at 20 °C (68 °F); torsion, 81 GPa (11.7×10^6 psi) at 20 °C (68 °F)
Impact strength. See Table 4.
Fatigue strength. See Table 5.

Structure

Crystal structure. Face-centered cubic
Microstructure. Nickel 200 is a solid-solution alloy. Its microstructure typically exhibits a minor amount of nonmetallic inclusions, principally oxides, which are unchanged by annealing. Prolonged exposure at 425 to 650 °C (800 to 1200 °F) causes precipitation of graphite, which is the reason the alloy is not recommended for service above 315 °C (600 °F).

Mass Characteristics

Density. 8.89 Mg/m^3 (0.321 lb/in.^{3}) at 20 °C (68 °F)

Thermal Properties

Liquidus temperature. 1466 °C (2635 °F)
Solidus temperature. 1435 °C (2615 °F)
Coefficient of thermal expansion. See Table 6.
Specific heat. 456 J/kg·K (0.106 Btu/lb·°F) at 21 °C (70 °F)
Thermal conductivity. See Table 6.

Electrical Properties

Electrical conductivity. Volumetric, 18.2% IACS at 21 °C (70 °F)
Electrical resistivity. See Table 6.

Magnetic Properties

Magnetic permeability. Ferromagnetic.
Curie temperature. 360 °C (680 °F)

Chemical Properties

General corrosion behavior. Nickel 200 is highly resistant to many corrosive media. Although most useful in reducing environments, it can be used also under oxidizing conditions that cause the formation of a passive oxide film. The alloy has excellent resistance to caustics, high-temperature halogens and hydrogen halides, and salts other than oxidizing halides. It is also well suited to food processing, in which product purity must be maintained. Nickel 201 (low-carbon nickel) should be used for applications involving temperatures above 315 °C (600 °F).

Resistance to specific agents. An outstanding characteristic of Nickel 200 is its resistance to caustic soda and other alkalies except ammonium hydroxide. Table 7 gives corrosion rates in caustic soda.

Table 1 Tensile properties(a) of Nickel 200

Form and condition	Tensile strength MPa	ksi	Yield strength (0.2% offset) MPa	ksi	Elongation, %
Rod and bar					
Hot finished	414-586	60-85	103-310	15-45	55-35
Cold drawn	448-758	65-110	276-690	40-100	35-10
Annealed	379-517	55-75	103-207	15-30	55-40
Plate					
Hot rolled	379-690	55-100	138-552	20-80	55-35
Annealed	379-552	55-80	103-276	15-40	60-40
Sheet					
Hard	621-793	90-115	483-724	70-105	15-2
Annealed	379-517	55-75	103-207	15-30	55-40
Strip					
Spring temper	621-896	90-130	483-793	70-115	15-2
Annealed	379-517	55-75	103-207	15-30	55-40
Tubing					
Stress relieved	448-758	65-110	276-621	40-90	35-15
Annealed	379-517	55-75	83-207	12-30	60-40
Wire					
Annealed	379-586	55-85	103-345	15-50	50-30
Spring temper	862-1000	125-145	724-931	105-135	15-2

(a) Values shown represent usual ranges for common section sizes. In general, values in the higher portions of the ranges are not obtainable with large section sizes, and exceptionally small or large sections may have properties outside the ranges.

Table 2 Typical tensile properties of annealed Nickel 200 as a function of temperature

Temperature °C	°F	Tensile strength MPa	ksi	Yield strength (0.2% offset) MPa	ksi	Elongation, %
20	68	462	67.0	148	21.5	47.0
93	200	458	66.5	154	22.3	46.0
149	300	460	66.7	150	21.7	44.5
204	400	458	66.5	139	20.2	44.0
260	500	465	67.5	135	19.6	45.0
316	600	456	66.2	139	20.2	47.0
371	700	362	52.5	117	17.0	61.5

Table 6 Thermal and electrical properties of annealed Nickel 200

Temperature °C	°F	Mean linear expansion(a) μm/m·K	μin./in.·°F	Thermal conductivity W/m·K	Btu/ft·h·°F	Electrical resistivity, nΩm
−253	−423	8.5	4.7	...	...	...
−184	−300	10.4	5.8	...	...	27
−200	−129	11.2	6.2	77.2	44.6	43
−100	−73	11.3	6.3	...	...	58
−18	0	...	...	72.1	41.7	80
21	70	...	...	...	...	95
93	200	13.3	7.4	67.1	38.8	126
204	400	13.9	7.7	61.3	35.4	188
316	600	14.4	8.0	56.3	36.5	273

(a) From 21 °C (70 °F) to temperature shown.

Fabrication Characteristics

Annealing temperature. 700 to 925 °C (1300 to 1700 °F)

Hot working temperature. 650 to 1230 °C (1200 to 2250 °F)

Table 3 Typical compressive strength and hardness of Nickel 200

Material condition	Compressive yield strength (0.2% offset) MPa	ksi	Hardness, HB
Hot rolled	159	23.0	107
Cold drawn 24%	400	58.0	177
Annealed	179	26.0	109

Table 4 Typical impact strength of Nickel 200

Material condition	Impact strength Izod J	ft·lb	Charpy V-notch J	ft·lb
Hot rolled	163	120	271	200
Cold drawn, stress relieved	163	120	277	204
Cold drawn, annealed	163	120	309	228

Table 5 Typical fatigue strength of Nickel 200

Cycles	Fatigue strength Cold drawn rod MPa	ksi	Annealed rod MPa	ksi
10^4	751	109	...	...
10^5	579	84	358	52
10^6	434	63	276	40
10^7	358	52	234	34
10^8	345	50	228	33

Table 7 Corrosion of Nickel 200 in caustic soda solutions

Environment	Temperature °C	°F	Corrosion rate μm/yr	mils/yr
Laboratory tests in 4% solution:	20	68		
Quiet immersion			1	0.05
Air-agitated immersion			1	0.05
Continuous alternate immersion			13	0.50
Intermittent alternate immersion			15	0.60
Spray test			1	0.05
Plant tests in 14% solution in first effect of multiple-effect evaporator	88	190	0.5	0.02
Plant tests in 23% solution in tank receiving liquor from evaporator	104	220	4.1	0.16
Plant tests in single-effect evaporator concentrating solution from 30 to 50%	82	179	2.5	0.10
Plant tests in evaporator concentrating to 50% solution			3	0.1
Laboratory tests during concentration from 32 to 52% (vacuum, 640-685 mm Hg)	85-91	185-196	33	1.3
Tests in storage tank containing 49-51% solution	55-75	131-167	0.5	0.02
Laboratory tests in 75% solution	121	250	25	1.0
	204	400	20	0.8
Plant tests in 70% electrolytic solution in receiving tank	90-115	194-239	3	0.1

Fig. 1 Typical rupture strength of annealed Nickel 201

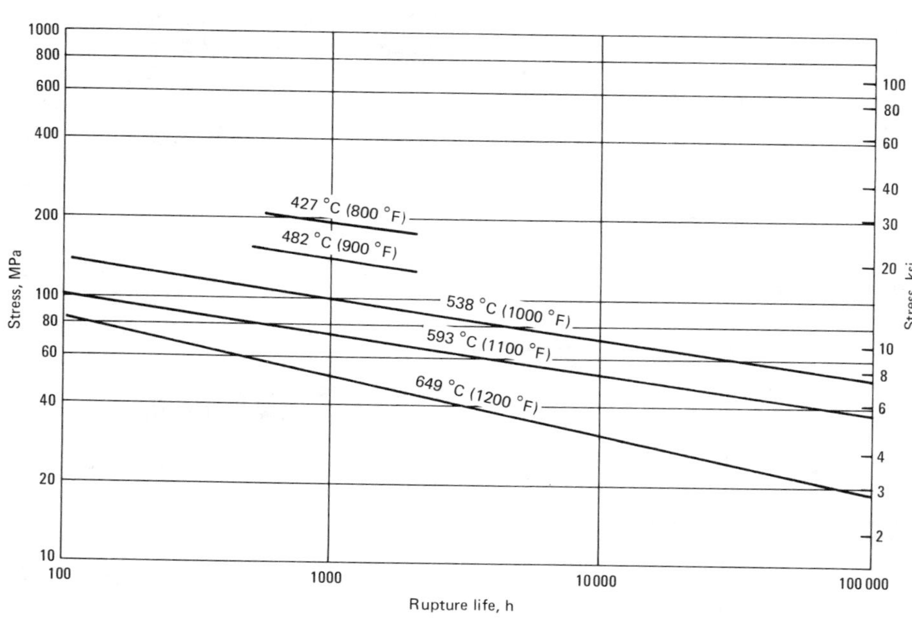

Nickel 201
99.5Ni-0.01C

Commercial Names

Trade name. Nickel 201
Common name. Low-carbon nickel
UNS number. N02201

Specifications

AMS. 5553
ANSI. H34.1, H34.7, H34.8, H34.15, H34.42
ASME. SB160, SB161, SB162, SB163
ASTM. B160, B161, B162, B163, B366

Chemical Composition

Composition limits. 99.0 min Ni + Co, 0.02 max C, 0.25 max Cu, 0.40 max Fe, 0.35 max Mn, 0.35 max Si, 0.010 max S

Applications

Typical uses. Caustic evaporators, combustion boats, plater bars, electronic parts

Mechanical Properties

Tensile properties. See Table 8.
Elastic modulus. Tension, 207 GPa (30×10^6 psi)
Creep-rupture characteristics. See Fig. 1 and 2.

Structure

Crystal structure. Face-centered cubic

Mass Characteristics

Density. 8.88 Mg/m^3 (0.321 lb/in.^{3}) at 20 °C (68 °F)

Thermal Properties

Liquidus temperature. 1446 °C (2635 °F)
Solidus temperature. 1435 °C (2615 °F)
Coefficient of thermal expansion. 13.3 μm/m·K (7.4 μin./in.·°F) at 21 to 93 °C (70 to 200 °F)
Specific heat. 456 J/kg·K (0.106 Btu/lb·°F) at 21 °C (70 °F)
Thermal conductivity. See Table 9.

Electrical Properties

Electrical conductivity. Volumetric, 20.5% IACS at 27 °C (80 °F)
Electrical resistivity. See Table 9.

Magnetic Properties

Magnetic permeability. Ferromagnetic

Curie temperature. 360 °C (680 °F)

Chemical Properties

General corrosion behavior. Nickel 201 has the same corrosion resistance as Nickel 200. Nickel 201, however, has a low carbon content and is not subject to embrittlement by precipitated carbon or graphite at high temperatures. It is therefore preferred to Nickel 200 for use at temperatures above 315 °C (600 °F).

Fabrication Characteristics

Annealing temperature. 650 to 927 °C (1200 to 1700 °F)
Hot working temperature. 650 to 1230 °C (1200 to 2250 °F)

Table 8 Typical tensile properties of annealed Nickel 201 as a function of temperature

Temperature °C	Temperature °F	Tensile strength MPa	Tensile strength ksi	Yield strength (0.2% offset) MPa	Yield strength (0.2% offset) ksi	Elongation, %
20	68	403	58.5	103	15.0	50
93	200	387	56.1	106	15.4	45
149	300	372	54.0	99	14.4	46
204	400	372	54.0	102	14.8	44
260	500	372	54.0	101	14.6	41
316	600	362	52.5	105	15.3	42
371	700	325	47.2	97	14.1	53
427	800	284	41.2	93	13.5	58
482	900	259	37.5	89	12.9	58
538	1000	228	33.1	83	12.1	60
593	1100	186	27.0	77	11.2	72
649	1200	153	22.2	70	10.2	74

Fig. 2 Typical creep strength of annealed Nickel 201

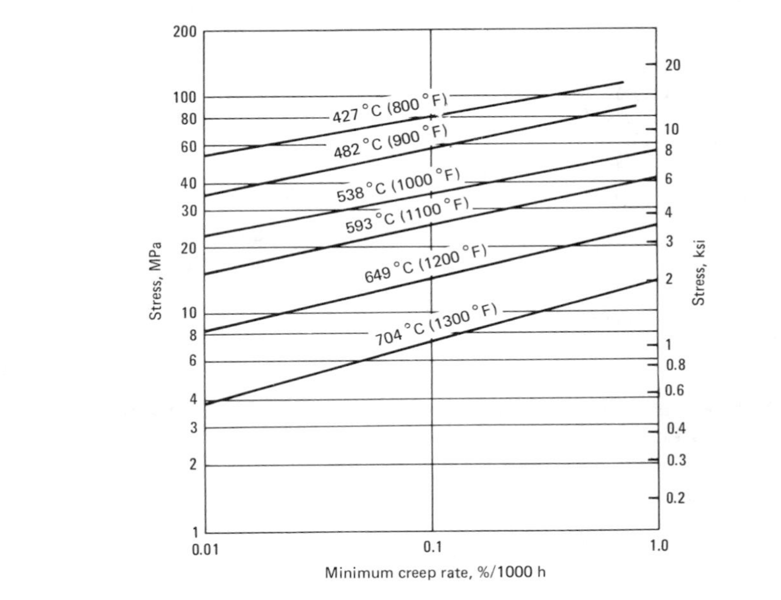

Table 9 Thermal and electrical properties of Nickel 201

Temperature °C	Temperature °F	Thermal conductivity W/m · K	Thermal conductivity Btu/ft · h · °F	Electrical resistivity, nΩ · m
−196	−320	...	...	16.6
−184	−300	95.5	55.2	...
−73	−100	...	...	48.2
−18	0	90.9	52.5	71.5
27	80	...	...	84.8
38	100	...	...	91.4
93	200	73.8	42.6	118.0
204	400	66.4	38.4	182.9
316	600	58.8	33.9	266.0
427	800	56.5	32.6	347.4
538	1000	59.1	34.1	385.7
649	1200	61.7	35.6	420.6
760	1400	64.2	37.1	455.5
871	1600	66.8	38.6	483.8
982	1800	69.2	40.0	512.0
1093	2000	...	...	523.7

Nickel 270
99.9Ni

Commercial Names

Trade name. Nickel 270
UNS number. N02270

Specifications

ASTM. F239

Chemical Composition

Composition limits. 99.97 min Ni, 0.02 max C, 0.005 max Fe, 0.001 max Cu, 0.001 max Mn, 0.001 max Si, 0.001 max S, 0.001 max Co, 0.001 max Cr, 0.001 max Mg, 0.001 max Ti

Applications

Typical uses. Cathode shanks, fluorescent lamps, hydrogen thyratrons, anodes and passive cathodes, heat exchangers, heat shields.

Mechanical Properties

Tensile properties. See Table 10.
Hardness. See Table 10.

Elastic modulus. Tension, 207 GPa (30×10^6 psi)

Structure

Crystal structure. Face-centered cubic

Mass Characteristics

Density. 8.88 Mg/m³ (0.321 lb/in.³) at 20 °C (68 °F)

Thermal Properties

Melting point. 1454 °C (2650 °F)

Table 10 Typical room-temperature tensile properties and hardness of Nickel 270

Form and condition	Tensile strength MPa	ksi	Yield strength (0.2% offset) MPa	ksi	Elongation, %	Hardness, HRB
Rod and bar, hot finished	345	50	110	16	50	40
Strip:						
Cold rolled	655	95	621	90	4	95
Annealed	345	50	110	16	50	35
Sheet, annealed	345	50	110	16	45	30

Table 11 Thermal and electrical properties of Nickel 270

Temperature °C	°F	Mean linear expansion(a) μm/m · K	μin./in. · °F	Thermal conductivity W/m · K	Btu/ft · h · °F	Electrical resistivity, nΩ · m
−196	−320	...	...	...	...	6.6
−129	−200	...	...	108	62.4	23.3
−18	0	...	...	91	52.6	59.8
27	80	...	...	...	...	74.8
93	200	13.3	7.4	79	45.6	106.4
204	400	13.7	7.6	70	40.4	169.6
316	600	14.4	8.0	62	35.8	254.3
427	800	15.1	8.4	59	34.1	329.2
538	1000	15.5	8.6	62	35.8	364.1
649	1200	15.8	8.8	65	37.6	395.6
760	1400	16.2	9.0	67	38.7	425.6
871	1600	16.6	9.2	70	40.4	448.8
982	1800	...	...	73	42.2	480.4
1093	2000	...	...	...	...	510.4

(a) From 21 °C (70 °F) to temperature shown.

Fig. 3 Recrystallization behavior of Nickel 270 (annealing time, 30 min)

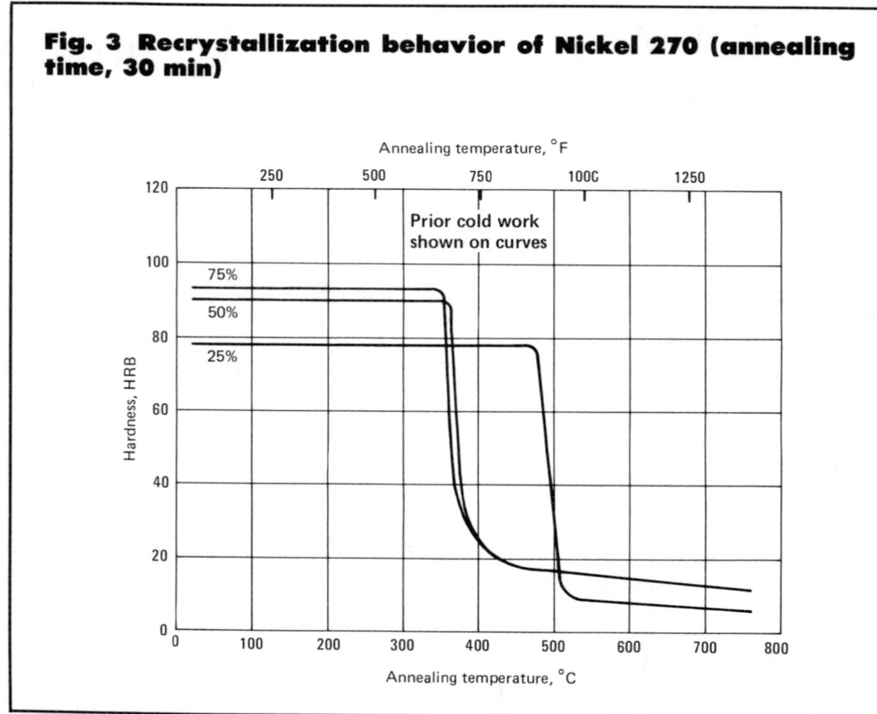

Coefficient of thermal expansion. See Table 11.
Specific heat. 460 J/kg·K (0.107 Btu/lb·°F) at 20 °C (68 °F)
Thermal conductivity. See Table 11.

Electrical Properties

Electrical conductivity. Volumetric, 23.0% IACS at 27 °C (80 °F)
Electrical resistivity. See Table 11.

Magnetic Properties

Magnetic permeability. Ferromagnetic
Curie temperature. 358 °C (676 °F)

Chemical Properties

General corrosion behavior. Nickel 270 has essentially the same corrosion resistance as Nickel 200 and Nickel 201. Nickel 270 is a high-purity product and therefore may be more susceptible to sulfur embrittlement under certain conditions.

Fabrication Characteristics

Recrystallization temperature. See Fig. 3.

Duranickel 301
95Ni-4.4Al-0.6Ti

Commercial Names

Trade name. Duranickel 301

Chemical Composition

Composition limits. 93.0 min Ni + Co, 0.25 max Cu, 0.60 max Fe, 0.50 max Mn, 0.30 max C, 1.0 max Si, 0.01 max S, 4.0 to 4.75 Al, 0.25 to 1.0 Ti

Applications

Typical uses. Duranickel 301 is used for applications that require the corrosion resistance of commercially pure nickel but the greater strength or spring properties. Examples are diaphragms, springs, clips, press components for extrusion of plastics, and molds for production of glass articles.

Mechanical Properties

Tensile properties. See Table 12.
Compressive properties. Compressive yield strength, hot rolled and aged material: 948 MPa (137.5 ksi) at 0.2% offset

Poisson's ratio. 0.31
Elastic modulus. Tension, 207 GPa (30×10^6 psi) at 27 °C (80 °F)
Fatigue strength. Hot rolled and aged material, rotating-beam: 352 MPa (51 ksi) at 10^8 cycles

Structure

Crystal structure. Face-centered cubic.
Microstructure. The alloy is age hardened by the precipitation of a gamma prime phase during heat treatment. Other phases that may be present are graphite and carbides.

Mass Characteristics

Density. 8.25 Mg/m^3 (0.298 lb/in.^{3}) at 20 °C (68 °F)

Thermal Properties

Liquidus temperature. 1438 °C (2620 °F)
Solidus temperature. 1399 °C (2550 °F)
Coefficient of thermal expansion. See Table 13.
Specific heat. Mean, 435 J/kg·K (0.101 Btu/lb·°F) at 21 to 100 °C (70 to 212 °F)
Thermal conductivity. See Table 13.

Electrical Properties

Electrical conductivity. Volumetric, 4.1% IACS at 21 °C (70 °F)
Electrical resistivity. See Table 13.

Magnetic Properties

Magnetic properties vs treatment. In the annealed condition the alloy is slightly magnetic at room temperature and nonmagnetic above about 49 °C (120 °F). Age hardening increases the magnetic properties slightly and raises the Curie temperature to about 93 °C (200 °F).
Curie temperature. Annealed, 16 to 49 °C (60 to 120 °F); aged, 93 °C (200 °F)

Chemical Properties

General corrosion behavior. The corrosion resistance of alloy 301 is essentially the same as that of Nickel 200. Alloy 301 has exceptional resistance to fluoride glasses, and is used for glass molds.

Fabrication Characteristics

Annealing temperature. 870 to 980 °C (1600 to 1800 °F)
Aging temperature. Annealed material: 582 to 593 °C (1080 to 1100 °F) for 16 h, furnace cool to 482 °C (900 °F) at a maximum rate of 9 °C (15 °F) per hour. Further cooling may be by air or furnace cooling or by water quenching.
Hot working temperature. 1038 to 1232 °C (1900 to 2250 °F)

Table 12 Typical tensile properties of age-hardened Duranickel 301

Temperature		Tensile strength		Yield strength (0.2% offset)		Elongation,
°C	°F	MPa	ksi	MPa	ksi	%
21	70	1276	185	910	132	28
316	600	1158	168	827	120	29
371	700	1124	163	807	117	27
427	800	1069	155	786	114	24
482	900	972	141	752	109	11
538	1000	814	118	683	99	7
593	1100	648	94	517	75	5
649	1200	476	69	372	54	4
704	1300	290	42	234	34	8
760	1400	172	25	97	14	60
816	1500	117	17	62	9	98

Table 13 Thermal and electrical properties of age-hardened Duranickel 301

Temperature		Mean linear expansion(a)		Thermal conductivity		Electrical resistivity,
°C	°F	μm/m · K	μin./in. · °F	W/m · K	Btu/ft · h · °F	nΩ · m
21	70	...	...	23.8	13.8	424
93	200	13.0	7.2	25.4	14.7	465
204	400	13.7	7.6	28.6	15.9	500
316	600	14.0	7.8	32.2	18.6	530
427	800	14.4	8.0	35.0	19.4	560
538	1000	14.8	8.2	38.2	22.1	580
649	1200	15.3	8.5	41.2	23.8	595
760	1400	15.8	8.8	44.1	25.4	610
871	1600	16.4	9.1	47.0	27.2	630
982	1800	...	...	49.3	28.5	650
1093	2000	...	...	51.6	29.8	670

(a) From 21 °C (70 °F) to temperature shown.

Monel 400
66.5Ni-31.5Cu

Commercial Names

Trade name. Monel 400
UNS number. N04400

Specifications

AMS. 4544, 4574, 4575, 4675, 4730, 4731, 7233
ANSI. H34.2, H34.6, H34.9, H34.15, H34.42
ASME. SB127, SB163, SB164, SB165
ASTM. B127, B163, B164, B165, B366, B564
UNS number. N04400
Government. QQ-N-281

Chemical Composition

Composition limits. 63.0 to 70.0 Ni + Co, 0.30 max C, 2.0 max Mn, 2.5 max Fe, 0.24 max S, 0.50 max Si, rem Cu

Applications

Typical uses. Valve and pump parts, propeller shafts, marine fixtures and fasteners, electronic components, chemical-processing equipment, gasoline and fresh-water tanks, petroleum-processing equipment, boiler feedwater heaters and other heat exchangers.

Mechanical Properties

Tensile properties. See Tables 14 and 15, and Fig. 4.
Compressive properties. See Table 15.
Hardness. Annealed bar, 110 to 150 HB
Poisson's ratio. 0.32
Elastic modulus. Tension and compression, 179 GPa (26×10^6 psi); torsion, 66 GPa (9.5×10^6 psi)
Impact strength. See Table 16.
Fatigue strength. Rod, rotating-beam: hot rolled, 290 MPa (42.0 ksi); cold drawn, 279 MPa (40.5 ksi); annealed, 231 MPa (33.5 ksi). All values at 10^8 cycles.
Creep-rupture properties. See Fig. 5 and 6.

Structure

Crystal structure. Face-centered cubic

Microstructure. Alloy 400 is a solid-solution, single-phase alloy. Its microstructure typically exhibits randomly dispersed nonmetallic inclusions such as sulfides or silicates.

Mass Characteristics

Density. 8.83 Mg/m^3 (0.319 lb/in.^{3}) at 20 °C (68 °F)

Thermal Properties

Liquidus temperature. 1349 °C (2460 °F)
Solidus temperature. 1299 °C (2370 °F)
Coefficient of thermal expansion. See Table 17.
Specific heat. 427 J/kg·K (0.099 Btu/lb·°F) at 21 °C (70 °F)
Thermal conductivity. See Table 17.

Electrical Properties

Electrical conductivity. Volumetric, 3.4% IACS at 21 °C (70 °F)
Electrical resistivity. See Table 17.

Magnetic Properties

Magnetic properties vs treatment. The Curie temperature of Monel 400 is near room temperature and is affected by normal variations in chemical composition. Therefore some lots of material may be magnetic at room temperature and others may not.
Curie temperature. −7 to +10 °C (20 to 50 °F)

Chemical Properties

General corrosion behavior. Monel 400, a nickel-copper alloy, is more resistant than nickel to corrosion under reducing conditions and more resistant than copper under oxidizing conditions. An important characteristic of the alloy is its general freedom from stress-corrosion cracking. It is highly resistant to seawater or brackish water, chlorinated solvents, glass-etching agents, many acids including sulfuric and hydrochloric, and nearly all alkalies.
Resistance to specific agents. Corrosion rates in strongly agitated and aerated seawater usually do not exceed 25.4 μm (1 mil) per year. Figure 7 shows corrosion rates in sulfuric acid; Fig. 8, rates in hydrochloric acid.

Fabrication Characteristics

Annealing temperature. 760 to 982 °C (1400 to 1800 °F)
Hot working temperature. 649 to 1177 °C (1200 to 2150 °F)

Table 14 Tensile properties(a) of Monel 400

Form and condition	Tensile strength, MPa	Tensile strength, ksi	Yield strength (0.2% offset), MPa	Yield strength (0.2% offset), ksi	Elongation, %
Rod and bar					
Annealed	517-621	75-90	172-345	25-50	60-35
Hot finished	552-758	80-110	276-690	40-100	60-30
Cold drawn, stress relieved	579-827	84-120	379-690	55-100	40-22
Plate					
Hot rolled	517-655	75-95	276-517	40-75	45-30
Annealed	483-586	70-85	193-345	28-50	50-35
Sheet					
Annealed	483-586	70-85	172-310	25-45	50-35
Hard	690-827	100-120	621-758	90-110	15-2
Strip					
Annealed	483-586	70-85	172-310	25-45	55-35
Spring temper	690-965	100-140	621-896	90-130	15-2
Tubing, cold drawn					
Annealed	483-586	70-85	172-310	25-45	50-35
Stress relieved	586-827	85-120	379-690	55-100	35-15
Wire					
Annealed	483-655	70-95	207-379	30-55	45-25
Spring temper	1000-1241	145-180	862-1172	125-170	5-2

(a) Values shown represent usual ranges for common section sizes. In general, values in the higher portions of the ranges are not obtainable with large section sizes, and exceptionally small or large sections may have properties outside the ranges.

Table 16 Impact strength of Monel 400

Material condition	Izod, J	Izod, ft·lb	Charpy V-notch, J	Charpy V-notch, ft·lb
Hot rolled	136-163+	100-120	298	220
Forged	102-156	75-115	...	...
Cold drawn	102-156	75-115	203	150
Annealed	122-163+	90-120	291	215

Table 15 Typical tensile and compressive properties of Monel 400

Material condition	Tension: Tensile strength, MPa	Tension: Tensile strength, ksi	Tension: Yield strength (0.01% offset), MPa	Tension: Yield strength (0.01% offset), ksi	Tension: Yield strength (0.2% offset), MPa	Tension: Yield strength (0.2% offset), ksi	Tension: Elongation, %	Compression: Yield strength (0.01% offset), MPa	Compression: Yield strength (0.01% offset), ksi	Compression: Yield strength (0.2% offset), MPa	Compression: Yield strength (0.2% offset), ksi
Hot rolled	579	84	255	37	283	41	39.5	228	33	262	38
Cold drawn, stress equalized	669	97	517	75	600	87	27.0	400	58	558	81
Cold drawn, annealed	538	78	193	28	228	33	44.0	131	19	193	28

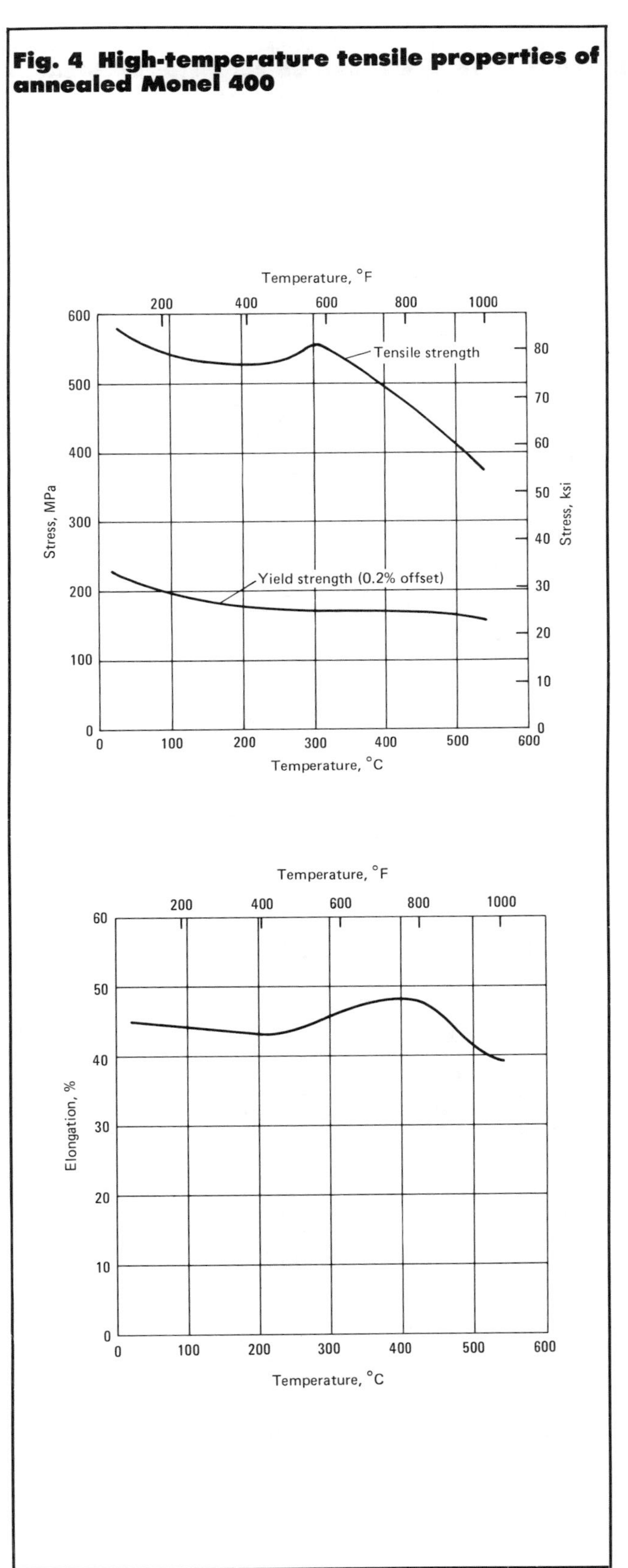

Fig. 4 High-temperature tensile properties of annealed Monel 400

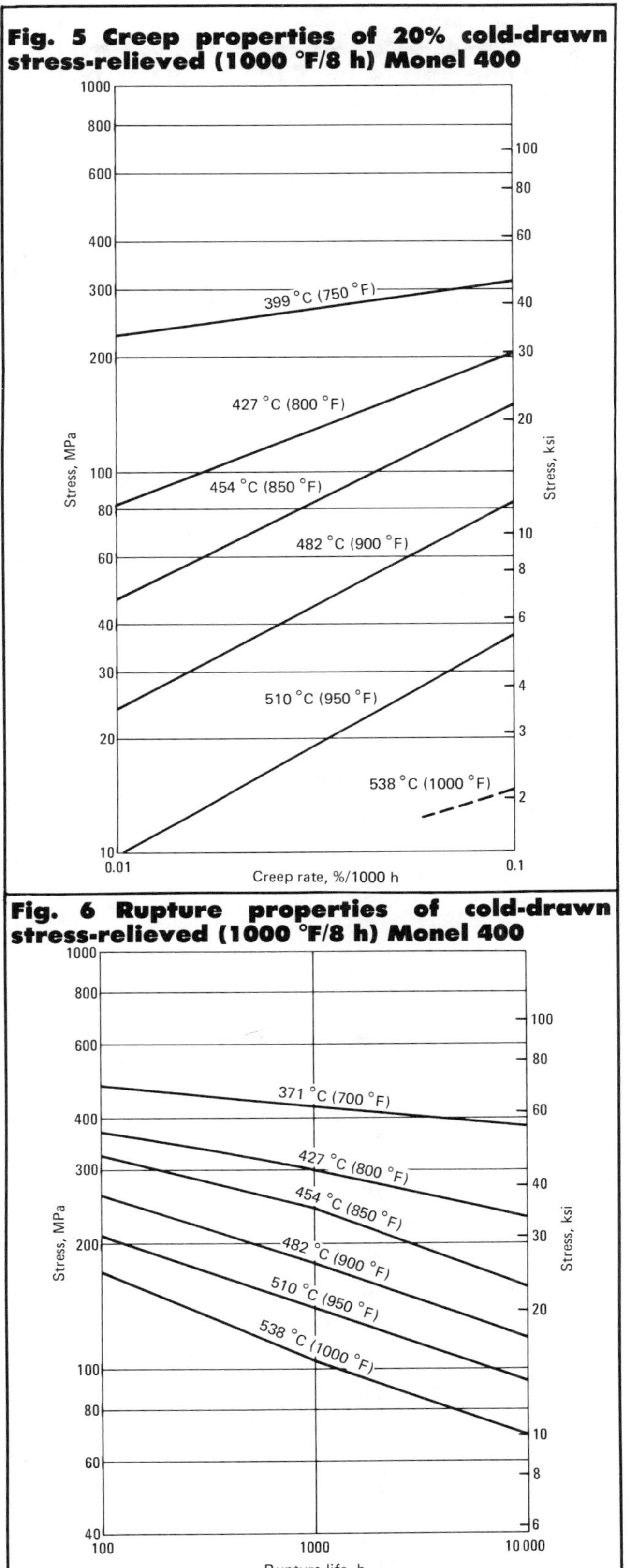

Fig. 5 Creep properties of 20% cold-drawn stress-relieved (1000 °F/8 h) Monel 400

Fig. 6 Rupture properties of cold-drawn stress-relieved (1000 °F/8 h) Monel 400

Fig. 7 Corrosion of Monel 400 in sulfuric acid (temperature 66 °C [86 °F]; velocity, 8.6 mm/s [17 ft/min])

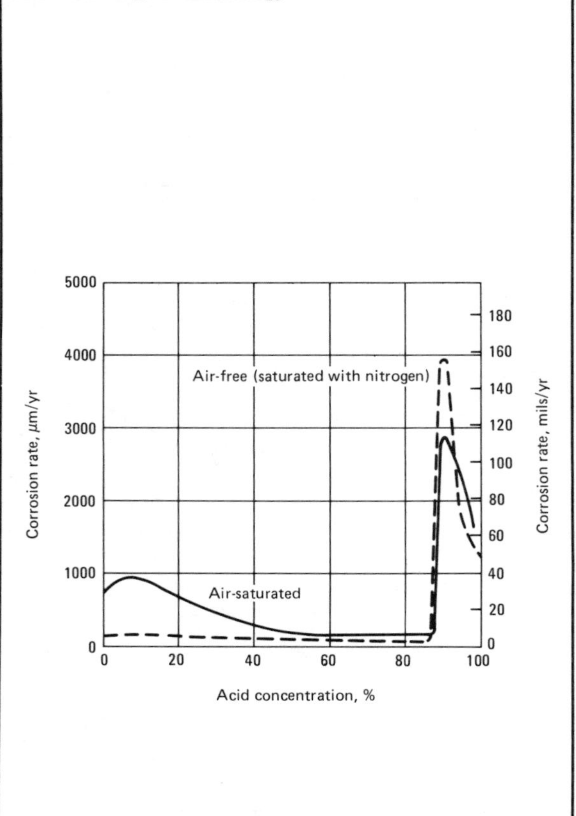

Fig. 8 Effect of temperature on corrosion of Monel 400 in 5% hydrochloric acid

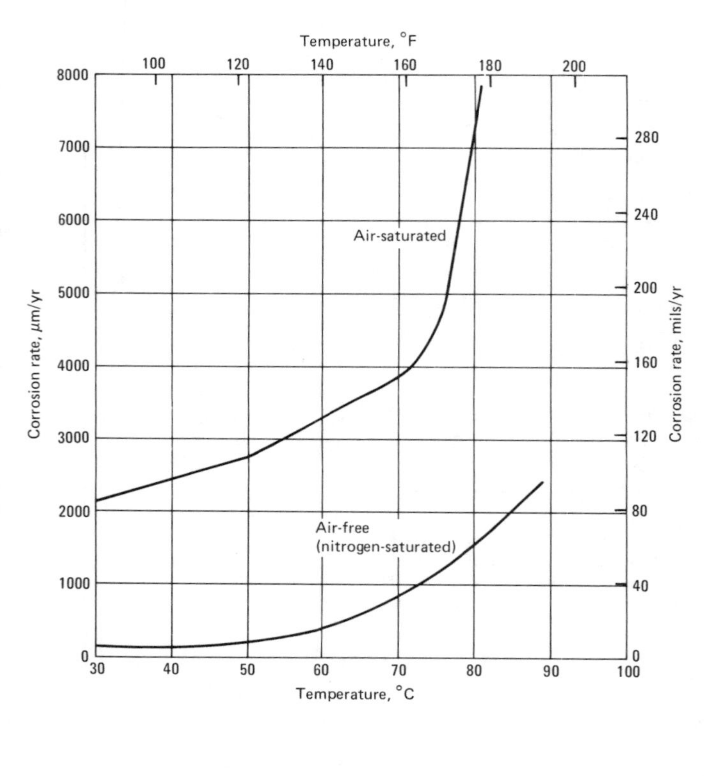

Table 17 Thermal and electrical properties of Monel 400

Temperature °C	Temperature °F	Mean linear expansion(a) μm/m · K	Mean linear expansion(a) μin./in. · °F	Thermal conductivity W/m · K	Thermal conductivity Btu/ft · h · °F	Electrical resistivity, nΩ · m
−196	−320	...	...	...	...	341
−184	−300	11.0	6.1	16.3	9.4	...
−129	−200	11.5	6.4	18.8	10.9	...
−73	−100	12.1	6.7	20.0	11.6	...
21	70	...	...	21.8	12.6	510
93	200	13.9	7.7	24.1	14.0	535
204	400	15.5	8.6	27.8	16.1	560
316	600	15.8	8.8	31.0	18.9	575
427	800	16.0	8.9	34.3	19.8	590
538	1000	16.4	9.1	38.1	22.0	610
649	1200	16.7	9.3	41.4	23.9	630
760	1400	17.3	9.6	44.9	25.9	650
871	1600	17.6	9.8	48.3	27.9	670
982	1800	18.0	10.0	51.9	30.0	690
1093	2000	18.5	10.3	...	...	710

(a) From 21 °C (70 °F) to temperature shown.

Monel R-405
66.5Ni-31.5Cu-0.04S

Commercial Names

Trade name. Monel R-405
UNS number. N04405

Specifications

AMS. 4674, 7234
ANSI. H34.9
ASME. SB164
ASTM. B164
Government. QQ-N-281

Chemical Composition

Composition limits. 63.0 min Ni + Co, 0.3 max C, 2.0 max Mn, 2.5 max Fe, 0.025 to 0.060 S, 0.5 max Si, 28.0-34.0 Cu

Applications

Typical uses. Monel R-405 is a free-machining version of Monel 400 designed for parts to be produced by automatic machining. Applications include screw-machine products, water-meter parts, valve-seat inserts, and fasteners for nuclear equipment.

Mechanical Properties

Tensile properties. See Table 18.
Compressive properties. See Table 18.
Poisson's ratio. 0.32
Elastic modulus. See Monel 400
Impact strength. Charpy V-notch: hot rolled, 254 J (187 ft·lb); cold drawn, 190 J (140 ft·lb); annealed, 266 J (196 ft·lb)
Fatigue strength. Rod, rotating-beam: hot rolled, 248 MPa (36 ksi); cold drawn, 252 MPa (36.5 ksi); annealed, 207 MPa (30 ksi). All values at 10^8 cycles

Structure

Crystal structure. Face-centered cubic
Microstructure. The relatively high sulfur content of Monel R-405 results in a substantial amount of Ni-Cu sulfides in the microstructure. The sulfides, which enhance machinability, appear as elongated stringers.

Mass Characteristics

Density. 8.83 Mg/m^3 (0.319 lb/in.^{3}) at 20 °C (68 °F)

Thermal Properties

Liquidus temperature. 1350 °C (2460 °F)
Solidus temperature. 1300 °C (2370 °F)
Coefficient of thermal expansion. Linear: 13.9 μm/m·K (7.7 μin./in.·°F) from 21 to 93 °C (70 to 200 °F); 15.7 μm/m·K (8.7 μin./in.·°F) from 21 to 260 °C (70 to 500 °F); 16.4 μm/m·K (9.1 μin./in.°F) from 21 to 538 °C (70 to 1000 °F).
Specific heat. 427 J/kg·K (0.009 Btu/lb·°F) at 20 °C (68 °F)
Thermal conductivity. See Monel 400.

Electrical Properties

Electrical conductivity. Volumetric, 3.4% IACS at 20 °C (68 °F)
Electrical resistivity. See Monel 400.

Magnetic Properties

Magnetic properties vs treatment. See Monel 400.
Curie temperature. −7 to +10 °C (19 to 50 °F)

Chemical Properties

General corrosion behavior. See Monel 400.

Fabrication Characteristics

Annealing temperature. 760 to 980 °C (1400 to 1800 °F)
Hot working temperature. Monel R-405 is not recommended for forging.

Table 18 Typical tensile and compressive properties of Monel R-405

Material condition	Tension: Tensile strength, MPa	ksi	Yield strength (0.01% offset), MPa	ksi	Yield strength (0.2% offset), MPa	ksi	Elongation, %	Compression: Yield strength (0.01% offset), MPa	ksi	Yield strength (0.2% offset), MPa	ksi
Hot rolled	524	76	228	33	248	36	39.5	179	26	234	34
Cold drawn, stress equalized	572	83	427	62	510	74	28.0	352	51	455	66
Cold drawn, annealed	503	73	172	25	193	28	44.5	159	23	179	26

Monel K-500
66.5Ni-29.5Cu-2.7 Al-0.6Ti

Commercial Names

Trade name. Monel K-500
UNS number. N05500

Specifications

AMS. 4676
Government. QQ-N-286

Chemical Composition

Composition limits. 63.0 min Ni + Co, 0.25 max C, 1.5 max Mn, 2.0 max Fe, 0.01 max S, 0.5 max Si, 2.30 to 3.15 Al, 0.35 to 0.85 Ti, 27.0 to 33.0 Cu

Applications

Typical uses. Monel K-500 is an age-hardenable Ni-Cu alloy used for applications that require the corrosion resistance of Monel 400 but greater strength. Examples are pump shafts and impellers, doctor blades, oil-well drill collars, springs, and valve trim.

Mechanical Properties

Tensile properties. See Table 19 and Fig. 9.
Compressive properties. See Table 19.
Hardness. See Table 19.
Poisson's ratio. 0.32
Elastic modulus. Tension, 179 GPa (26 × 10^6 psi); torsion, 66 GPa (9.5 × 10^6 psi)
Impact strength. Aged bar, Charpy V-notch: 50 J (36.9 ft·lb) at 20 °C (68 °F); 42 J (30 ft·lb) at −196 °C (−320 °F)
Fatigue strength. Rod, rotating-

beam: annealed, 262 MPa (38 ksi); hot rolled, 296 MPa (43 ksi); hot rolled and aged, 352 MPa (52 ksi). All values at 10^8 cycles

Creep-rupture characteristics. See Fig. 10 and 11.

Structure

Crystal structure. Face-centered cubic

Microstructure. Exposure to age-hardening temperatures results in precipitation of a gamma prime phase throughout the matrix.

Mass Characteristics

Density. 8.47 Mg/m^3 (0.305 lb/in.^{3}) at 20 °C (68 °F)

Thermal Properties

Liquidus temperature. 1350 °C (2460 °F)

Solidus temperature. 1315 °C (2400 °F)

Coefficient of thermal expansion. See Table 20.

Specific heat. 419 J/kg·K (0.097 Btu/lb·°F) at 21 °C (70 °F)

Thermal conductivity. See Table 20.

Electrical Properties

Electrical conductivity. Volumetric, 2.8% IACS at 21 °C) 70 °F)

Electrical resistivity. See Table 20.

Magnetic Properties

Magnetic permeability. Annealed and age-hardened material, 1.0018 at a field strength of 15.9 kA/m

Curie temperature. −134 °C (−210 °F)

Chemical Properties

General corrosion behavior. The corrosion resistance of Monel K-500 is essentially the same as that of Monel 400 except that, in the age-hardened condition, Monel K-500 is more susceptible to stress-corrosion cracking in some environments

Fabrication Characteristics

Annealing temperature. 870 to 980 °C (1600 to 1800 °F). Water quench

Solution temperature. Cold finished products, 1040 °C (1900 °F); other products, 980 °C (1800 °F). Water quench

Aging cycle. Annealed material: 595 to 605 °C (1100 to 1125 °F) for 16 h; furnace cool, at a rate of 8 to 14 °C (15 to 25 °F) per hour, to 480 °C (900 °F). Further cooling may be done at any rate.

Hot working temperature. 870 to 1150 °C (1600 to 2100 °F)

Table 19 Typical tensile properties, compressive properties and hardness of Monel K-500

Property	Hot rolled	Age-hardened
Tension		
Tensile strength, MPa (ksi)	690 (100)	1041 (151)
Yield strength (0.2% offset), MPa (ksi)	324 (47)	765 (111)
Elongation, %	42.5	30.0
Compression		
Yield strength (0.2% offset), MPa (ksi)	276 (40)	834 (121)
Yield strength (0.1% offset), MPa (ksi)	234 (34)	662 (96)
Hardness, HB	165	300

Fig. 9 High-temperature tensile properties of hot finished, age-hardened Monel K-500

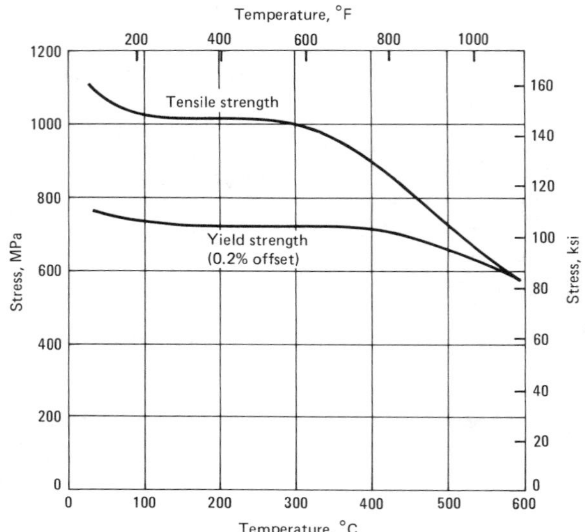

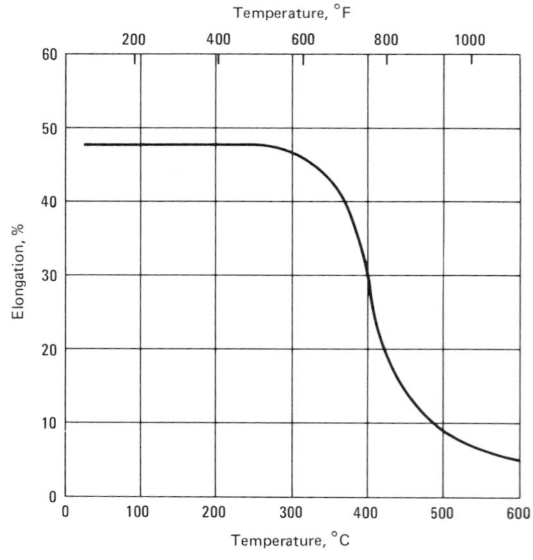

Table 20 Thermal and electrical properties of Monel K-500 as a function of temperature

Temperature °C	Temperature °F	Mean linear expansion(a) μm · K	Mean linear expansion(a) μin./in. · °F	Thermal conductivity W/m · K	Thermal conductivity Btu/ft · h · °F	Electrical resistivity, nΩ · m
−196	−320	11.2	6.2	. . .	. . .	550
−157	−250	11.7	6.5	12.4	7.2	. . .
−129	−200	12.2	6.8	13.3	7.7	. . .
−73	−100	13.0	7.2	14.9	8.6	. . .
21	70	. . .	. . .	17.4	10.0	615
93	200	13.7	7.6	19.6	11.3	618
204	400	14.6	8.1	22.5	13.0	628
316	600	14.9	8.3	25.7	14.8	640
427	800	15.3	8.5	28.6	16.5	648
538	1000	15.7	8.7	31.7	18.3	653
649	1200	16.4	9.1	34.6	20.0	658
760	1400	16.7	9.3	37.8	21.8	665
871	1600	17.3	9.6	40.7	23.5	678

(a) From 21 °C (70 °F) to temperature shown.

Fig. 10 Creep properties of Monel K-500 (cold drawn and aged)

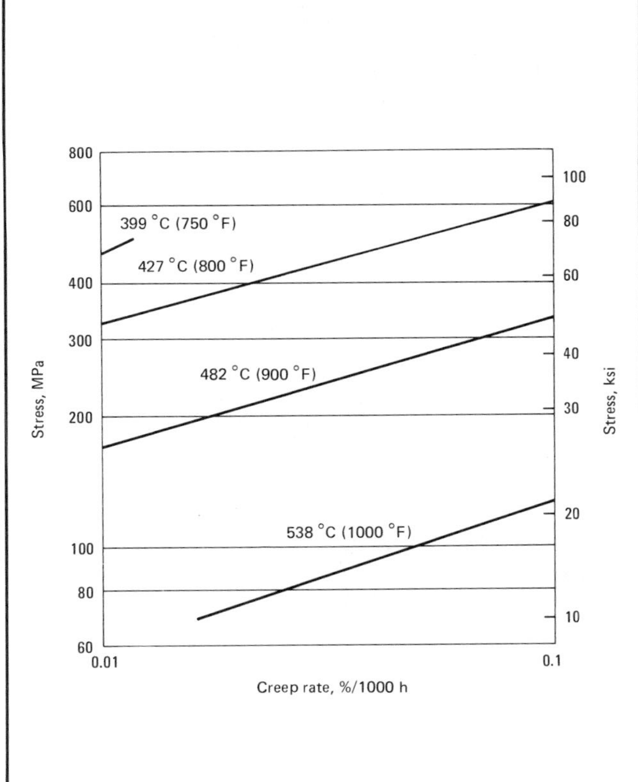

Fig. 11 Rupture life of hot finished aged Monel K-500

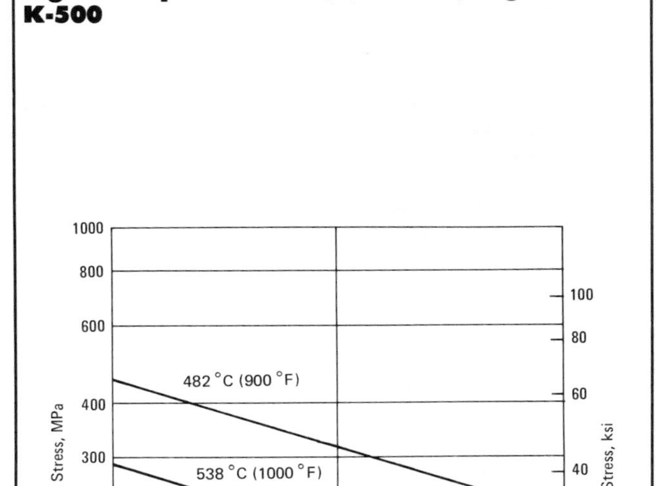

Monel 502
66.5Ni-28Cu-3Al-0.2Ti

Commercial Names

Trade name. Monel 502
UNS number. N05502

Specifications

AMS. 4677
Government. QQ-N-286

Chemical Composition

Composition limits. 63.0 min Ni + Co, 0.1 max C, 1.5 max Mn, 2.0 max Fe, 0.010 max S, 0.5 max Si, 2.5 to 3.5 Al, 0.5 max Ti, 27.0 to 33.0 Cu

Applications

Typical uses. Monel 502 is a modification of Monel K-500 that has better machinability than K-500. It is used for machined parts such as fasteners, pump and propeller shafts, and valve stems.

Mechanical Properties

Tensile properties. See Table 21 and Fig. 12.
Hardness. See Table 21.
Poisson's ratio. 0.32
Elastic modulus. Tension, 179 GPa (26 × 10^6 psi); torsion, 66 GPa (9.5 × 10^6 psi)

Structure

Crystal structure. Face-centered cubic
Microstructure. Like Monel K-500, Monel 502 is age hardened by precipitation of gamma prime. Monel 502, however, is lower in titanium and carbon to reduce formation of titanium carbides, which are abrasive to cutting tools.

Mass Characteristics

Density. 8.44 Mg/m³ (0.305 lb/in.³) at 20 °C (68 °F)

Thermal Properties

Liquidus temperature. 1350 °C (2460 °F)
Solidus temperature. 1315 °C (2400 °F)
Coefficient of thermal expansion. See Monel K-500.
Specific heat. 419 J/kg·K (0.098 Btu/lb·°F) at 20 °C (68 °F)
Thermal conductivity. See Monel K-500.

Electrical Properties

Electrical conductivity. Volumetric, 2.8% IACS at 21 °C (70 °F)
Electrical resistivity. See Monel K-500.

Magnetic Properties

Magnetic permeability. Age-hardened material, 1.002 at a field strength of 15.9 kA/m
Curie temperature. −134 °C (−210 °F)

Chemical Properties

General corrosion behavior. See Monel K-500.

Fabrication Characteristics

Annealing temperature. 760 to 870 °C (1400 to 1600 °F)
Solution temperature. 730 to 760 °C (1350 to 1400 °F). Water quench or rapid air cool
Aging cycle. 620 °C (1150 °F) for 2 h, furnace cool to 565 °C (1050 °F) and hold 4 h, furnace cool to 510 °C (950 °F) and hold 4 h, air cool
Hot working temperature. 925 to 1150 °C (1700 to 2100 °F)

Fig. 12 High-temperature tensile properties of cold drawn Monel 502 rod (annealed 1400 °F/½ h, water quenched and aged 1150 °F/2 h, furnace cooled to 1050 °F/4 h, furnace cooled to 950 °F/4 h, air cooled)

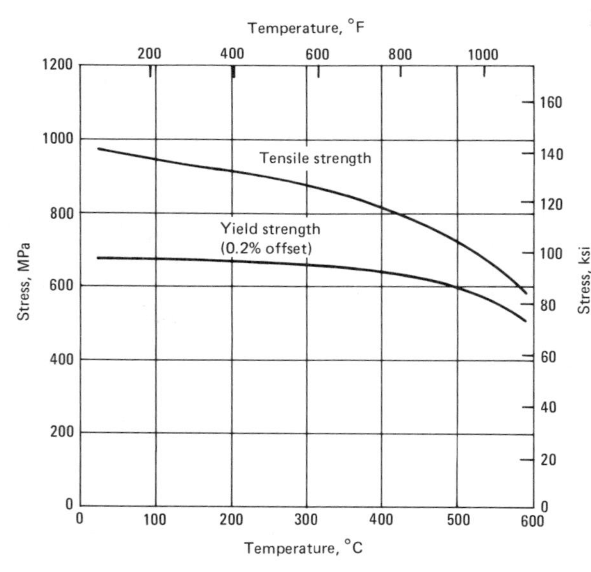

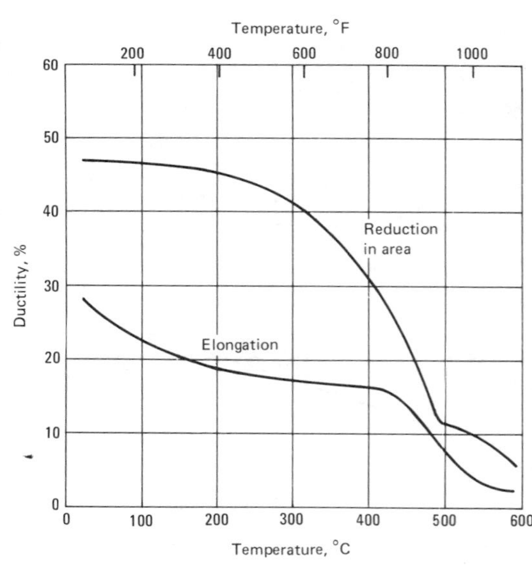

Table 21 Typical tensile properties and hardness of Monel 502

Material condition	Tensile strength MPa	ksi	Yield strength (0.2% offset) MPa	ksi	Elongation, %	Hardness
Hot rolled	586	85	255	37	47	74 HRB
Annealed	572	83	234	34	48	73 HRB
Annealed and aged	986	143	655	95	27	24 HRC

Table 22 Tensile properties(a) for Inconel 600

Form and condition	Tensile strength MPa	ksi	Yield strength (0.2% offset) MPa	ksi	Elongation, %
Rod and bar:					
Annealed	552-690	80-100	172-345	25-50	55-35
Cold drawn	724-1034	105-150	552-862	80-125	30-10
Hot finished	586-827	85-120	241-621	35-90	50-30
Plate:					
Hot rolled	586-758	85-110	241-448	35-65	50-30
Annealed	552-724	80-105	207-345	30-50	55-35
Sheet:					
Annealed	552-690	80-100	207-310	30-45	55-35
Hard	586-1034	120-150	621-862	90-125	15-2
Strip:					
Annealed	552-690	80-100	207-310	30-45	55-35
Spring temper	1000-1172	145-170	827-1103	120-160	10-2
Tubing:					
Hot finished	517-690	75-100	172-345	25-50	55-35
Cold drawn and annealed	552-690	80-100	172-345	25-50	55-35
Wire:					
Annealed	552-827	80-120	241-517	35-75	45-20
Spring temper	1172-1517	170-220	1034-1448	150-210	5-2

(a) Values shown represent usual ranges for common section sizes. In general, values in the higher portions of the ranges are not obtainable with large section sizes, and exceptionally small or large sections may have properties outside the ranges.

Fig. 13 Corrosion rates for Inconel 600 in hydrofluoric acid at 75 °C (167 °F)

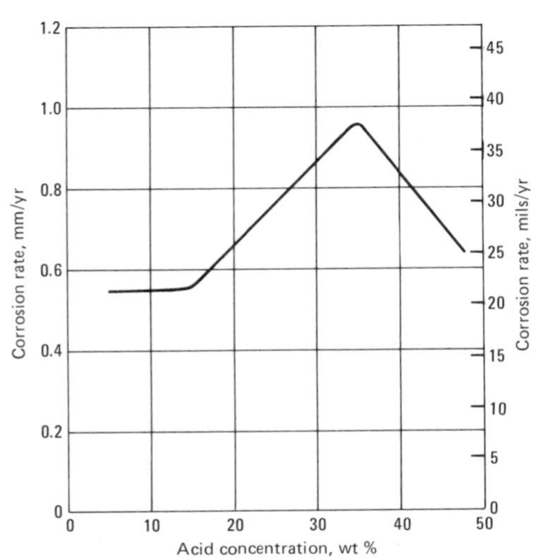

Inconel 600
76Ni-15.5Cr-8Fe

Commercial Names

Trade name. Inconel 600
UNS number. N06600

Specifications

AMS. 5540, 5580, 5665, 5687, 7232
ANSI. H34.3, H34.4, H34.10, H34.15, H34.25, H34.26, H34.42
ASME. SB163, SB166, SB167, SB168
ASTM. B163, B166, B167, B168, B366, B516, B517, B564
Government. QQ-W-390

Chemical Composition

Composition limits. 72.0 min Ni + Co, 14.0 to 17.0 Cr, 6.0 to 10.0 Fe, 0.15 max C, 1.0 max Mn, 0.015 max S, 0.50 max Si, 0.50 max Cu

Applications

Typical uses. Inconel 600 is used in a variety of applications involving temperatures from cryogenic to 1093 °C (2000 °F). Examples are chemical-processing vessels and piping, heat treating equipment, aircraft-engine and airframe components, electronic parts, and nuclear reactors.

Mechanical Properties

Tensile properties. See Tables 22, 23 and 24.
Compressive properties. See Table 23.
Hardness. See Table 24.
Tensile properties vs temperature. See Table 9 in the article "Wrought Superalloys".
Poisson's ratio. 0.29
Elastic modulus. Tension, 207 GPa (30×10^6 psi); torsion, 76 GPa (11×10^6 psi)
Impact strength. Plate, Charpy keyhole: 86.1 J (63.5 ft·lb) at 21 °C (70 °F); 88.8 J (65.5 ft·lb) at −79 °C (−110 °F); 82.4 J (60.8 ft·lb) at −196 °C (−320 °F)
Fatigue strength. Rotating-beam: annealed, 269 MPa (39.0 ksi); hot rolled, 279 MPa (40.5 ksi); cold drawn, 310 MPa (45.0 ksi). All values at 10^8 cycles and 21 °C (70 °F).
Creep-rupture characteristics. See Table 10 in the article "Wrought Superalloys".

Structure

Crystal structure. Face-centered cubic
Microstructure. Inconel 600 is a

stable, solid-solution material. The only precipitated phases in its microstructure are titanium nitrides, titanium carbides and chromium carbides.

Mass Characteristics

Density. 8.42 Mg/m³ (0.304 lb/in.³) at 20 °C (68 °F)

Thermal Properties

Liquidus temperature. 1415 °C (2575 °F)
Solidus temperature. 1355 °C (2470 °F)
Coefficient of thermal expansion. See Table 8 in the article "Wrought Superalloys".
Specific heat. 444 J/kg·K (0.103 Btu/lb·°F) at 21 °C (70 °F)
Thermal conductivity. See Table 8 in the article "Wrought Superalloys".

Electrical Properties

Electrical conductivity. Volumetric, 1.7% IACS at 21 °C (70 °F)
Electrical resistivity. 1030 nΩ·m at 21 °C (70 °F)

Magnetic Properties

Magnetic permeability. 1.010 at a field strength of 15.9 kA/m
Curie temperature. −124 °C (−192 °F)

Chemical Properties

General corrosion behavior. The high nickel content of Inconel 600 provides good resistance to corrosion under reducing conditions, and its chromium content, resistance under oxidizing conditions. The alloy is virtually immune to chloride stress-corrosion cracking.
Resistance to specific corroding agents. Inconel 600 has useful resistance to many acid solutions, both oxidizing and reducing. For corrosion rates in sulfuric acid and hydrofluoric acid, see Table 25 and Fig. 13, respectively. This alloy resists dilute hydrochloric acid but not concentrated or hot solutions. It is resistant to all concentrations of phosphoric acid at room temperature. It has poor resistance to nitric acid. Inconel 600 has excellent resistance to alkalies; Fig. 14 shows corrosion rates in boiling sodium hydroxide. Inconel 600 is unaffected by most neutral and alkaline salt solutions and resists many acid salts. It is one of the few materials suitable for use in hot, strong solutions of magnesium chloride, usually having a corrosion rate of about 25 μm (1 mil) per year.

Table 23 Typical tensile and compressive yield strengths of Inconel 600

Material condition	Tension 0.02% offset MPa	Tension 0.02% offset ksi	Tension 0.2% offset MPa	Tension 0.2% offset ksi	Compression 0.02% offset MPa	Compression 0.02% offset ksi	Compression 0.2% offset MPa	Compression 0.2% offset ksi
Hot rolled and annealed	268	38.9	303	43.9	276	40.0	309	44.8
Cold drawn and stress relieved	552	80.0	619	89.8	513	74.4	605	87.7
As extruded (tubing)	174	25.2	212	30.8	192	27.9	224	32.5

Table 24 Typical tensile properties and hardness of Inconel 600

Form and condition	Tensile strength MPa	Tensile strength ksi	Yield strength (0.2% offset) MPa	Yield strength (0.2% offset) ksi	Elongation, %	Hardness
Rod:						
As rolled	672	97.5	307	44.5	46	86 HRB
Annealed	624	90.5	210	30.4	49	75 HRB
Plate:						
As rolled	682	99.0	346	50.2	42	87 HRB
Annealed	639	92.7	199	28.9	49	75 HRB
Tubing:						
As drawn	993	144.0	916	132.8	8	34 HRC
Annealed	693	100.5	279	40.5	43	83 HRB

Fig. 14 Corrosion rates for Inconel 600 in boiling sodium hydroxide

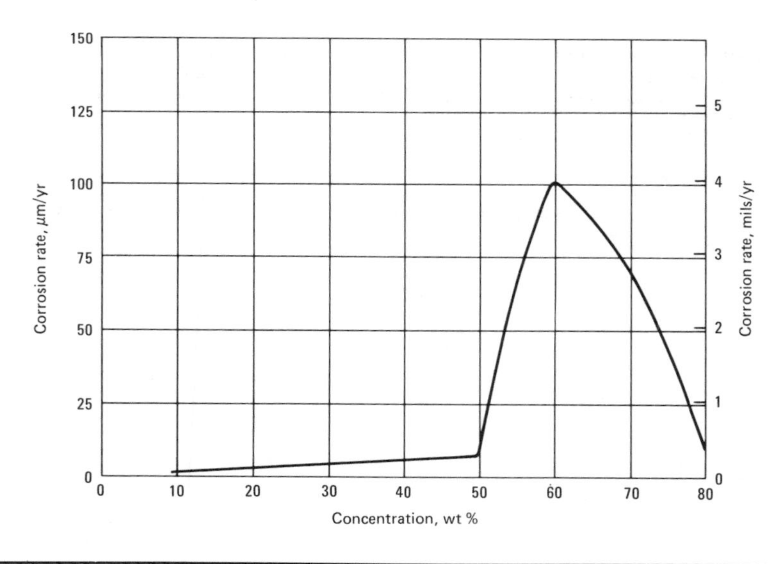

Fabrication Characteristics

Annealing temperature. Approx 1010 °C (1850 °F)
Hot working temperature. 870 to 1230 °C (1600 to 2250 °F)

Table 25 Corrosion rates for Inconel 600 in various concentrations of sulfuric acid at room temperature and at boiling temperature

Acid concentration, %	Room temperature, mm/yr	Room temperature, mils/yr	Boiling temperature, mm/yr	Boiling temperature, mils/yr
10	0.081	3.2	3.43	135
20	0.051	2.0	4.72	186
30	0.064	2.5	5.49	216
40	0.046	1.8	17.8	700
50	0.041	1.6	...	...
60	0.048	1.9	...	...
70	0.058	2.3	...	...
80	0.566	22.3	...	...
90	0.013	0.5	...	...
98	0.188	7.4	...	...

Table 26 Corrosion rates for Inconel 625 in sulfuric and hydrochloric acids at various concentrations

Sulfuric acid(a) Concentration, %	Corrosion rate mm/yr	Corrosion rate mils/yr	Hydrochloric acid(b) Concentration, %	Corrosion rate mm/yr	Corrosion rate mils/yr
15	0.188	7.40	5	1.803	71.0
50	0.432	17.0	10	2.057	81.0
60	0.711	28.0	15	1.651	65.0
70	1.626	64.0	20	1.270	50.0
80	2.286	90.0	25	0.965	38.0
			30	0.864	34.0
			Conc	0.381	15.0

(a) At 80 °C (176 °F). (b) At 66 °C (151 °F).

Fig. 15 Rotating-beam fatigue strength of hot rolled solution-treated Inconel 625 bar (15.9-mm diam [0.625-in. diam]) at elevated temperature. Average grain size, 0.10 mm (0.004 in.)

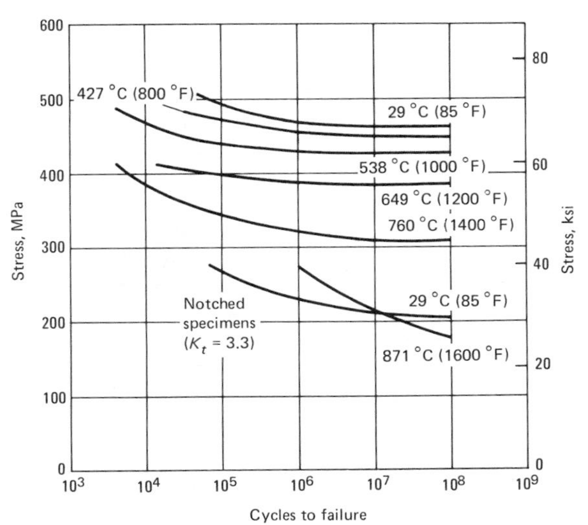

Inconel 625
61Ni-21Cr-9Mo-3.6Nb

Commercial Names

Trade name. Inconel 625
UNS number. N06625

Specifications

AMS. 5599, 5666, 5837
ANSI. H34.19, H34.20, H34.22
ASME. SB443, SB444, SB446
ASTM. B443, B444, B446

Chemical Composition

Composition limits. 20.0 to 23.0 Cr, 5.0 max Fe, 8.0 to 10.0 Mo, 3.15 to 4.15 Cb + Ta, 0.10 max C, 0.50 max Mn, 0.50 max Si, 0.015 max P, 0.015 max S, 0.40 max Al, 0.40 max Ti, 1.0 max Co, 58.0 min Ni

Applications

Typical uses. Chemical-processing equipment, aircraft-engine and airframe components, ship and submarine parts, nuclear reactors

Mechanical Properties

Tensile properties. See Table 9 in the article "Wrought Superalloys".
Poisson's ratio. Annealed material, 0.278 at 21 °C (70 °F)
Elastic modulus. Annealed material: tension, 208 GPa (30 × 10^6 psi) at 21 °C (70 °F); torsion, 81 GPa (11.8 × 10^6 psi) at 21 °C (70 °F)
Impact strength. As-rolled plate, Charpy keyhole: 66 J (48.7 ft·lb) at 29 °C (85 °F); 60 J (44.2 ft·lb) at −79 °C (−110 °F); 47 J (34.7 ft·lb) at −196 °C (−320 °F)
Fatigue strength. See Fig. 15.
Creep-rupture characteristics. See Table 10 in the article "Wrought Superalloys".

Structure

Crystal structure. Face-centered cubic
Microstructure. Inconel 625 is a solid-solution, matrix-stiffened alloy whose microstructure contains carbides. Sluggish precipitation of a gamma-prime phase occurs during exposure to intermediate temperatures. The gamma prime gradually transforms to orthorhombic Ni_3Cb after prolonged exposure.

Mass Characteristics

Density. 8.44 Mg/m^3 (0.305 $lb/in.^3$) at 20 °C (68 °F)

Thermal Properties

Liquidus temperature. 1350 °C (2460 °F)
Solidus temperature. 1290 °C (2350 °F)
Coefficient of thermal expansion. See Table 8 in the article "Wrought Superalloys".
Specific heat. 410 J/kg·K (0.095 Btu/lb·°F) at 21 °C (70 °F)
Thermal conductivity. See Table 8 in the article "Wrought Superalloys".

Electrical Properties

Electrical conductivity. Volumetric, 1.3% IACS at 21 °C (70 °F)
Electrical resistivity. 1290 nΩ·m at 21 °C (70 °F)

Magnetic Properties

Magnetic permeability. 1.006 at a field strength of 15.9 kA/m
Curie temperature. < -196 °C (< -320 °F)

Chemical Properties

General corrosion behavior. The high alloy content of Inconel 625 enables it to withstand a wide variety of corrosive environments. The alloy is almost completely resistant to mild environments such as the atmosphere, fresh water and seawater, neutral salts and alkaline media. In more severe environments, the combination of nickel and chromium provides resistance to oxidizing chemicals, and the combination of nickel and molybdenum provides resistance to reducing conditions. The molybdenum content also makes Inconel 625 highly resistant to pitting and crevice corrosion. The columbium stabilizes the alloy against sensitization and prevents intergranular corrosion. The high nickel content provides freedom from chloride stress-corrosion cracking.
Resistance to specific corroding agents. Table 26 gives corrosion rates in sulfuric and hydrochloric acids; Fig. 16 shows resistance to phosphoric acid. In boiling 65% nitric acid, Inconel 625 typically corrodes at a rate of 0.76 mm (30 mils) per year. In boiling 50% sodium hydroxide, the corrosion rate is 0.13 mm (5.0 mils) per year.

Fabrication Characteristics

Annealing temperature. 925 to 1205 °C (1700 to 2200 °F)
Hot working temperature. 1010 to 1175 °C (1850 to 2150 °F)

Fig. 16 Corrosion of Inconel 625 in boiling phosphoric acid solutions

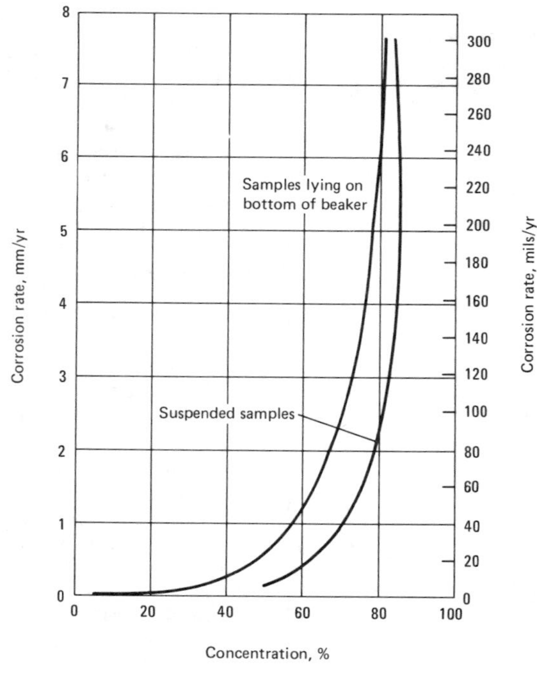

Inconel 671
52Ni-48Cr

Commercial Names

Trade name. Inconel 671

Chemical Composition

Nominal composition. 46 Cr, 0.05 C, 0.35 Ti, rem Ni

Applications

Typical uses. Power-generation equipment such as superheater-tube shields, soot-blower tubes, boiler splash plates, boiler-tube separators and hangers; high-temperature recuperators and waste incinerators; pressure vessels for pyrolysis of spent paper-mill sulfite pulping liquors.

Mechanical Properties

Tensile properties. Annealed plate: tensile strength, 862 MPa (125 ksi); yield strength, 483 MPa (70 ksi) at 0.2% offset; elongation, 25%
Hardness. Annealed plate, 95 HRB
Tensile properties vs temperature. See Fig. 17.
Creep-rupture characteristics. See Table 27.

Structure

Crystal structure. Matrix, face-cen-

Table 27 Rupture strength of Inconel 671 as a function of temperature

Temperature		Stress to rupture in 100 h		Stress to rupture in 1000 h	
°C	°F	MPa	ksi	MPa	ksi
649	1200	193	28.0	97	14.0
704	1300	110	16.0	57	8.2
760	1400	66	9.5	34	5.0
816	1500	41	6.0	21	3.0
871	1600	25	3.6	12	1.8
927	1700	16	2.3	9	1.3
982	1800	10	1.5	6	0.8

Table 28 Corrosion of Inconel 671 in mixtures of vanadium pentoxide and sodium sulfate at 899 °C (1650 °F)

Test mixture	Time, h	Weight loss, g/m²
80% V_2O_5, 20% Na_2SO_4	16	972
	150	6136
	300	7678
20% V_2O_5, 80% Na_2SO_4	16	9.0
	150	63.3
	300	182.1

tered cubic; alpha phase, body-centered cubic
Microstructure. The high chromium content exceeds the solubility limit in nickel, resulting in a two-phase microstructure.

Mass Characteristics

Density. 7.86 Mg/m^3 (0.284 $lb/in.^3$) at 20 °C (68 °F)

Thermal Properties

Liquidus temperature. 1350 °C (2460 °F)
Solidus temperature. 1305 °C (2385 °F)
Coefficient of thermal expansion. Linear, 11.8 μm/m·K (6.6 μin./in.·°F) from 26 to 93 °C (78 to 200 °F); 12.8 μm/m·K (7.1 μin./in.·°F) from 26 to 260 °C (78 to 500 °F);13.8 μm/m·K (7.7 μin./in.·°F) from 26 to 538 °C (78 to 1000 °F); 15.0 μm/m·K (8.3 μin./in.·°F) from 26 to 816 °C (78 to 1500 °F); 15.4 μm/m·K (8.6 μin./in.·°F) from 26 to 1038 °C (78 to 1900 °F)
Specific heat. 456 J/kg·K (0.106 Btu/lb·°F) at 21 °C (70 °F)

Electrical Properties

Electrical conductivity. Volumetric, 2.0% IACS at 21 °C (70 °F)
Electrical resistivity. 869 nΩ·m at 21 °C (70 °F)

Chemical Properties

General corrosion behavior. Inconel 671, a 50/50 nickel-chromium alloy, is designed for use in extremely corrosive environments. Its outstanding characteristic is its resistance to high-temperature corrosion, particularly in atmospheres containing sulfur and/or vanadium.
Resistance to specific corroding agents. Table 28 gives corrosion rates in mixtures of vanadium pentoxide and sodium sulfate.

Fabrication Characteristics

Annealing temperature. 1205 °C (2200 °F)

Fig. 17 High-temperature tensile properties of annealed (2200 °F/1 h, air cooled) Inconel 671

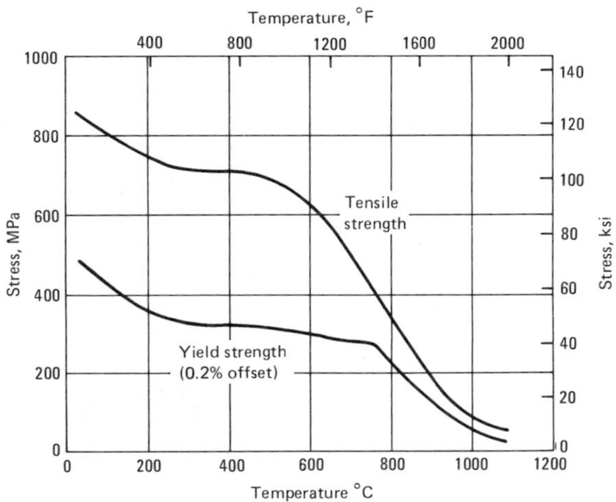

Inconel 690
60Ni-30Cr-9.5Fe

Commercial Names

Trade name. Inconel 690

Chemical Composition

Nominal composition. 60.0 Ni, 30.0 Cr, 9.5 Fe, 0.03 C

Applications

Typical uses. Applications involving nitric acid or nitric-plus-hydrochloric acid. Examples are tail-gas reheaters used in production of nitric acid and steam-heating coils in nitric-plus-hydrochloric acid solutions used for pickling stainless steels and reprocessing nuclear fuels. Inconel 690 is also useful for high-temperature service in gases containing sulfur.

Mechanical Properties

Tensile properties. See Table 29 and Fig. 18.
Hardness. See Table 29.
Poisson's ratio. 0.289
Elastic modulus. Tension, 210 GPa (30.5×10^6 psi)

Structure

Crystal structure. Face-centered cubic

Mass Characteristics

Density. 8.14 Mg/m^3 (0.29 $lb/in.^3$) at 20 °C (68 °F)

Thermal Properties

Liquidus temperature. 1375 °C (2510 °F)
Solidus temperature. 1345 °C (2450 °F)
Coefficient of thermal expansion. See Table 30.
Thermal conductivity. See Table 30.

Electrical Properties

Electrical conductivity. Volumetric, 1.5% IACS at 26 °C (78 °F)
Electrical resistivity. See Table 30.

Chemical Properties

General corrosion behavior. A high chromium content gives Inconel 690 good resistance to oxidizing chemicals such as nitric acid and to sulfur-containing gases at high temperature.
Resistance to specific corroding agents. Table 31 gives corrosion

Fig. 18 High-temperature tensile properties of Inconel 690 annealed at 1095 °C (2000 °F) for 1 h and water quenched

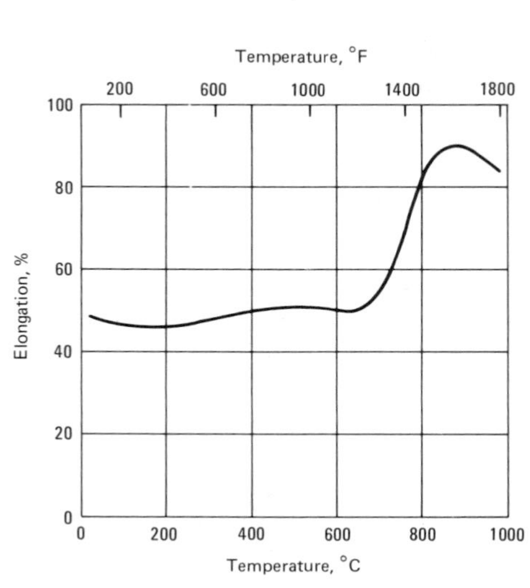

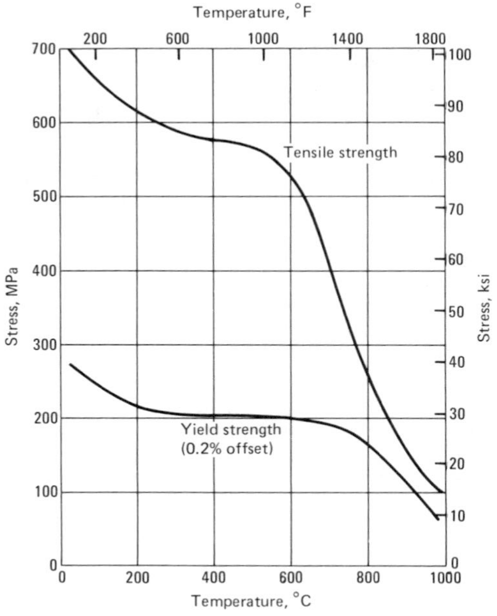

Table 29 Typical tensile properties and hardness of Inconel 690

Material condition	Tensile strength MPa	Tensile strength ksi	Yield strength (0.2% offset) MPa	Yield strength (0.2% offset) ksi	Elongation, %	Hardness, HRB
Tube, annealed	731	106	365	53	41	97
Strip, annealed	758	110	372	54	40	88
Rod, as rolled	765	111	434	63	40	90
Rod, annealed	710	103	317	46	49	90
Plate, as rolled	765	111	483	70	36	95

Table 30 Thermal and electrical properties of Inconel 690 as a function of temperature

Temperature °C	Temperature °F	Mean linear expansion(a) μm/m · K	Mean linear expansion(a) μin./in. · °F	Thermal conductivity W/m · K	Thermal conductivity Btu/ft · h · °F	Electrical resistivity, nΩ · m
26	78	...	...	13.9	8.0	148
93	200	13.5	7.5	15.0	8.7	161
204	400	...	...	17.0	42.1	181
316	600	15.3	8.5	18.8	10.9	202
427	800	...	...	20.5	11.8	224
538	1000	...	...	22.8	13.1	238
649	1200	16.2	9.0	24.2	14.0	238
760	1400	...	...	26.0	15.0	245
871	1600	...	...	27.8	16.1	252
982	1800	17.6	9.8	29.5	17.0	263
1093	2000	...	...	31.2	18.0	277
1204	2200	...	...	32.8	18.9	292

(a) From 21 °C (70 °F) to temperature shown.

rates in nitric-plus-hydrofluoric acid solutions.

Fabrication Characteristics

Annealing temperature. 980 to 1040 °C (1800 to 1900 °F)
Hot working temperature. 1040 to 1232 °C (1900 to 2250 °F)

Table 31 Corrosion rates for Inconel 690 in nitric-plus-hydrofluoric acid solutions at 60 °C (140 °F)

Solution	Corrosion rate mm/yr	Corrosion rate mils/yr
10% HNO_3, 3% HF	0.15	6.0
15% HNO_3, 3% HF	0.25	10.0
20% HNO_3, 3% HF	0.15	6.0

Incoloy 800
32.5Ni-21Cr-46Fe

Commercial Names

Trade name. Incoloy 800
UNS number. N08800

Table 32 Tensile properties(a) of Incoloy 800

Form and condition	Tensile strength MPa	Tensile strength ksi	Yield strength (0.2% offset) MPa	Yield strength (0.2% offset) ksi	Elongation, %
Rod and bar:					
Annealed	517-690	75-100	207-414	30-60	60-30
Hot finished	552-827	80-120	241-621	35-90	50-25
Cold drawn	690-1034	100-150	517-862	75-125	30-10
Plate:					
Hot rolled	552-758	80-110	207-448	30-65	50-25
Annealed	517-724	75-105	207-414	30-60	50-30
Sheet:					
Annealed	517-724	75-105	207-379	30-55	50-30
Strip:					
Annealed	517-690	75-100	207-379	30-55	50-30
Tubing:					
Hot finished	517-724	75-105	172-414	25-60	50-30
Cold drawn, annealed	517-690	75-100	207-414	30-60	50-30
Wire:					
Annealed	552-758	80-110	241-448	35-65	45-25
Spring temper	965-1207	140-175	896-1172	130-170	5-2

(a) Values shown represent usual ranges for common section sizes. In general, values in the higher portions of the ranges are not obtainable with large section sizes, and exceptionally small or large sections may have properties outside the ranges.

Specifications

AMS. 5766, 5871
ANSI. H34.15, H34.39, H34.40, H34.41, H34.42
ASME. SB163, SB407, SB408, SB409
ASTM. B163, B366, B407, B408, B409, B514, B515, B564

Chemical Composition

Composition limits. 30.0 to 35.0 Ni, 19.0 to 23.0 Cr, 0.10 max C, 1.50 max Mn, 0.015 max S, 1.0 max Si, 0.75 max Cu, 0.15 to 0.60 Al, 0.15 to 0.60 Ti, 39.5 min Fe

Applications

Typical uses. Heat treating equipment, petrochemical pyrolysis tubing and piping systems, sheathing for electrical heating elements, food-processing equipment.

Mechanical Properties

Tensile properties. See Tables 32 and 33.
Compressive properties. See Table 33.
Poisson's ratio. Annealed material, 0.339 at 24 °C (75 °F)
Elastic modulus. Annealed material: tension, 195 GPa (28.35×10^6 psi) at 24 °C (75 °F); torsion, 73 GPa (10.64×10^6 psi) at 24 °C (75 °F)
Impact strength. Annealed plate, Charpy keyhole: 122 J at 21 °C (70 °F); 122 J at −79 °C (−110 °F); 106 J at −196 °C (−320 °F); 99 J at −253 °C (−423 °F)
Fatigue strength. Rotating-beam: hot rolled, 352 MPa (51 ksi); cold drawn, 228 MPa (33 ksi); annealed, 214 MPa (31 ksi). All values at 10^8 cycles.

Structure

Crystal structure. Face-centered cubic
Microstructure. Incoloy 800 is a solid-solution alloy. Precipitated phases usually found in its microstructure and titanium nitrides, titanium carbides and chromium carbides. Small amounts of a gamma prime phase may form after prolonged exposure to temperatures of 565 to 620 °C (1050 to 1150 °F).

Mass Characteristics

Density. 7.94 Mg/m³ (0.287 lb/in.³) at 20 °C (68 °F)

Thermal Properties

Liquidus temperature. 1385 °C (2525 °F)
Solidus temperature. 1355 °C (2475 °F)
Specific heat. 502 J/kg·K (0.117 Btu/lb·°F) at 20 °C (68 °F)

Electrical Properties

Electrical conductivity. Volumetric, 1.7% IACS at 21 °C (70 °F)
Electrical resistivity. 989 nΩ·m at 21 °C (70 °F)

Magnetic Properties

Magnetic permeability. Annealed material, 1.0092 at a field strength of 15.9 kA/m
Curie temperature. −115 °C (−175 °F)

Chemical Properties

General corrosion behavior. A high chromium content gives Incoloy 800 good resistance to oxidation. It also resists many aqueous media and is relatively free from stress-corrosion cracking.
Resistance to specific corroding agents. Incoloy 800 has excellent resistance to nitric acid at concentrations up to about 70% and at temperatures up to the boiling point. It also has good resistance to organic acids such as formic, acetic and propionic. It resists a variety of oxidizing and nonoxidizing salts, but not halide salts. Corrosion rates in various media are given in Table 34.

Fabrication Characteristics

Annealing temperature. 980 to 1150 °C (1800 to 2100 °F)
Hot working temperature. 870 to 1205 °C (1600 to 2200 °F)

Table 33 Typical tensile and compressive properties of Incoloy 800

Material condition	Tension: Tensile strength MPa	Tension: Tensile strength ksi	Tension: Yield strength (0.02% offset) MPa	Tension: Yield strength (0.02% offset) ksi	Tension: Yield strength (0.2% offset) MPa	Tension: Yield strength (0.2% offset) ksi	Compression: Yield strength (0.02% offset) MPa	Compression: Yield strength (0.02% offset) ksi	Compression: Yield strength (0.2% offset) MPa	Compression: Yield strength (0.2% offset) ksi
Annealed bar	616	89.3	268	38.8	283	41.1	269	39.0	287	41.6
As-extruded tube	479	69.5	145	21.0	190	27.5	145	21.0	175	25.4

Table 34 Corrosion rates for Incoloy 800 in various media (laboratory tests at 80 °C [176 °F])

Environment	Test duration, days	Corrosion rate mm/yr	Corrosion rate mils/yr	Pitting resistance
Acetic acid (10%)	7	0.0003	0.01	No pitting
Acetic acid (10%)+ sulfuric acid (0.5%)	7	0.0006	0.02	No pitting
Acetic acid (10%)+ sodium chloride (0.5%)	42	0.0008	0.03	Incipient pits visible at 30X after 42 days
Aluminum sulfate (5%)	7	0.0003	0.01	No pitting
Ammonium chloride (5%)	42	0.0006	0.02	Pitting after 42 days
Ammonium hydroxide (5%)	7	0.0003	0.01	No pitting
Ammonium hydroxide (10%)	7	0.0003	0.01	No pitting
Ammonium sulfate (5%)	7	0.00	0.00	No pitting
Barium chloride (10%)	42	0.0008	0.03	Pitting after 42 days
Bromine water (saturated)	42	0.19	7.6	Pitting after 7 days
Calcium chloride (5%)	42	0.0003	0.01	Pitting after 42 days
Chromic acid (5%)	7	0.041	1.6	No pitting
Citric acid (10%)	7	0.00	0.00	No pitting
Copper sulfate (10%)	7	0.00	0.00	No pitting
Ferric chloride (5%)	42	11	420	Pitting after 7 days
Ferrous ammonium sulfate (5%)	7	0.002	0.08	No pitting
Lactic acid (10%)	7	0.001	0.04	No pitting
Methanol (absolute)	7	0.00	0.00	No pitting
Oxalic acid (5%)	7	0.003	0.12	No pitting
Oxalic acid (10%)	7	0.28	11.0	No pitting
Potassium ferricyanide (5%)	7	0.001	0.04	No pitting
Sodium bisulfite (5%)	7	0.0008	0.03	No pitting
Sodium carbonate	7	0.00	0.00	No pitting
Sodium chloride (10%)	42	0.0003	0.01	Incipient pits visible at 30X after 42 days
Sodium chloride (20%)	42	0.0086	0.34	Pitting after 7 days
Sodium hypochlorite (1%)	42	0.127	5.0	Pitting after 7 days
Sodium hypochlorite (5%)	42	0.2	8.0	Pitting after 7 days
Sodium sulfate (5%)	7	0.00	0.00	No pitting
Sodium sulfate (10%)	7	0.0006	0.02	No pitting
Sulfurous acid (5%)	7	1.09	43.0	No pitting
Tartaric acid (10%)	7	0.0006	0.02	No pitting
Zinc chloride (10%)	42	0.0003	0.01	Pitting after 42 days

Incoloy 801
32Ni-20.5Cr-44.5Fe-1.1Ti

Commercial Names

Trade name. Incoloy 801
UNS number. N08801

Specifications

AMS. 5552, 5742

Chemical Composition

Composition limits. 30.0 to 34.0 Ni, 19.0 to 22.0 Cr, 0.10 max C, 1.5 max Mn, 0.015 max S, 1.0 max Si, 0.5 max Cu, 0.75 to 1.5 Ti, rem Fe

Applications

Typical uses. Incoloy 801 is a modification of Incoloy 800 that is higher in titanium; it is used for hydrodesulfurizers in petrochemical processing and for other applications involving polythionic acid.

Mechanical Properties

Tensile properties. Annealed extruded tubing: tensile strength, 514 MPa (74.6 ksi); yield strength, 197 MPa (28.6 ksi) at 0.2% offset; elongation, 53%

Poisson's ratio. 0.413 at 26 °C (78 °F)

Elastic modulus. Tension, 207 GPa (30.07 × 10^6 psi) at 26 °C (78 °F); torsion, 73 GPa (10.64 × 10^6 psi) at 26 °C (78 °F)

Impact strength. Charpy V-notch: annealed, 324 J (239 ft·lb); aged, 127 J (93.7 ft·lb)

Fatigue strength. Annealed material, rotating-beam: 255 MPa (37 ksi) at 27 °C (80°F); 179 MPa (26 ksi) at 732 °C (1350 °F); 76 MPa (11 ksi) at 816 °C (1500 °F)

Creep-rupture characteristics. See Table 10 in the article "Wrought Superalloys".

Structure

Crystal structure. Face-centered cubic

Microstructure. Incoloy 801 is an austenitic material that can be strengthened by precipitation of a gamma prime phase during heat treatment. Other phases present are titanium nitrides, titanium carbides and chromium carbides.

Mass Characteristics

Density. 7.94 Mg/m^3 (0.287 lb/in.^{3}) at 20 °C (68 °F)

Thermal Properties

Liquidus temperature. 1385 °C (2525 °F)

Solidus temperature. 1355 °C (2475 °F)

Coefficient of thermal expansion. See "Wrought Heat-Resisting Alloys".

Specific heat. 452 J/kg·K (0.105 Btu/lb·°F) at 26 °C (78 °F)

Thermal conductivity See Table 8 in the article "Wrought Superalloys".

Electrical Properties

Electrical conductivity. Volumetric, 1.7% IACS at 26 °C (78 °F)

Electrical resistivity. 1012 nΩ·m at 26 °C (78 °F)

Chemical Properties

General corrosion behavior. In most environments, the corrosion resistance of Incoloy 801 is similar to that of Incoloy 800. However, Incoloy 801 has greater resistance to intergranular corrosion because its higher titanium content stabilizes the alloy against sensitization.

Resistance to specific corroding agents. Incoloy 801 is especially resistant to stress-corrosion cracking in solutions of polythionic acid.

Fabrication Characteristics

Annealing temperature. 940 °C (1725 °F)
Aging temperature. 650 to 730 °C (1200 to 1350 °F)
Hot working temperature. 870 to 1205 °C (1600 to 2200 °F)

Incoloy 825
42Ni-21.5Cr-30Fe-3Mo-2.2Cu

Commercial Names

Trade name. Incoloy 825
UNS number. N08825

Specifications

ASME. SB163, SB423, SB424, SB425
ASTM. B163, B423, B424, B425

Chemical Composition

Composition limits. 38.0 to 46.0 Ni, 19.5 to 23.5 Cr, 2.5 to 3.5 Mo, 1.5 to 3.0 Cu, 0.6 to 1.2 Ti, 0.05 max C, 1.0 max Mn, 0.03 max S, 0.5 max Si, 0.2 max Al, rem Fe

Applications

Typical uses. Phosphoric acid evaporators, pickling equipment, chemical-processing vessels and piping, equipment for recovery of spent nuclear fuel, propeller shafts, tank trucks.

Mechanical Properties

Tensile properties. See Table 35.
Compressive properties. Annealed bar with tensile yield strength of 396 MPa (57.5 ksi): compressive yield strength, 423 MPa (61.4 ksi) at 0.2% offset.
Elastic modulus. Tension, 195 GPa (28.3×10^6 psi) at 27 °C (80 °F)
Impact strength. Plate, Charpy keyhole: 107 J (78.9 ft·lb) at 20 °C (68 °F); 106 J (78.2 ft·lb) at −79 °C (−110 °F); 91 J (67.1 ft·lb) at −196 °C (−320 °F); 92 J (67.8 ft·lb) at −253 °C (−420 °F)

Fig. 19 Laboratory corrosion rates of Incoloy 825 in chemically pure sulfuric acid solutions

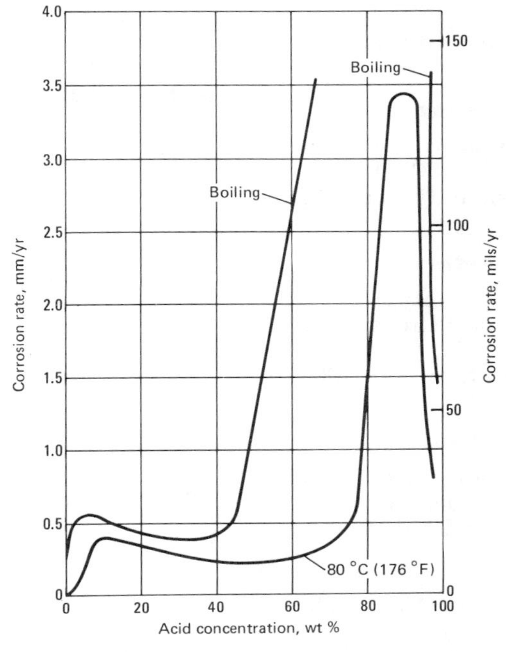

Table 35 Tensile properties of annealed Incoloy 825 as a function of temperature

Temperature °C	Temperature °F	Tensile strength MPa	Tensile strength ksi	Yield strength (0.2% offset) MPa	Yield strength (0.2% offset) ksi	Elongation, %
29	85	693	100.5	301	43.7	43
93	200	655	95.0	279	40.4	44
204	400	637	92.4	245	35.6	43
316	600	632	91.7	232	33.6	46
371	700	621	90.0	234	34.0	46
427	800	610	88.5	228	33.0	44
482	900	608	88.2	221	32.0	42
538	1000	592	85.9	229	33.2	43
593	1100	541	78.5	222	32.2	38
649	1200	465	67.5	213	30.9	62
760	1400	274	39.7	183	26.5	87
871	1600	135	19.6	117	17.0	102
982	1800	75	10.9	47	6.8	173
1093	2000	42	6.1	23	3.3	106

Structure

Crystal structure. Face-centered cubic
Microstructure. Incoloy 825 is a solid-solution austenitic material. Its microstructure usually contains titanium nitrides, titanium carbides and chromium carbides.

Mass Characteristics

Density. 8.14 Mg/m^3 (0.294 $lb/in.^3$) at 20 °C (68 °F)

Thermal Properties

Liquidus temperature. 1400 °C (2550 °F)
Solidus temperature. 1370 °C (2500 °F)
Coefficient of thermal expansion. See Table 36.
Thermal conductivity. See Table 36.

Electrical Properties

Electrical conductivity. Volumetric, 1.5% IACS at 26 °C (78 °F)
Electrical resistivity. See Table 36.

Magnetic Properties

Magnetic permeability. 1.005 at 21 °C (70 °F) and a field strength of 15.9 kA/m
Curie temperature. <−196 °C (<−320 °F)

Chemical Properties

General corrosion behavior. Incoloy 825 has exceptional resistance to seawater and to reducing chemicals such as sulfuric and phosphoric acids. Because it is stabilized against sensitization, Incoloy 825 resists intergranular corrosion. This alloy contains sufficient nickel to make it resistant to chloride stress-corrosion cracking. Its molybdenum content provides resistance to pitting. Its chromium content provides resistance to oxidizing media such as nitric acid, nitrates and oxidizing salts.
Resistance to specific corroding agents. Corrosion rates in sulfuric acid are shown in Fig. 19. Results of plant corrosion tests in phosphoric and nitric acid solutions are given in Tables 37 and 38. Table 39 gives corrosion rates in hydrochloric acid, and Table 40 gives rates in various organic acids.

Fabrication Characteristics

Annealing temperature. 925 to 980 °C (1700 to 1800 °F)
Hot working temperature. 870 to 1175 °C (1600 to 2150 °F)

Table 36 Thermal and electrical properties of Incoloy 825 as a function of temperature

Temperature °C	Temperature °F	Mean linear expansion(a) μm/m · K	Mean linear expansion(a) μin./in. · °F	Thermal conductivity W/m · K	Thermal conductivity Btu/ft · h · °F	Electrical resistivity, nΩ · m
26	78	...	...	11.1	6.4	1127
38	100	...	...	11.3	6.5	1130
93	200	14.0	7.7	12.3	7.1	1142
204	400	14.9	8.3	14.1	8.1	1180
316	600	15.3	8.5	15.8	9.1	1210
427	800	15.7	8.7	17.3	10.0	1248
538	1000	15.8	8.8	18.9	10.9	1265
649	1200	16.4	9.1	20.5	11.8	1267
760	1400	17.1	9.5	22.3	12.9	1272
871	1600	17.5	9.7	24.8	14.3	1288
982	1800	...	...	27.7	16.0	1300
1093	2000	...	...	...	...	1318

(a) From 27 °C (80 °F) to temperature shown.

Table 37 Plant corrosion tests of Incoloy 825 immersed in wet-process phosphoric acid solutions

Test solution	Temperature °C	Temperature °F	Duration of test, days	Corrosion rate mm/yr	Corrosion rate mils/yr
Recycle liquor from evaporator fume scrubber containing 15% H_3PO_4, 20% H_2SiF_6, 1% H_2SO_4	75-85	165-185	16	0.025	1.0
Solution containing 20% H_3PO_4 and 20% HF in tank	20-30	70-85	13	0.036	1.4
Slurry in digester tank. Mixture contains 20% H_3PO_4, 2% H_2SO_4, 1% HF, 40% H_2O, plus $CaSO_4$	75-95	170-200	117	0.02	0.7
Slurry containing 37% H_3PO_4 (27% P_2O_5) in acid transfer tank. Velocity 3 ft/s	65-90	150-190	46	0.02	0.7
Slurry containing 31.4% H_3PO_4, 1.6% H_2SO_4, 1.5% H_2SiF_6, 0.12% HF, plus $CaSO_4$ in filter tank	45-60	115-140	8.3	<0.003	<0.1
Thickener in evaporated acid containing 54% H_3PO_4, 1.7% HF, 2% H_2SO_4, 2% $CaSO_4$	50-65	125-150	51	0.01	0.5
Evaporator heated with hot gases in acid containing 53% H_3PO_4, 1-2% H_2SO_4, 1.5% HF plus Na_2SiF_6	120	250	42	0.15	6.0
In wet separator on top of concentrating drum in vapors from concentration of crude acid to 50-55% H_3PO_4 containing HF	105-150	225-300	21	0.79	31.0
Defluorinator in acid containing 75-80% H_3PO_4, 1% H_2SO_4, with some HF. Violent agitation	120-155	250-315	8	3.048	120.0

Table 39 Corrosion rates for Incoloy 825 in hydrochloric acid at three temperatures

Acid concentration, %	Temperature °C	Temperature °F	Corrosion rate mm/yr	Corrosion rate mils/yr
5	20	68	0.12	4.9
	40	104	0.45	17.8
	66	150	2.00	79
10	20	68	0.18	7.2
	40	104	0.47	18.6
	66	150	2.60	102
20	20	68	0.18(a)	7.3(a)
	40	104	0.44	17.2
	66	150	1.52	60

(a) 15% acid concentration.

Table 38 Plant corrosion tests of Incoloy 825 in nitric acid mixtures

Test conditions(a)	Temperature °C	Temperature °F	Duration of test, days	Corrosion rate mm/yr	Corrosion rate mils/yr
In evaporator during concentration of nitric acid solution saturated with potassium nitrate and containing chlorides.					
Liquid: 40-70% HNO_3, 0.2-0.02% Cl	105-115	220-240	4.2	0.10	4.0
Vapor: 50-10% HNO_3, 0.05-1.5% Cl	105-115	220-240	4.2	0.279	11.0
In evaporator during concentration of nitric acid solution from 35-45% nitric acid, saturated with zirconyl nitrate and containing 10-35% $ZrO(NO_3)_2$ crystals.					
Liquid	115-125	235-255	29	0.533	21.0
Vapor	115-125	235-255	29	0.660	26.0
In 40% nitric acid solution containing some nitrogen tetroxide and nitrous acid. Location in N_2O_2 absorption tower immediately below distributor.	30-40	85-105	15	<0.003	<0.1
In evaporator during concentration of 20% nitric acid solution containing 6% metal nitrates (iron, magnesium, lead and aluminum), 2% sulfate as metal sulfates.	70-90	160-190	52	0.01	0.4
Laboratory test in 53% nitric acid containing 1% hydrofluoric acid.					
Liquid	80	176	7	5.080	200.0
Vapor	80	176	7	2.18	86.0
In vapor during concentration of nitric acid solution containing 35-45% nitric acid, 3-20% chlorine as chlorides, and 10-20% metal nitrates (mainly zirconium).					
Liquid	115-125	240-260	21	0.330	13.0
Vapor	115-125	240-260	21	0.15	5.8
In evaporator during concentration of 36% nitric acid solution containing 30% potassium nitrate, some sodium, iron, calcium and magnesium nitrates, and 0.05-0.10% chlorine as chloride. Intermittently exposed to liquid and vapor.	65-80	150-180	8	0.01	0.5
In evaporator during concentration of nitric acid solution containing metal nitrates (mostly zirconium) and small amount of chlorides.					
Liquid at bottom of column (58% nitric acid, 5 ppm chlorides, 11-13% metal nitrates)	115-130	240-265	10	0.660	26.0
Liquid on 10th tray of column (2.21% nitric acid, 3-13 ppm chlorides, 0-25% metal nitrates)	105-130	225-265	10	0.12	4.6
In evaporator during concentration of raffinate solution containing 30-40% nitric acid and variable chlorides up to 2000 ppm Cl.					
Liquid	80(b)	175(b)	92	0.02	0.7
Vapor	80(b)	175(b)	92	0.028	1.1

(a) Specimens were immersed in solution unless otherwise stated. (b) Average.

DS Nickel
98Ni-2ThO₂

Commercial Names

Trade name. DS Nickel
Common names. Dispersion strengthened nickel; thoria dispersed nickel

Specifications

AMS. 5865

Chemical Composition

Composition limits. 1.80 to 2.60 ThO_2; 0.20 max Co; 0.15 max Cu; 0.05 max Fe; 0.05 max Cr; 0.05 max Ti; 0.02 max C; 0.0025 max S; rem Ni
Consequence of exceeding impurity limits. Sulfur contents in excess of above limit can cause brittleness and hot shortness

Applications

Typical uses. Components in combustion systems of advanced gas turbine engines; fixtures for high temperature tensile testing; specialized furnaces and heating elements
Precautions in use. Has poor oxidation resistance and should be coated for long-term high temperature surface stability. Strength depends on metallurgical structure, and the as-fabricated structure should be retained during secondary fabrication. Mildly radioactive; possession of this material may require licensing by the U.S. Nuclear Regulatory Commission.

Mechanical Properties

Tensile properties. See Table 41.
Poisson's ratio. 0.38 at 21 °C (70 °F)
Elastic modulus. Tension:

Temperature °C	Temperature °F	Elastic modulus GPa	Elastic modulus 10^6 psi
21	70	137	19.9
260	500	137	19.8
538	1000	99.3	14.4
816	1500	92.4	13.4
1093	2000	86.2	12.5

Creep-rupture characteristics:

Table 40 Plant corrosion tests of Incoloy 825 in organic acids

Test conditions	Temperature °C	Temperature °F	Duration of test, days	Corrosion rate mm/yr	Corrosion rate mils/yr
In vapors of 85% acetic acid, 10% acetic anhydride, 5% water, plus some acetone, acetonitrile, in vapor line just before condenser	115-135	240-275	875	0.008	0.3
In 99.9% acetic acid, less than 0.1% water in still	105	225	40	0.006	0.2
In mixture of 94% acetic acid, 1% formic acid, 5% high boiling esters	125	260	465	0.02	0.7
In mixture of 96.5-98% acetic acid, 1.5% formic acid, 1-1.5% water	125	255	262	0.15	6.0
In mixture of 91.5% acetic acid, 2.5% formic acid, 6.0 water	110-125	230-260	55	0.079	3.1
In mixture of 95% acetic acid, 1.5-3.0% formic acid, 0.5% potassium permanganate, balance water	110-145	230-290	55	0.038	1.5
In mixture of 40% acetic acid, 6% propionic acid, 20% butane, 5% pentane, 8% ethyl acetate, 5% methyl ethyl ketone, plus other esters and ketones	175	345	217	0.051	2.0
In liquid phthalic anhydride containing phthalic acid, some water, and small amounts of maleic acid, maleic anhydride, benzoic acid and naphthaquinones. On reflux plate of crude phthalic anhydride still	165-260	330-500	70	0.20	8.0

Table 41 Tensile properties of DS Nickel sheet

Temperature °C	Temperature °F	Tensile strength MPa	Tensile strength ksi	0.2% yield strength MPa	0.2% yield strength ksi	Elongation in 25 mm (1 in.), %
21	70	491	71.2	333	48.3	14
871	1600	157	22.8	148	21.5	4
982	1800	137	19.8	129	18.7	4
1093	2000	121	17.5	118	17.1	4
1260	2300	95	13.8	88	12.8	3

Temperature °C	Temperature °F	Time, h	Rupture strength MPa	Rupture strength ksi
649	1200	100	162	23.5
		1000	155	22.5
816	1500	100	131	19.0
		1000	121	17.5
982	1800	100	84.1	12.2
		1000	70.3	10.2
1093	2000	100	61.4	8.9
		1000	51.0	7.4

Structure

Crystal structure. Face-centered cubic. Crystallographic texture: sheet, [100] <001> cube; bar, <100>

Mass Characteristics

Density. 8.86 Mg/m^3 (0.32 lb/in.^{3}) at 22 °C (72 °F)

Thermal Properties

Melting point. 1453 °C (2647 °F)

Coefficient of thermal expansion:

Temperature °C	Temperature °F	μm/m·K	μin./in.·°F
20–300	68– 570	14.8	8.2
20–600	68–1110	15.7	8.7
20–900	68–1650	16.7	9.3
20–1100	68–2010	17.3	9.6
900–1100	1650–2010	20.1	11.2

Thermal conductivity:

Temperature °C	Temperature °F	Conductivity W/m·K	Conductivity Btu/ft·h·°F
260	500	55	32.1
538	1000	57.5	33.2
816	1500	62.4	36.1
1093	2000	69.5	40.2

Optical Properties

Color. Silvery

Chemical Properties

General corrosion behavior. Requires coating for oxidation protection at temperatures above 870 °C (1650 °F)

Fabrication Characteristics

Recrystallization temperature. Mill products are supplied in the "recrystallized" condition. This structure is very stable and resists recrystallization in normal deformation/annealing treatments. Annealing results in stress-relief only.

Annealing temperature. 980 to 1090 °C (1800 to 2000 °F) in nonoxidizing atmosphere recommended for stress relieving anneal

Hot working temperature. Hot working not recommended

Hot shortness temperature. 480 to 700 °C (900 to 1300 °F) if contaminated by sulfur

Maximum reduction between anneals. 20 to 25%

Suitable forming methods. Drawing, spinning, closed die forming, bending, machining and swaging. Ductility increases with increased strain rates.

Suitable joining methods. Mechanical fasteners, brazing, spot welding, diffusion bonding, gas metal-arc brazing using low-carbon NiCr wire. Conventional fusion welding procedures not recommended for highly stressed applications at elevated temperature.

Hastelloy B-2
68Ni-28Mo

Commercial Names

Trade name. HASTELLOY alloy B-2

Chemical Composition

Composition limits. 26 to 30 Mo; 2.00 max Fe; 1.00 max Co; 1.00 max Cr; 1.00 max Mn; 0.10 max Si; 0.040 max P; 0.030 max S; 0.02 max C; rem Ni

Applications

Typical uses. Suitable for most chemical process applications in the as-welded condition. Well suited for equipment handling hydrochloric acid in all concentrations and temperatures. Resistant to hydrogen chloride gas and sulfuric, acetic and phosphoric acids. Principal high-temperature uses are those in which a low coefficient of thermal expansion is required.

Precautions in use. Exposure to temperatures of 540 to 815 °C (1000 to 1500 °F) should be avoided because of a reduction in the ductility of the alloy. In oxidizing gases such as air, B-2 may be used at temperatures up to 540 °C (1000 °F). In reducing gases or in vacuum, the alloy may be used from 815 °C (1500 °F) to substantially higher temperatures. Ferric or cupric salts might develop when hydrochloric acid comes in contact with iron or copper, so Hastelloy B-2 should not be used with copper or iron piping in a system containing hydrochloric acid.

Mechanical Properties

Tensile properties. See Table 42.
Hardness. See Table 42.

Mass Characteristics

Density. 9.22 Mg/m³ (0.333 lb/in.³) at 22 °C (72 °F)

Thermal Properties

Coefficient of thermal expansion. Linear:

Temperature °C	°F	Coefficient µm/m·K	µin./in.·°F
20– 93	68–200	10.3	5.7
20–204	68–400	10.8	6.0
20–316	68–600	11.2	6.2
20–427	68–800	11.5	6.4
20–538	68–1000	11.7	6.5

Specific heat:

Temperature °C	°F	Specific heat J/kg·K	Btu/lb·°F
0	32	373	0.089
200	390	406	0.097
400	750	431	0.103
600	1100	456	0.109

Thermal conductivity:

Temperature °C	°F	Conductivity W/m·K	Btu/ft·h·°F
0	32	11.1	6.4
100	212	12.2	7.1
200	390	13.4	7.75
300	570	14.6	8.5
400	750	16.0	9.25
500	930	17.3	10.0
600	1100	18.7	10.8

Table 42 Average tensile properties for Hastelloy B-2

Temperature °C	°F	Tensile strength MPa	ksi	0.2% yield strength MPa	ksi	Elongation in 2 in. or 50 mm,%	Rockwell hardness
Sheet, 1.3 to 3 mm thick(a)							
RT	RT	965	140	525	76	53	22 HRC
204	400	885	128	450	65	50	
316	600	860	125	425	62	49	
427	800	860	125	415	60	51	
Sheet and plate, 2.5 to 9 mm thick(a)							
RT	RT	895	130	415	60	61	95 HRB
204	400	850	123	350	51	59	
316	600	820	119	325	47	60	
427	800	805	117	310	45	60	
Plate, 9 to 50 mm thick(a)							
RT	RT	905	131	405	59	61	94 HRB
204	400	870	126	360	52	60	
316	600	840	122	340	49	60	
427	800	820	119	315	46	61	

(a) Solution treated at 1065 °C (1950 °F) and rapidly quenched.

Thermal diffusivity:

Temperature °C	°F	Diffusion coefficient, $10^{-6}m^2/s$
0	32	3.2
100	212	3.4
200	390	3.6
300	570	3.8
400	750	4.0
500	930	4.2
600	1100	4.5

Electrical Properties

Electrical resistivity:

Temperature °C	°F	Resistivity, µΩ·m
0	32	1.37
100	212	1.38
200	390	1.38
300	570	1.39
400	750	1.39
500	930	1.41
600	1100	1.46

Optical Properties

Color. Silvery

Chemical Properties

General corrosion behavior. Excellent. Resists formation of grain boundary carbide precipitates in weld heat-affected zones. Excellent resistance to pitting and stress-corrosion cracking.

Resistance to specific corroding agents. Ferric or cupric salts may cause rapid corrosion failure. Ferric or cupric salts may develop when hydrochloric acid comes in contact with iron or copper. Limited tests indicate that corrosion resistance in boiling 20% HCl is not affected by cold reductions of up to 50%.

Hastelloy C-4
64Ni-16Cr-16Mo

Commercial Names

Trade name. HASTELLOY alloy C-4

Chemical Composition

Composition limits. 14 to 18 Cr; 14 to 17 Mo; 3.00 max Fe; 2.00 max Co; 1.00 max Mn; 0.70 max Ti; 0.15 max C; 0.08 max Si; 0.04 max P; 0.03 max S; rem Ni

Applications

Typical uses. Outstanding high

temperature stability; exhibits good ductility and corrosion resistance after long-time aging at 650 to 1040 °C (1200 to 1900 °F). Resists formation of grain-boundary precipitates in weld heat-affected zones, and is suitable for most chemical process applications in the as-welded condition. Has excellent resistance to stress-corrosion cracking and to oxidizing atmospheres up to 1040 °C (1900 °F).

Mechanical Properties

Tensile properties. Average, at room temperature, for material solution treated at 1065 °C (1950 °F) and quenched. Sheet: tensile strength, 785 MPa (114 ksi); yield strength, 400 MPa (58 ksi); elongation in 50 mm or 2 in., 54%. Plate: tensile strength, 785 MPa (114 ksi); yield strength, 345 MPa (50 ksi); elongation in 50 mm or 2 in., 60%

Hardness. At room temperature, for sheet heat treated at 1065 °C (1950 °F) and quenched: 91 HRB

Elastic modulus. In tension, average of three tests at each temperature for 12.7-mm (1/2-in.) thick plate heat treated at 1065 °C (1950 °F) and quenched:

Temperature °C	°F	Modulus GPa	10^6 psi
RT	RT	211	30.8
93	200	207	30.2
205	400	201	29.3
315	600	194	28.3
425	800	187	27.3
540	1000	179	26.2
650	1200	171	25.0
760	1400	162	23.7
870	1600	152	22.2
980	1800	141	20.6

Mass Characteristics

Density. 8.64 Mg/m³ (0.312 lb/in.³) at 20 °C (68 °F)

Thermal Properties

Coefficient of thermal expansion. Linear:

Temperature °C	°F	Coefficient µm/m·K	µin./in.·°F
20–93	68–200	10.8	6.0
20–205	68–400	11.9	6.6
20–315	68–600	12.6	7.0
20–425	68–800	13.0	7.2
20–540	68–1000	13.3	7.4
20–650	68–1200	13.5	7.5
20–760	68–1400	14.4	8.0
20–870	68–1600	14.9	8.3
20–980	68–1800	15.7	8.7

Specific heat:

Temperature °C	°F	Specific heat J/kg·K	Btu/lb·°F
0	32	406	0.097
100	212	426	0.102
200	390	448	0.107
300	570	465	0.111
400	750	477	0.114
500	930	490	0.117
600	1100	502	0.120

Thermal conductivity:

Temperature °C	°F	Conductivity W/m·K	Btu/ft·h·°F
23	74	10.0	5.8
100	212	11.4	6.6
200	390	13.2	7.7
300	570	14.9	8.7
400	750	16.6	9.7
500	930	18.4	10.7
600	1100	20.4	11.8

Thermal diffusivity:

Temperature °C	°F	Diffusion coefficient, 10^{-6} m²/s
23	74	2.8
100	212	3.1
200	390	3.3
300	570	3.7
400	750	4.0
500	930	4.3
600	1100	4.7

Electrical Properties

Electrical resistivity:

Temperature °C	°F	Resistivity, µΩ·m
23	74	1.25
100	212	1.25
200	390	1.26
300	570	1.27
400	750	1.28
500	930	1.29
600	1110	1.32

Optical Properties

Color. Silvery

Chemical Properties

General corrosion behavior. Exceptional resistance to a variety of chemical process environments, including hot contaminated mineral acids, solvents, chlorine, and chlorine-contaminated media (organic and inorganic, dry chlorine, formic and acetic acids, acetic anhydride, seawater and brine).

Fabrication Characteristics

Formability. Average Olsen cup depth, 13.2 mm (0.52 in.) for sheet either solution treated at 1065 °C (1950 °F) and quenched or aged 1000 h at 870 °C (1600 °F)

Hastelloy C-276 59Ni-15.5Cr-16Mo-3.75W-5.5Fe

Chemical Composition

Composition limits. 14.50 to 16.50 Cr; 15 to 17 Mo; 4.00 to 7.00 Fe; 3.00 to 4.50 W; 2.50 max Co; 1.00 max Mn; 0.35 max V; 0.04 max P; 0.03 max S; 0.08 max Si; 0.02 max C; rem Ni

Applications

Typical uses. Resists formation of grain boundary precipitates in weld heat-affected zones, and is suitable for most chemical process applications in the as welded condition. Excellent resistance to pitting, stress-corrosion cracking, and to oxidizing atmospheres to 1040 °C (1900 °F).

Mechanical Properties

Tensile properties. Room temperature, average for material solution treated at 1120 °C (2050 °F) and quenched. Sheet: tensile strength, 790 MPa (115 ksi); yield strength, 355 MPa (52 ksi); elongation, 61% in 50 mm or 2 in. Plate: tensile strength, 785 MPa (114 ksi); yield strength, 365 MPa (53 ksi); elongation, 59% in 50 mm or 2 in.

Hardness. Average: sheet, 90 HRB; plate, 87 HRB

Elastic modulus. Average, in tension:

Temperature °C	°F	Modulus GPa	10⁶ psi
RT	RT	205	29.8
204	400	195	28.3
316	600	188	27.3
427	800	182	26.4
538	1000	176	25.5

Structure

Crystal structure. Face-centered cubic

Mass Characteristics

Density. 8.89 Mg/m^3 (0.321 $lb/in.^3$) at 22 °C (72 °F)

Thermal Properties

Liquidus temperature. 1370 °C (2500 °F)
Solidus temperature. 1325 °C (2415 °F)
Coefficient of thermal expansion. Linear:

Temperature °C	°F	Coefficient μm/m·K	μin./in.·°F
24–93	75–200	11.2	6.2
24–205	75–400	12.0	6.7
24–315	75–600	12.8	7.1
24–425	75–800	13.2	7.3
24–540	75–1000	13.4	7.4
24–650	75–1200	14.1	7.8
24–760	75–1400	14.9	8.3
24–870	75–1600	15.9	8.8
24–925	75–1700	16.0	8.8

Specific heat. 427 J/kg·K (0.102 Btu/lb·°F) at room temperature
Thermal conductivity:

Temperature °C	°F	Conductivity W/m·K	Btu/ft·h·°F
−168	−270	7.2	4.2
−73	−100	8.6	5.0
−18	0	9.4	5.4
+38	+100	10.2	5.9
93	200	11.1	6.4
205	400	13.0	7.5
315	600	15.0	8.7
425	800	16.9	9.75
540	1000	19.0	11.0
650	1200	20.9	12.1
760	1400	23.0	13.3
870	1600	24.9	14.4
980	1800	26.7	15.4
1090	2000	28.2	16.3

Electrical Properties

Electrical resistivity. 1.30 μΩ·m at 24 °C (75 °F)

Optical Properties

Color. Silvery

Chemical Properties

General corrosion behavior. Exceptional resistance to ferric and cupric chlorides, hot contaminated mineral acids, solvents, chlorine and chlorine-contaminated media, dry chlorine, formic acid, acetic acid, acetic anhydride, seawater and brine.
Resistance to specific corroding agents. C-276 is one of the few materials that resists wet chlorine gas, hypochlorite and chlorine dioxide solutions.

Fabrication Characteristics

Weldability. All common methods can be used, although oxyfuel-gas welding is not recommended when the fabricated item is to be used in corrosion service. Avoid excessive heat input, especially with submerged-arc welding. Fluxes containing carbon or silicon should not be used with submerged-arc welding.

Hastelloy G

Commercial Names

Trade name. HASTELLOY alloy G

Chemical Composition

Composition limits. 21 to 23.5 Cr; 18.0 to 21.0 Fe; 5.5 to 7.5 Mo; 1.75 to 2.50 (Nb + Ta); 1.5 to 2.5 Cu; 1.0 to 2.0 Mn; 2.5 max Co; 1.0 max W; 1.0 max Si; 0.05 max C; 0.04 max P; 0.03 max S; rem Ni

Applications

Typical uses. Chemical applications, particularly those involving sulfuric and phosphoric acids, pulp digestion operations, dissolver vessels and attendant equipment for the dissolution of spent nuclear fuel elements.

Mechanical Properties

Tensile properties. See Table 43.
Compressive properties. Compressive yield strength, plate; 25-mm (1-in.) thick: 315 MPa (45.5 ksi); bar, 32 mm (1.25-in.). Diameter: 355 MPa (51.5 ksi)
Hardness. Sheet: 0% cold reduction, 84 HRB; 10% reduction, 97 HRB; 20% reduction, 28 HRC; 30% reduction, 31 HRC; 40% reduction, 34 HRC; 50% reduction, 36 HRC. In the aged condition:

Temperature °C	°F	Hardness
Aged for 1 h		
650	1200	84 HRB
705	1300	83 HRB
760	1400	84 HRB
815	1500	84 HRB
Aged for 4 h		
650	1200	86 HRB
705	1300	85 HRB
760	1400	86 HRB
815	1500	20 HRC
Aged for 16 h		
650	1200	86 HRB
705	1300	86 HRB
760	1400	94 HRB
815	1500	26 HRC
Aged for 50 h		
650	1200	88 HRB
705	1300	89 HRB
760	1400	21 HRC
815	1500	29 HRC
Aged for 100 h		
650	1200	88 HRB
705	1300	89 HRB
760	1400	25 HRC
815	1500	30 HRC

Elastic modulus:

Temperature °C	°F	Dynamic modulus of elasticity GPa	10⁶ psi
21	70	195	28
93	200	185	27
205	400	180	26
315	600	175	25
425	800	165	24
540	1000	160	23
650	1200	150	22
760	1400	145	21

Impact strength:

Temperature °C	°F	Charpy V-notch impact strength J	ft·lb
Aged for 16 h			
650	1200	181	134
705	1300	65	48
760	1400	27	20
815	1500	19	14
Aged for 100 h			
650	1200	49	36
705	1300	39	29
760	1400	11	8
815	1500	7	5

Fatigue strength. Plate, 13-mm (0.5-in.) thick, solution treated, 330 MPa (48 ksi) at 100 million cycles.

Rupture strength:

Temperature °C	°F	Average initial stress for rupture MPa	ksi
At 10 h			
650	1200	385	56
760	1400	195	28
870	1600	90	13
980	1800	41	6
At 100 h			
650	1200	310	45
760	1400	145	21
870	1600	62	9
980	1800	28	4
At 500 h			
650	1200	275	40
760	1400	125	18
870	1600	55	8
980	1800	21	3
At 1000 h			
650	1200	260	38
760	1400	110	16
870	1600	48	7
980	1800	14	2
At 2000 h			
650	1200	235	34
760	1400	105(a)	15(a)
870	1600	41(a)	6(a)
980	1800	14(a)	2(a)

(a) Extrapolated values

Mass Characteristics

Density. 8.31 Mg/m³ (0.300 lb/in.³) at 22 °C (70 °F)

Thermal Properties

Melting range. 1260 to 1340 °C (2300 to 2450 °F)

Coefficient of thermal expansion. Linear:

Temperature °C	°F	Mean coefficient µm/m·K	µin./in.·°F
21–93	70–200	13.5	7.5
21–205	70–400	13.9	7.7
21–315	70–600	14.2	7.9
21–425	70–800	14.9	8.3
21–540	70–1000	15.7	8.7
21–650	70–1200	16.4	9.1

Specific heat:

Temperature °C	°F	Specific heat J/kg·K	Btu/lb·°F
0	32	390	0.093
100	212	455	0.109
200	392	480	0.115
300	570	502	0.120
400	750	520	0.124
500	930	535	0.128
600	1110	553	0.132
700	1290	570	0.136
800	1470	586	0.140
900	1650	603	0.144
1000	1830	620	0.148

Thermal conductivity:

Temperature °C	°F	Conductivity W/m·K	Btu/ft·h·°F
25	77	10.1	5.8
100	212	11.2	6.5
200	392	12.8	7.4
300	570	14.3	8.3
400	750	15.9	9.2
500	930	17.5	10.1
600	1110	19.2	11.1
700	1290	20.8	12.0
800	1470	22.4	12.9
900	1650	24.0	13.9

Chemical Properties

Resistance to specific corroding agents. Outstanding resistance to hot sulfuric and phosphoric acids in the as-welded condition. Also resists mixed acids, fluosilicic acid, sulfate compounds, contaminated nitric acid, flue gases and hydrofluoric acid.

Table 43 Typical tensile properties of solution treated Hastelloy G

Temperature °C	°F	Tensile strength MPa	ksi	Yield strength MPa	ksi	Elongation, %
Sheet, 3.1 mm (0.125 in.) thick and under						
...	−320	840	122	450	65	48
...	−150	795	115	400	58	50
21	70	705	102	315	46	61
93	200	670	97	290	42	56
205	400	625	91	255	37	74
315	600	605	88	250	36	82
425	800	585	85	230	33	84
540	1000	565	82	230	33	83
650	1200	525	76	220	32	82
760	1400	425	62	220	32	61
Plate, 9.5 to 16 mm (0.375 to 0.625 in.) thick						
...	−320	840	122	460	67	60
...	−150	800	116	400	58	55
21	70	690	100	310	45	62
93	200	655	95	260	38	61
205	400	605	88	235	34	63
315	600	580	84	205	30	68
425	800	565	82	200	29	70
540	1000	525	76	195	28	73
650	1200	505	73	200	29	68
760	1400	415	60	195	28	57

Table 44 Typical mechanical properties of Hastelloy G-3 at room temperature

Form	Tensile strength MPa	ksi	Yield strength MPa	ksi	Elongation, %	Hardness, HRB
Sheet(a)	690	100	325	47	50	79
Sheet(b)	685	99	305	44	53	83
Plate(c)	740	107	365	53	56	87
Bar(d)	695	101	295	43	59	80
Weld metal(e)	690	100	450	65	46(f)	...

(a) 0.63 to 0.97 mm (0.025 to 0.038 in.) thick. (b) 1.4 to 4.8 mm (0.056 to 0.187 in.) thick. (c) 6.4 to 19 mm (0.25 to 0.75 in.) thick. (d) 13 to 25 mm (0.5 to 1.0 in.) diameter. (e) Shielded metal-arc welded. (f) In 57 mm or 2.25 in.

Fabrication Characteristics

Formability. Can be forged, hot-upset and extruded. Has been fabricated into welded and drawn tubing without difficulty.
Weldability. Can be welded by a variety of methods with no significant loss in strength or ductility.

Hastelloy G-3

Commercial Names

Trade name. HASTELLOY alloy G-3

Chemical Composition

Nominal composition. 22.2 Cr; 19.5 Fe; 7.0 Mo; 5.0 max Co; 1.9 Cu; 1.5 max W; 0.8 Mn; 0.4 Si; 0.3 Nb + Ta; 0.04 max P; 0.03 max S; 0.015 max C, rem Ni

Applications

Typical uses. Chemical applications, particularly those involving sulfuric and phosphoric acids; pulp digestion operations; dissolver vessels and attendant equipment for the dissolution of spent nuclear fuel elements

Mechanical Properties

Tensile properties. See Table 44.
Hardness. See Table 44.

Mass Characteristics

Density. 8.31 Mg/m³ (0.300 lb/in.³) at 21 °C (70 °F)

Chemical Properties

General corrosion behavior. Excellent resistance to hot sulfuric and phosphoric acids

Hastelloy N
70Ni-17Mo-7Cr

Chemical Composition

Composition limits. 15.00 to 18.00 Mo; 6.00 to 8.00 Cr; 5.0 max Fe; 0.80 max Mn; 0.50 max W; 0.35 max Cu; 0.20 max Co; 0.04 to 0.08 C; 0.020 max S; 0.015 max P; 0.010 max B; 0.50 max Al + Ti; 1.00 max Si; rem Ni

Applications

Typical uses. Developed specifically for primary-loop service in molten-salt reactors. Especially good for use with hot fluoride salts.

Mass Characteristics

Density. 8.930 Mg/m³ (0.320 lb/in.³) at 22 °C (72 °F)

Thermal Properties

Coefficient of thermal expansion. Linear:

Temperature °C	°F	Coefficient µm/m·K	µin./in.·°F
20–315	68–600	12.6	7.0
20–425	68–800	13.0	7.2
20–540	68–1000	13.3	7.4
20–650	68–1200	13.5	7.5
20–760	68–1400	14.2	7.9
20–870	68–1600	14.8	8.2
20–980	68–1800	15.3	8.5
20–1090	68–2000	15.7	8.7

Specific heat:

Temperature °C	°F	Specific heat J/kg·K	Btu/lb·°F
100	212	419	0.100
200	390	440	0.105
300	570	456	0.109
400	750	469	0.112
480	895	477	0.114
540	1005	486	0.116
570	1060	523	0.125
590	1095	565	0.135
620	1150	586	0.140
660	1220	582	0.139
680	1255	578	0.138
700	1290	578	0.138

Thermal conductivity:

Temperature °C	°F	Conductivity W/m·K	Btu/ft·h·°F
100	212	11.5	6.6
200	390	13.1	9.4
300	570	14.4	8.3
400	750	16.5	9.2
500	930	18.0	10.3
600	1110	20.3	11.8
700	1290	23.6	13.6

Electrical Properties

Electrical resistivity:

Temperature °C	°F	Resistivity, µΩ·m
RT	RT	1.20
705	1300	1.26
815	1500	1.24

Optical Properties

Color. Silvery

Chemical Properties

General corrosion behavior. Good oxidation resistance; also resists aging and embrittlement. Compares favorably with Hastelloy alloys B, C and W in various corrosive media.
Resistance to specific corroding agents. Exceptional resistance to hot fluoride salts; it can serve continuously at temperatures to 980 °C (1800 °F). In fluoride salts at 700 °C (1300 °F), corrosion is less than 0.025 mm/yr (1 mil/yr).

Hastelloy S
68Ni-15Cr-15Mo

Chemical Composition

Composition limits. 14.5 to 17.0 Cr; 14.0 to 16.5 Mo; 2.0 max Co; 3.0 max Fe; 0.20 to 0.75 Si; 0.3 to 1.0 Mn; 0.02 max C; 0.10 to 0.50 Al; 0.015 max B; 0.01 to 0.10 La; rem Ni

Applications

Typical uses. Developed for applications involving severe cyclic heating conditions where components must be capable of retaining strength, ductility, dimensional stability, and metallurgical integrity after long-time exposure.

Mechanical Properties

Tensile properties. Average at room temperature. Sheet: tensile strength, 890 MPa (129 ksi); yield strength, 495 MPa (72 ksi); elongation, 51% in 50 mm or 2 in. Plate: tensile strength, 850 MPa (123 ksi); yield strength, 385 MPa (56 ksi); elongation, 55% in 50 mm or 2 in.
Hardness. Solution heat treated: sheet, 52 HRA; plate, 57 HRA
Elastic modulus. Average, in tension, for sheet heat treated at 1065 °C (1950 °F) and air cooled:

Temperature °C	°F	Modulus GPa	10^6 psi
24	75	212	30.8
360	675	194	28.2
540	1000	182	26.4
650	1200	174	25.2
760	1400	166	24.1
815	1495	161	23.3
925	1700	151	21.9
1090	2000	132	19.2

Impact strength. Charpy V-notch, average for solution heat treated plate: 190 J (140 ft·lb)

Creep-rupture characteristics. See Tables 45 and 46.

Mass Characteristics

Density. 8.747 Mg/m^3 (0.316 lb/in.^{3}) at 22 °C (72 °F)

Thermal Properties

Liquidus temperature. 1380 °C (2516 °F)

Solidus temperature. 1335 °C (2435 °F)

Coefficient of thermal expansion. Linear:

Temperature °C	Temperature °F	Coefficient µm/m·K	Coefficient µin./in.·°F
20–93	68–200	11.5	6.4
20–205	68–400	12.2	6.8
20–315	68–600	12.8	7.1
20–425	68–800	13.1	7.3
20–540	68–1000	13.3	7.4
20–650	68–1200	13.7	7.6
20–760	68–1400	14.4	8.0
20–870	68–1600	14.9	8.3
20–980	68–1800	15.5	8.6
20–1090	68–2000	16.0	8.9

Specific heat:

Temperature °C	Temperature °F	Specific heat J/kg·K	Specific heat Btu/lb·°F
0	32	398	0.095
100	212	427	0.102
200	390	448	0.107
300	570	465	0.111
400	750	477	0.114
500	930	490	0.117
600	1110	498	0.119
700	1290	594	0.142
800	1470	590	0.141
900	1650	594	0.142
1000	1830	599	0.143
1100	2010	603	0.144

Thermal conductivity:

Temperature °C	Temperature °F	Conductivity W/m·K	Conductivity Btu/ft·h·°F
200	390	14.0	8.1
300	570	16.1	9.3
400	750	17.9	10.3
500	930	19.5	11.3
600	1110	21.0	12.2
700	1290	26.0	15.1
800	1470	26.0	15.1
900	1650	26.0	15.1
950	1740	27.0	15.7
1000	1830	28.0	16.2

Table 45 Average creep data for Hastelloy S

Temperature °C	Temperature °F	Creep, %	10 h MPa	10 h ksi	100 h MPa	100 h ksi	1000 h MPa	1000 h ksi
			Average initial stress to produce specified creep in:					
Sheet, 1.1 to 1.6 mm (0.045 to 0.063 in.) thick								
650	1200	0.2	310	45.0	217	31.5	145	21.0
		0.5	345	50.0	245	35.5	165	24.0
		1.0	390	56.5	276	40.0	186	27.0
730	1350	0.2	152	22.0	97	14.1	62	9.0
		0.5	172	25.0	112	16.2	72	10.4
		1.0	200	29.0	131	19.0	84	12.2
815	1500	0.2	70	10.2	41	5.9	...	...
		0.5	81	11.8	48	7.0	...	...
		1.0	95	13.8	58	8.4	...	...
Plate, 25 mm (1 in.) thick								
650	1200	0.2	310	45.0	186	27.0	117(a)	17.0(a)
		0.5	372	54.0	225	32.6	131	19.0
		1.0	386	56.0	234	34.0	143	20.8
705	1300	0.2	165	24.0	86	12.5	46(a)	6.7(a)
		0.5	200	29.0	114	16.5	62	9.0
		1.0	234	34.0	138	20.0	83	12.0
760	1400	0.2	90	13.0	45	6.5	23(a)	3.3(a)
		0.5	117	16.9	63	9.2	33	4.8
		1.0	143	20.7	81	11.8	46	6.7
815	1500	0.2	54	7.8	26	3.8	13(a)	1.9(a)
		0.5	69	10.0	39	5.7	21	3.0
		1.0	86	12.5	48	6.9	26	3.8
870	1600	0.2	32	4.7	15	2.2	7.6(a)	1.1(a)
		0.5	43	6.3	24	3.5	13	1.9
		1.0	52	7.6	28	4.1	15	2.2

(a) Extrapolated values.

Table 46 Stress-rupture data for Hastelloy S

Temperature °C	Temperature °F	10 h MPa	10 h ksi	100 h MPa	100 h ksi	1000 h MPa	1000 h ksi
		Average stress for rupture life of:					
Sheet, 1.1 to 1.6 mm (0.045 to 0.063 in.) thick							
650	1200	431	62.5	345	50.0	262	38.0
730	1350	269	39.0	194	28.2	139	20.2
815	1500	162	23.5	103	15.0	68	9.9
925	1700	66	9.6	40	5.8	...	...
Plate, 25 mm (1 in.) thick							
650	1200	552	80.0	400	58.0	269	39.0
705	1300	386	56.0	262	38.0	172	25.0
760	1400	262	38.0	169	24.5	107	15.5
815	1500	172	25.0	110	16.0	66	9.6
870	1600	114	16.5	68	9.8	37	5.4

Thermal diffusivity:

Temperature °C	Temperature °F	Diffusion coefficient, 10^{-6}m^2/s
100	212	3.8
200	390	3.8

(continued)

Temperature °C	°F	Diffusion coefficient, $10^{-6}m^2/s$
300	570	3.8
400	750	4.5
500	930	4.5
600	1110	5.1
700	1290	5.1
800	1470	5.1
900	1650	5.1
950	1740	5.1
1000	1830	5.8

Electrical Properties

Electrical resistivity. 1.28 μΩ·m at 25 °C (77 °F) for material aged 16 000 h at 650 °C (1200 °F)

Optical Properties

Color. Silvery

Chemical Properties

General corrosion behavior. Excellent oxidation resistance to 1090 °C (2000 °F)

Fabrication Characteristics

Weldability. Manual and automatic methods—including shielded metal-arc (covered electrodes), gas tungsten-arc and gas metal-arc welding—are all suitable.

Hastelloy W
62Ni-24.5Mo-5Cr-5.5Fe

Chemical Composition

Composition limits. Bars and forgings: 0.12 max C; 23 to 26 Mo; 4 to 6 Cr; 4 to 7 Fe; 0.60 max V; 2.50 max Co; 1.00 max Mn; 1.00 max Si; 0.040 max P; 0.030 max S; rem Ni

Applications

Typical uses. Primarily a high-temperature alloy for structural applications up to 760 °C (1400 °F). As a weld filler metal, it has excellent characteristics for joining dissimilar high-temperature metals.

Mechanical Properties

Tensile properties. Average at room temperature. Sheet, 0.109 in. thick: tensile strength, 850 MPa (123 ksi); yield strength, 370 MPa (53.5 ksi); elongation, 55% in 50 mm or 2 in. Weld metal: tensile strength, 875 MPa (127 ksi); elongation, 37% in 50 mm or 2 in. See also Table 47.

Table 47 Typical properties of solution treated Hastelloy W sheet

Temperature °C	°F	Tensile strength MPa	ksi	Yield strength MPa	ksi	Elongation in 50 mm or 2 in., %
425	800	725	105	260	38	56.0
650	1200	...	...	255	37	29.5
730	1350	465	67	...	...	16.0
815	1500	405	59	250	36	17.0
900	1650	353	52	220	32	14.5
980	1800	...	...	135	20	14.5
1065	1950	180	26	...	...	34.0

Mass Characteristics

Density. 9.03 Mg/m³ (0.325 lb/in.³) at 22 °C (72 °F)

Thermal Properties

Liquidus temperature. 1316 °C (2400 °F)

Coefficient of thermal expansion. Linear: 11.3 μm/m·K (6.28 μin./in.·°F) at 23 to 1000 °C (73 to 1832 °F)

Hastelloy X
49Ni-22Cr-18Fe-9Mo

Chemical Composition

Composition limits. 0.50 to 2.50 Co; 20.50 to 23.00 Cr; 8.00 to 10.00 Mo; 0.20 to 1.00 W; 17.00 to 20.00 Fe; 0.05 to 0.15 C; 1.00 max Si; 1.00 max Mn; 0.010 max B; 0.040 max P; 0.030 max S; rem Ni

Applications

Typical uses. Exceptional strength and oxidation-resistance up to 1200 °C (2200 °F); used in many industrial furnace applications because of its resistance to oxidizing, neutral and carburizing atmospheres; widely used for aircraft parts such as jet engine tailpipes, afterburners, turbine blades and vanes.

Precautions in use. Upper limit of usefulness just above 1200 °C (2200 °F)

Mechanical Properties

Tensile properties. Room temperature, average, for sheet solution treated at 1175 °C (2150 °F) and rapidly cooled: tensile strength, 785 MPa (114 ksi); yield strength, 360 MPa (52 ksi); elongation, 43% in 50 mm or 2 in.

Hardness. Average, for sheet solution treated at 1175 °C (2150 °F) and rapidly cooled: 89 HRB

Poisson's ratio. 0.320 at 22 °C (72 °F); 0.328 at −78 °C (−108 °F)

Elastic modulus. In tension, average, for sheet solution treated at 1175 °C (2150 °F) and rapidly cooled:

Temperature °C	°F	Modulus GPa	10^6 psi
25	76	196	28.5
100	212	193	28.0
200	390	185	26.9
300	570	179	26.0
400	750	172	25.0
500	930	164	23.8
600	1110	158	22.9
700	1290	150	21.8
800	1470	143	20.7
900	1650	134	19.5
1000	1830	126	18.3

Impact strength. Charpy V-notch, average, for plate solution treated at 1175 °C (2150 °F) and rapidly cooled:

Temperature °C	°F	Impact strength J	ft·lb
−196	−321	50	37
−157	−216	60	44
−78	−108	69	51
−29	−20	76	56
RT	RT	73	54
+815	+1500	79	58

Creep-rupture characteristics. See Tables 48 and 49.

Mass Characteristics

Density. 8.22 Mg/m³ (0.297 lb/in.³) at 22 °C (72 °F)

Thermal Properties

Melting range. 1260 to 1355 °C (2300 to 2470 °F)

Coefficient of thermal expansion. Linear:

Table 48 Average creep data for Hastelloy X

Temperature, °C	Temperature, °F	Creep, %	10 h, MPa	10 h, ksi	100 h, MPa	100 h, ksi	1000 h, MPa	1000 h, ksi
			Average initial stress to produce specified creep in:					
Sheet, solution heat treated(a)								
650	1200	0.5	276	40	186	27	121	17.5
		1.0	303	44	207	30	145	21
		2.0	331	48	227	33	155	22.5
760	1400	0.5	114	16.5	72	10.5	45	6.5
		1.0	131	19	90	13	62	9
		2.0	145	21	103	15	74	10.8
870	1600	0.5	54	7.8	34	4.9	21	3.1
		1.0	62	9	42	6.1	25	3.6
		2.0	72	10.5	50	7.2	30	4.3
980	1800	0.5	21	3.1	12	1.7	6	0.9
		1.0	25	3.6	13	1.9	7	1.0
		2.0	29	4.2	15	2.2	8	1.1
1090	2000	0.5	...	...	...	...	...	...
		1.0	6	0.8	...	...	...	...
		2.0	8	1.1	...	...	...	...
Plate and bar, solution heat treated(b)								
650	1200	1.0	331	48	220	32	151	22
		5.0	441	64	296	43	200	29
730	1350	1.0	186	27	131	19	89	13
		5.0	234	34	165	24	113	16.5
815	1500	1.0	103	15	71	10.4	50	7.3
		5.0	124	18	89	13	62	9
900	1650	1.0	55	8	37	5.4	25	3.7
		5.0	72	10.5	46	6.8	28	4.2
980	1800	1.0	28	4.2	17	2.6	11	1.6
		5.0	37	5.4	22	3.3	14	2.0

(a) Based on over 100 tests. (b) Based on over 90 tests for 1.0% creep; over 60 tests for 5.0% creep.

Table 49 Stress-rupture data for Hastelloy X

Temperature, °C	Temperature, °F	10 h, MPa	10 h, ksi	100 h, MPa	100 h, ksi	1000 h, MPa	1000 h, ksi	10 000 h, MPa	10 000 h, ksi
		Average rupture life strength for:							
Sheet, solution heat treated(a)									
650	1200	462	67	331	48	234	34	170	24.6
760	1400	221	32	155	22.5	109	15.8	77	11.1
870	1600	117	17	73	10.6	45	6.5	28	4.0
980	1800	45	6.5	26	3.7	14	2.1	8	1.2
1090	2000	17	2.4	8	1.2	4	0.6	...	...
Plate and bar, solution heat treated(b)									
650	1200	496	72	330	47.9	234	34	165	24
730	1350	248	36	172	25	124	18	86	12.5
815	1500	145	21	96	14	69	10	47	6.8
900	1650	83	12	52	7.5	32	4.7	21	3.0
980	1800	48	7.0	29	4.2	17	2.4	10	1.4
1065	1950	26	3.7	13	1.9	7	1.0(c)	...	...
1150	2100	12	1.7	5	0.7	2	0.3(c)	...	...

(a) Based on over 100 tests for sheet. (b) Over 150 tests for plate and bar. (c) Extrapolated.

Temperature °C	°F	Coefficient µm/m·K	µin./in.·°F
26–100	79–200	13.8	7.7
26–500	79–1000	14.9	8.4
26–600	79–1200	15.3	8.6
26–700	79–1350	15.7	8.8
26–800	79–1500	16.0	8.9
26–900	79–1650	16.3	9.1
26–1000	79–1800	16.6	9.2

Specific heat:

Temperature °C	°F	Specific heat J/kg·K	Btu/lb·°F
RT	RT	486	0.116
315	600	498	0.119
650	1200	582	0.139
870	1600	699	0.167
1095	2000	858	0.205

Thermal conductivity:

Temperature °C	°F	Conductivity W/m·K	Btu/ft·h·°F
20	70	9.7(a)	5.25(a)
100	200	11.1	6.3
300	500	14.7	8.2
600	1100	20.6	12.0
700	1300	22.8	13.3
800	1500	25.0	14.5
900	1700	27.4(a)	15.8(a)

(a) Extrapolated values.

Electrical Properties

Electrical resistivity. 1.18 µΩ·m at 22 °C (72 °F)

Magnetic Properties

Magnetic permeability. <1.002 at (200 oersteds) 16 000 A/m

Optical Properties

Color. Silvery

Chemical Properties

General corrosion behavior. Resistant to heat and oxidation; resistant to aqueous solutions at ambient temperatures

Fabrication Characteristics

Formability. Sheet, heat treated at 1177 °C (2150 °F), rapid cooled: typical Erichsen cup depth, 9.0 to 11.5 mm (0.35 to 0.45 in.)

Cast Nickel Alloys

By Donald E. Wenschhof
Huntington Alloys, Inc.

Robert B. Herchenroeder
High Technology Materials Division
Cabot Corporation

C. R. Bird
Stainless Foundry & Engineering, Inc.

CW-12M
55Ni-17Mo-16Cr-6Fe-4W

Commercial Names

Trade name. CW-12M-1, HASTELLOY alloy C; CW-12M-2, Chlorimet 3

Specifications

ASTM. A494, A743, A744

Chemical Composition

Composition limits. CW-12M-1: 0.12 max C; 1.0 max Mn; 1.0 max Si; 0.04 max P; 0.03 max S; 16.0 to 18.0 Mo; 4.5 to 7.0 Fe; 15.5 to 17.5 Cr; 3.75 to 5.25 W; 0.20 to 0.40 V; 2.50 max Co; rem Ni. CW-12M-2: 0.07 max C; 1.0 max Mn; 1.0 max Si; 0.04 max P; 0.03 max S; 17.0 to 20.0 Cr; 17.0 to 20.0 Mo; 3.0 max Fe; rem Ni

Applications

Typical uses. Chemical process equipment

Precautions in use. Except for very weak solutions, this alloy is not recommended for use above 50 °C (120 °F) in nitric or hydrochloric acid that is contaminated with metallic salts.

Mechanical Properties

Tensile properties. Minimum: tensile strength, 495 MPa (72 ksi); yield strength, 315 MPa (46 ksi); elongation, 25% in 50 mm or 2 in. Sheet, typical: tensile strength, 835 MPa (121 ksi); yield strength, 400 MPa (58 ksi); elongation, 47.5% in 50 mm or 2 in. Sand cast, typical: tensile strength, 550 MPa (80 ksi); yield strength, 345 MPa (50 ksi); elongation, 9% in 50 mm or 2 in. Investment cast, typical: tensile strength, 615 MPa (89 ksi); yield strength, 360 MPa (52 ksi); elongation, 11% in 25 mm or 1 in. See also Table 50 and Fig. 20 and 21.

Hardness. Typical: sand cast, 93 HB; investment cast, 96 HB

Elastic modulus. Sand cast, 180 GPa (26.0 × 10^6 psi); investment cast, 170 MPa (24.5 × 10^6 psi)

Impact strength. Typical, Izod V-notch: sand cast, 28 J (20.5 ft·lb); investment cast, 7 J (5 ft·lb). See also Table 50.

Structure

Crystal structure. Face-centered cubic alpha

Mass Characteristics

Density. 8.94 Mg/m^3 (0.323 lb/in.^{3}) at 25 °C (77 °F)

Patternmaker's shrinkage. 2%

Thermal Properties

Liquidus temperature. 1340 °C (2450 °F)

Solidus temperature. 1265 °C (2310 °F)

Minimum liquidation temperature. 1245 °C (2275 °F)

Coefficient of thermal expansion. Linear, 11.3 µm/m·K (6.3 µin./in.·°F) at 0 to 100 °C (32 to 212 °F); 15.3 µm/m·K (8.5 µin./in.·°F) at 0 to 1000 °C (32 to 1830 °F)

Specific heat. 385 J/kg·K (0.092 Btu/lb·°F) at room temperature

Thermal conductivity. 11.3 W/m·K (6.5 Btu/ft·h·°F) at 200 °C (392 °F)

Electrical Properties

Electrical conductivity. Volumetric, 1.3% IACS at 20 °C (68 °F)

Fig. 20 Statistical distribution of tensile strength and elongation at 815 °C (1500 °F) for investment cast CW-12M-1

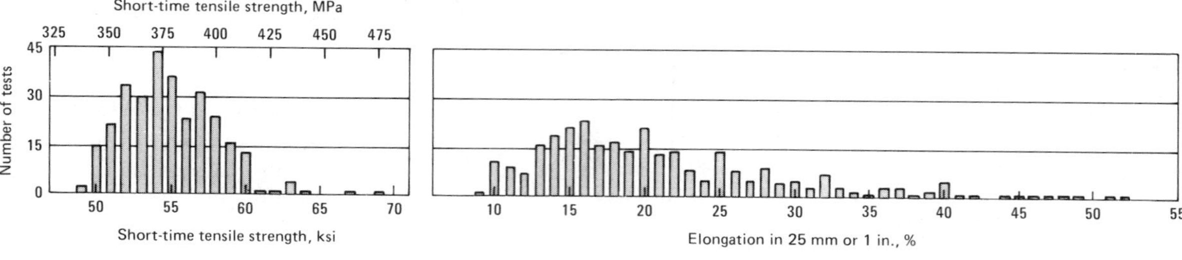

Data represent 299 as-cast test bars taken from 221 heats.

Electrical resistivity. 1.30 μΩ·m at 24 °C (75 °F)

Magnetic Properties

Magnetic permeability. Paramagnetic: over 1.000 and less than 1.001

Optical Properties

Color. Silvery

Chemical Properties

General corrosion behavior. Resists oxidizing agents such as wet chlorine, chlorine gas, hypochlorite, chlorine dioxide solutions, ferric chloride and nitric, hydrochloric and sulfuric acids at moderate temperatures or under oxidizing conditions; has excellent resistance to acetic acid, seawater and many corrosive organic acids and salts; has good high-temperature properties and is resistant to oxidizing and reducing atmospheres up to 1090 °C (2000 °F). This alloy should not be held for long periods of time in the range 500 to 800 °C (930 to 1470 °F), unless subsequently annealed, as these temperatures are likely to cause higher corrosion rates.

Resistance to specific corroding agents. See Table 51.

Fabrication Characteristics

Heat treatment. Solution treat at 1180 to 1230 °C (2150 °F), then water quench by spraying or immersion

Standard finishes. Solution heat treated and pickled

Joining. Oxyfuel-gas and carbon-arc welding are not recommended. All types of resistance welding are satisfactory. In shielded metal-arc welding, use AWS ENiCrMo-4 or ENiCrMo-5 electrode and dc reverse polarity, choke the arc short, and avoid overheating the weld and heat-affected zones.

Fig. 21 Selected mechanical and physical properties of CW-12M-1

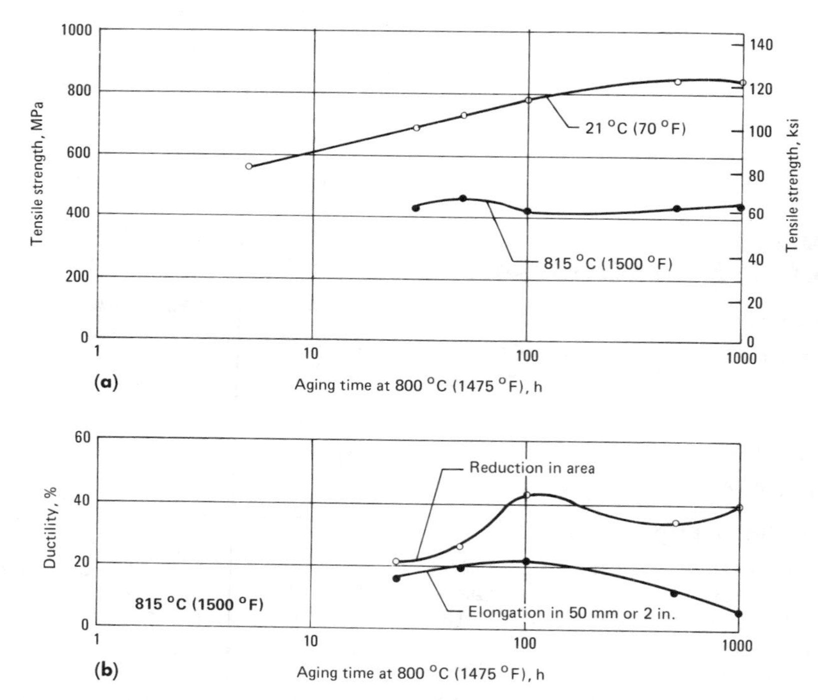

(a) Tensile strength, at room and elevated temperature, of aged investment cast test bars. (b) Elevated-temperature ductility of aged investment cast test bars. (c) 100-hour and 1000-hour rupture curves for aged investment cast test bars. (d) Mean coefficient of thermal expansion from 21 °C (70 °F) to the indicated temperature.

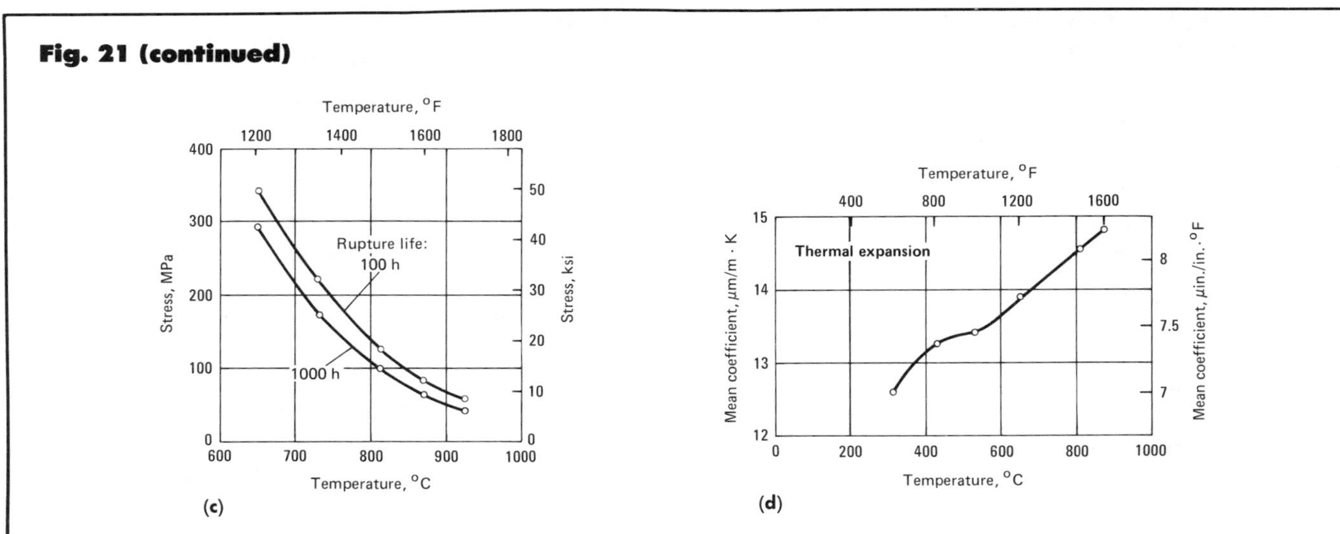

Table 50 Typical properties for CW-12M-1 and N-12M-1 at subzero temperatures

Temperature °C	Temperature °F	Tensile strength MPa	Tensile strength ksi	Yield strength MPa	Yield strength ksi	Elongation, %	Reduction in area, %	Charpy impact strength J	Charpy impact strength ft·lb
CW-12M-1									
−50	−58	925	134	450	65	38	33	34	25
−100	−150	985	143	545	79	25	25	30	22
−200	−330	1100	160	660	96	24	22	37	27
N-12M-1									
−50	−58	980	142	405	59	66	54	66	49
−100	−150	995	144	420	61	63	53	72	53
−200	−330	1190	172	570	83	58	45	72	53

CY-40
72Ni-16Cr-8Fe-2Si

Commercial Names

Common name. Nickel-chromium alloy
Trade name. Inconel

Specifications

ASTM. A494, A743, A744

Chemical Composition

Composition limits. 14.0 to 17.0 Cr; 11.0 max Fe; 3.0 max Si; 1.5 max Mn; 0.40 max C; 0.03 max P; 0.03 max S; rem Ni

Applications

Typical uses. For industrial applications that demand high strength, pressure tightness, and high resistance to destructive chemical action, mechanical wear and oxidation at elevated temperatures

Mechanical Properties

Tensile properties. Sand cast: tensile strength, 480 to 655 MPa (70 to 95 ksi); yield strength, 205 to 310 MPa (30 to 45 ksi); elongation, 30 to 10% in 50 mm or 2 in. Minimum: tensile strength, 480 MPa (70 ksi); yield strength, 195 MPa (28 ksi); elongation, 30% in 50 mm or 2 in.
Compressive properties. Compressive yield strength: 115 MPa (17 ksi) at 0.00% offset; 160 MPa (23 ksi) at 0.01% offset; 250 MPa (36 ksi) at 0.2% offset
Hardness. 160 to 190 HB
Elastic modulus. Tension, 155 GPa (22.7 × 10^6 psi)
Impact strength. Charpy, 81 J (60 ft·lb); Izod, 115 to 95 J (85 to 70 ft·lb)

Mass Characteristics

Density. 8.3 Mg/m^3 (0.30 lb./in.^{3}) at 20 °C (68 °F)
Solidification shrinkage. 21 mm/m (0.25 in./ft)

Thermal Properties

Melting temperature. 1370 to 1400 °C (2500 to 2550 °F)
Coefficient of thermal expansion. Linear, 11.5 μm/m·K (6.4 μin./in.·°F) at 25 to 100 °C (77 to 212 °F)

Magnetic Properties

Magnetic susceptibility. Nonmagnetic at 25 °C (77 °F)
Magnetic transformation temperature. −101 °C (−150 °F)

Optical Properties

Color. White

Fabrication Characteristics

Casting temperature. 1540 to 1620 °C (2800 to 2950 °F)
Preferred deoxidizer. Magnesium
Standard finishes. Machined

Table 51 Typical corrosion rates for CW-12M and N-12M in selected corrosive media

Concentration, %	Temperature °C	Temperature °F	N-12M µm/yr	N-12M mils/yr	CW-12M µm/yr	CW-12M mils/yr
Acetic acid						
10	65	150	152	6.0	5	0.2
10	Boiling		18	0.7	10	0.4
50	65	150	102	4.0	3	0.1
50	Boiling		10	0.4	3	0.1
99	65	150	13	0.5	3	0.1
99	Boiling		5	0.2	3	0.1
Chlorine, wet						
...	Room		12 140	478	3	0.1
Chromic acid						
2	Room		nil	nil	nil	nil
2	65	150	nil	nil	nil	nil
2	Boiling		710(b)	28(b)	51	2.0
20	Room		8	0.3	3	0.1
20	65	150	1650	65	127	5.0
20	Boiling		25 400	1000	1473	58.0
Hydrochloric acid						
2	Room		50	2.0	23	0.9
2	65	150	230	9.0	3	0.1
2	Boiling		76	3.0	1510	62
10	Room		50	2.0	50	2.0
10	65	150	180	7.0	560	22
10	Boiling		230	9.0	8430	332
20	Room		50	2.0	50	2.0
20	65	150	125	5.0	635	25
20	Boiling		610	24.0	10 338	407
37	Room		8	0.3	25	1.0
37	65	150	50	2.0	280	11
Hydrofluoric acid						
5	Room		100	4.0	25	1.0
25	Room		125	5.0	125	5.0
45	Room		76	3.0	150	6.0
Nitric acid						
10	Room		...	...	3	0.1
10	65	150	...	...	25	1.0
10	Boiling		...	...	180	7.0
30	Room		...	...	3	0.1
30	65	150	...	...	76	3.0
30	Boiling		...	...	3890	153
50	Room		...	...	25	1.0
50	65	150	...	...	280	11.0
50	Boiling		...	...	1200	473
70	Room		...	...	25	1.0
70	65	150	...	...	585	23
70	Boiling		...	...	19 810	780
Phosphoric acid (chemically pure)						
10	65	150	50	2.0	5	0.2
10	Boiling		25	1.0	15	0.6
30	65	150	20	0.8	3	0.1
30	Boiling		76	3.0	100	4.0
50	65	150	8	0.3	8	0.3
50	Boiling		76	3.0	100	4.0
85	65	150	10	0.4	8	0.3
85	Boiling		710	28.0	1140	45

(continued)

CZ-100
97Ni-1.5Si

Commercial Name

Trade name. Cast nickel

Specifications

ASTM. A494, A743, A744

Chemical Composition

Composition limits. 1.0 max C; 0.03 max P; 0.03 max S; 1.5 max Mn; 2.0 max Si; 1.25 max Cu; 3.0 max Fe; 95 min Ni

Applications

Typical uses. Equipment handling corrosives such as caustics; applications where it is necessary to avoid contamination of a product with metals such as copper and iron. Resistance to galling under conditions of boundary lubrication can be provided by adjustment of minor elements, including carbon.

Mechanical Properties

Tensile properties. Sand cast, typical: tensile strength, 310 to 415 MPa (45 to 60 ksi); yield strength, 140 to 205 MPa (20 to 30 ksi); elongation, 30 to 15% in 50 mm or 2 in.

Compressive properties. Compressive yield strength: 69 MPa (10 ksi) at 0.007% offset; 97 MPa (14 ksi) at 0.01% offset; 160 MPa (23 ksi) at 0.2% offset

Hardness. 80 to 125 HB

Elastic modulus. Tension, 150 GPa (21.5×10^6 psi)

Impact strength. Charpy: 81 J (60 ft·lb). Izod: 120 to 100 J (90 to 75 ft·lb)

Mass Characteristics

Density. 8.30 Mg/m^3 (0.30 lb/in.^{3}) at 20 °C (68 °F)

Solidification shrinkage. 21 mm/m (0.25 in./ft)

Thermal Properties

Liquidus temperature. 1425 to 1460 °C (2600 to 2660 °F)

Solidus temperature. 1360 to 1410 °C (2480 to 2565 °F)

Incipient melting temperature. 1345 °C (2450 °F)

Coefficient of thermal expansion. Linear, 13.0 µm/m·K (7.2 µin./in.·°F) at 24 to 100 °C (77 to 212 °F)

Electrical Properties

Electrical resistivity. 210 nΩ·m at 20 °C (68 °F)

Magnetic Properties

Magnetic susceptibility. Slightly magnetic at room temperature. Magnetic properties decrease with increasing silicon content.

Fabrication Characteristics

Casting temperature. Sand: 1510 to 1595 °C (2750 to 2900 °F)
Preferred deoxidizer. Magnesium
Standard finishes. Machined or sand blasted
Weldability. Solder with Pb- or Sn-base alloys, using an acid flux. Braze with P- and Pb-free alloys, using a fluoride flux and slightly reducing flame. Oxyfuel gas weld using unalloyed Ni filler metal, fluoride flux and slightly reducing flame. Shielded metal-arc weld using unalloyed Ni filler metal, fluoride flux and dc straight polarity.

Table 51 (continued)

Concentration, %	Temperature °C	Temperature °F	Penetration rate(s) for: N-12M µm/yr	N-12M mils/yr	CW-12M µm/yr	CW-12M mils/yr
Sodium hydroxide						
5	65	150	nil	nil	nil	nil
5	Boiling		nil	nil	nil	nil
50	65	150	nil	nil	nil	nil
Sulfuric acid						
2	Room		25	1.0	5	0.2
2	65	150	125	5.0	13	0.5
2	Boiling		25	1.0	280	11
10	Room		25	1.0	3	0.1
10	65	150	76	3.0	13	0.5
10	Boiling		50	2.0	840	33
25	Room		25	1.0	5	0.2
25	65	150	25	1.0	25	1.0
25	Boiling		50	2.0	1625	64
50	Room		10	0.4	25	1.0
50	65	150	25	1.0	100	4.0
50	Boiling		50	2.0	8430	332
60	Room		5	0.2	50	2.0
60	65	150	25	1.0	100	4.0
60	Boiling		180	7.0	24 330	958
80	Room		3	0.1	3	0.1
80	65	150	8	0.3	76	3.0
80	Boiling		>25 400	>1000	>25 400	>1000
96	Room		5	0.2	5	0.2
96	65	150	8	0.3	25	1.0
96	Boiling		>25 400	>1000	6450	254

(a) All data are steady-state, as calculated from a minimum of five 24-h test periods, unless otherwise indicated. (b) For the fifth 24-h test period, not steady-state rate.

M-35
65Ni-30Cu-1.5Si

Commercial Names

Trade name. Monel
Common name. Nickel-copper alloy

Specifications

ASTM. A494, A743, A744
Government. QQ-N-288 (composition A), QQ-N-288 (composition E)

Chemical Composition

Composition limits. Sand cast (composition A): 62 to 68 Ni; 26 to 33 Cu; 2.0 max Si; 2.5 max Fe; 1.5 max Mn; 0.35 max C; 0.50 max Al. Composition E contains 1.0 to 3.0 Nb + Ta. ASTM specifications allow 3.5 max Fe and do not specify a limit for Al; ASTM M-35-1 contains 1.25 max Si, and M-35-2 contains 2.0 max Si

Applications

Typical uses. Industrial applications that demand high strength, pressure tightness and high resistance to destructive chemical action or mechanical wear

Table 52 Minimum tensile properties for M-35

Alloy grade	Tensile strength MPa	Tensile strength ksi	Yield strength MPa	Yield strength ksi	Elongation, %
Comp A	450	65	225	32.5	25
Comp E	450	65	225	32.5	25
M35-1	450	65	170	25	25
M35-2	450	65	205	30	25

Mechanical Properties

Tensile properties. Sand cast, as cast: tensile strength, 450 to 620 MPa (65 to 90 ksi); yield strength, 220 to 275 MPa (32 to 40 ksi); elongation, 45 to 25% in 50 mm or 2 in. See also Table 52.
Compressive properties. Compressive yield strength, 130 MPa (19 ksi) at 0.00% offset; 160 MPa (23 ksi) at 0.01% offset; 200 MPa (29 ksi) at 0.2% offset
Hardness. 125 to 150 HB
Elastic modulus. Tension, 130 GPa (18.5 × 10^6 psi)
Impact strength. Charpy: 95 J (70 ft·lb). Izod: 110 to 88 J (80 to 65 ft·lb)

Mass Characteristics

Density. 8.63 Mg/m^3 (0.312 lb/in.^{3}) at 20 °C (68 °F)
Solidification shrinkage. 21 mm/m (0.25 in./ft)

Thermal Properties

Melting temperature. 1315 to 1345 °C (2400 to 2450 °F)
Coefficient of thermal expansion. Linear, 12.9 µm/m·K (7.2 µin./in.·°F) at 25 to 100 °C (77 to 212 °F)

Electrical Properties

Electrical resistivity. 533 nΩ·m at 20 °C (68 °F)

Magnetic Properties

Magnetic susceptibility. Slightly magnetic at 25 °C (77 °F)

Fabrication Characteristics

Casting temperature. 1480 to 1565 °C (2700 to 2850 °F)
Weldability. Shielded metal-arc weld using ENiCu-1 (preferred) or ENiCu-2 electrode, dc reverse polarity and a short arc length. Preheat not normally required. Postweld heat treatment is not used.

H Monel
65Ni-30Cu-3Si

Commercial Names

Common name. Nickel-copper alloy

Specifications

Government. QQ-N-288 (composition B)

Chemical Composition

Composition limits. Sand cast: 61 to 68 Ni; 27 to 33 Cu; 2.7 to 3.7 Si; 2.5 max Fe; 1.5 max Mn; 0.30 max C; 0.50 max Al

Applications

Typical uses. Industrial applications that demand nongalling and antiseizing characteristics, moderately high hardness, moderate machinability and resistance to corrosive attack

Mechanical Properties

Tensile properties. Sand cast: tensile strength, 690 to 895 MPa (100 to 130 ksi); yield strength, 415 to 550 MPa (60 to 80 ksi); elongation, 25 to 10% in 50 mm or 2 in.
Hardness. 240 to 290 HB

Mass Characteristics

Density. 8.5 Mg/m³ (0.308 lb/in.³) at 20 °C (68 °F)
Solidification shrinkage. 21 mm/m (0.25 in./ft)

Thermal Properties

Melting temperature. 1285 to 1315 °C (2350 to 2400 °F)
Coefficient of thermal expansion. Linear, 12.4 μm/m·K (7.0 μin./in.·°F) at 24 to 100 °C (77 to 212 °F)

Electrical Properties

Electrical resistivity. 583 nΩ·m at 20 °C (68 °F)

Magnetic Properties

Magnetic susceptibility. Nonmagnetic at room temperature

Fabrication Characteristics

Casting temperature. 1455 to 1315 °C (2650 to 2800 °F)
Preferred deoxidizer. Magnesium
Standard finishes. Machined or sand blasted
Weldability. Solder with Pb- or Sn-base alloys, using acid flux. Braze with P- and Pb-free alloys, using fluoride flux and a slightly reducing flame.

S Monel
63Ni-30Cu-4Si

Commercial Names

Common name. Nickel-copper alloy

Specifications

AMS. 4892, 4893
Government. QQ-N-288 (composition D)

Chemical Composition

Composition limits. Sand cast: 60 min Ni; 27 to 31 Cu; 3.5 to 4.5 Si; 2.5 max Fe; 1.5 max Mn; 0.25 max C; 0.50 max Al

Mechanical Properties

Tensile properties. Tensile strength, 760 to 1000 MPa (110 to 145 ksi); yield strength, 550 to 795 MPa (80 to 115 ksi); elongation, 4 to 1% in 50 mm or 2 in.
Compressive properties. Compressive yield strength, see Table 53.
Hardness. Sand cast: as cast, 275 to 350 HB; quenched, 225 to 260 HB; aged, 300 to 375 HB. As-cast material:

Temperature °C	°F	Hardness, HB
RT	RT	321
370	700	321
425	800	311
480	900	311
540	1000	321
565	1050	355
595	1100	293

Elastic modulus. Sand cast: as cast, 145 GPa (21 × 10⁶ psi); quenched, 140 GPa (20.5 × 10⁶ psi); aged, 150 GPa (21.5 × 10⁶ psi)
Impact strength. Sand cast. Charpy: as cast, 5.4 J (4 ft·lb). Izod: as cast, 12 to 4 J (9 to 3 ft·lb); aged, 7 to 1.4 J (5 to 1 ft·lb)

Mass Characteristics

Density. 8.36 Mg/m³ (0.302 lb/in.³) at 20 °C (68 °F)
Solidification shrinkage. 21 mm/m (0.25 in./ft)

Thermal Properties

Melting temperature. 1260 to 1290 °C (2300 to 2350 °F)
Coefficient of thermal expansion:

Temperature °C	°F	Linear coefficient μm/m·K	μin./in.·°F
21 to 100	70 to 212	12.2	6.8
21 to 315	70 to 600	14.8	8.2
21 to 455	70 to 850	15.1	8.4
21 to 540	70 to 1000	15.7	8.7

Electrical Properties

Electrical resistivity. 633 nΩ·m at 20 °C (68 °F)

Fabrication Characteristics

Casting temperature. 1455 to 1540 °C (2650 to 2800 °F)
Preferred deoxidizer. Magnesium
Standard finishes. Machined or sand blasted
Joining characteristics. Solder with Pb- or Sn-base solders, using an acid flux. Braze with P- and Pb-free alloys, using a fluoride flux and slightly reducing flame.
Heat treatment. To soften: 1 h at 870 °C (1600 °F) in air, air cool to 650 °C (1200 °F), then quench to room temperature. To harden: 4 to 6 h at 595 °C (1100 °F) in air, air or furnace cool

Table 53 Compressive yield strength of cast S Monel

Condition	Compressive yield strength at: 0% offset MPa	ksi	0.01% offset MPa	ksi	0.2% offset MPa	ksi
Sand cast, as cast	450	65	550	80	765	111
Quenched	205	30	280	41	420	61
Aged	460	67	625	91	860	125

Fig. 22 Selected mechanical properties of N-12M-1

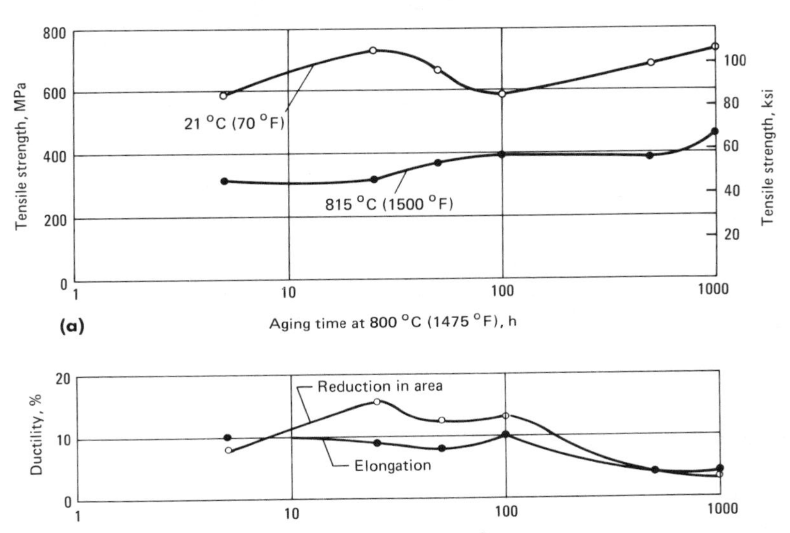

(a) Tensile strength, at room and elevated temperature, of aged investment cast test bars. (b) Elevated-temperature ductility of aged investment cast test bars.

N-12M
62Ni-30Mo-5Fe

Commercial Names

Trade name. N-12M-1, HASTELLOY alloy B; N-12M-2, Chlorimet 2

Specifications

ASTM. A494, A743, A744

Chemical Composition

Composition limits. N-12M-1: 0.12 max C; 1.0 max Mn; 1.0 max Si; 0.04 max P; 0.03 max S; 26.0 to 30.0 Mo; 4.0 to 6.0 Fe; 1.0 max Cr; 0.20 to 0.60 V; 2.50 max Co; rem Ni. N-12M-2: 0.07 max C; 1.0 max Mn; 1.0 max Si; 0.04 max P; 0.03 max S; 30.0 to 33.0 Mo; 3.0 max Fe; 1.0 max Cr; rem Ni

Applications

Typical uses. Chemical process equipment; high-temperature structural applications

Precautions in use. Not recommended for strongly oxidizing acids or salts. In oxidizing atmospheres, service temperatures should not exceed 760 °C (1400 °F); N-12M may be used at substantially higher temperatures in reducing atmospheres.

Mechanical Properties

Tensile properties. Average at room temperature. Sand cast: tensile strength, 620 MPa (90 ksi); yield strength, 345 MPa (50 ksi); elongation, 10% in 50 mm or 2 in. Investment cast: tensile strength, 585 MPa (85 ksi); yield strength, 370 MPa (54 ksi); elongation, 15% in 25 mm or 1 in. See also Fig. 22 and Table 50.

Hardness. Sand cast: 91 HB. Investment cast: 93 HB. See also Fig. 23.

Elastic modulus. Tension. Sheet, 180 GPa (26.4×10^6 psi); sand cast, 185 GPa (26.5×10^6 psi); investment cast, 195 GPa (28.5×10^6 psi)

Impact strength. Izod V-notch: sand cast, 25 J (18.5 ft·lb); investment cast, 17.6 J (13 ft·lb)

Mass Characteristics

Density. 9.24 Mg/m^3 (0.334 lb/in.^{3}) at 25 °C (77 °F)

Patternmaker's shrinkage. 2%

Thermal Properties

Liquidus temperature. 1370 °C (2495 °F)

Solidus temperature. 1300 °C (2375 °F)

Coefficient of thermal expansion. Linear: 10.0 μm/m·K (5.5 μin./in.·°F) at 0 to 100 °C (32 to 212 °F); 14.6 μm/m·K (8.1 μin./in.·°F) at 0 to 100 °C (32 to 1830 °F)

Specific heat. 380 J/kg·K (0.091 Btu/lb·°F)

Thermal conductivity. 2.2 W/m·K (7.1 Btu/ft·h·°F) at 200 °C (392 °F)

Electrical Properties

Electrical conductivity. Volumetric, 1.3% IACS at 20 °C (68 °F)

Electrical resistivity. 1.35 μΩ·m at 24 °C (75 °F)

Magnetic Properties

Magnetic permeability. Paramagnetic: over 1.000 and less than 1.001

Optical Properties

Color. Silvery

Chemical Properties

General corrosion behavior. Resists hydrochloric acid in all concentrations up to the boiling point; is also resistant to other nonoxidizing acids and salts, such as phosphoric

acid, sulfuric acid up to 60% concentration, cuprous chloride, and similar reducing chloride salts; has valuable high-temperature properties, retaining over two thirds of its room-temperature strength at 870 °C (1600 °F); in oxidizing atmospheres (not oxidizing solutions), may be used at up to 760 °C (1400 °F); in reducing atmospheres, may be used at substantially higher temperatures.
Resistance to specific corroding agents. See Table 51.

Fabrication Characteristics

Casting temperature. Chill casting, 1450 °C (2650 °F)
Heat treatment. Solution treated at 1177 °C (2150 °F), for N-12M-1 and 1120 °C (2050 °F) for N-12M-2, followed by either a rapid air cool or water quench. This heat treatment ensures maximum ductility, corrosion resistance and machinability.
Joining. Oxyfuel-gas and carbon-arc welding not recommended. All types of resistance welding are satisfactory. In metal-arc welding, use AWS ENiMo-1 electrode with reversed polarity, choke the arc short, and avoid overheating the weld.
Heat treatment. 1175 °C (2150 °F)
Annealing temperature. 1175 °C (2150 °F)

Illium 98
55Ni-28Cr-8.5Mo-5.5Cu

Chemical Composition

Nominal composition. 28 Cr; 8.5 Mo; 5.5 Cu; 1.0 Mn; 1.0 Si; 0.07 max C; rem Ni

Applications

Typical uses. A machinable high-strength casting alloy designed for hot sulfuric acid service. Because it has good resistance to sulfuric acid over the entire range of concentrations, Illium 98 is used in components for pumps, valves and various other types of process equipment.
Precautions in use. Not recommended for service in halogens, halogen acids or halogen salt solutions, except that Illium 98 is highly resistant to seawater and to many fluorine compounds.

Fig. 23 Corrosion of as-cast Illium 98 in sulfuric acid

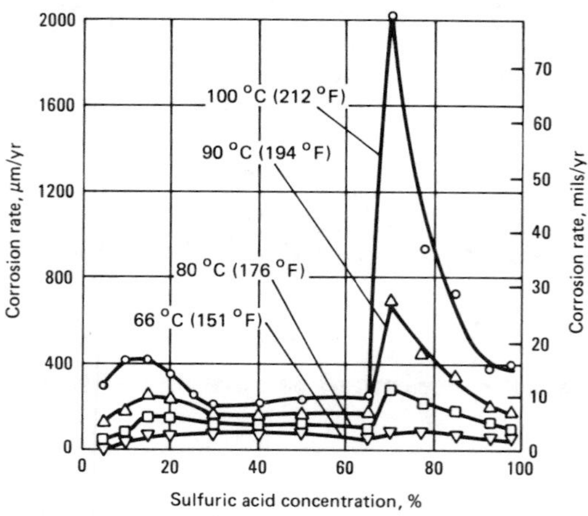

Mechanical Properties

Tensile properties. Typical. Tensile strength, 370 MPa (54 ksi); yield strength, 285 MPa (41 ksi); elongation, 18% in 50 mm or 2 in.; reduction in area, 22%
Hardness. 160 HB
Impact strength. Charpy V-notch: 270 J (200 ft·lb)

Structure

Crystal structure. Expanded face-centered cubic
Microstructure. Austenitic solid solution with minor isolated precipitation of heavy-metal carbides
Metallography. Macroetch by immersion in aqua regia. For microstructure, etch electrolytically with 10% oxalic acid.

Mass Characteristics

Density. 8.36 Mg/m^3 (0.302 $lb/in.^3$) at 25 °C (77 °F)
Patternmaker's shrinkage. 26 mm/m (5/16 in./ft)

Chemical Properties

General corrosion behavior. More resistant than Illium G to the same chemical group of reagents. Usable in acid reagents to higher temperature levels. Resists intergranular attack when properly heat treated.
Resistance to specific corroding agents. See Fig. 23 for resistance to sulfuric acid.

Fabrication Characteristics

Heat treatment. Anneal at 1120 °C (2150 °F) and water quench for best corrosion resistance.
Weldability. Shielded metal-arc welding preferred, using specially coated Illium 98 electrodes. Oxyfuel-gas welding not recommended.

Illium B
50Ni-28Cr-8.5Mo-5.5Cu-3.5Si

Chemical Composition

Nominal composition. 28 Cr; 8.5 Mo; 5.5 Cu; 3.5 Si; 1.0 Mn; 0.07 max C; rem Ni

Applications

Typical uses. Used in the form of machined castings where corrosion resistance is of primary importance but where resistance to erosion, wear and galling must also be considered. Typical uses include thrust and rotary bearings, cutting blades and pump impellers where high hardness is required in corrosive environments.
Precautions in use. Attacked by hot concentrated nitric acid and other highly oxidizing media. Not recommended for halogen compounds, but may satisfactorily resist seawater.

Fig. 24 Corrosion of Illium B in sulfuric acid at 100 °C

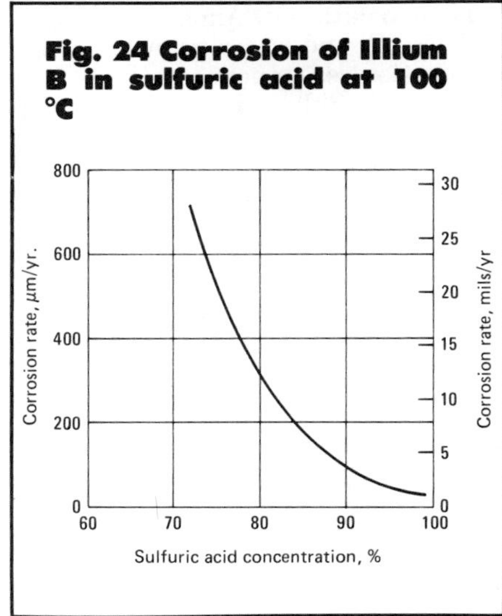

Mechanical Properties

Tensile properties. Typical. Tensile strength, 420 MPa (61 ksi); yield strength, 415 MPa (60 ksi); elongation, 1% in 50 mm or 2 in.
Hardness. 221 to 420 HB

Structure

Metallography. Macroetch by immersion in aqua regia. Microetch by immersion in a mixture of 92% HCl, 5% H_2SO_4 and 3% HNO_3
Microstructure. Two phase: complex silicon precipitate in an austenitic solid solution

Mass Characteristics

Patternmaker's shrinkage. 26 mm/m (5/16 in./ft)

Chemical Properties

General corrosion behavior. Extremely resistant to hot concentrated sulfuric acid in both reducing and moderately oxidizing environments. General corrosion resistance similar to Illium G except less resistant in strongly oxidizing media.
Resistance to specific corroding agents. See Fig. 24.

Fabrication Characteristics

Weldability. Shielded metal-arc welding using specially coated Illium B electrodes is preferred. Oxyfuel-gas welding not recommended.

Fig. 25 Isocorrosion chart for Illium G in sulfuric acid

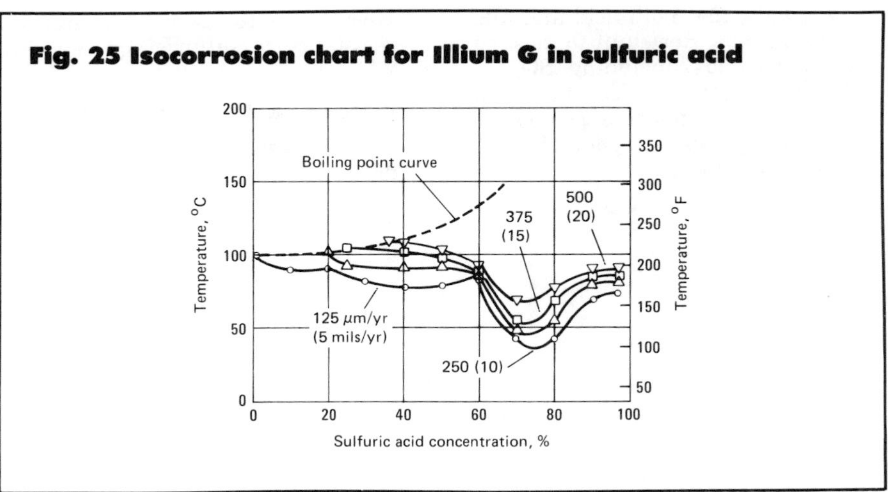

Illium G
56Ni-22.5Cr-6.5Mo-6.5Cu

Chemical Composition

Nominal composition. 22.5 Cr; 6.5 Cu; 6.5 Mo; 5.0 Fe; 1.0 Mn; 1.0 Si; 0.2 C; rem Ni

Applications

Typical uses. Provides superior corrosion resistance in a machinable high-strength casting alloy. Used extensively in dies for viscose cellophane extrusion. Also used in pumps, valves and other process equipment where severely corrosive environments are encountered.
Precautions in use. Generally not recommended for halogens, halogen acids or halogen salt solutions. However, the alloy is highly resistant to seawater and to fluorine compounds in oxidizing environments.

Mechanical Properties

Tensile properties. Typical. Tensile strength, 470 MPa (68 ksi); yield strength, 195 MPa (28 ksi) at 0.02% offset; 240 MPa (35 ksi) at 0.1% offset, 270 MPa (39 ksi) at 0.2% offset; elongation, 7.5% in 50 mm or 2 in.; reduction in area, 11.3%; proportional limit, 145 MPa (21.1 ksi)
Hardness. 168 HB
Elastic modulus. Tension, 170 GPa (24.3×10^6 psi)

Structure

Crystal structure. Expanded face-centered cubic
Microstructure. Austenitic solid solution with precipitate network of eutectic heavy-metal carbides

Mass Characteristics

Density. 8.58 Mg/m^3 (0.31 $lb/in.^3$)
Patternmaker's shrinkage. 2.6% (5/16 in./ft)

Thermal Properties

Liquidus temperature. 1340 °C (2440 °F)
Solidus temperature. 1250 °C (2290 °F)
Coefficient of thermal expansion. Linear:

Temperature °C	°F	Coefficient μm/m·K	μin./in.·°F
0–100	32–212....	12.2	6.8
0–200	32–390....	12.8	7.1
0–300	32–570....	13.3	7.4
0–400	32–750....	13.6	7.6
0–500	32–930....	14.0	7.8
0–600	32–1110...	14.4	8.0
0–700	32–1290...	14.9	8.3
0–800	32–1470...	15.3	8.5

Specific heat. 440 J/kg·K (0.105 Btu/lb·°F) at room temperature

Electrical Properties

Electrical resistivity. 1.235 μΩ·m at room temperature

Magnetic Properties

Magnetic permeability. 1.0011 at 26 °C (79 °F); 1.0017 at −30 °C (−22 °F)

Chemical Properties

General resistance to corrosion. Resists sulfuric, nitric, phosphoric, organic acids, mixed acids and many acid-salt mixtures; withstands the corrosion of both oxidizing and reducing agents, and both acid and alkaline solutions. Resistant to most sulfur compounds, including SO_2 and H_2S gases. Highly resistant to seawater and some fluorine compounds. Attacked by most compounds of chlorine or bromine.

Resistance to specific agents. For resistance to sulfuric acid, see Fig. 25.

Fabrication Characteristics

Machinability. 33% of B1112, cold drawn steel

Heat treatment. Anneal at 1120 °C (2150 °F) and water quench for best corrosion resistance

Joining. Shielded metal-arc welding, using specially coated Illium 98 electrodes, preferred. Oxyfuel-gas welding not recommended

Corrosion Resistance of Nickels and Nickel Alloys

By D. L. Graver
Huntington Alloys, Inc.

NICKEL is highly resistant to a variety of corrosive mediums. It can be readily alloyed with elements such as copper, chromium, molybdenum and iron to produce alloys that are resistant to corrosion throughout a wide range of environments.

The common nickel alloy families are as follows: commercially pure nickel; binary systems, such as Ni-Cu, Ni-Si and Ni-Mo; ternary systems, such as Ni-Cr-Fe and Ni-Cr-Mo; and more complex systems, such as Ni-Cr-Fe-Mo-Cu (with other possible additions). Table 1 gives nominal compositions of alloys representative of each of these alloy families, with nickel contents varying from 34 to 99.5%.

Nickel alloys are used extensively in a variety of chemical-processing applications. These applications include production of ethylene, ammonia, nitric acid, caustic-chlorine and hydrofluoric acid; refining of salt; and processing of pulp and paper.

Alloy Characteristics

The corrosion characteristics of nickels and nickel alloys—notably, their corrosion resistance in a variety of mediums, which accounts for their wide industrial use—are described in the following paragraphs.

Commercially Pure Nickels. This family is represented in Table 1 by Nickel 200 and Nickel 201. The latter is preferred for use at temperatures above 315 °C (600 °F) because its lower carbon content prevents graphitization and subsequent loss of ductility. The outstanding corrosion characteristics of Nickel 200 and Nickel 201 are their resistance to caustics; high-temperature halogens and hydrogen halides; salts other than oxidizing halides; and foods, in which these alloys are particularly suited for maintaining product purity. Other important characteristics are high thermal and electrical conductivities and magnetic and magnetostrictive properties.

Ni-Cu Alloys. Monel 400 and Monel K-500 differ in that the strength and hardness of the latter may be increased by age hardening. Although these alloys share many of the corrosion characteristics of commercially pure nickel, they exceed nickel in resistance to sulfuric and hydrofluoric acids and to brine. Handling of waters, including seawater and brackish water, is a major area of application. Monel 400 and Monel K-500 are immune to chloride-ion stress-corrosion cracking, which often is a factor in their selection.

Ni-Mo and Ni-Si Alloys. Hastelloy B, noted for its superior resistance to hydrochloric acid, also has very good resistance to many nonoxidizing acids. As with commercially pure nickel and Ni-Cu alloys, corrosion rates of Hastelloy B in acid solutions can be greatly increased by the presence of strong oxidizers in the acid.

Hastelloy D is a cast alloy with outstanding resistance to sulfuric acid. At 66 °C (150 °F), it has exhibited a maximum corrosion rate of 0.15 mm/yr (6 mils/yr) at acid concentrations of from 2 to 96% (see Table 2).

Ni-Cr-Fe Alloys. Inconel 600 and Incoloy 800 are used primarily for their oxidation resistance and strength at elevated temperatures. In aqueous environments, these materials often are used to combat chloride-ion stress-corrosion cracking. Inconel 600 frequently is substituted for commercially pure nickel to resist caustic soda and halo-

Table 1 Nominal chemical composition of nickels and nickel alloys

Alloy	C	Mn	S	Si	Cr	Ni	Mo	Cu	Ti	Fe	Others
						Percent by weight					
Nickel 200	0.08	0.18	0.005	0.18	...	99.5 (a)	...	0.13	...	0.2	...
Nickel 201	0.01	0.18	0.005	0.18	...	99.5 (a)	...	0.13	...	0.2	...
Monel 400	0.15	1.0	0.012	0.25	...	66.5 (a)	...	31.5	...	1.25	...
Monel K-500	0.13	0.75	0.005	0.25	...	66.5 (a)	...	29.5	0.60	1.00	2.73 Al
Hastelloy B	0.05 max	1.00	...	1.00	1.00	61.0	28.00	...	...	5.50	2.50 Co
Hastelloy D	0.12	0.90	...	9.25	1.00	82.0	...	3.00	...	2.00	1.50 Co
Inconel 600	0.08	0.5	0.008	0.25	15.5	76.0 (a)	...	0.25	...	8.00	...
Incoloy 800	0.05	0.75	0.008	0.50	21.0	32.5	...	0.38	0.38	46.0	0.38 Al
Hastelloy C-276	0.01	0.5	0.02	0.03	15.50	57.0	16.00	...	...	5.5	0.02 P, 3.75 W, 0.2 V, 1.25 Co
Inconel 625	0.05	0.25	0.008	0.25	21.5	61.0 (a)	9.0	...	0.2	2.5	0.2 Al, 3.65 Cb + Ta
Incoloy 825	0.03	0.50	0.015	0.25	21.5	42.0	3.0	2.25	0.90	30.0	0.10 Al
Hastelloy G	0.03	1.50	0.02	0.50	22.25	44.0	6.50	2.00	...	19.50	0.02 P, 0.50 W, 1.25 Co, 2.10 Cb + Ta
20 Cb-3	0.04	1.00	0.02	0.50	20.0	34.0	2.50	3.5	...	...	0.02 P, 0.50 Cb + Ta

(a) Includes cobalt.

Table 2 Corrosion rates in sulfuric acid

Acid concentration	Hastelloy B mm/yr	Hastelloy B mils/yr	Hastelloy D mm/yr	Hastelloy D mils/yr
Tested at 66 °C (150 °F)				
2	0.13	5	0.15	6
5	0.10	4	0.13	5
10	0.08	3	0.13	5
25	0.03	1	0.05	2
50	0.03	1	0.03	1
60	0.03	1	0.15	6
77	...	...	0.05	2
80	...	...	0.05	2
85	...	...	0.05	2
90	...	...	0.05	2
96	...	...	0.03	1
Tested in boiling acid solution				
2	0.03	1	0.10	4
5	0.03	1	0.18	7
10	0.05	2	0.33	13
25	0.05	2	0.23	9
50	0.05	2	0.28	11
60	0.18	7	0.20	8
80	...	...	0.91	36
85	...	...	2.31	91
90	...	...	4.85	191
96	...	...	2.18	86

gens at high temperature. A common application of Incoloy 800 is sheathing for electric heating elements.

Ni-Cr-Mo Alloys. Significant additions of molybdenum make Hastelloy C-276 and Inconel 625 highly resistant to pitting. These alloys retain high strength and oxidation resistance at elevated temperatures, but they are used in the chemical industry primarily for their resistance to a wide variety of aqueous corrodents. In many applications, these alloys are selected because they are considered the only materials of this type capable of withstanding the severe corrosion conditions encountered.

Ni-Cr-Fe-Mo-Cu Alloys. Carpenter 20 Cb-3, Hastelloy G and Incoloy 825 offer good resistance to pitting, intergranular corrosion, chloride-ion stress-corrosion cracking and general corrosion in a wide variety of both oxidizing and reducing environments. These alloys frequently are used in applications involving sulfuric or phosphoric acid.

Behavior of Nickel Alloys in Corrosive Environments

All of the alloys referred to in this article have excellent resistance to corrosion in rural, industrial and marine atmospheres. Of these materials, only Monel 400 has been used to a significant degree solely for its resistance to atmospheric attack. Because of its low corrosion rate and the attractive patina that develops on its surface, Monel 400 is used in architectural applications such as roofs, gutters and flashings.

Corrosion by Water. The major corrosion problems associated with industrial use of waters are crevice corrosion, pitting, stress-corrosion cracking and general corrosion. Many engineering materials have adequate resistance to general corrosion, but localized attack often is the life-limiting factor. In seawater, the high chloride content and fouling by marine organisms combine to produce severe crevice attack. Hastelloy C-276 and Inconel 625 are virtually immune to attack by seawater. Hastelloy G, Incoloy 825, Monel 400 and Monel K-500, although not immune to the detrimental effects of seawater, show good resistance to localized attack and excellent resistance to general corrosion. The other alloys listed in Table 1 are seldom used in seawater.

In distilled waters, fresh water and process waters, all of the alloys listed in Table 1 have good resistance to corrosion. Except for Carpenter 20 Cb-3, Incoloy 800 and 825 and Hastelloy G, these materials are virtually immune to chloride-ion stress-corrosion cracking, and the excepted alloys are sufficiently resistant to it to be frequently employed specifically for this characteristic.

Corrosion by Sulfuric Acid. Commercially pure nickel adequately resists unaerated sulfuric acid, but would be selected for service in such an environment only if its use were dictated by other conditions, such as the presence of contaminants, exposure to alternating environments, or simultaneous exposure to caustics (encountered in some heat-exchanger applications).

Monel 400 is widely used for handling sulfuric acid under reducing conditions. At room temperature, its corrosion rate in air-free acid at concentrations up to 85% is less than 0.25

mm/yr (10 mils/yr). Tests in boiling sulfuric acid produced corrosion rates of 0.086 mm/yr (3.4 mils/yr) at 5% acid concentration, 0.061 mm/yr (2.4 mils/yr) at 10% and 0.19 mm/yr (7.5 mils/yr) at 19%. At 95 °C (203 °F), corrosion rates in unaerated acid at concentrations of up to 60% were less than 0.51 mm/yr (20 mils/yr).

Hastelloy B and Hastelloy D have exceptional resistance to sulfuric acid; results of tests at various acid concentrations are given in Table 2.

Inconel 600 and Incoloy 800 are used in low-concentration sulfuric acid at room temperature. Although they are rarely used in sulfuric acid service under any other conditions, Incoloy 800 has been employed in 99% acid at 120 °C (250 °F).

Hastelloy C-276 and Inconel 625 both exhibit good resistance to sulfuric acid; however, neither would be selected on this basis alone.

Carpenter 20 Cb-3, Hastelloy G and Incoloy 825 have excellent resistance to sulfuric acid. Although the compositional differences among these alloys result in some variation in corrosion behavior, the alloys normally exhibit corrosion rates of less than 0.13 mm/yr (5 mils/yr) at all concentrations when solution temperature is below 50 °C (120 °F). Depending on composition, all three alloys exhibit maximum corrosion at acid concentrations between 60 and 80%.

Corrosion by Hydrochloric Acid. Nickel 200, Nickel 201, Monel 400 and Monel K-500 have room-temperature corrosion rates of below 0.25 mm/yr (10 mils/yr) in air-free hydrochloric acid at concentrations of up to 10%. Concentration of hydrochloric acid produced during hydrolysis of chlorides or chlorinated solvents usually is less than 0.5%; Nickel 200 and Monel 400 can withstand this environment at temperatures up to about 205 °C (400 °F). In air-saturated solutions, corrosion rate increases sharply. In boiling acid, Monel 400 has corroded at rates of 0.74 mm/yr (29 mils/yr) at 0.5% concentration, 1.07 mm/yr (42 mils/yr) at 1% and 1.12 mm/yr (44 mils/yr) at 5%. Rates for Nickel 200 are much higher.

Hastelloy B has outstanding resistance to hydrochloric acid, whereas Hastelloy D has moderate resistance. Hastelloy B corroded at a rate of 0.23 mm/yr (9 mils/yr) in 1, 2 and 5% HCl at 66 °C (150 °F). When acid concentration was increased to 37%, corrosion rate decreased to 0.05 mm/yr (2 mils/yr). In boiling HCl, corrosion rates were 0.05 mm/yr (2 mils/yr) at 1% concentration, 0.08 mm/yr (3 mils/yr) at 2%, 0.18 mm/yr (7 mils/yr) at 5%, 0.23 mm/yr (9 mils/yr) at 10%, 0.36 mm/yr (14 mils/yr) at 15% and 0.61 mm/yr (24 mils/yr) at 24%. Because chromium is rapidly attacked by hydrochloric acid, Inconel 600 and Incoloy 800 have little resistance to this acid. Because of their high molybdenum contents, both Hastelloy C-276 and Inconel 625 are resistant to all concentrations of hydrochloric acid at room temperature. At 66 °C (150 °F) in acid concentrations of from 5 to 37%, Hastelloy C-276 corrodes at rates of from 0.51 to 1.3 mm/yr (20 to 50 mils/yr). When tested at a 37% acid concentration at 66 °C (150 °F), Inconel 625 exhibited a corrosion rate of 0.38 mm/yr (15 mils/yr).

Carpenter 20 Cb-3, Hastelloy G and Incoloy 825, although normally not considered candidate materials for hydrochloric acid service, exhibit useful room-temperature resistance at acid concentrations of up to 15%. When tested at room temperature, Hastelloy G corroded at a rate of 0.25 mm/yr (10 mils/yr) at 10% acid concentration, and Incoloy 825 exhibited corrosion rates of 0.12 mm/yr (4.9 mils/yr) at 5%, 0.18 mm/yr (7.2 mils/yr) at 10% and 0.19 mm/yr (7.3 mils/yr) at 15%.

Corrosion by Phosphoric Acid. Neither commercially pure nickel nor Ni-Cu alloys are used extensively in applications involving hot, concentrated phosphoric acid. At low temperatures or low acid concentrations, these materials show minor corrosion. Monel 400, for example, exhibits corrosion rates below 0.25 mm/yr (10 mils/yr) for all acid concentrations when tested at temperatures below 95 °C (200 °F).

Even in boiling 10 to 85% phosphoric acid, Hastelloy B and Hastelloy D show low corrosion rates. Inconel 600 and Incoloy 800 are resistant to all concentrations of pure phosphoric acid at room temperature. Corrosion rates increase rapidly with temperature.

Several contaminants, such as chloride, fluoride and silica, are present during manufacture of wet-process phosphoric acid. Because these contaminants increase susceptibility to pitting and crevice corrosion, resistance to general corrosion is not as important as resistance to local attack. In evaporators handling wet acid, Hastelloy C-276 and Inconel 625 have proved useful.

Carpenter 20 Cb-3, Hastelloy G and Incoloy 825 have excellent resistance to all concentrations of phosphoric acid, even at boiling temperature. They also find application in the manufacture of superphosphoric acid.

Corrosion by Nitric Acid. Those alloys containing no chromium have poor resistance to nitric acid. Chromium-bearing nickel alloys show good resistance, with corrosion rates decreasing primarily as chromium content increases. Because boiling 65% nitric acid (Huey test) often is used to measure the resistance of an alloy to intergranular corrosion, it should be pointed out that alloying elements (such as columbium in 20 Cb-3, Inconel 625 or Hastelloy G and titanium in Incoloy 825) are added to preferentially modify carbide precipitation and to minimize intergranular corrosion.

Corrosion by Organic Acids. Both commercially pure nickel and nickel-copper alloys find limited use in monocarboxylic acids. In glacial acetic acid at 110 °C (230 °F), Monel 400 has shown a corrosion rate of 0.33 mm/yr (13 mils/yr). Again, aeration normally increases corrosion rate.

Hastelloy B and Hastelloy D have very good resistance to most organic acids. In either acetic or formic acid at 70 °C (160 °F), the highest corrosion rate was 0.1 mm/yr (4 mils/yr).

Inconel 600 and Incoloy 800 have excellent resistance to hot, long-chain organic acids. Fat-splitting towers for stearic, oleic, linoleic and abietic acids are commonly fabricated from Inconel 600. Both Hastelloy C-276 and Inconel 625 display excellent resistance to organic acids. Formic acid normally is considered the most corrosive monocarboxylic acid. In all concentrations of boiling acid, both Hastelloy C-276 and Inconel 625 corrode at rates of 0.025 to 0.05 mm/yr (1 to 2 mils/yr). These alloys are the preferred materials for construction of high-temperature distillation columns for glacial acetic acid.

Carpenter 20 Cb-3, Hastelloy G and Incoloy 825 are highly resistant to organic acids and are adequate for most applications involving them.

Corrosion by Alkalis. Commercially pure nickel is unsurpassed by any common engineering material in resistance to corrosion by bases. An exception is ammonium hydroxide, which forms complexes with nickel and copper. Nickel 200 is not subject to corrosion by anhydrous ammonia or ammonium hydroxide in concentrations up to 1%. However, ammonium hydroxide in higher concentrations causes rapid attack.

Nickel 200 and Nickel 201 have excellent resistance to sodium hydroxide and potassium hydroxide at all concentrations and at all temperatures (even molten). In sodium hydroxide or potassium hydroxide at concentrations of less than 50%, Nickel 200 and Nickel 201 exhibit negligible corrosion rates (usually less than 0.005 mm/yr, or 0.2 mils/yr, even in boiling solutions). As concentration and temperature increase, corrosion rates increase slowly. Although stress cracking of nickel in molten anhydrous caustic soda has been reported, long-term laboratory and plant exposures of stressed specimens have not revealed any susceptibility to cracking. The presence of chlorates or oxidizable sulfur compounds increases the corrosive effect of caustics on nickel.

Nickel-copper alloys are not as resistant as pure nickel to the corrosive effects of alkalies. In boiling alkalis in concentrations up to 50%, however, corrosion rates for Ni-Cu alloys are still below 0.025 mm/yr (1 mil/yr), and thus these less expensive materials are widely employed in such applications.

Hastelloy alloys B, C-276, D and G have excellent resistance to alkaline environments, but they are seldom employed unless other corrodents are involved. Also possessing excellent resistance to alkaline environments are 20 Cb-3 and Incoloy 825; however, they, too, are seldom used in such environments.

In certain applications involving high-temperature caustics where sulfur is present or high strength is required, Inconel 600 is used instead of Nickel 201. The chromium content of Inconel 600 provides greater resistance to sulfur embrittlement. Inconel 600, in common with all nickel alloys except commercially pure nickels, is subject to stress-corrosion cracking when brought in contact with high-temperature, high-strength caustics. Thus, equipment should be fully stress relieved prior to use, and operating stresses should be minimized.

Corrosion by Salts. Except for halide salts, the corrosivity of a salt is based primarily on its oxidizing strength and on whether it hydrolyzes to an acid or a base. For example, materials that are resistant to nitric acid most likely are resistant to nitrates, including both sodium nitrate and ferric nitrate. These nitrate salts have high oxidizing strength and will readily hydrolyze to form nitric acid.

Halide salts, particularly chlorides, tend to promote localized attack such as pitting, crevice corrosion and stress-corrosion cracking. In general, high molybdenum contents help to control pitting and crevice corrosion, and high nickel contents resist chloride-ion stress-corrosion cracking.

Nickel 200 and Monel 400 are not subject to stress-corrosion cracking in any of the chloride salts. They have excellent resistance to all of the nonoxidizing halides. Oxidizing acid chlorides, such as ferric chloride and cupric chloride, are very corrosive to these alloys. Hypochlorites can cause pitting. A mixed group of very reactive and corrosive salts—phosphorus oxychloride, phosphorus trichloride, nitrosyl chloride, benzyl chloride and benzoyl chloride—is commonly contained in equipment made of Nickel 200.

Nickel 200 and Monel 400 have good resistance to solutions of neutral and alkaline salts such as carbonates, sulfates, nitrates and acetates. Even under severe conditions of concentration, temperature, agitation and aeration, corrosion rates normally are less than 0.1 mm/yr (5 mils/yr). Nickel 200 tubing is being used successfully in sodium chloride and sodium sulfate evaporators, and nickel-clad steel is used in construction of rotary salt driers. Monel 400 is widely used in salt plants for evaporators, crystallizers, filters, piping and similar equipment. In solutions of acid salts such as zinc chloride, ammonium sulfate and ammonium chloride, both Nickel 200 and Monel 400 have good resistance, but Monel 400 is more widely employed.

Hastelloy B has excellent resistance to nonoxidizing salts. Cupric chloride and ferric chloride are extremely corrosive to this alloy, whereas ammonium, aluminum and zinc chlorides are relatively harmless. Hastelloy B has little resistance to nitrates, chromates and other oxidizing salts. Typical use of this Ni-Mo alloy has been in connection with aluminum chloride-type catalysts, such as those used in alkylation of benzene during production of styrene. Corrosion rates in strong, boiling magnesium chloride are less than 0.05 mm/yr (2 mils/yr). This alloy also is resistant to pitting attack in chloride solutions.

The resistance of Inconel 600 to salts is very similar to that of Nickel 200 and Monel 400; however, in oxidizing acid salts, Inconel 600 is superior. This resistance does not apply to oxidizing acid chlorides. Inconel 600 has excellent resistance to silver nitrate, as used in photographic processing, and to strong, hot magnesium chloride. In nitrosyl chloride at temperatures above 43 °C (110 °F), this alloy is preferred to Nickel 200.

Incoloy 800 is subject to pitting in strong chloride solutions. It is highly resistant, although not immune, to stress-corrosion cracking. In salts other than halides, Incoloy 800 exhibits excellent resistance to a wide variety of both oxidizing and nonoxidizing mediums.

Inconel 625 and Hastelloy C-276 are resistant to all classes of salts, including oxidizing chlorides. Carpenter 20 Cb-3, Hastelloy G and Incoloy 825 are less resistant to pitting than the higher molybdenum-containing alloys but much more resistant than Incoloy 800. These three alloys have excellent resistance to all classes of salts except the oxidizing halides.

Corrosion by Fluorine, Chlorine and Hydrogen Chloride. At room temperature, fluorine forms protective fluoride films on nickel, copper, magnesium and iron; thus, these metals are considered satisfactory for low-temperature service in fluorine. Nickel 201 and Monel 400 are preferred for high-temperature service in fluorine. All of the nickel-base alloys considered are resistant to dry chlorine and hydrogen chloride, most of them even at moderately high temperatures. Monel 400 is a standard material for trim on chloride cylinders and valves. Wet chlorine is successfully handled by Hastelloy C-276. Nickel 201 and Inconel 600 are the most widely used materials for service in chlorine and hydrogen chloride at elevated temperatures.

Corrosion-Resistant Nickel-Alloy Castings

By Warren M. Spear
International Nickel Company, Inc.
Adapted from Chapter 18
in *Steel Castings Handbook,*
5th Edition, published by the
Steel Founders' Society of
America

NICKEL-BASE ALLOY CASTINGS are widely used for handling corrosive media and are regularly produced by high alloy steel foundries. The principal alloys are identified by Alloy Casting Institute (ACI) designations, and are included in ASTM A743, A744 and A494. In addition, several specialized proprietary grades are often specified for severe corrosion applications. Cast nickel-base corrosion-resistant alloys may be classified as follows:

- Nickel
- Nickel-copper alloys
- Nickel-chromium-iron alloys
- Nickel-chromium-molybdenum alloys
- Nickel-molybdenum alloys
- Nickel-base proprietary alloys.

Compositions of these alloys are given in Table 1.

The cast nickel-base alloys, with the exception of some high silicon and proprietary grades, have wrought counterparts and frequently are specified as the cast components in systems built of both wrought and cast components made of a single alloy. Compositions of cast and equivalent wrought grades differ in minor elements because workability in wrought grades is a dominant factor, whereas castability and soundness are dominant factors in cast grades. The differences in minor elements do not result in significant differences in serviceability.

Nickel-base castings are employed most often in fluid handling systems where they are matched with equivalent wrought alloys. They also are quite commonly used for pump and valve components or where crevices and high-velocity effects require a superior material in a wrought stainless steel system. Because of high cost, nickel-base alloys are usually selected only for severe service conditions where maintenance or product purity is of great importance and where less costly stainless steels or other alternative materials are inadequate.

Cast Nickel

CZ-100 is the ACI designation for the standard grade of cast nickel. A higher carbon, higher silicon grade is occasionally specified for greater resistance to wear and galling. Cast nickel is unsurpassed in handling concentrated and anhydrous caustic at elevated temperatures. It is also used in applications where product contamination by elements other than nickel cannot be tolerated.

The minor elements in CZ-100 provide for excellent castability and the production of sound, pressure-tight castings. Usual practice is to aim for 0.75% C and 1.0% Si. Carbon is present as finely distributed spheroidal graphite. A maximum carbon content of 0.10% or less is occasionally specified where castings are welded into wrought nickel systems. Low carbon CZ-100, however, is a difficult material to cast and has no significant advantage over the higher carbon option under any known service conditions. CZ-100 is used in the as-cast condition.

Mechanical property requirements for CZ-100 are listed in Table 2. Cast nickel has excellent toughness, ther-

Table 1 Compositions of cast nickel-base corrosion-resistant alloys

Designation	C	Si	Mn	Cu	Fe	Ni	Cr	Mo	Other
	Composition limits(a), %								
Nickel									
CZ-100	1.0	2.0	1.5	1.25	3.0	Rem	...	...	...
Ni-Cu alloys									
M-35-1	0.35	1.25	1.5	26–33	3.5	Rem	...	...	...
M-35-2	0.35	2.0	1.5	26–33	3.5	Rem	...	...	...
QQ-N-288 Grade A	0.35	2.0	1.5	26–33	2.5	62-68	...	...	...
Grade B	0.30	2.7–3.7	1.5	27–33	2.5	61–68	...	...	...
Grade C	0.20	3.3–4.3	1.5	27–31	2.5	60 min	...	...	...
Grade D	0.25	3.5–4.5	1.5	27–31	2.5	60 min	...	...	...
Grade E	0.30	1.0–2.0	1.5	26–33	3.5	60 min	...	...	1–3 (Nb+Ta)
Ni-Cr-Fe alloy									
CY-40	0.40	3.0	1.5	...	11	Rem	14–17	...	
Ni-Cr-Mo alloy									
CW-12M-1	0.12	1.0	1.0	...	4.5–7.5	Rem	15.5–17.5	16–18	0.20–0.40 V, 3.75–5.25 W
CW-12M-2	0.07	1.0	1.0	...	3.0	Rem	17–20	17–20	...
Ni-Mo alloy									
N-12M-1	0.12	1.0	1.0	...	4–6	Rem	1.0	26–30	0.2–0.6 V
N-12M-2	0.07	1.0	1.0	...	3.0	Rem	1.0	30–33	...
Proprietary alloys									
Chlorimet 2	0.07	1.0	1.0	...	2.0	66 min	1.0	30-33	...
Chlorimet 3	0.07	1.0	1.0	...	3.0	60 min	17-20	17-20	...
Hastelloy B	0.12	1.0	1.0	...	4–6	Rem	1.0	26–30	0.2–0.6 V, 2.5 Co
Hastelloy C	0.12	1.0	1.0	...	4.5–7.0	Rem	15.5–17.5	16–18	0.2–0.6 V, 2.5 Co
Hastelloy D	0.12	8.5–10	0.5–1.25	2–4	2.0	Rem	1.0	...	1.5 Co
Illium 98(b)	0.05	1.0	1.0	5.5	...	Rem	28	8.5	...
Illium G(b)	0.20	1.0	1.0	6.5	5.0	Rem	22.5	6.5	...
Illium B(b)	0.05	3.5	1.0	5.5	...	Rem	28	8.5	...

(a) Where a single value is shown, that value is a maximum unless otherwise indicated. (b) Nominal composition.

mal resistance and heat-transfer characteristics.

CZ-100 can be readily repair welded. It can be joined to other castings or to wrought forms using any arc or gas welding process; filler metal is nickel rod or wire. Joints must be prepared very carefully for welding because small amounts of sulfur or lead will embrittle the welds.

The most common application for nickel castings is in the manufacture of caustic soda and in chemical processing with caustic. As temperature and caustic soda concentration increase, austenitic stainless steels are of limited usefulness. Ni-Cu (M-35) and Ni-Cr (CY-40) alloys often are applied at intermediate concentrations, and cast nickel is selected for higher caustic concentrations including fused anhydrous NaOH. Minor amounts of elements such as oxygen and sulfur can have profound effects on the corrosion rate of nickel in caustic. Detailed corrosion data should be obtained before making a final selection.

Nickel-Copper Alloys

Cast 70Ni-30Cu alloys (Monels) are listed in ASTM A494, A743 and A744 as M-35-1 and M-35-2, and in Federal Specification QQ-N-288 as compositions A, B, C, D, and E.

The low-silicon grades M-35-1 (1.0% Si) and M-35-2 (1.5% Si) and compositions A and E (1.5% Si) are commonly used in conjunction with wrought nickel-copper and copper-nickel alloys in pumps, valves and fittings. A higher silicon grade, composition B (3.5% Si), is used for rotating parts and wear rings because it combines corrosion resistance with high strength and wear resistance. Composition D (4.0% Si) is employed where exceptional resistance to galling is desired.

M-35-1, M-35-2, and QQ-N-288 compositions A and E are employed in the as-cast condition. Homogenization at 815 to 925 °C (1500 to 1700 °F) may slightly improve corrosion resistance, but under most corrosive conditions, alloy performance is not affected by the minor segregation present in as-cast metal.

At about 3.5% silicon, age hardening becomes possible. At approximately 3.8% Si, the solubility limit for silicon in the nickel-copper matrix is exceeded and hard, brittle silicides are formed. These effects are particularly evident in the high-silicon alloy composition D. The combination of aging during cooling to room temperature after casting, plus the hard silicides developed when the silicon content exceeds about 3.8% can cause considerable difficulty in machining. Softening of composition D is accomplished by solution heat treatment consisting of heating to 900 °C (1650 °F), holding at temperature 1 h

Table 2 Mechanical property requirements at room temperature for standard cast nickel-base alloys

Alloy designation	Min tensile strength MPa	ksi	Min yield strength MPa	ksi	Elongation, %	Hardness, HB
CZ-100	345	50	124	18	10	...
M-35-1	448	65	127	25	25	110–140
M-35-2	448	65	207	30	25	125–150
QQ-N-288, grade A	448	65	224	32.5	25	125–150
QQ-N-288, grade B	689	100	455	66	10	240–290
QQ-N-288, grade C(a)	825	120	550	80	10	250-300(b)
QQ-N-288, grade D(a)	...	...	...	...	...	300
QQ-N-288, grade E	448	65	221	32	25	125–150
CY-40	483	70	193	28	30	...
CW-12M-1	496	72	317	46	4	...
CW-12M-2	496	72	317	46	25	...
N-12M-1	524	76	317	46	6	...
N-12M-2	524	76	317	46	20	...

(a) Values are typical. (b) Minimum hardness requirement for solution treated and age hardened condition or cast and age hardened condition.

for each 25 mm (1 in.) of section, and oil quenching. Maximum softening is obtained by oil quenching from 900 °C, but it is apt to result in quench cracks in complicated castings or castings with large differences in section size.

In solution heat treating of complicated castings, it is advisable to charge them into a furnace below 315 °C (600 °F) and heat them to 900 °C (1650 °F) at a rate that will limit the maximum temperature difference within any casting to about 55 °C (100 °F). After soaking, castings should be transferred to a furnace held at 730 °C (1350 °F), allowed to equalize in temperature, and then oil quenched. Alternatively, the furnace may be cooled rapidly to 730 °C (1350 °F), the casting temperature equalized and the castings subsequently quenched in oil.

Solution treated castings are age hardened by placing them in a furnace at 315 °C (600 °F), heating uniformly to 600 °C (1100 °F), holding 4 to 6 h, and cooling in air.

Tensile properties of nickel-copper castings are controlled by the solution hardening effect of silicon or of silicon plus niobium. Increasing the copper content also has a minor strengthening effect. The tensile properties of M-35-1, M-35-2 and composition A are controlled by a carbon-plus-silicon relationship, and tensile properties of composition E are controlled by a silicon-plus-niobium relationship.

The combination of aging plus hard silicides in composition D results in an alloy with exceptional resistance to galling. As the silicon content is increased above 3.8%, the amounts of hard, brittle silicides in a tough nickel-copper matrix increase, ductility decreases sharply and tensile and yield strengths increase. Because of these effects, strength and ductility cannot be controlled readily, and thus minimum hardness is the only mechanical property specified for composition D.

The toughness of nickel-copper alloys decreases with increasing silicon content but all grades retain their room temperature toughness down to at least −195 °C (−320 °F).

Weldability of nickel-copper alloys decreases with increasing silicon content but is adequate up to at least 1.5% Si. Niobium can enhance weldability of nickel-copper alloys, particularly when small amounts of low-melting-point residuals are present. Careful raw-material selection and good foundry practice, however, have largely eliminated any difference in weldability between niobium-bearing and niobium-free grades.

The high-silicon compositions B (3.5% Si), and D (4.0% Si) are considered not weldable. They can be brazed or soldered, as can the low-silicon grades.

Principal advantages of 70Ni-30Cu alloys are high strength and toughness coupled with excellent resistance to reducing mineral acids, organic acids, salt solutions, food acids, strong alkalies and marine environments.

The most common applications for 70Ni-30Cu alloy castings are in equipment for handling hydrofluoric acid, salt water, neutral and alkaline salt solutions, and reducing acids.

Nickel-Chromium-Iron Alloy

The cast nickel-chromium-iron alloy CY-40 (Inconel) differs in composition from the parallel wrought grade in having higher carbon, manganese and silicon contents, which impart the required qualities of castability and pressure tightness.

CY-40 is used in the as-cast condition because the alloy is insensitive to intergranular attack of the type encountered in as-cast or sensitized stainless steels. A modified cast nickel-chromium-iron alloy for nuclear applications (0.12% max carbon) is usually solution treated as an extra precaution.

The minimum mechanical properties in Table 2 are for a typical composition of 0.20 C, 1.50 Si, 1.0 Mn, 15.5 Cr, 8.0 Fe, rem Ni. Lower carbon and silicon contents for nuclear-grade castings result in a lower yield strength, but do not lower the minimum tensile strength.

CY-40 is frequently used for elevated-temperature fittings in conjunction with a wrought alloy of similar composition. Typical elevated-temperature properties are listed in Tables 3 and 4.

CY-40 castings can be repair welded or joined to other components using any of the standard arc or gas welding processes. Rod and wire whose nickel and chromium contents match those of the castings should be used. Postweld heat treatment is not required after repair welding or fabrication unless residual stresses must be relieved.

CY-40 is commonly used to handle hot corrosives or corrosive vapors under moderately oxidizing conditions, where stainless steels might be subject to intergranular attack or stress corrosion cracking. In recent years, CY-40 has been used extensively for components in systems handling hot boiler feedwater in nuclear power plants because it provides a greater margin of safety over stainless steels.

Table 3 Typical elevated-temperature tensile properties of CY-40

Temperature °C	°F	Tensile strength MPa	ksi	Yield strength MPa	ksi	Elongation in 25 mm or 1 in., %
Room temperature		486	70.5	293	42	16
480	900	427	62	...	...	20
650	1200	372	54.5	...	...	21
730	1350	314	45.5	...	...	25
815	1500	186	27	...	...	34

Note: Data are typical for investment cast test bars with nominal analysis of 0.20 C and 1.50 Si.

Table 4 Typical stress-rupture properties of CY-40

Temperature °C	°F	Stress to rupture in 100 h MPa	ksi
650	1200	165	24
730	1350	103	15
815	1500	62	9
925	1700	38	5.5

Note: Data are typical for investment cast test bars with nominal analysis of 0.20 C and 1.50 Si.

Nickel-Chromium-Molybdenum Alloy

Two grades of cast Ni-Cr-Mo alloy (ACI CW-12M-1 and CW-12M-2) are used in severe corrosion service: most often, service involving combinations of acids at elevated temperatures. The alloy also is produced under the proprietary names Hastelloy C and Chlorimet 3.

The high chromium and molybdenum contents of CW-12M result in the precipitation of carbides and intermetallic compounds during cooling in the mold. Because these precipitates adversely influence corrosion resistance, ductility, and weldability, CW-12M should be solution treated at 1175 to 1230 °C (2150 to 2250 °F) and water or spray quenched.

The CW-12M grades have relatively high yield strengths due to the solution-hardening effect of chromium, molybdenum, and silicon in CW-12M-2, and similar effects plus those of tungsten and vanadium in CW-12M-1. Ductility is excellent up to the limits of solid solubility. Inadequate heat treatment or improper composition balance, however, may result in the formation of a hard brittle phase and result in a rapid loss of ductility. Careful control within the specified composition range is necessary. Carbon and sulfur contents should be kept as low as practicable.

CW-12M can be arc or gas welded, using wire or rod of matching composition. For best weldability, carbon content of the base metal should be as low as practicable. The usual practice is to solution treat and quench prior to repair welding. Heat treatment after welding generally is not necessary because the alloy is not subject to sensitization in the heat-affected zone.

CW-12M is probably the most common material for upgrading a system where service conditions are too demanding for standard or special stainless steels—most often, service involving combinations of acids and elevated temperatures. The cast alloy may be used in conjunction with similar wrought materials or it may be used to upgrade pump and valve components in a wrought stainless steel system.

Nickel-Molybdenum Alloy

Two grades of cast Ni-Mo alloy (ACI N-12M-1 and N-12M-2) are frequently used for handling hydrochloric acid in all concentrations at temperatures up to the boiling point. The alloy also is produced commercially under the proprietary names Hastelloy B and Chlorimet 2.

Slow cooling in the mold is detrimental to corrosion resistance, ductility and weldability of N-12M. Because of this, the alloy should be solution treated, at a temperature of at least 1180 °C (2150 °F) for N-12M-1 and 1120 °C (2050 °F) for N-12M-2, and water quenched.

The N-12M grades have a high yield strength due to the solution hardening effect of molybdenum (Table 2). Ductility is controlled by the carbon and molybdenum contents. For best ductility, carbon content should be as low as practicable and molybdenum content should be adjusted to avoid the formation of intermetallic phases.

N-12M can be arc or gas welded using wire or rod of matching composition. Castings should be solution treated and quenched prior to repair welding. Postweld heat treatment is not necessary because the alloy is not subject to sensitization in the heat-affected zone.

Applications of N-12M are specialized—mainly in the handling of hydrochloric acid. N-12M should not be used as a substitute for stainless steels in applications where the latter have proved inadequate; N-12M is not adequately resistant to most oxidizing solutions for which stainless steels are initially selected.

Nickel-Base Proprietary Alloys

In addition to the standard ACI corrosion-resistant nickel-base alloys, there are a number of proprietary alloys that are widely used in corrosive service. Many of the proprietary alloys have excellent general corrosion resistance and are most commonly used where stainless steels are inadequate. Others, such as Chlorimet 2, Hastelloy B and Hastelloy D are used in specialized applications and should not be considered substitutes for stainless steel. Producers should be consulted before specifying these alloys, particularly for applications where there is no history of use.

In chemical processing where a mixture of corrosive solutions is involved, or where small amounts of a contaminant are present, corrosion rates of many common corrosion-resistant materials vary widely. It is under these more severe chemical processing conditions that corrosion-resistant nickel-base alloys can be of greatest utility.

Industrial Applications of Nickel Plating*

NICKEL plating, a standard coating used for decorative purposes, is also widely used for its ability to protect other metals from wear and corrosion in industrial applications.

Corrosion protection and wear resistance are the two most important services that nickel plating provides to industry. However, nickel plating also is used extensively to salvage or repair worn or mismachined components and to improve or modify other surface properties such as magnetic characteristics, hardness, and the ability to withstand thermal shock. In addition, certain alloys (such as Inconel) are nickel plated as a preparation for brazing.

The properties of nickel plating depend on the type of solution used and on plating conditions.

There are four major methods of nickel plating metals and plastics: electroplating, electroless plating, brush plating (sometimes called selective plating) and immersion deposition. Nickel electroplating and electroless nickel plating are the two most popular processes; the latter provides some unique properties and advantages. Brush plating is valuable in salvage or repair of worn, damaged or mismachined parts and in production of electronic connectors or contacts. Immersion deposition is used primarily in preparing steel for ceramic or enamel coating.

Table 1 summarizes the typical applications of the various nickel plating methods. Table 2 gives the compositions and normal operating conditions for electroplating from the three most widely used types of plating baths.

The type of bath and the plating conditions can be selected to produce a deposit having properties within the ranges given in Table 2. When temperature, pH and current density are set for specific properties, those properties can be easily reproduced.

With proper preplating treatment, adhesion strength will at least equal the strength of the weaker metal—the electrodeposited nickel or the basis metal. Thus, separation of 630-MPa nickel plating from a 700-MPa steel substrate would occur in the nickel plating and not precisely at the interface between the two metals.

Deposition from a sulfamate bath is preferred when electrodeposited nickel is to be exposed to high temperatures. Nickel deposited from a sulfamate bath has much better thermal stability than nickel deposited from either a Watts or an all-chloride bath.

Plating by any of the methods discussed above can be either over-all or selective. In over-all plating, a coating of nickel is applied over the entire surface of the part. In selective plating, application of nickel is limited to only those areas where it is desired or needed—either by brush plating only those areas or by masking the part except in those areas prior to electroplating, electroless plating or immersion deposition. For example, selective electroplating is used for plating copper alloy electronic spring connectors, either to improve wear resistance, corrosion resistance or electrical characteristics or to provide a substrate for subsequent precious-metal plating of contact areas.

In a process related to nickel plating, called nickel electroforming, nickel is deposited as in electroplating but on a pattern or mandrel and in considerably greater thickness. Nickel electroforming is especially useful for producing intricate shapes, fine surface detail or very smooth inside surfaces. It can be used to join assemblies, and to make ultrathin parts—especially those whose shapes preclude forming and whose wall or web thicknesses preclude machining.

When the plated deposit is stripped from the pattern or mandrel, the result is a free-standing form or structural component made entirely of nickel. The thickness almost always exceeds 125 μm (0.005 in.); thickness can be uniform or varied, as required.

Nickel electroforming offers distinct

*Adapted from Nickel Topics, Vol 29, No. 2, 1976, The International Nickel Company, Inc.

Table 1 Applications of industrial nickel plating

Reason for plating	Type of plating	Usual thickness	Industry	Applications
To improve corrosion resistance	Electroplating	<125 μm (0.005 in.)	Food processing	Pots, kettles, vessels of all kinds
			Paper and pulp	Drying cylinders and rolls
			Textile	Tape condensers, calendar rolls
			Soap and caustic processing	Heating coils, pumps, and pipe
			Glass processing	Lehr rolls
			Chemical and nuclear	Internally and externally plated pipe, elbows, joints and other components
			Automotive	Hydraulic rams, cylinder liners, shock-absorber components
	Electroless plating	12 to 125 μm (0.0005 to 0.005 in.)	Food processing	Mustard pots and other vessels, base plates for ice cream machines
			Printing	Cylinder rolls used in offset printing
			Chemical and chemical equipment	Washers and fasteners, plastic extrusion dies, nitrogen gas cylinders, globe valves, drive shafts, gears, springs, chains, clamps, tank cars, meter yokes
To improve wear resistance	Electroplating	<125 μm (<0.005 in.)	Automotive	Coatings on pistons, cylinders, pump rods, rotary housing liners, gear shafts
			Printing	Cylinder rolls
	Electroless plating	<25 μm (<0.001 in.)	Printing	Cylinder rolls used in offset printing
			Automotive	Drive shafts, ball studs, transmission thrust washers and differential pinion cross shafts
For salvage and repair	Electroplating	Determined by extent of repair required	Heavy duty machinery and tools	Shafts and splines
	Brush plating	Tolerance of 1.2 μm (0.00005 in.) achievable with experience	Heavy duty machinery and tools	Molds, dies, shafts, housings and precision fitting of bearings
To prepare for enameling	Immersion deposition	Extremely thin	Consumer products and chemical-processing equipment	Appliances and tanks
To modify magnetic properties	Electroless plating	...	Data processing	Computer memory devices

advantages for producing many types of parts. For example, molds and dies—especially those for making plastic items such as nylon gears, ratchets, cams, synthetic fibers, toys and dolls—can be made at less cost than by hobbing from solid metal. Parts requiring precise inside dimensions are more accurately produced by electroforming than by cold forming, casting or machining. Among the parts in this category are waveguides, Pitot-static tubes for airspeed indicators, venturi nozzles and thrust chambers for rockets.

Nickel electroforming can be used to advantage whenever surface characteristics must be reproduced with a high degree of precision. Stamping dies for reproducing phonograph records can be made with such precision that the high fidelity of the master recording is retained in the stamped reproductions. Printing plates, bellows for flexible joints and certain pressure devices are so intricate in shape that they are difficult and very costly to make by conventional methods; nickel electroforming offers an economical alternative manufacturing method.

Electroforming is particularly suitable for making nickel foil ½ to 1 mil thick. Such foil is used in printed circuits for high-temperature service.

Woven wire screening has a tendency to fail by fretting wear at the points where the wires cross one another. Electroformed screening does not have this characteristic—all intersections are solid, and do not chafe in use. Besides, screening with as many as 1000 openings or more in an area 25 mm square (1000 openings or more per square inch) can be made without difficulty. Electric shaver heads are but one

Table 2 Compositions of, and operating conditions for, typical industrial-type electrolytes used in nickel electroplating

Constituent or condition	Watts	Type of plating bath(a) All-chloride	Sulfamate
$NiSO_4 \cdot 7H_2O$ (nickel sulfate), g/l (oz/gal)	300 (40)	...	...
$NiCl_2 \cdot 6H_2O$ (nickel chloride), g/l (oz/gal)	45 (6)	300 (40)	6 (0.8)
$Ni(NH_2SO_3)_2$ (nickel sulfamate), g/l (oz/gal)	...	...	300 (40)
H_3BO_3 (boric acid), g/l (oz/gal)	37.5 (5)	37.5 (5)	30 (4)
pH	2.5-4	2	3.5-4.1
Temperature, °C (°F)	45-60 (115-140)	50-70 (120-160)	27-60 (80-140)
Current density, A/m² (A/ft²)	250-1000 (25-100)	250-1000 (25-100)	250-1500 (25-150)
Hardness, HV	150	240	200
Tensile strength, MPa (ksi)	340-1030 (50-150)	480-930 (70-135)	550-1620 (80-235)
Elongation(b), %	25-30	8	...
Internal tensile stress, MPa (ksi)	<150 (<22)	240-340 (35-50)	3.5-14 (0.5-2)
Modulus of elasticity, GPa (10^6 psi)	160-170 (23-25)	210 (30)	145-170 (21-25)

(a) Wetting agents (for example, sodium lauryl sulfate) must be added to any of these baths; stress reducers (such as saccharin) usually are added to Watts and all-chloride baths. (b) In 50 mm or 2 in.

of the many screen-type products made by nickel electroforming.

Fine capillary tubing used in hypodermic needles seems an ideal application for nickel electroforming. Unfortunately, about 12% of the population exhibits an allergic reaction when nickel comes in contact with open lesions, which makes nickel electroforming questionable for manufacturing hypodermic needles. However, electroforming can be considered a satisfactory method of manufacture for capillary tubing used in other applications. Bellows, foils and meshes for communications and electronic equipment have been made by electroforming. In the aerospace industry, electroforming has been used to make rocket thrust chambers and to join certain types of assemblies. In medicine, hypodermic needles have been made by electroforming.

Corrosion Resistance. Most industrial nickel plating is done to prevent chemical reactions between metal parts and their environments. Usually the purpose is to prevent damage to the metal part by substances such as chemicals and petroleum products, as well as general corrosion in normal atmospheres; sometimes it is to preclude contamination of materials (such as foods and beverages) contained in or handled by the metal parts. Both nickel electroplating and electroless nickel plating are used to impart or enhance corrosion resistance. Electroplated coatings generally are thicker—in excess of 125 μm (0.005 in.)—and can be applied using a Watts bath, an all-chloride solution or a nickel sulfamate solution. The sulfamate solution provides coatings of higher tensile strength—from 550 to 1620 MPa (80 to 235 ksi). Electroplated parts can be easily assembled by welding and other methods. In weldments, care must be taken to prevent the plating from cracking in the heat-affected zone.

The coatings applied by electroless plating range from 12 to 125 μm (0.0005 to 0.005 in.) thick, depending on need, and can be deposited uniformly over all surfaces of a workpiece—internal surfaces and cavities as well as external surfaces. Electroless deposits are especially low in porosity. They also have lower magnetic susceptibility and electrical conductivity than electroplated coatings. There are two practical electroless nickel processes: one deposits nickel-phosphorus alloys (up to 8% P), and the other deposits nickel-boron alloys (1 to 8% B). Deposits increase in wear and corrosion resistance with increases in phosphorus or boron content. The phosphorus or boron also helps increase coating hardness. Nickel-phosphorus alloys can be electrodeposited from aqueous solutions. However, no method is available for electrodepositing nickel-boron alloys from aqueous solutions, and thus there is no alternative to electroless nickel plating for depositing these alloys.

Wear resistance is another major application of industrial nickel plating and one in which either electroplating or electroless plating can provide a range of properties. The automotive industries make substantial use of nickel plating in many applications involving metal-to-metal wear. Pistons and cylinder walls, journal bearings, gear shafts and hydraulic pump components are among the principal applications. In the printing industry, nickel plating is used to reduce wear on press rolls and cylinders.

In wear applications, hardness of nickel electroplating is critical and can be varied widely by co-depositing other elements, such as sulfur, cobalt, tungsten, phosphorus, boron, tin and iron, along with the nickel. This is done by adding salts of the alloying element and appropriate organic compounds to the electrolyte. Changes in plating conditions, pH, temperature and current density also alter hardness, which can range from 150 to 170 HB. In this type of application, however, galling can be a problem. Electrodeposited nickel galls when rubbed against nickel, chromium, steel and some phosphor bronzes. Galling can be prevented by an electroplated overlay of chromium.

Electroless nickel, on the other hand, is often used for its antigalling properties. Aluminum, titanium and stainless steels frequently are electroless nickel plated on one of two mating surfaces to prevent galling. Cast iron, titanium, beryllium, magnesium, plastics and ceramics also are frequently electroless nickel plated for wear or antigalling applications.

Electroless nickel coatings can be heat treated to increase tensile strength and provide hardness and wear properties comparable to those of chromium (see Table 3). Other properties can be attained through deposition of ternary alloys (nickel-phosphorus-copper alloys, for instance) or use of oxides, Teflon or powdered metals in conjunction with the electroless nickel or nickel alloy coatings.

Table 3 Taber wear index numbers of coatings

Coating	TWI(a)
Electrodeposited nickel (Watts type)	25
Typical electroless nickel	
As plated	17
Heat treated at 300 °C (572 °F)	10
Heat treated at 500 °C (932 °F)	6
Heat treated at 650 °C (1202 °F)	4
Electrodeposited chromium	2

(a) Taber Abrader CS-10 wheel under 1-kg load, average weight loss in milligrams per 1000 cycles.

Electrodeposited nickel can be strengthened and/or made more wear resistant by codepositing nonmetallic particles such as thoria, alumina or carbides to make a dispersion-strengthened coating.

Salvage and repair of mismachined, damaged or worn parts can be accomplished by electroplating or by brush plating. With either method, the part can be plated over its entire surface (over-all plating) or only in selected areas (selective plating). Also with either method, thickness is determined by the extent of repair required. There are no limitations on thickness, but application of very thick deposits may require intermediate machining to maintain smoothness.

In brush plating, the nickel is applied with a portable anode or stylus made of carbon, nickel or stainless steel, using a direct-current power source and an appropriate electrolyte, which is held in an absorbent material covering the portable anode or stylus. For plating, the anode or stylus is drawn or rubbed over the region to be plated, which is made the cathode in the electrical circuit. The circuit is closed when the anode and the part make contact. The nonconducting absorbent material not only holds the electrolyte during plating, but also prevents short circuiting between the stylus and the part being plated. This method permits plating in the field, without dismantling equipment, when only limited areas need to be plated or when time is not available for more complete repair.

Enameling and ceramic coating of steels and other metals can be preceded by immersion deposition. Different electrolytes are used for different basis metals. For steel, nickel chloride solutions containing boric acid are used at high temperature. Immersion deposited coatings are extremely thin and only moderately adherent, but improve adhesion of subsequently applied enamel and ceramic coatings.

Magnetic properties of nickel plated surfaces of various metals can be varied significantly by electroless plating with different quantities of phosphorus (or boron) in the deposits. The greater the phosphorus (or boron) content, the less the susceptibility to magnetic fields.

On the other hand, nickel-cobalt alloys plated on substrates provide increased susceptibility to magnetic fields, which is useful in computer memory banks and other data processing components. Typical magnetically hard films are cobalt-nickel alloys containing codeposited phosphorus, tungsten, chromium, molybdenum, arsenic, antimony, bismuth, copper, oxygen or hydrogen. Electrodeposition and electroless deposition are probably the only techniques capable of producing magnetic films with high coercive force and good squareness ratio.

Process Selection and Testing. Designers of parts that are to be plated with nickel can take advantage of characteristics peculiar to the different plating methods. For instance, in electroplating, areas of the part closer to the anode (high spots) receive thicker deposits of nickel than areas farther from the anode. In most instances, such uneven plating is undesirable. If so, it can be controlled to a considerable extent. Sometimes, however, it is advantageous for high spots to be plated more heavily, and in such instances designers can utilize this characteristic of nickel electroplating. Otherwise, design can be altered to overcome this feature, plating conditions can be adjusted or electroplating can be replaced with electroless nickel plating, which deposits the nickel evenly on all surfaces.

Testing procedures for nickel plating have been established by ASTM and are important in determining plating quality. Tests for quality control include thickness, porosity, corrosion-resistance and adhesion tests. Performance tests measure hardness, ductility, tensile strength and internal stress.

The best measure of over-all performance is service life. Accelerated testing for corrosion or wear can be misleading. However, once an acceptable service life has been established for a specific thickness and type of plating, the performance of other coatings can be extrapolated with reasonable accuracy through accelerated testing.

Materials for Corrosion-Resistant Fasteners

By Joseph S. Orlando
Technical Director
and
William Ballantine
Manager, Marketing Services
ITT Harper

CORROSION-RESISTANT metallic fasteners are those made of stainless steels and nonferrous alloys. This broad definition could include hundreds of alloys, but in practice the materials actually used are limited to several stainless steels and several copper alloys, plus a few nickel, aluminum and titanium alloys. Fasteners can and have been made from unusual materials (tantalum, for example), but this discussion is primarily limited to those corrosion-resistant materials used in commercial fasteners that are readily available as standard (see Table 1).

Stainless Steels

Over half of all industrial fasteners classified as corrosion resistant are made of stainless steels. This general designation covers austenitic, martensitic and ferritic stainless steels. Of all stainless steels, the 300 series austenitic types are the most popular for fastener use. Austenitic stainless steels are not hardenable by heat treatment and are nonmagnetic for all practical purposes. All alloys in this group have at least 8% nickel in addition to chromium. They offer a greater degree of corrosion resistance than martensitic and ferritic types, but offer a lesser degree of resistance to chloride stress-corrosion cracking.

Martensitic and ferritic stainless steels contain at least 12% chromium, but contain little or no nickel because it stabilizes austenite. Martensitic grades, such as types 410 and 416, are magnetic and can be hardened by heat treatment. Ferritic alloys, such as type 430, are also magnetic but cannot be hardened by heat treatment.

The fastener industry generally markets fasteners made of types 302, 303, 304 and 305 stainless steels as "18-8". These four alloys are similar in both corrosion resistance and mechanical properties. From the manufacturer's point of view, the choice of alloy depends on method of fastener production, which in turn depends on type and size of fastener and, to some extent, on production volume. Because no two manufacturers have identical equipment, the alloy selected for a given fastener will vary; as an indication, however, the alloys that a major fastener producer uses on orders for 18-8 are as follows:

- *Type 302* is used for machine and tapping screws.
- *Type 303* is used to make nuts machined from bar. It contains a small amount of sulfur, for improved machinability.
- *Type 304* is used for hot heading (examples: long bolts or large-diameter bolts beyond the range of cold heading equipment).
- *Type 305* is used for cold heading (examples: hex-head bolts and cold formed nuts).

Other 300-series stainless steels used in fasteners include the following:

- *Types 309 and 310* are higher in both nickel and chromium than the standard 18-8 alloys, and are used for high-temperature applications.
- *Types 316 and 317,* because they contain molybdenum, have better elevated-temperature strength and better resistance to pitting than 18-8 alloys.
- *Types 321 and 347* are similar to 18-8 alloys but are stabilized by addition of titanium (type 321) or niobium (type 347) to increase resistance to intergranular corrosion.

Ferritic and martensitic stainless steels for fasteners are largely limited to:

- *Types 410 and 416* are general-purpose corrosion and heat-resistant alloys; they are hardenable by heat treatment.
- *Type 430* has better corrosion- and heat-resistant qualities than type 410; it is not hardenable by heat treatment.

Copper Alloys

Silicon bronzes have tensile strengths higher than those of low-carbon steels and are resistant to corrosion by the atmosphere, by fresh water and seawater, and by gases and sewage. Silicon bronzes are the copper alloys most commonly used for fasteners. They are nonmagnetic and have excellent machining and forming characteristics. C65100 is a low-silicon alloy suitable for cold heading; C65500 is suitable for hot forged fasteners.

Aluminum bronzes have better mechanical properties than silicon bronzes, but are much less frequently used in fasteners. Because of its good machinability, C64200 is the aluminum bronze most often used for fasteners.

Brasses, once the most commonly used materials for corrosion-resistant fasteners, now are specified less frequently than steels and silicon bronzes, which have higher mechanical properties. Brasses are still used in various applications, including electrical communications equipment, builders' hardware and many other consumer and industrial products. C27000 is used for cold headed fasteners, and C36000 is used for fasteners milled from bar.

Table 1 Standard corrosion-resistant fastener alloys

Commercial name	UNS No.	ASTM specifications
Stainless steels		
17-4 PH	S17400	F593: stainless steel bolts, hex cap screws, and studs; F594: stainless steel nuts
Type 302	S30200	
Type 303	S30300	
Type 304	S30400	
Type 305	S30500	
Type 309	S30900	
Type 310	S31000	
Type 316	S31600	
Type 317	S31700	
Type 321	S32100	
Type 347	S34700	
Type 410	S41000	
Type 416	S41600	
Type 430	S43000	
Copper alloys		
ETP copper	C11000	F468: nonferrous bolts, hex cap screws, and studs for general use; F467: nonferrous nuts for general use
Yellow brass	C27000	
High leaded brass	C34200	
Free-cutting brass	C36000	
Naval brass, 63½%	C46200	
Naval brass, uninhibited	C46400	
Si-bearing aluminum bronze	C64200	
Low-silicon bronze B	C65100	
High-silicon bronze A	C65500	
Nickel alloys		
Monel 400	N04400	
Monel 405	N04405	
Monel K-500	N05500	
Inconel 600	N06600	
Aluminum alloys		
Aluminum 1100	A91100	
Alloy 2024	A92024	
Alloy 6061	A96061	
Titanium alloys		
Commercial-purity titanium		
ASTM grade 1	R50250	
ASTM grade 2	R50400	
ASTM grade 4	R50700	
Ti-6Al-4V (ASTM grade 5)	R56400	
Ti-0.2Pd (ASTM grade 7)	R52400	

Naval brasses are copper-zinc alloys containing small amounts of tin, which give them higher resistance to salt water and atmospheric corrosion. C46200 is used for cold headed fasteners and C46400 is used for hot forged fasteners and for fasteners milled from bar.

Nickel Alloys

Nickel-base alloys are characterized by good strength and good resistance to heat and corrosion. They are often specified for marine and chemical-plant uses.

- *Monel 400* is used for fasteners more often than any other nickel-base alloy.
- *Monel K-500* is heat treatable and, in effect, is a high-strength version (900-MPa, or 130-ksi, minimum tensile strength) of Monel 400.
- *Inconel 600* is used for fasteners that must retain both high strength and resistance to oxidation at temperatures as high as 870 °C (1600 °F).

Aluminum Alloys

Some aluminum alloys are used for industrial fasteners. They have good corrosion resistance and low weight. Typically, aluminum fasteners are used to join aluminum components.

- *2024-T4* is a heat treated alloy usually used for cold headed fasteners; its tensile strength is above 425 MPa (62 ksi).
- *6061-T6* is used for some nuts, both cold formed and machined.
- *1100* (commercial-purity aluminum) is used for some washers and rivets.

Titanium

Titanium and its alloys have excellent corrosion resistance and maintain their strength at moderately high temperatures. Most industrial titanium fasteners are made from commercial-purity titanium, and are used in chemical-equipment applications. Titanium aircraft fasteners, many of which are of proprietary design, are produced from titanium alloys of much higher strength.

Industry Standards

The American Society for Testing and Materials (ASTM) has a working committee, "F16-Fasteners", with a series of subcommittees each of which deals with development of specific fastener standards. Subcommittee 4 works with nonferrous and stainless steel fastener standards. ASTM specifications in Table 1 (ASTM F467, F468, F593 and F594) are the four standards initially created by Subcommittee 4. They can be referred to for design criteria applicable to corrosion-resistant fasteners.

Nonstandard Fastener Alloys

Previous sections of this article have dealt with those corrosion-resistant alloys that are used most often for fasteners whose designs are recognized as standard by the American National Standards Institute (ANSI). Not surprisingly, there are numerous other materials that are used, either for standard fasteners or for special parts, when dictated by strength considerations, corrosive conditions or temperature requirements. Some of these more specialized alloys are listed below.

- *Precipitation-hardening stainless steels,* such as 17-4 PH, 17-7 PH, PH 15-7 Mo, Custom 450 and Custom 455, are used to obtain higher strength than that available from 18-8 stainless steels.
- *Martensitic stainless steels* such as type 416 and type 420 are used to obtain better mechanical properties than can be achieved with types 410 and 430.
- *Carpenter 20Cb-3* is specified when greater corrosion resistance is required than can be offered by 18-8 stainless steels, such as for equipment handling hot sulfuric acid.
- *A-286,* a nonstandard stainless steel that has greater corrosion resistance than the 18-8 types, as well as good mechanical properties at elevated temperatures, has been used in applications requiring resistance to both heat and a corrosive substance, such as in specialized chemical-plant or petroleum-refinery applications.

Heat-
Resistant
Materials

Iron-Base Heat-Resistant Alloys

By the ASM Committee on
Heat-Resisting
Alloys and Refractory Metals*

WROUGHT IRON-BASE HEAT-RESISTANT ALLOYS are used for applications involving metal temperatures above about 370 °C (700 °F)—the approximate upper limit for use of plain carbon steels under continuous load. The heat-resistant alloys include low-alloy steels that have been used at 370 °C or higher and for which high-temperature data are available. Also included are the austenitic stainless steels, high-temperature steels and precipitation-hardening alloys that have useful strengths at temperatures from 540 to 650 °C (1000 to 1200 °F) and higher while maintaining adequate oxidation resistance. Carbon and low-alloy steels used at temperatures above ambient are dealt with in the article "Elevated-Temperature Properties of Constructional Steels", beginning on page 639 in Volume 1 (9th Edition) of this Handbook. The present article deals chiefly with the wrought stainless steels used for high-temperature applications. Wrought iron-base superalloys are covered elsewhere in this volume. Compositions of representative alloys dealt with in this article are presented in Table 1.

Production of Ingots

In recent years, melting and refining of wrought heat-resistant alloys have become more and more sophisticated. Traditionally, the most exotic melting techniques have been applied to high-strength, high-temperature-resistant, precipitation-hardening alloys, which usually contain large amounts of reactive elements. However, even for low-alloy steels and other lower-strength materials, innovations in melting have steadily increased as melting and casting techniques have been optimized.

Low-alloy steels generally are melted in electric-arc furnaces, although ½Mo, 1Cr-½Mo and 1¼Cr-½Mo steels are melted also by basic open-hearth procedures. In some instances where applications may be more stringent, such as for D6, D6a, D9, D11, H-11 and H-13, electric-arc-furnace melting may be followed by a further refining procedure such as vacuum arc refining or even electroslag refining. An ingot from the electric furnace is used as an electrode in these remelting operations. Vacuum degassing also may be used for some of the higher-alloy grades to remove gases, particularly hydrogen.

Ferritic stainless steels generally are electric-furnace melted, although sometimes the argon-oxygen-deoxidation (AOD) process is employed. E-Brite 26-1 and 29Cr-4Mo are vacuum induction melted. These methods are necessary to obtain the low carbon required for these two alloys.

Martensitic and semiaustenitic stainless steels are, for the most part, melted using the electric-furnace technique combined with vacuum arc remelting or electroslag remelting. Custom 450 and Custom 455 are vacuum induction melted.

The austenitic 300 series and types 202 and 216 stainless steels are undergoing a rapid transition to the AOD process for their manufacture. The old process for making stainless steel required the use of expensive low-carbon ferroalloys for producing these low-carbon alloys. With the AOD process, this is unnecessary, and cheap ferroalloys can be used. The ultimate product is not only cheaper but also of better quality. In the AOD process, the CO partial pressure over the bath is lowered with argon and thus the carbon-oxygen reaction is enhanced to remove carbon. One disadvantage is the high consumption of expensive argon, but plans are underway to replace argon with nitrogen. The AOD process probably will be used on other types of stainless steel, and on other types of highly alloyed materials discussed in this section, in the near future. Alloys that require rigid specifications, such as nuclear grades of type 304 and type 316, sometimes are melted by vacuum induction melting followed by vacuum arc refining or electroslag refining. The Nitronic alloys are electric-arc-furnace melted.

Precipitation-hardening iron-base alloys are electric-furnace or vacuum induction melted and then vacuum arc or electroslag refined. Sometimes, as required by specification, double vacuum induction melting may be employed for critical applications.

The yield of forgeable stock per ingot generally is higher with vacuum melting. With no change in composition except for an increase in purity, yield

*See page XI for committee list.

Table 1 Nominal compositions of wrought iron-base heat-resistant alloys

Designation	UNS number	Composition, %							
		C	Cr	Ni	Mo	N	Nb	Ti	Others
Ferritic stainless steels									
405	S40500	0.15 max	13.0	...	...	...	...	...	0.2 Al
406	...	0.15 max	13.0	...	...	...	...	...	4.0 Al
409	S40900	0.08 max	11.0	0.5	...	...	...	6 × C min	...
430	S43000	0.12 max	16.0	...	...	...	...	...	...
434	S43400	0.12 max	17.0	...	1.0	...	...	...	...
439	S43027	0.07 max	18.25	...	...	...	...	0.2 + 4 (C + N)	...
18 SR	...	0.05	18.0	0.5	...	...	...	0.40	2.0 Al
18Cr-2Mo	...	...	18.0	...	2.0	...	...	...	...
446	S44600	0.20 max	25.0	...	...	0.25	...	...	...
E-Brite 26-1	S44627	0.01 max	26.0	...	1.0	0.015 max	0.1	...	...
26-1Ti	...	0.04	26.0	...	1.0	...	...	10 × C min	...
29Cr-4Mo	...	0.01 max	29.0	...	4.0	0.02 max	...	...	...
Quenched and tempered martensitic stainless steels									
403	S40300	0.15 max	12.0	...	...	...	...	...	...
410	S41000	0.15 max	12.5	...	...	...	...	...	...
416	S41600	0.15 max	13.0	...	0.6(a)	...	...	...	0.15 min S
422	S42200	0.20	12.5	0.75	1.0	...	...	...	1.0 W, 0.22 V
H-46	...	0.12	10.75	0.50	0.85	0.07	0.30	...	0.20 V
Moly Ascoloy	...	0.14	12.0	2.4	1.80	0.05	...	...	0.35 V
Greek Ascoloy	...	0.15	13.0	2.0	...	...	...	...	3.0 W
Jethete M-152	...	0.12	12.0	2.5	1.7	...	...	...	0.30 V
Almar 363	...	0.05	11.5	4.5	...	...	...	10 × C min	...
431	S43100	0.20 max	16.0	2.0	...	...	...	...	...
Precipitation-hardening martensitic stainless steels									
Custom 450	...	0.05 max	15.5	6.0	0.75	...	8 × C min	...	1.5 Cu
Custom 455	...	0.03	11.75	8.5	...	...	0.30	1.2	2.25 Cu
15-5 PH	S15500	0.07	15.0	4.5	...	...	0.30	...	3.5 Cu
17-4 PH	S17400	0.04	16.5	4.25	...	...	0.25	...	3.6 Cu
PH 13-8 Mo	S13800	0.05	12.5	8.0	2.25	...	...	...	1.1 Al
Precipitation-hardening semiaustenitic stainless steels									
AM-350	S35000	0.10	16.5	4.25	2.75	0.10	...	...	...
AM-355	S35500	0.13	15.5	4.25	2.75	0.10	...	...	...
17-7 PH	S17700	0.07	17.0	7.0	...	...	...	...	1.15 Al
PH 15-7 Mo	S15700	0.07	15.0	7.0	2.25	...	...	...	1.15 Al

(continued)

may be increased 10% or more, depending on the alloy. Alloys for forgings used for gas-turbine rotors are manufactured on a large scale by the consumable-electrode vacuum arc process and the electroslag process, in both of which arc melting is accomplished directly in a water-cooled crucible. The ductility of the metal in the centers of forgings made from these ingots is higher than in forgings made from conventional ingots poured in iron molds. For this reason, the rotors forged from metal produced in this way can withstand higher bursting stresses.

The relative effects of the several methods of melting on mechanical properties of various alloys at elevated temperatures have not been determined statistically. In general, vacuum melting has improved the structure-sensitive properties such as fatigue strength, ductility and impact strength; its effect on tensile strength is negligible. The effects of directionality in conventional ingot designs are the same for vacuum and air melting.

Product Forms

Wrought heat-resistant alloys are manufactured in all the forms common to the metal industry: billet, bar, sheet, tubing and wire. Stainless steels are produced as tubing, wire, hot rolled sheet, plate, polished sheet, cold rolled strip, flat circles, hot rolled bar, cold finished bar, forging billet, tube rounds, structural shapes, pipe and rod. The elevated-temperature properties of any of these materials are influenced to some extent by the form of the product, depending largely on specific alloy characteristics such as oxidation resistance, type of oxide scale, thermal conductivity and thermal expansion. Duration and type of loading also may influence differences in properties among different product forms.

For alloys that form thin tenacious scales at elevated temperature, the stress-rupture properties of bar and sheet of the same alloy will be about the same. On the other hand, for alloys that are less resistant to oxidation, rupture values are likely to be significantly lower for sheet than for the same alloy in bar form because of the greater ratio of surface area to volume, which causes greater interaction between the environment and the substrate metal. In the case of oxidation, a fixed depth of oxidation (such as 3 to 5 mils) will more drastically affect sheet properties in 50 mil sheet than such a depth will affect 250 mil bar stock properties.

Heat-resistant alloys are cold worked by hammering, swaging and rolling. Cold worked products such as bolts and

Table 1 (continued)

Designation	UNS number	C	Cr	Ni	Mo	Composition, % N	Nb	Ti	Others
Austenitic stainless steels									
304	S30400	..... 0.08 max	19.0	10.0	...	...	...	...	...
304L	S30403	..... 0.03 max	19.0	10.0	...	...	...	...	...
304N	S30451	..... 0.08 max	19.0	9.25	...	0.13	...	...	...
309	S30900	..... 0.20 max	23.0	13.0	...	...	...	...	...
310	S31000	..... 0.25 max	25.0	20.0	...	...	...	...	...
316	S31600	..... 0.08 max	17.0	12.0	2.5	...	...	...	...
316L	S31603	..... 0.03 max	17.0	12.0	2.5	...	...	...	...
316N	S31651	..... 0.08 max	17.0	12.0	2.5	0.13	...	...	...
317	S31700	..... 0.08 max	19.0	13.0	3.5	...	...	...	...
321	S32100	..... 0.08 max	18.0	10.0	...	...	...	5 × C min	...
347	S34700	..... 0.08 max	18.0	11.0	...	...	10 × C min	...	...
19-9 DL	K63198	..... 0.30	19.0	9.0	1.25	...	0.4	0.3	1.25 W
19-9 DX	K63199	..... 0.30	19.2	9.0	1.5	...	...	0.55	1.2 W
17-14-CuMo	...	..... 0.12	16.0	14.0	2.5	...	0.4	0.3	3.0 Cu
202	S20200	..... 0.09	18.0	5.0	...	0.10	...	...	8.0 Mn
216	S21600	..... 0.05	20.0	6.0	2.5	0.35	...	...	8.5 Mn
21-6-9	S21900	..... 0.04 max	20.25	6.5	...	0.30	...	...	9.0 Mn
Nitronic 32	...	..... 0.10	18.0	1.6	...	0.34	...	...	12.0 Mn
Nitronic 33	...	..... 0.08 max	18.0	3.0	...	0.30	...	...	13.0 Mn
Nitronic 50	...	..... 0.06 max	21.0	12.0	2.0	0.30	0.20	...	5.0 Mn
Nitronic 60	...	..... 0.10 max	17.0	8.5	2.0	...	...	...	8.0 Mn, 0.20 V, 4.0 Si
Carpenter 18-18 Plus	...	..... 0.10	18.0	<0.50	1.0	0.50	...	...	16.0 Mn, 0.40 Si, 1.0 Cu

(a) Optional.

studs usually are stress relieved, but when tested at elevated temperature they retain portions of the additional tensile and yield strengths imparted by the cold reduction.

Allowable Design Stresses

The maximum allowable stresses of the ASME Boiler Code are based on high-temperature laboratory data, and also reflect consideration of past operating experience. The two sections of the code that specify maximum allowable stresses deal with power boilers and unfired pressure vessels.

The procedure used by the Ferrous Section of the ASME Boiler Code Subcommittee on Allowable Stresses, as a basis for their allowable-stress tables, has been published in Section I of the Code for Power Boilers and in Section VIII of the Code for Unfired Pressure Vessels. The subcommittee established design criteria for unfired vessels that operate under the ASME Boiler Code, as follows:

- In the temperature range where elastic properties predominate, which varies from about 315 to 540 °C (600 to 1000 °F) depending on the alloy, maximum allowable stress shall not exceed 1/4 of the tensile strength or 5/8 of the yield strength at 0.2% offset. For bolting steels, maximum stress shall not exceed 1/5 of the tensile strength or 1/4 of the yield strength.
- In the temperature range where plastic properties predominate, maximum allowable stress shall not exceed the stress for a creep rate of 1% per 100 000 h or the rupture strength for 100 000-h life.
- In the intermediate range where both plastic and elastic properties predominate, maximum allowable stress shall be determined by a smooth curve joining the curves for the other two ranges.

The procedure for establishing maximum allowable stresses for power boilers is the same as for unfired boilers except that in the plastic range the maximum allowable stress in a power boiler shall not exceed 60% of the average stress for rupture in 100 000 h or 80% of the minimum stress for rupture in the same period.

Curves for tensile, yield, creep and rupture strengths may be plotted to define the governing strength values as functions of temperature. Figure 1, an example of such treatment of data for 18-8 stainless steel, also illustrates another principle: that there is a wide difference between yield strength and tensile, creep, and rupture strengths for 18-8 up to quite high temperatures because of its low yield strength. For applications where yielding is not serious and where severe stress concentrations are not anticipated, some industries consider it safe to design up to 1/4 of the tensile strength as long as this stress does not exceed 90% of the yield strength or the applicable limitations on creep and rupture strengths.

In many service conditions the amount of deformation is not critical, and relatively high fractions of the rupture strength can be used in design. Under such conditions, with the combined uncertainties of actual stress, temperature and strength, it may be important that failure not occur without warning and that the metal retain high elongation and reduction in area throughout its service life. In the oil and chemical industries, for instance, many applications of tubing under high pressure require high long-time ductility, so that warning of impending rupture will be evident from bulging of the tubes.

Values of elongation and reduction in area obtained in rupture tests are

Fig. 1 Relationships between temperature and high-temperature strengths used by the ASME Boiler Code Committee to establish maximum allowable stresses in tension for type 18-8 austenitic stainless steel

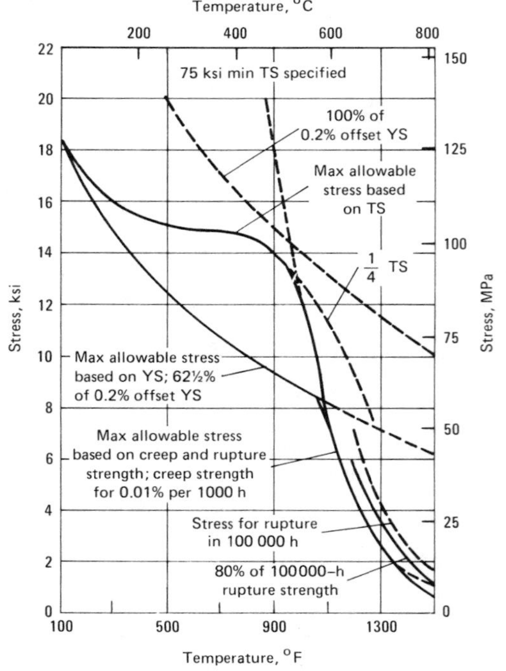

Fig. 2 Isochronous stress-strain curves for 17-7 PH sheet 1.3 mm (0.050 in.) thick, TH1050 condition

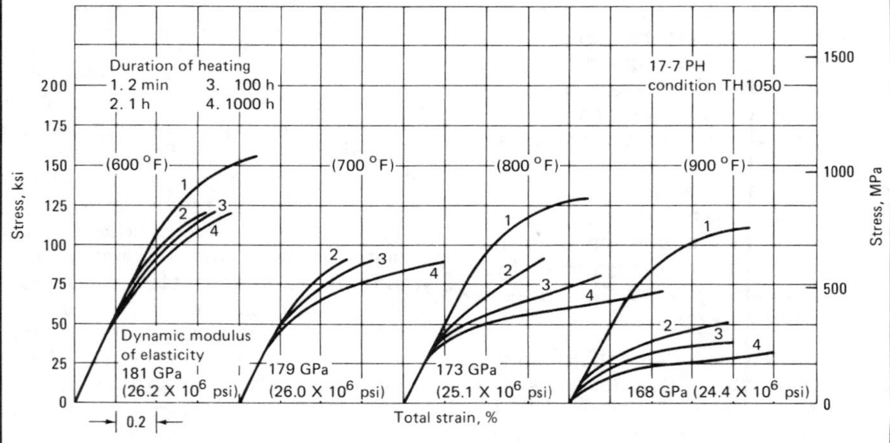

Room-temperature properties were: tensile strength, 1290 MPa (187 ksi); yield strength at 0.2% offset, 1225 MPa (178 ksi); dynamic modulus of elasticity, 200 GPa (29 × 10⁶ psi). Total strain was adjusted to the indicated modulus values.

used in judging the ability of metal to adjust to stress concentration. The requirements are not well defined and are controversial. Most engineers are reluctant to use alloys with elongations of less than 5%, and this limit sometimes is considerably higher. Low ductility in a rupture test almost always indicates high resistance to relaxation of stress by creep, and possible sensitivity to stress concentrations. Large changes in elongation with increasing fracture time usually are indicators of extensive changes in metallurgical structure or of surface corrosion.

Creep and Stress-Rupture

Methods of conducting creep and stress-rupture tests and of reporting and interpreting data are discussed in Appendices 2 and 3 of the article on wrought superalloys, in this volume.

Isochronous stress-strain curves such as those shown in Fig. 2 are useful in selection of design stresses for permissible total deformations during short and long periods of time. Because these data are taken from creep curves, extension due to thermal expansion is not included.

Many of the values for stress-rupture and creep given in the data compilations in this volume are typical or average values. There is a spread above and below these values caused by differences among heats of metal, methods of processing, and variations in conducting the standard test procedures. A typical spread in 100 000-h rupture strength is shown at upper left in Fig. 3 for type 347 stainless steel. At 590 °C (1100 °F), the test values range from 105 to 205 MPa (15 to 30 ksi), and at 650 °C (1200 °F) from 41 to 145 MPa (6 to 21 ksi). The average curve is well above the stress levels allowed by the ASME Power Boiler Code. The graph at lower left in Fig. 3 shows the total range in stresses for a creep rate of 1% per 100 000 h. On a percentage basis, the spread is about the same as in the upper left-hand chart. The stresses permitted by the ASME Power Boiler Code are below the average curve, except for low values for fine-grain material at 650 and 700 °C (1200 and 1300 °F). The results of an analysis of these tests are plotted in the upper right-hand chart in Fig. 3, classified according to grain size. Above 590 °C (1100 °F) the results fall into two distinct groups on the basis of grain size, with coarse-grain materials having higher rupture strengths.

The total range of rupture-strength values for fine- and coarse-grain type 347 stainless steel is shown at upper left in Fig. 3. However, when the total spread for coarse-grain material is considered alone (upper right), the range

Fig. 3 Rupture strength and creep properties of type 347 stainless steel compared with stresses permitted by the ASME Power Boiler Code

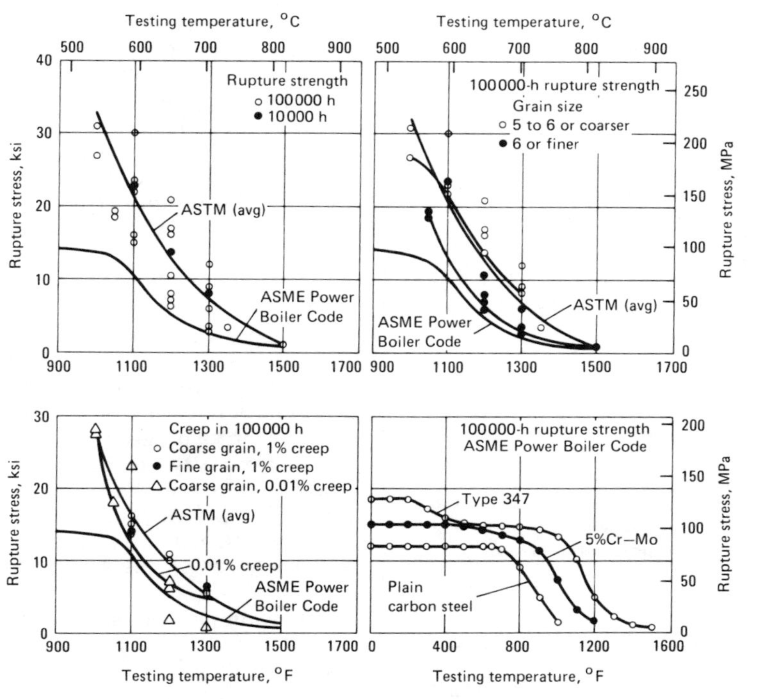

All charts except lower right are derived from ASTM STP 124, 1952.

Fig. 4 Creep and rupture characteristics of two austenitic stainless steels

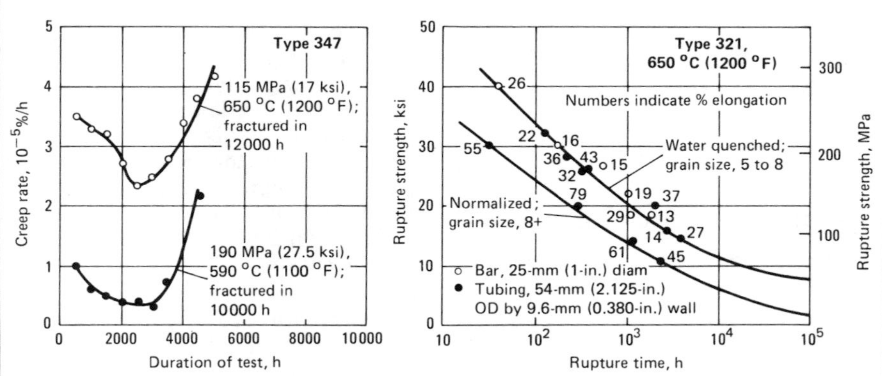

(Left) Effect of time on creep rate of type 347 stainless steel (ASTM STP 124). (Right) Rupture characteristics of type 321 stainless steel, cold worked and solution treated. Specimens indicated by lower curve were normalized at 955 °C (1750 °F), which resulted in grain sizes of ASTM No. 8 or finer. Specimens water quenched from between 1040 and 1120 °C (1900 and 2050 °F) exhibited grain sizes of ASTM No. 5 to 8.

in values at 650 °C (1200 °F) is 95 to 145 MPa (14 to 21 ksi) instead of 41 to 145 MPa (6 to 21 ksi).

Stresses allowed by the ASME Boiler Code are determined by, and vary considerably with, the type of steel. Allowable stresses for two heat-resistant steels are compared with those for carbon steel at lower right in Fig. 3.

Time has a greater effect on creep rate than on rupture strength for type 347 steel at 590 and 650 °C (1100 and 1200 °F), as shown at left in Fig. 4. The lower curve for a stress of 190 MPa (27.5 ksi) at 590 °C shows a decrease in creep rate for the first 3000 h to a minimum of 0.0036% per 1000 h. Shortly after 3000 h, the creep rate increased, and the specimen fractured at 10 000 h. The upper curve represents similar behavior at 650 °C and a stress of 115 MPa (17 ksi). Extrapolation of creep data of this type for longer periods of time may lead to gross inaccuracies because of variations in creep rate.

In central-station steam-generating equipment, many premature failures of type 321 superheater tubes have occurred at a metal temperature of 650 °C (1200 °F) and a maximum fiber stress of 34 MPa (5 ksi) in the tube wall. This stress is in accordance with the maximum allowable stress of the Boiler Code and is supposed to result in a tube life of at least 100 000 h. The tubes ranged approximately from 50 to 75 mm (2 to 3 in.) in outside diameter and had a maximum wall thickness of 13 mm (0.500 in.). They were produced by cold working (cold drawing or tube reducing), followed by solution heat treating at elevated temperature.

Although the tubes were designed for a life of 100 000 h, many exhibited considerable bulging (creep) within a few thousand hours and some actually ruptured in 30 000 h or less. All tubes that bulged or ruptured had an ASTM grain size of 8 or finer, whereas those that performed satisfactorily had a grain size of 8 or coarser. The contribution of grain size to performance is considered minor, however, and is overshadowed by the effects of, and rate of cooling from, the solution treating temperature.

The fine-grain tubes were solution treated at 950 °C (1750 °F) and cooled in air, whereas the coarser-grain tubes were solution treated at 1040 °C (1900 °F) or higher and water quenched. The rupture characteristics of both fine-grain and coarse-grain tubes are shown at right in Fig. 4.

Fig. 5 Effects of alternating loads on rupture life of type 403 stainless steel

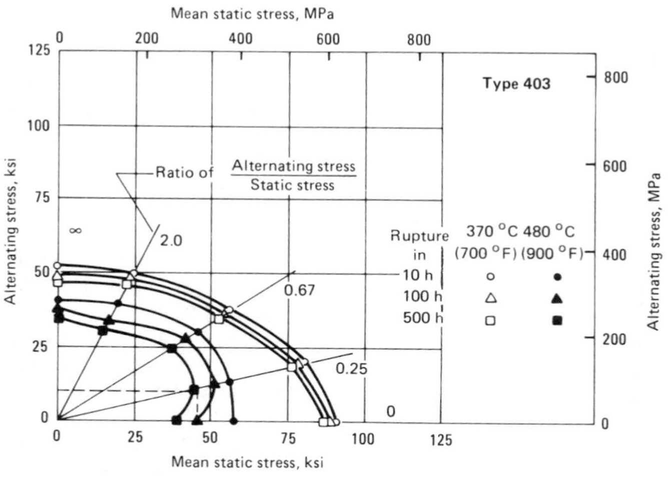

Relationship of static loads and superimposed alternating loads for type 403, oil quenched from 955 °C (1750 °F) and tempered at 540 °C (1000 °F). Axial alternating force up to ±2270 kg (±5000 lb) was produced by a 3600-rpm mechanical oscillator. A preload of 4540 kg (10 000 lb) was applied by calibrated springs and automatic follow-up.

Fig. 6 Room-temperature and high-temperature tensile properties of ferritic stainless steels

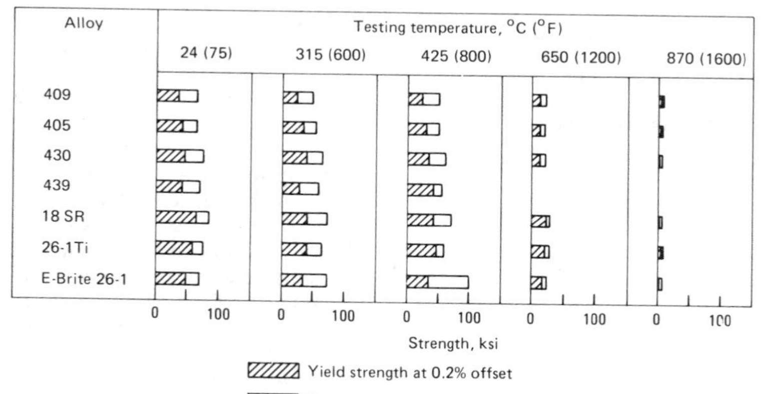

All alloys in the annealed condition: fast cooled from 815 to 925 °C (1500 to 1700 °F)

Both solution-treating temperature and quenching rate affect performance. Solution treating at a temperature of 950 °C (1750 °F) does not place sufficient titanium carbide in solution to exert the strengthening effect at 650 °C (1200 °F) that is produced by solution treating at a temperature of 1040 °C (1900 °F). Also, depending on section size, water quenching provides sufficiently rapid cooling to retain titanium carbide in solution. In larger sections, cooling in air may be too slow to produce this desirable result.

Dynamic Creep and Fatigue Properties

The relationship between static loads and superimposed fluctuating loads for type 403 stainless steel at two different temperatures is shown in Fig. 5. The points on the horizontal axes are the stresses that result in rupture under constantly applied static load at times indicated on the elliptical segments. The points of intersection of the elliptical segments with the vertical axes indicate the fatigue strengths for the indicated times when the mean stress is zero.

When a small fluctuating stress is superimposed on a static stress (for instance, $A = 0.25$), the static stress can be larger than the stress to rupture when no alternating stress is present, and the curve bulges to the right. This effect generally increases with time and temperature. A point in the bulge area also indicates longer life for a specimen with small alternating stress superimposed on the static stress.

For example, the stress-rupture life at 480 °C (900 °F) and 100 h for type 403 stainless steel is a little less than 310 MPa or 45 ksi (intercept on the horizontal axis of the 100-h rupture curve), but with an alternating stress of ±690 MPa (±10 ksi) superimposed on this static stress, the life is approximately 300 h, determined by the intersection of the dotted lines.

Overheating. An alloy is selected for high-temperature service on the basis of its performance at some specific limiting temperature at which the design stresses can be maintained for the life of the equipment and within the limit of some predetermined amount of deformation. Static and dynamic tests on the alloy at this limiting temperature are satisfactory criteria for design if the component is not overheated or overstressed in service. See Appendix 1 in "Wrought Superalloys" for additional information.

Relaxation of a bolted joint is a special case of tensile creep under gradually decreasing load with constant total elastic-plus-plastic deformation of the bolted assembly. The mechanical properties desired in a bolt are high yield strength, high relaxation resistance and satisfactory notch rupture strength for the expected life of the bolt.

Ferritic Stainless Steels

Many stainless steels of the 400 series have essentially ferritic structures at all temperatures. Types 405, 430, 434 and 446 form a certain amount of austenite when heated to high temperatures. Type 409 also may form some austenite, particularly if Ti is relatively low, while the other steels listed are completely ferritic at all temperatures.

The amount of chromium added for corrosion and oxidation resistance varies from 11% in type 409 to 29% in 29Cr-4Mo. Titanium is used to tie up carbon and nitrogen for structure control and resistance to intergranular corrosion. Molybdenum is used to improve corrosion resistance, whereas aluminum and silicon are added for resistance to oxidation.

Ferritic steels are melted in electric-arc furnaces with or without argon-oxygen decarburization. Basic open-hearth melting is also used successfully. Vacuum induction melting and electron beam hearth refining have been used for melting E-Brite 26-1 in order to control interstitial elements.

An important structural characteristic of ferritic stainless steels is precipitation of alpha prime, a chromium-rich ferrite, when the steel is exposed to temperatures in the range from 370 to 540 °C (700 to 1000 °F). This precipitation results in an increase in hardness and a drastic reduction in room-temperature toughness, which is known as 475 °C (885 °F) embrittlement. This embrittlement occurs in all ferritic grades that have chromium contents above approximately 13%, and its severity increases at higher chromium levels. This characteristic has to be taken into consideration for applications involving exposure to temperatures in the range from 370 to 540 °C (700 to 1000 °F), because subsequent room-temperature ductility will be severely impaired.

In the higher-chromium alloys such as type 446, 26-1 and 29-4, sigma phase is encountered at temperatures above 565 °C (1050 °F). Chi phase will also form in 26-1Ti and 29Cr-4Mo, if the Ti content is above 0.5%. Extensive formation of these phases can also result in severe embrittlement that persists up to at least 370 °C (700 °F).

Tensile and yield strengths of ferritic stainless steels in the annealed condition are shown in Fig. 6. At room temperature, these properties are nearly equivalent to those of austenitic stainless steels. At higher temperatures, however, ferritic steels are much lower in strength. The rupture strength and creep strength of types 430 and 446 are illustrated in Fig. 7. Long-time and short-time high-temperature strengths of ferritic steels are relatively low compared with those of austenitic steels.

Fig. 7 Creep and stress-rupture comparisons for selected iron-base heat-resistant alloys

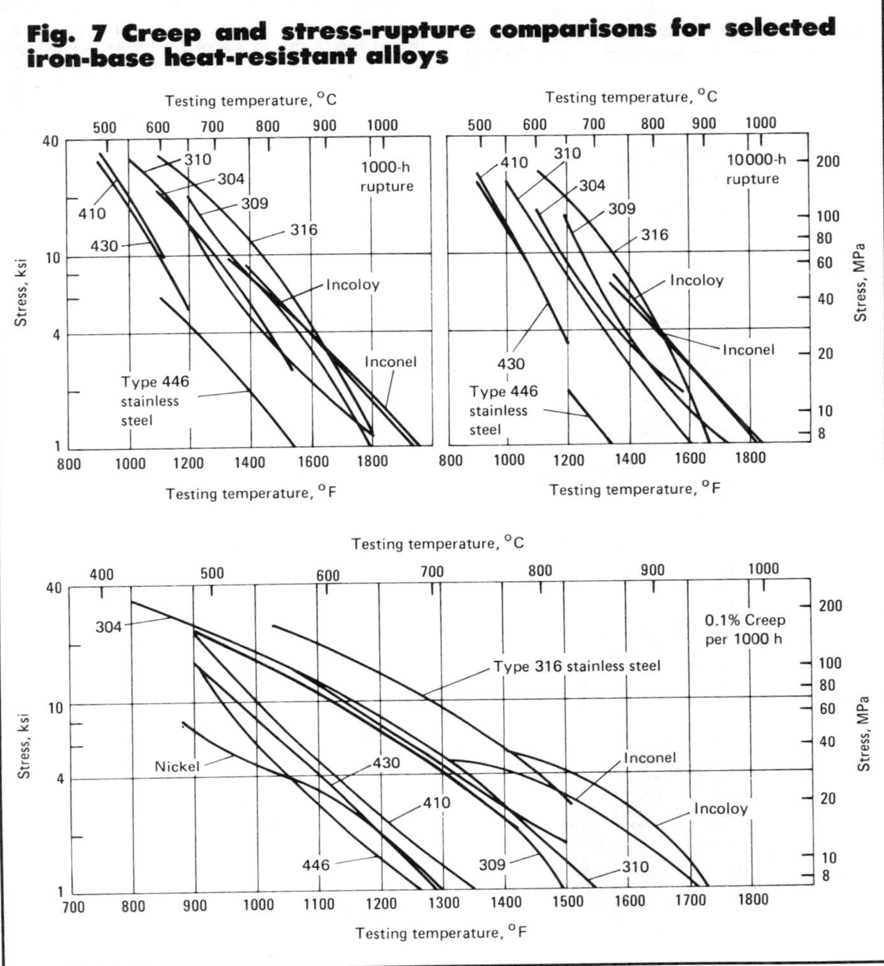

The main advantage of ferritic stainless steels for high-temperature use is their good oxidation resistance, which is comparable to that of austenitic grades. In view of their lower alloy content and lower cost, ferritic steels should be used in preference to austenitic steels, stress conditions permitting. Oxidation resistance of stainless steel is affected by many factors, including temperature, time, type of service (cyclic or continuous), and atmosphere. For this reason, selection of a material for a specific application should be based on tests that duplicate anticipated conditions as closely as possible. Figure 8 compares the oxidation resistance of type 430, type 446, and several martensitic and austenitic grades in 1000-h continuous exposure to water-saturated air at temperatures from 815 to 1095 °C (1500 to 2000 °F). Data on cyclic oxidation resistance of ferritic stainless steels in air containing 10% water vapor are presented in Table 2. At 705 and 815 °C (1300 and 1500 °F), all the alloys listed are resistant to oxidation. At 980 °C (1800 °F), the lower-alloy types 409, 430 and 304 exhibit high corrosion. At 1090 °C (2000 °F), only 18 SR and Inconel 601 have adequate oxidation resistance. The cycling oxidation resistances of E-Brite 26-1, type 310 and Incoloy 800 are compared in Fig. 9.

Type 409, the lowest-alloy stainless steel with a nominal chromium content

Fig. 8 1000-h oxidation resistance of selected stainless steels

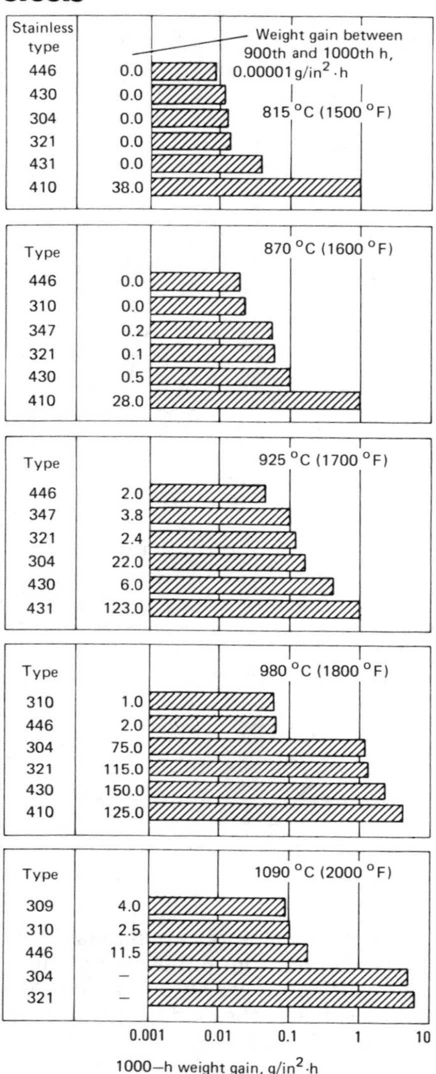

Table 2 Cyclic oxidation resistance of ferritic stainless steels
Specimens were exposed for 100 h in air containing 10% water vapor, and were cooled to room temperature every 2 h, then reheated to test temperature.

Alloy	Weight change (scale not removed), g/m², at: 705 °C (1300 °F)	815 °C (1500 °F)	980 °C (1800 °F)	1090 °C (2000 °F)
409	+0.1	+0.8	+1430(a)	−10 000(b)
430	+0.4	+1.3	−1660(c)	−10 000(b)
18 SR	+0.1	+0.3	+2.5	+7.4
304	+0.2	+1.7	−3400	−10 000(d)
309	+0.2	+2.7	−120	−910
210	+0.8	+3.1	+10	−150
Incoloy 800	+0.3	+3.2	+8.6	−560
Inconel 601	+0.1	+1.2	+10	−2.1

(a) Removed after 36 h. (b) Removed after 12 h. (c) Removed after 30 h. (d) Removed after 24 h.

of 11.0%, is used extensively because of its good fabricating characteristics, including weldability. Its best-known high-temperature applications are in automotive exhaust systems, where metal temperature in catalytic converters exceeds 540 °C (1000 °F). Type 409 is also used for exhaust ducting and silencers in gas turbines. Type 405 is used in stationary vanes and spacers in steam turbines and in various furnace components. Types 430 and 439 are used for heat exchangers, hot-water tanks, condensers and furnace parts. Type 18 SR, like type 446, is used in industrial ovens, blowers, exhaust systems, furnace equipment, annealing boxes, kiln liners and pyrometer tubes. E-Brite 26-1 and 26-1Ti are newer alloys that were developed primarily for corrosion service but that also can be used in applications similar to those of types 18 SR and 446.

Fig. 9 Cyclic oxidation behavior of three iron-base heat-resistant alloys at 980 °C (1800 °F)

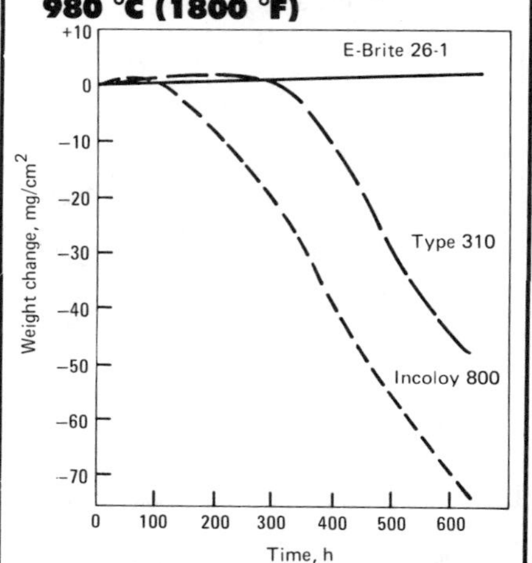

Specimens were alternately heated 15 min (to 980 °C) and cooled 5 min in air.

Quenched and Tempered Martensitic Stainless Steels

Quenched and tempered martensitic stainless steels are essentially martensitic, and harden when cooled from the austenitizing temperature. These alloys offer good combinations of mechanical properties, with usable strength up to 590 °C (1100 °F), and relatively good corrosion resistance. The strength levels at temperatures up to 590 °C that can be attained in these alloys through heat treatment are considerably higher than those attainable in ferritic stainless steels, but the martensitic alloys have inferior corrosion resistance.

These alloys normally are purchased in the annealed or the fully treated (hardened and tempered) condition. They are used in the hardened and tempered condition. For best long-time thermal stability, these alloys should be tempered at a temperature 110 to 165 °C (200 to 300 °F) above the expected service temperature.

Properties. Quenched and tempered martensitic stainless steels can be grouped according to increasing strength and heat resistance as follows:

- Group 1 (lowest strength and heat resistance): types 403, 410 and 416
- Group 2: Greek Ascoloy and type 431
- Group 3: Moly Ascoloy (Jethete M152)
- Group 4 (highest strength and heat resistance): H-46 and type 422.

An actual comparison of mechanical-property data is presented in Fig. 10 for some of these alloys. Data for type 410 are typical of group 1 alloys. Data for Greek Ascoloy are typical of data for type 431. Data for Moly Ascoloy are typical of data for group 3 alloys (the composition of Jethete M152 is very similar to that of Moly Ascoloy). Although H-46 and type 422 are similar in strength, their compositions are somewhat different; therefore, data are shown for both alloys.

The short-time tensile and rupture data shown in Fig. 10 were generated

Fig. 10 Comparison of mechanical properties of martensitic stainless steels

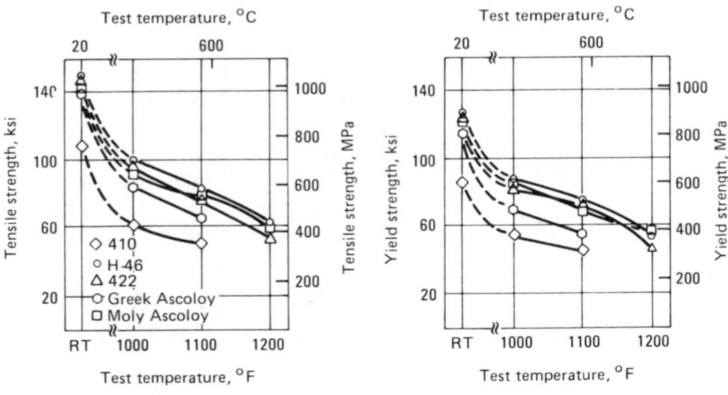

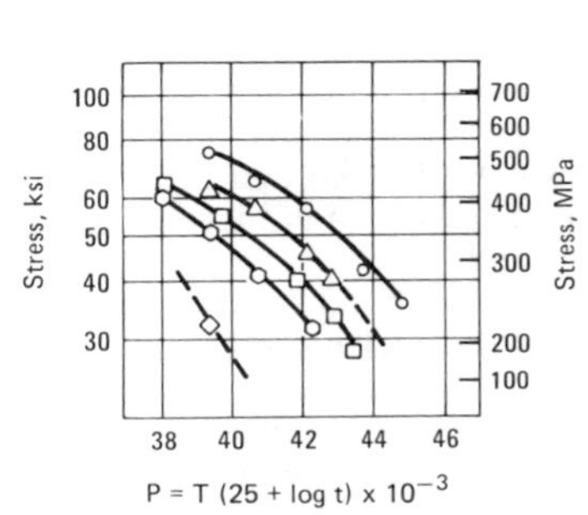

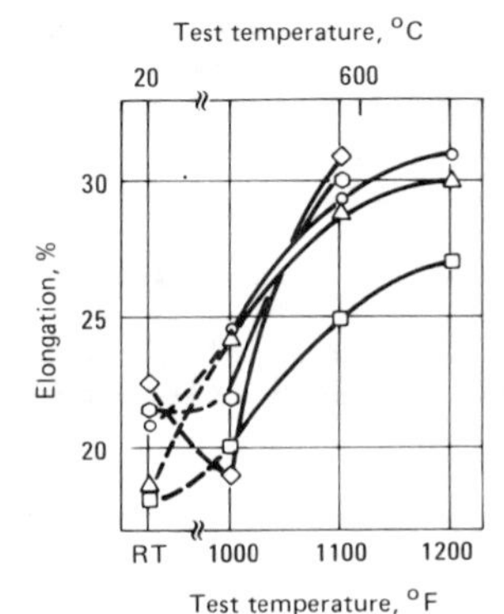

Heat treating schedules were as follows. Type 410: 1 h at 980 °C (1800 °F), oil quench; plus 2 h at 650 °C (1200 °F), air cool. H-46: 1 h at 1150 °C (2100 °F), air cool; plus 2 h at 650 °C, air cool. Type 422: 1 h at 1040 °C (1900 °F), oil quench; plus 2 h at 650 °C, air cool. Greek Ascoloy: 1 h at 955 °C (1750 °F), oil quench; plus 2 h at 650 °C, air cool. Moly Ascoloy: 30 min at 1050 °C (1925 °F), oil quench; plus 2 h at 650 °C, air cool.

in tests of material that had been given austenitizing treatments typical for the specific alloys tested. Because these alloys normally are used at service temperatures near 540 °C (1000 °F), although they may be used up to 590 °C (1100 °F), data are shown for a relatively high tempering temperature of 650 °C (1200 °F), which results in good thermal stability in these alloys at 540 °C (1000 °F).

Note that the group 1 alloys, of which type 410 is typical, show the lowest values of strength capability as a function of test temperature. Greek Ascoloy is considerably stronger than type 410, with a yield strength (0.2% offset) of 480 MPa (70 ksi) and a tensile strength of 585 MPa (85 ksi) at 540 °C (1000 °F). The tensile-strength capabilities of H-46, Moly Ascoloy and type 422 are fairly similar and are the highest in this group of alloys. Tensile-elongation data for all these alloys are similar—from about 20% elongation at 21 °C (70 °F) to about 30% at 650 °C (1200 °F).

Stress-rupture data for alloys typical of each subgroup are compared in Fig. 10 by means of a master stress-rupture plot. Note that the niobium-containing H-46 alloy has the highest stress-rupture capability, with type 422, Moly Ascoloy and Greek Ascoloy, in that order, having increasingly lower rupture capabilities. Type 410 has a very low stress-rupture capability and is the weakest of all the martensitic stainless alloys being considered. It should be noted that niobium-containing alloys such as H-46 usually show an advantage in stress-rupture capability (creep resistance) for short testing times (100 to 1000 h) but lose their strength advantage when tested for periods of about 10 000 h or more. The favorable effects of niobium additions on short-time stress-rupture properties are attributed to finely dispersed precipitation of NbX. The favorable effects tend to diminish as tempering temperature is increased, and a coarsely dispersed precipitate is formed. The effect of "tempering in service" is illustrated by the hardness data shown in Fig. 11 for H-46 and type 422. Note that H-46 alloy shows a larger hardness drop for extended thermal exposure at a testing temperature of 540 °C (1000 °F) than does type 422.

Quenched and tempered martensitic stainless steels find their greatest application in steam and gas turbines, where they are used in blading at temperatures up to 540 °C. Other uses include steam valves, bolts and miscellaneous parts requiring corrosion resistance and good strength up to 540 °C.

Type 410 is the basic, general-purpose steel, used for steam valves, pump shafts, bolts and miscellaneous parts requiring corrosion resistance and moderate strength up to 540 °C (1000 °F). Type 403 is similar to 410 in chemical composition. It is used extensively for steam-turbine rotor blades and gas-turbine compressor blades operating at temperatures up to 480 °C (900 °F). For this type of application the steel is tempered at 590 °C (1100 °F) or above, after which embrittlement is negligible in the service temperature range of 370 to 480 °C (700 to 900 °F).

A satisfactory heat treatment for these steels is austenitizing at 950 to 980 °C (1750 to 1800 °F), cooling rapidly in air or oil, and tempering. Cooling from the hot rolling temperature and tempering without intermediate austenitizing sometimes is practical but may result in a structure that contains free ferrite, which is detrimental to fatigue life. Warm or cold work after tempering sets up residual stresses that can be relieved by heating to approximately 620 °C (1150 °F).

Design curves for type 410 sheet are given in Fig. 12. These values are for special uses where heating rates are high.

Greek Ascoloy, type 431 and type 422 are variants of 410, modified by addition of such elements as nickel, tungsten, aluminum, molybdenum and vanadium. Nickel serves a useful purpose by causing the steel to be entirely aus-

Fig. 11 Approximate effects of time and stress on tempering of type 422 and H-46

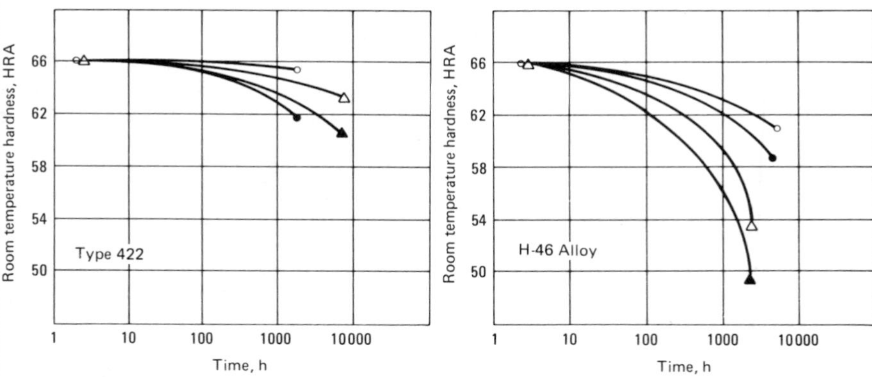

Circles indicate specimens heated to 1150 °C (2100 °F) and rapidly cooled, tempered 2 h at 705 °C (1300 °F), and tested (to fracture) at a temperature of 540 °C (1000 °F) and a stress of 380 MPa (55 ksi). Triangles indicate specimens heated to 980 °C (1800 °F) and rapidly cooled, tempered 2 h at 705 °C, and tested (to fracture) at 540 °C and 275 MPa (40 ksi). Open symbols represent data taken at the unstressed specimen shoulder; solid symbols represent data taken within the stressed gage length.

Fig. 12 Design curves for type 410 stainless steel sheet, showing effect of time at temperature on total deformation at specific stress levels

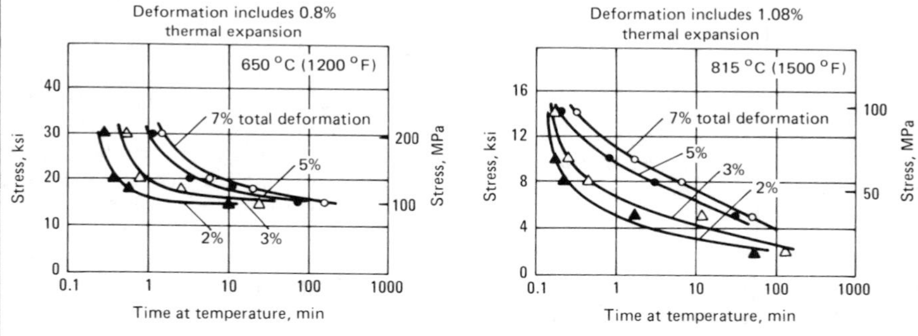

Design curves in chart at left represent a heating rate of 89 °C/s (160 °F/s) to 650 °C (1200 °F). Those at right represent a heating rate of 105 °C/s (190 °F/s) to 815 °C (1500 °F). Room-temperature properties of the sheet used in these tests were as follows: tensile strength, 650 to 695 MPa (94.5 to 101 ksi); yield strength at 0.2% offset, 555 to 565 MPa (80.7 to 82.3 ksi); and 9.6 to 16% elongation in 50 mm or 2 in. after air cooling from the normalizing temperature of 955 °C (1750 °F) (AFTR 6731, Part 4).

tenitic at conventional heating temperatures when the carbon and chromium contents are such that the structure would be two-phase if nickel were absent. The tempering temperature for Greek Ascoloy may be 55 °C (100 °F) or more higher than that for type 410 for equivalent strength and hardness. Type 422 develops the highest mechanical properties and at 650 °C (1200 °F) has a tensile strength equivalent to that of type 403 at 590 °C (1100 °F). The rupture strength of type 422 and 540 °C (1000 °F) is considerably higher than those of the other steels in this series (Fig. 13).

Types 430 and 446, which have nominal chromium contents of 16 and 25%, respectively, are not hardenable by conventional quenching and tempering treatments. Nitrogen to 0.25% max is added to type 446 for grain refinement. These grades generally are used in the annealed condition, but they can be cold worked to increase strength and decrease ductility. They have greater oxidation resistance than the 12% Cr steels and may be used without excessive scaling to temperatures of 840 °C (1550 °F) for type 430 and 1090 °C (2000 °F) for type 446 (Fig. 8). The rate of oxidation and the character of the scale are affected greatly by variations in the chemical and physical nature of the air and gases in the environment, and tests under actual service conditions are required for meaningful evaluation.

These alloys age harden, with a loss of ductility, when held for prolonged periods at 370 to 540 °C (700 to 1000 °F). A typical hardness-temperature curve for a 27% Cr alloy was obtained on a temperature-gradient bar (Fig. 14). The specimen was heated in a specially designed furnace so that one end only was subjected to a maximum and uniform temperature. The opposite end reached about 95 °C (200 °F), and the area between showed a uniform increase in temperature from the colder end to the hotter end. Thermocouples were inserted through the colder end of the bar to different depths so that temperatures could be measured at specified distances from the hot end.

The increase in hardness with time at 475 °C (885 °F) and the effect of increase in chromium content are shown in Fig. 15. The notch sensitivity is greatly increased by the increase in hardness. The embrittled condition can be alleviated and the original hardness restored by heating at 590 °C (1100 °F). This embrittlement is significant only for applications in the range from 425 to 540 °C (800 to 1000 °F).

Type 430 is used for heat-exchange equipment, condensers, and piping for nitric acid, and for furnace parts and retorts operating at temperatures up to 840 °C (1550 °F). It can be stamped, spun, and formed more easily than type 446.

Type 446 is used for furnace parts, oil burners, heat exchangers, kiln liners, glass molds, and stationary soot blowers in steam boilers. Its elevated-temperature strength is low, which is indicated by the creep and rupture-strength values given in Fig. 7. These data compare elevated-temperature properties of types 410, 430 and 446

Fig. 13 Tensile, yield, rupture and creep strengths for seven ferritic and martensitic stainless steels

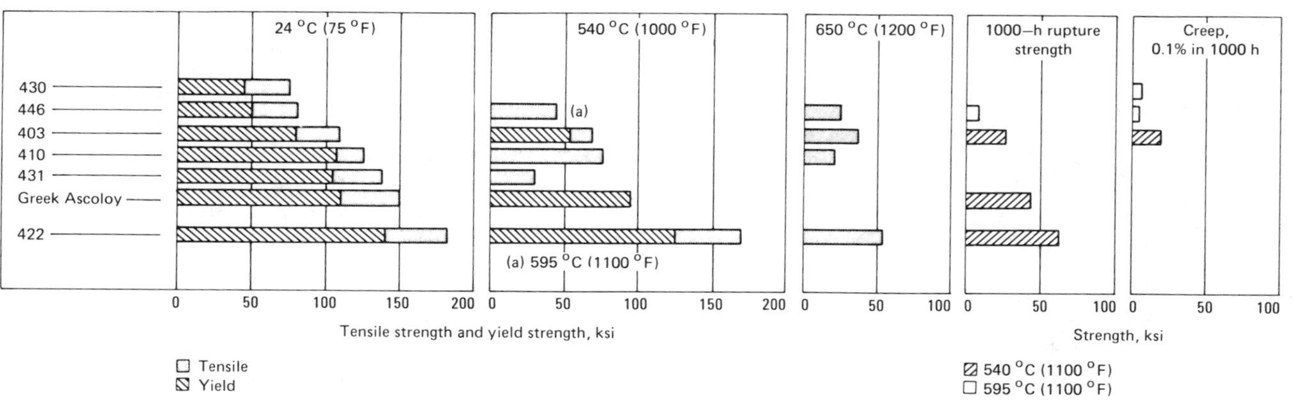

Types 430 and 446 were annealed. Type 403 was quenched from 870 °C (1600 °F) and tempered at 620 °C (1150 °F). Type 410, quenched from 955 °C (1750 °F) and tempered at 590 °C (1100 °F). Type 431, quenched from 1025 °C (1875 °F) and tempered at 590 °C. Greek Ascoloy, quenched from 955 °C and tempered at 590 °C. Type 422, quenched from 1040 °C (1900 °F) and tempered at 590 °C.

Fig. 14 Hardness values along temperature-gradient bar of type 446 stainless steel after exposure for indicated times

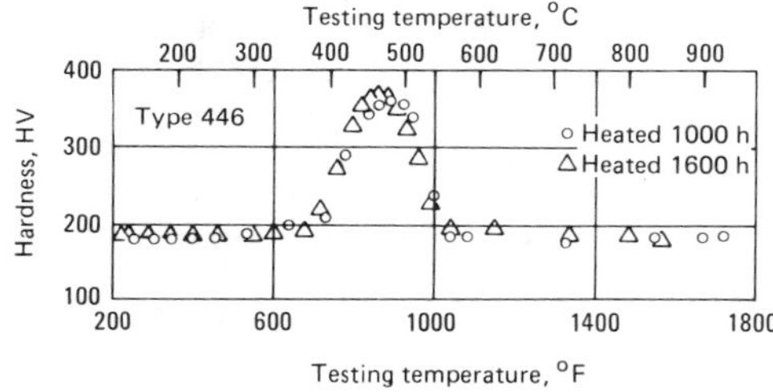

Composition: 0.19 C, 0.73 Mn, 0.54 Si, 27.31 Cr, 0.16 Ni, 0.15 N.

Fig. 15 Effect of time at 475 °C (885 °F) on age-hardening characteristics of chromium steels containing 12, 17 and 27% Cr

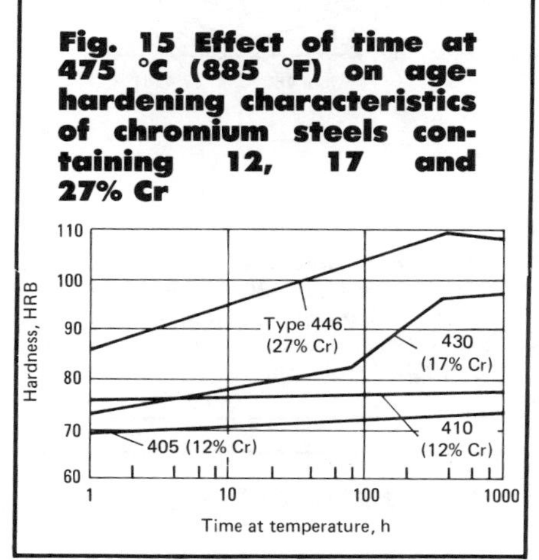

with those of several other materials, including the standard austenitic stainless steels, Incoloy, and Inconel, a nickel-base alloy.

Welding. The primary factor in welding of martensitic steels is their hardenability. The heat-affected zones surrounding the weld harden on cooling, setting up stresses that can give rise to cracking. If the filler metal is similar to the parent metal, the weld itself will also harden on cooling and become very brittle. These difficulties can be avoided by two precautions:

- Use austenitic filler metal, which will remain ductile and will absorb the stresses set up by hardening in the heat-affected zone.
- Preheat the work gradually before welding, and postheat after welding to avoid quenching stresses and cracks. Preheating should be carried out at 205 to 315 °C (400 to 600 °F), and postheating at 590 to 760 °C (1100 to 1400 °F).

Precipitation-Hardening Martensitic Stainless Steels

The precipitation-hardening martensitic stainless (maraging) steels fill an important position between the chromium-free 18% Ni maraging steels and the 12% Cr, low-nickel quenched and tempered martensitic stainless alloys. These PH alloys contain 12 to 16% Cr for corrosion resistance and scaling resistance at elevated temperatures and thus are intermediate in heat-resistance capability between the 18 Ni maraging steels and the 12% Cr, low-nickel quenched and tempered martensitic stainless alloys (such as type 422).

These alloys normally are purchased in the solution annealed condition. Depending on the application, they may be used in the annealed condition or in the annealed plus age hardened condi-

Fig. 16 Comparison of mechanical properties of precipitation-hardening martensitic stainless steels

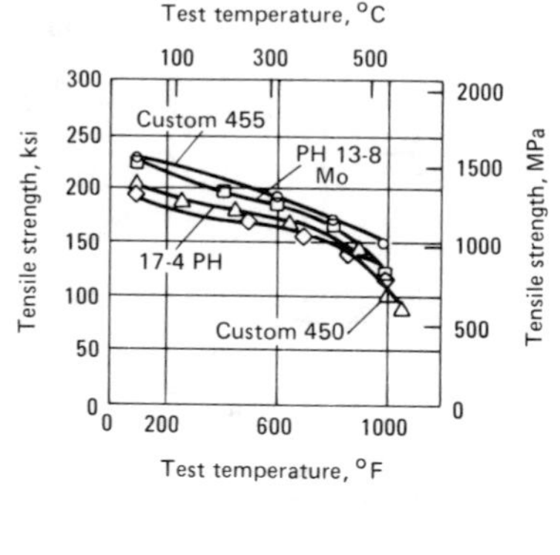

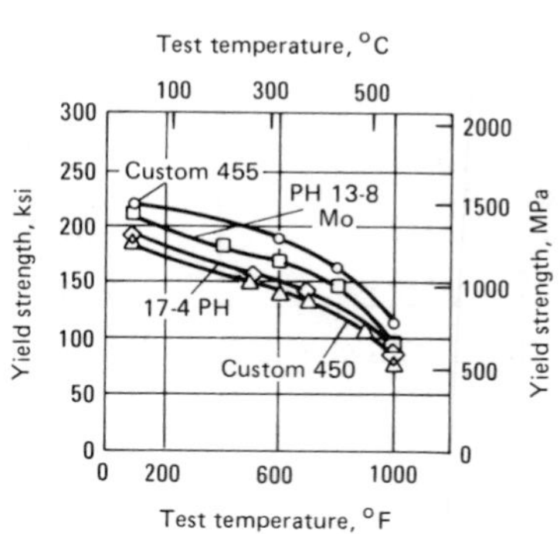

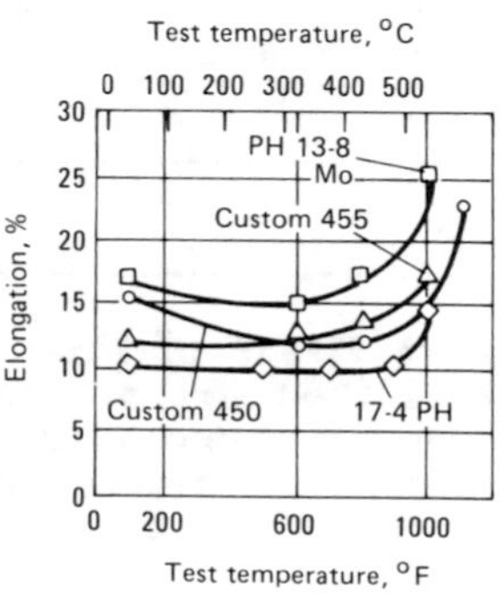

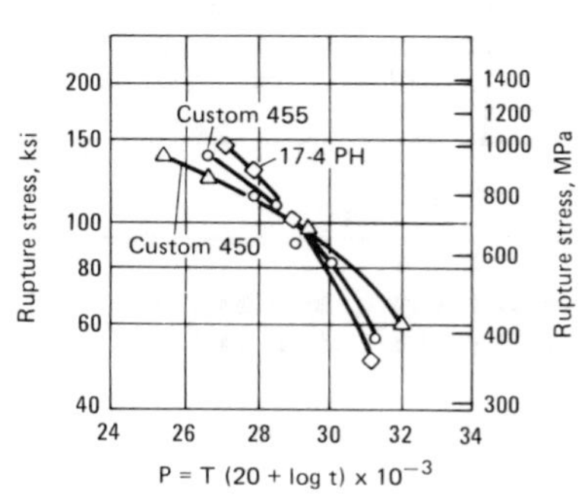

Heat treating schedules were as follows. Custom 450: 1 h at 1040 °C (1900 °F), water quench; plus 4 h at 480 °C (900 °F), air cool. 17-4 PH: 30 min at 1040 °C, oil quench; plus 4 h at 480 °C, air cool. Custom 455: 30 min at 815 °C (1500 °F), water quench; plus 4 h at 510 °C (950 °F), air cool. PH 13-8 Mo: oil quenched from 925 °C (1700 °F); plus 4 h at 540 °C (1000 °F), air cool.

tion. In some cases, material will be supplied in an "overaged" condition to facilitate forming of parts. The formed parts are then solution annealed following fabrication.

Properties. The following five maraging alloys are listed in Table 1 and can be divided into two groups on the basis of strength:

- Group 1 (lower strength): Custom 450, 17-4 PH and 15-5 PH
- Group 2 (higher strength): Custom 455 and PH 13-8 Mo.

Property data for these alloys are illustrated in Fig. 16. Data for 15-5 PH are not shown separately, because the properties of this alloy are very similar to those of 17-4 PH.

Short-time tensile data indicate that Custom 455 and PH 13-8 Mo have higher strengths than Custom 450, 17-4 PH and 15-5 PH. For all of these alloys, tensile and yield strengths drop rapidly at temperatures above 425 °C (800 °F), and tensile elongation is greater than 10% over the temperature range from ambient to 540 °C (1000 °F).

Stress-rupture data are compared in Fig. 16 by means of a master parameter plot. Data were developed at testing temperatures of 425 and 480 °C (800 and 900 °F) during time periods of 100 and 1000 h. Note that 17-4 PH appears to have better stress-rupture strength at 425 °C, whereas Custom 450 is superior in this respect at 480 °C. The only stress-rupture data available for any group 2 alloy are those for Custom 455, which appears to be intermediate in

Fig. 17 Effect of temperature on mechanical properties of Custom 450

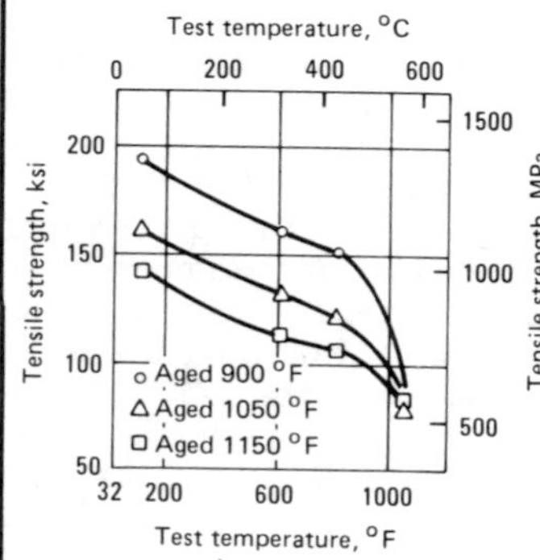

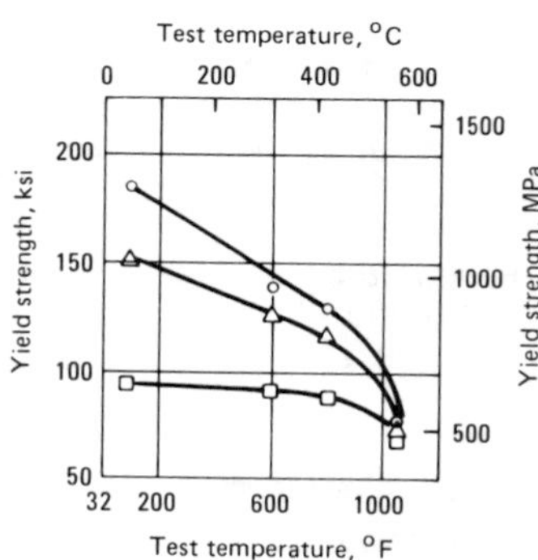

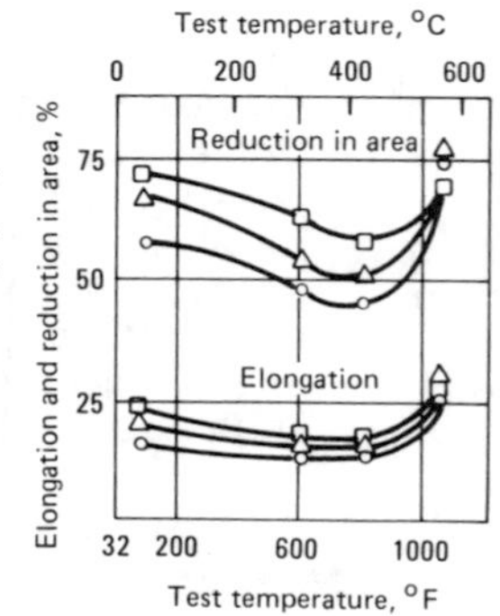

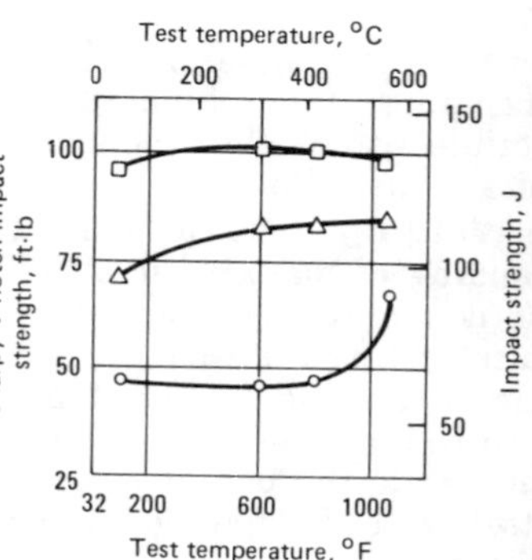

Material used for testing was round bar stock 25 mm (1 in.) in diameter that had been solution treated by heating 1 h at 1040 °C (1900 °F) and water quenching

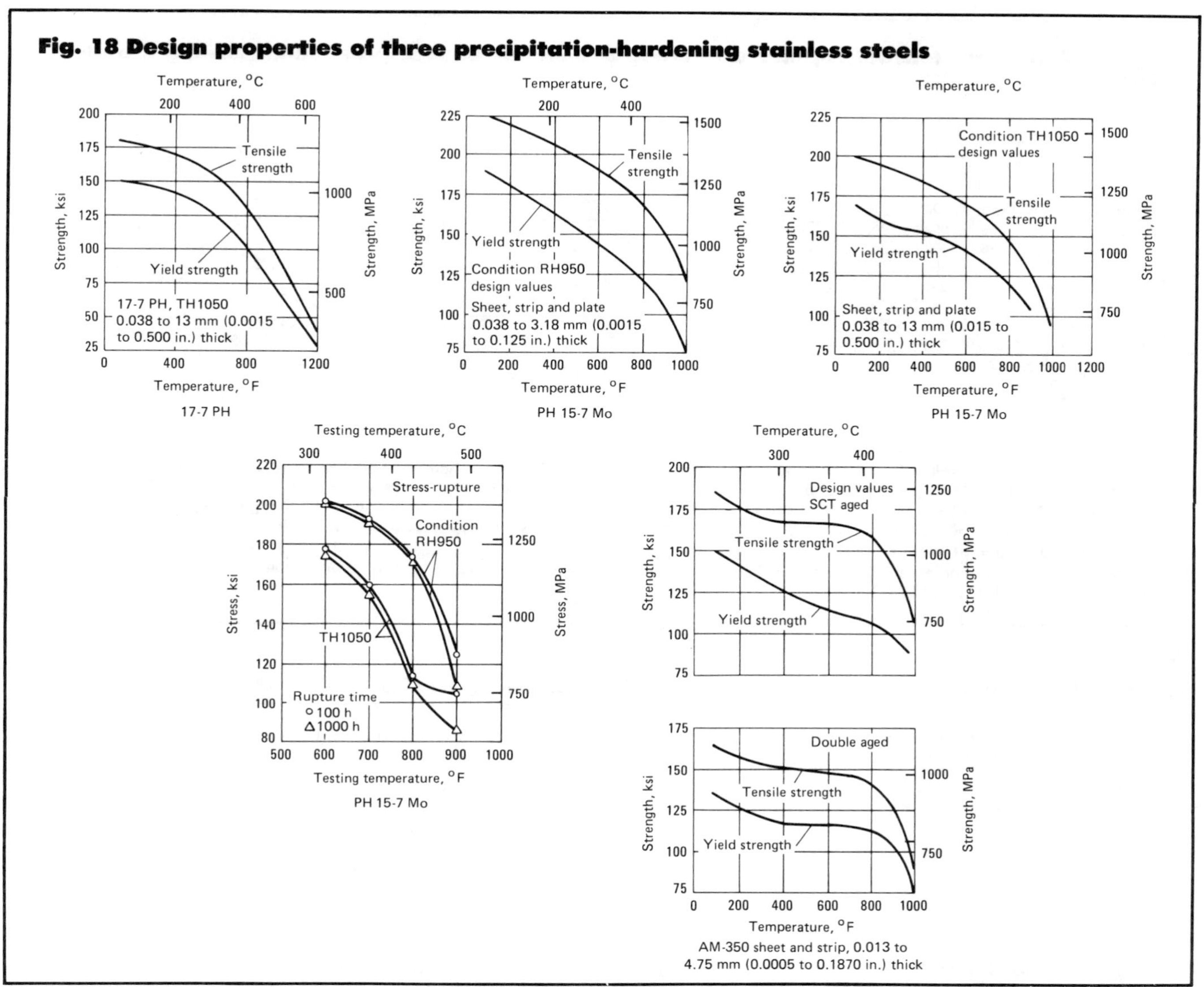

Fig. 18 Design properties of three precipitation-hardening stainless steels

rupture strength between 17-4 PH and Custom 450.

It is possible to produce a wide variety of useful properties in a given alloy by varying the aging temperature. An example of this can be seen in Fig. 17, where tensile-strength, yield-strength (0.2% offset), ductility and impact data are shown for Custom 450 at three different aging temperatures. Note that whereas aging at 480 °C (900 °F) can produce significant strengthening at testing temperatures as high as 450 °C (850 °F), it also results in lower values of toughness (as measured by Charpy V-notch testing) than aging at either of the two higher temperatures.

Precipitation-hardening martensitic stainless steels are used for industrial and military applications where resistance to corrosion and high mechanical properties at temperatures up to 425 °C (800 °F) are necessary. Typical uses include valve parts, ball bearings, forgings, turbine blades, mandrels, conveyor chain, miscellaneous hardware, and mechanical and structural components for aircraft.

Welding. All of these alloys are martensitic in the annealed condition and, because of their low carbon levels, are readily weldable with minimal danger of cracking. Any of the standard welding procedures, such as gas tungsten-arc, gas metal-arc, covered electrode and resistance welding, may be used. No preheating is required because the very low carbon contents of these alloys restrict the hardness of rapidly cooled metal and reduce the possibility of crack formation in weld metal and in heat-affected zones.

Precipitation-Hardening Semiaustenitic Stainless Steels

The precipitation-hardening semiaustenitic heat-resistant stainless steels are modifications of standard 18-8 austenitic stainless steels. Nickel contents are lower, and such elements as aluminum, copper, molybdenum and niobium are added.

Heat Treatment. Typical schedules for heat treating precipitation-hardening alloys are given in Table 3. These alloys are solution annealed above 1040 °C (1900 °F), and in this condition

Fig. 19 Short-time tensile, rupture and creep properties of precipitation-hardening stainless steels

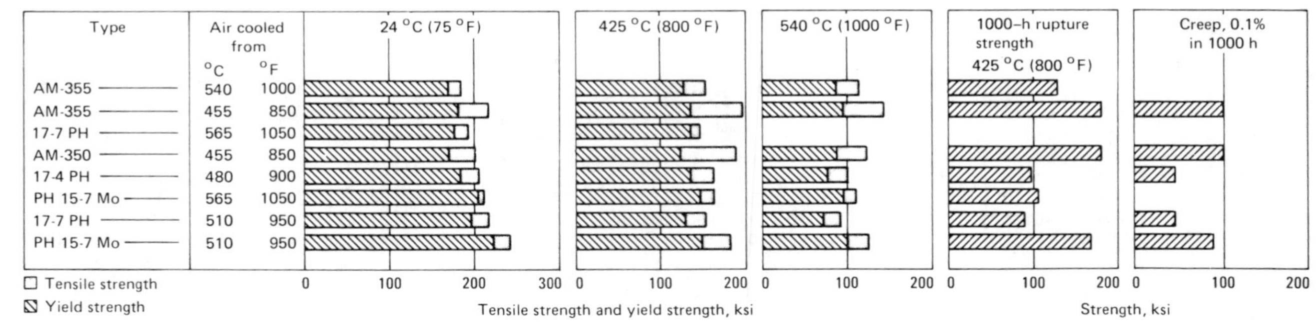

AM-355 was finish hot worked from a maximum temperature of 980 °C (1800 °F), reheated to 930 to 955 °C (1710 to 1750 °F), water quenched, treated at −75 °C (−100 °F) and aged at 540 and 455 °C (1000 and 850 °F). 17-7 PH and PH 15-7 Mo were solution treated at 1040 to 1065 °C (1900 to 1950 °F). 17-7 PH (TH1050) and PH 15-7 Mo (TH1050) were reheated to 760 °C (1400 °F), air cooled to 15 °C (60 °F) within 1 h and aged 90 min at 565 °C (1050 °F). 17-7 PH (RH950 and PH 15-7 Mo (RH950) were reheated to 955 °C (1750 °F) after solution annealing, cold treated at −75 °C (−100 °F) and aged at 510 °C (950 °F). 17-4 PH was aged at 480 °C (900 °F) after solution annealing. AM-350 was solution annealed at 1040 to 1065 °C (1900 to 1950 °F) and reheated to 930 °C (1710 °F), air cooled, treated at −75 °C (−100 °F) and then aged at 455 °C (850 °F).

Table 3 Heat treating schedules for precipitation-hardening semiaustenitic stainless steels

Alloy	Mill heat treatment (solution anneal)	Fabrication	Conditioning and hardening treatments	Aging or tempering treatment
17-7 PH	1065 °C (1950 °F), air cool	Forming, welding	10 min at 955 °C (1750 °F), air cool, 8 h at −75 °C (−100 °F)	1 h at 510, 565 or 620 °C (950, 1050 or 1150 °F)
			1½ h at 760 °C (1400 °F), air cool to 15 °C (60 °F), hold ½ h	1 h at 510, 565 or 620 °C (950, 1050 or 1150 °F)
15-7 Mo	1065 °C (1950 °F), air cool	Forming, welding	10 min at 955 °C (1750 °F), air cool, 8 h at −75 °C (−100 °F)	1 h at 510, 565 or 620 °C (950, 1050 or 1150 °F)
			1½ h at 790 °C (1450 °F), air cool to 15 °C (60 °F), hold ½ h	1 h at 510, 565 or 620 °C (950, 1050 or 1150 °F)
AM-350	1040 to 1080 °C (1900 to 1975 °F), air cool	Forming, welding	930 °C (1710 °F), air cool, 3 h at −75 °C (−100 °F)	3 h at 455 or 540 °C (850 or 1000 °F)
			3 h at 745 °C (1375 °F), air cool to 27 °C (80 °F) max	3 h at 455 °C (850 °F)
AM-355	3 h at 755 °C (1425 °F), oil or water quench to 27 °C (80 °F) max, 3 h at 580 °C (1075 °F), air cool	Machining and other	1040 °C (1900 °F), water quench, 3 h at −75 °C (−100 °F), reheat to 955 °C (1750 °F), air cool, 3 h at −75 °C (−100 °F)	3 h at 455 or 540 °C (850 or 1000 °F)

Table 4 Short-time tensile properties of heat treated AM-355 bar stock

Testing temperature		Tempering temperature		Yield strength 0.02% offset		Yield strength 0.2% offset		Tensile strength		Elongation in 50 mm or 2 in., %	Reduction in area, %
°C	°F	°C	°F	MPa	ksi	MPa	ksi	MPa	ksi		
21	70	455	850	980	142	1255	182	1490	216	19.0	38.5
		540	1000	1015	147	1180	171	1280	186	19.0	57.0
205	400	455	850	850	123	1125	163	1425	207	15.5	45.0
		540	1000	885	128	1050	152	1145	166	16.0	59.5
315	600	455	850	760	110	1050	152	1450	210	11.5	35.5
		540	1000	850	123	985	143	1095	159	14.0	49.0
425	800	455	850	675	98	960	139	1365	198	11.0	35.5
		540	1000	740	107	885	128	965	140	15.0	53.5
540	1000	455	850	450	65	670	97	995	144	16.0	57.0
		540	1000	485	70	660	96	795	115	19.0	65.0

Table 5 Typical mechanical properties of heat treated AM-355 at various temperatures

Testing temperature		Tensile properties: Yield strength(a)		Tensile properties: Tensile strength		Tensile properties: Modulus of elasticity		Compressive properties: Yield strength 0.02% offset		Compressive properties: Yield strength 0.2% offset		Compressive properties: Modulus of elasticity	
°C	°F	MPa	ksi	MPa	ksi	GPa	10^6 psi	MPa	ksi	MPa	ksi	GPa	10^6 psi
Room		1480	215	1615	234	200	29.0	840	122	1260	183	198	28.7
205	400	1280	186	1440	209	179	25.9	635	92	1075	156	185	26.8
315	600	1180	171	1395	202	170	24.7	585	85	1005	146	175	25.4
370	700	...	...	...	...	...	...	495	72	940	136	171	24.8
425	800	1025	149	1280	186	166	24.1	425	62	875	127	159	23.0
480	900	995	144	1200	174	164	23.8	440	64	820	119	148	21.5
540	1000	800	116	860	139	165	23.9	435	63	770	112	148	21.5

(a) At 0.2% offset.

Table 6 Stress-rupture of heat treated AM-350 and AM-355

Testing temperature		Tempering temperature		Stress to produce rupture in: 10 h		100 h		1000 h	
°C	°F	°C	°F	MPa	ksi	MPa	ksi	MPa	ksi
AM-350									
425	800	455	850	1280	186	1270	184	1255	182
		540	1000	925	134	895	130	885	128
480	900	455	850	1015	147	835	121	655	95
		540	1000	745	108	710	103	625	91
AM-355									
425	800	455	850	1305	189	1280	186	1240	180
		540	1000	940	136	925	134	910	132
480	900	455	850	995	144	835	121	675	98
		540	1000	770	112	725	105	675	98
540	1000	455	850	580	84	485	70	400	58
		540	1000	600	87	505	73	420	61

can be formed, stamped, stretched and otherwise cold worked to about the same extent as can 18-8 alloys, although they are less ductile and may require intermediate annealing.

All the semiaustenitic stainless steels also can be used in the cold worked condition, in either sheet or wire form. Cold working causes partial transformation of the rather unstable austenite to martensite due to plastic deformation. Aging or tempering is performed after cold working.

Mechanical properties of precipitation-hardening semiaustenitic stainless steels are shown graphically in Fig. 18. Tensile properties of AM-355 are presented in Tables 4 and 5.

Short-time tensile, rupture and creep properties of several precipitation-hardening alloys are compared in Fig. 19. Different hardening heat treatments may produce a wide variety of useful properties for the same alloy. For example, the 17-7 PH alloy treated to condition RH950 has higher room-temperature tensile strength than the same alloy treated to RH1050. There is no significant difference in properties at 425 °C (800 °F) for this alloy in these two conditions.

Compressive and tensile yield strengths are approximately equal for all precipitation-hardened steels. For sheet, yield strengths of specimens taken transverse and parallel to the direction of rolling may vary appreciably. The magnitude of this effect varies from grade to grade, and with heat treatment for a given grade.

Rupture strength at three temperatures as a function of rupture time is given in Fig. 20 for 17-7 PH in the TH1050 condition. These data indicate some notch strengthening at all three temperatures. Rupture strengths of AM-350 and AM-355 are presented in Table 6.

Welding. The precipitation-hardening stainless steels in either the annealed or the hardened condition can be

Fig. 20 Stress-rupture properties of 17-7 PH, condition TH1050

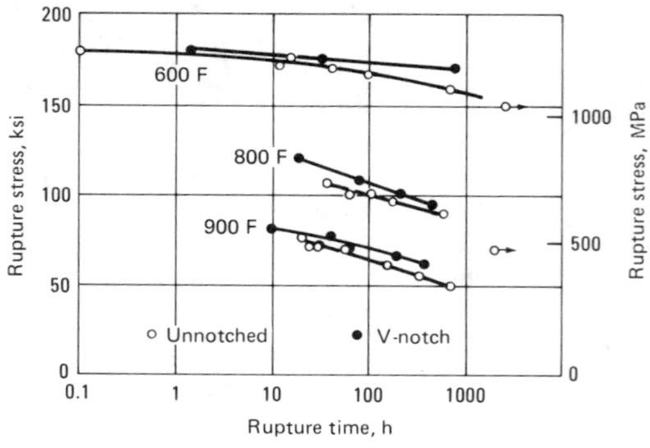

Average short-time properties

Temperature °C	°F	Tensile yield strength(a) (b) MPa	ksi	Compressive yield strength(a) MPa	ksi	Elongation in 50 mm or 2 in., %
24	75	1340	194	1525	221	6.7
315	600	1125	163	1280	186	4.8
425	800	950	138	1015	147	11.8
480	900	765	111	785	114	19.6

(a) At 0.2% offset. (b) Tensile strengths (low to high temperature) were: 1400, 1185, 1020 and 840 MPa (203, 172, 148 and 122 ksi). Hardness at 24 °C (75 °F) was 44 HRC.

Specimens were taken transverse to the rolling direction from sheet 915 by 3050 by 1.6 mm (36 by 120 by 0.063 in.) and were heat treated by air cooling from 760 to 260 °C (1400 to 500 °F), water quenching, holding 10 h at 15 °C (60 °F), aging 1½ h at 565 °C (1050 °F) and air cooling. V notches on both edges of specimens were 7.6 mm (0.300 in.) wide with 0.61-mm (0.024-in.) root radii; K_t= 3.1. Strain of 1 to 2% occurred within 10% of rupture life in smooth specimens.

Table 7 Typical mechanical properties of AM-350 sheet, 2.0 mm (0.078 in.) thick, GTA welded with AM-355 filler metal

Testing temperature(a) °C	°F	Tensile strength MPa	ksi	Yield strength(b) MPa	ksi	Elongation(c), %
Base metal(d)						
27	80	1470	213	1205	175	12.5
165	300	1345	195	1070	155	10.5
315	600	1340	194	960	139	6.5
370	700	1340	194	910	132	10.0
425	800	1290	187	855	124	11.5
480	900	1160	168	800	116	10.5
Transverse to weld(e)						
27	80	1315	191	1140	165	4.0
165	300	1315	191	1050	152	6.5
315	600	1260	183	925	134	5.0
370	700	1250	181	825	120	4.5
425	800	1185	172	780	113	5.0
480	900	1105	160	710	103	6.0

(a) Welded specimens were ground flush, and held 20 min at test temperature, prior to testing. (b) At 0.2% offset. (c) In 50 mm or 2 in. (d) Heat treated. (e) Heat treated after welding. All specimens broke in the weld.

welded by any of the processes used for welding 18-8 stainless steel. These steels are less ductile and more notch sensitive than conventional 18-8 grades and require more care to prevent stress concentrations at corners and notches.

The gas-shielded tungsten-arc process has been satisfactory for welding large assemblies of sheet and forgings that are subsequently aged. Postweld annealing is desirable.

A complete cycle of heat treatment including high-temperature solution annealing is desirable after welding. Strength and ductility in fully heat treated fusion welds are superior to these properties in material hardened before welding and not heat treated after welding. Joint efficiencies near 90% have been reported for welded joints fully heat treated after welding (Table 7).

The precipitation-hardening steels can be spot welded by the same technique used for 18-8 steels. It may be done before or after precipitation hardening. The minimum shear requirements for stainless steel at 1035 MPa (150 ksi) can be obtained.

Precipitation-hardening steels can be cold rolled and tempered. Work-hardening rates are higher than for type 301 stainless steel and can be varied by regulating the annealing temperature. Compared with cold rolled 301, cold rolled precipitation-hardened steels have higher ductility at a given strength level and less reduction in strength with increasing temperature.

Precipitation-hardening steels are used for industrial and military applications that require resistance to corrosion, and high mechanical properties at temperatures up to 425 °C (800 °F).

Typical uses of these steels include landing-gear hooks, poppet valves, fuel tanks, hydraulic lines, hydraulic fittings, compressor casings, miscellaneous hardware, and structural components for aircraft. The higher-carbon grade (0.13%) is used for compressor blades, spacers, frames and casings for gas turbines, oil-well drill rods, and rocket casings.

Austenitic Stainless Steels

Austenitic stainless steels comprise a group of iron-base alloys that contain 16 to 25% Cr and residual to 20% Ni. Some alloys may contain as much as 18% Mn. These stainless steels are not hardenable by heat treatment but can be hardened by cold work. The effect of cold work on elevated-temperature mechanical properties depends on the recrystallization temperature of the alloy, the amount of residual stress resulting from the cold work, and the duration of thermal exposure.

The austenitic stainless steels listed in Table 1 can be grouped into three categories, based primarily on composition, as follows:

- *300 series alloys,* which are essentially Cr-Ni and Cr-Ni-Mo austenitic stainless steels to which small amounts of other elements have been added.
- *19-9 DL, 19-9 DX* and *17-14-CuMo,* all of which contain 1.25 to 2.5 Mo and 0.3 to 0.55 Ti. Other elements used include 1.25 W and 3 Cu in 17-14-CuMo.
- *Cr-Ni-Mn alloys,* which include types 202 and 216; 21-6-9; Nitronics 32, 33, 50 and 60; and Carpenter 18-18 Plus. These alloys contain 5 to 18 Mn and 0.10 to 0.50 N.

Applications that take advantage of the heat-resisting capabilities of austenitic stainless steels include furnace parts, heat-exchanger tubing, steam lines, exhaust systems in reciprocating engines and gas turbines, afterburner parts, and similar parts that require strength and oxidation resistance.

Type 304 has good resistance to atmospheric corrosion and oxidation. Types 309 and 310 rank higher in these properties because of their higher nickel and chromium contents. Type 310 is preferred where intermittent heating and cooling are encountered, because it forms a more adherent scale than does type 309. Types 309 and 310 are used for parts such as firebox sheets, furnace linings and boiler baffles, thermocouple wells, aircraft cabin heaters and jet-engine burner liners.

Types 321 and 347 are for use primarily where solution treatment after welding is not feasible—for instance, steam lines and superheater tubes and exhaust systems in reciprocating engines and gas turbines that operate at temperatures from 425 to 870 °C (800 to 1600 °F). The low-carbon types 304L and 316L are used for similar applications but are more susceptible to intergranular attack during long exposure to high temperature.

Fig. 21 Effect of testing temperature on tensile properties of austenitic stainless steels

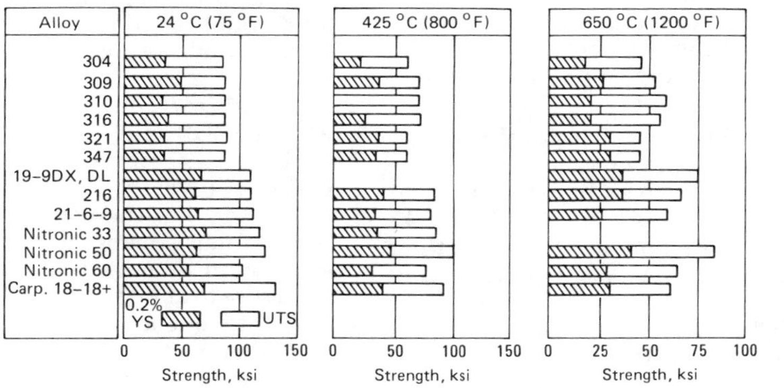

Heat treating schedules were as follows. Type 304: 1065 °C (1950 °F), water quench. Type 309: 1090 °C (2000 °F), water quench. Type 310: 1120 °C (2050 °F), water quench. Type 316: 1090 °C (2000 °F), water quench. Type 321: 1010 °C (1850 °F), water quench. Type 347: 1065 °C (1950 °F), water quench. 19-9 DX, DL: 705 °C (1300 °F), air cool. Type 216: 1065 °C (1950 °F), water quench. 21-6-9: 1065 °C (1950 °F), water quench. Nitronic 33: 1065 °C (1950 °F), water quench. Nitronic 50: 1090 °C (2000 °F), water quench. Nitronic 60: 1065 °C (1950 °F), water quench. Carpenter 18-18 Plus: 1065 °C (1950 °F), water quench.

Fig. 22 Stress-rupture plots for austenitic stainless steels

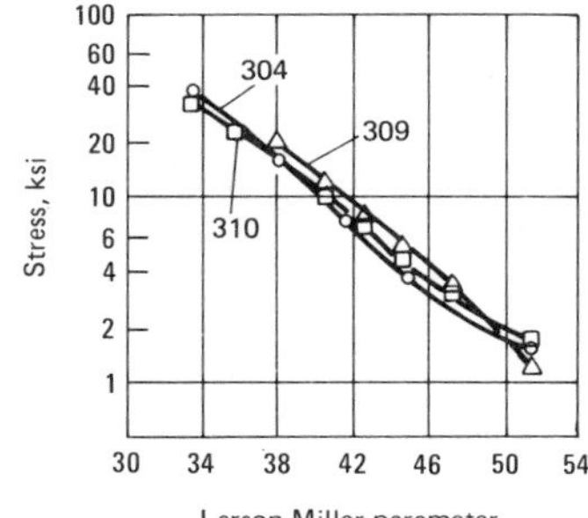

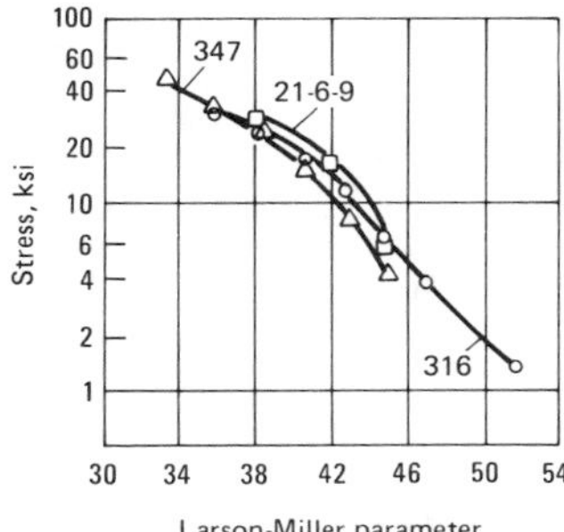

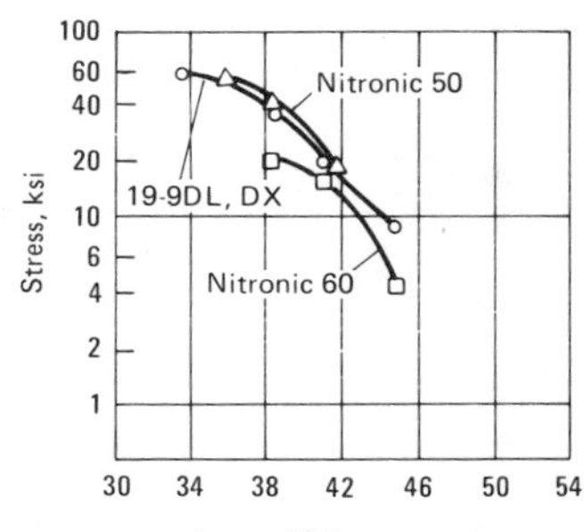

Heat treating schedules were as follows. Type 304: 1065 °C (1950 °F), water quench. Type 309: 1090 °C (2000 °F), water quench. Type 310: 1120 °C (2050 °F), water quench. Type 316: 1090 °C (2000 °F), water quench. Type 347: 1065 °C (1950 °F), water quench. 21-6-9: 1065 °C (1950 °F), water quench. 19-9 DX, DL: for tests above 705 °C (1300 °F), 1065 °C (1950 °F) and water quench, plus 705 °C and air cool; for tests below 705 °C, 705 °C and air cool. Nitronic 50: 1090 °C (2000 °F), water quench. Nitronic 60: 1065 °C (1950 °F), water quench. Larson-Miller parameter = T/1000 (20 + log t), where T is temperature in °R and t is time in h; all data taken from 1000-h tests.

Types 316 and 316L have higher mechanical properties than 304 and 321 and are more resistant to corrosion in some media, such as fatty acids at elevated temperature and mild sulfuric acid solutions.

Austenitic stainless steels are noted for high strength and for exceptional toughness, ductility and formability. As a class, they exhibit considerably better corrosion resistance than martensitic or ferritic steels and also have excellent strength and oxidation resistance at elevated temperatures.

Solution heat treatment of these alloys is done by heating at 955 to 1180 °C (1750 to 2150 °F) and cooling rapidly. Carbides that are dissolved at these temperatures may precipitate at grain boundaries on exposure to temperatures from 425 to 870 °C (800 to 1600 °F), causing chromium depletion in grain-boundary regions. In this condition, the metal is sensitive to intergranular corrosion. Precipitation of chromium carbides can be controlled by reducing carbon content, as in types 304L and 316L, or by adding the stronger carbide formers titanium and niobium, as in types 321 and 347. These alloys normally are purchased and used in the annealed condition.

Tensile Properties. Typical mechanical-property data for austenitic stainless steels are given in Fig. 21 and 22. Room-temperature tensile properties of annealed 300 series alloys are similar. At higher testing temperatures (425 and 650 °C; 800 and 1200 °F), types 321, 347 and 309 appear to have yield strengths somewhat higher than those of types 304, 310 and 316. Types 309, 310 and 316 have the highest tensile strengths at 650 °C.

Tensile and yield strengths of 19-9 DL and 19-9 DX are higher than those of any 300 series alloy. However, 19-9 DL and 19-9 DX are heat treated at 705 °C (1300 °F), compared with an average of 1065 °C (1950 °F) for 300 series alloys. Also, 19-9 DL and 19-9 DX normally are strengthened by controlled amounts of hot and cold work.

Tensile and yield strengths of Cr-Ni-Mn alloys are higher than those of 300 series alloys at both room and elevated temperatures. Carpenter 18-18 Plus exhibits the highest room-temperature tensile strength, whereas Nitronic 50 has the highest tensile strength at 650 °C (1200 °F).

Stress-Rupture Properties. Type 316 stainless steel has the highest stress-rupture capability of all the 300 series alloys (see Fig. 22). Type 347 appears to be somewhat weaker than type 316, and weakens with increasing temperature at a higher rate. Types 304, 309 and 310 are inferior to types 316 and 347 in stress-rupture capability.

The 19-9 DL and 19-9 DX alloys have rupture strengths superior to those of all 300 series alloys over the temperature range for which rupture data are available (540 to 815 °C, or 1000 to 1500 °F).

The Cr-Ni-Mn alloys have higher stress-rupture capabilities than 300 series alloys, with the following exceptions: type 316 is superior to 21-6-9, and both 316 and 347 are stronger than Nitronic 60.

The spread in rupture-strength capability among these alloys is greater at the lower testing temperatures (540 to 700 °C; 1000 to 1300 °F) and becomes progressively smaller as temperature is increased to approximately 980 °C (1800 °F), where all the alloys exhibit 1000-h rupture stresses of about 7 to 10 MPa (1 to 1.5 ksi). Types 304 and 310 have the lowest stress-rupture strengths.

Welding of austenitic stainless steels is not nearly as difficult as welding of martensitic stainless steels. From the standpoint of welding, austenitic stainless alloys can be considered to fall into two groups:

- *Nonstabilized alloys,* such as types 304, 309, 310 and 316, can be welded by all of the common welding processes. The weld zone will not harden on cooling, but the possibility of intergranular corrosion should be eliminated by reheating the welded structure to at least 1040 °C (1900 °F) and water quenching. Use of low-carbon filler metal will also help prevent intergranular corrosion. Filler metals whose compositions are balanced to form small amounts of ferrite are used to avoid hot cracking.
- *Stabilized alloys,* such as types 321 and 347, are resistant to harmful grain-boundary carbide precipitation and can be used in the as-welded condition. They are used primarily in applications that involve welding during fabrication and that require weldments to be placed in service without postweld annealing.

Wrought Superalloys

Edited by F. R. Morral

SEVERAL GROUPS of alloys are classified as "heat-resistant alloys", but the only group discussed in this article is the group known as "superalloys". Superalloys are iron-, cobalt- and nickel-base alloys that contain chromium for resistance to oxidation and hot corrosion and that contain other elements for strength at elevated temperatures. The high-temperature applications of superalloys are extensive, including components for aircraft, chemical-plant equipment and petrochemical equipment. The increasing significance of superalloys in today's commerce is typified by the fact that, whereas in 1950 only about 10% of the total weight of an aircraft turbojet engine was made of superalloys, by 1980 this figure had risen to about 50%.

In order to satisfy the requirements of the wide variety of uses for which superalloys have been employed, extensive property data have been generated—not only for the ordinary mechanical, physical, chemical, nuclear and magnetic properties, but also for several uncommon properties. Generally, superalloys must exhibit good resistance to corrosion, creep, fatigue, thermal fatigue, thermal shock, impact, cavitation and erosion, and in addition must have good fracture properties, forming characteristics and weldability. In many applications, components must be in service for prolonged periods of time, and therefore any variable such as temperature, corrosion, diffusion or solid-state reactions that might alter properties must be accounted for. Hot corrosion is particularly insidious: both gaseous corrodents such as sulfur compounds and liquid corrodents such as molten salts are responsible for attack in gas turbines. Impingement by flying particles may destroy inherent or artificial surface coatings that protect the alloy from corrosion. When this happens, the combination of erosion and hot corrosion often is more devastating than either effect would be alone. In certain respects, industrial experience does not agree with laboratory predictions—combinations of alloy and environment that, according to laboratory data, should lead to cracking after extended exposure at high temperatures have not done so in actual service.

The maximum service temperature for conventional superalloys in structural applications frequently does not exceed about 950 °C (1740 °F). In applications where the component does not bear a load, the temperature may exceed 1200 °C (2200 °F). In general, heating to any temperature which causes (*a*) melting, (*b*) solid solution of strengthening phases or (*c*) extensive oxidation or corrosion can be considered "overheating". Properties of a specific alloy at elevated temperatures are influenced to some extent by the form of the product. Duration and type of loading may emphasize the differences in properties of one product form versus another.

Wrought superalloys may be classified as follows:

- Conventional superalloys, as wrought
- Alloys prepared by powder metallurgy techniques in the following groups: (*a*) conventional-composition alloys, (*b*) oxide-dispersion-strengthened (ODS) alloys and (*c*) new alloy compositions.

Many recent developments in P/M plus hot isostatic pressing (HIP) technology, as well as the development of superplastic forging, are making materials selection for elevated-temperature service a more complicated and sophisticated endeavor. Instead of merely considering alloy composition, a materials engineer must decide whether P/M, P/M plus HIP, superplastic forging or some combination of these can impart the required properties. For instance, P/M plus HIP components have unusually fine grain sizes, which impart additional ductility and toughness, leading to improved resistance to low-cycle fatigue.

Current applications of superalloys

Fig. 1 Elevated-temperature corrosion of selected superalloys

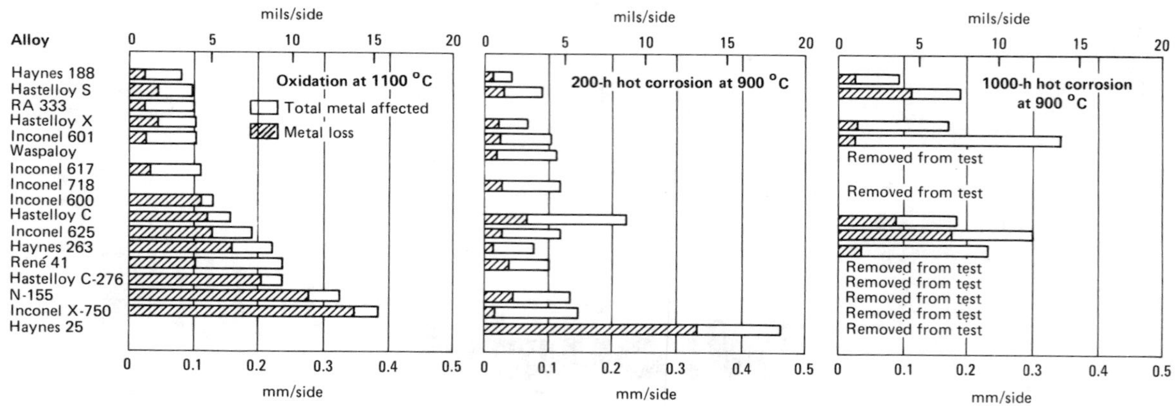

Left—100-h dynamic oxidation resistance at 1100 °C (2000 °F): tested in an environment created by burning No. 2 fuel oil (0.3 to 0.45% S) with an air-to-fuel ratio of 45 : 1 to 55 : 1; thermal shock frequency was 2 per hour. Center—200-h hot corrosion resistance at 900 °C (1650 °F): tested according to ASTM D-665 in an environment created by burning No. 2 fuel oil (0.3 to 0.45% S), to which 50 ppm salt were added, using an air-to-fuel ratio of 30 : 1; thermal shock frequency was 1 per hour. Right—1000-h hot corrosion resistance at 900 °C (1650 °F): conditions same as for center graph, except salt concentration was only 5 ppm. The following alloys were removed from test due to excessive attack, and are listed in order of decreasing resistance: N-155 (601 h), Inconel 718 (601 h), Waspaloy (700 h), René 41 (700 h), Inconel X-750 (700 h), Haynes 25 (142 h), Hastelloy C-276 (296 h); all seven were judged inferior to Inconel 601.

include the wide variety of uses listed below:

- Aircraft gas turbines:
  - Disks
  - Bolts
  - Shafts
  - Cases
  - Blades
  - Vanes
  - Burner cans
  - Afterburners
  - Thrust reversers
- Steam-turbine power plants:
  - Bolts
  - Blades
  - Stack-gas reheaters
- Reciprocating engines:
  - Turbochargers
  - Exhaust valves
  - Hot plugs
  - Precombustion cups (Riccardo-type diesel)
  - Valve-seat inserts
- Metal processing:
  - Hot work tools and dies
  - Cast dies
- Medical applications:
  - Dentistry
  - Prosthetic devices
- Space vehicles:
  - Aerodynamically heated skins
  - Rocket-engine parts
- Heat-treating equipment:
  - Trays
  - Fixtures
  - Conveyor belts
- Nuclear power systems:
  - Control-rod drive mechanisms
  - Valve stems
  - Springs
  - Ducting
- Chemical and petrochemical industries:
  - Bolts
  - Valves
  - Reaction vessels
  - Piping
  - Pumps

Superalloys often are diffusion or overlay coated in order to improve their corrosion resistance. The overlay coatings used are based on iron-chromium-aluminum-yttria, cobalt-chromium-aluminum-yttria and nickel-chromium-aluminum-yttria alloys. The diffusion coatings are based on reaction of aluminum with the substrate. Environments may be oxidizing or reducing, or may contain some reactive contaminant such as sodium or potassium.

An important consideration for coated superalloys is the reduction in incipient melting temperature of the system (coating/base metal) that may result from the change in composition caused by diffusion of coating components inward from the surface. Incipient melting reduces grain-boundary strength and ductility and thus reduces stress-rupture capabilities. Once an alloy has been heated above its incipient melting point, alloy properties cannot be restored by heat treatment. Generally, an alloy should not be used for a structural application at any temperature higher than the point about 125 °C (225 °F) below its incipient melting temperature. Oxidation behavior and strength will determine how close actual metal temperatures may approach this suggested upper limit.

An important difference between nickel and cobalt superalloys is related to the superior hot corrosion resistance claimed for cobalt-base alloys in atmospheres containing sulfur, sodium salts, halides, vanadium oxides and lead oxide, all of which can be found in fuel-burning systems. Nickel forms low-melting-point eutectics with nickel sulfide, and in sulfur-bearing gases attack of nickel alloys may be devastating. Generalizations about comparative oxidation and hot corrosion resistance of nickel- and cobalt-base alloys must be treated with some caution as

Fig. 2 Temperature capabilities and hot corrosion of several Nimonic alloys

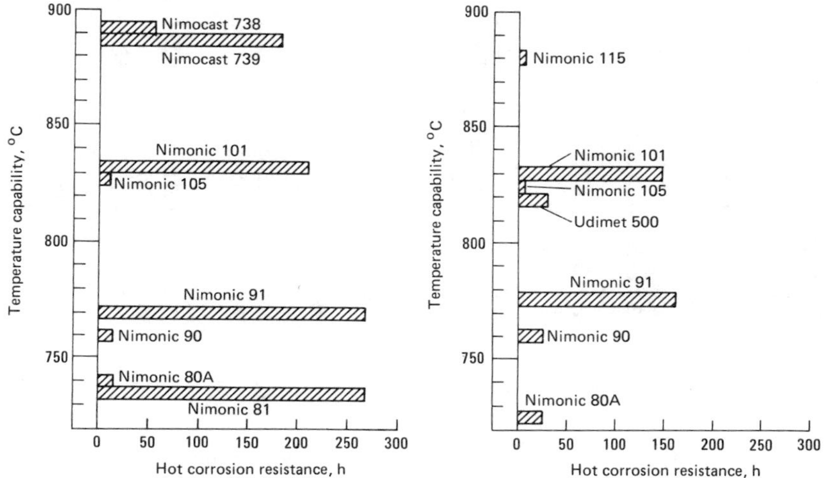

Temperature capabilities based on 10 000-h rupture life at 120 MPa (17.4 ksi). Hot corrosion resistance based on time to a weight loss of: left, 1 kg/m^2 (100 mg/cm^2) in continuously fed mixture of 75% Na_2SO_{42} and 25% NaCl at 900 °C (1650 °F); right, 180 g/m^2 (18 mg/cm^2) in continuously fed mixture of 75% Na_2SO_4 and 25% NaCl at 850 °C (1560 °F).

Fig. 3 Oxidation resistance of selected oxide dispersion strengthened alloys

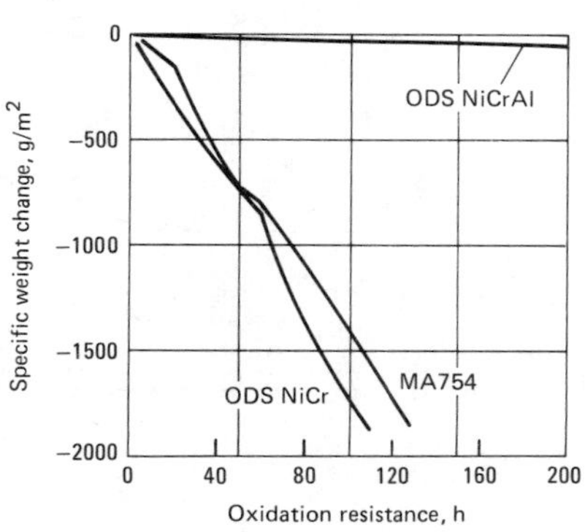

Tested in Mach 1 burner rig at 1200 °C (2200 °F); samples cooled to room temperature once each hour.

there are wide ranges of resistance within each alloy group, and seemingly small variations in service conditions can vastly affect results, as shown in Fig. 1. Figure 2 illustrates hot corrosion resistance of a selected group of nickel-base superalloys, and Fig. 3 presents hot corrosion data for selected ODS alloys.

Additional information about superalloys can be found in Ref 1 to 5. The effects of overheating and alloy stability are discussed in more detail in Appendix 1, while testing and assessment of mechanical properties are covered in Appendices 2 and 3.

Metallurgy of Superalloys

Iron-, cobalt- and nickel-base alloys can be strengthened by the following mechanisms: solid-solution strengthening, precipitation hardening and oxide-dispersion strengthening. Solid-solution and precipitation-hardening alloys are considered conventional alloys; the most widely used of these alloys are listed in Table 1. Oxide-dispersion-strengthened alloys, a new breed of superalloys that enjoy only limited commercial use, are listed in Table 2.

Table 3 reviews briefly the roles that the various alloying elements play in strengthening conventional alloys. The alloying elements can be classified as (*a*) solid-solution strengtheners, which affect only the matrix, (*b*) elements that form carbides or intermetallic compounds in grain boundaries or within grains, and (*c*) elements that form a coherent gamma-prime phase in nickel-base alloys. Residual elements also are important because they may have a deleterious effect on creep behavior.

Iron-Base Superalloys

Iron-base superalloys are nominally defined as those alloys that have iron as the major constituent and that are hardened by a carbide or intermetallic precipitate. Because many of the principles applicable to iron-base superalloys can be applied to other iron-containing systems, information will be included on those superalloys containing iron in relatively large amounts (defined for the present purposes as $\geq$10%). The increased emphasis on iron-base superalloys as lower-cost substitutes for nickel- and cobalt-base superalloys has resulted in development of moderate-strength, solid-solution alloys; these also are considered below.

Iron-base alloys with a face-centered-cubic matrix (often referred to as austenite) have a strong tendency to form topologically close-packed (tcp) phases such as sigma, mu, Laves and chi. On the other hand, fcc nickel-base alloys have a tendency toward precipitation of ordered geometrically close-packed (gcp) phases such as γ' [$Ni_3(Al, Ti)$] and η (Ni_3Ti); fcc cobalt-base alloys have a strong potential for solid-solution strengthening. Iron can tolerate, more readily than nickel, the formation of phases with abnormally short interatomic distances such as those which occur in Laves and sigma phases. This characteristic is of essential practical importance.

Various elements are added to perform one or more desirable functions. The most potent strengthening for alloys with fcc matrixes is provided by elements such as nickel, aluminum, titanium and niobium. These elements precipitate intermetallic phases such as ordered fcc γ' and ordered body-centered tetragonal γ'' from the matrix by suitable heat treatment. Elements such as iron and chromium may also be found in γ' and γ''. FCC alloys can be hardened by addition of carbon in relatively large amounts (~0.5%) to form a general carbide precipitate; nitrogen and phosphorus sometimes are added to enhance this effect. Carbon may promote formation of grain-boundary carbide phases such as $M_{23}C_6$ and M_6C to provide strength in these regions.

Elements in solid solution add some limited degree of strengthening; the most potent are interstitial in nature. These elements also alter the lattice

Table 1 Nominal compositions of wrought superalloys

Alloy	UNS number	Cr	Ni	Co	Mo	W	Nb	Ti	Al	Fe	C	Other
						Composition, %						
Iron-base solid-solution alloys												
16-25-6	...	16.0	25.0	...	6.00	...	...	...	...	50.7	0.06	1.35 Mn; 0.70 Si; 0.15 N
17-14CuMo	...	16.0	14.0	...	2.50	...	0.4	0.3	...	62.4	0.12	0.75 Mn; 0.50 Si; 3.0 Cu
19-9DL	K63198	19.0	9.0	...	1.25	1.25	0.4	0.3	...	66.8	0.30	1.10 Mn; 0.60 Si
Carpenter 20Cb-3	N08020	20.0	34.0	...	2.50	...	1.0 max	...	...	42.4	0.07 max	3.5 Cu
Incoloy 800	N08800	21.0	32.5	...	...	...	...	0.38	0.38	45.7	0.05	...
Incoloy 801	N08801	20.5	32.0	...	...	...	...	1.13	...	46.3	0.05	...
Incoloy 802	...	21.0	32.5	...	...	...	...	0.75	0.58	44.8	0.35	...
N-155	R30155	21.0	20.0	20.0	3.00	2.5	1.0	...	...	32.2	0.15	0.15 N; 0.02 La; 0.02 Zr
RA330	N08330	19.0	36.0	...	...	...	...	...	...	45.1	0.05	...
Cobalt-base solid-solution alloys												
Haynes 25 (L-605)	R30605	20.0	10.0	50.0	...	15.0	...	...	...	3.0	0.10	1.5 Mn
Haynes 188	R30188	22.0	22.0	37.0	...	14.5	...	...	...	3.0 max	0.10	0.90 La
S-816	R30816	20.0	20.0	42.0	4.0	4.0	4.0	...	...	4.0	0.38	...
Stellite 6B	...	30.0	1.0	61.5	...	4.5	...	...	...	1.0	1.0	...
UMCo-50	...	28.0	...	49.0	...	...	...	...	...	21.0	0.12 max	...
Nickel-base solid-solution alloys												
Hastelloy B	N10001	1.0 max	63.0	2.5 max	28.0	...	...	...	...	5.0	0.05 max	0.03 V
Hastelloy B-2	N10665	1.0 max	69.0	1.0 max	28.0	...	...	...	...	2.0 max	0.02 max	...
Hastelloy C	N10002	16.5	56.0	...	17.0	4.5	...	...	...	6.0	0.15 max	...
Hastelloy C-4	N06455	16.0	63.0	2.0 max	15.5	...	...	0.7 max	...	3.0 max	0.015 max	...
Hastelloy C-276	N10276	15.5	59.0	...	16.0	3.7	...	...	...	5.0	0.02 max	...
Hastelloy N	N10003	7.0	72.0	...	16.0	...	...	0.5 max	...	5.0 max	0.06	...
Hastelloy S	...	15.5	67.0	...	15.5	...	...	...	0.2	1.0	0.02 max	0.02 La
Hastelloy W	N10004	5.0	61.0	2.5 max	24.5	...	...	...	...	5.5	0.12 max	0.6 V
Hastelloy X	N06002	22.0	49.0	1.5 max	9.0	0.6	...	...	2.0	15.8	0.15	...
Inconel 600	N06600	15.5	76.0	...	...	...	...	...	...	8.0	0.08	0.25 max Cu
Inconel 601	N06601	23.0	60.5	...	...	...	...	...	1.35	14.1	0.05	0.5 max Cu
Inconel 604	...	16.0	74.0	...	...	...	2.25	...	...	7.5	0.02	0.03 max Cu
Inconel 617	...	22.0	55.0	12.5	9.0	...	...	...	1.0	...	0.07	...
Inconel 625	N06625	21.5	61.0	...	9.0	...	3.6	0.2	0.2	2.5	0.05	...
NA-224	...	27.0	48.0	...	...	6.0	...	...	...	18.5	0.50	...
Nimonic 75	...	19.5	75.0	...	...	...	...	0.4	0.15	2.5	0.12	0.25 max Cu
RA-333	N06333	25.0	45.0	3.0	3.0	3.0	...	...	...	18.0	0.05	...
Iron-base precipitation-hardening alloys												
A-286	K66286	15.0	26.0	...	1.25	...	...	2.0	0.2	55.2	0.04	0.005 B; 0.3 V
Discaloy	K66220	14.0	26.0	...	3.0	...	...	1.7	0.25	55.0	0.06	...
Haynes 556	...	22.0	21.0	20.0	3.0	2.5	0.1	...	0.3	29.0	0.10	0.50 Ta; 0.02 La; 0.002 Zr
Incoloy 903	...	0.1 max	38.0	15.0	0.1	...	3.0	1.4	0.7	41.0	0.04	...
Pyromet CTX-1	...	0.1 max	37.7	16.0	0.1	...	3.0	1.7	1.0	39.0	0.03	...

(continued)

(a) No longer active, but shown here for reference. (b) Also known as Rolls Royce C-263.

Table 1 (continued)

Alloy	UNS number	Cr	Ni	Co	Mo	W	Composition, % Nb	Ti	Al	Fe	C	Other
V-57	...	14.8	27.0	...	1.25	...	...	3.0	0.25	48.6	0.08 max	0.01 B; 0.5 max V
W-545	K66545	13.5	26.0	...	1.5	...	...	2.85	0.2	55.8	0.08	0.05 B
Cobalt-base precipitation-hardening alloys												
AR-213	...	19.0	0.5 max	65.0	...	4.5	...	...	3.5	0.5 max	0.17	6.5 Ta; 0.15 Zr; 0.1 Y
MP-35N	R30035	20.0	35.0	35.0	10.0	...	...	...	...	...	...	...
MP-159	...	19.0	25.0	36.0	7.0	...	0.6	3.0	0.2	9.0	...	...
Nickel-base precipitation-hardening alloys												
Astroloy	...	15.0	56.5	15.0	5.25	...	...	3.5	4.4	<0.3	0.06	0.03 B; 0.06 Zr
D-979	N09979; K66979(a)	15.0	45.0	...	4.0	4.0	...	3.0	1.0	27.0	0.05	0.01 B
IN 100	N13100	10.0	60.0	15.0	3.0	...	...	4.7	5.5	<0.6	0.15	1.0 V; 0.06 Zr; 0.015 B
IN 102	N06102	15.0	67.0	...	2.9	3.0	2.9	0.5	0.5	7.0	0.06	0.005 B; 0.02 Mg; 0.03 Zr
Incoloy 901	N09901	12.5	42.5	...	6.0	...	...	2.7	...	36.2	0.10 max	...
Inconel 706	N09706	16.0	41.5	...	...	...	...	1.75	0.2	37.5	0.03	2.9 (Nb + Ta); 0.15 max Cu
Inconel 718	N07718	19.0	52.5	...	3.0	...	5.1	0.9	0.5	18.5	0.08 max	0.15 max Cu
Inconel 751	...	15.5	72.5	...	...	...	1.0	2.3	1.2	7.0	0.05	0.25 max Cu
Inconel X750	N07750	15.5	73.0	...	...	...	1.0	2.5	0.7	7.0	0.04	0.25 max Cu
M252	N07252	19.0	56.5	10.0	10.0	...	...	2.6	1.0	<0.75	0.15	0.005 B
Nimonic 80A	N07080	19.5	73.0	1.0	...	...	...	2.25	1.4	1.5	0.05	0.10 max Cu
Nimonic 90	N07090	19.5	55.5	18.0	...	...	...	2.4	1.4	1.5	0.06	...
Nimonic 95	...	19.5	53.5	18.0	...	...	...	2.9	2.0	5.0 max	0.15 max	+B; +Zr
Nimonic 100	...	11.0	56.0	20.0	5.0	...	...	1.5	5.0	2.0 max	0.30 max	+B; +Zr
Nimonic 105	...	15.0	54.0	20.0	5.0	...	...	1.2	4.7	...	0.08	0.005 B
Nimonic 115	...	15.0	55.0	15.0	4.0	...	...	4.0	5.0	1.0	0.20	0.04 Zr
Nimonic 263(b)	...	20.0	51.0	20.0	5.9	...	...	2.1	0.45	0.7 max	0.06	...
Pyromet 860	...	13.0	44.0	4.0	6.0	...	...	3.0	1.0	28.9	0.05	0.01 B
Refractory 26	...	18.0	38.0	20.0	3.2	...	...	2.6	0.2	16.0	0.03	0.015 B
René 41	N07041	19.0	55.0	11.0	10.0	...	...	3.1	1.5	<0.3	0.09	0.01 B
René 95	...	14.0	61.0	8.0	3.5	3.5	3.5	2.5	3.5	<0.3	0.16	0.01 B; 0.05 Zr
René 100	...	9.5	61.0	15.0	3.0	...	...	4.2	5.5	1.0 max	0.16	0.015 B; 0.06 Zr; 1.0 V
Udimet 500	N07500	19.0	48.0	19.0	4.0	...	...	3.0	3.0	4.0 max	0.08	0.005 B
Udimet 520	...	19.0	57.0	12.0	6.0	1.0	...	3.0	2.0	...	0.08	0.005 B
Udimet 630	...	17.0	50.0	...	3.0	3.0	6.5	1.0	0.7	18.0	0.04	0.004 B
Udimet 700	...	15.0	53.0	18.5	5.0	...	...	3.4	4.3	<1.0	0.07	0.03 B
Udimet 710	...	18.0	55.0	14.8	3.0	1.5	...	5.0	2.5	...	0.07	0.01 B
Unitemp AF2-1DA	...	12.0	59.0	10.0	3.0	6.0	...	3.0	4.6	<0.5	0.35	1.5 Ta; 0.015 B; 0.1 Zr
Waspaloy	N07001	19.5	57.0	13.5	4.3	...	...	3.0	1.4	2.0 max	0.07	0.006 B; 0.09 Zr

(a) No longer active, but shown here for reference. (b) Also known as Rolls Royce C-263.

Table 2 Oxide dispersion strengthened superalloys (ODS)

Alloy	Ni	Cr	Y_2O_3	Ti	Al	C	Fe	Others
Inconel MA 754	rem	20	0.6	0.5	0.3	0.05	...	
Incoloy MA 956	...	20	0.5	0.5	4.5	...	rem	
Inconel MA 6000E	rem	15	1.1	2.5	4.5	0.05	...	2 Mo, 4 W, 2 Ta, 0.15 Zr, 0.1 B
HDA 8077	rem	16	...	...	4.0	...	...	...
IN 738 + Y_2O_3	rem	16	1.3	3.4	3.4	0.17	...	1.7 Mo, 1.7 Ta, 2.6 W, 0.9 Nb, 8.5 Co, 0.1 Zr

Table 3 Role of elements on superalloys

Effect(a)	Fe-base	Co-base	Ni-base
Solid solution strengtheners	Cr, Mo	Nb, Cr, Mo, Ni, W, Ta	Co, Cr, Fe, Mo, W, Ta
Fcc matrix stabilizers	C, W, Ni	Ni	...
Carbide Form: MC type	Ti	Ti	W, Ta, Ti, Mo, Nb
M_7C_3 type	...	Cr	Cr
$M_{23}C_6$ type	Cr	Cr	Cr, Mo, W
M_6C type	Mo	Mo, W	Mo, W
Carbonitrides: M(C,N) type	C, N	C, N	C, N
Promotes general precipitation of carbides	P	...	...
Forms γ′ Ni_3 (Al, Ti)	Al, Ni, Ti	...	Al, Ti
Retards formation of hexagonal η (Ni_3Ti)	Al, Zr	...	...
Raises solvus temperature of γ′	...	...	Co
Hardening precipitates and/or intermetallics	Al, Ti, Nb	Al, Mo, Ti(b) W, Ta	Al, Ti, Nb
Oxidation resistance	Cr	Al, Cr	Al, Cr
Improve hot corrosion resistance	La, Y	La, Y, Th	La, Th
Sulfidation resistance	Cr	Cr	Cr
Improves creep properties	B	...	B
Increases rupture strength	B	B, Zr	B(c)
Causes grain boundary segregation	...	...	B, C, Zr
Facilitates working	...	Ni_3Ti	...

(a) Not all these effects necessarily occur in a given alloy. (b) Hardening by precipitation of Ni_3Ti also occurs if sufficient Ni is present. (c) If present in large amounts, borides are formed.

Table 4 Effects of elements in iron-base superalloys

Elements	Effects
Aluminum	Forms γ′ Ni_3(Al, Ti); retards formation of hexagonal η Ni_3Ti
Titanium	Forms γ′ Ni_3(Al, Ti) and MC carbides
Niobium, Tantalum	Forms body-centered tetragonal γ″ and MC carbides
Carbon	Forms MC, M_7C_3, $M_{23}C_6$ and M_6C carbides; stabilizes fcc matrix
Phosphorus	Promotes general precipitation of carbides
Nitrogen	Forms M(C,N) carbonitrides
Chromium	Oxidation resistance; solid solution strengthening
Molybdenum; Tungsten	Solid solution strengthening; forms M_6C carbides
Nickel	Stabilize fcc matrix; form γ′, inhibit information of deleterious phases
Boron; Zirconium	Improve creep properties; retard formation of grain boundary η Ni_3Ti
Lanthanum	Enhances oxidation resistance

parameter of the fcc matrix, which can influence the strengthening effect of coherent precipitates because the value of the lattice parameter directly affects coherency strains between matrix and precipitate. Oxidation resistance is provided by judicious use of elements such as chromium and manganese, whereas nickel additions retard formation of the generally undesirable tcp phases. Use of boron or lanthanum in relatively limited amounts can greatly improve elevated-temperature properties.

Iron-base alloys of most importance for application temperatures above 540 °C (1000 °F) are those with a fcc matrix, because a close-packed lattice is more resistant to time-dependent deformation processes. Under many circumstances, the microstructure-property interactions in fcc iron-base superalloys can be considered analogous to those in nickel-base superalloys.

Table 4 lists several elements used in iron-base alloys together with their nominal functions. These effects are not necessarily the same for each alloy, nor are they necessarily produced simultaneously in a specific alloy. Also, under some circumstances deleterious effects may result from a given element.

The most important class of iron-base alloys comprises those alloys that are strengthened by precipitation of intermetallic compounds in fcc matrixes. The most common precipitate is γ′, as typified by A-286, V-57 and Pyromet 860. On the other hand, the iron-containing alloy Inconel 718 (usually considered a nickel-base superalloy, but which contains 18.5% iron, see Table 1) is hardened by γ″. HNM and the CRMD series represent alloys hardened by carbides, nitrides and carbonitrides; elements such as tungsten and molybdenum also may be added to give solid-solution strengthening. Hastelloy X (generally considered a nickel-base alloy but which contains 10% iron) is an example of a third class of alloys, which are essentially solid-solution hardened, but which may derive some strengthening from carbide precipitation when the alloy is cold worked and aged.

Iron-base superalloys hardened by intermetallic-compound precipitation have found primary usage in gas-turbine engines as blades, disks, casings and fasteners. For example, some gas-turbine engines use A-286 forgings for turbine disks and Incoloy 901 for turbine hubs. A-286 also has been used for turbine cases. Alloys relying on precipitation of carbides have not been widely employed in the United States, but the CRM alloys such as 18D may eventually be used in applications such as automotive gas turbines.

Solid-solution-hardened Hastelloy X is used in sheet form for applications such as burner cans in gas-turbine engines.

Alloy N-155 was one of the first sheet alloys to be used for gas-turbine combustors, transition ducts and afterburners, and still is used for these applications on some engines where stress, temperature and environment permit. This alloy also has been used extensively for furnace hardware and for industrial fans that operate in high-temperature environments.

The new alloy Haynes 556 has a basic composition similar to that of N-155. Because its content of minor elements (<1% concentration) is more closely controlled, Haynes 556 has improved oxidation resistance, hot corrosion resistance, weldability, high-temperature ductility, and more resistance to thermal shock and thermal fatigue than the older N-155.

High temperature annealing has been used to improve the high-temperature mechanical properties of Incoloy 800. This alloy is also commercially available as Incoloy 800H, a variation in which carbon content and grain size are controlled to improve creep-rupture strength. Most specifications for Incoloy 800H require a minimum average grain size of ASTM No. 5 and a minimum carbon content of 0.05%.

Cobalt-Base Superalloys

Cobalt solid-solution alloys can be subdivided into three simple groups on the basis of use: (*a*) alloys for use primarily at temperatures from 650 to 1150 °C (1200 to 2100 °F), including Haynes 25, Haynes 188, Haynes 556, UMCo-50 and S-816; (*b*) fastener alloys MP-35N and MP-159, for use to about 650 °C; and (*c*) wear-resistant Stellite 6B.

All alloys in the heat treated and softened condition have fcc crystal structures; however, alloys MP-35N and MP-159 develop controlled amounts of cph structure during the thermomechanical processing recommended before service applications. Stellite 6B heat treated between 650 and 1050 °C (1200 and 1900 °F) and Haynes 25 exposed for 1000 h or more at temperatures near 650 °C may partly transform to a cph structure.

None of the cobalt-base superalloys is a complete solid-solution alloy, because all contain secondary phases: carbides (M_6C, $M_{23}C_6$, M_7C_3 or MC) or intermetallic compounds. Aging causes additional second-phase precipitation, which generally results in some loss of room-temperature ductility.

Of the high-temperature group, alloy S-816 originally was used extensively in turbochargers and gas-turbine wheels, blades and vanes, but has been largely replaced by higher-strength, lower-density nickel-base alloys with improved resistance to adverse environments.

Haynes 25 is perhaps the best-known wrought cobalt-base alloy and has been widely used for hot sections of gas turbines, for nuclear-reactor components, for surgical implants and, in the cold worked condition, for fasteners and wear pads.

Haynes 188 is an alloy specially designed for sheet-metal components, such as combustors and transition ducts, in gas turbines. The basic composition, provided that lanthanum, silicon, aluminum and manganese contents are judiciously controlled, provides excellent qualities such as oxidation resistance at temperatures up to 1100 °C (2000 °F), hot corrosion resistance, creep resistance, room-temperature formability, and ductility after long-term aging at service temperatures. Figure 1 compares the dynamic oxidation resistance and hot corrosion resistance of Haynes 188 with those of several other alloys that have been used for similar applications. In these charts, Haynes 25, Haynes 188 and N-155 are the only high-cobalt superalloys; all of the others are nickel-base superalloys. Haynes 188, like Haynes 25, MP-35N and MP-159, can be work hardened to relatively high hardness and tensile strength; after 50% cold reduction, Haynes 188 has a tensile strength of 1690 MPa (245 ksi) at room temperature and 1585 MPa (230 ksi) at 540 °C (1000 °F).

UMCo-50, which contains about 21% iron, is not as strong as Haynes 25 or Haynes 188. It is not used extensively in the U.S., and especially not in gas turbine applications. In Europe, on the other hand, it is one of the alloys used extensively for furnace parts and fixtures.

MP-35N and MP-159 were specifically designed to be work hardened, and both alloys have high strength and relatively high ductility in the work-hardened condition. The combination of high strength and high ductility in these alloys is attributed to the formation of small platelets of cph structure in the work-hardened fcc matrix. MP-35N and MP-159 are used predominantly for fasteners; MP-159 has an approximate upper limit on service temperature of 650 °C (1200 °F).

The last group of high-temperature cobalt alloys consists of a single alloy, Stellite 6B. This alloy is characterized by high hot hardness and relatively good resistance to oxidation. The latter property is derived chiefly from the high chromium content (about 30%), whereas the hot hardness is obtained through formation of complex carbides of the Cr_7C_3 and $M_{23}C_6$ types. Stellite 6B is widely used for erosion shields in steam turbines, for wear pads in gas turbines and for bends in tube systems carrying particulate matter at high temperatures and high velocities.

Nickel-Base Superalloys

Nickel-base alloys generally have greater resistance to high temperatures than low-alloy steels and stainless steels. Nickel-base heat-resistant alloys contain 30 to 75% nickel and up to 30% chromium. Iron contents range from relatively small amounts in most Inconels, Nimonics and Hastelloys to about 35% in alloys such as Incoloy 901 and Inconel 706. Many nickel-base alloys contain small amounts of aluminum, titanium, niobium, molybdenum, and tungsten to enhance either strength or corrosion resistance.

The combination of nickel and chromium gives these alloys outstanding oxidation resistance. As a class, nickel-base superalloys exceed stainless steels in mechanical strength, especially at temperatures above 650 °C (1200 °F).

Heat-resistant nickel alloys are used in a wide range of applications that require high strength or resistance to oxidation and corrosion. Solid-solution alloys such as Inconel 600, Inconel 601, and RA 333 are frequently used for furnace parts and other heat treating equipment. Those alloys are also used in high-temperature chemical-processing equipment such as hydrocarbon reformers and crackers.

Power generation is another field in which nickel-base superalloys are used extensively. In nuclear power plants, applications include steam-generator tubing and structural components of reactor cores. In fossil-fueled plants, these alloys are used for superheater tubing, ash-handling systems, stack-gas scrubbers, and other parts that must resist heat or corrosion, or both.

Solid-solution nickel alloys normally are used in the annealed temper, the annealing treatment depending on the properties required by the application. In general, a relatively low annealing temperature of 870 to 980 °C (1600 to 1800 °F) is used to produce the highest tensile and fatigue strengths. A high-temperature anneal at about 1120 to 1200 °C (2050 to 2200 °F) usually produces optimum fatigue resistance and creep-rupture properties at service temperatures greater than 600 °C (1100 °F). High-temperature annealing has been used to improve the high-temperature mechanical properties of Incoloy 800. This alloy is also commercially available as Incoloy 800H, a variation in which carbon content and grain size are controlled to improve creep-rupture strength. Most specifications for Incoloy 800H require a minimum average grain size of ASTM No. 5 and a minimum carbon content of 0.05%.

Some solid-solution nickel alloys, notably Inconel 601, Inconel 617 and Inconel 625, are used in aerospace applications. Inconel 601 is used for jet-engine igniters and for gas-turbine components such as combustion-can liners and diffuser assemblies. The high stress-rupture strength of Inconel 617 at temperatures higher than 980 °C (1800 °F) makes it a useful material for various components of aircraft gas turbines. Inconel 625 is used for heat shields, aircraft ducting systems, exhaust systems, thrust reversers, fuel and hydraulic lines, spray bars and turbine shroud rings.

Precipitation-hardening nickel alloys contain aluminum, titanium or niobium to cause precipitation of a second phase during appropriate heat treatment. The precipitated phase, usually gamma prime or gamma double prime, substantially increases the strength and hardness of the alloy. For example, Inconel X-750, a precipitation-hardening version of Inconel 600, has a yield strength about three times that of Inconel 600 at 540 °C (1000 °F).

Precipitation-hardening Nimonic 80A shows a similar increase over its solid-solution counterpart, Nimonic 75.

Heat treatments for the precipitation-hardening alloys generally consist of a solution treatment at 970 to 1175 °C (1700 to 2150 °F) followed by one or more precipitation treatments at 600 to 815 °C (1100 to 1500 °F).

Most of these alloys utilize aluminum and titanium to promote precipitation hardening. In some alloys, niobium is used along with lesser amounts of aluminum and titanium. The niobium-hardened alloys, Inconel 718 for example, exhibit delayed responses to precipitation-hardening temperatures. Such alloys have significantly better weldability because the heat of welding does not induce hardening and consequent postweld cracking.

Incoloy 903, a precipitation-hardening iron-nickel-cobalt alloy, is used for its constant low coefficient of thermal expansion in addition to its high strength. The alloy typically exhibits a coefficient of expansion of about 7.2 μm/m·K (4.0 μin./in.·°F) from room temperature to around 425 °C (800 °F). This characteristic is useful in applications such as jet-engine parts that require close operating clearances over a range of temperatures.

Aerospace applications constitute the greatest field of use for precipitation-hardening superalloys. They are used in rocket engines and for such aircraft gas-turbine parts as blades, disks, rings, shafts, and various compressor and diffuser components. Other applications include bolts and springs in nuclear reactors.

Applications for precipitation-hardening alloys frequently involve forged components. Consequently, mechanical properties in the forged plus heat treated condition are often important. Also, because statistical methods frequently must be used to determine the reliability of a given part, statistical distributions of properties are of considerable help in designing parts, particularly critical aircraft parts.

Processing

The austenitic high-nickel alloys and nickel-base solid-solution alloys are generally electric-furnace melted. Inconels, Incoloys and RA alloys are electric-furnace melted, followed by argon-oxygen decarburization (AOD processing). Hastelloys generally are electric-furnace melted followed by electroslag remelting, although vacuum-induction melting followed by electroslag melting also may be employed. Some of the other alloys, such as Carpenter 20Cb-3, may be simply electric-arc-furnace melted.

Precipitation-hardening iron-base alloys are electric-furnace or vacuum-induction melted followed by vacuum-arc or electroslag remelting. Double vacuum-induction melting may be employed where critical applications are involved. In general, vacuum melting improves structure-sensitive properties such as fatigue strength, ductility and impact strength; its effect on tensile strength is negligible.

The precipitation-hardening nickel-base alloys generally are double vacuum-induction melted, followed by vacuum-arc or electroslag remelting. These alloys are usually employed in the most demanding applications with respect to temperature and stress, and their complicated processing reflects this fact. Vacuum-induction melting lowers the gas content (hydrogen, nitrogen and oxygen) and evaporates trace elements (such as Pb, Bi, Cd, Te, As, Sb, Ti and Se), whose presence can adversely affect mechanical properties and workability of ingots. Several of these elements, even in trace amounts, can adversely affect service life. Secondary remelting using vacuum-arc and electroslag processes further refines the metal by eliminating gases, nonmetallic and metallic impurities, and inclusions. Secondary remelting also produces larger ingots of uniform composition and dense homogeneous structure. Electroslag remelting is done by arc melting under a cover of slag, and removal of sulfur is quite easy compared with vacuum melting processes. High sulfur levels are injurious to the workability and properties of many wrought heat-resisting alloys, particularly high-nickel alloys.

Several of the precipitation-hardening nickel-base alloys are produced by powder metallurgy processes. Two processes are of commercial importance. The first is the rotating-electrode method, in which a consumable electrode is melted while being rotated at very high speed in a helium-filled chamber. The second method is inert-gas atomization, in which a prealloyed vacuum-cast ingot is vacuum remelted, and then atomized by a stream of high-purity argon gas. The objective in both cases is to produce powder of specified particle size (mesh) with an oxygen content not exceeding about

100 ppm. Billet is produced by (*a*) canning, evacuation, heating and extrusion, (*b*) canning followed by evacuation and hot isostatic pressing, and (*c*) canning into a glass or metallic container with the shape of the finished component followed by evacuation and hot isostatic pressing. The advantages of powder metallurgy for processing of superalloys are as follows:

- Rapid cooling of small powder particles eliminates segregation, hence the consolidated material is much more homogeneous than conventionally cast ingots.
- The fine grain structures improve hot workability and permit hot working of high titanium-plus-aluminum compositions, which otherwise would have to be used in the as-cast state.
- Powder metallurgy products may have superior tensile and fatigue strength at temperatures of 650 to 815 °C (1200 to 1500 °F) compared to conventionally processed superalloys.
- Powder metallurgy permits the production of components with dimensions and shape approaching those of the finished product, thus resulting in considerable increase in material yield and considerable savings in machining costs.
- Because gross ingot segregation can be removed, greater levels of strengthening elements can be added before deleterious phases are formed, which permits development of new alloys by P/M.

Fairly recent P/M processing techniques that have resulted in improved materials are (*a*) rapid solidification processing (RSP) of powders, (*b*) mechanical alloying of powders, (*c*) thermomechanical treatment (TMT), (*d*) thermoplastic (T/P) treatment, (*e*) Gatorizing and (*f*) hot isostatic pressing (HIP) of powder.

By RSP, unusual structures, as well as dispersions of fine intermetallics, are obtained. The very high solidification rates permit quite large amounts of elements to be held in supersaturated solution, and subsequent precipitation sequences and aging kinetics have been considerably altered, thus increasing the possibility of producing new alloys. Cooling rates in the range of 3×10^4 °C/s to 8×10^5 °C/s have been achieved with IN-100 powder.

Mechanical alloying has been very useful, particularly in production of ODS alloys, but it has been used to produce other types of alloys as well. In mechanical alloying, constituent powders are literally beaten into a homogeneous material.

Thermoplastic processing consists essentially of cold working the prealloyed powder, using techniques such as attrition and rolling, prior to compaction. On heating for compaction, the cold work in the powder causes recrystallization to a very fine grain size, which greatly reduces the flow strength of the material at normal superalloy hot working temperatures. The very fine grain sizes (1 to 1.5 μm) generated by HIP compaction of T/P processed material result in material that behaves superplastically.

In the Gatorizing process, alloy billets are worked to obtain enhanced ductility; then isothermally forged at temperatures of 1000 to 1200 °C (1830 to 2190 °F) and strain rates of about 5 m/min; and, finally, given suitable heat treatments to develop required properties.

HIP of prealloyed powders has produced strength levels comparable to

Table 5 Powder metallurgy processes applicable to superalloys

Alloy(s) and powder type	Consolidation route	Properties evaluated	Conclusions
MAR-M 200: argon atomized	HIP only	Hot working response-upset test	Intermediate extrusion step not necessary for superplastic response: can achieve superplasticity during HIP
IN-100: argon atomized	HIP + forge or HIP + extrude	Tensile; stress rupture; creep; low-cycle fatigue	Properties as good as those of the best conventional cast and wrought forgings
IN-100, IN-792, IN-744: argon atomized	HIP; extrusion; hot press + forge; HIP + isothermal forging; HIP + forge	Tensile; creep; stress-rupture	Workability of powders enhanced by prior T/P (thermoplastic processing); acceptable mechanical properties
Astroloy: argon atomized	Hot compaction + extrusion + forge; HIP + forge	Tensile; creep; creep-rupture; low-cycle fatigue	Higher mechanical properties than cast and extruded material
Low-carbon Astroloy: argon atomized and soluble gas (hydrogen) technique	HIP only	Tensile properties as a function of temperature	Strengths better than HIP + forged Astroloy, but not as good as conventional cast and wrought materials
Low-carbon Astroloy: soluble gas technique	HIP only	Tensile; fatigue; stress-rupture	Tensile and fatigue behavior equivalent to cast and wrought conditions; notched-bar stress-rupture test recommended to evaluate material integrity; high consolidation pressures (greater than 105 MPa or 15 ksi) required
Co-base Ni-Cr-Mo-C; Co-base Cr-Mo-Ti; nitrogen atomized or rotating electrode	Hot extrusion + hot rolling or hot swaging	Tensile properties as function of temperature	Strength levels achieved warrant further study

those of "coventional" forged alloys at reduced costs and has greatly enhanced fatigue resistance.

Advanced P/M processing techniques now applied to superalloys are expensive, but the expense can be justified in many instances because of either of two potential benefits: improvement in properties or substantial reduction in cost. Properties frequently can be improved through production of segregation-free billets, which is aided by many of the processes discussed in the previous paragraphs. The ability to produce segregation-free billets, in turn, enables advanced forging techniques and other specialized forms of thermomechanical treatment to achieve dramatic improvements in properties. Development of HIP and powder-canning techniques has made possible the production of "near-net-shape" components—that is, components requiring little or no machining to bring them to finished size and shape. Ingots and billets of homogeneous material are extruded, rolled, drawn, swaged, forged or otherwise compacted to near net shape, and often the resultant savings in both machining time and scrap loss more than compensate for the cost of the specialized preliminary processing. In addition, finished components often are more reliable than they would have been if processed by conventional techniques, and this improved reliability yields both design and economic benefits.

As indicated earlier, incorporation of a fine uniform dispersion of inert oxide particles in a metallic matrix frequently can lead to improved mechanical properties and surface stability at elevated temperatures. When materials such as ODS NiCrAl become available in sheet form, they are expected to have significantly higher temperature capabilities than conventional sheet alloys. In addition to solutions for the problems involved in developing a sheet production process, an acceptable degree of anisotropy and appropriate joining techniques that will not seriously degrade temperature capability also must be developed.

Table 5 describes some of the P/M processing that has been investigated for superalloys.

Table 6 compares data for HIP IN 100 and Gatorized IN 100, and Table 7 compares high-temperature properties of Udimet 700 and IN 738 for three conditions: ingot, cast and wrought; P/M, as-extruded; and P/M, grain coarsened.

Table 6 Typical properties comparison for as-HIP IN-100 and Gatorized IN-100

Property	Gatorized	As-HIP
Tensile strength at 21 °C (70 °F)		
Tensile strength, MPa (ksi)	1610 (233)	1490 (216)
Yield strength, MPa (ksi)	1120 (162)	1040 (151)
Ductility, %	26	10
Tensile strength at 704 °C (1300 °F)		
Tensile strength, MPa (ksi)	1225 (177)	1085 (157)
Yield strength, MPa (ksi)	1175 (170)	1010 (146)
Ductility, %	28	9
Stress-rupture at 732 °C (1350 °F) and 748 MPa		
Life, h	35	10
Elongation, %	12	Notched failure

Table 7 Typical hot tensile properties

Condition	Tensile strength MPa	ksi	Elongation, %	Secondary creep rates(a), %/h
Udimet 700 tested at 550 °C (1020 °F)				
Wrought	1280	185	...	...
P/M, as extruded	1400	203	30	...
P/M, grain coarsened	1190	172	25	...
Udimet 700 tested at 950 °C (1740 °F)				
Wrought	...	...	...	...
P/M, as extruded	428	62	19	187
P/M, grain coarsened	435	63	27	3.3
IN-738 tested at 550 °C (1020 °F)				
Wrought	1060	154	...	...
P/M, as extruded	1490	216	31	...
P/M, grain coarsened	1150	167	19	...
IN-738 tested at 950 °C (1740 °F)				
Wrought	...	...	...	...
P/M, as extruded	350	51	10	5400
P/M, grain coarsened	430	62	4	19.8

(a) At stress rate of 100 MPa (14.5 ksi).

Product Forms

Wrought heat-resistant alloys are manufactured in all mill forms common to the metal industry. Iron-base, cobalt-base and nickel-base superalloys are produced conventionally as bar, billet, extrusions, plate, sheet, strip, wire and forgings by primary mills. Inconels and Hastelloys also are available as rod, bar, plate, sheet, strip, tube, pipe, shapes, wire, forging stock and specialty items from secondary converters. An important application for P/M techniques has been HIP production of turbine disks made of IN-100, René 95 and Astroloy.

P/M Hastelloy X sheet has given good service in combustion components because of good resistance to oxidation and thermal fatigue, and alloys such as Haynes 188 and Nimonic 86 show good potential. Nevertheless, when they become available in sheet form, materials such as ODS NiCrAl potentially have a significantly higher temperature capability than conventional sheet alloys. Reproducible sheet processing procedures and joining techniques that will not seriously degrade the temperature capabilities of the alloys still have to be developed. The corrosion resistance (oxidation and hot corrosion) of a number of alloys was shown in Fig. 1. Figure 4 shows the criterion used to make the evaluation in laboratory and service samples.

Table 8 Physical properties of selected superalloys

Alloy	Density, Mg/m³	Melting temperatures Liquidus °C	Liquidus °F	Solidus °C	Solidus °F	Specific heat (a) J/kg	Specific heat (a) Btu/lb	Electrical conductivity, % IACS	Electric resistivity, nΩ·m	Magnetic permeability	Curie temperature °C	Curie temperature °F
Iron-base alloys												
Carpenter 20-Cb3	8.055	1425	2600	1370	2500	...	...	...	1040	...	...	...
Haynes 556	8.23	...	...	...	...	472	0.113	...	970	...	...	...
Incoloy 800	7.94	1385	2525	1355	2475	502	0.117	1.7	989	1.0092	...	...
Incoloy 801	7.94	1385	2525	1355	2475	452	0.105	1.7	1012	...	...	...
Incoloy 825	8.14	1400	2500	1370	2500	...	...	1.5	1127	1.005	< −196	< −520
Cobalt-base alloys												
Haynes 25 (L-605)	9.13	1410	2570	1329	2425	374	0.090	...	890	<1.00	...	...
Haynes 188	9.13	1398	2550	1302-1330	2375-2425	423(b)	0.101(b)	...	...	1.01	...	...
Stellite 6B	8.38	1354	2470	1265	2310	421	0.101	...	910	<1.2	...	...
UMCo 50	8.05	1395	2540	1380	2515			...	825	...	...	...
Nickel-base alloys												
Hastelloy B-2	9.21	...	...	...	...	389(b)	0.093(b)	...	1380(b)	...	...	...
Hastelloy C-4	8.64	...	...	...	...	426(b)	0.102(b)	...	1250	...	...	...
Hastelloy C-276	8.90	1371	2500	1323	2415	427	0.102	...	1330	...	...	...
Hastelloy N	8.93	...	...	...	...	419(b)	0.100(b)	...	1200(b)	...	...	...
Hastelloy S	8.76	1380	2516	1335	2435	427(b)	0.102(b)	...	...	...	...	...
Hastelloy W	9.03	1315	2400	...	...	...	...	...	...	...	...	...
Hastelloy X	8.23	1290	2350	1250	2280	486	0.116	...	1180	<1.002(c)	...	...
Inconel 600	8.42	1415	2575	1354	2470	444	0.103	1.7	1030	1.010	−124	−192
Inconel 617	...	...	...	1333	2430	...	...	...	...	...	...	...
Inconel 625	8.44	1350	2460	1290	2350	410	0.095	1.3	1290	1.006	−196	−320
Inconel 671	7.86	1350	2460	1305	2385	456	0.106	2.0	869	...	...	...
Inconel 690	8.03	1375	2510	1345	2450	...	...	1.5	148	...	...	...
Inconel X750	8.25	1425	2600	1393	2540	431	0.103	...	1215	1.0020	−143	−225
Nimonic 75	...	...	...	1380	2515	...	...	...	...	...	...	...
Nimonic 80A	...	...	...	1360	2480	...	...	...	...	...	...	...
Nimonic 90	...	...	...	1310	2390	...	...	...	...	...	...	...
Nimonic 100	...	...	...	...	...	...	...	...	...	...	...	...
Nimonic 105	...	...	...	1290	2256	...	...	...	...	...	...	...
René 41	8.25	1371	2500	1232	2250	452	0.108	...	1308	1.002	...	...
Udimet 500	8.14	1345	2450	1260	2300	...	...	...	1203	...	...	...
Udimet 700	7.92	1345	2450	1216	2220	...	...	...	...	...	...	...
Waspaloy	8.20	1355	2475	1339	2425	523(c)	0.125(c)	...	1240	...	...	...

(a) At room temperature. (b) At 100 °C (212 °F). (c) At 93 °C (200 °F).

Fig. 4 Schematic of metallographic technique for measuring hot corrosion damage

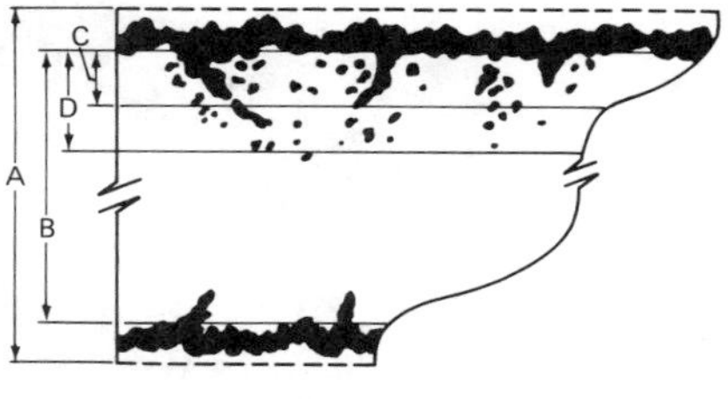

Metal loss per side = $(A\text{-}B)/2$
Continuous penetration per side = C
Maximum penetration per side = D
Total metal affected per side = $[(A\text{-}B)/2] + D$

Selection Based on Properties

The properties of individual commercial superalloys are given in the section of this volume immediately following this article. Table 8 summarizes physical properties of some alloys; Table 9 compares selected mechanical properties, and Table 10 compares rupture strengths. Of the mechanical properties, fatigue strength is probably of greatest importance; it has been estimated that about 90% of all engineering structures fail because of fatigue.

Figure 2 shows relative values of creep strength and hot corrosion resistance for some nickel-base superalloys between 700 and 900 °C, and shows the alloy temperature capabilities based on 10 000-h life at 120 MPa (17.4 ksi) under hot corrosion conditions. Many of the stress-rupture and creep data given in the data compilations following this article are typical or average values.

There is a spread above and below these values caused by differences among heats of metal, methods of processing, and variations in conducting standard tests that must be evaluated statistically whenever it is necessary to predict reliability of a new component or to design a component for a specified level of reliability.

For alloys that form thin tenacious scales at elevated temperature, the stress-rupture properties of bar and sheet of the same alloy will be about the same. On the other hand, for alloys

Table 9 Typical mechanical properties of cobalt-base and nickel-base superalloys

Temperature		Tensile strength		Yield strength		Elongation,
°C	°F	MPa	ksi	MPa	ksi	%
Cobalt-base alloys						
Haynes 25 (L-605) sheet						
21	70	1010	146	460	67	64
540	1000	800	116	250	36	59
650	1200	710	103	240	35	35
760	1400	455	66	260	38	12
870	1600	325	47	240	35	30
Haynes 188, sheet						
21	70	960	139	485	70	56
540	1000	740	107	305	44	70
650	1200	710	103	305	44	61
760	1400	635	92	290	42	43
870	1600	420	61	260	38	73
S-816, bar						
21	70	965	140	385	56	30
540	1000	840	122	310	45	27
650	1200	765	111	305	44	25
760	1400	650	94	285	41	21
870	1600	360	52	240	35	16
Nickel-base alloys						
Astroloy, bar						
21	70	1410	205	1050	152	16
540	1000	1240	180	965	140	16
650	1200	1310	190	965	140	18
760	1400	1160	168	910	132	21
870	1600	770	112	690	100	25
D-979, bar						
21	70	1410	204	1010	146	15
540	1000	1300	188	925	134	15
650	1200	1100	160	890	129	21
760	1400	7	104	655	95	17
870	1600	345	50	305	44	18
Hastelloy X, sheet						
21	70	785	114	360	52	43
540	1000	650	94	290	42	45
650	1200	570	83	275	40	37
760	1400	435	63	260	38	37
870	1600	255	37	180	26	50
IN-102, bar						
21	70	960	139	505	73	47
540	1000	825	120	400	58	48
650	1200	710	103	400	58	64
760	1400	440	64	385	56	110
870	1600	215	31	200	29	110
Inconel 600, bar						
21	70	620	90	250	36	47
540	1000	580	84	195	28	47
650	1200	450	65	180	26	39
760	1400	185	27	115	17	46
870	1600	105	15	62	9	80

(continued)

Table 9 (continued)

Temperature °C	°F	Tensile strength MPa	ksi	Yield strength MPa	ksi	Elongation, %
Inconel 601, sheet						
21	70	740	107	340	49	45
540	1000	725	105	150	22	38
650	1200	525	76	180	26	45
760	1400	290	42	200	29	73
870	1600	160	23	140	20	92
Inconel 625, bar						
21	70	855	124	490	71	50
540	1000	745	108	405	59	50
650	1200	710	103	420	61	35
760	1400	505	73	420	61	42
870	1600	285	41	475	40	125
Inconel 706, bar						
21	70	1300	188	980	142	19
540	1000	1120	163	895	130	19
650	1200	1010	147	825	120	21
760	1400	690	100	675	98	32
Inconel 718, bar						
21	70	1430	208	1190	172	21
540	1000	1280	185	1060	154	18
650	1200	1230	178	1020	148	19
760	1400	950	138	740	107	25
870	1600	340	49	330	48	88
Inconel 718, sheet						
21	70	1280	185	1050	153	22
540	1000	1140	166	945	137	26
650	1200	1030	150	870	126	15
760	1400	675	98	625	91	8
Inconel X 750, bar						
21	70	1120	162	635	92	24
540	1000	965	140	580	84	22
650	1200	825	120	565	82	9
760	1400	485	70	455	66	9
870	1600	235	34	165	24	47
M-252, bar						
21	70	1240	180	840	122	16
540	1000	1230	178	765	111	15
650	1200	1160	168	745	108	11
760	1400	945	137	715	104	10
870	1600	510	74	485	70	18
Nimonic 75, bar						
21	70	750	109	...	...	41
540	1000	635	92	...	...	41
650	1200	538	78	...	...	42
760	1400	290	42	...	...	70
870	1600	145	21	...	...	68
Nimonic 80A, bar						
21	70	1240	179	620	90	24
540	1000	1100	160	530	77	24
650	1200	1000	145	550	80	18
760	1400	760	110	505	73	20
870	1600	400	58	260	38	34

(continued)

that are less resistant to oxidation, rupture values are likely to be significantly lower for sheet than for the same alloy in bar form.

Short-time tensile and stress-rupture properties for several different product forms of Haynes 25 (L-605) are compared in Table 11. Some alloys are more sensitive than others to differences in form. Regardless of alloy, the difference in properties caused by product form is likely to be more significant for stress-rupture properties than for short-time tensile properties.

The data in Table 11 are from different heats of metal and are also affected by the different fabrication procedures for the products and by the shape of the test specimens, but they serve to illustrate the general relationship among product forms for a typical oxidation-resistant alloy.

The heat-resistant alloys are cold worked by hammering, swaging and rolling. Cold worked products such as bolts and studs usually are stress relieved but, when tested at elevated temperature, retain a portion of the additional tensile and yield strengths imparted by the cold reduction.

Short-time tensile properties for the cobalt-base alloy Haynes 25 in the form of sheet 1.4 mm (0.054 in.) thick, solution treated at 1220 °C (2225 °F), showed a gradual and significant increase in strength as the amount of subsequent cold reduction was increased. The rupture properties were not improved (Table 12).

REFERENCES

1. *The Superalloys:* J. Wiley, New York, 1972
2. *The Nimonic Alloys:* Crane, Russak and Co., New York, 1974
3. *Cobalt Superalloys—1970:* Centre d'Information, Brussels, Belgium, 1970
4. *High Temperature Alloys for Gas Turbines:* Applied Science Publishers Ltd., London, 1978
5. *Source Book on Materials for Elevated Temperature Applications:* American Society for Metals, Metals Park, OH, 1979
6. *Materials and Coatings to Resist High Temperature Corrosion,* edited by D. R. Holmes: Applied Scientific Publications, Essex, England, 1978

Table 9 (continued)

Temperature °C	°F	Tensile strength MPa	ksi	Yield strength MPa	ksi	Elongation, %
Nimonic 90, bar						
21	70	1240	180	805	117	23
540	1000	1100	160	725	105	23
650	1200	1030	150	685	99	20
760	1400	825	120	540	78	10
870	1600	430	62	260	38	16
Nimonic 105, bar						
21	70	1140	166	815	118	12
540	1000	1100	160	775	112	18
650	1200	1080	156	800	116	24
760	1400	965	140	655	95	22
870	1600	605	88	365	53	25
Nimonic 115, bar						
21	70	1240	180	860	125	25
540	1000	1090	158	795	115	26
650	1200	1120	163	815	118	25
760	1400	1080	157	800	116	22
870	1600	825	120	550	80	18
Pyromet 860, bar						
21	70	1300	188	835	121	22
540	1000	1250	182	840	122	15
650	1200	1110	161	850	123	17
760	1400	910	132	835	121	18
René 41, bar						
21	70	1420	206	1060	154	14
540	1000	1400	203	1010	147	14
650	1200	1340	194	1000	145	14
760	1400	1100	160	940	136	11
870	1600	620	90	550	80	19
René 95, bar						
21	70	1620	235	1310	190	15
540	1000	1540	224	1250	182	12
650	1200	1460	212	1220	177	14
760	1400	1170	170	1100	160	15
Udimet 500, bar						
21	70	1310	190	840	122	32
540	1000	1240	180	795	115	28
650	1200	1210	176	760	110	28
760	1400	1040	151	730	106	39
870	1600	640	93	495	72	20
Udimet 520, bar						
21	70	1310	190	860	125	21
540	1000	1240	180	825	120	20
650	1200	1170	170	795	115	17
760	1400	725	105	725	105	15
870	1600	515	75	515	75	20
Udimet 700, bar						
21	70	1410	204	965	140	17
540	1000	1280	185	895	130	16
650	1200	1240	180	855	124	16
760	1400	1030	150	825	120	20
870	1600	690	100	635	92	27

(continued)

Appendix 1 Overheating, Creep, and Alloy Stability

By Matthew J. Donachie, Jr.
Pratt & Whitney Aircraft

Heat-resistant alloys normally respond to heat treatment, and thus exposure of these alloys to elevated temperatures, with or without stress, can cause microstructural changes and resultant changes in properties. Generally, the higher the exposure temperature, the more rapid the structural change. As exposure temperature decreases, the type of microstructural degradation may change. At the highest exposure temperatures, an alloy may be subject to incipient melting. In addition, oxidation and surface corrosion will take place at all temperatures for which these alloys are normally specified.

This Appendix deals principally with the effects of microstructural changes, melting and corrosion on nickel-base and cobalt-base superalloys at temperatures above about 725 °C (1340 °F). It also touches briefly on microstructural changes affecting nickel-base, iron-base and iron-nickel superalloys, stainless steels and low alloy steels at temperatures below 700 °C (1300 °F). Accelerated oxidation (hot corrosion) induced by salts in marine and industrial environments at temperatures below 700 °C (1300 °F) is not considered. When times or temperatures exceed normal test or operating levels, alloys are exposed to substantially different operating environments from those ordinarily experienced, and may behave in a manner that could not have been predicted from normal test data. Additional information on behavior at elevated temperature may be found in failure analysis studies described in Volume 10 of the 8th Edition of this Handbook, as well as in those portions of the articles in this volume on superalloys and iron-base heat-resistant alloys that discuss creep, stress-rupture, high-temperature strength and corrosion resistance. Microstructural aspects of aging and exposure are described briefly in Volume 7 of the 8th Edition. For more detail, consult Ref 1 to 8.

Overheating

Overheating in the broad sense consists of exposing a metal to excessively high temperatures for short periods of

Table 9 (continued)

Temperature °C	°F	Tensile strength MPa	ksi	Yield strength MPa	ksi	Elongation, %
Udimet 710, bar						
21	70	1190	172	910	132	7
540	1000	1150	167	850	123	10
650	1200	1290	187	860	125	15
760	1400	1020	148	815	118	25
870	1600	705	102	635	92	29
Unitemp AF2-1DA, bar						
21	70	1290	187	1050	152	10
540	1000	1340	194	1080	157	13
650	1200	1360	197	1080	157	13
760	1400	1150	167	1010	146	8
870	1600	830	120	715	104	8
Waspaloy, bar						
21	70	1280	185	795	115	25
540	1000	1170	170	725	105	23
650	1200	1120	162	690	100	34
760	1400	795	115	675	98	28
870	1600	525	76	515	75	35

Table 10 Typical rupture strengths of selected superalloys

Temperature °C	°F	For stress rupture at: 100 h MPa	ksi	1000 h MPa	ksi
Incoloy 800					
650	1200	220	32	145	21
760	1400	115	17	69	10
870	1600	45	6.5	33	4.8
Incoloy 801					
650	1200	250	36	...	...
730	1350	145	21	...	...
815	1500	62	9	...	...
Incoloy 802					
650	1200	240	35	170	24
760	1400	145	21	105	15
870	1600	97	14	62	9
Inconel 600					
815	1500	55	8	39	5.6
870	1600	37	5.3	24	3.5
Inconel 601(a)					
540	1000	...	...	400	58
870	1600	48	7	30	4.3
980	1800	23	3.4	14	2.1
Inconel 617(b)					
815	1500	140	20	97	14
925	1700	62	9	...	5.5
980	1800	41	6	...	3.5

(continued)

(a) Solution treated 1150 °C (2100 °F). (b) Solution treated 1175 °C (2150 °F). (c) Heat treated to 980 °C (1800 °F) plus 720 °C (1325 °F) hold for 8 h, F.C. to 620 °C (1150 °F) hold for 8 h. (d) 730 °C (1350 °F) hold for 2 h. (e) Heat treat to 1150 °C (2100 °F) plus 840 °C (1550 °F) hold for 24 h, plus 705 °C (1300 °F) hold for 20 h. (f) Solution treated and aged. (g) Stress-relieved forging. (h) Heat treat to 1050 °C (1922 °F) hold for 1 h. (j) Heat treat to 1080 °C (1976 °F) hold for 8 h, plus 700 °C (1290 °F) hold for 16 h. (k) Heat treat to 1150 °C (2100 °F) hold for 4 h, plus 1050 °C (1920 °F) hold for 16 h, plus 850 °C (1560 °F) hold for 16 h. (m) Heat treat to 1190 °C (2175 °F) hold for 1.5 h, plus 1100 °C (2010 °F) hold for 6 h. (n) Heat treat to 1150 °C (2100 °F) hold for 2 h, W.Q., plus 800 °C (1475 °F) hold for 8 h.

time. Allowable metal temperatures for wrought heat-resistant alloys in structural applications generally do not exceed about 950 °C (1740 °F). In applications where the component does not bear a load, allowable temperatures may exceed 1200 °C (2200 °F). In general, any temperature can be considered to be in the overheating range when it (*a*) causes melting, (*b*) causes strengthening phases to dissolve in the matrix, or (*c*) causes extensive oxidation or corrosion. Results of overheating depend on the maximum temperature reached by the metal.

Nickel-base and cobalt-base superalloys generally have incipient melting temperatures above 1200 °C (2200 °F). Table 13 gives melting temperatures for some representative wrought superalloys. Figure 5 shows the microstructures of two typical wrought superalloys before and after incipient melting. Incipient melting reduces grain-boundary strength and ductility. It thus reduces alloy rupture capabilities. Once an alloy has exceeded its incipient melting point, normal properties cannot be restored by any known heat treatment. In cast alloys, incipient melting may occur at temperatures substantially below the temperatures predicted from alloy composition. This behavior results from alloy segregation to grain boundaries and interdendritic areas during solidification. In wrought alloys, incipient melting takes place at a temperature much closer to the general alloy melting temperature (solidus), and overheating actually may cause significant portions of the structure to melt.

Overheating may deplete alloying elements that provide oxidation resistance. Oxidation processes are thus accelerated on return of the alloy to normal operating temperatures. Even if prolonged exposure to overtemperature does not result in mechanical failure, it frequently will cause excessive surface corrosion. This, in turn, can reduce the strength of the alloy. Even if an alloy is coated for oxidation resistance, the coating will be degraded by surface attack and eventually mechanical properties of the alloy will be affected. The alloy under a coating may also degrade rapidly because of excessive interdiffusion of alloying elements, if temperatures exceed the normal operating range. As a rule of thumb, alloys used in structural applications should not be exposed at temperatures within about 125 °C (225 °F) of their incipient melting temperatures. The strength

Table 10 (continued)

Temperature °C	°F	For stress rupture at: 100 h MPa	100 h ksi	1000 h MPa	1000 h ksi
Inconel 625(a)					
650	1200	440	64	370	54
815	1500	130	19	93	13.5
870	1600	72	10.5	48	7
Inconel 718(c)					
540	1000	...	...	951	138
595	1100	860	125	760	110
650	1200	690	100	585	85
Inconel 751(d)					
815	1500	200	29	125	185
870	1600	120	175	69	10
Inconel X750(e)					
540	1000	...	...	827	120
870	1600	83	12	45	6.5
925	1700	58	8.4	21	3.1
N-155, bar(f)					
650	1200	360	52	295	43
730	1350	195	28	150	22
870	1600	97	14	66	9.5
N-155(g)					
650	1200	380	55	290	42
N-155, sheet(f)					
980	1800	39	5.6	20	2.9
Nimonic 75(h)					
815	1500	38	5.5	24	3.5
870	1600	23	3.4	15	2.2
925	1700	14	2.1	10	1.5
980	1800	...	...	7.6	1.1
Nimonic 80A(j)					
540	1000	...	...	825	120
815	1500	185	27	115	17
870	1600	105	15	...	...
Nimonic 90(j)					
815	1500	240	35	155	22.5
870	1600	150	22	69	10
925	1700	69	10	...	...
Nimonic 105(k)					
815	1500	325	47	225	32
870	1600	210	30.2	135	19
Nimonic 115(m)					
815	1500	425	62	315	46
870	1600	315	46	205	30
925	1700	205	30	130	18.5
Nimonic 263(n)					
815	1500	170	24.5	105	15
870	1600	93	13.5	46	6.7
925	1700	45	6.5	...	...

(a) Solution treated 1150 °C (2100 °F). (b) Solution treated 1175 °C (2150 °F). (c) Heat treated to 980 °C (1800 °F) plus 720 °C (1325 °F) hold for 8 h, F.C. to 620 °C (1150 °F) hold for 8 h. (d) 730 °C (1350 °F) hold for 2 h. (e) Heat treat to 1150 °C (2100 °F) plus 840 °C (1550 °F) hold for 24 h, plus 705 °C (1300 °F) hold for 20 h. (f) Solution treated and aged. (g) Stress-relieved forging. (h) Heat treat to 1050 °C (1922 °F) hold for 1 h. (j) Heat treat to 1080 °C (1976 °F) hold for 8 h, plus 700 °C (1290 °F) hold for 16 h. (k) Heat treat to 1150 °C (2100 °F) hold for 4 h, plus 1050 °C (1920 °F) hold for 16 h, plus 850 °C (1560 °F) hold for 16 h. (m) Heat treat to 1190 °C (2175 °F) hold for 1.5 h, plus 1100 °C (2010 °F) hold for 6 h. (n) Heat treat to 1150 °C (2100 °F) hold for 2 h, W.Q., plus 800 °C (1475 °F) hold for 8 h.

and oxidation resistance of the alloy and the operating environment will determine how close actual metal temperatures may approach the suggested upper limit.

An important factor to consider when dealing with coated superalloys is the reduction in incipient melting temperature of the system (coating/base metal) that may result from the change in composition brought about by diffusion. For example, for aluminide coatings on an alloy such as U-700, which has an incipient melting temperature of about 1215 °C (2220 °F), incipient melting may occur in the inner diffusion zone at temperatures between 1175 and 1190 °C (2150 and 2175 °F). Incipient melting in coated systems often leads to accelerated degradation of the coating.

At temperatures that cause neither incipient melting nor surface degradation, alloy strength may still be reduced because strengthening phases are taken into solution. Wrought nickel-base alloys frequently are strengthened by a dispersion of the phase γ', Ni_3 (Al, Ti), an fcc intermetallic compound. In wrought alloys, the γ' phase can be taken into solution at temperatures of about 1175 °C (2150 °F) or less (see Table 14). Exposure at or near the solution temperature will reduce the amount of γ' phase and thereby reduce alloy strength. Prolonged operation at temperatures within the solution range is inadvisable, although occasional excursions into this range may be tolerated. If γ' phase is dissolved, it can be reprecipitated as fine particles by subsequent aging; original property levels can be reasonably recovered, assuming there was no other damage due to stress or oxidation. However, properties will not be recovered if slow cooling is used after extensive solution has occurred or if the material is held at a high temperature so that coarse γ' forms while fine γ' dissolves. Figure 6 compares the life of a nickel-base superalloy when it is solution treated only and when it is fully heat treated. Under high-stress applications, strength is significantly reduced if a γ'-hardened alloy has been exposed to a solution treatment temperature and not re-aged. During long-time low-stress applications within the aging-temperature range, a lesser reduction of strength occurs.

Carbide phases in heat-resistant alloys behave somewhat like γ', but subsequent reprecipitation is not as easily controlled. Carbides are taken into solution or agglomerated during overtemperature exposure, and there can be substantial variations in the amount, form and distribution of the resulting carbide structure. In Hastelloy X, a solid-solution hardened high-nickel austenitic alloy, there are large differences in the volume fractions and structure of carbides between normal-temperature and overtemperature exposures. The volume fraction of M_6C is about 12 vol % after about 7500 h at 980 °C (1800 °F). During overtemperature exposure, carbide form and distribution are considerably changed due to agglomeration and rounding particles. As a result, mechanical properties are reduced, as shown in Fig. 7, for room-temperature strength after exposure at 1040 °C (1900 °F) in air. Similar results were obtained from a series of exposures at about 1100 °C (2000 °F).

Extensive carbide precipitation frequently can occur in alloys when there is a change from the original carbide that was present in the mill-annealed condition. Figure 8 shows Haynes 188, a cobalt-base wrought alloy, before and after exposure for 6000 h at 870 °C (1600 °F). During exposure, M_6C carbides dissolved and were replaced predominantly by $M_{23}C_6$ carbides and, to a lesser extent, by Laves phase. Significant losses in ductility resulted from the precipitation shown in Fig. 8.

In alloys where extensive carbide precipitation or agglomeration occurs, it is frequently possible to recover the original carbide distribution, and a major portion of alloy properties, by solution heat treatment. For example, it is claimed that Haynes 188 can recover original room-temperature ductility by heat treatment at 1175 °C (2150 °F) for 15 min.

Table 11 Typical short-time tensile and stress-rupture properties of different mill forms of Haynes 25

Product form	Tensile strength MPa	Tensile strength ksi	Yield strength MPa	Yield strength ksi	Elongation, %
At 540 °C (1000 °F)					
Bar	750	109	...	...	71
Sheet	690	100	...	...	68
Tube	710	103	315	46	12
At 650 °C (1200 °F)					
Bar	670	97	...	...	37
Sheet	515	75	...	...	25
Tube	620	90	275	40	12
At 815 °C (1500 °F)					
Bar	455	66	...	...	24
Sheet	345	50	...	...	15
Tube	345	50	170	25	25

	Stress for rupture in: 100 h		450 h		1000 h	
	MPa	ksi	MPa	ksi	MPa	ksi
At 790 °C (1350 °F)						
Bar	255	37	...	...	...	...
Tube	255	37	...	...	...	...
At 815 °C (1500 °F)						
Bar	150	22	...	...	117	17
Sheet	150	22	...	...	121	17.5
At 870 °C (1600 °F)						
Bar	...	...	...	...	86	12.6
Sheet	...	...	82	12	...	...

Note: Bar properties were determined from standard 13-mm (0.5-in.) diameter tension-test bars. Sheet stock was 1.0 to 1.8 mm (0.040 to 0.070 in.) thick. Tube stock was 9.5-mm (0.375-in.) OD by 0.7-mm (0.028-in.) wall thickness, and was tested in full section.

Microstructural Degradation in Normal Operation Above 700 °C (1300 °F)

Whereas some alloys are exposed to extremely high temperatures in furnace and petrochemical applications, alloys for gas-turbine applications generally are exposed to the most demanding combinations of high temperature and stress. The normal operating regime of turbine blades and vanes is about 725 to 850 °C (1340 to 1560 °F) for wrought alloys, and up to 1050 °C (1920 °F) for cast alloys. Within these ranges, microstructural changes readily occur with time at temperature. Furthermore, when stress is applied, the changes may be accelerated. The principal changes are (*a*) breakdown of primary carbides and formation of secondary carbides, (*b*) agglomeration of primary geometrically close-packed (gcp) strengthening phases such as γ', and (*c*) formation of topologically close-packed (tcp) phases, such as sigma, Laves and mu. Processes described under (*a*) and (*b*) are an extension of the normal strengthening process. These

Table 12 Effect of cold rolling on elevated-temperature properties of Haynes 25 sheet

Cold reduction, %	Temperature °C	Temperature °F	Tensile strength MPa	Tensile strength ksi	0.2% yield strength MPa	0.2% yield strength ksi	Rupture time, h, at 130 MPa (19 ksi)
0	540	1 000	690	100	...	...	...
0	870	1 600	310	45	...	...	30
10	540	1 000	725	105	525	76	...
10	870	1 600	425	62	325	47	35
15	540	1 000	930	135	745	108	...
15	870	1 600	485	70	345	50	30
20	540	1 000	1 070	155	930	135	...
20	870	1 600	515	75	415	60	15

Note: Sheet 14 mm (0.054 in.) thick was solution treated at 1220 °C (2225 °F) before cold rolling.

Table 13 Incipient melting temperatures of selected wrought superalloys

Alloy	Incipient melting temperature °C	°F
Hastelloy X	1250	2280
Haynes 25 (L-605)	1329	2425
Haynes 188	1302	2375
Incoloy 800	1357	2475
Incoloy 825	1370	2500
Inconel 617	1333	2430
Inconel 625	1288	2350
Inconel X-750	1393	2540
Nimonic 80A	1360	2480
Nimonic 90	1310	2390
Nimonic 105	1290	2354
René 41	1232	2250
Udimet 500	1260	2300
Udimet 700	1216	2220
Waspaloy	1329	2425

Table 14 Solution treatments for selected wrought nickel-base superalloys

Alloy	Solution temperature(a) °C	°F	Time, h
Inconel X-750	1150	2100	4
Nimonic 90	1080	1975	8
Nimonic 105	1125-1150	2060-2100	4
Udimet 500	1175	2150	2
Udimet 700	1175	2150	4
Waspaloy	1080	1975	4

(a) All materials air cooled after solution treatment.

Fig. 5 Effect of incipient melting on microstructure of two nickel-base superalloys

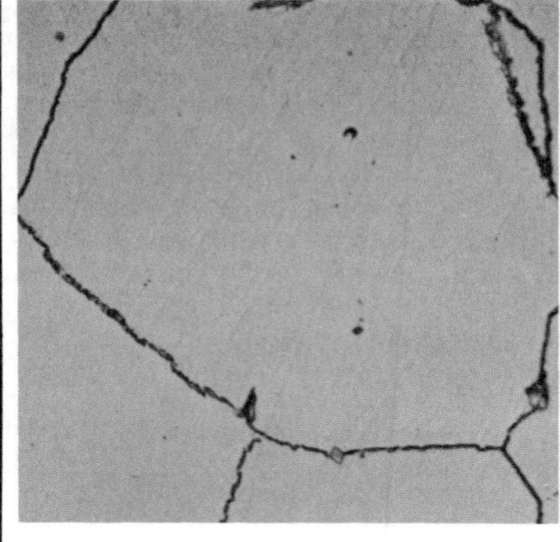

Inconel 617: No melting (500×, unetched)

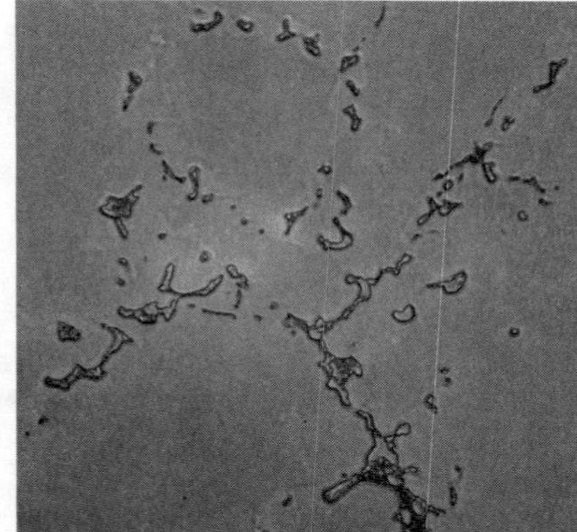

Inconel 617: Incipient melting (500×, unetched)

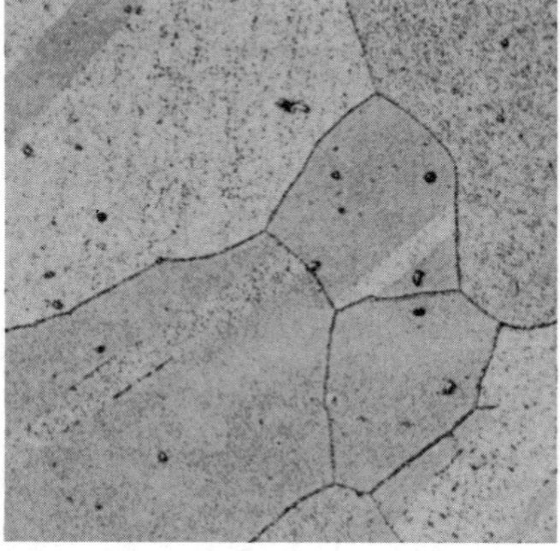

Udimet 700: No melting (500×, lactic acid + HCl + HNO_3 etch)

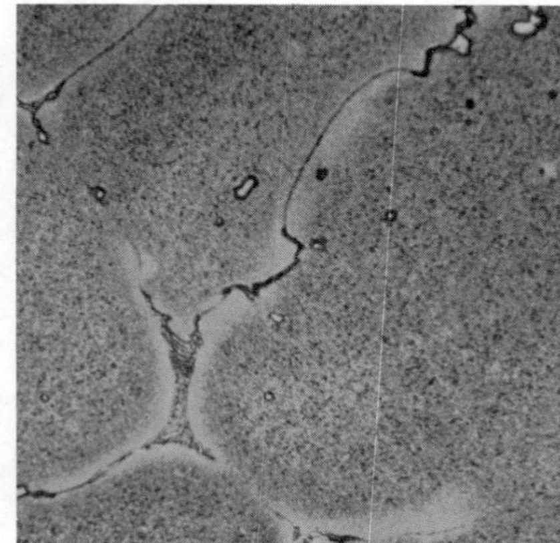

Udimet 700: Incipient melting (500×, lactic acid + HCl + HNO_3 etch)

reactions are recognized during the design of components and allowances are made for their effects on alloy strength at moderate times. Carbide transition and γ′ agglomeration do reduce strength with time but generally are not as detrimental as formation of tcp phases. Figure 9 shows the microstructure of U-700 nickel-base alloy before and after exposure at 870 °C (1600 °F). During exposure, the γ′ became agglomerated and $M_{23}C_6$ carbides formed. Generally, design parameters take into account γ′ agglomeration if it is experienced in the moderate times used to test alloys (most often, 20 to 1000 h). However, for significantly longer times at normal temperatures, γ′ agglomeration may reduce alloy strength below predicted values.

Changes in carbide phases also adversely affect strength, although initially there may be increases in strength as additional carbides precipitate. In the wrought cobalt-base solid-solution alloy Haynes 25 (L-605), carbide precipitation at 815 °C (1500 °F) is responsible for alloy hardening in both early and late stages of exposure. In the late stages, $M_{23}C_6$, M_6C and Laves phase participate in strengthening. In Hastelloy X, extensive precipitation occurs at 705 to 790 °C (1300 to 1450 °F). As a result, sigma, mu and a dense intragranular secondary M_6C carbide can occupy as much as 27 vol % of the structure. As primary M_6C carbides coalesce, there is a continual reduction in strength; formation of small secondary carbides and tcp phases enhances strength but also reduces ductility.

Figure 10 compares the strength of U-700 with and without sigma formation. As sigma forms, creep-rupture strength of the alloy is reduced. Furthermore, room-temperature ductility is greatly reduced and ductility at temperature may be reduced.

Design procedures (Ref 14 to 17) are now available to assist in minimizing detrimental secondary (tcp) phase formation in nickel-base and cobalt-base superalloys. Every element is assigned an electron vacancy number (N_v). The matrix composition after normal precipitation is then determined. The electron vacancy number for the alloy (matrix) is computed by summing the products of electron vacancy number times atomic fraction for each element present. For nickel alloys, if the N_v of the alloy exceeds about 2.4, tcp phase

Fig. 6 Stress-rupture plot for a γ′-hardening nickel alloy

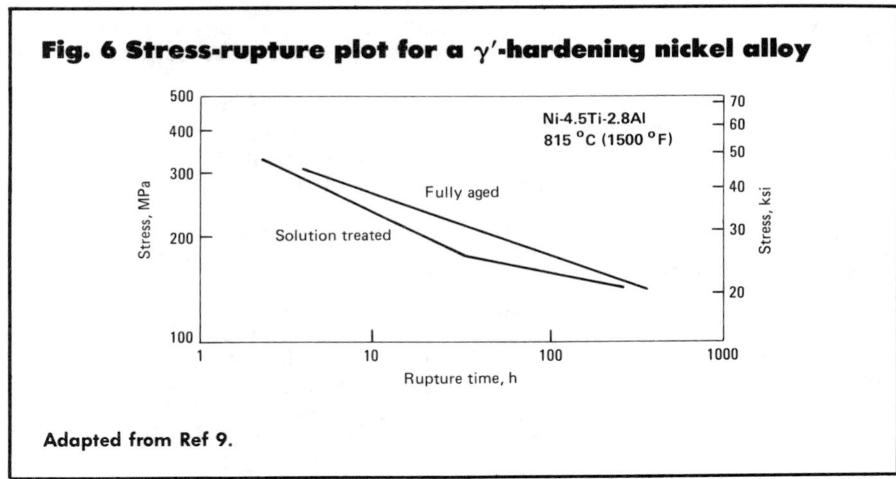

Adapted from Ref 9.

Fig. 7 Room-temperature tensile strength vs exposure time at 1040 °C (1900 °F) in air for Hastelloy X

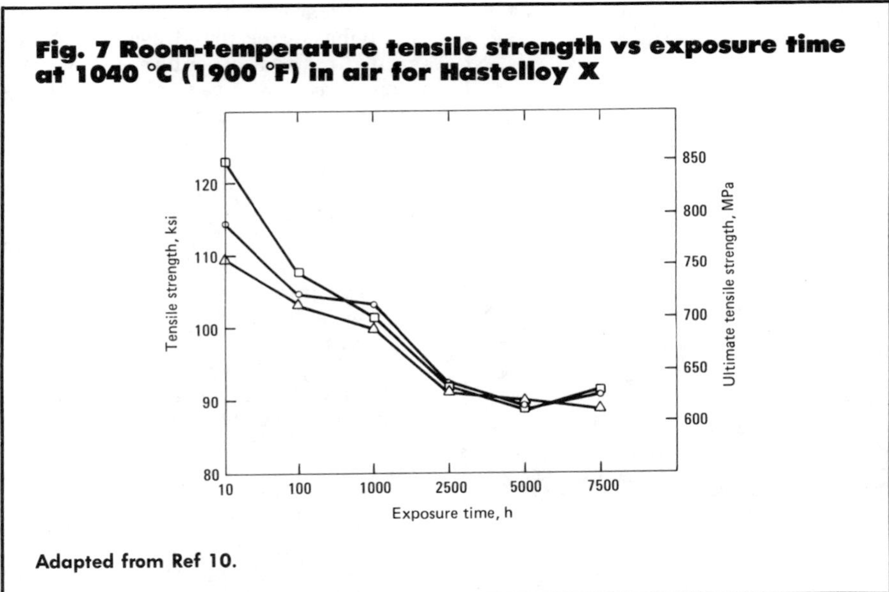

Adapted from Ref 10.

formation is likely. For cobalt alloys, the corresponding N_v is about 2.6. By proper adjustment of alloy composition, it is often possible to maintain base-alloy strength while reducing susceptibility to tcp phase formation. Application of the N_v concept thus should improve long-time creep-rupture behavior of alloys by reducing microstructural degradation at normal operating temperatures.

Formation of tcp phases is not restricted to nickel-base and cobalt-base alloys. Solid-solution high-nickel austenitic alloys such as Hastelloy X also may show a tendency to tcp phase formation above 725 °C (1340 °F). A Ni-Cr-Mo sigma phase was identified in Hastelloy X exposed for more than 2500 h at 705 °C (1300 °F) and 790 °C (1450 °F). Mu phase was also identified in specimens aged at temperatures between 705 and 955 °C (1300 and 1750 °F). However, mu was most prominent in specimens aged at 870 and 955 °C (1600 and 1750 °F). Figure 11 shows the microstructure of Hastelloy X after a 2232-h creep test at 760 °C (1400 °F).

An aspect of microstructural and creep degradation that may become important in future applications is the reported ability to heat treat creep-exposed alloys to recover creep-rupture strength. Most of this work has been done on British alloys (Ref 18). U.S. engineers and others also reportedly have developed procedures for recovering creep damage in wrought nickel-base alloys. During creep exposure, alloys gradually form cavities along grain boundaries, and these cavities lead to intergranular failure. In addition, there is carbide breakdown and γ′-phase coarsening such as that discussed above. When these processes have not proceeded for an excessively long time, it is claimed that complete recovery of strength is possible by using a solution heat treatment. However, ductility at rupture is reduced. Figure 12 illustrates the effect of one such treatment on the creep behavior of Nimonic 105 at 870 °C (1600 °F). Nimonic 115, roughly equivalent in strength to U-700, is the most advanced wrought alloy for which data are available. Results were less conclusive for Nimonic 115 than for Nimonic 105. It has been suggested that hot isostatic pressing (HIP) may significantly accelerate the healing process, at least for creep cavities that do not extend to a free surface.

Microstructural Degradation in Normal Operation Below 700 °C (1300 °F)

For alloys normally used at temperatures of 425 to 725 °C (800 to 1340 °F), melting by overtemperature is not a problem; microstructural changes are important. Hot corrosion and other accelerated oxidation is important for some alloys, and oxidation can be significant in alloy steels. The alloys used at these low temperatures may be grouped with respect to alloy stability as follows: (*a*) steels with up to 10% alloy; (*b*) stainless steels; (*c*) precipitation-hardening high-nickel austenitic alloys and certain nickel-base precipitation-hardening alloys (A-286, Incoloy 901, V-57, Waspaloy and Astroloy, for instance). The latter alloys generally are considered microstructurally stable at the temperatures for which they are employed as disks. Their properties are almost exclusively determined by prior heat treatment. If nickel-base alloys such as Waspaloy, Astroloy and René 95 are exposed for prolonged times at metal temperatures that exceed about 650 °C (1200 °F), the strengthening phase may coarsen slightly. The same effect may occur above 600 °C (1100 °F) for the iron-base or iron-nickel austenitic alloys. However, unless the alloys are operated at temperatures well into or above their aging-temperature ranges, no signifi-

Fig. 8 Microstructure of Haynes 188

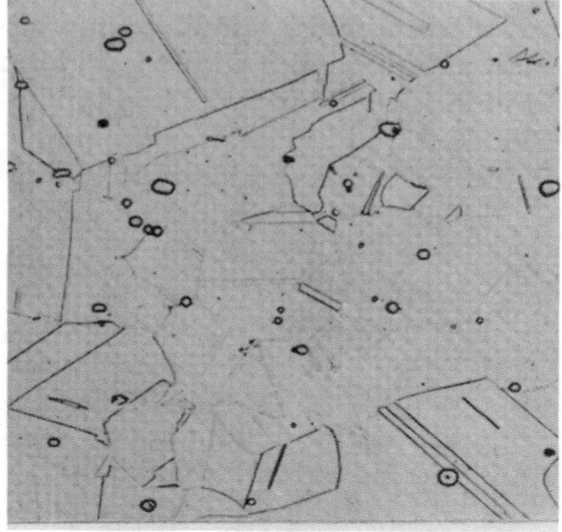

Solution treated

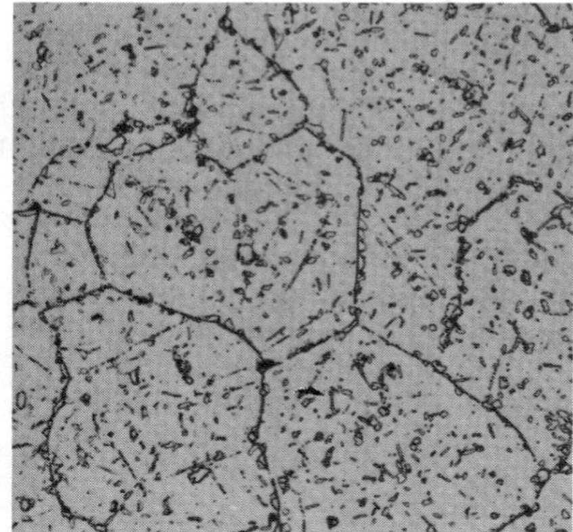

After 6244 h at 870 °C (1600 °F)

500×; etched in HCl + H_2O_2 (From Ref 11)

Fig. 9 Microstructure of Udimet 700

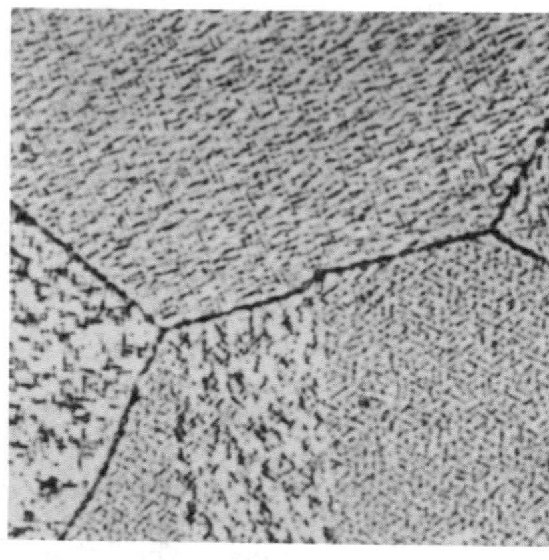

Solution treated and aged

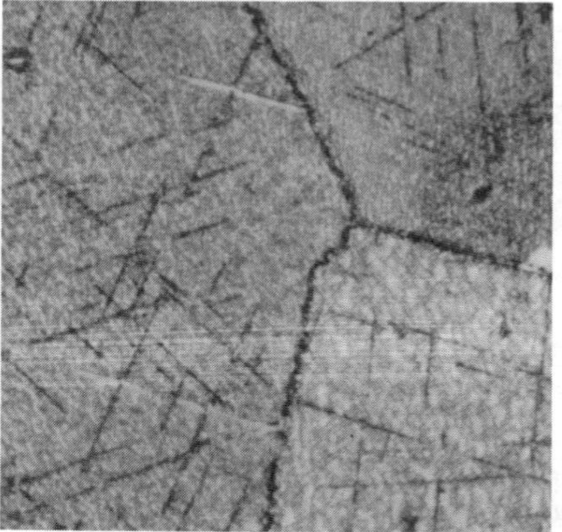

After 500 h at 870 °C (1600 °F)

100×; etched in Kalling's reagent (From Ref 12)

cant microstructural effects will be noted. Carbide precipitation at dislocations may be significant when operating times at 480 to 650 °C (900 to 1200 °F) exceed 10 000 h. However, there are no published data to support the existence of significant effects of thermal exposure on creep-rupture behavior although notch behavior may become important. At these temperatures, surface oxidation is not generally a problem, although long-term exposure to certain highly corrosive environments may produce surface attack such as hot corrosion (sulfidation).

Inconel 718 deserves some mention because it often is used at temperatures above its final aging temperature of 620 °C (1150 °F). Although some minor coarsening of the γ'' phase may take place, no detrimental effects normally occur at temperatures up to 650 °C (1200 °F). If overheating to above 700 °C (1300 °F) should occur, the strengthening γ'' phase is degraded; significant sigma-phase precipitation can then occur, with resultant losses in strength.

With respect to carbide changes and corrosive attack, austenitic stainless steels may be expected to behave similarly to the precipitation-hardening nickel-base, iron-base and iron-nickel austenitic alloys described above. Precipitated strengthening phases of the γ' type are not present in austenitic stainless steels. Extensive changes in carbide phases can occur in austenitic stainless steels and may increase resistance to creep. Properties will be determined almost exclusively by prior heat treatment. In special tests, type 316 stainless steel, solution treated and aged, crept much faster than solution-treated material that was not aged. This behavior occurred because strain aging (dislocation-enhanced $Cr_{23}C_6$ precipitation) was very effective in retarding creep in solutioned material. (No excess carbon was available for strain aging in previously aged material.) Carbide precipitation during exposure could also influence fatigue life. In general, if excess carbon is available after prior heat treatment, then carbide precipitation will take place in stainless steels and will influence mechanical properties.

In addition to carbide changes, austenitic stainless steels form tcp phases. At temperatures of 425 to 725 °C (800 to 1340 °F), the kinetics of sigma formation are quite slow. Chi and Laves phases may form more rapidly than sigma phase. The time-temperature precipitation behavior of type 316 stainless steel is shown in Fig. 13. At temperatures between 650 and 725 °C (1200 and 1340 °F), Laves phases formed in 10 to 100 h. The higher the solution temperature, the sooner $Cr_{23}C_6$ and Laves formed at lower temperatures. Chi and sigma phases may form with extended exposure, the time required for them to form depends on composition and prior processing.

With respect to the properties of austenitic stainless steels, the formation of tcp phases can be significant. Figure 14 shows the effects of Laves formation on the low-temperature impact strength of type 316 stainless steel after long-time exposure at 650 °C (1200 °F). The potential effects of tcp phases on stress-rupture behavior are illustrated in Fig. 15 where test results for two 18-8 stainless steels (with differing carbon contents) are plotted. High carbon and nitrogen contents stabilized the alloy

Fig. 10 Log stress vs log rupture life at 815 °C (1500 °F) for Udimet 700

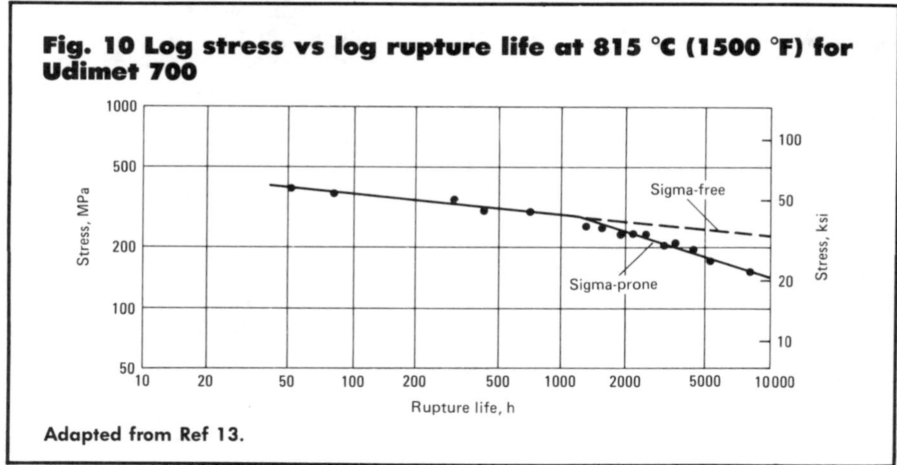

Adapted from Ref 13.

Fig. 11 Microstructure of Hastelloy X after creep exposure for 2232 h at 760 °C (1400 °F)

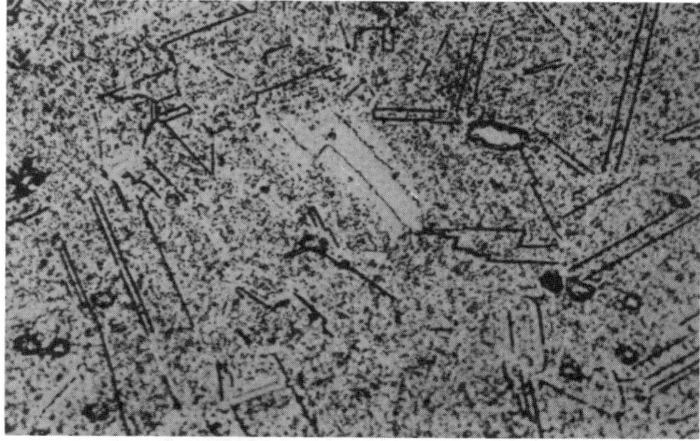

500×; etched in 10% chromic acid (From Ref 10).

Table 15 Influence of prior exposure on rupture strength of type 410 stainless steel

Exposure and test temperature		Stress for 1000-h rupture life			
		Unexposed		After exposure	
°C	°F	MPa	ksi	MPa	ksi
480	900	167	24.2	179	26.0
565	1050	81	11.8	74	10.8
650	1200	34	4.9	29	4.2

against sigma formation. Sigma formed in the alloy having lower carbon and nitrogen contents, with a resultant decrease in load-carrying capability.

In general, the only microstructural changes in ferritic and heat treatable stainless steels are due to tempering and temper embrittlement processes. Occasionally, other phases form in some of the precipitation-hardening steels (such as Laves in Pyromet X-15). Generally, tcp phase formation is not a problem with ferritic and heat treatable stainless steels, although some high-chromium ferritic grades may show tendencies for formation of sigma phase.

Ferritic stainless steels such as type 405 may be subject to "885 °F embrittlement" if exposed for long times at temperatures below 480 °C (900 °F). Martensitic grades such as type 410 also may be subject to embrittlement at such temperatures. Composition seems to influence the effects of exposure on 885 °F embrittlement. At constant chromium content, ferritic and heat treatable stainless steels with higher nickel and lower silicon and niobium contents may be less susceptible.

In addition to this specific type of embrittlement, overtempering is a potential problem for heat treatable stainless steels. As a rule of thumb, for 100 000-h life, these steels should not be used at temperatures greater than 165 °C (300 °F) below their tempering temperatures. For 30 000-h life, the limit is about 110 °C (200 °F) below the tempering temperatures. For ferritic and heat treatable stainless steels, the results of exposure vary with alloy and temperature.

Table 15 compares stress-rupture properties for type 410 before and after long-time exposure. At 565 °C (1050 °F) and above, modest decreases in load-carrying ability result from thermal exposures of 100 000 h in the unstressed condition. Additional information on stainless steels can be found in other articles in this volume of the Handbook.

Low-alloy steels are peripheral in interest to iron-nickel and cobalt-base superalloys and stainless steels at high temperatures. However, some generalizations are desirable. Low-alloy steels are susceptible to microstructural changes in the 425 to 725 °C (800 to 1340 °F) range, particularly when used in the normalized or quenched and tempered condition. Long-time exposure can cause spheroidization, graphitization and overtempering. In general, the finer the original microstructure, the more rapid the trend to an equilibrium spheroidized or graphitized structure. Consequently, the finer the original microstructure, the more rapid the degradation of creep-rupture strength.

As a general rule, a normalized and tempered or quenched and tempered steel will have the highest short-time load-carrying ability but may be inferior to a full annealed structure for long exposure times. Actual behavior depends on composition and temperature. Information on low-alloy steels is given in greater detail in Volume 1 of the 9th Edition of this Handbook.

REFERENCES

1. Microstructure of Precipitation

Fig. 12 Effect of one reheat treatment on creep of Nimonic 105 at 870 °C (1600 °F)

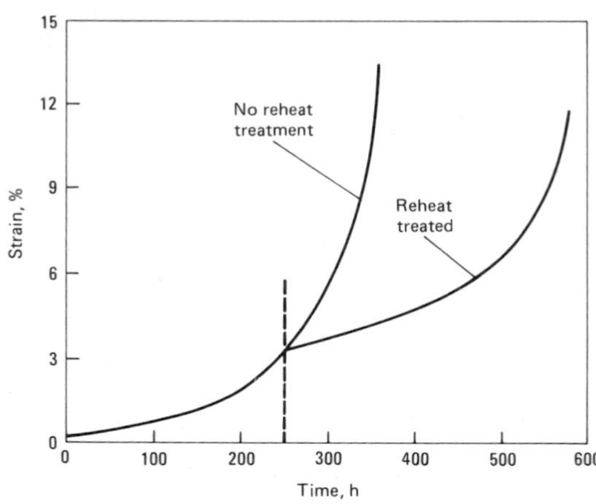

Curve at right is for specimen reheat treated after 250 h creep exposure (From Ref 18).

Fig. 13 Time-temperature-precipitation diagram for type 316L stainless steel

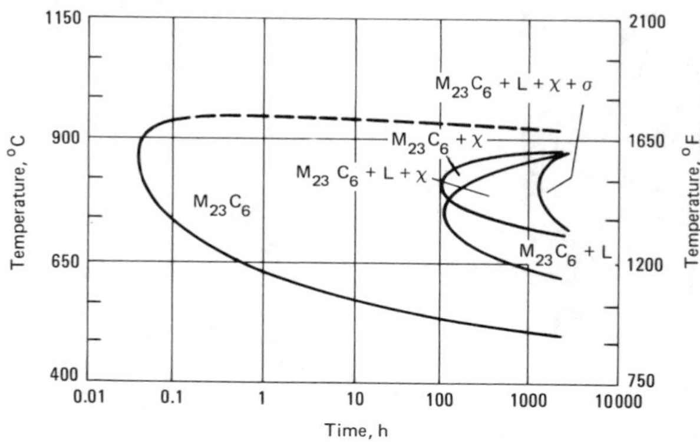

Diagram shows formation of $M_{23}C_6$ carbides, Laves phase (L), chi phase (χ) and sigma phase (σ) for material solution treated 1.5 h and water quenched from various temperatures. Data from Ref 19.

Strengthened Nickel-Base Superalloys, by J. L. Johnson and M. J. Donachie, Jr.: American Society for Metals Report System Paper C 6-18.1, 1966

2. Microstructure of Nickel-Based Superalloys, by G. P. Sabol and R. Stickler: *Physica Status Solidi,* Vol 35, 1969, p 11-52
3. The Microstructure of Superalloys, by P. S. Kotval: *Metallography,* Vol 1, 1969, p 251-285
4. *Cobalt Base Superalloys—1970,* by C. P. Sullivan *et al.:* Cobalt Information Center, Brussels, 1970
5. *The Superalloys,* edited by C. T. Sims and W. C. Hagel: J. Wiley and Sons, New York, 1972
6. Phase Stability in Superalloys, by R. Stickler: in *High Temperature Materials in Gas Turbines,* Elsevier, New York, 1974, p 115-149
7. *High Temperature Alloys for Gas Turbines,* edited by D. Coutsouradis, *et al.:* Applied Science Publishers Ltd., London, 1978
8. *Source Book on Materials for Elevated Temperature Applications,* edited by E. F. Bradley: American Society for Metals, Metals Park, OH, 1979
9. Aging Characteristics of Ni-Cr Alloys Containing Appreciable Amounts of Ti and Al, by N. Rogen and N. J. Grant: *Proceedings of ASTM,* Vol 58, 1958, p 697
10. Oxidation and Structural Stability Investigations, Vol 1 of *Evaluation Study of Hastelloy X as a Nuclear Cladding,* by W. L. Clark, Jr. and G. W. Titus: Nuclear Div., Aerojet—General Corp., San Ramon, CA, June 1968
11. Haynes Alloy No. 188, by R. B. Herchenroeder *et al.: Cobalt,* No. 54, Mar 1972, p 3
12. M. S. Thesis, by J. Johnson, Rensselaer Polytechnic Institute, Troy, NY, 1965
13. The Effect of Phase Instability on the High Temperature Stress Rupture Properties of Representative Nickel Base Superalloys, by D. Moon and F. Wall: *Proceedings Symposium on Structural Stability in Superalloys,* AIME, 1968, p 115
14. Prediction of Sigma Type Phase Occurrence from Composition in Austenitic Superalloys, by L. Woodyatt, C. Sims and H. Beattie: *Transactions of AIME,* Vol 236, p 519-527
15. "Strengthening Mechanisms in Nickel-Base Superalloys": International Nickel Co., New York, 1970
16. A Modified System for Predicting σ Formation, by R. G. Barrow and J. B. Newkirk: *Metallurgical Transactions,* Vol 3, 1972, p 2889-2893
17. Prediction of Sigma Phase Formation in High Temperature Alloys, by W. Wallace: *DME/NAE Quarterly Bulletin,* No. 1974, Vol 3, 1974
18. Recovery of Mechanical Properties in Nickel Alloys by Re-Heat-Treatment, by R. Hart and H. Gayter: *Journal of the Institute of Metals,* Vol 96, 1968, p 338-344

19. Phase Instabilities During High Temperature Exposure of 316 Austenitic Stainless Steel, by B. Weiss and R. Stickler: *Metallurgical Transactions,* Vol 3, 1972, p 851

20. The Effect of Composition and Structure on the Creep-Rupture Properties of 18-8 Stainless Steels, by F. C. Monkman, P. E. Price and N. J. Grant: *Transactions of ASM,* Vol 48, 1956, p 418

21. SIGMA SAFE: A Phase Diagram Approach to the Sigma Phase Problem in Ni Base Superalloys, by E. S. Machlin and J. Shao: *Metallurgical Transactions,* Vol. 9A, 1978, p 561-568

22. Sigma Phase Formation in Conventional and P/M Nickel-Base Superalloys, by C. C. Law, *et al.:* *Metal Science,* Vol. 13, 1979, p 627

Fig. 14 Effect of formation of Laves phase on impact strength of type 316 stainless steel at −185 °C (−365 °F)

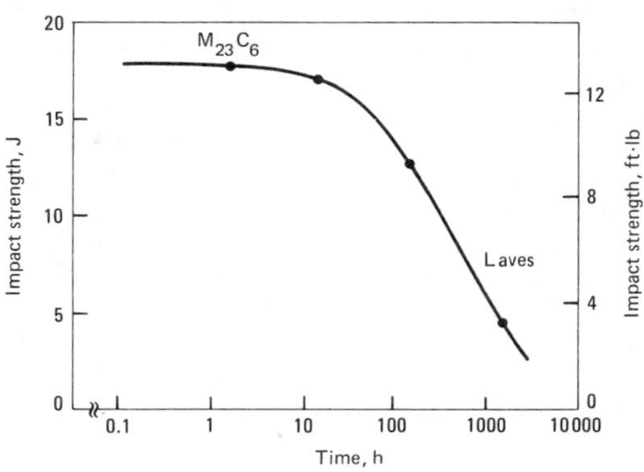

Material was aged various times at 650 °C (1200 °F).

Fig. 15 Log stress vs log rupture life at 700 °C (1300 °F) for 18-8 stainless steel

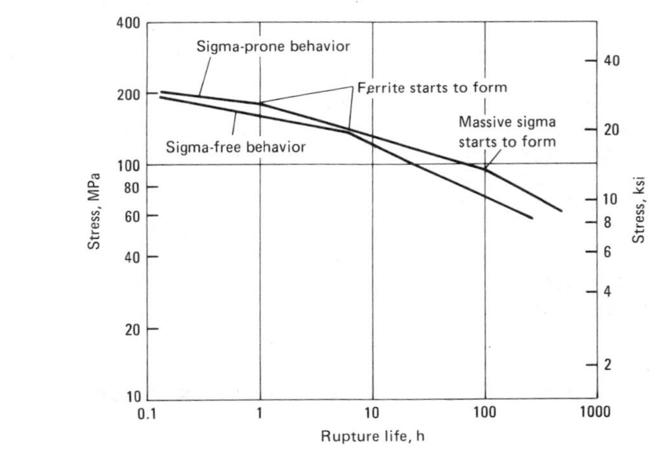

Adapted from Ref 20.

Appendix 2
Testing of Superalloys

Edited by F. R. Morral

Short-Time Tensile Tests

The short-time tensile values given in the data compilations were obtained by tests conforming to ASTM E21. The tensile-test data for Inconel "X" sheet in Fig. 16 indicate the effect of rate of heating on rupture strength. The specimens were machined from annealed Inconel "X" sheet 1.3 mm (0.050 in.) thick, with axes parallel to the direction of rolling. They were loaded at room temperature to the desired stress level by dead weights and heated at a constant rate by an electric current until failure occurred. A heating rate of 0.11 °C (0.2 °F) per second gave results equivalent to those obtained by the ASTM E21 procedure. When the rate was raised to 55 °C (100 °F) per second the rupture strength was increased more than 100%. These data have limited application in missile design.

Elastic properties at elevated temperature in the short-time tension test usually are recorded as yield strength at 0.02 or 0.2% offset, thus involving some limited plastic flow. Elastic strains usually are included in the total deformation in creep and stress-rupture tests but not in creep derived from secondary creep rate. Yield strengths are used to a limited extent in selection of allowable design stresses.

Stress-Rupture Tests

Standard creep and stress-rupture tests are covered by ASTM E139, E150 and, for notched specimens, E292. In many applications, the amount of deformation that occurs in service is not critical, and relatively high fractions of the rupture strength can be used in design. Under such conditions with the combined uncertainties of actual stress, temperature and strength, it may be important that failure not occur without warning and that the metal retain high elongation and reduction in area throughout its service life. In the oil and chemical industries, for instance, many applications of tubing under high pressure require high long-time ductility, so that warning of impending rupture will be evident from bulging of the tubes.

Values of elongation and reduction in area obtained in rupture tests are used in judging the ability of metal to adjust to stress concentration. The requirements are not well defined and are controversial. Most engineers are reluctant to use alloys with elongations of less than 5%, and this limit sometimes is considerably higher. Low ductility in a rupture test almost always indicates high resistance to relaxation of stress by creep, and possible sensitivity to stress concentrations. Large changes in elongation with increasing fracture time usually are indicators of extensive changes in metallurgical structure or of surface corrosion.

Although there is at present no real substitute for conventional long-time stress-rupture tests for determination of rupture ductility and rupture strength, parameter tests are valuable tools. In the time required to develop one series of long-time test data on one heat of material, parameter tests permit testing of enough heats to determine the variation in rupture strength for a specific alloy. This variation in strength is normally considerably larger than the error in parameter testing. In addition, parameter tests can be used to check the rupture strength of material for critical applications to ensure that the strength is typical of the specified material.

Rupture of Notched Specimens. Notched specimens are used principally as a qualitative alloy-selection tool for comparing the suitability of materials for components that may contain deliberate or accidental stress concentrations. The rupture life of notched specimens is an indication of the ability of a material to deform locally without cracking under multiaxial stresses. The most common practice is to use a circumferential 60° V-notch in round specimens with a cross-sectional area at the base of the notch one-half that of the section. However, size and shape of test specimens should be based on requirements necessary for obtaining representative samples of the material being investigated. In a notch test, the material being tested most severely is the small volume at the root of the notch. Therefore, surface effects and residual stresses can be very influential. The notch radius must be carefully machined or ground, because it can have a pronounced effect on test results. The root radius is generally 0.13 mm (0.005 in.) or less, and should be measured using an optical comparator or other equally accurate means. Size effects, stress-concentration factors introduced by notches, methods of preparing notches, grain size and hardness are all known to effect notch-rupture life.

Notch rupture properties can be obtained by using individual notched and unnotched specimens or by using a specimen with a combined unnotched or notched test section. The ratio of rupture strength of notched specimens to that of unnotched specimens varies with (*a*) notch shape and acuity, (*b*) specimen size, (*c*) rupture life (and therefore stress level), (*d*) testing temperature, and (*e*) heat treatment and

Fig. 16 Rupture strength of Inconel X-750 sheet under conditions of rapid heating at a constant rate

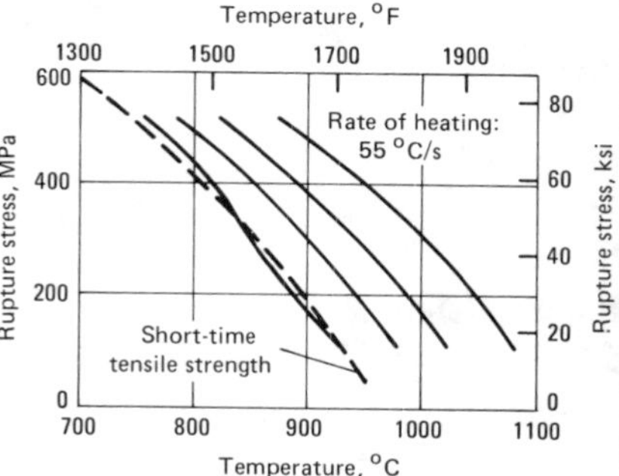

Specimens were machined from annealed sheet, 1.3 mm (0.050 in.) thick, with specimen axes parallel to rolling direction.

Fig. 17 Three general types of notch effects in stress-rupture tests

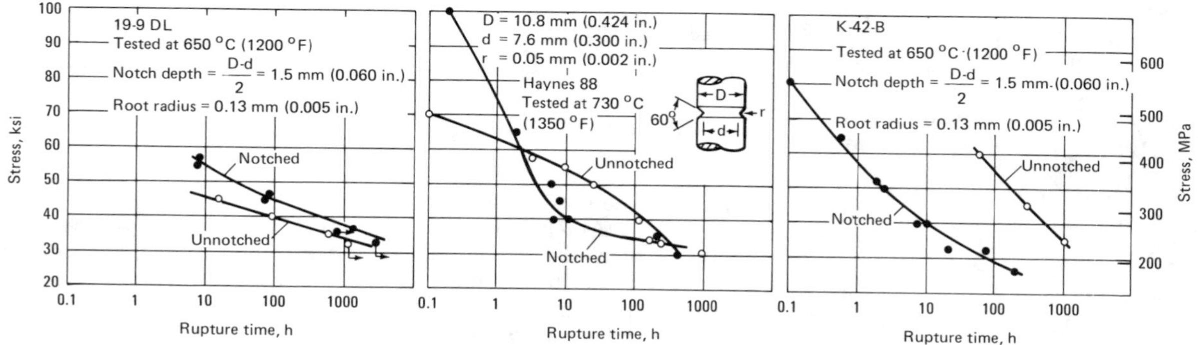

Left—Notch strengthening in 19-9 DL heat treated 50 h at 650 °C (1200 °F) and air cooled. Center—Mixed behavior in Haynes 88 heat treated 1 h at 1150 °C (2100 °F), air cooled, and worked 40% at 760 °C (1400 °F). Right—Notch weakening in K-42-B heat treated 1 h at 955 °C (1750 °F), water quenched, reheated 24 h at 650 °C and air cooled.

processing history. To avoid introducing large experimental errors, notched and unnotched specimens must be machined from adjacent sections of the same piece of material, and the gage sections must be machined to very accurate dimensions. For the combination specimen, the diameter of the unnotched section and the diameter at the root of the notch should be the same within ±0.025 mm (±0.001 in.).

Notch sensitivity in creep rupture is influenced by various factors, including both material and test conditions. The presence of a notch may increase life, have no effect, or decrease life. When the presence of a notch increases life over the entire range of rupture time, as shown at left in Fig. 17, the alloy is said to be notch strengthened—that is, the notched specimen can withstand higher nominal stresses than the unnotched specimen. Conversely, when the notch rupture strength is consistently below the unnotched rupture strength, as shown at right in Fig. 17, the alloy is said to be notch sensitive, or notch weakened. Many investigators have defined a notch-sensitive condition as one for which the notch strength ratio is below unity. However, this ratio is unreliable and can vary according to class of alloy and rupture time.

Certain alloys and test conditions show notch strengthening at high nominal stresses (short rupture times) and notch weakening at lower nominal stresses (longer rupture times), with the result that the stress-rupture curve for notched specimens crosses the curve for unnotched specimens as nominal stress is reduced. The center graph in Fig. 17 shows that Haynes 88 becomes notch sensitive under high nominal stresses in a rupture time of about 2 h, and that the material becomes notch strengthened again at lower nominal stresses at a rupture time of approximately 400 h. This same phenomenon has been observed in many superalloys and is illustrated in a different manner in Fig. 18. The "notch ductility trough" varies with alloy composition. For example, A-286 is notch sensitive at 540 °C (1000 °F) whereas Inconel X is notch sensitive at 650 °C (1200 °F). A given alloy may show notch weakening at some temperatures and notch strengthening at others. In general, notch sensitivity appears to increase as temperature is reduced.

Changes in heat treatment of some

Fig. 18 Notch rupture strength ratio vs temperature at four different rupture times for Inconel X-750

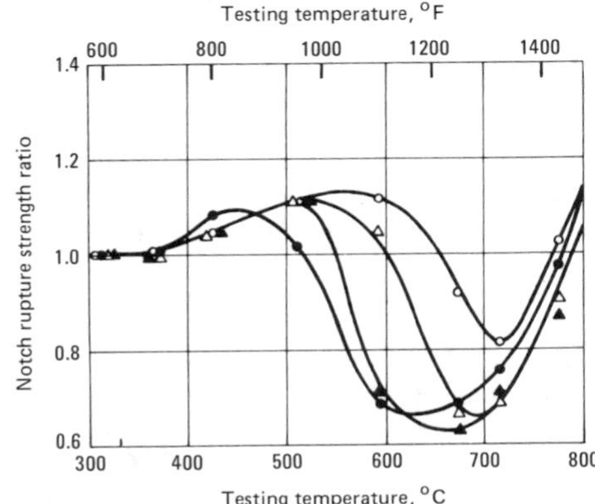

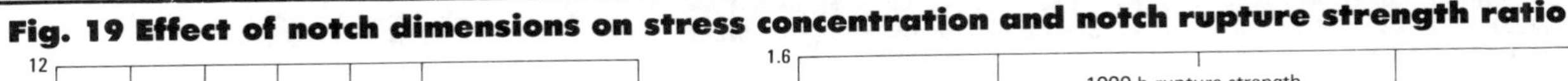

Fig. 19 Effect of notch dimensions on stress concentration and notch rupture strength ratio

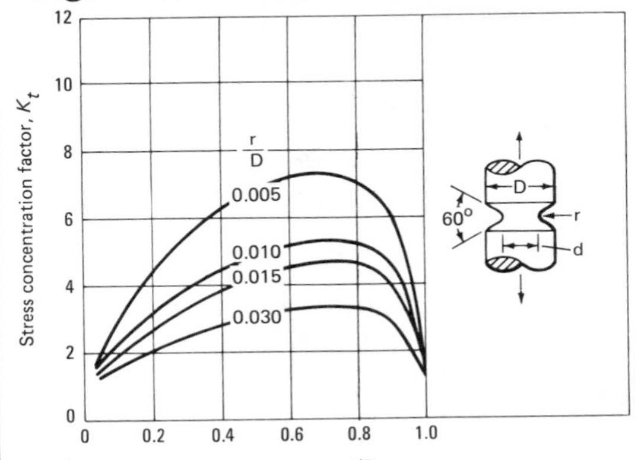

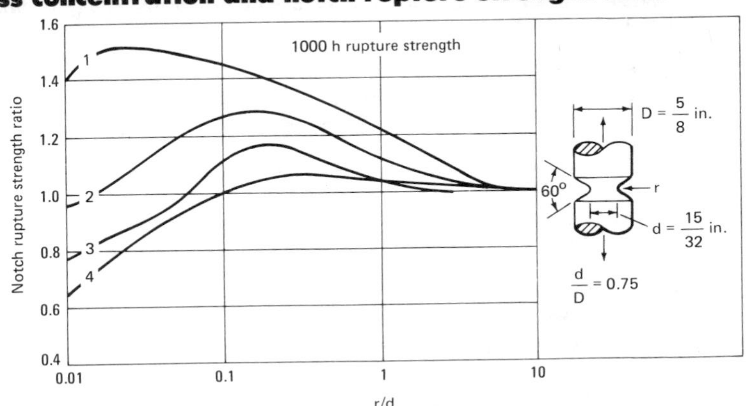

Left—Variation of stress concentration factor with ratio of minor to major diameter and with ratio of root radius to major diameter for a notched bar stressed in tension within the elastic range. Right—Variation of notch rupture strength ratio for 1000 h life with ratio of root radius to minor diameter. Curve 1 is for 12Cr-3W steel heated 3 h at 900 °C (1650 °F) and air cooled. Grain size, ASTM No. 12; hardness, 215 HV; unnotched rupture ductility, 40%; test temperature, 540 °C (1000 °F). Curve 2 is for Refractaloy 26 oil quenched from 1010 °C (1850 °F); reheated 20 h at 815 °C (1500 °F) and air cooled; reheated 20 h at 650 °C (1200 °F) and air cooled; reheated 20 h at 815 °C and air cooled; and finally reheated 20 h at 650 °C and air cooled. Grain size, ASTM No. 7 to 8; hardness, 330 HV; unnotched rupture ductility, 7%; test temperature, 650 °C. Curve 3 is for Refractaloy 26 oil quenched from 1175 °C (2150 °F); reheated 20 h at 815 °C and air cooled; reheated 20 h at 730 °C (1350 °F) and air cooled; and finally reheated 20 h at 650 °C and air cooled. Grain size, ASTM No. 2 to 3; hardness, 325 HV; unnotched rupture ductility, 10%; test temperature, 650 °C. Curve 4 is for Refractaloy 26 oil quenched from 980 °C (1800 °F); reheated 44 h at 730 °C and air cooled; and finally reheated 20 h at 650 °C and air cooled. Grain size, ASTM 7 to 8; hardness, 375 HV; unnotched rupture ductility, 3%; test temperature, 650 °C.

Table 16 Effect of a notch on rupture time of Waspaloy

Condition	Hours to failure at 730 °C (1350 °F) and 360 MPa (52 ksi) Unnotched bar	Notched bar
Solution heat treated 4 h at 1080 °C (1975 °F), air cooled, aged 16 h at 760 °C (1400°F), air cooled	76.0	1.5
Solution heat treated 4 h at 1080 °C (1975 °F), air cooled, stabilized 4 h at 845 °C (1550 °F), air cooled, aged 16 h at 760 °C (1400 °F), air cooled	82.8	150(a)
Solution heat treated 4 h at 1080 °C (1975 °F), air cooled, stabilized 4 h at 870 °C (1600 °F), air cooled, aged 16 h at 760 °C (1400 °F), air cooled	87.4	150(a)
Solution heat treated 4 h at 1080 °C (1975 °F), air cooled, stabilized 1 h at 980 °C (1800 °F), air cooled, aged 16 h at 760 °C (1400 °F)	1.9	46.6

(a) No failure; test discontinued after 150 h.

Fig. 20 Rupture strength as a funcion of time for notched and unnotched bars of different grain size

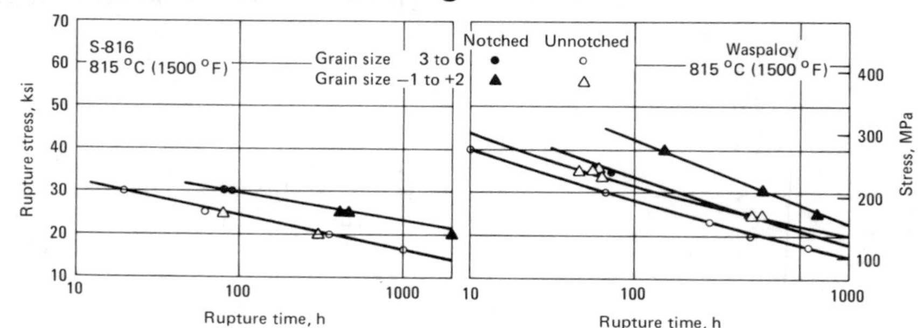

S-816 was heated to 1175 °C (2150 °F) and water quenched, reheated to 760 °C (1400 °F), held 12 h and air cooled. Waspaloy was heated to 1080 °C (1975 °F), held 4 h and air cooled; reheated to 840 °C (1550 °F), held 4 h and air cooled; and finally reheated to 760 °C (1400 °F), held 16 h and air cooled.

Smaller grain sizes were produced by cold reducing the S-816 1%, and the Waspaloy 1.25%, by cold rolling at 24 °C (75 °F) and then heat treating.

Diameter of specimens was 12.7 mm (0.5 in.), diameter at base of notch was 8.9 mm (0.35 in.), root radius was 0.1 mm (0.004 in.) and notch angle was 60°. Data are a composite of results from two laboratories.

alloys may alter notch sensitivity drastically. For example, single low-temperature aging of some alloys may produce very low rupture ductilities because the structure is not sufficiently stabilized. Consequently, exposure of such materials for prolonged rupture times will further reduce rupture ductility because of continued precipitation of particles that enhance notch sensitivity. On the other hand, multiple aging usually stabilizes the structure, and thus reduces notch sensitivity.

The notch configuration itself can have a profound effect on test results, particularly in notch-sensitive alloys. Most studies on notch configuration present results in terms of the elastic stress-concentration factor. The design criterion for the weakening effect of notches at normal and low temperatures is that of complete elasticity. The design stress is the yield stress divided by the elastic stress-concentration factor K_t (Fig. 19, left). The value of the peak axial (design) stress depends on the configuration of the notch.

There is no similar relationship for the effect of notches at elevated temperatures. The metallurgical effects that influence the behavior of notched material are complex and include composition, fabrication history and heat treatment. Effects of several heat treatments on rupture time of Waspaloy are shown in Table 16.

For ductile metals, the ratio of rupture strength of notched specimens to that of unnotched specimens usually increases to some maximum as the stress-concentration factor is increased. For very insensitive alloys, there may be little further change. Metals that are more notch-sensitive may undergo a reduction in ratio as the notch sharpness (stress-concentration factor) is increased beyond the maximum, and may show notch weakening for even sharper notches. Very notch-sensitive alloys may undergo little or no notch strengthening, even for very blunt notches (low stress-concentration factor), and undergo progressive weakening as notch sharpness increases.

Relationships betweeen notch configuration and the ratio of rupture strengths of notched and unnotched specimens are shown at right in Fig. 19. In curve 1, for an alloy with an unnotched rupture ductility of 40%, the notch-strengthening factor increases as the notch is increased in sharpness (decrease in ratio r/d). In curve 2, for an alloy with unnotched rupture ductility of 7%, the notch-strength factor increases with increasing notch sharpness, reaches a peak and then drops to a notch-strength reduction factor less than unity. For an alloy with a still-lower unnotched rupture ductility of 3% (curve 4), the notch-strength factor is only slightly greater than unity for large radii of curvature, and becomes less than unity, and continues to decrease, as the notches become sharper.

The effects of grain size on notched and unnotched rupture strength are shown in Fig. 20. The coarse grain sizes (ASTM −1 to +2) were obtained by reheating bars in which small strains had been introduced by cold reducing them 1 to 1.25%. Notches had a strengthening effect on both S-816 and Waspaloy when tested at 815 °C (1500 °F). There was no measured difference in the effect of grain size on either the notched or unnotched specimens of S-816. On the other hand, the coarse-grained Waspaloy specimens showed a longer rupture time at the same rupture stress for both notched and unnotched specimens.

The rupture time for Discaloy at 650 °C (1200 °F) increases with increasing hardness up to about 290 HV for notched specimens (K_t, 3.9) and up to 330 HV for unnotched specimens, as shown in Fig. 21. Ductility, as measured by elongation values for un-

Fig. 21 Variation of rupture time at 650 °C (1200 °F) with initial hardness for Discaloy

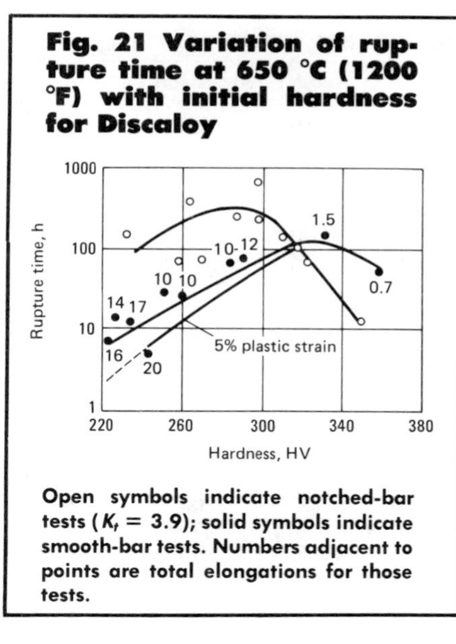

Open symbols indicate notched-bar tests (K_t = 3.9); solid symbols indicate smooth-bar tests. Numbers adjacent to points are total elongations for those tests.

Fig. 22 Effect of notches on rupture life of three superalloys

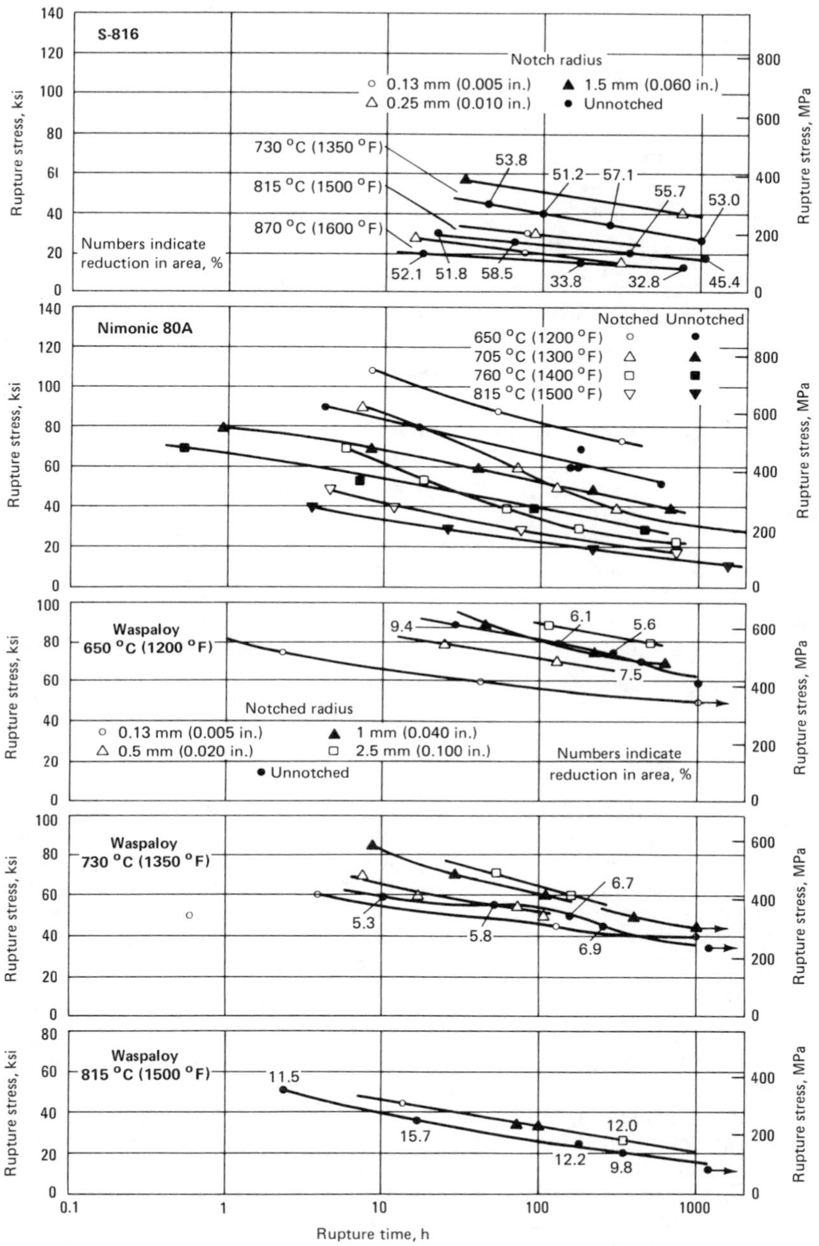

notched bars, decreases with increasing hardness. The peak in rupture time at 650 °C (1200 °F) corresponds to a rupture elongation of 1.5%. The continual reduction in rupture elongation with increasing hardness indicates that the alloy exhibits time-dependent notch sensitivity. Notched bars exhibit a strengthening effect at lower hardnesses and higher ductilities; for specimens of higher hardness and lower ductility, rapid notch weakening is apparent. For this particular alloy at this temperature, 5% rupture elongation indicates the point at which no notch strengthening or weakening takes place; this point also is indicated by the crossover of the two curves in Fig. 20 at about 318 HV. For other alloys, this crossover may occur at rupture ductilities as low as 3% or as high as 25%. Obviously, alloys with lower rupture ductilities will be more notch sensitive.

The effects of notches on rupture life of three alloys are shown in Fig. 22. Nimonic 80A, S-816, Inconel 751 (formerly Inconel X-550) and Waspaloy show various degrees of notch sensitivity. Nimonic 80A shows a notch-strengthening effect at 650 °C (1200 °F), but is notch weakened in about 100 h at 705 °C (1300 °F) and in about 40 h at 760 °C (1400 °F). Nimonic 80A again becomes notch strengthened at 815 °C (1500 °F), which illustrates that the alloy exhibits a notch ductility trough between 705 and 760 °C (1300 and 1400 °F). Using 0.13 mm (0.005 in.) as the standard notch radius, Waspaloy and Inconel 751 both exhibited notch weakening at the lower test temperatures 650 and 730 °C (1200 and 1350 °F) and notch strengthening at the higher test temperatures 815 and 870 °C (1500 and 1600 °F). Similar tests on Inconel X-750 with rupture ductilities of 10% reduction in area at 730 °C (1350 °F) and 24% at 815 °C (1500 °F) indicated notch strengthening at both temperatures. These results support the theory that materials with high rupture ductilities under the initial test conditions will be less notch sensitive in long exposure times than materials with low initial rupture ductilities. Alloy S-816 further illustrates this theory in that the alloy

Table 17 Specified requirements for selected superalloys used in aircraft gas turbines

Alloy	ASM No.	Hardness, HB	Test temperature °C	Test temperature °F	First test Stress MPa	First test Stress ksi	First test Min time, h	Accelerated tests(a) Stress MPa	Accelerated tests(a) Stress ksi	Elongation, % min(b)
Iron-base solid-solution alloys										
19-9 DL	5722	228–285	650	1200	275	40	100	414	60	15
16-25-6	5727	235–293	650	1200	275	40	100	310	45	15
N-155	5768	192–241	732	1350	165	24	100	275	40	10
Cobalt-base solid-solution alloy										
Haynes 25 (L-605)	5759	275 max	816	1500	165	24	24	207	30	10
S-816	5765	248–331	732	1350	262	38	100	345	50	8
Nickel-base solid-solution alloy										
Hastelloy X	5754	241 max	816	1500	103	15	24	172	25	10
Iron-base precipitation-hardening alloys										
Discaloy	5733	248–321	650	1200	414	60	15	448	65	5,3(c)
A-286	5735	248–341	650	1200	483	70	23	517	75	5,3(c)
A-286	5737	277–363	650	1200	448	65	23	483	70	5,3(c)
Nickel-base precipitation-hardening alloys										
Incoloy 901	5660	302–388	650	1200	552	80	23	586	85	5
D-979	5746	340–418	650	1200	655	95	23	690	100	5,3(c)
Waspaloy	5707	321–403	732	1350	517	75	23	552	80	8
Inconel 706	5703	285 min	650	1200	690	100	23	724	105	4
Inconel 718	5663	331 min	650	1200	690	100	23	724	105	4
Inconel X-750	5668	262–341	732	1350	362	52	23	414	60	5
Udimet 500	5753	285 min	900	1650	172	25	24	207	30	5

Note: Studies have indicated the above rupture elongations are not high enough to alleviate a notch sensitive condition at lower test temperatures 540 °C (1000 °F).
(a) The first test may continue to failure after the minimum time, or the stress may be increased and maintained to failure. In either case, the elongation shall be as specified. (b) In 50 mm or 2 in. (c) Rupture in 48 h or less, 5% minimum; more than 48 h, 3% minimum.

has very high rupture ductilities at each test temperature and does not show signs of notch weakening at any test temperature.

The curves for the alloys with various notch radii show that, in general, notch sensitivity increases with increased notch severity at the lower test temperatures. This effect is especially evidenced in Waspaloy: at 650 and 730 °C (1200 and 1350 °F) and for a notch radius of 0.13 mm (0.005 in.), the alloy is highly notch sensitive, but it shows notch strengthening at these same test temperatures at a blunter notch radius of 2.5 mm (0.10 in.). The effect of the radius in the root of the notch is minimal for Inconel 751 at 730 °C (1350 °F) in that the material is notch sensitive under all conditions of notch severity at prolonged rupture times. The material does show slight notch strengthening, but only with the larger radii and very short rupture times. Thus the larger radii would compensate for notch sensitivity in Waspaloy but not in Inconel 751 at 730 °C (1350 °F). At the highest test temperature, 815 °C (1500 °F), Waspaloy was notch strengthened under all conditions of notch severity. The notch radius did not have any effect on alloy S-816, as evidenced by notch strengthening at all notch severities and test temperatures, but again high rupture ductilities enhanced notch strengthening.

Correlation With Laboratory Tests. In steam and gas turbines, the blade-root fastenings always introduce stress raisers in both the blades and the rotor. Optimum design of blade-root fastenings depends not only on the creep strength of the blade material but also on the stress distribution at the root and any possible embrittlement of the material. Alloys that embrittle can be used only if the root indentation is given the most favorable form. This can be determined by photoelastic methods, pull-out tests and simulated service tests. The embrittlement of an alloy as determined by rupture tests of notched bars has been correlated with that determined by rupture tests of blade-root fastenings. An indication of the probable life of a blade fastening can be gained from rupture tests of bars that have a notch similar to that represented by the blade root.

Material specifications for alloys used in critical components of aircraft gas turbines require stress-rupture tests, and in some instances stress-rupture tests of notched specimens, for quality control. Some of the requirements are given in Table 17.

Dynamic Creep and Fatigue Properties

Metals are tested for fatigue at elevated temperatures by methods similar to those used for room-temperature tests (see ASTM E466). Testing time must be considered in analyzing the results for design purposes because creep stresses can be the most important cause of failure at high temperatures. As shown in Table 18, unnotched specimens of S-816 failed at 730 °C (1350 °F) at the same stress level for both static and dynamic loads; at 815 °C (1500 °F), dynamic loading brought on a decided increase in rupture stress (fatigue strength) compared with static conditions. At all temperatures, the

Table 18 Comparison of creep-rupture and fatigue strengths for notched and unnotched specimens of S-816

Temperature		Static rupture strength				Fatigue strength			
		Unnotched		Notched(a)		Unnotched		Notched(a)	
°C	°F	MPa	ksi	MPa	ksi	MPa	ksi	MPa	ksi
24	75	1010	147(b)	...	...	385	56	165	24
595	1100	625	91(c)	...	...	315	46	150	22
650	1200	450	65(c)	...	...	295	43	145	21
730	1350	290	42	405	59	290	42	130	19
815	1500	170	25	255	37	255	37	130	19
900	1650	97	14	...	...	195	28	...	...

(a) Circular, 60° V-notch, D = 9.5 mm (0.375 in.), d = 6.4 mm (0.25 in.), r = 0.25 mm (0.010 in.), K_t = 3.4. (b) Tensile strength. (c) Typical values; all other test specimens from same heat.

notch was more damaging under dynamic stress than under static stress.

Low-Cycle Fatigue

Low-cycle fatigue (LCF) is a term used to describe fatigue failures that occur in less than about 10 000 stress cycles. In low-cycle fatigue, cyclic strains usually are high enough to initiate cracking within a relatively few stress cycles, and the majority of the applied stress cycles act to propagate the fatigue crack. By contrast, in high-cycle fatigue, most of the applied stress cycles occur before a crack is initiated; consequently, crack-initiation mechanisms rather than crack-growth mechanisms exert the dominant control over the fatigue process.

A whole new type of mechanical and thermal fatigue testing (Ref 1 to 3) is needed to deal with the phenomenon of low-cycle fatigue. This type of testing is done using closed-loop, servo-controlled, electrohydraulic testing systems that can impose cyclic strain-controlled axial loads on laboratory-size specimens. A large number of computer-controlled test systems are now in common use. Typically, LCF test specimens have either a cylindrical test section or, for tests involving diametral strain control, an hourglass section. The nominal diameter is 6.5 mm (0.25 in.), and the cylindrical specimens have a uniform gage length of 12.5 or 25 mm (0.5 or 1.0 in.). Strain-controlled LCF testing was introduced in an attempt to simulate the fatigue-crack-initiation process that occurs in real parts. The primary variables (stress, strain, temperature and environment) are directly controlled, measured and recorded. Low-cycle fatigue tests must be strain controlled because of the shape of the stress-strain hysteresis loop at large strains. Under these circumstances, a slight variation in the stress-strain response of the material (cyclic hardening or softening) could produce very drastic changes in the hysteresis loop (strain range) if the test were run under stress control. Under strain control, however, changes in the hysteresis loop caused by cyclic hardening or softening are considerably smaller.

Cyclic inelastic deformation at the roots of strain concentrations eventually produces microscopic discontinuities at the surface outcroppings of crystallographic slip bands. These tiny surface features are called intrusions and extrusions. With continued cycling, they become the paths of first-stage fatigue cracking, which occurs along crystallographic slip planes. Eventually, crack progression reaches the point at which it is controlled by the self-induced stress-strain field at the crack front. Stage II cyclic crack growth then begins, and the plane of cracking aligns itself normal to the direction of cyclic tensile stresses. Stage II cracking is well suited for treatment by fracture mechanics principles. This mode of cracking usually is transgranular and essentially independent of crystallographic orientation. Stage II growth continues until the specimen cross section can no longer carry the imposed load. If the material is ductile, final fracture often occurs by tearing. If the material is brittle, the fatigue crack grows to a critical size and rapid fracture ensues in accordance with fracture mechanics principles. The preceding describes low-cycle fatigue at temperatures below the creep limit. At higher temperatures, this process can be appreciably altered as a result of grain-boundary cracking.

Prediction of LCF Life. Low-cycle fatigue tests have been conducted only since the 1950's. Test temperatures as low as −269 °C (−452 °F) and higher than 1650 °C (3000 °F) have been investigated. Test environments such as hard vacuum, hot flowing sodium and nuclear radiation have been used in attempts to study possible deleterious effects on LCF resistance of engineering materials.

Advanced design concepts and the need for more efficient high-temperature equipment require accurate analyses of proposed designs in order to guarantee performance and reliability of a new structure. With respect to LCF, several life-prediction approaches have been developed over the past two decades; these prediction methods account for the interaction of creep and other time-dependent effects with fatigue at high temperatures. Some of the many methods for predicting LCF life both below and in the creep range (Ref 4 to 17) are listed in Table 19. A thorough assessment of the methods discussed in Ref 4 to 17 is presented in Ref 18. The most significant currently used life-prediction approaches are discussed briefly below.

Universal slopes was one of the first practical methods to be developed for predicting LCF life. Its practicality arises from two sources: it is based on a relationship between total strain range and life, and it relates tensile properties to the constants in the fatigue equation. Thus, a knowledge of the tensile properties of a material enables a designer to estimate its LCF behavior. This method has achieved widespread use.

The 10% rule for estimating the effect of creep on LCF is based directly on the universal slopes equation. A value equal to 10% of the cyclic life computed by the equation is considered to be the lowest life that can be expected when creep-induced intergranular cracking is present. The primary utility of both the universal slopes equation and the 10% rule is in the early stages of design, when fatigue properties of a material of interest may not have been measured but its tensile properties are available.

The time and cycle fraction rule considers fatigue and creep damage to be two distinct processes contributing to the cyclic deterioration of a material, but that damages from the two processes can be summed linearly. Failure is assumed to occur when the sum of the damage fractions reaches unity. The time and cycle fraction rule has seen considerable use since its adoption in Code Case 1592 of the ASME Boiler and Pressure Vessel Code (1974). Although the rule is relatively universal in applicability, it is difficult to apply accurately to practical problems. This disadvantage stems from the requirement that stresses be known accurately in order to assess creep damage.

The frequency-modified approach, in some respects, resembles the universal slopes method, because it relates total strain range to low-cycle fatigue life. The principal distinction is that the frequency-modified approach introduces a frequency term into both the elastic and inelastic strain-range components. It is tacitly assumed that these frequency terms can properly account for all time-dependent influences on high-temperature cyclic life, including thermally activated creep, environment-surface interactions and metallurgical transformations. If these factors are present in the tests used to establish the constants in frequency-modified equations, then these factors will also be accounted for in a design that does not have to operate under conditions appreciably beyond those imposed in the laboratory tests. A significant drawback to the frequency-modified approach is that large quantities of test data are required to establish the values of constants in the equations.

Strainrange partitioning is based on the fundamental concept that there are two distinct modes of inelastic strain: time-dependent creep and time-independent plasticity. Low-cycle fatigue life can be strongly affected by the magnitudes of these two types of strain as well as by the manner in which they are combined within a cycle. There are only four basic combinations of these two strains within a completely reversed strain cycle: tensile and compressive plasticity, tensile and compressive creep, tensile plasticity with compressive creep, and tensile creep with compressive plasticity. By conducting laboratory tests featuring these types of generic partitioned strain ranges, four relationships of partitioned strainrange versus life can be established. These relationships represent bounds on cyclic life. By use of this method, a lower limit and upper limit of life can be ascertained immediately. In principle, any conceivable strain cycle can be partitioned into components of the four generic strainrange types. By use of a life-fraction damage rule, and by correlating strain components with their basic life relationships, it is possible to predict the life associated with very complex strain cycles. The predicted life, which lies between known bounds, is determined by a process similar to interpolation. The inherent accuracies associated with interpolation are obviously better than those associated with extrapolation (which usually is involved in the other methods). Hence, the accuracy of LCF life predictions made by strainrange partitioning is very high.

In addition to the potential for high-accuracy life predictions, the method is general and therefore is applicable to any strain cycle, whether it be rapid or slow, at high temperatures or low, with isothermal or variable temperatures within the cycle, or with periods of constant strain or of constant stress. For many metallurgically stable materials, life relationships for the four generic strainrange types are essentially independent of temperature.

It may ultimately be possible to express strainrange-partitioning relationships in terms of both total strain range and inelastic strain range. This would greatly increase the usefulness of the method, especially when inelastic strain ranges are small and less accurately known than total strain ranges.

The damage-rate approach places primary emphasis on two phenomenological variables: inelastic strain rate, $\dot{\epsilon}_p$, and the instantaneous inelastic strain, ϵ_p. In the original development of the equations relating cyclic life, N_f, to $\dot{\epsilon}_p$ and ϵ_p, it was assumed that the major portion of the life of a low-cycle fatigue specimen is spent in extending pre-existent microcracks into macrocracks. Also, it was assumed that the greater the inelastic strain imposed per cycle, the faster the crack would grow. Likewise, the lower the rate of straining, the greater the extent of crack growth within a cycle. The result of carrying these hypotheses to their logical conclusion is a life-prediction equation

Table 19 Summary of methods for predicting low cycle fatigue life

Method	Temperature regime	Primary variable(s)	Form of equation
Manson-Coffin law	Below creep range	Plastic strain	$\Delta\epsilon_p = AN_f{}^{a_1}$
Hysteresis energy	Below creep range	Plastic energy	$\Sigma\Delta W_p = f(N_f)$
Universal slopes	Below creep range	Total strain	$\Delta\epsilon_t = \frac{3.5\sigma_u}{E} N_f{}^{-0.12} + D^{0.6}N_f{}^{-0.6}$
10% rule	In creep range	Total strain	$\Delta\epsilon_t = \frac{3.5\sigma_u}{E} (10N_f)^{-0.12} + D^{0.6}(10N_f)^{-0.6}$
Time and cycle fraction rule	In creep range	Stress and total strain	$\sum \frac{\Delta t}{t_r} + \sum \frac{n}{N_f} = 1$
Frequency-modified approach	In creep range	Total strain	$\Delta\epsilon_t = AN_f{}^{a_1}\nu^{b_1} + BN_f{}^{a_2}\nu^{b_2}$
Time-cycle diagram	In creep range	Total strain	Plot of log (t_f) vs log (N_f)
Strainrange partitioning	In creep range	Creep and plastic strains	$\Delta\epsilon_{pp} = A_1N_{pp}{}^{\alpha_1}$ $\Delta\epsilon_{cc} = A_3N_{cc}{}^{\alpha_3}$ $\Delta\epsilon_{pc} = A_2N_{pc}{}^{\alpha_2}$ $\Delta\epsilon_{cp} = A_4N_{cp}{}^{\alpha_4}$
Damage rate	In creep range	Inelastic strain and strain rate	$N_f = \left(\frac{m+1}{4A}\right)\left(\frac{\Delta\epsilon_p}{2}\right)^{(1-m)} \left(\dot{\epsilon}_p\right)^{(1-k)}$

that is shown in its simplest form in Table 19. Here, the equation is appropriate for representing results of continuous strain-cycling tests. As the cycle is made more complex by introduction of hold times or other forms of nonsymmetrical loading, the life equation can be expanded to include terms that account for complications in both the tensile and compressive halves of the cycle. Consequently, the damage-rate approach provides a procedure for dealing with unbalanced strain cycles, which are of extreme importance in low-cycle thermal fatigue.

Frequently, the material constants needed to apply this life-prediction method are themselves functions of one of the primary variables, the inelastic strain rate. Thus, closed-form solutions for specific problems seldom can be obtained. Nevertheless, numerical solutions are possible.

The damage-rate approach can correlate laboratory creep-fatigue life within a factor of approximately two for instances in which a sufficient data base has been generated on which to evaluate the material constants. This method seems to have an inherent limitation for practical problems involving low strains and nominally elastic creep-fatigue interactions. Here, it may be impossible to determine the magnitudes of inelastic strains and their rates of change with time, thus making it impossible to use the damage-rate approach.

REFERENCES

1. Handbook for Fatigue Testing: STP 465, American Society for Testing and Materials, 1975
2. Manual on Low-Cycle Fatigue Testing: STP 465, American Society for Testing and Materials, 1969
3. *Fatigue in the Creep Range—Material Testing, Documentation and Interpretation,* by G. R. Halford: Feb 1975. [Position paper prepared under the auspices of the ASME Working Group on Creep Fatigue for use with Code Case 1592.]
4. *Behavior of Materials Under Conditions of Thermal Stress,* by S. S. Manson: Heat Transfer Symposium, University of Michigan, University of Michigan Press 1952, (Also NACA TN 2933, July 1953)
5. *A Study of the Effects of Cyclic Thermal Stresses on a Ductile Metal,* by L. F. Coffin, Jr.: Trans. ASME, 76, 1954, p 931-950
6. *Cyclic Plastic Strain Energy and Fatigue of Metals,* by J. Morrow: in STP 378, American Society for Testing and Materials, 1965, p 45-84
7. The Energy Required for Fatigue, by G. R. Halford: *J. Materials,* Vol. 1, No. 1, 1966, p 3-18
8. Fatigue: A Complex Subject—Some Simple Approximations, by S. S. Manson: *Exp. Mech.,* Vol. 5, No. 7, 1965, p 193-226
9. *A Method of Estimating High-Temperature Low-Cycle Fatigue Behavior of Materials,* by S. S. Manson and G. R. Halford: Proc., Intl. Conf. on Thermal and High-Strain Fatigue, Metals and Metallurgy Trust, London, 1967, p 154-170
10. Application of a Method of Estimating High-Temperature Low-Cycle Fatigue Behavior of Materials, by G. R. Halford and S. S. Manson: *ASM Trans.,* Vol. 61, No. 1, 1968, p 94-102
11. *The Role of Creep in High-Temperature, Low-cycle Fatigue,* by S. S. Manson, G. R. Halford, and D. A. Spera: Chapter 12 in *Advances in Creep Design,* edited by A. T. Smith and A. M. Nicolson, Appl. Sci. Publ. Ltd., London, 1971, p 229-249
12. Comparison of Experimental and Theoretical Thermal-Fatigue Lives for Five Nickel-Base Alloys, by D. A. Spera: in *STP 520,* ASTM, 1973, p 648-656
13. Fatigue at High Temperatures—Prediction and Interpretation, by L. F. Coffin, Jr.: James Clayton Memorial Lecture, International Conf. on Creep and Fatigue in Elevated Temperature Application, Institution of Mechanical Engineers, London, 1974
14. Considerations of Creep-Fatigue Interaction in Design Analysis, by J. R. Ellis and E. P. Esztergar: symposium on *Design for Elevated Temperature Environment,* ASME, 1971, p 29-43
15. Creep-Fatigue Analysis by Strainrange Partitioning, by S. S. Manson, G. R. Halford, and M. H. Hirschberg: symposium on *Design for Elevated Temperature Environment,* ASME, 1971, p 12-28
16. The Challenge to Unity Treatment of High Temperature Fatigue—A Partisan Proposal Based on Strainrange Partitioning, by S. S. Manson: in *STP 520,* ASTM, 1973, p 744-775
17. "Strainrange Partitioning—A Tool for Characterizing High-Temperature Low-Cycle Fatigue", by M. H. Hirschberg and G. R. Halford: NASA TN, in press
18. "Interpretive Report on Interaction of Creep and Fatigue at Elevated Temperatures", contributed by Manson, Coffin, Carden, and Severud: Oak Ridge National Laboratories, 1976

Appendix 3
Time-Temperature Parameters

By R. M. Goldhoff (deceased)
General Electric Company

Use of time-temperature parameters (TTP's) for presenting and extrapolating high-temperature creep-rupture data has been practiced for many years. For reasons that will be evident, time-temperature parameters have not become the ultimate tools for reliable long-time data prediction, but in the course of their development they have provided much useful information and have aided in practical decision making. Furthermore, because of continuing interest in their development, they have the best potential among all available techniques for improving the technology of reliable prediction of creep-rupture data for superalloys.

A TTP is basically a function correlating the creep-rupture test variables of stress (load), temperature and time, which normally are recorded for every standard uniaxial, isothermal, constant-load test performed throughout the world. When properly developed, these correlations can be used (*a*) to represent creep-rupture data in a compact form, allowing for analytical representation and interpolation of data not experimentally determined; (*b*) to provide a simple means of comparing the behavior of materials and of rating them in a relative manner; or (*c*) to extrapolate experimental data to time ranges ordinarily difficult to evaluate directly because of test limitations.

Much literature has been devoted to descriptions of experiences with these uses, and summaries are presented in the proceedings of recent conferences

Fig. 23 Characteristic appearance of creep-rupture correlations using four different time-temperature parameters

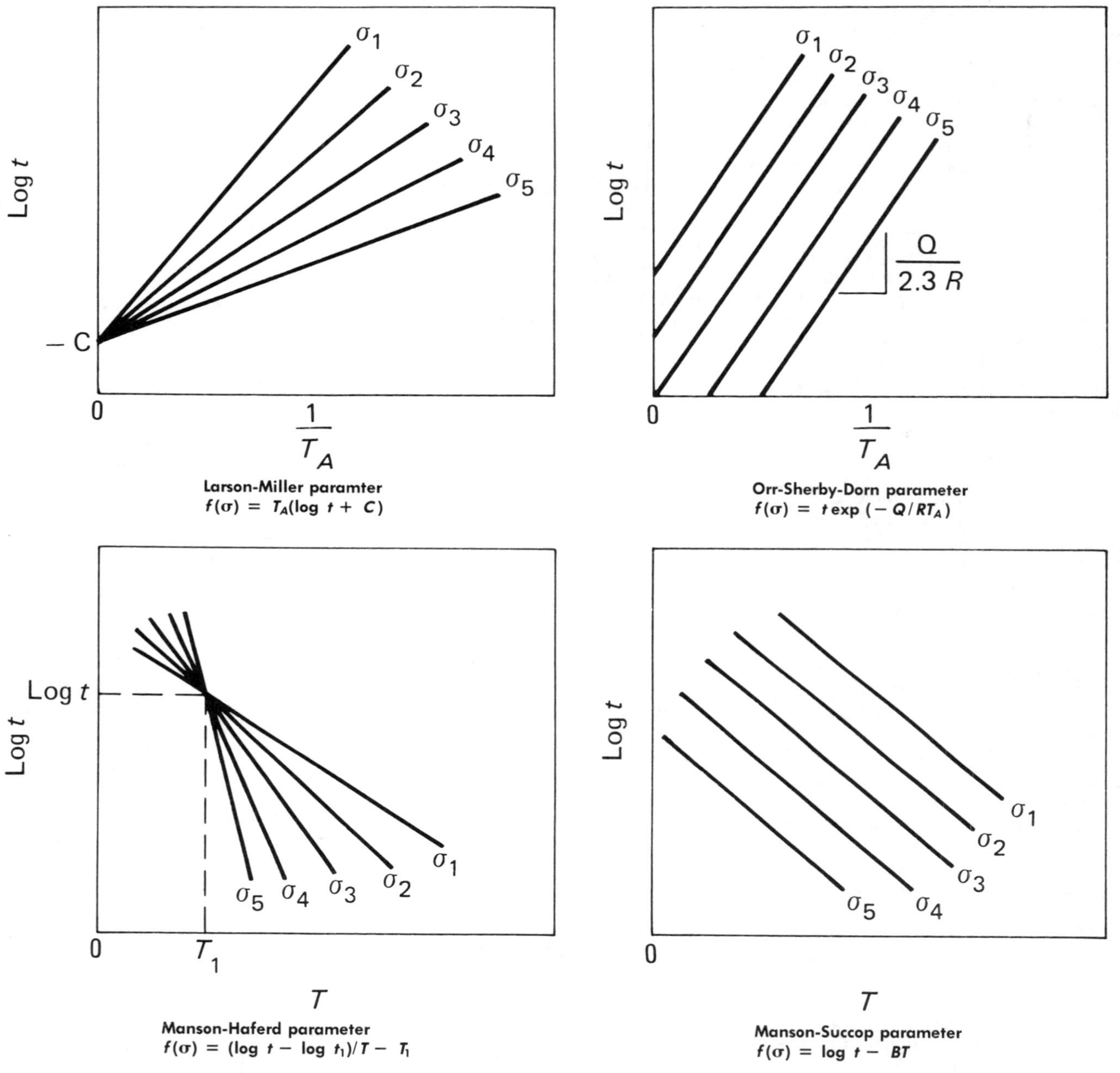

Larson-Miller paramter
$f(\sigma) = T_A(\log t + C)$

Orr-Sherby-Dorn parameter
$f(\sigma) = t \exp(-Q/RT_A)$

Manson-Haferd parameter
$f(\sigma) = (\log t - \log t_1)/T - T_1$

Manson-Succop parameter
$f(\sigma) = \log t - BT$

In the equations given above for each parameter, σ is applied stress, t is time, T is temperature in °C or °F, T_A is temperature in K or °R, Q is the activation energy, R is the gas constant, and B and C are numerical constants characteristic of the material and its metallurgical condition.

on the subject (Ref 3 and 7). Comprehensive treatments of the history of TTP development are given in Ref 2 and 3 and are recommended for those having further interest.

Conceptually, the TTP has a physical basis in chemical rate theory in that the Arrhenius equation, with suitable but not rigorous assumptions, is used to derive a number of useful expressions. Grounes (Ref 4) has reviewed this thoroughly, showing over thirty existing equations that have been proposed. By far the most common and illustrative of these are the four linear parameters described in Fig. 23. Here they are shown by name, form of equation, type of graphical plot they suggest and final-

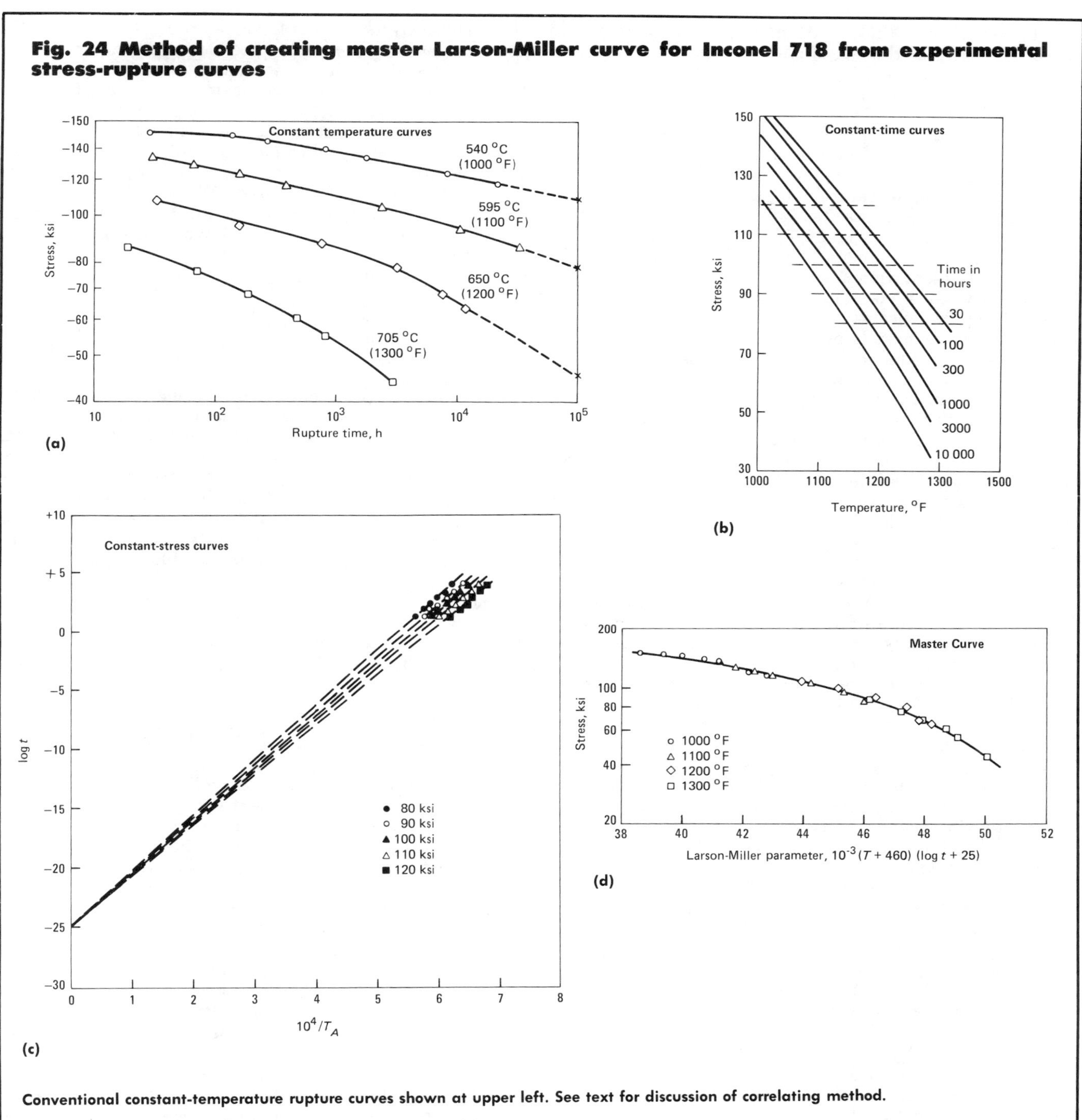

Fig. 24 Method of creating master Larson-Miller curve for Inconel 718 from experimental stress-rupture curves

Conventional constant-temperature rupture curves shown at upper left. See text for discussion of correlating method.

ly a schematic of that plot. The reasons for the graphs will become apparent but first it will help to give an actual example to illustrate the technique and its uses. In Fig. 24(a), an actual set of stress-rupture data (log stress vs log time to rupture) is shown for the nickel-base alloy Inconel 718. Next the data is replotted in the form shown in Fig. 23 for the Larson-Miller parameter. This is done by first using constant-time intercepts, shown in Fig. 24(b) by the dashed lines, to arrive at the constant-time curves shown in Fig. 24(b). Next, constant-stress curves are plotted as shown in Fig. 24(c) using again the intercepts shown by the dashed lines in Fig. 24(b). By extending the data in Fig. 24(c), a plausible set of convergent isostress lines meeting on the ordinate at a value of $\log t = -25$ can be obtained. Of course, such data could have been obtained by direct experiment also. Referring again to Fig. 23, it can be established that the Larson-Miller equation for this set of data is:

$$P = f(\sigma) = (T + 460)(\log t_R + 25) \quad \text{(Eq 1)}$$

where T is temperature in °F and t_R is time to rupture, in hours. At this stage, for each data point in the original set of stress, time and temperature data, the proper value can be substituted in Eq 1 and plotted as shown in Fig. 24(d). This is the compact parameter form of graphical representation known as the "master curve". It is possible, of course, to find the optimum constant(s) for any of the parameter models by computational data-fitting techniques (Ref 4). In either event it is now possible to see that a TTP is an analytical expression containing both time and temperature such that any combination of these two variables that produces the same numerical value of parameter number will cause failure at the same stress level. Thus the parameter is a means of interchanging time and temperature in the analysis of the creep-rupture process. Note then that the expression for time in the parameter in Eq 1 may also be the time to reach a fixed amount of creep strain (such as 1%), and indeed such expressions will accommodate the normal minimum-creep-rate data from evaluation of engineering creep curves when such values are substituted for time in a pertinent equation.

Returning to description of the uses of parameters and referring to Fig. 24(d), it is immediately recognized that the master curve can be described analytically by some relation between stress and the chosen parameter (Ref 5). It is most useful in this form for engineers who need to interpolate creep-rupture data. For example, assuming the validity of the master curve, isothermal rupture curves between 540 and 700 °C (1000 and 1300 °F) can be reproduced for parameter values included in the span of the master curve. Again viewing the master curve in Fig. 24(d), it is evident that the line drawn could have been determined by fewer data points—in fact, as few as three: the points on either extreme and one midway on the curve. In practice, for comparing materials for acceptance or in an alloy development program, three or four well-chosen points at various stresses and temperatures will produce such a curve in only several hundred hours of testing. This is an excellent method for comparing heats within a specification over a period of time or alternatively as a first cut in selecting the few best performing materials out of many prepared in a new-material development program. Although these are useful procedures, by far the most needed and controversial employment of parameters is in extrapolation. To identify the stress required for failure of Inconel 718 in 10^5 h at temperatures of 1000, 1100 and 1200 °F, simply substitute in Eq 1 the value 10^5 for t_R and, consecutively, the values 1000, 1100 and 1200 for T and solve for the parameter numbers, which are approximately 43 800, 46 900 and 49 900. Now proceed from the abscissa of the graph in Fig. 24(d) at each of these values to the master curve and thence to the ordinate to arrive at approximate stresses of 110, 78 and 46 ksi. These values are shown as dashed extensions on the curves in Fig. 24(a). Obviously, this is a very simple way of arriving at necessary design inputs without having to wait for eleven years of testing time to elapse. But we must consider the assumptions that are made and ask how sure we can be that these answers are correct.

First consider Fig. 23, in which common forms of linear parameters are shown. This suggests, and correctly so, that among many available data sets one or more of the types shown will better fit the data and hence should be used. Note that the word "linear" is used to describe these common forms because linear isostress lines are used to construct the graphical presentations. Practically, of course, inherent scatter in the test data, metallurgical characteristics of the material (discussed later) and extent of the experimental data may have to be taken into account when deciding which method to use. The most important factors affecting this decision are the degree of accuracy needed and the end use for which the procedure is being employed. Most complete data sets do not conform to the simple requirement of linearity. That is to say, the techniques described up to this point may not produce the linear isostress lines and convergence or parallelism so neatly shown in Fig. 23. This then requires much more sophisticated mathematical treatment and perhaps even less certainty in the results. These problems are treated in detail in the references.

Up to now it has been assumed that large quantities of data, such as shown in Fig. 23, are available to be treated as described, but this is most emphatically not the case. Because of this and the pressing need for a parameter extrapolation estimate, the universalized constant parameter has come into use. In these cases, experience on many materials over extended periods has indicated that on the average certain constants in the linear parameters are most statistically useful. For example, (refer to Fig. 23), the following values of constants have been suggested:

Parameter	Constant(s)
Larson-Miller	$C = 20$
Manson-Haferd	$t_1 = 10$; $T = 600$
Manson-Succop	$B = 0.01$

Thus it is reasonable in certain instances to make comparisons and extrapolations for incomplete sets of data by assuming the validity of one of these universalized parameter methods. In so doing, the analyst is well advised of the probability of inaccuracy. Further to this point, because of observations that Larson-Miller evaluations were too liberal and Manson-Haferd evaluations too conservative, the following "compromise" parameter was devised by Manson (Ref 3):

$$f(\sigma) = \log t + \tfrac{1}{40}(\log t)^2 - \frac{40\,000}{T + 460} \quad \text{(Eq 2)}$$

This, too, is a universal parameter that may provide the analyst with a better fit for the data at hand.

By now it is clear that the TTP concept of interchanging time and temperature is not without problems and in certain cases is suspect for some materials, at least in some ranges of time and temperature. The most serious lack is the inability to predict metallurgical phase instabilities, which may cause, for example, curvature in isothermal stress-rupture lines beyond the point at which data are available. Naturally, the parameter cannot represent effects other than those apparent in the short-time data involved in its derivation. Note, too, that optimizing a constant in a parametric method ensures only that a best fit of short-time data is obtained; it does not guarantee the reliability of extrapolation beyond that range. Of the types of instabilities occurring in metallurgical systems, those in which precipitates nucleate and grow in a discontinuous way are the most disturbing. The occurrence of "sigma" phase is perhaps the best known of these. Such precipitations usually are accompanied by changes in fracture mode occurring discontinuously over certain time and temperature ranges, which disturbs the "family" relationship of curves needed to justify the parameter usage. On the other hand, some metallurgical changes (such as grain growth and overaging [softening]) will occur after short times at higher temperatures much as they occur after longer times at lower temperatures. Here the parameter methods remain tractable. The obvious point to be made is that all possible information concerning the high-temperature physical metallurgy of the alloy in question should be sought out ahead of analysis and judgment used in parameter application. All sets of data should be further examined to ensure that testing stresses above the yield point are not included, that temperatures inducing such surface problems as excessive oxidation have not been used (such data should be discarded) and that sufficient data are available to justify parameter use (master curves should not be extrapolated).

It has become obvious that the metallurgical complexity of engineering materials precludes a strict theoretical basis for the derivation of correlating parameters. The tendency now is to seek expressions in which material performance itself reveals the best form of parameter to use (Ref 1, 6 and 7). Despite the inherent problems in existing TTP's and the pressures for development of new ones, current technology dictates continued use of TTP's. What is required is complete cooperation between the metallurgist and the analyst in making objective decisions on how and where to use the techniques. Finally, it is more than clear that the parameter does not obviate the need for generating data, rather it justifies the continued accumulation of long-time data for analysis of these complex engineering materials.

REFERENCES

1. *Characterization of Materials for Service at Elevated Temperatures:* American Society of Mechanical Engineers, New York, 1978
2. The Evaluation of Elevated Temperature Creep and Rupture Strength Data: An Historical Perspective, by R. M. Goldhoff: in *Characterization of Materials for Service at Elevated Temperatures,* American Society of Mechanical Engineers, New York, 1978, p 247-265
3. Time-Temperature Parameters—A Re-evaluation and Some New Approaches, by S. S. Manson: in *Time-Temperature Parameters for Creep Rupture Analysis:* American Society for Metals, Metals Park, OH 1969
4. A Reaction-Rate Treatment of the Extrapolation Methods in Creep Testing, by M. Grounes: *Journal of Basic Engineering (Transactions of ASME, Series D),* 1969
5. *Numerical Methods for Creep and Rupture Analysis,* by J. B. Conway: Gordon and Breach, New York, 1967
6. Towards the Standardization of Time-Temperature Parameter Usage in Elevated Temperature Data Analysis, by R. M. Goldhoff *et al.: ASTM Journal of Testing and Evaluation,* Vol 2, No. 5, Sept. 1974
7. *Development of a Standard Methodology for Correlation and Extrapolation of Elevated Temperature Creep and Rupture Data,* edited by R. M. Goldhoff: Electric Power Research Institute, Palo Alto, California, Vols 1 and 2, April 1979
8. *Characterization of Materials for Service at Elevated Temperature,* edited by G. V. Smith: American Society of Mechanical Engineers, June 1978

Properties of Superalloys

Properties of Nickel-Base Superalloys

B-1900, B-1900 + Hf

Chemical Composition

Typical composition. 64 Ni; 8.0 Cr; 10.0 Co; 6.0 Mo; 4.0 Ta; 6.0 Al; 1.0 Ti; 0.10 C; 0.015 B; 0.10 Zr. B-1900 + HF contains 1.5 Hf

Applications

Typical uses. For high temperature applications, jet engines, turbines

Mechanical Properties

Tensile properties. See Table 1.
Elastic modulus:

Temperature °C	Temperature °F	Modulus of elasticity GPa	Modulus of elasticity 10^6 psi
21	70	215	31.0
93	200	210	30.5
205	400	205	29.7
315	600	200	28.9
425	800	195	28.0
540	1000	185	27.0
650	1200	180	26.0
760	1400	170	24.9
870	1600	165	23.7
980	1800	155	22.4

Table 1 Typical mechanical properties of B-1900 alloy

Temperature °C	Temperature °F	Tensile strength MPa	Tensile strength ksi	Yield strength MPa	Yield strength ksi	Elongation, %
21	70	970	141	825	120	8
540	1000	1010	146	870	126	7
650	1200	1010	147	925	134	6
760	1400	950	138	805	117	4
870	1600	795	115	695	101	4
980	1800	550	80	415	60	7
1090	2000	270	38	195	28	11

Creep-rupture characteristics. 650 MPa (94 ksi) at 760 °C (1400 °F) from bars machined from blades: B-1900, 1.6%; B-1900 + Hf, 4.8% **Stress-rupture characteristics.** 650 MPa (94 ksi) at 760 °C (1400 °F): bars machined from blades: B-1900, 19.2 h; B-1900 + Hf, 86.4 h; 6.4 mm (0.250 in.) cast to size, 81 h; 4.1 mm (0.160 in.) cut from blade, 19 h; 1.3 mm (0.050 in.) cut from thin wall, 10 h. 200 MPa (29 ksi) at 980 °C (1800 °F): 6.4 mm (0.250 in.) cast to size, 34 h; 4.1 mm (0.160 in.) cut from blade, 32 h; 1.3 mm (0.050 in.) cut from thin wall, 13 h

Temperature °C	Temperature °F	Typical rupture strength MPa	Typical rupture strength ksi
At 100 h			
815	1500	505	73
870	1600	385	56
925	1700	260	38
980	1800	170	25
1040	1900	105	15
1090	2000	62	9
At 1000 h			
815	1500	380	55
870	1600	255	37
925	1700	120	25
980	1800	105	15
1040	1900	62	9
1090	2000	34	5

Structure

Microstructure. fcc matrix containing one or more carbides. Additions of hafnium alter the MC carbide from script to a more equiaxed morphology

Mass Characteristics

Density. 8.22 Mg/m^3 (0.297 lb/in.^{3}) at 20 °C (68 °F)

Thermal Properties

Coefficient of thermal expansion. Linear:

Temperature °C	Temperature °F	Mean coefficient μm/m·K	Mean coefficient μin./in.·°F
93	200	11.7	6.5
205	400	12.2	6.75
315	600	12.6	7.0
425	800	13.0	7.2
540	1000	13.3	7.4
650	1200	13.7	7.6
760	1400	14.2	7.9
870	1600	14.9	8.3
980	1800	15.8	8.8
1090	2000	16.2	9.0

Thermal conductivity:

Temperature °C	°F	Conductivity W/m·K	Btu/ft·h·°F
93	200	10.2(a)	5.9
205	400	11.7	6.8
315	600	13.1	7.6
425	800	14.7	8.5
540	1000	16.3	9.4
650	1200	18.1	10.5
760	1400	20.0	11.6
870	1600	21.9	12.7
980	1800	23.0	13.3

(a) Extrapolated values.

Chemical Properties

General corrosion behavior. Excellent resistance to oxidation and hot-corrosion

Resistance to specific corroding agents. Poor resistance to hot corrosion when coated with Na_2CO_3, $NaNO_3$ or Na_2SO_4

Fabrication Characteristics

Weldability. Impossible to make fissure-free fusion welds

IN-100
60Ni-10Cr

Specifications

AMS. Investment cast: 5397

Chemical Composition

Typical composition. 60 Ni; 10 Cr; 15.0 Co; 3.0 Mo; 5.5 Al; 4.7 Ti; 1.0 V; 0.18 C; 0.014 B; 0.06 Zr

Applications

Typical uses. For jet engine parts, turbine blades for operating temperatures up to 1040 °C (1900 °F)

Mechanical Properties

Tensile properties. See Table 2.

Elastic modulus:

Temperature °C	°F	Modulus of elasticity GPa	10^6 psi
21	70	215	31.2
93	200	210	30.7
205	400	205	29.9
315	600	200	28.9
425	800	195	28.1
540	1000	190	27.1
650	1200	180	26.1
760	1400	175	25.1
870	1600	160	23.5
980	1800	150	21.9

Table 2 Typical mechanical properties of IN-100

Temperature °C	°F	Tensile strength MPa	ksi	Yield strength MPa	ksi	Elongation, %
21	70	1010	147	850	123	9
540	1000	1090	158	885	128	9
650	1200	1110	161	890	129	6
760	1400	1070	155	860	125	6.5
870	1600	885	128	695	101	6
980	1800	565	82	370	54	6
1090	2000	380(a)	55(a)	240(a)	35(a)	...

(a) Extrapolated values.

Stress-rupture characteristics. Typical:

Temperature °C	°F	Rupture strength MPa	ksi
At 100 h			
760	1400	625	91
816	1500	505	73
870	1600	380	55
925	1700	260	38
980	1800	170	25
1040	1900	110	16
1090	2000	62	9
At 1000 h			
760	1400	515	75
816	1500	380	55
870	1600	255	37
925	1700	170	25
980	1800	105	15
1040	1900	59	8.5

Mass Characteristics

Density. 7.75 Mg/m³ (0.280 lb/in.³) at 20 °C (68 °F)

Thermal Properties

Coefficient of thermal expansion. Linear:

Temperature °C	°F	Mean coefficient μm/m·K	μin./in.·°F
93	200	13.0	7.2
205	400	13.0	7.2
315	600	13.1	7.3
425	800	13.5	7.5
540	1000	13.9	7.7
650	1200	14.4	8.0
760	1400	14.4	8.3
870	1600	15.5	8.8
980	1800	16.8	9.3
1090	2000	18.1	10.1

Specific heat:

Temperature °C	°F	Specific heat J/kg	Btu/lb·°F
540	1000	480	0.115
650	1200	500	0.12
760	1400	525	0.125
870	1600	545	0.13
980	1800	585	0.14
1090	2000	605	0.145

Thermal conductivity:

Temperature °C	°F	Conductivity W/m·K	Btu/ft·h·°F
540	1000	17.3	10.0
650	1200	18.9	10.9
760	1400	21.7	12.5
870	1600	25.2	14.6
980	1800	28.8	16.6

Table 3 Typical mechanical properties of IN-162

Temperature °C	°F	Tensile strength MPa	ksi	Yield strength MPa	ksi	Elongation, %
21	70	1010	146	815	118	7
540	1000	1020	148	795	115	6.5
650	1200	1090	158	855	124	5.5
760	1400	1010	146	850	123	5.5
870	1600	825	120	710	103	6
980	1800	585	85	450	65	5.5

IN-162
73Ni-10Cr

Chemical Composition

Typical composition. 73 Ni; 10 Cr; 4 Mo; 2 W; 2 Ta; 1 Nb; 6.5 Al; 1.0 Ti; 0.12 C; 0.02 B; 0.10 Zr

Applications

Typical uses. Gas turbine blades and vanes

Mechanical Properties

Tensile properties. See Table 3.
Elastic modulus:

Temperature °C	Temperature °F	Modulus of elasticity GPa	Modulus of elasticity 10^6 psi
21	70	195	28.5
93	200	195	28.0
205	400	190	27.3
315	600	185	26.6
425	800	180	25.8
540	1000	170	24.9
650	1200	165	24.0
760	1400	160	23.0
870	1600	150	21.8
980	1800	140	20.4

Stress-rupture characteristics. Typical:

Temperature °C	Temperature °F	Rupture strength MPa	Rupture strength ksi
At 100 h			
760	1400	620	90
815	1500	505	73
870	1600	340	49
925	1700	240	35
980	1800	165	24
1040	1900	108	17
At 1000 h			
760	1400	525	76
815	1500	370	54
870	1600	255	37
925	1700	170	25
980	1800	110	16
1040	1900	90	13

Mass Characteristics

Density. 8.08 Mg/m³ (0.292 lb/in.³) at 20 °C (68 °F)

Thermal Properties

Coefficient of thermal expansion. Linear:

Temperature °C	Temperature °F	Mean coefficient μm/m·K	Mean coefficient μin./in.·°F
93	200	12.2	6.75
205	400	12.9	7.15
315	600	13.4	7.45
425	800	13.8	7.65
540	1000	14.1	7.85
650	1200	14.5	8.05
760	1400	15.0	8.35
870	1600	15.5	8.6
980	1800	16.6	9.2

IN-731

Chemical Composition

Typical composition. 9.5 Cr; 10.0 Co; 2.5 Mo; 5.5 Al; 4.6 Ti; 1.0 V; 0.18 C; 0.015 B; 0.06 Zr; 0.2 max Mn; 0.5 max Fe, rem Ni

Mechanical Properties

Tensile properties. See Table 4.
Stress-rupture characteristics. Typical:

Temperature °C	Temperature °F	Rupture strength MPa	Rupture strength ksi
At 100 h			
815	1500	505	73
925	1700	250	38
980	1800	165	24
1040	1900	110	16
At 1000 h			
815	1500	365	53
925	1700	160	23.5
980	1800	105	15
1040	1900	48	7

Mass Characteristics

Density. 7.75 Mg/m³ (0.280 lb/in.³) at 20 °C (68 °F)

Table 4 Typical mechanical properties of IN-731

Temperature °C	Temperature °F	Tensile strength MPa	Tensile strength ksi	Yield strength MPa	Yield strength ksi	Elongation, %
21	70	835	121	725	105	6.5
540	1000	...	...	...	...	...
650	1200	895	130	745	108	5
760	1400	915	133	770	112	4.5
870	1600	750	108	585	88	3.5
980	1800	525	76	360	52	6.5
1090	2000	275	40	170	25	7.5

IN-738

Chemical Composition

Typical composition. 61 Ni; 16.0 Cr; 8.5 Co; 1.7 Mo; 2.6 W; 1.7 Ta; 0.9 Nb; 3.4 Al; 3.4 Ti; 0.17 C; 0.01 B; 0.10 Zr

Mechanical Properties

Tensile properties. See Table 5.
Elastic modulus:

Temperature °C	Temperature °F	Modulus of elasticity GPa	Modulus of elasticity 10^6 psi
21	70	200	29.2
93	200	195	28.3
205	400	190	27.6
315	600	185	26.8
425	800	180	26.0
540	1000	175	25.4
650	1200	170	24.3
760	1400	160	23.2
870	1600	150	21.9
980	1800	140	20.3

Stress-rupture characteristics. Typical:

Temperature °C	Temperature °F	Rupture strength MPa	Rupture strength ksi
At 100 h			
760	1400	595	86
815	1500	455	66
870	1600	315	46
925	1700	215	31
980	1800	130	19
At 1000 h			
760	1400	475	69
815	1500	335	48
870	1600	215	31.5
925	1700	140	20
980	1800	83	12

Mass Characteristics

Density. 8.10 Mg/m³ (0.293 lb/in.³) at 20 °C (68 °F)

Table 5 Typical mechanical properties of IN-738

Temperature °C	°F	Tensile strength MPa	ksi	Yield strength MPa	ksi	Elongation, %
21	70	1100	159	915	133	5
540	1000	1000	145	795	115	5
650	1200	1060	154	815	118	6
760	1400	965	140	795	115	6.5
870	1600	770	112	550	80	11
980	1800	455	66	345	50	13

Thermal Properties

Coefficient of thermal expansion. Linear:

Temperature °C	°F	Mean coefficient μm/m·K	μin./in.·°F
93	200	11.6	6.45
205	400	12.2	6.75
315	600	12.9	7.15
425	800	13.6	7.55
540	1000	14.0	7.75
650	1200	14.5	8.05
760	1400	14.8	8.25
870	1600	15.4	8.55
980	1800	15.9	8.85

Specific heat:

Temperature °C	°F	Specific heat J/kg	Btu/lb·°F
21	70	420	0.10
93	200	460	0.11
205	400	500	0.12
315	600	525	0.125
425	800	545	0.13
540	1000	565	0.135
650	1200	585	0.14
760	1400	625	0.15
870	1600	670	0.16
980	1800	710	0.17
1090	2000	710	0.17

Thermal conductivity:

Temperature °C	°F	Conductivity W/m·K	Btu/ft·h·°F
205	400	11.8	6.8
315	600	13.7	7.9
425	800	15.6	9.0
540	1000	17.7	10.2
650	1200	19.7	11.4
760	1400	21.5	12.4
870	1600	23.3	13.5
980	1800	25.3	14.6
1090	2000	27.2	15.7

Chemical Properties

General corrosion behavior. Excellent hot-corrosion resistance

IN-792

Chemical Composition

Typical composition. 61 Ni; 12.4 Cr; 9.0 Co; 1.9 Mo; 3.8 W; 3.9 Ta; 3.1 Al; 4.5 Ti; 0.12 C; 0.02 B; 0.10 Zr

Mechanical Properties

Tensile properties. See Table 6.
Stress-rupture characteristics. Typical:

Temperature °C	°F	Rupture strength MPa	ksi
At 100 h			
760	1400	690	100
815	1500	515	75
870	1600	365	53
925	1700	255	37
980	1800	165	24
At 1000 h			
760	1400	545	79
815	1500	380	55
870	1600	260	38
925	1700	170	24.5
980	1800	105	15

Mass Characteristics

Density. 8.25 Mg/m^3 (0.298 $lb/in.^3$) at 20 °C (68 °F)

Chemical Properties

General corrosion behavior. Excellent hot-corrosion resistance

IN-939

Chemical Composition

Composition limits. 0.13 to 0.17 C; 22.2 to 22.8 Cr; 18.5 to 19.5 Co; 1.8 to 2.2 W; 0.9 to 1.1 Nb; 1.3 to 1.5 Ta; 3.6 to 3.8 Ti; 1.8 to 2.0 Al; 0.05 to 0.14 Zr; 0.004 to 0.014 B; 0.005 max N; 0.005 max S; 0.5 max Fe; 0.2 max Si; 0.2 max Mn; 0.00005 max Pb; rem Ni

Applications

Typical uses. For engines operating in a marine environment for extended periods and industrial gas turbines operating on fuels that are less pure than distillate fuel

Mechanical Properties

Tensile properties. See Table 7.
Elastic modulus:

Temperature °C	°F	Modulus of elasticity GPa	10^6 psi
20	68	200	28.8
100	212	195	28.0
200	390	190	27.4
300	570	185	26.6
400	750	180	26.0
500	930	170	24.9
600	1110	165	24.1
700	1290	155	22.8
800	1470	150	21.6
900	1650	140	20.2
1000	1830	125	18.2

Stress-rupture characteristics. Typical:

Temperature °C	°F	Rupture strength MPa	ksi
At 100 h			
730	1350	600	87
815	1500	410	59
870	1600	285	41
925	1700	180	26
980	1800	105	15
At 300 h			
730	1350	530	77

(continued)

Table 6 Typical mechanical properties of IN-792

Temperature °C	°F	Tensile strength MPa	ksi	Yield strength MPa	ksi	Elongation, %
21	70	1170	170	1060	154	4
540	1000	...	...	1100	160	5.5
650	1200	1090	158	1100	160	6.5
760	1400	995	144	995	144	4
870	1600	840	122	660	96	8

Temperature °C	°F	Rupture Strength MPa	ksi
815	1500	360	52
870	1600	240	35
925	1700	145	21
980	1800	83	12
At 1000 h			
730	1350	455	66
815	1500	305	44
870	1600	200	29
925	1700	115	17
980	1800	7	1
At 3000 h			
815	1500	260	38
870	1600	165	24
925	1700	90	13
At 10 000 h			
815	1500	220	32
870	1600	140	20
925	1700	76	11

Mass Characteristics

Density. 8.16 Mg/m³ (0.295 lb/in.³)

Thermal Properties

Liquidus temperature. 1315 °C (2400 °F)
Solidus temperature. 1235 °C (2255 °F)
Coefficient of thermal expansion. Linear:

Temperature °C	°F	Mean coefficient GPa	10⁶ psi
20–100	68–212	77	11.2
20–200	68–390	86	12.5
20–300	68–570	89	12.9
20–400	68–750	91	13.2
20–500	68–930	92	13.4
20–600	68–1110	94	13.6
20–700	68–1290	95	13.8
20–800	68–1470	97	14.0
20–900	68–1650	98	14.2

Specific heat:

Temperature °C	°F	Specific heat J/kg	Btu/lb·°F
20	68	416	0.099
100	212	435	0.104
200	390	460	0.109
300	570	485	0.115
400	750	510	0.122
500	930	534	0.128
600	1110	559	0.134
700	1290	583	0.139
800	1470	608	0.145
900	1650	633	0.151
1000	1830	658	0.157

Thermal conductivity:

Temperature °C	°F	Conductivity W/m·K	Btu/ft·h·°F
20	68	11.3	6.5
100	212	12.6	7.3
200	390	14.2	8.2
300	570	15.6	9.0
400	750	17.1	9.9
500	930	18.5	10.7
600	1110	20.0	11.6
700	1290	21.5	12.4
800	1470	22.9	13.2

Electrical Properties

Electrical resistivity:

Temperature °C	°F	Resistivity, μΩ·m
20	68	1.23
100	212	1.24
200	390	1.27
300	570	1.31
400	750	1.34
500	930	1.36
600	1110	1.37
700	1290	1.38
800	1470	1.40

Table 7 Typical mechanical properties of IN-939

Temperature °C	°F	Tensile strength MPa	ksi	Yield strength MPa	ksi	Elongation, %	Reduction in area, %
20	68	950–965	138–140	770–800	112–116	3.0–3.8	9.0
200	390	960	139	770	112	4.9	5.0
400	750	905	131	715	104	4.6	10.0
600	1110	890	129	685	99	5.2	13.0
650	1200	865–930	125–135	690	100	5–6	6–9
700	1290	915	133	695	101	4	7
732	1350	905–970	132–141	665–690	96–100	4–8	7–12
760	1400	1020	148	690	100	8	12
815	1500	865	125	620	90	13	26
925	1700	505	73	275	40	23	40

Inconel 713C

Specifications

AMS. Investment cast: 5391

Chemical Composition

Typical composition. 74 Ni; 12.5 Cr; 4.2 Mo; 2.0 Nb; 6.1 Al; 0.8 Ti; 0.12 C; 0.012 B; 0.10 Zr

Applications

Typical uses. Developed specifically for integral rotor castings

Mechanical Properties

Tensile properties. See Table 8.
Elastic modulus:

Temperature °C	°F	Modulus of elasticity GPa	10⁶ psi
21	70	205	29.9
93	200	205	29.5
205	400	200	28.7
315	600	195	28.0
425	800	190	27.2
540	1000	180	26.2
650	1200	170	25.1
760	1400	165	24.2
870	1600	155	22.6
980	1800	150	21.4

Stress-rupture characteristics. Typical:

Temperature °C	°F	Rupture strength MPa	ksi
At 100 h			
760	1400	570	83
815	1500	415	60
870	1600	290	42
925	1700	205	30
980	1800	145	21
1040	1900	83	12
1090	2000	45	6.5
At 1000 h			
705	1300	605(a)	88(a)
760	1400	460	65
815	1500	305	44
870	1600	195	28
925	1700	125	18
980	1800	90	13
1040	1900	52(a)	7.5(a)

(a) Extrapolated values.

Mass Characteristics

Density. 7.91 Mg/m³ (0.286 lb/in.³) at 20 °C (68 °F)

Thermal Properties

Melting range. 1260 to 1290 °C (2300 to 2350 °F)

Table 8 Typical mechanical properties of Inconel 713C

Temperature °C	°F	Tensile strength MPa	ksi	Yield strength MPa	ksi	Elongation, %
21	70	850	123	740	107	8
540	1000	860	125	705	102	10
650	1200	870	126	715	104	7
760	1400	940	136	745	108	6
870	1600	725	105	495	72	14
980	1800	470	68	305	44	20

Coefficient of thermal expansion. Linear:

Temperature °C	°F	Mean coefficient μm/m·K	μin./in.·°F
93	200	10.6	5.9
205	400	11.9	6.6
315	600	12.6	7.0
425	800	13.1	7.3
540	1000	13.5	7.5
650	1200	14.0	7.8
760	1400	14.8	8.2
870	1600	15.5	8.6
980	1800	16.4	9.1
1090	2000	17.1	9.5

Specific heat:

Temperature °C	°F	Specific heat W/m·K	Btu/lb·°F
21	70	420	0.10
93	200	460	0.11
205	400	500	0.12
315	600	525	0.125
425	800	545	0.13
540	1000	565	0.135
650	1200	585	0.14
760	1400	625	0.15
870	1600	670	0.16
980	1800	690	0.165
1090	2000	710	0.17

Thermal conductivity:

Temperature °C	°F	Conductivity W/m·K	Btu/ft·h·°F
93	200	10.9	6.3
205	400	12.2	7.0
315	600	13.8	8.0
425	800	15.4	8.9
540	1000	17.0	9.8
650	1200	18.6	10.7
760	1400	20.4	11.8
870	1600	22.3	12.9
980	1800	24.3	14.0
1090	2000	26.4	15.3

Chemical Properties

General corrosion behavior. Excellent resistance to oxidation

Inconel 713LC

Chemical Composition

Typical composition. 75 Ni; 12.0 Cr; 4.5 Mo; 2.0 Nb; 5.9 Al; 0.6 Ti; 0.05 C; 0.010 B; 0.10 Zr

Applications

Typical uses. Rotor castings (addition of hafnium allows larger and more complex rotor castings to be made)

Mechanical Properties

Tensile properties. See Table 9.
Elastic modulus:

Temperature °C	°F	Modulus of elasticity GPa	10^6 psi
21	70	195	28.6
93	200	195	28.1
205	400	185	27.3
315	600	185	26.6
425	800	180	25.8
540	1000	170	25.0
650	1200	165	24.0
760	1400	160	23.1
870	1600	150	21.6
980	1800	135	19.7

Stress-rupture characteristics. Typical:

Temperature °C	°F	Rupture strength MPa	ksi
At 100 h			
760	1400	550	80
815	1500	430	62
870	1600	295	43
925	1700	200	29
980	1800	140	20
At 1000 h			
760	1400	415	60
815	1500	310	45
870	1600	205	30
925	1700	130	19
980	1800	90	13

Mass Characteristics

Density. 8.0 Mg/m^3 (0.289 $lb/in.^3$) at 20 °C (68 °F)

Thermal Properties

Coefficient of thermal expansion:

Temperature °C	°F	Mean coefficient μm/m·K	μin./in.·°F
93	200	10.1	5.6
205	400	12.6	7.0
315	600	13.7	7.6
425	800	14.9	8.3
540	1000	15.8	8.75
650	1200	16.0	8.9
760	1400	16.2	9.0
870	1600	16.7	9.3
980	1800	17.0	9.8
1090	2000	18.9	10.5

Specific heat:

Temperature °C	°F	Specific heat J/kg	Btu/lb·°F
21	70	440	0.105
93	200	460	0.11
205	400	500	0.12
315	600	525	0.125
425	800	545	0.13
540	1000	565	0.135

(continued)

Table 9 Typical mechanical properties of Inconel 713LC

Temperature °C	°F	Tensile strength MPa	ksi	Yield strength MPa	ksi	Elongation, %
21	70	895	130	750	109	15
540	1000	895	130	760	110	11
650	1200	1080	157	785	114	11
760	1400	950	138	760	110	11
870	1600	750	109	580	84	12
980	1800	470	68	305	44	22

Temperature °C	°F	Specific heat J/kg	Btu/lb·°F
650	1200	585	0.14
760	1400	625	0.15
870	1600	670	0.16
980	1800	690	0.165
1090	2000	710	0.17

Thermal conductivity:

Temperature °C	°F	Conductivity W/m·K	Btu/ft·h·°F
93	200	10.7	6.2
205	400	12.1	7.0
315	600	13.7	7.9
425	800	15.3	8.8
540	1000	16.7	9.6
650	1200	18.3	10.6
760	1400	20.0	11.6
870	1600	21.7	12.6
980	1800	23.1	13.3
1090	2000	25.3	14.6

M-22

Chemical Composition

Typical composition. 71 Ni; 5.7 Cr; 2.0 Mo; 11.0 W; 3.0 Ta; 6.3 Al; 0.13 C; 0.60 Zr

Mechanical Properties

Tensile properties. See Table 10.
Stress-rupture characteristics. Typical:

Temperature °C	°F	Rupture strength MPa	ksi
At 100 h			
650	1200	725	105
705	1300	670	97
760	1400	600	87
815	1500	515	75
870	1600	395	57
925	1700	285	41
980	1800	200	29
1040	1900	140	20
1090	2000	83	12
At 1000 h			
760	1400	545	79
815	1500	385	56
870	1600	285	41
925	1700	195	28
980	1800	130	19
1040	1900	83	12
1090	2000	41	6

Table 10 Typical mechanical properties of M-22

Temperature °C	°F	Tensile strength MPa	ksi	Yield strength MPa	ksi	Elongation, %
21	70	730	106	685	99	5.5
540	1000	780	113	730	106	4.5
650	1200	835	121	765	111	5
760	1400	910	132	770	112	5
870	1600	885	128	675	98	4.5
980	1800	545	79	360	52	5.5

Mass Characteristics

Density. 8.63 Mg/m³ (0.312 lb/in.³) at 20 °C (68 °F)

Thermal Properties

Coefficient of thermal expansion. Linear:

Temperature °C	°F	Mean coefficient µm/m·K	µin./in.·°F
93	200	12.4	6.9
205	400	12.4	6.9
315	600	12.8	7.1
425	800	13.1	7.3
540	1000	13.3	7.4
650	1200	13.7	7.6
760	1400	14.2	7.9
870	1600	14.8	8.2
980	1800	15.5	8.6

MAR-M 200
MAR-M 200(DS)
60Ni-9Cr

Commercial Name

Trade name. SM200 (obsolete)

Chemical Composition

Composition limits. Typical. 60 Ni; 9.0 Cr; 10.0 Co; 12.0 W; 1.0 Nb; 5.0 Al; 2.0 Ti; 0.15 C(a); 0.015 B; 0.05 Zr. Range: 1.75 to 2.25 Ti; 11.5 to 13.5 W; 9.0 to 11.0 Co; 8.0 to 10.0 Cr; 4.75 to 5.25 Al; 0.01 to 0.02 B; 0.12 to 0.17 C; 0.75 to 1.25 Nb; 0.03 to 0.08 Zr; 1.50 max Fe; 0.20 max Mn; 0.20 max Si; rem Ni

(a) In MAR-M200(DS), 0.13 C.

Applications

Typical uses. Turbine blades and vanes in aircraft gas turbine engines

Mechanical Properties

Tensile properties. See Table 11.
Elastic modulus:

Temperature °C	°F	Modulus of elasticity GPa	10⁶ psi
MAR-M 200			
21	70	220	31.6
93	200	215	31.2
205	400	215	30.9
315	600	195	28.6
425	800	190	27.7
540	1000	185	26.7
650	1200	175	25.6
760	1400	170	24.5
870	1600	160	23.3
980	1800	145	21.0
MAR-M 200(DS)			
21	70	130	19.0
93	200	130	18.5
205	400	125	17.8
315	600	120	17.2
425	800	115	16.7
540	1000	110	16.0
650	1200	105	15.2
760	1400	100	14.3
870	1600	90	13.1
980	1800	79	11.5

Stress-rupture characteristics. Typical:

Temperature °C	°F	Rupture strength MPa	ksi
MAR-M 200 at 100 h			
760	1400	635	92
815	1500	495	72
870	1600	385	56
925	1700	285	41
980	1800	180	26
1040	1900	120	17.5
MAR-M 200 at 1000 h			
760	1400	580	84
815	1500	415	60
870	1600	295	43
925	1700	200	29
980	1800	130	18.5
1040	1900	83	12

(continued)

Table 11 Typical mechanical properties of MAR-M 200 and MAR-M 200(DS)

Temperature °C	Temperature °F	Tensile strength MPa	Tensile strength ksi	Yield strength MPa	Yield strength ksi	Elongation, %
MAR-M 200						
21	70	930	135	840	122	7
540	1000	945	137	850	123	5
650	1200	950	138	855	124	4
760	1400	930	135	840	122	3.5
870	1600	840	122	760	110	4
980	1800	550	80	470	68	4.5
1090	2000	325	47	...	...	...
MAR-M 200(DS)						
21	70	1000	145	860	125	10
540	1000	1010	147	875	127	10
650	1200	1020	148	890	129	9
760	1400	1050	152	925	134	4.5
870	1600	915	133	780	113	4.5
980	1800	655	95	620	90	8

Temperature °C	Temperature °F	Rupture strength MPa	Rupture strength ksi
MAR-M 200(DS) at 100 h			
760	1400	725	105
815	1500	580	84
870	1600	450	65
925	1700	315	40
980	1800	200	29
1040	1900	140	20
MAR-M 200(DS) at 1000 h			
760	1400	660	96
980	1800	140	20
1040	1900	97	14

Mass Characteristics

Density. 8.53 Mg/m^3 (0.308 lb/in.^{3}) at 20 °C (68 °F)

Thermal Properties

Coefficient of thermal expansion. Linear, MAR-M 200:

Temperature °C	Temperature °F	Mean coefficient μm/m·K	Mean coefficient μin./in.·°F
205	400	11.9	6.6
315	600	12.4	6.9
425	800	12.8	7.1
540	1000	13.1	7.3
650	1200	13.5	7.5
760	1400	14.0	7.8
870	1600	14.8	8.2
980	1800	15.8	8.8
1090	2000	17.0	9.8

Specific heat. MAR-M 200:

Temperature °C	Temperature °F	Specific heat J/kg	Specific heat Btu/lb·°F
21	70	400	0.095
93	200	400	0.095
205	400	395	0.095
315	600	420	0.10
425	800	440	0.105
540	1000	420	0.10
650	1200	460	0.11
760	1400	480	0.115
870	1600	500	0.12
980	1800	525	0.125
1090	2000	565	0.135

Thermal conductivity. MAR-M 200:

Temperature °C	Temperature °F	Conductivity W/m·K	Conductivity Btu/ft·h·°F
21	70	12.7	7.3
93	200	13.0	7.5
205	400	13.5	7.8
315	600	13.8	8.0
425	800	15.1	8.7
540	1000	15.2	8.8
650	1200	17.3	10.0
760	1400	14.0	8.1
870	1600	21.6	12.5
980	1800	24.9	14.4
1090	2000	29.7	17.0

Chemical Properties

General corrosion behavior. Oxidation resistant to 1040 °C (1900 °F)

MAR-M 246

Chemical Composition

Composition limits. Typical. 60 Ni; 9.0 Cr; 10.0 Co; 2.5 Mo; 10.0 W; 1.5 Ta; 5.5 Al; 1.5 Ti; 0.15 C; 0.015 B; 0.05 Zr. Range. 9.0 to 11.0 W; 9.0 to 11.0 Co; 8.0 to 10.0 Cr; 5.25 to 5.75 Al; 2.25 to 2.75 Mo; 1.25 to 1.75 Ta; 1.25 to 1.75 Ti; 0.13 to 0.17 C; 0.03 to 0.08 Zr; 0.01 to 0.02 B; 1.0 max Fe; 0.20 max Mn; 0.20 max Si; 0.10 max Cu; 0.015 max S; rem Ni

Applications

Typical uses. Turbine vanes, nozzles, jet engine components

Mechanical Properties

Tensile properties. See Table 12.
Hardness. 36 to 42 HRC
Elastic modulus:

Temperature °C	Temperature °F	Modulus of elasticity GPa	Modulus of elasticity 10^6 psi
21	70	205	29.8
93	200	200	29.3
205	400	195	28.5
315	600	190	27.6
425	800	185	26.8
540	1000	180	25.8
650	1200	170	24.8
760	1400	165	23.7
870	1600	155	22.6
980	1800	150	21.7
1090	2000	145	21.1

Stress-rupture characteristics. Typical:

Temperature °C	Temperature °F	Rupture strength MPa	Rupture strength ksi
At 100 h			
760	1400	675	98
815	1500	540	76
870	1600	440	63
925	1700	295	43
980	1800	195	28
1040	1900	130	19
At 1000 h			
760	1400	595	86
815	1500	435	62
870	1600	290	42
925	1700	185	27
980	1800	125	18

Mass Characteristics

Density. 8.44 Mg/m^3 (0.305 lb/in.^{3}) at 20 °C (68 °F)

Thermal Properties

Incipient melting temperature. 1220 °C (2230 °F)

Table 12 Typical mechanical properties of MAR-M 246

Temperature °C	°F	Tensile strength MPa	ksi	Yield strength MPa	ksi	Elongation, %
21	70	965	140	860	125	5
540	1000	1000	145	860	125	5
650	1200	1030	150	860	125	5
760	1400	1030	150	860	125	5
870	1600	860	125	690	100	5
980	1800	550	80	380	55	8
1090	2000	345	50	...	...	...

Liquidus temperature. 1360 °C (2475 °F)
Coefficient of thermal expansion. Linear:

Temperature °C	°F	Mean coefficient µm/m·K	µin./in.·°F
93	200	11.3	6.3
205	400	13.0	7.2
315	600	13.3	7.4
425	800	14.1	7.85
540	1000	14.8	8.2
650	1200	14.9	8.3
760	1400	15.6	8.65
870	1600	16.0	8.9
980	1800	16.8	9.3
1090	2000	18.6	10.35

Thermal conductivity:

Temperature °C	°F	Conductivity W/m·K	Btu/ft·h·°F
425	800	16.8	9.7
540	1000	18.9	10.9
650	1200	21.0	12.1
760	1400	23.0	13.3
870	1600	25.2	14.6
980	1800	27.2	15.7
1090	2000	30.0	17.3

Chemical Properties

General corrosion behavior. High oxidation resistance

MAR-M 247

Commercial Name

Trade name. M-M-0011 alloy

Chemical Composition

Typical composition. 60 Ni; 10.0 Co; 10.0 W; 8.25 Cr; 5.5 Al; 3.0 Ta; 1.5 Hf; 1.0 Ti; 0.70 Mo; 0.50 Fe; 0.15 C; 0.05 Zr; 0.015 B; 0.20 max Mn; 0.20 max Si; 0.015 max S; rem Ni

Mechanical Properties

Tensile properties. See Table 13.
Stress-rupture characteristics. Typical(a):

Temperature °C	°F	Rupture strength MPa	ksi
At 100 h			
760	1400	690	100
870	1600	450	65
925	1700	290	42
980	1800	185	27
1040	1900	125	18
At 1000 h			
870	1600	290	42
925	1700	195	28
980	1800	125	18

(a) Test bar machined from blade.

Mass Characteristics

Density. 8.53 Mg/m^3 (0.308 lb/in.^{3}) at 21 °C (70 °F)

Table 13 Typical tensile properties of MAR-M 247

Temperature °C	°F	Tensile strength MPa	ksi	Yield strength MPa	ksi	Elongation, %
21	70	970	141	805	117	7.9
540	1000	1025	149	815	118	8.0
650	1200	1050	152	805	117	7.0
760	1400	1010	147	825	120	6.0
870	1600	800	116	670	97	5.0
980	1800	525	76	330	48	7.5
1090	2000	250	36	180	26	10.5

MAR-M 421

Chemical Composition

Typical composition. 61 Ni; 15.8 Cr; 9.5 Co; 2.0 Mo; 3.8 W; 2.0 Nb; 4.3 Al; 1.8 Ti; 0.15 C; 0.015 B; 0.05 Zr. Range. 15.2 to 16.0 Cr; 9.0 to 10.0 Co; 4.1 to 4.5 Al; 3.6 to 4.0 W; 1.8 to 2.2 Nb; 1.8 to 2.2 Mo; 1.6 to 1.9 Ti; 0.12 to 0.17 C; 0.03 to 0.07 Zr; 0.012 to 0.017 B; rem Ni

Applications

Typical uses. Turbine rotors and discs, jet engine components

Mechanical Properties

Tensile properties. See Table 14.
Hardness. 32 to 39 HRC
Elastic modulus:

Temperature °C	°F	Modulus of elasticity GPa	10^6 psi
21	70	205	29.4
205	400	195	28.1
425	800	185	26.5
650	1200	170	24.4
760	1400	160	23.2
870	1600	150	21.9
1090	2000	140	20.4

Stress-rupture characteristics. Typical:

Temperature °C	°F	Rupture strength MPa	ksi
At 100 h			
760	1400	600	81
815	1500	450	65
870	1600	310	46
925	1700	195	28
980	1800	125	18
1040	1900	69	9.5
1090	2000	34	5
At 1000 h			
760	1400	435	63
815	1500	305	44
870	1600	215	31
925	1700	140	20
980	1800	83	12
1040	1900	48	7
1090	2000	28	4

Mass Characteristics

Density. 8.08 Mg/m^3 (0.292 lb/in.^{3}) at 20 °C (68 °F)

Thermal Properties

Coefficient of thermal expansion:

Table 14 Typical mechanical properties of MAR-M 421

Temperature °C	°F	Tensile strength MPa	ksi	Yield strength MPa	ksi	Elongation, %
21	70	1080	157	930	135	4.5
540	1000	1010	147	815	118	3
650	1200	965	140	820	119	4
760	1400	950	138	860	125	2.5
870	1600	750	109	650	94	6
980	1800	380	55	270	39	22

Temperature °C	°F	Mean coefficient μm/m·K	μin./in.·°F
425	800	14.6	8.1
540	1000	14.9	8.3
650	1200	15.1	8.4
760	1400	15.7	8.7
870	1600	16.4	9.1
980	1800	17.8	9.9
1090	2000	19.8	11.0

Thermal conductivity:

Temperature °C	°F	Conductivity W/m·K	Btu/ft·h·°F
425	800	17.4	10.1
540	1000	19.1	11.0
650	1200	20.0	11.6
760	1400	24.5	14.2
870	1600	26.9	15.5
980	1800	29.8	17.2
1090	2000	32.0	18.5

MAR-M 432

Chemical Composition

Composition limits. Typical. 50 Ni; 15.5 Cr; 20.0 Co; 3.0 W; 2.0 Ta; 2.0 Nb; 2.8 Al; 4.3 Ti; 0.15 C; 0.015 B; 0.05 Zr. Range. 19.5 to 20.5 Co; 15.3 to 15.8 Cr; 4.2 to 4.5 Ti; 2.9 to 3.3 W; 2.7 to 3.0 Al; 1.8 to 2.2 Nb; 1.8 to 2.2 Ta; 0.12 to 0.18 C; 0.03 to 0.07 Zr; 0.012 to 0.020 B; rem Ni

Applications

Typical uses. For integrally cast turbine wheels, jet engine components, turbine blades

Mechanical Properties

Tensile properties. See Table 15.
Hardness. 42 to 48 HRC
Elastic modulus:

Temperature °C	°F	Modulus of elasticity GPa	10^6 psi
26	78		29.7
66	150		29.4
260	500		27.9
540	1000		25.5
815	1500		22.4
1090	2000		17.1

Stress-rupture characteristics. Typical:

Temperature °C	°F	Rupture strength MPa	ksi
At 100 h			
760	1400	620	90
815	1500	435	63
870	1600	295	40
925	1700	205	30
980	1800	140	20
1040	1900	83	12
At 1000 h			
760	1400	485	70
815	1500	330	48
870	1600	215	31
925	1700	150	22
980	1800	97	14
1040	1900	52	7.5

Table 15 Typical mechanical properties of MAR-M 432

Temperature °C	°F	Tensile strength MPa	ksi	Yield strength MPa	ksi	Elongation, %
21	70	1240	180	1070	155	6
540	1000	1100	160	910	132	...
650	1200	1090	158	910	132	5.0
760	1400	1080	156	910	132	3.5
870	1600	730	106	605	88	8
980	1800	370	54	285	41	21

Mass Characteristics

Density. 8.16 Mg/m³ (0.295 lb/in.³) at 20 °C (68 °F)

Thermal Properties

Coefficient of thermal expansion. Linear:

Temperature °C	°F	Mean coefficient μm/m·K	μin./in.·°F
315	600	14.1	7.85
425	800	14.4	8.0
540	1000	14.9	8.3
650	1200	15.5	8.6
760	1400	15.7	8.7
870	1600	16.6	9.2
980	1800	17.5	9.7
1040	2000	19.3	10.7

Chemical Properties

Resistance to specific corroding agents. Resistant to sulfidation

MC-102

Chemical Composition

Typical composition. 64 Ni; 20 Cr; 6 Mo; 2.5 W; 0.6 Ta; 6 Nb; 0.30 Mn; 0.25 Si; 0.04 C

Mechanical Properties

Tensile properties. See Table 16.
Stress-rupture characteristics. Typical:

Temperature °C	°F	Rupture strength MPa	ksi
At 100 h			
650	1200	540	78
705	1300	370	54
760	1400	260	38
815	1500	195	28
870	1600	145	21
925	1700	105	15
At 1000 h			
650	1200	415	60
705	1300	260	38
760	1400	185	27
815	1500	145	21
870	1600	105	15
925	1700	62	9

Mass Characteristics

Density. 8.84 Mg/m³ (0.319 lb/in.³) at 21 °C (70 °F)
Coefficient of thermal expansion. Linear:

Table 16 Typical mechanical properties of MC-102

Temperature °C	°F	Tensile strength MPa	ksi	Yield strength MPa	ksi	Elongation, %
21	70	675	98	605	88	5
540	1000	655	95	540	78	9
650	1200	640	93	525	76	9
760	1400	605	88	530	77	4
870	1600	395	57	250	35	11
980	1800	250	36	130	19	12

Temperature °C	°F	Mean coefficient μm/m·K	μin./in.·°F
93	200	12.8	7.1
205	400	13.1	7.3
315	600	13.3	7.4
425	800	14.4	8.0
540	1000	14.9	8.3
650	1200	15.6	8.7
760	1400	16.2	9.0
870	1600	16.7	9.3
980	1800	17.3	9.6

Nimocast 75

Chemical Composition

Typical composition. 73 Ni; 20 Cr; 5 Fe; 0.4 Ti; 0.4 Mn; 0.4 Si; 0.2 Al; 0.10 C

Mechanical Properties

Tensile properties. Tensile strength, 500 MPa (72 ksi); yield strength, 180 MPa (26 ksi); elongation, 39% in 50 mm or 2 in.

Mass Characteristics

Density. 8.44 Mg/m^3 (0.305 lb/in.^{3}) at 21 °C (70 °F)

Coefficient of thermal expansion. Linear:

Temperature °C	°F	Mean coefficient μm/m·K	μin./in.·°F
93	200	12.8	7.1
205	400	13.6	7.6
315	600	14.1	7.8
425	800	14.6	8.1
540	1000	14.9	8.3
650	1200	15.6	8.7
760	1400	16.2	9.0
870	1600	16.7	9.3
980	1800	17.3	9.6

Nimocast 80

Chemical Composition

Typical composition. 71 Ni; 20 Cr; 2.4 Ti; 1.3 Al; 5 Fe; 0.4 Mn; 0.4 Si; 0.07 C

Mechanical Properties

Tensile properties. Tensile strength, 730 MPa (106 ksi); yield strength, 520 MPa (75 ksi); elongation, 39% in 50 mm or 2 in.

Stress-rupture characteristics. Typical:

Temperature °C	°F	Rupture strength MPa	ksi
At 100 h			
650	1200	285	41
705	1300	200	29
760	1400	145	21
At 1000 h			
650	1200	200	29
705	1300	145	21
760	1400	105	15

Mass Characteristics

Density. 8.17 Mg/m^3 (0.295 lb/in.^{3}) at 21 °C (70 °F)

Thermal Properties

Coefficient of thermal expansion. Linear:

Temperature °C	°F	Mean coefficient μm/m·K	μin./in.·°F
93	200	12.8	7.1
205	400	13.5	7.5
315	600	14.1	7.8
425	800	14.6	8.1
540	1000	14.9	8.3
650	1200	15.6	8.7
760	1400	16.5	9.2
870	1600	17.5	9.7
980	1800	18.7	10.4

Nimocast 90

Chemical Composition

Typical composition. 53 Ni; 20 Cr; 17.5 Co; 5 Fe; 2.4 Ti; 1.3 Al; 0.4 Mn; 0.4 Si; 0.09 C

Mechanical Properties

Tensile properties. See Table 17.

Stress-rupture characteristics. Typical:

Temperature °C	°F	Rupture strength MPa	ksi
At 100 h			
650	1200	350	51
705	1300	260	38
760	1400	200	29
815	1500	160	23
870	1600	125	18
At 1000 h			
650	1200	305	44
705	1300	220	32
760	1400	160	23
815	1500	110	17
870	1600	83	12

Mass Characteristics

Density. 8.18 Mg/m^3 (0.296 lb/in.^{3}) at 21 °C (70 °F)

Thermal Properties

Coefficient of thermal expansion. Linear:

Temperature °C	°F	Mean coefficient μm/m·K	μin./in.·°F
93	200	12.3	6.8
205	400	13.3	7.4
315	600	14.1	7.8
425	800	14.6	8.1
540	1000	14.8	8.2
650	1200	15.4	8.6
760	1400	16.2	9.0
870	1600	17.1	9.5
980	1800	18.1	10.1

Nimocast 242

Chemical Composition

Typical composition. 57 Ni; 20.5 Cr; 10.0 Co; 10.5 Mo; 0.2 Al; 0.3 Ti; 1.0 Fe; 0.3 Mn; 0.3 Si; 0.34 C

Mechanical Properties

Tensile properties. Tensile strength, 460 MPa (67 ksi); yield strength, 300 MPa (44 ksi); elongation, 8% in 50 mm or 2 in.

Table 17 Typical mechanical properties of Nimocast 90

Temperature °C	Temperature °F	Tensile strength MPa	Tensile strength ksi	Yield strength MPa	Yield strength ksi	Elongation, %
21	70	700	102	520	75	14
540	1000	595	86	420	61	15
650	1200	560	81	410	60	17
760	1400	490	71	390	57	14
870	1600	260	38	110	16	31

Stress-rupture characteristics. Typical:

Temperature °C	Temperature °F	Rupture strength MPa	Rupture strength ksi
At 100 h			
760	1400	170	25
815	1500	110	16
870	1600	90	13
925	1700	70	10
980	1800	45	6.5
At 1000 h			
760	1400	110	16
815	1500	83	12
870	1600	59	8.5
925	1700	43	6.2

Mass Characteristics

Density. 8.40 Mg/m^3 (0.303 lb/in.^{3}) at 21 °C (70 °F)

Thermal Properties

Coefficient of thermal expansion. Linear:

Temperature °C	Temperature °F	Mean coefficient μm/m·K	Mean coefficient μin./in.·°F
93	200	12.5	7.0
205	400	13.1	7.3
315	600	13.7	7.6
425	800	14.1	7.8
540	1000	14.4	8.0
650	1200	14.9	8.3
760	1400	15.6	8.7
870	1600	16.3	9.1
980	1800	17.0	9.5

Nimocast 263

Chemical Composition

Typical composition. 51 Ni; 20 Cr; 20 Co; 5.8 Mo; 2.2 Ti; 0.5 Al; 0.5 Fe; 0.5 Mn; 0.06 C; 0.008 B; 0.04 Zr

Mechanical Properties

Tensile properties. Tensile strength, 730 MPa (106 ksi); yield strength, 510 MPa (74 ksi); elongation, 18% in 50 mm or 2 in.

Mass Characteristics

Density. 8.36 Mg/m^3 (0.302 lb/in.^{3}) at 21 °C (70 °F)

Thermal Properties

Coefficient of thermal expansion. Linear:

Temperature °C	Temperature °F	Mean coefficient μm/m·K	Mean coefficient μin./in.·°F
93	200	11.0	6.1
205	400	12.1	6.7
315	600	12.7	7.1
425	800	13.1	7.3
540	1000	13.6	7.6
650	1200	14.3	8.0
760	1400	15.1	8.4
870	1600	16.3	9.1
980	1800	17.9	10.0

NX 188(DS)

Chemical Composition

Typical composition. 74 Ni; 18.0 Mo; 8.0 Al; 0.04 C

Applications

Typical uses. Jet engine vanes

Mechanical Properties

Tensile properties. See Table 18.
Rupture strength. At 100 h: 180 MPa (26 ksi) at 980 °C (1800 °F); 97 MPa (14 ksi) at 1090 °C (2000 °F)

Mass Characteristics

Density. 8.19 Mg/m^3 (0.296 lb/in.^{3}) at 20 °C (68 °F)

René 77

Chemical Composition

Typical composition. 58 Ni; 14.6 Cr; 15.0 Co; 4.2 Mo; 4.3 Al; 3.3 Ti; 0.07 C; 0.016 B; 0.04 Zr

Applications

Typical uses. Jet engine applications

Mechanical Properties

Tensile properties. See Table 19.
Rupture strength. Typical. At 100 h: 310 MPa (45 ksi) at 870 °C (1600 °F); 130 MPa (19 ksi) at 980 °C (1800 °F). At 1000 h: 215 MPa (31.5 ksi) at 870 °C (1600 °F); 62 MPa (9 ksi) at 980 °C (1800 °F)

Mass Characteristics

Density. 7.91 Mg/m^3 (0.286 lb/in.^{3}) at 20 °C (68 °F)

Chemical Properties

Resistance to specific corroding agents. Resists hot sulfidation corrosion

René 80

Chemical Composition

Typical composition. 60 Ni; 14.0 Cr; 9.5 Co; 4.0 Mo; 4.0 W; 3.0 Al; 5.0 Ti; 0.17 C; 0.015 B; 0.03 Zr

Applications

Typical uses. Turbine blades

Mechanical Properties

Tensile properties. See Table 20.
Elastic modulus. Tension, 210 GPa (30.2×10^6 psi)
Rupture strength. Typical. At 100 h: 350 MPa (51 ksi) at 870 °C (1600 °F); 165 MPa (24 ksi) at 980 °C (1800 °F). At 1000 h: 240 MPa (35 ksi) at 870 °C (1600 °F); 105 MPa (15 ksi) at 980 °C (1800 °F)

Mass Characteristics

Density. 8.16 Mg/m^3 (0.295 lb/in.^{3}) at 20 °C (68 °F)

Chemical Properties

General corrosion behavior. Excellent hot-corrosion resistance

Table 18 Typical mechanical properties of NX 188(DS)

Temperature		Tensile strength		Yield strength		Elongation,
°C	°F	MPa	ksi	MPa	ksi	%
21	70	1040	151	960	139	5
650	1200	1090	158	1050	152	3
760	1400	1190	173	1140	166	1.5
870	1600	1210	176	1190	172	2
980	1800	595	86	585	85	3
1090	2000	395	57	345	50	3

Table 19 Typical mechanical properties of René 77

Temperature		Tensile strength		Yield strength		Elongation,
°C	°F	MPa	ksi	MPa	ksi	%
21	70	1020	148	795	115	7
540	1000	1010	147	730	106	11
650	1200	1050	152	715	104	12
760	1400	940	136	690	100	...

Table 20 Typical mechanical properties of René 80

Temperature		Tensile strength		Yield strength		Elongation,
°C	°F	MPa	ksi	MPa	ksi	%
21	70	1030	149	855	124	5.2
540	1000	1030	149	730	106	7
650	1200	1030	149	725	105	8
760	1400	995	144	715	104	9.5
870	1600	705	102	530	77	12.0

René 100

Chemical Composition

Typical composition. 60 Ni; 9.5 Cr; 15 Co; 3 Mo; 5.5 Al; 4.2 Ti; 1.0 V; 0.18 C; 0.014 B; 0.06 Zr

Applications

Typical uses. Turbine blades operating to 1040 °C (1900 °F)

Mechanical Properties

Tensile properties. As cast: tensile strength, 1010 MPa (147 ksi); yield strength, 850 MPa (123 ksi); elongation, 9% in 50 mm or 2 in. At 815 °C (1500 °F): tensile strength, 995 MPa (144 ksi); yield strength; 815 MPa (118 ksi); elongation, 6% in 50 mm or 2 in.

Hardness. As cast: 30 to 44 HRC

Mass Characteristics

Density. 7.75 Mg/m³ (0.280 lb/in.³) at 20 °C (68 °F)

TAZ-8A

Chemical Composition

Typical composition. 68 Ni; 6.0 Cr; 4.0 Mo; 4.0 W; 8.0 Ta; 2.5 Nb; 6.0 Al; 0.12 C; 0.004 B; 1.0 Zr

Applications

Typical uses. Low-stress components, stator vanes, turbine buckets, after-burner liners, transition ducts

Mechanical Properties

Tensile properties. See Table 21.

Stress-rupture characteristics. Typical:

Temperature		Rupture strength(a)	
°C	°F	MPa	ksi
At 100 h			
650	1200	760	110
705	1300	620	90
760	1400	470	68
815	1500	365	53
870	1600	270	39
925	1700	205	30
980	1800	145	21
1040	1900	105(b)	15(b)
1090	2000	69	10
1150	2100	41	6
At 1000 h			
650	1200	655	95
705	1300	495	72
760	1400	380	55
815	1500	275	40
870	1600	200	29
925	1700	150	22
980	1800	97	14
1040	1900	69	10
1090	2000	41	6

(a) Extrapolated values. (b) Actual values.

Mass Characteristics

Density. 8.63 Mg/m³ (0.312 lb/in.³) at 20 °C (68 °F)

Thermal Properties

Thermal conductivity:

Temperature		Conductivity	
°C	°F	W/m·K	Btu/ft·h·°F
205	400	10.9	6.3
315	600	12.3	7.1
425	800	13.8	8.0
540	1000	15.5	9.0
650	1200	17.4	10.1
760	1400	19.4	11.2
870	1600	21.6	12.5
980	1800	23.9(a)	13.8(a)
1090	2000	26.2(a)	15.1(a)

(a) Extrapolated values.

Chemical Properties

General corrosion behavior. High oxidation resistance

TRW-NASA VI A

Chemical Composition

Typical composition. 61 Ni; 6.1 Cr; 7.5 Co; 2.0 Mo; 5.8 W; 9.0 Ta; 0.5 Nb; 5.4 Al; 1.0 Ti; 0.13 C; 0.02 B; 0.13 Zr; 0.5 Re; 0.4 Hf

Applications

Typical uses. Blades

Mechanical Properties

Tensile properties. See Table 22.
Stress-rupture characteristics. Typical:

Temperature °C	°F	Rupture strength MPa	ksi
At 100 h			
705	1300	860(a)	125(a)
760	1400	725(a)	105(a)
815	1500	550(a)	80(a)
870	1600	395(a)	57(a)
925	1700	330	44
980	1800	215	31
1040	1900	145	21
1090	2000	90(a)	13(a)
At 1000 h			
760	1400	585	85
815	1500	420	61
860	1600	305	44
925	1700	215	31
980	1800	140	20

(a) Extrapolated values.

Mass Characteristics

Density. 8.77 Mg/m³ (0.317 lb/in.³) at 20 °C (68 °F)

Udimet 500
52Ni-18Cr

Chemical Composition

Typical composition. 52 Ni; 18.0 Cr; 19.0 Co; 4.2 Mo; 3.0 Al; 3.0 Ti; 0.07 C; 0.007 B; 0.05 Zr

Applications

Typical uses. Jet engine components

Mechanical Properties

Tensile properties. See Table 23.
Stress-rupture characteristics. Typical:

Temperature °C	°F	Rupture strength MPa	ksi
At 100 h			
760	1400	450	65
815	1500	330	48
860	1600	230	33
925	1700	145	21
980	1800	90	13
At 1000 h			
705	1300	460	67
760	1400	395	50
815	1500	240	35
860	1600	165	24
925	1700	90	13

Table 21 Typical mechanical properties of TAZ-8A

Temperature °C	°F	Tensile strength MPa	ksi	Elongation, %
21	70	885	128	5
760	1400	885	128	3
870	1600	760	110	4
980	1800	550	80	5
1090	2000	340	49	4

Table 22 Typical mechanical properties of TRW-NASA VI A

Temperature °C	°F	Tensile strength MPa	ksi	Yield strength MPa	ksi	Elongation, %
21	70	1050	152	940	136	4
650	1200	1140	165	945	137	4
760	1400	1100	159	945	137	4.5
870	1600	870	126	770	112	2.5
980	1800	595	86	520	75	...
1090	2000	340	49	310	45	5

Table 23 Typical mechanical properties of Udimet 500

Temperature °C	°F	Tensile strength MPa	ksi	Yield strength MPa	ksi	Elongation, %
21	70	930	135	815	118	13
540	1000	895	130	725	105	13
650	1200	885	128	705	102	13
760	1400	855	124	705	102	9
870	1600	660	96	600	87	8.5
980	1800	130	19	...	...	50

Mass Characteristics

Density. 8.02 Mg/m³ (0.290 lb/in.³) at 20 °C (68 °F)

Coefficient of thermal expansion. Linear, 13.3 μm/m·K (7.4 μin./in.·°F) at 20 to 93 °C (68 to 200 °F)

Chemical Properties

General corrosion behavior. Excellent hot-corrosion resistance

Udimet 700

Chemical Composition

Composition limits. 17.0 to 20.0 Co; 13.0 to 17.0 Cr; 4.5 to 5.7 Mo; 3.7 to 4.7 Al; 3.0 to 4.0 Ti; 0.15 max C; rem Ni

Applications

Typical uses. Turbine blades, discs and combustion chambers

Mechanical Properties

Creep-rupture characteristics. Typical. At 760 °C (1400 °F), 585 MPa (85 ksi). From bars machined from blades: U-700, 3.8%; U-700 + Hf, 9.1%
Stress-rupture characteristics. Typical. 585 MPa (85 ksi) at 760 °C (1400 °F): bars machined from blades: U-700 48.5 h; U-700 + Hf, 68.6 h. 6.4 mm (0.250 in.) cast to size, 90 h; 4.1 mm (0.160 in.) cut from blade, 49 h; 1.3 mm (0.050 in.) cut from thin wall, 15 h. 330 MPa (48 ksi) at 870 °C (1600 °F): 6.4 mm (0.250 in.) cast to size, 71 h; 4.1 mm (0.160 in.) cut from blade, 66 h; 1.3 mm (0.050 in.) cut from thin wall, 18 h

Table 24 Mechanical properties of Udimet 710

Temperature °C	Temperature °F	Tensile strength MPa	Tensile strength ksi	Yield strength MPa	Yield strength ksi	Elongation, %
21	70	1075	156	895	130	8
650	1200	1110	161	835	121	12
760	1400	1055	153	820	119	18
870	1600	730	106	630	91	23
980	1800	415	60	295	43	27
1090	2000	240	35	170	25	25

Udimet 710

Chemical Composition

Typical composition. 55 Ni; 18 Cr; 15 Co; 5 Ti; 3 Mo; 2.5 Al; 1.5 W; 0.13 C; 0.08 Zr; 0.02 B

Applications

Typical uses. Land-based turbine blades, high temperature applications to 980 °C (1800 °F)

Mechanical Properties

Tensile properties. See Table 24.

Stress-rupture characteristics. Typical:

Temperature °C	Temperature °F	Rupture strength MPa	Rupture strength ksi
At 100 h			
705	1300	730(a)	106(a)
760	1400	560	81
815	1500	420	61
870	1600	305	44
925	1700	220	32
980	1800	150	22
1040	1900	76	11
At 1000 h			
650	1200	750(a)	109(a)
705	1300	600(a)	87(a)
760	1400	440	64
815	1500	325	47
870	1600	215	31
925	1700	130	19
980	1800	76	11

(a) Extrapolated values.

Thermal Properties

Thermal conductivity:

Temperature °C	Temperature °F	Conductivity W/m·K	Conductivity Btu/ft·h·°F
21	70	11.4(a)	6.6(a)
93	200	12.1	7.0
205	400	13.5	7.8
315	600	15.0	8.1
425	800	16.6	9.6
540	1000	18.1	10.5
650	1200	19.6	11.3
760	1400	21.2	12.2
870	1600	22.8	13.2
980	1800	24.2	14.0

(a) Extrapolated values.

WAZ-20(DS)

Chemical Composition

Typical composition. 72 Ni; 20.0 W; 6.5 Al; 0.20 C; 1.50 Zr

Applications

Typical uses. Jet engine discs and vanes

Mechanical Properties

Tensile properties. See Table 25.

Stress-rupture characteristics. Typical:

Temperature °C	Temperature °F	Rupture strength(a) MPa	Rupture strength(a) ksi
At 100 h			
650	1200	930	135
705	1300	760	110
760	1400	605	88
815	1500	450	65
870	1600	345	50
925	1700	255	37
980	1800	185	27
1040	1900	130	19
1090	2000	86	12.5
1150	2100	48	7
At 1000 h			
980	1800	105	15

(a) Values at 100 h are extrapolated.

Mass Characteristics

Density. 9.02 Mg/m^3 (0.326 $lb/in.^3$) at 20 °C (68 °F)

Table 25 Typical mechanical properties of WAZ-20(DS)

Temperature, °C	Tensile strength MPa	Tensile strength ksi	Elongation, %
21	895	130	13
650	825	120	5
760	825	120	4
870	775	113	5
980	485	70	10
1090	310	45	12

Properties of Cobalt-Base Superalloys

AR-13

Chemical Composition

Typical composition. 58 Co; 21 Cr; 11.0 W; 3.5 Al; 2.5 Fe; 2.0 Ta; 1.0 Ni; 0.50 Mn; 0.45 C; 0.1 Y

Mechanical Properties

Tensile properties. See Table 26.

Structure

Microstructure. fcc matrix (austenite) solid solution

Mass Characteristics

Density. 8.43 Mg/m^3 (0.305 lb/in.^{3}) at 20 °C (68 °F)

Table 26 Typical mechanical properties of AR-13

Temperature		Tensile strength		Yield strength		Elongation,
°C	°F	MPa	ksi	MPa	ksi	%
21	70	600	87	530	77	1.5
650	1200	475	69	385	56	4.5
760	1400	420	61	330	48	4.5
870	1600	290	42	275	40	21.0

Table 27 Typical mechanical properties of AR-213

Temperature		Tensile strength		Yield strength		Elongation,
°C	°F	MPa	ksi	MPa	ksi	%
21	70	1120	162	625	91	14
650	1200	960	139	455	66	28
760	1400	485	70	385	56	47
870	1600	315	46	220	32	55

AR-213

Chemical Composition

Typical composition. 66.0 Co; 19.0 Cr; 6.5 Ta; 4.7 W; 3.5 Al; 0.18 C; 0.15 Zr; 0.1 Y

Applications

Typical uses. For gas turbine components

Mechanical Properties

Tensile properties. See Table 27.
Hardness. Age-hardened: 47 HRC

Structure

Microstructure. fcc matrix (austenite) solid solution

Mass Characteristics

Density. 8.51 Mg/m^3 (0.307 lb/in.^{3}) at 20 °C (68 °F)

Chemical Properties

Resistance to specific corroding agents. Sulfidation and oxidation resistant

AR-215

Chemical Composition

Typical composition. 64.0 Co; 19.0 Cr; 7.5 Ta; 4.5 W; 4.3 Al; 0.35 C; 0.13 Zr; 0.17 Y

Mechanical Properties

Tensile properties. See Table 28.

Structure

Microstructure. fcc matrix (austenite) solid solution

Mass Characteristics

Density. 8.47 Mg/m³ (0.306 lb/in.³) at 20 °C (68 °F)

Table 28 Typical mechanical properties of AR-215

Temperature		Tensile strength		Yield strength		Elongation,
°C	°F	MPa	ksi	MPa	ksi	%
21	70	690	100	485	70	4
650	1200	570	83	315	46	12
870	1600	275	40	215	31	59

FSX-414

Chemical Composition

Typical composition. 52 Co; 29 Cr; 10 Ni; 7.5 W; 1.0 Fe; 0.25 C; 0.010 B

Applications

Typical uses. Gas turbine vanes

Mechanical Properties

Tensile properties. See Table 29.
Stress-rupture characteristics. Typical:

Temperature		Rupture strength	
°C	°F	MPa	ksi
At 100 h			
815	1500	160	23
870	1600	110	16
925	1700	83	12
980	1800	55	8
1040	1900	34	5
1090	2000	24	3.5
At 1000 h			
760	1400	165	24
815	1500	115	16.5
870	1600	83	12
925	1700	55	8
980	1800	34	5
1040	1900	21	3

Structure

Microstructure. fcc matrix (austenite) solid solution

Mass Characteristics

Density. 8.30 Mg/m³ (0.300 lb/in.³) at 20 °C (68 °F)

Table 29 Typical mechanical properties of FSX-414

Temperature		Tensile strength		Yield strength		Elongation,
°C	°F	MPa	ksi	MPa	ksi	%
21	70	740	107	440	64	11
540	1000	540	78	240	35	15
650	1200	485	70	215	31	15
760	1400	400	58	195	28	18
870	1600	310	45	165	24	23

Table 30 Creep-rupture data for Haynes 25 solution heat treated sheet

Temperature		Stress		Creep rate, % per hour at:	
°C	°F	MPa	ksi	50 h	100 h
816	1500	97	14	0.0033	0.0033
		130	19	0.022	0.015
		130	19	0.027	0.007
		150	22	0.065	0.030
870	1600	55	8	0.0047	0.0015
		69	10	0.0085	0.0037
		69	10	0.0044	0.0035
		83	12	0.012	0.006
		97	14	0.024	0.030
925	1700	41	6	0.0029	0.0047
		55	8	0.0058	0.0093
		55	8	0.011	0.019
		69	10	0.063	...
		69	10	0.032	0.108
980	1800	28	4	0.0030	0.0024
		34	5	0.0044	0.0050
		41	6	0.020	...

Haynes 25

Commercial Name

Trade name. HAYNES alloy No. 25

Specifications

AMS. Solution heat treated sheet, 5537-C; solution heat treated bar and forgings, 5759-D; welding wire bright annealed, 5796; coated welding electrodes, 5797
Government. MIL-E-17496D, Type 3NIL (Class 2)

Chemical Composition

Composition limits. 19.00 to 21.00 Cr; 14.00 to 16.00 W; 9.00 to 11.00 Ni; 3.00 max Fe; 1.00 to 2.00 Mn; 1.00 max Si; 0.05 to 0.15 C; 0.030 P; 0.030 max S; rem Co

Applications

Typical uses. Jet engine parts, such as turbine blades, combustion chambers, afterburner parts and turbine rings. Used extensively in industrial furnaces for muffles and liners in critical spots. Combines good formability with excellent high-temperature properties.

Mechanical Properties

Tensile properties. Typical short-time data for solution heat treated material. Sheet: tensile strength, 930

Table 31 Typical oxidation resistance for Haynes 25, Haynes 188 and Hastelloy X

Temperature		Oxidation rate, weight loss, g/m²			Oxidation rate, metal loss					
					Haynes 188		Hastelloy X		Haynes 25	
°C	°F	Haynes 188	Hastelloy X	Haynes 25	μm/side	mils/side	μm/side	mils/side	μm/side	mils/side
870	1600	10	15	21	1.3	0.05	1.8	0.07	2.3	0.09
980	1800	15	24	85	1.8	0.07	2.8	0.11	9.4	0.37
1090	2000	39	54	424	4.1	0.16	6.4	0.25	46.5	1.83
1150	2100	92	92	1061	9.9	0.39	10.7	0.42	116.1	4.57

Note: Intermittent exposure for 100 h in dry air based on descaled weight change.

MPa (135 ksi); yield strength, 450 MPa (65 ksi); elongation, 60% in 50 mm or 2 in. Bar: tensile strength, 805 to 1030 MPa (117 to 150 ksi); yield strength, 285 to 485 MPa (41 to 70 ksi); elongation, 87% in 50 mm or 2 in. Plate: tensile strength, 950 to 985 MPa (139 to 143 ksi); yield strength, 460 to 470 MPa (66 to 68 ksi); elongation, 56 to 60% in 50 mm or 2 in.

Compressive properties. Average room temperature value for sheet, solution heat treated, cold reduced 20%, ages 2 h at 593 °C (1100 °F): yield strength, 1210 MPa (176 ksi): proportional limit, 840 MPa (122 ksi)

Elastic modulus. Average in tension for solution heat treated sheet:

Temperature °C	°F	Elastic modulus GPa	10^6 psi
25	77	225	32.6
100	212	222	32.2
200	392	214	31.0
300	572	204	29.6
400	752	197	28.6
500	932	188	27.3
600	1112	181	26.3
700	1292	174	25.2
800	1472	163	23.7
900	1652	154	22.4
1000	1832	146	21.2

Tangent, in compression. Room temperature solution heat treated sheet; cold reduced 20%; aged 2 h at 590 °C (1100 °F): 250 GPa (36.2 × 10^6 psi)

Creep-rupture characteristics. See Table 30.

Impact strength. Typical Charpy V-notch for solution heat treated plate:

Temperature °C	°F	Mean Charpy J	V-notch strength ft·lb
−195	−321	148	109
−138	−216	182	134
−78	−108	212	156
−29	−20	243	179
20	68	262	193
260	500	297	219
540	1000	273	201
650	1200	230	170
760	1400	194	143
870	1600	163	120
980	1800	144	106

Fatigue strength. Solution heat treated sheet. At 816 °C (1500 °F): 275 MPa (40 ksi) at 10^7 cycles; 215 MPa (31 ksi) at 10^8 cycles. At 980 °C (1800 °F): 145 MPa (21 ksi) at 10^6 cycles; 105 MPa (15 ksi) at 10^7 cycles; 90 MPa (13 ksi) at 10^8 cycles

Mass Characteristics

Density. 9.130 Mg/m³ (0.330 lb/in.³) at 22 °C (72 °F)

Thermal Properties

Liquidus temperature. 1410 °C (2570 °F)

Solidus temperature. 1329 °C (2425 °F)

Coefficient of thermal expansion. Linear:

Temperature °C	°F	Average coefficient μm/m·K	μin./in.·°F
21–935	70–200	12.3	6.8
21–205	70–400	12.9	7.2
21–315	70–600	13.5	7.6
21–425	70–800	13.9	7.8
21–540	70–1000	14.4	8.0
21–650	70–1200	14.8	8.2
21–760	70–1400	15.4	8.6
21–870	70–1600	16.2	9.1
21–980	70–1800	16.9	9.4
21–1090	70–2000	17.7(a)	9.8(a)

(a) Extrapolated data.

Specific heat. 384 J/kg·K (0.092 Btu/lb·°F) at 28 to 100 °C (80 to 212 °F)

Thermal conductivity:

Temperature °C	°F	Conductivity W/m·K	Btu/ft·h·°F
20	70	9.8(a)	5.6(a)
100	100	11.2(a)	6.5(a)
200	300	13.0	7.5
300	500	14.7	8.5
400	700	16.6	9.6
500	900	18.5	10.7
600	1100	20.4	11.8
700	1300	22.4	12.9
800	1500	24.4	14.1
900	1700	26.5	15.3

(a) Extrapolated data.

Electrical Properties

Electrical resistivity:

Temperature °C	°F	Resistivity, nΩ·m
24	75	890
240	460	970
450	840	990
650	1200	1060
740	1360	1070
815	1500	1080
950	1740	970
1000	1830	950
1080	1980	1000

Magnetic Properties

Magnetic permeability. <1.0 at 9 kA/m (116 oersteds)

Optical Properties

Color. Silvery

Chemical Properties

General corrosion behavior. Resists oxidation and carburization to 1065 °C (1900 °F)

Resistance to specific corroding agents. Tests show a penetration rate of less than 1 mil/year in wet chlorine at room temperature. In addition, the alloy is resistant to hydrochloric and nitric acids at certain

Table 32 Average short-time tensile properties for Haynes 188 (a)

Temperature °C	Temperature °F	Tensile strength MPa	Tensile strength ksi	Yield strength MPa	Yield strength ksi	Elongation, %
Sheet, 0.8 to 1.7 mm (0.030 to 0.65 in.) thick						
−184	−300	1310	190.2	715	103.5	48
−129	−200	1145	166.1	645	93.7	47
−73	−100	1080	156.1	565	82.2	47
+20	+68	960	139.4	480	69.5	56
315	600	800	116.3	335	48.5	71
540	1000	740	107.2	300	43.8	70
650	1200	710	103.1	305	43.9	61
760	1400	635	92.0	290	42.2	43
870	1600	420	60.7	260	38.0	73
980	1800	250	36.8	160	23.5	72
1090	2000	135	19.3	79	11.5	47
1150	2100	97	14.0	57	8.2	37
1200	2200	77	11.1	48	7.0	35
Sheet, 2.8 to 3.3 mm (0.109 to 0.130 in.) thick						
−184	−300	1280	186.1	685	99.6	58
−129	−200	1150	167.0	630	91.4	62
−73	−100	1090	158.4	565	82.0	65
+20	+68	945	136.8	465	67.6	61
315	600	810	117.6	330	47.5	75
540	1000	755	109.3	310	45.3	78
650	1200	730	106.2	300	43.6	73
760	1400	660	95.5	285	41.6	54
870	1600	435	63.2	279	40.4	81
980	1800	260	37.7	170	24.7	86
1090	2000	140	20.4	88	12.7	75
1150	2100	100	14.6	60	8.7	61
1200	2200	83	12.1	47	6.8	48
Plate, 6.4 mm (0.250 in.) thick						
+20	+68	965	139.7	450	65.2	52
315	600	845	122.9	310	45.3	61
540	1000	810	117.8	320	46.4	59
760	1400	630	91.5	280	40.6	58
870	1600	405	59.0	260	37.8	69
1090	2000	125	18.2	76	11.0	79

(a) Test conditions: below 871 °C (1600 °F), 0.005 in./in./min strain rate to 0.6% offset and 0.500 in./min crosshead velocity to failure. At 871 °C (1600 °F) and above, 0.050 in./min crosshead velocity to 0.6% offset and 0.500 in./min crosshead velocity to failure. Product heat treated at 1177 °C (2150 °F), water quenched. (b) In 28 mm or 1[1]/8 in.

concentrations and temperatures. See also Table 31.

Haynes 188 40Co-22Cr-22Ni-14.5W

Commercial Name

Trade name. HAYNES alloy No. 188

Chemical Composition

Composition limits. 20.00 to 24.00 Cr; 20.00 to 24.00 Ni; 13.00 to 16.00 W; 3.00 max Fe; 1.25 max Mn; 0.20 to 0.50 Si; 0.05 to 0.15 C; 0.03 to 0.15 La; rem Co

Applications

Typical uses. Transition ducts, combustor cans; spray bars, flameholders, liners in jet engines. Excellent strength, ductility and oxidation resistance allow Haynes 188 to meet critical high temperature requirements for gas turbine applications, as well as those of the aircraft, chemical and nuclear fields.

Mechanical Properties

Tensile properties. See Table 32.

Hardness:

Temperature °C	Temperature °F	Average hardness, HRA, after indicated time at temperature 200 h	500 h
980	1800	62	61
925	1700	63	63
870	1600	63	64
815	1500	64	63
760	1400	64	63
705	1300	64	64
650	1200	62	62
590	1100	62	61
540	1000	61	61
480	900	62	61

Creep-rupture characteristics. See Table 33.

Structure

Microstructure:

Condition(a)	Structure
Solution annealed	fcc matrix; primary M_6C; LaRich compound (La_xM_y) associated with M_6C
Aged 980 °C (1800 °F)	fcc matrix; primary M_6C, La_xM_y; secondary M_6C
Aged 870 °C (1600 °F)	fcc matrix; primary M_6C, La_xM_y; secondary M_6C, $M_{23}C_6$
Aged 760 °C (1400 °F)	fcc matrix; primary M_6C, La_xM_y; secondary M_{23},C_6
Aged 650 °C (1200 °F) and lower	Annealed constituents

(a) Samples aged 200 and 500 h at temperature.

Mass Characteristics

Density. 9.130 Mg/m³ (0.330 lb/in.³) at 22 °C (72 °F)

Thermal Properties

Liquidus temperature. 1398 °C (2550 °F)

Incipient melting temperature. 1302 to 1330 °C (2375 to 2425 °F)

Coefficient of thermal expansion. Linear:

Temperature °C	Temperature °F	Mean coefficient μm/m·K	μin./in.·°F
21–−240	70–−400	9.7	5.4

(continued)

Table 33 Average creep stress for Haynes 188, 0.8 to 2.0 mm (0.03 to 0.08 in.) sheet

Temperature °C	Temperature °F	Creep elongation, %	Average stress to creep elongation in: 1 h MPa	1 h ksi	10 h MPa	10 h ksi	100 h MPa	100 h ksi	1000 h MPa	1000 h ksi
760	1400	0.2	205	30	140	20	93	13.5	...	...
		0.5	240	35	170	24.5	115	17.0	86	12.5
		1.0	260	38	185	27	130	19.0	93	13.5
870	1600	0.2	97	14	66	9.5	43	6.3	...	...
		0.5	120	17.5	83	12	59	8.5	40	5.8
		1.0	130	19	93	13.5	68	9.8	48	7.0
980	1800	0.2	50	7.5	31	4.5	17	2.4	...	...
		0.5	70	10	46	6.6	26	3.8	14	2.1
		1.0	80	11.5	50	7.2	30	4.4	17	2.4
1090	2000	0.2	26	3.8	12	1.8	...	...	...	...
		0.5	33	4.8	19	2.7	7	1.0	...	...
		1.0	38	5.5	22	3.2	8	1.2	...	...

Note: Heat treated at 1232 °C (2150 °F); water quenched.

Temperature °C	Temperature °F	Mean coefficient μm/m·K	μin./in.·°F
21– –130	70– –200	10.4	5.8
21– –20	70– 0	11.1	6.2
21– +40	70– +100	11.5	6.4
21– 93	70– 200	11.9	6.6
21– 205	70– 400	12.2	7.0
21– 315	70– 600	13.3	7.4
21– 425	70– 800	14.0	7.8
21– 540	70– 1000	14.8	8.2
21– 650	70– 1200	15.5	8.6
21– 760	70– 1400	16.3	9.0
21– 870	70– 1600	17.0	9.4
21– 980	70– 1800	17.7	9.9
21– 1090	70– 2000	18.5	10.3

Specific heat:

Temperature °C	Temperature °F	Specific heat J/kg·K	Btu/lb·°F
0	32	398	0.095
100	212	423	0.101
200	410	444	0.106
300	572	465	0.111
400	752	486	0.116
500	932	502	0.120
600	1112	523	0.125
700	1292	540	0.129
800	1472	557	0.133
900	1652	573	0.137
1000	1832	590	0.141
1100	2012	607	0.145
1200	2192	624	0.149

Thermal conductivity:

Temperature °C	Temperature °F	Conductivity W/m·K	Btu/lb·°F
38	100	10.8	6.2
93	200	13.4	7.7
149	300	13.2	7.6
205	400	14.4	8.3
260	500	15.2	8.8
315	600	16.1	9.3
370	700	17.0	9.8
425	800	18.0	10.4
480	900	19.0	11.0
540	1000	19.9	11.5
595	1100	21.0	12.1
650	1200	21.9	12.7
705	1300	23.0	13.3
760	1400	24.0	13.9
815	1500	25.1	14.5

Thermal diffusivity:

Temperature °C	Temperature °F	Diffusivity μm²/s	μin.²/s
300	572	3.6	0.006
400	752	4.0	0.006
500	932	4.5	0.007
600	1112	4.8	0.007
700	1292	5.2	0.008
765	1409	5.3	0.008
800	1472	5.3	0.008
900	1652	5.1	0.008
1000	1832	5.5	0.009
1100	2012	5.8	0.009
1200	2192	5.9	0.009

Electrical Properties

Electrical resistivity. 922 nΩ·m at 21 °C (70 °F)

Magnetic Properties

Magnetic permeability. 1.01 at 16 kA/m (200 oersteds) and 20 °C (68 °F)

Optical Properties

Color. Silvery

Chemical Properties

General corrosion behavior. Excellent oxidation resistance results in part from minute additions of lanthanum which modifies the protective oxide scale to make the alloy extremely resistant to diffusion when exposed to temperatures through 1150 °C (2100 °F).

Resistance to specific corroding agents. See Table 31.

Haynes 556 29Fe-22Cr-21Ni-19Co-3Mo-3W-1Ta

Commercial name

Trade name. HAYNES alloy No. 556

Specifications

Government. MIL-STD-163

Chemical Composition

Composition limits. 19 to 22.5 Ni; 16 to 21 Co; 21 to 23 Cr; 2.5 to 4.0 Mo; 2.0 to 3.5 W; 0.5 to 2.0 Mn; 0.2 to 0.5 Si; 0.5 to 0.15 C; 0.30 to 1.25 Ta; 0.30 max Nb; 0.005 to 0.10 La; 0.10 to 0.50 Al; 0.001 to 0.10 Zr; 0.10 to 0.30 N; 0.02 max B; 0.04 max P; 0.015 max S; rem Fe

Applications

Typical uses. Suited for structural applications up to 1100 °C (2000 °F), such as gas turbines, pollution control equipment, incinerators, industrial fans, and furnace hardware and structures. Ideal for stressed components where oxidation resistance, creep resistance, and high-

Table 34 Average tensile properties for Haynes 556 sheet

Temperature °C	°F	Tensile strength MPa	ksi	0.2% yield strength MPa	ksi	0.02% yield strength MPa	ksi	Elongation, %
4.6-mm (0.180-in.) thickness								
20	68	815	118.4	400	58.1	325	46.9	58
315	600	675	98.1	255	37.1	205	30.1	62
540	1000	650	94.3	240	34.6	205	29.6	69
650	1200	585	85.4	220	31.7	180	26.5	62
760	1400	490	71.2	230	33.4	180	26.1	52
870	1600	335	48.6	205	29.8	190	27.5	50
980	1800	210	30.3	140	20.2	105	15.4	56
1090	2000	80	11.9	50	7.3	41	5.9	61
2.3-mm (0.090-in.) thickness								
20	68	820	119.0	400	57.9	350	51.0	53
315	600	690	99.9	265	38.2	255	36.9	58
540	1000	670	97.3	245	35.4	225	32.5	63
650	1200	595	86.2	240	34.6	220	31.9	54
760	1400	440	63.5	225	32.7	210	30.8	45
870	1600	300	43.3	195	28.3	165	23.7	49
980	1800	175	25.1	120	17.7	105	15.2	49
1.3-mm (0.050-in.) thickness								
20	68	820	119.0	390	56.5	330	48.2	52
315	600	685	99.3	300	43.4	230	33.3	53
540	1000	670	97.0	240	35.1	230	33.4	58
650	1200	615	88.9	245	35.5	225	32.8	55
760	1400	460	66.4	225	32.7	205	29.5	47
870	1600	310	45.0	190	27.8	170	25.0	45
980	1800	185	26.6	105	15.4	110	16.2	39
1090	2000	97	14.1	50	7.2	59	8.5	34
0.8-mm (0.032-in.) thickness								
20	68	800	116.0	380	55.1	325	46.8	55
315	600	670	97.3	240	35.1	195	28.4	58
540	1000	655	94.7	225	32.4	195	28.2	63
650	1200	555	80.4	215	31.0	190	27.8	53
760	1400	440	64.1	210	30.5	185	26.9	44
870	1600	290	42.0	185	26.9	170	24.5	40
980	1800	190	27.7	110	16.1	125	18.2	43
1090	2000	115	16.8	59	8.5	66	9.5	39
0.5-mm (0.019-in.) thickness								
20	68	838	121.6	405	58.9	340	49.0	51
315	600	698	101.2	290	41.7	245	35.6	48
540	1000	665	96.6	270	38.9	245	35.2	54
650	1200	605	88.0	265	38.5	240	34.6	45
760	1400	415	60.1	230	33.7	205	29.6	36
870	1600	260	37.6	155	22.7	155	22.5	33
980	1800	175	25.6	100	14.7	105	15.0	37
1090	2000	100	14.5	51	7.4	38	5.5	34

Note: Average of two tests at each temperature. Samples solution treated at 1175 °C (2150 °F) and rapidly cooled.

temperature ductility are also important. Haynes 556 is available as billet, round bar, plate, sheet and wire.

Mechanical Properties

Tensile properties. See Table 34.
Hardness. See Table 35.
Elastic modulus:

Temperature °C	°F	Modulus in tension GPa	10⁶ psi
20	68	203	29.5
93	200	199	28.8
150	300	194	28.2
204	400	190	27.6
260	500	186	27.0
315	600	181	26.3
370	700	177	25.7
425	800	173	25.1
480	900	169	24.5
540	1000	165	23.9
590	1100	161	23.3
650	1200	156	22.6
705	1300	151	21.9
760	1400	145	21.1
815	1500	141	20.5
870	1600	137	19.9
925	1700	133	19.3
980	1800	129	18.7

Creep-rupture characteristics. See Table 36.

Structure

Crystal structure. Face-centered cubic
Microstructure. In the annealed condition has MC, M_6C, and sometimes $M_{23}C_6$ carbides in an austenitic-type matrix. Depending on aging temperature, additional M_6C and $M_{23}C_6$, as well as Laves phase or M_7N_3, may precipitate. Grain size depends on thermomechanical history and cannot be related to annealing temperature alone.

Mass Characteristics

Density. 8.230 Mg/m³ (0.297 lb/in.³) at 22 °C (72 °F)

Thermal Properties

Coefficient of thermal expansion. Linear:

Temperature °C	°F	Mean coefficient(a) μm/m·K	μin./in.·°F
20–95	68–200	14.0	7.8
20–205	68–400	14.9	8.3
20–315	68–600	15.5	8.6
20–425	68–800	15.8	8.8
20–590	68–1000	16.2	9.0
20–650	68–1200	16.7	9.3
20–760	68–1400	17.1	9.5
20–870	68–1600	17.5	9.7
20–980	68–1800	17.8	9.9

(a) Accuracy: ±0.4 μm/m·K (±0.2 μin./in.·°F)

Specific heat:

Temperature °C	°F	Specific heat J/kg·K	Btu/lb·°F
0	32	439	0.1049
100	212	472	0.1128

(continued)

Temperature °C	Temperature °F	Specific heat J/kg·K	Specific heat Btu/lb·°F
200	392	494	0.1180
300	572	510	0.1218
400	752	523	0.1250
500	932	534	0.1275
600	1112	541	0.1292
700	1292	545	0.1303
800	1472	553	0.1322
900	1652	619	0.1480
1000	1832	690	0.1649
1100	2012	690	0.1650

Enthalpy:

Temperature °C	Temperature °F	Enthalpy, kJ/kg
0	32	0
75	168	33.919
224	435	105.375
498	928	247.960
837	1540	432.320
956	1750	506.890
1047	1922	569.970
1138	2080	632.270

Thermal conductivity:

Temperature °C	Temperature °F	Conductivity W/m·K	Conductivity Btu/ft·h·°F
24	75	11.2	6.4
200	392	14.5	8.4
394	740	17.9	10.3
600	1112	20.8	12.0
695	1290	22.3	12.9
795	1460	22.6	13.1
900	1650	26.7	15.4
980	1800	29.6	17.1
1100	2012	30.9	17.8

Thermal diffusivity:

Temperature °C	Temperature °F	Diffusivity, $10^{-6}m^2/s$
24	75	3.05
200	392	3.55
394	740	4.13
600	1112	4.67
695	1290	4.94
795	1460	4.95
900	1650	5.22
986	1800	5.18
1100	2012	5.42

Electrical Properties

Electrical resistivity:

Temperature °C	Temperature °F	Resistivity, μΩ·m
24	75	0.970
32	90	0.970
40	104	0.980
121	250	1.010
180	356	1.040
330	626	1.100
412	774	1.120
423	790	1.120
579	1070	1.160
713	1315	1.190
828	1520	1.210
858	1575	1.210
926	1700	1.220
1066	1950	1.250
1210	2210	1.270

Table 35 Effect of cold work and heat treatment on room temperature hardness of Haynes 556 sheet

Heat treatment Temperature °C	°F	Time, min	Average hardness, HRA, after cold reduction of 10%	20%	30%	40%	50%
As-cold reduced		0	65	67	68	70	72
815	1500	10	63	67	69	70	71
		30	64	67	68	70	71
870	1600	10	64	66	69	70	68
		30	63	66	68	68	67
925	1700	10	63	65	68	66	64
		30	63	65	67	64	64
980	1800	10	62	64	63	63	64
		30	62	64	63	64	62
1100	2000	10	57	59	60	61	61
		30	57	58	60	61	61

Note: Hardness of material prior to cold reduction and subsequent heat treatment was 59 HRA. Thickness before cold reduction was 2.3 mm (0.090 in.).

Table 36 Typical stress-rupture data for annealed 0.73-mm (0.029-in.) thick Haynes 556 sheet

Temperature °C	Temperature °F	Life, h	Stress MPa	Stress ksi	Elongation, %
650	1200	394.4	310	45.0	43
760	1400	135.9	170	25.0	60
815	1500	46.0	140	20.0	67
815	1500	71.5	125	18.0	71
870	1600	68.3	83	12.0	60
925	1700	153.2	52	7.5	31
980	1800	108.0	31	4.5	24
1040	1900	123.0	19	2.7	42
1090	2000	82.2	12	1.7	85

Chemical Properties

General corrosion behavior. Improved oxidation and hot corrosion resistance compared to N-155

Fabrication Characteristics

Machinability. Compared to other high-temperature alloys, machinability is good to excellent; can be machined by standard techniques for this type of alloy; can be welded by common welding techniques.

Recrystallization temperature. An interrelated function of prior cold deformation, temperature and time. Material cold reduced 50% may recrystallize at temperatures as low as 840 °C (1500 °C); material cold reduced 10% generally requires greater than 980 °C (1800 °F). To avoid secondary precipitation of carbides and to achieve a product with good formability, 1175 °C (2150 °F) is the suggested heat treatment temperature.

Solution temperature. Microconstituents dissolve at various temperatures; some such as MC carbides remain in the structure through 1230 °C (2250 °F)

Aging temperature. Haynes 556 is normally put into service in the annealed condition. Precipitation of secondary phases, such as carbides, will occur at all temperatures below the annealing temperature, but reaction kinetics are slow below about 700 °C (1300 °F).

Hot working temperature. Best results are obtained in the temperature range 1180 to 980 °C (2150 to 1800 °F); however the alloy has been "hot" worked at temperatures as low as 650 °C (1200 °F).
Hot shortness temperature. 1230 °C (2250 °F)

MAR-M 302

Commercial Name

AISI designation. Type 684
Trade name. SM 302 (obsolete)

Specifications

AMS. Investment castings: 5384
ASTM. A-567 castings

Chemical Composition

Composition limits. Typical. 58.0 Co; 21.5 Cr; 10.0 W; 9.0 Ta; 0.85 C; 0.005 B; 0.20 Zr. Range. 20 to 23 Cr; 9 to 11 W; 8 to 10 Ta; 0.78 to 0.93 C; 0.75 to 1.5 Fe; 0.10 to 0.40 Si; 0.10 to 0.30 Zr; 0.10 max Mn; 0.010 max B; rem Co

Applications

Typical uses. For turbine vanes, nozzle guide vanes and buckets in gas turbines. For service to 1150 °C (2100 °F)

Mechanical Properties

Tensile properties. See Table 37.
Stress-rupture characteristics. Typical, at 100 h:

Temperature		Rupture strength	
°C	°F	MPa	ksi
815	1500	250	36
870	1600	185	27
925	1700	140	20
980	1800	97	14
1040	1900	69	10
1090	2000	41	6

Structure

Microstructure. fcc matrix (austenite) solid solution

Mass Characteristics

Density. 9.21 Mg/m³ (0.333 lb/in.³) at 20 °C (68 °F)

Thermal Properties

Coefficient of thermal expansion. Linear:

Temperature		Mean coefficient	
°C	°F	µm/m·K	µin./in.·°F
205	400	12.4	6.9
315	600	13.0	7.2
425	800	13.3	7.4
540	1000	13.7	7.6
650	1200	14.0	7.8
760	1400	14.4	8.0
870	1600	14.9	8.3
980	1800	15.7	8.7
1090	2000	16.6	9.2

Thermal conductivity:

Temperature		Conductivity	
°C	°F	W/m·K	Btu/ft·h·°F
21	70	18.7	10.8
93	200	19.2	11.1
205	400	20.2	11.7
315	600	21.2	12.2
425	800	21.7	12.5
540	1000	22.2	12.8
650	1200	22.3	12.9
760	1400	22.6	13.1
870	1600	23.2	13.4
980	1800	24.2	14.0

Table 37 Typical mechanical properties of MAR-M 302

Temperature		Tensile strength		Yield strength		Elongation,
°C	°F	MPa	ksi	MPa	ksi	%
21	70	930	135	690	100	2
540	1000	795	115	505	73	...
650	1200	785	114	450	65	...
760	1400	705	102	385	56	8
870	1600	450	65	310	45	11
980	1800	275	40	215	31	15
1090	2000	150	22	150	22	21

MAR-M 322

Chemical Composition

Composition limits. Typical. 61.0 Co; 21.5 Cr; 9.0 W; 4.5 Ta; 0.75 Ti; 1.0 C; 2.25 Zr. Range. 20 to 23 Cr; 8 to 10 W; 4 to 5 Ta; 2 to 2.5 Zr; 0.90 to 1.1 C; 0.65 to 0.85 Ti; 1.5 max Fe; 0.20 max Mn; 0.20 max Si; rem Co

Applications

Typical uses. For turbine vanes and blades, jet engine components

Mechanical Properties

Tensile properties. See Table 38.

Table 38 Typical mechanical properties of MAR-M 322

Temperature		Tensile strength		Yield strength		Elongation,
°C	°F	MPa	ksi	MPa	ksi	%
21	70	825	120	625	91	3.2
540	1000	655	95	415	60	6
650	1200	655	95	415	60	6
760	1400	625	91	380	55	6.5
870	1600	550	80	345	50	12

Structure

Microstructure. fcc matrix (austenite) solid solution

Mass Characteristics

Density. 8.91 Mg/m³ (0.322 lb/in.³) at 20 °C (68 °F)

Chemical Properties

General corrosion behavior. Oxidation resistant to 1090 °C (2000 °F)

MAR-M 509

Chemical Composition

Composition limits. Typical. 55.0 Co; 23.5 Cr; 10.0 Ni; 7.0 W; 3.5 Ta; 0.20 Ti; 0.60 C; 0.50 Zr. Range. 21.0 to 24.0 Cr; 9.0 to 11.0 Ni; 6.5 to 7.5 W; 3.0 to 4.0 Ta; 0.55 to 0.65 C; 0.4 to 0.6 Zr; 0.15 to 0.25 Ti; 1.50 max Fe; 0.4 max Si; 0.1 max Mn; 0.01 max B; 0.015 max S; rem Co

Applications

Typical uses. For aircraft and industrial turbines

Mechanical Properties

Tensile properties. See Table 39.
Hardness. 24 to 34 HRC
Elastic modulus:

Temperature °C	°F	Modulus of elasticity GPa	10^6 psi
21	70	225	32.7
205	400	215	30.9
425	800	195	28.4
650	1200	180	25.8
760	1400	165	23.9
870	1600	155	22.5
980	1800	135	19.8

Stress-rupture characteristics. Typical:

Temperature °C	°F	Rupture strength MPa	ksi
At 100 h			
760	1400	345	50
815	1500	250	36.5
870	1600	180	26
925	1700	135	19.5
980	1800	105	15
1040	1900	79	11.5
1090	2000	52	7.5
At 1000 h			
760	1400	260	38
815	1500	195	28
870	1600	140	20
925	1700	105	15
980	1800	79	11.5
1040	1900	55	8
1090	2000	34	5

Structure

Microstructure. fcc matrix (austenite) solid solution with blocky, script and eutectic carbides

Mass Characteristics

Density. 8.85 Mg/m^3 (0.320 lb/in.^{3}) at 20 °C (68 °F)

Thermal Properties

Incipient melting temperature. 1290 °C (2350 °F)
Liquidus temperature. 1400 °C (2550 °F)
Softening temperature. 1340 °C (2450 °F)
Coefficient of thermal expansion. Linear:

Temperature °C	°F	Mean coefficient μm/m·K	μin./in.·°F
315	600	14.6	8.1
425	800	15.3	8.5
540	1000	15.9	8.85
650	1200	16.2	9.0
760	1400	16.7	9.3
870	1600	17.2	9.55
980	1800	17.6	9.8
1090	2000	18.2	10.1

Thermal conductivity:

Temperature °C	°F	Conductivity W/m·K	Btu/ft·h·°F
425	800	25.2	14.6
540	1000	27.9	16.1
650	1200	31.1	18.0
760	1400	34.3	19.8
870	1600	37.6	21.7
980	1800	41.2	23.8
1090	2000	44.6	25.8

Table 39 Typical mechanical properties of MAR-M 509

Temperature °C	°F	Tensile strength MPa	ksi	Yield strength MPa	ksi	Elongation, %
21	70	785	114	570	83	4
540	1000	570	83	400	58	6
650	1200	560	81	370	54	7
760	1400	570	83	365	53	10
870	1600	350	51	290	42	20
980	1800	215	31	180	26	26

Table 40 Typical tensile properties for Stellite-6B (a)

Temperature °C	°F	Tensile strength MPa	ksi	Yield strength MPa	ksi	Elongation, %
2-mm (0.063-in.) sheet						
20	68	1010	146	635	92	11
815	1500	510	74	310	45	17
870	1600	385	56	270	39	18
980	1800	230	33	140	20	36
1090	2000	140	20	76	11	44
1150	2100	90	13	55	8	22
13-mm (1/2-in.) plate						
20	68	1020	148	605	88	7
540	1000	915	133	405	59	9
675	1250	795	115	420	61	9
16-mm (5/8-in.) bar						
20	68	1060	154	640	93	17(b)
315	600	1020	148	515	75	30(b)
540	1000	890	129	460	67	28(b)
815	1500	515	75	325	47	38(b)
870	1600	400	58	260	38	34(b)

(a) All specimens solution heat treated at 1232 °C (2250 °F) and rapid air cooled; limited number of tests. (b) Elongation in 25 mm or 1 in.

Stellite 6B
60Co-30Cr-4.5W-1.1C-2Fe-2Ni

Commercial Name

Trade name. HAYNES STELLITE alloy No. 6B

Chemical Composition

Composition limits. 28.00 to 32.00 Cr; 3.50 to 5.50 W; 3.00 max Fe; 3.00 max Ni; 2.00 max Mn; 2.00 max Si; 1.50 max Mo; 0.90 to 1.40 C; rem Co

Applications

Typical uses. Wear plates and bars, bushings and sleeves for shafts operating in hot or possibly corrosive atmospheres where lubrication is difficult or impossible

Mechanical Properties

Tensile properties. See Table 40.

Table 41 Typical creep-rupture characteristics for Stellite 6B (a)

Temperature, °C	Stress, MPa	Stress, ksi	Initial elongation, %	Life, h	Time (h) for total elongation of: 0.5%	1.0%	2.0%	Elongation of rupture, %
2-mm (0.063-in.) sheet								
540	415	60	0.70	192.8(b)	...	...	...	0.8
650	345	50	0.45	361.4	0.5	113.8	...	3.0
760	240	35	0.35	59.3	0.4	3.8	16.3	5.1
815	170	25	0.35	70.6	0.2	4.3	16.9	4.7
870	130	19	0.10	57.9	0.5	2.2	11.1	4.3
925	83	12	0.19	104.0	1.8	20.9	89.9	2.6
980	55	8	0.05	113.4	5.1	22.7	57.6	5.5
1090	18	2.6	0.004	116.7	5.4	...	...	13.3
16-mm (5/8-in.) bar								
540	310	45	0.79	330.8(b)	...	...	...	0.83
	450	65	1.03	329.9	...	...	...	...
650	275	40	0.06	164.6(b)	...	...	...	0.44
	310	45	0.70	367.1	...	15.9	117.5	11.6
	345	50	0.54	282.3(b)	...	163.6	...	1.73
730	240	35	0.37	137.7(b)	1.57	11.3	55.7	3.22
	275	40	0.17	13.6	...	...	...	16.0
	310	45	0.27	62.7	1.03	4.75	9.8	16.9
870	105	15	0.40	130.9	0.03	1.77	7.6	13.3
	125	18	0.42	52.7	0.05	0.68	2.9	11.0
	130	19	0.22	55.9	0.20	0.70	2.4	20.0

(a) Specimens solution heat treated 1232 °C (2250 °F), rapid air cooled prior to testing. (b) Test discontinued before rupture.

Table 42 Typical tensile properties for Stellite 6K (a)

Temperature, °C	Temperature, °F	Tensile strength, MPa	Tensile strength, ksi	Yield strength, MPa	Yield strength, ksi	Elongation, %
2-mm (0.063-in.) sheet						
20	68	1220	177	710	103	4
650	1200	1010	146	...	...	8
815	1500	485	70	310	45	17
980	1800	235	34	130	19	28
1090	2000	115	17	59	8.6	53
13-mm (1/2-in.) and 23-mm (7/8-in.) plate						
20	68	1010	146	745	108	1
316	600	785	114	560	81	2
540	1000	730	106	515	75	3

(a) All specimens solution heat treated at 1232 °C (2250 °F) and rapid air cooled; limited number of tests.

Compressive properties. Compressive strength, 2399 MPa (348 ksi)

Hardness. 2-mm (0.063-in.) sheet, 39 HRC; 13-mm (0.5-in.) plate, 38 HRC

Impact strength. Charpy(a):

Temperature, °C	Temperature, °F	Type of test	Longitudinal, J	Longitudinal, ft·lb
3-mm (1/2-in.) plate				
RT	RT	Unnotched	97.6	72
		Notched	8.1	6
540	1000	Unnotched	110	81
		Notched	20.3	15
675	1250	Unnotched	160	116
		Notched	20.3	15
815	1500	Unnotched	170	126
		Notched	20.3	15

(a) Solution heat treated at 1230 °C (2250 °F), rapid air cooled.

Creep-rupture characteristics. See Table 41.

Mass Characteristics

Density. 8.380 Mg/m^3 (0.303 lb/in.^{3}) at 22 °C (72 °F)

Thermal Properties

Liquidus temperature. 1355 °C (2470 °F)

Solidus temperature. 1265 °C (2310 °F)

Coefficient of thermal expansion. Linear:

Temperature, °C	Temperature, °F	Mean coefficient, µm/m·K	Mean coefficient, µin./in.·°F
0–100	32–212	13.9	7.7
0–200	32–392	14.1	7.8
0–300	32–572	14.5	8.0
0–400	32–752	14.7	8.2
0–500	32–932	15.0	8.3
0–600	32–1112	15.3	8.5
0–700	32–1292	15.8	8.8
0–800	32–1472	16.3	9.1
0–900	32–1652	16.9	9.4
0–1000	32–1832	17.4	9.7

Specific heat. 421 J/kg·K (0.101 Btu/lb·°F)

Thermal conductivity. 14.7 W/m·K (8.5 Btu/ft·h·°F) at 22 °C (72 °F)

Electrical Properties

Electrical resistivity. 910 nΩ·m at 22 °C (72 °F)

Magnetic Properties

Magnetic permeability. <1.2 at 16 kA/m (200 oersteds) at 22 °C (72 °F)

Optical Properties

Color. Silvery

Reflectivity. 57 to 70%

Fabrication Characteristics

Hot hardness:

Temperature, °C	Temperature, °F	Hardness(a), HB
540	1000	226
650	1200	203
760	1400	167
870	1600	102

(a) Mutual indentation method.

Stellite 6K
59Co-31Cr-4.5W-1.6C-2Fe-2Ni

Commercial Name

Trade name. Haynes Stellite alloy No. 6K

Chemical Composition

Composition limits. 28.00 to 32.00 Cr; 3.50 to 5.50 W; 3.00 max Fe; 3.00 max Ni; 2.00 max Si; 2.00 max Mn; 1.40 to 1.90 C; 1.50 max Mo; rem Co

Applications

Typical uses. Machine knives for cutting rubber, plastics, wood, leather, food products, synthetic fibers and paper

Mechanical Properties

Tensile properties. See Table 42.
Compressive properties. Compressive strength, sheet: 2240 MPa (325 ksi)
Hardness. 2-mm (0.063-in.) sheet, 46 HRC; 13-mm (0.5-in.) and 23-mm (7/8-in.) plate, 45 HRC

Mass Characteristics

Density. 8.38 Mg/m^3 (0.303 lb/in.^{3}) at 22 °C (72 °F)

Thermal Properties

Liquidus temperature. 1317 °C (2403 °F)
Solidus temperature. 1257 °C (2295 °F)
Coefficient of thermal expansion. Linear:

Temperature °C	Temperature °F	Mean coefficient µm/m·K	Mean coefficient µin./in.·°F
0–100	32–212	13.8	7.7
0–200	32–392	13.8	7.7
0–300	32–572	13.8	7.7
0–400	32–752	13.8	7.7
0–500	32–932	13.8	7.7
0–600	32–1112	14.0	7.8
0–700	32–1292	14.2	7.9
0–800	32–1472	14.5	8.1
0–900	32–1652	14.9	8.3
0–1000	32–1832	15.5	8.6

Magnetic Properties

Magnetic permeability. <1.2 at 16 kA/m (200 oersteds) at 22 °C (72 °F)

Optical Properties

Color. Silvery
Reflectivity. 57 to 70%

Table 43 Typical tensile properties and hardness for UMCo-50 alloy

Temperature °C	Temperature °F	Tensile strength MPa	Tensile strength ksi	0.2% offset yield strength MPa	0.2% offset yield strength ksi	Elongation, %	Hardness, DPH
UMCo-50, as-cast							
25	70	550	80.0	315	45.6	8	250
700	1290	195	28.6	150	21.4	19	163
900	1650	130	18.6	110	15.7	9	72
1000	1830	79	11.4	69	10.0	18	...
UMCo-50, wrought							
25	70	925	134	610	88.5	10	350
500	930	885	128	570	83.0	23	...
700	1290	325	47	225	32.8	21	210
900	1650	155	22.8	150	21.4	12	...
1000	1830	79	11.4	59	8.6	18	...

Table 44 Typical resistance of wrought UMCo-50

Temperature °C	Temperature °F	Stress to cause creep elongation of: 1% MPa	1% ksi	0.5% MPa	0.5% ksi
In 10 h					
850	1560	61	8.9	55	8.0
900	1650	35	5.1	30	4.3
950	1740	23	3.3	19	2.7
1000	1830	16	2.3	14	2.0
In 100 h					
800	1470	61	8.9	53	7.7
850	1560	34	5.0	30	4.3
900	1650	21	3.1	18	2.6
950	1740	14	2.1	12	1.8
In 1000 h					
750	1380	64	9.3	58	8.4
800	1470	34	5.0	30	4.3
850	1560	21	3.1	18	2.6
900	1650	14	2.1	12	1.7
In 10 000 h					
700	1290	72	10.4	63	9.2
750	1380	37	5.3	32	4.6
800	1470	21	3.1	19	2.7
850	1560	14	2.1	12	1.8

UMCo-50
50Co-28Cr-22Fe

Commercial Name

Trade name. Haynes Alloy No. 150; Esco Alloy 75; Illium H; MO-RE 5

Chemical Composition

Composition limits. 48 to 52 Co; 27 to 29 Cr; 0.5 to 1.0 Mn; 0.5 to 1.0 Si; 0.05 to 0.12 C; 0.02 max P; 0.02 max S; rem Fe

Applications

Typical uses. Furnace grates, trays, and rolls; skids and rails in heat treating furnaces; slag-notch rings and tundishes

Mechanical Properties

Tensile properties. See Table 43.
Hardness. See Table 43.
Elastic modulus. Tension, 215 GPa (31.5×10^6 psi)
Impact strength. Charpy, 95 J (70

ft·lb) at 20 °C (68 °F)

Creep-rupture characteristics. See Table 44.

Mass Characteristics

Density. 8.050 Mg/m³ (0.29 lb/in.³)

Thermal Properties

Liquidus temperature. 1395 °C (2540 °F)

Solidus temperature. 1380 °C (2515 °F)

Coefficient of thermal expansion. Linear, 16.8 μm/m·K (9.33 μin./in.·°F) at 20 to 1000 °C (68 to 1830 °F)

Thermal conductivity. 8.9 W/m·K (5.1 Btu/ft·h·°F) at 20 °C (68 °F)

Electrical Properties

Electrical resistivity. 825 nΩ·m at 25 °C (77 °F)

Chemical Properties

General corrosion behavior. Resistant to attack by dilute H_2SO_4 and by boiling HNO_3; rapidly attacked by HCl. High degree of oxidation resistance, e.g., comparable to 25Cr-20Ni steel in air up to 1200 °C (2190 °F). High degree of hot corrosion resistance to oxidizing-sulfidizing environments as well as in combustion products from sulfur-bearing fuel oils. Resistant to molten copper but rapidly attacked by molten aluminum.

Fabrication Characteristics

Hot working temperature. 900 to 1100 °C (1650 to 2010 °F)

X-40/X-45

Specifications

AMS. Investment castings: 5382

ASTM. Investment castings: A567

Chemical Composition

Typical composition. 54.0 Co; 25.5 Cr; 10.5 Ni; 7.5 W; 0.75 Mn; 0.75 Si; 0.50 C (a)

(a) 0.25 C for X-45.

Mechanical Properties

Tensile properties. See Table 45.

Stress-rupture characteristics. Typical:

Temperature °C	°F	Rupture strength MPa	ksi
At 100 h			
650	1200	385	56
760	1400	260	38
870	1600	140	20
980	1800	76	11
At 1000 h			
650	1200	350	51
760	1400	230	33
870	1600	110	16
980	1800	68	9.8

Structure

Microstructure. fcc matrix (austenite) solid solution with blocky, script and eutectic carbides

Mass Characteristics

Density. 8.60 Mg/m³ (0.311 lb/in.³) at 20 °C (68 °F)

Thermal Properties

Coefficient of thermal expansion. Linear:

Temperature °C	°F	Coefficient μm/m·K	μin./in.·°F
315	600	14.0	7.8
425	800	14.6	8.1
540	1000	15.1	8.4
650	1200	15.8	8.75
760	1400	16.6	9.2

Thermal conductivity:

Temperature °C	°F	Conductivity W/m·K	Btu/ft·h·°F
21	70	11.8(a)	6.8(a)
93	200	13.5	7.8
205	400	15.1	8.7
315	600	17.9	10.3
425	800	19.1	11.0
540	1000	21.6	12.5
650	1200	22.8(a)	13.2(a)

(a) Extrapolated values.

Table 45 Typical mechanical properties of X-40

Temperature °C	°F	Tensile strength MPa	ksi	Yield strength MPa	ksi	Elongation, %
21	70	745	108	525	76	9
540	1000	550	80	275	40	17
650	1200	515	75	260	38	12
760	1400	485	70	...	...	10
870	1600	325	47	...	...	16
980	1800	200	29	...	...	31

Heat-Resistant Castings

By the ASM Committee on Cast Corrosion-Resisting and Heat-Resisting Alloys*

CASTINGS are classified as heat resistant if they are capable of sustained operation while exposed, either continuously or intermittently, to operating temperatures that result in metal temperatures in excess of 650 °C (1200 °F). Alloys used in castings for such service include iron-chromium ("straight chromium"), iron-chromium-nickel, iron-nickel-chromium, nickel-base and cobalt-base alloys. In application of heat-resistant alloys, considerations include: (*a*) resistance to corrosion at elevated temperatures; (*b*) stability (resistance to warping, cracking or thermal fatigue); and (*c*) creep strength (resistance to plastic flow).

Many alloys of the same general types are used also for their resistance to corrosive media at temperatures below 650 °C (1200 °F), and castings intended for such service are classified as corrosion resistant. Although there is usually a distinction between heat-resistant alloys and corrosion-resistant alloys, based on carbon content, the line of demarcation is vague — particularly for alloys used in the range from 480 to 650 °C (900 to 1200 °F).

Commercial applications of heat-resistant castings include metal-treatment furnaces, gas turbines, aircraft engines, military equipment, oil-refinery furnaces, cement-mill equipment, petrochemical furnaces, chemical-process equipment, power-plant equipment, steel-mill equipment, turbochargers, and equipment used in manufacturing glass and synthetic rubber. Alloys of the iron-chromium and iron-chromium-nickel groups are of the greatest commercial importance.

General Properties

General characteristics of the five types of cast heat-resistant alloys are discussed below. For Fe-Cr, Fe-Cr-Ni and Fe-Ni-Cr alloys, Table 1 lists designations and compositions, Table 2 gives typical room-temperature properties and Table 3 presents corrosion rates in

Table 1 Compositions of ACI heat-resistant casting alloys

ACI designation	UNS number	ASTM specifications(a)	Composition, %(b) C	Cr	Ni	Si (max)
HA	...	A217	0.20 max	8 to 10	...	1.00
HC	J92605	A297, A608	0.50 max	26 to 30	4 max	2.00
HD	J93005	A297, A608	0.50 max	26 to 30	4 to 7	2.00
HE	J93403	A297, A608	0.20 to 0.50	26 to 30	8 to 11	2.00
HF	J92603	A297, A608	0.20 to 0.40	19 to 23	9 to 12	2.00
HH	J93503	A297, A608	0.20 to 0.50	24 to 28	11 to 14	2.00
HI	J94003	A297, A567, A608	0.20 to 0.50	26 to 30	14 to 18	2.00
HK	J94224	A297, A351, A567, A608	0.20 to 0.60	24 to 28	18 to 22	2.00
HL	J94604	A297, A608	0.20 to 0.60	28 to 32	18 to 22	2.00
HN	J94213	A297, A608	0.20 to 0.50	19 to 23	23 to 27	2.00
HP	...	A297	0.35 to 0.75	24 to 28	33 to 37	2.00
HP-50WZ (c)	...	...	0.45 to 0.55	24 to 28	33 to 37	2.50
HT	J94605	A297, A351, A567, A608	0.35 to 0.75	13 to 17	33 to 37	2.50
HU	...	A297, A608	0.35 to 0.75	17 to 21	37 to 41	2.50
HW	...	A297, A608	0.35 to 0.75	10 to 14	58 to 62	2.50
HX	...	A297, A608	0.35 to 0.75	15 to 19	64 to 68	2.50

(a) ASTM designations are same as ACI designations. (b) Rem Fe in all compositions. Manganese content: 0.35 to 0.65% for HA, 1% for HC, 1.5% for HD and 2% for the other alloys. Phosphorus and sulfur contents: 0.04% max for all but HP-50WZ. Molybdenum is intentionally added only to HA, which has 0.90 to 1.20% Mo; maximum for other alloys is set at 0.5% Mo. HH also contains 0.2% max N. (c) Also contains 4 to 6% W, 0.1 to 1.0% Zr, and 0.035% max S and P.

*See page X for committee list.

Table 2 Typical room-temperature properties of ACI heat-resistant casting alloys

Alloy	Condition	Tensile strength MPa	Tensile strength ksi	Yield strength MPa	Yield strength ksi	Elongation, %	Hardness, HB
HC	As cast	760	110	515	75	19	223
	Aged(a)	790	115	550	80	18	...
HD	As cast	585	85	330	48	16	90
HE	As cast	655	95	310	45	20	200
	Aged(a)	620	90	380	55	10	270
HF	As cast	635	92	310	45	38	165
	Aged(a)	690	100	345	50	25	190
HH, type 1	As cast	585	85	345	50	25	185
	Aged(a)	595	86	380	55	11	200
HH, type 2	As cast	550	80	275	40	15	180
	Aged(a)	635	92	310	45	8	200
HI	As cast	550	80	310	45	12	180
	Aged(a)	620	90	450	65	6	200
HK	As cast	515	75	345	50	17	170
	Aged(b)	585	85	345	50	10	190
HL	As cast	565	82	360	52	19	192
HN	As cast	470	68	260	38	13	160
HP	As cast	490	71	275	40	11	170
HT	As cast	485	70	275	40	10	180
	Aged(b)	515	75	310	45	5	200
HU	As cast	485	70	275	40	9	170
	Aged(c)	505	73	295	43	5	190
HW	As cast	470	68	250	36	4	185
	Aged(d)	580	84	360	52	4	205
HX	As cast	450	65	250	36	9	176
	Aged(c)	505	73	305	44	9	185

(a) Aging treatment: 24 h at 760 °C (1400 °F), furnace cool. (b) Aging treatment: 24 h at 760 °C (1400 °F), air cool. (c) Aging treatment: 48 h at 980 °C (1800 °F), air cool. (d) Aging treatment: 48 h at 980 °C (1800 °F), furnace cool.

Table 3 Approximate rates of corrosion for ACI heat-resistant casting alloys in air and in flue gas

Alloy	Oxidation rate in air, mils/yr(a) 870 °C (1600 °F)	980 °C (1800 °F)	1090 °C (2000 °F)	Corrosion rate, mils/yr(a)(b), in flue gas with sulfur content of: 0.12 g/m³ Oxidizing	0.12 g/m³ Reducing	2.3 g/m³ Oxidizing	2.3 g/m³ Reducing
HB	25−	250−	500−	100+	500−	250−	500
HC	10	50	50	25−	25+	25	25−
HD	10−	50−	50−	25−	25−	25−	25−
HE	5−	25−	35−	25−	25−	25−	25−
HF	5−	50+	100	50+	100+	50+	250−
HH	5−	25−	50	25−	25	25	25−
HI	5−	10+	35−	25−	25−	25−	25−
HK	10−	10−	35−	25−	25−	25−	25−
HL	10+	25−	35	25−	25−	25−	25−
HN	5	10+	50−	25−	25−	25	25
HP	25−	25	50	25−	25−	25−	25−
HT	5−	10+	50	25	25−	25	100
HU	5−	10−	35−	25−	25−	25−	25
HW	5−	10−	35	25	25−	50−	250
HX	5−	10−	35−	25−	25−	25−	25−

(a) Data based on 100-h tests. To convert to μm/yr, multiply by 25. (b) At 980 °C (1800 °F).

Sources: Alloy Casting Institute; A. Brasunas, J. T. Gow and O. E. Harder, "Resistance of Fe-Ni-Cr Alloys to Corrosion in Air at 1600 to 2200 F", ASTM Symposium on Materials for Gas Turbines, June 1946, p 129 to 152

air and in flue gas. Compositions of low-iron nickel-base and low-iron cobalt-base alloys are given in Tables 4 and 5, respectively. Other data for cast heat-resistant alloys are presented in the data compilations that directly follow this article.

Fe-Cr alloys contain 10 to 30% Cr and little or no nickel. These alloys are useful chiefly for resistance to oxidation; they have low strength at elevated temperatures. Use of these alloys is restricted to conditions, either oxidizing or reducing, that involve low static loads and uniform heating. Chromium content depends on anticipated service temperature.

Fe-Cr-Ni alloys contain more than 13% Cr and more than 7% Ni (always more chromium than nickel). These austenitic alloys ordinarily are used under oxidizing or reducing conditions similar to those withstood by the ferritic iron-chromium alloys, but in service they have greater strength and ductility than the straight chromium alloys. They are used, therefore, to withstand greater loads and moderate changes of temperature. These alloys also are used in the presence of oxidizing and reducing gases that are high in sulfur content.

Fe-Ni-Cr alloys contain more than 25% Ni and more than 10% Cr (always more nickel than chromium). These austenitic alloys are used for withstanding reducing as well as oxidizing atmospheres, except where sulfur content is appreciable. (In atmospheres containing 0.05% or more hydrogen sulfide, for example, Fe-Cr-Ni alloys are recommended.) In contrast with Fe-Cr-Ni alloys, Fe-Ni-Cr alloys do not carburize rapidly or become brittle and do not take up nitrogen in nitriding atmospheres. These characteristics become enhanced as nickel content is increased, and in carburizing and nitriding atmospheres casting life increases with nickel content. Austenitic Fe-Ni-Cr alloys are used extensively under conditions of severe temperature fluctuations such as those encountered by fixtures used in quenching and by parts that are not heated uniformly or that are heated and cooled intermittently. In addition, these alloys have characteristics that make them suitable for electrical-resistance heating elements.

Nickel-base alloys contain about 50% Ni and appreciable amounts of chromium, cobalt and refractory metals, but little or no iron; they may also

Table 4 Compositions of nickel-base heat-resistant casting alloys

Alloy designation	Nominal composition, %											
	C	Ni	Cr	Co	Mo	Fe	Al	B	Ti	W	Zr	Others
B-1900	0.1	64	8	10	6	...	6	0.015	1	...	0.10	4Ta(a)
Hastelloy X	0.1	50	21	1	9	18	...	...	...	1	...	...
IN-100	0.18	60.5	10	15	3	...	5.5	0.01	5	...	0.06	1V
IN-738X	0.17	61.5	16	8.5	1.75	...	3.4	0.01	3.4	2.6	0.1	1.75Ta, 0.9Nb
IN-792	0.2	60	13	9	2.0	...	3.2	0.02	4.2	4	0.1	4Ta
Inconel 713C	0.12	74	12.5	...	4.2	...	6	0.012	0.8	...	0.1	2Nb
Inconel 713LC	0.05	75	12	...	4.5	...	6	0.01	0.6	...	0.1	2Nb
Inconel 718	0.04	53	19	...	3	18	0.5	...	0.9	...	...	0.1Cu, 5Nb
Inconel X-750	0.04	73	15	...	...	7	0.7	...	2.5	...	...	0.25Cu, 0.9Nb
M-252	0.15	56	20	10	10	...	1	0.005	2.6	...	...	...
MAR-M 200	0.15	59	9	10	...	1	5	0.015	2	12.5	0.05	1Nb(b)
MAR-M 246	0.15	60	9	10	2.5	...	5.5	0.015	1.5	10	0.05	1.5Ta
MAR-M 247	0.15	59	8.25	10	0.7	0.5	5.5	0.015	1	10	0.05	1.5Hf, 3Ta
NX 188 (DS)	0.04	74	...	...	18	...	8	...	...	...	...	...
René 77	0.07	58	15	15	4.2	...	4.3	0.015	3.3	...	0.04	...
René 80	0.17	60	14	9.5	4	...	3	0.015	5	4	0.03	...
René 100	0.18	61	9.5	15	3	...	5.5	0.015	4.2	...	0.06	1V
TRW-NASA VIA	0.13	61	6	7.5	2	...	5.5	0.02	1	6	0.13	0.4Hf, 0.5Nb, 0.5Re, 9Ta
Udimet 500	0.1	53	18	17	4	2	3	...	3	...	...	...
Udimet 700	0.1	53.5	15	18.5	5.25	...	4.25	0.03	3.5	...	...	...
Udimet 710	0.13	55	18	15	3	...	2.5	...	5	1.5	0.08	...
Waspaloy	0.07	57.5	19.5	13.5	4.2	1	1.2	0.005	3	...	0.09	...
WAZ-20 (DS)	0.20	72	...	...	...	...	6.5	...	...	20	1.5	...

(a) B-1900 + Hf also contains 1.5% Hf. (b) MAR-M 200 + Hf also contains 1.5% Hf.

Table 5 Compositions of cobalt-base heat-resistant casting alloys

Alloy designation	Nominal composition, %										
	C	Co	Cr	Ni	Al	B	Fe	Ta	W	Zr	Others
AiResist 13	0.45	62	21	...	3.4	...	...	2	11	...	0.1Y
AiResist 213	0.20	64	20	0.5	3.5	...	0.5	6.5	4.5	0.1	0.1Y
AiResist 215	0.35	63	19	0.5	4.3	...	0.5	7.5	4.5	0.1	0.1Y
Haynes 21	0.25	64	27	3	...	...	1	...	...	...	5 Mo
Haynes 25; L-605	0.1	54	20	10	...	...	1	...	15	...	...
Haynes 151(a)	0.48	65	20	...	...	0.03	...	...	12.8	...	3 max Fe + Ni
J-1650	0.20	36	19	27	...	0.02	...	2	12	...	3.8 Ti
MAR-M 302	0.85	58	21.5	...	...	0.005	0.5	9	10	0.2	...
MAR-M 322	1.0	60.5	21.5	...	...	...	0.5	4.5	9	2	0.75 Ti
MAR-M 509	0.6	54.5	23.5	10	...	...	...	3.5	7	0.5	0.2 Ti
MAR-M 918	0.05	52	20	20	...	...	...	7.5	...	0.1	...
NASA Co-W-Re	0.40	67.5	3	...	...	...	...	...	25	1	2 Re, 1 Ti
S-816	0.4	42	20	20	...	...	4	...	4	...	4 Mo, 4 Nb, 1.2 Mn, 0.4 Si
V-36	0.27	42	25	20	...	...	3	...	2	...	4 Mo, 2 Nb, 1 Mn, 0.4 Si
WI-52	0.45	63.5	21	...	...	...	2	...	11	...	2 Nb+Ta
X-40	0.50	57.5	22	10	...	...	1.5	...	7.5	...	0.5 Mn, 0.5 Si

(a) Obsolete alloy, included for reference purposes.

contain aluminum and titanium. Originally, the high-chromium nickel-base alloys were developed for oxidation resistance, and those high in molybdenum for chemical corrosion resistance. Alloys with no more than 20% Cr became known as "superalloys" when strengths were remarkably increased by adding aluminum and titanium. Nickel-base heat-resistant alloys have high-temperature mechanical properties superior to those of other heat-resistant alloys. They are precision investment cast under vacuum in many configurations, but principally as turbine airfoils (blades and vanes). Nickel-base heat-resistant alloys are more costly than iron-base alloys but less expensive than cobalt-base alloys.

ASTM A560 describes two grades of chromium-nickel casting alloys: 50Cr-50Ni and 60Cr-40Ni. These two alloys (not strictly considered nickel-base alloys) are cast into tube supports and other firebox fittings for certain stationary and marine boilers. The chief attribute of the Cr-Ni alloys is their resistance to hot-slag corrosion in boilers that fire oil high in vanadium content. Hot slag high in V_2O_5 content is extremely destructive to most other heat-resistant alloys. For example, iron-base alloys frequently last less than one-fourth as long as chromium-nickel alloys in regions of a firebox where the components are continually covered with molten slag containing V_2O_5.

Cobalt-base alloys contain about 50% or more cobalt plus appreciable amounts of chromium and refractory metals. The cobalt-base alloys were developed in the 1940's from Vitallium for turbocharger applications and became known as "superalloys" because of their superior high-temperature strengths compared with those of the Fe-Ni-Cr alloys then available. Cobalt-base alloys have good high-temperature mechanical properties and are precision investment cast in many configurations. They are not as strong as nickel-base alloys in short-time tests but are competitive in strength and corrosion resistance with nickel-base alloys at high temperatures and for long periods of operation. At present, the major uses of these alloys are furnace fixtures and turbine vanes.

Manufacture

Iron-base alloys can be cast from heats melted in electric-arc furnaces that have either acid or basic linings. When melting is done in acid-lined furnaces, however, chromium losses are high and silicon content is difficult to control, and thus acid-lined furnaces are seldom used. All heat-resistant alloys can be melted in high-frequency induction furnaces. Initial melting also can be done in consumable-arc, electron beam or other furnaces with appropriate atmosphere control. Heat-resistant alloys that contain appreciable amounts of aluminum, titanium or other reactive metals are melted by induction or electron beam processes under vacuum or a protective atmosphere prior to casting.

Iron-base alloy castings can be made by the static method, the centrifugal method or the investment process. The centrifugal method is used extensively in production of tubular parts such as radiant tubes for furnaces.

With the exception of a small tonnage of Fe-Cr alloys, iron-base heat-resistant alloy castings usually are not heat treated before being shipped. Because the 12Cr and 18Cr alloys are hardenable, castings of these alloys sometimes require full annealing for removal of casting stresses.

Nickel-base and cobalt-base alloy castings generally are made by investment casting. Cobalt-base alloys usually are melted and cast in air, although some of the more advanced alloys such as MAR-M 509 must be melted and cast under vacuum. Most nickel-base and cobalt-base castings are small, about 1 kg (2 lb) or less. Large nickel-base alloy castings, up to 1 m (3 ft) in diameter and about 0.3 m (1 ft) in length, have been produced.

For many applications, both nickel-base and cobalt-base alloy castings must be diffusion coated with a material high in silicon or aluminum to enable the alloy to resist the service atmosphere. Cobalt-base alloys are used as cast — with or without a diffusion coating, but without any other thermal treatment. Nickel-base alloys may be used as cast; in the cast, coated and aged condition; or in the cast, solution treated, coated and aged condition. The last condition applies more generally to advanced nickel-base alloys. Turbine vanes and blades cast in nickel-base alloys usually are shipped in the as-cast or solution-treated condition. Turbine manufacturers perform coating and aging treatments on machined castings as part of their manufacturing operation.

Metallurgical Structures of Iron-Base Alloys

The structures of Fe-Cr-Ni and Fe-Ni-Cr alloys must be wholly austenitic, or mostly austenitic with some ferrite, if these alloys are to be used for heat-resistant service. Depending on chromium and nickel contents, the structures of these iron-base alloys can be austenitic (stable), ferritic (stable, but also soft, weak and ductile) or martensitic (unstable); therefore, chromium and nickel levels should be selected to achieve good strength at elevated temperatures combined with resistance to carburization and hot-gas corrosion.

A fine dispersion of carbides or intermetallic compounds in an austenitic matrix increases high-temperature strength considerably. For this reason, iron-base heat-resistant alloys are higher in carbon content than corrosion-resistant alloys of comparable Cr and Ni contents. By holding at temperatures where carbon diffusion is rapid (such as above 1200 °C) and then rapidly cooling, a high and uniform carbon content is established and up to about 0.20% C is retained in the austenite. Some chromium carbides are present in the structures of alloys with carbon contents greater than 0.20%, regardless of solution treatment.

Castings develop considerable segregation as they freeze. In standard grades either in the as-cast condition or after rapid cooling from a temperature near the melting point, much of the carbon is in supersaturated solid solution. Subsequent reheating precipitates excess carbides. The lower the reheating temperature, the slower the reaction and the finer the precipitated carbides. Fine carbides increase creep strength and decrease ductility. Intermetallic compounds such as Ni_3Al, if present, have a similar effect.

Reheating material containing precipitated carbides in the range from 980 to 1200 °C (1800 to 2200 °F) will agglomerate and spheroidize the carbides, which reduces creep strength and increases ductility. Above 1100 °C (2000 °F), so many of the fine carbides are dissolved or spheroidized that this strengthening mechanism loses its importance. For service above 1100 °C, certain proprietary alloys of the Fe-Ni-Cr type have been developed. Alloys for this service contain tungsten to form tungsten carbides, which are more stable than chromium carbides at these temperatures.

Aging at a low temperature, such as 760 °C (1400 °F), where a fine, uniformly dispersed carbide precipitate will form, confers a high level of strength that is retained at temperatures up to those where agglomeration changes the character of the carbide dispersion (overaging temperatures). Solution heat treatment or quench annealing, followed by aging, is the treatment generally employed to attain maximum creep strength.

Ductility usually is reduced when strengthening occurs; but in some alloys the strengthening treatment corrects an unfavorable grain-boundary network of brittle carbides, and both properties benefit. However, such treatment is costly and may warp castings excessively. Hence, this treatment is applied to heat-resistant castings only for the small percentage of applications where the need for premium performance justifies the high cost.

Carbide networks at grain boundaries are generally undesirable in iron-base heat-resistant alloys. Grain-boundary networks usually occur in very-high-carbon alloys or in alloys that have cooled slowly through the high-temperature ranges where excess carbon in the austenite is rejected as grain-boundary networks rather than as dispersed particles. These networks confer brittleness in proportion to their continuity.

Carbide networks also provide paths for selective attack in some atmospheres and in certain molten salts. Therefore, it is advisable in some salt-bath applications to sacrifice the high-temperature strength imparted by high carbon content and gain resistance to intergranular corrosion by specifying that carbon content be no greater than 0.08%.

Iron-chromium alloys, also known as "straight chromium" alloys, contain either 9 or 28% Cr. HC and HD alloys are included among the straight chromium alloys, although they contain low levels of nickel.

HA alloy (9Cr-1Mo), a heat treatable material, contains enough chromium to provide good resistance to oxidation at temperatures up to about 650 °C (1200 °F). The 1% molybdenum is present to provide increased strength. HA alloy castings are widely used in oil-refinery service. A higher-chromium modification of this alloy (12 to 14% Cr) is widely used in the glass industry.

HA alloy has a structure that is essentially ferritic; carbides are present in pearlitic areas or as agglomerated particles, depending on prior heat treatment. Hardening of the alloy occurs on cooling in air from temperatures above 815 °C (1500 °F). In the normalized-and-tempered condition, the alloy exhibits satisfactory toughness throughout its useful temperature range.

HC alloy (28% Cr) resists oxidation and the effects of high-sulfur flue gases at temperatures up to 1100 °C (2000 °F). It is used for applications where strength is not a consideration, or where only moderate loads are involved, at temperatures of about 650 °C (1200 °F). It is also used where appreciable nickel cannot be tolerated, as in very-high-sulfur atmospheres, or where nickel may act as an undesirable catalyst and destroy hydrocarbons by causing them to crack.

HC alloy is ferritic at all temperatures. Its ductility and impact strength are very low at room temperature, and its creep strength is very low at elevated temperatures unless some nickel is present. In a variation of HC alloy that contains more than 2% Ni, substantial improvement in all three of these properties is obtained by increasing the nitrogen content to 0.15% or more.

HC alloy becomes embrittled when heated for prolonged periods at temperatures between 400 and 550 °C (750 and 1025 °F), and it shows low resistance to impact. The alloy is magnetic and has a low coefficient of thermal expansion, comparable to that of carbon steel. It has about eight times the electrical resistivity, and about half the thermal conductivity, of carbon steel. Its thermal conductivity, however, is roughly double the value for austenitic iron-chromium-nickel alloys.

HD alloy (28Cr-5Ni) is very similar in general properties to HC, except that its nickel content gives it somewhat greater strength at high temperatures. The high chromium content of this alloy makes it suitable for use in high-sulfur atmospheres.

HD alloy has a two-phase, ferrite-plus-austenite structure that is not hardenable by conventional heat treatment. Long exposure at 700 to 900 °C (1300 to 1650 °F), however, may result in considerable hardening and severe loss of room-temperature ductility through formation of sigma phase. Ductility may be restored by heating uniformly to 980 °C (1800 °F) or higher, and then cooling rapidly to below 650 °C (1200 °F).

Iron-Chromium-Nickel Alloys. Ferrous alloys in which chromium content exceeds nickel content are made in compositions ranging from 20Cr-10Ni to 30Cr-20Ni.

HE alloy (28Cr-10Ni) has excellent resistance to corrosion at elevated temperatures. Because of its higher chromium content, it can be used at higher temperatures than HF alloy and is suitable for applications up to 1100 °C (2000 °F). This alloy is stronger and more ductile at room temperature than the straight-chromium alloys.

In the as-cast condition, HE alloy has a two-phase, austenite-plus-ferrite structure containing carbides. HE castings cannot be hardened by heat treatment; however, as with HD castings, long exposure to temperatures near 815 °C (1500 °F) will promote formation of sigma phase and consequent embrittlement of the alloy at room temperature. The ductility of this alloy can be improved somewhat by quenching from about 1100 °C (2000 °F).

Castings of HE alloy have good machining and welding properties. Thermal expansion is about 50% greater than that of either carbon steel or the Fe-Cr alloy HC. Thermal conductivity is much lower than for HD or HC, but electrical resistivity is about the same. HE alloy is weakly magnetic.

HF alloy (20Cr-10Ni) is the cast version of 18-8 stainless steel, which is widely used for its outstanding resistance to corrosion. HF alloy is suitable for use at temperatures up to 870 °C (1600 °F). When this alloy is used for resistance to oxidation at elevated temperatures, it is not necessary to keep the carbon content at the low level specified for corrosion-resistant castings. Molybdenum, tungsten, niobium and titanium sometimes are added to the basic HF composition to improve elevated-temperature strength.

In the as-cast condition, HF alloy has an austenitic matrix that contains interdendritic eutectic carbides and, occasionally, a lamellar constituent presumed to consist of alternating platelets of austenite and carbide or carbonitride. Exposure at service temperatures usually promotes precipitation of finely dispersed carbides, which increases room-temperature strength and causes some loss of ductility. If improperly balanced, as-cast HF may be partly ferritic. HF is susceptible to embrittlement due to sigma-phase formation after long exposure at 760 to 815 °C (1400 to 1500 °F).

HH Alloy (26Cr-12Ni). Alloys of this nominal composition comprise about one-third of total production of iron-base heat-resistant castings. HH alloy is used extensively in high-temperature applications because of its combination of relatively high strength and good resistance to oxidation at temperatures up to 1100 °C (2000 °F).

For iron-base alloys of this nominal composition, composition balance is critical. The limits for nickel and chromium contents define a region that encompasses portions of the alpha-plus-gamma, gamma and gamma-plus-sigma regions of the Fe-Cr-Ni system at 900 °C (1650 °F). Consequently, an imbalance in the levels of ferrite-promoting elements over levels of austenite-promoting elements may result in substantial amounts of ferrite. Ferrite improves ductility, but decreases strength at high temperatures.

To achieve maximum strength at high temperature, HH alloy must be wholly austenitic. If a balance is maintained between ferrite-promoting elements (such as chromium and silicon) and austenite-promoting elements (such as nickel, carbon and nitrogen), the desired austenitic structure can be obtained. In commercial HH alloy castings, with the usual carbon, nitrogen, manganese and silicon contents, the ratio of chromium to nickel necessary for a stable austenitic structure is expressed by the inequality:

$$\frac{\%\,Cr - 16\,(\%\,C)}{\%\,Ni} < 1.7 \qquad \text{(Eq 1)}$$

Silicon and molybdenum have definite effects on formation of sigma. A silicon content in excess of 1% is equivalent to a chromium content three times as great, and any molybdenum content is equivalent to a chromium content four times as great.

When HH alloy is heated between 650 and 870 °C (1200 and 1600 °F), a loss in ductility may be produced by either of two changes within the alloy: precipitation of carbides, and transformation of ferrite to sigma. When composition is balanced so that the structure is wholly austenitic, only carbide precipitation normally occurs. In partly ferritic alloys, both carbides and sigma phase may form.

Figure 1, which shows the relationship of carbon, chromium and nickel in wholly austenitic HH alloy, is useful in balancing composition to obtain an austenitic structure.

Before HH alloy is selected as a material for heat-resistant castings, it is advisable to consider the relation between chemical composition and operating-temperature range. For castings that are to be exposed continuously at temperatures appreciably above 870 °C (1600 °F), there is little danger of severe embrittlement from either precipitation of carbide or formation of sigma phase, and composition should be 0.50% max C (0.35 to 0.40% preferred), 10 to 12% Ni and 24 to 27% Cr. On the other hand, castings to be used at temperatures from 650 to 870 °C (1200 to 1600 °F) should have compositions of 0.40% max C, 11 to 14% Ni and 23 to 27% Cr. For applications involving either of these temperature ranges (650 to 870 °C, or appreciably above 870 °C), composition should be balanced to provide an austenitic structure. For service from 650 to 870 °C, for example, a combination of 11% Ni and 27% Cr is likely to produce sigma phase and its associated embrittlement, which occurs most rapidly around 870 °C. It is preferable, therefore, to avoid using the maximum chromium content with the minimum nickel content. The mechanical properties of HH alloy at room temperature vary considerably with composition, as shown in the data compilations following this article.

Short-time tensile testing of fully austenitic HH alloys shows that tensile strength and elongation depend on carbon and nitrogen contents. The nomographic charts in Fig. 2 may be used to determine the properties of these alloys at 870 and 980 °C (1600 and 1800 °F).

For maximum creep strength, HH alloy should be fully austenitic in structure. In design of load-carrying castings, data concerning creep stresses should be used with an understanding of the limitations of such data. An extrapolated limiting creep stress for 1% elongation in 10 000 h cannot necessarily be sustained for that length of time without structural damage. Stress-rupture testing is a valuable adjunct to creep testing and a useful aid in selecting section sizes to obtain appropriate levels of design stress.

Because HH alloys of wholly austenitic structure have greater strength at high temperatures than partly ferritic alloys of similar composition, measurement of ferrite content is recommended. Although a ratio calculated from Eq 1 that is less than 1.7 indicates wholly austenitic material, ratios greater than 1.7 do not constitute quantitative indications of ferrite content. It is possible, however, to measure ferrite content by magnetic analysis after quenching from about 1100 °C (2000 °F). The magnetic permeability of HH alloys increases with ferrite content. This measurement of magnetic permeability, preferably after holding 24 h at 1100 °C and then quenching in water, can be related to creep strength, which also depends on structure.

HH alloys often are evaluated by measuring percentage elongation in room-temperature tension testing of specimens that have been held 24 h at 760 °C (1400 °F). Such a test may be misleading, because there is a natural tendency for engineers to favor compositions that exhibit the greatest elongation after this particular heat treatment. High ductility values are often measured for alloys that have low creep resistance, but, conversely, low ductility values do not necessarily connote high creep resistance.

HI alloy (28Cr-15Ni) is similar to HH but contains more nickel and chromium. The higher chromium content makes HI more resistant to oxidation than HH, and the additional nickel serves to maintain good strength at high temperatures. Exhibiting adequate strength, ductility and corrosion resistance, this alloy has been used extensively for retorts operating with an internal vacuum at a continuous temperature of 1175 °C (2150 °F). It has an essentially austenitic structure that contains carbides and that, depending on the exact composition balance, may or may not contain small amounts of ferrite. Service at 760 to 870 °C (1400 to 1600 °F) results in precipitation of finely dispersed carbides, which increases strength and decreases ductility at room temperature. At service temperatures above 1100 °C (2000 °F), however, carbides remain in solution and room-temperature ductility is not impaired.

Fig. 1 Nomograph for determining C, Cr and Ni contents that produce wholly austenitic structure in HH alloy

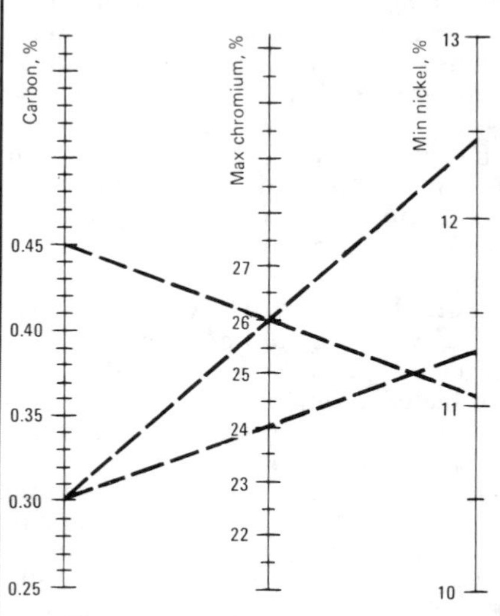

Source: J. T. Gow and O. E. Harder, *Trans ASM,* Vol 30, 1942, p 855

HK alloy (26Cr-20Ni) is somewhat similar to wholly austenitic HH alloy in general characteristics and mechanical properties. Although less resistant to oxidizing gases than HC, HE or HI, HK alloy contains enough chromium to ensure good resistance to corrosion by hot gases, including sulfur-bearing gases, under both oxidizing and reducing conditions. The high nickel content of this alloy helps make it one of the strongest heat-resistant casting alloys at temperatures above 1040 °C (1900 °F). Accordingly, HK alloy castings are widely used for stressed parts in structural applications at temperatures up

Fig. 2 Nomographs relating short-time tensile properties of HH alloy to C and N contents

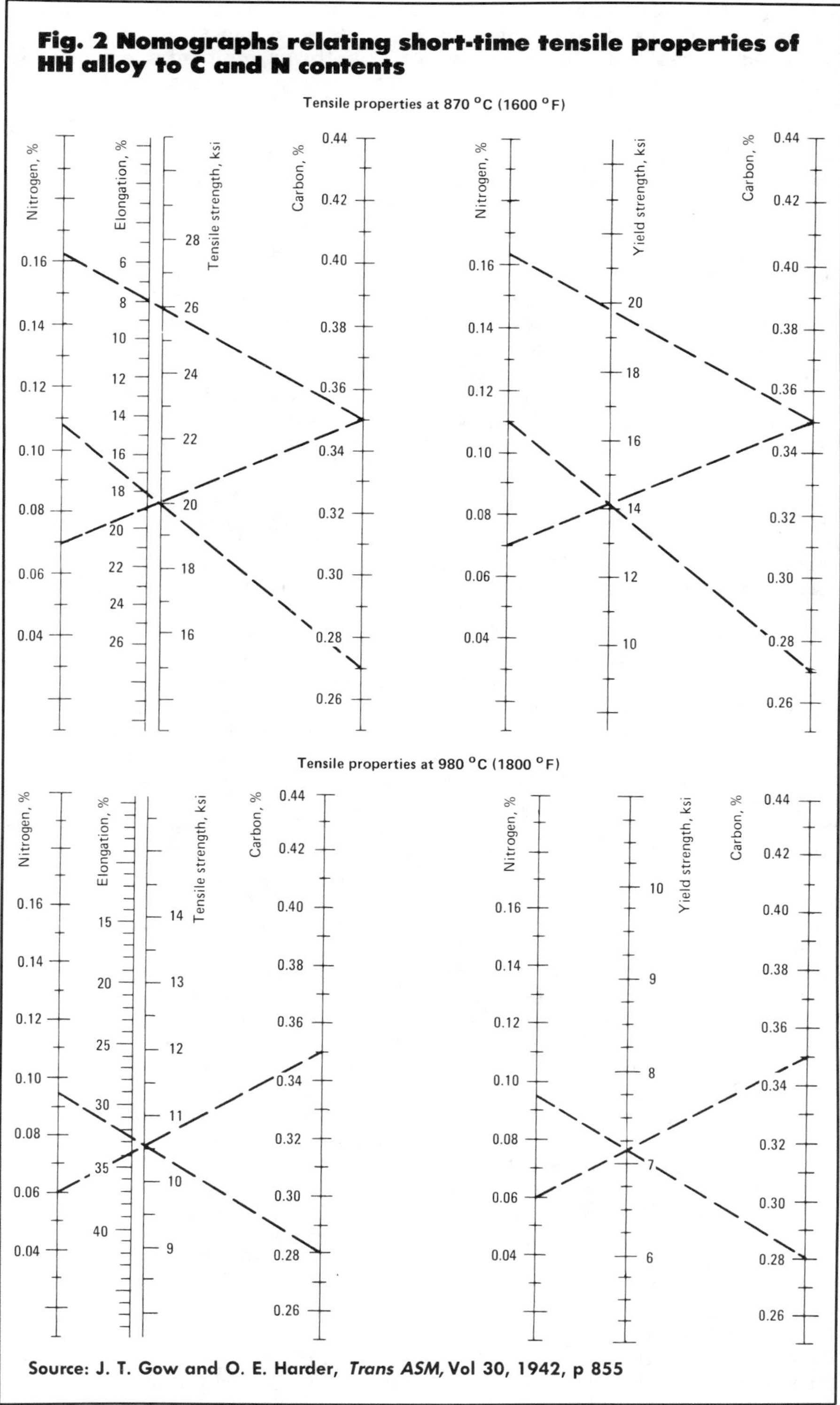

Source: J. T. Gow and O. E. Harder, *Trans ASM,* Vol 30, 1942, p 855

to 1150 °C (2100 °F). As normally produced, HK is a stable austenitic alloy over its entire range of service temperatures. The as-cast microstructure consists of an austenitic matrix containing relatively large carbides as either scattered islands or networks. After the alloy has been exposed to service temperature, fine granular carbides precipitate within the grains of austenite and, if the temperature is high enough, undergo subsequent agglomeration. These fine, dispersed carbides contribute to creep strength. A lamellar constituent that resembles pearlite, but that is presumed to be carbide or carbonitride platelets in austenite, also is frequently observed in HK alloy.

Unbalanced compositions are possible within the standard composition range for HK alloy, and hence some ferrite may be present in the austenitic matrix. Ferrite will transform to brittle sigma phase if the alloy is held for more than a short time at about 815 °C (1500 °F), with consequent embrittlement on cooling to room temperature. Direct transformation of austenite to sigma phase can occur in HK alloy in the range 760 to 870 °C (1400 to 1600 °F), particularly at lower carbon levels (0.20 to 0.30%). The presence of sigma phase can cause considerable scatter in property values at intermediate temperatures.

Minimum creep rate and average rupture life of HK are influenced strongly by variations in carbon content. Under the same conditions of temperature and load, alloys with higher carbon contents have lower creep rates and longer lives than lower-carbon compositions. Room-temperature properties after aging at elevated temperatures are affected also: the higher the carbon, the lower the residual ductility. For these reasons, three grades of HK alloy with carbon ranges narrower than that indicated in Table 1 are recognized: HK-30, HK-40 and HK-50. In these designations, the number indicates the midpoint of a 0.10% carbon range. HK-40 is widely used for high-temperature processing equipment in the petroleum and petrochemical industries.

Figure 3 indicates the statistical spread in room-temperature mechanical properties obtained for an HK alloy. These data were obtained in a single foundry and are based on 183 heats of the same alloy.

HL alloy (30Cr-20Ni) is similar to HK; its higher chromium content gives it greater resistance to corrosion by hot gases, particularly those containing appreciable amounts of sulfur. Because essentially equivalent high-temperature strength can be obtained with either HK or HL, the superior corrosion resistance of HL makes it especially useful for service in which excessive

Fig. 3 Statistical spread in mechanical properties of HK alloy

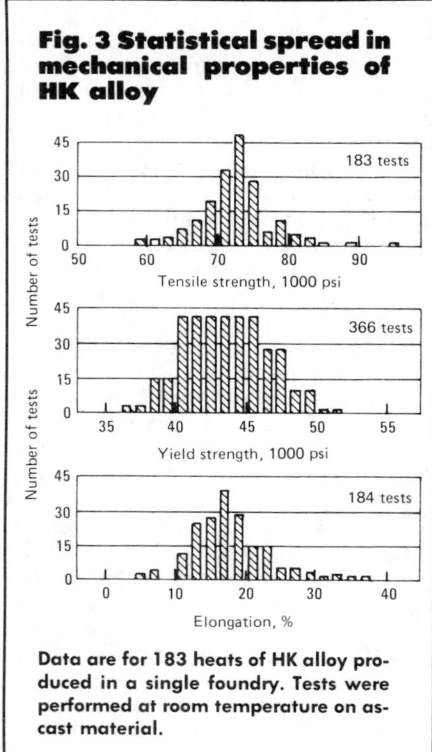

Data are for 183 heats of HK alloy produced in a single foundry. Tests were performed at room temperature on as-cast material.

scaling must be avoided. The as-cast and aged microstructures of HL alloy, as well as its physical properties and fabricating characteristics, are similar to those of HK.

Iron-nickel-chromium alloys generally have structures more stable than those of iron-base alloys in which chromium is the predominant alloying element. There is no evidence of an embrittling phase change in Fe-Ni-Cr alloys that would impair their ability to withstand prolonged service at elevated temperature. Experimental data indicate that composition limits are not critical; therefore, production of castings from these alloys does not require the close composition control necessary for making castings from Fe-Cr-Ni alloys.

The following general observations should be considered in selection of Fe-Ni-Cr alloys: (*a*) as nickel content is increased, the ability of the alloy to absorb carbon from a carburizing atmosphere decreases; (*b*) as nickel content is increased, tensile strength at elevated temperatures decreases somewhat, but resistance to thermal shock and thermal fatigue increases; (*c*) as chromium content is increased, resistance to oxidation and to corrosion in chemical environments increases; (*d*) as carbon content is increased, tensile strength at elevated temperatures increases; and (*e*) as silicon content is increased, tensile strength at elevated temperatures decreases, but resistance to carburization increases somewhat.

HN alloy (25Ni-20Cr) contains enough chromium for good high-temperature corrosion resistance. HN has mechanical properties somewhat similar to those of the much more widely used HT alloy, but has better ductility. It is used for highly stressed components in the temperature range from 980 to 1100 °C (1800 to 2000 °F). In several specialized applications (notably, brazing fixtures), it has given satisfactory service at temperatures from 1100 to 1150 °C (2000 to 2100 °F). HN alloy is austenitic at all temperatures: its composition limits lie well within the stable austenite field. In the as-cast condition it contains carbide areas, and additional fine carbides precipitate on aging. HN is not susceptible to sigma-phase formation, and increases in its carbon content are not especially detrimental to ductility.

HP, HT, HU, HW and *HX alloys* together constitute about one-third of total production of heat-resistant alloy castings. When used for fixtures and trays for heat treating furnaces, which are subjected to rapid heating and cooling, these five high-nickel alloys have exhibited excellent service life. Because these compositions are not as readily carburized as Fe-Cr-Ni alloys, they are used extensively for parts of carburizing furnaces. Because they form an adherent scale that does not flake off, castings of these alloys are also used in enameling applications where loose scale would be detrimental.

In many respects there are no sharp lines of demarcation among HP, HT, HU, HW and HX alloys as far as service applications are concerned.

HP alloy (35Ni-26Cr) is related to HN and HT alloys but is higher in alloy content. It contains the same amount of chromium but more nickel than HK, and the same amount of nickel but more chromium than HT. This combination of elements makes HP resistant to both oxidizing and carburizing atmospheres at high temperatures. It has creep-rupture properties from 980 to 1100 °C (1800 to 2000 °F) that are comparable to, or better than, those of HK-40 and HN alloys.

HP alloy is austenitic at all temperatures, and is not susceptible to sigma-phase formation. Its microstructure consists of massive primary carbides in an austenitic matrix; in addition, fine secondary carbides are precipitated within the austenite grains upon exposure to elevated temperature.

HT alloy (35Ni-17Cr) contains nearly equal amounts of iron and alloying elements. Its high nickel content enables it to resist the thermal shock of rapid heating and cooling. In addition, HT is resistant to high-temperature oxidation and carburization and has good strength at the temperatures ordinarily used for heat treating steel. Except in high-sulfur gases, and provided that limiting creep-stress values are not exceeded, it performs satisfactorily in oxidizing atmospheres at temperatures up to 1150 °C (2100 °F) and in reducing atmospheres at temperatures up to 1100 °C (2000 °F).

HT alloy is widely used for highly stressed parts in general heat-resistant applications. It has an austenitic structure containing carbides in amounts that vary with carbon content and thermal history. In the as-cast condition, it has large carbide areas at interdendritic boundaries; but fine carbides precipitate within the grains after exposure to service temperature, causing a decrease in room-temperature ductility. Increases in carbon content may decrease the high-temperature ductility of the alloy. Silicon contents above about 1.6% provide additional protection against carburization, but at some sacrifice in elevated-temperature strength. HT can be made still more resistant to thermal shock by addition of up to 2% niobium.

HU alloy (39Ni-18Cr) is similar to HT, but its higher chromium and nickel contents give it greater resistance to corrosion by either oxidizing or reducing hot gases, including those that contain sulfur in amounts up to 2.3 g/m^3 (see Table 3). Its high-temperature strength and resistance to carburization are essentially the same as those of HT, and thus its superior corrosion resistance makes it especially well suited for severe service involving high stress and/or rapid thermal cycling in combination with an aggressive environment.

HW alloy (60Ni-12Cr) is especially well suited for applications where wide and/or rapid fluctuations in temperature are encountered. In addition, HW exhibits excellent resistance to carburization and high-temperature oxidation. HW alloy has good strength at steel-treating temperatures, although it is not as strong as HT. HW performs satisfactorily at temperatures up to about

1120 °C (2050 °F) in strongly oxidizing atmospheres and up to 1040 °C (1900 °F) in oxidizing or reducing products of combustion, provided that sulfur is not present in the gas. The generally adherent nature of its oxide scale makes HW suitable for enameling-furnace service, where even small flakes of dislodged scale could ruin the work in process.

HW alloy is widely used for intricate heat treating fixtures that are quenched with the load, and for many other applications (such as furnace retorts and muffles) that involve thermal shock, steep temperature gradients and high stresses. Its structure is austenitic and contains carbides in amounts that vary with carbon content and thermal history. In the as-cast condition, the microstructure consists of a continuous interdendritic network of elongated eutectic carbides. Upon prolonged exposure at service temperatures, the austenitic matrix becomes uniformly peppered with small carbide particles except in the immediate vicinity of eutectic carbides. This change in structure is accompanied by an increase in room-temperature strength, but no change in ductility.

HX alloy (66Ni-17Cr) is similar to HW but contains more nickel and chromium. Its higher chromium content gives it substantially better resistance to corrosion by hot gases (even sulfur-bearing gases), which permits it to be used in severe service applications at temperatures up to 1150 °C (2100 °F). However, it has been reported that HX alloy decarburizes rapidly at temperatures from 1100 to 1150 °C (2000 to 2100 °F). High-temperature strength, resistance to thermal fatigue and resistance to carburization are essentially the same as for HW; hence HX is suitable for the same general applications where corrosion must be minimized. The as-cast and aged microstructures of HX, as well as its mechanical properties and fabricating characteristics, are similar to those of HW.

Fig. 4 Stress-rupture curves for 1000-h life of selected nickel-base alloys

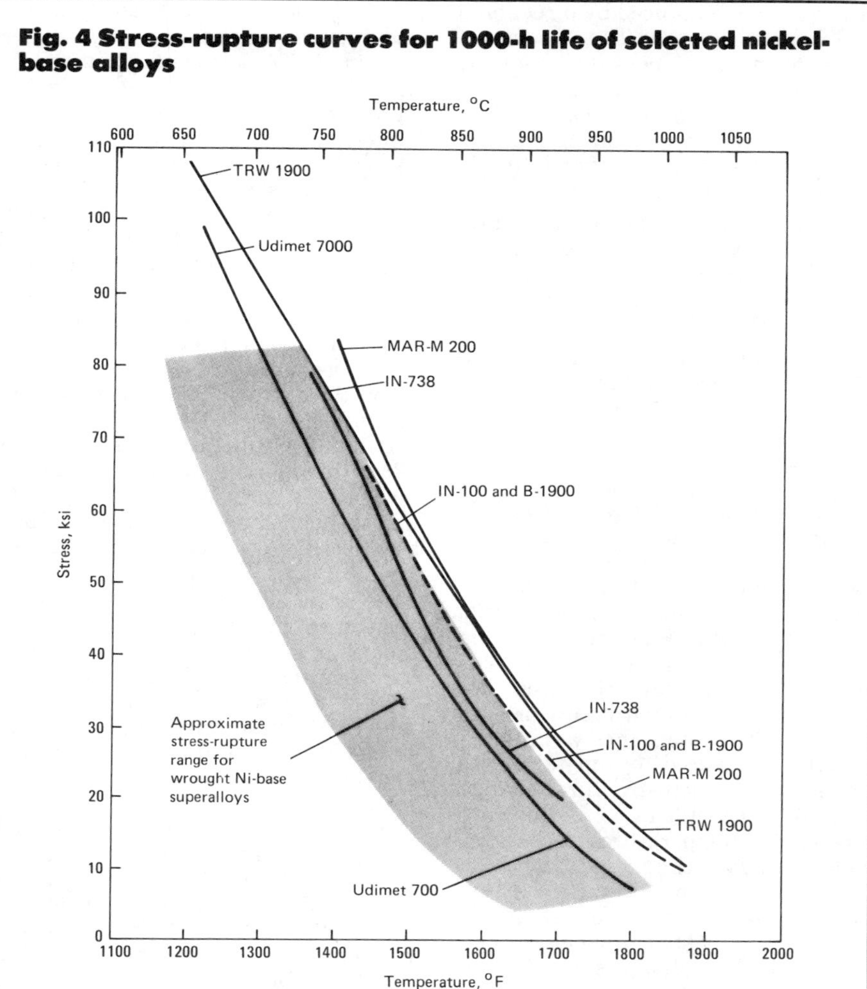

To convert stress values to MPa, multiply by 6.8948.

Metallurgical Structures of Nickel-Base Alloys

Nickel-base heat-resistant casting alloys generally contain substantial levels of aluminum and titanium (see Table 4). These elements strengthen the austenitic matrix through precipitation of $Ni_3(Al, Ti)$, an ordered fcc compound referred to as "gamma prime" (γ'). Various ratios of aluminum and titanium are used in the different nickel-base heat-resistant alloys; generally, titanium atoms can replace aluminum atoms up to a ratio of 3 Ti to 1 Al without altering the ordered fcc crystallographic structure of γ'. When excess titanium is present, Ni_3Ti, an ordered cph compound known as "eta phase" (η), precipitates. Because γ' is coherent with the matrix, precipitation of this phase has a greater strengthening effect than precipitation of η.

In addition to the strengthening imparted by γ' precipitation, solid-solution strengthening is conferred by addition of refractory elements, and grain-boundary strengthening by additions of boron, zirconium, carbon and hafnium. Hafnium also enhances grain-boundary ductility. Stress-rupture curves for various nickel-base alloys are shown in Fig. 4.

The strength of these alloys is complemented by superior corrosion resistance, which is conferred by chromium and aluminum (titanium may be more favorable than aluminum under hot-corrosion conditions). Coatings are used on most nickel alloys for temperatures exceeding about 815 °C (1500 °F) in order to provide adequate protection from oxidation and corrosion at these temperatures.

Nickel-base heat-resistant alloy castings are produced by investment casting under vacuum, and improvements in properties have been made not only through control of composition but also through more precise control of microstructure. A significant advance in microstructure control was the development of a columnar grain structure produced by directional solidification. Single crystals of nickel-base superalloys have been produced by directional solidification as well. The absence of grain boundaries permits elements such as carbon, zirconium and boron to be deleted from the composition. The resulting increase in melting point in turn provides improved flexibility in alloy composition and heat treatment.

Extensive use of nickel alloy castings essentially began with Inconel 713, and now alloys are available that can be used at temperatures up to about 1040 °C (1900 °F).

In addition to creep strength and corrosion resistance, two other properties — stability, and resistance to thermal fatigue — are important considerations in selection of nickel-base heat-resistant casting alloys. Thermal-fatigue resistance is partly controlled by composition, but it is also significantly affected by grain-boundary area and alignment relative to applied stresses. Crystallographic orientation of grains also influences thermal stresses, because the modulus of elasticity, which directly influences thermal stresses, varies with grain orientation. Stability of property values is directly influenced by metallurgical stability: any microstructural changes that take place during long-time exposure at high temperatures under stress cause attendant changes in properties. For instance, if the γ' phase coarsens, strength decreases. Also, potentially deleterious topologically close-packed (TCP) secondary phases, such as sigma, Laves and mu, may form. Coarsening of γ' can be controlled to some degree by adjusting alloy additions. Formation of TCP phases is controlled by adjusting the composition of the matrix to minimize the electron vacancy number (N_v). A high N_v indicates a tendency toward formation of TCP phases. In general, an N_v value below 2.4 indicates minimal formation of deleterious phases; however, this relationship varies with base-alloy composition.

Inconel 713C and *713LC* are closely related investment casting alloys used principally for low-pressure turbine airfoils in gas turbines. Intended for operation at intermediate temperatures from 790 to 870 °C (1450 to 1600 °F), these alloys generally are used in uncooled airfoil designs.

IN-738X is an investment casting alloy similar in strength to Inconel 713C and Udimet 700 but with outstanding sulfidation resistance. It is used principally for latter-stage turbine airfoils and for hot-corrosion-prone applications such as industrial and marine engines.

Udimet 700, although primarily a wrought alloy, is also used in investment cast high-pressure turbine blades. In cast form, it is similar in strength to Inconel 713C but offers better hot-corrosion resistance. It is designed for operation at intermediate temperatures from 730 to 900 °C (1350 to 1650 °F). A stability-controlled version of U-700 is known as René 77.

IN-100 is designed for use at metal temperatures up to about 980 °C (1800 °F) in cooled and uncooled airfoils. A stability-controlled version of IN-100 is known as René 100.

B-1900, to which 1% Hf usually is added to improve ductility and thermal-fatigue resistance, is designed for use at metal temperatures up to about 980 °C (1800 °F) in cooled and uncooled airfoils.

René 80 offers excellent corrosion resistance in sulfur-bearing environments. It is designed for use at metal temperatures up to about 950 °C (1750 °F).

IN-792 is designed for use in applications similar to those of René 80. It is one of the most sulfidation-resistant nickel alloys available.

MAR-M 246 and *MAR-M 247* are designed for use at metal temperatures of about 980 to 1010 °C (1800 to 1850 °F) in cooled and uncooled airfoils.

D.S. MAR-M 200 + Hf is produced by unidirectional solidification and is designed for metal temperatures of about 1010 to 1040 °C (1850 to 1900 °F). It is used in cooled airfoils.

Other alloys (such as Udimet 500) are occasionally used in turbine airfoil applications, and Inconel 718 has been cast into large static structures for gas turbines.

Metallurgical Structures of Cobalt-Base Alloys

Cobalt-base heat-resistant casting alloys were first used in highly stressed gas-turbine blades during World War II. Although the initial alloy was used at temperatures no higher than those at which the Fe-Cr-Ni and Fe-Ni-Cr heat-resistant alloys were used, it gave satisfactory service under high stress and far surpassed the older alloys in creep and rupture properties. Compositions of selected cobalt-base casting alloys are given in Table 5. Stress-rupture properties of commonly used cobalt-base alloys are shown in Fig. 5.

Most high-temperature cobalt-base alloys contain appreciable amounts of carbon and derive their strength not only from solid-solution hardening by such elements as tungsten and chromium but also from carbide precipitation, which becomes less effective above about 815 °C (1500 °F). Nickel often is added to stabilize the high-temperature form of cobalt (face-centered cubic).

With the advent of vacuum melting, nickel-base alloys strengthened by precipitation of γ' phase promptly surpassed cobalt alloys in strength, and very few new cobalt-base casting alloys were developed for use in gas turbines after 1952. Generally, X-40 is used for latter-stage turbine airfoils of older gas turbines, and MAR-M 509, WI-52 and MAR-M 302 are employed for first-stage, and occasionally second-stage, turbine vanes in gas turbines. In terms of strength, cobalt alloys typically cannot compete with nickel alloys except at temperatures above 980 °C (1800 °F). However, because of the ease with which they can be repaired by welding, and because of their high chromium contents, which give them good corrosion resistance at these high temperatures, cobalt alloys find extensive use in high-pressure turbine vanes. Additional uses include cast components in burner cans and seals, and, for some cobalt alloys, heat-resistant castings for furnace hardware.

Five commonly used cobalt-base alloys are described below. All are investment cast, usually from air-melted stock. None of the commercial alloys is directionally solidified.

In application of cobalt-base alloys, stability and thermal-fatigue resistance must be considered, because cobalt-base alloys generally are less resistant to thermal fatigue than nickel-base alloys of comparable strength. Stability can be inferred from the electron vacancy number (N_v). However, N_v numbers that delineate cutoff points for TCP-phase formation tend to be near 2.6, which is higher than corresponding numbers for nickel-base alloys. All cobalt-base alloys exhibit aging on exposure as a result of the strengthening effects of changes in the carbide structure.

Haynes 21 and *X-40* are used for turbine vanes in some gas turbines. They have good corrosion resistance and strength at temperatures from 700 to 815 °C (1300 to 1500 °F).

WI-52 is used for turbine vanes in some gas turbines. It can be used at temperatures up to about 950 °C (1750

Fig. 5 Stress-rupture curves for 1000-h life of selected cobalt-base alloys

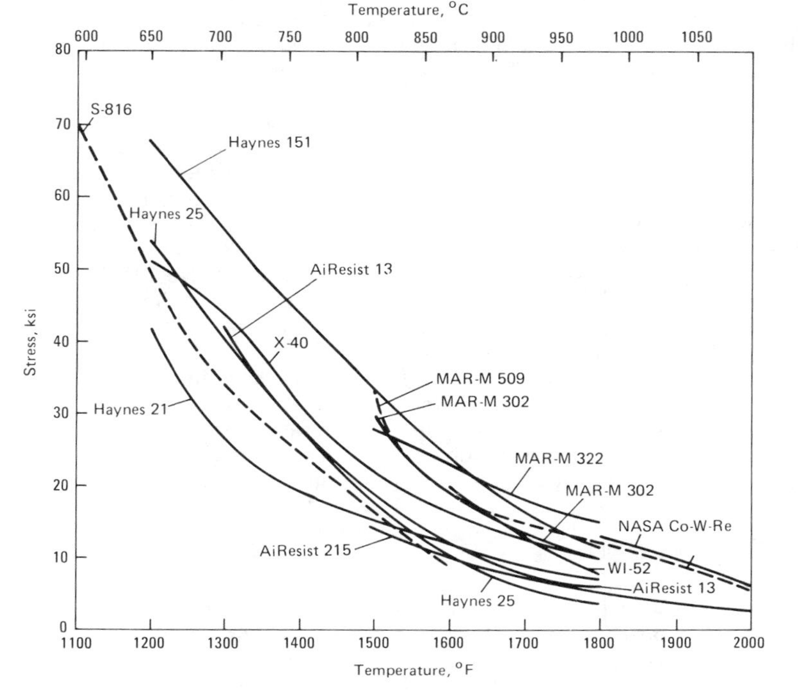

To convert stress values to MPa, multiply by 6.8948.

Table 6 Machining data for ACI heat-resistant casting alloys

ACI designation	Typical hardness, HB	Rough turning(a) Speed, sfm(b)	Rough turning(a) Feed, ipr(c)	Finishing Speed, sfm(b)	Finishing Feed, ipr(c)	Drilling speed(d), sfm(b)
HA	220	40 to 50	0.010 to 0.030	80 to 100	0.005 to 0.010	35 to 70
HC	220	40 to 50	0.025 to 0.035	80 to 100	0.010 to 0.015	40 to 60
HD	190	40 to 50	0.025 to 0.035	80 to 100	0.010 to 0.015	40 to 60
HE	270	30 to 40	0.020 to 0.025	60 to 80	0.005 to 0.010	30 to 60
HF	190	25 to 35	0.020 to 0.025	50 to 70	0.005 to 0.010	20 to 40
HH	200	25 to 35	0.015 to 0.020	50 to 70	0.005 to 0.010	20 to 40
HI	200	25 to 35	0.015 to 0.020	50 to 70	0.005 to 0.010	20 to 40
HK	190	25 to 35	0.020 to 0.025	50 to 70	0.005 to 0.010	20 to 40
HL	190	30 to 40	0.020 to 0.025	60 to 80	0.005 to 0.010	30 to 60
HN	160	35 to 45	0.020 to 0.025	70 to 90	0.005 to 0.010	40 to 60
HP	...	35 to 45	0.020 to 0.025	70 to 90	0.005 to 0.010	40 to 60
HT	200	40 to 45	0.025 to 0.035	80 to 90	0.005 to 0.010	40 to 60
HU	190	40 to 45	0.025 to 0.035	80 to 90	0.010 to 0.015	40 to 60
HW	200	40 to 45	0.025 to 0.035	80 to 90	0.010 to 0.015	40 to 60
HX	185	40 to 45	0.025 to 0.035	80 to 90	0.010 to 0.015	40 to 60

(a) Single-point high speed steel tools usually are ground to 4 to 10° side and back rake, 4 to 7° side relief, 7 to 10° end relief, 8 to 15° end cutting-edge angle, 10 to 15° side cutting-edge angle, and 1/32- to 1/8-in. nose radius. (b) To convert to m/s, multiply by 0.005. (c) To convert to mm/rev, multiply by 25. (d) Recommended drilling feeds are as follows: for drill diameters up to 1/8 in., 0.001 to 0.002 ipr; 1/8 to 1/4 in., 0.002 to 0.004 ipr; 1/4 to 1/2 in., 0.004 to 0.007 ipr; 1/2 to 1 in., 0.007 to 0.015 ipr; over 1 in., 0.015 to 0.025 ipr. Tapping speeds recommended for HA, HC, HD, HE and HL are 10 to 25 sfm; for HF, HH, HI, and HK, 10 to 20 sfm; and for HN, HT, HU, HW and HX, 5 to 15 sfm.

°F) but must be coated for corrosion resistance.

MAR-M 302 is vacuum cast into turbine vanes for some gas turbines. This alloy is used at temperatures up to about 980 °C (1800 °F).

MAR-M 509 is vacuum cast into high-pressure turbine vanes and seals for gas turbines. The alloy has good corrosion resistance and strength from 815 to 1010 °C (1500 to 1850 °F). MAR-M 509 generally is coated to provide the required corrosion resistance.

Applications

In terms of tonnage, the most important use of heat-resistant castings is in metallurgical and other industrial furnaces. Iron-base alloys are most often used for this service, although significant amounts of nickel-base and cobalt-base alloys are used also. Other major applications for heat-resistant castings include turbochargers, gas turbines, power-plant equipment, and equipment used in the manufacture of glass, cement, synthetic rubber, chemicals and petroleum products.

Alloy Selection. Heat-resistant alloys are selected on the basis of structural integrity in a specific application. Strength, creep resistance and corrosion resistance are prime factors influencing alloy selection. Next in importance is castability, though it is difficult to obtain a quantitative evaluation of this factor. The ability of the alloy to fill the mold must be assessed, as must the statistical distributions of product porosity and critical dimensions.

Because many castings must be machined to final dimensions, machinability of the as-cast metal often must be evaluated. Quantitative machining data are available for many cast heat-resistant alloys. Typical recommendations of speed, feed and depth of cut are summarized in Table 6 for iron-base casting alloys. These recommendations must be considered starting points for evaluation of a particular machining operation; some adjustment of one or more of these factors will invariably be required to achieve optimum tool life.

When total manufacturing cost and projected service life of a cast article have been determined, it is then possible to evaluate unit operating cost. Cost per hour of service life is the ultimate criterion in selection of a heat-resistant alloy. On this basis an alloy that is more expensive in initial cost frequently provides lower cost per hour of service than a less-expensive type.

Example 1. Use of Cast Heat-Resistant Alloys in a Glass Fiber Plant (Fig. 6). Data on use of cast heat-resistant alloys in manufacture of glass

Fig. 6 Service life of cast heat-resistant alloy bushings used in equipment for forming glass fibers (Example 1)

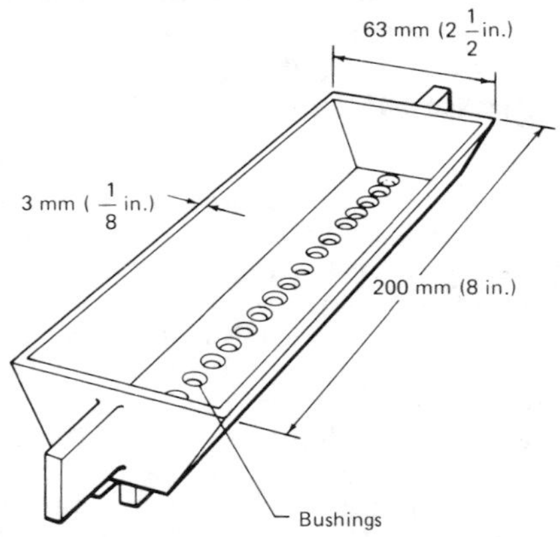

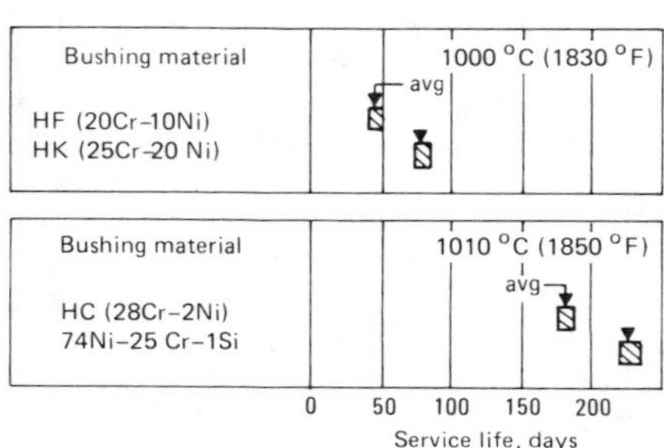

Corrosion and erosion of the orifices were the main criteria of failure in bushings exposed to molten glass on one side and air on the other.

fiber are given in Fig. 6. In this application, cast bushings are mounted in the forehearth of a glass-melting tank. Molten glass is fed by gravity to each bushing, flows through forming tips at the bottom of the bushing and is mechanically drawn into elongated fibers. The diameter of each glass fiber is determined by the size of the hole in the bushing tip, the speed of pull, and the temperature and type of glass used. The bushing must be kept hot, so it is resistance heated by means of water-cooled clamps attached to terminals at each end.

The cast bushings are subjected to corrosion and erosion resulting from passage of molten glass at high temperature and from the oxidizing effects of the surrounding air. At 1000 °C (1830 °F), HF alloy with low chromium and nickel contents had an average life of only 46 days. It was replaced by HK alloy, thereby increasing average bushing life to 77 days. This improvement was effected with only a moderate increase in alloy content. When the equipment was used at 1010 °C (1850 °F), however, it was necessary to use bushings made of an alloy rich in nickel; average life of these bushings was 229 days, as shown in Fig. 6. (HT and HW were tried experimentally but were not used in production, and thus no performance data on these alloys are available.)

Example 2. Use of Cast Heat-Resistant Alloys in a Cement Plant (Fig. 7). The cast heat-resistant alloys that are used for various components of a burning layout in a cement mill are given in Fig. 7. Significantly, more than one alloy can be used successfully in most of the twenty types of components listed. Environmental conditions that provide the basis for selection are listed for each component.

Casting Design

Iron-Base Alloys. For most applications, ACI heat-resistant alloys are sand cast, although shell molding is also used. Sections in thicknesses of 4.8 mm (3/16 in.) and greater can be cast satisfactorily, and somewhat thinner sections also may be castable depending on casting design and pattern equipment. Dimensional tolerances for rough castings are influenced by quality of pattern equipment. In general, over-all dimensions and locations of cored holes can be held to 5.2 mm/m (1/16 in./ft).

More specific information on casting design is presented in Fig. 8. Figure 9 compares dimensional variations encountered in casting belt links in green sand molds and in shell molds: variations are about the same for both types of molds. Both distributions are skewed toward the plus side of the tolerance range, which is not particularly unusual for pilot runs and which can be corrected by making adjustments in the dimensions of the pattern. The bottom portion of Fig. 9 shows two stages in the development of link A, the second stage being one that incorporated a pattern correction to bring the variation in casting dimensions from the high side of the tolerance range to the middle of the range. In this instance, removing approximately 0.25 mm (0.010 in.) from all surfaces of the pilot pattern cost about 1% of the cost of a new pattern, and resulted in a production pattern that yielded castings consistently within design tolerances.

Figure 10 provides a comparison of recommended tolerances for two typical shapes cast in sand molds and in shell molds. Alloy composition does not significantly affect these tolerances for heat-resistant alloy castings.

The relations between process variables and casting shape are of fundamental importance for all castings. High-alloy castings are similar to steel castings in many of these design-process relations.

Example 3. Design-Process Relations for a Complex Casting (Fig. 11 and 12). A shell molded valve body is shown in Fig. 11, along with a record of the dimensional variations encountered in its production. These data illustrate the magnitude and frequency of dimensional variations likely to be encountered in casting a relatively complex form. This valve body is an example of a casting for which design revisions were necessary to suit processing requirements. Two of the revisions involved blending at section changes, one dealt with ribs, and another required a reduction in the mass of a chilled boss. Each change was related to the mode of solidification applicable to the casting shape. The gating and risering needed for a casting of this over-all shape are shown in Fig. 12 (top).

The section changes shown in view A in Fig. 12 were not expected to be a problem but nevertheless resulted in hot tears. Why hot tears developed here can be understood by studying the direction of freezing, which could have been predicted by the location of the six risers. Because freezing progressed generally from points E1 and E2 (Fig. 12) diagonally toward the four side risers and the two end risers, shrinkage stresses were concentrated across these sections, and a localized hot zone was created at the 1/16-in. fillet. Had the changes in section size been oriented parallel to the freezing direction, stresses would not have developed and hot tears would not have occurred. The blend was redesigned with a 15° taper, as shown in view A, and hot tears were not observed during subsequent production of 245 castings.

The four flanged outlets shown in view B in Fig. 12 obviously required risering with directional freezing toward each riser. The original design provided for a sharply changing section, which resulted in hot tears in 20%

Fig. 7 Typical cement-mill burning layout (Example 2)

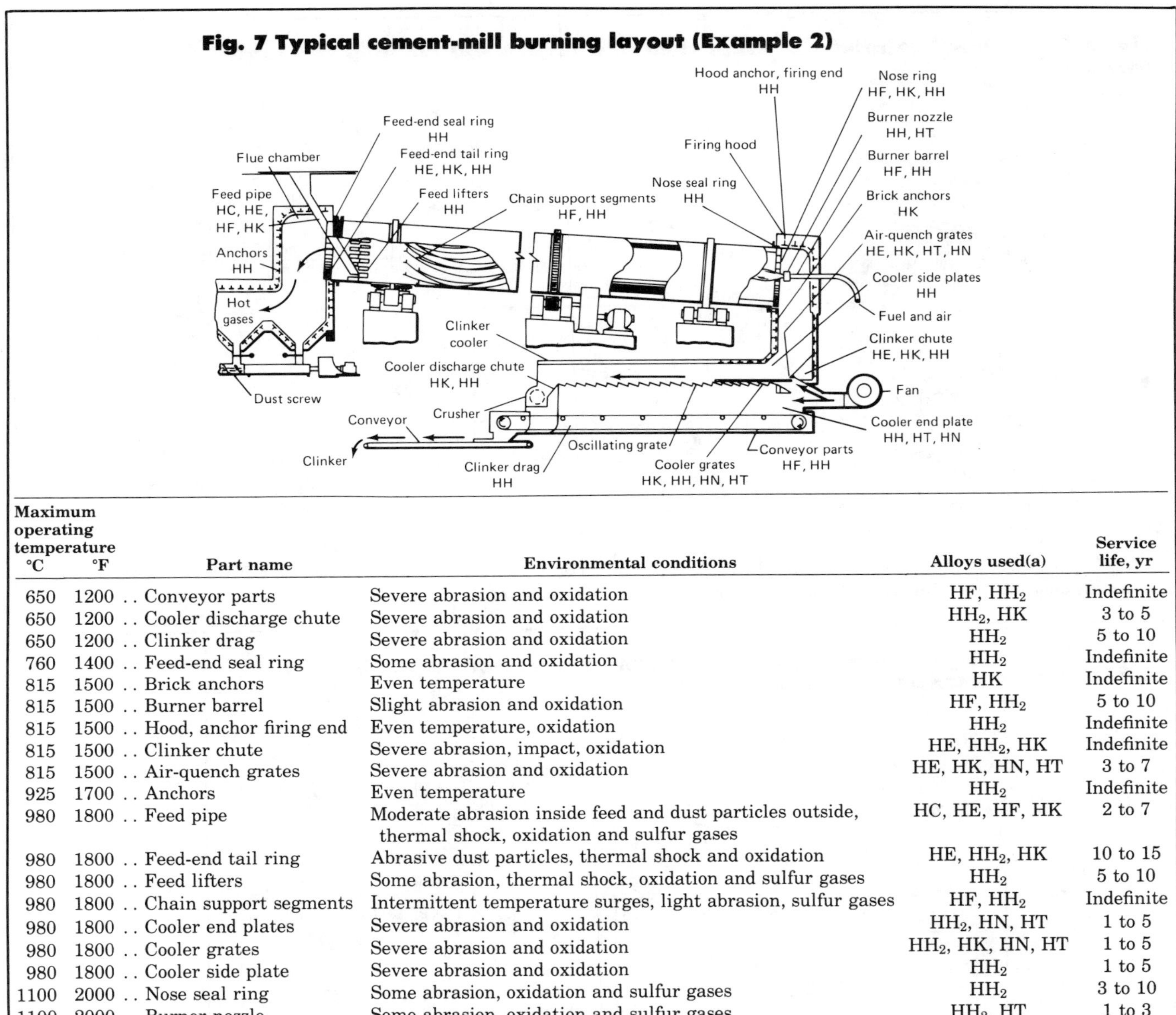

Maximum operating temperature °C	°F	Part name	Environmental conditions	Alloys used(a)	Service life, yr
650	1200	.. Conveyor parts	Severe abrasion and oxidation	HF, HH_2	Indefinite
650	1200	.. Cooler discharge chute	Severe abrasion and oxidation	HH_2, HK	3 to 5
650	1200	.. Clinker drag	Severe abrasion and oxidation	HH_2	5 to 10
760	1400	.. Feed-end seal ring	Some abrasion and oxidation	HH_2	Indefinite
815	1500	.. Brick anchors	Even temperature	HK	Indefinite
815	1500	.. Burner barrel	Slight abrasion and oxidation	HF, HH_2	5 to 10
815	1500	.. Hood, anchor firing end	Even temperature, oxidation	HH_2	Indefinite
815	1500	.. Clinker chute	Severe abrasion, impact, oxidation	HE, HH_2, HK	Indefinite
815	1500	.. Air-quench grates	Severe abrasion and oxidation	HE, HK, HN, HT	3 to 7
925	1700	.. Anchors	Even temperature	HH_2	Indefinite
980	1800	.. Feed pipe	Moderate abrasion inside feed and dust particles outside, thermal shock, oxidation and sulfur gases	HC, HE, HF, HK	2 to 7
980	1800	.. Feed-end tail ring	Abrasive dust particles, thermal shock and oxidation	HE, HH_2, HK	10 to 15
980	1800	.. Feed lifters	Some abrasion, thermal shock, oxidation and sulfur gases	HH_2	5 to 10
980	1800	.. Chain support segments	Intermittent temperature surges, light abrasion, sulfur gases	HF, HH_2	Indefinite
980	1800	.. Cooler end plates	Severe abrasion and oxidation	HH_2, HN, HT	1 to 5
980	1800	.. Cooler grates	Severe abrasion and oxidation	HH_2, HK, HN, HT	1 to 5
980	1800	.. Cooler side plate	Severe abrasion and oxidation	HH_2	1 to 5
1100	2000	.. Nose seal ring	Some abrasion, oxidation and sulfur gases	HH_2	3 to 10
1100	2000	.. Burner nozzle	Some abrasion, oxidation and sulfur gases	HH_2, HT	1 to 3
1200	2200	.. Nose ring	Extreme abrasion, oxidation and sulfur gases	HF, HH_2, HK	3 to 5

(a) HH_2 is the type 2, wholly austenitic grade of iron-base 26Cr-12Ni alloy.

of the castings. It was recognized that these hot tears were caused by the sharpness of the section change closest to the riser and that the blend should be more gradual. The temperature gradient across each outlet was too great; a reduction in temperature gradient was needed to avoid hot tears.

The importance of this temperature gradient can be appreciated when it is realized that the cylindrical part of the outlet was freezing considerably in advance of the flange. This, combined with the predictable occurrence of retarded cooling in the corner at the junction of the cylinder and flange, produced an insufficient thickness of solidified skin at the corner to resist hot tearing due to the stresses caused by differential shrinkage of these adjacent portions.

The flange was redesigned with more generous, blended fillets, as shown at right in view B, and no more castings were rejected for hot tears in this area.

Inspection of the eight ribs (view C) revealed about 5% rejects due to cold shuts. One common foundry solution to the problem of cold shuts is to pour the metal at a higher temperature in order to keep it fluid long enough so that it will run into the narrow openings in the mold. Although the pouring temperature was already high, it was decided that pouring at a still higher temperature was worth the calculated risk. But this resulted in excessive evolution of gas from the mold and of gas dissolved in the metal, which produced porosity at random locations throughout the casting. Thus, the alternative

Fig. 8 Dimensional relations in sand, shell and investment molding for heat-resistant alloy castings

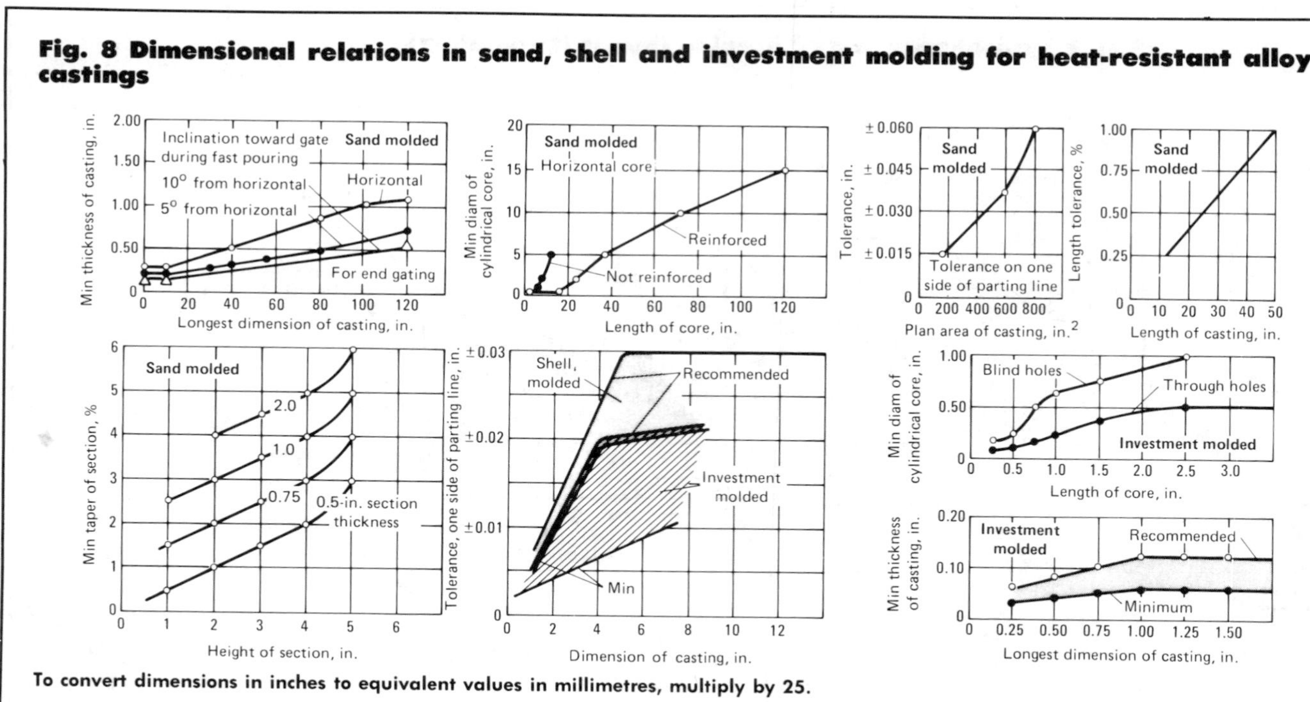

To convert dimensions in inches to equivalent values in millimetres, multiply by 25.

Fig. 9 Dimensional variations encountered in casting two models of conveyor-belt links

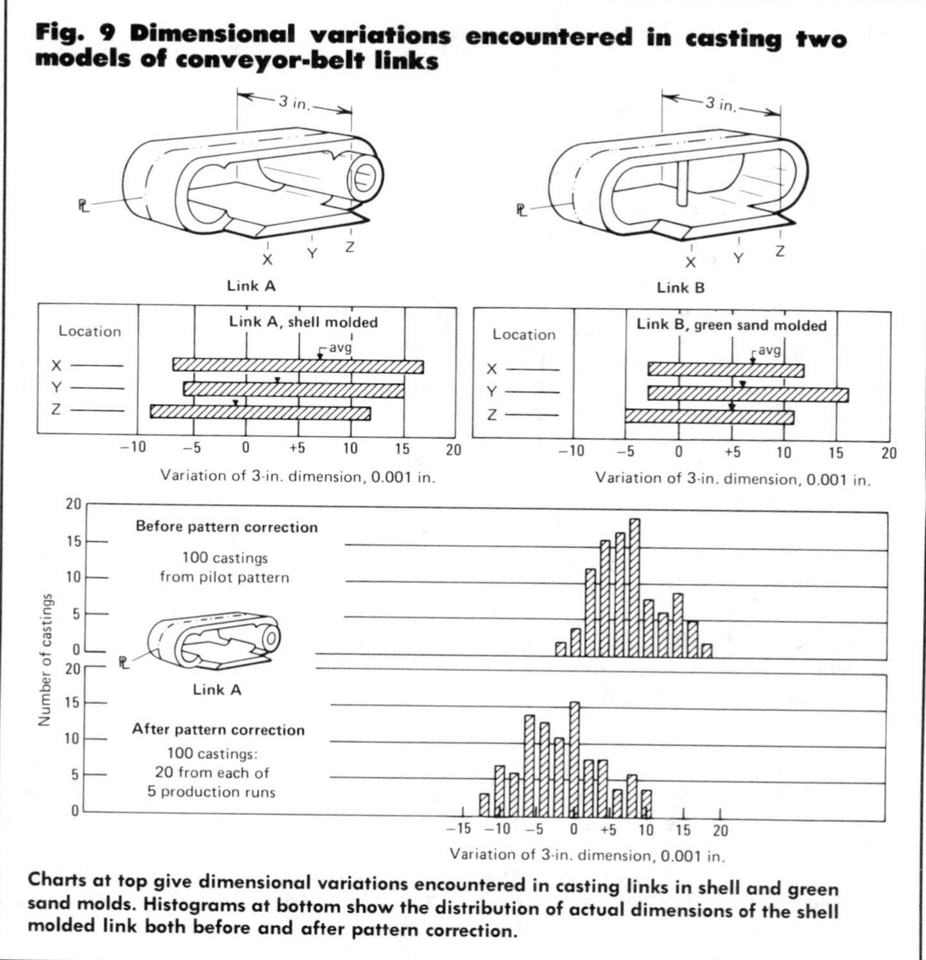

Charts at top give dimensional variations encountered in casting links in shell and green sand molds. Histograms at bottom show the distribution of actual dimensions of the shell molded link both before and after pattern correction.

Fig. 10 Recommended tolerances for casting two simple shapes

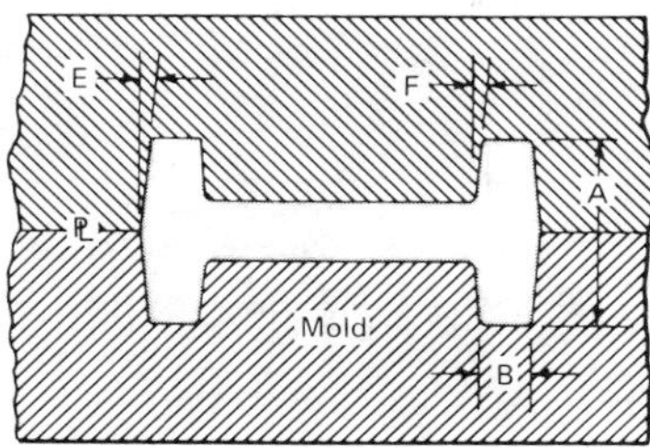

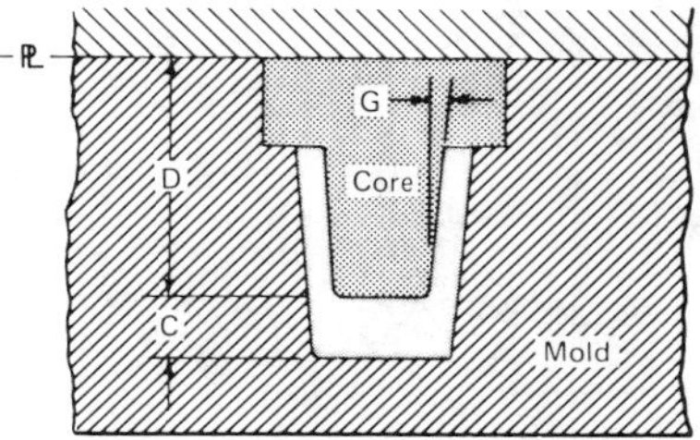

Dimension	Tolerance for casting in: Sand molds mm	in.	Shell molds mm	in.
A (across parting line)	...	...	0.25 + 0.5%	0.010 + 0.5%
B (within one part of mold)	...	...	0.5%	0.5%
C (between core and mold):				
For C of 25 mm (1 in.) or less	0.8	1/32	0.25 + 0.5%	0.010 + 0.5%
For C of 25 to 250 mm (1 to 10 in.)	1.5	1/16	...	...
D (length of core supported at one end)	375 max	15 max	4 × core diameter	
E (outside draft)	1° minimum		1° minimum	
F (draft in recesses)	1° minimum		1° minimum	
G (draft on cores)	1° minimum		1/2° minimum	

Fig. 11 Dimensional variations encountered in production of a shell molded valve body (Example 3)

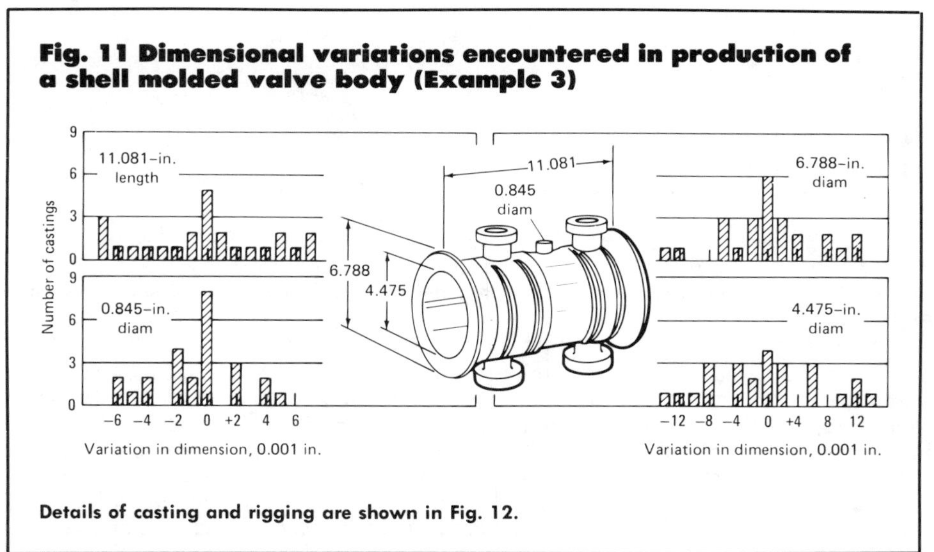

Details of casting and rigging are shown in Fig. 12.

Fig. 12 Details of casting and rigging design for a complex valve body (Example 3)

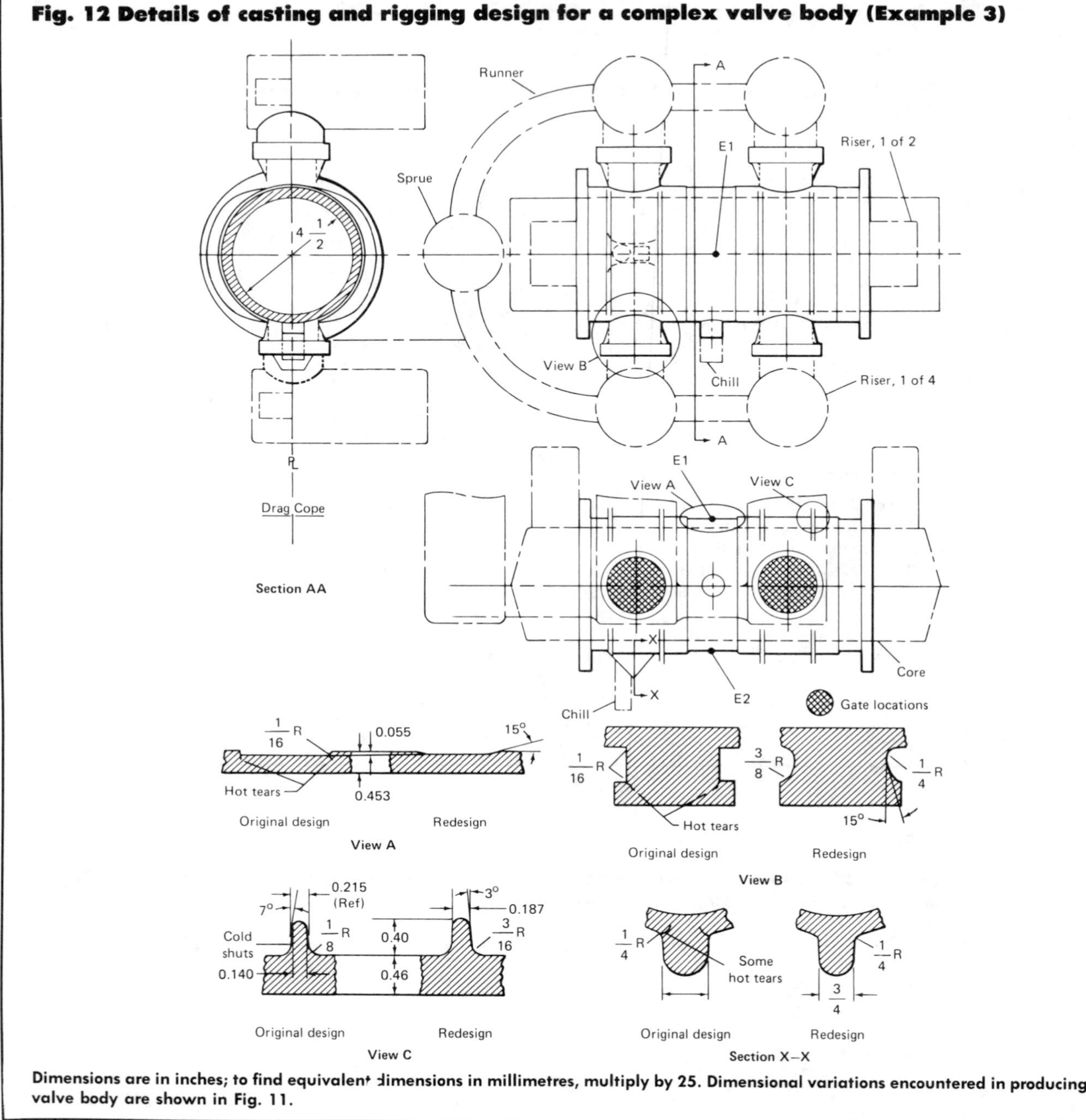

Dimensions are in inches; to find equivalent dimensions in millimetres, multiply by 25. Dimensional variations encountered in producing this valve body are shown in Fig. 11.

was to enlarge the rib in the areas where cold shuts were observed.

It is evident in comparing these two designs that the increase in width of the rib was very near the minimum required to avoid cold shuts. For this reason, the 4.75-mm (0.187-in.) rib width probably reflects a minimum for ribs of this height in a shell molded 17 Cr steel casting of comparable design and rigging.

The boss shown in section X-X in Fig. 12 is an example of the conflict that can exist between part function and required foundry procedures. The chilled face of this boss was to be drilled to receive a butterfly valve. This required that the face be acceptable as cast. Thus, if a riser were added or a rib were used to feed the boss from a nearby riser, a machining operation would have been required to remove the riser or rib. The boss was at first believed small enough to respond to chilling. Nevertheless, the total mass of metal cooled more slowly than had been anticipated, and therefore tended to freeze after surrounding areas had solidified. This late freezing isolated the boss from feed metal, which resulted in shrinkage cracks at the base of the boss (see section X-X). The solution was to reduce the mass of the boss so that chilling would be effective and freezing would begin in this boss and continue into adjacent areas.

Properties of Cast Heat-Resistant Alloys

Adapted from data compiled for the Alloy Casting Institute Division of the Steel Founders' Society of America by Ernest A. Schoefer

HA 9Cr-1Mo

Commercial Name

UNS number. J82090

Specifications

ASTM. A217 (C12)

Chemical Composition

Composition limits. 0.2 max C; 0.35 to 0.65 Mn; 1.0 max Si; 0.04 max P; 0.04 max S; 0.9 to 1.2 Mo; 8 to 10 Cr; rem Fe

Applications

Typical uses. Widely used in the oil refining industry. Also, in fan blades, furnace rollers, Lehr rolls, refinery fittings and trunnions

Mechanical Properties

Tensile properties. Annealed: tensile strength, 655 MPa (95 ksi); 0.2% yield strength, 450 MPa (65 ksi); elongation, 23% in 50 mm or 2 in. Normalized, tempered (a): tensile strength, 740 MPa (107 ksi); 0.2% yield strength, 560 MPa (81 ksi); elongation, 21% in 50 mm or 2 in.; reduction of area, 56%. Minimum for ASTM A217: tensile strength, 620 MPa (90 ksi); 0.2% yield strength, 415 MPa (60 ksi); elongation, 18% in 50 mm or 2 in.; reduction of area, 35%

Hardness. Annealed: 180 HB, normalized 966 °C (1825 °F), tempered at 677 °C (1250 °F): 220 HB

Elastic modulus. Tension, 200 GPa (29×10^6 psi)

Impact strength. Room temperature. Charpy keyhole: 43 J (32 ft·lb)

Creep-rupture characteristics. (b) Limiting creep stress, 110 MPa (16 ksi) for 0.0001%/h at 538 °C (1000 °F); stress to rupture: 10 h, 310 MPa (45 ksi); 100 h: 255 MPa (37 ksi); 1000 h: 185 MPa (27 ksi)

(a) Normalized at 966 °C (1825 °F), tempered 677 °C (1250 °F). (b) For constant temperature; for cyclic temperature, lower values would apply.

Structure

Microstructure. Ferritic with carbides in pearlite areas or agglomerated particles depending on prior heat treatment

Mass Characteristics

Density. 7.53 Mg/m³ (0.279 lb/in.³) at 20 °C (68 °F)

Patternmaker's shrinkage. ¼ in./ft

Thermal Properties

Melting point. Approx 1510 °C (2750 °F)

Coefficient of thermal expansion. Linear:

Temperature °C	°F	Mean coefficient µm/m·K	µin./in.·°F
21–100	70–212	11	6.1
21–316	70–600	12	6.5
21–538	70–1000	13	7.1
21–650	70–1200	14	7.5

Specific heat. 461 J/kg·k (0.11 Btu/lb·°F) at 21 °C (70 °F)

Thermal conductivity:

Temperature °C	°F	Conductivity W/m·K	Btu/ft·h·°F
100	212	25.9	15.0
316	600	26.7	15.4
538	1000	27.1	15.7
650	1200	27.3	15.8

Electrical Properties

Electrical resistivity. 700 nΩ·m at 21 °C (70 °F)

Magnetic Properties

Magnetic permeability. Ferromagnetic

Chemical Properties

General corrosion behavior. Resistant to high-temperature air, flue gases, petroleum, steam

Fabrication Characteristics

Heat treatment. For maximum softness, anneal by heating to approximately 885 °C (1625 °F) and furnace cooling at approximately 10 °C/h (18 °F/h). For improved strength, normalize by heating to 995 °C (1825 °F), air cooling to below 700 °C (1300 °F), and tempering at 675 °C (1250 °F)
Machinability. Most machining operations can be performed on castings of alloy HA. The work-hardening rate of this grade is much lower than that of the iron-chromium-nickel types, but it is advisable in all cases that the tool be kept continously entering into the metal. Slow feeds, deep cuts and powerful, rigid machines are necessary for best results. Work should be firmly mounted and supported, and tool mountings should provide maximum stiffness. Both high speed steel and carbide tools may be used successfully. Chips are stringy
Weldability. Shielded metal-arc, inert-gas arc and oxyacetylene gas methods. Shielded metal arc is preferred for high-temperature applications. Electrodes of similar composition (AWS E505-18) are recommended for arc welding. Before welding, heat castings to 230 to 290 °C (450 to 550 °F). After welding, heat castings to 650 to 700 °C (1200 to 1300 °F), depending on original draw temperature, hold long enough to ensure uniform heating, and rapidly air cool

HC, HC-30
28Cr

Commercial Name

UNS number. HC: J92605. HC-30: J92613

Specifications

ASTM. A297 (grade HC); A608
SAE. 70446

Chemical Composition

Composition limits. HC: 0.5 max C; 1.0 max Mn; 2.0 max Si; 0.04 max P; 0.04 max S; 0.5 max Mo (a); 26 to 30 Cr, 4 max Ni; rem Fe. HC-30: 0.35 max C; 1.0 max Mn; 2.0 max Si; 0.04 max P; 0.04 max S; 0.5 max Mo (a); 26 to 30 Cr; 4 max Ni; rem Fe

(a) Molybdenum not intentionally added.

Applications

Typical uses. Applications where strength is not a consideration or for moderate load bearing service around 650 °C (1200 °F). Also used where appreciable nickel cannot be tolerated (e.g., very high sulfur atmospheres) or where nickel tends to crack hydrocarbons through catalytic action. Castings used in boiler baffles, electrodes, furnace grate bars, gas outlet dampers, kiln parts, lute rings, rabble blades and holders, recuperators, salt pots, soot blower tubes, support skids, tuyeres. HC-30 is used for tubing in pressure applications at high temperatures.

Mechanical Properties

Tensile properties. At room temperature: minimum tensile strength, 379 MPa (55 ksi). As cast: tensile strength, 483 MPa (70 ksi) (a); 758 MPa (110 ksi) (b); 0.2% yield strength, 448 MPa (65 ksi) (a); 517 MPa (75 ksi) (b); elongation, 2%(a); 19%(b) in 50 mm or 2 in. Aged 24 h at 760 °C (1400 °F), furnace cooled: tensile strength, 793 MPa (115 ksi) (b); 0.2% yield strength, 552 MPa (80 ksi) (b); elongation, 18% in 50 mm or 2 in. See also Fig. 1.

Temperature °C	°F	Minimum tensile strength(a) MPa	ksi	Minimum elongation(a), %
760	1400	36.5	5.3	40
870	1600	20.4	2.96	50
980	1800	11.0	1.6	40

(a) From ASTM 608

Elastic modulus. Tension, 200 GPa (29 × 10^6 psi)
Creep-rupture characteristics. See Table 1.

(a) Under 1.0% nickel, low nitrogen. (b) Over 2.0% nickel with 0.15% min nitrogen.

Structure

Microstructure. Ferritic at all temperatures and for this reason is not hardenable by heat treatment. Ductility and impact strength are very low at room temperature. Creep strength very low at elevated temperature, unless some nickel is present

Fig. 1 Selected properties of alloy HC

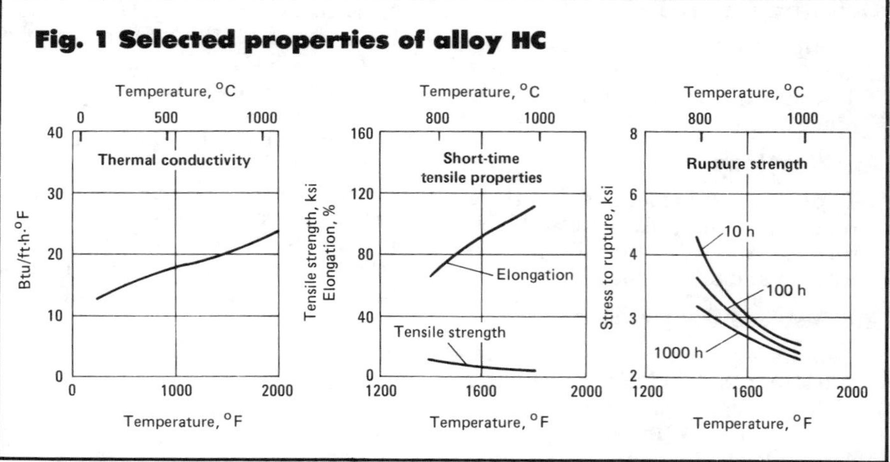

Table 1 Representative creep-rupture properties for alloy HC(a)

Temperature		Limiting stress for 0.0001%/h creep rate		Stress to rupture in: 10 h		100 h		1000 h	
°C	°F	MPa	ksi	MPa	ksi	MPa	ksi	MPa	ksi
760	1400	8.96	1.3	31.7	4.6	22.8	3.3	15.5	2.3
870	1600	5.17	0.75	13.8	2.0	11.7	1.7	8.96	1.3
980	1800	2.48	0.36	7.6	1.1	5.86	0.85	4.27	0.62

(a) For constant temperature; for cyclic temperature lower values would apply. Values represent test samples containing over 2.0% nickel with 0.15% min nitrogen.

Fig. 2 Results of 100-h corrosion tests of alloy HC in gases

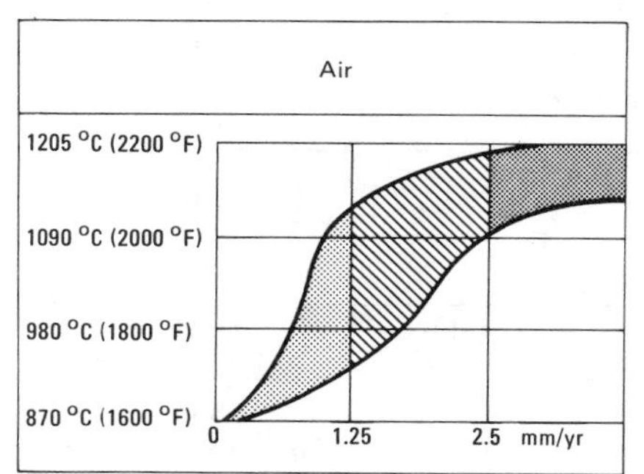

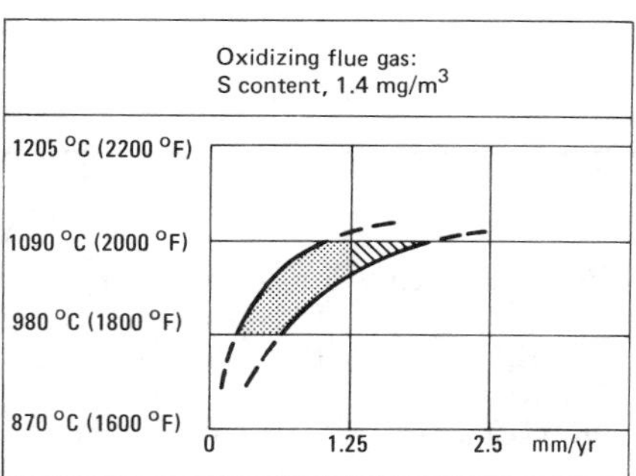

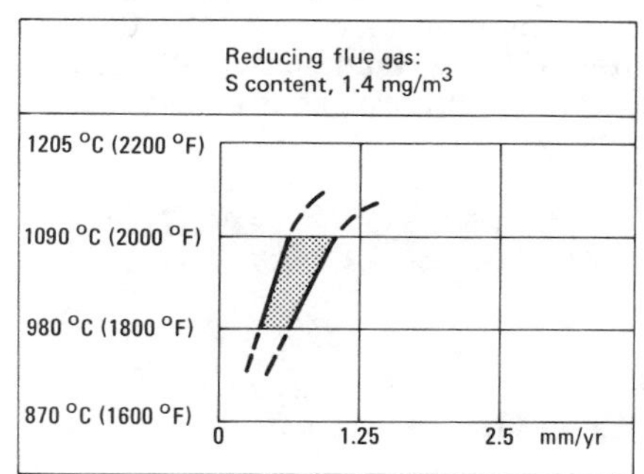

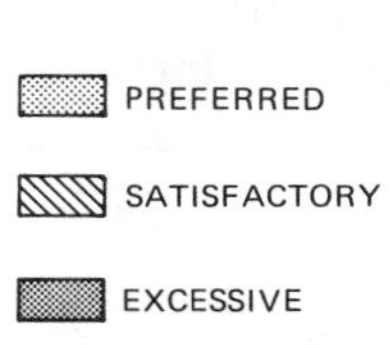

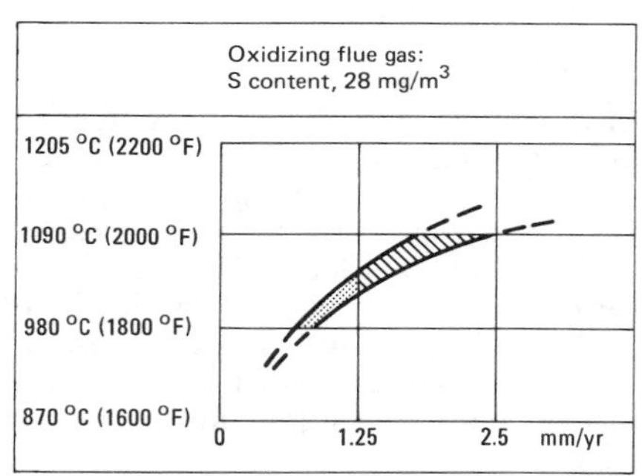

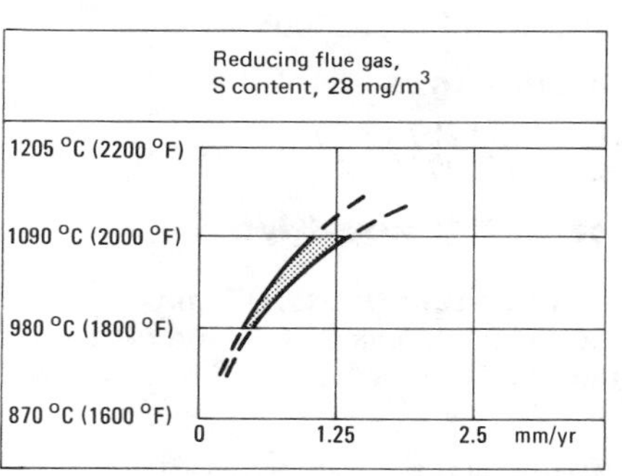

Mass Characteristics

Density. 7.53 Mg/m^3 (0.272 $lb/in.^3$)
Patternmaker's shrinkage. 1/32 in./ft

Thermal Properties

Melting point. 1496 °C (2725 °F)
Coefficient of thermal expansion. Linear:

Temperature		Mean coefficient	
°C	°F	μm/m·K	μin./in.·°F
21– 538	70–1000	11.3	6.3
21– 650	70–1200	11.5	6.4
21– 760	70–1400	11.9	6.6
21– 870	70–1600	12.6	7.0
21– 980	70–1800	13.3	7.4
21–1093	70–2000	13.9	7.7
650– 871	1200–1600	15.7	8.7
650– 980	1200–1800	16.7	9.3

Specific heat. 502 J/kg·K (0.12 Btu/lb·°F) at 21 °C (72 °F)
Thermal conductivity:

Temperature		Conductivity	
°C	°F	W/m·K	Btu/ft·h·°F
100	212	21.8	12.6
538	1000	30.9	17.9
816	1500	35.1	20.3
1093	2000	41.9	24.2

Electrical Properties

Electrical resistivity. 770 nΩ·m at 21 °C (70 °F)

Magnetic Properties

Magnetic permeability. Ferromagnetic

Chemical Properties

General corrosion behavior. Provides excellent resistance to oxidation and high-sulfur containing flue gases up to 1095 °C (2000 °F).
Resistance to specific corroding agents. Poor resistance to neutral salts. Poor resistance to molten aluminum and zinc, and good resistance to molten magnesium. See also Fig. 2.

Fabrication Characteristics

Machining. Most operations can be performed satisfactorily. Tool should be kept entering into the metal. Slow feeds, deep cuts and powerful, rigid machines are necessary. High-speed steel and carbide tools may be successfully used. Chips are tough and stringy; chip curler and breaker tools are recommended. Good lubrication and cooling essential; cutting fluid should flood both tool and work.

Weldability. Shielded metal arc, inert-gas arc and oxyacetylene-gas methods can be used. Shielded metal arc welding is generally preferred for high-temperature applications. Lime coated electrodes (AWS E446-15) are recommended for arc welding. Castings should be heated to 204 to 704 °C (400 to 1300 °F) before welding. Each bead should be peened before depositing next bead. Heat castings to 843 °C (1550 °F) after welding; hold sufficiently long to ensure uniform heating; air cool rapidly

Fig. 3 Selected properties of alloy HD

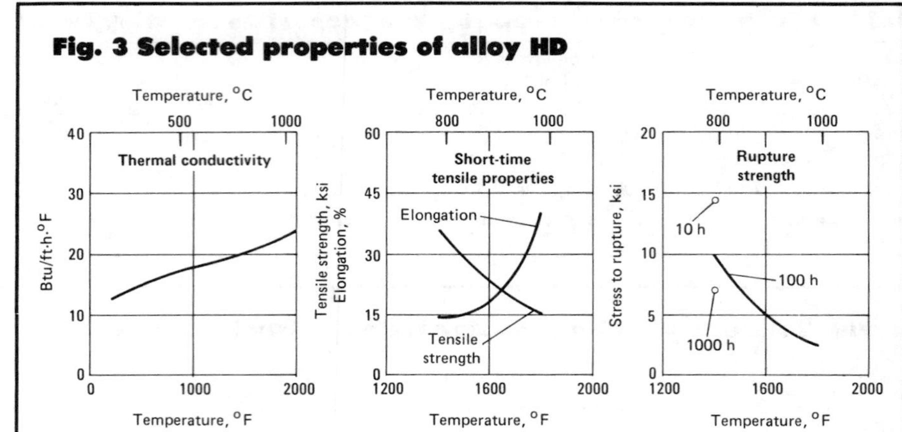

HD
28Cr-5Ni

Commercial Name

UNS number. J93005

Specifications

ASTM. A297 (HD)
SAE. 70327

Chemical Composition

Composition limits. 0.5 max C; 1.5 max Mn; 2.0 max Si; 0.04 max P; 0.04 max S; 0.5 max Mo(a); 26 to 30 Cr; 4 to 7 Ni, rem Fe

(a) Molybdenum not intentionally added.

Applications

Typical uses. Brazing furnace parts, cracking equipment, furnace blowers, gas burner parts, holding pots, kiln parts, pouring spouts, rabble shoes and arms, and recuperator sections

Mechanical Properties

Tensile properties. Room temperature. As cast: tensile strength, 586 MPa (85 ksi); 0.2% yield strength, 330 MPa (48 ksi); elongation, 16% in 50 mm or 2 in. Minimum for ASTM A297: tensile strength, 517 MPa (75 ksi); 0.2% yield strength, 240 MPa (35 ksi); elongation, 8% in 50 mm or 2 in. See also Fig. 3.

Hardness. As cast: 190 HB

Elastic modulus. Tension, 186 GPa (27 × 10⁶ psi)

Creep-rupture characteristics(a). Representative long-time values at 760 °C (1450 °F): limiting creep stress, 24.1 MPa (3.5 ksi) for 0.0001%/h. Stress to rupture in: 10 h, 96.5 MPa (14 ksi); 100 h, 69 MPa (10 ksi); 1000 h: 48.2 MPa (7.0 ksi)

(a) For constant temperature; lower values would apply for cyclic temperature.

Structure

Microstructure. Two-phase: ferrite plus austenite nonhardenable by customary heat treating procedure. Long exposure to temperatures at 704 to 816 °C (1300 to 1500 °F), however, may result in hardening of the alloy accompanied by severe loss of room temperature ductility through the formation of sigma phase. Ductility may be restored by heating the alloy to a uniform temperature of 982 °C (1300 °F) or higher and then cooling rapidly to below 649 °C (1200 °F)

Mass Characteristics

Density. 7.58 Mg/m³ (0.274 lb/in.³) at 21 °C (70 °F)

Patternmaker's shrinkage. 7/32 in./ft

Thermal Properties

Melting point. Approx 1480 °C (2700 °F)

Coefficient of thermal expansion. Linear:

Temperature °C	°F	Mean coefficient µm/m·K	µin./in.·°F
21–538	70–1000	13.9	7.7
21–650	70–1200	14.4	8.0
21–760	70–1400	14.9	8.3
21–870	70–1600	15.5	8.6
21–980	70–1800	16.0	8.9
21–1090	70–2000	16.6	9.2
650–870	1200–1600	18.5	10.3
650–980	1200–1800	19.1	10.6

Specific heat. 502 J/kg·K (0.12 Btu/lb·°F) at 21 °C (70 °F)

Thermal conductivity.

Temperature °C	°F	Conductivity W/m·K	Btu/ft·h·°F
100	212	21.8	12.6
538	1000	30.9	17.9
816	1500	35.1	20.3
1093	2000	41.9	24.2

Electrical Properties

Electrical resistivity. 810 nΩ·m at 21 °C (70 °F)

Magnetic Properties

Magnetic permeability. Ferromagnetic

Chemical Properties

General corrosion behavior. Resists corrosion in air, combustion gases, flue gases, high-sulfur, molten copper and copper alloys, and molten neutral salts

Resistance to specific corroding agents. Poor resistance to molten magnesium. See also Fig. 4.

Fabrication Characteristics

Machinability. Most machining operations can be performed satisfactorily on castings. The tool should be

Fig. 4 Results of 100-h corrosion tests of alloy HD in gases

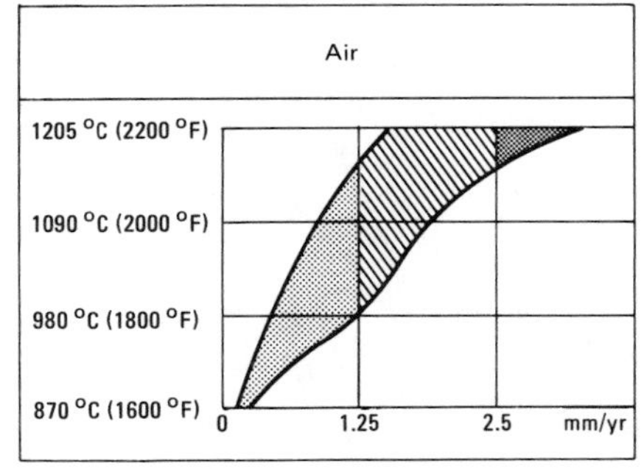

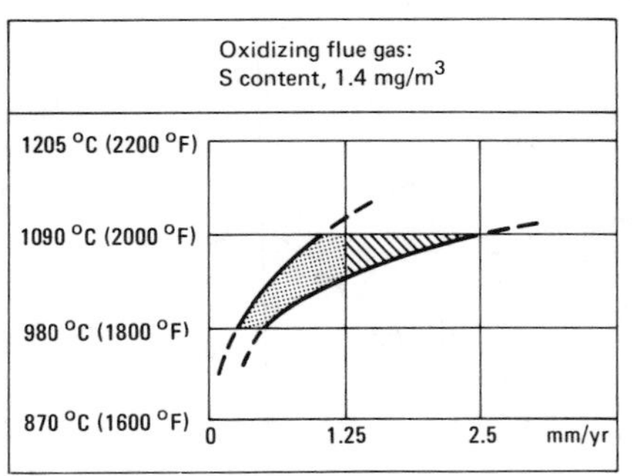

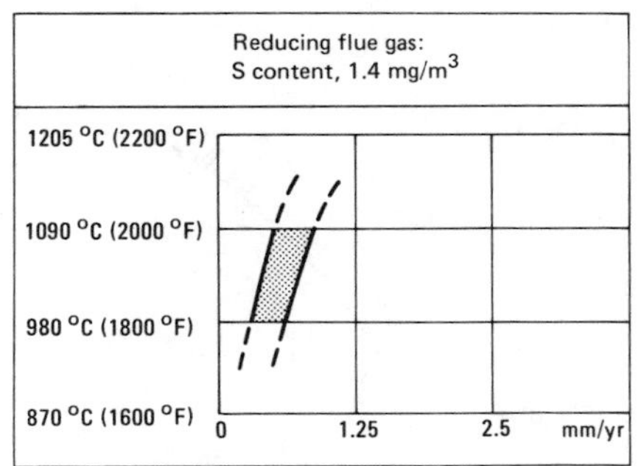

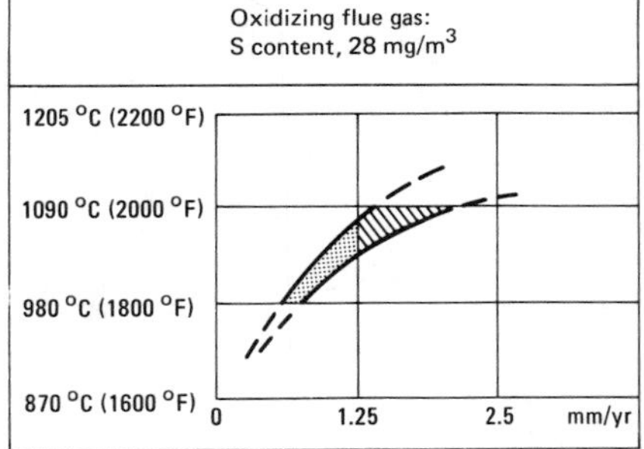

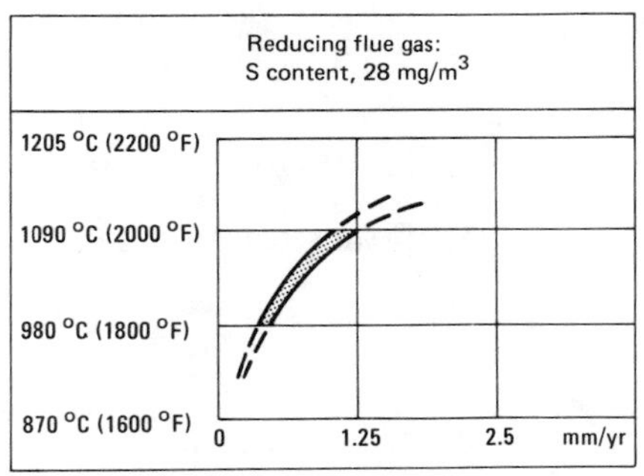

kept continuously entering into the metal to avoid work hardening the surface from rubbing or scraping. For best results, slow feeds, deep cuts and powerful rigid machines are necessary. Both high-speed steel and carbide tools may be successfully used. Chips are tough and stringy; chip curler and breaker tools are recommended. Good lubrication and cooling are essential. Cutting fluid should flood both the tool and the work

Weldability. Shielded metal arc, inert-gas arc and oxyacetylene gas methods. Shielded metal arc welding preferred for high-temperature applications. Lime coated electrodes of similar composition (AWS E446-15) are recommended for arc welding. Preweld heat treatment is not usually required. No heat treatment usually required after welding except to relieve welding stresses

HE
28Cr-10Ni

Commercial Name

UNS number. J93403

Specifications

ASTM. A297 (HE)
SAE. 70312

Chemical Composition

Composition limits. 0.2 to 0.5 C; 2.0 max Mn; 2.0 max Si; 0.04 max P; 0.04 max S; 0.5 max Mn(a); 26 to 30 Cr; 8 to 11 Ni; rem Fe

(a) Molybdenum not intentionally added.

Applications

Typical uses. Used extensively in ore-roasting equipment. Uses include billet skids, burner nozzles, dampers, furnace chains and conveyors, furnace door frames, oil burner parts, rabble arms and blades, recuperators, rotating shafts, soot blower elements, steam generator parts, tube supports

Mechanical Properties

Tensile properties. Representative values at room temperature. As cast: tensile strength, 655 MPa (95 ksi); 0.2% yield strength, 310 MPa (45 ksi); elongation, 20% in 50 mm or 2 in. Aged 24 h at 760 °C (1400 °F), furnace cooled: tensile strength, 620 MPa (90 ksi); 0.2% yield strength, 379 MPa (55 ksi); elongation, 10% in 50 mm or 2 in. Minimum for ASTM A297: tensile strength, 586 MPa (85 ksi); 0.2% yield strength, 276 MPa (40 ksi); elongation, 9% in 50 mm or 2 in.

Hardness. As cast: 200 HB. Aged 24 h at 760 °C (1400 °F), furnace cooled: 270 HB

Elastic modulus. Tension, 172 GPa (25×10^6 psi)

Impact strength. Charpy keyhole, 13.6 J (10 ft·lb) at 21 °C (70 °F)

Creep-rupture characteristics. See Table 2.

Structure

Microstructure. As cast, two-phase austenite plus ferrite containing carbides

Mass Characteristics

Density. 7.67 Mg/m^3 (0.277 lb/in.^{3}) at 21 °C (70 °F)

Patternmaker's shrinkage. 9/32 in./ft

Thermal Properties

Melting point. Approx 1454 °C (2650 °F)

Coefficient of thermal expansion. Linear:

Fig. 5 Results of 100-h corrosion tests of alloy HE in gases

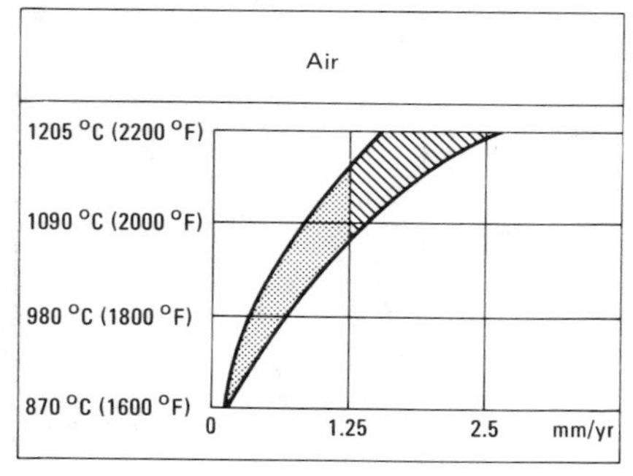

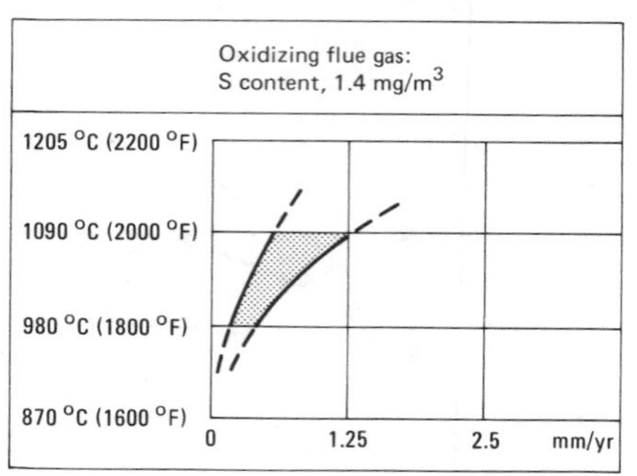

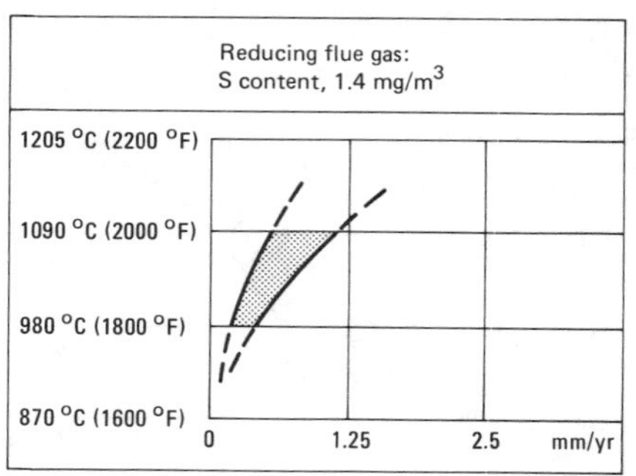

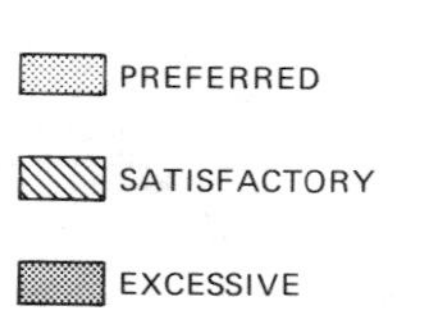

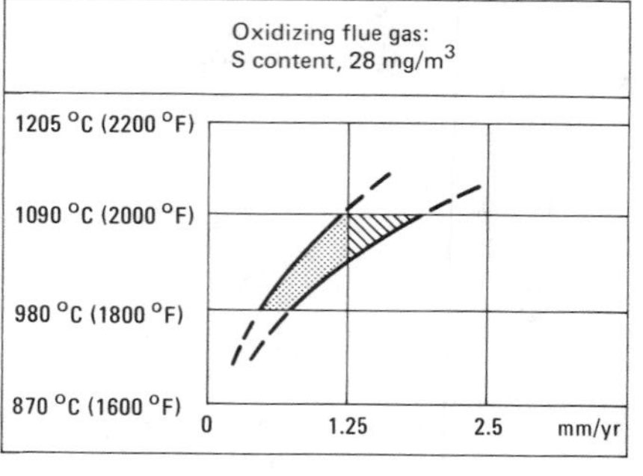

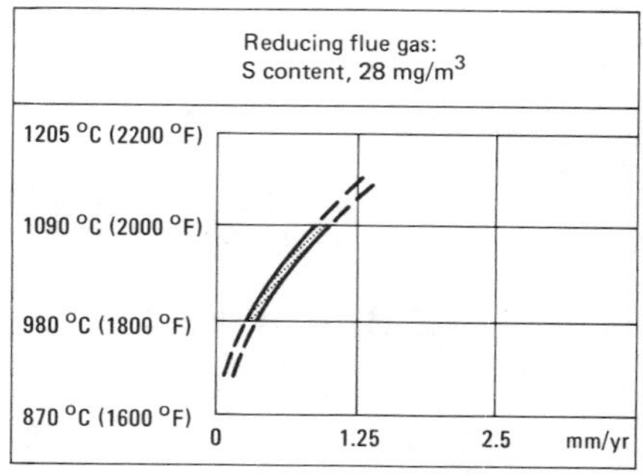

Temperature °C	°F	Mean coefficient µm/m·K	µin./in.·°F
21–538	70–1000	17.3	9.6
21–650	70–1200	17.8	9.9
21–760	70–1400	18.4	10.2
21–870	70–1600	18.9	10.5
21–980	70–1800	19.4	10.8
21–1090	70–2000	19.9	11.1
650–870	1200–1600	21.9	12.2
650–980	1200–1800	22.5	12.5

Specific heat. 586 J/kg·K (0.14 Btu/lb·°F) at 21 °C (70 °F)
Thermal conductivity:

Temperature °C	°F	Conductivity W/m·K	Btu/ft·h·°F
100	212	14.7	8.5
316	600	18.2	10.5
538	1000	21.5	12.4
650	1200	23.4	13.5
760	1400	25.3	14.6
870	1600	27.5	15.9
980	1800	29.2	16.9
1090	2000	31.5	18.2

Electrical Properties

Electrical resistivity. 850 nΩ·m at 21 °C (70 °F)

Table 2 Representative long-time values of alloy HE creep-rupture characteristics(a)

Temperature °C	°F	Limiting stress for 0.0001%/h creep rate MPa	ksi	Stress to rupture in 100 h MPa	ksi
760	1400	27.6	4.0	75.8	11.0
870	1600	16.5	2.4	36.5	5.3
980	1800	9.65	1.4	17.2	2.5
1090	2000	2.76	0.4	...	...

(a) For constant temperature; lower values would apply for cyclic temperature.

Magnetic Properties

Magnetic susceptibility. Weakly magnetic
Magnetic permeability. 1.3 to 2.5

Chemical Properties

General corrosion behavior. Excellent corrosion resistance at high temperatures
Resistance to specific corroding agents. Good resistance to very-high-sulfur-content gases (300 to 500 grains sulfur per 100 ft^3 of gas) at high temperatures. See Fig. 5.

Fabrication Characteristics

Machinability. Most machining operations can be satisfactorily used on castings. The tool must be kept continuously entering into the metal to avoid work hardening the surface from rubbing or scraping. Slow feeds, deep cuts and powerful, rigid machines are necessary. High-speed steel and carbide tools may be successfully used. Chips are tough and stringy; chip curler and breaker tools are recommended. Good lubrication and cooling are essential. Cutting fluid must flood both the tool and the work
Weldability. Shielded metal arc, inert-gas arc and oxyacetylene-gas methods. Shielded metal arc welding is generally preferred for high-temperature applications. Lime coated electrodes of similar composition (AWS E312-15) are recommended for arc welding. Preweld and postweld heat treating are not required

HF
20Cr-10Ni

Commercial Name

UNS number. J92603

Specifications

ASTM. A297 (HF)
SAE. 70308
Government. MIL-S-17509

Chemical Composition

Composition limits. 0.2 to 0.4 C; 2.0 max Mn; 2.0 max Si; 0.04 max P; 0.04 max S; 0.5 max Mo(a); 19 to 23 Cr; 9 to 12 Ni; rem Fe

(a) Molybdenum not intentionally added.

Applications

Typical uses. Suitable for applications requiring high strength and corrosion resistance at 650 to 870 °C (1200 to 1600 °F). Used extensively in oil refinery and heat treating furnaces. Uses include arc furnace electrode arms, annealing boxes and trays, baskets, brazing channels, burner tips, burnishing rolls, conveyor belts and chains, fan housings, furnace rails, gas burner rings, hardening retorts, hearth plates, Lehr rolls, pier caps, soaking pit dampers, tempering baskets, and wear plates

Mechanical Properties

Tensile properties. Representative values at room temperature. As cast: tensile strength, 634 MPa (92 ksi); 0.2% yield strength, 310 MPa (45 ksi); elongation, 38% in 50 mm or 2 in. Aged 24 h at 760 °C (1400 °F), furnace cooled: tensile strength, 689 MPa (100 ksi); 0.2% yield strength, 345 MPa (50 ksi); elongation, 25% in 50 mm or 2 in. Minimum for ASTM A297: tensile strength, 483 MPa (70 ksi); 0.2% yield strength, 241 MPa (35 ksi); elongation, 25% in 50 mm or 2 in. See also Fig. 6.

Hardness. As cast, 165 HB. Aged 24 h at 760 °C (1400 °F), furnace cooled: 190 HB. See also Fig. 7.
Elastic modulus. Tension, 193 GPa (28×10^6 psi). See also Fig. 8.
Creep-rupture characteristics. See Table 3 and Fig. 9.

Fig. 6 Effect of temperature on short-time tensile properties of alloy HF

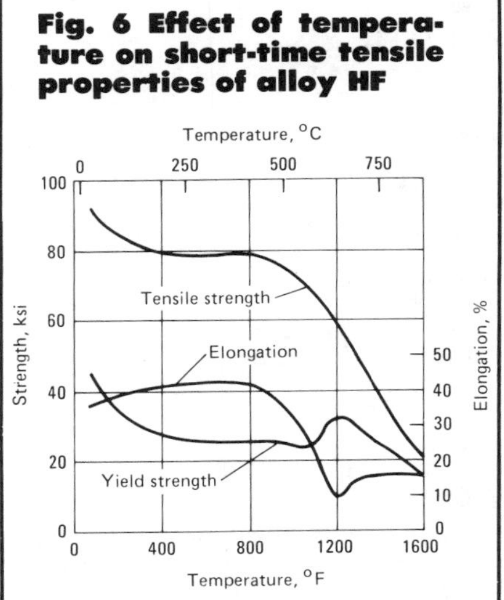

Structure

Microstructure. As cast: austenitic matrix containing interdendritic eutectic carbides, occasional unidentified lamellar constituent

Mass Characteristics

Density. 7.75 Mg/m³ (0.280 lb/in.³)
Patternmaker's shrinkage. 9/32 in./ft

Fig. 8 Effect of temperature on approximate elastic modulus of alloy HF

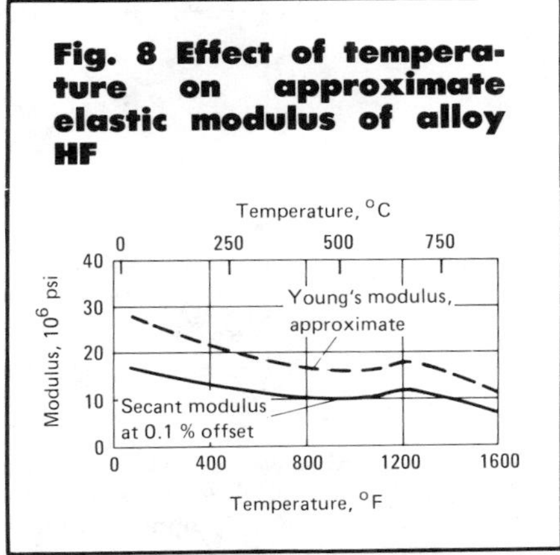

Fig. 7 Effect of temperature on hardness of alloy HF

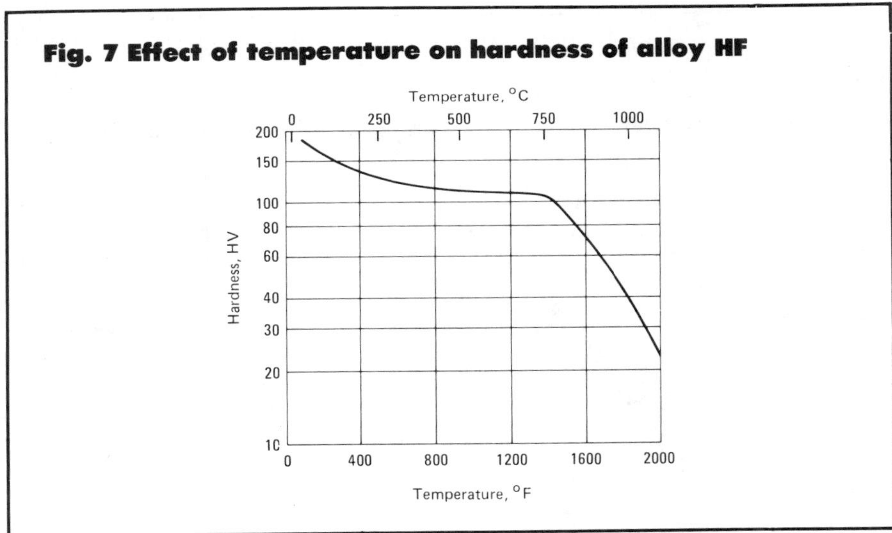

Table 3 Creep-rupture data(a) for alloy HF

Temperature °C	Temperature °F	Limiting stress for: Minimum creep rate of 0.0001% MPa	ksi	1% total creep in 100 000 h MPa	ksi	Stress to rupture in: 100 h MPa	ksi	1000 h MPa	ksi	10 000 h MPa	ksi	100 000 h MPa	ksi
649	1200	124	18	77.9	11.3	228	33.0	172	25	114	16.5	75.8	11
760	1400	46.9	6.8	30.3	4.4	93.1	13.5	62.7	9.1	42.1	6.1	27.6	4.0
871	1600	26.9	3.9	9.65	1.4	49.6	7.2	30.3	4.4	18.6	2.7	11.7	1.7

(a) Representative values for constant temperature; for cyclic temperature, lower values would apply. (b) Extrapolated values.

Fig. 9 Creep-rupture properties of alloy HF

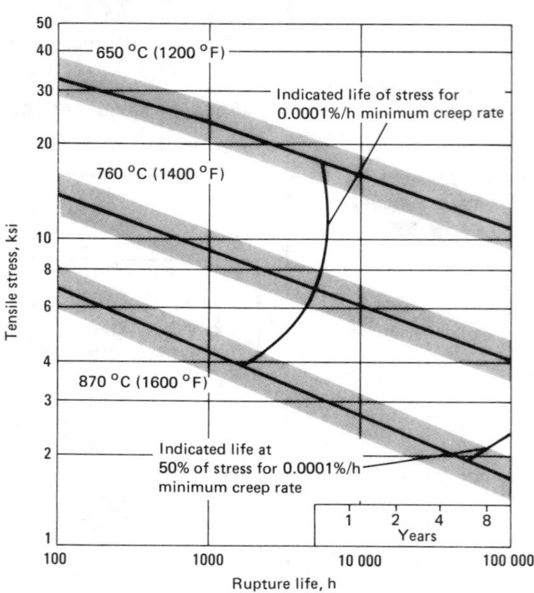

The scatter bands shown are set arbitrarily at ±20% of the stress for the central tendency line. Such a range usually embraces test data for similar alloy compositions, but should not be considered statistically significant confidence limits. Scatter of values may be much wider, particularly at the longer times and higher temperatures.

Thermal Properties

Melting point. Approx 1400 °C (2550 °F)

Coefficient of thermal expansion. Linear:

Temperature °C	°F	Mean coefficient µm/m·K	µin./in.·°F
20– 100	68–212	13.0	7.2
20– 538	68–1000	17.8	9.9
20– 650	68–1200	18.2	10.1
20– 760	68–1400	18.5	10.3
20– 871	68–1600	18.7	10.4
20– 982	68–1800	18.9	10.5
20–1093	68–2000	19.6	10.9
650– 871	1200–1600	19.9	11.1

Specific heat. 502 J/kg·K (0.12 Btu/lb·°F) at 21 °C (70 °F)

Electrical Properties

Electrical resistivity. 800 nΩ·m at 21 °C (70 °F)

Magnetic Properties

Magnetic permeability. 1.00

Chemical Properties

General corrosion behavior. Often selected because of its superior corro-

Fig. 10 Results of 100-h corrosion tests of alloy HF in gases

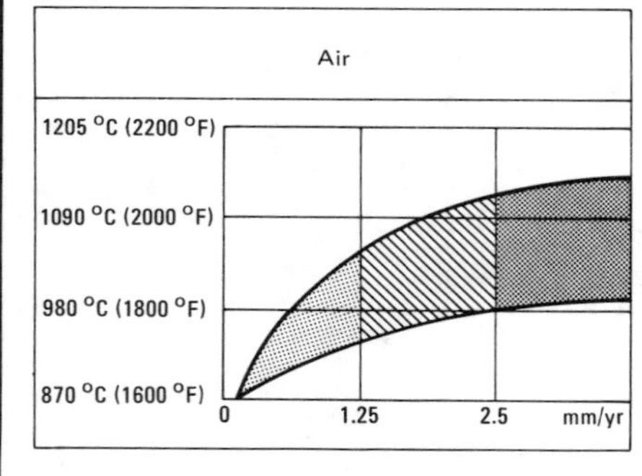

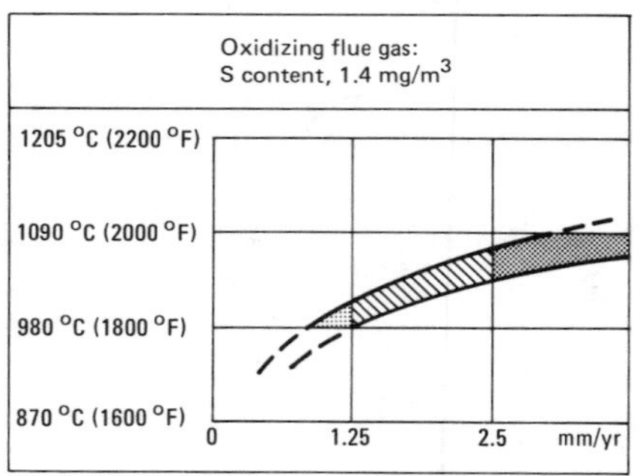

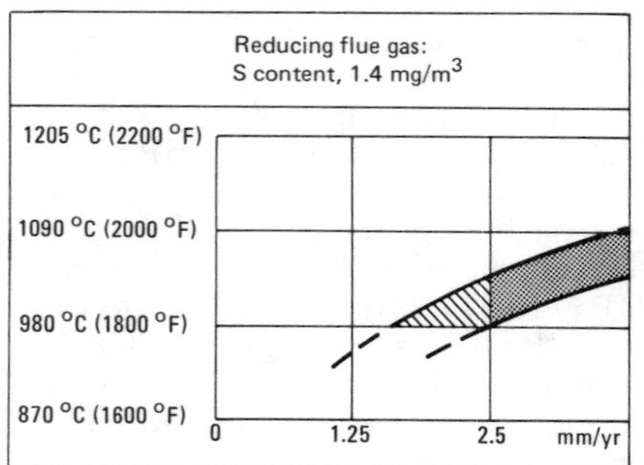

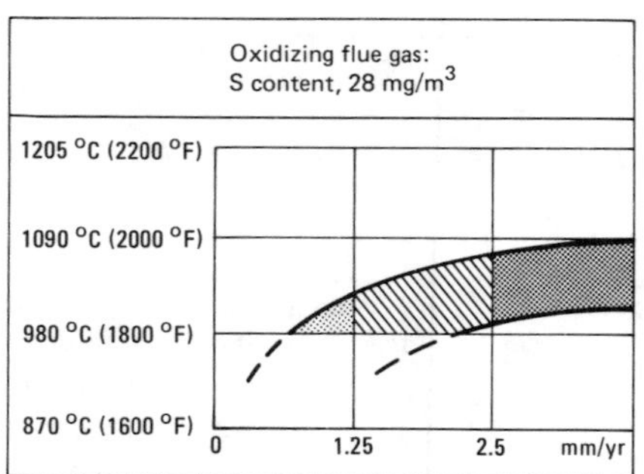

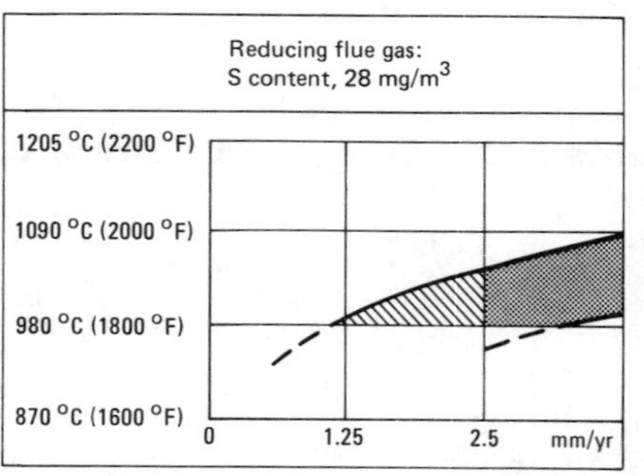

PREFERRED
SATISFACTORY
EXCESSIVE

sion resistance to air, conbustion gases, flue gases oxidizing and reducing, and steam. When used for resistance to oxidation at elevated temperatures, it is not necessary to keep the carbon content at the lowest level specified for corrosion-resistant castings

Resistance to specific corroding agents. Gases: See Fig. 10. Molten metals: good resistance to molten tin to 345 °C (650 °F); poor resistance to molten zinc. Good resistance to molten antimony to 705 °C (1300 °F; good resistance to molten cadmium to 410 °C (775 °F).

Fabrication Characteristics

Heat treatment. Castings are normally used in the as-cast condition. The alloy cannot be hardened by heat treatment but if service conditions involve repeated heating and cooling, improved performance may be obtained by heating castings at 1035 °C (1900 °F) for 6 h followed by furnace cooling prior to placing in service

Machinability. See alloy HE

Weldability. Shielded metal arc, inert-gas arc, oxyacetylene-gas methods can be used. Shielded metal arc is preferred for high-temperature applications

HH
26Cr-12Ni

Commercial Name

UNS number. J93503

Specifications

ASTM. A297 (HH), A447
SAE. 70309

Chemical Composition

Composition limits. 0.20 to 0.50 C; 2.0 max Mn; 2.0 max Si; 0.04 max P; 0.04 max S; 0.05 max Mo(a); 0.2 max N; 24 to 28 Cr; 11 to 14 Ni; rem Fe

Consequence of exceeding impurity limits. Improper balance of nickel and chromium can result in the presence of substantial amounts of ferrite in the metal. Ferrite decreases strength at high temperatures. Transformation to sigma at intermediate temperatures can cause

(a) Molybdenum not intentionally added.

Table 4 Typical tensile properties at room temperature for alloy HH

Type	Tensile strength MPa	ksi	0.2% yield strength MPa	ksi	Elongation in 50 mm or 2 in., %
As cast					
I	586	85	345	50	25
II	552	80	276	40	15
Aged 24 h at 760 °C (1400 °F), furnace cooled					
I	593	86	379	55	11
II	634	92	310	45	8
Minimum for ASTM A297					
I, II	517	75	241	35	10
Minimum for ASTMA447 after aging					
I	552	80	...	...	9
II	552	80	...	...	4

Fig. 11 Effect of temperature on short time tensile properties and Charpy impact strength of wholly austenitic alloy HH

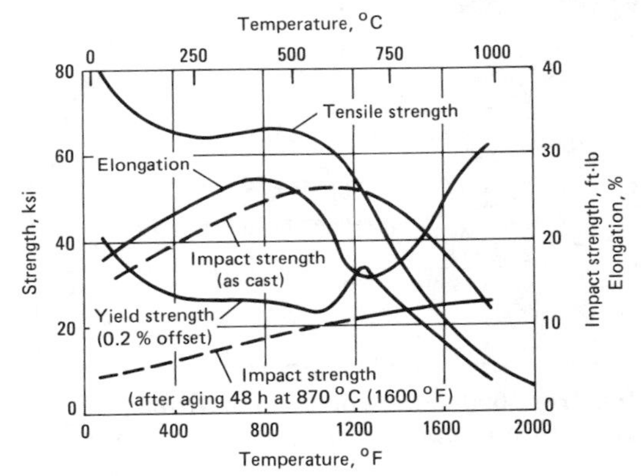

Fig. 12 Effect of temperature on hardness of wholly austenitic alloy HH

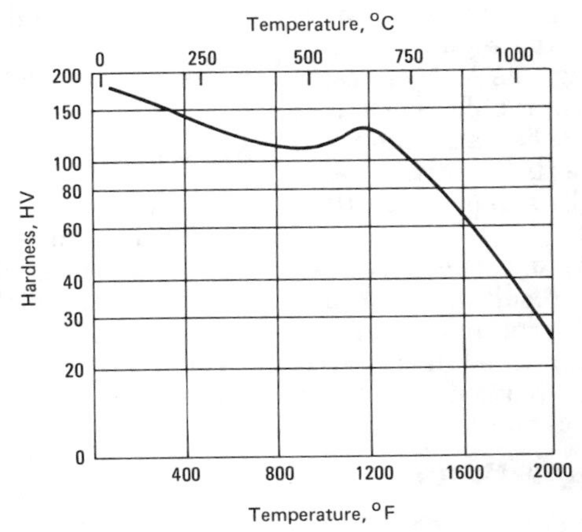

Table 5 Creep-rupture data(a) for alloy HH

Temperature, °C	Limiting stress for: Minimum creep rate (b) of 0.0001%/h		Limiting stress for: 1% total creep in 100 000 h		Stress to rupture in: 10 h		100 h		1000 h		10 000 h		100 000 h	
	MPa	ksi	MPa	ksi	MPa	ksi	MPa	ksi	MPa	ksi	MPa	ksi	MPa	ksi
Type I														
649	...	...	...	...	...	...	...	...	...	...	...	...	...	...
760	20.7	3.0	...	...	...	...	96.5	14.0	44.8	6.5	...	...	...	...
871	11.7	1.7	...	...	...	...	44.1	6.4	26.2	3.8	...	...	...	...
982	7.6	1.1	...	...	32	4.7	21.4	3.1	14.5	2.1	...	...	...	...
1093	2.1	0.3	...	...	...	...	10.3	1.5	...	...	...	...	...	...
Type II														
649	124	18	65.5	9.5	...	...	241	35	152	22	96.5	14.0	62.1	9.0(b)
760	43	6.3	13.8	2.0	...	...	96.5	14	55.2	8.0	33.1	4.8	19.3	2.8(b)
871	26.9	3.9	7.6	1.1	...	...	46.9	6.8	26.2	3.8	14.8	2.15	8.3	1.2(b)
982	14.5	2.1	...	...	...	...	22.1	3.2	11.4	1.65	5.9(b)	0.86(b)	3.0(b)	0.44(b)
1093	5.5	0.8	...	...	...	...	9.7	1.4	4.7	0.68	2.34(b)	0.34(b)	1.0(b)	0.15(b)

(a) Representative long-time values for constant temperature; for cyclic temperature, lower values would apply. (b) Extrapolated data.

Fig. 13 Effect of temperature on approximate modulus of elasticity of alloy HH

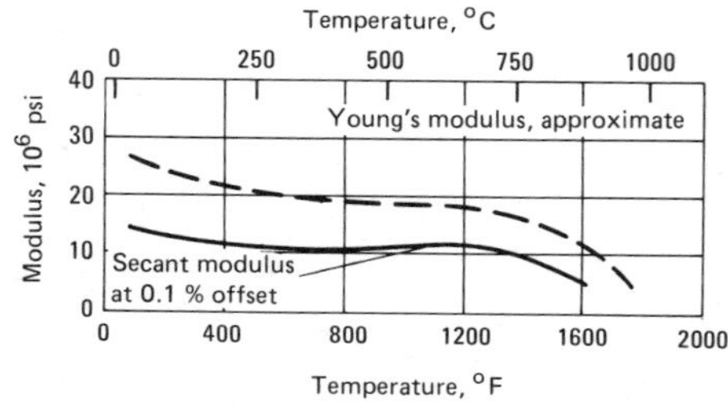

castings to suffer from brittleness or cracking within certain temperature ranges

Applications

Typical uses. Annealing trays, billet skids, burner nozzles, carburizing boxes, convection tube supports, dampers, exhaust manifolds, flue gas stacks, grate supports, hardening trays, kiln nose ring segments, muffles, normalizing discs, pier caps, quenching trays, rabble arms and blades, radiant tubes and supports, refractory supports, retorts, roller hearths and rails, stoker parts, tube hangers

Precautions in use. Because embrittlement can result from service environment absorption of carbon, alloy HH is seldom used in carburizing applications, particularly those involving thermal shock

Mechanical Properties(a)

Typical tensile properties. See Table 4 and Fig. 11.

Hardness. As cast: Type I, 185 HB; Type II, 180 HB. Aged 24 h at 760 °C (1400 °F), furnace cooled: Types I and II; 200 HB. See also Fig. 12.

Elastic modulus. Tension, 186 GPa (27×10^6 psi). See Fig. 13.

Impact strength. Charpy keyhole, see Fig. 11.

Creep-rupture characteristics. See Table 5 and Fig. 14.

(a) Type I, partially ferritic; Type II, wholly austenitic.

Structure

Microstructure. Alloy HH is basically austenitic and holds considerable carbon in solid solution, but carbides, ferrite (soft, ductile and magnetic) and sigma (hard, brittle and nonmagnetic) may also be present in the microstructure. The amounts of the various structural constituents present depend on composition and thermal history. In fact, two distinct grades of material can be obtained within the stated chemical composition range of the type alloy HH. These grades are defined as type I wholly austenitic and type II partially ferritic in ASTM A447.

Near 870 °C (1600 °F) the partially ferritic alloys tend to embrittle from the development of sigma phase, while around 760 °C (1400 °F) carbide precipitation may cause comparable loss of ductility. Such possible embrittlement suggests that 927 to 1093 °C (1700 to 2000 °F) is the best service temperature range, but this is not critical for steady temperature conditions in the absence of unusual thermal or mechanical stresses.

A serious cause of embrittlement is absorption of carbon from the service environment. Hence, type alloy HH is seldom used for carburizing applications. High silicon content (over 1.5%) will fortify the alloy against carburization under mild conditions but will promote ferrite formation and possible sigma embrittlement.

The partially ferritic (type I) alloy HH is adapted to operating conditions which are subject to changes in temperature level and applied stress. A plastic extension in the weaker, ductile ferrite under changing load tends to occur more readily than in the stronger austenitic phase, thereby reducing unit stresses and stress concentrations and permitting rapid adjustment to suddenly applied overloads without cracking. Where load and temperature conditions are comparatively constant, the wholly austenitic (type II) alloy HH provides the highest creep strength and permits use of maximum design stress. The stable austenitic alloy is also favored for cyclic temperature service that might induce sigma phase formation in the partially ferritic type.

Mass Characteristics

Density. 7.72 Mg/m^3 (0.279 $lb/in.^3$)

Thermal Properties

Melting point. Approximately 1370 °C (2500 °F)

Coefficient of thermal expansion. Linear:

Temperature °C	°F	Mean coefficient µm/m·K	µin./in.·°F
21– 538	70–1000	17.1	9.5
21– 650	70–1200	17.5	9.7
21– 760	70–1400	17.8	9.9
21– 870	70–1600	18.4	10.2
21– 980	70–1800	18.9	10.5
21–1090	70–2000	19.3	10.7
650– 870	1200–1600	20.5	11.4
650– 980	1200–1800	21.1	11.7

Specific heat. 502 J/kg·K (0.12 Btu/lb·°F) at 21 °C (70 °F).

Thermal conductivity:

Temperature °C	°F	Conductivity W/m·K	Btu/ft·h·°F
100	212	14.2	8.2
316	600	17.5	10.1
538	1000	20.8	12.0
650	1200	22.5	13.0
760	1400	24.4	14.1
870	1600	26.5	15.3
980	1800	28.2(a)	16.3(a)
1090	2000	30.3(a)	17.5(a)

(a) Estimated.

Electrical Properties

Electrical resistivity. 750 to 850 nΩ·m at 21 °C (70 °F)

Magnetic Properties

Magnetic permeability. 1.0 to 1.9.

Chemical Properties

Resistance to specific corroding agents. Alloy HH has been employed successfully in the following corrosive environments; air, ammonia, carburizing gas, combustion gases, flue gases oxidizing and reducing, high sulfur gases, molten cyanide, steam and tar. Fair corrosion resistance to tempering, neutral, cyaniding and high speed salts. Good resistance to molten lead. See Fig. 15.

Fig. 14 Creep-rupture properties of alloy HH

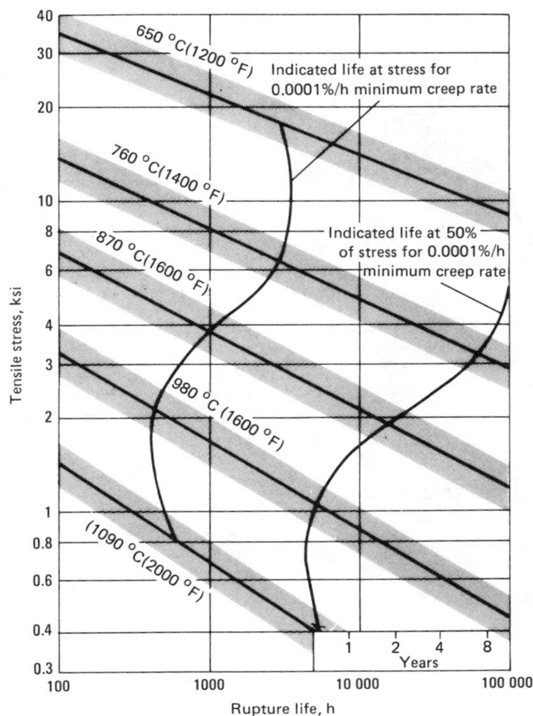

The scatter bands shown are set arbitrarily at ±20% of the stress for the central tendency line. Such a range usually embraces test data for similar alloy compositions, but should not be considered statistically significant confidence limits. Scatter of values may be much wider, particularly at the longer times and higher temperatures.

Fabrication Characteristics

Heat treatment. Castings are used in the as-cast condition. The alloy cannot be hardened by heat treatment. For alloys of medium carbon content (about 0.30%), in applications involving thermal fatigue from rapid heating and cooling, improved performance sometimes may be obtained by heating castings at 1040 °C (1900 °F) for 12 h followed by furnace cooling prior to placing in service

Machinability. Most machining operations can be performed satisfactorily on alloy HH castings. It is important in all cases that the tool be kept continuously entering the metal in order to avoid work-hardening the surface from rubbing or scraping. Slow feeds, deep cuts and powerful, rigid machines are necessary for best results. Work should be firmly mounted and supported, and tool mountings should provide maximum stiffness. Both high speed steel and carbide tools may be used successfully

Good lubrication and cooling are essential. The low thermal conductivity of the alloy makes it most important to have the cutting fluid flood both the tool and the work. Sulfo-chlorinated petroleum oil containing active sulfur and about 8 to 10% fatty oil is recommended for high speed steel tools. Water-soluble cutting fluids are primarily coolants and are most useful for high speed operation with carbide tools

Weldability. HH castings can be welded by shielded metal-arc, inert-gas, and oxyacetylene gas methods. Shielded metal-arc welding is preferred in high temperature applications. Lime-coated electrodes of similar composition (AWS E309-15, special high carbon) generally are used in shielded metal-arc welding. Bare wire from 0.9 to 2.4 mm (0.035 to 0.094 in.) diameter (AWS ER 309, special high carbon) is used for filler metal in the gas metal arc (GMAW) and gas tungsten-arc (GTAW) processes

Fig. 15 Results of 100-h corrosion tests of alloy HH in gases

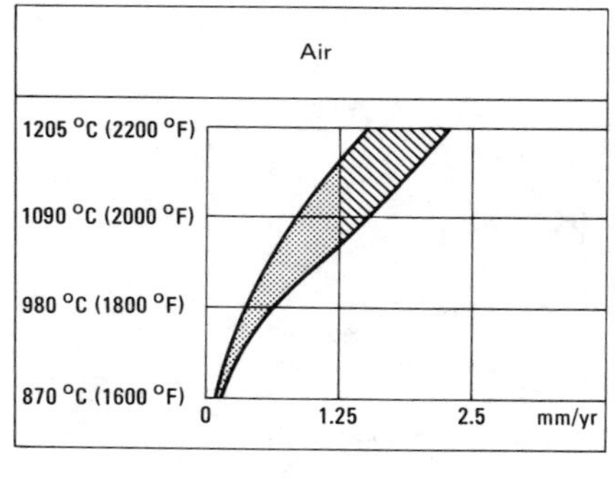

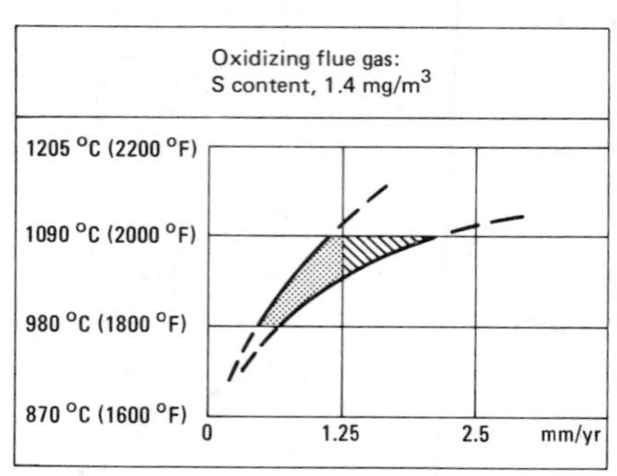

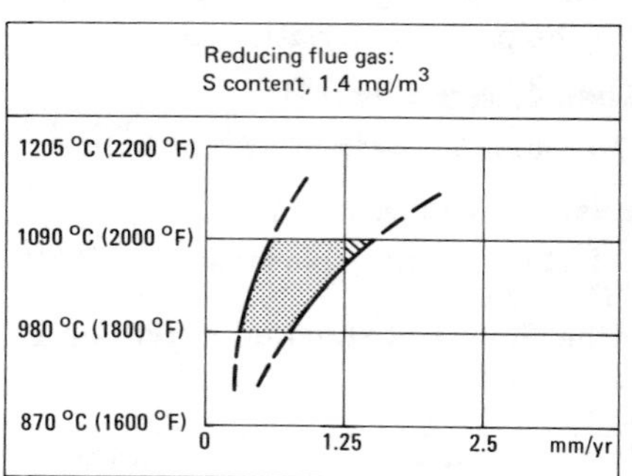

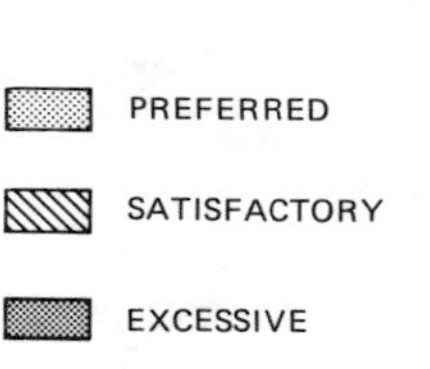

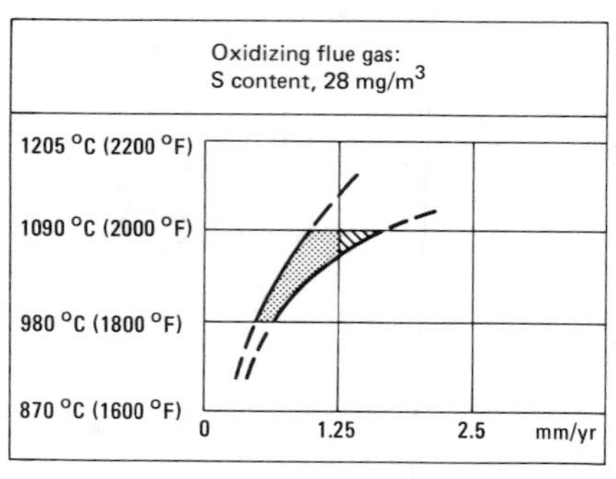

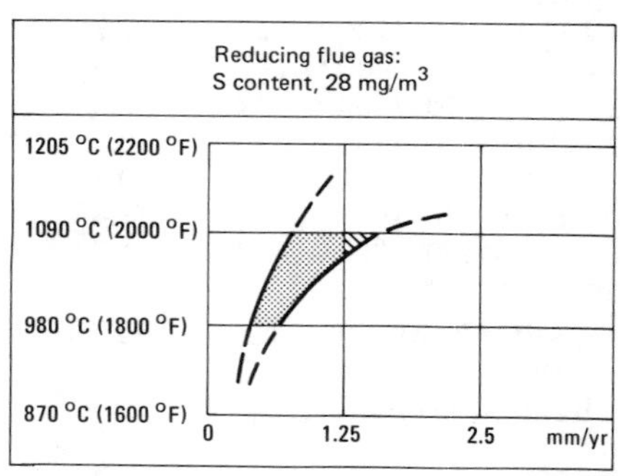

HI
28Cr-15Ni

Commercial Name

UNS number. J94003

Specifications

ASTM. A297 (HI)

Chemical Composition

Composition limits. 0.2 to 0.5 C; 2.0 max Mn; 2.0 max Si; 0.04 max P; 0.04 max S; 0.5 max Mo(a); 26 to 30 Cr; 14 to 18 Ni; rem Fe

(a) Molybdenum not intentionally added.

Applications

Typical uses. Used extensively for retorts operating with an internal vacuum at continuous temperature of 1177 °C (2150 °F). Castings used as billet skids, brazing fixtures, conveyor rollers, furnace rails, hearth plates, lead pots, pier caps, and tube spacers

Mechanical Properties

Tensile properties. Representative values at room temperature. As cast: tensile strength, 552 MPa (80 ksi); 0.2% yield strength, 310 MPa (45 ksi); elongation, 12% in 50 mm or 2 in. Aged 24 h at 760 °C (1400 °F), furnace cooled: tensile strength, 621 MPa (90 ksi); 0.2% yield strength, 448 MPa (65 ksi); elongation, 6% in 50 mm or 2 in. Minimum for A297: tensile strength, 483 MPa (70 ksi); 0.2% yield strength, 241 MPa (35 ksi); elongation, 10% in 50 mm or 2 in.

Hardness. As cast: 180 HB. Aged 24 h at 760 °C (1400 °F), furnace cooled: 200 HB

Elastic modulus. 186 GPa (27×10^6 psi)

Table 6 Creep-rupture data(a) for alloy HI

Temperature		Limiting stress for 0.0001%/h creep rate		Stress to rupture in: 100 h		1000 h	
°C	°F	MPa	ksi	MPa	ksi	MPa	ksi
760	1400	45.5	6.6	89.6	13	58.6	8.5
871	1600	24.8	3.6	51.7	7.5	33.0	4.8
982	1800	13.1	1.9	28.3	4.1	17.9	2.6
1093	2000	5.5	0.8	13.1	1.9	8.62	1.25
1177	2150	1.03	0.15	...	...	...	...

(a) Representative long-time values for constant temperature; for cyclic temperature lower values would apply.

Creep-rupture characteristics. See Table 6.

Structure

Microstructure. Austenitic containing carbides and, depending on the composition balance, small amounts of ferrite. Aging at 760 to 871 °C (1400 to 1600 °F) is accompanied by precipitation of finely dispersed carbides which tend at room temperature to increase the mechanical strength and decrease the ductility. At service temperature about 1093 °C (2000 °F) however, such carbides remain in solution; room temperature ductility is not impaired

Fig. 16 Results of 100-h corrosion test of alloy HI in gases

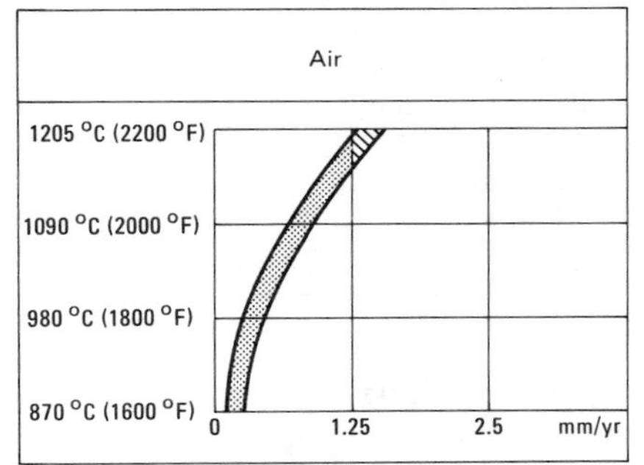

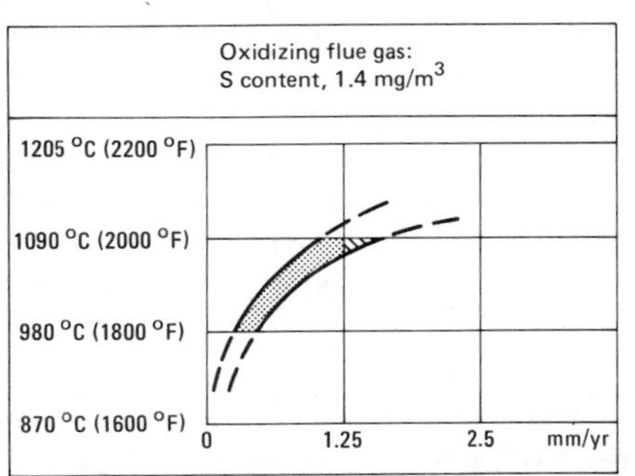

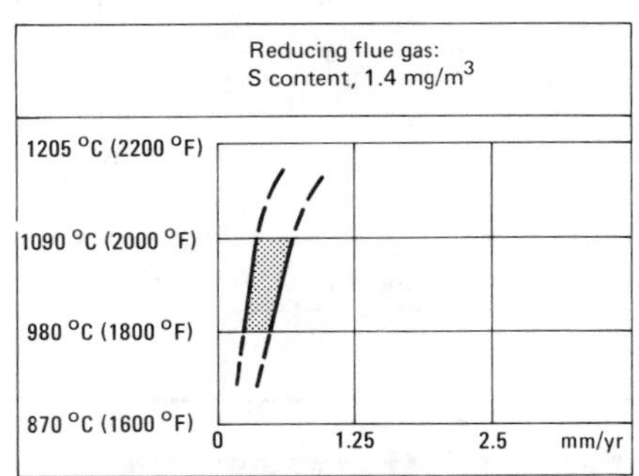

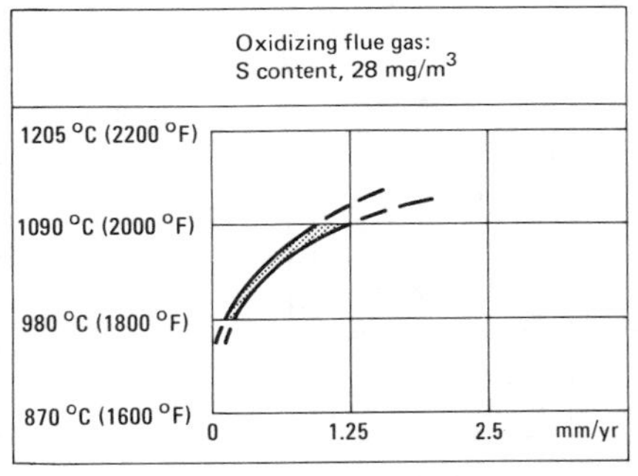

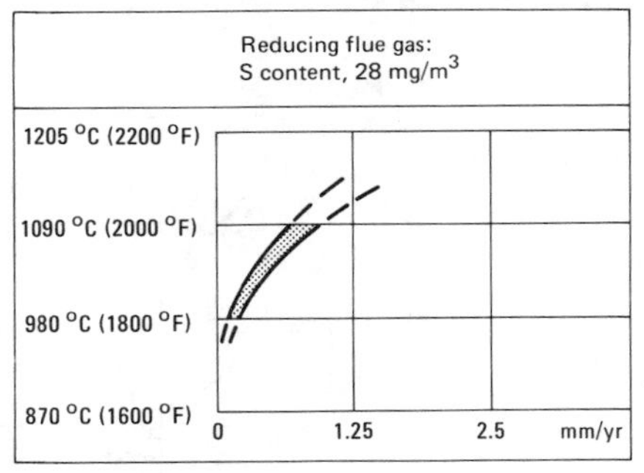

Mass Characteristics

Density. 7.72 Mg/m^3 (0.279 lb/in.^{3}) at 20 °C (68 °F)
Patternmaker's shrinkage. 5/16 in./ft

Thermal Properties

Melting point. 1232 °C (2250 °F)
Coefficient of thermal expansion. Linear:

Temperature °C	°F	Mean coefficient µm/m·K	µin./in.·°F
21–538	70–1000	17.8	9.9
21–650	70–1200	18.0	10.0
21–760	70–1400	18.2	10.1
21–870	70–1600	18.5	10.3
21–980	70–1800	18.9	10.5
21–1090	70–2000	19.4	10.8
650–870	1200–1600	19.8	11.0
650–980	1200–1800	21.6	12.0

Specific heat. 502 J/kg·K (0.12 Btu/lb·°F) at 20 °C (68 °F)
Thermal conductivity:

Temperature °C	°F	Conductivity W/m·K	Btu/ft·h·°F
100	212	14.2	8.2
316	600	17.5	10.1
538	1000	20.8	12.0
650	1200	22.5	13.0
760	1400	24.4	14.1
870	1600	26.5	15.3
980	1800	28.2	16.3
1090	2000	29.8	17.5

Electrical Properties

Electrical resistivity. 850 nΩ·m at 21 °C (70 °F)

Magnetic Properties

Magnetic permeability. 1.0 to 1.7

Chemical Properties

General corrosion behavior. More resistant to oxidation than similar type HH. Also resists air, reducing flue gases, and molten lead
Resistance to specific corroding agents. Fair resistance to tempering salts, good resistance to cyaniding salts and poor resistance to high speed salts. Good resistance to molten lead; poor resistance to molten zinc. See Fig. 16.

Fabrication Characteristics

Machinability. See alloy HE
Weldability. Shielded metal arc, inert-gas arc and oxyacetylene-gas methods. Shielded metal arc welding is more satisfactory for high-temperature applications. Type 310 bare electrodes should be used for gas welding and flame should be adjusted to be slightly rich in acetylene. Lime coated electrodes of similar composition (AWS E310-15 HC) are recommended for arc welding. No preweld or postweld treatment is required

HK
26Cr-20Ni

Commercial Name

UNS number. J94224

Specifications

AMS. Sandcast: 5365; investment cast: 5366
ASTM. Cast: HK, A297; HK-30 and HK-40, A608 and A351; HK-40 and HK-50, A567
SAE. 70310

Chemical Composition

Composition limits. (a) 0.2 to 0.6 C;

Table 7 Creep-rupture values(a) for alloy HK (grade HK-40)

Temperature, °C	Minimum creep rate of 0.0001%, MPa	ksi	1% total creep h in 100 000 h, MPa	ksi	Stress to rupture in: 100 h, MPa	ksi	1000 h, Mpa	ksi	10 000 h, MPa	ksi	100 000 h, MPa	ksi
760	70.3	10.2	43.4	6.3(b)	108	15.6	82.7	12.0	60.7	8.8	42.7(b)	6.2(b)
871	41.2	6.0	17.2	2.5(b)	63.4	9.2	41.4	6.0	26.2	3.8	17.2(b)	2.5(b)
982	17.2	2.5	6.20	0.90(b)	32.8	4.75	19.3	2.8	11.7	1.7	6.89(b)	1.0(b)
1038	9.66	1.4	2.69	0.39(b)	22.1	3.2	13.1	1.9	7.58(b)	1.1(b)	4.55(b)	0.66(b)
1093	4.48	0.65	1.57	0.23(b)	15.2	2.2	8.62	1.25	4.96(b)	0.72(b)	2.90(b)	0.42(b)

(a) Representative long-time values for constant temperature; for cyclic temperature lower values would apply. (b) Extrapolated values.

Fig 17 Effect of temperature on short-time tensile properties of alloy HK-40

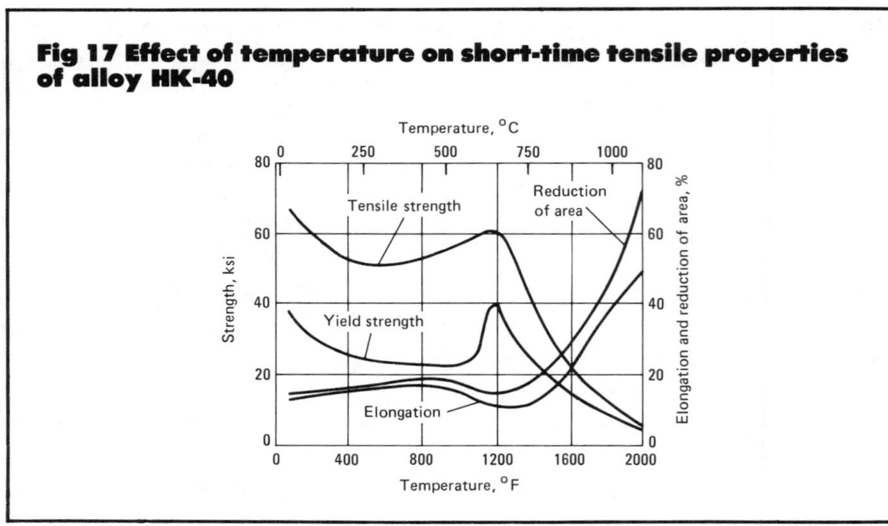

Fig. 18 Effect of temperature on hardness of alloy HK-40

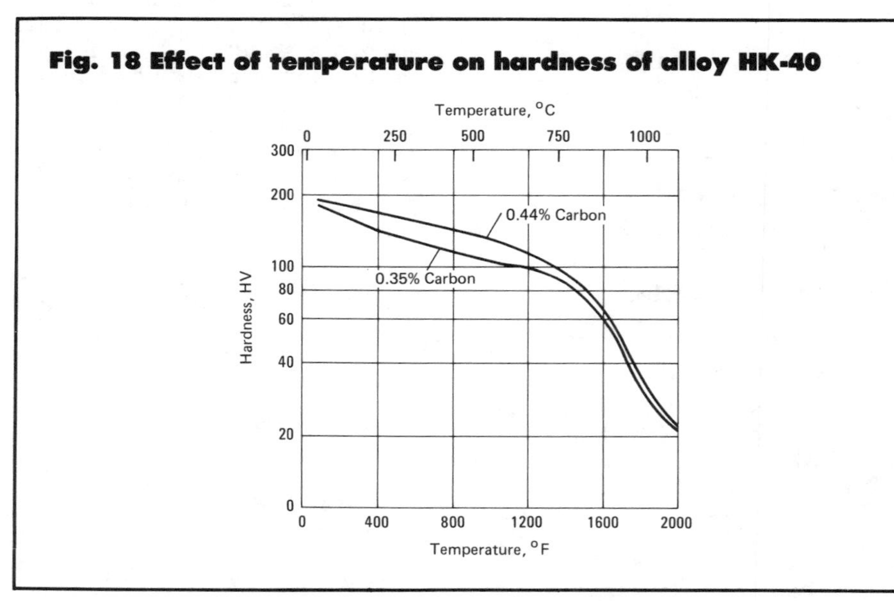

2.0 max Mn; 2.0 max Si; 0.04 max P; 0.04 max S; 0.5 max Mo(b); 24 to 28 Cr; 18 to 22 Ni; rem Fe

(a) HK-30, HK-40 and HK-50 are three grades in which the number indicates the midpoint of a ±0.05 C range. They also have narrower carbon and lower composition ranges. (b) Molybdenum not intentionally added.

Applications

Typical uses. Because of its high temperature strength, HK is widely used for stressed parts in structural applications up to 1150 °C (2100 °F). Castings used for billet skids, brazing fixtures, calcining tubes, cement kiln nose segments, conveyor rolls, furnace door arches and lintels, heat treating trays and fixtures, pier caps, rabble arms and blades, radiant tubes, reformer tubes, retorts, rotating shafts, skid rails, sprockets, stack dampers

Mechanical Properties

Tensile properties. Representative values at room temperature. As cast: tensile strength, 517 MPa (75 ksi); 0.2% yield strength, 345 MPa (50 ksi); elongation, 17% in 50 mm or 2 in. Aged 24 h at 760 °C (1400 °F), air cooled: tensile strength, 586 MPa (85 ksi); 0.2% yield strength, 345 MPa (50 ksi); elongation, 10% in 50 mm or 2 in. Minimum for ASTM A297: tensile strength, 448 MPa (65 ksi); 0.2% yield strength, 241 MPa (35 ksi); elongation, 10% in 50 mm or 2 in. See also Fig. 17.

Hardness. As cast: 170 HB. Aged 24 h at 760 °C (1400 °F), air cooled: 190 HB. See also Fig. 18.

Poisson's ratio. 0.30; see also Fig. 19.

Elastic modulus. In tension at 21 °C (70 °F): equiaxed grains, 186 GPa (27 × 10^6 psi); columnar grains, 138 GPa (20 × 10^6 psi), see also Fig. 19 for variation with temperature

Impact strength. Charpy keyhole, 29 J (22 ft·lb)

Creep-rupture characteristics. See Table 7 and Fig. 20.

Structure

Microstructure. Alloy HK is a stable austenite over its applied temperature range. As cast, it contains massive carbides present as scattered

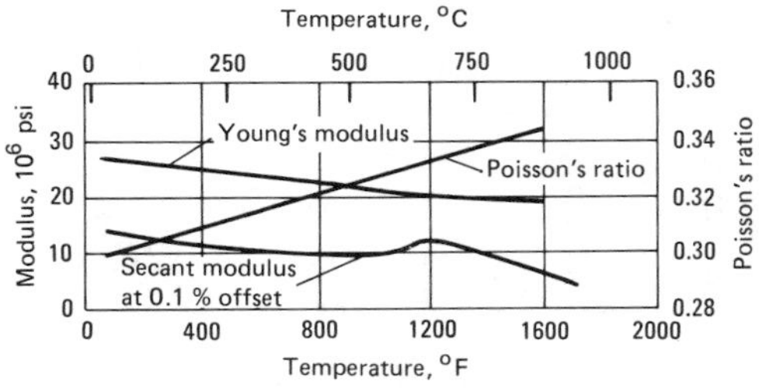

Fig. 19 Effect of temperature on elastic moduli and Poisson's ratio of equiaxed grain alloy HK-40

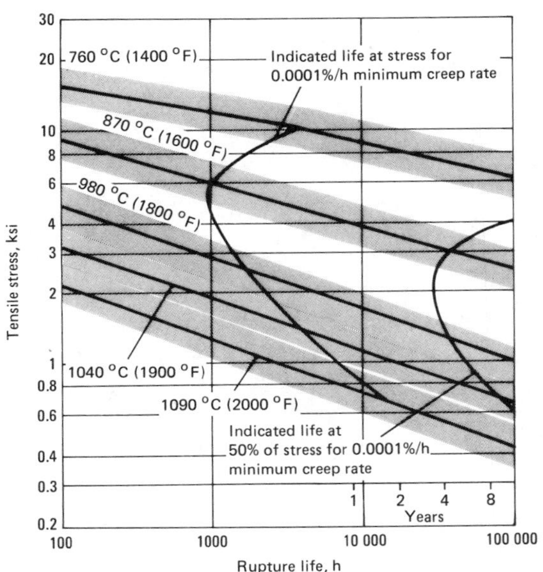

Fig. 20 Creep-rupture properties of alloy HK-40

The scatter bands shown are set arbitrarily at ±20% of the stress for the central tendency line. Such a range usually embraces test data for similar alloy compositions, but should not be considered statistically significant confidence limits. Scatter of values may be much wider, particularly at the longer times and higher temperatures.

islands or networks. After aging at service temperature, fine granular carbides precipitate, giving the alloy high creep strength. Unbalanced compositions are possible, resulting in ferrite which can transform sigma if the austenite is held at 815 °C (1500 °F) for more than a short time

Mass Characteristics

Density. 7.75 Mg/m^3 (0.280 $lb/in.^3$) at 21 °C (70 °F)

Patternmaker's shrinkage. 5/16 in./ft

Thermal Properties

Melting point. 1230 °C (2550 °F)

Coefficient of thermal expansion. Linear:

Temperature °C	Temperature °F	Mean coefficient µm/m·K	Mean coefficient µin./in.·°F
21– 538	70–1000	16.9	9.4
21– 650	70–1200	17.3	9.6
21– 760	70–1400	17.6	9.8
21– 870	70–1600	18.0	10.0
21– 980	70–1800	18.4	10.2
650– 870	1200–1800	20.5	11.4
650– 980	1200–2000	20.7	11.5

Specific heat. 502 J/kg·K (0.12 Btu/lb·°F) at 21 °C (70 °F)

Thermal conductivity:

Temperature °C	Temperature °F	Conductivity W/m·K	Conductivity Btu/ft·h·°F
100	212	13.7	7.9
316	600	16.9	9.8
538	1000	20.4	11.8
650	1200	22.3	12.9
760	1400	24.6	14.2
870	1600	27.2	15.7
980	1800	29.6(a)	17.1(a)
1090	2000	32.2(a)	18.6(a)

(a) Estimated values.

Electrical Properties

Electrical resistivity. 900 nΩ·m at 21 °C (70 °F)

Magnetic Properties

Magnetic permeability. 1.02

Chemical Properties

General corrosion behavior. Alloy HK offers good resistance to corrosion by hot gases, including sulfur-bearing gases, in both oxidizing and reducing conditions (although HC, HE, and HI are more resistant in oxidizing gases). It is also used in air, ammonia, hydrogen, and molten neutral salts

Resistance to specific corroding agents. Good resistance to tempering, neutral, cyaniding and high speed salts. See Fig. 21.

Fabrication Characteristics

Machinability. See "Machinability" for alloy HE

Heat treatment. Castings are normally used in the as-cast condition. Alloy HK cannot be hardened by heat treatment

Weldability. Shielded metal arc, inert-gas arc and oxyacetylene-gas methods. Shielded metal arc welding is more satisfactory for high-temperature applications. Lime coated electrodes of similar composition (AWS E310-15 high carbon) are generally used for shielded metal-arc welding. Bare wire from 0.8 to 2.4 mm (0.035 to 0.094 in.) in diameter is used for filler metal in gas metal arc and gas tungsten arc processes. Preweld or postweld heat treating not required

Fig. 21 Results of 100-h corrosion tests of alloy HK-40 in gases

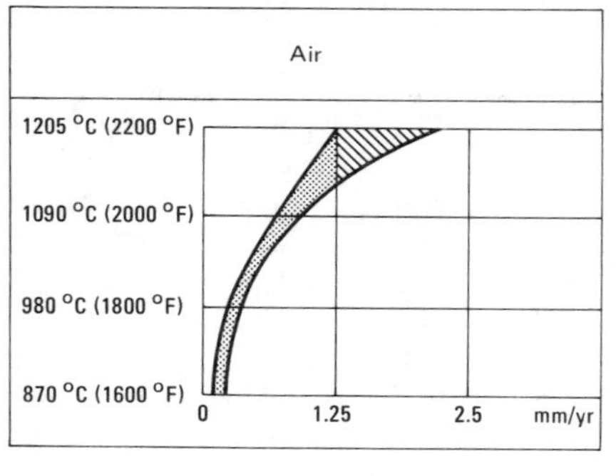

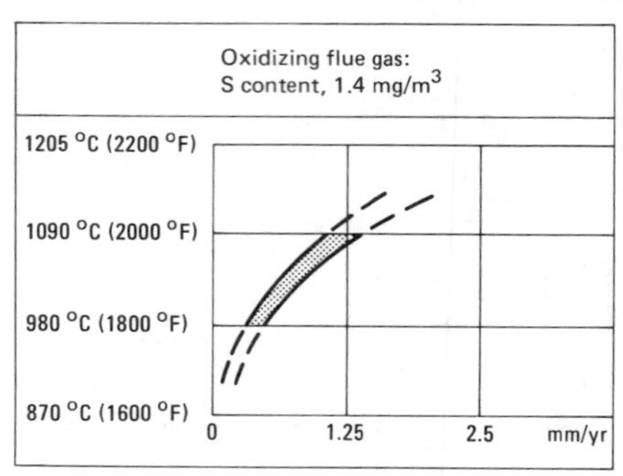

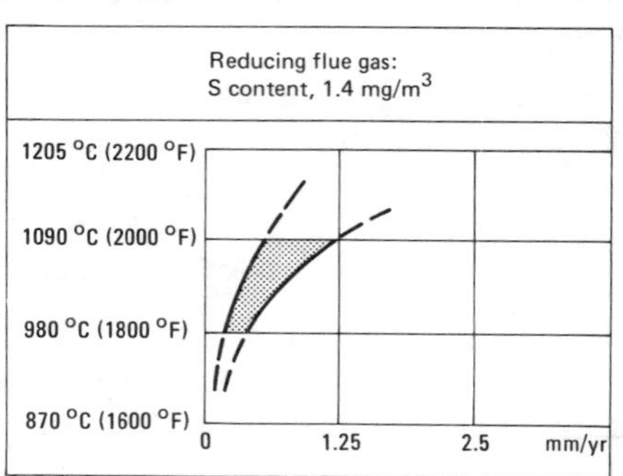

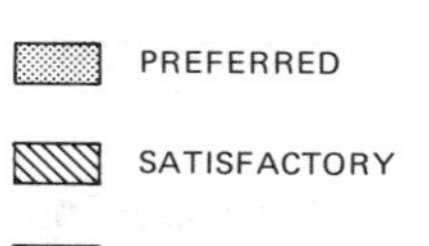

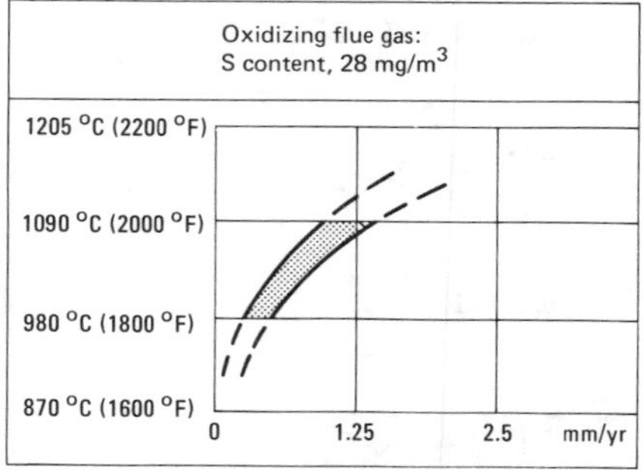

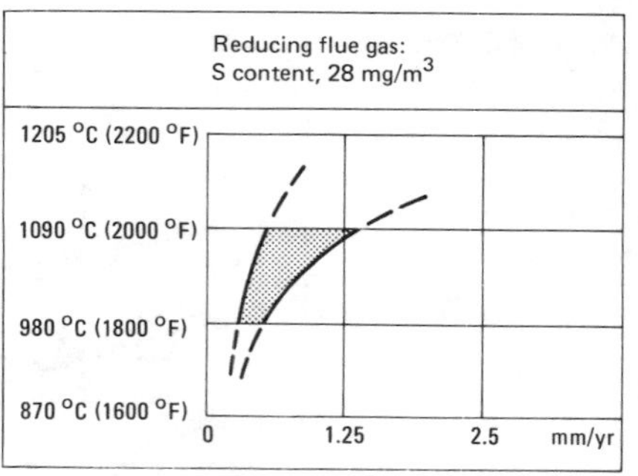

HL, HL-30
30Cr-20Ni

Commercial Name

UNS number. HL: J94604. HL-30: J94613

Specifications

ASTM. A297 (HL); A608 (HL-30)
SAE. Cast: 70310A

Chemical Composition

Composition limits. 0.2 to 0.6 C; 2.0 max Mn; 2.0 max Si; 0.04 max P; 0.04 max S; 0.5 max Mo(a); 28 to 32 Cr; 18 to 22 Ni; rem Fe. For HL-30, carbon content is 0.25 to 0.35; manganese content is 1.50 max. For HL-40, carbon content is 0.35 to 0.45; manganese content is 1.50 max

(a) Molybdenum not intentionally added.

Applications

Typical uses. Especially useful for severe service where excessive scaling must be avoided. Castings used for carrier fingers, enameling furnace fixtures, furnace skids for slab and bars, radiant tubes, stack dampers. HL-30 and HL-40 are used for pressure pipe applications at high temperatures

Table 8 Creep-stress rupture of alloy HL(a)

Temperature °C	Temperature °F	Limiting stress for 0.0001%/h creep rate MPa	Limiting stress for 0.0001%/h creep rate ksi	Stress to rupture in 100 h MPa	Stress to rupture in 100 h ksi
760	1400	48	7.0	103	15
871	1600	30	4.3	63	9.2
982	1800	15	2.2	36	5.2

(a) Representative values, long time for constant temperatures, for cyclic temperature lower values would apply.

Mechanical Properties

Tensile properties. Room temperature. As cast: 565 MPa (82 ksi); 0.2% yield strength, 359 MPa (52 ksi); elongation, 19% in 50 mm or 2 in. Minimum for ASTM A297: tensile strength, 448 MPa (65 ksi); 0.2% yield strength, 241 MPa (35 ksi); elongation, 10% in 50 mm or 2 in.
Hardness. As cast: 192 HB
Elastic modulus. Tension, 200 GPa (29×10^6 psi)
Creep-rupture characteristics. See Table 8.

Structure

Microstructure. Similar to type HK both as cast and aged

Mass Characteristics

Density. 7.72 Mg/m³ (0.279 lb/in.³)

Patternmaker's shrinkage. 5/16 in./ft

Thermal Properties

Melting point. Approx 1427 °C (2600 °F)
Coefficient of thermal expansion. Linear:

Temperature °C	Temperature °F	Mean coefficient μm/m·K	Mean coefficient μin./in.·°F
21– 538	70–1000	16.6	9.2
21– 650	70–1200	16.9	9.4
21– 760	70–1400	17.3	9.6
21– 870	70–1600	17.5	9.7
21– 980	70–1800	17.8	9.9
21–1090	70–2000	18.2	10.1
650– 870	1200–1600	18.9	10.5
650– 980	1200–1800	19.3	10.7

Fig. 22 Results of 100-h corrosion tests of alloy HL in gases

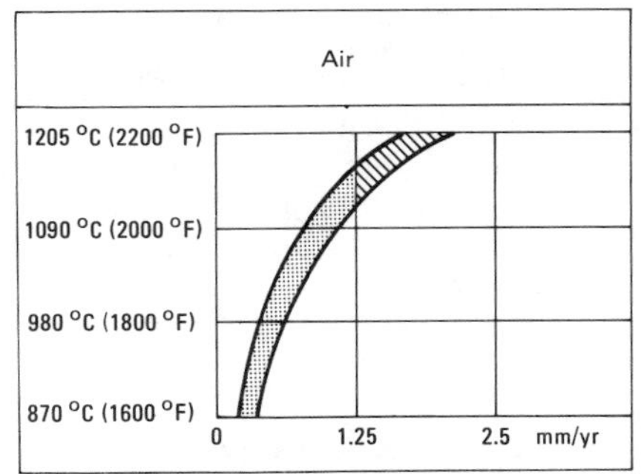

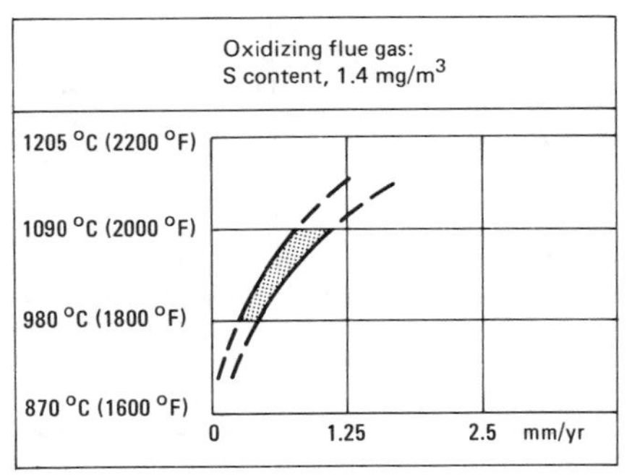

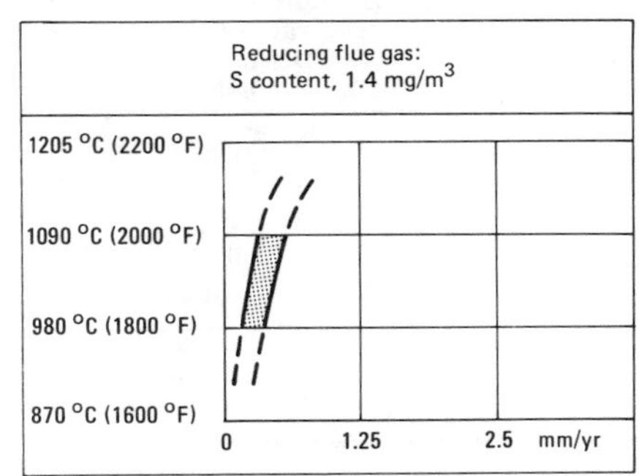

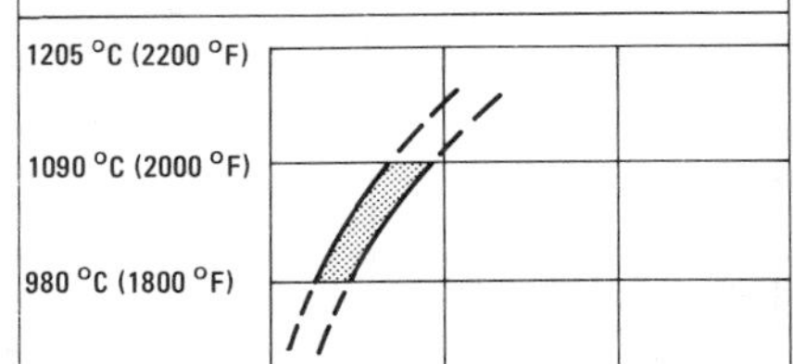

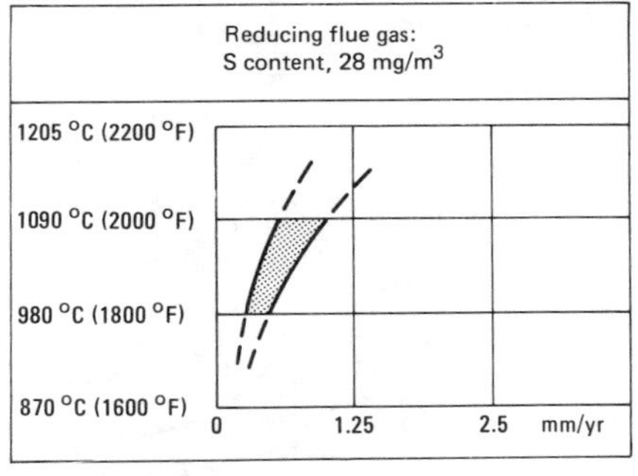

Specific heat. 502 J/kg·K (0.12 Btu/lb·°F) at 20 °C (68 °F)
Thermal conductivity:

Temperature °C	°F	Conductivity W/m·K	Btu/ft·h·°F
100	212	14.2	8.2
316	600	17.7	10.2
538	1000	21.1	12.2
650	1200	23.2	13.4
760	1400	25.4	14.7
870	1600	28.2	16.3
980	1800	30.6	17.7
1090	2000	33.4	19.3

Electrical Properties

Electrical resistivity. 940 nΩ·m at 21 °C (70 °F)

Magnetic Properties

Magnetic permeability. 1.01

Chemical Properties

General corrosion behavior. Especially useful for severe service where excessive scaling must be avoided. Resists corrosion by hot gases, particularly those with high sulfur levels, better than the similar type HK. Also resists corrosion by air

Resistance to specific corroding agents. See Fig. 22.

Fabrication Characteristics

Machinability. See alloy HE
Weldability. Shielded metal arc, inert-gas arc and oxyacetylene-gas methods. Shielded metal arc welding is more satisfactory for high-temperature applications. Bare wire for 0.8 to 1.6 mm (0.035 to 0.062 in.) in diameter (AWS ER 310 HC) is used for filler metal in the gas metal-arc and gas tungsten arc welding processes. Lime coated electrodes of similar composition (AWS E310-15HC) are generally used for shielded metal arc welding. Preweld or postweld heat treatment not required.

HN
25Ni-20Cr

Commercial Name

UNS number. J94213

Specifications

ASTM. A297 (HN)

Chemical Composition

Composition limits. 0.2 to 0.5 C; 2.0 max Mn; 2.0 max Si; 0.04 max P; 0.04 max S; 0.5 max Mo(a); 19 to 23 Cr; 23 to 27 Ni; rem Fe

(a) Molybdenum not intentionally added.

Applications

Typical uses. Brazing fixtures, chain, furnace beams and parts, pier caps, radiant tubes and tube supports, sill plate brackets, torch nozzles, trays, tubes

Mechanical Properties

Tensile properties. Representative values, as cast: tensile strength, 469 MPa (68 ksi); 0.2% yield strength, 262 MPa (38 ksi); elongation, 13% in 50 mm or 2 in. See also Fig. 23.
Hardness. As cast: 160 HB. See also Fig. 24.
Elastic modulus. Tension, 186 GPa (27×10^6 psi). See also Fig. 25.
Creep-rupture characteristics. See Table 9 and Fig. 26.

Structure

Microstructure. Austenitic at all temperatures, well within the stable austenite field. In the as-cast condition carbide areas are present and

Table 9 Representative long-time(a) creep-rupture values for alloy HN

Temperature, °C	Limiting stress for: Minimum creep rate of 0.0001%/h, MPa	ksi	1% total creep in 100 000 h, MPa	ksi	Stress to rupture in: 100 h, MPa	ksi	1000 h, MPa	ksi	10 000 h, MPa	ksi	100 000 h, MPa	ksi
871	43	6.3	21	3.0(b)	76	11.0	51	7.4	33	4.8	22	3.2
982	17	2.4	8	1.1(b)	39	5.6	23	3.4	14	2.1	9.0	1.3
1038	11	1.6	3	0.45(b)	32	4.6	14	2.1	6.6	0.96(b)	3.0	0.44
1093	7	1.04	1.2	0.17(b)	20	2.9	9	1.25	3.6	0.52(b)	1.5	0.22

(a) For constant temperature; for cyclic temperature lower values would apply. (b) Extrapolated values.

Fig. 23 Effect of temperature on short-time tensile properties of alloy HN

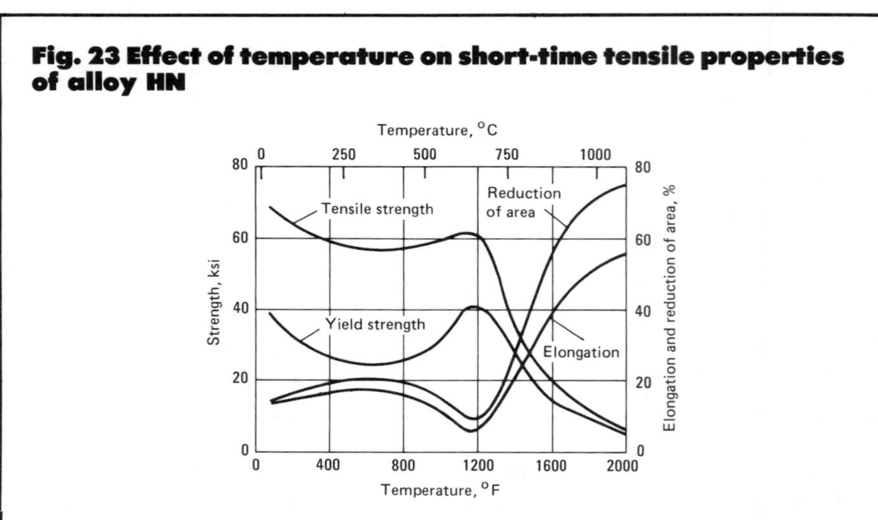

Fig. 24 Effect of temperature on hardness of alloy HN

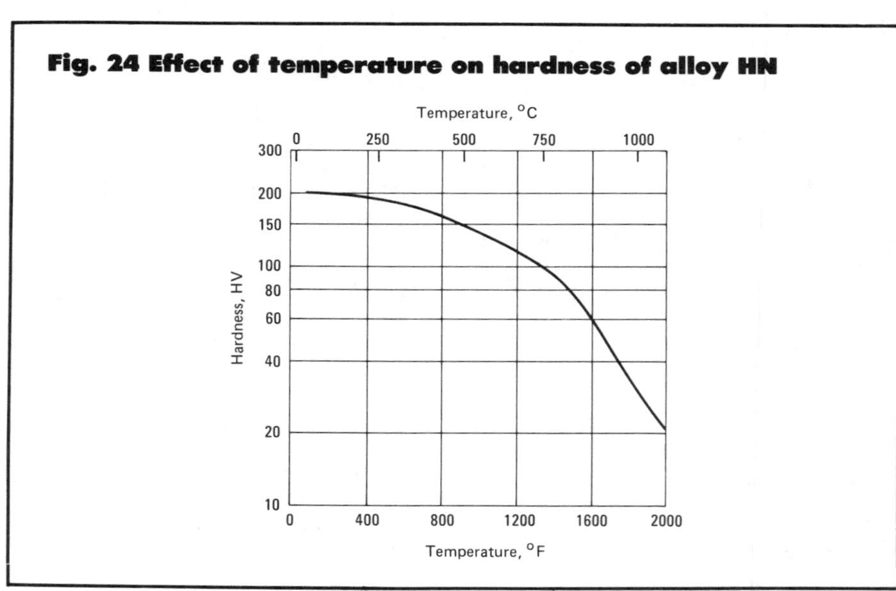

additional fine carbides precipitate on aging. The alloy is not susceptible to sigma phase formation; nor is increased carbon content especially detrimental to ductility

Mass Characteristics

Density. 7.83 Mg/m^3 (0.283 lb/in.^{3}) at 20 °C (68 °F)

Patternmaker's shrinkage. 5/16 in./ft

Thermal Properties

Melting point. Approx 1370 °C (2500 °F)

Coefficient of thermal expansion. Linear:

Temperature °C	°F	Mean coefficient µm/m·K	µin./in.·°F
21– 538	70–1000	16.7	9.3
21– 650	70–1200	17.1	9.5
21– 760	70–1400	17.2	9.7
21– 870	70–1600	17.8	9.9
21– 980	70–1800	18.1	10.1
21–1090	70–2000	18.3	10.2
650– 980	1200–1800	19.8	11.0

Specific heat. 461 J/kg·K (0.11 Btu/lb·°F) at 21 °C (70 °F)

Thermal conductivity:

Temperature °C	°F	Conductivity W/m·K	Btu/ft·h·°F
100	212	13	7.5
316	600	16	9.2
538	1000	19	11.0
650	1200	21	12.1
760	1400	23	13.2
870	1600	25	14.5
980	1800	27(a)	15.7(a)
1090	2000	29(a)	17.0(a)

(a) Estimated values.

Electrical Properties

Electrical resistivity. 991 nΩ·m at 21 °C (70 °F)

Magnetic Properties

Magnetic permeability. 1.10

Chemical Properties

General corrosion behavior. Re-

Fig. 25 Effect of temperature on approximate modulus of elasticity of alloy HN

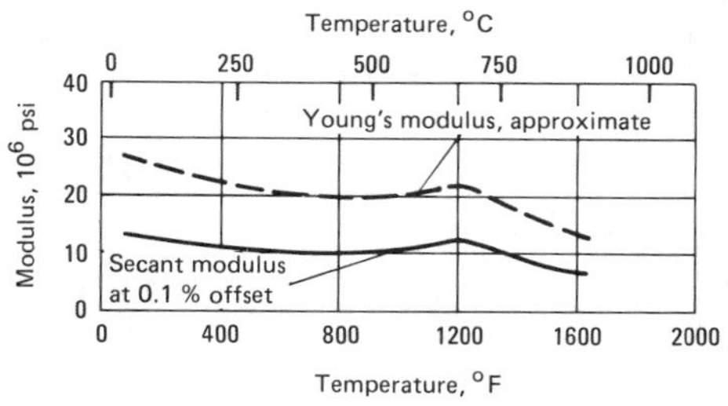

Fig. 26 Creep-rupture properties of alloy HN

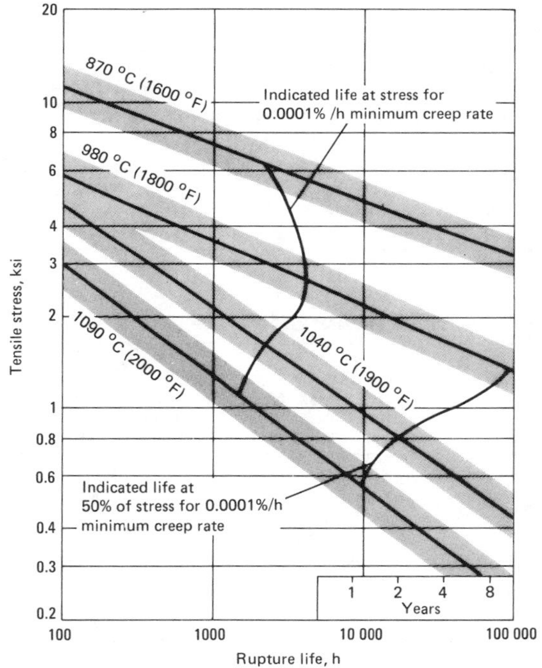

The scatter bands shown are set arbitrarily at ±20% of the stress for the central tendency line. Such a range usually embraces test data for similar alloy compositions, but should not be considered statistically significant confidence limits. Scatter of values may be much wider, particularly at the longer times and higher temperatures.

sistant to air, flue gases oxidizing and reducing.
Resistant to specific corroding agents. See Fig. 27.

Fabrication Characteristics

Machinability. See alloy HE
Weldability. Shielded metal arc, inert-gas and oxyacetylene-gas methods. Electric arc processes are most widely used; oxyacetylene flame also produces satisfactory welds. Lime coated electrodes of AWS E-330-15 with special high carbon are generally used for shielded metal arc welding. Bare wire 0.8 to 1.6 mm (0.035 to 0.063 in.) in diameter (AWS Er 330 high carbon), is used for filler in gas metal arc, gas tungsten-arc and oxyacetylene gas processes. Flame should be adjusted to be very rich in acetylene in this last process. Preweld or postweld heat treating not required

HP
35Ni-26Cr

Commercial Name

UNS number. J95705

Specifications

ASTM. A297 (HP)

Chemical Composition

Composition limits. 0.035 to 0.75 C; 20 max Mn; 2.0 max Si; 0.04 max P; 0.04 max S; 0.5 max Mo(a); 24 to 28 Co; 33 to 37 Ni; rem Fe
(a) Molybdenum not intentionally added.

Applications

Typical uses. Castings used as ethylene pyrolysis heaters, heat treat fixtures, radiant tubes, refinery tubes

Mechanical Properties

Tensile properties. As cast: tensile strength, 490 MPa (71 ksi); 0.2% yield strength, 276 MPa (40 ksi); elongation, 11.5% in 50 mm or 2 in. See also Fig. 28.
Poisson's ratio. 0.3 at 21 °C (70 °F)
Elastic modulus. Tension, 186 GPa (27×10^6 psi). See also Fig. 29.
Creep-rupture characteristics. See Table 10 and Fig. 30.

Structure

Microstructure. Austenitic structure at all temperatures, thus not susceptible to embrittlement from sigma formation. Microstructure consists of massive primary carbides in an austenitic matrix together with fine carbides which are precipitated within the austenite grains after aging at elevated temperatures

Mass Characteristics

Density. 7.86 Mg/m³ (0.284 lb/in.³) at 21 °C (70 °F)
Patternmaker's shrinkage. 5/16 in./ft

Fig. 27 Results of 100-h corrosion tests of alloy HN in gases

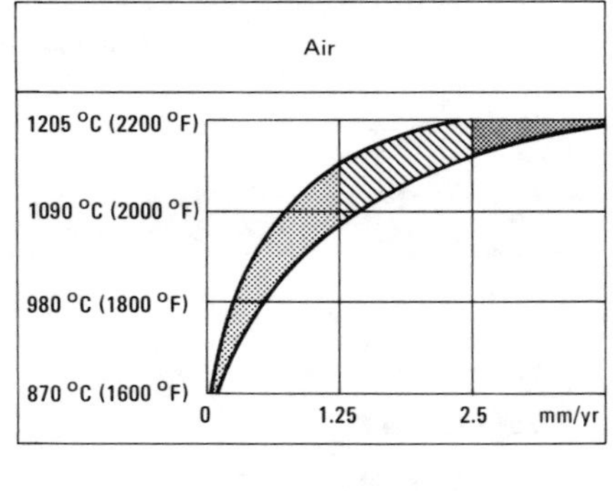

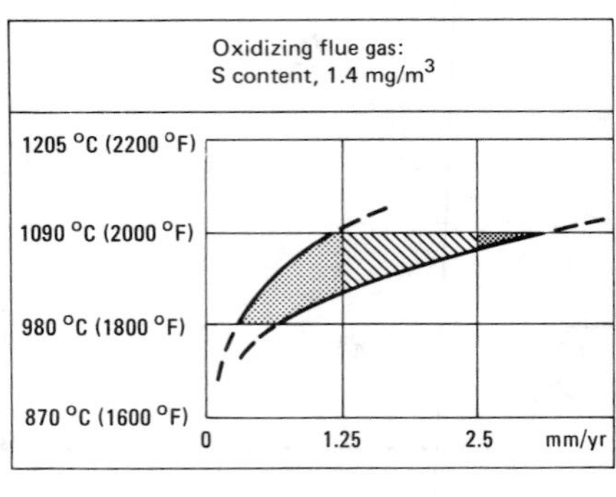

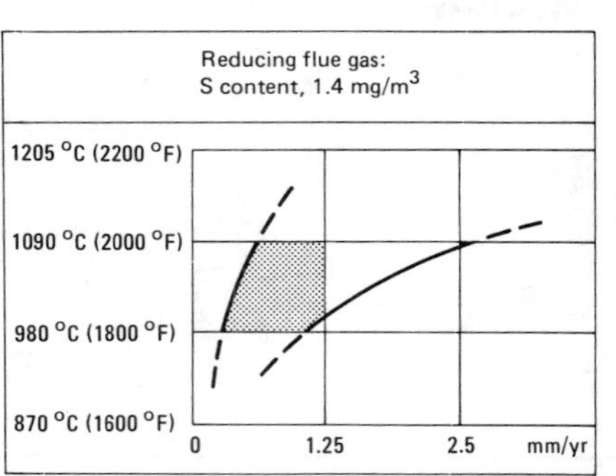

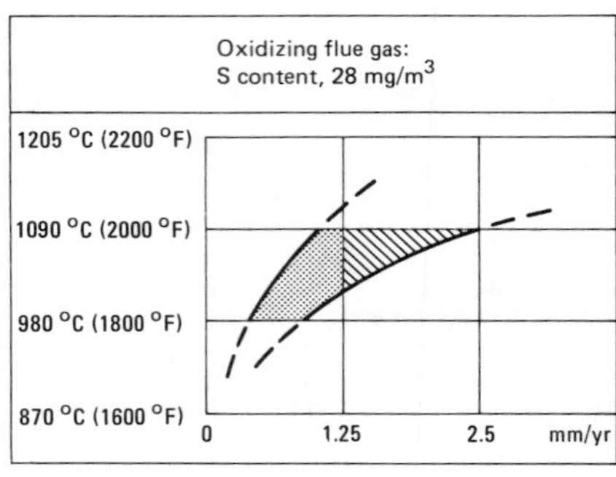

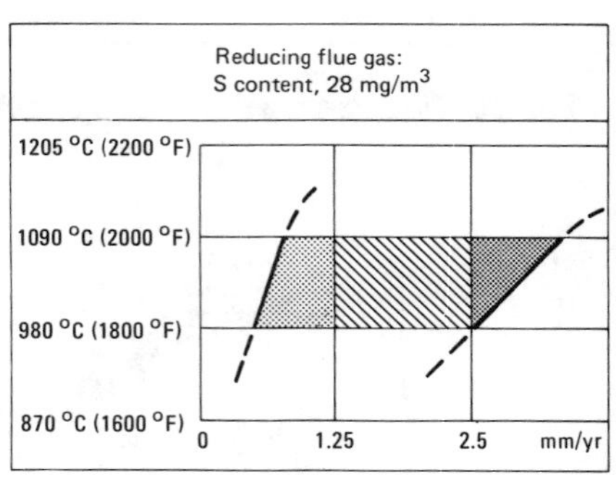

Table 10 Representative(a) long-time creep-rupture values for alloy HP

Temperature, °C	Limiting stress for: Minimum creep rate of 0.0001%/h, MPa	ksi	Limiting stress for: 1% total creep in 100 000 h, MPa	ksi	Stress to rupture in: 100 h, MPa	ksi	1000 h, MPa	ksi	10 000 h, MPa	ksi	100 000 h, MPa	ksi
871	40	5.8	34	4.9	69	10.0	52	7.5	35	5.1	23(b)	3.3(b)
982	19	1.8	14	2.1	41	5.9	25	3.6	15	2.2	7.6(b)	1.1(b)
1093	6.9	1.0	2.8	0.4	19	2.8	10	1.5	4.1(b)	0.6(b)	...	...

(a) For constant temperature; for cyclic temperature, lower values would apply. (b) Extrapolated values.

Thermal Properties

Melting point. Approx 1343 °C (2450 °F)

Solidus temperature. 1340 °C (2450 °F)

Coefficient of thermal expansion. Linear:

Temperature °C	°F	Mean coefficient µm·K	µin./in.·°F
21–358	70–1000	16.6(a)	9.2(a)
21–650	70–1200	17.1	9.5
21–760	70–1400	17.6(a)	9.8(a)
21–870	70–1600	18.0(a)	10.0(a)
21–980	70–1800	18.5(a)	10.3(a)
21–1090	70–2000	19.1(a)	10.6(a)
650–870	1200–1600	20.5(a)	11.4(a)
650–980	1200–1800	21.4(a)	11.9(a)
870–1090	1600–2000	23.6(a)	13.1(a)

(a) Estimated.

Specific heat. 461 J/kg·K (0.11 Btu/lb·°F) at 21 °C (70 °F)

Thermal conductivity:

Temperature °C	°F	Conductivity W/m·K	Btu/ft·h·°F
100	212	12.9	7.5
316	600	15.9(a)	9.2(a)
538	1000	19.0(a)	11.0(a)
650	1200	20.9(a)	12.1(a)
760	1400	22.8(a)	13.2(a)
870	1600	25.1(a)	14.5(a)
980	1800	27.2(a)	15.7(a)
1090	2000	29.4(a)	17.0(a)

(a) Estimated.

Electrical Properties

Electrical resistivity. Estimated value, 1020 nΩ·m at 21 °C (70 °F)

Magnetic Properties

Magnetic permeability. 1.02 to 1.25

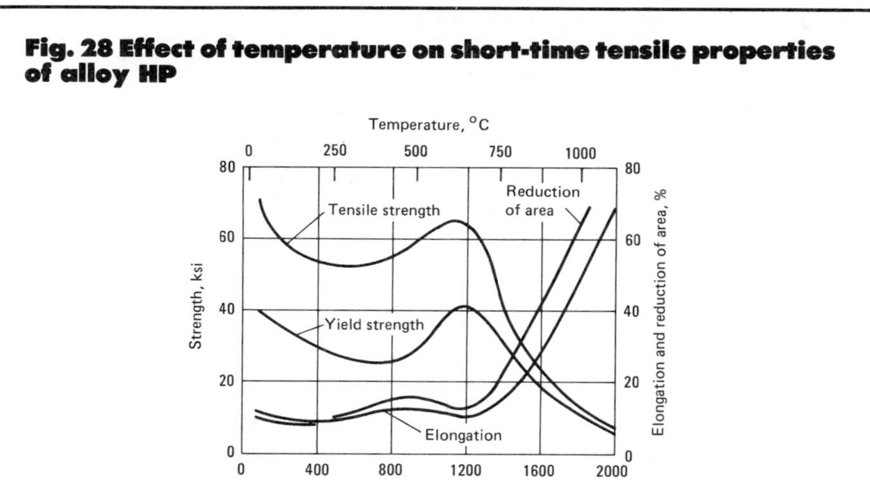

Fig. 28 Effect of temperature on short-time tensile properties of alloy HP

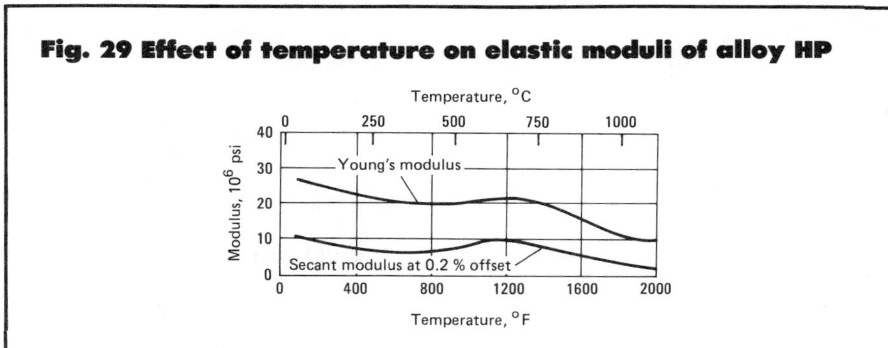

Fig. 29 Effect of temperature on elastic moduli of alloy HP

Chemical Properties

General corrosion behavior. Resistant to air, carburizing atmospheres and flue gases oxidizing and reducing

Resistant to specific corroding agents. See Fig. 31.

Fabrication Characteristics

Machinability. See alloy HE.

Weldability. Shielded metal arc, inert-gas arc and oxyacetylene-gas methods, with arc electric processes being most widely used. Lime coated electrodes (AWS E310-15 special high carbon) are used for shielded metal arc welding. Preweld or postweld heat treating not required

HP-50WZ
26Cr-35Ni

Chemical Composition

Composition limits. 0.45 to 0.55 C; 24 to 28.0 Cr; 33.0 to 37.0 Ni; 4.0 to 6.0 W; 0.1 to 1.0 Zr; 2.0 max Mn; 2.5 max Si; 0.035 max S; 0.035 max P; rem Fe

Applications

Typical uses. In high temperatures up to 1205 °C (2200 °F)

Mechanical Properties

Tensile properties. 269 MPa (39 ksi) at room temperature; 169 MPa (24.5 ksi) at 650 °C (1200 °F); 52 MPa (7.5 ksi) at 1075 °C (2000 °F)

Stress-rupture characteristics. 7 MPa (1.0 ksi) at 1205 °C (2200 °F) in 94.2 and 236 h. See also Fig. 32.

Creep-rupture characteristics. Limiting creep stress, 8.3 MPa (1.2 ksi) for 0.001%/h at 1075 °C (2000 °F); 2.1 MPa (0.3 ksi) for 0.001%/h at 1205 °C (2200 °F)

Chemical Properties

General corrosion behavior. Good resistance to oxidation and carburization at high temperatures

HT
35Ni-17Cr

Commercial Name

UNS number. HT: J94605; HT-30: J94603; HT-50: J94805

Specifications

ASTM. Cast: A297 (HT)
SAE. 70330

Chemical Composition

Composition limits. 0.35 to 0.75 C; 2.0 max Mn; 2.5 max Si; 0.04 max P; 0.04 max S; 0.5 max Mo(a); 15 to 19 Cr; 33 to 37 Ni; rem Fe. For HT-50, 0.40 to 0.60 C; 1.50 max Mn; 0.50 to 2.0 Si; 15 to 19 Cr; 33 to 37 Ni; 0.04 max P; 0.04 max S; 0.50 max Mo (a); rem Fe

(a) Molybdenum not intentionally added.

Applications

Typical uses. Widely used for general heat resistant applications in highly stressed parts. Air ducts, brazing trays, carburizing containers, chain, cyanide pots, dampers, dippers, door frames, enameling bars and supports, fan blades, feed screws, gear spacers, glass molds, glass rolls, hearth plates, heat treating fixtures and trays, idler drums, kiln nose rings, lead pots, malleablizing baskets, muffles, oil burner nozzles, point bars, radiant tubes, resistor guides, retorts, roller rails, rolling mill guides, salt pots, tube supports. HT-30 and HT-50 are used for pressure pipe applications at high temperatures.

Mechanical Properties

Tensile properties. Representative values at room temperature. As cast: tensile strength, 483 MPa (70 ksi); 0.2% yield strength, 276 MPa (40 ksi); elongation, 10% in 50 mm or 2 in. Aged 24 h at 760 °C (1400 °F), air cooled: tensile strength, 517 MPa (75 ksi); 0.2% yield strength, 310 MPa (45 ksi); elongation, 5% in 50 mm or 2 in. Minimum for ASTM A297: tensile strength, 448 MPa (65 ksi). See also Fig. 33.

Hardness. As cast: 180 HB. Aged 24 h at 760 °C (1400 °F), air cooled: 200 HB. See also Fig. 34.

Elastic modulus. 186 GPa (27×10^6 psi). See also Fig. 35.

Impact strength. Charpy keyhole, 5.4 J (4 ft·lb)

Creep-rupture characteristics. See Table 11 and Fig. 36.

Table 11 Representative(a) long-time creep-rupture values for alloy HT

Temperature, °C	Limiting stress for 0.0001%/h creep rate MPa	ksi	Stress to rupture (b) in: 100 h MPa	ksi	1000 h MPa	ksi	10 000 h MPa	ksi	100 000 h MPa	ksi
760	55	8.0	110	16	83	12	58	8.4	39	5.6(c)
871	31	4.5	61	8.9	40	5.8	26	3.7	17	2.4(c)
982	14	2.0	30	4.4	19	2.7	12	1.7	7.2	1.05(c)
1093	3	0.5	14	2.1	8.9	1.3	...	...	...	...
1177	1	0.15	...	...	...	...	...	...	...	...

(a) For constant temperature; for cyclic temperature lower values would apply. (b) Type HT-45 (0.45 to 0.50 C, 1.0 to 1.6 Si. (c) Extrapolated values.

Fig. 30 Creep-rupture properties of alloy HP

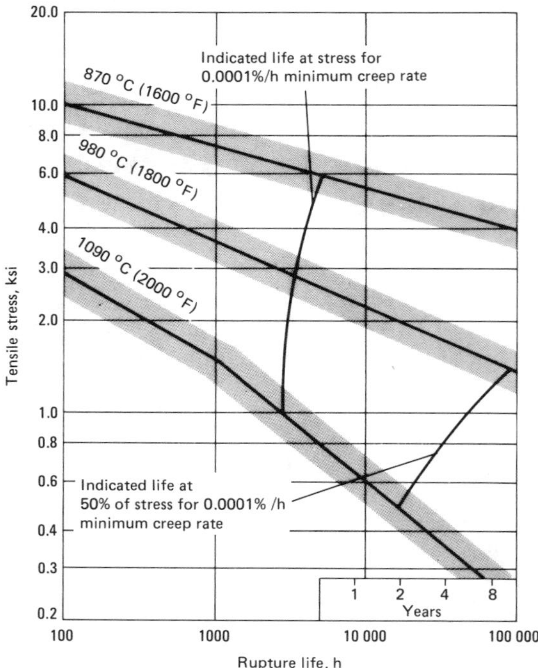

The scatter bands shown are set arbitrarily at ±20% of the stress for the central tendency line. Such a range usually embraces test data for similar alloy compositions, but should not be considered statistically significant confidence limits. Scatter of values may be much wider, particularly at the longer times and higher temperatures.

Structure

Microstructure. Austenite containing various amounts of carbides depending on the carbon content and thermal history. In the as-cast condition large carbide areas are present at the grain boundaries, but fine carbides precipitate within the grains after exposure at service temperatures with subsequent decrease in room temperature ductility. Increased carbon content does not significantly affect the high temperature ductility of the alloy; this characteristic makes it especially useful for carburizing fixtures or containers. Additional protection against carburization is obtained with silicon contents above about 1.6%, but at some sacrifice of hot strength

Mass Characteristics

Density. 7.92 Mg/m³ (0.286 lb/in.³)
Patternmaker's shrinkage. 5/16 in./ft

Thermal Properties

Melting point. Approx 1340 °C (2450 °F)
Coefficient of thermal expansion. Linear:

Temperature °C	°F	Mean coefficient µm/m·K	µin./in.·°F
20– 93	68–200	14.2	7.9
20– 204	68–400	14.6	8.1
20– 316	68–600	15.1	8.4
20– 427	68–800	15.2	8.6
20– 358	68–1000	16.0	8.9
20– 650	68–1200	16.2	9.1
20– 760	68–1400	16.7	9.3
20– 870	68–1600	17.3	9.6
20– 980	68–1800	17.6	9.8
20–1090	68–2000	18.0	10.0
650– 870	1200–1600	19.4	10.8
650– 980	1200–1800	19.8	11.0

Specific heat. 461 J/kg·K (0.11 Btu/lb·°F)
Thermal conductivity:

Temperature °C	°F	Conductivity W/m·K	Btu/ft·h·°F
100	212	12.1	7.0
316	600	15.4	8.9
538	1000	18.7	10.8
650	1200	20.6	11.9
760	1400	22.3	12.9
870	1600	24.2	14.0
980	1800	26.5(a)	15.3(a)
1090	2000	28.2(a)	16.3(a)

(a) Estimated values.

Electrical Properties

Electrical resistivity. 1000 nΩ·m at 21 °C (70 °F)

Magnetic Properties

Magnetic permeability. 1.10 to 2.00

Chemical Properties

General corrosion behavior. Resists air, carburizing gas, flue gases oxidizing and reducing, molten metals, and salts. Except in high-sulfur gases, it performs satisfactorily up to 1150 °C (2100 °F) in oxidizing atmospheres and up to 1095 °C (2000 °F) in reducing atmospheres, provided

Fig. 31 Results of 100-h corrosion tests of alloy HP in gases

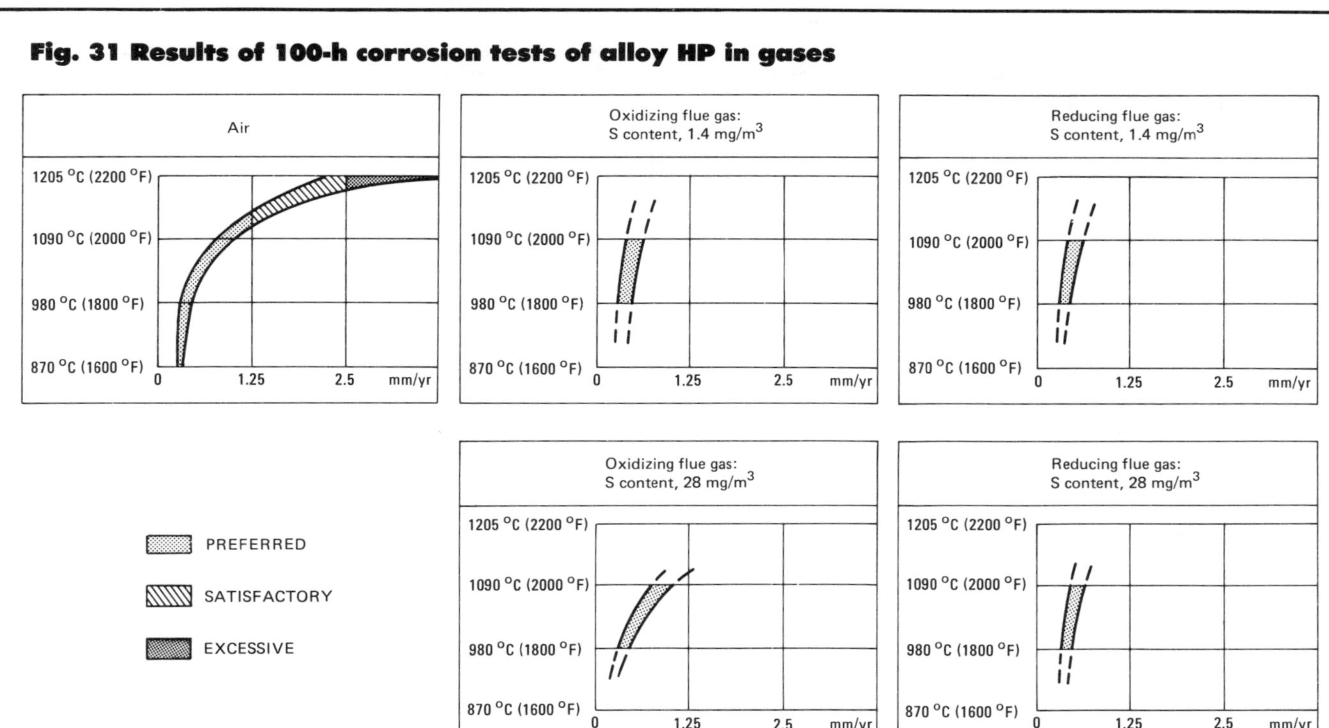

Fig 32 Stress-rupture curves for HP-50WZ

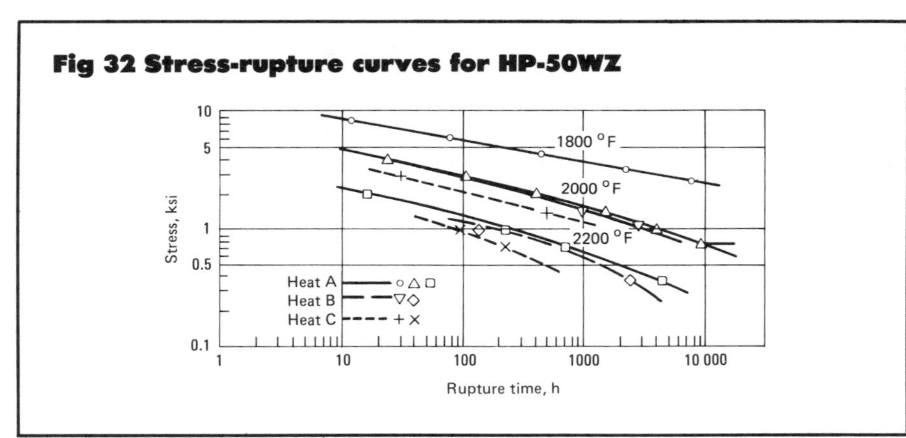

that limiting creep stress values are not exceeded

Resistance to specific corroding agents. Good resistance to tempering and cyaniding salts; fair resistance to high speed salts and fair resistance to neutral salts with proper control. Molten metals: good resistance to lead, tin (to 345 °C or 650 °F) and cadmium (to 410 °C or 775 °F); poor resistance to zinc, babbitt, soft solder, antimony and type metal. See also Fig. 37.

Fabrication Characteristics

Machinability. See alloy HE

Heat treatment. Castings are normally used in the as-cast condition. This alloy cannot be hardened by heat treatment but for applications involving thermal fatigue from repeated rapid heating and cooling, improved performance may be obtained by heating castings at 1038 °C (1900 °F) for 12 h followed by furnace cooling prior to placing them in service

Weldability. Castings of this type have good welding properties if proper techniques are employed. Thermal expansion is about one third greater than for carbon steel or Fe-Cr alloy types HC or HD. Electrical resistance is over six times that of carbon steel and is characterized by a low temperature, coefficient of resistivity 0.00017 °F in the range from 20 to 500 °C (68 to 930 °F). Welding inert-gas arc, and oxyacetylene gas methods. Oxyacetylene welding is more satisfactory than arc welding for high-temperature applications of this alloy. Type 330 bare filler rods should be used for gas welding, and the flame should be adjusted to be very rich in acetylene

HU
39Ni-19Cr

Commercial Name

UNS number. HU: J95405. HU-50: J95404

Fig 33 Effect of temperature on short-time tensile properties and Charpy impact strength of alloy HT

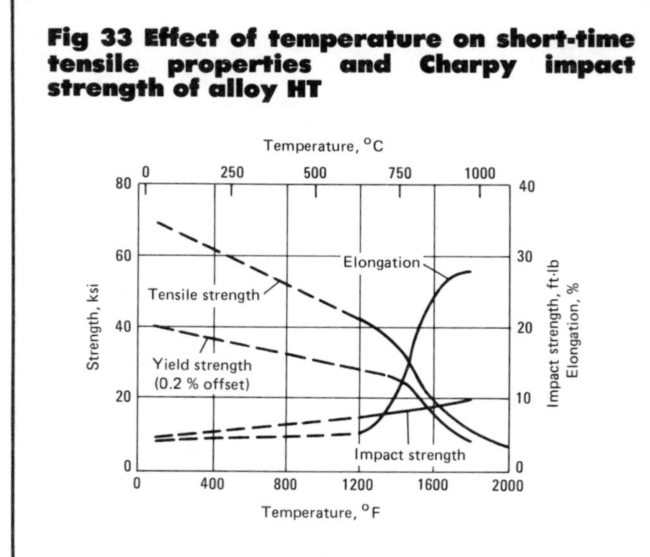

Fig. 34 Effect of temperature on hardness of alloy HT

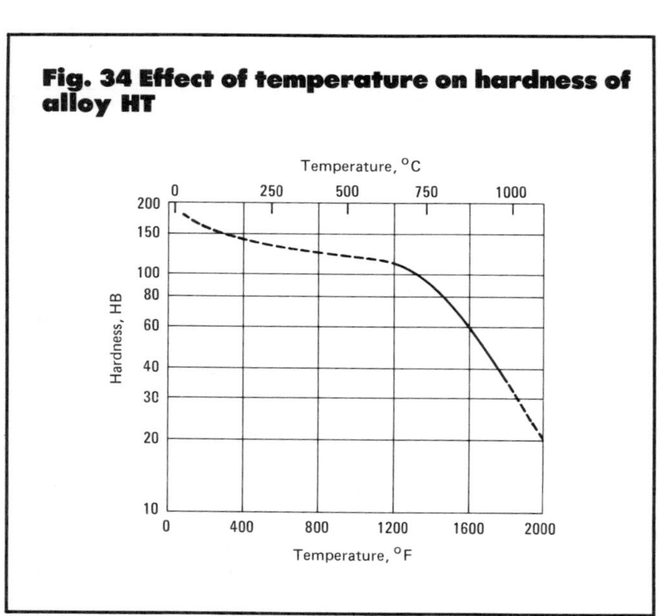

Fig. 35 Effect of temperature on modulus of elasticity of alloy HT

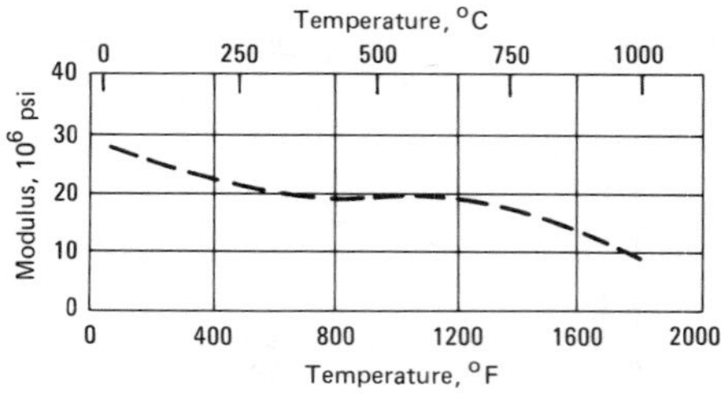

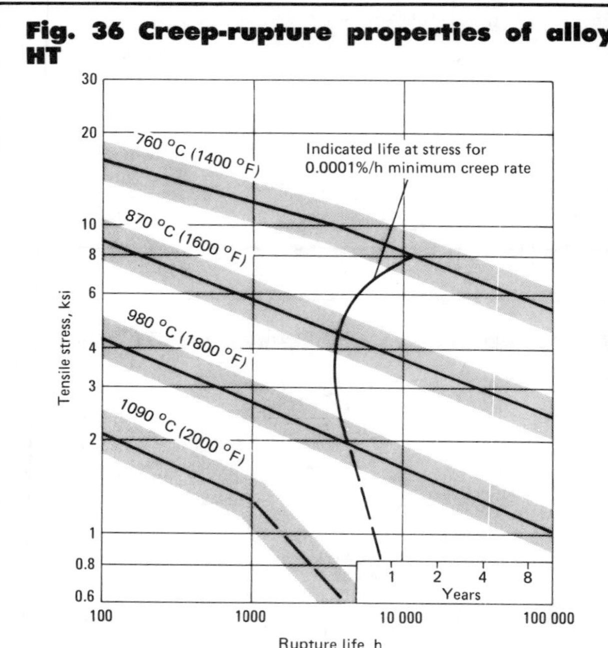

Fig. 36 Creep-rupture properties of alloy HT

The scatter bands shown are set arbitrarily at ±20% of the stress for the central tendency line. Such a range usually embraces test data for similar alloy compositions, but should not be considered statistically significant confidence limits. Scatter of values may be much wider, particularly at the longer times and higher temperatures.

Table 12 Representative creep-rupture values for alloy HU(a)

Temperature		Limiting stress for 0.0001%/h creep rate		Stress to rupture in: 100 h		1000 h		10 000 h	
°C	°F	MPa	ksi	MPa	ksi	MPa	ksi	MPa	ksi
760	1400	58.6	8.5	103	15	...	...	...	...
870	1600	34.5	5.0	55	8.0	35.9	5.2	22.8	3.3
980	1800	15.2	2.2	31	4.5	19.9	2.9	12.4	1.8
1090	2000	4.1	0.6	...	...	...	...	...	...

(a) For constant temperature, for cyclic-temperature, lower values would apply.

Fig. 37 Results of 100-h corrosion tests of alloy HT in gases

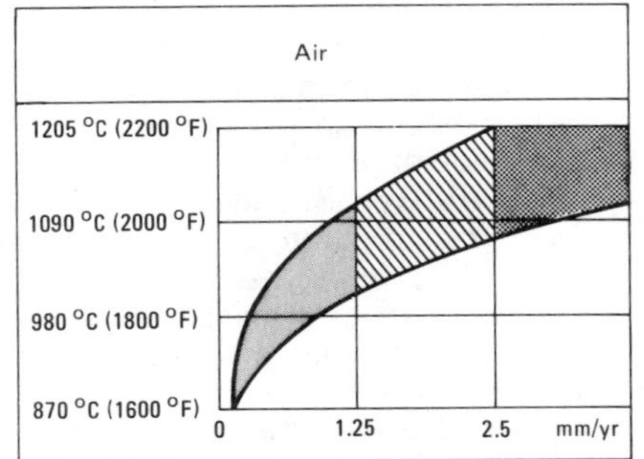

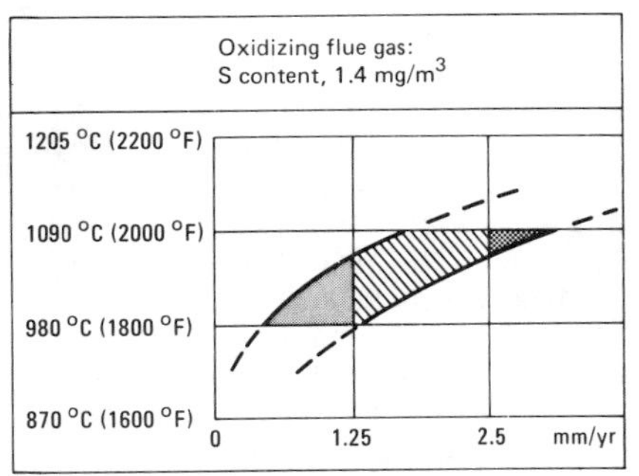

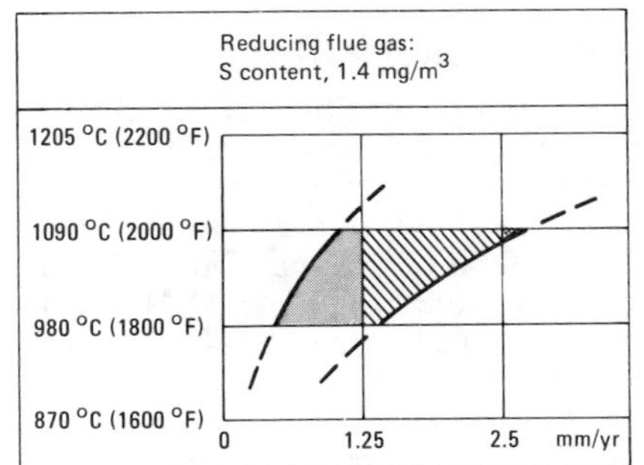

PREFERRED

SATISFACTORY

EXCESSIVE

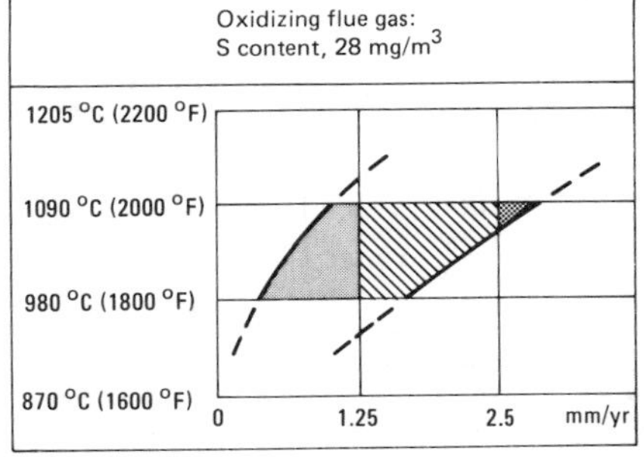

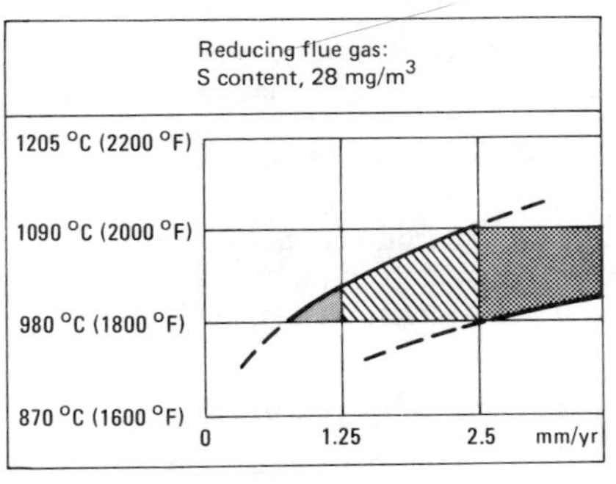

Fig. 38 Results of 100-h corrosion tests of alloy HU in gases

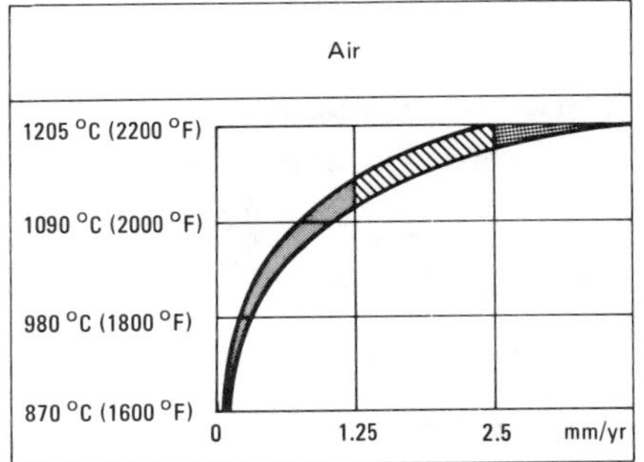

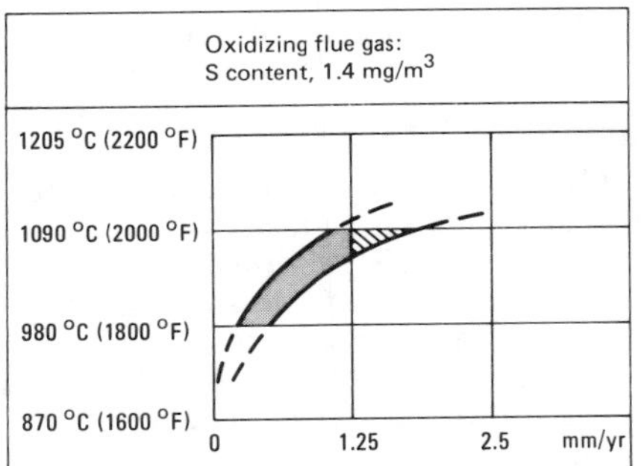

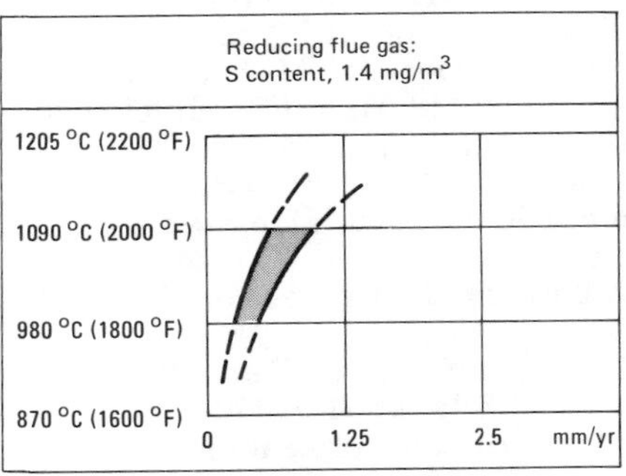

PREFERRED

SATISFACTORY

EXCESSIVE

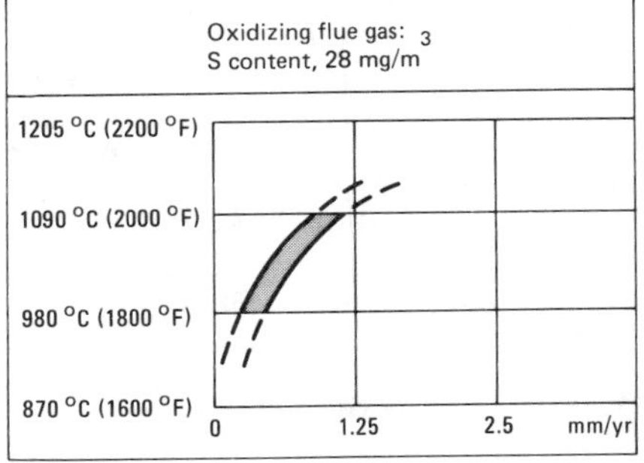

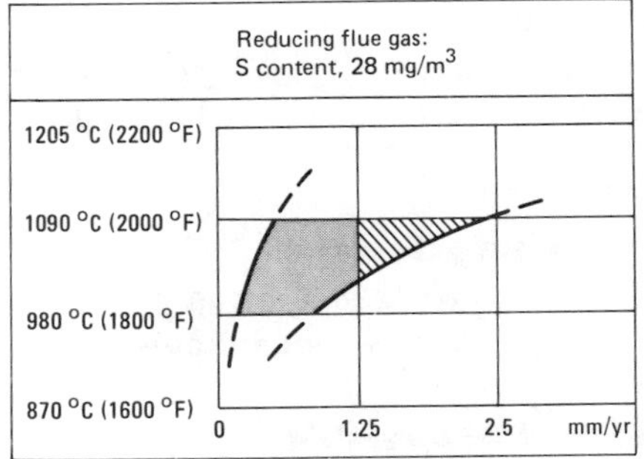

Specifications

ASTM. A297 (HU)
SAE. 70331

Chemical Composition

Composition limits. 0.35 to 0.75 C; 2.0 max Mn; 2.5 max Si; 0.04 max P; 0.04 max S; 0.5 max Mo(a); 17 to 21 Cr; 37 to 41 Ni; rem Fe. For HU-50; 0.40 to 0.60 C; 1.5 max Mn; 0.50 to 2.0 Si; 17 to 21 Cr; 37 to 41 Ni; 0.04 max P; 0.04 max S; 0.50 max Mo(a)

(a) Molybdenum not intentionally added.

Applications

Typical uses. Especially suited for severe service conditions involving high stress rapid thermal cycling. Articulated trays, burner tubes, carburizing retorts, conveyor screws and chains, cyanide pots, dipping baskets, furnace rolls, lead pots, muffles, pouring spouts, radiant tubes, resistor guides

Mechanical Properties

Tensile properties. At room temperature. As cast: tensile strength, 483 MPa (70 ksi); 0.2% yield strength, 276 MPa (40 ksi); elongation, 9% in 50 mm or 2 in. Aged 48 h at 980 °C (1800 °F), air cooled: tensile strength, 503 MPa (73 ksi); 0.2% yield strength, 296 MPa (43 ksi); elongation, 5% in 50 mm or 2 in. Minimum for ASTM A297: tensile strength, 448 MPa (65 ksi); elongation, 4% in 50 mm or 2 in.
Hardness. As cast: 170 HB. Aged 48 h at 980 °C (1800 °F), air cooled: 190 HB
Elastic modulus. Tension, 186 GPa (27×10^6 psi)
Impact strength. Charpy keyhole, 5.4 J (4 ft·lb) at 21 °C (70 °F)
Creep-rupture characteristics. See Table 12.

Structure

Microstructure. Austenite with varying amounts of carbides depending on carbon content and thermal history

Mass Characteristics

Density. 8.04 Mg/m^3 (0.290 $lb/in.^3$)
Patternmaker's shrinkage. 5/16 in./ft

Thermal Properties

Melting point. Approx 1340 °C (2450 °F)
Coefficient of thermal expansion. Linear:

Temperature °C	Temperature °F	Mean coefficient µm/m·K	Mean coefficient µin./in.·°F
21– 358	70–1000	15.8	8.8
21– 650	70–1200	16.2	9.0
21– 760	70–1400	16.6	9.2
21– 870	70–1600	16.9	9.4
21– 980	70–1800	17.3	9.6
21–1090	70–2000	17.5	9.7
650– 870	1200–1600	18.9	10.5
650– 980	1200–1800	19.1	10.6

Specific heat. 461 J/kg·K (0.11 Btu/lb·°F)
Thermal conductivity:

Temperature °C	Temperature °F	Conductivity W/m·K	Conductivity Btu/ft·h·°F
100	212	12.1	7.0
316	600	15.4	8.9
538	1000	18.7	10.8
650	1200	20.6	11.9
760	1400	22.3	12.9
870	1600	24.2	14.0
980	1800	26.5	16.3

Electrical Properties

Electrical resistivity. 1050 nΩ·m at 21 °C (70 °F)

Magnetic Properties

Magnetic permeability. 1.10 to 2.00

Chemical Properties

General corrosion behavior. Often selected for its superior corrosion resistance. Resists corrosion by oxidizing and reducing hot gases, particularly those containing high sulfur levels, better than type HT. Also resists air, carburizing gases, combustion gases, molten cyanide, and molten lead.
Resistance to specific corroding agents. Good resistance to tempering and cyaniding salts; fair resistance to neutral salts with proper control. Good resistance to molten lead, molten tin to 345 °C (650 °F), and molten cadmium to 410 °C (775 °F); poor resistance to molten antimony, babbitt, soft solder and type metal. See also Fig. 38.

Fabrication Characteristics

Machinability. See alloy HE
Heat treatment. Castings are normally used in the as-cast condition. The alloy cannot be hardened by heat treatment, but for applications involving thermal fatigue from repeated rapid heating and cooling, improved performance may be obtained by heating castings at 1040 °C (1900 °F) for 12 h followed by furnace cooling prior to placing in service
Weldability. Shielded metal arc, inert-gas arc and oxyacetylene-gas methods. Arc processes most widely used but oxyacetylene flame provides satisfactory welds also. Lime coated electrodes of similar composition (AWS E330-a5, special high silicon, high carbon) are generally used for shielded metal arc welding. Bare wire 0.8 to 1.6 mm (0.035 to 0.063 in.) in diameter (AWS ER30 high carbon), is used for filler metal in gas metal arc, gas tungsten arc and oxyacetylene gas processes. Flame should be adjusted to very rich acetylene in last method. Preweld or postweld heat treatments not required

HW
60Ni-12Cr

Specifications

ASTM. A297 (HW)
SAE. 70334

Chemical Composition

Composition limits. 0.35 to 0.75 C; 2.0 max Mn; 2.5 max Si; 0.04 max P; 0.04 max S; 0.5 max Mo(a); 10 to 14 Cr; 58 to 62 Ni; rem Fe

(a) Molybdenum not intentionally added.

Applications

Typical uses. When resistance to carburization and tolerance of temperature fluctuations are needed. Cyanide pots, electric heating elements, enameling tools, gas retorts, hardening fixtures, hearth plates, lead pots, and muffles

Mechanical Properties

Tensile properties. At room temperature. As cast: tensile strength, 469 MPa (68 ksi); 0.2% yield strength, 248 MPa (36 ksi); elongation, 4% in 50 mm or 2 in. Aged 48 h at 980 °C (1800 °F), furnace cooled: tensile strength, 579 MPa (84 ksi);

Table 13 Representative(a) creep-rupture values for alloy HW

Temperature °C	Temperature °F	Limiting stress for 0.0001%/h creep rate, MPa	Limiting stress for 0.0001%/h creep rate, ksi	Stress to rupture (b) in: 10 h, MPa	10 h, ksi	100 h, MPa	100 h, ksi	1 000 h, MPa	1 000 h, ksi
760	1400	41.3	6.0	110	16	69.0	10	53.8	7.8
870	1600	20.7	3.0	56.5	8.2	41.3	6.0	31.0	4.5
980	1800	9.7	1.4	29.6	4.3	24.8	3.6	17.9	2.6

(a) For constant temperature; for cyclic temperature lower values would apply. (b) Extrapolation of stress values to longer times than those shown for rupture life should not be attempted.

Fig. 39 Results of 100-h corrosion tests of alloy HW in gases

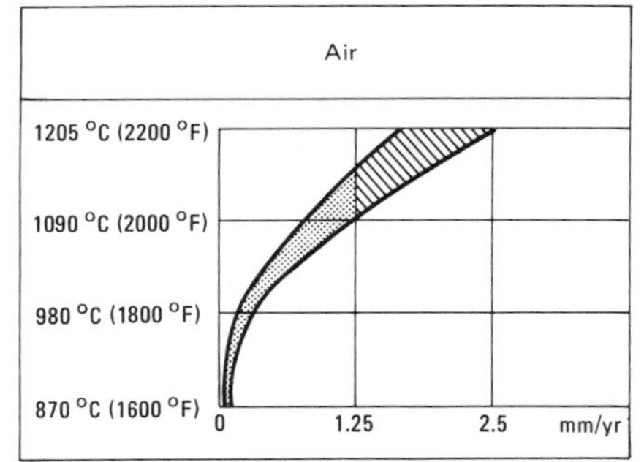

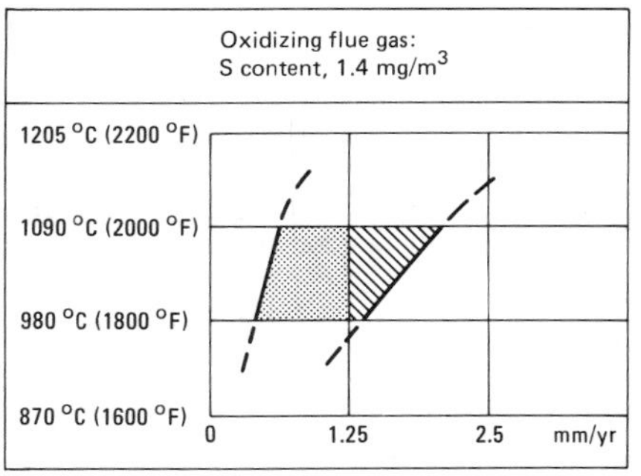

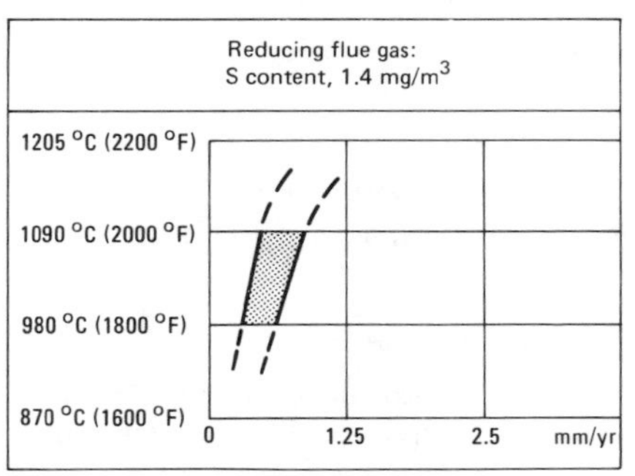

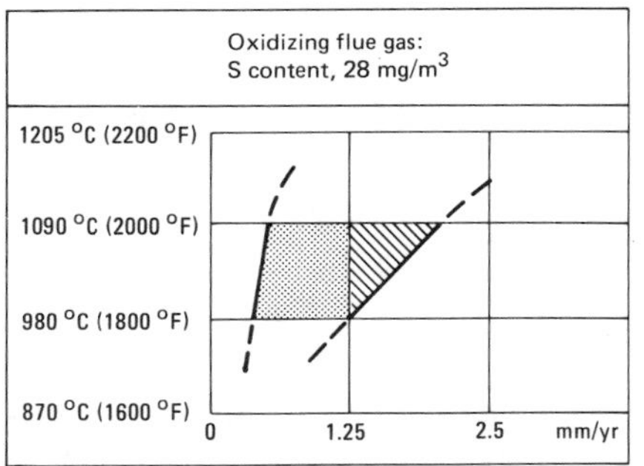

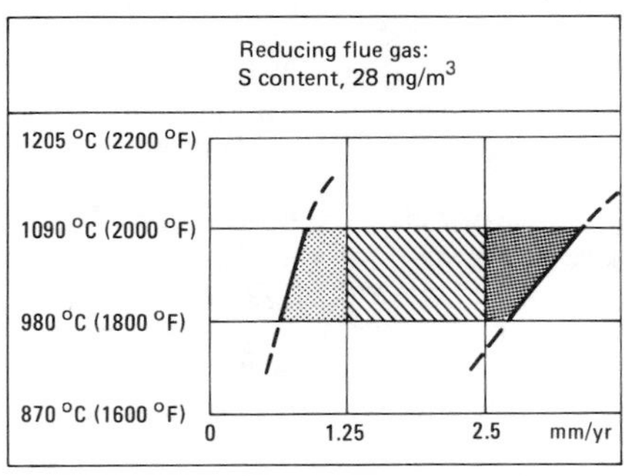

0.2% yield strength, 359 MPa (52 ksi); elongation, 4% in 50 mm or 2 in. Minimum for ASTM 297; tensile strength, 414 MPa (60 ksi)

Hardness. As cast: 185 HB. Aged 48 h at 980 °C (1800 °F), furnace cooled: 205 HB

Elastic modulus. Tension, 172 GPa (25 × 10^6 psi)

Creep-rupture characteristics. See Table 13.

Structure

Microstructure. Austenite with varying amounts of carbides depending on the carbon content and thermal history. In the as-cast alloy, the microstructure consists of a continuous interdendritic network of massive and elongated eutectic carbides

Mass Characteristics

Density. 8.14 Mg/m³ (0.294 lb/in.³) at 21 °C (70 °F)

Patternmaker's shrinkage. 9/32 in./ft

Thermal Properties

Melting point. Approx 1287 °C (2350 °F)

Coefficient of thermal expansion:

Temperature °C	Temperature °F	Mean coefficient µm/m·K	Mean coefficient µin./in.·°F
20–93	68– 200	12.6	7.0
20–204	68– 400	12.9	7.22
20–316	68– 600	13.4	7.45
20–427	68– 800	13.9	7.7
20–358	68–1000	14.3	7.95
20–650	68–1200	14.8	8.2
20–760	68–1400	15.2	8.47
20–870	68–1600	15.7	8.74
20–980	68–1800	16.2	9.07
20–1090	68–2000	16.7	9.28
650–870	1200–1600	18.0	10.0
650–980	1200–1800	18.6	10.33

Specific heat. 461 J/kg·K (0.11 Btu/lb·°F) at 21 °C (70 °F)

Thermal conductivity:

Temperature °C	°F	Conductivity W/m·K	Btu/ft·h·°F
100	212	12.5	7.2
316	600	15.6	9.0
538	1000	19.2	11.1
650	1200	21.1	12.2
760	1400	23.0	13.3
870	1600	25.1	14.5
980	1800	27.2(a)	15.7(a)
1090	2000	29.4(a)	17.0(a)

(a) Estimated values.

Electrical Properties

Electrical resistivity. 1.12 μΩ·m at 21 °C (70 °F)

Magnetic Properties

Magnetic permeability. 16.0

Chemical Properties

General corrosion behavior. Resistant to air, carburizing gases, combustion gases, molten cyanide, and molten lead. Performs satisfactorily to 1120 °C (2050 °F) in strongly oxidizing atmospheres and to 1040 °C (1900 °F) in oxidizing or reducing products of combustion which do not contain sulfur.
Resistance to specific corroding agents. Excellent resistance to tempering and cyaniding salts; fair resistance to neutral salts, with proper control. Excellent resistance to molten lead; good resistance to molten tin, to 345 °C (650 °F) and good resistance to molten cadmium to 410 °C (775 °F). Poor resistance to antimony, babbitt, soft solder and type metal. See also Fig. 39.

Fabrication Characteristics

Machinability. See alloy HE.
Weldability. Shielded metal arc, inert-gas arc and oxyacetylene-gas methods. Electric arc most widely used though oxyacetylene produces satisfactory welds also. Lime coated electrodes of similar composition (AWS EN:Cr-1 of AWS EN:Cr Fel) are generally used for shielded metal arc welding. Bare Inconel wire and stainless flux should be used for gas welding; and flame should be adjusted to be very rich acetylene. Preweld or postweld heat treating not required.

HX
66Ni-17Cr

Specifications

ASTM. A297(HX)
SAE. 70335

Chemical Composition

Composition limits. 0.35 to 0.75 C; 2.0 max Mn; 2.5 max Si; 0.04 max P; 0.04 max S; 0.05 max Mo(a); 15 to 19 Cr; 64 to 68 Ni; rem Fe
(a) Molybdenum not intentionally added.

Applications

Typical uses. Autoclaves, brazing furnace rails and doors, calciner tubes, carburizing boxes, cyanide pots, enameling tools, heating elements, hearth plates, heat treating trays and fixtures, lead pots, muffles, retorts, roller hearths, salt bath electrodes, salt pots, shaker hearths. Severe service applications at temperatures up to 1150 °C (2100 °F), situations where corrosion must be minimized.

Mechanical Properties

Tensile properties. Room temperature. As cast: tensile strength, 448 MPa (65 ksi); 0.2% yield strength, 248 MPa (36 ksi); elongation, 9% in 50 mm or 2 in. Aged 48 h at 980 °C (1800 °F), air cooled: tensile strength, 503 MPa (73 ksi); 0.2% yield strength, 303 MPa (44 ksi) elongation, 9% in 50 mm or 2 in. Minimum for ASTM A297: tensile strength, 414 MPa (60 ksi)
Hardness. As cast: 176 HB. Aged 48 h at 980 °C (1800 °F), air cooled: 185 HB
Elastic modulus. Tension, 172 GPa (25 × 10^6 psi)
Creep-rupture characteristics. See Table 14.

Structure

Microstructure. Continuous interdendritic network of massive and eutectic carbides. After aging at service temperatures, the austenitic matrix becomes uniformly peppered with small carbide particles except in the immediate vicinity of the eutectic carbides.

Mass Characteristics

Density. 8.14 Mg/m^3 (0.294 lb/in.^{3}) at 21 °C (70 °F)
Patternmaker's shrinkage. 9/32 in./ft

Thermal Properties

Melting point. Approx 1287 °C (2350 °F)
Coefficient of thermal expansion. Linear:

Temperature °C	°F	Mean coefficient μm/m·K	μin./in.·°F
21–358	70–1000	14.0	7.8
21–650	70–1200	14.6	8.1
21–760	70–1400	15.3	8.5
21–870	70–1600	15.8	8.8
21–980	70–1800	16.6	9.2
21–1090	70–2000	17.1	9.5
650–870	1200–1600	19.3	10.7
650–980	1200–1800	20.3	11.3

Specific heat. 461 J/kg·K (0.11 Btu/lb·°F) at 21 °C (70 °F)
Thermal conductivity:

Temperature °C	°F	Conductivity(a) W/m·K	Btu/ft·h·°F
100	212	12.5	7.2
316	600	15.6	9.0
538	1000	19.2	11.1
650	1200	21.1	12.2
760	1400	23.0	13.3
870	1600	25.1	14.5
980	1800	27.2	15.7
1090	2000	29.4	17.0

(a) Estimated values.

Electrical Properties

Electrical resistivity. 1160 nΩ·m at 21 °C (70 °F)

Magnetic Properties

Magnetic susceptibility. Slightly magnetic
Magnetic permeability. 2.0

Chemical Properties

General corrosion behavior. Often selected because of its superior corrosion resistance. Resistant to air carburizing gases, combustion gases, flue gases, hydrogen, molten cyanide, molten lead, and molten neutral salts at temperatures up to 1150 °C (2100 °F).
Resistance to specific corroding agents. Excellent resistance to tempering and cyaniding salts; fair resistance to neutral salts, with proper control. Excellent resistance to molten lead; good resistance to molten tin to 345 °C (650 °F) and molten cadmium to 410 °C (775 °F). See also Fig. 40.

Fabrication Characteristics

Machinability. See alloy HE
Weldability. See alloy HW

Table 14 Representative(a) creep-rupture data for alloy HX

Temperature		Limiting stress for 0.0001%/h creep rate		Stress to rupture (b) in: 10 h		100 h		1000 h	
°C	°F	MPa	ksi	MPa	ksi	MPa	ksi	MPa	ksi
760	1400	44	6.4	125	18	90	13	...	...
870	1600	22	3.2	70	10	46	6.7	28	4.0
980	1800	11	1.6	37	5.4	24	3.5	14	2.0
1090	2000	4	0.6	17	2.5	12	1.7	6	0.9

(a) For constant temperature; for cyclic temperature lower values would apply. (b) Extrapolation of stress values to longer times than those shown for rupture life should not be attempted.

Fig. 40 Results of 100-h corrosion tests of alloy HX in gases

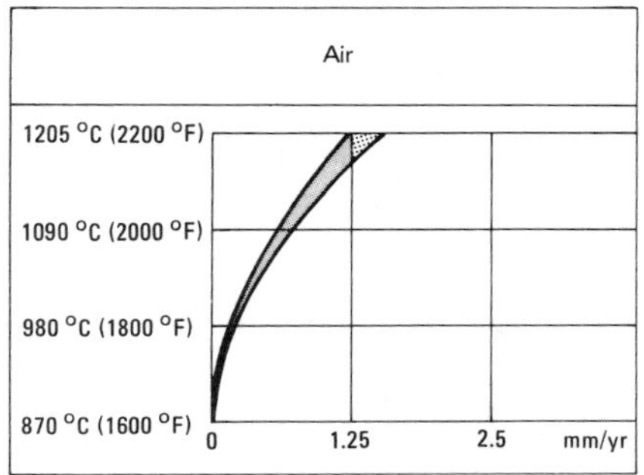

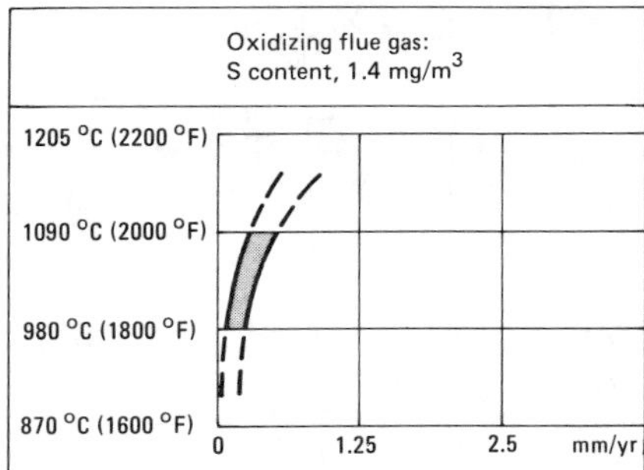

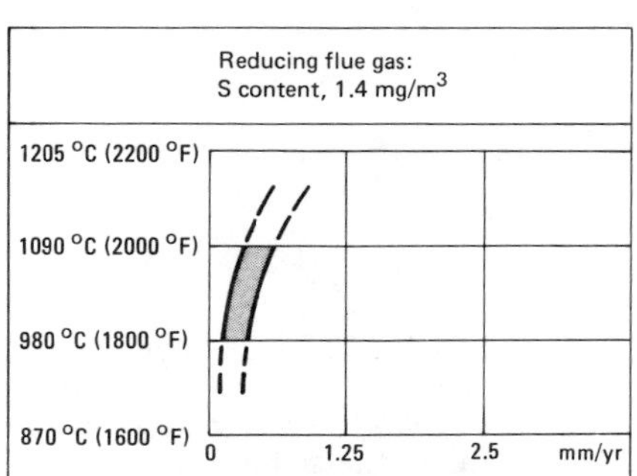

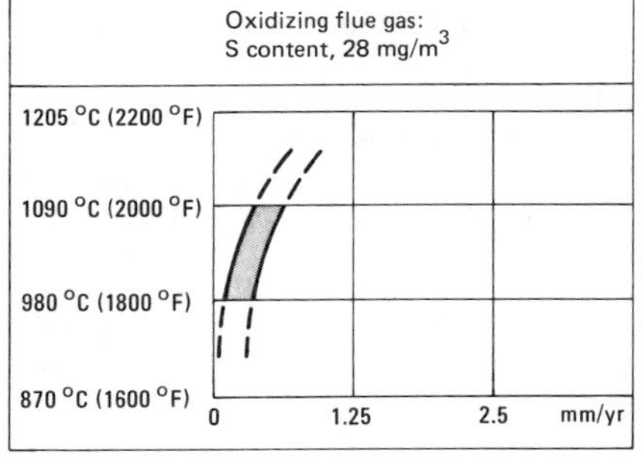

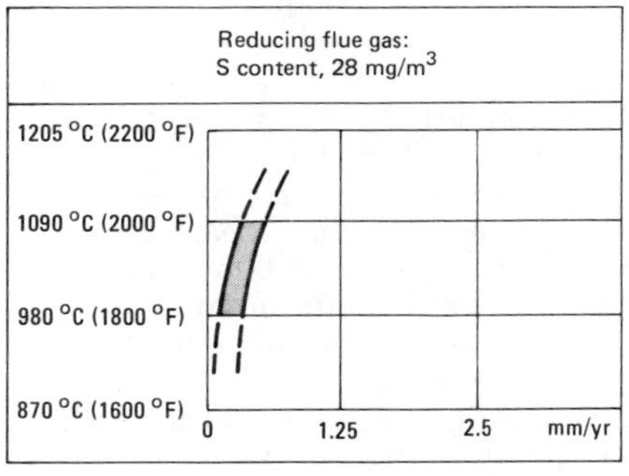

Refractory Metals and Alloys

REFRACTORY METALS once were limited to use in lamp filaments, electron-tube grids, heating elements and electrical contacts, but they have since gained applications in the aerospace, nuclear, electronics and chemical-process industries. A majority of the total tonnage of refractory metals produced is now used in aerospace applications, and much of the development of specific manufacturing techniques has been in that area.

Selection of a specific alloy from the refractory-metal group often is based on fabricability rather than on strength or corrosion resistance. Niobium, tantalum and their alloys are the most easily fabricated refractory metals. They can be formed, machined and joined by conventional methods. They are ductile in the pure state and have high interstitial solubilities for carbon, nitrogen, oxygen and hydrogen. Because of such high solubilities, these embrittling contaminants normally do not present problems in fabrication. However, tantalum and niobium dissolve sufficient amounts of oxygen at elevated temperatures to destroy ductility at normal operating temperatures. Therefore, elevated-temperature fabrication of these metals is used only when necessary. Protective coatings or atmospheres are mandatory unless some contamination can be tolerated. The allowable level of contamination, in turn, determines the maximum permissible exposure time in air at elevated temperature.

Molybdenum, molybdenum alloys, tungsten and tungsten alloys require special techniques for fabrication. Fabrication involving mechanical working should be performed below recrystallization temperature. These materials have limited solubilities for carbon, nitrogen, oxygen and hydrogen. Because the residual levels of these elements required to prevent embrittlement are impractically low, the microstructure must be controlled to ensure a sufficiently low ductile-to-brittle transition temperature (DBTT).

Based chiefly on experience gained in aerospace programs, applications for refractory metals now encompass almost every type of industry. Table 1 summarizes the commercially significant uses of these metals. Table 2 lists nominal compositions of the refractory-metal alloys that are now commercially prominent.

Resistance of refractory metals to corrosion by liquid metals and aggressive acid solutions can cut maintenance and downtime if high initial cost can be accepted. Systems for containing liquid metals such as lithium and cesium at high temperature have been fabricated of Nb-1Zr alloy tubing; tantalum and tantalum-clad steel process equipment has performed well in high-temperature sulfuric acid service.

Most refractory metals and alloys are available as wire. Tungsten wire, for example, which comes in diameters as small as 0.0102 mm (0.0004 in.), is used as fiber reinforcement in composite materials in which the matrix is any one of various ductile alloys.

In the nuclear field, tungsten crucibles that are pressed and sintered, shear spun, chemical vapor deposited, or plasma sprayed and sintered are used in recovery of uranium and plutonium from spent reactor fuel.

In the chemical-process industries, tantalum and (to a lesser extent) molybdenum have been used for many years. The severe corrosion problems accompanying many recently developed chemical processes have given impetus to greater use of refractory alloys. Recently, chemical equipment has been fabricated from steel plate explosively clad with tantalum. Forming and welding methods have been developed for fabrication of the clad plate into reactor vessels, tanks and other types of chemical equipment. Explosive bonding produces a metallurgical bond at the tantalum/steel interface. Bond efficiency is over 98%, and bond shear strength exceeds the ASME minimum acceptable value for clad material.

Electronic Applications

Tantalum packages for hermetic sealing of wet electrolytic tantalum capacitors are made by cold drawing a can and a cap or header. A glass-to-metal seal in the header insulates a tantalum lead wire that passes through the header and connects to the capacitor element. The lead supports a sintered P/M tantalum slug that is spot welded to it. After the slug has been inserted and the can has been filled with electrolyte, the header is resistance welded to the tantalum can, forming a hermetic seal.

Headers are formed in progressive dies. Cans, approximately 9.52 mm (0.375 in.) in diameter and 19.0 mm (0.750 in.) in length usually are made either by drawing on a transfer press or by spinning, but some are vapor deposited on removable mandrels.

Other electronic components in which refractory metals are employed include a composite x-ray target consisting of a forged or spun molybdenum substrate with a plasma-sprayed optical track of W-Re alloy.

In vacuum metallizing equipment, evaporation boats are commonly fabricated by coating formed molybdenum substrates with plasma-sprayed refractory oxide insulation.

Chemical vapor deposition has proved useful for fabricating free-standing refractory-metal parts for advanced electronics applications. Tungsten emitters of preferred crystal orientation, tungsten collectors, and sandwich insulators ($Nb-Al_2O_3-Nb$, for example) are used in nuclear thermionic-conversion devices.

Production

Primary production of refractory metals starts with consolidation by melting or powder metallurgy techniques. Hot forging or extrusion is used for breaking down ingots or powder compacts into sheet bar and solid rounds for processing into sheet, plate, foil tubing and bar products. Table 3 gives typical mill-processing temperatures.

Industrial practices for niobium and tantalum now include vacuum electron beam melting. High-purity ingots are electron beam melted and then often remelted in vacuum or an inert atmosphere to ensure homogeneity. Tantalum alloys and niobium alloys can be extruded, open-hammer forged, upset, pierced and drawn, or closed-die forged.

Machining

Compared with familiar structural alloys, refractory metals are considered difficult to machine. However, they can be machined using modifications of conventional machining practices.

Table 1 Typical applications of refractory metals

Application	Material
Aerospace and nuclear industries	
Solid-propellent rockets, 2650-2750 °C flame temperature	Molybdenum, tungsten
Solid-propellent rockets, 3425-3550 °C flame temperature	Tungsten (silver infiltrated)
Lifting and guidance structures for glide re-entry vehicles	Cb-752 (silicide coated), T-111, T-222, B-66, FS-85
Leading edges and nose caps for hypersonic flight vehicles	Cb-752, SCb-291
Thrust chambers	C-103 (silicide-coated)
Radiation skirts	C-103 (aluminide-coated)
Rocket nozzles	SCb-291 (silicide-coated), Ta-10W (Ta-Hf clad)
Counterweights (aircraft, inertial-guidance controls)	Tungsten alloys
Liquid-metal containers and piping	Nb-1Zr
Radiation shields	Tungsten alloys
Thermal shields	C-129Y (aluminide-coated)
Porous ionizer plates (ion engine)	Tungsten
Heat shields and cesium vapor inlet tubes (ion engine)	Tantalum
Honeycomb structures	Molybdenum, Cb-752, Ta-10W
Electronics industry	
Circuits	Tantalum
Capacitor packages	Tantalum foil
Rectifiers, railway signals	Tantalum
Battery chargers	Tantalum
High-performance capacitance powder	Tantalum
Lead wire	Tantalum
Transducers	Molybdenum, tungsten
Electron tube parts:	
Heaters	Tungsten
Supports	Molybdenum
Cathodes	Tantalum
Anodes	Molybdenum, tungsten
Superconducting wire	Nb-Ti, Nb-Sn, Nb-Zr
X-ray targets	Tungsten, molybdenum, composite W-Mo
Electrodes (Hg switches)	Molybdenum, tungsten
Thin-film substrates	Tungsten, molybdenum
Electrical contacts	Tungsten, W-Ag, W-Cu
Heat sinks	Tungsten, molybdenum
Backing wafers, semiconductors	Tungsten, molybdenum

(continued)

Equipment for machining refractory metals must be rigid and powerful to ensure optimum results. Carbide tools are often mandatory for acceptable tool life and cutting properties.

Niobium alloys and tantalum alloys are readily machined using high speed steel or carbide tools. Machining and grinding characteristics vary from those of soft copper to those of annealed stainless steel.

Molybdenum is machined using carbide tools of the same configurations as those used for machining 1040 steel, the machining characteristics of these two metals being similar. Machining speeds for molybdenum alloys (TZM, for example) are about 40% higher than those for type 302 stainless steel. Finish grinding of molybdenum requires heavy coolant flow and use of alundum wheels to prevent heat checking. Tool configurations and grinding techniques are similar to those for grinding cast iron; conventional machines with standard feeds and speeds are satisfactory.

Turning is a problem only with tungsten. Carbide tools ground with negative back rake, 15° lead and 0° side rake are mandatory. All turning is done at room temperature.

For grinding tungsten, wheels of 60-

Table 1 (continued)

Application	Material
Process industries	
Heating and cooling coils	Tantalum
Shell and tube heat exchangers	Tantalum
Condensers	Tantalum
Tantalum-clad steel vessels	Tantalum
Distillation towers	Tantalum
Valves for hot H_2SO_4 service	Molybdenum, tantalum
Expansion joints (bellows)	Tantalum
Glass-processing equipment	Tantalum
Crucibles, all sizes to 1 m diam by 1.3 m high (3 ft diam by 4 ft high)	Tungsten, tantalum
Spinnerettes, textile industry	Tantalum
Thermocouple-protection tubes	Tantalum-coated copper or steel
Thermowells	Tantalum-clad copper, tantalum
Spargers, funnels, jet ejectors	Tantalum
Bayonet heaters	Tantalum
Pumps for HCl service at 200 kPa and 150 °C (30 psi and 300 °F)	Tantalum (exposed parts)
Special equipment	
Furnace parts:	
Heating elements, shields, boats, trays, platens, fixtures	Tungsten, molybdenum, tantalum
Susceptors (induction furnace)	Tungsten
Extrusion dies	Tungsten, molybdenum
Piercing points, hot punches	Tungsten, molybdenum
Cups	Tungsten, molybdenum, tantalum
Fasteners (nuts, screws, studs, rivets)	Tantalum, molybdenum
Die casting molds, cores	Molybdenum, tungsten
Vacuum metallizing coils, boats	Tungsten, molybdenum
Springs	Tungsten, molybdenum, tantalum
Boring bars	Tungsten, molybdenum
Surgical implants	Tantalum
Instruments	Tantalum
Electroplating equipment	Tantalum
Thermocouples	W, W-Re alloys
Cathode, plasma-generator	W-1Ni

grit silicon carbide or 46-grit alumina are recommended, and normal precautions, extra-light pressures and heavy coolant flow are required.

Tungsten and molybdenum must be punched and sheared above their ductile-to-brittle transition temperatures. Sheets over 1.3 mm (0.050 in.) thick must have excess thickness of 1.6 to 3.2 mm (1/16 to 1/8 in.) to allow for belt sanding to final dimensions. Cutting can be done using abrasive (60-grit SiC) cutoff wheels. Tungsten and molybdenum are also suitable for electrical-discharge machining (EDM) and electrochemical machining (ECM).

Electrochemical machining can also be used for shaping niobium and tantalum alloys, but it produces hydrogen embrittlement which must be removed by high-temperature vacuum annealing or inert-atmosphere annealing. For example, T-222 is held for 90 min at 1175 °C (2150 °F) in an argon atmosphere after ECM.

Photoetching and chemical blanking have been used on molybdenum. In these processes, photographic masking is followed by etching in a solution of HNO_3 and HF.

Vibration radiusing using loose abrasive and frequencies of 23 to 30 Hz can eliminate sharp corners on all refractory metals. The abrasive action rounds edges, eliminating the need for hand filing and polishing.

Table 2 Compositions of commercially important refractory alloys

Designation	Nominal composition, %
Molybdenum alloys	
Mo-0.5Ti	Mo-0.5Ti-0.02W
TZM	Mo-0.5Ti-0.1Zr-0.02W
Mo-30W	Mo-30W
Niobium alloys	
Nb-1Zr	Nb-1Zr
FS-85	Nb-27.5Ta-11W-1Zr
SCb-291	Nb-10Ta-10W
Cb-752	Nb-10W-2.5Zr
B-66	Nb-5Mo-5V-1Zr
C-103	Nb-10Hf-1Ti
C-129Y	Nb-10W-10Hf-0.15Y
Tantalum alloys	
"63" Metal	Ta-2.5W-0.15Nb
Ta-10W	Ta-10W
T-111	Ta-8W-2Hf
T-222	Ta-10W-2.5Hf-0.01C
Tungsten alloys	
W-ThO_2	W-1ThO_2; W-2ThO_2
W-Mo alloys	Various Mo contents; W-2Mo and W-15Mo are most common
W-Re alloys	Various Re contents up to 26%; W-1.5Re, W-3Re and W-25Re are most common
Doped W (a)	50 ppm Si, 90 ppm K, 15 ppm Al, 35 ppm O

(a) See also Table 15.

Forming

Sheet metal or tubing of tantalum or niobium alloys usually are formed in the annealed condition. Their forming behavior is similar to that of mild steel, except that they are more prone to galling, seizing and tearing. In thicknesses from 0.1 to 1.5 mm (0.004 to 0.060 in.), tantalum and niobium can be readily formed, blanked, punched, stamped or deep drawn at room temperature in steel dies (6% *t* clearance). Sheet must have a homogeneous, fine grain size (generally, ASTM No. 5 or finer) for satisfactory results. Coarse-grain sheet is likely to fail by localized necking during severe forming.

Table 3 Mill-processing temperatures for refractory metals

Metal or alloy	Forging Temperature(a) °C	Forging Temperature(a) °F	Forging Typical total reduction, %	Extrusion Temperature(a) °C	Extrusion Temperature(a) °F	Extrusion Typical reduction ratio	Rolling Temperature(a) °C	Rolling Temperature(a) °F	Rolling Typical total reduction (between anneals), %
Nb	980 to 650	1800 to 1200	50 to 80	1090 to 650	2000 to 1200...	10:1	315 to 205	600 to 400...	50 breakdown
							20	70	90 finish
Nb-1Zr	1200 to 980	2200 to 1800	50 to 80	1200 to 980	2200 to 1800...	10:1	315 to 205	600 to 400...	50 breakdown
							20	70	80 finish
FS-85	1320 to 980	2400 to 1800	50	1320 to 980	2400 to 1800...	4:1	370 to 205	700 to 400...	40 breakdown
							20	70	50 to 65 finish
SCb-291	1200 to 980	2200 to 1800	30	1320 to 980	2400 to 1800...	4:1	370 to 260	700 to 500...	50 breakdown
							20	70	60 to 75 finish
Cb-752	1200 to 980	2200 to 1800	30	1320 to 980	2400 to 1800...	4:1	370 to 260	700 to 500...	50 breakdown
							20	70	60 to 75 finish
B-66	1290 to 980	2350 to 1800	50	1320 to 980	2400 to 1800...	4:1	1200 to 1090	2200 to 2000 .	50 breakdown
							20	500 to 70 ...	25 to 50 finish
C-103	1320 to 980	2400 to 1800	50	1320 to 980	2400 to 1800...	8:1	205	400......	50 breakdown
							20	70	60 to 70 finish
C-129Y	1320 to 980	2400 to 1800	50	1320 to 980	2400 to 1800...	4:1	425	800......	50 breakdown
							20	70	60 to 70 finish
Ta	<500	<930	50 to 80	1090	2000	10:1	370 to 260	700 to 500...	80 breakdown
	20	70	Finish				20	70	90 finish
Ta-10W	1260 to 980	2300 to 1800	50	1650 to 1425	3000 to 2600...	10:1	370 to 260	700 to 500...	80 breakdown
	1090 to 815	2000 to 1500	Finish				20	70	90 finish
T-222	1260 to 1200	2300 to 2200	50	2040 to 1650	3700 to 3000...	10:1	370 to 260	700 to 500...	75 breakdown
							20	70	50 to 75 finish
Mo	1320 to 1150	2400 to 2100	50	1760 to 1370	3200 to 2500...	8:1	1200	2200	50 breakdown
	925 to 815	1700 to 1500	Finish				870	1600	90 to 75 finish
Mo-0.5Ti	1425 to 1260	2600 to 2300	50	1820 to 1480	3300 to 2700...	8:1	1200	2200	50 breakdown
	1320 to 1150	2400 to 2100	Finish				870	1600	75 finish
TZM	1480 to 1320	2700 to 2400	50	1820 to 1540	3300 to 2800...	8:1	1350 to 1200	2460 to 2200 .	50 breakdown
	1370 to 1200	2500 to 2200	Finish				1000 to 980	1830 to 1800 .	60
							315	600......	10 finish
W	1820 to 1590	3300 to 2900	20	1925 to 1650	3500 to 3000...	9:1	1450 to 1400	2640 to 2550 .	50 breakdown
	1320 to 1010	2400 to 1850	Finish				1370 to 980	2500 to 1800 .	90 finish

(a) Where a range is given, the higher temperature is typical starting temperature and the lower temperature is minimum working temperature for that process.

Both tungsten and molybdenum must have highly worked, fibrous microstructures to ensure adequate formability. Molybdenum disulfide and graphite lubricants facilitate forming. Sheet 0.5 mm (0.020 in.) or less in thickness is bent or rolled at 21 to 93 °C (70 to 200 °F) for molybdenum and at 425 to 540 °C (800 to 1000 °F) for tungsten. Deep drawing of heavier sheets of molybdenum and tungsten is done at 425 °C (800 °F) and 925 °C (1700 °F), respectively.

For conventional forming of molybdenum sheet, as well as for blanking, punching and shearing with heated dies, the following temperature-thickness relationships apply:

Thickness, mm	Temperature °C	Temperature °F
0.5	21	70
0.5-1.0	93-165	200-325
1.0	480-540	900-1000

Despite difficulties in fabrication, tungsten is used in more applications than any other refractory metal except tantalum. Many tungsten parts are die formed or deep drawn.

Fabricators can shear, draw or form tungsten by many techniques if they understand its directional and recrystallization properties. Thin sections can be deformed into simple shapes from room temperature to about 95 °C (200 °F). Heavier sections, however, require higher forming temperatures:

Thickness, mm	Temperature °C	Temperature °F
0.25-0.4	205-260	400-500
0.4 to 1.0	540	1000
1.0	1260-1590	2300-2900

Punching and shearing must be done hot in accordance with the same temperature-thickness relationships. Material 1.3 mm (0.050 in.) or greater in thickness should be sheared to within 1.6 to 3.2 mm (1/16 to 1/8 in.) of final dimensions, and parts should be finished by edge grinding.

Table 4 gives blanking pressures,

shear strengths and approximate blanking temperatures for tungsten blanks of various sizes. Shear strengths were derived from strain-gage tests on disks about 32 and 50 mm (1¼ and 2 in.) in diameter.

In roll, open-die and closed-die forming of tungsten and molybdenum, rolls and dies are heated to 425 to 540 °C (800 to 1000 °F). Otherwise, conventional techniques prevail.

The triaxial compressive forces of high-energy-rate extrusion can successfully produce symmetrical shapes in the crack-sensitive refractory metals. Sprayed glass is the preferred lubricant.

By form spinning, shear spinning (flow turning) and extrusion spinning, employed singly or in combination, refractory-metal plates, sheets or tubular blanks can be fabricated into configurations that are impractical to produce by conventional forming processes. Often, the only alternative is a combination of open-die forging and machining, which is comparatively expensive. Spinning involves relatively low tooling and finishing costs and short setup times; it also can produce parts within relatively tight dimensional tolerances.

All refractory metals can be spun in air if careful control of temperature and time can be maintained. Generally, the metal to be formed is mounted on a heavy spinning lathe and one or (preferably) two rollers. An oxypropane torch is used to heat the workpiece to the following temperatures:

Metal	Temperature °C	°F
Niobium:		
Protected	400 to 620	750 to 1150
Unprotected	425 max	800 max
Tantalum:		
Protected	480 to 650	900 to 1200
Unprotected	480 max	900 max
Molybdenum	480 to 1065	900 to 1950
Tungsten	760 to 1315	1400 to 2400

Many specialized processes are employed for fabrication of complex refractory-metal components. Typical examples include plasma coating, vapor deposition and photoetching.

Small tubing of all refractory metals except tungsten, currently a specialty item, is generally available. It is made by a variety of techniques, depending on quality, quantity and size. Most heat exchangers require tubing 1.6 to 13 mm (1/16 to ½ in.) in diameter and 0.25 to 0.75 mm (0.010 to 0.030 in.) in wall thickness.

Table 4 Blanking characteristics of tungsten sheet

Blank thickness mm	in.	Blank diameter mm	in.	Blanking temperature °C	°F	Blanking pressure Mg	lb	Shear strength MPa	ksi
1.5	0.060	51	2	1000	1830	13.0	28 750	480	70
2.3	0.090	51	2	1050	1920	14.8	32 500	405	59
3.2	0.125	51	2	1100	2010	19.3	42 500	380	55
1.5	0.060	32	1.250	950	1740	6.8	15 000	440	64
2.3	0.090	32	1.250	1000	1830	9.3	20 500	400	58
3.2	0.125	32	1.250	1100	2010	10.0	22 000	390	57

Methods of producing tubes of niobium, molybdenum and tantalum alloys include gas tungsten-arc welding and drawing; extruding a tube shell and reducing (rocking) it to finished size; and cupping and drawing.

Tungsten tubing produced by chemical vapor deposition (CVD) can range from 0.025 to 300 mm (0.001 to 12 in.) in inside diameter, 0.10 to 3.2 mm (0.004 to 0.125 in.) in wall thickness, and up to 1.8 m (6 ft) in length. Other processes, such as extrusion, cost more and impose greater limitations on size. CVD tungsten tubing has lower mechanical properties than as-worked wrought tubing because of the inherently low ductility of its columnar microstructure. After recrystallization, however, CVD tungsten displays superior properties. The CVD technique is also used for making tubing of molybdenum, W-Re alloys and other refractory metals in diameters of 3.2 mm (1/8 in.) and less.

Tubing of any refractory metal can be made by extrusion. Tungsten tubing has been extruded from billets at ram speeds of 0.13 to 0.2 m/s (5 to 8 in./s). Seamless tungsten tubing has been produced experimentally by a filled-billet technique. Refractory-metal tubing is also made by sinking with a deformable mandrel, by tube drawing with a removal hardened mandrel, and by plug drawing.

A low-temperature, back-extrusion – spinning technique and a floating-mandrel extrusion technique have been used for producing tantalum and molybdenum tube shells. Tantalum tubing for chemical-process applications is either welded and drawn or seamless. The material has good drawing properties and can be reduced in area up to 60% without intermediate annealing. Niobium alloys are similarly processed.

Joining

All refractory metals can be joined by electron beam (EB) welding, gas tungsten-arc welding (GTAW) or resistance welding. Two major problems are encountered: chemical changes due chiefly to atmospheric contamination, and microstructural changes resulting from thermal cycling. The latter changes include grain growth and different stages of precipitation hardening (solution, precipitation and overaging). Preheating and postheating generally are required to minimize deleterious effects arising from precipitation hardening as well as from the residual stresses normally induced by welding. Although recrystallization and grain growth are unavoidable in weldments of wrought tungsten and molybdenum, proper choice of welding process and procedure can localize these effects. Electron beam welding has proved effective in achieving full weld penetration with an extremely narrow heat-affected zone.

All refractory metals suffer losses in ductility and increases in ductile-to-brittle transition temperature when welded, but niobium and tantalum alloys are less affected than molybdenum and tungsten alloys. Tantalum and niobium alloys generally retain greater than 75% joint efficiency after gas tungsten-arc welding. Preheating is not required, but postweld annealing can restore large amounts of ductility and toughness to commercial alloys. Table 5 summarizes recommended postweld annealing treatments, and Table 6 lists recommended welding conditions.

Although pure niobium shows no evidence of an aging reaction, Nb-1Zr undergoes abrupt losses in strength and ductility when treated at 815 to 980 °C (1500 to 1800 °F) for up to 500 h. Welds are subject to such embrittlement but can be restored to a ductile condition by postweld vacuum annealing at 1040 to 1200 °C (1900 to 2200 °F) for 3 h. This treatment produces overaging, preventing embrittlement on subsequent heating at a lower temperature.

In contrast to welds in niobium and tantalum, which retain good ductility, welds in molybdenum and tungsten are brittle (less than 50% joint efficiency), and thus these metals are difficult to join. Before welding, molybdenum and tungsten must be preheated above their ductile-to-brittle transition temperatures to prevent fracture. Sections in thicknesses of 0.6 mm (0.025 in.) and less demand special attention in this respect and, at best, present serious cracking problems. Welds are always brittle, and depend for joint efficiency on the reinforcing effect of the weld bead.

Tungsten is the most difficult refractory metal to join for satisfactory high-temperature service. Welding, and especially the EB process, offers the best compromise for joining tungsten for service at high temperatures. Mechanical joints are unsatisfactory unless molybdenum fasteners are used. Diffusion bonding is impractical because of severe tooling problems. Brazing for relatively low-temperature applications is done using precious metals (silver, palladium and platinum alloys) and transition metals (nickel and manganese alloys) as filler metals.

Table 7 lists typical brazing filler metals, and their maximum service temperatures for all four refractory-metal systems. Molybdenum brazing has received much attention. Brazed molybdenum honeycomb configurations are used for structural and heat-shield applications at temperatures from 1370 to 1650 °C (2500 to 3000 °F). Low-temperature brazing processes have been developed for TZM. The high remelt temperatures of the filler metals listed in Table 7 permit relatively high service temperatures.

Resistance welding is feasible, but some problems with electrode sticking can arise. RWMA class I copper electrodes show the least susceptibility to sticking. Projection welding can result in relatively high mechanical properties.

Coatings

Surface protection is the most significant obstacle to widespread use of refractory metals in high-temperature oxidizing environments. The existing ceiling of about 1650 °C (3000 °F) is dictated by coating limitations: coatings have insufficient life at reduced pressures (below about 13 kPa, or 100 torr) and high temperatures (about 1370 °C, or 2500 °F) in oxidizing atmospheres and give unreliable protection, particularly at edges and corners.

Table 8 summarizes coating systems of current importance. Aluminide and silicide coatings with various modifications are available commercially. Much controversy exists on application methods; however, slurry techniques are usually preferred, because pack cementation processes can be used only for small parts.

Silicide coatings are used more often than aluminide coatings, and have been selected for radiative heat shields and leading edges on aerospace vehicles. Successful simulations and actual flight tests have lessened the concern over possible catastrophic failure of refractory-metal components due to localized defects in coatings. The vanadium-modified niobium disilicide coating system has proved outstanding.

Because operating temperatures must exceed 1650 °C (3000 °F) before use of tantalum and tungsten alloys can be justified for aerospace applications, coatings for niobium and molybdenum are more highly developed. Surface protection at temperatures above 1650 °C requires more complicated approaches, including new coating systems, new application techniques, and fresh concepts in materials design.

Some promising systems for protection of tantalum and tungsten from 1650 to 2200 °C (3000 to 4000 °F) include roll cladding with Ta-Hf alloys; slurry-type coatings of iridium-base al-

Table 5 Recommended postweld annealing treatments for selected refractory alloys

Alloy	Annealing temperature(a): GTAW welds °C	GTAW welds °F	EB welds °C	EB welds °F
B-66	Not annealed		1040	1900
D-43	1315	2400	1315	2400
FS-85, C-129Y	1315	2400	1200	2200
Cb-752	1200	2200	1315	2400
SCb-291	1200	2200	Not annealed	
T-111, T-222	1315	2400	1315	2400
Ta-10W	Not annealed		Not annealed	

(a) One hour at temperature.

Table 6 Typical conditions for welding 0.9-mm (0.035-in.) refractory-metal sheet

Alloy	Gas tungsten-arc welds: Speed, in./min (a)	Clamp spacing, in.(a)	Current, A(b)	Arc gap, in.(a)	Electron beam welds: Speed, in./min (a)	Clamp spacing, in.(a)	Deflection, in.(a)(c)	Voltage, kV	Current, mA
B-66	15	3/8	86	0.06	25	3/16	0.050	150	3.2
C-129Y	30	3/8	110	0.06	50	1/2	0.050	150	4.1
Cb-752	30	3/8	87	0.06	15	3/16	0.050	150	3.3
FS-85	15	3/8	90	0.06	50	3/16	0.050	150	4.4
T-111	15	3/8	115	0.06	15	1/2	0.050	150	3.8
T-222	30	1/4	190	0.06	15	1/2	0.050	150	4.5
Ta-10W	7.5	1/4	118	0.06	15	1/2	0.050	150	3.8

(a) To convert inch values to equivalent values in mm, multiply by 25. (b) Direct current, straight polarity. (c) Beam deflection at 60 cycles parallel to weld direction.

Table 7 Typical brazing filler metals and service temperatures

Filler metal	Maximum service temperature °C	°F
For niobium alloys		
Si-Cr-Ni	980	1800
4Be-48Zr-48Ti	925	1700
Zr-6Be-19Nb	925	1700
Ti-0.5Si	1370	2500
Zr-0.1Be-16Ti-25V	1200	2200
V-35Nb	1200	2200
Ti-50Zr	1650	3000
Titanium	1760	3200
Ti-33Cr	1370	2500
Ti-3Al-11Cr-13V	1650	3000
For tantalum alloys		
Hf-7Mo	2090	3800
V-20Nb-20Ta	1870	3400
Ti-15Ta-25V	1650	3000
Ta-10Hf-(15-70)Nb	2200	4000
Nb-(30-50)Hf	2200	4000
Ta-10Hf	2200	4000
Nb-1.3B	1925	3500
Copper	980	1800
For molybdenum alloys		
Ti-3Be-25Cr	1590	2900
Pt-Mo	1650	3000
Zr-Ti	1230	2250
Cu-Au	815	1500
Ni-Cu	1200	2200
V-35Nb	1200	2200
Ti-30V	1370	2500
Ti-13Ni-25Cr	1760	3200
Co-10Ni-15W-20Cr	1320	2400
For tungsten		
Ag-Mn	870	1600
V-Nb-Ta	1925	3500
V-Ti-Ta	1925	3500
W-25Os	2200	4000
W-3Re-50Mo	2200	4000
Mo-5Os	1925	3500
Niobium	1650	3000
Tantalum	2200	4000

Table 8 Coatings for refractory metals

Coating designation	Method of application	Developer	Applicable substrate	Temperature limit(a) °C	°F
Aluminide coatings					
LB-2 (Al-Cr-Si)	Fused slurry	General Electric;	Nb	1425	2600
		McDonnell Douglas	Ta	1650	3000
Al-Si-Cr	Fused slurry	Sylvania	Nb	1425	2600
Sn-Al	Slurry dip or spray	GT&E	Mo	1480	2700
Ag-Si-Al	Hot dip	Sylvania	Nb	1540	2800
NAA-85 (Al_2O_3 + Al)	Slurry fusion	North American Rockwell	Nb	1425	2600
Silicide coatings					
Cr-Ti-Si (multilayered)	Vacuum pack and vacuum slip pack	TRW	Nb	1480	2700
			Ta	1480	2700
W Modified	Plasma spraying	TRW	W	1980	3600
Disil (Si + V, Cr, Ti)	Fluidized bed	Boeing	Nb	1540	2800
			Ta	1540	2800
L-7 ($MoSi_2$)	Slip pack	McDonnell Douglas	Mo	1650	3000
			W	1980	3600
PFR (Si + additives)	Pack cementation; fluidized bed	Pfaudler	Nb	1650	3000
PFR-5 ($MoSi_2$ + Cr	Pack cementation	Pfaudler	Mo	1650	3000
R (Si-20Cr-5Ti)	Fusion	Sylvania	Nb	1650	3000
NS-4, TNV-7 (complex silicides)	Vacuum and high-pressure pack	Solar	Nb	1650	3000
Durak KA	Pack cementation	Chromizing	Nb	1425	2600
Durak B	Pack cementation	Chromizing	Mo	1650	3000
N-2 (Si + Cr, Al, B)	Pack cementation	Chromalloy	Nb	1425	2600
W-3	Pack cementation	Chromalloy	Mo	1650	3000

(a) Maximum temperature at which coating will give one hour of protection at atmospheric pressure.

loys such as Ir-30Rh; and duplex and triplex silicide-base coating systems that combine slurry, slip, chemical vapor deposition and pack cementation processes.

Molybdenum

Initially, molybdenum alloys were considered rather specialized aerospace materials, but today they are also used in a wide range of more conventional applications in the thermal-processing, electronic, nuclear, automotive, chemical, glass, specialty-alloy and metal-working industries. This wide range of application reflects the important characteristics of molybdenum: high melting point, high modulus of elasticity, high strength at elevated temperatures, high thermal conductivity, high resistance to corrosion, low specific heat, relatively low density (compared with those of tantalum and tungsten) and low coefficient of expansion.

Molybdenum generally is produced by conventional cold-press-and-sinter P/M methods, although vacuum arc-melted molybdenum and molybdenum-tungsten alloys are available. Molybdenum can be worked in air at temperatures up to 540 °C (1000 °F) without difficulty, but it must be protected if exposed to air at higher temperatures or it will oxidize very rapidly. Its relatively high ductile-to-brittle transition temperature of roughly -20 to +100 °C (0 to 200 °F) requires that fabrication be done at elevated temperature, which adds to cost and inconvenience. Wrought material offers the best ductility at normal temperatures.

Applications

In its most common application,

molybdenum is used as an alloying element in cast irons, steels, heat-resistant alloys and corrosion-resistant alloys to improve hardenability, toughness, abrasion resistance, corrosion resistance, and strength and creep resistance at elevated temperatures. In its pure form, molybdenum is used in a wide range of industries in tools and components that can perform satisfactorily at high temperatures or under severe abrasive or corrosive conditions.

In the electrical and electronic industries, molybdenum is used in cathodes, cathode supports for radar devices, current leads for thoria cathodes, magnetron "end hats", and mandrels for winding tungsten filaments, and is also employed as a filler metal for brazing tungsten. Molybdenum resistance-heating elements are used in electric furnaces that operate at temperatures up to 2200 °C (4000 °F).

Molybdenum is important in the missile industry, where it is used for high-temperature structural parts including nozzles, leading edges of control surfaces, support vanes, struts, re-entry cones, heat-radiation shields, heat sinks, turbine wheels and pumps. These alloys are particularly well-suited for use in airframes because of their high stiffness, high recrystallization temperature, retention of mechanical properties after thermal cycling, and good creep strength.

Mo-0.5Ti has been used in many aerospace applications, but TZM is preferred where higher hot strength is needed.

In the metalworking industry, molybdenum is used for die-casting cores, for hot work tools such as piercer points and extrusion and isothermal forging dies, for boring bars, tool shanks and chill plates, and for tips on resistance-welding electrodes. It is also used for cladding, for truing of grinding wheels, for molds, and for thermocouples.

Molybdenum has also been useful in the nuclear, chemical, glass and metallizing industries. Service temperatures for molybdenum alloys in structural applications are limited to about 1650 °C (3000 °F). Pure molybdenum has good resistance to hydrochloric acid and is comparable to tantalum for acid service in chemical-process industries.

Precautions in Use. Because unprotected molybdenum oxidizes rapidly at temperatures above 500 °C (930 °F) in oxidizing atmospheres, it is not suitable for continued service under such conditions unless it is protected by an adequate coating. Consumable-arc melting either in vacuum or in an inert atmosphere is required. Molten tin, aluminum, nickel, iron and cobalt severely attack molybdenum, as do molten oxidizing salts such as potassium nitrate and potassium carbonate.

Mechanical Properties

Mechanical properties greatly depend on the amount of working performed below the recrystallization temperature and on the ductile-to-brittle transition temperature (DBTT). The minimum recrystallization temperature for molybdenum is 900 °C (1650 °F).

Corrosion Resistance

Molybdenum has particularly good resistance to corrosion by mineral acids provided that oxidizing agents are not present. It is relatively inert in carbon dioxide, hydrogen, ammonia and nitrogen atmospheres up to about 1100 °C (2000 °F), and in reducing atmospheres containing hydrogen sulfide. It has excellent resistance to corrosion by iodine vapor, bromine and chlorine up to clearly defined temperature limits, and good resistance to attack by several liquid metals including bismuth, lithium, magnesium, potassium and sodium. In inert atmospheres it is unaffected up to at least 1750 °C (3180 °F) by refractory oxides such as alumina, zirconia, beryllia, magnesia and thoria. It is subject to attack by fused caustic alkalis, but not by aqueous caustic solutions.

Protective coatings for molybdenum fall into two major classifications. When stresses are not too high, ceramic coatings are effective and long-lasting. At higher stresses, metallic coatings such as gold, nickel, platinum, rhodium and silver are recommended. Metal coatings may be applied by cladding, electroplating, flame spraying, liquid-phase diffusion (dipping), vapor-phase deposition or pack cementation.

Fabrication of Parts

Molybdenum can be drilled, turned, spun, milled, shaped, broached, ground, threaded, sawed, forged, bent, rolled, punched, stamped, sheared, deep drawn or extruded. Both machining and fabrication can be performed using conventional processes, equipment and tools. Machining can be done using cutting tools of tungsten carbide or even high speed steel, usually with soluble-oil coolants. In addition, both electrical discharge machining and electrochemical machining work well with molybdenum.

Molybdenum and its alloys can be joined by brazing, fusion welding, resistance welding, and a few of the more exotic welding methods such as percussion and flash welding. Good results have also been obtained with simple pressure welding.

Molybdenum Alloys

There are two basic alloys of molybdenum in use today: TZM and Mo-30W. Nominal compositions of these alloys are given in Table 2.

TZM is used for high-temperature structural applications and in tooling for hot die forging. It has a higher recrystallization temperature, higher creep strength and higher tensile strength than pure molybdenum.

Mo-30W has outstanding resistance to corrosive attack by liquid metals, especially liquid zinc. Its melting point is 2830 °C (5125 °F), which is higher than the melting point of either TZM or pure molybdenum.

Niobium (Columbium)

By Chun T. Wang
Teledyne Wah Chang Albany

Commercially pure (electron beam welded) niobium is ductile and is easy to fabricate at room temperature by conventional forming practices. Because it is relatively light in weight and high in elevatedtemperature strength, it is used extensively in aerospace applications. Alloy C-103 has been widely used for rocket components requiring moderate strength at temperatures of about 1100 to 1370 °C (2000 to 2500 °F). Nb-1Zr is used in nuclear applications because it has a low thermal-neutron absorption cross section, good corrosion resistance, and good resistance to radiation damage. It is used extensively for liquid-metal systems operating at temperatures from 980 to 1200 °C (1800 to 2200 °F). Nb-1Zr combines moderate strength with excellent fabricability, and as a result it is also used for parts in sodium or magnesium vapor lamps.

Vapor deposition of Nb-1Zr or niobium on the inside surface of type 316 stainless steel tubing improves the performance of the tubing in many chemical-process applications without degrading the mechanical properties of the stainless steel. An intermediate layer of pure niobium under Nb-1Zr improves adherence to the steel substrate.

C-129Y and Cb-752 have shown higher elevated-temperature tensile and creep strengths than C-103 while maintaining good fabricability, coatability and thermal stability. They are used as leading edges, nose caps for hypersonic flight vehicles, rocket nozzles, and guidance structure for reentry vehicles. Cb-132M is a superstrength alloy, specifically for use in turbine applications at temperatures above 1100 °C (2000 °F). FS-85 and B-66 exhibit high strength at temperatures above 1200 °C (2200 °F), yet remain sufficiently ductile for forming and welding operations. They are used as lifting and guidance structures for reentry vehicles, nozzles and gas-turbine parts.

Nb-46.5Ti is a widely used superconducting material. Nb-55Ti is used, in large quantities, for fasteners for airplane structures.

Niobium generally is consolidated and refined by electron beam melting and may be further purified by electron beam zone refining. Powder metallurgy techniques or arc melting also may be used. Purity achieved by these processes is shown in Table 9. Niobium alloys generally are produced by arc melting of high-purity niobium with alloying elements.

Table 9 Typical purity of niobium produced by various methods

Impurity element	Typical concentration, ppm, in niobium: Electron beam melted	Electron beam zone refined	P/M consolidated
C	50	100	50
O	150	50	500
N	50	50	100
H	5	...	1
Al	20	<20	20
B	...	<1	...
Cr	...	<20	...
Cu	30	<30	30
Fe	30	<30	30
Mg	<10	<10	<10
Mn	10	<10	10
Mo	100	40	100
Ni	60	<20	60
Pb	20	40	...
Si	50	<50	100
Sn	20	...	...
Ta	<1000	<1000	...
Ti	<150	<20	20
V	...	<300	...
W	<300	<150	...
Zn	...	<20	...
Zr	100	100	<500

Chemical Properties

Niobium forms an oxide coating in most acid environments. This coating provides excellent corrosion resistance, especially to nitric and hydrochloric acids. Strong alkaline solutions and hydrofluoric acid attack niobium severely. Table 10 presents typical data on corrosion of niobium in various acids and alkalis. At elevated temperatures the metal reacts with halogens, oxygen, nitrogen, carbon, hydrogen and sulfur. It forms high-melting-point refractory compounds with light elements such as carbon, boron, silicon and nitrogen.

Niobium can be used in contact with liquid lithium, sodium and sodium-potassium eutectic at temperatures well above 800 °C (1470 °F). Addition of 1% Zr increases the resistance of niobium to embrittlement caused by oxygen absorbed from the liquid metal.

Table 10 Corrosion of niobium in aqueous media

Solution	Temperature °C	Temperature °F	Duration of test, days	Loss in weight(a) mg/m²·d	Loss in weight(a) oz/ft²·yr	Condition of specimen at end of test
20% HCl	21	70	82	2.5	0.03	No change
Concentrated HCl	21	70	82	6	0.072	Slight etch; not embrittled
	100	212	67	234	2.79	Brittle
Concentrated HNO_3	100	212	67	Nil	Nil	No change
Aqua regia	22	72	6	Nil	Nil	No change
H_2SO_4:						
20% by vol	21	70	3650	0.2	0.0025	No change
25% by vol	21	70	3650	0.3	0.0036	No change
Concentrated (98%)	21	70	3650	5.6	0.067	Partial embrittlement (18.3% drop in toughness)
	50	122	67	48	0.57	Brittle
	100	212	32	1 131	13.5	Brittle
	150	302	2	12 470	149.3	Brittle
	175	347	1	83 200 +	995 +	Completely dissolved
	100	212	42	464	5.56	Pitted and brittle
85% H_3PO_4	21	70	82	0.7	0.0084	No change
	100	212	31	193	2.32	Brittle
20% tartaric acid	22	72	82	Nil	Nil	No change
10% oxalic acid	21	70	82	33	0.40	Brittle
NH_4OH	21	70	82	Nil	Nil	No change
20% Na_2CO_3	100	212	50	74	0.88	Brittle
5% NaOH	21	70	31	66	0.79	Action at surface of liquid
	100	212	5	1 086	13.0	Brittle
5% KOH	21	70	31	442	5.3	Action at surface of liquid
	100	212	5	2 744	32.8	Brittle
30% H_2O_2	21	70	61	11	0.13	Oxide film; not embrittled

(a) Original specimen dimensions: thickness, 0.2 mm (0.008 in.); surface area, 26 cm² (4 in.²). 75% of specimen surface was immersed in the liquid.

Mechanical Properties

Mechanical properties of niobium are highly dependent on purity, particularly on concentration of interstitial elements. In general, mechanical properties reflect a compromise in which fabricability is enhanced at the expense of high-temperature strength. Most niobium alloys display good forming and welding qualities and moderate strength. Detailed mechanical-property data for niobium and specific niobium alloys are presented in the data compilations immediately following this article. It should be noted that these data are for high-purity commercial-grade recrystallized materials. Structural properties are good up to 1650 °C (3000 °F), but protective coatings are required above 425 °C (800 °F) to minimize interaction with the atmosphere.

Fabrication of Parts

Niobium and its alloys are readily machined, varying in machinability from that of copper to that of stainless steel. Special attention should be paid to tool angles and lubrication, because niobium and its alloys gall easily. Lathe turning works best with high speed steel tools. Carbide tooling can be used only for light cuts from 0.25 to 0.4 mm (0.010 to 0.015 in.) deep. Tooling recommendations are given in Table 11.

Niobium metals can be easily forged, rolled or otherwise worked directly from ingot at room temperature. After 50 to 95% reduction in area, annealing is required. Recrystallization occurs in one hour at a temperature of 1200 °C (2200 °F). Niobium alloys usually are broken down at temperatures above recrystallization temperatures and reworked at lower or room temperatures.

Niobium can be joined by electron beam (EB) welding, resistance welding or gas tungsten-arc welding (GTAW). Ductile weldments of niobium and of niobium alloys can be made if proper care is taken to avoid atmospheric contamination in the fusion and heat-affected zones. Postweld annealing increases ductility and toughness.

Niobium and its alloys also can be joined by brazing. Table 12 gives brazing filler metals and maximum service temperatures for brazed joints.

Coatings

Niobium and its alloys oxidize readily in air at elevated temperatures, and surface protection is necessary for operation at high temperatures in oxidizing environments. The coatings generally used are various modifications of aluminide or silicide. Available coatings, methods of coating application and maximum service temperatures are listed in Table 13.

Table 11 Tooling recommendations for machining niobium

Approach angle	15 to 20°
Side rake	30 to 35°
Side and end clearance	5°
Plan relief angle	15 to 20°
Nose radius	0.50 to 0.75 mm (0.020 to 0.030 in.)
Cutting speed:	
High speed steel tools	0.3 to 0.4 m/s (60 to 80 ft/min)
Carbide tools	1.3 to 1.5 m/s (250 to 300 ft/min)
Feed:	
Roughing	0.22 to 0.30 mm/rev (0.009 to 0.012 in./rev)
Finishing	0.13 mm/rev max (0.005 in./rev max)
Depth of cut	0.75 to 3.2 mm (0.030 to 0.125 in.)

Table 12 Typical filler metals for brazing niobium and its alloys

Filler metal	Maximum service temperature °C	°F
Si-Cr-Ni	980	1800
4Be-48Zr-48Ti	925	1700
6Be-19Nb-75Zr	925	1700
Ti-0.5Si	1370	2500
Zr-25V-15Ti-0.1Be	1200	2200
V-35Nb	1200	2200
Ti-33Cr	1370	2500
Ti-3Al-11Cr-13V	1650	3000
Ti-50Zr	1650	3000
Ti	1760	3200

Tantalum

By Mortimer Schussler
Fansteel Inc.

Tantalum is available as powder metallurgy and electron-beam-melted products. Welding of P/M products is not recommended because of the porosity that forms in the heat-affected zone; however, P/M tantalum has superior deep-drawing properties. EB-melted sheet can be used for various welded products, including welded-and-drawn tubing for heat-exchanger bundles.

Applications

Tantalum can be used for structural applications at service temperatures from 1370 to 1980 °C (2500 to 3600 °F), but for any exposure to an oxidizing environment at temperatures above 480 °C (900 °F) it requires a protective coating. It exhibits exceptional resistance to corrosion by acids (except HF and fuming H_2SO_4).

Tantalum's dense, dielectric oxide film makes it useful for miniature capacitors. Currently, the largest use of tantalum is in electrolytic capacitors; sintered P/M anodes are used in solid and wet electrolyte capacitors, and to a lesser extent precision tantalum foil is used in foil capacitors. Tantalum also is used extensively in chemical-process equipment such as heat exchangers, condensers, thermowells and lined vessels. Notably, it is used for condensing, reboiling, preheating and cooling of nitric, hydrochloric, bromic and sulfuric acids, and combinations of these acids with many other chemicals. Spinnerettes for extruding man-made textile fibers constitute another important use.

Because of its high melting point, tantalum is used for heating elements, heat shields and other components in vacuum furnaces. Tantalum and some tantalum alloys have been used in specialized aerospace and nuclear applications. Tantalum has been used in prosthetic devices in contact with body fluids and as an alloying element in superalloys. Tantalum carbide is an important constituent in complex cemented carbides used for cutting steel.

Tantalum-clad parts have been used in numerous process-equipment applications. For instance, steel bellows for liquid-metal service have been clad with tantalum on all surfaces.

Corrosion Characteristics

Tantalum oxidizes in air above 300 °C (570 °F). It is attacked by hydrofluoric acid, fuming sulfuric acid, and strong alkalis. Salts that hydrolyze to form hydrofluoric acid or strong alkalis also attack tantalum. It may be embrittled by hydrogen if it is the cathodic member of a galvanic couple exposed in an acid environment or if it is exposed to a hydrogen-containing atmosphere at elevated temperature. Other agents that can attack tantalum include bromine plus methanol, and halogen gases—fluorine at or above room temperature, chlorine at 250 °C (480 °F), bromine at 300 °C (570 °F), and iodine at somewhat higher temperatures.

Tantalum has excellent resistance to corrosion by most acids, by most aqueous salt solutions and by organic chemicals. It also has good resistance to many corrosive gases and liquid metals.

General Fabrication Practice

Consolidation of tantalum powder is usually accomplished by electron beam or arc melting in vacuum. Consolidation by cold pressing followed by self-resistance sintering is used in some applications. Tantalum products are made by cold working from ingot and can be given dull, matte or bright surface finishes.

Machinability of fully recrystallized, unalloyed tantalum is similar to that of soft copper, whereas machinability of tantalum-base alloys is similar to that of annealed stainless steel. Machining of tantalum and its alloys should be done with high speed steel or cemented carbide tools, using chlorinated hydrocarbons, light oils or water-soluble oils for lubrication. Tantalum and tantalum alloys can be turned, bored, drilled, tapped, reamed, shaped, milled, sawed or ground to desired tolerances and surface finishes.

Cleaning. To avoid contamination of tantalum it is mandatory that the material be chemically clean prior to any heating operation such as annealing or welding. This involves thorough degreasing (detergent or solvent); chemical etching in mixed acids, such as a solution containing 40% HNO_3, 10 to 20% HF, up to 25% H_2SO_4, remainder water; hot- and cold-water rinsing (deionized water is recommended); and spot-free drying. The etching solution may be strengthened by adding HF, or weakened by adding water, to achieve the amount of stock removal necessary for ensuring cleanness of parts.

Annealing Practice. Tantalum often is used in the fully recrystallized condition to achieve best fabricability. Occasionally it is used in the stress-relieved or the cold worked condition. Recrystallization temperature depends on purity, amount of cold work, and prior history. Recrystallization treatment of popular-thickness mill products of unalloyed tantalum ordinarily is done at temperatures from 1000 to 1250 °C (1830 to 2280 °F).

Forming Methods. Operations such as spinning, deep drawing, bulging, bending, blanking, punching and stretch forming can be performed using conventional methods, equipment and tooling.

Joining. Tantalum and its alloys can be joined by gas tungsten-arc welding (GTAW), gas metal-arc welding (GMAW), resistance welding, electron beam (EB) welding or laser welding. Fusion welding must be done in vacuum or in high-purity inert gas (argon or helium); resistance spot or seam welding can be done in air or under water.

Silver brazing alloys, copper brazing alloys and several specially developed refractory-metal brazing alloys can be used for brazing tantalum to itself or to dissimilar metals such as stainless steels. Brazing is done in vacuum or in an inert atmosphere.

Tantalum can also be bonded to mild steel and to copper and other nonferrous alloys by explosive cladding or by roll bonding.

Table 13 Application methods and temperature limits for protective coatings for niobium

Coating designation	Method of application	Maximum service temperature °C	°F	Developer
Aluminide coatings				
NAA-85	Slurry	1425	2600	North American Rockwell
LB-2 (Cr-Si)	Slurry	1425	2600	McDonnell Douglas
(Ti-Si)	Slurry	1425	2600	GT&E
(Sn)	Slurry	1480	2700	GT&E
(Ag-Si)	Slurry	1540	2800	GT&E
Silicide coatings				
Disil	Fluidized bed	1540	2800	Boeing
L-7	Slip pack	1650	3000	McDonnell Douglas
PFR-30,32	Pack cementation	1650	3000	Pfaudler
(Cr,Ti)	Vacuum pack cementation	1480	2700	TRW
(Cr,B)	Pack cementation	1540	2800	LTV
R-512A (Cr)	Slurry	1650	3000	GT&E
R-512E (Cr-Fe)	Slurry	1650	3000	GT&E
(Cb,V)	Fluidized bed	1700	3100	Boeing

Coatings

The greatest obstacle to use of tantalum and its alloys at elevated temperatures in oxidizing environments is the difficulty of providing the material with adequate surface protection. Coatings on tantalum generally have insufficient life in service involving pressures below 1 atm and temperatures above 1370 °C (2500 °F). Coatings also can raise the ductile-to-brittle transition temperature of the substrate and can give unreliable oxidation protection, particularly at edges and corners.

Fused-slurry silicide coatings (such as Si-20Cr-20Fe) and metallic-sintered-slurry-plus-silicide coatings are most popular. Aluminide-base coatings applied in slurry form also have been developed.

Tantalum Alloys

Tantalum alloys have good fabrica-

Table 14 Processing characteristics of tantalum metals

	Unalloyed tantalum	"63" Metal	Ta-10W	T-111	T-222
Forging					
Temperature, °C:					
Start	<500	<500	1100-1200	1100-1200	1150-1250
Finish	20	20	800	800	800
Typical total reduction, %	75	75	75	75	75
Extrusion					
Temperature, °C	1100	1100	1425-1650	1425-1650	1650
Typical reduction ratio	4:1	4:1	4:1	4:1	3:1
Rolling					
Temperature, °C:					
Breakdown	20	20	500	500	500
Finish	20	20	20	20	20
Typical total reduction between anneals, %	80	80	75	75	75
Stress-relieving temperature, °C	900	1000	1100	1100	1100
Recrystallization temperature, °C	980-1200	1200-1300	1300-1650	1400-1650	1400-1650
Ductile-to-brittle transition temperature, °C:					
Wrought	<−250	<−250	<−250	<−250	<−196
Recrystallized	<−250	<−250	<−250	<−250	<−196

tion characteristics, high melting points and relatively good mechanical properties, and thus they are well-suited for high-temperature structural applications. Their use in aerospace applications has increased despite high cost and high density.

High yield and ultimate strength accompanied by loss in ductility results from increases in interstitial contents (oxygen, nitrogen, carbon and hydrogen). Embrittlement of tantalum and its alloys can occur if contamination by these elements is sufficiently severe.

Commercial tantalum alloys are discussed individually, below. Because of their relatively high strengths, ingot breakdown by forging or extrusion must be done at quite high temperatures. Mill-processing temperatures, recrystallization temperatures and ductile-to-brittle transition temperatures for tantalum alloys are listed in Table 14. Tables 15 and 16 present some typical properties of tantalum and Ta-base alloys at both room and elevated temperatures.

Ta-2.5W-0.15Nb, an electron-beam-melted, solid-solution alloy, is similar to commercially pure tantalum in fabricating characteristics, welding characteristics and corrosion resistance, but has about twice the yield strength at 200 °C (400 °F). Its applications include heat exchangers, valves, piping, fittings, gaskets and linings for chemical-process towers.

Ta-10W is the oldest Ta alloy, and the most widely used for aerospace parts. It is stronger than pure tantalum and retains ductility and general fabricability at low temperatures.

In the aerospace industry Ta-10W has been used either bare or coated at temperatures up to 2480 °C (4500 °F). Applications include hot-gas metering valves, rocket-engine extension skirts, complex manifold assemblies, and fasteners. In the chemical-process industries, it is used in machined valves, internal seats and plugs of large valves, liners requiring abrasion and corrosion resistance, disks used in patching glass-lined steel vessels, and tubing for some nuclear applications.

Ta-7.5W (P/M) is consolidated by powder metallurgy techniques. It has a lower tungsten content, but higher interstitial contents, than Ta-10W, and as a result has higher strength at room temperature. Cold drawn Ta-7.5W wire (83% reduction) has a yield strength of about 1100 MPa (160 ksi). Applications include springs and other elastic parts in gas chlorinators.

T-111 (Ta-8W-2Hf) is an essentially solid-solution alloy that is consolidated by EB melting followed by consumable-electrode vacuum-arc melting. It provides higher strength, especially stress-rupture strength and resistance to creep, than Ta-10W, and has comparable fabricability and low-temperature properties. It is also more resistant to some liquid metals, such as potassium, sodium, NaK eutectic, lithium, cesium and mercury, than either unalloyed tantalum or Ta-10W. T-111 is used as a containment material. Applications include handling of liquid metals used as coolants in high-temperature nuclear reactors and as heat-transfer and working fluids in compact nuclear reactor systems such as the Rankine cycle satellite power system.

T-222 (Ta-10W-2.5Hf-0.01C) also is consolidated by EB melting followed by consumable-electrode vacuum-arc melting. It is comparable to T-111 in fabricability and low-temperature properties; but, as a result of higher tungsten and hafnium contents and addition of the interstitial element carbon, it has higher strength. Applications for T-222 are generally the same as those for T-111.

T-222 currently offers the best combination of properties for high-temperature (1980 °C, or 3600 °F), radiation-cooled structures; however, suitable protective coatings have not yet been developed.

Astar 811C (Ta-8W-1Re-0.7Hf-0.025C), which is consolidated by double consumable-electrode vacuum-arc melting, can be considered a semicommercial or developmental alloy. It was developed to surpass T-111 and T-222 in both creep resistance and fabricability. It is intended for high-temperature structural applications and for containment of liquid alkali metals.

Tungsten

By Chester W. Dawson
and
Heinz G. Sell
Westinghouse Electric Company

Tungsten's high melting point makes it the obvious choice for structural applications at very high temperatures. Design engineers must first

Table 15 Typical mechanical properties of tantalum and tantalum-base alloys

Temperature °C	°F	Tensile strength MPa	ksi	Yield strength MPa	ksi	Elongation, %	Modulus of elasticity GPa	10^6 psi
Commercially pure Ta, EB melted								
20	70	205	30	165	24	40	185	27
200	390	190	27.5	69	10	...	...	...
750	1380	140	20	41	6	...	160	23
1000	1830	90	13	...	...	...	...	...
Commercially pure Ta, P/M								
20	70	310	45	220	32	30	185	27
"63" Metal, EB melted								
20	70	345	50	230	33	40	195	28
200	390	315	46	195	28	33	...	...
750	1380	180	26	83	12	22	...	...
1000	1830	125	18	69	10	20	...	...
Ta-10W, EB melted								
20	70	550	80	460	67	25	205	30
200	390	515	75	400	58	...	...	...
750	1380	380	55	275	40	...	150	22
1000	1830	305	44	205	30	...	...	...
Ta-7.5W (P/M product), wire								
20	70	1030	150	1010	146	6	200	29
Ta-7.5W (P/M product), sheet								
20	70	1170	169	875	127	7	200	29
T-111, EB plus arc melted								
20	70	690	100	585	85	29	200	29
1980	3596	90	13	85	12.6	36	...	...
T-222, EB plus arc melted								
20	70	805	117	800	116	28	200	29
200	390	620	90	515	75	...	...	...
750	1380	565	82	330	48	...	...	...
1000	1830	550	80	275	40	...	...	...
1980	3595	97	14	97	14	...	...	...

Table 16 Typical properties of tantalum and tantalum-base alloys

Alloy	Consolidation practice	Hardness, HV	Density Mg/m³	lb/in.³	Melting point °C	°F
Commercially pure Ta	EB melted	110	16.6	0.600	3000	5430
	P/M	120	16.6	0.600	3000	5430
"63" Metal	EB melted	130	16.6	0.602	3010(a)	5440(a)
Ta-10W	EB melted	245	16.8	0.608	3030	5490
Ta-7.5W:						
Wire	P/M	326	16.7	0.606	3025(a)	5477(a)
Sheet	P/M	404	16.7	0.606	3025(a)	5477(a)
T-111	EB plus arc melted	...	16.7	0.604	2980(a)	5400(a)
T-222	EB plus arc melted	280	16.7	0.604	...	...

(a) Estimated.

overcome or accommodate its high density, brittleness at normal temperatures, poor oxidation resistance and poor fabricability.

Tungsten generally is produced by conventional cold-press-and-sinter P/M methods, although vacuum-arc-melted W-Mo alloys are available. Ingots of arc-melted and EB-melted tungsten and W-25 Re alloy also have been produced.

Tungsten produced by powder metallurgy is available as wrought bar, sheet or wire. These products are available in either commercially pure (undoped) tungsten or commercially doped (AKS) tungsten to which small amounts of aluminum, potassium and silicon have been added. These additives improve the recrystallization and creep properties of tungsten, which are especially important when tungsten is used for incandescent lamp filaments. Wrought P/M stock can be zone refined by electron beam melting to produce single crystals higher in purity than the commercially pure product. EB zone-melted tungsten single crystals are of commercial interest for applications where single crystals with very high electrical resistance ratios are needed. A comparative impurity analysis of the three grades of tungsten is given in Table 17.

Tungsten Alloys

Three types of tungsten alloys are produced commercially: $W\text{-}ThO_2$, W-Mo, and W-Re alloys.

$W\text{-}ThO_2$ alloy is a dispersed second-phase alloy containing 1 to 2% thoria. The thoria dispersion enhances thermionic electron emission, which improves starting characteristics of GTAW welding electrodes. It also increases the efficiency of electron-discharge tubes and imparts creep strength to wire at temperatures above one-half the absolute melting point of tungsten.

W-Mo Alloys. Molybdenum forms a continuous solid solution with tungsten, the solidus of which is about 20 °C (36 °F) below the liquidus at 50% W. W-Mo alloys are used mostly for improved machinability where strength somewhat lower than those of tungsten and $W\text{-}ThO_2$ can be tolerated.

W-Re Alloys. Rhenium is soluble in tungsten up to 26%, above which the embrittling sigma phase begins to form. The W-1.5Re and W-3Re alloys are used to improve resistance to cold fracture in lamp filaments, especially for lamps exposed to vibrations. These alloys also contain the AKS dopants to improve creep strength in filament wires. Undoped W-1.5Re and W-3Re are used in thermocouple applications where strength is not a primary factor.

Alloy Structures

The high melting point and low va-

Table 17 Typical purity of the three commercial grades of tungsten

Impurity element	Concentration, ppm, in tungsten: Electron beam zone refined	Undoped	Doped
Fe	1	10	11
Ni	2	5	5
Si	5	21	47
Al	<2	<5	15
K	<1	12	91
O	10	27	36
C	20	31	24

Fig. 1 Recrystallized microstructure of undoped tungsten wire

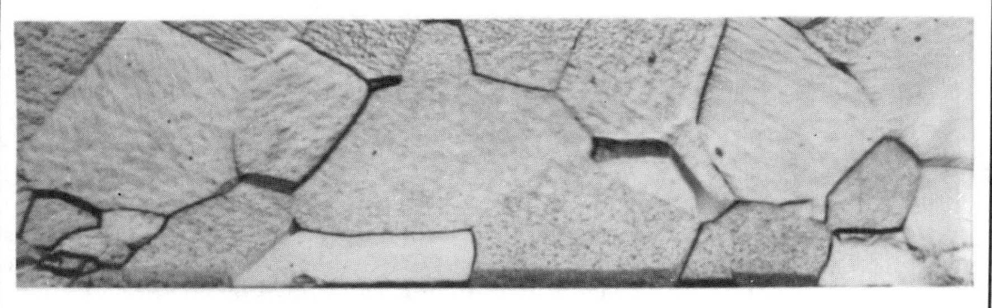

Fig. 2 Recrystallized microstructure of doped tungsten wire

por pressure of tungsten constitute the basis for many applications of tungsten and tungsten alloys.

Tungsten metals can be divided into three distinct groups on the basis of recrystallization behavior. The first group consists of EB-melted, zone-refined, arc-melted and other very pure forms of tungsten by itself or alloyed with rhenium or molybdenum. These materials exhibit equiaxed grain structures upon primary recrystallization. Recrystallization temperature and grain size both decrease with increasing deformation.

The second group, consisting of commercial-grade or undoped powder metallurgy tungsten, demonstrates tungsten's sensitivity to purity. These materials also exhibit equiaxed grain structures (Fig. 1), but their recrystallization temperatures are higher and they do not necessarily exhibit decreases in recrystallization temperature and grain size with increasing deformation. In EB-melted tungsten wire, the recrystallization temperature can be as low as 900 °C (1650 °F) or less, whereas in commercially pure (undoped) tungsten it can be as high as 1200 to 1400 °C (2200 to 2550 °F).

The third group of materials consists of AKS doped tungsten, doped tungsten alloyed with rhenium, and undoped tungsten alloyed with more than 1% ThO_2. These materials are characterized by higher recrystallization temperatures (>1800 °C or 3270 °F) and uniquely different recrystallized grain structures (Fig. 2). The structure of heavily drawn wire or heavily rolled sheet consists of very long, interlocking grains. This structure is more readily found in AKS doped tungsten and in doped tungsten alloyed with 1 to 5% Re. The K dopant is smeared out in the direction of rolling or drawing, and volatilizes when heated into a linear array of submicron-size bubbles. These bubbles pin grain boundaries in the manner of a dispersion of second-phase particles. The recrystallization temperature will rise, and the interlocking structure will become more pronounced, as the rows of bubbles become finer and longer with increasing deformation. Higher concentrations of rhenium (7 to 10%) destroy this effect. In W-2ThO_2, occurrence of the elongated, interlocking structure depends on thermomechanical treatment and on the fineness of the thoria dispersion.

Addition of 1.5% or more ThO_2 raises the recrystallization temperature of tungsten much as the potassium dopant raises it, but ThO_2 additions generally result in a much finer grain structure.

Rhenium in amounts up to about 5% inhibits recrystallization. In greater amounts, it lowers resistance to recrystallization.

Electrical Properties

Tungsten's electrical resistivity and temperature coefficient of electrical resistivity both are strongly affected by purity and deformation. The effects of recovery annealing on these two properties for commercially pure tungsten wire are shown in Table 18. The product of resistivity and temperature coefficient is a nearly constant value that is independent of the degree of residual cold work.

Addition of rhenium, molybdenum or ThO_2 increases the resistivity of tungsten wire but has no appreciable effect on its temperature coefficient.

Thermocouples in which tungsten is one of the thermoelements are used extensively at very high temperatures. Tungsten-molybdenum thermocouples, for example, can be used at temperatures up to 2200 °C (4000 °F) if maintained in a protective envelope or a reducing atmosphere.

Table 18 Effect of annealing on electrical resistivity of drawn tungsten wire

Annealing temperature °C	°F	Electrical resistivity, μΩ·m	Temperature coefficient	Matthiessen's rule(a)
As drawn		617	0.355	219
400	750	591	0.376	222
600	1110	543	0.415	225
800	1470	523	0.433	226
1000	1830	518	0.440	228
1200	2200	500	0.432	216
2500	4530	484	0.481	233

(a) Product of specific resistance and temperature coefficient.

Fig. 3 Variation of DBTT with annealing temperature for undoped tungsten

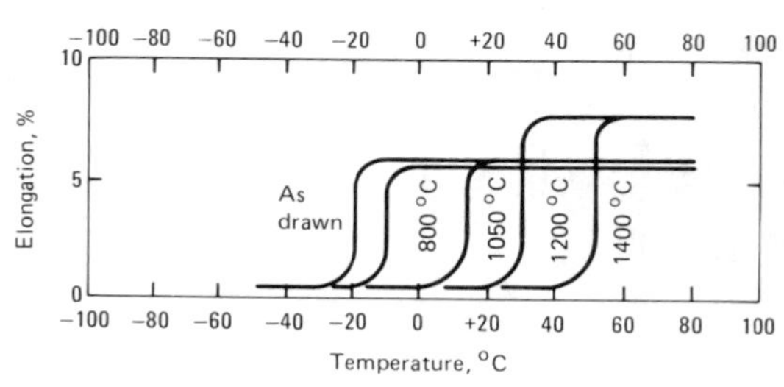

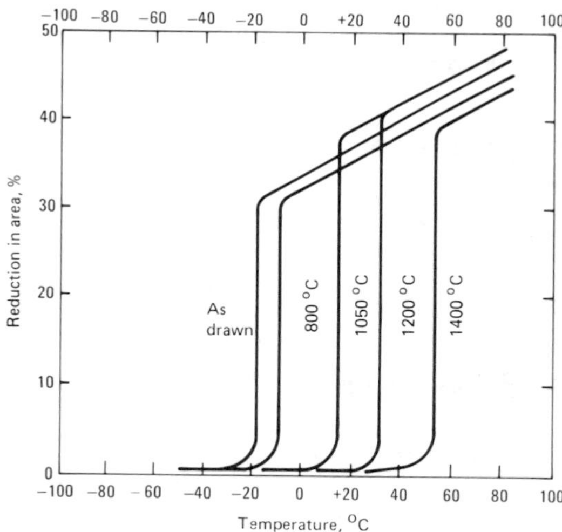

Data are for 10-min recovery annealing of heavily worked 0.75-mm-diam (0.030-in.-diam) wire.

Mechanical Properties

Tungsten has high tensile strength and good creep resistance. Above 2200 °C (4000 °F) tungsten has twice the tensile strength of the strongest tantalum alloys but only about 10% more density. However, its high density, poor low-temperature ductility and strong reactivity in air limit its usefulness.

Maximum service temperatures for tungsten range from 1925 to 2480 °C (3500 to 4500 °F), but surface protection is required for use in air at these temperatures.

Wrought tungsten (as cold worked) has high strength, strongly directional mechanical properties and some room-temperature toughness. But recrystallization occurs rapidly above 1370 °C (2500 °F) and produces a grain structure that is crack-sensitive at all temperatures.

Recrystallized tungsten undergoes a ductile-to-brittle transition above 200 °C (400 °F). Only by heavy warm or cold working is the ductile-to-brittle transition temperature (DBTT) lowered to below room temperature (Fig. 3). Annealing raises the DBTT of cold worked tungsten until it approaches that of recrystallized material.

The exact DBTT is influenced by many factors, including grain size, strain rate and impurity levels. The DBTT decreases as grain size decreases unless the grains are larger than 1 mm in diameter. DBTT also drops with increases in strain rate, but it climbs rapidly as impurity levels increase. Like all brittle metals, tungsten is very notch-sensitive. Therefore, removal of even minute surface flaws by grinding, oxidizing or electrolytic polishing prior to service improves ductility and lowers DBTT.

Alloying can have a beneficial effect on DBTT; the effect of rhenium in producing a ductile alloy is the best-known example. AKS doping or alloying with a dispersion of thoria retards recrystallization and thereby improves the ductility of annealed wire. In addition, a fine dispersion of thoria causes a decrease in grain size, which promotes a reduction in DBTT.

Below the DBTT, recrystallized tungsten fails by a combination of cleavage and grain-boundary fracture. Near the DBTT, fracture by cleavage increases. At higher temperatures, usually above 500 °C (930 °F), grain-boundary and ductile fracture predominate. Generally, grain-boundary fracture predominates in commercially pure tungsten, and ductile fracture in AKS doped tungsten.

Aside from its use in abrasive and wear-resistant tools and as an alloying element, tungsten's primary commercial application is in filaments for incandescent lamps. Thoria particles, and the potassium bubble dispersion that occurs in AKS doped tungsten, impede the annealing process that progressively eliminates substructure. This allows tungsten to retain hardness and tensile strength at temperatures higher than those at which commercially pure or refined tungsten is softened and weakened. It also improves the creep resistance of tungsten wire at elevated temperatures. Upon recrystallization, a nonsag, interlocking grain structure forms. This structure gives tungsten wire added creep resistance at high temperatures, which allows tungsten filaments in incandescent lamps to burn at high temperatures without sagging.

Alloying with rhenium also improves the tensile strength of AKS doped or undoped tungsten. Although small additions of less than 5% Re cause softening of W-Re alloys, hardness again increases when solid-solution strengthening becomes the overriding factor. Alloying with molybdenum has a softening effect that is proportional to molybdenum content.

Tungsten is not as anisotropic in elastic behavior as are some other cubic metals, but its stress-strain curve does vary somewhat with crystallographic direction.

Chemical Properties

At room temperature, tungsten is generally resistant to most chemicals but can be easily dissolved with a solution of nitric and hydrofluoric acids. Murakami's reagent is the most commonly used metallographic etchant, although hot hydrogen peroxide and solutions of nitric, hydrofluoric, sulfuric and phosphoric acids are also available. At higher temperatures, tungsten becomes more prone to attack. At about 250 °C (480 °F), it reacts rapidly with phosphoric acid and with chlorine. It begins to oxidize readily at 500 °C (930 °F), and at 1000 °C (1830 °F) it reacts with many gases, including water vapor, iodine, bromine and carbon monoxide. Above 1000 °C it begins to form compounds with various metals.

SELECTED REFERENCES

- "Aerospace Structural Metals Handbook"—Code 5201 (1963), Code 5206 (1966), Code 5208 (1967), Code 5209 (1971), Code 5211 (1973): Department of Defense, Mechanical Properties Data Center, Traverse City, MI
- "Aerospace Structural Metals Handbook, Volume 11A—Nonferrous Heat Resistant Alloys", by J. G. Sessler and V. Weiss: AFML-TR-68-115, Air Force Materials Laboratory, Wright-Patterson AFB, 1969
- *Columbium and Tantalum,* by F. T. Sisco and E. Epremian: John Wiley & Sons, New York, 1963
- "Creep Behavior of Tantalum Alloy T-222 at 1365 to 1700 K," by R. H. Titran: NASA Technical Note TN D-7673, June 1974
- "The Engineering Properties of Columbium and Columbium Alloys", by F. F. Schmidt and H. R. Ogden: DMIC Report 188, Battelle Memorial Institute, Columbus, OH, 1963
- "The Engineering Properties of Tantalum and Tantalum Alloys", by F. F. Schmidt and H. R. Ogden: DMIC Report 189, Battelle Memorial Institute, Columbus, OH, 1963
- "The Engineering Properties of Tungsten and Tungsten Alloys": DMIC Report 191, Battelle Memorial Institute, Columbus, OH, 1962
- Fabricating the Refractory Metals, by D. R. Mash, D. W. Bauer and M. Schussler: *Metal Progress,* Vol 99, No. 2, 3 and 4, Feb, Mar, Apr, 1971
- "Generation of Long Time Creep Data on Refractory Alloys at Elevated Temperatures", by K. D. Sheffler: ER-7442, TRW Inc., Jan 1970
- "The High Temperature Specific Heat of Body-Centered Cubic Refractory Metals", by M. Hoch: GEMP-696, General Electric Company, 1969
- "Physical and Mechanical Properties of Tungsten and Tungsten-Base Alloys": DMIC Report 127, Battelle Memorial Institute, Columbus, OH, 1960
- "Physical Metallurgy of Tungsten and Tungsten Base Alloys", by H. G. Sell, G. H. Keith, R. C. Koo, R. H. Schnitzel and R. Corth: Technical Report WADD-TR-60-37, Parts I-III, Armed Services Technical Information Agency, 1961
- "The Spectral Emissivity and Optical Properties of Tungsten", by R. D. Larrabee: Technical Report 328, Massachusetts Institute of Technology, 1957
- "State of the Science for Tungsten Alloys", by S. Foldes: General Electric Technical Information Series 62-LMC-197, 1962
- *Tungsten*: Metallwerk Plensee AG. and Co., 1971
- *Tungsten and Its Compounds,* by G. D. Rieck: Pergamon Press Inc., New York, 1967
- *Tungsten, Its History, Geology, Ore Dressing, Metallurgy, Chemistry, Analysis, Applications and Economics,* by K. C. Li and C. Y. Wang: Reinhold Publishing, 1955
- *Tungsten, Its Metallurgy, Properties and Applications,* by C. J. Smithells: Chemical Publishing Co., 1953
- *Vanadin·Niob·Tantal,* by R. Kieffer and H. Braun: Springer-Verlag, Berlin, Germany, 1963
- *Tungsten: Sources, Metallurgy, Properties and Applications,* by S. Y. H. Yih and C. T. Wang: Plenum Press, 1979

Machinable Tungsten Alloys

By Jerome F. Kuzmick
Powder Tech Company

MACHINABLE TUNGSTEN ALLOYS are of interest for applications requiring material of high density or mass, such as counterweights for self-winding wristwatches and instruments, counterbalance weights for jet aircraft and helicopters, gyroscope rotors and similar inertial components, tool shanks for chatter-free machining tools, and shielding for use against x-rays and gamma rays. Because of the high melting point of tungsten, these components must be made by powder-metallurgy techniques.

Pure tungsten powder is difficult to sinter to 100% of theoretical density, and the sintered product is difficult to machine because of its brittle character. Shortly before World War II, machinable, fairly ductile tungsten-base alloys were developed (Ref 1). These alloys are a classic example of the application of liquid-phase sintering to the production of powder-metallurgy parts. In this case, the basic metal is tungsten, and the liquid phase in which tungsten is partly soluble is primarily nickel. In the original heavy-metal alloys, it was found that addition of copper was desirable because it lowered the melting temperature of the liquid phase, thus lowering the sintering temperature. The resulting tungsten-nickel-copper alloy had good mechanical properties, fair ductility and good machinability.

Subsequently developed were tungsten-nickel-iron alloys that had greater ductility than the tungsten-nickel-copper materials (Ref 2). It was also found that the tungsten-nickel-iron alloys could be sintered to practically theoretical density with higher percentages of tungsten, so that materials of even higher specific gravity could be produced.

Table 1 Classification of machinable tungsten alloys by composition, density, and hardness

Class	Tungsten content, %	Density, Mg/m^3	Hardness, HRC	Available in type
1	89-91	16.85-17.25	30-36	I
1	89-91	16.85-17.25	32 (max)	II, III
2	91-94	17.15-17.85	33 (max)	II, III
3	94-96	17.75-18.35	34 (max)	II, III
4	96-98	18.25-18.85	35 (max)	II, III

Classification of Alloys

The specifications for machinable, high-density tungsten-base alloys usually divide them into four classes based on composition (Table 1) and three types based on tensile properties (Table 2).

Class 1 alloys may be basically tungsten-nickel-copper or tungsten-nickel-iron alloys. Tungsten-nickel-copper alloys of this class typically contain 90 W, 6 to 7 Ni and 3 to 4 Cu. Minor additions of other metals, such as molybdenum or cobalt, may be made to modify properties such as hardness. Class 1 tungsten-

Table 2 Classification of machinable tungsten alloys by tensile properties

Type	Tensile strength, ksi	0.2% yield strength, ksi	Elongation, %
I	130	105	1.5
II	94	75	2.0
III	60	...	1.0

nickel-iron alloys usually contain 90 W, 5 to 7 Ni and 3 to 5 Fe.

Class 2, 3 and 4 alloys are usually tungsten-nickel-iron alloys with tungsten weight percentages in the range shown in Table 1, remainder Ni-Fe in a ratio of 7Ni:3Fe to 5Ni:5Fe. Sometimes a portion of the iron may be replaced with copper.

Methods of Manufacture

The heavy-metal alloys usually are produced from a mixture of elemental, high-purity, fine-particle-size metal powders. The tungsten powder is about 2 to 3 μm in average particle size and is 99.99% pure. Fine high-purity nickel powder such as carbonyl nickel, fine electrolytic copper powder, and fine high-purity iron powder such as carbonyl iron are used. The powders are blended in a powder blender or ball mill for sufficient time to produce a homogeneous mixture and to achieve an apparent density compatible with the molding operation. If molding is to be done by isostatic pressing, no binder is required. If molding is to be done in a steel or carbide die in a hydraulic or mechanical press, the powder is coated with paraffin or other suitable organic binder. Molding pressures of about 70 to 140 MPa (10 to 20 ksi) are used. The molded compact must be of such size as to allow for considerable shrinkage during the sintering operation, which usually is on the order of 20% lineal shrinkage, or more than 50% by volume. Because of the high shrinkage involved, most parts produced from these alloys require finish machining if close dimensional tolerances are required.

Sintering. The molded parts are usually sintered in box-type electric sintering furnaces either by plunging or stoking. The furnaces must have molybdenum or tungsten heating elements, because sintering temperatures range from about 1425 to 1650 °C (2600 to 3000 °F) depending on the exact composition of the alloy. In some instances, vacuum furnaces are used for sintering these materials, but the usual operation utilizes dry hydrogen or dissociated ammonia for the sintering atmosphere. Sintering times at temperature range from about 20 minutes for small parts to several hours for large blanks. Weights of parts may range from a few grams to 20 kilograms or more. During sintering, rapid densification of the compact occurs, as the fine tungsten particles are dissolved in the liquid phase and then are precipitated on the larger tungsten particles. This causes the compact to shrink, and produces a very dense structure with rounded tungsten-rich grains considerably greater in diameter than the original tungsten particles.

The blanks are cooled to room temperature in the cooling chamber of the furnace and then are removed.

Tensile bars and other test blanks usually are sintered from each powder mix and tested for mechanical and physical properties prior to approval of that mix for production.

Hot Pressing. Some very large parts are produced by hot pressing

Table 3 Typical mechanical properties of commercial machinable tungsten alloys

W-Ni-Cu alloy, class 1	
Density	17.0 Mg/m^3 (0.614 lb/in.^{3})
Tensile strength	785 MPa (114 ksi)
Yield strength(a)	605 MPa (88 ksi)
Elongation(b)	4%
Hardness	27 HRC
Proportional limit	205 MPa (30 ksi)
Modulus of elasticity	275 GPa (40 × 10^6 psi)
Coefficient of thermal expansion	5.5 μm/m · °C (3.0 μin./in. · °F)
Magnetic properties	Virtually nonmagnetic
W-Ni-Fe alloy, class 1	
Density	17.0 Mg/m^3 (0.614 lb/in.^{3})
Tensile strength	895 MPa (130 ksi)
Yield strength(a)	615 MPa (89 ksi)
Elongation(b)	16%
Hardness	27 HRC
Proportional limit	260 MPa (38 ksi)
Modulus of elasticity	275 GPa (40 × 10^6 psi)
Coefficient of thermal expansion	5.4 μm/m · °C (3.0 μin./in. · °F)
Magnetic properties	Slightly magnetic
W-Ni-Fe alloy, class 3	
Density	18.0 Mg/m^3 (0.650 lb/in.^{3})
Tensile strength	925 MPa (134 ksi)
Yield strength(a)	655 MPa (95 ksi)
Elongation(b)	6%
Hardness	29 HRC
Proportional limit	350 MPa (51 ksi)
Modulus of elasticity	310 GPa (45 × 10^6 psi)
Coefficient of thermal expansion	5.3 μm/m · °C (2.9 μin./in. · °F)
Magnetic properties	Slightly magnetic
W-Ni-Fe alloy, class 4	
Density	18.5 Mg/m^3 (0.667 lb/in.^{3})
Tensile strength	795 MPa (115 ksi)
Yield strength(a)	690 MPa (100 ksi)
Elongation(b)	3%
Hardness	32 HRC
Proportional limit	450 MPa (65 ksi)
Modulus of elasticity	345 GPa (50 × 10^6 psi)
Coefficient of thermal expansion	5.0 μm/m · °C (2.6 μin./in. · °F)
Magnetic properties	Slightly magnetic

(a) 0.2% offset. (b) In 25 mm or 1 in.

Table 4 Additional properties of machinable tungsten alloys

Property	Minimum	Average
Type I		
Modulus of rupture (flexure), MPa (ksi)	1380 (200)	1585 (230)
Proportional limit, MPa (ksi)	310 (45)	425 (62)
Modulus of elasticity, GPa (10^6 psi)	205 (30)	305 (44)
Modulus of rigidity, GPa (10^6 psi)	130 (19)	...
Angle of twist at rupture, degrees	80	100
Shear strength, MPa (ksi)	895 (130)	...
Electrical conductivity, % IACS	13.5	14
Type II		
Modulus of rupture, (flexure), MPa (ksi)	1240 (180)	1515 (220)
Proportional limit, MPa (ksi)	...	170 (25)
Modulus of elasticity, GPa (10^6 psi)	170 (25)	275 (40)
Modulus of rigidity, GPa (10^6 psi)	130 (19)	132 (19.2)
Angle of twist at rupture, degrees	160	166
Shear strength, MPa (ksi)	550 (80)	560 (81)
Electrical conductivity, % IACS	13	14
Type III (a)		
Modulus of rupture (flexure), MPa (ksi)	690 (100)	...

(a) This type used almost exclusively for radiation shielding; properties other than modulus of rupture not available.

rather than by cold pressing and sintering. This usually is done by leveling the powder mix in a graphite mold and heating in an induction coil while light pressure, sufficient to compact the mix to the required density at temperatures similar to the sintering temperatures mentioned above, is applied to the assembly. Hot pressed alloys of this type usually are more brittle and lower in strength than the cold pressed and sintered materials. Also, the graphite mold may cause a carburized layer to form on the surface of the blank, which is then difficult to remove in machining.

Machining and Finishing. Tungsten heavy alloys can be machined by the usual methods, such as turning, boring, shaping, drilling and tapping. For small runs steel tools may be used, but generally carbide tools of the type used for machining cast iron are recommended.

Although these alloys have good corrosion resistance, for special applications they are sometimes plated with nickel or cadmium, or coated with corrosion-resistant paints.

Properties

Minimum mechanical properties of machinable heavy tungsten alloys are given in the specifications under which these materials are purchased. There are three specifications in general use: MIL-T-21014, ASTM B459 and AMS 7725.

Tables 3 and 4 give typical mechanical and physical properties of tungsten heavy-metal alloys as commercially produced.

REFERENCES

1 U. S. Patent 2,183,359 to Smithells. Also British Patent 447,567

2 Development of Ductile Tungsten Base Heavy Metal Alloys, by J. F. Kuzmick: *Modern Developments in Powder Metallurgy,* Vol 3, Metal Powder Industries Federation, 1965

Properties of Refractory Metals

Niobium Alloys

By Chun T. Wang
Sr. Research Metallurgist
Teledyne Wah Chang Albany

Niobium

Commercial Name

Common name. Columbium, commercial high-purity

Chemical Composition

Composition limits. See Table 9 of the article "Refractory Metals and Alloys".
Consequence of exceeding impurity limits. High content of interstitial impurities increases strength, but decreases ductility

Applications

Typical uses. Superconductor, aerospace and nuclear applications
Precautions in use. At temperatures above 650 °C (1200 °F), too soft to be used as a structural material. Absorbs oxygen when exposed to atmosphere at moderately high temperature.

Mechanical Properties

Tensile properties. Tensile strength, 170 to 205 MPa (25 to 30 ksi); yield strength, 75 to 105 MPa (11 to 15 ksi); elongation, 30-35%
Hardness. 40 to 80 HV
Elastic modulus. Tension, 105 GPa (15.2×10^6 psi)
Fatigue strength. 140 MPa (20 ksi) at 1 million cycles

Mass Characteristics

Density. 8.66 Mg/m^3 (0.310 lb/in.^{3}) at 20 °C (68 °F)

Thermal Properties

Liquidus temperature. 2470 °C (4475 °F)
Coefficient of thermal expansion. Linear, 7.1 μm/m·K (3.94 μin./in.·°F)
Specific heat. 268 J/kg·K (0.064 Btu/lb·°F) at 0 °C (32 °F)
Thermal conductivity. 649 W/m·K (32 Btu/ft·h·°F) at 0 °C (32 °F)

Electrical Properties

Electrical resistivity. 16 nΩ·m at 0 °C (32 °F)

Chemical Properties

Resistance to specific corroding agents. Nb can be used in contact with Li, Na, and Na-K eutectic provided the oxygen content is low.

Fabrication Characteristics

Formability. Conventional sheet-metal or tube working processes can be applied.
Weldability. Nb of commercial purity can be joined by electron-beam, resistance, or gas tungsten-arc welding.
Annealing temperature. 980 to 1250 °C (1800 to 2280 °F)

Nb-1Zr

Commercial Names

UNS number. Commercial grade, R04261; reactor grade, R04251
Trade name. Wah Chang WC-1Zr, Fansteel 80
Common name. Cb-1Zr

Specifications

ASTM. B391, B392, B393, B394

Chemical Composition

Composition limits. Commercial grade: 98.5 min Nb; 0.8 to 1.2 Zr; 0.0100 max C; 0.0300 max N; 0.0300 max O; 0.0020 max H; 0.01 max Hf; 0.01 max Fe; 0.005 max Mo; 0.005 max Ni; 0.005 max Si; 0.2 max Ta; 0.05 max W. For reactor grade, Fe is 0.005 max, O is 0.015 max, Ta is 0.1 max and W is 0.03 max

Consequence of exceeding impurity limits. Increasing interstitial content decreases ductility of the material.

Applications

Typical uses. For thermal barriers, high temperature parts; nuclear applications; liquid metal containers; sodium or magnesium vapor lamp parts

Mechanical Properties

Tensile properties. Recrystallized: typical tensile strength, 241 MPa (35 ksi); yield strength, 138 MPa (20 ksi); elongation, 20% in 25.4 mm or 1 in. See also Table 1.

Shear strength. See Table 2.

Elastic modulus. Tension, 68.9 GPa (10×10^6 psi)

Impact strength. See Table 3.

Creep-rupture properties. See Fig. 1.

Mass Characteristics

Density. 8.59 Mg/m^3 (0.31 lb/in.^{3})

Thermal Properties

Liquidus temperature. 2407 °C (4365 °F)

Coefficient of thermal expansion. Linear, 7.54 μm/m·K (4.19 μin./in.·°F) at 20 to 400 °C (68 to 750 °F)

Specific heat. 270 J/kg·K (0.065 Btu/lb·°F) at 20 °C (70 °F)

Thermal conductivity. 41.9 W/m·K (24.2 Btu/ft·h·°F) at 25 °C (77 °F)

Electrical Properties

Electrical resistivity. 14.7 nΩ·m at 0 °C (32 °F)

Chemical Properties

Resistance to specific corroding agents. Especially resistant to liquid metals

Fabrication Characteristics

Machinability. 80% of C36000, free-cutting brass

Forgeability. 75% at 650 to 980 °C (1200 to 1800 °F)

Formability. Extrusion: reduction ratio of 10:1 at 1066 °C (1950 °F).

Table 1 Temperature effect on tensile properties of Nb-1Zr

Temperature °C	Temperature °F	Tensile strength MPa	Tensile strength ksi	Yield strength MPa	Yield strength ksi	Elongation, %
21	70	345	50	255	37	15
1090	2000	185	27	165	24	...
1650	3000	83	12	69	10	...

Table 2 Shear strength for Nb-1Zr rivets

Fastener type	Diameter mm	Diameter in.	Shear strength at 20 °C (70 °F) MPa	ksi	Shear strength at 870 °C (1600 °F) MPa	ksi
Huck rivet	3.18	0.125	265	38.5	220	32.0
			300	43.5	220	32.0
Deutsch rivet	3.18	0.125	240	35.0	180	26.0
			230	33.0	...	...
DuPont explosive	3.18	0.125	185	27.0	90	13.0
			180	26.0	90	13.0

Table 3 Charpy impact strength of Nb-1Zr

Condition	Temperature °C	Temperature °F	Impact energy J	Impact energy ft·lb	Impact fracture
Unnotched					
As rolled	24	75	210	156	None
	−73	−100	180	133	Partial
Stress relieved 1 h 899 °C (1650 °F)	24	75	175	129	None
	−73	−100	170	126	None
Recrystallized 1 h 1205 °C (2200 °F)	24	75	174	128	None
	−73	−100	164	121	None
Notched					
As rolled	24	75	>81	>60	Partial(a)
	−73	−100	93	69	Partial
Stress relieved 1 h 899 °C (1650 °F)	24	75	160	119	Partial
	−73	−100	129	95	Partial
Recrystallized 1 h 1205 °C (1650 °F)	24	75	126	93	Partial
	−73	−100	156	116	None

(a) Specimen stopped hammer, 60 ft·lb range.

Fig. 1 Stress-rupture properties of Nb-1Zr

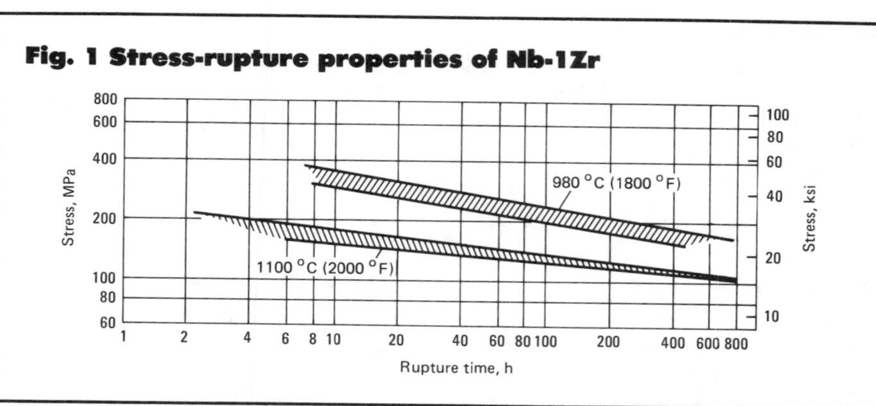

Rolling: 85% reduction at 204 to 315 °C (400 to 600 °F) and at finish. Readily formable by conventional metal forming processes.
Weldability. It can be joined by electron beam welding, resistance welding and gas tungsten-arc welding
Recrystallization temperature. 980 to 1205 °C (1800 to 2200 °F)
Hot working temperature. 1090 to 1200 °C (2000 to 2200 °F)
Stress relief temperature. 1 h at 900 to 980 °C (1650 to 1800 °F)

C-103
89Nb-10Hf-1Ti

Commercial Names

Trade name. WC 103

Chemical Composition

Composition limits. 0.0100 max C; 9 to 11 Hf; 0.7 to 1.3 Ti; 0.7 max Zr; 0.0300 O; 0.0300 max N; 0.0020 max H; 0.5 max W; 0.5 max Ta; rem Nb

Applications

Typical uses. Thrust chambers and radiation skirts for rocket and aircraft engines; guidance structure for glide re-entry vehicles; piping or container of chromic and other acids; thermal shields; etc.
Precautions in use. For elevated temperature applications, aluminide or silicide coatings should be used. Not recommended for use in hydrofluoric acid or strong alkaline solutions.

Mechanical Properties

Tensile properties. Typical. Cold rolled: tensile strength, 725 MPa (105 ksi); yield strength, 670 MPa (97 ksi); elongation, 4.5% in 50 mm or 2 in. Recrystallized: tensile strength, 405 MPa (59 ksi); yield strength, 310 MPa (45 ksi); elongation, 26% in 50 mm or 2 in. See also Table 4.
Hardness. 230 HV
Elastic modulus. Tension: 87 GPa (12.6×10^6 psi) at 21 °C (70 °F); 43 GPa (6.3×10^6 psi) at 1370 °C (2500 °F); 25 GPa (3.6×10^6 psi) at 1480 °C (2700 °F); 10 GPa (1.5×10^6 psi) at 1650 °C (3000 °F)
Creep-rupture properties. See Fig. 2.

Table 4 Typical tensile properties of arc-cast C-103 sheet

Temperature °C	Temperature °F	Direction	Tensile strength MPa	Tensile strength ksi	Yield strength(a) MPa	Yield strength(a) ksi	Elongation, %(b)
0.75 mm (0.03 in.) thick, cold rolled							
RT	RT	L	725	105	660	96	4.5
		T	745	108	640	93	4
1100	2000	L	235	34	160	23	39
		T	215	31	185	27	35
1370	2500	L	90	13	76	11	87
		T	90	13	76	11	80
1 mm (0.04 in.) thick, stress relieved 1 h at 870 °C (1600 °F)							
RT	RT	L	640	93	605	88	9
1100	2000	L	180	26	125	18	63
1370	2500	L	76	11	69	10	>75
1480	2700	L	55	8	48	7	>73
0.75 mm (0.03 in.) thick, recrystallized 1 h at 1315 °C (2400 °F)							
RT	RT	L	405	50	345	50	26
1100	2000	L	185	27	125	18	45
1370	2500	L	83	12	69	10	>70
1480	2700	L	62	9	55	8	>70
1650	3000	L	34	5	28	4	>70

(a) At 0.2% offset. (b) In 25 mm or 1 in.

Fig. 2 Larson-Miller plot for recrystallized C-103

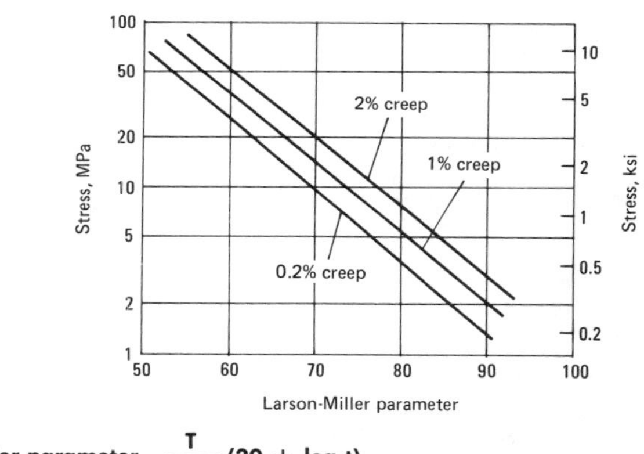

Larson-Miller parameter - $\frac{T}{1000}$ (20 + log t).

Mass Characteristics

Density. 8.87 Mg/m^3 (0.32 lb/in.^{3}) at 25 °C (77 °F)

Thermal Properties

Liquidus temperature. 2348 °C (4260 °F)
Coefficient of thermal expansion. Linear, 8.10 µm/m·K (4.5 µin./in.·°F) at 20 to 1200 °C (68 to 2200 °F)
Specific heat. 340 J/kg·K (0.082 Btu/lb·°F) at 21 °C (70 °F)
Thermal conductivity. 41.9 W/m·K (24.2 Btu/ft·h·°F) at 25 °C (77 °F)

Chemical Properties

General corrosion behavior. A protective oxide forms on C103 in most acid media, providing excellent corrosion resistance. However, the alloy is severely attacked by hydrofluoric acid and strong alkaline solutions.
Resistance to specific corroding agents. Excellent resistance to nitric

acid of all concentrations and to dilute hydrochloric acid

Fabrication Characteristics

Forgeability. 60% total reduction at 1200 to 930 °C (2200 to 1700 °F)
Hot formability. Extrusion: reduction ratio is 10:1 at 1200 °C (2200 °F). Rolling: 50% reduction at 425 °C (800 °F); 60 to 80% reduction at finish
Weldability. Good gas tungsten arc weldability
Recrystallization temperature. 1040 to 1315 °C (1900 to 2400 °F)
Stress relief temperature. 1 h at 870 °C (1600 °F)

C-129Y
80Nb-10W-10Hf-0.1Y

Commercial Name

Trade name. WC-129Y

Chemical Composition

Composition limits. 9 to 11 W; 9 to 11 Hf; 0.05 to 0.3Y; 0.5Ta; 0.5 max Zr; 0.015 max C; 0.025 max O; 0.015 max N; 0.0015 max H
Consequence of exceeding impurity limits. Increasing interstitials content decreases material ductility.

Applications

Typical uses. For high temperature applications, space vehicles, missiles; leading edges, nose caps for hypersonic flight vehicles, rocket nozzles; guidance structure for glide re-entry vehicles, etc.
Precautions in use. Interstitial contamination should be avoided during welding. A post weld annealing at 1205-1315 °C (2200 -2400 °F) for 1 h is recommended. For elevated temperature applications, silicide or aluminide coatings are required.

Mechanical Properties

Tensile properties. Tensile strength, 620 MPa (90 ksi); yield strength, 517 MPa (75 ksi); elongation, 25% in 25 mm or 1 in. See also Fig. 3.
Hardness. Recrystallized: 220 HV
Elastic modulus. Tension, 112 GPa (16.2×10^6 psi). See Fig. 4.

Fig. 3 Tensile properties of C-129Y sheet

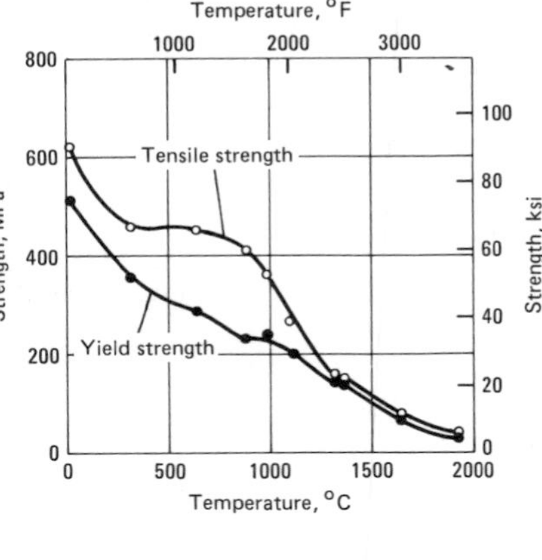

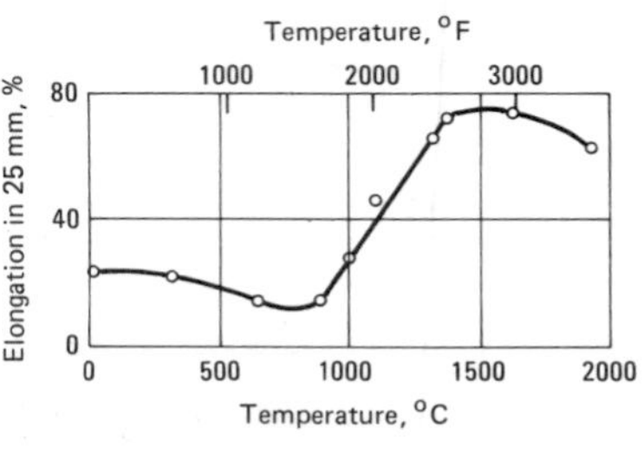

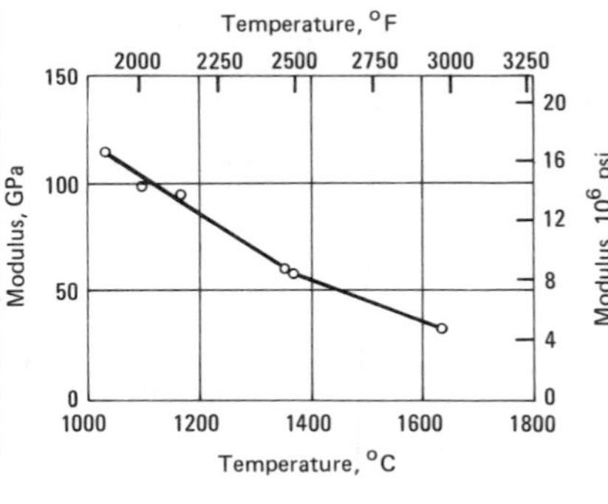

Fig. 4 Static modulus of elasticity for C-129Y

Creep-rupture properties. See Fig. 5 and 6.

Mass Characteristics

Density. 9.50 Mg/m³ (0.343 lb/in.³)

Thermal Properties

Liquidus temperature. 2400 °C (4350 °F)
Coefficient of thermal expansion. Linear, 6.88 μm/m·K (3.82 μin./in.·°F) at 20 to 1100 °C (70 to 2010 °F)

Fig. 5 Creep curves for C-129Y sheet in vacuum

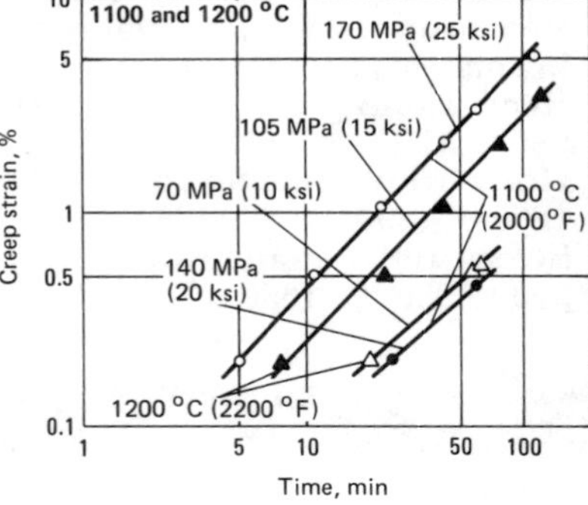

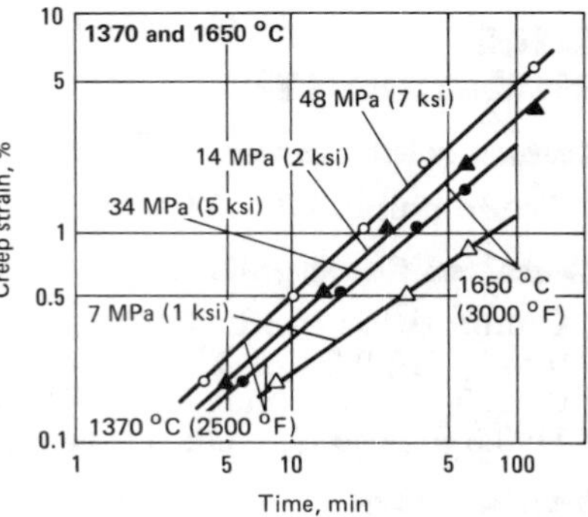

C-129Y sheet, 1 mm (0.04 in.) thick, was annealed 1 h at 1315 °C (2400 °F) and tested in vacuum at 13 MPa (10^{-4} torr).

Specific heat. 268 J/kg·K (0.064 Btu/lb·°F) at 1095 °C (2000 °F)
Thermal conductivity. 69.6 W/m·K (40 Btu/ft·h·°F)

Chemical Properties

General corrosion behavior. Good elevated temperature properties combined with heat and oxidation resistance. Excellent corrosion resistance to most acid media. However, C-129Y is severely attacked by hydrofluoric acid and strong alkaline solutions.

Fabrication Characteristics

Machinability. 75% of C36000, free-cutting brass
Forgeability. 50% at 930 to 1205 °C (1705 to 2200 °F)
Formability. Extrusion: reduction ratio of 4:1 at 1205 °C (2200 °F). Rolling: reduction of 50% at 430 °C (805 °F), 60 to 70% reduction at finish.
Weldability. Weldments exhibit

ductility as low as −170 °C (−275 °F) if prevention measures from atmosphere contamination are taken
Recrystallization temperature. 1315 °C (2400 °F)
Annealing temperature. 980 to 1315 °C (1800 to 2400 °F).
Hot working temperature. 980 to 1315 °C (1800 to 2400 °F)
Stress relief temperature. 1 h at 870 °C (1800 °F)

Fig. 6 Secondary creep rate vs stress for C-129Y sheet, cold worked 50%

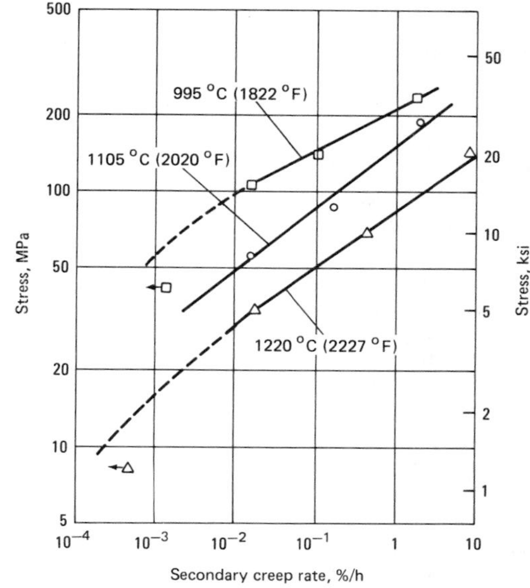

Cb-752
Nb-10W-2.5Zr

Commercial Name

Trade name. Haynes alloy Cb-752

Chemical Composition

Composition limits. 9 to 11 W; 2 to 3 Zr; 0.015 max C; 0.02 max O; 0.01 max N; 0.001 max H
Consequence of exceeding impurity limits. Increasing interstitials content decreases ductility of the metal

Applications

Typical uses. Guidance structure for glide re-entry vehicles, jet engine structure, thermal radiation and ducting for space power systems.
Precautions in use. Resistance to high temperature oxidation is poor, and protective coatings are required for elevated temperature applications

Mechanical Properties

Tensile properties. Annealed sheet: tensile strength, 540 MPa (78 ksi); yield strength, 400 MPa (58 ksi); elongation, 20% minimum in 50 mm or 2 in. At 1200 °C (2200 °F): tensile strength, 195 MPa (28 ksi); yield strength, 150 MPa (22 ksi); elongation, 25% minimum at 50 mm or 2 in. See also Fig. 7.
Shear strength. 427 MPa (62 ksi)
Compressive properties. Compressive strength, 345 MPa (50.9 ksi)
Bearing properties. Bearing strength, ultimate, 703 MPa (102 ksi); bearing strength, yield, 627 MPa (91 ksi)
Hardness. 180 KHN; annealed at 1370 °C (2500 °F), 1 h

Fig. 7 Effect of test temperature on average tensile properties of duplex annealed Cb-752 sheet

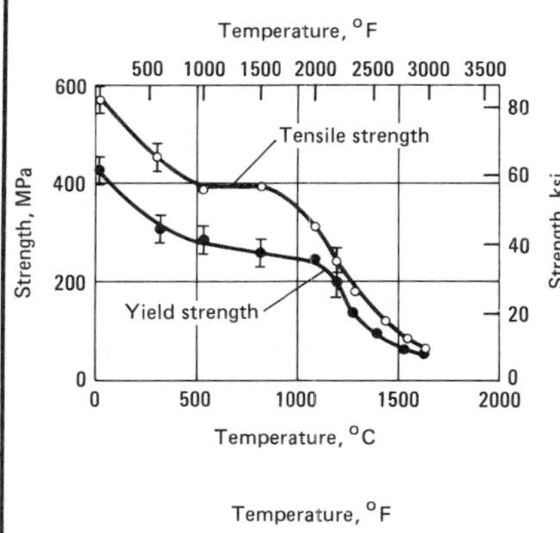

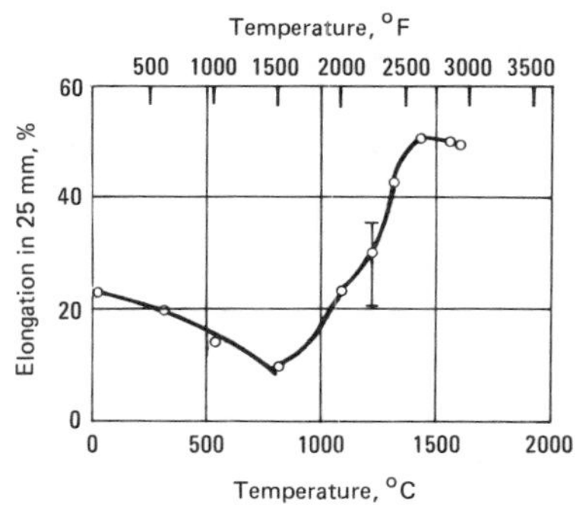

Elastic modulus. Tension, 110 GPa (16×10^6 psi)
Fatigue strength. 276 MPa (40 ksi) at 1 million cycles
Creep-rupture properties. See Fig. 8.

Mass Characteristics

Density. 9.03 Mg/m^3 (0.326 $lb/in.^3$) at 25 °C (77 °F)

Thermal Properties

Liquidus temperature. 2430 °C (4400 °F)
Coefficient of thermal expansion. Linear, 7.4 μm/m·K (4.1 μin./in.·°F) at 20 to 1205 °C (68 to 2200 °F)
Specific heat. 281 J/kg·K (0.067 Btu/lb·°F) at 538 °C (1000 °F); temperature coefficient, 0.0335 J/kg·K per K at 0 to 538 °C (32 to 1000 °F). See also Fig. 9.
Thermal conductivity. 48.7 W/m·K (28 Btu/ft·h·°F) at 760 °C (1400 °F); temperature coefficient, 0.0219 W/m·K per K at 204 to 760 °C (400 to 1400 °F)

Chemical Properties

General corrosion behavior. Excellent corrosion resistance to most acid media, but severely attacked by hydrofluoric acid and strong alkaline solutions
Resistance to specific corroding

agents. Good corrosion resistance to oxygen-free liquid metals, e.g. potassium, lithium, at elevated temperatures 982 to 1204 °C (1800 to 2200 °F) for some 4000 h

Fabrication Characteristics

Machinability. 80% of C36000, free-cutting brass
Forgeability. 50% at 930 to 1205 °C (1705 to 2200 °F)
Formability. The alloy can be formed using most of the conventional methods. Primary ingot breakdown must be done above the recrystallization temperature. Subsequent working may be accomplished in range of room temperature to 430 °C (800 °F).
Weldability. Fusion welding by GTA and EB processes can be accomplished readily, if prevention from atmosphere contamination is taken care of.
Recrystallization temperature. 1200 to 1315 °C (2200 to 2400 °F).
Annealing temperature. 1200 to 1315 °C (2200 to 2400 °F)
Solution temperature. 1425 to 1540 °C (2600 to 2800 °F)
Aging temperature. 1100 °C (2000 °F)
Hot working temperature. 980 to 1315 °C (1800 to 2400 °F)
Stress relief temperature. 1 h at 870 to 1090 °C (1800 to 2000 °C)

Fig. 8 Total creep curves for Cb-752 sheet

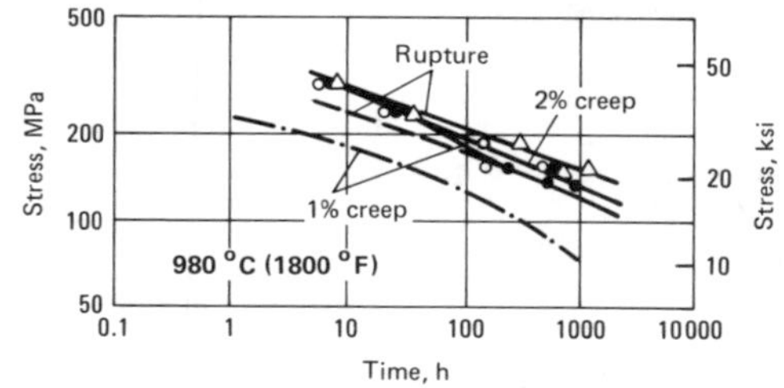

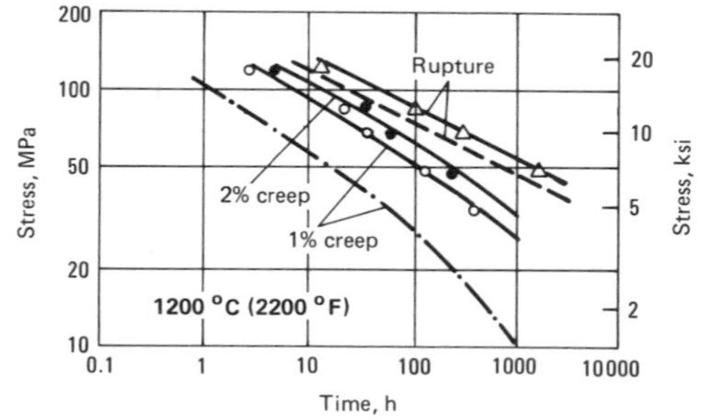

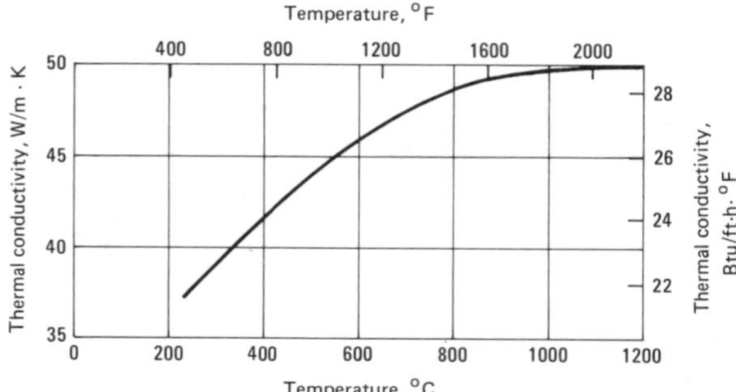

Points represent material duplex annealed, then aged 1 h at 1600 °C (2900 °F). Broken lines are for material duplex annealed only.

B-66
89Nb-5Mo-5V-1Zr

Chemical Composition

Composition limits. 4.5-5.5 Mo; 4.5-5.5V; 0.85-1.3 Zr; 0.02 max C; 0.02 max N; 0.03 max O; rem Nb
Consequence of exceeding impurity limits. Increasing interstitial contents decrease ductility of the material

Applications

Typical uses. Space vehicles, nuclear reactors, lifting and guidance structures for glide re-entry vehicles
Precautions in use. A protective coating should be applied for high temperature service

Fig. 9 Thermal properties of Cb-752

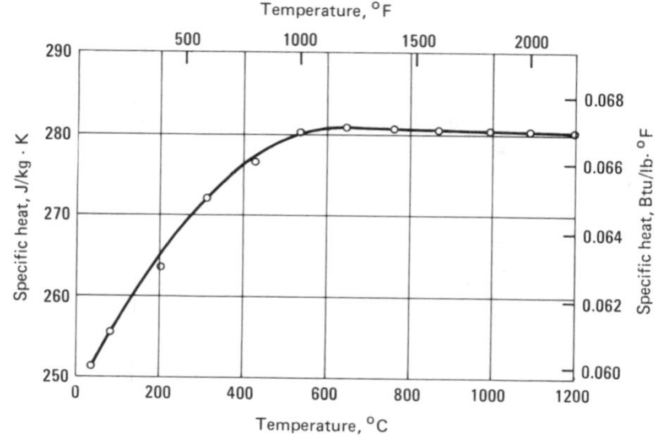

Mechanical Properties

Tensile properties. See Table 5.
Hardness. At room temperature: 228 HV. See also Fig. 10.
Elastic modulus. Tension, 105 GPa (15.3×10^6 psi) at 21 °C (70 °F); 83 GPa (12×10^6 psi) at 1100 °C (2000 °F). See also Fig. 11.
Bend ductility. See Fig. 12.
Creep-rupture properties. See Fig. 13.

Mass Characteristics

Density. 8.45 Mg/m^3 (0.305 lb/in.^{3}) at 20 °C (68 °F)

Thermal Properties

Liquidus temperature. 2370 °C (4300 °F)
Coefficient of thermal expansion. Linear, 8.51 μm/m·K (4.73 μin./in.·°F) at 20 to 1370 °C (68 to 2500 °F)

Chemical Properties

General corrosion behavior. Good oxidation resistance at high temperatures

Fabrication Characteristics

Forgeability. 50% total reduction at 930 to 1290 °C (1705 to 2355 °F)
Formability. Extrusions: Reduction ratio is 4:1 at 1205 °C (2200 °F); Rolling: Total reduction is 50% at 1205 to 1090 °C (2200 to 1995 °F); 25 to 50% reduction at finish from 20 to 250 °C (70 to 480 °F). The alloy can be formed, punched, blanked or sheared at room temperature, but with greater ease at slightly elevated temperatures.
Weldability. The alloy can be welded by tungsten-arc gas welding or electron beam welding
Recrystallization temperature. 1205 to 1370 °C (2200 to 2500 °F)
Annealing temperature. 1205 to 1370 °C (2200 to 2500 °F)
Hot working temperature. 1205 to 1315 °C (2200 to 2400 °F)
Stress relief temperature. 1100 °C (2000 °F); 1 h at temperature is typical

Table 5 Typical tensile properties of B-66

Temperature		Tensile strength		Yield strength		Elongation,
°C	°F	MPa	ksi	MPa	ksi	%
-100	−148	885	128	745	108	12
+23	+73	795	115	625	91	14
1090	2000	450	65	400	58	28

Fig. 10 Hot hardness of B-66

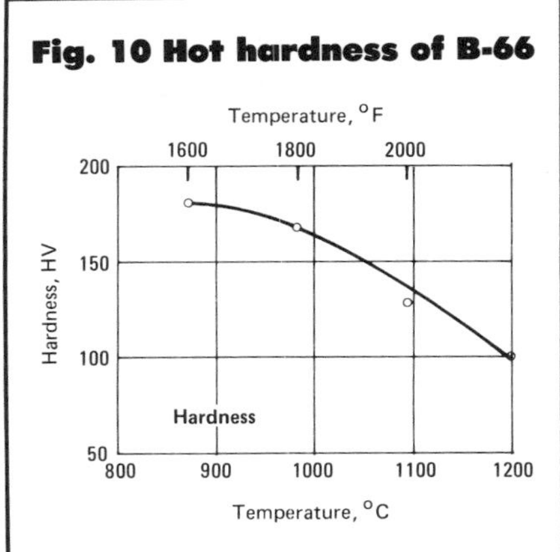

Fig. 11 Modulus of elasticity at elevated temperature for B-66

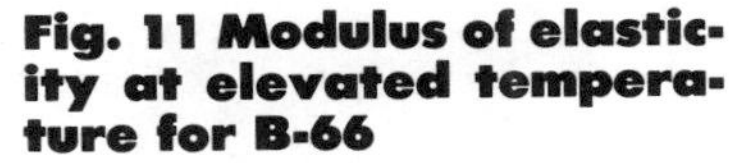

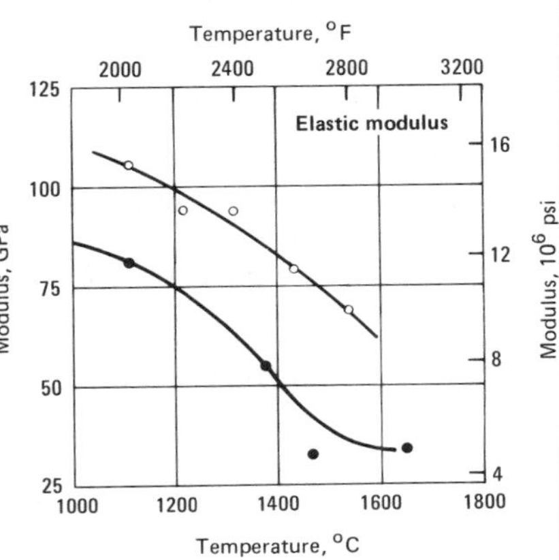

Open symbols are for tests done at a strain rate of 24%/min; solid symbols, for tests done at 0.5%/min.

Fig. 12 Bend ductility of as-received B-66 alloy sheet

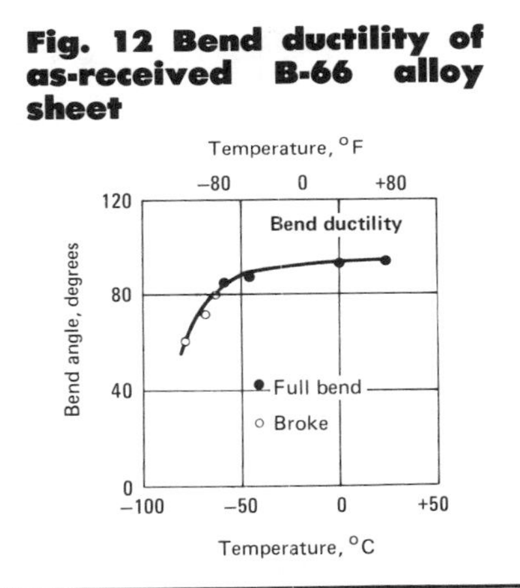

Consequence of exceeding impurity limits. Increasing interstitials decreases ductility of the alloy

Applications

Typical uses. Moderate strength alloy suitable for gas turbine blading
Precautions in use. Protective coating should be applied for service at moderately high temperatures in oxidizing environments

Cb-132M
Cb-20Ta-15W-5Mo-1.5Zr-0.12C

Chemical Composition

Composition limits. 0.025 max O; 0.001 max H; 0.01 max N

Mechanical Properties

Tensile properties. Tensile strength, 670 MPa (97 ksi); yield strength, 570 MPa (83 ksi); elongation, 5% in 25.4 mm or 1 in.
Hardness. 297 HV
Fatigue strength. 240 MPa (35 ksi) at 10 million cycles.

Mass Characteristics

Density. 10.7 Mg/m^3 (0.385 lb/in.^{3}) at 25 °C (77 °F)

Thermal Properties

Coefficient of thermal expansion. Linear, 10.1 μm/m·K (5.6 μin./in.·°F) at 20 to 110 °C (70 to 200 °F)

Electrical Properties

Electrical resistivity. 0.354 nΩ·m at 115 °C (240 °F)

Chemical Properties

General corrosion behavior. Excellent corrosion resistance to most acid media, but severely attacked by hydrofluoric acid and strong alkaline solutions
Resistance to specific corroding

agents. Resists liquid metal corrosion; a candidate for turbine components in turboalternator systems that use liquid metals such as potassium

Fabrication Characteristics

Formability. Because of the low ductility of recrystallized material, Cb-132M is somewhat difficult to fabricate. In the as-processed condition, working temperatures should stay below the recrystallized temperatures.
Weldability. Satisfactory bonds can be obtained at 1205 °C (2200 °F) by diffusion bonding, which is facilitated by using a vanadium foil intermediate layer acting as an interstitial sink.
Recrystallization temperature. 1760 °C (3200 °F)
Annealing temperature. 1370 to 1650 °C (2500 to 3000 °F)
Hot working temperature. 1370 to 1650 °C (2500 to 3000 °F)

Fig. 13 Creep and stress-rupture curves in vacuum for B-66 alloy

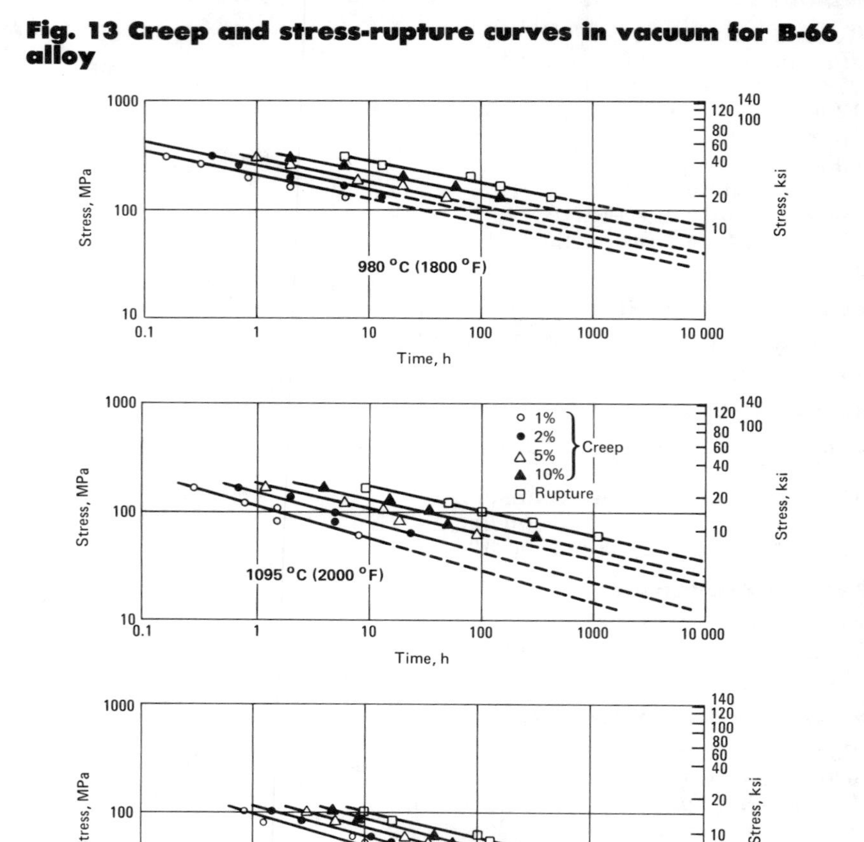

FS-85
Cb-28Ta-10W-1Zr

Chemical Composition

Composition limits. 0.01 max C; 26 to 29 Ta; 10 to 12 W; 0.6 to 1.1 Zr; 0.03 max O; 0.015 max N; 0.001 max H

Mechanical Properties

Tensile properties. See Table 6 and Fig. 14.
Creep-rupture properties. See Table 7.
Elastic modulus:

Temperature °C	°F	Modulus of elasticity GPa	10^6 psi
RT	RT	140	20
980	1800	125	18
1090	2000	125	18
1200	2200	110	16
1540	2800	105	15
1590	2900	83	12
1650	3000	83	12

Mass Characteristics

Density. 10.61 Mg/m^3 (0.383 lb/in.^{3})

Thermal Properties

Liquidus temperature. 2590 °C (4695 °F)
Coefficient of thermal expansion. Linear, 9.0 μm/m·K (5.0 μin./in.·°F) at 20 to 1315 °C (68 to 2400 °F)

Fabrication Characteristics

Forgeability. 50% at 930 to 1290 °C (1705 to 2355 °F)
Formability. Extrusion: reduction ratio of 4:1 at 1205 °C (2200 °F). Rolling: 40% reduction at 205 to 370 °C (400 to 700 °F); 50 to 65% reduction at finish
Recrystallization temperature. 1090 to 1370 °C (2000 to 2500 °F)
Stress relief temperature. 1 h at 1010 °C (1850 °F)

SCb-291
80Nb-10Ta-10W

Chemical Composition

Composition limits. 0.0060 C; 80.0 Nb; 9 to 11 Ta; 9 to 11 W; 0.0090 max O; 0.0100 max N
Consequence of exceeding impurity limits. High interstitial contents decrease ductility

Applications

Typical uses. Rocket nozzles and other aerospace applications

Mechanical Properties

Tensile properties. See Tables 8 and 9.
Creep-rupture properties. Rupture strength of cold rolled EB-

Table 6 Room-temperature tensile properties of FS-85 Sheet 0.8 mm (0.030 in.) thick strain rate: 2%/min

Condition	Tensile strength, MPa	ksi	Yield strength(a) MPa	ksi	Elongation(b), %	Reduction in area, %
Stress relieved	830	120	735	106	11	47
Recrystallized 1 h at 1260 °C (2300 °F)	590	85	475	69	22	54
Recrystallized and Cr-Ti-Si coated	580	84	470	68.4	18	32

(a) 0.2% offset. (b) In 25 mm or 1 in.

Table 7 Creep and stress-rupture data for FS-85 sheet

Temperature, °C	°F	Stress, MPa	ksi	Secondary creep rate, %/h	Rupture time, h
1 mm (0.04 in.) sheet, cold worked 50%					
1090	2000	115	17	0.0321	>158
		135	19.8	0.171	...
		140	20	0.0363	...
		185	27	0.606	...
1200	2200	69	10	<0.0845	...
		83	12	0.119	...
		97	14	0.246	...
		110	16	0.525	...
		125	18	1.06	...
		130	19	1.32	10.42
		140	20	2.20	...
1320	2400	90	13	1.85	9.98
1430	2600	69	10	1.88	9.29
1.5 mm (0.063 in.) sheet, cold worked 94%					
980	1800	240	35	0.930	6.51
1090	2000	140	20	0.462	23.78
		185	27	...	2.33
1200	2200	130	19	5.13	2.72
1320	2400	90	13	3.99	4.93
		90	13	3.24	6.12
		83	12(a)	...	11.21
1430	2600	69	10	4.14	>4.7
		69	10(a)	3.45	5.01

(a) Cold worked 94% and annealed 1 h at 1320 °C (2400 °F).

Table 8 Typical room-temperature tensile properties of electron-beam-melted SCb-291

Product and condition	Tensile strength MPa	ksi	Yield strength(a) MPa	ksi	Elongation(b), %	Reduction in area, %
Forged bar	550	80	455	66	31	78
Annealed bar	455	66	350	51	20	93
As-rolled sheet	825	120	...	...	8	...
Annealed sheet	605	88	510	74	22	...

(a) At 0.2% offset. (b) In 50 mm or 2 in.

melted sheet: 54 MPa (7.8 ksi) for rupture in 2.5 h at 1500 °C (2730 °F); 34 MPa (5 ksi) for rupture in 2.4 h at 1630 °C (2970 °F); 26 MPa (3.8 ksi) for rupture in 2.1 h at 1760 °C (3200 °F). See also Fig. 15.

Mass Characteristics

Density. 9.6 Mg/m³ (0.347 lb/in.³)

Thermal Properties

Liquidus temperature. 2600 °C (4710 °F)

Coefficient of thermal expansion. Linear, 14.04 μm/m·K (7.8 μin./in.·°F) at 20 to 400 °C (68 to 750 °F)

Fabrication Characteristics

Forgeability. 60% at 930 to 1205 °C (1705 to 2200 °F)

Formability. Good. Extrusion: reduction ratio of 10:1 at 1205 °C (2200 °F). Rolling: 50% reduction at 260 to 371 °C (500 to 700 °F), 60 to 70% reduction at finish.

Recrystallization temperature. 1150 to 1400 °C (2400 to 2550 °F)

Stress relief temperature. 1 h at 1095 °C (2000 °F)

Molybdenum Alloys

By Russell W. Burman
Market Development Manager
Amax Specialty Metals Corporation

Mo-0.5Ti
Mo-0.5Ti-0.02C

Commercial Names

UNS number. R03620

ASTM designation. Molybdenum alloy 362

Specifications

ASTM. B384, B385, B386, B387

Chemical Composition

Composition limits. 0.010 to 0.040 C; 0.010 max Fe; 0.001 max N; 0.005 max Ni; 0.003 max O; 0.010 max Si; 0.40 to 0.55 Ti; rem Mo

Mechanical Properties

Tensile properties. Typical. Tensile strength: 895 MPa (130 ksi) at 21 °C (70 °F); 415 MPa (60 ksi) at 1100 °C (2000 °F); 76 MPa (11 ksi) at 1650 °C (3000 °F). Yield strength: 825 MPa (120 ksi) at 21 °C; 345 MPa (50 ksi) at 1100 °C; 48 MPa (7 ksi) at 1650 °C. Elongation: 10% at 21 °C

Elastic modulus. Tension: 315 GPa (46 × 10⁶ psi) at 21 °C (70 °F); 180 GPa (26 × 10⁶ psi) at 1100 °C (2000 °F)

Fig. 14 Effect of temperature on the typical tensile properties of FS-85

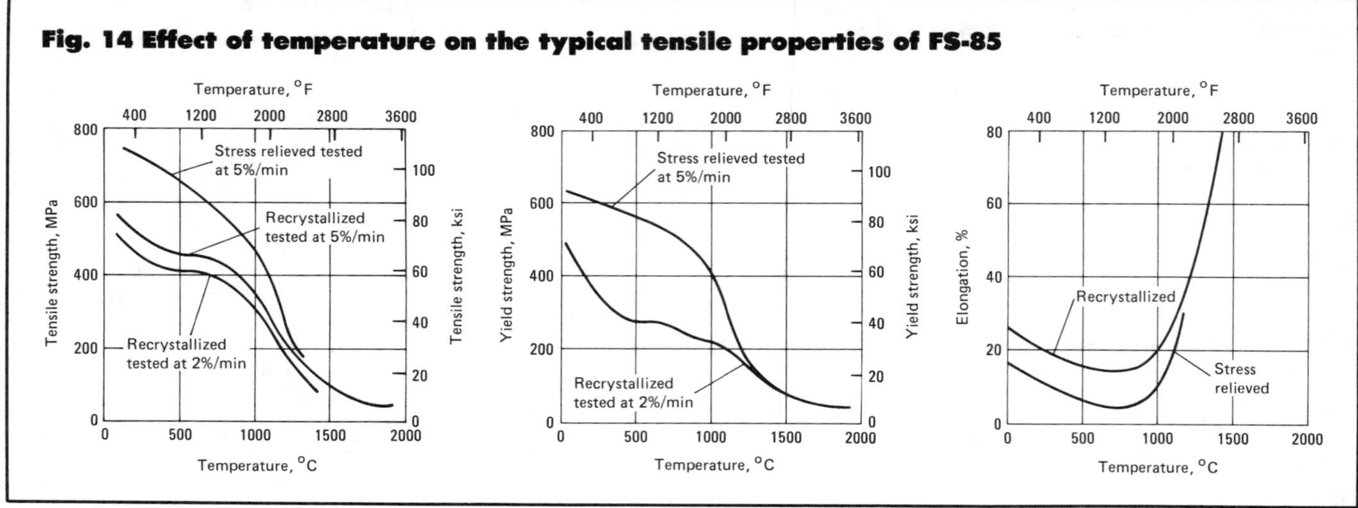

Mass Characteristics

Density. 10.2 Mg/m³ (0.367 lb/in.³)

Thermal Properties

Liquidus temperature. 2610 °C (4730 °F)

Coefficient of thermal expansion. Linear, 6.1 μm/m·K (3.41 μin./in.·°F) at 20 to 1000 °C (68 to 1850 °F)

Fabrication Characteristics

Recrystallization temperature. 1315 to 1425 °C (2400 to 2600 °F)

Stress relief temperature. 1 h at 1100 to 1200 °C (2000 to 2200 °F)

Table 9 Average tensile properties of annealed electron-beam-melted SCb-291 sheet at various temperatures

Temperature, °C	°F	Tensile strength MPa	ksi	Yield strength(a) MPa	ksi	Elongation(b), %
RT	RT	517	75	414	60	25
430	800	414	60	345	50	22
540	1000	359	52	276	40	23
650	1200	331	48	255	37	22
870	1600	310	45	241	35	20
980	1800	269	39	207	30	22
1095	2000	221	32	165	24	24
1205	2200	186	27	138	20	22
1315	2400	145	21	103	15	25
1370	2500	124	18	90	13	24
1425	2600	110	16	76	11	24
1540	2800	83	12	59	8.5	22
1650	3000	69	10	48	7	23
1760	3200	55	8	41	6	24
1870	3400	45	6.5	33	4.8	25

TZC
Mo-1Ti-0.3Zr

Chemical Composition

Nominal composition. 1.25 Ti; 0.3 Zr; 0.15 C; rem Mo

Applications

Typical uses. Aerospace equipment and components

Mechanical Properties

Tensile properties. Stress relieved: tensile strength, 990 MPa (144 ksi); yield strength, 725 MPa (105 ksi); elongation, 22% in 50 mm or 2 in.; reduction in area, 36%. At 1090 °C (2000 °F): tensile strength, 640 MPa (93 ksi). At 1315 °C (2400 °F): tensile strength, 415 MPa (60 ksi)

Table 10 Typical tensile properties of TZM

Temperature °C	°F	Tensile strength MPa	ksi	Yield strength(a) MPa	ksi	Elongation, % (b)
Stress-relieved condition						
21	70	965	140	860	125	10
1100	2000	490	71	435	63	...
1650	3000	83	12	62	9	...
Recrystallized material						
21	70	550	80	380	55	20
1100	2000	505	73	...	...	...
1315	2400	369	53.5	...	...	...

(a) At 0.2% offset. (b) In 50 mm or 2 in.

TZM
Mo-0.5Ti-0.1Zr

Commercial Names

UNS number. Arc-cast, R03630; P/M, R03640
ASTM designation. Arc-cast: molybdenum alloy 363. P/M: molybdenum alloy 364

Specifications

ASTM. B384, B385, B386, B387

Chemical Composition

Composition limits. Arc-cast: 0.40 to 0.55 Ti; 0.06 to 0.12 Zr; 0.01 to 0.04 C; 0.010 max Fe; 0.010 max Si; 0.005 max Ni; 0.001 max N; 0.0030 max O; 0.0005 max H. For P/M products, limit on N is 0.002 max, on O is 0.030 and on Si is 0.005 max; all other limits remain the same.

Applications

Typical uses. For heat engines, heat exchangers, nuclear reactors, radiation shields, extrusion dies, boring bars

Mechanical Properties

Tensile properties. See Table 10.
Elastic modulus. Tension, 315 GPa (46 × 10^6 psi) at 21 °C (70 °F); 205 GPa (30 × 10^6 psi) at 1100 °C (2000 °F)

Mass Characteristics

Density. 10.16 Mg/m^3 (0.367 lb/in.^{3}) at 20 °C (68 °F)

Thermal Properties

Liquidus temperature. 2620 °C (4750 °F)
Coefficient of thermal expansion. Linear, 4.9 μm/m·K (2.7 μin./in.·°F) at 20 to 40 °C (68 to 100 °F)
Thermal conductivity. See Fig. 16.

Fig. 16 Thermal conductivities of selected refractory metals

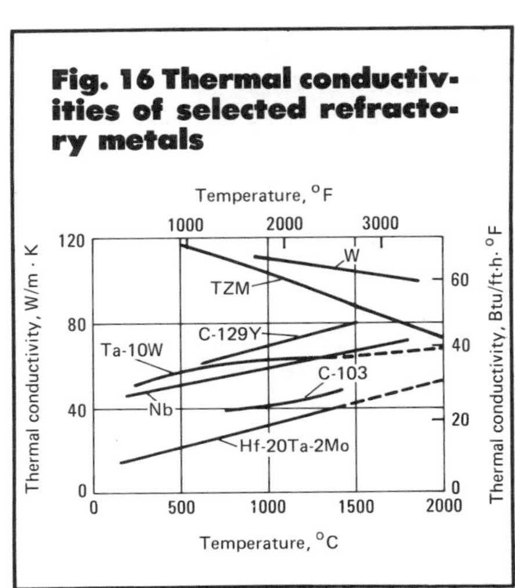

Fabrication Characteristics

Recrystallization temperature. 1425 to 1600 °C (2600 to 2900 °F)
Stress relief temperature. 1 h at 1100 to 1260 °C (2000 to 2300 °F)

Fig. 15 Time for 0.2% creep in electron-beam-melted SCb-291

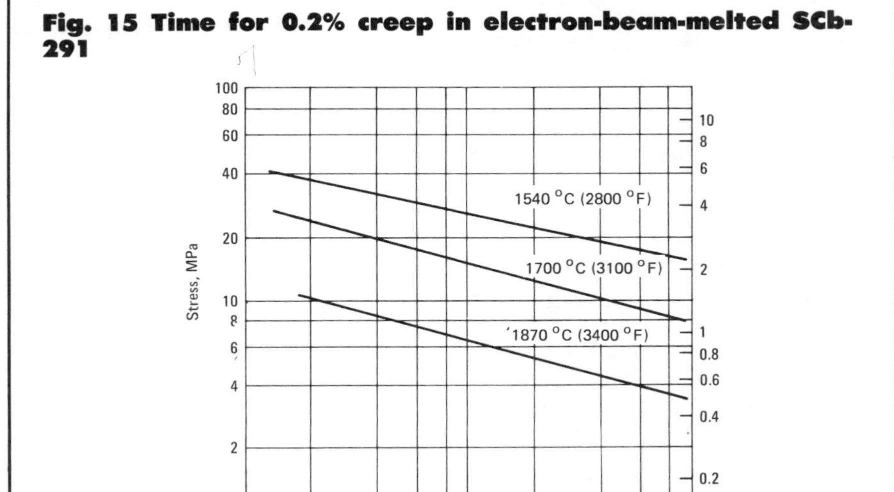

Tantalum Alloys

By Mortimer Schussler
Senior Scientist
Fansteel Inc.

"61" Metal

Chemical Composition

Composition limits. 7.5 W; rem Ta

Applications

Typical uses. Springs and other elastic parts for gas chlorinators or severe acid corrosive conditions

Mechanical Properties

Tensile properties. Cold drawn wire, 83% reduction: tensile strength, 1140 MPa (165 ksi); yield strength, 1100 MPa (160 ksi); elongation, 5% in 25 mm or 1 in.
Hardness. As cold worked: 35 HRC

Mass Characteristics

Density. 16.7 Mg/m^3 (0.606 lb/in.^{3})

Thermal Properties

Melting point. Approximately 3025 °C (5475 °F)

Fabrication Characteristics

Consolidated by powder metallurgy techniques.

"63" Metal
Ta-2.5W-0.15Nb

Chemical Composition

Composition limits. 2.0 to 3.0 W; 0.5 max Nb; 50 ppm max

Applications

Typical uses. Heat exchangers, linings for towers, valves, piping, fittings and gaskets

Mechanical Properties

Tensile properties. At room temperature: tensile strength, 345 MPa (50 ksi); yield strength, 228 MPa (33 ksi); elongation, 40% in 25.4 mm or 1 in.; At recrystallization: tensile strength, 385 MPa (56 ksi); yield

strength, 235 MPa (34 ksi); elongation, 38%
Hardness. Recrystallized: 54 HR30T

Mass Characteristics

Density. 16.6 Mg/m³ (0.602 lb/in.³)

Thermal Properties

Melting point. 3005 °C (5440 °F) est.

Fabrication Characteristics

Consolidated by electron beam melting.
Weldability. Excellent, for electron beam welding
Recrystallization temperature. 1200 to 1300 °C (2200 to 2375 °F)

Ta-10W

Commercial Name

Trade name. Fansteel 60 Metal

Chemical Composition

Nominal composition. 10W; rem Ta

Applications

Typical uses. Used in aerospace applications up to 2480 °C (4500 °F), such as hot-gas metering valves, rocket engine extension skirts, complex manifold assemblies, and fasteners. In chemical process industries, applications include machined solid valves, internal seats and plugs of large valves, liners requiring abrasion and corrosion resistance, disks used in patching glass-lined steel vessels, and tubing in some nuclear applications.

Mechanical Properties

Tensile properties:

Temperature, °C	Tensile strength MPa	ksi
21	550	80
1090	275	40
1650	69	10
Recrystallized		
21	650	94

See also Fig. 17.

Fig. 17 Elevated temperature tensile properties of Ta-10W sheet

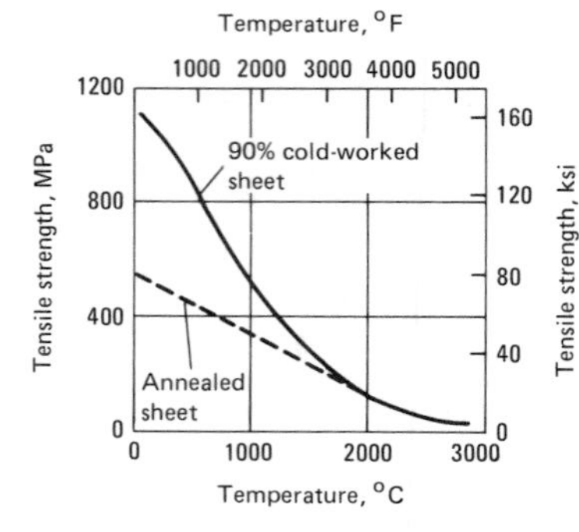

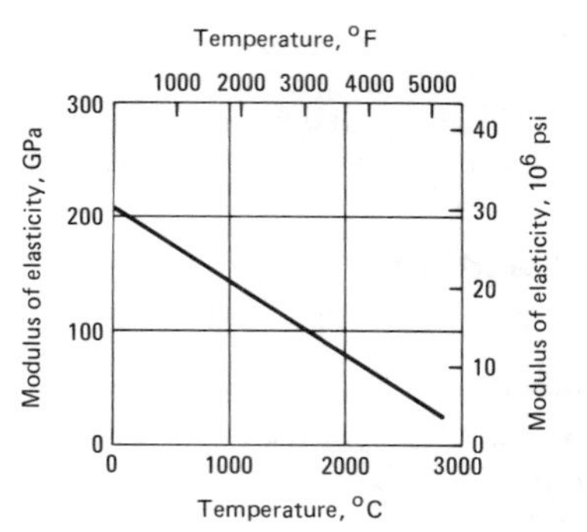

Tensile properties:

Temperature, °C	Yield strength MPa	ksi	Elongation, %
21	460	67	25
1090	240	35	...
1650	62	9	...
Recrystallized			
21	560	81	35

Hardness. Recrystallized: 78 HR30T
Impact strength. See Fig. 18.
Fatigue strength. See Fig. 19.
Creep-rupture characteristics. See Fig. 20.
Stress-rupture characteristics. See Fig. 21.

Fig. 18 Charpy-keyhole impact curve for electron-beam-melted Ta-10W

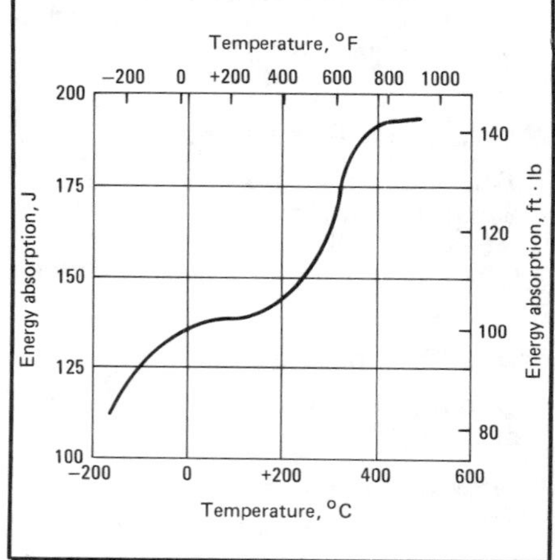

Fig. 19 Completely reversed sheet-bending fatigue properties of cold rolled Ta-10W sheet (0.025 in.) at room temperature

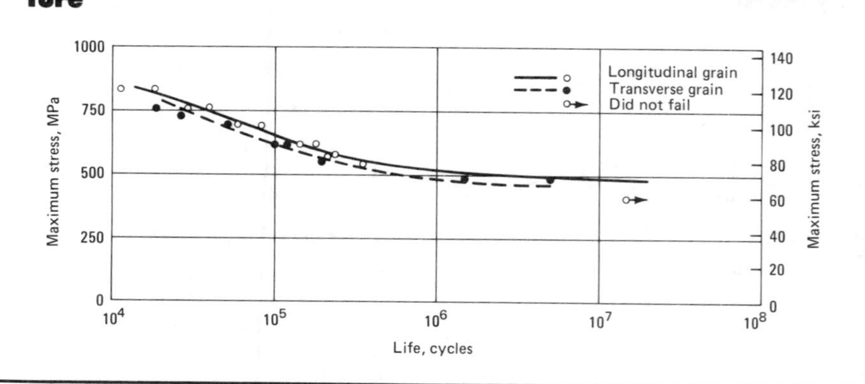

Mass Characteristics

Density. 16.8 Mg/m³ (0.608 lb/in.³)

Thermal Properties

Melting point. 3035 °C (5495 °F)
Coefficient of thermal expansion. Linear, 6.7 μm/m·K (3.74 μin./in.·°F) at 20 to 1650 °C (68 to 3000 °F)
Thermal expansion:

Temperature °C	°F	$\Delta l/l$(a), mm/m
0	32	0
200	392	1.1
400	750	2.3
600	1110	3.6
800	1470	4.9
1000	1830	6.3
1200	2190	7.8
1400	2550	9.3
1600	2910	10.9
1800	3270	12.6
2000	3630	14.4
2200	3990	16.3
2400	4350	18.3
2600	4710	20.4
2800	5070	22.6

(a) Approximate total expansion from 20 °C to indicated temperature.

Thermal conductivity:

Temperature °C	°F	Conductivity(a) W/m·K	Btu/ft·h·°F
1400	2550	57	33
1600	2910	54	31
1800	3270	51	29
2000	3630	48	28
2200	3990	45	26
2400	4350	42	24
2600	4710	39	23
2800	5070	36	21

(a) Approximate value at indicated temperature.

Electrical resistivity:

Temperature °C	°F	Resistivity(a), nΩ·m
0	32	170
200	392	230
400	750	300
600	1110	350
800	1470	410
1000	1830	470
1200	2190	520
1400	2550	570
1600	2910	620
1800	3270	670
2000	3630	710
2200	3990	750
2400	4350	790

(a) Approximate value at the indicated temperature.

Fig. 20 Time for 0.2% creep at various temperatures

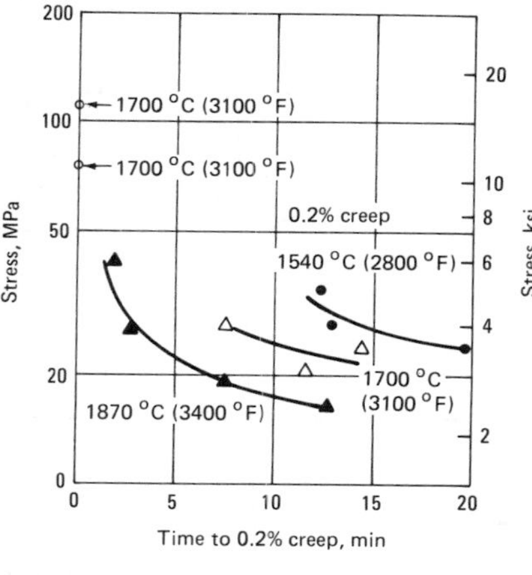

Fabrication Characteristics

Weldability. Electron beam welded
Recrystallization temperature. 1300 to 1650 °C (2375 to 3000 °F)
Stress relief temperature. 1 h at 1090 to 1230 °C (2000 to 2250 °F)

T-111
Ta-8W-2Hf

Chemical Composition

Composition limits. 7.0 to 9.0 W; 2.0 to 2.8 Hf; 0.003 C; rem Ta

Applications

Typical uses. A containment material in applications including handling liquid metals used as coolants in high temperature nuclear reactors, as heat transfer and working fluids in power generation that operate in conjunction with compact nuclear reactors.

Mechanical Properties

Tensile properties. At room temperature: tensile strength, 690 MPa (100 ksi); yield strength, 586 MPa (85 ksi); elongation, 29% in 25.4 mm or 1 in. At 1320 °C (2400 °F): tensile strength, 255 MPa (37 ksi); yield strength, 165 MPa (24 ksi). At 1930 °C (3500 °F): tensile strength, 90 MPa (13 ksi); yield strength, 90 MPa (13 ksi). See also Fig. 22.
Hardness. See Fig. 23.
Elastic modulus. Tension, 200 GPa (29 × 10⁶ psi)
Stress-rupture characteristics. See Fig. 24.

Mass Characteristics

Density. 16.7 Mg/m³ (0.604 lb/in.³)

Thermal Properties

Melting point. 2982 °C (5400 °F) est.

Chemical Properties

Resistance to specific corroding agents. Improved resistance to some liquid metals, such as potassium, sodium, NaK, lithium, cesium and mercury, over unalloyed tantalum and Ta-10W

Fabrication Characteristics

Consolidated by electron beam melting followed by consumable-electrode vacuum-arc melting.
Formability. Good ductility for forming
Recrystallization temperature. 1400 to 1650 °C (2550 to 3000 °F)
Stress relief temperature. 1 h at 1090 to 1320 °C (2000 to 2400 °F)

T-222
Ta-10W-2.5Hf-0.01C

Chemical Composition

Composition limits. 0.01C, 9.64W; 2.4Hf; rem Ta

Applications

Typical uses. For high temperature, re-entry space vehicles, rocket reaction chambers, nozzle parts, liquid metal systems

Mechanical Properties

Tensile properties. See Table 11 and Fig. 25.
Elastic modulus. Tension, 200 GPa (29 × 10⁶ psi)
Creep characteristics. See Fig. 26 and 27.

Structure

Microstructure. Consolidated by electron-beam melting, followed by consumable electrode, vacuum arc melting

Mass Characteristics

Density. 16.7 Mg/m³ (0.604 lb/in.³)

Thermal Properties

Liquidus temperature. 3020 °C (5480 °F)

Fabrication Characteristics

Formability. Good ductility at low temperatures
Weldability. Good
Recrystallization temperature. 1425 to 1650 °C (2600 to 3000 °F)
Stress relief temperature. 1 h at 1090 to 1320 °C (2000 to 2400 °F)

Fig. 21 Typical stress-rupture data of Ta-10W sheet (tests in vacuum)

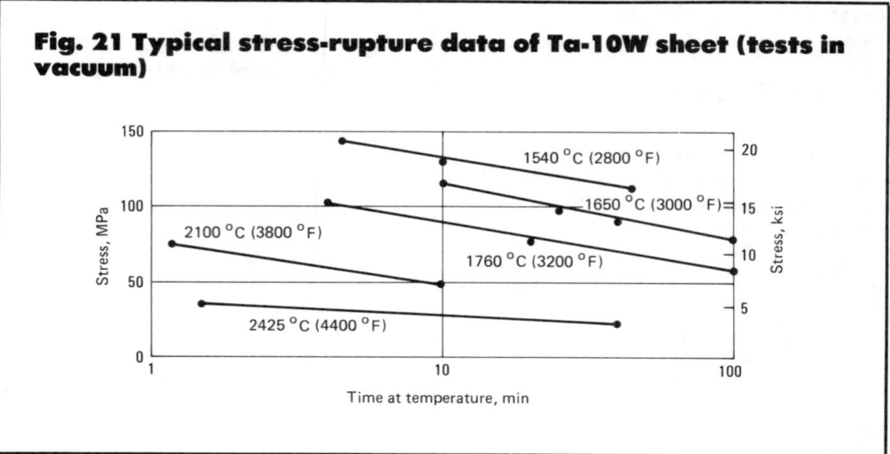

Fig. 22 Tensile properties vs temperature for cold rolled sheet of T-111

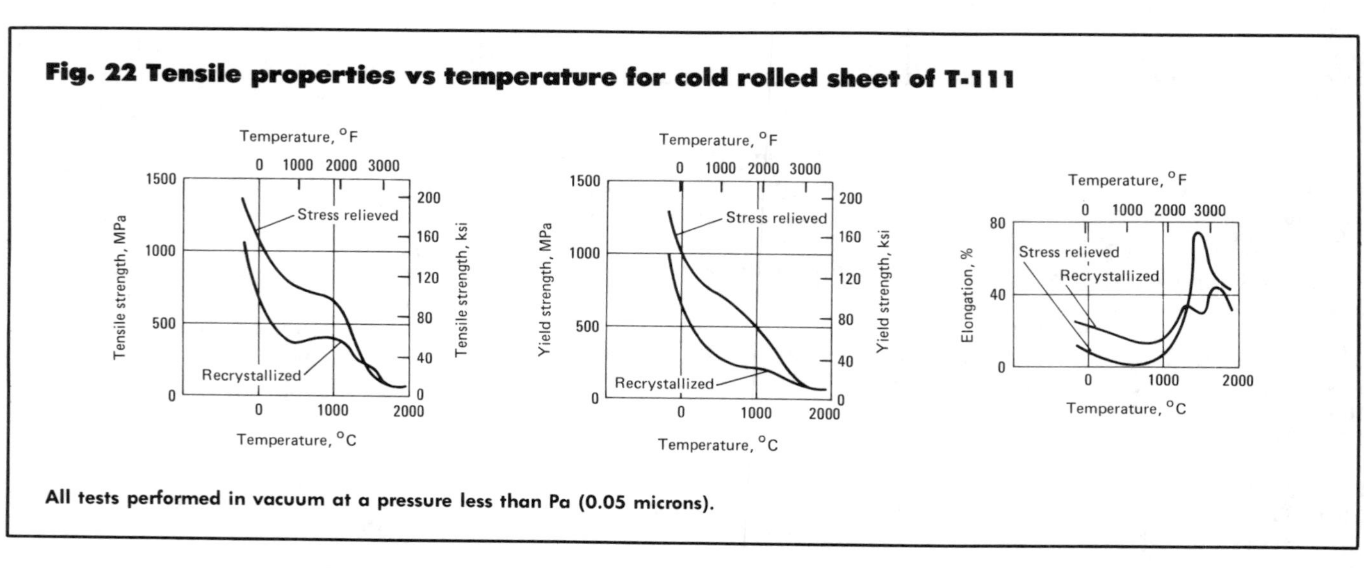

All tests performed in vacuum at a pressure less than Pa (0.05 microns).

Fig. 23 Hardness vs annealing temperatures for T-111

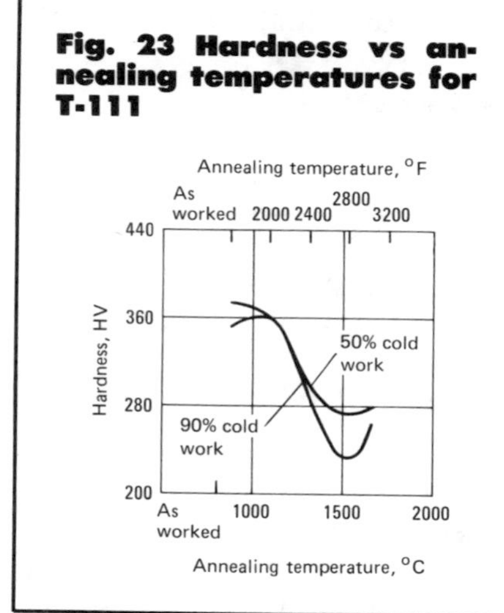

Fig. 24 Stress rupture properties of T-111 alloy at 1200 °C (2400 °F) in vacuum

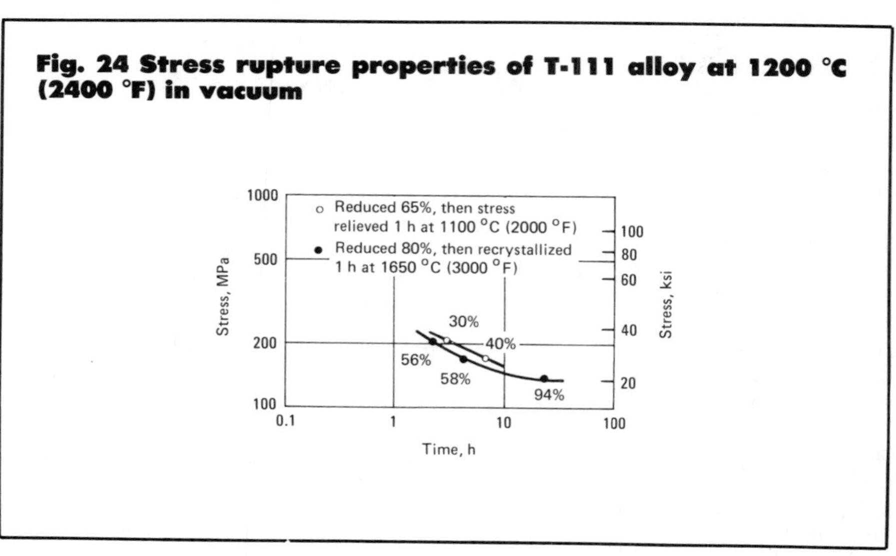

Fig. 25 Effect of test temperature on tensile properties of T-222 alloy sheet

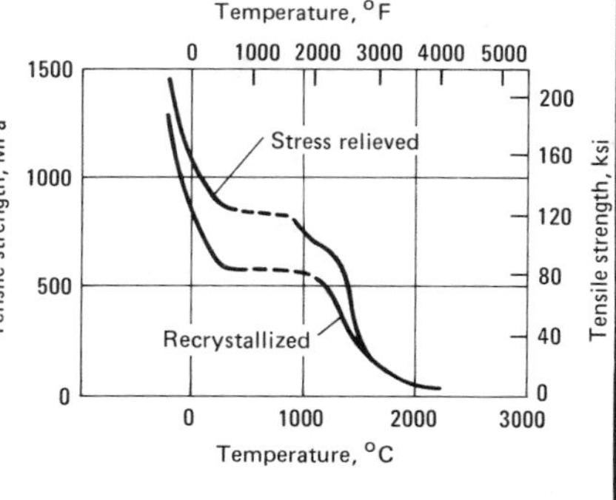

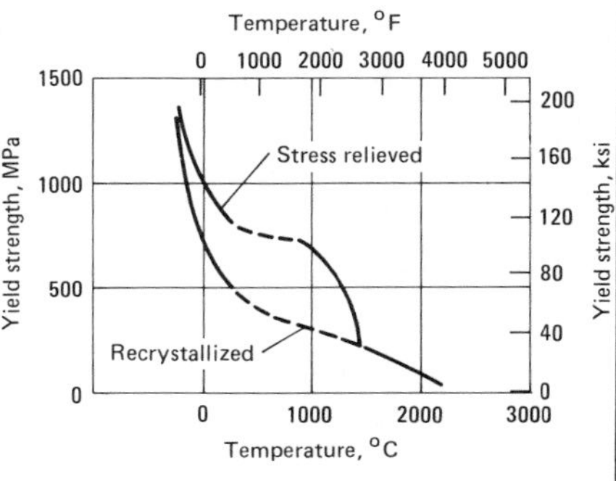

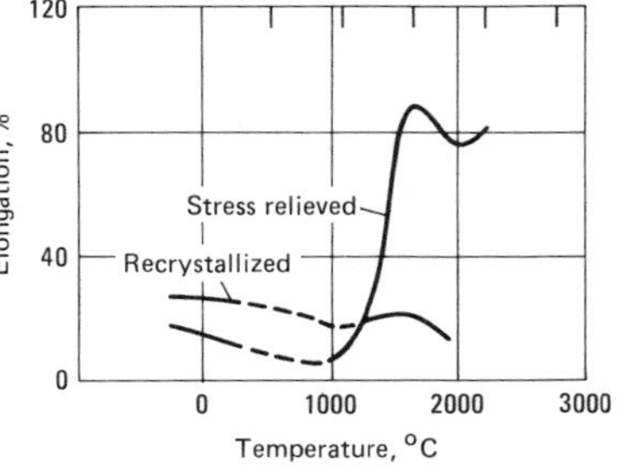

Table 11 Typical mechanical properties of T-222

Temperature °C	Temperature °F		Tensile strength MPa	Tensile strength ksi	Yield strength MPa	Yield strength ksi	Elongation, %
-195	-320	....	1280	185	1200	175	28
20	68	....	805	117	800	116	28
24	75	....	765	111	725	105	30
1650	3000	....	165	24.2	165	24	36

Fig. 26 Creep rupture curves for T-222 alloy at 2200 °F, 2400 °F and 3000 °F

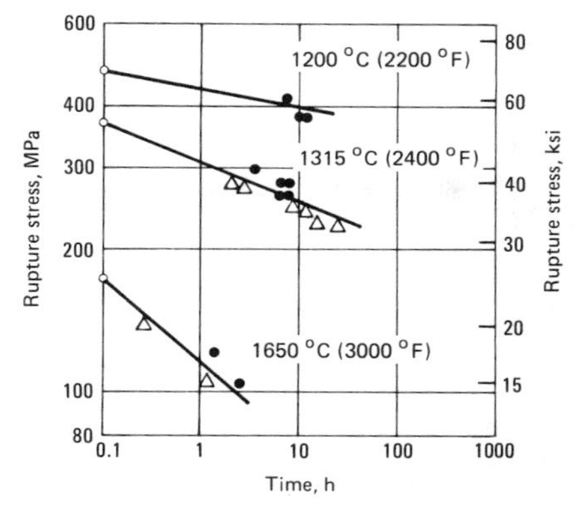

Fig. 27 Minimum creep rate curves for T-222 alloy at 2200 °F, 2400 °F and 3000 °F

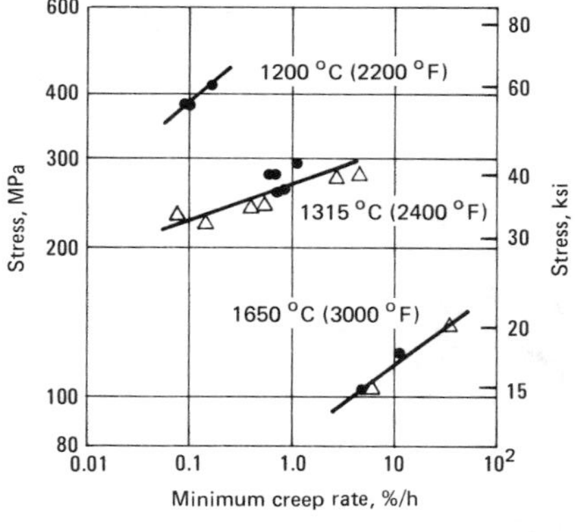

Tungsten Alloys

By Chester W. Dawson
and Heinz G. Sell
Westinghouse Lamp Division

Specific electrical resistivity of Mo-W alloys

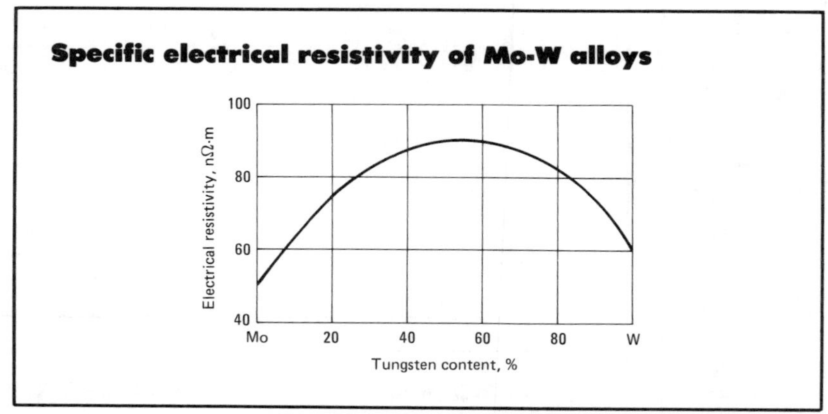

Room temperature mechanical properties of Mo-W alloys

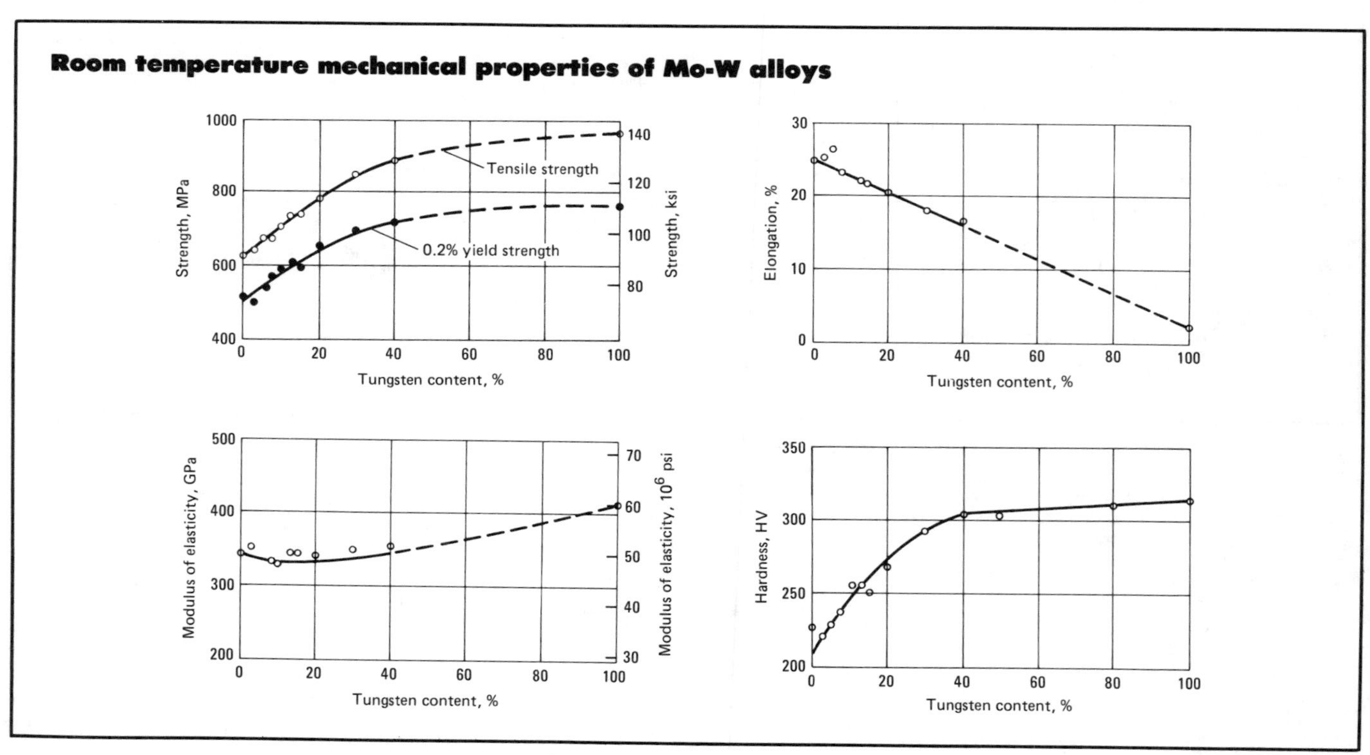

Specific electrical resistivity of W-Re alloys

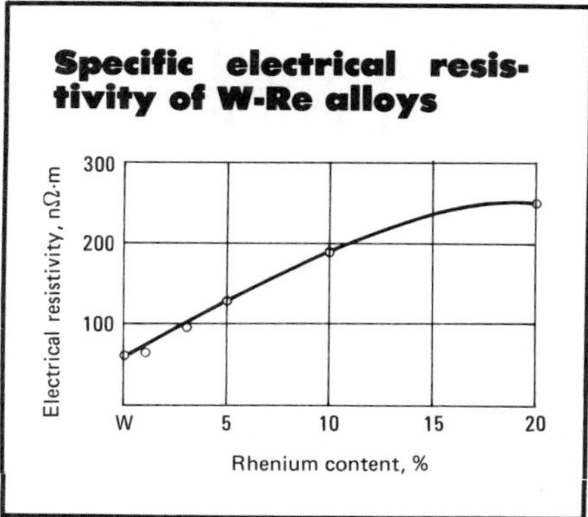

Recrystallization behavior of Undoped W bar

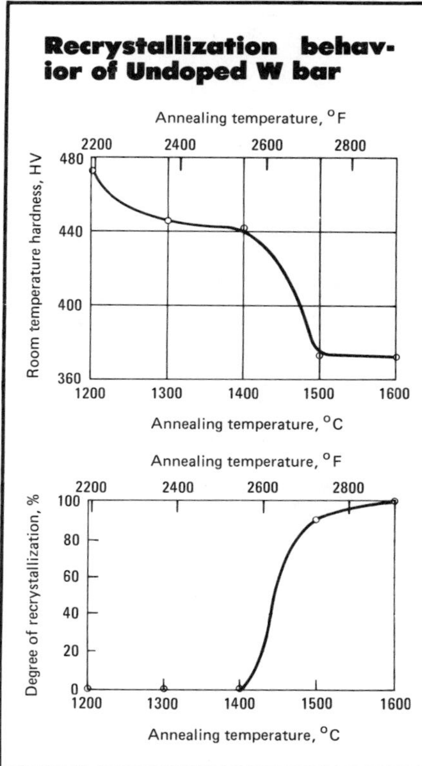

Thermal conductivity of Undoped W

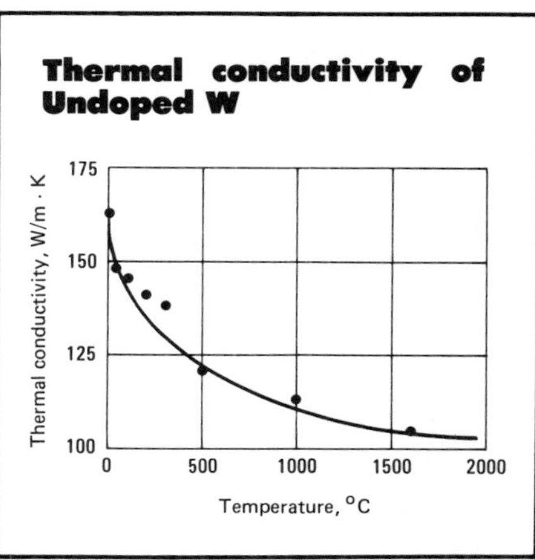

Creep curves for coiled tungsten wires at 2500 °C

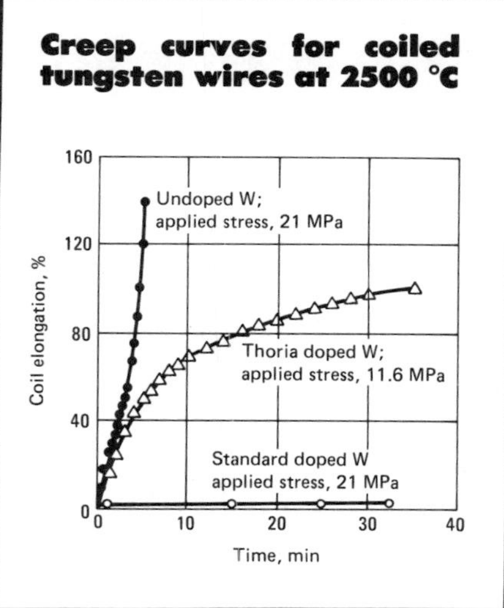

Electrical resistivity of Doped W and W-Re alloys

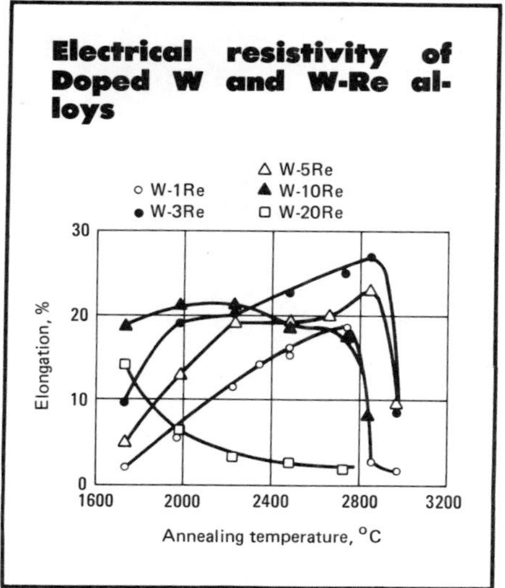

Room temperature ductility of annealed wire for five W-Re alloys

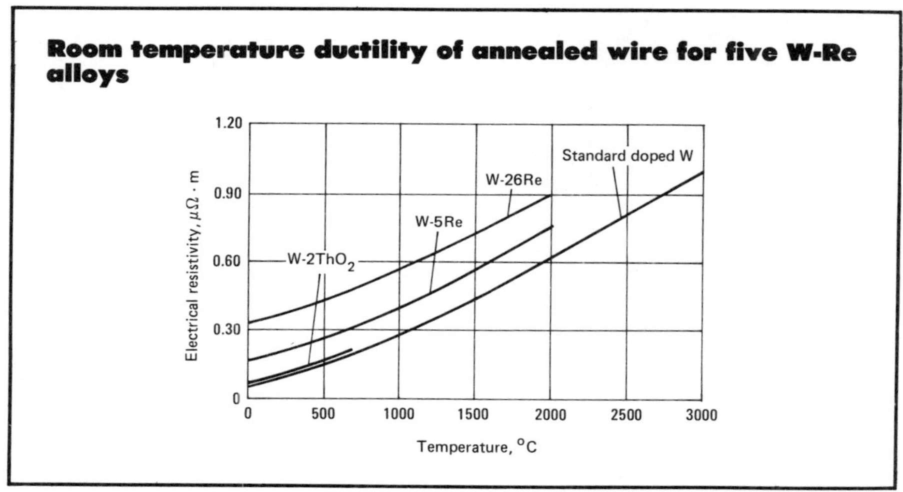

Short-time tensile strength of five W-Re alloys

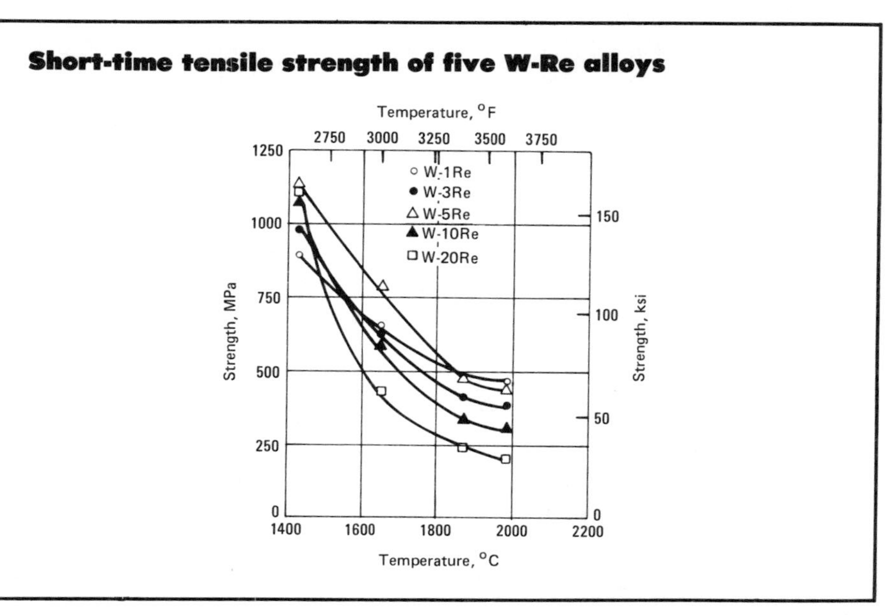

Titanium and Titanium Alloys

Introduction to Titanium and Its Alloys

By the ASM Committee on Titanium and Titanium Alloys*

TITANIUM was first identified as a metallic element by Gregor in England in 1791. Several years later, Klaproth in Germany named the metal after the Titans of Greek mythology. Production of high-purity titanium, because of the strong tendency of this metal to react with oxygen and nitrogen, proved difficult, and it was not until almost the middle of the 20th century (1938–1940) that a commercially attractive process was developed by Dr. Wilhelm Kroll. The Kroll process involves reduction of titanium tetrachloride with magnesium in an inert-gas atmosphere. The titanium so produced is called "titanium sponge" or "sponge metal" because of its porous, spongy appearance. Dr. Kroll interested the U.S. Bureau of Mines in his process, and a pilot plant was set up at the Bureau's Boulder City, NV station. The properties of titanium determined by the Bureau, and by others from samples provided by the Bureau, excited commercial interest, and the titanium industry began to take form shortly after the end of World War II. This commercial interest was prompted by the relatively low density of titanium (between those of aluminum and iron), its excellent corrosion resistance and its wide occurrence in mineral deposits.

*See page XI for committee list.

Mining and Production

The titanium production processes in use today are (*a*) the basic Kroll process, which uses magnesium as the reductant and (*b*) a modified process that uses sodium. Electrolytic reduction of molten salts has been studied on laboratory and pilot-plant scales. Such a process has already been demonstrated to be technically feasible and may become economically feasible in the future.

Titanium ore deposits are of two types: one based on rutile and the other on ilmenite. Rutile deposits of commercial significance occur as beach sands in Australia, South Africa, India and the United States. Australian rutile has been the primary source of ore for the United States titanium production. Ilmenite deposits are found widely throughout the world; leading producers are the United States, USSR, Norway, Canada and Malaysia. Ilmenite is the chief titanium ore used in the USSR. Estimated titanium content of the earth's minerals indicates that it is the fourth most abundant metal of commercial significance. Only aluminum, iron and magnesium are more abundant. However, despite the abundance of the metal, production of titanium is comparatively expensive because of the complexity of the reduction process, which involves producing titanium tetrachloride from the ore, purifying it, and reducing the purified tetrachloride with magnesium or sodium in an inert-atmosphere batch process.

The price of titanium as it is now produced is intimately related to the cost of magnesium and sodium, whether purchased or recycled. Manufacture of either magnesium or sodium is energy intensive. However, the reaction of $TiCl_4$ with magnesium or sodium is exothermic, so little or no energy need be supplied at this stage of the process.

Important Metal Characteristics

Titanium and its alloys are used primarily in two areas of application where the unique characteristics of these metals justify their selection: corrosion-resistant service and strength-efficient structures. Titanium has the ability to passivate, and thereby exhibit a high degree of immunity to attack by most mineral acids and chlorides. The combination of high strength, stiffness, good toughness and low density provided by various titanium alloys at very low to moderately elevated temperatures allows weight savings in aerospace structures and other high-performance applications. For these two diverse areas, selection criteria differ markedly. Corrosion applications normally utilize low-strength "unalloyed" titanium mill products fabricated into tanks, heat exchangers or reactor vessels for chemical-processing, desalination or power-generation plants. In contrast, high-performance applications typically utilize high-strength titanium alloys in a very selective manner depending on factors such as thermal environment, loading parameters, available product forms, fabrication characteristics, and inspection and/or reliability requirements. As a result of their specialized usage, alloys for high-performance applications normally are processed to more stringent and costly requirements than "unalloyed" titanium for corrosion service.

Selection for Corrosion Resistance. Economic considerations normally determine whether titanium alloys will be used for corrosion service. Capital expenditures for titanium equipment generally are higher than for equipment fabricated from competing materials such as stainless steel, brass, bronze, copper nickel or carbon steel. As a result, titanium equipment must yield lower operating costs, longer life or reduced maintenance to justify selection based on lower total-life-cycle cost. Commercially pure titanium satisfies the basic requirements for corrosion service. Unalloyed titanium normally is produced to requirements of ASTM B265, B338 or B367 in grades 1, 2, 3 and 4; these grades vary in oxygen and iron contents, which control strength level and corrosion behavior, respectively. For certain corrosion applications, Ti-0.2Pd (ASTM grades 7, 8 and 11) may be preferred over unalloyed grades 1, 2, 3 and 4.

Due to its unique corrosion behavior, titanium is used extensively in prosthetic devices such as heart-valve parts and load-bearing leg-bone replacements or splints. In general, body fluids are chloride brines that have pH values from 7.4 into the acidic range and that also contain a variety of organic acids and other components—media to which titanium is totally immune. Of the grades available, ASTM grade 2 is normally used for low-stress applications, whereas Ti-6Al-4V is normally employed for applications requiring higher strength.

Selection for Strength Efficiency. Historically, titanium alloys have been widely used instead of iron or nickel alloys in aerospace applications because titanium saves weight in highly loaded components that operate at low to moderately elevated temperatures. Many titanium alloys have been custom designed to have optimum tensile, compressive and/or creep strength at selected temperatures, and at the same time to have sufficient workability to be fabricated into mill products suitable for a specific application. During the life of the titanium industry, various compositions have had transient usage—Ti-4Al-3Mo-1V, Ti-7Al-4Mo and Ti-8Mn, for example. Ti-6Al-4V is unique in that it combines attractive properties with inherent workability (which allows it to be produced in all types of mill products, in both large and small sizes), good shop fabricability (which allows the mill products to be made into complex hardware), and the production experience and commercial availability that lead to reliable and economic usage. Thus Ti-6Al-4V has become the standard alloy against which other alloys must be compared when selecting a titanium alloy (or custom designing one) for a specific application. Ti-6Al-4V also is the standard alloy selected for castings that must exhibit superior strength.

Rotating components such as jet-engine blades and gas turbine parts require titanium alloys that maximize strength efficiency and metallurgical stability at elevated temperatures. These alloys also must exhibit low creep rates along with predictable behavior in stress rupture and low-cycle fatigue. To reproducibly provide these properties, stringent user requirements are specified to ensure controlled homogeneous microstructures and total freedom from melting imperfections such as alpha segregation, high-density or low-density tramp inclusions, and unhealed ingot porosity or pipe. Originally, alloys such as Ti-2Fe-2Cr-2Mo and Ti-1.5Fe-2.7Cr were used for engine applications; these alloys were replaced by Ti-6Al-4V in the 1950's and by Ti-8Al-1Mo-1V in the 1960's. Currently, Ti-2.25Al-11Sn-5Zr-1Mo (IMI679), Ti-6Al-2Sn-4Zr-2Mo and Ti-6Al-2Sn-4Zr-6Mo are also used in production engine applications.

Aerospace pressure vessels similarly require optimized strength efficiency; required auxiliary properties include weldability and predictable fracture toughness at cryogenic to moderately elevated temperatures. To provide this combination of properties, stringent user specifications require controlled microstructures and freedom from melting imperfections. For cryogenic applications, the interstitial elements oxygen, nitrogen and carbon are carefully controlled to improve ductility and fracture toughness. For these applications, the basic titanium alloy Ti-6Al-4V or (Ti-6Al-4V-ELI), processed to either the annealed or the solution treated and aged (STA) condition, is widely used. Ti-5Al-2.5Sn-ELI is an attractive alternative.

Aircraft structural applications plus high-performance automotive and marine applications require high strength efficiency, which normally is achieved by judicious alloy selection combined with close control of mill processing. However, when the design includes redundant structures, when operating environments are not severe, when there are constraints on the fabrication methods that can be used for specific components or when there are low operational risks, selection of the appropriate alloy and process must take these factors into account. Ti-6Al-4V and Ti-6Al-6V-2Sn are the basic alloys currently used for high-performance applications. Strength efficiency, fatigue-crack-growth rate and fracture toughness, plus manufacturing considerations such as welding and forming requirements, normally provide the criteria that determine the alloy, microstructure (alpha-beta or beta), heat treatment (some variant of either annealing or solution treating and aging), and level of process control. For lightly loaded structures (where titanium is normally selected because it offers greater resistance to temperature effects than aluminum), commercial availability of required mill products and ease of fabrication normally dictate selection. Here, one of the grades of unalloyed titanium is usually chosen.

Optic system support structures are a little-known but very important structural application for titanium castings. Complex castings are used in surveillance and guidance systems for aircraft and missiles to support the optics where wide temperature variations are encountered in service. The chief reason for selecting titanium for this application is the fact that the thermal expansion coefficient of titanium most closely matches that of the optics.

Lastly, a category exists where optimum strength efficiency is dictated by

Fig. 1 Titanium mill products shipped from U.S. mills from 1950 to 1978

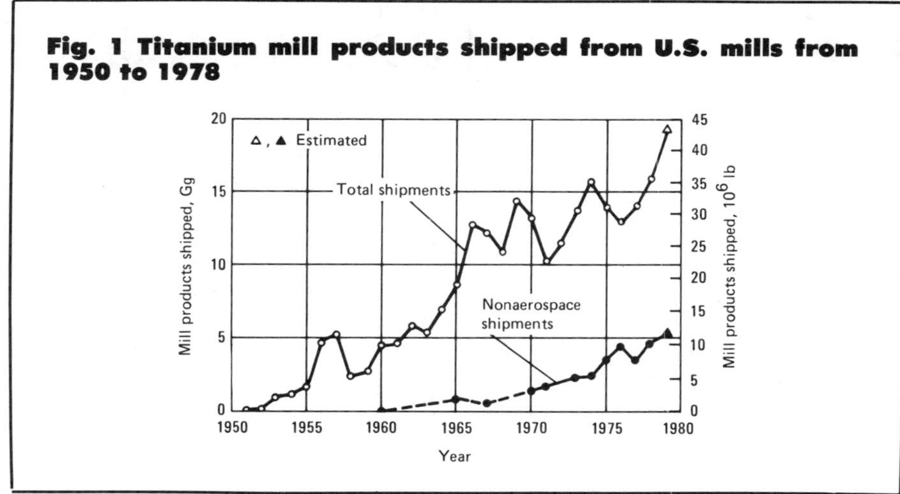

severe fabrication-imposed restraints, such as in high-strength foil and tubing. For this class of applications, the basic Ti-6Al-4V composition has been modified to Ti-3Al-2.5V, which is used in either the annealed or the cold worked and stress relieved condition.

Growth of Titanium Markets

The first actual production application of titanium was in the nacelles and firewalls of the DC-7—the last commercial aircraft powered by a reciprocating engine (initial flight, 1952). Soon after, newly developed titanium alloys were used to make compressor disks and blades for second-generation jet engines and airframes for military and commercial jet aircraft. These production applications resulted in rapid growth of the titanium industry between 1952 and 1957 (see Fig. 1). A market correction occurred in 1958 when military expenditures were transferred from development of aircraft to development of ballistic missiles. However, this low point has been followed by relatively steady growth. Periods of decreased production in 1968 and 1970–71 resulted from shifts in military spending, a general economic downturn and cancellation of programs such as development of the supersonic transport aircraft (SST).

Figure 1 shows total and nonaerospace tonnages of titanium mill products (bar, billet, plate and sheet, for example) shipped from U.S. mills during the period from 1950 to 1978. Data on tonnages of ingot melted would be roughly double these figures. The sponge produced in the United States is insufficient for domestic needs, and significant amounts are imported from Japan and the USSR. In the past, these overseas sources were able to sell sponge at lower prices than domestic producers, which made expansion of domestic facilities unattractive in spite of the need for additional capacity. The situation has changed in recent years because of worldwide inflationary pressures, and the cost differential between domestic and imported sponge has all but disappeared. The consumers of imported sponge are primarily those metal producers who do not have their own spongemaking facilities.

Because of the excellent corrosion resistance of titanium, both in marine service and in many aggressive chemical environments, it was predicted that large tonnages of this metal would find use in such applications. However, for a variety of reasons, such extensive application on the basis of corrosion resistance was slow to develop. As shown in Fig. 1, it was not until about 1965 that nonaerospace uses accounted for a significant fraction of titanium production. However, steady growth is now forecast for these applications because titanium and its alloys have been proved to have long service lives, which makes them cost effective in many applications where less expensive but less resistant materials were previously used.

Current estimated titanium sales (1979) are distributed as follows: 35% for commercial aircraft, 28% for industrial (corrosion) applications and 37% for military aircraft and missiles. Aircraft applications include both airframes and jet-engine parts. It has been said that the large, high-thrust, fan-type jet engines, which power commercial passenger aircraft such as the Boeing 747, Douglas DC-10, Aerospace Industries A300 AirBus, and Lockheed L-1011, would not be feasible without materials such as titanium alloys, which have high ratios of strength to density. Continued growth is forecast for nonaerospace marine and chemical applications of titanium, perhaps accompanied by a slight downturn in aerospace applications. This combination, however, is expected to produce continued over-all growth in total production and use of titanium and its alloys.

About 1960, the first commercial titanium castings were produced for chemical pumps and valves. Roughly ten years later, castings for aerospace applications became a production reality. By the late 1970's, sales of titanium castings paralleled sales for the titanium industry as a whole—about 65 to 70% for aerospace applications and 30 to 35% for commercial applications. Total casting sales are small; in 1979, they represented only about 2% of total titanium sales. However, it has been estimated that the same parts produced from wrought titanium would have accounted for about 14% of total sales, which emphasizes the chief advantage of castings—that of producing parts that are very close to the required final shape.

Titanium Alloy Systems

Titanium undergoes an allotropic transformation at about 885 °C (1625 °F), changing from a close-packed hexagonal crystal structure (alpha phase) to a body-centered cubic crystal structure (beta phase). The transformation temperature is strongly influenced by the interstitial elements oxygen, nitrogen and carbon (alpha stabilizers), which raise the transformation temperature, and by metallic impurity or alloying elements, which may either raise or lower the transformation temperature.

Figure 2 is the phase diagram for the titanium-rich portion of the titanium-oxygen system, illustrating the high solubility of oxygen in both the alpha and beta phases and its strong effect in raising the transformation temperature.

Figure 3 is the phase diagram for the titanium-molybdenum system, an ex-

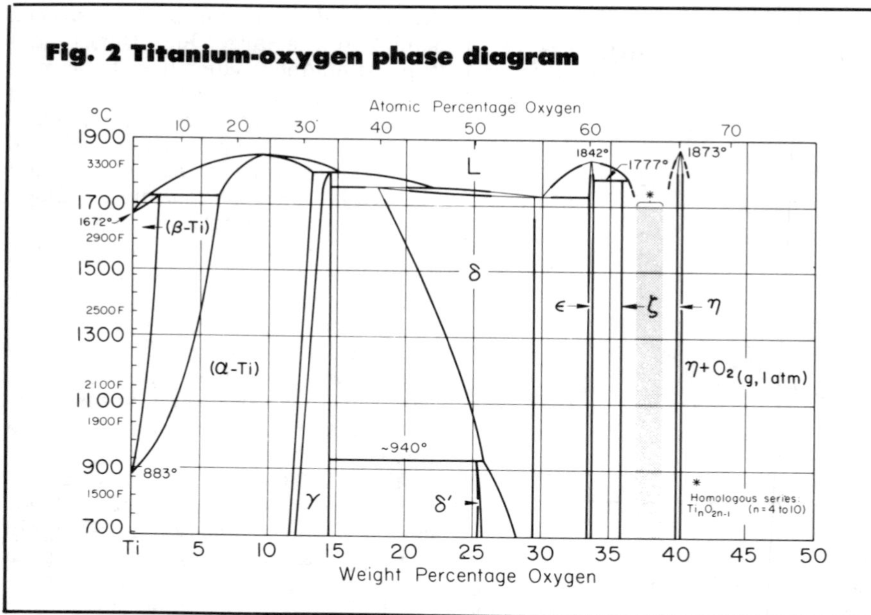

Fig. 2 Titanium-oxygen phase diagram

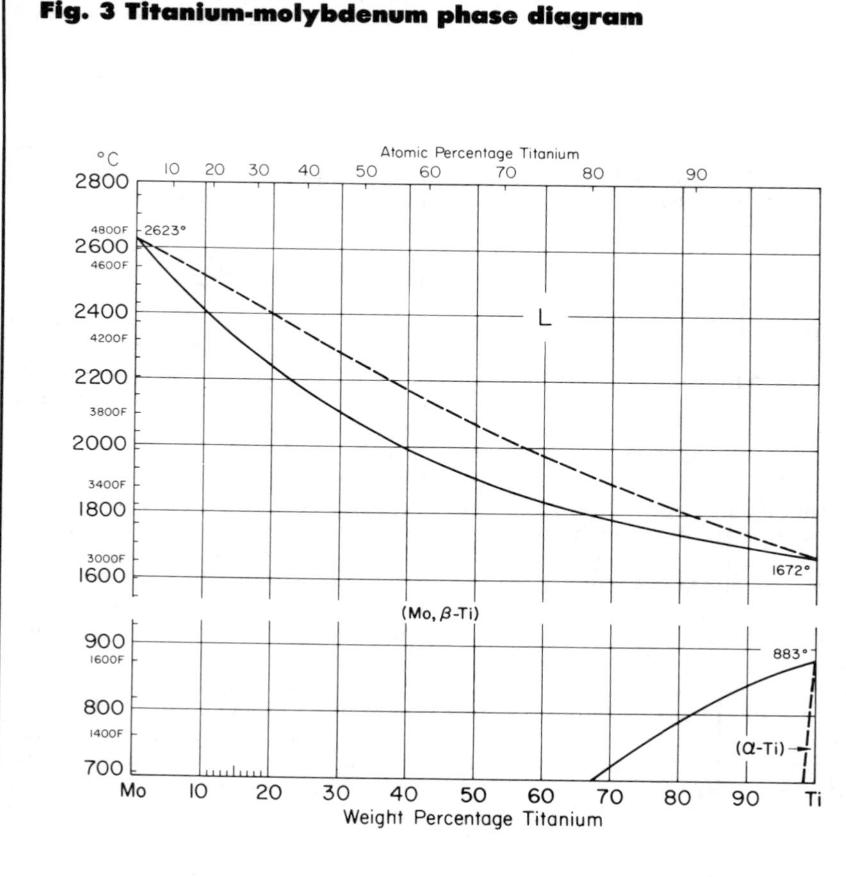

Fig. 3 Titanium-molybdenum phase diagram

ample of a beta isomorphous system. Tantalum, vanadium and niobium have similar phase relations with titanium. Titanium does not form intermetallic compounds with the beta isomorphous elements.

The titanium-chromium phase diagram (see Fig. 4) typifies eutectoid systems, which also are formed with iron, copper, nickel, palladium, cobalt, manganese and certain other transition metals. These elements have low solubility in alpha titanium and decrease the transformation temperature. They usually are added to alloys in combination with one or more of the beta isomorphous elements to stabilize the beta phase and prevent or minimize formation of intermetallic compounds, which might occur during service at elevated temperature.

Zirconium and hafnium are unique in that they are isomorphous with both the alpha and beta phases of titanium, and for this reason, they have been called the sister elements to titanium. Tin and aluminum have significant solubility in both alpha and beta phases. Aluminum increases the transformation temperature; tin lowers it slightly. Aluminum, tin and zirconium are commonly used together in alpha and near-alpha alloys. In alpha-beta alloys, these elements are distributed approximately equally between the alpha and beta phases. In beta alloys, they stabilize and strengthen the beta phase and modify transformation characteristics. Almost all commercial titanium alloys contain one or more of these three elements because they are soluble in both alpha and beta phases, and particularly because they improve creep strength in the alpha phase.

Many more elements are soluble in beta titanium than in alpha. Beta isomorphous alloying elements are preferred because they do not form intermetallic compounds. However, iron, chromium, manganese and other compound formers sometimes are used in beta-rich alpha-beta alloys or in beta alloys, because they are strong beta stabilizers and improve hardenability and response to heat treatment. Nickel, molybdenum and palladium improve corrosion resistance of unalloyed titanium in certain media. The trend in recent years has been to design and develop alloys for specific applications that demand a particular set of mechanical properties. Alloying elements are then selected to achieve the goals set for the new alloy by taking advan-

Table 1 Summary of commercial and semicommercial grades and alloys of titanium

Designation	Tensile strength (min) MPa	ksi	0.2% yield strength (min) MPa	ksi	Impurity limits, wt % N (max)	C (max)	H (max)	Fe (max)	O (max)	Nominal composition, wt % Al	Sn	Zr	Mo	Others
Unalloyed grades														
ASTM Grade 1	240	35	170	25	0.03	0.10	0.015	0.20	0.18	...	...	...	...	...
ASTM Grade 2	340	50	280	40	0.03	0.10	0.015	0.30	0.25	...	...	...	...	...
ASTM Grade 3	450	65	380	55	0.05	0.10	0.015	0.30	0.35	...	...	...	...	...
ASTM Grade 4	550	80	480	70	0.05	0.10	0.015	0.50	0.40	...	...	...	...	...
ASTM Grade 7	340	50	280	40	0.03	0.10	0.015	0.30	0.25	...	...	...	...	0.2Pd
Alpha and near-alpha alloys														
Ti Code 12	480	70	380	55	0.03	0.10	0.015	0.30	0.25	...	...	...	0.3	0.8Ni
Ti-5Al-2.5Sn	790	115	760	110	0.05	0.08	0.02	0.50	0.20	5	2.5	...	...	...
Ti-5Al-2.5Sn-ELI	690	100	620	90	0.07	0.08	0.0125	0.25	0.12	5	2.5	...	...	...
Ti-8Al-1Mo-1V	900	130	830	120	0.05	0.08	0.015	0.30	0.12	8	...	...	1	1V
Ti-6Al-2Sn-4Zr-2Mo	900	130	830	120	0.05	0.05	0.0125	0.25	0.15	6	2	4	2	...
Ti-6Al-2Nb-1Ta-0.8Mo	790	115	690	100	0.02	0.03	0.0125	0.12	0.10	6	...	...	1	2Nb, 1Ta
Ti-2.25Al-11Sn-5Zr-1Mo	1000	145	900	130	0.04	0.04	0.008	0.12	0.17	2.25	11.0	5.0	1.0	0.2Si
Ti-5Al-5Sn-2Zr-2Mo(a)	900	130	830	120	0.03	0.05	0.0125	0.15	0.13	5	5	2	2	0.25Si
Alpha-beta alloys														
Ti-6Al-4V(b)	900	130	830	120	0.05	0.10	0.0125	0.30	0.20	6.0	...	...	...	4.0V
Ti-6Al-4V-ELI(b)	830	120	760	110	0.05	0.08	0.0125	0.25	0.13	6.0	...	...	...	4.0V
Ti-6Al-6V-2Sn(b)	1030	150	970	140	0.04	0.05	0.015	1.0	0.20	6.0	2.0	...	...	0.75Cu, 6.0V
Ti-8Mn(b)	860	125	760	110	0.05	0.08	0.015	0.50	0.20	...	...	...	...	8.0Mn
Ti-7Al-4Mo(b)	1030	150	970	140	0.05	0.10	0.013	0.30	0.20	7.0	...	...	4.0	...
Ti-6Al-2Sn-4Zr-6Mo(c)	1170	170	1100	160	0.04	0.04	0.0125	0.15	0.15	6.0	2.0	4.0	6.0	...
Ti-5Al-2Sn-2Zr-4Mo-4Cr(a)(c)	1125	163	1055	153	0.04	0.05	0.0125	0.30	0.13	5.0	2.0	2.0	4.0	4.0Cr
Ti-6Al-2Sn-2Zr-2Mo-2Cr(a)(b)	1030	150	970	140	0.03	0.05	0.0125	0.25	0.14	5.7	2.0	2.0	2.0	2.0Cr, 0.25Si
Ti-10V-2Fe-3Al(a)(c)	1170	170	1100	160	0.05	0.05	0.015	2.5	0.16	3.0	...	...	...	10.0V
Ti-3Al-2.5V(d)	620	90	520	75	0.015	0.05	0.015	0.30	0.12	3.0	...	...	...	2.5V
Beta alloys														
Ti-13V-11Cr-3Al(c)	1170	170	1100	160	0.05	0.05	0.025	0.35	0.17	3.0	...	...	...	11.0Cr, 13.0V
Ti-8Mo-8V-2Fe-3Al(a)(c)	1170	170	1100	160	0.05	0.05	0.015	2.5	0.17	3.0	...	...	8.0	8.0V
Ti-3Al-8V-6Cr-4Mo-4Zr(a)(b)	900	130	830	120	0.03	0.05	0.020	0.25	0.12	3.0	...	4.0	4.0	6.0Cr, 8.0V
Ti-11.5Mo-6Zr-4.5Sn(b)	690	100	620	90	0.05	0.10	0.020	0.35	0.18	...	4.5	6.0	11.5	...

(a) Semicommercial alloy; mechanical properties and composition limits subject to negotiation with suppliers. (b) Mechanical properties given for annealed condition; may be solution treated and aged to increase strength. (c) Mechanical properties given for solution treated and aged condition; alloy not normally applied in annealed condition. Properties may be sensitive to section size and processing. (d) Primarily a tubing alloy; may be cold drawn to increase strength.

tage of effects determined in earlier studies.

Commercial Grades and Alloys

There are several grades of unalloyed titanium. The primary difference between grades is oxygen and iron content. Grades of higher purity (lower interstitial content) are lower in strength, hardness and transformation temperature than those higher in interstitial content. The high solubility of the interstitial elements oxygen and nitrogen makes titanium rather unique among metals, and also creates problems not of concern in most other metals. For example, heating titanium in air at high temperature results not only in oxidation but also in solid-solution hardening of the surface as a result of inward diffusion of oxygen. A surface-hardened zone of "alpha-case" (or "air contamination layer") is formed. Normally, this layer is removed by machining, chemical milling or other mechanical means prior to placing a part in service, because the presence of alpha-case reduces fatigue strength and ductility.

Table 1 lists the commercial and semicommercial titanium grades and alloys currently available, which are subdivided into four groups: commercially pure grades, alpha and near-alpha alloys, alpha-beta alloys and beta alloys.

Ti-6Al-4V is the most widely used titanium alloy, accounting for about 45% of total titanium production. Unalloyed grades comprise about 30% of production, and all other alloys combined comprise the remaining 25%.

AMS specifications have been prepared for all commercial titanium alloys and some semicommercial alloys (see Table 2). Some alloys and product forms also are covered by MIL specifications, which are cross referenced in Table 2. Table 3 summarizes grades and alloys covered by ASTM specifications, and gives a cross reference to AMS specifications. ASTM specifications are used chiefly for nonaerospace applications where commercially pure titanium is used for corrosion resistance and Ti-6Al-4V (ASTM grade 5) is used for strength and erosion resistance. Typical applications are heat exchangers, process tubing, reaction vessels, fittings, pump and valve components, and gas compressors.

Table 2 AMS specifications for titanium and titanium alloys

AMS No.	Mill form	Condition	Alloy	Similar MIL specification
4900	Plate, sheet, strip	Annealed	Unalloyed; 55-ksi YS	MIL-T-9046
4901	Plate, sheet, strip	Annealed	Unalloyed; 70-ksi YS	MIL-T-9046
4902	Plate, sheet, strip	Annealed	Unalloyed; 40-ksi YS	MIL-T-9046
4906	Sheet, strip; continuously rolled	Annealed	Ti-6Al-4V	...
4907	Plate, sheet, strip	Annealed	Ti-6Al-4V-ELI	MIL-T-9046
4908	Sheet, strip	Annealed	Ti-8Mn; 110-ksi YS	MIL-T-9046
4909	Plate, sheet, strip	Annealed	Ti-5Al-2.5Sn-ELI	MIL-T-9046
4910	Plate, sheet, strip	Annealed	Ti-5Al-2.5Sn	MIL-T-9046
4911	Plate, sheet, strip	Annealed	Ti-6Al-4V	MIL-T-9046
4915	Plate, sheet, strip	Single annealed	Ti-8Al-1Mo-1V	MIL-T-9046
4916	Plate, sheet, strip	Duplex annealed	Ti-8Al-1Mo-1V	MIL-T-9046
4917	Plate, sheet, strip	Solution treated	Ti-13V-11Cr-3Al	MIL-T-9046
4918	Plate, sheet, strip	Annealed	Ti-6Al-6V-2Sn	MIL-T-9046
4921	Bar, forgings, rings	Annealed	Unalloyed; 70-ksi YS	MIL-T-9047
4924	Bar, forgings, rings	Annealed	Ti-5Al-2.5Sn-ELI; 90-ksi YS	MIL-T-9047
4926	Bar, rings	Annealed	Ti-5Al-2.5Sn; 110-ksi YS	MIL-T-9047
4928	Bar, forgings	Annealed	Ti-6Al-4V; 120-ksi YS	MIL-T-9047
4930	Bar, forgings, rings	Annealed	Ti-6Al-4V-ELI	MIL-T-9047
4935	Extrusions	Annealed	Ti-6Al-4V	...
4936	Extrusions	...	Ti-6Al-6V-2Sn	...
4941	Tubing, welded	Annealed	Unalloyed; 40-ksi YS	...
4942	Tubing, seamless	Annealed	Unalloyed; 40-ksi YS	...
4943	Tubing, seamless	Annealed	Ti-3Al-2.5V	...
4944	Tubing, seamless hydraulic	Cold worked and stress relieved	Ti-3Al-2.5V	...
4951	Wire, welding	...		...
4953	Wire, welding	Annealed	Ti-5Al-2.5Sn	...
4954	Wire, welding	...	Ti-6Al-4V	...
4955	Wire, welding	...	Ti-8Al-1Mo-1V	...
4956	Wire, welding	...	Ti-6Al-4V-ELI	...
4965	Bar, forgings, rings	Precipitation heat treated	Ti-6Al-4V	...
4966	Forgings	Annealed	Ti-5Al-2.5Sn; 110-ksi YS	MIL-F-83142
4967	Bar, forgings	Annealed	Ti-6Al-4V	MIL-T-9047
4970	Bar, forgings	Precipitation heat treated	Ti-7Al-4Mo	MIL-T-9047
4971	Bar, forgings, rings	Annealed	Ti-6Al-6V-2Sn	MIL-T-9047, MIL-F-83142
4972	Bar, rings	Solution treated and stabilized	Ti-8Al-1Mo-1V	...
4973	Forgings	Solution treated and stabilized	Ti-8Al-1Mo-1V	...
4974	Bar, forgings	Precipitation heat treated	Ti-11Sn-5Zr-2.3Al-1Mo-0.21Si	...
4975	Bar, rings	Precipitation heat treated	Ti-6Al-2Sn-4Zr-2Mo	MIL-T-9047
4976	Forgings	Precipitation heat treated	Ti-6Al-2Sn-4Zr-2Mo	...
4977	Bar, wire	Solution treated	Ti-11.5Mo-6Zr-4.5Sn	MIL-T-9047
4978	Bar, forgings, rings	Annealed	Ti-6Al-6V-2Sn; 140-ksi YS	MIL-T-9047, MIL-F-83142
4979	Bar, forgings, rings	Precipitation heat treated	Ti-6Al-6V-2Sn	MIL-T-9047, MIL-F-83142
4980	Bar, wire	Solution treated at 745 °C (1375 °F)	Ti-11.5Mo-6Zr-4.5Sn	...
4981	Bar, forgings	Precipitation heat treated	Ti-6Al-2Sn-4Zr-6Mo	MIL-T-9047

Fig. 4 Titanium-chromium phase diagram

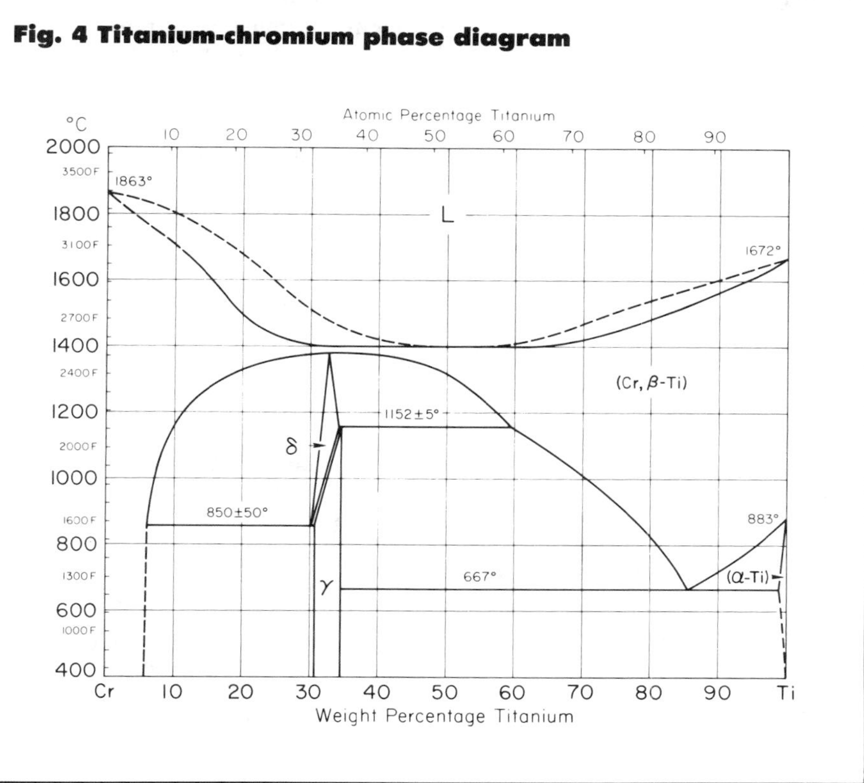

Selection of an unalloyed grade of titanium, an alpha or near-alpha alloy, an alpha-beta alloy or a beta alloy depends on desired mechanical properties, service requirements, cost considerations, and the other factors that enter into any material-selection process.

Unalloyed titanium usually is selected for its excellent corrosion resistance, especially in applications where high strength is not required. Yield strengths of "unalloyed" (commercially pure) grades (see Table 1) vary from less than 170 MPa (25 ksi) to over 485 MPa (70 ksi) simply as a result of variation in the interstitial and impurity levels. Oxygen and iron are the primary variants in these grades; strength increases with increasing oxygen and iron contents.

Alpha and Near-Alpha Alloys. Alpha alloys that contain aluminum, tin and/or zirconium are preferred for high-temperature and cryogenic applications. Alpha-rich alloys generally are more resistant to creep at high temperature than alpha-beta or beta alloys. The extra-low-interstitial alpha alloys (ELI grades) retain ductility and toughness at cryogenic temperatures, and 5Al-2.5Sn-ELI has been used extensively in such applications.

Unlike alpha-beta and beta alloys, alpha alloys cannot be strengthened by heat treatment. Generally, alpha alloys are annealed or recrystallized to remove residual stresses induced by cold working. Alpha alloys have good weldability because they are insensitive to heat treatment. They generally have poorer forgeability and narrower forging-temperature ranges than alpha-beta or beta alloys, particularly at temperatures below the beta transus. This poorer forgeability is manifested by a greater tendency for center bursts or surface cracks to occur, which means that small reduction steps and frequent reheats must be incorporated in forging schedules.

Alpha alloys that contain small additions of beta stabilizers (Ti-8Al-1Mo-1V or Ti-6Al-2Nb-1Ta-0.8Mo, for example) sometimes have been classed as "super-alpha" or "near-alpha" alloys. Although they contain some retained beta phase, these alloys consist primarily of alpha and behave more like conventional alpha alloys than alpha-beta alloys.

Alpha-beta alloys contain one or more alpha-stabilizers or alpha-soluble elements plus one or more beta stabilizers. These alloys retain more beta phase after final heat treatment than do near-alpha alloys, the specific amount depending on the quantity of beta stabilizers present and on heat treatment.

Alpha-beta alloys can be strengthened by solution treating and aging. Solution treating usually is done at a temperature high in the two-phase alpha-beta field, and is followed by quenching in water, oil or other suitable quenchant. As a result of quenching, the beta phase present at the solution treating temperature may be retained or may be partly transformed during cooling by either martensitic transformation or nucleation and growth. The specific response depends on alloy composition, solution treating temperature (beta-phase composition at the solution temperature), cooling rate and section size. Solution treatment is followed by aging, normally at 480 to 650 °C (900 to 1200 °F), to precipitate alpha and produce a fine mixture of alpha and beta in the retained or transformed beta phase. Transformation kinetics, transformation products and specific response of a given alloy can be quite complex, and a detailed review of the subject is beyond the scope of this article.

Solution treating and aging can increase the strength of alpha-beta alloys 30 to 50%, or more, over the annealed or overaged condition. Response to solution treating and aging depends on section size; alloys relatively low in beta stabilizers (Ti-6Al-4V, for example) have poor hardenability and must be quenched rapidly to achieve significant strengthening. For Ti-6Al-4V, the cooling rate of a water quench is not rapid enough to significantly harden sections thicker than about 25 mm (1 in.). As the content of beta stabilizers increases, hardenability increases; Ti-5Al-2Sn-2Zr-4Mo-4Cr, for example, can be through hardened with relatively uniform response throughout sections up to 150 mm (6 in.) thick. For some alloys of intermediate beta-stabilizer content, the surface of a relatively thick section can be strengthened, but the core may be 10 to 20% lower in hardness and strength. The strength that can be achieved by heat treatment is also a function of the volume fraction of beta phase present at the solution treating temperature. Alloy composi-

Table 3 ASTM specifications for titanium and titanium alloys

Specification	Grade	Alloy	Min 0.2% yield strength(a) MPa	ksi	Similar AMS specification(b)
Plate, sheet and strip					
B265	1	Unalloyed	170	25	...
	2	Unalloyed	280	40	4902
	3	Unalloyed	380	55	4900
	4	Unalloyed	480	70	4901
	5	Ti-6Al-4V	830	120	4911
	6	Ti-5Al-2.5Sn	790	115	4910
	7	Ti-0.2Pd	280	40	...
	10	Ti-4.5Sn-11.5Mo-6Zr	620	90	4977
	11	Ti-0.2Pd	170	25	...
Seamless and welded pipe					
B337	1	Unalloyed	170	25	...
	2	Unalloyed	280	40	4941(c), 4942(d)
	3	Unalloyed	380	55	...
	7	Ti-0.2Pd	280	40	...
	9	Ti-3Al-2.5V	480	70	4943
	9	Ti-3Al-2.5V	720(e)	105(e)	4943
	10	Ti-11.5Mo-6Zr-4.5Sn	620(f)	90(f)	4977, 4980
	11	Ti-0.2Pd	170	25	...
Seamless and welded tube for condensers and heat exchangers					
B338	1	Unalloyed	170	25	...
	2	Unalloyed	280	40	...
	3	Unalloyed	380	55	...
	7	Ti-0.2Pd	280	40	...
	9	Ti-3Al-2.5V	720	105	4943, 4944
	10	Ti-11.5Mo-6Zr-4.5Sn	620	90	4977, 4980
	11	Ti-0.2Pd	170	25	...
Bar and billet					
B348	1	Unalloyed	170	25	...
	2	Unalloyed	280	40	...
	3	Unalloyed	380	55	...
	4	Unalloyed	480	70	4921(g)
	5	Ti-6Al-4V	830	120	4928(g)
	6	Ti-5Al-2.5Sn	790	115	4926(g)
	7	Ti-0.2Pd	280	40	...
	10	Ti-4.5Sn-11.5Mo-6Zr	620	90	4977, 4980
	11	Ti-0.2Pd	170	25	...
Castings					
B367	C-1	Unalloyed	170	25	...
	C-2	Unalloyed	280	40	...
	C-3	Unalloyed	380	55	...
	C-4	Unalloyed	480	70	...
	C-5	Ti-6Al-4V	830	120	...
	C-6	Ti-5Al-2.5Sn	720	105	...
	C-7A	Ti-0.2Pd	170	25	...
	C-7B	Ti-0.2Pd	280	40	...
	C-8A	Ti-0.2Pd	380	55	...
	C-8B	Ti-0.2Pd	480	70	...
Forgings					
B381	F1	Unalloyed	170	25	...
	F2	Unalloyed	280	40	...
	F3	Unalloyed	380	55	...
	F4	Unalloyed	480	70	4921
	F5	Ti-6Al-4V	830	120	4928
	F6	Ti-5Al-2.5Sn	790	115	4966
	F7	Ti-0.2Pd	280	40	...
	F11	Ti-0.2Pd	170	25	...

(a) Annealed. (b) Interstitial and impurity levels, and mechanical property requirements, may show minor differences compared with ASTM specifications. (c) Welded tubing. (d) Seamless tubing. (e) Cold worked and stress relieved. (f) Solution treated. (g) AMS specifications cover bar and forgings but not billet.

tion, solution temperature and aging conditions must be carefully selected and balanced to produce the desired mechanical properties in the final product.

Although the ability of alpha-beta alloys to be precipitation hardened has been studied in laboratory programs since the early days of the titanium industry, there have been relatively few production applications of solution treated and aged alloys. This situation appears to be changing, because alloys such as Ti-6Al-2Sn-4Zr-6Mo, Ti-5Al-2Sn-2Zr-4Mo-4Cr and certain high hardenability beta alloys have been developed specifically to be age hardened for improved strength—about 30 to 40% above that of annealed alloys.

Beta alloys are richer in beta stabilizers and leaner in alpha stabilizers than alpha-beta alloys. They are characterized by high hardenability, with beta phase completely retained on air cooling of thin sections or water quenching of thick sections. Beta alloys have excellent forgeability, and in sheet form can be cold formed more readily than high-strength alpha-beta or alpha alloys. After solution treating, beta alloys are aged at temperatures of 450 to 650 °C (850 to 1200 °F) to partially transform the beta phase to alpha. The alpha forms as finely dispersed particles in the retained beta, and strength levels comparable or superior to those of aged alpha-beta alloys can be attained. The chief disadvantages of beta alloys in comparison with alpha-beta alloys are higher density, lower creep strength and lower tensile ductility in the aged condition. Although tensile ductility is lower, the fracture toughness of an aged beta alloy generally is higher than that of an aged alpha-beta alloy of comparable yield strength.

In the solution treated condition (100% retained beta), beta alloys have good ductility and toughness, relatively low strength, and excellent formability. Solution treated beta alloys begin to precipitate alpha phase at slightly elevated temperatures, and thus are unsuitable for elevated-temperature service without prior stabilization or overaging treatment.

Beta alloys, despite the name, actually are metastable, because cold work at ambient temperature or heating to a slightly elevated temperature can cause partial transformation to alpha. The chief advantages of beta alloys are high hardenability, excellent forgeability, and good cold formability in the solution treated condition.

Relation of Properties to Processing for Wrought Titanium Alloys

By the ASM Committee on Titanium and Titanium Alloys*

TITANIUM metal passes through four major steps during processing from ore to finished product: reduction of titanium ore to a porous form of titanium metal called "sponge"; melting of sponge to form ingot; primary fabrication, in which ingots are converted into general mill products; and secondary fabrication of finished shapes from mill products. At each of these steps, mechanical and physical properties of titanium in the finished shape may be affected by any of several factors, or by a combination of factors. Among the most important are (*a*) amounts of specific alloying elements and impurities, (*b*) melting process used to make ingot, (*c*) method for mechanically working ingots into mill products and (*d*) the final step employed in working, fabrication or heat treatment.

Because the properties of titanium are so readily influenced, processors must exercise great care in controlling the conditions under which processing is carried out. At the same time, this characteristic of titanium makes it possible for the titanium industry to serve a wide range of applications with a minimum number of grades or alloys. By varying thermal or mechanical processing, or both, a broad range of special properties can be produced in commercially pure titanium and titanium alloys.

Raw Materials

Control of raw materials is extremely important in producing titanium and its alloys, because there are many elements of which small amounts can have major effects on the properties of these metals in finished form. The raw materials used in producing titanium

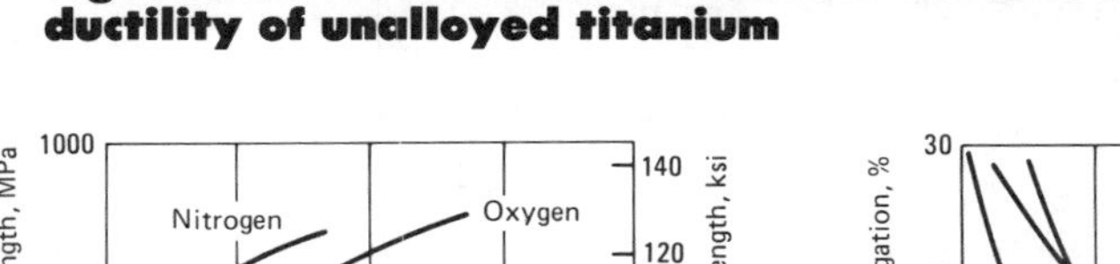

Fig. 1 Effects of interstitial-element content on strength and ductility of unalloyed titanium

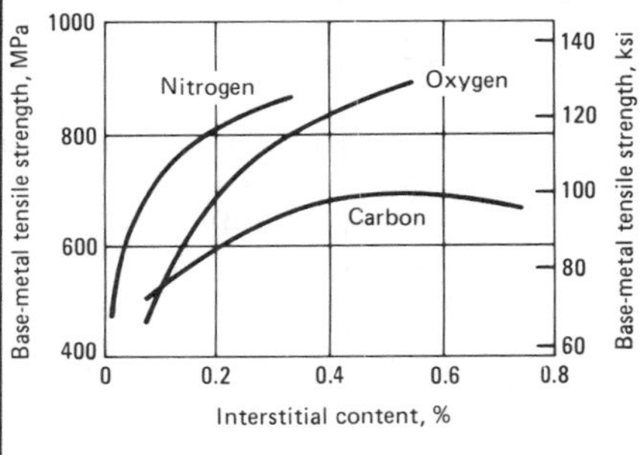

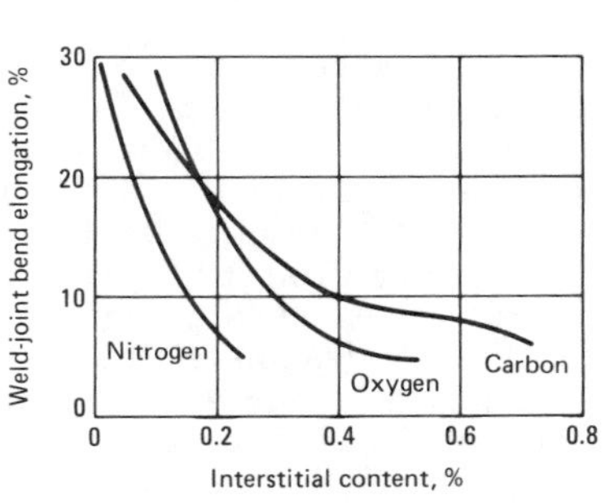

*See page XI for committee list.

are: titanium in the form of sponge metal, alloying elements, and reclaimed titanium scrap (usually called "revert").

Titanium sponge must meet stringent specifications for control of ingot composition. Most importantly, sponge must not contain hard, brittle and refractory titanium oxide, titanium nitride or complex titanium oxynitride particles that, if retained through subsequent melting operations, could act as crack initiation sites in the final product.

Carbon, nitrogen, oxygen, silicon and iron are commonly found as residual elements in sponge. These elements must be held to acceptably low levels because they raise the strength and lower ductility of the final product (see Fig. 1).

Titanium sponge is manufactured by first chlorinating rutile ore and then reducing the resulting $TiCl_4$ with either sodium or magnesium metal. Sodium-reduced sponge is leached with acid to remove the NaCl by-product of reduction. Magnesium-reduced sponge may be leached, inert-gas swept, or vacuum distilled to remove the excess $MgCl_2$ by-product. Vacuum distilling results in lower residual levels of magnesium, hydrogen and chlorine, but increases cost. Modern melting techniques remove volatile substances from sponge, so that ingot of high quality can be produced regardless of which method is used for production of sponge. Electrolytic methods have been used to produce titanium sponge on a pilot-plant scale.

Alloying Elements. Purity of alloying elements added to titanium during melting is as important as purity of sponge, and must be controlled with the same degree of care to avoid undesirable residual elements—especially those that can form refractory or high-density inclusions in the titanium matrix.

Basically, oxygen and iron contents determine strength levels of commercially pure titanium (ASTM and ASME grades 1, 2, 3 and 4) and the differences in mechanical properties between extra-low interstitial (ELI) grades and standard grades of titanium alloys. (Table 1 illustrates this effect.) In higher strength grades, oxygen and iron are intentionally added to the residual amounts already in the sponge to provide extra strength. On the other hand, carbon and nitrogen usually are held to minimum residual levels to avoid embrittlement.

Table 1 Tensile properties of annealed titanium sheet as influenced by oxygen and iron contents

Material	Maximum impurity content, %		Minimum tensile strength		Minimum yield strength(a)	
	Oxygen	Iron	MPa	ksi	MPa	ksi
Unalloyed Ti, grade 1	0.18	0.20	240	35	170	25
Unalloyed Ti, grade 2	0.25	0.30	345	50	275	40
Unalloyed Ti, grade 3	0.35	0.30	450	65	380	55
Unalloyed Ti, grade 4	0.40	0.50	655	95	485	70
Ti-6Al-4V	0.20	0.30	925	134	870	126
Ti-6Al-4V-ELI	0.13	0.25	900	130	830	120
Ti-5Al-2.5Sn	0.20	0.50	830	120	780	113
Ti-5Al-2.5Sn-ELI	0.12	0.25	690	100	655	95

(a) At 0.2% offset.

Reclaimed scrap makes production of ingot titanium more economical than production solely from sponge. If properly controlled, addition of scrap (commonly referred to as "revert") is fully acceptable and can be used even in materials for critical structural applications, such as rotating components for jet engines.

All forms of scrap can be remelted—machining chips, cut sheet, trim stock and chunks. To be utilized properly, scrap must be thoroughly cleaned and carefully sorted by alloy and by purity before being remelted. During cleaning, surface scale must be removed, because adding titanium scale to the melt could produce refractory inclusions or excessive porosity in the ingot. Machining chips from fabricators who use carbide tools are acceptable for remelting only if all carbide particles adhering to the chips are removed; otherwise, hard high-density inclusions could result. Improper segregation of alloy revert would produce off-composition alloys and could potentially degrade the properties of the resulting metal.

Titanium Ingot

Double melting is considered necessary for all applications to ensure an acceptable degree of homogeneity in the resulting product. Triple melting is used to achieve better uniformity. Triple melting also reduces oxygen-rich or nitrogen-rich inclusions in the microstructure to a very low level by providing an additional melting operation to dissolve them.

Melting Practice. Most titanium and titanium alloy ingot is melted twice in an electric-arc furnace under vacuum—a procedure known as the "double consumable-electrode vacuum-melting process". In this two-stage process, titanium sponge, revert and alloy additions are initially mechanically consolidated and then are melted together to form ingot. Ingots from the first melt are used as the consumable electrodes for second-stage melting. Processes other than consumable-electrode arc melting are used in some instances for first-stage melting of ingot for noncritical applications. Usually, all melting is done under vacuum, but in any event the final stage of melting must be done by the consumable-electrode vacuum-arc process.

Segregation and other compositional variations directly affect the final properties of mill products. Melting technique alone does not account for all segregation and compositional variations, and thus cannot be correlated with final properties.

Melting in a vacuum reduces the hydrogen content of titanium and essentially removes other volatiles. This tends to result in high purity in the cast ingot. However, anomalous operating factors such as air leaks, water leaks, arc-outs, or even large variations in power level affect both soundness and homogeneity of the final product.

Still another factor is ingot size. Normally, ingots are 650 to 900 mm (26 to 36 in.) in diameter and weigh 3600 to 6800 kg (8000 to 15 000 lb). Larger ingots are economically advantageous to use and are important in obtaining refined macrostructures and microstructures in very large sections, such as billets with diameters of 400 mm (16 in.) or greater. Ingots up to 1000 mm

Fig. 2 Representative microstructural inhomogeneities in titanium metals

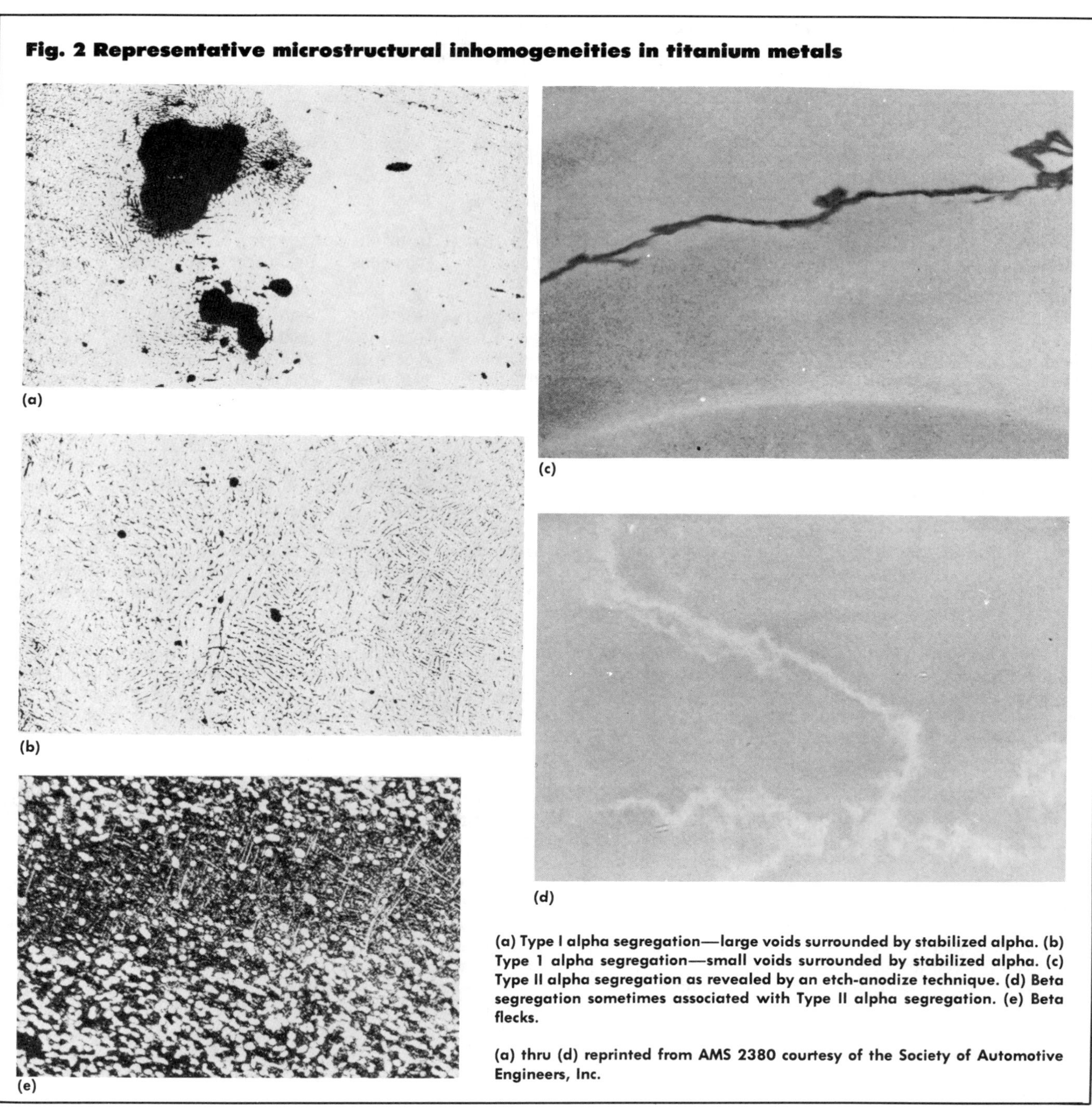

(a) Type I alpha segregation—large voids surrounded by stabilized alpha. (b) Type 1 alpha segregation—small voids surrounded by stabilized alpha. (c) Type II alpha segregation as revealed by an etch-anodize technique. (d) Beta segregation sometimes associated with Type II alpha segregation. (e) Beta flecks.

(a) thru (d) reprinted from AMS 2380 courtesy of the Society of Automotive Engineers, Inc.

(40 in.) in diameter and weighing more than 9000 kg (20 000 lb) have been melted successfully, but there appear to be limitations on the improvements that can be achieved by producing large ingots due to increasing tendency for segregation with increasing ingot size.

Segregation in titanium ingot must be controlled because it leads to several different types of imperfections that cannot be readily eliminated by homogenizing heat treatments or combinations of heat treatment and primary mill processing.

Type I imperfections, usually called "high interstitial defects", are regions of interstitially stabilized alpha phase that have substantially higher hardness and lower ductility than the surrounding material, and also exhibit a higher beta transus temperature. They arise from very high nitrogen or oxygen concentrations in sponge, master alloy or revert. Type I imperfections frequently, but not always, are associated with voids or cracks (see Fig. 2a and b).

Table 2 Standard forging temperatures for several titanium metals

Alloy	Beta transus °C	Beta transus °F	Forging temperatures: Ingot breakdown °C	Ingot breakdown °F	Intermediate °C	Intermediate °F	Finish °C	Finish °F
Commercially pure titanium								
Grades 1 to 4	900-955	1650-1750	955-980	1750-1800	900-925	1650-1700	815-900	1500-1650
Alpha and near-alpha alloys								
Ti-5Al-2.5Sn	1030	1890	1120-1175	2050-2150	1065-1095	1950-2000	1010-1040	1850-1900
Ti-6Al-2Sn-4Zr-2Mo-0.08Si	995	1820	1095-1150	2000-2100	1010-1065	1850-1950	955-980	1750-1800
Ti-8Al-1Mo-1V	1040	1900	1120-1175	2050-2150	1065-1095	1950-2000	1010-1040	1850-1900
Alpha-beta alloys								
Ti-8Mn	800	1475	925-980	1700-1800	845-900	1550-1650	815-845	1500-1550
Ti-6Al-4V	995	1820	1095-1150	2000-2100	980-1040	1800-1900	925-980	1700-1800
Ti-6Al-6V-2Sn	945	1735	1040-1095	1900-2000	955-1010	1750-1850	870-940	1600-1725
Ti-7Al-4Mo	1005	1840	1120-1175	2050-2150	1010-1065	1850-1950	955-980	1750-1800
Beta alloy								
Ti-13V-11Cr-3Al	720	1325	1120-1175	2050-2150	1010-1065	1850-1950	925-980	1700-1800

Although type I imperfections sometimes are referred to as "low-density inclusions", they often are of higher density than is normal for the alloy.

Type II imperfections, sometimes called "high-aluminum defects", are abnormally stabilized alpha-phase areas that may extend across several beta grains (see Fig. 2c). Type II imperfections are caused by segregation of metallic alpha stabilizers, such as aluminum, and contain an excessively high proportion of primary alpha having a microhardness only slightly higher than that of the adjacent matrix. Type II imperfections sometimes are accompanied by adjacent stringers of beta—areas low in both aluminum content and hardness. This condition, shown in Fig. 2(d), is generally associated with closed solidification pipe into which alloy constituents of high vapor pressure migrate, only to be incorporated into the microstructure during primary mill fabrication. Stringers normally occur in the top portions of ingots and can be detected by macroetching or anodized blue etching. Material containing stringers usually must undergo metallographic review to ensure that the indications revealed by etching are not artifacts.

Beta flecks, another type of imperfection, are small regions of stabilized beta in material that has been alpha-beta processed and heat treated. In size, they are equal to or greater than prior beta grains (see Fig. 2e). Beta flecks are either devoid of primary alpha or contain less than some specified minimum level of primary alpha. They are caused by localized regions either abnormally high in beta-stabilizer content or abnormally low in alpha-stabilizer content. Beta flecks are attributed to microsegregation during solidification of ingots of alloys that contain strong beta stabilizers. They are most often found in products made from large-diameter ingots. Beta flecks also may be found in beta-lean alloys such as Ti-6Al-4V that have been heated to a temperature near the beta transus during processing.

Type I and type II imperfections are not acceptable in aircraft-grade titanium because they degrade critical design properties. Beta flecks are not considered harmful in alloys lean in beta stabilizers if they are to be used in the annealed condition. On the other hand, they constitute regions that incompletely respond to heat treatment, and for this reason microstructural standards have been established for allowable limits on beta flecks in various alpha-beta alloys. Beta flecks are more objectionable in beta-rich alpha-beta alloys than in leaner alloys.

Primary Fabrication

Primary fabrication includes all operations that convert ingot into general mill products—billet, bar, plate, sheet, strip, extrusions, tube and wire. These mill products can be readily utilized in secondary manufacture of parts and structures.

Originally, primary fabrication processes were designed around the capabilities of the steel mill equipment available at that time. As the titanium industry matured, special furnace equipment was introduced to improve temperature control during heating; special presses and mills were developed for proper hot and cold working of various alloys, and conditioning equipment was procured to improve surface quality. In general, all mill products today are produced to much higher quality standards than those specified in the aerospace industry in the 1950's. This improvement reflects the advances in titanium technology that have taken place over the last two decades.

Primary fabrication is very important in establishing final properties, because many secondary fabrication operations may have little or no effect on metallurgical characteristics. Some secondary fabrication processes, such as forging and ring rolling, do impart sufficient reduction to play the major role in establishing material properties.

Reduction to Billet. Generally, the first breakdown of production ingot is a press cogging operation done in the beta temperature range. Modern processes utilize substantial amounts of working below the beta transus to produce billets with refined structures. These processes are carried out at temperatures high in the alpha region to allow greater reduction and improved grain refinement with a minimum of surface rupturing. Where maximum fracture toughness is required, beta processing (or alpha-beta processing followed by beta heat treatment) is generally preferred. Table 2 gives standard forging-temperature ranges for manufacture of billet stock.

Table 3 Variation of typical room-temperature tensile properties with section size for four titanium alloys

Section size(a) mm	in.	Tensile strength MPa	ksi	Yield strength MPa	ksi	Elongation(b), %	Reduction in area, %
6Al-4V(c)							
25-50	1-2	1015	147	965	140	14	36
102	4	1000	145	930	135	12	25
205	8	965	140	895	130	11	23
330	13	930	135	860	125	10	20
6Al-4V-ELI(c)							
25-50	1-2	950	138	885	128	14	36
102	4	885	128	827	120	12	28
205	8	885	128	820	119	10	27
330	13	870	126	795	115	10	22
6Al-6V-2Sn(c)							
25-50	1-2	1105	160	1035	150	15	40
102	4	1070	155	965	145	13	35
205	8	1000	145	930	135	12	25
8Al-1Mo-1V							
25-50	1-2(d)	985	143	905	131	15	36
102	4(e)	910	132	840	122	17	35
205	8(f)	1000	145	895	130	12	23
6Al-2Sn-4Zr-2Mo + Si(g)							
25-50	1-2	1000	145	930	135	14	33
102	4	1000	145	930	135	12	30
205	8	1035	150	940	136	12	28
330	13	1000	145	825	120	11	21

(a) Properties are in longitudinal direction for sections 50 mm (2 in.) or less, and in transverse direction for sections 100 mm (4 in.) or more, in section size. (b) In 50 mm or 2 in. (c) Annealed 2 h at 700 °C (1300 °F) and air cooled. (d) Annealed 1 h at 900 °C (1650 °F), air cooled, then heated 8 h at 600 °C (1100 °F) and air cooled. (e) Annealed 1 h at 1010 °C (1850 °F), air cooled, then heated to 566 °C (1050 °F). (f) Annealed 1 h at 1010 °C (1850 °F) and oil quenched. (g) Annealed 1 h at 954 °C (1750 °F), air cooled, then heated 8 h to 600 °C (1100 °F) and air cooled.

Table 4 Typical rolling temperatures for several titanium metals

Alloy	Bar °C	Bar °F	Plate °C	Plate °F	Sheet °C	Sheet °F
Commercially pure titanium						
Grades 1 to 4	760-815	1400-1500	760-790	1400-1450	705-760	1300-1400
Alpha and near-alpha alloys						
Ti-5Al-2.5Sn	1010-1065	1850-1950	980-1040	1800-1900	980-1010	1800-1850
Ti-6Al-2Sn-4Zr-2Mo	955-1010	1750-1850	955-980	1750-1800	925-980	1700-1800
Ti-8Al-1Mo-1V	1010-1040	1850-1900	980-1040	1800-1900	980-1040	1800-1900
Alpha-beta alloys						
Ti-8Mn	...	...	705-760	1300-1400	705-760	1300-1400
Ti-4Al-3Mo-1V	925-955	1700-1750	900-925	1650-1700	900-925	1650-1700
Ti-6Al-4V	955-1010	1750-1850	925-980	1700-1800	900-925	1650-1700
Ti-6Al-6V-2Sn	900-955	1650-1750	870-925	1600-1700	870-900	1600-1650
Ti-7Al-4Mo	955-1010	1750-1850	925-955	1700-1750	925-955	1700-1750
Beta alloy						
Ti-13V-11Cr-3Al	955-1065	1750-1950	980-1040	1800-1900	730-900	1350-1650

Some billets intended for further forging, rolling or extrusion go through a grain-refinement process. This technique, developed in the early 1970's, utilizes the fact that titanium recrystallizes when it is heated above the beta transus. By starting with grain-refined billet, secondary fabricators may be able to produce forgings that meet strict requirements with respect to macrostructure, microstructure and mechanical properties, without extensive hot working below the beta transus.

Final tensile properties of alpha-beta alloys are strongly influenced by the amount of processing in the alpha-beta field—both below the beta transus temperature and after recrystallization. Such processing increases the strength of high-alpha grades in large section sizes. With modern processing techniques, billet and forged sections readily meet specified tensile properties prior to final forging. Table 3 shows how billet and forging section size affects room-temperature tensile properties of various titanium alloys.

Rolling of Bar, Plate and Sheet. Roll cogging and hot roll finishing of bar, plate and sheet are now standard operations, and special rolling and auxiliary equipment have been installed by the larger titanium producers to allow close control of all rolling operations. Rolling processes used by each manufacturer are proprietary and in some respects unique, but because all techniques must produce the same specified structures and mechanical properties, a high degree of similarity exists among the processes of all manufacturers.

A representative range of temperatures used for hot rolling of titanium metals is presented in Table 4. Rolling at these temperatures produces end products with the desired grain structures.

Bars up to about 100 mm (4 in.) in diameter are unidirectionally rolled, and their properties commonly reflect total reduction in the alpha-beta range. For example, a round bar 50 mm (2 in.) in diameter rolled from a Ti-6Al-4V billet 100 mm square typically is 140 to 170 MPa (20 to 25 ksi) lower in tensile strength than rod 7.8 mm (5/16 in.) in diameter rolled on a rod mill from a billet of the same size at the same rolling temperatures. For bars about 50 to 100 mm in diameter, strength does not decrease with section size, but transverse ductility and notched stress-rupture strength at room temperature do become lower. In diameters greater than about 75 to 100 mm (3 to 4 in.), annealed Ti-6Al-4V bars usually do not meet prescribed limits for stress-rupture at room temperature—1170 MPa (170 ksi) min to cause rupture of a notched specimen in 5 h—unless the material is given a special duplex anneal. Transverse ductility is lower in bars about 65 to 100 mm (2½ to 4 in.) in diameter because it is not possible to obtain the preferred texture throughout bars of this size.

Plate and sheet commonly exhibit the same tensile properties in both the transverse and longitudinal directions relative to the final rolling direction. With the precise control systems now available, proper texturing and directionality can be obtained in alpha-beta sheet by unidirectional rolling. These characteristics favorably affect tensile properties of Ti-6Al-4V sheet in various gages (see Table 5). Other properties, such as fatigue resistance, also are improved by this type of rolling.

Directionality in properties is observed only as a slight drop in transverse ductility of plate greater than 25 mm (1 in.) thick. Military, AMS and customer specifications all prescribe lower minimum tensile and yield strengths as plate thickness increases. For forming applications, some customers specify a maximum allowable difference between tensile strengths in the transverse and longitudinal directions.

Table 5 Tensile properties of unidirectionally rolled Ti-6Al-4V sheet

Gage mm	Gage in.	Tensile strength MPa	Tensile strength ksi	Yield strength MPa	Yield strength ksi	Elongation(a), %	Tensile modulus GPa	Tensile modulus 10^6 psi
Longitudinal direction								
0.737	0.029	945	137	870	126	7.0	100	14.5
1.016	0.040	970	141	855	124	6.5	106	15.4
1.168	0.046	915	133	860	125	6.5	105	15.2
1.524	0.060	985	143	925	134	6.5	104	15.1
1.778	0.070	995	144	915	133	8.0	105	15.3
Transverse direction								
0.737	0.029	1105	160	1061	154	7.5	130	18.8
1.016	0.040	1195	173	1105	160	7.5	145	21.1
1.168	0.046	1225	178	1165	169	7.5	140	20.2
1.524	0.060	1125	163	1090	158	8.0	125	18.2
1.778	0.070	1095	159	1055	153	9.5	135	19.5

(a) In 50 mm or 2 in.

Secondary Fabrication

Secondary fabrication refers to manufacturing processes such as die forging, extrusion, hot and cold forming, machining, chemical milling, and joining, all of which are used for producing finished parts from mill products. Each of these processes may strongly influence properties of titanium and its alloys, either alone or by interacting with effects of processes to which the metal has previously been subjected.

Die Forging. One of the main purposes of die forging is to obtain a combination of mechanical properties that generally does not exist in bar or billet. Tensile strength, creep resistance, fatigue strength and toughness all may be better in forgings.

A recent program evaluated the effects of different thermomechanical processing schedules on the mechanical properties and the corresponding structures of three titanium alloys: Ti-6Al-4V, Ti-6Al-6V-2Sn and Ti-6Al-2Sn-4Zr-6Mo. Table 6 summarizes four thermomechanical schedules that produced optimum combinations of proper-

Table 6 Thermomechanical schedules for producing various combinations of properties in Ti-6Al-4V forgings

Initial microstructure	Blocker forging temperature range	Finish forging temperature range	Finish forging reduction	Cooling after forging	Heat treated condition	Final microstructure
Best combinations of properties						
...	Alpha-beta	Alpha-beta	...	Air cooled	Annealed	6% equiaxed alpha plus fine platelet alpha
Grain-boundary alpha	Alpha-beta	Alpha-beta	...	Air cooled	Annealed	26% elongated partly broken up grain-boundary primary alpha plus fine platelet alpha
Grain-boundary alpha	Alpha-beta	Alpha-beta	...	Water quenched	Annealed	23% elongated partly broken up primary alpha plus very fine platelet alpha
...	Beta	Alpha-beta	10%	Air cooled	Annealed	63% fine elongated primary alpha plus fine platelet alpha
Subnormal properties						
Spaghetti alpha	Alpha	Alpha	...	Air cooled	STOA	25% blocky primary alpha plates plus very fine platelet alpha
...	Beta	Alpha-beta	10%	Water quenched	STOA	43% coarse elongated primary alpha plates plus very fine platelet alpha
...	Beta	Beta	...	Slow cooled	Annealed	92% alpha basket-weave structure

All data from Ref 1.

Fig. 3 Microstructures corresponding to different combinations of properties in Ti-6Al-4V forgings

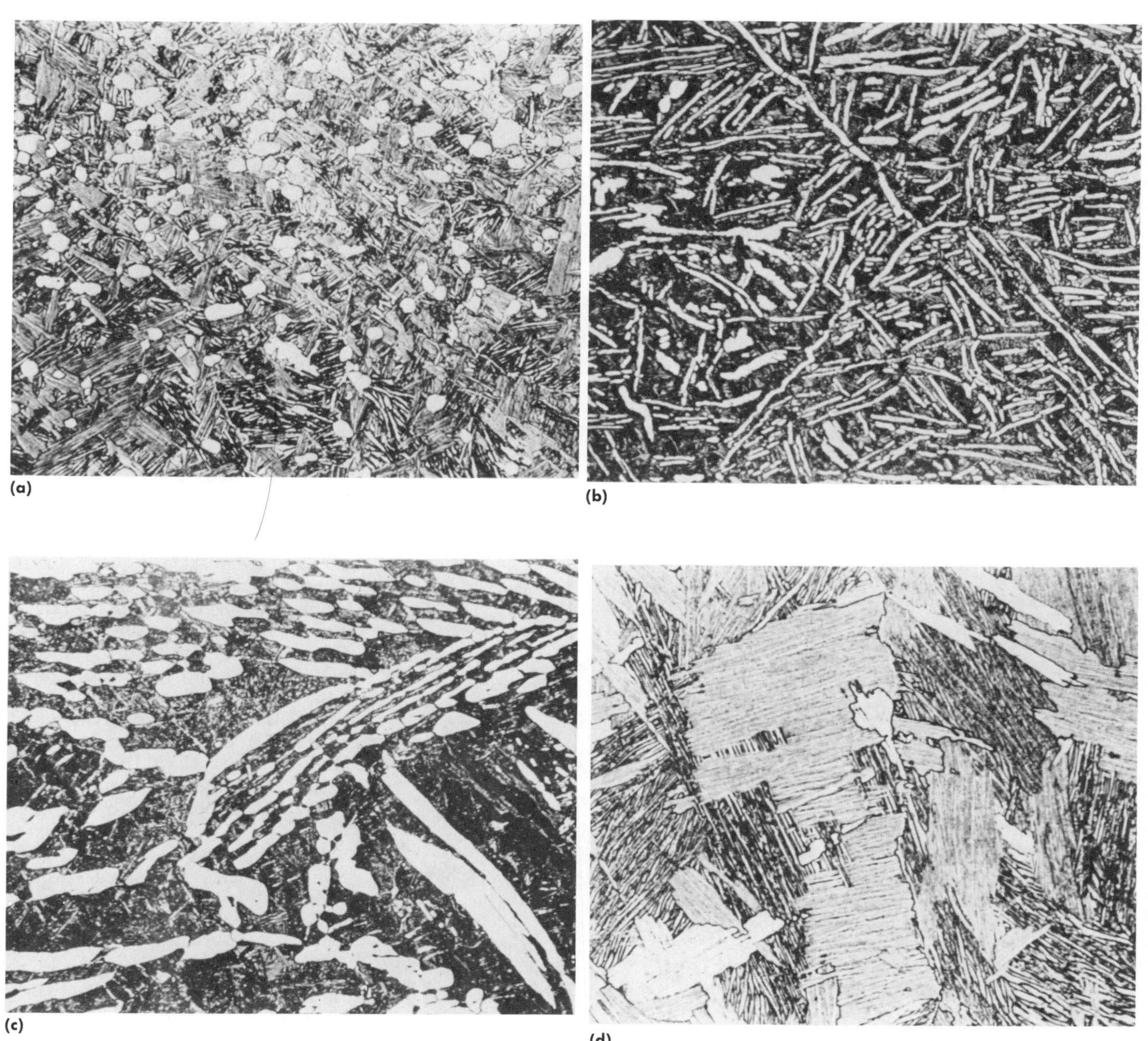

(a) (b) (c) (d)

(a) 6% equiaxed primary alpha plus fine platelet alpha in Ti-6Al-4V alpha-beta forged, then annealed 2 h at 705 °C (1300 °F) and air cooled. (b) 23% elongated, partly broken up primary alpha plus grain-boundary alpha in Ti-6Al-4V, alpha-beta forged and water quenched, then annealed 2 h at 705 °C and air cooled. (c) 25% blocky (spaghetti) alpha plates plus very fine platelet alpha in Ti-6Al-4V alpha-beta forged from a spaghetti-alpha starting structure, then solution treated 1 h at 955 °C (1750 °F) and reannealed 2 h at 705 °C. (d) 92% alpha basket-weave structure in Ti-6Al-4V beta forged and slow cooled, then annealed 2 h at 705 °C.

Structures in (a) and (b) produced excellent combinations of tensile properties, fatigue strength and fracture toughness. Structure in (c) produced very poor combination of mechanical properties. Structure in (d) produced good fracture toughness, but poor tensile properties and fatigue resistance. (Ref 1)

Table 7 Typical tensile, bend and hardness data for as-welded titanium and several titanium alloys

Material condition	Tensile strength MPa	Tensile strength ksi	Yield strength MPa	Yield strength ksi	Elongation, %	Minimum bend radius	Hardness Knoop	Hardness Rockwell
Ti Grade 1								
Unwelded sheet	315	46	215	31	50.4	0.7*t*	140	63.5 HRB
Single-bead weld	345	50	255	37	37.5	1.0*t*	140	55.8 HRB
Multiple-bead weld	365	53	270	39	37.7	...	...	...
Transverse weld	325	47(a)	...	...	...	...	...	...
Ti Grade 2								
Unwelded sheet	460	67	325	47	26.2	2.9*t*	165	80.6 HRB
Single-bead weld	505	73	380	55	18.3	2.9*t*	175	83.1 HRB
Multiple-bead weld	510	74	385	56	13.3	...	...	...
Transverse weld	475	69(a)	...	...	...	...	...	...
Ti Grade 3								
Unwelded sheet	545	79	395	57	25.9	1.9*t*	175	94.4 HRB
Single-bead sheet	605	88	475	69	15.5	4.7*t*	220	92.4 HRB
Multiple-bead weld	615	89	480	70	14.7	...	...	...
Transverse weld	560	81(a)	...	...	...	...	...	...
Ti Grade 4								
Unwelded sheet	660	96	530	77	22.3	3.2*t*	215	23.4 HRC
Single-bead weld	695	101	580	84	16.4	5.6*t*	240	21.2 HRC
Multiple-bead weld	710	103	585	85	16.0	...	...	...
Transverse weld	660	96(a)	...	...	...	...	...	...
Ti-5Al-2.5Sn-ELI								
Unwelded sheet	850	123	805	117	15.7	3.8*t*	265	33.2 HRC
Single-bead weld	920	133	770	112	9.8	5.9*t*	310	28.0 HRC
Multiple-bead weld	935	136	820	119	7.5	...	...	...
Transverse weld	850	123(a)	...	...	...	...	...	...
Ti-6Al-2Cb-1Ta-1Mo								
Unwelded sheet	895	130	855	124	9.7	2.8*t*	275	29.6 HRC
Single-bead weld	930	135	800	116	5.9	7.7*t*	300	27.7 HRC
Multiple-bead weld	945	137	815	118	5.7	...	...	...
Transverse weld	890	129(a)	...	...	...	...	...	...
Ti-3Al-2.5V								
Unwelded sheet	705	102	670	97	15.2	4.0*t*	230	23.6 HRC
Single-bead weld	705	102	600	87	12.7	5.4*t*	250	19.6 HRC
Multiple-bead weld	745	108	625	91	11.2	...	...	...
Transverse weld	710	103(a)	...	...	...	...	...	...
Ti-6Al-4V								
Unwelded sheet	1000	145	945	137	11.0	2.6*t*	320	32.2 HRC
Single-bead weld	1060	154	920	133	3.5	10.5*t*	350	35.9 HRC
Multiple-bead weld	1090	158	945	137	3.2	...	...	...
Transverse weld	1015	147(a)	...	...	...	...	...	...
Ti-8Al-1Mo-1V								
Unwelded sheet	1060	154	1020	148	15.0	2.9*t*	325	36.0 HRC
Single-bead weld	1085	157	930	135	5.5	7.0*t*	345	35.2 HRC
Multiple-bead weld	1115	162	960	139	3.2	...	...	...
Transverse weld	1060	154(a)	...	...	...	...	...	...
Ti-6Al-6V-2Sn								
Unwelded sheet	1060	154	1005	146	9.8	2.8*t*	350	34.0 HRC
Single-bead weld	1295	188	1255	182	0.3	25.6*t*	420	46.8 HRC
Multiple-bead weld	1280	186	...	...	0.1	...	...	...
Transverse weld	1103	160(a)	...	...	...	...	...	...
Ti-13V-11Cr-3Al								
Unwelded sheet	965	140	910	132	13.9	2.7*t*	300	30.6 HRC
Single-bead weld	950	138	925	134	11.6	2.7*t*	320	30.1 HRC
Multiple-bead weld	925	134	875	127	9.1	...	...	...
Transverse weld	950	138(a)	...	...	...	...	...	...

(a) Fracture occurred in base metal.

ties in Ti-6Al-4V test forgings: excellent tensile strength; and good to excellent notch fatigue strength, low-cycle fatigue strength and fracture toughness. Also included in Table 6 are three schedules that produced subnormal properties. The microstructures shown in Fig. 3 correspond to two of the schedules that produced good combinations of properties and two that produced inferior combinations. This evaluation program demonstrated that control of thermomechanical processing can control the microstructures and corresponding final properties of forgings.

Figure 4 summarizes the results of an extensive study of alpha-beta forging versus beta forging for several titanium alloys. Although yield strength after beta forging was not always as high as that after alpha-beta forging, values of notch tensile strength and fracture toughness were consistently higher for beta-forged material.

In another series of tests, beta forging produced improved notch tensile strength, notch fatigue strength, creep strength and fracture toughness in Ti-6Al-4V, Ti-8Al-1Mo-1V and Ti-6Al-2Sn-4Zr-2Mo. Tensile strength, elongation and creep stability were about the same for beta-forged material as for alpha-beta-forged material, but yield strength was slightly lower, and reduction in area significantly lower, for beta-forged material. Also, improper beta forging further degraded only ductility; other tensile properties were unaffected, and fracture toughness was slightly lower than that of properly beta-forged material but still almost twice that of alpha-beta-forged material.

An early report on Ti-8Al-1Mo-1V compared the mechanical properties obtained by alpha-beta processing with those obtained by beta processing. Advantages were found in both processes and the differences in properties clearly illustrated the influence of thermomechanical processing.

One aircraft manufacturer reported achieving a 60% increase in fatigue strength at 10^7 cycles by forging at temperatures about 30 °C (50 °F) above the beta transus. The same report showed that fine alpha grain size in conventional alpha-beta forgings was associated with high fatigue strength at 10^8 cycles.

Extrusion is used to make rodlike products, as an alternative mill process to rolling. Properties are affected by processing conditions in much the same

Fig. 4 Comparison of mechanical properties of alpha-beta forged and beta forged titanium alloys

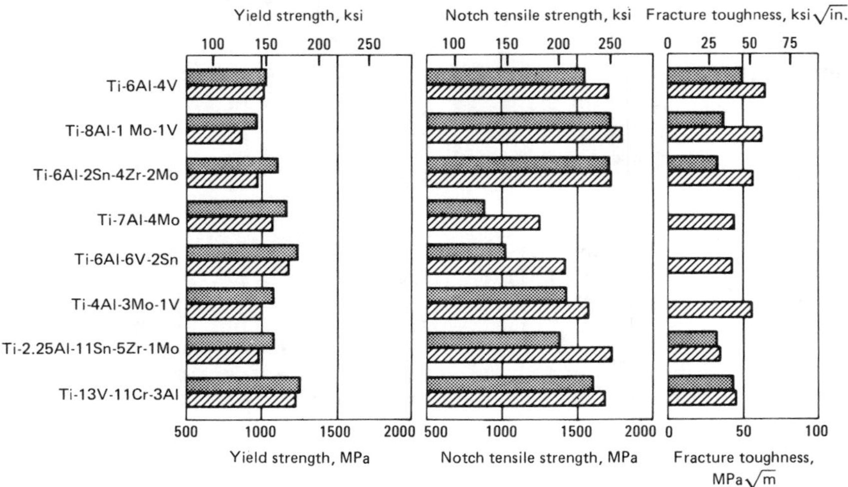

Shaded bars represent alpha-beta forged material; striped bars, beta forged material. (Ref 2)

Table 8 Typical properties of welded Ti-6Al-6V-2Sn after postweld heat treatment

Heat treatment	Tensile strength MPa	ksi	Yield strength MPa	ksi	Elongation, %	Bend radius (min)
None						
Base metal	1060	154	1010	146	9.8	2.8*t*
Single-bead weld	1300	188	1250	182	0.3	25.6*t*
Furnace cool from 830 °C (1525 °F)						
Base metal	1030	149	970	141	8.1	...
Single-bead weld	1050	152	990	144	3.7	15.5*t*
Furnace cool after 6 h at 700 °C (1300 °F) in vacuum						
Base metal	1120	163	1040	151	9.8	2.0*t*
Single-bead weld	1170	169	1100	159	3.8	12.2*t*
Air cool after ½ h at 700 °C						
Base metal	1050	152	1000	145	9.8	...
Single-bead weld	1230	178	1160	168	1.3	14.5*t*

way as they are for rolled or forged products. One of the more unique applications of extrusion has been in the production of tapered wing spars for a military aircraft.

Forming. Titanium and titanium alloy sheet and plate are strain hardened by cold forming. This normally increases tensile and yield strengths, and causes a slight drop in ductility. Titanium metals exhibit a high degree of springback in cold forming. To overcome this characteristic, titanium must be extensively overformed or, as is done most frequently, hot sized after cold forming.

Hot forming does not greatly affect final properties. Forming at temperatures from 595 to 815 °C (1100 to 1500 °F) allows the material to slip more readily and simultaneously stress relieves the deformed material; it also minimizes the degree of springback. The true net effect in any forming operation depends on total deformation and actual temperature during forming. Titanium metals also tend to creep at elevated temperature; holding under load at the forming temperature (creep forming) is another alternative for achieving the desired shape without having to compensate for extensive springback.

In all forming operations, titanium and its alloys are susceptible to the Bauschinger effect—a drop in compressive yield strength in one loading direction accompanied by an increase in tensile strength in another direction due to strain hardening. The Bauschinger effect is most pronounced at room temperature: plastic deformation (1 to 5% tensile elongation) at room temperature always introduces a significant loss in compressive yield strength, regardless of the initial heat treatment or strength of the alloys. At 2% tensile strain, for instance, the compressive yield strengths of Ti-4Al-3Mo-1V and Ti-6Al-4V drop to less than half the values for solution-treated material. Increasing the temperature reduces the Bauschinger effect; subsequent full thermal stress relieving completely removes it.

Temperatures as low as the aging temperature will remove most of the Bauschinger effect in solution-treated titanium alloys. Heating or plastic deformation at temperatures above the normal aging temperature for solution-treated Ti-6Al-4V will cause overaging to occur and, as a result, all mechanical properties will decrease.

Joining. Adhesive bonding, brazing, mechanical fastening, metallurgical bonding and welding are all used routinely and successfully to join titanium and its alloys. The first three processes do not affect the properties of these metals as long as joints are properly designed. Metallurgical bonding includes all solid-state joining processes in which diffusion or deformation play the major role in bonding the members together.

Because these processes are performed below the beta transus, metallurgical effects, either normally caused by heating at that temperature or resulting from contamination, should be anticipated. Properly processed joints have the same properties as the base metal and, because bonding is carried out at a temperature high in the alpha-beta field, material properties appear similar to those resulting from high-temperature annealing. With most alloys, a final low-temperature anneal will produce properties characteristic of typical annealed material.

Welding has the greatest potential

for affecting material properties. In all types of welds, contamination by interstitial impurities such as oxygen and nitrogen must be closely controlled to maintain useful ductility in the weldment. Alloy composition, welding procedure and subsequent heat treatment are highly important in determining the final properties of welded joints. Tables 7 and 8 review mechanical properties for representative alloys and types of welds. The data can be summarized as follows:

- Welding generally increases strength and hardness.
- Welding generally decreases tensile and bend ductility.
- Welds in unalloyed titanium grades 1, 2 and 3 do not require postweld treatment unless the material will be highly stressed in a strongly reducing atmosphere. In such event, stress relieving or annealing may prove useful.
- Welds in beta-rich alloys such as Ti-6Al-6V-2Sn have a high likelihood of fracturing with little or no plastic straining. Weld ductility can be improved by postweld heat treatment consisting of slow cooling from a high annealing temperature.
- Rich beta-stabilized alloys can be welded, and such welds exhibit good ductility.

Electron-beam and laser welds are made without filler metal and weld beads have high ratios of depth to width. This combination allows excellent welds to be made in heavy sections, with properties very close to those of the base metal.

Welding must be done under strict environmental controls to avoid pick-up of interstitials that can embrittle the weld metal. Small and moderate-size weldments ordinarily are enclosed within environmentally controlled chambers during welding. Larger weldments are made with the aid of portable chambers that only partly enclose the components, or with the aid of "trailers", both of which maintain a protective atmosphere on both front and back sides of the weld until it has cooled below about 480 °C (1000 °F).

Powder metallurgy (P/M) products having properties equal to or exceeding those of other forms are now manufactured on a production basis. Table 9 compares room-temperature properties of titanium and several titanium alloys in wrought, cast and powder forms. The processes for manufacture of titanium powders are slow and costly, however, and this has been the primary reason for slow growth of powder metallurgy as a means of manufacturing titanium parts.

One of the most important considerations in manufacture of a titanium powder metallurgy product is control of oxygen content, because oxygen has the same undesirable effects on properties of P/M parts as it has on those of wrought products. Powders must be

Table 9 Comparison of typical room-temperature properties of wrought, cast and P/M titanium products

Product and condition	Tensile strength MPa	Tensile strength ksi	Yield strength MPa	Yield strength ksi	Elongation, %	Reduction in area, %	Impact strength (a) J	Impact strength (a) ft·lb
Unalloyed Ti								
Wrought bar, annealed	550	80	480	70	18	33	35	26
Cast bar, as cast	635	92	510	74	20	31	26	19
P/M compact, annealed(b)	480	70	370	54	18	22	...	...
Ti-5Al-2.5Sn-ELI								
Wrought bar, annealed	815	118	710	103	19	34	...	...
Cast bar, as cast	795	115	725	105	10	17	...	...
P/M compact, annealed and forged(c)	795	115	715	104	16	27	...	...
Ti-6Al-4V								
Wrought bar, annealed	1000	145	925	134	16	34	22	16
Cast bar								
As cast	1025	149	880	128	12	19	19	14
Annealed	1015	147	890	129	10	16	...	...
Solution treated and aged(d)	1180	171	1085	157	6	11	...	...
P/M compact								
Annealed(b)	825-855	120-124	740-785	107-114	5-8	8-14	...	...
Annealed and forged(c)	925	134	840	122	12	27	...	...
Solution treated and aged(d)	965	140	895	130	4	6	...	...
Ti-6Al-6V-2Sn								
Wrought bar, annealed	1125	163	1055	153	16	38	20	15
Cast bar, as cast	1105	160	965	140	6	11	14	10
P/M compact, annealed(b)	965	140	840	122	5	5	...	...

All data from Ref. 3. (a) Charpy, at −40 °C (−40 °F), (b) About 94% dense. (c) Almost 100% dense. (d) Aging treatment not specified.

Table 10 Typical room-temperature tensile properties of titanium P/M compacts

Powder manufacturing process	Oxygen content, ppm	Tensile strength MPa	Tensile strength ksi	Yield strength MPa	Yield strength ksi	Elongation, %
Ti-6Al-4V						
Mechanical attrition	1750	1000	145	940	136	1.5
Rotating electrode	900	1000	145	925	134	7.5
Hydride-to-hydride	1570	1025	149	970	141	2.0
Ti-5Al-2.5Sn						
Rotating electrode	980	905	131	905	131	4.0
Gas attrition	3530	895	130	...	...	...

All data from Ref 4 for compacts hot pressed to 1380 MPa (100 tons/in.^{2}) at 1010 °C (1850 °F).

handled very carefully, because they have a very high affinity for oxygen. Table 10, which shows room-temperature tensile properties of compacts made from alloy powders of varying oxygen content, illustrates the importance of low oxygen content in obtaining satisfactory ductility in hot-pressed powder compacts.

REFERENCES

1. "Improved Manufacturing Methods for Producing High Integrity More Reliable Titanium Forgings", by R. B. Sparks and J. R. Long: AFML TR-73-301, Wyman Gordon Company, Worcester, Massachusetts, Feb 1974
2. "Effect of Alpha plus Beta and Beta Forging on the Fracture Toughness of Several High Strength Titanium Alloys", by G. H. Heitman, J. E. Coyne and R. P. Galipean: Wyman Gordon Company, Worcester, Massachusetts, Oct 1967
3. "The Titanium Industry in the mid 1970's": MCIC Report 75-26, Metals and Ceramics Information Center, Battelle Columbus Laboratories, Columbus, Ohio, June 1975
4. "Properties of Powdered Titanium Alloys", by G. Friedman: NASA CR-72568, Nuclear Metals Division, Whittaker Corp., Concord, Massachusetts, May 1969

Properties of Titanium and Titanium Alloys

By the ASM Committee on Titanium and Titanium Alloys*

Unalloyed Ti Grade 1

Commercial Names

Trade name. RMI 30; Ti-35A
Common names. Unalloyed titanium; commercially pure titanium; unalloyed titanium of 25 ksi min yield strength
UNS number. R50250

Specifications

ASTM. Bar and billet: B348. Forgings: B381, grade F-1. Sheet, strip and plate: B265. Pipe: B337. Tubing: B338. Castings: B367, grade C-1
Government. Extruded bar, rod and special shaped sections: MIL-T-81556, type I, composition A. Welding rod and wire: MIL-R-81588, type I, composition A

Chemical Composition

Composition limits. 0.03 max N; 0.10 max C; 0.015 max H(a); 0.18 max O; 0.20 max Fe; 0.05 max others (each); 0.3 max others (total); rem Ti. Minimum titanium content, 99.175

(a) Deviations from max H values: bar, 0.0125; billet and castings, 0.0100.

*See page XI for committee list.

Fig. 1 Thermal expansion of unalloyed titanium

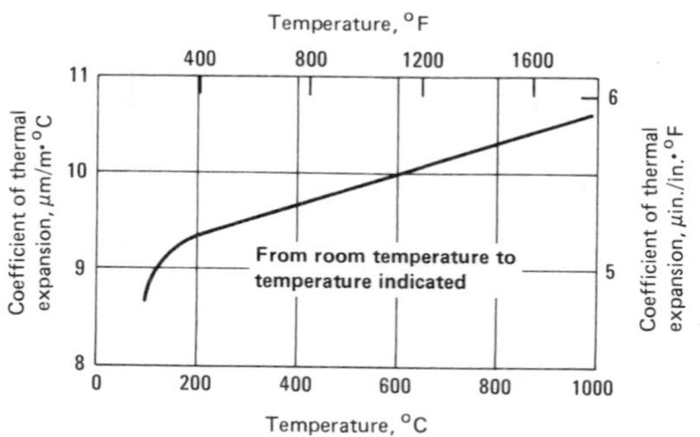

Data are for mean coefficient of linear expansion from room temperature to indicated temperature for titanium of undetermined grade (but probably not of high purity).

Table 1 Typical elevated temperature tensile properties for unalloyed titanium, grade 1

Temperature °C	Temperature °F	Tensile strength MPa	Tensile strength ksi	Yield strength(a) MPa	Yield strength(a) ksi	Elongation, %
204	400	193-207	28-30	110-124	16-18	40-50
316	600	152-179	22-26	83-103	12-15	30-35
427	800	124-138	18-20	76-90	11-13	25-30

(a) 0.2% offset.

Table 2 Corrosion rates for unalloyed titanium (99.2% Ti) in selected media

Corrodent	Concentration, %	Temperature °C	Temperature °F	Corrosion rate µm/yr	Corrosion rate mils/yr
Acetic acid	5, 25, 75	100	212	Nil	Nil
	50, 99.5	100	212	0.25	0.01
Aluminum chloride, aerated	25	25	77	<2	<0.1
	5, 10	60	140	<2.5	<0.1
	10	100	212	<2.5	<0.1
Ammonium chloride	1, 10, saturated	20-100	68-212	<13	<0.5
Ammonium sulfate	5	25	77	Nil	Nil
	Saturated + 5% H_2SO_4	25	77	25	1
Aqua regia (3:1)	100	25	77	Nil	Nil
	100	77	170	890	35
Calcium chloride	28	Boiling		Nil	Nil
	5, 10, 20	100	212	<25	<1
Calcium hypochlorite	Saturated	25	77	Nil	Nil
	2, 6	100	212	1.3	0.05
Chlorine					
Saturated with H_2O	...	25	77	125	5
More than 0.013% H_2O	...	79	175	Nil	Nil
Dry	...	32	90	Rapid	Rapid
Copper nitrate	Saturated	25	77	Nil	Nil
Cupric chloride	20, 40	Boiling		Nil	Nil
Ferric chloride	10, 20	25	77	Nil	Nil
	5	60	140	Nil	Nil
	10 to 40	Boiling		Nil	Nil
	30	93	200	Nil	Nil
	10 to 30	100	212	<13	<0.5
	5 + 10% NaCl	100	212	<13	<0.5
Ferric sulfate	10	25	77	Nil	Nil
Ferrous sulfate	Saturated	25	77	Nil	Nil
Hydrochloric acid	5	35	95	<50	<2
	10	35	95	1000	40
	20	35	95	4400	175
Hydrochloric acid plus copper sulfate	10 + 0.05	65	150	<50	<2
	10 + 0.1	65	150	<25	<1
	10 + 0.2, 0.25 or 0.5	65	150	Nil	Nil
	10 + 1	65	150	<25	<1
Hydrogen sulfide	Saturated water	25	77	<125	<5
Lactic acid	10 to 85	100	212	<125	<5
	10 to 100	Boiling		<125	<5
Lead acetate	Saturated	25	77	Nil	Nil
Magnesium chloride	5 to 40	Boiling		Nil	Nil
	5 to 40	100	212	<125	<5
Nitric acid	5	100	212	<25	<1
	10	100	212	<50	<2
	40 to 50, 69.5	100	212	<25	<1
	65	175	347	<125	<5
	40	200	392	<1250	<50
	70	270	518	<1250	<50
	20	290	554	300	12
Phosphoric acid	5 to 30	25	77	<50	<2
	35 to 85	25	77	<1250	<50
	85	38	100	1000	40
	5 to 35	60	140	<1250	<50
	10	79	175	1250	50
	5	100	212	<1250	<50
Seawater	...	25	77	Nil	Nil
Silver nitrate	50	25	77	Nil	Nil
Sulfuric acid	15	25	77	Nil	Nil
	1	60	140	Nil	Nil
	3	60	140	1.3	0.05
	5	60	140	730	29
Zinc chloride	Saturated	25	77	Nil	Nil
	10	Boiling		Nil	Nil
	20	100	212	<125	<5

Consequence of exceeding impurity limits. May result in raising yield strength above maximum permitted or in lowering elongation or reduction in area below minimums

Applications

Typical uses. Applications requiring high ductility for fabrication but relatively low strength in service; other uses where maximum ease of formability is required and where low iron and interstitial content might enhance corrosion resistance. Weldable.

Precautions in use. Hydrogen embrittlement of titanium can occur in pickling solutions (or other hydrogenating solutions) at room temperature and at elevated temperatures during air exposure or in exposures to reducing atmospheres. Elevated temperature atmospheric exposure also results in oxygen and nitrogen contamination which increases in severity with increasing temperature and time of exposure. Violent oxidation reactions can occur between titanium and liquid oxygen or between titanium and red fuming nitric acid.

Mechanical Properties

Tensile properties. Minimum: tensile strength, 241 MPa (35 ksi); 0.2% yield strength, 172 MPa (25 ksi); elongation in 50 mm or 2 in., 24%; reduction in area, 30%. Maximum yield strength, 310 MPa (45 ksi). See also Table 1.

Hardness. 120 HB, 70 HRB. Maximum hardness: grade C-1, 190 HB

Elastic modulus. Tension, 103 GPa (15.0×10^6 psi)

Impact strength. 135 J/kg (100 ft·lb) at 20 °C (68 °F)

Coefficient of friction. Titanium sliding on titanium: 0.8 at 40 m/min (125 ft/min); 0.68 at 300 m/min (1000 ft/min)

Structure

Crystal structure. Alpha phase: cph below 882 °C (1620 °F). Typical values for unit cell parameters: *a*, 0.2950 nm; *c*, 0.4683 nm; *c/a*, 1.587; all at 25 °C (77 °F). Beta phase: bcc above 882 °C. Typical value for unit cell parameter: 0.329 nm at 900 °C (1652 °F). Impurity elements (commonly oxygen, nitrogen, carbon and iron) change unit cell dimensions

Microstructure. 100% alpha. As amounts of impurity elements in-

Fig. 2 Specific heat of unalloyed titanium

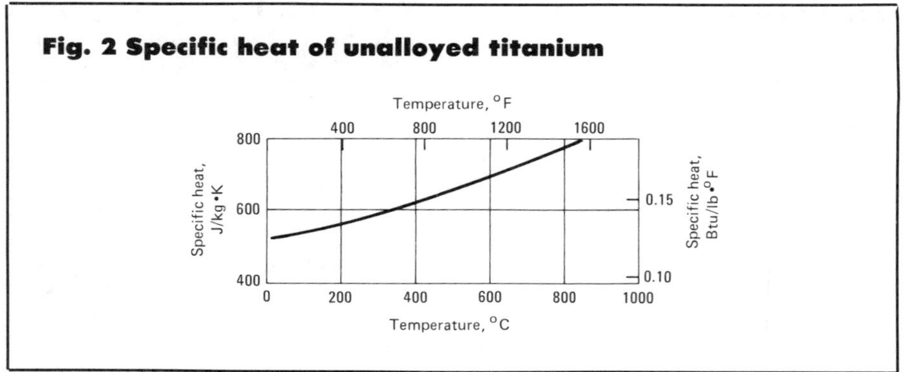

crease, small but increasing amounts of beta are observed metallographically, usually at alpha grain boundaries. Annealed unalloyed titanium may have an equiaxed or acicular alpha microstructure. Acicular alpha occurs by transformation of beta to alpha on cooling through the transformation temperature range. Platelet width decreases with cooling rate. Equiaxed alpha can only be produced by recrystallization of material that has been extensively worked in the alpha phase. The presence of acicular alpha, therefore, is an indication that the material has been heated to a temperature above the beta transus. Beta cannot be retained at low temperatures in unalloyed titanium, except in small quantities in materials containing beta stabilizing contaminants such as iron.

Mass Characteristics

Density. 4.50 Mg/m^3 (0.163 lb/in.^{3}) at 20 °C (68 °F)

Thermal Properties

Liquidus temperature. 1670 ± 5 °C (3040 ± 9 °F)
Phase-transformation temperature. Grade 1 unalloyed titanium transforms from beta to alpha phase at about 882 °C (1620 °F)
Coefficient of thermal expansion. Linear, 8.35 ± 0.15 μm/m·K (4.64 μin./in.·°F) at 15 °C (60 °F). See also Fig. 1.
Specific heat. 520 J/kg (0.124 Btu/lb·°F) at 15 °C (60 °F). For values for unalloyed titanium at elevated temperatures, see Fig. 2.

Electrical Properties

Electrical conductivity. Volumetric, 3.6% IACS at 20 °C (68 °F)
Electrical resistivity. 420 nΩ·m at 20 °C (68 °F). See also Fig. 7.

Hydrogen overvoltage. The value reported for unalloyed titanium is −0.6 V at a current density of 10 A/m^2. This is close to the overvoltages accepted for cadmium and mercury (−1 V) and is a commonly used value

Magnetic Properties

Magnetic susceptibility. Paramagnetic: 3.98 × 10^{-8}/kg
Magnetic permeability. 1.00005 at 1.6 kA/m

Optical Properties

Emissivity. Total: 0.31 at 27 °C (81 °F), 0.33 at 777 °C (1430 °F) and 0.40 at 1027 °C (1880 °F). Room temperature values generally are between 0.13 and 0.31, depending on surface oxidation state

Nuclear Properties

Thermal neutron cross section. Average for naturally occurring titanium: 5.6 b

Chemical Properties

General corrosion behavior. Though highly reactive, titanium is extremely resistant to corrosion in many aggressive environments. For example, it is highly resistant to seawater and to nitric acid. Resistance to general corrosion has been ascribed to a thin, inert film that forms rapidly on the surface when titanium is exposed to air and to passive films that form in certain aggressive media.
Resistance to specific corroding agents. See Table 2.

Fabrication Characteristics

Machinability. Requires a rigid tooling set-up with sharp tools and adequate cooling (flood or mist). Deep cuts at medium speeds are preferred.

Annealing temperature. Full anneal: 700 °C (1300 °F), 30 min to 2 h, air cool
Stress relieving temperature. 480 to 600 °C (900 to 1100 °F), 30 to 60 min, air cool
Hot working temperature. Breakdown forging: 870 to 925 °C (1600 to 1700 °F). Finishing: 800 to 870 °C (1450 to 1600 °F)
Hot forming temperature. Light forming operations, 260 to 400 °C (500 to 750 °F); heavy forming operations, 480 to 760 °C (900 to 1400 °F)

Unalloyed Ti Grade 2

Commercial Names

Trade names. RMI 40; A-40; Ti-50A
Common names. Unalloyed titanium; commercially pure titanium; unalloyed titanium of 40 ksi min yield strength
UNS number. R50400

Specifications

AMS. Sheet, strip and plate: 4902. Welding wire: 4951. Welded tube: 4941. Seamless tube: 4942
ASTM. Bar and billet: B348. Forgings: B381, grade F-2. Sheet, strip and plate: B265. Pipe: B337. Tube: B338. Castings: B367, grade C-2
Government. Bar and forging stock: MIL-T-9047, composition 1. Sheet, strip and plate: MIL-T-9046, type I, composition A. Premium quality forgings: MIL-F-83142, composition 1. Extruded bar, rod and special shaped sections: MIL-T-81556, type I, composition B. Welding rod and wire: MIL-R-81588, type I, composition B

Chemical Composition

Composition limits. 0.03 max N; 0.10 max C; 0.015 max H(a); 0.25 max O; 0.30 max Fe; 0.05 max others (each); 0.30 max others (total); rem Ti. Minimum titanium content 98.885

(a) Deviations from max H values: bar, 0.0125; billet and castings, 0.0100.

Consequence of exceeding impurity limits. May result in raising yield strength above maximum permitted or lowering elongation or reduction in area below minimum

Table 3 Typical elevated temperature tensile properties for unalloyed titanium, grade 2

Temperature °C	Temperature °F	Tensile strength MPa	Tensile strength ksi	Yield strength(a) MPa	Yield strength(a) ksi	Elongation(b), %	Reduction in area, %
205	400	207-228	30-33	138-152	20-22	38-45	55-65
315	600	172-207	25-30	90-124	13-18	30-38	70-80
425	800	131-186	19-27	76-103	11-15	24-28	70-80
540	1000	103-138	15-20	62-76	9-11	30-35	70-80

(a) 0.2% offset. (b) In 50 mm or 2 in.

Fig. 3 Charpy V-notch impact strength of grades 2, 3 and 4 unalloyed titanium

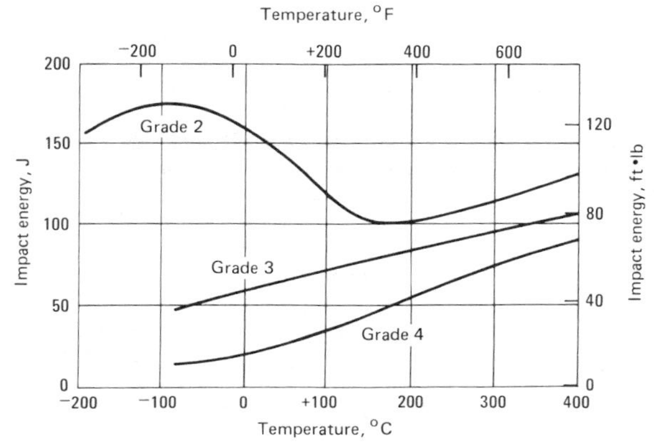

Applications

Typical uses. Aerospace airframe and engine components such as engine shrouds, casings and airframe hot gas ducting; corrosion resistant hardware such as sparger tubes for the chemical process industry; marine applications (such as seawater heat exchangers); other applications requiring high ductility for fabrication but relatively low strength in service

Precautions in use. Hydrogen embrittlement of titanium can occur in hydrogenating solutions at room temperature and during exposure to reducing atmospheres at elevated temperatures. Elevated temperature atmospheric exposure also results in surface contamination by oxygen and nitrogen (oxygen preferentially), which can be extended subsurface to greater depths with increasing time or temperature. Violent oxidation reactions can occur between titanium and liquid oxygen or between titanium and red fuming nitric acid.

Mechanical Properties

Tensile properties. Minimum: tensile strength, 345 MPa (50 ksi); 0.2% yield strength, 276 MPa (40 ksi); elongation, 20% in 50 mm or 2 in.; reduction in area, 30%. Maximum: yield strength, 448 MPa (65 ksi). See also Table 3 and Fig. 10.

Hardness. Typical: 200 HB, 80 HRB. Maximum, grade C-2: 210 HB

Elastic modulus. Tension, 103 GPa (15×10^6 psi)

Impact strength. See Fig. 3.

Coefficient of friction. Titanium sliding on titanium: 0.8 at 40 m/min (125 ft/min); 0.68 at 300 m/min (1000 ft/min)

Structure

Crystal structure. cph below about 890 °C (1635 °F). bcc above about 913 °C (1675 °F). Transition temperature is largely influenced by amounts and kinds of impurity elements present

Mass Characteristics

Density. 4.5 Mg/m^3 (0.163 lb/in.^{3}) at 20 °C (68 °F)

Thermal Properties

Liquidus temperature. 1665 ± 5 °C (3029 ± 9 °F)

Phase-transformation temperature. 890 to 913 °C (1635 to 1675 °F), depending on impurity content

Coefficient of thermal expansion. See Fig. 1.

Specific heat. See Fig. 2.

Thermal conductivity. 16 W/m·K (9.5 Btu/ft·h·°F) at room temperature

Electrical Properties

Electrical conductivity. Volumetric, 3.6% IACS at 20 °C (68 °F)

Electrical resistivity. See Fig. 7.

Hydrogen overvoltage. −0.6 V at a current density of 10 A/m^2

Magnetic Properties

Magnetic susceptibility. Paramagnetic: approx 4×10^{-8} kg

Magnetic permeability. 1.00005 at 1.6 kA/m

Optical Properties

Emissivity. Values between 0.13 and 0.31 depending on surface oxidation state

Nuclear Properties

Thermal neutron cross section. Average absorption of naturally occurring titanium: 5.6 b

Chemical Properties

Effects of specific corroding agents. See Table 2; see also Table 7 and Fig. 11 and 12.

Fabrication Characteristics

Machinability. Requires a rigid tooling set-up with sharp tools and adequate cooling (flood or mist). Deep cuts at medium speed are generally preferred

Formability. Minimum bend radius: less than 1.8 mm (0.07 in.) thick, 3*t*; 1.8 to 4.8 mm (0.07 to 0.187 in.) thick, 4*t*

Annealing temperature. Full anneal: 700 °C (1300 °F), 30 min to 2 h, air cool

Stress relieving temperature. 480 to 600 °C (900 to 1100 °F), 30 to 60 min, air cool. Lower temperatures for longer times have been used with success

Hot working temperature. Breakdown forging: 870 to 925 °C (1600 to 1700 °F). Finishing: 800 to 870 °C (1450 to 1600 °F)

Hot forming temperature. Light forming operations, 260 to 400 °C (500 to 750 °F); heavy forming operations, 480 to 760 °C (900 to 1400 °F)

Unalloyed Ti Grade 3

Commercial Names

Trade names. A-55; RMI 55; Ti-65A

Common names. Unalloyed titanium; commercially pure titanium; unalloyed titanium of 55 ksi min yield strength

UNS number. R50550

Specifications

AMS. Sheet, strip and plate: 4900

ASTM. Bar and billet: B348. Forgings: B381, grade F-3. Sheet, strip and plate: B265. Pipe: B337. Tubing: B338. Castings: B367, grade C-3

Government. Bar and forging stock: MIL-T-9047, composition 1. Sheet, strip and plate: MIL-T-9046, type I, composition C. Premium quality forgings: MIL-F-83142, composition 1. Extruded bar, rod and special shaped sections: MIL-T-81556, type I, composition C

Chemical Composition

Composition limits. 0.05 max N; 0.10 max C; 0.015 max H(a); 0.35 max O; 0.30 max Fe; 0.05 max others (each); 0.30 max others (total); rem Ti. Minimum titanium content 98.885

(a) Deviations from max H values: bar, 0.0125; billet and castings, 0.0100.

Applications

Typical uses. Jet engine shrouds, cases, airframe skins, firewalls, and other hot-area equipment for aircraft and missiles; corrosion resistant equipment for marine and chemical processing industries. Other applications requiring good fabricability, weldability and intermediate strength in service

Precautions in use. Hydrogen embrittlement can occur in hydrogenating solutions at room temperature and in reducing atmospheres at elevated temperature. Violent oxidation reactions can occur between titanium and liquid oxygen or between titanium and red fuming nitric acid

Mechanical Properties

Tensile properties. Minimum: tensile strength, 450 MPa (65 ksi); 0.2% yield strength, 380 MPa (55 ksi); elongation in 50 mm or 2 in., 18%. Maximum: 0.2% yield strength, 550 MPa (80 ksi). Typical: tensile strength, 550 MPa (80 ksi); 0.2% yield strength, 450 MPa (65 ksi); elongation, 25%; reduction in area, 50%. See also Table 4.

Shear properties. See Table 4.

Bearing properties. See Table 4.

Compressive yield strength. Typical at room temperature, 450 MPa (65 ksi) at 0.2% offset. See also Table 4.

Hardness. 225 HB, 90 HRB. Maximum hardness, 235 HB for grade C-3

Poisson's ratio. 0.33

Elastic modulus. Tension and com-

Table 4 Typical mechanical properties for unalloyed titanium, grade 3, at various temperatures

Temperature		Tensile strength		Yield strength(a)		Compressive yield strength		Shear strength		Bearing strength		Bearing yield strength(b)	
°C	°F	MPa	ksi	MPa	ksi	MPa	ksi	MPa	ksi	MPa	ksi	MPa	ksi
−253	−423	1210	175	...	...	...	...	...	...	...	...	...	...
−196	−320	1140	165	...	...	...	...	...	...	...	...	...	...
−54	−65	690	100	510	74	...	...	...	...	...	...	...	...
21	70	550	80	450	65	450	65	380	55	680	99	565	82
93	200	440	64	340	49	315	46	315	46	580	84	455	66
204	400	310	45	200	29	200	29	240	35	435	63	345	50
316	600	250	36	140	20	200	29	195	28	345	50	240	35
427	800	200	29	115	17	180	26	160	23	290	42	220	32
538	1000	165	24	90	13	140	20	110	16	180	26	145	21

(a) 0.2% offset. (b) 2% permanent set.

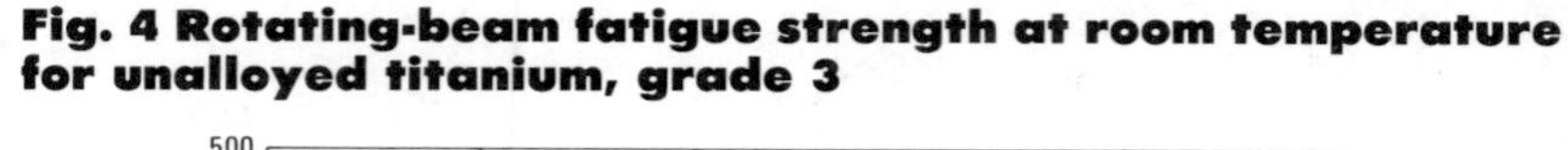

Fig. 4 Rotating-beam fatigue strength at room temperature for unalloyed titanium, grade 3

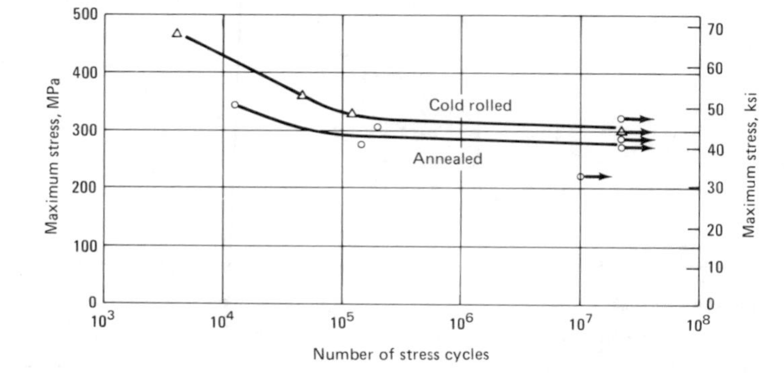

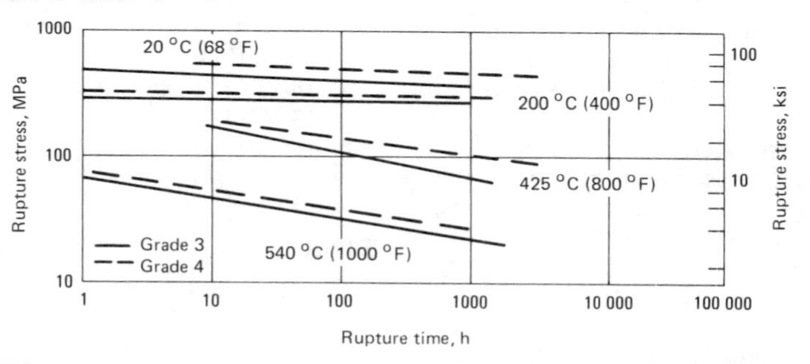

Fig. 5 Typical creep-rupture data for unalloyed titanium, grades 3 and 4

Fig. 6 Thermal conductivities of unalloyed titanium, grades 3 and 4

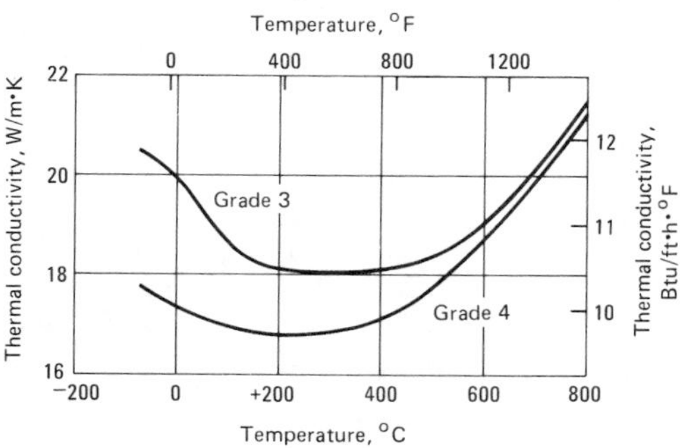

Fig. 7 Electrical resistivity of unalloyed titanium

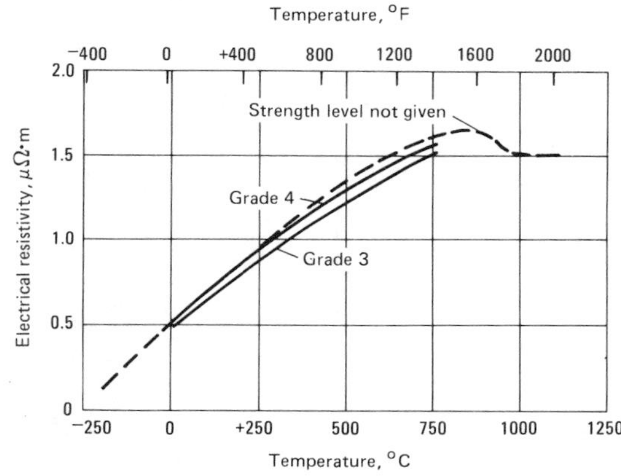

pression, 107 GPa (15.5 × 10^6 psi); shear, 41 GPa (6.0 × 10^6 psi)
Impact strength. See Fig. 3.
Fatigue strength. See Fig. 4.
Creep-rupture characteristics. See Fig. 5.
Coefficient of friction. Titanium sliding on titanium: 0.8 at 40 m/min (125 ft/min); 0.68 at 300 m/min (1000 ft/min)

Structure

Crystal structure. cph below about 900 °C (1650 °F). bcc above about 920 °C (1685 °F)

Mass Characteristics

Density. 4.50 to 4.54 Mg/m^3 (0.163 to 0.164 lb/in.^{3}) at 20 °C (68 °F)

Thermal Properties

Liquidus temperature. 1660 ± 10 °C (3020 ± 18 °F)
Phase-transformation temperature. 900 to 920 °C (1650 to 1685 °F), depending on impurity content
Coefficient of thermal expansion. See Fig. 1.
Specific heat. See Fig. 2.
Thermal conductivity. See Fig. 6.

Electrical Properties

Electrical conductivity. Volumetric, 3.6% IACS
Electrical resistivity. See Fig. 7.
Hydrogen overvoltage. −0.6 V at a current density of 10 A/m^2

Magnetic Properties

Magnetic permeability. 1.00005 at 1.6 kA/m

Optical Properties

Emissivity. 0.13 to 0.31, depending on surface oxidation state

Nuclear Properties

Thermal neutron cross section. Average absorption of naturally occurring titanium: 5.6 b

Chemical Properties

Effects of specific corroding agents. See Table 2.

Fabrication Characteristics

Machinability. Requires a rigid tooling set-up with sharp tools and adequate cooling (flood or mist). Deep cuts at medium speed are preferred
Formability. Bend radius: less than 1.8 mm (0.07 in.) thick, 3*t*; 1.8 to 4.8 mm (0.07 to 0.187 in.) thick, 4*t*
Annealing temperature. Full anneal: 700 °C (1300 °F), 30 min to 2 h, air cool
Stress relieving temperature. 480 to 600 °C (900 to 1100 °F), 30 to 60 min, air cool. Lower temperatures for longer times have been used with success
Hot working temperature. Breakdown forging: 870 to 925 °C (1600 to 1700 °F). Finishing: 800 to 870 °C (1450 to 1600 °F)
Hot forming temperature. Light forming operations, 260 to 400 °C (500 to 750 °F); heavy forming operations, 480 to 760 °C (900 to 1400 °F)

Unalloyed Ti Grade 4

Commercial Names

Trade names. A-70; RMI 70; Ti-75A
Common names. Unalloyed titanium; commercially pure titanium; unalloyed titanium of 70 ksi min yield strength
UNS number. R50700

Specifications

AMS. Forgings, bar and rings: 4921B. Sheet, strip and plate: 4901E
ASTM. Bar and billet: B348. Forgings: B381, grade F-4. Sheet, strip and plate: B265. Pipe: B337. Tubing: B338. Castings: B367, grade C-4
Government. Bar and forging stock: MIL-T-9047E composition 1. Sheet, strip and plate: MIL-T-9046H type I,

Table 5 Typical mechanical properties of unalloyed titanium, grade 4

Temperature		Tensile strength		Yield strength(a)		Compressive yield strength		Shear strength		Bearing strength		Bearing yield strength(b)	
°C	°F	MPa	ksi	MPa	ksi	MPa	ksi	MPa	ksi	MPa	ksi	MPa	ksi
−253	−423	1275	185	...	...	...	...	...	...	...	...	...	...
−196	−320	1205	175	...	...	...	...	...	...	...	...	...	...
−54	−65	760	110	580	84	...	...	...	...	...	...	...	...
21	70	630	90	550	80	550	80	450	65	830	120	695	101
93	200	495	72	415	60	385	56	380	55	700	102	560	81
204	400	345	50	250	36	250	36	290	42	530	77	390	57
316	600	275	40	165	24	240	35	220	32	415	60	295	43
427	800	220	32	145	21	220	32	185	27	345	50	270	39
538	1000	185	27	110	16	165	24	140	20	215	31	170	26

(a) 0.2% offset. (b) 2% permanent set.

Fig. 8 Variation of elastic modulus with temperature for unalloyed titanium, grade 4

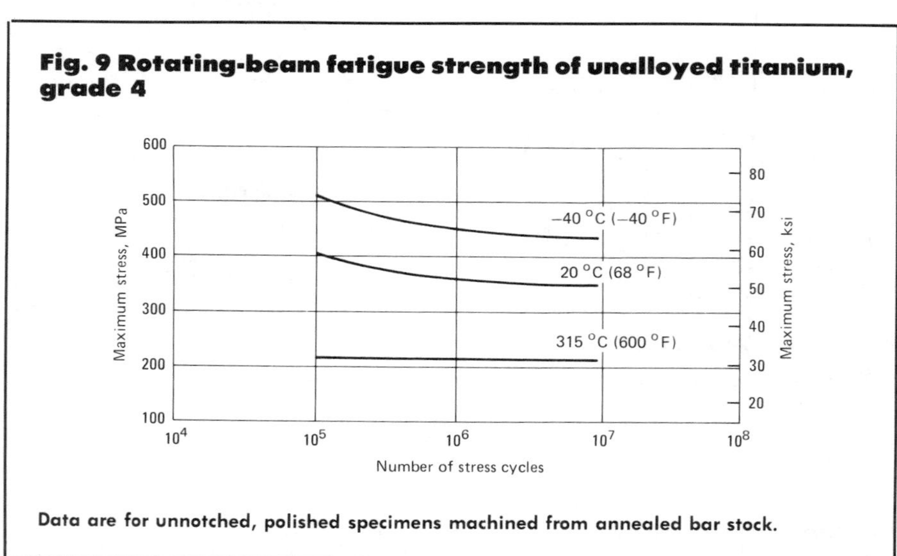

Fig. 9 Rotating-beam fatigue strength of unalloyed titanium, grade 4

Data are for unnotched, polished specimens machined from annealed bar stock.

composition B. Premium quality forgings: MIL-F-83142A composition 1. Extruded bar, rod and special shaped sections: MIL-T-81556, type I, composition D

Chemical Composition

Composition limits. 0.05 max N; 0.10 max C; 0.015 max H(a); 0.40 max O; 0.50 max Fe; 0.05 max others (each); 0.30 max others (total); rem Ti. Minimum titanium content, 98.635% (determined by difference)

(a) Deviations from max H values: bar, 0.0125; billet and castings, 0.0100.

Applications

Typical uses. Firewalls, bulkheads, compressor cases, shrouds and tailpipes for jet engines; pumps, valves, piping, and tanks in chemical processing equipment. Other applications requiring intermediate strength in service where simple parts can be made by cold forming, complex parts by hot forming, or by combinations of forming and machining and where good weldability might be a requirement

Precautions in use. Hazardous reactions can occur with liquid oxygen or red fuming nitric acid. Hydrogen embrittlement can occur during exposure to hydrogenating solutions at room temperature and in air at elevated temperatures.

Mechanical Properties

Tensile properties. Minimum: tensile strength, 550 MPa (80 ksi); yield strength, 480 MPa (70 ksi); elongation in 50 mm or 2 in., 15%. Maximum: yield strength, 655 MPa (95 ksi). Typical: tensile strength, 620 MPa (90 ksi); 0.2% yield strength, 550 MPa (80 ksi); elongation, 20%; reduction in area, 40%. See also Table 5.

Shear properties. See Table 5.

Bearing properties. See Table 5.

Compressive yield strength. Typical at room temperature, 550 MPa (80 ksi) at 0.2% offset. See also Table 5.

Hardness. 265 HB, 100 HRB. Maximum hardness, 245 HB

Poisson's ratio. 0.33
Elastic modulus. Tension and compression, 107 GPa (15.5 × 10^6 psi); shear, 41 GPa (6.0 × 10^6 psi). See also Fig. 8.
Impact strength. See Fig. 3.
Fatigue strength. See Fig. 9.
Creep-rupture characteristics. See Fig. 5.
Coefficient of friction. Titanium sliding on titanium: 0.8 at 40 m/min (125 ft/min); 0.68 at 300 m/min (1000 ft/min)

Structure

Crystal structure. cph below about 905 °C (1665 °F). bcc above about 950 °C (1740 °F)

Mass Characteristics

Density. 4.500 to 4.540 Mg/m^3 (0.163 to 0.164 lb/in.^{3}) at 20 °C (68 °F)

Thermal Properties

Liquidus temperature. 1660 ± 10 °C (3020 ± 18 °F)
Phase-transformation temperature. 905 to 950 °C (1665 to 1740 °F), depending on impurity content
Coefficient of thermal expansion. See Fig. 1.
Specific heat. See Fig. 2.
Thermal conductivity. See Fig. 6.

Electrical Properties

Electrical conductivity. Volumetric, 3.6% IACS
Electrical resistivity. See Fig. 7.
Hydrogen overvoltage. −0.6 V at a current density of 10 A/m^2

Magnetic Properties

Magnetic permeability. 1.00005 at 1.6 kA/m

Optical Properties

Emissivity. Values between 0.13 and 0.31 depending upon surface oxidation state

Nuclear Properties

Effects of neutron irradiation. Irradiation under specified exposures results in increased yield strength and hardness but no change in impact strength. Density appears to be unaffected, but electrical resistivity is decreased.
Thermal neutron cross section. Average absorption for naturally occurring titanium: 5.6 b

Chemical Properties

Resistance to specific corroding agents. See Table 2.

Fabrication Characteristics

Machinability. Requires a rigid tooling set-up with sharp tools and adequate cooling (flood or mist). Deep cuts at medium speed are preferred.
Formability. Bend radius: less than 1.8 mm (0.07 in.) thick, 4*t*; 1.8 to 4.8 mm (0.07 to 0.187 in.) thick, 5*t*
Annealing temperature. Full anneal: 700 °C (1300 °F), 30 min to 2 h, air cool
Stress relieving temperature. 480 to 600 °C (900 to 1100 °F), 30 to 60 min, air cool. Lower temperatures for longer times have been used with success.
Hot working temperature. Breakdown forging: 870 to 925 °C (1600 to 1700 °F). Finishing: 800 to 870 °C (1450 to 1600 °F)
Hot forming temperature. Light forming operations, 260 to 400 °C (500 to 750 °F); heavy forming operations, 480 to 760 °C (900 to 1400 °F)

Table 6 ASTM composition limits for titanium-palladium alloys

	Composition limits (ASTM B265), %							
Grade	N, max	C, max	H, max	O, max	Fe, max	Pd	Others, max Each	Others, max Total
7	0.03	0.10	0.015(a)	0.25	0.30	0.12-0.25	0.05(b)	0.30(b)
C-7A....	0.03	0.10	0.010	0.18	0.30	0.12 min	0.10	0.40
C-7B....	0.03	0.10	0.010	0.25	0.30	0.12 min	0.10	0.40
F-7	0.03	0.10	0.015	0.25	0.30	0.15-0.25	0.05	0.30
C-8A....	0.05	0.10	0.010	0.35	0.30	0.12 min	0.10	0.40
C-8B....	0.05	0.10	0.010	0.40	0.50	0.12 min	0.10	0.40
11	0.03	0.10	0.015(a)	0.18	0.20	0.12-0.25	0.05(b)	0.30(b)

(a) Compositions of grades 7 and 11 specified in ASTM B348 are the same as those specified in B265 except that limits on hydrogen are lower. Hydrogen limit in bar, 0.0100%; in billet, 0.0125%. (b) Compositions of grades 7 and 11 specified in ASTM B337 and B338 are the same as those specified in B265, except that limits on other elements are not defined.

Ti-Pd Alloys

Commercial Names

Trade names. RMI 0.2 Pd; Ti Tech 0.2 Pd; MMA-1942; Ti-0.2Pd
Common names. P-D alloy; palladium alloy
UNS numbers. Grade 7, R52400; grade 11, R52250

Specifications

ASTM. Bar and billet: B348, grade 7 and grade 11. Forgings: B381, grade F-7. Sheet, strip and plate: B265, grade 7 and grade 11. Pipe: B337, grade 7 and 11. Tube: B338, grade 7 and grade 11. Castings: B367, grades C-7A, C-7B, C-8A and C-8B

Chemical Composition

Composition limits. See Table 6.
Consequence of exceeding impurity limits. Exceeding impurity limits may negate beneficial increase in corrosion resistance afforded by palladium additions. Palladium additions of less than specified minimums are less effective in promoting an improved corrosion resistance. Excess palladium (above specified range) is not cost effective

Applications

Typical uses. Those requiring excellent corrosion resistance in chemical processing or storage applications where the media is mildly reducing or fluctuates between oxidizing and reducing. The palladium-containing alloys extend the range of titanium's application in hydrochloric, phosphoric and sulfuric acid solutions. Characteristics of good fabricability, weldability, and strength level are similar to those of corresponding unalloyed titanium grades (see table below). Material selection based on properties of unalloyed grades is applicable.

Ti-Pd grades	Corresponding grades of unalloyed Ti
7	2
F-7	F-2
C-7A	C-1
C-7B	C-2
11	1
C-8A	C-3
C-8B	C-4

Precautions in use. Hydrogen embrittlement can occur in hydrogenat-

Table 7 Comparative corrosion rates for Ti-Pd, grade 7, and unalloyed titanium, grade 2

Corrodent	Concentration, %	Temperature, °C	Temperature, °F	Corrosion rate, Grade 7, mm/yr	Corrosion rate, Grade 7, mils/yr	Corrosion rate, Grade 2, mm/yr	Corrosion rate, Grade 2, mils/yr
Aluminum chloride	10	100	212	<0.025	<1	<0.025	<1
	25	100	212	0.025	1	50	2020
Chlorine (wet)	...	Room		<0.025	<1	<0.025	<1
Citric acid	50	Boiling		<0.025	<1	0.4	17
Hydrochloric acid, (N_2 saturated)	3	190	374	0.025	1	>28	>1120
	5	190	374	0.1	4	>28	>1120
	10	190	374	8.8	350	>28	>1120
	15	190	374	40	1620	...	...
Hydrochloric acid, (O_2 saturated	3	190	374	0.13	5	>28	>1120
	5	190	374	0.13	5	>28	>1120
	10	190	374	9.2	368	>28	>1120
Sodium chloride	Brine	93	200	<0.025	<1	...	...
	10	190	374	<0.025	<1	...	...
	23(a)	Boiling		...	...	Nil	Nil
Sulfuric acid, (N_2 saturated)	1	100	212	...	...	7	282
	1	190	374	0.13	5	...	...
	5	100	212	...	...	26.5	1060
	5	190	374	0.13	5	...	...
	10	190	374	1.5	59	...	...
Formic acid	50	Boiling		0.075	3	3.6	143
Hydrochloric acid	5	Boiling		0.18	7	>10	>400
Oxalic acid	1	Boiling		1.13	45	45	1800
Phosphoric acid	50	70	158	1.8	71	10	405
	10	Boiling		3.2	127	11	439
Sulfuric acid	5	Boiling		0.5	20	48	1920

(a) Acidified: pH 1.2.

ing solutions at room temperature and at elevated temperatures during exposure to air or reducing atmospheres. These conditions also produce oxygen and nitrogen contamination which increases in severity with increasing temperature and time of exposure. Violent oxidation reactions can occur between Ti-Pd alloys and liquid oxygen or red or white fuming nitric acid.

Mechanical Properties

Essentially the same as equivalent grades of unalloyed titanium

Structure

Crystal structure. The small amount of palladium added does not significantly modify the crystal structure of unalloyed titanium. Transformation temperatures are similarly unchanged. For Ti-Pd, grade 7, the alpha-beta transformation temperature is 890 to 913 °C (1635 to 1675 °F), depending on impurity content.

Microstructure. Only alpha soluble amounts of palladium are added to make titanium-palladium alloys; therefore, microstructures are essentially the same as for equivalent grades of unalloyed titanium. Titanium-palladium intermetallic compounds formed in this system have not been reported to occur with normal heat treatments.

Mass Characteristics

Density. 4.500 to 4.540 Mg/m³ (0.163 to 0.164 lb/in.³) at 20 °C (68 °F)

Other Physical Properties

Essentially the same as for equivalent grades of unalloyed titanium

Chemical Properties

Resistance to specific corroding agents. See Table 7.

Fabrication Characteristics

Essentially the same as for equivalent grades of unalloyed titanium

Ti-0.3Mo-0.8Ni

Commercial Names

Trade name. Ti Code 12

Specifications

ASTM. Bar and billet: B348. Forgings: B381, grade F-12. Sheet, strip and plate: B265. Pipe: B337. Tubing: B338

Chemical Composition

Composition limits. 0.03 max N; 0.08 max C; 0.015 max H(a); 0.25 max O; 0.30 max Fe; 0.2–0.4 Mo; 0.6–0.9 Ni; 0.05 max others (each); 0.30 max others (total); rem Ti

(a) Deviations from max H values: bar, 0.0125; billet, 0.0100.

Applications

Typical uses. Applications are similar to those for unalloyed titanium, where corrosion resistance is important. Compared to unalloyed titanium, grades 1 to 4, Ti-0.3Mo-0.8Ni has better corrosion resistance and higher strength. This combination allows Ti-0.3Mo-0.8Ni to be used in harsher environments and/or in thinner sections. The alloy is particularly resistant to crevice corrosion in hot brines.

Precautions in use. Similar to unalloyed titanium.

Mechanical Properties

Tensile properties. See Table 8 and Fig. 10. Typical properties for welds

Fig. 10 Minimum tensile strength of low-strength titanium metals

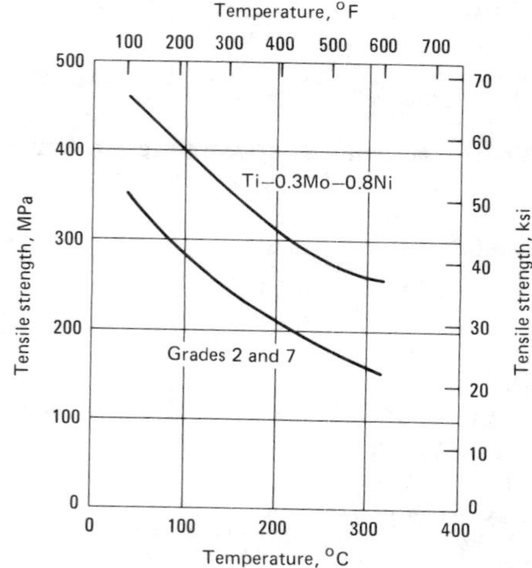

Table 8 Typical tensile properties of Ti-0.3Mo-0.8Ni

Temperature °C	°F	Tensile strength MPa	ksi	0.2% yield strength(a) MPa	ksi	Elongation, %
25	77	510	74	415	60	33
240	400	345	50	250	36	37
316	600	325	47	205	30	32

Fig. 11 Crevice corrosion of Ti-0.3Mo-0.8Ni and grade 2 unalloyed Ti in saturated NaCl solution

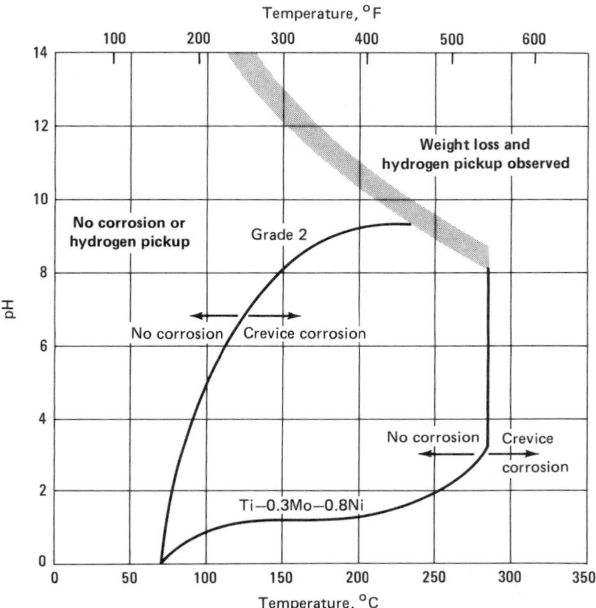

Shaded band represents transition zone between active and passive behavior.

Fig. 12 Corrosion of titanium metals in boiling nitric acid

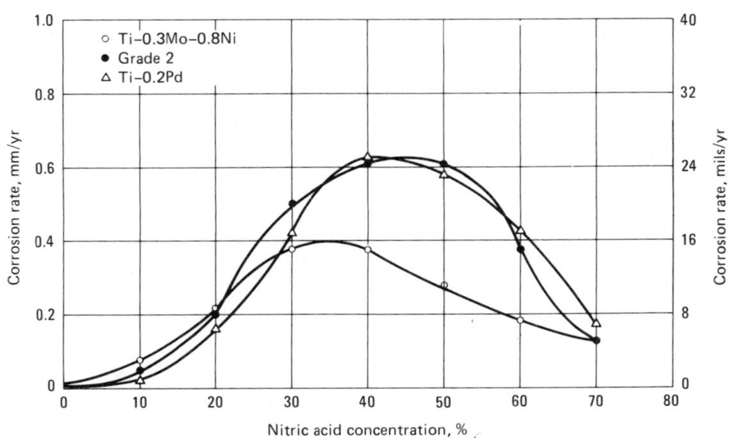

Solution replaced with fresh solution every 24 h; total time of exposure, 480 h.

Table 9 Creep properties of Ti-0.3Mo-0.8Ni

Applied stress MPa	ksi	Creep strain(a), %/h
At 25 °C (77 °F)		
138	20.0	0
207	30.0	9.2×10^{-8}
290	42.0	3.6×10^{-5}
331	48.0	3.94×10^{-5}
At 150 °C (300 °F)		
108	15.6	0
179	26.0	7.2×10^{-6}
221	32.0	4.8×10^{-6}
At 315 °C (600 °F)		
48	7.0	3.3×10^{-6}
83	12.0	8.2×10^{-6}
103	15.0	1.4×10^{-5}
124	18.0	1.5×10^{-5}
138	20.0	1.86×10^{-5}

(a) Best fit linear rate over test period of 5000 to 10 000 h.

in 13 mm (0.5 in.) plate: tensile strength, 550 MPa (80 ksi); elongation, 12%

Poisson's ratio. 0.33

Fatigue strength. Reverse bending, 1.3 mm (0.05 in.) thick sheet: smooth specimen, 345 MPa (50 ksi) at 10^6 to 10^7 cycles; notched specimen (K_t = 3.5), 160 MPa (25 ksi) at 10^5 to 10^7 cycles

Elastic modulus. Typically 110 GPa (16×10^6 psi), but may vary up to 10% about this value depending on direction and processing variables

Impact strength. Typical, 15 J (11 ft·lb)

Creep properties. See Table 9.

Structure

Microstructure. Either equiaxed or acicular alpha with minor amounts of beta. Acicular alpha microstructures are found primarily in welds or heat-affected zones.

Mass Characteristics

Density. 4.52 Mg/m³ (0.163 lb/in.³) at 20 °C (68 °F)

Thermal Properties

Coefficient of thermal expansion. Linear: 9.5 μm/m·K (5.3 μin./in.·°F) at 0 to 315 °C (32 to 600 °F)

Specific heat. 540 J/kg·K (0.13 Btu/lb·°F)

Thermal conductivity. 19.0 W/m·K (11.0 Btu/ft·h·°F)

Electrical Properties

Electrical resistivity. 520.7 nΩ·m at 26 °C (79 °F)

Chemical Properties

General corrosion resistance. Intermediate between unalloyed Ti and Ti-0.2Pd.

Resistance to specific corroding agents. In a series of crevice corrosion tests, Ti-0.3Mo-0.8Ni was completely resistant in 500-h exposures to the following boiling solutions: saturated $ZnCl_2$ at pH of 3.0; 10% $AlCl_3$; $MgCl_2$ at pH of 4.2; 10% NH_4Cl at pH of 4.1; saturated NaCl, and saturated NaCl + Cl_2, both at pH of 1.0; and 10% Na_2SO_4 at pH of 1.0. In a similar test in boiling 10% $FeCl_3$, crevice corrosion was observed in metal-to-Teflon crevices after 500 h. Ti-0.3Mo-0.8Ni also exhibits the following typical corrosion rates:

Environment	Corrosion rate mm/yr	mils/yr
Wet Cl_2 gas	0.00089	0.035
5% NaOCl + 2% NaCl + 4% NaOH(a)	0.06	2.4
70% $ZnCl_2$	0.005-0.0075	0.2-0.3
50% citric acid	0.013	0.5
10% sulfamic acid	11.6	455
45% formic acid	nil	nil
88 to 90% formic acid	0-0.56	0-22
90% formic acid(b)	0.56	2.2
10% oxalic acid	104	4100

(a) No crevice corrosion in metal-to-metal or metal-to-Teflon crevices. (b) Anodized specimens. For corrosion behavior in saturated brine, see Fig. 11; in boiling nitric acid, see Fig. 12.

Fabrication Characteristics

Machining. Requires rigid setup, sharp tools, heavy feeds, slow speeds and copious amounts of a nonchlorinated coolant

Welding. Similar to unalloyed titanium

Annealing temperature. 760 to 790 °C (1400 to 1450 °F)

Stress-relieving temperature. 480 to 600 °C (900 to 1050 °F) for 1 to 2 h; air cool

Ti-5Al-2.5Sn

Commercial Names

Trade names. Standard grade: A-110AT; MMA-5137; RMI 5Al-2.5Sn. High-purity grade: usually the name of the standard grade with the suffix ELI (for extra-low interstitial content) as in RMI 5Al-2.5Sn-ELI, but also in code (A-95 AT, for example).

Common names. Standard grade: alpha alloy; aluminum-tin alloy; 5-2½ alloy; and combinations of these terms. High-purity grade: 5-2½-ELI alloy; low-temperature alpha alloy; low-temperature aluminum-tin alloy; high-purity 5-2½ alloy; and combinations of these terms

UNS numbers. Ti-5Al-2.5Sn: R54520. Ti-5Al-2.5Sn-ELI: R54521

Specifications

AMS. Standard grade: forgings, 4966; bar and rings, 4926; sheet, strip and plate, 4910; welding wire, 4953. High-purity grade: forgings, bar and rings, 4924; sheet, strip and plate, 4909

ASTM. Standard grade: bar and billet: B348, grade 6. Forgings: B381, grade F-6. Sheet, strip and plate: B265, grade 6. Castings: B367, grade C-6. High-purity grade: Ti-5Al-2.5Sn ELI, not defined in ASTM specifications

Government. Standard grade: bar and forging stock: MIL-T-9047, composition 2. Sheet, strip and plate: MIL-T-9046, type II, composition A. Premium quality forgings: MIL-F-83142, composition 2. Extrusions: MIL-T-81556, type II, composition B. Welding rod and wire: MIL-R-81588, type II, composition A. High-purity grades: bar and forging stock, MIL-T-9047, composition 3. Sheet, strip and plate: MIL-T-9046, type II, composition B. Premium quality forgings: MIL-F-83142, composition 3. Extrusions: MIL-T-81556, type II, composition B. Welding rod and wire: MIL-R-81588, type II, composition B

Fig. 13 Rotating-beam fatigue strength of Ti-5Al-2.5Sn

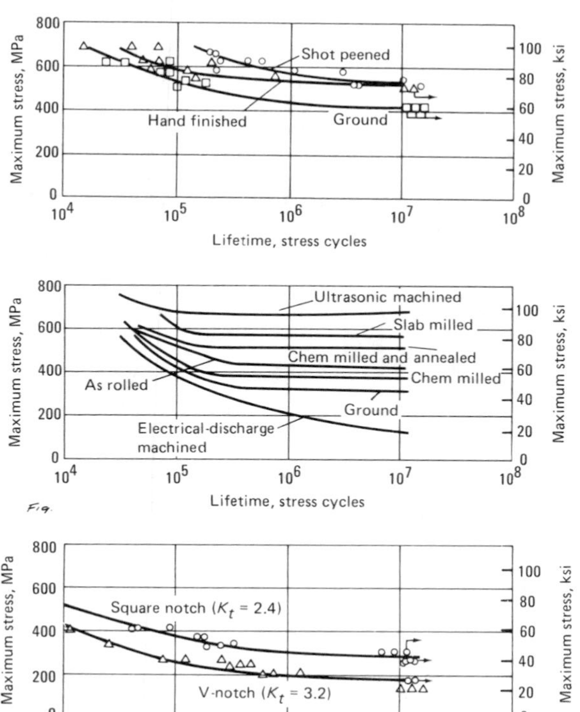

Top and center, fatigue strengths for different types of surface finish; bottom, notch fatigue strength for two different types of notches.

Table 10 Composition limits for Ti-5Al-2.5Sn

Specification and grade	N, max	C, max	H, max	O, max	Fe, max	Al	Sn	Others, max Each	Others, max Total
ASTM									
B265, grade 6	0.05	0.10	0.020	0.20	0.50	4.0-6.0	2.0-3.0	0.05	0.30
B348, grade 6	0.05	0.10	0.0125(a)	0.20	0.50	4.0-6.0	2.0-3.0	0.05	0.30
B367, grade C-6	0.05	0.10	0.0100	0.20	0.50	4.0-6.0	2.0-3.0	0.10	0.40
B381, grade F-6	0.05	0.10	0.020	0.30	0.50	4.0-6.0	2.0-3.0	0.05	0.30
Military									
MIL-T-9046, MIL-T-81556									
Std grade	0.05	0.08	0.020	0.20	0.50	4.5-5.75	2.0-3.0	...	0.40
ELI	0.035	0.05	0.0125	0.12	0.25	4.5-5.75	2.0-3.0	...	0.30
MIL-T-9047, MIL-F-83142									
Std grade	0.05	0.08	0.020	0.20	0.50	4.5-5.75	2.0-3.0	...	0.40
ELI	0.035	0.05	0.0125	0.12	0.25	4.7-5.6	2.0-3.0	0.10Mn	0.30

(a) Bar only; for billet, 0.0100 max H.

Table 11 Tensile property limits at room temperature for Ti-5Al-2.5Sn

Specification	Min tensile strength MPa	Min tensile strength ksi	Min yield strength MPa	Min yield strength ksi	Min reduction in area, %
Plate, sheet and strip					
AMS 4910 (Std)	827	120	779	113	...
ASTM B265, grade 6	827	120	793	115	...
MIL-T-9046					
Std grade, up to 1.5 in. thick	827	120	779	113	...
Std grade, 1.5 to 4.0 in. thick	793	115	758	110	...
ELI, all thicknesses	689	100	655	95	...
Billet, bar, forgings and rings					
AMS 4924 (ELI)	689	100	620	90	20(b)
AMS 4926 and 4966 (Std)	793	115	758	110	25
ASTM B348, grade 6	827	120	793	115	25
ASTM B381, grade F-6	827	120	793	115	25
MIL-T-9047					
Std grade	793	115	758	110	25
ELI	689	100	620	90	20(b)
Castings					
ASTM B367, grade C-6	793	115	724	105	...

(a) 0.2% offset. (b) 15% for material 2 to 4 in. thick.

Chemical Composition

Composition limits. See Table 10.

Applications

Typical uses. Standard grade: gas turbine engine casings and rings, aerospace structural members in hot spots (near engines and wing leading edges), and chemical processing equipment that requires both better elevated-temperature strength than unalloyed titanium and excellent weldability. Other applications requiring good weld fabricability and intermediate strength at service temperatures up to 480 °C (900 °F). High-purity grade: pressure vessels for liquified gases and other applications requiring better ductility and toughness (at somewhat lower strength) than the standard grade, particularly in hardware for service to cryogenic temperatures

Precautions in use. Hydrogen embrittlement of Ti-5Al-2.5Sn (and its ELI modification) can occur in hydrogenating solutions at room temperature, in air or reducing atmospheres at elevated temperatures, and even in pressurized hydrogen at cryogenic temperatures. Oxygen and nitrogen contamination can occur in air at elevated temperatures; such contamination increases in severity with increasing exposure time and temperature. Ti-5Al-2.5Sn is susceptible to hot-salt (particularly chloride) stress-corrosion cracking and to accelerated crack propagation in aqueous solutions at ambient temperatures. Environments in which Ti-5Al-2.5Sn is to be used should be carefully controlled to ensure against material degradation. Ti-5Al-2.5Sn-ELI is more tolerant of environments capable of degrading properties, but over-all immunity is not ensured by using the high-purity alloy. Ti-5Al-2.5Sn should never come in contact with liquid oxygen.

Mechanical Properties

Tensile properties. Typical. Tensile strength: standard, 862 MPa (125 ksi); ELI, 779 MPa (113 ksi). 0.2% yield strength: standard, 827 MPa (120 ksi); ELI, 717 MPa (104 ksi). Elongation in 50 mm or 2 in.: standard, 15%; ELI, 17%. Reduction in area, 40%. Minimum elongation: castings, 8%; all other forms, 10%. See Table 11 for minimum requirements.

Compressive yield strength. Typical, room temperature, standard grade: 896 MPa (130 ksi) at 0.2% offset

Hardness. 36 HRC. Grade C-6 castings, standard grade: 335 HB max

Poisson's ratio. 0.35

Elastic modulus. Tension, 107 to 110 GPa (15.5 to 16.0 × 10^6 psi); shear, 48 GPa (7 × 10^6 psi)

Impact strength. At room temperature: standard grade, 23 J (17 ft·lb); ELI, 43 J (32 ft·lb)

Fatigue strength. See Fig. 13.

Plane-strain fracture toughness. See Table 12.

Creep-rupture characteristics. Ti-5Al-2.5Sn will creep at room temperature at stresses above the proportional limit (about 80% of the yield stress). See Table 13.

Structure

Microstructure. Either acicular or equiaxed alpha (cph) depending on prior processing. Acicular alpha is the form observed after thermal excursions above the beta transus. Equiaxed alpha results from working the metal below the beta transus followed by annealing in the alpha field. Commonly there is a very small amount of beta (bcc) in microstructures of Ti-5Al-2.5Sn containing high iron. Equiaxed alpha is the form

Table 12 Fracture toughness of Ti-5Al-2.5Sn

Heat treatment variable(a)	Test temperature K	Test temperature °F	Stress intensity, K_{Ic} MPa $\sqrt{m}$	Stress intensity, K_{Ic} ksi $\sqrt{in.}$	Specimen, orientation(b) and type(c)	Yield strength(d) MPa	Yield strength(d) ksi
Standard grade							
Air cooled	295	72	71.4	65	LT-CT	876	127
	77	−320	53.8	49	LT-CT	1338	194
	20	−423	51.6	47	LT-B	1482	215
	20	−423	50.5	46	LS-B	...	...
Furnace cooled	295	72	65.9	60	LT-CT	882	128
	77	−320	57.1	52	LT-CT	1379	200
	20	−423	47.2	43	LT-B	1517	220
	20	−423	52.7	48	LS-B	...	...
ELI grade							
Air cooled	295	72	118.7	108(e)	LT-CT	703	102
	77	−320	111.0	101	LT-CT	1179	171
	20	−423	91.2	83	LT-B	1303	189
	20	−423	106.6	97	LS-B	...	...
Furnace cooled	295	72	115.4	105(e)	LT-CT	682	99
	77	−320	82.4	75	LT-CT	1179	171
	20	−423	68.1	62	LT-B	1303	189
	20	−423	80.2	73	LS-B	...	...

(a) Air cooled or furnace cooled from annealing temperature. (b) Orientation notation per ASTM E399-74. (c) CT, compact tension specimen; B, bend specimen. (d) 0.2% offset. (e) Invalid toughness values (not 100% plane-strain conditions).

Table 13 Typical creep properties of Ti-5Al-2.5Sn

Temperature °C	Temperature °F	0.1% MPa	0.1% ksi	0.2% MPa	0.2% ksi	0.5% MPa	0.5% ksi
		Stress for total deformation in 1000 h of					
315	600	83	12	190	27	385	56
370	700	69	10	175	25	330	48
425	800	45	6.5	100	15	235	34
540	1000	2.8	0.4	5	0.7	23	3.3

most frequently encountered in mill products of either standard Ti-5Al-2.5Sn or Ti-5Al-2.5Sn-ELI

Mass Characteristics

Density. 4.460 Mg/m^3 (0.161 lb/in.^{3}) at 20 °C (68 °F)

Thermal Properties

Liquidus temperature. 1590 ± 20 °C (2895 ± 35 °F)
Phase-transformation temperature. Beta phase to alpha on cooling, 1040 to 1090 °C (~1900 to ~2000 °F). On heating, alpha to beta, 955 to 985 °C (~1750 to ~1805 °F)
Coefficient of thermal expansion. Linear: 9.4 μm/m·K (5.2 μin./in.·°F) at 20 °C (68 °F)
Specific heat. 523 J/kg·K (0.125 Btu/ft·°F) at 20 °C (68 °F)
Thermal conductivity. 7.4 to 7.8 W/m·K (4.3 to 4.5 Btu/ft·h·°F) at 20 °C (68 °F)

Electrical Properties

Electrical conductivity. Volumetric, 2.5% IACS at 20 °C (68 °F)
Electrical resistivity. At 20 °C (68 °F): standard grade, 1.57 μΩ·m; ELI grade, 1.80 μΩ·m

Magnetic Properties

Magnetic permeability. 1.00005 at 1.6 kA/m

Optical Properties

Emissivity. Not substantially different from unalloyed titanium (e.g., 0.13 to 0.31 at 27 °C). Values dependent on the surface oxidation state

Nuclear Properties

Effect of neutron irradiation. Low temperature tensile properties (at 17 K, or 30 °R) deteriorate (strength increases, ductility decreases) with exposure to irradiation of 1 × 10^{17} nvt fast neutrons

Chemical Properties

General corrosion behavior. Similar to grades 2 and 3 of unalloyed titanium. The standard alloy and its high-purity modification (Ti-5Al-2.5Sn-ELI) are highly resistant to many reactive substances (such as seawater and nitric acid solutions). Deep sea exposures (about 1500 m, or 1 mile depth) produced no corrosion in 751 days (rate estimated at <2.5 μm/yr or <0.1 mils/yr). In seawater flowing at about 36 m/s (120 ft/s), the alloy exhibited a corrosion rate of 5.5 μm/yr (0.22 mils/yr)
Resistance to specific corroding agents. Especially susceptible to corrosion by solid salt (such as chlorides) at elevated temperatures (hot-salt stress corrosion). Exposure to salt causes pitting and cracking in 100 h at 315 °C (600 °F) at stress levels above about 207 MPa (30 ksi). Stress-corrosion at lower temperatures is possible for longer exposure times at higher stress levels. Accelerated crack propagation at room temperature in the presence of a preexisting crack is possible in seawater, chloride solutions and other active solutions

Fabrication Characteristics

Machinability. Rigid tooling plus liberal cooling via flood or mist systems are requirements. Deep cuts at slow speed are preferred
Annealing temperature. Full anneal: 720 to 845 °C (~1325 to ~1550 °F), 10 min to 4.25 h (shorter times at the higher temperatures), air cool
Stress-relieving temperature. 540 to 650 °C (~1000 to ~1200 °F), 1 to 4 h, air cool
Hot working temperature. Breakdown forging, 1010 to 1040 °C (1850 to ~1900 °F). Finishing, 900 to 955 °C (~1650 to ~1750 °F)

Ti-8Al-1Mo-1V

Commercial Names

Common names. Ti-8-1-1; Ti-811
Trade names. RMI-8Al-1Mo-1V; HA-8116; UTA8DV
UNS number. R54810

Specifications

AMS. 4915, 4916, 4955, 4972, 4973
AWS. A5.16

Government. MIL-F-8312, MIL-T-9046, MIL-T-9047, MIL-T-81556, MIL-T-81588

Chemical Composition

Nominal composition. 8.0 Al; 1.0 Mo; 1.0 V; rem Ti

Applications

Typical uses. Sheet and forgings for high speed aircraft structural components (primarily skin); turbine parts; forgings for compressor disks, plates, and hubs; cargo flooring. Other applications where light, high strength, highly weldable material with low density is required

Precautions in use. Similar to Ti-5Al-2.5Sn

Mechanical Properties

Tensile properties. Typical. Tensile strength, 965 MPa (140 ksi); 0.2% yield strength, 827 MPa (120 ksi); elongation, 14% in 50 mm or 2 in.

Elastic modulus. Tension, 120 GPa (17.4×10^6 psi); compression, 117 GPa (17.0×10^6 psi)

Impact strength. 33.9 J (25 ft·lb) at room temperature

Creep-rupture characteristics. At elevated temperatures, duplex annealed Ti-8-1-1 has creep strength superior to that of other titanium alloys. At 500 °C (930 °F), stress for indicated deformation in 1000 h: 0.05%, 68.9 MPa (10 ksi); 0.1%, 97.9 MPa (14.2 ksi); 0.2%, 157 MPa (22.8 ksi); 216 MPa (31.3 ksi)—20-mm diam forged test specimens

Mass Characteristics

Density. 4.37 Mg/m^3 (0.158 $lb/in.^3$) at 20 °C (68 °F)

Electrical Properties

Electrical resistivity. 1.99 μΩ·m at room temperature

Fabrication Characteristics

Machinability. Similar to Ti-6Al-4V but does not work harden

Formability. Bend radius, 6.0*t* (min)

Weldability. Fusion welded using conventional welding processes and equipment provided contamination of bead and heat-affected zone by interstitials such as oxygen, hydrogen and nitrogen is prevented. Resistance welding: readily welded by spot, seam or stitch welding processes

Fig. 14 Constant-life fatigue diagram for duplex annealed Ti-6Al-2Sn-4Zr-2Mo sheet, 1 mm (0.04 in.) thick

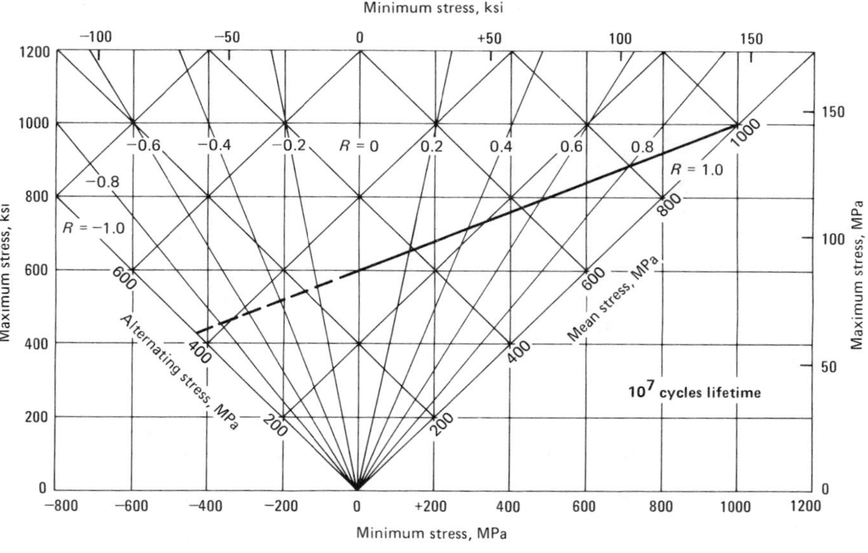

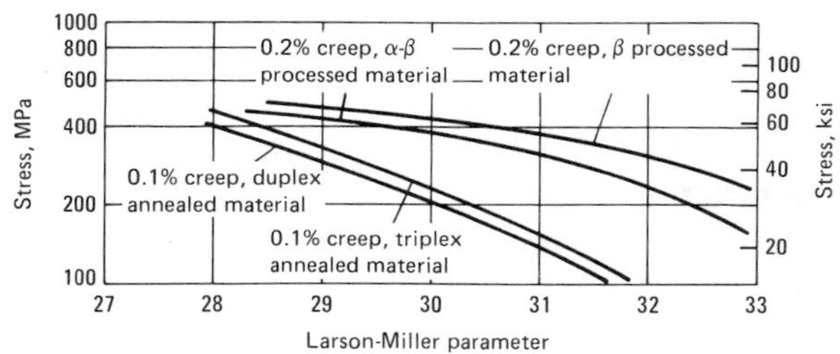

Fig. 15 Master creep curves for Ti-6Al-2Sn-4Zr-2Mo

Larson-Miller parameter equals $10^{-3}\,T(20 + \log t)$, where T is temperature in °R and t is time in hours.

Ti-6Al-2Sn-4Zr-2Mo

Commercial Names

Trade name. Ti-6242

Common name. Ti-6-2-4-2

UNS number. R54620

Specifications

AMS. 4975, 4976

Government. MIL-F-83142, MIL-T-9046, MIL-T-9047, MIL-T-81915

Chemical Composition

Composition limits. 5.5 to 6.5 Al; 1.8 to 2.2 Sn; 3.6 to 4.4 Zr; 1.8 to 2.2 Mo; 0.25 max Fe; 0.05 max C; 0.05 max N; 0.12 max O; 0.0125 max H(a); 0.10 max others (each); 0.30 max others (total); rem Ti

(a) For bar and billet, 0.0100 max H; for forgings and sheet, 0.0150 max H.

Applications

Typical uses. Forgings and flat-rolled products used in gas turbine engine and air-frame applications where high strength and toughness, excellent creep resistance, and stress stability at temperatures up to 595 °C (1100 °F) are required

Precautions in use. Similar to Ti-5Al-2.5Sn

Mechanical Properties

Tensile properties. See Table 14.

Table 14 Typical mechanical properties of Ti-6Al-2Sn-4Zr-2Mo sheet

Temperature °C	°F	Tensile strength MPa	ksi	0.2% yield strength MPa	ksi	Elongation, %	0.2% compressive yield strength MPa	ksi	Shear strength MPa	ksi	Elastic modulus Tension GPa	10^6 psi	Compression GPa	10^6 psi
Longitudinal														
Room temperature	...	1010	146	990	144	3	1075	156.3	690	100	120	16.9	130	18.6
204	400 ...	890	129	760	110	10.8	800	116	...	...	110	15.9	120	17.5
371	700 ...	830	120	650	94	11.5	695	101	...	...	100	14.4	105	15.4
540	1000 ...	710	103	570	83	17.8	635	92	...	...	85	12.2	100	14.1
Transverse														
Room temperature	...	1020	148	1010	146	2.7	1165	169	695	101	120	17.9	140	19.9
204	400 ...	890	129	770	112	10.6	860	125	...	...	115	16.7	125	18.3
371	700 ...	830	120	670	97	21.6	750	109	...	...	105	15	115	16.5
540	1000 ...	720	104	590	86	16.5	695	101	...	...	95	13.5	105	15.2

Compressive properties. See Table 14.
Elastic modulus. See Table 14.
Fatigue strength. See Fig. 14.
Plane-strain fracture toughness. Average for sheet: 148 MPa$\sqrt{m}$ (135 ksi$\sqrt{in.}$)
Creep-rupture characteristics. See Fig. 15.

Mass Characteristics

Density. 4.54 Mg/m^3 (0.164 lb/in.^{3})

Thermal Properties

Coefficient of thermal expansion. Linear, 7.7 μm/m·K (4.3 μin./in.·°F) at 205 °C (400 °F); 8.1 μm/m·K (4.5 μin./in.·°F) at 315 to 540 °C (600 to 1000 °F)
Specific heat. 460 J/kg·K (0.110 Btu/lb·°F) at 100 °C (212 °F)
Thermal conductivity. 7.1 W/m·K (4.1 Btu/ft·h·°F) at 100 °C (212 °F)

Electrical Properties

Electrical resistivity. 1.9 μΩ·m at room temperature

Fabrication Characteristics

Weldability. 100% joint efficiencies in annealed material and limited weld zone ductility; relatively high fracture toughness in the fusion and heat affected zones. Fusion welding by either TIG or MIG process can be performed on any of the heat treated conditions of Ti-6242. Weldment properties inferior to those of Ti-6Al-4V

Ti-6Al-2Nb-1Ta-0.8Mo

Commercial Names

Common name. Ti-621/0.8
Trade name. RMI 6Al-2Cb-1Ta-0.8Mo
UNS number. R56210

Specifications

AWS. A5.16
Government. MIL-T-9046, MIL-T-9047, MIL-T-81588

Chemical Composition

Composition limits. 5.5 to 6.2 Al; 1.5 to 2.5 Nb; 0.5 to 1.5 Ta; 0.5 to 1.0 Mo; 0.12 max Fe; 0.03 max C; 0.02 max N; 0.10 max O; 0.125 max H; 0.10 max others (each); 0.40 max others (total); rem Ti

Applications

Typical uses. Plate for naval shipbuilding applications, submersible hulls, pressure vessels, and other high toughness applications
Precautions in use. Similar to Ti-5Al-2.5Sn

Mechanical Properties

Tensile properties. Minimum for plate, as-rolled from above beta transus temperature. Tensile strength, 830 MPa (120 ksi); 0.2% yield strength, 760 MPa (110 ksi); elongation, 10%; reduction in area, 20%
Compressive yield strength. 815 MPa (118 ksi)
Hardness. 30 HRC
Poisson's ratio. 0.31
Elastic modulus. Tension, 117 GPa (17.0 × 10^6 psi); compression, 126 GPa (18.3 × 10^6 psi); tangent, 21 GPa (3.10 × 10^6 psi); secant, 86 GPa (12.5 × 10^6)
Impact strength. Charpy V-notch: 37 J (27 ft·lb) at 0 °C (32 °F)

Mass Characteristics

Density. 4.48 Mg/m^3 (0.162 lb/in.^{3})

Thermal Properties

Incipient melting temperature. 1650 °C (3000 °F)
Phase-transformation temperature. 1110 ± 14 °C (1860 ± 25 °F) beta transus
Coefficient of thermal expansion. Linear, 9 μm/m·K (5 μin./in.·°F) at 22 to 650 °C (72 to 1200 °F)
Thermal conductivity. 6.4 W/m·K (3.7 Btu/ft·h·°F) at 20 °C (70 °F)

Magnetic Properties

Alloy is not magnetic

Fabrication Characteristics

Formability. Recommended forging temperatures: blocking, 1110 to 1065 °C (1850 to 1950 °F); finishing, 980 to 1110 °C (1800 to 1850 °F)
Weldability. Can be welded by GTAW or GMAW processes
Annealing temperature. 900 °C (1650 °F); 1 h at temperature; air cool
Solution temperature. 1110 °C (1850 °F); 1 h at temperature, water quench
Aging temperature. 590 °C (1100 °F); 2 h at temperature; air cool

Ti-2.25Al-11Sn-5Zr-1Mo

Commercial Names

Trade name. Ti-679
UNS number. R54790

Specifications

AMS. 4974
Government. MIL-F-83142, MIL-T-9047

Chemical Composition

Composition limits. 2.00 to 2.50 Al; 10.50 to 11.50 Sn; 4.00 to 6.00 Zr; 0.80 to 1.20 Mo; 0.15 to 0.27 Si; 0.12 max Fe; 0.15 max O; 0.040 max C; 0.040 max N; 0.008 max H; rem Ti

Applications

Typical uses. Jet engine blades and wheels, large bulkhead forgings, other applications requiring high temperature creep strength plus stability and short time strength
Precautions in use. Similar to Ti-5Al-2.5Sn

Mechanical Properties

Tensile properties. See Table 15 and Fig. 16.
Elastic modulus. Dynamic, tensile: 108 GPa (15.7 × 10^6 psi) at 20 °C (68 °F); 94.4 GPa (13.7 × 10^6 psi) at 300 °C (580 °F)

Mass Characteristics

Density. 4.816 Mg/m^3 (0.174 $lb/in.^3$)

Thermal Properties

Coefficient of thermal expansion. Bar: linear, 10.3 μm/m·K (5.7 μin./in.·°F) at 20 to 300 °C (68 to 570 °F)
Thermal conductivity. Bar: 6.9 W/m·K (4.0 Btu/ft·h·°F) at 20 °C (68 °F); 10.5 W/m·K (6.1 Btu/ft·h·°F) at 300 °C (570 °F)

Electrical Properties

Electrical resistivity. Bar: 1.62 μΩ·m at 20 °C (68 °F); 1.81 μΩ·m at 300 °C (570 °F)

Chemical Properties

Resistance to specific corroding agents. Superior to Ti-6Al-4V, Ti-8Al-1Mo-1V and Ti-5Al-5Sn-5Zr in resistance to hot-salt stress-corrosion cracking at 290 to 480 °C (550 to 900 °F). Critical stress: about 525 MPa (76 ksi) at 290 °C and 275 MPa (40 ksi) and 480 °C for cracking in fused NaCl

Fig. 16 Typical short-time tensile properties of Ti-2.25Al-11Sn-5Zr-1Mo forgings

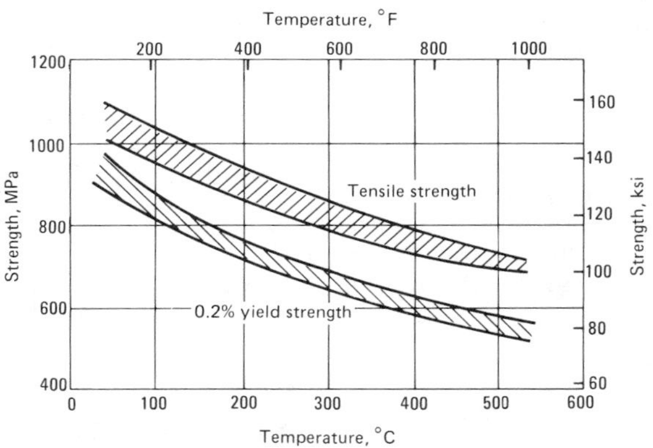

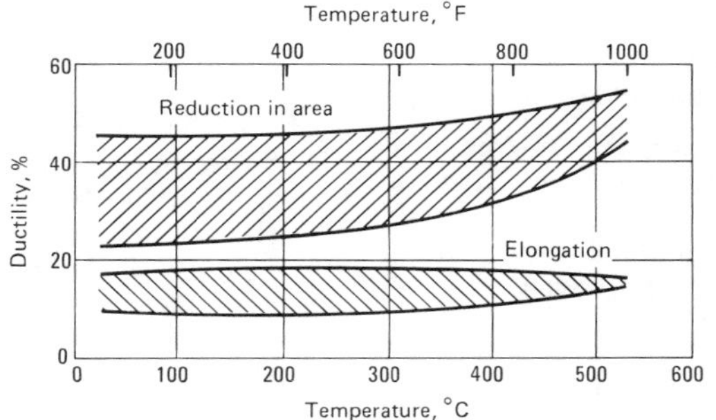

Specimens were taken from forged compressor wheels with a maximum thickness of about 50 mm (2 in.) that were solution treated 1 h at 900 °C (1650 °F), air cooled, aged 24 h at 500 °C (930 °F) and air cooled.

Table 15 Variation of tensile properties with heat treatment for forged Ti-2.25Al-11Sn-5Zr-1Mo

Condition(a)	Tensile strength MPa	Tensile strength ksi	Yield strength MPa	Yield strength ksi	Elongation, %	Reduction in area, %
As forged	1040	151	940	136	14	50
Furnace cooled	990	144	920	133	11	26
FC and aged(b)	990	144	910	132	11	23
Air cooled	1100	160	985	143	16	47
AC and aged(b)	1155	167	1030	150	17.5	44
Water quenched	1220	177	1045	151	12	43
WQ and aged(b)	1375	200	1240	180	10	33

(a) All samples (except as forged) were solution treated at 900 °C (1650 °F). (b) Aged 24 h at 500 °C (930 °F).

Ti-5Al-5Sn-2Zr-2Mo-0.25Si

Commercial Names

Common name. Ti-5522S
UNS number. R54560

Specifications

Government. MIL-T-9046, MIL-T-9047

Chemical Composition

Composition limits. 4.5 to 5.5 Al; 4.5 to 5.5 Sn; 1.75 to 2.25 Zr; 1.75 to 2.25 Mo; 0.20 to 0.30 Si; 0.15 max Fe; 0.04 max C; 0.03 max N; 0.13 max O; rem Ti

Applications

Typical uses. Forged billet and bar, special products available in plate and sheet. Products supplied as forged, hot rolled and annealed, solution annealed, or solution annealed and stabilized.

Mechanical Properties

Tensile properties. Room temperature, minimum values: tensile strength, 900 MPa (130 ksi); yield strength, 830 MPa (120 ksi); elongation, 10%; reduction in area, 25%. At 535 °C (1000 °F), minimum values: tensile strength, 689 MPa (100 ksi); yield strength, 517 MPa (75 ksi); elongation, 15%; reduction in area, 35%
Hardness. 32 to 38 HRC
Elastic modulus. Tension, 110 to 117 GPa (16 to 17 × 10^6 psi)
Plane-strain fracture toughness. 82.4 MPa$\sqrt{m}$ (75 ksi$\sqrt{in.}$) for forging with yield strength of 890 MPa (129 ksi)
Creep-rupture characteristics. Min rupture stress: 1170 MPa (170 ksi) for rupture in 5 h at room temperature, notched specimen. Min creep strength: 345 MPa (50 ksi) for 0.1% creep in 100 h at 510 °C (950 °F)

Mass Characteristics

Density. 4.52 Mg/m^3 (0.163 lb/in.^{3})

Thermal Properties

Phase-transformation temperature. 980 °C (1800 °F)
Coefficient of thermal expansion. Linear, 10.2 μm/m·K (5.71 μin./in.·°F) at 20 to 815 °C (68 to 1500 °F)

Chemical Properties

General corrosion behavior. Similar to unalloyed titanium and to other near-alpha and alpha-beta titanium alloys

Fabrication Characteristics

Annealing temperature. 650 to 790 °C (1200 to 1450 °F); ½ to 2 h at temperature; vacuum cool
Stabilization anneal. 590 °C (1100 °F); 2 to 8 h
Solution temperature. 960 to 970 °C (1765 to 1785 °F); ¼ to 4 h at temperature; air cool

Ti-6Al-4V
Ti-6Al-4V-ELI

Commercial Names

Trade names. C-120AV; HA-6510
UNS number. Ti-6Al-4V, R56400; Ti-6Al-4V-ELI, R56401

Specifications

AMS. Standard grade: 4906, 4911, 4928, 4934, 4935, 4954, 4965, 4967. ELI grade: 4907, 4930, 4956
ASTM. Bar and billet: B348. Castings: B367. Forgings: B381. Sheet, strip and plate: B265
AWS. A5.16
Government. MIL-T-9046, MIL-T-9047, MIL-T-8884, MIL-T-14557, MIL-T-14558, MIL-T-81556, MIL-T-81588, MIL-F-83142

Chemical Composition

Composition limits. Ti-6Al-4V: 5.5 to 6.75 Al; 3.5 to 4.5 V; 0.05 max N; 0.10 max C; 0.015 max H; 0.40 max Fe; 0.20 max O; 0.1 max others (each); 0.5 max others (total); rem Ti

Applications

Typical uses. Ti-6Al-4V is the most widely used titanium alloy. It is specially processed to both standard and ELI compositions to provide mill-annealed or beta-annealed structures, and is sometimes solution treated and aged. It is used for aircraft gas turbine disks and blades; is extensively used, in all mill product forms, for airframe structural components and other applications requiring strength at temperatures up to 315 °C (600 °F); and is also used for high-strength prosthetic implants and chemical-processing equipment.
Precautions in use. Rolled sheet may exhibit strong directionality of physical and mechanical properties. In other ways, similar to Ti-5Al-2.5Sn.

Mechanical Properties

Tensile properties. Annealed bar: tensile strength, 895 MPa (130 ksi); yield strength, 825 MPa (120 ksi); elongation, 10%; reduction in area, 20%. Solution treated aged bar and forgings, 2.54 to 5.08 cm (1 to 2 in.) thick: tensile strength, 1035 MPa (150 ksi); yield strength, 965 MPa (140 ksi); elongation, 8%; reduction in area, 20%. See also Fig. 17.
Compressive yield strength. Annealed bar: 860 MPa (125 ksi). Solution treated and aged bar: 1070 MPa (155 ksi)
Hardness. 36 to 39 HRC
Poisson's ratio. 0.33
Elastic modulus. Tension, 110 GPa (16 × 10^6 psi); compression, 117 GPa (17 × 10^6 psi)

Impact strength:

Temperature °C	Temperature °F	Fracture energy J	Fracture energy ft·lb
−185	−300	20.3	15
−70	−94	24.4	18
−20	−4	25.7	19
90	194	29.8	22
205	400	54.2	40
315	600	74.5	55

Fatigue strength. See Fig. 18.
Plane-strain fracture toughness. At 25 °C (77 °F): annealed plate, 74.6 MPa$\sqrt{m}$ (68 ksi$\sqrt{in.}$); solution treated and aged, 42.9 MPa$\sqrt{m}$ (39 ksi$\sqrt{in.}$)
Creep-rupture characteristics. See Fig. 19.
Velocity of sound. 4.987 km/s at 25 °C (77 °F)

Structure

Crystal structure. cph, $a_0 = 2.925 \times 10^{-8}$ m at 25 °C (77 °F). $c = 4.644 \times 10^{-8}$ m at 25 °C (77 °F)
Slip planes. $\{10\bar{1}0\}$ $\langle 11\bar{2}0\rangle$, $\{10\bar{1}1\}$ $\langle 11\bar{2}0\rangle$, $\{0001\}$, $\langle 11\bar{2}0\rangle$ at 25 °C (77 °F)
Twinning planes. $\{10\bar{1}2\}$, $\{11\bar{2}1\}$, $\{11\bar{2}2\}$ at 25 °C (77 °F)
Microstructure. Predominantly cph alpha phase with some bcc beta phase at room temperature. Morphology and relative amounts of alpha and beta depend upon thermal

Fig. 17 Typical tensile properties of Ti-6Al-4V

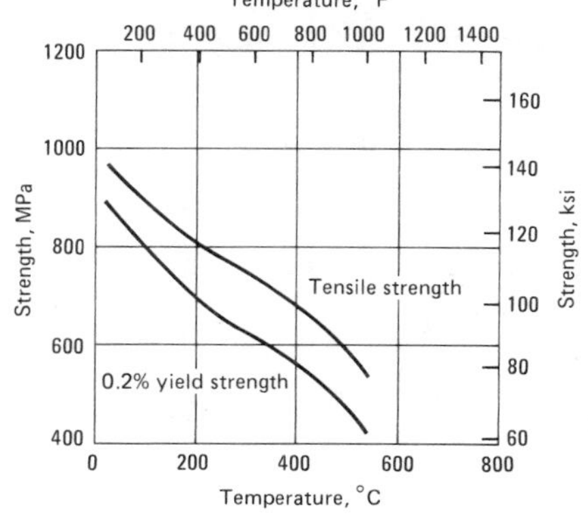

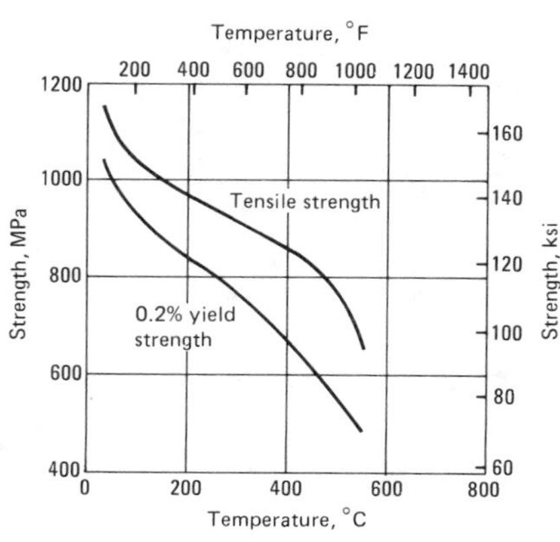

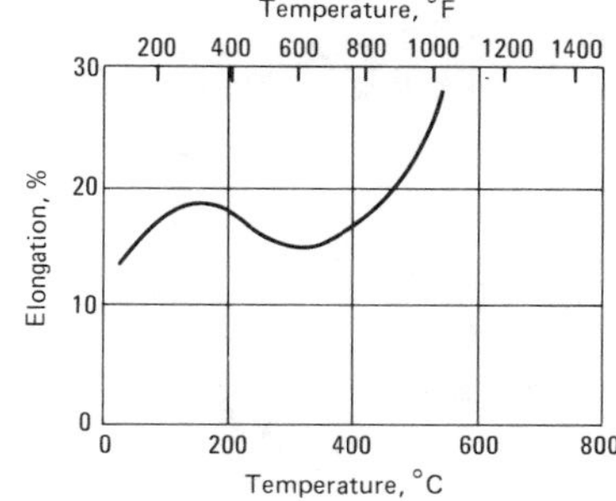

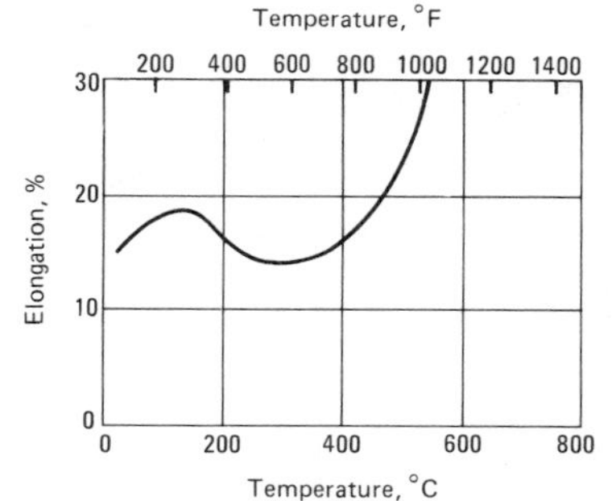

Left: Mill annealed condition. Right: STA condition—1 h at 955 °C (1750 °F), water quench, age 4 h at 525 °C (975 °F) and air cool.

Fig. 18 Typical rotating-beam fatigue curves for Ti-6Al-4V bar stock

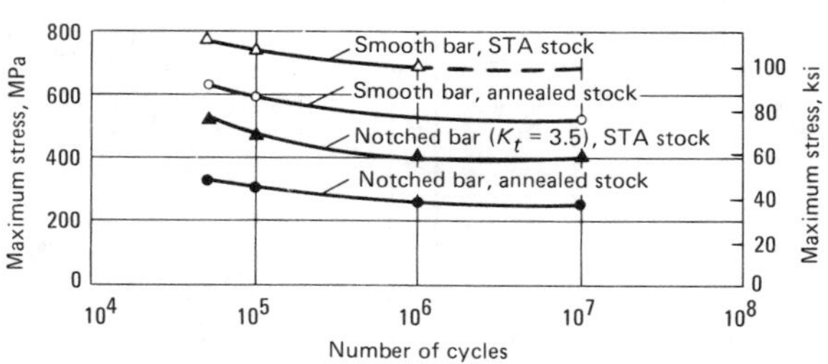

history. Martensitic structure can also be produced.

Mass Characteristics

Density. 4.43 Mg/m³ (0.160 lb/in.³) at 25 °C (77 °F)

Thermal Properties

Liquidus temperature. 3020 ± 25 °F (1660 ± 14 °C)
Solidus temperature. 2920 ± 20 °F (1604 ± 11 °C)
Phase-transformation temperature. Beta transus, 955 to 1010 °C (1750 to 1850 °F)
Coefficient of thermal expansion. See Fig. 20.
Specific heat:

Temperature °C	Temperature °F	Specific heat J/kg·K	Specific heat Btu/lb·°F
20	68	580	0.14
205	400	610	0.15
425	800	670	0.16
650	1200	760	0.18
870	1600	930	0.22

Thermal conductivity:

Temperature °C	Temperature °F	Conductivity W/m·K	Conductivity Btu/ft·h·°F
Mill annealed			
20	68	6.6	3.8
93	200	7.3	4.2
205	400	9.1	5.3
315	600	10.6	6.1
425	800	12.6	7.3
540	1000	14.6	8.4
650	1200	17.5	10.1
Solution treated and aged			
20	68	6.8	3.9
93	200	7.5	4.3
205	400	8.5	4.9
315	600	9.6	5.5
425	800	10.9	6.3
540	1000	12.6	7.3
650	1200	14.1	8.1

Electrical Properties

Electrical resistivity. 1.71 μΩ·m at room temperature

Magnetic Properties

Magnetic permeability. 1.00005 at 1.6 kA/m

Chemical Properties

General corrosion behavior. Same as commercially pure titanium
Resistance to specific corroding agents. Not attacked by 10% boiling

Fig. 19 Creep-rupture data for Ti-6Al-4V bar

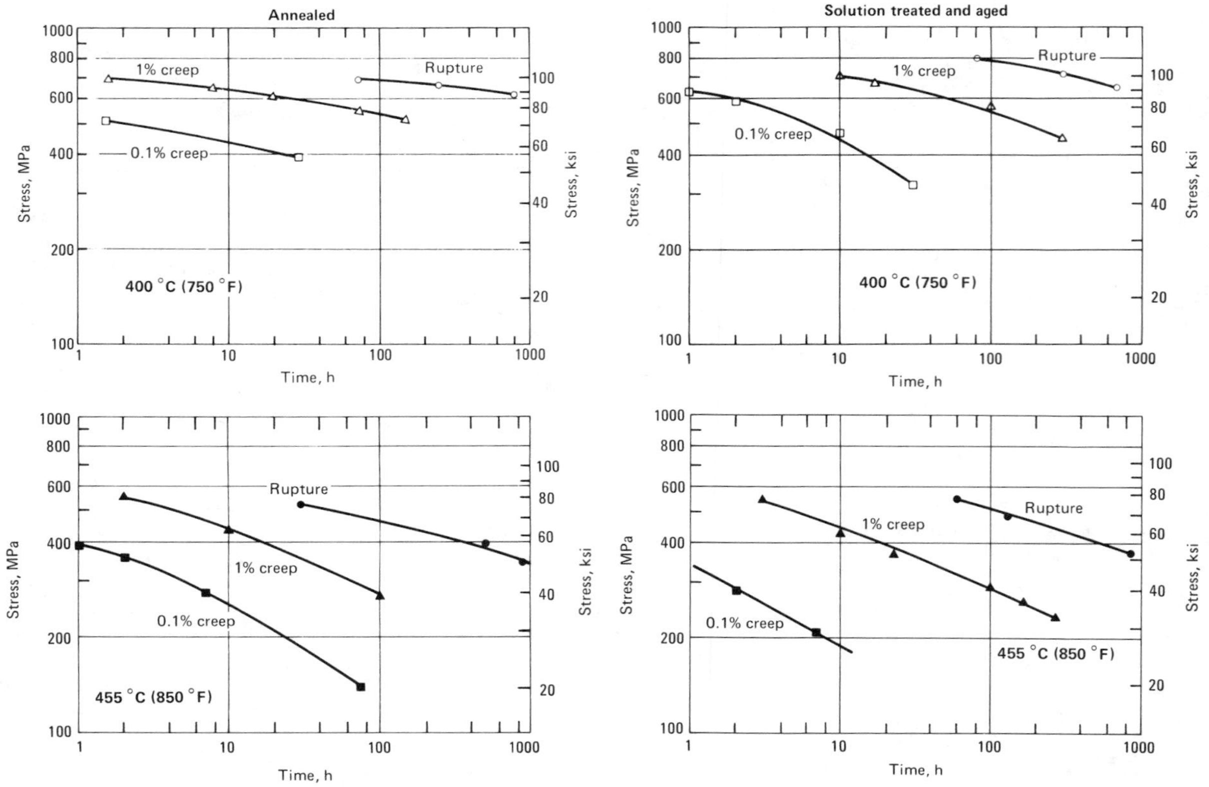

Fig. 20 Mean coefficient of thermal expansion for Ti-6Al-4V

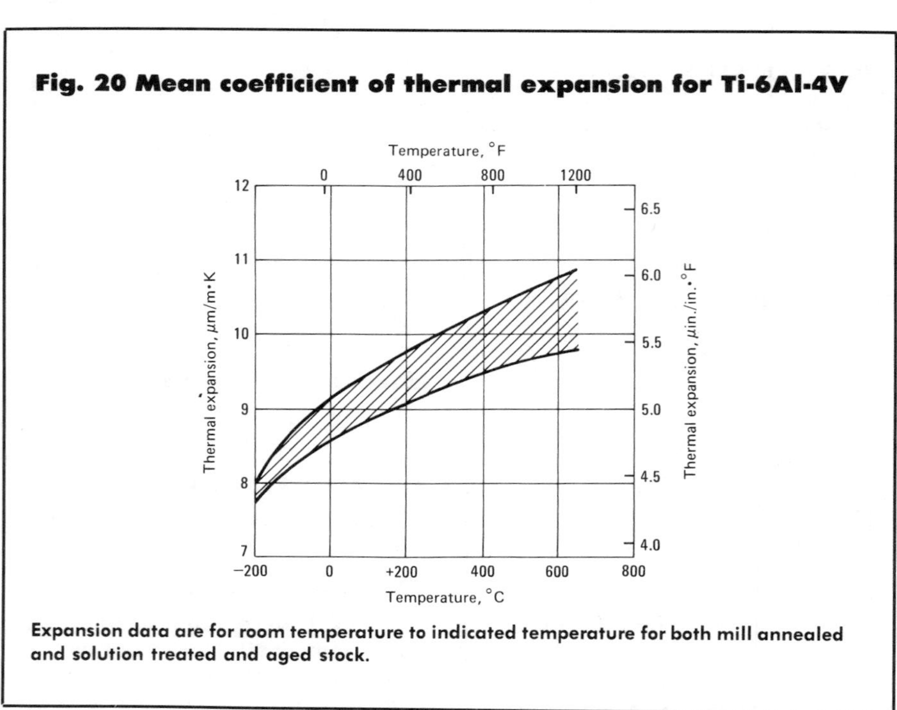

Expansion data are for room temperature to indicated temperature for both mill annealed and solution treated and aged stock.

Fig. 21 Threshold stress for solid salt stress-corrosion cracking in Ti-6Al-4V

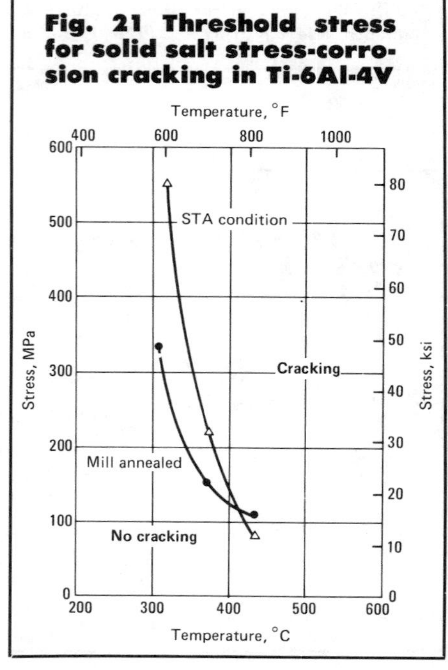

HCl; 1.8 mm/yr (70 mils/yr) corrosion rate in boiling 15% H_2SO_4. See also Fig. 21.

Fabrication Characteristics

Machinability. Same as commercially pure titanium
Annealing temperature. 705 to 785 °C (1300 to 1450 °F)
Solution temperature. 900 to 955 °C (1650 to 1750 °F)
Aging temperature. 540 °C (1000 °F)
Hot working temperature. 870 to 1010 °C (1650 to 1850 °F)

Ti-6Al-6V-2Sn

Commercial Names

Common name. Ti-6-6-2
UNS number. R56620

Specifications

AMS. 4918, 4936, 4971, 4978, 4979
Government. MIL-F-83142, MIL-T-9046, MIL-T-9047, MIL-T-81556

Applications

Typical uses. Applications requiring high strength at temperatures up to 315 °C (600 °F). In the forms of sheet, light-gage plate, extrusions and small forgings, this alloy is used for airframe structures where strength higher than that of Ti-6Al-4V is required. Usage is generally limited to secondary structures, because attractiveness of higher strength efficiency is minimized by lower fracture toughness and fatigue properties.
Precautions in use. Susceptible to tensile embrittlement following long exposures above 315 °C (600 °F). In other ways, similar to Ti-5Al-2.5Sn.

Mechanical Properties

Tensile properties. At room temperature, solution treated and aged: tensile strength, 1100 MPa (160 ksi); yield strength, 1035 MPa (150 ksi); elongation, 12%; reduction in area, 18%. See also Fig. 22.
Elastic modulus. See Fig. 22.
Fatigue strength. See Fig. 23.
Plane-strain fracture toughness. Solution treated and aged: 60 MPa$\sqrt{m}$ (55 ksi$\sqrt{in.}$) at 22 °C (71 °F); 99 MPa$\sqrt{m}$ (90 ksi$\sqrt{in.}$) at 93 °C (200 °F); 143 MPa$\sqrt{m}$ (130 ksi$\sqrt{in.}$) at 205 °C (400 °F)

Fig. 22 Typical tensile properties and tensile modulus of Ti-6Al-6V-2Sn, STA condition

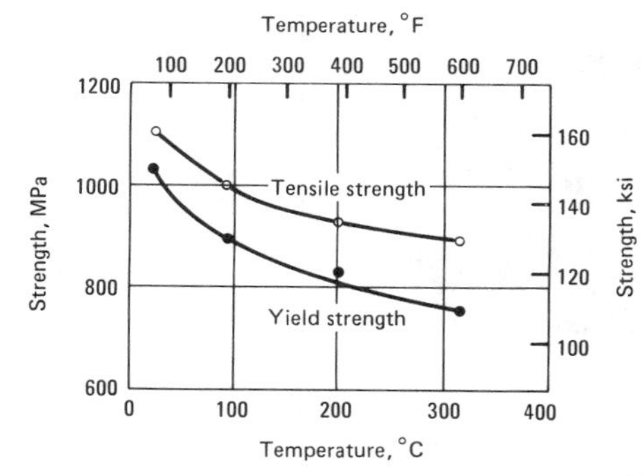

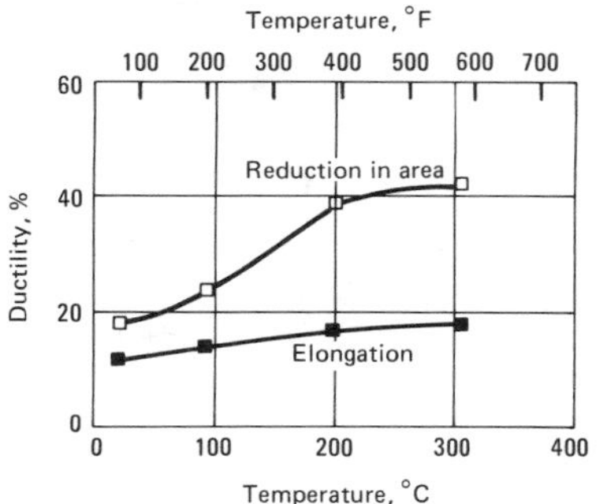

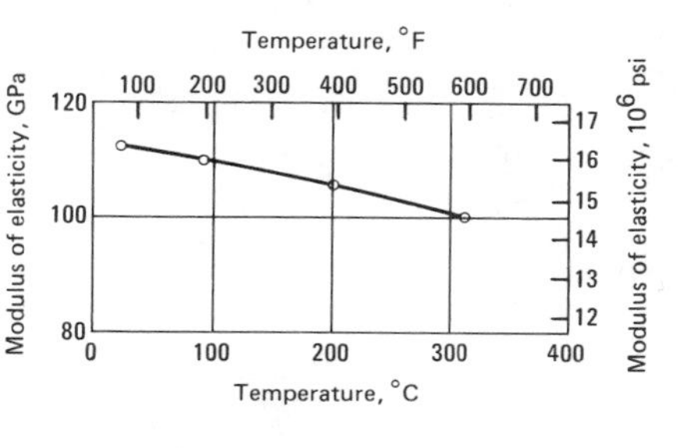

STA condition: 1 h at 870 °C (1600 °F), water quench, age 4 h at 565 °C (1050 °F).

Fig. 23 Typical fatigue behavior of Ti-6Al-6V-2Sn

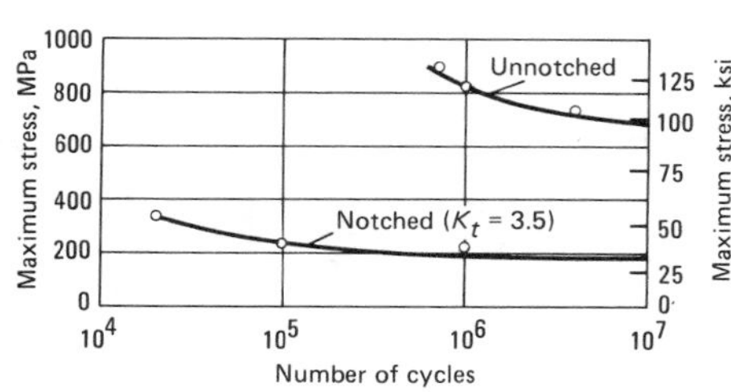

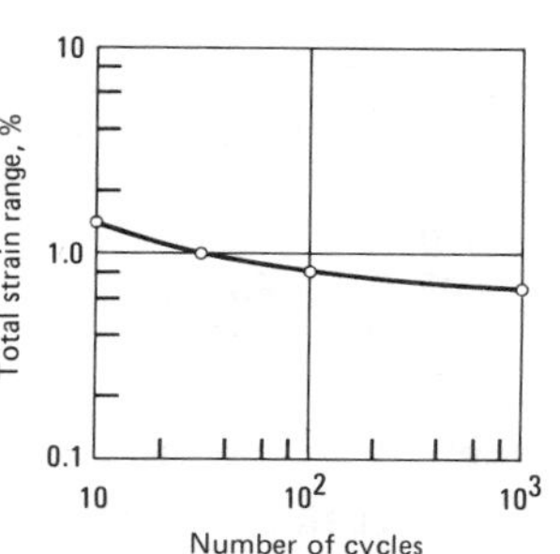

Left: low-cycle axial fatigue ($R = 0$) for stock annealed 2 h at 705 °C (1300 °F) and furnace cooled. Right: high-cycle axial fatigue ($R = 0.1$) for STA condition.

Creep-rupture characteristics. At 315 °C (600 °F). For 0.5% total creep strain: in 100 h, 800 MPa (116 ksi); in 500 h, 770 MPa (110 ksi). For 1.0% total creep strain: in 100 h, 820 MPa (120 ksi); in 500 h, 800 MPa (116 ksi)

Structure

Crystal structure. Alpha phase: cph; beta phase: bcc

Microstructure. Equiaxed primary alpha in a transformed beta matrix

Mass Characteristics

Density. 4.58 Mg/m^3 (0.165 lb/in.^{3}) at 21 °C (70 °F)

Thermal Properties

Liquidus temperature. 1649 °C (3000 °F)

Solidus temperature. 1627 °C (2940 °F)

Phase-transformation temperature. 946 °C (1735 °F)

Coefficient of thermal expansion. Linear, annealed:

Temperature °C	°F	Average coefficient µm/m·K	µin./in.·°F
93	200	9.0	5.0
205	400	9.2	5.1
315	600	9.3	5.2
425	800	9.5	5.3

Specific heat. Annealed:

Temperature °C	°F	Specific heat kJ/kg·K	Btu/lb·°F
93	200	670	0.160
150	300	674	0.161
205	400	682	0.163
260	500	687	0.164
315	600	691	0.165
370	700	699	0.167
425	800	703	0.168
480	900	712	0.170

Thermal conductivity. Annealed:

Temperature °C	°F	Conductivity W/m·K	Btu/ft·h·°F
93	200	6.6	3.8
205	400	8.12	4.7
315	600	9.86	5.7
425	800	11.9	6.9

Electrical Properties

Electrical resistivity. 1.57 µΩ·m at room temperature

Fabrication Characteristics

Annealing temperature. 730 °C (1350 °F)

Solution temperature. 845 to 875 °C (1550 to 1600 °F)

Aging temperature. 593 °C (1100 °F)

Hot working temperature. 875 to 925 °C (1600 to 1700 °F)

Fig. 24 Creep-rupture properties of annealed Ti-8Mn sheet

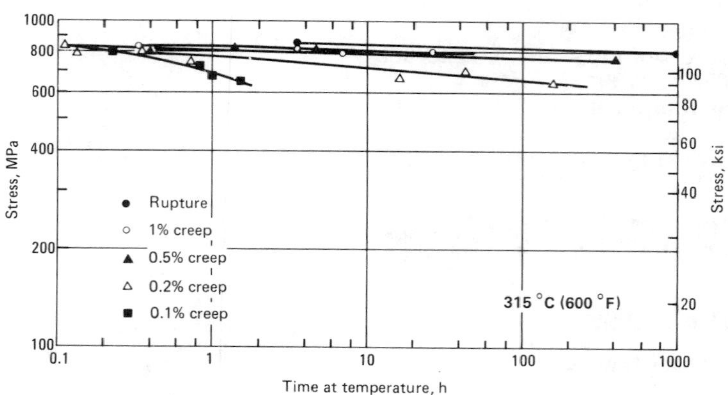

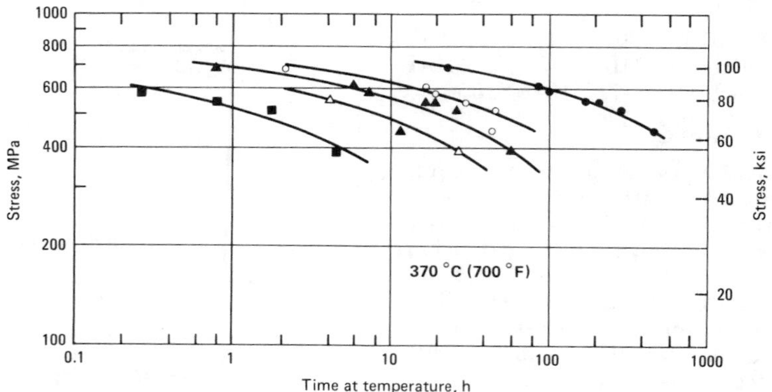

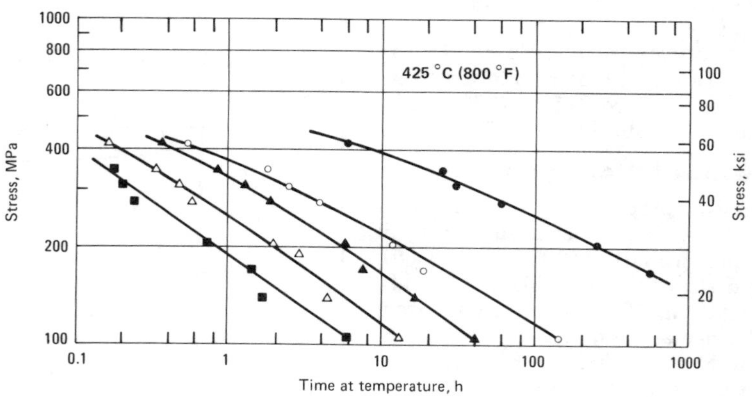

Ti-8Mn

Commercial Names

Trade name. C-110M
UNS number. R56080

Specifications

AMS. 4908
Government. MIL-T-90406

Chemical Composition

Composition limits. 6.5 to 9.0 Mn; 0.20 max C; 0.20 max O; 0.07 max N; 0.015 max H; rem Ti

Applications

Typical uses. Sheet, strip and plate in limited current usage for aircraft sheet components, structural parts
Precautions in use. Similar to Ti-5Al-2.5Sn; not weldable

Mechanical Properties

Tensile properties. Typical. Annealed sheet, at 24 °C (75 °F): tensile strength, 965 MPa (140 ksi); 0.2% yield strength, 860 MPa (125 ksi); elongation, 15%
Compressive yield strength. Annealed sheet at 24 °C (75 °F): 875 MPa (127 ksi)
Hardness. Annealed: 33 to 36 HRC
Elastic modulus. Tension, annealed sheet. Static: 114 GPa (16.5 × 10^6 psi). Dynamic: 110 GPa (15.9 × 10^6 psi) at 93 °C (200 °F)
Fatigue strength. Direct stress test (A = 0.6, R = 0.25), smooth specimen of annealed sheet: 585 MPa (85 ksi) at 10^7 cycles
Creep-rupture characteristics. See Fig. 24.

Structure

Microstructure. Alpha-beta

Mass Characteristics

Density. 4.70 Mg/m³ (0.170 lb/in.³) at 20 °C (68 °F)

Thermal Properties

Liquidus temperature. 1635 °C (2970 °F)
Solidus temperature. 1300 °C (2370 °F)
Phase-transformation temperature. 550 to 750 °C (1020 to 1380 °F). Beta to alpha-plus-beta on cooling
Coefficient of thermal expansion. Linear, 11 μm/m·K (6 μin./in.·°F) at 20 to 540 °C (68 to 1000 °F)
Specific heat. 495 J/kg·K (0.118 Btu/lb·°F) at room temperature
Thermal conductivity. 10.9 W/m·K (6.3 Btu/ft·h·°F)

Electrical Properties

Electrical resistivity. 0.92 μΩ·m at room temperature

Magnetic Properties

Magnetic permeability. 1.00005 at 1.6 kA/m

Chemical Properties

General corrosion behavior. Similar to unalloyed Ti
Effect of specific corroding agents. Resistance to strong mineral acids somewhat lower than unalloyed Ti

Fabrication Characteristics

Machinability. Similar to unalloyed Ti
Annealing temperature. 705 °C (1300 °F)
Aging temperature. 480 to 510 °C (900 to 950 °F); 1 to 8 h at temperature
Hot-working temperature. 260 to 315 °C (500 to 600 °F)

Ti-7Al-4Mo

Commercial Names

Trade names. C-135AMo; Ti(6 to 7) Al-(3 to 4)Mo; RMI 7Al-4Mo; HA-7146
UNS number. R56740

Specifications

AMS. 4970

Chemical Composition

Composition limits. 6.5 to 7.3 Al; 3.5 to 4.5 Mo; 0.10 max C; 0.013 max H; 0.30 max Fe; 0.05 max N; 0.20 max O; rem Ti

Applications

Typical uses. Bar, forgings and forging stock in limited current usage for jet engine disks, compressor blades and spacers, and for sonic horns
Precautions in use. Similar to Ti-5Al-2.5Sn

Mechanical Properties

Tensile properties. Solution treated and aged: tensile strength, 1170 MPa (170 ksi); 0.2% yield strength, 1035 MPa (150 ksi); elongation, 10%; reduction in area, 20%. See also Fig. 25.
Hardness. 32 to 38 HRC
Elastic modulus. See Fig. 25.
Impact strength:

Temperature °C	Temperature °F	Impact strength J	Impact strength ft·lb
24	75	17	13
93	200	19	14
200	390	27	20
325	615	39	29
540	1000	47	35

Fatigue strength. Rotating beam tests, R = −1: smooth specimen, 670 MPa (97 ksi) at 10^7 cycles; notched specimen (K_t = 3.9), 200 MPa (29 ksi) at 10^7 cycles
Creep-rupture characteristics. See Fig. 26.

Structure

Microstructure. Alpha-beta

Mass Characteristics

Density. 4.480 Mg/m³ (0.162 lb/in.³) at 20 °C (68 °F)

Thermal Properties

Liquidus temperature. 1650 °C (3000 °F)
Phase-transformation temperature. 900 to 1018 °C (1815 to 1865 °F) for beta to alpha-plus-beta on cooling
Coefficient of thermal expansion. Linear: 9.7 μm/m·K (5.4 μin./in.·°F) at 20 to 450 °C (68 to 850 °F)
Specific heat. 515 J/kg·K (0.123 Btu/lb·°F) at 20 °C (68 °F)
Thermal conductivity:

Temperature °C	Temperature °F	Conductivity W/m·K	Conductivity Btu/ft·h·°F
20	68	6.1	3.5
315	600	10.4	6.0
427	800	12.1	7.0
540	1000	13.8	8.0

Electrical Properties

Electrical resistivity. 1.7 μΩ·m at 24 °C (75 °F)

Chemical Properties

General corrosion behavior. Similar to 99.2% Ti (see grade 1)
Effects of specific corroding agents. Resistance to strong mineral acids somewhat less than 99.2% Ti

Fig. 25 Tensile properties and elastic modulus of solution treated and aged Ti-7Al-4Mo

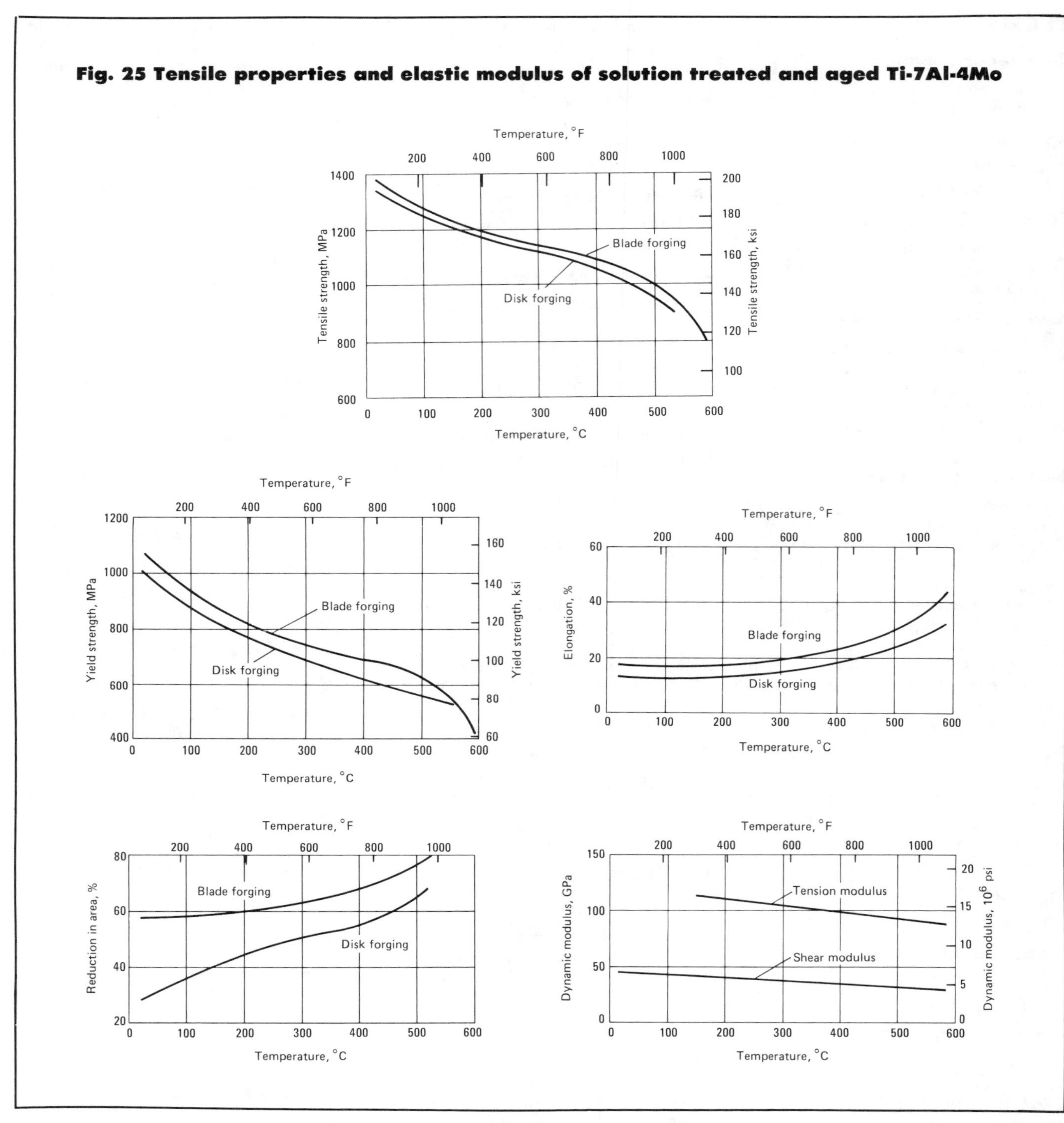

Fabrication Characteristics

Machinability. Similar to unalloyed Ti

Annealing temperature. 790 °C (1450 °F)

Solution temperature. 930 to 980 °C (1700 to 1800 °F); ½ to 2 h at temperature; water quench

Aging temperature. 540 to 650 °C (1000 to 1200 °F), 2 to 8 h at temperature, air cool

Hot-working temperature. 900 to 1000 °C (1650 to 1830 °F)

Fig. 26 Master creep and stress-rupture curves for Ti-7Al-4Mo

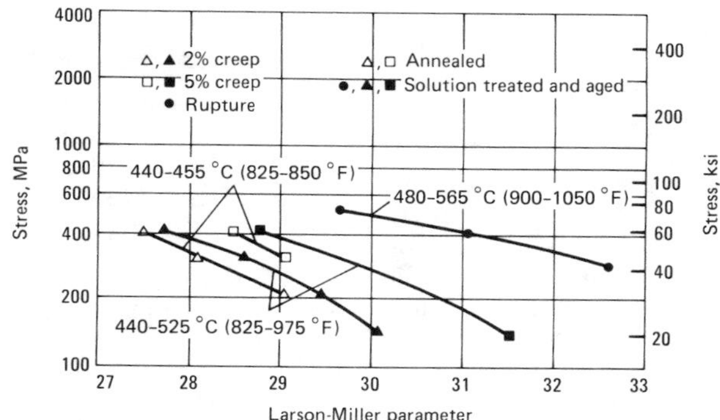

Specimens were taken from compressor-wheel forgings forged below the beta transus. Open symbols refer to material annealed 1 h at 790 °C (1450 °F), then furnace cooled to 565 °C (1050 °F) and air cooled to room temperature. Solid symbols refer to material annealed as above, then solution treated ½ h at 850 °C (1560 °F), air cooled, aged 24 h at 550 °C (1020 °F) and air cooled.

Fig. 27 Low-cycle axial fatigue curves for Ti-6Al-2Sn-4Zr-6Mo

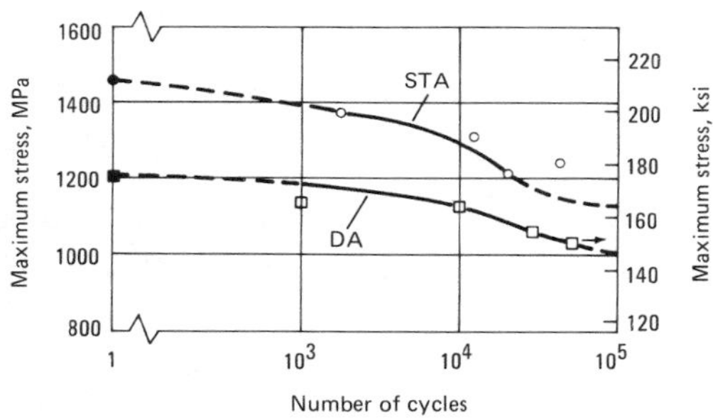

STA (solution treated and aged) condition: 1 h at 870 °C (1600 °F), water quench, age 8 h at 595 °C (1100 °F) and air cool. DA (duplex annealed) condition: 15 min at 870 °C, air cool, then 8 h at 540 °C (1000 °F) and air cool. All fatigue tests conducted at a stress ratio of R = 0.1. Open symbols indicate fatigue tests; solid symbols, tension tests.

Ti-6Al-2Sn-4Zr-6Mo

Commercial Names

Trade name. Ti-6246
Common name. Ti-6-2-4-6
UNS number. R56260

Chemical Composition

Composition limits. 5.5 to 6.5 Al; 3.6 to 4.4 Zr; 1.8 to 2.2 Sn; 5.5 to 6.5 Mo; 0.04 max C; 0.04 max N; 0.12 max O; 0.15 max Fe; rem Ti

Applications

Typical uses. Forgings in intermediate temperature range sections of gas turbine engines, particularly in disk and fan blade components of compressors. Available as billet and bar for forging stock, sheet and plate. Should be considered for long-time load-carrying applications at temperatures up to 400 °C (750 °F) and short-time load-carrying applications at temperatures up to 540 °C (1000 °F)

Precautions in use. Similar to Ti-5Al-2.5Sn

Mechanical Properties

Tensile properties. Typical, at room temperature: tensile strength, 965 MPa (140 ksi); yield strength, 1034 MPa (150 ksi); elongation, 10% in 50 mm or 2 in.; reduction in area, 20%

Compressive yield strength. 1270 MPa (184 ksi)

Hardness. 36 to 42 HRC; 350 to 450 HB

Elastic modulus. Triplex annealed. Tension: 121 GPa (17.5 × 10^6 psi) at 20 °C (68 °F); 114 GPa (16.6 × 10^6 psi) at 93 °C (200 °F); 97 GPa (14.0 × 10^6 psi) at 315 °C (600 °F); 76 GPa (11.0 × 10^6 psi) at 540 °C (1000 °F). Compression: 119 GPa (17.2 × 10^6 psi) at 20 °C (68 °F); 117 GPa (17.0 × 10^6 psi) at 93 °C (200 °F); 106 GPa (15.4 × 10^6 psi) at 315 °C (600 °F); 86 GPa (12.5 × 10^6 psi) at 540 °C (1000 °F)

Impact strength. Charpy V-notch. Forged and heat treated: 8.1 to 15 J (6 to 11 ft·lb)

Fatigue strength. See Fig. 27.

Plane-strain fracture toughness. 21 to 28 MPa$\sqrt{m}$ (19 to 26 ksi$\sqrt{in.}$)

Creep-rupture characteristics. See Fig. 28.

Structure

Microstructure. Two phase, alpha-beta

Mass Characteristics

Density. 4.65 Mg/m³ (0.168 lb/in.³) at 20 °C (68 °F)

Thermal Properties

Melting range. 1595 to 1675 °C (2900 to 3050 °F)

Phase-transformation temperature. 932 ± 8 °C (1710 ± 15 °F)

Coefficient of thermal expansion. Linear, STA condition:

Temperature °C	°F	Average coefficient µm/m·K	µin./in.·°F
20	68	8.6	4.8
93	200	9.4	5.2
205	400	9.9	5.5
315	600	10.3	5.7
425	800	10.4	5.8
540	1000	10.4	5.8

Thermal conductivity. STA condition:

Temperature °C	°F	Conductivity W/m·K	Btu/ft·h·°F
20	68	7.7	4.4
93	200	7.9	4.6
205	400	9.3	5.4
315	600	10.7	6.2
425	800	12.1	7.0
540	1000	13.5	7.8

Chemical Properties

General corrosion behavior. Oxidation characteristics in air. Good resistance at 205, 315 and 425 °C (400, 600 and 800 °F), both long and short exposures. Low resistance at 538 °C (1000 °F), both long and short exposures. Resistance to crack propagation in salt water reported better after beta forging than after alpha-beta forging

Fabrication Characteristics

Machinability. Rigid set-up, slow speed, heavy feed, sharp tools, adequate coolant. Readily cut with saw or abrasive wheel.

Formability. Hot forging recommended for moderate to complex structures

Weldability. Electron beam weld for sheet or plate; inertia weld for forgings

Annealing temperature. Mill anneal: 705 to 730 °C (1300 to 1350 °F), 2 h at temperature; air cool. Single anneal: 870 °C (1600 °F); 1/4 h at temperature; air cool. Duplex anneal: same as single anneal plus: 705 °C (1300 °F), 1/4 h at temperature, air cool. Triplex anneal same as duplex anneal plus: 590 °C (1100 °F); 2 h at temperature; air cool

Solution temperature. 815 to 926 °C (1500 to 1700 °F)

Aging temperature. 540 to 730 °C (1000 to 1350 °F)

Hot-working temperature. 840 to 930 °C (1550 to 1700 °F)

Stress relieving temperature. 480 to 650 °C (900 to 1200 °F), 1 to 4 h at temperature

Fig. 28 Typical creep curves for Ti-6Al-2Sn-4Zr-6Mo

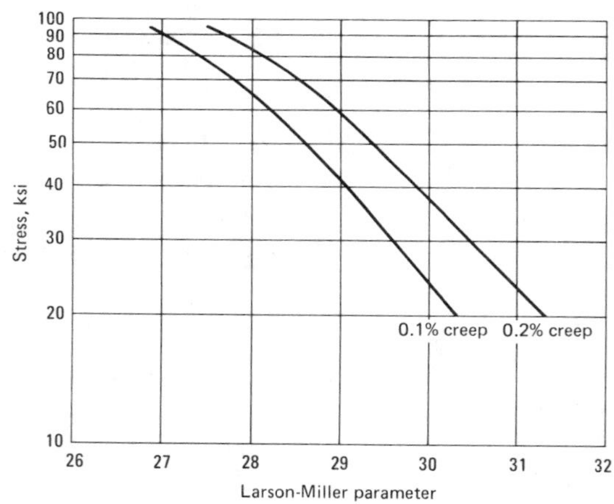

Larson-Miller parameter equals $10^{-3}T(20 + \log t)$, where T is temperature in °R and t is time in hours.

Ti-5Al-2Sn-2Zr-4Mo-4Cr

Commercial Names

Common name. Ti-17

Chemical Composition

Composition limits. 4.5 to 5.5 Al; 1.5 to 2.5 Sn; 1.5 to 2.5 Zr; 3.5 to 4.5 Mo; 3.5 to 4.5 Cr; 0.08 to 0.13 O; 0.04 max N; 0.0125 max H; 0.30 max Fe; rem Ti

Applications

Typical uses. Heavy section forgings up to 6 in. thick used for gas turbine engine components and other applications requiring high tensile strength and good fracture toughness

Precautions in use. Similar to Ti-5Al-2.5Sn

Mechanical Properties

Tensile properties. Forging at 24 °C (75 °F): tensile strength, 1205 MPa (175 ksi); 0.2% yield strength, 1125 MPa (163 ksi); elongation, 10%; reduction in area, 25%

Elastic modulus. Tension, 116 GPa (16.8×10^6 psi) at 25 °C (77 °F)

Fatigue strength. Forgings, axial fatigue ($R = 0$), smooth specimen: at room temperature and 315 °C (600 °F), 450 MPa (65 ksi) at 10^5 cycles

Creep-rupture characteristics. See Fig. 29.

Structure

Microstructure. Mixture of alpha and beta according to thermal history

Mass Characteristics

Density. 4.65 Mg/m³ (0.168 lb/in.³) at 24 °C (75 °F)

Thermal Properties

Phase-transformation temperature. 870 °C (1600 °F) beta transus

Coefficient of thermal expansion. Linear:

Temperature °C	°F	Average coefficient µm/m·K	µin./in.·°F
100	212	9.0	5.0
200	390	9.2	5.1
300	570	9.4	5.2
400	750	9.5	5.3

Fabrication Characteristics

Solution temperature. Alpha-beta forgings: 815 to 855 °C (1500 to 1575 °F) for 4 h plus 775 to 815 °C (1425 to 1500 °F). Beta forgings: latter step only

Aging temperature. 621 °C (1150 °F)

Hot-working temperature. Alpha-beta forgings: 815 to 855 °C (1500 to 1575 °F); beta forgings: 900 to 925 °C (1650 to 1700 °F)

Fig. 29 Master creep-rupture curves for forged Ti-5Al-2Sn-2Zr-4Mo-4Cr

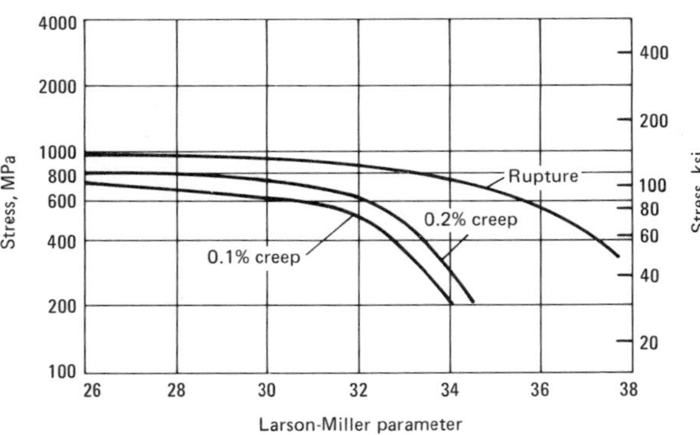

Larson-Miller parameter equals 10^{-3} $T(25 + \log t)$, where T is temperature in °R and t is time in hours.

Fig. 30 Creep-rupture data for Ti-6Al-2Sn-2Zr-2Cr-2Mo-0.25Si

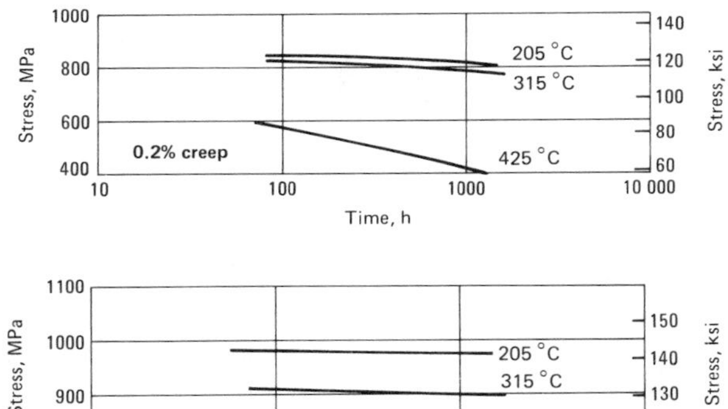

Ti-6Al-2Sn-2Zr-2Cr-2Mo-0.25Si

Commercial Names

Common name. Ti-6-22-22-S

Chemical Composition

Composition limits. 5.25 to 6.25 Al; 1.75 to 2.25 Sn; 1.75 to 2.25 Zr; 1.75 to 2.25 Cr; 1.75 to 2.25 Mo; 0.20 to 0.27 Si; 0.14 max O; 0.03 max N; 0.04 max C; 0.25 max Fe; rem Ti

Applications

Typical uses. Heavy section forgings for applications requiring high strength and fracture toughness

Precautions in use. Possible hot-salt stress-corrosion cracking at elevated temperatures. Should never come in contact with liquid oxygen.

Mechanical Properties

Tensile properties. At 20 °C (68 °F): tensile strength, 1160 MPa (168 ksi); yield strength, 1070 MPa (155 ksi); elongation, 18%; reduction in area, 25%

Compressive yield strength. At 25 °C (77 °F): 1170 MPa (170 ksi)

Elastic modulus. At 25 °C (77 °F): tension, 123 GPa (17.9 × 10^6 psi). At 50 °C (120 °F): compression, 125 GPa (18.2 × 10^6 psi)

Impact strength. 20.3 J (15 ft·lb) at 24 °C (75 °F)

Fatigue strength. Axial, tension-tension (R = 0.1): smooth specimen, 500 MPa (73 ksi) at 10^7 cycles; notched specimen (K_t = 30), 290 MPa (42 ksi) at 10^7 cycles

Creep-rupture characteristics. See Fig. 30.

Structures

Microstructure. Two-phase alpha-beta distributed according to processing temperature

Mass Characteristics

Density. 4.57 Mg/m^3 (0.165 lb/in.^{3}) at 24 °C (75 °F)

Thermal Properties

Coefficient of thermal expansion. Linear: 9.18 μm/m·K (5.1 μin./in.·°F) at 20 to 425 °C (68 to 800 °F)

Fabrication Characteristics

Solution temperature. 950 °C (1740 °F); 1 h at temperature; air cool or faster method

Aging temperature. 540 °C (1000 °F); 8 h at temperature; air cool

Ti-10V-2Fe-3Al

Commercial Names

Common name. Ti-10-2-3

Chemical Composition

Composition limits. 2.6 to 3.4 Al; 9.0 to 11.0 V; 1.8 to 2.2 Fe; 0.13 max O; 0.05 max C; 0.05 max N; 0.015 max H; 0.30 max others (total); rem Ti

Applications

Typical uses. Applications up to 315 °C (600 °F) where medium to high strength and high toughness are required in bar, plate or forged sections up to 125 mm (5 in.) thick. The combination of high strength and high toughness available with this alloy is superior to any other commercial titanium alloy. For applications requiring uniformity of tensile properties at surface and center locations.

Precautions in use. Mechanical properties depend heavily on prior

Fig. 31 Tensile properties vs temperature for round bars of Ti-10V-2Fe-3Al, STOA condition

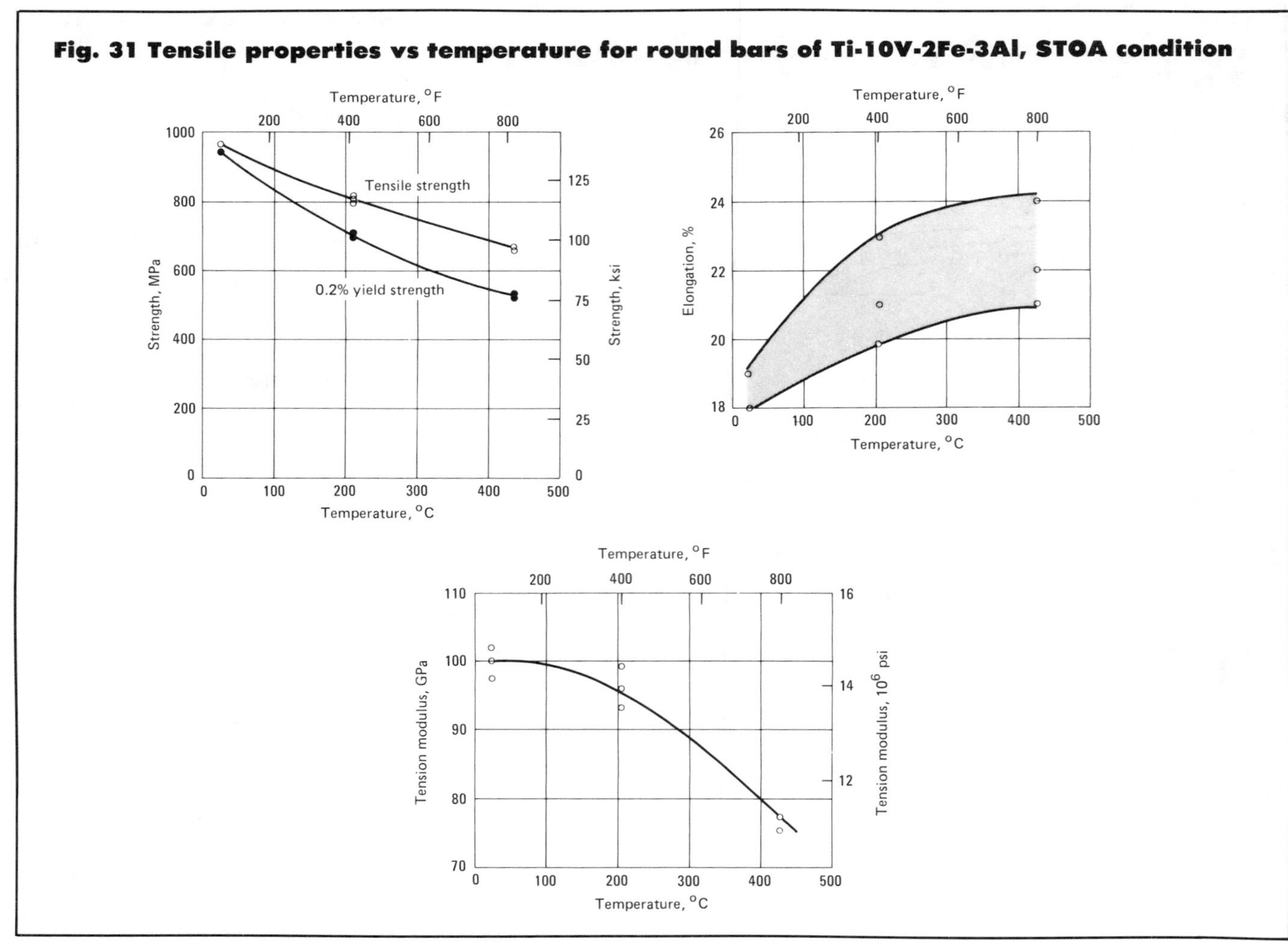

Table 16 Typical room-temperature tensile properties of a Ti-10V-2Fe-3Al airframe forging

Condition(a)	Tensile strength MPa	Tensile strength ksi	0.2% yield strength MPa	0.2% yield strength ksi	Elongation(b), %	Reduction in area, %
Longitudinal direction, 15 mm (0.6 in.) section thickness						
STA	1275	185	1200	174	11	25
STOA	980	142	940	136	22	56
Transverse direction, 15 mm (0.6 in.) section thickness						
STA	1260	183	1200	174	9	20
STOA	950	138	895	130	21	56
Transverse direction, 56 mm (2.2 in.) section thickness						
STA	1270	184	1195	173	7	33
STOA	970	141	910	132	21	56
Short transverse direction, 56 mm (2.2 in.) section thickness						
STA	1280	186	1200	174	8	21
STOA	950	138	890	129	19	55

(a) STA condition: Solution treat 1 h at 760 °C (1400 °F), water quench and age 8 h at 510 °C (950 °F). STOA condition: Solution treat 1 h at 730 °C (1350 °F), air cool and age 8 h at 580 °C (1075 °F). (b) In 50 mm or 2 in.

thermomechanical processing. Heat treatment should be tailored to give desired properties for each specific application. Overaging recommended to provide intermediate strength levels with enhanced toughness and reduced scatter in property values.

Mechanical Properties

Tensile properties. See Table 16 and Fig. 31.
Elastic modulus. Tension, 110 GPa (16×10^6 psi) at room temperature
Impact strength. Typical, STOA condition: 30 J (22 ft·lb) at room temperature
Plane-strain fracture toughness. See Fig. 32.
Fatigue properties. See Fig. 33.
Creep-rupture properties. See Fig. 34.

Structure

Microstructure. Mixture of cph al-

Fig. 32 Plane-strain fracture toughness vs tensile strength for Ti-10V-2Fe-3Al forgings

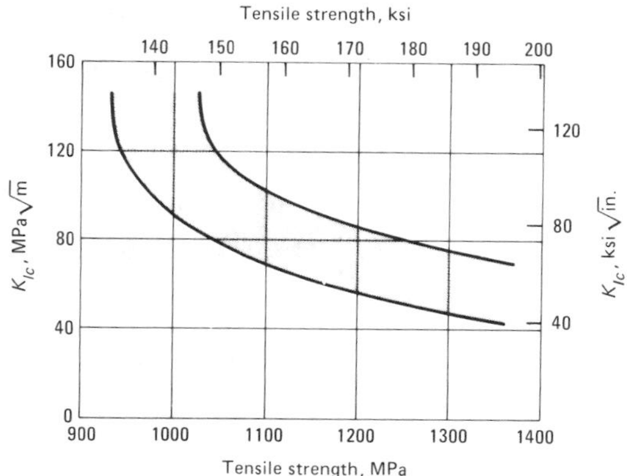

Data represent a composite of fracture toughness values for beta-forged die forgings, beta-forged block forgings, and beta forged plus alpha-beta-forged die forgings.

Fig. 33 Axial fatigue of Ti-10V-2Fe-3Al bar stock in the STOA condition

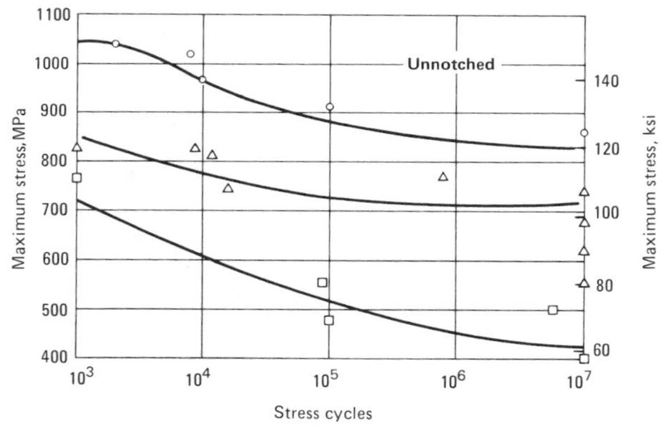

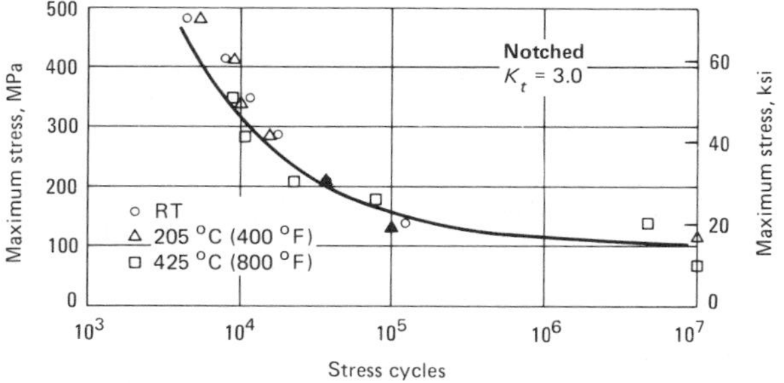

Specimens were taken from round bars 75 mm (3 in.) in diameter that had been solution treated 1 h at 760 °C (1400 °F), furnace cooled, overaged 8 h at 565 °C (1050 °F) and air cooled. Tests were conducted at a stress ratio of R = 0.1 and a frequency of 20 Hz.

pha and bcc beta that varies with mechanical and thermal history; beta predominates in all instances.

Mass Characteristics

Density. 4.65 Mg/m^3 (0.168 lb/in.^{3})

Thermal Properties

Coefficient of thermal expansion. Linear, average: plate, 9.7 μm/m·K (5.4 μin./in.·°F) for 20 to 425 °C (68 to 800 °F); forgings, 8.6 μm/m·K (4.8 μin./in.·°F) for 425 to 1150 °C (800 to 2100 °F)

Phase-transformation temperature. Beta transus, 795 °C (1460 °F)

Chemical Properties

Resistance to specific corroding agents. Excellent resistance to atmospheric and seawater corrosion. Resists stress-corrosion cracking in salt water ($K_{Iscc} > 0.7\ K_{Ic}$).

Fabrication Characteristics

Weldability. Good, partly because of fine grain sizes developed during welding.

Machinability. Slightly more difficult to machine than Ti-6Al-4V, but better than all-beta titanium alloys.

Hot working temperature. 760 to 815 °C (1400 to 1500 °F)

Annealing temperature. 705 °C (1300 °F)

Solution treating temperature. 760 to 775 °C (1400 to 1425 °F) for STA condition; 730 to 745 °C (1350 to 1375 °F) for STOA condition.

Aging temperature. 495 to 580 °C (925 to 1075 °F). Typical: 8 h at 510 °C (950 °F) for STA condition; 8 h at 580 °C (1075 °F) for STOA condition. Overaging (STOA condition) recommended to enhance toughness and ductility as well as to reduce scatter in mechanical property values. Overaging at 480 °C (900 °F) avoids development of the embrittling omega phase, and yet allows the development of intermediate strength levels.

Ti-3Al-2.5V

Commercial Names

Common names. Ti-3-2½; Half 6-4; "Tubing" alloy

Specifications

AMS. 4943

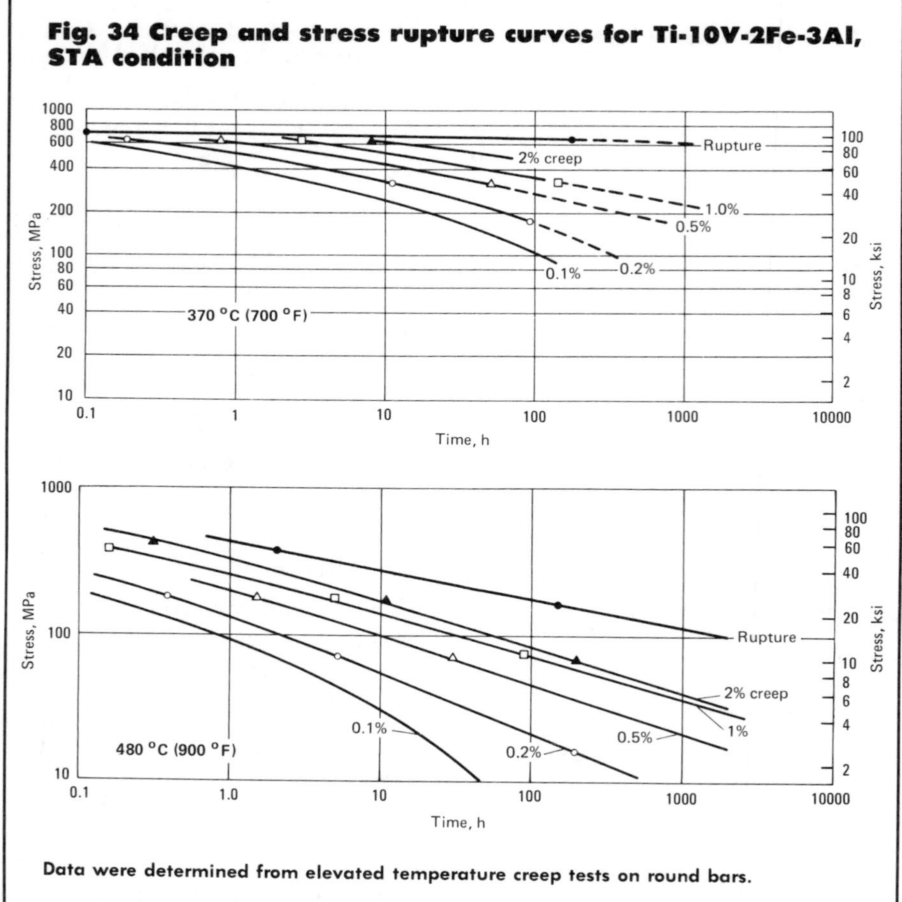

Table 17 Typical tensile properties of Ti-3Al-2.5V

Temperature °C	°F	Tensile strength MPa	ksi	Yield strength MPa	ksi	Elongation(a), %
Cold worked and stress relieved						
Room temperature		895	130	760	110	19
150	300	785	114	640	93	17
250	480	715	104	585	85	15
Annealed						
Room temperature		655	95	560	81	29
150	300	565	82	455	66	25
250	480	490	71	380	55	23

(a) In 50 mm or 2 in.

Chemical Composition

Composition limits. 2.5 to 3.5 Al; 2.0 to 3.0 V; 0.12 max O; 0.02 max N; 0.05 max C; 0.30 max Fe; 0.015 max H; rem Ti

Applications

Typical uses. Seamless tubing for aircraft hydraulic and ducting applications; weldable sheet; mechanical fasteners. Normally used in the cold-worked and stress-relieved condition

Mechanical Properties

Tensile properties. See Table 17.
Elastic modulus. Tension, 103 GPa (15×10^6 psi)
Impact strength. 54 J (40 ft·lb) at 21 °C (70 °F)

Mass Characteristics

Density. 4.48 Mg/m^3 (0.162 lb/in.^{3}) at 24 °C (75 °F)

Thermal Properties

Solidus temperature. 1704 °C (3100 °F)
Phase-transformation temperature. 935 °C (1715 °F)
Coefficient of thermal expansion. Linear: 9.9 μm/m·K (5.5 μin./in.·°F) at 21 to 315 °C (70 to 600 °F)

Fabrication Characteristics

Annealing temperature. 760 °C (1400 °F)
Solution temperature. 913 °C (1675 °F)
Aging temperature. 510 °C (950 °F)
Hot-working temperature. Range 871 to 927 °C (1600 to 1700 °F)

Ti-13V-11Cr-3Al

Commercial Names

Trade names. Crucible B-120VCA; OMC VCA; RMI 13V-11Cr-3Al; Teledyne Tel-Ti 13V-11Cr-3Al; Timet Ti-13V-11Cr-3Al
Common names. 13-11-3; B-120; VCA
UNS number. R58010

Specifications

AMS. 4917, 4959
Government. MIL-F-83142; MIL-T-9046; MIL-T-9047; MIL-T-81588

Chemical Composition

Composition limits. 12.5 to 14.5 V; 10 to 12 Cr; 2.4 to 4.0 Al; 0.35 max Fe; 0.08 max N; 0.10 max C; 0.015 max H; 0.20 max O; rem Ti

Applications

Typical uses. Missile applications such as solid rocket motor cases where extremely high strengths are required for short periods of time and for other structural applications in advanced manned and unmanned airborne systems. Springs for airframe applications.

Mechanical Properties

Tensile properties. Typical, see Table 18.
Shear strength. Typical, annealed bar: 760 MPa (110 ksi)

Fig. 35 Constant-life fatigue diagrams for Ti-13V-11Cr-3Al, STA condition, longitudinal orientation

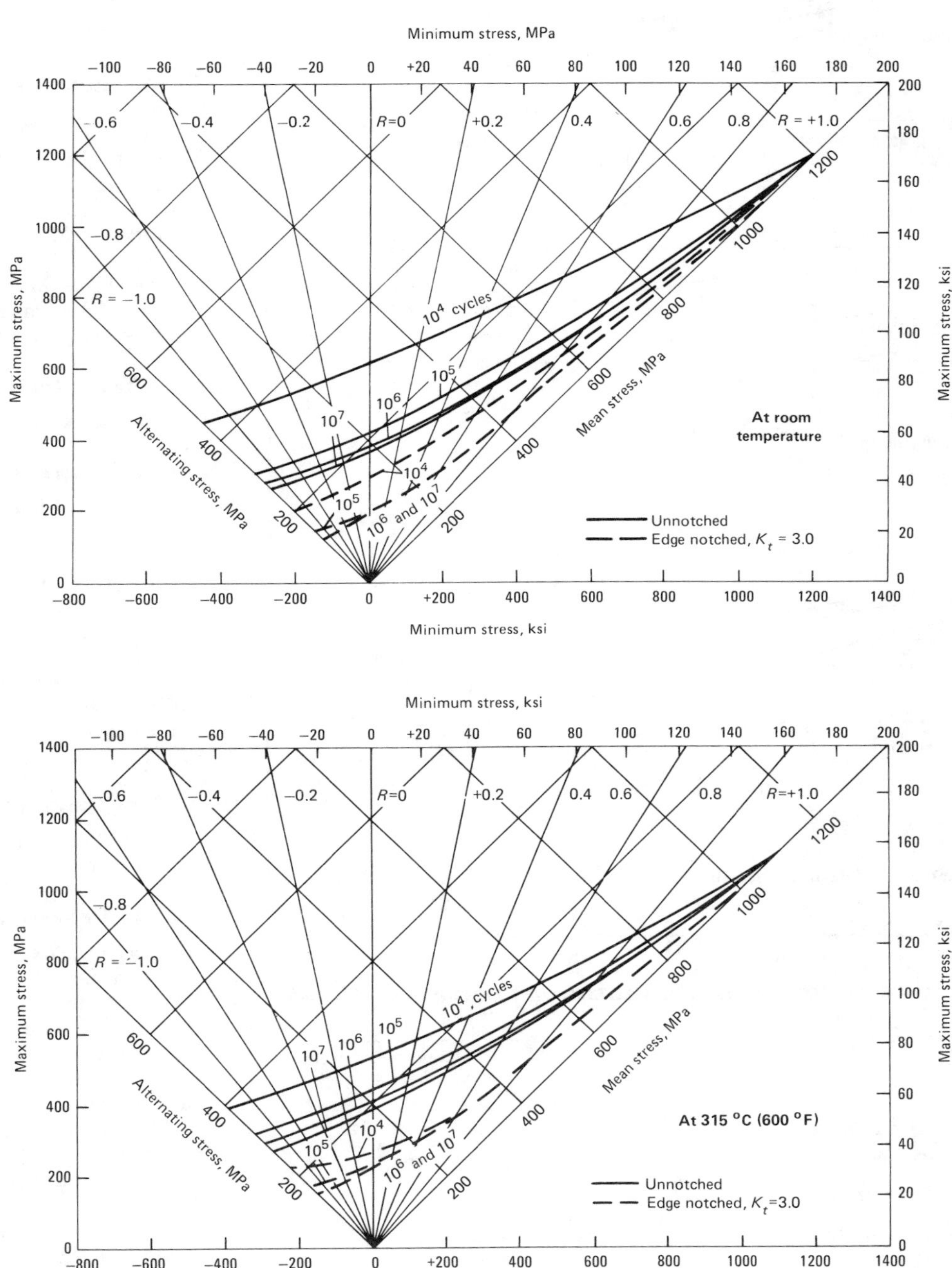

Data are for axial fatigue of edge-polished sheet specimens of material solution treated and aged to room-temperature tensile strength of 1203 MPa (174.5 ksi). Corresponding yield strength was 1080 MPa (156.7 ksi); at 315 °C (600 °F), the tensile strength was 1078 MPa (156.3 ksi) and the yield strength was 876 MPa (127.0 ksi). Tests were conducted at a speed of 60 Hz.

Fig. 36 Creep curves for Ti-13V-11Cr-3Al

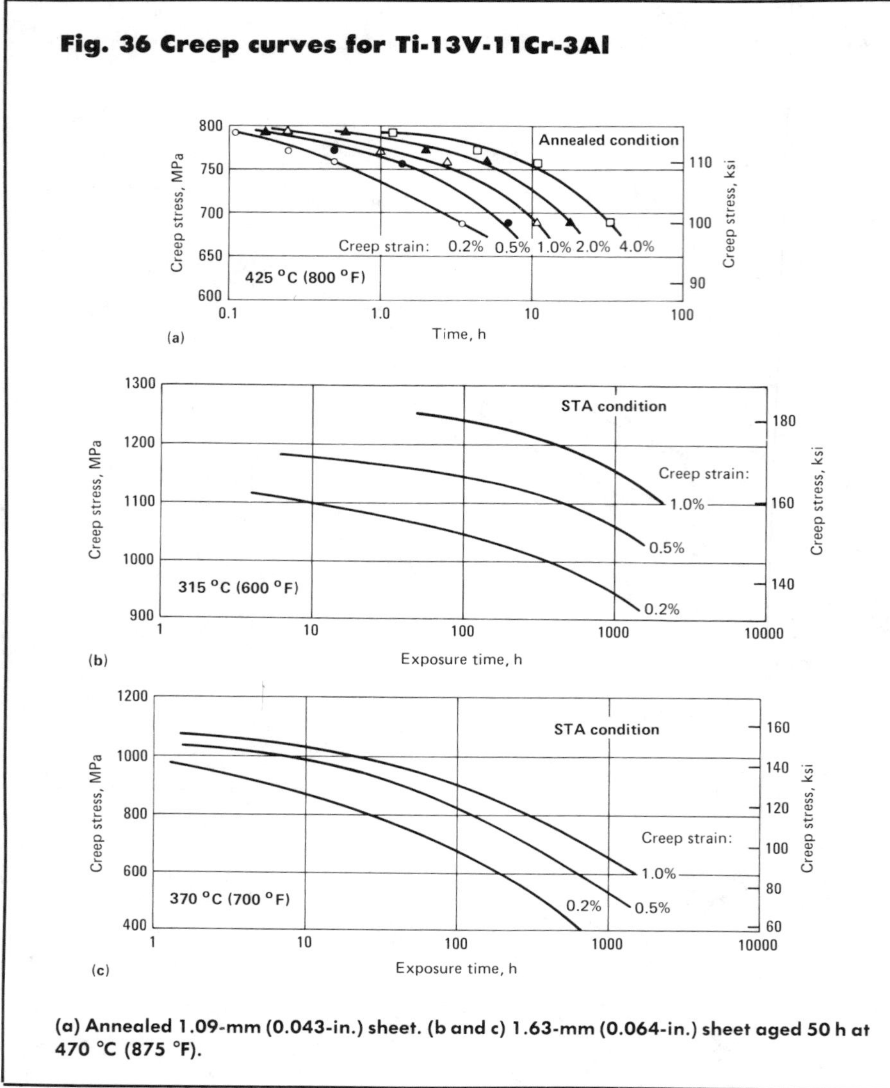

(a) Annealed 1.09-mm (0.043-in.) sheet. (b and c) 1.63-mm (0.064-in.) sheet aged 50 h at 470 °C (875 °F).

Table 18 Typical room temperature tensile properties of Ti-13V-11Cr-3Al

Diameter, in.	Condition	Tensile strength MPa	Tensile strength ksi	0.2% yield strength MPa	0.2% yield strength ksi	Elongation, %
Annealed or annealed and aged						
0.750	64 h at 480 °C (900 °F) + 1 h at 565 °C (1050 °F)	1290	187	1215	176	5.0
0.575	Annealed	995	144	985	143	22.5
	48 h at 480 °C (900 °F)	1365	198	1205	175	8.0
	72 h at 480 °C (900 °F)	1460	212	1295	188	6.0
0.257	Annealed	1035	150	995	144	23.3
	48 h at 480 °C (900 °F)	1420	206	1290	187	10.0
	72 h at 480 °C (900 °F)	1475	214	1345	195	6.7
Cold worked or cold worked and aged						
0.062	Cold drawn, 92%	1750	254	...	...	4.0
	92% + 24 h at 370 °C (700 °F)	2000	290	...	...	8.0
	92% + 24 h at 425 °C (800 °F)	2150	312	...	...	4.0

Compressive properties. 0.2% compressive yield strength at room temperature: annealed, 905 MPa (131 ksi); aged 50 h at 425 °C (800 °F), 995 MPa (144 ksi); aged 100 h at 425 °C (800 °F), 1010 MPa (155 ksi)
Poisson's ratio. 0.304 at 21 °C (70 °F)
Elastic modulus in tension. See Table 19.
Fatigue strength. See Fig. 35.
Plane-strain fracture toughness. Forging aged 20 h at 480 °C (900 °F) to give 1170 MPa (170 ksi) yield strength: 41 MPa$\sqrt{m}$ (45.8 ksi $\sqrt{in.}$). Forging aged 15 h at 480 °C (900 °F) to give 1138 MPa (165 ksi) yield strength: 23 MPa$\sqrt{m}$ (25 ksi$\sqrt{in.}$)
Creep strength. See Fig. 36.

Mass Characteristics

Density. 4.84 Mg/m^3 (0.175 lb/in.^{3})

Thermal Properties

Coefficient of thermal expansion. Linear, 10.6 μm/m·K (5.9 μin./in.·°F) at 20 to 540 °C (68 to 1000 °F)
Specific heat. Instantaneous: 545 J/kg·K (0.13 Btu/lb·°F) at 95 °C (200 °F); 670 J/kg·K (0.16 Btu/lb·°F) at 315 °C (600 °F)
Thermal conductivity. 6.9 W/m·K (4.0 Btu/ft·h·°F) at 21 °C (70 °F); 14.2 W/m·K (8.2 Btu/ft·h·°F) at 425 °C (800 °F)

Electrical Properties

Electrical resistivity. 1.78 μΩ·m at room temperature

Chemical Properties

General corrosion behavior. Excellent resistance to corrosion; protective coating not required. Exhibits immunity to general or pitting corrosion during long time storage of completed assemblies without protective coatings, which is an important quality for advanced airborne systems.
Resistance to specific corroding agents. Immune to pitting or general corrosion in the atmosphere, seawater and other natural environments; at ambient temperatures, inert to oxidizing media, inhibited reducing acids, alkalis and metallic chlorides.

Fabrication Characteristics

Machinability. Machines best in aged condition; requires slower speed than other titanium alloys

Table 19 Elastic modulus in tension for Ti-13V-11Cr-3Al

Temperature °C	°F	Solution-treated material GPa	Solution-treated material $\times 10^6$ psi	Aged material GPa	Aged material $\times 10^6$ psi
−54	−65	102	14.8	112	16.2
21	70	101	14.7	110	16.0
205	400	96.5	14.0	107	15.5
315	600	91.0	13.2	103	15.0
425	800	85.5	12.4	99.3	14.4
540	1000	80.0	11.6	94.5	13.7

Weldability. Readily weldable by various methods (fusion and resistance welding); high base metal strength combined with strong, ductile weldments produces highly efficient, lightweight assemblies
Annealing temperature. Solution-treating: 1775 ± 4 °C (1425 ± 25 °F); hold at temperature for 10 to 30 min; air cool
Aging temperature. Solution-treated material aged to optimum combination of strength and ductility: 426 to 482 °C (800 to 900 °F); 10 to 100 h at temperature depending on wrought configuration

Table 20 Room temperature tensile property limits for Ti-8Mo-8V-2Fe-3Al

Form(a)	Tensile strength MPa	Tensile strength ksi	Yield strength MPa	Yield strength ksi	Elongation, %	Reduction in area, %
Bar stock	1180-1295	171-188	1125-1240	163-180	3.5-10.5	4.0-24
Closed die forging						
Test forging	1220-1310	177-190	1185-1270	172-184	3-5	5.5-11.5
"Navaho" test forging	1215-1305	176-189	1105-1235	160-179	2.5-10.5	4.0-16.2
Hand-mill sheet	1070-1380	155-200	965-1380	140-200	3-12	...
Simulated strip	825-1415	120-205	965-1380	140-200	4-18	...
Flat-rolled products	785-1035	114-150	710-810	103-126	12-24	36-51

(a) Specific property values vary with section size, orientation in product, and heat treatment.

Fig. 37 Fatigue strength of Ti-8Mo-8V-2Fe-3Al sheet, STA condition

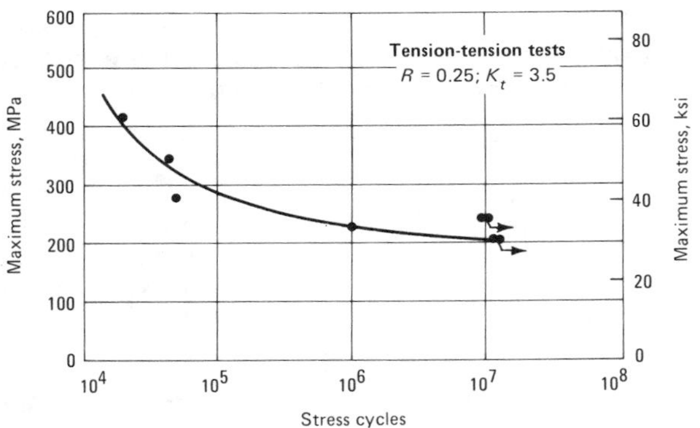

Data are for 1.5-mm (0.060-in.) thick sheet solution treated 10 min at 790 °C (1450 °F), air cooled, and aged 8 h at 480 °C (900 °F).

Ti-8Mo-8V-2Fe-3Al

Commercial Names

Common name. 8-8-2-3 beta alloy
UNS number. R58820

Specifications

Government. MIL-F-83142; MIL-T-9046; MIL-T-9047

Chemical Composition

Composition limits. 7.5 to 8.5 Mo; 7.5 to 8.5 V; 1.6 to 2.4 Fe; 2.6 to 3.4 Al; 0.10 to 0.16 O; 0.05 max Ni; 0.05 max C; 0.01 max S; 0.015 max H; 0.10 max others (each); 0.30 max others (total); rem Ti

Applications

Typical uses. Rod and wire for fastening applications; sheet, strip and forgings for aerospace structures. Should be considered for use in process tanks, tubing and other chemical processing equipment for use at elevated temperatures.

Mechanical Properties

Tensile properties. See Table 20.
Elastic modulus. Tension: 98.5 to 114 GPa (14.3 to 16.6 × 10^6 psi). Specific values vary with section size, orientation in product and heat treatment.
Fatigue strength. At 10^7 cycles. Rotating-beam tests: closed die forging, 482 to 517 MPa (70 to 75 ksi); forged bar, 503 MPa (73 ksi). Notched sheet (K_t = 3.5), tension-tension tests (R = 0.25): 207 to 241 MPa (30 to 35 ksi). See also Fig. 37.

Structure

Microstructure. Typical: Solution annealed condition, bcc beta; solution treated plus aged condition, beta plus beta transformed to alpha

Mass Characteristics

Density. 4.84 Mg/m³ (0.175 lb/in.³)

Chemical Properties

General corrosion behavior. Good resistance expected in media such as seawater, inhibited reducing acids, and metallic chlorides, although confirming data are sparse.

Fabrication Characteristics

Weldability. No data reported; this material expected to have limited weldability similar to other commercial beta titanium alloys.

Annealing temperature. 790 °C (1450 °F); 15 min recommended for forgings, plate, bar; 5 min recommended for sheet; air cool

Stress-relieving temperature. Material aged at 510 °C (950 °F): re-age at temperature for 4 h. Aged at 595 °C (1100 °F): re-age at temperature for 1 h

Ti-3Al-8V-6Cr-4Zr-4Mo

Commercial Names

Trade names. RMI 3Al-8V-6Cr-4Mo-4Zr; Teledyne Tel-Ti 3Al-8V-6Cr-4Mo-4Zr

Common names. Beta C alloy; 38-6-44

UNS number. R58640

Specifications

Government. MIL-F-83142; MIL-T-9046; MIL-T-9047

Chemical Composition

Composition limits. 3.0 to 4.0 Al; 7.5 to 8.5 V; 5.5 to 6.5 Cr; 3.5 to 4.5 Mo; 3.5 to 4.5 Zr; 0.30 max Fe; 0.05 max C; 0.03 max N; 0.14 max O; rem Ti

Applications

Typical uses. Fasteners, springs, torsion bars, foil used in making cores for sandwich structures

Precautions in use. Highly susceptible to hydrogen pick-up during heating, pickling and chemical milling.

Mechanical Properties

Tensile properties. See Table 21.

Shear strength. Typical: 620 to 655 MPa (90 to 95 ksi) for solution annealed condition; 620 to 965 MPa (90 to 140 ksi) after various aging or overaging heat treatments

Compressive properties. Compressive yield strength, forged billet, solution annealed and aged at 565 °C (1050 °F): 1070 MPa (155 ksi) (transverse); 1110 MPa (161 ksi) (longitudinal)

Elastic modulus. At room temperature: tension, 103 GPa (15.0×10^6 psi); torsion, 41 GPa (6.0×10^6 psi)

Creep-rupture properties. Stress-rupture strength, typical: 965 MPa (140 ksi) for 100 h life at 315 °C (600 °F). Creep strength, typical: 690 MPa (100 ksi) for 0.2% creep in 100 h at 315 °C (600 °F)

Fig. 38 Variation of tensile properties with aging treatment for upset forged Ti-3Al-8V-6Cr-4Mo-4Zr

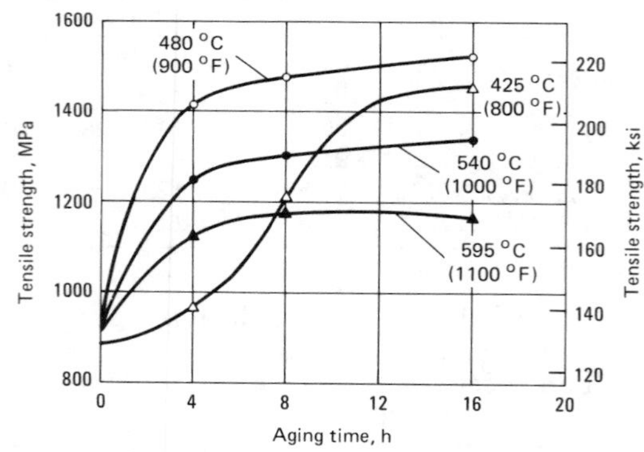

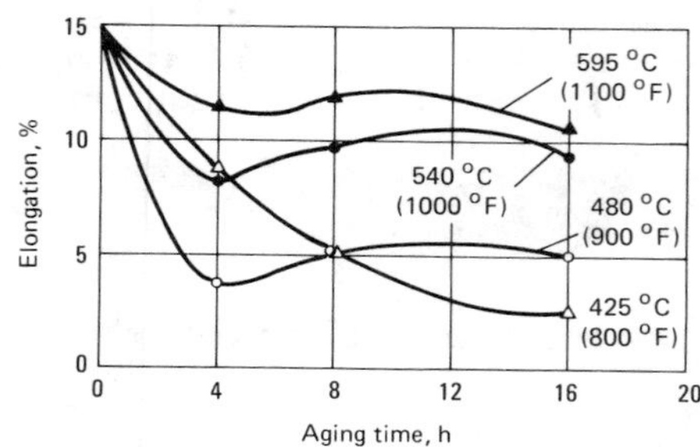

Forgings upset to 16 mm (0.625 in.) from 64 mm (2.5 in.) at 815 °C (1500 °F), followed by solution annealing at 815 °C.

Table 21 Room temperature longitudinal tensile properties for Ti-3Al-8V-6Cr-4Mo-4Zr

Form(a)	Tensile strength MPa	Tensile strength ksi	Yield strength MPa	Yield strength ksi	Elongation, %	Reduction in area, %
Forged billet(b)	1220	177	1151(c)	167(c)	10.0	18.2
Forging	1103-1267	160-184	1089-1192(c)	158-173(c)	4.5-8.0	12-20
Bar and wire(d)	1103-1220	160-177	993-1124	144-163	13-18	33-50

(a) All material in solution treated and aged condition. (b) Solution treated 15 min at 815 °C (1500 °F); air cooled aged 12 h at 565 °C (1050 °F), air cooled. (c) 0.2% offset. (d) Aged 6 h at 565 °C (1050 °F), air cool.

Structure

Crystal structure. Metastable bcc beta alloy at room temperature. Aging causes precipitation of finely divided cph alpha in the beta matrix.

Microstructure. Solution treated: equiaxed beta grains. STA condition: equiaxed beta with fine alpha precipitate in matrix.

Mass Characteristics

Density. 4.82 Mg/m³ (0.174 lb/in.³)

Thermal Properties

Liquidus temperature. 1650 °C (3000 °F)

Solidus temperature. 1555 °C (2830 °F)

Table 22 Linear thermal expansion for Ti-3Al-8V-6Cr-4Mo-4Zr

Temperature range °C	Temperature range °F	Mean coefficient on heating(a) μm/m·K	Mean coefficient on heating(a) μin./in.·°F
20-38	68-100	8.44	4.69
20-93	68-200	8.65	4.81
20-150	68-300	8.82	4.94
20-205	68-400	9.05	5.03
20-260	68-500	9.25	5.14
20-315	68-600	9.41	5.23
20-370	68-700	9.54	5.30
20-425	68-800	9.61	5.34
20-480	68-900	9.68	5.38

(a) Material forged, solution treated, and then aged 12 h at 565 °C (1050 °F).

Beta transus temperature. 793 ± 14 °C (1460 ± 25 °F)
Coefficient of thermal expansion. See Table 22.

Chemical Properties

Resistance to specific corroding agents. Limited data indicate reasonably low susceptibility to accelerated crack propagation in salt water

Fabrication Characteristics

Weldability. No data reported; this material expected to have limited weldability similar to other commercial beta titanium alloys.
Annealing temperature. 790 °C (1460 °F) or above; 15 to 30 min at temperature; air cool or water quench
Solution temperature. 815 °C (1500 °F) or temperatures up to 927 °C (1700 °F) depending on form used, processing history and mechanical properties expected; 30 min at temperature; water quench or air cool
Aging temperature. Varies with preferred strength level and associated mechanical properties (see Fig. 38); 6 to 12 h at temperature; air cool

Ti-11.5Mo-6Zr-4.5Sn

Commercial Names

Trade name. Beta III

Specifications

AMS. 4977; 4980
ASTM. Bar and billet: B348. Sheet, strip and plate: B265. Pipe: B337. Tube: B338
Government. MIL-F-83142; MIL-T-9046; MIL-T-9047

Table 24 Creep properties of Ti-11.5Mo-6Zr-4.5Sn

Temperature °C	Temperature °F	Creep stress MPa	Creep stress ksi	Plastic creep strain(a), %
260	500	895	130	0.07
315	600	795	115	0.16
370	700	240	35	0.05
		515	75	0.17
425	800	70	10	0.04
		160	23	0.20

(a) For 100 h at temperature under load.

Table 23 Typical mechanical properties of Ti-11.5Mo-6Zr-4.5Sn

Aging temperature °C	Aging temperature °F	Tensile strength MPa	Tensile strength ksi	Yield strength(a) MPa	Yield strength(a) ksi	Elongation(b), %	Reduction in area, %
Rivet wire(c)							
As solution treated		993	144	792	115	24	65
480	900	1365	198	1269	184	15	36
510	950	1303	189	1213	176	18	38
540	1000	1186	172	1124	163	21	44
565	1050	1089	158	1034	150	25	56
590	1100	986	143	945	137	27	65
Bar, 13.6 mm (0.522 in.) diam(c)							
As solution treated		855	124	752	109	21	72
480	900	1386	201	1317	191	11	33
540	1000	1165	169	1096	159	17	63
590	1100	1041	151	1007	146	17	67
Plate, 12.7 and 25.4 mm (0.5 and 1.0 in.) thick(c)(d)							
As solution treated		896	130	827	120	22	62
480	900	1351	196	1262	183	3	6.4
510	950	1289	187	1200	174	5	10.7
540	1000	1255	182	1179	171	4.8	12.3
590	1100	1041	151	979	142	11	24
Sheet, 1.6 mm (0.063 in.) thick							
Solution treated, 720 °C (1325 °F)							
Air cooled		972	141	882	128	17	45
Water quenched		841	122	738	107	20	52
480	900	1413	205	1317	191	7	29
590	1000	1158	168	1089	158	8	45
Solution treated, 770 °C (1425 °F)							
Air cooled		896	130	834	121	18	45
Water quenched		827	120	745	108	21	48
480	900	1310	190	1234	179	6	35
590	1000	1138	165	1062	154	8	42
Lab foil specimen(e)							
As solution treated, 760 °C (1400 °F)							
0.010 in. thick		1000	145	958	139	8.0	...
0.005 in. thick		979	142	924	134	8.5	...
0.002 in. thick		1014	147	958	139	6.5	...
480	900						
0.010 in. thick		1282	186	1248	181	6.7	...
0.005 in. thick		1510	219	1413	205	4.5	...
0.002 in. thick		1586	230	1538	223	2.0	...
540	1000						
0.010 in. thick		1158	168	1082	157	6.5	...
0.005 in. thick		1262	183	1186	172	8.2	...
0.002 in. thick		1344	195	1276	185	4.0	...

(a) 0.2% offset. (b) In 2 in. or 4*d*, where *d* is diameter of reduced section of tensile test specimen. (c) Solution treated 730 to 790 °C (1350 to 1450 °F), water quenched, aged 8 h. (d) Longitudinal properties. (e) Solution treated, descaled and pickled, aged 8 h.

Chemical Composition

Composition limits. 10.0 to 13.0 Mo; 4.50 to 7.50 Zr; 3.75 to 5.25 Sn; 0.10 max C; 0.020 max H; 0.05 max N; 0.18 max O; 0.35 max Fe; 0.01 max others (each); 0.04 max others (total); rem Ti

Applications

Typical uses. Aircraft fasteners (especially rivets) and sheet metal parts where cold formability and strength potential can be used to greatest advantage. Possible use in plate and forging applications where high strength capability, deep hardenability and resistance to stress corrosion are required and somewhat lower aged ductility can be accepted
Precautions in use. Similar to Ti-10V-2Fe-3Al

Mechanical Properties

Tensile properties. See Table 23.

Elastic modulus. Tension, 70 GPa (10×10^6 psi) at 20 °C (68 °F)
Plane-strain fracture toughness. Transverse: as solution treated, 163 MPa$\sqrt{m}$ (148 ksi$\sqrt{in.}$); aged 8 h at 510 °C (950 °F), 66 MPa$\sqrt{m}$ (60 ksi$\sqrt{in.}$); aged 8 h at 595 °C (1100 °F), 95 MPa$\sqrt{m}$ (87 ksi$\sqrt{in.}$)
Creep-rupture characteristics. See Table 24.

Mass Characteristics

Density. 5.07 Mg/m^3 (0.183 lb/in.^{3})

Electrical Properties

Electrical resistivity. 1.56 μΩ·m at room temperature

Chemical Properties

General corrosion behavior. Virtually immune to aqueous stress corrosion; almost no loss of fracture toughness in salt solutions. Limited tests show resistance to hot salt stress corrosion but not immunity

Fabrication Characteristics

Formability. Minimum room-temperature bend radius for 1.6 mm (0.063 in.) sheet. Solution treated at 720 °C (1325 °F): air cooled, 3.5*t*; water quenched, 1.7*t*. Solution treated at 775 °C (1425 °F): air cooled, 3.0*t*; water quenched, 1.7*t*. Olsen cup test, mill solution treated 1.65 mm (0.065 in.) thick sheet: cup height at failure, 8.38 mm (0.33 in.); load at failure, 5490 kg (12 100 lb)

Titanium Castings

By J. R. Newman
Vice President
TiTech International, Inc.

ALL TITANIUM CASTINGS have compositions based on those of the common wrought alloys. There are no commercial titanium alloys developed strictly for casting applications. This is unusual, because in other metallic systems alloys have been developed specifically as casting alloys, often to overcome certain problems such as poor castability of a wrought-alloy composition. No unusual problems regarding castability or fluidity have been encountered in any of the titanium metals cast to date.

The major reason for selecting a titanium casting instead of a wrought titanium product is cost. This cost advantage may be attained through increased design flexibility, better utilization of available metal or reduction in the cost of machining or forming parts.

Titanium castings are unlike castings of other metals in that they are equal or nearly equal in strength to their wrought counterparts. Strength guarantees in most specifications for titanium castings are the same as for wrought forms. Typical ductilities of cast products, as measured by elongation and reduction in area, are lower than typical values for wrought products of the same alloys. Fracture toughness and crack-propagation resistance equal or exceed those of corresponding wrought material. The fatigue strength of cast titanium is inferior to that of wrought titanium. However, results of ongoing research suggest that fatigue strength of cast titanium can be enhanced by further processing and heat treatment.

Titanium castings, like wrought titanium products, are used primarily in three areas of application: aerospace products, marine service and industrial (corrosion) service. Commercially pure titanium (ASTM grade 1, 2 or 3) is used for the vast majority of corrosion applications, whereas Ti-6Al-4V is the dominant alloy for aerospace and marine applications. Ti-6Al-2Sn-4Zr-2Mo-Si is being selected with increasing frequency for elevated-temperature service. Castings also have been supplied in alloys Ti-5Al-2.5Sn, Ti-8Al-1Mo-1V and Ti-6Al-6V-2Sn, as well as in several European alloys.

Developing the art and science of casting titanium presented formidable problems. The melting process used today (skull melting, or melting within a pivotable water-cooled copper crucible) is an adaptation of the established process for ingot melting, the consumable-electrode vacuum-arc process. Development efforts for titanium castings centered primarily around finding acceptable and economical techniques for containing the molten reactive metal.

Most formidable was the problem of finding a suitable mold system to receive the molten titanium. Conventional foundry sands and volatile binders were unacceptable because they reacted violently with the titanium, a powerful solvent when molten. Another problem was the changeable viscosity of molten titanium when poured from a rapidly cooling melt.

The earliest titanium castings were made in machined graphite molds at the U.S. Bureau of Mines. Machined graphite is an excellent mold material from a technical standpoint because it has high thermal conductivity and is relatively inert; however, it is severely limited for production of complex shapes.

Air Force funding in the early 1960's aided the development of a rammed-graphite molding technique. In this procedure graphite powder is mixed with a petroleum-base binder and rammed around a pattern made of wood, plastic or metal. Molds must be cured at high temperature to completely reduce the binder to carbon. The process is effective and is being used in production of industrial castings such as those for pump and valve parts.

In 1969 a modified rammed-graphite system was developed. This system employed water-soluble binders, which reduced mold curing to merely a low-temperature drying cycle. Dimensional stability was better than with the older rammed-graphite technique, and this led (in 1970) to the first use in production aircraft of titanium castings made using modified rammed-graphite molds.

Precision "lost wax" investment casting of titanium was also developed in the late 1960's. Several companies have developed proprietary "nonreactive" shell systems, and some of these systems have been patented. Essential shell ingredients for the various processes range from pyrolitic graphite, to a tungsten facing, to special oxide layers. There is little in the way of design complexity, tolerance or surface finish obtainable in other metals that cannot now be achieved in titanium.

By use of hot isostatic pressing, internal soundness of titanium castings can be improved to the point that no porosity or small voids can be detected.

The lack of fluidity of molten titanium was overcome early in the casting development period by employing a centrifuge to accelerate the molten metal up to 60 to 70 g's. Centrifuge design is such that the molten titanium is literally pumped along runner systems into mold cavities.

Molten titanium cannot be superheated to increase fluidity as can other metals that are contained in ceramic crucibles. In consumable-electrode melting, where the molten metal drips into a water-cooled copper crucible, application of more melting power simply increases the rate at which the electrode is consumed and decreases the skull—the thickness of titanium frozen against the water-cooled copper crucible. The temperature of the mass of molten titanium is little affected.

It was not until the mid-1970's that the technology was advanced enough for titanium castings to be widely accepted in applications utilizing their full potential. By then, applications included not only industrial pumps and valves but also airframe parts.

Titanium castings still represent a very small portion of the titanium industry—about 1% (by weight) of total industry shipments. However, because a one-pound titanium casting typically replaces a wrought product made from seven pounds of billet, it can be expected that use of titanium castings will grow rapidly during the 1980's. In the late 1970's, casting shipments (as reported by the U.S. Department of Commerce) averaged 30 000 to 35 000 pounds per month, whereas wrought-product shipments averaged close to 3.5 million pounds per month. Typical production (in pounds) of titanium ingot, mill products and castings for 1969, 1973 and 1979 were as follows:

Fig. 1 Sales trend for the U.S. titanium casting industry

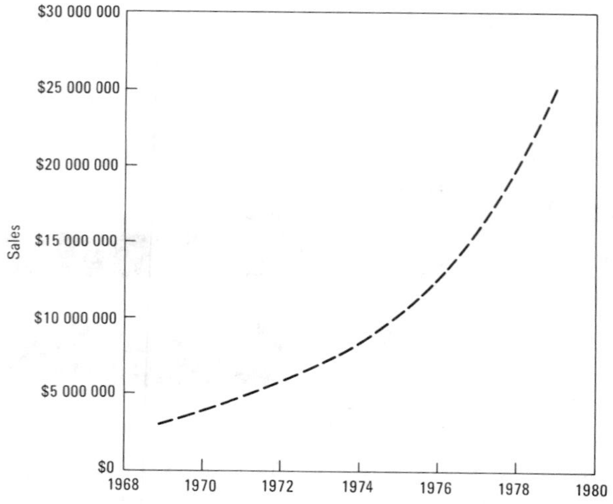

Product form	Typical production, lb
1969	
Ingot	56 972 000
Mill products	31 882 000
Castings	78 000
1973	
Ingot	58 080 000
Mill products	29 156 000
Castings	254 000
1979	
Ingot	74 520 000
Mill products	42 243 000
Castings	369 000

Total industry sales of titanium castings for 1969 through 1979 are presented graphically in Fig. 1.

Whereas in the late 1960's virtually all titanium castings were used in corrosion applications, by the late 1970's aerospace applications had become dominant and the distribution began to approach that for wrought titanium products. It is estimated that in 1979 aerospace applications accounted for more than 60% of total casting sales.

Applications of Titanium Castings

Titanium castings are now used extensively in the aerospace industry and to lesser but increasing measure in the chemical-process, marine and other industries.

Current aerospace applications include major structural fittings weighing over 45 kg (100 lb) each, and small switch guards weighing less than 30 g (1 oz) each, for the space shuttle; wings, engine components, brake components, optical-sensor housings, ordnance and other parts for military aircraft and missiles; and engine and brake components for commercial passenger aircraft. Additional aerospace applications, including rotor hubs for helicopters, flap tracks for fighters, gas-turbine compressors, and various missile and ordnance parts—are under active development.

In the chemical-process industry, components for pumps, valves and compressors are made of cast titanium. Marine applications include water-jet inducers for hydrofoil propulsion and seawater valve balls for nuclear submarines. Titanium castings are also used in various other industrial applications, such as well-logging hardware for the petroleum industry, special automotive parts, boat deck hardware and medical implants.

Mechanical Properties

Cast titanium alloys are equal or nearly equal in strength to wrought alloys of the same compositions. However, typical ductilities are below the typical values for comparable wrought alloys, but still above the guaranteed minimum values for the wrought metals. Because castings of Ti-6Al-4V have been used in aerospace applications, the most extensive data have been developed for this alloy.

Tensile Properties and Fracture Toughness. Typical room-temperature tensile properties are given in Table 1 for cast commercially pure titanium and for three cast titanium alloys. Figure 2 shows a distribution of room-temperature tensile properties for Ti-6Al-4V castings from one producer. Figure 3 shows typical elevated-temperature tensile properties. Figure 4 compares plane-strain fracture-toughness values for Ti-6Al-4V castings with values for Ti-6Al-4V plate and for other wrought titanium alloys.

In a comparison of elevated-temperature properties for cast and wrought Ti-6Al-2Sn-4Zr-2Mo, the results of triplicate tensile tests on bars cut from duplex annealed castings were quite uniform—typically less than 14 MPa (2 ksi) variation at each test temperature. The strength of cast material dropped off at a slightly greater rate with temperature than did published values for forged material. Yield strengths for cast material were essentially equivalent to those for bar stock at all temperatures. Ductility of cast material was generally lower than for wrought material. These data are shown graphically in Fig. 5.

Fatigue Properties. Measurements of fatigue-crack growth rates have been made on double-cantilever-beam specimens of cast Ti-6Al-4V. The tests were carried out in air at 24 °C (75 °F) on a closed-loop electrohydraulic fatigue machine using a sinusoidal wave form in load control at 10 Hz. Crack-growth rates at three values of R (ratio of minimum stress to maximum stress) were investigated—at R values of 0.1, 0.3 and 0.5. In this series of tests, the crack-growth rate for cast material was shown to be lower than that for rolled material at all values of R. Figure 6 (for $R = 0.5$) is typical of crack-growth behavior at all three stress ratios.

A great number of individual fatigue evaluations have been made. Figure 7 is a compilation of data from various sources. All results are for specimens cut from castings; and all specimens, except for those from one source, have notches or machined holes to introduce stress concentrations.

Table 1 Typical room-temperature tensile properties of several cast titanium alloys

Alloy	Condition	Tensile strength MPa	Tensile strength ksi	Yield strength(a) MPa	Yield strength(a) ksi	Elongation(b), %	Reduction in area, %
Commercially pure titanium	As-cast or annealed ..	550	80	450	65	17	32
Ti-6Al-4V	As-cast or annealed .	1035	150	890	129	10	19
Ti-6Al-2Sn-4Zr-2Mo	Duplex annealed .	1035	150	895	130	8	16
Ti-5Al-2.5Sn ELI	Annealed...	805	117	745	108	11	...

(a) At 0.2% offset. (b) In 50 mm or 2 in.

Fig. 2 Typical distribution of tensile properties for as-cast Ti-6Al-4V

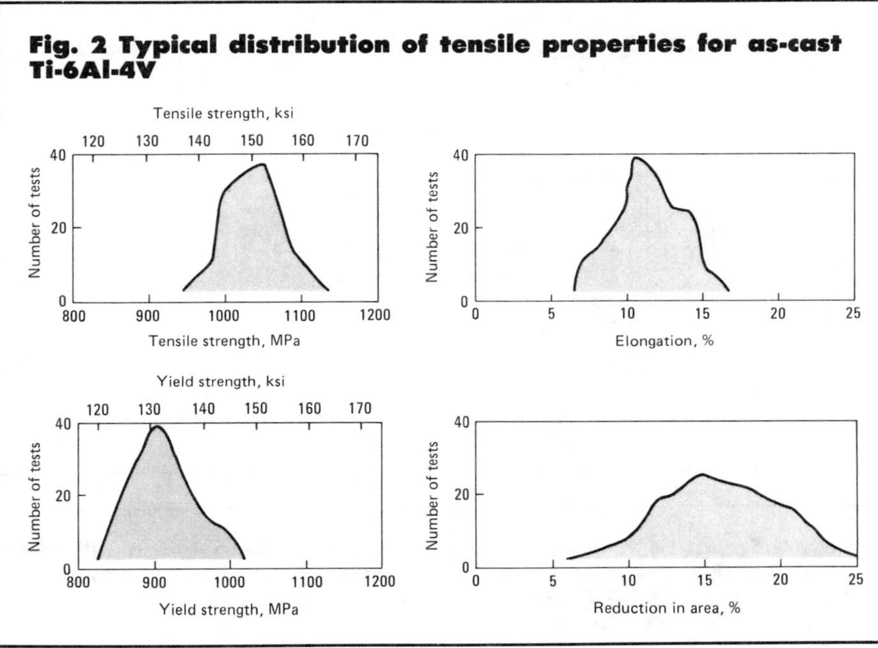

Little improvement in fatigue resistance is gained by hot isostatic pressing of properly cast material. On the other hand, results of ongoing research suggest that substantial improvement in fatigue life can be obtained by beta heat treating and overaging of cast titanium alloys.

A comparison of push-pull fatigue properties at 260 °C (500 °F) for as-cast Ti-6Al-4V and for forged and annealed Ti-6Al-4V showed very minor differences in fatigue strength. In this study, data were obtained for both smooth and notched specimens at three different stress ratios: $A = \infty$, $A = 0.67$ and $A = 0.25$. (A is the ratio of alternating stress to mean stress. A value of infinity for A indicates fully reversed stress, and is equivalent to $R = -1$.)

Impact Strength. Notched Charpy impact tests on cast Ti-6Al-4V were conducted in an AFML study. The average impact strength of the cast material was 26 J (19.2 ft·lb), which compares favorably with the typical value of 24 J (17 ft·lb) for forged material (AMS 4928).

Effect of Weld Repair. Weld repair is common foundry practice. However, because titanium can become embrittled due to pickup of oxygen, hydrogen and other contaminants during welding, weld repair of titanium castings must be carefully executed. The results of a fatigue study indicate that properly executed weld repair will not degrade the fatigue resistance of cast titanium. Welding also has been evaluated with regard to its effect on creep properties. This study demonstrated that welding does not drastically affect the creep properties of cast Ti-6Al-4V. The rupture times for welded and unwelded

Fig. 3 Typical elevated-temperature tensile properties of cast Ti-6Al-4V

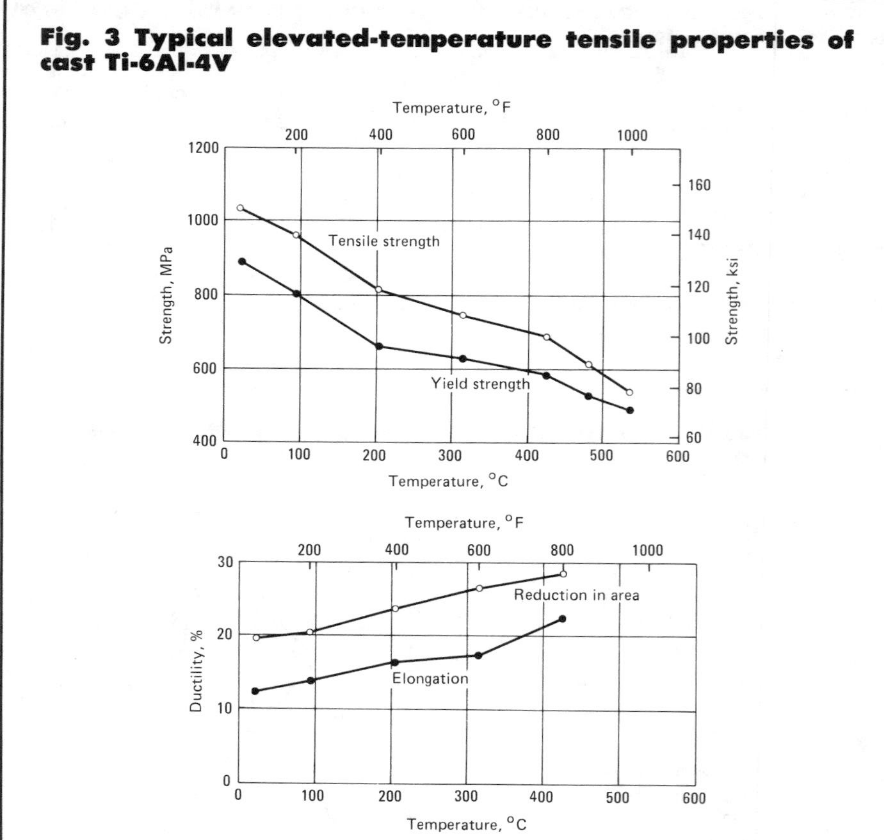

bars are similar at 315 °C (600 °F) and at 650 °C (1200 °F). Strain rates at 2% creep strain and 650 °C are the same for welded and unwelded bars.

Hot Isostatic Processing

Hot isostatic processing (HIP) of titanium castings became a production reality in the late 1970's. The HIP schedule that has become the industry standard is 2 h at 900 °C (1650 °F) under argon pressurized to 105 MPa (15 000 psi).

Initially, hot isostatic processing served as an excellent means of salvaging parts that had been rejected after radiographic inspection. The economic feasibility of deliberate plans to HIP routine parts was questionable due to the cost per cycle, which typically was $2700 for a chamber load approximately 400 mm in diameter by 1525 mm long (16 in. in diameter by 60 in. long). However, as the cost of titanium increases, metal utilization plays an increasingly greater role in pricing. Whereas conventional foundry practice calls for extensive risering to meet rigid aerospace acceptance standards, risering can be eliminated or greatly reduced if HIP is used to close shrinkage voids by diffusion bonding. This results

Fig. 4 Fracture toughness of Ti-6Al-4V castings compared to Ti-6Al-4V plate and to other Ti alloys

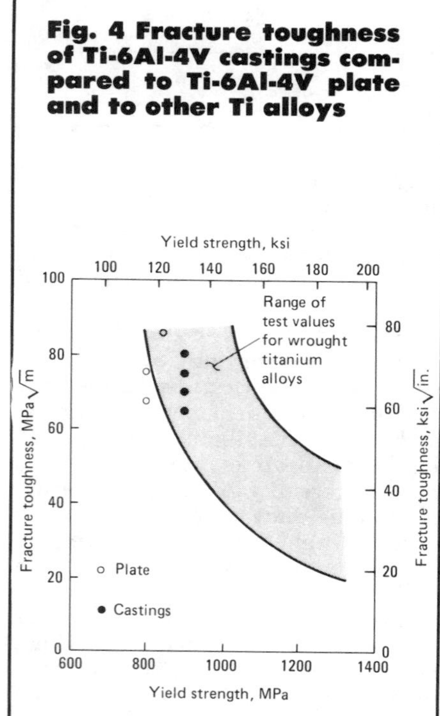

Fig. 5 Comparison of short-time tensile properties for wrought and cast forms of Ti-6Al-2Sn-4Zr-2Mo

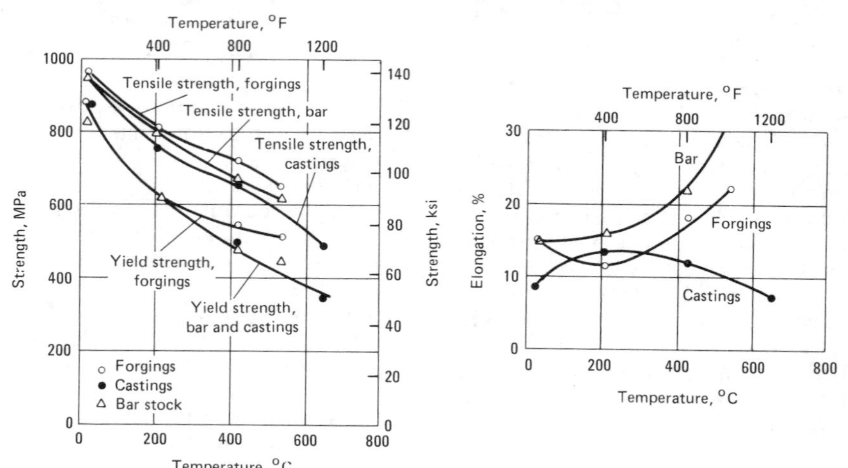

Test bars were cut from castings that were duplex annealed: 1 h at 900 °C (1650 °F) and air cool, plus 1 h at 790 °C (1450 °F) and air cool. Published data on forgings are for material annealed and aged: 1 h at 900 °C and air cool, plus 8 h at 595 °C (1100 °F) and air cool. Published data for 57-mm-diam (2¼-in.-diam) bar stock are for material annealed and aged: 1 h at 955 °C (1750 °F) and air cool, plus 8 h at 595 °C and air cool.

Fig. 6 Comparison of fatigue-crack-growth behavior of Ti-6Al-4V castings and hot rolled plate

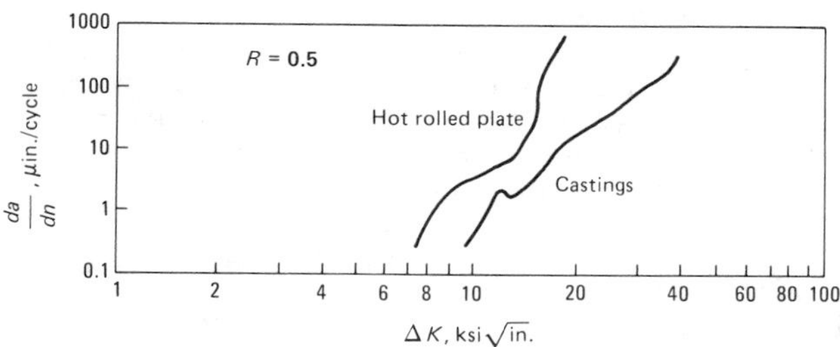

Fatigue testing was performed at 24 °C (75 °F) in air at 10 Hz.

Fig. 7 Notch fatigue strength of as-cast Ti-6Al-4V

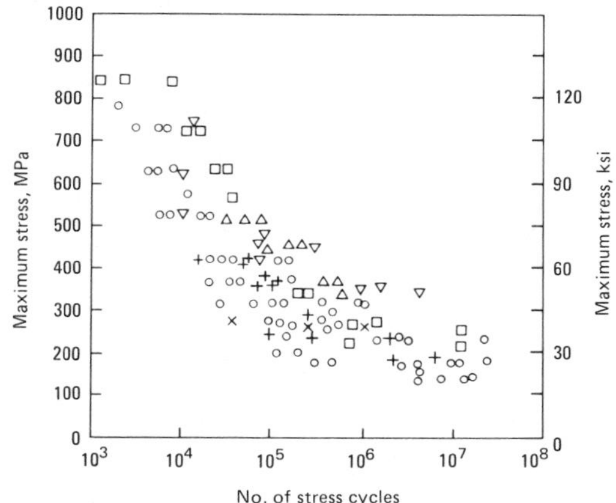

Each symbol represents fatigue data from a different source. Stress ratio, *R*, typically was +0.1; stress concentration factor, K_t, was mostly 3.0, but a few tests were run at K_t = 1.0.

Fig. 8 Cost comparison for one design of titanium alloy aircraft fitting machined from blocks, forgings and castings

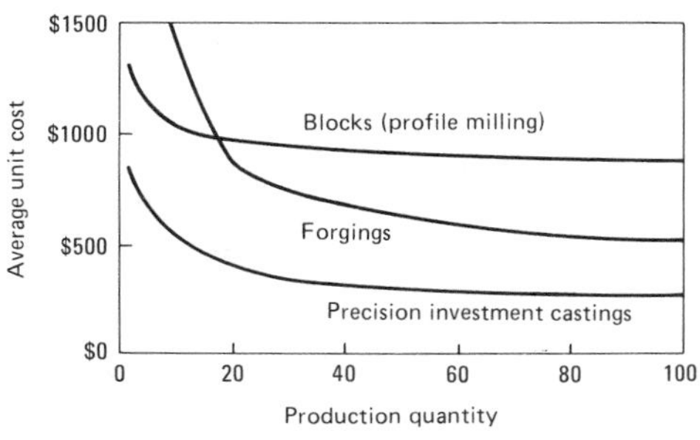

in greater yield: more metal from each melt in the form of usable castings and less in the form of revert (remelt stock).

For certain casting configurations, adequate feeding by use of conventional risering is virtually impossible, and, in order to meet aerospace nondestructive inspection standards, shrinkage voids are closed by welding. In such instances, hot isostatic processing becomes a means of avoiding weld repair and its attendant extra handling and NDT costs.

From a technical viewpoint, hot isostatic processing is not necessarily considered to be a heat treatment but rather a process for healing internal voids. Air Force and privately funded studies have shown that hot isostatic processing alone does little, if anything, to enhance mechanical properties of Ti-6Al-4V castings.

In one AFML report, the following conclusions were drawn:

- Shrinkage voids 1.5 to 3.0 mm (0.060 to 0.120 in.) in diameter do not affect tensile strength or ductility.
- Gas voids 0.6 to 1.9 mm (0.025 to 0.075 in.) in diameter result in equivalent strength, and a reduction of 2 to 5% in elongation, compared to void-free material.
- Castings containing typical repair welds no larger than 13 mm (0.5 in.) are 35 to 40 MPa (5 to 6 ksi) greater in strength, and 2 to 3% lower in elongation, than defect-free material.
- Hot isostatic processing results in ductilities comparable to those of defect-free material and in strengths that are on the low side of the scatter band for defect-free material.
- With the exception of a few data points, low-cycle fatigue life of both hot isostatic processed material and material containing a small amount of shrinkage or gas voids was equivalent to the mean curve for defect-free material generated in an earlier Air Force report.
- High-cycle fatigue properties of defect-free material, hot isostatic processed material and material containing internal imperfections are essentially equivalent.

With castings of marginal to substandard quality, hot isostatic processing raises the lower limit of data scatter and raises the degree of confidence in the reliability of cast products.

HIP is considered by many to be a

process that simplifies the problem of defining a standard for internal casting quality. Hot isostatic processed parts also are esthetically more acceptable. At the same time, use of HIP ensures that subsurface microporosity will be healed and therefore will not become exposed on a subsequently machined or polished surface to mar the finish or to act as a possible site for fatigue-crack propagation.

Cost Comparisons

A major aircraft manufacturer made an in-depth study in which costs of precision titanium alloy castings were compared with costs of parts machined from forgings and blocks of both titanium and aluminum. On the average, metal-removal (machining) costs constitute about 60% of total fabrication cost of an airplane. The use of precision investment castings reduces machining cost to about 5% of total part cost compared to as much as 70 to 80% for the same part made from a forging or hogged out of a block.

Figure 8 illustrates the relation of cost to number of units produced for a specific design of aircraft fitting. For all quantities, it was least expensive to produce the fitting as an investment casting.

For a series of 16 parts from one model of commercial aircraft, the average cost of a single part was $749 when 200 units were hogged from titanium alloy blocks. The average cost for investment castings in the same production quantity was $227. This represented a savings of $835 100 for each 100 airplanes constructed if the parts were made from castings.

For a series of eight parts from another model, average cost was $316 when the parts were machined from aluminum forgings, but only $179 when the parts were made from titanium investment castings. This represented a savings of $109 800 for each 100 airplanes.

Corrosion Resistance of Titanium

By L. C. Covington
Chief Corrosion Engineer
Henderson Technical Laboratories
Timet Division, Titanium Metals
Corporation of America

TITANIUM and its alloys are used chiefly for their desirable mechanical properties (most notably, high strength-to-weight ratio), but they also have outstanding resistance to corrosion. These metals have been used in a wide variety of aerospace, marine and chemical applications, and have been found to be immune to corrosion-related failure in most environments (Ref 1, 2).

Unalloyed titanium is highly resistant to corrosion by many natural environments, including seawater, body fluids, and fruit and vegetable juices. Titanium exposed continuously to seawater for about 18 years has undergone only superficial discoloration. Wet chlorine, molten sulfur, many organic compounds (including acids and chlorinated compounds) and most oxidizing acids have essentially no effect on this metal. Titanium is used extensively for handling salt solutions (including chlorides, hypochlorides, sulfates and sulfides), wet chlorine gas and nitric acid solutions.

On the other hand, hot concentrated low-pH chloride salts (such as boiling 30% $AlCl_3$ and boiling 70% $CaCl_2$) corrode titanium. Warm or concentrated solutions of HCl, H_2SO_4, H_3PO_4 and oxalic acid also are damaging (Ref 3). In general, all acidic solutions that are reducing in nature corrode titanium unless they contain inhibitors. Strong oxidizers, including anhydrous red fuming nitric acid and 90% H_2O_2, also cause attack. Ionizable fluoride compounds, such as NaF and HF, activate the surface and can cause rapid corrosion; dry chlorine gas is especially harmful.

Titanium has limited oxidation resistance in air at temperatures above about 650 °C (1200 °F), and chlorides or hydroxides deposited on its surface can accelerate oxidation. Exposure to liquid or gaseous oxygen, nitrogen tetroxide or red fuming nitric acid can cause titanium to react violently under impact loading.

Titanium has been used to contain liquid or supercritical hydrogen at cryogenic temperatures, but above −100 °C (−150 °F) hydrogen may severely embrittle titanium. The potential for embrittlement is enhanced where hydrogen flow rates are high or where coatings on the titanium become damaged.

In unalloyed titanium and many titanium alloys, weld zones are just as resistant to corrosion as the base metal. Other fabrication processes (such as bending, forming or machining) also appear to have no influence on basic corrosion resistance.

Passivation and Inhibition

Although titanium is chemically reactive, the thin oxide film that forms on titanium surfaces in most corrodents is relatively impervious and therefore quite protective. When titanium is not corrosion resistant, it is because the film is not fully protective. Reducing conditions, very powerful oxidizing environments and the presence of fluoride ions diminish the protective nature of the oxide film, but its stability and integrity can be improved substantially by adding inhibitors to the environment.

Most acidic solutions (except those containing soluble fluorides) can be inhibited by the presence of even small amounts of oxidizing agents and heavy-metal ions. Thus, titanium can be used in certain industrial process solutions (including hydrochloric and sulfuric acids) that otherwise would be corrosive. Nitric and chromic acids, and dissolved salts of iron, nickel, copper and chromium, are especially effective inhibitors. Attack by red fuming nitric acid and chlorine gas can be inhibited by small amounts of water.

Metallic ions and oxygen from the air apparently adsorb on the titanium surface (Ref 4), whereas strong oxidizing conditions (such as nitric acid, air at moderately high temperatures, and anodic treatments) promote resistance through growth of the oxide film. Dissolved oxygen is an important inhibitor in hot or mildly reducing chloride solutions, but if its supply is restricted, as in deep crevices, corrosion may be accelerated.

Galvanic Corrosion

Coupling of titanium to dissimilar metals usually does not accelerate corrosion of the titanium except in reducing environments, where titanium does not become passivated. Under reducing conditions, it has a galvanic potential similar to that of aluminum and will undergo accelerated corrosion when coupled to more noble metals.

The table given below is one version of the galvanic series in seawater. In this environment, titanium exhibits a potential of about −0.1 V versus a saturated calomel reference cell, which places it high on the passive (noble) end of the series.

Galvanic series in seawater
Cathodic end (noble metals)

Platinum
Gold

Silver
Titanium

Cr-Ni stainless steels, passive

Straight Cr stainless steels, passive

Ni-Cu alloys (Monels)

Ni-Cr-Fe alloys (Inconels), passive
Nickel, passive
Silver solder
Tin bronzes
Copper nickels

Silicon bronzes
Copper
Red and yellow brasses
Aluminum bronzes

Ni-Cr-Fe alloys (Inconels), active
Nickel, active

High-Zn yellow brasses (>30% Zn)
Manganese bronzes

Tin
Lead

Cr-Ni stainless steels, active

Straight Cr stainless steels, active

Cast iron
Wrought iron
Low-carbon steel

2*xxx* and 7*xxx* aluminum alloys

Cadmium

Alclad aluminum alloys
6*xxx* aluminum alloys

Galvanized steel
Zinc
Magnesium alloys
Magnesium

Anodic end (active metals)

In most environments, titanium is the cathodic member of any galvanic couple. It may accelerate corrosion of the other member but in most instances will itself be unaffected. If the surface area of the titanium exposed to the environment is small in relation to the exposed surface area of the other metal, the effect of the titanium on the corrosion rate of the other metal will be negligible; but if the exposed area of titanium greatly exceeds that of the other metal, severe corrosion of the other metal may result.

Because titanium is nearly always the cathodic member of any galvanic couple, hydrogen may be evolved at its surface in an amount proportional to the galvanic current flow. This may result in formation of surface hydride films that generally are stable and cause no problems. At temperatures above 75 °C (170 °F), however, the hydrogen may diffuse into the metal, causing embrittlement. In some environments, titanium hydride is unstable and decomposes or reacts, with a resultant loss of metal.

Alloying Additions

Alloying titanium with other metals has pronounced effects on its mechanical, physical and chemical properties.

Anodic control of the corrosion reaction predominates when titanium is exposed to reducing acids such as hydrochloric or sulfuric. Alloying with elements that reduce anodic activity therefore should improve corrosion resistance. This can be accomplished by using alloying elements that: (*a*) shift the corrosion potential of the alloy in the positive direction (cathodic alloying), (*b*) increase the thermodynamic stability of the alloy and thus reduce the ability of the titanium to dissolve anodically or (*c*) increase the tendency of titanium to passivate (Ref 5). The first group includes noble metals such as platinum, palladium, and rhodium. The second includes nickel, molybdenum and tungsten. The third group includes zirconium, tantalum, chromium and possibly molybdenum.

Cathodic Alloying. Considerable work has been done on the use of noble metals as alloying additions in titanium (Ref 5 to 7). An outgrowth of this work has been the development of Ti-0.2Pd, which has considerably greater resistance to corrosion in reducing environments than that of unalloyed titanium.

Alloying for Thermodynamic Stability. In studying corrosion of titanium in aqueous salt solutions, Griess (Ref 8) noted that titanium alloys containing nickel, molybdenum or palladium were more resistant to nonoxidizing acid solutions than commercially pure titanium. He concluded that they also should be more resistant to crevice corrosion.

An alloy containing 2% nickel was developed (Ref 9) and recommended for service in hot brine environments where crevice corrosion is sometimes a problem. Subsequent studies (Ref 10) confirmed that this alloy has much better resistance to crevice corrosion than the unalloyed metal. However, the nickel addition has detrimental effects that diminish its over-all value. Ti-2Ni is very susceptible to hydrogen embrittlement (Ref 11) and is subject to severe edge cracking during rolling, which makes it difficult to produce. The increased susceptibility to hydrogen embrittlement is most likely due to the formation of Ti_2Ni, which absorbs hydrogen much more readily than does unalloyed titanium (Ref 12).

Passivation Alloying. Various studies have demonstrated that corrosion resistance of titanium is improved by addition of molybdenum (Ref 5, 8). The principal problem with Ti-Mo alloys is the difficulty of obtaining uniform distribution of the molybdenum in large ingots. Because titanium and molybdenum have such widely different melting points, molybdenum is difficult to dissolve and may segregate as high-density inclusions when large amounts are added.

The commercial alloy Ti Code 12, which contains 0.3% Mo and 0.8% Ni, combines some of the favorable properties of nickel and molybdenum additions while avoiding the negative aspects. This alloy has excellent resistance to pitting and crevice corrosion in high-temperature brines, which sometimes attack commercially pure titanium, and also has better resistance to oxidizing environments such as nitric acid. It resists corrosion in reducing environments such as HCl and H_2SO_4 better than commercially pure titanium but not as well as the titanium-palladium alloy.

Crevice Corrosion

Titanium is subject to crevice corrosion in brine solutions containing oxygen, because the oxygen in the crevice is consumed faster than it can diffuse in from the bulk solution (Ref 8, 13). As a result, the corrosion potential of metal in the crevice becomes more electronegative than the potential of metal exposed to the bulk solution. Metal in the crevice acts as the anode and dissolves under the influence of the resulting galvanic current. This produces an excess of positive ions at the anode, which is balanced by the migration of chloride ions into the crevice. The titanium chlorides formed in the crevice are unstable and tend to hydrolize, forming small amounts of HCl. This reaction is very slow at first, but in the very restricted volume of the crevice it can reduce the pH to values as low as 1, which reduces the potential still further until corrosion becomes severe.

Although crevice corrosion of titanium is observed most often in hot chloride solutions, it also occurs in iodide, bromide and sulfate solutions. Susceptibility increases with increasing temperature, increasing concentration of chloride ions, decreasing concentration of dissolved oxygen and decreasing pH. In solutions with neutral pH, crevice corrosion of titanium has not been observed at temperatures below 120 °C (250 °F). At lower pH values, crevice corrosion sometimes is encountered at temperatures below 120 °C.

Pitting Corrosion

Pitting corrosion is a form of localized corrosion closely related to crevice corrosion (Ref 14). Both are observed mainly on passive metals such as aluminum, stainless steels and titanium (Ref 15).

Pitting initiates at imperfections in the oxide film. Aggressive ions such as Cl^- concentrate at these sites (Ref 15) until they are able to displace the oxygen in the passive film. A small crevice is soon formed by an insoluble corrosion product, TiO_2, which fills and covers the pit, thus restricting diffusion into the growing pit and permitting acid conditions to develop.

During fabrication and installation of titanium equipment, the titanium must be handled with enough care to avoid contaminating it with embedded iron particles. Pitting failures of titanium tubing have been traced to scratches in which traces of iron were detected; the failures were attributed to smearing of iron particles into the passive film of TiO_2 until they had penetrated it.

The difference in corrosion potential between low-carbon steel and unalloyed titanium is nearly 0.5 V in saturated brine at temperatures near the boiling point. This difference is sufficient to establish an electrochemical cell in which the iron is consumed at the anode. Before the iron is completely consumed, however, a pit begins to grow in the titanium. Once the pit becomes established, acid conditions develop in the pit. Acid conditions prevent the passive film from re-forming, and corrosion continues until the titanium is perforated.

Anodizing as a final operation in fabrication and installation of titanium equipment helps remove extraneous iron particles and thickens the passive film so that corrosion and pitting are much less likely.

Erosion-Corrosion and Cavitation

For most materials there are critical velocities beyond which protective films are swept away and accelerated corrosion attack occurs. This accelerated attack is known as erosion-corrosion. The critical velocity differs greatly from one material to another and may be as low as 0.6 to 0.9 m/s (2 to 3 ft/s). For titanium, the critical velocity in seawater (Ref 16) is more than 27 m/s (90 ft/s). Numerous erosion-corrosion tests (Ref 1, 17) have shown titanium to have outstanding resistance to this form of attack.

Erosion-corrosion can be greatly aggravated by the presence of abrasive particles (such as sand) in a flowing fluid. Titanium (Ref 18) exhibited superior resistance to this type of attack in seawater containing fine sand that flowed through conventional titanium condenser tubes at 1.8 m/s (6 ft/s).

Cavitation is a phenomenon that occurs in flowing liquids, wherein the relative motion between the liquid and a surface across which it flows is great enough to locally reduce the pressure below the vapor pressure of the liquid. At the reduced pressure, bubbles form in the liquid. When the liquid containing cavitation bubbles flows into a region of higher pressure, the bubbles collapse, inflicting severe, highly localized forces on the surface against which they collapse. This can produce deep, rounded pits in the surface of almost any solid — even glass.

Cavitation resistance tests performed in Boston Harbor (Ref 19) proved titanium to be one of the metals most resistant to this type of damage.

Stress-Corrosion Cracking

Unalloyed titanium generally is immune to stress-corrosion cracking unless it has a high oxygen content (0.3% or more). For this reason stress-corrosion cracking is of little concern in the chemical process industries where unalloyed titanium is most commonly used. On the other hand, certain alloys of titanium used principally in the aerospace industry are subject to stress-corrosion cracking.

Early investigators (Ref 20) were unable to crack titanium alloys in boiling 42% Mg/Cl_2, and therefore concluded that they were immune to stress-corrosion cracking. One of the first reported instances of stress-corrosion cracking of unalloyed titanium occurred in red fuming nitric acid. It was also found that a pyrophoric surface deposit was formed on exposure to this acid. There was no evidence that these two phenomena are related, but adding 1.5 to 2.0% water to the acid completely inhibited both reactions.

Since then, stress-corrosion cracking has been demonstrated in hot dry sodium chloride, methanol, HCl solutions, seawater, chlorinated solvents, nitrogen tetroxide, mercury and cadmium.

One of the important variables affecting susceptibility to stress-corrosion cracking is alloy composition. Aluminum additions increase susceptibility to stress-corrosion cracking; alloys containing more than 6% Al generally are susceptible to stress corrosion. Additions of tin, manganese and cobalt are detrimental, whereas zirconium appears to be neutral. Beta stabilizers such as molybdenum, vanadium and niobium are beneficial. Susceptibility to stress-corrosion cracking also can be affected by heat treatment (Ref 21).

Accelerated Crack Propagation in Seawater

Titanium is known to be highly resistant to corrosion by seawater. However, for certain alloys, components containing very sharp notches or cracks exhibit accelerated crack propagation and thus lose resistance to fracture when exposed to seawater. This phenomenon was first discovered in laboratory tests using a technique developed by Brown at the Naval Research Laboratory (Ref 22). Failure of titanium due to loss of fracture resistance appears to be similar to delayed fracture of high-strength steels containing sharp notches or cracks on exposure to various liquid environments; it is not considered a form of stress-corrosion cracking.

Exposure to seawater does not appear to diminish service life of titanium alloys, such as Ti-8Mn and Ti-5Al-2.5Sn, that exhibit this phenomenon in laboratory testing. These two alloys have been employed successfully in aircraft during the past 15 years without reported failures. Apparently, the conditions leading to accelerated crack propagation (primarily, the existence of a crack) have not been encountered in service.

Accelerated crack propagation in seawater can be avoided by proper alloy selection. Alloys containing more than 6% aluminum are particularly susceptible. Additions of tin, manganese, cobalt and oxygen are detrimental, whereas beta stabilizers such as molybdenum, niobium and vanadium tend to reduce or eliminate susceptibility to this phenomenon. Unalloyed titanium is not susceptible unless it contains more than about 0.3% oxygen.

Hot Salt Corrosion

Titanium and titanium alloys can be damaged by halogenated compounds at temperatures above 260 °C (500 °F). Chloride salts — especially sodium chloride — are very detrimental. Residual salts cause surface pitting, or even cracking of certain alloys under high tensile loads. Although rarely encountered in service, cracking of titanium parts due to hot salt corrosion has been encountered by fabricators during stress-relieving operations. Responsibility has been traced to vapors of chlorinated cleaning fluids that were not completely removed prior to thermal processing, chloride traces from other process fluids (including tap water) and even salt residues from fingerprints.

Since the original finding that hot halogenated salts can damage titanium, the phenomenon has been studied extensively. Although much has been learned about the reaction, relative susceptibility, and related variables, it is generally agreed that laboratory tests do not simulate service conditions well and thus do not correctly predict field performance.

The extent of damage by salts is directly related to temperature, exposure time and tensile-stress level. Processing history, alloy composition, salt composition and other environmental conditions also have important effects (Ref 23).

Susceptibility to hot salt corrosion appears to be influenced considerably by processing and alloy additions. Therefore, control of these factors should make it possible to avoid the phenomenon in service.

Liquid-Metal Embrittlement

Some titanium alloys crack under tensile stress when in contact with liquid cadmium, mercury or silver-base brazing alloys. This type of embrittlement differs from stress-corrosion cracking although there are some similarities. Liquid-metal embrittlement appears to result from diffusion along grain boundaries and formation of brittle phases, which in turn produce the loss of ductility.

Titanium also can be embrittled by contact with certain solid metals (cadmium and silver, for example) when the titanium is under tensile stress. The failure mechanism is not completely understood, although many investigators believe it is similar to liquid-metal embrittlement. Service failures have occurred in cadmium-plated titanium alloys at temperatures as low as 65 °C (150 °F), and in silver-brazed titanium parts at temperatures above 315 °C (600 °F).

Silver-plated components should not be used in contact with titanium under stress at temperatures above 230 °C (450 °F). Cadmium-plated parts such as interference-fit fasteners or press-fit bushings should not be used in contact with titanium at any temperature. Other cadmium-plated parts and fasteners should not be used in contact with titanium at temperatures above 230 °C.

REFERENCES

1. D. W. Stough, F. W. Fink and R. S. Peoples, "The Corrosion of Titanium", TML Report No. 57, Battelle Memorial Institute, Oct 1956, 184 pages
2. D. Schlain, "Corrosion Properties of Titanium and Its Alloys", Bulletin 619, U.S. Bureau of Mines, 1964, 228 pages
3. L. B. Golden, I. R. Lane, Jr. and W. L. Acherman, Corrosion Resistance of Titanium, Zirconium and Stainless Steel in Mineral Acids, *Industrial and Engineering Chemistry,* Vol 44, Aug 1952, p 1930-1939
4. J. R. Cobb and H. H. Uhlig, Resistance of Titanium to Sulfuric and Hydrochloric Acids Inhibited by Ferric and Cupric Ions, *Journal of the Electrochemical Society,* Vol 99, Jan 1952, p 13-15
5. N. D. Tomashov, R. M. Altovsky, and G. P. Chernova, Passivity and Corrosion Resistance of Titanium and Its Alloys, *Journal of the Electrochemical Society,* Vol 108 (No. 2), Feb 1961, p 113-119
6. M. Stern and H. Wissenberg, The Influence of Noble Metal Alloy Additions on the Electrochemical and Corrosion Behavior of Titanium, *Journal of the Electrochemical Society,* Vol 106 (No. 9), Sept 1959, p 759-764
7. J. B. Cotton, in *Platinum Metals Review,* Vol 11, 1967, p 50-52
8. J. C. Griess, Crevice Corrosion of Titanium in Aqueous Salt Solutions, *Corrosion,* Vol 24, 1968, p 96-109
9. N. G. Feige and T. J. Murphy, "Corrosion Resistance of Titanium and Ti-2Ni in Hot Brine Environments", paper presented at NACE meeting, Houston, TX, March 1969
10. A. J. Sedriks, J. A. S. Green, and D. L. Novak, Electrochemical Behavior of Ti-Ni Alloys in Acidic Chloride Solutions, *Corrosion,* Vol 28, Apr 1972, p 137
11. L. C. Covington and N. G. Feige, A Study of Factors Affecting the Hydrogen Uptake Efficiency of Titanium in NaOH Solutions, paper presented at the Symposium on

Localized Corrosion, ASTM Annual Meeting, June 1971, Atlantic City, NJ
12. "Investigation of Galvanically Induced Hydriding of Titanium in Saline Solutions", Report to Office of Saline Water, U.S. Dept. of the Interior, Battelle Memorial Institute, Richland, WA, Sept 1970
13. E. G. Bohlmann and F. A. Posey, in *Proceedings of the First International Symposium on Water Desalination,* Vol 1, p 306-325
14. B. F. Brown, The Concept of the Occluded Corrosion Cell, *Corrosion,* Vol 26 (No. 8), Aug 1970, p 249-250
15. H. H. Uhlig, Distinguishing Characteristics of Pitting and Crevice Corrosion, *Materials Protection and Performance,* Vol 12 (No. 2), Feb 1973, p 42-44
16. G. J. Danek, Jr., The Effect of Sea-Water Velocity on the Corrosion Behavior of Metals, *Naval Engineers Journal,* Vol 78 (No. 5), Oct 1966, p 763-769
17. J. A. Davis, "The Effect of Velocity on the Sea Water Corrosion Behavior of High Performance Ship Materials", paper No. 78, Corrosion/74, Chicago, IL, March 4-8, 1974
18. J. B. Cotton and B. P. Downing, Corrosion Resistance of Titanium to Sea Water, *Transactions of the Institute of Marine Engineers,* Vol 69, Aug 1957, p 314
19. W. L. Williams, The Titanium Program at the U.S. Naval Experiment Station, *Journal of the American Society of Naval Engineers,* Vol 62, Nov 1950, p 865-869
20. W. K. Boyd, "Stress Corrosion Cracking of Titanium Alloys—An Overview", paper presented at the International Symposium on Stress Corrosion Mechanisms in Titanium Alloys, Jan 27-29, 1971, Georgia Institute of Technology, Atlanta, GA
21. A. J. Hatch, H. W. Rosenberg and E. F. Erbin, Effects of Environment on Cracking in Titanium Alloys", *Stress Corrosion Cracking of Titanium,* STP 397, American Society for Testing and Materials, Philadelphia, 1966
22. B. F. Brown, "A New Stress-Corrosion Cracking Test Procedure for High-Strength Alloys", paper presented at ASTM 68th Annual Meeting, June 13-18, 1965
23. V. C. Petersen and H. B. Bomberger, "The Mechanism of Salt Attack on Titanium Alloys", paper presented at ASTM meeting, Seattle, WA, Oct 31-Nov 5, 1965

Tool Materials

Tool Steels

By the ASM Committee on Tooling Materials*

A TOOL STEEL is any steel used to make tools for cutting, forming or otherwise shaping a material into a part or component adapted to a definite use. The earliest tool steels were simple, plain carbon steels, but beginning in 1868, and to a greater extent early in the 20th century, many complex, highly alloyed tool steels were developed. These complex alloy tool steels, which contain, among other elements, relatively large amounts of tungsten, molybdenum, manganese and chromium, make it possible to meet increasingly severe service demands and to provide greater dimensional control and freedom from cracking during heat treatment. Many alloy tool steels are also widely used for machinery components and structural applications where particularly severe requirements must be met, such as high-temperature springs, ultrahigh-strength fasteners, special-purpose valves, and bearings of various types for elevated-temperature service.

In service, most tools are subjected to extremely high loads that are applied rapidly. They must withstand these loads a great number of times without breaking and without undergoing excessive wear or deformation. In many applications, tool steels must provide this capability under conditions that develop high temperatures in the tool. No single tool material combines maximum wear resistance, toughness, and resistance to softening at elevated temperatures. Consequently, selection of the proper tool material for a given application often requires a trade-off to achieve the optimum combination of properties.

Most tool steels are wrought products, but precision castings can be used to advantage in some applications. The powder metallurgy (P/M) process also is used in making tool steels, and provides (*a*) more uniform carbide size and distribution in large sections and (*b*) special compositions that are difficult or impossible to produce by melting and casting and then mechanically working the cast product.

For typical wrought tool steels, raw materials (including scrap) are carefully selected, not only for alloy content but also for qualities that ensure cleanness and homogeneity in the finished product. Tool steels are generally melted in small-tonnage electric-arc furnaces to economically achieve composition tolerances, good cleanness and precise control of melting conditions. Special refining and secondary remelting processes have been introduced to satisfy particularly difficult demands regarding tool steel quality and performance. The medium-to-high alloy contents of many tool steels require careful control of forging and rolling, which often results in a large amount of process scrap. Semifinished and finished bars are given rigorous in-process and final inspection. This inspection can be so extensive that both ends of each bar may be inspected for macrostructure (etch quality), cleanness, hardness, grain size, annealed structure and hardenability, and that the entire bar may be subjected to magnetic and ultrasonic inspections for surface and internal discontinuities. It is important that finished tool steel bars have minimal decarburization within carefully controlled limits, which requires that annealing be done by special procedures under closely controlled conditions.

Such precise production practices and stringent quality controls contribute to the high cost of tool steels, as do the expensive alloying elements they contain. Insistence on quality in the manufacture of these specialty steels is justified, however, because tool steel bars generally are made into complicated cutting and forming tools worth many times the cost of the steel itself. Although some standard construction-al alloy steels resemble tool steels in composition, they are seldom used for expensive tooling because, in general, they are not manufactured to the same rigorous quality standards as are tool steels.

The performance of a tool in service depends on proper design of the tool, accuracy with which the tool is made, selection of the proper tool steel and application of the proper heat treatment. A tool can perform successfully in service only when all four of these requirements have been fulfilled.

With few exceptions, all tool steels must be heat treated to develop specific combinations of wear resistance, resistance to deformation or breaking under high loads, and resistance to softening at elevated temperatures. A few simple shapes may be obtained directly from tool steel producers in correctly heat treated condition. However, most tool steels first are formed or machined to produce the required shape and then are heat treated either by the tool manufacturer or by the ultimate user.

*See page XII for committee list.

Table 1 Composition limits of principal types of tool steels

Designations AISI	SAE	UNS	Composition(a), % C	Mn	Si	Cr	Ni	Mo	W	V	Co
Molybdenum high speed steels											
M1	M1	T11301	0.78-0.88	0.15-0.40	0.20-0.50	3.50-4.00	0.30 max	8.20-9.20	1.40-2.10	1.00-1.35	...
M2	M2	T11302	0.78-0.88; 0.95-1.05	0.15-0.40	0.20-0.45	3.75-4.50	0.30 max	4.50-5.50	5.50-6.75	1.75-2.20	...
M3, class 1	M3	T11313	1.00-1.10	0.15-0.40	0.20-0.45	3.75-4.50	0.30 max	4.75-6.50	5.00-6.75	2.25-2.75	...
M3, class 2	M3	T11323	1.15-1.25	0.15-0.40	0.20-0.45	3.75-4.50	0.30 max	4.75-6.50	5.00-6.75	2.75-3.75	...
M4	M4	T11304	1.25-1.40	0.15-0.40	0.20-0.45	3.75-4.75	0.30 max	4.25-5.50	5.25-6.50	3.75-4.50	...
M6	...	T11306	0.75-0.85	0.15-0.40	0.20-0.45	3.75-4.50	0.30 max	4.50-5.50	3.75-4.75	1.30-1.70	11.00-13.00
M7	...	T11307	0.97-1.05	0.15-0.40	0.20-0.55	3.50-4.00	0.30 max	8.20-9.20	1.40-2.10	1.75-2.25	...
M10	...	T11310	0.84-0.94; 0.95-1.05	0.10-0.40	0.20-0.45	3.75-4.50	0.30 max	7.75-8.50	...	1.80-2.20	...
M30	...	T11330	0.75-0.85	0.15-0.40	0.20-0.45	3.50-4.25	0.30 max	7.75-9.00	1.30-2.30	1.00-1.40	4.50-5.50
M33	...	T11333	0.85-0.92	0.15-0.40	0.15-0.50	3.50-4.00	0.30 max	9.00-10.00	1.30-2.10	1.00-1.35	7.75-8.75
M34	...	T11334	0.85-0.92	0.15-0.40	0.20-0.45	3.50-4.00	0.30 max	7.75-9.20	1.40-2.10	1.90-2.30	7.75-8.75
M36	...	T11336	0.80-0.90	0.15-0.40	0.20-0.45	3.75-4.50	0.30 max	4.50-5.50	5.50-6.50	1.75-2.25	7.75-8.75
M41	...	T11341	1.05-1.15	0.20-0.60	0.15-0.50	3.75-4.50	0.30 max	3.25-4.25	6.25-7.00	1.75-2.25	4.75-5.75
M42	...	T11342	1.05-1.15	0.15-0.40	0.15-0.65	3.50-4.25	0.30 max	9.00-10.00	1.15-1.85	0.95-1.35	7.75-8.75
M43	...	T11343	1.15-1.25	0.20-0.40	0.15-0.65	3.50-4.25	0.30 max	7.50-8.50	2.25-3.00	1.50-1.75	7.75-8.75
M44	...	T11344	1.10-1.20	0.20-0.40	0.30-0.55	4.00-4.75	0.30 max	6.00-7.00	5.00-5.75	1.85-2.20	11.00-12.25
M46	...	T11346	1.22-1.30	0.20-0.40	0.40-0.65	3.70-4.20	0.30 max	8.00-8.50	1.90-2.20	3.00-3.30	7.80-8.80
M47	...	T11347	1.05-1.15	0.15-0.40	0.20-0.45	3.50-4.00	0.30 max	9.25-10.00	1.30-1.80	1.15-1.35	4.75-5.25
Tungsten high speed steels											
T1	T1	T12001	0.65-0.80	0.10-0.40	0.20-0.40	3.75-4.00	0.30 max	...	17.25-18.75	0.90-1.30	...
T2	T2	T12002	0.80-0.90	0.20-0.40	0.20-0.40	3.75-4.50	0.30 max	1.00 max	17.50-19.00	1.80-2.40	...
T4	T4	T12004	0.70-0.80	0.10-0.40	0.20-0.40	3.75-4.50	0.30 max	0.40-1.00	17.50-19.00	0.80-1.20	4.25-5.75
T5	T5	T12005	0.75-0.85	0.20-0.40	0.20-0.40	3.75-5.00	0.30 max	0.50-1.25	17.50-19.00	1.80-2.40	7.00-9.50
T6		T12006	0.75-0.85	0.20-0.40	0.20-0.40	4.00-4.75	0.30 max	0.40-1.00	18.50-21.00	1.50-2.10	11.00-13.00
T8	T8	T12008	0.75-0.85	0.20-0.40	0.20-0.40	3.75-4.50	0.30 max	0.40-1.00	13.25-14.75	1.80-2.40	4.25-5.75
T15	...	T12015	1.50-1.60	0.15-0.40	0.15-0.40	3.75-5.00	0.30 max	1.00 max	11.75-13.00	4.50-5.25	4.75-5.25
Chromium hot work steels											
H10	...	T20810	0.35-0.45	0.25-0.70	0.80-1.20	3.00-3.75	0.30 max	2.00-3.00	...	0.25-0.75	...
H11	H11	T20811	0.33-0.43	0.20-0.50	0.80-1.20	4.75-5.50	0.30 max	1.10-1.60	...	0.30-0.60	...
H12	H12	T20812	0.30-0.40	0.20-0.50	0.80-1.20	4.75-5.50	0.30 max	1.25-1.75	1.00-1.70	0.50 max	...
H13	H13	T20813	0.32-0.45	0.20-0.50	0.80-1.20	4.75-5.50	0.30 max	1.10-1.75	...	0.80-1.20	...
H14		T20814	0.35-0.45	0.20-0.50	0.80-1.20	4.75-5.50	0.30 max	...	4.00-5.25	...	...
H19		T20819	0.32-0.45	0.20-0.50	0.20-0.50	4.00-4.75	0.30 max	0.30-0.55	3.75-4.50	1.75-2.20	4.00-4.50
Tungsten hot work steels											
H21	H21	T20821	0.26-0.36	0.15-0.40	0.15-0.50	3.00-3.75	0.30 max	...	8.50-10.00	0.30-0.60	...
H22	...	T20822	0.30-0.40	0.15-0.40	0.15-0.40	1.75-3.75	0.30 max	...	10.00-11.75	0.25-0.50	...
H23	...	T20823	0.25-0.35	0.15-0.40	0.15-0.60	11.00-12.75	0.30 max	...	11.00-12.75	0.75-1.25	...
H24	...	T20824	0.42-0.53	0.15-0.40	0.15-0.40	2.50-3.50	0.30 max	...	14.00-16.00	0.40-0.60	...
H25	...	T20825	0.22-0.32	0.15-0.40	0.15-0.40	3.75-4.50	0.30 max	...	14.00-16.00	0.40-0.60	...
H26	...	T20826	0.45-0.55(b)	0.15-0.40	0.15-0.40	3.75-4.50	0.30 max	...	17.25-19.00	0.75-1.25	...
Molybdenum hot work steels											
H42	...	T20842	0.55-0.70(b)	0.15-0.40	...	3.75-4.50	0.30 max	4.50-5.50	5.50-6.75	1.75-2.20	...

(a) All steels except group W contain 0.25 max Cu, 0.03 max P and 0.03 max S; group W steels contain 0.20 max Cu, 0.025 max P and 0.025 max S. Where specified, sulfur may be increased to 0.06 to 0.15% to improve machinability of group H, M and T steels. (b) Available in several carbon ranges. (c) Contains free graphite in the microstructure. (d) Optional. (e) Specified carbon ranges are designated by suffix numbers.

Table 1 (continued)

Designations AISI	SAE	UNS	C	Mn	Si	Cr	Composition(a), % Ni	Mo	W	V	Co
Air-hardening medium-alloy cold work steels											
A2	A2	T30102	0.95-1.05	1.00 max	0.50 max	4.75-5.50	0.30 max	0.90-1.40	...	0.15-0.50	...
A3	...	T30103	1.20-1.30	0.40-0.60	0.50 max	4.75-5.50	0.30 max	0.90-1.40	...	0.80-1.40	...
A4	...	T30104	0.95-1.05	1.80-2.20	0.50 max	0.90-2.20	0.30 max	0.90-1.40	...	...	...
A6	...	T30106	0.65-0.75	1.80-2.50	0.50 max	0.90-1.20	0.30 max	0.90-1.40	...	...	...
A7	...	T30107	2.00-2.85	0.80 max	0.50 max	5.00-5.75	0.30 max	0.90-1.40	0.50-1.50	3.90-5.15	...
A8	...	T30108	0.50-0.60	0.50 max	0.75-1.10	4.75-5.50	0.30 max	1.15-1.65	1.00-1.50	...	...
A9	...	T30109	0.45-0.55	0.50 max	0.95-1.15	4.75-5.50	1.25-1.75	1.30-1.80	...	0.80-1.40	...
A10	...	T30110	1.25-1.50(c)	1.60-2.10	1.00-1.50	...	1.55-2.05	1.25-1.75	...	...	...
High-carbon, high-chromium cold work steels											
D2	D2	T30402	1.40-1.60	0.60 max	0.60 max	11.00-13.00	0.30 max	0.70-1.20	...	1.10 max	1.00 max
D3	D3	T30403	2.00-2.35	0.60 max	0.60 max	11.00-13.50	0.30 max	...	1.00 max	1.00 max	...
D4	...	T30404	2.05-2.40	0.60 max	0.60 max	11.00-13.00	0.30 max	0.70-1.20	...	1.00 max	...
D5	D5	T30405	1.40-1.60	0.60 max	0.60 max	11.00-13.00	0.30 max	0.70-1.20	...	1.00 max	2.50-3.50
D7	D7	T30407	2.15-2.50	0.60 max	0.60 max	11.50-13.50	0.30 max	0.70-1.20	...	3.80-4.40	...
Oil-hardening cold work steels											
O1	O1	T31501	0.85-1.00	1.00-1.40	0.50 max	0.40-0.60	0.30 max	...	0.40-0.60	0.30 max	...
O2	O2	T31502	0.85-0.95	1.40-1.80	0.50 max	0.35 max	0.30 max	0.30 max	...	0.30 max	...
O6	O6	T31506	1.25-1.55(c)	0.30-1.10	0.55-1.50	0.30 max	0.30 max	0.20-0.30	...	...	...
O7	...	T31507	1.10-1.30	1.00 max	0.60 max	0.35-0.85	0.30 max	0.30 max	1.00-2.00	0.40 max	...
Shock-resisting steels											
S1	S1	T41901	0.40-0.55	0.10-0.40	0.15-1.20	1.00-1.80	0.30 max	0.50 max	1.50-3.00	0.15-0.30	...
S2	S2	T41902	0.40-0.55	0.30-0.50	0.90-1.20	...	0.30 max	0.30-0.60	...	0.50 max	...
S5	S5	T41905	0.50-0.65	0.60-1.00	1.75-2.25	0.35 max	...	0.20-1.35	...	0.35 max	...
S6		T41906	0.40-0.50	1.20-1.50	2.00-2.50	1.20-1.50	...	0.30-0.50	...	0.20-0.40	...
S7		T41907	0.45-0.55	0.20-0.80	0.20-1.00	3.00-3.50	...	1.30-1.80	...	0.20-0.30(d)	...
Low-alloy special-purpose tool steels											
L2	...	T61202	0.45-1.00(b)	0.10-0.90	0.50 max	0.70-1.20	...	0.25 max	...	0.10-0.30	...
L6	L6	T61206	0.65-0.75	0.25-0.80	0.50 max	0.60-1.20	1.25-2.00	0.50 max	...	0.20-0.30(d)	...
Low-carbon mold steels											
P2	...	T51602	0.10 max	0.10-0.40	0.10-0.40	0.75-1.25	0.10-0.50	0.15-0.40	...	...	...
P3	...	T51603	0.10 max	0.20-0.60	0.40 max	0.40-0.75	1.00-1.50	...	...	...	...
P4	...	T51604	0.12 max	0.20-0.60	0.10-0.40	4.00-5.25	...	0.40-1.00	...	...	...
P5	...	T51605	0.10 max	0.20-0.60	0.40 max	2.00-2.50	0.35 max	...	...	...	...
P6	...	T51606	0.05-0.15	0.35-0.70	0.10-0.40	1.25-1.75	3.25-3.75	...	...	...	...
P20	...	T51620	0.28-0.40	0.60-1.00	0.20-0.80	1.40-2.00	...	0.30-0.55	...	...	...
P21	...	T51621	0.18-0.22	0.20-0.40	0.20-0.40	0.20-0.30	3.90-4.25	...	...	0.15-0.25	1.05-1.25Al
Water-hardening tool steels											
W1	W108,W109, W110,W112	T72301	0.70-1.50(e)	0.10-0.40	0.10-0.40	0.15 max	0.20 max	0.10 max	0.15 max	0.10 max	...
W2	W209,W210	T72302	0.85-1.50(e)	0.10-0.40	0.10-0.40	0.15 max	0.20 max	0.10 max	0.15 max	0.15-0.35	...
W5	...	T72305	1.05-1.15	0.10-0.40	0.10-0.40	0.40-0.60	0.20 max	0.10 max	0.15 max	0.10 max	...

(a) All steels except group W contain 0.25 max Cu, 0.03 max P and 0.03 max S; group W steels contain 0.20 max Cu, 0.025 max P and 0.025 max S. Where specified, sulfur may be increased to 0.06 to 0.15% to improve machinability of group H, M and T steels. (b) Available in several carbon ranges. (c) Contains free graphite in the microstructure. (d) Optional. (e) Specified carbon ranges are designated by suffix numbers.

Classification and Characteristics

Table 1 gives composition limits for the tool steels most commonly used in 1978. Each group of tool steels of similar composition and properties is identified by a capital letter; within each group, individual tool steel types are assigned code numbers. Table 2 identifies tool steel types that have been dropped from active listings because they are no longer commonly used.

Specifications. Tool steels are produced to various standards including several ASTM specifications. Reference 1 contains much useful information that essentially represents the normal manufacturing practices of most of the tool steel producers. Frequently, more stringent chemical and/or metallurgical standards are invoked by the individual producers to achieve certain commercial goals. Where appropriate, standard specifications for tool steels—ASTM A600, A681 and A686—may be used as a basis for procurement. ASTM A600 sets forth standard requirements for both tungsten and molybdenum high speed steels; A681 is applicable to hot work, cold work, shock-resisting, special-purpose and mold steels; A686 covers water-hardening tool steels. In many instances, however, tool steels are purchased by trade name because the user has found that a particular tool steel from a particular producer gives better performance in the specific application than a tool steel of the same AISI type classification purchased from another source.

Table 2 Compositions of tool steels no longer in common use

Type	Composition, % C	W	Mo	Cr	V	Others
High speed steels						
M8	0.80	5.00	5.00	4.00	1.50	1.25 Nb
M15	1.50	6.50	3.50	4.00	5.00	5.00 Co
M35	0.80	6.00	5.00	4.00	2.00	5.00 Co
M45	1.25	8.00	5.00	4.25	1.60	5.50 Co
T3	1.05	18.00	...	4.00	3.00	...
T7	0.75	14.00	...	4.00	2.00	...
T9	1.20	18.00	...	4.00	4.00	...
Hot work steels						
H15	0.40	...	5.00	5.00	...	...
H16	0.55	7.00	...	7.00	...	...
H20	0.35	9.00	...	2.00	...	...
H41	0.65	1.50	8.00	4.00	1.00	...
H43	0.55	...	8.00	4.00	2.00	...
Cold work steels						
D1	1.00	...	1.00	12.00	...	...
D6 (a)						
A5	1.00	...	1.00	1.00	...	3.00 Mn
Shock-resisting steels						
S3	0.50	1.00	...	0.74	...	...
S4	0.55	...	...	...	...	2.00 Si, 0.80 Mn
Mold steel						
P1	0.10	...	...	...	...	...
Special-purpose tool steels						
L1	1.00	...	...	1.25	...	...
L3	1.00	...	...	1.50	0.20	...
L4	1.00	...	...	1.50	0.25	0.60 Mn
L5	1.00	...	0.25	1.00	...	1.00 Mn
L7	1.00	...	0.40	1.40	...	0.35 Mn
F1	1.00	1.25	...	...	...	...
F2	1.25	3.50	...	...	...	...
F3	1.25	3.50	...	0.75	...	...
Water-hardening tool steels						
W3	1.00	...	...	...	0.50	...
W4	0.60/1.40(b)	...	...	0.25	...	...
W6	1.00	...	...	0.25	0.25	...
W7	1.00	...	...	0.50	0.20	...

(a) Now included with D3 in Table 1. (b) Various carbon contents were available.

High Speed Steels

High speed steels are tool materials developed largely for use in high speed cutting-tool applications. There are two classifications of high speed steels: molybdenum high speed steels (group M) and tungsten high speed steels (group T). Group M steels constitute about 95% of all high speed steel produced in the U. S.

Group M and group T high speed steels are equivalent in performance; the main advantage of group M steels is lower initial cost (approximately 40% lower than that of similar group T steels). This difference in cost results from the fact that the atomic weight of molybdenum is about one-half that of tungsten; on a weight-percent basis, only about one-half as much molybdenum as tungsten is required to provide the same atom ratio.

Molybdenum high speed steels and tungsten high speed steels are similar in many other respects, including hardenability. Typical applications for group M and group T steels include cutting tools of all kinds. Some grades are satisfactory for cold work applications, such as cold-header die inserts, thread-rolling dies, punches and blanking dies. Steels of the M40 series are used to make cutting tools for machining modern, very tough, high-strength steels.

For die inserts and punches, high speed steels frequently are under hardened—that is, quenched from austenitizing temperatures lower than those recommended for cutting-tool applications—as a means of increasing toughness.

Molybdenum high speed steels contain molybdenum, tungsten, chromium, vanadium, cobalt and carbon as principal alloying elements. Group M steels have slightly greater toughness than group T steels at the same hardness. Otherwise, mechanical properties of the two groups are similar.

Increasing the carbon and vanadium contents of group M steels increases wear resistance; increasing the cobalt content improves red hardness but concurrently lowers toughness. Type M2 has unusually high resistance to softening at elevated temperatures (see Fig. 1) as a result of high alloy content.

Fig. 1 Variation of hardness with tempering temperature for four typical tool steels

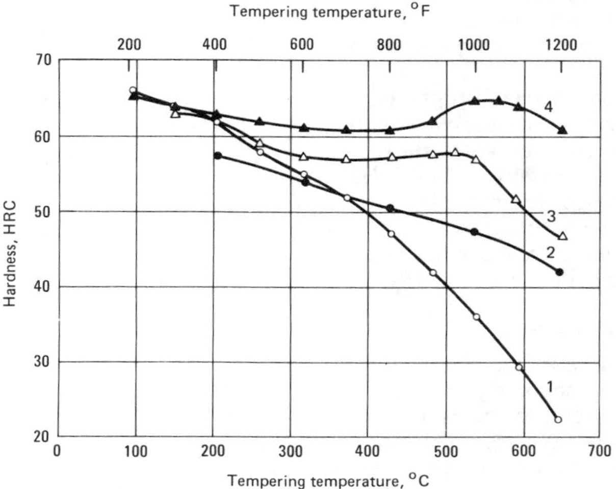

Curves are for 1 h at temperature. Curve 1 illustrates low resistance to softening at elevated temperature, such as is exhibited by group W and group O tool steels. Curve 2 illustrates medium resistance to softening, such as is exhibited by type S1 tool steel. Curves 3 and 4 illustrate high and very high resistance to softening, respectively, such as are exhibited by the secondary-hardening tool steels A2 and M2.

Because group M steels readily decarburize and are easily damaged due to overheating under adverse austenitizing environments, they are more sensitive than group T steels to hardening conditions—particularly austenitizing temperature and atmosphere. This is especially true of the high-molybdenum, low-tungsten compositions.

Group M high speed steels are deep hardening. They must be austenitized at temperatures lower than those for hardening group T steels to avoid incipient melting. Group M high speed steels can develop full hardness when quenched from temperatures of 1175 to 1230 °C (2150 to 2250 °F). Type M10, which usually has slightly lower hardenability than other molybdenum high speed steels, must be oil quenched if section size is larger than about 40 to 50 mm (about 1½ to 2 in.).

The maximum hardness that can be obtained in group M tool steels varies with composition. For those with lower carbon contents—types M1, M2, M10 (low-carbon composition), M30, M33, M34 and M36—maximum hardness is usually 65 HRC. For higher carbon contents—including types M3, M4 and M7—maximum hardness is about 66 HRC. A hardness of 66 HRC also can be developed in the lower-carbon, high-cobalt type M6. Maximum hardness of the higher-carbon cobalt-containing steels—types M41, M42, M43, M44 and M46—is 69 to 70 HRC. However, few industrial applications exist for steels of the M40 series at this maximum hardness. Usually, the tempering temperature is adjusted to provide a hardness of 66 to 68 HRC.

Tungsten high speed steels contain tungsten, chromium, vanadium, cobalt and carbon as the principal alloying elements. Type T1 was developed partly as a result of the work of Taylor and White, who in the early 1900's found that certain steels with over 14% W, about 4% Cr and about 0.3% V resisted softening at temperatures high enough to cause the steel to emit radiation in the red part of the visible spectra—or in other words, they exhibited red hardness. In its earliest form, type T1 contained about 0.68 C, 18 W, 4 Cr and 0.3 V. By 1920, the vanadium content had been increased to about 1.0%. Carbon content was gradually increased over a 30-year period to its present level of 0.75%.

Group T tool steels are characterized by high red hardness and wear resistance. They are so deep hardening that sections up to 75 mm (3 in.) in thickness or diameter can be hardened to 65 HRC or more by quenching in oil or molten salt. The high alloy and high carbon contents produce a large number of hard, wear-resistant carbides in the microstructure, particularly in those types containing more than 1.5% V and more than 1.0% C. Type T15 is the most wear-resistant steel of this group.

The combination of good wear resistance and high red hardness makes Group T high speed steels suitable for many high-performance cutting-tool applications; their toughness allows them to outperform cemented carbides in delicate tools and interrupted-cut applications. Group T tool steels are used primarily for cutting tools such as bits, drills, reamers, taps, broaches, milling cutters, and hobs. These steels are also used for making dies, punches, and high-load, high-temperature structural components such as aircraft bearings and pump parts.

Group T tool steels are all deep hardening when quenched from their recommended hardening temperatures of 1200 to 1300 °C (2200 to 2375 °F). They are seldom used to make hardened tools with section sizes greater than 75 mm (3 in.). Even very large cutting tools, such as drills 75 and 100 mm (3 and 4 in.) in diameter, have relatively small effective sections for hardening because metal has been removed to form the flutes. Some large-diameter solid tools are made from group T high speed steels; these include broaches and cold extrusion punches as large as 100 to 125 mm (4 to 5 in.) in diameter. For such tools, surface hardness is of primary importance.

As shown in Fig. 2, the difference between surface hardness and center hardness varies with bar size. The data in Fig. 2 are given more to indicate the general trend than to provide specific values of hardness variation. Section size and total mass of a given tool often have an effect on its response to a given hardening treatment that is equal to or greater than the effect of the grade of tool steel selected. For tools of extremely large diameter or heavy section, it is relatively common practice to use an accelerated oil quench to provide full hardness. This practice may yield values of Rockwell C hardness only one or two points higher than those obtainable through hot-salt quenching or air cooling, which ordinarily produce full hardness in tools smaller than about 75 mm (3 in.), but at such high hardnesses a one- or two-point increase in Rockwell hardness may prove quite significant.

Maximum hardness of tungsten high speed steels varies with carbon content, and to a lesser degree with alloy content. A hardness of at least 64.5 HRC can be developed in any high speed steel. Those types that have high carbon contents and hard carbides, such as T15, may be hardened to 67 HRC.

Hot Work Steels

Many manufacturing operations involve punching, shearing or forming of metals at high temperatures. Hot work steels (group H) have been developed to withstand the combinations of heat, pressure and abrasion associated with such operations. Table 3 gives data on resistance to softening after 100 h at temperatures from 480 to 760 °C (900 to 1400 °F) for four of these steels.

Generally, group H tool steels have medium carbon contents (0.35 to 0.45%), and chromium, tungsten, molybdenum and vanadium contents totaling 6 to 25%. They are divided into three subgroups: chromium hot work steels (types H10 to H19), tungsten hot work steels (types H21 to H26) and molybdenum hot work steel (type H42).

Chromium hot work steels (types H10 to H19) have good resistance to heat softening because of their medium chromium content and the addition of carbide-forming elements such as molybdenum, tungsten and vanadium. The low carbon and low total alloy contents promote toughness at the normal working hardnesses of 40 to 55 HRC. Higher tungsten and molybdenum contents increase hot strength but slightly reduce toughness. Vanadium is added to increase resistance to washing (erosive wear) at high temperatures. An increase in silicon content improves oxidation resistance at temperatures up to 800 °C (1475 °F). The most widely used types in this group are H11, H12, H13 and, to a lesser extent, H19.

All of the chromium hot work steels are deep hardening. H11, H12 and H13 may be air hardened to full working hardness in section sizes up to 150 mm (6 in.); other group H steels may be air hardened in section sizes up to 300 mm (12 in.). The air-hardening qualities and balanced alloy contents of these steels result in low distortion during hardening. Chromium hot work steels are especially well adapted to hot die work of all kinds, particularly dies for extrusion of aluminum and magnesium, as well as die-casting dies, forging dies, mandrels and hot shears. Most of these steels have alloy and carbon contents low enough that tools made from them can be water cooled in service without cracking.

H11 tool steel is used to make certain highly stressed structural parts, particularly in aerospace technology. Material for such demanding applications is produced by vacuum-arc remelting of air-melted electrodes. Vacuum-arc remelting provides extremely low residual gas contents, excellent microcleanness and a high degree of structural homogeneity.

The chief advantage of H11 over conventional high-strength steels is its ability to resist softening during continued exposure to temperatures up to 540 °C (1000 °F) and at the same time provide moderate toughness and ductil-

Fig. 2 Variation of surface and center hardness with bar diameter for four high speed steels

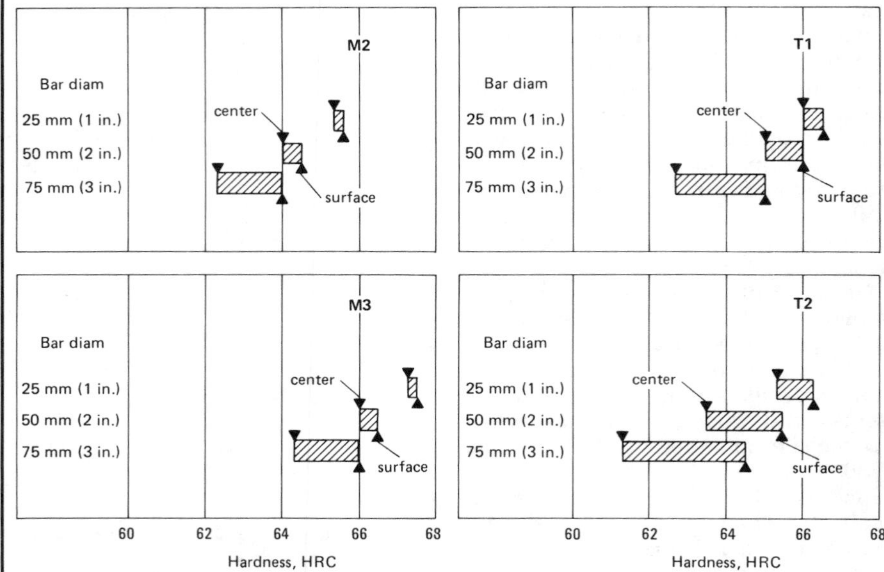

Steels M2 and M3 were oil quenched from 1205 °C (2200 °F) and 1230 °C (2250 °F), respectively. Steels T1 and T2 were oil quenched from 1290 °C (2350 °F).

Table 3 Resistance of four hot work steels to softening at elevated temperatures

Type	Original hardness, HRC	480 °C 900 °F	540 °C 1000 °F	600 °C 1100 °F	650 °C 1200 °F	700 °C 1300 °F	760 °C 1400 °F
		Hardness(a), HRC, after 100 h at:					
H13	50.2	48.7	45.3	29.0	22.7	20.1	13.9
	41.7	38.6	39.3	27.7	23.7	20.2	13.2
H21	49.2	48.7	47.6	37.2	27.4	19.8	15.2
	36.7	34.8	34.9	32.6	27.1	19.8	14.9
H23	40.8	40.0	40.6	40.8	38.6	33.2	25.8
	38.9	38.9	38.0	38.0	37.1	32.5	25.6
H26	51.0	50.6	50.3	47.1	38.4	26.9	21.3
	42.9	42.4	42.3	41.3	34.9	26.4	21.1

(a) At room temperature.

ity at room-temperature tensile strengths of 1720 to 2070 MPa (250 to 300 ksi). In addition, because of its secondary hardening characteristic, H11 can be tempered at high temperatures, resulting in nearly complete relief of residual hardening stresses, which is necessary for maximum toughness at high strength levels. Other important advantages of H11, H12 and H13 steels for structural and hot work applications include ease of forming and working, good weldability, relatively low coefficient of thermal expansion, acceptable thermal conductivity and above-average resistance to oxidation and corrosion.

Tungsten Hot Work Steels. The principal alloying elements of tungsten hot work steels (types H21 to H26) are carbon, tungsten, chromium and vanadium. The higher alloy contents of these steels make them more resistant to high-temperature softening and washing than H11 and H13 hot work steels. However, high alloy content also makes them more prone to brittleness at normal working hardnesses (45 to 55 HRC) and makes it difficult for them to be safely water cooled in service.

Although tungsten hot work steels can be air hardened, they are usually quenched in oil or hot salt to minimize scaling. When air hardened, they exhibit low distortion. Tungsten hot work steels require higher hardening temperatures than chromium hot work steels, which makes them more likely to scale when heated in an oxidizing atmosphere.

Although these steels have much greater toughness, in many characteristics, they are similar to high speed steels; in fact, type H26 is a low-carbon version of T1 high speed steel. If tungsten hot work steels are preheated to operating temperature before use, breakage can be minimized. These steels have been used to make mandrels and extrusion dies for high-temperature applications such as extrusion of brass, nickel alloys and steel, and are also suitable for use in hot forging dies of rugged design.

Molybdenum Hot Work Steel. There is only one active molybdenum hot work steel: type H42. This alloy contains molybdenum, chromium, vanadium and carbon, with varying amounts of tungsten. It is similar to tungsten hot work steels, having almost identical characteristics and uses. Although its composition resembles those of various molybdenum high speed steels, type H42 has a low carbon content and greater toughness. The principal advantage of type H42 over tungsten hot work steels is its lower initial cost. Type H42 is more resistant to heat checking than tungsten hot work steels but, in common with all high-molybdenum steels, requires greater care in heat treatment—particularly with regard to decarburization and control of austenitizing temperature.

Cold Work Steels

Cold work tool steels, because they do not have the alloy content necessary to make them resistant to softening at elevated temperature, are restricted in application to those uses that do not involve prolonged or repeated heating above 200 to 260 °C (400 to 500 °F). There are three categories of cold work steels: air-hardening steels (group A); high-carbon, high-chromium steels (group D); and oil-hardening steels (group O).

Air-hardening medium-alloy cold work steels (group A) contain enough alloying elements to enable them to achieve full hardness in sections up to about 100 mm (4 in.) in diameter on air cooling from the austenitizing temperature. (Type A6 through hardens in sections as large as a cube 175 mm, or 7 in., on a side.) Because they are air-hardening, group A tool steels exhibit minimum distortion and the highest safety (least tendency to crack) in hardening. Manganese, chromium and molybdenum are the principal alloying elements used to provide this deep hardening. Types A2, A3, A7, A8 and A9 contain a high percentage of chromium (5%), which provides moderate resistance to softening at elevated temperatures (see curve 3 in Fig. 1 for a plot of hardness versus tempering temperature for type A2).

Types A4, A6 and A10 are lower in chromium content (1%) and higher in manganese content (2%). They can be hardened from temperatures about 100 °C (200 °F) lower than those required for the high-chromium types, further reducing distortion and undesirable surface reactions during heat treatment.

To improve toughness, silicon is added to type A8 and both silicon and nickel are added to types A9 and A10. Because of the high carbon and silicon contents of type A10, graphite is formed in the microstructure; as a result, A10 has much better machinability when in the annealed condition, and somewhat better resistance to galling and seizing when in the fully hardened condition, than other group A tool steels.

Typical applications for group A tool steels include shear knives, punches, blanking and trimming dies, forming dies and coining dies. The inherent dimensional stability of these steels makes them suitable for gages and precision measuring tools. In addition, the extreme abrasion resistance of type A7 makes it suitable for brick molds, ceramic molds and other highly abrasive applications.

The complex chromium or chromium-vanadium carbides in group A tool steels enhance the wear resistance provided by the martensitic matrix. Therefore, these steels perform well under abrasive conditions at less than full hardness. Although cooling in still air is adequate to produce full hardness in most tools, very massive sections should be hardened by cooling in an air blast or by interrupted quenching in hot oil.

High-carbon, high-chromium cold work steels (group D) contain 1.50 to 2.35% carbon and 12% chromium; with the exception of type D3, they also contain 1% molybdenum. All group D tool steels except type D3 are air hardening, and attain full hardness when cooled in still air. Type D3 is almost always quenched in oil (small parts can be austenitized in vacuum and then gas quenched); therefore, tools made of D3 are more susceptible to distortion and are more likely to crack during hardening.

Group D steels have high resistance to softening at elevated temperatures. These steels also exhibit excellent resistance to wear—especially type D7, which has the highest carbon and vanadium contents. All group D steels—particularly the higher-carbon types D3, D4 and D7—contain massive carbides that make them susceptible to edge brittleness.

Typical applications for group D steels include long-run dies for blanking, forming, thread rolling and deep drawing; dies for cutting laminations; brick molds; gages; burnishing tools; rolls; and shear and slitter knives.

Oil-hardening cold work steels (group O) have high carbon contents, plus enough other alloying elements so that small to moderate sections can attain full hardness when quenched in oil from the austenitizing temperature.

Group O tool steels vary in type of alloy, as well as in alloy content, even though they are similar in general characteristics and are used for similar applications. Type O1 contains manganese, chromium and tungsten. Type O2 is alloyed primarily with manganese. Type O6 contains silicon, manganese and molybdenum; it has a high total carbon content that includes free carbon as well as sufficient combined carbon to enable the steel to achieve maximum as-quenched hardness. Type O7 contains manganese and chromium, and has a tungsten content higher than that of type O1.

The most important service-related property of group O steels is high resistance to wear at normal temperatures, a result of high carbon content. On the other hand, group O steels have low resistance to softening at elevated temperatures.

The ability of group O steels to harden fully on relatively slow quenching yields lower distortion and greater safety (less tendency to crack) in hardening than is characteristic of the water-hardening tool steels. Tools made from these steels can be successfully repaired or renovated by welding if proper procedures are followed. In addition, graphite in the microstructure of type O6 greatly improves the machinability of annealed stock and helps reduce galling and seizing of fully hardened stock.

Group O steels are used extensively in dies and punches for blanking, trimming, drawing, flanging and forming. Surface hardnesses of 56 to 62 HRC, obtained through oil quenching followed by tempering at 175 to 315 °C (350 to 600 °F), provide a suitable combination of mechanical properties for most dies made from type O1, O2 or O6. Type O7, which has lower hardenability but better general wear resistance than any other group O tool steel, is more often used for tools requiring keen cutting edges. Oil-hardening tool steels are also used for machinery components (such as cams, bushings and guides) and for gages (where good dimensional stability and wear properties are needed).

The hardenability of group O steels can be measured effectively by the Jominy end-quench test. Hardenability bands for group O steels are shown in Fig. 3. Variation of hardness with diameter is shown in Fig. 4 for center, surface and ¾ radius locations in oil-quenched bars of group O steels.

Fig. 3 End-quench hardenability bands for group O tool steels

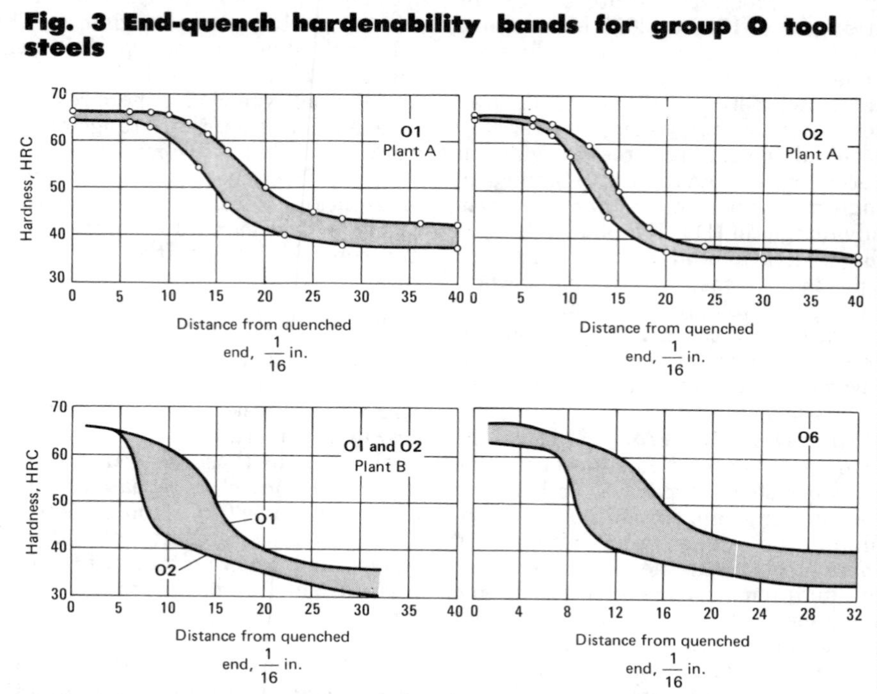

Hardenability bands from plant B represent the data from five heats each for O1 and O2 tool steels. Data from plant A were determined only on the basis of average hardness, not as hardenability bands. Data for O6 is for a spheroidized prior structure. O1 and O6 steels were quenched from 815 °C (1500 °F); O2 from 790 °C (1450 °F).

At normal hardening temperatures, group O steels retain greater amounts of undissolved carbides and thus do not harden as deeply as do steels that are lower in carbon but similar in alloy content. On the other hand, group O steels attain higher surface hardness. Raising the hardening temperature increases grain size, increases solution of alloying elements and dissolves more of the excess carbide, thereby increasing hardenability. However, raising the hardening temperature can have an adverse effect on certain mechanical properties—most notably ductility toughness—and also can increase the likelihood of cracking during hardening.

Shock-Resisting Steels

The principal alloying elements in shock-resisting (group S) steels are manganese, silicon, chromium, tungsten and molybdenum, in various combinations. Carbon content is about 0.50% for all group S steels, which produces a combination of high strength, high toughness and low-to-medium wear resistance. Group S steels are used primarily for chisels, rivet sets, punches, driver bits and other applications requiring high toughness and resistance to shock loading. Types S1 and S7 are also used for hot punching and shearing, which require some heat resistance.

Group S steels vary in hardenability from shallow hardening (S2) to deep hardening (S7). In these steels of intermediate alloy content, hardenability is controlled to a greater extent by actual composition than by the incidental effects of grain size and melting practice, which are so important for group W steels. Group S steels require relatively high austenitizing temperatures to achieve optimum hardness; consequently, undissolved carbides are not a factor in control of hardenability. Type S2 is normally water quenched; types S1, S5 and S6 are oil quenched; type S7 is normally cooled in air, except for large sections, which are oil quenched.

Because group S steels exhibit excellent toughness at high strength levels, they often are considered for nontooling or structural applications. The nominal mechanical properties of S1, S5 and S7, in both annealed and hard-

Fig. 4 Variation of as-quenched hardness with bar diameter for four oil-hardening tool steels

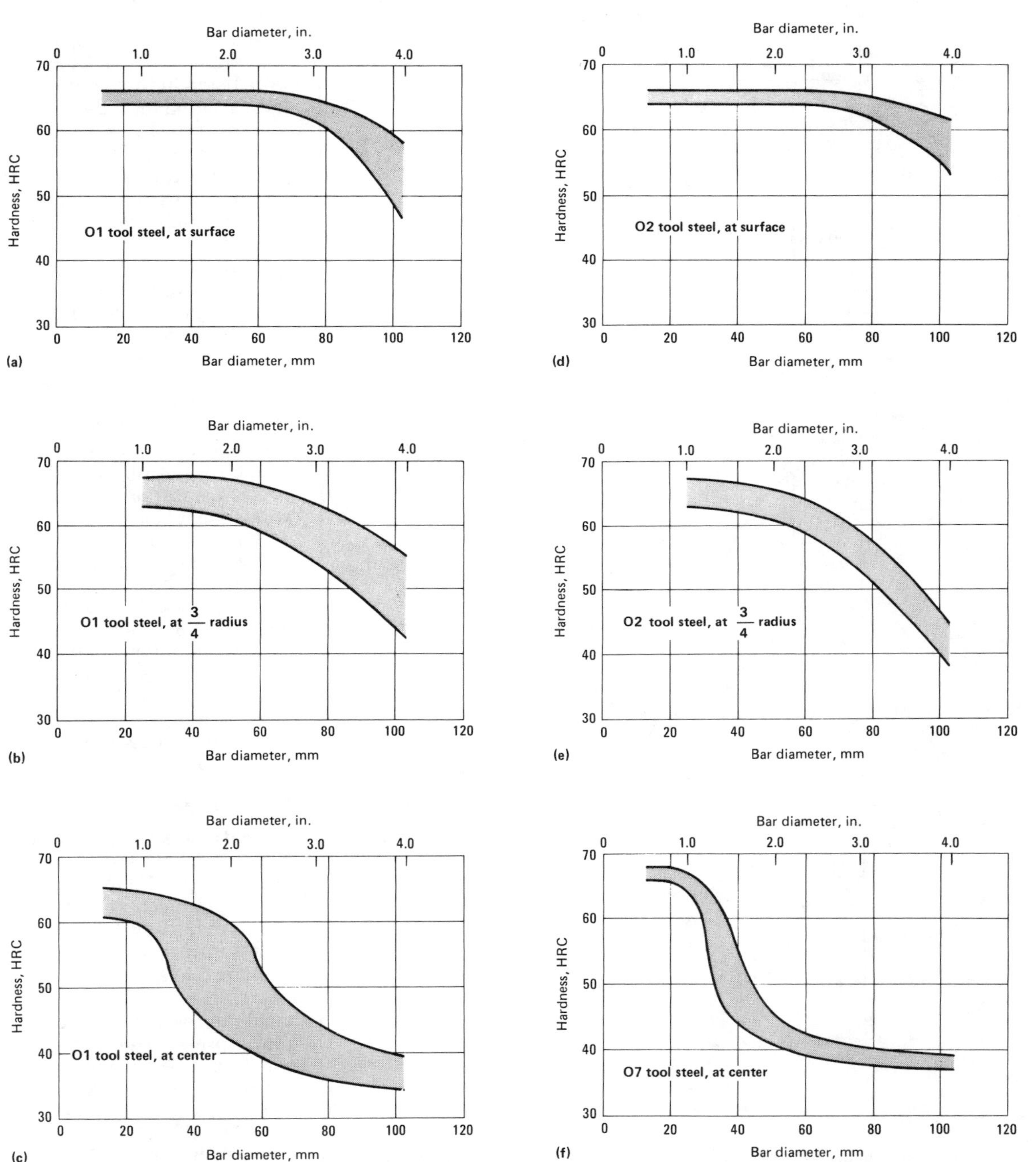

Data for (a), (b), (d) and (e) are from tests on five heats of each steel from Plant A. Data for (c) are from tests on 23 heats of O1 from Plant B. Data for (f) are from tests on eight heats (source unknown). Information on number of heats and source of data not available for (g), (h) and (j). Center hardness data not available for type O2, surface and ¾-radius data not available for type O7. Type O1 austenitized at 815 °C (1500 °F) in Plant A, and at 775 °C (1425 °F) in Plant B. Type O2 austenitized at 790 °C (1450 °F). Austenitizing temperatures for types O6 and O7 not available.

Fig. 4 (continued)

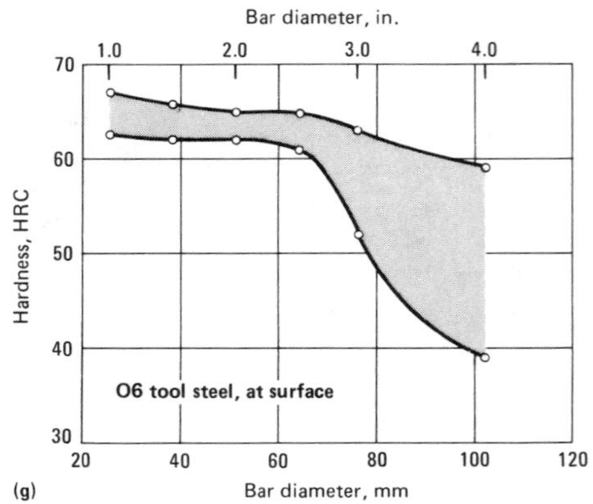

(g)

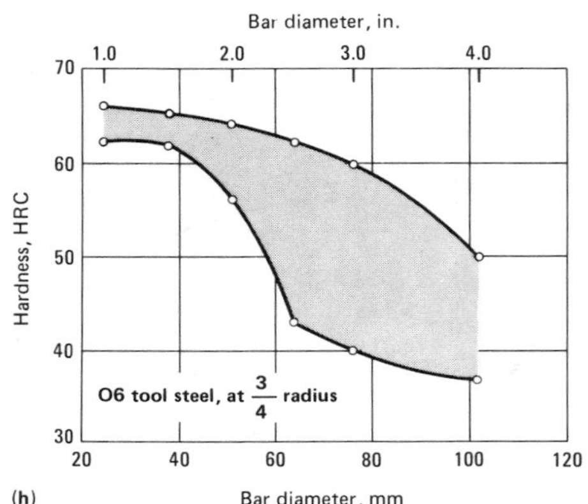

(h)

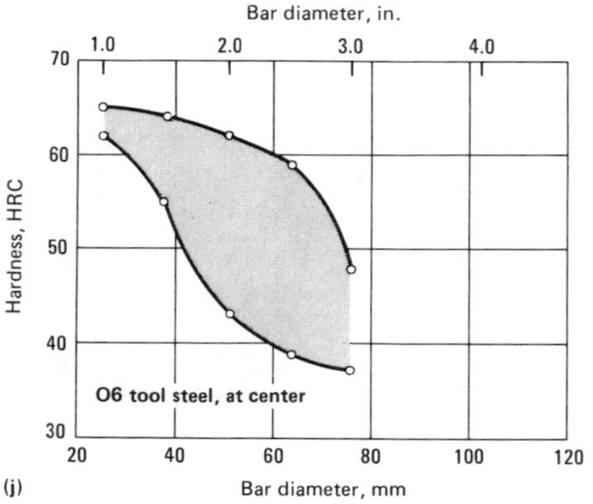

(j)

Data for (a), (b), (d) and (e) are from tests on five heats of each steel from Plant A. Data for (c) are from tests on 23 heats of O1 from Plant B. Data for (f) are from tests on eight heats (source unknown). Information on number of heats and source of data not available for (g), (h) and (j). Center hardness data not available for type O2, surface and 3/4-radius data not available for type O7. Type O1 austenitized at 815 °C (1500 °F) in Plant A, and at 775 °C (1425 °F) in Plant B. Type O2 austenitized at 790 °C (1450 °F). Austenitizing temperatures for types O6 and O7 not available.

ened and tempered conditions, are presented in Table 4.

Low-Alloy, Special-Purpose Steels

The low-alloy, special-purpose (group L) tool steels contain small amounts of chromium, vanadium, nickel and molybdenum. At one time, seven steels were listed in this group, but because of falling demand, only types L2 and L6 remain. Type L2 is available in several carbon contents from 0.50 to 1.10%; its principal alloying elements are chromium and vanadium, which make it an oil-hardening steel of fine grain size. Type L6 contains small amounts of chromium and molybdenum, plus 1.50% nickel for increased toughness.

Although both L2 and L6 are considered oil-hardening steels, large sections of L2 are often quenched in water. A type L2 steel containing 0.50% carbon is capable of attaining about 57 HRC as oil quenched, but it will not through harden in sections more than about 13 mm (0.5 in.) thick. Type L6, which contains 0.70% carbon, has an as-quenched hardness of about 64 HRC; it can maintain a hardness above 60 HRC throughout sections 75 mm (3 in.) thick.

Group L steels generally are used for machine parts such as arbors, cams, chucks and collets, and for other special applications requiring good strength and toughness. Nominal mechanical properties of annealed and of hardened-and-tempered L2 and L6 steels are given in Table 4.

Mold Steels

Mold steels (group P) contain chromium and nickel as principal alloying elements. Types P2 to P6 are carburizing steels produced to tool steel quality standards. They have very low hardness and low resistance to work hardening in the annealed condition. These factors make it possible to produce a mold impression by cold hubbing. After the impression is formed, the mold is carburized, hardened and tempered to a surface hardness of about 58 HRC. Types P4 and P6 are deep hardening; in type P4, full hardness in the carburized case can be achieved by cooling in air.

Types P20 and P21 normally are supplied heat treated to 30 to 36 HRC—a condition in which they can be machined readily into large, intricate dies and molds. Because these steels are prehardened, no subsequent high-temperature heat treatment is re-

Table 4 Nominal room-temperature mechanical properties of group L and group S tool steels

Type	Condition	Tensile strength MPa	Tensile strength ksi	0.2% yield strength MPa	0.2% yield strength ksi	Elongation(a), %	Reduction in area, %	Hardness, HRC	Impact energy J	Impact energy ft·lb
L2	Annealed	710	103	510	74	25	50	96 HRB	...	...
	Oil quenched from 855 °C (1575 °F) and single tempered at:									
	205 °C (400 °F)	2000	290	1790	260	5	15	54	28(b)	21(b)
	315 °C (600 °F)	1790	260	1655	240	10	30	52	19(b)	14(b)
	425 °C (800 °F)	1550	225	1380	200	12	35	47	26(b)	19(b)
	540 °C (1000 °F)	1275	185	1170	170	15	45	41	39(b)	29(b)
	650 °C (1200 °F)	930	135	760	110	25	55	30	125(b)	92(b)
L6	Annealed	655	95	380	55	25	55	93 HRB	...	...
	Oil quenched from 845 °C (1550 °F) and single tempered at:									
	315 °C (600 °F)	2000	290	1790	260	4	9	54	12(b)	9(b)
	425 °C (800 °F)	1585	230	1380	200	8	20	46	18(b)	13(b)
	540 °C (1000 °F)	1345	195	1100	160	12	30	42	23(b)	17(b)
	650 °C (1200 °F)	965	140	830	120	20	48	32	81(b)	60(b)
S1	Annealed	690	100	415	60	24	52	96 HRB	...	...
	Oil quenched from 930 °C (1700 °F) and single tempered at:									
	205 °C (400 °F)	2070	300	1895	275	...	...	57.5	249(c)	184(c)
	315 °C (600 °F)	2030	294	1860	270	4	12	54	233(c)	172(c)
	425 °C (800 °F)	1790	260	1690	245	5	17	50.5	203(c)	150(c)
	540 °C (1000 °F)	1680	244	1525	221	9	23	47.5	230(c)	170(c)
	650 °C (1200 °F)	1345	195	1240	180	12	37	42	...	...
S5	Annealed	725	105	440	64	25	50	96 HRB	...	...
	Oil quenched from 870 °C (1600 °F) and single tempered at:									
	205 °C (400 °F)	2345	340	1930	280	5	20	59	206(c)	152(c)
	315 °C (600 °F)	2240	325	1860	270	7	24	58	232(c)	171(c)
	425 °C (800 °F)	1895	275	1690	245	9	28	52	243(c)	179(c)
	540 °C (1000 °F)	1520	220	1380	200	10	30	48	188(c)	139(c)
	650 °C (1200 °F)	1035	150	1170	170	15	40	37	...	...
S7	Annealed	640	93	380	55	25	55	95 HRB	...	...
	Fan cooled from 940 °C (1725 °F) and single tempered at:									
	205 °C (400 °F)	2170	315	1450	210	7	20	58	244(c)	180(c)
	315 °C (600 °F)	1965	285	1585	230	9	25	55	309(c)	228(c)
	425 °C (800 °F)	1895	275	1410	205	10	29	53	243(c)	179(c)
	540 °C (1000 °F)	1820	264	1380	200	10	33	51	324(c)	239(c)
	650 °C (1200 °F)	1240	180	1035	150	14	45	39	358(c)	264(c)

(a) In 50 mm or 2 in. (b) Charpy V-notch. (c) Charpy unnotched.

quired, and distortion and size changes are avoided. However, when used for plastic molds, type P20 sometimes is carburized and hardened after the impression has been machined. Type P21 is an aluminum-containing precipitation-hardening steel and is supplied prehardened to 32 to 36 HRC. After machining and low-temperature aging, type P21 can reach 38 to 40 HRC in sections as large as it is practical to produce.

All group P steels have low resistance to softening at elevated temperatures, except for P4 and P21, which have medium resistance. Group P steels are used almost exclusively in low-temperature die-casting dies and in molds for injection or compression molding of plastics. Plastic molds often require very massive steel blocks up to 750 mm (30 in.) thick and weighing as much as 9 Mg (10 tons). Because these large die blocks must meet stringent requirements for soundness, cleanness and hardenability, electric-furnace melting, vacuum degassing and special deoxidation treatments have become standard practice in the production of group P tool steels. In addition, ingot casting and forging practices have been refined to achieve a high degree of homogeneity.

Water-Hardening Tool Steels

Water-hardening (group W) tool steels contain carbon as the principal alloying element. Small amounts of chromium and vanadium are added to most of the group W steels—chromium to increase hardenability and wear resistance, and vanadium to maintain fine grain size and thus enhance toughness. Group W tool steels are made with various nominal carbon contents (from about 0.60 to 1.40%); the most popular grades contain approximately 1.00% carbon.

Group W tool steels are very shallow hardening, and consequently develop a fully hardened zone that is relatively thin, even when quenched drastically. Sections more than about 13 mm (1/2 in.) thick generally have a hard case over a strong, tough and resilient core.

Group W steels have low resistance to softening at elevated temperatures. They are suitable for cold heading, striking, coining and embossing tools; woodworking tools; hard metal-cutting tools, such as taps and reamers; wear-resistant machine-tool components; and cutlery.

Group W steels are made in as many as four different grades or quality levels for the same nominal composition. These quality levels have been given various names by different manufacturers, and range from a clean carbon tool steel with precisely controlled hardenability, grain size, microstructure and annealed hardness to a grade less carefully controlled but satisfactory for noncritical low-production applications.

The Society of Automotive Engineers defines four grades of plain carbon tool steels as follows:

- *Special (grade 1)* is the highest-quality water-hardening tool steel. Hardenability is controlled, and composition held to close limits. Bars are subjected to rigorous testing to ensure maximum uniformity in performance.
- *Extra (grade 2)* is a high-quality water-hardening tool steel that is controlled for hardenability and is subjected to tests that ensure good performance in general applications.
- *Standard (grade 3)* is a good-quality water-hardening tool steel that is not controlled for hardenability and that is recommended for applications where some latitude in uniformity can be tolerated.
- *Commercial (grade 4)* is a commercial-quality water-hardening tool steel that is neither controlled for hardenability nor subjected to special tests.

Limits on manganese, silicon and chromium generally are not required for "special" and "extra" grades; the following Shepherd hardenability limits are prescribed instead:

Hardenability classification	Radial depth of hardening (*P*), 1/64 in.	Minimum fracture grain size (*F*)
Carbon content, 0.70 to 0.95%		
Shallow	10 max	8
Regular	9 to 13	8
Deep	12 min	8
Carbon content, 0.95 to 1.30%		
Shallow	8 max	9
Regular	7 to 11	9
Deep	10 to 16	8

(See "Testing of Tool Steels" on page 434 of this article for more information on Shepherd hardenability.)

The combined manganese, silicon and chromium contents of standard and commercial grades should not exceed 0.75%. Generally, both manganese and silicon are limited to 0.35% max in all standard and commercial grades; chromium is limited to 0.15% max in standard grades and to 0.20% max in commercial grades.

The ability of a group W tool steel to perform satisfactorily in many applications depends on depth of the hardened zone. Depth of hardening in these steels is controlled mainly by austenitic grain size, melting practice, alloy content, amount of excess carbide present at the quenching temperature and, to a lesser extent, initial structure of the steel prior to austenitizing for hardening.

Typical results in the Shepherd PF test indicate an increase in *P* value of 0.80 mm (2/64 in.) for every increase in austenitic grain size of one ASTM number for the same grade. Increased amounts of undissolved carbides at the hardening temperature will reduce hardenability. This is doubly important in hypereutectoid grades, which are deliberately quenched to retain carbides undissolved at the austenitizing temperature, in order to increase wear resistance. A fine lamellar microstructure prior to hardening, such as that obtained by normalizing, will result in fewer undissolved carbides at the normal austenitizing temperature than a prior spheroidized microstructure. Fewer carbides present at the austenitizing temperature promotes deeper hardening because more carbon is dissolved in the austenite and there are fewer carbides to act as nucleation sites for nonmartensitic transformation products. Thus, normalized bars have deeper hardenability than spheroidized bars of the same grade.

Fig. 5 Relation of bar diameter and depth of hardened zone for shallow, medium and deep hardening grades of W1 tool steel containing 1% C

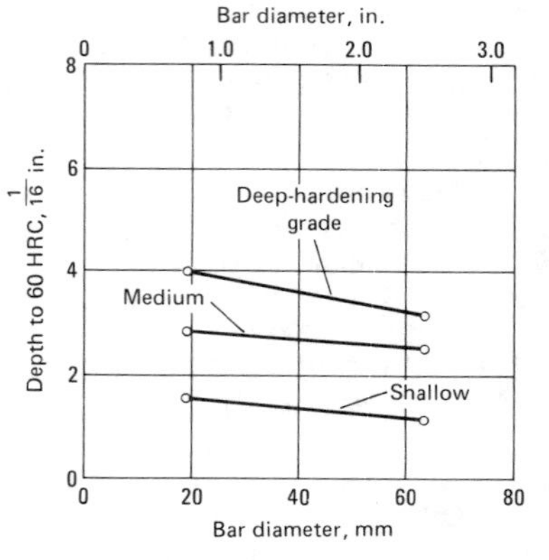

Addition of vanadium frequently decreases hardenability under normal hardening conditions due to formation of many fine carbides that not only act as nucleation sites for nonmartensitic transformation products, but also refine the austenitic grain size. Austenitizing at higher-than-normal temperatures will dissolve these excess carbides and thus increase the hardenability.

Group W steels with carbon contents lower than that of the eutectoid composition often have greater hardenability than hypereutectoid grades. Grain coarsening resulting from the higher austenitizing temperatures used for hypoeutectoid grades is one cause of this, but the main cause is the absence of excess carbides at the austenitizing temperature.

Figure 5 shows a typical relationship between bar diameter and case depth (60 HRC or above) for three W1 tool steels that are equal in carbon content (1% C) but differ in hardenability. Hardenability is varied by adjusting manganese and silicon contents and altering deoxidation procedure. This relationship illustrates the need for precise specification of hardenability in the selection of these grades: group W tool steels purchased without hardenability requirements could vary widely enough in this property to cause severe

Table 5 Normalizing and annealing temperatures of tool steels

Type	Normalizing temperature(a) °C	Normalizing temperature(a) °F	Annealing(b) Temperature °C	Annealing(b) Temperature °F	Rate of cooling, max °C/h	Rate of cooling, max °F/h	Hardness, HB
Molybdenum high speed steels							
M1, M10	Do not normalize		815 to 870	1500 to 1600	22	40	207 to 235
M2	Do not normalize		870 to 900	1600 to 1650	22	40	212 to 241
M3, M4	Do not normalize		870 to 900	1600 to 1650	22	40	223 to 255
M6	Do not normalize		870	1600	22	40	248 to 277
M7	Do not normalize		815 to 870	1500 to 1600	22	40	217 to 255
M30, M33, M34, M36, M41, M42, M46, M47	Do not normalize		870 to 900	1600 to 1650	22	40	235 to 269
M43	Do not normalize		870 to 900	1600 to 1650	22	40	248 to 269
M44	Do not normalize		870 to 900	1600 to 1650	22	40	248 to 293
Tungsten high speed steels							
T1	Do not normalize		870 to 900	1600 to 1650	22	40	217 to 255
T2	Do not normalize		870 to 900	1600 to 1650	22	40	223 to 255
T4	Do not normalize		870 to 900	1600 to 1650	22	40	229 to 269
T5	Do not normalize		870 to 900	1600 to 1650	22	40	235 to 277
T6	Do not normalize		870 to 900	1600 to 1650	22	40	248 to 293
T8	Do not normalize		870 to 900	1600 to 1650	22	40	229 to 255
T15	Do not normalize		870 to 900	1600 to 1650	22	40	241 to 277
Chromium hot work steels							
H10, H11, H12, H13	Do not normalize		845 to 900	1550 to 1650	22	40	192 to 229
H14	Do not normalize		870 to 900	1600 to 1650	22	40	207 to 235
H19	Do not normalize		870 to 900	1600 to 1650	22	40	207 to 241
Tungsten hot work steels							
H21, H22, H25	Do not normalize		870 to 900	1600 to 1650	22	40	207 to 235
H23	Do not normalize		870 to 900	1600 to 1650	22	40	212 to 255
H24, H26	Do not normalize		870 to 900	1600 to 1650	22	40	217 to 241
Molybdenum hot work steels							
H41, H43	Do not normalize		815 to 870	1500 to 1600	22	40	207 to 235
H42	Do not normalize		845 to 900	1550 to 1650	22	40	207 to 235
High-carbon high-chromium cold work steels							
D2, D3, D4	Do not normalize		870 to 900	1600 to 1650	22	40	217 to 255
D5	Do not normalize		870 to 900	1600 to 1650	22	40	223 to 255
D7	Do not normalize		870 to 900	1600 to 1650	22	40	235 to 262
Medium-alloy air-hardening cold work steels							
A2	Do not normalize		845 to 870	1550 to 1600	22	40	201 to 229
A3	Do not normalize		845 to 870	1550 to 1600	22	40	207 to 229
A4	Do not normalize		740 to 760	1360 to 1400	14	25	200 to 241
A6	Do not normalize		730 to 745	1350 to 1375	14	25	217 to 248
A7	Do not normalize		870 to 900	1600 to 1650	14	25	235 to 262
A8	Do not normalize		845 to 870	1550 to 1600	22	40	192 to 223
A9	Do not normalize		845 to 870	1550 to 1600	14	25	212 to 248
A10	790	1450	765 to 795	1410 to 1460	8	15	235 to 269
Oil-hardening cold work steels							
O1	870	1600	760 to 790	1400 to 1450	22	40	183 to 212
O2	845	1550	745 to 775	1375 to 1425	22	40	183 to 212
O6	870	1600	765 to 790	1410 to 1450	11	20	183 to 217
O7	900	1650	790 to 815	1450 to 1500	22	40	192 to 217

(a) Time held at temperature varies from 15 min for small sections to 1 h for large sizes. Cooling is done in still air. Normalizing should not be confused with low-temperature annealing. (b) The upper limit of ranges should be used for large sections and the lower limit for smaller sections. Time held at temperature varies from 1 h for light sections to 4 h for heavy sections and large furnace charges of high-alloy steel. (c) For 0.25Si type, 183 to 207 HB; for 1.00Si type, 207 to 229 HB. (d) Temperature varies with carbon content: 0.60 to 0.75 C, 816 °C (1500 °F); 0.75 to 0.90 C, 788 °C (1450 °F); 0.90 to 1.10 C, 871 °C (1600 °F); 1.10 to 1.40 C, 871 to 927 °C (1600 to 1700 °F). (e) Temperature varies with carbon content: 0.60 to 0.90 C, 738 to 788 °C (1360 to 1450 °F); 0.90 to 1.40 C, 760 to 788 °C (1400 to 1450 °F).

Table 5 (continued)

Type	Normalizing temperature(a) °C	Normalizing temperature(a) °F	Annealing(b) Temperature °C	Annealing(b) Temperature °F	Rate of cooling, max °C/h	Rate of cooling, max °F/h	Hardness, HB
Shock-resisting steels							
S1	Do not normalize		790 to 815	1450 to 1500	22	40	183 to 229(c)
S2	Do not normalize		760 to 790	1400 to 1450	22	40	192 to 217
S5	Do not normalize		775 to 800	1425 to 1475	14	25	192 to 229
S7	Do not normalize		815 to 845	1500 to 1550	14	25	187 to 223
Mold steels							
P2	Not required		730 to 815	1350 to 1500	22	40	103 to 123
P3	Not required		730 to 815	1350 to 1500	22	40	109 to 137
P4	Do not normalize		870 to 900	1600 to 1650	14	25	116 to 128
P5	Not required		845 to 870	1550 to 1600	22	40	105 to 116
P6	Not required		845	1550	8	15	183 to 217
P20	900	1650	760 to 790	1400 to 1450	22	40	149 to 179
P21	900	1650		Do not anneal			
Low-alloy special-purpose steels							
L2	871 to 900	1600 to 1650	760 to 790	1400 to 1450	22	40	163 to 197
L3	900	1650	790 to 815	1450 to 1500	22	40	174 to 201
L6	870	1600	760 to 790	1400 to 1450	22	40	183 to 212
Carbon-tungsten special-purpose steels							
F1	900	1650	760 to 800	1400 to 1475	22	40	183 to 207
F2	900	1650	790 to 815	1450 to 1500	22	40	207 to 235
Water-hardening steels							
W1, W2	790 to 925(d)	1450 to 1700(d)	740 to 790(e)	1360 to 1450(e)	22	40	156 to 201
W5	870 to 925	1600 to 1700	760 to 790	1400 to 1450	22	40	163 to 201

(a) Time held at temperature varies from 15 min for small sections to 1 h for large sizes. Cooling is done in still air. Normalizing should not be confused with low-temperature annealing. (b) The upper limit of ranges should be used for large sections and the lower limit for smaller sections. Time held at temperature varies from 1 h for light sections to 4 h for heavy sections and large furnace charges of high-alloy steel. (c) For 0.25Si type, 183 to 207 HB; for 1.00Si type, 207 to 229 HB. (d) Temperature varies with carbon content: 0.60 to 0.75 C, 816 °C (1500 °F); 0.75 to 0.90 C, 788 °C (1450 °F); 0.90 to 1.10 C, 871 °C (1600 °F); 1.10 to 1.40 C, 871 to 927 °C (1600 to 1700 °F). (e) Temperature varies with carbon content: 0.60 to 0.90 C, 738 to 788 °C (1360 to 1450 °F); 0.90 to 1.40 C, 760 to 788 °C (1400 to 1450 °F).

processing difficulties or actual tool failures.

With the very fast cooling rate required for hardening of the W grades, there is a greater chance that the tool will crack during hardening. Consequently, most manufacturers prefer to use tool steels that can be hardened satisfactorily by quenching in oil or cooling in air in order to attempt to avoid the expense involved if a tool cracks during heat treatment.

Typical Heat Treatments and Properties

Condensed information on processing and service characteristics of tool steels is presented in Table 5 to 7. This information is essential in understanding the problems involved in selection, processing and application of tool steels.

More detailed heat treating information for each of these steels is available in the heat treating volume of Metals Handbook. Additional detailed information on resistance to softening at elevated temperatures is summarized in Fig. 1, which presents curves of hardness versus tempering temperature. Similar curves for most of the tool steels covered in this article are presented on pages 251 to 292 in Ref 2.

Technical representatives of tool steel producers can supply more specific information on the properties developed by specific heat treatments in the steels produced by their companies. They should be consulted as to the type of steel and heat treatment best suited to meet all service requirements at the least over-all cost.

Physical properties—density, thermal expansion and thermal conductivity—of selected tool steels are given in Tables 8 and 9.

Testing of Tool Steels

Because of the difficulty of obtaining reliable correlation between the properties of tool steels as measured by laboratory tests and the performance of these steels in service or in fabrication, most frequently these properties are presented as general comparisons rather than specific data.

Performance in Service

The basic properties of tool steels that determine their performance in service are resistance to wear, deformation and breakage; toughness; and in many instances, resistance to softening at elevated temperatures. Often, these characteristics can be measured by, or inferred from, direct measurement of hardness. Hardness of tool steels is most commonly measured and reported on the Rockwell C scale (HRC) in the United States and on the Vickers scale (diamond pyramid hardness, or HV) in the United Kingdom and Europe. It is significant that conversion from HRC to HV, or vice versa, is not linear (see Fig. 6). For example, an increase from 67 to 68 HRC corresponds to a 40-point

Table 6 Hardening and tempering of tool steels

Type	Rate of heating	Hardening: Preheat temperature °C	Preheat temperature °F	Hardening temperature °C	Hardening temperature °F	Time at temperature, min	Quenching medium(a)	Tempering temperature °C	Tempering temperature °F
Molybdenum high speed steels									
M1, M7, M10	Rapidly from preheat	730 to 845	1350 to 1550	1177 to 1219	2150 to 2225(b)	2 to 5	O, A or S	540 to 595(c)	1000 to 1100(c)
M2	Rapidly from preheat	730 to 845	1350 to 1550	1190 to 1230	2175 to 2250(b)	2 to 5	O, A or S	540 to 595(c)	1000 to 1100(c)
M3, M4, M30, M33, M34	Rapidly from preheat	730 to 845	1350 to 1550	1200 to 1230(b)	2200 to 2250(b)	2 to 5	O, A or S	540 to 595(c)	1000 to 1100(c)
M6	Rapidly from preheat	790	1450	1180 to 1200(b)	2150 to 2200(b)	2 to 5	O, A or S	540 to 595(c)	1000 to 1100(c)
M36	Rapidly from preheat	730 to 845	1350 to 1550	1220 to 1250(b)	2225 to 2275(b)	2 to 5	O, A or S	540 to 595(c)	1000 to 1100(c)
M41	Rapidly from preheat	730 to 845	1350 to 1550	1190 to 1220(b)	2175 to 2220(b)	2 to 5	O, A or S	540 to 595(d)	1000 to 1100(d)
M42	Rapidly from preheat	730 to 845	1350 to 1550	1190 to 1210(b)	2175 to 2210(b)	2 to 5	O, A or S	510 to 595(d)	950 to 1100 (d)
M43	Rapidly from preheat	730 to 845	1350 to 1550	1190 to 1220(b)	2175 to 2220(b)	2 to 5	O, A or S	510 to 595(d)	950 to 1100(d)
M44	Rapidly from preheat	730 to 845	1350 to 1550	1200 to 1230(b)	2190 to 2240(b)	2 to 5	O, A or S	540 to 625(d)	1000 to 1160(d)
M46	Rapidly from preheat	730 to 845	1350 to 1550	1190 to 1220(b)	2175 to 2225(b)	2 to 5	O, A or S	525 to 565(d)	975 to 1050(d)
M47	Rapidly from preheat	730 to 845	1350 to 1550	1180 to 1200(b)	2150 to 2200(b)	2 to 5	O, A or S	525 to 595(d)	975 to 1100(d)
Tungsten high speed steels									
T1, T2, T4, T8	Rapidly from preheat	815 to 870	1500 to 1600	1260 to 1300(b)	2300 to 2375(b)	2 to 5	O, A or S	540 to 595(c)	1000 to 1100(c)
T5, T6	Rapidly from preheat	815 to 870	1500 to 1600	1270 to 1300(b)	2325 to 2375(b)	2 to 5	O, A or S	540 to 595(c)	1000 to 1100(c)
T15	Rapidly from preheat	815 to 870	1500 to 1600	1200 to 1260(b)	2200 to 2300(b)	2 to 5	O, A or S	540 to 650(d)	1000 to 1200(d)
Chromium hot work steels									
H10	Moderately from preheat	815	1500	1010 to 1040	1850 to 1900	15 to 40(e)	A	540 to 650	1000 to 1200
H11, H12	Moderately from preheat	815	1500	995 to 1020	1825 to 1875	15 to 40(e)	A	540 to 650	1000 to 1200
H13	Moderately from preheat	815	1500	995 to 1040	1825 to 1900	15 to 40(e)	A	540 to 650	1000 to 1200
H14	Moderately from preheat	815	1500	1010 to 1070	1850 to 1950	15 to 40(e)	A	540 to 650	1000 to 1200
H19	Moderately from preheat	815	1500	1090 to 1200	2000 to 2200	2 to 5	A or O	540 to 705	1000 to 1300
Tungsten hot work steels									
H21, H22	Rapidly from preheat	815	1500	1090 to 1200	2000 to 2200	2 to 5	A or O	595 to 675	1100 to 1250
H23	Rapidly from preheat	845	1550	1204 to 1260	2200 to 2300	2 to 5	O	650 to 815	1200 to 1500
H24	Rapidly from preheat	815	1500	1090 to 1230	2000 to 2250	2 to 5	O	565 to 650	1050 to 1200
H25	Rapidly from preheat	815	1500	1150 to 1260	2100 to 2300	2 to 5	A or O	565 to 675	1050 to 1250
H26	Rapidly from preheat	870	1600	1180 to 1260	2150 to 2300	2 to 5	O, A or S	565 to 675	1050 to 1250

(a) O = oil quench; A = air cool; S = salt bath quench; W = water quench; B = brine quench. (b) When the high-temperature heating is carried out in a salt bath, the range of temperatures should be about 14 °C (25 °F) lower than given in this line. (c) Double tempering recommended for not less than 1 h at temperature each time. (d) Triple tempering recommended for not less than 1 h at temperature each time. (e) Times apply to open-furnace heat treatment. For pack hardening, a common rule is to heat 30 min/inch of cross section of the pack. (f) Preferable for large tools to minimize decarburization. (g) Carburizing temperature. (h) After carburizing. (j) Carburized case hardness. (k) P21 is a precipitation-hardening steel having a thermal treatment which involves solution treating and aging rather than hardening and tempering. (m) Recommended for large tools and tools with intricate sections.

Table 6 (continued)

Type	Rate of heating	Hardening: Preheat temperature °C	Preheat temperature °F	Hardening temperature °C	Hardening temperature °F	Time at temperature, min	Quenching medium(a)	Tempering temperature °C	Tempering temperature °F
Molybdenum hot work steels									
H41, H43	Rapidly from preheat	730 to 845	1350 to 1550	1090 to 1190	2000 to 2175	2 to 5	O, A or S	565 to 650	1050 to 1200
H42	Rapidly from preheat	730 to 845	1350 to 1550	1120 to 1220	2050 to 2225	2 to 5	O, A or S	565 to 650	1050 to 1200
High-carbon, high-chromium cold work steels									
D1, D5	Very slowly	815	1500	980 to 1020	1800 to 1875	15 to 45	A	205 to 540	400 to 1000
D3	Very slowly	815	1500	925 to 980	1700 to 1800	15 to 45	O	205 to 540	400 to 1000
D4	Very slowly	815	1500	970 to 1010	1775 to 1850	15 to 45	A	205 to 540	400 to 1000
D7	Very slowly	815	1500	1010 to 1070	1850 to 1950	30 to 60	A	150 to 540	300 to 1000
Medium-alloy air-hardening cold work steels									
A2	Slowly	790	1450	925 to 980	1700 to 1800	20 to 45	A	175 to 540	350 to 1000
A3	Slowly	790	1450	955 to 980	1750 to 1800	25 to 60	A	175 to 540	350 to 1000
A4	Slowly	675	1250	815 to 870	1500 to 1600	20 to 45	A	175 to 425	350 to 800
A6	Slowly	650	1200	830 to 870	1525 to 1600	20 to 45	A	150 to 425	300 to 800
A7	Very slowly	815	1500	955 to 980	1750 to 1800	30 to 60	A	150 to 540	300 to 1000
A8	Slowly	790	1450	980 to 1010	1800 to 1850	20 to 45	A	175 to 595	350 to 1100
A9	Slowly	790	1450	980 to 1020	1800 to 1875	20 to 45	A	510 to 620	950 to 1150
A10	Slowly	650	1200	790 to 815	1450 to 1500	30 to 60	A	175 to 425	350 to 800
Oil-hardening cold work steels									
O1	Slowly	650	1200	790 to 815	1450 to 1500	10 to 30	O	175 to 260	350 to 500
O2	Slowly	650	1200	760 to 800	1400 to 1475	5 to 20	O	175 to 260	350 to 500
O6	Slowly	...	...	790 to 815	1450 to 1500	10 to 30	O	175 to 315	350 to 600
O7	Slowly	650	1200	790 to 830 845 to 885	W: 1450 to 1525 O: 1550 to 1625	10 to 30	O or W	175 to 290	350 to 550
Shock-resisting steels									
S1	Slowly	...	...	900 to 955	1650 to 1750	15 to 45	O	205 to 650	400 to 1200
S2	Slowly	650(f)	1200(f)	845 to 900	1550 to 1650	5 to 20	B or W	175 to 425	350 to 800
S5	Slowly	760	1400	870 to 925	1600 to 1700	5 to 20	O	175 to 425	350 to 800
S7	Slowly	650 to 705	1200 to 1300	925 to 955	1700 to 1750	15 to 45	A or O	205 to 620	400 to 1150

(a) O = oil quench; A = air cool; S = salt bath quench; W = water quench; B = brine quench. (b) When the high-temperature heating is carried out in a salt bath, the range of temperatures should be about 14 °C (25 °F) lower than given in this line. (c) Double tempering recommended for not less than 1 h at temperature each time. (d) Triple tempering recommended for not less than 1 h at temperature each time. (e) Times apply to open-furnace heat treatment. For pack hardening, a common rule is to heat 30 min/inch of cross section of the pack. (f) Preferable for large tools to minimize decarburization. (g) Carburizing temperature. (h) After carburizing. (j) Carburized case hardness. (k) P21 is a precipitation-hardening steel having a thermal treatment which involves solution treating and aging rather than hardening and tempering. (m) Recommended for large tools and tools with intricate sections.

Table 6 (continued)

Type	Rate of heating	Hardening: Preheat temperature °C	Preheat temperature °F	Hardening temperature °C	Hardening temperature °F	Time at temperature, min	Quenching medium(a)	Tempering temperature °C	Tempering temperature °F
Mold steels									
P2	...	900 to 925(g)	1650 to 1700(g)	830 to 845(h)	1525 to 1550(h)	15	O	175 to 260	350 to 500
P3	...	900 to 925(g)	1650 to 1700(g)	800 to 830(h)	1475 to 1525(h)	15	O	175 to 260	350 to 500
P4	...	970 to 995(g)	1775 to 1825(g)	970 to 995(h)	1775 to 1825(h)	15	A	175 to 480	350 to 900
P5	...	900 to 925(g)	1650 to 1700(g)	845 to 870(h)	1550 to 1600(h)	15	O or W	175 to 260	350 to 500
P6	...	900 to 925(g)	1650 to 1700(g)	790 to 815(h)	1450 to 1500(h)	15	A or O	175 to 230	350 to 450
P20	...	870 to 900(h)	1600 to 1650(h)	815 to 870	1500 to 1600	15	O	480 to 595(j)	900 to 1100(j)
P21(k)	Slowly	Do not preheat		705 to 730	1300 to 1350	60 to 180	A or O	510 to 550	950 to 1025
Low-alloy special-purpose steels									
L2	Slowly	...	...	W: 790 to 845 O: 845 to 925	W: 1450 to 1550 O: 1550 to 1700	10 to 30	O or W	175 to 540	350 to 1000
L3	Slowly	...	...	W: 775 to 815 O: 815 to 870	W: 1425 to 1500 O: 1500 to 1600	10 to 30	O or W	175 to 315	350 to 600
L6	Slowly	...	...	790 to 845	1450 to 1550	10 to 30	O	175 to 540	350 to 1000
Carbon-tungsten special-purpose steels									
F1, F2	Slowly	650	1200	790 to 870	1450 to 1600	15	W or B	175 to 260	350 to 500
Water-hardening steels									
W1, W2, W3	Slowly	565 to 650(m)	1050 to 1200(m)	760 to 815	1400 to 1550	10 to 30	B or W	175 to 345	350 to 650

(a) O = oil quench; A = air cool; S = salt bath quench; W = water quench; B = brine quench. (b) When the high-temperature heating is carried out in a salt bath, the range of temperatures should be about 14 °C (25 °F) lower than given in this line. (c) Double tempering recommended for not less than 1 h at temperature each time. (d) Triple tempering recommended for not less than 1 h at temperature each time. (e) Times apply to open-furnace heat treatment. For pack hardening, a common rule is to heat 30 min/inch of cross section of the pack. (f) Preferable for large tools to minimize decarburization. (g) Carburizing temperature. (h) After carburizing. (j) Carburized case hardness. (k) P21 is a precipitation-hardening steel having a thermal treatment which involves solution treating and aging rather than hardening and tempering. (m) Recommended for large tools and tools with intricate sections.

increase on the HV scale, whereas increases from 57 to 58 HRC and from 49 to 50 HRC correspond, respectively, to 20-point and 10-point increases on the HV scale.

For a given tool steel at a given hardness, wear resistance may vary widely depending on the wear mechanism involved and the heat treatment used. It is important to note also that among tool steels with widely differing compositions but identical hardnesses, wear resistance may vary widely under identical wear conditions.

For practical purposes, the resistance to elastic deformation (modulus of elasticity) of all tool steels in all conditions is about 200 GPa (30×10^6 psi) at room temperature. This decreases uniformly to about 185 GPa (27×10^6 psi) at 260 °C (500 °F) and about 150 GPa (22×10^6 psi) at 540 °C (1000 °F).

Except for special grades, compositions and heat treatments of most tool steels are selected to provide very high resistance to plastic deformation. This leaves the metal with very little ability to absorb deformation; in other words, it leaves the metal very brittle. Therefore, it is difficult to determine reliable values of strength at maximum hardness by tensile testing, even when specially designed clamping fixtures are used to provide accurate alignment. Compression tests have been used to some extent to measure resistance to deformation. Bending tests using either three- or four-point supports can provide useful comparative information on tool steels with high hardness levels, but the results often are difficult to evaluate. Torsion tests have been used effectively to measure the toughness of tool steels, particularly those to be used in drills and other tools loaded in torsion during service.

The amount of energy absorbed when a notched bar of fully hardened tool

Table 7 Processing and service characteristics of tool steels

Adapted from Ref 1

AISI designation	Resistance to decarburization	Hardening and tempering: Hardening response	Amount of distortion(a)	Resistance to cracking	Approximate hardness(b), HRC	Fabrication and service: Machinability	Toughness	Resistance to softening	Resistance to wear
Molybdenum high speed steels									
M1	Low	Deep	A or S, low; O, medium	Medium	60-65	Medium	Low	Very high	Very high
M2	Medium	Deep	A or S, low; O, medium	Medium	60-65	Medium	Low	Very high	Very high
M3 (class 1 and class 2)	Medium	Deep	A or S, low; O, medium	Medium	61-66	Medium	Low	Very high	Very high
M4	Medium	Deep	A or S, low; O, medium	Medium	61-66	Low to medium	Low	Very high	Highest
M6	Low	Deep	A or S, low; O, medium	Medium	61-66	Medium	Low	Highest	Very high
M7	Low	Deep	A or S, low; O, medium	Medium	61-66	Medium	Low	Very high	Very high
M10	Low	Deep	A or S, low; O, medium	Medium	60-65	Medium	Low	Very high	Very high
M30	Low	Deep	A or S, low; O, medium	Medium	60-65	Medium	Low	Highest	Very high
M33	Low	Deep	A or S, low; O, medium	Medium	60-65	Medium	Low	Highest	Very high
M34	Low	Deep	A or S, low; O, medium	Medium	60-65	Medium	Low	Highest	Very high
M36	Low	Deep	A or S, low; O, medium	Medium	60-65	Medium	Low	Highest	Very high
M41	Low	Deep	A or S, low; O, medium	Medium	65-70	Medium	Low	Highest	Very high
M42	Low	Deep	A or S, low; O, medium	Medium	65-70	Medium	Low	Highest	Very high
M43	Low	Deep	A or S, low; O, medium	Medium	65-70	Medium	Low	Highest	Very high
M44	Low	Deep	A or S, low; O, medium	Medium	62-70	Medium	Low	Highest	Very high
M46	Low	Deep	A or S, low; O, medium	Medium	67-69	Medium	Low	Highest	Very high
M47	Low	Deep	A or S, low; O, medium	Medium	65-70	Medium	Low	Highest	Very high
Tungsten high speed steels									
T1	High	Deep	A or S, low; O, medium	High	60-65	Medium	Low	Very high	Very high
T2	High	Deep	A or S, low; O, medium	High	61-66	Medium	Low	Very high	Very high
T4	Medium	Deep	A or S, low; O, medium	Medium	62-66	Medium	Low	Highest	Very high
T5	Low	Deep	A or S, low; O, medium	Medium	60-65	Medium	Low	Highest	Very high
T6	Low	Deep	A or S, low; O, medium	Medium	60-65	Low to medium	Low	Highest	Very high
T8	Medium	Deep	A or S, low; O, medium	Medium	60-65	Medium	Low	Highest	Very high
T15	Medium	Deep	A or S, low; O, medium	Medium	63-68	Low to medium	Low	Highest	Highest

(a) A, air cool; B, brine quench; O, oil quench; S, salt bath quench; W, water quench. (b) After tempering in temperature range normally recommended for this steel. (c) Carburized case hardness. (d) After aging at 510 to 550 °C (950 to 1025 °F). (e) Toughness decreases with increasing carbon content and depth of hardening.

Table 7 (continued)

Adapted from Ref 1

AISI designation	Resistance to decarburization	Hardening and tempering: Hardening response	Amount of distortion(a)	Resistance to cracking	Approximate hardness(b), HRC	Fabrication and service: Machinability	Toughness	Resistance to softening	Resistance to wear
Chromium hot work steels									
H10	Medium	Deep	Very low	Highest	39-56	Medium to high	High	High	Medium
H11	Medium	Deep	Very low	Highest	38-54	Medium to high	Very high	High	Medium
H12	Medium	Deep	Very low	Highest	38-55	Medium to high	Very high	High	Medium
H13	Medium	Deep	Very low	Highest	38-53	Medium to high	Very high	High	Medium
H14	Medium	Deep	Low	Highest	40-47	Medium	High	High	Medium
H19	Medium	Deep	A, low; O, medium	High	40-57	Medium	High	High	Medium to high
Tungsten hot work steels									
H21	Medium	Deep	A, low; O, medium	High	36-54	Medium	High	High	Medium to high
H22	Medium	Deep	A, low; O, medium	High	39-52	Medium	High	High	Medium to high
H23	Medium	Deep	Medium	High	34-47	Medium	Medium	Very high	Medium to high
H24	Medium	Deep	A, low; O, medium	High	45-55	Medium	Medium	Very high	High
H25	Medium	Deep	A, low; O, medium	High	35-44	Medium	High	Very high	Medium
H26	Medium	Deep	A or S, low; O, medium	High	43-58	Medium	Medium	Very high	High
Molybdenum hot work steels									
H42	Medium	Deep	A or S, low; O, medium	Medium	50-60	Medium	Medium	Very high	High
Air-hardening medium-alloy cold work steels									
A2	Medium	Deep	Lowest	Highest	57-62	Medium	Medium	High	High
A3	Medium	Deep	Lowest	Highest	57-65	Medium	Medium	High	Very high
A4	Medium to high	Deep	Lowest	Highest	54-62	Low to medium	Medium	Medium	Medium to high
A6	Medium to high	Deep	Lowest	Highest	54-60	Low to medium	Medium	Medium	Medium to high
A7	Medium	Deep	Lowest	Highest	57-67	Low	Low	High	Highest
A8	Medium	Deep	Lowest	Highest	50-60	Medium	High	High	Medium to high
A9	Medium	Deep	Lowest	Highest	35-56	Medium	High	High	Medium to high
A10	Medium to high	Deep	Lowest	Highest	55-62	Medium to high	Medium	Medium	High
High-carbon, high-chromium cold work steels									
D2	Medium	Deep	Lowest	Highest	54-61	Low	Low	High	High to very high
D3	Medium	Deep	Very low	High	54-61	Low	Low	High	Very high
D4	Medium	Deep	Lowest	Highest	54-61	Low	Low	High	Very high
D5	Medium	Deep	Lowest	Highest	54-61	Low	Low	High	High to very high
D7	Medium	Deep	Lowest	Highest	58-65	Low	Low	High	Highest
Oil-hardening cold work steels									
O1	High	Medium	Very low	Very high	57-62	High	Medium	Low	Medium
O2	High	Medium	Very low	Very high	57-62	High	Medium	Low	Medium
O6	High	Medium	Very low	Very high	58-63	Highest	Medium	Low	Medium
O7	High	Medium	W, high; O, very low	W, low; O, very high	58-64	High	Medium	Low	Medium

(a) A, air cool; B, brine quench; O, oil quench; S, salt bath quench; W, water quench. (b) After tempering in temperature range normally recommended for this steel. (c) Carburized case hardness. (d) After aging at 510 to 550 °C (950 to 1025 °F). (e) Toughness decreases with increasing carbon content and depth of hardening.

Table 7 (continued)

Adapted from Ref 1

AISI designation	Resistance to decarburization	Hardening and tempering: Hardening response	Amount of distortion(a)	Resistance to cracking	Approximate hardness(b), HRC	Fabrication and service: Machinability	Toughness	Resistance to softening	Resistance to wear
Shock-resisting steels									
S1	Medium	Medium	Medium	High	40-58	Medium	Very high	Medium	Low to medium
S2	Low	Medium	High	Low	50-60	Medium to high	Highest	Low	Low to medium
S5	Low	Medium	Medium	High	50-60	Medium to high	Highest	Low	Low to medium
S6	Low	Medium	Medium	High	54-56	Medium	Very high	Low	Low to medium
S7	Medium	Deep	A, lowest; O, low	A, highest; O, high	45-57	Medium	Very high	High	Low to medium
Low-alloy special-purpose steels									
L2	High	Medium	W, low; O, medium	W, high; O, medium	45-63	High	Very high(c)	Low	Low to medium
L6	High	Medium	Low	High	45-62	Medium	Very high	Low	Medium
Low-carbon mold steels									
P2	High	Medium	Low	High	58-64(c)	Medium to high	High	Low	Medium
P3	High	Medium	Low	High	58-64(c)	Medium	High	Low	Medium
P4	High	High	Very low	High	58-64(c)	Low to medium	High	Medium	High
P5	High	...	W, high; O, low	High	58-64(c)	Medium	High	Low	Medium
P6	High	...	A, very low; O, low	High	58-61(c)	Medium	High	Low	Medium
P20	High	Medium	Low	High	28-37	Medium to high	High	Low	Low to medium
P21	High	Deep	Lowest	Highest	30-40(d)	Medium	Medium	Medium	Medium
Water-hardening steels									
W1	Highest	Shallow	High	Medium	50-64	Highest	High(e)	Low	Low to medium
W2	Highest	Shallow	High	Medium	50-64	Highest	High(e)	Low	Low to medium
W5	Highest	Shallow	High	Medium	50-64	Highest	High(e)	Low	Low to medium

(a) A, air cool; B, brine quench; O, oil quench; S, salt bath quench; W, water quench. (b) After tempering in temperature range normally recommended for this steel. (c) Carburized case hardness. (d) After aging at 510 to 550 °C (950 to 1025 °F). (e) Toughness decreases with increasing carbon content and depth of hardening.

Fig. 6 Relation of Vickers and Rockwell hardness scales

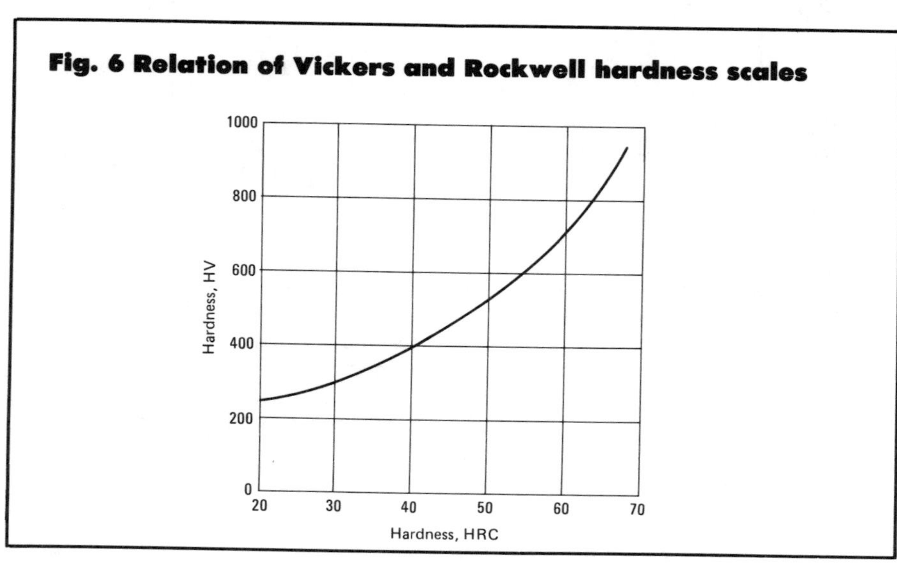

steel (except for certain grades) is broken in impact (Charpy test) is so small that it is very difficult to measure the differences in toughness that may make it possible to predict service performance. Attempts have been made to perform impact tests on unnotched tool steel bars, but excessive deformation of supporting fixtures makes it very difficult to obtain reproducible results. Torsion impact testing yields useful, reproducible data on the effects of variations in composition and heat treatment of tool steels; however, it is difficult to correlate the results of such testing with service experience. Fatigue tests have been useful for research, but only in some instances have the results correlated well with field experience.

In general, the ability of tool steels to

Table 8 Density and thermal expansion of selected tool steels

Type	Density		Thermal expansion									
			μm/m·K from 20 °C to					μin./in. °F from 68 °F to				
	Mg/m³	lb/in.³	100 °C	200 °C	425 °C	540 °C	650 °C	200 °F	400 °F	800 °F	1000 °F	1200 °F
W1	7.84	0.282	10.4	11.0	13.1	13.8(a)	14.2(b)	5.76	6.13	7.28	7.64(a)	7.90(b)
W2	7.85	0.283	...	...	14.4	14.8	14.9	...	...	8.0	8.2	8.3
S1	7.88	0.255	12.4	12.6	13.5	13.9	14.2	6.9	7.0	7.5	7.7	7.9
S2	7.79	0.281	10.9	11.9	13.5	14.0	14.2	6.0	6.6	7.5	7.8	7.9
S5	7.76	0.280	...	...	12.6	13.3	13.7	...	...	7.0	7.4	7.6
S6	7.75	0.279	...	...	12.6	13.3	...	...	...	7.0	7.4	...
S7	7.76	0.280	...	12.6	13.3	13.7(a)	13.3	...	7.0	7.4	7.6(a)	7.4
O1	7.85	0.283	...	10.6(c)	12.8	14.0(d)	14.4(d)	...	5.9(c)	7.1	7.8(d)	8.0(d)
O2	7.66	0.277	11.2	12.6	13.9	14.6	15.1	6.2	7.0	7.7	8.1	8.4
O7	7.8	0.282	...	...	...	...	...	...	...	...	...	...
A2	7.86	0.284	10.7	10.6(c)	12.9	14.0	14.2	5.96	5.91(c)	7.2	7.8	7.9
A6	7.84	0.283	11.5	12.4	13.5	13.9	14.2	6.4	6.9	7.5	7.7	7.9
A7	7.66	0.277	...	...	12.4	12.9	13.5	...	...	6.9	7.2	7.5
A8	7.87	0.284	...	...	12.0	12.4	12.6	...	...	6.7	6.9	7.0
A9	7.78	0.281	...	...	12.0	12.4	12.6	...	...	6.7	6.9	7.0
D2	7.70	0.278	10.4	10.3	11.9	12.2	12.2	5.8	5.7	6.6	6.8	6.8
D3	7.70	0.278	12.0	11.7	12.9	13.1	13.5	6.7	6.5	7.2	7.3	7.5
D4	7.70	0.278	...	...	12.4	...	...	...	...	6.9	...	...
D5	...	...	...	...	...	12.0	...	...	...	...	6.7	...
H10	7.81	0.281	...	...	12.2	13.3	13.7	...	...	6.8	7.4	7.6
H11	7.75	0.280	11.9	12.4	12.8	12.9	13.3	6.6	6.9	7.1	7.2	7.4
H13	7.76	0.280	10.4	11.5	12.2	12.4	13.1	5.8	6.4	6.8	6.9	7.3
H14	7.89	0.285	11.0	...	...	...	...	6.1	...	...	...	...
H19	7.98	0.288	11.0	11.0	12.0	12.4	12.9	6.1	6.1	6.7	6.9	7.2
H21	8.28	0.299	12.4	12.6	12.9	13.5	13.9	6.9	7.0	7.2	7.5	7.7
H22	8.36	0.302	11.0	...	11.5	12.0	12.4	6.1	...	6.4	6.7	6.9
H26	8.67	0.313	...	...	...	12.4	...	...	...	...	6.9	...
H42	8.15	0.295	...	...	...	11.9	...	...	...	...	6.6	...
T1	8.67	0.313	...	9.7	11.2	11.7	11.9	...	5.4	6.2	6.5	6.6
T2	8.67	0.313	...	...	...	...	...	...	...	...	...	...
T4	8.68	0.313	...	...	...	11.9	...	...	...	...	6.6	...
T5	8.75	0.316	11.2	...	...	11.5	...	6.2	...	...	6.4	...
T6	8.89	0.321	...	...	...	...	...	...	...	...	...	...
T8	8.43	0.305	...	...	...	...	...	...	...	...	...	...
T15	8.19	0.296	...	9.9	11.0	11.5	...	...	5.5(c)	6.1	6.4	...
M1	7.89	0.285	...	10.6(c)	11.3	12.0	12.4	...	5.9(c)	6.3	6.7	6.9
M2	8.16	0.295	10.1	9.4(c)	11.2	11.9	12.2	5.6	5.2(c)	6.2	6.6	6.8
M3, class 1	8.15	0.295	...	...	11.5	12.0	12.2	...	...	6.4	6.7	6.8
M3, class 2	8.16	0.295	...	...	11.5	12.0	12.8	...	...	6.4	6.7	7.1
M4	7.97	0.288	...	9.5(c)	11.2	12.0	12.2	...	5.3(c)	6.2	6.7	6.8
M7	7.95	0.287	...	9.5(c)	11.5	12.2	12.4	...	5.3(c)	6.4	6.8	6.9
M10	7.88	0.255	...	...	11.0	11.9	12.4	...	...	6.1	6.6	6.9
M30	8.01	0.289	...	...	11.2	11.7	12.2	...	...	6.2	6.5	6.8
M33	8.03	0.290	...	...	11.0	11.7	12.0	...	...	6.1	6.5	6.7
M36	8.18	0.296	...	...	...	...	...	...	...	...	...	...
M41	8.17	0.295	...	9.7	10.4	11.2	...	...	5.4	5.8	6.2	...
M42	7.98	0.288	...	...	...	...	...	...	...	...	...	...
M46	7.83	0.283	...	...	...	...	...	...	...	...	...	...
M47	7.96	0.288	10.6	11.0	11.9	...	12.6	5.9	6.1	6.6	...	7.0
L2	7.86	0.284	...	...	14.4	14.6	14.8	...	...	8.0	8.1	8.2
L6	7.86	0.284	11.3	12.6	12.6	13.5	13.7	6.3	7.0	7.0	7.5	7.6
P2	7.86	0.284	...	...	13.7	...	...	...	...	7.6	...	...
P5	7.80	0.282	...	...	...	...	...	...	...	...	...	...
P6	7.85	0.284	...	...	...	...	...	...	...	...	...	...
P20	7.85	0.284	...	...	12.8	13.7	14.2	...	...	7.1	7.6	7.9

(a) From 20 °C to 500 °C (68 °F to 930 °F). (b) From 20 °C to 600 °C (68 °F to 1110 °F). (c) From 20 °C to 260 °C (68 °F to 500 °F). (d) From 38 °C (100 °F).

Table 9 Thermal conductivity of selected tool steels

Temperature °C	°F	Thermal conductivity W/m·K	Btu/ft·h·°F
Type W1			
100	200	48.3	27.9
260	500	41.5	24.0
400	750	38.1	22.0
540	1000	34.6	20.0
675	1250	29.4	17.0
815	1500	24.2	14.0
Type H11			
100	200	42.2	24.4
260	500	36.3	21.0
400	750	33.4	19.3
540	1000	31.5	18.2
675	1250	30.1	17.4
815	1500	28.6	16.5
Type H13			
215	420	28.6	16.5
350	660	28.4	16.4
475	890	28.4	16.4
605	1120	28.7	16.6
Type H21			
100	200	27.0	15.6
260	500	29.8	17.2
400	750	29.8	17.2
540	1000	29.4	17.0
675	1250	29.1	16.8
Type T1			
100	200	19.9	11.5
260	500	21.6	12.5
400	750	23.2	13.4
540	1000	24.7	14.3
Type T15			
100	200	20.9	12.1
200	500	24.1	13.9
400	750	25.4	14.7
540	1000	26.3	15.2
Type M2			
100	200	21.3	12.3
200	500	23.5	13.6
400	750	25.6	14.8
540	1000	27.0	15.6
675	1250	28.9	16.7

Fig. 7 Hot hardness of H11 and T15 tool steels

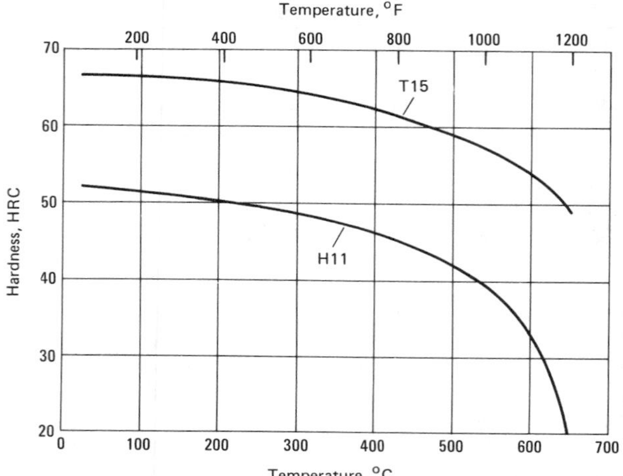

Type H11 has high resistance to softening at elevated temperatures; T15 has the highest resistance to softening. For these tests, H11 was air cooled from 1010 °C (1850 °F) and tempered 2 + 2 h at 565 °C (1050 °F); T15 was oil quenched from 1230 °C (2250 °F) and tempered 2 + 2 h at 550 °C (1025 °F). After hot hardness testing at 650 °C, T15 had a room temperature hardness of 63.4 HRC.

withstand rapid application of high loads without breaking increases with decreasing hardness. With hardness held constant, wide differences can be observed among tool steels of different compositions, or among steels of the same nominal composition made by different melting practices or heat treated according to different schedules.

The ability of a tool steel to resist softening at elevated temperatures is related to (*a*) its ability to develop secondary hardening and (*b*) the amount of special phases, such as excess alloy carbides, in the microstructure. Useful information on the ability of tool steels to resist softening at elevated temperatures can be obtained from tempering curves such as those in Fig. 1. Hardness testing at elevated temperatures (see Fig. 7) also can provide useful information.

Fabrication

The properties that influence the fabricability of tool steels include: machinability; grindability; weldability; hardenability; and extent of distortion, safety (freedom from cracking) and tendency to decarburize during heat treatment.

Machinability of tool steels can be measured by the usual methods applied to constructional steels. Results are reported as percentages of the machinability of water-hardening tool steels (see Table 10); 100% machinability in tool steels is equivalent to about 30% machinability in constructional steels, for which 100% machinability would be that of a free-machining constructional steel such as B1112.

Improving the machinability of a tool steel by altering either composition or preliminary heat treatment can be very important if a large amount of machining is required to form the tool and a large number of tools is to be made.

Grindability. One measure of grindability is the ease with which the necessary excess stock on heat treated tool steel can be removed using standard grinding wheels. The grinding ratio (grindability index) is the volume of metal removed per volume of wheel wear. The higher the grindability index the easier the metal is to grind. The index is valid only for specific sets of grinding conditions. Table 11 gives grinding ratios for several high speed steels. It should be noted that the grindability index does not indicate susceptibility to cracking during or after grinding, ability to produce the required surface (and subsurface) stress distribution, or ease of obtaining the required surface smoothness.

Weldability. The ability to construct, alter or repair tools by welding without causing the material to crack may be an important factor in selection of a tool material, especially if the tool is large. It is only rarely of importance in selecting materials for small tools. Weldability is largely a function of composition, but welding method and procedure also influence weld soundness. Generally, tool steels that are deep hardening and classified as having relatively high safety in hardening

Table 10 Approximate machinability ratings for annealed tool steels

Type	Machinability rating
O6	125
W1, W2, W5	100(a)
A10	90
P2, P3, P4, P5, P6	75 to 90
P20, P21	65 to 80
L2, L6	65 to 75
S1, S2, S5, S6, S7	60 to 70
H10, H11, H13, H14, H19	60 to 70(b)
O1, O2, O7	45 to 60
A2, A3, A4, A6, A8, A9	45 to 60
H21, H22, H24, H25, H26, H42	45 to 55(b)
T1	40 to 50
M2	40 to 50
T4	35 to 40
M3 (class 1)	35 to 40
D2, D3, D4, D5, D7, A7	30 to 40
T15	25 to 30
M15	25 to 30

(a) Equivalent to approximately 30% of the machinability of B1112. (b) For hardness range 150 to 200 HB.

Table 11 Typical grinding ratios for high speed steels

Type	Hardness, HRC	Grinding ratio(a) 32A46-H8VBE	32A60-H8VBE	32A80-H8VBE
T15	65.7	0.49	0.62	0.51
M44	67.7	0.97	0.99	0.88
M41	68.7	1.2	1.6	1.4
M43	67.5	1.4	2.2	1.7
M42	68.8	4.8	6.5	3.8
M2	64.9	6.1	7.2	6.7
M1	64.9	7.8	8.0	11.9

(a) For the following conditions: work, 152 mm (6 in.) long by 38 mm (1.5 in.) wide; wheel size, 200 mm (8 in.) in diameter by 13 mm (0.5 in.) wide; wheel speed (idling), 30 m/s (6000 ft/min); table speed, 0.3 m/s (60 ft/min); unit crossfeed, 1.27 mm (0.050 in.) after each table traverse; unit downfeed, 0.025 mm (0.001 in.) after each complete crossfeed; total downfeed, 0.25 mm (0.010 in.) preceded by four unit downfeeds to break wheel in after dressing with a diamond tool; grinding fluid, 1.25% water emulsion of general-purpose soluble oil.

are among the more readily welded tool steel compositions.

Hardenability includes both the maximum hardness obtainable when the quenched steel is fully martensitic and the depth of hardening obtained by quenching in a specific manner. In this context, depth of hardening must be defined—generally as a specific value of hardness or a specific microstructural appearance. As a very general rule, maximum hardness of a tool steel increases with increasing carbon content; increasing the austenitic grain size and the amount of alloying elements reduces the cooling rate required to produce maximum hardness (increases the depth of hardening). The Jominy end-quench test, which is applied extensively to measure hardenability of constructional steels (see pages 471 to 525 in Volume 1 of the 9th Edition of this Handbook) has limited application to tool steels. This test gives useful information only for oil-hardening grades. Air-hardening grades are so deep hardening that the standard Jominy test is not sufficient to evaluate hardenability.

An air hardenability test has been developed that is based on the principles involved in the Jominy test, but which uses only still air cooling and a 150 mm (6 in.) diameter end block to produce the very slow cooling rates of large sections. Such tests provide useful information for research but are of limited use in devising production heat treatments. By contrast, water-hardening grades of tool steel are so shallow hardening that the Jominy test is not sensitive enough. Special tests, such as the Shepherd PF test, are useful for research and for special applications of water-hardening tool steels.

In the Shepherd PF test, a bar 19 mm (3/4 in.) in diameter, in the normalized condition, is brine quenched from 740 °C (1450 °F) and fractured; the case depth (penetration, *P*) is measured in 0.4-mm (1/64-in.) intervals, and the fracture grain size of the case (*F*) is determined by comparison with standard specimens. A *PF* value of 6–8 indicates a case depth of 6/64 in. (2.4 mm) and a fracture grain size of 8. Fine-grain water-hardening tool steels are those with fracture grain sizes (*F* values) of 8 or more. Deep-hardening steels of this type have *P* values of 12 or more; medium-hardening steels, 9 to 11; and shallow-hardening steels, 6 to 8.

Distortion and Safety in Hardening. Minimal distortion in heat treating is important for tools that must remain within close size limits. For a more detailed discussion of this factor, see the article in this volume on distortion in tool steels. In general, the amount of distortion and the tendency to crack increase as the severity of quenching increases.

Resistance to decarburization is an important factor in determining whether or not a protective atmosphere is required during heat treating. In a decarburizing atmosphere, the rate of decarburization increases rapidly with increasing austenitizing temperature, and for a given austenitizing temperature, the depth of decarburization increases directly with holding time. Some types of tool steel decarburize much more rapidly than others under the same conditions of atmosphere, austenitizing temperature and time.

Machining Allowances

The standard machining allowance is the recommended total amount of stock that the user should remove from the as-supplied mill form to provide a surface free from imperfections that might adversely affect response to heat treatment or the ability of tools to perform properly.

The decarburization resulting from oxidation at the exposed surfaces during forging and rolling of the tool steel is a major factor in determining the amount of stock that should be removed. Although extra care is used in producing tool steels, scale, seams and other surface imperfections may be present, and if so, must be removed.

Table 12 gives the standard machining allowances for various sizes of hot rolled square and flat bars. Similar tables are available for other shapes and other methods of forming and finishing in ASTM specifications A600, A681 and A686.

Besides the standard machining allowance, sufficient additional stock must be provided to allow for cleanup of any decarburization and distortion that may occur during final heat treatment. The amount of this allowance varies with the type of tool steel, the type of heat treating equipment and the size and shape of the tool.

Group W and group O tool steels are considered highly resistant to decarburization. Group M steels, cobalt-containing group T steels, group D steels and types H42, A2 and S5 are rated as poor in resisting decarburization.

Decarburization during final heat

Table 12 Standard machining allowances for hot rolled square and flat bars

Specified width		Machining allowances(a)			
		Top and bottom surfaces		Edges	
mm	in.	mm	in.	mm	in.
Specified thickness, up to 12.7 mm (½ in.)					
0 to 12.7	0 to ½	0.64	0.025	0.64	0.025
>12.7 to 25.4	>½ to 1	0.64	0.025	0.89	0.035
>25.4 to 50.8	>1 to 2	0.76	0.030	1.02	0.040
>50.8 to 76.2	>2 to 3	0.89	0.035	1.27	0.050
>76.2 to 101.6	>3 to 4	1.02	0.040	1.65	0.065
>101.6 to 127.0	>4 to 5	1.14	0.045	2.03	0.080
>127.0 to 152.4	>5 to 6	1.27	0.050	2.41	0.095
>152.4 to 177.8	>6 to 7	1.40	0.055	2.67	0.105
>177.8 to 203.2	>7 to 8	1.52	0.060	3.05	0.120
>203.2 to 228.6	>8 to 9	1.52	0.060	3.30	0.130
>228.6 to 304.8	>9 to 12	1.52	0.060	3.56	0.140
Specified thickness, >12.7 to 25.4 mm (>½ to 1 in.)					
>12.7 to 25.4	>½ to 1	1.14	0.045	1.14	0.045
>25.4 to 50.8	>1 to 2	1.14	0.045	1.27	0.050
>50.8 to 76.2	>2 to 3	1.27	0.050	1.52	0.060
>76.2 to 101.6	>3 to 4	1.40	0.055	1.90	0.075
>101.6 to 127.0	>4 to 5	1.52	0.060	2.41	0.095
>127.0 to 152.4	>5 to 6	1.65	0.065	2.92	0.115
>152.4 to 177.8	>6 to 7	1.78	0.070	3.30	0.130
>177.8 to 203.2	>7 to 8	1.90	0.075	3.81	0.150
>203.2 to 228.6	>8 to 9	1.90	0.075	3.94	0.155
>228.6 to 304.8	>9 to 12	1.90	0.075	3.94	0.155
Specified thickness, >25.4 to 50.8 mm (>1 to 2 in.)					
>25.4 to 50.8	>1 to 2	1.65	0.065	1.65	0.065
>50.8 to 76.2	>2 to 3	1.65	0.065	1.78	0.070
>76.2 to 101.6	>3 to 4	1.78	0.070	2.16	0.085
>101.6 to 127.0	>4 to 5	1.78	0.070	2.67	0.105
>127.0 to 152.4	>5 to 6	1.90	0.075	3.18	0.125
>152.4 to 177.8	>6 to 7	2.03	0.080	3.68	0.145
>177.8 to 203.2	>7 to 8	2.03	0.080	4.19	0.165
>203.2 to 228.6	>8 to 9	2.41	0.095	4.32	0.170
>228.6 to 304.8	>9 to 12	2.54	0.100	4.32	0.170
Specified thickness, >50.8 to 76.2 mm (>2 to 3 in.)					
>50.8 to 76.2	>2 to 3	2.16	0.085	2.16	0.085
>76.2 to 101.6	>3 to 4	2.16	0.085	2.54	0.100
>101.6 to 127.0	>4 to 5	2.16	0.085	3.05	0.120
>127.0 to 152.4	>5 to 6	2.16	0.085	3.43	0.135
>152.4 to 177.8	>6 to 7	2.29	0.090	3.94	0.155
>177.8 to 203.2	>7 to 8	2.54	0.100	4.32	0.170
>203.2 to 228.6	>8 to 9	2.54	0.100	4.83	0.190
>228.6 to 304.8	>9 to 12	2.54	0.100	4.83	0.190
Specified thickness, >76.2 to 101.6 mm (>3 to 4 in.)					
>76.2 to 101.6	>3 to 4	2.92	0.115	2.92	0.115
>101.6 to 127.0	>4 to 5	2.92	0.115	3.18	0.125
>127.0 to 152.4	>5 to 6	2.92	0.115	3.56	0.140
>152.4 to 177.8	>6 to 7	2.92	0.115	4.32	0.170
>177.8 to 203.2	>7 to 8	3.18	0.125	4.83	0.190
>203.2 to 228.6	>8 to 9	3.18	0.125	4.83	0.190
>228.6 to 304.8	>9 to 12	3.18	0.125	4.83	0.190

(a) Minimum allowance per side for machining prior to heat treatment. Maximum decarburization limit, 80% of machining allowance.

Table 13 Out-of-roundness distortion in large-diameter bars of M2S tool steel

Bar diameter		Production method	Typical out-of-roundness(a)	
mm	in.		mm	in.
75	3	P/M	0.008	0.0003
		Conventional	0.020	0.0008
125	5	P/M	0.013	0.0005
		Conventional	0.033	0.0013
190	7.5	P/M	0.015	0.0006
		Conventional	0.051	0.0020

(a) Maximum diameter minus minimum diameter after normal hardening treatment.

treatment is undesirable because it alters the composition of the surface layer, thereby changing the response to heat treatment of this layer and usually affecting adversely the properties resulting from heat treatment. Decarburization can be controlled or avoided by heat treating in a salt bath or in a controlled atmosphere or vacuum furnace. When heat treating is done in vacuum, a vacuum of 100 to 200 μm Hg is satisfactory for most tools if the furnace is in good operating condition and has a very low leak rate. However, it is recommended that a vacuum of 50 to 100 μm Hg be used wherever possible.

If special heat treating equipment is not available, appreciable decarburization can be avoided by wrapping the tool in stainless steel foil. Type 321 stainless steel foil can be used at austenitizing temperatures up to about 1000 °C (1850 °F); either type 309 or type 310 foil is required at austenitizing temperatures from 1000 to 1200 °C (1850 to 2200 °F).

Powder Metallurgy Steels

In recent years, tool steels with improved properties have been produced by the powder metallurgy (P/M) process. In this process, a bath of prealloyed molten metal is gas atomized and quenched to produce a fine powder. The particles of this powder are screened and loaded into a steel container, which is then evacuated, and the particles are hot isostatically pressed to full density. The resulting compact is rolled or forged to size on conventional steel-mill equipment or, in some instances, is used as compacted to make tools.

P/M tool steels have two major advantages: complete freedom from mac-

Fig. 8 Comparison of response to hardening for P/M and conventionally produced bars of M2S (HC) tool steel

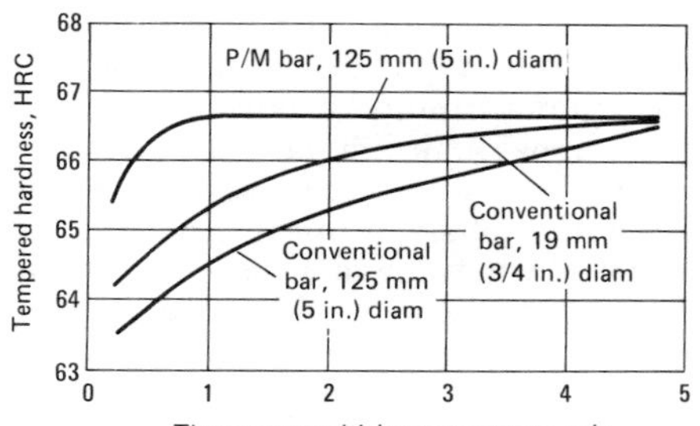

Hardness at mid-radius was evaluated for bars oil quenched from 1200 °C (2200 °F) and tempered 2 + 2 + 2 h at 550 °C (1025 °F).

rosegregation and uniform distribution of extremely fine carbides. These characteristics provide deeper hardening and faster response to hardening (see Fig. 8). The latter is important, particularly for molybdenum high speed steels, which tend to decarburize rapidly at austenitizing temperatures. P/M products also show less out-of-roundness distortion in large-diameter bars (see Table 13).

When sulfur is added to P/M tool steels, they exhibit a very fine homogeneous distribution of sulfides. This uniform sulfide distribution promotes better machinability. After heat treating, the refined microstructure of P/M tool steels promotes superior grindability (see Fig. 9) and greater toughness (see Fig. 10 and Table 14) compared to conventionally processed tool steels.

As of 1979, the following AISI types of high speed steels were available as P/M tool steels: M2, M2 with high sulfur and high carbon, M3 class 2 with sulfur, M4, M35 with sulfur, M42 and T15. P/M steels can be substituted for their conventional counterparts in all applications, and are particularly advantageous when heavy sections are required.

The freedom from gross segregation provided by the P/M process makes it possible to readily fabricate new higher-alloy tool steels. One type now available, which contains 1.50 C, 3.75 Cr, 3.00 V, 10 W, 5.25 Mo and 9.00 Co, is reported to have the highest hot hardness of any high speed steel. It has been used in cutting tools for critical applications such as machining of certain aero-

Table 14 Mechanical properties of P/M and conventional tool steels

		Toughness			
		Charpy C-notch impact value		Bend fracture strength	
Grade	Hardness, HRC	J	ft·lb	MPa	ksi
T15, P/M	67	19	14	4670	678
T15, conventional	66	5	4	2150	312
M2, conventional	65	23	17	3820	554

Fig. 9 Comparison of grinding ratios for P/M and conventionally produced tool steels

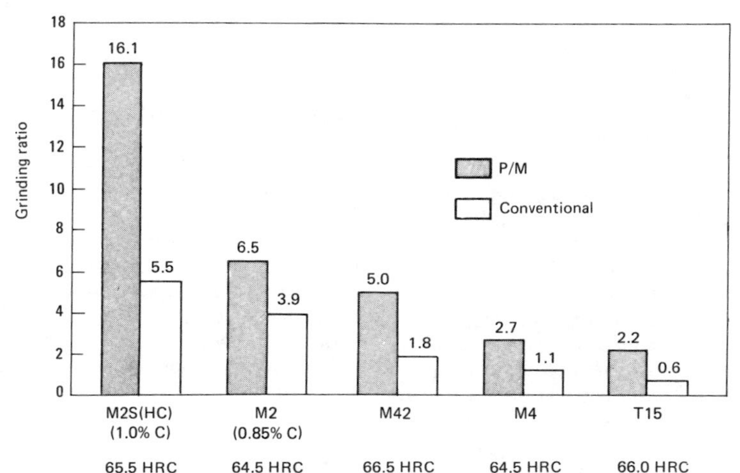

Grinding ratio is measured as the volume of metal removed divided by the volume of grinding wheel worn away. (High grinding ratios mean more efficient metal removal.)

Fig. 10 Comparison of bend fracture strength for P/M and conventionally produced M2S(HC) tool steel

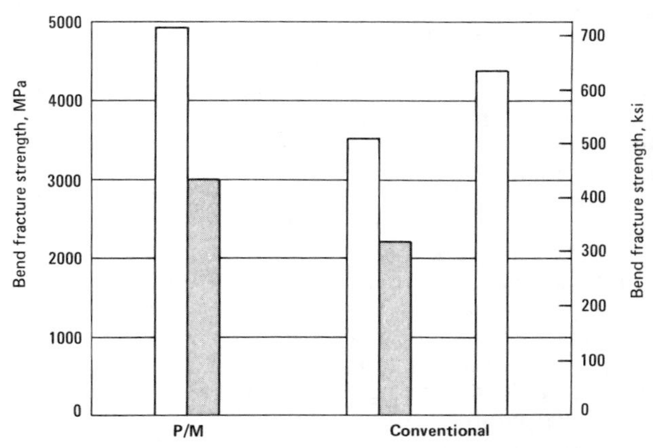

Material was hardened and tempered to 65 HRC. Unshaded bars are for longitudinal direction; shaded bars for transverse.

Fig. 11 Comparison of hot hardness for cast and wrought H13 tool steel

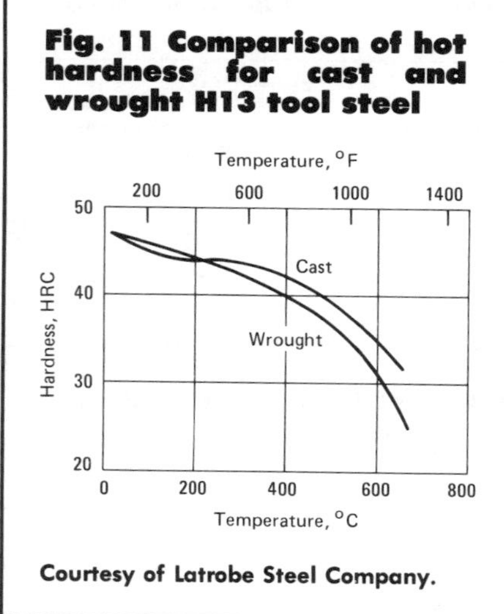

Courtesy of Latrobe Steel Company.

space alloys and cutting applications where the highest speeds and feeds are required. Another type, CPM 10V, which contains 2.45 C, 5.25 Cr, 9.75 V and 1.30 Mo, is designed for extreme wear resistance in cold and warm work tooling. Its microstructure consists chiefly of a uniform distribution of hard, wear resistant vanadium carbides in a tool steel matrix. The wear resistance of CPM 10V at ordinary temperatures typically exceeds that of T15 high speed steel.

Precision-Cast Hot Work Tools

Precision casting of tools to nearly finished size offers important cost advantages through reductions in waste and machining. Casting is particularly advantageous when patternmaking costs can be distributed over a large number of tools.

Experience with cast forging and extrusion dies has shown that cast tools are more resistant to heat checking; minute cracks do occur, but they grow at much lower rates than in wrought material of the same grade and hardness. Slower propagation of thermal-fatigue cracks generally extends die life significantly. Mechanical testing of cast and wrought H13 indicates that yield and tensile strengths are virtually identical from room temperature to 600 °C (1100 °F), but that ductility is moderately lower in cast material. Hot hardness of cast H13 is higher than that of wrought H13 at temperatures above about 300 °C (600 °F); this hardness advantage increases with temperature, as illustrated in Fig. 11, and measures about eight points on the Rockwell C scale at 650 °C (1200 °F).

Because cast dies exhibit uniform properties in all directions, no problem of directionality (anisotropy) exists. Dimensional control of castings is very consistent after an initial die is made and any necessary corrections are incorporated in the pattern. Reasonable finishing allowances are 0.25 to 0.38 mm (0.010 to 0.015 in.) on the impression faces, 0.8 to 1.5 mm (1/32 to 1/16 in.) at the parting line of the mold, and 1.6 to 3.2 mm (1/16 to 1/8 in.) on the back and outside surfaces. The hot work tool steels most commonly cast include H12, H13, H21 and H25.

Surface Treatments

In many applications, service life of high speed steel tools can be increased by surface treatments.

Oxide coatings, provided by treatment of the finish-ground tool in an alkali-nitrate bath or by steam oxidation, prevent or reduce adhesion of the tool to the workpiece. Oxide coatings have doubled tool life—particularly in machining of gummy materials such as soft copper and non-free-cutting low-carbon steels.

Plating of finished high speed steel tools with 0.0025 to 0.0125 mm (0.1 to 0.5 mil) of chromium also prolongs tool life by reducing adhesion of the tool to the workpiece. Chromium plating is relatively expensive, and precautions must be taken to prevent tool failure in service due to hydrogen embrittlement.

Carburizing is not recommended for high speed steel cutting tools because the cases on such tools are extremely brittle. However, carburizing is useful for applications such as cold work dies that require extreme wear resistance and that are not subjected to impact or highly concentrated loading. Carburizing is done at 1035 to 1065 °C (1900 to 1950 °F) for short periods of time (10 to 60 min) to produce a case 0.05 to 0.25 mm (0.002 to 0.010 in.) deep. The carburizing treatment also serves as an austenitizing treatment for the whole tool. A carburized case on high speed steels has a hardness of 65 to 70 HRC, but does not have the high resistance to softening at elevated temperatures exhibited by normally hardened high speed steel.

Nitriding successfully increases the life of all types of high speed steel cutting tools. However, gas nitriding in dissociated ammonia produces a case that is too brittle for most applications. Liquid nitriding for about 1 h at 565 °C (1050 °F) provides a light case, increasing both surface hardness and resistance to adhesion. For nitrided high speed steel taps, drills and reamers used in machining annealed steel, fivefold increases in life have been reported, with average increases of 100 to 200%. Obviously, if this nitrided case is removed when the tool is reground, the tool must then be retreated, which reduces the cost advantage of the process.

In addition, special surface-treatment processes, such as aerated nitriding baths, improve resistance to adhesive wear without producing excessive brittleness. Sulfur-containing nitriding baths provide a high-sulfur surface layer for additional resistance to seizing.

Sulfide Treatment. A low-temperature (190 °C; 375 °F) electrolytic process using sodium and potassium thiocyanate provides a seizing-resistant iron sulfide layer. This process can be used as a final treatment for all types of hardened tool steels without much danger of overtempering.

Maraging Steels

Certain high-nickel maraging steels are being used for special noncutting tool applications; 18Ni(250) is the type most frequently used. However, for the most demanding applications, the higher-strength 18Ni(300) is often preferred. For applications requiring maximum abrasion resistance, any of the maraging steels can be nitrided.

Maraging steels achieve full hardness—nominally 50 HRC for 18Ni(250), 54 HRC for 18Ni(300) and 58 HRC for 18Ni(350)—by a simple aging treatment, usually 3 h at about 480 °C (900 °F). Because hardening does not depend on cooling rate, full hardness can be developed uniformly in massive sections, with almost no distortion. Decarburization is of no concern in these alloys, because they contain very little carbon and because their aging temperature is relatively low. However, if the long-time service temperature exceeds the aging temperature, maraging steels overage with a significant drop in hardness.

The 18Ni(300) grade is used for aluminum die-casting dies and cores, aluminum hot forging dies, dies for molding plastics, and various support tooling used in extrusion of aluminum. In die casting of aluminum, maraging steel dies can be used at higher hardness than is possible with dies made of H13 tool steel because maraging steel is not as prone to heat checking. Because the aging process results in very little size change, it is possible to machine the intricate impressions for plastic molding dies to final size prior to final hardening.

For molding extremely abrasive types of plastics, the higher surface hardness provided by 18Ni(350) maraging steel is desirable.

REFERENCES

1. "Tool Steels", (a Steel Products Manual): American Iron and Steel Institute, March 1978
2. *Source Book on Industrial Alloys and Engineering Data:* American Society for Metals, Metals Park, OH, 1978, p 251-292

SELECTED REFERENCES

- *The Metallurgy of Tool Steels,* by P. Payson: John Wiley & Sons, Inc., 1962
- *Metallurgy and Heat Treatment of Tool Steels,* by R. Wilson: McGraw-Hill, London, 1975
- *Tool Steels,* by G. A. Roberts and R. A. Cary: American Society for Metals, 1980
- *Tool Steel Simplified,* Revised Ed., by F. R. Palmer *et al.*: Chilton Book Co., Radnor, PA, 1978

Superhard Tool Materials

By the ASM Committee on Tooling Materials*

SUPERHARD TOOL MATERIALS are exceptionally hard—far harder than any tool steel. Some superhard tool materials, such as hard alloys and cemented carbides, exhibit largely metallic characteristics, and others, such as ceramic tool materials, are considered nonmetallic.

Cemented carbides are the most widely used group of superhard tool materials. Their chief applications are in metalworking tools, mining tools and wear-resistant parts. Steel bonded carbides have a heat treatable matrix, and are used extensively to make forming and stamping tools. Ceramics, boron nitride and diamond are used chiefly to make cutting tools. Diamond is also used for making small-diameter wiredrawing dies.

To take full advantage of the wear resistance of cemented carbides and other superhard tool materials in metalcutting applications, new types of machine tools having the required power and rigidity have been developed. The degree of improvement that can be obtained by using superhard tools in older existing equipment may be limited by a lack of sufficient power and rigidity.

*See page XII for committee list.

Classification of Superhard Tool Materials

At present (1980), there is no universally accepted system for classifying carbides and other superhard tool materials. The systems most often employed by both producers and users to describe superhard tool materials are discussed in the following paragraphs. Each of the systems has inherent strengths and weaknesses in describing specific materials, and for this reason close cooperation between user and producer is often the most fruitful means of selecting the proper grade, especially in circumstances where previous experience on which to base the choice is lacking.

SAE Classification. SAE Recommended Practice J1072 sets forth one system for the identification and classification of ceramic, cemented carbide and other cermet products. This system uses alphanumeric designations consisting of:

- SAE Standard number, if applicable
- Basic classification, which identifies the type of ceramic or cermet
- One or more suffixes that define specific characteristics of the material.

Table 1 summarizes the code used in constructing the alphanumeric designation. The basic classification is a five-digit number. The first digit indicates the type of compound; the second, the type of binder; the third, the predominating metallic element in the compound; and the fourth and fifth digits, other metallic elements in the compound. Suffixes each consist of a letter identifier followed by one to three digits that provide a quantitative definition of the property (see Table 1 for a descriptive summary of the suffix system). As an example of the SAE classification system, a material designated 23200A060C920D150E123F322G230 would be a straight grade of tungsten carbide containing 6% cobalt binder and having the following properties: hardness, 92.0 HRA; specific gravity, 15.0; grain size, predominantly 1 μm and at least 80% of the grains 3 μm or less in diameter; porosity, appearance

Table 1 SAE J1072 system for classification of superhard tool materials

Basic classification	
Material classification	Designation
Material compound	
Nitride	1
Carbide	2
Oxide	3
Other	9(a)
Binder material	
None	0
Nickel	1
Iron	2
Cobalt	3
Other	9(a)
Base metal	
None	0
Niobium	1
Tungsten	2
Titanium	3
Tantalum	4
Chromium	5
Aluminum	6
Other	9(a)

Suffixes	
Material property	Identifier(b)
Binder metal quantity (wt% to nearest 0.1%)	A
Base metal quantity (wt% to nearest 0.1%)	B
Hardness(c) (HRA to nearest 0.1)	C
Specific gravity(c) (to nearest 0.1)	D
Grain size(c) (maximum amount of each type)	E
Apparent porosity(c) (the first digit indicates the amount of type A, the second the amount of type B and the third the amount of type C porosity)	F
Transverse rupture strength(d) (minimum, in ksi)	G
Other properties (written description required)	Z

(a) Material in this category shall be described by suffix Z. (b) Complete description consists of the letter identifier followed by one to three digits that express a quantitative value for the specific property. (c) Determined according to procedures outlined in SAE J439. (d) Determined according to procedures outlined in ASTM B406.

Table 2 C-grade classification system for cemented carbides

C grade	Application category
Machining of cast iron, nonferrous and nonmetallic materials	
C-1	Roughing
C-2	General-purpose machining
C-3	Finishing
C-4	Precision finishing
Machining of carbon and alloy steels	
C-5	Roughing
C-6	General-purpose machining
C-7	Finishing
C-8	Precision finishing
Wear-surface applications	
C-9	No shock
C-10	Light shock
C-11	Heavy shock
Impact applications	
C-12	Light impact
C-13	Medium impact
C-14	Heavy impact
Miscellaneous applications	
C-15	Hot weld-flash removal, light cuts
C-15A	Hot weld-flash removal, heavy cuts
C-16	Rock bits
C-17	Cold header dies
C-18	Wear at elevated temperatures and/or resistance to chemicals
C-19	Radioactive shielding, counterbalances and kinetic-energy devices

corresponding to types A-3, B-2 and C-2 of SAE J439; and transverse rupture strength, 230 ksi (1585 MPa). This material corresponds to the material described on line 3 of Table 8 in the section of this article on properties of the cemented carbides.

Clearly, the SAE system makes it possible to be very precise in specifying the material desired. However, even if the designation were shortened to include only composition and grain size, a combination of 12 numbers and letters would be required to identify a specific grade.

C-Grade System. Some U.S. users and producers use an informal application-oriented system of classification to assist in selecting proper grades of cemented carbide. This "C-grade" system does not require use of trade names to identify specific carbide products. C-grade identifications (see Table 2) are not really sufficient for procurement purposes because they fail to reflect the characteristics that significantly influence selection of the proper carbide material. Definition of work materials involved in classifying the machining grades of carbide is not precise, nor is there universal agreement on meanings of the various terms used to describe specific application categories.

ISO Classification. The International Organization of Standardization has issued ISO Recommendation R513 entitled "Application of Carbides for Machining by Chip Removal". Table 3 summarizes the basis for ISO classification of carbides according to their use in machining.

In the ISO system, P grades are used to machine ferrous metals that generate long chips; M grades are recommended for ferrous metals that generate long or short chips and for certain nonferrous metals; and K grades are preferred for ferrous metals that generate short chips, for many nonferrous metals and for nonmetallic materials. In each category, low numbers are used for high speed, low feed machining, and higher numbers for lower speeds and higher feeds. Increasing numbers also indicate increasing toughness and decreasing wear resistance.

The British Hard Metal Association System is a three-digit system, with values from zero to nine for each digit. The first digit describes the wear resistance, in terms of hardness; the second digit designates transverse rupture strength; and the third digit represents crater resistance, in terms of the amount of titanium carbide and tantalum carbide present in the material. The specified values corresponding to each digit in the BHMA system are shown in Table 4.

This system has the advantage of brevity, but the fundamental question of the degree to which different cemented carbides having the same BHMA classification are equivalent in service remains unanswered.

Cemented Carbides

Cemented carbides are made by a powder metallurgy process in which finely divided compounds of refractory metals and carbon are bonded together to form a compacted solid of high strength and hardness. The first cemented carbide to be produced was tungsten carbide with a cobalt binder. Over the years, this original material has been modified in many ways to produce a variety of cemented carbides that can be used in a wide range of applications (see Table 5). These modifications consist mainly of varying the amount of metal used as the binder,

Table 3 ISO R513 classification of carbides according to use for machining

Designation(a)	Groups of application: Material to be machined	Use and working conditions
P 01	Steel, steel castings	Finish turning and boring; high cutting speeds, small chip section, accuracy of dimensions and fine finish, vibration-free operation
P 10	Steel, steel castings	Turning, copying, threading and milling; high cutting speeds, small or medium chip sections
P 20	Steel, steel castings Malleable cast iron with long chips	Turning, copying, milling, medium cutting speeds and chip sections; planing with small chip sections
P 30	Steel, steel castings Malleable cast iron with long chips	Turning, milling, planing, medium or low cutting speeds, medium or large chip sections, and machining in unfavorable conditions(b)
P 40	Steel Steel castings with sand inclusion and cavities	Turning, planing, slotting, low cutting speeds, large chip sections with the possibility of large cutting angles for machining in unfavorable conditions(b) and work on automatic machines
P 50	Steel Steel castings of medium or low tensile strength, with sand inclusion and cavities	For operations demanding very tough carbide: turning, planing, slotting, low cutting speeds, large chip sections, with the possibility of large cutting angles for machining in unfavorable conditions(b) and work on automatic machines
M 10	Steel, steel castings, manganese steel Grey cast iron, alloy cast iron	Turning, medium or high cutting speeds. Small or medium chip sections
M 20	Steel, steel castings, austenitic or manganese steel, grey cast iron	Turning, milling. Medium cutting speeds and chip sections
M 30	Steel, steel castings, austenitic steel, grey cast iron, high temperature resistant alloys	Turning, milling, planing. Medium cutting speeds, medium or large chip sections
M 40	Mild free cutting steel, low tensile steel Nonferrous metals and light alloys	Turning, parting off, particularly on automatic machines
K 01	Very hard grey cast iron, chilled castings of over 85 scleroscope hardness, high silicon aluminum alloys, hardened steel, highly abrasive plastics, hard cardboard, ceramics	Turning, finish turning, boring, milling, scraping
K 10	Grey cast iron over 220 HB malleable cast iron with short chips, hardened steel, silicon aluminum alloys, copper alloys, plastics, glass, hard rubber, hard cardboard, porcelain, stone	Turning, milling, drilling, boring, broaching, scraping
K 20	Grey cast iron up to 220 HB nonferrous metals: copper, brass, aluminum	Turning, milling, planing, boring, broaching, demanding very tough carbide
K 30	Low hardness grey cast iron, low tensile steel, compressed wood	Turning, milling, planing, slotting, for machining in unfavorable conditions(b) and with the possibility of large cutting angles
K 40	Soft wood or hard wood Nonferrous metals	Turning, milling, planing, slotting, for machining in unfavorable conditions(b) and with the possibility of large cutting angles

(a) In each letter category, low designation numbers are for high speeds and light feeds, higher numbers for slower speeds and/or heavier feeds. Also, increasing designation numbers imply increasing toughness and decreasing wear resistance of the cemented carbide materials. (b) Unfavorable conditions include: shapes that are awkward to machine; material having a casting or forging skin; material having variable hardness; and machining that involves variable depth of cut, interrupted cut or moderate to severe vibrations.

varying the structure (grain size) of the carbide and substituting other metallic carbides for part of the tungsten carbide. Titanium carbide and tantalum carbide have been the metallic carbides used most widely in complex grades, but because of limited world supply of tantalum, niobium carbides are replacing tantalum carbides in increasing

Table 4 British Hard Metal Association classification of cemented carbides for machining applications

Adapted from Ref 4

Value	Hardness HV	Hardness HRA	Transverse rupture strength MPa	Transverse rupture strength tons/in.^{2}	Composition index(a)
0	...	...	...	...	<0.6
1	Up to 1300	Up to 88.9	Up to 1240	Up to 80	0.6–2.2
2	1300–1450	88.9–90.1	1240–1380	80–90	2.2–4.2
3	1450–1550	90.1–90.9	1380–1550	90–100	4.2–7.2
4	1550–1600	90.9–91.3	1550–1725	100–110	7.2–10.6
5	1600–1650	91.3–91.7	1725–1860	110–120	10.6–14.2(b)
6	1650–1700	91.7–92.0	1860–2000	120–130	14.2–17.7(b)
7	1700–1750	92.0–92.4	2000–2170	130–140	17.7–21.2(b)
8	1750–1800	92.4–92.7	2170–2340	140–150	21.2–24.6(b)
9	>1800	>92.7	>2340	>150	>24.6(b)

(a) Number listed is the sum of % TiC + 0.4(% TaC); rem WC. (b) 30% TaC max.

Fig. 1 Indexable cemented carbide inserts and methods of attachment

amounts. Cemented carbides containing only tungsten carbide are commonly referred to as "straight grades", whereas those containing other metallic carbides in addition to tungsten carbide are called "complex grades".

Straight grades of cemented carbides are used for cutting tools, drawing dies, forming-die inserts, punches and many other types of tools. Complex grades are used chiefly to make cutting tools and cutting-tool inserts for machining plain carbon and low-alloy steels. In these applications, tools made of complex grades exhibit less wear, because the crater formed on the top rake of the tool due to chip adherence is much smaller than that developed on tools made of straight grades. Also, the complex grades have better resistance to deformation at the temperatures developed along the cutting edge.

Cemented carbide is very hard, and holds its high hardness at temperatures well above those at which high speed steels begin to soften. Cemented carbide cutting tools originally were constructed by copper brazing small compacts into heavy steel shanks. These tools had to be sharpened by regrinding frequently for maximum tool efficiency. Although regrinding of brazed-tip tools is still done in many instances, it is much more common for cemented carbide to be used as relatively small indexable inserts that are mechanically held in a steel holder. Several types of inserts and methods of attachment are shown in Fig. 1. These inserts are indexed until all available cutting edges have become worn, and then are reground or recycled.

Manufacture by Powder Metallurgy. The conventional means of making cemented carbide tools is a powder metallurgy process in which finely divided tungsten carbide powders are blended with cobalt powders, compacted by cold pressing to the desired shape (with allowance for shrinkage during sintering), and sintered in vacuum or under controlled atmosphere at a temperature high enough to melt, or at least partly melt, the binder. One of the main limitations of this conventional process is the requirement that the parts be of uniform cross section along the direction of pressing, and be of limited length-to-diameter ratio, so that the compacting pressure can be applied relatively uniformly.

Large parts and parts that cannot be conveniently compacted in dies generally are made by cold isostatic pressing. The parts may or may not be presintered, and may or may not be machined to shape, before being liquid-phase sintered to final density. Large parts occasionally are compacted by hot pressing in graphite dies.

Hot isostatic pressing is commonly used to increase the soundness of sintered carbide parts. Hot isostatic pressing improves the surface finish and increases the average transverse rupture, compression and impact strengths of carbide tools. It is an essential step in the production of certain tools such as compressor plungers, grinder spindles, anvils and dies for very-high-pressure apparatus, cold extrusion punches and swaging mandrels. Hot isostatic pressing also is preferred for parts that require a nearly perfect surface finish. Drawing dies and mandrels; extrusion punches; Sendzimer-mill rolls; strip-mill rolls; wire-flattening rolls; liners and plungers for pumps and compres-

Table 5 Typical application of cobalt-bonded cemented carbides

Grade	Grain size	Application
Straight grades		
97WC-3Co	Medium	Machining of cast iron, nonferrous metals and nonmetallic materials; excellent abrasion resistance and low shock resistance; the most wear resistant of the straight WC-Co grades; maintains a sharp cutting edge and makes long finishing cuts to close tolerances possible; also used for fine wire dies and small nozzles
94WC-6Co	Fine	Machining nonferrous and high-temperature alloys
94WC-6Co	Medium	General-purpose machining of work materials other than steel; also used for small and medium size compacting dies, coating dies, burnishing rings and nozzles
94WC-6Co	Coarse	Machining of cast iron, nonferrous metals and nonmetallic materials; also used for small wire-drawing dies, compacting dies, small drawing dies and caps and rings. The hardest grade used in mining applications where impact is encountered, as in rotary percussive bits
90WC-10Co	Fine	Machining steel and milling high-temperature metals (including titanium and its alloys) at low feeds and speeds: face mills, end mills, form tools, cutoff tools and screw-machine tools
90WC-10Co	Coarse	Primarily used for mining roller bits and percussive drilling bits
84WC-16Co	Fine	Primarily used for mining and metal-forming components
84WC-16Co	Coarse	Metal-forming and mining components: medium and large dies where great toughness is required, blanking dies for punch presses, and large mandrels
75WC-25Co	Medium	Metal-forming components for heavy impact applications, such as heading dies, cold extrusion dies, and punches and dies for blanking heavy stock
Complex grades		
71-74.5WC-10-12.5TiC-11-12.0TaC-4.5Co	Medium	Finishing, semifinishing and light roughing operations on plain carbon and alloy steels and alloy cast irons
72-73WC-7-8TiC-11.5-12TaC-8-8.5Co	Medium	Tough, wear-resistant grade for heavy-duty roughing cuts. Successfully withstands high temperatures encountered in heavy-duty machining, interrupted turning, scale cuts and milling of plain carbon and alloy steels and alloy cast irons
64TiC-28WC-2TaC-2Cr_3C_2-4Co	Medium	High-speed finishing of steels and cast irons
57WC-27TaC-16Co	Coarse	Cutting hot flash formed in the manufacture of welded tubing; also used to make dies for hot extrusion of aluminum wirebar and tubing

sors; burnishing rolls; and balls for burnishing, sizing and valve applications are among the types of tools and other parts that need exceptionally good surface finishes for optimum performance.

Parts that are long and slender, such as rod stock for circuit-board drills, are difficult to produce by either compaction in dies or isostatic pressing. Such parts are produced by extruding a mixture of carbide powders, metallic binders and a suitable organic vehicle, followed by sintering.

Properties of Cemented Carbides

The basic mechanical and physical properties of refractory metal carbides used in the production of cemented carbide tools are shown in Table 6. A knowledge of these basic properties of the refractory carbides is useful in understanding, generally, why cemented carbides resist wear at elevated temperatures and have high moduli of elasticity and high densities. However, specific properties of individual grades depend not only on the composition of the carbide but also on its particle size and on the amount and type of binder.

Table 7 lists accepted test methods for determining basic properties of cemented carbides and other superhard materials. Although experience is the ultimate means of evaluating these materials for specific applications, certain of the tests identified in Table 7 are helpful in narrowing the choice of materials for applications where there is only limited and unsatisfactory experience or where previous experience does not exist.

The compositions and properties of nine straight grades and four complex grades of cobalt-bonded carbide are given in Table 8. Because properties are influenced by both composition and structure, both characteristics must be specified to define a specific grade. Representative microstructures of the 13 grades identified in Table 8 are shown in Fig. 2.

Several properties of straight grades of tungsten carbide are plotted in Fig. 3 against cobalt content. These graphs illustrate not only the effect of composition but also the importance of structure.

Microstructure. Varying the structure of cemented carbides can improve tool performance in specific applications. The photomicrographs in Fig. 2

Table 6 Properties of refractory metal carbides(a)

Carbide	Hardness(b), HV	Crystal system	Melting point °C	Melting point °F	Theoretical density, Mg/m^3	Modulus of elasticity GPa	Modulus of elasticity 10^6psi
TiC	3200	Cubic	3065 ± 15	5550 ± 30	4.92	448	65
VC	2950	Cubic	2730 ± 75	4950 ± 150	5.48	434	63
HfC	2700	Cubic	3925 ± 50	7100 ± 100	12.67	...	...
ZrC	2600	Cubic	3440 ± 20	6225 ± 40	6.56	474	68.8
NbC	2400	Cubic	3500 ± 75	6330 ± 135	7.82	~290	~42
Cr_3C_2	2280	Orthorhombic	~1900	~3440	6.68	386	56
WC	2080	Hexagonal	~2800	~5030	15.8	669	97
Mo_2C	1950	Hexagonal	2490-2520	4510-4570	9.12	227	33
TaC	1790	Cubic	3915 ± 50	7080 ± 100	14.50	276	40

(a) Data from Ref 2 unless otherwise indicated. (b) Data from Ref 3.

Table 7 Test methods for determining properties of cemented carbides

Property	ASTM/ANSI	CCPA	ISO	SAE
Abrasive wear resistance	B611	P112	(a)	...
Apparent grain size	B390	M203	...	J439
Apparent porosity	B276	M201	4505	J439
Axial load fatigue	...	...	(a)	...
Coefficient of sliding friction	...	P111	...	...
Coercive force	...	...	3326	...
Compressive strength	E9(b)	P104	4506	...
Density	B311	P101	3369	J439
Diametral compression testing	B485	P115	...	...
Electrical resistivity	B421	P107	...	...
Fracture toughness	(a)	...	...	...
Hardness, HRA	B294	P103	3738	J439
Hardness, HV	E92	...	3878	...
Linear thermal expansion	B95	P108	...	...
Magnetic permeability	A342	P109	...	...
Metallographic preparation of samples	(a)	...	...	...
Microstructure	B657	M202	4499	...
Poisson's ratio	E132	P105	...	...
Powder sampling and testing	...	...	4884	...
Sampling and testing	...	...	4889	...
Tensile testing	B437(c)	P113	...	...
Thermal shock resistance	...	P110	...	...
Transverse rupture strength	B406	P102	3327	...
Young's modulus	E111	P106	3312	...

(a) In preparation. (b) A procedure derived from ISO 4506 is in preparation. (c) Being withdrawn.

(a) through (j) show microstructures of the nine straight grades whose properties are given in Table 8. These microstructures and properties correspond to the applications summarized for the first nine grades listed in Table 5.

For straight tungsten carbides of comparable grain size (WC particle size), increasing cobalt content increases transverse strength and toughness but decreases hardness, compressive strength, elastic modulus and abrasion resistance. If we compare, for example, medium-grain carbides having 3, 6 and 25% cobalt (microstructures a, b and j in Fig. 2), we find the 3% Co grade to have the greatest hardness and abrasion resistance—properties that make it well suited for wiredrawing dies and for cutting tools used in machining of cast iron and other abrasive or gummy materials. The 6% Co grade has moderate values for all properties, and is a good general-purpose carbide material. The 25% Co grade has the greatest toughness, and is used for applications involving heavy impact. Because of its relatively low hardness and abrasion resistance, it is not used for cutting tools. Similar parallels in properties and uses can be drawn for the fine-grain grades containing 6, 10 and 16% cobalt (micrographs b, e and g in Fig. 2), and for the coarse-grain grades containing 6, 10 and 16% cobalt (micrographs c, f and h in Fig. 2).

Another set of comparisons can be drawn for the grades containing 6% cobalt. All three grades—fine, medium and coarse—are used for cutting tools, but the applications to which they are applied involve different machining conditions and different work materials. The fine-grain material (Fig. 2b) is used for finish to medium-rough machining of ductile, gray and chilled irons and of austenitic stainless steels, high-temperature alloys and nonmetallic materials; the medium-grain material (Fig. 2c) for light to heavy machining of these same wrought work materials; and the coarse-grain grade (Fig. 2d) for heavy to extremely heavy rough machining of such materials. The medium-grain material is widely employed for general-purpose machining because its properties have been found to offer a good practical balance between hardness and toughness. The coarse-grain grade, which has the lowest hardness and abrasion resistance and the best toughness of the three grades, is used where a combination of moderate hardness and high toughness is needed. Similar comparisons can be made for the grades that contain 10 and 16% Co. In general, decreasing grain size improves abrasion resistance and makes it easier to retain the edge on a cutting tool; increasing grain size improves toughness and makes the cemented carbide more suitable for die applications.

For complex grades (the last four grades listed in Tables 5 and 8), comparisons similar to those drawn for the straight grades are not as readily made. Variations in carbide type, as well as in binder content, affect properties, which in turn influence suitability for specific types of service. In the microstructures (k through p in Fig. 2), the WC particles are angular, whereas those of TiC and TaC are more rounded.

The first two complex grades (microstructures k and m) contain about the same amount of WC, but the latter contains twice as much binder. The lower-cobalt grade is used for lighter-duty cutting.

The complex grade high in TiC (microstructure n) is relatively low in transverse strength and high in resistance to abrasion and cratering. It is used extensively for high-speed, light-duty finishing.

The complex grade highest in cobalt content and in TaC (microstructure p) is preferred for hot work tools, in both cutting and shaping of metals.

Hardness. Following the practice of

Fig. 2 Typical microstructures of cemented carbides

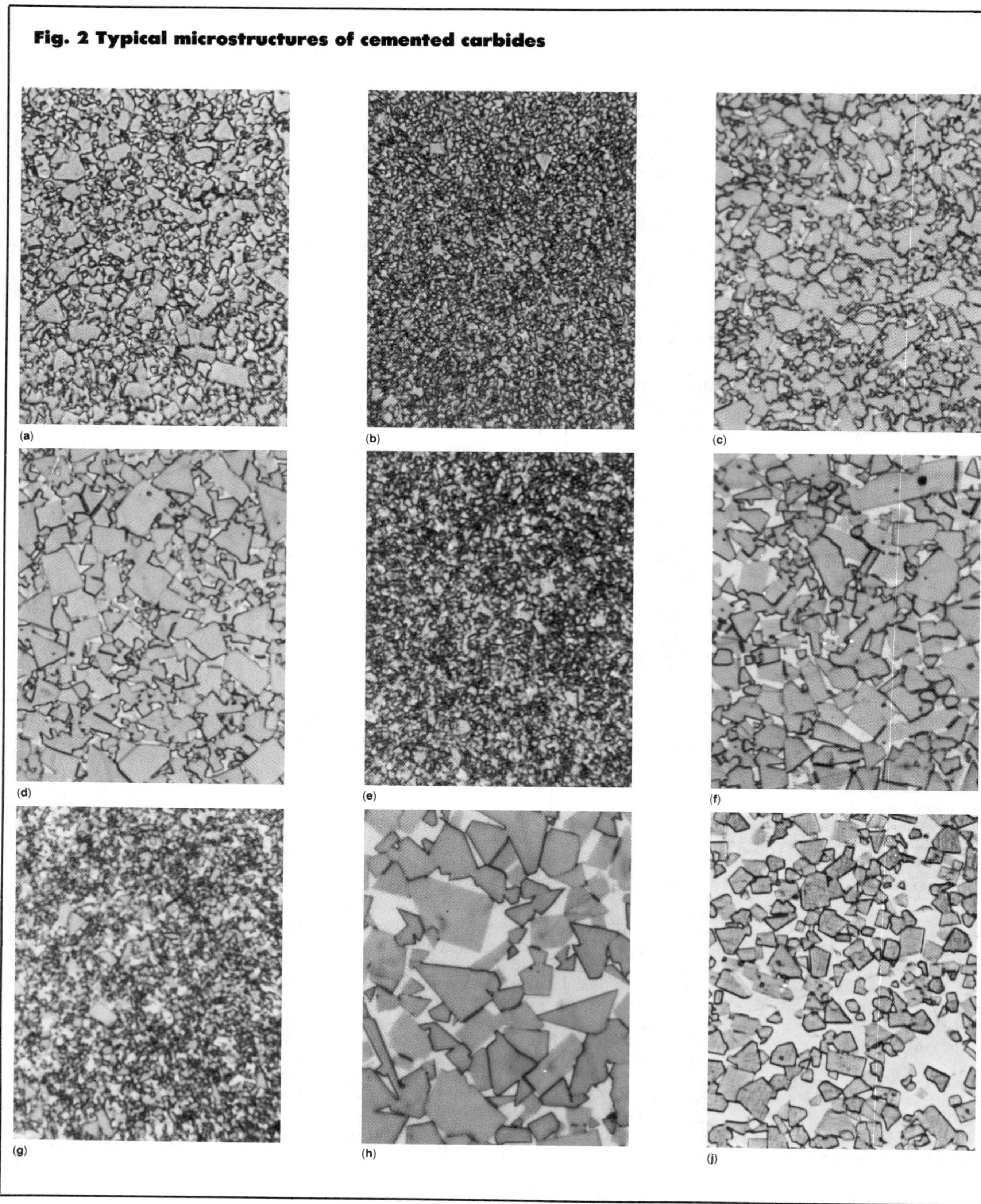

(a) (b) (c) (d) (e) (f) (g) (h) (j)

Fig. 2 (continued)

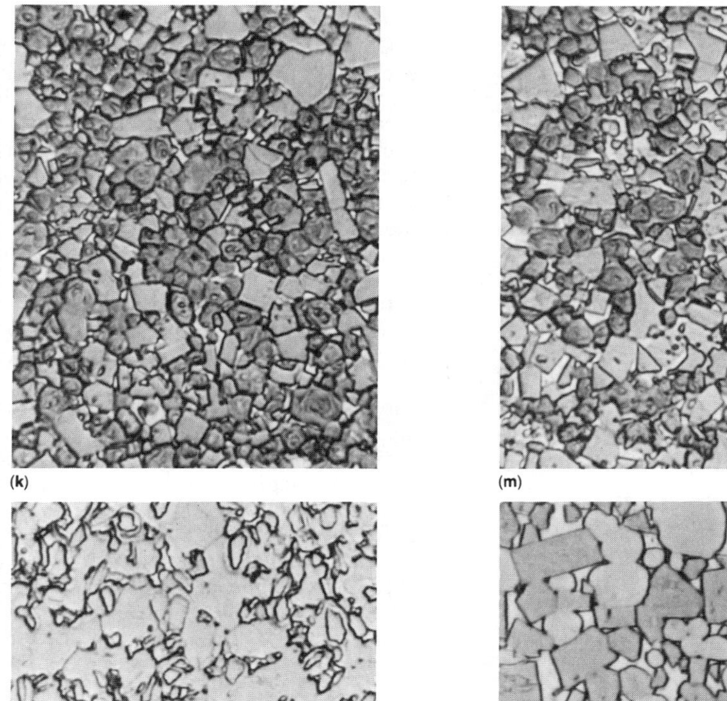

All specimens etched in Murakami's reagent, 1500 X. (a) 97WC-3Co, medium grain. (b) 94WC-6Co, fine grain. (c) 94WC-6Co, medium grain. (d) 94WC-6Co, coarse grain. (e) 90WC-10Co, fine grain. (f) 90WC-10Co, coarse grain. (g) 84WC-16Co, fine grain. (h) 84WC-16Co, coarse grain. (j) 75WC-25Co, medium grain. (k) 71WC-12.5TiC-12TaC-4.5Co, medium grain. (m) 72WC-8TiC-11.5TaC-8.5Co, medium grain. (n) 64TiC-28WC-2TaC-2Cr_2C_3-4Co, medium grain. (p) 57WC-27TaC-16Co, coarse grain. For corresponding properties, see Table 8.

the producers and users of cemented carbides in the United States, Rockwell A hardness is reported in Table 8 and Fig. 3. Figures 4 and 5 correlate Rockwell A values with Vickers (diamond pyramid hardness) and Knoop values for these materials. For very hard materials such as cemented carbides, no single curve will provide accurate conversion among hardness values taken by different methods. However, it is clear that variations of a single point on the HRA scale can be very significant, because of the great degree of nonlinearity in the conversions, especially at the high hardness end of the scales. For instance, an increase in hardness from 92 HRA to 93 HRA is equivalent to an increase from about 1700 HV to about 2000 HV, a difference of 300 points on the Vickers scale. On the other hand, an increase from 75 HRA to 76 HRA is equivalent to an increase from 500 HV to about 520 HV, a difference of only about 20 points on the Vickers scale.

In hardness testing of superhard tool materials, the diamond point of the indentor can be damaged easily, particularly during testing at high loads. Because of the low ductility of cemented carbides, a fragment may flake away under the indentor, causing the point to chip or break off. Frequent calibration using a fully hardened steel test block and periodic visual inspection of the indentor point is necessary to ensure that tests are being made with an undamaged indentor.

Figure 6 graphically depicts changes in Rockwell A hardness as temperature is increased. The reduction in hardness between room temperature and about 800 °C for 94WC-6Co with a very fine structure is 7 points on the Rockwell A scale—from 93 to 86 HRA. Lower hot hardness generally signifies lower resistance to deformation at high temperature. Nevertheless, cemented carbides with similar hot hardness values sometimes show very significant differences in resistance to deformation at high temperature in service.

Abrasion Resistance. Most producers of cemented carbides and other superhard tool materials measure abrasion resistance by subjecting the materials to a test in which a sample is held against a rotating wheel for a fixed number of revolutions while the sample and wheel are immersed in a water slurry containing sharp aluminum oxide particles. Comparative rankings are reported, usually on the basis of a wear rating based on the reciprocal of volume loss. Two standards apply to abrasive wear testing and the method of reporting results: ASTM B611 and CCPA P112. Not all producers have adopted the ASTM method of abrasive wear testing, so values of abrasion resistance cited by producers vary widely. Because of this variance, it is almost impossible to make valid comparisons among test results reported by different producers. It also is fallacious to use abrasion resistance as a measure of wear resistance of superhard tool materials when they are used for cutting steel or other materials—abrasion resistance in a standard test does *not* correspond directly to wear resistance in machining operations.

Values of comparative abrasion resistance are listed in Table 8 and Fig. 3. These are relative values only, and are based on a value of 100 for the most abrasion-resistant grade. Comparative abrasion resistance is lowered as cobalt

Fig. 3 Variation of properties with cobalt content and grain size for straight grades of cemented carbide

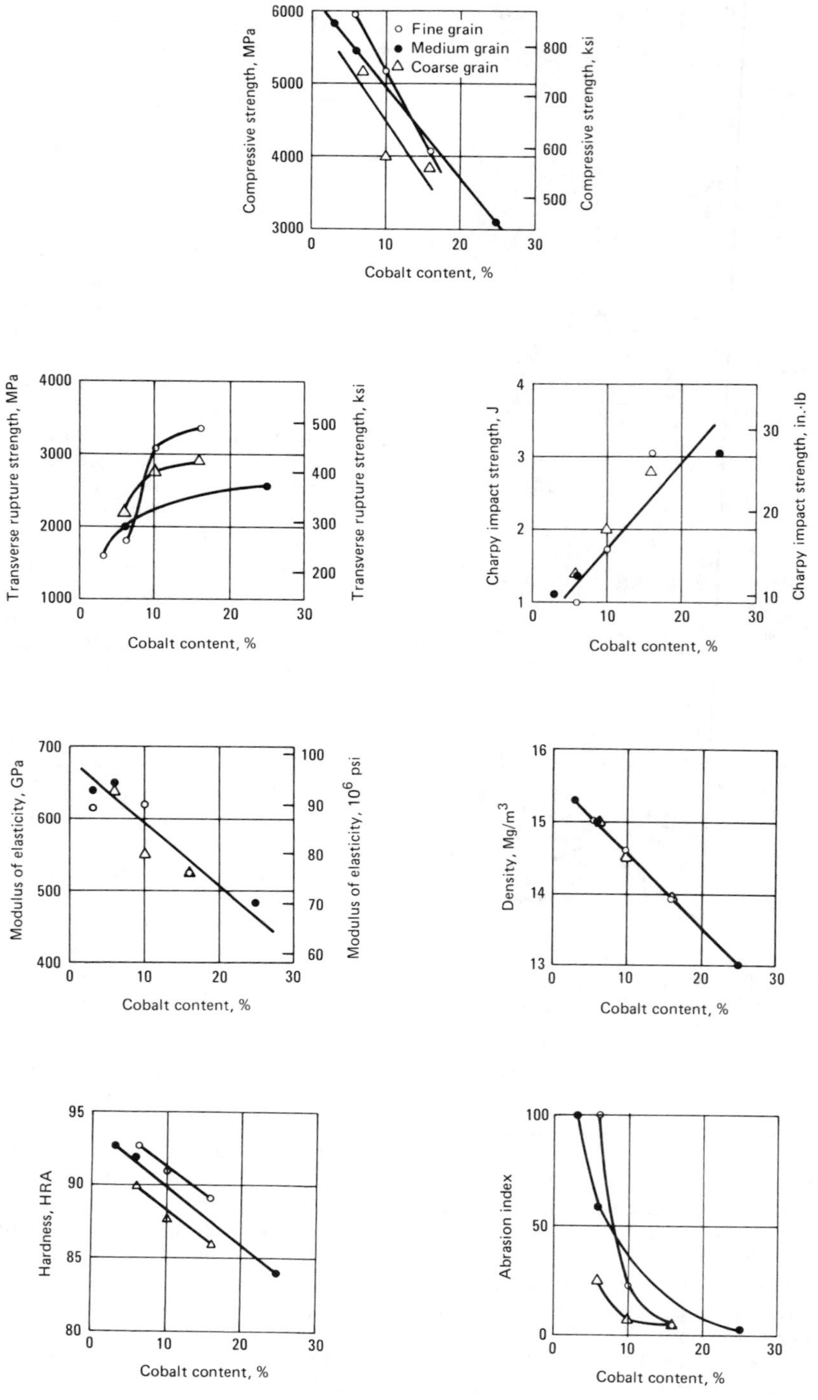

content or grain size is increased. However, abrasion resistance is lower for complex carbides than for straight WC grades having the same cobalt content.

Corrosion resistance of cemented carbides is fairly good, and they may be employed advantageously in certain corrosive environments for applications where outstanding wear resistance is required. Resistance to water, to oils and other cutting fluids used in machining, and to alkaline attack is excellent; but the cobalt matrix in many cemented carbides is subject to attack by acids. When the cobalt is attacked, accelerated wear develops because of the rapid crumbling of unsupported carbide particles. Corrosion of the cobalt binder also may cause a drastic reduction in strength.

Cemented carbides used as tool materials begin to oxidize when heated in air above about 500 to 600 °C (900 to 1100 °F). However, as measured by weight changes or shape distortion, the rate of oxidation of grades containing large amounts of titanium carbide is much lower than that of straight grades at temperatures as high as 1000 °C (1800 °F).

Toughness. Cemented carbides are brittle materials; usually they will show less than 0.2% elongation in a tensile test.

Values of Charpy impact strength for cemented carbides have little value and may be misleading (Ref 5). The energy absorbed during impact testing of very hard materials consists mainly of energy absorbed in elastically bending the specimen and energy absorbed by the testing machine. The portion of total absorbed energy that is a measure of the toughness of a material—namely, the energy of plastic work and the energy necessary to create new surfaces—is only a few percent of the total energy measured.

Coated Cemented Carbides

Coated cemented carbides are materials consisting of a substrate having a composition similar to a conventional cemented carbide onto which a thin coating of very hard material is deposited, usually by chemical vapor deposition.

Coated cemented carbides have a combination of wear and breakage superior to uncoated carbides. Coated cemented carbides became commercially

Table 8 Properties of representative cobalt-bonded cemented carbides

Nominal composition	Grain size	Hardness, HRA	Density		Transverse strength		Compressive strength		Proportional limit, compression		Modulus of elasticity	
			Mg/m^3	lb/in.^{3}	MPa	ksi	MPa	ksi	MPa	ksi	GPa	10^6 psi
97WC-3Co	Medium	92.5–93.2	15.3	0.55	1590	230	5860	850	2410	350	641	93
94WC-6Co	Fine	92.5–93.1	15.0	0.54	1790	260	5930	860	2550	370	614	89
	Medium	91.7–92.2	15.0	0.54	2000	290	5450	790	1930	280	648	94
	Coarse	90.5–91.5	15.0	0.54	2210	320	5170	750	1450	210	641	93
90WC-10Co	Fine	90.7–91.3	14.6	0.53	3100	450	5170	750	1590	230	620	90
	Coarse	87.4–88.2	14.5	0.52	2760	400	4000	580	1170	170	552	80
84WC-16Co	Fine	89	13.9	0.50	3380	490	4070	590	970	140	524	76
	Coarse	86.0–87.5	13.9	0.50	2900	420	3860	560	700	100	524	76
75WC-25Co	Medium	83–85	13.0	0.47	2550	370	3100	450	410	60	483	70
71WC-12.5TiC-12TaC-4.5Co	Medium	92.1–92.8	12.0	0.43	1380	200	5790	840	1170	170	565	82
72WC-8TiC-11.5TaC-8.5Co	Medium	90.7–91.5	12.6	0.45	1720	250	5170	750	1720	250	558	81
64TiC-28WC-2TaC-2Cr_2C_3-4.0Co	Medium	94.5–95.2	6.6	0.24	690	100	4340	630	...	...	...	...
57WC-27TaC-16Co	Coarse	84.0–86.0	13.7	0.49	2690	390	3720	540	1170	170	441	64

(a) Based on a value of 100 for the most abrasion-resistant grade.

Fig. 4 Correlation of diamond pyramid and Rockwell A hardness values for very hard materials

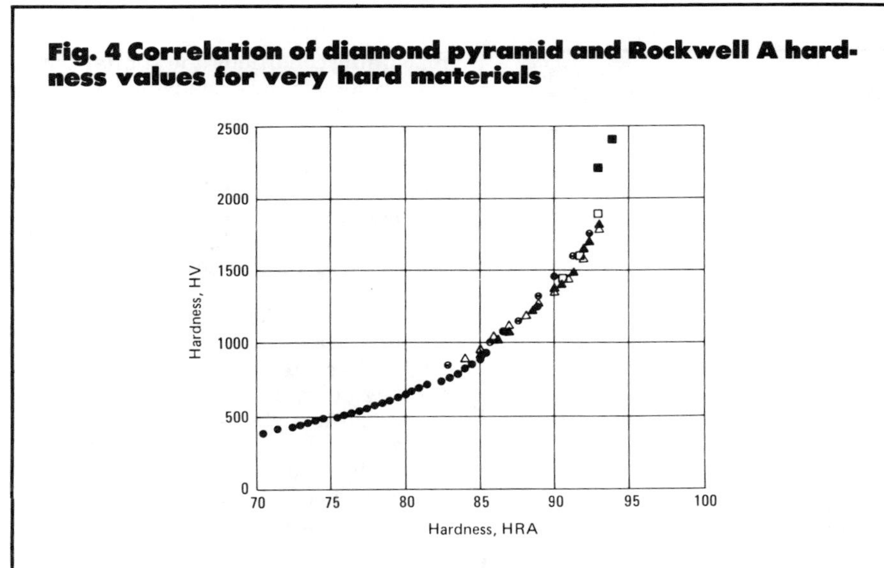

available in 1970. In the United States, about 35% of the cemented carbide tools used for metalcutting in 1979-1980 were coated. Acceptance of coated carbides for metalcutting continues to increase steadily as improvements in the coating itself and in substrate materials are introduced. Coatings now are more uniform in thickness and have less porosity than earlier coatings; adhesion to the substrate has been improved, and undesirable interface reactions suppressed; and substrates designed to be coated are being used, which has resulted in improved resistance to breakage and thermal deformation, the two most important substrate properties.

Coated carbides are becoming more widely used because, in high-production machining operations, they permit cutting speeds to be significantly increased. For instance, compared to uncoated carbide tools, the same tool life can be obtained at as much as 50% greater cutting speed with a TiC coated tool, and as much as 90% greater cutting speed with an Al_2O_3 coated tool.

Selection of the proper coating should be made, however, not merely on the basis of cutting speed, but also on tool wear mode (edge wear vs flank wear vs cratering). Tool configuration, relief angles and type of cut (roughing or finishing) also influence selection of the proper combination of substrate material and coating.

Titanium carbide coatings were the first coatings used on cemented carbide and are still the most widely used. TiC usually is deposited to a thickness of about 0.005 mm (0.0002 in.), but commercial titanium carbide coated tool materials vary greatly in thickness of coating and type of bond between coating and substrate. These variations can cause significant differences in metalcutting performance.

Some manufacturers use several different substrates and may offer as many as three or four grades of TiC-coated carbide that differ significantly in performance characteristics. A photomicrograph of the cross section of a typical cemented carbide coated with TiC is shown in Fig. 7. The TiC layer is at the top.

Commercial grades of TiC-coated carbide do not have the superior breakage resistance of uncoated heavy-duty roughing grades of cemented carbide, such as WC-TaC-8TiC-8.5Co, a roughing grade low in TiC and relatively high in cobalt. Coated grades are not as wear resistant as the most wear resis-

Table 8 (continued)

Nominal composition	Tensile strength MPa	Tensile strength ksi	Impact strength J	Impact strength in.·lb	Relative abrasion resistance(a)	Thermal expansion µm/m·°C at 200 °C	Thermal expansion µm/m·°C at 1000 °C	Thermal expansion µin./in.·°F at 400 °F	Thermal expansion µin./in.·°F at 1800 °F	Thermal conductivity, W/m·K	Electrical conductivity, % IACS
97WC-3Co	...	...	1.13	10	100	4.0	...	2.2	...	121	5.3
94WC-6Co	...	...	1.02	9	100	4.3	5.9	2.4	3.3	...	...
	1450	210	1.36	12	58	4.3	5.4	2.4	3.0	100	7.8
	1520	220	1.36	12	25	4.3	5.6	2.4	3.1	121	10.0
90WC-10Co	...	...	1.69	15	22	...	...	...	...	...	...
	1340	195	2.03	18	7	5.2	...	2.9	...	112	11.4
84WC-16Co	...	...	3.05	27	5	...	...	...	...	...	...
	1860	270	2.83	25	5	5.8	7.0	3.2	3.9	88	9.2
75WC-25Co	1380	200	3.05	27	3	6.3	...	3.5	...	71	9.8
71WC-12.5TiC-12TaC-4.5Co	...	...	0.79	7	11	5.2	6.5	2.9	3.6	35	4.3
72WC-8TiC-11.5TaC-8.5Co	...	...	0.90	8	13	5.8	6.8	3.2	3.8	50	5.2
64TiC-28WC-2TaC-2Cr_2C_3-4.0Co	...	...	...	...	8	...	...	...	...	...	...
57WC-27TaC-16Co	...	...	2.03	18	3	5.9	7.7	3.3	4.3	...	...

(a) Based on a value of 100 for the most abrasion-resistant grade.

Fig. 5 Correlation of Knoop and Rockwell A hardness values for very hard materials

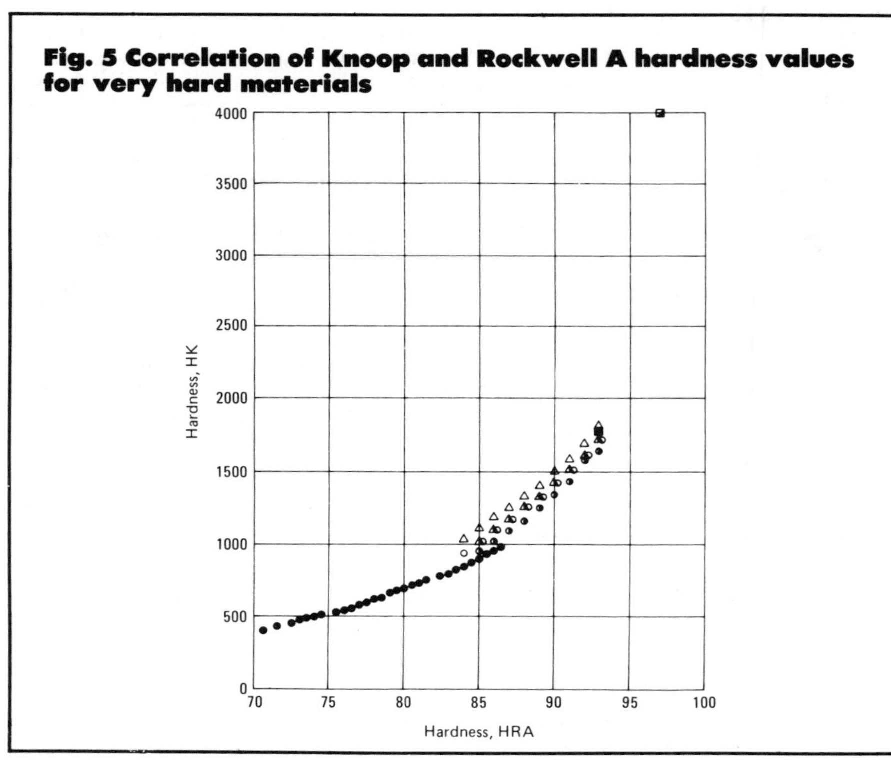

tant uncoated grades, such as 64TiC-32WC-4Co, a grade high in TiC and low in cobalt. Nevertheless, TiC-coated carbides perform well under a wide variety of machining conditions, often producing two to three times as many parts as can be produced using uncoated finishing or general-purpose grades of comparable breakage resistance. This comparison presumes the use of similar machining conditions and similar criteria for determining the point at which the tool edge no longer can be used.

Aluminum oxide coatings are rapidly gaining acceptance for efficient metal removal at high cutting speeds. Carbide tools coated with aluminum oxide possess an edge strength much higher than that of solid ceramic cutting tools.

Commercial Al_2O_3-coated cemented carbides are available as single-coated or double-coated products. The former consists of an oxide coating 0.005-mm thick that is metallurgically bonded to a specially designed cemented carbide substrate. Double-coated products consist of a thin Al_2O_3 coating over a TiC-coated cemented carbide substrate.

Titanium nitride coatings are claimed to impart superior resistance to crater formation and flank wear in certain metal-cutting applications. TiN-coated products are easily recognized by their gold color.

Multiple coatings consisting of successive thin layers of titanium carbide, titanium carbonitride and titanium nitride are being used in increasing quantities. Multiple coatings are claimed to combine the desirable qualities of both nitride and carbide coatings, and impart to the tools better resistance to edge wear, flank wear and cratering, a combination that cannot be obtained with a single-layer coating.

Fig. 6 Hot hardness of cemented carbides

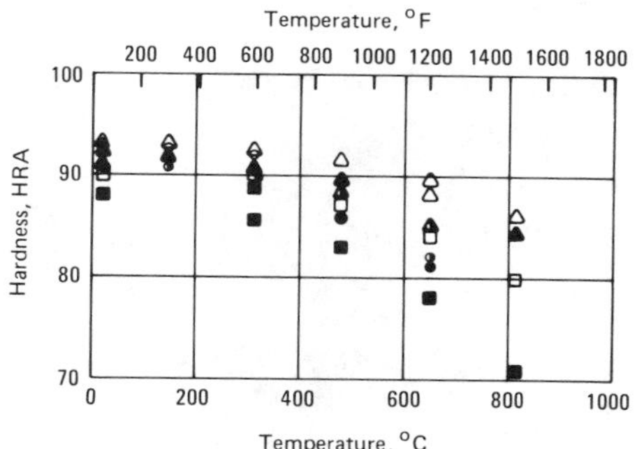

Symbol	Composition	Grain size
Key: ○	94WC-6Co	Fine
●	94WC-6Co	Coarse
△	94WC-6Co	Very fine
▲	90WC-10Co	Very fine
□	85WC-15Co	Very fine
■	85WC-15Co	Coarse
◑	79WC-8TiC-4TaC-9Co	Medium
◭	72WC-8TiC-11.5TaC-8.5Co	Medium

Fig. 7 Cross section of TiC-coated cemented carbide

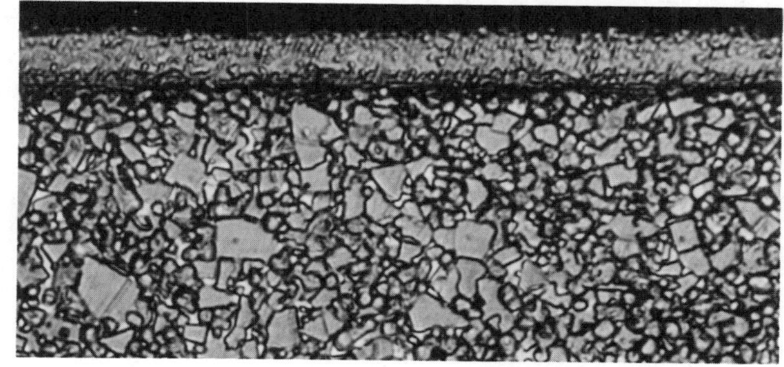

Etched with Murakami's reagent, 1500 X.

Nickel-Bonded Titanium Carbide

A satisfactory method of cementing titanium carbide using molybdenum carbide and nickel has been developed. The resultant material has good crater resistance, low coefficient of friction and low thermal conductivity. Although penetration hardness is high, abrasion resistance is lower than that of a cobalt-bonded tungsten carbide having equal or lower penetration hardness.

Typically, cemented titanium carbides contain 8 to 25% Ni, 8 to 15% Mo_2C and 60 to 80% TiC. Occasionally, tungsten carbide, cobalt, titanium nitride and other additives may be present in smaller amounts.

The properties and typical classifications for machining uses of four types of cemented titanium carbides are presented in Table 9. A typical structure is shown in Fig. 8.

Additional information on nickel-bonded titanium carbides may be found in Ref 6.

Table 9 Properties and uses of nickel-bonded titanium carbides

Property	High TiC, plus Mo_2C; low nickel	High TiC, plus Mo_2C; low nickel	TiC, plus Mo_2C; intermediate nickel	Lower TiC, plus Mo_2C; high nickel
Grain size	Fine	...	Fine	Fine
Hardness: HRA	93.3	93.0	91.7	90.5
HV	1970	1890	1600	1440
Density: Mg/m^3	5.50	5.63	5.71	5.82
$lb/in.^3$	0.198	0.203	0.206	0.210
Transverse strength: MPa	1170	1380	1720	1890
ksi	170	200	250	275
Compressive strength: MPa	3585	3450	3270	2960
ksi	520	500	475	430
Tensile strength: MPa	970	1100	1170	1240
ksi	140	160	170	180
Modulus of elasticity: GPa	462	448	414	379
10^6 psi	67	65	60	55
Impact strength: J	0.79	0.90	1.02	1.24
in.·lb	7	8	9	11
Thermal expansion(a): μm/m·°C	7.5	7.8	8.4	9.1
μin./in.·°F	4.2	4.3	4.7	5.1
Thermal conductivity(b): W/m·K	16.7	16.7	16.7	16.7
Btu/ft·h·°F	9.6	9.6	9.6	9.6
Typical classification for machining use: C-grade	C-8	C-7	C-6	C-6
ISO	P01	P10	P20	P30

(a) At 21 to 650 °C (70 to 1200 °F). (b) At 100 to 300 °C (200 to 575 °F).

Steel-Bonded Carbide

Steel-bonded carbide is a P/M tool material intermediate in wear resistance between tool steels and cemented carbides. Steel-bonded carbide consists of 40 to 55 vol % titanium carbide homogeneously dispersed in a steel matrix.

It is customary to follow tool steel practice and make the entire tool from steel-bonded carbide. The tool can be joined to a supporting member either of

Table 10 Compositions and selected properties of three principal grades of steel-bonded titanium carbide

Characteristic	Grade C	Grade CM	Grade SK
Composition:			
Titanium carbide	45 vol%	45 vol%	40 vol%
Steel matrix	0.6C-3Cr-3Mo	0.85C-10Cr-3Mo	0.40C-5Cr-4Mo-0.50Ni
Hardness:			
Annealed HRC	40	45	37
Hardened(a) HRC	70	69	65
Max service temperature: °C	200	540	540
°F	400	1000	1000
Density: Mg/m^3	6.60	6.45	6.80
lb/in.^{3}	0.238	0.232	0.245
Transverse strength: MPa	2070	2140	2070
ksi	300	310	300
Modulus of elasticity: GPa	305	305	270
10^6 psi	44	44	39
Thermal expansion: μm/m·°C	7.83(b)	8.3(c)	9.47(c)
μin./in.·°F	4.35	4.6	5.26
Electrical conductivity, % IACS	3.2	2.8	3.0

(a) Grade C: austenitized at 950 °C (1750 °F), oil quenched and tempered 1 h at 190 °C (375 °F). Grade CM: austenitized at 1100 °C (2000 °F), oil quenched (gas quenched if heat treated in vacuum), and double tempered at 525 °C (975 °F). Grade SK: austenitized at 1000 °C (1850 °F), oil quenched (gas quenched if heat treated in vacuum), and double tempered at 525 °C. (b) At 21 to 200 °C (70 to 400 °F). (c) At 21 to 540 °C (70 to 1000 °F).

Fig. 8 Microstructure of a typical nickel-bonded titanium carbide

Intermediate Ni content; hardness, 91.7 HRA. Etched with Murakami's reagent, 1500 X.

Fig. 9 Microstructures of grade C steel-bonded carbide

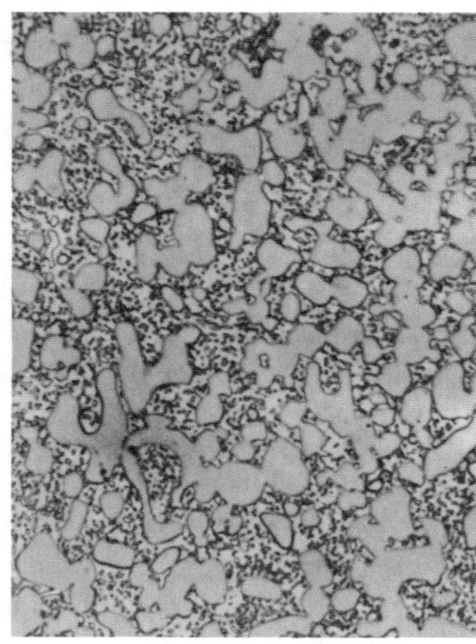

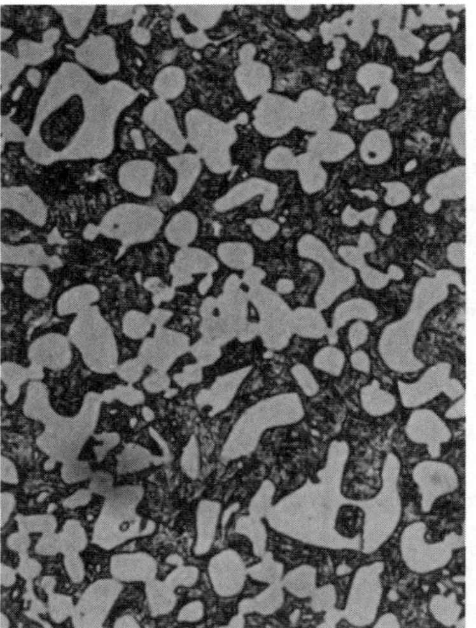

Nominal composition: 45 vol % TiC in steel matrix (0.6C-3Cr-3Mo). Left: annealed condition—rounded grains of TiC in soft, spheroidized cementite-ferrite matrix. Right: hardened condition—rounded grains of TiC in tempered martensite matrix. (Nital etch, 750 X.)

the same material or of steel. This may be done by mechanical fastening, adhesive bonding or brazing. Brazing under vacuum using AWS BNi-8 (a nickel-manganese brazing filler metal) provides a ductile, high-strength joint with a remelt temperature in excess of the temperature recommended for heat treating grade C steel-bonded carbide.

The properties and recommended heat treatments for grades of steel-bonded titanium carbide commonly used for tools are given in Table 10; representative microstructures of grade C are shown in Fig. 9.

Grade C is used for progressive stamping dies; lamination dies; dies for drawing, bending and curling sheet metal and wire; tube rolls; gages and fixtures. Grade CM is used to make tools for cold and warm forming of heavy-gage stock. Grade SK is used to make hot work rolls and forging dies, including dies for forging hot powder metals. Generally, steel-bonded carbide is not recommended for cutting tools to be used to machine ferrous metals because the hardness drops off too rapidly at the high temperatures developed at the cutting edge. Interface temperatures developed during cutting of nonferrous metals are not too high, on the other hand, and several grades of steel-bonded carbide have been used for machining these materials.

Steel-bonded carbide performs well in severe wear applications involving sliding friction. Its success has been attributed to exceptionally hard, extremely fine, rounded grains of titani-

Table 11 Typical properties of cast Tantung G

Property	Chill cast	Refractory mold cast
Melting temperature:		
°C	1150–1200	
°F	2100–2200	
Casting temperature:		
°C	1370	
°F	2500	
Density:		
Mg/m^3	8.3	8.3
$lb/in.^3$	0.30	0.30
Thermal expansion:		
μm/m·°C	4.2	4.2
μin./in.·°F	2.3	2.3
Thermal conductivity:		
W/m·K	26.8	26.8
Btu/ft·h·°F	15.5	15.5
Hardness: HRC	60–63	53–58
Transverse strength:		
MPa	2240	1030–1200
ksi	325	150–175
Modulus of elasticity:		
GPa	265	...
10^6 psi	41	...
Tensile strength:		
MPa	585–620	450
ksi	85–90	65
Compressive strength:		
MPa	2760	2930
ksi	400	425
Impact strength:		
J	6.1	6.1
ft·lb	4.5	4.5

um carbide exposed in slight relief at the surface. To provide this condition, the surface preparation of the tool after heat treatment requires lapping with a coarse compound (about 30 μm particle size) to remove any grinding marks and smeared metal, then lapping with a medium compound (15 μm particle size) and finally polishing with a fine abrasive (6 μm particle size) to a mirror finish. Aluminum oxide is satisfactory as the abrasive.

In addition to the three grades shown in Table 10, other grades of steel-bonded carbide are available that have special-alloy matrices to provide corrosion resistance or nonmagnetic properties. Also, some grades can be surface treated to improve the wear resistance of the matrix in applications involving extremely fine abrasives. These surface treatments include nitriding and boriding.

Reference 7 contains additional information on steel-bonded carbide.

Table 12 Mutual indentation hot hardness of cast Tantung G

Temperature °C	Temperature °F	Hardness, HB(a)	Equivalent hardness(b) HRA	HRC	HRB
RT	RT	654	81.3	60.1	...
425	800	479	75.7	49.8	...
650	1200	479	75.7	49.8	...
870	1600	267	63.8	27.1	104
980	1800	114	...	...	66.7

(a) 3000 kg load, applied for 30 s. (b) Converted values.

Cast Co-Cr-W-Nb-C Alloys

Developed early in the 20th century to provide tools with better capabilities than high speed steel, cast cobalt-chromium-tungsten-niobium-carbon alloys are still produced. In general, as cutting tools, they bridge the gap between high speed steels and cemented carbides. Several cast Co-Cr-W-Nb-C wear-resistant parts have been used in machinery applications as well.

These alloys are produced by electric or induction melting under a protective atmosphere, and for cutting-tool applications they are preferably cast in chill graphite molds. However, they may be cast in investment, shell or sand molds to produce special and intricate shapes.

Two of these cast tool alloys are Tantung G, which has a nominal composition of 47Co-30Cr-15W-3Nb-2.5C-2Mo-0.5B, and Tantung 144, which has a composition of 45Co-27Cr-18W-5Nb-2Mn-2.3C-0.7B. Typical properties of Tantung G are shown in Table 11. For comparison, chill cast Tantung 144 has a hardness of 61 to 65 HRC, a transverse strength of 2070 MPa (300 ksi) and an elastic modulus of 295 GPa (43 × 10^6 psi).

Tantung G is recommended for general-purpose cutting tools and parts for wear applications, and is more widely used than Tantung 144. Tantung 144 has higher hardness than Tantung G and was developed for use where resistance to abrasion is paramount and where there is little or no shock or impact.

The typical microstructure of chill cast Tantung G is shown in Fig. 10(a) and its replica electron micrograph in Fig. 10(b). The much coarser grain size of investment cast Tantung G is shown in Fig. 10(c).

The good resistance of Tantung G and Tantung 144 to foods—especially those containing acetic acid—makes them well-suited for food-processing equipment, especially parts requiring good resistance to both abrasion and corrosion.

Water containing chlorine and hypochlorites may produce some corrosion and pitting of alloys G and 144. In addition, these alloys are attacked by strong acid solutions, by alkalis and by solutions of some heavy-metal salts such as ferric chloride, ferric sulfate and cupric chloride.

When heated in air, Tantung G and Tantung 144 both tarnish on short-time exposures at 400 °C (750 °F) and lose appreciable weight at 750 °C (1380 °F) or higher. Scaling may be progressive above 1000 °C (1830 °F). Hot-hardness data for Tantung G are shown in Table 12.

Use of cast Co-Cr-W-Nb-C cutting tools should be considered:

- Where relatively low surface speeds cause build-up with cemented carbides
- Where machines lack the power or rigidity to use cemented carbides effectively
- Where higher production is desired than is possible with high speed steel tools
- For multiple-tool operations where surface speed of one or more operations falls between the recommended speeds for high-speed steel and carbide tools
- For short runs on automatic equipment where form grinding of carbide tools is excessively costly
- For machining rough surfaces of castings where the surfaces contain abrasive material such as residual sand, surface oxides, slag or refractory particles.

Tools made of cast Co-Cr-W-Nb-C alloys usually are not recommended for light, very fast finishing cuts.

Typical wear applications for these alloys include: wear strips for belt sanders; dies for extruding copper, for extruding molybdenum tubing and for hot swaging tungsten rod; burnishing rolls; internal chuck jaws; drill bushings; and knives for slicing fruits, vegetables and meat.

Additional information on cast Co-Cr-W-Nb-C alloys may be found in Ref 8.

Ceramics

Ceramic cutting-tool materials use aluminum oxide as the base material. They are available as indexable inserts that can be used in the same holders as those used for cemented-carbide inserts. However, because ceramics are more brittle than carbides, extra care must be used to ensure that the inserts are firmly seated. Also, tool overhang should be kept at 50 to 100% of toolholder thickness.

Ceramic inserts may be made by cold pressing and sintering or by hot pressing. Early in their development, it was recognized that the retention of fine grain size was important, and various procedures (including additions of other compounds) have been developed to hold final grain size to a maximum of 2 to 5 μm.

The many available ceramic tools fall into three general groups:

- *Group A-1.*
  Al_2O_3 with up to about 10% of other oxides or carbides, primarily those of titanium, magnesium, molybdenum, chromium, nickel or cobalt. The mixtures are cold pressed and sintered
- *Group A-2.*
  Essentially pure Al_2O_3, hot pressed
- *Group A-3.*
  Al_2O_3 plus 25 to 30% of a refractory carbide such as titanium carbide, hot pressed.

Typical properties of these groups are shown in Table 13. Also, the properties of a specific example of a Group A-1 material have been included, for which the microstructure is shown in Fig. 11 and for which hot-hardness data are given in Table 14. Data in Table 15 for Groups A-2 and A-3 ceramics indicate that hardness greater than 90 HRA is retained, and transverse rupture strength remains unchanged, at temperatures up to about 1000 °C (1800 °F).

The comparative abrasion resistance of the example of a Group A-1 ceramic tool material in Table 13 is 5, based on the results of the slurry-type wear test described previously in the section on cemented carbides, using a value of 100 for the material most resistant in that test. As was pointed out for cemented carbides, this test does not measure the wear resistance of actual tools.

Because ceramic tools are predominantly oxides, they are not subject to the oxidation that limits the usefulness of cemented carbides at high temperatures in air.

Ceramic tool materials retain their resistance to wear and deformation at

Fig. 10 Typical microstructures of Co-Cr-W-Nb-C alloy Tantung G

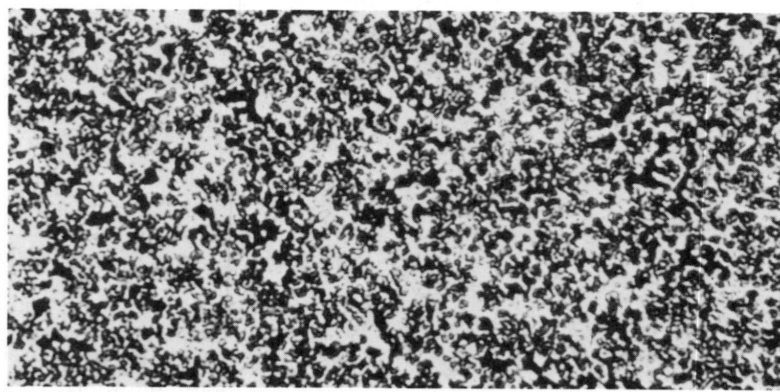

(a) Chill cast: 6:1 $HCl-H_2O$ etch, 1000 X.

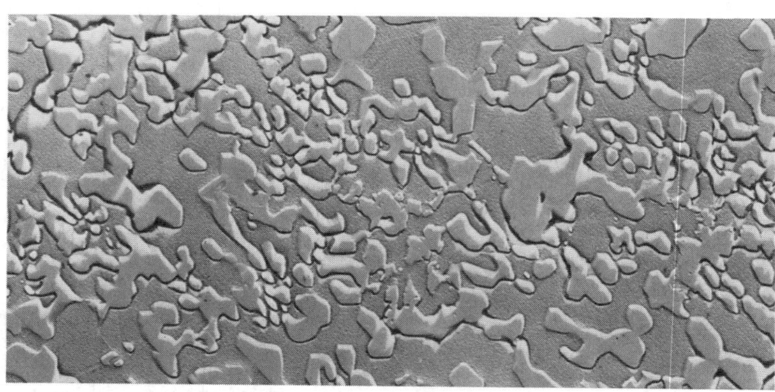

(b) Replica electron micrograph of chill cast structure: $HNO_3-H_2SO_4-CH_3OH$ electrolytic etch, 10 000 X.

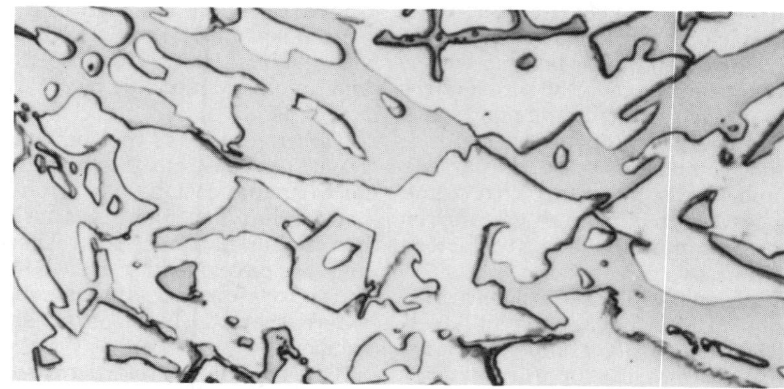

(c) Investment cast in refractory mold: 6:1 $HCl-H_2O$ etch, 1000 X.

Table 13 Typical properties of ceramic tool materials

Property	Group A-1 General	Group A-1 Example(a)	Group A-2	Group A-3
Hardness: HRA	93–94	93–94	93–94	93–94
Density:				
Mg/m^3	3.96–3.98	4.1	4.0	4.24
$lb/in.^3$	0.142–0.143	0.148	0.144	0.153
Transverse strength:				
MPa	480–690	620	640	760
ksi	70–100	90	92.5	110
Compressive strength:				
MPa	3790–4480	2140(b)	4140	3930–4070
ksi	550–650	310(b)	600	570–590
Modulus of elasticity:				
GPa	390	400	390	...
10^6 psi	57	58	57	...
Impact strength:				
J	...	0.23	...	...
in.·lb	...	2	...	...
μm/m·°C	...	6.1(d)	7.2	7.7
Thermal expansion(c): μin./in.·°F	...	3.4(d)	4.0	4.3
Thermal conductivity:				
At room temperature:				
W/m·K	29	...	...	17–21
Btu/ft·h·°F	17	...	...	10–12
At 100 °C (212 °F):				
W/m·K	22	...	29	...
Btu/ft·h·°F	13	...	17	...
At 450 °C (850 °F):				
W/m·K	11	...	...	...
Btu/ft·h·°F	6.5	...	...	...
At 600 °C (1100 °F):				
W/m·K	...	...	...	14.7
Btu/ft·h·°F	...	...	...	8.4

(a) $89Al_2O_3$-11TiO, cold pressed and sintered. See Table 14 for hot hardness data. (b) Proportional limit. (c) At 21 to 200 °C (70 to 400 °F). (d) 8.3 μm/m·°C (4.6 μin./in.·°F) at 21 to 980 °C (70 to 1800 °F).

Fig. 11 Microstructure of $89Al_2O_3$-11TiO grade A-1 ceramic tool material

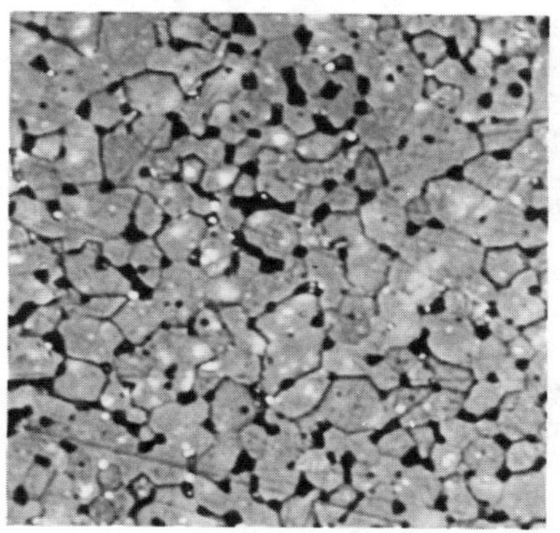

Phosphoric acid etch, 750 X. For material properties, see A-1 example in Table 13.

Table 14 Hot hardness of an $89Al_2O_3$-11TiO ceramic tool material

See group A-1 example in Table 13 for other properties

Temperature °C	Temperature °F	Hardness, HRA
RV	RT	94.1
150	300	94.0
315	600	93.0
480	900	91.0
650	1200	86.5

much higher temperatures than the best cemented carbides. Consequently, ceramic tools generally can cut for acceptable periods of time at speeds much higher than those possible with cemented carbides. Coolants should not be used when cutting with ceramic tools.

The lower transverse rupture and shear strengths of the ceramic tools require that the feed be reduced and that negative-rake tools be used when cutting materials other than low-strength materials such as hard rubber. The edges of ceramic tools must be very carefully ground to prevent formation of cracks, and regrinding of ceramic tools generally is not worthwhile.

The most extensive use of ceramic tools in the automotive industry has been in machining cast iron at cutting speeds up to 10 m/s (2000 ft/min). Ceramic tools are also being used effectively for cutting low-carbon steels at similar speeds and can be used to finish machine heat treated steel parts with hardnesses up to 60 HRC, which previously had to be ground.

Although Al_2O_3 tool materials do not react with ferrous work materials, they react chemically with magnesium, aluminum, titanium and zirconium. Therefore, Al_2O_3 tools are not recommended for cutting these four nonferrous metals or their alloys.

Additional information on ceramic tool materials may be found in Ref 9 and 10.

Polycrystalline Cubic Boron Nitride

In 1973, polycrystalline composite cubic boron nitride (CBN) cutting tools were introduced. By use of high-pressure, high-temperature processes, a layer of CBN is bonded to a cemented carbide substrate. The CBN is held together primarily by CBN-CBN intercrystalline bonds.

Most frequently, composite CBN tools are supplied in the same shapes as those used for cemented carbides. In many cases, CBN inserts can be brazed into a steel holder using the same procedures employed for cemented carbides, as long as special care is taken to avoid overheating the structure and to prevent molten flux from contacting the CBN layer.

Because the cost of a CBN tool is several times that of a cemented carbide tool, it is usually necessary to regrind CBN tools with diamond wheels and reuse them to minimize the cost per tool edge.

Some selected properties of CBN crystals are given in Table 16. The hardness of the polycrystalline CBN layer on a tool insert is 97 HRA (4000 HK).

Because of their high hardness, their ability to hold this hardness and the fact that they resist oxidation at much

Table 15 Typical elevated-temperature hardness and strength of groups A-2 and A-3 ceramic tool materials

See Table 13 for other typical properties

Temperature		Hardness, HV		Transverse rupture strength			
				Group A-2		Group A-3	
°C	°F	Group A-2	Group A-3	MPa	ksi	MPa	ksi
RT	RT	2100	2400	690	100	735	107
480	900	2000	2000	690	100	715	104
650	1200	1950	1850	...	...	...	...
815	1500	1850	1700	...	...	...	...
980	1800	1700	1500	700	101	700	102
1200	2200	1400	1400	610	90	690	100

higher temperatures than carbides, composite CBN tools are used effectively to cut difficult-to-machine superalloys at speeds several times higher than those possible with cemented carbides. The best conditions for their use are in the self-induced thermal machining mode.

In addition to nickel-base and cobalt-base high-temperature alloys, other ferrous materials—including chilled cast iron, Meehanite cast iron, tool steels (M2, M42, D2, A2, S5 and O1) and other steels (1055, 8620, 52100 and 4140) hardened to 50 to 70 HRC—have been machined successfully with CBN tools.

Additional information on polycrystalline composite cubic boron nitride cutting tools may be found in Ref 11.

Diamond

Diamond, the tetrahedral form of carbon, is the hardest and most scratch-resistant material known. Its scratch hardness number is 10 on Moh's scale. It will scratch all other materials and be scratched by none. Similarly, its penetration hardness is 5 000 to 12 000 HK, roughly twice that of the next hardest material.

These characteristics make diamond very attractive as a tool material. However, industrial-grade natural single-crystal diamonds are expensive even in small sizes. In addition, diamonds are very brittle and cleave easily along certain crystallographic planes. Also, diamond starts to oxidize rapidly at 650 °C (1200 °F), and at atmospheric pressure reverts to graphite above 1500 °C (2700 °F).

These properties and the fact that carbon dissolves rapidly in iron at high temperatures make diamond unsatisfactory for machining ferrous alloys.

However, diamond tools are used effectively in machining high-silicon cast aluminum alloys, copper and its alloys, sintered cemented tungsten carbides, rubber impregnated with silica glass, glass-fiber/plastic and carbon/plastic composites, and high-alumina ceramics.

Diamonds are used extensively in resin-bonded and metal-bonded grinding wheels. In addition, diamond dies are used for drawing fine wire. Other common industrial uses of diamonds are as tools for dressing abrasive grinding wheels, as diamond cutoff saws and laps (when dispersed in a metal matrix) and as a polishing abrasive when in very finely divided form.

In 1965, polycrystalline diamond compacts were made by sintering diamond powder under pressures in the 10 GPa (million psi) range at high temperatures. The process was based in part on techniques developed earlier to produce small individual diamonds. This process fuses diamond powder into a cohesive solid with a random crystallographic orientation. The process not only extends the use of diamond tools by providing larger sizes at reasonable cost but also, due to the random orientation, provides uniform wear resistance in all directions and eliminates the planes of easy cleavage that frequently result in premature failure of single-crystal diamond cutting tools.

In 1973, a vastly improved laminated tool material consisting of polycrystalline diamond bonded to a cemented carbide substrate was introduced. These laminated tool blanks (inserts) are a product of high-pressure, high-temperature technology in which the hardness and resistance to cleavage of polycrystalline diamond is combined with the toughness and bondability of cemented carbide to produce an extremely wear-resistant and impact-resistant composite structure. The laminated tool blanks are characterized by a 0.5 to 1.5 mm (0.02 to 0.06 in.) thick polycrystalline diamond layer bonded to a cemented carbide base. Insert blanks are available in a wide range of shapes and sizes that can be brazed and finished into many cutting-tool and drawing-tool configurations.

The relative abrasion resistance and Knoop hardness values of laminated diamond, natural diamond and cemented 94WC-6Co tool blanks are compared below:

Tool material	Hardness, HK	Relative abrasion resistance
Laminated diamond/ carbide composite	5500–8000	250
Natural diamond single crystal	8000–12 000	96 to 245
Cemented 94WC-6Co	1800–2200	2

The relative abrasion resistance is based on the time, in minutes, required to generate a specific size of wear land on the material when turning a siliceous hard rubber commonly used as a coating for steel rolls in the paper industry.

References 11, 12 and 13 contain additional information on diamond as a tool material.

Table 16 Selected properties of cubic boron nitride

Crystal structure	Zinc blende (F $\bar{4}$3m)
Density	3.48 Mg/m^3 (0.125 lb/in.^{3})
Hardness:	
At 20 °C (70 °F)	4000 HV
At 1000 °C (1800 °F)	~4000 HV
Temperature for dislocation mobility	1300-1400 °C (2400-2550 °F)
Melting point (triple point)	3500 K
Thermal conductivity (theoretical)	13 W/m·K (7.5 Btu/ft·h·°F)
Thermal stability:	
Limit of oxidation resistance in air	~1300 °C (~2400 °F)
Metastable reversion temperature	~1500 °C (~2700 °F)
Thermal expansion, 21 to 500 °C (70 to 900 °F)	4.8 μm/m·°C (2.7 μin./in.·°F)

REFERENCES

1. *World Directory and Handbook of Hard Metals,* 2nd Ed., by Kenneth J. A. Brooks: Engineers' Digest Limited, London, 1979
2. "Engineering Properties of Ceramics", by J. F. Lynch, C. G. Rederer and W. H. Duckworth: Technical Report AFML-TR 66-52, Air Force Materials Laboratory, 1966
3. Some Plain Talk About Carbides, by H. S. Kalish: *Manufacturing Engineering and Management,* July 1973
4. "A System of Classification of Hard Metal Grades for Machining": Technical Publication No. 1, The British Hard Metal Association, Sheffield, England, 1967
5. An Analysis of Charpy Impact Testing as Applied to Cemented Carbide, by R. C. Lueth: in *Instrumented Impact Testing,* STP 563, American Society for Testing and Materials, 1974
6. Where Solid Titanium Carbide Stands, by H. S. Kalish: *American Machinist,* 7 Jan 1974, p 50-52
7. "Machinable Carbides for High Performance Tooling and Wear Parts", by S. E. Tarkan and M. K. Mal: Technical Paper MR 73-927, Society of Manufacturing Engineers
8. "Tantung: The Premiere Cast Alloy": Bulletin 72-1, VR/Wesson Div. of Fansteel, Inc., 2 June 1972
9. "Ceramic Tools", by E. D. Whitney: Technical Paper TE 73-205, Society of Manufacturing Engineers
10. Cutting Performance and Practical Merits of Carbide Ceramics, by K. Ogawa, M. Furukawa and Y. Hara: *Nippon Tungsten Review,* Vol C, 1973
11. "Borazon and Diamond Cutting Tools", by R. E. Hanneman and L. E. Gibbs: Technical Information Series, Report No. 73 CRD 182, General Electric Co., June 1973
12. Some Experiments to Compare Diamond and Diamond Compact Cutting Tools, by M. Casey and J. Wilks: *Sixteenth International Tool Design and Research Conference,* McMillan Press Ltd., Sept 1975
13. "The Characteristics and Performance of COMPAX® Diamond Tools in Machining Applications", by M. D. Dennis and J. D. Christopher: Technical Paper MR 75-986, Society of Manufacturing Engineers

Distortion in Tool Steels

Edited by Daniel S. Zamborsky
Corporate Metallurgist
The Warner and Swasey Company

DISTORTION in tool steel parts includes all irreversible changes in size and shape that result from processing, from heat treatment, and from temperature variations and loading in service. A basic understanding of distortion is important for two reasons. First, finishing operations for correcting distortion not only are expensive but also may destroy some desirable properties and introduce others that are undesirable. Second, most tool steel parts must interact with other parts in service, and excessive distortion may prevent them from interacting in the intended manner.

This article has been condensed from the more complete discussion of this complicated subject provided by Bernard S. Lement in his book *Distortion in Tool Steels* (Ref 1). The compositions and basic properties of the tool materials mentioned in this article are discussed elsewhere in this volume.

Changes in size or shape of tool steel parts may be either reversible or irreversible. Reversible changes are those caused by stressing in the elastic range or by temperature variations that neither cause changes in the metallurgical structure nor induce stresses that exceed the elastic range. Under such conditions, the initial dimensional values can be restored by a return to the original state of stress or temperature.

The upper limit of reversible dimensional change in a tool steel is determined by the stress required to initiate deformation (that is, the elastic limit corresponding to a preselected value of plastic strain), the elastic deformation per unit stress (modulus of elasticity), the effect of temperature on these properties, the coefficient of thermal expansion and the temperature-time combinations at which stress relief and phase changes occur.

For practical purposes the modulus of elasticity of all tool steels, regardless of composition or heat treatment, is 210 GPa (30×10^6 psi) at room temperature. Therefore, if a tool steel part deforms excessively under service loading but returns to its original dimensions when the load is removed, a change in grade or type of tool steel or in heat treatment will not be useful. To counteract excessive elastic distortion it is necessary to (*a*) reduce the applied stress by increasing the section size or (*b*) use a tool material with a higher modulus of elasticity (such as cemented tungsten carbide).

Irreversible changes in size or shape of tool steel parts are those caused by stresses that exceed the elastic limit or by changes in metallurgical structure (most notably, phase changes). Such irreversible changes sometimes can be corrected by thermal processing (annealing, tempering or cold treating) or by mechanical processing to remove excess material or to redistribute residual stresses.

Nature and Causes of Distortion

Distortion is a general term encompassing all irreversible dimensional changes. There are two main types: size distortion, which involves expansion or contraction in volume or linear dimensions without changes in geometrical form; and shape distortion, which entails changes in curvature or angular relations, as in twisting (warpage) or bending, and nonsymmetrical changes in dimensions. Usually, both size distortion and shape distortion (illustrated schematically in Fig. 1) occur during any heat treating operation.

Size distortion is the result of a net change in specific volume due to a change in metallurgical structure dur-

Fig. 1 Size and shape distortion in hardening

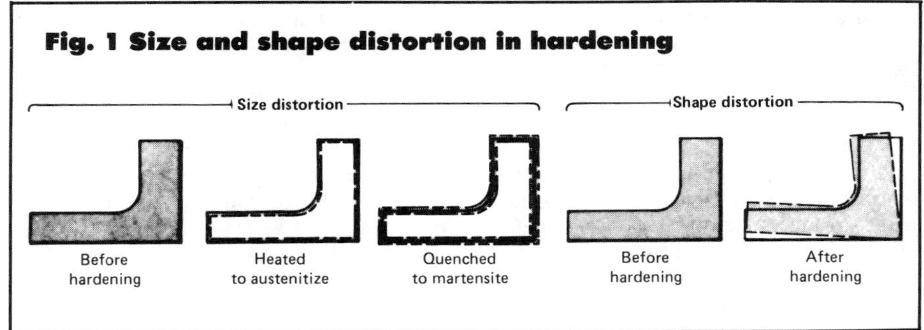

ing heat treatment. Shape distortion results from either residual or applied stresses. Residual stresses resulting from heat treatment are caused by nonuniform thermal gradients within the metal during heating and cooling, by nonuniform changes in metallurgical structure, and by nonuniformity in the metal itself, such as that due to segregation or ingot pattern.

Changes in metallurgical structure during heat treatment of tool steels are produced by the three steps described below.

The first step involves heating an annealed structure (usually consisting of ferrite and spheroidal carbides) to a temperature of about 800 °C (1450 °F) or higher to change the ferrite to austenite and to dissolve all or most of the spheroidal cementite in that austenite. For plain carbon or low-alloy tool steels, austenitizing results in a contraction in volume that decreases with increasing amounts of carbon in solution. This can be approximated as follows:

$$V_{SA} = -4.64 + 2.21(\%C) \qquad \text{(Eq 1)}$$

where V_{SA} is the volume change in percent that occurs when spheroidite transforms to austenite. By use of this equation it can be estimated that, if heated to a temperature high enough to dissolve all of the carbon in the austenite, a 0.50% C tool steel would exhibit a volume change of −3.53%, a common type containing 1% C would exhibit a change of −2.43%, and a very-high-carbon type containing 1.5% C would exhibit a change of −1.33%. However, tool steels having carbon contents higher than that of the eutectoid composition are normally austenitized at temperatures only high enough to dissolve the eutectoid amount of carbon. Under these circumstances, 1% C and 1.5% C tool steels would exhibit changes in volume of −2.77 and −2.53%, respectively, after austenitizing. These percentages are less than that calculated directly from Eq 1 because an allowance must be made for the volume occupied by undissolved carbides, which is about 3.5% for the 1.0% C steel and about 12% for the 1.5% C steel.

The second step involves cooling fast enough to cause the austenite to transform to martensite. The steel expands on transformation, the amount of expansion being in inverse proportion to the amount of carbon in solution in the austenite. Mathematically:

$$V_{AM} = 4.64 - 0.53(\%C) \qquad \text{(Eq 2)}$$

where V_{AM} is the percent volume change that occurs when austenite transforms to martensite. By use of Eq 2, it can be estimated that a 0.5% C tool steel would exhibit a volume increase for this transformation of 4.37% and that 1.0 and 1.5% C steels would exhibit increases of 4.07% and 3.71%, respectively, if austenitized at the normal austenitizing temperature (only 0.8% C, the eutectoid amount of carbon, in solution and again allowing for the volume occupied by undissolved carbides).

Equations 1 and 2 can be used to calculate the net change in dimensions in a tool steel when it is heat treated to transform it from an annealed to a fully hardened (martensitic) state. For the examples referred to above, normal heat treatment would produce net volume increases of −3.53 + 4.37 = 0.84% in the 0.5% C tool steel, −2.77 + 4.07 = 1.30% in the 1.0% C steel and −2.53 + 3.71 = 1.18% in the 1.5% C steel. Net changes in linear dimensions would be about one-third the corresponding net changes in volume.

The third step involves reheating the freshly formed martensite to relatively low temperatures (tempering) to increase toughness and reduce lattice stress. Tempering produces various changes in metallurgical structure, depending on temperature and time at temperature.

After very long times at room temperature or shorter times at temperatures up to 200 °C (400 °F), the high-carbon martensite in plain carbon and low-alloy tool steels transforms to low-carbon martensite (about 0.25% C) plus epsilon carbide, with an accompanying contraction in volume. At higher tempering temperatures, 200 to 430 °C (400 to 800 °F), the tempered structure is ferrite plus cementite.

Transformation of the maximum amount of austenite to martensite on quenching usually requires continuous cooling to below the martensite-finish temperature (M_f), which for a eutectoid tool steel is about −50 °C (−60 °F). To prevent cracking, it is common practice to remove the tool from the quenching medium and begin tempering it while it is still too warm to hold comfortably in the bare hands (about 60 °C, or 140 °F). Under these conditions, even in plain carbon and low-alloy steels, a substantial proportion of the structure (up to 10% or more) may still be austenite. On tempering at increasing temperatures in the range 120 to 260 °C (250 to 500 °F), increasing amounts of this retained austenite transform to bainite plus epsilon carbide with an accompanying expansion in volume.

Most alloying elements lower the M_f temperature. Consequently, more austenite is retained at room temperature in the more highly alloyed tool steels. In addition, cementite reacts with carbide-forming elements during tempering above 370 °C (700 °F) to form alloy carbides, which induces an additional expansion in volume. The formation of alloy carbides during tempering is characteristic of tool steels containing large amounts of carbide-forming elements such as chromium, molybdenum and tungsten.

Size Distortion in Tool Steels

Typical volume percentages of martensite, retained austenite and undissolved carbides are given in Table 1 for four different tool steels quenched from their usual austenitizing temperatures.

Typical changes in linear dimensions for several tool steels are given in Table 2. As shown in this table, some tool steels such as A10 show very little size change when hardened and tempered over the entire range from 150 to 600 °C (300 to 1100 °F).

Other types, such as the M2 and M41 high speed steels, expand about 0.2% (2 mm/m, or 0.002 in./in.) when hardened and tempered in the usual range of 540 to 595 °C (1000 to 1100 °F) to develop full secondary hardness. Although the information in Table 1 is useful in comparing size distortion in several tool steels, the factor of shape distortion makes it impossible to use these data alone to predict dimensional changes of a particular tool made from any of these steels.

Shape Distortion in Tool Steels

In considering shape distortion, it is important to recognize that the strength of any tool steel decreases rapidly above about 600 °C (1100 °F). At the austenitizing temperature, the yield strength is so low that plastic deformation often occurs simply from the stresses exerted on the part by its own weight. Therefore, long parts, large parts and parts of complex shape must be properly supported at critical locations to prevent sagging at the hardening temperature.

Rapid heating increases shape distortion, especially in large tools and in complex tools containing both light and heavy sections. If the rate of heating is high, light sections will increase in temperature much faster than heavy sections. Likewise, the outer surfaces will increase in temperature much faster than the interior, especially in moderate-to-heavy sections. Differences in thermal expansion due to the differences in temperature will be enough to set up large stresses in the material. Under these stresses, the hotter regions will deform plastically, thereby relieving the thermally induced stress.

On continued heating, the hotter portions will begin to level off at the furnace temperature, while the cooler portions will continue to increase in temperature. This produces a decrease in thermal differential, which in turn causes at least a partial reversal in thermal stress because of the plastic deformation that had taken place when the temperature differential was high. This may or may not cause the part to undergo further plastic deformation, but if it does, the deformation will most likely be lesser in extent than the deformation that took place when the temperature differential was high, and will most likely be in a different direction.

Slow heating minimizes distortion by keeping temperature differentials low throughout the heating cycle. Ideally, all heat treatment of tool steel parts should start from a cold furnace to provide the greatest freedom from shape distortion during heating. Starting from a cold furnace is neither very practical nor energy efficient unless heat treating is being done in a vacuum furnace. For heat treating in fused salt or an atmosphere furnace, preheating the parts at an intermediate temperature prior to heating them to the austenitizing temperature provides a useful compromise.

On cooling to form martensite, large temperature differences between surface and interior, and between light and heavy sections, can cause severe shape distortion. This problem is most likely to arise if the hardenability of the steel is so low that a fast cooling rate is required to obtain full hardness. In that event, it may be best to substitute a high-hardenability, air-hardening tool steel, especially when making a large or complex part.

However, if lower-hardenability steels that require liquid quenching are used, fixturing and pressure die quenching will help minimize distortion. Long symmetrical parts should be fixtured, and should be quenched in the vertical position and agitated in the vertical direction while completely submerged in the quenching medium.

Special Techniques for Controlling Shape Distortion

Besides being reduced through control of rates of heating and cooling, as discussed above, shape distortion can

Table 1 Microconstituents in four tool steels after hardening

Steel	Hardening treatment	As-quenched hardness, HRC	Martensite, vol %	Retained austenite, vol %	Undissolved carbides, vol %
W1	790 °C (1450 °F), 30 min; WQ	67.0	88.5	9	2.5
L3	840 °C (1550 °F), 30 min; OQ	66.5	90	7	3.0
M2	1225 °C (2235 °F), 6 min; OQ	64	71.5	20	8.5
D2	1040 °C (1900 °F), 30 min; AC	62	45	40	15

Table 2 Typical dimensional changes in hardening and tempering

Tool steel	Hardening treatment °C	°F	Quenching medium	Total change in linear dimensions, % after quenching	Total change in linear dimensions, %, after tempering at: 150 °C (300 °F)	205 °C (400 °F)	260 °C (500 °F)	315 °C (600 °F)	370 °C (700 °F)	425 °C (800 °F)	480 °C (900 °F)	510 °C (950 °F)	540 °C (1000 °F)	565 °C (1050 °F)	595 °C (1100 °F)
O1	816	1500	Oil	0.22	0.17	0.16	0.18	...	...	...	...	...	...	...	...
O1	788	1450	Oil	0.18	0.09	0.12	0.13	...	...	...	...	...	...	...	...
O6	788	1450	Oil	0.12	0.07	0.10	0.14	0.10	0.00	−0.05	−0.06	...	−0.07	...	...
A2	954	1750	Air	0.09	0.06	0.06	0.08	0.07	...	0.05	0.04	...	0.06	...	...
A10	788	1450	Air	0.04	0.00	0.00	0.08	0.08	0.01	0.01	0.02	...	0.01	...	0.02
D2	1010	1850	Air	0.06	0.03	0.03	0.02	0.00	...	−0.01	−0.02	...	0.06	...	...
D3	954	1750	Oil	0.07	0.04	0.02	0.01	−0.02	...	...	...	...	...	...	...
D4	1038	1900	Air	0.07	0.03	0.01	−0.01	−0.03	...	−0.4	−0.03	...	0.05	...	...
D5	1010	1850	Air	0.07	0.03	0.02	0.01	0.00	...	0.3	0.03	...	0.05	...	...
H11	1010	1850	Air	0.11	0.06	0.07	0.08	0.08	...	0.3	0.01	...	0.12	...	...
H13	1010	1850	Air	−0.01	...	...	...	...	...	...	0.00	...	0.06	...	...
M2	1210	2210	Oil	−0.02	...	...	...	...	...	...	...	−0.06	0.10	0.14	0.16
M41	1210	2210	Oil	−0.16	...	...	...	...	...	...	...	−0.17	0.08	0.21	0.23

be reduced by quenching locally instead of quenching the entire part, or by using flame, induction, electron beam or laser methods to harden only that portion of the tool that must be hardened.

Special hardening procedures such as martempering and austempering may be useful for controlling distortion. In martempering, parts are quenched in hot molten salt fast enough to avoid transformation to high-temperature transformation products such as ferrite or pearlite. The parts are held at a bath temperature in the range from slightly above to slightly below the M_s just long enough to equalize the temperature, then they are removed from the bath and are allowed to cool in air to room temperature. Slow cooling through the martensitic transformation range does not induce as much distortion as normal quenching to martensite. Martempered tools must be given the usual tempering treatment.

Austempering can be used to reduce distortion if a hardness no higher than 57 HRC is acceptable for the application. In austempering, parts are quenched in hot molten salt as in martempering, but, instead of being held a relatively short time, they are held long enough at a temperature above the M_s (usually about 230 °C, or 450 °F) to permit the austenite to transform to lower bainite (about 1 h for plain carbon or low-alloy tool steels). When air cooled to room temperature, austempered tools exhibit lower-than-normal shape distortion and require no subsequent tempering.

Controlling out-of-roundness is important for certain precision applications, such as class C and D cutting hobs made of high speed steels. Class C and D hobs must be held to close size limits because they are not ground to size after heat treatment, but rather are used in the unground condition.

Normal size distortion in hardening and tempering can be accommodated by making the tool slightly oversize or slightly undersize, as required, and then heat treating. High speed steel bars, however, have been observed to go out-of-round as much as 0.05 mm (0.002 in.) when conventionally processed. The pattern shown at left in Fig. 2 is characteristic, and appears to be related to the initial shape of the cast ingot and to the specific primary-mill processing used to break down the ingot. By changing steelmaking, forging and rolling procedures, it has been possible to reduce out-of-roundness to the smaller differential pattern shown at right in Fig. 2, where the difference between high and low points is only 0.005 mm (0.0002 in.). High speed steel bars made this way are marketed by a few tool steel producers as "close tolerance hob stock". An even better method of combating out-of-roundness is to use high speed tool steel bars made from hot isostatically pressed powders, which maintain the best possible symmetry during conventional heat treatment.

Fig. 2 Typical diametral size changes during heat treatment for high speed steel bars

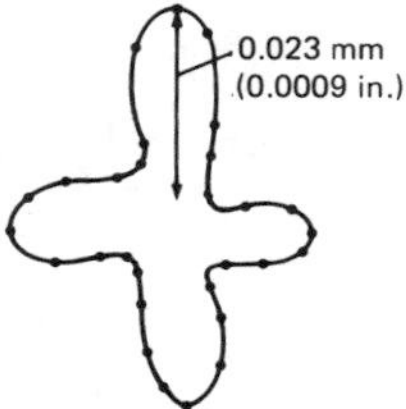

Conventional process

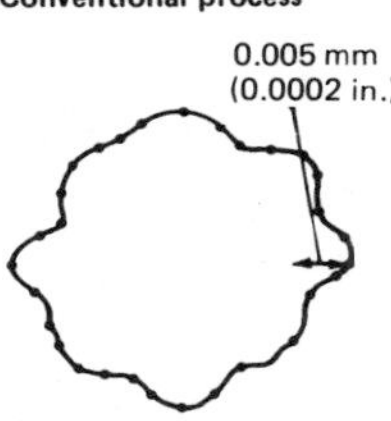

Special process

Drawings produced by calculation from precision measurements of diameter. Charts are plots on polar coordinates depicting variations in diameter after heat treatment for a bar that was round within ±1.25 μm (± 0.00005 in.) before heat treatment. (Courtesy of Latrobe Steel Co., Latrobe, Pennsylvania)

Stabilization involves reducing the amount of retained austenite that can slowly transform and thus produce distortion if the material is heated or subjected to stress. Stabilization also reduces internal (residual) stress, which in turn makes distortion in service due to stress relaxation less likely to occur. Stabilization is most important for tools that must retain their size and shape over long periods of time.

If the tool steel chosen provides the required hardness after tempering at a relatively high temperature, it is possible to reduce the amount of retained austenite and the internal stress by multiple tempering. Initial tempering reduces internal stress and conditions the retained austenite so that it can transform to martensite on cooling from the tempering temperature. Usually, a second or third retempering is necessary to reduce the internal stress set up by the transformation of retained austenite.

Single or repeated cold treatment to a temperature below M_f will cause most of the retained austenite to transform to martensite in plain carbon or low-alloy tool steels that must be tempered at low temperatures to achieve the hardness required for the application. Cold treatment may be applied either before or after the first temper, but if the tools tend to crack because of the additional stress induced by dimensional expansion during cold treatment, it is generally prudent to apply cold treatment after the tools have been tempered the first time. When cold treatment is applied after the first temper, the amount of retained austenite that transforms on cold treatment may be considerably less than would be expected because some of the austenite may be stabilized by tempering prior to cold treating. Cold treatment is usually done in a commercial refrigeration unit capable of attaining −70 to −95 °C (−100 to −140 °F). Tools must be retempered promptly after the treatment returns to room temperature following cold treatment to reduce internal stress and increase the toughness of the newly formed martensite.

For some tools, a small percentage of retained austenite is desirable to improve toughness and provide a favorable internal stress pattern that will help the tool withstand service stresses. For these tools, little or no stabilization may be preferred, and a full stabilizing treatment may actually result in tools that are unfit or only marginally able to perform their required functions.

REFERENCE

1. *Distortion in Tool Steels,* by B. S. Lement: American Society for Metals, 1959

Selection of Materials for Cutting Tools

CUTTING TOOLS include all of the various styles of cutters used in machining of metals, plastics, woods and other machinable structural materials. The most common styles include single-point tools, drills, reamers, taps, threading dies, milling cutters, end mills, broaches, saws and hobs. For many of these styles, the actual cutting edge is on a detachable portion of the tool called an "insert". Inserts generally are made of materials different from those of the tool holders in which they are affixed.

Because of the necessity for a cutting tool to physically cut away material from a workpiece in order to create a usable part, three properties required at the cutting edge are of paramount importance in selecting material for a solid tool or insert. High resistance to elevated temperatures (hot hardness) enables the tool to resist deformation and the cutting edge to remain sharp for longer periods of time. High wear resistance permits tools to last longer, because the abrasive and adhesive actions that take place during cutting do not readily dull the cutting edge. High toughness helps avoid chipping of the cutting edge and enables the tool to absorb the high loads necessary to force the cutting edge through tough, strong and "gummy" metals. To a certain extent, materials having exceptionally high hot hardness and wear resistance do not have good toughness. This means that a compromise may have to be reached in selecting the best tool material for a given cutting application. Generally, high speed steels have good wear resistance, hot hardness superior to other classes of tool steels, and good toughness at high hardness. Nonmetallic materials such as cemented carbides and ceramics have outstanding hot hardness and wear resistance, but are decidedly inferior to tool steels in toughness.

The tool materials discussed in this article are those that have proved most versatile in many different machining applications. Compositions and specific properties of these materials can be found in the articles in this volume entitled "Tool Steels" and "Superhard Tool Materials".

Single-Point Cutting Tools

Single-point cutting tools are those types of cutters having essentially only one cutting edge and/or one corner in contact with the workpiece throughout a given cutting cycle. Single-point cutting tools are used for turning, boring, shaping, planing and threading. In some instances, two or more single-point tools may be mounted in tandem, so that two or more cuts can be taken simultaneously.

A substantial portion of all single-point tools used in the U.S. consist of superhard tool bits (inserts) mounted in tool holders usually made of carbon steel or alloy steel: some sources estimate that more than 40% of the single-point tools are insert-type tools. The remainder consists largely of solid high speed steel and carbide-tipped steel tools. For both classes of tool materials, tool performance depends not only on the specific tool material but also on the material being cut, tool shape, machining speed, cutting angles, and type and amount of cutting fluid, or lubricant. For most operations, a recommended combination of tool material, machining speed and cutting fluid can be obtained from references such as manufacturers' literature and *Machining Data Handbook* (Metcut Research Associates, Cincinnati, Ohio) for each combination of workpiece material and hardness. It must be recognized, however, that the recommendations of tool material, machining speed and cutting fluid are not necessarily optimum for the specific machining job. This often must be determined experimentally. Recommendations are intended only as a starting point that will enable the user to find more quickly a combination that will provide reasonably satisfactory service, or to estimate more reliably plant capacity and manufacturing costs.

Actual determination of the proper tool material, type of tool, shape at the cutting edge and specific machining conditions is a very complex undertaking, and must include consideration of workpiece characteristics (including metallurgical condition), the machine(s) to be used, and proposed machine setups. Only in this way can tools be properly selected to achieve maximum production rate, optimum tool life and minimum over-all machining cost.

High speed steels are used for single-point cutting tools in the form of both solid tool bits and inserts. Molybdenum types M2 and M4 generally are recommended for solid tool bits used for general-purpose machining of metals whose hardness is below about 250 HB. For machining harder metals, M42 and T15 high speed steels, which have better hot hardness, are generally preferred. Types T4 and T5 are relatively common alternatives for types M2 and M4 in single-point tools for machining cast irons and copper alloys.

Type T15 is the predominant tool steel used for inserts. Insert-type tools generally are recommended for machining materials that are relatively hard or gummy or that for other reasons are among the more difficult to machine. In some instances, insert-type tools are expected to last somewhat longer than solid tool bits in a given machining application. In others, machining time is saved by using insert-type tools under more demanding operating conditions than would be imposed on solid bits. All of these reasons make it necessary to manufacture high speed steel inserts of a type such as T15, which has superior edge-retention qualities.

Cemented carbides are used largely for inserts. Single-point carbide tools are generally preferred for high-volume production machining, where productivity is significantly enhanced by high machining speeds, relatively deep cuts and relatively high feed rates, and where it is highly desirable to change tool bits only at infrequent intervals.

With single-point carbide tools it is most important to establish optimum machining speed, because this factor has the greatest effect on tool life. Depth of cut has the smallest effect, and it is this factor that usually is adjusted to give the desired metal-removal rate; however, it is important to avoid setting the depth of cut too low. More of the heat generated during machining is concentrated at the nose of the insert when the depth of cut is below about 10 times the feed rate, and this can have an adverse effect on tool life. For optimum tool life, it is best to set both depth of cut and feed rate as high as is practicable to avoid having to set an excessively high machining speed.

Selecting the proper tool shape is almost as important as selecting the right machining conditions. For this, the recommendations of the producers of carbide inserts usually give the most satisfactory results because the shape of the insert and the rake angles built into the insert and tool holder are designed to give best performance under specific conditions predetermined by the producer.

Although selection of machining conditions and tool shape have a greater effect on tool life than does the type of carbide chosen, specific types of carbide are preferred for single-point turning of certain metals. Straight grades of tungsten carbide are intended primarily for use in machining cast iron and nonferrous metals. When used for machining steel, the straight grades tend to crater rapidly, and thus it is often better to select a coated carbide or a complex grade containing titanium carbide instead.

The properties of straight grades are varied by adjusting carbide particle size and amount of cobalt binder. The low-cobalt grades are more wear-resistant, but also more prone to breakage, than grades higher in cobalt content. In selecting the proper grade for roughing, resistance to breakage and deformation are the most important properties. For finishing, on the other hand, wear resistance is the key property because of the higher machining speeds used and the closer tolerances held on the completed workpiece.

Coated carbides are available with coatings of titanium carbide, titanium nitride or multiple layers of these compositions. A coating of aluminum oxide, either as a single layer or as a layer over titanium carbide-nitride, also can be used. Coated inserts are superior to uncoated inserts in resistance to abrasive wear and cratering. Ceramic-coated inserts are best at resisting chemical reactions between tool and workpiece, which are most likely to occur at high machining speeds.

Many inserts on the market have substrates that are not conventional cutting grades of carbide, but are grades specially formulated to provide resistance to edge deformation and breakage. This results in an insert that performs well over an exceptionally broad range of machining conditions.

Ceramic inserts, such as those of solid aluminum oxide, permit very high machining speeds to be used with no sacrifice in tool life. They also produce finer finishes than those obtained with other insert materials. Ceramics are weaker than carbides or coated carbides, and can be used only for applications where impact loading is low. The selection principles applicable to carbides are also applicable to ceramics, but with the following limitations:

- The entire setup—part, fixturing, machine tool and cutting tool—must be exceptionally rigid to avoid variable shock loading.
- Tool shape should be chosen to protect the cutting edge from stress. High negative rake angles, strong insert shapes, large nose radii and large lead angles are preferred.
- Use of an insert without a hole, which requires use of a tool holder with a clamp, is also preferred.

Diamond inserts may be either single-crystal natural diamonds or carbide substrates under a layer of randomly oriented fine-grain polycrystalline synthetic diamond. Single-crystal natural diamond inserts have outstanding wear resistance, but they cannot withstand high shock loading. They also may vary by as much as an order of magnitude in performance, depending on crystal orientation with respect to the cutting edge. Diamond-carbide composite inserts combine excellent wear resistance with good resistance to shock loading.

Diamond tools are preferred for machining soft, abrasive nonferrous metals, and abrasive nonmetallic materials such as cemented tungsten carbide, unfired ceramics, filled plastics, rubber, carbon and graphite. Recommendations generally applicable to diamond tools include:

- Use positive-rake tooling.
- Avoid built-up edge by using a high cutting speed, greater positive (or less negative) rake and/or a reduced feed rate.

- Provide for regrinding to increase the economic advantage. (Note: certain types of complex inserts cannot be reground.)

Boron nitride bonded to a cemented carbide substrate offers exceptionally high wear resistance and edge life in the machining of high-temperature alloys and hardened ferrous metals. Next to diamond, cubic boron nitride is the hardest known material. This accounts for its cutting properties, and use of a cemented carbide substrate gives the insert satisfactory resistance to shock.

Drills

Edited by M. J. McGinty
Cleveland Twist Drill,
an Acme Cleveland Company

Drilling of holes for assembly of metal parts is performed principally with twist drills made of high speed steel. Carbide drills of several designs are sometimes used for drilling cast iron, high-silicon aluminum alloys and abrasive materials, as well as certain ceramic and plastic materials of high hardness. For drilling holes of large diameter—about 25 mm (1 in.) or greater—indexable carbide insert-style drills are sometimes used. The number of holes drilled with all types of carbide drills is still a small percentage of the total number of holes drilled. Use of carbon and low-alloy steels for drills used in wood or on low-production jobs in metals is outside the scope of this article.

High Speed Steel Drills

Materials for drills must have good resistance to elevated temperatures in order to withstand the heat generated in drilling holes at high production rates. In this requirement drills are similar to single-point cutting tools. However, because of their shape, drills must have greater edge strength and toughness to resist breaking. High speed steels M1, M2, M7 and M10, which have the highest strength and toughness, are used for most of the drills manufactured in the United States. Each of these steels has a specific range of carbon normally associated with it. Steels with carbon contents at the upper end of the range normally are used for drills requiring less toughness and more resistance to abrasion. Drills subject to shock during use often are produced from steels with carbon contents at the lower end of the normal range, to provide better toughness. Some manufacturers also use high speed steels with lower carbon contents for relatively high-cost large-diameter drills, because the greater toughness of these steels provides a greater factor of safety against breakage.

A small percentage of all holes made in metals is drilled with carbide drills or with drills made from high speed steels other than those so far discussed. In recent years there has been a trend toward drilling metal components at high hardness levels—in many instances, after heat-treatment. This trend has been accompanied by a greater demand for drilling of high-alloy heat-resistant metals, which are more difficult to machine than ordinary low-alloy steels. The requirements of such applications for additional hot hardness in tool materials have been met by using high speed steels containing cobalt, such as M33, M42 and T15.

Drills are more complicated in design than tool bits, and in drilling difficult-to-machine materials, more success has been attained by improving the mechanical design of the drill and the rigidity of the machine setup than through selection of tool material. Careful selection of the most suitable drill design, of the most rigid machine setup, of the most suitable coolant and of proper speeds and feeds have helped greatly in overcoming the difficulties in drilling hard and heat-resistant materials.

For example, tests in which 4340 steel was drilled at 55 HRC showed that drills of standard construction made of M33, T15 and M3 high speed steels would barely drill one hole to a depth of 13 mm (½ in.) under any conditions of feed, speed and coolant. When the tooling setup was made more rigid by decreasing overhang of drill and spindle and when the flute lengths of the drills were decreased, the M33, M3 and T15 drills were able to make from two to four holes. When drill construction was made more rugged by using a very heavy-web, low-helix design, M3 and T15 drills produced an average of 19 to 20 holes, and M33 drills averaged 26 holes. With drills of rugged design and with rigid machine setups, the cobalt-type high speed steels, with their better resistance to softening at elevated temperatures, are useful for drilling the more difficult-to-drill materials. Rugged-design drills made from M33, M42 and T15 have been used commonly in the United States, and those from M35 in Europe, for drilling titanium alloys, stainless steels, high-strength steels like 4340 at HRC 36 to 50, and heat-resistant alloys at 45 HRC and below. Whether or not one of the special grades of high speed steel is needed for drilling a difficult-to-machine material cannot be resolved solely on the basis of drill-life testing. If costs of making holes in difficult-to-drill materials are to be minimized, consideration must be given to drill speed and penetration, original tool cost, and tool life. The total cost of making holes includes cost of labor, original cost of drills, and cost of downtime and regrinding.

Fig. 1 Over-all comparative cost index for high speed steel drills

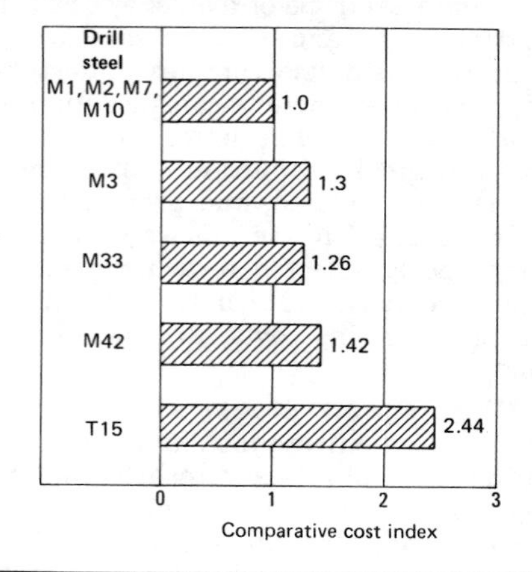

The relative initial costs of these highly alloyed high speed steels have varied widely in recent years as changing political situations around the world have influenced the supply of raw materials used in the manufacture of these alloys. In addition, the relative importance of labor costs for machining and grinding these steels, and for removing and replacing the drills in service, has increased markedly. Over-all costs of drills made from several types of high speed steel, based on data available in 1979, are compared in Fig. 1.

Relative performance of several types of high speed steel in drilling heat-resisting alloy A-286 is illustrated in Fig. 2. In this instance, the ratio of the performance of M7 drills to that of T15 drills is 34:47, or 0.72, and the corresponding cost indexes are in the ratio of 1:2.44, or 0.41. On this basis, the cost

Fig. 2 Performance of high speed steel drills in drilling A-286 at a hardness of 30 HRC

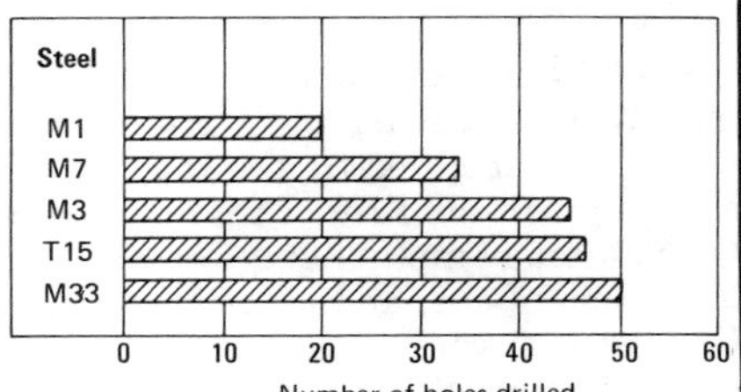

Drill design, type C aircraft; diameter, 6.35 mm (0.250 in.); over-all length, 64 mm (2½ in.); flute length, 35 mm (1⅜ in.); drill-point included angle, 118°; lip relief, 16°; depth of hole, 13 mm (½ in.); cutting speed, 0.2 m/s (40 ft/min); feed, 0.13 mm/rev (0.005 in./rev); cutting fluid, sulfurized cutting oil diluted 1:1 with light machine oil; test end point, gross wear on drill point.

of drilling holes in A-286 under the drilling conditions given in Fig. 2 will be lower if M7 drills are used.

Cemented Carbides for Drills

Drilling of holes in concrete, glass and various ceramic materials with carbide-tipped "masonry" drills is a well-known application for cemented carbide tool materials. Solid cemented carbide twist drills of special design are being used in large quantities for drilling printed circuit boards made of glass-filled epoxy that is abrasive and comparatively low in strength. Increasing use of graphite-fiber composites, particularly for reducing structural weight in aircraft applications, is leading to increasing use of specialized solid carbide drills because of the highly abrasive nature of the material. Carbide twist drills are now used in production drilling of cast iron, aluminum and other nonferrous metals. A special type of carbide-tipped drill known as a "gun drill", which has a hollow body through which coolant is conducted to the drill tip, is finding use where holes of exceptionally good straightness, size and parallelism are needed, or where use of a drill of this style eliminates subsequent operations such as honing, lapping and reaming. Carbide-tipped gun drills have been used successfully in drilling a wide variety of materials, including brass, bronze, aluminum, cast iron, stainless steel, high-strength valve steel, titanium, and cobalt-base alloys.

Carbide-tipped die drills have been used successfully on some difficult-to-machine heat-resistant materials, especially where the machine setup is very rigid. For example, in drilling 6.35-mm (¼-in.) holes in 4340 steel heat treated to a hardness of 55 HRC, the best drill life that could be obtained under any conditions with high speed steel drills was 26 holes for M33 drills. Carbide-tipped die drills produced an average of 90 holes in the same steel. In these tests, tool overhang was reduced to a minimum, which resulted in a very rigid setup.

Whether or not cemented carbide should be selected as a material for drills depends on the type of material being drilled and on the number of holes to be drilled. The bodies of most carbide-tipped twist drills and some types of straight-flute die drills are made of hardened high speed steel to provide strong support for the carbide tip and abrasion-resistant margins for guiding purposes.

Straight tungsten carbides containing up to 6% cobalt are used successfully in tipped twist drills for drilling cast iron and nonferrous metals. Grades of this type having higher transverse rupture strength and slightly lower hardness (90.5 to 91.5 HRA) are preferred for tips of straight-flute die drills that are used for drilling heat-resistant alloys and high-strength steel heat treated to 50 HRC or more.

Indexable-insert carbide drills approximately 25 mm (1 in.) or more in diameter have been developed in which mechanically held carbide inserts are clamped to a body having one, two or even more flutes. Such indexable-insert drills generally have very short flute lengths for rigidity because there is a general imbalance in cutting forces from flute to flute and because a large amount of cutting torque is required for drills this large. They are run at carbide turning speeds and lend themselves well to numerical-control lathe operations. These coolant fed insert drills generally are used for drilling holes up to two diameters deep.

Coated tungsten carbides, and complex cemented carbides that contain titanium and tantalum carbides in addition to tungsten carbide, are preferred for the inserts in indexable-insert drills used for drilling steels.

Reamers

Edited by M. J. McGinty
Cleveland Twist Drill,
an Acme Cleveland Company

Reamers are cutting tools used for finishing holes that must meet stringent finish and size-tolerance requirements that cannot be satisfied by simple drilled holes. Reamers are available in a wide variety of sizes and designs that suit specific applications and work materials.

Reamer design must be the first consideration. After this has been determined, consideration can be given to selection of tool material.

High speed steel, with its high resistance to softening at elevated temperatures, has replaced carbon steel for all except hand reamers. Reamers remove only small amounts of material. Accordingly, they do not need deep flutes for easy passage of chips and coolant, as do drills, and can be designed as heavy, rigid tools. Moreover, less toughness is required for reamers than for drills.

In choosing materials for machine reamers, high hardness and high abrasion resistance are the most important properties. The general-purpose high speed steels (M1, M2, M7, M10 and T1) at high hardness levels, and the high-vanadium types (M3, M4 and T15), have been used successfully. The latter steels have greater resistance to abrasion than the lower-vanadium types. The very high resistance to softening at elevated temperatures provided by cobalt-containing high speed steels such as M33, M42, M6 and T5 is less often required in reaming than in drilling, because heat is more easily controlled in reaming.

Solid cemented carbide reamers and carbide-tipped reamers have been used widely and successfully in reaming almost all metals. General-purpose and harder grades of carbide are used most commonly. Carbide-tipped reamers are made with hardened high speed steel bodies that furnish support for the carbide tips and guide the reamer in the hole. Large shell reamers, made with inserted blades tipped with carbide, are constructed most often with medium-carbon alloy steel bodies.

In applications in which high speed steel and cemented carbide reamers are used, the high hardness levels of these cutting-tool materials demand optimum operating conditions. High hardness is associated with brittleness, so that poor conditions—such as a sloppy machine spindle, flimsy support, or misalignment of the reamer with respect to the drilled hole—result in early tool failure due to flaking and chipping.

Taps

Edited by Dino J. Emanuelli
Greenfield Tap and Die Division
TRW, Inc.

Taps are tools used for cutting or forming internal screw threads. Selection of tool materials for taps usually is done by the tap manufacturer. Three families of materials generally are considered: (*a*) carbon and alloy tool steels, (*b*) high speed steels and (*c*) cemented carbides.

Carbon and alloy tool steels are suitable for taps used for hand tapping or other low-speed, light-duty applications. The major reason for selecting this family of tap materials is low cost. Taps used for nonprecision maintenance and repair work usually are made from these grades.

High speed steels are used for taps that must cut efficiently at high speeds and thus must resist softening under the extreme heat generated at the cutting edge of the tool. Most tap manufacturers use M1, M2, M7 and M10 for general purpose taps, because these materials exhibit good toughness, good grindability and relatively low cost.

For special tap designs that require greater resistance to abrasion, type M3 (class 2) or type M4 is generally used. Where requirements include high abrasion resistance and/or very high resistance to softening at elevated temperatures, types T15 and M42 are the most popular choices. Although there has been a slight increase in the number of applications requiring T15 or M42 steel, most regular and special taps are made from the general-purpose types M1, M2, M7 and M10. In many instances, the higher material and tap-grinding costs involved in using high speed steels other than the general-purpose types cannot be justified by an observed improvement in performance. However, use of T15 and M42 steels has proved worthwhile in tapping high-strength alloys and high-temperature alloys and in applications that require extremely high tapping speeds.

Many supplementary surface treatments have been adopted by tap manufacturers for use on high speed steel ground-thread taps to improve tap life and performance. Three common surface treatments are nitriding, oxide coating and hard chromium plating. Nitriding is recommended for counteracting abrasive wear in tapping ferrous, nonferrous and nonmetallic materials that have highly abrasive qualities. Oxide treatments improve tap life and performance in tapping ferrous materials that tend to gall, such as certain steels and stainless steels. Hard chromium plating increases surface hardness and decreases friction between tap-thread surfaces and the material being tapped, and thus results in better chip flow and increased tap life.

Cemented carbides can be considered for extremely abrasive applications, such as tapping of filled plastics and certain grades of cast iron. However, because carbide taps are expensive and highly susceptible to breaking and chipping, there are few tapping applications for which they can be justified solely on the basis of lowest over-all cost.

End Mills

Edited by Thomas Ribich
and
Gary Testen
The Weldon Tool Company

An end mill is a shank-type milling cutter with cutting edges on its periphery and end surface. The shank may be either straight or tapered. These tools are available in diameters ranging from 0.8 to 75 mm (1/32 to 3 in.), and may have one or more cutting teeth (most have 2, 4 or 6 teeth). Larger sizes are made as special items on customer request. End teeth may be of the noncenter type or may be designed to cut to the axis of the tool to permit plunge cutting. The peripheral cutting teeth may be produced without a helix (end mills having such teeth are called straight-flute end mills) or with either a right-hand or left-hand helix. The most common helix angle is approximately 30°. The direction of cutting rotation may be right hand or left hand and either hand of cut may be combined with the same or opposite hand of helix. The type of flute, direction of cutting rotation and helix angle have no direct bearing on the choice of tool materials for end mills.

The length of unsupported cutting flutes may vary from short to extra long: standard ratios of cutting length to diameter are about 1½ to 1 for the shortest "stub end mills" and about 6 to 1 for the longest "extra long end mills".

Tool Life. Most end mills fail ultimately by wearing out. The rate of wear is determined by the tool material, the machinability of the workpiece, the volume of metal being removed (dimensions of cut, and cutter feed rate), the general configuration of the tool, the condition of the cutting edges and the lubricant. However, the factor most critical to end-mill life is the length of unsupported cutting edge that is under extreme stress during use and that can suffer premature catastrophic failure if it cannot withstand the stress due to interrupted cutting action and deflective cutting forces.

End-Mill Materials. The majority of end mills larger than 16 mm (5/8 in.) in diameter are made from high speed steel bars, which in the annealed state may be easily machined to shape, heat treated and finished by grinding. End mills smaller than 16 mm in diameter are commonly made from hardened cylindrical blanks of high speed steel into which the flutes are ground.

End mills are made not only from high speed steel bars produced by conventional electric furnace melting and rolling but also from those produced by powder metallurgy processes. For some end mills, blanks are made by cold isostatic pressing of high speed steel powders into the desired blank shape. The blanks are then sintered, annealed, hardened and ground.

The majority of workpieces machined with high speed steel end mills have hardnesses below 300 HB. For these applications, end mills made from the more popular and less costly general-purpose tool steels such as M1, M2, M7 and M10 generally have proved satisfactory on the basis of both economy and performance.

For work materials that are difficult to machine and that have hardnesses from 350 to 450 HB, end mills made of cobalt high speed steels such as T15, M33 and M42 are most effective. Figure 3(a) shows the improvement in tool life at cutting speeds above 0.25 m/s (50 ft/min) that was achieved when T15 rather than M7 was used to end mill 4340 steel at 49 HRC. Figure 3(b) illustrates the typical rapid reduction in tool life that occurs with increasing

Fig. 3 Typical life of four-flute end mills made of M7 and T15 high speed steels

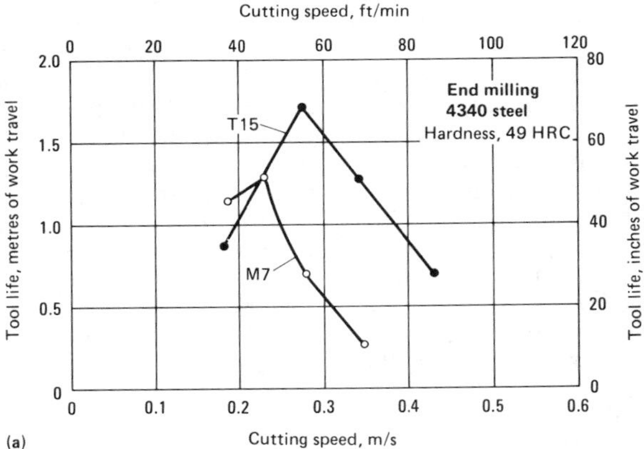

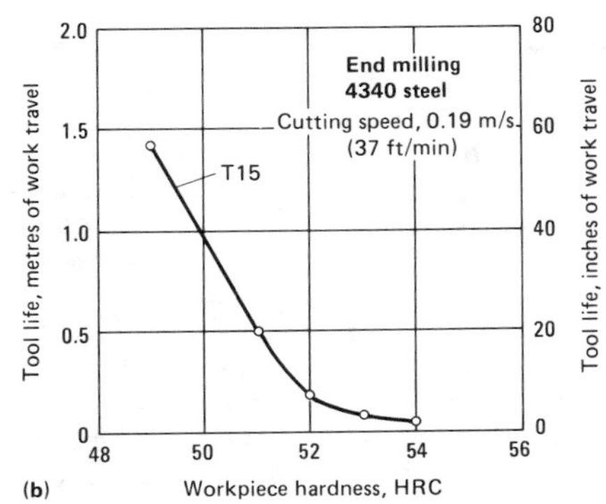

	Chart (a)	Chart (b)
Cutting conditions		
End-mill diameter	19 mm (3/4 in.)	19 mm (3/4 in.)
Feed per tooth	25 μm (0.001 in.)	13 μm (0.0005 in.)
Depth of cut	6.4 mm (1/4 in.)	6.4 mm (1/4 in.)
Width of cut	19 mm (3/4 in.)	19 mm (3/4 in.)
Soluble-oil ratio	1:20	1:20
Wear land	0.4 mm (0.016 in.)	0.4 mm (0.016 in.)
Chip load per tooth	25 μm (0.001 in.)	13 μm (0.0005 in.)
Tool angles		
Helix angle	35° rh, rh cut	35° rh, rh cut
Radial rake	15°	15°
Clearance angle	45°, 1.5 mm (0.060 in.) wide	45°, 1.5 mm (0.060 in.) wide
Side clearance	5° primary, 10° secondary	5° primary, 10° secondary
End clearance	5° primary, 20° secondary	5° primary, 20° secondary

workpiece hardness in end milling of 4340 steel with a T15 mill at an intermediate cutting speed.

Cemented carbide end mills have the ability to withstand higher cutting temperatures and greater abrasion than high speed steel end mills. Carbide end mills are used most often for machining nonferrous alloys; nonmetallic materials; work materials with hardnesses exceeding 450 HB; and materials with highly abrasive scaled surfaces, such as sand castings.

Small end mills made of cemented carbide are ground from solid cylindrical blanks using diamond grit wheels. Intermediate-size carbide end mills—up to 25 mm (1 in.) in diameter—usually are made from preformed fluted blanks. Finishing of preformed blanks is done by grinding with diamond grit wheels. Most large end mills (> 25 mm in diameter) and some smaller end mills have cemented carbide tips brazed or mechanically held in position on high speed steel bodies.

The straight grade of tungsten carbide containing 6% cobalt is used most commonly for solid carbide end mills and for brazed-tip tools, both of which are used for cutting cast iron, titanium alloys, nonferrous metals, stainless steels and plastics. A complex grade containing 72 to 73% WC, 7 to 8% TiC, 11.5 to 12% TaC and 8 to 8.5% cobalt binder is used for end mills in steel-cutting applications.

Base-price ratios for typical end-mill materials are given below; the ratios are based on the standard mill price for M1.

End-mill material	Base-price ratio
M1	1.0
M2	1.16
M7	1.0
M10	1.0
M33	1.5
M42	1.6
T15	2.5
T15 P/M	2.7
Cemented carbide	18.0

Milling Cutters

Edited by V. Edward Simms, Jr.
and
Norman F. Nau
Union/Butterfield Division
Litton Industrial Products, Inc.

Milling cutters that are relatively small and of complex configuration generally are made of high speed steels. The cutting teeth on tools of this type often have helical cutting edges, deep radial and/or axial gashes, irregular profiles or thin web sections. These complex configurations can be most readily obtained by machining the tool as an integral unit from annealed high speed steel, heat treating it to a suitable hardness, and then grinding it to size.

The useful lives of most milling cutters end as a result of gradual wear rather than catastrophic failure. Consequently, the ability of high speed steel to resist abrasion at elevated temperatures is important. High speed steel milling cutters do not break unless cutting conditions are very abusive or unless the structure and hardness of the workpiece are not uniform.

There are many special applications and operating conditions for which milling cutters are constructed of materials other than solid high speed steel. For economy, large-diameter cutters, and cutters of simple configuration used in high-production applications, often comprise high speed steel blade inserts attached to low-cost alloy steel bodies. A cutter of this type has blades of high speed steel, which are clamped to the body and which form the cutting edges. The blades can be removed and replaced with new blades as needed. Blade design and clamping requirements often limit over-all cutter design to straight cutting edges and to sizes suitable for accepting the clamping mechanism. Face mills, side mills, and large counterbores or end-feeding tools can be readily adapted to such a design.

Because of their high cost, cemented carbides are used predominantly in those milling cutters into which carbide tips can be brazed or carbide blades inserted. Carbide, because it is not readily machinable, is difficult to use in cutters of complex shape. Nevertheless, by overlapping tips of various sizes, a variety of milling-cutter configurations can be designed. Small-diameter tools such as end mills and slotting cutters can be constructed completely of carbide. Configurations as complex as those used for high speed steel end mills and slotting cutters often can be reproduced in carbide. However, on these types of milling cutters, where tool deflection is often a major problem, small-diameter carbide tools are likely to break.

In a design that is finding widespread use in high-production face milling and shell milling, solid carbide inserts are clamped to an alloy steel body. Classified as an indexable-blade milling cutter, this type of cutter is unique because its inserts can be indexed to two or more positions without affecting the concentricity or contour of the cutting profile of the tool, thus providing new cutting edges without regrinding. The size and configuration needed to accommodate the inserts and clamping mechanisms somewhat restrict over-all design of the tool, but many ingenious styles are available to extend applications of this type of tool.

For general-purpose cutting of steel, the grades containing complex carbides are widely used in all types of milling cutters. Other types of workpiece materials, such as cast irons, various brasses, aluminum alloys and fiber composites, are most productively cut using the straight tungsten carbide grade containing 6% Co. This straight tungsten carbide also performs well in cutting stainless steels of the 300 series.

Tool steel milling cutters used for cutting plain carbon and low-alloy steels, cast irons and nonferrous alloys, where workpiece hardness does not exceed 30 HRC, normally are made of high speed steels such as M1, M2, M7 and M10. T1 cutters are available from most cutter manufacturers, but they are high in initial cost.

At intermediate hardness levels from 30 to 35 HRC, the high-vanadium grades M3 and M4 provide increased tool performance. As the hardness of the workpiece increases beyond 35 HRC, tool life of the general-purpose grades drops very rapidly. For workpiece hardness levels from 35 to about 45 HRC, cobalt high speed steels such as M42 and T15 are recommended. Addition of cobalt to high speed steel increases its resistance to softening at elevated temperatures and permits an appreciable increase in cutting speed. No high speed steel, however, is recommended for cutting metals with hardnesses much above 45 HRC; for these high-hardness materials, carbide tools must be used. In general, carbide milling cutters can be used at cutting speeds three to six times those permitted with high speed steel cutters.

As a group, nickel and cobalt high-temperature alloys are very difficult to machine, and best results are obtained by using cobalt-bearing high speed steel cutters, low cutting speeds and heavier-than-normal feed rates. Carbide cutters normally are not recommended because they are prone to chipping of the cutting edge.

Other factors to be considered in selecting tool materials for milling cutters are rigidity of the machining setup, milling-machine horsepower and facilities for sharpening of cutting edges. A rigid setup is particularly important when carbide milling cutters are used. Looseness in the milling machine can result in vibrations that shorten cutter life by causing chipping or even breakage. Changing from high speed steel cutters to carbide cutters to permit an increase in cutting speed is feasible only if the additional horsepower required is available. Finally, grinding of carbide is more difficult than grinding of high speed steel and requires diamond wheels and operators trained in using them.

Hobs

Edited by V. Edward Simms, Jr.
and
Norman F. Nau
Union/Butterfield Division
Litton Industrial Products, Inc.

A hob is a type of milling cutter used to generate a repeating form about a center, such as in cutting of gear teeth, spline teeth and serrations. The hob and workpiece must rotate and mesh in a specific timed relationship, and thus the cutting teeth on the hob follow a helical or thread pattern around the periphery of the tool. This is the feature that distinguishes a hob from a milling cutter.

Whereas milling cutters usually are used under conditions involving a relatively heavy chip load per tooth—frequently as heavy as tool strength, machine rigidity and surface finish requirements will allow—hobs are operated under conditions involving a very low chip load per tooth because of the generating action of each hob tooth. As a consequence, strength in hob materials is not as important a consideration as abrasion resistance.

For hobs, as for milling cutters, service life usually is terminated by wear; hobs do not break unless operated under abusive conditions.

The majority of hobs are made from high speed steels, although hobs for special applications have been manufactured from cemented tungsten carbide, cast cobalt-chromium-tungsten alloys, and even some low-alloy tool steels. Virtually all types of high speed steel have been used, but the most widely used is type M2. This steel is available in two carbon levels. Because of its higher hardenability and better abrasion resistance, the high-carbon M2 has become the more popular of the two. Some manufacturers prefer to use M1 or even T1 for small-diameter hobs because those alloys provide improved grindability. M1, T1 and the low-carbon version of M2 work well in general-purpose applications and, when properly heat treated, are similar in cutting ability. Bar stock for hobs generally is made with a free-machining addition that enables the manufacturer to produce unground hobs with greater accuracy and a smoother finish. There is no evidence that free-machining additions degrade either tool life or tool performance.

General-purpose types like M2 offer good combinations of edge strength and wear resistance, but high speed steels such as M3 and M4, which contain more carbon and vanadium, are used for hobbing harder and more abrasive materials. For applications that require increased resistance to softening at elevated temperatures, cobalt-bearing grades such as M42 and T15 are more suitable.

A recent innovation has been the production of high speed steel by powder metallurgy. Steels manufactured in this manner exhibit improved grindability and dimensional stability. Because of these properties, use of P/M high speed steels for hobs and cutters is increasing. A broad range of grades is available, and the same application guidelines apply to them as apply to wrought tool steels.

Selection of Hob Material

Constructional steels in the annealed, cold drawn, or quenched-and-tempered condition that have hardnesses below about 300 HB can be readily cut with hobs made from one of the general-purpose high speed steels such as M2. As hardness increases above 300 HB, it is often more economical to use hobs made from a higher-vanadium grade such as M3 or M4. For workpiece hardnesses from 350 to 475 HB, cobalt-bearing types such as M42 and T15 may be required. Experience in cutting steels of high hardness with carbide hobs has been generally unsatisfactory.

Hardnesses of about 440 to 475 HB represent the upper limit for practical hobbing. Even at these hardnesses, hobbing is expensive and tools must operate at low speeds.

Cast irons usually are hobbed with tools made from standard general-purpose high speed steels. If the iron contains an appreciable amount of free carbide, it is more economical to use hobs made from one of the more abrasion-resistant steels (M3 or M4, for instance). For cutting heat treated cast iron, the same hob steel should be used that would be required for hobbing steel of the same hardness. Carbide hobs have been used to a limited extent for hobbing abrasive grades of cast iron.

Brasses and bronzes normally are cut with hobs made from one of the general-purpose grades of high speed steel. The harder, more abrasive bronzes require hobs made of higher-vanadium steels. Carbide hobs are useful for cutting these materials if warranted by production quantities.

Aluminum alloys may be cut with hobs made of any high speed steel or of carbide. Use of carbide greatly increases the number of gears cut between sharpenings and at the same time permits higher cutting speeds to be used.

Abrasive nonmetallic materials such as laminated phenolic plastics and fibers are advantageously cut with carbide hobs. Because some of these materials are extremely abrasive, even high-vanadium high speed steel hobs may wear out rapidly, whereas carbide hobs will successfully cut large numbers of parts. For best results with carbide hobs, a machine capable of running at high speeds is desirable. In cutting these nonmetallics, however, carbide hobs will produce good results even at normal cutting speeds.

Nonabrasive plastics such as nylons are satisfactorily cut with high speed steel hobs.

Factors in Selection. Besides composition, hardness, and microstructure of the material being hobbed, several other factors may influence choice of hob material. Some of these factors are:

- Form to be cut (gear, spline, special shape)
- Available hobbing machines (rigidity of setup, attainable cutting speed)
- Quantity of identical parts to be produced
- Dimensional tolerance of workpieces
- Cost.

In some instances, a special hob material such as carbide might be warranted for the material being cut, but unless rigid machines capable of high speeds are available, hobs made of such a material would not be a wise choice and it would be better to use cobalt-bearing high speed steel hobs.

For large quantities, and especially for close dimensional tolerances, one of the more highly alloyed high speed steels or a cemented carbide might be the best choice because of their high resistance to wear.

Cost of hobs is influenced mainly by tolerance class and by costs of material, fabrication and sharpening. Carbide hobs not only are high in initial cost but also require diamond wheels for resharpening, which are expensive. High-vanadium high speed steels are more costly to sharpen than standard grades.

Tolerance class has a marked influence on the cost of a hob. Unnecessarily close tolerances result in needlessly high tool cost for two reasons: (*a*) initial cost is greater and (*b*) unground hobs provide a longer tooth length for more regrinds, and frequently give better tool life, than ground hobs. Hob tolerances are classified as follows:

- Class A—precision ground hobs
- Class B—commercial quality ground hobs
- Class C—accurate unground hobs
- Class D—commercial quality unground hobs.

The influences of hob tolerance and tool material on the cost of a hob are shown in the following table:

Hob class	Cost ratio (a)			
	M2	M3	T15	Carbide tipped
A	1.9	2.4	4.2	6.0
B	1.6	2.0	3.5	(b)
C	1.2	1.5	2.7	(b)
D	1.0	1.3	2.2	(b)

(a) Based on the cost of an unground M2 high speed steel 6-diametral-pitch, 20°-pressure-angle, single-thread hob about 90 mm ($3\frac{1}{2}$ in.) in both diameter and length and having a 32-mm ($1\frac{1}{4}$-in.) bore. (b) Carbide hobs are made as class A hobs only.

Selection of Materials for Shearing and Slitting Tools

Edited by L. A. Hauser
Universal Cyclops Specialty Steel Div.

MOST SHEAR BLADES are solid, one-piece blades made of tool steel. However, some are composite tools that consist of tool-material inserts in heat treated medium-carbon or low-alloy steel backings.

Cold Shearing and Slitting of Metals

Blade materials recommended for cold shearing of various metals are presented in Table 1, and blade materials for rotary slitting are given in Table 2. The composition and thickness of the material being sheared are the most important factors in the selection of a blade material. Other significant factors are cost, availability, heat treating characteristics, and previous experience in similar applications.

Table 1 Recommended blade materials for cold shearing flat metals

Material to be sheared	Blade material(s) for work-metal thickness of: 6 mm (1/4 in.) or less	6 to 13 mm (1/4 to 1/2 in.)	13 mm (1/2 in.) and over
Carbon and low-alloy steels up to 0.35% C	D2, A2, CPM 10V	A2, A9	S2, S5, S6, S7
Carbon and low-alloy steels, 0.35% C and over	D2, A2, CPM 10V	A9, S5, S7	S2, S5, S6, S7
Stainless steels and heat-resisting alloys	D2, A2, CPM 10V	A2, A9, S2	S2, S5, S6, S7
High-silicon electrical steels	D2, T15, CPM 10V, cemented carbide inserts(a)	S2, S5, S7	(b)
Copper and aluminum alloys	D2, A2	A2	S2, S5, S6, S7
Titanium alloys	D2	...	...

(a) Carbide inserts usually are brazed to heat treated medium-carbon or low-alloy steel backings. (b) Seldom sheared in these thicknesses.

Tool materials vary in toughness and wear resistance, and the metals being sheared vary in hardness and resistance to shearing. If the material to be sheared is very thin and of relatively low hardness, the shear-blade material can be low in toughness but must have optimum wear resistance. For shearing material of greater thickness and higher hardness, it may be necessary to

Table 2 Recommended blade materials for rotary slitting of flat metals

Material to be sheared	Blade material for work-metal thickness of: 4.5 mm (3/16 in.) or less	4.5 to 6.5 mm (3/16 to 1/4 in.)	6.5 mm (1/4 in.) or more
Carbon, alloy and stainless steels	D2, CPM 10V	D2, A2, A9	A9, S5, S6, S7
High-silicon electrical steels	D2, M2, CPM 10V cemented carbide inserts	D2	...
Copper and aluminum alloys	A2, D2, CPM 10V	A2, D2	A2, S5, S6, S7
Titanium alloys	D2, A2, CPM 10V	...	...

Fig. 1 Comparison of wear and life of different tool steels in cold shearing of steel

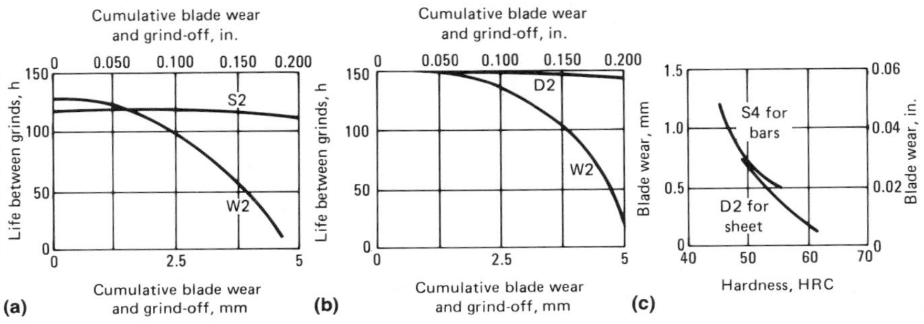

(a) Cold shearing of 19-mm-thick (3/4-in.-thick) low-carbon steel plate. (b) Cold shearing of 6-mm-thick (1/4-in.-thick) low-carbon steel plate. (a) and (b) As more and more of the edge is worn away and ground off in resharpening, the life of W2 between grinds continuously decreases until all of the hardened shell is gone and the blade must be scrapped or rehardened. (c) Effect of hardness of cutting edge on wear of shear blades.

decrease blade hardness, or change to a less wear-resistant blade material or to a shock-resistant tool steel having a hard case over a tough core, to obtain the toughness needed to resist edge chipping. For example, D2 tool steel is recommended for shearing any metal substantially less than 6 mm (1/4 in.) thick, but for thicknesses of 6 mm and greater, tool steels of greater toughness must be used (successively, A2, A9 and the shock-resisting types S2, S5 and S6).

Effect of Hardness on Wear. The rate at which a blade wears in cold shearing depends chiefly on its carbon content and hardness. As shown in Fig. 1(a) and (b), shallow-hardening steels such as W2 may equal D2 in performance until several sharpenings have ground the hardened stock away; then hardness is lower, and life between grinds decreases accordingly. In these tests, shallow-hardening W2 was superior to deeper-hardening S2 until about 1.5 mm (0.060 in.) had been removed, because of the higher initial surface hardness of W2.

Use of a low-hardness blade for cold shearing will result in short blade life. A blade made of S4 or S5 tool steel with a hardness of 44 HRC wore three times as fast as one with a hardness of 54 HRC when other conditions were equal. A similar comparison of blade wear as a function of hardness for S4 and D2 blades is given in Fig. 1(c).

Recommendations on hardness of blades for cold shearing cannot always be made without knowledge of operational details. For example, a D2 blade performed satisfactorily at 61 HRC in one application, but blades at this hardness broke under similar operating conditions in a different plant. Blades made of D2 tool steel at 58 to 60 HRC usually perform satisfactorily in shearing mild steel up to 6 mm (1/4 in.) thick, and in many instances, D2 shear blades have been used successfully at 60 to 62 HRC. However, in shearing metals such as high-strength low-alloy steels, hardness of D2 blades must be kept below 58 HRC to prevent breakage.

The shock-resisting group S tool steels are used at hardness levels from 50 to 58 HRC. Blades with hardnesses near the higher end of this range are recommended for shearing steels 6 to 13 mm (1/4 to 1/2 in.) thick and for nonferrous metals. For shearing harder or thicker metals, blades with hardnesses nearer the low end of this range are used in order to compensate for increased shock loading. Table 3 gives some typical hardnesses of blades used in ten different plants for cold shearing of specific ferrous and nonferrous products.

Hot Shearing of Metals

Shearing is done at elevated temperatures when the work material is thick and resistant to shearing or when hot shearing is otherwise desirable as part of the manufacturing process. The strong secondary hardening of group H tool steels provides sufficient resistance to softening to make them useful for shear blades operating at temperatures up to 425 °C (800 °F).

Type H11 tool steel is satisfactory for most hot shearing operations. The slightly more expensive types H12 and H13 also have been recommended. No data are available that prove any of these three steels to be superior to the other two. More costly steels such as H21 and H25 are recommended only when H11 has been tried and found to be inadequate. In many instances, however, higher-alloy tool steels are used for shear blades. For example, one large mill used H25 as the standard blade material for hot shearing aluminum at 150 to 425 °C (300 to 800 °F), as indicated in Table 3.

Hardness of blades for hot shearing varies considerably with conditions such as thickness and temperature of the metal being sheared, and type and condition of available equipment. However, hardness is usually kept within the range from 38 to 48 HRC. In one steel mill, H12 tool steel at 45 to 47 HRC is used for slab shear blades 1.45 m (57 in.) long, 125 mm (5 in.) high, 75.4 mm (2.969 in.) thick at the top and 73.0 mm (2.875 in.) thick at the bottom. In another mill, slabs and blooms are

Table 3 Service data for shear blades

Type of shear	Material sheared	Thickness of material sheared mm	in.	Blade steel	Blade hardness, HRC	Blade service before regrinding
Cold shearing of steel						
Sheet metal	Low-carbon steel	5	Up to 3/16	W2	58 to 60	30 000 cuts
Sheet metal	Low-carbon steel	5	Up to 3/16	A2	58 to 60	55 000 cuts
Sheet metal	Low-carbon steel	5	Up to 3/16	D2	58 to 60	100 000 cuts
Bar shear	1025 & 1040 steel	25	1	W2	58 to 60	20 000 cuts
Bar shear	1025 & 1040 steel	25	1	L6	...	40 000 cuts
Bar shear	1025 & 1040 steel	25	1	S5	...	100 000 cuts
Sheet and strip	1010 steel	5	Up to 3/16	D2	58 to 60	150 000 cuts
Sheet and strip	Stainless steel	12	0.478	D2	58 to 60	65 000 cuts
Sheet	2 to 5% silicon steel	0.8	0.032	D2	58 to 60	45 000 cuts
Slitter	Carbon, silicon and galvanized steels	0.16 to 4	0.005 to 0.160	D2	58 to 60	1 week(a)
Slitter	Stainless steel	0.8 to 1.5	0.030 to 0.060	D2	58 to 60	15 000 ft
Blade 1.5 m by 100 mm by 25 mm (60 by 4 by 1 in.)	Stainless and silicon steels	2 to 2.5	0.080 to 0.100	D2	60 to 62	2 weeks(b)
Hot shearing of steel						
Slab shear	Various steels	...	...	4340 mod	...	30 000 to 40 000 tons
Cold shearing of copper alloys						
Sheet	Brass	0.16 to 5	0.005 to 0.187	S1	54 to 58	5 000 cuts
Slab	Brass	25 to 60	1.0 to 2.25	S1	54 to 58	25 000 cuts
Hot shearing of copper alloys						
Slab	Brass	Up to 45	Up to 1.75	H11	42	15 000 cuts
Hot shearing of aluminum alloys						
Automatic flying cutoff shear	Aluminum alloys at 150 to 250 °C (300 to 500 °F)	7	0.280	H25	43 to 46	10 000 cuts
75-mm (3-in.) shear	Aluminum alloys at 315 to 425 °C (600 to 800 °F)	76 to 140	3 to 3.5	H25	43 to 46	17 000 to 20 000 cuts

(a) Blades were reground weekly, with about 0.025 mm (0.001 in.) of stock being removed. (b) Maximum. Blades usually were changed weekly.

sheared with blades made from a steel containing 0.35 C, 0.90 Mn, 1.00 Cr, 1.25 Mo and 0.25 V, hardened to 38 to 42 HRC.

The temperature of the metal being sheared influences blade life. Higher-alloy blades may be required if the metal being sheared has high strength at elevated temperatures.

Hard faced blades are satisfactory, and in some plants are used exclusively, for hot shearing. Table 4 summarizes data on blade life in one steel plant where hard faced blades are used for most hot shearing operations. For more information on hard facing alloys, see the article that begins on page 563 in this volume.

Shear-Blade Life

Service data on life of shear blades are scarce, because maintenance programs employed in most high-production mills call for removal and redressing of blades, regardless of condition, during scheduled shutdowns. Available data report blade life in terms of number of cuts, linear footage (in slitting), tonnage or time between redressings. The service data in Table 3 encompass all of these variables.

The number of cuts per edge before regrinding is the most common basis of evaluation. For different types of shearing, the number of cuts reported has varied from 5000 to more than 2 million. Therefore, meaningful comparisons can be made only when conditions are nearly identical. Even when identical blades are used for the same shearing operation, life may vary by as much as 100%. This is substantiated in Fig. 2, which presents data on performance of 33 blades with carbide inserts used for shearing small rods. Wide variation in blade life also is characteristic of hot and cold shearing of nonferrous metals (see Fig. 3).

Fig. 2 Statistical distribution of life for shear blades with tungsten carbide inserts

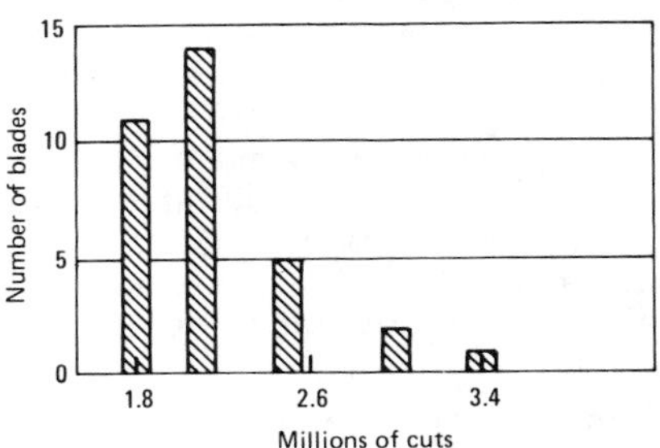

Variation in life of 33 identical shear blades used for the same operation. Blades were made of carbon steel with cemented tungsten carbide inserts (15 to 30% Co) for the cutting edges. Material sheared was 4620 steel rod, 8.7 mm (0.343 in.) in diameter, at 149 to 187 HB.

Table 4 Life of hard faced blades for hot shearing of steel in a specific steel mill

Type of shear	Steel for blade body	Hard facing alloy(a)	Blade life, tons of steel sheared
Billet, 300 by 300 mm (12 by 12 in.)	1030 cast	1B new, 2B repair, 3A ribs	29 000
Billet, 250 by 250 mm (10 by 10 in.)	1030 forged	2B	5 800
Billet duplex	1030 cast	None; inserts of H21 or M2	26 000
Billet	H21	None	3 584
Bloomer, 1 m (40 in.)	1045 plate	1B new, 2B repair	7 680
Bloomer, 1.1 m (44 in.)	1045 plate	1B	87 000
Slab, 900 mm (36 in.)	1045 plate	1B	71 000
Plate	6150 (mod)	None	6 000
Rail, 250 by 250 mm (10 by 10 in.)	1030 cast	1B new, 2B repair, 3A ribs	12 960
Rail billet	1045 plate	2B	5 400 10 800

(a) Nominal compositions. Alloy 1B: 0.5 C, 0.9 Si, 4.75 Cr, 1.2 W, 1.4 Mo, rem Fe. Alloy 2B: 0.75 C, 0.5 Mn, 0.65 Si, 4 Cr, 1 V, 1.2 W, 8 Mo, rem Fe. Alloy 3A: 3 C, 1 Si, 28 Cr, 4 Mo, rem Fe.

Fig. 3 Variation in life of identical blades used for shearing nonferrous metals

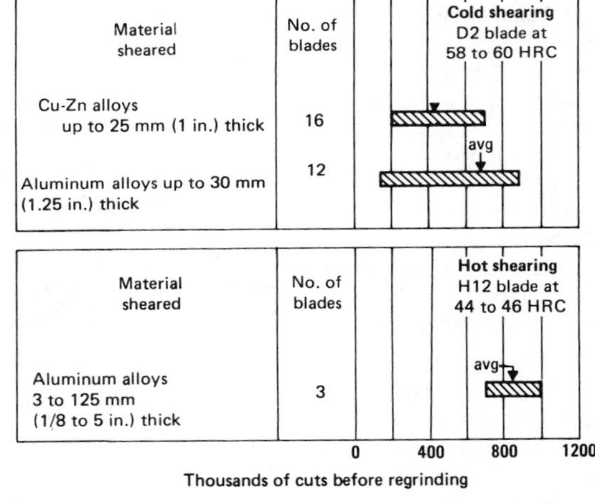

Tools for Slitting Very Thin Metallic and Nonmetallic Materials

By John W. Beckhard
President
Bach & Company

Selection of tool materials for shearing and slitting papers, films, foils and textiles is complicated by the vast array of materials to be cut and by the equally numerous types of tools and methods employed for such operations.

Tool Descriptions

It is common to use the term *machine knives* to identify those tools that are most commonly used to slit, score, shear, trim, perforate or otherwise cut very thin stock. There are two main categories of machine knives—circular knives and straight knives—each of which in turn can be classified according to knife design and application.

Circular Knives. *Score cutters* exert pressure on the material to be slit by crushing it against a surface (usually a roll or sleeve on a shaft) that is harder than the cutter. Such knives usually have a double bevel at the cutting edge.

Shear cutters, which utilize the scissor-cutting principle, comprise pairs of opposing knives through which the material to be slit is passed.

Burst cutters either penetrate the material with a progressive motion (as in core cutting) or pass through one or more layers of material without contacting another tool surface. In burst cutting, the knife may be used in conjunction with a grooved roll, or it may be entirely free-cutting in a manner similar to razor-blade slitting.

Single knives, slicers and *disc knives* are used for cutting rolls of textile, foam or film in a progressive fashion without unwinding (and rewinding) the roll.

Straight Knives. *Trimmer knives* and *guillotine knives* are used for cutting one or more layers of material by virtue of a bevel that exerts a downward force through the material.

Sheeter knives and *cutoff knives* are used in pairs for cutting roll stock into sheet or for cutting sheet to length. One or both of the knives in a pair can be mounted on a rotating drum or reciprocating holder. The lower (bed) knife frequently is stationary.

Specialty Knives. There are many other applications involving cutting of thin materials for which there is a wide variety of specially designed knives that do not fit into any of the foregoing categories. The selection of materials for specialty knives is not included in this article.

Materials for Machine Knives

Unlike materials for metal-slitting applications, materials for knives used in cutting papers, films and foils usually are selected on the basis of cost and wear resistance, without consideration of toughness. This allows wider latitude in selection of tool materials, which often results in better performance with fewer compromises. The final choice must be influenced by specific production requirements, maintenance considerations and the projected life of the tool being used. It is general practice in this field to standardize on one or a few tool materials for a given application or type of tool to take advantage of the economy of mass production or the ready availability of raw material in the required form.

Score cutters are most commonly made of 52100 steel, because this material is widely available in bar form and is well suited to mass production and easy maintenance by the user. Score cutters are hardened to 60 to 62 HRC, a hardness level that is adequate for most operations and that will not result in scoring of an opposing platen sleeve or hardened roll. Such sleeves and rolls are made of 52100 or carburized steel, hardened to not less than 60 HRC. Alternative materials for score cutters are O7 and D2 tool steels, both of which are available as bar stock. These steels are heat treated to 60 HRC or higher, the same hardness as that specified for 52100 steel cutters.

Score cutters generally should be made of bar stock, not of sheet. The more uniform directional properties of bar stock yield more consistent service life. The angle and radius of the double bevel are important determinants of performance, as is the pressure with which the knife cuts against the platen sleeve. Group M and group T high speed steels are not recommended: score cutting cannot make full use of their high hardness and resistance to softening at elevated temperatures, and their higher cost is not justified. In addition, brittle, high-hardness tool materials should not be used for score cutters because they will give unsatisfactory service.

Finally, because most score cutters are dry ground by the user, materials must be selected from only those steels that will retain their hardness under these conditions. A properly adjusted score cutter will slice about 350 km (1-million feet) of paper between sharpenings.

Shear cutters can be produced from a much wider array of alloys, the choice being influenced by such factors as tool design, material to be cut, machine design, and maintenance limitations. The entire range of tool steels (including the popular 52100 and other low-alloy types plus the higher-alloy group D, group M and group T steels), as well as specialty tool steels and cemented carbides, can be considered. For standard applications, 52100 and O1 are selected most often, chiefly because they are readily available as sheet, bar stock or tubing. Most types of paper can be cut efficiently and economically with these tool steels. Cutting of coated and impregnated papers sometimes results in premature wear of low-alloy tool steel cutters, in which case D2 is a suitable alternative. Films and foils—whether plain, laminated, coated or metallized—usually can be cut more efficiently with knives made from tool steels of higher alloy content, such as group D or group M tool steels, or specialty tool steels such as CPM 10V, hardened to 62 to 64 HRC.

In many instances, other considerations also have strong influences on selection of materials for cutting tools and on tool performance, including the design of the knife or machine, the surface finish of the tool, and other limitations imposed by maintenance or production practices. High-alloy tools are necessarily more sensitive to misuse and require greater skills in setup, operation, and maintenance.

Certain films, such as polyester coated with iron oxide or chromium dioxide (magnetic tape), are particularly destructive to knife blades. For these applications, expense becomes secondary to performance, and group M and group T high speed steels (including the cobalt-containing types M42 and T15) are used to obtain maximum hardness and wear resistance. High speed steels are preferred because they seem better able to preserve the high degrees of accuracy and surface finish required for such knives. CPM 10V also is well-suited for these applications, offering a higher degree of wear resistance than most of the alternative materials.

Burst cutters are, by design and function, fairly thin knives. Material selection thus is restricted to alloys that are readily available in thin sheet, a group that includes 52100 steel, 1075 steel and razor-blade stock. Except in those applications where resistance to elevated temperature is required (as in core cutting with a stationary blade), use of high speed steels is rarely economical.

Single-knife cutters are used to reduce wide rolls of paper, foam or textile to narrower rolls without unwinding and rewinding. These knives have long, thin, one-sided or two-sided bevels, are kept sharp by means of one or more grinding wheels situated on the machine, and are activated either automatically (for grinding at specific intervals) or manually (for grinding as required).

Under normal circumstances, such knives are made from L2, L6 or D2 tool steel. Certain applications of single knives, including slicing of foam or impregnated fabrics, cause the thin bevel to heat up considerably as it penetrates the material being slit. In these cases, an alloy that has higher resistance to elevated temperatures, such as M2 or T1 tool steel, may be required. The choice may be restricted by availability of tool material or by production limitations. For instance, some single-blade slicer knives are 750 mm (30 in.) or more in diameter and would require heat treating facilities not generally available. Slicer knives are also particularly sensitive to flutter and straightness problems, so that selection of tool steel must be based on the ability to keep these factors within acceptable bounds.

Group D, group M and group T tool steels should be selected for slicer knives only where it is possible to use grinding wheels specifically intended for grinding materials other than standard group L tool steels. Large-diameter D2 knives, for instance, respond well to on-machine grinding with cubic boron nitride wheels, but these wheels represent a considerable investment and must be set accurately to achieve optimum performance from both wheel and knife.

Recent developments in circular-knife design include (*a*) coating or impregnation of low-alloy or carbon steels with wear-resistant materials and (*b*) use of laminated steels produced by cladding low-cost cores on one or both sides, depending on bevel configuration, with high-alloy, wear-resistant alloys such as group D and group M tool steels. Knives tipped with cemented carbides have become popular for certain types of service, especially in shear slitting of coated films. The advantages of extremely long life frequently outweigh the expense and difficulties encountered in proper setup and maintenance of such tools.

Straight Knives. The choice of material for a straight knife is governed by several factors, including the cross section and length of the tool, production and maintenance facilities, and the cost of the knife itself.

A standard paper-trimmer knife consists of a carbon steel backing and an insert that provides the actual cutting edge. The backing, or carrier, remains soft after heat treatment. This allows the manufacturer or user of the knife to drill the necessary mounting holes without having to compensate for distortion during heat treatment, and gives the knife a certain amount of shock resistance. The insert material can range from O1 to M2 to one of the group T high speed steels. For most applications, O1 is ideal with respect to initial cost, ease of maintenance, and adequate performance between resharpenings. For cutting certain materials, such as coated papers, films and foils, use of a more highly alloyed insert frequently can be justified in terms of quality of cut and service life, even though it may double the original cost of the knife. Trimmer knives also are available in solid steel, in which case the material is selectively hardened or induction hardened to produce a zone in which hardness diminishes from the cutting edge to the back of the blade. Care must be taken to ensure that a solid knife has uniform hardness along the entire length of the cutting edge and that the hardened zone is sufficiently deep to allow for numerous resharpenings. Such knives are considerably cheaper than insert-type knives, but are also necessarily limited to steels that can be locally hardened.

Sheeter and cutoff knives are similar in function to trimmer knives and frequently are made with cutting-edge inserts of M2 high speed steel hardened to 62 to 64 HRC. Under optimum conditions, life of these blades is very long—50 to 100 million cuts between resharpenings.

Some sheeter or cutoff blade designs are too narrow in cross section to allow for use of inserts; such knives are most often made of solid 52100, O1 or D2 tool steel.

Selection of Materials for Blanking and Piercing Dies

Edited by Bruce Wright
Research Engineer
Buick Motor Division
General Motors Corporation

BLANKING AND PIERCING DIES, as discussed in this article, include the punches, dies and related components used to blank, pierce and shape metallic and nonmetallic sheet and plate in a stamping press. The primary measure of the performance of die materials in blanking and piercing service is the number of acceptable parts produced.

Sectional views of the blanking dies and the blanking and piercing punches used for making simple parts are shown in Fig. 1. More complex parts require notching and compound dies.

A common indication of tool deterioration is production of a burr along the sheared edge of the workpiece. When tools are new, there is minimum clearance between punch and die, and cutting edges are sharp. Under these conditions, the break in the stock starts at the underside (the side not in contact with the punch), because there the stock is subjected to the greatest tensile stress from stretching of the outer fiber. As more and more parts are produced, the cutting edges of the punch and die become rounded by wear and the stress distribution in the stock is changed. Stress on the underside is reduced, breaking at that point is delayed, and deformation accompanied by work hardening occurs. When breaking starts, it nucleates from both sides simultaneously, and a burr develops on both the slug and the surrounding area of the sheet from which it was cut. The height of this burr increases with increasing tool wear. Acceptable burr height varies with the application but usually is between 0.025 and 0.125 mm (0.001 and 0.005 in.).

Materials for Specific Tools

The compositions and properties of the common tool materials referred to in this article are presented elsewhere in this volume.

Fig. 1 Sectional views of typical tools used for blanking and piercing simple shapes

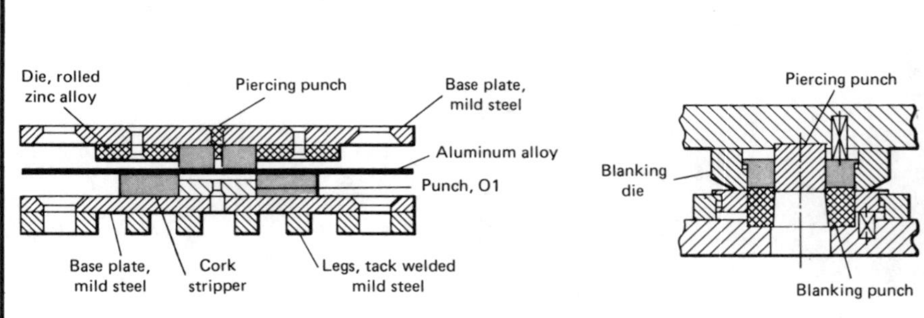

Tools at left are for short-run production of parts similar to parts 1 and 2 in Fig. 2 from relatively thin-gage sheet metal; tools at right are for longer runs.

Table 1 Typical punch and die materials for blanking 1.3-mm (0.050-in.) sheet

For sketches of typical parts, see Fig. 2

Work material	Tool material for production quantity of: 1000	10 000	100 000	1 000 000	10 000 000
Part 1 and similar 75-mm (3-in.) parts					
Aluminum, copper and magnesium alloys	Zn (a), O1, A2	O1, A2	O1, A2	D2, CPM 10V	Carbide
Carbon and alloy steel, up to 0.70% C, and ferritic stainless steel	O1, A2	O1, A2	O1, A2	D2, CPM 10V	Carbide
Stainless steel, austenitic, all tempers	O1, A2	O1, A2	A2, D2	D4, CPM 10V	Carbide
Spring steel, hardened, 52 HRC max	A2	A2, D2	D2	D4, CPM 10V	Carbide
Electrical sheet, transformer grade, 0.6 mm (0.025 in.)	A2	A2, D2	A2, D2	D4, CPM 10V	Carbide
Paper, gaskets, and similar soft materials	W1 (b)	W1 (b)	W1 (c), A2 (d)	W1 (d), A2 (d)	D2, CPM 10V
Plastic sheet, not reinforced	O1	O1	O1, A2	D2, CPM 10V	Carbide
Plastic sheet, reinforced	O1 (e), A2	A2 (f)	A2 (f)	D2 (f), CPM 10V	Carbide
Part 2 and similar 300-mm (12-in.) parts					
Aluminum, copper and magnesium alloys	Zn (a), 4140 (g)	4140 (h), A2	A2	A2, D2, CPM 10V	Carbide
Carbon and alloy steel, up to 0.70% C, and stainless steels up to quarter hard	4140 (h), A2	4140 (h), A2	A2	A2, D2, CPM 10V	Carbide
Stainless steel, austenitic, over quarter hard	A2	A2, D2	D2	D2, D4, CPM 10V	Carbide
Spring steel, hardened, 52 HRC max	A2	A2, D2	D2	D2, D4, CPM 10V	Carbide
Electrical sheet, transformer grade, 0.6 mm (0.025 in.)	A2	A2, D2	A2, D2	D2, D4, CPM 10V	Carbide
Paper, gaskets, and similar soft materials	4140 (j)	4140 (j)	A2	A2	D2, CPM 10V
Plastic sheet, not reinforced	4140 (j)	4140 (h), A2	A2	D2, CPM 10V	Carbide
Plastic sheet, reinforced	A2 (e)	A2 (e)	D2 (e)	D2 (e), CPM 10V	Carbide
Part 3 and similar 75-mm (3-in.) parts					
Aluminum, copper and magnesium alloys	O1, A2	O1, A2	O1, A2	A2, D2, CPM 10V	Carbide
Carbon and alloy steel, up to 0.70% C, and ferritic stainless steel	O1, A2	O1, A2	O1, A2	A2, D2, CPM 10V	Carbide
Stainless steel, austenitic, all tempers	A2	A2, D2	A2, D2	D2, D4, CPM 10V	Carbide
Spring steel, hardened, 52 HRC max	A2	A2, D2	D2, D4	D2, D4, CPM 10V	Carbide
Electrical sheet, transformer grade, 0.6 mm (0.025 in.)	A2	A2, D2	D2, D4	D2, D4, CPM 10V	Carbide
Paper, gaskets and other soft materials	W1 (b)	W1 (b)	W1 (k), A2	W1 (k), A2	D2, CPM 10V
Plastic sheet, not reinforced	O1	O1	A2	A2, D2, CPM 10V	Carbide
Plastic sheet, reinforced	O1 (m)	A2 (f)	A2 (f)	D2 (f), CPM 10V	Carbide
Part 4 and similar 300-mm (12-in.) parts					
Aluminum, copper and magnesium alloys	A2	A2	A2, D2	A2, D2, CPM 10V	Carbide
Carbon and alloy steel, up to 0.70% C, and ferritic stainless steel	A2	A2	A2, D2	A2, D2, CPM 10V	Carbide
Stainless steel, austenitic, up to quarter hard	A2	A2	A2, D2	D2, D4, CPM 10V	Carbide
Stainless steel, austenitic, over quarter hard	A2	D2	D2	D2, D4, CPM 10V	Carbide
Spring steel, hardened, 52 HRC max	A2	A2, D2	D2	D2, D4, CPM 10V	Carbide
Electrical sheet, transformer grade, 0.6 mm (0.025 in.)	A2	A2, D2	D2	D2, D4, CPM 10V	Carbide
Paper, gaskets, and other soft materials	W1 (b)	W1 (b)	W1 (n)	W1, A2	D2, CPM 10V
Plastic sheet, not reinforced	A2	A2	A2	A2, D2, CPM 10V	Carbide
Plastic sheet, reinforced	A2 (f)	A2 (f)	D2 (f)	D2 (f), CPM 10V	Carbide

Note: Although carbide is recommended in this table only for 10 million pieces, it should usually be considered also for runs of 1 to 10 million pieces. (a) Zn refers to a die made of zinc alloy plate and a punch of hardened tool steel. (b) For punching up to 10 000 parts, the W1 punch and die would be left soft and the punch peened to compensate for wear if necessary. (c) For punching 10 000 to 1 000 000 pieces, the W1 punch can be soft so that it can be peened to compensate for wear, or it can be hardened and ground to size. (d) Of the two alternatives listed, A2 tool steel is preferred if compound tooling is to be used for quantities of 10 000 to 1 000 000. (e) This O1 punch may have to be cyanided 0.1 to 0.2 mm (0.004 to 0.008 in.) deep to make even 1000 pieces. (f) For the application indicated, the punch and die should be gas nitrided 12 h at 540 to 565 °C (1000 to 1050 °F). (g) Soft. (h) Working edges are flame hardened in this application. (j) May be soft or flame hardened. (k) For punching 10 000 to 1 000 000 pieces, the punch would be W1, left soft so that it can be peened to compensate for wear, and the die would be O1, hardened. (m) Cyaniding of the punch is advisable, even for 1000 pieces. (n) For punching 10 000 to 1 000 000 pieces, the W1 die would be hardened and the W1 punch would be soft, so that it can be peened to compensate for wear.

Punches and Dies. Typical materials for punches and dies used for blanking parts of different sizes and degrees of severity from several different work materials about 1.3 mm (0.050 in.) thick, in various quantities, are given in Table 1. (Sketches of typical parts are presented in Fig. 2.) Typical materials for the punches and dies used to shave several work materials of this same thickness in various quantities are given in Table 2.

Tables 1 and 2 may be used to select punch and die materials for parts made of sheet thicker or thinner than the 1.3 mm used in the examples. For sheet of greater thickness, use the punch and die material recommended for the next

Fig. 2 Typical parts of varying severity that are commonly produced by blanking and piercing

Dimensions are in inches; to find equivalent metric values (mm), multiply listed values by 25. Parts 1 and 2 are relatively simple parts, and require dies similar to those illustrated in Fig. 1. Parts 3 and 4 are more complex, and require notching and compound or progressive dies.

Fig. 3 Typical hardnesses for tool steel perforator punches

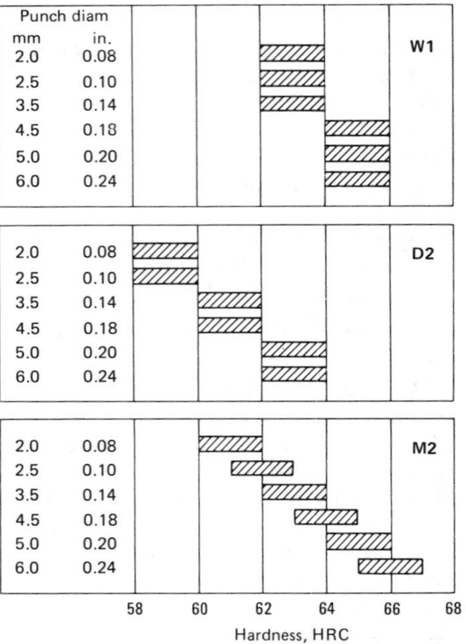

Regardless of material, punches should be tempered back to 56 to 60 HRC if they are to be subjected to heavy shock or used to pierce thick material.

Table 2 Typical punch and die materials for shaving 1.3-mm (0.050-in.) sheet

Work material	Tool material for production quantity of: 1000	10 000	100 000	1 000 000
Aluminum, copper and magnesium alloys	O1 (a)	A2	A2	D4 (b), CPM 10V
Carbon and alloy steel, up to 0.30% C, and ferritic stainless steel	A2	A2	D2	D4 (b), CPM 10V
Carbon and alloy steel, 0.30 to 0.70% C	A2	D2	D2	D4 (b), CPM 10V
Stainless steel, austenitic, up to quarter hard	A2	D2	D4 (b)	D4 (b), CPM 10V
Stainless steel, austenitic, over quarter hard, and spring steel hardened to 52 HRC max	A2	D2	D4 (b)	M2 (b), CPM 10V

(a) Type O2 is preferred for dies that must be made by broaching. (b) On frail or intricate sections, D2 should be used in preference to D4 or M2. Carbide shaving punches may also be practical for this quantity.

greater production quantity than the quantity actually to be made (the column to the right of the actual production quantity in the table). Similarly, for sheet of lesser thickness, use the punch and die material recommended for the next lower production quantity (shift one column to the left of the actual production quantity).

Typical materials for perforator punches used on several different work materials are given in Table 3. The usual limiting slenderness ratio (punch diameter to sheet thickness) for piercing aluminum, brass and steel is 2.5:1 for unguided punches and 1:1 for guided punches. The limiting slenderness ratio for piercing spring steel and stainless steel ranges from 3:1 to 1.5:1 for unguided punches and from 1:1 to 0.5:1 for accurately guided punches. Typical hardnesses for these perforator punches are given in Fig. 3.

Table 4 gives typical materials for perforator bushings of all three types (punch holder, guide or stripper, and perforator or die). These recommendations are particularly applicable to precision bushings—for instance, where the outside diameter is ground to a tolerance of −0, +0.008 mm (−0, +0.0003 in.) and is concentric with the inside diameter within 0.005 mm (0.0002 in.) TIR. The hardness of W1 bushings should be 62 to 64 HRC, and that of D2 bushings, 61 to 63 HRC.

Die plates and die parts that hold inserts normally are made of gray iron, alloy steel or tool steel. For stamping thick sheet or hard materials, either class 50 gray iron, or 4140 steel heat treated to a hardness of 30 to 40 HRC, should be used. For long-run die plates for stamping thick or hard materials, steels such as 4340 and H11 are preferred when inserts are pressed into the die plates, and 4340 is nearly always

Table 3 Typical materials for perforator punches

Work material	Punch material for production quantity of: 10 000	100 000	1 000 000
Punch diameters up to 6.4 mm (1/4 in.)			
Aluminum, brass, carbon steel, paper and plastics	M2	M2, CPM 10V	M2, CPM 10V
Spring steel, stainless steel, electrical sheet and reinforced plastics	M2	M2, CPM 10V	M2, CPM 10V
Punch diameters over 6.4 mm (1/4 in.)			
Aluminum, brass, carbon steel, paper and plastics	W1	W1	D2, CPM 10V
Spring steel, stainless steel, electrical sheet and reinforced plastics	M2	M2, CPM 10V	M2, CPM 10V

Table 4 Typical materials for perforator bushings

Work material	Bushing material for production quantity of: 10 000	100 000	1 000 000
Aluminum, brass, carbon steel, paper and plastics	W1 (a)	W1 (a)	D2
Spring steel, stainless steel, electrical sheet and reinforced plastics	D2	D2	D2 or carbide

(a) When bushings are of a shape that cannot be ground after hardening, an oil-hardening or air-hardening steel is recommended to minimize distortion.

Fig. 4 Relative life of three steel dies and one carbide die

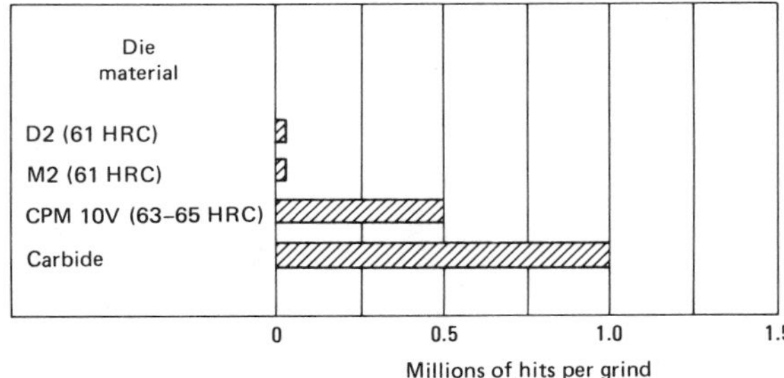

Die life was obtained under the same operating conditions, blanking 3.25% Si electrical steel sheet 0.4 mm (0.014 in.) thick. Dies were reground when they had worn sufficiently to produce a burr 0.13 mm (0.005 in.) high.

used when inserts are screwed in. Die plates for stamping thin or soft sheet may be made of class 25 or class 30 gray iron or of mild steel.

Secondary Tooling. Punch holders and die shoes for carbide dies are made of high-strength gray iron or low-carbon steel plate. Yokes for retaining carbide sections usually are made of O1 tool steel hardened to 55 to 60 HRC. Backup plates for carbide tools are preferably made of O1 hardened to 48 to 52 HRC. Strippers ordinarily can be made of low-carbon or medium-carbon steel (1020 or 1035) plate. Where a hardened plate is used for medium-production work, 4140 flame hardened, W1 conventionally hardened or W1 cyanided and oil quenched are often preferred. Hardened strippers for carbide dies and high-production D2, D4 or CPM 10V dies are made of O1 or A2, hardened to 50 to 54 HRC.

Custom-made hardened guides and locator pins usually are made of W1 or W2 for most medium- or long-run dies, or of alloy steels such as 4140 for low-cost short-run dies. Commercial guide pins often are made of SAE 1117, then carburized, hardened and finished to a surface roughness of 0.4 μm (15 μin.) rms.

Applications of Specific Materials

The following additional information is provided to assist in the selection of tool materials for blanking punches and dies.

Rolled zinc alloy, or tooling plate, is available in the form of 6.4-mm (1/4-in.) plate from the principal suppliers of zinc-base die casting alloys. Dies of this material are sheared in with a flame-hardened O1 punch, and strippers of 9.5-mm (3/8-in.) sheet cork are invariably used with them.

Tools of hot rolled low-carbon steel plate (0.10 to 0.20% C) may be used for short runs of small parts, provided these tools have been surface hardened, either by carburizing to a depth of 0.25 to 0.50 mm (0.010 to 0.020 in.) or by cyaniding to a depth of 0.1 to 0.2 mm (0.004 to 0.008 in.). Because of distortion in heat treatment, use of this material is limited to blanking of small, symmetrical shapes.

For long-run blanking of soft materials, various sizes of aircraft-quality 4140 steel plate have been used. In this application, 4140 normally is flame hardened to about 50 HRC. Flame hardening the working edge of a large die has an advantage over through hardening in that very little warping or change of size occurs. However, tools with inside or outside corners may have soft spots after flame hardening and, if so, will perform poorly.

The tool steels in Table 1 are assumed to have been hardened and tempered by conventional methods to their maximum usable hardness (58 to 61 HRC).

In addition to the tool steels listed in Table 1, type O6 has given satisfactory service in multiple-stage progressive dies, and type A10, because of its low austenitizing temperature, high dimensional accuracy and good dimensional stability, is often used to make large dies for stamping laminations.

In some applications, M2 high speed steel tools may produce smaller burrs than D2 tools (for equal numbers of parts). In addition, steel-bonded carbides and high–vanadium carbide powder metallurgy tool materials such as CPM 10V should be considered for critical applications.

Cemented tungsten carbide tooling should be considered where production life must be four or more times that possible with D4 tool steel. Partial or complete carbide inserts in tool steel dies may be considered for lower quantities, especially where close tolerances and minimum burr height are desired, or where tool life between resharpenings needs to be extended. However, brazed inserts are hazardous, and the cost of dovetailed or mechanically held inserts approaches that of complete carbide dies.

Composition and hardness of carbides frequently used in blanking and piercing dies are as follows:

Composition, %			Hardness,
W	C	Co	HRA
75.1	4.9	20.0	86
78.9	5.1	16.0	86
81.7	5.3	13.0	88
88.3	5.7	6.0	91

The first material listed above should be used where shock is appreciable. The second combines toughness and wear resistance, and is preferred for heavy-duty service such as piercing of silicon steel. Where close tolerances must be held in piercing silicon steel laminations, the third material is useful. The fourth material is best for guides and guide rolls and for applications involving very light shock.

The data in Fig. 4 show that the difference in wear life between two different tool steels at the same hardness is negligible compared with the difference between the average life of conventional tool steel dies and the life of a carbide die or a CPM 10V die.

Selection of Materials for Press Forming Dies

Edited by Bruce Wright
Research Engineer
Buick Motor Division
General Motors Corporation

PRESS FORMING is a process in which sheet metal is made to conform to the contours of a die and punch—largely by bending or moderately stretching it, or both. The suitability of a tool material for a press forming die is determined by the number of parts that can be produced using that die. This number is influenced by size and shape of the part, type and thickness of the metal being formed, lubrication practice, and the allowable variation in dimensions.

Table 1 presents nominal compositions of the tool materials most often used for press forming dies. Tool materials are usually selected on the basis of providing adequate die life at minimum cost. However, the final choice often depends on availability rather than on a small difference in die life or cost.

Die Life. Wear determines the useful performance of a press forming die. Total wear is affected primarily by the length of the production run and the severity of the forming operation. This total wear may be produced by abrasion or adhesion (galling), or both.

The amount of wear on a given die during forming is proportional to the total accumulated distance over which the sheet metal slides against the die at a given pressure between the surfaces in contact. Thin, soft or weak sheet metals exert the least pressure and thus cause the least wear; thick, moderately hard or strong metals cause the most rapid wear. The rate of wear for each combination of work metal and die metal may vary considerably depending on surface characteristics, speed of forming, and die lubrication. In situations where wrinkles form in the parts, high localized pressures are developed on the tools because of the ironing that takes place at these locations, and prohibitively high rates of abrasive wear and galling are almost always encountered.

Typical Tool Materials. Tooling for the part shown in Fig. 1 consists of a punch and a lower die. In operation, the

Fig. 1 Cross section of die used for small part of mild severity

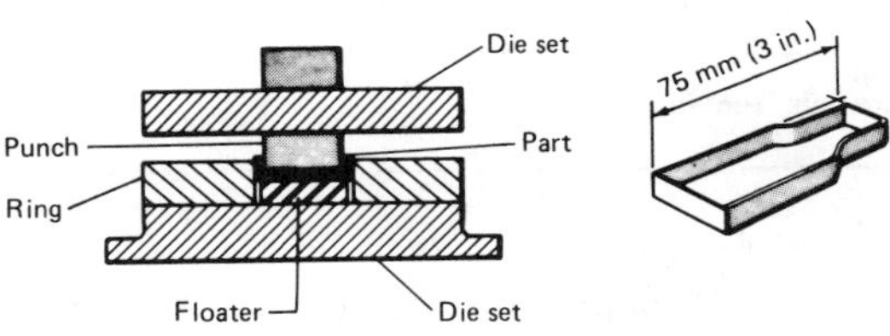

For typical die materials, see Table 2.

Table 1 Tool materials commonly used for press forming dies

Material	Nominal composition
Tool steels	
W1, water-hardening	0.60 to 1.40 C
S1, oil-hardening, shock-resisting	0.50 C; 1.50 Cr; 2.50 W
O1, oil-hardening	0.90 C; 1.00 Mn; 0.50 Cr; 0.50 Mo
A2, air-hardening, medium-alloy	1.00 C; 5.00 Cr; 1.00 Mo
A4, low-temperature air-hardening	1.00 C; 2.00 Mn; 1.00 Cr; 1.00 Mo
D2, high-carbon, high-chromium	1.50 C; 12.00 Cr; 1.00 Mo; 1.00 V
D3, high-carbon, high-chromium	2.25 C; 12.00 Cr
D5, high-carbon, high-chromium	1.50 C; 12.00 Cr; 1.00 Mo; 3.00 Co
D7, high-carbon, high-chromium	2.35 C; 12.00 Cr; 1.00 Mo; 4.00 V
M2, molybdenum high speed	0.80 C; 4.00 Cr; 5.00 Mo; 6.00 W; 2.00 V
M4, molybdenum high speed	1.30 C; 4.00 Cr; 4.50 Mo; 5.50 W; 4.00 V
Other ferrous alloys	
Hot rolled carbon steel	Steels 1010 to 1018
Unalloyed cast iron, 185 to 225 HB	3% total C; 1.6 Si; 0.7 Mn (or equivalent)
Alloy cast iron, 200 to 250 HB	3% total C; 0.7% combined C; 1.6 Si; 0.4 Cr; 0.4 Mo (or equivalent)
Cast carbon steel, 185 to 225 HB	0.75 C
Cast alloy steel, 200 to 235 HB	0.45 C; 1.10 Cr; 0.40 Mo
4140 alloy steel	0.40 C; 0.60 Mn; 0.30 Si; 1 Cr; 0.20 Mo
4140 modified (Nitralloy)	0.35 C; 1.20 Cr; 0.20 Mo; 1.00 Al
Nonferrous alloys	
Zinc alloy	4 Al; 3 Cu; 0.03 Mg; rem Zn
Aluminum bronze, 270 to 300 HB	13 Al; 4 Fe; rem Cu
Plastics	
Polyester-glass	50% polyester plastic, 50% glass in the form of cloth, strand or chopped fibers
Epoxy-glass	50% epoxy plastic, 50% glass as above
Polyester-metal	Polyester plastic reinforced with metal powder
Epoxy-metal	Epoxy plastic reinforced with metal powder
Nylon-metal	Polyamide plastic reinforced with metal powder
Polyester or epoxy-glass-metal	Polyester or epoxy plastic with both glass and metal as above

Fig. 2 Cross section of die used for large part of mild severity

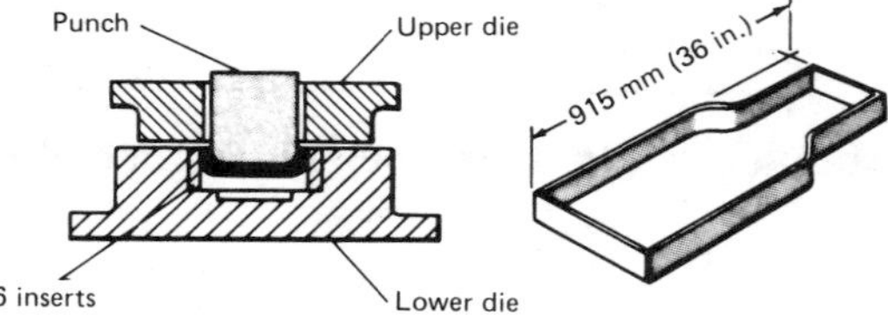

For typical die materials, see Table 3.

punch pushes the blank through the lower die, which causes wear of the lower die. The metal closely envelops the punch, with little sliding, and in that event a punch generally produces about ten times as many parts as a lower die made of the same material. However, at areas where the part shrinks against the punch during forming, wear (and possibly galling) of the punch surface will occur, particularly when the forming is done in single-action dies. For a small die and punch, the cost of steel is of minor importance, and type D2 tool steel may be used for production quantities as low as 10 000. If galling occurs during preproduction trials, the tool can be nitrided. Typical materials for lower dies used in press forming small parts similar to that shown in Fig. 1 are given in Table 2.

Tooling for the part in Fig. 2 consists of a punch, an upper die and a lower die. Without the upper die, excessive wrinkling would occur at the shrink flanges. As for the part shown in Fig. 1, a less wear-resistant material is required for the punch and upper die than for the lower die. Under conditions for which the tooling is typically made of tool steel (see Table 3), the tooling is in the form of inserts in a lower die made of cast iron, as indicated in Fig. 2, and the punch is made of a cast tool steel such as D2. For example, a cast iron die with A2 or D2 inserts at points of greatest wear is typical for production quantities of 10 000 to 100 000 pieces. When this part must be held to close tolerances over lengthy production runs, type D2 tool steel inserts should be used at all surfaces subject to wear.

Typical lower-die materials for press forming large parts similar to that shown in Fig. 2 are given in Table 3. For quantities less than 100 000 pieces, the entire lower die is typically made of the material indicated in the selection table, without inserts. The punch is made of a less wear-resistant material, which usually is the same as the lower-die material in the first column to the left of the quantity being considered.

Tables 2 and 3 may be used to select lower-die materials for parts made of sheet thicker or thinner than the 1.3 mm used for the examples, or for parts of greater or lesser severity than those shown in Fig. 1 and 2. For parts of greater severity or sheet of greater thickness, use the die material recommended for the next greater production quantity than the quantity actually to be made (the column to the right of the actual production quantity in the table). Similarly, for parts of lesser severity or sheet of lesser thickness, use the die material recommended for the next lower production quantity (shift to the next column to the left of the actual production quantity).

Selection for Galling Resistance. As indicated previously, galling, which is cold welding of the metal being

Fig. 3 Data on die wear and die life

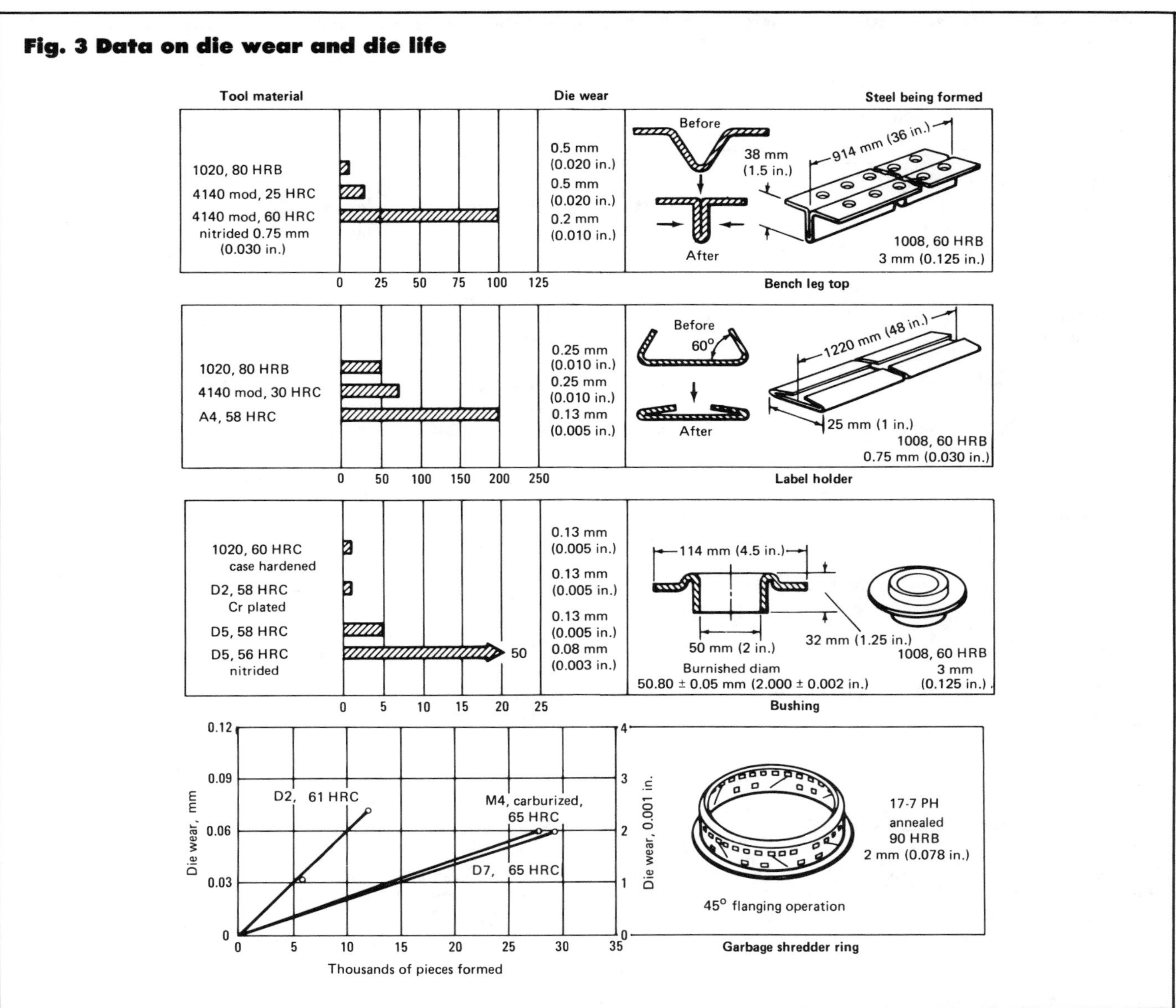

formed to that in the dies, drastically reduces the number of parts that can be made using a particular set of dies. Galling is caused by attempts to stretch sheet metal beyond practical limits, by inadequate lubrication, by poor tool fitting, and by rough finishes on tool surfaces. Therefore, when galling is encountered, the tool fit and the thickness of the metal being formed should be checked first to determine whether clearance is adequate. If clearance is considered adequate, lubrication practice should be reviewed before considering a change in die materials. Galling is less likely if the die materials and the metal being formed are dissimilar in hardness, chemical composition and/or surface characteristics. For instance, effective combinations are: (*a*) aluminum bronze tools for forming carbon steel and stainless steel, (*b*) tool steel tools for forming aluminum and copper alloys, and (*c*) carbide tools for forming carbon steel, stainless steel and aluminum.

Aluminum bronzes have excellent resistance to galling and are desirable for dies in applications where best finish is required on carbon steel or stainless steel parts. However, for medium-to-high production (10 000 to 100 000 parts), the use of inserts permits easy reconditioning of worn tools.

Nitriding minimizes or prevents galling of dies made of alloy steels or alloy tool steels (such as A2 or D2) that contain chromium and molybdenum. However, the nitrided surfaces may spall off at radii smaller than about 3 mm (1/8 in.), especially from dies having very complex contours.

Hard chromium plating usually eliminates galling of mild steel, alloy steel and tool steel dies, and it is often used for severe duty. For operations involving high local pressures, hardened alloy steels or tool steels will be less likely to yield plastically and to cause cracking of the hard chromium plating. With dies for complex parts, hard chromium plating may spall off at radii smaller than about 6 mm (1/4 in.).

For some press forming operations,

Table 2 Typical lower-die materials for forming a small part of mild severity from 1.3-mm (0.050-in.) sheet

For die cross section and part shape, see Fig. 1

Metal being formed	Quality requirements: Finish	Tolerance, mm	Tolerance, in.	Lubrication(b)	Lower-die materials(a) for total production quantity of: 100	1 000	10 000	100 000	1 000 000
1100 aluminum, brass, copper(c)	None	None	None	Yes	Epoxy-metal, mild steel	Polyester-metal, mild and 4140 steel	Polyester-glass(d), mild and 4140 steel	O1, 4140	A2, D2
1100 aluminum, brass, copper(c)	None	±0.1	±0.005	Yes	Epoxy-metal, mild and 4140 steel	Polyester-metal, mild and 4140 steel	Polyester-glass(d), mild and 4140 steel	4140, O1, A2, D2	A2, D2
1100 aluminum, brass, copper(c)	Best	±0.1	±0.005	Yes	Epoxy-metal, mild steel	Polyester-metal, mild and 4140 steel	Polyester-glass(d), mild and 4140 steel	4140, O1, A2	A2, D2
Magnesium or titanium(e)	Best	±0.1	±0.005	Yes	Mild steel	Mild and 4140 steel	A2	A2	A2, D2
Low-carbon steel, to 1/4 hard	None	None	None	Yes	Mild and 4140 steel	Mild and 4140 steel	4140, mild steel chromium plated, D2	A2	D2
Type 300 stainless, to 1/4 hard	None	None	None	Yes	Mild and 4140 steel	Mild and 4140 steel	Mild and 4140 steel	A2, D2	D2
Low-carbon steel	Best	±0.1	±0.005	Yes	Mild and 4140 steel	Mild and 4140 steel	Mild and 4140 steel	A2, D2, nitrided D2	D2, nitrided D2
High-strength aluminum or copper alloys	Best	±0.1	±0.005	No(f)	Mild and 4140 steel	Mild and 4140 steel	Mild steel chromium plated and 4140	Cr plated O1; A2	D2, nitrided D2
Type 300 stainless, to 1/4 hard	None	±0.1	±0.005	Yes	Mild and 4140 steel	Mild and 4140 steel	Mild steel and 4140	Cr plated O1; A2	D2
Type 300 stainless, to 1/4 hard	Best	±0.1	±0.005	Yes	Mild and 4140 steel	Mild and 4140 steel	Mild steel chromium plated, D2	D2, nitrided D2	D2, nitrided D2
Heat-resisting alloys	Best	±0.1	±0.005	Yes	Mild and 4140 steel	Mild and 4140 steel	Mild steel chromium plated, D2	D2, nitrided D2	D2, nitrided D2
Low-carbon steel	Good	±0.1	±0.005	No(f)	Mild and 4140 steel	Mild and 4140 steel	Mild steel chromium plated	D2, nitrided D2	D2, nitrided D2

(a) Description of die materials is given in Table 1. When more than one material for the same conditions of tooling is given, the materials are listed in order of increasing cost; however, final choice often depends on availability rather than on small differences in cost or performance. Where mild steel is recommended for forming fewer than 10 000 pieces, the dies are not heat treated. For forming 10 000 pieces and more, such dies should be carburized and hardened. Where 4140 is recommended for fewer than 10 000 pieces, it should be pretreated to a hardness of 28 to 32 HRC. Flame hardening of high wear areas is recommended for quantities greater than 10 000 pieces. (b) Specially applied lubrication, rather than mill oil. (c) Soft. (d) With inserts. (e) Heated sheet. (f) Use lubrication to make 1 to 100 parts.

dies made from tool steels other than those discussed above may be desirable. For example, shock-resisting tool steels such as S1, S5 and S7 may be used for die components subjected to severe impact in service. H11 and H13, possibly nitrided for greater wear resistance, also may be used for such components. In press forming operations requiring significantly greater wear life than is routinely attained with D2 or nitrided D2, it may be necessary to specify a more wear-resistant cold work tool steel such as A7, D3, D4 or D7, or a high speed steel such as M2, M4 or T15. The example in Fig. 3 that pertains to flanging of 17-7 PH stainless steel illustrates the improvement in die life that can be expected by changing from D2 to the more highly alloyed D7 and M4 steels. Cost generally determines whether or not it is desirable to change to an alternative material, although toughness may also be a determining factor. Costs to be considered include not only material costs but also tool fabrication costs and the cost of periodic resharpening.

Significant advances have been made in recent years in the area of high-alloy wear-resistant tool steels made by powder metallurgy (P/M) techniques. For example, P/M high speed steels, hot isostatically pressed to full density, offer greater ease of fabrication and significantly improved toughness in service compared to conventional ingot-cast steels of the same compositions. New grades that could not have been produced economically by conventional steelmaking practices have been introduced through the use of powder metallurgy. One such alloy is Crucible CPM 10V (2.45 C, 5.0 Cr, 9.75 V, 1.25 Mo), which is an air-hardening cold work tool steel designed specifically for tooling applications requiring long wear life and good toughness. Due to its high vanadium carbide content, CPM 10V will outwear any commercially available cold work tool steel and even the most wear-resistant high speed steel. It also can be a cost-effective alternative to carbide in applications where breaking or chipping of carbide is a problem or where the full potential of carbide is either not realized or not required.

Where maximum resistance to galling and wear is required, cemented car-

Table 3 Typical lower-die materials for forming a large part of mild severity from 1.3-mm (0.050-in.) sheet

For die cross section and part shape, see Fig. 2

Metal being formed	Quality requirements: Finish	Tolerance, mm	Tolerance, in.	Lubrication(b)	Lower-die materials(a) for total production quantity of: 100	1 000	10 000	100 000	1 000 000
1100 aluminum, brass, copper(c)	None	None	None	Yes	Epoxy-metal, polyester-metal, zinc alloy	Polyester-metal, zinc alloy	Epoxy or polyester-glass(d), zinc alloy	Alloy cast iron	Cast iron or A2(e)
1100 aluminum, brass, copper(c)	None	±0.1	±0.005	Yes	Epoxy-metal, polyester-metal, zinc alloy	Polyester-metal, zinc alloy	Alloy cast iron	Alloy cast iron	Alloy cast iron
1100 aluminum, brass, copper(c)	Best	±0.1	±0.005	Yes	Epoxy-metal, polyester-metal, zinc alloy	Polyester-metal, zinc alloy	Alloy cast iron	Alloy cast iron	Alloy cast iron, A2(e)
Magnesium or titanium(f)	Best	±0.1	±0.005	Yes	Cast iron, zinc alloy	Cast iron, zinc alloy	Cast iron	Alloy cast iron	Alloy cast iron, A2(e)
Low-carbon steel, to 1/4 hard	None	None	None	Yes	Epoxy-metal, polyester-metal, zinc alloy	Epoxy-glass, polyester-glass, zinc alloy	Epoxy or polyester-glass(d), cast iron	Alloy cast iron	
Type 300 stainless, to 1/4 hard	None	None	None	Yes	Epoxy-metal, polyester-metal, zinc alloy	Epoxy-glass, polyester-glass, zinc alloy	Epoxy or polyester-glass(d), alloy cast iron	A2(e)	D2(e)
Low-carbon steel	Best	±0.1	±0.005	Yes	Zinc alloy	Epoxy-glass, polyester-glass, zinc alloy	Alloy cast iron	D2; nitrided A2(e)	D2, nitrided D2(e)
High-strength aluminum or copper alloys	Best	±0.1	±0.005	No(g)	Zinc alloy	Polyester-glass, zinc alloy	Alloy cast iron	Alloy cast iron	Nitrided A2(e), nitrided D2(e)
Type 300 stainless, to 1/4 hard	None	±0.1	±0.005	Yes	Zinc alloy	Zinc alloy	Alloy cast iron	D2; nitrided A2(e)	D2(e), nitrided D2(e)
Type 300 stainless, to 1/4 hard	Best	±0.1	±0.005	Yes	Zinc alloy	Zinc alloy	Alloy cast iron	Nitrided D2	Nitrided D2(e)
Heat-resisting alloys	Best	±0.1	±0.005	Yes	Zinc alloy	Zinc alloy	Alloy cast iron	Nitrided D2	Nitrided D2(e)
Low-carbon steel	Good	±0.1	±0.005	No(g)	Zinc alloy	Zinc alloy	Alloy cast iron	Nitrided D2	Nitrided D2(e)

(a) Description of die materials is given in Table 1. When more than one material for the same conditions of tooling is given, the materials are listed in order of increasing cost; however, final choice often depends on availability rather than on small differences in cost or performance. Where mild steel is recommended for forming fewer than 10 000 pieces, the dies are not heat treated. For forming 10 000 pieces and more, such dies should be carburized and hardened. Where 4140 is recommended for fewer than 10 000 pieces, it should be pretreated to a hardness of 28 to 32 HRC. Flame hardening of high-wear areas is recommended for quantities greater than 10 000 pieces. (b) Specially applied lubrication, rather than mill oil. (c) Soft. (d) With inserts. (e) Use as inserts in cast iron body. (f) Heated sheet. (g) Use lubrication to make 1 to 100 parts.

bides have traditionally been recognized as the ultimate tooling materials. However, because of the high cost of these materials and their tendency to be brittle in service, carbides frequently are used only for inserts in critical die areas. These inserts usually are made of a straight grade of tungsten carbide containing about 6% cobalt binder, but higher cobalt contents can be specified to provide greater resistance to shock. The more recently developed steel-bonded carbides offer greater ease of fabrication and very often can be demonstrated to be cost-effective substitutes for the more costly cemented carbides with cobalt binder.

Selection of Materials for Deep Drawing Dies

Edited by Bruce Wright
Research Engineer
Buick Motor Division
General Motors Corporation

DEEP DRAWING is a process in which sheet metal is formed into round or square cup-shape parts by making it conform to a punch as it is drawn through a die. In conventional deep drawing, successive draws are made in the same direction. The types of dies and other tooling used for conventional deep drawing are illustrated in Fig. 1.

Occasionally, redrawn shells must have a wrinkle-free sidewall of uniform thickness, or a section in the bottom of the cup that is sharply raised, usually by forming in two operations. Such operations are difficult, impossible or uneconomical to perform by conventional single-action drawing, but are easily done by reverse redrawing. Figure 2 shows typical tooling for reverse redrawing of thin-wall shells.

For economy in manufacture, a drawn part should always be produced in the fewest steps possible. Ironing—that is, thinning the walls of the part being drawn by using a reduced clearance between punch and die—is used almost universally in multioperation deep drawing. Ironing helps produce

Table 1 Typical lubricants for deep drawing

Metal being drawn	Severity of drawing: 10% or less	25% average	50% or more
Aluminum and aluminum alloys	Straight mineral oil, 100-s viscosity (a); mineral oil with approximately 10% lard oil	Straight mineral oil, 200- to 250-s viscosity (a); mineral oil with approximately 15% lard oil	Mineral oil with EP additives—sulfur and others; coating of soap or wax dried on blanks (or shells) prior to drawing (or redrawing)
Copper and copper alloys	5% soap solution; lard and soap emulsion	10% soap solution with stearic or oleic acid; lard oil and mineral oil with stearic acid	Lard oil blended with 50% mineral oil, coating of soap or wax dried on blanks or draws prior to draw or redraw
Carbon steel	Mineral oil, 250- to 350-s viscosity (a); 5% soap solution	Emulsions of lard oil, mineral oil and sulfonated oils	Phosphate coating impregnated with dried soap or wax
Stainless steel	Castor oil and soap emulsion	Castor oil with fillers such as mica or zinc oxide	Boiled linseed oil with mica or lithopone; phosphoric acid etch with dried soap or wax film

Note: When more than one lubricant is given, they are listed in order of increasing effectiveness.
(a) Saybolt viscosity at 40 °C (100 °F).

Table 2 Sheet metals that require similar drawing-die materials

Type of sheet metal	Maximum hardness	Metals that require similar drawing-die materials
Drawing quality aluminum and copper alloys	64 HRF(a)	All aluminum and clad aluminum alloy sheet, copper and alloys, zinc and alloys, silver, pewter and Monel
Drawing quality steel	70 HRB	Carbon steel, grades 1008 to 1020
	75 HRB	Carbon steel, grades 1021 to 1030
Austenitic stainless steel	95 HRB	301, 302, 304, 305, 308, 310, 316, 317 steel; 410 and 430 carbon steel clad with stainless steel; copper clad with stainless steel; magnesium drawn at 200 to 300 °C (400 to 600 °F) with no ironing of sides; 17-4 PH, 17-7 PH and PH 15-7 Mo stainless steels

(a) Roughly equivalent to 58 HB (500-kg load) or 24 HR30T.

Fig. 1 Tooling components used in conventional drawing operations

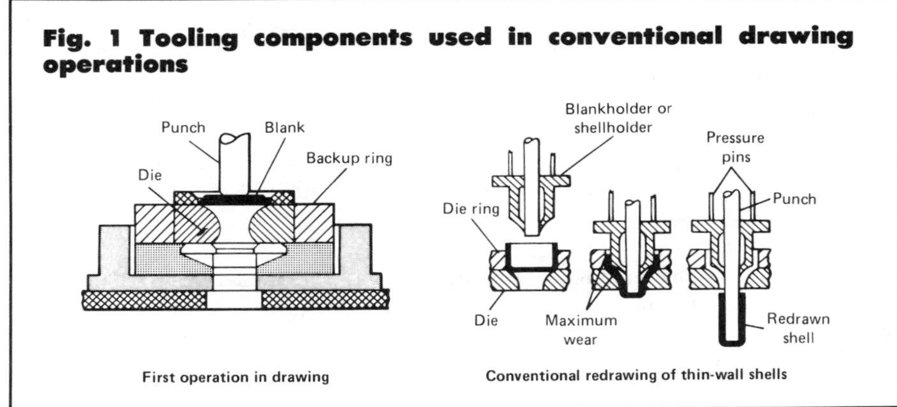

deep draws and uniform wall thickness in the fewest operations. Each operation is designed for maximum practical reduction of the metal being drawn. Accordingly, the information given in this article is predicated on use of reductions near the maximum of about 35%.

The selection of material for a drawing die is aimed at production of the desired quality and quantity of parts with the least possible tooling cost per part. In small dies—for instance, those for making parts up to 75 mm (3 in.) across—performance is the predominant consideration. Material cost is a minor factor, because the cost of even the more highly alloyed tool steels is probably less than 5% of total die cost. In dies for parts larger than about 200 mm (8 in.), material cost is more important, and in a die for a 300 mm (12 in.) part, it may amount to nearly one-half of total die cost, even when the tool consists of a tool steel insert in a flame-hardened alloy cast iron die.

Die Performance

The performance of a drawing die is determined by the total amount of wear (abrasive and adhesive) that occurs during a production run. Wear of a given die material is determined largely by its hardness, the type and thickness of the sheet metal being drawn, sharpness of die radii, lubrication, and the construction and surface finish of the die. The amount of wear on die radii can vary by a factor of 20 between the sharpest and most liberal radii. In drawing square cups, formation of wrinkles at the corners, accompanied by high localized pressures, may produce prohibitively high rates of wear.

Lubrication. Correct lubrication of the sheet metal is essential if friction, wear and galling are to be held to the lowest possible levels during deep drawing. In fact, deep drawing is impossible if the sheet metal is not lubricated. In actual practice, die materials are selected after trials employing one or more candidate production lubricants. If excessive wear or galling occurs, a better lubricant is usually applied. For extremely difficult draws, the best lubricants usually are applied at the outset.

Fig. 2 Typical tooling used in reverse redrawing of thin-wall shells

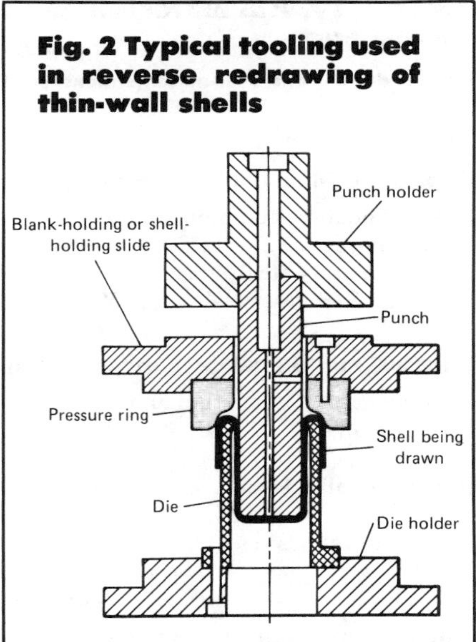

Table 1 gives typical lubricants used for different work metals and severities of drawing. Lubricants are marketed under proprietary names, but any supplier of lubricants can recommend commercial compounds fitting the descriptions in Table 1.

Table 2 lists three basic types of sheet metal, and for each basic type gives several other types of sheet metal that require the same tooling materials for drawing parts of equivalent severity. As indicated in this table, soft nonferrous alloys require die materials different from those used in drawing of steel sheet. Nevertheless, cold worked nonferrous alloys about 1/4 hard cause almost as much die wear as drawing quality steel. The rates of die wear caused by ferritic stainless steels are about the same as those caused by carbon or alloy steels of equivalent hardness.

Materials for Specific Tools

Draw Rings. Table 3 gives typical materials for draw rings (both dies and backup rings) used in drawing and ironing cups of various diameters and lengths from the three basic types of sheet metal listed in Table 2. The data

Table 3 Typical materials for draw rings used in drawing and ironing both round and square parts

For part designs and over-all dimensions, see Fig. 3

Metal to be drawn	Total number of parts to be drawn: 10 000	100 000	1 000 000
Cups up to 76 mm (3 in.) across, drawn from 1.6-mm (0.062-in.) sheet (parts 1, 2 and 3)			
Drawing quality aluminum and copper alloys	W1; O1	O1; A2	A2; D2
Drawing quality steel	W1; O1	O1; A2	A2; D2
300-series Stainless steel	W1 chromium plated; aluminum bronze	Nitrided A2; aluminum bronze	Nitrided D2 or D3; cemented carbide
Cups 305 mm (12 in.) or more across, drawn from 1.6-mm (0.062-in.) sheet (parts 4 and 5)			
Drawing quality aluminum and copper alloys	Alloy cast iron(a)	Alloy cast iron (a); A2 inserts(b)	A2 or D2 inserts(b)
Drawing quality steel	Alloy cast iron(a)	Alloy cast iron(c); A2 inserts (b)	A2 or D2 inserts(b)
300-series stainless steel	Alloy cast iron(d); aluminum bronze inserts(b)	A2 or aluminum bronze inserts(b)	Nitrided A2 or D2 inserts(b)
Square cups similar to part 6, drawn from 1.6-mm (0.062-in.) sheet			
Drawing quality aluminum and copper alloys(e)	W1	O1; A2	A2; D2
Drawing quality steel(e)	W1	O1; A2	A2; D2; nitrided A2 or D2
300-series stainless steel(f)	W1; aluminum bronze	Nitrided A2; aluminum bronze	Nitrided A2 or D2
Large pans similar to part 7; drawn from 0.8-mm (0.031-in.) sheet			
Drawing quality aluminum and copper alloys	Alloy cast iron(a)	Alloy cast iron(a); A2 corner inserts(b)	Nitrided A2 or D2 inserts(b)
Drawing quality steel	Alloy cast iron(a)	Alloy cast iron(a); A2 corner inserts(b)	Nitrided A2 or D2 inserts(b)
300-series stainless steel	Alloy cast iron(d); aluminum bronze	Nitrided A2 or aluminum bronze inserts(b)	Nitrided A2 or D2 inserts(b)

(a) Wearing surfaces flame hardened. (b) In flame hardened alloy cast iron. (c) Quenched and tempered for part 4; flame hardened for part 5. (d) Flame hardened on wearing surfaces to not over 420 HB. (e) For drawing aluminum, copper and steel, the tool material would be used as corner inserts. (f) For drawing stainless steel, inserts would be used for all wear surfaces.

Table 4 Typical materials for draw rings used in making part 4 from flat rolled steel of six different thicknesses

For part design and over-all dimensions, see Fig. 3

Thickness of steel, mm	in.	Total number of parts to be drawn: 1000	10 000	100 000	1 000 000
0.4	0.015	Alloy cast iron(a)	Alloy cast iron(a)	Alloy cast iron(a)	Alloy cast iron(b), O1, A2
0.8	0.031	Alloy cast iron(a)	Alloy cast iron(a)	Alloy cast iron(b)	A2, D2
1.6	0.062	Alloy cast iron(a)	Alloy cast iron(b)	Alloy cast iron(b), A2	A2, D2
3.2	0.125	Alloy cast iron(b)	Alloy cast iron(b)	A2, D2	D2
6.4	0.250	A2	A2	D2	D2
12.7	0.500	A2 (c)	A2 (c)	D2 (c)	D2 (c)

Note: Where tool steels are recommended, they are used as inserts in flame-hardened alloy cast iron.
(a) Flame hardening not necessary. (b) Wearing surfaces flame hardened. (c) In drawing 12.7 mm (0.500 in.) plate with A2 or D2 inserts, press speed is slower than for thinner stock and the plate is phosphate coated.

in Table 3 are given for both round and square cups drawn from stock 1.5 mm (0.062 in.) thick in three typical production quantities. Similar data for a large square cup and a large pan are given also. Design dimensions for all seven parts referred to in Table 3 are given in Fig. 3. The square parts have liberal corner radii consistent with favorable die life. Compositions and properties of the tool materials listed in Table 3, and of other tool materials mentioned in

Fig. 3 Seven typical deep drawn parts

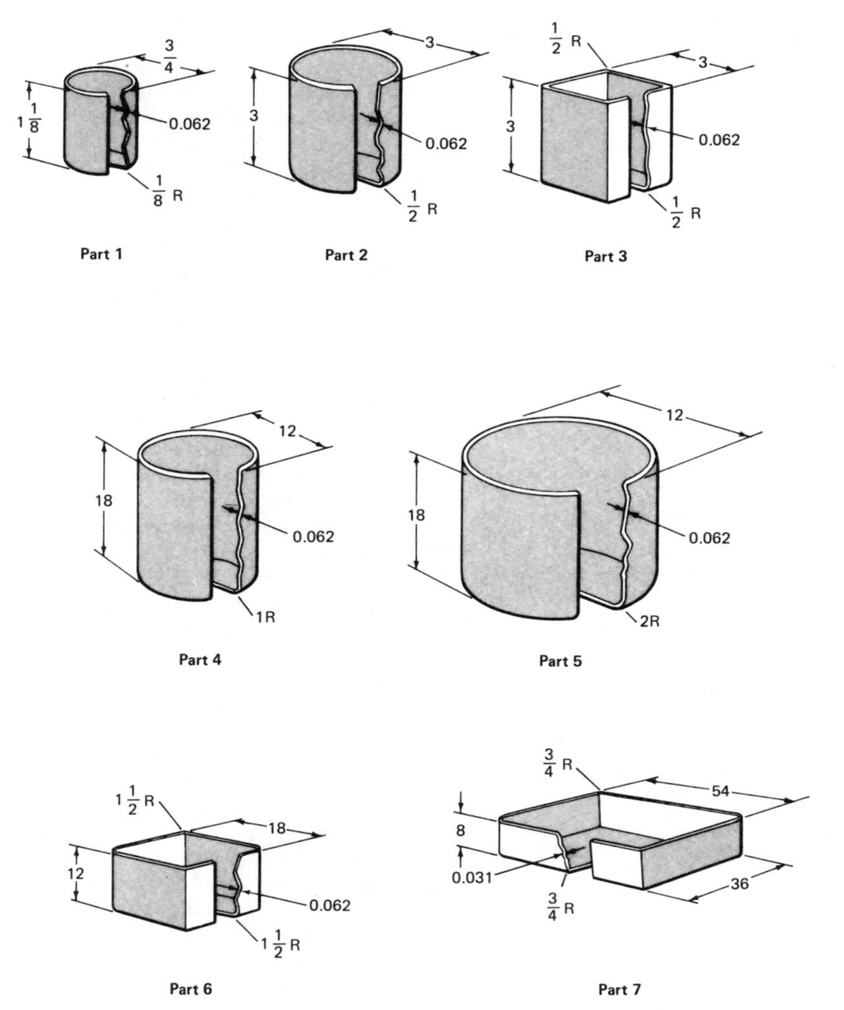

Dimensions are given in inches; to find equivalent metric dimensions (mm), multiply listed dimensions by 25. Corner radii comply with standard commercial practice. For typical deep drawing materials see Table 3.

this article, are given elsewhere in this volume. Compositions and properties of cast irons, and characteristic properties of carburized and nitrided parts, are discussed in Volume 1 of the 9th Edition of this Handbook; properties of aluminum bronzes are presented in Volume 2.

The effect on material selection of changing the thickness of the sheet metal being drawn is illustrated in Table 4. Here, it may be seen that tool materials of increasingly greater wear resistance are required as either thickness of the work metal or total quantity of parts is increased.

Punches and Blankholders. Typical materials for punches and for blankholders (pressure pads) or shellholders (pressure sleeves) are given in Table 5. The materials listed in Table 5 are for punches and blankholders used in drawing and ironing round and square steel cups similar to parts 2 through 7 in Fig. 3.

More wear-resistant materials are required not only for the tools used in drawing and ironing harder or thicker stock or for those used for longer runs, but also for tools used to achieve greater percent reductions during ironing. Table 6 lists typical tool steels used in punches and dies for short-, medium- and long-run production at four different levels of reduction in ironing.

Typical materials for punches and dies used in reverse redrawing of steel cups are given in Table 7.

Selection to Combat Specific Service Problems

Wear—most notably galling—is the most common sign of deterioration in deep drawing tools. Wear can be reduced by selection of a harder and more wear-resistant material, by application of a surface coating such as chromium plating to the finished tools, or by use of a surface treatment such as carburizing or carbonitriding. The information given in the following paragraphs is intended to supplement the basic information given in Tables 3 to 7.

Galling. The common causes of galling of deep-drawing tooling are:

- Attempts to stretch sheet metal beyond practical limits
- Poor tool fit-up, with poor alignment or insufficient die clearance for the sheet thickness
- Excessive wrinkling
- Insufficient or otherwise inadequate lubrication
- Use of tool steels that are susceptible to galling without applying a surface coating to the tools or using a lubricant of superior lubricating qualities
- Rough finishes on tool surfaces.

For short runs, dies made of carburized hot rolled steel or hardened alloy steel will, in many instances, produce parts equal in quality to those drawn over most tool steel dies. Exceptions may be encountered in ironing to severe reductions or in drawing metals that tend to gall, such as austenitic stainless steel. These exceptions may be of little consequence, however, because tool steel dies also may become galled under the same circumstances. Longest die life can be expected when die surfaces have a very fine finish, with final surface scratches parallel to the direction of drawing.

Die materials for resistance to galling can be selected on the following basis:

- For parts drawn from carbon steel or nonferrous alloy sheet, the die material may be selected without regard to galling and then, as a finishing

Table 5 Typical materials for punches and blankholders

For part designs and over-all dimensions, see Fig. 3

Die component	Total number of parts to be drawn 10 000	100 000	1 000 000
For round steel cups like part 2			
Punch(a)	Carburized 4140; W1	W1; carburized S1	A2; D2
Blankholder(b)	W1; O1	W1; O1	W1; O1
For square steel cups like part 3			
Punch(a)	Carburized 4140; W1	W1; carburized S1	A2; D2
Blankholder(b)	W1; O1	W1; O1	W1; O1
For round steel cups like parts 4 and 5			
Punch(a)	Alloy cast iron(c)	O1 (d)	A2 (c); D2 (c)
Blankholder(b)	Alloy cast iron(c)	Alloy cast iron(e)	O1; A2
For square steel cups like parts 6 and 7			
Punch(a)	Carburized 4140 (f)	W1; O1 (d)	Nitrided A2; D2 (d)
Blankholder(b)	Alloy cast iron(c)	W1; O1	O1; A2

(a) Chromium plating is optional on punches, to reduce friction between part and punch and thus facilitate removal of the part. Cast iron, however, should not be plated. (b) Also applies to shellholder, pressure pad or pressure sleeve. (c) Flame hardening not necessary. (d) The punch holder is flame hardened alloy cast iron with a nose insert of the indicated tool steel. (e) For part 4, this blankholder is quenched and tempered; for part 5, it is flame hardened. (f) The punch holder is alloy cast iron with a nose insert of the indicated steel.

Table 6 Typical tool steels for punches and dies to iron soft steel sheet at various reductions

Ironing reduction, %	Total quantity of shells(a) to be ironed 1000	10 000	100 000	1 000 000
Ironing punches(b)				
Up to 25	W1	O1	A2	A2; S1 carburized
25 to 35	W1	A2	A2; S1 carburized	D2
35 to 50	A2	A2; S1 carburized	D2	D2
Over 50	D2	D2	D2	D2
Ironing dies				
Up to 25	W1 (c)	O1	O1	D2
25 to 35(d)	W1 (c)	O1	D2	D2
35 to 50(d)	O1	D2	D2	D2
Over 50(d)	D2	D2	D2	D2

(a) Steel sheet up to 75 HRB, or softer metals. (b) All tool steel punches should be plated with chromium 0.005 to 0.010 mm (0.0002 to 0.0004 in.) thick for easier removal of the part from the punch. (c) W1 is quenched on the inside and tempered to 60 HRC min for these applications. (d) Draw rings must be inserted in shrink rings for ironing at reduction greater than 25% and for quantities of more than 10 000 parts.

operation, the punch and die should be either nitrided or chromium plated. If a tool steel such as A2, D2, D3 or D4, which contain chromium and molybdenum, has been selected, the smoothly ground tools should be nitrided, and then polished or buffed.

- For parts drawn from stainless steel or from high-nickel alloy steel, the draw-ring material with best resistance to galling is aluminum bronze. The second choice is D2, D3 or D4, smoothly ground, nitrided and polished. The third choice is alloy cast iron, quenched and tempered to 400 to 420 HB.

Chromium plating is used to extend service life of tool steel draw rings. On punches, its primary function is to reduce frictional forces and facilitate removal of parts from the punch after the sidewalls have been ironed tight to the punch. Chromium plating usually improves punch life somewhat less than would changing the punch material to the next best tool steel.

For successful tool performance, chromium plating must always be deposited on a surface harder than 50 HRC; preferably, plating thickness should be 0.005 to 0.01 mm (0.0002 to 0.0004 in.)—never less than 0.002 mm (0.0001 in.). This gives the required hardness and reduction of friction without excessive spalling or chipping at corners. Chromium-plated dies should be heated to 150 to 205 °C (300 to 400 °F) for a minimum of 3 h immediately after plating to minimize the possibility of hydrogen embrittlement.

Combined Operations. Over the past three decades, combined operations have become progressively more widely used. Among the more popular combined operations are one that combines drawing and coining, and another that combines successive or tandem drawing (or ironing) operations. This latter combination is called double drawing or double ironing. Advancements in combined operations have paralleled advancements in die materials, such as better selection of drawing steels, improvements in engineering and construction of tools, and especially in surface treatments such as those using zinc phosphate with emulsified soap.

These operations have increased production by doubling reductions and decreasing the number of operations, but at the same time have required capital investment in larger presses. In addition, tool steels of greater resistance to compression and heat have become necessary for drawing and ironing tools.

Double drawing and double ironing operations are successive operations in one tooling setup, with two dies placed in tandem so that a punch forces the cup through one die and then directly through the second die while the cup is still hot. Punches are longer than those used in conventional deep drawing and, because of their slenderness, preferably are made of S1 tool steel. Die materials are much the same as in single operations except that when temperatures are high in the second operation, selection is confined to tool steels such as A2 and D2. These more temper-resistant

Table 7 Typical punch and die material for reverse redrawing of steel

Die component	Total quantity of parts(a) to be redrawn 1000	10 000	100 000	1 000 000
Small thick-wall cups				
Die and pressure ring	O1	O1 (b)	A2 (c)	D2 (c)
Punch(d)	4140, 6150	O1, A2	D2	D3
Medium and large thin-wall cups				
Die and pressure ring	1018 (e), 4140	4140 (f), O1	A2 (c)	D2 (c)
Punch(d)	W1	A2	D2	D2, D3

(a) No specific finish or tolerance requirements. (b) Dies are polished and chromium plated. (c) A2 and D2 should be nitrided. (d) All punches used for making more than 1000 pieces should be heat treated to 60 to 62 HRC, polished and chromium plated. (e) Carburized, hardened and polished to a fine finish. (f) 4140 or 6150 may be used if carburized and highly polished.

steels can better withstand the effects of the higher temperatures developed by increased plastic deformation of the work.

Cemented Carbides. For long runs, inserts of cemented carbide are widely used in deep drawing dies. In dies up to 200 mm (8 in.) across for continuous production of over one million drawn parts, carbide in many instances has proved to be the most economical die material. Such dies have maintained size in drawing 500 000 parts with 60% reductions, and have made as many as one million parts with reductions greater than 40% when the steel to be drawn was surface treated with zinc phosphate and soap. However, cemented carbide dies do not provide satisfactory service with inferior lubricants. Also, carbide dies are not superior to dies made of a tool steel such as D2 in complex deep drawing operations (for example, those that combine drawing with coining or forming and in which the reduction in drawing is greater than 40%).

The cemented carbides most often used for deep drawing inserts are straight tungsten grades, of normal particle size, that contain about 9% cobalt. Steel-bonded carbides also are being used for deep drawing tools.

Plastics are the most economical tooling materials for short runs, especially when a prototype part is available as a pattern for lay-up of plastic-impregnated glass cloth facing, which is backed with chopped glass fibers impregnated with 50% resin. Among resins, polyester, epoxy, phenolic resin and nylon have been used. The plastic dies that exhibit longest life are those constructed so that the wearing surface is faced with glass cloth that has had most of the plastic material forced out under pressure before and during curing.

Except for very short runs, plastics should not be selected as blankholder materials where burred edges of the blank slide over the plastic surface and produce severe wear or gouging.

Zinc Alloy Tools. Because they are relatively soft, zinc alloy tools should be used only in drawing (without ironing) small quantities of large-diameter thin-wall parts. Zinc alloy tools work best for drawing well-lubricated stock into parts 300 mm (12 in.) or more in diameter under circumstances where wrinkling is not likely to occur.

Selection of Materials for Shear Spinning Tools

By James E. Denton
Manager of Materials Technology
Cincinnati Milacron, Inc.

SPINNING is a technique for chipless forming of a generally circular or conical thin-wall object from a flat or preformed blank by forcing the rotating blank against a rotating mandrel by means of a forming tool or roller. Strictly speaking, spinning includes such applications as the forming of cooking utensils and musical instruments from light-gage sheet metal, in which a mandrel may or may not be used and in which the forming tool may even be made of wood. In this discussion we will consider only applications in which power-driven rollers simultaneously form the contours of the part and reduce the thickness of the work metal by ironing the blank against the mandrel. This type of spinning is generally referred to as shear spinning. Figure 1 shows a typical tooling arrangement for shear spinning a cone from a flat disk.

Tooling for shear spinning consists of a mandrel of appropriate shape and multiple rollers (tool rings). Frequent-

Fig. 1 Typical tooling arrangement for shear spinning a cone from a flat disk

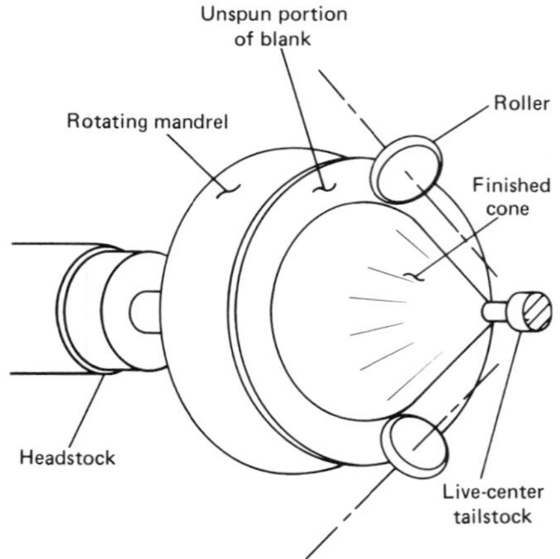

Circular metal disk is plastically deformed in a horizontal direction by rollers that move parallel to the mandrel surface and exert pressures up to 2800 MPa (400 000 psi) against the blank. Relationship of blank thickness, t, to wall thickness of finished cone, t_1, is given by the equation $t_1 = t \sin\alpha$, where α is one-half the apex angle of the cone. Maximum diameter of the cone equals diameter of blank.

Fig. 2 Three types of shear spinning rollers

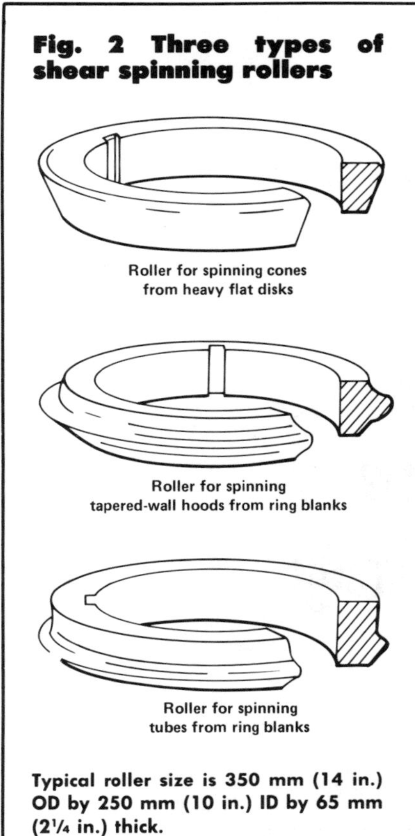

Typical roller size is 350 mm (14 in.) OD by 250 mm (10 in.) ID by 65 mm (2¼ in.) thick.

ly, two or three rollers are positioned equidistant around the machine headstock to reduce reaction forces on the headstock and on the center that holds the blank in position on the mandrel. This arrangement also reduces or eliminates bending stresses on the mandrel.

Tooling Materials

Demands on the mandrel material are not severe, but obviously it must be higher in compressive strength than the material being formed. For most applications, alloy steels such as 4140 hardened to 50 to 55 HRC, or hardened alloy ductile irons, are satisfactory mandrel materials. For high-volume production, mandrels made of type A2 tool steel are typical. More exotic mandrel materials seldom are required.

Demands on materials for spinning rollers are considerably more severe. Typical failure mechanisms for spinning rollers include abrasive wear, adhesive wear, surface fatigue and spalling, thermal cracking, and breakage. Typical materials selected for spinning rollers—in order of increasing resistance to abrasion—include A2, D2 and D3 tool steels.

Types A2 and D2 are tempered in their secondary-hardening range of 510 to 540 °C (950 to 1000 °F) to about 61 HRC; type D3 is tempered at 200 °C (400 °F) to 63 HRC. Titanium parts frequently must be formed hot; for such applications, M2 and M4 high speed steel rollers are preferred.

Proper design of roller contours reduces sliding-contact velocities and consequent thermal shock in service. Figure 2 shows typical roller designs for various types of shear spinning.

In order of increasing preference, spinning rollers may be made from static castings, centrifugal castings, upset forgings and ring-rolled forgings. Static castings typically contain undesirable interdendritic shrinkage and carbide segregation. Centrifugal castings usually are free of shrinkage porosity but may contain grain-boundary carbides. Hot working refines the carbide network in upset and ring-rolled forgings, but the isotropic grain orientation in a ring-rolled forging results in best tool life.

Surface Integrity and Finish

Extremely high hertzian contact stresses are encountered by spinning rollers, and for this reason surface integrity of finished rollers is of paramount importance. It is essential to prevent carburization or decarburization of rollers during heat treatment. Decarburization results in a soft surface that is low in compressive strength. On the other hand, a carburized surface would be too brittle for this application. High-chromium tool steels such as D2 and D3 are easily carburized; furnace atmospheres that would be neutral to low-chromium steels are carburizing to high-chromium steels of the same carbon content.

If finished rollers are to have the required surface characteristics, they must be ground and polished before being put into service. Grinding must be carefully controlled to prevent metallurgical damage and to minimize residual tensile stresses in the surface. It is advisable to stress relieve rollers between rough and finish grinding operations by retempering them at the original tempering temperature. Retempering for stress relief also is advisable after rough regrinding of worn rollers.

Finish grinding always should be followed by polishing. One manufacturer polishes spinning rollers by turning them in a lathe and using a felt pad saturated with 30-micron diamond lapping compound to produce a finish of 0.2 μm (8 μin.) rms.

Finish turning of fully hardened spinning rollers using a ceramic cutting tool has been used as a successful alternative to grinding and polishing. Machining with ceramic tools induces significant residual compressive stresses in the surface of the rollers and results in long service life.

Selection of Materials for Metalworking Rolls

By Roll Manufacturers Institute Committee*

ROLLS are used to reduce the cross section of metal or to change its shape, or both. Cylindrically shaped, rolls are placed in a mill housing supported on journal bearings against which screw-down pressure is exerted. Rolls rotate in opposite directions, the metal bar, plate or sheet passing between them longitudinally (Fig. 1a). Cross rolling is used for making seamless tube or for straightening round bar (Fig. 1b). Cross rolling is also used for widening billet (to make wide sheet or plate products), as well as for producing a more homogeneous microstructure.

Metal may be rolled hot or cold; it may be reduced in cross section or changed in shape and elongated. Coolants and lubricants may be used to reduce friction, maintain roll surfaces and control roll shape.

Roll Parts. The principal parts of a roll are the large central working portion called the body, the smaller-diameter ends known as necks or journals, and the driving ends commonly known as wabblers. In modern mills the wabblers may be replaced by keyways, splines or flat spade ends. The body of the roll is the portion that actually rolls the metal. The necks or journals are the sections on which the bearings are mounted. The wabblers or drive ends are the sections through which driving torque is transmitted.

Fig. 1 Typical arrangements of metalworking rolls

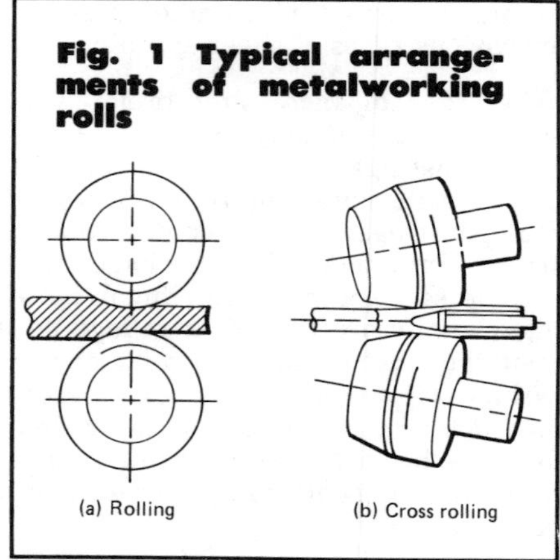

Fig. 2 Principal parts of metalworking rolls

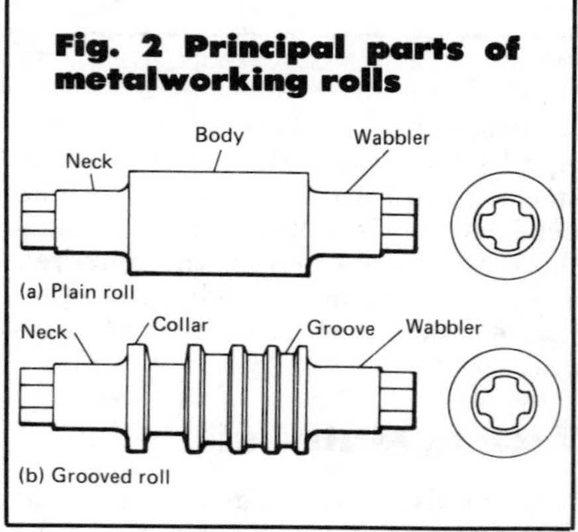

Figure 2(a) shows a plain-bodied roll used in a two-high mill for rolling flat products. Figure 2(b) shows a grooved roll for rolling shapes. In grooved rolls, the metal is reduced by passing it through the grooves or passes; the ridges separating the passes are known as collars. Figure 3 shows a plain roll used for rolling flat products in a four-high mill.

Types of Rolls. Rolls may be classified as to type of product rolled (blooming-mill, plate- and slab-mill, billet- and rod-mill, structural-mill, pipe-mill, foil-mill and strip-mill rolls). Rolls also may be classified according to their

*T. H. Caddy, *Chairman,* Superintendent, Metallurgical Heat Treatment & Forge, Mesta Machine Co.; J. M. Dugan, Vice President—Research, Blaw-Knox Co.; J. A. McKinnon, Chief Metallurgist, Mackintosh-Hemphill Div., Gulf + Western; E. W. Schane, Manager, Metallurgy, Corporate Quality Control, Wean United, Inc.; D. J. Tarney, Manager, Metallurgical & Technical Services, National Roll Div., General Steel Industries, Inc.; Dr. R. B. Corbett, Technical Director, Roll Manufacturers Institute

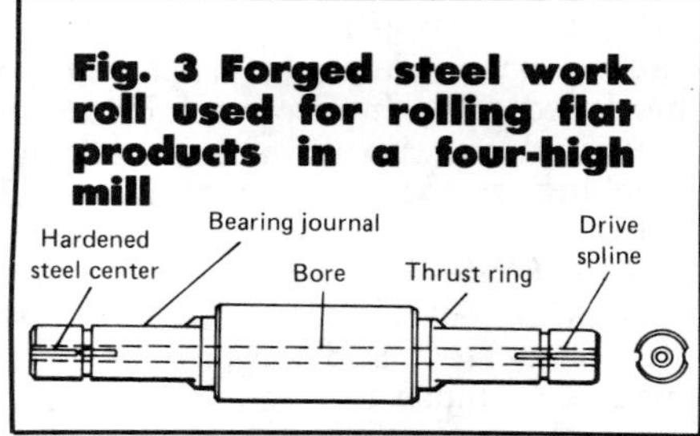

Fig. 3 Forged steel work roll used for rolling flat products in a four-high mill

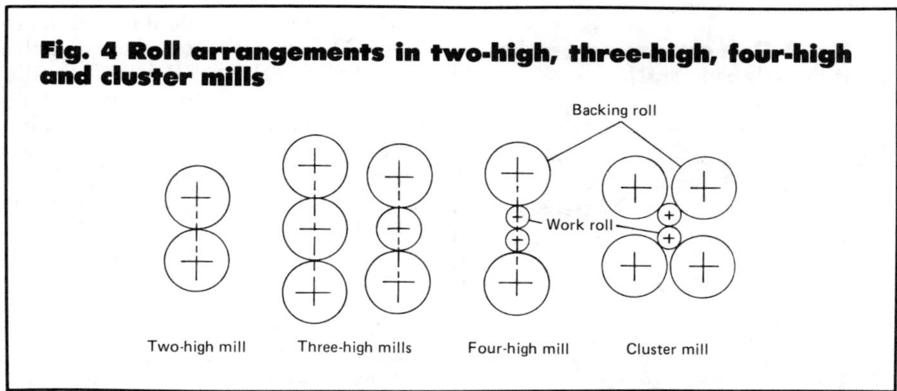

Fig. 4 Roll arrangements in two-high, three-high, four-high and cluster mills

position in a series of mills (breakdown, roughing and rundown, intermediate, pony, strand, leader and finishing rolls). Further roll classification may be functional (scale-breaker, piercing, straightener, skin-pass and calender rolls).

Arrangement of Rolls in Mill Housings. Rolls may be placed in the mill housing in two-high, three-high, four-high or cluster arrangements (Fig. 4) or in the more complicated arrangement of the Sendzimir ("Z") mill (Fig. 5). Except in two-high and three-high mills, not all rolls contact the work metal: some merely support the rolls that work the metal. Regardless of roll arrangement, any roll that contacts the work metal is known as a work roll; any roll whose primary purpose is to support the work rolls and keep them from deflecting under load is known as a backing roll.

Roll Requirements

Resistance to Wear. A major cause of roll deterioration in service is wear that impairs the surface and enlarges the work-metal section beyond specified tolerances. Wear results from friction and abrasion caused by the relative motion of metal and roll surfaces while pressed against each other under high loading. Wear, scoring and indentation of the roll-body working area will mark the metal being rolled and adversely affect its shape and mechanical tolerance. There is no general index for determining wear resistance of rolls. For lack of a better index, hardness is used, although composition of the roll must be considered also.

Selection of the proper hardness specification is based on type of rolling, on experience and on operating requirements. Hardness of the roll body must be high enough to maintain the surface for economical operation. Roll hardness is limited by requirements for ductility, shock, strength, and resistance to fire cracking and surface heat checking.

Roll hardness is commonly measured by the C-model Scleroscope, a portable instrument that rapidly determines hardness with minimum marking of the roll face. Rockwell C hardness tests sometimes are used for rolls that can be brought to, and tested on, a standard Rockwell machine.

Approximate relationships among Scleroscope, Rockwell and Vickers hardness numbers are given in ASTM E140 and A427. It should be noted that forged roll Scleroscope numbers (HFRSc) are higher than regular Scleroscope numbers (HSc), as explained in ASTM E448. Other testers for determining hardness of rolls are commercially available, and the approximate relationships of values indicated by these testers to values indicated by Scleroscope and other testers are being determined.

Strength. Rolls must have sufficient strength to withstand fracture from bending and the torsional and shearing stresses to which they are subjected in operation. High strength usually is needed more in rolls for roughing stands than in rolls for finishing stands. Adequate strength for a given application depends on composition, heat treatment, and freedom from internal defects.

Size, heat treatment, design and processing conditions have important effects on the residual stresses that are present to some extent in all rolls. In service, rolls having high residual stresses may be more easily ruptured by unusual shocks or by rapid or uneven heating on the mill. Subsurface strength is necessary to resist spalling, which frequently originates below the surface and proceeds outward until a piece breaks loose from the surface.

Bite is necessary for grabbing the metal and starting it through the mill, and is particularly important for large reductions. Ordinary bite is a function of roll diameter, varying inversely with diameter. Bite can be increased by roughening the roll face. The roughening operation, commonly known as "ragging", can be accomplished by shot blasting, by grinding, or by cutting longitudinal grooves in, or welding ridges on, the roll face.

Surface finish on rolls for flat rolled products is specified either to produce desirable characteristics in operation of the mill or to transfer a pattern or finish to the product being rolled. For transfer of a pattern or finish, depending on specific requirements, rolls may be smooth turned, ground and polished to a high luster, shot blasted to a matte finish, or embossed or stamped. Where a high-quality finish is important, the roll surface must be ground and then polished or lapped to a high luster, and must be free from surface blemishes. Smoothness of roll surfaces usually is measured and designated in microinches root mean square (rms). High roll hardness is necessary to preserve quality of finish for profitable production.

Rigidity of rolls is essential for obtaining minimum deflection (or spring) across the face of the roll. This is particularly important in rolling flat products such as plate, sheet and strip, to ensure more uniform thickness tolerances. For a given diameter, a steel roll is more rigid than a cast iron roll, but the hardness of the steel roll has no effect on its rigidity because deflections depend solely on modulus of elasticity, which is the same for hard steel as for soft steel. Short rolls of large diameter will deflect less than long rolls of smaller diameter. Crowning of rolls (making the diameter slightly larger at the center of the face than at the ends) will compensate for spring and produce a flatter rolled product; but because a

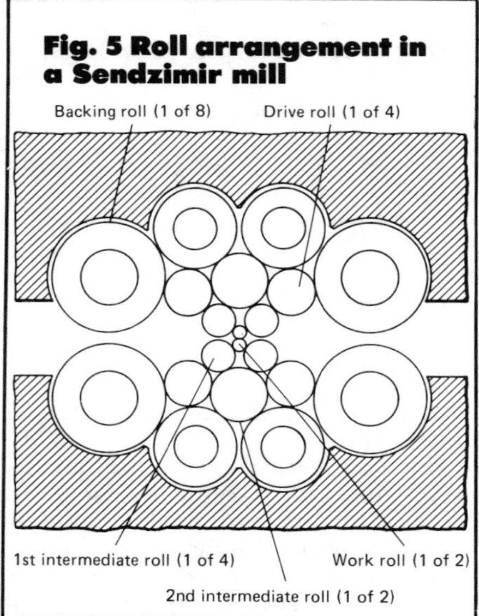

Fig. 5 Roll arrangement in a Sendzimir mill

Table 1 Minimum safe ratios of neck diameter to body diameter

Application	Ratio of neck diameter to body diameter
Blooming mills	0.50 to 0.55
Sheet and tin mills	0.76 to 0.80
Strip mills:	
Work rolls	0.60 to 0.65
Backing rolls	0.55 to 0.65
Merchant bar mills	0.60 to 0.67
Plate mills:	
Middle rolls	0.60 to 0.70
Top and bottom rolls	0.65 to 0.72
Structural mills	0.57 to 0.65
Continuous bar mills	0.70
Universal mills, vertical rolls	0.57
Cold rolling brass:	
Heavy reduction	0.775
Finishing reduction	0.72 to 0.75

specific crown is correct for only one combination of thickness, hardness and section width, crowning is a limited remedy.

Roll Design. Rolls are designed by engineering companies and builders of rolling mills except for pass and groove designs on grooved rolls, which generally are engineered in the user's roll shop. The proportions of rolls are based on application and mill design. The width of the metal to be rolled, or the length of the billet where cross rolling is required, determines the width of the body face. Body diameter is selected to provide the required bite and pass angle to accomplish reduction, and to provide sufficient mass to resist roll deflection and breakage. Rolls of smaller diameter result in less spread of the work metal and require less rolling pressure, separating force and power for a given reduction. In design of grooved rolls, deep grooves should be placed as far as possible from the center, in a location where the bending moment is at a minimum.

The size of a roll generally is designated by body diameter and body length, in that order; for example, a 600-by-1200-mm (24-by-48-in.) roll would have a body diameter of 600 mm and a body length of 1200 mm. For rolls used in processing shapes, the body diameter given is the nominal, or pitch, diameter.

Journal or neck dimensions are determined by imposed bending loads and by bearing design. The abrupt change in diameter from roll body to roll neck intensifies bending and torsional stresses at this location. To prevent breakage, the neck diameter should be as large a proportion of the body diameter as is feasible. Minimum safe ratios of neck diameter to body diameter are given in Table 1 for rolls used in various types of rolling mills; safe ratios vary with type of bearing, type of mill and conditions of service. In any event, neck diameter should never be smaller than 50% of body diameter.

Misuse. Rolls can be broken or damaged easily in service. Among the principal causes of failure are mill wrecks, overloads, bruises, cobbles and stickers. The work metal may wrap around the roll or enter the wrong pass, or fire cracking may occur. The metal being rolled may be too cold or of uneven temperature. Severe damage may result if a mill is stalled or stopped with hot work metal between the rolls. After any mill wreck, rolls should be removed from the mill and overheated areas cooled under controlled conditions. The roll may be heated too rapidly after initial installation or after a mill shutdown, with resulting rupture by expansion forces due to unfavorable heat gradients.

In cold rolling, any laminations in the work metal may break or fold and pass through in multiple thicknesses. Slippage of strip or bar in the rolls may cause high local frictional heat and heat checks on the roll face. Sometimes actual failure of the roll from these mishaps does not take place until much later. Use of improper techniques in grinding or regrinding of rolls can cause serious damage, especially on hardened rolls. Ineffective lubrication of journals may result in heat checking, cracking or excessive scoring of roll necks.

Roll Materials. The three main classes of material for rolls are cast iron, cast steel and forged steel. Cemented carbides are used occasionally, particularly for the small rolls used in bar-mill finishing stands.

Cast Iron Rolls

Cast iron rolls are used in the as-cast condition or after stress relief. Some high-alloy iron rolls are heat treated by holding at high temperature followed by several lower-temperature treatments. Cast irons used for rolls are metastable and may be white or gray depending on composition, inoculation (if any), cooling rate and other factors. Because of the number of elements present, determination of transformation diagrams is complicated.

Development of proper roll specifications to meet widely varying rolling requirements is an extremely complicated, technical undertaking; for example, when specifying radial hardness penetration, roll manufacturers must consider the requirements dictated by the design of each particular mill. Because of these factors, each roll must be more or less tailored for its intended use, and close cooperation between manufacturer and user is necessary for obtaining maximum roll life and performance.

In American practice, cast iron rolls are classified as *(a)* chilled iron rolls, *(b)* grain rolls, *(c)* sand iron rolls, *(d)* ductile iron rolls and *(e)* composite rolls.

Chilled iron rolls (hardness, 50 to 90 HSc) have a definitely formed, clear, homogeneous chilled white iron body surface and a fairly sharp line of demarcation between the chilled surface and the gray iron interior portion of the body. Clear chilled iron rolls can be made in unalloyed or alloyed grades, as shown in Table 2. The depth of chill is measured visually as the distance between the finished surface of the body and the depth at which the first graphitic specks appear. Below this, there is an area consisting of a mixture of white and gray iron known as mottle, which gradually becomes more gray and more graphitic, until it merges with the main gray iron structure of the roll interior.

Table 2 Applications of cast iron rolls

Type of roll	Applications
Chilled iron rolls:	
Unalloyed (50–72 HSc)	Hot and cold rolls for sheet mills, tin mills, two-high and three-high plate mills, and jobbing mills; wet and dry work rolls for four-high hot strip mills and for intermediate and finishing stands in rod, merchant, sheet, bar and skelp mills
Alloy iron (60–90 HSc)	Hot rolls for sheet and strip mills in ferrous, nonferrous, rubber, plastic and paper industries, two-high and three-high plate mills, and universal mills; work rolls for four-high hot strip mills and for finishing stands in sheet, bar, skelp, strip and merchant mills; cold rolls for finishing ferrous and nonferrous sheet and strip
Grain rolls (40–90 HSc):	
Mild hard	Light-duty roughing rolls for small merchant and bar mills
Medium hard	Intermediate rolls for merchant and bar mills and for large structural mills
Hard	Finishing rolls for merchant, bar and structural mills; flat finishing rolls for sheet, bar and skelp mills; sizing, high-mill, reeler and welding rolls for tube mills
Sand iron rolls (35–45 HSc)	Mild-duty rolls for roughing stands in small mills and finishing stands in large structural mills
Ductile iron rolls (50–80 HSc)	Roughing and intermediate rolls for bar and merchant mills and for tube mills and various other uses

Alloy chilled iron rolls have hardnesses that range from 60 to 90 HSc and that are controlled by carbon and alloy contents. Customary maximum percentages of alloying elements are 1.25 Mo, 1.00 Cr and 5.5 Ni. Many different combinations are used to produce desired properties. Rolls of this type, particularly in the harder grades, are used chiefly for rolling flat work, both hot and cold. The softer, machinable grades are used for rolling rod and small shapes.

Grain rolls are "indefinite chill" iron rolls (hardness, 40 to 90 HSc) that have an outer chilled face on the body. There is finely divided graphite at the surface, which gradually increases in amount and in flake size, with a corresponding decrease in hardness, as distance from the surface increases. These rolls have high resistance to wear and good finishing qualities, to considerable depths. The harder grades are used for hot and cold finishing of flat rolled products, and the softer grades are for deep sections (even with small rolls). Alloying elements such as chromium, nickel, and molybdenum usually are added, either singly or in combination, to develop specific levels of hardness and toughness similar to those of chilled iron rolls.

Sand iron rolls (no chill; hardness, 35 to 45 HSc) are cast in sand molds, in contrast to chilled iron rolls and grain rolls, the bodies of which are cast directly against chills. In a sand iron roll, the metal in the grooves of the body may be mildly hardened by use of cast iron ring inserts set in the sand mold. Sand iron rolls are used chiefly for intermediate and finishing stands on mills that roll large shapes. They are also used for roughing operations in primary mills.

Ductile iron rolls (hardness, 50 to 65 HSc) are made of iron of restricted composition to which magnesium or rare earth metals are added under controlled conditions to cause the graphite to form, during solidification, as nodules instead of the flakes common to gray iron. The resulting iron has strength and ductility properties between those of gray iron and those of steel.

Composite rolls, sometimes called double-pour rolls (hardness: bodies, 70 to 90 HSc; necks, 40 to 50 HSc) are rolls in which the body surface is made of a richly alloyed, hard cast iron resistant to wear, and the necks, wabblers, and central areas of the body are of a tougher and softer material. The metals are bonded firmly together during casting to form an integral structure that produces a wearing surface of high hardness, along with a tougher body and neck. Composite rolls are thus better able to withstand impact and thermal stresses. The outer rolling surface may be of either chilled or grain iron. The chief application of composite rolls in rolling of steel has been work rolls for four-high hot and cold strip mills and for plate mills; in rolling of nonferrous metals, the chief application has been rolls for hot breakdown and cold reduction of sheet and strip.

Manufacture. Most cast iron for rolls is melted in coreless electric induction furnaces or electric arc furnaces, although air furnaces fired by pulverized coal or open-hearth furnaces fired by gas or oil may be used. Cast iron rolls are cast vertically with the metal entering the bottom of the mold tangentially. The resulting rotation of the molten iron concentrates all slag products in the center, resulting in cleaner outer metal. The necks are cast in sand molds. Depending on the specific application, the body may be cast into (*a*) cast iron chill rings lightly coated or (*b*) sand molds containing cast iron half chill rings.

In casting of composite (double-poured) rolls, molten metal of the desired composition for the shell is held in the mold until a solidified outer layer of the proper thickness has been formed. The still-molten metal within the core of the body and the necks is then replaced by introducing metal of the desired core composition through the same pouring basin. The resulting roll has machinable necks and a hard, wear-resistant roll face.

An alternative technique is to use a slide-gate valve located at the very bottom of the mold cavity. When the shell metal has solidified to sufficient thickness, the slide-gate valve is opened to drain out the remaining liquid metal. The mold is then filled with the core metal. This process is also referred to as the "flush cast" process.

Another method used for making composite rolls is vertical or horizontal centrifugal casting (spin casting). Shell material of the exact weight is poured against a metal chill mold protected by

Table 3 Applications of cast steel rolls

Carbon, %	Applications
0.50 to 0.65	Applications where strength is the prime and only requirement
0.70 to 0.85	Blooming mills; roughing stands in jobbing, plate and sheet mills; muck mills
0.90 to 1.05	Blooming mills; slab mills; roughing stands in continuous bar mills; backing rolls
1.10 to 1.25	Blooming and slab mills where breakage is not great; piercing mills; roughing stands in billet, bar, rail and structural mills
1.35 to 1.55	Intermediate stands for rail mills; structural, continuous billet and continuous bar mills
1.60 to 1.80	Intermediate stands for continuous bar and billet mills; middle rolls for three-high mills
1.85 to 2.05	Middle rolls for rail and structural mills; finishing mills where housing design is too limited for iron rolls
2.10 to 2.60	Finishing rolls for unusual conditions
2.65 and up	Special applications

a refractory coating. Solidification occurs as the metal is subjected to a "G" force of 60 or greater.

Cast Steel Rolls

Differentiation between cast iron rolls and cast steel rolls cannot be made strictly on the basis of carbon content. Iron rolls usually are of compositions that produce free graphite in unchilled portions; steel rolls do not exhibit free graphite.

The harder cast alloy steel rolls have hardnesses equivalent to those of the softer cast iron rolls, and the superior strength of cast steel rolls often makes them preferable.

Composition. Alloy steel rolls have almost entirely superseded carbon steel rolls. Compositions of most alloy steel rolls are within the following limits: 0.40 to 2.60 C; less than 0.12 S, usually 0.06 max; less than 0.12 P, usually 0.06 max; up to 1.25 Mn; up to 1.50 Cr; up to 1.50 Ni; and up to 0.60 Mo. Higher carbon contents increase hardness and wear resistance. Some rolls have higher alloy contents, but these usually are for special purposes.

Manufacture. Steel rolls are cast from steel made in open-hearth or electric furnaces. The steel is cast in a sand mold in which suitable metal chills may be placed for more rapid solidification of certain portions of the roll. For all steel rolls, various complicated heat treatments, including induction hardening of wear surfaces, are specified to meet service requirements.

Applications. Cast steel rolls are graded according to carbon content. The general applications of these rolls are listed in Table 3. This table does not constitute a rigid classification, because conditions vary widely from mill to mill. Adjustments in carbon and alloy content are commonly made to suit individual conditions.

Forged Steel Rolls

Hardened forged steel rolls are used principally for cold rolling various metals in the form of coiled sheet and strip. Extremely high pressures are used in cold rolling, and forged rolls have sufficient strength, surface quality and wear resistance for cold rolling operations. Forged rolls sometimes are employed in nonferrous hot mills in preference to iron rolls because of their higher bending strength and resistance to metal pickup.

Type and Design. Forged steel rolls generally are flat-bodied (or plain-bodied) rolls designed to close dimensional tolerances and concentricity. They vary widely in size from a few kilograms to as much as 45 000 kg (50 tons). During manufacture, holes are bored through the centers of larger rolls for heat treatment and inspection purposes. New design developments include tapered journals with drilled holes to accommodate a special type of roller bearing, and somewhat greater use of fully hardened bearing journals for direct roller-bearing contact. Forged rolls have been specified for work rolls, backing rolls, auxiliary rolls and special rolls.

Manufacture. Forged steel rolls are made from basic electric-furnace or acid open-hearth steel and are cast into corrugated big-end-up ingot molds of suitable size and shape, so that after forge reduction the desired structure and grain refinement are obtained. New developments are use of vacuum melting by the consumable-electrode process and vacuum degassing of ingots by vacuum casting methods. In vacuum melting processes, the size of the ingot is limited, whereas in vacuum casting no such limitations exist. Vacuum melting produces the cleanest steel, and is of some value in the manufacture of rolls for foil or other products requiring blemish-free surfaces, although the additional cost is often prohibitive.

Composition. The most commonly used composition for forged steel rolls, sometimes known as regular roll steel, averages 0.85 C, 0.30 Mn, 0.30 Si, 1.75 Cr and 0.10 V. About 0.25% Mo sometimes is added to this basic composition, and the chromium content may be varied to obtain specific characteristics. For rolling nonferrous metals, a forged steel containing 0.40 C and 3.00 Cr is preferred. In Sendzimir mills, the work rolls and first and second intermediate supporting and drive rolls usually are made from high-carbon high-chromium steel with 1.50 or 2.25% C and 12.00% Cr (D1 or D4). For more severe service, work rolls of M1 molybdenum high speed steel are used. The new alloy CPM 10V, produced through application of the P/M process to high-alloy tool steel, has wear resistance approaching that of carbide, which makes it attractive for some special forged steel rolls. The composition of CPM 10V is 2.45 C, 5.25 Cr, 10.0 V and 1.30 Mo.

Hardness. Selection of the proper hardness for the body of the roll is essential for successful service performance. The hardness range varies with the specific application and is developed with the cooperation of mill operators. Most forged rolls are heat treated to high hardness, but they may be processed to lower values for specific purposes. Because of their high hardness, hardened steel rolls require careful handling in shipping, storage, mill service and grinding.

Hardness of work rolls for rolling thin strip averages about 95 HSc; lower hardnesses are employed for rolling thicker strip. In temper and finishing mills, work-roll hardness sometimes is higher than 95 HSc, and for special applications such as foil rolls, up to 100 HSc. In nonferrous rolling, especially in aluminum plate mills, work-roll hardness generally ranges from 60 to 80 HSc. Hardness of backing rolls varies from 55 to 95 HSc; values on the

Fig. 6 Sleeve-type backing roll

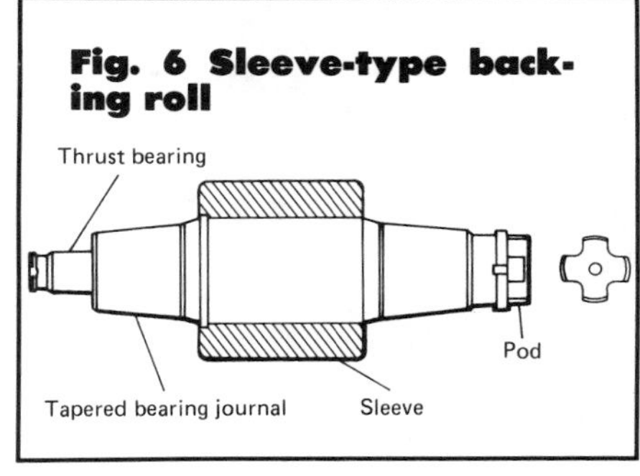

high side of this range are specified for rolls in small mills and foil mills.

For Sendzimir mills, customary hardness is 61 to 64 HRC for D1 and D4 steel work rolls and 64 to 66 HRC for high speed steel work rolls. Customary hardness of intermediate rolls is 58 to 62 HRC.

Only the body section of a forged roll is hardened. Journals usually are not hardened, except those for direct-contact roller-bearing designs, for which a minimum hardness of 80 HSc is specified. In normal practice, the journals of forged rolls range in hardness from 30 to 50 HSc.

Sleeve Rolls

Use of forged and hardened sleeve-type backing rolls in certain hot strip and cold reduction mills has become a common practice because such rolls are more economical. A sleeve roll is shown in Fig. 6. Sleeves are forged from high-quality alloy steel. Chromium-molybdenum-vanadium and nickel-chromium-molybdenum-vanadium compositions are generally used. Sleeves are heat treated by liquid quenching in either oil or water and are tempered to hardnesses of 50 to 85 HSc, depending on application.

The mandrel over which the sleeve is slipped may be made from a cast roll that has been worn below its minimum usable diameter, from a new casting made specifically for use as a mandrel, or from an alloy steel forging.

The outside diameter of the mandrel and the inside diameter of the sleeve are accurately machined or ground for a shrink fit. Mounting is accomplished by heating the sleeve to obtain the required expansion and then either slipping the sleeve over the mandrel or inserting the mandrel in the sleeve. This operation is performed with the mandrel in a vertical position. A locking device prevents lateral movement of the sleeve. Final machining is done after the sleeve is mounted.

Forged sleeves provide the hard, dense, spall-resistant surface required for the severe service encountered in hot and cold reduction mills. Another economic advantage of this type of roll is that the mandrel may be resleeved four or five times.

Miscellaneous Forged Rolls

Auxiliary rolls such as leveler rolls and pinch rolls are employed in processing and handling equipment associated with rolling mills. These rolls are characterized by their long, slender shape. They are made from forged or rolled bars of 52100 steel or carburizing grades, and are processed to a hardness of approximately 95 HSc. Rolls used for various types of straightening machines generally are of sleeve design. One roll may be concave in body shape and the mating roll straight. Standard compositions for forged steel rolls may be employed, and bodies may be hardened to 85 to 90 HSc.

Cemented Tungsten Carbide Rolls

Cemented tungsten carbide rolls have been used for rolling metals under a wide variety of conditions and in many types of rolling mills. They are used for both cold and hot rolling, and are made in all sizes from 6 to 400 mm (1/4 to 16 in.) in diameter. Rolls for slitters and trimmers also have been made of cemented tungsten carbide.

In flat rolling of wire and sheet, cemented tungsten carbide produces an extremely fine surface finish. In sheet rolling, the rigidity and dimensional stability of cemented carbide rolls result in more uniform stock thickness across the sheet width. Carbide rolls can be run at higher speeds for longer wear lives than are possible with cast iron or steel rolls.

For hot rolling of rod—both smooth rod and concrete-reinforcing rod (rebar)—cemented carbide rolls are used where increased speed and life, close size control and better surface finish are desired. These benefits result from the higher hardness, better wear and corrosion resistance, higher hot hardness and greater compressive strength of cemented tungsten carbide. It is in this application that CPM 10V forged steel rolls are being found to perform well and to be more cost efficient than carbide rolls.

Compositions of cemented tungsten carbides used in metalworking rolls have been varied greatly in amount of cobalt or nickel binder, in tungsten carbide grain size, and in types and amounts of other carbide additions. These variations have been made to achieve longer roll life by increasing wear and thermal-shock resistance, and to reduce the corrosion rate of the binder.

Use of cemented tungsten carbide rolls for applications varying from conventional cold rolling of flat sheet and wire to continuous hot rolling of rod in a wide range of sizes is continuing to expand, and technology of design and composition for these rolls is changing rapidly. Therefore, consultation with experienced carbide manufacturers is advised in selecting materials for specific applications.

Selection of Materials for Coining Dies

Edited by Robert McCreery
Vice President, Metallurgical Engineering
Teledyne—Portland Forge
and
Joseph Kozol
Director, Technical Services
The Franklin Mint

IN COLD COINING operations, the surface metal being worked is free to flow only to the small extent required by the coining operation. Flow to a larger extent constitutes cold extrusion, and information on selection of materials for press dies used in gross transfer of metal from one part of the die to another is covered in the article in this volume "Selection of Materials for Cold Extrusion Tools".

Coining dies may fail by wear, deformation due to compression (called "sinking"), or cracking. With low coining pressures and soft work metal, wear failures predominate. With some combinations of die metal and work metal, dies may fail by adhesion (wear caused by metal pickup).

Failure of dies from deformation or cracking is usually caused by: (*a*) coining of extremely intricate designs, (*b*) attempts to coin large areas that confine the metal and build up excessive pressure, or (*c*) coining of oversize slugs.

Constraints due to the pattern being produced may limit die life and cause premature cracking. If the obverse and reverse artwork of a decorative medal are not aligned properly, metal flow will be restricted and the die will not fill properly. As a result, excess tonnage (pressure) must be used to obtain fill, which sharply reduces die life. Stress raisers such as straight lines and sharp edges, which often are present in designs for decorative medals, also reduce die life unless the tonnage can be lowered. Low tonnage requirements often can be achieved by striking softer blanks, provided the blank is not so soft that a fin is extruded on coining.

Decorative Coining

Selection of tool steels for fabrication of dies used for striking high-quality coins and medals requires consideration of several important properties and characteristics. Among these are machinability, hardenability, distortion in hardening, hardness, wear resistance and toughness. In dies used for decorative coining, materials that can be through hardened to produce a combination of good wear resistance, high hardness and high toughness are preferred. Specific properties and compositions of the tool materials referred to in this article are presented elsewhere in this volume.

A smooth, polished background surface on the die is required for striking proof-type coins and medals. Massive undissolved carbides or nonmetallic inclusions make it more difficult to obtain this smooth background. Special processing and inspection should be required for tool steels to be used for coining dies (particularly in large sections), because any such imperfections can be troublesome. The stringent controls ordinarily applied to tool steels may not be sufficient to ensure that the required die surface condition will be obtainable.

Typical Die Materials. For dies up to 50 mm (2 in.) in diameter, consumable-electrode vacuum-melted or electroslag remelted 52100 steel provides the clean microstructure necessary for development of critical polished die surfaces. When heat treated to a hardness of 59 to 61 HRC, 52100 steel provides optimum die life. This steel also is suitable for photochemical etching, a process used in place of mechanical die sinking for engraving many low-relief dies. L6 tool steel at a hardness of 58 to 60 HRC is suitable for dies up to 100 mm (4 in.) in diameter. It can be through hardened, has enough toughness for long-life applications, and is suitable for photochemical etching of low-relief patterns. Air-hardening tool steels are preferred for coining and embossing dies greater than 100 mm (4 in.) in diameter. One of the chief reasons for choosing air-hardening tool steels is their low degree of distortion during heat treatment. Type A6 is a nondeforming, deep-hardening tool steel that often is used for large dies that must be hardened to 59 to 61 HRC. Air-hardening hot work steels such as type H13 are used at a hardness of 52 to 54 HRC for applications requiring especially high toughness.

For dies containing high-relief impressions, lowest die cost is obtained by machining the impressions directly into the dies when the number of pieces to be coined is less than the anticipated die life. For longer runs that require two or more identical dies, it is less expensive to produce the impressions by hubbing. Hubbing is done by cutting the pattern into a male master plug (hub), hardening this hub, and pressing the hardened hub into a die block to make the coining impression. Highly alloyed tool steels are relatively difficult to hub. When coining dies are made from these steels, it may be necessary to form the impression by hot hubbing, or by hubbing in several stages with intermediate anneals between stages.

Table 1 gives typical materials used to make the punches and dies for coining small pieces such as the 13-mm-diam (½-in.-diam) emblem shown in the accompanying sketch. The choice of tool material often depends less on the alloy to be coined than on the way the tools are made and on the type of stamping equipment to be used.

O1 and A2 tool steels are alternative choices for machined dies in production quantities up to about 100 000 pieces. The small additional cost of A2 is often justified because A2 gives longer life, especially when aluminum alloys, alloy steels, stainless steels or heat-resisting alloys are being coined.

Production of coins and medallions, as described in the article beginning on page 83 in Volume 4 of the 8th Edition of this Handbook, frequently involves quantities greatly in excess of 100 000 pieces. Coins are usually produced on high-speed mechanical presses using dies containing impressions that have relatively low relief above the background plane. Dies for this type of operation must be easily hubbed, inexpensive, wear resistant and made of nondeforming materials. W1 often is selected for small dies, and 52100 for either small dies or large dies. Average die life can be expected to range from 200 000 to more than 1 000 000 strikes, depending on the type of coinage alloy and on coin diameter.

Dies for Coining Silverware. Probably the greatest amount of industrial coining is done with drop hammers in the silverware industry, in producing highly embossed designs on surfaces such as the teaspoon handle shown in Fig. 1. Water-hardening steels such as W1 are almost always used for making such coining dies, whether the product is made of silver, a copper alloy or stainless steel. Water-hardening grades are selected because die blocks made of these steels can be reused repeatedly. After a die block fails—either by shallow cracking of the hardened shell or by wear of the high points of the impressed pattern—the block is annealed, the impression is machined off, and a new impression is hubbed before the die is rehardened. Dies made of deep-hardening tool steels such as O1, A2 and D2 are not reused (as are W1 dies), because they fail by deep cracking.

For ordinary designs requiring close reproduction of dimensions, dies may be made of A2, or of the high-carbon high-chromium steels D2, D3 and D4, to obtain greater compression resistance. For coining of designs with deep configurations and either coarse or sharp details, where dies usually fail by cracking, a deep-hardening carbon tool steel may be used at lower hardness, or O1, or S5 or S6 may be selected. In some instances, it may be desirable to select an air-hardening type such as A2, which would provide improved dimensional stability and wear resistance. A hot work steel such as H11, H12 or H13 may prove to be best where extreme toughness is the predominant requirement. When die failure occurs by rapid wear, a higher-hardness steel, or a more highly alloyed wear-resistant steel such as A2, may solve the problem.

For articles coined on drop hammers from series 300 austenitic stainless steels, it has sometimes been found advantageous to use steels of the S1, S5, S6 and L6 types, oil quenched and

Fig. 1 A deeply coined teaspoon handle

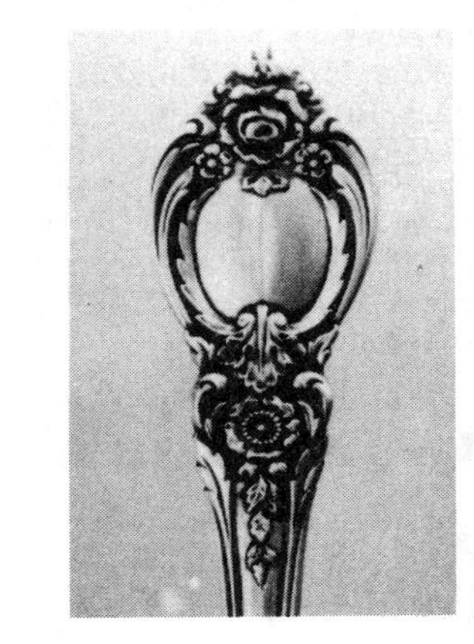

Table 1 Typical materials for dies to coin small emblems

Type of tool	Tool material(a) for striking a total quantity of: 1000	10 000	100 000
Machined dies for use on drop hammers	W1	W1	O1(b), A2
Machined dies for use on presses	O1	O1, A2	O1, A2
Hubbed dies for use on drop hammers	W1	W1	W1(c)
Hubbed dies for use on presses	O1	O1, A2	A2, D2(d)

(a) For coining the emblem from aluminum, copper, gold or silver alloys, or from low-carbon, alloy or stainless steel. (b) O1 recommended only for coining low-carbon steel and alloys of copper, gold or silver. (c) The average life of W1 dies in coining alloys of copper, gold or silver softer than 60 HRB would be about 40 000 ± 10 000 pieces. Life of W1 dies in coining harder materials would be about half as great; therefore, more than one set of dies would be needed for 100 000 parts or more. (d) Hot hubbed.

tempered to 57 to 59 HRC. Because the carbon contents of these grades are between 0.50 and 0.70%, they are less resistant to wear than W1, A2 or D2, but are tougher and more resistant to chipping and splitting. If necessary, the wear resistance of S5 tool steel dies can be improved slightly by carburizing to a depth of 0.13 to 0.25 mm (0.005 to 0.010 in.).

Coining in Progressive Dies

Tool steels recommended for coining a cup-shape part to final dimensions in the last stages of progressive stamping are shown in Table 2. This press coining operation involves partial confinement of the entire cup within the die. This produces high radial die pressures and thus requires pressed-in inserts on long runs, to prevent die cracking. Quantities up to about 10 000 can be made with the steels given in the table, without danger of failure by cracking; the D2 steel listed for quantities greater than 10 000 pieces would be used in the form of an insert pressed into the die plate.

The punch material can be the same as the die material, except that O1 should be substituted for W1 in applications where W1 might crack during quenching.

The coining illustrated in the sketch accompanying Table 2 is typical of the coining stage for articles stamped from strip material through progressive forming operations employing die and punch inserts for each stage. Frequently, the inserts are near, or even below, the minimum size that provides the amount of die stock required by good practice. Dies often cannot be any larger, or they will not fit in the over-all space available, as shown in the sketch in Table 2. In such instances, hot work steels give better life than W1, O1, A2 or S2. The separate pieces of the punch body and pilot in the tooling setup illustrated in Table 2 might be made of H12, at 49 to 52 HRC—a compromise between lower hardnesses that result in scoring deterioration and higher hardnesses that lead to failure by splitting. Scoring of the pilot part of the punch is best prevented by hard chromium plating 0.008 to 0.01 mm (0.0003 to 0.0004 in.) thick that has been baked at least 3 h at 150 to 200 °C (300 to 400 °F) to minimize hydrogen embrittlement.

In the coining die, type H12 hot work tool steel at 45 to 48 HRC would probably be more resistant to splitting stresses than any of the cold coining die steels. For the kickout pin, an L6 tool steel at a hardness of 40 to 45 HRC is recommended.

H11, H12, H13, H20 and H21 at or near their full hardness of 50 to 54 HRC often perform well in coining dies having circular grooves, beads, thin sections, or any configuration that demands improved resistance to breakage and that can tolerate some sacrifice in wear resistance.

Working Hardnesses

The normal working hardnesses of the tool steels listed in Tables 1 and 2 are as follows:

W1	59 to 61 HRC
O1	58 to 60 HRC
A2	56 to 58 HRC
D2	56 to 58 HRC

D2 might be used at 60 to 62 HRC for coining small aluminum parts.

Coining of Gears and Similar Parts

Figure 2 shows a coining application involving a degree of severity so great that the risk of coining die breakage and wear in normal service is extremely high. For this steel forging, the working surfaces of the gear teeth were cold coined 0.025 to 0.13 mm (0.001 to 0.005 in.) for sizing to final dimensions; simultaneously, the central section was cold coined to impart the concentricity specified for the part. The recommended tool steel for such an application is A2 when 1000 to 10 000 gears are to be coined, and D2 when the production quantity is 100 000. If a similar part made of a stainless steel or heat-resistant alloy were to be coined, several D2 die sets might be required to complete a run of 100 000 because of the more severe die wear encountered with stainless steels and heat-resistant alloys.

P/M Steels for Coining Dies

The application of hot isostatic processing to P/M production of high speed steels and special high-alloy steels has expanded the range of tool steel grades available for long-run coining dies. Dramatic increases in toughness and grindability have been achieved (see discussion under "Powder Metallurgy Steels" in the article "Tool Steels" in this volume). Type M4 is an excellent example. When made by P/M processing, M4 has approximately twice the toughness and two to three times the grindability of conventionally pro-

Table 2 Typical tool steels for coining a preformed cup to final size on a press

Metal to be coined	Die material(a) for total quantity of: 1000	10 000	100 000
Aluminum and copper alloys	W1	W1	D2
Low-carbon steel	W1	O1	D2
Stainless steel, heat-resisting alloys and alloy steels	O1	A2	D2

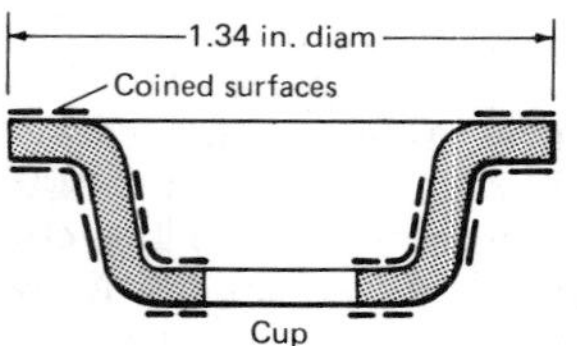

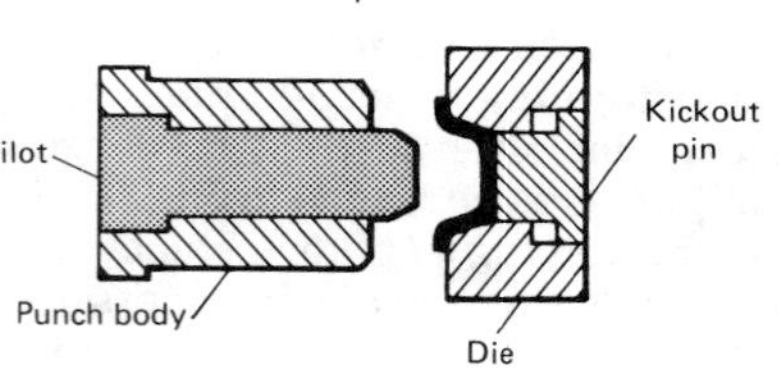

(a) For quantities over 10 000, the materials given are for die inserts. All selections shown are for machined dies. The same material would be used for the punch, except that O1 should be substituted for W1 in applications where W1 might crack during heat treating.

Fig. 2 Severely coined bevel gear

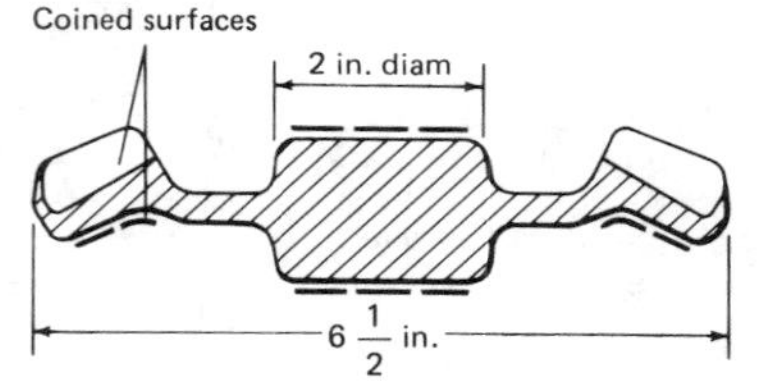

Forged bevel gear (20°, 26-tooth, 4-pitch) coined 0.025 to 0.13 mm (0.001 to 0.005 in.) for sizing to dimensions and for improving concentricity to 0.90 mm (0.035 in.) TIR maximum runout at the pitch line.

cessed M4. Consequently, P/M M4 heat treated to 63 to 64 HRC has better toughness, wear resistance and compressive strength than conventionally processed D2 at 62 HRC.

CPM 10V is one of the specialty grades created for cold work and warm work tools used in high-wear applications. CPM 10V is more wear resistant than even the most wear-resistant high speed steels, and has toughness and compressive strength equivalent to those of D2.

Cemented Carbide Coining Dies

Cemented carbides are occasionally used to make coining dies, but generally only for light coining of small pieces in very large production quantities. The successful application of cemented carbides for this service depends to a great extent on the design of the die (or die insert), and to an even greater extent on the design of the hardened tool steel supporting and backup members that surround the carbide dies or inserts. It is most important that the supporting and backup members counteract any tensile stresses imposed on the carbide by the coining operation and that they ensure minimum movement of the die parts.

For light-load applications with minimal shock or impact loading, cemented tungsten carbide containing at least 13% cobalt is used. For applications involving greater shock loading, higher cobalt contents (up to 25%) are required.

Selection of Materials for Cold Heading Tools

Edited by Ralph G. Barton
President
Rico Machine Co., Inc.

THE MANY TYPES of fasteners that are used in large quantities are manufactured in cold heading machines. Most of the tools used in these machines require maximum surface hardness for wear resistance, along with maximum strength and toughness to enable them to withstand high service pressures without breaking. Proper selection of materials for these tools is essential to profitable production.

Materials used for tools in single-die, two-die-three-blow and multistation cold heading machines are listed in Table 1.

Tool Steel Dies

Solid cold heading dies made of W1 and W2 tool steels are used for short production runs. These tools are hardened by flush quenching, which provides the desired combination of high hardness, high strength and high toughness. Most long production runs require dies made of the more highly alloyed tool steels M1, M2 and D2, or of cemented carbides. A new, wear-resistant P/M tool steel, CPM 10V, has also given excellent performance and is an economical alternative to cemented carbides on long runs. These materials are most commonly used as inserts in H13 tool steel cases; hardness of such cases usually is held to about 48 HRC to provide the best combination of back-

Table 1 Typical materials for tools in cold heading machines

Tool	Material
Cutoff quill (die)	M4, or cemented carbide insert
Cutoff knife	M4, or cemented carbide insert
Upset, cone or spring punch	W1, S1, M1, or cemented carbide insert
Cone-punch knockout pin	M2, or CPM 10V
Backing plug	O1
Finish-punch case	H13
Finish-punch insert	M1, M2, CPM 10V, or cemented carbide
Die case	H13
Die insert	M1, D2, M2, CPM 10V, or cemented carbide
Die knockout pin	M2, or CPM 10V

Table 2 Typical cemented carbides for cold heading tools

Composition, %		Hardness,	Transverse rupture strength	
WC	Co	HRA	MPa	ksi
75	25	84.5	2760	400
80	20	84.7	3030	440
85	15	87.0	3240	470
88	12	88.0	2960	430

up strength and freedom from hazardous breakage. However, for applications involving high-interference fits between insert and case, the greater high-strength fracture toughness of maraging steels such as 18Ni(300) makes them very desirable as materials for cases, although at some sacrifice in wear resistance.

Cemented Carbide Tools

As a rule of thumb, cemented carbide tools properly designed and utilized should provide roughly ten times the life of steel tools. Carbide tools also tend to improve product quality as a result of their greater dimensional stability and consistency throughout long production runs. Thus, although tungsten carbide tooling is higher in initial cost than tool steel tooling, the longer life and superior dimensional integrity of carbide result in far lower cost per thousand pieces produced.

The carbides most useful for cold heading tools are straight tungsten carbides with cobalt binders. Compositions, hardnesses and rupture strengths of the cemented tungsten carbides most commonly used are presented in Table 2. Toughness generally is proportional to cobalt content, as well as to the coarseness of the grain structure. The materials in Table 2 have medium to coarse grain structures. Grain size and cobalt content can be adjusted to provide the levels of hardness, strength and wear resistance necessary for best performance. Once a baseline of performance has been established for a specific carbide material in a given application, composition and microstructure can be adjusted to improve tool life.

Cold heading tools for which cemented carbides are used include knives and quills, upset hammers (both standard and sliding), dies and finish punches. Carbide dies are of the insert type, with a hardened steel case supporting the carbide insert. Solid carbide dies have been used, but unsupported dies of this type are extremely rare.

For best results, cemented carbide tools should be used in cold heading machines that are in good condition, with tight rams, minimum vibration and accurate alignment.

Reworking and Resharpening

Although cemented carbide cold heading tools can be reworked and reshaped on trial runs, recommended practice is to use steel tooling (which can be altered more easily) for trial runs and then switch to carbide when design has been finalized.

When steel knives and quills are used, they should be removed from the machine once per shift and ground slightly to sharpen the edges and thus ensure good, straight cutoffs. This practice requires that each machine be shut down for 10 to 30 min each shift, but it eliminates a variety of cold heading problems. On the other hand, cemented carbide knives and quills can be used for many days, and sometimes weeks, without being resharpened. When they do become dull or slightly chipped, they can be resharpened with diamond wheels.

The 25% Co grade of straight tungsten carbide is widely used for cold heading work and has the distinct advantage of being machinable; it can be bored, drilled and turned with carbide or diamond tools. The new tool steel CPM 10V provides comparable wear resistance and better toughness and is more readily machinable and reworkable. For applications that require greater wear resistance and that involve less shock, the 12% and 15% Co grades of straight tungsten carbide are used. These grades must be ground or lapped with diamond.

Selection of Materials for Cold Extrusion Tools

Edited by George E. Ferber
Vice President, Sales
Braun Engineering Company

IN COLD EXTRUSION, neither the tooling nor the work is preheated. However, the heat generated by plastic deformation of the workpiece under steady and nearly uniform pressure may be sufficient to require tool steels with relatively high resistance to softening at elevated temperatures.

In the cold extrusion process, backward displacement from a closed die progresses in the direction opposite that of punch travel, as shown at left in Fig. 1. Parts often are cuplike in shape and have wall thicknesses equal to the clearance between punch and die.

In forward extrusion, the metal is forced in the direction of punch travel (Fig. 1, center). One end of the die recess is just large enough to receive the starting slug, and the other end has a small orifice of the shape required for the final part.

Fig. 1 Backward, forward and combined forward and backward displacement in cold extrusion

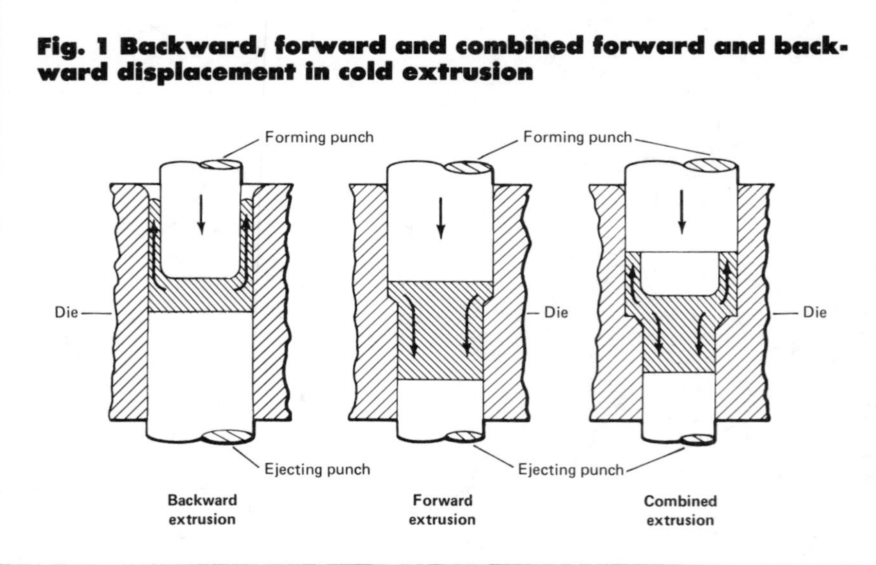

Sometimes the two methods of extrusion are combined so that some of the metal flows backward and some forward, as shown at right in Fig. 1.

Compressive strength of the punch and tensile strength of the die are among the most important factors influencing the selection of materials for cold extrusion tools. Because the die is invariably prestressed in compression by the pressure of inner and outer shrink rings, the principal requirement for a satisfactory die is a combination of tensile yield strength and degree of compressive prestressing that will withstand repeated tensile loading in service. Punches require sufficient compressive strength to enable them to resist upsetting without being hazardously brittle. Thus, almost without exception and particularly for extrusion of steel, the primary tools in contact with the work must be made from steels that will through harden in the section sizes involved.

The primary mechanism of tool deterioration in cold extrusion is abrasive and adhesive (galling) wear of both punch and die. Wear of the tools causes the finished parts to exhibit progressive changes in dimensions or contour and progressive increases in surface roughness. For those cold extrusion processes involving drawing, in which the slug is extended to several times its original length, the die wears much more rapidly than the punch because the metal slides farther over the die surface.

Lubrication, at least theoretically, is the prime factor controlling tool wear, because there is no metal-to-metal contact between the tools and the work when lubrication is ideal. In practice, the stresses imposed on tooling can vary by more than 100% with changes in lubrication. Use of an insufficient amount or an inferior type of lubricant usually increases downtime for polishing out score marks or adhesive pickup on dies and punches and, therefore, results in shorter die life and greater scrap loss.

Punch and Die Materials

Cold extrusion of a part from 1018 or 1021 steel requires about 10% more extrusion pressure than extrusion of the same part from 1010 steel. For low-alloy steels, forming loads are about 20 to 30% greater than those for 1010; this conclusion is based on extensive experience in cold extrusion of 8620, and less experience with 4130, 5120, 5130 and 4027. Medium-carbon steels such as 1030 and 1040 also require forming loads about 20 to 30% higher than those for 1010 steel.

Dies made with W1 tool steel generally are satisfactory for extruding the softer alloys of aluminum. Steels such as A2 and D2 are preferred for tools used in extrusion of the stronger aluminum alloys, because enough heat sometimes is generated to soften tools made of W1. In extrusion of aluminum, tool wear is roughly proportional to the yield strength of the work metal. Thus, the common impact extrusion alloys 1100, 6061, 2014 and 7075 cause progressively more wear on tools in the order listed. Compositions and properties of the tool materials discussed in this article are presented elsewhere in this volume.

Table 1 Typical tool steels for backward extrusion of parts 1 and 2
For designs of parts, see Fig. 2

Metal to be extruded(a)	Total quantity of parts to be extruded: 5 000	50 000
Punch material(b)		
Aluminum alloys	A2	A2, D2, M4 (c)
Carbon steel, up to 0.40% C	A2	D2, M2 (b), M4 (c)
Carburizing grades of alloy steel	A2	M2 (d), M4 (c)
Die material(b)		
Aluminum alloys	W1 (e)	W1 (e)
Carbon steel, up to 0.40% C	O1, A2	A2 (f)
Carburizing grades of alloy steel	O1, A2	A2 (f)
Knockout material(b)		
Aluminum alloys	A2	D2
Carbon steel, up to 0.40% C, and carburizing grades of alloy steel	A2	A2, D2

(a) For part 1, starting with a solid slug; for part 2, starting with part 1. In aluminum, part 2 can be made directly from a cylindrical blank. (b) Where two or more tool materials are recommended for the same conditions, they are given in order of cost, with the less or least expensive shown first. (c) First choice in automotive parts processing. (d) Liquid nitrided. (e) The 1.00% C grade is recommended. (f) Gas nitrided on the inside surface only.

Table 2 Typical tool steels for drawing part 3
For design of part, see Fig. 2

Metal to be drawn(a)	Total quantity of parts to be extruded: 5 000	50 000(b)
Punch material(c)		
Aluminum alloys	A2	D2, M4 (d)
Carbon and alloy steel, up to 0.40% C	O1	A2, M4 (d)
Die material(c)		
Aluminum alloys	W1 (e)	W1 (e), A2
Carbon and alloy steel, up to 0.40% C	O1, A2	A2 (f), D2

(a) Starting with part 2 (Table 1) for steel. In aluminum, the part would be made in one backward extrusion from a cylindrical slug. (b) For quantities greater than about 100 000 parts in steel, carbide punches and dies should be considered, especially if close tolerances must be maintained. (d) First choice in automotive parts processing. (e) The 1.00% C grade is recommended. (f) Gas nitriding is recommended on the inside surface. F2 tool steel may be used in place of A2.

Tables 1 to 6 list typical tool steels used in tools for cold extrusion of steels and aluminum alloys, in two quantities, for the series of hypothetical parts shown in Fig. 2. These simple parts are seldom encountered in practice; however, the principles described can be related to actual production components of comparable severity.

Backward-Extrusion Tools. Table 1 gives typical tool steels used for punches, dies and knockouts for backward extrusion of parts 1 and 2. Extrusion of part 2 using part 1 as the starting slug is similar in severity to extrusion of part 1 from a cylindrical slug. (If made in aluminum, part 2 can be extruded in a single step directly from a cylindrical blank.)

The recommendations of die materials for extrusion of 50 000 parts from steel are conservative. A D2 punch might achieve a total life of 300 000 pieces with 60 000 between redressings, an O1 die a total of 160 000 with 40 000 between redressings, and an A2 die a total of 200 000 with 70 000 between redressings.

In one extrusion department, there was a marked difference in punch wear in extruding steel parts similar to parts 1 and 2—up to four times more wear for the part similar to part 2.

Drawing Tools. Table 2 shows typical punch and die steels used for draw-

Fig. 2 Seven hypothetical cold extruded parts

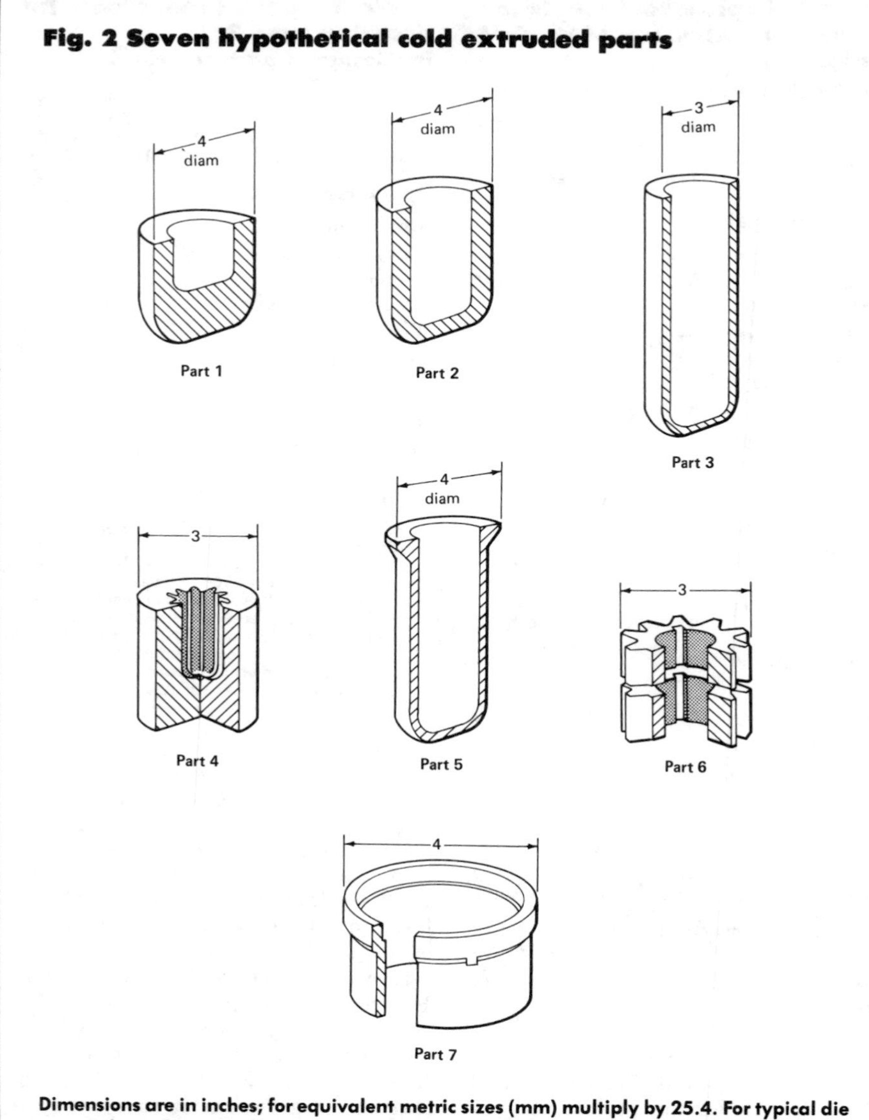

Dimensions are in inches; for equivalent metric sizes (mm) multiply by 25.4. For typical die materials used in extruding these parts from low-carbon steel and aluminum, see Tables 1 to 6.

ing part 3. When made of steel, this part is assumed to be drawn from part 2; when made of aluminum, it is assumed to be drawn directly from a cylindrical slug. During this operation, the metal in the part is extended twice as much as in part 2, and sliding and wear on the die are much more severe than on the punch.

In drawing part 3 from steel, an O1 punch will make 5000 to 10 000 pieces per redressing, and an A2 punch about 50 000 pieces. Dies of O1 might make 5000 steel parts, and dies of normally hardened A2 might make 30 000. Neither D2 dies nor gas nitrided A2 dies would be likely to make more than about 50 000 parts.

In drawing part 3 from aluminum, an A2 punch may be expected to have a total life of 75 000 pieces, and a D2 punch, 200 000 pieces; a die of either W1 or A2 might make 500 000 pieces.

Table 3 presents typical recommendations of tool steels for tooling used in making part 4. This selection problem entails ordinary requirements for punch and die, but the cold extrusion operation involved is considered severe because of the shape of the part. This drawing operation is comparable to backward extrusion of part 1, except that the punch loads are higher in the drawing operation (and would be higher still if the part were backward extruded).

Table 3 Typical tool steels for drawing part 4
For design of part, see Fig. 2

Metal to be drawn(a)	Total quantity of parts to be drawn: 5 000	50 000
Punch material(b)		
Aluminum alloys	A2	D2, M4 (c)
Carbon steel, 1010	M2	M2, M4 (c)
Carbon steel, 1020 to 1040, and carburizing grades of alloy steel	M2 (d)	M2 (d), T15, M4 (c)
Die material(b)		
Aluminum alloys	A2	A2, D2
Carbon and alloy steel	A2	A2 (d), D2

(a) In steel, a part would be made in two operations with an intermediate process anneal (see text). In aluminum, it would be made in one backward extrusion. (b) Where two or more tool materials are recommended for the same conditions, they are given in order of cost, with the less or least expensive shown first. (c) First choice in automotive parts processing. (d) Nitriding treatment is recommended.

In steel, part 4 normally would be made by first forming a primary cup to the depth shown, then process annealing the cup at about 650 °C (1200 °F). To actually form the part, a splined punch would enter and bottom out in the hole of the annealed primary cup, then a ring at the top of the punch would engage the rim of the cup and push forward into the die to reduce the outside diameter of the cup and form the internal splines. This is sometimes termed a "push-draw" operation.

Forward-Extrusion Tools. Table 4 lists the tool steels typically used in forward extrusion of part 5 from part 2. (In aluminum, this part would be made from a cylindrical slug.) In forward extrusion, the work metal moves over the die several times farther than over the punch, and normal wear of the die is more rapid. In this instance, no knockout is required, because the part is stripped from the punch on the return stroke.

Table 5 gives the tool steels typically used in forward extrusion of part 6. Because of the shape of the die in this operation, there may be: (*a*) local resid-

Table 4 Typical tool steels for forward extrusion of part 5
For design of part, see Fig. 2

Metal to be extruded(a)	Total quantity of parts to be extruded 5 000	50 000(b)
Punch material(c)		
Aluminum alloys	A2	D2, M4 (d)
Carbon steel, 1010	A2	D2, M4 (d)
Carbon steel, 1020 and 1040, and carburizing grades of alloy steel	A2	M2 (e)
Die material(c)		
Aluminum alloys	W1 (f)	A2, D2
Carbon and alloy steel, up to 0.40% C	A2	A2 (g)

(a) Starting with part 2 (Table 1) for steel. Aluminum would be extruded from a cylindrical slug. (b) For quantities greater than about 100 000 parts in steel, carbide punches and dies should be considered, especially if close tolerances must be maintained. (c) Where two tool materials are recommended for the same conditions, they are given in order of cost, with the less expensive shown first. (d) First choice in automotive parts processing. (e) Nitrided. (f) The 1.00% C grade is recommended. (g) Liquid nitrided.

Table 5 Typical tool steels for forward extrusion of part 6
For design of part, see Fig. 2

Metal to be extruded	Total quantity of parts to be extruded 5 000	50 000
Punch material(a)		
Aluminum alloys	A2	D2, M4 (b)
Carbon and alloy steels, up to 0.40% C	A2, D2	M2 (c), M4 (b)
Die material(a)		
Aluminum alloys	A2	A2
Carbon and alloy steels, up to 0.40% C	(d)	(d)

(a) Where two tool materials are recommended for the same conditions, they are given in order of cost, with the less expensive shown first. (b) First choice in automotive parts processing. (c) Liquid nitrided. (d) No tool steel can be recommended without qualification. Medium-carbon alloy tool steels such as H12, H21 and 6F5 have given the best results.

ual stresses at corners, resulting from heat treating, (*b*) discontinuous loading stresses across corners, and (*c*) extra frictional surface that increases the pressure required to extrude the part. In making this part from steel, the stresses involved are high enough to preclude use of standard tool steels, except perhaps H12 or H21, at low hardness. Lower-alloy, less wear-resistant steels that resist splitting have given good results when quenched and tempered to 55 or 56 HRC. These nonstandard steels, designated 6F, 6G or 6H in this article, should not be carburized.

Fig. 3 Terminology for tools used in backward extrusion of steel parts

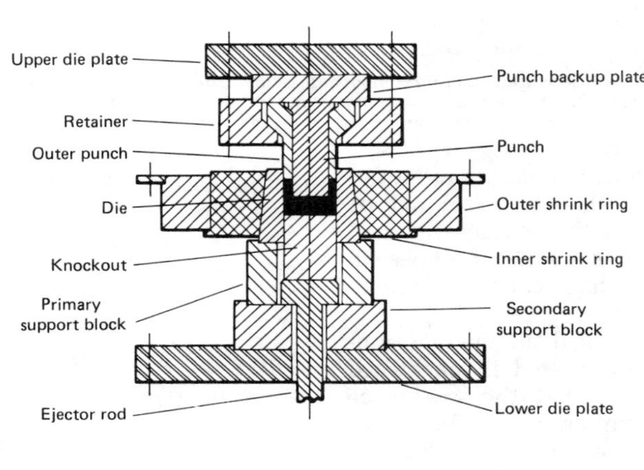

Table 6 Typical tool steels for forward extrusion of part 7
For design of part, see Fig. 2

Metal to be extruded(a)	Total quantity of parts to be extruded 5 000	50 000
Punch material(b)		
Aluminum alloys	A2	D2, M4 (c)
Carbon and alloy steel, up to 0.40% C, and series 300 stainless steels	A2	D2, M4 (c)
Die material(b)		
Aluminum alloys	A2	D2
Carbon and alloy steel, up to 0.40% C, and series 300 stainless steels	(d)	(d)
Knockout material(b)		
Aluminum alloys, steels and series 300 stainless steels	1020 (e) O1 (f)	1020 (e) O1 (f)

(a) Starting with a ring-shaped blank. (b) Where two tool materials are recommended for the same conditions, they are given in order of cost, with the less expensive shown first. (c) First choice in automotive parts processing. (d) No tool steel can be recommended without qualification. Medium-carbon alloy tool steels such as H12, H21, 6F5 and 6H2 have given the best results. (e) Or other low-carbon or low-alloy steels for knockout pins. (f) Knockout heads.

Table 6 gives the tool steels typically used in forward extrusion of part 7 from a ring-shape blank. The pressure required for extrusion of such thin-wall shapes in ferrous alloys is so high that the dies used are made of lower-alloy nonstandard steels such as those described in the preceding paragraph.

Secondary Tooling Components

Table 7 lists the constructional and tool steels used for the secondary tooling components for cold extrusion of parts 1 to 7 (see Fig. 2) in steel. (Tooling for forward or backward extrusion of aluminum consists of only a die, a die holder, a punch and an ejector.) The steels listed in Table 7 reflect the moderate to high extrusion severity involved in producing parts 1 to 7.

Die terminology is presented in Fig. 3, which illustrates a typical tooling setup used for backward extrusion of steel parts. There are some variations in tooling setup from plant to plant. For

Table 7 Typical steels for secondary tooling components used in extruding steel parts

Part number	Related table	Type of operation	Upper die plate	Punch backup plate	Inner shrink ring	Outer shrink ring	Primary support block	Secondary support block	Lower die plate
1 and 2	Table 1	Backward	1040	S1	S4, S5	1040	W2	W2	1040
3	Table 2	Drawing	1040	S1	S4, S5	1040	W2	W2	1040
4	Table 3	Drawing	W2	8620 (a)	4340 (a)	W2	S4, S5	D4	W2
5	Table 4	Forward	1040	S1	S4, S5	1040	W2	W2	1040
6	Table 5	Forward	W2	8620 (a)	4340 (a)	W2	S4, S5	D4	W2
7	Table 6	Forward	W2	8620 (a)	4340 (a)	W2	S4, S5	D4	W2

Note: Terminology for the various tooling components is defined in Fig. 3; part designs are shown in Fig. 2.
(a) Or other alloy steel having hardenability appropriate for the component.

example, an outer punch (sometimes called a punch sleeve) would not be used in some tooling setups for cold extrusion of parts 1 and 2 because an oversize slug might cause a "crack-up".

Tooling for forward extrusion is similar to the tooling used in backward extrusion, except that the workpiece does not bottom on a knockout (although a knockout may be incorporated in the tooling to eject the workpiece from the die). Tooling for drawing-type cold extrusion differs from that shown in Fig. 3, chiefly by incorporating a stripper and a delivery hole for ejection of the drawn part.

The inner shrink ring has a highly finished bore with a taper of 1° to 2°; finishes of 0.2 to 0.25 μm (8 to 10 μin.) rms are usual. The die is press fitted into the inner shrink ring and thereby is subjected to compressive prestressing, which counteracts the tensile stresses developed in the die extrusion. Inner shrink rings are made from low-alloy or medium-alloy steels that have been quenched and tempered to about 50 HRC to enable them to withstand service stresses without undergoing permanent deformation.

Outer shrink rings usually are made from normalized low-carbon or medium-carbon steel such as 1040 or 1045, and are shrunk or pressed onto the inner shrink ring. Design of and materials chosen for support blocks depend on the severity of the cold forming operation, the type of press equipment to be used, and the vertical components of the required extrusion pressures. Tool steels or alloy steels must have sufficient hardenability: support blocks must be hardened deep enough to enable them to absorb service stresses without undergoing permanent deformation.

Materials for upper and lower die plates are selected to resist the deforming pressures inherent in a particular tooling design. Punch retainers may have to be made from deep-hardening steels such as 4140, depending on design. Steels such as 6F5, hardened to 56 to 58 HRC, have been used successfully for outer punches.

Uses of Specific Materials

For primary tools used in forward or backward extrusion, water-hardening tool steels such as W1 and W2 are generally limited to applications where they will harden throughout most of the section, or at least to applications where they will not deform or upset in use as a result of low core hardness.

Oil-hardening tool steels (particularly O1) are used for secondary components and sometimes for short-run punches and dies. O1 steel has sufficient hardenability to develop nearly full hardness in section thicknesses up to about 50 mm (2 in.). Therefore, when quenched and tempered to a working hardness of 58 to 62 HRC at the surface, it will be only slightly lower in hardness in the center of a 50-mm (2-in.) section.

Shock-resisting tool steels S1, S4, S5 and S7 are used for secondary extrusion tools at 57 to 60 HRC. S7 is the most readily available grade and may be substituted for S4 or S5 if either of these grades cannot be obtained.

A2 air-hardening tool steel is a deep-hardening, wear-resistant steel that exhibits minimal dimensional change during hardening. It is the preferred tool steel for cold extrusion dies for all except the highest extrusion pressures and longest runs. It is slightly more expensive than O1 but is more wear resistant, especially when nitrided. A2 is used extensively in short and medium runs for both punches and dies because of its high tensile strength. For most die applications, it is quenched from 950 to 980 °C (1750 to 1800 °F) and tempered at 495 to 525 °C (925 to 975 °F) to obtain a hardness of 56 to 58 HRC.

High-Carbon, High-Chromium Tool Steels. D2 is the most popular high-carbon, high chromium tool steel for large punches used in backward or forward extrusion of steel or aluminum. D2 also is frequently used for dies employed in extrusion or drawing of aluminum. It is used for tools subjected to compressive loads. Although it usually is not used as a die material for forward or backward cold extrusion of steel, it is sometimes used for drawing-type extrusion tools.

Type D2 tool steel is usually double tempered to a working hardness of 57 to 59 HRC. Occasionally, it is tempered back to approximately 55 HRC to avoid punch breakage, albeit at the expense of shorter punch life; in such applications, A2 should be considered an alternative material. The use of hardnesses above 60 HRC is not recommended because a D2 die that hard might shatter explosively when subjected to a high service load.

Types D3 (2.25 C, 12.0 Cr) and D4 (2.25 C, 12.0 Cr, 1.0 Mo) at 62 to 64 HRC are better in wear resistance and compressive strength than D2, and may be used in applications where these advantages outweigh the liabilities of greater brittleness and extra cost of fabrication. One plant reports good results from use of D7 punches (2.35 C, 12.0 Cr, 1.0 Mo, 4.0 V) for long runs in extruding parts such as 1, 2 and 7 (see Fig. 2) in steel. However, for small punches and long runs, the D-series tool steels generally are inferior to M2 high speed steel.

High Speed Steels. M2 is preferred for small punches on short to moderately long runs, and for long runs when the only available press equipment and tooling design result in breakage of carbide tools. M2 at 64 to 66 HRC fre-

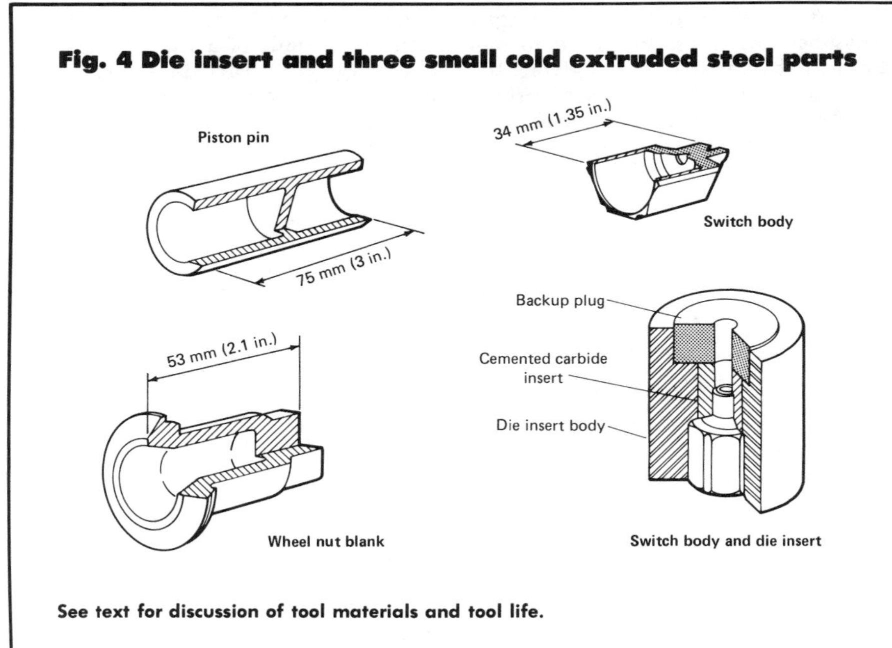

Fig. 4 Die insert and three small cold extruded steel parts

See text for discussion of tool materials and tool life.

quently is specified for punches for moderate runs on operations where the pressure of extrusion would deform D2. The higher-carbon, higher-vanadium M4 high speed steel, heat treated to 62 to 64 HRC, is the first choice for long-run cold extrusion punches processing automotive parts. The difficulty in grinding that previously limited the usefulness of this M4 steel has been overcome by using certain recently developed grinding-wheel formulations. However, if an M4 punch breaks in service, it can be replaced with M2 at the same hardness. T15 high speed steel is sometimes preferred for punches of very large size that must withstand high extrusion pressures.

The wheel nut in Fig. 4 is an example of a part design requiring that a punch form the outside area of the part in such a way that the tensile stresses are imposed on the totally unsupported punch. Such a punch made of nitrided M3 high speed steel has been run extensively in production and has averaged 25 000 to 30 000 pieces.

The piston pin in Fig. 4 is in extensive production. The compressive forces on the punch are only about 1380 MPa (200 ksi), and for small quantities such steels as W1 can be used for dies and M2 or M3, both nitrided, for punches. For longer production runs, the die material would be changed to A2. However, for long and continuous production of this part, the preferred punch and die material to be considered is cemented carbide—justified by ten to twenty times more production (an average of 1 000 000 pieces) per punch and a 90 to 95% reduction in the number of press shutdowns for tooling changes.

Cemented Carbides. The grades of cemented carbides used most frequently in cold extrusion punches and dies are shown in Table 8. In general, the quantity of parts to be produced and the required dimensional tolerances are more important than tool size in making a decision to use carbide.

As a specific example, it is estimated that carbide tooling could not be justified for extruding part 5 (see Fig. 2) in steel in quantities less than 100 000 pieces. Much larger quantities would be required (perhaps 500 000 or more) to justify use of carbide tooling for extruding part 5 from aluminum.

In addition to its use in cold extrusion dies and die inserts, cemented carbide has been used extensively for cold extrusion punches for many years, producing small parts for automotive, farm-equipment and other high-production industries. Typical of such parts are bearing cups, valve lifters, wrist pins and spark-plug bodies. Larger parts now are being produced by cold extrusion. Primarily, these are cuplike shapes that either are used as is or are subsequently punched out to produce bushings. Another trend encouraging the selection of carbides is the need to extrude advanced alloys that require very high forming pressures either in cold extrusion or in warm extrusion of slugs preheated to about 500 °C (about 1000 °F).

Table 8 Cemented carbides used most frequently for cold extrusion punches and die inserts

Type of service	Composition, % Tungsten carbide	Composition, % Cobalt binder	Grain size
Punches			
High impact	84	16	Fine
Medium impact	88	12	Fine
Dies and die inserts			
High impact	75	25	Medium coarse
Medium impact	84	16	Medium fine
Light impact, maximum wear	88	12	Fine

The first and most critical requirement of punches is high compressive yield strength. Strain-gage measurements made during back extrusion of large cups have revealed an average cross-sectional compressive load in excess of 2300 MPa (330 ksi). Because this value is an average across the entire punch, it can be assumed that localized forces may be as much as 50% greater due to temporary misalignment, variations in force during the working stroke, and impact at initial contact with the work material. The compressive yield strength of the high-impact grades of carbide used for these punches is about 1½ times that of fully hardened tool steels.

Another critical requirement is high stiffness, or high modulus of elasticity. This property is most critical for punches having length-to-diameter ratios of 4:1 or greater. Carbides of the high-impact type have 2½ times the elastic modulus of high-alloy tool steels.

Because of the overhang involved in cold extrusion punches and the imbalance in forces that can occur during backward or forward extrusion, the punch material must be capable of resisting high bending stress. The high modulus of rupture, or high breaking strength, of high-impact cemented carbides provides them with the necessary resistance to bending.

In addition, because cold extrusion presses are operated at high rates, punches must have good shock resistance to withstand high impact at the point of contact with the work metal. Good fatigue resistance is required in punches if they are to endure the tens of thousands, or even hundreds of thousands, of strokes expected during the service life of such tooling. Fatigue resistance is inversely related to mechanical hysteresis, and, given the same deflection, high-impact carbide tool materials have approximately one-eighth the mechanical hysteresis of steel. This, combined with low deflection as a result of a high modulus of elasticity, gives carbide a mechanical hysteresis advantage over steel of about 18:1 or 20:1 for a given loading on the tool.

Selection of Tool Materials for Drawing Wire, Bar and Tubing

Edited by L. C. Shaheen
Director, Mechanical Development
Superior Tube Company

SELECTION of tool materials for cold drawing metal into continuous forms such as wire, bar and tubing depends primarily on the size, composition, shape, stock tolerance and quantity of the metal being drawn. Cost of the tool material is also important and may be decisive.

Dies and mandrels used for cold drawing are subjected to severe abrasion. For this reason, most wire, bar and tubing is drawn through dies having diamond or cemented tungsten carbide inserts, and tube mandrels usually are fitted with carbide nibs. Small quantities, odd shapes and large sizes are more economically drawn through hardened tool steel dies.

The compositions and properties of the tool materials mentioned in this article are discussed elsewhere in this volume.

Wiredrawing Dies

Table 1 gives recommended materials for wiredrawing dies. For round wire, dies made of diamond or cemented tungsten carbide are always recommended, without regard to the composition or quantity of the metal being drawn. For short runs or special shapes, hardened tool steel is less costly, although carbide gives superior performance in virtually any application.

Die Life. In a wiredrawing die, both the approach angle and the bearing area (see Fig. 1) are subjected to severe abrasion. Normal die life is defined as the length or mass of metal drawn through a die that causes the bearing area of the die to increase from minimum to maximum size. Factors that influence die wear, both singly and collectively, are drawing speed, composition of the metal being drawn, wire temperature, reduction per pass, and hardness of the die material. Figure 2 presents data that illustrate the effects of some of these factors on wear of both natural diamond and tungsten carbide dies.

Wear often begins as an angular ring on the approach angle of the die. Die life may be increased by as much as 200% if the die is removed and repolished at the first appearance of this ring; otherwise, die wear will accelerate. Redressing should never shorten the length of the bearing area to less than 30% of the product diameter.

Diamond Dies. The use of diamond dies is restricted only by limitations on the sizes of available industrial diamonds and by cost, which is extremely high for diamonds in larger sizes. These tools can outperform cemented tung-

Fig. 1 Cross section of typical wire die

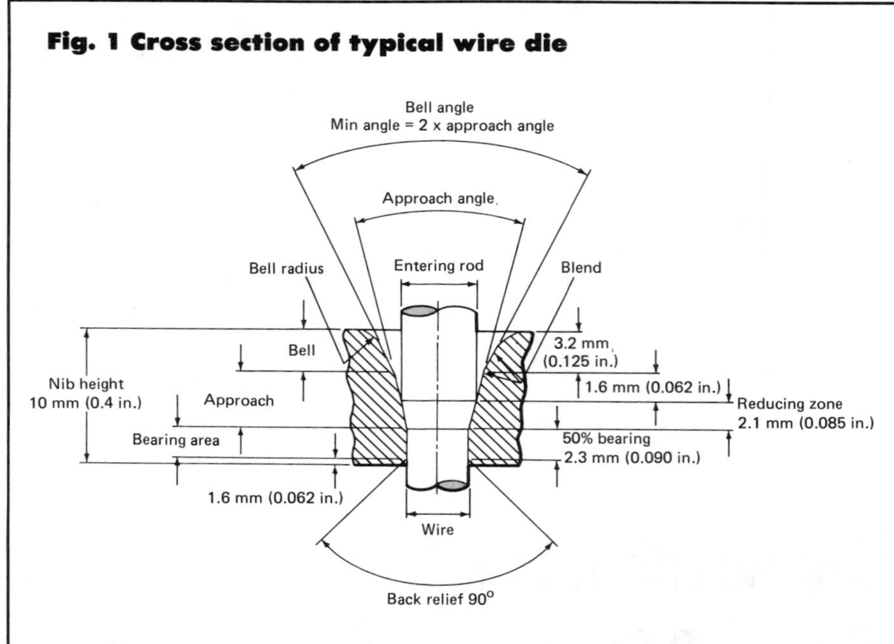

For drawing 5.5-mm (0.218-in.) diameter rod to 4.6-mm (0.180-in.) diameter wire (17% reduction per pass).

Table 1 Recommended materials for wiredrawing dies

Metal to be drawn	Wire size mm	Wire size in.	Recommended die material: Round wire	Recommended die material: Special shapes
Carbon and alloy steels	<1.57	<0.062	Diamond, natural or synthetic	CPM 10V, M2 or cemented tungsten carbide
	>1.57	>0.062	Cemented tungsten carbide	
Stainless steels; titanium, tungsten, molybdenum and nickel alloys	<1.57	<0.062	Diamond, natural or synthetic	CPM 10V, M2 or cemented tungsten carbide
	>1.57	>0.062	Cemented tungsten carbide	
Copper	<2.06	<0.081	Diamond, natural or synthetic	CPM 10V, D2 or cemented tungsten carbide
	>2.06	>0.081	Cemented tungsten carbide	
Copper alloys and aluminum alloys	<2.5	<0.100	Diamond, natural or synthetic	CPM 10V, D2 or cemented tungsten carbide
	>2.5	>0.100	Cemented tungsten carbide	
Magnesium alloys	<2.06	<0.081	Diamond, natural or synthetic	
	>2.06	>0.081	Cemented tungsten carbide	

sten carbide dies by 10 to 200 times, depending on the alloy being drawn, and thus are cost effective despite their high unit cost.

Diamond dies are available in two types: natural and synthetic. Natural diamond dies, which have been used the longest, are made from single crystals and produce exceptional surface finishes in the drawn product.

Performance can be erratic; one out of every four or five dies breaks after only 30% of its expected life because of crystal flaws or an unfavorable orientation of natural cleavage planes in the single crystal. In addition to this limitation, supplies of the large natural diamond crystals needed for drawing dies are becoming scarce and extremely expensive.

Synthetic diamond dies are manufactured from polycrystalline synthetic diamond and cemented tungsten carbide fused together under high temperature and pressure. Minute, randomly oriented diamond and carbide crystals are bonded together to form a die blank that is high in hardness and wear resistance and that is uniform in all directions. The fine-grain isotropic structure of the material greatly diminishes cracking and nonuniform wear. As a result, comparative die-life tests have shown that synthetic diamond dies give longer and more consistent service than natural diamond dies (see Table 2). Unfortunately, the surface finish generated by synthetic diamond dies is dull and more striated than that generated by natural diamond dies. As a result, some users employ synthetic material for breakdown operations and natural diamond for finishing passes.

Table 3 gives ratios of initial cost of natural diamond dies to initial cost of synthetic diamond dies for hole sizes in the range typical of diamond tooling. Generally, natural diamond is less expensive than synthetic diamond for smaller hole sizes. For a hole size of 0.66 mm (0.0259 in.), the two materials are about the same in cost. For larger sizes, synthetic diamond is more economical. Synthetic diamond dies are generally more cost effective throughout the entire size range when both initial cost and total die life are considered together.

Cemented tungsten carbide is economical for wiredrawing dies in most applications above the range of size where diamond can be used. The softer cemented carbides, which contain about 8% cobalt, are less brittle

Fig. 2 Wear of natural diamond and cemented tungsten carbide dies used in wiredrawing

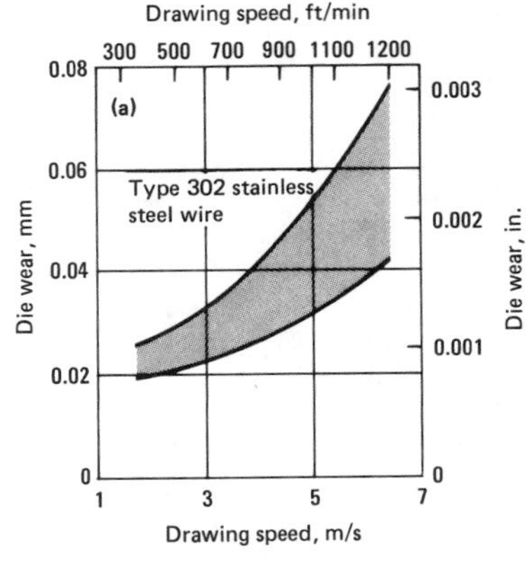

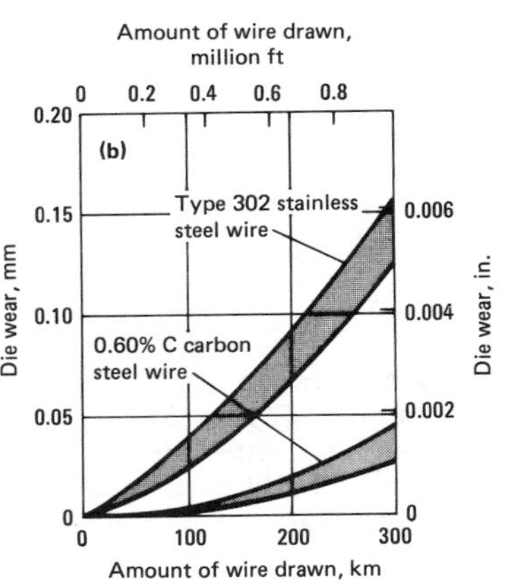

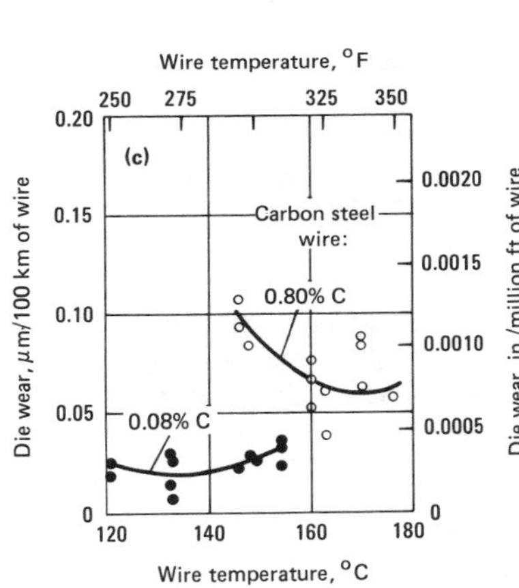

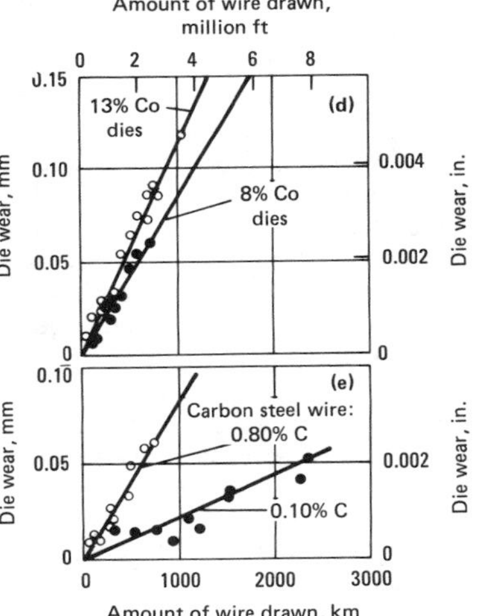

(a) Effect of drawing speed on die wear 302 stainless steel wire. Wire was 1.4 mm (0.55 in.) in diameter, was lime coated, and had a tensile strength of 1030 MPa (150 ksi). A total of 7.6 km (255 000 ft) was drawn using an oil lubricant and 20% reduction in area per die. (b) Die wear in drawing 302 stainless steel with tensile strength of 965 MPa (140 ksi) and 0.60% C steel wire with tensile strength of 900 MPa (130 ksi). Drawn in oil from 0.84 to 0.50 mm (0.033 to 0.0199 in.) in diameter; 20% reduction in area per pass. Both alloys were lime coated. (c) Effect of temperature on die wear. Temperature of wire at exit side of die measured by temperature-indicating crayons. Steel wire (0.80% C) drawn to 0.8 mm (0.033 in.) at 10 m/s (2000 ft/min), using a special lubricant. (d) Influence of cobalt content of cemented tungsten carbide dies on rate of die wear in drawing 0.80%-C patented wire at 76 to 360 m/s (255 to 1200 ft/min). Drawn from 2.8 to 1.0 mm (0.110 to 0.041 in.) in diameter, using a special lubricant. (e) Effect of wire composition on wear of 8%-Co cemented tungsten carbide dies. Steel wire of different carbon contents drawn using a proprietary lubricant.

and can withstand greater stock reductions without breaking, but wear more rapidly than lower-cobalt grades (Fig. 2). If not damaged or broken, carbide dies can be progressively reworked to accommodate larger wire sizes. Diamond dies also can be reworked, but greater numbers of reworkings are expected from carbide dies. Data on life of carbide dies used for drawing various sizes of steel wire, reported by two different mills, are shown in Table 4. Data on life of carbide dies used in drawing stainless steel wire were as follows:

Wire size	Life (a)
1.27 mm (0.050 in.)	810 kg (1800 lb)
0.64 mm (0.025 in.)	450 kg (1000 lb)

(a) Weight of wire drawn per die.

Tool steel used for wiredrawing dies should have near-maximum hardness (62 to 64 HRC) for reductions below about 20%. For greater reductions, because of the possibility of breakage, hardness should be decreased to 58 to 60 HRC, even though the rate of wear will increase.

Die breakage is usually caused by abnormal reductions, lack of mechanical support for the insert, inadequate lubrication, or use of a tool material that is too hard and brittle for the amount of reduction and speed. Some wear resistance is always sacrificed to minimize breakage.

Drawing Bars and Tubing

Table 5 shows die and mandrel materials recommended for drawing bars and tubing. Diamond is virtually never used in larger sizes; cemented tungsten carbide is recommended for three-fourths of all applications. Tool steels are rarely used to make tools for drawing commercial quality round bars less than 90 mm (3.5 in.) in diameter. Cemented tungsten carbide is used to draw stainless steel tubes as large as 280 mm (11 in.) in outside diameter.

Common sizes are usually drawn in sufficient quantities to warrant the investment in carbide dies. In addition, carbide bar or tube dies can be reworked to the next larger size. Die life after reworking is substantially the same as for the first run. In drawing steel bars, it is possible to increase normal die life by properly planning the sequence of compositions to be drawn. For example, in drawing 0.45% carbon steel bars 25.40 mm (1.000 in.) in diameter, a minus tolerance of 0.08 mm (0.003 in.) is allowed, but for this grade it is necessary to allow for a 0.05-mm

Table 2 Die-life comparison of natural and synthetic diamond

Metal	Wire diameter mm	in.	Die-life ratio, synthetic : natural
Aluminum	0.64-2.5	0.025-0.100	3:1
Copper	0.38-2.0	0.015-0.080	10:1
Nickel 200	0.33-1.5	0.013-0.057	10:1
Tungsten	0.18-0.64	0.007-0.025	4:1
Molybdenum	0.18-1.0	0.007-0.040	5:1
Brass-plated steel	0.15-0.97	0.006-0.038	4:1
Type 304 stainless steel	0.41-1.6	0.016-0.062	6:1
Type 304L stainless steel	0.71-1.5	0.028-0.060	3:1
Type 302 stainless steel	0.36-0.71	0.014-0.028	3:1

(0.002-in.) elastic expansion of the bar after it passes through the die. When the die is worn to maximum size at the bearing area, it will still be only 25.35 mm (0.998 in.) in diameter. It is then possible to draw 0.20% carbon steel bars, which expand less because of their lower yield strength. After the limit of tolerance has been reached for this grade (a diameter of 25.38 mm, or 0.999 in., at the bearing area), the die can be used for drawing a still lower carbon steel, such as a low-carbon free-machining grade that expands even less, until the diameter of the bearing area reaches 25.40 mm (1.000 in.). The dies then can be reworked to the next usable size. In many instances, the planning of drawing sequences is more complicated than described above. Bell angle, approach angle, back relief and amount of subsequent straightening all affect as-drawn size because they influence the amount of elastic growth that occurs, and thus must be taken into account when planning drawing sequences.

Because of their high hardness, carbide dies become progressively less expensive to rework as the worn size approaches the next usable size. This is not necessarily true for tool steel dies, because they can be annealed to soften them for reworking, and then rehardened.

There has been some hesitation in using carbide for large dies because of cost. If a large carbide die is broken, it represents an appreciable monetary loss; however, carbide dies up to 280 mm (11 in.) in diameter have been used successfully and economically. One mill reported that steel dies produced an average of 10 km (32 000 ft) of 180-mm (7-in.) oil-well casing before being permanently withdrawn from service. Carbide dies averaged 122 km (400 000 ft) of the same casing, and one carbide die produced 180 km (590 000 ft). In this example, the die cost per unit length drawn was about one-fifth as much for carbide as for steel.

When complex shapes are to be drawn, selection of die material is somewhat uncertain. In short runs less than 300 m (1000 ft), tool steels generally are more economical. For longer runs, carbide is usually more economical, unless sharp edges, which may cause the carbide to chip, are involved. In that event, tool steel dies must be used, even though they may have to be replaced more frequently because of wear. A proprietary P/M tool steel, CPM 10V, is another alternative to cemented carbide. CPM 10V has toughness equivalent to the conventional tool steels D2 and M2, and has substantially superior wear resistance in drawing-die service.

Adjustable sectional dies are widely used for drawing rectangular and square bars where sharp corners are required on the finished material. The die is composed of side and end blocks of solid cemented tungsten carbide. The sizes of the end blocks determine the position of the bar drawn; the positions of the side blocks determine the width. A special die holder is required, using hardened and ground back-up plates to ensure proper alignment. This type of die is not suitable for drawing rectangular bars with corner radii.

Mandrels. Either carbide or hardened tool steel is satisfactory for mandrels used in drawing tubes, but carbide is more economical for tubes less than 125 mm (5 in.) in diameter. Carbide nibs are available in lengths sufficient for this purpose. Mandrel nibs are available in either a braze-type design or shell design that permits mechanical attachment to the shank for easy replacement. Carbide tips are also recommended for mandrels used in drawing

Table 3 Relative cost of natural and synthetic diamond dies

Hole size mm	in.	Relative cost(a)
0.051	0.0020	0.60
0.064	0.0025	0.60
0.15	0.0060	0.60
0.23	0.0092	0.60
0.26	0.0102	0.65
0.29	0.0116	0.70
0.33	0.013	0.70
0.37	0.0146	0.80
0.42	0.0164	0.95
0.47	0.0184	0.75
0.52	0.0205	0.90
0.59	0.0231	0.95
0.66	0.0259	1.0
0.74	0.0292	1.2
0.83	0.0327	0.80
0.93	0.0368	0.90
1.04	0.0411	1.1
1.18	0.0463	1.4
1.32	0.0521	1.6
1.48	0.0581	1.9
1.66	0.0653	1.4
1.87	0.0736	1.5
2.10	0.0827	1.6
2.36	0.0931	1.0
2.62	0.1031	1.1
2.95	0.1161	1.2

(a) Ratio of initial cost of natural diamond dies to initial cost of synthetic diamond dies.

Table 4 Life of 9%-Co cemented tungsten carbide dies used in drawing steel wire

Diameter mm	in.	Life (approx), tonnes/die or tons/die
4.1	0.162	100
3.2	0.125	70
2.3	0.0915	40
1.8	0.072	35
0.5	0.018	1
2.7-4.7	0.105-0.187	50-75(a)
0.9-2.7	0.035-0.105	10-20(b)
0.4-0.9	0.016-0.035	7½-10(c)
0.15-0.4	0.005-0.016	1½-2(d)

(a) Drawing speed, 1.25-3.75 m/s (250-750 ft/min). (b) Drawing speed, 3.75-9.0 m/s (750-1800 ft/min). (c) Drawing speed, 7.5-10.0 m/s (1500-2000 ft/min). (d) Drawing speed, 10.0-15.0 m/s (2000-3000 ft/min).

of shapes, but with reservations on the use of carbide similar to those that apply to the use of carbide in dies for drawing of shapes. Tool steel mandrels are recommended for drawing of tubes over 125 mm (5 in.) in inside diameter.

Tool Breakage. The most frequent cause of die breakage in bar and tube drawing is a die design inappropriate for the percentage reduction. Excessive die hardness also frequently leads to breakage, particularly of dies for drawing thin-wall tubing. Lack of lubrication, excessive drawing speeds and other extreme conditions of operation also contribute to die breakage.

The permissible hardness for tool steel dies varies inversely with speed of drawing and percentage reduction. For reductions above 20% and for high speeds, hardness should be kept between 58 and 60 HRC. For lesser reductions and lower speeds, hardness may be increased to 60 to 64 HRC.

As tool size increases, closer attention should be given to grade selection for carbide dies. Several types of cemented tungsten carbide that have been used to make dies for drawing bars and tubes of different sizes are given in Table 6. Final selection of grade is usually made after consultation with the supplier.

Table 5 Recommended tool materials for drawing bars, tubing and complex shapes

Metal to be drawn	Round bars and tubing(a): Common commercial sizes: Bar and tube dies	Round bars and tubing(a): Common commercial sizes: Tube mandrels(b)	Round bars and tubing(a): Maximum commercial size(c): dies and mandrels	Complex shapes: dies and mandrels(a)(b)
Carbon and alloy steels	Tungsten carbide	W1 or carbide	D2 or CPM 10V	CPM 10V or carbide
Stainless steels, titanium, tungsten, molybdenum and nickel alloys	Diamond or carbide(d)	D2 or carbide	D2, M2 or CPM 10V(a)	F2 or carbide(e)
Copper, aluminum and magnesium alloys	W1 or carbide	W1 or carbide	D2 or CPM 10V	O1, CPM 10V or carbide

(a) Tool steels for both dies and mandrels are usually chromium plated. (b) "Carbide" indicates use of cemented carbide nibs fastened to steel rods. (c) 10-in. OD by 3/4-in. wall. (d) Under 1.5 mm (0.062 in.), diamond; over 1.5 mm (0.062 in.), tungsten carbide. (e) Recommendations for large tubes or complex shapes apply to stainless steel only.

Table 6 Types of tungsten carbide recommended for dies for drawing bars, tubes and shapes

Diameter(a), mm	Diameter(a), in.	Composition(b), %: Co	Composition(b), %: TaC + TiC	Hardness, HRA
3 to 16	1/8 to 5/8	2.5 to 6.5	0 to 3	93 to 91
16 to 50	5/8 to 2	6.5 to 15	0 to 2	92 to 86
Over 50	Over 2	15 to 30	0 to 5	88 to 83

(a) Inside diameter for dies, outside diameter for mandrels, diameter of the circumscribed circle for complex shapes. (b) Remainder WC for all sizes.

The finish on the working surface of the die is extremely important when drawing bar and tubing. Polishing may increase die life by 200% compared to similar unpolished dies. Chromium plating 0.08 to 0.25 mm (0.003 to 0.010 in.) thick is usually recommended for hardened steel dies and mandrels.

Selection of Materials for Closed-Die Hot Forging Tools

Edited by William Wilson
Director of Research
A. Finkl & Sons Co.
and
Sanjay N. Shah
Research Group Leader
Wyman-Gordon Company

THE CLOSED-DIE FORGING TOOLS discussed in this article are restricted to die blocks, die inserts and trimming tools used for hot forging in vertical presses and hammers. In hammer forging, the hammer—whether a gravity, steam, air-drop or counterblow hammer—strikes a sudden blow, imposing a shock load on the forging dies. In press forging, the working pressure is applied as a fast push rather than a blow, so that the dies are subjected to less shock. However, because a press is much slower acting than a hammer, the dies in a press absorb more heat from the hot blank during the forging cycle.

The size and shape of the part being forged influence the force and energy required to reshape the hot plastic metal—from the initial shape of the forging stock (usually round, square or flat) to that of the finished forging. The force and energy required are further influenced by the composition of the metal being forged. For example, as the alloy content of a steel increases, hot strength and subsequent resistance to flow also increase, and more energy is required to forge the same shape. Similarly, some copper alloys and aluminum alloys are considerably easier to forge than others.

Causes of Die Failure

The basic causes of premature die failure are excessive force, abrasion, and excessive temperature, as discussed in the three subsections that follow. In addition, dies may break in a

Table 1 Nominal compositions of nonstandard tool steels for die blocks

Steel(a)	C	Mn	Si	Cr	Ni	Mo	V
	Composition, %						
6G	0.55	0.80	0.25	1.00	...	0.45	0.10(b)
6F2	0.55	0.85	0.25	1.00	1.00	0.40	0.10(b)
6F3	0.55	0.60	0.85	1.00	1.80	0.75	0.10(b)

(a) Neither AISI nor SAE has assigned type numbers to these tool steels. (b) Optional.

brittle manner if used cold, and thus preheating to 260 to 300 °C (500 to 600 °F) is recommended. This may be accomplished by placing "warmers" (pieces of hot steel) between the dies or by installing gas-fired or electrical heating devices to maintain temperature during idle periods.

Excessive force, either general or localized, can result in excessive die wear and possibly breakage. Both wear and breakage can be minimized by selecting die steel and hardness carefully; using blocks of adequate size; controlling stock weight and forging temperature; and ensuring proper application of working pressures, correct metal flow from preform to blocker to finisher, and proper seating of dies in the hammer or press.

Abrasion is inherent in forging, and there is no practical way of preventing it. Abrasion caused by flowing and spreading of hot metal in the die impressions increases in intensity as the impressions increase in complexity and severity. The degree of abrasion also is influenced by the hot strength of the metal being forged. Abrasive action may be further aggravated if scale is present on the metal being forged.

Although abrasion cannot be eliminated, its effects can be minimized by good die design, correct combinations of die composition and hardness, and forging techniques that include proper heating, descaling (if necessary) and correct die lubrication.

Excessive temperature is the largest single cause of early die wear, and resistance to heat is a major consideration in selecting die steels. Typical temperatures for forging various metals are as follows:

Magnesium alloys	370 °C (700 °F)
Aluminum alloys	425 °C (800 °F)
Copper alloys	815 °C (1500 °F)
Tool steels	1040 °C (1900 °F)
Stainless steels and heat-resisting alloys	1200 °C (2200 °F)
Carbon and alloy steels	1260 °C (2300 °F)

Actual temperatures may vary 55 °C (100 °F) or more from those listed, depending on composition. Because forging may be done at temperatures from 370 to 1260 °C (700 to 2300 °F), the working temperatures to which die materials may be subjected also cover a broad range. The temperature within die blocks used for forging steel usually reaches 150 to 300 °C (300 to 600 °F), although thin projections or plugs may reach 400 to 550 °C (750 to 1000 °F). In forging aluminum and magnesium alloys, it is desirable to heat the dies and to maintain die temperature near the forging temperature of the metal to minimize extraction of heat from the forging due to contact with the dies.

During forging of steel and other heavy metals, die surfaces may reach temperatures higher than the die temperatures mentioned above, and heat checking (see Fig. 1) may result. Heat checking is the development of minute surface cracks, mainly in corners or on projections within the die cavity. It results from high surface temperatures that cause greater thermal expansion at working surfaces than in the interior of the die block. The difference in expansion causes plastic deformation at the surface, which results in tensile stress and cracking on cooling. Once surface cracks are started, working pressures (especially impact) will cause them to grow, and if crack growth is allowed to continue, the die block will break.

Heat checking is greatest in forging of steel and heat-resisting alloys, considerably less in forging of copper alloys, and very minor in forging of aluminum or magnesium alloys. Because of the lower forging temperatures involved, considerations other than heat checking govern the selection of a die steel for nonferrous alloys.

Continuous production may cause some parts of the die to reach excessively high temperatures unless adequate cooling is provided. Water sprays should not be used to cool hot spots;

Table 2 Normal hardness ranges for prehardened die blocks of various tool steels

Hardness, HB	Brinell indentation diameter(a), mm	Hardness, HRC	Scleroscope hardness
444 to 477	2.80 to 2.90	47 to 50	63 to 66
388 to 429	2.95 to 3.10	42 to 46	56 to 61
341 to 375	3.15 to 3.30	37 to 40	50 to 54
302 to 331	3.35 to 3.50	32 to 36	45 to 48
269 to 293(b)	3.55 to 3.70	28 to 31(b)	40 to 43(b)

(a) Die block hardness often specified as the Brinell indentation diameter for a 3000-kg test. (b) Hardened and tempered die blocks also available at lower hardnesses.

Fig. 1 Gross heat checking in a D-6ac forging die due to excessive temperature

Heat checking occurred after an undetermined number of 225-kg (500-lb) Ni-base alloy preforms had been forged from an average temperature of 1100 °C (2000 °F).

Fig. 2 Six hypothetical parts of progressively increasing severity

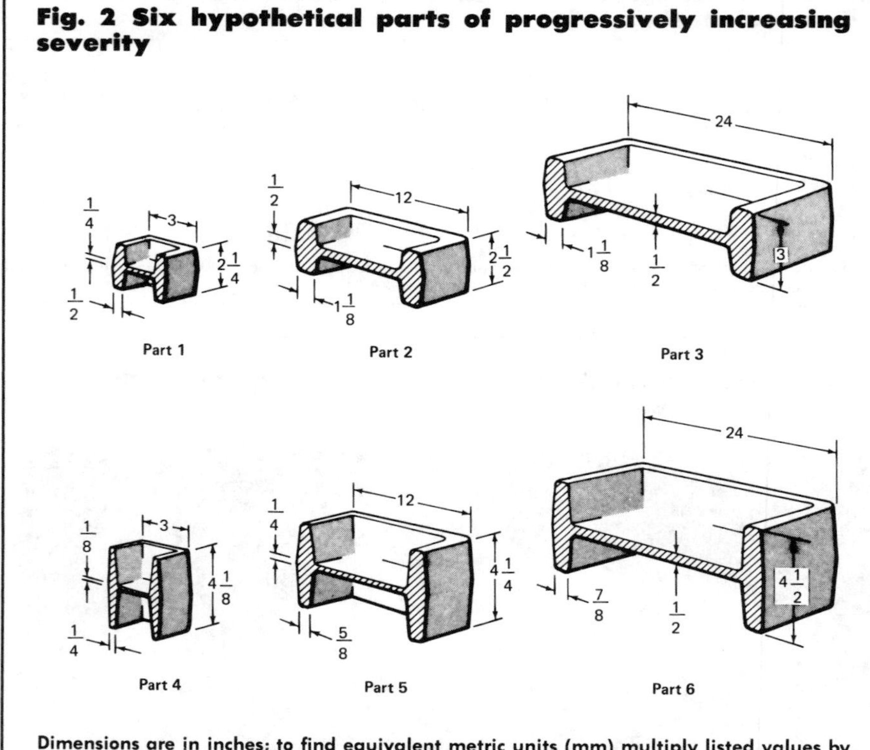

Dimensions are in inches; to find equivalent metric units (mm) multiply listed values by 25. For typical die materials and hardness ranges, see Table 3.

Fig. 3 Typical die design for automobile axle housings

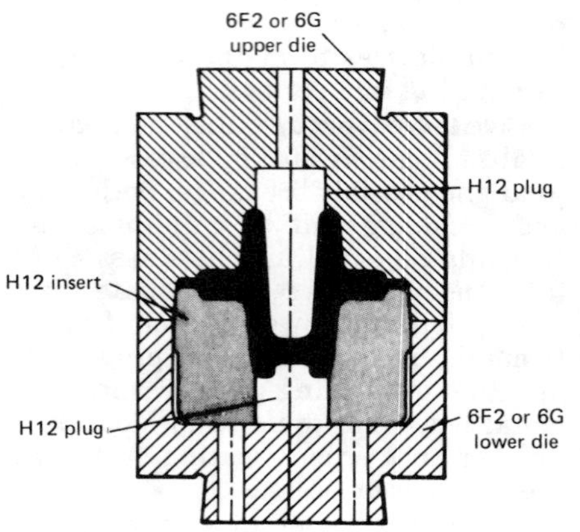

Plug-type inserts (unshaded) would always be recommended for making forgings of the severity illustrated above (black area represents section of an automotive axle housing). In this instance, an H12 steel plug in the upper die section is used in combination with a nearly complete H12 insert in the lower die section.

they introduce very high thermal stresses, which can lead to early die failure. If the forging technique being used makes prevention of excessive die temperatures impractical, dies or portions of dies should be constructed of tool steel grades more resistant to heat.

Die Materials

The compositions and properties of standard grades and types of tool steel discussed below are presented elsewhere in this volume. The nominal compositions of three nonstandard low-alloy tool steels that are widely used for prehardened forging-die blocks are presented in Table 1.

Prehardened die blocks suitable for making forging dies are available in a range of compositions and hardnesses. Prehardened tool steels also are available in other forms for making small die blocks, die inserts and trimming tools. All steels available in a prehardened condition also are available in the annealed condition for ease of machining; once machined to the desired contours, they can be hardened and finished by methods ordinarily used for standard tool steels.

Table 2 lists five hardness ranges in which prehardened die blocks are commercially available. Hardness ranges are tabulated in Brinell, Rockwell and Scleroscope scales.

Hardness measurements are usually made with a Brinell tester, because almost all die blocks are too large for testing on a standard Rockwell machine and because the Brinell test is less sensitive to the minute structural dissimilarities that are present in massive blocks. A carbide ball is used for Brinell tests on hardened die blocks because it has better resistance to deformation at high hardness levels and better dimensional stability than a steel ball.

In many forging shops, hardness is measured with a Scleroscope because of its portability; often, Scleroscope readings are converted and reported in some other hardness scale.

Prehardened die-block steels usually are purchased on the basis of hardness and proprietary name. Typical tool steels, and their Brinell hardnesses, for die blocks and die inserts used in hammer and press forging are shown in Table 3. In this table, recommended die materials are listed for the six hypothetical shapes of increasing severity illustrated in Fig. 2. Information is included for two production quantities and four types of work metal.

The recommendations for steel and hardness level in Table 3 are based on lowest over-all cost (which includes both material and fabrication costs), and on avoiding breakage. Lower-alloy tool steels at lower hardnesses are acceptable when relatively few parts are to be made from carbon or low-alloy steel. Somewhat higher hardness is required in dies for forging stainless steels and heat-resisting alloys.

The relatively expensive, high-alloy H11 and H12 tool steels are desirable for forging copper alloys, because copper alloys are quite resistant to flow at their maximum forging temperatures. Copper oxide is abrasive, and copper alloy forgings usually are made to closer tolerances than steel forgings; both of these characteristics demand high wear resistance in the die material.

As shown in Table 3, higher hardnesses can be used for press-forging dies than for hammer-forging dies because the former are not subjected to

Table 3a Typical tool steels for die blocks and die inserts

For illustration of forged parts, see Fig. 2

Work metals	Typical die materials and hardness ranges — Hammer forging: 100 to 10 000 parts	Hammer forging: 10 000 parts or more	Press forging: 100 to 10 000 parts	Press forging: 10 000 parts or more
For making parts of severity no greater than part 1				
Carbon and alloy steels	6F2 or 6G at 341 to 375 HB	6F2 or 6G at 388 to 429 HB	6F2 or 6G at 388 to 429 HB	6F3 at 369 to 388 HB or H11 or H12 at 388 to 405 HB(a)
Stainless steels and heat-resisting alloys	6F2 or 6G at 388 to 429 HB	6F2 or 6G at 388 to 429 HB	6F2 or 6G at 388 to 429 HB	H11 or H12 insert at 477 to 543 HB or H26 at 514 to 577 HB(b)
Aluminum and magnesium alloys	6F2 or 6G at 302 to 331 HB	6F2 or 6G at 341 to 375 HB	6F2 or 6G at 341 to 375 HB	6F3 at 375 to 405 HB or H11 or H12 at 448 to 477 HB(a)
Copper and copper alloys	6F2 or 6G at 341 to 375 HB or H11 or H12 at 405 to 433 HB	H11 or H12 at 405 to 448 HB	6F2 or 6G at 341 to 375 HB or H11 or H12 at 477 to 514 HB	H11 or H12 at 477 to 514 HB
For making parts of severity no greater than part 2				
Carbon and alloy steels	6F2 or 6G at 341 to 375 HB	6F2 or 6G at 341 to 375 HB	6F2 or 6G at 388 to 429 HB	6F3 at 369 to 388 or H11 or H12 at 388 to 405 HB(a)
Stainless steels and heat-resisting alloys	6F2 or 6G at 341 to 375 HB	6F2 or 6G at 341 to 375 HB with 6F3 insert at 405 to 448 HB	...	...
Aluminum and magnesium alloys	6F2 or 6G at 302 to 331 HB	6F2 or 6G at 341 to 375 HB or H11 or H12 at 405 to 448 HB(a)	6F2 or 6G at 341 to 375 HB	6F2 or 6G at 341 to 375 HB with 6F3 insert at 405 to 448 HB or H11 or H12 at 448 to 477 HB(a)
Copper and copper alloys	6F2 or 6G at 341 to 375 HB	6F2 or 6G at 341 to 375 HB or H11 or H12 at 405 to 448 HB(a)	6F2 or 6G at 341 to 375 HB	H11 or H12 at 477 to 514 HB
For making parts of severity no greater than part 3				
Low-alloy steels, stainless steels and heat-resisting alloys	6F2 or 6G at 302 to 331 HB	6F2 or 6G at 302 to 331 HB with insert of same steel at 341 to 375 HB	...	...
Aluminum, magnesium and copper alloys	6F2 or 6G at 269 to 293 HB	6F2 or 6G at 302 to 331 HB	...	...

(a) Recommended for long runs—for example, 50 000 forgings. (b) Recommended for forging higher-alloy heat-resisting materials, such as nickel-base and cobalt-base alloys.

impact. Because of the longer times in contact with hot work metal, press-forging dies must be made of a steel having greater resistance to softening at elevated temperatures.

Die designs that incorporate plug-type inserts are frequently employed for forging of shapes of greater severity. Fig. 3 shows the plug-type inserts used in closed-die forging of an automotive axle housing.

The effects of forging size and shape on die life are given in Fig. 4. The chart at upper left in Fig. 4 presents life data for dies used in making eight different parts of stainless steel and heat-resisting alloys; all of these parts were forged in the same plant. A difference in the type or grade of alloy being forged can cause some variation in die life, but in this study, the shape of the part was the largest factor affecting die life. For example, three die sets had an average life of only 1500 pieces in forging a gear

Table 3b Typical tool steels for die blocks and die inserts

For illustration of forged parts, see Fig. 2

Work metals	Typical die materials and hardness ranges			
	Hammer forging		Press forging	
	100 to 10 000 parts	10 000 parts or more	100 to 10 000 parts	10 000 parts or more
For making parts of severity no greater than part 4				
Carbon and alloy steels	6F2 or 6G at 341 to 375 HB, solid or with H11 or H12 plug(a) at 369 to 388 HB	6F2 or 6G at 341 to 375 HB with H11 or H12 plug at 369 to 388 HB or H11 or H12 at 405 to 433 HB	6F2 or 6G at 388 to 429 HB, solid or with H11 or H12 plug(a) at 405 to 433 HB	6F2 or 6G at 388 to 429 HB with H11 or H12 plug at 405 to 433 HB
Stainless steels and heat-resisting alloys	6F2 or 6G at 341 to 375 HB, solid or with H11 or H12 plug(a) at 429 to 448 HB	6F2 or 6G at 341 to 375 HB with H11 or H12 insert at 429 to 448 HB	6F2 or 6G at 388 to 429 HB, solid or with H11 or H12 plug(a) at 429 to 448 HB	6F2 or 6G at 341 to 375 HB with H11 or H12 plug at 429 to 448 HB
Aluminum and magnesium alloys	6F2 or 6G at 341 to 375 HB or H11 or H12 at 405 to 433 HB	6F2 or 6G at 341 to 375 HB with H11 or H12 plug at 405 to 433 HB or H11 or H12 at 405 to 433 HB	6F2 or 6G at 341 to 375 HB or H11 or H12 at 405 to 433 HB	6F2 or 6G at 341 to 375 HB with H11 or H12 plug at 429 to 448 HB
Copper and copper alloys	H11 or H12 at 405 to 433 HB	H11 or H12 at 405 to 433 HB	H11 or H12 at 405 to 433 HB	6F2 or 6G at 341 to 375 HB with H11 or H12 plug at 429 to 448 HB
For making parts of severity no greater than part 5				
Carbon and alloy steels	6F2 or 6G at 302 to 331 HB	6F2 or 6G at 302 to 331 HB, solid or with 6F3 plug(b) at 369 to 388 HB	6F2 or 6G at 341 to 375 HB	6F3 at 369 to 388 HB with H11 or H12 plug at 369 to 388 HB
Stainless steels and heat-resisting alloys	6F2 or 6G at 302 to 331 HB	6F2 or 6G at 302 to 331 HB with H11 or H12 plug at 369 to 388 HB	...	...
Aluminum and magnesium alloys	6F2 or 6G at 269 to 293 HB	6F2 or 6G at 269 to 293 HB with plug of same steel at 302 to 331 HB	6F2 or 6G at 341 to 375 HB	6F2 or 6G at 341 to 375 HB with H11 or H12 plug(c) at 429 to 448 HB
Copper and copper alloys	6F2 or 6G at 302 to 331 HB	6F2 or 6G at 302 to 331 HB with H11 or H12 plug at 405 to 448 HB	6F2 or 6G at 341 to 375 HB	H11 or H12 at 477 to 514 HB
For making parts of severity no greater than part 6				
Low-alloy steels, stainless steels and heat-resisting alloys	6F2 or 6G at 269 to 293 HB(d)	6F2 or 6G at 269 to 293 HB with plug of same steel at 341 to 375 HB	...	...
Aluminum, magnesium and copper alloys	6F2 or 6G at 269 to 293 HB	6F2 or 6G at 269 to 293 HB	...	...

(a) Recommended for 1000 to 10 000 forgings. (b) Recommended for long runs—for example, 50 000 forgings. (c) For long runs—for example, 50 000 forgings—a solid block made from H11 or H12 tool steel at 477 to 514 HB is recommended. (d) For quantities over 1000, a plug of the same material at 341 to 375 HB is recommended.

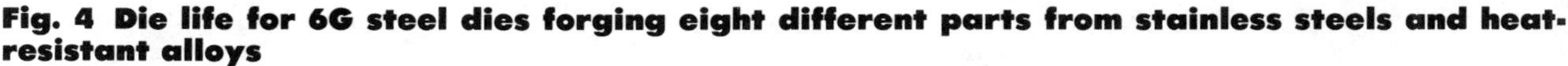

Fig. 4 Die life for 6G steel dies forging eight different parts from stainless steels and heat-resistant alloys

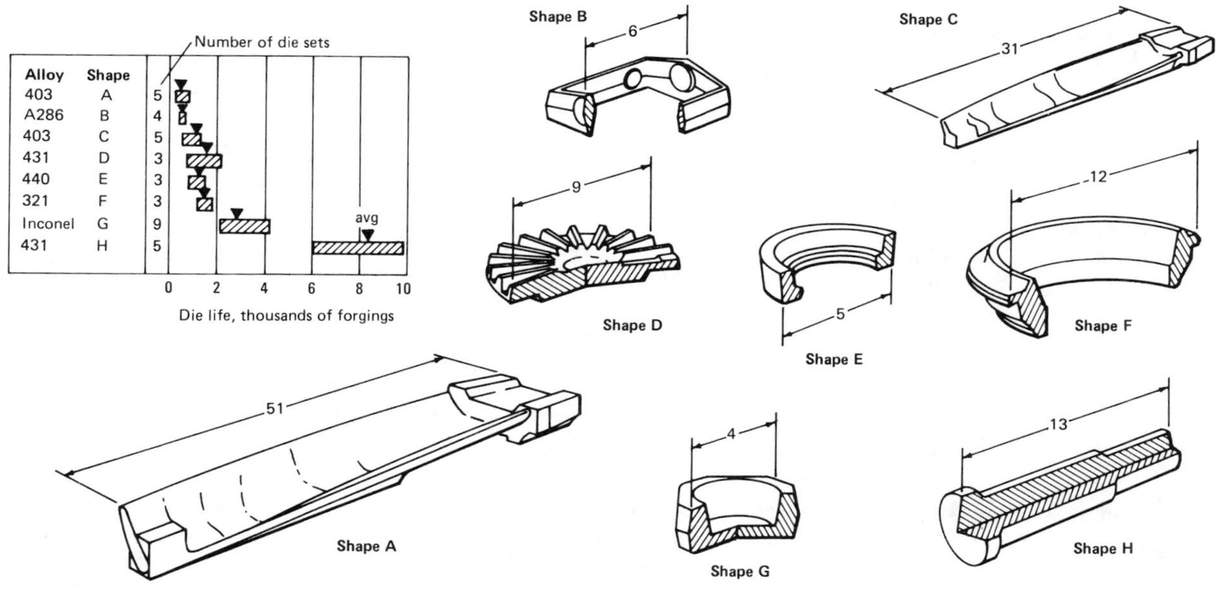

Die hardness, 341 to 375 HB. Chart at upper left gives both average life and life range for the indicated number of die sets evaluated. Shapes A to H are arranged in approximate order of decreasing severity. Dimensions are in inches; to find equivalent metric units (mm) multiply listed values by 25.

blank (shape D) from type 431 stainless steel, whereas eight die sets had an average life of 8000 pieces in forging a simpler part (shape H) from the same steel.

Tools for Trimming

Usually, trimming of metal flash at temperatures below 150 °C (300 °F) is referred to as cold trimming, and trimming at 500 °C (1000 °F) or higher is termed hot trimming. For the purpose of selecting materials for trimming tools, trimming at temperatures between 150 and 500 °C is also considered hot trimming.

Whether trimming is done hot or cold depends chiefly on whether the trim is to be normal or close and on the composition of the metal being trimmed. Table 4 gives typical materials for both hot and cold trimming tools. Carbon and alloy steels may be trimmed either hot or cold, but close trimming is generally done hot. Stainless steels and heat-resisting alloys are usually trimmed hot, whereas nonferrous alloys may be trimmed either hot or cold.

Cold Trimming. Punches for cold trimming carbon and alloy steel forgings are commonly made from discarded die blocks. Because of this practice, prehardened low-alloy die steels are the materials most widely used for punches and are considered satisfactory for many different types of punches. Although high-carbon alloy tool steels such as A2 have been used successfully as blade materials for normal cold trimming of carbon and alloy steel forgings, D2 is usually a better choice because of its longer life.

Hardened and tempered 6150 steel has been used successfully in punches for normal cold trimming of nonferrous forgings. Other alloy steels similar to 6150 in hardenability are also satisfactory for this application. For close trimming, the edge-holding properties of D2 make it more desirable as a punch material. It is also recommended for blades used in close cold trimming of nonferrous forgings.

Hot Trimming. Prehardened die steels are satisfactory punch materials for hot trimming of carbon, alloy and stainless steel forgings. Punch materials for hot trimming of nonferrous forgings are less critical, and 1020 steel, either as rolled or as annealed, is widely used for reasons of economy.

Hard faced carbon steels are often preferred for use as trimming edges (see Table 4). One advantage of hard faced blades is that chipped or broken edges are easily repaired. The cobalt-base alloy indicated in Table 4 (type 4A) is extremely high in resistance to shock, heat and abrasion. Most forging shops prefer to use this alloy (or a similar one) for facing all trimming blades, either hot or cold. Information on alloy 4A and other hard-facing alloys may be found elsewhere in this volume.

Table 4 Typical materials for trimming dies

Material to be trimmed	Cold trimming: Normal trim, Punch	Cold trimming: Normal trim, Blade	Cold trimming: Close trim, Punch	Cold trimming: Close trim, Blade	Hot trimming(a): Punch	Hot trimming(a): Blade
Carbon and alloy steels	6F2 or 6G, at 341 to 375 HB	D2 at 54 to 56 HRC	Generally hot trim		6F2 or 6G at 341 to 375 HB	Hard facing alloy 4A on 1035 steel(b); or D2 at 58 to 60 HRC
Stainless steels and heat-resisting alloys	Generally hot trim		Generally hot trim		6F2 or 6G at 388 to 429 HB	D2 at 58 to 60 HRC
Aluminum, magnesium and copper alloys	6150 at 461 to 477 HB	Hard facing alloy 4A on 1020 steel(b); or O1 at 58 to 60 HRC	D2 at 58 to 60 HRC	D2 at 58 to 60 HRC	1020 soft	Hard facing alloy 4A on 1020 steel(b)

(a) Both normal and close trimming. (b) Hard facing alloy 4A has nominal composition as follows: 1 C, 30 Cr, 3 Ni, 4.5 W, 60 Co, rem Fe. For greater detail, refer to the article on hard facing alloys in this volume.

Tools for Isothermal Forging

By James M. Marder
Principal Engineer
TRW Materials Technology

Forging of materials with substantial high-temperature strength, such as titanium alloys and nickel-base superalloys, is most economically performed by an isothermal forging process. In isothermal forging, the dies are carefully maintained at the same temperature as the workpiece throughout the press stroke. This procedure eliminates chill effects and, therefore, results in excellent dimensional precision, reduced forging loads, and fewer preliminary forging steps than would be required in conventional hot die forging.

The temperatures typically used for isothermal forging of titanium alloys (870 to 980 °C, or 1600 to 1800 °F) and of nickel-base superalloys (925 to 1100 °C, or 1700 to 2000 °F) impose severe hot-strength, creep and oxidation-resistance requirements on the die materials. Dies for isothermal forging of titanium alloys typically are made of cast nickel-base superalloys that were initially developed for blades in gas turbine engines. Of these superalloys, IN-100 has been used with great success for isothermal forging dies. However, in the most severe environments of high temperature and stress, excessive creep has been a major cause of failure for IN-100 dies. In severe environments, replacement of IN-100 by an alloy of even higher creep strength, TRW-NASA VIA, has proved effective.

Isothermal forging of nickel-base superalloys often is performed with dies made of the molybdenum alloy TZM. The susceptibility of TZM dies to oxidation at working temperatures requires that forging be performed in vacuum. The large capital investment and relatively low production rate attendant to forging under vacuum can be justified only for forging alloys that cannot be forged in any other manner or for which considerable savings in critical or expensive raw materials can be realized.

Selection of Materials for Hot Upset Forging Tools

Edited by George Linck
Ford Motor Co. (retired)

SELECTION of materials for hot upset forging tools depends on material to be forged, size of parts to be forged, complexity of die shape and design, production rate and quantity, and the forging and heating equipment to be used. The tools discussed in this article are restricted to header dies, gripper dies and auxiliary tools used in forging machines or upsetters operating in a horizontal plane.

Forgings made by the hot upset method vary in size from 13-mm (1/2-in.) bolts to 305-mm (12-in.) flanged pipe sections for the oil industry. Production rate varies with size of forging and amount of automation. Use of automatic feeds allows rates as high as 7200 pieces per hour in such applications as boltmaking, and such rates require tool materials that will serve continuously at high temperatures for long periods of time. For medium-size parts such as automotive forgings, which are produced at rates of 120 to 150 pieces per hour, the dies cool off enough between blows so that die materials with lower hot strength can be used. The high-alloy, high-hardness upsetting tools used in the bolt industry are not suitable for medium-size automotive upset forgings because such high-hardness tools are too susceptible to breakage. Dies for still larger upset forgings made at lower rates may require higher strength at forging temperature and higher alloy content because of the longer sustained contact between the hot workpiece and the tools.

Complexity of die shape also influences selection of die steels for hot upset forging. Sharp corners and edges greatly increase stress concentration, and thin sections may be subject to extreme loads and high thermal stresses. Internal punches and mandrels are subjected to high impact loads and sliding abrasive wear, and often are designed to be replaceable because of their short life. Replaceable inserts may be used for areas of gripper dies subject to short life and for parts requiring close tolerances.

The material to be forged is important in selecting a die material. At forging temperature, carbon and low-alloy steels have lower strength than stainless steels and heat-resisting alloys and can be forged with less costly

Table 1 Nominal compositions of nonstandard tool steels used in tools for hot upset forging

Type(a)	Composition, % C	Mn	Si	Cr	Ni	Mo	V
6G	0.55	0.80	0.25	1.00	...	0.45	0.10(b)
6F2	0.55	0.75	0.25	1.00	1.00	0.30	0.10(b)
6F3	0.55	0.60	0.85	1.00	1.80	0.75	0.10(b)
6F4	0.20	0.70	0.25	...	3.00	3.35	...
6H1	0.55	...	...	4.00	...	0.45	0.85
6H2	0.55	0.40	1.10	5.00	1.50	1.50	1.00

(a) UNS, AISI and SAE have not assigned type numbers to these tool steels. (b) Optional.

Table 2 Typical tool materials for hot upset forging

Material forged	Tool material types and hardness ranges for total production quantity of: 100 — Gripper die	100 — Heading tool	1000 to 10 000 — Gripper die	1000 to 10 000 — Heading tool	50 000 and up — Gripper die	50 000 and up — Heading tool
For parts of maximum outside upsetting severity (part 1)(a)						
Carbon and low-alloy steels	4150 at 38 to 42 HRC or 4340 insert at 38 to 42 HRC	W1 with 0.70 C at 42 to 46 HRC or 4340 insert at 38 to 42 HRC	6H1 or H11 at 46 to 50 HRC or 4340 insert at 38 to 42 HRC	6H1 or H11 at 44 to 48 HRC or 6G (b) insert at 41 to 45 HRC	6H1 or H11 at 46 to 50 HRC or 4340 insert at 38 to 42 HRC	H11 at 46 to 50 HRC or 6H2 at 52 to 56 HRC or 6G (b) insert at 41 to 45 HRC
Stainless steels and heat-resistant alloys (up to type 310)	6G (b) insert at 38 to 42 HRC	6F3 insert at 42 to 46 HRC	6F3 insert at 42 to 46 HRC	H11 at 46 to 50 HRC or same for insert	6H2 at 52 to 56 HRC or H11 insert at 44 to 48 HRC	6H2 at 52 to 56 HRC or H11 insert at 48 to 52 HRC
For parts requiring both upsetting and piercing (parts 2 and 3)						
Carbon and alloy steels	4340 insert at 38 to 42 HRC or 6G (b) insert at 36 to 40 HRC	4340 insert at 42 to 46 HRC or 6G (b) insert at 41 to 45 HRC	4340 insert at 38 to 42 HRC or 6G (b) insert at 36 to 40 HRC	H11 at 42 to 46 HRC or H11 insert at 46 to 48 HRC	H11 or H11 insert at 42 to 46 HRC	H12 or M10 at 50 to 52 HRC or H11 insert at 46 to 50 HRC
Stainless steels and heat-resistant alloys (up to type 310)	(c)	(c)	(c)	(c)	(c)	(c)

(a) All heads are round and made in one blow with relative dimensions shown. (b) 6F2 die steel may be used interchangeably with 6G. (c) The same tool materials are recommended for upsetting part 2 as are shown for part 1. Part 3 is too severe to be made from a stainless steel or heat-resistant alloys.

Fig. 1 Multiple-pass upset die and header for making rear axle stem pinions

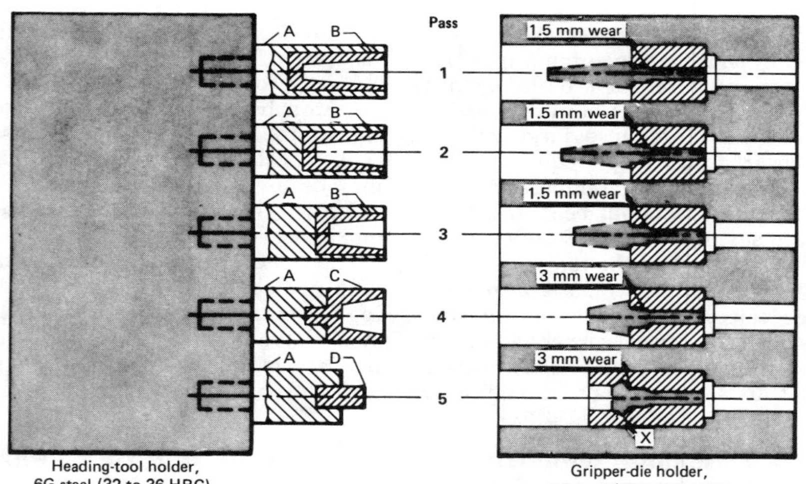

Header die inserts: Insert holders (A) and sleeve inserts (B) are made from 4340 at 38 to 40 HRC. Cap inserts (C) are made from 4340 at 36 to 38 HRC. Punch inserts (D) are made from 6F3 at 44 to 46 HRC.
Gripper die inserts: First-, second- and third-pass inserts are made from 4340, 6G or 6F2 at 36 to 40 HRC. Steel 6F4 is used for fourth and fifth passes because of better weldability. Point X is built up by welding after 3 mm (1/8 in.) of wear. Inserts are tempered at 540 °C (1000 °F) after welding. Inserts are replaced when wear reaches the extent indicated at the shoulder area in each pass. (Wear at shoulder is given in inches; for equivalent value in millimetres, multiply by 25.)

tool steels. In upset forging of heat-resisting steels and titanium alloys, even the best die steels may have a short life.

The compositions and properties of the AISI grades and types of tool steel discussed in this article are presented elsewhere in this volume. The nominal compositions of six nonstandard tool steels used to make tools for hot upset forging are shown in Table 1.

Heading Tools and Gripper Dies

Tools for hot upset forging generally are constructed in the form of insert dies. An example of a multiple-pass upset die and header and of the inserts that are used with them is shown in Fig. 1. Table 2 summarizes the typical materials used for hot upset forging tools for making parts of the various degrees of severity shown in Fig. 2. Part 1 in Fig. 2 represents straight upset forging; parts 2 and 3 represent two degrees of severity for pierced and upset forgings. Each of the three shapes is made in one blow. For all three, the original unsupported length of stock required to make the part is 2.5 times the diameter of the starting stock.

In hot upset forging of simple flanged shapes from low-carbon and alloy carburizing steels, the heading tool usually wears faster than the gripper die. For example, in hot upsetting of hexagonal bolt heads on 1020, 1045 or 4140 steel shanks 13 to 25 mm (0.5 to 1 in.) in diameter, H13 tool steel is used for both the heading tool and the gripper die, but the heading tool has greater hardness. Under the same wear conditions, the higher hardness could have been expected to provide a lower rate of wear, but Fig. 3 shows that forging of about 28 000 pieces produced the same amount of wear (0.15 mm, or 0.006 in.) in the heading tool as forging of about 40 000 pieces produced in the softer gripper tool.

The reverse may be true under certain severe conditions of forming. For instance, the unsymmetrical steering-gear forging shown in Fig. 4 was made with the segment stock all gathered on one side in the gripper die, and this resulted in greater metal movement, with correspondingly heavier die loads and increased abrasion, in that area.

The quantity of water used as a coolant for hot upset forging dies ordinarily does not affect selection of a die steel for forging a given part. In the bolt industry, however, water-cooled tools made of tungsten-bearing tool steels are susceptible to cracking and heat checking at high production rates (on the order of 100 pieces per minute). At high production rates such as these, the hot bolt stock is in nearly constant contact with the dies, allowing them little chance to cool, and under these conditions water cooling imposes severe thermal shock. Low-carbon (0.58 to 0.65% C) M10 high speed steel heat treated to 58 HRC performs well in such applications, as does T1 heat treated to the same hardness.

When a lower-alloy tool steel such as 6G or 6F2 is used for inserts, wear resistance can be improved by adjusting water flow to keep die temperatures below 200 °C (400 °F). At such temperatures, little improvement can be gained by changing to a different low-alloy die steel.

Lubrication of dies impairs speed and is not used extensively for upset forging of steel. Some lubrication may be used in deep punching and piercing, but only enough to prevent the workpiece from sticking to the punch. Resistance to wear and abrasion can be improved by use of die lubricants, but there is no known correlation between lubrication and die performance that would make lubrication a factor in die-steel selection.

Table 3 Materials for typical auxiliary tools

Type of tool	Material(a)
Tools for hot shearing of flash	8630, hard faced(b)(c)
	H11 at 46 to 50 HRC
	H21 at 50 to 52 HRC(d)
Tools for cold shearing of flash	1040, hard faced(b)
	6G or 6F2 at 48 to 50 HRC
	D2 at 56 to 58 HRC
Auxiliary punches	6F3 at 44 to 48 HRC
Bending tools	8630, hard faced(b)
	6F3 at 40 to 44 HRC
	H11 at 46 to 50 HRC
Kicker rods	6F3 at 44 to 48 HRC
	M10 at 56 to 58 HRC
Hot marking tools	M2 at 58 to 60 HRC

(a) Where more than one tool material is recommended for a specific purpose, they are arranged in order of increasing cost. (b) Hard faced with an alloy of 1.10 C, 30 Cr, 3 Ni, 4.5 W, 60 Co, rem Fe; hardness, 48 HRC minimum. (c) Not recommended if trimming of flash is incorporated in the upsetting tools. (d) Preferred for trimming extra-heavy flash.

Fig. 2 Three typical parts made by hot upset forging

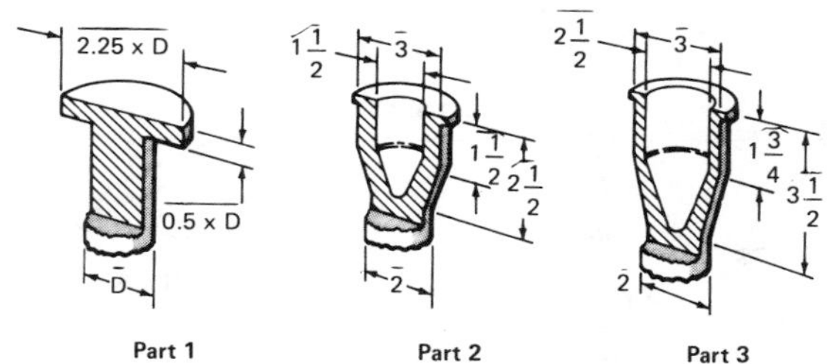

Dimensions for parts 2 and 3 are given in inches; to find equivalent metric dimensions (mm), multiply listed dimensions by 25. Corner radii comply with standard commercial practice. For typical header tool and gripper die materials, see Table 2.

Auxiliary Tools

Typical materials and hardnesses for auxiliary hot upset forging tools are presented in Table 3. Resistance to softening at elevated temperatures is important in selection of materials for hot shearing tools. Without this property, early deformation and ultimate rupture would occur. Repeated contact between the tool and the stock being sheared conducts considerable heat to the cutting edge, softening it by tempering. This problem is more acute with larger stock sizes because of the greater time of contact and volume of material sheared. High-alloy steels with high hot strength and resistance to shearing will require the highest-performance tool materials. Hard faced alloy constructional steels (8630 and 4340, for example) are satisfactory for most shearing and biting applications for small and medium-size forgings.

More care must be taken in selecting tool steels for combined trimming and upsetting dies. If trimming tools fail when placed in the same die set as upsetting tools, costly downtime could result. Also, damage to trimming tools could result in improper trimming, severely reducing the life of subsequent die impressions and producing defective parts.

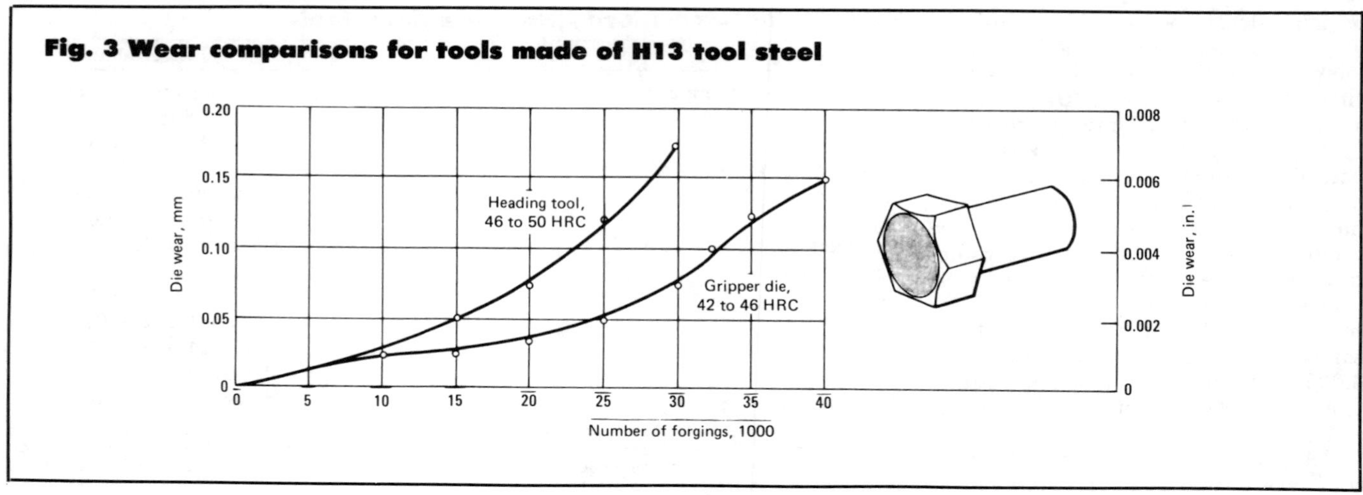

Fig. 3 Wear comparisons for tools made of H13 tool steel

Fig. 4 Wear comparisons between tools made of 6F4 and H14 tool steels

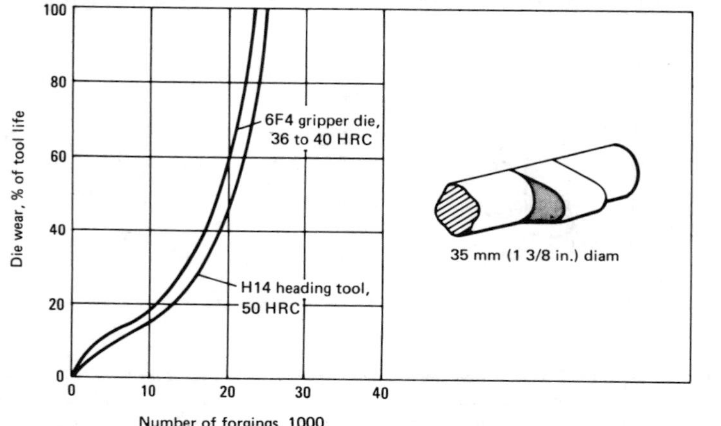

Data are based on 438 000 steering-gear shafts of the configuration shown that were forged from 5132 steel.

If flash is trimmed in combined tooling, the tools usually are not hard faced because of the possibility that the hard facing will chip or spall off and the debris will damage the die set. Such trimming tools are usually made from H11 or H21 tool steel. No change in trimming-tool requirements is normally required to accommodate changes in size or material for most upset parts. H21 is especially recommended for extra-heavy flashes.

Hard faced carbon steel (such as 1040) is satisfactory for most cold trimming operations. Die steels such as 6G and 6F2 are also used. Type D2 tool steel lasts longer than 6G or 6F2, but is more costly.

Selection of Materials for Hot Extrusion Tooling

Edited by John F. McGraw
Metallurgist
Tool and Alloy Metallurgy
Carpenter Technology Corp.

TOOLING for hot extrusion must operate under severe conditions of temperature, pressure and abrasive wear. Fundamentally, the extrusion process consists of forcing material in a plastic condition through a suitable die under high pressure to form a long, continuous shape. Some of the softer metals such as lead and aluminum are sufficiently plastic to be extruded at or near room temperature, but most other metals and alloys are extruded only at elevated temperatures.

Typical billet temperatures for hot extrusion are given in Table 1.

The process of hot extrusion usually is performed in a horizontal hydraulic extrusion press. Rated capacities of such presses range from several hundred Mg to 13 Gg (several hundred to 13 000 tonnes). Presses with capacities of 1.5 to 2.3 Gg (1600 to 2500 tons) are the most common, and such presses can produce the complete range of aluminum alloy extruded shapes and tubes in all standard sizes. Table 2 gives recommended press capacities for making aluminum alloy extrusions from billets ranging in diameter from 100 to 275 mm (4 to 11 in.). Specific mechanical setups are varied as required for satisfactory extrusion of different products from a given work metal. One of the chief variations is in container bore, which affects extrusion pressure. For instance, if a 2.3-Gg (2500-ton) extrusion press is selected, varying the diameter of the container bore would produce varying extrusion pressures as illustrated below.

Bore diameter mm	in.	Extrusion pressure MPa	psi
160	6 3/8	1100	157 000
185	7 1/4	855	122 000
210	8 1/4	660	94 000
235	9 1/4	520	75 000
285	11 1/4	360	51 000

The prevailing commercial hot extrusion process, based on tonnage of product, is single-charge direct extrusion with butt discard. The essential parts of the tooling for this process are described below and illustrated in Fig. 1.

- A *container,* which receives the hot billet to be extruded. The container may be heated and almost always is fitted with a liner. The liner must resist the abrasive action of the billet during extrusion and should maintain relatively high hardness at elevated temperatures. The container and liner should be made of materials having high fracture toughness and good resistance to low-cycle fatigue.
- A *ram,* which operates within the liner and transmits pressure from the press cylinder to the billet. The ram must sustain high cyclic compressive loading. The material used

Table 1 Typical billet temperatures for hot extrusion

Material	Billet temperature °C	°F
Lead alloys	90 to 260	200 to 500
Magnesium alloys	340 to 430	650 to 800
Aluminum alloys	340 to 510	650 to 950
Copper alloys	650 to 1100	1200 to 2000
Titanium alloys	870 to 1040	1600 to 1900
Nickel alloys	1100 to 1260	2000 to 2300
Steels	1100 to 1260	2000 to 2300

Table 2 Recommended press capacities for extrusion of aluminum alloys

Shape circle diameter, mm	Shape circle diameter, in.	Shape thickness, min, mm	Shape thickness, min, in.	Required extrusion pressure, MPa	Required extrusion pressure, psi	Container bore diameter, mm	Container bore diameter, in.	Required press capacity, Gg	Required press capacity, tons
Low- and intermediate-strength alloys									
0 to 75	0 to 3	1.15	0.045	620	90 000	110	4 1/4	0.6	650
75 to 100	3 to 4	1.25	0.050	520	75 000	160	6 3/8	1.0	1150
100 to 150	4 to 6	1.60	0.062	520	75 000	185	7 1/4	1.4	1550
150 to 200	6 to 8	2.00	0.078	520	75 000	235	9 1/4	2.3	2500
200 to 250	8 to 10	2.80	0.109	350	50 000	285	11 1/4	2.3	2500
High-strength alloys									
0 to 75	0 to 3	1.25	0.050	1035	150 000	110	4 1/4	0.9	1060
75 to 100	3 to 4	1.60	0.062	1035	150 000	160	6 3/8	2.1	2300
100 to 150	4 to 6	2.00	0.078	830	120 000	185	7 1/4	2.3	2500
150 to 200	6 to 8	2.80	0.109	690	100 000	235	9 1/4	3.1	3400
200 to 250	8 to 10	4.80	0.188	585	85 000	285	11 1/4	3.8	4200

Table 3 Typical materials and hardnesses for tools used in hot extrusion

Tooling application	For tools used in extruding: Aluminum and magnesium — Tool material	Aluminum and magnesium — Hardness, HRC	Copper and brass — Tool material	Copper and brass — Hardness, HRC	Steel — Tool material	Steel — Hardness, HRC
Dies, for both shapes and tubing	H11, H12, H13	47-51	H11, H12, H13	42-44	H13	44-48
			H14, H19, H21	34-36	Cast H21 inserts	51-54
Dummy blocks, backers, bolsters, and die rings	H11, H12, H13	46-50	H11, H12, H13	40-44	H11, H12, H13	40-44
			H14, H19	40-42	H19, H21	40-42
			Inconel 718	...	Inconel 718	...
Mandrels	H11, H13	46-50	H11, H13	46-50	H11, H13	46-50
Mandrel tips and inserts	T1, M2	55-60	Inconel 718	...	H11, H12, H13	40-44
					H19, H21	45-50
Liners	H11, H12, H13	42-47	A-286, V-57	...	H11, H12, H13	42-47
Rams	H11, H12, H13	40-44	H11, H12, H13	40-44	H11, H12, H13	40-44
Containers	4140, 4150, 4340	35-40	4140, 4150, 4340	35-40	H13	35-40

Table 4 Typical compositions of superalloys used for extrusion tools

Alloy	C	Mn	Si	Cr	Ni	Mo	Nb	Ti	Al	Fe	V	B
						Composition, %						
A-286	0.05	1.35	0.50	15.00	26.00	1.25	...	2.00	0.20	53.6	0.30	0.015
V-57	0.08	0.35	0.75	14.80	27.00	1.25	...	3.00	0.25	52.0	0.30	0.010
Inconel 718	0.04	0.20	0.30	18.60	53.00	3.10	5.0	0.90	0.40	18.5	...	...

for the ram should have good resistance to upsetting, to work hardening, to thermal shock and to rapidly applied stress.

- A *dummy block,* which is inserted between the ram and the billet to absorb the heat and erosion that otherwise would be imposed directly on the ram. The dummy block should resist indentation by the billet at temperatures approaching the billet temperature.
- A *die* or *die assembly,* through which the billet is pushed to form the extruded bar or shape. A die assembly typically consists of the following component parts (see Fig. 2): (*a*) a die containing the extrusion cavity, (*b*) a die ring (die holder), (*c*) a die backer, (*d*) one or more bolsters, (*e*) one or more sub-bolsters and (*f*) a tool carrier (die slide or housing). The die itself must have high toughness combined with resistance to wear and softening at elevated temperatures. The other components of the die assembly should be high in both strength and toughness.
- A *mandrel,* which is used for production of hollow products. Mandrels

Fig. 1 Typical tooling for extrusion of seamless tube

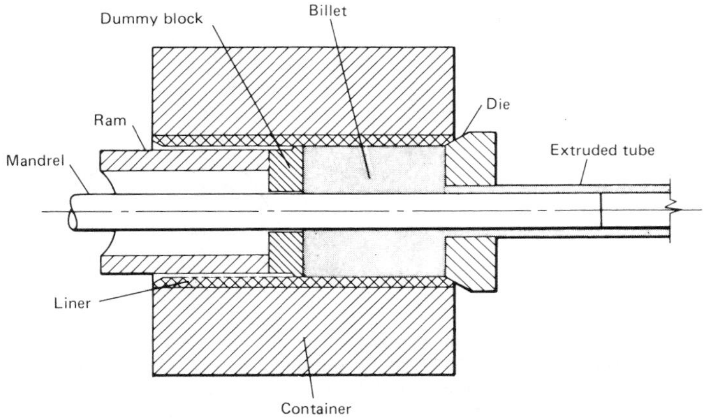

In this arrangement, the mandrel is attached to the press piercer and operates through a hollow ram.

Fig. 2 Typical extrusion die assembly, showing relative positions of components in a tool carrier

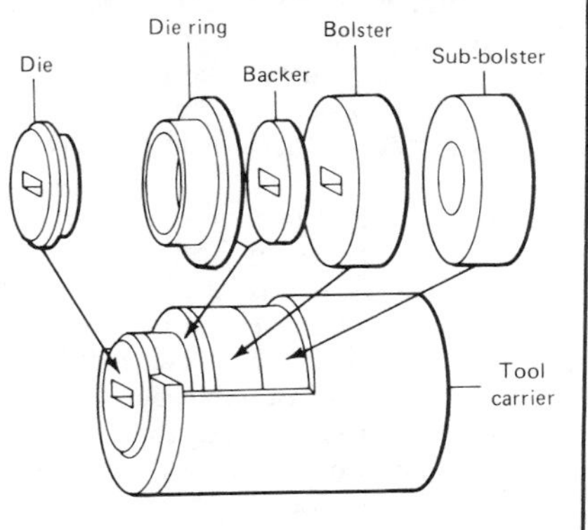

Fig. 3 Typical die life in extrusion of austenitic stainless steel tubes

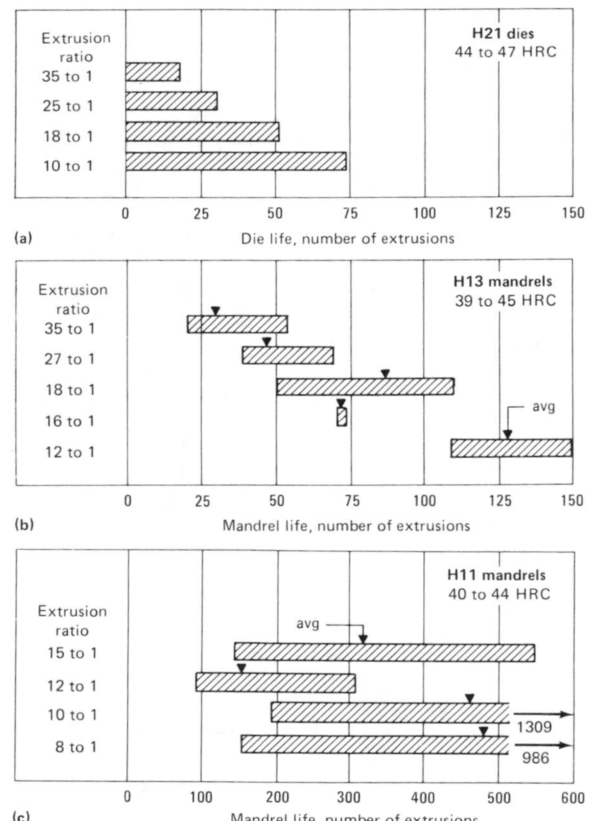

Life of dies and mandrels at several different extrusion ratios in extrusion of 300-series stainless steel tubes at 1230 to 1260 °C (2250 to 2300 °F) using a glass lubricant. Data are summarized from records on (a) 22 dies, (b) 25 mandrels and (c) 53 mandrels.

Fig. 4 Life of H21 extrusion dies at two hardnesses

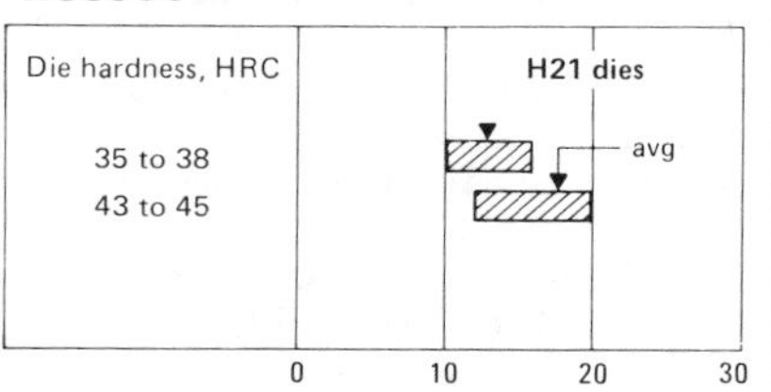

Data are for extrusion of irregular shapes from type 410 stainless steel at an extrusion ratio of 18.5 to 1. The extrusions were about 10 m (32 ft) long, and had a cross-sectional area of about 1075 mm^2 (1.35 in.^{2}).

should be high in hot hardness, abrasion resistance, fracture toughness and yield strength.

With the ram retracted, a hot billet or slug is placed in the container. A dummy block is inserted between the ram and billet; then the hot billet is pushed into the container liner and advanced under high pressure against the die. The metal is squeezed through the die opening, assuming the desired shape, and is severed from the remaining stub by sawing or shearing.

Causes of Die Failure

Die breakage accounts for most die failures. Many instances of die breakage are caused by attempts to extrude "cold" billets; however, other variations in the process also can result in die breakage. Excessive die hardness or excessive carbon content can cause breakage, particularly during extrusion of complex shapes. Cracking or splitting may be caused by insufficient preheating, by carburization, by drastic cooling, by excessively high temperature, by mechanical factors such as misalignment and poor support, or by improper steel selection or heat treatment.

Wear, wash, or loss of size from erosion may be caused by low hardness or a decarburized die surface. Flow or closure of dies may be caused by excessive heat in service or by low hardness. Excessive heat often can be minimized by using coolants where applicable or by rotation of tooling in sets. Increasing the extrusion speed also may be necessary to avoid excessive heat buildup. Heat checking results from excessive heat in service, excessive hardness, drastic quenching, and grinding stresses, and is often more severe if the alloy has become either carburized or decarburized during processing.

Periodic stress relief of rams and other high-hardness tools used without preheating will increase tool life and decrease the likelihood of downtime due to tool failure. Storage of rams at about 65 °C (150 °F) will help minimize failures. Frequent inspection of rams by ultrasonic and magnetic-particle methods is essential for safe operation.

Fig. 5 Life of H21 dies in extrusion of copper alloys

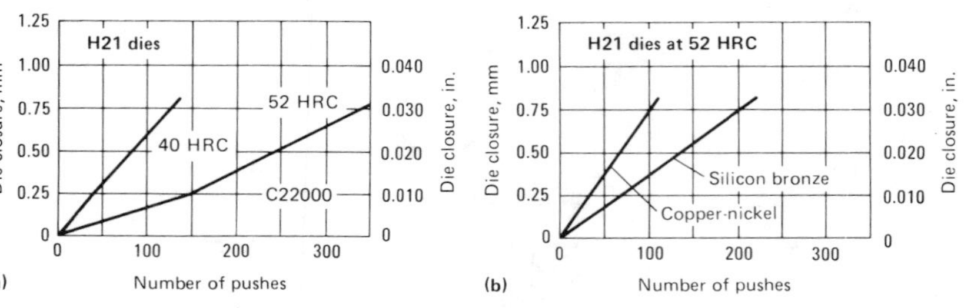

Die life under different conditions of extrusion, measured as die closure of 0.8 mm (0.03 in.). (a) For extrusion of C22000 (90Cu-10Zn) tubing, 92 mm (3⅝ in.) in OD by 8 mm (5/16 in.) in wall thickness; data are based on eight dies of each hardness. (b) For extrusion of silicon bronze at 830 °C (1520 °F) and copper nickel at 1090 °C (2000 °F); data are based on eight dies for each alloy extruded.

Tool Materials

Table 3 lists typical materials and hardnesses for tools used in hot extrusion. Hot extrusion of aluminum and magnesium is in many respects similar, the major difference being the pressure required. Often, the same tooling materials can be used for extrusion of either aluminum or magnesium. The compositions and properties of the tool steels mentioned in Table 3 are discussed elsewhere in this volume. Typical compositions of the superalloys used for some hot extrusion tools are given in Table 4. Compositions of the AISI alloy steels in Table 3 are given in Volume 1 (9th Edition) of this Handbook.

Dies for extrusion of aluminum alloys and copper alloys generally are made from H11, H12 or H13. For extrusion of copper alloys, some companies specify tungsten hot work steels such as H14, H19 and H21. For extrusion of steel, H13 solid dies or H13 dies with cast H21 inserts are often used.

Dummy blocks, backers, bolsters and die rings are routinely made from H11, H12 and H13. For extrusion of copper, brass and steel, H14, H19 and H21 sometimes are used. Inconel 718 and other superalloys occasionally are used for dummy blocks; use of these alloys often results in extremely long tool life.

Mandrels generally are made of either H11 or H13, regardless of the material being extruded. Most mandrel tips and inserts for extrusion of aluminum are made of T1 or M2. Inconel 718 mandrel tips and inserts are commonly used in the extrusion of copper and brass, whereas H11, H12, H13, H19 or H21 tips and inserts may be used in extrusion of steel.

Liners used in extrusion of aluminum or steel usually are made of H11, H12 or H13. Liners for extrusion of copper and brass normally are made of A-286 or V-57.

Rams generally are made of H11, H12 or H13.

Containers for extrusion of aluminum or copper products usually are made of either 4140, 4150 or 4340 alloy steel. Containers for extrusion of steel also can be made from alloy steels; however, H13 generally is preferred.

Die Life. To supplement the general information in Table 3, Fig. 3 shows the effect of extrusion ratio (cross-sectional area of the billet divided by cross-sectional area of the extrusion) on life of H21 dies and H11 and H13 mandrels in extruding tubes of austenitic stainless steels of the 300 series at about 1260 °C (2300 °F). With a relatively low extrusion ratio of about 10 to 1, die life is relatively short but mandrel life is relatively long. However, even with a simple shape such as a tube, die life and mandrel life are short at high extrusion ratios.

Figure 4 presents similar data on die life for extruding an irregular shape in type 410 stainless steel at an intermediate extrusion ratio of 18.5 to 1. Comparison of the data in Fig. 3 and Fig. 4 shows that die life for a given die steel and hardness is much shorter if the extrusions being made are of irregular shape.

Die life is further reduced if, as is common, the hardness of the die must be reduced by about five points HRC to reduce the probability of breakage.

The information in Fig. 5(a) shows that, in extrusion of C22000 tubing, increasing the hardness of H21 dies from 40 HRC to 52 HRC increased die life by a factor of about three. Figure 5(b) indicates that die life for a given material and hardness is significantly influenced by extrusion temperature; the life of H21 dies (hardness, 52 HRC) in extrusion of silicon bronze at 830 °C (1520 °F) was about twice the life of similar dies in extrusion of copper nickel at 1090 °C (2000 °F).

Special Materials. In addition to the typical materials listed in Table 4, special insert materials and surface treatments have been specified (particularly for tools used in extruding complex shapes) where better resistance to wear at higher temperatures is required. Special insert materials include special grades of cemented tungsten carbide, nickel-bonded titanium carbides and aluminum oxide ceramics. Special surface treatments include nitriding, aluminide coating, and application of proprietary materials by vapor deposition or sputtering.

Detailed information on these special materials is provided in several other articles in this volume.

Selection of Materials for Die-Casting Dies

By Walter Smith
Die Cast Dies, Inc.

MATERIALS for die-casting dies must have good resistance to thermal shock and good resistance to softening at elevated temperatures. Resistance to softening is necessary because dies must be able to withstand the erosive action of molten metal under high injection velocity. Other properties that influence selection of materials for die-casting dies are hardening characteristics, machinability, resistance to heat checking, and weldability.

Performance of die-casting dies is directly related to casting temperature, injection pressure, thermal gradients within the dies, and frequency of exposure to high temperature. These variables are the principal ones used in the selection tables in this article. Tool steels of relatively high alloy content are required for components in direct contact with molten metal, as indicated in Table 1.

Die hardness is less critical for die casting of zinc than for die casting of alloys of higher casting temperature. Consequently, pre-hardened insert steels (29 to 34 HRC) are often used to make tooling for die casting of zinc.

Table 1 Recommended materials for die-casting dies and die inserts

Components	Typical material(s) and hardness
Cavity inserts for Al and Mg castings	H13 hardened to 45 to 48 HRC
Cavity inserts for Zn castings	P20 prehardened to 300 HB
Cavity inserts for long-run Zn castings	H13 hardened to 44 to 46 HRC
Cavity inserts for Cu castings	H20, H21, H22 hardened to 44 to 48 HRC
Holder blocks	4140 prehardened to 300 HB

Hot work tool steels are almost always used to make tooling for casting higher-melting-point alloys, such as aluminum, magnesium and copper. For casting aluminum and magnesium alloys, H13 steels are hardened to about 44 to 48 HRC; for casting copper alloys, H20, H21 and H22 steels generally are used at hardnesses of 38 to 45 HRC.

Injection components for both zinc and aluminum die-casting dies are subjected to considerable contact with molten metal and to severe erosion. Recommended materials and heat treatments for such dies are given in Table 2.

Slides, Guides, Cores and Pins. The moving parts of a die-casting die must have the same general characteristics as those of the stationary parts, and also must provide resistance to wear. Table 3 lists materials recommended for slides, guides, cores and ejector pins. Wear of such parts can be alleviated by one or more of the following preventive measures:

- Use of dissimilar hardnesses for contacting die parts
- Use of optimum hardness for hot work applications, but with one or

both surfaces of the contacting parts nitrided
- Use of dissimilar materials (where incoming molten metal does not touch the moving die component)
- Use of a lubricant on contacting surfaces
- Interruption of the continuity of contacting surfaces by shallow notching or dimpling of the slide or gib to retain lubricant
- Maintenance of proper fit between wearing surfaces
- Provision of smoothly polished surfaces on both members.

The degree of erosion (washing away of die material by the jet action of an incoming stream of molten metal) is a function of gating and casting technique as well as of die material. Impingement of molten metal against a core or die face at the gate inlet often can be lessened through proper die design.

Nitriding has been used to improve the resistance of die materials to erosion. Although nitriding usually minimizes erosion in early service, it decreases ultimate service life. Spalling of a nitrided surface, which causes undercuts at the spalled area, results in tearing of the casting when it is withdrawn from the die. Spalled die parts are very difficult to repair and almost always must be replaced.

Cavity configurations affect service life of dies. If design is improper, gross cracking and excessive heat checking can result. Among the design faults that lead to low cavity life are:

- Cavity too large for die block—not enough body in die
- Impression too close to the water-cooling chamber
- Inappropriate location of gates, vents, offsets, undercuts, deep narrow recesses, cores and inserts
- Cavity contour irregular with sharp corners and knife edges
- Improper section balance—that is, locating large areas of molten metal close to smaller areas without providing satisfactory transition.

Maximum die life is obtained when parts being cast are symmetrical in shape, thin and uniform in cross section, and small in volume of cast metal.

Die Lubricants. Most commercial die lubricants or mold-release agents reduce soldering or sticking of the casting to the die. The use of an effective lubricant results in easier cleaning of dies, less wear and longer runs between die polishes.

However, some of these die lubricants attack the die steel as well as the particles of the casting alloy adhering to the die surface. If used frequently, such lubricants will shorten die life.

Preheating and Cooling. Water cooling of cores and slides must be considered where significant wear is likely. Die-casting dies used for casting aluminum or copper are water cooled in service and should be preheated to about 175 °C (350 °F) before a run is started. Preheating should be done with water flowing slowly through the die to prevent the serious damage that would result from introduction of cold water into a hot die.

Whenever possible, dies should be preheated with the slides removed (however, this practice is not often followed because equipment design makes it impractical to remove the slides). The mass of a slide is generally less than that of a die, so that heating the die and slide together causes the slide to expand sooner than the guide, thus reducing clearance.

Ejection components are not subjected to high temperatures. Consequently, hot rolled 1020 steel in the as-received condition is used for making ejector boxes, rails, ejector plates, support posts and support blocks for tooling employed in die casting of aluminum and zinc parts.

Trim dies are used in secondary operations and are made of wear-resistant material, as indicated in Table 4.

Table 2 Typical materials for injection components

Component	Metal being cast	Material and condition
Sprue spreader	Zinc	H13 hardened to 250 to 290 HB and nitrided
Sprue bushing	Zinc	Nitralloy hardened to 250 to 300 HB and nitrided
Nozzle and adapter	Zinc	H13 hardened to 46 to 48 HRC and nitrided
Shot sleeve	Aluminum	H13 hardened to 46 to 48 HRC and nitrided
Shot pad	Aluminum	H13 hardened to 46 to 48 HRC and nitrided
Plunger tip	Aluminum	Beryllium copper hardened to 38 to 42 HRC

Table 3 Materials for slides, guides, cores and ejector pins

Component	Material(s)	Condition
Slide carrier	4130, 4140, 6150	Hardened to 46 to 50 HRC
Slide lock	4140, 6150	Hardened to 46 to 50 HRC
Leader pin	1117	Carburized and hardened to 58 to 62 HRC
Guide bushing	1018	Carburized and hardened to 58 to 62 HRC
Guide block	4140	Carburized and hardened to 56 to 60 HRC
Guide plate	4140	Carburized and hardened to 56 to 60 HRC
Ejector pin	H13	Hardened to 34 to 40 HRC and nitrided
Return (surface) pin	H13	Hardened to 34 to 40 HRC and nitrided
Core	H13	Hardened to 48 to 55 HRC

Table 4 Typical materials for trim dies

Component	Material	Condition
Base, holder	1020(a)	As received
Trim ring, plate	A2	Hardened to 56 to 58 HRC
Punch, pad	1020(a)	As received
Guide pin, guide bushing	C1117	Carburized and hardened to 58 to 62 HRC
Slide block	A2	Hardened to 56 to 58 HRC
Cam	1020(a)	Carburized and hardened to 46 to 48 HRC

(a) Hot rolled.

Selection of Materials for Powder-Compacting Tools

By James J. McCarthy
Metallurgical Engineer
Bethlehem Steel Corp.

THE COMPACTING PROCESS most widely used to convert metallic powders into components is the closed-die process. The basic tooling for this process includes a precision-machined die, an upper and lower punch and, if the part is to be hollow, a core rod. These components are attached to the platens of a mechanical or hydraulic press by means of adapters, which often enclose the basic tooling and give it backup support.

Production rates of 500 to 2500 parts per hour, depending on part configuration and size, are not uncommon. Because many of the powders subjected to compaction are abrasive in nature, the tools that contact the powder must be wear resistant to withstand the attritional action involved.

The pressures used in compaction of powders vary from 70 to 830 MPa (5 to 60 tons/in.^{2}, or 10 to 120 ksi), and tools must have sufficient tensile and compressive strengths to withstand these pressures (plus some margin of safety to allow for possible overloading) without suffering permanent deformation. Usually, dies encounter tensile stresses in use, whereas punches are subjected to compressive stresses; frequently, these stresses are nonuniform because the powder being compacted does not fill the die uniformly.

Besides having sufficient mechanical strength, tools must be tough enough to withstand substantial impact loads. Also, where pressing is done hot, as in hot isostatic pressing, tools must be able to resist softening at the elevated temperatures involved.

Highly polished tool surfaces often are desired, which necessitate materials with low nonmetallic-inclusion content and/or porosity. As an extra benefit, this requirement results in improved fatigue strength, which also helps minimize premature tool failure.

Two other properties of interest—particularly to fabricators of tool components—are machinability and resistance to distortion in heat treatment. With greater use of unconventional metal-removal processes such as electrical discharge machining, the difficulties usually associated with machining intricate and complex shapes have been significantly lessened. Resistance to distortion in heat treatment is important when close tolerances must be maintained.

The materials generally used for these tools are (*a*) cemented carbides and (*b*) tool steels with or without special surface treatments.

Dies. In the most common type of die construction, wear-resistant inserts or liners are held in place by clamping or shrink fitting. Die inserts for compaction of carbide, ceramic or ferrite powder most frequently are the medium- or coarse-grain 94WC-6Co grades of cemented carbide. Cemented tungsten carbide containing 12 to 16% cobalt may be used to make inserts for compacting metal powders in medium-to-long production runs.

The elastic moduli of carbides are considerably higher than those of steels—a fact that should be considered when designing composite steel and carbide die assemblies. Because carbide will deflect only 33 to 40% as much as steel, the steel portion generally should be designed with enough stiffness to support three times the expected loading in order to match the deflection of the carbide. Shrink-fit allowance should be about 1.0 mm/m (0.0010 in./in.). Shrink rings and similar supporting parts of the tooling can be made from medium-carbon alloy steel quenched and tempered to about 42 to 46 HRC. It is especially important that supporting parts for carbide tools provide sufficient support; otherwise, the carbide tools are likely to break in service.

Cemented carbides are relatively expensive, and shaping of parts to the required form must be done either by electrical discharge machining or by specialized methods of grinding.

Wear-resistant tool steel inserts are sometimes used instead of carbide inserts. Tool steel inserts are tougher and easier to fabricate than carbide inserts. Crucible CPM 10V is frequently chosen for medium-to-long production runs because it has wear resistance approaching that of carbide. Other wear-resistant tool steels, usually D2 or a high speed steel such as M2 or M4, have been used for short-run applications. Tool steel inserts generally are heat treated to a working hardness of 62 to 64 HRC. For increased wear resistance, a nitrided case may be specified for dies made of CPM 10V or D2. For certain part designs, a solid die rather than an insert die is a more practical choice; an air-hardening 5%-Cr tool steel such as A2 is generally used for such applications.

Punches. The stresses imposed on punches during service are such that toughness is a much more important material requirement than wear resistance, although wear resistance cannot be ignored. Type A2, and sometimes the shock-resisting type S7, is preferred for punches. (Wear-resisting grades such as D2 and Crucible CPM 10V often lack the required toughness, particularly for solid punches.) In type A2, which is deep hardening on air quenching, an as-quenched hardness of 60 HRC can be developed in the center of a section 125 mm (5 in.) square (even though a solid punch this large would seldom be used), whereas for type S7 the maximum section size in which such hardness can be obtained is 65 mm ($2^1/_2$ in.). If sizes larger than 125 mm are required, Type A2 should be oil quenched from the austenitizing temperature to about 540 °C (1000 °F), then air quenched to 65 °C (150 °F) before tempering. S7 can be carburized or nitrided for added wear resistance.

For applications in which A2 or S7 punch faces become severely abraded, a more wear-resistant grade such as Crucible CPM 10V, D2, D3 or M2 should be considered. Cemented carbides are too brittle to perform successfully as punches or punch-face inserts.

Core Rods. Both toughness and wear resistance are important criteria in selection of core-rod materials, but generally the primary consideration is wear resistance. For particularly abrasive conditions, CPM 10V has been used successfully, as have D2, M2 and A2 tool steels that have been nitrided or coated with tungsten carbide.

Operational Factors. Die working surfaces and core rods should be polished or lapped to a mirror-like surface finish, and final polishing should be done in a direction parallel to the axis of the tool. The faces and lands of the punches also should be given a fine finish. An exceptionally smooth surface finish reduces friction, thereby reducing some of the load on the tooling. It also makes it easier to eject the compacts, and eliminates minute scratches and other stress raisers that could lead to premature fatigue failure.

Hard chromium plating is sometimes recommended to improve the life of tool steel punches and core rods, particularly when abrasive powders are involved. Some users claim that nitrided or chromium-plated die parts have up to ten times the wear resistance of untreated tool steel die parts; others claim that chromium plating is not very effective. Both nitrided and chromium-plated die parts are subject to chipping or flaking, especially at sharp edges. When this is a problem, a diffused surface layer such as that produced by Chromalizing may prove an effective alternative.

Selection of Materials for Molds for Plastics and Rubbers

Edited by Charles C. Davis, Jr.
Consultant

IN MOLDS for parts made of plastic materials (including rubbers), the type of material to be molded is a major factor governing the choice of mold material. Plastic materials basically are of two major types—thermoplastic materials (thermoplastics) and thermosetting materials (thermosets).

Thermoplastics can be remelted after full melting and solidification. Thermosets cannot be remelted after they have been heat cured. Thermoplastics and thermosets are further classified into seven distinct groups, as shown in Table 1. Some of these plastics require mold materials that resist abrasion, corrosion, heat, or high compression loads.

There are several methods for molding plastics and several for forming the molds. A detailed discussion of these methods may be found in Ref 1.

This article covers typical mold materials, and methods of producing molds, for injection, blow, transfer and compression molding of the chief groups of plastic materials. The following definitions describe the basic steps in each molding process, and the principal tooling involved.

Injection molding is a method of forming plastic objects from granular or powdered thermoplastics by (*a*) using heat and pressure to plasticize the material in a chamber and then (*b*) forcing part of the fluid mass into a cooler chamber, where it solidifies. Injection molding requires pressures of 70 to 140 MPa (10 000 to 20 000 psi).

Blow molding is a method of forming hollow objects in which a thermoplastic in the form of a molten or softened tube (parison) is inserted into a cool mold. The parison is then pressurized at low internal pressure so that it expands against the sides of the mold, where it solidifies. A major increase in the use of blow molding took place early in the 1960's with general public acceptance of plastic bottles made from high-density polyethylene. Blow molding requires pressures of only 0.2 to 0.7 MPa (25 to 100 psi). The primary advantage of blow molding is the ability to form re-entrant curves.

Transfer molding is a method of forming plastic objects from granular, powdered or preformed thermosets by softening the material in a heated chamber and then forcing essentially the entire mass into another hot chamber, where it cures.

Compression molding is a method of forming objects from either thermosets or thermoplastics by placing the material in a heated mold cavity open at the top, and then simultaneously applying heat and compressing the material with a "force".

Cavity refers to the female portion of the mold, which forms the outer surface of the molded article. (The cavity often is called the *die*.)

Plunger refers to the ram or piston used to displace the fluid or semifluid material in transfer molding and injection molding. (In compression molding, the ram or piston usually is called the *force*.)

Table 1 Classification of plastic materials

Classification of plastic materials
Thermoplastics and unpolymerized thermosets
Group 1—General-purpose plastics and rubbers
Acetal
Acrylic
Acrylo-butadiene-styrene (ABS)
Acrylo-nitrile and copolymers
Cellulose acetate/butyrate
Polycarbonate
Polyester (polyethylterpthalate) (PET)
Polyethylene
Polyphenyl oxide
Polypropylene
Polystyrene and copolymers
Structural foams
Thermoplastic rubbers and elastomers
Group 2—Fluid plastics
Nylon
Urethane(a)
Group 3—Corrosive and high-temperature thermoplastics
Vinyl chloride
Fluorocarbon
Polysulfone
Thermosets
Group 4—General-purpose high-pressure and abrasive plastics
Alkyd
Melamine formaldehyde
Phenol formaldehyde
Urea formaldehyde
Group 5—Plastics requiring high-temperature curing
Silicone resins
Group 6—Rubbers
Thermosetting rubbers
Group 7—Low-pressure plastics(b)
Epoxy
Polyester

(a) Two-component "RIM" type. (b) Abrasive when filled.

Fig. 1 Hypothetical shapes typical of molded plastic products

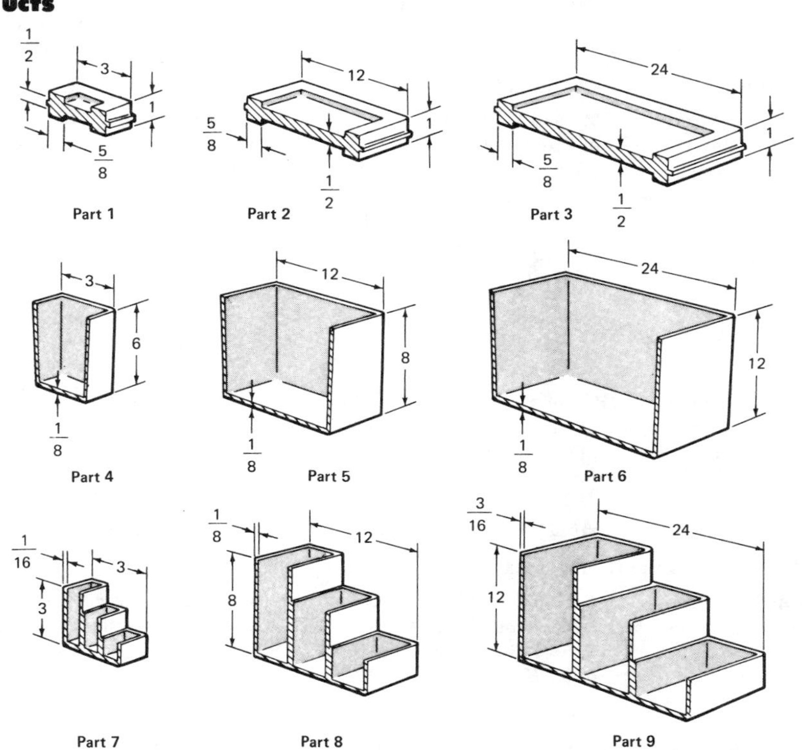

Dimensions are in inches; for equivalent metric dimensions (mm), multiply listed values by 25. See Tables 2 and 3 for typical mold materials used in producing parts similar in general configuration and degree of severity to those shown above.

Selection of Mold Materials

Table 2 lists typical materials used for making machined molds. The choice of mold material depends on the type of plastic to be molded, and on the quantity and shape of the parts to be made. The mold materials given in Table 2 are for several hypothetical parts representing different degrees of molding severity; these hypothetical parts are illustrated in Fig. 1. Table 3 presents typical materials used for making hubbed or cast molds.

To supplement the information in Tables 2 and 3, Table 4 lists compositions and properties of the beryllium copper alloys used for plastic molds, which are prepared in most instances by pressure or ceramic mold casting but sometimes by machining, especially for simple shapes.

Machined Molds. Large cavities usually are produced by machining. For these large cavities, steels that require high-temperature heat treatment and rapid cooling may exhibit an unacceptable amount of distortion; it may be necessary to use prehardened die steels, or steels (such as nitriding steels or maraging steels) that require only a relatively low-temperature final heat treatment to develop acceptable wear resistance.

Hubbed Molds. In many instances, small multiple-cavity molds can be produced at lowest cost by hubbing. This involves pressing a male master plug, known as the hub, into the metal block. (This process is also called "hobbing".) If forming a mold by hubbing is to be economical, the mold steel must be very soft and have a low rate of work hardening. Such steels must be carburized and heat treated to provide the required wear resistance and surface hardness.

Satisfactory hubbing of mold cavities depends not only on the steel to be hubbed but also on selection of the proper steel for the hub. Steels for master hubs should have good machinability and workability in the annealed condition, high compressive strength, and resistance to abrasion in the heat treated condition. Hub steels also should exhibit minimum distortion and size change in heat treatment, minimum scaling, and the ability to be polished to a high finish.

For relatively simple hubs containing sharp detail but no feather edges (which are susceptible to edge wear, edge breakdown, or loss of detail), S1 is usually recommended. Hardnesses of 59 to 61 HRC are recommended, and slight carburization of the surface during heat treatment is often beneficial.

Table 2 Typical materials for machined molds

Material to be molded(a)	Mold materials(b) for making parts shown in Fig. 1 in various production quantities — Parts 1, 4 and 7: 10 000 to 100 000	Parts 1, 4 and 7: 1 000 000 to 10 000 000	Parts 2, 5 and 8: 10 000 to 100 000	Parts 2, 5 and 8: 1 000 000 to 10 000 000	Parts 3, 6 and 9: 10 000 or more
Mold materials for thermoplastics					
Group 1: general-purpose plastics and rubbers	P20 or P21, prehardened(c); 414L, prehardened	O1, 53 to 57 HRC; P6 or P20, carburized; S7, 51 to 57 HRC; 420 stainless, 45 to 50 HRC	P20 or P21, prehardened(c); 414L, prehardened	P6, carburized, 54 to 58 HRC; P20 or P21, prehardened(c); 414L, prehardened	P20 or P21, prehardened(c); 414L, prehardened
Group 2: fluid plastics	P6 or P20, carburized, 54 to 58 HRC	O1, 53 to 57 HRC; S7, 51 to 57 HRC	P6, carburized, 54 to 58 HRC	P6, carburized, 54 to 58 HRC; H13 (d), 48 to 52 HRC; S7, 51 to 57 HRC; 420 stainless, 45 to 50 HRC	P20 or P21, prehardened(c); 414L, prehardened
Group 3: corrosive and high-temperature plastics	P20 or P21, prehardened(c) and nickel plated(e); 414L, prehardened; 420 stainless, 45 to 50 HRC	O1, 53 to 57 HRC, nickel plated(e); 420 stainless, 45 to 50 HRC; S7, 51 to 57 HRC	P20 or P21, prehardened(c) and nickel plated(e); 414L prehardened; 420 stainless, 45 to 50 HRC	P6, carburized, 54 to 58 HRC, nickel plated(e); 420 stainless, 45 to 50 HRC	P20 or P21, prehardened(c) and nickel plated(e); 414L, prehardened
Mold materials for thermosets					
Group 4: general-purpose plastics	L2, 53 to 57 HRC; P20, carburized, 54 to 58 HRC; S7, 51 to 57 HRC	L2, carburized, 53 to 57 HRC; A2, 53 to 57 HRC; P20, carburized, 54 to 58 HRC; S7, 51 to 57 HRC	P20 or P6, carburized, 54 to 58 HRC	P20 or P6, carburized, 54 to 58 HRC	P20 or P6, carburized, 50 to 55 HRC
Group 5: plastics requiring high-temperature curing	H13, 48 to 52 HRC; S7, 51 to 57 HRC	H13, 48 to 52 HRC; S7, 51 to 57 HRC	P4, carburized, 52 to 56 HRC; H13, 48 to 52 HRC	P4, carburized, 52 to 56 HRC; H13, 48 to 52 HRC; S7, 51 to 57 HRC	P4, carburized, 52 to 56 HRC
Group 6: rubbers	Class 30 gray iron(f)	Class 30 gray iron; 1020, soft, chromium plated(g); A2, 53 to 57 HRC(h)	1020, soft, chromium plated(g)	1020, soft, chromium plated(g)	1020, soft, chromium plated(g)
Group 7: low-pressure and abrasive plastics	P20, carburized, 54 to 58 HRC; L2, 53 to 57 HRC; S7, 51 to 57 HRC	P20, carburized, 54 to 58 HRC; L2, 53 to 57 HRC; S7, 51 to 57 HRC	P20, carburized, 54 to 58 HRC; L2, flame hardened	P20, carburized, 54 to 58 HRC; L2, flame hardened; S7, 51 to 57 HRC	P20 or P6, carburized, 50 to 55 HRC

(a) See Table 1. (b) Where more than one mold material is given for a specific set of conditions, they are arranged in order of preference unless otherwise noted. (c) Hardness of prehardened steels should be 300 HB minimum. (d) Preferred for molding parts 5 and 8. (e) Recommended thickness of plating, 0.005 to 0.025 mm (0.0002 to 0.0010 in.). (f) Cast 356-T6 aluminum recommended for quantities up to 10 000 parts. (g) Recommended thickness of plating, 0.005 to 0.015 mm (0.0002 to 0.0005 in.). (h) Provides increased resistance to handling.

Most hubs fall into this general category. Therefore, S1 is the tool steel most commonly used in hubbing.

Steels O1, A2 and D2, which have high hardness and good abrasion resistance, are used for high-production hubs of simple design for which long life and resistance to bulging are important considerations but for which high toughness is relatively unimportant. A hardness in the range from 60 to 63 HRC is recommended for this type of hubbing. Hubs of intricate or complex design must be notch tough and often are made of 6F5 tool steel. Such hubs are heat treated to hardnesses of 58 to 59 HRC. However, 6F6 tool steel is more suitable for hubs of complex design that incorporate feather edges, which require exceptional resistance to brittleness and edge failure. Intentional carburization during heat treatment of this type of steel is often helpful.

L6 tool steel is used principally for inexpensive hubs that are pressed into the mold material with relatively low hubbing pressures.

Cast Molds. Many small-cavity and multiple-cavity molds are pressure cast using a hub as the pattern for the die cavity.

Hubs used for pressure casting of beryllium copper molds must resist softening at the elevated temperatures involved. Therefore, a hot work steel such as H13, is usually chosen for this application.

Alternative Mold Materials. In addition to the materials listed in Table 2, maraging steels are useful for making large molds—especially those used for molding abrasive plastics. Maraging steels require only simple aging at about 480 °C (900 °F) for several hours to achieve full hardness, which ranges from 50 to 58 HRC depending on the specific grade. In maraging steels,

Table 3 Typical materials for hubbed or cast molds

Material to be molded(a)	Mold materials(b) for making parts shown in Fig. 1 in production quantities of 10 000 or more: Part 1	Part 4	Part 7
Mold materials for thermoplastics			
Group 1: general-purpose plastics and rubbers	P1 (c), P3 or P5, all carburized(d), 54 to 58 HRC	Beryllium copper(e), 36 to 42 HRC; P5 or P1 (c), carburized(d), 54 to 58 HRC	Beryllium copper(f), 36 to 42 HRC; P1, carburized(d), 54 to 58 HRC
Group 2: fluid plastics	P3 (g), carburized(d), 54 to 58 HRC	P5, carburized(d), 54 to 58 HRC	P1, carburized(d), 54 to 58 HRC; beryllium copper(f), 36 to 42 HRC
Group 3: corrosive and high-temperature plastics	Beryllium copper(e), 36 to 42 HRC, nickel plated(h); P1 (c) or P3 (g), carburized(d), 54 to 58 HRC, nickel plated(h)	Beryllium copper(e), 36 to 42 HRC, nickel plated(h); P5 or P1 (c), carburized(d), 54 to 58 HRC, nickel plated(h)	Beryllium copper(f), 36 to 42 HRC, nickel plated(h); P1, carburized(d), 54 to 58 HRC, nickel plated(h)
Mold materials for thermosets			
Groups 4 and 7: general-purpose plastics	P3, carburized(d), 54 to 58 HRC	P3 or P5, carburized(d), 54 to 58 HRC	P1, carburized(d), 54 to 58 HRC
Group 5: plastics requiring high-temperature curing	P4, carburized(d), 56 to 60 HRC	P4 (j), carburized(d), 56 to 60 HRC	P4 (j), carburized(d), 56 to 60 HRC
Group 6: rubbers	1020, chromium plated(k); P1, chromium plated(k)	1020, chromium plated(k); P1, chromium plated(k)	P1, chromium plated(k)

(a) See Table 1. (b) Where more than one mold material is given for a specific set of conditions, they are arranged in order of preference unless otherwise noted. (c) P1 not recommended for quantities of 1 000 000 or more. (d) Carburizing recommended for mold accuracy in molds for less than 1 000 000 parts, and for increased abrasion resistance in molds for 1 000 000 parts or more. (e) Beryllium copper not recommended for quantities of 1 000 000 or more. (f) Beryllium copper not recommended for quantities greater than about 100 000 unless periodic replacement can be tolerated. (g) P5 also recommended for quantities of 1 000 000 or more. (h) Recommended thickness of plating, 0.005 to 0.025 mm (0.0002 to 0.0010 in.). (j) Hubbing probably is not a satisfactory means of producing mold cavities in P4 steel for parts of this depth. (k) Recommended thickness of plating, 0.005 to 0.015 mm (0.0002 to 0.0005 in.).

Table 4 Properties and applications of beryllium coppers used for cast molds

Alloy	Density Mg/m^3	Density lb/in.^{3}	Thermal conductivity(a) W/m·K	Thermal conductivity(a) Btu/ft·h·°F	Hardness Solution treated	Hardness Aged	Pouring temperature °C	Pouring temperature °F	Characteristics and applications
C82500(b)	8.26	0.298	93 to 114	54 to 67	63 HRB	40 HRC	1000 to 1120	1850 to 2050	High strength and hardness, for investment, sand and ceramic cast molds; excellent fluidity
C82600(c)	8.16	0.295	86 to 107	50 to 63	75 HRB	42 HRC	1000 to 1090	1850 to 2000	High strength, hardness and wear resistance; for pressure, sand and ceramic cast molds; very high fluidity
C82800(d)	8.09	0.292	86 to 107	50 to 63	85 HRB	43 HRC	1000 to 1080	1850 to 1975	High hardness and wear resistance; more brittle than C82500 or C82600; for pressure, sand and ceramic cast molds; very high fluidity

(a) At 20 °C (68 °F). (b) 97.2Cu-2Be-0.5Co-0.25Si. (c) 97Cu-2.4Be-0.5Co. (d) 96.6Cu-2.6Be-0.5Co-0.3Si.

hardening does not depend on fast cooling, and thus full hardness can be developed in massive sections. For enhanced abrasion resistance, maraging steel cavities can be nitrided.

Corrosion can cause significant problems for plastic molders, ranging from corrosion caused by molding of corrosive plastics such as polyvinyl chloride to plugging of water-cooling lines due to rust. Chromium plating has typically been used on molds for polyvinyl chloride, but such plating cannot prevent rusting of water lines. In recent years, stainless mold steels have become available. The stainless mold steels offer the advantages of standard mold steels such as P20 with the additional advantage of being highly resistant to corrosion. Two grades are manufactured as mold steel: type 414L and type 420. Type 414L is available prehardened to about 300 HB, which makes it ideal for injection molds; type 420 can be heat treated to 50 HRC, which makes it suitable for both injection and compression molds.

As an alternative to the materials listed in Table 3, small molds or mold inserts made by powder metallurgy should be considered where small articles are to be mass produced in high volume from abrasive plastics. The composition of the most successful P/M material for this purpose is 94.5 to 98.5 Fe, 0.25 to 0.75 C, 1.0 to 3.0 Ni, nil Cu and 2.0 max others.

Electroforming of molds by deposition of a relatively thin layer of hard nickel, followed by softer nickel or copper, on a pattern known as a "plating master" is very effective for certain types of work, particularly when the mold cavity includes delicate detail.

To extend life at points of high wear,

such as gate inserts, machinable steel-bonded carbides can be shaped into the insert, and then heat treated to hardnesses on the order of 68 HRC.

Influence of Service on Selection

Except in molding of unreinforced and unfilled thermoplastics, mold wear is one of the most troublesome problems with which a molder has to contend. Case depth (for case hardening steels), surface hardness of the mold, abrasive characteristics of the material being molded and the type of molding process are the most important factors influencing mold wear and mold life. Wear is lowest in blow molding, because the molding pressure is low and the plastic simply expands against the mold wall. Therefore, molds for blow molding can be made of relatively soft materials, such as annealed low-carbon steels.

In transfer molding, on the other hand, mold wear is relatively high, because the plastic is confined to channels under high pressure and flows at high velocity from the transfer pot into the cavity. Any abrasive material erodes the metal as it flows and in a short time enlarges the channels, causing leakage and loss of pressure. In addition to the fact that mold wear, general or local, is a function of mold hardness and the abrasiveness of the plastic being molded, there are other reasons why harder steels give more satisfactory life for extended runs.

For a given set of molding conditions, the molder's prime maintenance cost is the downtime necessary to maintain clean parting lines and to eliminate excessive flashing. Flashing, or cavity overflow, is caused by closing an empty mold on small particles of flash left over from the previous cycle. The areas directly adjacent to the cavity opening eventually become indented and distorted, and the mold must be reworked or repaired by welding or plating, or the damaged components must be completely replaced. Harder steels are not as readily indented as softer steels, and thus give longer service before they have to be reworked.

Resistance to parting-line indentation is roughly proportional to the hardness of the mold material. However, a hardness greater than that recommended for a given steel in a particular application should not be used. It has been stated that more molds fail by cracking than are worn out by use.

Special resistance to corrosion is required for molding certain plastics, and may be desirable if common steels rust due to sweating of the mold surfaces. Types 414L and 420 can meet these requirements.

Other Factors in Selection

In addition to service-related requirements, several other factors must be considered in selecting materials for molds.

The mold must be free from all surface defects. One pore or nonmetallic inclusion at the surface of a finished mold cavity may render the mold useless because the imperfection will be reproduced as a surface flaw on every plastic article produced in that mold. Therefore, although the compositions of tool steels used for plastic molds are very similar to those of less expensive constructional steels, the higher quality (freedom from large inclusions and other imperfections) provided by the special melting and casting practices used in producing tool steels imparts a most important material characteristic.

Despite special care in all phases of mill processing, there are still inherent sources of internal flaws in mold steels. For example, large tool steel ingots cool and solidify more slowly than small ones, and thus have more center porosity. This porosity may become partly welded shut during subsequent forging and hot rolling, but it is impossible to obtain as much reduction with bars of large cross section as with bars of smaller cross section and thus the former are more likely to contain internal voids. For molds larger than about 150 mm (6 in.) in thickness or 0.125 m^2 (200 $in.^2$) in cross section, individually upset forged blocks are recommended, rather than pieces cut from bar stock. This adds to the cost and the lead time required to obtain the mold steel, but for critical applications it may be the only available means of obtaining the required freedom from porosity.

The probability that a flaw will be present on the surface of the finished mold cavity can be further reduced by subjecting each mold block to ultrasonic inspection prior to machining.

REFERENCE

1. *Plastic Mold Engineering Handbook,* 3rd Ed., edited by J. Harry DuBois and Wayne I. Pribble: Van Nostrand-Reinhold, 1977

Selection of Materials for Thread-Rolling Dies

Edited by R. A. Cary
Vice President and Technical Director
Teledyne VASCO

THE PROPERTIES that are most significant in selecting materials for thread-rolling dies are hardness, toughness and wear resistance. Hardness and toughness must be high enough to enable the dies to withstand the forces exerted on them in service; good wear resistance is necessary because the prime cause for removing thread-rolling dies from service is spalling of the crests of the die threads, which is allowed to continue until the contours of the threads being rolled no longer meet dimensional or functional requirements.

Die performance, which is measured by the number of satisfactory parts produced, depends not only on the mechanical properties of the die material, but also on (*a*) the hardness, size and dimensional accuracy of the blanks being threaded, (*b*) the pitch of the threads produced, (*c*) whether the die is flat or circular, (*d*) accuracy of the die setup, (*e*) the inherent stiffness of the thread-rolling machine, (*f*) design of die-entrance threads, (*g*) speed of rolling, (*h*) number of revolutions per blank and (*i*) surface condition of the material being worked (the surfaces of blanks should be free of oxide, scale and other unwanted surface layers, such as one that has been inadvertently carburized).

The materials most commonly used to make thread-rolling dies are M1 and M2 high speed tool steels; D2 high-carbon, high-chromium tool steel; and A2 medium-alloy cold work tool steel. The steel chosen should be adequately annealed before hardening and should have a sufficiently uniform carbide-particle distribution.

In most applications, D2, M1 and M2 are about equal in performance, whereas the service life of A2 dies is somewhat lower. In general, D2, M1 and M2 should be selected for long production runs and for rolling larger parts, coarser threads and alloys of higher hardness. The hardness ranges given below have been found satisfactory for most applications:

Die material	Recommended hardness, HRC Flat dies	Circular dies
For threading aluminum, copper or soft steel blanks		
A2	57 to 60	56 to 58
D2	60 to 62	58 to 60
M2	58 to 60	58 to 60
For threading ferritic steel (hardness, >95 HRB) or austenitic stainless steel blanks		
A2	57 to 59	56 to 58
D2	59 to 61	58 to 60
M2	59 to 61	58 to 60

Hardness values indicated above should be achieved by double tempering after quenching. Early failure is more likely if double tempering is not used.

If diameter and lead tolerances of the rolled part permit, the dies can be ground or machined before hardening. The recommended steels for that practice, given in descending order of preference, and their average distortion during heat treatment necessary for obtaining the required hardness, are as follows:

Type of steel	Approx average distortion, %
A2	0.04
D2	0.05
M1 or M2	0.11

If tolerances require lower average distortions, dies must be ground after heat treatment. Die life usually is not decreased if grinding is done after heat treatment, as long as proper (nonabusive) grinding techniques are used.

Ordinarily, the surfaces of screw threads are about 20 to 40% rougher than the corresponding die surfaces. The specified surface roughness of dies is usually attained easily, regardless of the tool steel selected. Required finishes for fine-pitch threads can be obtained most economically by using A2, and either machining or grinding the die threads.

As a short-term solution for a deficient die setup, which is the cause of most early die failures, die hardness can be reduced to increase toughness. However, a better solution is achieved by providing good setups, because reduction of hardness generally decreases die life. For both flat and circular dies, insufficient hardness can cause failure by upsetting, sinking or flattening of thread crests, whereas excessive hardness can cause the die threads to crack off at the base.

A compromise that provides an optimum combination of hardness and toughness in the die material is essential in many applications. Where abrasive wear is the prime consideration, higher hardness may be justified, but this may lead to the other types of failure mentioned above.

Flat dies are used for producing most standard threaded fasteners and most wood screws. The stationary half of a pair of standard flat thread-rolling dies is shown in Fig. 1. Dies are made to roll the maximum standard length of thread for the specific screw size and can be used to roll screws of any thread length up to the maximum. Because the same die can be used for threading fasteners of different lengths as required until it fails or wears out, die material should be selected for maximum production except where special threads are called for and production quantities are small.

Flat dies made of D2 are usually ground before hardening, because D2 is susceptible to grinding cracks if improperly ground after hardening.

A circular thread-rolling die is also shown in Fig. 1. Circular dies usually are ground after hardening. A2 is preferred for short production runs on all except the materials most difficult to thread, because its grindability is good and its wear resistance is adequate. More expensive steels such as D2, M1 and M2 are justified for long runs and for work materials that are difficult to roll.

Fig. 1 Typical flat and circular thread-rolling dies

Examples

Specific evaluations of thread-rolling dies are given in the following examples. Although all the circular dies in these examples were made of A2 and all the flat dies of D2, it should not be assumed that A2 is the best steel for circular dies or that D2 is the best for flat dies, either for these examples or for other similar applications.

Seventy-two thread-rolling dies, made by hobbing D2 steel before hardening, ultimately failed in accordance with the die-life distribution shown in Fig. 2. These data were for dies of the same design and tool steel, all used for rolling the same threads in the same screw-blank material. It is clear that even under these conditions, in which many of the factors influencing die life were controlled, a wide spread and skewed distribution in life were obtained. This spread and distribution are not unusual for tools that must operate close to the endurance limit. It serves to emphasize, however, that the effects of changes in source of supply, tool material or heat treatment on the life of thread-rolling dies usually cannot be evaluated by testing only a few tools in production.

Fig. 2 Life of D2 tool steel dies for threading steel bolts

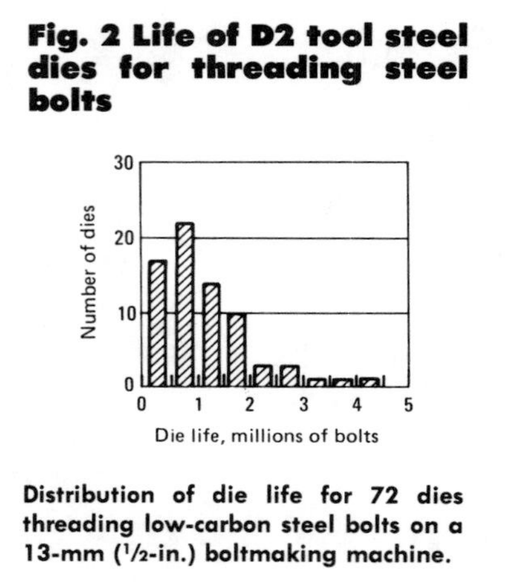

Distribution of die life for 72 dies threading low-carbon steel bolts on a 13-mm (1/2-in.) boltmaking machine.

Figure 3(a) shows the approximate average life of flat D2 tool steel dies used for rolling 1/4-20 threads in blanks of 1022 steel of various hardnesses. Some of the tests were run on hardened blanks, including those yielding the two data points showing the highest hardness. At high blank hardnesses, die life becomes short and inconsistent. The shaded portion of the curve indicates this spread and inconsistency in die life. A hardness of about 33 HRC for the blank is the limit for thread rolling with normal die steels and die processing. M42 high speed steel has been used in dies for rolling threads in higher-hardness blanks.

Figure 3(b) shows minimum and maximum die life versus blank hardness for a large number of circular dies made of A2 tool steel. These dies were used mainly in rolling fine-pitch threads in screws made mostly from free-cutting brass or aluminum alloys. The data include some high-production runs where conditions of setup and blank material were nearly ideal. Furthermore, dies that roll fine threads in small parts have longer average life than those that roll coarse threads in

Fig. 3 Variation of life of thread-rolling dies with blank hardness and diameter

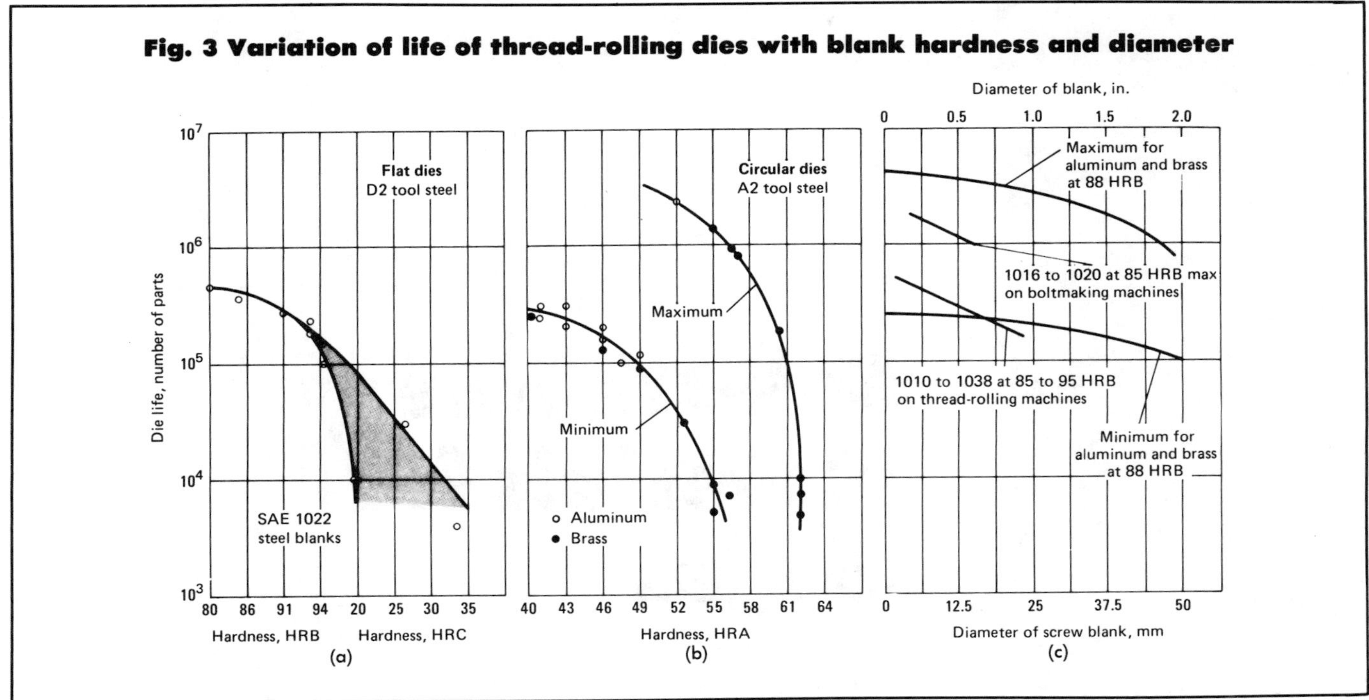

large parts. Consequently, the average life expected for all types of circular dies made of A2 tool steel is less than the mean between the two curves in Fig. 3(b).

Figure 3(c) shows the relation between die life and diameter of threads rolled in aluminum and brass blanks with circular dies and in steel blanks with flat dies. The upper and lower curves are for circular A2 dies, most of which were ground after hardening and were used primarily for rolling fine-pitch threads—the majority of them special threads. For blanks having a hardness of 88 HRB, no detectable difference in die life was found regardless of whether the blanks were aluminum or brass.

The curve for 1016 to 1020 steels gives average life in a single setup of several hundred D2 dies that had special entering-thread contours; dies were machined and hardened, and were used in the as-hardened condition.

The other curve represents the relation between size of blank and approximate average life expected from flat dies in a single setup when rolling threads in 1010 to 1038 steels on reciprocating thread-rolling machines.

Unusual conditions can alter die life considerably. For example, rolling threads in a 16-mm (5/8-in.) 1040 hot rolled steel blank that was nonuniform in diameter, scaled, and unannealed gave a die life of only 45 000 parts.

Selection of Materials for Gages

Edited by T. A. Newmeyer
L. S. Starrett Co.

SELECTION OF MATERIALS for gages depends to a large extent on tolerance to be checked, number of items to be gaged, composition and hardness of the material being gaged, size and complexity of the item and cost of the gage material. Abrasive wear is the predominant factor in determining the useful life of production gages. Therefore, gage surfaces that contact the workpiece must be hard enough to provide adequate resistance to abrasion. The actual hardness required depends on the hardness and abrasive characteristics of the workpiece surfaces, number of pieces to be gaged and the tolerance to be checked. Production gages that must be used in hostile environments may have to be corrosion resistant. Also, if the gage must operate over a range of temperatures, thermal expansion must be considered when making final material selection.

Gage Materials

Typical materials and hardnesses for the more common types of production gages are given in Table 1. In this table, it should be noted that feeler gages, which are thin flexible strips of precisely controlled thickness, must be tempered back to lower hardness so that they will not break during use.

Production Gages. Compositions and basic properties of most of the materials used for gages may be found elsewhere in this volume. To supplement the information given in Table 1, actual wear comparisons are shown in Fig. 1 to 3 for various plug-gage and ring-gage materials used to evaluate gray iron, steel and aluminum parts. The ability of cemented tungsten carbide to increase the useful life of these gages is clearly shown in these illustrations. Straight tungsten carbide grades of normal particle size and containing 6 to 9% cobalt (hardness, 90 to 92 HRA) are used most frequently for gage applications. Steel-bonded carbides also are acceptable for gage service.

Carbide should be used as a gage material only after careful consideration of its advantages and disadvantages. Based on material cost and number of pieces gaged, economy should be the first consideration. Because carbide will outwear steel by a factor of 20 to 100, the gage should have high use and close tolerance requirements in order to justify the higher cost. In constant

Fig. 1 Comparison of wear resistance for several plug-gage materials used for gaging parts made of class 30 gray iron and of two steels

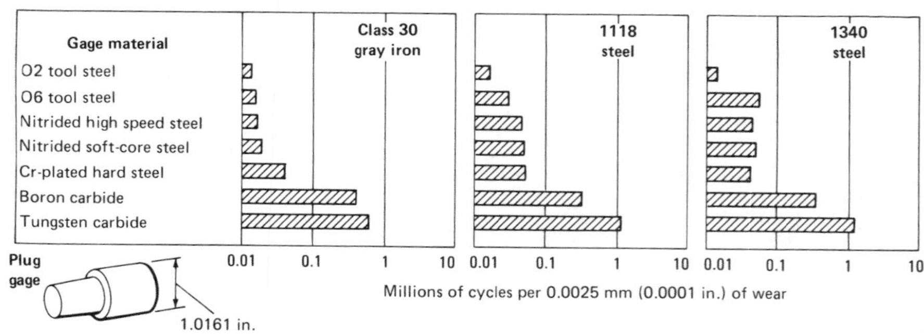

Table 1 Typical materials and hardnesses for common types of production gages

	Gaging tolerance 0.01 mm (0.0005 in.)				Gaging tolerance 0.05 mm (0.002 in.)			
	Part hardness up to 350 HB		Part hardness over 350 HB		Part hardness up to 350 HB		Part hardness over 350 HB	
Types of gage	Gage material(a)	Hardness, HRC	Gage material(a)	Hardness, HRC	Gage material(a)	Hardness, HRC	Gage material(a)	Hardness, HRC
Occasional gaging of dimensions up to 100 mm (4 in.)								
Gage blocks, gaging pins, anvils, buttons	W1, O1, O2	61-64						
Cylindrical ring and plug gages	1212 (b), W1, O1, O2	61-64						
Threaded ring and plug gages	W1, O1, O2	61-64						
Height and length gages	W1, O1, O2	61-64						
Spline gages	W1, O1, O2	61-64						
Snap gages(c), thread rolls	O1, O2, L7	61-64						
Feeler gages	W1, O1, O2	45-52						
Alignment bars	1212 (b)	...						
Frequent, long-term gaging of dimensions up to 12 mm (1/2 in.)								
Gage blocks, gaging pins, anvils, buttons	L7 (d) D2 (e)	61-64 57-64	D2 (e) M2 (f) Carbide(g)	57-64 62-65 ...				
Cylindrical ring and plug gages	M2 (f)	62-65	M2 (f) Carbide(g)	62-65 ...				
Threaded ring and plug gages	A2 (e)	56-64	M2	62-65				
Height and length gages	M2	62-65	M2	62-65				
Spline gages	O6	61-64	A2 (e)	56-64				
Snap gages(c), thread rolls	L7 (d)	61-64	L7 (d)	61-64				
Feeler gages	L7 (d)	45-50	D2	45-50				
Frequent, long-term gaging of dimensions from 12 to 100 mm (1/2 to 4 in.)								
Gage blocks, gaging pins, anvils, buttons	A2 (e) D2 (e) M2	56-64 57-64 62-65	Carbide(g)	...	A2 D2 M2	62-64 62-64 62-65	M2 (f)	62-65
Cylindrical ring and plug gages	M2 (f)	62-65	Carbide(g)	...	D2 M2	62-64 62-65	M2 (f)	62-65
Threaded ring and plug gages	A2 (e) D2 (e)	56-64 57-64	M2 (f)	62-65	A2	62-64	M2	62-65
Height and length gages	A2 (e) D2 (e)	56-64 57-64	M2	62-65	A2	62-64	M2	62-65
Spline gages	O6	61-64	A2 (e) D2 (e)	56-64 57-64	O6	61-64	A2, D2	62-64
Snap gages(c), thread rolls	L7 (d)	61-64	D2 (e) M2	57-64 62-65	L7	61-64	D2 M2	62-64 62-65
Alignment bars	1212 (b) 8620 (b)	... ...	8620 (b) 4140 (h)	... ...	1212 (b) 8620 (b)	... ...	8620 (b) 4140 (h)	... ...

(a) Where more than one tool material is listed for a specific set of conditions, the last material listed is usually preferred for large sections. (b) Carburized, with a case not more than 1/5 the section thickness and having a minimum surface hardness equivalent to 61 HRC. (c) Snap-gage bodies generally are made of stress-relieved cast iron, ASTM A48, class 20, 30 or 35. (d) 52100 steel has proved a satisfactory substitute for L7. (e) For close tolerances, this steel must be tempered in the secondary hardening range for maximum stability. (f) Liquid nitriding after full hardening is recommended to produce a surface hardness of about 1100 HV. (g) Cemented tungsten carbide is usually selected, but chromium carbide or boron carbide also can be used. (h) Heat treated to 26 to 30 HRC, then gas nitrided for 24 h.

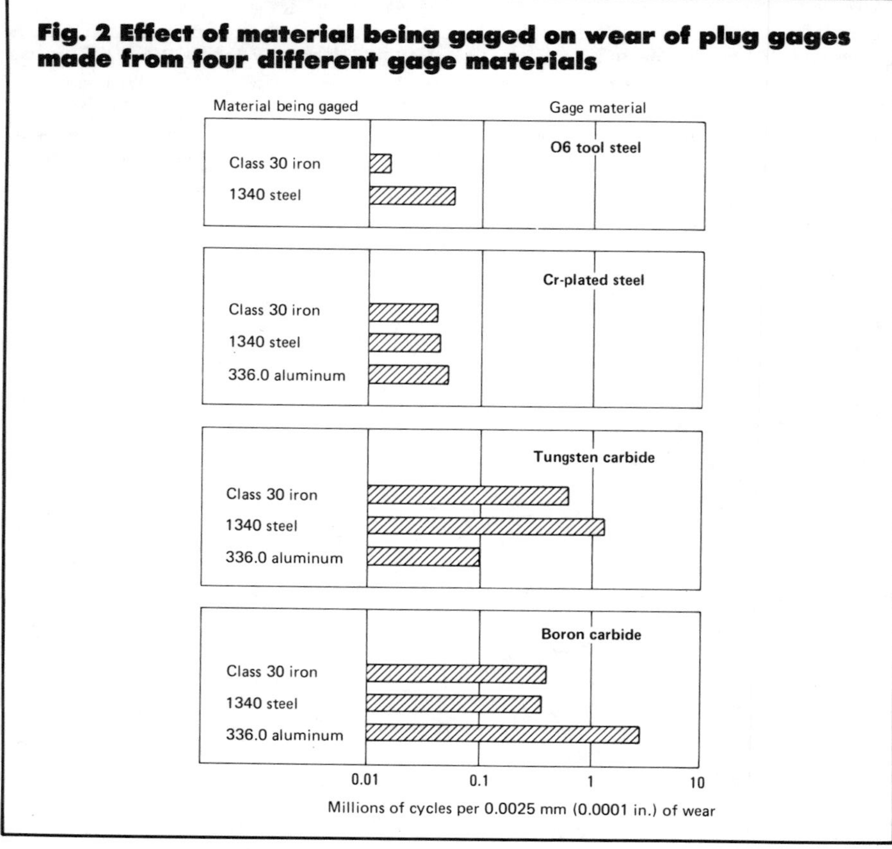

Fig. 2 Effect of material being gaged on wear of plug gages made from four different gage materials

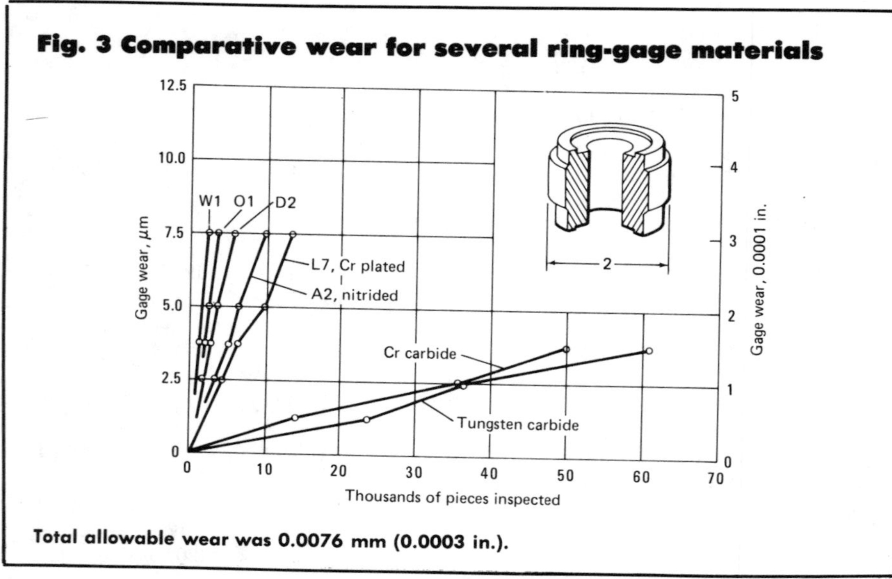

Fig. 3 Comparative wear for several ring-gage materials

Total allowable wear was 0.0076 mm (0.0003 in.).

use, steel gages sometimes will wear out in a few hours or a few days. For such applications, carbide is often the most practical and economical material based on total gage cost per piece gaged. If the gage must be altered from time to time because of changes in product dimensions or tolerances, it must be kept in mind that a diamond grinding wheel and diamond polishing compound are necessary for reworking of a carbide gage.

The lower coefficient of expansion of carbide, compared with that of steel, must also be considered when tolerances are very small (0.0025 mm; 0.001 in.). A change in room temperature may cause the product to expand or contract out of tolerance, whereas the gage would not change the same amount.

Carbide is more brittle than steel, and thus sharp corners and thin sections need more care and protection from misuse in carbide gages than in gages made of hardened and tempered tool steel. With few exceptions, any gage that will not break or chip when made from D4 steel can be made from carbide. In one plant that uses large numbers of small gages, the loss of carbide gages through breakage was so great that carbide was replaced by alloy tool steel as a gage material.

Precision Gages. Making gages to extra-close tolerances often requires special gage materials. In applications that require extremely close gage tolerances, fine surface finishes, exceptional wear characteristics and outstanding dimensional stability, boron carbide and even jewels make excellent gage elements. Gage elements have been made of jewel materials such as synthetic sapphire, ruby and diamond. Because of fragility, susceptibility to chipping, and high cost, use of these gage materials is restricted to highly specialized applications, such as high-precision gaging or gaging of extremely abrasive products.

Combination Gages. Frames, bodies and bases of combination gages commonly are made of cast iron. The cast iron used for gages is generally a class 20, 30 or 35 gray iron that has been stress relieved at 450 to 480 °C (850 to 900 °F).

Aluminum is also used in combination gages for handles, bodies and bases. Aluminum handles and bodies may be made either of soft grades such as 1100 or, if more strength is required, of harder grades such as 2017 or 2024. The light weight of aluminum combination gages such as plug gages, male thread gages and ring gages reduces inspector fatigue. Aluminum bases cast from alloys such as 355 or 356 are used for large gages where dimensions are not held to close tolerances, such as gages to check automobile body dimensions and window openings.

Materials used for wear surfaces in combination gages must have the same properties as those required for the solid gages discussed above. Wear inserts generally are made from hardened tool steel or cemented carbide. However, when the insert is a relatively simple shape that is not very susceptible to distortion and cracking in heat treatment, it may be possible to use an inexpensive tool steel instead.

Inspection fixtures are gages, but belong to a different category from those discussed so far. They are used mainly for checking dimensions and contours of large stampings such as automobile body components.

Plastics or plastic-faced cast iron composites are frequently used for inspection fixtures (checking fixtures) because they are lower in cost and lighter in weight than solid cast iron. Use of plastics is limited because they are less stable, less wear resistant and higher in thermal expansion than metals. Because tolerances for most stampings are about ± 0.5 mm (± 0.020 in.), plastics are well suited for many of these applications.

The coefficient of thermal expansion of the type of resin used in inspection fixtures is about 67 μm/m·K (37 μin./in.·°F), compared with about 12 μm/m·K (6.5 μin./in.·°F) for cast iron. However, if a glass cloth filler is used in the ration of four parts glass to one part resin, this very wide difference in coefficient of thermal expansion between resin and cast iron is reduced considerably. The change in coefficient of thermal expansion is not directly proportional to the amount of glass filler. For instance, in a filled plastic, adding about 80% glass reduces the coefficient by only about 40% of the difference between the coefficient of the resin and that of the glass.

Plastic materials used in inspection fixtures must be thoroughly cured for stability. Cast iron must be stress relieved.

Neither all-plastic nor composite construction is practical for all inspection fixtures, but because of cost and weight advantages, they should both be considered. Although composite plastic-gray iron construction generally is preferred for reasons of economy, no substitute has been found for hardened steel as a material for the wear parts of inspection fixtures.

Master gages require high accuracy, good wear resistance and maximum stability. They are used in checking other gages and are expected to hold their accuracy over long periods of time. Because one of the most important characteristics of master gages is dimensional stability, relative freedom from retained austenite and residual stress are necessary in the materials from which they are made.

No one material has the ideal combination of properties desired for master gages, although high-carbon high-chromium steels such as D2 and D4 have most of the desired characteristics. These steels are difficult to machine, but they exhibit low dimensional change in heat treating. After tempering in the secondary hardening range and subzero treating (the same treatment used for close-tolerance production gages), master gages made from these steels have good dimensional stability, particularly when given a final stress-relief temper.

Wear resistance is an important (although secondary) consideration in selecting materials for master gages. The high wear resistance of properly heat treated D2 and D4 tool steels can be further increased by gas nitriding at 515 °C (960 °F). For master gages used only in modern constant-temperature rooms, solid cemented carbide is an attractive gage material.

Surface Treatments

Chromium plating of simple gages increases gage life, affords some resistance to atmospheric corrosion (without a nickel underplate) and is useful in salvaging worn gages. Gages in which the wear areas are situated along edges or at corners should not be chromium plated, because the plating will flake away with continued use. This applies particularly to formed pin gages. No gage subjected to impact near an edge or corner should be chromium plated either. If plated gages are used, plating should be done at 38 to 60 °C (100 to 140 °F), and the gages should be baked immediately for 2 h or more at 150 °C (300 °F) to remove hydrogen and thus combat embrittlement.

Corrosion of gages may become a problem where humidity is high (above 60%) or where industrial fumes are present. A common practice for protection of carbon steel and low-alloy steel gages is to coat them lightly with phosphate so that they will retain a film of oil. This does not affect gage dimensions. Corrosion also can be avoided by using a stainless steel such as type 440C (1% C, 17% Cr), which can be hardened, tempered in the secondary hardening range, subzero treated and given a final stress-relief temper. When carbon or low-alloy steels are used, they may be electroplated with nickel and then chromium. Carbide gages do not require protection against corrosion.

Selection of Tool Materials for Structural Components

By James E. Denton
Manager of Materials Technology
Cincinnati Milacron, Inc.

TOOL MATERIALS have unique characteristics that make them particularly suitable for many demanding structural (nontool) applications. Many tool steels are nondeforming in heat treatment—that is, they undergo little or no size and shape distortion. They also exhibit good to excellent abrasion resistance, hot hardness, toughness, shock resistance, hardenability and dimensional stability. In addition to these properties, tool steels have fine grain size and good microcleanness.

Approximately half of the total U. S. production of cemented tungsten carbide is used to make rock drills and valve seats for the mining, petroleum and natural gas industries. For these applications, the high abrasion resistance of tungsten carbide is its most useful property. Its high density, 15 Mg/m^3 (0.54 $lb/in.^3$), is useful in applications where high inertia is required. Gyroscope rotors, flywheels, vibration dampers, centrifugal clutches and governors are parts that take advantage of the high density of tungsten carbide. The high modulus of elasticity, 620 GPa (90×10^6 psi), is useful in devices that require minimum deflection under load, such as boring bars.

Although necessarily higher in cost, tool materials frequently are preferable to constructional alloy steels because they provide longer life and increased reliability.

Abrasion Resistance

Typical examples of machinery components that are subject to high abrasion include guides, cams, plain sliding bearings, bushings, cams and cam followers. These components frequently function under conditions involving high contact stress, high surface velocity, marginal lubrication and abrasive environments. Materials appropriate for these types of applications, in order of increasing wear resistance, include W1, O1, A2, D2, M2, T15 and cemented carbides.

The spool in a hydraulic valve was changed from carburized 8620 to A2 tool steel. After three years in service it was found that the valve with the A2 spool was more precise in its response than the valve with the 8620 spool, and exhibited a rate of hydraulic-fluid leakage 70% lower. This improvement in performance was due to retention of sharp edges at the port shoulders of the spool and to lack of wear on the closely fitted diameters.

In the textile industry, D2 tool steel has proved very effective for thread guides.

Hot Hardness

Some mechanical components are required to operate continuously at elevated temperatures that would result in ordinary constructional alloy steels being tempered to hardness levels below their useful limit. High speed steels have been found to perform very satisfactorily in such applications. Roller bearings for service in gas turbine engines must operate in the region of 540 °C (1000 °F); type M42 tool steel is commonly used in such bearings. Valves and valve seats in plastic injection molding equipment must withstand the plasticizing temperature of the resin being molded. For some polymers this may be as high as 300 °C (575 °F). Type M2 tool steel is frequently used in this application.

Toughness

Many machinery components are designed to withstand repeated flexing. These spring-type applications include split collets, leaf and reed springs, torsion bars, and critical aircraft parts such as landing-gear components and helicopter rotors. Type L2 and type L6 tool steels are among the only standard grades that contain nickel. This imparts high toughness, which makes these steels leading candidates for use in flexural applications.

The hot work die steels H11, H11 Mod, H12 and H13 exhibit several properties that are important in airframe and landing-gear applications. These grades show secondary hardening maxima in their tempering curves, and typically develop the combination of maximum hardness, maximum tensile strength, high fracture toughness and maximum fatigue strength on being tempered at about 565 °C (1050 °F). These steels are susceptible to hydrogen embrittlement and must be protected with a corrosion-resistant coating, even in mild atmospheric environments. Similarly, they must be protected against oxidation if exposed to air for prolonged times at temperatures above 400 °C (750 °F).

Shock Resistance

Although they also have remarkable toughness, the group S shock-resisting steels are perhaps more notable for their resistance to chipping. Types S5 and S7 are used in applications involving impact loading, such as clutch teeth, ratchets, dogs and detents.

Safety in Heat Treatment

The exceptional hardenability of tool steels permits them to be quenched less drastically than constructional-grade alloy steels. This provides an increased degree of safety in heat treatment—that is, less likelihood of cracking—for parts such as templates and manifold plates, which have intricate shapes incorporating many stress raisers at corners, edges, fillets and holes. Air-hardening grades such as A2 are frequently selected for these applications. Several grades of tool steel, including O2, O6, A7 and L6, can be hardened from a relatively low austenitizing temperature of about 790 °C (1450 °F). For parts that are not amenable to precision finishing after hardening—such as gears, splines and couplings—the low distortion resulting from the reduction in austenitizing temperature is advantageous.

Dimensional Stability

Precision measuring instruments such as gage blocks, toolmakers' squares, straight edges, templates and micrometer anvils must have extreme dimensional stability. Dimensional inaccuracies may arise from two major sources: wear resulting from use, and changes in size resulting from transformation of retained austenite. Type O2 tool steel has proved very effective in minimizing both of these effects. The presence of uncombined carbon in the graphitic O6 grade is extraordinarily effective in promoting dimensional stability. For both O2 and O6, multiple subzero treatments followed by retempering are beneficial in transforming or stabilizing retained austenite.

Wear-Resistant Materials

Hard Facing Materials

HARD FACING is the process of applying, by welding, plasma spraying or flame plating, a layer, edge or point of wear-resistant metal onto a metal part to increase its resistance to abrasion, erosion, galling, hammering or other form of wear. Hard facing may be applied to new parts to improve their resistance to wear during service, or to worn parts for the purpose of restoring them to serviceable condition. It is frequently used in applications where systematic lubrication against abrasion is not feasible or is inadequate to give the desired service life. In general the wear-resistant coating is applied only to those critical surfaces of components where wear is maximum. Worn parts can be satisfactorily refaced or rebuilt many times before replacement becomes mandatory.

The economic success of the process often depends on selective application of relatively expensive hard facing alloys to comparatively inexpensive base metals. Heavy or bulky parts that are difficult and costly to move often can be repaired or rebuilt in the field or in the plant where they are installed, by welding with portable equipment.

Material Classifications

Most hard facing alloys are marketed as proprietary materials. They are classified here in five major groups (1 to 5), primarily according to total alloy content (elements other than iron), with subdivisions based on major alloying elements (see Table 1). Usually, both wear resistance and cost increase as the group number increases. Choice of form and type of alloy depends on the application and the welding process to be used.

Forms. Alloys for hard facing usually are available as bare cast or tubular rod, covered solid or tubular electrodes, solid wire, and powder. The availability of a particular alloy in a given form depends on the ability of the alloy to be cast, shaped into wire or tubing, or produced in powder form. Alloys that do not lend themselves to economical production by any of these methods may be inserted in granular form into a low-carbon steel tube during the process of roll forming the tube from strip. The continuous tube can be cut to rod lengths for "stick" electrodes or it can be coiled for use as electrode wire in semiautomatic or automatic welding.

Many alloys are available in more than one form. The bare cast and bare tubular rod forms are used in gas tungsten-arc welding, covered electrodes in shielded metal-arc welding, solid wire in gas metal-arc or submerged-arc welding, tubular (alloy-cored) wire in open-arc welding, and powder in plasma-arc welding.

Alloy Composition. The alloys in group 1A are low-alloy steels that, with few exceptions, contain chromium as the principal alloying element. The total alloy content (including carbon) is between 2 and 6%. These alloys often are used as buildup materials for support of harder, more highly alloyed hard facing alloys.

The iron-base alloys in group 1B are

Table 1 Classification of hard facing materials by alloy groups

Group	Total alloy content, %	Principal alloying elements
Low-alloy ferrous materials		
1A........	2 to 6	Cr, Mo, Mn
1B........	6 to 12	Cr, Mo, Mn
High-alloy ferrous materials		
2A.......	12 to 25	Cr, Mo
2B.......	12 to 25	Mo, Cr
2C.......	12 to 25	Mn, Ni
2D.......	30 to 37	Mn, Cr, Ni
3A.......	25 to 50	Cr, Ni, Mo
3B.......	25 to 50	Cr, Mo
3C.......	25 to 50	Co, Cr
Nickel-base and cobalt-base alloys		
4A.......	50 to 100	Co, Cr, W
4B.......	50 to 100	Ni, Cr, B
4C.......	50 to 100	Cr, Ni, Co
Carbides		
5........	75 to 96	WC, alone or in combination with other carbides such as TiC and TaC, all in a metal matrix.

similar to those in group 1A except that they contain higher total alloy contents (6 to 12%), and in some instances carbon contents of 2% or more. Many tool steels and several alloy cast irons are included in this group.

Alloys in group 1 have the greatest shock resistance of all hard facing alloys except austenitic manganese steels, and have better wear resistance than low-carbon and medium-carbon steels, which are the base metals to which they are usually applied. These alloys are less expensive than the other hard facing alloys, and are extensively used where machinability is necessary and only moderate improvement over the wear properties of the base metal is required.

The alloys in group 2A are chromium-containing alloys with total alloy contents of 12 to 25%. Many of these alloys have appreciable molybdenum contents. This group also includes certain medium-alloy cast irons.

Molybdenum is the principal alloying element in nearly all group 2B alloys, most of which also contain appreciable amounts of chromium. These and the group 2C steels have total alloy contents between 12 and 25%.

Group 2C alloys are austenitic manganese steels. Although manganese content predominates, each of these alloys contains an appreciable amount of nickel or molybdenum as an austenite stabilizer.

The hard facing alloys in groups 2A and 2B are more wear resistant, less shock resistant and more expensive than those in group 1. Groups 2C and 2D are highly shock resistant, but have limited wear resistance unless subjected to work hardening. Group 2D alloys have total alloy contents of 30 to 37%, and carbon contents ranging from less than 0.10% to more than 1.0%.

Group 3 alloys have total alloy contents from 25 to 50%. They are high-chromium alloys, many of which contain nickel or molybdenum, or both. Carbon content ranges from about 1.75% to more than 5%. Group 3B alloys contain appreciable amounts of molybdenum and chromium, and group 3C, of cobalt and chromium. Groups 3A, 3B and 3C alloys are characterized by massive hypereutectic alloy carbides that impart wear resistance and some degree of resistance to corrosion and heat. They are more expensive than the alloys in groups 1 and 2.

Cobalt-base and nickel-base alloys with total content of nonferrous metals from 50 to 99% are classified in group 4. The cobalt-base alloys (group 4A) generally are rated as the most versatile of the hard facing materials. They resist heat, abrasion, corrosion, impact, galling, oxidation, thermal shock, erosion and metal-to-metal wear. Some of these alloys retain useful hardness up to 825 °C (1500 °F) and resist oxidation temperatures up to 1100 °C (2000 °F). The nickel-base alloys (group 4B) are most effective for service involving both corrosion and wear. They are superior to other hard facing alloys where wear is caused by metal-to-metal contact, as in bearings. They retain useful hardness up to about 650 °C (1200 °F) and resist oxidation at temperatures up to 875 °C (1600 °F).

Group 5 materials consist of hard granules of carbide distributed in a metal matrix; they are extremely important for severe abrasion and cutting applications. Historically, tungsten carbides were used exclusively. Recently, however, carbides of certain other elements—notably titanium, tantalum and chromium—have been used with satisfactory results. Various matrix metals are employed, including iron, carbon steel, nickel-base alloys, cobalt-base alloys, and bronzes. Group 5 materials provide maximum abrasion resistance under service conditions involving low or moderate impact.

In addition to the materials in Table 1, copper-base alloys comparable in composition to bronzes are used as matrix metals for carbides or as overlays on less-expensive base metals, notably low-carbon steels. The overlays sometimes serve as bearing materials and provide resistance to corrosion and cavitation damage. They offer poor resistance to corrosion by sulfur compounds, to abrasive wear and to creep at elevated temperatures.

Alloy Selection

In order to select the correct hard facing alloy for a particular wear application, a thorough analysis of anticipated service conditions must be carried out, and the factors that could potentially cause severe material degradation must be established. Many investigators acknowledge the following types of wear:

Adhesive wear
- Mild or oxidative wear
- Severe or metallic wear

Abrasive wear
- Low-stress scratching abrasion
- High-stress grinding abrasion
- Gouging abrasion

Erosive wear
- Impingement erosion
- Cavitation erosion

Fretting

Adhesive wear is a term generally used to describe wear due to sliding of two metallic components against each other where no abrasives are intended to be present (for example, wear due to sliding caused by slippage of roller bearings, or to sliding between a valve and a seat). A quantitative relationship of the factors involved in adhesive wear can be expressed as follows:

$$V = \frac{KLD}{H}$$

where V is volume wear, L is load, D is sliding distance, H is hardness and K is wear coefficient.

When the applied load is low enough, an oxide film usually is generated as a result of frictional heating accompanying sliding. The oxide film prevents formation of metallic bonds between the asperities on the sliding surfaces, resulting in low wear rates. This form of wear is called mild wear or oxidative wear. In general, moving components should be designed so that mild wear conditions prevail.

On the other hand, if the applied load is high, metallic bonds are formed between the surface asperities of the mating materials, and the resulting wear rates are extremely high. This form of wear is called severe wear or metallic wear. The load at which there is a transition from mild to severe wear is called the transition load.

A form of wear called galling is a special form of severe wear. Galling occurs when the wear debris is larger than clearances between mating surfaces, and seizure of the moving component results.

Quite often, even a small amount of sliding can produce galling of mating surfaces. For example, in a high-pressure gate valve, high seating stresses may result in formation of metallic bonds between the surface asperities of mating materials. When the valve is actuated, material is transferred from one point to another due to adhesion or locking of surface asperities. The resulting surface damage prevents relatively leak-free reseating of the valve, and valve failure is said to have occurred.

In situations where lubrication is not possible, hard facing is recommended for minimizing adhesive wear. One example is automotive exhaust valves, which are subjected to extreme temperatures at which lubricants are not stable. Nickel-base hard facing alloys usually have very high transition loads compared with those of cobalt-base hard facing alloys. On the other hand, cobalt-base alloys are much more resistant to galling than nickel-base or iron-base alloys.

Abrasive Wear. Low-stress scratching abrasion is defined as wear resulting from the cutting action of sliding abrasives stressed below their crushing strength. In this form of abrasion, wear scars usually show scratches and the amount of subsurface deformation is minimal.

High-stress grinding abrasion is wear under conditions of stress high enough to crush the abrasives. Such stresses usually are high enough to also cause plastic deformation of any ductile constituents of the wear material. Typical examples of high-stress grinding abrasion are those that occur in ball and rod mills.

The term gouging abrasion is used to describe high-stress abrasion by which sizable grooves or gouges are created on the wear surface. These gouges may result from sliding followed by impact such as that encountered in gyratory crushers, where high stresses (high even on a macroscale) are transferred through large chunks of abrasives.

For many materials, abrasion resistance increases with increasing hardness. However, abrasion resistance is not determined solely by material hardness, but is strongly influenced by microstructural condition. In general, abrasion resistance of hard facing alloys increases as carbide content increases. As a result, in extremely abrasive situations, alloys with large amounts of carbides are recommended. Whether or not heat treatment following application of a hard facing material improves wear resistance depends on the specific abrasion-resistant alloy. For example, heat treatment does not improve abrasion resistance of cobalt-base alloys.

The deposition process used has a significant influence on alloy dilution and on size of carbides. For example, in depositing Stellite 6, the gas tungsten-arc process creates finer carbides than the oxyfuel-gas process. Thus, in abrasive-wear tests, wear resistance of oxyfuel-gas deposits are superior to those of gas tungsten-arc deposits.

Erosive Wear. Wear in which loss of material occurs due to the cutting action of moving particles carried in a fluid stream is termed impingement erosion. Usually, resistance of materials to impingement erosion varies with the angle of impingement. At low impingement angles ($<15°$), hard facing alloys with large amounts of carbides (hypereutectic alloys), such as Stellite 1, are recommended. At high impingement angles ($>80°$), hard facing alloys with large amounts of matrix (hypoeutectic alloys), such as Stellite 21, are recommended.

In fluid-handling valves and pumps, a phenomenon known as cavitation erosion often occurs as a result of turbulent flow. In this type of wear, air bubbles caused by the turbulence collapse near the metal surface, and the resultant shock waves detach material from the surface. Typical examples include fluid-flow valves with large pressure differentials and high-velocity pumps. A similar type of damage occurs on the leading edges of steam-turbine blades. This damage is caused by impact of small droplets of water that have condensed in the flowing steam. Cobalt-base hard facing alloys such as Stellite 6 and Stellite 21 have been found to be the best suited for resisting cavitation erosion and liquid-impingement erosion.

Fretting is the term used to describe material loss due to very-small-amplitude vibrations at mechanical connections, such as riveted joints or joints made with other fasteners. This type of wear is a combination of oxidation and abrasive wear. Oscillation of two metallic surfaces produces tiny metallic fragments, which oxidize to form abrasive particles. Subsequent wear proceeds by both adhesion and abrasion, and the wear debris is continually converted into abrasive oxides. Most cobalt-base hard facing alloys are recommended for resisting this form of wear.

Other Considerations. In addition to the aforementioned types of wear, selection of hard facing alloys is strongly influenced by impact, corrosion, oxidation, hot corrosion and thermal stability.

In general, impact resistance of hard facing alloys decreases as carbide content is increased. Because abrasion resistance increases with increasing carbide content, a compromise often must be made between abrasion resistance and impact resistance. In applications where impact resistance is extremely important, austenitic manganese steels are used.

Quite often, wear is accompanied by corrosion from acids or alkalis such as those encountered in the chemical-process or petroleum industries or in flue-gas scrubbers. Very few of the iron-base hard facing alloys have the necessary corrosion resistance in such media. As a result, nickel-base or cobalt-base hard facing alloys are often recommended when both corrosion resistance and wear resistance are required. For example, a tool steel knife used to cut tomatoes in a food-processing plant will last many times longer if the edge is hard faced with a cobalt-base alloy.

Oxidation resistance and hot-corrosion resistance of iron-base alloys are generally poor. For the most part, nickel-base alloys containing borides do not contain sufficient chromium in the matrix to resist oxidation. Hence, nickel- or cobalt-base alloys containing Laves phase or carbides are typically recommended for applications where wear resistance and resistance to oxidation or hot corrosion are required.

Thermal stability, or the ability to retain strength at elevated temperatures, is important in wear applications such as hot forging dies or valves for service at 870 °C (1600 °F), as well as in coal gasification applications. Iron-base alloys with martensitic structures lose hardness at elevated temperatures. In general, thermal stability of a hard facing alloy increases with its tungsten or molybdenum content. In applications requiring elevated-temperature strength and wear resistance, cobalt-base alloys or Laves phase alloys are strongly recommended.

Once the service conditions for a particular hard facing application have been characterized, consideration should be given to interactions between candidate alloys and the base metal, and thus the process to be used for depositing the hard facing alloy becomes an important consideration. A general guide for selection of hard facing alloys, based on service conditions, is given in Table 2. In general, the following steps should be taken:

- Analyze service conditions to determine the types of wear resistance and environmental resistance required.
- Select a few candidate hard facing alloys.
- Analyze compatibility of these alloys with the base metal, including consideration of thermal stresses and possible cracking.
- Field test parts hard faced with candidate alloys.

- Select optimum hard facing alloy, considering both cost and wear life.
- Select hard facing process, considering deposition rate, degree of dilution, deposition efficiency and overall cost. Over-all cost should include cost of consumables and cost of processing.

Selection of Deposition Method

For a given hard facing application, selection of the most suitable deposition process and technique may be as important as selection of the alloy. Along with service requirements, the physical characteristics of the workpiece, the metallurgical properties of the base metal, the form and composition of the hard facing alloy, the property and quality requirements of the deposit, the skill of the operator and the cost of the hard facing operation must be considered in selecting an appropriate process.

Workpiece Factors. The size, shape and weight of the workpiece always influence selection of deposition process. Very large, heavy parts that require hard facing or buildup coatings, and that are difficult or impossible to transport, usually require selection of a process for which the equipment can be readily moved to the site of the workpiece. In such applications, the hard facing operation is most likely to be manual or semiautomatic, particularly when hard facing of relatively inaccessible areas is involved. In contrast, parts that can be readily transported and that are to be processed in large quantities can be hard faced most effectively and economically by machine or automatic methods. Shielded metal-arc and open-arc welding, for which portable equipment is readily available, lend themselves to field applications, whereas submerged-arc welding, gas tungsten-arc welding, gas metal-arc welding, plasma-arc welding, plasma spraying and flame plating are better suited to in-plant hard facing.

Properties of Base Metal. The chemical composition, melting-temperature range, and expansion and contraction characteristics are the principal attributes of the base metal that influence selection of a deposition process. The susceptibility of the base metal to thermal cracking, oxidation, or contamination at elevated temperatures also may need to be considered.

Table 2 General guide to selection of hard facing alloys

Service conditions	Hard facing materials
Metal-to-metal sliding, high contact stresses	Stellite 1, Tribaloy alloys
Metal-to-metal sliding, low contact stresses	Low-alloy hard facing steels
Metal-to-metal sliding combined with corrosion or oxidation	Cobalt-base or nickel-base alloys; selection of specific alloy depends on environmental conditions
Low-stress abrasion; particle-impingement erosion at low angles	High-alloy cast irons
Severe low-stress abrasion; cutting-edge retention	Materials containing high proportions of carbides
Cavitation erosion; liquid-impingement erosion	Cobalt-base alloys
Heavy impact	High-alloy manganese steels
Heavy impact combined with corrosion or oxidation	Stellite 21, Stellite 6
Gouging abrasion	Austenitic manganese steels
Galling	Stellite 21, Stellite 6, Tribaloy T-400, Tribaloy T-800
Thermal stability and/or creep resistance at high temperatures	Cobalt-base alloys; carbide-type nickel-base alloys

Thus, when rapid heating could cause thermal cracking of the base metal, adequate preheat and a process that provides a moderate heating rate, preferably without sacrifice in efficiency, should be selected. In addition, the cooling rate from the deposition temperature may have to be controlled, and residual stresses may have to be reduced by subsequent stress relieving.

Form and Composition of Hard Facing Alloy. The physical and metallurgical properties of a hard facing alloy determine the forms in which it is available. Some of the harder, more brittle alloys cannot be produced in the form of drawn wire, and therefore they are produced as mixtures of powder and are inserted in a carbon steel tubular wire if they are to be deposited by gas metal-arc welding, open-arc welding or submerged-arc welding, which require continuous electrode wire. A variety of forms, such as wire, bare cast rod and bare tubular rod, can be used as filler metals in hard facing by gas tungsten-arc welding.

Hard facing materials in powder form are by far the most widely used, regardless of whether or not the materials are capable of being produced in other forms. In part, this is due to the prominence of plasma-arc welding, plasma spraying and flame plating as deposition processes. In other respects, however, it is due to the fact that custom-tailored compositions can be produced merely by mixing appropriate combinations of standard powders. The ability to work the hard facing alloy into some wrought form (such as welding wire) thus has no bearing on the ability to produce a particular hard facing composition.

Properties and Quality Requirements of the Deposit. The properties and quality of hard facing deposits depend primarily on composition of the hard facing alloy. Other influential factors are base-metal composition, the deposition process and technique employed, the number of layers deposited, and deposition characteristics of the hard facing material. The extent of base-metal "dilution" of the hard faced surface during deposition varies with the process and the number of layers. Dilution is interalloying of the hard facing alloy and the base metal, and usually is expressed as the percentage of base metal in the hard facing deposit. A dilution of 10% means that the deposit contains 10% base metal and 90% hard facing alloy. As dilution in-

Table 3 Characteristics of welding processes used in hard facing

Welding process	Mode of application	Form of hard facing alloy	Weld-metal dilution, %	Deposition rate, kg/h	Deposition rate, lb/h	Minimum deposit thickness, mm	Minimum deposit thickness, in.
Oxyfuel-gas	Manual	Bare cast rod; tubular rod	1 to 10	0.5 to 2.5	1 to 6	0.8	1/32
	Manual	Powder	1 to 10	0.5 to 7	1 to 15	0.8	1/32
	Automatic	Extra-long bare cast rod; tubular wire	1 to 10	0.5 to 2.5	1 to 6	0.8	1/32
Shielded metal-arc	Manual	Flux-covered cast rod; flux-covered tubular rod	15 to 25	0.5 to 2.5	1 to 6	3	1/8
Open-arc	Semiautomatic	Alloy-cored tubular wire	15 to 25	2.3 to 11	5 to 25	3	1/8
	Automatic	Alloy-cored tubular wire	15 to 25	2.3 to 11	5 to 25	3	1/8
Gas tungsten-arc	Manual	Bare cast rod; tubular rod	10 to 20	0.5 to 3.5	1 to 8	2.3	3/32
	Automatic	Various forms	10 to 20	0.5 to 3.5	1 to 8	2.3	3/32
Submerged-arc	Semiautomatic	Bare tubular wire	20 to 60	4.5 to 9	10 to 20	3	1/8
	Automatic, single wire	Bare tubular wire	30 to 60	4.5 to 11	10 to 25	3	1/8
	Automatic, multiwire	Bare tubular wire	15 to 25	11 to 27	25 to 60	5	3/16
	Automatic, series arc	Bare tubular wire	10 to 25	11 to 16	25 to 35	5	3/16
Plasma-arc	Automatic	Powder	5 to 30	0.5 to 7	1 to 15	0.8	1/32
Plasma transferred arc	Automatic	Powder	5 to 20	2.5 to 6.5	6 to 14	1.5	1/16

creases, the hardness, wear resistance and other desirable properties of the deposit are reduced. Sometimes, in order to control composition, a buffer layer is deposited between the base metal and the hard facing alloy. In addition to minimizing dilution, a buffer layer is often used to counteract the adverse effects of differences in coefficient of expansion between the hard facing alloy and the base metal.

One criterion for selection of a hard facing process should be the ability of the process to limit dilution while achieving a high deposition rate, and thus to avoid the necessity for a deposit thickness greater than that required for service. Typical dilution percentages, deposition rates, and minimum deposit thicknesses for different deposition processes, and various forms, compositions and modes of application of hard facing alloys, are given in Table 3.

As shown in Table 3, dilution may vary from 1 to 60%, and the minimum practical deposit thickness of a single layer of hard facing alloy ranges from 0.8 to 5 mm (1/32 to 3/16 in.), depending on the deposition process and technique. On parts that require several layers, deposits may be built up to 25 mm (1 in.) or more in thickness. For these parts, care must be taken in selecting the hard facing alloy, because very highly alloyed deposits are likely to spall if applied in thicknesses of more than 6 mm (1/4 in.). When deposits less than 0.8 mm (1/32 in.) thick are required, a process such as metal spraying usually is more efficient than any welding process.

Operator Skill. It is essential to relate the quality requirements of the deposit to both deposition process and operator skill. For example, although manual gas tungsten-arc welding can be used to obtain high-quality deposits on relatively small areas, and in grooves and recesses, and although thin layers can be deposited with dilution as low as 10%, relatively high welder skill and close control of the welding operation are necessary. In contrast, automatic submerged-arc welding requires a minimum of welder skill and results in high deposition rates, but penetration is deep, dilution is greater, and consequently it may be necessary to make use of a buffer layer or to deposit two or more layers of the hard facing alloy to obtain the full properties of the alloy.

In general, earthmoving and mining equipment can be hard faced adequately in the field by workers of lesser talents. Selection of process usually is based on maximum deposition rate, and high dilution rates seldom have significant effects on the suitability of the coating for service. Hard facing of valves, on the other hand, requires highly skilled technicians and precise control of the operation, and frequently is automated. Excessive dilution or contamination of the deposit can result in failure of the part during service.

Cost. Depending on form and composition, initial cost of hard facing alloys may vary by a factor of as much as 30 to 1. Equipment and operating costs for appropriate deposition processes also vary considerably. These variables, together with the metallurgical factor of base-metal compatibility, obviously must be coordinated as part of the selection process.

Austenitic Manganese Steel

THE ORIGINAL AUSTENITIC MANGANESE STEEL, containing about 1.2% C and 12% Mn, was invented by Sir Robert Hadfield in 1882. Hadfield's steel is unique in that it combines high toughness and ductility with high work-hardening capacity and, usually, good resistance to abrasion. Consequently, Hadfield's steel rapidly gained acceptance as a very useful engineering material.

Applications. Hadfield's austenitic manganese steel is still used extensively, with minor modifications in composition and heat treatment, primarily in the fields of earthmoving, mining, quarrying, oil-well drilling, steelmaking, manufacturing of cement and clay products, railroading, dredging and lumbering. Austenitic manganese steel is used chiefly in equipment for handling and processing earthen materials (such as rock crushers, grinding mills, dredge buckets, power-shovel buckets and teeth, and pumps for handling gravel and rocks) and in a multitude of associated applications. Another important use is in railway trackwork at frogs, switches and crossings, where multiple impacts at intersections are especially severe.

Because austenitic manganese steel resists metal-to-metal wear, it is used in sprockets, pinions, gears, wheels, conveyor chain, wear plates and shoes. Nonmagnetic parts required for lifting-magnets, induction furnaces and special electrical equipment are an expanding field of application.

Austenitic manganese steel has certain properties that tend to restrict its use. It is difficult to machine and usually has a yield strength of only 345 to 415 MPa (50 to 60 ksi). Consequently, it is not well suited for parts that require close-tolerance machining or that must resist plastic deformation when highly stressed in service. However, hammering, pressing, cold rolling or explosion shocking of the surface raises the yield strength to provide a hard surface on a tough core structure.

Table 1 Standard composition ranges for austenitic manganese steel castings

ASTM A128 grade	Composition, %						
	C	Mn	Cr	Mo	Ni	Si (max)	P (max)
A	1.05–1.35	11.0 min	...	...	...	1.00	0.07
B-1	0.9–1.05	11.5–14.0	...	...	...	1.00	0.07
B-2	1.05–1.2	11.5–14.0	...	...	...	1.00	0.07
B-3	1.12–1.28	11.5–14.0	...	...	...	1.00	0.07
B-4	1.2–1.35	11.5–14.0	...	...	...	1.00	0.07
C	1.05–1.35	11.5–14.0	1.5–2.5	...	...	1.00	0.07
D	0.7–1.3	11.5–14.0	...	...	3.0–4.0	1.00	0.07
E-1	0.7–1.3	11.5–14.0	...	0.9–1.2	...	1.00	0.07
E-2	1.05–1.45	11.5–14.0	...	1.8–2.1	...	1.00	0.07
F	1.05–1.35	6.0–8.0	...	0.9–1.2	...	1.00	0.07

Composition

Many variations of the original Hadfield's steel have been proposed, often in unexploited patents, but only a few have been adopted as significant improvements. These usually involve variations of carbon and manganese, with or without additional alloys such as chromium, nickel, molybdenum, vanadium, titanium and bismuth. The most common of these compositions, as listed in ASTM A128, are given in Table 1.

Fig. 1 Solubility of carbon in 13% Mn steels (Ref 1)

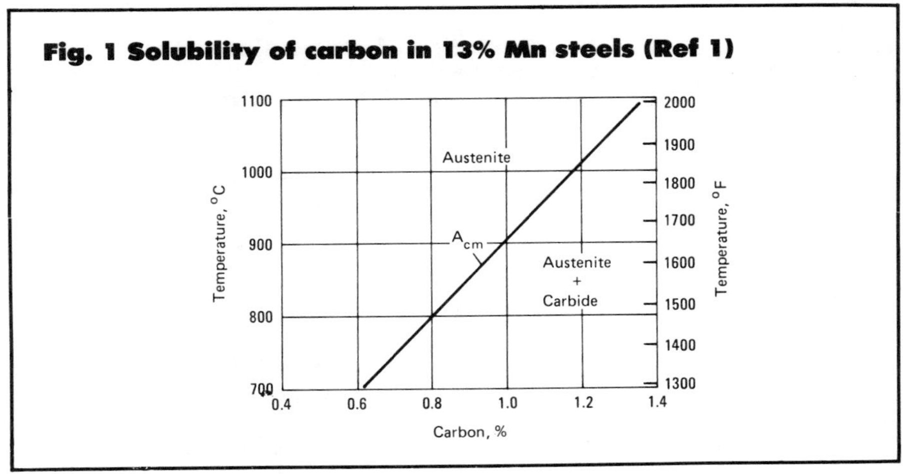

Fig. 2 Variation of M_s temperature with carbon and manganese contents (Ref 2)

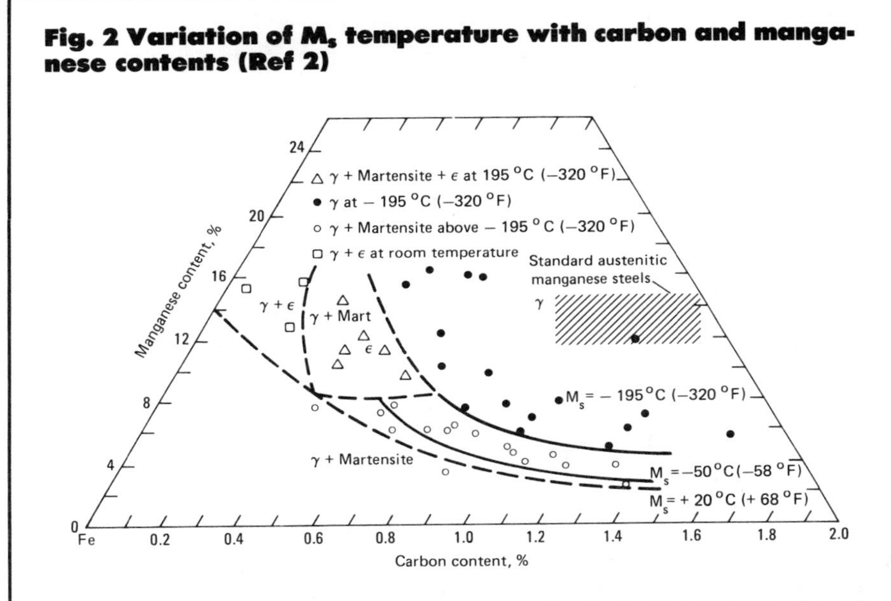

The available assortment of wrought grades is smaller and usually approximates ASTM composition B-3. Some wrought grades contain about 0.8% C and either 3% Ni or 1% Mo. Large heat orders are usually required for production of wrought grades, whereas cast grades and their modifications are more easily obtained in small lots. A manganese steel foundry may have several dozen modified grades on its production list.

Modified grades usually are produced to meet special requirements by application, section size, casting size, cost and weldability considerations.

Carbon and Manganese. The ASTM A128 compositions in Table 1 do not permit any austenite transformation when the alloys are water quenched from above the A_{cm}, but this does not preclude lower ductility in heavy sections of the higher carbon steels, because quenching is slowed by the heavy sections. The effect is due to formation of carbides along grain boundaries, and in some degree, affects nearly all commerical castings except the very smallest. Figure 1 shows A_{cm} temperatures for 13% Mn steels containing between 0.6 and 1.4% C. Figure 2 illustrates the effects of carbon and manganese contents on the M_s temperature of a homogeneous austenite with all carbon and manganese in solid solution.

The mechanical properties of austenitic manganese steel vary with both carbon and manganese contents. Figure 3 illustrates the variations that occur in cast 13% manganese steel as the carbon content is varied from 0.8 to 1.7%. Over this span, there is a gradual increase in yield strength, whereas tensile strength and ductility reach a maximum at about 1.2% carbon and then decrease steadily. As carbon content is increased above about 1.1%, it becomes increasingly difficult to retain all of the carbon in solution in the austenite, which largely accounts for the reductions in tensile strength and ductility. Nevertheless, because abrasion resistance tends to increase as carbon content increases up to about 1.4% and possibly higher, carbon contents higher than 1.1% often are preferred even though the ductility of the steel may be lowered. Carbon levels over 1.4% are seldom used because of the difficulty of obtaining an austenite structure sufficiently free of grain boundary carbides to avoid undesirably low values of both strength and ductility. The effect also may be observed in manganese steels containing less than 1.1% carbon because segregation may result in local variations of ±0.2% from the average carbon content determined by chemical analysis. The 0.7% C minimum of grades D and E-1 may be used to minimize carbide precipitation in heavy castings or in weldments and similar low-carbon contents are specified for welding filler metal. Carbides form in castings that are cooled slowly in the molds. In fact, carbides form in practically all as-cast grades containing more than 1.0% C, regardless of mold cooling rates. They often form in heavy-section castings during heat treatment if quenching is even somewhat ineffective in producing rapid cooling throughout the entire section thickness. Carbides can form during welding or during service at temperatures above about 275 °C (530 °F).

If carbon and manganese are lowered together, for instance to 0.53% C with 8.3% Mn or 0.62% C with 8.1% Mn, the speed of hardening by cold work is increased and strain induced martensite may be produced. However, this does not provide enhanced abrasion resistance (at least to high-stress grinding abrasion) as is often hoped.

Manganese contributes the vital austenite-stabilizing effect of delaying transformation (but not eliminating it). Thus, in a simple steel that contains 1.1% Mn, isothermal transformation at 370 °C (700 °F) begins about 15 s after

Fig. 3 Variation of properties with carbon content for austenitic manganese steel containing 12.6 to 13% manganese

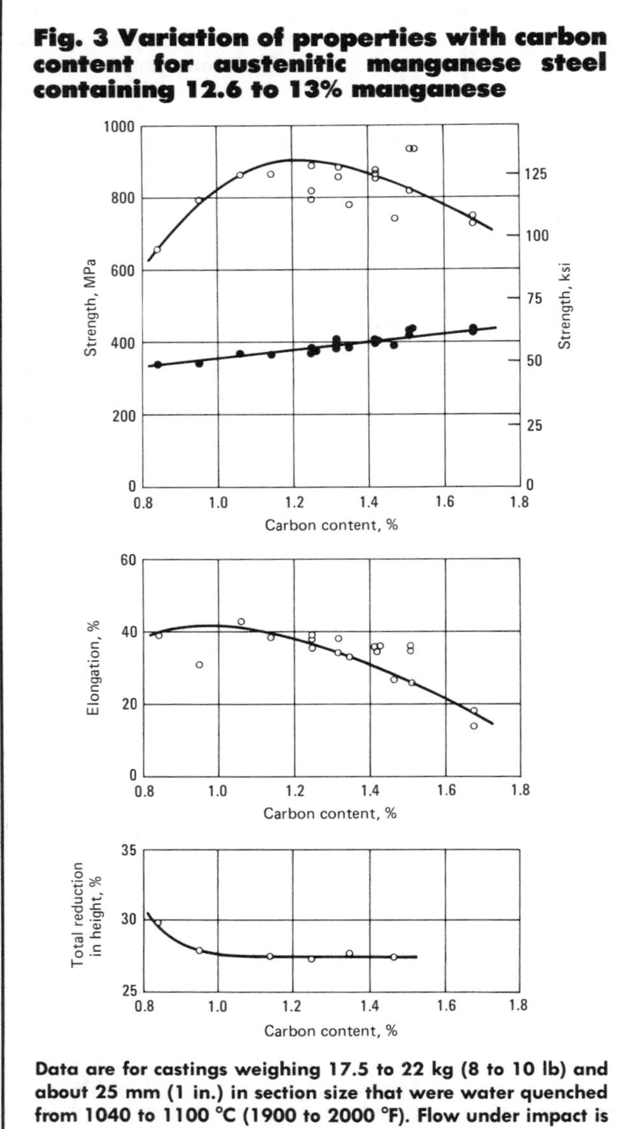

Data are for castings weighing 17.5 to 22 kg (8 to 10 lb) and about 25 mm (1 in.) in section size that were water quenched from 1040 to 1100 °C (1900 to 2000 °F). Flow under impact is the total reduction in height sustained by a cylindrical specimen 25 mm in both diameter and length after absorbing 20 blows of 680 J (500 ft·lb) each.

Fig. 4 Variation of properties with manganese content for austenitic manganese steel containing 1.15% carbon

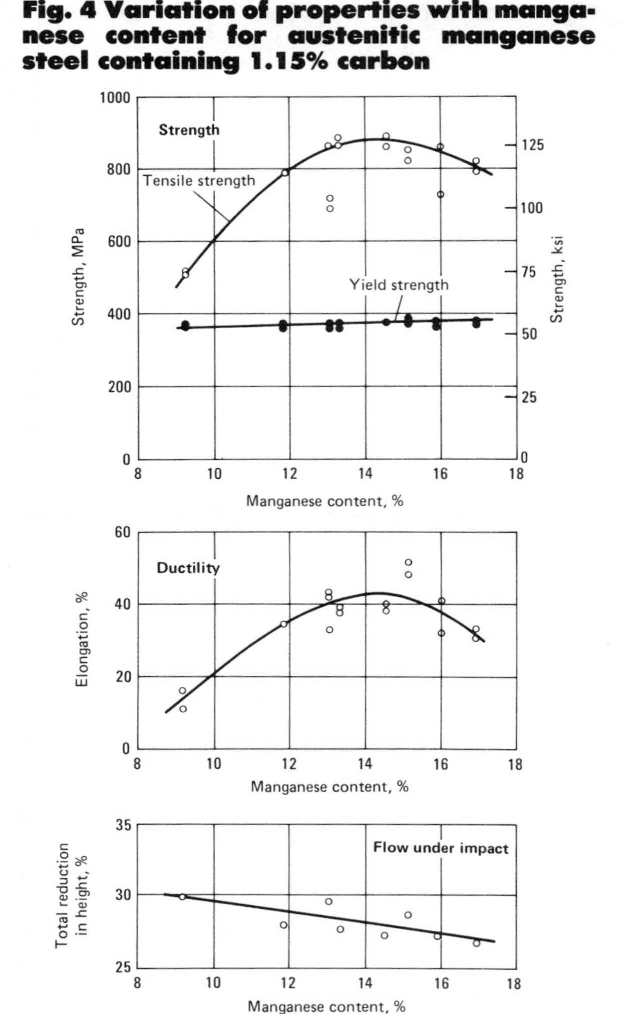

Data are for castings weighing 17.5 to 22 kg (8 to 10 lb) and about 25 mm (1 in.) in section size that were water quenched from 1040 to 1100 °C (1900 to 2000 °F). Flow under impact is the total reduction in height sustained by a cylindrical specimen 25 mm in both diameter and length after absorbing 20 blows of 680 J (500 ft·lb) each.

the steel is quenched to that temperature, whereas in a 13% Mn steel, transformation at 370 °C does not begin until after 48 h (Ref 1). Below 260 °C (500 °F), phase changes and carbide precipitation are so sluggish that for all practical purposes they may be neglected, in the absence of deformation, if manganese content exceeds 10%.

Figure 4 illustrates the influence of manganese content on strength and ductility of cast austenitic steel that has been solution treated and water quenched. It confirms the observations of many investigators, including Sir Robert Hadfield (Ref 4), who studied the influence of manganese contents up to about 22%. Manganese content has little effect on yield strength. In tensile testing, ultimate strength and ductility increase fairly rapidly with increasing manganese content up to about 12% and then tends to level off, although small improvements normally continue up to about 13% Mn.

Silicon and Phosphorus. As noted in Table 1, silicon and phosphorus are present in all ASTM A128 grades of austenitic manganese steel. Silicon is seldom added except for steelmaking purposes. Silicon contents exceeding 1% are not usual, because foundries do not like to have it pyramid in melts containing returned scrap. Silicon contents of 1 to 2% might be used to modally increase yield strength, but other elements are preferred for this effect. Loss of strength is abrupt above 2.2% Si and Mn steel containing more than 2.3% Si may be worthless.

Since about 1960, low-phosphorus ferromanganese has been available, which has enabled steelmakers to greatly reduce phosphorus levels in manganese steel. The preferred prac-

Fig. 5 Effects of nickel, molybdenum and chromium contents on tensile properties of cast manganese steel

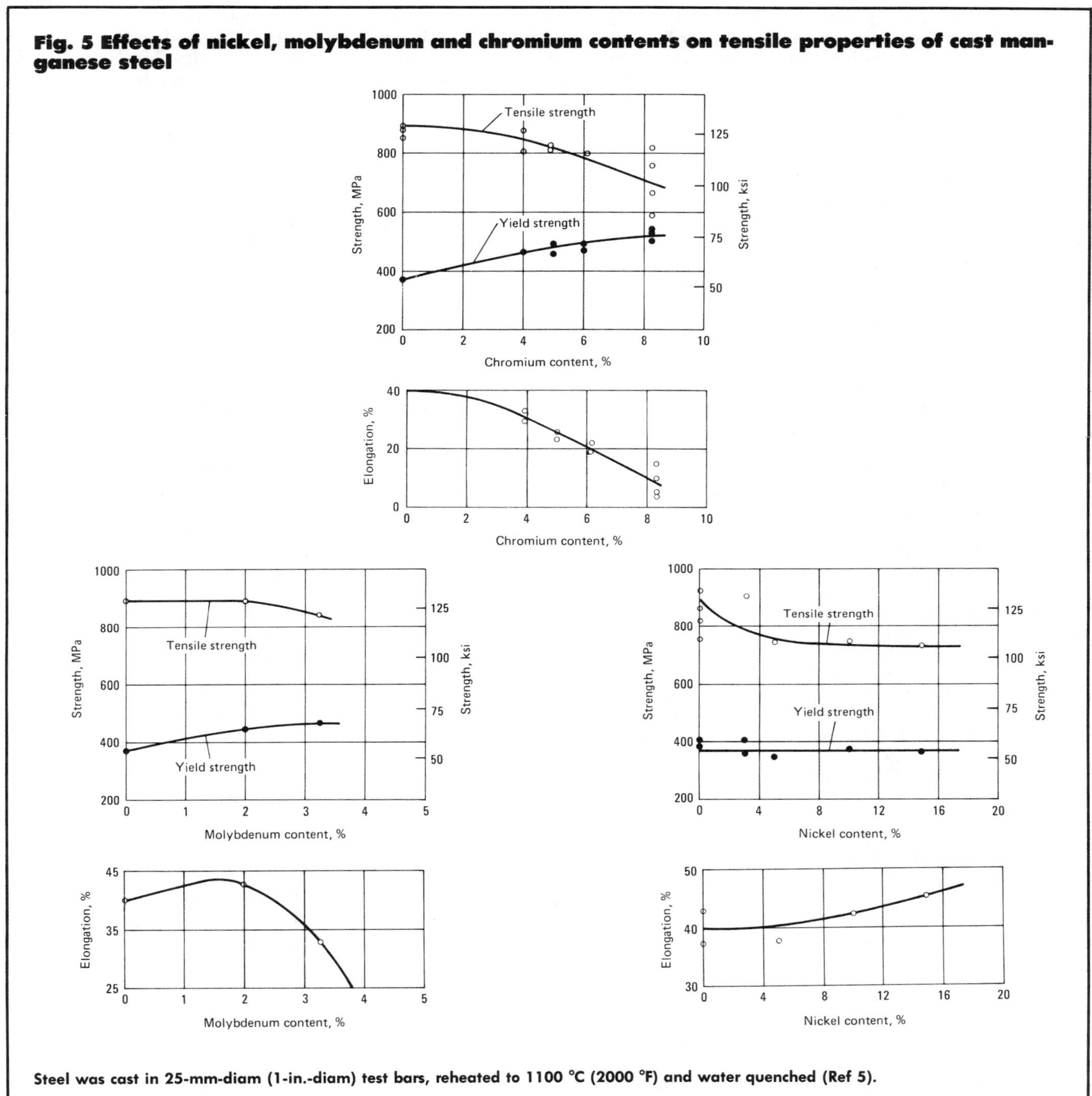

Steel was cast in 25-mm-diam (1-in.-diam) test bars, reheated to 1100 °C (2000 °F) and water quenched (Ref 5).

tice is to hold phosphorus content below 0.04% even though a level of 0.07% is permitted by ASTM A128. Levels above 0.06%, which formerly were prevalent, contribute to hot shortness and low elongation at very high temperatures, and frequently are the cause of hot tears in castings and underbead cracking in weldments. In general, phosphorus in manganese steel tends to lower the ductility; the effect is more critical at or below ambient temperatures. It is particularly advantageous to keep phosphorus at the lowest possible level in the grades that are welded and in manganese steel welding electrodes.

Special-Purpose Additives. Other elements are also added to certain grades of manganese steel to achieve desired properties. As can be seen in Table 1, chromium is added to grade C, molybdenum to grades E-1, E-2 and F, and nickel to grade D. Figure 5 summarizes the effects of these elements on the mechanical properties of a steel containing about 15% Mn and from 1.1

to 1.2% C. Several other elements not listed in Table 1 (vanadium, copper, bismuth and titanium, for instance) are added to manganese steel for unique applications.

Chromium levels from 1.8 to 2.2% (and occasionally as high as 5%) are employed in manganese steel to moderately raise yield strength. Its effect on resistance to abrasion is unproven and appears to be inconsistent. Chromium reduces ductility by increasing the number of embrittling carbides in the austenite—particularly those containing more than 2.5% chromium.

Molybdenum additions, usually 0.5 to 2%, generally are made to improve the toughness and resistance to cracking of castings in the as-cast condition, and to raise the yield strength (and possibly toughness) of heavy-section castings in the solution treated and quenched condition. These effects occur because molybdenum in manganese steel is distributed partly in solution in the austenite and partly in primary carbides formed during solidification of the steel. The molybdenum in solution effectively suppresses formation of both embrittling carbide precipitates and pearlite, even when the austenite is exposed to temperatures above 275 °C (530 °F) during service or during welding. The molybdenum in primary carbides tends to change distribution of the carbides from continuous envelopes around the austenite dendrites to less-harmful nodular carbides, especially when molybdenum content exceeds about 1.5%.

Grade E-2, which contains about 2% Mo, may be given a special heat treatment to develop a structure of finely dispersed carbides in the austenite, which provides unusually high yield strength and, under some conditions, an improvement in abrasion resistance. Molybdenum is added to the lean-manganese steel, grade F, to partly suppress embrittlement in both as-cast and heat-treated conditions.

Nickel, in amounts up to 4% or more, stabilizes the austenite because it remains in solid solution. It is particularly effective in suppressing precipitates of carbide platelets, which can form between about 300 and 550 °C (570 and 1020 °F). Therefore, the presence of nickel helps retain nonmagnetic qualities in the steel, especially in its decarburized skin, and slows the rate of work hardening, thus providing high ductilities in lower-carbon grades. Abrasion resistance is lowered, but yield strength is not significantly affected. Nickel is used primarily in the lower-carbon or weldable grades of cast manganese steel and in wrought manganese steel products (including welding electrodes). In wrought products, nickel often is used in conjunction with molybdenum.

Because vanadium is a strong carbide former, it is added to manganese steels to substantially increase yield strength. However, a comparable decrease in ductility occurs. Vanadium is used on a limited basis in special-purpose compositions, such as a Mn-Ni-Mo-V austenitic alloy that can be age-hardened to provide yield strengths of over 700 MPa (100 ksi). Tests in a jaw crusher demonstrated that the abrasion resistance of this steel is not as great as that of the regular Hadfield types (Ref 6).

Like nickel, copper in amounts of 1 to 5% is sometimes used in austenitic manganese steels to stabilize the austenite. The effects of copper on mechanical properties have not been clearly established. Scattered reports indicate it may have an embrittling effect, which may be due to the limited solubility of copper in austenite.

Other elements, such as bismuth and titanium, also are added to standard manganese steels. Bismuth can improve machinability.

Titanium can reduce carbon in the austenite by forming very stable carbides. The resulting properties may simulate those of a lower carbon grade. Titanium may also somewhat neutralize the effect of too much phosphorus; some European practice is apparently based on this idea.

Sulfur. The sulfur content in manganese steel seldom influences its properties, because the scavenging effect of manganese operates to eliminate sulfur by fixing it in the form of innocuous rounded inclusions of manganese sulfide. Elongation of these inclusions in wrought steels may contribute to directional properties; in cast steel, such inclusions are harmless.

As-Cast Properties

Although austenitic manganese steels in the as-cast condition generally are considered too brittle for normal use, Table 2 demonstrates that there are exceptions to this rule. Mechanical properties are listed for five grades of as-cast austenitic manganese steels in various thicknesses. These data indicate that lowering carbon content to less than 1.1% and/or adding about 1.0% Mo or about 3.5% Ni results in commercially acceptable as-cast ductilities in light and moderate section thicknesses. These data also apply to weld deposits, which normally are left in the as-deposited condition and therefore are essentially equivalent to material in the as-cast condition.

Adjustments in composition, by limiting carbide embrittlement and austenite transformation, reduce or eliminate cracking of manganese steel castings during cooling in the molds or reheating for solution treatment. The steel generally is poured at temperatures just high enough to avoid misruns in the castings and excessive skulling of the metal in the ladle. This practice helps ensure rapid solidification of the metal in the molds, which in turn helps prevent excessively coarse grain size. The final microstructure in most castings is not fully austenitic but rather contains carbide precipitates, pearlite and bainite in an austenitic matrix.

Commercial use of castings in the as-cast condition results in cost and energy savings and eliminates the problems of (*a*) decarburization of thin castings during solution treatment and (*b*) warpage during water quenching. Full ductility or toughness is not required in certain applications of solution treated and quenched manganese steels. For example, as-cast manganese steels have been used successfully for pans on pan conveyors and for other light-section applications where solution treated and water quenched castings were prone to severe warpage.

Of the steels listed in Table 2, the 6Mn-1Mo grades are most susceptible to embrittlement because of their low manganese levels. However, even these grades can be used in applications where 1% elongation (either determined in a tensile test or estimated from a transverse bend test) is considered adequate ductility. For example, the as-cast 6Mn-1Mo grade containing 0.8 to 1.0% C has been used successfully in grinding-mill liners.

Heavy Sections. As section thickness increases, the rate at which castings cool in sand molds also slows, which increases the opportunity for embrittlement to occur by carbide precipitation. Shapes that tend to develop high residual stresses, such as cylinders and cones, can be particularly affected. These stresses most probably result from volume changes accompa-

Table 2 Typical mechanical properties of as-cast austenitic manganese steels

Composition, %				Form	Section size		0.2% yield strength		Tensile strength		Elongation,	Reduction in area,	Impact strength(a)		Hardness,
C	Mn	Si	Other		mm	in.	MPa	ksi	MPa	ksi	%	%	J	ft·lb	HB
Plain manganese steels															
0.85	11.2	0.57	. . .	Round	25	1	. . .	. . .	440	64	14.5	. . .	. . .	. . .	. . .
0.95	13.0	0.51	. . .	Round	25	1	. . .	. . .	420	61	14	. . .	. . .	. . .	. . .
1.11	12.7	0.54	. . .	Round	25	1	360	52	450	65	4	. . .	. . .	. . .	. . .
1.27	11.7	0.56	. . .	Round	25	1	. . .	. . .	360	52	2	. . .	. . .	. . .	. . .
1.28	12.5	0.94	. . .	Keel block	100	4	. . .	. . .	330(b)	48(b)	1(b)	. . .	3.4	2.5	245
1.36	20.2	0.6	. . .	Y-block	50	2	. . .	. . .	425(b)	62(b)	1(b)	. . .	. . .	. . .	283
1Mo manganese steels															
0.61	11.8	0.17	1.10 Mo	Round	25	1	315	46	710	103	27.5	23	. . .	. . .	163
0.75	13.9	0.58	0.90 Mo	Round	25	1	340	49	740	107	39.5	30	. . .	. . .	183
0.83	11.6	0.38	0.96 Mo	Round	25	1	345	50	695	101	30	29	. . .	. . .	163
0.89	14.1	0.54	1.00 Mo	Round	25	1	360	52	690	100	29.5	22	. . .	. . .	196
1.16	13.6	0.60	1.10 Mo	Round	25	1	400	58	560	81	13	15	. . .	. . .	185
0.93	13.6	0.67	0.96 Mo	Plate	25	1	365	53	510	74	11	16	72	53	188
0.99	12.6	0.6	0.87 Mo	Plate	25	1	. . .	. . .	460(b)	67(b)	6(b)	. . .	. . .	. . .	. . .
0.98	12.6	0.6	0.87 Mo	Plate	50	2	. . .	. . .	435(b)	63(b)	4(b)	. . .	. . .	. . .	. . .
0.95	12.6	0.6	0.87 Mo	Plate	100	4	345	50	385	56	4	4	. . .	. . .	. . .
1.30	13.1	0.78	0.99 Mo	Keel block	100	4	. . .	. . .	435(b)	63(b)	2(b)	. . .	8	6	230
1.33	19.8	0.6	0.99 Mo	Y-block	50	2	. . .	. . .	505(b)	73(b)	2.5(b)	. . .	. . .	. . .	231
2Mo manganese steels															
0.52	14.3	1.47	2.4 Mo	Round	25	1	370	54	600	87	15.5	13	. . .	. . .	220
0.70	13.6	0.63	2.0 Mo	Round	25	1	360	52	785	114	41	29	. . .	. . .	180
0.75	14.1	0.99	2.0 Mo	Round	25	1	365	53	745	108	34.5	27	. . .	. . .	183
0.91	14.1	0.60	2.0 Mo	Round	25	1	395	57	705	102	27.5	21	. . .	. . .	196
1.24	14.1	0.64	3.0 Mo	Round	25	1	440	64	600	87	7.5	10	. . .	. . .	235
1.40	12.5	0.62	2.1 Mo	Round	25	1	420	61	550	80	3.5	5	. . .	. . .	228
1.34	12.0	0.43	2.2 Mo	Keel block	50	2	415	60	435	63	3.5	7	. . .	. . .	235
3.5Ni manganese steels															
0.75	13.0	0.95	3.65 Ni	Round	25	1	295	43	655	95	36	26	. . .	. . .	150
0.80	13.5	0.53	3.61 Ni	Round	25	1	. . .	. . .	530	77	26	. . .	. . .	. . .	. . .
0.91	13.3	0.53	3.38 Ni	Round	25	1	. . .	. . .	510	74	24	. . .	. . .	. . .	. . .
6Mn-1Mo alloys															
0.90	5.8	0.37	1.46 Mo	Mill liner	100	4	325	47	340	49	2	. . .	9	7	181
1.00	6.0	0.43	1.03 Mo	Keel block	100	4	330	48	365	53	2	3	. . .	. . .	195
0.89	6.3	0.6	1.20 Mo	Plate	100	4	. . .	. . .	330(b)	48(b)	1(b)	. . .	. . .	. . .	. . .
1.27	6.1	0.42	1.07 Mo	Keel block	50	2	365	53	400	58	1	1	3	2	273

(a) Charpy V-notch. (b) Properties converted from transverse bend tests on 6-by-13-mm (1/4-by-1/2-in.) bars cut from castings and broken by center loading across 25-mm (1-in.) span.

nying the carbide precipitation and austenite transformation that occur during normal cooling of castings.

Figure 6 shows the volume changes that occur during isothermal decomposition of a 1.25C-12.8Mn steel at temperatures between 850 and 500 °C (1560 and 930 °F), the principal range in which embrittlement occurs when a casting is cooled in its mold or reheated for reaustenitization. Between 850 and about 700 °C (1560 and 1300 °F), only carbides are precipitated, principally as envelopes around austenite grains and as lamellar-type patches within grains. The lamellar carbide patches have the appearance of coarse pearlite, but actually they are carbide plates in austenite. Below 700 °C, and particularly between 650 and 550 °C (1200 and 1020 °F), pearlite nodules, nucleated by previously precipitated carbides, grow relatively rapidly.

Transgranular acicular carbides also tend to precipitate below about 600 °C (1110 °F), especially in austenite containing more than about 1.1% C. This precipitation can continue down to about 300 °C (570 °F) in a 1.2C-12Mn steel. It may be followed by transformation of some of the carbon-depleted austenite to martensite as the temperature approaches ambient.

Heat Treatment

Heat treatment strengthens austenitic manganese steel so that it can be used safely and reliably in a wide variety of engineering applications. Solution annealing and quenching, the standard treatment that produces normal tensile properties and the desired toughness, involves austenitizing followed quickly by water quenching. Figure 7 shows the microstructures of a 75-mm (3-in.) section of austenitic manganese steel in the as-cast condition and after solution annealing and quenching. Variations of this treatment can be used to enhance specific desired proper-

Fig. 6 Change in length of an austenitic 1.25C-12.8Mn steel during isothermal transformation (Ref 7)

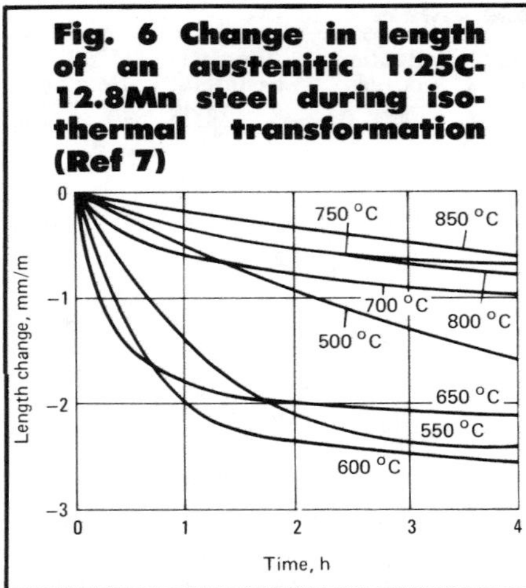

Fig. 7 Typical structures of as-cast and heat treated ASTM A128, grade B-3, manganese steel

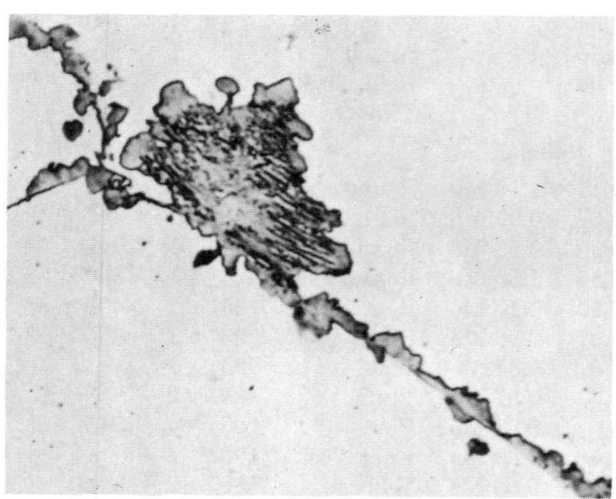

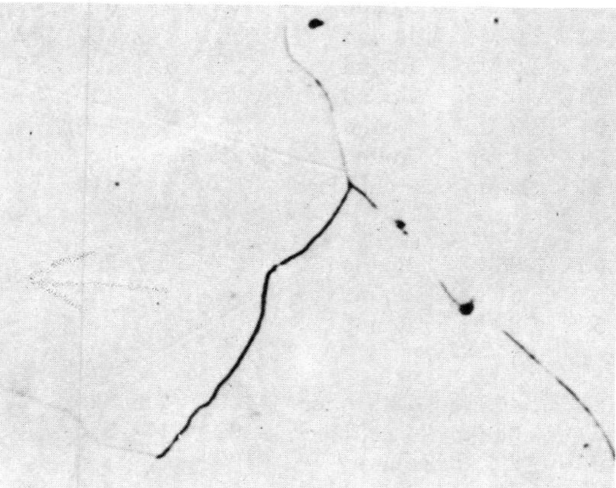

Top: As-cast material 75 mm (3 in.) thick, which has large carbides along grain boundaries. Bottom: 75-mm-thick material heated to 1120 °C (2050 °F) and water quenched. Both samples etched in 2½% nital, rinsed in methanol and re-etched in 15% HCl; magnification, 500×.

ties such as yield strength and abrasion resistance. Usually, a fully austenitic structure, essentially free of carbides and reasonably homogeneous with respect to carbon and manganese, is desired in the as-quenched condition, although this is not always attainable in heavy sections or in steels containing carbide-forming elements such as chromium, molybdenum, vanadium and titanium. If carbides exist in the as-quenched structure, it is desirable for them to be present as relatively innocuous particles or nodules within the austenite grains rather than as continuous envelopes at grain boundaries.

Procedures. Full solution of carbides requires that the solution treating temperature exceed A_{cm} by about 30 to 50 °C (50 to 90 °F). Soaking for 1 to 2 h at temperature is usually adequate. Although it might appear that very high solution temperatures should permit carbon contents of 1.4 to 1.5%, three factors make it impractical to use very high temperatures: (*a*) incipient melting occurs in areas of carbon segregation, (*b*) scaling and decarburization become excessive and (*c*) commercial quenching rates are limited in their ability to retain high carbon concentrations in solution.

Commercial heat treatment of manganese steel castings normally involves heating slowly to 1010 to 1090 °C (1850 to 2000 °F), soaking for 1 to 2 h at temperature and then quenching in agitated water. There is some tendency toward growth of austenite grains during soaking, although final austenite grain size in castings is determined largely by pouring temperature and solidification rate.

For grade E-2 manganese steel (see Table 1), a modified heat treatment often is specified or recommended. This treatment consists of heating castings to about 600 °C (1100 °F) and soaking them 8 to 12 h at temperature, which causes substantial amounts of pearlite to form in the structure. The castings then are further heated to about 980 °C (1800 °F) to reaustenitize the structure. This step converts the pearlitic areas to fine-grain austenite containing a dispersion of small carbide particles, which remain undissolved as long as the austenitizing temperature does not exceed about 1000 °C (1850 °F). Quenching then results in a dispersion-hardened austenite, which is characterized by higher yield strength, higher hardness and lower ductility than would be obtained if the same steel were given a full solution treatment at a higher austenitizing temperature. When this dispersion-hardening heat treatment is used, relatively high carbon contents are permissible, which in turn can improve abrasion resistance.

Precautions. Speed of quenching is important, but it is difficult to increase it beyond the rate of heat transfer from a hot surface to agitated water or the

Fig. 8 Cooling curves for austenitic manganese steel of various thicknesses

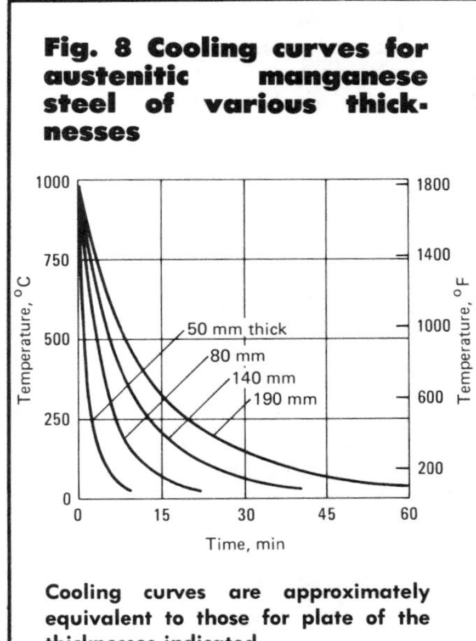

Cooling curves are approximately equivalent to those for plate of the thicknesses indicated.

rate fixed by the thermal conductivity of the metal. As a result, heavy-section castings have lower mechanical properties at the center than do thinner castings. Figure 8 illustrates the cooling rates that can be expected when metal plates of four different thicknesses are quenched in water. Table 3 lists average properties observed in castings of 1.11C-12.7Mn-0.5Si-0.043P steel water quenched from about 980 °C (1800 °F), which cooled the castings at the rates shown in Fig. 8.

Residual stresses from quenching, coupled with the lower properties of heavy sections, establish the usual maximum thickness of commercial castings at about 125 to 150 mm (5 to 6 in.), although castings with sections up to 400 mm (16 in.) thick have been produced.

The relatively high austenitizing temperature leads to marked surface decarburization by furnace gases and to some loss of manganese. Surface decarburization may extend as much as 3 mm (1/8 in.) below the casting surface. Thus, the skin may be partly martensitic at times and usually exhibits properties less desirable than those of the underlying metal. This characteristic is not significant in parts subject to abrasion, such as those used in crushing or grinding, because in these applications the skin is removed by normal wear. Tensile deformation in service sometimes produces numerous cracks in this inferior skin, which terminate where they reach the tough austenite of normal composition. Service performance is not seriously affected except under critical fatigue conditions or in very light sections; in such instances, premature failure may result. If considered necessary, a proprietary grade (equivalent to ASTM grade B-2 to which 6% Cr has been added) that is less prone to surface decarburization can be specified. Under certain conditions, sections such as wrought sheets may be protected with covers, metal envelopes, organic coatings or inorganic coatings, thereby reducing the effects of decarburizing atmospheres used in heat treatment.

For applications that require nonmagnetic properties, alteration of the skin by furnace gases requires attention. If the affected skin, which usually is magnetic, is quite shallow, it may be possible to remove it by pickling. In heavy sections, where skin thickness may approach 3 mm (1/8 in.), the altered layer should be removed by grinding if a surface permeability below 1.3 is required.

Table 3 Average mechanical properties of 12.7Mn-1.1C-0.5Si-0.043P castings water quenched from 1040 °C (1900 °F)

Tension tests were performed on specimens 6.40 mm (0.252 in.) in diameter and 25 mm (1 in.) in gage length

Plate thickness, mm	Plate thickness, in.	Type of grain	Tensile strength, MPa	Tensile strength, ksi	Elongation(a), %	Reduction in area, %	Impact strength(b), J	Impact strength(b), ft·lb
50	2	Coarse	635	92	37.0	35.7	137	101
		Fine	820	119	45.5	37.4	134	99
83	3 1/4	Coarse	620	90	25.0	34.5	133	98
		Fine	765	111	36.0	33.0	115	85
140	5 1/2	Coarse	545	79	22.5	25.6	115	85
		Fine	705	102	32.0	28.3	100	74
190	7 1/2	Coarse	455	66	18.0	25.1	77	57
		Fine	725	105	33.5	29.2	66	49

(a) In 25 mm or 1 in. (b) Izod V-notch.

Mechanical Properties After Heat Treatment

As the section size of manganese steel increases, tensile strength and ductility decrease substantially in specimens cut from heat treated castings. This occurs because, except under specially controlled conditions, heavy sections do not solidify fast enough in the mold to prevent coarse grain size, a condition that is not altered by heat treatment. As shown in Table 3, fine-grain specimens may exhibit strength and elongation as much as 30% greater than those of coarse-grain specimens.

Grain size also accounts chiefly for the differences between cast and wrought manganese steels (the latter usually are of fine grain size). For cast grade B-2, the standard deviations for tensile strength and elongation are about 69 MPa (10 ksi) and 9%, respectively. The midrange values of 825 MPa (120 ksi) and 40% apply to sound, medium-grain cast specimens that have been properly heat treated. The scatter bands for this grade extend from 620 to 1035 MPa (90 to 150 ksi) for tensile strength and from 13 to 67% for elongation.

Mechanical properties vary with section size. Tensile strength, tensile elongation, reduction in area and impact strength are substantially lower in 100-mm-thick (4-in.-thick) sections than in 25-mm-thick (1-in.-thick) sections. Because section thicknesses of production castings often are from 100 to 150 mm (4 to 6 in.), this factor is an important consideration in specification of the proper grade.

Notched-bar impact test values can be exceptionally high. Charpy test specimens are sometimes bent and dragged through the machine rather than being fractured. Sometimes, observed values are biased because of incorrect preparation of specimens. Notches should be cut by precision grinding to minimize work hardening at the apex of the notch.

Austenitic manganese steel remains tough at subzero temperatures above the M_s temperature. It is apparently immune to hydrogen embrittlement, although embrittlement has been produced in steels with low carbon contents (less than about 0.02%) and high manganese contents.

Resistance to crack propagation is high and is associated with very sluggish progressive failures. Because of

Fig. 9 True stress vs engineering strain for manganese steel, cast ferritic tank armor of similar tensile strength and a high-strength gray iron

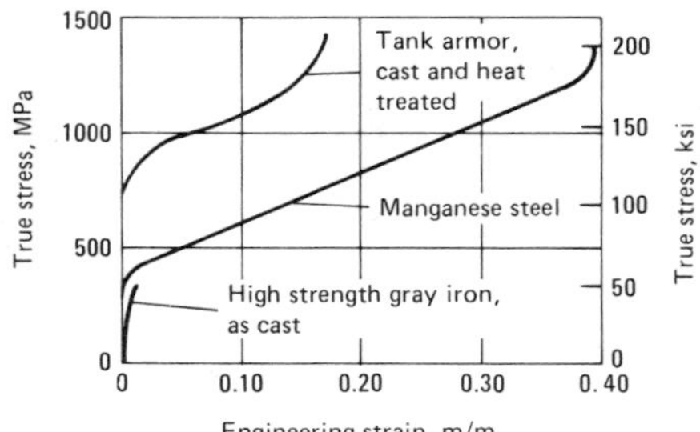

Alloy	C	Mn	Si	Cr	Other
Tank armor	0.29	1.30	0.52	0.37	0.36 Mo
Manganese steel	1.22	13.08	0.33	0.09	0.05 Al
Gray iron	2.79	0.75	1.32	0.10	...

this, any fatigue cracks that develop can be monitored, and the affected part(s) removed from service, before complete failure occurs—a capability that is a distinct advantage in railway trackwork. The fatigue limit of austenitic manganese steel has been reported as 270 MPa (39 ksi).

Yield strength and hardness vary only slightly with section size. Hardness of most grades is about 200 HB after solution annealing and quenching, but this value has little significance for estimating machinability or wear resistance. Hardness increases so rapidly due to work hardening during machining or in wear service that austenitic manganese steels must be evaluated on some basis other than hardness.

The true tensile characteristics of manganese steel are better revealed by the stress-strain curves in Fig. 9, which compare manganese steel with gray iron and with a tough ferritic steel of about the same nominal tensile strength. The low yield strength is significant and may prevent selection of this alloy where slight or moderate deformation is undesirable, unless the usefulness of the parts in question can be restored by grinding. However, if deformation is immaterial, the low yield values may be considered temporary—that is, deformation will produce a new, higher yield strength corresponding to the amount of strain that is absorbed locally.

Work Hardening

The approximate ranges of tensile properties produced in constructional alloy steels by heat treatment are developed in austenitic manganese steels by work hardening. In a tension test, yielding signifies the beginning of work hardening, and elongation is associated with its progress. Little or no reduction in area occurs in austenitic manganese steels by necking, because work hardening is greatest at the point of greatest deformation. The increase in strength due to cold work stops elongation, and deformation then occurs elsewhere until the hardening and reduction in area are substantially equalized throughout the specimen.

Manganese steels are unequalled in their ability to work harden, exceeding even the metastable austenitic stainless steels in this feature. For example, a standard grade of manganese steel containing 1.0 to 1.4% C and 10 to 14% Mn can work harden from an initial level of 220 HV to a maximum of more than 900 HV. After extended service, the hardness at the wearing surfaces of railway frogs typically ranges from 495 to 535 HB. Maximum attainable hardness depends on many factors, including specified composition, service limitations, method of work hardening and preservice hardening procedures. It appears that rubbing under heavy pressure can produce higher values of maximum attainable hardness than can be produced by simple impact.

Service Limitations. In some cases, abrasion may remove surface metal before it can attain maximum hardness. In other instances, work hardening raises the elastic limit to the point where succeeding impact blows cannot cause more plastic flow and thus merely bounce off. However, if a rotating tool is applied with enough force and does not cut, accentuated work hardening can be expected. This is a common occurrence during drilling. A sharp twist drill of superior steel can drill 13% Mn steel provided a deep enough cut is taken, but cutting ceases if the drill becomes dull. When this happens, it frequently is futile to continue drilling, even with a sharp drill, because the bottom of the hole has become so hard that the cutting edge cannot penetrate it.

The low yield strength of manganese steel is sometimes a disadvantage in service. For example, plastic deformation due to impact, as in railway frogs and crossings, increases yield strength to levels more resistant to flow, but the associated changes in dimensions are undesirable. In trackwork, low spots develop at critically pounded locations, resulting in a bumpy roadway and eventually requiring rebuilding with weld deposits.

Low yield strength also can be a disadvantage when manganese steel is used in light armor and similar applications. Because much energy is absorbed during work hardening, manganese steel sheet is an effective light armor against slow-moving projectiles. However, manganese steel is relatively ineffective against high-velocity projectiles that shear through the armor with little accompanying deformation. Prehardened alloy steels and other steels of higher yield strength are preferred for armor against high-velocity projectiles. Most of the impact blows encountered in industrial service are of low velocity, and for this service, manganese steel is acceptable. It is the preferred choice where high shock resistance, toughness and absorption of energy are required.

Work-Hardening Methods. Work hardening usually is induced by impact, as from hammer blows. Light blows, even if they are of high velocity, cause shallow deformation with only superficial hardening even though the resulting surface hardness ordinarily is high. Heavy impact produces deeper hardening, usually with lower values

Fig. 10 Plastic flow and work hardening of a manganese steel and an air-hardening steel under repeated impact

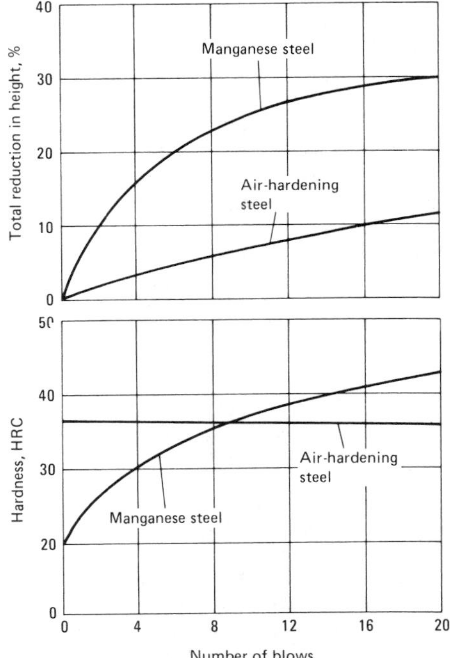

Specimens 25 mm (1 in.) in both diameter and length were struck repeatedly on one end by blows with an impact energy of 680 J (500 ft·lb). Composition and heat treatment of the manganese steel: 1.17C-12.8Mn-0.46Si; water quenched from 1010 °C (1850 °F). Composition and heat treatment of the air-hardening steel: 0.74C-0.88Mn-0.30Si-0.75Ni-1.40Cr-0.38Mo; air cooled from 900 °C (1650 °F), reheated to 700 °C (1300 °F) and air cooled.

of surface hardness. The course of flow under impact and the associated increase in hardness are illustrated in Fig. 10, which compares a standard 12% Mn steel with an air-hardening Cr-Ni-Mo alloy steel. Less well known is the fact that abrasion itself can produce work hardening.

Explosion hardening was developed as a substitute for hammer or press hardening that would achieve hardening with less deformation. Pentaerythritol tetranitrate in the form of plastic explosive sheet is cemented to the surface of the steel and then detonated. Several explosions may be required to attain the desired hardness. The use of plastic explosive permits hardening of areas like trackwork flangeways and unsupported sections that cannot be hammered satisfactorily. Figure 11 compares depth and intensity of hardening from three different explosion treatments and from one hammer-peening operation applied to railway trackwork. Explosion hardening is now considered a satisfactory but expensive substitute for hammer or press hardening of solid-base manganese steel trackwork castings. On open-base and other "unsupported" castings, it provides hardening to a depth previously unattainable.

Fig. 11 Hardening patterns in manganese steel railway crossings

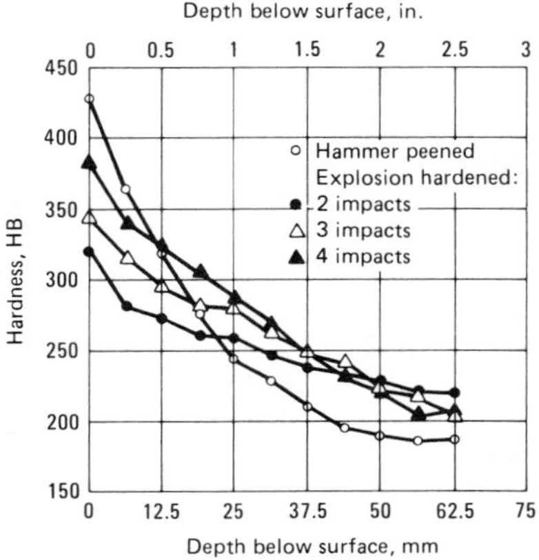

Measurements were made on sections taken 19 mm (3/4 in.) from the flangeway on each sample crossing.

Explosion treatment also has been applied to parts subject to abrasive wear. Initial reports were favorable, but have been reversed by subsequent experience; there is no solid evidence that explosion hardening is advantageous for service involving grinding or gouging abrasion.

Studies have been conducted in an attempt to discover why explosion hardening is accompanied by insignificant deformation in spite of the fact that work hardening is usually associated with plastic flow and deformation. These studies have centered on the premise that a different hardening mechanism is involved in explosion hardening. Results indicate that among the mechanisms by which Hadfield's steel hardens are creation of stacking faults, transformation to epsilon martensite, transformation to alpha martensite, and multiple twinning. However, there are differences of opinion as to whether martensite can be produced by cold working—differences that have resulted from the fact that most workers have studied only one alloy rather than a series of high-manganese steels. As carbon and/or manganese contents are reduced, the austenite becomes less stable and reaches a point where phase transformation can take place.

Fig. 12 Flow under repeated impact for several manganese steels and for rail steel at different hardnesses

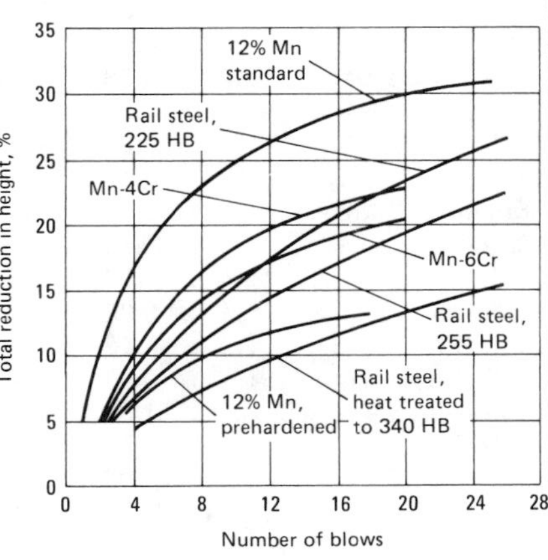

Specimens 25 mm (1 in.) in both diameter and length were struck repeatedly by blows with an impact energy of 680 J (500 ft·lb).

Preservice Hardening. The low flow resistance and consequent low yield strength of manganese steel can be increased by several methods; preservice hardening by deformation is preferred. Special equipment for hammering or pressing is usually employed to induce deep-seated hardening. The depth of the affected zone usually ranges up to about 25 mm (1 in.). The superficial hardening that can be produced by high-velocity shot blasting is seldom satisfactory for service requirements.

Addition of alloying elements such as vanadium, chromium, silicon and molybdenum also is an effective means of raising yield strength, but vanadium and chromium reduce ductility. The relative effects of alloying and of prehardening by deformation are com-

Table 4 Typical mechanical properties, toughness and work-hardening capacity of experimental manganese steels

Steel	Carbon content, %	Tensile strength MPa	Tensile strength ksi	Yield strength MPa	Yield strength ksi	Elongation, %	Merit No.(a)	Brinell indentation diameter, mm, for load, kg, of: 1000	1500	2500	3000	Meyer exponent
20Mn-Bi	0.60	855(b)	123.8(b)	310	45.3	63	780	2.70(b)	3.10	3.90	4.40(b)	2.22
13Mn-3Ni	0.89	795	115.4	310	45.3	54	624	...	...	...	...	...
20Mn-0.07S	0.62	730(b)	105.8(b)	300	43.2	55	582	2.92(b)	3.58(b)	4.37	4.65	2.26
20Mn-Ni-Cu-Bi-S	0.38	615(b)	89.0(b)	245	35.4	60	534	3.16(b)	3.70(b)	4.80	4.98(b)	2.28
Type 304 stainless steel	0.08	555(b)	80.3(b)	295(b)	43.1(b)	65	522	2.70	3.20	...	4.45	2.17
20Mn-0.02S	0.67	765	110.9	320	46.3	46	500	2.97	3.50	4.30	4.65	2.42
13Mn-Zr	1.14	745	108.4	325	47.4	45	489	2.80	3.45	4.20	4.50	2.40
Standard 13 Mn	1.12	825	120.0	360	52.0	40	480	2.80	3.25(b)	...	4.24(b)	2.60
13Mn-Bi	1.15	785(b)	113.6(b)	360	52.0	41	466	3.10	3.40	4.20	4.50	2.57
13Mn-Se	1.07	746	108.3	335	48.7	43	465	2.90	3.35	4.15	4.60	2.35
13Mn-1Mo	0.95	800	116.0	400	58.2	40	465	...	...	...	...	...
13Mn-Pb	1.12	760	110.4	330	48.0	42	459	2.80	3.40	4.30	4.60	2.17
CF-8 stainless steel	0.05	500	72.4	220	31.7	60	435	3.20(b)	3.65(b)	4.60	5.00(b)	2.47
13Mn-S	1.07	750(b)	109.0(b)	340	49.7	32	425	2.80	3.40	4.10	4.40	2.38
20Mn-Bi	0.55	650(b)	94.0(b)	280	40.9	43	404	2.95	3.51	4.45	4.75(b)	2.30
13Mn-0.98C	0.98	760	110.0	345	50.0	35	385	...	...	...	...	...
13Mn-Ag	1.14	760(b)	110.3(b)	345	49.9	33	364	2.90	3.40	4.20	4.60	2.43
20Mn-0.9C	0.90	650	94.0	340	49.0	34	320	2.90	3.40	4.30	4.50	2.50

(a) Merit No. is the product of tensile strength and elongation, and is an approximate measure of toughness. (b) Average of two or more tests.

pared in Fig. 12. A manganese steel that combines high yield strength with high ductility has recently been developed. This alloy (U.S. Patent 3,075,838), which contains 0.6 C, 13 Mn, 3 Cr, 3 Ni, 2 Mo and 0.7 V, has not yet been given a commercial designation. It has been considered for applications that require strong, nonmagnetic steel and do not require substantial resistance to corrosion. The high-yield-strength manganese steel would be used as a replacement for ordinary manganese steel or nonmagnetic stainless steel. The new steel has a yield strength of 620 MPa (90 ksi) and 20 to 30% elongation. Compared with standard manganese steel, it is similar in abrasion resistance, more expensive, much more resistant to deformation under repeated impact or static stresses and more difficult to machine.

Prehardening by deformation is particularly recommended for such operations as deep drawing of military helmets and cold forming of strip for use in body armor. The requirements of these operations necessitate careful control of the martensitic skin formed during solution annealing and quenching, because tensile deformation may produce cracks that render these thin products unsuitable for the intended use. The skin may not be thick enough to result in a noticeably defective product during manufacture but may contain minute cracks that lower resistance to further deformation. Solution and distribution of carbides are important in such applications.

Hardness Measurement. Methods of measuring and reporting the hardness of manganese steel differ from those used on other metals. The reasons underlying this can explain in part the unsubstantiated notions that manganese steel must work harden to develop abrasion resistance and that a specific instance of poor performance in abrasion service occurred because the metal did not work harden. Preservice work hardening often is advocated, but may produce variable results because of other factors involved. Composition of the metal, abrasive conditions, speed of work hardening and maximum hardness attainable in the steel are variables that all influence performance.

The common Brinell test reports hardness in terms of load divided by the curved area of the indentation, whereas the Meyer test (which uses the same indentor) reports hardness as load divided by the area of the indentation measured in the plane of the original surface. If a series of hardness numbers are obtained with various loads, it will be found that, for an annealed metal indented with a ball, Meyer hardness will increase with load, whereas for fully work-hardened metal, Meyer hardness is constant. The behavior of annealed metal—metal free of plastic strain or residual stress—can be used as a measure of the extent of work hardening.

This measurement technique generally employs a 10-mm-diam Brinell ball and a series of loads. Test load is plotted against diameter of the corresponding impression on logarithmic scales. The result is expected to be a straight line that fits the equation:

$$P = Ad^n \quad \text{(Eq 1)}$$

where P is the applied load, d is the diameter of the impression, A is a constant for a given indenting ball size, and n is a measure of the tendency of the metal to strain harden. The exponent n, also called the Meyer index or Meyer exponent, is the slope of a plot of P versus d on log-log paper. It also can be expressed as:

$$n = \tan \phi \quad \text{(Eq 2)}$$

where ϕ is the angle between the plot of impression diameters and its horizontal coordinate. Meyer exponents for some austenitic manganese steels are given in Table 4; exponents for types 304 and CF8 stainless steels are included in the table for comparison.

Reheating

Before manganese steel parts are reheated in the field, the effects of such reheating must be seriously considered. Unlike ordinary structural steels, which become softer and more ductile when reheated, manganese steels suffer reduced ductility when reheated enough to induce carbide precipitation or some transformation of the austen-

Fig. 13 Structure of a typical reheated manganese steel

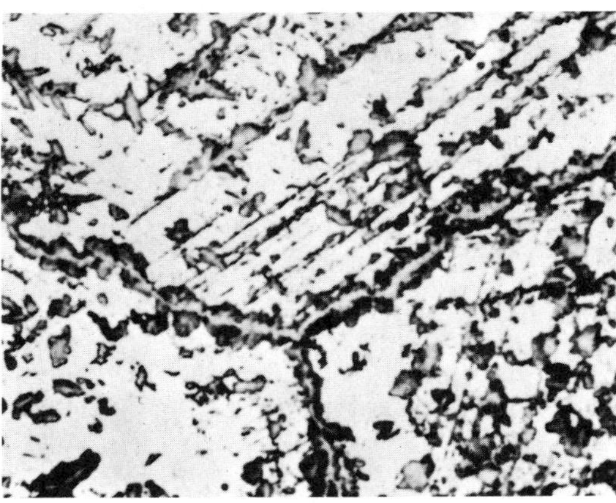

ASTM A128, grade B-3, steel annealed at 1120 °C (2050 °F), water quenched and reheated above 315 °C (600 °F). Structure is austenite with extensive carbide precipitation along grain boundaries and within grains. Specimen was etched in 2.5% nital, rinsed in ethanol and re-etched in 15% HCl; magnification, 500×.

Fig. 14 Time-temperature relationship for embrittlement of 13Mn-1.2C-0.5Si steel

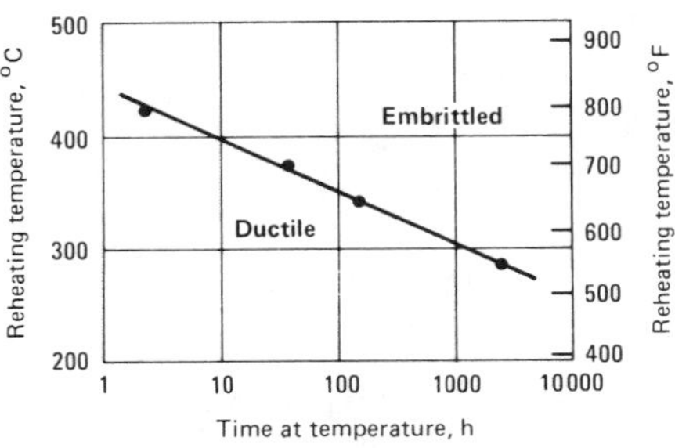

Prior to reheating, the alloy was annealed 2 h at 980 °C (2000 °F) and water quenched.

Fig. 15 Embrittlement from reheating manganese steel

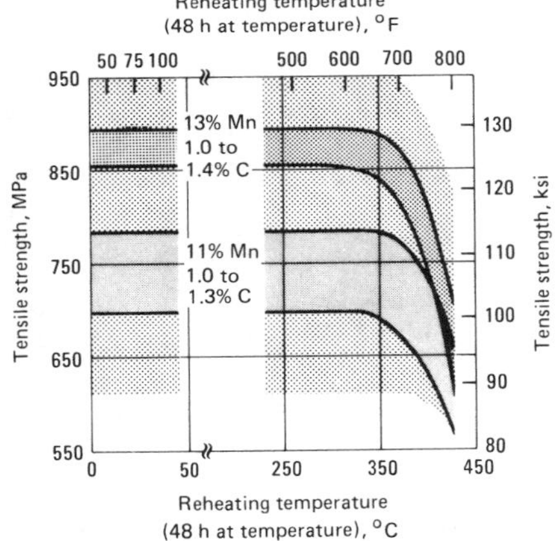

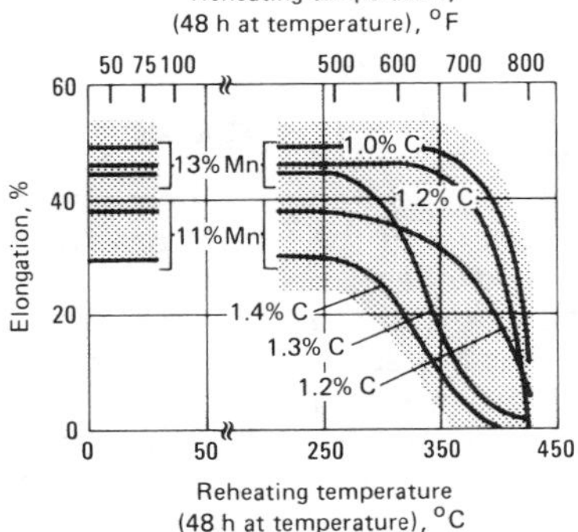

Cast bars 25 mm (1 in.) in diameter were reheated 48 h at the temperatures indicated, after solution annealing and quenching.

ite. Figure 13 illustrates the microstructural effects of such heating. As a general rule, manganese steels should never be heated above 260 °C (500 °F), either intentionally or accidentally, unless such heating can be followed by standard solution annealing and quenching.

Time, temperature and composition are variables in the embrittlement process. At lower temperatures, embrittlement takes longer to develop. The time-temperature relationship in 13Mn-1.2C-0.5Si steel is illustrated in Fig. 14, which presents data based on metallographic examination for structural changes that indicate the beginning of embrittlement. At 260 °C (500 °F), transformation requires more than 1000 h; reheating to as high as 425 °C (800 °F), even with close control of temperature, may be done for no longer than 1 h if transformation is to be avoided. Figure 15 shows the effect of composition on magnitude of embrittlement.

Because large castings sometimes are mounted with backings of molten lead or zinc, the temperature of such castings should be carefully controlled during reheating. The time-temperature relationship should also be given due consideration for parts that must be welded.

When 12 to 14% Mn steels are to be heated above about 290 °C (550 °F) during service or welding, it is recommended that carbon content be held below 1.0%, which will suppress embrittlement for at least 48 h at temperatures up to 370 °C (700 °F). Addition of 1.0% Mo will suppress embrittlement completely at temperatures up to 480 °C

Table 5 Mechanical properties of three reheated austenitic manganese steels

Condition	Time at temper-ature, h	Tensile strength(a) MPa	Tensile strength(a) ksi	Yield strength(a) MPa	Yield strength(a) ksi	Elonga-tion(a), %	Reduction in area(a), %	Hardness(b), HB	Charpy V-notch impact strength(a) J	Charpy V-notch impact strength(a) ft·lb
12% Mn steel										
Solution treated	...	615	89	340	49	28	31	164	129	95.5
Reheated at 370 °C (700 °F)	0.5	560	81	325	47	26	25	175	117	86.2
	2	600	87	315	46	27	30	168	137	100.7
	10	670	97	330	48	31	30	177	112	82.5
Reheated at 480 °C (900 °F)	0.5	620	90	330	48	29	31	177	138	101.7
	2	565	82	325	47	24	20	171	115	84.5
	10	460	67	345	50	6	6	177	12	8.5
Reheated at 590 °C (1100 °F)	0.5	395	57	325	47	4	7	173	12	8.8
	2	475	69	350	51	1	1	182	5	3.8
	10	540	78	340	49	1	1	180	3	2.3
12Mn-1Mo steel										
Solution treated	...	595	86	360	52	26	27	187	110	81.5
Reheated at 370 °C (700 °F)	0.5	635	92	350	51	32	31	171	129	95.2
	2	595	86	360	52	25	29	182	116	85.3
	10	625	91	350	51	31	31	163	115	84.5
Reheated at 480 °C (900 °F)	0.5	640	93	350	51	32	39	177	120	88.5
	2	625	91	350	51	28	32	187	112	82.5
	10	585	85	350	51	27	31	183	136	100.5
Reheated at 590 °C (1100 °F)	0.5	540	78	345	50	21	26	182	103	76.0
	2	405	59	350	51	5	6	182	19	13.8
	10	525	76	380	55	1	2	227	5	4.0
12Mn-3.5Ni steel										
Solution treated	...	635	92	290	42	39	38	151	135	99.7
Reheated at 370 °C (700 °F)	0.5	695	101	330	48	40	33	149	144	106
	2	685	99	325	47	40	35	146	127	93.7
	10	805	117	340	49	50	37	163	126	93
Reheated at 480 °C (900 °F)	0.5	685	99	340	49	35	31	142	72	52.8
	2	485	70	325	47	17	20	150	57	42.2
	10	385	56	330	48	5	6	148	14	10.2
Reheated at 590 °C (1100 °F)	0.5	460	67	325	47	14	11	146	30	21.8
	2	435	63	330	48	7	6	142	15	10.8
	10	395	57	315	46	4	3	145	6	4.2

(a) Average of two determinations. (b) 1000-kg load; average of six determinations.

(900 °F) and will partly suppress it at temperatures of 480 to 600 °C (900 to 1100 °F). If carbon content is held below about 0.9%, addition of 3.5% Ni will completely suppress embrittlement up to 480 °C and will partly suppress it above this temperature. These rules can be expected to apply during heating periods of up to 100 h. For periods of 1000 h or more, limiting temperatures are substantially lower.

Table 5 lists properties of three 1.0% C manganese steels that were reheated for periods of up to 10 h at temperatures up to 600 °C (1100 °F). At 370 °C (700 °F), none of the three steels was embrittled; in fact, there were indications that the treatment had effected slight improvements in ductility. At 480 °C (900 °F), the plain manganese steel and the Mn-Ni steel were embrittled, but there was little change in the properties of the Mn-Mo steel. At 600 °C, all three steels were embrittled, but the plain manganese steel was more severely affected than the others.

Embrittlement, as revealed by microstructural investigation, was caused by formation of acicular carbide and pearlite in the austenite grains in the plain manganese steel, by carbide nodules surrounded by pearlite within the austenite grains in the Mn-Mo steel, and by an envelope of proeutectoid cementite around each grain in the Mn-Ni steel. If the Mn-Ni steel had been lower in carbon content, it might have been substantially less susceptible to embrittlement. In another investigation, a 0.9C-14.3Mn-1.75Si-3.4Ni steel did not become significantly embrittled when it was heated for 1½ h at 480 °C (900 °F), and its ductility was reduced by only 17 to 20% when it was heated for 1½ h at 600 to 760 °C (1100 to 1400 °F).

Wear Resistance

Compared with most other abrasion-resistant ferrous alloys, manganese steels are superior in toughness and moderate in cost, and it is primarily for these reasons that they are selected for a wide variety of abrasive applications. They usually are less resistant to abrasion than martensitic white irons and martensitic high-carbon steels, but often are more resistant than pearlitic white irons and pearlitic steels.

The type of abrasion to which a manganese steel is exposed has a major influence on how well it wears. Manganese steels have excellent resistance to metal-to-metal wear, as in sheave wheels, crane wheels and mine-car wheels; moderately good resistance to gouging abrasion, as in equipment for handling or crushing rock; intermediate resistance to high-stress (grinding) abrasion, as in ball-mill and rod-mill liners; and relatively low resistance to low-stress abrasion, as in equipment for handling loose sand or sand slurries.

Metal-to-Metal Contact. In applications involving metal-to-metal contact, work hardening of manganese steel is a distinct advantage because it decreases the coefficient of friction and confers resistance to galling if temperatures are not excessive. Compressive loads, rather than impact, provide the deformation required, producing a smooth, hard surface that has good resistance to wear but that does not abrade the contacting part. Sheaves, rails and castings for railway trackwork are common applications of this type. Manganese steel also has been used in some water-lubricated bearings.

Railway center plates, which are the bearing surfaces where trucks swivel under freight cars, provide a good example of the merit of manganese steel for metal-to-metal frictional wear. Initially lubricated, these plates soon are operating dry and are accessible to airborne grit. When plates are made of carbon steel, the mating surfaces become rough from galling (adhesive wear); the increased friction prevents easy motion, the trucks do not swivel properly on curves, and the lateral pressure is so accentuated that early wheel-flange wear is induced. Poor swiveling may accentuate thrust loads at the ends of wheel bearings, thus developing excessive frictional heat, and center plates and mating bowls on trucks may wear severely. Many years ago, substitution of 13% Mn steel center plates on heavily loaded ore cars demonstrated their superiority. More recently, various service tests have demonstrated that manganese steel not only wears less than carbon steel but also develops a low-friction polished surface. The advantages of manganese steel center plates are most evident for cars that are very heavily loaded.

Field tests that involve intermittent metal-to-metal contact provide additional evidence that manganese steel can render distinctive service in such applications. Locomotive chafing irons and plates of manganese steel lasted for an average mileage of 447 000, compared with 84 000 for carbon steel parts. Manganese steel wheels on a 6-ton gas locomotive were reported to have exhibited no appreciable wear after three years, whereas the flanges on chilled cast steel wheels had worn about one-third of the way through in eight months and had caused more wear of rails. A steel mill has reported double life from manganese steel crane wheels, with freedom from spalling and from the diametral wear that characterizes carbon steel wheels.

Fig. 16 Relative wear ratios of ferrious alloys in jaw-crusher tests

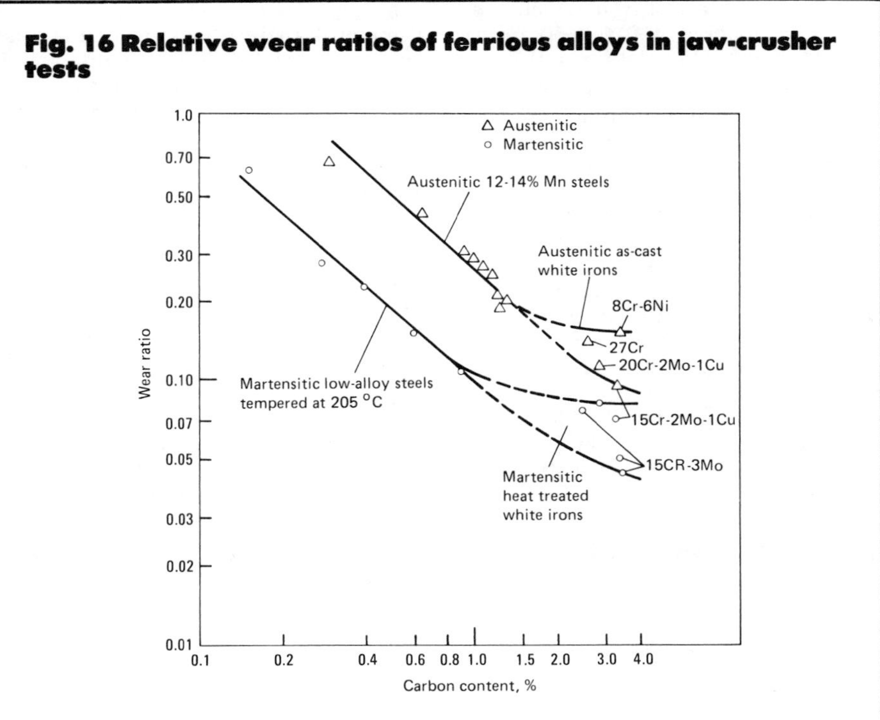

Adapted from Ref 6.

Manganese steel wear plates in a blooming-mill housing were in good condition after 16 years, whereas carbon steel plates required replacement every two years. Manganese steel wear plates on ore-bridge haulage drums were reported good as new after 5½ years of service, whereas ordinary cast steel plates wore out in about three years. Manganese steel sheave wheels are expected to last about four times as long as cast carbon steel wheels, and do not groove and cause excessive rope wear as do wheels made of alloys that do not work harden. In steel-mill applications, welded manganese steel overlays on large mill-coupling boxes, pinions, spindles and other items working under heavy impact loads are satisfactory.

The concept that manganese steel has poor wear resistance unless it has been work hardened is not a valid generalization. The misunderstanding has probably developed because, where much impact and attendant work hardening are present, 12% Mn steel is so clearly superior to other metals that its performance is attributed to surface hardening. However, controlled abrasion tests have indicated that there are circumstances under which the abrasion resistance of austenitic manganese steel is modified little by preservice work hardening, and others under which this steel will outwear harder pearlitic white cast irons without work hardening.

Abrasion. In applications that involve heavy blows or high compressive and structural stresses, the very hard and abrasion-resistant martensitic cast irons may wear more slowly than manganese steel. However, these irons usually fail by early fracture with a considerable portion of the original cross section unworn, whereas manganese steel may wear almost to paper thinness before fracturing.

Pearlitic white cast iron, which has a hardness of about 400 to 450 HB, is equally brittle but less resistant to wear. Comparative tests on log washer lugs indicated that manganese steel was about 25% worn out with no breakage, whereas in the same period white iron lugs wore to the point of uselessness with 14% breakage. In clay-crusher rolls, manganese steel lasted two to three times as long as white or chilled iron. In grinding-barrel liners, cast irons lasted two to three years compared with ten years for manganese steel. Part of this superiority of manganese steel over white cast iron is attributed to greater freedom from breakage and spalling, but some is probably due to better intrinsic wear resistance.

Manganese steel chain, with endless links cast in interlocking molds, also provides resistance to wear, lasting three to nine times as long as heat treated steel chain in certain applications. Manganese steel is valuable in conveyors as well as in dragline chain subjected to abrasion and used for carrying heavy loads.

Manganese steel is not satisfactorily resistant to wear by a stream of air-borne abrasive particles (impingement erosion), such as in sandblasting or gritblasting equipment, and consequently should not be selected for such service.

Abrasion Testing. The abrasion resistance of austenitic 12 to 14% Mn steels with various carbon contents has been compared with the resistance of other steels and white irons in a jaw-crusher abrasion test (see Fig. 16). The wear rate for a quenched and tempered low-carbon, low-alloy steel, (ASTM 517, type B at 269 HB), which was used as a comparative standard in each test, is shown also. When the relative wear rate (wear ratio) of each test material is plotted against increasing carbon content on a log-log scale, results for austenitic steels and irons tend to fall on a descending straight line, and results for martensitic steels and irons fall on a parallel line below the line for the austenitic alloys. A decrease in wear ratio represents a proportionate increase in abrasion resistance. Thus, Fig. 16 strongly supports the conclusions that abrasion resistance of both austenitic and martensitic steels improves with increasing carbon content and that, for a given carbon content, martensitic steels have better abrasion resistance than austenitic steels. However, martensitic steels and white irons have limited resistance to gouging abrasion due to their lack of toughness. The wear ratios of pearlitic steels, if plotted on Fig. 16, would lie above those of the austenitic steels. There is considerable scatter in the rates for pearlitic steels due to their wide variation in hardness for any given carbon content.

The same trends have been observed in ore-milling tests and in sand-slurry abrasion tests. Generally, austenitic 12 to 14% Mn steels have higher wear rates, and therefore poorer abrasion resistance, than pearlitic or martensit-

Table 6 Relative resistance of various materials to high-stress grinding abrasion by wet quartz sand

Material	Hardness, HB	Abrasion factor(a)
Cemented tungsten carbide	...	0.17
Martensitic Ni-Hard cast iron	550 to 750	0.25 to 0.60
Martensitic 4150 steel	715	0.60±
Bainitic 4150 steel	512	0.75±
Austenitic 12% Mn steel	200	0.75 to 0.85
Pearlitic 0.85% C steel	220 to 350	0.75 to 0.85
Alloyed white cast irons	400 to 600	0.70 to 1.00
Unalloyed white cast irons	400±	0.90 to 1.00
1020 steel (standard)	107	1.00
Gray cast irons	200±	1.00 to 1.50
Ferritic ingot iron	90	1.40

(a) Particle size of abrasive, 50 to 55 AFS grain fineness number; compressive stress on specimen about 370 kPa (54 psi), as determined at the American Brake Shoe Co. laboratories. Ratio of weight loss of sample to weight loss of the standard material, 1020 steel. High factors indicate poor wear resistance.

Table 7 Relative wear rates of various materials tested as 125-mm (5-in.) grinding balls in a ball mill

Material	Nominal composition, % C	Mn	Cr	Mo	Ni	Hardness(a), HB	Relative wear rate(b)	Order of toughness(c)
Martensitic Cr-Mo white iron	2.8	1.0	15.0	3.0	...	740	89	7
Martensitic high-Cr white iron	2.7	1.0	26.0	...	...	705	98	8
Martensitic high-carbon Cr-Mo steel (type 3)	1.0	0.8	6.0	1.0	...	615	100	6
Martensitic high-carbon Cr-Mo steel (type 2)	0.7	0.7	2.0	0.4	...	560	110	5
Martensitic Ni-Cr white iron	3.2	0.8	2.0	...	4.0	650	112	9
Austenitic 6Mn-1Mo steel	0.9	6.0	...	1.0	...	490	114	3
Martensitic medium-carbon Cr-Mo steel (type 1)	0.4	1.5	0.8	0.4	...	560	120	2
Pearlitic high-carbon Cr-Mo steel (type a)	0.8	0.8	2.5	0.4	...	380	127	4
Austenitic 12Mn steel	1.2	12.0	...	...	...	410	138	1

(a) Hardnesses are converted from average of HRC readings on the worn surfaces of the balls. Hardnesses below the cold worked surface normally are lower than those given. (b) Relative wear rates are based on a nominally assigned factor of 100 for the high-carbon martensitic steel (type 3). Factors greater than 100 indicate higher rates of wear and factors less than 100 indicate proportionately lower rates of wear. (c) Order of toughness as listed here is a qualitative value, based partly on results of laboratory tests and partly on the basis of resistance to breakage and spalling that the respective materials have shown in service.

ic steels of equivalent carbon content in low-stress slurry abrasion; a 6Mn-1Mo composition usually is intermediate in wear resistance between pearlitic and martensitic steels.

Rankings of various metals in abrasion tests are given in Tables 6 to 8. A considerable amount of test data on abrasion resistance of a wide variety of austenitic manganese steels, constructional steels and cast irons has been accumulated. Some of these data are summarized and discussed in the article "Wear Resistance" on pages 597 to 638 in Volume 1 of this Handbook (9th Edition).

Corrosion

Manganese steel is not corrosion resistant. It rusts readily; and where corrosion and abrasion are combined, as they frequently are in mining and manufacturing environments, the metal may deteriorate or be dissolved at a rate only slightly lower than that of carbon steel. If the toughness or nonmagnetic nature of manganese steel is essential for a marine application, protection by galvanizing usually is satisfactory.

Effects of Temperature

The excellent properties of 13% Mn steel between −45 and +205 °C (−50 and +400 °F) make it useful for all ambient-temperature applications, even in arctic climates. It is not recommended for hot wear applications because of structural instability between 260 and 870 °C (500 and 1600 °F). At higher temperatures, it may lack the strength and ductility necessary to withstand severe welding stresses, and thus welding must be done under closely controlled conditions. It is not resistant to oxidation, and its creep-rupture properties are inferior to those of Fe-Cr-Ni austenites. Thus, it is unsuitable for structural applications in the red-heat range.

Tensile properties from −195 to +98 °C (−320 to +208 °F) for a wrought steel containing 1.40% C, 12.1% Mn and 0.12% Si are given in Table 9. Elongation is above 25% at temperatures as low as about −100 °C (−150 °F), and at −75 °C (−100 °F) decrease in strength and ductility is not significant for most uses.

At −75 °C, cast manganese steels retain from 50 to 85% of their room-temperature impact resistance (see Table 10). They are considerably more brittle at liquid-air temperature (−185 °C, or −300 °F), but at all atmospheric temperatures encountered by railway trackwork and mining and construction equipment they have outstanding toughness that provides a valuable safety factor in comparison with ferritic steels at subzero temperatures.

Associated with the embrittlement produced by reheating above 260 °C (500 °F) are changes in physical properties stemming from the same transformations that cause the loss in toughness. Because both composition and time at temperature influence these changes, erratic behavior and a considerable range in such properties as ther-

Table 8 Relative gouging abrasion resistance in the Esco jaw-crusher test

Material	Hardness prior to test, HB	Abrasion factor(a)
Martensitic Cr-Mo white iron	750	0.088
A2 tool steel	653	0.127
0.50C-3Cr tool steel	653	0.164
4340 steel(b)	555	0.262
Austenitic 12Mn-1.16C steel	217	0.279
Martensitic 0.18C-Cr-Ni steel(b)	461	0.386
4340 steel(c)	321	0.788
Martensitic T-1 steel plate (tempered)	269	1.000 ± 3%
Pearlitic mild steel plate	197	1.198

(a) Ratio of weight loss of sample to weight loss of the standard material, martensitic T-1 steel plate. Tests were performed in a specially modified small commercial jaw crusher (Esco Laboratories Test). High factors indicate poor wear resistance. (b) Quenched and tempered at 205 °C (400 °F). (c) Quenched and tempered at 650 °C (1200 °F).

Table 9 Results of individual mechanical tests on wrought 1.4C-12Mn steel at various temperatures

Temperature °C	Temperature °F	Fracture stress (a) MPa	Fracture stress (a) ksi	Natural strain, Ao/A	Tensile strength MPa	Tensile strength ksi	Yield strength MPa	Yield strength ksi	Elongation (b), %	Reduction in area, %	Hardness(c), HV
98	208	1580	229	0.414	1040	151	435	63	56.0	34.0	518
91	196	1830	265	0.489	1120	163	400	58	68.0	38.7	...
23	74.5	1600	232	0.431	1040	151	460	67	48.5	35.1	524
23	74.5	1740	253	0.468	1090	158	295	43	62.5	37.7	...
22	72.5	1580	229	0.416	1040	151	365	53	46.8	34.2	...
−30	−22	1470	213	0.351	1030	150	485	70	37.5	29.6	470
−97	−143	1320	192	0.267	1010	147	225	62	25.0	23.5	...
−110	−166	1240	180	0.277	950	138	525	76	25.0	23.3	438
−150	−236	1060	154	0.120	945	137	695	101	12.2	11.3	317
−195	−320	995	144	0.050	945	137	855	124	4.0	4.9	251
−195	−320	950	138	0.033	940	136	805	117	4.0	3.2	...

(a) A uniform strain rate of 0.13 mm (0.005 in.) per minute was used for all tests. Fracture stress is taken as load at fracture divided by area at fracture; with no necking, fracture load is maximum load. (b) In 25 mm or 1 in. (c) 2-kg load at tensile break; tested at room temperature.

Table 10 Charpy V-notch impact strength (Ref 8)

Composition, %				Impact strength (a)			
				At 24 °C (75 °F)		At −73 °C (−100 °F)(b)	
C	Mn	Si	Ni	J	ft·lb	J	ft·lb
1.03	12.9	0.52	...	128	94.5	71	52.5
1.18	13.0	0.50	...	144	106	79	58.5
1.19	14.6	0.50	...	141	104	79	58.5
0.84	12.5	0.48	3.46	136	100	108	80
1.17	12.7	0.53	3.56	142	104.5	119	88

(a) Average of duplicate tests. Temperature ±3 °C (±5 °F). V-notches machine ground to avoid work hardening. Specimens water quenched from A_{cm} before machining. (b) Held in solid CO_2 and acetone until just before testing.

Table 11 Effects of temperature on physical properties of 13% Mn steel (Ref 9)

Temperature °C	Temperature °F	Mean apparent specific heat, J/kg·K	Mean coefficient of thermal expansion from 0 °C, μm/m·K	Electrical resistivity, nΩ·m	Thermal conductivity, W/m·K
0	32	494	...	6.65	13.2
50	120	510	...	7.11	14.0
100	212	527	18.01	7.57	14.9
150	302	553	...	8.02	15.7
200	390	573	19.37	8.47	16.5
250	480	590	...	8.89	17.4
300	570	603	20.71	9.31	18.0
350	660	607	...	9.69	18.6
350 to 650	660 to 1200	(a)	(a)	(a)	(a)
700	1290	...	20.49	11.53	21.8
750	1380	...	...	11.80	21.8
800	1470	...	21.86	12.11	22.2
850	1560	...	...	12.40	22.4

Composition: 1.22 C, 13.0 Mn, 0.22 Si, 0.038 P, 0.010 S, 0.07 Ni, 0.03 Cr, 0.07 Cu, 0.004 Al, 0.038 As. Condition: wrought steel, heated to 1050 °C (1925 °F) and air cooled.
(a) Values depend on time at temperature and on amount of transformation.

mal and electrical conductivity may be expected above 315 °C (600 °F), as shown in Table 11.

Thermal-expansion characteristics of austenitic manganese steels are similar to those of other austenitic materials. The expected change in length on heating is about 1½ times that of ferritic steels. A coefficient of linear thermal expansion of 18 μm/m·°C (10 μin./in.·°F) generally is precise enough near room temperature (Fig. 17). Transformation to pearlite and precipitation of carbide influence values of the expansion coefficient in the range from 370 to 760 °C (700 to 1400 °F).

Magnetic Properties

The untransformed austenite of 13% Mn steel is virtually nonmagnetic, with a permeability of about 1.03 or less. This permits use of the material where a strong, tough, nonmagnetic metal is required, as in magnet cover plates, collector shoes for traveling cranes, stator core parts for generators and motors, liner plates for storage bins holding materials that are handled by lifting magnets, magnetic-separator parts, instrument-testing devices and furnace parts located in the magnetic fields of induction furnaces.

The changes that occur in composition of the surface during heat treatment may produce a magnetic skin, one that is either a martensitic surface layer or a magnetic oxide. Permeability values of 1.3± have been obtained on specimens that have this magnetic surface layer. Frequently, this layer does no harm, but if necessary it may be removed by grinding or pickling or may be prevented by suitable (although often expensive) corrective measures during heat treatment.

Cast or wrought manganese steel probably is the most economical material for strong nonmagnetic parts if machining is not required. Cost of fabrication and anticipated operating temperature are decisive factors in selection; operating temperature must not exceed 260 °C (500 °F).

A 20% Mn steel containing bismuth (U.S. Patent 3,010,823) has been developed for nonmagnetic parts that require machining. Laboratory tests indicate machinability comparable to that of the more expensive type 304 stainless steel that this material was designed to replace. It can be lathe turned (horizontal force: 14 to 39 kg at 1.35 m/s, or 30 to 85 lb at 265 ft/min), drilled and tapped. This steel has not yet been exploited commercially.

Lack of magnetism is a disadvantage when austenitic manganese steel is used in components of material-handling systems that depend on magnetic separators to remove tramp iron from the process stream before it enters crushers, grinders and other machinery. When there is a possibility that manganese steel parts may become detached from working equipment and fall into the process stream, it is advisable to cast mild steel inserts in the manganese steel parts. The inserts must be large enough to provide the level of ferromagnetism necessary for magnetic separators to detect and remove the lost parts so they cannot enter working machinery and cause damage.

Fig. 17 Thermal expansion of a 13% Mn steel

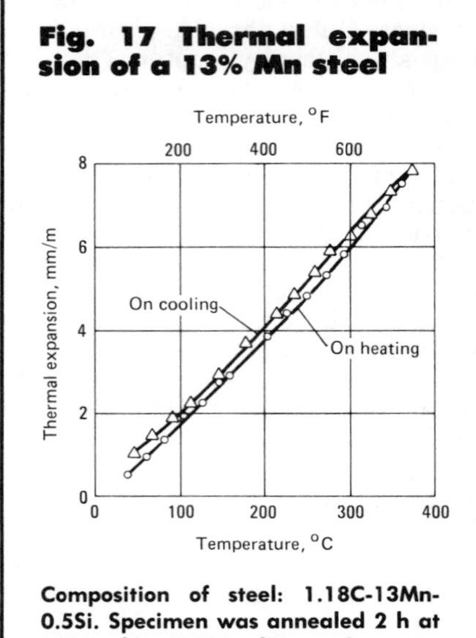

Composition of steel: 1.18C-13Mn-0.5Si. Specimen was annealed 2 h at 1100 °C (2000 °F) and water quenched.

Table 12 Compositions and tensile properties of 14Mn-Ni steels meeting MIL-S-17758(Ships)

Element or property	Alloy(a) Type 1	Type 2
Composition limits		
Carbon	0.70 to 0.90	0.70 to 0.90
Manganese	13 to 15	13 to 15
Silicon	0.50 to 1.00	0.50 to 1.00
Nickel	3.0 to 3.5	1.75 to 2.25
Chromium	0.50 max	0.50 max
Molybdenum	...	0.35 to 0.55
Phosphorus	0.07 max	0.07 max
Sulfur	0.05 max	0.05 max
Minimum tensile properties		
Tensile strength, MPa (ksi)	690 (100)	690 (100)
Yield strength(b) MPa (ksi)	345 (50)	345 (50)
Elongation, % in 50 mm or 2 in.	18	18
Elongation plus reduction in area, %	56	56

(a) These grades are intended primarily for nonmagnetic applications, and have maximum surface permeability of 1.2, measured with a "Go/No go" low-mu permeability indicator, as described in MIL-I-17214 and MIL-N-17387. The comparable casting specification, MIL-S-17249(Ships), includes only the standard 1.00 to 2.35% C grade. (b) 0.2% offset.

Alloy Modifications

The most common alloying elements are chromium, molybdenum and nickel, each of which contributes an additional austenite stabilizing effect. Added to the usual carbon level, about 1.15%, both chromium and molybdenum increase yield strength (Fig. 5) and flow resistance under impact; chromium is less expensive for a given increase, and chromium grades (ASTM A128, grade C, for instance) are probably the most common modifications. The common "B" grades often contain some chromium. The 2% Cr of the C grade does not significantly lower toughness in light sections; higher levels have an effect similar to that of raising carbon content. Chromium has been used up to 6% for some applications, sometimes in combination with copper, but these grades no longer receive much attention. Chromium enhances resistance to atmospheric corrosion. It is also used up to 18% in low-carbon electrodes for welding manganese steel.

Molybdenum and nickel will both tend to prevent the deterioration of properties of 13% Mn steel with reduced carbon contents (lowered to enhance ductility after slow cooling or reheating). Such lowered carbon in castings as well as weld deposits will reduce yield strength unless compensated; molybdenum can compensate, but nickel does not. The carbon ranges of grades "D" and "E-1" are wide, encompassing narrower ranges that are suitable for selected applications: low carbon for minimum carbide precipitation and high carbon for abrasion resistance. Thus, low carbon specifications are used for heavy sections, good as-cast properties, and welding electrodes.

The 1% Mo grades (ASTM A128, grade E-1 and AWS A5.13, grade EFeMn-B) are resistant to the reheating effect that limits the usefulness of the standard B-2, B-3 and B-4 grades. Grade E-1 is adapted to heavy casting with thick sections and to parts such as rolls and impact crushers that are frequently reheated during weld buildup and overlays.

The 2% Mo grade (ASTM A128, grade E-2) has a higher yield strength and is exploited for crusher service, usually with a duplex heat treatment. This entails partial grain refinement (U.S. Patent 1,975,746) by pearlitizing near 590 °C (1100 °F) for 12 h and then water quenching from 980 °C (1800 °F). This "dispersion treatment" results in many small carbides well distributed in the austenite. It is reported that this material gave a record of 370 to 470 thousand tons of highly siliceous ore crushed in 7-ft secondary cone crushers vs 230 to 390 thousand tons for the standard Hadfield's steel.

In terms of service life, the effect of carbon was shown as 653.6 h for 1.10 to 1.20%; 758.4 h for 1.25 to 1.35%; and 848.4 for 1.40 to 1.50%. Tensile properties of specimens removed from worn cone-crusher parts ranged from 440 to 485 MPa (64 to 70 ksi) yield strength, 695 to 850 MPa (100 to 125 ksi) tensile strength, and 15 to 25% elongation.

ASTM A128, type F, has reduced manganese (6.0 to 8.0%) to make the austenite less stable, but this requires compensation with 1% Mo to gain acceptable properties. Work hardening is expected to be more rapid than for the 13% Mn grades, at a sacrifice of toughness. It has found favor for use in scoop lips, ball-mill end liners, discharge grates, and grizzly screens in siliceous ore milling. One record indicates about 45% longer life as ball-mill discharge grates in comparison with the pearlitic Cr-Mo steel previously used. Average properties reported were 415 MPa (60 ksi) yield strength 585 MPa (85 ksi) tensile strength and 12% elongation at 193 HB and 1.2 to 1.4% carbon.

The F grade is not adapted to heavy sections or to service involving temperatures above 315 °C (600 °F). It has poor weldability and should be avoided if a casting is to be rebuilt or hardfaced.

The 3.5% nickel grade, ASTM A128, type D, has been applied to resist reheating embrittlement, for welding electrodes, for some trackwork, for nonmagnetic parts, and for naval applications (Table 12). Its stability when reheated is shown in Table 6.

Welding

Many of the common applications of austenitic manganese steel involve welding, either for fabrication or for repair. Consequently, it is important to understand that this material is unusually sensitive to the effects of reheating, often becoming embrittled to the point of losing its characteristic toughness. Oxyfuel-gas welding is so likely to produce embrittlement that it is not accepted as a practical method of welding this alloy. When properly done, electric-arc welding is the preferred method of joining manganese steels.

Arc Welding. Electrodes for arc welding austenitic manganese steel are commercially available in many compositions. They may be used for surfacing, for repair welds, and for joining manganese steel to itself or to carbon steels. Some of them are shown in Table 13. They have a lowered carbon content to minimize carbon precipitation as they cool from welding heat. Though formulated to avoid embrittlement of the deposited filler metal, proper welding procedures still must be used to avoid damage in the heat affected zone.

Electrodes of high manganese content, containing insignificant amounts of other alloying elements, are available also. Usually, these electrodes are recommended only for build-up of worn areas, because they are inherently lower in toughness than more highly alloyed grades. High-alloy, low-manganese electrodes generally have not been accepted as equivalent to high-manganese types.

Factors that are frequently overlooked are the losses in carbon, man-

Table 13 Compositions of electrodes for weld surfacing with austenitic manganese steel

Classification(a)	Composition, %(b)							
	C	Mn	Si	Ni	Cr	Mo	P	Total others
EFeMn-A:								
Bare	0.55-0.90	12.5-16.0	0.4-1.3	2.75-5.0	0.50	...	0.07	1.0
Covered	0.5-0.9	11.0-16.0	0.3-1.3	2.75 min	0.50	...	0.07	1.0
EFeMn-B:								
Bare	0.65-0.90	12.5-16.0	0.4-1.3	...	0.50	0.6-1.4	0.07	1.0
Covered	0.5-0.9	11.0-16.0	0.3-1.3	...	0.50	0.6-1.4	0.07	1.0
EFeMn-C:								
Bare	0.55-0.90	12.0-16.0	0.2-0.8	1.00 min	2.50-5.0	...	0.035	1.0
Covered	0.55-0.90	12.0-16.0	0.2-0.8	1.00 min	2.5-5.0	...	0.035	1.0

(a) AWS A5.13. (b) Single values are maximum values except where otherwise noted.

ganese and silicon that occur during welding. Although many electrode manufacturers compensate for these losses, improper welding techniques, such as use of excessive arc length and excessive puddling, may cause additional losses. The result is inferior properties in the weld deposit.

Carbon steels frequently are welded to high-manganese steel using austenitic stainless steel electrodes. Because the deposit tends to be a mixture or hybrid of the base and the filler metal, it can have quite different properties. Often it is air-hardening, producing a martensitic zone as the weld cools. The ductility of martensite is low, but the strength is high and weldments are often satisfactory; the chief adverse factor may be cracks in the martensite. Cross-weld tensile properties of low-carbon 14Mn-1Mo steel plate welded to 1045 steel with EFeMn-A electrodes were 435 MPa (63 ksi) yield strength, 650 MPa (94 ksi) ultimate tensile strength, and 11% elongation with fracture in the 1045 steel. These properties are superior to those of many weldments of carbon steel.

Filler metal from 0.75% carbon, 15% Mn, 3.5% Ni and 4.0% Cr seems to be superior to that of EFeMn-A and EFeMn-B.

The grades with chromium above 14% are also useful for joining manganese steel and buildup of worn parts but because of their low carbon content have sacrificed abrasion resistance. However, they are more machinable than the higher carbon grades. If used to restore dimensions of crusher parts they should be overlaid with an effective hard-facing alloy.

Precautions. The primary consideration in welding austenitic manganese steel is minimum heating of the parent metal to avoid embrittling transformations or carbide precipitation. This precludes preheating. Under the most favorable circumstances, some precipitation is expected, and the resulting heat-affected zones seldom attain the toughness of normal parent metal. Because manganese steel work hardens in service, it may be assumed that any worn area requiring repair or rebuilding will have a work-hardened surface. This surface must be removed before welding in order to prevent cracking in the heat-affected zone.

The low heat conductivity and high thermal expansion of manganese steel also cause difficulties, combining to produce steep thermal gradients and high residual stresses. Weld beads are subjected to tension as they cool; to minimize cracking, it is desirable to peen them while they are hot, producing plastic flow and changing the stress to compression. This hammering should be done promptly after 150 to 225 mm (6 to 9 in.) of weld bead has been deposited.

Machining

Manganese steels are so tough, and work harden at the point of a cutting tool to such an extent, that frequently they are considered commercially unmachinable. However, these steels are regularly cut by adhering to generally accepted procedures. In addition, a new, more highly machinable grade of manganese steel has been developed and may be helpful in appropriate applications.

Procedures. Although details of practice and tool design differ, there is general agreement on the following procedures for machining manganese steels:

- Machine tools should be rigid and in good condition. Any factors that encourage chatter are undesirable.
- Tools should be sharp. Dull tools cause excessive work hardening of the cut surface and accentuate the difficulty in machining.
- Low speeds of about 9 to 12 m/min (30 to 40 ft/min) should be used. High speeds are likely to create red hot chips and to cause rapid tool breakdown.
- Both cobalt high speed steel and cemented carbide tools can be used. The latter are preferred.
- Liberal use of a good grade of sulfur-bearing cutting oil is beneficial.
- In castings, holes should be formed by cores in the foundry, rather than by machining, whenever possible.

Various sources provide statements in favor of both positive-rake and negative-rake tools and both dry cutting and use of liquid coolants. Because high temperatures at the cutting edge are a large part of the problem, effective cooling seems desirable. Negative-rake tools are likely to require more force and thus to produce more heat. However, the thinner edge of a positive-rake tool is more vulnerable to heat. Comparative machining data are presented in Table 14.

Machinability is increased by the embrittlement that develops on reheating between about 550 and 650 °C (1000 and 1200 °F). Although not usually practicable, such a treatment may be useful if the part can subsequently be properly toughened. Milling generally is not considered practicable.

New Machinable Grade. A 20Mn-0.6C steel was developed specifically for improved machinability. Table 15 gives the mechanical properties of this material. To obtain improved machinability, the yield strength was deliberately reduced from 360 MPa (52 ksi) to a value between 240 and 310 MPa (35 and 45 ksi), and yet the ultimate tensile strength exceeds 620 MPa (90 ksi) and elongation in small castings may reach 40%. Heat treatment of this steel involves water quenching from 1040 °C (1900 °F). As-cast properties are lower

Table 14 Feed forces required in lathe turning of austenitic manganese steels(a)

	Negative 7° rake				Flat tool				Positive 6° rake			
	Horizontal		Vertical		Horizontal		Vertical		Horizontal		Vertical	
Type of steel	kg	lb	kg	lb	kg	lb	kg	lb	kg	lb	kg	lb
1.12C-13Mn steel	68	150	161	355	79	175	161	355	116	255	170	375
3Ni-13Mn steel	63	140	156	345	...	...	...	...	127	280	172	380
1Mo-13Mn steel	82	180	168	370	...	...	...	...	111	245	175	385
1.12C-13Mn leaded steel(b)	59	130	152	335	50	110	141	310	127	280	172	380
Wrought type 304 stainless steel (c)	70	155	118	260	91	200	127	280	Welded to tool			
Cast CF-8 stainless steel(c)	68	150	116	255	...	...	...	...	229	505	261	575

(a) Specimens were 32-mm-diam (1.25-in.-diam) bars, toughened by water quenching. Roughing cuts 2.5 mm deep were taken using complex-carbide tools containing about 15% TiC + TaC (predominantly TiC) and about 7 to 10% Co. New cutting edges were used for each positive or negative rake. Cutting speed was 0.19 to 0.20 m/s (37 to 39 ft/min). (b) Recovery of 0.02% Pb from 0.35% added in ladle. The effect on machinability is inconclusive. (c) Stainless steels suffer in the comparison at this speed. They are more machinable at higher speeds and permit certain operations, such as drilling of 6.4-mm-diam (1/4-in.-diam) holes, that are very difficult with austenitic manganese steel. The type 304 stainless steel was cold finished.

Table 15 Typical room-temperature properties of machinable Mn steel

Type	Treatment	Tensile strength MPa	Tensile strength ksi	Yield strength MPa	Yield strength ksi	Elongation, %	Reduction in area, %	Hardness, HB	Magnetic permeability
Standard 13% Mn	Toughened	825	120	360	52	40	35	200	1.01
Machinable grade A, 20 Mn-0.6 C	Toughened	640 to 855	93 to 124	275 to 310	40 to 45	39 to 65	26 to 44	159 to 170	1.003
	As cast	380 to 580	55 to 84	275 to 305	40 to 44	13 to 22	24	159 to 170	1.003

Table 16 Force requirements for single-point lathe turning of austenitic manganese steel

		Feed force(a)				
		Horizontal		Vertical		Friction
Type	Condition	kg	lb	kg	lb	coefficient
Standard 13% Mn	As cast	54 to 68	120 to 150	125	275	0.64
	Toughened	70	155	135	295	0.76
Machinable grade A (20% Mn)	As cast	16 to 29	35 to 65	91 to 100	200 to 220	0.31 to 0.48
	Toughened	18 to 39	40 to 85	98 to 102	215 to 225	0.33 to 0.57

(a) Depth of cut, 3 mm (0.1 in.) on radius; feed, 0.16 mm/rev (0.0062 in./rev); turning speed, 1.35 m/s (265 ft/min); 6° positive-rake tool.

but are probably adequate for many applications.

This nonmagnetic modified grade can be lathe turned, drilled, tapped and threaded; even holes 6.4 mm (1/4 in.) in diameter can be drilled and tapped in this metal. In some machine shops, it is rated only slightly more difficult to drill than plain 1020 steel, and the quality of the tapped threads is considered very good. Typical machining data for this steel are presented in Table 16. Wear resistance has been sacrificed for machinability and this grade has significantly less abrasion resistance than the various types in ASTM A128.

REFERENCES

1. The Equilibrium Diagram of Iron-Manganese-Carbon Alloys of Commercial Purity, by E. C. Bain, E. S. Davenport and W. S. N. Waring: *Transactions of AIME,* Vol 100, 1932, p 228
2. "Work Hardening and Martensite Formation in Austenitic Manganese Alloys", by C. H. Shih, B. L. Averbach and M. Cohen: Massachusetts Institute of Technology research report, 1953
3. Static and Dynamic Tension Tests on Austenitic Manganese Steel, by H. Krainer: *Archiv Eisenhuttenwesen,* Vol 11, 1937, p 279
4. *Manganese Steel,* by Hadfields Ltd.: Oliver and Boyd, London, 1956
5. Austenitic Manganese Steel Welding Electrodes, by H. S. Avery and H. J. Chapin: *The Welding Journal,* Vol 33, 1954, p 459
6. Gouging Abrasion Test for Materials Used in Ore and Rock Crushing, Part II, by F. Borik and W. G. Scholz: *Journal of Materials,* Vol 6, No. 3, Sept 1971, p 590
7. Some Decomposition Structures of Austenitic Manganese Steels, by R. Castro and P. Garnier: *Revue de Metallurgie,* Vol 55, Jan 1958, p 17

8. Austenitic Manganese Steel, by H. S. Avery: American Brake Shoe Company, 1949

9. "The Physical Properties of a Series of Steels—Part II", Special Report No. 23: from the Alloy Steels Research committee of the British Iron and Steel Institute, Sept 1946

Nonferrous Alloys for Wear Applications

By T. C. Du Mond
Consultant
and
P. A. Tully
Manager, Research and Development
Ampco Metal Division
Ampco-Pittsburgh Corporation
and
Keith Wikle
Manager, Technical Services
Kawecki Berylco Industries, Inc.

NONFERROUS ALLOYS most widely used for wear-resistant service are cobalt-base materials. Only wrought cobalt-base alloys are discussed in this article; hard facing alloys are the subject of a separate article in this volume. Cobalt-base materials are not the only class of nonferrous wear-resistant materials. For some types of applications, copper-beryllium alloys and certain aluminum bronzes are frequently specified—primarily in cast form and primarily for bearing applications.

Cobalt-Base Alloys

Wrought cobalt-base wear alloys are nearly identical in chemical composition with their hard facing alloy counterparts but with subtle differences in boron, silicon or manganese levels. Equally subtle differences in mill practice are used in producing the two classes of wear alloys. The significant difference between hard facing alloys and other cobalt-base wear alloys is microstructure, which varies with casting, sintering and rolling practices.

Wrought cobalt-base wear resistant alloys are produced to standard production specifications designed to develop consistent microstructures and properties. The greatest quantity of these alloys is processed on typical hot and cold rolling equipment. However, there is a strong trend toward use of cobalt-base wear resistant alloys in powder form. Powders are used either to produce powder-metallurgy parts or to be applied as coatings by various commonly accepted methods.

Most wrought cobalt-base wear resistant materials are proprietary and are not included under technical society, national standard or government specifications. Compositions of the five principal cobalt-base wear alloys are given in Table 1.

The primary alloys for severe wear applications are Stellite 6B, Stellite 6K and Haynes 25. These alloys have excellent resistance to most types of wear, as indicated in Table 2. The wear resistance is inherent and not the result of special heat treatment or of methods used to produce high surface hardness.

Cobalt-base wear alloys possess a number of other properties that significantly widen their range of application:

- Good resistance to impact and thermal shock
- Good resistance to heat and oxida-

Table 1 Compositions of principal cobalt-base wear alloys

Alloy	C	Cr	Ni	Mo	W	Fe	Mn	Si	Co
	Chemical composition, %(a)								
Stellite 6B	0.9-1.4	28-32	3.0	1.5	3.5-5.5	3.0	2.0	2.0	Rem
Stellite 6K	1.4-2.2	28-32	3.0	1.5	3.5-5.5	3.0	2.0	2.0	Rem
Haynes 25 (L-605)	0.05-0.10	19-21	9-11	...	14-16	3.0	1.0-2.0	1.0	Rem(b)
Tribaloy T-400(c)	0.06	8.5	(d)	28.5	...	(d)	...	2.6	Rem
Tribaloy T-800(c)	0.06	17.5	(d)	28.5	...	(d)	...	3.4	Rem

(a) Where a single value is shown instead of a range (except for the Tribaloy alloys), the value is a maximum limit. (b) Also contains 0.03 max P and 0.03 max S. (c) Nominal composition. (d) Ni + Fe content is 3.0 max.

Table 2 Typical wear data for selected alloys

Alloy	Condition	Volume loss, mm^3	Average wear coefficient(a)	Coefficient of friction(b)
Abrasive wear at room temperature(c)				
Stellite 6B	Mill annealed (38 HRC)	8.2	1.6×10^{-4}	...
Stellite 6K	Mill annealed (46 HRC)	13.3	3.2×10^{-4}	...
Haynes 25	Mill annealed (24 HRC)	53.0	6.7×10^{-4}	...
1090 steel	Hardened and tempered (55 HRC)(d)	37.2	2.7×10^{-3}	...
Type 316 stainless steel	As received (86 HRB)	81.4	6.7×10^{-4}	...
Type 304 stainless steel	As received (92 HRB)	102.1	1.0×10^{-3}	...
Adhesive wear at room temperature(e)				
Stellite 6B	Mill annealed (38 HRC)	0.293	1.2×10^{-5}	...
	Cold reduced 10% (44 HRC)	0.347	1.7×10^{-5}	...
Stellite 6K	Mill annealed (46 HRC)	0.561	2.4×10^{-5}	...
Haynes 25	Mill annealed (24 HRC)	0.285	0.8×10^{-5}	...
1090 steel	Hardened and tempered (55 HRC)(d)	0.293	2.0×10^{-5}	...
Adhesive wear at 540 °C (1000 °F)				
Tribaloy T-400		0.12(f)	4.0×10^{-5}(f)	0.25(f)
		0.07(g)	1.0×10^{-6}(g)	0.37(g)
		0.11(h)	1.0×10^{-6}(h)	0.44(h)
Tribaloy T-800		0.37(f)	3.0×10^{-5}(f)	0.19(f)
		0.03(g)	1.0×10^{-5}(g)	0.42(g)
		2.1(h)	1.0×10^{-5}(h)	0.62(h)
Stellite 6		0.07(f)	1.0×10^{-6}(f)	0.24(f)
		1.7(g)	1.7×10^{-5}(g)	0.62(g)
		2.3(h)	2.3×10^{-5}(h)	0.66(h)

(a) Wear coefficient, K, calculated from $K = Vh/PL$, where V is volume loss in mm^3, h is diamond pyramid (Vickers) hardness, P is test load in kg and L is sliding distance in mm. For elevated-temperature wear tests, hot hardness at the wear test temperature was used to calculate the wear coefficient. (b) Average final friction measurements on wear test specimens. (c) Rubber wheel test using dry sand abrasive. (d) Austenitized 1 h at 870 °C (1600 °F), water quenched, and tempered 4 min at 480 °C (900 °F). (e) Average of two or more tests of an alloy block rubbing against the edge of a rotating ring made of 4620 steel, case hardened to a surface hardness of 63 HRC. (f) 45-kg (100-lb) test load. (g) 90-kg (200-lb) test load. (h) 136-kg (300-lb) test load.

tion: high temperatures have little effect on hardness, toughness and dimensional stability. The alloys are highly resistant to atmospheric oxidation at ordinary temperatures, and have good resistance to oxidation at elevated temperatures.

- Good corrosion resistance: the alloys resist attack by a variety of corrosive media. The combination of wear resistance and corrosion resistance makes Stellite 6B, for example, particularly useful in applications as food-handling machinery and chemical-process equipment.
- High hot-hardness: wrought cobalt-base alloys retain high hardness even at red heat. Once cooled back to room temperature they recover full original hardness.

Stellite 6B is the most widely used cobalt-base wear alloy. Half-bushings and half-sleeves made of Stellite 6B can be used where abrasive particles such as fly ash, coke, metal powders, shale or cement dust tend to collect on bearing surfaces. The ability of Stellite 6B parts to withstand the abrasive effects of hard, sharp particles makes them especially useful in screw conveyors, rock crushers, rollers, tile-making machines, and cement-plant and steel-mill equipment. Figure 1 illustrates the use of Stellite 6B erosion shields on the leading edges of critical aircraft structures such as helicopter rotors.

In wear applications, Stellite 6B resists seizing and galling, and also resists erosive wear. Its low coefficient of friction allows sliding contact with other metals without damage through metal pickup. Stellite 6B can be used in applications where no lubricants can be tolerated or where lubrication cannot be maintained. Even under constant erosive conditions, Stellite 6B exhibits long service life. Its outstanding resistance to cavitation damage makes the alloy well suited to such applications as erosion shields in steam turbines. In one turbine installation, Stellite 6B shields have blades for more than 20 years of continuous service.

Stellite 6K contains more carbon than Stellite 6B and is intended for applications where the part must retain a keen edge for comparatively long periods of time. Other properties of Stellite 6K are similar to properties of Stellite 6B, including good resistance to high temperatures and corrosion. Applications of Stellite 6K range from blades for custom-made hunting knives to dish-trimming knives used in the manufacture of ceramic ware.

Haynes 25 is essentially a high-strength, high-temperature alloy but can be considered for wear applications in situations where high heat and severe corrosive conditions exist. Its chemical composition differs from those of Stellite 6B and Stellite 6K primarily in a substantially lower chromium con-

Fig. 1 Exploded view of section through a helicopter rotor having an erosion shield of Stellite 6B

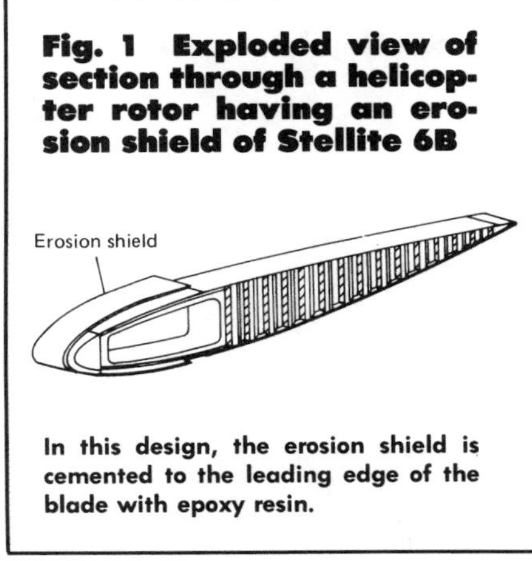

In this design, the erosion shield is cemented to the leading edge of the blade with epoxy resin.

Table 3 Amsler wear test results for six aluminum bronzes, one nickel tin bronze and one aluminum brass

Alloy	Nominal composition	Typical hardness, HB	Wear rate(a) for load stress of: 23 ksi	27.5 ksi	Dynamic coefficient of friction for load stress of: 23 ksi	27.5 ksi
Lubricated with Mobil DTE-25						
AMS 4881	Cu-11Al-4.7Fe-5Ni	277	0.40	0.90	0.120	0.140
C61300	Cu-7Al-3.5Fe-0.3Sn	183	0.50	1.60	0.110	0.135
C62400	Cu-10.7Al-3.3Fe	200	1.10	1.30	0.120	0.140
C62500	Cu-13Al-4.2Fe	302	0.20	0.40	0.100	0.120
C86300	Cu-6.3Al-3Fe-25Zn-3.8Mn	223	0.85	0.90	0.125	0.140
C91700	Cu-12Sn-1.6Ni	106	2.50	4.30	0.130	0.140
C95400	Cu-10.7Al-4Fe	180	0.40	0.70	0.140	0.112
C95500	Cu-10.7Al-4Fe-3.7Ni-1Mn	228	0.25	0.60	0.105	0.140
Unlubricated						
AMS 4881		277	1.66	2.40	0.32	0.22
C61300		183	2.48	3.82	0.52	0.42
C62400		200	1.32	2.96	0.27	0.22
C62500		302	1.42	3.18	0.24	0.22
C86300		223	5.96	5.73	0.38	0.22
C91700		106	Erratic	5.18	0.88	0.69
C95400		180	1.76	4.85	0.28	0.22
C95500		228	1.20	2.20	0.42	0.26

(a) Wear rate, based on weight loss in mg for each 1000 revolutions of the test specimen, for a bronze disk 38 mm (1.5 in.) in diameter by 9.5 mm (0.375 in.) thick rotating against a similar specimen of 4340 steel hardened and tempered to 43 HRC under test conditions involving 110% slip and 4 mm (0.16 in.) axial slide. Wear surfaces were 0.6 μm (25 μin.) rms surface finish, or better.

tent, higher nickel and tungsten contents considerably lower carbon.

Tribaloy Alloys. Two cobalt-base intermetallic compounds, used primarily as powder additions to other compositions, also are available in compacted form at densities approaching theoretical values. These Laves-phase compounds, known as Tribaloy alloys, are increasingly being adopted for wear applications. Tribaloy T-400 combines excellent mechanical wear resistance with good corrosion resistance. Tribaloy T-800 has a higher chromium content than T-400, which gives Tribaloy T-800 greater resistance to oxidation and corrosion than the leaner alloy.

Tribaloy intermetallic materials maintain their surface properties over a wide temperature range, and are particularly well suited for applications where lubrication is a problem. Tribaloy alloys are used in such products as piston rings, bushings, cams, seals, pump-compressor parts, thrust washers, valves and vanes. Typical high-temperature adhesive wear data may be found in Table 2.

Powder Alloys. Many of the alloys available for hard facing also are produced as powders. Some of the powders are used for spray-coating processes for protecting metal surfaces, but an increasingly high percentage of powders are pressed and sintered approximately to final shape or into dense solid forms that then can be machined to desired shape.

Of all of the hard facing alloys capable of being formed into usable parts, Stellite 3PM and equivalents are best suited to the production of wear resistant parts by the powder metallurgy method. Stellite 3PM is a cobalt-chromium-tungsten P/M alloy that attains a transverse rupture strength of 965 MPa (140 ksi) at 97% of theoretical density. In one set of adhesive wear tests, Stellite 3PM gave the best results of eight potential P/M wear alloys. Tests showed only 0.10 cm^3 volume loss under the following conditions: 2000 revolutions at 80 rev/min for a test specimen of Stellite 3PM held with a force of 13.6 kg (30 lb) against 4620 steel case hardened to a surface hardness of 62 HRC.

Highly alloyed powders are difficult to compact into powder metallurgy shapes because they develop oxide coatings. Special techniques have been developed that permit pressing and sintering to a density approaching 100% of theoretical.

Among products made from powders are hard facing welding rod; ball and race blanks for ball bearings; aerospace bushings, spacers, and counterweights; automotive valve-seat inserts; and wear parts such as blade inserts for surgical scissors.

Although Tribaloy powders can be converted into final shapes through conventional powder metallurgy techniques or hot isostatic pressing, the powders frequently are used to enhance wear resistant properties of other metals. Up to 25% of all Tribaloy powder produced is blended with powders of bronze, stainless steel, aluminum, iron and low alloy steel to permit the use of P/M products of these alloy types under wear conditions that would destroy the unenhanced materials.

Aluminum Bronzes

The aluminum bronze family of alloys ranges from soft, ductile alpha alloys such as C60800, C61300, and C61400 to proprietary die alloys that are very hard and brittle. The hardness of aluminum bronzes in various forms ranges from 30 HRB to 40 HRC (67 to 375 HB), and there is a similarly large range of tensile and compressive strengths. Proper selection of an alloy from this family is the key to successful wear service, and is based on many complex factors.

As a group, the aluminum bronzes are not considered self-lubricating, and are therefore used where an adequate reliable source of lubrication can be maintained. The alloys are recommended for applications involving high loads and moderate to low speeds. Natural surface oxidation during service forms a film that helps provide galling and seizing resistance in applications where the lubricating film might become temporarily marginal. General corrosion resistance is good, and the high strength and hardness result in relatively low conformability and embeddability. Abrasion resistance, particularly in the case of the hard die alloys, generally is superior to that of other copper alloys.

Some typical wear applications for aluminum bronzes are:

Alloy	Typical hardness	Applications
C61300, C61400	30-82 HRB	Wear plates, gibs, press guides
C62400, C95400	76-98 HRB	Gears, bushings, wear plates, mill slippers, rollers
C63000	90-100 HRB	Valve guides, bushings, rollers
C62500	27-32 HRC	Rollers, cams, bending dies
Die alloys	32-40 HRC	Automotive die inserts, dies for drawing stainless steel sinks, tube-bending dies

Fig. 2 Wear of four lubricated copper alloy bearings

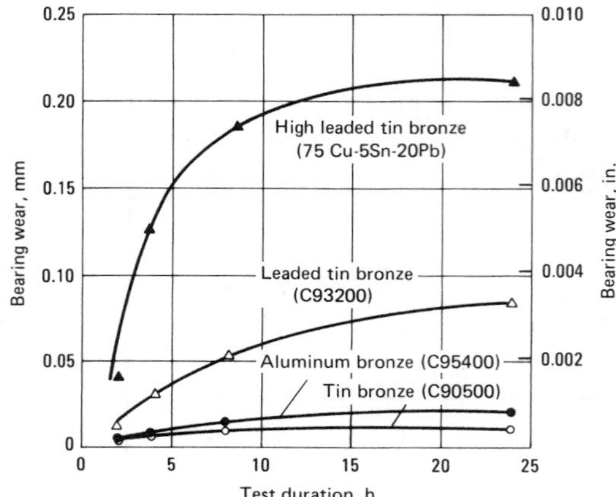

Bearing hardnesses: C95400, 91 HRB, C90500, 67 HRB; C93200, 52 HRB and 75-5-20, 5 HRB. Shaft material: 1045 steel at 30 HRC min, ground and polished. Lubricant: Sun Oil Co. Prestige 741 EP. Test conditions: bearing load, 6.9 MPa (1000 psi); bearing speed, 0.1 m/s (20 ft/min).

Wear Tests. The wear resistance and effective coefficient of friction for aluminum bronzes, as is true for most alloys, depend so much on specific conditions that generalizations can be misleading. One type of laboratory evaluation of wear involves the use of Amsler wear tests. In Amsler tests, thick disks of materials rotate against each other in a variety of modes, including counterrotating slip, corotating slip, or one specimen rotating against a stationary specimen. A combination of rotational and axial slide may also be used. Tests are run dry or lubricated, and abrasive or corrosive substances may also be introduced. Wear rates are measured by weight loss and/or dimensional changes in the specimens. Careful visual examination of the wear surfaces for fretting, galling, pitting or scoring completes the evaluation. Friction coefficients are calculated from driving torque and load during the test runs, generally after an initial "wear-in" period during which torque values become stable.

Table 3 summarizes results of some Amsler wear tests on six aluminum bronzes and two other copper alloys. A complete laboratory evaluation of relative wear characteristics might require a much larger variety of testing modes and materials in order to be of real value. Figure 2 compares wear characteristics of aluminum bronze C95400 with those of three other cast copper-base bearing alloys.

Beryllium Copper Alloys

Prior to the mid-1960's, few beryllium copper components were designed into equipment for service under conditions of sliding and rolling metal-to-metal contact. However, since that time, beryllium copper components have been produced for submarine telephone equipment, jet aircraft landing gear, wind tunnel apparatus and molds for producing plastic parts. There also has been tremendous growth in available forms of alloys for wear applications. They can now be cast and forged to shape and fabricated from tube, rod, bar or sheet.

During the late 1960's, beryllium copper alloys were used mostly to replace bronzes and low-alloy steels in applications where those alloys gave marginal service. Beryllium coppers were seldom specified in original bearing or bushing design until the 1970's. Beryllium coppers are among the hardest and strongest of all copper-base metals. Properly lubricated beryllium copper surfaces are more wear resistant than those of other copper-base alloys and many ferrous alloys. Wear rates for one alloy sleeve bearing are compared with those of aluminum-nickel bronze in Fig. 3.

In addition, beryllium coppers exhibit excellent corrosion resistance in industrial and marine environments, have high electrical conductivity, are nonmagnetic and nonsparking, and resist anelastic behavior.

Specific Applications. Beryllium copper wear parts are almost always used with a lubricant, and preferably in a moist environment. These conditions result in an extremely low wear rate as shown in Fig. 4. In most applications, a beryllium copper component is mated with a steel component. In typical practice, the less critical part is beryllium copper. Because it is softer than a hardened steel part, the beryllium copper part wears faster and is replaced more often.

Higher cost (two to three times the cost of other copper-base alloys) is often offset by a low wear rate. For example, in aircraft landing gear, beryllium copper bushings normally last two to three times longer than nickel bronze bushings. This is important in scheduling the aircraft to operate for longer periods—before a costly overhaul stop is required for replacement of bushings in the nose landing gear and main landing gear.

Beryllium coppers also are being used by a number of heavy equipment

Fig. 3 Comparative wear data at three bearing loads for copper alloys C17200 and C63000

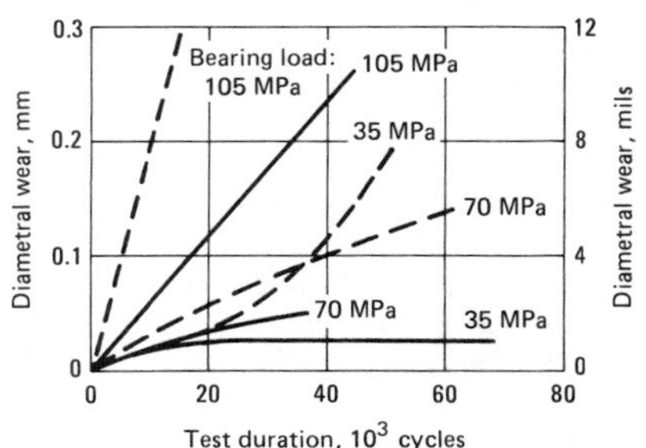

Solid lines are for C17200, beryllium copper; broken lines are for C63000, aluminum nickel bronze. Shaft material: chromium-plated 4340 steel. Lubrication: AF17 (MIL-G-21164), applied at start of test only for 35 MPa test, and applied every 100 cycles in 70 and 105 MPa tests. Test conditions: bearing loads, 35, 70 and 105 MPa (5000, 10 000 and 15 000 psi); oscillation amplitude, ±45°; oscillation frequency, 2 Hz; velocity, 0.16 m/s (31.4 ft/min). Direction of loading reversed every 100 cycles of oscillation.

Fig. 4 Wear data for two beryllium coppers showing influence of moisture on wear rate

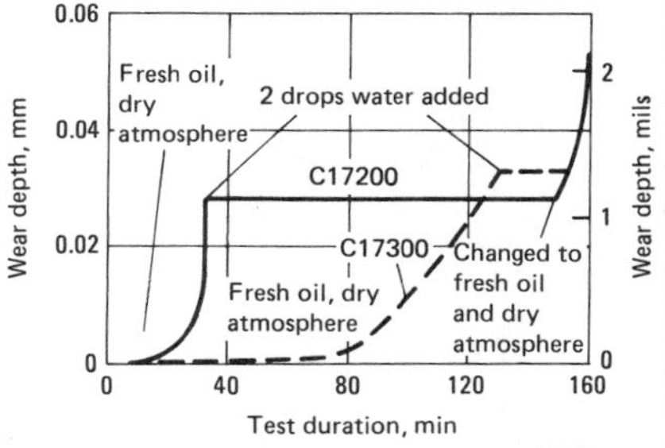

Results of Sinclair wear test in which copper test pieces were run against a 52100 steel disk hardened to 60 HRC. Bearing load: 415 MPa (60 ksi). Lubricant: highly refined white mineral oil. Test velocity: 0.08 m/s (16 ft/min). Wear proceeds rapidly in an artificially dried atmosphere, but slows to almost nil when two drops of water are added.

Fig. 5 Wear of three beryllium copper alloys illustrating the effect of CoBe in a hard matrix

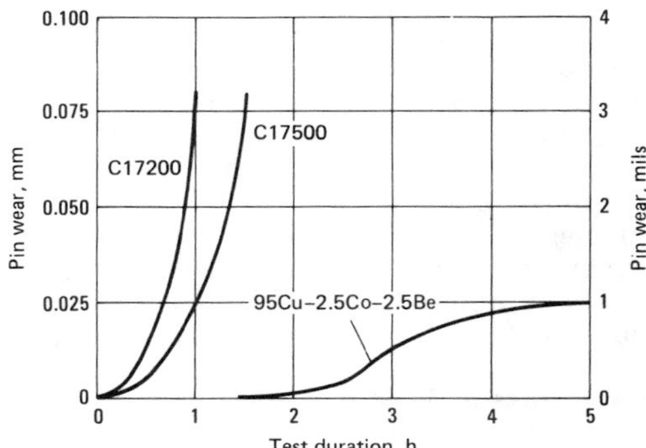

Pin-on-rotating-disk tests comparing C17200 at 40 HRC, C17500 at 98 HRB and 95Cu-2.5Co-2.5Be at 41 HRC. Test conditions: dry lubricant MIL-G-81322; bearing load, 415 MPa (60 ksi); velocity, 0.08 m/s (15 ft/min).

manufacturers. A manufacturer of coal mining machinery uses a beryllium copper rod 55 mm (2 ⅛ in.) in diameter as a pinion gear axle in a planetary gear. The gear free-wheels on the beryllium copper axle with no need for a bushing.

In the steel industry, a threaded nut made of beryllium copper is being used with a large diameter threaded steel screw that extrudes a 150-mm (6-in.) diameter rod of clay to plug the tap hole of blast furnaces. The beryllium copper nut replaced an aluminum bronze nut that lasted an average of only six months in this abrasive environment. Over 20 of the beryllium copper nuts have been placed in service. None has failed, and the oldest has been in service in the clay gun for 15 months.

A manganese bronze air compressor valve body had been giving about 1000 h of service. The valve body and a steel disk that shut off flow through the valve came in contact with a hammering action, up to 1500 times per minute. A valve body made of C17200 has not shown any signs of wear after 4000 h of service.

In a sand-casting foundry, C17200 replaced carbon steel in the wear plates of core boxes and was found to last 20 times longer.

C17200 (98Cu-1.9Be-0.2Co), which has a tensile strength of 470 to 1460 MPa (68 to 212 ksi) is usually specified for wear applications. When well lubricated, no other copper-base alloy has equal load carrying capacity as a bushing material. It resists vibration-induced fretting as well as wear in the presence of molten aluminum, zinc, cast iron, and non-chloride plastics. It is more resistant to sand blasting than low-alloy steels.

Sleeve bearing tests simulating airplane landing gear operation have shown that C17200 can sustain operating bearing loads up to about 700 MPa (100 ksi) when lubricated. Customarily, bearing stresses up to 415 MPa (60 ksi) are normal on beryllium copper aircraft bushings; overloads can be as high as 655 MPa (90 ksi). Even at such high stresses, no seizing or galling occurs as long as there is adequate lubrication, and wear rate is minimal for service lives exceeding 60 000 oscillatory cycles.

To get the most favorable wear rate from C17200 bushings, a complex BeO-containing oxide surface film must be developed and maintained. The thin, tenacious film probably is formed as a result of chemical reaction between the beryllium copper bearing and the lubricating grease under the conditions prevailing at the bearing interface. Breakthrough of the film in the absence of grease leads to galling and seizure.

Because well-designed lubrication systems are costly, design engineers usually seek applications where good lubrication is relatively easy to obtain. Typical situations that provide natural conditions for good lubrication are hammering action, slow speed reverse oscillation, and sliding motion between mating surfaces.

A linear actuator (jackscrew) for an aircraft hatch positioner, operating at bearing stresses over 205 MPa (30 ksi) at a low wear rate as well as at a high mechanical efficiency (high work output versus work input), is a good illustration of the optimum choice of design elements for a beryllium copper/steel wear couple. In this application, a C17200 beryllium copper nut (40 HRC) works against a nitrided AISI 4340 jackscrew (case hardness, 56 HRC) with a 29-mm (1-1/8-in.) ACME 32-64 thread. A MIL-G-27617 high-temperature, Teflon-filled grease is used to obtain a coefficient of friction of 0.01 to 0.03. This combination provides high performance, low maintenance and long service life.

Techniques for Enhancing Wear Resistance. Self-lubrication of components has been accomplished by placing graphite in the beryllium copper surface. Wear properties can be augmented by casting beryllium coppers rather than machining components from wrought alloys. Wear properties also can be increased by orders of magnitude simply by oxidizing the surface of the alloy.

Structures with greater wear resistance than normal wrought C17200 have been developed by plasma depositing or detonation depositing atomized C17200 powder on suitable substrates. There is a thin BeO film on the surface of each spherical particle before plasma deposition. Thus, after deposition, a network of BeO films exists throughout the coating structure. Wear resistance superior to that of conventional cast and wrought C17200 also can be developed by hot compaction extrusion of atomized powders. Here again, the components contain a network of BeO films. These experiments all tend to show that a complex surface-oxide film containing BeO, or a substrate BeO film network, aid formation of a wear resistant film in the presence of a suitable lubricant.

Wear resistance can be further augmented by incorporating a dispersion of a hard phase of cobalt beryllide (CoBe) intermetallic particles in the 99.5Cu-1.9Be matrix. Figure 5 shows that a liberal amount of CoBe phase improves the wear properties of standard beryllium copper alloys.

Materials for Special Applications

Magnetically Soft Materials

By the ASM Committee on Magnetically Soft Materials*

MAGNETICALLY SOFT MATERIALS are ferromagnetic materials that have little or no retentivity—that is, if they are magnetized in a magnetic field and then are removed from that field, they lose most, if not all, of the magnetism they exhibited while in the field.

The most important characteristics of magnetically soft materials are: (*a*) low hysteresis loss (easy domain movement during magnetization); (*b*) low eddy-current loss from electric currents induced by flux changes; (*c*) high magnetic permeability, and sometimes constant permeability at low field strengths; (*d*) high magnetic saturation induction; and (*e*) minimum or definite change in permeability with temperature in special applications. Cost, availability and ease of processing are other factors that influence final choice of material.

Magnetically soft materials made in large quantities include high-purity iron, low-carbon steels, silicon steels, iron-nickel alloys, iron-cobalt alloys, and ferrites.

Impurities. Ferromagnetic properties depend on crystal structure and composition. Carbon, sulfur, nitrogen and oxygen are especially deleterious to ferromagnetic properties because they distort the lattice of the crystal structure and even in small amounts may greatly interfere with easy movement of magnetic domains, which is the basis of such properties.

A similar disturbance caused by carbon and nitrogen, known as "aging", occurs when low solubility at room temperature causes the excess solute to precipitate slowly as small particles within grains. In irons and iron-silicon alloys, carbon content preferably should be less than 0.003% for best permeability, low hysteresis loss and minimal aging. Figure 1 shows the relationship between carbon content and hysteresis loss for iron. Hysteresis loss is similarly related to sulfur and oxygen contents.

Grain Size. For most applications, grain size should be as large as possible for nonoriented materials (Fig. 2a). In oriented grades of silicon steel (see Fig. 2b), optimum magnetic properties are usually obtained with grain sizes of 2 to 10 mm depending on the degree of crystal orientation: increases in grain size above 10 mm are accompanied by significant increases in both domain-wall spacing and eddy-current losses.

Grain Orientation. All ferromagnetic crystals are magnetically anisotropic—that is, they have different magnetic properties in different crystallographic directions. In nickel, the direction of easiest magnetization is the cube diagonal <111>; in iron it is

Fig. 1 Relationship between carbon content and hysteresis loss for unalloyed iron

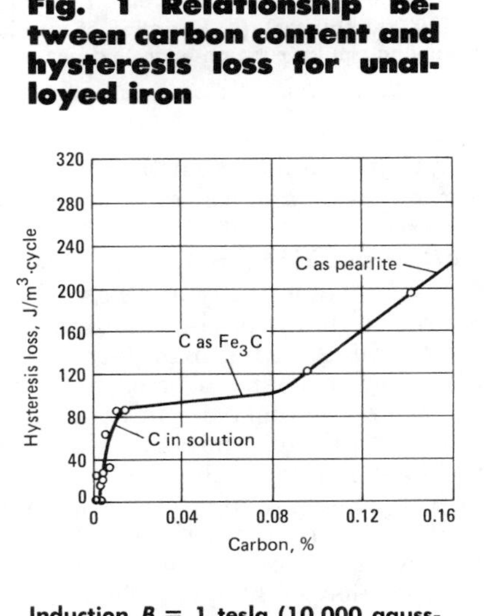

Induction B = 1 tesla (10 000 gausses).

*See page XII for committee list.

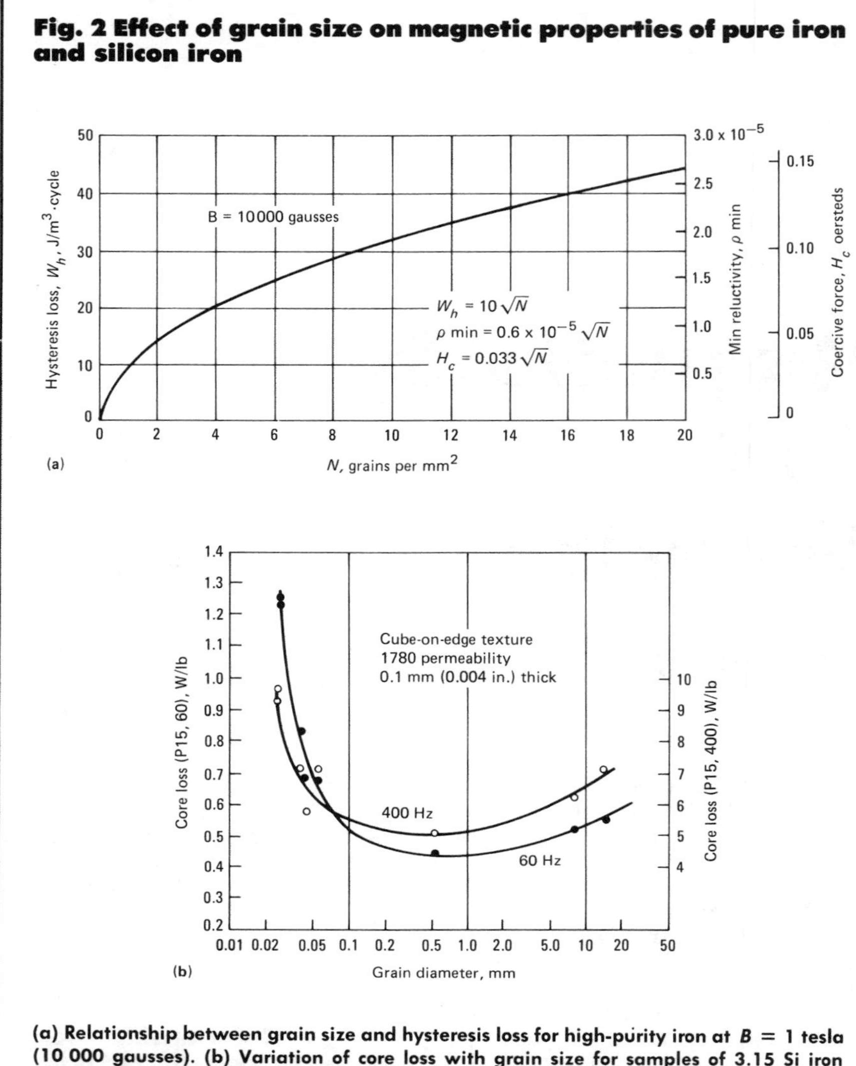

Fig. 2 Effect of grain size on magnetic properties of pure iron and silicon iron

(a) Relationship between grain size and hysteresis loss for high-purity iron at B = 1 tesla (10 000 gausses). (b) Variation of core loss with grain size for samples of 3.15 Si iron having similar cube-on-edge textures and chemical purity.

Fig. 3 Effect of alloying elements on electrical resistivity of iron

the cube edge <100>. Crystal orientation, therefore, is a basic factor in the determination and control of magnetic properties.

Alloying Elements. Pure iron is very soft magnetically and has a high saturation induction, but requires special handling during manufacture due to its low mechanical strength. Pure iron is used extensively in dc applications and in some ac relays. However, its low electrical resistivity makes it unsuitable for use in ac circuits, which constitute a great majority of all industrial applications of magnetic materials. Addition of alloying elements to iron increases its electrical resistivity and results in alloys that can be used in ac circuits. Figure 3 shows the changes in resistivity that result from additions of silicon, aluminum and other elements to iron.

Addition of silicon in sufficient amounts eliminates the allotropic transformation in iron. Consequently, Si-Fe alloys can be annealed at high temperature to promote grain growth, thus facilitating development of preferred grain orientation. However, room-temperature saturation induction is reduced by alloy additions other than cobalt (see Fig. 4).

High-Purity Irons

For experimental uses, 99.99% pure iron can be produced; by high-temperature annealing for prolonged times, a maximum permeability of about 100 000 with a hysteresis loss of about 10^{-5} J/cm^3 per cycle at a flux density of 1.0 tesla (10 000 gausses) has been obtained (Fig. 5).

The saturation induction of iron, based on a density of 7.878 g/cm^3, is given as $4\pi I_s$ = 2.1580 ± 0.001 tesla (21 580 ± 10 gausses), where I_s is the intensity of magnetization or magnetic moment per unit volume. The electrical resistivity is 98 nΩ·m (59 Ω·circ mil/ft) at 20 °C (68 °F), and the temperature coefficient is 0.0065 per °C (0.0036 per °F).

Commercial irons with purities of 99.6 to 99.8% are available in a variety of shapes. These irons have a saturation induction of about 2.15 teslas (21 500 gausses), a specific gravity of 7.85 and a resistivity of 107 nΩ·m (64 Ω·circ mil/ft) at 20 °C (68 °F).

Low-Carbon Steels

For many applications that require less than superior magnetic properties, low-carbon steels (type 1010, for example) are used. Frequently, higher-than-normal phosphorus and manganese contents are used to increase electrical resistivity. Such steels are not purchased to magnetic specifications. Although low-carbon steels exhibit power losses higher than those of silicon steels, they have better permeability at

Fig. 4 Effect of alloying elements on room-temperature saturation induction of iron

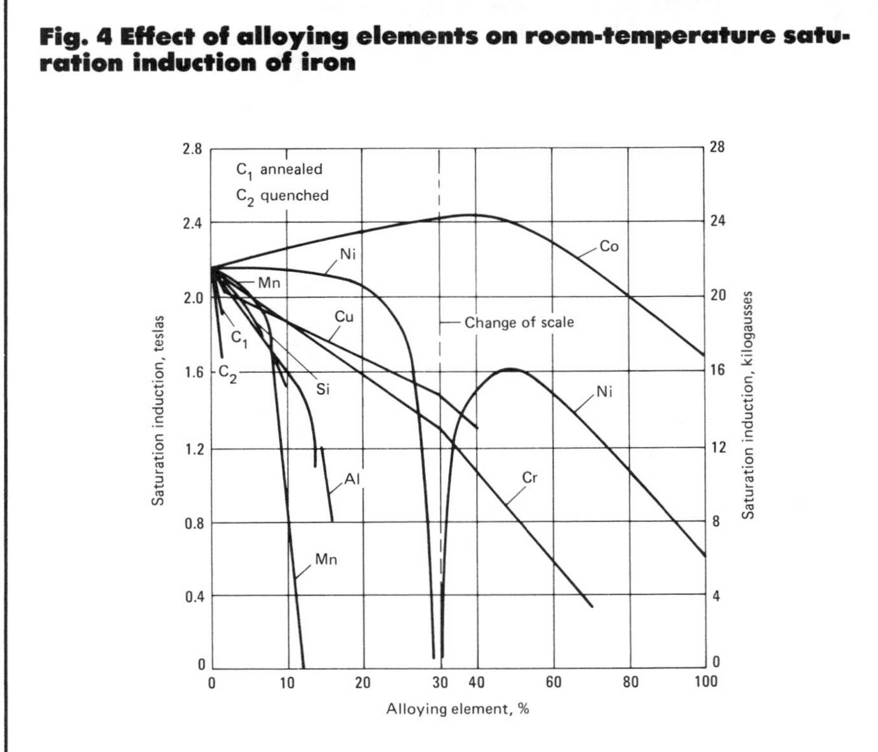

Fig. 5 Partial hysteresis loops and *B-H* curves for two types of iron

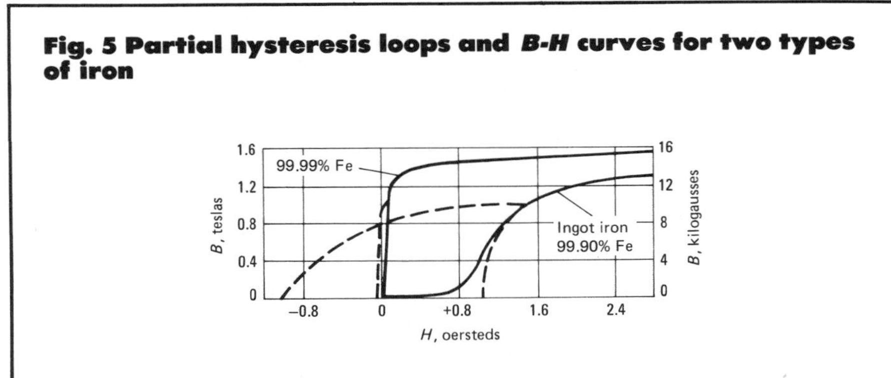

high flux density. This combination of magnetic properties, coupled with low price, makes low-carbon steels especially suitable for applications such as fractional-horsepower motors, which are used intermittently.

Nonoriented Silicon Steels

Except for saturation induction, the magnetic properties of iron containing a small amount of silicon are better than those of pure iron. Few commercial steels contain more than 3.5% silicon, because at levels above 4% the steel becomes brittle and difficult to process with cold rolling methods.

The commercial grades of silicon steel in common use (0.5 to 3.5% Si) are made mostly in electric or basic-oxygen furnaces. Nonoriented grades are melted with careful control of impurities; better grades have sulfur contents of about 0.01% or less. Continuous casting and vacuum degassing may be employed. After hot rolling, the hot bands are annealed, pickled and cold rolled to final thickness as continuous coils.

Semiprocessed grades of strip are not sufficiently decarburized for general use, and so decarburization and annealing to develop potential magnetic quality must be done by the user. This procedure is practical for small laminations accessible to the annealing atmosphere. Fully processed grades are strand annealed in moist hydrogen at about 825 °C (1520 °F) to remove carbon. The final annealing operation is very important and is carried out at a higher temperature—up to 1100 °C (2000 °F) for continuous strip—to cause grain growth and development of magnetic properties. Use of a protective atmosphere is vital. The steel frequently is coated with organic or inorganic materials after annealing, to reduce eddy currents in lamination stacks.

The vast majority of finished nonoriented silicon steel is sold in either full-width coils (860 to 1220 mm; 34 to 48 in.) or slit-width coils, but some is sold as sheared sheets. All coils are sampled and tested according to ASTM A343, and graded as to quality.

The historical trend of improvement in core loss for the best grade of 0.35-mm (0.014-in.) silicon steel is given in Fig. 6.

Tables 1 and 2 give properties specified by ASTM and AISI for standard grades of electrical steel. The AISI designations were adopted in 1946 to eliminate the wide variety in nomenclature formerly used. When originally adopted, the AISI designation number approximated ten times the maximum core loss, in watts per pound, exhibited by 29-gage samples when tested at a flux density of 1.5 teslas (15 000 gausses) and a magnetic circuit frequency of 60 Hz.

More specific information is given by the present ASTM designation, as follows: the first two digits indicate thickness in mm × 100; following these digits is a letter (F, S, G or H) that indicates material and test conditions; the last three digits indicate maximum core loss (in watts per pound) × 100. In the present designation, code letter F denotes fully processed nonoriented electrical steel whose core loss is determined at 1.5 teslas and 60 Hz on a 50/50 Epstein specimen consisting of as-sheared test strips, 50% sheared parallel to the grain and 50% sheared across the grain. Code letter S indicates semiprocessed nonoriented electrical steel whose core loss is determined at 1.5 teslas and 60 Hz on a 50/50 Epstein specimen that has been annealed 1 h at 845 °C (1550 °F). Code G denotes fully processed grain-oriented electrical steel

Table 1 Magnetic and thermal properties of electrical steel sheet and strip

AISI type	Thickness mm(a)	ASTM designation	Maximum core loss at 60 Hz(b) Induction, 1.5 teslas W/lb	W/kg	Induction, 1.7 teslas W/lb	W/kg	Saturation induction, teslas	Thermal conductivity W/m·K
Fully processed								
...	0.64	64F610	6.10	13.45	...	...	...	...
...	0.47	47F475	...	...	...	...	...	...
M-47	0.64	64F490	4.90	10.80	...	...	2.11	37.7
	0.47	47F400	4.00	10.14	...	...	2.11	37.7
M-45	0.64	64F360	3.60	7.94	...	...	2.07	25.1
	0.47	47F305	3.05	6.72	...	...	2.07	25.1
M-43	0.64	64F270	2.70	5.95	...	...	2.04	20.9
	0.47	47F230	2.30	5.07	...	...	2.04	20.9
M-36	0.64	64F240	2.40	5.29	...	...	2.02	18.8
	0.47	47F205	2.05	4.52	...	...	2.02	18.8
	0.36	36F190	1.90	4.19	...	...	2.02	18.8
M-27	0.64	64F225	2.25	4.96	...	...	2.02	18.8
	0.47	47F190	1.90	4.19	...	...	2.02	18.8
	0.36	36F180	1.80	3.97	...	...	2.02	18.8
M-22	0.64	64F218	2.18	4.81	...	...	2.00	18.8
	0.47	47F185	1.85	4.08	...	...	2.00	18.8
	0.36	36F168	1.68	3.70	...	...	2.00	18.8
M-19	0.64	64F208	2.08	4.59	...	...	1.99	16.7
	0.47	47F174	1.74	3.84	...	...	1.99	16.7
	0.36	36F158	1.58	3.48	...	...	1.99	16.7
M-15	0.47	47F168	1.68	3.70	...	...	1.98	16.7
	0.36	36F145	1.45	3.20	...	...	1.98	16.7
M-6	0.35	35G066	0.66	1.46	...	...	2.00	16.7
	0.35	35H094	...	...	0.94	2.07	2.00	16.7
M-5	0.30	35G058	0.58	1.28	...	...	2.00	16.7
	0.30	30H083	...	...	0.83	1.83	2.00	16.7
M-4	0.27	27G053	0.53	1.17	...	...	2.00	16.7
	0.27	27H076	...	...	0.76	1.68	2.00	16.7
M-3	0.27	27H071(c)	...	...	0.76	1.57	2.00	16.7
...	0.35	35P076(d)	...	...	0.76	1.68	2.01	16.7
...	0.30	30P079(d)	...	...	0.70	1.54	2.01	16.7
...	0.27	27P066(d)	...	...	0.66	1.46	2.01	16.7
Semiprocessed								
M47	0.64	64S350	3.50	7.72	...	...	...	...
	0.47	47S300	3.00	6.61	...	...	...	...
M45	0.64	64S280	2.80	6.17	...	...	...	...
	0.47	47S250	2.50	5.51	...	...	...	...
M43	0.64	64S230	2.30	5.07	...	...	...	...
	0.47	47S200	2.00	4.41	...	...	...	...
M36	0.64	64S213	2.13	4.70	...	...	...	...
	0.47	47S188	1.88	4.14	...	...	...	...
M27	0.64	64S194	1.94	4.28	...	...	...	...
	0.47	47S178	1.78	3.92	...	...	...	...

(a) 0.64 mm is equivalent to 24-gage sheet; 0.47 mm to 26-gage sheet; and 0.36 mm to 29-gage sheet. (b) Standard tests on all nonoriented sheets (M-47 to M-19 inclusive) are made on samples cut half with and half across the rolling direction; grain-oriented grades (ASTM designations G0 and H0) are tested in the rolling direction only. (c) Unofficial designation. (d) Unofficial designations of tentative grades of high-permeability type.

whose core loss is determined at 1.5 teslas and 60 Hz on a parallel-grain Epstein specimen stress relieved 1 h at 790 °C (1450 °F). Code H indicates fully processed grain-oriented electrical steel whose core loss is determined at 1.7 teslas and 60 Hz on a parallel-grain Epstein specimen stress relieved 1 h at 790 °C.

Table 2 lists some typical applications. Fully processed grades require only stress-relief annealing after fabrication or stamping. Semiprocessed grades must be decarburized by the customer to develop full magnetic properties. Suppliers' recommendations on annealing should be followed carefully.

Best permeability at high induction is obtained in steels with lower silicon contents. Low core loss is obtained with higher silicon contents, larger grains, lower impurity levels and thinner gages. The absence of preferred crystal orientation is actually helpful, because isotropic properties are desired in punchings for rotating machinery.

Table 2 Silicon contents, densities and applications of electrical steel sheet and strip

AISI type	Nominal Si + Al content, %	Assumed density, Mg/m^3	Characteristics and applications
Lamination steel			
. . .	0	7.85	High magnetic saturation; magnetic properties may not be guaranteed; intermittent-duty small motors
Nonoriented electrical steels			
M47	1.05	7.80	Ductile, good stamping properties, good permeability at high inductions; small motors, ballasts, relays
M45	1.85	7.75	Good stamping properties, good permeability at moderate and high inductions, good core loss; small generators, high-efficiency continuous-duty rotating machines, ac and dc
M43	2.35	7.70	
M36	2.65	7.70	Good permeability at low and moderate inductions, low core loss; high reactance cores, generators, stators of high-efficiency rotating machines
M27	2.80	7.70	
M22	3.20	7.65	Excellent permeability at low inductions, lowest core loss; small power transformers, high-efficiency rotating machines
M19	3.30	7.65	
M15	3.50	7.65	
Oriented electrical steels			
M6	3.15	7.65	Grain-oriented steel has highly directional magnetic properties with lowest core loss and highest permeability when flux path is parallel to rolling direction; heavier thicknesses used in power transformers, thinner thicknesses generally used in distribution transformers. Energy savings improve with lower core loss
M5	3.15	7.65	
M4	3.15	7.65	
M3	3.15	7.65	
High-permeability oriented steel			
. . .	2.9–3.15	7.65	Low core loss at high operating inductions

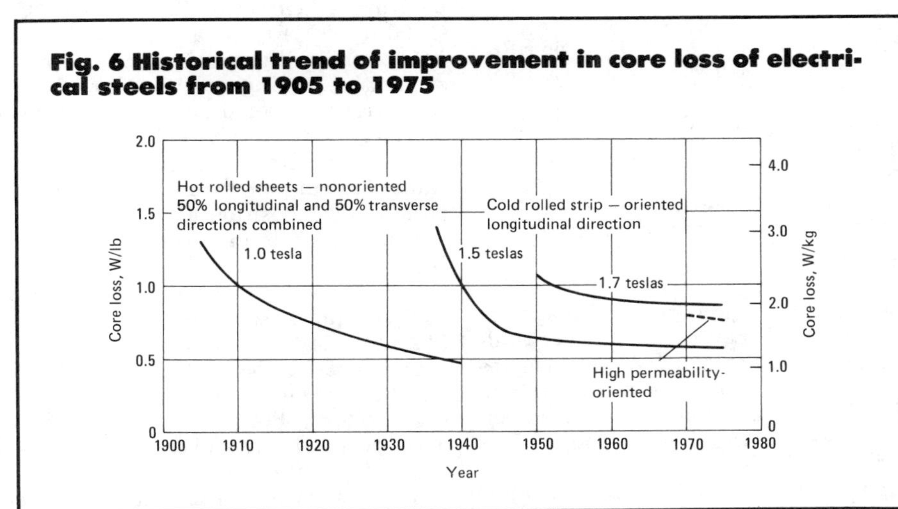

Fig. 6 Historical trend of improvement in core loss of electrical steels from 1905 to 1975

Oriented Silicon Steels

Grain size is as important in silicon steel as it is in iron with regard to core losses and low-flux-density permeability. For high-flux-density permeability, however, crystallographic orientation is the deciding factor. Like iron, silicon steels are more easily magnetized in the direction of the cube edge, <100>, as shown in Fig. 7 and 8. For special compositions, rolling and heat treating techniques are used to promote secondary recrystallization in the final anneal at about 1175 °C (2150 °F) or higher, which results in a well-developed texture with the cube edge parallel to the rolling direction {110} <001>. Conventional oriented grades contain about 3.15% Si.

About 1970, improved {110} <001> texture was developed by modification of composition and processing. The improved high-permeability material usually contains about 2.9 to 3.2% Si. Conventional grain-oriented 3.15% silicon steel has grains about 3 mm (0.12 in.) in diameter. The high-permeability silicon steel tends to have grains about 8 mm (0.31 in.) in diameter. Ideally, grain diameter should be less than 3 mm to minimize excess eddy-current effects from domain-wall motion. Special coatings provide electrical insulation and induced tensile stresses in the steel substrate. These induced stresses lower core loss and minimize noise in transformers.

Figure 9 compares variations in flux density and core loss, with respect to direction of rolling, for oriented and nonoriented silicon steels. These curves indicate the advantage of using oriented steels in a manner such that the flux is parallel to the rolling direction. Corresponding B−H curves and half hysteresis loops are given in Fig. 10.

High-grade silicon electrical steel does not age significantly as received from the mill, because its carbon content has been reduced to about 0.003% or less. With higher carbon contents, core loss can increase with time because of carbide precipitation. Also, silicon steel may age appreciably if not correctly heat treated in a manner that completely stabilizes its physical structure.

Iron-Aluminum Alloys

Although aluminum and silicon have similar effects on electrical resistivity and some magnetic properties of iron, aluminum is seldom substituted for silicon because of the resulting difficulties in fabrication. Aluminum is used most commonly as small (<0.5%) additions to the better grades of nonoriented silicon steel to increase electrical resistivity and thereby reduce eddy currents. Ternary alloys of iron, silicon and aluminum have high resistivity and good permeability at low flux density (see Tables 3 and 4). Increases in silicon and aluminum reduce saturation induction. At low flux densities, the magnetic properties of these alloys can be made to approach those of some

Table 3 Magnetic and physical properties of alloys with moderately high permeability at low field strength and high electrical resistance

Alloy(a)	Permeability Initial	Permeability Max	B value at max permeability	Hysteresis loss, ergs per cm^3 per cycle	Residual induction, gausses	Coercive force, oersteds	Saturation, gausses	Resistivity, μΩ·cm	Specific gravity
45 Permalloy	2 500	25 000	...	...	...	0.25	16 000	45	8.17
Sinimax	2 200	50 000	5400	400	5 500	0.06	11 000	90	7.70
Thermenol	6 000	60 000	1500	...	2 070	0.018	6 100	162	6.58
Monimax	3 000	60 000	6200	800	8 900	0.06	14 500	80	8.27
High Permalloy 49, A-L 4750, Armco 48, Hipernik	5 000	70 000	4500	300	10 000	0.50	16 000	48	8.25
4-79 Moly Permalloy, Hymu 80	20 000 min	90 000 min	4000	200	4 000 to 5 500	0.03	7 000 to 7 800	58	8.74
Mumetal	20 000 min	100 000	2000	...	2 300	0.05	6 500	60	8.58
1040 alloy	20 000 min	100 000	2000	200	2 400	0.20	6 000	56	8.76
Supermalloy	55 000 min	300 000 min	4000	20	4 000 to 5 500	0.006	6 800 to 7 800	65	8.77
Metglas	...	...	...	...	...	0.075	16 100	140	7.05

(a) Compositions and heat treatments are given in Table 4.

Table 4 Compositions, heat treatments and mechanical properties of alloys with moderately high permeability at low field strength and high electrical resistance

Alloy	Nominal composition(a)	Heat treatment(b)	Tensile strength MPa	Tensile strength ksi	Elongation, %	Hardness
4-79 Moly Permalloy, Hymu 80	79Ni, 4Mo	H_2 anneal, 2050 °F	545	79	40	58 HRB
45 Permalloy	45Ni	Anneal, 1920 °F	...	...	...	...
1040 alloy	72Ni, 14Cu, 11Fe, 3Mo	H_2 anneal, 2050 °F	...	...	...	...
High Permalloy 49 A-L 4750	47 to 50Ni	H_2 anneal, 2150 °F	495	72	45	58 HRB
Mumetal	77Ni, 5Cu, 2.75Cr	H_2 anneal, 2050 °F	440	64	27	60 HRB
Supermalloy	79Ni, 5Mo	H_2 anneal, 2370 °F	...	...	...	...
Thermenol	16Al, 3.5Mo	Anneal 1920 °F, fc to 1290 °F, oil quench	450	65	3	30 HRC

(a) Remainder Fe plus deoxidizers. (b) fc, furnace cool; H_2 anneal, hydrogen-atmosphere anneal.

iron-nickel alloys (see Tables 5 and 6).

Iron-Nickel Alloys

The effects of nickel content in iron-nickel alloys on saturation induction after annealing and on initial permeability are illustrated in Fig. 11 and 12, respectively. Various amounts of other elements, particularly molybdenum, often are added to these alloys in order to develop or accentuate specific characteristics.

In iron-nickel alloys, as in all magnetic materials, magnetic properties are controlled by saturation magnetization and magnetic anisotropy energies. Of the various types of magnetic anisotropy energy, the magnetocrystalline and magnetostrictive anisotropies are the most important in this system. Two broad classes of alloys have been developed in the Fe-Ni system. The high-nickel alloys (about 79% Ni) have high initial permeability (Fig. 12) but low saturation induction (approximately 0.9 tesla, or 9000 gausses), whereas the low-nickel alloys (about 50% Ni) are lower in initial permeability but higher in saturation induction (about 1.6 teslas, or 16 000 gausses).

The data plotted in Fig. 13 are from early laboratory studies and illustrate the effects of both composition and heat treatment on initial permeability (μ_0). Values of μ_0 above 12 000 are now obtained commercially in 50% Ni alloys, and values above 60 000 are obtained for 79% Ni alloys containing 4% Mo.

To obtain high initial permeability, both magnetocrystalline anisotropy (K_1) and magnetostrictive anisotropy (λ_s) must be minimized. Cooling rate—or, in other words, the degree of atomic ordering achieved in the critical temperature range from 760 to 315 °C (1400 to 600 °F), has a profound influence on the ability to minimize K_1. Consequently, cooling of this alloy from heat treating temperatures must be precisely controlled to obtain $K_1 = 0$. For commercially practical cooling rates of about 100 °C/min (180 °F/min), optimum composition for achieving $K_1 = 0$ and $\lambda_s = 0$ is about 4% Mo and 80% Ni.

Because $K_1 = 0$ and $\lambda_s = 0$ for commercial alloys such as Supermalloy, Moly Permalloy and Hymu 80, all of which contain about 4% Mo and 80% Ni, grain orientation is not critical for these alloys. Purity, however, influences permeability and core loss. Interstitial impurities such as carbon and nitrogen must be minimized by special melting procedures and by careful final annealing of laminations and other core configurations. Oxygen and sulfur are also objectionable. Best magnetic properties in Fe-Ni alloys are obtained by annealing in pure dry hydrogen (dew point less than -50 °C, or -58 °F) at 1000 to 1200 °C (1830 to 2200 °F) for several hours to reduce carbon, nitrogen, sulfur and oxygen contents. Sulfur contents higher than several ppm and carbon in excess of 20 ppm are detrimental to final annealed magnetic properties.

Most applications of 50% Ni alloys are based on requirements for high saturation induction. Nickel content is not critical near the middle of the iron-nickel series (50% Ni) and may be varied from 45 to 60%, but for highest saturation induction it should be held close to 50% (Fig. 11). Although $K_1 = 0$, the value of λ_{100}, which is the magnetostrictive constant in the <100> easy direction of magnetization, is close to zero for these alloys. Therefore, the initial permeability is still reasonable and in fact reaches a small maximum (Fig. 12). In some applications, such as converters, a high squareness ratio is desired—that is, a high ratio of remanence to saturation induction, Br/Bs. I. 50% Ni alloys, where the magnetocrystalline anisotropy is not zero, excellent squareness can be achieved by careful development of a cube texture during the final high-temperature anneal. In applications of Fe-Ni alloys it is thus important to control magnetic anisotropy energies, purity and texture, depending on what combination of properties is desired.

Trade names, nominal compositions, magnetic properties and typical applications of various iron-nickel and iron-cobalt alloys are given in Tables 7 and 8.

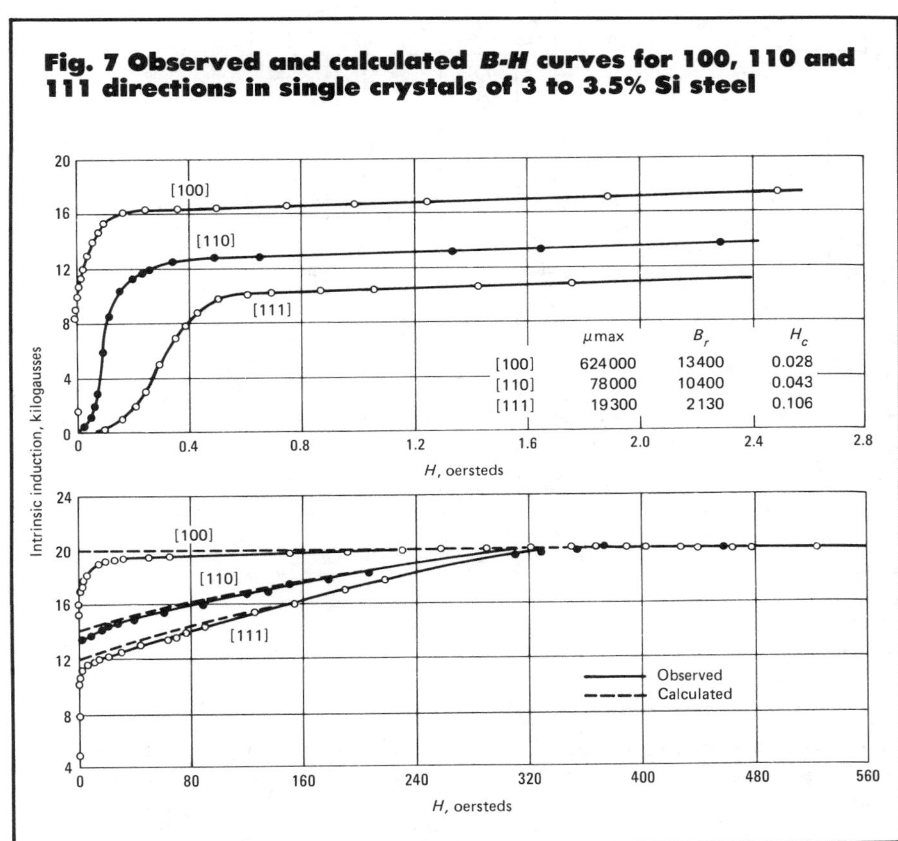

Fig. 7 Observed and calculated *B-H* curves for 100, 110 and 111 directions in single crystals of 3 to 3.5% Si steel

Amorphous Materials

Suitable alloys of iron prepared in amorphous, noncrystalline form have the attractive combination of high permeability and high volume resistivity.

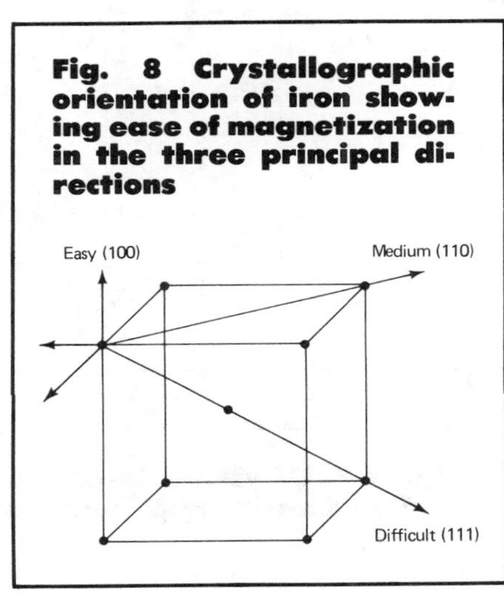

Fig. 8 Crystallographic orientation of iron showing ease of magnetization in the three principal directions

Table 5 Typical magnetic properties of alloys with high permeability at higher field strength, annealed for optimum magnetic properties

Material	Permeability Initial	Permeability Max	Induction, gausses at maximum permeability	Hysteresis loss, ergs per cm³ per cycle	Maximum induction(a), gausses	Coercive force, oersteds	Residual induction, gausses	Saturation value, gausses	Resistivity, μΩ·cm
0.5% Si steel	280	3 000	8 000	2300	10 000	0.90	...	20 500	28
Ingot iron	150	5 000	8 000	2700	10 000	1.00	7 700	21 400	10
1.75% Si steel	280	5 000	6 000	2100	10 000	0.80	...	20 000	37
2V Permendur, 49% Co, 2% V	800	5 000	13 000	...	15 000	1.00	10 000	23 500	40
3.0% Si steel	290	8 000	8 000	1600	10 000	0.70	...	20 100	47
Grain-oriented 3.0% Si steel	1400	50 000	10 000	400	15 000	0.09	12 000	20 100	50
Supermendur, 49% Co, 2% V	800	70 000	20 000	1500	21 000	0.23	21 400	24 000	40
50% Ni iron	3500	100 000	5 500	250	10 000	0.05	9 000	15 500	50
Grain-oriented 50% Ni iron	500	200 000	8 000	450	15 000	0.02	9 000	16 000	50

(a) Maximum induction for hysteresis loss, coercive force and residual induction measurements.

Fig. 9 Comparative flux densities and core losses for nonoriented M-19 and oriented M-6 electrical steels as a function of the direction of applied field

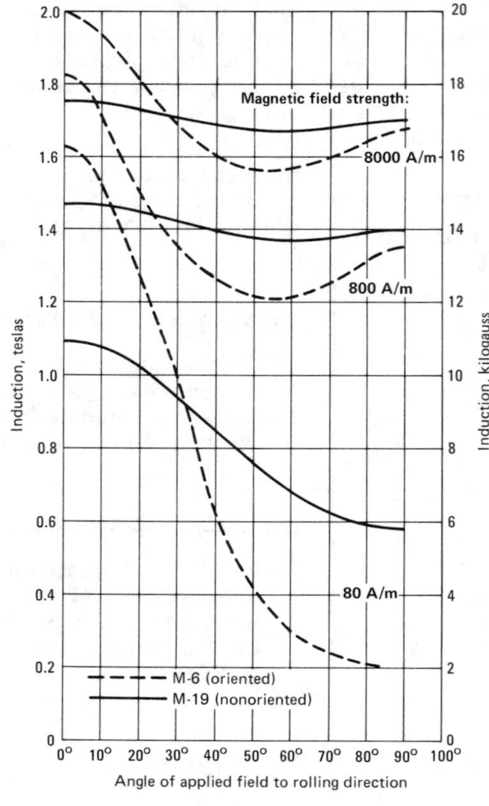

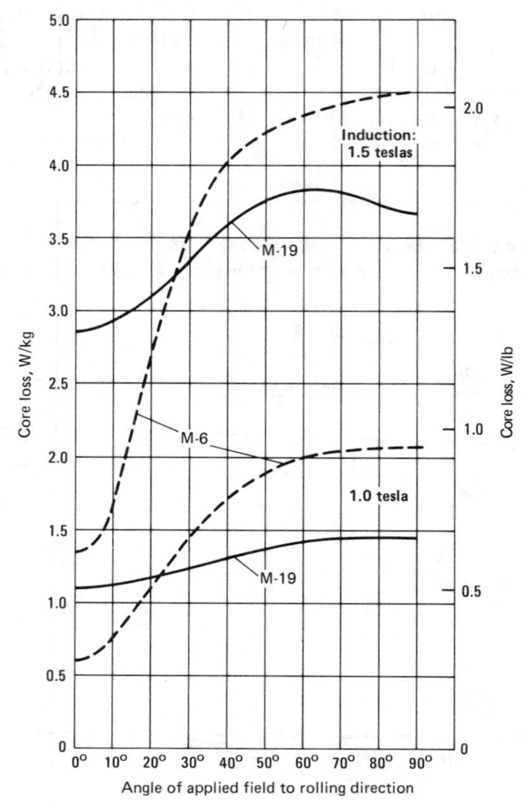

Fig. 10 Half hysteresis loops and dc magnetization curves for grain-oriented M-6 and cold rolled nonoriented M-19 steels

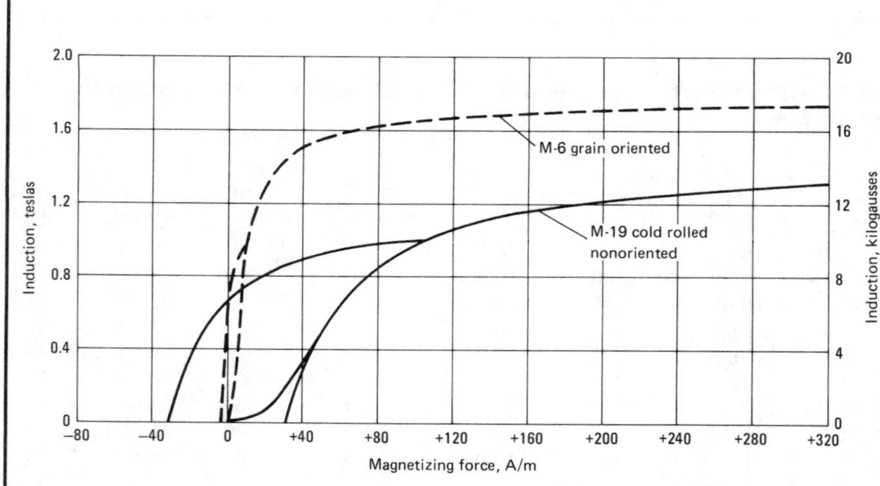

In the preferred method of production, the metal is rapidly quenched from the melt onto cooled rotating drums to form long ribbons about 40 μm (1.6 mils) thick. Attractive compositions contain 40 to 80% Fe with various additions of carbon, cobalt, boron, silicon, nickel and phosphorus. These materials are characterized by low hysteresis loss and low coercive force. However, Curie temperature is limited to about 400 °C (750 °F), and magnetic saturation to about 1.6 teslas (16 000 gausses).

Potential applications include substitution for conventional nickel-iron alloys in electronic devices. A major goal is to utilize the low core loss of amorphous material, which is less than one-half that of conventional grain-oriented silicon steel 0.26 mm (0.010 in.) thick, in distribution transformers (provided that production costs are competitive).

Fig. 11 Magnetic saturation of iron-nickel alloys at various field strengths

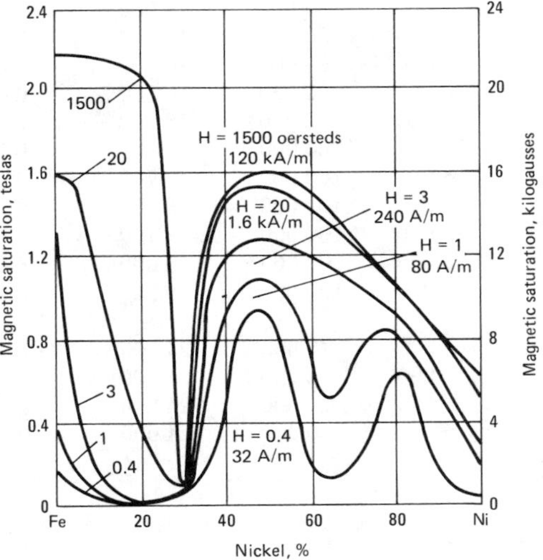

All samples were annealed at 1000 °C (1830 °F) and cooled in the furnace.

Fig. 12 Initial permeability at 2 mT (20 gausses) for annealed Fe-Ni alloys

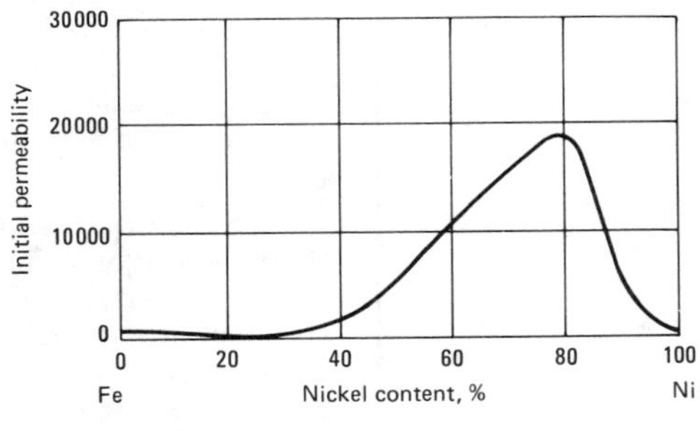

Table 6 Properties of alloys with high permeability at higher field strength, annealed for optimum magnetic properties

Material	Density, g/cm³	Tensile strength MPa	Tensile strength ksi	Elongation, %	Hardness, HRB
0.5% Si steel	7.75	380	55	32	43
1.75% Si steel	7.75	450	65	28	63
3.0% Si steel	7.65	565	82	20	100
50% Ni iron	8.20	400	58	27	45
2V Permendur, 49% Co, 2% V	8.15	550	80	5	77
Grain-oriented 3.0% Si steel	7.65	344	50	17	75
Grain-oriented 50% Ni iron	8.25	...	...	...	...
Ingot iron	7.88	275	40	40	30
Supermendur, 49% Co, 2% V	8.15	...	...	...	...

Iron-Cobalt Alloys With High Magnetic Saturation

Pure iron has a saturation induction of 2.16 teslas (21 600 gausses). Higher values require cobalt additions of up to 65%; the highest values (about 2.42 teslas, or 24 000 gausses) are obtained with cobalt contents of about 35% (Fig. 4). Use of alloys containing 25 to 50% Co is limited by low resistivity and high hysteresis loss, the high cost of cobalt and the brittleness of alloys containing more than 30% Co. However, with small additions of vanadium and special treatment in processing, 50% Co alloys can be cold rolled commercially to any gage, and strip is ductile enough to be punched and sheared. The 27% Co alloy is more easily fabricated, and less subject to degradation by stresses, than the 50% Co alloy. Furthermore, proper annealing of the 27% Co alloy produces magnetic properties suitable for both dc and ac applications.

Alternating-current applications require low eddy-current and hysteresis losses. Eddy-current losses can be minimized by a proper combination of composition and thickness.

Austenitic Stainless Steels

Austenitic stainless steels usually are not considered to be magnetic materials. However, increases in tensile strength due to cold working of these alloys are accompanied by increases in intrinsic permeability. This phenomena is illustrated graphically in Fig. 14 for nine austenitic stainless steels.

Ferrites

Ferrites for high-frequency applications are ceramics with characteristic spinel-magnetic structures ($M{\cdot}Fe_2O_4$, where M is a metal) and usually comprise solid solutions of iron oxide and one or more oxides of other metals such as manganese, zinc, magnesium, copper, nickel and cobalt. They are unique among magnetic materials in their outstanding magnetic properties at high frequencies, which result from very high resistivities ranging from about 10^8 Ω·cm to as high as 10^{14} Ω·cm. Hence at frequencies where eddy-current losses for metals become excessive, fer-

Table 7 Typical magnetic properties for various Fe-Ni and Fe-Co alloys

Material	Nominal composition(a)	Typical anneal(b)	Permeability At B=20	Permeability Max	Saturation induction, gausses	Coercivity, H_c, oersteds	Retentivity, B_r, gausses
45 Permalloy	45Ni	1920 °F	2 500	30 000	16 000	0.20	8000
		H_2, 2150 °F	4 000	50 000	16 000	0.06	8000
4750 alloy	47-50Ni	H_2, 2050 °F	4 000	50 000	16 000	0.07	8000
Carpenter 49 alloy	47-50Ni	H_2, 2050 °F	4 000	50 000	16 000	0.07	8000
Conpernik	50Ni	...	1 500	2 000	16 000	...	...
Orthonol	50Ni(c)	H_2, 1825 °F	...	60 000	15 600	0.20	14500
78 Permalloy	78Ni	1920 °F	8 000	100 000	10 700	0.05	6000
4-79 Moly Permalloy	79Ni, 4Mo	2000 °F, Q	20 000	100 000	8 700	0.03	5000
Hymu 80	79Ni, 4Mo	2000 °F, Q	20 000	100 000	8 700	0.05	5000
Supermalloy	79Ni, 5Mo	H_2, 2375 °F, Q	75 000	800 000	8 000	0.006	5000
Mumetal	77Ni, 5Cu, 2.75Cr	2050 °F	20 000	100 000	6 500	0.05	3000
Permendur	50Co	1470 °F	800	5 000	24 500	2.00	14000
2V Permendur	49Co, 2V	1470 °F	800	8 000	24 000	1.2	14000
Hiperco 2.7	37Co, 0.6Cr	...	650	10 000	24 200	1.00	13000
Supermendur	49Co, 2V	...	...	60 000	24 000	0.20	21500
2-81 Moly Permalloy powder	81Ni, 2Mo	1200 °F	125	130	...	...	...
Carbonyl iron powder	...	...	60	150	...	...	...

(a) Remainder iron plus deoxidizer. (b) H_2, annealed in hydrogen; Q, quenched or controlled cooled. (c) Grain oriented.

Fig. 13 Relative initial permeability at 2 mT (20 gausses) for Fe-Ni alloys given various heat treatments

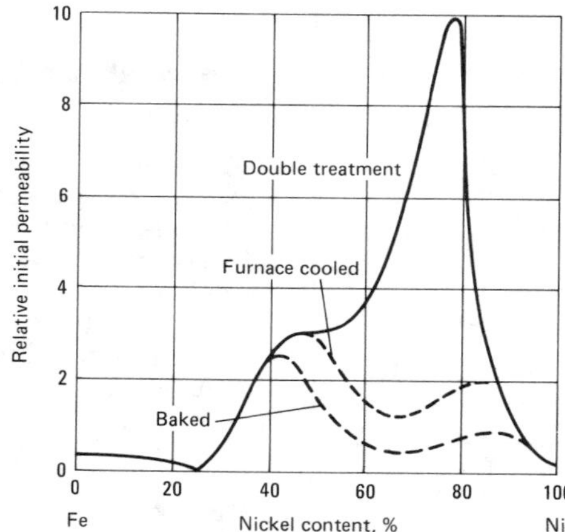

Treatments were as follows: furnace cooled—1 h at 900 to 950 °C (1650 to 1740 °F), cooled at 100 °C/h (180 °F/h); baked—furnace cooled plus 20 h at 450 °C (840 °F); double treatment—furnace cooled plus 1 h at 600 °C (1110 °F) and cooled at 1500 °C/min (2700 °F/min).

rites make ideal soft magnetic materials. Because ferrites have inherently high corrosion resistance, parts made of these materials normally do not require protective finishing.

Disadvantages of ferrites include low magnetic saturation, low Curie temperature, and relatively poor mechanical properties compared with those of metals. Ferrites are produced from powdered raw materials by mixing, calcining, ball milling, pressing to shape, and firing to the desired magnetic properties. The final product is hard, brittle and unmachinable, and thus close dimensional tolerances must be obtained by grinding.

Many different types of ferrites are available for magnetic use. They can be classified into three general types: square-loop ferrites for computer memories, linear ferrites for transformers and for inductors in filters, and microwave ferrites for microwave devices. In recent years, due to increasing usage of semiconductors for computer memories, square-loop ferrites have decreased in importance.

Microstructure and composition have much stronger influences on the magnetic properties of ferrites than on those of metals. Hence, properties of finished ferrite parts can vary drastically with purity and structure of raw materials, with the nature of binders used and with the ceramic-processing technique employed. In general, lithium ferrites, Mn-Mg-Zn ferrites and Mn-Mg-Di ferrites are used for computer memories. Lithium ferrite is higher in Curie temperature and saturation magnetization, but lower in switching speed, than Mn-Mg-Zn and Mn-Mg-Di ferrites. Linear ferrites comprise Mn-Zn and Ni-Zn ferrites. Mn-Zn ferrite is higher in saturation magnetization, but lower in resistivity, than Ni-Zn ferrite. Mn-Zn ferrite is preferred for frequencies up to about 1 MHz. For microwave applications, Ni-Zn, Mg-Mn-Al and Mg-Mn-Cu ferrites are used, as well as garnets of the type $M_{3+x}Fe_{5-x}O_{12}$ (where M = Y + Al or Y + Gd + Al).

Table 8 Applications of high-permeability Fe-Ni and Fe-Co alloys

Material	Typical applications
45 Permalloy 4750 alloy Carpenter 49 alloy Nicalloy Audioalloy	Good over-all characteristics, high μ and low losses; audio transformers, coils, relays
Conpernik	Nearly constant μ up to B = 200 gausses; choke coils
Blendalloy Orthonol Permanite	Rectangular hysteresis loop; magnetic amplifier coils, contact rectifiers
78 Permalloy	High μ_0, low H_c, low resistivity; sensitive d-c relays
4-79 Moly Permalloy Hymu 80	Very high μ_0, low losses; audio coils, transformers, magnetic shields
Supermalloy	Highest μ_0, low losses; high-efficiency a-c coils, pulse transformers, magnetic amplifier coils
Mumetal	Similar to 4-79 Moly Permalloy
Permendur	High μ at very high flux densities; d-c electromagnets, pole tips
2V Permendur	Similar to Permendur, but ductile; receiver diaphragms
Hiperco	Similar to 2V Permendur; high-flux-density motors, transformers
Supermendur	Similar to 2V Permendur, but much higher magnetic quality

Note: μ, permeability; μ_0, initial permeability; H_c, coercivity, in oersteds; B, flux density.

Fig. 14 Correlation of increased tensile strength from cold working and the permeability of cold worked austenitic stainless steels

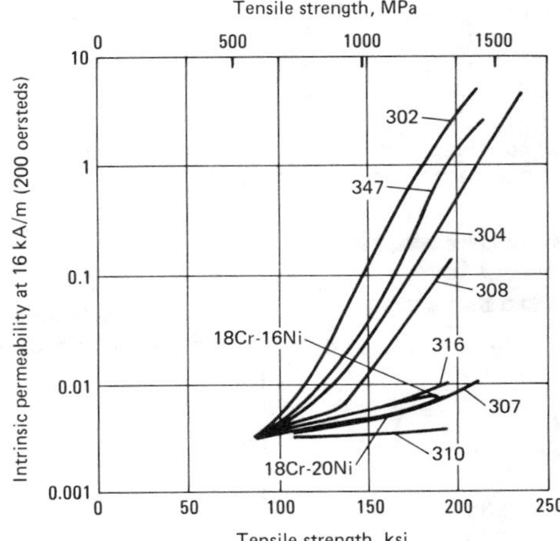

Annealed hot rolled strips 2.4 to 3.2 mm (0.095 to 0.125 in.) thick before cold reduction. For normal permeability values, add unity to the numbers given on vertical scale.

Fig. 15 Variation of induction with temperature for iron, at four different values of magnetizing force

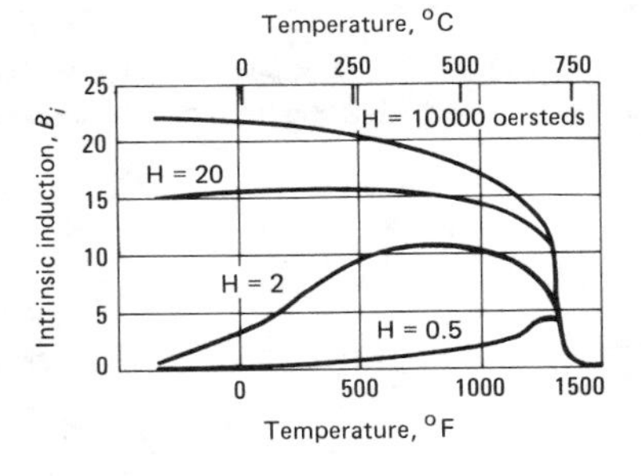

Constant Permeability With Changing Temperature

In all magnetic materials, magnetic properties change with temperature. Proper selection and preparation of materials, and proper circuit design, can minimize these changes.

Change in flux density with temperature for iron tested at four different values of magnetizing force is plotted in Fig. 15. Operation of a device at a flux density of 1.5 teslas (15 000 gausses) would be only slightly affected by variations in operating temperature near ambient. There is a similar minimized temperature effect for all materials, except that the flux density for optimum operation depends on the materials, given the proper flux density and temperature range. Great changes occur at temperatures approaching the Curie temperature (Fig. 15).

For many reasons, it is not always possible to operate a material at the best flux density for temperature stability. One way of obtaining better temperature stability is to use magnetic materials in insulated powder form, such as pressed Permalloy powder cores, which have good temperature stability due to the presence of many built-in air gaps. Another method involves use of Fe-Ni alloys, such as Isoperm or Conpernik, that have been drastically cold rolled and then under-annealed to produce a partly strained alloy less sensitive to temperature changes.

In a third method, two alloy powders with opposite temperature coefficients are combined for use in the desired temperature range. Special Permalloy powder cores are combined with small amounts of Fe-Ni-Mo powder having a low Curie temperature and a negative temperature coefficient near room temperature. The 30% Ni irons are of the low-Curie-temperature type.

Alloys for Magnetic Temperature Compensation

The variation in magnetic permeability of iron and iron alloys that occurs with temperature can seriously affect the accuracy of various electrical measuring instruments. To compensate for such changes, a certain amount of the magnetic flux can be shunted around the moving part of the instrument by using an alloy with high negative magnetic temperature coefficient between −18 and +100 °C (0 and 212 °F). The amount of shunted flux therefore decreases with increases in ambient temperature, forcing more flux through the moving part than would otherwise be possible. Nearly complete

Table 9 Variations in permeability of Fe-Ni magnetic alloys with temperature

Alloy(a)	Permeability(b), for H = 46 oersteds, at: 45 °C (−50 °F)	−15 °C (0 °F)	10 °C (50 °F)	25 °C (77 °F)	40 °C (100 °F)	65 °C (150 °F)	95 °C (200 °F)	120 °C (250 °F)
29.0% Ni	92	70	25	3	2	...	...	...
29.5% Ni	120	102	74	46	27	4	...	...
31.0% Ni	156	140	120	110	98	73	45	15
32.5% Ni	212	202	191	180	170	145	120	85

(a) Remainder is iron in all alloys. (b) Temperature-permeability properties can be varied by manufacturing procedure and thermal treatment at finished size. Above values are average and representative of temperature-permeability properties available in a commercial product.

Table 10 Curie temperature for some alloys used for temperature compensation

Material	Curie temperature °C	°F
Supermendur, 49% Co, 2% V	930	1710
2V Permendur, 49% Co, 2% V	930	1710
Ingot iron	770	1420
0.5% Si steel	755	1390
3% Si steel, grain oriented	730	1350
50% Ni iron, grain oriented	500	930
50% Ni iron	480	900
4-79 Moly Permalloy	455	850
Mumetal	400	750
Monimax	400	750
Sinimax	290	550
30% Ni iron compensator alloy	−75 to +150	−100 to +300

Fig 16 Effect of nickel content on the permeability-temperature characteristics of annealed iron-nickel temperature-compensator alloys

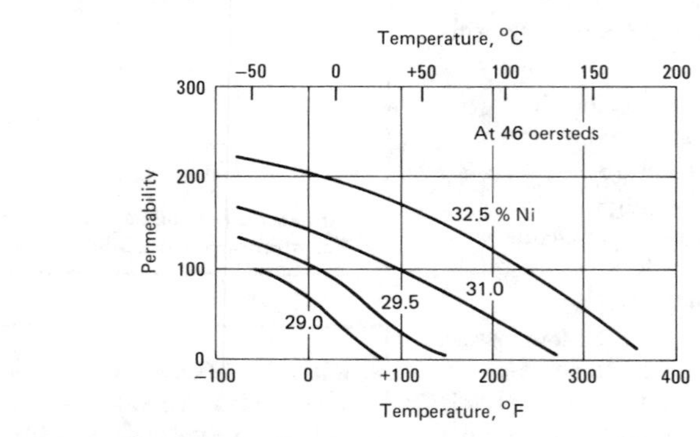

Fig. 17 Variation in permeability index of P/M iron as a result of changes in forming pressure

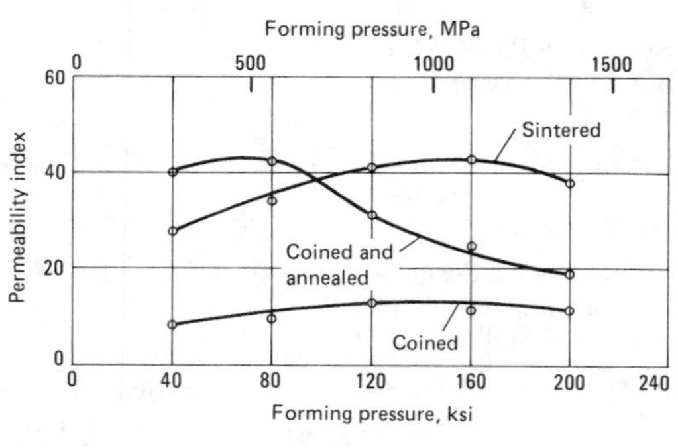

compensation for temperature changes can be made by correct design of parts. A watt-hour meter is one example. Another example is an automobile speedometer, where a temperature compensator compensates for the change in pole strength of the permanent magnet and the change in electrical resistivity of the aluminum drag cup, over a temperature range.

Iron-nickel alloys are frequently used commercially for this purpose. Table 9 and Fig. 16 show how some typical Fe-Ni alloys vary in permeability with temperature at $H = 46$ oersteds. Table 10 lists the Curie temperatures, ranging from below room temperature to 930 °C (1710 °F), for 12 alloys that may be used for temperature compensation.

Magnetic induction and losses decrease, and permeability at low to intermediate induction increases, as operating temperature increases. However, temperature instability becomes greater as the operating temperature approaches the Curie temperature, or nonmagnetic point.

Compressed Powdered Iron

For applications in which complicated magnetic circuits would otherwise require considerable machining, it may be helpful to press iron powder in a mold and sinter the part in vacuum or in a reducing atmosphere. Figures 17, 18 and 19 show anticipated effects of sintering temperature and sintering time on magnetic permeability of powdered iron, expressed as a percentage of the permeability of annealed low-carbon steel, for a constant magnetizing force. Use of P/M techniques to make magnetically soft components

Table 11 Direct-current magnetic properties of relay steels and alloys after annealing

Metal	Magnetizing force, H_{max} from B = 10 000	H_c, coercive force(a), oersteds	B_r, kilogausses from B = 10 000	Permeability, μ Initial	Permeability, μ Maximum	Flux density at max permeability, kilogausses	Saturation induction, kilogausses	Resistivity, μΩ·cm
Low-carbon iron and steel								
Low-carbon iron	2.0 to 5.8	0.80 to 1.70	7.0 to 8.5	200(b) 500 to 1000(c)	2 200 to 5 500	6.4 to 7.5	21.5	10
1010 steel	3.0	1.00 to 2.00	8.4	200(b)	3 800	7.5	21.0	12
Silicon steels								
1% Si	1.7 to 3.2	0.40 to 0.80	2.8 to 8.5	400(b), 650(c)	1 700 to 6 000	5.0 to 9.0	20.6	23
2.5% Si	0.8 to 1.90	0.13 to 0.70	2.3 to 6.7	900(b)	1 800 to 11 000	4.0 to 8.0	20.0	41
3% Si	1.30 to 2.00	0.23 to 0.65	3.9 to 7.6	550(b)	7 500 to 10 000	4.7 to 8.0	18.4 to 20.0	48
3% Si, grain oriented	0.22	0.05 to 0.10	6.6 to 8.6	2500	55 000 to 60 000	8.0 to 10.0	19.7	48
Stainless steels								
Type 430(d)	....	3.00 to 4.00	8.0 to 9.5	230(c)	1 100 to 1 600	6.0 to 8.2	14.7	60
Type 416(e)	....	4.00 to 6.00	9.0 to 11.0	200(c)	800 to 1 000	9.0 to 10.5	15.0	57
Type 410(e)	....	4.50 to 7.50	8.0 to 12.0	110 to 180(c)	800 to 1 000	6.0 to 7.0	16.0	57
Type 443(f)	....	6.50 to 7.50	3.2 to 3.8	60(c)	450 to 550	2.5 to 3.5	12.0	68
Type 446(g)	....	3.00 to 4.00	4.5 to 5.0	100(c)	800 to 900	5.7 to 5.9	12.0	61
Nickel irons								
50% Ni(h)	0.20 to 0.50	0.06 to 0.14	5.5 to 8.0	2500 to 3500(b)	30 000 to 120 000	5.0 to 7.0	16.0	48
78% Ni	0.20	0.05	8.0	8000(b)	100 000	5.0	10.7	16
77% Ni (Cu, Cr)(i)	0.10	....	4.0	20 000	150 000	2.0	7.5	60
79% Ni (Mo)(j)	....	....	3.2 to 3.5	....	70 000 to 75 000	3.2 to 4.0	8.0	58

(a) From B = 10 000. (b) At B = 20. (c) At B = 200. (d) Coercive force from saturation is 3.50 to 4.50 oersteds. (e) Coercive force from saturation is 5.00 to 7.00 oersteds. (f) Coercive force from saturation is 7.30 oersteds. (g) Coercive force from saturation is 3.50 to 5.00 oersteds. (h) Coercive force from saturation is 0.15 oersteds. (i) Coercive force from B = 5000 is 0.02 oersteds. (j) For 79% Ni iron with molybdenum, coercive force from B = 5000 is 0.010 to 0.12 oersteds.

can eliminate all machining operations and may save as much as 50% of total cost.

The densities of most iron-base P/M parts range from 6.2 to 7.0 Mg/m^3. High densities and superior magnetic properties can be obtained by using higher molding pressures or higher sintering temperatures, or by making the parts from high-purity iron-base powders (with or without small additions of alloying elements such as silicon and phosphorus). The cost of such parts is higher and, unless magnetic flux density exceeds 1 tesla (10 000 gausses), performance is no better than that of less-expensive iron-base P/M parts. This is especially true in magnetic circuits with short flux paths.

Selection of a rolled or sintered iron for a specific application depends on die cost, required finish, tolerances and number of parts. For example, design of soft iron pole pieces in the magnetic circuits of small panel-type motors permits the use of sintered P/M parts.

Heat Treatment

Magnetic materials such as iron-nickel alloys in the cold rolled condition must be annealed to develop the desired grain structure and magnetic properties.

Annealing conditions required for silicon steels depend on processing carried out by the supplier and on cost-versus-performance factors. Semiprocessed grades must be annealed near 840 °C (1550 °F), after stamping of laminations, for removal of carbon and development of magnetic properties. Fully processed grades of nonoriented or grain-oriented steels require annealing in the range 750 to 875 °C (1375 to 1600 °F) only for removal of fabrication stresses. For wide laminations used in a flat condition, as in large power transformers, low-stress strip requiring no annealing at all is available.

Table 11 lists 15 magnetically soft materials and their electrical properties after annealing. Table 12 gives nominal compositions, recommended annealing temperatures and resulting mechanical properties for these alloys.

Stress Effects

Magnetic properties such as power (core) loss and permeability are very sensitive to stresses. Plastically strained materials usually must be annealed unless the volume of strained material is only a small fraction of the total. In oriented silicon steels, a compressive stress of as little as 3450 to 6900 kPa (500 to 1000 psi) can increase core loss by 50 to 100%. Thus, great care must be exercised in punching and assembling electrical steels to ensure that stresses are relieved before the unit is assembled and that new stresses are not introduced during assembly. In some instances, however, small tensile stresses induced by coatings can improve the properties of grain-oriented silicon steel.

Selection for Specific Applications

The volume of magnetically soft materials used for more specialized applications is small compared to that used for heavy rotating equipment and transformers. Materials used in industry range from commercial low-carbon steel sheet and strip to high-silicon and grain-oriented silicon steels and special

Table 12 Nominal compositions, recommended annealing cycles and typical mechanical properties of relay steels and alloys after annealing

Metal	Nominal composition C	Si	Ni	Cr	Recommended annealing cycle: Temperature °C	°F	Time, h	Cooling method	Mechanical properties: Tensile strength MPa	ksi	Yield strength MPa	ksi	Elongation, %	Reduction area, %	Hardness, HRB
Low-carbon iron and steel															
Low-carbon iron	0.03	...	...	...	815 to 1150	1500 to 2100	1 to 4	Furnace	275	40	165	24	40	75	40
1010 steel	0.10	...	...	...	815 to 980	1500 to 1800	1 to 6	Furnace	310	45	170	25	32	...	42
Silicon steels															
1% Si	0.06	1.10	...	...	785 to 1150	1450 to 2100	1 to 6	Furnace	345	50	...	...	22	...	50
2.5% Si	0.05	2.50	...	...	785 to 1150	1450 to 2100	1 to 6	Furnace	385	56	295	43	8	8	65
3% Si	0.03	3.10	...	...	800 to 1150	1475 to 2100	1 to 3	Furnace	515	75	...	...	10	...	80
3% Si, grain oriented	0.01	3.00	...	...	800 to 1205	1475 to 2200	4	Furnace	415	60	...	...	...	...	65
Stainless steels															
Type 430	0.10	...	...	17.00	760 to 815	1400 to 1500	2	Still air	515	75	310	45	30	65	83
Type 416	0.12	...	...	13.00	815 to 900	1500 to 1650	2	Still air	515	75	275	40	30	60	83
Type 410	0.12	0.40	0.50	12.50	815 to 900	1500 to 1650	2	Still air	515	75	275	40	35	70	83
Type 443(a)	0.18	0.60	0.40	21.00	815 to 900	1500 to 1650	2	Water quench	525	76	330	48	32	65	86
Type 446	0.16	0.50	0.50	29.00	815 to 900	1500 to 1600	1 to 3	Water quench	525	76	370	54	31	59	78
Nickel irons															
50% Ni	0.03	0.35	50.0	...	1065 to 1120	1950 to 2050	1 to 4	Furnace	450	65	145	21	38	68	60
78% Ni	...	...	78.00	...	1065 to 1120	1950 to 2050	1 to 3	(b)	515	75	97	14	35	...	60
76% Ni (Cu, Cr)(c)	0.02	...	76.00	2.75	1065 to 1120	1950 to 2050	4	Furnace	...	...	...	...	...	...	...
79% Ni (Mo)(d)	0.05	0.15	79.00	...	1150 to 1175	2100 to 2150	4	Furnace	545	79	150	22	64	70	62

Note: Where there is a spread of several hundred degrees for annealing temperature, the lower temperature is intended for relief of stresses induced in fabricating fully processed material. The higher temperature is for full annealing. (a) Also 1.00% Cu. (b) Furnace cool to 610 °C (1125 °F), then oil quench. (c) Also 5% Cu. (d) Also 4.00% Mo.

Fig. 18 Variation in permeabilty index of P/M iron with sintering temperature

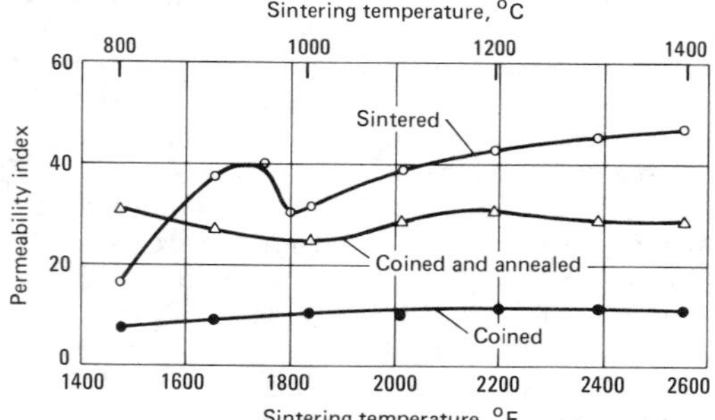

The magnetic permeability (for a constant magnetizing force) is shown as a percentage of the permeability of annealed hot rolled low-carbon steel.

alloys. Tables 13 to 16 can be used as guides in selecting compositions, grades and gages for several specific applications; these tables also take into account factors such as cost, availability, punchability, temperature and corrosion resistance.

Product requirements in terms of cost and plant standardization should be studied to determine which materials are most suitable. More than one alloy often may be suitable for a given set of conditions. Many of these materials are similar, and slight upgrading or downgrading will not have significant effects.

As stated previously, grain-boundary disturbances induce greater eddy-current and hysteresis losses, and the ideal material would be, of course, one grain thick and properly oriented. By extension, the ideal lamination would be made from material of the thinnest possible gage. However, in practice, some compromise is necessary because the thinnest gages of material are very costly.

In considering the relationships between electrical properties and gage, it might appear practical to downgrade to a cheaper and lower-efficiency material and use a thinner gage, thereby retaining about the same electrical properties. For example, a stack of laminations of 29-gage M-43 and the same size stack of 26-gage M-36 might be equal in electrical properties, although M-36 is more efficient and slightly more expensive. However, selection on such a basis does not take into consideration all aspects of cost.

Table 13 Recommended materials for motors and generators(a)

Type of motor	Material thickness mm	in.	Material
Starting motors			
Automotive	0.35	0.014	1008
	0.46	0.0185	1008
	0.63	0.025	1008
Medium, 1 to 100 hp	0.35	0.014	M-36
	0.46	0.0185	M-43
	0.63	0.025	1008
Large, 100 hp min	0.35	0.014	M-19
	0.46	0.0185	M-36, M-27
	0.63	0.025	M-43, M-36
Motors and generators for intermittent operation(b)			
Miniature	0.46	0.0185	M-50, M-43
	0.63	0.025	1008, M-50
Gyros	0.35	0.014	M-15
	0.46	0.0185	M-15
Selsyns	0.35	0.014	M-15, 45 to 50% Ni iron
Fractional, 1/4 hp	0.63	0.025	1008
Fractional, 1/2 hp	0.63	0.025	M-43
Fractional, 3/4 hp	0.63	0.025	M-36
Medium and large	0.46	0.0185	M-43, M-36, M-27
	0.63	0.025	M-43, M-36, M-27
Motors and generators for continuous operation			
Fractional, 1/4 hp	0.63	0.025	1008
Fractional, 1/2 hp	0.63	0.025	M-43
Fractional, 3/4 hp	0.63	0.025	M-36
Medium, 1 to 100 hp	0.35	0.014	M-22, M-19
	0.46	0.0185	M-36, M-27
Large, 100 to 5000 hp	0.35	0.014	M-19, M-15
	0.46	0.0185	M-27, M-19, M-15
	0.63	0.025	M-27, M-19
Large, over 5000 hp	0.35	0.014	M-15, M-6
	0.46	0.0185	M-19, M-15
Frequency	**Thickness µm**	**mils**	**Material**
High-frequency motors			
To 400 cycles	180	7	3% Si steel(c)
	380	15	M-19, M-15
800 to 1200 cycles	125	5	3% Si steel(c)
	180	7	3% Si steel(c)
Servo motors	125	5	3% Si steel(c)
Synchronous motors	100 to 355	4 to 14	45 to 50% Ni iron

(a) Where more than one grade of silicon sheet is shown, they are listed in the order of increasing cost and efficiency as a result of electrical properties. (b) 1008 steel is used in 0.76 mm (0.030 in.) for all applications in this category. (c) Cold rolled, nonoriented.

While there is a gradual increase in price per pound as the M number decreases and efficiency increases, other elements of cost must be considered. Most important of these is the cost of making laminations from the strip or sheet. A punch press running at constant speed will produce the same number of laminations regardless of material thickness. Hence, it would require nearly twice as much machine time to produce a stack of given size from 29-gage as from 24-gage sheet. Annealing, stacking and handling costs also increase as the gage becomes thinner. Therefore, where electrical efficiency is critical and thin-gage material is used, the major cost is in making the stack with thinner laminations. It is false economy to downgrade the material and sacrifice electrical properties for only slight savings in material cost. Recommendations in the selection tables reflect the factors described above.

Motors and Generators

Table 13 recommends alloys for motors and generators. For small starting motors that operate infrequently and for short periods of time, there is little emphasis on electrical efficiency. For this reason, the cheapest core material is recommended, regardless of gage. The most common gage for such applications is 24, although thinner gages are used occasionally. For heavier starting motors (up to 100 hp), 29-gage M-36 is recommended for greater efficiency. Increasing the thickness to 26 gage, or using a lower grade, will decrease cost and electrical efficiency. Laminations for these larger motors are seldom made of 24-gage material, but when the heavy gage is used, 1008 steel is recommended. For the largest starting motors, efficiency is of greater importance and upgrading of material is justified.

Table 13 also gives recommended materials and thicknesses for ac and dc motors and generators for intermittent use. For small motors, only the heavier gages and cheaper materials are suggested. Most of these motors are used for highly competitive light-duty consumer items; electrical efficiency is a secondary consideration.

For gyros and selsyns, which are specialized applications demanding more efficiency, higher-grade materials in thinner gages are recommended. Large industrial motors, even in intermittent operation, require more efficiency than small motors, and it is for this reason that both intermediate gages and intermediate materials are recommended for large motors.

Compositions and gages of sheets for motors and generators for continuous operation are also recommended in Table 13. Upgrading of material for the more rigorous service is recommended, but the same principles of selection should be followed.

In high-speed rotating machinery, yield strength of the magnetic rotor material may be the decisive factor in selection. For instance, in an alternator designed for 27 600 rpm at 15% overspeed, stresses at the base of the rotor teeth were calculated to be 189 MPa (27.4 ksi) for the overspeed condition.

Table 14 Recommended materials for transformers

Type	Material thickness mm	in.	Material
Continuous duty(a)			
Distribution	0.27	0.011	M-3, M-4
	0.30	0.012	M-5
	0.35	0.014	M-6
Power	0.30	0.012	M-5
	0.35	0.014	M-6
Voltage regulator	0.30	0.012	M-5
	0.35	0.014	M-15
	0.63	0.025	M-22
Welding transformer	0.30	0.012	M-5
	0.35	0.014	M-6
	0.63	0.025	M-43, M-36, M-27

Application	Standard electrical steels	Other alloys
Special application transformers		
Instrument	M-15, M-6 0.30 to 0.63 mm (0.012 to 0.025 in.)	Vanadium Permendur 70 to 80% Ni iron, 45 to 50% Ni iron
Radio, power	M-27, M-22, M-19 0.30 to 0.35 mm (0.012 to 0.014 in.)	Vanadium Permendur 70 to 80% Ni iron
Radio, audio	M-19, M-17, M-15, M-6, M-5 0.35 to 0.46 mm (0.014 to 0.0185 in.)	45 to 50% Ni iron
Radar pulse transformers	25 to 100 μm (1 to 4 mil): oriented 3% Si steel; 125 to 180 μm (5 to 7 mil): nonoriented 3% Si steel	Oriented 45 to 50% Ni iron, 4-79 Moly Permalloy, Supermalloy, Supermendur, Monimax, Sinimax, 13 to 100 μm (0.5 to 4 mil)
Chokes, power	M-22, M-19 M-15, M-6	...
Chokes, radio	...	Carbonyl irons, ferrites
Ballasts	M-27, M-22 0.46 to 0.63 mm (0.0185 to 0.025 in.)	...
Miscellaneous bell-ringing and toy	1008	...

(a) For core laminations for welding transformers, M-27 and M-22 are recommended in 0.46-mm (0.0185-in.) sheet.

Because ingot iron, the material best suited electrically, has a yield strength of less than 138 MPa (20 ksi), both 1010 and 1015 steel were tested. The 1010 hot rolled steel, with a yield strength of 221 MPa (32 ksi), was selected.

Transformers

Tables 14 to 16 recommend magnetically soft materials for nonrotating equipment, the most important of which are transformers.

In transformers, one of the important considerations is weight. The volume of a transformer is closely proportional to that of the core, and the weight of core and coil are usually about equal. For minimum weight, material with the highest possible flux density must be used in the core. Overvoltage requirements must be considered. Regular oriented grades M-6 to M-4 have flux densities of about 1.8 to 18.4 teslas (18 000 to 18 400 gausses) in a field of 10 oersteds. This limits design flux density to about 1.7 teslas (17 000 gausses). Somewhat higher flux density is attainable with the high-permeability grades designated "PO" by ASTM nomenclature in Table 1.

Although core weight is important, values of energy losses for transformers are receiving more and more attention as costs of generating electrical power escalate. This factor is placing more and more emphasis on the importance of reducing core loss by lowering operating inductions or through use of more costly grades of core material, which are thinner and which may have higher permeability.

Noise produced in transformers results ultimately from magnetostriction of the core material, which varies with operating flux density, silicon content, operating strain and type of surface insulation. Flux density is important in magnetostriction and joint noise. For typical designs of power transformers using grain-oriented material, a 10% reduction in flux density will reduce noise about 3 dB.

Interlaminar insulation is necessary for high electrical efficiency in the magnetic core, whether the application is static or rotating. For small cores used in communications and in fractional-horsepower motors, an oxide surface on the laminations insulates the core adequately. Insulations of AISI types C-1, C-2, C-3, C-4 and C-5 are used for more rigorous requirements. Types C-1 and C-3 are organic and cannot be successfully applied to laminations before annealing. They are unsuitable for electrical equipment operated at high temperatures or for power transformers with certain types of coolants. However, they improve the punchability of the sheet steel.

Inorganic types C-4 and C-5 are used when insulation requirements are severe and when annealing temperatures up to 790 °C (1450 °F) must be withstood. Typical values of interlaminar resistance for these two types are between 3 and 100 Ω·cm per lamination under a pressure of 2070 kPa (300 psi). These coatings also can be made to impart residual tensile stresses in the steel substrate, which improves magnetic properties as described above.

Core insulation is required to be sufficiently thin and uniform so as to have no more than 1.0% effect on the lamination factor (solidity of the core). To calculate the required insulation for most operations at power frequency, the square of the resistivity, in ohm-centimetres per lamination, should at least equal the square of the width of the magnetic path, in inches. This usually ensures a negligible interlaminar loss that is less than 1.0% of the core loss.

Selection for Fabrication. Lami-

Table 15 Recommended materials for applications based on specialized properties

Application	Special property	Silicon steels(a)	Other alloys(b)
Filters	High incremental permeability, low loss	M-22, M-19, M-15	Powdered Moly Permalloy, 45 to 50% Ni iron
Loading coils	Constant permeability with changing temperature	....	81-2 brittle Moly Permalloy, stabilized 81-2 brittle Moly Permalloy
Magnetic amplifiers	Rectangular hysteresis loop	Oriented silicon steel M-6	HYRA 49, Orthonol ½ to ¼ mil, 4-79 Moly Permalloy
Magnetic switch storage device	....	....	⅛ to ½ mil Moly Permalloy, ferrites
Ultrasonic	High magnetostriction	....	Vanadium Permendur, nickel-iron alloys
Magnetic shields	High permeability in weak fields	M-6	45 to 50% Ni iron

(a) The various grades of silicon steel sheet presented here are listed in order of increasing cost. (b) No preference in either economy or operating characteristics is intended by the order in which other alloys are listed.

Fig. 19 Variation in permeability index of P/M iron with duration of sintering

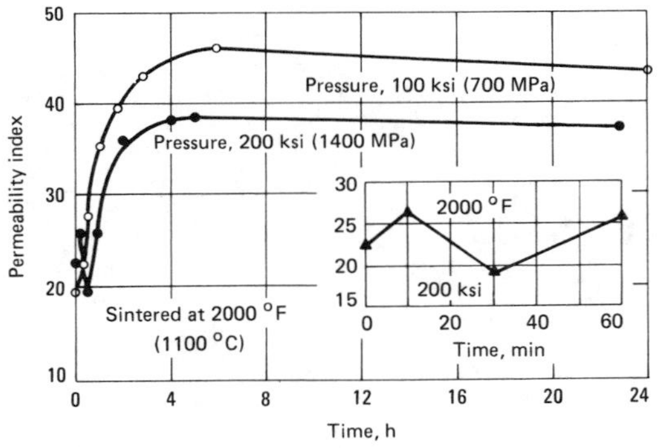

nations usually are fabricated by punching the required shape from flat or coiled sheet, or from coiled strip. Selection of grade and gage is based primarily on electrical requirements. Differences in punchability may be considered in final selection. Studies of electrical sheet indicate that punchability decreases as silicon content increases, but test results are not conclusive. High-silicon alloys are inherently more abrasive, but they are also more brittle and can be punched with less roll and drag at the edges than can the more ductile low-silicon grades. Typical data on the relative wear of steel and carbide dies in punching various grades of electrical sheet having different coatings are given in the article in this volume on selection of material for blanking dies.

Optimum punchability also depends on factors other than steel composition or method of manufacture. If it is difficult to punch a fully processed sheet, it may be advisable to use a semiprocessed grade with magnetic properties not fully developed by the supplier. In this situation, the laminations must be annealed to produce the desired properties. Laminations for small motors may require annealing at temperatures from 790 to 870 °C (1450 to 1600 °F). For grain-oriented grades, purchased in the semiprocessed condition, annealing temperatures are as high as 1200 °C (2200 °F).

The advantage of the better punchability of semiprocessed electrical sheet is offset somewhat by the need for better control of annealing conditions during fabrication and by adhesion of laminations during annealing. It is impractical to grade steel for electrical properties after laminations have been fabricated. Therefore, the supplier must grade the product by annealing samples cut from the sheet before it is shipped. The lower cost of the semiprocessed material may influence the designer's decision.

Table 16 Applications of Fe-Ni and Fe-Co magnetically soft alloys

Application	Recommended material
Cores for computers and radar pulse transformers; special transformers	4-79 Moly Permalloy, Supermalloy
Shields; special choke coils and transformers for audio and carrier circuits	Mumetal, 4-79 Moly Permalloy
Shields; audio chokes and transformers; relays; pole pieces for communication use	45 to 50% Ni iron
Pole tips and magnet yokes	Permendur
Telephone diaphragms	Vanadium, Permendur

Corrosion Resistance. In specific applications, corrosion resistance limits the choice of magnetically soft materials. Vanadium Permendur and Mumetal have high magnetic quality and good corrosion resistance. Supermalloy and 4-79 Moly Permalloy have fair corrosion resistance. The iron-nickel alloys have only mild resistance to corrosion, and silicon steels corrode readily. Consequently, many of these materials must be protected from corrosion by painting, canning, plating, potting or molding; where none of these can be done, materials of inferior magnetic quality but better corrosion resistance, such as ferritic stainless steels, may be used. In such applications, allowance must be made for higher core losses, lower saturation (compared with silicon steels), lower permeability, and a different specific gravity of the core material. Nonhardening types such as 430 and 405 develop the best magnetic properties of all stainless steels, and these steels, along with type 416 because of its machinability, are most often used where a corrosion-resistant core material is needed.

Field Frame Materials

The magnetic properties of cast low-carbon steels are adequate for yoke structures or pole pieces in dc machinery. Field structures of complicated shape with mounting lugs or feet can be produced more economically as castings than as fabricated structures unless the proper size of commercial steel tubing is available and can be welded in place.

Automotive field frames made of steel tubing cost about 6% more than rolled and welded frames. The total cost of the completely fabricated frame is about the same as the cost of the tubing alone. Steels such as 1211, 1010, 1015 and 1020 are commonly used for these applications and are annealed to improve magnetic properties by increasing permeability and narrowing the hysteresis loop. The limitations of these steels should be clearly understood; for more nearly ideal properties, annealed ingot iron or commercially pure iron should be used.

SELECTED REFERENCES

- *Soft Magnetic Materials,* by R. Ball: Heyden & Son Ltd., London, 1979. [Handbook of soft magnetic materials with comprehensive commercial data.]
- *Magnetism and Metallurgy of Soft Magnetic Materials,* by C. W. Chen: North Holland, New York, 1977. [General text on soft magnetic materials.]
- *Magnetic Materials and Their Applications,* by C. Heck: Crane, Russak & Co., Inc., New York, 1974. [Comprehensive treatment of all magnetic materials in terms of component applications.]
- *Ferromagnetic Materials—A Handbook on the Properties of Magnetically Ordered Substances,* edited by E. P. Wohlfarth: Elsevier, New York, 1980. [Volume 1 contains an article on "Amorphus Ferromagnets" by F. E. Luborsky. Volume 2 contains articles on "Soft Magnetic Metallic Materials" by G. Y. Chin and J. H. Wernick, "Ferrites for Non-Microwave Applications" by P. L. Slick and "Microwave Ferrites" by J. Nicolas.]
- "Steel Products Manual—Flat Rolled Electrical Steel": American Iron and Steel Institute, March 1978. [Properties of low-carbon steel and silicon steel as established by ASTM.]

Materials for Permanent Magnets

By the ASM Committee on Permanent Magnets*

THE TERM "permanent magnet" is used to describe materials that are normally used in a single magnetic state and that have sufficiently high resistance to demagnetizing fields and sufficiently high magnetic flux output to provide useful and stable magnetic fields. This implies insensitivity to temperature effects, mechanical shock and demagnetizing fields. This article does not consider magnetic memory or recording materials in which the magnetic state is altered during use. It does include, however, hysteresis alloys used in motors.

Permanent magnet materials include a variety of metals, intermetallics and ceramics. Commonly included are certain steels, Alnico, Cunife, Fe-Co alloys containing V or Mo, Pt-Co, hard ferrites, and cobalt–rare earth alloys. Each type of magnet material possesses unique magnetic and mechanical properties, corrosion resistance, temperature sensitivity, fabrication limitations and cost. These factors provide designers with a wide range of options in designing magnetic parts.

Permanent magnet materials are based on the cooperation of atomic and molecular moments within a magnet body to produce a high magnetic induction. This induced magnetization is retained because of a strong resistance to demagnetization. These materials are classified ferromagnetic, and do not include diamagnetic or paramagnetic materials. The natural ferromagnetic elements are iron, nickel and cobalt. Other elements, such as manganese or chromium, can be made ferromagnetic by alloying to induce proper atomic spacing. Ferromagnetic metals combine with other metals or with oxides to form ferrimagnetic substances; ceramic magnets are of this type. Although scientific literature lists many magnetic substances, relatively few have gained commercial acceptance because of the commercial requirement for low cost and high efficiency.

Permanent magnet materials are marketed under a variety of trade names and designations throughout the world. The United States designations will be used here; other designations are listed in Table 1.

Permanent magnet materials are developed for their chief magnetic characteristics: high induction, high resistance to demagnetization, and maximum energy content. Magnetic induction is limited by composition; the highest saturation induction is found in binary Fe-Co alloys. Resistance to demagnetization is conditioned less by composition than by shape or crystal anisotropies and the mechanisms that subdivide materials into microscopic regions. Precipitations, strains and other material imperfections, and fine particle technology are all used to obtain a characteristic resistance to demagnetization.

Maximum energy content is most important because permanent magnets are used primarily to produce a magnetic flux field (which is a form of potential energy). Maximum energy content and certain other characteristics of materials used for magnets, are best described by a hysteresis loop. Hysteresis is measured by successively applying magnetizing and demagnetizing fields to a sample and observing the related magnetic induction.

Fundamentals of Magnetism

For understanding a permanent magnet, Faraday's concept of representing a magnetic flux field by "lines of force" is very useful. The lines of force radiate outward from a "north pole" and return at a "south pole". The lines of force can be revealed by a powder pattern made by sprinkling iron powder on a paper placed above a bar magnet. The number of lines per unit area is the magnetic induction and is designated B. Induction consists of lines of force due to the magnetic field

*See page XIII for committee list.

and lines of magnetization due to the ferromagnetism of the magnet:

$$B = H + B_i \quad \text{(Eq 1)}$$

where H is the magnetic field strength and B_i is the intrinsic induction.

Magnetic Hysteresis. A hysteresis loop is a common method of characterizing a permanent magnet. The intrinsic induction is measured as the magnetizing field is changed (see Fig. 1). Starting with a virgin state of the material at the origin O, induction increases along curve I to the point marked $+S$ as the field is increased from zero to maximum. The point $+S$ is the point at which induction no longer increases with higher magnetizing field, and is known as the saturation induction. When the magnetizing field is reduced to zero in permanent magnets, most of the induction is retained. In Fig. 1, when the field is reduced through zero to $-S$, the induction decreases from $+S$ to B_r to $-S$. At zero field, there is a remanent magnetization in the sample, defined as B_r; the value of B_r approaches the saturation induction in well prepared permanent magnet materials.

If the field is increased again in the positive direction, the induction passes through $-B_r$ to $+S$ as shown, and not through the origin. Thus, there is a hysteresis between the magnetization and demagnetization curves, and this plot is called the hysteresis loop. The two halves of the loop are generally symmetrical and form a major loop, which represents the maximum energy content, or the amount of magnetic energy that can be stored in the material. Innumerable minor loops can be measured within the major loop, measurements being made to show the effects of lesser fields on magnets under operating conditions.

Demagnetization. The particular value of the demagnetizing field is called the intrinsic coercive force H_{ci}. Figure 1 includes the normal demagnetization curve derived from the intrinsic curve. The field required to reduce induction B to zero is the normal coercive force H_c. The important practical features of the curve for application to permanent magnet materials are the numerical values of B_r and H_c, and the area within the hysteresis loop.

Because a permanent magnet most often is used to provide a flux field in a space outside itself, the material rests within its own field, which is a self-demagnetizing field. Therefore, for

Table 1 Principal magnet designations and their origins

Designation	Magnet type(a)	Country	Company(b)
Acier Co.	Co steel	France	Giffey-Pretre
Acier Cr. Co.	CoCr steel	France	Giffey-Pretre
Acero al carbono	Carbon steel	Spain	Echevarria
Acero al wolframio	W steel	Spain	Echevarria
Alcomax	Anisotropic Alnico	England	PMA
Alloy 1751	CoPt	U.S.	...
Alni	Isotropic Alnico (no cobalt)	England	PMA
Alnico	Alnico	U.S.	Arnold, Crucible, Hitachi Magnetics, Indiana, Simonds, Thomas and Skinner, Westinghouse, General Magnet, Permanent Magnet Co.
		Canada	Canadian GE
		India	Elpro
		Germany	Magnetfabrik
		Italy	Sampas, Elett. Lombarda
Alnico	Isotropic Alnico (with cobalt)	England	PMA
Alnicomax	Anisotropic Alnico	Italy	Elett. Lombarda
Arnox	Barium ferrite	U.S.	Arnold
Bismanol	MnBi	U.S.	...
Bonded Feroba	Organic-barium ferrite	England	PMA
Caslox I	Cobalt ferrite	England	...
Caslox II, III	Barium ferrite	England	...
Ceramagnet	Barium ferrite	U.S.	Stackpole
Ceramic magnet	Barium ferrite	U.S.	Steward, Allen-Bradley
CHJ	CoCr steel	Japan	Hitachi Metals, Ltd.
Chrome steel	Cr steel	U.S.	Bethlehem, Crucible, Indiana, Permanent Magnet Co., Simonds, Thomas and Skinner
		England	PMA
		Italy	Elett. Lombarda
Coalni	Isotropic Alnico	Italy	Centro
Coalnimax	Anisotropic Alnico	Italy	Centro
Cobalt platinum	CoPt	U.S.	Engelhard
Cobalt steel	Co steel	U.S.	Engelhard
		England	...
		Italy	...
Coercimax	Anisotropic Alnico	Italy	Centro
Columax	Anisotropic Alnico	U.S.	Thomas and Skinner
		England	PMA
Comalloy	CoMoFe	England	...
Comol	CoMoFe	U.S.	Simonds
CoNiAl	Isotropic Alnico	Austria	...
Coramag	Co rare-earth	France	Ugimag
Coramax	Co rare-earth	Japan	Sumitomo Special Metals
Crucore	Co rare-earth	U.S.	Crucible
CS 1	Cunife	Japan	Sumitomo
CS 2	Cunico	Japan	Sumitomo
CS 3	Vicalloy	Japan	Sumitomo
Cunico	CuNiCo	U.S.	Arnold, Indiana
Cunife	CuNiFe	U.S.	Arnold, Hoskins, Indiana

(continued)

(a) Alnico refers to the FeNiAl or the FeCoNiAl type alloys. (b) PMA refers to the Permanent Magnet Association whose members include Edgar Allen, Darwins, English Steel Magnet, H. J. Foster, Jessop-Saville, Swift Levick, Marrison and Catherall, Murex, Neil, Sheffield Magnet, Turton and Matthews, and Watson and Saville.

Table 1 (continued)

Designation	Magnet type(a)	Country	Company(b)
E.S.D.	Elongated single domain (see Lodex)	U.S.	...
Ergit	Steel	Hungary	...
F310	Isotropic barium	U.S.	Steward
F510, 620	Oriented barium ferrite	U.S.	Steward
Fama 700	Isotropic Alnico	Sweden	...
Ferbalite	Barium ferrite	...	Lignes Teleg. et Telep.
Feroba	Barium ferrite	England	PMA
Ferriflex	Flexible barium ferrite	France	Ugimag
Ferrimag	Barium ferrite	U.S.	Crucible
Ferrolite	Barium ferrite	...	Lignes Teleg. et Telep.
Ferroxdure	Barium ferrite	France Holland Italy	Giffey-Pretre Philips Sampas, Centro
FXD	Barium ferrite	Japan	Sumitomo
Gaussit	Vicalloy	Germany	...
Gecalloy	Fe or FeCo powder	England	...
Genox	Barium ferrite	U.S.	General Magnet
Gumox	Flexible barium ferrite	Germany	Magnetfabrik
Ha	Steel	Belgium	Henricot
Hicorex	Co rare-earth	U.S.	Hitachi Magnetics
Hicorex	Co rare-earth	Japan	Hitachi Metals, Ltd.
HLRA	Co rare-earth	England	Magnetic Polymers
Honda alloy (new KS)	27 Co, 18 Ni, 7 Ti, 4 Al, bal Fe	Japan	...
Hycomax	Anisotropic Alnico	England	PMA
Hyflux	Anisotropic	U.S.	Indiana
Hynical	12 Al, 32 Ni, 56 Fe	England	...
Hynico	Isotropic Alnico	England	PMA
Hysterloy	Alnico (Hysteresis material)	U.S.	Permag
Incor	Co rare-earth	U.S.	Indiana General
Indalloy	CoMoFe	U.S.	Indiana General
Indox	Barium ferrite	U.S.	Indiana General
Iron Ti	FeTi	Japan	...
K-	Steel	Germany	...
Katos oxide	Cobalt ferrite	Japan	...
Koerox	Barium ferrite	Germany	Krupp
Koerzit	Alnico	Germany	Krupp
Koerzit T, H	Vicalloy	Germany	Krupp
Koroseal	Flexible barium ferrite	U.S.	Goodrich
KS	Co steel	Japan	Sumitomo
Lanthanet	Co rare-earth	Japan	Tohoku Metals
Lodex	Elongated iron-cobalt	U.S.	Hitachi Magnetics
M-01, 05	Barium ferrite	U.S.	Allen-Bradley
Magnadur	Barium ferrite	England	Mullard
Magnetoflex	CuNiFe	Germany	...
Magnico	Alnico	USSR	...

(continued)

(a) Alnico refers to the FeNiAl or the FeCoNiAl type alloys. (b) PMA refers to the Permanent Magnet Association whose members include Edgar Allen, Darwins, English Steel Magnet, H. J. Foster, Jessop-Saville, Swift Levick, Marrison and Catherall, Murex, Neil, Sheffield Magnet, Turton and Matthews, and Watson and Saville.

practical applications, a magnet designer is interested primarily in the second quadrant of the hysteresis loop, called the demagnetization curve (see Fig. 2). This curve represents the resistance to demagnetization and, in an affirmative sense, the ability of a material to establish a magnetic field in an air gap or adjoining magnetic material.

Magnetic Energy. The maximum magnetic energy available for use outside the magnet body is proportional to the largest rectangle that fits inside the normal demagnetization curve. It is indicated by the product $(B_dH_d)_{max}$ and is usually cited as the figure of merit for determining the quality of permanent magnet materials.

A characteristic useful in selecting permanent magnet materials subjected to varying demagnetizing conditions is the permeability at the operating point: $\mu = B_d/H_d$. For example, a straight-line demagnetization curve where $B_r = H_c$ would have the ideal permeability of 1.0; a magnet of such a material would recover spontaneously all flux when a partial demagnetizing field is removed. The corresponding intrinsic curve would be flat out to the knee, and the material would retain maximum energy. Cobalt–rare earth magnets come closest to ideal permanent magnet behavior.

Figure 3 is the product of B and H at each point along the demagnetization curve, plotted against B. On the demagnetization curve, each value of B or H involves the other as a coordinate variable. The maximum value of their product, designated as $(B_dH_d)_{max}$, represents the maximum magnetic energy that a unit volume of the material can produce in an air gap. The most efficient design for a magnet is that which employs the magnet at the flux density corresponding to the $(B_dH_d)_{max}$ value.

The amount of total external magnetic flux available from a magnet operating in an open-circuit condition depends on its shape. This relation is shown in Fig. 4 for one specific shape. The permeability μ is the ratio of the total external permeance B_d to that of the permeance of the space occupied by the magnet, H_d, and is equal to the slope of the demagnetization curve.

An enlarged plot of the first and second quadrants of the intrinsic induction curve is given in Fig. 5. The intrinsic induction curve is primarily of interest to the materials scientist, who is concerned about the effect of composition and processing on the various intrinsic parameters of the material: B_{is}, B_r, and H_{ci}. The applications engineer, on the other hand, is concerned about the flux density in an air gap due to both B_i and H. Accordingly, he is likely to be more interested in the induction curve and in the values of B_r, H_c and $(BH)_{max}$.

Lines of constant energy product (B_dH_d) usually are plotted in the second quadrant area and, as illustrated in Fig. 6, are a series of hyperbolic curves superimposed on the rectangular B-H grid of the demagnetization curves. The maximum values of external energy are therefore readily available in relation to the demagnetization curve. In this form, the grid constitutes an efficient guide for the design engineer. In practice, a magnet with a fixed air gap would have one fixed B_d/H_d operating point on the demagnetization curve corresponding to the material being used. For variable air gaps, such as are produced by relative movement between the armature and field poles of electrical machinery, the external energy available at the air gap changes continuously, resulting in a so-called minor loop with minimum and maximum values. In practice, the minor loop is plotted on the demagnetization curve to determine location of the loop on the curve, and to evaluate the extent of flux variation within the minor loop cycle. Efficient design of equipment using permanent magnets, such as magnetos, small generators and motors, requires that the minor loop operate near the $(B_dH_d)_{max}$ point.

Magnetically soft materials differ from permanent magnet materials not only in their higher permeabilities, but also, and more significantly, in their much lower resistance to demagnetization. The best magnetically soft materials have H_c values of virtually zero. The hysteresis loop of such a material retraces itself through or near the origin point with each cycle.

Conversely, permanent magnet materials have wide hysteresis loops, characterized by high values of H_{ci}, which range from about 100 to over 20 000 Oe.

Commercial Permanent Magnet Materials

Table 2 lists most of the permanent magnet materials commercially available in the U.S. and their nominal compositions. Magnetic properties are

Table 1 (continued)

Designation	Magnet type(a)	Country	Company(b)
Maxalco	Alnico	Italy	Sampas
Mishima metal	29 Ni, 14 Al, bal Fe	Japan	...
MK	Alnico	Japan	Mitsubishi
Mo	FeCrMo	Japan	...
MRC	Co rare-earth	Japan	Mitsubishi
MS	Steel	Switzerland	...
MT	90.5 Fe, 8 Al, 1.5 C	Japan	Mitsubishi
MVC	FeCoV	Japan	Mitsubishi
New KS (Honda alloy)	27 Co, 18 Ni, 7 Ti, 4 Al, bal Fe Alnico 12	Japan	...
NiAl	Isotropic Alnico	France	Giffey-Pretre
NiAl	Isotropic Alnico	Belgium	Henricot
NiAlCo	Alnico	Austria	...
NiAlCo	Isotropic Alnico	France	Giffey-Pretre
NiAlCo	Isotropic Alnico	France	Ugimag
Nipermag	31 Ni, 13 Al, 0.5 Ti, bal Fe	U.S.	...
		England	...
NKS	Alnico	Japan	Sumitomo
Oerstit 30 to 90	Steel	Germany	T.E.W.
Oerstit 90 to 700	Alnico	Germany	T.E.W.
Ox	Barium ferrite	Germany	Magnetfabrik
Oxilit	Bonded barium ferrite	Germany	T.E.W.
Oxit	Barium ferrite	Germany	T.E.W.
P-6	45 Fe, 45 Co, 6 Ni, 4 V	U.S.	Hitachi Magnetics Carpenter Technology
Permet	45 Cu, 25 Ni, 30 Co (sintered)	...	...
PF	Fe or FeCo powder	France	...
Placo	CoPt	Germany	Magnetfabrik
Placovar	CoPt	U.S.	Hamilton Watch
Plastiform	Bonded barium ferrite	U.S.	Leyman
Plasto ferrite	Bonded barium ferrite	France	Ugimag
Platinex	CoPt	U.S.	Bishop
		England	Johnson Matthey
Prac	Bonded Alnico	Germany	Magnetfabrik
Raeco	Co rare-earth	U.S.	Raytheon
Reco	Isotropic Alnico	Holland	Philips
		England	Mullard
Recoma	Co rare-earth	Switzerland	Brown, Boveri & Cie
Remco	Co rare-earth	U.S.	Electron Energy
Rarenet	Co rare-earth	Japan	Shinetsu Chemical Industries
		Spain	Echevarria
		Belgium	Henricot
Remalloy	CoMoFe	U.S.	Arnold, Simonds
Remendur	CoFeV	U.S.	Wilbur Driver, Carpenter Technology
RM, RN	Rubber barium ferrite	Japan	Sumitomo

(continued)

(a) Alnico refers to the FeNiAl or the FeCoNiAl type alloys. (b) PMA refers to the Permanent Magnet Association whose members include Edgar Allen, Darwins, English Steel Magnet, H. J. Foster, Jessop-Saville, Swift Levick, Marrison and Catherall, Murex, Neil, Sheffield Magnet, Turton and Matthews, and Watson and Saville.

Table 1 (continued)

Designation	Magnet type(a)	Country	Company(b)
Safe-Nialco	Isotropic Alnico	Spain	Hamsa
Safe-Supernialco	Anisotropic Alnico	Spain	Hamsa
Samoomag	Co rare-earth	England	Magnetic Developments
Silmanal	87 Ag, 9 Mn, 4 Al	U.S.	...
Simonds No. 73 to 3500	Steel	U.S.	Simonds
Spinal	Isotropic barium ferrite	France	Ugimag
Spinalor	Anisotropic barium ferrite	France	Ugimag
Sprox	Flexible barium ferrite	Germany	Magnetfabrik
Sura	Isotropic and Anisotropic Alnico	Sweden	Surahammars Bruks
Ticonal	Anisotropic Alnico	Holland	Philips
		England	Mullard
		France	Giffey-Pretre, Ugimag
		Spain	Echevarria
		Belgium	Henricot
Tromalit	Bonded Alnico	France	Ugimag
Tungsten steel	Steel	U.S.	Bethlehem, Crucible, Indiana, Permanent Magnet Co., Simonds, Thomas and Skinner
		England	PMA
		Italy	Elett. Lombarda
Ugimax	Anisotropic Alnico	France	Allevard-Ugine
Vacomax	Cobalt ferrite	Germany	Vacuumschmelze Gmbh
Vectolite	Cobalt ferrite	U.S.	...
VS 30	FeCoCrV	Germany	Krupp
Vicalloy	FeCoV	U.S.	Arnold, Carpenter Technology, Thomas and Skinner, Wilbur Driver
Westro-Alpha-Beta	Oriented barium ferrite	U.S.	Westinghouse
YBM	Barium ferrite	Japan	Hitachi Metals, Ltd.
YCM	Alnico	Japan	Hitachi Metals, Ltd.
YRM	Flexible barium ferrite	Japan	Hitachi Metals, Ltd.

(a) Alnico refers to the FeNiAl or the FeCoNiAl type alloys. (b) PMA refers to the Permanent Magnet Association whose members include Edgar Allen, Darwins, English Steel Magnet, H. J. Foster, Jessop-Saville, Swift Levick, Marrison and Catherall, Murex, Neil, Sheffield Magnet, Turton and Matthews, and Watson and Saville.

given in Table 3. Figure 6 presents demagnetization curves associated with the materials listed in Table 2. Physical and mechanical properties are summarized in Table 4. Generally, the production of permanent magnet materials is controlled to achieve magnetic characteristics, and other properties are allowed to vary according to the manufacturing process used. The selection of materials and the design of permanent magnets for particular applications is a well-defined engineering art; design assistance is available from most major producers.

Magnet Steels

Until about 1930, all the commercial permanent magnet materials were quench-hardening steels. Up to about 1910, plain carbon steels containing about 1.5% carbon were the principal magnet alloys. Alloy steels with up to 6% W were then developed, and later, high-carbon steels with 1 to 6% Cr came into use. Coercive forces for this group of alloys ranged from 40 to 70 Oe. The most significant improvement in the quench-hardening steels came in 1917, when the Japanese introduced a cobalt steel containing 36% Co and having a coercive force as high as 250 Oe.

The coercive force in all of these martensitic magnet steels is due to the difficulty of domain boundary movement, resulting from the combined effects of nonmagnetic inclusions, internal strains, lattice defects, and inhomogeneities or voids. When any of these steels is heated into the austenite range, the carbides dissolve. When cooled, the alloy transforms back to ferrite and carbide particles; the size of the carbide particles depends on the rate of cooling or on the time and temperature of annealing below the transformation temperature. Quenching to produce martensite results in the greatest magnetic and mechanical hardness, probably through a combination of the effect of the large number of inclusions and other lattice defects, and the high degree of strain introduced. The more carbon, the higher H_{ci}, but the lower the saturation moment. Any reheating relieves the strains developed during processing and, if continued, will cause growth of precipitated particles. Both effects result in a decrease in H_{ci} and an increase in B_{is}. The composition and processing of each of these steels is controlled to provide the optimum combination of H_{ci}, B_r, stability, ease of processing and cost.

Fabrication of magnet steels is similar to that of tool steels: hot-rolling, forging, or casting and then hardening by quenching from temperatures above 850 °C (1560 °F) to retain a critical proportion of austenite. During hot-rolling or forging—usually between 950 and 1000 °C (1740 and 1830 °F)—massive carbides may form, preventing optimum magnetic properties from developing. Austenizing at 1200 °C (2190 °F) before hardening is then necessary. In order to be machined or cold formed, the steels may be softened by heating for a short time below the eutectoid temperature of about 730 °C (1345 °F), usually after hot rolling but before hardening. Because some deterioration of magnetic properties results, softening should be avoided. A typical manufacturing cycle for cobalt steel consists of solution treatment at 1150 °C (2100

Fig. 1 Major hysteresis loop for a permanent magnet material

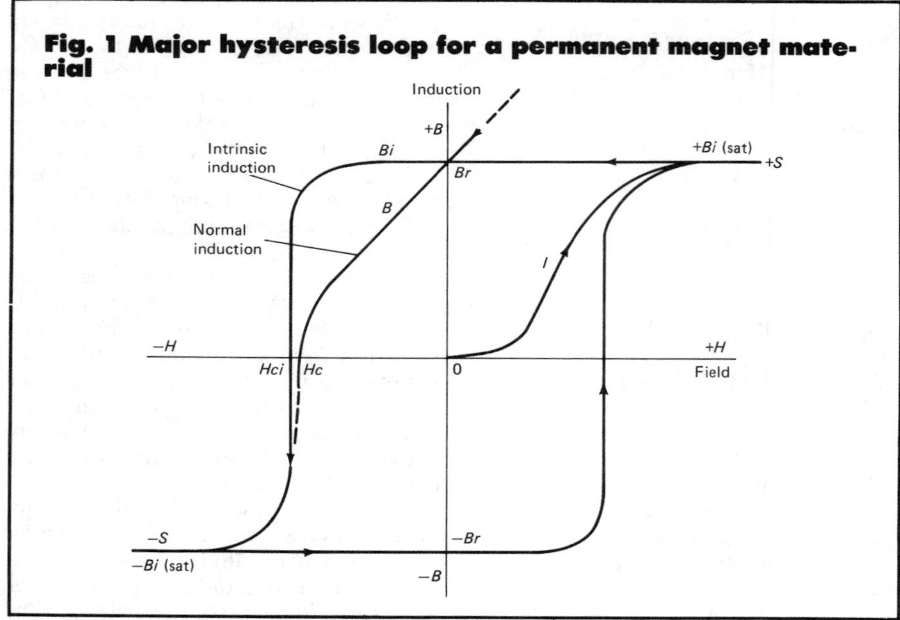

Fig. 3 Typical energy-product curve for a permanent magnet material

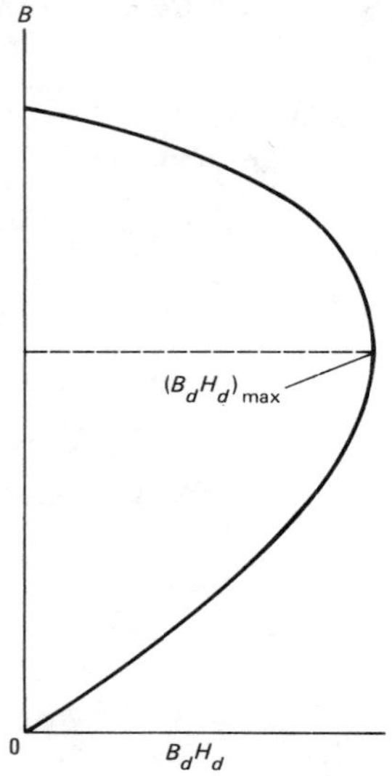

Fig. 2 Normal demagnetization curve for a permanent magnet material

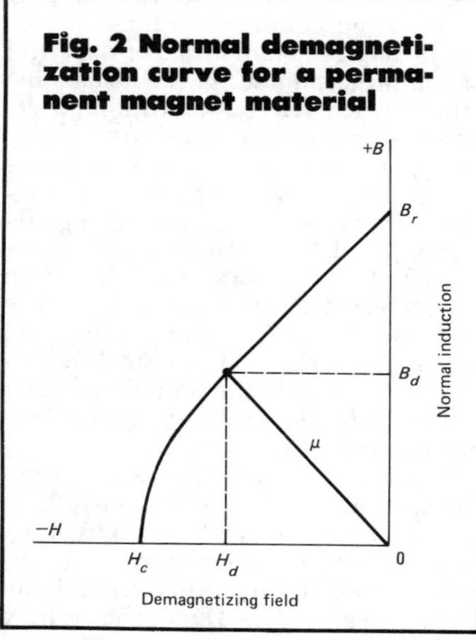

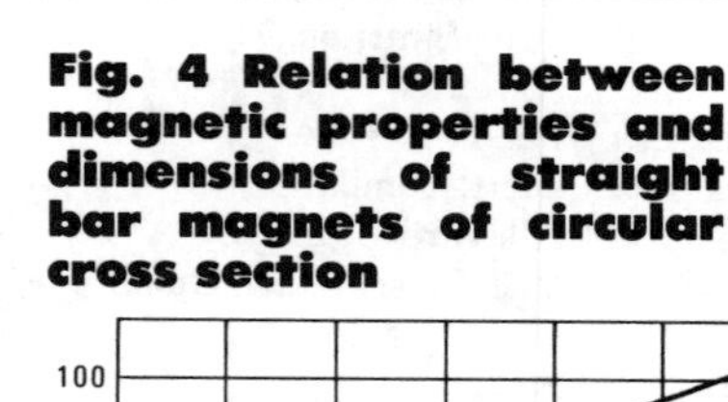

Fig. 4 Relation between magnetic properties and dimensions of straight bar magnets of circular cross section

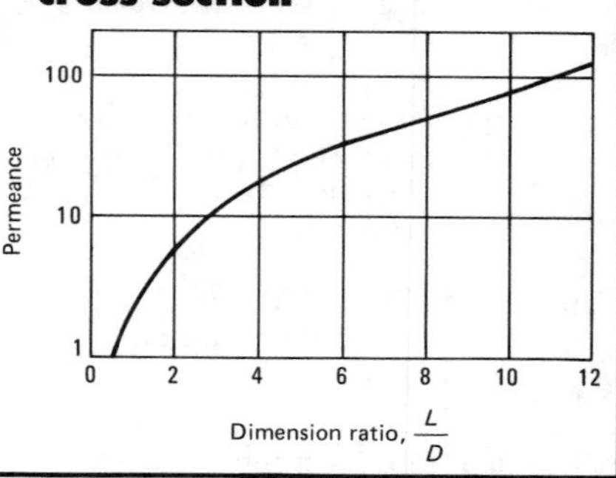

°F) to dissolve the carbides, followed by an air quench; heating at 780 °C (1435 °F) to convert all austenite to ferrite; reheating rapidly to 1000 °C (1830 °F) for 3% to 15% Co steels (or to 950 °C (1740 °F) for 35% Co steels) to precipitate carbides; and quenching to harden the steel and to obtain the optimum number and size of retained austenite regions. After hardening and before magnetizing, the steels may be ground to meet dimensional tolerances. The more intricate shapes are frequently cast, especially if made of 17% or 36% Co steel. Cast magnets are slightly higher in coercive force and lower in residual induction than their wrought counterparts.

Compositions and properties are shown in Tables 2, 3 and 4 for six typical magnet steels. Two carbon steels are included, although carbon steels are now virtually obsolete for magnets. Magnetic properties and cost increase as alloy content increases. Demagnetization curves for steel magnets are shown in Fig. 6.

Magnet steels are unsuitable for use where ambient temperature exceeds 100 °C (212 °F). The steels are more susceptible to aging than other permanent magnet materials. This aging effect, which causes some loss of energy content, is greatest immediately after hardening and becomes negligible after a few weeks. Magnet steels are also more susceptible to loss of energy content from mechanical shock, but after several impacts, the flux remains nearly stable. To ensure the stability of magnet steels, it is customary to subject them to stabilizing treatments before use. These usually consist of heating the magnets for 24 h or more at 100 °C (212 °F) and partially demagnetizing them with ac fields; sometimes mechanical shock treatments also are used.

The use of magnet steels has declined steadily since 1930, when the Alnico alloys were introduced. The relative ease with which magnet steels can be shaped, their low cost, and their use as replacement parts in existing equipment account for their continued use.

Magnet Alloys

The first advance away from magnet steels came in 1931, with the development of a series of ternary alloys of iron and cobalt plus molybdenum or tungsten.

Remalloy. The modern permanent magnet alloys Remalloy and Comol are representative of the ternary alloys. They can be hot rolled and are machinable. Typical properties are given in

Fig. 5 Intrinsic magnetization curve *Bi* in the first and second quadrants compared with the curve for *B*

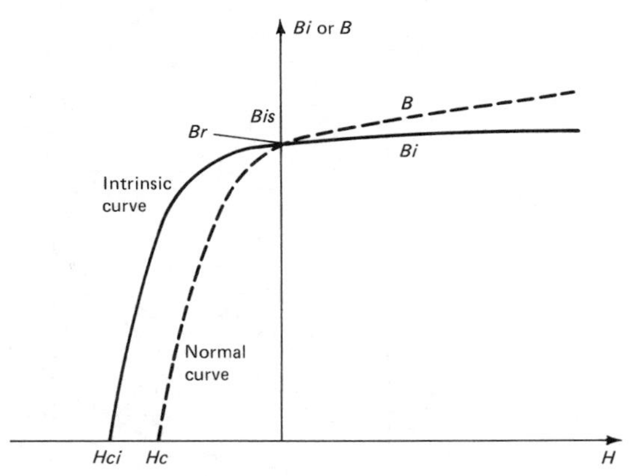

Fig. 6 Demagnetization curves for selected permanent magnet materials

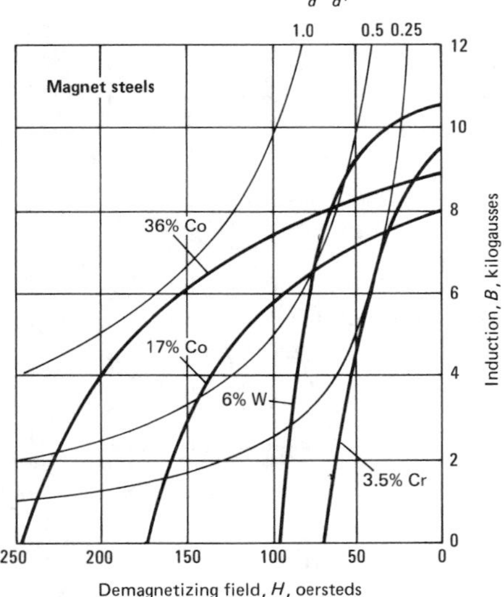

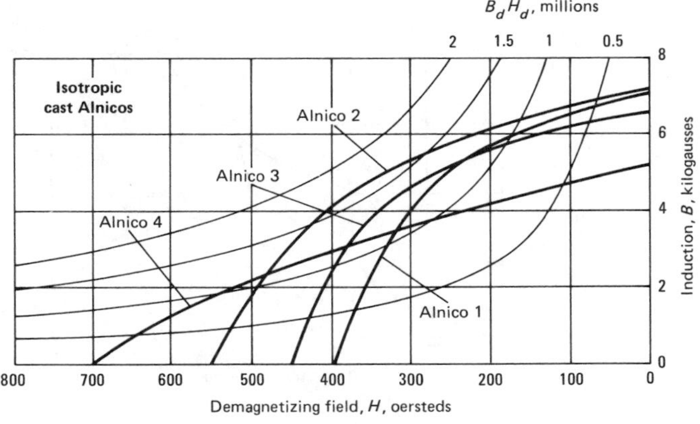

Tables 3 and 4 and the demagnetization curve is shown in Fig. 6.

Although the energy product of Remalloy is about the same as that of 36% Co steel, it costs less and is essentially stable with time. Remalloy was the forerunner of the whole series of dispersion-hardening, carbon-free, permanent magnet materials from which modern magnets are made.

The commercial alloys are isotropic and contain 17 to 20 Mo and 12 Co. The material is solution treated by quenching from a temperature of 1200 to 1300 °C (2190 to 2370 °F). It is then aged at 685 to 700 °C (1265 to 1290 °F) for about 1½ h. This material can be cast and forged, and may also be made by sintering of P/M compacts. There are some problems in obtaining large cross sections.

Cunife, Cunico and Vicalloy. The other three principal alloys intermediate between magnet steels and Alnicos are Cunife, Cunico and Vicalloy. All can be swaged, drawn, rolled, machined and punched. For most shapes, Cunico is isotropic, but in very thin strips it has slightly superior properties at right angles to the rolling direction. Although Cunico can also be cast in some shapes, its best magnetic properties are developed in wire or rod 0.4 to 25 mm (0.015 to 1 in.) in diameter, and in strip with a maximum thickness of about 3 mm (0.125 in.).

Commercial Cunife contains approximately 20 Fe, 20 Ni and 60 Cu. This composition is in the two-phase region of the phase diagram. The material is quenched from about 1000 °C (1830 °F) to give a homogeneous fcc structure. The quenched specimens are already fully magnetic and contain FeNi-rich clusters, 5 to 10 nm in cluster size, in a Cu-rich matrix. It is then aged at 650 °C (1200 °F) to develop the amounts of both phases to optimum proportions. It is usually cold worked in stages to produce maximum directional properties in the final shape. The nature of the phase diagram, the periodicity of the microstructure and x-ray diffraction effects all support the view that the magnetic structure develops by spinodal decomposition. The coercive force can be accounted for on the basis of shape anisotropy of FeNi-rich magnetic regions in a Cu-rich nonmagnetic matrix. The material is extremely anisotropic, with the superior magnetic properties in the direction of rolling, a factor that must be considered in magnet design.

Fig. 6 (continued)

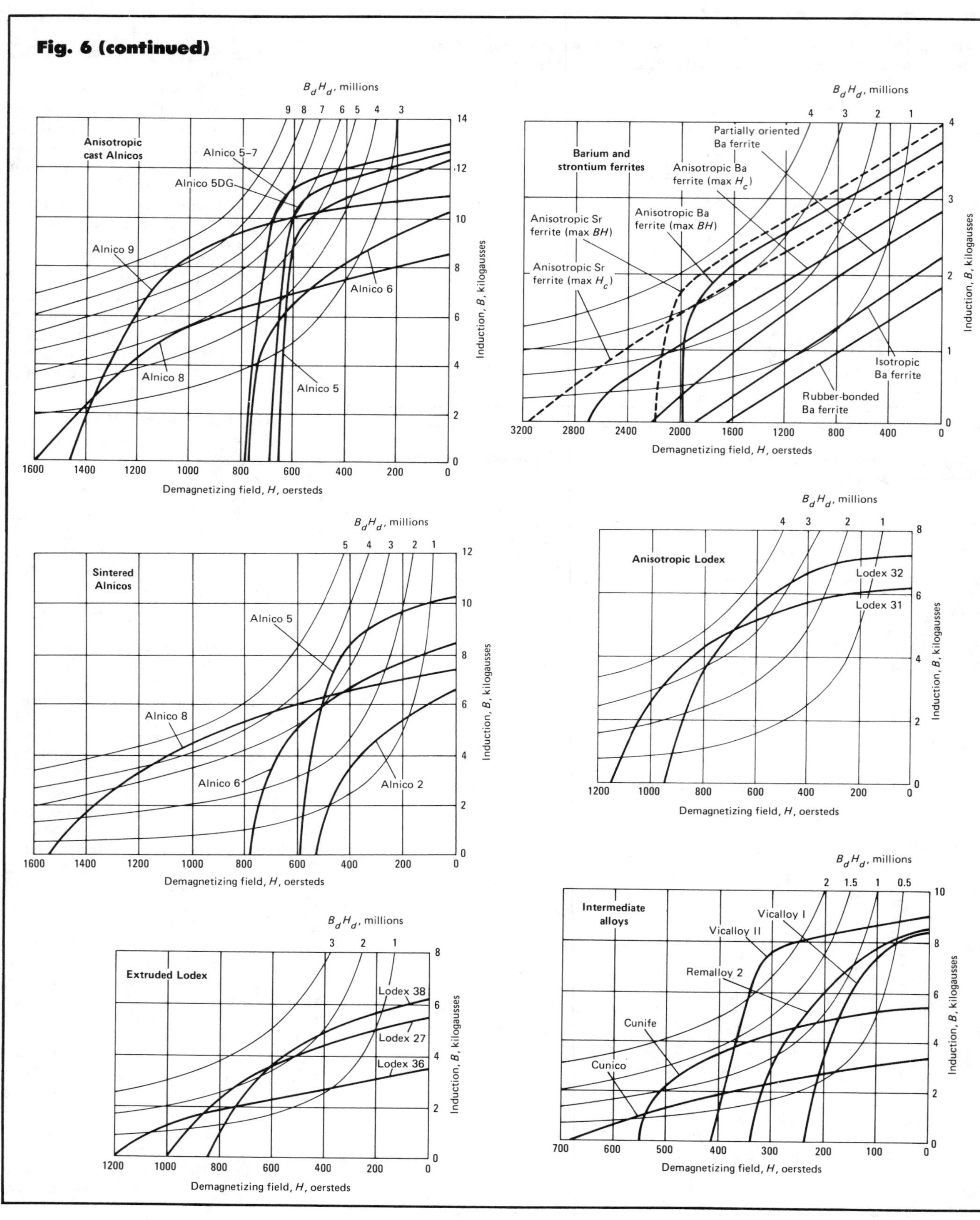

Fig. 6 (continued)

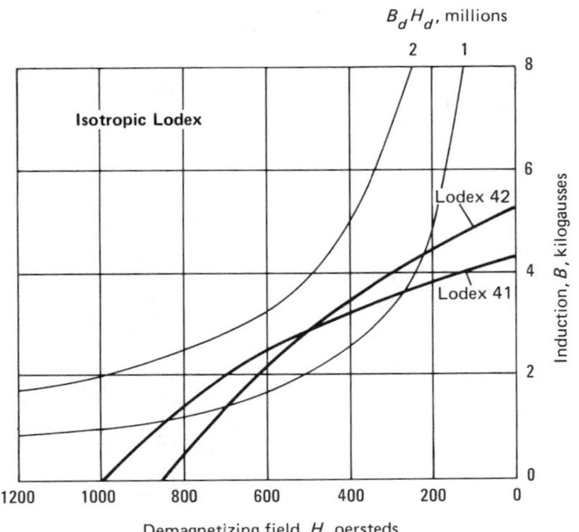

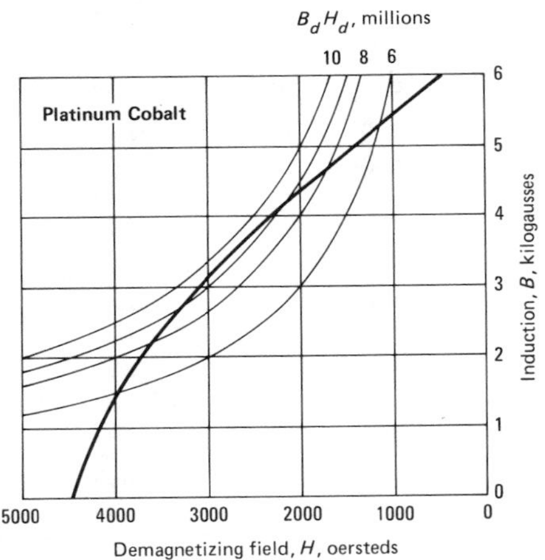

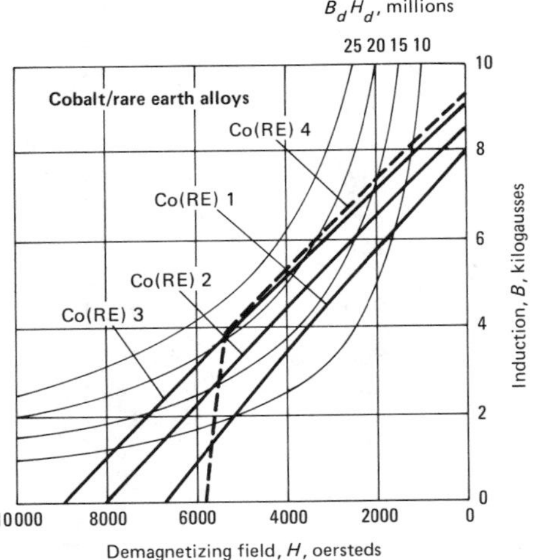

Cold working produces a crystallographic texture which is developed strongly by subsequent annealing. The spinodal decomposes along definite crystallographic planes, and interaction of the spinodal with the crystal texture gives rise to the strong magnetic texture. An additional contribution to magnetic texture may arise from straightforward deformation of the partially decomposed spinodal phase. Aging in a magnetic field has no influence on the magnetic properties because the Curie temperature is too far below the decomposition temperature.

The mechanical softness in this alloy system permits easy cold reduction and working, thus leading to many applications in the form of wire or tape. Optimum properties are obtained only after 95% or more cold reduction.

Commercial Cunico contains about 30% Co, 20% Ni, and 50% Cu and, as expected, is similar to Cunife in both processing and the origin of properties. It has a somewhat lower B_r, but higher H_c than Cunife. Typically, Cunico is solution treated at 1150 °C (2100 °F), quenched, and then aged at 700 °C (1290 °F) to develop the critical two-phase structure. If it is worked, it must be solution treated and aged again—apparently, working destroys the magnetic structure in this alloy. This alloy is too hard to work in the hardened state, and it does not develop any magnetic texture, probably because only weak crystal textures are developed during rolling. Annealing in a magnetic field produces a small anisotropy. Cunico can be cast, but its best properties are developed in wire or rod less than 20 mm in diameter.

Vicalloy alloys can be cast into large billets and rolled into thin sheets. The Vicalloy alloys contain about 50% Co, 10% to 14% V, and remainder iron. In some alloys, Cr replaces part of the vanadium. Vicalloy I, which contains about 10% vanadium, is made without subjecting it to cold deformation; this results in isotropic magnetic properties. Vicalloy II, which contains about 14% vanadium, is subjected to 90% or more reduction during mill processing, resulting in highly directional properties. Vicalloy I is made by quenching from about 1200 °C (2190 °F), forming to shape, and then aging 8 h at 600 °C (1110 °F). Vicalloy II is cast, worked to a convenient size, cold reduced 90% or more, and then aged 8 h at 600 °C (1110 °F). In the solution treated or cold worked conditions, Vicalloy alloys may

Table 2 Nominal compositions, Curie temperatures and magnetic orientations of selected permanent magnet materials

Designation	Nominal composition	Approximate Curie temperature °C	°F	Magnetic orientation(a)
3½% Cr steel	Fe-3.5Cr-1C	745	1370	No
6% W steel	Fe-6W-0.5Cr-0.7C	760	1400	No
17% Co steel	Fe-17Co-8.25W-2.5Cr-0.7C	...	...	No
36% Co steel	Fe-36Co-3.75W-5.75Cr-0.8C	890	1630	No
Cast Alnico 1	Fe-12Al-21Ni-5Co-3Cu	780	1440	No
Cast Alnico 2	Fe-10Al-19Ni-13Co-3Cu	810	1490	No
Cast Alnico 3	Fe-12Al-25Ni-3Cu	760	1400	No
Cast Alnico 4	Fe-12Al-27Ni-5Co	800	1475	No
Cast Alnico 5	Fe-8.5Al-14.5Ni-24Co-3Cu	900	1650	Y,H
Cast Alnico 5DG	Fe-8.5Al-14.5Ni-24Co-3Cu	900	1650	Y,H,C
Cast Alnico 5-7	Fe-8.5Al-14.5Ni-24Co-3Cu	900	1650	Y,H,C
Cast Alnico 6	Fe-8Al-16Ni-24Co-3Cu-2Ti	860	1580	Y,H
Cast Alnico 7	Fe-8Al-18Ni-24Co-4Cu-5Ti	840	1540	Y,H
Cast Alnico 8	Fe-7Al-15Ni-35Co-4Cu-5Ti	860	1580	Y,H
Cast Alnico 9	Fe-7Al-15Ni-35Co-4Cu-5Ti	...	...	Y,H,C
Cast Alnico 12	Fe-6Al-18Ni-35Co-8Ti	...	...	No
Sintered Alnico 2	Fe-10Al-17Ni-12.5Co-6Cu	610	1490	No
Sintered Alnico 4	Fe-12Al-28Ni-5Co	800	1475	No
Sintered Alnico 5	Fe-8.5Al-14.5Ni-24Co-3Cu	900	1650	Y,H
Sintered Alnico 6	Fe-8Al-16Ni-24Co-3Cu-2Ti	860	1580	Y,H
Sintered Alnico 8	Fe-7Al-15Ni-35Co-4Cu-5Ti	860	1580	Y,H
Bonded ferrite A	$BaO-6Fe_2O_3$ + organics	450	840	No,P
Bonded ferrite B	$BaO-6Fe_2O_3$ + organics	450	840	No
Sintered ferrite 1	$BaO-6Fe_2O_3$	450	840	No,P
Sintered ferrite 2	$BaO-6Fe_2O_3$	450	840	Y,A
Sintered ferrite 3	$BaO-6Fe_2O_3$	450	840	Y,A
Sintered ferrite 4	$SrO-6Fe_2O_3$	460	860	Yes
Sintered ferrite 5	$SrO-6Fe_2O_3$	460	860	Yes
Lodex 30	9.9Fe-5.5Co-77.0Pb-8.6Sb	980	1800	Y,A
Lodex 31	16.0Fe-9.0Co-67.5Pb-7.5Sb	980	1800	Y,A
Lodex 32	19.2Fe-10.8Co-63.0Pb-7.0Sb	980	1800	Y,A
Lodex 33	21.9Fe-12.3Co-59.2Pb-6.6Sb	980	1800	Y,A
Lodex 36	9.9Fe-5.5Co-77Pb-8.6Sb	980	1800	No,E
Lodex 37	16Fe-9Co-67.5Pb-7.5Sb	980	1800	No,E
Lodex 38	19.2Fe-10.8Co-63Pb-7.0Sb	980	1800	No,E
Lodex 40	9.9Fe-5.5Co-77Pb-8.6Sb	980	1800	No,P
Lodex 41	16Fe-9Co-67.5Pb-7.5Sb	980	1800	No,P
Lodex 42	19.2Fe-10.8Co-63.0Pb-7.0Sb	980	1800	No,P
Lodex 43	21.9Fe-12.3Co-59.2Pb-6.6Sb	980	1800	No,P
P-6 alloy	45Fe-45Co-6Ni-4V	...	...	No
Cunife	20Fe-20Ni-60Cu	410	770	Y,R
Cunico	29Co-21Ni-50Cu	860	1580	No
Vicalloy I	39Fe-51Co-10V	855	1570	No
Vicalloy II	35Fe-52Co-13V	855	1570	Y,R
Remalloy 1	17Mo-12Co-71Fe	900	1650	No
Remalloy 2	20Mo-12Co-68Fe	900	1650	No
Platinum cobalt	76.7Pt-23.3Co	480	900	No
Cobalt rare earth 1	Co_5 Sm	725	1340	Y,A
Cobalt rare earth 2	Co_5Sm	725	1340	Y,A
Cobalt rare earth 3	Co_5 Sm	725	1340	Y,A
Cobalt rare earth 4	$(Co, Cu, Fe)_7$ Sm	...	...	Y,A

(a) Y, yes; H, orientation developed during heat treatment; C, columnar crystal structure developed; P or E, some orientation developed during pressing or extrusion; R, orientation developed by rolling or other mechanical working; A, orientation developed predominantly by magnetic alignment of powder prior to compacting but alignment influenced by pressing forces also.

Table 3 Nominal magnetic properties of selected permanent magnet materials

For nominal compositions, see Table 2; for mechanical and physical properties, see Table 4.

Designation	H_c, Oe	H_{cf}, Oe	B_r, G	B_{is}, G	$(BH)_{max}$, MG·Oe	B_d, G	H_d, Oe	Required magnetizing field, Oe	Permeance coefficient at $(BH)_{max}$	Average recoil permeability
3½% Cr steel	66	...	9 500	...	0.29	...	...	...	...	...
6% W steel	74	...	9 500	...	0.33	...	...	...	...	...
17% Co steel	170	...	9 500	...	0.65	...	...	...	...	...
36% Co steel	240	...	9 750	...	0.93	...	...	...	...	...
Cast Alnico 1	440	455	7 100	10 500	1.4	4 500	305	2 000	14	6.8
Cast Alnico 2	550	580	7 250	10 900	1.6	4 500	350	2 500	12	6.4
Cast Alnico 3	470	485	7 000	10 000	1.4	4 300	320	2 500	13	6.5
Cast Alnico 4	730	770	5 350	8 600	1.3	3 000	420	3 500	8.0	4.1
Cast Alnico 5	620	625	12 500	13 500	5.25	10 200	525	3 000	18	4.3
Cast Alnico 5DG	650	655	12 900	14 000	6.1	10 500	580	3 500	17	4.0
Cast Alnico 5-7	730	735	13 200	14 000	7.4	11 500	640	3 500	17	3.8
Cast Alnico 6	750	...	10 500	13 000	3.7	7 100	525	4 000	13	5.3
Cast Alnico 7	1050	...	8 570	9 450	3.7	...	...	5 000	8.2	...
Cast Alnico 8	1600	1 720	8 300	10 500	5.0	5 060	950	8 000	5.0	3.0
Cast Alnico 9	1450	...	10 500	...	8.5	...	...	7 000	7.0	...
Cast Alnico 12	950	...	6 000	...	1.7	3 150	540	5 000	5.6	...
Sintered Alnico 2	525	545	6 700	11 000	1.5	4 300	345	2 500	12	6.4
Sintered Alnico 4	700	760	5 200	...	1.2	3 000	400	3 500	...	7.5
Sintered Alnico 5	600	605	10 400	12 050	3.60	7 850	465	3 000	18	4.0
Sintered Alnico 6	760	790	8 800	11 500	2.75	5 500	500	4 000	12	4.5
Sintered Alnico 8	1550	1 675	7 600	9 400	4.5	4 600	1000	8 000	5.0	2.1
Bonded ferrite A	1940	...	2 140	...	1.0	1 160	...	12 000	1.3	1.1
Bonded ferrite B	1150	...	1 400	...	0.4	...	...	8 000	1.2	1.1
Sintered ferrite 1	1800	3 450	2 200	...	1.0	1 100	900	10 000	1.2	1.2
Sintered ferrite 2	2200	2 300	3 800	...	3.4	1 850	1650	10 000	1.1	1.1
Sintered ferrite 3	3000	3 650	3 200	...	2.5	1 600	1600	10 000	1.1	1.1
Sintered ferrite 4	2200	2 300	4 000	...	3.7	2 150	1700	12 000	1.2	1.05
Sintered ferrite 5	3150	3 590	3 550	...	3.0	1 730	1730	15 000	1.0	1.05
Lodex 30	1250	1 470	4 000	4 400	1.6	2 200	750	6 000	3.4	1.5
Lodex 31	1140	1 180	6 300	7 000	3.4	4 400	770	6 000	5.3	1.9
Lodex 32	940	960	7 350	8 300	3.5	5 400	650	5 000	8.2	2.6
Lodex 33	865	875	8 000	9 200	3.2	5 850	545	5 000	10.5	3.0
Lodex 36	1210	1 380	3 500	4 400	1.5	1 850	800	5 000	2.0	2.0
Lodex 37	1000	1 080	5 450	7 000	2.1	3 150	670	5 000	5.8	3.0
Lodex 38	850	890	6 200	8 300	2.2	3 700	600	5 000	7.0	3.5
Lodex 40	1100	1 400	2 700	4 400	0.8	1 400	600	5 000	2.0	2.5
Lodex 41	990	1 100	4 350	7 000	1.4	2 400	600	5 000	3.8	3.2
Lodex 42	845	920	5 300	8 300	1.4	2 750	510	5 000	7.6	3.5
Lodex 43	710	750	6 000	9 200	1.3	3 300	400	5 000	10	3.8
P-6 alloy	58	...	14 000	19 000	0.5	10 500	48	300	220	23
Cunife	550	555	5 400	5 900	1.5	4 000	325	2 500	12	3.7
Cunico	680	750	3 400	4 500	0.8	1 950	390	3 000	5.0	3.2
Vicalloy I	240	242	8 400	12 900	0.9	5 600	160	1 000	...	...
Vicalloy II	415	420	9 050	...	2.3	7 000	325	2 000	...	...
Remalloy 1	250	...	9 700	14 200	1.0	6 100	155	1 000	40	13
Remalloy 2	340	345	8 550	...	1.2	5 400	220	2 000	...	...
Platinum cobalt	4450	5 400	6 450	...	9.2	3 500	2700	20 000	1.2	1.2
Cobalt, rare earth 1	9000	20 000	9 200	9 800	21	...	...	30 000	...	...
Cobalt, rare earth 2	8000	>25 000	8 600	...	18	4 400	4100	30 000	...	1.05
Cobalt, rare earth 3	6700	>15 000	8 000	...	15	4 000	3700	30 000	...	1.1
Cobalt, rare earth 4	5700	6 500	9 400	...	21	4 600	4600	>15 000	...	...

Table 4 Nominal mechanical and physical properties of selected permanent magnet materials

See Table 2 for compositions, Curie temperatures and magnetic orientations; see Table 3 for nominal magnetic properties.

Designation	Density, Mg/m^3	Tensile strength MPa	Tensile strength ksi	Transverse modulus of rupture MPa	Transverse modulus of rupture ksi	Hardness, HRC	Coefficient of linear expansion, μm/m·K	Electrical resistivity, nΩ·m	Maximum service temperature °C	Maximum service temperature °F
3½% Cr steel	7.77	...	...	...	...	60-65	12.6	290	...	...
6% W steel	8.12	...	...	...	...	60-65	14.5	300	...	...
17% Co steel	8.35	...	...	...	...	60-65	15.9	280	...	...
36% Co steel	8.18	...	...	...	...	60-65	17.2	270	...	...
Cast Alnico 1	6.9	28	4.1	96	14	45	12.6	750	540	1004
Cast Alnico 2(a)	7.1	21	3.1	52	7.5	45	12.4	650	540	1004
Cast Alnico 3	6.9	83	12	157	23	45	13.0	600	480	896
Cast Alnico 4	7.0	63	9.1	167	24	45	13.1	750	590	1094
Cast Alnico 5(a)(b)	7.3	37	5.4	73	11	50	11.4	470	540	1004
Cast Alnico 5DG	7.3	36	5.2	62	9.0	50	11.4	470	...	...
Cast Alnico 5-7	7.3	34	4.9	55	8.0	50	11.4	470	540	1004
Cast Alnico 6(a)	7.4	157	23	314	46	50	11.4	500	540	1004
Cast Alnico 7	7.3	108	16	...	...	60	11.4	580	...	...
Cast Alnico 8	7.3	64	9.3	...	...	56	11.0	500	540	1004
Cast Alnico 9	7.3	48	6.9	55	8.0	56	11.0	...	...	...
Cast Alnico 12	7.4	275	40	343	50	58	11.0	620	480	896
Sintered Alnico 2	6.8	451	65	480	70	43	12.4	680	480	896
Sintered Alnico 4	6.9	412	60	588	85	...	13.1	680	590	1094
Sintered Alnico 5	7.0	343	50	392	57	44	11.3	500	540	1004
Sintered Alnico 6	6.9	382	55	755	110	44	11.3	530	540	1004
Sintered Alnico 8	7.0	...	...	382	55	43	...	...	...	...
Bonded ferrite A(c)	3.7	4.4	0.63	...	...	...	94	~10^{13}	95	203
Sintered ferrite 1(d)	4.8	49	7.1	...	...	...	10	~10^{13}	400	752
Sintered ferrite 2	5.0	...	...	...	...	...	10	~10^{13}	400	752
Sintered ferrite 3	4.5	...	...	...	...	...	18	~10^{13}	400	752
Sintered ferrite 4	4.8	...	...	...	...	...	...	10^{13}	400	752
Sintered ferrite 5	4.5	...	...	...	...	...	...	10^{13}	...	...
Lodex 30	10.1	...	...	31	4.5	...	18	1200	200	392
Lodex 31	9.6	6.9	1.0	31	4.5	...	18	1200	200	392
Lodex 32	9.3	6.9	1.0	31	4.5	...	18	1200	200	392
Lodex 33	9.2	...	...	31	4.5	...	18	1200	200	392
Lodex 36	10.2	...	...	108	16	...	18	1200	200	392
Lodex 37	9.7	...	...	108	16	...	18	1200	200	392
Lodex 38	9.6	...	...	108	16	...	18	1200	200	392
Lodex 40	10.2	...	...	27	3.9	...	18	1200	200	392
Lodex 41	10.1	6.9	1.0	27	3.9	...	18	1200	200	392
Lodex 42	9.8	6.9	1.0	27	3.9	...	18	1200	200	392
Lodex 43	9.4	...	...	27	3.9	...	18	1200	200	392
P-6 alloy	7.9	2160	313	1 180	170	65	11	300	...	...
Cunife	8.6	686	99	...	...	95 HRB	12	180	350	662
Cunico	8.3	588	85	...	...	95 HRB	14	240	500	932
Vicalloy I	8.2	2060	299	...	...	62	7	630	450	842
Remalloy 1	8.2	882	128	...	...	60	9.3	450	500	932
Platinum cobalt	15.5	1370	199	1 570	230	26	11	280	350	662
Cobalt rare earth(e)	8.2	3430	498	13 730	1 990	50	511; 131	500	250	482

(a) Specific heat: 460 J/kg·K (0.11 Btu/lb·°F). (b) Thermal conductivity: 25 W/m·K (170 Btu·in./ft^2·h·°F) at room temperature. (c) Thermal conductivity: 0.62 W/m·K (4.3 Btu·in./ft^2·h·°F). (d) Thermal conductivity: 5.5 W/m·K (38 Btu·in./ft^2·h·°F). (e) Specific heat: J/kg·K (0.09 Btu/lb·°F). Thermal conductivity: 15 W/m·K (104 Btu·in./ft^2·h·°F).

be cold formed or machined. After aging, the alloys no longer have sufficient ductility. The remarkable features of these alloys are their extreme hardness, tensile strength and sensitivity to strain. Typical properties of Cunife, Cunico and Vicalloy alloys are given in Tables 3 and 4; demagnetization curves are presented in Fig. 6.

Alnico Alloys

Alnico alloys are one of the major classes of permanent magnet materials. The Alnicos vary widely in composition and in preparation, to give a broad spectrum of properties, costs, and workability. Alnico alloys are sold under a variety of names throughout the world (see Table 1). As a group, Alnico alloys are brittle and hard, and can be machined only by surface grinding, electrical discharge machining or electrochemical milling. They resist atmospheric corrosion well up to 500 °C (930 °F). Magnetic properties are negligibly affected by vibration or shock. Generally, Alnico is superior to other permanent magnet materials in resisting temperature effects on magnetic performance. Typical compositions and properties are summarized in Tables 2, 3 and 4; demagnetization curves are shown in Fig. 6.

Alnicos 1 through 4 are isotropic and generally lower in cost than the other Alnico alloys, but are much lower in quality. The remainder of the alloys are generally anisotropic. Maximum properties, and therefore greatest economy, are obtained when the device is designed to make use of oriented material. Optimum properties are achieved by casting and heat treating.

Magnets made by sintering Alnico powders or by bonding are used where small or intricate shapes to precise tolerances are required. Sintered Alnico is produced by blending powders, pressing and sintering just below the melting temperature in an oxygen-free atmosphere. The sintered alloys have mechanical properties superior to those of cast Alnicos, but the magnetic properties generally are slightly lower. Sintered magnets are given the same heat treatments as cast.

In general, optimum properties are developed by solution treating at 1100 °C (2010 °F), where the alloy is in equilibrium as a bcc phase, followed by cooling at a rapid, but critical, rate. Between 900 and 800 °C (1650 and 1470 °F), the alloy separates into two nonequilibrium bcc phases. One is almost pure iron and the other, a weakly magnetic phase, is roughly FeNiAl. The resulting microstructure is typical of a structure resulting from spinodal decomposition. Various combinations of H_c and B_r can be obtained.

Various other elements are often added to Alnicos. For example, to lessen the deleterious effect of carbon, carbide stabilizers such as Ti or Nb may be added. Ti and Cu increase H_c at the expense of B_r, but Nb increases H_c without decreasing B_r. Increasing the basic cobalt content by about 20% or more results in a major improvement in quality, because such large amounts of cobalt make it possible to develop a preferred orientation by heat treating the material in a magnetic field. Less striking improvements in H_c and B_r are obtained in nonoriented samples.

The ability of the magnetic field to influence the orientation (and thus the anisotropy) of the decomposing phase originates in the mechanism of decomposition. In processing the most commonly used alloy in this family, Alnico 5, for instance, the molten alloy is cast, then solution treated above 1250 °C (2280 °F). If Zr and/or Si is present, the alloy is solution treated at 900 to 925 °C (1650 to 1700 °F). The alloy is then placed in a magnetic field and cooled at a controlled rate from the solution treating temperature. Finally, the alloy is aged at 600 to 500 °C (1110 to 930 °F). An Alnico casting ordinarily can be shaped only by grinding or electrolytic machining, although hot working and certain other very specialized processing techniques are possible. Final finishing generally is done by grinding.

The critical phenomenon in this process is the spinodal decomposition of the high-temperature α-phase into a FeCo-rich α-phase and a NiAl-rich α'-phase. In the high-cobalt Alnico alloys, heat treating in a magnetic field appears to favor decomposition of parallel compositional waves and to suppress transverse waves. This effect is strongest if aging is carried out just below the intersection, on the phase diagram, of the Curie temperature and the temperature where the spinodal instability sets in. This condition limits the number of alloys that respond to magnetic aging. The addition of cobalt to FeNiAl promotes magnetic aging by moving the Curie temperature and decomposition temperature closer together.

Because the magnetic structure is influenced by crystallographic orientation during its formation, careful development of the proper <100> crystal texture is required to achieve the best properties in the Alnicos. In commercial alloys, the magnetic properties of Alnico 5DG, which is a partly oriented material, are significantly better than those of standard Alnico 5; those of Alnico 5-7, which has almost perfect orientation, are even better.

In the construction of magnets from crystal oriented Alnico, designs are limited by the possible shapes in which a properly oriented magnetic field can be established during heat treatment. (Magnetic fields are easy to create, for instance, in the shapes of straight lines, circles or arcs of circles.) This is not a major limitation on component design, however, because it can generally be overcome by using segmented magnets or by secondary fabrication of simple cast shapes, or both.

Lodex

Elongated single-domain iron-cobalt magnets sold under the trademark Lodex are fine-particle powder metallurgy (P/M) magnets that derive their properties predominantly from shape anisotropy. Properties are comparable to those of some Alnicos.

Lodex magnets are made by the following sequence: electrodepositing iron and cobalt into mercury to form elongated structures; aging to remove dendritic branches from these structures; adding antimony to form a protective monolayer and improve the magnetic properties of the particles; adding a lead-tin alloy to provide the matrix metal; pressing the slurry into large blocks in a magnetic field to align the particles and remove some of the liquid mercury; heating the blocks in vacuum to remove the remainder of the mercury by distillation; grinding the resulting porous cake to a powder of about 100-mesh particle size; and pressing the powder in a die to final shape and size. Anisotropic magnets are made by compacting the powder in a magnetic field; for isotropic magnets, powders are pressed without a field. Variations in magnetic properties may be obtained by varying the fraction of magnetic particles in the nonmagnetic matrix. Variations in H_c, B_r, and $(BH)_{max}$ with packing fraction follow the variations predicted for shape anisotropy. Mechanical forces enhance particle alignment in a direction transverse to the pressing direction because the easy magnetic axis is along the long axis of the original dendrites. Lodex permanent magnets also can be formed by extrusion. In this case, partial orientation develops along the long axis of the extrusion.

Lodex having a ratio of 60Fe:40Co has the highest M_s of the available Lodex alloys. H_{ci} and B_{is} also are maximum at this 60Fe:40Co ratio. However, because of small contributions from crystal anisotropy and packing effects that vary with iron-cobalt ratio, the value of $(BH)_{max}$ actually is quite constant for compositions from about 80Fe:20Co to 40Fe:60Co. Both Alnico and Lodex permanent magnets derive their high H_c predominantly from shape anisotropy of an FeCo phase.

Lodex permanent magnets are used extensively where the application requires close dimensional and magnetic tolerances, high uniformity and an intricate shape—such as in hearing-aid receivers and microphones, meters, electrical instruments, precision relays, reed switches and electronic speedometers. Typical properties of commercial grades are given in Tables 3 and 4 and in Fig. 6. These properties are far below the maximum properties theoretically obtainable.

Platinum-Cobalt

Although Pt-Co magnets are expensive, they are useful in certain applications. Pt-Co is isotropic, ductile, easily machined, resistant to corrosion and high temperatures, and has magnetic properties superior to all except the rare earth/cobalt alloys. Best magnetic properties are obtained at an atomic ratio of 50Pt:50Co. Above about 820 °C (1510 °F) the alloy has a disordered fcc structure. Below this temperature, ordering develops a slightly tetragonal structure. To process this alloy to develop a $(BH)_{max}$ of 9 megagauss-oersteds, the following treatment is used: heat to 1000 °C (1830 °F) to fully disorder, cool at a controlled rate to room temperature, and age at 600 °C (1110 °F) for about 5 h. The final structure is only partly ordered; its physical structure is without distinction until overaged to the completely ordered structure. Values of $(BH)_{max}$ of more than 10 megagauss-oersteds have been achieved by fabricating parts using powder metallurgical techniques and heat treating with essentially the same treatment as that used for cast parts. Typical properties are given in Tables 3 and 4 and the demagnetization curve is shown in Fig. 6.

Semihard Alloys

Semihard magnetic alloys exhibiting square hysteresis loops (B_r/B_s>0.85) and medium coercive forces (H_c = 10 to 50 Oe) have been developed for memory and reed switch applications. For such uses, the alloys must be sufficiently ductile to be drawn to fine wire or rolled to thin tape. Remendur, a low-vanadium alloy of the Vicalloy family, is characterized by a very high residual induction. It is, however, magnetostrictive and thus undesirable for components subjected to complex stresses.

Commercial semihard alloys containing gold or niobium are listed in Tables 2, 3 and 4. The gold-cobalt-iron material was the first of a family of precipitation alloys based on the 90Co-10Fe binary, which exhibits an almost complete lack of magnetostrictive behavior. Similar magnetic behavior is observed when Be, Ti or Mo is added to the 90Co-10Fe binary alloy.

Cobalt/Rare Earth Alloys

The newest family of permanent magnet materials is based on combinations of cobalt and the lighter rare earth (lanthanide) metals. Permanent magnets made of powder metallurgy Co-RE alloys have the highest magnetic properties of all known materials. Yttrium, lanthanum, cerium, praseodymium, neodymium and samarium alloys apparently are the best choices for commercial permanent magnet applications. An alloy need not contain only a single lanthanide metal; mixtures often are used. Likewise, a portion of the cobalt can be replaced with copper and iron to obtain desired magnetic characteristics.

Sintered Co_5Sm magnets, with minor additions of praseodymium or other elements, are the highest-quality permanent magnet materials available. In addition, $Co_{17}(RE)_2$ compounds have properties of considerable interest and are being investigated. $(Co,Fe)_{17}(RE)_2$ compositions are higher in crystal anisotropy than $Co_5(RE)$ compounds and have unrealized potential energy products of up to 60 megagauss-oersteds—the highest projected value for any known material. Typical properties of commercially available materials are given in Tables 3 and 4, and demagnetization curves are presented in Fig. 6. Because these alloys are new, no standard designations have been developed for them. The four compositions listed in Table 2 (arbitrarily numbered 1 to 4) represent the range of magnetic qualities available from the many producers of these materials.

It was the discovery of the large magnetocrystalline anisotropy of Co_5Y that initiated the interest in $Co_5(RE)$ alloys as permanent magnet materials. An energy product of nearly 28 megagauss-oersteds was achieved in a single particle of Co_5Y. However, no investigator has reported the achievement of a usefully high coercive force in a $Co_5(RE)$ permanent magnet containing a substantial amount of yttrium.

Cerium is by far the most abundant, and potentially the least expensive, of all rare earth elements. Procedures for preparing large quantities of Co_5Ce have been described in the literature. However, because cerium-rich mischmetal (MM) is at present less expensive than cerium metal, and because $Co_5(MM)$ has some permanent magnet properties superior to those of Co_5Ce, very little effort has been made to develop permanent magnets with Co_5Ce as the basic magnetic phase. Most development efforts involving cerium metal have been conducted with precipitation-hardening Ce-Co-Cu alloys.

The relative abundance of lanthanum, coupled with the magnetic properties that have been reported for Co_5La, appear to make this phase very attractive for some permanent magnet applications. Indeed, some excellent permanent magnets, especially in terms of intrinsic coercive force, have been prepared using $Co_5La_{0.5}Sm_{0.5}$ as the basic magnetic component. Preparation of Co_5La by standard melting practice requires an annealing step. The $Co_{13}La$ peritectic isotherm, which extends well beyond the stoichiometric Co_5La composition (83.3 at.% Co), interferes with formation of Co_5La. Unless cooling rate is carefully controlled, very little Co_5La will be observed in the cast alloy.

The relative abundance of neodymium, coupled with the high saturation and high Curie temperature of Co_5Nd, appear to make Co_5Nd desirable for some magnetic applications. The relatively small anisotropy constant of Co_5Nd, and the fact that it changes sign just below room temperature, apparently have discouraged efforts at developing this material. However, it has been reported that dramatic increases in the intrinsic coercive forces of Co_5Nd and Co_5Di (Di is the symbol for didymium, a commercial alloy composed principally of neodymium and praseodymium in the ratio of about 3 to 1) can be achieved by sintering these powders with praseodymium-rich and samarium-rich cobalt alloy sintering additives. It was observed that the properties of the sintered magnets were very sensitive to composition and even more sensitive to sintering temperature.

Among the binary $Co_5(RE)$ phases, Co_5Pr has the highest potential energy product. Moreover, praseodymium is more than eight times as plentiful as samarium in bastnasite, the major source of rare earth metals in the United States. However, in spite of the potential advantages of Co_5Pr, development of Co_5Pr permanent magnets has been neglected in favor of those based on Co_5Sm and $Co_5Pr_xSm_{(1-x)}$.

Sintered Co-RE materials are hard and brittle, very much like the Alnicos. Magnets made of these materials often are pressed to final shape to eliminate machining; magnets are also produced by cutting or slicing them from rounds or other simple shapes.

Co-RE materials have coercive forces much higher than other permanent magnet materials. Nevertheless, they can be satisfactorily magnetized in fields lower than those necessary to achieve saturation induction. Virgin magnets—that is, magnets that have never been exposed to a magnetizing field after final heat treatment—can be magnetized in fields of about 15 000 Oe.

To obtain anisotropic properties, tooling and dies are designed to compact powders in orienting fields in a manner similar to that used for ferrites. Bonded magnets are produced simply by compacting aligned powder mixed with plastic or soft metal binders. Magnetic quality of bonded magnets is lower than that of sintered magnets.

In general, Co-RE magnets cost more than any other type except platinum-cobalt. Co-RE magnets have replaced most Pt-Co magnets, particularly in microwave applications, where not only the lower price but also the higher coercivity of Co-RE magnets are advantageous. The unique combination of properties in these magnets (high H_c, high $(BH)_{max}$, recoil permeability of one, and temperature stability) has led to a variety of new designs in traveling-wave tubes, watches, motors and generators, couplers and magnetic bearings.

Ferrites

Hard ferrites, also known as ceramic permanent magnet materials, are predominantly complex oxides. The ferrites most commonly used for magnets are $BaO \cdot 6Fe_2O_3$ or $SrO \cdot 6Fe_2O_3$, although various additives such as SiO_2 or Bi_2O_3 are beneficial. These substances have coercive forces two to eight times the coercive forces obtained with Alnicos. Both sintered and bonded ferrite magnets are used extensively.

A typical sintered ferrite magnet is made by mixing BaO or $BaCO_3$ (or Sr compounds) with Fe_2O_3. A slight excess of Ba^{++} is usual, and a substance to act as a lubricant and aid pressing is commonly present. This mixture is compacted in a die and sintered at 1200 to 1300 °C (2190 to 2370 °F). Sintering develops a typical ceramic hardness machinable only by diamond grinders and slicing tools. Shrinkage during sintering is higher than for sintered metal magnets. About 15% shrinkage occurs, giving a sintered magnet with a density about 90% of theoretical.

Nonoriented ferrites are almost isotropic. To obtain anisotropic magnets, with a resultant twofold to threefold improvement in magnetic properties, the original powdered mixture is prefired at temperatures above 900 °C (1650 °F) to form the complex oxide compound, which is then ground to a powder of about 1 μm particle size. This powder is compacted in a die within an orienting field and finally sintered at 1200 to 1300 °C (2190 to 2370 °F). The same powder may be used to prepare bonded magnets by mixing it with an appropriate proportion of a polymer and pressing, injection molding, or extruding to final shape. Magnetic properties generally are inferior to sintered magnets, but dimensional tolerances can be held close and processing costs kept low.

Ceramic permanent magnet materials have high electrical resistivities and are poor conductors of heat. Although ferrites are not affected by high temperatures or atmospheric corrosion, the magnetic properties are more temperature-dependent than they are with other permanent magnet materials. Coercive forces decrease with lowering temperatures; flux density B_d decreases at a rate of 0.19%/°C with rising temperature. The ferrites permanently lose a portion of their magnetic properties at low subzero temperatures, the amount of loss depending on the L/D ratio. For example, the permanent loss at −57 °C (−70 °F) is 3% when the L/D ratio is 0.09, but when the L/D ratio is 0.50, no loss occurs. Typical properties are given in Tables 3 and 4 and in the demagnetization curves in Fig. 6. Although the $(BH)_{max}$ of the ferrites is relatively low, their low cost and high coercive force make them attractive for applications such as door closers, motors, speakers, etc.

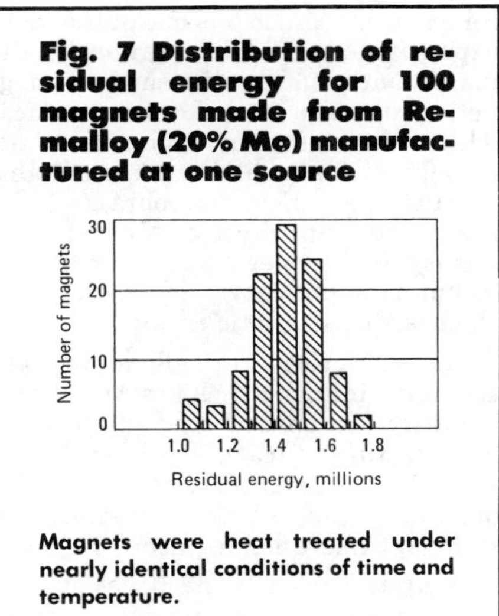

Fig. 7 Distribution of residual energy for 100 magnets made from Remalloy (20% Mo) manufactured at one source

Magnets were heat treated under nearly identical conditions of time and temperature.

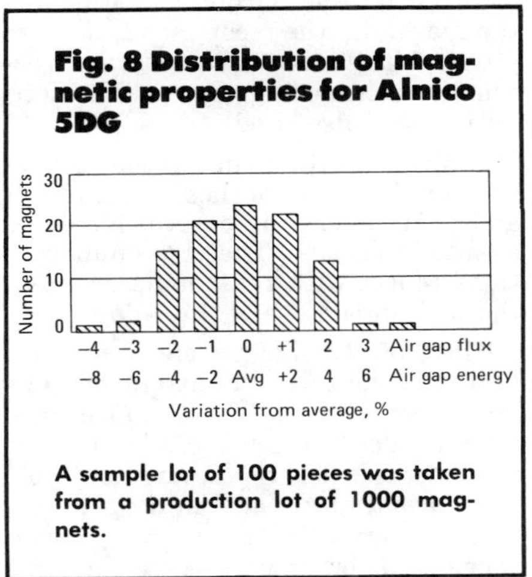

Fig. 8 Distribution of magnetic properties for Alnico 5DG

A sample lot of 100 pieces was taken from a production lot of 1000 magnets.

Selection and Application

Permanent magnets are superior to electromagnets for many uses because they maintain their fields without an expenditure of electrical power and without the generation of heat.

Tables 3 and 4 and Fig. 6 give nominal properties only. Even under the most carefully controlled manufacturing conditions, some variation from these nominal values must be expected and considered in practical application. Figures 7 and 8 are examples of variations in energy values for two magnet materials.

Cost per pound is seldom considered in selecting a magnet material. Cost per unit of magnetic energy is more significant and is more often the basis of comparison. Alnico 5 is one of the more expensive of the Alnico group by the pound, but because of its superior magnetic properties is the most economical Alnico alloy for many applications. Alnico 6 and Alnico 5DG are both slightly, but not significantly, more costly by the pound than Alnico 5. Alnico 2 costs about 35% less per pound than Alnico 5, but is significantly more costly on the basis of magnetic energy.

Magnet prices for all alloys are strongly influenced by size, shape, quantity purchased and method of manufacture (cast, sintered or wrought). A significant increase in price may also be associated with a stringent tolerance requirement.

Usage. Tables 5 and 6 list specific applications of permanent magnet materials. For each application, the primary material recommendation and the reason for the recommendation are given. Alternative materials and reasons for considering the specific alternatives are also cited.

Alnico 5 is the prime recommendation for about half of these applications and is also shown in several instances as an alternative. These recommendations reflect with reasonable accuracy the widespread use of this alloy.

Fig. 9 Hysteresis loss vs magnetizing force for various permanent magnet materials

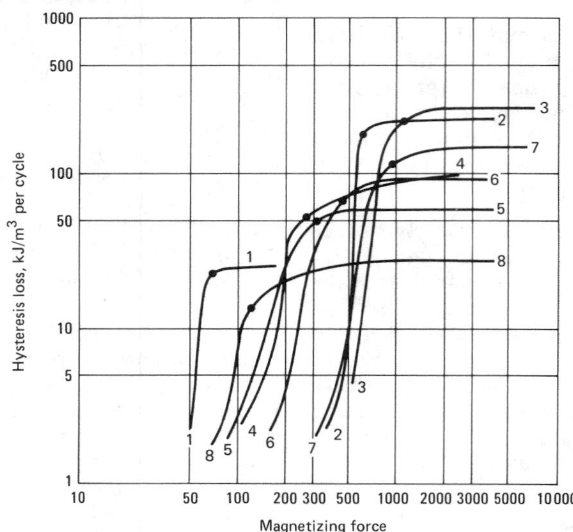

Dots indicate points of maximum efficiency. 1, P-6 alloy; 2, cast Alnico 5; 3, cast Alnico 6; 4, Vicalloy; 5, 17% cobalt steel; 6, 36% cobalt steel; 7, cast Alnico 2; 8, 3½% chrome steel.

The 3.5% Cr magnet steels are still used to a considerable extent for telephone bell ringers. There is also some residual demand for the 3.5% Cr and 6% W steels for replacements in old machines that are still in use. However, the magnet steels are rarely, if ever, considered for new designs. Virtually all new equipment requiring magnets is designed around the Alnicos, ferrites or cobalt rare-earth materials.

Considering the Alnicos only, it is estimated that ferrite is used for 90% of the Alnico applications. Considering all magnet applications, it is estimated that Alnico 5 is used for more than 50% of the total.

Special applications and conditions may indicate the superiority of one material over another. For example, if greater ductility with a higher H_c value is desired, Cunico might be used in place of Cunife. If lighter weight with high coercive force is needed, Vectolite might be considered.

Hysteresis Applications

In most applications, permanent magnets are used to supply flux in an air gap, and operate only in the second quadrant of the hysteresis loop. In a few specialized applications, notably hysteresis torque devices, it is necessary to consider the entire hysteresis loop in evaluating a magnetic material.

In hysteresis torque devices, the driving force is provided by a rotating magnetic field. The rotor is usually a thin-wall cylinder magnet that is not premagnetized but that can be magnetized by the rotating field. Immediately after starting, the induced poles rotate within the rotor, subjecting the material to repeated transversals of the hysteresis loop. Hysteresis causes the induced rotor poles to lag behind those of the applied field, producing an accelerating torque. The torque developed is proportional to the area of the hysteresis loop through which the magnet is driven.

If torque were the only requirement, the hysteresis material chosen would be a magnet with the largest possible hysteresis loop (and therefore the largest hysteresis loss). However, two other factors must be considered. Because the applied field must be capable of magnetizing the rotor, there is a practical limit to the coercive force that can be used.

Materials that are very high in coercive force seldom are used for these applications. For example, Alnico 6 develops the largest hysteresis loss, as indicated in Fig. 9, which shows hysteresis loss as a function of magnetizing force for various permanent magnet materials. Therefore, Alnico 6 would produce the greatest amount of torque. However, to produce this torque would require a magnetizing force of more than 1000 Oe—a magnetizing force that is not practical for most applications.

Because the efficiency of the motor is determined by the magnetic characteristics of the rotor, the shape of the hysteresis loop is important. A convenient measure of this shape is the energy factor, η. The energy factor is defined as the ratio of the area of the hysteresis loop to the area of a rectangle drawn through the extremes (peak B and H) of the curve and ranges from 0.5 to 0.75. Two materials may have the same intercepts in terms of induction at a given magnetizing force, yet have different energy factors.

In operation, materials are not magnetized to saturation, but operate with-

Table 5 Dynamic applications of permanent magnet materials

Application	Typical operating point H_d	B_d	Typical design conditions: Area of air gap, mm²	Length of air gap, mm	Recommended material	Primary reason for selection	Alternate material	Condition or reason favoring selection of the alternate material
Aircraft magnetos, military and civilian	250–450	6500–7200	660	to 0.25	Cast Alnico 5	Maximum energy per unit volume, shape permitting	Cast Alnico 6	Where demagnetization forces are large
	550–600	7000–10 000	900	to 0.25	Cast Alnico 5; Cast Alnico 5DG			
Truck magnetos	150–375	5000–6000	650	0.13 to 0.25	Cast Alnico 5	Compactness and reliability	Cast Alnico 2 or 3	Where space is available for a larger volume of material of lower magnetic energy and cost
Tractor magnetos	150–375	5000–6000	650	0.13 to 0.20	Cast Alnico 3	Adequate magnetic energy at lower cost than Alnico 5	...	...
Magnetos for medium and large marine and stationary engines	200–425	5250–6000	650	0.25 to 0.75	Cast Alnico 3	Same as above	Cast Alnico 2	Where a material of higher energy is required
Outboard engines with magnet welded to steel flywheel	150–450	3000–4000	400	0.25 to 0.75	Cast Alnico 3	Lowest-cost Alnico and best resistance to temperature effects	Cast Alnico 2 or 4	Quite resistant to temperature effects, greater magnetic energy
Outboard engines with magnet mechanically fixed in flywheel	150–450	3000–4000	400	0.25 to 0.50	Cast Alnico 5	Optimum material where fabrication temperatures are low	Cast Alnico 6	(a)
Outboard engines with flywheel die cast around magnet	150–450	3000–4000	400	0.25 to 0.50	Cast Alnico 2	Highest-energy Alnico with magnetic energy unaffected by temperature	Cast Alnico 3	The lowest-cost Alnico; resists effects of temperature as well as Alnico 2
Small dc motors, field	1100–3000	1000–1700	200 to 3000	0.4 to 1.6	Nonoriented barium ferrite	If area available to take advantage of low-cost material; high coercive force	Oriented barium ferrite, cast Alnico 5 or 6	Higher magnetic energy for smaller spaces
Small dc motors, rotor	400–550	7000–10 000	50 to 500	0.4 to 0.75	Cast Alnico 5	Rotor size requires high-energy material	Cast Alnico 6	Greater strength and resistance to demagnetization
Synchronous hysteresis motors	125–200	3500–8000	25 to 160	to 0.25	36% Co steel	Shape favors fabrication from wrought material	3.5% Cr or 17% Co steel	Availability and cost
Torque drives	0–2875	1750–2100	4.809 to 19.356	0.5 to 3.2	Barium ferrite	If area available to take advantage of low-cost material	Cast Alnico 5	Where design space dictates a high-energy material

(a) If space is available for lower-cost and lower-energy material, where greater resistance to demagnetization is required, or greater strength.

in a minor hysteresis loop. The actual loop is determined not only by the material, but also by the intensity of the magnetizing field. If only a small field is available, greater torque is developed by a low-energy material that is operated near its saturation value than by a higher-energy material that is little affected by the small applied field.

There is one value of magnetizing force for each material at which it operates most efficiently—that is, at which it produces maximum torque for a given magnetizing force. In Fig. 9, this point is marked with a dot on the curve for each material. The efficiency of a hysteresis material is the ratio of hysteresis loss to magnetomotive force required. Peak efficiencies of commonly used materials are:

Table 6 Static applications of permanent magnet materials

Application	Typical operating point H_d	B_d	Typical design conditions: Area of air gap, mm^2	Length of air gap, mm	Recommended material	Primary reason for selection	Alternative material	Condition or reason favoring selection of the alternative material
Holding device, air gap or direct contact	525	10 000	650 to 22 500	Up to 1.3	Cast Alnico 5	Highest energy available for external field for pulling	Oriented barium ferrite(a)	Critical weight or where economical to use larger volumes of lower-cost ferrites
Loud speakers	500	10 000	6000 to 20 000	1.3	Cast Alnico 5 or Alnico 5DG	Greatest flux density per unit volume per unit cost	Oriented barium ferrite	For magnets above about 0.8 kg
Ammeters and voltmeters with external magnet	400	10 500	100 to 200	2.5 to 4	Cast Alnico 5	Same as above	Cast Alnico 6 or 2	Alnico 6 for demagnetization resistance; Alnico 2 for a lower-cost material
Same with core magnet	550	7 000	100 to 200	...	Cast Alnico 6	*L/D* ratio is generally small	Cast Alnico 5	Where *L/D* ratio is not too restrictive
Watt-hour meters with cast magnets	320	11 000	100	3	Cast Alnico 5	For very high flux density	Cast Alnico 2	Where design allows larger volume of lower-cost material
Same with wrought magnets	60	7 500	350	3	17% Co steel	Large volume of lower-cost material provides adequate flux density	...	...
Flowmeters	550	7 000	70	3	Cast Alnico 5	Greatest flux density per unit volume	Cast Alnico 6(b)	Where large demagnetization forces are present
Tachometers, speedometers	260	4 400	30	4.5	Wrought Cunife	Shape favors a material that can be cold punched	3.5% Cr steel	Lower-cost material, but requires hot punching
Telephone receivers	525	2 590	107	1.0	Remalloy	Small volume with high flux density	Alnico 3	Larger volume at a lower point on the demagnetization curve
Temperature controls	300	4 500	50 to 150	1.5 to 3.2	Sintered Alnico 2	Size and shape favor sintered product; magnetic strength adequate	...	...
Polarized relays	550	9 000	20	2.5	Cast Alnico 5	Usually require highest-energy material available	3.5% Cr steel	For a larger-volume magnet at lower cost
Magnetic signs, games, toys and novelties	250	4 300	100 to 1000	Up to 2.5	Sintered Alnico 2	Nonfunctional design and sizes allow sintering. Lower-cost Alnico offers adequate magnetic energy	Sintered Alnico 3 or 4	Lower cost

(a) If the device can be designed for direct contact and large-area magnets, unoriented barium ferrite will be favored. (b) Barium ferrite is the preferred alternative where larger volumes of the lower-cost material can be used.

Alloy	Efficiency
P-6 alloy	0.330
Cast Alnico 5	0.304
Cast Alnico 6	0.202
Vicalloy	0.197
17% Co steel	0.158
36% Co steel	0.142
Cast Alnico 2	0.124
3.5% Cr steel	0.117

Stabilization and Stability

There is an important group of permanent magnet applications where the accuracy or performance of the device is drastically affected by very small changes (1% or less) in the strength of the magnet. These applications include braking magnets for watt-hour meters, magnetron magnets, special torque motor magnets, and most dc panel and switchboard instrument magnets. Operation of these devices requires extreme accuracy over a moderate range of conditions, or moderate accuracy over an extreme range of conditions.

If the nature and magnitude of the conditions are known, it often is possi-

Fig. 10 Irreversible changes in H_c after aging for various permanent magnet materials

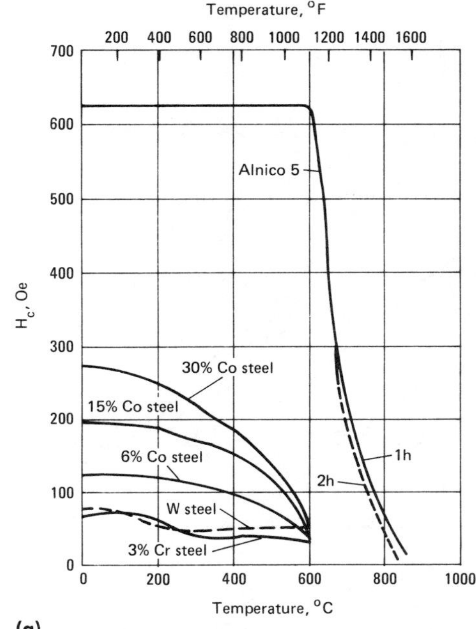

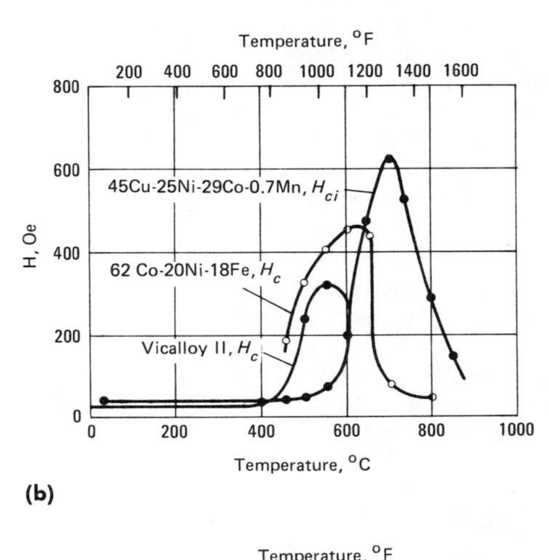

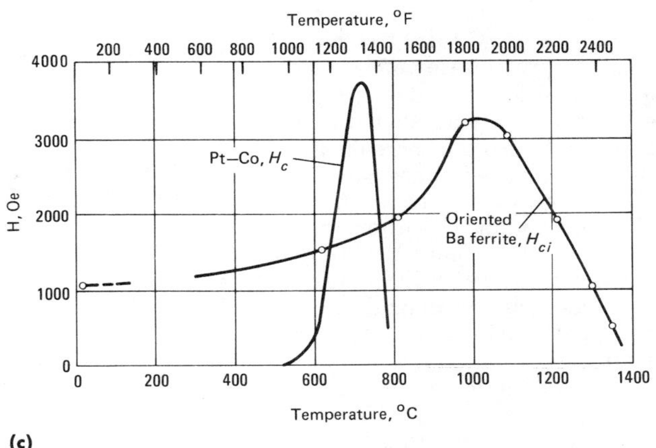

(a) Steels were aged ½ h at temperature after being austenitized and quenched. Alnico 5 was aged 1 and 2 h, as indicated, after an optimum heat treatment. (b) Vicalloy wire was cold reduced 92.5% and aged 2 h at temperature. 62Co-20Ni-18Fe alloy was oil quenched from 1000 to 1050 °C (1830 to 1925 °F) and aged 1 h at temperature. 45Cu-25Ni-29Co-0.7Mn alloy was quenched from 1140 °C (2080 °F) and aged 1 h at temperature. (c) Pt-Co was quenched from 1000 °C and aged ½ h at temperature. Oriented barium ferrite was prefired at 1000 °C, milled 94 h and pressed, then aged 1 h at temperature.

ble to predict the flux change. It also may be possible, by exposing the magnet to certain influences in advance, to render the magnet insensitive to subsequent changes in service. For many years, permanent magnets in instruments have exhibited long-term stability on the order of one part per thousand (0.1%). More recently, investigations in conjunction with inertial guidance systems for space vehicles have shown that long-term stability of the order of one to 10 ppm (0.0001 to 0.001%) can be achieved. This incredible stability of a magnetic field achieved with modern permanent magnets contrasts sharply with the stability of very early permanent magnets, in which both structural and magnetic changes caused a significant loss of magnetization with time.

Fig. 11 Flux loss in bars dropped one metre onto hardwood floors

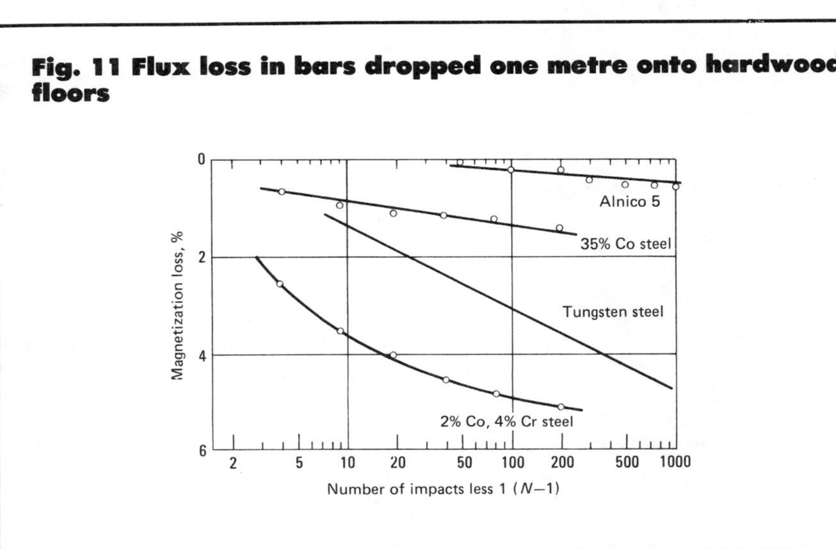

Irreversible Changes

Losses in magnetization with time can be classified as either reversible or irreversible. Irreversible changes are defined as changes where the affected properties remain altered after the influence responsible for the change has been removed. For example, if a magnet loses field strength under the influence of elevated temperature, and if the field strength does not return to its original value when the magnet is cooled to room temperature, the change is considered irreversible.

Changes in Metallurgical State. Irreversible changes begin to occur at different temperatures for different alloys. These changes usually depend on both time and temperature, and thus short exposures above the recommended temperatures may be tolerated. These changes may take the form of growth of the precipitate phase, such as in Alnico, Cunife and Cunico; precipitation of another phase, such as γ precipitation in Alnico; an increase in the amount of an ordered phase, such as in PtCo; stress-relief effects, such as in quenched steels and Vicalloy; an increase in grain size, as in $BaO{\cdot}6Fe_2O_3$; oxidation, as occurs with metals, or reduction, as occurs with oxides; radiation damage; cracking; or changes in dimensions.

Typical permanent changes in H_c due to changes in metallurgical structure are shown in Fig. 10. Results for the steels and Alnico 5 were obtained on samples that initially were in the optimum magnetic state. The remainder of the materials were aged, starting from their quenched, as-cast, or as-prepared state, as indicated in the figure. The temperature at which changes in properties first become noticeable corresponds to the beginning of metallurgical changes; this temperature corresponds closely to the maximum temperature to which each material can be exposed, even after aging to the optimum magnetic state.

Irreversible metallurgical changes often can be counteracted, and original properties restored, by a suitably chosen thermal treatment. For example, if Alnico 5 has become degraded by exposure to 700 °C (1290 °F), it may be solution treated at 1300 °C (2370 °F), cooled in a magnetic field and aged at 600 °C (1110 °F) to re-attain the optimum metallurgical structure.

A nuclear environment is known to cause changes in metallurgical structure, and thus may cause changes in magnetic properties. Permanent magnet materials tested were not affected by neutron irradiation at levels below about 3×10^{17} neutrons/cm^2. Results of later work at levels up to 10^{20} n/cm^2 showed some degradation. The Alnicos are not affected by radiation up to 5×10^{20} n/cm^2 at neutron energies greater than 0.4 eV, and up to 2×10^{19} n/cm^2 for neutron energies greater than 2.9 MeV. Radiation effects were found to be independent of temperature, but high temperatures tended to counteract radiation effects.

Changes in magnetic state may be caused by temperature effects, such as ambient temperature changes or statistical local temperature fluctuations within the material; mechanical effects, such as mechanical shock or acoustical noise; or magnetic field effects, such as external fields, circuit reluctance changes or magnetic surface contacts. In all of these situations, the loss in magnetization may be restored by remagnetizing.

Fig. 12 Temperature dependence of saturation magnetization or remanence for various permanent magnet materials

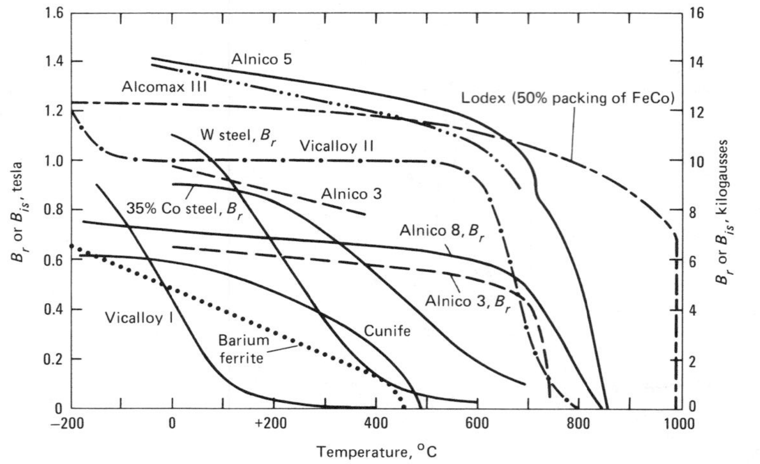

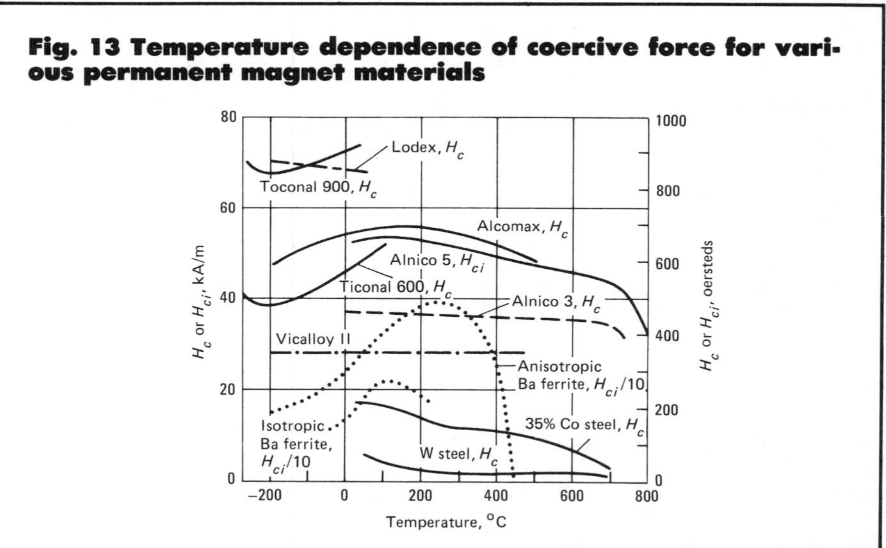

Fig. 13 Temperature dependence of coercive force for various permanent magnet materials

Table 7 Magnetization changes on cooling below room temperature (20°C)(a)

Material	Dimensional ratio, L/D	% irreversible loss at room temperature after exposure to: −190 °C	−60 °C	Reversible temperature coefficient, % remanence change per °C
Alnico 2	5.29	0	0	−0.025
	3.69	0	0	−0.021
	2.66	0	0	−0.018
	1.77	0	0	−0.009
	0.94	0	0	−0.014
Alnico 5	8.00	0	0	−0.022
	5.36	4.6	1.4	−0.012
	3.63	9.0	2.5	−0.002
	2.72	6.2	3.6	+0.010
	1.84	7.9	2.1	+0.016
	0.94	8.5	3.4	+0.007
Alnico 6	8.00	0	0	−0.045
	6.03	1.8	0.4	−0.020
	3.57	8.5	1.3	−0.007
	2.70	10.1	4.1	+0.007
	1.78	10.5	4.2	+0.022
	0.89	7.9	3.1	+0.046
Alnico 8	5.62	0	0	−0.013
	2.85	0.5	0.1	+0.003
	1.91	0.7	0.3	+0.015
Barium ferrite	1.01	1.3	0.5	+0.033
(isotropic)	0.50	4.0	0(b)	−0.19
	0.28	...	1.3	−0.19
Barium ferrite	0.09	...	2.4	−0.19
([*BH*] max anisotropic)	1.20	...	0(b)	−0.19
Barium ferrite	0.50	...	...	−0.19
([H_c] max anisotropic)	0.40	...	0(b)	−0.19
Platinum cobalt	31.67	0	0	−0.015
	16.02	0.3	0	−0.015
	10.63	0.2	0	−0.015
	5.56	0.2	0	−0.015
Rare earth cobalt 1	...	...	...	−0.050
Rare earth cobalt 2	...	...	...	−0.045
Rare earth cobalt 3	...	...	...	−0.053

(a) R. Parker and R. Studders, *Permanent Magnets and Their Application,* John Wiley and Sons, Inc., New York (1959), p 345-348. (b) In the case of the low-temperature irreversible loss occurring in oriented barium ferrite, only the smallest dimension ratio resulting in no irreversible loss at −60 °C is shown. This is the recommendation of a major producer, to avoid catastrophic loss. The minimum dimension ratio will depend upon the lowest temperature to be encountered.

Fig. 14 Demagnetization curves for Alcomax and for oriented barium ferrite at various temperatures

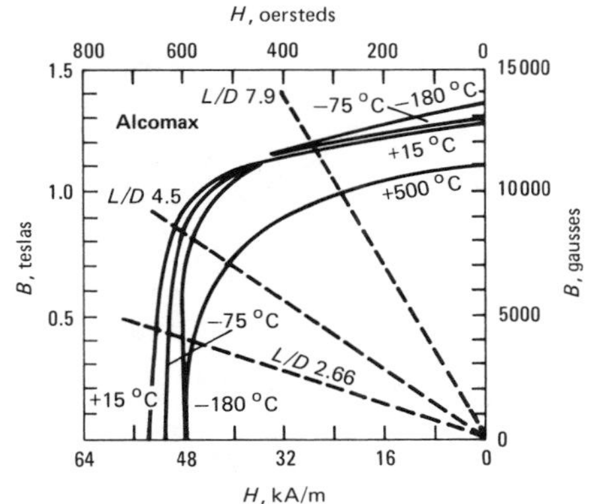

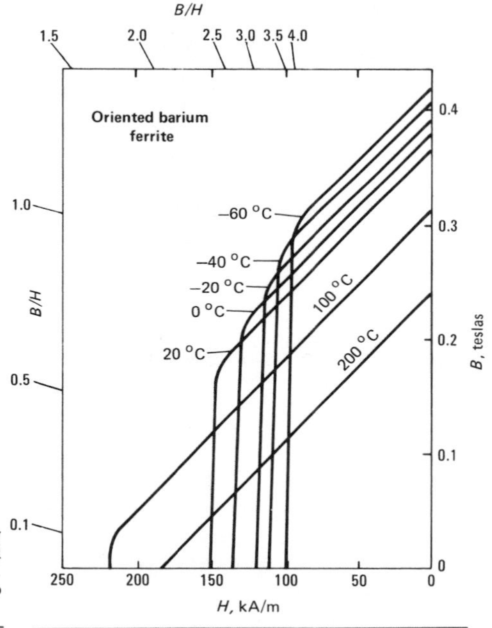

Mechanical shock and vibration add energy to a permanent magnet, and decrease the magnetization in the same manner as discussed for the case of thermal energy. The only difference is that energy imparted thermally to the magnet is precisely kT, whereas the energy imparted mechanically usually is not known. Thus, repetitive shocks or continual vibration should decrease the magnetization by the same logarithmic relations as for thermal effects, but where time is replaced, for example, by number of impacts. Figure 11 shows some typical data. The nonlinear curve for 2Co-4Cr steel may result from a combination of both the large normal decrease with time at room temperature and the decrease due to impacts.

Little work has been done regarding stabilization to minimize mechanical effects because it is seldom found necessary after thermal and field stabilization. There is limited information that suggests that both thermal and alternating field exposure will minimize, but not entirely eliminate, the change in magnetization due to shock.

Reversible Changes

A loss in magnetization caused by a disturbing influence, such as temperature or an external magnetic field, is considered reversible if the original properties of the magnet return when the disturbing influence is removed.

Temperature Effects. The properties of a magnet vary with temperature in a manner that often can be predicted. The variation of B_{is} with temperature can be calculated from theory, provided detailed knowledge of the crystallographic and magnetic structure of the magnetic phase is available. In many other instances, such information is not yet available, but direct measurements of B_{is} vs T have been

Table 8 Magnetization losses on heating above room temperature (20 °C)

Material	Dimensional ratio (L/D)	Temperature, °C (°F) 100 (212) I	II	200 (390) I	II	300 (570) I	II	400 (750) I	II	500 (930) I	II
Alnico 2	8.00	2.0	98	3.1	94	4.2	90	6.1	86	8.2	80
	3.62	3.1	98	4.0	92	6.9	88	8.6	84	12.0	78
	2.00	3.5	97	4.7	91	7.4	89	10.7	85	13.1	81
Alnico 5	8.00	0.1	99.9	0.2	96	0.4	93.6	0.7	91.2	1.2	88.0
	4.68	0.4	99.6	0.8	96.3	1.1	93.8	1.7	91.1	2.0	88.2
	2.00	0.5	99.4	1.7	96.6	2.1	94.1	2.6	92.2	3.0	88.6
Alnico 6	20.00	0.1	98.2	0.2	95.6	0.4	93.0	0.8	89.7	1.8	86.5
	4.12	0.5	98.7	0.9	95.6	1.2	92.7	2.0	89.4	3.0	85.2
	2.00	0.7	99.1	1.2	97.2	1.5	94.2	2.1	90.5	3.3	86.0
Alnico 8	5.62	0.7	98.8	...	...	...	...	...	...	...	...
	2.85	0.7	99.0	...	...	...	...	...	...	...	...
	1.91	0.9	99.4	...	...	...	...	...	...	...	...
	1.01	1.0	99.8	...	...	...	...	...	...	...	...
Barium ferrite (all grades)	All	0	85	0	68	0	50	...	...	...	...
Platinum cobalt	31.67	0	97.9	...	...	...	...	...	...	...	...
	16.02	0	98.5	...	...	...	...	...	...	...	...
	10.63	0	98.8	...	...	...	...	...	...	...	...
	5.56	0	97.9	...	...	...	...	...	...	...	...
Rare earth cobalt	10	0	...	0.2	...	...	...	...	...	...	...
	2	0.1	...	0.5	...	...	...	...	...	...	...
	1	0.8	...	2.0	...	...	...	...	...	...	...
	0.5	0.9	...	2.5	...	...	...	...	...	...	...
	0.25	1.8	...	5.5	...	...	...	...	...	...	...

Note: Column I: Percent irreversible remanence loss at room temperature after heating to indicated temperature. Column II: Percent of initial room temperature remanence found stable at indicated temperature.

made. Examples of such curves for many permanent magnet materials are shown in Fig. 12. The curves in Fig. 12 include some that represent changes in B_r with T, which depend extensively on properties of the system, such as particle size, sample shape, demagnetizing fields and domain structures. Changes with temperature must be determined experimentally for the specific magnet and configuration of interest. Typical reversible and irreversible changes are given in Tables 7 and 8 as a function of length-to-diameter ratio for a number of permanent magnet materials.

Changes in H_{ci} with temperature can be predicted from the changes with temperature of anisotropy and magnetization. This assumes knowledge of the physical origin of all anisotropies contributing to H_{ci}. Experimental results are shown for many magnets in Fig. 13. For a case where uniaxial anisotropy predominates, as in $BaO \cdot 6Fe_2O_3$, quite good agreement between calculated and experimental results is obtained. In a case where shape anisotropy is dominant, as for Lodex elongated particles with various Co contents, calculated and experimental results also are in good agreement, especially when the small crystal anisotropy contributions are considered. The case of Alnico is similar to that of Lodex, but crystal anisotropy is more in evidence. In addition, there is greater uncertainty as to the effect of the so-called nonmagnetic phase, especially at lower temperatures where the nonmagnetic phase may contribute appreciable magnetization. In the case of steels, the temperature dependence based on the inclusion mechanism is difficult to predict.

Demagnetization curves may change in both shape and peak values with changes in temperature. Families of demagnetization curves at various temperatures are shown in Fig. 14 for Alnico and barium ferrite.

Time Effects at Constant Temperature. In ferromagnetic materials, the intensity of magnetization does not instantly attain its equilibrium value when the applied field is suddenly changed. This time dependence may be due to eddy current effects or to reversible or irreversible magnetic viscosity. In general, eddy current effects are important only for a very short time—normally, less than a second after a change in the applied field. Such effects are not considered here. "Reversible" magnetic viscosity has been shown to be due to ionic diffusion in the crystal lattice and thus has a time-temperature dependence characteristic of diffusion processes. The time constant is

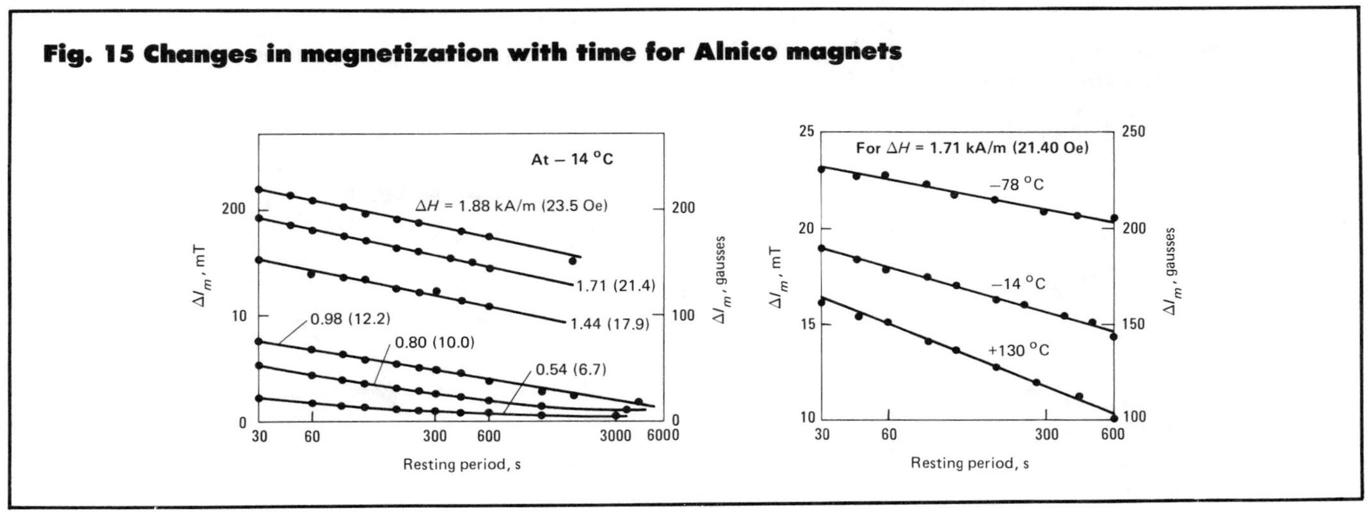

Fig. 15 Changes in magnetization with time for Alnico magnets

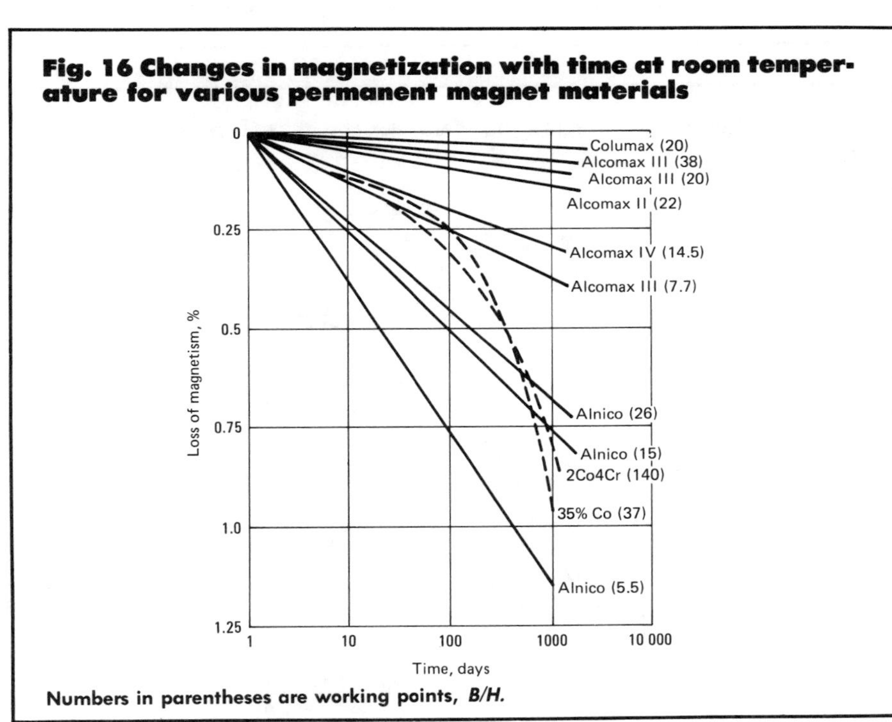

Fig. 16 Changes in magnetization with time at room temperature for various permanent magnet materials

Numbers in parentheses are working points, *B/H*.

Table 9 Recommended magnetizing fields, demagnetizing methods and maximum service temperature

Permanent magnet material	Magnetizing field, Oe	Demagnetizing method	Maximum service temperature, °C
Steels, Cunife, Vicalloy, Remalloy	1 000	ac field	100
Alnico 3 to 6	3 000	ac field	500
Alnico 8 and 9	6 000	ac field	500
Ceramic 1 to 8	10 000	Curie temperature	300
ESD Fe-Co	3 000	ac field	100
Pt-Co	20 000	Curie temperature	325
Co(RE)	15 000 to 50 000(a)	Heat treatment	300

(a) Depending on previous magnetic history.

$$\tau = \tau_\infty \exp (E/RT)$$

where τ_∞ is the time constant at infinite temperature and E is the activation energy, normally 0.1 to 1 eV. The time constant appears to be important only in magnetically soft materials, and only at high frequencies.

"Irreversible" magnetic viscosity is important to the stability of permanent magnets. Irreversible magnetic viscosity is due to the influence of thermal fluctuations on magnetization or the domain process responsible for magnetization. The effect of thermal agitation has been considered in terms of the energy required to activate irreversible domain processes. The time-temperature dependence of magnetization was shown to be given by

$$M(t) = S \ln t$$

where $S = \lambda N M_s k T$. Here, N is the number of blocks, or regions of magnetization M_s per unit volume; λ is the constant probability density of energy E of all these blocks; and k is Boltzmann's constant. Because these factors are all relatively independent of temperature (except near the Curie temperature), S is nearly directly proportional to T. The results of experiments are in agreement with this equation, as shown in Fig. 15. Aging at room temperature results in losses in magnetization for many materials, as shown in Fig. 16.

Effects of Temperature Variations. Various permanent magnet materials undergo changes in magnetization as the temperature is cycled above and below room temperature, see Fig. 17. For a long bar operating above $(BH)_{max}$, as in Fig. 17(a), the change in

Fig. 17 Temperature variation for Alcomax III bars of various length-to-diameter ratios and working points

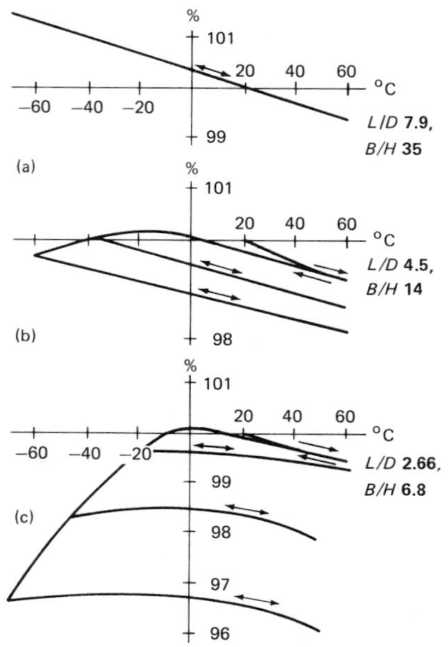

Fig. 18 Effect of stress on the magnetization curves for Vicalloy

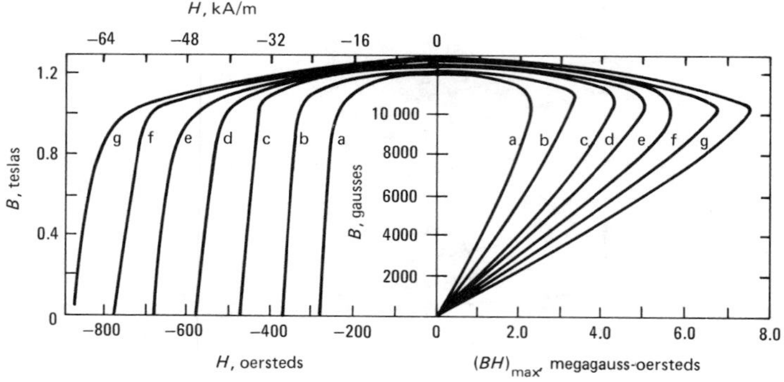

Applied stresses: (a), 0; (b), 500 MPa (73 ksi); (c), 990 MPa (144 ksi); (d), 1490 MPa (216 ksi); (e), 1990 MPa (289 ksi); (f), 2490 MPa (361 ksi); (g), 2990 MPa (434 ksi).

Fig. 19 Effect of stress on the magnetization curves for Cunife

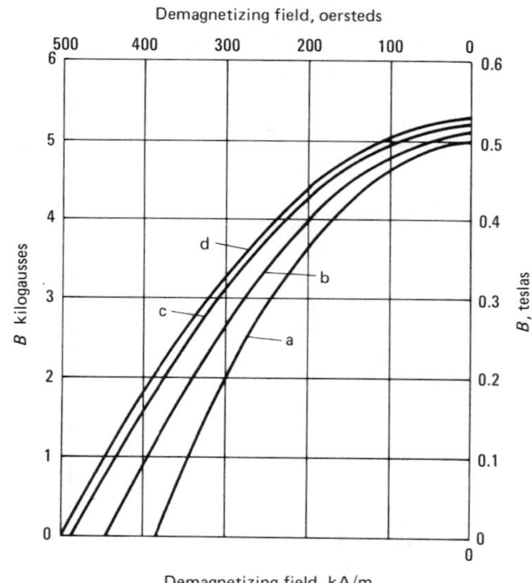

Applied stresses: (a), 0; (b), 385 MPa (56 ksi); (c), 570 MPa (83 ksi); (d), 635 MPa (92 ksi).

M is reversible. For a shorter bar operating below $(BH)_{max}$, as in Fig. 17(b), the first cooling cycle results in a substantial loss in magnetization. After the initial low-temperature exposure, the changes in M are reversible, but at a level below the initial magnetization. Results on an even shorter bar are shown in Fig. 17(c). These data suggest that by proper choice of dimensions, a reversible coefficient of approximately zero could be achieved over a limited range of temperature.

Design Considerations

Stability can have a significant influence on choice of magnet material, as well as on component shape and magnetic circuit arrangement. For example, the rather drastic change in coercive force of oriented barium ferrite with temperature (Fig. 13) requires special considerations in design. Here, the lowest permeance coefficient (B/H) that can be used is established by stability considerations rather than by magnetic circuit analysis.

For the more widely used permanent magnet materials, reversible changes in magnetization are encountered by cooling below room temperature (Table 7). Because the reversible remanence changes are closely approximated by a straight line, a reversible temperature coefficient is listed. The values of the coefficient are very small and may be of different sign for different magnet shapes. Consequently, it is often possible to carefully design magnet shape to yield very small variations in remanence with temperature. Similar changes may result upon heating above room temperature (Table 8). It is important to distinguish between irreversible losses and reversible changes. It is common practice prior to use to cycle a magnet between the temperature extremes to be encountered in service. Nearly all of the irreversible loss is encountered in one temperature cycle, but in some instances four or five cycles may be necessary.

In applications that are extremely

sensitive to magnetization changes, it is very common to use a temperature compensating circuit to counteract reversible changes over the operating temperature range. Temperature-sensitive iron-nickel alloys are used as magnetic shunts for this purpose. A shunt is mounted beside the permanent magnet and simply diverts flux from the air gap as the temperature decreases. Temperature compensation by shunting requires overdesign of the magnet to allow for the loss in flux through the shunt at low operating temperatures.

Exposure at Very High Temperatures. There is considerable interest in using permanent magnets at temperatures approaching the Curie temperature of the permanent magnet material. The anisotropic Alnico 5 and Alnico 6 have been considered for use at 500 to 700 °C (930 to 1290 °F). At these temperatures, metallurgical effects as well as irreversible and reversible temperature effects are present. Alnico 5 exposed to 700 °C (1290 °F) for 20 h resulted in the reduction of $(BH)_{max}$ and H_c to approximately half. It is possible to program such changes into equipment and devices to allow permanent magnets to function for a limited time at extreme temperatures.

Stress Effects. Some magnets subjected to tension or compression show large changes in properties. This is especially true of Vicalloy, as shown in Fig. 18, and Cunife, as shown in Fig. 19. The changes are reversible, often even after considerable deformation has occurred. The changes are due to the contribution that stress makes to the total anisotropy of the system.

Magnetization Prior to Use. Magnets are magnetized in applied fields supplied by dc or pulsed-current electromagnets. Where practical, saturating magnetizing fields are recommended to gain full use of magnetic potential energy (see Table 9). Magnets are demagnetized by heating to Curie temperature or by applying an ac or dc field to reduce the measured induction to zero. Partial demagnetization may be needed to reduce the flux density B_d to some calibrated level, or to prestabilize against anticipated magnetic losses. These losses can occur due to external demagnetizing forces, such as in electric motors, or by temperature cycles. Partial demagnetization is accomplished by initial exposure to the operating environment or by applying an ac field equivalent to about twice the amount of knockdown anticipated.

Handling of all permanent magnets is best done in the nonmagnetized condition. There is less risk of attracting dirt and chips, of snapping magnetic objects onto the magnet with possible injury, and of partial demagnetization due to mechanical shock.

Electrical Resistance Alloys

Edited by the ASM Committee on Electrical Resistance*

ELECTRICAL RESISTANCE ALLOYS include both the types used in instruments and control equipment to measure and regulate electrical characteristics and those used in furnaces and appliances to generate heat. In the former applications, properties near ambient temperature are of primary interest; in the latter, elevated-temperature characteristics are of prime importance. In common commercial terminology, electrical resistance alloys used for control or regulation of electrical properties are called *resistance alloys,* and those used for generation of heat are referred to as *heating alloys.* A third class of electrical resistance materials, used in applications where heat generated in a metal resistor is converted to mechanical energy are termed *thermostat metals.* All three classes of electrical resistance alloys are discussed in this article.

Resistance Alloys

The primary requirements for resistance alloys are uniform resistivity, stable resistance (no time-dependent aging effects), reproducible temperature coefficient of resistance, and low thermoelectric potential, versus copper. Properties of secondary importance are coefficient of expansion, mechanical strength, ductility, corrosion resistance, and ability to be joined to other metals by soldering, brazing or welding.

Nominal compositions and physical properties of metals and alloys used to make resistors for instruments and controls are listed in Table 1.

Resistance alloys must be ductile enough so that they can be drawn into wire as fine as 0.01 mm (0.0004 in.) in diameter or rolled into narrow ribbon from 0.4 to 50 mm (1/64 to 2 in.) wide and from 0.025 to 6.4 mm (0.001 to 0.25 in.) thick.

Alloys must be strong enough to withstand fabrication operations, and it must be easy to procure an alloy that has consistently reproducible properties. For instance, successive batches of wire must have closely similar electrical characteristics: if properties vary from lot to lot, resistors made of wire from different batches may cause a given model of instrument to exhibit widely varying performance under identically reproduced conditions or may cause large errors in a given instrument when a resistor from one batch is used as a replacement part for a resistor from another batch.

Coefficients of expansion of both the resistor and the insulator on which it is wound must be considered, because stresses can be established that will cause changes in both resistance and temperature coefficient of resistance. It is equally important that consideration be given to the choice between single-layer and multiple-layer wound resistors, because of the difference in rate of heat dissipation between the two styles.

In design of primary electrical standards of very high accuracy, cost of resistance material is not a consideration. For ordinary production components, however, cost may be the deciding factor in material selection.

Resistors

Resistors for electrical and electronic devices may be divided into two arbitrary classifications on the basis of permissible error: those employed in precision instruments in which over-all error is considerably less than 1%, and those employed where less precision is needed. The choice of alloy for a specific resistor application depends on the variation in properties that can be tolerated.

In many electronic devices, resistors whose error in resistance value is 5 to 10% are entirely satisfactory. Most resistors for this classification are made of carbon. Carbon resistors are not discussed in this article. Here, we are concerned chiefly with metallic resistors such as wirewound precision resistors and potentiometers, resistance thermometers, and ballast resistors.

Some applications of resistance

*See page XIII for committee list.

Table 1 Typical properties of electrical resistance alloys

Basic composition, %	Resistivity(a), nΩ·m(b)	TCR, ppm/°C(c)	Thermo-electric potential vs Cu, μV/°C	Coefficient of thermal expansion(d), μm/m·°C	Tensile strength(a) MPa	Tensile strength(a) ksi	Density(a) Mg/m³	Density(a) lb/in.³
Radio alloys								
98Cu-2Ni	50	+1350 (25-105 °C)	−13 (25-105 °C)	16.5	205-410	30-60	8.9	0.32
94Cu-6Ni	100	+550 (25-105 °C)	−13 (25-105 °C)	16.3	240-585	35-85	8.9	0.32
89Cu-11Ni	150	+430 (25-105 °C)	−25 (25-105 °C)	16.1	240-515	35-75	8.9	0.32
78Cu-22Ni	300	+160 (25-105 °C)	−36 (0-75 °C)	15.9	345-690	50-100	8.9	0.32
Manganins								
87Cu-13Mn	480	±15 (15-35 °C)	+1 (0-50 °C)	18.7	275-620	40-90	8.2	0.30
83Cu-13Mn-4Ni	480	±15 (15-35 °C)	−1 (0-50 °C)	18.7	275-620	40-90	8.4	0.31
85Cu-10Mn-4Ni (e)	380	±10 (20-45 °C)	−1.5 (0-50 °C)	18.7	345-690	50-100	8.4	0.31
Constantans								
57Cu-43Ni	490	±20 (25-105 °C)	−43 (25-105 °C)	14.9	410-930	60-135	8.9	0.32
55Cu-45Ni	500	±40 (20-1000 °C)	−42 (0-75 °C)	14.9	455-860	66-125	8.9	0.32
Nickel-chromium-aluminum alloys								
75Ni-20Cr-3Al-2(Cu, Fe or Mn)	1330	±20 (−55-+105 °C)	−0.1 (25-105 °C)	12.6	690-1380	100-200	8.1	0.29
72Ni-20Cr-3Al-5Mn	1355	±20 (−55-+105 °C)	−0.1 (25-105 °C)	13	690-1380	100-200	7.1	0.26
Nickel-base alloys								
94Ni-3Mn-2Al-1Si	315	+2400 (20-100 °C)		12.3	550-1035	80-150	8.5	0.31
80Ni-20Cr	1125	+85 (−55-+100°C)	+5 (0-100 °C)	13	690-1380	100-200	8.4	0.31
78.5Ni-20Cr-1.5Si	1080	+85 (25-105 °C)	+3.9 (25-105 °C)	13.5	690-1380	100-200	8.3	0.30
76Ni-17Cr-4Si-3Mn	1330	±20 (−55-+150 °C)	−1 (20-100 °C)	15	900-1380	130-200	7.8	0.28
71Ni-29Fe	200	+4500 (25-105 °C)	−40 (25-105 °C)	15	480-1035	70-150	8.4	0.31
68.5Ni-30Cr-1.5Si	1180	+90 (25-105 °C)	−1.2 (25-105 °C)	12.2	825-1380	120-200	8.1	0.29
60Ni-16Cr-24Fe	1120	+150 (25-105 °C)	+0.9 (25-105 °C)	13.5	655-1200	95-175	8.4	0.30
35Ni-20Cr-45Fe	1015	+400 (25-105 °C)	−1.1 (25-105 °C)	15.6	550-1200	80-175	8.1	0.29
Iron-chromium-aluminum alloys								
73.5Fe-22Cr-4.5Al	1350	±50	−3.0 (0-100 °C)	11	690-965	100-140	7.25	0.262
73Fe-22Cr-5Al	1390	±50	−2.8 (0-100 °C)	11	690-965	100-140	7.15	0.258
72.5Fe-22Cr-5.5Al	1450	±50	−2.6 (0-100 °C)	11	690-965	100-140	7.1	0.256
81Fe-15Cr-4Al	1250	±50	−1.2 (0-100 °C)	11	620-900	90-130	7.43	0.268
Pure metals								
Aluminum (99.99+)	26.55	+4290(a)	−3.4 (0-50 °C)	23.9(a)	50-110	7-16	2.70	0.098
Copper (99.99)	16.73	+4270 (0-50 °C)	0	16.5(a)	115-130	17-19	8.96	0.324
Gold (99.999+)	23.50	+4000 (0-100 °C)	+0.2 (0-100 °C)	14.2(a)	130	19	19.32	0.698
Iron (99.94)	970	+5000(a)	+12.2 (0-100 °C)	11.7(a)	180-220	26-32	7.87	0.284
Molybdenum (99.9)	52	+3300(a)	+6.9 (0-100 °C)	4.9	690-2140	100-310	10.22	0.369
Nickel (99.8)	80	+6000 (20-35 °C)	−22 (0-75 °C)	15	345-760	50-110	8.90	0.322
Platinum (99.99+)	106	+3920 (0-100 °C)	+7.6 (0-100 °C)	8.9(a)	125	18	21.45	0.775
Silver (99.99)	16	+4100(a)	−0.2 (0-100 °C)	19.7	125	18	10.49	0.379
Tantalum (99.96)	135	+3820 (0-100 °C)	−4.3 (0-100 °C)	6.5(a)	690-1240	100-180	16.6	0.600
Tungsten (99.9)	55	+4500(a)	+3.6 (0-100 °C)	4.3(a)	1825-4050	265-590	19.25	0.695

(a) At 20 °C (68 °F). (b) To convert to Ω·circ mil/ft, multiply by 0.6015. (c) Temperature coefficient of resistance is $(R - R_0)/R_0\,(t - t_0)$, where R is resistance at t °C and R_0 is resistance at the reference temperature t_0 °C. (d) At 25 to 105 °C. (e) Shunt manganin.

materials require devices with large thermal coefficients of resistance, either positive or negative. A device of this type is called a thermistor. Thermistors are made almost exclusively of ceramic semiconductor materials.

Precision resistors (those with less than 1% error) require careful material selection. The ideal material for a precision resistor should have a thermal coefficient of resistance equal to zero for the temperature range over which the resistor will operate. In addition, to ensure freedom from thermoelectric effects, it should have a small or negligible thermoelectric potential versus copper, which is the material normally used for the connecting conductor. Temperature differentials may exist among various junctions between a resistance wire and a connecting wire, resulting in a network of thermocouples that can cause parasitic electromotive forces in the circuit; this effect is especially critical in precise dc circuits. In an apparatus where extreme precision is required, it is advisable to make the connecting wires of the same material as the resistors or to design the apparatus so that all dissimilar-metal junctions are at the same operating temperature.

Selection of a material for, and specific dimensions of, a precision resistor must include consideration of equipment size and heat-dissipation characteristics. Temperature excursions from the ambient or from a specified operating temperature may be undesirable, because they may cause net changes in resistance that will affect the stability or accuracy of the instrument. The magnitude of the change in resistance can be calculated using the temperature coefficient of resistance. For example, a resistor made of a low-resistivity material could be several times larger than one made of a higher-resistivity material and yet achieve the same total resistance. The large resistor would have a much greater surface area and therefore could dissipate much more heat, and thus, despite its low resistivity, would attain a lower steady-state temperature than would be possible for a small, high-resistivity resistor operating under the same conditions. Alloys used for precision resistors generally have resistivities ranging from 500 to 1350 μΩ·m (300 to 800 Ω·circ mil/ft).

Resistance thermometers are commonly made of copper, nickel or platinum; these devices are precision resistors whose resistance change with temperature is stable and reproducible over specified ranges of temperature. For resistance thermometers, the larger the temperature coefficient of the material, the greater the accuracy and ease of measurement. Temperature coefficients of relatively pure metals are greatly affected by small amounts of impurities. In fact, one of the most sensitive tests of the purity of a metal is measurement of its temperature coefficient of resistivity, which decreases sharply with increasing impurity or alloy content.

Ballast resistors are used extensively in industrial circuits to maintain constant currents over long periods of time. In such an application, a ballast resistor must be able to dissipate energy in such a way as to control current over a wide range of voltages. Wires with the proper temperature coefficient of resistance can be made to change resistance rapidly with changes in current, due to self heating, in such a manner that the current in the circuit will remain nearly constant even when there are fluctuations in voltage across the circuit. Because ballast resistors operate at elevated temperatures, mechanical properties are important also. Typical materials used in ballast resistors are pure iron, pure nickel, and nickel-iron alloys such as 71Ni-29Fe (see Table 1).

Reference resistors and virtually all other applications of resistance alloys demand temperature coefficients of resistance lower than ±20 ppm/°C (±20 μΩ/Ω·°C). This requirement stems from the fact that, for these applications, resistance errors resulting from the small changes in ambient temperature that are continually taking place cannot be tolerated. In the most demanding of these applications, resistors often are mounted in thermally insulated containers and are carefully maintained at a temperature slightly above the maximum anticipated ambient temperature.

Fig. 1 Change in resistance of a 10-kΩ resistor with time

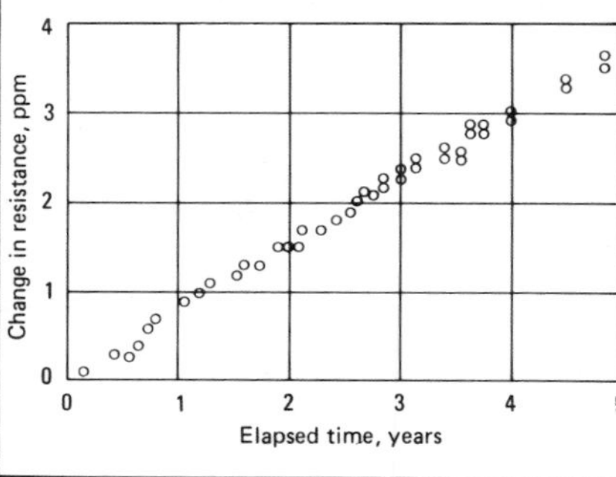

The most important requirement of a resistor used as a reference standard is that its value be predictable within narrow limits over long periods of time. Many reference resistors exhibit a nearly linear change in resistance with time. Hence, resistance between dates of calibration can be determined by interpolation; resistance at future points in time can be determined by extrapolation, but undue reliance should not be placed on extrapolated values. Figure 1 shows the change in resistance with time for a 10-kΩ resistor made of a Ni-Cr-Al-Cu alloy.

Stability, or the ability to maintain a specific value of resistance within narrow limits over a long period of time, is an important requirement of materials for precision resistors and reference resistors. Principal sources of instability are: (*a*) relief of residual stresses during service; (*b*) time-dependent or time-temperature-dependent metallurgical changes, such as precipitation, of a second phase; and (*c*) corrosion or oxidation.

Residual stresses often are relieved at room temperature over long periods of time through a process known as stress relaxation. Stress relaxation alters the resistance of a coil at a rate of change that increases with the original level of residual stress. For this reason, only carefully preannealed wires are used for precision resistors. Stresses induced during winding, weaving or other operations in fabrication of resistors from preannealed wire must be kept to a minimum. Thorough annealing of finished resistors is not always possible, because the wires may be enameled or may be coated with a textile insulation of only moderate resistance to heat. Either type of coating limits to about 140 °C (285 °F) the temperature that can be used for stress relieving finished resistors.

Fig. 2 Change in resistance of manganin resistors upon aging at room temperature

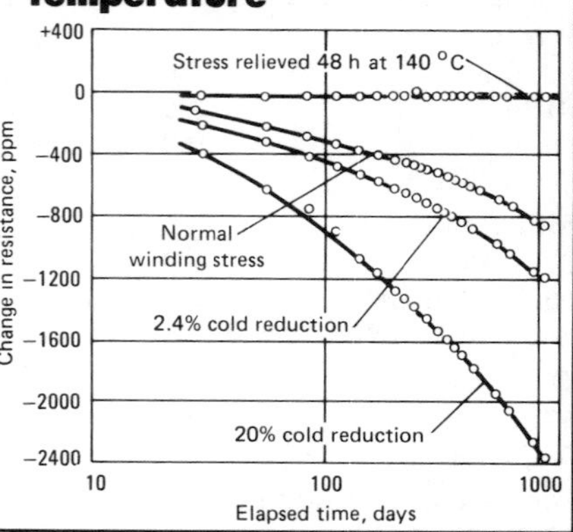

Figure 2 shows the effect of residual stress on the stability of a manganin alloy subjected to different amounts of cold work. The top curve illustrates that a low-temperature stress-relieving treatment substantially eliminates the stresses that would, in time, have been eliminated due to natural relaxation at room temperature.

Resistors represented by the top curve were stress relieved at 140 °C for 48 h to stabilize their resistance within about 20 ppm of the nominal value. Resistors not stress relieved, as represented by the other curves, continue to change in resistance almost indefinitely. For most modern, hermetically sealed precision resistors annealed at 140 °C, the change in resistance does not exceed 10 ppm/yr, and for many it does not exceed 5 ppm/yr. However, resistors made of manganin that are used as reference standards require greater stability, and stress relief at 140 °C is not adequate. One-ohm resistors of the best grade (the double-wall type) are treated as follows. A coil of wire is wound on a steel mandrel and annealed at about 500 °C (930 °F) in a protective atmosphere for 6 h or more. The coil is removed and slipped over an insulated tube of the same diameter as the mandrel, and then is hermetically sealed using a second tube slightly greater in diameter. In most resistors of this type, the change in resistance does not exceed 1 ppm/yr.

The second factor affecting stability of precision resistors is the metallurgical stability of the alloy being used as the resistance element; any metallurgical change will be detrimental. All resistance alloys are single-phase solid-solution alloys; thus, the changes in resistance that occur are relatively small but not insignificant. Changes in resistance are caused by internal changes such as long-range order-disorder reactions in 71Ni-29Fe alloys, short-range order or clustering in quaternary nickel-chromium alloys, and even minor ordering in manganin alloys. Accordingly, resistance of these alloys is affected by heat treatment and by rates of cooling from heat treating temperatures. Power resistors that can operate as high as 300 °C (570 °F) can in effect be heat treated during service. The net effect during service can be an increase in resistance for nickel-chromium alloys, a decrease for manganin and either an increase or decrease for nickel-iron alloys.

The third factor affecting stability of resistors is corrosion and/or oxidation. Corrosion of the resistance element will decrease its effective cross section, resulting in a corresponding increase in resistance. If the corrosive attack is selective, changes will occur in temperature coefficient of resistance and thermal emf as well as in resistivity. These corrosive effects may be minimized by protecting the wire with an enamel or plastic coating. One relatively common source of corrosive attack, but one that is often overlooked, is flux residue at soldered or brazed joints. Another less obvious cause of instability is the presence of tin-containing solder. Intergranular stress corrosion, believed to originate during thermal stress-relieving treatments, may cause open circuits.

Combinations of these three factors—residual stresses, metallurgical instability, and corrosion or oxidation—account for the complex changes in resistance that often occur in resistors.

The ease with which alloys can be soldered, brazed or welded is an important consideration in selection of materials for precision resistors. Improperly brazed or soldered joints frequently cause resistance instability in the circuit. Metals to be soldered must be cleaned prior to tinning so that solder can completely wet the surfaces and maintain electrical continuity. For copper-nickel alloys this is relatively simple, because protective oxide coatings are not formed on these alloys. Nickel-chromium alloys must be tinned immediately after cleaning and before an inherent protective oxide forms.

The resistance value of a resistor may change if the hydrostatic pressure on the resistance element is changed; for manganin this change is about 22.5 pΩ/Ω·Pa (0.155 μΩ/Ω·psi). Sealed resistors may also be affected by changes in external pressure. In a double-wall one-ohm resistor, for example, a change in pressure on the inner tube will cause a change in tube diameter, thus altering the length of wire wound on the tube. The magnitude of the resistance change depends in part on the thickness of the wall, and for commercial resistors is typically less than the hydrostatic pressure coefficient (PCR) of manganin. Unsealed resistors wound on mica cards containing air bubbles may have pressure coefficients several times greater than that predicted from the hydrostatic pressure coefficient of the alloy. This effect is important only if there is a large change in pressure, which would be most likely if there were a large change in elevation above sea level.

Types of Resistance Alloys

Copper-nickel resistance alloys, generally referred to as *radio alloys,* have very low resistivities and moderate temperature coefficients of resistance (TCR), as shown in the first four listings in Table 1. Resistivity of radio alloys increases, and TCR decreases, as nickel content increases. Thermal emf is negative with respect to copper, the magnitude being directly proportional to nickel content. All radio alloys can be readily soldered or brazed. Those with 12 and 22% nickel have high enough resistance to permit welding. Because of their high copper contents, radio alloys have low resistance to oxidation and thus are restricted to applications involving low operating temperatures. They are used chiefly for resistors that carry relatively high currents, and for this reason rapid dissipation of heat from the surface of the resistor is desirable. In this application, resistor temperature may vary over a wide range, but temperature changes are relatively unimportant.

Fig. 3 Variation of resistance with temperature for four precision resistor alloys

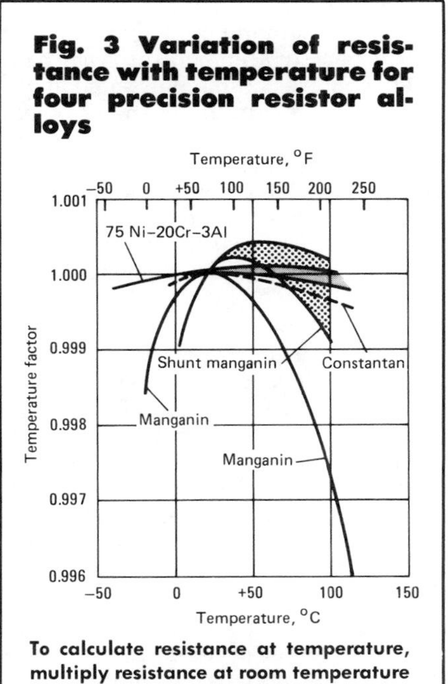

To calculate resistance at temperature, multiply resistance at room temperature by the temperature factor.

Copper-manganese-nickel resistance alloys, generally referred to as *manganins,* have been adopted almost universally for precision resistors, slide wires and other resistive components with values of 1 kΩ or less, and are also used for components with values up to 100 kΩ.

Originally, manganin was the name of a specific alloy, but the term is now generic and covers several different compositions (see Table 1). All manganins are moderate in resistivity (from 380 to 480 nΩ·m, or 230 to 290 Ω·circ mil/ft) and low in TCR (less than ±15 ppm/°C).

Manganins are stable solid-solution alloys. The electrical stability of these alloys, verified by several decades of experience, is such that their resistance values change no more than about 1 ppm per year when the material is properly heat treated and protected. Manganin-type alloys are characterized by rather steep, parabolic relations between resistance and temperature (see Fig. 3). This severely restricts the range of temperature over which resistance is stable, thus limiting the use of manganins to devices for which operating temperatures are both stable and predictable. For some applications, the maximum of the parabola (peak, or peak temperature) is kept near room temperature by controlling composition. This minimizes the effects of small changes in ambient temperature. The temperature coefficient of commercial manganin is generally less than ±10 ppm/°C for an interval of 10 °C (18 °F) on either side of the peak.

When instruments are designed for operation above ambient temperature, the chemical composition of the manganin is chosen so that the peak will occur in the operating temperature range. So-called "shunt manganin", which carries high currents and consequently gets hot in use, usually has a peak temperature from 45 to 65 °C (115 to 150 °F).

Manganins are susceptible to selective oxidation or preferential corrosive attack. This may occur during heat treatment, wire manufacture or coil fabrication. Selective oxidation results in formation of a copper-rich (manganese-depleted) zone on the wire. This copper-rich sheath has the effect of greatly increasing the temperature coefficient of resistance and raising the peak temperature well beyond the range where any precision resistor would ordinarily be used.

The resistivity of manganin—roughly 500 nΩ·m (300 Ω·circ mil/ft) at 25 °C—is adequate for most instrumentation purposes. The thermoelectric potential versus copper is very low, usually less than −2 μV/°C from 0 to 100 °C.

Constantan, like manganin, has become a generic term for a series of alloys that have moderate resistivities and low temperature coefficients of resistance. Nominally, constantans are 55Cu-45Ni alloys, but specific compositions vary from about 50Cu-50Ni to about 65Cu-35Ni. The most favored composition is 57Cu-43Ni. Most constantans contain small but significant amounts of iron, manganese and cobalt.

The temperature coefficient of resistance of constantan is very low and parabolic like that of manganin, but remains flat over a much wider range (Fig. 3). Other properties are given in Table 1; specific property values vary somewhat with composition. Constantans are considerably more resistant to corrosion than manganins.

Use of constantans as electrical resistance alloys is restricted largely to ac circuits, because thermoelectric potential versus copper is quite high for these materials (about 40 μV/°C at room temperature). However, if the circuit voltage is high enough to overshadow thermoelectric effects, constantans may be used in dc circuits as well.

Nickel-Chromium-Aluminum Resistance Alloys. Nickel-chromium alloys containing small amounts of other metals—usually aluminum plus either copper, manganese or iron—have resistivities about 2½ to 3½ times that of manganin. Ni-Cr-Al resistance alloys have been adopted almost universally for the construction of wirewound precision resistors having resistance values of about 100 kΩ, and are also used for resistors with values as low as about 100 Ω. The temperature coefficients of resistance of these alloys are vastly superior to those of manganin and constantan, being less than ±20 ppm/°C between −50 and +150 °C (−60 and +300 °F). The difference in TCR between the hot region (25 to 150 °C) and the cold region (−50 to +25 °C) is about 16 ppm/°C for constantan, but only about 5 ppm/°C for the original quaternary Ni-Cr-Al alloys and only 1 ppm/°C for the new quaternary alloys. The high resistivity and low TCR of Ni-Cr-Al alloys are obtained by an order-disorder type of heat treatment at approximately 540 °C (1000 °F). Therefore, if desired, the temperature coefficient for small-diameter wires can be decreased to ±5 ppm/°C or less without resorting to melt selection, which is required for alloys that do not respond to heat treatment. Electrical stability of quaternary Ni-Cr-Al alloys is excellent—1 to 10 ppm/yr or less. Their thermoelectric potential versus copper also is excellent—about 1 μV/°C at temperatures from 0 to 100 °C.

As indicated in Table 1, the mechanical properties of Ni-Cr-Al alloys are higher than those of manganin and constantan. Wires made of Ni-Cr-Al alloys are available in diameters as small as 0.0102 mm (0.0004 in.), whereas wires of copper-base alloys such as constantan are seldom produced in diameters smaller than 0.025 mm (0.001 in.). Because the resistance of a wire varies inversely with the square of its diameter, it is possible with small-diameter Ni-Cr-Al wires to produce miniature resistors that are exceedingly high in resistance.

The Ni-Cr-Al alloys resist oxidation better than other commercial electrical resistance alloys. This is an advantage in resistors that are not covered with enamel, teflon or other coatings. It is a disadvantage for making acceptable soldered or brazed joints, because it necessitates greater care in joint preparation. However, suitable soldered or silver-brazed connections can be made readily using appropriate fluxes.

Other Precision Resistance Materials. In high-resistance precision resistors, where TCR limits are less stringent, 80Ni-20Cr alloys may be used. 80Ni-20Cr alloys have temperature coefficients about four times the nominal value for Ni-Cr-Al alloys. Other similar alloys, such as Ni-Cr-Fe alloys, are used primarily to accomplish specific design objectives, because they permit designers to vary wire diameter and wire coatings in order to accommodate design constraints such as severe space limitations. All of these precision resistance alloys have low thermoelectric potentials versus copper at temperatures from 0 to 100 °C.

Aside from electrical resistance and temperature coefficient, several other design factors must be taken into consideration in selection of precision resistance metals for specific applications. For example, pure nickel (99.8%) is used in precision instruments that require a high positive temperature coefficient; pure copper is used for compensating resistors in precision instruments where the temperature coefficient of moving coils must be matched; and pure platinum is used for heating elements in thermocouple-type ac-to-dc converters as well as for precision resistance bulbs. Aluminum is seldom used as a resistor material, but because of its favorable ratio of weight to resistance it is often used in windings for the moving coils of permanent-magnet electrical instruments.

Alloys such as 71Ni-29Fe and pure metals such as nickel and platinum have low resistivities and high temperature coefficients. For these metals, it is important that operating temperature be specified carefully, because the temperature coefficient varies quite sharply with temperature (see Table 2).

Although their high temperature coefficients eliminate them from use in ordinary precision resistors, pure metals are useful in other applications such as temperature-measurement instruments and ballast devices.

Semiprecision Resistance Alloys. The alloys in Table 1 other than those previously discussed are used chiefly for rheostats and potentiometers. These semiprecision resistance alloys are not made to the rigid specifications that apply to manganins, constantans, or quaternary Ni-Cr-Al alloys. For these alloys, long-term stability generally is not critical, and such factors as thermal emf versus copper and temperature coefficient of resistance are relatively unimportant.

Semiprecision resistance alloy wires are produced in diameters ranging from a lower limit of 0.013 mm (0.0005 in.)—a lower limit of 0.02 mm (0.0008 in.) for certain Cu-Ni alloys—to an approximate maximum of 6.35 mm (¼ in.). Above 0.078 mm (0.0031 in.), wires are made only in standard B&S gages; below 0.078 mm, both standard B&S gages and intermediate sizes are available. Also, ribbon and strip usually are available in thicknesses from 0.013 to 6.35 mm (0.0005 to ¼ in.). Ribbon and strip in thicknesses near the lower end of the foregoing range usually are available only as cold rolled stock, and those in thicknesses near the extreme upper end of the range usually are available only as hot rolled stock.

Dimensional tolerances are seldom specified for applications in which resistivity and electrical resistance per unit length, rather than physical dimensions, are of primary importance. Typical tolerances on resistance per unit length are as follows:

Hot rolled ribbon rods . . . . ±8%
Cold rolled ribbon. . . . . ±5%
Cold drawn round wire:
- Finer than 0.05 mm (0.002 in.). . . . . ±10%
- 0.05 to 0.1 mm (0.002 to 0.004 in.) . . . . ±8%
- 0.1 to 0.57 mm (0.004 to 0.0226 in.) . . . . ±5%
- 0.57 mm (0.0226 in.) and heavier . . . . ±3%

Closer tolerances are available.

Thermostat Metals

A thermostat metal is a composite material (usually in the form of sheet or strip) that consists of two or more materials bonded together, of which one may be a nonmetal. Because the materials bonded together to form the composite differ in thermal expansion, the curvature of the composite is altered by changes in temperature: this is the fundamental characteristic of any thermostat metal. A thermostat metal is, therefore, a complete, self-contained transducing system capable of transforming heat directly into mechanical energy for control, indicating or monitoring purposes.

In applications such as circuit breakers, thermal relays, motor overload protectors, and flashers, the change in temperature necessary for operation of the element is produced by the passage of current through the element itself—in other words, the change is produced by I^2R heating. In certain other applications, any increase in the temperature of the thermostat element caused by I^2R heating is objectionable, and a thermostat metal with low electrical resistivity is required.

For circuit breakers and similar devices, there are thermostat metals that differ in electrical resistivity but that are similar in other properties. This allows a manufacturer to design a complete series of circuit breakers of different ratings in which the thermostat elements are all of the same size but have different electrical resistances. Resistivity is varied by incorporating a layer of a low-resistivity metal between outer layers of two other metals that have high resistivities and that differ widely in expansion coefficient.

In one series of commercial thermostat metals with resistivities ranging from 165 to 780 nΩ·m (100 to 470 Ω·circ mil/ft) at 24 °C (75 °F), high-purity nickel is used for the intermediate layer. In a series with resistivities from 33 to 165 nΩ·m (20 to 100 Ω·circ mil/ft), high-conductivity copper alloys are employed for the intermediate layer.

The use of a manganese-copper-nickel alloy having a resistivity of 1745 nΩ·m (1050 Ω·circ mil/ft) for one of the outer layers has extended the practical upper resistivity limit of thermostat metals to 1620 nΩ·m (975 Ω·circ mil/ft) at 24 °C.

Tolerances on resistivity at a standard temperature vary from ±3 to ±10%, depending on the type of thermostat metal and its resistivity.

About 30 different alloys are used to make over 50 different thermostat metals. Most of these 30 alloys are nickel-iron, nickel-chromium-iron, chromium-iron, high-copper and high-manganese alloys.

Thermostat metals are available as strip or sheet in thicknesses ranging from 0.13 to 3.2 mm (0.005 to 0.125 in.) and widths from 0.5 to 300 mm (0.020 to 12 in.). They are easily formed into the required shapes. Thermostat metals usually are selected on the basis of the temperature range in which they are required to operate. They are available for various operating ranges between −185 and +540 °C (−300 and +1000 °F). Properties and typical bimetal combinations for several temperature ranges are given in Table 3.

Table 2 TCR values for 71Ni-29Fe and pure nickel determined using various reference temperatures

Reference temperature, °C	Temperature range, °C	TCR, ppm/°C
71Ni-29Fe alloy		
20	20 to 100	4500
25	25 to 100	4300
Nickel (99.9% purity)		
0	0 to 100	6730
20	20 to 100	6150
25	25 to 100	6000
Nickel (99% purity)		
0	0 to 100	5250
20	20 to 100	4750
25	25 to 100	4620

Heating Alloys

Resistance heating alloys are used in many varied applications—from small household appliances to large industrial furnaces. In appliances, heating elements are designed for intermittent short-term service at about 100 to 1090 °C (200 to 2000 °F). In industrial furnaces, elements often must operate continuously at temperatures as high as 1300 °C (2350 °F) for furnaces used in metal treating industries, 1700 °C (3100 °F) for kilns used for firing ceramics, and occasionally as high as 2000 °C (3600 °F) for special applications.

The primary requirements of materials used for heating elements are high melting point, high electrical resistivity, reproducible temperature coefficient of resistance, good oxidation resistance in furnace environments, absence of volatile components, and resistance to contamination. Other desirable properties are good elevated-temperature creep strength, high emissivity, low thermal expansion and low modulus (both of which help minimize thermal fatigue), good resistance to thermal shock, and good strength and ductility at fabrication temperatures.

Table 4 gives physical and mechani-

Table 3 Properties of thermostat metals frequently selected for some common service temperatures

Temperature range of maximum sensitivity		Composition		Resistivity at 24 °C (75 °F)		Flexivity(a)	
°C	°F	High-expanding side	Low-expanding side	nΩ·m	Ω·circ mil/ft	μm/m·°C	μin./in.·°F
−20 to +150	0 to 300	75Fe-22Ni-3Cr	64Fe-36Ni	780	470	26.3	14.6
−20 to +200	0 to 400	75Fe-22Ni-3Cr	Pure Ni	160	95	8.3	4.6
		72Mn-18Cu-10Ni	64Fe-36Ni	1120	675	38.5	21.4
120 to 290	250 to 550	67Ni-30Cu-1.4Fe-1Mn	60Fe-40Ni	565	340	16.6	9.2
150 to 450	300 to 850	66.5Fe-22Ni-8.5Cr	50Fe-50Ni	580	350	11.2	6.2

(a) At 40 to 150 °C (100 to 300 °F). See ASTM B106 for standard test method for determining flexivity of thermostat metals.

Table 4 Typical properties of resistance heating materials

Basic composition	Resistivity(a), nΩ·m(b)	Average change in resistance(c), %, from 20 °C to: 260 °C	540 °C	815 °C	1095 °C	Thermal expansion, μm/m·°C, from 20 °C to: 100 °C	540 °C	815 °C	Tensile strength MPa	ksi	Density Mg/m³	lb/in.³
Nickel-chromium and nickel-chromium-iron alloys												
78.5Ni-20Cr-1.5Si (80-20)	1080	+4.5	+7.0	+6.3	+7.6	13.5	15.1	17.6	690 to 1380	100 to 200	8.41	0.30
77.5Ni-20Cr-1.5Si-1Nb	1080	+4.6	+7.0	+6.4	+7.8	13.5	15.1	17.6	690 to 1380	100 to 200	8.41	0.30
68.5Ni-30Cr-1.5Si (70-30)	1180	+2.1	+4.8	+7.6	+9.8	12.2	...	...	825 to 1380	120 to 200	8.12	0.29
68Ni-20Cr-8.5Fe-2Si	1165	+3.9	+6.7	+6.0	+7.1	...	12.6	...	895 to 1240	130 to 180	8.33	0.30
60Ni-16Cr-22.5Fe-1.5Si	1120	+3.6	+6.5	+7.6	+10.2	13.5	15.1	17.6	655 to 1205	95 to 175	8.25	0.30
35Ni-30Cr-33.5Fe-1.5Si	1055	+7.95	+14.9	+18.0	+22.0	14.6	17.5	16.0	895 to 1380	130 to 200	7.90	0.29
35Ni-20Cr-43.5Fe-1.5Si	1015	+8.0	+15.4	+20.6	+23.5	15.7	15.7	...	550 to 1205	80 to 175	7.95	0.29
35Ni-20Cr-42.5Fe-1.5Si-1Nb	1015	+8.0	+15.4	+20.6	+23.5	15.7	15.7	...	550 to 1205	80 to 175	7.95	0.29
Iron-chromium-aluminum alloys												
83.5Fe-13Cr-3.25Al	1120	+7.0	+15.5	...	...	10.6	...	...	515 to 1035	75 to 150	7.30	0.26
81Fe-14.5Cr-4.25Al	1245	+3.0	+9.7	+16.5	...	10.8	11.5	12.2	550 to 1170	80 to 170	7.28	0.26
79.5Fe-15Cr-5.2Al	1370	+1.9	+5.5	+8.9	+9.6	11.3	12.6	...	550 to 895	80 to 130	7.12	0.26
73.5Fe-22Cr-4.5Al	1355	+0.3	+2.9	+4.3	+4.9	10.8	12.6	13.1	725 to 1205	105 to 175	7.15	0.26
72.5Fe-22Cr-5.5Al	1455	+0.2	+1.0	+2.8	+4.0	11.3	12.8	14.0	760 to 1205	110 to 175	7.10	0.26
Pure metals												
Molybdenum	52	+110	+238	+366	+508	4.8	5.8	...	690 to 2160	100 to 313	10.2	0.369
Platinum	105	+85	+175	+257	+305	9.0	9.7	10.1	345	50	21.5	0.775
Tantalum	125	+82	+169	+243	+317	6.5	6.6	...	345 to 1240	50 to 180	16.6	0.600
Tungsten	55	+91	+244	+396	+550	4.3	4.6	4.6	3380 to 6480	490 to 940	19.3	0.697
Nonmetallic heating-element materials												
Silicon carbide	995 to 1995	−33	−33	−28	−13	4.7	...	...	28	4	3.2	0.114
Molybdenum disilicide	370	+105	+222	+375	+523	9.2	...	...	185	27	6.2	0.212
$MoSi_2$ + 10% ceramic additives	270	+167	+370	+597	+853	13.1	14.2	14.8	...	...	5.6	0.202
Graphite	9100	−16	−18	−13	−8	1.3	...	...	1.8	0.26	2.3	0.057

(a) At 20 °C (68 °F). (b) To convert to Ω·circ mil/ft, multiply by 0.6015. (c) Changes in resistance may vary somewhat, depending on cooling rate.

cal properties, and Table 5 presents recommended maximum operating temperatures, for resistance heating materials. Of the four groups of materials listed in these tables, the first group (Ni-Cr and Ni-Cr-Fe alloys) serves by far the greatest number of applications.

The ductile wrought alloys in the first group have properties that enable them to be used at both low and high temperatures in a wide variety of environments. The Fe-Cr-Al compositions (second group) are also ductile alloys. They play an important role in heaters for the higher temperature ranges, which are constructed to provide more effective mechanical support for the element. The pure metals that comprise the third group have much higher melting points. All of them except platinum are readily oxidized and are restricted to use in nonoxidizing environments. They are valuable for a limited range of application, primarily for service above 1370 °C (2500 °F). The cost of platinum prohibits its use except in small, special furnaces.

The fourth group, nonmetallic heating-element materials, are used at still higher temperatures. Silicon carbide can be used in oxidizing atmospheres at temperatures up to 1650 °C (3000 °F); two varieties of molybdenum disilicide are effective up to maximum temperatures of 1700 and 1800 °C (3100 and 3270 °F) in air. Molybdenum disilicide heating elements are gaining increased acceptance for use in industrial and

Table 5 Recommended maximum furnace operating temperatures for resistance heating materials

Basic composition, %	Approximate melting point °C	°F	Maximum furnace operating temperature in air °C	°F
Nickel-chromium and nickel-chromium-iron alloys				
78.5Ni-20Cr-1.5Si (80-20)	1400	2550	1150	2100
77.5Ni-20Cr-1.5Si-1Nb	1390	2540		
68.5Ni-30Cr-1.5Si (70-30)	1380	2520	1200	2200
68Ni-20Cr-8.5Fe-2Si	1390	2540	1150	2100
60Ni-16Cr-22.5Fe-1.5Si	1350	2460	1000	1850
35Ni-30Cr-33.5Fe-1.5Si	1400	2550		
35Ni-20Cr-43.5Fe-1.5Si	1380	2515	925	1700
35Ni-20Cr-42.5Fe-1.5Si-1Nb	1380	2515		
Iron-chromium-aluminum alloys				
83.5Fe-13Cr-3.25Al	1510	2750	1050	1920
81Fe-14.5Cr-4.25Al	1510	2750		
79.5Fe-15Cr-5.2Al	1510	2750	1260	2300
73.5Fe-22Cr-4.5Al	1510	2750	1280	2335
72.5Fe-22Cr-5.5Al	1510	2750	1375	2505
Pure metals				
Molybdenum	2605	4730	400(a)	750(a)
Platinum	1770	3216	1500	2750
Tantalum	2975	5390	500(a)	930(a)
Tungsten	3375	6116	300(a)	570(a)
Nonmetallic heating-element materials				
Silicon carbide	2410	4370	1600	2900
Molybdenum disilicide	2080	3775	1700 to 1900	3100 to 3270
$MoSi_2$ + 10% ceramic additives	1800	3270		
Graphite	3650 to 3695(b)	6610 to 6690(b)	400(c)	400(c)

(a) Recommended atmospheres for these metals are a vacuum of 10^{-4} to 10^{-5} mm Hg, pure hydrogen, and partly combusted city gas dried to a dew point of +4 °C (+40 °F). In these atmospheres the recommended temperatures would be:

	Vacuum	Pure H_2	City gas
Mo	1650 °C (3000 °F)	1760 °C (3200 °F)	1700 °C (3100 °F)
Ta	2480 °C (4500 °F)	Not recommended	Not recommended
W	1650 °C (3000 °F)	2480 °C (4500 °F)	1700 °C (3100 °F)

(b) Graphite volatilizes without melting at 3650 to 3695 °C (6610 to 6690 °F). (c) At approximately 400 °C (750 °F) (threshold oxidation temperature), graphite undergoes a weight loss of 1% in 24 h in air. Graphite elements can be operated at surface temperatures up to 2205 °C (4000 °F) in inert atmospheres.

laboratory furnaces. Among the desirable properties of molybdenum disilicide elements are excellent oxidation resistance, long life, constant electrical resistance, self-healing ability, and resistance to thermal shock.

Nickel-Chromium and Nickel-Chromium-Iron Alloys. The resistivities of Ni-Cr and Ni-Cr-Fe alloys are high, ranging from 1015 to 1180 nΩ·m (610 to 710 Ω·circ mil/ft) at 25 °C. Figure 4 shows that the resistance changes more rapidly with temperature for 35Ni-20Cr-45Fe than for any other alloy in this group. The curve for 35Ni-30Cr-35Fe is similar, but slightly lower. The other four curves, which are for alloys with substantially higher nickel contents, reflect relatively low changes in resistance with temperature. For these alloys, rate of change reaches a peak near 540 °C (1000 °F), goes through a minimum at about 760 to 870 °C (1400 to 1600 °F) and then increases again. For Ni-Cr alloys, the change in resistance with temperature depends on section size and cooling rate. Figure 5 presents values for a typical 80Ni-20Cr alloy. The maximum change (curve A) occurs with small sections, which cool rapidly from the last production heat treatment. The smallest change occurs for heavy sections, which cool slowly. The average curve (curve B) is characteristic of medium-size sections.

Iron-Chromium-Aluminum Alloys. Fe-Cr-Al heating alloys are higher in electrical resistivity, and lower in density, than Ni-Cr and Ni-Cr-Fe alloys. Resistivity of Fe-Cr-Al alloys depends on both aluminum and chromium contents, with aluminum being predominant (see Fig. 6).

These alloys have excellent resistance to oxidation at elevated temperatures, because reaction with atmospheric oxygen forms a protective layer of relatively pure alumina. At about 1200 °C (2200 °F), this oxide consists of approximately 94.5% Al_2O_3, 3.5% Cr_2O_3 and 2% Fe_2O_3. This gray-white protective skin has extremely high dielectric strength. The electrical resistivity of aluminum oxide is 10^{12} Ω·m at room temperature, and at about 1100 °C (2000 °F) it is still 10^4 Ω·m. Under normal operating conditions, deterioration of the oxide surface layer, and the resulting aluminum depletion, are fairly slow provided that there is no contact with certain refractories at temperatures above 980 °C (1800 °F). The time required for a 10% change in resistance varies from 75 to 100% of heater life (time to burnout), depending on the particular melt, the size of the heater, and the operating temperature.

Tensile strength of Fe-Cr-Al alloys is relatively low, as shown in Table 6. Because of this low strength, the weight of the lower terminal in straight-wire testing causes the wire to stretch; consequently, life tests on Fe-Cr-Al alloys are conducted using U-shape specimens. Life of heaters made from Fe-Cr-Al alloys is also influenced by the fact that these alloys exhibit large increases in resistance with time as a result of their growth characteristics, which are discussed in a later section of this article.

Iron-chromium-aluminum alloys undergo a metallurgical change that causes brittleness after cyclic exposure to high temperatures. As a result, when heaters made of these alloys fail they cannot be repaired but must be replaced.

Figure 7 presents data on life of 74.5Fe-20Cr-5Al-0.5Co wire 0.40 and 0.29 mm (0.0159 and 0.0113 in.) in diameter at a series of temperatures. With Fe-Cr-Al alloys, as with Ni-Cr alloys, life of a heater of given size decreases as temperature increases.

Pure metals used as heating alloys (see Table 4) have very low resistivities and very high temperature coefficients of resistance. The tensile strengths of

Fig. 4 Variation of resistance with temperature for six Ni-Cr and Ni-Cr-Fe alloys

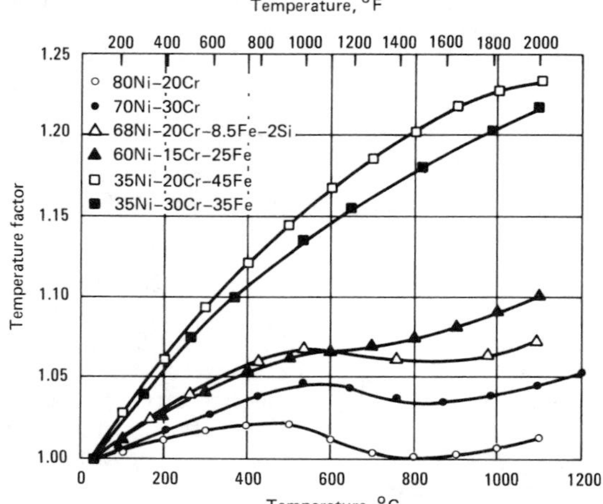

To calculate resistance at temperature, multiply resistance at room temperature by the temperature factor.

Fig. 5 Variation of resistance with temperature for 80Ni-20Cr heating alloy

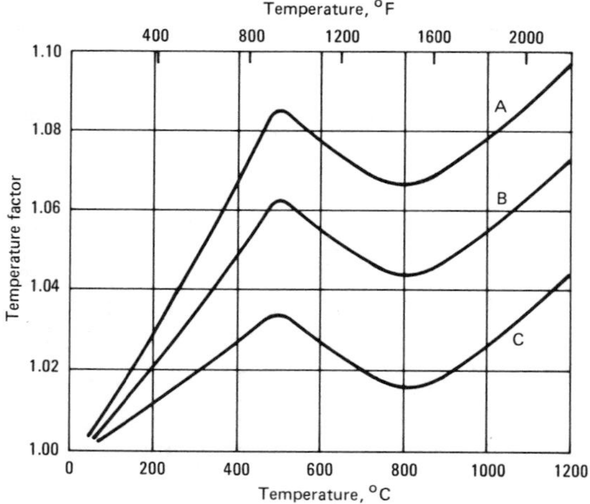

Curve A is for a specimen cooled rapidly after the last production heat treatment. Curve C is for a specimen cooled slowly after the last production heat treatment. Curve B represents the average value for material as delivered by the producer. To calculate resistance at temperature, multiply the resistance at room temperature by the temperature factor.

molybdenum and tungsten are quite high, even at elevated temperatures, and that of tantalum is medium (see Table 6). Above 800 °C (1500 °F), all three of these refractory metals are substantially stronger than Ni-Cr, Ni-Cr-Fe and Fe-Cr-Al alloys at any given temperature.

Platinum can be used for temperatures up to 1500 °C (2750 °F) because it has excellent resistance to oxidation in air. Molybdenum, tantalum and tungsten must be kept below 400, 500 and 300 °C (750, 930 and 570 °F), respectively, because they are subject to catastrophic oxidation in air at moderate and elevated temperatures.

Nonmetallic Materials. Of the nonmetallic heating materials, graphite and silicon carbide are much higher in resistivity than wrought metallic heating materials. The temperature coefficient of resistance of these materials is negative. Molybdenum disilicide elements have relatively low resistivity at room temperature and a very high positive temperature coefficient of resistance.

All of these materials have low tensile strengths. However, because the resistivity of silicon carbide is so high, silicon carbide elements are made in large cross sections in order to reduce resistance to reasonable values, and as a result these elements can withstand relatively high mechanical loads. Graphite has poor oxidation resistance and should not be used above 400 °C (750 °F). Silicon carbide, however, has excellent oxidation resistance and can be used at temperatures up to 1600 or 1650 °C (2900 or 3000 °F).

Two materials composed of about 90% molybdenum disilicide and 10% refractory oxides have been produced. One has a maximum operating temperature of 1700 °C (3100 °F), while the other serves at temperatures up to 1800 °C (3270 °F).

Pure molybdenum disilicide is too brittle for practical use as an electric heating element. Addition of metallic and ceramic binding agents reduces this brittleness to a practical level. Sintering in the presence of a liquid phase then produces a material that is essentially free of porosity. The high resistance to thermal shock of $MoSi_2$ has enabled $MoSi_2$ elements to undergo, without damage, a test in which they were cycled from room temperature to 1650 °C (3000 °F) through 20 000 cycles.

The excellent high-temperature performance of molybdenum disilicide elements is brought about by the chemical reaction that takes place on the element surface.

Above 980 °C (1800 °F), the material reacts with oxygen to form a silicon dioxide (quartz glass) coating, which protects the base material against chemical attack including further oxidation. This film is "self-healing": surface cracks developed by mechanical damage are covered by a new coating of quartz glass when the element is again heated above 980 °C in air.

Design of Resistance Heaters

Regardless of which heating alloy is selected, design of the heating element

Table 6 Elevated-temperature tensile strength of selected resistance heating materials

Heating material	Tensile strength at: 425 °C (800 °F) MPa	ksi	650 °C (1200 °F) MPa	ksi	870 °C (1600 °F) MPa	ksi	1100 °C (2000 °F) MPa	ksi
Nickel-chromium and nickel-chromium-iron alloys								
68.5Ni-30Cr-1.5Si	735	107	675	98	205	30	75	11
78.5Ni-20Cr-1.5Si	715	104	620	90	170	25	75	11
68Ni-20Cr-8.5Fe-1.5Si	760	110	655	95	195	28	75	11
Iron-chromium-aluminum alloys								
79.5Fe-15Cr-5.2Al	480	70	205	30	48	7	...	...
73.5Fe-22Cr-4.5Al	525	76	165	24	14	2	...	...
72.5Fe-22Cr-5.5Al	550	80	345	50	52	7.5	26	3.8
Pure metals								
Tungsten	560	81	525	76	395	57	295	43
Molybdenum	620	90	585	85	365	53	235	34
Tantalum	315	46	315	46	280	41	195	28

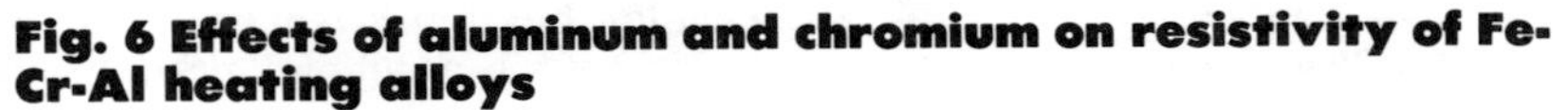

Fig. 6 Effects of aluminum and chromium on resistivity of Fe-Cr-Al heating alloys

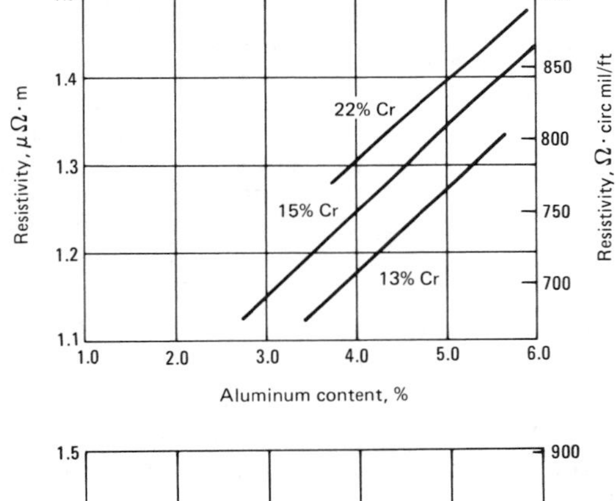

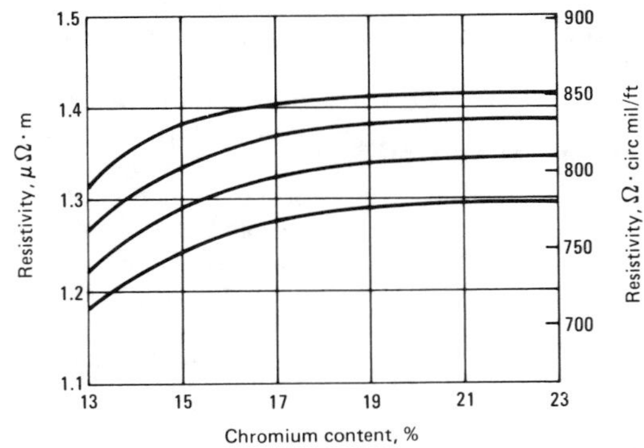

is important. One of the most important rules is to allow for unhindered expansion and contraction so as to avoid concentration of stresses as the temperature changes.

For service at lower temperatures, particularly from 400 to 600 °C (750 to 1100 °F), formed heating elements are used in ovens. In this construction, a heater support is made of two high-alloy rods spaced approximately 300 mm (12 in.) apart in a frame made of angle sections. The rods, whose length is determined by rated electrical input, contain spool insulators around which is wound a ribbon element made of a heating alloy. In a similar alternative construction, the ribbon element is replaced by a continuous helical coil of 5 gage or smaller wire.

Ribbon sizes for oven heaters range from 0.09 to 0.20 mm (0.0035 to 0.008 in.) thick, and from 9.5 to 16 mm (3/8 to 5/8 in.) wide. Oven heaters are rated to give maximum output at a watt density of approximately 8 kW/m^2 (5 W/in.^{2}). (Watt density is obtained by dividing total power input to the elements by total surface area of the heater.) For 110- or 220-V oven heaters operating under normal conditions, expected life of Ni-Cr elements in air is three to five years.

For furnace temperatures up to 1175 °C (2150 °F), sinuous loop elements generally are formed from ribbon having a width-to-thickness ratio of about 12 to 1 and dimensions varying from 0.76 to 3.2 mm (0.030 to 0.125 in.) in thickness and from 13 to 38 mm (1/2 to 1 1/2 in.) in width. Round rods of resistance material also may be formed into elements. Rod-type elements have been used by several furnace manufacturers.

Dimensional relationships of loops are important in achieving the desired combination of uniform furnace temperature and long heater life. Recommended loop dimensions for various locations within a furnace are shown in Table 7 for several different ranges of operating temperature. Ribbon size and watt density are correlated with operating temperature in Table 8. When designed in accordance with these recommendations, Ni-Cr elements have life expectancies of up to seven years at temperatures of 540 to 925 °C (1000 to 1700 °F), two to five years at 980 to 1100 °C (1800 to 2000 °F), and one to two years at 1100 to 1175 °C (2000 to 2150 °F).

Iron-chromium-aluminum heating elements used for temperatures up to 1300 °C (2350 °F) also may be made from ribbon. However, Fe-Cr-Al ribbon must be formed into short, sinuous loops having a loop spacing of 50 mm (2 in.) or less, and requires better loop support than Ni-Cr ribbon.

Refractory Supports. Ceramic refractories that come in contact with heating elements may influence selection of heating alloys. Below 1000 °C (1825 °F), protective oxides that form

Fig. 7 Total life vs temperature for two sizes of 74.5Fe-20Cr-5Al-0.5Co wire

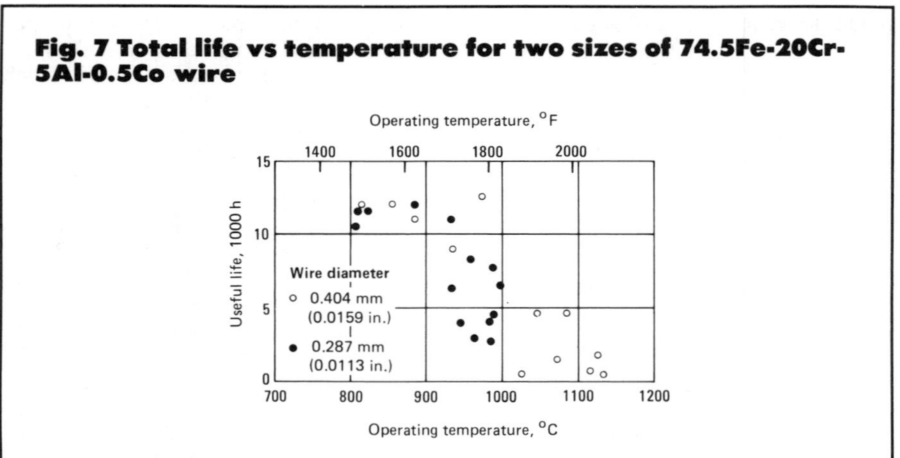

Table 7 Typical length and spacing of loops in Ni-Cr and Ni-Cr-Fe heating elements

Maximum operating temperature °C	Maximum operating temperature °F	Loop spacing(a) for ribbon width of: Up to 19 mm (3/4 in.) mm	in.	Over 19 mm (3/4 in.) mm	in.	Maximum loop length(b) mm	in.
Sidewall heaters							
540-760	1000-1400 ...	50-75	2-3	65-75	2.5-3	450	18
760-1100	1400-2000 ...	50-75	2-3	65-75	2.5-3	450	18
1100-1175	2000-2150 ...	65-75	2.5-3	75	3	300	12
Roof heaters							
540-760	1000-1400 ...	50-75	2-3	50-75	2-3	300	12
760-1100	1400-2000 ...	50-75	2-3	50-75	2-3	300	12
1100-1175	2000-2150 ...	50-75	2-3	50-75	2-3	300	12
Floor heaters							
540-760	1000-1400 ...	40 min	1.5 min	55 min	2.25 min	450	18
760-1100	1400-2000 ...	40 min	1.5 min	55 min	2.25 min	450	18
1100-1175	2000-2150 ...	55 min	2.25 min	55 min	2.25 min	450	18

(a) Loop spacing is the lineal distance between centers for two adjacent bends on the same side of the element. (b) Loop length is the over-all lineal distance from one side of the element to the other. For elements over 300 mm in loop length, the loops must be separated or supported by additional insulators. Two end supports or one hanger and a bottom guide separator are sufficient for elements under 300 mm in length.

Table 8 Typical ribbon size and electrical capacity of Ni-Cr and Ni-Cr-Fe heating elements

Maximum operating temperature °C	Maximum operating temperature °F	Size of ribbon: Min thickness mm	in.	Width mm	in.	Watt density kW/m²	W/in.²
540-760	1000-1400	0.75	0.030	Any	Any	21.5	14
760-925	1400-1700	1.8	0.070	13-40	1/2-1 1/2	18.5	12
925-1100	1700-2000	2.3	0.090	19-40	3/4-1 1/2	15.5	10
1100-1175	2000-2150	2.5	0.100	19-40	3/4-1 1/2	12.5	8

on the surfaces of Ni-Cr, Ni-Cr-Fe and Fe-Cr-Al alloys do not react with ceramic oxides, including refractory grades of SiO_2, Al_2O_3, CaO, Na_2O, MgO, K_2O and ZrO_2. Above 1000 °C, pure MgO, Al_2O_3 and ZrO_2 are recommended. Many ordinary refractory-grade materials become conductive at such temperatures; the sodium and potassium contents of the refractory material should be low to prevent this. Sulfur-containing refractories should not be used with Ni-Cr, Ni-Cr-Fe or Fe-Cr-Al alloys. Refractories also must be as low as possible in ferric oxide, if they are to be used with Fe-Cr-Al resistance elements.

Use of molybdenum, tungsten or platinum heating elements at temperatures above 1200 °C (2200 °F) necessitates use of pure oxide refractories. High-purity alumina (99%) and magnesia are the most satisfactory. Zirconia becomes conductive above 1300 °C (2350 °F); silica decomposes and embrittles platinum at about 1200 °C. Consequently, neither zirconia nor silica can be used at or above these temperatures.

In cyclic temperature applications, some alloys elongate continuously and at the same time continuously decrease in cross-sectional area. Iron-chromium-aluminum alloys do this even in the absence of applied external force. A popular but unproved explanation of this growth is as follows: first, the alloy oxidizes at elevated temperature; on cooling, the high compressive strength of the oxide layer forces the weaker core to elongate; on reheating, the oxide is weak in tension so that it cracks, and reoxidation takes place; thus growth continues in a cyclic manner. Growth is detrimental because it causes large changes in resistance, so suitable steps must be taken in the design of high-temperature elements to minimize cyclic growth.

Iron-chromium-aluminum heating alloys must be adequately supported for use at temperatures near 1300 °C (2350 °F).

Nickel-chromium alloys, which exhibit little or no growth, do not require special support at their maximum operating temperatures, which are 175 °C lower than those of Fe-Cr-Al alloys (1375 °C vs. 1200 °C). Although all electrical heating alloys have low tensile properties at high temperatures, Ni-Cr and Ni-Cr-Fe alloys have higher strength at elevated temperature than do Fe-Cr-Al alloys. For example, 80Ni-20Cr tested at 1100 °C (2000 °F) and a strain rate of 3.3 mm/m··s (0.2 in./in.··min) exhibited yield and tensile strengths of 41 and 48 MPa (6 and 7 ksi), respectively. An Fe-Cr-Al alloy tested under identical conditions exhibited values of only 12 and 16 MPa (1.8 and 2.3 ksi).

One way of establishing allowable stresses in heating elements is to define allowable stress as the load that will produce 0.1% creep in 1000 h or 1% creep in 10 000 h. For 80Ni-20Cr, this load is 1.4 MPa (200 psi) at 1100 °C (2000 °F).

Fabrication of Heaters

Annealed wire, rod or ribbon of any common nickel-chromium heating alloy can be bent at room temperature around a mandrel whose diameter equals the diameter or thickness of the stock. Some difficulty has been experienced in making such bends in 80Ni-20Cr alloys, because these alloys tend to strain age during forming if they have been heated in the range 100 to 200 °C (200 to 400 °F) between forming stages. In a sheathed heater helix, strain aging can lead to nonuniform stretching during forming of the helix, which in turn can lead to hot spots in service.

Iron-chromium-aluminum alloys are harder at room temperature, and somewhat more difficult to form, than Ni-Cr alloys. They can be shaped into heating elements by techniques much like those used for making elements from Ni-Cr alloys, but heavy-gage material should be preheated to 300 °C (about 550 °F) to facilitate forming of sharp radii.

Joining. Nickel-chromium and nickel-chromium-iron alloys are readily joined by welding. Preferred processes include resistance welding, gas-shielded metal-arc welding and oxyfuel gas welding. Filler metal of essentially the same composition as the base alloy should be used. These alloys can be brazed, but heating elements are seldom joined in this way because brazing introduces a zone whose melting temperature is lower than that of the base metal.

Iron-chromium-aluminum alloys are weldable, but they should be welded as quickly as possible to prevent grain growth. Postheating, which sometimes is done for stress relief, must be done at a temperature low enough to prevent grain growth.

Sheathed Heaters

Nickel-chromium (80Ni-20Cr) and nickel-chromium-iron (60Ni-16Cr-22.5Fe-1.5Si and 35Ni-20Cr-43.5Fe-1.5Si) alloys are extensively used as heating elements in sheathed heaters, where compacted granular magnesium oxide provides electrical insulation between a wire element and a metallic sheath. This construction permits operation at high watt density without rapid degradation. Various grades of magnesium oxide are available for use at different temperature levels.

Insulation resistance and heater life are affected by factors such as chemical and physical characteristics of the insulation, operating temperature, and atmospheric conditions within the heater sheath. Penetration of oxidation products into the insulating layer of magnesium oxide may occur, causing a reduction in insulation resistance with prolonged cyclic testing or use. This can result in excessive electrical leakage to the metallic sheath and failure if the sheath has inadequate resistance to cyclic oxidation. In the preparation of element wire and fabrication of heaters, care must be taken to ensure uniformity of cross section and to avoid surface damage that could shorten service life.

In the design of sheathed heaters, it must be recognized that the operating temperature of the internal heating element is considerably higher than that of the external sheath. Selection of heater material is based on expected operating temperature and environment. Approximate temperature limitations for fused and unfused magnesium oxide insulation used in sheathed heaters are as follows:

Insulation purity	Max operating temperature °C	°F
Fused MgO		
	98	180
98%	0	0
	87	160
96%	0	0
Unfused MgO		
	70	130
96%	0	0
	67	125
94%	5	0

Except for immersion heaters, where operating temperatures are low, sheathing is not recommended for the standard composition Fe-Cr-Al alloys.

Service Life of Heating Elements

Maximum service temperature is one of the most important factors governing service life of heating elements. Whether the temperature is constant or intermittent also has a marked effect. Data from accelerated laboratory testing (Fig. 8) illustrate the effects of temperature on 80Ni-20Cr heating elements. The graph shows a rapid decrease in element life as temperature increases above 1120 °C (2050 °F).

Fig. 8 Variation in useful life with temperature in cyclic testing of 80Ni-20Cr heating elements

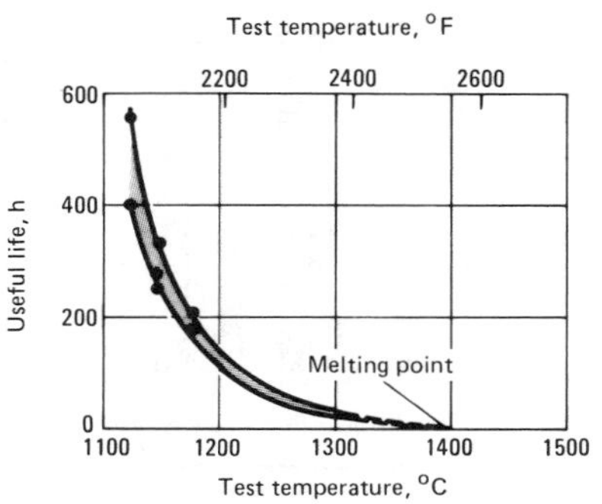

Life is defined as time to burnout or to 10% change in electrical resistance, whichever occurs first. Tests were conducted in accordance with ASTM B76; specimens were at temperature for 2 min out of every 4-min cycle.

Despite the consistency of such laboratory data, life of a heating element in service cannot be predicted accurately. Even under closely controlled test conditions there will be considerable variation in element life, as shown by the data in Fig. 9, for 56 identical specimens of 80Ni-20Cr tested at 1175 °C (2150 °F). In actual service, where conditions of operation are less closely controlled, the variation in heating-element life is likely to be considerably greater. In this article, figures given for the life of an alloy are mean values obtained from tests of a large number of identical specimens.

The life of a heating element increases with ribbon thickness or wire diameter, as shown in Fig. 10. In some applications, heaters may be required to operate for 10 to 15 years without element failure. Predictions, based on data from accelerated life tests, that elements will achieve such extended service lives are not particularly reliable. Often, maximum heater temperatures must be lowered considerably from those normally given in data sheets derived from accelerated life tests in order to ensure exceptionally long element life.

Oxidation resistance of alloys used for heating elements is critical. In addition to the inherent oxidation resistance necessary for any alloy used at elevated temperatures, heating alloys

Fig. 9 Distribution of life to burnout in 56 identical tests of 80Ni-20Cr elements at 1175 °C (2150 °F)

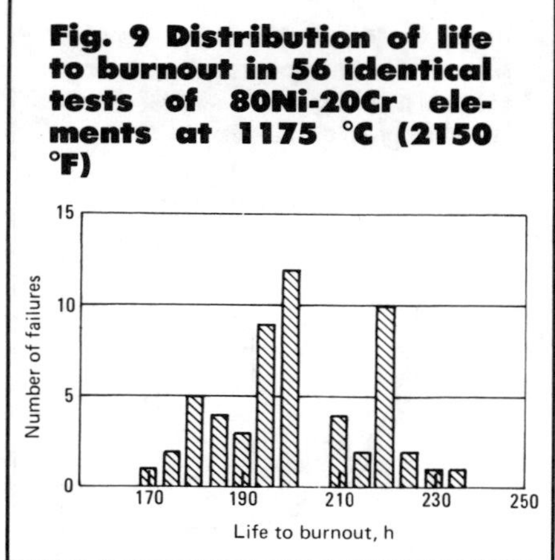

Fig. 10 Variation of heater life with section thickness or wire diameter

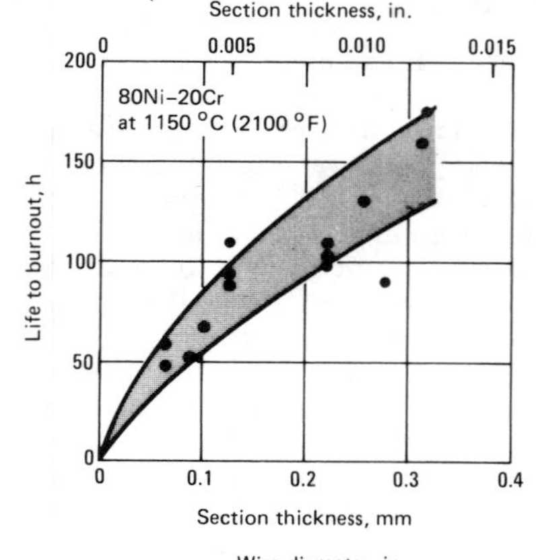

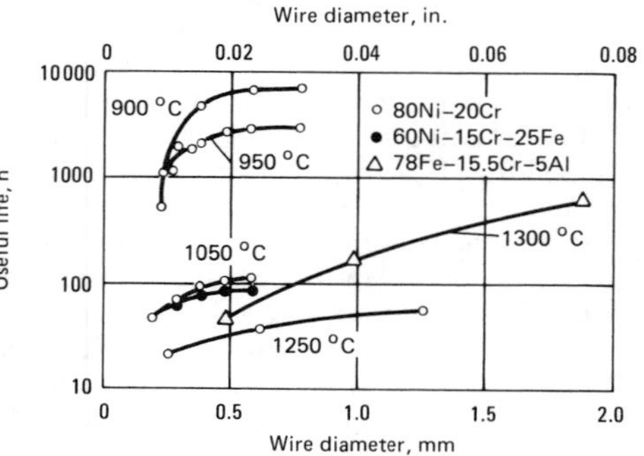

Life is defined as time to burnout or to 10% change in resistance, whichever occurs first.

also must have adherent oxides that resist spalling during temperature cycling. Because elements are heated electrically to attain temperature, oxidation and spalling in a localized area results in a local increase in resistance with consequent increase in local temperature, which creates a hot spot and shortens the life of the element. Because localized oxidation or spalling increases total resistance only slightly, the current through the element remains essentially constant, and I^2R heating causes an increase in temperature only in the region of increased resistance.

The effect of composition on the life of heating-element alloys is evaluated by both static and cyclic testing. In static testing, the element is heated to an elevated temperature and held for a prescribed time, and the weight gain or loss due to oxidation is measured. In cyclic testing, the element is alternately (*a*) heated to an elevated temperature and held for a prescribed time and (*b*) cooled to ambient temperature and held for an equal time. The weight loss in percent, or total testing time in hours (total life), is used as a measure of the quality of the alloy. Either low weight loss or long life is desired. For heating elements used with a fixed voltage, life is defined as the time in hours to burnout or to a 10% increase in total resistance, whichever occurs first. Any increase in total resistance results in a decrease in power and consequently a decrease in temperature. Usually, a 9% decrease in power is sufficient to cause "economic failure" in a constant-voltage appliance.

The upper graph in Fig. 11 compares two alloys that were continuously oxidized in air at 1175 °C (2150 °F). Cylindrically coiled strips 0.13 by 9.5 by 250 mm (0.005 by 0.375 by 10 in.) were used. Oxidation was evaluated by determining the thickness of the oxide layer. The lower graph in Fig. 11 presents test results for the same two alloys oxidized under similar conditions except that heating was intermittent.

Chemical composition of heating-element alloys in the first two groups in Table 4 is carefully controlled to achieve resistance to intermittent oxidation in oxidizing environments. Figure 12 presents data on weight loss of Ni-Cr-Fe alloys as nickel-plus-chromium content is increased from just over 20% to almost 100%. In practice, nickel-plus-chromium content is maintained above 50% to inhibit the oxidation and spalling that lead to large weight changes in high-iron compositions. Figure 13 shows the weight gain in Ni-Cr alloys after continuous oxidation for 100 h at 1100 °C (2000 °F). Increasing the chromium content of Ni-Cr alloys causes a substantial decrease in oxidation rate. Minimum weight gain is obtained at about 30% Cr.

Minor chemical constituents are important in governing life at elevated temperatures. Close control of minor elements, especially silicon, has substantially increased life of Ni-Cr heating alloys. An example of the effect of silicon on life of 80Ni-20Cr alloys is shown in Fig. 14.

The five graphs in Fig. 15 show the effects of manganese and silicon contents on rate of oxidation of 80Ni-20Cr alloys under the conditions of time and temperature indicated. Specimens for these tests were first polished with emery papers through No. 0000. They were tested in oxygen at a pressure of 0.1 atm (76 mm Hg), which constitutes accelerated testing. Such tests performed under controlled conditions give some useful information, but they cannot necessarily be correlated with element life for any specific application. Element life is related to oxide stability and adherence under actual conditions of use; life is affected by furnace atmo-

Fig. 11 Typical rates of oxidation at 1175 °C (2150 °F) for two common heating alloys

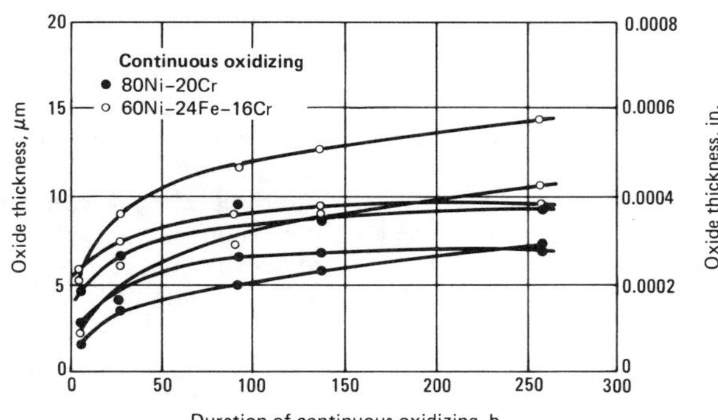

In both continuous tests (upper graph) and intermittent tests (lower graph), cylindrically coiled strips 0.13 by 9.5 by 250 mm (0.005 by 3/8 by 10 in.) were heated in air. In the intermittent oxidizing tests, power was cycled 7.5 min on and 7.5 min off.

Fig. 12 Intermittent oxidation of Ni-Cr, Ni-Cr-Fe and stainless steel heating elements in air

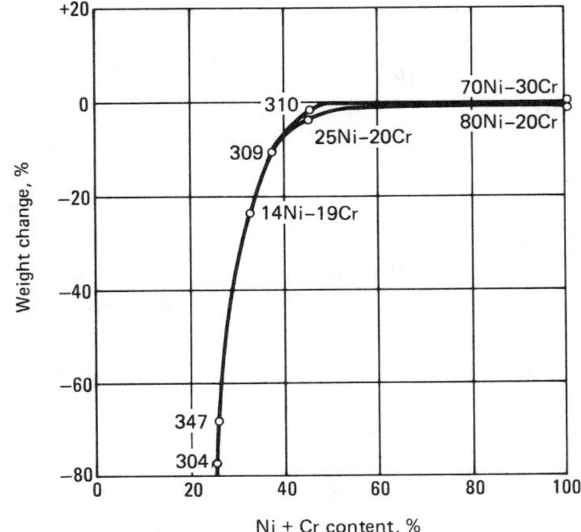

Weight change determined after 400 h in a cyclic test at 980 °C (1800 °F) where the power was cycled 15 min on and 15 min off.

sphere, temperature cycling, and contact with other materials (such as refractories).

Addition of 1% Nb to 80Ni-20Cr increases resistance to preferential oxidation of chromium, commonly called "green rot". Green rot occurs when Ni-Cr alloys are exposed to environments with low partial pressures of oxygen at temperatures from 870 to 1040 °C (1600 to 1900 °F); maximum oxidation rate occurs at 955 °C (1750 °F). Green rot is most common in 90Ni-10Cr alloys used in thermocouples. However, it can occur in 80Ni-20Cr alloys exposed for long periods of time. Although green rot in 70Ni-30Cr alloys is theoretically possible, it has not occurred during testing in which such alloys have been exposed for up to eight years.

Addition of niobium to a 35Ni-20Cr-45Fe alloy stabilizes the carbon by forming niobium carbides. If carbon is tied up in this manner, heating elements remain ductile when stressed during service.

Atmospheres

Based on element temperature, Table 9 rates serviceabilities of various heating-element materials as good, fair or not recommended for the temperatures and atmospheres indicated. Element temperatures are always higher than furnace control temperatures; the difference depends on watt-density loading on the element surface. Thus, when furnaces are operated near maximum element temperature in the more active atmospheres, watt-density loading should be lower and element cross-sectional area should be higher.

With the exception of molybdenum, tantalum, tungsten and graphite, commonly used resistor materials have satisfactory life in air and in most other oxidizing atmospheres.

Oxidizing Atmospheres. Nickel-chromium and nickel-chromium-iron alloys are the most widely used heating materials in electric heat treating furnaces. The 80Ni-20Cr alloys are more commonly used than the 60Ni-20Cr-20Fe or the 35Ni-20Cr-45Fe types. In fact, most electric-furnace manufacturers provide 80Ni-20Cr elements as standard, both because they permit a wider range of furnace temperatures and because it is usually more economical to stock only a limited number of heater materials. The 80Ni-20Cr alloys permit a wider range of operating temperatures because they have the great-

Fig. 13 Continuous oxidation of Ni-Cr heating alloys held 100 h at 1100 °C (2000 °F)

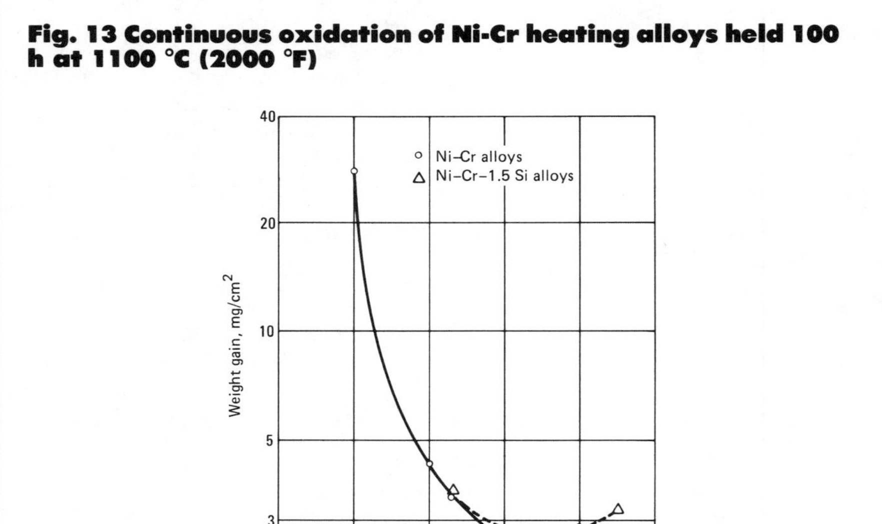

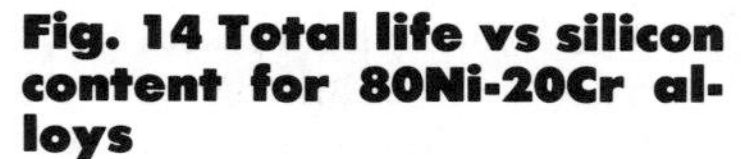

Fig. 14 Total life vs silicon content for 80Ni-20Cr alloys

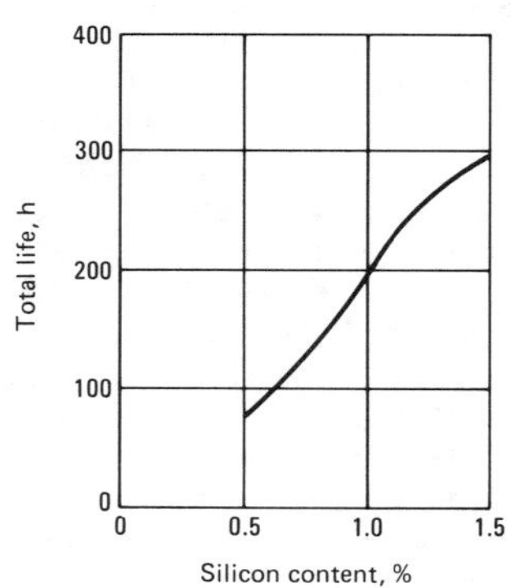

Total life determined for 0.64-mm-diam (0.025-in.-diam) wire in intermittent tests in air at 1175 °C (2150 °F).

est resistance to oxidation, and therefore can be used at higher temperatures than other Ni-Cr and Ni-Cr-Fe alloys. Heating elements of lower nickel content are required for certain special applications, such as where an oxidizing atmosphere contaminated with sulfur, lead or zinc is present.

The iron-chromium-aluminum alloys are widely used in furnaces operating at 800 to 1300 °C (1500 to 2350 °F). In general, Ni-Cr heating elements are unsuitable above 1150 °C (2100 °F) because the oxidation rate in air is too great and the operating temperature is too close to the melting point of the alloy, although some Ni-Cr elements have been used at element temperatures up to 1200 °C (2200 °F). The Fe-Cr-Al elements generally are recommended for operation in air; reducing atmospheres have adverse effects on elements that have not been preoxidized in air.

For temperatures above 1300 °C (2350 °F), silicon carbide or molybdenum disilicide elements are employed in industrial furnaces. Here again, maximum life of heating elements is obtained in air. These nonmetallic materials give fair service life in slightly reducing atmospheres at temperatures up to 1300 °C for SiC and 1500 °C for $MoSi_2$. Because they can be used in both oxidizing and slightly reducing atmospheres, they are used more commonly than Fe-Cr-Al elements, which are recommended only for service in air or inert atmospheres.

Platinum has been used in some small laboratory furnaces up to 1480 °C (2700 °F) in air. Because of the high cost of platinum, it is used only in special applications where silicon carbide cannot be worked into the furnace design. Platinum is restricted to service in air and cannot be used in reducing atmospheres. Although the initial cost of platinum is high, it has a high salvage value.

Elements made of 90% molybdenum disilicide and 10% refractory oxide mixtures perform well at continuous temperatures of 1700 and 1800 °C (3100 and 3270 °F) (depending on type) in air and in other oxidizing or inert atmospheres.

Carburizing Atmospheres. Unpurified exothermic (type 102) and purified exothermic (type 202) atmospheres are less harmful than endothermic (type 301) or charcoal (type 402) atmospheres, which are higher in carbon potential. The higher-carbon-potential atmospheres have a tendency to carburize Ni-Cr alloys, especially at higher temperatures. Chromium is a strong carbide former and may pick up enough carbon to lower the melting point of the alloy, causing localized fusion in the heating element. For this reason, in reducing atmospheres of high carbon potential, it is safer to limit the operating temperature of 80Ni-20Cr to about 1000 °C (1850 °F).

Unprotected heating elements made of Ni-Cr alloys are not recommended for use at more than 30 volts in enriched endothermic carburizing or carbonitriding (type 309) atmospheres. Short life of heating elements in these types of atmospheres may be caused by carbon deposits on the element or refractory and by carburization of the element alloy. Carbon deposits may also short out extension wires and terminals, and cause them to melt. In recent years, a coated Ni-Cr heating element has been developed for use in carburizing atmospheres; in this element, the alloy is protected with a high-temperature ceramic coating that resists carburization. The element is designed to operate at low voltage (8 to 10 volts) to prevent arcing at the terminals in carbon-impregnated brickwork.

In recent times, when more and more carburizing furnaces are being converted from fossil fuels to electricity, it has been found that molybdenum disilicide can operate safely in both carburizing and reducing atmospheres at element temperatures up to 1500 °C (2700 °F). This makes it possible to eliminate the radiant tubes often used to protect metallic heating elements, thereby greatly increasing the efficiency of these furnaces by allowing faster recovery when a cold charge is placed in the furnace. Radiant tubes form a thermal barrier that slows heat trans-

Table 9 Comparative life of heating-element materials in various furnace atmospheres

See Table 10 for atmosphere compositions.

Element material	Relative life and maximum operating temperature in: Oxidizing: (air)	Reducing: dry H_2 or type 501	Reducing: type 102 or 202	Reducing: type 301 or 402	Carburizing: type 307 or 309	Reducing or oxidizing, with sulfur	Reducing, with lead or zinc	Vacuum
Nickel-chromium and nickel-chromium-iron alloys								
80Ni-20Cr	Good to 1150 °C	Good to 1175 °C	Fair to 1150 °C	Fair to 1000 °C	Not recommended(a)	Not recommended	Not recommended	Good to 1150 °C
60Ni-16Cr-24Fe	Good to 1000 °C	Good to 1000 °C	Good to fair to 1000 °C	Fair to poor to 925 °C	Not recommended	Not recommended	Not recommended	...
35Ni-20Cr-45Fe	Good to 925 °C	Good to 925 °C	Good to fair to 925 °C	Fair to poor to 870 °C	Not recommended	Fair to 925 °C	Fair to 925 °C	...
Iron-chromium-aluminum alloys								
Fe, 23Cr, 4.5Al, 1Co	Good to 1150 °C	Fair to poor to 1150 °C(b)	Not recommended	Not recommended	Not recommended	Fair in oxidizing atmosphere	Not recommended	...
37Cr, 7.5Al, rem Fe	Good to 1320 °C	Fair to poor to 1300 °C(b)	Not recommended	Not recommended	Not recommended	Fair in oxidizing atmosphere	Not recommended	...
Pure metals								
Molybdenum	Not recommended(c)	Good to 1650 °C	Not recommended	Not recommended	Not recommended	Not recommended	Not recommended	Good to 1650 °C
Platinum	Good to 1400 °C	Not recommended	Not recommended	Not recommended	Not recommended	Not recommended	Not recommended	...
Tantalum	Not recommended	Not recommended	Not recommended	Not recommended	Not recommended	Not recommended	Not recommended	Good to 2500 °C
Tungsten	Not recommended	Good to 2500 °C(d)	Not recommended	Not recommended	Not recommended	Not recommended	Not recommended	Good to 1650 °C
Nonmetallic heating element materials								
Silicon carbide	Good to 1600 °C	Fair to poor to 1200 °C	Fair to 1375 °C	Fair to 1375 °C	Not recommended	Good to 1375 °C	Good to 1375 °C	Not recommended
Graphite	Not recommended	Fair to 2500 °C	Not recommended	Fair to 2500 °C	Fair to poor to 2500 °C	Fair to 2500 °C in reducing	Fair to 2500 °C	...
Molybdenum disilicide	Good to 1800 °C	...	...	...	...	...	...	...

Inert atmosphere of argon or helium can be used with all materials. Nitrogen recommended only for the nickel-chromium group. Temperatures listed are element temperatures, not furnace temperatures.

(a) Special 80Ni-20Cr elements with ceramic protective coatings designated for low voltage (8 to 16 V) can be used. (b) Must be oxidized first. (c) Special molybdenum heating elements with $MoSi_2$ coating can be used in oxidizing atmospheres. (d) Good with pure H_2 only.

Fig. 15 Effects of silicon and manganese contents on continuous oxidation of 80Ni-20Cr alloys

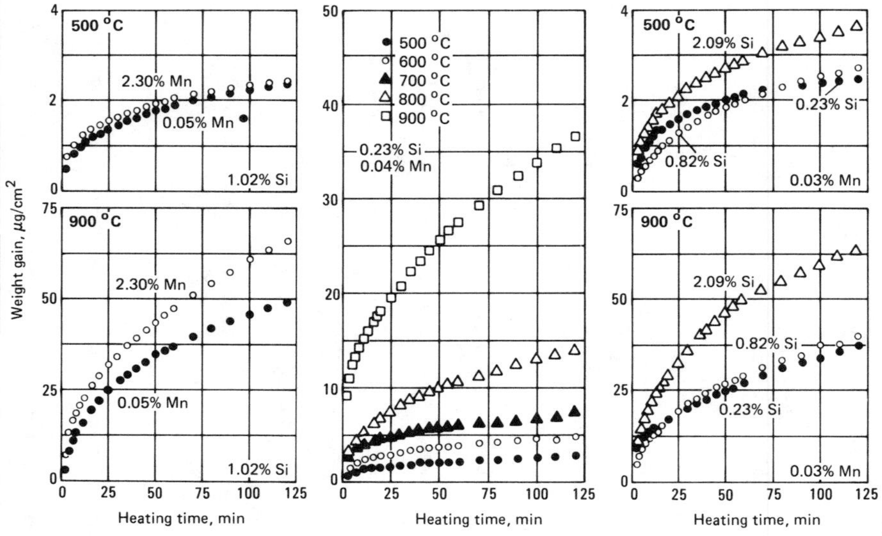

Alloys can be identified in the tabulation below, and in the graphs, by manganese and silicon contents. Specimens were first polished with emery papers through No. 0000 and then tested in pure oxygen at 0.1 atm (76 mm Hg), which constitutes accelerated testing.

Mn	Si	Cr	Ni	Fe	Al	C	Useful life, h
2.30	1.02	20.00	rem	0.25	0.13	0.06	105
0.05	1.02	20.00	rem	0.25	0.15	0.06	106
0.04	0.23	20.00	rem	0.25	0.20	0.06	63
0.03	2.09	20.00	rem	0.25	0.17	0.06	124
0.03	0.82	20.00	rem	0.25	0.15	0.06	178

Table 10 Types and compositions of standard furnace atmospheres

See Table 9 for comparative life of heating elements in these atmospheres.

Type	Description	Composition, vol %: N_2	CO	CO_2	H_2	CH_4	Typical dew point, °C	°F
Reducing atmospheres								
102(a)	Exothermic unpurified	71.5	10.5	5.0	12.5	0.5	+27	+80
202	Exothermic purified	75.3	11.0	...	13.0	0.5	−40	−40
301	Endothermic	45.1	19.6	0.4	34.6	0.3	+10	+50
402	Charcoal	64.1	34.7	...	1.2	...	−29	−20
501	Dissociated ammonia	25	...	...	75	...	−51	−60
Carburizing atmospheres								
307	Endothermic + hydrocarbon	No standard composition					...	...
309	Endothermic + hydrocarbon + ammonia	No standard composition					...	...

(a) This atmosphere, refrigerated to obtain a dew point of +4 °C (+40 °F), is widely used.

fer from the heating elements to the charge.

Reducing Atmospheres. In conventional heat treating terminology, a reducing atmosphere is one that will reduce iron oxide on steel. Reducing atmospheres are of several types, as shown in Table 10.

With the exception of dry hydrogen and dissociated ammonia (type 501), all atmospheres listed as reducing in Table 9 are oxidizing to Ni-Cr and Fe-Cr-Al alloys. Even hydrogen or dissociated ammonia will selectively oxidize chromium in a Ni-Cr alloy unless the gas is extremely dry. The type of oxide produced in "reducing" atmospheres is entirely different from that produced in air. The oxide produced in air is a green-to-black, impervious type that retards further oxidation of the underlying metal. It is usually a combination of Cr_2O_3 and $NiO{\cdot}Cr_2O_3$. The oxide produced on Ni-Cr elements in reducing atmospheres is green and porous, and allows the atmosphere to continue to attack the base metal. This type of attack, frequently referred to as "green rot", takes place over a limited temperature range—870 to 1040 °C (1600 to 1900 °F)—in any atmosphere that is oxidizing to chromium and reducing to nickel, and occurs as particles or stringers of Cr_2O_3 surrounding metallic nickel.

Among the listed reducing atmospheres, type 501 has the smallest effect on Ni-Cr heating elements. At temperatures above 1100 °C (2000 °F) a Ni-Cr element may have better life in dry hydrogen than in air, because oxidation in air occurs more rapidly at elevated temperatures. Wet hydrogen, on the other hand, will cause oxidation, and, around 950 °C (1750 °F), green rot will occur.

Graphite heating elements have been used for laboratory applications at temperatures near 1370 °C (2500 °F) in atmospheres free from O_2, CO_2 and H_2O. Silicon carbide elements give fair life in some reducing atmospheres; however, the maximum operating temperature is lower than for operation in air.

The poorest life for silicon carbide elements is obtained in hydrogen or dissociated ammonia. All atmospheres, including air, must be relatively dry; wet atmospheres shorten the life of silicon carbide elements. Silicon carbide is not recommended for use in carburizing atmospheres because it absorbs carbon, thus reducing electrical resistance and overloading the power supply.

Molybdenum disilicide heating elements can be safely used in carbon monoxide environments at 1500 °C (2730 °F), in dry hydrogen at 1350 °C (2460 °F), and in moist hydrogen at 1460 °C (2660 °F). As a rule of thumb, any combination of temperature and atmosphere that does not attack silica glass is compatible with molybdenum disilicide.

Atmosphere Contamination. Sulfur, if present, will appear as hydrogen sulfide in reducing atmospheres and as sulfur dioxide in oxidizing atmospheres. Sulfur contamination usually comes from one or more of the following sources: high-sulfur fuel gas used to generate the protective atmosphere; residues of sulfur-base cutting oil on the metal being processed; high-sulfur refractories, clays or cements used for sealing carburizing boxes; and the metal being processed in the furnace. Sulfur is destructive to Ni-Cr and Ni-Cr-Fe heating elements. Pitting and blistering of the alloy occur in oxidizing atmospheres, and a Ni-S eutectic that melts at 645 °C (1190 °F) may form in any type of atmosphere. The higher the nickel content, the greater the attack. Therefore, if sulfur is present and cannot be eliminated, 35Ni-20Cr-43.5Fe-1.5Si elements are preferred over those made of other Ni-Cr-Fe alloys.

Lead and zinc contamination of a furnace atmosphere may come from the work being processed. This is a common occurrence in sintering furnaces for processing powder metallurgy parts. In the presence of a reducing atmosphere, lead will vaporize from leaded bronze powders (such as those used to make sintered bronze bushings) and attack the heating elements, forming lead chromate. Metallic lead vapors are even more harmful than sulfur to Ni-Cr alloys, and will cause severe damage to a heating element in a matter of hours if unfavorable conditions of concentration and temperature exist. Higher-nickel alloys are affected more than lower-nickel alloys. Elements made of 35Ni-20Cr-43.5Fe-1.5Si give satisfactory life for sintering lead-bearing bronze powders at 845 °C (1550 °F) in reducing atmospheres; 80Ni-20Cr elements give poor life in this application.

Zinc contamination results from zinc stearate used as a lubricant and binder when P/M compacts are pressed. The zinc stearate volatilizes when the compacts are heated and may carburize the heating element. (Brazing of nickel silvers, which contain at least 18% Zn, also results in a high concentration of zinc vapors in the furnace atmosphere.) Zinc vapors, which alloy with Ni-Cr heating elements and result in poor life, may be eliminated at the higher sintering temperatures by using a separate burn-off furnace at 650 °C (1200 °F), with heating elements protected by full muffle, by sheathing, or by a high-temperature ceramic protective coating. If these precautions are not feasible, silicon carbide elements (which are not affected by sulfur, lead or zinc contamination) should be used at both low and high temperatures when contamination is anticipated.

Molybdenum disilicide heating elements should not be used in the presence of uncombined chlorine or other halogens.

Vacuum Service. For vacuum heating, 80Ni-20Cr elements have been used at temperatures up to 1150 °C (2100 °F). The 80Ni-20Cr alloys generally are not satisfactory much above 1150 °C (2100 °F), because the vapor pressure of chromium is high enough for chromium to vaporize from the elements, resulting in poor life, contamination of the material being processed, and loss of vacuum. Because of this, watt density must be kept low, especially at higher temperatures.

In vacuum heating, the estimated maximum operating temperature at which weight loss by evaporation from refractory-metal heating elements will not exceed 1% in 100 h is:

Metal	Temperature °C	Temperature °F
Tungsten	2550	4620
Tantalum	2400	4350
Molybdenum	1900	3470
Platinum	1600	2910

Molybdenum disilicide heating elements are not suitable for use in high vacuum.

Properties of Electrical Resistance Alloys

80Ni-20Cr

Commercial Name

Common name. 80-20 alloy

Specifications

ASTM. B344

Chemical Composition

Composition limits. 19 to 21 Cr; 1 max Fe; 2.5 max Mn; 0.15 max C; 0.75 to 1.6 Si; 0.01 max S; rem Ni

Applications

Typical uses. Electric heating elements for household appliances and industrial furnaces; fine-wire resistors for electronic applications

Precautions in use. Avoid sulfur-bearing and reducing atmospheres at high temperatures

Table 1 Variation of tensile properties of 80Ni-20Cr wire with temperature

Annealed wire 0.72 mm (0.0285 in.) in diameter

Temperature °C	Temperature °F	Tensile strength MPa	Tensile strength ksi	Elongation, %
20	68	860	125	30
425	800	725	105	22
650	1200	620	90	20
870	1600	205	30	17
1040	1900	97	14	15

Mechanical Properties

Tensile properties. Annealed: tensile strength, 655 MPa (95 ksi); elongation, 25 to 35% in 50 mm or 2 in.; reduction in area, 55%. Hard: tensile strength, 1140 MPa (165 ksi); elongation, 0 to 1% in 50 mm or 2 in.; reduction in area, 0 to 1%. See also Table 1.
Hardness. Annealed: 85 to 90 HRB. Hard: 100 to 105 HRB
Elastic modulus. Tension, 215 GPa (31×10^6 psi)

Structure

Crystal structure. fcc

Mass Characteristics

Density. 8.4 Mg/m^3 (0.30 lb/in.^{3}) at room temperature

Thermal Properties

Liquidus temperature. 1400 °C (2550 °F)
Coefficient of thermal expansion. Linear, 17.3 μm/m·K (9.6 μin./in.·°F) at 20 to 1000 °C (68 to 1830 °F)
Specific heat. 450 J/kg·K (0.107 Btu/lb·°F) at room temperature
Thermal conductivity. 134 W/m·K (74 Btu/ft·h·°F) at 100 °C (212 °F)

Electrical Properties

Electrical conductivity. Volumetric, 1.6% IACS at 20 °C (68 °F)
Electrical resistivity. 1.079 μΩ·m at 20 °C (68 °F); temperature coefficient, 90 μΩ/Ω·K at −65 to +150 °C (−85 to +300 °F). See also Fig. 1.

Chemical Properties

General corrosion behavior. Highly resistant to oxidizing atmospheres up to 1200 °C (2200 °F). It is subject to corrosion in sulfur-bearing atmospheres at elevated temperatures and in certain reducing atmospheres at around 925 °C (1700 °F)

Fabrication Characteristics

Weldability. Can be soldered, brazed, or welded using oxyfuel-gas, carbon-arc or resistance methods
Heat treatment. Annealing only
Annealing temperature. 870 to 1040 °C (1600 to 1900 °F)
Reduction between anneals. 65%
Hot working temperature. 1205 °C (2220 °F)

Fig. 1 Electrical resistance vs temperature for 80Ni-20Cr

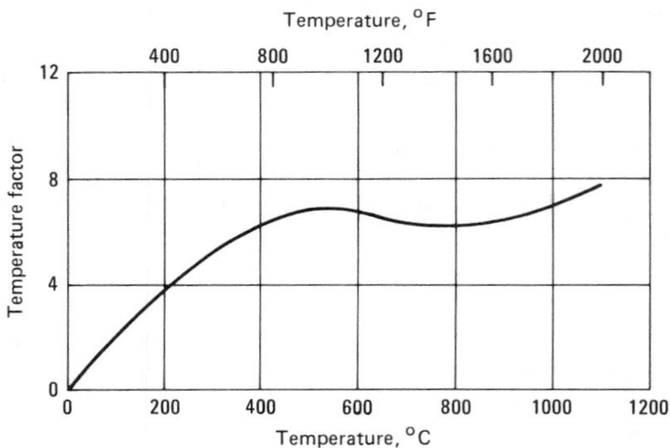

To find resistance at any given temperature, multiply resistance at room temperature by the temperature factor.

Fig. 2 Electrical resistance vs temperature for 60Ni-24Fe-16Cr

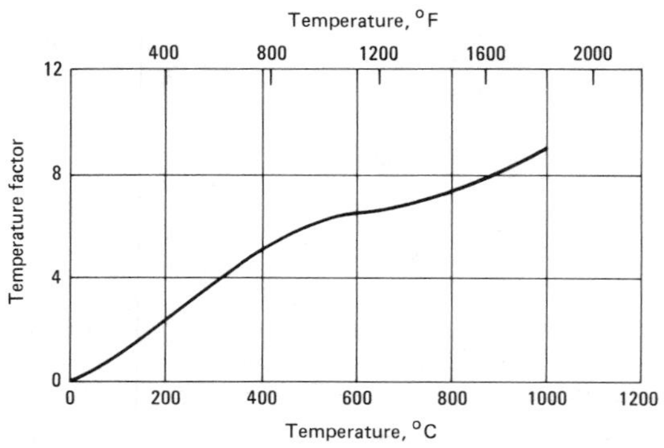

To find resistance at any given temperature, multiply resistance at room temperature by the temperature factor.

60Ni-24Fe-16Cr

Commercial Name

Common name. 60Ni-16Cr alloy

Specifications

ASTM. B344

Chemical Composition

Composition limits. 57 min Ni + Co; 14 to 18 Cr; 1.0 max Mn; 0.75 to 1.6 Si; 0.15 max C; 0.01 max S; rem Fe

Applications

Typical uses. Heat-resisting elements for heating devices such as toasters, percolators, waffle irons, flat irons, ironing machines, heater pads, hair driers, permanent wave equipment, hot water heaters, carburizing and annealing containers for sheet metal. Electrical usage in high-resistance rheostats for electronic equipment, oxidized wire with high-resistance coating for close-wound rheostats, potentiometers and thermocouples. Corrosion-resisting usage in dipping baskets for acid pickling and cyanide hardening, automatic pickling-machine parts, filters, enameling racks, containers for molten salts

Mechanical Properties

Tensile properties. Sand cast: ten-

sile strength, 450 MPa (65 ksi); elongation, 2% in 50 mm or 2 in.; reduction in area, 2%. Annealed: tensile strength, 725 MPa (105 ksi); elongation, 30% in 50 mm or 2 in.; reduction in area, 47%. See also Table 2.
Hardness. Sand cast: 92 HRB. Annealed: 83 HRB

Structure

Microstructure. fcc
Metallography. Rough polish with emery paper through 000; use levigated alumina on polishing cloth to finish. Etching for grain size: Marble's reagent preferably or aqua regia (3 HCl, 1 HNO_3); sometimes less HCl is used and sometimes it is diluted to 50% with water, depending on the worked or annealed condition of the alloy. Use aqua regia for microstructure. Upon completion of polishing, this alloy often exhibits a cold worked film obscuring the true structure, and hence must be repolished on the last stage and re-etched. This is common to other stainless austenitic materials. Usual method of reporting grain size—grains per sq mm

Mass Characteristics

Density. 8.25 Mg/m^3 (0.298 $lb/in.^3$) at 20 °C (68 °F)
Patternmaker's shrinkage. 2.0%

Thermal Properties

Liquidus temperature. 1425 °C (2600 °F)
Coefficient of thermal expansion. Linear, 17.0 μm/m·K (9.4 μin./in.·°F) at 20 to 1000 °C (68 to 1830 °F)
Specific heat. 450 J/kg·K (0.107 Btu/lb·°F) at 20 °C (68 °F)
Thermal conductivity. 13.4 W/m·K (7.7 Btu/ft·h·°F) at 100 °C (212 °F)

Electrical Properties

Electrical conductivity. Volumetric, 1.5% IACS at 20 °C (68 °F)
Electrical resistivity. 1.12 μΩ·m at 20 °C (68 °F); temperature coefficient, 150 μΩ/Ω·K at 20 to 100 °C (68 to 212 °F). See also Fig. 2.

Chemical Properties

Corrosion testing. Strauss test, using changes in both weight and electrical resistance as criteria for evaluation
Resistance to specific corroding agents:

Corrosive agent	Resistance
Acid, acetic	Resistant
Acid, hydrochloric	Moderate general attack
Acid, lactic	Resistant
Acid, nitric	Resistant
Acid, phosphoric	Moderate general attack when hot
Acid, sulfuric	Resistant
Air	Resistant
Alcohol, ethyl	Resistant
Alcohol, methyl	Resistant
Ammonia	Resistant
Ammonium nitrate	Resistant
Carbon dioxide	Resistant
Carbon tetrachloride	Resistant
Copper sulfate	Moderate intergranular attack
Foodstuffs	Resistant
Fruit products	Resistant
Hydrogen sulfide	Severe general attack when hot
Lead and its compounds	Severe attack when hot
Milk	Resistant
Mineral oils	Resistant
Motor fuel	Resistant
Oxygen	Resistant
Petroleum products	Resistant when cold
Photographic solutions	Resistant
Potassium hydroxide	Resistant
Potassium nitrate	Resistant
Sodium carbonate	Resistant
Sodium hydroxide	Resistant
Sodium nitrate	Resistant
Sugar	Resistant
Sulfur	Severe general attack when hot
Sulfur dioxide	Severe general attack when hot
Water, distilled	Resistant
Water, rain	Resistant

Table 2 Nominal tensile properties of 60Ni-24Fe-16Cr at elevated temperatures

Temperature		Cast material			Wrought material		
		Tensile strength		Elongation,	Tensile strength		Elongation,
°C	°F	MPa	ksi	%	MPa	ksi	%
20	68	450	65	2.0	725	105	30
540	1000	325	47	2.1	365	53	12
650	1200	295	43	2.3	260	38	13
760	1400	255	37	2.5	200	29	23
870	1600	150	22	3.7	130	19	32
980	1800	105	15	6.0	62	9	45

Fabrication Characteristics

Formability. Suited to forming by hot and cold rolling, forging, drawing, pressing and bending
Weldability. Soft solder with 50Pb-50Sn, using HCl plus Zn flux. Silver-braze (silver solder) with low-melting-point filler metals, using borax flux and a neutral flame. Braze with low brass filler metal, using borax flux. Oxyfuel-gas weld with 60Ni-24Fe-16Cr filler metal, using no flux and a neutral or carburizing flame
Casting temperature. Sand and ingot. 1510 to 1540 °C (2750 to 2800 °F)
Annealing temperature. 760 to 1090 °C (1400 to 2000 °F)
Reduction between anneals. 75% max in area
Standard finishes. Castings are sandblasted, forgings pickled, and wrought mill products given various finishes.

35Ni-45Fe-20Cr

Commercial Name

Common name. 35Ni-20Cr alloy

Specifications

ASTM. B344

Chemical Composition

Composition limits. 34 to 37 Ni; 18 to 21 Cr; 1.0 max Mn; 0.25 max C; 0.03 max S; rem Fe

Applications

Typical uses. Although 35Ni-45Fe-20Cr has lower electrical resistivity than 60Ni-24Fe-16Cr near room temperature, it has a higher temperature coefficient. It is normally used at temperatures up to 815 °C (1500 °F) for heavy-duty rheostats, low priced electrical appliances and resistors operating in cracked gas atmospheres.

Mechanical Properties

Tensile properties. Sand cast: tensile strength, 425 MPa (62 ksi); elongation, 2% in 50 mm or 2 in.; reduction in area, 2%. Annealed: tensile strength, 705 MPa (102 ksi); elongation, 30% in 50 mm or 2 in.; reduction in area, 45%. See also Table 3.

Structure

Microstructure. fcc
Metallography. Same as 60Ni-24Fe-16Cr

Mass Characteristics

Density. 7.95 Mg/m^3 (0.287 $lb/in.^3$) at 20 °C (68 °F)
Patternmaker's shrinkage. 2.0%

Thermal Properties

Liquidus temperature. 1425 °C (2600 °F)
Coefficient of thermal expansion. Linear, 28.4 μm/m·K (15.8 μin./in.·°F) at 20 to 500 °C (68 to 930 °F)
Specific heat. 460 J/kg·K (0.110 Btu/lb·°F) at 20 °C (68 °F)
Thermal conductivity. 13 W/m·K (7.5 Btu/ft·h·°F) at 100 °C (212 °F)

Electrical Properties

Electrical conductivity. Volumetric, 1.7% IACS at 20 °C (68 °F)
Electrical resistivity. 1.0 μΩ·m at 20 °C (68 °F); temperature coefficient, 310 μΩ/Ω·K at 20 to 500 °C (68 to 932 °F). See also Fig. 3.

Chemical Properties

Corrosion testing. Same as 60Ni-24Fe-16Cr

Fabrication Characteristics

Formability. Suited to forming by hot and cold rolling, forging, drawing, pressing and bending.
Weldability. Soft solder, silver braze and braze with the same methods as for 60Ni-24Fe-16Cr alloy. Oxyfuel-gas weld with 35Ni-45Fe-20Cr filler metal, using no flux, and a neutral or carburizing flame.
Casting temperature. 1525 to 1550 °C (2775 to 2825 °F)
Annealing temperature. 760 to 1090 °C (1400 to 2000 °F)

Table 3 Nominal tensile properties of 35Ni-45Fe-20Cr at elevated temperatures

Temperature		Cast material: Tensile strength		Cast: Elongation,	Wrought material: Tensile strength		Wrought: Elongation,
°C	°F	MPa	ksi	%	MPa	ksi	%
20	68	425	62	2	705	102	32
540	1000	295	43	4	345	50	22
650	1200	250	36	6	250	36	20
760	1400	235	34	8.5	185	27	26
870	1600	130	19	20	125	18	28
980	1800	83	12	25	62	9	30

Fig. 3 Electrical resistance vs temperature for 35Ni-45Fe-20Cr

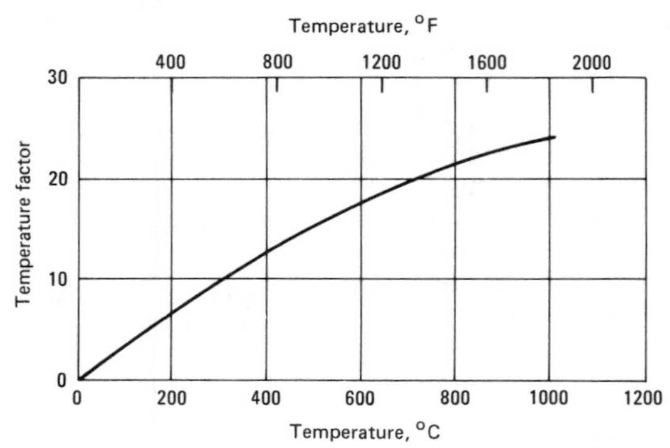

To find resistance at any given temperature, multiply resistance at room temperature by the temperature factor.

Reduction between anneals. 75% max
Hot working temperature. 980 to 1260 °C (1800 to 2300 °F)
Standard finishes. See 60Ni-24Fe-16Cr.

Constantan 45Ni-55Cu

Chemical Composition

Composition limits. Variable; several specific compositions—all having an approximate nominal composition of 42 to 45 Ni, rem Cu—are commonly supplied. Composition limits of both major and minor constituents are varied to suit the specific application.

Applications

Typical uses. This alloy has about the highest electrical resistivity, the lowest temperature coefficient of resistance, and the highest thermal emf against platinum of any of the copper-nickel alloys. Because of the first two of these properties, it is used for electrical resistors; and because of the latter property, for thermocouples.
Precautions in use. Maximum temperature for resistor use is 500 °C (930 °F); maximum temperature for thermocouple use is 900 °C (1650 °F)

Mechanical Properties

Tensile properties. Wrought: tensile strength, annealed, 415 MPa (60 ksi); cold worked, 930 MPa (135 ksi). Cast: tensile strength, 380 MPa (55 ksi); yield strength, 145 MPa (21 ksi); elongation, 32% in 50 mm or 2 in.; reduction in area, 4%

Shear strength. 295 MPa (43 ksi)
Hardness. 75 to 85 HB, 48 to 54 HRB
Impact resistance. Charpy: 41 J (30 ft·lb)

Mass Characteristics

Density. Wrought: 8.9 Mg/m^3 (0.32 lb/in.^{3}) at 20 °C (68 °F). Cast: 8.6 Mg/m^3 (0.31 lb/in.^{3})
Patternmaker's shrinkage. 2.1%

Thermal Properties

Liquidus temperature. 1290 °C (3354 °F)
Solidus temperature. 1220 °C (2225 °F)
Coefficient of thermal expansion. Linear, 14.9 μm/m·K (8.3 μin./in.·°F) at 20 to 100 °C (68 to 212 °F); 16.3 μm/m·K (9.0 μin./in.·°F) at 20 to 500 °C (68 to 930 °F); 18.8 μm/m·K (10.4 μin./in.·°F) at 20 to 1000 °C (68 to 1830 °F)
Specific heat. 395 J/kg·K (0.094 Btu/lb·°F) at 20 °C (68 °F)
Thermal conductivity. 21 W/m·K (12.3 Btu/ft·h·°F) at 20 °C (68 °F)

Electrical Properties

Electrical resistivity. 500 nΩ·m at 20 °C (68 °F); temperature coefficient, ±20 μΩ/Ω·K at 20 to 150 °C (68 to 300 °F). The coefficient may be either positive or negative, depending on small variations in composition and on variations in the amount of cold work. In any event, the value of the coefficient is small.
Thermoelectric potential. The basic alloy is modified by additions of manganese and iron to give slightly different emf characteristics in thermocouple service, as specified by different pyrometer manufacturers. For more detailed information on this subject, see the article in this volume entitled "Thermocouples for Industrial Applications".

Fabrication Characteristics

Formability. Suited to forming by hot and cold rolling, forging, drawing, pressing and bending
Weldability. Can be welded, brazed and soldered by conventional methods
Casting temperature. Minimum: 1350 °C (2460 °F)
Annealing temperature. 870 to 980 °C (1600 to 1800 °F)
Reduction between anneals. 80% max
Hot working temperature. 870 to 1120 °C (1600 to 2050 °F)

Molybdenum Disilicide $MoSi_2$, $MoSi_2$ + 10% ceramic additives

Chemical Composition

Composition limits. 90% $MoSi_2$; 10% metal and ceramic additives

Applications

Typical uses. Electric resistance heating elements in industrial and laboratory furnaces. Normally supplied in hairpin shape consisting of terminals twice the diameter of the hot zone. Most common hot zone diameters are 3, 6, and 9 mm. Will operate in oxidizing atmospheres at temperatures of 1700 °C for $MoSi_2$, and 1800 °C for $MoSi_2$ + 10% ceramic additives. Also usable at lower temperatures in reducing atmospheres. For maximum recommended service temperatures, see Table 4.
Precautions in use. Avoid elemental chlorine, other halogens and high vacuum environments

Mechanical Properties

Tensile strength. 195 MPa (28 ksi) at room temperature
Compressive strength. 2350 MPa (340 ksi) at room temperature
Hardness. Knoop: 1280 at 20 °C (68 °F), decreasing to 640 at 700 °C (1290 °F)
Elastic modulus. Tension, 410 GPa (59 × 10^6 psi) at room temperature
Bending strength. 345 MPa (50 ksi) at 20 °C (68 °F), 135 MPa (19 ksi) at 1100 °C (2010 °F)

Structure

Crystal structure. Pure $MoSi_2$: body-centered tetragonal

Mass Characteristics

Density. 5.6 Mg/m^3 (0.202 lb/in.^{3}) at room temperature

Thermal Properties

Incipient melting temperature. 2050 °C (3720 °F). Decomposes before melting at approximately 1740 °C (3165 °F) for $MoSi_2$, and 1825 °C (3315 °F) for $MoSi_2$ + 10% additives
Coefficient of thermal expansion. Linear, 7.1 to 8.8 μm/m·K (3.9 to 4.9 μin./in.·°F) at 20 to 1500 °C (68 to 2730 °F)
Specific heat. 420 J/kg·K (100 Btu/lb·°F) at 20 °C (68 °F)

Table 4 Maximum service temperatures for $MoSi_2$ heating elements

Atmosphere	Temperature °C	°F
Air	1700(a)	3100(a)
Nitrogen	1590	2900
Argon, helium	1450	2730
Dry hydrogen	1350	2460
Moist hydrogen(b)	1460	2660
Carbon dioxide	1590	2900
Carbon monoxide	1450	2730
Sulfur dioxide	1590	2900
Cracked, partly burnt ammonia(c)	1400	2550
Methane	1350	2460

(a) For "1700 °C" material; for "1800 °C" material, maximum service temperature is 1800 °C (3270 °F). (b) Dew-point 15 °C (60 °F). (c) Approximately 8% H_2.

Electrical Properties

Electrical resistivity. For $MoSi_2$ + 10% ceramics:

Temperature °C	°F	1700 °C material μΩ·m	1800 °C material μΩ·m
20	68	465	440
1370	2500	4800	4570
1650	3000	6060	5750

Thermoelectric potential. In microvolts vs Pt: $5.13\,T + 1.88 \times 10^{-2}\,T^2 - 5.22 \times 10^{-6}\,T^3$
Temperature of superconductivity. 1.3 K
Emissivity. 0.34 at 370 °C; 0.60 at 1370 °C; 0.75 to 0.80 at 1500 °C in air due to formation of SiO_2 coating

Chemical Properties

General corrosion behavior. In normal use, material has a SiO_2 glass coating and any combination of temperature and atmosphere that is harmful to quartz glass will deteriorate the elements

Fabrication Characteristics

Sintering temperature. 1620 °C (2950 °F)
Hot working temperature. $MoSi_2$, 1450 to 1700 °C (2640 to 3090 °F); $MoSi_2$ + 10% additives, 1500 to 1800 °C (2730 to 3270 °F)

Electric-Contact Materials

By the ASM Committee on Electric-Contact Materials*

IF AN IDEAL ELECTRICAL CONTACT MATERIAL could be found, it would have high electrical conductivity to minimize the heat generated during passage of current; high thermal conductivity to dissipate both the resistive and arc heat developed; high reaction resistance to all environments in which it was to be used to avoid formation of insulating oxides, sulfides and other compounds; and immunity to arcing damage on the making and breaking of electrical contact. The force required to close a contact made of this material would be low, as would the electrical resistance between mating members. The melting point of the material would be high enough to limit arc erosion, metal transfer, and welding or sticking, but it would also be low enough to increase resistance to re-ignition in switching. (When the melting point is high, contacts continue to heat gas in the contact gap after the current drops to zero, thus facilitating re-ignition.) The vapor pressure would be low to minimize arc erosion and metal transfer. Hardness would be high to provide good wear resistance, and yet ductility would be high enough to ensure ease of fabrication. Purity of the material would be maintainable at a level that ensures consistent performance. Neither the material, nor any process step necessary to fabricate it, would present an environmental hazard. Finally, the material would be available at low cost in any desired form.

Because no metal has all the desired properties, a wide variety of contact materials is required to accomplish the objectives of different contact applications. The economic choice of materials is usually a compromise between the various processing variables and the application requirements. Load conditions, service requirements, and ambient conditions present during the life of the unit must be considered in the selection of contact materials.

The electrical characteristics of the circuit and the current must be considered in selecting contact materials. Proper choice of materials depends not only on intended service application, but on whether the current is alternating or direct, whether the circuit voltage or amperage is high or low, and whether the voltage at contacts during interruption of the circuit is high or low. Whether the load is inductive, capacitive or resistive, or is a motor load, is also important. Allowance should be made for overload, where such a condition can be expected, and attention should be given to the method of arc suppression. Consideration must also be given to potential hazards during service life of the electrical contact. For example, failure of a railroad signal to open might cause a fatal accident, although failure to close would only be troublesome. For such applications, carbon-to-metal contacts are employed because they cannot become welded together, even under most adverse conditions.

Among mechanical factors, it is important to consider: nature of the contact force; frequency of operation; speed of opening and closing of the contact; manner in which the contact is made (wipe, slice or butt); degree of chatter on opening or closing; size of contact gap when the contact is fully opened; and method of operation—whether by cam, simple lever or push button, electromagnet, or bimetal thermostatic control.

Of the environmental factors that influence selection of a contact material, most important are type of atmosphere, ambient temperature, and whether contacts are exposed to ambi-

*See page XIII for committee list.

ent atmosphere or enclosed in a hermetically sealed, artificial environment such as might be specified for use in outer space. Atmospheric pressure and minor contaminants that are present affect contact life and, in some instances, may cause a sharp decrease in service life. Gases may cause tarnishing, as does the presence of humidity, airborne salts, dust and organic vapors.

The required service life for contacts varies according to the application. This may range from a few operations, on missiles and detection systems, to 100 million cycles in automotive vibrators or 40 years in telephone relays. Likewise, the dormant time between successive openings or closings may vary widely—from milliseconds or microseconds in vibrators to months or years in alarms and similar safety devices.

Failure Modes of Make-and-Break Contacts

In an electric make-and-break switching device, the contact points, which usually consist of a pair of thin shaped slabs, perform the actual duty of making, carrying and breaking the current. Make-and-break contacts differ from sliding contacts in that the moving member of the switching device travels perpendicular to the contact surfaces. As a result, arcs generated during opening and closing actions always strike and consequently damage the conducting surfaces.

Arcing, except in a circuit with an extremely low potential or low current, is a major factor—if not the main factor—causing failure of contact points.

When a pair of contacts opens in a live circuit, an arc is often generated between the contact pair, which remains until they are separated by a certain gap. Relatively less severe arcing occurs when the contacts close. Arcing also occurs when the moving contact bounces away from the stationary contact during closing. The arc causes contact erosion by blowing away the molten metal droplets, vaporizing the material, and transforming the metal to ion jets. Sometimes the material vaporizes from one contact and then condenses onto the other contact, thereby altering the surface configuration of both contacts. This is known as material transfer.

Oxidation of the contact surfaces, which may be accelerated by the heat from arcing, is a serious problem, because most metallic oxide films are nonconductive or semiconductive. The oxide film may easily increase contact resistance. In high-current circuits, this may cause excessive contact heating. In low-voltage and low-current circuits, the oxide films can grow so thick that they completely insulate the contact surfaces before the contact bodies erode. This happens more frequently when a pair of contacts operates in a hostile environment such as a polluted industrial atmosphere. Condensed organic polymers also play a role in precious metal contacts at light loads. These polymers come from monomers which evaporate from resins and are polymerized on the active catalytic metal surfaces of contacts.

Welding. To make a pair of contacts more conductive, a mechanical load is always applied on the contact pairs. Theoretically, the load could make two rigid contact surfaces touch at no more than three points. However, the touching points at both surfaces yield either elastically or plastically, resulting in larger areas of contact. These constricted regions carry the current through the contacts and form regions of high current density. Heat is generated in these areas and, if the temperature becomes high enough, the two contact points eventually are welded together.

Another kind of welding occurs as contacts close. The arcs generated during closing and bouncing of a moving contact melt a small portion of both contacts. On reclosure, solidification of the molten material welds the contact pair. Occasionally, the strength of the weld exceeds the opening force of the switching device, resulting in catastrophic failure of the entire electrical system because the contacts fail to open on command.

Bridge Formation. When a pair of contacts opens, the contact area gradually decreases because of the gradual lessening of contact pressure. The continuous opening action causes the contact areas to reach a stage at which the current density of the constricted areas is so great that it melts the material in these regions. Continuous separation of the contact points now pulls the molten metal, forming a current-carrying bridge. The temperature of the molten bridge continues to rise as the contact points pull apart. It may become high enough to evaporate the material and finally break the circuit. This "bridge" phenomenon during the opening of a pair of contacts slightly damages the surfaces of the contacts and evaporates some of the bridge material. This generally results in pitting of one contact surface and buildup of material on the other; an uneven continuous transfer may eventually erode one of the contacts. Furthermore, the surface asperities from the continuous bridge formation may interlock the contact pair and interfere with their mechanical separation.

Property Requirements for Make-and-Break Contacts

The four failure modes discussed above determine the requirements of materials for make-and-break contacts. The most important requirements are listed below. In selecting a material, it is often necessary to reach a compromise that provides adequate properties without jeopardizing essential qualities of the component as a whole, such as reliability, life and cost.

- **Electrical conductivity:** Because the conduction of electricity between the pair of contacts depends on only a few constricted spots, the higher the electrical conductivity, the less the amount of heat that will be generated by high current density in these spots.
- **Thermal properties:** High melting and boiling points decrease evaporation loss caused by high arcing heat. High thermal conductivity disperses the heat rapidly and quenches the arc.
- **Chemical properties:** Contact materials should be corrosion resistant so that insulating films (either oxides or other compounds) do not form easily when the contacts operate in a hostile environment.
- **Mechanical properties:** The major loads applied to a contact pair are the closing force and the impact between movable and stationary contact points during closing. An induced relative movement between two contact surfaces always exists when closing. In some devices, such as certain types of relays, a wiping motion is purposely designed into the device to destroy any oxide films that form. However, friction between wiping surfaces produces wear of the contacts upon repeated opening and

closing. Generally, hard materials are more resistant to wear. However, hard materials often have high contact resistances and low thermal conductivities, both of which contribute to a greater tendency to contact welding. Hard materials also have high tensile strengths, which may or may not be advantageous in electrical contact applications.

- **Fabrication properties:** Contact materials should have the capability of being welded, brazed or otherwise joined to backing materials. In addition, they should have sufficient malleability to enable them to be shaped, or they should be capable of being formed by P/M techniques.

None of the materials used for make-and-break contacts meets all of the criteria for an ideal contact material. For instance, silver has the best electric and thermal conductivity and good oxidation resistance, but its resistance to arcing and mechanical wear is low. Tungsten resists arcing and withstands mechanical wear, but it has poor conductivity and poor corrosion resistance. Properties of a contact material can usually be improved by combining metals, either by alloying or by powder metallurgy, but improvement often is achieved at the expense of other properties. For example, the alloying element that increases the hardness of silver also decreases its conductivity.

Sliding Contacts

The applications of sliding contacts are usually quite different from those of make-and-break contacts. Friction, contact temperature, mechanical considerations and wear also are different.

The fundamental difference between make-and-break contacts and sliding contacts is that sliding contacts require films on the contact faces to facilitate sliding without seizure or galling; shear must take place within this film with only minor disturbance of both materials. A lubricant of some kind is always necessary. This can be provided by graphite if there is moisture present—such as in an environment having a dew point of about −20 °C (−4 °F) or higher. Alternatively, lubrication can be provided by very thin oil films, although excessive oil vapor causes over-filming. It can also be provided by molybdenum disulfide, and other chalcogenides of molybdenum, tungsten and niobium. Oxygen, sulfur and other contaminants cause increased filming.

In applications in air, a drop in voltage can result from an equilibrium between oxidation and filming (which tend to increase the drop) and fretting or film breakdown and cleaning action (which tend to decrease the drop). In the absence of lubricants, fretting and oxidation are most important. In inert or reducing gases, oxidation is largely eliminated, and the voltage drop decreases until counteracted by mechanical factors. Noble metals that are properly lubricated also minimize voltage drops in air.

Brush contacts generally contain an appreciable amount of metal if they are intended for use in low-voltage (<24 V) applications. Large quantities of brush contacts are used in automotive and related industries as starter brushes and auxiliary motor brushes; copper-graphite is the principal material. Silver-graphite brushes are used primarily in instruments and in outer space applications. Some silver-graphite brushes are used in seam welders and similar equipment.

Oxidation of sliding contacts is similar to that of make-and-break contacts, except that the surface disturbed by friction oxidizes more rapidly. In most applications in air, the metal surface generates a film that is a complex mixture of graphite, oxide, sulfide and water, which tends to decrease the conducting area.

The surfaces generated on metal-graphite brushes as they wear are effective cleaning agents in that they abrade films and keep larger areas available for conduction. Even so, it is sometimes advisable to have additional abrasive material in the brushes to prevent over-filming in critical atmospheres.

Because the major factor in friction is the shear strength of any film that is present, the composition of this film, as affected by atmospheric contaminants, is important. Table 1 lists common materials that affect friction in sliding contacts and the mechanism by which they affect contact friction.

Table 1 Effects of atmospheric contaminants on friction between carbon brushes and copper

Contaminant	Effect
CCl_4	Friction becomes high and erratic
Cl_2	Small amounts, friction decreases; large amounts, friction increases
SO_2	Friction increases
H_2S	Small amounts, friction decreases; large amounts (detectable by smell), friction increases
Silicones	With current, friction increases; without current, friction decreases
Steam, H_2O	Friction may increase or decrease
Tobacco smoke	Friction increases
Oil vapor	Small amounts, friction decreases; large amounts, friction increases

Brush Materials. Considering the range of commercially available metal powders, graphites, other lubricants, and processing variables, there are unlimited possibilities for development of suitable brush materials. However, only a limited number of commercial grades have been developed; a list of grades compiled by one supplier appears in Table 2.

More brush contacts are made from copper and its alloys than from any other class of material. In applications where copper metals undergo substantial oxidation, silver metals may be used. Tungsten or, more rarely, molybdenum is used where a high melting point is required. Platinum, palladium and gold are used where reliable closure with low force is required. Brushes clad or electroplated with precious metals, and brushes made of sintered alloys, are important for general applications in power switching relays. Silver/cadmium oxide is the most widely used contact material for medium- to high-energy applications. Recently, aluminum has become more popular for insert-type contacts because of its good machinability, formability and light weight. However, it is not as good as some other metals in contact properties—for instance, corrosion can be a limiting factor. When contact properties are important, aluminum should be plated or clad with copper or silver. More recently, tin platings have been accepted for certain applications. Protective lubricants have been added to the joints in some instances.

Interdependence Factors. When

Table 2 Properties of metal-graphite electrical contact materials

Composition, % Metal	Graphite	Approximate density, Mg/m^3	Electrical conductivity, % IACS	Applications
Copper-graphite materials				
30	70	2.60	0.11	Alternators; small auxiliary motors; low-metal, long-life brushes
30	70	2.50	2	Automotive auxiliary and appliance motors
36	64	2.75	3	Automotive heaters and blower motors
40	60	2.75	4	Automotive and other small auxiliary starting motors
40	60	2.75	2.5	Automotive heaters and ac motors
50	50	3.05	0.73	Automotive alternators
50	50	2.97	6	Automotive auxiliary and appliance motors
50	50	3.18	0.83	Industrial truck motors
62	38	3.65	3	Automotive starters. Excellent grade for low-humidity applications; excellent filming properties
65	35	3.15	3	Starters
75	25	3.25	0.51	ac wound motors and rotary converters
95	5	6.30	34	Collector roll brushes
92	8	7.30	41	High-current-carrying brush material for grounding applications
96	4	7.75	42	Automotive starters
Silver-graphite materials				
90	10	7.5	42	Instruments, fuel pumps
50	50	3.4	2	Welding machines, motors
85	3(a)	7.8	45	Antenna motors(b)
75	20(c)	5.2	12	Antenna motors and generators(b)

(a) Plus 12 MoS_2. (b) With altitude protection. (c) Plus 5 MoS_2.

contacts are attached to a carrier, which is usually a copper alloy, the properties of the carrier material and the properties of the interface between contact and carrier (that is, the area of bond and the conductivity across the interface) are critical to ultimate performance. The contact carrier serves as a heat sink as well as a structural member and electrical conductor. The over-all efficiency of the system depends on the contact, the contact carrier, and the method of attachment, all of which affect the size of the contact required for a specific application. To conserve precious metal, the contact materials, carrier material, and method of attachment must be optimized. Some high-strength, high-conductivity copper alloys are used for carriers because of their structural properties and resistance to softening at brazing temperatures.

The attachment method that provides minimum interface alloying, minimum softening of the carrier, and maximum bond area generally produces the best combination of properties for the contact system as a whole. The methods that best satisfy these criteria are percussion welding and diffusion bonding. Percussion welding can be utilized for round, square, or rectangular contacts and does not require a special backing for attachment purposes. This is especially helpful in minimizing the cost of Ag-CdO contacts, because the fine silver backing adds about 20% to the cost. Percussion welding does not soften the carrier as brazing does, and can be controlled to provide a high level of reliability.

Commercial Contact Materials

Commercial materials for electrical contacts are divided into two categories based on their manufacturing methods: (*a*) wrought materials, which include both pure metals and alloys; and (*b*) composite materials, which include powder metallurgy products and internally oxidized silver alloys.

Copper Metals

High electrical and thermal conductivities, low cost, and ease of fabrication account for the wide use of copper in electrical contacts. The main disadvantage of copper contacts is low resistance to oxidation and corrosion. In many applications, the voltage drop resulting from the film developed by normal oxidation and corrosion is acceptable. In some circuit breaker applications, the contacts are immersed in oil to prevent oxidation. In other applications, such as in drum controllers, sufficient wiping occurs to maintain fairly clean surfaces, thus providing a circuit of low resistance. In some applications, such as knife switches, plugs and bolted connectors, contact surfaces are protected with grease or coatings of silver, nickel or tin. In power circuits, where oxidation of copper is troublesome, contacts frequently are coated with silver. Vacuum-sealed circuit breakers use oxygen-free copper contacts (wrought or powder metal) for optimum electrical properties.

In air, copper does not provide high resistance to arcing, welding or sticking. Where these characteristics are important, copper-tungsten or copper-graphite mixtures are used. However, when used in a helium atmosphere, a Cu-CdO contact performs similarly to an Ag-CdO contact. Copper alloys are used for high currents in vacuum interrupters.

Pure copper is relatively soft, anneals at low temperatures, and lacks the spring properties sometimes desired. Some copper alloys, harder than pure copper and having much better spring properties, are listed in Table 3. The annealing temperature of copper can be increased by additions of 0.25% Zn, 0.5% Cr, 0.03 to 0.06% Ag (10 to 20 oz per ton) or small amounts of finely dispersed metal oxides, such as Al_2O_3, with little loss of conductivity. On the other hand, improved spring properties are obtained only at the expense of elec-

Table 3 Properties of copper metals used for electrical contacts

UNS number	Solidus temperature °C	Solidus temperature °F	Electrical conductivity, % IACS	Hardness OS035 temper	Hardness H02 temper	Tensile strength OS035 temper MPa	Tensile strength OS035 temper ksi	Tensile strength H02 temper MPa	Tensile strength H02 temper ksi
C11000	1065	1950	100	40 HRF	40 HRB	220	32	290	42
C16200	1030	1886	90	54 HRF	64 HRB(a)	240	35	415(a)	60(a)
C17200	865	1590	15 to 33(b)	60 HRB(c)	93 HRB(d)	495(c)	72(c)	655(d)	95(d)
C23000	990	1810	37	63 HRF	65 HRB	285	41	395	57
C24000	965	1770	32	66 HRF	70 HRB	315	46	420	61
C27000	905	1660	27	68 HRF	70 HRB	340	49	420	61
C50500	1035	1900	48	60 HRF	59 HRB	276	40	365	53
C51000	975	1785	20	28 HRB	78 HRB	340	49	470	68
C52100	880	1620	13	80 HRF	84 HRB	400	58	525	76

(a) H04 temper. (b) Depends on heat treatment. (c) TB00 temper. (d) TD02 temper.

Table 4 Properties of silver metals used for electrical contacts

Alloy	Solidus temperature °C	Solidus temperature °F	Electrical conductivity, % IACS	Hardness, HR15T Annealed	Hardness, HR15T Cold worked	Tensile strength Annealed MPa	Tensile strength Annealed ksi	Tensile strength Cold worked MPa	Tensile strength Cold worked ksi	Density, Mg/m³	Elongation in 50 mm or 2 in., % Annealed	Elongation in 50 mm or 2 in., % Cold worked
99.9Ag	960	1760	104	30	75	17	25	310	45	10.51	55	5
99.55Ag-0.25Mg-0.2Ni	...	...	70	61	77	207	30	345	50	10.34	35	6
99.47Ag-0.18Mg-0.2Ni-0.15Cu	...	...	75	64	84	...	...	...	...	10.38	...	...
99Ag-1Pd	...	...	79	44	76	179	26	324	47	10.14	42	3
97Ag-3Pd	977	1790	58	45	77	186	27	331	48	10.53	37	3
97Ag-3Pt	982	1800	45	45	77	172	25	324	47	10.17	37	3
92.5Ag-7.5Cu	821	1510	88	65	81	269	39	455	66	10.34	35	5
90Ag-10Au	971	1780	40	57	76	200	29	317	46	11.03	28	3
90Ag-10Cu	777	1430	85	70	83	276	40	517	75	10.31	32	4
90Ag-10Pd	999	1830	27	63	80	234	34	365	53	10.57	31	3
86.8Ag-5.5Cd-0.2Ni-7.5Cu	...	...	43	72	85	276	40	517	75	10.10	43	3
85Ag-15Cd	877	1610	35	51	83	193	28	400	58	10.17	55	5
77Ag-22.6Cd-0.4Ni	...	...	31	50	85	241	35	469	68	10.31	55	4
75Ag-24.5Cu-0.5Ni	...	...	75	78	85	310	45	552	80	10.00	32	4
72Ag-28Cu	777	1430	84	79	85	365	53	552	80	9.95	20	5
60Ag-23Pd-12Cu-5Ni	...	...	11	86	93	517	78	758	110	10.51	22	3

trical conductivity. Precipitation-hardened alloys, dispersion hardened alloys, and powder metal mixtures can provide a wide range of mechanical and electrical properties.

Applications. Copper-base metals are commonly used in plugs, jacks, sockets, connectors and sliding contacts. Because of tarnish films, the contact force and amount of slide must be kept high to avoid excessive contact resistance and high levels of electrical noise. Yellow brass (C27000) is preferred for plugs and terminals because of its machinability. Phosphor bronze (C50500 or C51000) is preferred for thin socket and connector springs and for wiper-switch blades because of its strength and wear resistance. Nickel silver is sometimes preferred over yellow brass for relay and jack springs because of its high modulus of elasticity and strength, and also for its resistance to tarnishing and better appearance. Sometimes, copper alloy parts are nickel plated to improve surface hardness, reduce corrosion and improve appearance. However, nickel carries a thin but hard oxide film that has high contact resistance; very high contact force and long slide are necessary to rupture the film. To maintain low levels of resistance and noise, copper metals should be plated or overlaid with a precious metal.

Silver Metals

Silver, in pure or alloyed form, is the most widely used material for a considerable range of make-and-break contacts (1 to 600 A). Mechanical properties and hardness of pure silver are improved by alloying, but its thermal and electrical conductivities are adversely affected. Fig. 1 shows the effect of different alloying elements on the hardness and electrical resistivity of silver. Properties of the principal silver metals used for electrical contacts are given in Table 4. Silver is widely used in contacts that remain closed for long periods of time and, in the form of electroplate, is widely used as a coating for connection plugs and sockets. It is also used on contacts subject to occasional sliding, such as in rotary switches, and to a limited extent for low-resistance sliding contacts, such as slip rings.

Electrical and Thermal Conductivity. Silver has the highest electrical and thermal conductivities of all metals at room temperature and, as a result, will carry high currents without excessive heating, even when dimensions of the contacts are only moderate. Although good thermal conductivity is desired once the contact is in service, such conductivity increases the difficulty of assembly welding.

Fig. 1 Hardness and electrical resistivity versus alloy content for silver alloy contacts

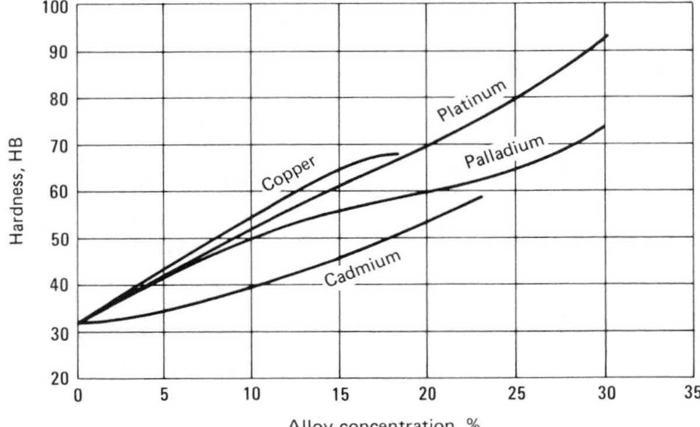

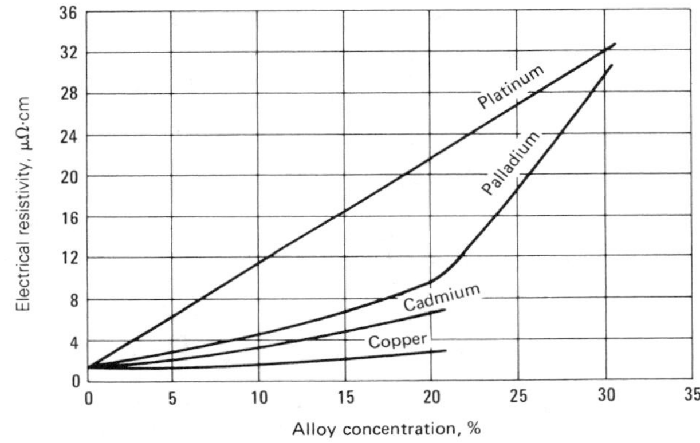

In component assemblies, migration of silver through electrical insulation may cause failure of the insulation. When in contact with certain materials such as phenol fiber and under electric potential, silver migrates ionically through or across the insulating material, producing thread-like connections that lower the resistance across the insulation. This reduces insulating qualities, and the reduction is even greater if moisture is present in the atmosphere. Insulators must be designed with care to avoid this hazard.

Oxidation Resistance. Silver is used instead of copper chiefly because of its resistance to oxidation in air. In general, silver oxide is not a problem on silver contacts, whether or not the contacts make and break the circuit. However, silver oxide can be produced by exposure to ozone, as well as by other methods. This oxide has high resistivity, is decomposed slowly on heating at about 175 °C (350 °F), is decomposed rapidly at about 350 °C (650 °F) and is removed by arcing.

Silver is vulnerable to attack by sulfur or sulfide gases in the presence of moisture. The resulting sulfide film may produce significant contact resistance, particularly where contact force, voltage, or current is low. Direct current brings silver ions from the matrix into the sulfide where they form connecting bridges. Therefore, particularly at high direct current, the film becomes somewhat conducting. The resistance of a silver sulfide film decreases as temperature increases—Ag_2S decomposes slowly at 360 °C (680 °F) and more rapidly at higher temperatures. In addition, the film may increase erosion and entrap dust.

Limitations of Silver Contacts. Silver will provide a fairly long contact life for make-and-break contacts and will handle up to 600 A. In pure silver contacts, difficulties sometimes arise from transfer of metal from one electrode to the other, which leads to the formation of buildups on one contact surface and holes in the other. When used in dc circuits, silver contacts are subject to ultimate failure by mechanical sticking as a direct result of metal transfer. The direction of transfer is generally from the positive contact to the negative, but under the influence of arcing, the direction may be reversed. With high currents or inductive loads, it may be desirable to shunt the load with a resistance-capacitance (RC) protection network to reduce erosion.

When arcing produces a glow discharge in air, the rate of erosion of silver is unusually high because of a chemical interaction with air to form $AgNO_2$.

For low resistance and low noise levels, the design of the contact device must provide sufficient force and slide to break through any silver sulfide film and maintain film-free metal-to-metal contact at the interface. Connectors should have high slide force and several newtons normal force. Rotary switches that have up to 490 mN normal force and considerable slide should have a protective coating of grease to reduce sulfiding and to remove abrasive particles. In low-noise transmission circuits, silver should not be used on relay and other butting contacts that have less than 196 mN force; other precious-metal coatings, such as gold or palladium, should be used instead of silver.

A silver sulfide film has a characteristic voltage drop of several tenths of a volt. Where this drop is tolerable, silver contacts will provide reliable contact closure. Failure to close, however, may be greater than with other precious metal contacts because of impacted dirt, with a sulfide film acting as a dirt catcher.

For many applications, silver is too soft to give acceptable mechanical wear. Alloying additions of copper, cadmium, platinum, palladium, gold and other elements are effective in increasing the hardness and modifying the contact behavior of silver.

Fine silver (99.9 Ag) has the highest electrical and thermal conductivities of all metals. It has a high current capacity, which limits heat generation, and a high thermal conductivity, which allows contacts to readily dissipate the heat generated by arcing. Silver also has good oxidation resistance, and therefore a low resistance and low voltage drop across the contact interface can be maintained.

The low boiling and melting points of

silver are disadvantages. Fine silver contacts tend to weld easily, and usually have high erosion loss. The low hardness and low mechanical strength of fine silver result in high rates of mechanical wear.

Fine silver contacts are used in low-current applications such as switches and relays in appliances and automotive products.

Because fine silver is very ductile, it can be fabricated into many designs, including contacts in the form of solid, tubular, and composite rivets and solid buttons.

Silver-Copper Alloys. Copper additions improve the hardness of silver appreciably and slightly decrease its conductivity. However, copper decreases the corrosion resistance of silver; hence, the oxidized film increases contact resistance. Switching devices that have silver-copper contacts should have a high closing force and large wiping action to break down the oxide films.

Silver-copper alloys are used in place of fine silver where electrical, mechanical, and atmospheric conditions are compatible. For the same application, silver-copper alloys usually cost less than fine silver.

Addition of a small amount of nickel to a silver-copper alloy (as in Ag-24.5Cu-0.5Ni, for example) makes the oxide film brittle, so that switching devices can use less closing force.

Silver-Cadmium Alloys. Cadmium improves the arc quenching ability of silver, and also increases its resistivity and mechanical strength. Silver-cadmium alloys are more resistant to arc erosion and welding than fine silver and silver-copper alloys.

Ag-22.6Cd-0.4Ni: Because of the nickel addition, the oxidation film of this alloy is also brittle. Ag-22.6Cd-0.4Ni is used in electrical gages and automotive voltage regulators, where the closing force is light and where a stable resistance and low transfer rate are required. It is also used to make positive contact and retard material transfer when paired with Ag-3Pd alloy in polarized low-voltage circuits.

Ag-15 Cd: This alloy exhibits low welding tendencies at the contact interface and is the material most commonly used to switch light or medium current in ac or dc circuits such as line starters, solenoid relays for automotive starters, and other devices subjected to high surge current.

Ag-5.5Cd-0.2Ni-7.5Cu: This alloy has excellent resistance to corrosion and good spring properties. It is used to make current-carrying spring contacts in television tuners, collector rings and rf switches.

Silver-Platinum Alloys. A small addition of platinum increases the hardness, wear resistance and corrosion resistance of silver, but concurrently decreases its electrical conductivity. Silver-platinum alloys are used in switching devices having low closing force where cost is not the main concern.

Silver-Palladium Alloys. Palladium improves the wear resistance of silver, but also decreases its conductivity. Silver-palladium alloys are less susceptible to oxidation than fine silver. Ag-3Pd alloy is used as the negative contact paired with Ag-22.6Cd-0.4Ni in low-voltage dc circuits. Silver and palladium form a complete solid solution, and their alloys have very good fabricability.

Silver-Gold Alloys. Gold increases hardness and improves oxidation resistance of silver. The tarnish films on contact surfaces are more stable than those of any other alloy. Ag-10Au is primarily used in ac and dc relays with current capacities less than 0.5 A where high reliability is essential. This alloy is very ductile and can be fabricated in the same manner as fine silver.

Multi-component Alloys. Ag-0.25Mg-0.20Ni and Ag-0.18Mg-0.20Ni-0.15Cu have similar properties. In low-current dc applications (voltage regulators, thermal gages, and relays), these materials provide low transfer characteristics. Mechanical properties can be improved by internally oxidizing the alloying elements into oxides.

Ag-23Pd-12Cu-5Ni has high hardness (good resistance to wear), better tarnish resistance, and a higher melting point than fine silver. It is limited to light current applications. Because of its high hardness, the alloy is used as brush contacts in potentiometers and other sliding applications. This alloy is also made into disks for composite rivets.

Gold Metals

Pure gold has unsurpassed resistance to oxidation and sulfidation, but a low melting point and susceptibility to erosion limit its use in electrical contacts to situations where the current is not more than 0.5 A. Although oxide and sulfide films do not form on gold, a carbonaceous deposit is sometimes formed when a gold contact is operated in the presence of organic vapors. The resistance of this film may be several ohms.

When gold is used in contact with palladium or rhodium, very low contact resistances have been reported.

The low hardness of gold can be increased by alloying with copper, silver, palladium and platinum, but usage is necessarily restricted to low-current applications because of the low melting point.

Properties of gold and its alloys are listed in Table 5. If low tarnish rates and low contact resistance are to be preserved, the gold content should not be less than about 70%.

Fine Gold. The unique property of fine gold (99.9 Au) as a contact material is its superb tarnish resistance in air. Only platinum is more tarnish resistant, but fine gold is less costly than platinum. Pure gold is very soft and susceptible to mechanical wear, metal transfer and welding. Pure gold electroplated or roll bonded over a base metal substrate is used in dry circuit connectors and relays to improve reliability. It is widely used in computers and telecommunications equipment where reliability is a major concern.

Au-26.2Ag-1.8Ni. Silver and nickel increase the hardness of gold, thereby increasing resistance to mechanical wear and deformation. Au-26.2Ag-1.8Ni resists welding and transfer better than pure gold and is used in devices that carry less than 0.5 A current where high reliability is required. The alloy is ductile and has good fabricability.

Au-25Ag-6Pt. Both silver and platinum increase the hardness of gold. Au-25Ag-6Pt is employed in low-current and low-closing-force relays such as those used in telecommunication systems, where high reliability is required. Under conditions of erosion, contacts have long life if the current is limited to 0.4 A. The alloy is also highly satisfactory for use in sliding contacts, such as in rotary switches or low-pressure slip rings, because it has good wear resistance and maintains low contact resistance. It is less susceptible to polymer formation than palladium and, where this is important, its greater cost may be justified.

Au-25Ag and Au-50Ag. Silver increases the hardness of gold, but decreases its tarnish resistance. Gold-silver alloys are used where a higher

Table 5 Properties of gold metals used for electrical contacts

Alloy	Solidus temperature °C	Solidus temperature °F	Electrical conductivity, %IACS	Hardness, HR15T Annealed	Hardness, HR15T Cold worked	Tensile strength Annealed MPa	Tensile strength Annealed ksi	Tensile strength Cold worked MPa	Tensile strength Cold worked ksi	Density, Mg/m^3
99Au	1085	1985	74	40	65	...	...	...	...	19.36
90Au-10Cu	932	1710	16	76	91	400	58	705	102	17.18
75Au-25Ag	1029	1885	17	50	77	...	...	...	...	15.96
72.5Au-14Cu-8.5Pt-4Ag-1Zn	954	1750	10	88	96	...	...	...	...	16.11
72Au-26.2Ag-1.8Ni	...	...	14	61	81	230	33	345	50	15.56
71Au-5Ag-9Pt-15Cu	...	...	8	88.5	75(a)	700	101	1170	170	16.02
69Au-25Ag-6Pt	1029	1885	10	70	84	275	40	415	60	15.92
50Au-50Ag	...	...	...	...	...	...	...	...	...	13.59

(a) Rockwell 15N.

degree of reliability is required than can be obtained with silver-base alloys.

Au-5Ag-9Pt-15Cu provides good tarnish resistance as well as high hardness and strength. It is used as a contact where a large wipe is required. It is also used as brush contacts against slip rings made of Au-26.2Ag-1.8Ni.

Au-10Cu. Copper increases the hardness of gold with only a small sacrifice in corrosion resistance. Au-10Cu is used in low-voltage dc devices such as alternators or voltage regulators, and as a positive contact paired with a platinum-iridium negative contact. Under light closing forces, this combination provides a low transfer rate and good anti-welding characteristics.

Precious Metals of the Platinum Group

Platinum and palladium are the two most important metals of the platinum group. These metals have a high resistance to tarnishing, and therefore provide reliable contact closure for relays and other devices having contact forces of less than 490 mN. Their high melting points, low vapor pressure, and resistance to arcing make them suitable for contacts that close and open the load, particularly in the range up to 1 A. The low electrical and thermal conductivities of these metals, as well as their cost, generally exclude them from use at currents above about 5A.

Palladium has an arcing limit only slightly less than that of platinum and gives comparable performance in relays for telephones and similar services handling 1 A or less. Palladium is a satisfactory substitute for platinum in these applications.

Chemical Properties. Platinum has a high resistance to corrosion, including resistance to oxidation, sulfidation and salt water. It will not form a stable oxide at any temperature.

Palladium is resistant to oxidation at ordinary temperatures. If heated above 350 °C (660 °F), it will oxidize slowly to form an oxide that is stable at room temperature. However, the oxide is decomposed promptly on heating to 800 °C (1470 °F) or by arcing. The oxide is not considered to be a significant factor in the reliability of closure of telephone-type relays.

The presence of organic vapors in the contact area can seriously influence the life and reliability of electrical contacts, particularly the low-force precious metal contacts universally employed in high-reliability low-noise circuits. The damaging organic vapors may arise from coil forms, wire coatings, insulation, soldering flux, potting and sealing compounds, and other organics in associated electrical equipment, as well as from external sources.

Organic contamination may produce two distinctly different forms of contact damage: activation and polymer formation.

Activation is the development of a carbon deposit on the contacting surfaces, formed by the decomposition of the organic contaminant in the arc. This deposit markedly increases arc erosion. The carbon deposits decrease the current needed to sustain an arc and prolong the arcing time. A 95% reduction in contact life may result from activation brought about by the presence of organic vapors. Activation can be reduced or eliminated by using insulating materials that are not sources of organic vapors, by adequate ventilation and, perhaps, by absorbing the organic vapors in a getter such as carbon.

Polymer formation is the development of a polymer-like insulating brown powder on contacts in dry circuits (those not carrying current on make or break), and may lead to transient open circuits. The insulating brown powder is believed to result from the adsorption of the organic vapor on the contact surface, followed by its polymerization by the friction associated with contact operation. The sliding motion both forms the polymer and pushes it outside the slide area, where it builds up as a brown powder. A transient open circuit occurs when some of the built-up powder falls into the contact area.

Controlled experiments have shown that the type of contact metal influences the amount of polymer formed. The greatest amount of polymer is formed on the platinum metals; lesser amounts are formed on gold and some base metals; polymer does not form on silver.

Elimination of materials that give rise to organic vapors is a possible solution to polymer formation, but one that is difficult to carry out. From a practical standpoint, the problem has been solved in telephone circuits by cladding one of a mating pair of palladium contacts with a very thin layer of gold. In dry circuits, the one gold surface significantly reduces polymer formation, although in working circuits, the gold soon wears off and exposes the palladium base.

Erosion and Sticking. The arcing current limit for platinum group metals is about 1 A, and contact life is long if the current is kept below this value. With currents higher than 1 A or with inductive loads, it may be desirable to shunt the load or contact with a resistance-capacitance network to reduce erosion, and to reduce failures caused by snagging of pits and buildups. In general, for equal volumes of

Table 6 Properties of platinum and palladium metals used for electrical contacts

Alloy	Solidus temperature, °C	Solidus temperature, °F	Electrical conductivity(a), % IACS	Hardness, HR15T, Annealed	Hardness, HR15T, Cold worked	Tensile strength, Annealed, MPa	Tensile strength, Annealed, ksi	Tensile strength, Cold worked, MPa	Tensile strength, Cold worked, ksi	Density, Mg/m³	Elongation(a) in 50 mm or 2 in., %
99.9Pt	1770	3220	15	60	73	138	20	241	35	21.45	35
95Pt-5Ru	1775	3230	5	84	89	414	60	793	115	20.57	18
92Pt-8Ru	...	...	4	86	91	483	70	896	130	20.27	15
90Pt-10Ir	1780	3240	7	87	92	379	55	620	90	21.52	12
89Pt-11Ru	1815	3300	4	91	96	586	85	1034	150	19.96	12
86Pt-14Ru	1843	3350	3	93	99	655	95	1172	170	19.06	10
85Pt-15Ir	1787	3250	6	90	95	517	75	827	120	21.52	12
80Pt-20Ir	1808	3290	5	93	97	689	100	1000	145	21.63	12
75Pt-25Ir	1819	3310	5	95	98	862	125	1172	170	21.68	10
73.4Pt-18.4Pd-8.2Ru	...	...	4	90	92	517	75	862	125	17.77	12
65Pt-35Ir	1899	3450	4	97	99	965	140	1344	195	21.80	8
99.9Pd	1554	2830	16	62	78	193	28	324	47	12.17	28
95Pd-5Ru	1593	2900	8	79	89	372	54	517	75	12.00	15
89Pd-11Ru	1649	3000	6	85	92	483	70	689	100	12.03	13
72Pd-26Ag-2Ni	1382	2520	4	82	90	469	68	689	100	11.52	13
60Pd-40Ag	1338	2440	4	65	91	372	54	689	100	11.30	28
60Pd-40Cu	1199	2190	8	82	92	565	82	1331	193	10.67	20
35Pd-9.5Pt-9Au-14Cu-32.5Ag	1085	1985	5	90	94	689	100	1034	150	11.63	18

(a) For material in annealed condition.

contact metal, the life of platinum or palladium contacts is about ten times the life of silver contacts.

Resistance and Noise. Palladium is used almost universally on relays and relay-type switch contacts in telephone systems within the United States for talking circuit transmission. In this service, palladium is essentially noise-free, and is used in preference to platinum or gold alloys because it is more economical.

In a few isolated instances, where the palladium talking circuit contacts have been subject to vibration in service, noise troubles have developed because of polymer formation. In these few instances, the difficulty has been met by the use of gold alloys, which greatly reduces the production of polymer.

Fine Platinum. Platinum has higher melting and boiling points (1799 °C, 3270 °F, and 4530 °C, 8186 °F, respectively) than those of gold (1063 °C, 1945 °F and 2970 °C, 5380 °F), and has excellent corrosion resistance.

Fine platinum (99.9 Pt) provides a very low and consistent surface resistance at a wide range of temperatures. Contacts made with platinum remain clear in hostile environments. It is usually used in light-force relays with a current capacity of up to 2 A when reliability is the most important factor.

Properties of platinum and its alloys are listed in Table 6. The effect of alloying on the hardness and electrical resistivity of platinum is shown in Fig. 2.

Platinum-Iridium Alloys. Platinum and iridium form a complete solid solution. The physical properties, melting point, hardness and mechanical strength of Pt-Ir alloys increase almost linearly with the amount of iridium in platinum, without affecting the corrosion resistance of the latter. This group of alloys is used in low-current ac and dc circuits when the mechanical forces are high and high wear resistance and strength are required.

Fabrication of platinum-iridium alloys becomes difficult when the iridium content is high. These alloys are usually used in the form of disks to make composite rivet faces.

Platinum-Ruthenium Alloys. Ruthenium forms a solid solution with platinum up to 79% Ru. Ruthenium increases the hardness, strength and wear resistance of platinum, but high ruthenium content makes the alloy brittle. To ensure good fabricability, the ruthenium content should not exceed 15%.

Platinum-ruthenium alloys have good tarnish resistance and cost less than platinum-iridium alloys. Pt-Ru alloys are used as positive contacts in low-voltage dc circuits, almost always paired with negative contacts made of tungsten. Pt-Ru alloys can be made into rivets or disks for composite rivets.

Pt-18.4Pd-8.2Ru. This ternary alloy has properties similar to those of platinum-ruthenium and platinum-iridium alloys, but costs less. It is used in low-voltage dc circuits as the positive contact, paired with tungsten as the negative contact. This alloy has poor fabricability and can be used in the form of disks for composite rivets.

Fine Palladium. The oxidation resistance of fine palladium (99.9 Pd) is second only to that of platinum. The boiling point (3950 °C, 7142 °F) and melting point (1552 °C, 2825 °F) are slightly lower than those of platinum. However, palladium costs about one-fourth as much as platinum, and usually replaces platinum when cost is the main concern. Table 6 gives properties of palladium metals. Fig. 3 shows the effect of alloying elements on palladium.

Palladium is used in light-closing-force and low-current applications. It has very good fabricability and can be easily made into rivets or composite rivets.

Palladium-Ruthenium Alloys. Palladium forms only a limited solid solution with ruthenium. Ruthenium increases the hardness of palladium

Fig. 2 Hardness and electrical resistivity versus alloy content for platinum contacts

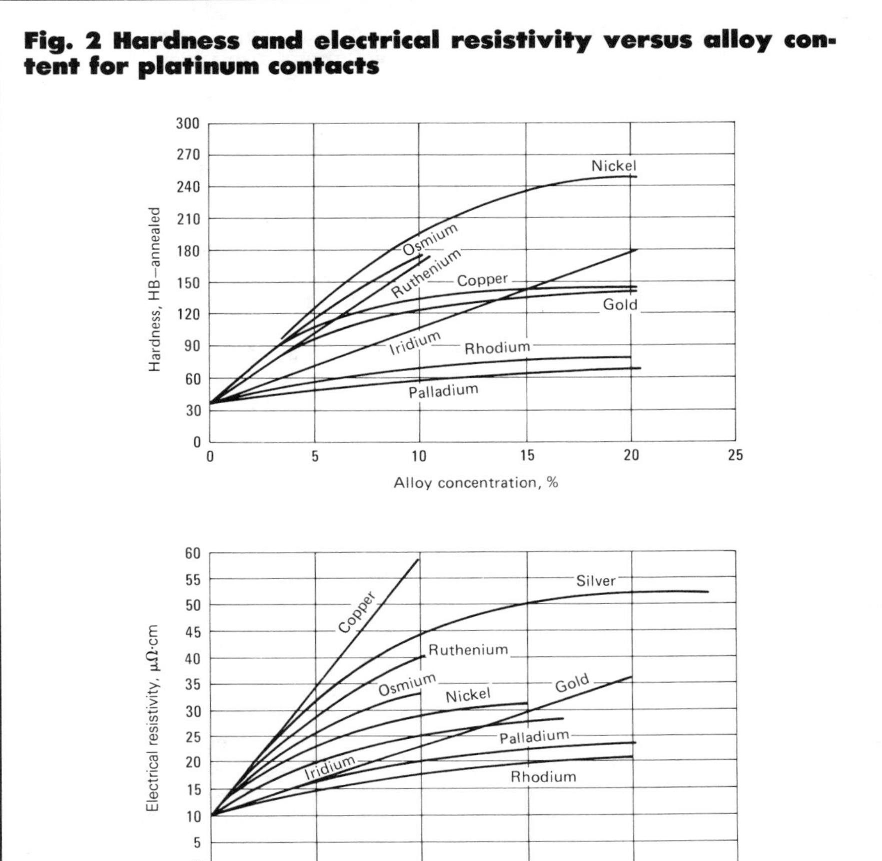

without sacrificing its corrosion resistance. Palladium-ruthenium alloys are used in low-current relays where closing forces are high. These alloys are used mostly as positive contacts paired with tungsten in dc circuits. In some applications, they are used to replace platinum-ruthenium alloys when cost is important. Practically all of the palladium-ruthenium alloys are used in the form of rivets because of their poor fabricability.

Pd-40Cu. Corrosion resistance of Pd-40Cu is good, but it is inferior to that of other precious metals. The alloy has high hardness and is used primarily as brush contacts and in instruments and gages. Because of its very poor headability, it can be used only in disk form to make composite rivets.

Pd-40Ag. Silver improves the hardness of palladium. Pd-40Ag costs less than fine palladium, but has the same corrosion resistance against a sulfiding atmosphere. It is used in high-closing-force contacts with less than 1 A current. It has a fair headability, but is used mostly in disk form for composite rivets.

Multiple-component Alloys. Table 6 lists multiple-component alloys that are designed to increase mechanical properties and decrease cost, with some sacrifice of corrosion resistance.

Pd-9.5Pt-9.0Au-32.5Ag alloy is used for brushes and slide contacts. It has a modulus of elasticity of 115 GPa (17×10^6 psi) and a proportional limit of 930 MPa (135 ksi), which are the highest for precious metal contacts.

Pd-26Ag-2Ni is used in ac or dc contact devices where operation frequency is high, such as in business machines and computers. In dc circuits, it is also used as a positive contact paired with tungsten.

Precious Metal Overlays

Silver, gold, rhodium and, to a lesser extent, platinum and palladium are employed in clad and electroplated contacts. Electrodeposition and cladding compete for the same applications. Clad overlays are favored because of their lower porosity, but electrodeposits are slightly less expensive. Electrodeposits frequently have higher hardness than the annealed wrought material; rhodium, which has a deposit hardness in excess of 600 HV in the annealed condition, is an outstanding example. High hardness accounts for the superior wear resistance of electroplated rhodium where rubbing or wiping occurs. Even at a thickness of 0.13 to 0.50 μm (5 to 20 μin.) over silver or base-metal contacts, rhodium improves wear resistance and minimizes tarnishing.

Electroplated gold is employed on silver contacts to minimize tarnishing. Nickel underlayers (barrier coats) are used to prevent migration of silver through the gold plate. Recent studies indicate that migration of silver along nickel grain boundaries is rapid at high temperatures. Hence, a nickel barrier coat is questionable for high-temperature applications. Other work in this area has disclosed that palladium can be substituted for gold as a protective coating for extension of shelf life. Electroplated gold also is used on palladium contacts to minimize polymer formation in dry circuits. However, gold electrodeposits on both silver and palladium contacts soon wear off if the contacts wipe, rub or arc.

Electroplated silver is sometimes applied to copper-base materials to make less expensive components. Electroplated silver is slightly harder than annealed wrought silver. Palladium and platinum electroplated on silver have improved tarnish resistance. Platinum, palladium and gold electroplates are used on silver to prevent the development of conducting filaments in insulating supports.

Tungsten and Molybdenum

Most tungsten and molybdenum contacts are made in the form of composites with silver or copper as the other principal component. Tungsten, which was one of the earliest metals other than copper and silver adopted for electrical contact applications, has the highest boiling point (5930 °C, or 10 700 °F) and melting point (3110 °C, or 5625 °F) of all metals; it also has very high hardness at both room and elevated temperatures. Therefore, as a contact material, it offers excellent resis-

Fig. 3 Hardness and electrical resistivity versus alloy content for palladium contacts

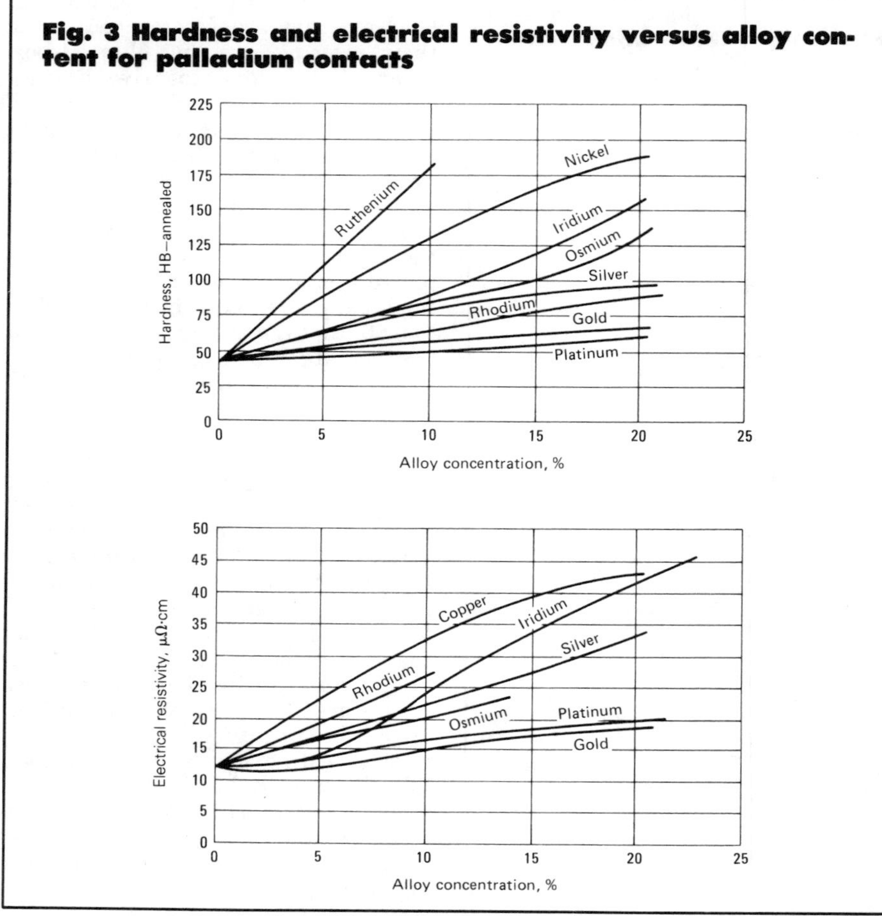

tance to mechanical wear and electrical erosion. Its main disadvantages are low corrosion resistance and high electrical resistance. After a short period of operation, an oxidized film will build up on tungsten contacts, resulting in very high contact resistance. Considerable force is required to break through the film, but high pressure and considerable impact cause little damage to the underlying metal because of its high hardness. Tungsten contacts are used in switching devices with closing forces of more than 19.6 N and in circuits with high voltages and currents not more than 5 A, such as automotive ignitions, vibrators, horns, voltage regulators, magnetos, and electric razors. In low-voltage dc devices, tungsten is always used as the negative contact, and is paired with a positive contact made of precious metal.

Tungsten rods or strips that are consolidated by swaging or rolling from sintered powder compacts have very poor ductility. They cannot be cold worked, in contrast to other contact materials. Tungsten disks are usually cut from rods or punched from strips and then brazed directly to functional parts such as breaker arms, brackets or springs.

Properties such as grain size, grain configuration, and the degree of fibrous structure, which affect contact behavior, are controlled by using special swaging methods and annealing cycles. Tungsten disks usually are supplied with a ground finish but they can also be electrochemically polished to obtain high-luster surfaces.

The high boiling and melting points of molybdenum—5560 °C (10 040 °F) and 2610 °C (4730 °F), respectively—are second only to those of tungsten and rhenium. Molybdenum is not used as widely as tungsten because it oxidizes more readily and erodes faster on arcing than tungsten. Nevertheless, because the density of molybdenum (10.2 Mg/m^3 or 0.369 $lb/in.^3$) is about half that of tungsten (19.3 Mg/m^3 or 0.697 $lb/in.^3$), use of molybdenum is advantageous where mass is important. Its cost by volume is also lower.

In addition to its use in make-and-break contacts, molybdenum is widely used for mercury switches because it is not attacked, but only wetted, by mercury.

Like tungsten, molybdenum strips and sheets are made by swaging or rolling sintered powder compacts. Disks made from rods or sheets are brazed to blanks or other structural components. Table 7 lists the properties of tungsten and molybdenum, and Fig. 4 and 5 show the effect of temperature or diameter on various properties.

Aluminum

In recent years, because of its good electrical and mechanical properties, ready availability and favorable cost, aluminum has gained importance as a conductor material. It has replaced copper in many applications, such as rectangular, tubular and channel bus conductors. It has advantages over copper in density, mass, electrical conductivity, availability and cost; it is lighter, and the same mass will conduct more current for the same voltage drop.

Aluminum 1350 is preferred for contact materials because of its high conductivity (61.8% IACS), but it is low in strength and, for some designs, requires additional support. Where strength and resistance to joint relaxation are important, heat treated 6101 is better suited and is used to a considerable extent, although there is some sacrifice in electrical conductivity (57 to 60% IACS for 6101).

As a contact metal, aluminum is generally poor because it oxidizes readily. Where aluminum is used in contacting joints, it should be plated or clad with copper, silver or tin. Aluminum should never be used for power applications where arcing is present. For instance, if aluminum contacts were substituted for silver in a motor starter, an explosion due to noninterruption of current on motor-starter de-energization would probably occur on load interruption.

Composite Materials

There are three major groups of composite contact materials made by powder metallurgy methods: refractory and carbide-base, silver-base, and copper-base.

Table 7 Typical properties of tungsten and molybdenum(a)

Tungsten	
Hardness	70 HRA, 385 HV
Modulus of elasticity	
At 20 °C (68 °F)	405 GPa (59 × 10^6 psi)
At 1000 °C (1830 °F)	325 GPa (47 × 10^6 psi)
Density	19.3 Mg/m^3 (0.697 lb/in.^{3})
Melting point	3410 °C (6170 °F)
Boiling point	5900 °C (10 650 °F)
Specific heat (Fig. 8)	
At 20 °C (68 °F)	140 J/kg (0.033 Btu/lb·°F)
Thermal conductivity (Fig. 9)	
At 20 °C (68 °F)	130 W/m·K (75 Btu/ft·h·°F)
Coefficient of linear thermal expansion (Fig. 10)	
At 20 °C (68 °F)	4.43 μm/m·°C (2.46 μin./in.·°F)
Specific resistance (Fig. 11)	
At 20 °C (68 °F)	5.5 nΩ·m
Electrical conductivity	
At 20 °C (68 °F)	31% IACS
Molybdenum	
Hardness	58 HRA, 210 HV
Modulus of elasticity	
At 20 °C (68 °F)	325 GPa (47 × 10^6 psi)
At 1000 °C (1830 °F)	270 GPa (39 × 10^6 psi)
Density	10.22 Mg/m^3 (0.369 lb/in.^{3})
Melting point	2622 °C (4750 °F)
Boiling point	4800 °C (8672 °F)
Specific heat (Fig. 8)	
At 20 °C (68 °F)	270 J/kg (0.065 Btu/lb·°F)
Thermal conductivity (Fig. 9)	
At 20 °C (68 °F)	155 W/m·K (89 Btu/ft·h·°F)
Coefficient of linear thermal expansion (Fig. 10)	
At 20 °C (68 °F)	5.53 μm/m·°C (3.07 μin./in.·°F)
Specific resistance (Fig. 11)	
At 20 °C (68 °F)	5.2 nΩ·m
Electrical conductivity	
At 20 °C (68 °F)	33% IACS

(a) Some of the physical properties of tungsten and molybdenum vary considerably with cross sectional area and grain structure.

Manufacturing Methods

Because manufacturing methods affect properties of materials with the same composition, the most common methods of producing composite electrical contact materials are discussed below.

Infiltration is used exclusively for making refractory metal and carbide-base composite contact materials. Metal powder or carbide powder is first blended to the desired composition with or without a small amount of binder to impart green strength, then is pressed and sintered into a skeleton of the required shape. Silver or copper is then infiltrated into the pores of the skeleton. This method produces the most densified composites, generally 97% or more of theoretical density. Complete densification is not possible because of the presence of some closed pores in the sintered skeleton. After infiltration, the contact is sometimes chemically or electrochemically etched so that only pure silver appears on the surface. The contact thus treated has better corrosion resistance and performs better in the early stages of use.

Press-Sinter. For small refractory-metal contacts (not exceeding about 25 mm, or 1 in., in diameter), a high-density material can be obtained by pressing a blended powder of exact final composition into shape and then sintering it at the melting temperature of the low-melting-point component (liquid-phase sintering). In some cases, an activating agent such as nickel, cobalt, or iron is added to improve the sintering effect on the refractory metal particles. For this process, powders of much finer particle size are required so that more bonding surface exists. However, the skeleton formed by this process is weaker than that formed by the infiltration process. Formation of the skeleton usually shrinks the apparent volume of the refractory portion of the composition, thus bleeding out the molten component onto the surface of the finished contact.

Press-Sinter-Re-press. The press–sinter–re-press process is used for all categories of contact materials, especially those in the silver-base category. Blended powders of the correct composition are compacted to the required shape and then sintered. Afterward, the material is further densified by a second pressing (re-pressing). Sometimes the properties can be modified by a second sintering or annealing. The versatility of this process makes it applicable for contacts of any configuration and of any material. However, it is difficult to obtain material with as high a density as is obtained with other processes. Material thus produced also may have weak bonding between particles.

Press-Sinter-Extrude. Blended powder of final composition is pressed into an ingot and sintered. The ingot is then extruded into wires, slabs or other desired shapes. The extruded material may be subsequently worked by rolling, swaging or drawing. Material made by this method is usually fully dense.

The press-sinter-extrude process is used mostly for silver-base composites. Other processes used for manufacturing silver-base composite contacts are direct extrusion or direct rolling of loose powder. Although they appear to be uncommon, they are economically feasible if the equipment is properly designed and built.

Preoxidize-Press-Sinter-Extrude. This method is used exclusively for making silver/cadmium oxide (Ag-CdO) material. Alloys are reduced to small particles in the shape of flakes, slugs or shredded foil. These particles are oxidized and then consolidated with the press-sinter-extrude process. Material made by this method is more uniform than the same material made by conventional internal oxidation. Me-

Fig. 4 Variation of properties with temperature for tungsten and molybdenum

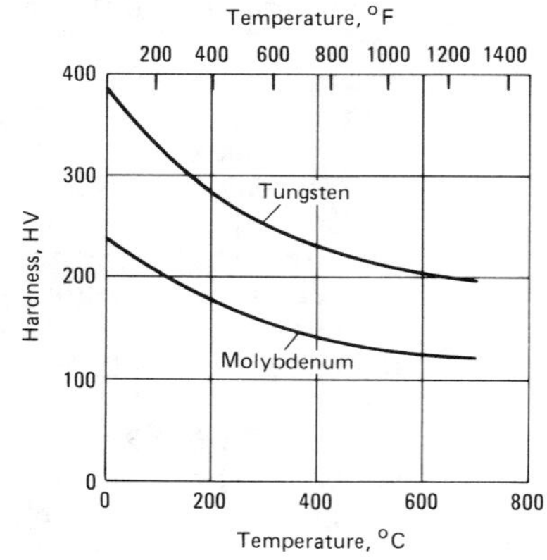

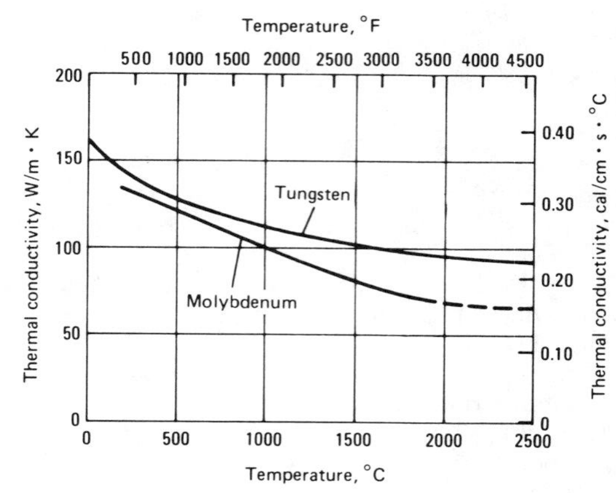

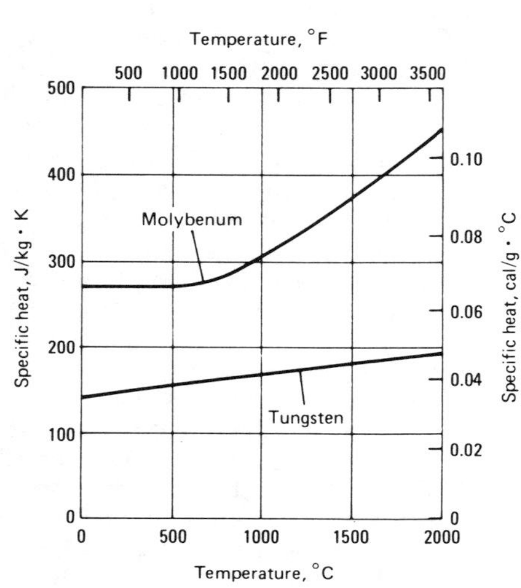

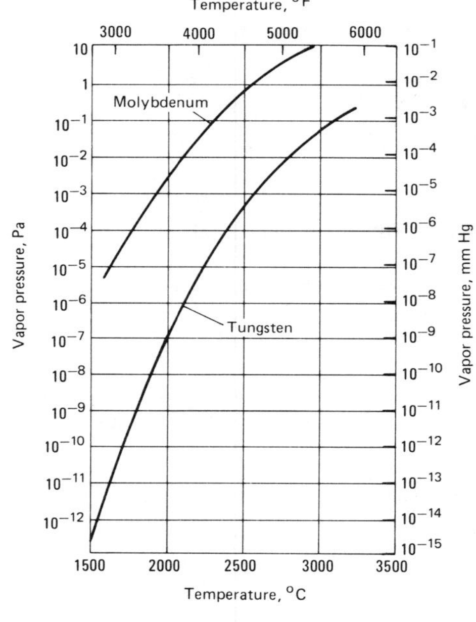

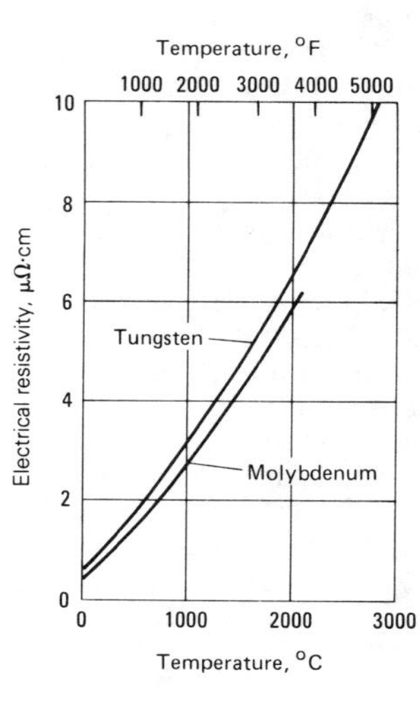

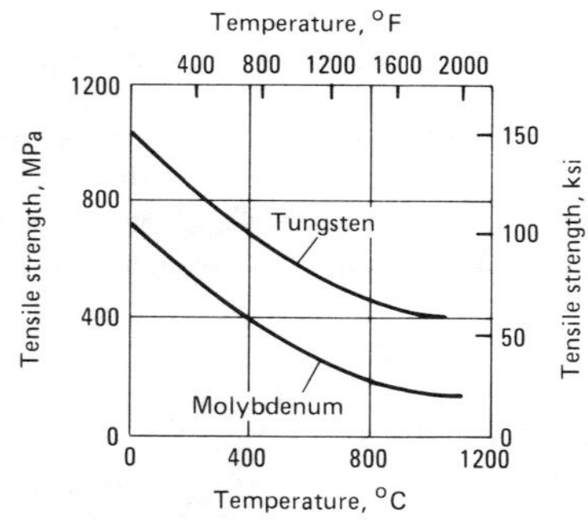

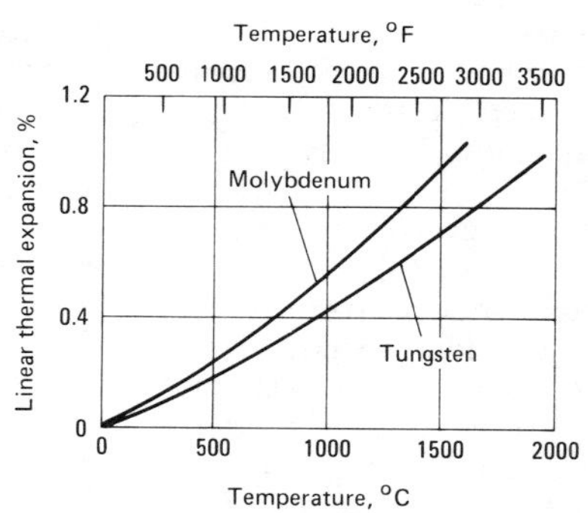

Table 8 Comparison of Ag-CdO material made by different methods

Properties	Press–sinter–re-press	Press-sinter-extrude	Internal oxidation	Preoxidize-press-sinter-extrude
Performance characteristics				
Resistance to arc erosion	3	2	1	1
Resistance to sticking and welding	1	1	2	2
Low contact resistance and temperature rise	1	1	1	1
Arc interruption	3	2	1	1
Resistance to corrosion	1	1	1	1
Material characteristics				
High mechanical properties	3	2	2	1
Resistance to annealing	3	2	2	1
Electrical and thermal conductivity	2	1	1	1
Flexibility of composition	2	2	2	1
Uniform cadmium oxide distribution	1	1	3	1

Note: 1 indicates that under most conditions this is the preferred material; 2 indicates that under most conditions the material is preferable to 3, but not as good as 1; 3 indicates that the material may be acceptable, but under typical operating conditions it is not as good as 1 or 2.

Fig. 5 Variation of tensile strength with diameter for tungsten and molybdenum rod

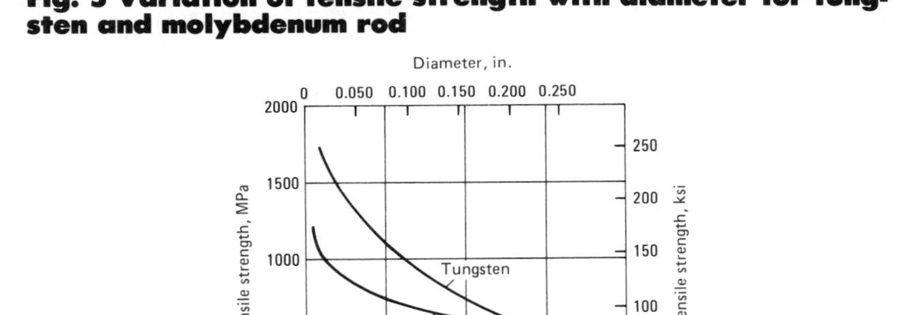

chanical properties are superior to those of the same material made by the press–sinter–re-press method.

Coprecipitation. Conventional blending or mechanical mixing of silver and cadmium oxide powders begins by dissolving the proper amounts of silver and cadmium metals in nitric acid. Compounds of silver and cadmium coprecipitate from the solution when the pH value of the solution is changed by adding either hydroxide or carbonate solutions. During subsequent calcination at about 500 °C (930 °F), the compound mixture decomposes to form a mixture of silver and cadmium oxide. Alkali-metal content can be controlled in the ppm range by adequate washing. Controlled amounts of sodium, potassium and lithium may enhance electrical life. Excessive amount of these elements can lead to rapid erosion, restrike, and generally poor electrical life. Depending on device design, the range may be from 10 to 300 ppm. Contacts are consolidated from this mixture by conventional P/M methods. The microstructure of contacts made by this method displays a finer particle size and a more uniformly dispersed CdO phase than material made by conventional blending. The fine particle dispersion results in good contact welding resistance, presumably because of the formation of slag-like inclusions.

For comparisons of Ag-CdO material manufactured by different methods, see Table 8.

Material Classification

Refractory Metal and Carbide-Base Composites. Refractory metals and their carbides are distinguished by high melting and boiling points, and high hardness, but poor electrical and thermal conductivities and poor oxidation resistance. In pure elemental form, refractory metals perform well only under low-current conditions. Forming a composite can compensate for these drawbacks. In a material made by infiltration, the function of the infiltrant (silver or copper) is twofold. First, because silver or copper does not alloy with tungsten, molybdenum, or carbides, the conductivity of the composite depends strictly on the volume percentage of infiltrant. Second, during arcing, the high temperature melts the infiltrant; consequently, the heat of fusion absorbs (quenches) a portion of the heat generated by the arc. Theoretically, the skeleton, which is made of a high-melting element, will not begin to melt until all the low-melting component evaporates. The refractory skeleton also prevents molten infiltrant from flowing by capillary action. Because of this, erosion loss of the contact is low. Properties of the contact vary with the composition of the composite. A composite with high skeletal composition has high hardness and better wear resistance, but lower current-carrying capacity. On the other hand, a high-silver composite possesses high electrical and thermal conductivities, and undergoes lower temperature rise, but is softer.

There is a lower limit for the composition of the skeleton material. Generally, when the amount of refractory or carbide is less than about 30 vol %, it is difficult to form a sound and uniform skeleton to accommodate the amount of silver. For practical purposes, the skeleton material should amount to a minimum of 50 wt % for tungsten and molybdenum, and 35 wt % for tungsten carbide. Any composite containing lesser amounts than these limits should be made by the press–sinter–re-press method, and should be considered a silver-base composite in which the function of the refractory material is to reinforce the silver matrix. For convenience of tabulation, all composites consisting of silver and refractory metal or carbide are listed in Table 9.

Tungsten, tungsten carbide, and molybdenum powders are the most commonly used materials for making skeletons for infiltrated contacts. Com-

Table 9 Properties of composites for electrical make-and-break contacts

Nominal composition, %	Manufacturing method(a)	Density, Mg/m³ Calculated	Density, Mg/m³ Typical	Electrical conductivity, % IACS	Hardness, Rockwell	Tensile strength MPa	Tensile strength ksi	Modulus of rupture MPa	Modulus of rupture ksi	Data source(b)	Application examples
Molybdenum-silver											
90Ag-10Mo	PSR	10.47	10.38	65–68	35–40HRB	...	...	...	...	G	Air conditioner controls
80Ag-20Mo	PSR	10.44	10.36	59–62	38–42HRB	...	...	...	...	G	Light and medium duty applications, automotive circuit breakers
75Ag-25Mo	PSR	10.42	10.33	58–61	44–47HRB	...	...	...	...	G	
70Ag-30Mo	PSR	10.41	10.31	56–60	46–48HRB	...	...	...	...	G	
65Ag-35Mo	PSR	10.39	10.30	55–64	49–55HRB	...	...	...	...	G	Automatic circuit protectors, starting switches
60Ag-40Mo	PSR	10.38	10.28	55–62	55–62HRB	...	...	...	...	G	
50Ag-50Mo	INF	10.35	10.10–10.24	45–52	70–80HRB	...	...	758	110	C,G,A	Air and oil circuit breakers, arcing tips, traffic signal relays, home circuit breakers
	PSR	10.35	10.14	50	65HRB	...	...	552	80	C,G,A	
45Ag-55Mo	INF	10.33	10.10–10.32	44–58	75–82HRB	...	...	...	...	G,A	
40Ag-60Mo	INF	10.32	10.10–10.22	42–49	80–90HRB	...	...	...	...	C,G,A	Aircraft switches, breaker arcing tips, electric razors, air and oil circuit breakers
	PSR	10.32	10.12	45	50–68HRB(c)	...	...	676	98	C,G,A	
35Ag-65Mo	INF	10.30	10.00–10.08	40–45	82–92HRB	...	...	...	...	G,A	
30Ag-70Mo	INF	10.29	10.00–10.31	35–45	85–95HRB	414	60	931	135	C,G,A	Air circuit breakers, low-erosion arcing tips
25Ag-75Mo	INF	10.27	10.27	31–34	93–97HRB	414	60	958	139	C,G	
20Ag-80Mo	INF	10.26	10.23–10.26	28–32	96–98HRB	407	59	965	140	C,G	Arcing contacts, heavy duty electrical applications
15Ag-85Mo	INF	10.24	10.18	28–31	97–102HRB	...	...	...	...	G	
10Ag-90Mo	INF	10.23	10.13	27–30	97–102HRB	...	...	...	...	G	Semiconducting material
Silver/cadmium oxide											
97.5Ag-2.5CdO	PSR	10.42	10.21	85	22HRF(c)	110(c)	16(c)	...	...	C	Aircraft circuit breakers, aircraft relays, automotive relays, truck controls, snap switches, contactors, motor controllers, circuit breakers, governor relays.
	PSE	10.42	10.42	95	37HRF(c) 60HRF(d)	131(c) 172(d)	19(c) 25(d)	...	...	E,C	
95Ag-5CdO	PSR	10.35	9.50–10.14	80–90	32HRF(c)	110(c)	16(c)	...	...	C,A	
	PSE	10.35	10.35	92	40HRF(c) 70HRF(d)	131(c) 172(d)	19(c) 25(d)	...	...	E,C	
	IO	10.35	10.35	80	40HRF(c) 75HRF(d)	186(c) 241(d)	27(c) 35(d)	...	...	E,C	
	PPSE	10.35	10.35	85	70HRF(c) 90HRF(d)	207(c) 248(d)	30(c) 36(d)	...	...	E,C	
90Ag-10CdO	PSR	10.21	9.30–9.80	72–85	42HRF(c)	103(c)	15(c)	...	...	C,A	
	PSE	10.21	10.21	84–87	46HRF(c) 80HRF(d)	172(c) 228(d)	25(c) 33(d)	...	...	E,G,C,T	
	IO	10.21	10.21	75	45HRF(c) 81HRF(d)	186(c) 262(d)	27(c) 38(d)	...	...	E,C	
	PPSE	10.21	10.21	82	71HRF(c) 90HRF(d)	269(c) 317(d)	39(c) 46(d)	...	...	E,C	
87Ag-13CdO	...	...	9.20	43	56HRF(c)	...	...	...	...	A	
86.7Ag-13.3CdO	IO	10.11	10.11	68	48HRF(c) 84HRF(d)	200(c) 262(d)	29(c) 38(d)	...	...	E,C	
86.5Ag-13.5CdO	PPSE	10.11	10.11	75	70HRF(c) 90HRF(d)	276(c) 324(d)	40(c) 47(d)	...	...	E,C	

(continued)

(a) PSR, press–sinter–re–press; INF, press-sinter-infiltrate; PS, press-sinter; PSE, press-sinter-extrude; IO, internal oxidation; PPSE, preoxidize-press-sinter-extrude. (b) A: Advance Metallurgy, Inc., McKeesport, PA. C: Contacts, Materials, Welds, Inc., Indianapolis, IN. E: Engelhard Industries, Plainville, MA. G: Gibson Electric Inc., Delmont, PA. S: Stackpole Carbon Co., St. Marys, PA. T: Texas Instruments Inc., Attleboro, MA. (c) Annealed. (d) Cold worked.

Table 9 (continued)

Nominal composition, %	Manu-facturing method(a)	Density, Mg/m³ Calcu-lated	Typical	Electrical conductivity, % IACS	Hardness, Rockwell	Tensile strength MPa	ksi	Modulus of rupture MPa	ksi	Data source(b)	Application examples
Silver/cadmium oxide (continued)											
85Ag-15CdO.....	PSR	10.06	8.60– 9.58	55–75	35HRF(c)	83(c)	12(c)	...	...	E,C,A	
	PSE	10.06	9.90– 10.06	55–75	57HRF(c) 80HRF(d)	193(c) 241(d)	28(c) 35(d)	...	...	E,T,C,G	
	IO	10.06	10.06	65	50HRF(c) 85HRF(d)	207(c) 269(d)	30(c) 39(d)	...	...	C	Pressure and temperature controls
	PPSE	10.06	10.06	72	70HRF(c) 90HRF(d)	276(c) 331(d)	40(c) 48(d)	...	...	C	
83Ag-17CdO.....	IO	10.01	10.01	62	52HRF(c) 88HRF(d)	214(c) 276(d)	31(c) 40(d)	...	...	C,E	Aircraft circuit breakers, aircraft relays, truck controls, contactors, circuit breakers, governor relays
	PPSE	10.01	10.01	70	70HRF(c) 90HRF(d)	276(c) 352(d)	40(c) 51(d)	...	...	C,E	
80Ag-20CdO.....	PPSE	9.93	9.93	68	70HRF(c) 90HRF(d)	276(c) 345(d)	40(c) 50(d)	...	...	C,E	
75Ag-25CdO.....	PPSE	9.79	9.79	60	...	...	...	...	...	C,E	
Silver-graphite											
99.75Ag-0.25C ...	PSR	10.41	9.70– 10.40	95–103	33–45HRF(c) 70–73HRF(d)	172(c) 255(d)	25(c) 37(d)	...	...	C,A,S	Automotive regulators, low voltage make-and-break contacts, sliding contacts
99.5Ag-0.5C	PSR	10.31	9.60– 10.30	92–102	26–44HRF(c) 69–72HRF(d)	169(c) 252(d)	24.5(c) 36.5(d)	...	...	C,A,S	
99.25Ag-0.75C ...	PSR	10.22	10.21	90–100	39HRF(c) 70HRF(d)	165(c) 247(d)	24(c) 35.8(d)	...	...	C	
99Ag-1C	PSR	10.13	9.40– 10.12	87–99	24–36HRF(c) 68–69HRF(d)	162(c) 241(d)	23.5(c) 35(d)	...	...	C,A,S	
98.5Ag-1.5C	PSR	9.96	10.04	97	33HRF(c) 66HRF(d)	152(c) 231(d)	22(c) 33.5(d)	...	...	C,A	Mate with other contact materials in circuit breakers
98Ag-2C	PSR	9.79	9.15– 9.57	82–90	22HRF(c) 65HRF(d)	...	...	...	...	C,A,S	
97Ag-3C	PSR	9.46	8.80	55–62	20HRF(c) 60HRF(d)	...	...	...	...	A,S	
95Ag-5C	PSR	8.88	8.30– 8.68	55–62	25HRF(d)	...	...	...	...	C,A,S	
	PSE	8.88	8.84	75	40HRF(d)	...	...	...	...	C	
93Ag-7C	PSR	8.37	7.80	50–57	15HRF(c) 45HRF(d)	...	...	...	...	C,A,S	
90Ag-10C	PSR	7.69	6.30– 7.20	43–53	13HRF(c) 30HRF(d)	...	...	...	...	C,A,S	
Silver-iron											
90Ag-10Fe	PSR	10.16	9.60– 10.25	87–92	48HRF(c) 81HRF(d)	214(c) 272(d)	31(c) 39.5(d)	...	...	C,A	Wall switches, thermostat controls
Silver-nickel											
99.7Ag-0.3Ni	...	10.49	...	100	53HR15T(c) 79HR15T(d)	...	...	...	...	T	...
95Ag-5Ni........	PSR	10.41	9.80– 10.41	80–95	32HRF(c) 84HRF(d)	165(c)	24(c)	...	...	C,A,S	Appliance switches
90Ag-10Ni	PSR	10.31	9.70– 10.32	75–90	35HRF(c) 89HRF(d)	172(c)	25(c)	...	...	C,S,A,E	Low rating line starters
85Ag-15Ni	PSR	10.22	9.50– 10.02	66–80	40HRF(c) 93HRF(d)	186(c)	27(c)	...	...	C,A,S, G,E	Circuit breakers
80Ag-20Ni	PSR	10.13	9.30– 9.50	63–75	52–59HRF(c) 80HRF(d) (continued)	...	...	...	...	E,A,S	

(a) PSR, press–sinter–re–press; INF, press-sinter-infiltrate; PS, press-sinter; PSE, press-sinter-extrude; IO, internal oxidation; PPSE, preoxidize-press-sinter-extrude. (b) A: Advance Metallurgy, Inc., McKeesport, PA. C: Contacts, Materials, Welds, Inc., Indianapolis, IN. E: Engelhard Industries, Plainville, MA. G: Gibson Electric Inc., Delmont, PA. S: Stackpole Carbon Co., St. Marys, PA. T: Texas Instruments Inc., Attleboro, MA. (c) Annealed. (d) Cold worked.

Table 9 (continued)

Nominal composition, %	Manufacturing method(a)	Density, Mg/m³ Calculated	Density, Mg/m³ Typical	Electrical conductivity, % IACS	Hardness, Rockwell	Tensile strength MPa	Tensile strength ksi	Modulus of rupture MPa	Modulus of rupture ksi	Data source(b)	Application examples
Silver-nickel (continued)											
75Ag-25Ni	PSR	10.05	9.20	59	61HRF(c)	...	...	...	...	S	Circuit breakers, disconnect switches
70Ag-30Ni	PSR	9.96	9.40–9.53	55–56	42HRF(c) 87HR(d)	...	...	...	...	C,S,G	
65Ag-35Ni	PSR	9.88	9.00	49	26HR30T(c)	...	...	...	...	S	
60Ag-40Ni	PSR	9.80	8.90–9.60	44–47	40HR30T(c) 92HR30T(d)	241(c) 414(d)	35(c) 60(d)	...	...	C,S,A,G	
	PSE	9.80	9.60	60	46HR30T(c)	...	...	...	...	S	
55Ag-45Ni	PSR	9.71	8.80	41	25HR30T(c)	...	...	...	...	S	
50Ag-50Ni	PSR	9.63	9.00	38	50HR30T(c)	...	...	...	...	S	
45Ag-55Ni	PSR	9.56	8.50	35	30HR30T(c)	...	...	...	...	S	
40Ag-60Ni	PSR	9.48	8.80	32	35HR30T(c) 97HR(d)	...	...	...	...	S	Circuit breakers
	PSE	9.48	9.30	40	68HR30T(c)	...	...	...	...	S	Transformer protectors, contractors, relays
35Ag-65Ni	PSR	9.40	8.60	30	40HR30T(c)	...	...	...	...	S	
30Ag-70Ni	PSR	9.32	8.50	27	40HR30T(c)	...	...	...	...	S	
25Ag-75Ni	PSR	9.25	8.20	24	40HR30T(c)	...	...	...	...	S	
20Ag-80Ni	PSR	9.17	8.00	21	35HR30T(c)	...	...	...	...	S	
Tungsten carbide-silver											
65Ag-35WC	INF	11.86	11.53–11.85	55–60	50–65HRB	272	39.5	483	70	C,G	Aircraft contactors, lighting relays, low-voltage switches, circuit breakers
	PSR	11.86	11.10–11.80	50–60	50–62HRB	...	...	...	...	C,G	
60Ag-40WC	PSR	12.09	11.40–11.92	46–55	60–70HRB	...	...	...	...	A,G	
58Ag-42WC	PSR	12.17	11.86–11.97	50–55	75–85HRB	...	...	...	...	C	
50Ag-50WC	INF	12.56	12.12–12.50	43–52	75–85HRB	276	40	793	115	C,G,A	
40Ag-60WC	INF	13.07	12.70–12.92	40–47	90–100HRB	379	55	827	120	C,G,A	Heavy-duty circuit breakers
38Ag-62WC	INF	13.18	12.92–13.29	35–38	90–100HRB	552	80	...	...	C	
35Ag-65WC	INF	13.35	12.90–13.18	30–37	95–105HRB	...	...	...	...	G,A	Semiconducting material
Tungsten-silver											
90Ag-10W	PSR	11.00	10.30–11.20	90–95	20–33HRB	...	...	...	...	C,A,G	Controls, automatic circuit protectors, wall switches
85Ag–15W	PSR	11.27	10.60–11.30	85–90	25–38HRB	...	...	...	...	A,G	Current-carrying contacts in circuit breakers, light-duty contactors
80Ag–20W	PSR	11.55	10.90–11.70	80–85	30–43HRB	...	...	...	...	A,G	
70Ag–30W	PSR	12.16	12.00	72–80	40–47HRB	...	...	...	...	G	
60Ag–40W	PSR	12.84	12.10–12.60	60–65	50–60HRB	...	...	...	...	A,G	
35Ag-65W	INF	14.92	14.20–14.77	45–53	80–93HRB	...	...	827	120	C,G	Automotive starting switches, circuit breakers
	PS	14.92	13.90–14.20	47–50	85–87HRB	...	...	...	...	C,G	
	PSR	14.92	14.65–14.74	47–50	55–65HRB(c)	...	...	572	83	C	

(continued)

(a) PSR, press–sinter–re–press; INF, press-sinter-infiltrate; PS, press-sinter; PSE, press-sinter-extrude; IO, internal oxidation; PPSE, preoxidize-press-sinter-extrude. (b) A: Advance Metallurgy, Inc., McKeesport, PA. C: Contacts, Materials, Welds, Inc., Indianapolis, IN. E: Engelhard Industries, Plainville, MA. G: Gibson Electric Inc., Delmont, PA. S: Stackpole Carbon Co., St. Marys, PA. T: Texas Instruments Inc., Attleboro, MA. (c) Annealed. (d) Cold worked.

Table 9 (continued)

Nominal composition, %	Manufacturing method(a)	Density, Mg/m³ Calculated	Density, Mg/m³ Typical	Electrical conductivity, % IACS	Hardness, Rockwell	Tensile strength MPa	Tensile strength ksi	Modulus of rupture MPa	Modulus of rupture ksi	Data source(b)	Application examples
Tungsten-silver (continued)											
30Ag-70W	INF	15.42	15.02	40–50	85–93HRB	...	...	...	...	G	Motor starters, aircraft equipment, circuit breakers, contactors, computers, arcing tips
27.5Ag-72.5W	INF	...	15.56	49	90HRB	483	70	896	130	C	
	PSR	...	15.44	...	58–68HRB(c)	...	...	586	85	C	
25Ag-75W	INF	15.96	15.25–15.40	40–50	85–95HRB	...	...	...	...	A,G	
20Ag-80W	INF	16.53	16.18	35–40	91–100HRB	...	...	...	...	G	
15Ag-85W	INF	17.14	16.60–17.05	32–41	90–100HRB	448	65	758	110	A,G,C	Motor governors
10Ag-90W	PSR	17.81	17.25	29–35	95–105HRB	379	55	758	110	G	Semiconducting material
Tungsten carbide-copper											
50Cu	INF	11.39	11.00–11.27	42–47	90–100HRF	...	...	1103	160	C,A	Arcing contacts in oil switches, wiping shoes in power transformers
44Cu	INF	11.77	11.64	43	99HRF	...	...	1241	180	C	
30Cu	INF	12.78	12.65	30	38HRC	...	...	...	...	...	
Tungsten-copper											
75Cu-25W	PSR	10.37	9.45–10.00	50–79	35–60HRB	...	...	414	60	C,A,G	Current-carrying contacts
70Cu-30W	...	10.70	10.45	76	59–66HRB	...	...	...	...	G	
65Cu-35W	...	11.06	11.40	72	63–69HRB	...	...	...	...	G	
60Cu-40W	...	11.45	11.76	68	69–75HRB	...	...	...	...	G	Oil circuit breakers, arcing tips
50Cu-50W	INF	12.30	11.90–11.96	45–63	60–81HRB	...	...	...	...	A,G	
44Cu-56W	INF	12.87	12.76	55	79HRB	434	63	827	120	C	
40Cu-60W	INF	13.29	12.80–12.95	42–57	75–86HRB	...	...	...	...	A,G	Oil circuit breakers, reclosing devices, arcing tips, tap change arcing tips, contactors
35Cu-65W	INF	13.85	13.35	54	83–93HRB	...	...	...	...	G	
32Cu-68W	INF	14.20	13.95	50	90HRB	...	...	896	130	C	
30Cu-70W	INF	14.45	13.85–14.18	36–51	86–96HRB	...	...	1000	145	C,G,A	Circuit breaker runners, arcing tips, tap change arcing tips
26Cu-74W	INF	14.97	14.70	46	98HRB	621	90	1034	150	...	
25Cu-75W	INF	15.11	14.50	33–48	90–100HRB	...	...	...	...	G,A	Vacuum switches, arcing tips, oil-circuit breakers
20Cu-80W	INF	15.84	15.20	30–40	95–105HRB	758	110	...	...	C,G	
13.4Cu-86.6W	INF	16.71	16.71	33	20HRC	621	90	1034	150	C	
10.4Cu-89.6W	INF	17.22	17.22	30	30HRC	765	111	1138	165	C	
Tungsten-graphite-silver											
48Ag-51.75W-0.25C	PSR	13.21	13.38	65	55HRB	...	...	552	80	C	Circuit breakers, arcing tips
46Ag-53W-1C	PSR	13.58	12.85	55	85HRB	...	...	...	...	C	
45Ag-50W-5C	PSR	11.00	10.60	37–43	45–55HRB	...	...	621	90	A	
Complex composite contacts											
88Ag-10Ni-2C	PSR	9.63	9.37	70	26HRF(c) 64HRF(d)	...	...	...	...	C,A	Sliding contacts
25Ag-50Fe-25Cu	PSR	8.67	8.52	21	84HRF(c) 94HRF(d)	...	...	...	...	C	Circuit breakers

(a) PSR, press–sinter–re–press; INF, press-sinter-infiltrate; PS, press-sinter; PSE, press-sinter-extrude; IO, internal oxidation; PPSE, preoxidize-press-sinter-extrude. (b) A: Advance Metallurgy, Inc., McKeesport, PA. C: Contacts, Materials, Welds, Inc., Indianapolis, IN. E: Engelhard Industries, Plainville, MA. G: Gibson Electric Inc., Delmont, PA. S: Stackpole Carbon Co., St. Marys, PA. T: Texas Instruments Inc., Attleboro, MA. (c) Annealed. (d) Cold worked.

posites with tungsten skeletons have the best arc-interrupting and arc-resisting characteristics and the best arc-erosion resistance. Their antiwelding properties are moderate. High-energy devices usually use silver-infiltrated composites having a tungsten skeleton.

Composites with tungsten carbide skeletons have better resistance to welding, better anticorrosion properties, and more stable contact resistance compared with other infiltrated composites. Devices that handle switching arcs usually use composites based on tungsten carbide skeletons.

Composites with molybdenum skeletons have relatively low contact resistance and behave well in circuit-interrupting devices. For the same current-carrying capacity, a molybdenum-base composite costs less than the other two, but the antiwelding and anticorrosion properties of molybdenum-base composites are inferior to those with tungsten or tungsten carbide skeletons.

For a combination of properties, or sometimes for a special requirement, a skeleton made of a mixture of tungsten and tungsten carbide is used. The blended powder contains either the mixture of tungsten and tungsten carbide or a mixture of tungsten and graphite. In the latter case, the graphite and part of the tungsten react to form tungsten carbide during sintering.

Most infiltrated composite contacts use silver as the infiltrant because of its excellent thermal and electrical conductivities, as well as its superb oxidation resistance. Copper infiltrant, which costs less but has very poor corrosion resistance, is used for composites that operate in noncorrosive environments such as oil, vacuum, or inert atmospheres.

Fig. 6 Contact resistance versus force for fine silver and Ag-CdO contacts

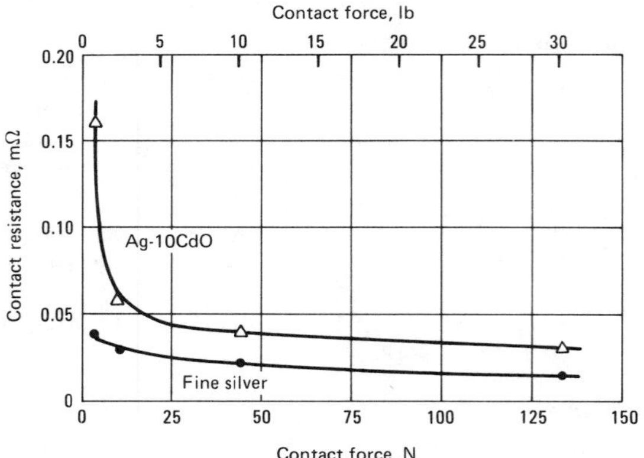

Unarced contacts were 12.7 mm (1/2 in.) in diameter with a 38-mm (1 1/2-in.) spherical radius. Resistance measurements were made with ac current at 50 A and 60 Hz.

Silver-base Composites. The main advantage of a silver composite over a silver alloy is that the conductivity of a silver composite depends strictly on the percentage of silver by volume. An alloying element in solution greatly decreases the conductivity of silver. For instance, the volume of silver in Ag-15CdO composite is less than that in Ag-15Cd alloy, yet the electrical conductivity of the former (65% IACS) is much greater than that of the latter (35% IACS).

In silver composites, the second phase forms discrete particles that are dispersed in the silver matrix. The dispersed phase improves the matrix in two ways. First, it increases the hardness of the composite material in a manner similar to dispersion hardening. Second, in the region where contact welding occurs, the second phase particles can form oxide particles, which behave as slag inclusions and greatly reduce sticking or welding.

Silver-base composites can be divided into two types: type 1 uses a pure element or carbide as the dispersed phase; type 2 uses oxides as the dispersed phase. In both types, the hardness increases and the conductivities decrease as the volume fraction of dispersed phase increases, and vice versa.

In type 1, the dispersed phase functions as a hardener and improves the mechanical properties of the silver matrix. The dispersed phase also promotes improved electrical performance such as antiwelding properties. Elements used include tungsten, tungsten carbide, molybdenum, nickel, iron, graphite, and mixtures of these materials.

Silver composites (made by the press–sinter–re-press method using tungsten, tungsten carbide, and molybdenum as the dispersed phases) show electrical conductivities similar to those of infiltrated composites of the same components. However, their mechanical properties are inferior because the dispersed phases do not form a refractory skeleton.

Composites with nickel as the dispersed phase resist mechanical deformation or peening under impact and possess good antiwelding properties. Silver-nickel composite contacts can be used as both members of a contact pair. Sometimes, a silver-nickel composite is used as the moving contact operating against a stationary contact of a different composite such as silver-graphite.

Composites of silver-iron exhibit good antiwelding and good wear characteristics when used in creep-type thermostat devices. These materials have poor corrosion resistance.

Graphite in silver-base composites serves as a good lubricant, reducing the damage caused by frictional forces. Silver-graphite composites are used chiefly as sliding or brush contacts. These materials have high resistance to welding and are also used as make-and-break contacts. In circuit breakers, they are usually paired with silver-nickel composites.

Type 2 silver-base composite contacts use oxide as the dispersed phase. Although contact materials incorporating several different oxides have been patented, only CdO is used commercially. CdO strengthens the silver matrix and reduces the tendency to weld. Cadmium oxide decomposes at about the melting point of silver. The decomposition reaction absorbs much of the heat from arcing, minimizing evaporative loss of the silver matrix. Because of this, CdO is the most efficient of all the elements and compounds used as dispersed phases for decreasing erosion loss and welding tendency.

A substantial portion of the Ag-CdO composites are made by internal oxidation. Alloys of silver and cadmium are fabricated into the required shape and are heated in oxygen or air at about 800 °C (1472 °F) for some time. Oxygen diffuses into the alloy, forming CdO particles uniformly dispersed in the silver matrix.

Multiple-component Composites. There is no ideal material to meet all conditions for contact applications. If required by manufacturers of switching devices, contact manufacturers can offer composite materials consisting of as many as four or five components. Most of these composites serve only special purposes. They are not universally accepted and generally cost more. Two common three-component composites are listed in Table 9.

Fig. 7 Contact erosion characteristics of silver and silver-tungsten contacts

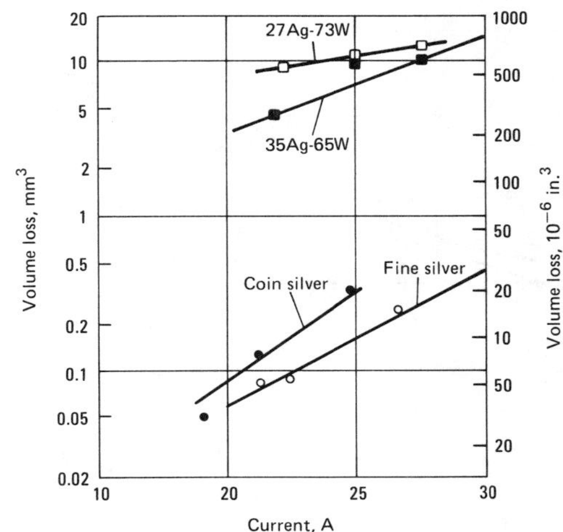

Test conditions were 115 V, 60 Hz, and 1.0 power factor for 100 000 operations at 60 operations per minute. Closing and opening speeds were 38 mm/s (1 1/2 in./s). Closing force was 980 mN and opening force, 735 mN.

Composition and Properties of Composite Contacts

Data published by contact manufacturers usually include density, hardness, and electrical conductivity. These data provide designers of electrical devices with a basic concept of the behavior of a composite contact. Characteristics that relate directly to failure modes such as arc erosion or material transfer are usually described in a qualitative manner. Very few quantitative data pertaining to these characteristics have been published because these properties depend on several test parameters. For instance, the factors affecting arc erosion rate include the following:

Mechanical

- Opening force and opening speed
- Closing force and closing speed
- Bouncing of the movable contact
- Wiping distance
- Gap between opposing contacts

Electrical

- Current—both amperage and whether ac or dc
- Voltage
- Power factor (inductive/capacitive)

Fig. 8 Contact welding characteristics of silver and silver-tungsten contacts

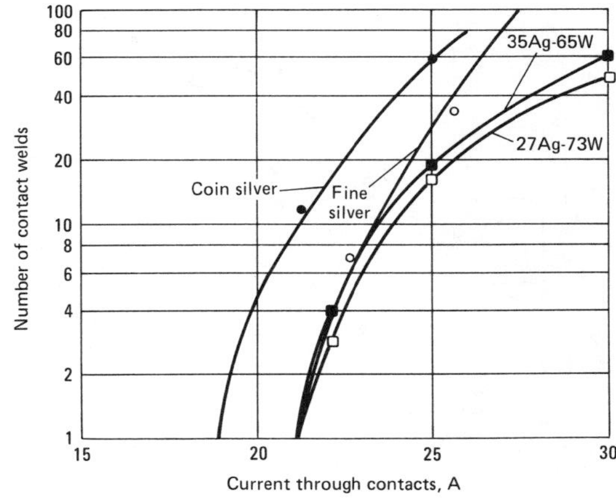

Operation characteristics are the same as for Fig. 7.

Because each variable can greatly affect the arc erosion rate of a composite, it is virtually impossible to define a universal test to evaluate erosion rate. Published data on erosion rate and welding frequency usually are collected under very specific conditions. They are valid only for qualitative description in a specific set of circumstances and cannot be extrapolated to suit other applications. The only means of learning how a composite will perform in a specific application is to test it extensively in the device in which it will be used.

Figures 6 to 9 show one way of expressing the characteristics of composites. Another way is to compare the properties of a group of materials such as in Tables 8 and 9.

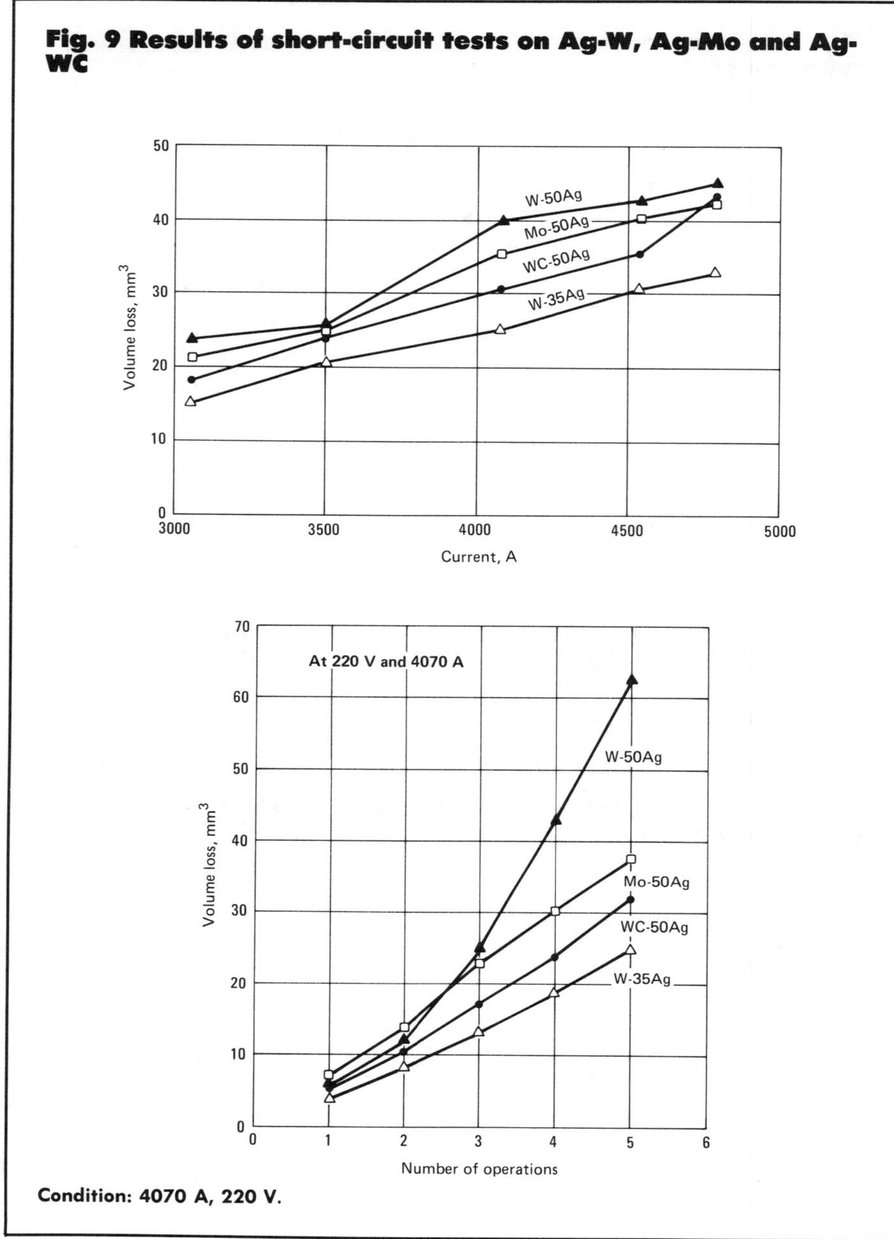

Fig. 9 Results of short-circuit tests on Ag-W, Ag-Mo and Ag-WC

Condition: 4070 A, 220 V.

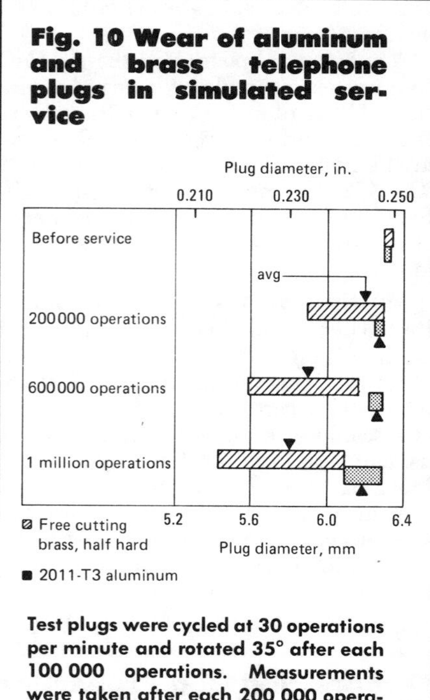

Fig. 10 Wear of aluminum and brass telephone plugs in simulated service

Test plugs were cycled at 30 operations per minute and rotated 35° after each 100 000 operations. Measurements were taken after each 200 000 operations.

Life Tests

Example 1. Wear of Telephone Plugs (Fig. 10). A simulated service test, representing the service of telephone plug bodies, has provided a comparison of the wear properties of 2011-T3 aluminum and half-hard brass rubbing against nickel-silver.

The outside diameter of each sleeve was measured at intervals of 60° around the circumference. After the original diameter of approximately 6.32 mm (0.2490 in.) was measured, the plug was mounted on a test fixture designed to simulate normal mating of the plug and corresponding jack. Upon entering the jack, the sleeve rubbed against a nickel-silver spring contact. The test lasted for one-million operations at a rate of about 30 per minute. To get more uniform wear, the plugs were rotated about 35° after each 100 000 operations. There was considerably less wear on the aluminum alloy plugs than on the brass, as indicated in Fig. 10.

Example 2. Life Tests Using an ASTM Microcontact Tester (Fig. 11). Effect of voltage on contact resistance and contact area for 100 000 operations, using 0.38-mm (0.015-in.) diam 80Pt-20Ir contacts with a load of 100 mA, is shown in Fig. 11. Tests were made using an ASTM microcontact tester that allowed selection of any make-or-break contact force up to 49 mN. Contacts were protected from dust by a glass cover during testing. In these tests, the make force was 10 mN and the break force was 5 mN. Ten readings were taken at each interval, resulting in the spread shown in Fig. 11.

Low noninductive voltages had little effect on contact operation. There appeared to be more consistency in the 3 and 6 V readings, and they finished at lower resistance.

Example 3. Life Tests Using a Movable-Coil Relay. An accelerated life test was conducted on five contact materials mounted in a movable-coil type of relay. These contact materials were solid silver, solid 80Pt-20Ir, and gold, ruthenium and rhodium plated on the platinum-iridium alloy. The moving contacts were two 20-mm (0.80-in.)

Fig. 11 Effect of voltage on contact resistance and contact area for 80Pt-20Ir contacts 0.4 mm (0.015 in.) thick

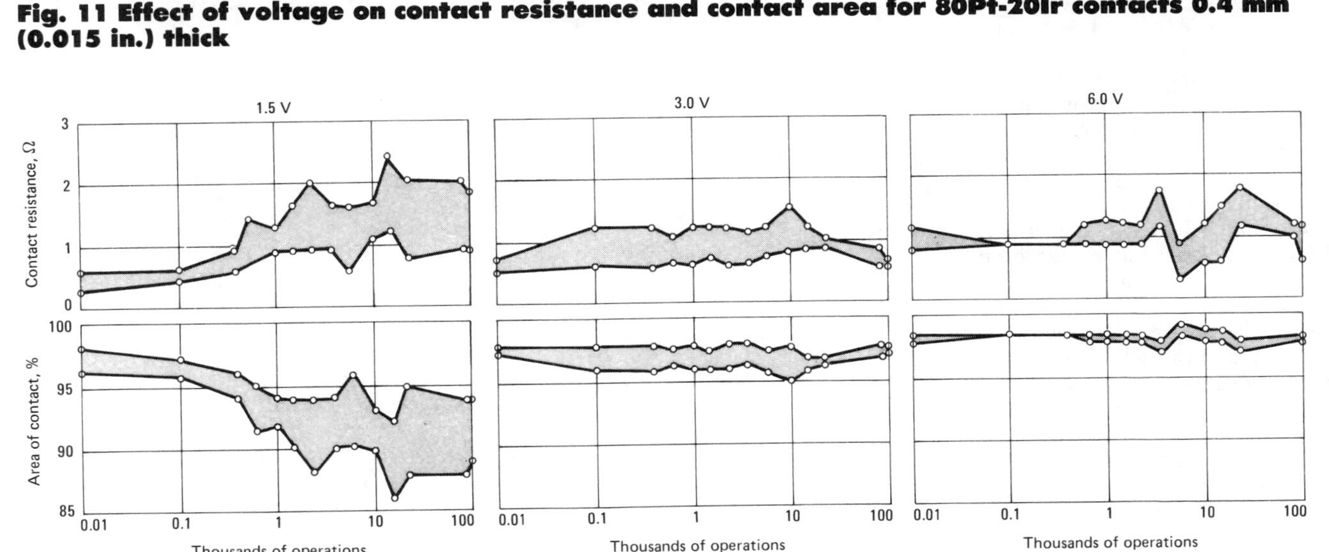

(Upper row of graphs) Effect of voltage on contact resistance at a load of 100 mA. Contacts were held in a clamp-type holder and tested in a closed jar. Shaded area represents spread for ten readings.
(Lower row of graphs) Effect of voltage on contact area for tests same as above. Initial contact area was 96 to 98%.

Fig. 12 Effect of time on maximum temperature of Ag-W contacts, compared with similar contacts having silver-enriched surfaces

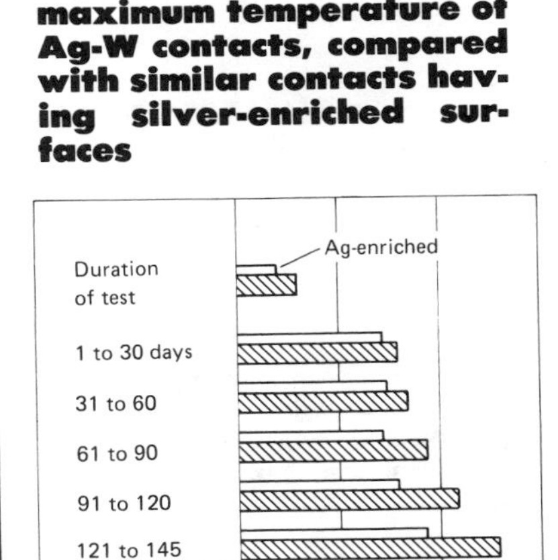

See text for description of test.

diameter flat disks mounted one on each side of a phosphor bronze strip. The stationary contacts were pointed wire having a 0.075-mm (0.003-in.) radius at the point, mounted in adjusting screws, one on each side of the moving contact. The contacts were operated with a force of about 980 μN on one side and no mechanical load on the other side. The electrical load was purely resistive: 1 mA at 9 V dc. Contact resistances were measured using an ohmmeter that operated on 1.5 V, 30 mA, and a force of 29 μN. Results of this test support these conclusions:

- If sticking were not a problem, silver would be the best material, having low contact resistance up to 12-million operations. However, silver began to stick after 800 000 operations.
- Gold-plated platinum-iridium had low contact resistance, and there was no sticking until about 3.5-million operations.
- The three platinum-family metals were similar, all of them having higher contact resistance than silver or gold from the start. However, platinum-iridium did not increase in resistance as much as the other two, and finished with the lowest contact resistance of the three platinum metals.

An examination of the contacts at the end of the test revealed that there had been no arcing and that the load was not sufficient to erode the contacts. Thus, failure (high contact resistance) occurred solely because of wear.

Although silver and gold were superior for applications involving frequent make-and-break, platinum-iridium would still be chosen for use where the relay would be idle for long periods of time, as in a burglar alarm.

Example 4. Life Tests in Circuit Breakers (Fig. 12). Ag-W contacts and the same kind of contacts having silver-enriched surfaces were tested for about five months in circuit breakers to determine how the average maximum temperature of the contacts changed during this simulated service. The current through the contact was 20 A at 120 V, 60 Hz, ac. The breakers were mounted in an enclosure where the temperature was 40 °C (105 °F); outside the enclosure it was 25 °C (77 °F). The breakers were operated in the closed position for 8½ h per day from Monday through Friday but remained open overnight and on weekends (one make and one break per day). The temperature values shown in Fig. 12 are averages of six readings.

The maximum temperatures for the Ag-W contacts changed almost linearly from 78 to 127 °C (172 to 260 °F) during the period. For the first three months, there was no appreciable change for the silver-enriched contacts. The temperature increased significantly during the fourth and fifth months, reaching about 95 °C (203 °F) at the end of the test. This is 23 °C (42 °F) higher than for the first period, but 32 °C (57 °F) lower than the temperature developed in the same period for the contacts that were not silver enriched.

Effect of Atmosphere. Life tests involving butt contacts have been made for several materials in atmospheres of helium, hydrogen and air under different operating conditions. Extensive data are presented in Fig. 13. The contacts were subjected to five operations per second, the closing being made magnetically to a closed force of 195 mN; the opening was accomplished by a spring to a gap of 0.8 mm (1/32 in.).

The criterion of failure was a weld or an open circuit, whichever developed first. If failure did not occur in 3.88-million operations, the test was discontinued.

Not only can the effect of the different atmospheres be determined for any one of the five conditions used, but comparisons can be made of direct and alternating current, variations in current, and variations in frequency. The current conditions were as follows:

Volts	Amperes	Frequency, Hz
28 dc.........	1	...
28 dc.........	8	...
115 ac.........	5	60
115 ac.........	5	400
115 ac.........	30	400

For the direct-current test, none of the ten materials gave long life in air when the current was 8 A. Six materials lasted for the full 3.88-million operations in air when the current was 1 A: fine silver, tungsten, molybdenum, 90Ag-10CdO, 50Ag-50W and 89Pt-11Ru. Tungsten was the only material to last the full test period when helium was used as the test medium and the direct current was 8 A. Copper, molybdenum and tungsten lasted the full test period when hydrogen was the test medium and the direct current was 8 A.

The alloy containing 97Ag-3Pt had long life when air and alternating current at 5 A were used, regardless of frequency. The lives of 77Ag-22.4Cd-0.6Ni, 97Ag-3Pt and 90Ag-10CdO, in a hydrogen atmosphere, were increased considerably by increasing the frequency from 60 to 400 Hz. In general, the life of a material was decreased by increasing the current. Fine silver showed a definite superiority over all other materials at 30 A. Copper was a fairly strong rival at 30 A, except in air.

Fig. 13 Effect of atmosphere and current characteristics on life of contact materials

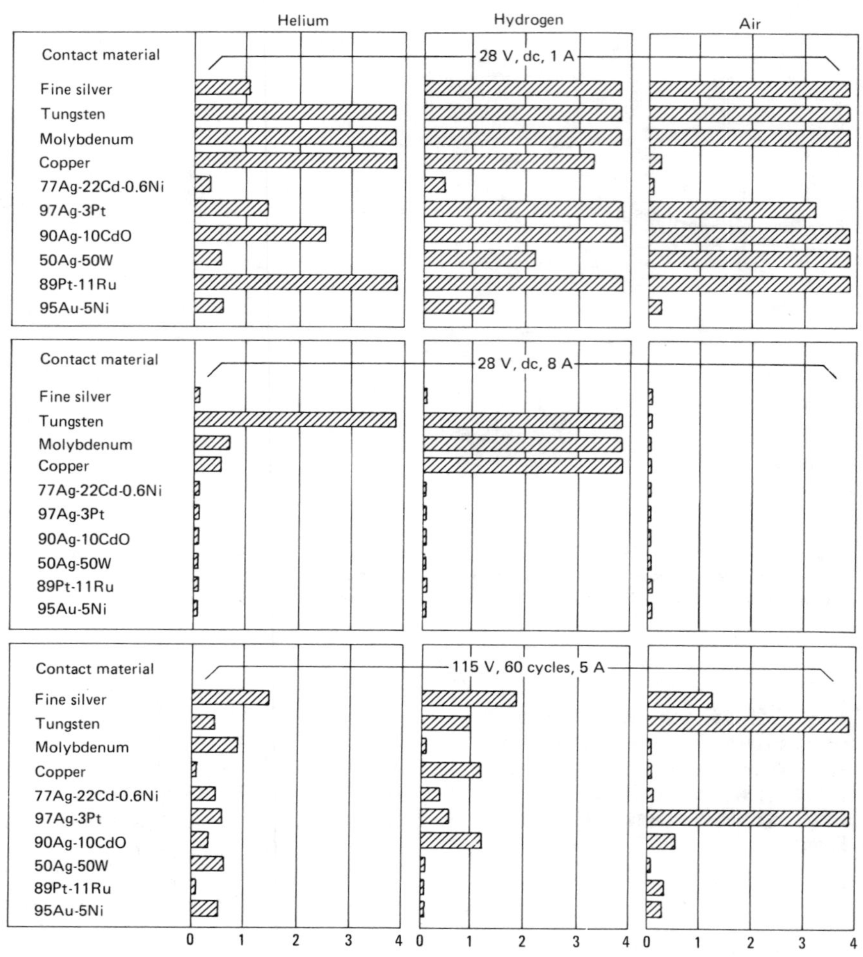

Tests were performed with a purely resistive load and a rate of five operations per second. Action was entirely butting, the contact being closed magnetically under a force of 20 g and opened by a spring to 0.8 mm (1/32 in.) width of gap. Welding or open circuit was the criterion of failure. Tests were discontinued after 3.88 million operations.

Erosion and Wear. Three simulated service tests, the results of which are shown in Fig. 14 and 15, are related to erosion and wear. The following contact materials are dealt with in Fig. 14: fine silver, 90Ag-10Cu, 77Ag-21Cu-2Ni, 77Ag-22.7Cd-0.3Ni, palladium, 35Ag-65W and 27Ag-73W.

Alternating current at 115 V and 60 Hz was employed, with the test lasting 100 000 operations at the rate of 60 per minute. The contacts were butt-type with closing and opening speeds of 38 mm (1½ in.) per second. The closing force was 980 mN, and the opening force 735 mN.

The number of contact welds was far less for fine silver and for the silver-tungsten sintered products than for palladium. The first two are about the same, but palladium can carry only about 40% as much current through the contact for the same number of contact welds.

The volume loss for fine silver at 25 A is about the same as that of palladium at 12 A. The volume loss for 27Ag-73W and 35Ag-65W is about 100 times that of fine silver at the same amperage.

In Fig. 15, comparison is made of copper, tungsten, 40Cu-60W, and 32Cu-68W, when tested for loss in volume for 30 operations in 10C transformer oil at 240 V, 60 Hz, 1400 A. The volume loss for 40Cu-60W was about two thirds

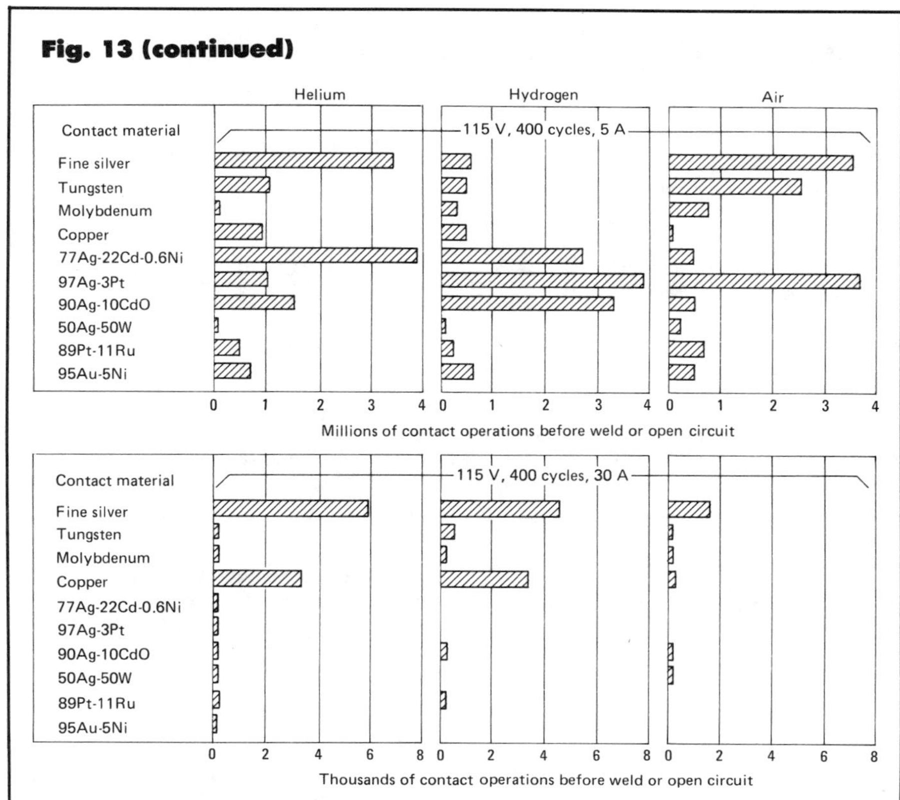

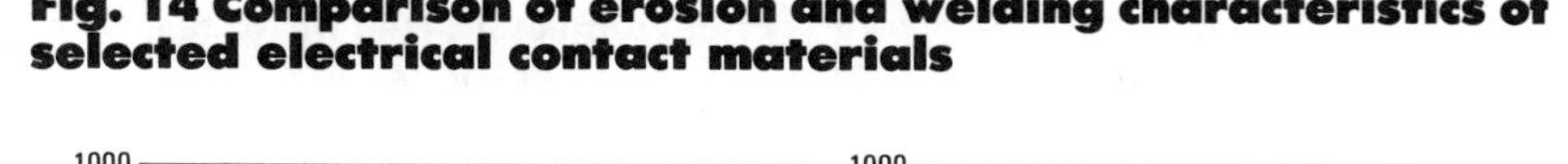

Fig. 14 Comparison of erosion and welding characteristics of selected electrical contact materials

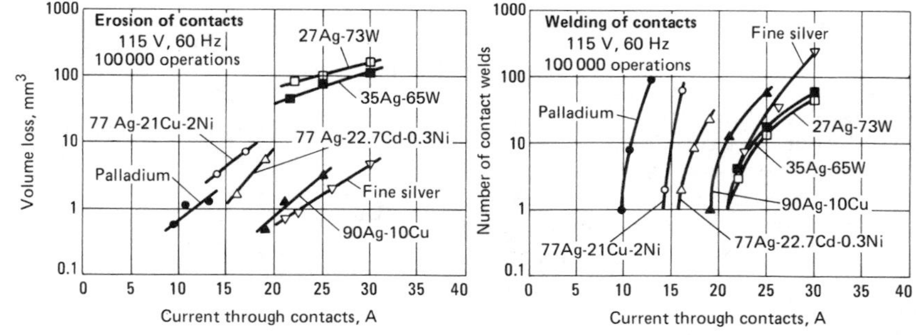

All contacts were butting, with a closing force of 980 mN, an opening force of 735 mN, and closing and opening speeds of 38 mm/s (1½ in./s). Contacts operated at the rate of 60 per minute.

that for 32Cu-68W, and both were far below that for tungsten or copper.

In the second test, silver, Ag-CdO and silver-tungsten were used to determine the number of operations required for failure in short circuit in a circuit breaker operating at 12 V dc, 7.5 A, interrupting a 300-A short circuit. A substantial increase in life is realized with a proper polarized combination: Ag-W positive and Ag-Fe negative.

Life of Polarized Contacts. Some information on life of polarized contacts has already been given in connection with erosion and wear (Fig. 15). It was shown that the polarized combination of 20Ag-80W or 35Ag-65W positive and 90Ag-10Fe negative, operating at 12 V dc, 7.5 A, with short-circuit interruption of 300 A, had a longer life than a similar combination where the tungsten content of the sintered contact was as low as 50%.

Test results are shown in Fig. 16 for a device operating with various contact materials in a dc circuit, operating at 12 V, 1.04 A, where polarized contact materials were used, with different materials as positive and negative terminals. The effect of the polarity of the contact materials is clearly demonstrated. Instead of plotting the number of operations obtained with each polarized contact combination, the time required to complete 100 cycles is used. Because sustained arcing affects the temperature rise of the contacts, and also the bimetal element used to make and break the circuit, the time required to complete 100 cycles is a good indication of the effectiveness of different contact combinations, as well as the effect of polarity on a given combination.

Of the materials and combinations tested, 90Pd-10Ru positive and 90Pt-10Ir negative gave the shortest time (12 min) for the completion of 100 operations; 90Pt-10Ir positive and 72Pd-26Ag-2Ni negative required the longest time (62 min). It takes less time for 100 cycles when the palladium alloy is positive and the platinum alloy is negative than for the reversed polarity. The time for 100 cycles for the combination 90Pd-10Ru positive and 90Pt-10Ir negative is about one third that for the same combination with reversed polarity.

Figure 17 shows that polarized contacts, using 50Ag-50WC at both terminals for 24 V dc and 1500 A, have considerably longer life than some other combinations of sintered silver products. Results are also given in this same figure for automotive voltage regulators for 15 V dc, 1.0 A and 15 V dc, 1.8 A. Tungsten-tungsten had a very long life of about 70-million operations when the current was 1.0 A. However, when the current was increased to 1.8 A, the life was reduced to 3-million operations. Positive 90Ag-10Ni and 97Ag-3Pd negative was the best combination tested for the higher amperage, but the poorest of all combinations for the lower amperage.

Fig. 15 Contact materials compared for erosion and wear characteristics

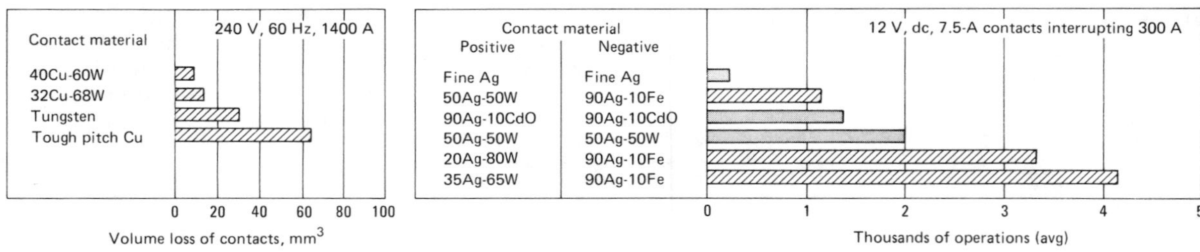

(Left) Four contact materials compared for volume loss after 30 operations in 10 c transformer oil at a power factor of 70%, maximum arc time of 1/2 cycle, one operation per minute, using a closing force of 13 N (3 lb) and an average opening speed of 2.4 m/s (8 ft/s). (Right) Life of polarized contact pairs and nonpolarized pairs based on the number of short circuits observed.

Fig. 16 Effect of composition and polarity on life of some platinum and palladium alloy contacts actuated by a bimetal element

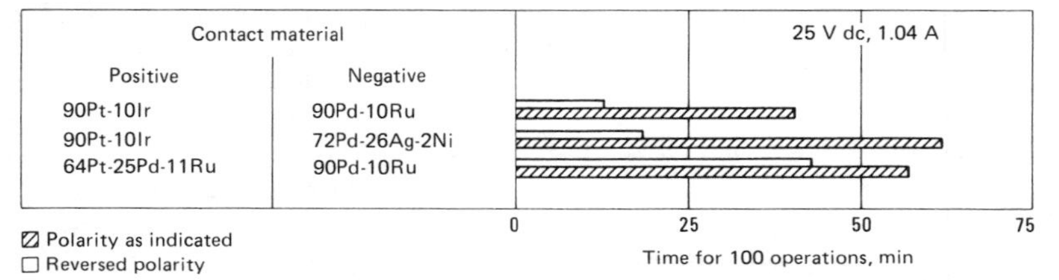

Sustained arcing affects the bimetal, thus indicating effectiveness of contact combinations by time for 100 operations.

Recommended Contact Materials

Fixed and stationary contacts for operations at low frequency are made of less expensive materials such as copper metals and aluminum metals. The ordinary two-prong domestic appliance plug is this type in its simplest form. Plug connectors are manually connected and disconnected. There is little trouble from arcing, pitting and welding because the current is controlled at some other contact where it will be interrupted before the plug is engaged or removed. The life of the spring is the life of the contact; it is the spring that usually fails. Therefore, spring bronze is used for plugs frequently connected and disconnected that do not need the highest conductivity. Where higher conductivity is needed silver on copper is better.

For bolted connectors, low contact resistance is important. Silver-coated copper is ideal in a noncorrosive environment. For wet-cell battery terminals, where corrosion is severe, lead against lead is used. Many combinations of metals are used for bolted connectors, with factors such as temperature, corrosivity, conductivity and cost affecting selection.

Power Circuits. (Tables 10, 11 and 12) Materials selected for contacts in power circuits usually have high electrical and thermal conductivity, and high resistance to arc erosion and to welding or sticking. Because these contacts are relatively large, precious metals like platinum are seldom used except in extremely corrosive atmospheres. Bare copper, copper faced or plated with silver, and aluminum faced or plated with silver are commonly used for stationary contacts carrying high current. Carbon, occasionally mixed with copper or silver in a sin-

Table 10 Recommended materials for fixed or stationary contacts for power circuits

Materials are listed in order of decreasing preference

Alloy	Advantages(a)
Plug connectors (1 to 100 A) 1 to 10 000 operations	
Brass	(b, c)
Bronze	(c, d, e)
Sn on Cu	(f, g, h)
Plug connectors (100 to 100 000 A) 1 to 10 000 operations	
Cu	(b, f)
Ag	(f, g, h, j)
Ag on Cu	(b, f, g, h, j)
Ag on Al	(b, g, h, j)
Cr-Cu	(d, e)
Blade connectors (10 to 100 000 A) 10 to 10 000 operations	
Cu	(b, f)
Ag on Cu	(f, g, h)
Bolted connectors (100 to 1000 A) 1 to 100 operations	
Brass	(b, c)
Bronze	(e, k)
Sn on Cu	(f, g, h)
Bolted connectors (1000 to 1 000 000 A) 1 to 100 operations	
Cu	(b, f)
Ag on Cu	(f, g, h)
Sn on Cu	(f, g)
Ag on Al	(b, g, h)

(a) With fixed or stationary contacts, failure is ultimately caused by deterioration of the contact surface. (b) Low-cost material. (c) Material easy to fabricate. (d) Wear resistance. (e) Material provides lower contact resistance. (f) Electrical conductivity. (g) Surface oxidation resistance. (h) Material provides lower contact resistance. (j) Material permits use of a low contact force. (k) Material has higher strength.

Fig. 17 Effect of composition, combination and electrical characteristics on contact life of various materials

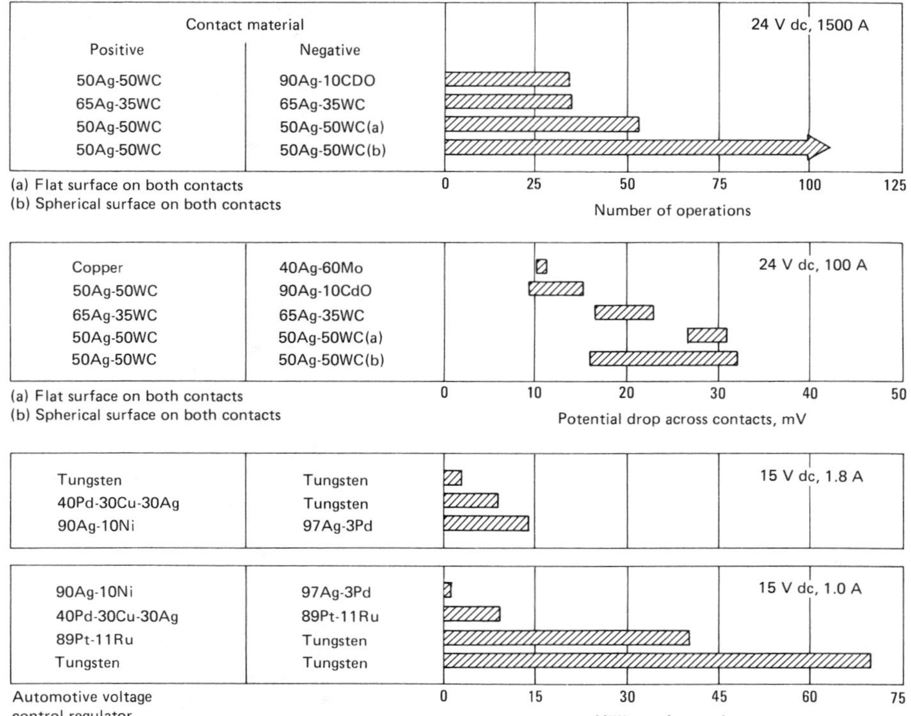

24 V dc, 1500 A: A comparison of several polarized contact material combinations for their susceptibility to welding when subjected to an overload of 1500 A at 24 V. The 50Ag-50WC combination with a spherical surface shows considerably longer life than some other combinations of silver sintered products; it did not weld in 100 operations. 24 V dc, 100 A: Shown is the scatter of contact potential after 200 000 operations for several samples of different alloy combinations. Under conditions of relatively high current, the contact force used in these tests was approximately 2.2 to 2.8 N (8 to 10 oz). 15 V dc, 1.8 A: A comparison of the life of three contact material combinations intended for use in automotive voltage regulators. At this level of amperage, tungsten with tungsten failed because of insulation resulting from oxide formation after 3-million operations. Palladium-copper-silver with tungsten failed as a result of sticking. Silver-nickel with silver-palladium failed from metal transfer. 15 V dc, 1.0 A: Life comparison for four contact material combinations intended for use in automotive voltage regulators. Silver-nickel with silver-palladium made the poorest showing and failed from sticking. Palladium-copper-silver with platinum-ruthenium, and platinum-ruthenium with tungsten, failed from oxide insulation. Tungsten with tungsten, which failed from oxide insulation at 1.8 A, survived about 70-million operations.

tered product, is commonly used for brushes against copper commutators or slip rings for sliding contacts. Occasionally, silver or silver-plated copper is used. Sintered products consisting of tungsten, tungsten carbide or molybdenum mixed with copper or silver are used for spark gaps, and for moving contacts that are required to interrupt high currents at high voltage with heavy arcing.

Liquid metals are seldom used in power circuits, but mercury and the sodium-potassium eutectic are used against solid contacts where a circuit must be completed between a stationary part and one involving an unusual type of motion, or where currents of more than 100 000 A must be "collected" from rotating equipment, as is accomplished with the liquid eutectic.

Brushes. A cursory examination of Table 12 shows that sliding power contacts have a wide range of application. Although the list is not complete, it covers machines with current ranges from 1 to 10 000 A. Even with this broad application range, there is one condition that must be satisfied in any successful sliding contact. Somewhere between the two surfaces that are moving relative to each other, there must be a film or region in shear that prevents seizing or welding of the clean surfaces. Trouble ensues when the film disappears in plating generators because of low humidity, or when the film can no longer be formed because of lack of oxygen and moisture, as in aircraft flying at high altitudes. Under normal conditions, this film is inherent.

Lubricants added to the brush material also help in preventing cold welds. In certain power stations operating at room temperature in the wintertime, the relative humidity must be kept above 25% to prevent rapid wear and dusting of brushes. In aircraft applications, adjuvants are helpful in maintaining this film. The identity of such films has not been clearly established. Graphite is most effective, but it requires the presence of considerable moisture and oxygen.

There are no simple rules for selecting materials for brushes, but the usual practice is to start with a material in the correct resistance range. If this is unsuitable, the reasons for failure, which might be poor commutation, overfilming, high wear, or arcing, must be considered and brush properties improved.

For many functions, it is imperative to select contact materials that do not contribute unduly to high-frequency radio noise, as may be generated by sliding contacts and brushes. Such noise may be minimized by having low or uniform contact resistance between brushes and rings, a condition that can be effected through the use of graphite, silver-graphite, gold-graphite or other noble metal brushes against alloys of silver, gold or platinum, or against graphite. A good expedient is to use two or more brushes against a single ring to have parallel circuits, the ultimate being fiber brushes, molybdenum, or some other metal that gives the effect of many separately supported parallel contacts. In most applications, silver-graphite brushes containing from 5 to 50% graphite are used against silver rings. At high altitude, silver-graphite must have a protecting adjuvant to prevent the rapid wear that may ensue in dry rarefied air.

Circuit Breakers. Contact materials recommended for use in air in cir-

Table 11 Recommended materials for sliding contacts for power circuits
Materials are listed in order of decreasing preference

Alloy	Advantages	Cause of failure
Power brushes (10 to 10 000 A) continuous slide		
Electrographite	(a, b, c)	(d)
Carbon graphite	(a, b, c)	(d)
Fractional horsepower brushes (1 to 10 A) continuous slide		
Electrographite	(a, b, c)	(e)
Carbon graphite	(a, b, c)	(e)
Resin bonded	(a, b, c)	(e)
Aviation brushes (10 to 1000 A) continuous slide		
Electrographite	(a, b, c, f)	(g)
Carbon graphite plus BaF, CdI, MoS_2 for altitude performance	(a, b, c, f)	(g)
Automotive starter brushes (10 to 1000 A) continuous slide		
Copper graphite	(a, b, h, j)	(k)
Automotive generator brushes (10 to 100 A) continuous slide		
Electrographite	(a, b, c)	(m)
Carbon graphite	(a, b, c)	(m)
Automotive auxiliary brushes (1 to 10 A) continuous slide		
Copper graphite	(a, b, h, j)	(n)
Plating generator brushes (100 to 10 000 A) continuous slide		
Copper graphite	(a, b, h, j)	(p)
Alternating-current and slip ring brushes (1 to 1000 A) continuous slide		
Electrographite	(a, b)	(q)
Carbon graphite	(a, b)	(q)
Copper graphite	(a, b, h, j)	(q)

(continued)

(a) Low-cost material. (b) Wear resistance. (c) High contact resistance for commutation. (d) Wear, poor commutation. (e) Wear, arcing. (f) Suitable for operation in dry air or at altitude. (g) Wear, dusting, poor commutation. (h) Electrical conductivity. (j) Lower contact resistance. (k) Wear, high resistance. (m) Wear, arcing, poor performance. (n) Wear, poor performance. (p) Wear, dusting, grooving of commutators. (q) Wear, sparking, grooving. (r) Wear, arc erosion. (s) Surface oxidation resistance. (t) Higher strength. (u) Higher annealing temperature. (v) Ease of fabrication. (w) Wear. (x) Arc-erosion resistance. (y) Liquids practically never, solids by arc erosion. (z) Less sticking and welding tendency.

cuit breakers having a maximum current rating of 800 A and a maximum voltage rating of 600 V are as follows:

Stationary contact	Moving contact
Silver-tungsten	Silver-tungsten
Silver-WC	Silver-WC
Silver-molybdenum	Silver-WC
Silver-nickel	Silver-molybdenum
Fine silver	Silver-tungsten
Silver-molybdenum	Silver-tungsten

Vibrators have severe requirements as contact materials because of the high localized temperature caused by the rapidly repeated making and breaking of contact. As a result, only a small amount of current at low voltage can be handled if the operation is continuous and reasonable life is expected. Contact materials recommended for vibrators are listed in the first group of Table 13.

Automobile horns usually operate for only a short length of time at each operation. They carry 10 to 20 A with a closing force of 6 to 7 lb. Tungsten is almost universally used as the contact material because of the combination of good conductivity and high melting point. The high pressure is needed to break down an oxide film that develops.

Tungsten is also used for contacts in radio vibrators that operate in a protected environment with low current and pressure for comparatively long periods of time. Silver and silver alloy points are used in buzzers that operate with very low current for short periods of time. Platinum or gold is used in critical applications, such as fire alarms, where reliability is essential.

Voltage Regulators. A wide range of contact materials is used in automotive voltage regulators because even minor differences in alternator field circuits or in suppression devices can have a marked effect on contact life. Generally, regulator contacts are polarized, the substance with the higher melting point being specified for the side that ordinarily loses material. A most successful contact combination is a tungsten negative contact against platinum-ruthenium positive contact. Other commonly used materials are tungsten against tungsten at higher voltage with good environmental protection; silver alloys against palladium alloys where cost is important; and gold against platinum-iridium where high current and low induced voltage prevail. In these combinations, the material listed first is used for the positive contact.

Most regulator contacts fail through development of an open circuit because of oxides or other contaminants that come from the contact arc, hydrocarbon vapors or dust particles. Some failures for the same equipment at a different level of field current are caused by welding. Hundreds of different contact material combinations have been used in voltage regulators.

Switches. Two classes of snap-action switches must be considered. The manual type is found in the walls of homes and offices and on electric ranges, ovens and other similar appliances. Lower-cost contact materials are used for manual switches because substantial slide or wipe of contact surfaces and high contact force can be tolerated.

The other type is a precision snap-action switch that may be operated electromagnetically or mechanically by precision rotating or sliding cams with little movement or low operating forces. This type of switch is used on equipment such as machine tools, precision controls and thermostats.

Bronze is unsuitable for precision snap-action switches because it corrodes too readily. In circuits that are sensitive to resistance and in which the voltage is low (1 V or less), either silver, gold or platinum alloys are used. In more than 80% of heat thermostats, mercury-molybdenum or mercury-platinum switches are used; the remainder are of materials such as fine silver, 90Ag-10CdO, 90Pt-10Ir, tung-

Table 11 (continued)

Alloy	Advantages	Cause of failure
Commutators (10 to 10 000 A) continuous slide		
Cu	(a, h)	(r)
Ag on Cu	(h, j, s)	(r)
Silver-bearing Cu	(b, h, t)	(r)
Zirconium-bearing Cu	(b, h, t, u)	(r)
Slip rings (10 to 10 000 A) continuous slide		
Stainless steel	(b, t)	(r)
Silver copper	(h, j, s)	(r)
Bronze	(a, b, v)	(r)
Tool steel	(a, b, t)	(r)
Wire against slider or trolley wheel (10 to 1000 A) continuous slide or roll		
Bronze wire against bronze wheels	(a, b)	(w)
Ag-Cu against Ag-W	(b, h, j, s, x)	(w)
Cd-Cu wire against Cu-C sliders	(h, s)	(w)
Cd-Sn-Cu wire against hard Cu	(h, r)	(w)
Steel against cast iron	(a, b)	(w)
Liquid to collector assembly (10 to 10 000 A) continuous dip		
Mo against Hg	(b, h, x)	(y)
Steel against Hg	(a, b, t)	(y)
Bearings and swivels (10 to 10 000 A) intermittent slide		
Brass	(a)	(r)
Steel	(s)	(r)
Cu	(h)	(r)
Ag-graphite	(b, z)	(r)
Ag	(h, j, s)	(r)
Bronze	(b, t)	(r)
Graphite	(x)	(r)

(a) Low-cost material. (b) Wear resistance. (c) High contact resistance for commutation. (d) Wear, poor commutation. (e) Wear, arcing. (f) Suitable for operation in dry air or at altitude. (g) Wear, dusting, poor commutation. (h) Electrical conductivity. (j) Lower contact resistance. (k) Wear, high resistance. (m) Wear, arcing, poor performance. (n) Wear, poor performance. (p) Wear, dusting, grooving of commutators. (q) Wear, sparking, grooving. (r) Wear, arc erosion. (s) Surface oxidation resistance. (t) Higher strength. (u) Higher annealing temperature. (v) Ease of fabrication. (w) Wear. (x) Arc-erosion resistance. (y) Liquids practically never, solids by arc erosion. (z) Less sticking and welding tendency.

sten, 35Pd-30Ag-14Cu-10Pt-10Au-1Zn, 90Ag-10Fe and 69Au-25Ag-6Pt.

Mercury switches are well suited for use in thermostats because they promise absolute reliability in making and breaking of contact. However, they must be kept in the desired position.

Fine silver is used extensively if the voltage is as high as 20 V and the current is 1 A or higher. Where a current of 10 to 30 A is controlled and the voltage is higher than 110 V, Ag-CdO is used, especially if the making and breaking of electrical contact is slow.

The precious metals, particularly gold alloys, are recommended where low voltage and so-called dry circuits are involved. It is good practice to have multiple contacts when there is a need for high reliability in making electrical contact.

It has been demonstrated experimentally that reliability can be increased as much as 2700% by using two contacts rather than one, and with the pressure force on each contact being only half of the force on the single contact.

Telecommunications Equipment. In electrical contacts for use in telecommunications equipment, the contact resistance must be low enough to ensure satisfactory circuit operation and to prevent excessive transmission losses. The contact resistance also must be constant to avoid noise modulation in the transmitted signal. Contact separation should not become unreliable because of welding, snagging or excessive surface roughness due to arcing. The contacts should not become excessively worn due to mechanical action or eroded due to arcing.

The types of contacts most frequently used in telecommunications equipment are

- Connector contacts that close with considerable slide and high force but are not required to be changed often—for example, a multiwire cable plug and its socket.
- Sliding contacts that operate with considerable slide but with low force to make operation easier and to reduce mechanical wear from frequent operation. The contact in a rotary switch is an example.
- Butting contacts that generally close with light force and without much slide. The low force and slide minimize power requirements and wear, thereby permitting operation at a high rate. Telephone relays feature this kind of contact.

Table 14 lists materials recommended for telecommunications equipment. The most frequently used metals are platinum, palladium, iridium, ruthenium, gold, silver, copper, nickel and tungsten. The precious metals, because of their tarnish resistance, provide greater reliability of closure and sometimes greater resistance to arc erosion. The base metals, where they can be used, are more economical and provide greater freedom from welding and snagging of pits and buildups arising from arcing.

Microcontacts are considered to be those contacts having a closure force of 49 mN or less and carrying currents measured in milliamperes at voltages of 120 V or less. Materials recommended for microcontacts are given in Table 15.

Silver metals are generally used as microcontacts that have more than 10 mN force and preferably some wipe. Other metals plated with silver are also used, as in magnetically operated microcontacts. In such devices, the fixed contact is a magnet that must be capable of being plated with a metal that can carry the current. The moving contact is iron or a magnetically soft alloy, which also must be plated. These contacts are used in meter-movement re-

Table 12 Recommended materials for make-and-break contacts for power circuits

Alloy	Advantages	Cause of failure
Tap changers (10 to 100 A) no-load make-and-break, 100 to 100 000 V, 10 000 max operations		
Bronze wiper against brass	(a)	(b)
Cu-Cd against Cu-Cd	(c, d, e)	(b)
Brass or bronze against brass or bronze	(a)	(b)
Tap changers (100 to 100 000 A) no-load make-and-break 100 to 100 000 V, 10 000 max operations		
Cu	(f)	(b)
Cu-Cd	(f, h, j)	(b)
Ag on Cu	(c, h, j)	(b)
Ag-Ni	(f, h)	(b)
Ag-Cu-Ni	(f, h)	(b)
Cr-Cu	(c, f, j)	(b)
Tap changers (10 to 10 000 A) load make-and-break, 100 to 100 000 V, 10 000 max operations		
Cu-W	(e, k, m)	(n)
Cu-WC	(e, k, m)	(n)
Ag-graphite	(c, d, g, p)	(n)
Ag-W	(c, d, e, g, k, m)	(n)
Ag-Ni	(a, j, h, f)	(n)
Contactors and motor starters (10 to 10 000 A) alternating current, 10 to 100 V, 10 000 000 max operations		
Ag	(c, d, g, j, p)	(n)
Ag-CdO	(e, f, h, k, m)	(n)
Ag-Cd	(e, g, j)	(n)
Cu	(a, c, j)	(n)
Cu-W, Ag-W, Ag-Ni	(e, f, h, k, m)	(n)
Contactors and motor starters (10 to 10 000 A) direct current, 1 to 1000 V, 10 000 000 max operations		
Ag-CdO	(e, f, h, k, m)	(n)
Ag-Cd	(e, j, k)	(n)
Ag	(c, d, g, j)	(n)
Ag-Ni	(e, h)	(n)
Ag-W or Ag-WC	(e, f, h, k, m)	(n)
Air circuit breakers (10 to 100 A) current-carrying and arcing, 10 to 1000 V, 1 000 000 max operations		
Ag-Cu-Ni	(a, c)	(q)
Ag-Ni	(e, h, k)	(q)
Ag-graphite	(c, d, g)	(q)
Ag-W	(e, f, h, k, m)	(q)
Ag-CdO	(c, d, e, g, k, m, p)	(q)
Air circuit breakers (100 to 1000 A) current-carrying and arcing, 10 to 10 000 V, 100 000 max operations		
Ag-W	(e, h, k, m)	(q)
Ag-CdO	(c, d, e, j, g, p)	(q)
Ag-WC	(c, g, h, k)	(q)

(continued)

(a) Low-cost material. (b) Surface deterioration, wear. (c) Electrical conductivity. (d) Lower contact resistance. (e) Less sticking and welding tendency. (f) Higher strength. (g) Surface oxidation resistance. (h) Wear resistance. (j) Ease of fabrication. (k) Arc-erosion resistance. (m) Resistance to transfer and pitting. (n) Sticking, arc erosion. (p) Allows low contact force. (q) Arc-erosion, material transfer and pitting, or overheating.

lays that have an actuating current of as little as 2 mA. When the soft iron on the pointer is deflected so that it enters the field of the magnetic contact, magnetic attraction causes the contacts to close with a force of 18 to 40 mN, and with some impact and wipe. Silver is used in armature-type sensitive relays less frequently than palladium, which is better than silver in such low-force low-current applications. Palladium is used, for example, for brushes in micropotentiometers.

Platinum-iridium (usually 80Pt-20Ir) is the material most commonly used where the contact force is less than 10 mN. Contacts made from 80Pt-20Ir are used in meter-movement relays, both load-current-contact-aiding (LCCA) and sensitive types. In both types, the contacts must close with extremely low force. With the LCCA type, the load current through the contact also passes through an extra winding on the moving coil and thereby increases the closure force. Initial contact often is made with less than 10 μN of force. Although hardness in a contact material is an advantage, it can also be a disadvantage. For instance, 70Pt-30Ir and 65Pt-35Ir are too hard for most microcontact applications.

Gold has become more widely used in microcontacts since the development of sealed relays. It is a less potent catalyst than members of the platinum family and therefore does not collect as much polymer on contacting surfaces. Like silver, it is easily plated and is used in this form on magnetic contacts. Gold is also plated over palladium in sealed relays to decrease the polymer effect. In miniature slip rings and micropotentiometers, gold alloys are used as brushes, rings and wire.

Availability

Silver, gold, platinum, palladium and most of their ductile alloys are available as stamped contacts. Except for material from which contact disks are produced, a variety of stock sizes is not maintained because, in general, no two applications are identical. For disks, material in strip form, varying in thickness from 0.25 mm to 1 mm (0.010 to 0.040 in.) is available.

Except for tungsten and molybdenum, contact materials are ductile enough so that they can be produced in all contact forms. Tungsten and molybdenum and some of the P/M materials have lower ductility and are available

Table 12 (continued)

Alloy	Advantages	Cause of failure
Air circuit breakers (100 to 100 000 A) current-carrying only, 100 to 100 000 V, 10 000 max operations		
Cu	(a, c)	(q)
Ag on Cu	(c, d, g, p)	(q)
Ag-Ni	(e, h)	(q)
Ag-graphite	(c, d, g, h, p)	(q)
Air circuit breakers (1000 to 1 000 000 A) arcing tip only, 100 to 100 000 V, 10 000 max operations		
Cu-W	(a)	(q)
Ag-W, Ag-WC, Ag-Mo	(h, k, m)	(q)
Magnet steel	(a)	(q)
Oil circuit breakers (10 to 10 000 A) current-carrying and arcing, 1000 to 100 000 V, 100 000 max operations		
Cu-W	(a)	(q)
Ag-W	(d, e, g)	(q)
Ag-Mo	(c, g, p)	(q)
Oil circuit breakers (100 to 100 000 A) current-carrying only, 1000 to 1 000 000 V, 10 000 max operations		
Cu	(a, f)	(q)
Ag on Cu	(c, d, g)	(q)
Ag-Ni	(f, h)	(q)
Oil circuit breakers (100 to 1 000 000 A) arcing tip only, 1000 to 1 000 000 V, 10 000 max operations		
Cu-W	(a)	(q)
Cu-WC	(k)	(q)
Ag-W	(c, d, g)	(q)
Ag-Mo	(c, d, e, g)	(q)
Spark gaps (10 to 1 000 000 A) 100 to 1 000 000 V, 1 000 000 max operations		
W	(k)	(q)

(a) Low-cost material. (b) Surface deterioration, wear. (c) Electrical conductivity. (d) Lower contact resistance. (e) Less sticking and welding tendency. (f) Higher strength. (g) Surface oxidation resistance. (h) Wear resistance. (j) Ease of fabrication. (k) Arc-erosion resistance. (m) Resistance to transfer and pitting. (n) Sticking, arc erosion. (p) Allows low contact force. (q) Arc-erosion, material transfer and pitting, or overheating.

in fewer forms. Contact manufacturers have displayed exceptional ingenuity in producing contacts clad or surfaced with precious metal in appropriate areas. Such coated forms have led to substantial savings.

The commercially available forms of common electrical contact materials are listed in Table 16.

Silver, gold, platinum, palladium and nearly all the alloys of these metals, as well as tungsten, molybdenum and the various sintered products of silver and the refractory metals, can be used to produce steel-backed contacts. Steel-backed contacts have been made in the form of screws, rivets, or buttons for projection welding.

Powder metallurgy materials are available with final densities up to 99% of theoretical and with high-conductivity surfaces as inserts or overlays. Both P/M and wrought materials may be attached to appropriate carriers by brazing, welding or diffusion bonding even though they contain cadmium oxide. For percussion welding of Ag-CdO contacts, backing is not required. For resistance welding, a fine silver backing is needed.

P/M contacts are available with a silver matrix for air and oil applications, and with a copper matrix for oil applications only. The second phase may be tungsten, molybdenum, graphite, tungsten carbide, cadmium oxide or zinc oxide.

Materials with a high content of refractory metal (50% or more) are usually made by infiltration. In this process, the refractory metal powder is pressed into the desired size and shape with a controlled porosity. The compact is sintered at high temperature in a reducing atmosphere, and then molten silver or copper is infiltrated into the porous sintered compact.

Compositions of a lower refractory content are made by blending powders, pressing them to a desired size and shape, sintering the pressed compact at a high temperature, and re-pressing to size the parts and to increase the density of the compact.

Although parts usually are molded to final shape, they are sometimes finished by machining or grinding to obtain special shapes or unusually close tolerances. When only a few parts are needed, they may be machined from bars to save the cost of expensive dies for pressing a P/M compact.

Sintered materials are available in sizes ranging from rectangles or disks about 0.8 mm (1/32 in.) thick by 3 mm (1/8 in.) square or 3 mm in diameter to bars 200 mm (8 in.) long. Most are available in widths up to 75 mm (3 in.) and thicknesses up to 12.5 mm (1/2 in.). For some materials, these dimensions may be exceeded.

In small sizes (up to about 12.5 mm square), the higher refractory compositions are frequently made with serrated surfaces coated with excess silver or silver-base brazing alloy so that they can be attached easily to a backing by welding or by resistance brazing in a welding machine. Larger sizes are generally attached to backing by silver-alloy brazing. The materials are often pre-tinned with silver-base brazing alloy in a controlled atmosphere for good wetting of the contact material.

Many of the low-refractory contact materials are fabricated as disks, rectangles and special-contour facings that also are attached to backing by silver-alloy brazing. Most materials with 90% or more silver are ductile enough to allow fabrication as rivets, and to allow assembly by staking or spinning. Among silver-graphite materials, those containing more than 0.5% graphite cannot be satisfactorily cold headed into rivets.

Ag-CdO mixtures are available as round, rectangular or special-shape facings, and also as rivets.

Cost

Copper is the least expensive elemental material used as electrical contacts, followed by silver, palladium, gold and platinum in that order.

Sometimes, it is cheaper to use a steel-backed contact. Silver, gold, platinum, palladium and their alloys all have been used to produce steel-backed contacts in the form of screws, rivets, or buttons for projection welding.

Table 13 Recommended materials for light power and engineering contacts of current range of 0.1 to 30 A

Materials are listed in order of decreasing preference

Alloy	Advantages	Cause of failure	Applications
Electromagnetic vibrators (0.1 to 30 A)(a)			
W-0.5Mo	(b, c, d, e, f)	(g)	Automotive voltage regulators, bells, buzzers, horns, radio vibrators
Pd-Ag-Ni-W			
Pt-Ru-W			
Ag-Mo			
Au-Pt-Ir			
Thermomechanical thermostats (0.1 to 30 A)(h)			
Ag	(f, j, k, m, n, q)	(p)	Household heating and cooling, cooking, electric blankets
Ag-Pd	(e, k, m, n)		
Ag-Cd	(b, c)		
Hg-Pt	(k)		
Hg-Mo	(b, c, e, k)		
Ag-CdO	(b, c, e)		
Pt-Ir	(d, k, m, q)		
Au	(k, m, q)		
Manual or electromagnetic snap-action switches (0.1 to 30 A)(r)			
Bronze	(f, n, s)	(t)	Lighting, appliances, engineering equipment
Ag-CdO	(b, c, e)		
Ag	(f, j, k, m, n, q)		
Pt	(d, k, m, q)		
Au	(k, m, q)		
Electromagnetic relays (0.1 to 30 A)(u)			
Ag-CdO	(b, c, e)	(v)	Control systems, engineering equipment, lighting, appliances
Ag	(j, k, m, q)		
Ag-Cu	(b, d, f)		
Ag-Cu-Ni	(d, f)		
Pd	(k, m, q)		

(a) Fast action with force of 0.137 to 4.45 N (1/2 to 16 oz). Contact, butting plus slight wipe. 1 to 110 V; 1 billion max operations. (b) Less sticking and welding tendency. (c) Arc-erosion resistance. (d) Wear resistance. (e) Resistance to transfer and pitting. (f) Low-cost material. (g) Wear, welding transfer, and arcing (open). (h) Slow or fast action with 2.7 N (10 oz) max force. Contact, butting to considerable wipe. 300 V max; 100 000 max operations. (j) Electrical conductivity. (k) Lower contact resistance. (m) Surface oxidation resistance. (n) Ease of fabrication. (p) Arcing, welding and contamination by dust. (q) Allows low contact force. (r) Fast action with 0.2 to 4.45 N (1 to 16 oz) force. Contact, butting to wipe. 300 V max; 1 million max operations. (s) Spring properties. (t) Arcing (open), and welding. (u) Fast action with 0.2 to 4.45 N (1 to 16 oz) force. Contact, butting to considerable wipe. 300 V max; 1 million max operations. (v) Wear (open), some welding (closed).

Table 14 Recommended contact materials for telecommunication equipment

Voltage, V	Current, A	Closed force, g	Make and break with v and A	Expected life, max no. of operations	Materials used(a)	Advantages	Usual cause of failure
				Connectors			
Plug and jack							
...	...	500	No	10^5	Brass against Ni-brass springs	(b, c, d)	(e)
					Ni-plated brass against Ag-plated brass springs	(f, g)	(e)
Vacuum tube plug and socket							
...	...	High	No	10^2	Ni-plated brass against phosphor bronze springs	(b, c, d)	...
...	...	High	No	10^2	Ni-plated brass against Ag-plated phosphor bronze springs	(f, g)	...
Multicontact connector							
...	...	150	No	10^3	Brass against phosphor bronze springs	(b, c, d)	(h)
...	...	150	No	10^3	Au-plated phosphor bronze against phosphor bronze springs	(f, g, j)	(h)

(continued)

(a) Materials are listed in the order of decreasing preference. (b) Low-cost material. (c) Ease of fabrication. (d) Spring properties. (e) Wear and surface deterioration. (f) Surface oxidation resistance. (g) Lower contact resistance. (h) Surface deterioration. (j) Wear resistance. (k) Surface deterioration, resistance, and noise. (m) Resistance and noise. (n) Wear, resistance, and noise. (p) Allows low contact force. (q) Electrical conductivity. (r) Dust, polymer, carbonaceous deposits, and erosion sticking. (s) Reduced polymer formation. (t) Material provides improved resistance to metal transfer and pitting.

Table 14 (continued)

Voltage, V	Current, A	Closed force, g	Make and break with v and A	Expected life, max no. of operations	Materials used(a)	Advantages	Usual cause of failure
				Connectors			
Multicontact connector (continued)							
...	...	150	No	10^3	Au-plated copper against phosphor bronze springs	(c, f, g)	(h)
				Sliding contacts			
Electromechanical rotary switches (telephone type)							
50 max	0.1 max	20 to 65	Not usually	10^6	Phosphor bronze wiper against brass or phosphor bronze terminal	(b, d)	(k)
50 max	0.1 max	20 to 65	Not usually	10^6	69Au-25Ag-6Pt overlay on phosphor bronze terminal	(f, g, j)	(k)
50 max	0.1 max	20 to 65	Not usually	10^6	69Au-25Ag-6Pt overlay on phosphor bronze or brass for both wiper and terminal	(f, g, j)	(k)
Manual rotary switches							
50 max	0.1 max	20 to 65	Not usually	10^6	Ag-plated brass	(b)	(m)
50 max	0.1 max	20 to 65	Not usually	10^6	90Ag-10Cu rotor blades against 90Ag-5Cu-5Zn or against 90Ag-5Cu-5Cd clips	(j)	(m)
50 max	0.1 max	20 to 65	Not usually	10^6	69Au-25Ag-6Pt overlay on brass for both blades and clips	(f, g, j)	(m)
Slip rings and brushes							
25 max	0.01 max	25	No	...	Ag	(b)	(n)
25 max	0.01 max	25	No	...	90Ag-10Cu	(j)	(n)
25 max	0.01 max	25	No	...	70Au-30Ag	(f, g, j)	(n)
25 max	0.01 max	25	No	...	Pd-Ag-Cu	(f, g, p)	(n)
25 max	0.01 max	25	No	...	Coin silver	(b, c, f, q)	(n)
25 max	0.01 max	25	No	...	Au-Ag-Pt	(f, g, p)	(n)
25 max	0.01 max	25	No	...	Au-Pt	(f, g, p, q)	(n)
				Butting contacts			
Sensitive relays							
50 max	0 to 1	1 to 5	No	10^9	Pd	(b)	(r)
50 max	0 to 1	1 to 5	No	10^9	69Au-25Ag-6Pt	(s)	(r)
50 max	0.4 max	1 to 5	Yes	10^7	Pd	(b)	(r)
50 max	0.4 max	1 to 5	Yes	10^7	69Au-25Ag-6Pt	(s)	(r)
50 max	0.4 to 1	1 to 5	Yes	10^7	Pd	(b)	(r)
50 max	0.4 to 1	1 to 5	Yes	10^7	Pt	(f, g)	(r)
Telegraph relays							
± 130	0.060	1 to 5	Yes	10^8	W against 60Pd-40Cu	(t)	Sticky
General-purpose relays							
50 max	1 max	5 to 50	No	10^9	Pd	(b)	(r)
50 max	1 max	5 to 50	No	10^9	69Au-25Ag-6Pt	(s)	(r)
50 max	0.4 max	5 to 50	Yes	10^8	Pd	(b)	(r)
50 max	0.4 max	5 to 50	Yes	10^8	69Au-25Ag-6Pt	(s)	(r)
50 max	0.4 to 1	5 to 50	Yes	10^8	Pd	(b)	(r)
50 max	0.4 to 1	5 to 50	Yes	10^8	Pt	(f, g)	(r)
10 to 50	1 max	20 to 50	No	10^9	Ag	(b)	(r)
10 to 50	0.4 max	20 to 50	Yes	10^7	Ag	(b)	(r)
Switches							
50 max	0.4 max	50 to 250	Yes	10^6	69Au-25Ag-6Pt	(j)	Dust
50 max	0.4 max	50 to 250	Yes	10^6	Pt-Ru or Pt-Ir	(j)	Dust
50 max	1 max	5 to 50	Yes	10^6	Pd	(b)	Dust
50 max	1 max	5 to 50	Yes	10^6	Pt	(f, g)	Dust

(a) Materials are listed in the order of decreasing preference. (b) Low-cost material. (c) Ease of fabrication. (d) Spring properties. (e) Wear and surface deterioration. (f) Surface oxidation resistance. (g) Lower contact resistance. (h) Surface deterioration. (j) Wear resistance. (k) Surface deterioration, resistance, and noise. (m) Resistance and noise. (n) Wear, resistance, and noise. (p) Allows low contact force. (q) Electrical conductivity. (r) Dust, polymer, carbonaceous deposits, and erosion sticking. (s) Reduced polymer formation. (t) Material provides improved resistance to metal transfer and pitting.

Table 15 Recommended materials for microcontacts

Current, microamperes	Voltage, V	Closed force, g	Slide or wipe	Expected operations, millions	Materials used(a)	Advantages	Usual cause of failure
Meter-movement relays							
Magnetic contacts							
100 max	120 max, ac or dc	2 to 4	Small	2 to 5	Plated bright Au	(b, c, d, e)	(g)
					Plated bright Ag	(b, c, e, f)	(g)
Load-current-aiding type							
5 to 25	75 to 120, dc	½ to 1	Small	2 to 5	Pt-Ir	(b, c, d, e, h)	(g)
Sensitive contacts							
200 max	6 max, dc	0.001 to 0.1	Small	2 to 20	Pt-Ir	(b, c, d, e, j)	(g)
Armature-type sensitive relays							
3000 max	120 max, ac or dc	1 to 5	Small	0.1 to 1.0	Pd	(b, c, d, k)	(g)
					Au-plated Pd	(b, c, d, e, f)	
					Au	(b, c, d, e, f)	
Micropotentiometers							
Brushes							
100 max	120 max, ac or dc	0.1 to 5	Intermittent	...	69Au-25Ag-6Pt	(b, c, d)	(n)
					75Au-18.5Ni-5.5Zn-1Cu	(b, c, d, e, h)	
					60Pd-40Cu	(h, m)	
					35Pd-30Ag-14Cu-10Pt-10Au-1Zn	(h)	
Miniature slip rings							
Wire							
1 max	120 max, ac or dc	1 to 5	Continuous	...	90Pt-10Rh	(b, c, d)	(p)
					70Au-16Cu-7Ag-5Pt-1Pd-1Zn	(b, c, d, e, h)	
Brush rings							
1 max	120 max, ac or dc	1 to 5	Continuous	...	75Au-18.5Ni-5.5 Zn-1Cu	(h)	(p)
					69Au-25Ag-6Pt	(b,c,d)	
					35Pd-30Ag-14Cu-10Pt-10Au-1Zn	(h)	

(a) Materials are listed in the order of decreasing preference. (b) Surface oxidation resistance. (c) Lower contact resistance. (d) Allows low contact force. (e) Reduced polymer formation. (f) Electrical conductivity. (g) Black deposit and wear. The black or brown deposit has high resistance and actually separates the contacts. (h) Wear resistance. (j) Less sticking and welding tendency. (k) Arc-erosion resistance. (m) Low cost material. (n) Wear, noise, brown deposit. (p) Wear, noise.

Table 16 Commercially available forms of electrical contact materials

Alloy	Product form										Mig method	
	Solid rivet	Wire	Strip	Tape	Disks	Attached(a)	Composite weld disks(b)	Clad(c)	Rings	Brushes	Melting	P/M
100 Ag	x	x	x	x	x	x	x	x	x	...	x	...
100 Pd	x	x	x	x	x	x	x	x	x	...	x	...
100 Au	x	x	x	x	x	...	...	x	...	...	x	...
100 Ru	...	...	x	...	x	x	...	...	...	...	x	x
100 Ir	...	x	x	...	x	x	...	...	...	...	x	x
100 Pt	x	x	x	...	x	x	x	x	...	x	x	x
100 Os	...	...	...	...	x	x	...	...	...	...	...	x
100 Rh	...	x	x	...	x	...	...	...	Plate	...	x	...
92.5Ag-7.5Cu	x	x	x	...	x	x	x	x	...	...	x	...
90Ag-10Cu	x	x	x	x	x	x	x	x	x	...	x	...
75Ag-24.5Cu-0.5Ni	x	x	x	x	x	x	x	...	x	...	x	...
72Ag-28Cu	...	x	x	...	...	...	...	...	...	x	x	...
99Ag-1Pd	x	x	x	x	x	x	x	x	...	x	x	...
97Ag-3Pd	x	x	x	x	x	x	x	x	...	x	x	...
90Ag-10Pd	x	x	x	x	x	x	x	x	...	...	x	...
90Ag-10Au	x	x	x	...	x	x	...	...	...	...	x	...
97Ag-3Pt	x	x	x	...	x	x	...	...	x	...	x	...
85Ag-15Cd	x	x	x	...	x	x	x	...	...	...	x	...
95Ag-5CdO	x	x	x	x	x	x	x	...	...	...	x	x
90Ag-10CdO	x	x	x	x	x	x	x	...	...	...	x	x
85Ag-15CdO	x	x	x	x	x	x	x	...	...	...	x	x
90Ag-10Fe	x	...	x	...	x	x	x	...	...	...	...	x
90Ag-10W	x	...	...	...	x	x	x	...	...	...	...	x
50Ag-50WC	...	...	...	...	x	x	x	...	...	...	...	x
65Ag-35WC	...	...	...	...	x	x	x	...	...	...	...	x
75Ag-25Zn	...	...	x	...	x	...	...	...	...	...	x	...
85Ag-15Ni	...	...	...	...	x	x	x	...	x	...	...	x
70Ag-30Ni	...	...	...	...	x	x	x	...	...	...	...	x
70Ag-30Mo	...	...	...	...	x	x	x	...	...	...	...	x
97Ag-3 Graphite	...	...	...	...	x	x	x	...	...	...	...	x
95Ag-5 Graphite	...	...	...	...	x	x	x	...	...	x	...	x
60Pd-40Ag	x	x	x	...	x	x	x	...	...	...	x	...
60Pd-40Cu	x	x	x	...	x	x	x	...	...	x	x	...
95Pd-5Ru	x	x	x	...	x	x	x	...	...	...	x	...
75Au-25Ag	x	x	x	...	x	x	x	...	...	...	x	...
90Au-10Cu(Coin)	x	x	x	...	x	x	x	x	x	...	x	...
95Pt-5Ru	...	...	x	...	x	x	x	...	...	...	x	...
90Pt-10Ru	...	...	x	...	x	x	x	...	...	...	x	...
90Pt-10Rh	...	...	x	...	x	x	x	...	...	...	x	...
90Pt-10Ir	...	x	x	...	x	x	x	...	...	...	x	...
85Pt-15Ir	...	x	x	...	x	x	x	...	...	...	x	...
80Pt-20Ir	...	x	x	...	x	x	x	...	...	...	x	...
75Pt-25Ir	...	x	x	...	x	x	x	...	...	...	x	...
65Pt-35Os	...	...	...	...	x	x	x	...	...	...	...	x
69Au-25Ag-6Pt	x	x	x	...	x	x	x	...	x	x	x	...
60Ru-35Ir-5Pt	...	...	...	...	x	x	x	...	...	...	x	...

(a) Contact disks attached to screws, rivets, blades and bars. (b) Composite welding-type buttons produced for resistance welding attachment. Backings of nickel, Monel and steel. (c) Clad materials including overlay, throughlay, edgelay, strip of precious metal on (or in) base metal.

Thermocouples for Industrial Applications

By T. P. Wang
Director of Research
AMAX Specialty Metals Corporation
Alloy Division
and
E. D. Zysk
Technical Director
Carteret Operation
Engelhard Industries Division
Engelhard Minerals & Chemicals Corporation

ACCURATE MEASUREMENT of temperature is one of the most common and vital requirements in science, engineering and industry. Measurement of temperature is generally thought to be one of the simplest and most accurate measurements that can be made. This is a misconception. Unless proper techniques are employed, highly inaccurate readings can occur and either useless data can be generated or materials can be misprocessed. Also, under certain conditions, it may be difficult or impossible to obtain accurate temperature measurements regardless of whether or not proper techniques are employed.

Seven types of instruments, under appropriate conditions and within specific operating ranges, may be used for measurement of temperature: thermocouple thermometers, radiation pyrometers, resistance thermometers, liquid-in-glass thermometers, filled-system thermometers, optical pyrometers and bimetal thermometers. The success of any temperature-measuring system depends not only on the capacity of the system but also on how well the user understands the principles, advantages and limitations of its application.

The thermocouple thermometer is by far the most widely used device for measurement of temperature. Its favorable characteristics include good accuracy, suitability over a wide temperature range, fast thermal response, ruggedness, high reliability, low cost and great versatility of application.

Essentially, a thermocouple thermometer is a system consisting of (*a*) a temperature-sensing element called a thermocouple, which produces an electromotive force (emf) that varies with temperature, (*b*) a device for sensing emf, which may include a printed scale for converting emf to equivalent temperature units, and (*c*) an electrical conductor (extension wires) for connecting the thermocouple to the sensing device. Although any combination of two dissimilar metals and/or alloys will generate a thermal emf, only seven thermocouples are in common industrial use today. These seven have been chosen on the basis of such factors as mechanical and chemical properties, stability of emf, reproducibility and cost. They will be discussed individually after a review of the principles and practice of measuring temperature by use of thermocouples.

Principles of Thermocouple Thermometers

The principle on which thermocouples depend was discovered in 1821 by Seebeck, who found that when two dissimilar metals are joined in a closed circuit an electromotive force is generated if the two junctions are maintained at different temperatures. This *thermal emf* induces an electric current to flow continuously through the circuit, and is termed *Seebeck emf* in honor of its discoverer.

Figure 1(a) is a schematic diagram of two electrical conductors, A and B, whose two junctions are exposed to dif-

Fig. 1 Schematic and equivalent circuit diagrams of a typical thermocouple system

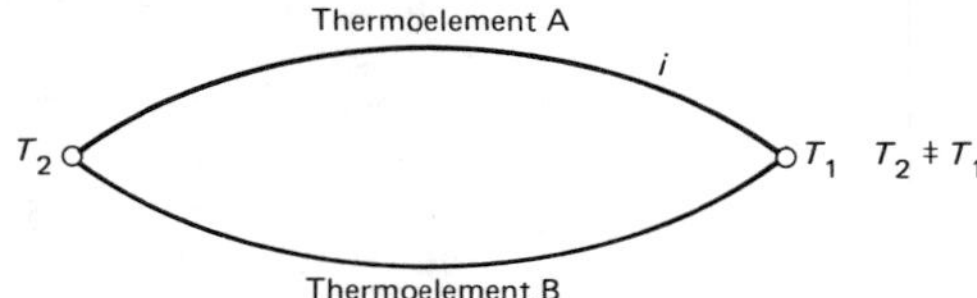

(a) Schematic diagram of a thermocouple

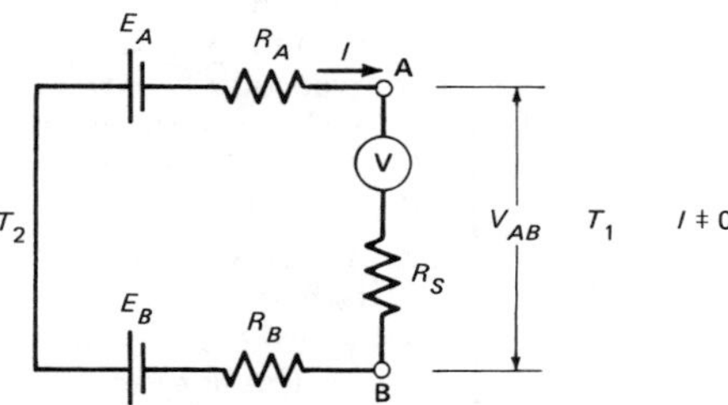

(b) Thermocouple in a voltmeter circuit

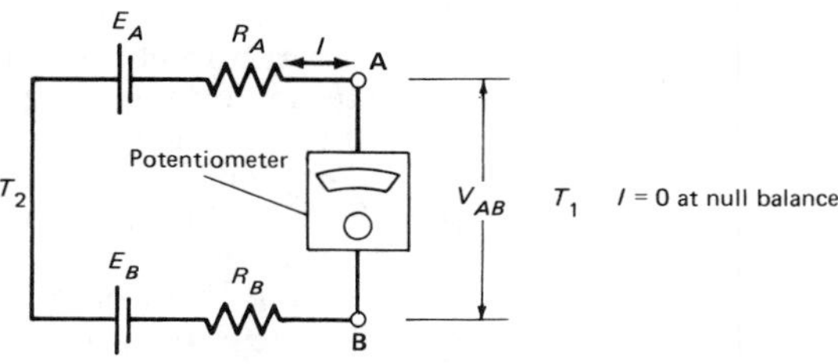

(c) Thermocouple in a potentiometer circuit

Adapted from Ref 1.

ferent temperatures, T_1 and T_2. The thermal emf generated in this circuit, E_{AB}, is expressed as follows:

$$E_{AB} = f[A, B, (T_2 - T_1)] \quad \text{(Eq 1)}$$

The thermal emf E_{AB} is a vector quantity. Its magnitude and direction depend on the material characteristics of A and B as well as the temperature difference between the hot and cold junctions, $T_2 - T_1$, providing that A and B are homogeneous in composition.

A circuit diagram describing a thermocouple in a voltmeter circuit is shown in Fig. 1(b). Thermoelement A is represented by a battery E_A and resistance R_A. E_A is the emf output of thermoelement A with reference to a certain standard, and R_A is the resistance of thermoelement A. Similarly, E_B is the emf output of thermoelement B with reference to the same standard and R_B is the resistance of thermoelement B.

The voltage drop between terminals A and B is given by the following equation:

$$V_{AB} = E_A - E_B - I(R_A + R_B + R_S) \quad \text{(Eq 2)}$$

where R_S is the resistance of a large resistor in series with the thermoelements to minimize the effect of the resistance of the thermoelements. If $E_A - (E_A + V_{AB})$ is positive, the thermoelectric current I will flow continuously from A to B at the cold junction. In this case, A is termed the *positive thermoelement* and B the *negative thermoelement* of the thermocouple.

In Fig. 1(c), a potentiometer is connected across the terminals in place of the voltmeter. A bucking voltage is applied at the potentiometer until it is equal in magnitude and opposite in direction to the thermoelectric voltage E_{AB}. At null balance, there is no current flow. All the IR terms in Eq 2 become zero. Under this condition,

$$V_{AB} = E_{AB} = E_A - E_B \quad \text{(Eq 3)}$$

Then, the measured emf at the potentiometer V_{AB} is the thermal emf of the thermocouple AB. It may be observed from Eq 3 that thermal emf is a bulk property. It is independent of the resistance, and hence of the diameter, of the wire.

The thermoelectric property of an electrical conductor is usually expressed in millivolts against a common reference. Platinum 67, because of its purity and excellent oxidation resistance at elevated temperatures, is the United States thermometric reference standard. This standard can be obtained from the National Bureau of Standards.

The change in emf with temperature for two electrical conductors A and B, which are, respectively, positive and negative with reference to platinum, is shown in Fig. 2(a). The emf values of both conductors change linearly with temperature. If A and B are joined as a couple and the two ends are exposed to temperatures T_1 and T_2, the emf of the couple AB at temperatures between T_1 and T_2 is equal to the algebraic difference in emf between the positive thermoelement A vs Pt and the negative thermoelement B vs Pt at temperatures in the same range. This is shown in Fig. 2(b) and expressed in the following equation:

$$E_{AB}\Big|_{T_1}^{T_2} = E_A\Big|_{T_1}^{T_2} - E_B\Big|_{T_1}^{T_2} \quad \text{(Eq 4)}$$

where $E_{AB}\Big|_{T_1}^{T_2}$ is the emf of thermocouple AB between T_1 and T_2, in millivolts

$E_A\Big|_{T_1}^{T_2}$ is the emf of thermoelement A vs Pt between T_1 and T_2, in millivolts

$E_B\Big|_{T_1}^{T_2}$ is the emf of thermoelement B vs Pt between T_1 and T_2, in millivolts.

Fig. 2 Thermal emf of two thermoelements with respect to platinum and to each other

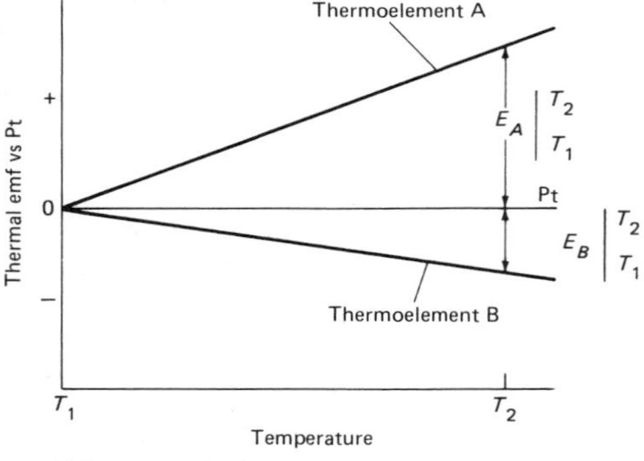

(a) Thermal emf of thermoelements A and B versus Pt

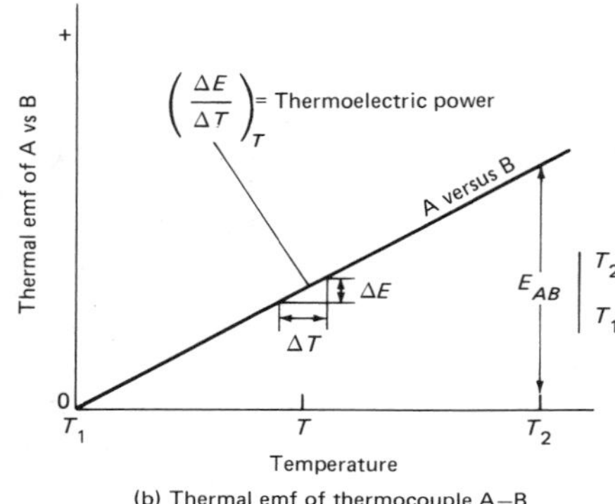

(b) Thermal emf of thermocouple A–B

Thermoelectric power at a given temperature T is defined as the rate of change of thermal emf with respect to temperature. The thermoelectric power of the thermocouple AB at temperature T is the slope of its emf/temperature curve,

$$\frac{\Delta E}{\Delta T}$$

as shown in Fig. 2(b). The thermoelectric power of a thermoelement at a given temperature T is the rate of change of its emf, referenced to Pt, with respect to temperature. In using a thermocouple for temperature measurement, it is essential that the thermoelectric power of the thermocouple be fairly large and uniform within the applicable temperature range.

In case the emf/temperature relationship of the thermocouple AB is well established we may determine the *temperature difference* between T_2 and T_1 by measuring with a potentiometer the generated thermal emf E_{AB}. Note that a thermocouple really does not measure the temperature of the hot junction T_2 but measures the temperature difference $T_2 - T_1$.

The ice point (0 °C or 32 °F) is set universally as the reference cold junction (T_1) for all established emf tables. If we use the ice point as our reference cold junction T_1, then the measured emf does correspond to the temperature of the hot junction T_2.

Measurement of Temperature by a Thermocouple. A setup for measurement of temperature by use of a thermocouple is illustrated schematically in Fig. 3. The welded junction of thermocouple PN is inserted into an electric furnace the temperature of which is to be measured. The ice-point cold junction is provided by two mercury U-tubes embedded in a Dewar flask packed with shaved ice. The legs of the thermocouple are inserted into the mercury U-tubes and connected to the positive and negative terminals of a potentiometer by insulated copper wires. The termperature of the furnace then can be obtained by measuring the emf generated by the thermocouple and referenced to the established emf table for that particular thermocouple. Commercially available automatic compensating cold junctions can be used in place of the above-mentioned mercury U-tubes to achieve a 0 °C reference junction. These may be built into an indicating or recording instrument used to measure the emf developed by the thermocouple or external of the measuring instrument.

In the absence of an ice junction, the thermocouple wires may be connected directly to the terminals of the potentiometer. The ambient temperature of the terminals is measured by a thermometer and converted to emf in millivolts from the emf table. The total emf generated by the thermocouple between the hot junction and the ice point is the sum of the emf thus measured by the potentiometer and this ambient-temperature correction factor. The temperature of the hot junction can be obtained by referring this total emf to the established table.

For additional information, see ASTM E563, "Standard Recommended Practice for Preparation and Use of Freezing Point Reference Baths".

Preparation of the Measuring Junction. The two dissimilar thermoelements must be joined at the temperature-measuring junction to form the thermocouple. The joint must have good thermal and electrical conductivity without adversely affecting the mechanical and electrical properties of the thermocouple wires at this joint.

Prior to being joined, the thermoelements are straightened to facilitate insertion into hard-fired ceramic insulators. In this operation, care should be taken to avoid excessive cold working of the wires, which has a deleterious effect on the emf of the couple. After being cut to the desired length, the thermocouple wires are cleaned carefully (to remove lubricant residue, fingerprints and other contaminants) with a suitable solvent such as methyl ethyl ketone, Freon TF or isopropyl alcohol prior to joining.

For applications below about 500 °C (about 1000 °F), base-metal thermocouple wires may be silver brazed using boraz as a flux. Above this temperature, thermocouple junctions usually are prepared by welding. Noble-metal thermocouples should always be joined by welding. Thermocouples are usually welded using gas, electric-arc, resis-

Fig. 3 Schematic diagram of the experimental setup for measuring temperature using a thermocouple and an ice-point reference junction

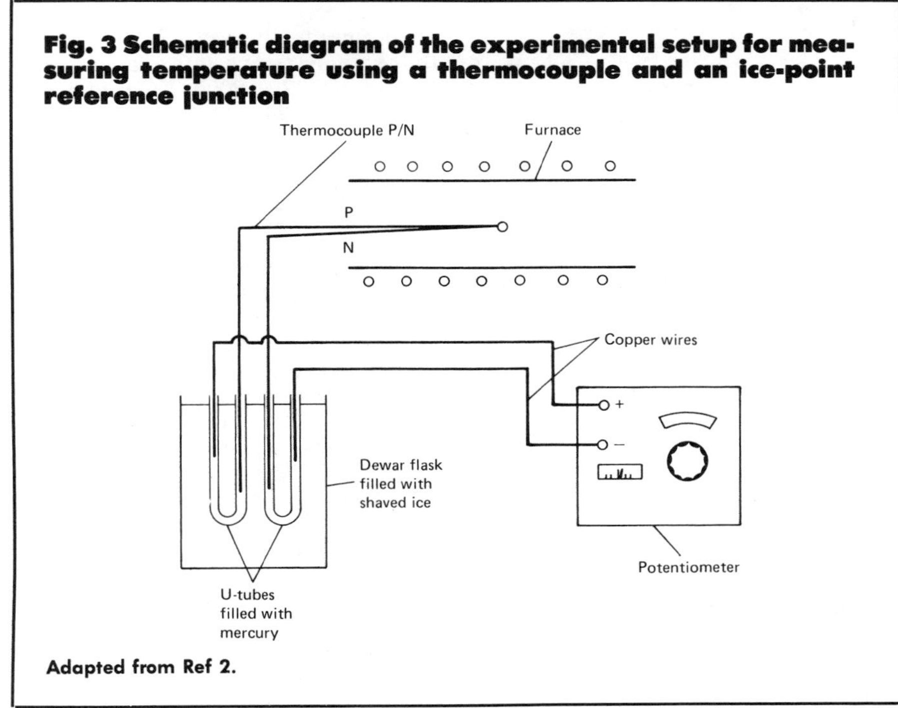

Adapted from Ref 2.

Fig. 4 Thermal emf of standard thermoelements

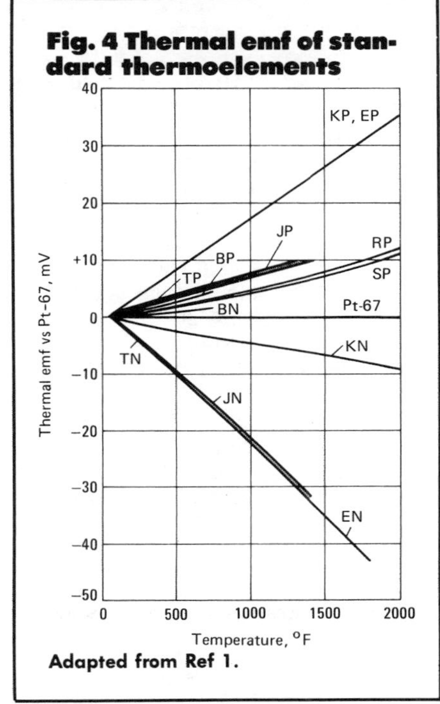

Adapted from Ref 1.

tance, tungsten-inert gas and plasma-arc processes. In gas welding, a neutral flame is required (preferably oxidizing for noble metals). Prior to gas or arc welding, the ends of types E, J, K and T thermocouple wires are first twisted one and a half turns.

Effecting a hot junction in a sheathed thermocouple requires a higher degree of skill, special equipment and considerable care. After the sheath has been stripped away, joining usually is done by gas tungsten-arc or plasma-arc welding. A clean, dry and well-lighted work area is required to produce a finished element of good integrity. An oven capable of continuous operation at 90 °C (200 °F) should be available for storage of unsealed sheathed thermocouples during unavoidable delays in forming of junctions. Use of such an oven will minimize a pickup of airborne moisture and other contaminants.

Thermocouple Materials

Commercially available thermocouples are grouped according to material characteristics (base metal or noble metal) and standardization. At present, four base-metal thermocouples and theree noble-metal thermocouples have been standardized and given letter designations by ANSI (American National Standards Institute), ASTM (American Society for Testing and Materials) and ISA (Instrument Society of America). Among the remaining thermocouples in use, some have not been assigned letter designations because of limited usage, and some are being considered for standardization.

Standard Thermocouples

The base compositions, melting points and electrical resistivities of the individual thermoelements of the seven standard thermocouples are presented in Table 1. Maximum operating temperatures, and limiting factors in environmental conditions, are listed also.

The relations between emf and temperature for the individual thermoelements with reference to Platinum 67 and for the seven standard thermocouples are shown in Fig. 4 and 5, respectively. Tolerances for initial calibration of standard thermocouples (those meeting established tables within a specified tolerance) are listed in Table 2.

Type J. The type J thermocouple is widely used, primarily because of its versatility and low cost. In this couple, the positive thermoelement is iron and the negative thermoelement is constantan, a 44Ni-55Cu alloy. As shown in Fig. 4, the emf of iron is positive with reference to platinum, but the emf of constantan is the most negative with respect to platinum among all thermoelements. The thermoelectric power of the type J couple as a whole is about 55 μV/°C (30 μV/°F) over the temperature range from 0 to 750 °C (32 to 1380 °F), a value higher than that of any other standard couple except the type E couple.

The commercial grade of iron used as the positive leg of the type J thermocouple contains small amounts of carbon, cobalt, manganese and silicon. Therefore, its emf can vary significantly from one heat to another. Accordingly, the emf of this iron is shown in Fig. 4 as a shaded band instead of a single line. However, constantan of a slightly different composition can be obtained commercially to match the iron wire on hand so that the thermocouple as a whole meets established emf/temperature requirements. As shown in Table 2, the initial calibration tolerance (to established table values) for standard-grade type J couples is 2.2 °C (4 °F) or ±3/4% of temperature, whichever is greater over the range from 0 to 750 °C (32 to 1380 °F).

Type J couples can be used in both oxidizing and reducing atmospheres at temperatures up to about 760 °C (1400 °F). They find extensive use in heat treating applications in which they are

Table 1 Properties of standard thermocouples

Type	Thermo-elements	Base composition	Melting point, °C	Resistivity, nΩ·m	Recommended service	Max temperature °C	Max temperature °F
J	JP	Fe	1450	100	Oxidizing or reducing	760	1400
	JN	44Ni-55Cu	1210	500			
K	KP	90Ni-9Cr	1350	700	Oxidizing	1260	2300
	KN	94Ni-Al, Mn, Fe, Si, Co	1400	320			
T	TP	OFHC Cu	1083	17	Oxidizing or reducing	370	700
	TN	44Ni-55Cu	1210	500			
E	EP	90Ni-9Cr	1350	700	Oxidizing	870	1600
	EN	44Ni-55Cu	1210	500			
R	RP	87Pt-13Rh	1860	196	Oxidizing or inert	1480	2700
	RN	Pt	1769	104			
S	SP	90Pt-10Rh	1850	189	Oxidizing or inert	1480	2700
	SN	Pt	1769	104			
B	BP	70Pt-30Rh	1927	190	Oxidizing, vacuum or inert	1700	3100
	BN	94Pt-6Rh	1826	175			

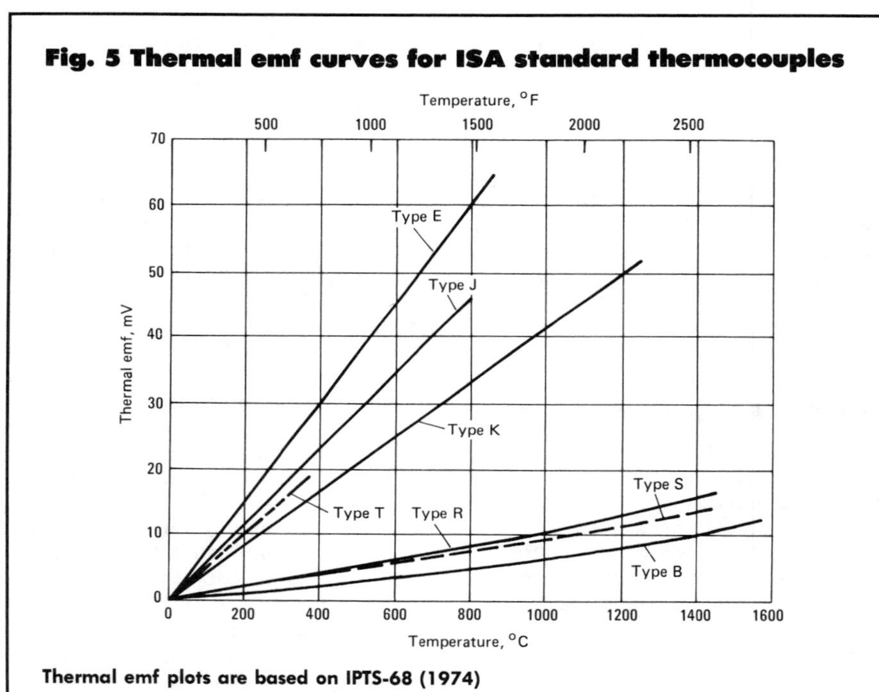

Fig. 5 Thermal emf curves for ISA standard thermocouples

Thermal emf plots are based on IPTS-68 (1974)

exposed directly to the furnace atmosphere. The No. 8 gage (3.25-mm- or 0.128-in.-diam) type J couple can have a useful service life of about 1000 hours at 760 °C in an oxidizing atmosphere. The same size couple can be used at higher temperatures in a reducing atmosphere. Smaller-gage type J couples are also available for laboratory use or applications where quicker heat-sensing response is desired. As thermocouple wire diameter decreases, the recommended upper temperature limit also decreases, as shown in Table 3.

Type K thermocouples, like type J couples, are also widely used in industrial applications. The positive thermoelement is a 90Ni-9Cr alloy; its thermal emf versus platinum is the most positive. The negative thermoelement is a 94% Ni alloy containing silicon, manganese, aluminum, iron and cobalt as alloying constituents; its emf is negative with respect to platinum. The thermoelectric power of the type K couple is close to 40 μV/°C (22 μV/°F) over the extended temperature range from 0 to 1100 °C (32 to 2000 °F); at temperatures up to 1250 °C (2300 °F), it is still close to 35 μV/°C (19 μV/°F).Commercial type K couples are Chromel-Alumel (trademark of Hoskins Manufacturing), T_1-T_2 (trademarks of Driver-Harris) and Tophel-Nial (trademark of AMAX Specialty Metals, formerly Wilbur B. Driver).

Type K thermocouples can be used up to 1250 °C in oxidizing atmospheres. As shown in Table 2, initial calibration of standard-grade type K couples should be within ±2.2 °C (±4 °F) or ±3/4% of table values (whichever is greater) up to 1260 °C. The maximum operating temperature of No. 8 gage wire (3.25-mm or 0.128-in. diam) is 1260 °C (2300 °F). For smaller-diameter wire, recommended upper temperature limits are lower, as shown in Table 3.

Type K couples should not be used in elevated-temperature service in reducing atmospheres or in environments containing sulfur, hydrogen or carbon monoxide. At elevated temperatures in oxidizing atmospheres, uniform oxidation takes place, and the oxide formed on the surface of the positive (90Ni-10Cr) thermoelement is a spinel, $NiO \cdot Cr_2O_3$. However, in reducing atmospheres, preferential oxidation of chromium takes place, forming only Cr_2O_3. The presence of this greenish oxide (commonly known as "green rot") depletes the chromium content, causing a very large negative shift of emf (up to −2 MV) and rapid deterioration of the thermoelement.

The type K positive thermoelement is susceptible to short-range ordering (changes from a random to an ordered atomic structure in localized regions) on aging at about 500 °C (930 °F). This causes a change in emf of about +0.2 MV, which is equivalent to a change of 5 °C (9 °F). This change can be essentially eliminated by preaging the thermoelement at 500 °C. However, the short-range ordering is a reversible process. The type K positive thermoelement will change back to its initial disordered condition if heated to 800 °C (1470 °F) or above.

Besides the above considerations, type K couples are quite versatile. They are the only standard base-metal thermocouples that can be used for sensing temperatures from 900 to 1260 °C (1650 to 2300 °F).

Type T thermocouples are used extensively for cryogenic measurements. The positive thermoelement is copper, the thermal emf of which is positive with respect to platinum. Commercial-grade OFHC copper (C10100), unlike commercial grades of iron, is of high purity and is very homogeneous. Its emf is quite uniform from lot to lot. The constantan (44Ni-55Cu) used for the negative thermoelement of the type T couple has the same base composition as that of the constantan used in the type J couple, but is slightly different in minor alloying constituents. Because of this, the two types of constantan have different emf characteristics. As manufactured, the copper-constantan couple has an emf output that conforms to the emf table for type T thermocouples to within ±1 °C at temperatures from 0 to 350 °C (32 to 660 °F). From −200 to 0 °C (−330 to +32 °F), the tolerance on the initial calibration is ±1% of temperature. Type T couples can be used in either oxidizing or reducing atmospheres. They should not be used above 370 °C (700 °F) because of the poor oxidation resistance of copper.

Type E. The positive thermoelement of the type E thermocouple is 90Ni-9Cr, the same as that of the type K thermocouple; and the negative element is 44Ni-55Cu, the same as that of the type T couple. Among all standard thermoelements, these two are the most positive and most negative with respect to Platinum 27. Therefore, the thermoelectric power of the type E couple is the highest among all standard couples. Type E couples are used primarily for power-generation applications such as thermopiles.

The recommended maximum operating temperature for type E thermocouples is 870 °C (1600 °F). Like type K thermocouples, type E couples should be used only in oxidizing atmospheres, because their use in reducing atmospheres results in preferential oxidation of chromium ("green rot").

Type S. The type S thermocouple serves as the interpolating instrument for defining the International Practical Temperature Scale of 1968 (amended in 1975) from the freezing point of antimony (630.74 °C, or 1103.33 °F) to the freezing point of gold (1064.43 °C, or 1883.97 °F). It is characterized by a high degree of chemical inertness and stability at high temperatures in oxidizing atmospheres. The materials used in the legs of this thermocouple, Pt-10Rh and platinum, both are ductile and can be drawn into fine wire (as small as 0.025 mm or 0.001 in. in diameter for special applications). For general-purpose use, wire 0.51 mm (0.020 in.) in diameter is commonly used.

The thermoelectric output of the type S thermocouple is about 6 μV/°C (3.3 μV/°F) at temperatures from 0 to 100 °C (32 to 212 °F) and about 11.5 μV/°C (6.4 μV/°F) at 1000 °C (1830 °F). Type S couples can be used in intermittent service up to 1750 °C or 3180 °F (the melting point of platinum is 1769 °C or 3216 °F) and can be used continuously up to 1500 °C (2730 °F) if properly protected. Because of its low emf output, this thermocouple is not used for measuring subzero temperatures. Type S couples that match standard emf/temperature values within ±1.5 °C (±0.83 °F) or ±0.25% (whichever is greater), and under special conditions within ±0.6 °C (±0.33 °F) or ±0.1%, can be obtained from reliable sources.

The type S couple is widely used in industrial laboratories as a standard for calibration of base-metal thermocouples and other temperature-sensing instruments. It is commonly used for controlling processing of steel, glass and many refractory materials. It

Table 2 Initial calibration tolerances for thermocouples when the reference junction is at 0 °C

Adapted from ANSI MC96.1

Thermocouple type	Temperature range, °C	Initial calibration tolerance: Standard (whichever is greater)	Initial calibration tolerance: Special (whichever is greater)
T	0 to 350	±1 °C or ±0.75%	±0.5 °C or 0.4%
J	0 to 750	±2.2 °C or ±0.75%	±1.1 °C or 0.4%
E	0 to 900	±1.7 °C or ±0.5%	±1 °C or ±0.4%
K	0 to 1250	±2.2 °C or ±0.75%	±1.1 °C or ±0.4%
R or S	0 to 1450	±1.5 °C or ±0.25%	±0.6 °C or ±0.1%
B	800 to 1700	±0.5%	...
T(a)	−200 to 0 °C	±1 °C or ±1.5%	(b)
E(a)	−200 to 0 °C	±1.7 °C or ±1%	(b)
K(a)	−200 to 0° C	±2.2 °C or ±2%	(b)

(a) Thermocouples and thermocouple materials are normally supplied to meet the limits of error specified in the table for temperatures above 0 °C. The same materials, however, may not fall within the subzero limits of error given in the second section of the table. If materials are required to meet the subzero limits, the purchase order must so state. Selection of materials usually will be required. (b) Little information is available to justify establishment of special limits of error for subzero temperatures. Limited experience suggests the following limits for types E and T thermocouples: Type E −200 to 0 °C ± 1 °C or ± 0.5%; Type T, 200 to 0 °C ± 0.5 °C or ± 0.8%. These limits are given only as a guide for discussion between purchaser and supplier. Due to the characteristics of the materials, subzero limits of error for type J thermocouples and special subzero limits for type K thermocouples are not listed.

Table 3 Recommended upper temperature limits for protected thermocouples of various wire sizes

Adapted from ANSI MC96.1

Type of thermocouple	No. 8 AWG: 3.25 mm (0.128 in.) °C	°F	No. 14 AWG: 1.63 mm (0.064 in.) °C	°F	No. 20 AWG: 0.81 mm (0.032 in.) °C	°F	No. 24 AWG: 0.51 mm (0.020 in.) °C	°F	No. 28 AWG: 0.33 mm (0.013 in.) °C	°F
T	...	...	370	700	260	500	200	400	200	400
J	760	1400	590	1100	480	900	370	700	370	700
E	870	1600	650	1200	540	1000	430	800	430	800
K	1260	2300	1090	2000	980	1800	870	1600	870	1600
R, S	...	...	...	...	...	...	1480	2700	...	...
B	...	...	...	...	...	...	1700	3100	...	...

(Upper temperature limit)

should be used in air or in oxidizing or inert atmospheres. It should not be used unprotected in reducing atmospheres in the presence of easily reduced oxides, atmospheres containing metallic vapors such as lead or zinc, or atmospheres containing nonmetallic vapors such as arsenic, phosphorus or sulfur. It should not be inserted directly into metallic protection tubes and is not recommended for service in vacuum at high temperatures except for short periods of time. Because the negative leg of this couple is fabricated from high-purity platinum (approximately 99.99% for commercial couples and 99.995%+ for special grades), special care should be taken to protect the couple from contamination by the insulators used as well as by the operating environment.

Type R. The type R thermocouple (Pt-13Rh/Pt) has characteristics similar to those of the type S couple. In 1922 it was found that the British Pt-10Rh alloy had a higher emf than the U.S. version but was unstable due to the presence of 0.34% Fe. In order to produce an alloy free from iron (and therefore stable) but with an emf that met the calibration of existing instruments (in other words, having an emf output equivalent to that of the couple using impure Pt-10Rh element), it was necessary to increase the rhodium content to 13%. The emf output of the type R thermocouple is slightly higher than that of the type S couple. End-use applications for type S couples also apply to type R.

Type B thermocouples (Pt-30Rh/Pt-6Rh) may be used in still air or inert atmospheres for extended periods at temperatures up to 1700 °C or 3100 °F and intermittently up to 1760 °C or 3200 °F (Pt-6Rh leg melts at approximately 1826 °C or 3319 °F). Because both of its legs are platinum-rhodium alloys, the type B couple is less sensitive than type R or type S to pickup of trace impurities from insulators or from the operating environment. Under corresponding conditions of temperature and environment, type B thermocouples exhibit less grain growth and less drift in calibration than type R or type S thermocouples.

The type B couple also is suitable for short-term use in vacuum at temperatures up to about 1700 °C (3100 °F); its emf stability varies with temperature, time at temperature and degree of vacuum. It should not be used in reducing atmospheres, or in those containing metallic or nonmetallic vapors, unless suitably protected with ceramic protection tubes. It should never be inserted directly into a metallic primary protection tube.

This couple has a very small emf in the normal reference range from 0 to about 100 °C (32 to about 212 °F), and particularly from 0 to 50 °C (120 °F). Errors arising because of uncertainties in the temperature of the reference junction, or as a result of that temperature being ignored, are relatively small for measurements of high temperature (over 1000 °C or 1830 °F). The thermoelectric power of the type B thermocouple is 9.1 μV/°C (5.1 μV/°F) at 1000 °C (1830 °F) and 11.3 μV/°C (6.3 μV/°F) at 1400 °C (2550 °F).

Fig. 6 Thermal emf of 19 alloy and 20 alloy vs Pt-67

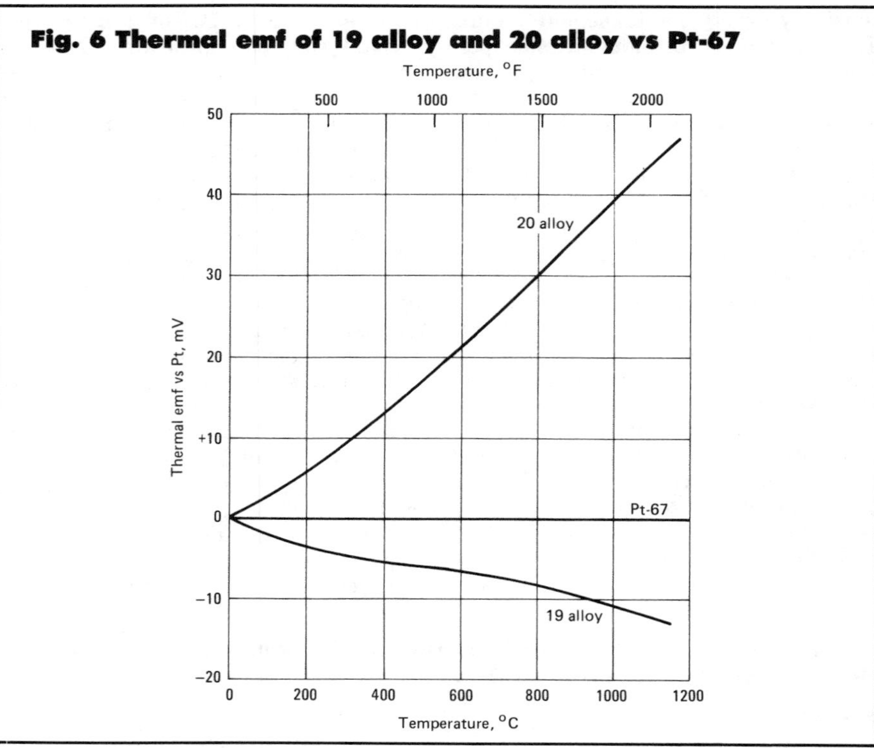

Nonstandard Thermocouples

19 Alloy/20 Alloy. The 19 alloy/20 alloy thermocouple was developed for temperature-sensing and control applications at elevated temperatures in hydrogen or in reducing atmospheres. The positive thermoelement is the 20 alloy, which has a nominal composition of 82Ni-18Mo. The negative thermoelement is the 19 alloy, the nominal composition of which is 99Ni-1Co. Values of emf versus platinum for 19 alloy and 20 alloy are given in Fig. 6. The emf of the 19 alloy/20 alloy thermocouple is somewhat larger than that of the type K couple. Physical, electrical and mechanical properties of the 19 alloy/20 alloy thermocouple are listed in Table 4.

19 alloy/20 alloy thermocouples can be used in hydrogen or in reducing atmospheres over the entire range from 0 to 1260 °C (32 to 2300 °F) with excellent performance. Hotchkiss (Ref 3) showed that, after exposure at about 950 °C (1750 °F), a type K couple was out of calibration by about 2 mV (about 50 °C, or 90 °F) in the negative direction whereas a 19 alloy/20 alloy thermocouple remained essentially in calibration. The oxidation resistance of 19 alloy/20 alloy is not good when compared with that of the type K couple. 19 alloy/20 alloy thermocouples should not be used in oxidizing atmospheres above about 650 °C (1200 °F).

Nicrosil/Nisil. The Nicrosil/Nisil thermocouple (trademark of AMAX Specialty Metals) was developed for oxidation resistance and emf stability superior to those of type K thermocouples at elevated temperatures. The positive thermoelement is Nicrosil (nominal composition, 14 Cr, 1.4 Si, 0.1 Mg, rem Ni), and the negative thermoelement is Nisil (nominal composition,

Table 4 Properties of two nonstandard thermocouples: 19 alloy/20 alloy and Nicrosil-Nisil

	19 alloy	20 alloy	Nicrosil	Nisil
Nominal composition	Ni-1Co	Ni-18Mo	Ni-14Cr-1.4Si	Ni-4.4Si-0.1Mg
Melting point, °C	1450	1425	1410	1400
Specific gravity	8.9	9.1	8.52	8.70
Thermal conductivity, W/m·K at 20 °C	50	15	130	230
Coefficient of thermal expansion, μm/m·°C (20 to 100 °C)	13.6	11.9	13.3	12.1
Magnetic susceptibility	Magnetic	Magnetic	Nonmagnetic	Nonmagnetic
Resistivity, nΩ·m at 20 °C	80	1650	930	370
Temperature coefficient of resistance, μΩ/Ω·°C (20 to 100 °C)	3050	290	100	900
Tensile strength, MPa (ksi)	415(60)	895(130)	760(110)	655(95)
Yield strength, MPa (ksi)	170(25)	515(75)	415(60)	380(55)
Elongation, %	240(35)	240(35)	205(30)	240(35)

Fig. 7 Thermal emf of Nicrosil, Nisil and type K thermoelements vs Pt-67

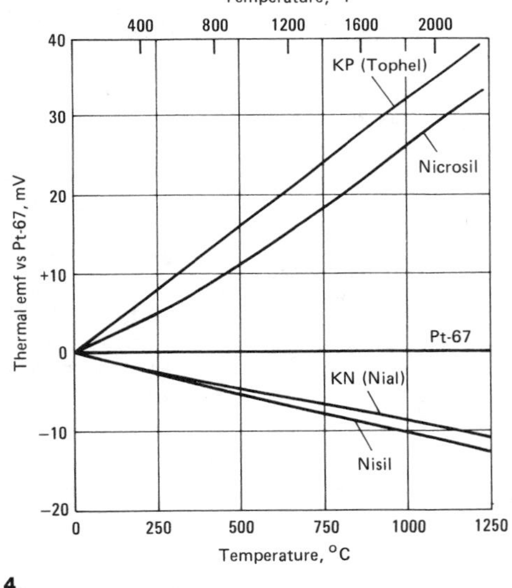

Adapted from Ref 4.

4.4 Si, 0.1 Mg, rem Ni). Reference 9 presents emf values and various physical properties of Nicrosil/Nisil thermocouples. These couples have been shown (Ref 6 and 7) to have longer life and better emf stability than type K thermocouples at elevated temperatures in air, both in the laboratory and in several industrial applications (see Fig. 7). At present (1980), an ASTM committee is evaluating Nicrosil/Nisil to determine its suitability for standardization.

Iridium-Rhodium. Three iridium-rhodium thermocouples are commercially available: 60Ir-40Rh/Ir, 50Ir-50Rh/Ir and 40Ir-60Rh/Ir. Of these three combinations, 60Ir-40Rh/Ir appears to be preferred at this time. Properties of iridium-rhodium couples are given in Table 5 and Fig. 8.

Iridium-rhodium couples are suitable for use for limited periods of time in air or other oxygen-carrying atmospheres at temperatures up to about 2000 °C (3600 °F), and generally are used for such service at temperatures above the range in which types R, S and B thermocouples are employed. They can be used in inert atmospheres and in vacuum, but not in reducing atmospheres (easily reduced oxides in contact with iridium or with Ir-Rh alloys are sources of contamination). These couples have been used for short periods of time at temperatures up to only 60 °C (110 °F) below the melting point of the alloy leg—that is, up to 2180 °C (3960 °F) for 60Ir-40Rh, up to 2140 °C (3880 °F) for 50Ir-50Rh, and up to 2090 °C (3790 °F) for 40Ir-60Rh.

After being hot worked, iridium thermoelements have a fibrous structure and are reasonably ductile. However, annealing causes the structure to become equiaxed, with resultant decreases in ductility and handleability. This should be considered if any preinstallation fabrication is anticipated.

Compensating extension wires are available for iridium-rhodium thermocouples—copper for the positive leg and stainless steel for the negative leg.

Platinum-Molybdenum. The Pt-5Mo/Pt-0.1Mo thermocouple is used for measuring temperatures from 1100 to about 1500 °C (2000 to about 2700 °F) under neutron radiation (type K couples are employed for temperatures up to 1100 °C). Basic characteristics of Pt-5Mo/Pt-0.1Mo couples are given in Table 6. Platinum alloys containing rhodium are not suitable for use under neutron radiation, because the rhodium is slowly transmuted to palladium.

The output of the Pt-5Mo/Pt-0.1Mo couple is high and increases with temperature in a uniform manner (see Fig. 9). Detailed emf/temperature tables (at 1 °C intervals) are available from suppliers. Compensating lead wires, good from 0 to 70 °C (32 to 160 °F), are available for Pt-5Mo/Pt-0.1Mo thermocouples—copper for the positive leg and Cu-1.6Ni for the negative leg.

Platinel. The Platinel thermocouple (trademark of Engelhard Minerals and Chemicals Corp.), an all-noble-metal combination, was metallurgically designed to approximate the emf/temperature characteristics of the type K couple. Actually, two combinations have been produced: Platinel I and Platinel II (see Table 7). Both have negative legs of 65Au-35Pd. The positive leg of the Platinel I couple is 83Pd-14Pt-3Au, and the positive leg of the Platinel II couple consisits of 55Pd-31Pt-14Au. The Platinel II couple has superior high-temperature fatigue properties and appears to be preferred over Platinel I.

Table 5 Properties of iridium-rhodium thermocouples
Adapted from Ref 5

	60Ir-40Rh versus Ir	50Ir-50Rh versus Ir	40Ir-60Rh versus Ir
Nominal operating temperature range, in:			
Wet hydrogen	(a)	(a)	(a)
Dry hydrogen	(a)	(a)	(a)
Inert atmosphere	2100 °C (3812 °F)	2050 °C (3722 °F)	2000 °C (3632 °F)
Oxidizing atmosphere	(a)	(a)	(a)
Vacuum	2100 °C (3812 °F)	2050 °C (3722 °F)	2000 °C (3632 °F)
Approximate microvolts per degree:			
Mean over nominal operating range	5.3/°C (2.9/°F)	5.7/°C (3.2/°F)	5.2/°C (2.9/°F)
At top temperature of normal range	5.6/°C (3.1/°F)	6.2/°C (3.5/°F)	5.0/°C (2.8/°F)
Melting temperature, nominal:			
Positive thermoelement	2250 °C (4082 °F)	2202 °C (3996 °F)	2153 °C (3907 °F)
Negative thermoelement	2443 °C (4429 °F)	2443 °C (4429 °F)	2443 °C (4429 °F)
Stability with thermal cycling	Fair	Fair	Fair
Ductility (of more brittle thermoelement) after use	Poor	Poor	Poor

(a) Not recommended.

Table 6 Properties of platinum molybdenum thermocouples
Adapted from Ref 5

	Pt-5Mo versus Pt-0.1Mo
Nominal operating temperature range, in:	
Reducing atmosphere (nonhydrogen)	(a)
Wet hydrogen	(a)
Dry hydrogen	(a)
Inert atmosphere (helium)	1400 °C (2552 °F)
Oxidizing atmosphere	(a)
Vacuum	(a)
Maximum short-time temperature	1550 °C (2822 °F)
Approximate microvolts per degree:	
Mean, over nominal operating range	29/°C (51.2/°F)
At top temperature of normal range	30/°C (54/°F)
Melting temperature, nominal:	
Positive thermoelement	1788 °C (3250 °F)
Negative thermoelement	1770 °C (3218 °F)
Stability with thermal cycling	Good
High-temperature tensile properties	Fair
Stability under mechanical working	Good
Ductility (of most brittle thermoelement) after use	Good
Resistance to handling contamination	Fair
Recommended extension wire 70 °C (158 °F) max:	
Positive conductor	Cu
Negative conductor	Cu-1.6Ni

(a) Not recommended.

Figure 10 compares thermal emf of Platinel thermocouples with that of the type K couple. The emf match with type K is good at elevated temperatures, but some departure occurs at lower temperatures. In an application involving measurement of turbine inlet temperatures in an aircraft engine, the connection between the thermocouple and type K extension wire is effected at 800 °C (1470 °F) where the emf match is excellent. In this application, only about 13 mm (½ in.) of Platinel II wire is used, and the remainder is type K. Other base-metal extension wires capable of matching the emf of the Platinels very closely over the range from 0 to 160 °C (320 °F) are also available.

Platinel couples can be used unprotected (insulators only) in air to 1200 °C (2190 °F) for extended periods of time and to 1300 °C (2370 °F) for shorter periods. Platinel II aged in commercial hydrogen for 1000 h at 1000 °C (1830 °F) showed reasonably good stability. Drift did not exceed 0.75%. It is recommended that precautions usually followed with platinum-rhodium thermocouples also be observed with Platinels. In particular, it should be noted that phosphorus, sulfur and silicon have deleterious effects on life of Platinel thermocouples.

Tungsten-Rhenium. Three tungsten-rhenium thermocouples are commercially available: W/W-26Re, doped W-3Re/W-25Re and W-5Re/W-26Re. All three couples have been used at temperatures up to 2760 °C (5000 °F), but they usually are employed only below 2315 °C (4200 °F) due to temperature limitations of ceramic insulators.

Early use of tungsten-rhenium thermocouples, particularly W/W-26Re couples, indicated that these couples might be capable of measuring high temperatures with reasonable accuracy; but one serious drawback was immediately observed. The tungsten leg, when heated to its recrystallization temperature of 1200 °C (2200 °F), became embrittled, an effect that was not experienced with the opposite leg (W-25Re or W-26Re). Early research showed that addition of 10% rhenium to the tungsten element did much to retain ductility after recrystallization. However, although large additions of rhenium to tungsten solved this problem, they also lowered the emf of the thermocouple. Consequently, other techniques intended to retain room-temperature ductility were employed, including special processing and doping combined with addition of 5% or less rhenium to the tungsten element.

The "dope", in the form of potassium, silicon and aluminum compounds, is added during preparation of the tungsten powder. With the exception of potassium, these additives are volatile and are almost eliminated during processing. Doping, however, assists in formation of a microstructure characterized by large, elongated grains

Fig. 8 Thermal emf of iridium-rhodium/iridium thermocouples

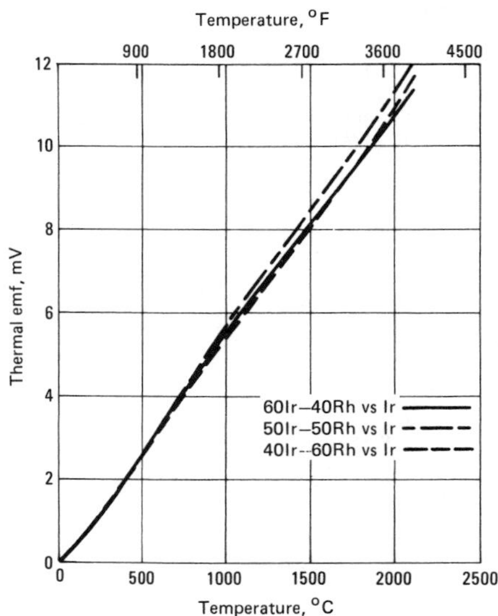

Adapted from Ref 5.

Fig. 9 Thermal emf of Pt-5Mo/Pt-0.1Mo thermocouples

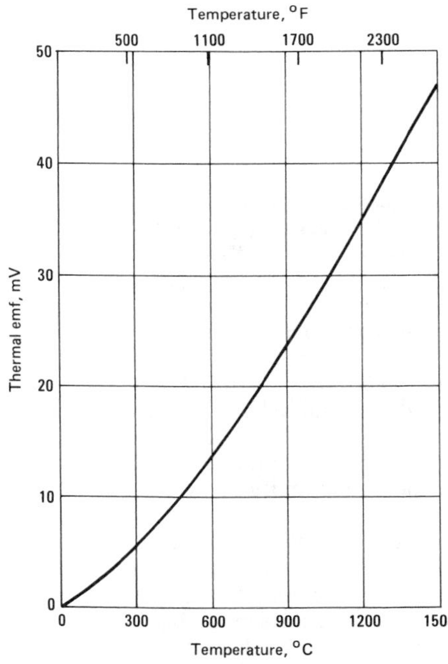

Adapted from Ref 5.

Fig. 10 Thermal emf of Platinel thermocouples compared with that of type K thermocouple

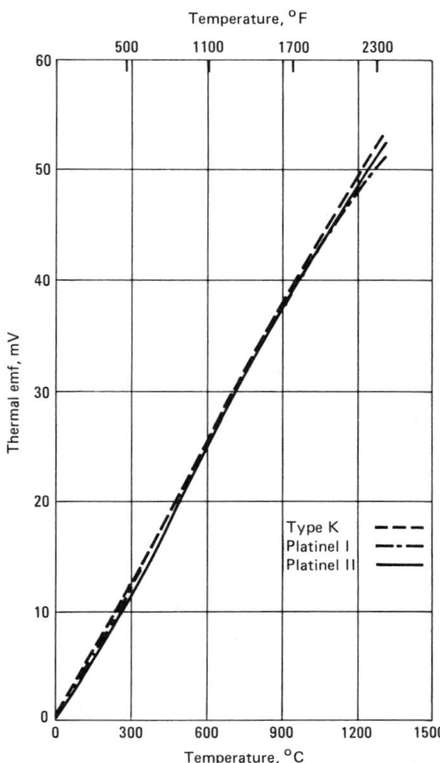

Adapted from Ref 5.

Fig. 11 Thermal emf of tungsten-rhenium thermocouples

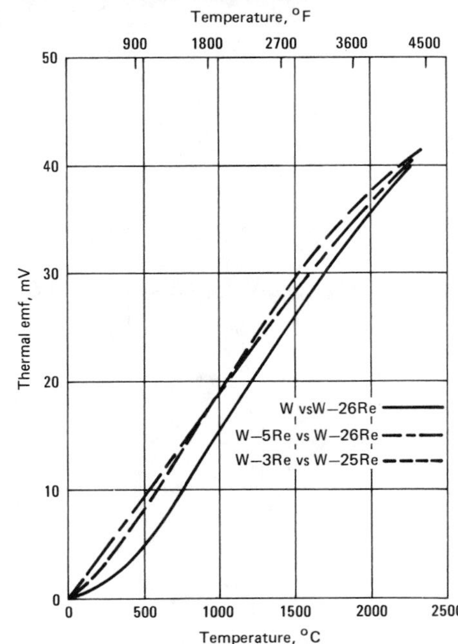

Adapted from Ref 5. For calibration procedures see Ref 21 and ASTM E452.

whose boundaries make relatively small angles with the wire axis. This structure is similar to that found in the well-known "nonsag" tungsten filament wire. Within recent years, microvoids or "bubbles" containing potassium "plated" on the inside surface were found decorating the boundaries of these elongated grains. It is now accepted that these voids promote formation of the desired elongated grains.

The thermal emf values of the three W-Re combinations are compared in Fig. 11, and other pertinent properties are shown in Table 8. All three thermocouple combinations are supplied as matched pairs guaranteed to meet the emf outputs given in producer-developed tables within ±1% (see Ref 21 and ASTM E452 for calibration procedure). Compensating extension wires are available for each combination.

Important factors controlling the use of W-Re thermocouples at high temperatures include: (*a*) insulation, sheaths and protection tubes (choice of insulation, sheaths and protection tubes depends on operating temperature and environment); (*b*) diameter of thermoelements (larger diameters for higher temperatures); and (*c*) atmosphere (vacuum, high-purity hydrogen, high-purity inert atmospheres required). There is evidence that selective vaporization of rhenium occurs at temperatures on the order of 1900 °C (3450 °F) and higher when bare, unprotected W-Re couples are used in vacuum. This is not a problem, however, when these couples are protected with suitable refractory-metal sheaths.

For swaged-type thermocouples, maximum service temperature is affected by the diameter of the thermocouple wire as well as by the thickness of the ceramic insulating material (wire-to-wire and wire-to-sheath resistance). The problem here is mainly "shunt" error. At high temperatures, the resistivity of ceramic insulating materials decreases exponentially with temperature. Therefore, at a sufficiently high temperature, the insulation shunt resistance between thermocouple wires becomes comparable with wire resistance, and shunting results. The error in thermocouple reading may be positive or negative. Increasing the

Fig. 12 Computation of thermal emf compensated to 0 °C (32 °F) for a normal industrial thermocouple setup

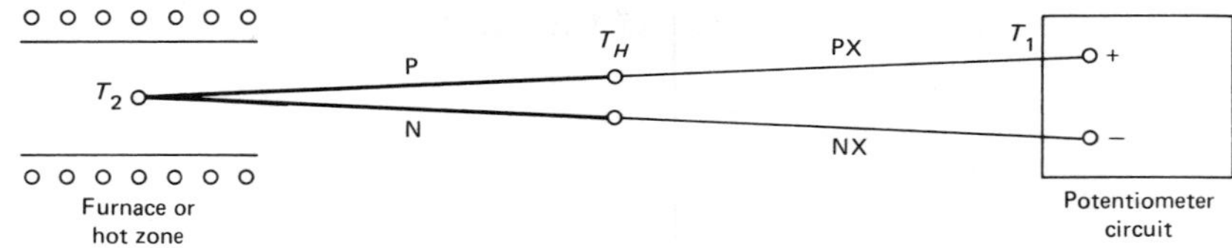

$$E_{\text{assembly}}\Big|_{T_1}^{T_2} = E_{PX}\Big|_{T_1}^{T_H} + E_P\Big|_{T_H}^{T_2} + E_N\Big|_{T_2}^{T_H} + E_{NX}\Big|_{T_H}^{T_1} = (E_P - E_N)\Big|_{T_H}^{T_2} + (E_{PX} - E_{NX})\Big|_{T_1}^{T_H} \quad \text{(Eq 5)}$$

$$E_{PN}\Big|_{T_1}^{T_2} = (E_P - E_N)\Big|_{T_1}^{T_2} = (E_P - E_N)\Big|_{T_H}^{T_2} + (E_P - E_N)\Big|_{T_1}^{T_H} \quad \cdots\cdots \text{(Eq 6)}$$

$$(E_P - E_N)\Big|_{T_1}^{T_H} = (E_{PX} - E_{NX})\Big|_{T_1}^{T_H} \quad \text{(Eq 7)}$$

For the extension wire: $E_P = E_{PX}\Big|_{T_1}^{T_H}$ and $E_N = E_{NX}\Big|_{T_1}^{T_H}$

For an alternative extension wire: $E_P \neq E_{PX}\Big|_{T_1}^{T_H}$ and $E_N \neq E_{NX}\Big|_{T_1}^{T_H}$

Adapted from Ref 9. In this example, the emf of the assembly is calculated from Eq 5; the remaining equations define terms used in Eq 5. In the sketch and equations, P and N designate elements in the thermocouple wire, and PX and NX designate elements in the extension wire. T_2 = hot-junction temperature; T_1 = cold-junction temperature; and T_H = head-junction temperature (205 °C, or 400 °F).

insulation thickness will also help to increase maximum allowable operating temperature.

Thermocouple Extension Wires (Ref 8, 9)

Thermocouple extension wires, also known as extension wires or lead wires, are electrical conductors used for connecting the thermocouple wires to the temperature measuring and control instrument. Extension wires usually are supplied in cable form, with positive and negative wires electrically insulated from each other. The chief reasons for using extension wires are economy and mechanical flexibility.

- *Economy.* Base-metal thermoelements, which cost less than $10 per pound in 1980, are always used as extension wires for the noble-metal thermocouple wires, which in 1980 cost about $700 per troy ounce. For base-metal thermocouples, use of extension wires permits periodic replacement of the thermocouple, which is exposed to elevated temperatures, without replacing the insulated extension-wire cables.
- *Mechanical Flexibility.* Insulated solid or stranded wires in sizes from 14 to 20 gage are used as extension wires. This lends mechanical sturdiness and flexibility to the thermocouple circuitry while permitting the use of larger-diameter (usually 3.2 mm, or 1/8 in.) base-metal thermocouples for improved oxidation resistance and service life, or smaller-diameter (usually 0.51 mm, or 0.020 in.) noble-metal thermocouple wire to save cost.

Table 7 Platinel thermocouples

Adapted from Ref 5

	Platinel II	Platinel I
Nominal operating temperature range, in:		
Reducing atmosphere (nonhydrogen)	(a)	(a)
Wet hydrogen	(a)	(a)
Dry hydrogen(b)	1010 °C (1850 °F)	1010 °C (1850 °F)
Inert atmosphere	1260 °C (2300 °F)	1260 °C (2300 °F)
Oxidizing atmosphere	1260 °C (2300 °F)	1260 °C (2300 °F)
Vacuum	(a)	(a)
Maximum short-time temperature (<1 h)	1360 °C (2480 °F)	1360 °C (2480 °F)
Approximate microvolts per degree:		
Mean, over nominal operating range (100 to 1000 °C)	42.5/°C (23.5/°F)	41.9/°C (23.3/°F)
At top temperature of normal range (1000 to 1300 °C)	35.5/°C (19.6/°F)	33.1/°C (18.4/°F)
Melting temperature, nominal:		
Positive thermoelement—solidus	1500 °C (2732 °F)	1580 °C (2876 °F)
Negative thermoelement—solidus	1426 °C (2599 °F)	1426 °C (2599 °F)
Stability with thermal cycling	Good	Good
High-temperature tensile properties	Fair	Fair
Ductility (of most brittle thermoelement) after use	Good	Good
Recommended extension wire at approximately 800 °C (1472 °F):		
Positive conductor	Type KP	Type KP
Negative conductor	Type KN	Type KN

(a) Not recommended. (b) High-purity alumina insulators are recommended.

Circuitry of Thermocouple Wires and Extension Wires. A schematic diagram and circuitry of thermocouple and extension wires are shown in Fig. 12. One end of the positive extension wire PX and one end of the negative extension wire NX are joined to the positive thermoelement P and the negative thermoelement N, respectively, at the head junction. The temperature at the head junction is usually less than 205 °C (400 °F). The other ends of PX and NX are connected to the positive and the negative terminals, respectively, of the measuring or control instrument.

By substituting in the equation for Kirchoff's law and rearranging terms, the emf of the thermocouple and extension wire assembly between hot- and cold-junction temperatures T_2 and T_1 can be shown to be equal to the sum of the emf of the thermocouple PN between T_2 and T_H (the head-junction temperature) and the emf of extension wire PX-NX between T_H and T_1 (see Eq 5, in Fig. 12).

The emf of the thermocouple PN between T_2 and T_1, without the use of any extension wire, can be expressed as the sum of the emf of the couple PN between T_2 and T_H and the emf of the couple between T_H and T_1 (see Eq 6, in Fig. 12).

In order that the thermocouple and extension wire assembly generates the same emf as that of the thermocouple PN between the hot-junction temperature T_2 and the cold-junction temperature T_1, the following condition must be met:

$$(E_P - E_N)\Big|_{T_1}^{T_H} = (E_{PX} - E_{NX})\Big|_{T_1}^{T_H} \quad \text{(Eq 7)}$$

where E_P is the emf of the positive thermoelement versus platinum, E_N is the emf of the negative thermoelement versus platinum, E_{PX} is the emf of the positive extension wire versus platinum, E_{NX} is the emf of the negative extension wire versus platinum, T_1 is the cold-junction temperature and T_H is the head-junction temperature. This expression means that the emf of the thermocouple PN within the temperature range T_1 to T_H must be the same as the emf of the extension-wire couple PX-NX within the same temperature range. It is not necessary, however, for the emf of the extension-wire couple to match that of the thermocouple over the entire range from T_1 to T_2. Extension wires of the same composition as that of the thermocouple wire are termed *extension wires*. Wires of different alloys used as extension wires to develop the same emf as that of the thermocouple wire from 0 to 200 °C (32 to 400 °F) are called *alternate* or *compensating extension wires*.

Extension wires are used for base-metal thermocouples. Alternate extension wires are used for noble-metal thermocouples. Base compositions, physical properties, ranges of application and initial calibration tolerances of the extension wires for standardized thermocouples are listed in Table 9.

Error Analysis. The error introduced by incorporation of extension wire in the thermocouple circuitry can be expressed as follows:

$$\Delta E_{\text{assembly}}\Big|_{T_1}^{T_2} = \Delta E_{PX-NX}\Big|_{T_1}^{T_H} - \Delta E_{PN}\Big|_{T_1}^{T_H} + \Delta E_{PN}\Big|_{T_1}^{T_2} \quad \text{(Eq 8)}$$

where

$\Delta E_{\text{assembly}}\Big|_{T_1}^{T_2}$ is the deviation (of initial calibration) of the thermocouple and extension wire assembly from established emf table between temperatures T_1 and T_2,

$\Delta E_{PX-NX}\Big|_{T_1}^{T_H}$ is the deviation of the extension wire between T_1 and T_H,

$\Delta E_{PN}\Big|_{T_1}^{T_H}$ is the deviation of the thermocouple between T_1 and T_H, and

$\Delta E_{PN}\Big|_{T_1}^{T_2}$ is the deviation of the thermocouple between T_1 and T_2.

Equation 8 shows that, besides the initial calibration error of the thermocouple over the temperature range T_1 to

Table 8 Properties of tungsten-rhenium thermocouples
Adapted from Ref 5

	W versus W-26Re	W-3Re versus W-25Re	W-5Re versus W-26Re
Nominal operating temperature range, in:			
Dry hydrogen	2760 °C (5000 °F)	2760 °C (5000 °F)	2760 °C (5000 °F)
Inert atmosphere	2760 °C (5000 °F)	2760 °C (5000 °F)	2760 °C (5000 °F)
Vacuum (a)	2760 °C (5000 °F)	2760 °C (5000 °F)	2760 °C (5000 °F)
Maximum short-time temperature	3000 °C (5430 °F)	3000 °C (5430 °F)	3000 °C (5430 °F)
Approximate microvolts per degree:			
Mean, over nominal operating range 0 °C to 2316 °C (32 °F to 4200 °F)	16.7/°C (9.3/°F)	17.1/°C (9.5/°F)	16.0/°C (8.9/°F)
At top temperature of normal range 2316 °C (4200 °F)	12.1/°C (6.7/°F)	9.9/°C (5.5/°F)	8.8/°C (4.9/°F)
Melting temperature, nominal:			
Positive thermoelement	3410 °C (6170 °F)	3360 °C (6080 °F)	3350 °C (6062 °F)
Negative thermoelement	3120 °C (5648 °F)	3120 °C (5648 °F)	3120 °C (5648 °F)
Stability with thermal cycling	Good	Good	Good
High-temperature tensile properties	Good	Good	Good
Stability under mechanical working	Fair	Fair	Fair
Ductility (of most brittle thermoelement) after use	Poor	Poor to good depending on atmosphere or degree of vacuum	Poor to good depending on atmosphere or degree of vacuum
Resistance to handling contamination	Good	Good	Good
Extension wire	Available	Available	Available

(a) Preferential vaporization of rhenium may occur when bare (unsheathed) couple is used at high temperatures and high vacuum. Check vapor pressure of rhenium at operating temperature and vacuum before using bare couple.

Table 9 Properties of thermocouple extension wires
Adapted from ANSI MC96.1

Thermocouple type	Extension wire	Base composition	Resistivity, nΩ·m	Melting point, °C	Temperature range, °C	Initial calibration tolerance
J	JPX	Fe	100	1450	0 to 200	±2.2 °C
	JNX	45Ni-55Cu	500	1210		
K	KPX	90Ni-10Cr	700	1350	0 to 200	±2.2 °C
	KNX	95Ni-AlSiMg	320	1210		
T	TPX	Cu	17	1083	−60 to +100	±1.0 °C
	TNX	45Ni-55Cu	500	1210		
E	EPX	90Ni-10Cr	700	1450	0 to 200	±1.7 °C
	ENX	45Ni-55Cu	500			
R or S	SPX	Cu	17	1083	0 to 200	±5 °C (±57 μV)
	SNX	Cu-1Ni-0.3Mn	45	1100		
B(a)	BPX(b)	Cu-2Mn	150	1100	0 to 200	±33 μV
	BNX	Cu	17	1083		

(a) Cu/Cu extension wire can be used if head-junction temperature is 100 °C or less. (b) Proprietary alloy. Can be used up to 300 °C with initial calibration of +50 μV at this temperature.

T_2, an additional term is introduced when extension is used. This term is equal to the difference of initial calibration of the extension-wire couple and the thermocouple between the cold-junction temperature T_1 and head-junction temperature T_H. This additional term can be minimized by judiciously choosing a pair of extension wires or alternate extension wires the initial emf calibration of which closely matches that of the thermocouple wire between T_1 and T_H.

Color Coding of Thermocouple Wires and Extension Wires (Ref 5)

For many years the Instrument Society of America has coordinated an effort to standardize color coding of thermocouple and extension wires in the United States. The main objective has been to establish uniformity in designation of various types of thermocouples and extension wires to provide, by means of insulation color, identification of wires by type or composition as well as by polarity when used as part of a thermocouple system. The present color designations, as indicated in ANSI MC96.1(1975), are given in Tables 10, 11 and 12. Color coding is not uniform throughout the world and presently is being evaluated by the International Electrotechnical Commission in an attempt at world standardization.

Thermocouple Calibration

The temperature/emf relationship for a specific thermocouple combination is a definite physical property and thus does not depend on details of the apparatus or method used for determining this relationship. Consequently, thermocouples can be calibrated by any of several methods, the choice of which depends on type of thermocouple, temperature range, accuracy required, size of wires, apparatus available and personal preference.

Calibration of a thermocouple is achieved through determination of its electromotive force (emf) at a series of known temperatures, which when coupled with a standardized means of interpolation will give values of emf over the entire temperature range in which it will be used. A standard thermometer that indicates temperatures on a universally acceptable scale is required, as well as a means of measuring the emf of the thermocouple and a controlled heat source wherein the thermocouple and the standard can be brought to the same temperature.

Only the basic points of calibration

Table 10 Color coding of duplex insulated thermocouple wire

Adapted from ANSI MC96.1

Type	Thermocouple Positive wire	Negative wire	Color of insulation Over-all(a)	Positive(a)	Negative
T	TP	TN	Brown	Blue	Red
J	JP	JN	Brown	White	Red
E	EP	EN	Brown	Purple	Red
K	KP	KN	Brown	Yellow	Red

(a) A tracer color of the positive wire code color may be used in the over-all braid.

Table 11 Color coding of single conductor insulated thermocouple extension wires

Adapted from ANSI MC96.1

Type	Extension wire type Positive	Negative	Color of insulation Positive	Negative(a)
T	TPX	TNX	Blue	Red-blue trace
J	JPX	JNX	White	Red-white trace
E	EPX	ENX	Purple	Red-purple trace
K	KPX	KNX	Yellow	Red-yellow trace
R or S	SPX	SNX	Black	Red-black trace
B	BPX	BNX	Gray	Red-gray trace

(a) The color identified as a trace may be applied as a tracer, braid, or by any other readily identifiable means.

NOTE OF CAUTION: In the procurement of random lengths of single conductor insulated extension wire, it must be recognized that such wire is commercially combined in matching pairs to conform to established calibration curves. Therefore, it is imperative that all single conductor insulated extension wire be procured in pairs, at the same time, and from the same source.

Table 12 Color coding of duplex insulated thermocouple extension wires

Adapted from ANSI MC96.1

Type	Extension wire type Positive	Negative	Color of insulation Over-all	Positive	Negative(a)
T	TPX	TNX	Blue	Blue	Red
J	JPX	JNX	Black	White	Red
E	EPX	ENX	Purple	Purple	Red
K	KPX	KNX	Yellow	Yellow	Red
R or S	SPX	SNX	Green	Black	Red
B	BPX	BNX	Gray	Gray	Red

(a) A tracer having the color corresponding to the positive wire code color may be used on the negative wire color code.

techniques will be described in this review; the reader is directed to other more detailed sources of information, such as Ref 5 and Ref 12.

Temperature Scales (Ref 5 and 10). Meaningful measurement of temperature requires a scale with appropriate units, just as measurement of length requires a yardstick or metre stick with all of its subdivisions. The ideal temperature scale is known as the thermodynamic scale. However, measurement of temperature on this scale (using a gas thermometer) is extremely difficult even under laboratory conditions. For many years prior to 1927, the need for a more practical temperature scale had been apparent.

In 1927, such a scale, named the International Temperature Scale (ITS 27), was adopted by the Seventh General Conference on Weights and Measures. Among other advantages, this scale served to unify the existing national temperature scales (Germany, UK, USA, etc.). The scale was revised in 1948, and in a 1960 modification the word "Practical" was inserted in the name of the Scale, which now became the International Practical Temperature Scale. The Scale was revised again in 1968, and was amended in 1975.

The present scale, the "International Practical Temperature Scale of 1968 (amended in 1975)", or "IPTS 68 (amended 1975)", was designed in such a way that the temperature measured on it closely approximates the thermodynamic temperature; the difference is within the limits of the present accuracy of measurement.

The IPTS 68 (amended 1975) is based on the assigned values of the temperatures of 13 reproducible equilibrium states (defining fixed points) and on standard instruments calibrated at these temperatures. Interpolation is provided by formulas used to establish the relations between indications on standard instruments and values of International Practical Temperature.

The IPTS 68 uses both International Practical Kelvin Temperature, symbol T_{68}, and International Practical Celsius Temperature, symbol t_{68}. The relation between T_{68} and t_{68} is the same as that between T and t on the Thermodynamic Scale—that is, $t_{68} = T_{68} - 273.15$ K. The units of T_{68} and t_{68} are the kelvin symbol, K, and the degree Celsius symbol, °C, as in the case of thermodynamic temperature T and Celsius temperature t. The standard instruments used are:

- Platinum resistance thermometer 13.81 K to 630.74 °C
- Pt-10Rh/Pt thermocouple 630.74 °C to 1064.43 °C (gold point)
- Above 1064.43 °C, defined in terms of the Planck radiation law using 1064.43 as a reference temperature (Optical Pyrometer).

Methods of Thermocouple Calibration. Initial calibration of a thermocouple can be done by any of the following methods:

- Freezing-point calibration
- Direct thermoelement emf measurement vs platinum
- Thermoelement comparison method
- Calibration of thermocouples by comparison methods.

In the freezing-point method of calibration, the emf output of the thermocouple as a whole is measured during the cooling cycle of molten pure metals. In the second and third methods, the emf of both the positive and negative thermoelements are individually measured versus platinum or another calibrated standard.

Freezing-Point Calibration. In the calibrication of a thermocouple at freezing points, the thermocouple (properly protected) is slowly immersed in the molten metal. The metal is brought essentially to a uniform temperature at the beginning of freezing by

Fig. 13 Experimental setup for direct measurement of the emf of thermoelements vs platinum

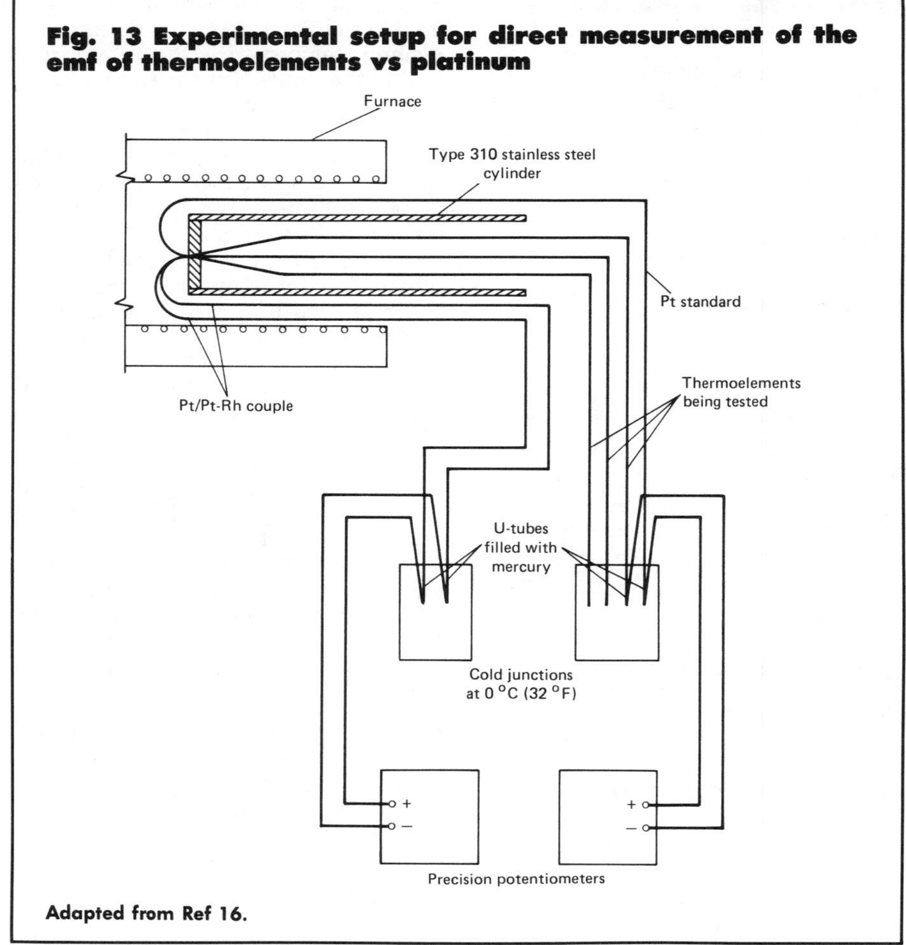

Adapted from Ref 16.

holding its temperature constant at about 10 °C (18 °F) above the freezing point for several minutes and then cooling slowly, or by agitating the metal with the thermocouple protection tube just before freezing begins. The emf of the thermocouple is observed at regular intervals of time. These values are plotted, and the emf corresponding to the flat portion of the cooling curve is the emf at the freezing point of the metal.

Metals of sufficient purity that may be used in freezing-point calibrations are tin (freezing point, 231.9681 °C), indium (156.634 °C), cadmium (321.108 °C), lead (327.502 °C), zinc (419.58 °C), antimony (630.755 °C), aluminum (660.46 °C), silver (961.93 °C), gold (1064.43 °C), copper (1084.88 °C), nickel (1455 °C), palladium (1554 °C) and platinum (1769 °C). The freezing points of zinc, silver and gold are primary reference points used to define the IPTS 68 between 630.755 °C and 1064.43 °C; the freezing points of the other metals listed above are secondary reference points. Of all these metals, antimony and tin have marked tendencies to undercool before freezing, but such undercooling will not be excessive if the liquid metal is stirred.

A Pt-10Rh/Pt thermocouple may be calibrated using values obtained at the freezing point of antimony (630.755 °C), the freezing point of silver (961.93 °C) and the freezing point of gold (1064.43 °C).

The emf developed by a homogeneous thermocouple at the freezing point of a metal is constant and reproducible if all of the following conditions are fulfilled:

- The thermocouple is protected from contamination.
- The thermocouple is immersed in the freezing-point sample sufficiently far to eliminate heating or cooling of the junction by heat flow along the wires and protection tube.
- The reference junctions are maintained at a constant and reproducible temperature.
- The freezing-point sample is of sufficient purity.
- The metal is maintained at an essentially uniform temperature during freezing.

Freezing points can be reproduced under industrial conditions within 0.1 to 5 °C for calibrations between the ice point and the melting point of platinum. Because of difficulty in testing, fixed points at temperatures above the freezing point of copper (1084.88 °C) usually are expressed as melting points rather than freezing points. Complete units including freezing-point sample, crucible and heating source are available commercially.

For additional information on the freezing-point method of calibrating thermocouples, the reader should consult Ref 11 to 14.

Direct emf Measurement Versus Pt. The method used for direct measurement of thermoelement emf versus platinum at a fixed temperature is illustrated in Fig. 13. The thermocouple wire specimens, a primary platinum standard and a calibrated reference-grade type R couple (Pt-13Rh/Pt) are welded together at one end. The multiple couple is in turn welded into a heat sink, and the whole assembly is inserted halfway into a 2-m (6-ft) horizontal electric furnace to a depth of approximately 1 m (3 ft).

Precision potentiometers are used so that measurement of emf versus platinum for a test specimen, and measurement of temperature by the calibrated platinum thermocouple, can be made simultaneously. The test specimens, the calibrated platinum couple and the platinum standard are inserted into mercury U-tubes embedded in Dewar flasks packed with shaved ice and are electrically connected to two precision potentiometers as shown in Fig. 13. Simultaneous measurements can be made as soon as the furnace temperature and the cold-junction temperature reach equilibrium.

Consider the case of emf measurements of Tophel, a type K positive thermoelement, at 980 °C (1800 °F). It is not necessary that the furnace temperature as measured by the calibrated Pt/Rh couple be exactly 980 °C. In actual practice, all that is needed is that the furnace temperature be within ±2.8 °C (±5 °F) of the desired temperature. The emf of the thermoelement at the desired test temperature can be computed with the following equation:

$$E_D = E_M - (T_M - T_D)\zeta \quad \text{(Eq 9)}$$

where E_D is the corrected specimen emf vs Pt at the desired temperature T_D, E_M is the measured emf at the measuring temperature T_M, and ζ is the thermoelectric power of the thermoelement vs Pt at T_D. As an example of the use of Eq 9, the following data were generated for a sample of Tophel:

Emf of Pt/Rh couple, mV	10.247
	10.248
	10.248
Corresponding temperature, T_M, °F	1801.4
Measured emf, E_M, mV	31.999
	32.000
	32.000
Thermoelectric power, mV/°F	0.017

Note:

E_D = Corrected emf of sample vs Pt at 1800 °F

$= 32.000\ \text{mV} - (1801.4 - 1800) \times 0.017\ \text{mV/°F}$

$= 32.000\ \text{mV} - 0.024\ \text{mV}$

$= 31.976\ \text{mV}$

This value is in excellent agreement with National Bureau of Standards (NBS) calibrations on both ends of a single coil of Tophel: 31.970 mV and 31.980 mV.

Comparison Method. In industrial practice, a thermoelement is calibrated at several fixed temperatures against a thermocouple standard of the same alloy calibrated by NBS. ASTM E207 describes the preferred standard method. The emf of the test specimen can be obtained by the following general equation:

$$\text{emf vs Pt} = \Delta\, emf_{\text{specimen vs std}} + emf_{\text{std vs Pt}} \qquad \text{(Eq 10)}$$

Figure 14(a) shows the emf vs Pt curve for a Tophel standard, a type K positive thermoelement. The type K positive thermoelement has a large thermoelectric power vs Pt. For example, its thermoelectric power vs Pt at 980 °C (1800 °F) is 31 μV/°C (17 μV/°F). An error of 34 μV would have been introduced by the temperature measurement error of only 1.1 °C (2 °F) in the case of direct measurement of emf vs Pt. However, this is not so in the comparison method, as can be readily observed from Fig. 14(b).

Figure 14(b) shows values of Δ emf vs NBS nominal emf as ordinates against temperature as abscissa. The emf deviations of the NBS-calibrated Tophel standard from NBS nominal emf values for the type K positive thermoelement are plotted with a heavy line. The deviation of the test specimen from NBS nominal emf at any test temperature can be obtained graphically by plotting measured Δ emf on the chart or by use of the following equation:

$$\Delta\, emf_{\text{specimen vs NBS nominal emf}} = \Delta\, emf_{\text{specimen vs std}} + (emf_{\text{std vs Pt}} - \text{NBS nominal emf vs Pt}) \qquad \text{(Eq 11)}$$

Fig. 14 Thermal emf plots for KP thermoelements illustrating the comparison method of emf measurement

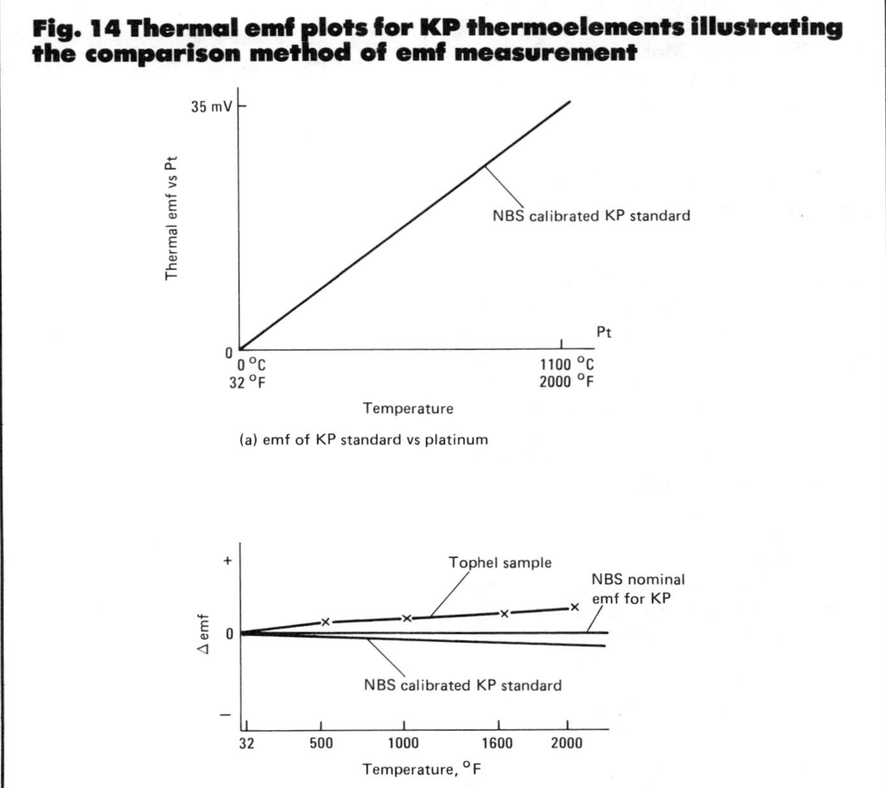

(a) emf of KP standard vs platinum

(b) Δemf of Tophel sample and KP standard thermoelements versus NBS nominal emf for type KP

Adapted from Ref 2.

Equation 11 is a general equation applicable to both the positive and negative thermoelements. If we calculate Δ emf for both thermoelements and substitute these values in Eq 4, then:

$$\Delta\, emf_{\text{couple}} = \Delta E_P - \Delta E_N \qquad \text{(Eq 12)}$$

Calibration of Thermocouples by Comparison Methods. Calibration of a thermocouple by comparing it to a working standard is sufficiently accurate for most purposes and can be done conveniently in most industrial and technical laboratories. The emf of the thermocouple being calibrated is measured at selected calibration points, the temperature of each point being measured by a standard thermocouple (usually one calibrated by NBS) or other standard thermometer. Test points are selected on the basis of thermocouple type, temperature range to be covered, accuracy required and end use.

The accuracy obtained with this technique depends on the ability of the observer to bring the junction of the thermocouple to the same temperature as that of the sensing portion of the standard used, such as the measuring junction of a standard thermocouple or the sensitive portion of a resistance or liquid-in-glass thermometer. The accuracy obtained is further limited by the accuracy of the standard. The method of bringing both measuring junctions to the same temperature depends on type of thermocouple, type of standard and method of heating.

Potentiometric instruments or high-impedance electronic instruments are used to measure emf, thus eliminating instrument loading as a contributor of significant error.

Additional information relative to

this calibration method and attainable accuracies may be found in ASTM E220, "Standard Method for Calibration of Thermocouples By Comparison Techniques", and in ANSI MC96.1.

Reference Tables for Thermocouples

Practical use of thermocouples requires that the selected thermocouple meet an established or standardized temperature/emf relationship within acceptable tolerance limits. Because the thermocouple in a thermoelectric thermometer system is replaced periodically due to drift, failure or other reasons, conformance to an established temperature/emf relationship is necessary in order to permit interchangeability when commercially available readout equipment is used. Such widely acceptable reference tables have S and T thermocouples and are available in NBS Monographs 124 (Cryogenic) and 125 (Standard Couples), ANSI MC96.1 and ASTM E230. Less detailed versions of these tables (at intervals of 10 °C, or 18 °F) usually may be obtained from producers or distributors of these thermocouples.

For other nonstandard thermocouples, including those that do not have letter designations, tables usually are developed by producers and are available from either producers or suppliers. Additionally, temperature/emf values for three W/Re combinations and for Nicrosil/Nisil have been published "for information" by ASTM in the standards book containing standards related to thermocouples, and one combination (W-3Re/W-25Re) has values published in ASTM E696.

All tables, in order to gain wide acceptance, must conform to an internationally recognized temperature scale. At this time, the scale is IPTS 68, and the latest published tables should conform to it. However, a large quantity of control or measurement instruments still in use are in compliance with IPTS 48, and replacement thermocouples for these instruments are purchased on this scale. The difference in the two scales may or may not be significant depending on the application. In this regard, particular attention should be paid to types S, R and B thermocouples for use above 1000 °C (1830 °F). See Ref 10 for differences arising from use of either scale.

Fig. 15 Thermal emf plots for a type K thermocouple illustrating the method of evaluating emf deviation

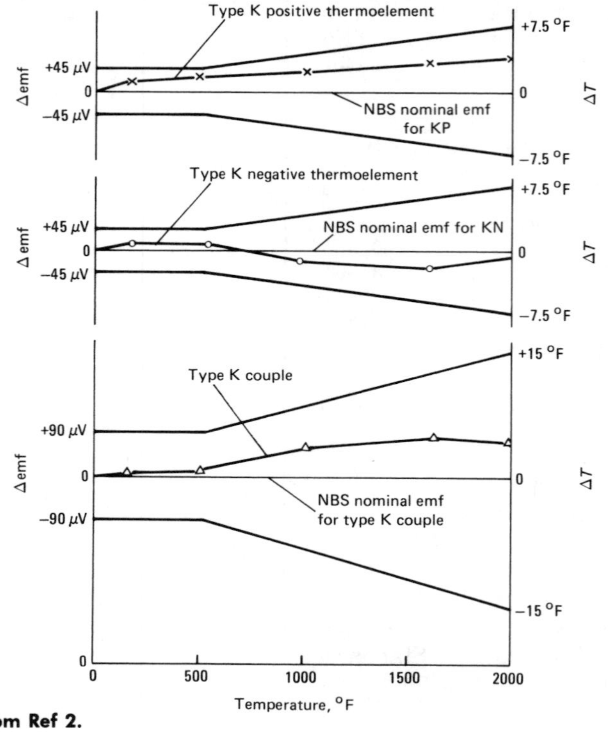

Adapted from Ref 2.

Initial Calibration Tolerances. Table 2 lists manufacturers' tolerances for initial calibration of all standardized thermocouples. For example, a brand new type K couple could be in error by as much as ±4.2 °C (±7.5 °F) when used for temperature measurement at 540 °C (1000 °F). The deviation in emf of a thermocouple from the standard table value is equal to the algebraic difference of the individual emf deviations of the thermoelements, as shown in the following equation:

$$\Delta E_{couple} = \Delta E_P - \Delta E_N \qquad \text{(Eq 13)}$$

where ΔE_{couple} is the emf deviation of the couple from table value, in millivolts; ΔE_P is the emf deviation of the positive thermoelement from NBS nominal value, in millivolts; and ΔE_N is the emf deviation of the negative thermoelement from NBS nominal value, in millivolts.

The deviations in initial calibration of a typical type K couple from NBS table values are illustrated in Fig. 15. The corresponding deviation of initial calibration expressed in temperature is obtained as follows:

$$\Delta T = \frac{\Delta E_{couple}}{\text{Th.p.}} \qquad \text{(Eq 14)}$$

where ΔE is the emf deviation of the couple at a certain temperature and Th.p. is the thermoelectric power of the couple at the same temperature.

Change of Calibration During Service

Any thermocouple can be subject to failure (of a type that creates an open circuit) during service. Failure can be caused by localized melting of the thermoelements as a result of overheating, by vibration resulting in fatigue failure, or by gradual reduction of wire diameter through high-temperature oxidation. Prior to failure, the emf calibration of a thermocouple will change, primarily as a result of the individual or combined changes in chemical composition, homogeneity and structure that take place in the thermoelements. The magnitudes and directions of these changes are dependent on temperature, time, wire diameter and environmental conditions.

Fig. 16 Changes in thermal emf of a type K thermocouple resulting from long-time exposure in air at temperatures up to 1100 °C (2000 °F)

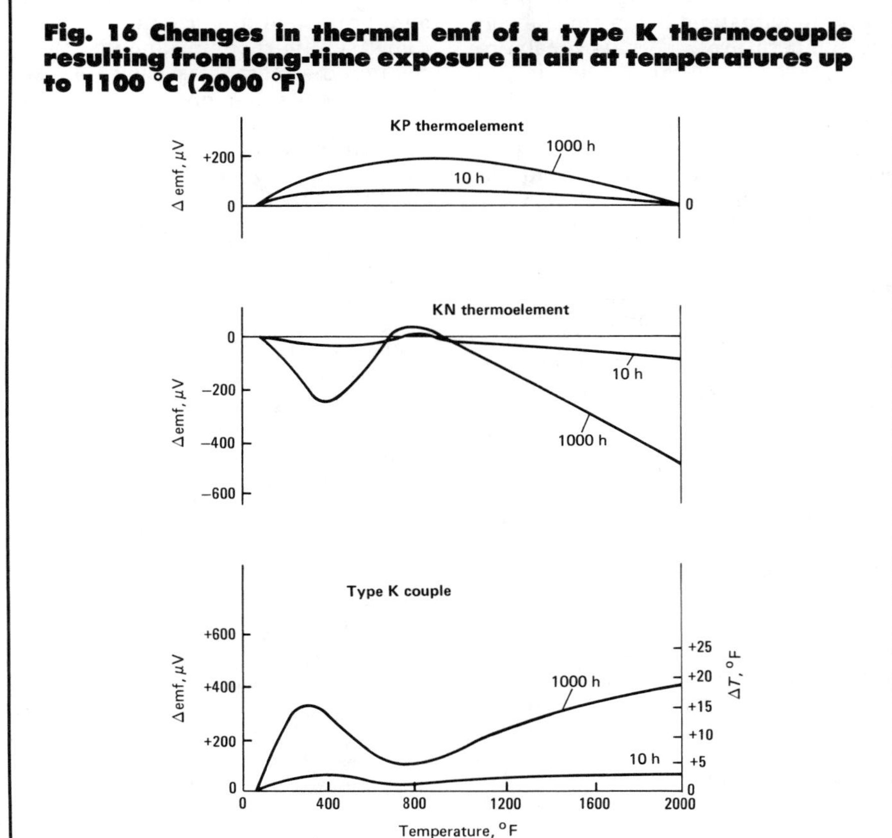

Effect of Environment on Base-Metal Thermocouples. The change in emf, as a function of test temperature, of a 3.25-mm-diam (0.128-in.-diam) type K thermocouple on exposure to air at 1083 °C (2000 °F) is shown in Fig. 16. After 10 h of exposure, the emf of the couple had changed about 3 °C (5 °F) at 1083 °C in the positive direction. The change had increased to +10 °C (+18 °F) after exposure for 1000 h. At lower test temperatures, the magnitude of the change is smaller. A decrease in silicon and chromium contents of the positive thermoelement through preferential oxidation causes a net change in emf. Similarly, the change in emf of the negative thermoelement is attributed to preferential oxidation of its alloy constituents Si, Mn, Al and Fe (Ref 15 and 16).

The oxidation resistance and emf stability of type J couples are inferior to those of type K couples, and type J couples should not be used above 760 °C (1400 °F).

Type K couples are recommended for use in inert atmospheres at elevated temperatures only for short intervals. Type J couples, on the other hand, are stable and can perform better in inert environments than in air.

Type K thermocouples are not recommended for use in reducing or hydrogen-bearing atmospheres. Type J couples are stable and can be expected to perform well at temperatures up to 760 °C (1400 °F).

Effect of Environment on Bare Pt-Rh Thermocouples. Pt-10Rh/Pt, Pt-13Rh/Pt and Pt-30Rh/Pt-6Rh thermocouples can be used with very good results continuously in air or in oxidizing atmospheres (to 1500 °C, or 2730 °F, for types S and R; to 1700 °C, or 3090 °F, for type B) and intermittently to temperatures approaching the melting point of platinum (1769 °C, or 3216 °F) for types S and R and to 1780 °C (3235 °F) for type B. For these couples in these atmospheres, life is governed by the temperature of operation, partial pressure of oxygen, rate of change of the atmosphere in the vicinity of the hot junction, and method of mounting of the thermocouple.

It is generally agreed that volatile oxides of platinum and rhodium are formed when these metals are heated at high temperatures and are the principal cause of the loss of metal. Experience indicates that slightly more rhodium than platinum volatilizes in air, which results in a negative drift of the couple after long periods of operation. A negative drift of 6 to 9 °C (11 to 16 °F) has been reported for a thermocouple that was in continuous use in air at 1290 °C (2350 °F) for over three years.

Bare Pt-Rh thermocouples can be used in inert atmospheres such as argon, helium or nitrogen with very good results.

As far as can be ascertained, reducing gases such as carbon monoxide and hydrogen do not have adverse effects on types S, R and B couples directly, but it is suspected that these gases reduce impurity oxides such as silica, which is usually present in alumina. The silicon reduced from the silica is known to unite with platinum to form a low-melting eutectic (830 °C, 1530 °F). Close contact of these couples by easily reduced oxides of any metal should not be permitted.

Type S, type R and type B thermocouples have been used for short periods of time in vacuum. Long-time exposure to vacuum is not recommended.

It has been reported that unstable hydrocarbons crack in contact with hot platinum-group metals, causing damage to these metals in the form of a fine intergranular precipitate of carbon.

Halogen gases have harmful effects on platinum-group metals at high temperatures.

Direct contact between bare couples and compounds of easily reduced metals such as lead, bismuth and antimony should be avoided at high temperatures, because such contact results in formation of low-fusing platinum alloys.

Unprotected platinum thermocouples are attacked by phosphorus, arsenic, sulfur and vapors of metals such as zinc and lead. This attack generally results in brittleness and hot shortness.

All contact between bare couples and caustic alkalis, nitrates, cyanides, alkaline earths and the hydroxides of barium and lithium should be avoided, because these substances attack platinum at red heat.

Judicious use of insulators and protection tubes will eliminate problems with many of the contaminants listed above. A variety of ceramics, some of which are gastight, are available, but it should be kept in mind that no single ceramic will suit all applications.

Certain precautions should be followed when platinum-group metals are in contact with ceramics in reducing atmospheres. It is generally known that when platinum or platinum alloys in contact with silica are heated in reducing atmospheres above 1200 °C (2200 °F), platinum silicides with low melting points (as low as 830 °C or 1530 °F) are formed at the grain boundaries, which results in embrittlement of the wire. It has been shown that this attack also occurs at and above 1100 °C (2000 °F) when platinum and platinum alloys are adjacent to but not in contact with silica-bearing materials.

In experiments conducted at about 1100 °C, in which silica was present in the alumina insulation and carbon and sulfur were also present (residue of drawing compound on wire), thermocouple wires failed due to melting. Based on failure analysis, it was hypothesized that a volatile compound of SiS_2 is first formed, which serves to transport silicon present in the insulator or protection tube to the platinum or platinum alloy. This compound decomposes in contact with the hot platinum, and the liberated sulfur recombines with additional silicon in the refractory.

It is quite obvious that all traces of lubricating oils, drawing compounds or other sulfur-bearing compounds should be removed from the thermocouple assembly, because silica is present in varying amounts in all commercial refractories (particularly mullite and sillimanite). Fractures in wires, caused by platinum silicide, generally present a melted appearance, and the fracture surface contains a number of glazed areas.

Insulation and Protection

To operate properly, thermocouple wires must be electrically insulated from one another at all points other than the measuring junction and must be protected from the operating environment.

Insulation. For cryogenic applications (below 0 °C and as low as about 4K) varnish or varnish-type coatings are used to insulate thermoelements from one another. The coating usually is selected on the basis of good electrical resistance, ease of application and ability to withstand flexing at the very low temperatures. Formvar, polyurethane, teflon, Pyre-ML-Polyimide (DuPont) and GE 7031 (General Electric) have been used for this purpose. In particular, the polyimide coating has not only good electrical resistance but also excellent flexing strength at very low temperatures.

At the lower temperatures above ambient, fiberglass, asbestos, rubber, fabrics, enamels and various plastics are used for insulation. Each type has its own limitations, and a knowledge of these limitations is essential if accurate and reliable measurements are to be made. It is important that these types of insulation be selected only after consideration of exposure temperatures, heating rates, number of temperature cycles, mechanical handling, moisture, routing of wires and chemical deterioration.

At temperatures above approximately 300 °C (570 °F), hard-fired ceramic insulators are used on most bare thermocouple elements. Such insulators are available with single, double or multiple bores, and in a variety of shapes, diameters and lengths. The thermocouple supplier should be consulted on the type or types of insulation that are available for each specific application. The hard-fired ceramic insulators that are used with base-metal thermocouples are mullite, aluminum oxide and steatite. Steatite is the most commonly used material for fish-spline insulators.

Platinum-rhodium thermocouples (types R, S and B) for use below 1000 °C (1830 °F) may be insulated with quartz, mullite, sillimanite or porcelain. Mullite and sillimanite have been used in industrial applications involving oxidizing atmospheres and temperatures from 1000 to 1400 °C (1830 to 2550 °F), but 99% Al_2O_3 is preferable for such service. Because both of these materials contain silica in various proportions, care should be taken to prevent promotion of a reducing atmosphere via carbonaceous impurities (such as residual lubricant on thermoelements). For all laboratory uses, for industrial uses in slightly reducing atmospheres (above 1000 °C), for critical applications and for all uses of type B couples to around 1750 °C (3180 °F), pure, sintered, dense alumina (99.5% min Al_2O_3) is recommended. This insulation should be of one-piece, full-length construction to provide maximum protection from contamination.

For iridium-rhodium and tungsten-rhenium thermocouples, choice of insulation depends on temperature of use as well as environment. Hard-fired insulators of high-purity alumina may be used to approximately 1800 °C (3270 °F). From 1800 °C to approximately 2300 °C (4170 °F), beryllium oxide (melting point: 2570 °C, or 4650 °F) should be considered. However, when beryllium oxide is used, certain safety precautions are necessary.

When hard-fired, dense, beryllia insulators are used, dimensional changes should be considered in design of the temperature-measuring system if it is to be used at or above approximately 2150 °C (3900 °F). At this temperature, beryllia undergoes a phase change. The problem is not serious in swaged thermocouples when crushable beryllia is used.

Thoria, which has a melting point higher than that of beryllia, has been used at temperatures up to about 2500 °C (4500 °F). However, the low electrical resistivity of this ceramic material limits its applications at very high temperatures. Hafnia has been used on an experimental basis with some success.

In addition to conventional thermocouple assemblies with hard-fired ceramic insulators, sheathed, compacted, ceramic-insulated thermocouples are in common use. Magnesium oxide generally is used as the insulating material. A more detailed discussion of this type of construction is presented in a later section on metal sheathed thermocouples.

Protection. Closed-end tubes made of metal, porcelain, mullite, sillimanite, quartz or pyrex-glass may be used to prevent contamination of thermocouple sensing elements by the environment and to provide mechanical protection and support. Such tubes are called protection tubes. In some instances, two concentric tubes are employed. A protection tube must be large enough in inside diameter to accommodate an insulated matched couple (positive and negative thermoelements joined at the hot end). However, larger-diameter tubes may be used (*a*) for strength, (*b*) to permit insertion of a checking thermocouple alongside the service thermocouple and (*c*) to provide an adequate diameter-to-length ratio. Metallic protection tubes are generally available in pipe sizes of ½ in., ¾ in. and 1 in.

Bare, insulated, base-metal thermocouples may be inserted directly into base-metal protection tubes. For noble-metal thermocouples, however, a ceramic protection tube generally is employed between the couple and the base-metal protection tube. For severe operating environments at elevated temperatures, platinum or platinum-rhodium protection tubes may be used. Bare but insulated noble-metal thermocouples may be inserted directly into these tubes. In any event, protection tubes must be internally clean and free of sulfur-bearing compounds, oils and easily reduced oxides.

A wide range of metal and ceramic protection tubes is available commercially (see for example Table 13). This allows for the selection of a particular protection tube for a specific application.

Steel protection tubes may be used at temperatures up to about 500 °C (930 °F). Stainless steels of the 18-8 variety may be used at up to 800 °C (1470 °F), and stainless steels of higher alloy content at up to about 1000 °C (1830 °F). The 80Ni-20Cr alloys and certain Ni-Cr-Fe alloys may be used to around 1100 °C (2000 °F), with the latter having better resistance to sulfur. It should be remembered that high-nickel alloys should not be used in sulfur-containing atmospheres at temperatures above 400 °C (750 °F).

Protection tubes for platinum-rhodium thermocouples (types R, S and B) have been made of quartz for service at temperatures up to about 1000 °C (1830 °F), and of mullite for service up to around 1650 °C (3000 °F). Both materials have good resistance to thermal shock. However, in order to ensure long life and emf stability, fused alumina tubes or insulators are preferable for such couples at temperatures above 1200 °C (2200 °F). Fused alumina tubes are more expensive than mullite tubes and have lower resistance to thermal shock. Double ceramic tubes, comprising fused-alumina primary tubes and mullite secondary tubes, are used in certain applications.

Metal protection tubes made of iridium, tantalum, tungsten and molybdenum, and of Ir-Rh, Nb-1Zr, W-26Re and Mo-50Re alloys, have been used to protect tungsten-rhenium thermocouples at high temperature. The noble metal iridium is the only known metal that may be used in air unprotected for short periods of time at temperatures up to approximately 2100 °C (3800 °F) without undergoing catastrophic failure. Iridium-rhodium alloys may be used under similar circumstances up to 2000 °C (3600 °F). Experience has shown that, within their recommended temperature ranges, Ir-Rh alloys have better oxidation resistance than that of iridium.

Table 13 Maximum service temperatures for protection tubes
Adapted from ANSI MC96.1

Materials	Maximum service temperature °C	°F
Carbon steel	540	1000
Wrought iron	700	1300
Cast iron	700	1300
304 stainless steel	870	1600
316 stainless steel	870	1600
Chrome iron (446)	980	1800
Nickel	980	1800
Inconel	1150	2100
Porcelain	1650(a)	3000(a)
Silicon carbide	1650	3000
Sillimanite	1650(a)	3000(a)
Aluminum oxide	1760(a)	3200(a)

(a) Horizontal tubes should receive additional support above 1480 °C (2700 °F).

The refractory metal tubes noted above must always be used in inert atmospheres or in a good vacuum. Of these, the tantalum and Nb-1Zr tubes, because of their excellent cold workability, have found extensive use in swaged-type W-Re thermocouples. They are presently being used at temperatures up to approximately 2100 °C (3800 °F).

The Mo-50Re alloy has some interesting possibilities as a material for protective sheaths. This alloy has some cold workability and, more important, is still ductile at room temperature after exposure to temperatures above its recrystallization temperature. At these high temperatures, cleanness of both wire and tubing (thermowell or swaged sheath) is very important. It has been found that carbon present in the tubing (possibly lubricant residue) can react with beryllium oxide insulators at temperatures below 2000 °C (3600 °F).

Protecting wells are employed for thermocouples used in liquids and gases at high pressure. These wells are made of metal, and may be turned and drilled from bar stock or built up by welding. Materials such as stainless steels (18-8), carbon steel and 14% chromium iron are used to fabricate these wells depending on end use.

The foregoing paragraphs describe the procedures used to insulate and protect conventional thermocouples ("bare-wire" thermocouples). Sheathed thermocouple elements ("swaged-type" thermocouples) are fabricated from commercially available sheathed thermocouple wires. Fabricating such thermocouples successfully requires special equipment, special precautions and more skill than is usually required for fabricating conventional bare-wire thermocouple assemblies. This type of thermocouple assembly is described briefly in the following section.

Thermocouple Assemblies

Conventional Thermocouples. Some typical thermocouple assemblies employed in industrial applications are shown in Fig. 17. In the assembly shown at the top of this figure, a closed-end pipe protection tube may be substituted for the nipple and ceramic protection tube in base-metal thermocouple applications. For additional details, see also ANSI MC96.1, "Temperature Measurement Thermocouples", and suppliers' literature.

Metal-Sheathed Thermocouples. In metal-sheathed couples, the wires are insulated from each other and from the sheath by means of compressed pure refractory oxide powder. The resulting assembly (thermocouple wires, oxide powder and integral sheath) is flexible enough to be formed around a diameter equal to four times that of the assembly, without damage.

Fabrication of a metal-sheathed thermocouple is simple and begins with matched thermocouple wires surrounded by a partly sintered ceramic material held within a metal tube. By swaging, drawing or any other

Fig. 17 Typical industrial thermocouples insulated with hard-fired ceramics

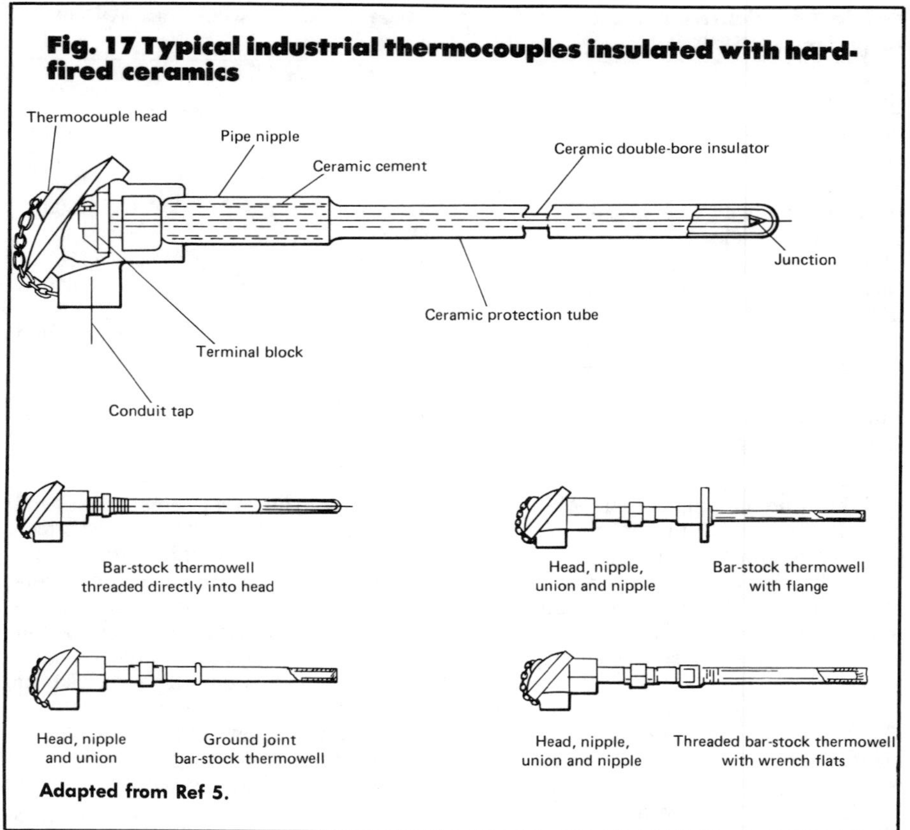

Adapted from Ref 5.

mechanical-reduction process, the assembly is reduced in diameter. As a result of this working, the insulation is first broken into powder and then is compacted around the wires while the assembly is elongated.

An assembly produced in this fashion should have a minimum insulation resistance of 100 MΩ at 500 V dc for sizes larger than 1.6 mm (1/16 in.) in outside diameter. This requires some care during fabrication and use of dry, uncontaminated compacted ceramic. Because of the hygroscopic nature of powdered ceramics—especially MgO—moisture can be absorbed through the exposed ends of the sheath by capillary action. For this reason, the metal-sheathed couple or cable should be purchased with the ends closed by welding or suitably sealed in some other manner. Under certain circumstances, organic seals may not be suitable for this purpose. It has been reported that cable sealed with an organic material leaked when shipped by air freight, with a resulting decalibration of the thermocouple. The following precautions should be exercised when handling compacted ceramic-insulated thermocouples, in order to preserve the integrity of the insulation:

- Never leave an end of a sheathed couple exposed for more than 2 or 3 min; seal ends immediately. Use appropriate seal, depending on method of shipping.
- Expose ends only in areas of low relative humidity.
- Store sheathed assemblies in an area that is warm (above 38 °C, or 100 °F) and dry (relative humidity less than 25%).

Sheaths are selected to suit specific end-use requirements. The materials that have been used for this purpose include: types 304, 310, 316, 321, 347 and 440; stainless steels; platinum alloys; Hastelloy X; copper; aluminum; Inconel 600; Inconel 702; and tantalum and niobium alloys.

Depending on temperature and application, magnesia, alumina, beryllia or thoria may be used for the insulation. Grounded (to sheath) or ungrounded junctions may be supplied as required, the former having faster response time in temperature sensing.

Among the advantages of compacted sheathed thermocouple construction are small dimensions (as small as 0.5 mm, or 0.020 in., OD) and flexibility. In addition, the assembly is completely resistant to thermal shock, to which more conventional assemblies comprising hard-fired insulators and outer refractory ceramic sheaths are prone.

Criteria for Selection of Thermocouples for Industrial Applications

No thermocouple meets all requirements of temperature measurements over the entire range from cryogenic through 2700 °C (4900 °F). However, each of the previously discussed standard or nonstandard thermocouples possesses characteristics most desirable for a particular application. The following criteria should be given careful consideration during selection and design of thermocouple systems:

Performance requirements
- Accuracy in temperature measurement
- EMF stability
- Service life

Operating environment
- Temperature range
- Time at temperature
- Temperature gradient
- Thermal cycling
- Effect of pressure or vacuum
- Nuclear radiation
- Chemical compatibility

Cost and availability
- Initial and replacement cost of thermocouple (parts and labor)
- Initial and replacement cost of thermocouple extension wire (parts and labor)
- Initial and replacement cost of thermocouple accessories (parts and labor)
- Downtime
- Delivery time (immediate or extended)

Design selection
- Thermocouple and extension wires (types)
- Temperature/emf relationship and temperature range
- Sensitivity of couple
- Available wire diameter
- Insulation and protection (types)
- Bare-wire versus sheathed construction, with proper end sealing for sheathed construction
- Assembly configuration and type of measuring junction
- Chemical, physical and mechanical properties (electrical resistance,

temperature coefficient of resistivity, coefficient of expansion, thermal conductivity, density, specific heat).

Cryogenic Applications. With the exception of types J, E, T and K, standard thermocouples developed for use at moderate or high temperatures are too low in sensitivity at cryogenic temperatures to be of any practical value in cryogenic applications. Of these four, type E, and to a lesser extent type T, are suitable for general low-temperature service down to −200 °C (−330 °F).

Advocated for applications at still lower temperatures (below 50 K, and possibly as low as 4 K) is a thermocouple consisting of gold plus a trace amount of iron versus either the KP or the EP thermoelement (90Ni-9Cr; see Table 1). Actually, three different Au-Fe thermoelements have been used: Au−0.02 at.% Fe, Au−0.03 at.% Fe and Au−0.07 at.% Fe. Of these three, the latter may have wider application.

To ensure a high emf in the cryogenic range as well as reproducibility from lot to lot, the Au-Fe alloys must be carefully prepared. High-purity (99.999%) gold is used, and trace amounts of iron are added. For example, only 57 ppm (by weight) of iron are added to the gold in making Au−0.02 at.% Fe, 85 ppm in making Au−0.03 at.% Fe and 200 ppm in producing Au−0.07 at.% Fe.

If standard thermocouples (types E and T) are intended for cryogenic use (to −200 °C), the supplier should be notified of this intention in order to facilitate selection of materials. If a Au-Fe/KP or Au-Fe/EP couple is being considered, it should be kept in mind that the gold-bearing thermoelement must be made to special order because it is not a stocked item. Wires used in cryogenic applications generally are fine-gage wires (0.10 to 0.15 mm, or 0.004 to 0.006 in., in diameter).

Good Thermocouple Practice

After the proper thermocouple and extension wire have been selected, care must be taken in installation of the thermocouple system to ensure that errors are not introduced which can affect service. Following are some precautions that should be observed:

- Avoid cold working of thermocouple and extension wire. Excessive deformation can adversely affect accuracy of the thermocouple. Severe bending, flexing or hammering of the thermocouple wire should be avoided. If cold working does occur, heat treatment should be considered to remove its effects. The degree of cold work that may be tolerated without annealing will depend on the end use and accuracy required.
- Extraneous junctions should be monitored—as when connecting lead wires in a thermocouple circuit and when connecting the circuit to a recorder. The solution is to maintain a uniform ambient temperature in the vicinity of these junctions and at the measuring device.
- Provide adequate protection. Generally, thermocouples must be equipped with suitable protection in the form of wells or protection tubes to guard the immersed portion against physical damage or contamination.

A protecting tube is a tube designed to enclose a temperature sensing device and protect it from the deleterious effects of the environment. It may provide for attachment to a connection head, but it is not primarily designed for pressure-tight attachment. A thermowell is a pressure-tight receptacle adapted to receive a temperature sensing element and is provided with external threads or other means for pressure-tight attachment to a vessel. There are many varieties of these tubes and wells, in various metals, alloys and refractory materials, available on the market today to meet special requirements. Adequate protection may also be obtained through the use of metal sheathed, mineral insulated thermocouples. In the latter case, the sheath is an integral part of the thermocouple assembly. On the debit side, it should be stated that the protecting tube interferes with ideal temperature measurement and control. It decreases the sensitivity of measurement (speed of response) and increases installation space and cost.

- Select largest practical wire size for a particular end use. The largest practical wire size should be used, consistent with end use requirements such as rapid response, flexibility, and available space. Heavy gage thermocouples have greater long-term stability at high temperatures than thermocouples of a lighter gage but also have a slower response. Speed of response, or rate at which the thermocouple detects temperature changes, may be of vital concern in many applications; particularly where these changes occur rapidly. However, many factors of heat transfer affect the speed of response of a thermocouple and the mass of the couple is only one of them.
- Thermocouples should be located properly to achieve maximum benefits. The thermocouple should be placed so that the measured temperature is representative of the equipment or medium that is being studied. For example, stagnant areas (not at representative temperature) or exposure to direct flame impingement (unless desired) could result in erroneous readings. If the thermocouple is immersed in a fluid (liquid or gas), the depth of immersion should be sufficient to minimize heat transfer away from the measuring junction. For many applications, a minimum immersion depth of ten times the outside diameter of the protection tube is considered adequate to prevent serious temperature errors (readings on low side). Also to be considered is the probability of radiation heat transfer to a bare thermocouple junction from the environment. In this case, radiation shields should be used.
- When measuring high temperatures, install the thermocouple vertically whenever possible to prevent sagging of the protection tube. However, care should be taken to properly support the thermocouple within the tube. This is particularly true for the noble metal thermocouples when used at elevated temperatures (greater than 600 °C). In this case, the thermocouple assembly may be supported by resting the bead of the thermocouple and the insulator on the bottom of the tube (on high purity alumina powder when a metal tube is used).
- Make sure that the protecting tube or well extends far enough beyond the outer surface of the vessel or heat source to bring the connecting head to approximately ambient temperature (particularly with type K using alternate extension wire and types R or S).
- When making thermocouples, clean the free ends of the wire well before fastening them to the connecting head, and be sure that they are inserted with proper polarity as identified on the terminal block.

Maintenance of Thermocouples

Scheduled maintenance can be beneficial. The life and reliability of a thermocouple measuring system can be improved, and the likelihood of catastrophic failure is reduced if recommended calibration and maintenance procedures are followed. In addition, any gradual aging or drift can be determined by periodic observation and recording of thermocouple behavior. Based on information obtained in the maintenance program, scheduled replacement of the thermocouple probe can be made (*a*) before it has deteriorated beyond acceptable limits or (*b*) before failure. The portion of the Metal Treating Institute specification MTI2000, "Quality Assurance Specifications for Performance of Heat Treating Processes", relating to temperature measurement is one example of a planned maintenance approach.

The personnel employed in the maintenance program should be familiar with the operating procedures upon which the system is based. Of primary importance, the equipment used for maintenance should be in good working order.

The following items are generally considered a good basis for planned maintenance:

- Thermocouples should be checked regularly at intervals determined by experience. For example, checking base metal thermocouples once a month may be sufficient for some applications but not for others. Exceptions could vary greatly.
- If a thermocouple must be removed for examination, carefully reinsert it so as not to change the depth of immersion. Most important of all, do not decrease immersion.
- It is preferable to check thermocouples in place. However, rather than checking thermocouples, it may be preferable to replace thermocouples after a predetermined average life has been achieved. This would ensure nearly perfect operation without the periodic problem of checking thermocouples.
- A type K thermocouple should not be exposed to temperatures greater than 760 °C (1400 °F) if it is to be used for accurate temperature measurement below 570 °C (1000 °F).
- "Burned out" protection tubes should not be used, otherwise the thermocouple may be damaged or ruined. In particular, old, previously utilized, protection tubes and insulators should not be used with new noble metal thermocouples because of possible contamination (especially true for types R and S thermocouples).
- Contacts must be kept clean if switches are used in the thermocouple circuit.
- When recording or indicating potentiometers are connected in parallel for operation from a single thermocouple, the circuit should be analyzed carefully to determine possible effects of one instrument on the other.

Troubleshooting. The prime requisite in troubleshooting is a good knowledge and understanding of how the temperature-measuring system operates and why it fails to operate. Familiarity with the previous history of operation of a particular system can be important in quickly determining and correcting a problem.

In general, troubleshooting entails systematic checking of the system one section at a time. Readings should be taken with independent instruments or you can substitute components that were previously found to be in good working order. This is continued until the problem has been isolated. If an independent source of heat is used to stimulate the system, it may not be possible to produce the same temperature gradients found in normal operation, and the emf results obtained with aged wire will not correspond with those obtained in actual operation.

Because it would be difficult, if not impossible, to discuss all potential operating problems, several common ones have been selected to get an idea of what may be expected in actual practice.

- First make sure that the correct extension or compensating extension wire has been used. This is particularly true for installations that may have a number of different types of thermocouples. In this case, it is not unusual to find extension wire designed for one type of thermocouple to be used with another. Also, it is not unusual to find that the initially purchased extension wire or thermocouples were wrong. A common mistake is to use compensating extension wire intended for the type R or type S thermocouple with the type B couple.
- The positive extension wire must be connected to the positive thermoelement and the negative extension wire must be connected to the negative thermoelement. A very large temperature error will be introduced if the polarity is reversed. Fig. 18 shows that a temperature error of about 200 °C (355 °F) is observed when a type K couple is used in measuring a temperature of 982 °C (1800 °F) with its extension wire in a reversed position.
- Check the polarity of spliced extension wire and make certain that the splice will not be subject to intermittent or high resistance contact. The system may be installed by electricians or mechanics who may not appreciate the seriousness of polarity reversal.
- Make sure that the thermocouple is suitable for the instrument used and that the connections at the terminal block are tight.
- If the extension wire color coding differs from the standard U. S. color coding shown in Table 10, 11 and 12 (see also ANSI Standard MC96.1, Temperature Measurement Thermocouples), particularly in new installations, check to see that this particular extension wire is correct for this thermocouple.
- When there is doubt about the type of a thermocouple used, there are several ways by which this can be determined quickly. This can be done visually as for example distinguishing the noble metal thermocouples (types R, S and B) from the base metal ones (types J, K, E and T); copper leg of the type T thermocouple. With the use of a magnet, the positive legs of the types E, J and K couples can be distinguished (magnetic). Other checks consist of making up a thermocouple of the lead wire and checking the output at a fixed temperature; this is also true for distinguishing whether the couple is type R or S. Checks on lead wire may not be necessary where standard color coding is clearly distinguishable but may be useful in old installations where the color coding has become faded.
- Checking the resistance of a thermocouple circuit will indicate immedi-

Fig. 18 Effect of reversed polarity at the head junction on emf output of a type K thermocouple assembly

$$E_t = E_{PN}\Big|_{T_H}^{T_2} + E_{PXNX}\Big|_{T_1}^{T_H}$$

Case 1: Correct polarity

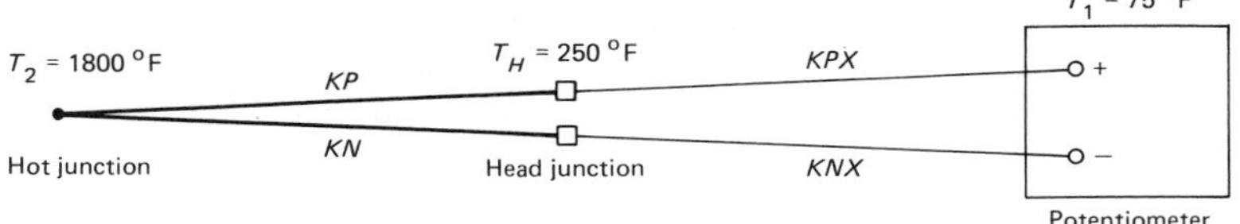

$$E_t = E_{KPKN}\Big|_{250\,°F}^{1800\,°F} + E_{KPXKNX}\Big|_{75\,°F}^{250\,°F} + E_{KPXKNX}\Big|_{32\,°F}^{75\,°F}$$

$$E_t = (40.62 - 4.97) + (4.97 - 1.00) + 1.00 \text{ mV}$$

$$E_t = 40.62 \text{ mV; equivalent to } 1800\,°F$$

Case 2: Reversed polarity

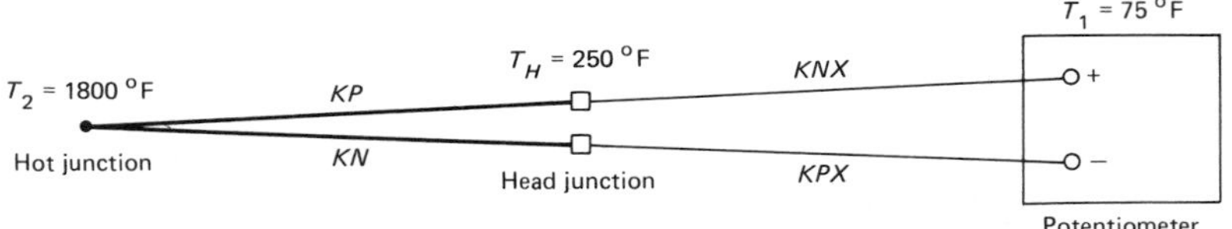

$$E_t = E_{KPKN}\Big|_{250\,°F}^{1800\,°F} - E_{KPXKNX}\Big|_{75\,°F}^{250\,°F} + E_{KPXKNX}\Big|_{32\,°F}^{75\,°F}$$

$$E_t = (40.62 - 4.97) - (4.97 - 1.00) + 1.00 \text{ mV}$$

$$E_t = 32.68 \text{ mV; equivalent to } 1445\,°F$$

Adapted from Ref 2.

ately whether it is in good condition. Low resistance may be equated to the probability of good performance. High resistance may be an indication that the thermocouple is nearing the end of its useful life or that there is a loose connection. In particular, this is a good test in installations where a large number of couples are connected to a readout device through switches.

REFERENCES

1. Temperature Sensors, by T. P. Wang: *Instrument and Control Systems,* Vol 40, 1967, p 100
2. "EMF Measurements", by T. P. Wang: Technical Paper MF77-958, Society of Manufacturing Engineers, 1977
3. *Protective Atmospheres,* by A. G. Hotchkiss and H. M. Webber: Wiley, New York, 1953, p 1295
4. "The Nicrosil Versus Nisil Thermocouple Properties and Thermoelectric Reference Data", by N. A. Burley, R. L. Powell and G. W. Burns: NBS Monograph 161, National Bureau of Standards, 1978
5. *Manual on the Use of Thermocouples in Temperature Measurement:* ASTM STP 470B (revised 1980), American Society for Testing and Materials, 1980
6. Nicrosil-Nisil Thermocouples in Production Furnaces in the 538 °C (1000 °F) to 1177 °C (2150 °F) Range, by T. P. Wang and C. D. Starr: *ISA Transactions,* Vol 18, No. 4, 1979, p 83
7. Electromotive Force Stability of Nicrosil-Nisil, by T. P. Wang and C. D. Starr: *ASTM Journal of Testing and Calibration,* Vol 8, No. 4, July 1980, p 192
8. A New Stable Nickel-Base Thermocouple, by C. D. Starr and T. P. Wang: *ASTM Journal of Testing and Evaluation,* Vol 21, 1976, p 42
9. The HI BX, A New Type B Thermocouple Extension Wire, by T. P. Wang and C. D. Starr: *ISA Transactions,* Vol 16, No. 3, 1977, p 85
10. The International Practical Temperature Scale of 1968, Amended Edition of 1975: *Metrologia,* Vol 12, 1976
11. "Methods of Testing Thermocouple Materials", by W. F. Roeser and S. T. Lomberger: NBS Circular 590, National Bureau of Standards, 1958
12. The Freezing Points of High Purity Metals as Precision Temperature Standards, by E. H. McClaren: in *Temperature, Its Measurement and Control in Sci-*

ence and Industry, Vol 3, Part 1, Reinhold, New York, 1962, p 185

13. W. G. Trabolt: in *Temperature, Its Measurement and Control in Science and Industry,* Vol 3, Part 2, Reinhold, New York, 1962, p 45
14. "Thermocouple Reference Table Based on IPTS 68": NBS Monograph 125, National Bureau of Standards, 1973
15. "The EMF Stability of Type K Thermocouple Alloys", by T. P. Wang, A. J. Gottlieb and C. D. Starr: Society of Automotive Engineers, 1969
16. Effect of Oxidation on Stability of Thermocouples, by C. D. Starr and T. P. Wang: Transactions of ASTM, Vol 63, 1963

ADDITIONAL REFERENCES

- Newer Thermocouple Materials, by E. D. Zysk and A. R. Robertson: in *Temperature, Its Measurement and Control in Science and Industry,* Vol 4, Part 3, Instrument Society of America, 1967
- Platinum Metal Thermocouples, by E. D. Zysk: in *Temperature, Its Measurement and Control in Science and Industry,* Vol 3, Part 2, Reinhold, New York, 1962
- "Noble Metals in Thermometry", by E. D. Zysk: Engelhard Industries Technical Bulletin, Vol 5, No. 3, Dec 1964
- "Calibration of Refractory Metal Thermocouples", by E. D. Zysk and D. A. Toenshoff: Paper # 12, 11-4-66, Instrument Society of America, Oct 1966
- "Thermocouple Thermometers": PMC Standard No. 8-10-1963, Scientific Apparatus Makers Association, Process Measurement and Control Section, 1963
- "Precision Measurement and Calibration Temperature": NBS Special Publication, Vol II, National Bureau of Standards, 1968

Alloys for Structural Applications at Subzero Temperatures

By James E. Campbell
Metallurgical Consultant

ALL STRUCTURAL METALS undergo changes in properties when cooled from room temperature to temperatures below 0 °C or 32 °F (temperatures in the "subzero" range). The greatest changes in properties occur when the metal is cooled to very low temperatures near the boiling points of liquid hydrogen and liquid helium. However, even at the less-severe subzero temperatures encountered in arctic regions, where the lowest temperature recorded has been −71 °C (−96 °F), carbon steels become embrittled. To avoid brittle fracture of structures, pressure vessels and vehicles in cold regions, certain low-alloy steels can be used that retain high degrees of notch toughness and crack toughness (fracture toughness) at the lowest exposure temperatures.

The effects of subzero temperatures also must be considered in selection of materials for aircraft, missiles and space vehicles that are exposed to the temperatures of upper altitudes and outer space. These structures are "weight limited", so they must be fabricated from materials with high strength-to-weight ratios both at room temperature and at subzero temperatures. At the same time, these materials are required to retain high levels of fracture toughness at all exposure temperatures for "fail safe" service. Certain titanium alloys, aluminum alloys and cold worked stainless steels have been used successfully in fabricating weight-limited structures according to state-of-the-art design.

A major requirement of materials for liquefaction equipment and for containment and transport of liquefied hydrocarbon gases and liquefied elemental gases is toughness at the boiling temperature of the liquid. For the convenience of the reader, the boiling points of ammonia; common liquefied hydrocarbon gases; and the elemental gases oxygen, argon, nitrogen, neon, hydrogen and helium are given in Table 1. Temperatures below −150 °C (−238 °F) often are identified as cryogenic temperatures. Materials selection depends on the lowest exposure temperature to be encountered in service. Aluminum alloys, low-alloy steels, 9Ni steels and austenitic stainless steels have been used successfully for liquefaction, containment and transport of these liquids.

Requirements are more critical in selecting materials for welded structures such as liquid oxygen, liquid hydrogen and liquid helium tankage and associated piping and fittings for rockets and launch vehicles. Among these requirements are minimum weight, high toughness of the base metal at cryogenic temperatures, and high strength and toughness of the welded joints. Materials that have been used successfully for these structures include aluminum alloys, titanium alloys, cold worked austenitic stainless steels, and a high-nickel alloy (for spherical pressure vessels in liquid oxygen).

Certain metals, alloys and compounds become superconductors at temperatures below about −260 °C (−436 °F). To achieve these tempera-

Table 1 Boiling points of liquefied gases

Liquefied gas	K	Boiling temperature(a) °C	°F
Ammonia	239.8	−33.3	−27.9
Propane	230.8	−42.3	−44.1
Propylene	226.1	−47.0	−52.6
Carbon dioxide	194.6(b)	−78.5(b)	−109.3(b)
Acetylene	189.1	−84.0	−119.2
Ethane	184.8	−88.3	−126.9
Ethylene	169.3	−103.8	−154.8
Methane	111.7	−161.4	−258.5
Oxygen	90.1	−183.0	−297.4
Argon	87.4	−185.7	−302.3
Nitrogen	77.3	−195.8	−320.4
Neon	27.2	−245.9	−410.6
Hydrogen	20.4	−252.7	−422.9
Helium	4.2	−268.9	−452.1

(a) At 1 atm. (b) Sublimation temperature.

Fig. 1 Yield and tensile strengths at subzero temperatures for two aluminum alloys

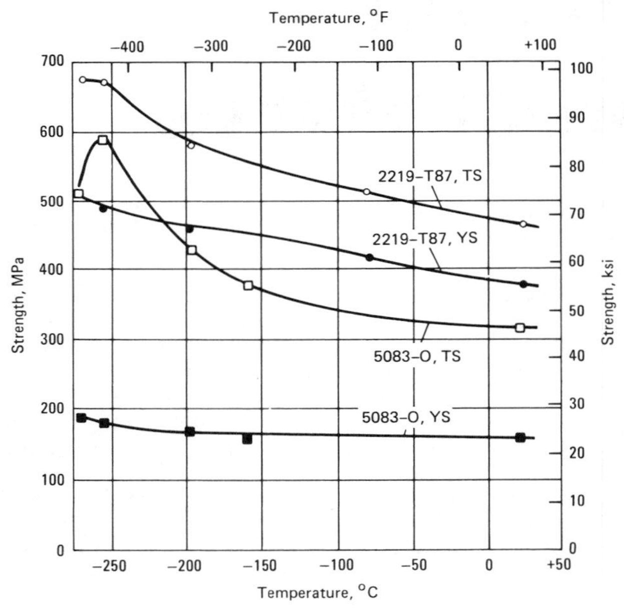

Open symbols are tensile strengths; solid symbols, yield strengths.

tures, all superconducting devices must be cooled with liquid helium. Therefore, structural materials selected for cryogenic components of superconducting machinery, magnets, and transmission systems must be suitable for use at liquid-helium temperature. Furthermore, these components are subjected to high stresses in service, and thus safeguards must be employed to minimize service failures. In order to obtain the required strength and toughness along with a reasonable degree of fabricability, certain austenitic stainless steels and high-nickel alloys (superalloys) usually are designated for highly stressed components of structures that will be cooled with liquid helium.

Because of several extensive research programs conducted over the past 25 years to obtain information on properties of structural materials at subzero temperatures, the state of the art has advanced to the point at which materials can be selected for most critical components with reasonable confidence. There are still applications involving exposure of materials to extreme conditions where current materials fall short of requirements. However, a large volume of information has been accumulated to provide a basis for selection of materials (metals and alloys) for subzero applications.

In this article, typical room-temperature and subzero-temperature mechanical properties are presented for:

- Aluminum alloys
- Copper and copper alloys
- Nickel and high-nickel alloys, including superalloys
- Ferritic alloy steels
- Stainless steels
- Titanium and titanium alloys.

In general, yield strengths and tensile strengths of structural alloys increase as the exposure temperature is decreased. This trend is illustrated in Fig. 1, which gives the yield and tensile strengths of aluminum alloys 2219-T87 and 5083-O at subzero temperatures. These alloys are used extensively for major cryogenic applications. The yield strength of 5083-O increases only slightly at very low temperatures. Yield strengths of some austenitic stainless steels are slightly lower at −78 °C (−108 °F) than at room temperature, but then increase at still lower temperatures. Results of tension tests at liquid-helium temperature may indicate slight drops in yield and tensile strengths for some alloys compared with the values at −253 °C (−423 °F). This effect may be a valid trend in tensile properties, but also may be one of the vagaries of testing in liquid helium. The phenomenon of serrated yielding occurs in many metals during plastic deformation at −253 and −269 °C (−423 and −452 °F), and it is more obvious the lower the testing temperature. It is characterized by interrupted yielding at localized points in the specimens as they are being loaded, and may lead to scatter in tensile-strength and yield-strength data.

In addition to tensile-property data, this article presents typical data on fracture toughness, fatigue-crack-growth rates and fatigue strengths at room temperature and at subzero temperatures for base metal and weld

Table 2 Nominal compositions of aluminum alloys

Alloy designation	Si	Cu	Mn	Mg	Cr	Zn	Ti	Zr	Others
	Nominal composition, %								
Wrought alloys									
1100	···	0.12	···	···	···	···	···	···	···
2014	0.8	4.4	0.8	0.5	···	···	···	···	···
2024	···	4.4	0.6	1.5	···	···	···	···	···
2219	···	6.3	0.3	···	···	···	0.06	0.18	0.1V
3003	···	0.12	1.2	···	···	···	···	···	···
5083	···	···	0.7	4.4	0.15	···	···	···	···
5456	···	···	0.8	5.1	0.12	···	···	···	···
6061	0.6	0.28	···	1.0	0.20	···	···	···	···
7005	···	···	0.45	1.4	0.13	4.5	0.04	0.14	···
7039	0.1	0.05	0.25	2.8	0.20	3.0	0.05	···	0.2Fe
7075	···	1.6	···	2.5	0.23	5.6	···	···	···
Cast alloys									
355	5.0	1.2	···	0.5	···	···	···	···	···
C355	5.0	1.3	···	0.5	···	···	···	···	···
356	7.0	···	···	0.3	···	···	···	···	···
A356	7.0	···	···	0.3	···	···	···	···	···

Table 3 Typical tensile properties of aluminum 1100 weldments
Data from Ref 1-3

Temperature		Tensile strength		Yield strength		Elongation, %	Reduction in area, %	Notch tensile strength(a)	
°C	°F	MPa	ksi	MPa	ksi			MPa	ksi
Weld in 1100-O bar, longitudinal specimens									
24	75	93.1	13.5	48	6.9	46	88	···	···
−78	−108	114	16.5	50	7.2	50	86	···	···
−196	−320	191	27.7	62	9.0	56	81	···	···
−253	−423	328	47.6	65	9.4	54	60	···	···
Weld in 1100-H12 bar, longitudinal specimens									
24	75	110	16	96	14	24	76	···	···
−78	−108	121	17.5	103	15	27	77	···	···
196	−320	186	27	117	17	46	75	···	···
Weld in 1100-H112 plate, transverse specimens(b)									
24	75	80	11.6	42	6.1	26	···	123	17.8
−196	−320	157	22.8	55	8.0	31	···	214	31.0

(a) K_t = 16. (b) Gas metal-arc welded; 1100 filler metal; as welded.

metal. Data are also presented for sheet, plate, bar, forgings and castings when available, and for different orientations, because all of these variables may affect results.

The numerical values presented in the tables have been averaged from available property data from the literature and represent state-of-the-art information. Because only limited amounts of test data are available for most metals and alloys, the numerical values represent only "composite" averages and are not statistical mean values or design values (except for Tables 9 and 13).

The data in the tables represent significant properties that may be used in selecting candidate alloys for subzero-temperature applications. This information, along with text discussion of the alloys that have been used successfully for such applications, will aid in avoiding use of alloys that are not suitable. Information is not included for unsuitable alloys or for experimental alloys. Some of the alloys indicated in the tables may not be available "off the shelf", but all of them have been produced in production heats. Some alloys, such as ferritic alloy steels, are suitable for use only in the upper portions of the subzero range. These limitations will be noted in the text.

Fabrication of these alloys usually requires fusion welding, and the welds often are lower in strength than the base metal. Subzero-temperature properties of welded joints are presented where available. These data represent only typical values, there being much variability in welding processes and processing. No attempt has been made to present processing details.

Appropriate reference numbers are listed with each table and figure. More complete information on the metals, alloys and weldments may be obtained from the original references. However, the tables and figures present enough detailed information to minimize the need to refer to the original documents.

Aluminum Alloys

Aluminum alloys represent a very important class of structural metals for subzero-temperature applications. Aluminum and aluminum alloys have face-centered-cubic (fcc) crystal structures. Most fcc metals retain good ductility at subzero temperatures. Aluminum can be strengthened by alloying and heat treatment while still retaining good ductility along with adequate toughness at subzero temperatures. Nominal compositions of aluminum alloys that are most often considered for subzero service are presented in Table 2.

Tensile Properties. Typical tensile properties of aluminum alloys and aluminum alloy weldments at room temperature and at subzero temperatures are presented in Tables 3 to 22. Data are presented for various product forms, tempers and orientations because these variables influence properties.

The tensile properties of high-purity aluminum are not presented here, because this material has very low strength even at −269 °C (−452 °F). However, high-purity aluminum may be useful as a stabilizer for airborne superconducting magnets and in superconducting power cables. The stabilizer, which is part of the superconductor composite, serves to stabilize superconducting conditions for the magnet or cable. Magnets with aluminum-stabilized coils are much lighter than comparable magnets with copper-stabilized coils. The aluminum may be strengthened by cold working. In these applications, the aluminum is subjected to

Table 4 Typical tensile properties of aluminum alloy 2014

Data from Ref 2, 4-10

Temperature		Tensile strength		Yield strength		Elongation,	Reduction in area,	Notch tensile strength(a)		Young's modulus	
°C	°F	MPa	ksi	MPa	ksi	%	%	MPa	ksi	GPa	10^6 psi
Sheet, T6 temper, longitudinal orientation											
24	75	490	71.1	440	63.6	9.6	...	445	64.7	70	10.2
−78	−108	515	74.7	475	68.8	9.6	...	420	60.7	74	10.8
−196	−320	580	84.0	510	74.2	11.4	...	390	56.9	81	11.7
−253	−423	685	99.7	565	82.0	12.2	...	510	74.2	83	12.1
−269	−452	670	97.0	570	82.5	10.4	...	...	...	...	...
Sheet, T6 temper, transverse orientation											
24	75	485	70.0	430	62.4	9.6	...	410	59.7	72	10.4
−78	−108	510	74.1	450	65.4	9.8	...	370	53.9	74	10.7
−196	−320	575	83.1	495	71.5	11.4	...	430	62.4	80	11.6
−253	−423	685	99.2	545	78.7	14.0	...	480	69.4	82	11.9
Sheet, T62 temper, longitudinal orientation											
24	75	475	69.2	425	61.5	10.8	...	...	...	...	...
−78	−108	500	72.6	465	67.1	9.3	...	...	...	...	...
−196	−320	570	82.6	505	73.1	13	...	...	...	...	...
−253	−423	650	94.2	540	78.3	14	...	...	...	...	...
Sheet, T62 temper, transverse orientation											
24	75	475	69.0	420	61.2	11	...	...	...	...	...
−78	−108	495	71.8	455	65.9	11.5	...	...	...	...	...
−196	−320	570	82.8	500	72.4	12.2	...	...	...	...	...
−253	−423	665	96.4	530	76.9	14.5	...	...	...	...	...
Plate, T62 temper, longitudinal orientation											
24	75	475	69.0	425	62.0	9	...	...	...	...	...
−196	−320	570	83.0	495	72.0	11	...	...	...	...	...
−253	−423	650	94.0	525	76.0	...	...	...	...	...	...
Plate, T651 temper, longitudinal orientation											
24	75	465	67.6	430	62.2	11.2	33	570	82.8	...	...
−196	−320	580	84.0	525	76.0	12.0	22	685	99.6	...	...
−253	−423	660	95.6	555	80.2	15.0	23	710	103	...	...
−269	−452	660	95.4	565	81.8	12.8	20	710	103	...	...
Plate, T651 temper, transverse orientation											
24	75	480	69.5	430	62.7	8.8	16	550	79.8	...	...
−196	−320	560	81.5	515	74.4	9.0	12	575	83.1	...	...
−253	−423	665	96.4	535	77.4	11.0	15	...	...	...	...
−269	−452	668	96.9	585	84.8	10.2	12	645	93.7	...	...
Bar, T6 temper, longitudinal orientation											
24	75	515	74.4	460	66.4	16	33	...	...	72	10.5
−78	−108	520	75.4	455	66.1	13	28	...	...	81	11.8
−196	−320	580	84.2	500	72.2	15	29	...	...	79	11.5
−269	−452	815	118	705	102	...	...	...	...	82	11.9
Forging, T6 temper, longitudinal orientation											
24	75	465	67.6	410	59.8	12.5	24	...	...	...	...
−78	−108	510	73.8	460	67.0	14	23	...	...	...	...
−196	−320	610	88.8	535	77.3	11	22	...	...	...	...

(a) For sheet in T6 temper, K_t = 19; for plate in T61 temper, K_t = 16.

high stresses in service and are part of the over-all stressed structure.

Tensile properties of aluminum 1100 weldments are presented in Table 3. This alloy has relatively low strength in the annealed condition (O temper) and in the H (cold worked) tempers at room and subzero temperatures. In weldments, both base metal and weld metal retain good ductility at very low temperatures. Aluminum 1100 is available as tubing, sheet, plate and bar for noncritical components of cryogenic structures.

As shown in Table 4, aluminum alloy 2014-T6 has relatively high strength at room temperature and at subzero temperatures. It retains about the same ductility and notch tensile strength at liquid-hydrogen temperature as at room temperature. Values of Young's

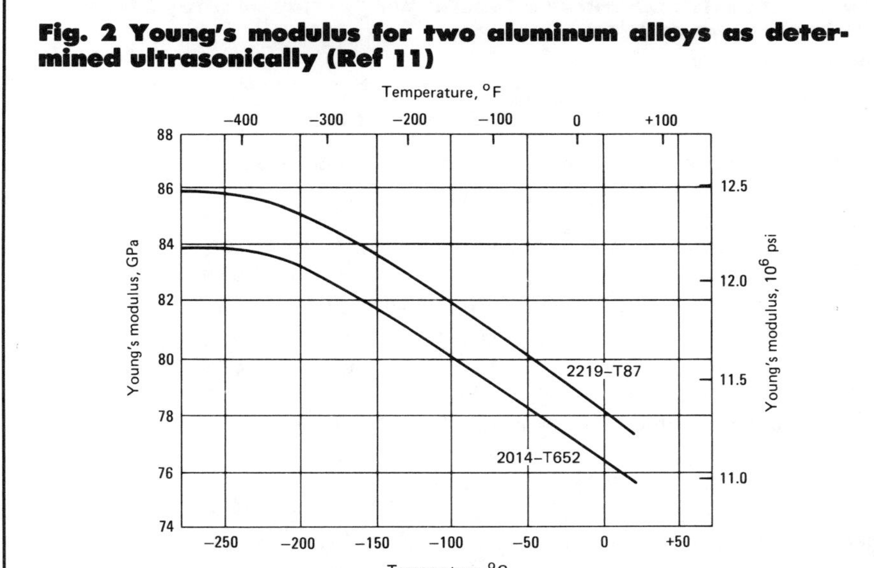

Fig. 2 Young's modulus for two aluminum alloys as determined ultrasonically (Ref 11)

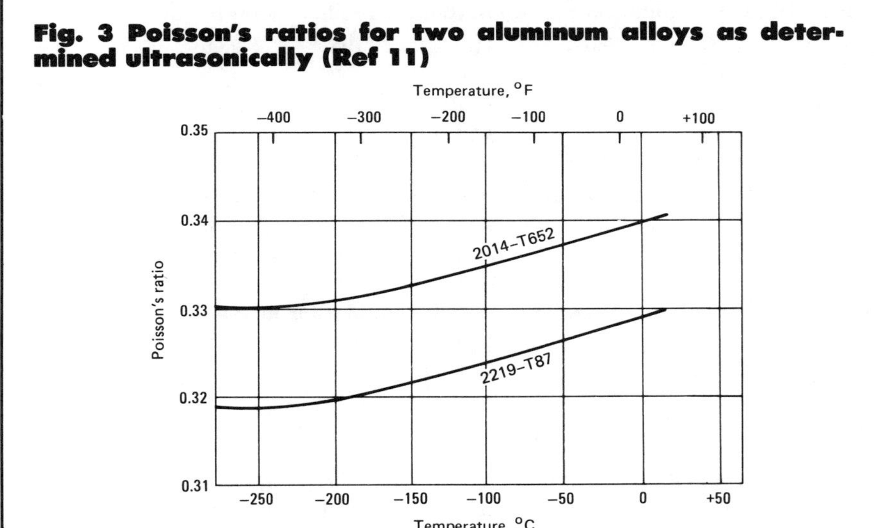

Fig. 3 Poisson's ratios for two aluminum alloys as determined ultrasonically (Ref 11)

modulus and Poisson's ratio for alloy 2014-T652, determined by an ultrasonic method, are plotted in Fig. 2 and 3.

Alloy 2014-T6 plate was employed in making very large welded tanks for the propellants (liquid oxygen and liquid hydrogen) used in the Saturn launch vehicle. These tanks were 6.6 m (21 ft, 8 in.) in diameter and were assembled by fusion welding (Ref 4); tensile properties of the welded joints are presented in Table 5. Strength and ductility of the weld metal were substantially lower than the strength and ductility of the base metal. Therefore, the welded areas of the tanks were made thicker than the remaining areas of the tank shells, which were milled out to produce a waffle grid pattern or integral ribs. As a result of this design, the tanks had high strength and stiffness with minimum weight.

Aluminum alloy 2024 (Table 6) has relatively high strength at both room and subzero temperatures in the T3 temper (solution treated and cold worked), the T4 temper (solution treated and naturally aged) and the T8 temper (solution treated, cold worked and artificially aged). Fusion welded joints in this alloy are less satisfactory than such joints in alloy 2014. Alloy 2024 is used primarily in aircraft and aerospace structures and is joined by mechanical fasteners.

Tensile properties of aluminum alloy 2219 in several forms and tempers are presented in Fig. 1 and Table 7. Values of Young's modulus and Poisson's ratio, determined by ultrasonic testing, are given in Fig. 2 and 3. This alloy has somewhat lower strength than 2014-T6 but better toughness at room and subzero temperatures. Tensile properties of alloy 2219 weldments are presented in Table 8. Alloy 2219 was originally developed for structural applications requiring strength at elevated temperatures. However, because of its favorable properties at cryogenic temperatures, alloy 2219-T87 plate has been used for the liquid oxygen and liquid hydrogen tanks for the space shuttle. These tanks are 8.4 m (27.6 ft) in diameter. The segments of the tanks are joined by fusion welding while assembled in very large fixtures (Ref 20). The external tank of the space shuttle contains one liquid oxygen tank and one liquid hydrogen tank joined by an intertank skirt of aluminum alloy 7075-T6 plate. Because of the space shuttle program, sufficient testing has been conducted on the 2219-T87 alloy to permit an evaluation of the design-allowable properties for sheet, plate and weldments, as presented in Table 9.

Aluminum alloy 3003 is used in fabrication of brazed heat exchangers and other equipment in gas-liquefaction plants. It is available as tubing (including finned tubing), pipe, sheet and plate. It is readily joined by brazing or welding. Tensile properties of alloy 3003 as annealed, as processed to two H tempers, and as welded are given in Table 10. For this alloy, ease of fabrication, along with good ductility and notch toughness at room and subzero temperatures, are the most significant properties.

Results of several studies on the tensile properties of aluminum alloy 5083 plate and weldments are presented in Tables 11 and 12. Average yield and tensile strengths are plotted in Fig. 1 for 5083-O plate. Values of Young's modulus and Poisson's ratio for alloy 5083, determined by ultrasonic testing, are plotted in Fig. 4 and 5. This alloy is not heat treatable; for maximum toughness, it is used in the annealed (O) condition. It is readily weldable, and the yield strength of the weld metal is nearly equal to that of the base metal.

A ship of current design for transport of liquefied natural gas (LNG) contains five spherical tanks made of aluminum alloy 5083-O built into the ship's structure (Ref 32). These tanks are about 36.5 m (120 ft) in diameter. Each tank has a capacity of 25 000 m^3 (883 000 ft^3) of LNG. The tanks are fabricated by butt welding preformed segments of 5083-O aluminum alloy plate in spherical fixtures. The equatorial rings for these tanks are machined from plate 195 mm (7.7 in.) thick. Because of the danger of explosion that would exist if the gas should escape through leaks in the tanks, an extensive evaluation program was conducted to determine the properties, including fracture properties, of thick 5083-O plate and of welds in such plate. These studies indicated that both base metal and weld metal had sufficient toughness at room temperature and LNG temperature to resist development of through cracks during the lifetime of the vessels. It has been projected that 70 LNG carriers with capacities of 25 000 to 130 000 m^3 (883 000 to 4 590 000 ft^3) will be in service in the 1980's. Land-based LNG storage tanks at distribution terminals also have been fabricated from alloy 5083-O. Design stresses for 5083-O sheet, plate and weldments at room temperature and at −162 °C (−260 °F) are presented in Table 13.

Aluminum alloy 5456 is another non-heat-treatable alloy that has good welding characteristics and good ductility and toughness at cryogenic temperatures. It is an alternative to alloy 5083. Tensile properties of 5456 base metal and weld metal are presented in Tables 14 and 15.

Aluminum alloy 6061 is usually used in the T6 temper. It is weldable and may be reheat treated after welding, although this practice is not recommended because it lowers weld ductility considerably. One significant application for 6061-T6 is weld-fabricated housings for pumps and motors used in pumping liquefied natural gas at distribution terminals. The pumps are submerged in LNG in service. For applications such as this, it is important that full-penetration welds be obtained at all welded joints. Alloy 6061 is available in all standard forms. Tensile properties of 6061 base metal and weld metal are presented in Tables 16 and 17. Specimens in T6 temper have higher strength and lower ductility than those in T4 temper. Specimens from

Table 5 Typical as-welded tensile properties of alloy 2014 weldments, as welded

Data from Ref 2, 5, 9

Temperature °C	Temperature °F	Tensile strength MPa	Tensile strength ksi	Yield strength MPa	Yield strength ksi	Elongation, %
Sheet in T6 temper, longitudinal orientation, 2319 filler metal(a)						
24	75	380	54.9	290	42.1	2.3
−73	−100	390	56.5	285	41.2	1.6
−196	−320	440	63.8	330	48.1	1.4
−253	−423	500	72.4	450	65.0	1.4
Sheet in T6 temper, longitudinal orientation, 4043 filler metal(a)						
24	75	350	50.6	240	34.6	1.7
−196	−320	410	59.6	395	57.2	0.7
−253	−423	480	69.4	...	...	0.6
Sheet in T6 temper, longitudinal orientation, 4043 filler metal(b)						
24	75	320	46.3	260	37.8	2.8
−196	−320	420	60.9	325	46.8	2.0
−253	−423	415	60.0	350	51.0	1.2
Sheet in T6 temper, transverse orientation, 4043 filler metal(b)						
24	75	355	51.5	260	37.7	2.0
−196	−320	420	61.1	330	47.6	2.1
−253	−423	410	59.4	360	52.2	1.2
Plate in T6 temper, longitudinal orientation, 2319 filler metal(b)						
24	75	285	41.0	195	28.6	2.4
−196	−320	345	49.9	265	38.3	1.5
Plate in T6 temper, longitudinal orientation, 4043 filler metal(b)						
24	75	255	37.2	160	23.1	2.4
−196	−320	310	45.0	200	29.2	1.6
Plate in T62 temper, longitudinal orientation, 2319 filler metal(a)						
24	75	330	47.8	180	26.0	15
−196	−320	385	55.5	230	33.1	10
−253	−423	435	63.4	300	43.5	...
Plate in T62 temper, longitudinal orientation, 4043 filler metal(a)						
24	75	315	46.0	200	28.7	10
−196	−320	380	55.4	245	35.4	8
−253	−423	435	63.2	260	37.8	...

(a) Gas tungsten-arc welding process. (b) Gas metal-arc welding process.

sheet, plate, bar and forgings, all in T6 temper, have about the same strength and ductility at the same subzero temperature. Strengths of weldments reheat treated to T6 temper are lower than the corresponding strengths of the base metal in T6 temper.

Aluminum alloy 7005 is produced commercially only in the form of extrusions, but tension tests at subzero temperatures have also been performed on base-metal and weld-metal specimens from weldments of 7005 plate. Results of these tests are presented in Table 18. The 7005-T6351 plate and the 7005-T5351 extrusions have higher strength than 6061-T6 plate, but about the same ductility at the same temperatures. As-welded strengths of 7005-T6351 plate exceed the strengths of 6061 plate reheat treated to the T6 temper.

Aluminum alloy 7039 was originally developed as an armor alloy. It is readily welded by inert-gas arc processes. Tensile tests have shown that alloy 7039 plate in the T6 temper retains good ductility and notch toughness at cryogenic temperatures. Specimens from weldments also retain good ductility and notch toughness. These properties are presented in Tables 19 and 20. Alloy 7039-T6 has been recommended for cryogenic pressure vessels.

Aluminum alloy 7075 is representative of the high-strength nonweldable 7000 series alloys. It is primarily an aircraft and aerospace alloy and has relatively low fracture toughness in the T6 temper. Normally it is not used in applications involving cryogenic temperatures. The choice of 7075-T6 plate for the intertank skirt between the liq-

Table 6 Typical tensile properties of alloy 2024

Data from Ref 2, 4, 6, 8, 9

Temperature °C	Temperature °F	Tensile strength MPa	Tensile strength ksi	Yield strength MPa	Yield strength ksi	Elongation, %	Reduction in area, %	Notch tensile strength(a) MPa	Notch tensile strength(a) ksi	Young's modulus GPa	Young's modulus 10^6 psi
Sheet in T3 temper, longitudinal orientation											
24	75	470	67.9	325	47.4	18	...	415	60.2	70	10.2
−78	−108	485	70.2	335	48.9	21	...	420	61.2	72	10.5
−196	−320	600	87.0	420	60.9	22	...	525	76.2	75	10.9
−253	−423	760	110	505	73.1	17	...	610	88.8	79	11.4
Sheet in T3 temper, transverse orientation											
24	75	455	65.8	305	43.9	18	...	435	62.9	71	10.3
−78	−108	465	67.8	305	44.5	21	...	435	62.8	73	10.6
−196	−320	575	83.4	385	56.1	22	...	515	74.7	76	11.0
−253	−423	740	107	475	69.0	18	...	600	86.8	79	11.5
Sheet in T4 temper, longitudinal orientation											
24	75	465	67.7	295	42.8	19	...	405	59.0	74	10.7
−78	−108	480	69.8	300	43.7	22	...	420	60.7	74	10.7
−196	−320	585	84.9	375	54.1	27	...	495	71.9	77	11.2
−253	−423	740	107	505	73.3	16	...	610	88.3	80	11.6
Sheet in T4 temper, transverse orientation											
24	75	465	67.1	285	41.5	20	...	395	57.5	72	10.4
−78	−108	470	68.0	295	42.7	24	...	405	58.9	74	10.7
−196	−320	565	81.8	370	53.6	19	...	470	68.2	76	11.0
−253	−423	670	97.1	465	67.5	10	...	590	85.4	80	11.6
Plate in T351 temper, longitudinal orientation											
24	75	465	67.1	345	50.3	22	28	500	72.8	...	...
−253	−423	740	107	530	77.2	22	20	660	95.6	...	...
Plate in T4 temper, longitudinal orientation											
24	75	465	67.7	365	53.3	17	17	480	69.3	72	10.5
−78	−108	480	69.6	375	54.7	17	4	510	73.7	73	10.6
−196	−320	555	80.8	460	66.5	11	11	580	84.3	75	10.9
−253	−423	650	94.4	555	80.6	8	9	670	97.4	80	11.6

(continued)

(a) For sheet in T3 and T4 tempers and plate in T351 and T4 tempers, K_t = 6.3; for bar in T351 temper, K_t = 13.3.

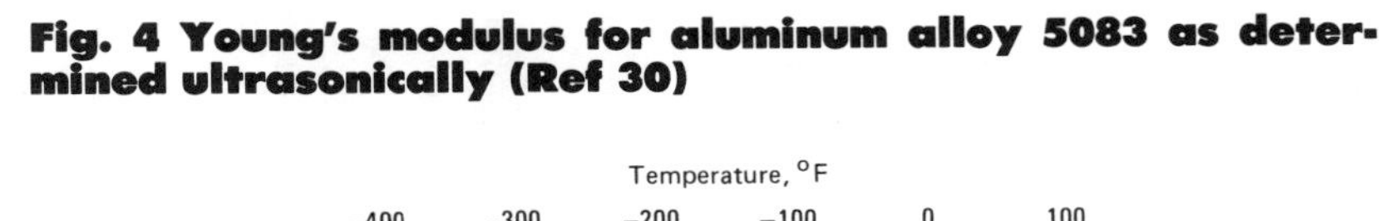

Fig. 4 Young's modulus for aluminum alloy 5083 as determined ultrasonically (Ref 30)

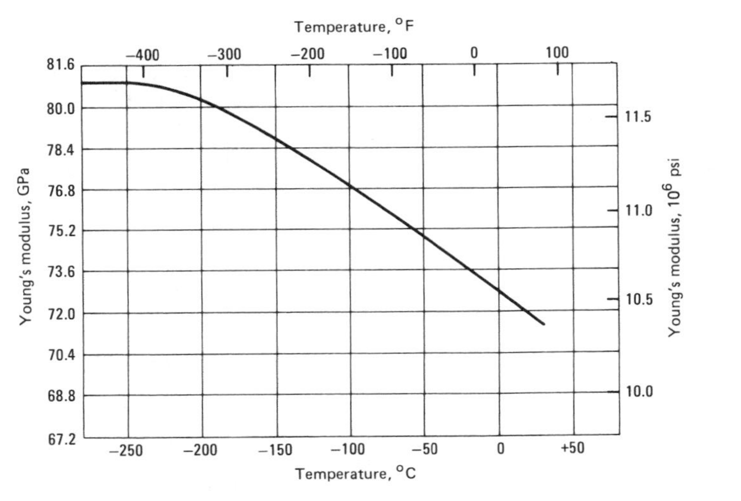

uid oxygen and liquid hydrogen sections of the external tank of the space shuttle was probably based on the high strength of this alloy. Mechanical fasteners are used in joints between the skirt and the tanks. Tensile properties of 7075 sheet and plate at subzero temperatures are presented in Table 21. Processing of alloy 7075 to the T7351 temper improves its ductility and notch toughness at cryogenic temperatures.

Other high-strength 7000 series aluminum alloys (7079 and 7178) also have been evaluated at subzero temperatures, but they are not generally weldable and have relatively low toughness at subzero temperatures.

Aluminum alloy castings often are specified for pump parts and for other components that are more readily produced by casting than by other fabricating processes. A variety of alloys, casting methods and heat treatments can be applied in producing aluminum al-

Table 6 (continued)

Temperature °C	°F	Tensile strength MPa	ksi	Yield strength MPa	ksi	Elongation, %	Reduction in area, %	Notch tensile strength(a) MPa	ksi	Young's modulus GPa	10^6 psi
Plate in T4 temper, transverse orientation											
24	75	460	66.9	315	45.5	16	15	525	76.2	73	10.6
−78	−108	460	66.4	310	45.2	16	19	535	77.4	73	10.6
−196	−320	555	80.6	350	50.7	15	12	620	89.9	77	11.1
−253	−423	645	93.6	480	69.4	8	9	655	95.0	79	11.5
Plate in T851 temper, longitudinal orientation											
24	75	495	72.0	455	65.8	8	17	...	...	...	...
−78	−108	535	77.8	490	71.3	6	14	...	...	...	...
−196	−320	690	99.8	575	83.3	8	13	...	...	...	...
−269	−452	715	104	625	90.7	9	14	...	...	...	...
Plate in T851 temper, transverse orientation											
24	75	490	70.8	445	64.4	7	...	...	...	...	...
−78	−108	525	76.0	475	69.2	7	...	...	...	...	...
−196	−320	605	87.7	545	79.0	7	...	...	...	...	...
Bar in T351 temper, longitudinal orientation											
24	75	480	68.4	365	53.0	19	...	520	75.2	...	...
−196	−320	610	88.5	480	69.3	19	...	625	90.6	...	...
Bar in T351 temper, transverse orientation											
24	75	470	69.4	325	47.4	17	...	495	72.1	...	...
−196	−320	595	86.1	430	62.1	14	...	595	86.3	...	...
Bar in T4 temper, longitudinal orientation											
24	75	490	70.8	365	52.6	20	30	...	...	...	...
−78	−108	500	72.8	365	52.9	21	26	...	...	...	...
−196	−320	615	89.4	470	68.1	19	18	...	...	...	...
−253	−423	750	109	585	85.0	18	19	...	...	...	...
−269	−452	740	107	525	76.0	20	21	...	...	...	...
Bar in T86 temper, longitudinal orientation											
24	75	510	74.2	495	71.6	10	27	...	...	...	...
−78	−108	550	80.0	525	76.3	9	23	...	...	...	...
−196	−320	635	91.8	590	85.6	11	21	...	...	...	...
−253	−423	725	105	645	93.2	15	24	...	...	...	...

(a) For sheet in T3 and T4 tempers and plate in T351 and T4 tempers, K_t = 6.3; for bar in T351 temper, K_t = 13.3.

Fig. 5 Poisson's ratio for aluminum alloy 5083 as determined ultrasonically (Ref 30)

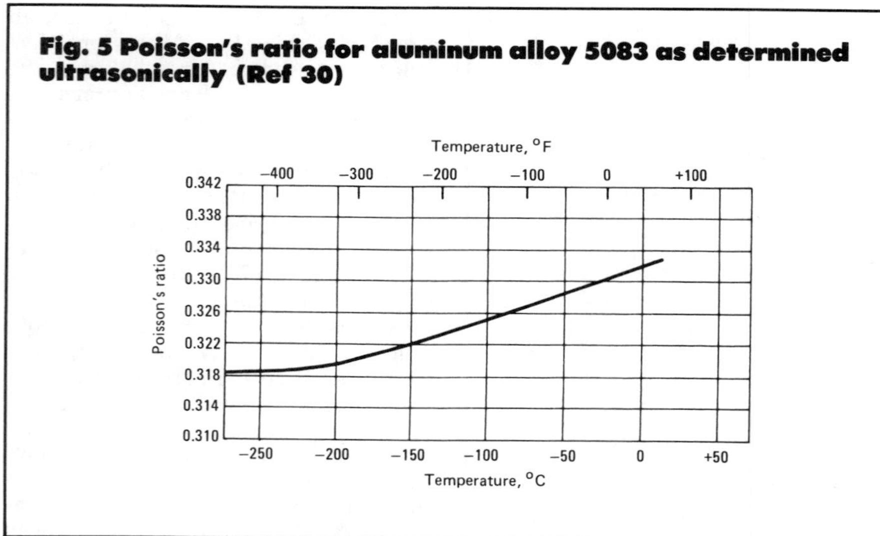

loy castings for subzero applications. Some of these alloys and casting methods are identified in Table 22, along with tensile properties at room and subzero temperatures. These data indicate that, with suitable designs and quality-control standards, cast aluminum components can be produced that are satisfactory for subzero applications.

Fracture Toughness. Data on fracture toughness of several aluminum alloys at room and subzero temperatures are summarized in Table 23. The room-temperature yield strengths for the alloys in this table range from 142 to 536 MPa (20.6 to 77.7 ksi), and room-temperature plane-strain fracture-toughness values for both bend and compact tension (CT) specimens range from 22.3 to 39.9 MPa $\sqrt{m}$ (20.3 to 36.3

Table 7 Typical tensile properties of aluminum alloy 2219

Data from Ref 2, 4, 7-10, 12-18

Temperature		Tensile strength		Yield strength		Elongation,	Reduction in area,	Notch tensile strength(a)		Young's modulus	
°C	°F	MPa	ksi	MPa	ksi	%	%	MPa	ksi	GPa	10^6 psi
Sheet in T6 temper, longitudinal orientation											
24	75	450	65.4	360	52.2	10	...	420	60.6	...	...
−78	−108	480	69.5	380	55.2	9	...	445	64.8	...	...
−196	−320	565	82.3	440	64.0	12	...	505	73.2	...	...
−253	−423	665	96.4	545	79.2	15	...	545	78.8	...	...
−269	−452	650	94.2	475	69.1	12	...	550	80.1	...	...
Sheet in T6 temper, transverse orientation											
24	75	455	66.1	360	52.1	10	...	...	...	...	...
−78	−108	485	70.3	375	54.6	10	...	...	...	...	...
−196	−320	570	82.8	430	62.4	12	...	...	...	...	...
−253	−423	655	94.7	535	77.8	14	...	...	...	...	...
−269	−452	660	95.9	475	68.7	11	...	...	...	...	...
Sheet in T62 temper, longitudinal orientation											
24	75	415	60.2	290	41.9	10	...	330	47.9	70	10.2
−78	−108	460	66.5	345	50.2	11	...	...	...	...	...
−196	−320	545	78.8	370	53.6	14	...	400	58.1	77	11.1
−253	−423	650	94.5	390	56.3	23	...	465	67.4	81	11.7
Sheet in T62 temper, transverse orientation											
24	75	405	58.8	280	40.6	12	...	320	46.1	...	...
−78	−108	440	63.9	315	46.0	12	...	...	...	...	...
−196	−320	525	76.4	360	52.0	14	...	385	56.0	...	...
−253	−423	635	91.8	385	55.6	17	...	400	58.2	...	...
Sheet in T81 temper, longitudinal orientation											
24	75	445	64.8	345	50.4	9	...	365	52.6	69	9.94
−78	−108	480	69.8	375	54.3	9	...	350	50.9	72	10.5
−196	−320	565	82.1	425	61.3	11	...	415	60.0	80	11.6
−253	−423	665	96.8	475	69.0	13	...	465	67.8	82	11.9
Sheet in T81 temper, transverse orientation											
24	75	450	65.2	340	49.5	10	...	365	52.6	69	10.0
−78	−108	490	70.8	365	53.2	10	...	320	46.7	74	10.7
−196	−320	565	81.6	415	60.0	10	...	390	56.9	79	11.5
−253	−423	675	97.6	465	67.6	12	...	420	61.2	81	11.8
Sheet in T87 temper, longitudinal orientation											
24	75	470	67.9	380	55.4	10	...	320	46.5	73	10.6
−78	−108	505	73.1	410	59.5	10	...	380	54.8	72	10.5
−196	−320	580	84.4	460	67.0	12	...	425	61.8	80	11.6
−253	−423	670	97.5	485	70.6	15	...	460	67.0	81	11.8
Sheet in T87 temper, transverse orientation											
24	75	475	68.6	380	55.4	9	...	...	...	73	10.6
−78	−108	505	73.4	405	58.6	8	...	...	...	82	11.9
−196	−320	590	85.8	460	66.4	9	...	...	...	83	12.1
−253	−423	685	99.6	490	71.3	13	...	...	...	81	11.7
Plate in T81 temper, longitudinal orientation											
24	75	445	64.3	345	50.2	9	18	...	...	...	...
−196	−320	575	83.5	435	63.3	10	17	...	...	...	...
−253	−423	650	94.6	485	70.0	13	25	...	...	...	...
Plate in T81 temper, transverse orientation											
24	75	440	63.9	335	48.9	6	9	...	...	...	...
−196	−320	510	74.1	400	57.9	3	7	...	...	...	...

(continued)

(a) K_t = 10 for sheet in T6 temper; K_t = 19 for sheet in T62 and T81 tempers; K_t = 22 for sheet in T87 temper; K_t = 16 for plate in T851 and T87 tempers.

Table 7 (continued)

Temperature °C	Temperature °F	Tensile strength MPa	Tensile strength ksi	Yield strength MPa	Yield strength ksi	Elongation, %	Reduction in area, %	Notch tensile strength(a) MPa	Notch tensile strength(a) ksi	Young's modulus GPa	Young's modulus 10^6 psi
Plate in T81 temper, S-T orientation											
24	75	400	58.3	335	48.6	...	8	...	...	...	...
−196	−320	485	70.6	380	55.4	3	3	...	...	...	...
Plate in T851 temper, longitudinal orientation											
24	75	465	67.6	370	53.8	11	27	545	79.4	...	...
−78	−108	490	71.4	395	57.6	11	28	580	84.3	...	...
−196	−320	570	82.5	440	63.8	14	30	650	94.5	...	...
−253	−423	660	95.6	475	68.8	16	28	715	104	...	...
−269	−452	660	95.7	485	70.3	15	26	705	102	...	...
Plate in T851 temper, transverse orientation											
24	75	460	66.4	355	51.2	10	22	530	77.0	...	...
−78	−108	490	71.0	380	55.0	10	22	560	81.5	...	...
−196	−320	570	83.0	420	61.1	12	24	625	90.5	...	...
−253	−423	665	96.7	465	67.5	16	25	665	96.5	...	...
−269	−452	660	95.7	480	69.8	13	20	665	96.5	...	...
Plate in T87 temper, longitudinal orientation											
24	75	465	67.8	380	55.2	11	26	565	82.3	71	10.3
−78	−108	515	74.8	420	61.0	14	27	570	82.8	...	...
−196	−320	585	84.5	460	66.6	13	25	630	91.5	81	11.8
−253	−423	675	97.6	490	71.0	14	21	705	102	79	11.5
−269	−452	675	97.8	510	74.2	15	23	690	100	...	...
Plate in T87 temper, transverse orientation											
24	75	465	67.6	385	55.8	9	17	...	...	73	10.6
−78	−108	505	73.4	410	59.2	10	20	...	...	...	...
−196	−320	595	86.1	460	66.8	11	17	...	...	79	11.5
−253	−423	685	99.4	490	71.3	12	17	...	...	80	11.6
Plate in T87 temper, S-T orientation											
24	75	455	66.2	375	54.4	6	7	...	...	...	...
−196	−320	530	76.6	455	65.8	3	6	...	...	...	...
−253	−423	545	79.0	505	73.6	1	...	...	...	...	...
Forging in T852 temper, radial orientation											
24	75	445	64.7	335	48.3	13	25	...	...	...	...
−196	−320	545	78.8	395	57.4	15	36	...	...	...	...
Forging in T852 temper, tangential orientation											
24	75	450	65.0	345	50.1	10	17	...	...	...	...
−196	−320	540	78.4	400	57.9	14	25	...	...	...	...
Forging in T852 temper, S-T orientation											
24	75	460	66.7	370	53.4	4	6	...	...	...	...
−196	−320	535	77.5	425	61.6	4	6	...	...	...	...
Extrusion in T851 temper, longitudinal orientation											
24	75	470	68.2	360	52.4	...	...	...	...	...	...
−196	−320	565	81.8	415	60.0	...	...	...	...	...	...
−269	−452	665	96.3	460	66.9	...	...	...	...	...	...

(a) K_t = 10 for sheet in T6 temper; K_t = 19 for sheet in T62 and T81 tempers; K_t = 22 for sheet in T87 temper; K_t = 16 for plate in T851 and T87 tempers.

Table 8 Typical tensile properties of alloy 2219 weldments using 2319 filler metal

Data from Ref 2, 5, 12, 15, 19

Temperature, °C	Temperature, °F	Tensile strength, MPa	Tensile strength, ksi	Yield strength, MPa	Yield strength, ksi	Elongation, %	Reduction in area, %	Notch tensile strength(a), MPa	Notch tensile strength(a), ksi
Sheet in T62 temper, longitudinal orientation(b)(c)									
24	75	315	45.4	210	30.5	3	...	325	47.3
−196	−320	370	54.0	275	39.7	2	...	390	56.5
−253	−423	435	63.2	345	50.0	6	...	410	59.2
Sheet in T62 temper, longitudinal orientation(c)(d)									
24	75	415	60.5	300	43.5	8	...	410	59.5
−196	−320	520	75.2	355	51.8	8	...	505	73.0
−253	−423	565	81.6	405	58.5	4	...	545	78.7
Sheet in T87 temper, transverse orientation(b)(c)									
24	75	320	46.2	220	31.8	2	...	370	53.4
−78	−108	335	48.5	240	34.6	2	...	445	64.6
−196	−320	465	67.2	255	37.0	3	...	475	69.1
Sheet in T87 temper, longitudinal orientation(b)(e)									
24	75	310	45.3	200	29.3	2	...	...	...
−196	−320	400	58.3	285	41.5	3	...	...	...
−253	−423	480	69.9	275	40.1	2	...	...	...
Sheet in T87 temper, transverse orientation(b)(c)									
24	75	275	39.8	165	24.2	4	...	...	...
−196	−320	390	56.4	210	30.6	6	...	...	...
−253	−423	470	68.4	245	35.3	5	...	...	...
Plate in T62 temper, longitudinal orientation(d)(e)									
24	75	395	57.3	275	40.2	7	7	...	...
−78	−108	415	60.4	280	40.5	6	8	...	...
−196	−320	475	68.9	320	46.6	5	6	...	...
−269	−452	495	72.0	355	51.5	3	5	...	...
Plate in T87 temper, longitudinal orientation(b)(e)									
24	75	280	40.9	170	24.6	5	27	...	...
−78	−108	285	41.4	170	25.0	4	...	...	...
−196	−320	395	57.0	205	29.6	8	18	...	...
−253	−423	425	61.9	270	38.8	4	...	...	...
Plate in T87 temper, longitudinal orientation(b)(c)									
24	75	275	40.0	165	23.8	6	27	265	38.6
−78	−108	285	41.0	170	24.8	4	...	...	...
−196	−320	410	59.3	190	27.8	9	22	325	46.8
Plate in T87 temper, transverse orientation(b)(c)									
24	75	275	40.2	165	24.0	6	...	...	...
−196	−320	390	56.8	195	28.5	7	...	...	...
−253	−423	465	67.8	230	33.5	6	...	...	...
Plate in T87 temper, longitudinal orientation(c)(f)									
24	75	310	44.7	190	27.8	...	...	...	...
−196	−320	335	48.9	225	32.5	...	...	...	...

(a) For sheet in T62 temper longitudinal orientation and T87 temper transverse orientation, $K_t = 30$; for plate in T87 temper longitudinal orientation, $K_t = 13$. (b) As welded. (c) Gas tungsten-arc welding process. (d) Reheat treated to T62. (e) Gas metal-arc welding process. (f) Aged 24 h at 165 °C (325 °F).

ksi $\sqrt{\text{in.}}$). This range in numerical values is not as impressive as actual service performance.

Of the alloys listed in Table 23, 5083-O has substantially greater toughness than the others. Because this alloy is too tough for obtaining valid K_{Ic} data, the values shown for 5083-O were converted from J_{Ic} data. The fracture toughness of this alloy increases as exposure temperature decreases. Of the other alloys, which were all evaluated in various heat treated conditions, 2219-T87 has the best combination of strength and fracture toughness, both at room temperature and at −196 °C (−320 °F), of all the alloys that can be readily welded.

Alloy 6061-T651 has good fracture toughness at room temperature and at −196 °C (−320 °F), but its yield strength is lower than that of alloy 2219-T87. Alloy 7039 also is weldable and has a good combination of strength and fracture toughness at room temperature and at −196 °C (−320 °F). Alloy 2124 is similar to 2024 but with a higher-purity base and special processing for improved fracture toughness. Tensile properties of 2124-T851 at subzero temperatures can be expected to be similar to those for 2024-T851.

Several other aluminum alloys, including 2214, 2419, 7050 and 7475, have been developed in order to obtain room-temperature fracture toughness superior to that of other 2000 and 7000 series alloys. Information on subzero properties of these alloys is limited, but it is expected that these alloys also would have improved fracture toughness at subzero temperatures as well as at room temperature.

Fatigue-Crack-Growth Rates. Data on rates of fatigue-crack growth in aluminum alloys at subzero temperatures are limited, but the available data are plotted for alloy 3003-O in Fig. 6 and for alloy 5083-O in Fig. 7. For both alloys, fatigue-crack-growth rates at subzero temperatures are lower than those at room temperature at the same ΔK levels. The equations of the type $da/dN = C(\Delta K)^n$ in Fig. 6 and 7 can be used to calculate fatigue-crack-growth rates at any ΔK value within the limited range shown.

Fatigue Strength. Results of axial and flexural fatigue tests at 10^6 cycles on aluminum alloy specimens at room temperature and at subzero temperatures are presented in Table 24. These data indicate that, for a fatigue life of 10^6 cycles, fatigue strength is higher at subzero temperatures than at room temperature for each alloy. This trend is not necessarily valid for tests at higher stress levels and shorter fatigue lives, but at 10^6 cycles results are consistent with the effect of subzero temperatures on tensile strength.

Table 9 Design-allowable tensile properties of aluminum alloy 2219-T87 and weldments

Data from Ref 2, 13

Temperature °C	Temperature °F	Reliability factor	Tensile strength MPa	Tensile strength ksi	Yield strength MPa	Yield strength ksi
Sheet and plate(a), longitudinal orientation						
24	75	Mean	474	68.7	391	56.7
		99/95(b)	443	64.3	354	51.4
−196	−320	Mean	578	83.8	454	65.9
		99/95(b)	537	77.9	410	59.4
−253	−423	Mean	685	99.3	498	72.2
		99/95(b)	625	90.5	454	65.8
Sheet and plate(a), transverse orientation						
24	75	Mean	476	69.0	388	56.2
		99/95(b)	445	64.6	350	50.7
−196	−320	Mean	585	84.8	454	65.9
		99/95(b)	543	78.7	412	59.8
−253	−423	Mean	692	100.4	495	71.8
		99/95(b)	649	94.1	446	64.7
Weldment, gas tungsten-arc welded, 2319 filler, as welded						
24	75	Mean	288	41.8	170(c)	24.6(c)
		99/95(b)	235	34.1	...	...
−196	−320	Mean	375	54.4	197(c)	28.6(c)
		99/95(b)	314	45.6	...	...
−253	−423	Mean	434	62.9	208(c)	30.2(c)
		99/95(b)	349	50.6	...	...
Weldment, gas metal-arc welded, 2319 filler, as welded						
24	75	Mean	296	42.9	...	...
		99/95(b)	239	34.7	...	...
−196	−320	Mean	386	56.0	...	...
		99/95(b)	292	42.3	...	...
−253	−423	Mean	432	62.7	...	...
		99/95(b)	316	45.8	...	...

(a) Thickness range, 0.81 to 102 mm (0.032 to 4.00 in.). (b) Values correspond to 99% probability at 95% confidence level. (c) "Apparent" mean values.

Copper and Copper Alloys

Copper and copper alloys have fcc crystal structures similar to those of aluminum and retain a high degree of ductility and toughness at subzero temperatures. Nominal compositions of copper alloys that might be considered for use at subzero temperatures are presented in Table 25.

Copper was once used extensively for heat exchangers and cold box components in air liquefaction and separation systems. A shortage of the skilled welders required for fabricating copper equipment and the development of brazed aluminum equipment were factors that caused copper to be replaced by aluminum for many of these components. At present, the major use of copper at subzero temperatures is for the stabilizer components of the windings in superconducting magnets, solenoids and power cables. High-purity high-conductivity copper is usually specified for this application. The purpose of such a stabilizer is to stabilize the superconducting condition in the windings. Composites of the superconducting filaments and the copper stabilizer may be produced by extrusion. Copper alloys containing tin are bonded to niobium in several processes for producing superconducting Nb_3Sn by thermal diffusion, and then extruded. In another application, bronze foil is used in wrapping superconducting power cable.

Tensile Properties. Available data on the tensile properties of copper and copper alloys at subzero temperatures are summarized in Table 26. Supple-

Fig. 6 Fatigue-crack-growth rates for aluminum alloy 3003-O sheet at room temperature and at −101 °C (−150 °F) (Ref 39)

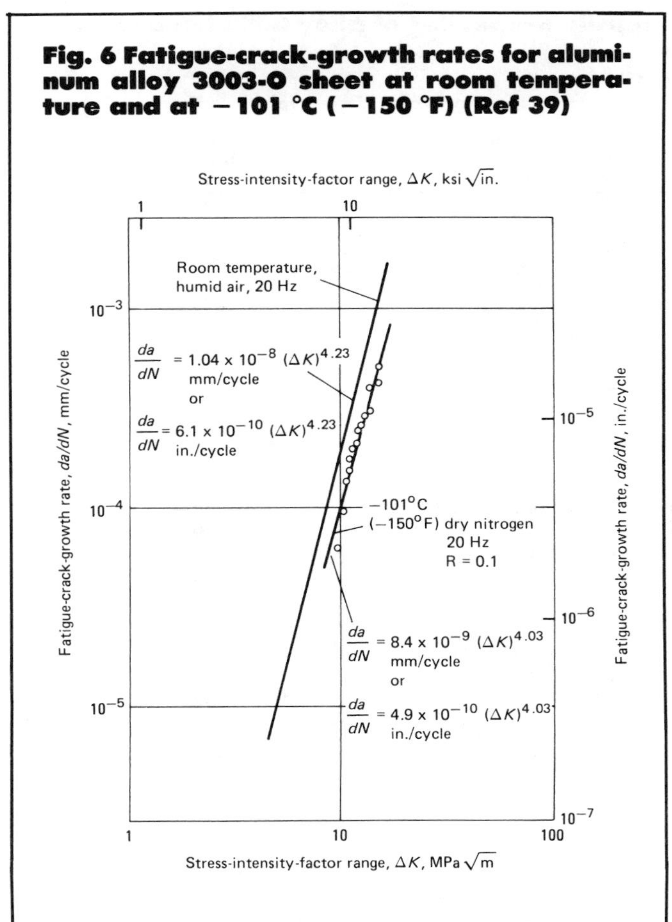

Fig. 7 Fatigue-crack-growth rates for aluminum alloy 5083-O plate (Ref 27)

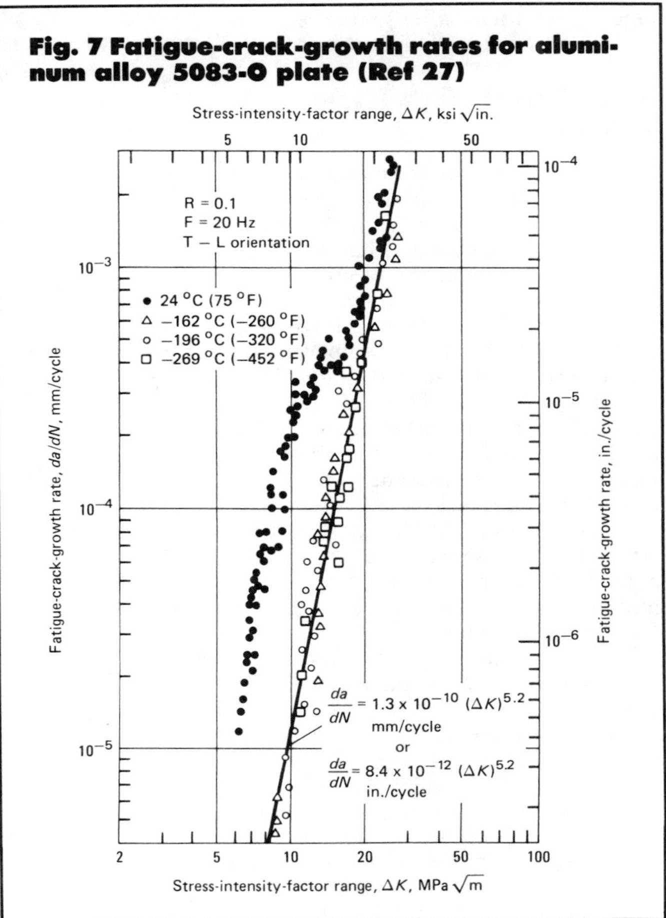

Fig. 8 Young's modulus for copper and two copper-nickel alloys as determined ultrasonically (Ref 48)

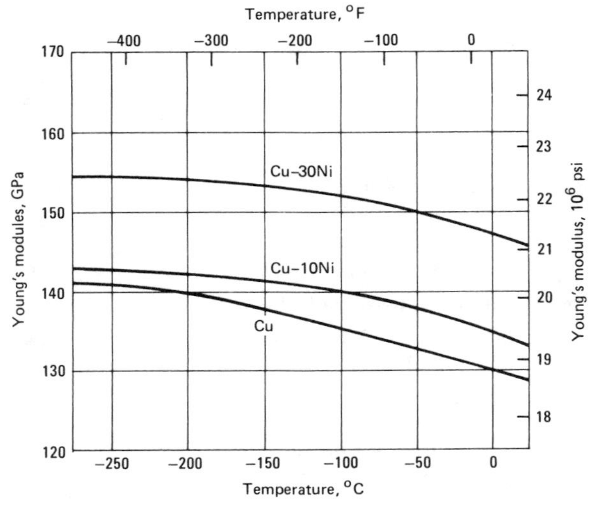

mentary data on values of Young's modulus for copper and two copper-nickel alloys, obtained by ultrasonic testing, are plotted in Fig. 8.

Many copper alloys appear to have good strength, ductility and notch toughness even at liquid-helium temperature. Some are used for valves and tubular fittings in cryogenic equipment. Both spring cold drawn phosphor bronze A and the copper-beryllium alloy in the HT condition are spring alloys. These alloys exhibit increases in yield strength and Young's modulus at very low temperatures, and thus their spring properties are better at cryogenic temperatures than at room temperature.

Fatigue Strength. Results of flexural fatigue testing of several higher-strength copper alloys at subzero temperatures are presented in Table 27. These data constitute fatigue-strength values at 10^6 cycles. Fatigue strengths of both unnotched and notched specimens increase as testing temperature is decreased, which is an added advantage when these alloys are used for springs at subzero temperatures.

Nickel and High-Nickel Alloys

Nickel is another fcc metal that retains good ductility and toughness at subzero temperatures. Unalloyed nickel is low in strength and has only limited usage at subzero temperatures. However, several nickel-base alloys, including some superalloys, have been evaluated at cryogenic temperatures. Compositions of selected high-nickel alloys are given in Table 28. Some of these alloys exhibit excellent combinations of strength, ductility and toughness over the entire range of subzero temperatures. These alloys have been selected for some of the most critical structural components in recent designs for large superconducting motors and generators. They also are suitable for service at elevated temperatures and may be used in applications involving exposure to both subzero and elevated temperatures.

Tensile Properties. Typical tensile properties of nickel and high-nickel alloys, and of high-nickel alloy weldments, at room temperature and at subzero temperatures are presented in Tables 29 and 30. Yield strengths at room temperature for the heat treated alloys range from 703 to 1172 MPa (102 to 170 ksi), whereas the maximum yield strength at −269 °C (−452 °F) is 1406 MPa (204 ksi). All of the heat treated alloys in Table 29, and some of the cold rolled alloys, retain good ductility at the lowest testing temperature. Notch tensile data were not available for some of these alloys, but the available data indicate that they also retain good notch toughness at the lowest testing temperatures. Ductility of heat treated weldments is lower, but is not affected by extended exposure to subzero temperatures. Values of Young's modulus at subzero temperatures for Inconel 718, Inconel X-750 and Inconel 600 are given in Fig. 9 and 10. Values of Poisson's ratio at subzero temperatures for Inconel 600 and Inconel X-750 are presented in Fig. 11.

The torque tube, the damper shield, the outer rotor shell and the stub shafts of a prototype 5 MV·A superconducting generator were made of Inconel X-750 (Ref 58). Larger superconducting generators are being built by several contractors using high-nickel alloys for many critical components. One of these

Table 10 Typical tensile properties of alloy 3003 base metal and weldment(a)

Data from Ref 1, 3, 9, 19, 21

Temperature		Tensile strength		Yield strength		Elongation,	Reduction in area,	Notch tensile strength(b)	
°C	°F	MPa	ksi	MPa	ksi	%	%	MPa	ksi
Plate, O temper									
24	75	110	16.1	41	6.0	40	...	...	...
−78	−108	140	20.1	48	7.0	43	...	...	...
−196	−320	230	33.7	59	8.5	46	...	...	...
−253	−423	365	53.0	69	10.0	48	...	...	...
Plate, H12 temper									
24	75	130	19.0	110	16.0	...	...	...	...
−78	−108	150	22.0	125	18.0	...	...	...	...
−196	−320	235	34.0	160	23.0	...	...	...	...
Bar, H14 temper									
24	75	160	22.9	145	21.1	17	68	...	...
−78	−108	175	25.3	155	22.3	18	59	...	...
−196	−320	250	36.6	180	25.9	32	56	...	...
−269	−452	400	58.1	210	30.1	32	49	450	65.1
Plate, H112 temper(c)									
24	75	110	16.1	52	7.6	24	67	155	22.7
−78	−108	135	19.3	57	8.3	26	66	...	...
−196	−320	230	33.7	74	10.8	31	52	...	...
−269	−452	350	51.1	130	18.5	28	25	275	39.8

(a) Orientation of 3003-O plate and 3003-H14 bar is longitudinal. (b) K_t = 16. (c) Gas metal-arc welded, 1100 filler, as welded.

Table 11 Typical tensile properties of alloy 5083

Data from Ref 2, 3, 9, 22-28

Temperature		Tensile strength		Yield strength		Elongation,	Reduction in area,	Notch tensile strength(a)	
°C	°F	MPa	ksi	MPa	ksi	%	%	MPa	ksi
Plate, O temper, longitudinal orientation									
24	75	310	45.3	155	22.8	20	32	335	48.8
−162	−260	380	54.8	150	21.7	33	39	385	56.2
−196	−320	425	61.7	165	24.1	33	34	410	59.2
−253	−423	585	85.2	175	25.2	32	24	410	59.3
−269	−452	515	74.6	180	26.4	28	22	430	62.3
Plate, O temper, transverse orientation									
24	75	315	45.5	160	22.9	19	26	325	46.8
−162	−260	385	55.7	155	22.8	29	30	370	53.9
−196	−320	430	62.0	175	25.6	28	26	390	56.3
Plate, H113 temper, longitudinal orientation									
24	75	335	48.6	235	34.2	15	23	425	61.4
−196	−320	465	67.1	275	39.6	31	31	485	70.4
−253	−423	620	90.0	305	43.9	30	24	500	72.8
−269	−452	590	85.8	280	40.5	29	28	510	73.7
Plate, H113 temper, transverse orientation									
24	75	345	50.1	235	34.3	16	24	...	...
−196	−320	460	66.4	280	40.3	24	25	...	...

(a) K_t = 14.

generators is rated at 18 million watts and has a rotor 4 m (13 ft) long. It is half the size of a conventional generator of the same capacity. Spherical pressure vessels 63.5 cm (25 in.) in diameter were produced from Inconel 718 to contain supercritical oxygen for the Apollo vehicle (Ref 59). This was a weight-limited application requiring 100% reliability.

Invar 36 is used where minimum thermal expansion and contraction are required. Other low-expansion alloys also have been evaluated for subzero service and can be used as alternatives to Invar 36.

Fracture Toughness. Available information on fracture toughness of several high-nickel alloys and weldments at room temperature and at subzero temperatures is presented in Table 31. Because of the high toughness of most of these alloys, all of the fracture-toughness data except that for Inconel 718 solution treated and double aged (STDA) were obtained by the J-integral method and converted to K_{Ic} (J) by means of the equation

$$\left[K_{Ic}(J)\right]^2 = \frac{E\,J_{Ic}}{1-\nu^2}$$

where E is Young's modulus and ν is Poisson's ratio. The $K_{Ic}(J)$ values are, therefore, estimates of the linear elastic plane-strain critical stress-intensity factors based on the J-integral results.

Fracture toughness of one heat of Inconel X-750 that had been vacuum-induction melted and vacuum-arc remelted (VIM-VAR process) was abnormally low because of carbide precipitation at the grain boundaries. This precipitation probably occurred because of insufficient breakdown of the billet during forging. The condition was not evident from results of tensile tests, but it was evident from metallographic examination. Fracture-toughness data for the other heats of Inconel X-750 are more representative and they normally retain a high degree of toughness at temperatures as low as −269 °C (−452 °F). Fracture toughness of the fusion zones and heat-affected zones of welds in heat treated weldments, however, tend to be lower than that of the base metal.

Fatigue-Crack-Growth Rates. Data on rates of fatigue-crack growth for several high-nickel alloys are presented in Table 32. These data are for the constants C and n in the equation

$$da/dN = C(\Delta K)^n$$

where da/dN is the crack-growth rate and ΔK is the stress-intensity-factor range for constant-load-amplitude fatigue tests on precracked specimens. It

Table 12 Typical tensile properties of gas metal-arc weldments in aluminum alloy 5083

Data from Ref 2, 3, 24, 28, 29

Temperature		Tensile strength		Yield strength		Elongation,	Reduction in area,	Notch tensile strength(a)	
°C	°F	MPa	ksi	MPa	ksi	%	%	MPa	ksi
Weld in O temper plate, 5183 filler metal, longitudinal specimen									
24	75	290	42.4	155	22.7	15	28	315	45.5
−78	−108	305	43.9	145	20.7	31	44	345	49.9
−162	−260	270	53.7	175	25.4	24	31	370	53.6
−196	−320	395	57.5	180	26.3	20	21	365	52.9
−253	−423	445	64.6	185	26.7	12	13	345	49.8
−269	−452	380	55.3	175	25.2	27	37	370	53.9
Weld in O temper plate, 5183 filler metal, transverse specimen									
24	75	300	43.7	160	23.2	12	...	340	49.3
−196	−320	420	61.2	180	26.2	18	...	395	57.6
Weld in O temper plate, 5356 filler metal, longitudinal specimen									
24	75	290	41.7	145	21.2	15	22	...	...
−78	−108	290	42.4	150	21.4	13	23	...	...
−196	−320	410	59.6	170	24.8	21	21	...	...
Weld in H113 temper plate, 5183 filler metal, longitudinal specimen									
24	75	295	42.5	145	20.9	16	30	305	44.5
−196	−320	420	60.8	175	25.1	21	25	365	52.8
−253	−423	445	64.2	185	27.1	11	14	355	51.5

(a) For O temper plate, 5183 filler metal, K_t = 14 to 16; for H113 temper plate, K_t = 14.

Table 13 Design stresses for 5083-O aluminum alloy sheet, plate and weldments

Data from Ref 2, 31

Plate thickness		Service temperature		Minimum tensile strength		Minimum yield strength		Design stress(a)	
mm	in.	°C	°F	MPa	ksi	MPa	ksi	MPa	ksi
1.3 to 38.1	0.051 to 1.50	24	75	276	40.0	124	18.0	69	10.0
		−162	−260	340	49.3	134	19.4	85	12.3
38.11 to 76.2	1.501 to 3.00	24	75	269	39.0	117	17.0	67	9.7
		−162	−260	332	48.1	126	18.3	83	12.0
76.21 to 127	3.001 to 5.00	24	75	262	38.0	110	16.0	66	9.5
		−162	−260	326	46.8	118	17.2	79	11.5
127.01 to 177.8	5.001 to 7.00	24	75	255	37.0	103	15.0	63	9.2
		−162	−260	314	45.6	111	16.1	74	10.7
177.81 to 203.2	7.001 to 8.00	24	75	248	36.0	96.5	14.0	62	9.0
		−162	−260	306	44.4	105	15.1	70	10.1

(a) Equal to 1/4 tensile strength or 2/3 yield strength, whichever is lower; applies to weldments in 5083-O sheet and plate welded using 5183 filler metal.

Table 14 Typical tensile properties of alloy 5456

Data from Ref 2, 3-5, 7, 9, 24

Temperature °C	Temperature °F	Tensile strength MPa	Tensile strength ksi	Yield strength MPa	Yield strength ksi	Elongation, %	Reduction in area, %	Notch tensile strength(a) MPa	Notch tensile strength(a) ksi	Young's modulus GPa	Young's modulus 10^6 psi
Sheet, O temper, longitudinal orientation											
24	75	350	50.9	185	26.5	17	...	330	48.1	...	...
−253	−423	585	85.1	210	30.5	25	...	305	44.5	...	...
Plate, O temper, longitudinal orientation											
24	75	340	49.0	160	23.2	22	31	350	50.9	...	...
−78	−108	340	49.1	165	23.6	26	43	...	...	...	...
−196	−320	455	66.0	180	26.1	34	35	410	59.6	...	...
−253	−423	585	84.5	195	28.6	25	22	415	59.9	...	...
−269	−452	580	84.4	205	29.5	31	24	420	60.9	...	...
Sheet, H321 temper, longitudinal orientation											
24	75	395	57.4	270	39.5	14	...	330	47.6	...	...
−196	−320	530	76.6	320	46.7	27	...	365	52.7	...	...
−253	−423	665	96.4	365	52.6	22	...	390	56.9	...	...
Sheet, H321 temper, transverse orientation											
24	75	400	57.9	275	39.6	18	...	330	48.2	...	...
−196	−320	500	72.5	335	48.7	24	...	365	53.1	...	...
−253	−423	615	88.9	375	54.1	15	...	390	56.4	...	...
Plate, H321 temper, longitudinal orientation											
24	75	355	51.2	235	34.1	14	11	410	59.7	...	...
−78	−108	380	55.0	235	34.3	20	31	...	...	...	...
−196	−320	490	71.3	275	39.9	26	28	455	66.2	...	...
−253	−423	635	92.2	305	43.9	18	17	490	71.2	...	...
−269	−452	640	92.6	320	46.5	24	25	525	75.8	...	...
Sheet, H343 temper, longitudinal orientation											
24	75	395	57.3	325	46.8	8	...	395	57.0	72	10.4
−78	−108	400	58.2	345	49.9	10	...	395	57.2	74	10.7
−196	−320	510	73.9	375	54.7	12	...	445	64.4	74	10.7
−253	−423	590	85.9	410	59.8	9	...	470	68.5	78	11.3
−269	−452	595	86.5	395	57.0	9	...	...	...	...	...
Sheet, H343 temper, transverse orientation											
25	75	405	58.4	295	43.1	10	...	395	57.3	70	10.2
−78	−108	400	58.1	300	43.8	11	...	390	56.9	74	10.8
−196	−320	500	72.4	355	51.7	12	...	410	59.4	73	10.6
−253	−423	565	82.1	390	56.5	8	...	420	61.2	81	11.7
Plate, H343 temper, longitudinal orientation											
24	75	385	56.1	305	44.4	10	...	365	53.3	...	...
−78	−108	385	55.6	310	44.6	12	...	360	52.2	...	...
−196	−320	495	72.1	360	52.3	14	...	420	61.2	...	...
−253	−423	515	74.8	380	54.8	7	...	...	...	...	...

(a) For sheet in O and H343 tempers, K_t = 6.3; for plate in H343 temper, K_t = 13; for plate in O and H321 tempers, K_t = 14 to 16; for sheet in H321 temper, K_t = 30.

is not applicable for negative stress ratios. However, the constants in Table 32 permit calculation of the fatigue-crack-growth rate at any ΔK value within the limits of the ΔK range.

All of the data in Table 32 were obtained at the same stress ratio *(R)* and at nearly the same frequency. Variations in these conditions may affect fatigue-crack-growth rates. However, for these alloys, fatigue-crack-growth rates at subzero temperatures are either equal to or lower than the rates at room temperature for the same ΔK values. For design purposes, the use of room-temperature fatigue-crack-growth-rate data for subzero applications is feasible.

Fatigue Strength. Results of fatigue-life tests at 10^6 cycles on axial and flexural specimens of several high-nickel alloys at room temperature and at temperatures are presented in Table 33. These data indicate that, at 10^6 cycles, the fatigue strengths of these alloys are higher at subzero temperatures than at room temperature. Furthermore, specimens with smoother surface finishes had higher fatigue strengths at each testing temperature.

Table 15 Typical as-welded tensile properties of alloy 5456 weldments

Data from Ref 2, 3, 24

Temperature °C	Temperature °F	Tensile strength MPa	Tensile strength ksi	Yield strength MPa	Yield strength ksi	Elongation, %	Reduction in area, %	Notch tensile strength(a) MPa	Notch tensile strength(a) ksi
Plate in O temper, longitudinal orientation, 5556 filler metal(b)									
24	75	320	46.7	155	22.7	15	27	310	44.6
−196	−320	415	60.3	180	26.1	15	17	350	50.9
−253	−423	430	62.4	200	28.8	9	11	330	47.5
Sheet in H321 temper, longitudinal orientation, 5556 filler metal(c)									
24	75	360	51.9	225	32.6	13	...	375	54.4
−196	−320	440	63.7	260	37.7	16	...	410	59.7
−253	−423	405	59.1	340	49.5	4	...	410	59.7
Sheet in H321 temper, transverse orientation, 5556 filler metal(c)									
24	75	355	51.7	210	30.3	8	...	355	51.8
−196	−320	435	63.1	245	35.8	10	...	395	57.1
−253	−423	430	62.1	380	55.0	5	...	385	55.7
Plate in H321 temper, longitudinal orientation, 5556 filler metal(b)									
24	75	310	44.6	155	22.5	13	23	...	...
−78	−108	310	45.2	155	22.5	16	28	...	...
−196	−320	405	59.0	180	26.2	14	18	...	...
Sheet in H343 temper, longitudinal orientation, 5556 filler metal(c)									
24	75	340	49.6	200	28.7	5	...	...	...
−78	−108	330	47.5	205	30.0	5	...	...	...
−196	−320	420	61.2	240	34.8	7	...	...	...
−253	−423	405	58.5	270	39.5	4	...	...	...
−269	−452	445	64.2	265	38.3	4	...	...	...
Plate in H343 temper, longitudinal orientation, 5356 filler metal(c)									
24	75	285	41.6	180	26.3	6	...	...	...
−78	−108	315	45.7	180	25.9	8	...	...	...
−196	−320	350	51.1	215	31.0	6	...	...	...
−253	−423	380	55.0	230	33.0	4	...	...	...
Plate in H343 temper, longitudinal orientation, 5356 filler(b)									
24	75	270	39.1	175	25.6	4	...	...	...
−78	−108	270	39.0	185	26.7	6	...	...	...
−196	−320	320	46.3	210	30.5	5	...	...	...
−253	−423	320	46.2	230	33.1	4	...	...	...

(a) For plate in O temper, K_t = 14; for sheet in H321 temper, K_t = 30. (b) Gas metal-arc welding process. (c) Gas tungsten-arc welding process.

Ferritic Alloy Steels

Ferritic steels exhibit a transition in toughness from ductile, high-energy fracture at temperatures above the transition temperature to brittle, low-energy fracture at temperatures below the transition temperature. This is characteristic of metals with body-centered-cubic (bcc) crystal structures. The usual method of measuring transition temperature in ferritic steels is the Charpy V-notch impact test, in which specimens are tested over a range of subzero temperatures (ASTM method A370). The transition temperatures determined by this method do not indicate at what temperatures brittle fracture will occur in structural compo-

Fig. 9 Young's modulus for Inconel 718 as determined ultrasonically (Ref 56)

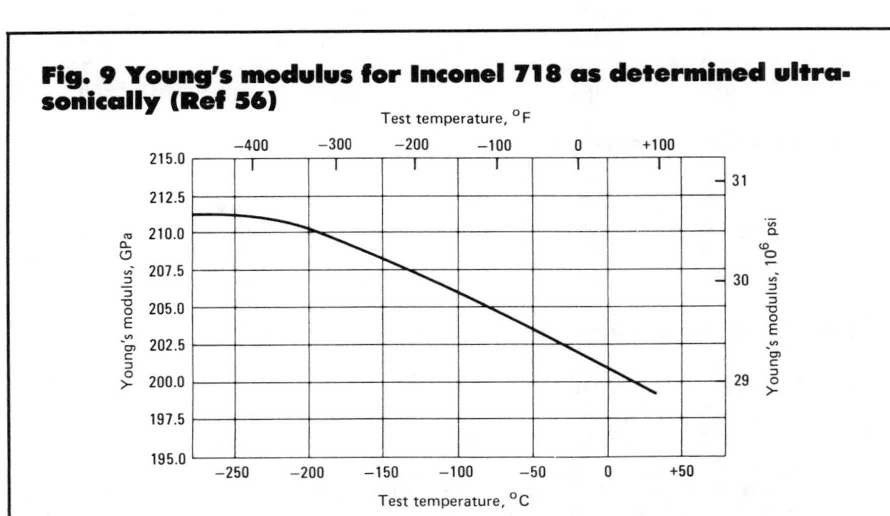

Table 16 Typical tensile properties of alloy 6061

Data from Ref 2, 4, 5, 8

Temperature °C	Temperature °F	Tensile strength MPa	Tensile strength ksi	Yield strength MPa	Yield strength ksi	Elongation, %	Reduction in area, %	Notch tensile strength(a) MPa	Notch tensile strength(a) ksi
Sheet, T4 temper, longitudinal orientation									
24	75	280	40.7	185	26.8	19	...	285	41.0
−78	−108	305	44.2	220	32.0	20	...	290	42.4
−196	−320	410	59.4	235	33.9	31	...	355	51.2
−253	−423	600	86.8	325	47.2	31	...	425	62.0
Sheet, T4 temper, transverse orientation									
24	75	275	40.1	170	24.4	20	...	270	39.3
−78	−108	300	43.5	200	29.1	21	...	290	41.8
−196	−320	400	57.7	205	30.0	32	...	350	51.1
−253	−423	635	92.3	295	42.6	34	...	430	62.6
Sheet, T6 temper, longitudinal orientation(b)									
24	75	320	46.3	290	42.2	12	...	340	49.6
−78	−108	350	50.8	310	44.8	12	...	370	53.5
−196	−320	425	61.8	340	49.6	18	...	430	62.3
−253	−423	495	72.0	365	52.6	26	...	515	74.6
Sheet, T6 temper, transverse orientation									
24	75	325	46.8	280	40.9	12	...	335	48.5
−78	−108	355	51.3	305	44.5	11	...	320	46.7
−196	−320	425	61.4	330	47.5	19	...	400	58.3
−253	−423	505	73.2	350	50.9	25	...	490	70.8
Plate, T6 temper, longitudinal orientation									
24	75	320	46.5	295	43.0	16	36	305	44.0
−78	−108	360	52.5	325	46.8	21	44	...	...
−196	−320	445	64.2	360	52.3	25	40	...	...
−253	−423	525	76.2	375	54.3	26	36	420	61.0
Plate, T651 temper, longitudinal orientation									
24	75	310	44.9	290	42.2	16	50	475	69.2
−196	−320	400	58.3	335	48.9	23	48	575	83.4
−269	−452	485	70.1	380	55.0	26	42	620	89.9
Plate, T651 temper, transverse orientation									
24	75	330	48.0	290	41.9	14	42	465	67.8
−78	−108	345	50.1	315	45.5	12	...	...	...
−196	−320	400	58.3	325	46.9	19	39	555	80.5
−269	−452	485	70.2	370	53.5	22	34	600	87.2
Bar, T6 temper, longitudinal orientation									
24	75	325	46.8	290	41.8	18	54	...	...
−78	−108	345	49.9	295	42.6	20	52	...	...
−196	−320	425	61.5	340	49.2	24	47	...	...
−253	−423	520	75.7	370	53.7	30	45	...	...
−269	−452	570	83.0	400	58.0	22	39	...	...
Forging, T6 temper, longitudinal orientation									
24	75	310	44.9	275	39.9	20	52	435	63.3
−78	−108	350	50.6	300	43.7	20	51	465	67.2
−196	−320	435	63.0	325	47.3	24	42	540	78.6
−253	−423	500	72.4	380	54.9	25	39	560	81.5

(a) For sheet in T4 and T6 tempers and plate in T6 temper, K_t = 6.3; for plate in T651 temper, K_t = 16; for forgings in T6 temper, K_t = 6 to 8. (b) Young's modulus for sheet in T6 temper at 24 °C (75 °F) is 68 GPa (9.9 × 10^6 psi); at −196 °C (−320 °F), 79 GPa (11.4 × 10^6 psi); at −253 °C (−423 °F), 87 GPa (12.6 × 10^6 psi).

Table 17 Typical tensile properties of gas metal-arc welded alloy 6061 weldments

Data from Ref 2, 5, 9, 19

Temperature °C	°F	Tensile strength MPa	ksi	Yield strength MPa	ksi	Elongation, %	Reduction in area, %	Notch tensile strength(a) MPa	ksi
Weld in T4 temper sheet, 4043 filler metal, longitudinal specimen(b)									
24	75	250	35.9	140	20.0	12	...	...	...
−196	−320	375	54.4	180	25.9	19	...	...	...
Weld in T4 temper sheet, 4043 filler metal, transverse specimen(b)									
24	75	235	34.2	135	19.7	12	...	...	...
−196	−320	375	54.5	195	28.5	16	...	...	...
Weld in T4 temper sheet, 4043 filler metal, longitudinal specimen(c)									
24	75	305	44.1	275	40.2	4	...	...	...
−196	−320	395	57.6	325	47.3	5	...	...	...
Weld in T4 temper sheet, 4043 filler metal, transverse specimen(c)									
24	75	300	43.6	260	38.0	4	...	...	...
−196	−320	400	57.9	330	48.2	6	...	...	...
Weld in T6 temper sheet, 4043 filler metal, longitudinal specimen(b)									
24	75	220	32.2	160	23.2	5	...	235	34.1
−196	−320	325	47.4	200	28.8	10	...	260	38.0
−253	−423	450	65.5	220	32.0	10	...	290	41.8
Weld in T6 temper sheet, 4043 filler metal, transverse specimen(b)									
24	75	215	31.3	170	24.8	5	...	...	...
−196	−320	325	47.2	195	28.4	8	...	...	...
−253	−423	415	59.9	235	34.3	6	...	...	...
Weld in T6 temper plate, 4043 filler metal, longitudinal specimen(b)									
24	75	215	31.0	145	20.9	6	19	235	34.0
−78	−108	240	34.6	165	23.6	6	19	265	38.6
−196	−320	305	44.0	180	25.8	6	12	275	39.6
−269	−452	340	49.1	260	37.6	4	9	275	39.9
Weld in T6 temper plate, 5356 filler metal, longitudinal specimen(b)									
24	75	225	32.7	155	22.6	8	31	325	46.9
−78	−108	255	37.1	170	24.7	9	36	345	50.1
−196	−320	325	47.0	190	27.3	14	39	375	54.1
−269	−452	400	57.7	245	35.3	14	24	370	53.3
Weld in T6 temper plate, 5356 filler metal, longitudinal specimen(d)									
24	75	280	40.5	200	29.3	10	33	...	...
−78	−108	320	46.4	240	35.1	12	44	400	57.8
−196	−320	395	57.1	235	33.9	20	29	460	66.4
−269	−452	475	69.1	305	44.5	19	24	420	60.8

(a) For sheet in T6 temper, K_t = 30; for plate in T6 temper, K_t = 16. (b) As welded. (c) Aged to T6. (d) Heat treated to T6.

nents or threaded fasteners, but they do provide a means of rating the notch toughness of steels at subzero temperatures. The method provides a means for setting minimum standards for steels intended for use at subzero temperatures. These minimum standards are presented in ASTM A20 (standard specification for pressure vessels). ASTM specifications A203, A353, A442, A516, A517, A537, A553, A612, A645, A662 and A724 describe steel plates with minimum Charpy V-notch energy or lateral expansion requirements at testing temperatures from −26 to −196 °C (−15 to −320 °F). ASTM specifications A353, A553 and A645 describe alloy steel plates with minimum Charpy energy requirements at testing temperatures from −170 to −196 °C (−275 to −320 °F).

The following ASTM specifications describe other steel products for subzero service:

- A333 Seamless and welded steel pipe
- A334 Seamless and welded carbon and alloy steel tubes
- A350 Forged or rolled carbon and alloy steel flanges, forged fittings, and valves
- A352 Ferritic steel castings
- A420 Piping and fittings of wrought carbon steel and alloy steel
- A522 Forged or rolled 8 and 9% nickel steel flanges, fittings and valves
- A671 Electric-fusion-welded steel pipe

In addition to the steels listed above, a number of proprietary ferritic steels have been developed by several steel producers to meet certain requirements for service at subzero temperatures as low as about −112 °C (−170 °F).

Table 18 Typical tensile properties of aluminum alloy 7005 base metal and weldments

Data from Ref 2, 9, 33

Temperature °C	Temperature °F	Tensile strength MPa	Tensile strength ksi	Yield strength MPa	Yield strength ksi	Elongation, %	Reduction in area, %	Notch tensile strength(a) MPa	Notch tensile strength(a) ksi
Plate, T6351 temper, longitudinal orientation									
24	75	375	54.7	330	47.7	19	51	555	80.2
−78	−108	430	63.4	360	51.8	17	42	605	87.4
−196	−320	510	74.2	390	56.5	18	34	660	95.7
−269	−452	600	86.8	430	62.4	18	27	670	97.2
Plate, T6351 temper, transverse orientation									
24	75	370	54.0	325	47.0	18	46	555	80.4
−78	−108	435	63.0	360	52.3	16	38	605	87.4
−196	−320	510	73.9	400	57.7	17	28	640	93.0
−269	−452	590	85.4	425	61.9	17	26	660	95.5
Plate, T6351 temper(b)									
24	75	330	48.0	220	32.0	12	...	...	...
−196	−320	385	56.0	290	42.0	4	...	...	...
−269	−452	415	60.0	330	48.0	4	...	...	...
Extrusion, T5351 temper, longitudinal orientation									
24	75	425	62.0	380	55.0	15	43	595	86.2
−78	−108	465	67.8	405	58.6	14	30	635	92.1
−196	−320	580	83.9	465	67.5	17	27	685	99.1
−253	−423	705	102	505	73.4	18	28	735	107
−269	−452	675	97.6	520	75.6	17	22	735	107

(a) K_t = 16. (b) Gas metal-arc welded using 5039 filler metal; as welded.

Fig. 10 Young's modulus for Inconel 600 and Inconel X-750 as determined ultrasonically (Ref 57)

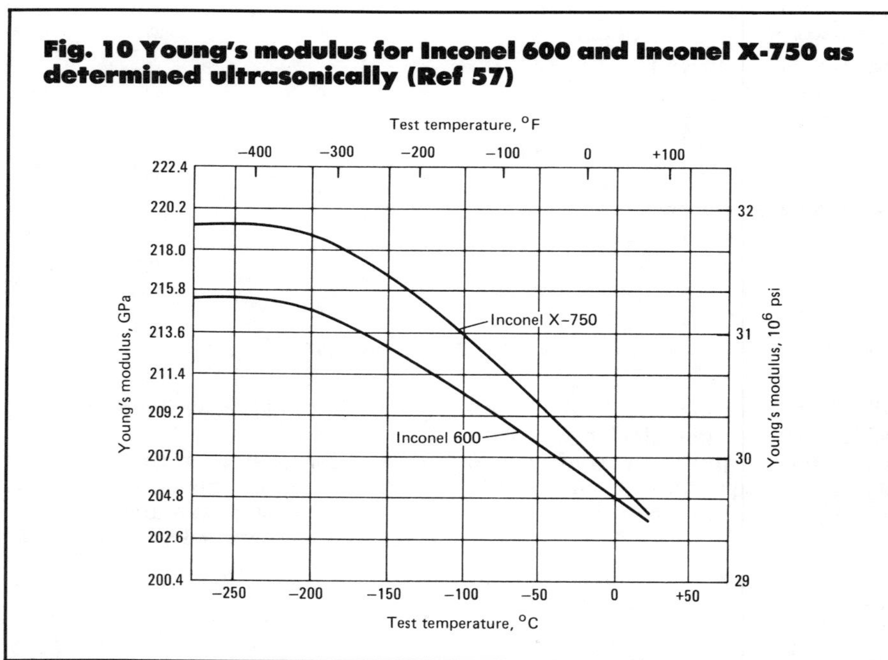

ASTM specification A203 covers ferritic alloy steel plates with nominal nickel contents of 2.25 and 3.50%. For 1-in. plate, minimum Charpy impact requirements are applicable for the 2.25% Ni grades at −68 °C (−90 °F) and for the 3.50% Ni grades at −101 °C (−150 °F).

ASTM specification A645 covers ferritic alloy steel plates containing 4.75 to 5.25% Ni and 0.20 to 0.35% Mo. For 1-in. plate, minimum Charpy impact requirements are designated at −170 °C (−275 °F) for hardened, temperized and reversion-annealed plate.

Double normalized and tempered 9% nickel steel is covered by ASTM A353, and quenched and tempered 8% and 9% nickel steels are covered by ASTM A553 (types I and II). For quenched and tempered material, the minimum lateral expansion in Charpy V-notch impact tests is 0.38 mm (0.015 in.). Charpy tests on 9% Ni steel (type I) are conducted at −195 °C (−320 °F); for 8% Ni steel (type II), they are conducted at −170 °C (−275 °F).

Carbon and alloy steel castings for subzero-temperature service are covered by ASTM Standard Specification A757. These specifications recognize the effect of nickel content in reducing the transition temperatures of nickel steels. Chemical compositions of these steels are presented in Table 34.

In order to ensure that a given steel structure will have a reasonable degree of toughness at minimum service temperature, it is recommended that the appropriate ASTM specification (or ASME SA20) be applied in designating the steels to be used. Minimum strength levels also are given in these specifications.

For applications involving exposure to temperatures from 0 to −196 °C (+32 to −320 °F), the ferritic nickel steels usually are considered first, if they have sufficient corrosion resistance. Such applications include storage tanks for liquefied hydrocarbon gases and structures and machinery designed for use in cold regions.

Tensile Properties. Typical tensile properties of 5% and 9% Ni steels at room temperature and at subzero temperatures are presented in Table 35. Yield and tensile strengths increase as testing temperature is decreased. These steels remain ductile at the lowest testing temperatures. Minimum tensile properties of all the ASTM steels for subzero applications are designated in ASTM Standard Specifications. Values of Young's modulus and Poisson's ratio at subzero temperatures for ferritic nickel steels are plotted in Fig. 12 and 13.

Fracture Toughness. Ferritic nickel steels are too tough at room temperature for valid fracture-toughness (K_{Ic}) data to be obtained on specimens of reasonable size (Ref 63), but limited

Table 19 Typical tensile properties of aluminum alloy 7039

Data from Ref 2, 4, 7, 23, 33, 34

Temperature °C	°F	Tensile strength MPa	ksi	Yield strength MPa	ksi	Elongation, %	Reduction in area, %	Notch tensile strength(a) MPa	ksi	Young's modulus GPa	10^6 psi
Sheet in T6 temper, longitudinal orientation											
24	75	455	66.0	410	59.8	11	...	470	68.1	68	9.82
−78	−108	510	73.9	460	66.7	12	...	445	64.5	70	10.2
−196	−320	575	83.2	495	72.0	15	...	345	50.2	79	11.4
−253	−423	665	96.2	535	77.4	14	...	395	57.3	79	11.4
Sheet in T6 temper, transverse orientation											
24	75	460	66.7	400	58.1	10	...	450	65.1	70	10.1
−78	−108	505	73.4	445	64.5	10	...	410	59.8	69	9.96
−196	−320	580	84.3	490	71.2	14	...	305	44.0	73	10.6
−253	−423	675	97.6	520	75.7	10	...	330	47.8	76	11.0
Sheet in T61 temper, longitudinal orientation											
24	75	400	58.0	325	47.3	14	...	...	...	...	...
−196	−320	495	71.7	365	53.1	18	...	...	...	...	...
Sheet in T61 temper, transverse orientation											
24	75	410	59.3	330	47.6	13	...	...	...	...	...
−196	−320	530	77.2	385	55.7	16	...	...	...	...	...
Plate in T6 and T61 tempers, longitudinal orientation											
24	75	450	65.2	400	57.7	13	...	515	74.4	...	...
−78	−108	490	71.0	440	63.6	18	...	550	79.8	...	...
−196	−320	575	83.5	480	69.6	14	...	565	81.7	...	...
−253	−423	660	95.9	525	76.0	13	...	560	81.5	...	...
Plate in T6 and T61 tempers, transverse orientation											
24	75	460	66.8	390	56.6	12	...	510	74.3	...	...
−78	−108	495	71.7	435	63.2	11	...	525	76.5	...	...
−196	−320	570	82.5	475	68.8	11	...	500	72.8	...	...
−253	−423	645	93.7	515	74.8	10	...	480	69.6	...	...
Plate in T6 and T61 tempers, S-T orientation											
24	75	465	67.2	385	55.5	8	...	515	75.0	...	...
−78	−108	495	71.8	410	59.5	6	...	450	65.6	...	...
−196	−320	520	75.4	460	66.8	2	...	300	43.3	...	...
−253	−423	276	39.3	175	25.5	2	...	160	23.3	...	...

(continued)

(a) K_t = 19 for sheet in T6 temper; K_t = 10 for plate in T6 and T61 tempers; K_t = 15 to 16 for plate in T6151 and T6351 tempers; K_t = 6.3 for extrusions in T6 and T61 tempers.

fracture-toughness data have been obtained on these steels at subzero temperatures by the J-integral method. Results of these tests are presented in Table 36. The 5% Ni steel retains relatively high fracture toughness at −162 °C (−260 °F), and the 9% Ni steel retains relatively high fracture toughness at −196 °C (−320 °F). These temperatures approximate the minimum temperatures at which these steels may be used.

Fatigue-Crack-Growth Rates. Constants for determining rates of fatigue-crack growth for ferritic nickel steels are presented in Table 37. For the 3.5% Ni steel, fatigue-crack-growth rates at −78 °C (−108 °F) and at −101 °C (−150 °F) are about the same as the rate at room temperature. However, in the stress-intensity-factor range (ΔK) above about 20 MPa$\sqrt{m}$ (18 ksi$\sqrt{in.}$), fatigue-crack-growth rates are substantially greater at −196 °C (−320 °F) than at room temperature. For the 5% Ni steel specimens, fatigue-crack-growth rates also were substantially higher at −196 °C (−320 °F) than at room temperature or at −162 °C (−260 °F) in the stress-intensity-factor range above 30 MPa$\sqrt{m}$ (27 ksi$\sqrt{in.}$). For the 9% Ni steel, fatigue-crack-growth rates were substantially higher at −269 °C (−452 °F) than at higher testing temperatures in the stress-intensity-factor range above 24 MPa$\sqrt{m}$ (22 ksi$\sqrt{in.}$). The fatigue-crack-growth-rate data for the ferritic nickel steels are consistent with results of other tests.

Stainless Steels

Austenitic stainless steels have been used extensively for subzero applications to −269 °C (−452 °F). These steels contain sufficient amounts of nickel and manganese to depress the M_s temperature into the subzero range. Thus they retain face-centered-cubic crystal structures on cooling from hot working or annealing temperatures. Yield and tensile strengths of austenitic stainless steels increase substantially as testing temperature is decreased, and these steels retain good ductility and toughness at −269 °C (−452 °F). Most austenitic stainless steels may be readily fabricated by welding, but sometimes the welding heat causes sensitization that reduces corrosion re-

Table 19 (continued)

Temperature °C	Temperature °F	Tensile strength MPa	Tensile strength ksi	Yield strength MPa	Yield strength ksi	Elongation, %	Reduction in area, %	Notch tensile strength(a) MPa	Notch tensile strength(a) ksi	Young's modulus GPa	Young's modulus 10^6 psi
Plate in T6151 temper, longitudinal orientation											
24	75	415	60.4	345	50.2	14	36	545	78.8	...	...
−129	−200	470	68.3	365	52.8	15	33	560	81.5	...	...
−196	−320	560	81.4	415	60.5	16	26	595	86.2	...	...
−269	−452	650	94.2	475	69.0	14	22	615	89.1	...	...
Plate in T6151 temper, transverse orientation											
24	75	405	59.1	335	48.7	14	32	535	77.5	...	...
−129	−200	465	67.1	360	52.2	14	30	555	80.4	...	...
−196	−320	540	78.6	410	59.2	16	25	560	81.3	...	...
−269	−452	625	90.8	455	66.0	12	15	585	85.0	...	...
Plate in T6351 temper, longitudinal orientation											
24	75	465	67.2	390	56.5	14	32	610	88.6	...	...
−78	−108	520	75.4	430	62.1	14	23	635	91.9	...	...
−196	−320	610	88.4	465	67.8	17	20	590	85.4	...	...
−269	−452	710	103	525	76.4	18	19	615	89.5	...	...
Plate in T6351 temper, transverse orientation											
24	75	460	66.5	390	56.6	13	33	605	88.0	...	...
−78	−108	500	72.8	425	61.4	12	23	585	84.7	...	...
−196	−320	600	87.1	475	69.0	13	19	520	75.1	...	...
−269	−452	695	101	535	77.7	13	15	555	80.6	...	...
Extrusion in T6 and T61 tempers, longitudinal orientation											
24	75	485	70.3	425	61.6	14	...	680	98.4	...	...
−196	−320	645	93.6	535	77.6	12	...	745	108	...	...
Extrusion in T6 and T61 tempers, transverse orientation											
24	75	455	66.0	390	56.8	16	...	645	93.9	...	...
−196	−320	575	83.5	470	67.9	11	...	665	96.2	...	...
Extrusion in T6 and T61 tempers, S-T orientation											
24	75	435	63.2	355	51.3	12	...	635	91.8	...	...
−196	−320	540	78.5	445	64.3	6	...	575	83.2	...	...

(a) K_t = 19 for sheet in T6 temper; K_t = 10 for plate in T6 and T61 tempers; K_t = 15 to 16 for plate in T6151 and T6351 tempers; K_t = 6.3 for extrusions in T6 and T61 tempers.

Fig. 11 Poisson's ratios for Inconel 600 and Inconel X-750 as determined ultrasonically (Ref 57)

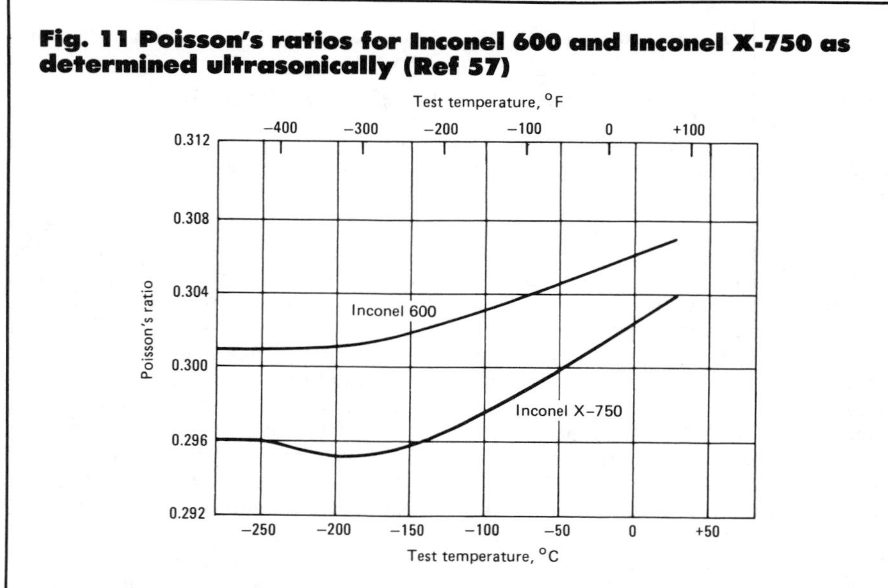

sistance in the weld area. Strength of austenitic steels can be increased by cold rolling or cold drawing. Cold working at −196 °C (−320 °F) is more effective in increasing strength than cold working at room temperature. For metallurgically unstable stainless steels such as 301, 304 and 304L, plastic deformation at subzero temperatures causes partial transformation to martensite, which increases strength. For some cryogenic applications, it is desirable to use a stable stainless steel such as type 310.

Compositions of austenitic stainless steels of interest are presented in Table 38. Small amounts of nitrogen increase the strengths of these steels. Manganese additions are used in some steels to replace some of the nickel. Maximum strength is obtained for A-286 alloy by solution treating and aging. Type 416

Table 20 Typical tensile properties of weldments in aluminum alloy 7039, longitudinal specimens
Data from Ref 2, 23, 34, 35

Temperature °C	Temperature °F	Tensile strength MPa	Tensile strength ksi	Yield strength MPa	Yield strength ksi	Elongation, %	Reduction in area, %	Notch tensile strength(a) MPa	Notch tensile strength(a) ksi
Sheet in T6 temper, 5183 filler metal(b)(c)									
24	75	365	53.2	240	34.8	7	...	405	58.7
−196	−320	390	56.7	300	43.4	4	...	435	63.0
Sheet in T61 temper(d)(e)									
24	75	315	46.0	175	25.6	10	...	...	...
−196	−320	420	61.2	210	30.3	14	...	...	...
Plate in T6 temper, 5183 filler metal(c)(f)									
24	75	295	42.6	170	24.5	10	...	355	51.5
−196	−320	420	60.8	205	29.7	10	...	405	58.9
Plate in T6 temper, 5039 filler metal(e)(f)									
24	75	345	50.3	215	31.5	11	...	335	48.7
−196	−320	440	64.0	265	38.6	7	...	330	47.6
Plate in T61 temper, 5039 filler metal(c)(g)									
24	75	355	51.8	215	31.1	9	23	360	52.1
−196	−320	400	58.3	250	36.1	9	12	375	54.2
−253	−423	375	54.5	...	...	2	4	330	47.9
Plate in T6151 temper, 5039 filler metal(c)(d)									
24	75	355	51.8	235	34.3	10	...	405	58.9
−196	−320	420	61.2	280	40.5	6	13	410	59.7

(a) $K_t = 6.3$. (b) Naturally aged 15 to 30 days. (c) Gas metal-arc welding process. (d) As welded. (e) Gas tungsten-arc welding process. (f) Naturally aged 15 days. (g) Naturally aged 27 days.

Fig. 12 Young's modulus for ferritic nickel alloys as determined ultrasonically (Ref 30)

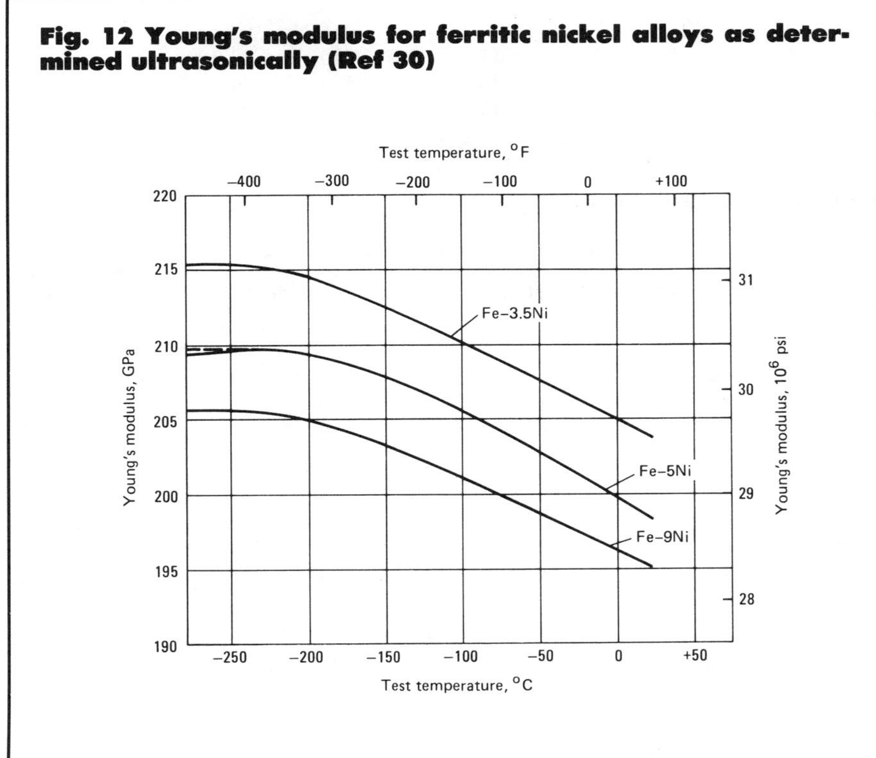

is a ferritic chromium stainless steel that is usually used in the quenched-and-tempered condition. It is included in this series because there are applications in rotating pumps and other machinery, in which a magnetic material is needed to activate counters.

Types 301 and 310 have been used in the form of extra-hard cold rolled sheet to provide high strength in such applications as the liquid oxygen and liquid hydrogen tanks for the Atlas and Centaur rockets. Joining was done by butt fusion welding, and reinforcing strips were spot welded to the tank along the weld joint. In another method for producing high-strength cylindrical tanks, welded preform tanks are fabricated from annealed type 301 stainless steel, submerged in liquid nitrogen while in a cylindrical die, and expanded (cryoformed) by pressurizing until the preform fits the die. The amount of strengthening depends on the amount of plastic deformation incurred in expanding the preform to the size of the die. Strengthening results from the dual effects of cold working of the austenite and partial transformation of the

Table 21 Typical tensile properties of aluminum alloy 7075

Data from Ref 5-7, 33, 36

Temperature		Tensile strength		Yield strength		Elongation,	Reduction in area,	Notch tensile strength(a)		Young's modulus	
°C	°F	MPa	ksi	MPa	ksi	%	%	MPa	ksi	GPa	10^6 psi
Sheet, T6 temper, longitudinal orientation											
24	75	555	80.8	500	72.4	10	...	470	68.4	72	10.5
−73	−100	585	85.1	525	76.2	10	...	...	...	74	10.8
−196	−320	680	98.6	610	88.7	12	...	295	42.5	76	11.0
−256	−423	705	114	705	102	7	...	280	40.4	86	12.5
Sheet, T6 temper, transverse orientation											
24	75	555	80.6	485	70.6	12	...	415	60.5	73	10.6
−73	−100	575	83.2	510	74.1	9	...	...	...	70	10.2
−196	−320	675	98.0	580	84.2	10	...	290	41.9	73	10.6
−253	−423	785	114	685	99.3	9	...	260	37.4	...	...
Plate, T6 temper, longitudinal orientation											
24	75	605	87.4	545	79.2	15	24	790	115	73	10.6
−78	−108	635	92.4	560	81.3	11	18	800	116	79	11.4
−196	−320	715	104	675	97.7	9	13	770	112	...	...
−253	−423	825	120	745	108	10	15	800	116	79	11.4
Plate, T651 temper, longitudinal orientation											
24	75	580	84.2	530	77.1	10	14	715	104	...	...
−78	−108	595	86.2	595	86.0	10	11	680	98.4	...	...
−196	−320	705	102	650	94.4	7	10	565	81.8	...	...
−253	−423	770	112	710	103	5	...	...	...	...	...
−269	−452	825	120	770	112	6	9	555	80.2	...	...
Plate, T651 temper, transverse orientation											
24	75	585	85.0	525	76.3	8	18	655	95.2	...	...
−78	−108	610	88.6	555	80.4	6	12	575	83.2	...	...
−196	−320	695	101	640	92.6	4	8	490	71.0	...	...
Plate, T7351 temper, longitudinal orientation											
24	75	525	76.2	455	66.2	10	22	645	93.4	...	...
−78	−108	575	83.4	500	72.3	10	17	640	93.0	...	...
−196	−320	675	98.2	570	82.5	11	14	580	84.4	...	...
−269	−452	760	110	605	88.1	11	12	650	94.0	...	...

(a) For sheet in T6 temper, K_t = 30; for plate in T6 temper, K_t = 6.3; for plate in T651 and T7351 tempers, K_t = 16.

Table 22 Typical tensile properties of cast aluminum alloys

Data from Ref 6, 9, 17, 37

Temperature		Tensile strength		Yield strength		Elongation,	Reduction in area,	Notch tensile strength(a)	
°C	°F	MPa	ksi	MPa	ksi	%	%	MPa	ksi
355-T6 sand cast									
24	75	225	32.7	180	26.3	4	3	...	...
−196	−320	410	59.3	320	46.5	2	4	...	...
−253	−423	440	63.6	390	56.8	2	2	...	...
355-T6 permanent mold cast									
24	75	345	50.0	300	43.6	3	5	...	...
−78	−108	345	50.3	315	45.5	4	4	...	...
−196	−320	410	59.7	365	53.2	3	3	...	...
−253	−423	490	70.8	420	60.8	2	2	...	...
355-T61 permanent mold cast									
24	75	325	47.4	240	34.8	5	5	...	...
−78	−108	345	49.8	255	36.7	5	7	...	...
−196	−320	385	55.6	340	49.0	3	6	...	...
−253	−423	480	69.9	365	53.2	2	3	...	...

(continued)

(a) K_t = 6.3 for A356-T6, cast. All other values, K_t = 16.

Table 22 (continued)

Temperature °C	°F	Tensile strength MPa	ksi	Yield strength MPa	ksi	Elongation, %	Reduction in area, %	Notch tensile strength(a) MPa	ksi
C355-T61, premium strength									
24	75	300	43.6	210	30.3	6	9	365	52.6
−78	−108	335	48.4	230	33.2	7	8	390	56.6
−196	−320	375	54.4	270	39.4	5	6	430	62.7
356-T4, sand cast									
24	75	215	31.1	135	19.8	4	6	220	31.6
−78	−108	250	36.6	160	23.4	4	6	260	37.6
−196	−320	280	40.8	190	27.2	3	3	290	42.2
356-T6, sand cast									
24	75	265	38.6	225	32.6	2	3	260	37.4
−78	−108	295	43.1	245	35.8	2	2	275	40.0
−196	−320	325	47.5	270	39.2	2	2	305	44.0
356-T6, permanent mold cast									
24	75	255	36.8	215	31.1	1	...	295	43.0
−78	−108	290	42.0	235	34.1	3	5	285	41.1
−196	−320	315	45.7	250	36.5	3	4	310	45.0
356-T7, sand cast									
24	75	260	37.8	230	33.7	2	2	240	34.5
−78	−108	285	41.4	235	34.4	2	2	270	38.8
−196	−320	310	45.1	270	38.8	1	0	295	43.1
356-T71, sand cast									
24	75	200	28.8	140	20.2	5	...	220	32.0
−78	−108	220	32.2	155	22.2	4	5	205	29.6
−196	−320	260	37.4	175	25.3	3	2	235	34.4
356-T7, permanent mold cast									
24	75	195	28.4	150	21.4	4	7	245	35.3
−78	−108	225	32.6	170	24.3	4	5	255	37.2
−196	−320	255	37.3	175	25.6	3	4	275	39.9
A356-T6, cast									
24	75	210	30.6	165	23.8	3	2	200	29.2
−253	−423	330	48.1	290	42.0	2	2	250	36.3
A356-T61, permanent mold cast									
24	75	270	39.4	210	30.8	4	...	330	47.8
−78	−108	290	41.9	225	32.6	4	6	330	47.7
−196	−320	340	49.4	245	35.8	4	6	365	52.6
A356-T61, premium strength									
24	75	285	41.6	210	30.2	9	10	355	51.4
−78	−108	330	48.2	240	34.8	9	10	380	55.2
−196	−320	355	51.7	260	38.0	4	4	410	59.8
−269	−452	455	66.0	330	48.0	7	9	495	71.9
A356-T62, permanent mold cast									
24	75	280	40.9	255	36.7	2	6	320	46.2
−78	−108	310	45.2	275	39.6	3	5	345	49.8
−196	−320	335	48.6	285	41.4	3	5	400	57.9
−253	−423	380	55.0	310	45.3	3	3	435	63.3
A356-T7, sand cast									
24	75	255	37.1	210	30.5	4	7	310	44.9
−78	−108	275	40.0	220	31.7	4	5	285	41.0
−196	−320	315	45.6	245	35.2	3	5	305	44.0
A356-T7, permanent mold cast									
24	75	195	28.2	150	21.4	5	9	255	36.9
−78	−108	245	35.4	180	25.8	6	8	305	43.9
−196	−320	295	42.7	195	28.5	6	7	325	47.1

(a) K_t = 6.3 for A356-T6, cast. All other values, K_t = 16.

Table 23 Fracture toughness of aluminum alloy plate

Data from 2, 13, 16, 27, 38

Alloy and condition	Room temperature yield strength MPa	ksi	Specimen design	Orientation	Fracture toughness, K_{Ic} or $K_{Ic}(J)$ at: 24 °C (75 °F) MPa$\sqrt{m}$	ksi$\sqrt{in.}$	−196 °C (−320 °F) MPa$\sqrt{m}$	ksi$\sqrt{in.}$	−253 °C (−423 °F) MPa$\sqrt{m}$	ksi$\sqrt{in.}$	−269 °C (−452 °F) MPa$\sqrt{m}$	ksi$\sqrt{in.}$
2014-T651	432	62.7	Bend	T-L	23.2	21.2	28.5	26.1	...	...	...	...
2024-T851	444	64.4	Bend	T-L	22.3	20.3	24.4	22.2	...	...	...	...
2124-T851(a)	455	66.0	CT	T-L	26.9	24.5	32.0	29.1	...	...	...	...
	435	63.1	CT	L-T	29.2	26.6	35.0	31.9	...	...	...	...
	420	60.9	CT	S-L	22.7	20.7	24.3	22.1	...	...	...	...
2219-T87	382	55.4	Bend	T-S	39.9	36.3	46.5	42.4	52.5	48.0	...	...
			CT	T-S	28.8	26.2	34.5	31.4	37.2	34.0	...	...
	412	59.6	CT	T-L	30.8	28.1	38.9	32.7	...	...	...	...
5083-O	142	20.6	CT	T-L	27.0(b)	24.6(b)	43.4(b)	39.5(b)	...	...	48.0(b)	43.7(b)
6061-T651	289	41.9	Bend	T-L	29.1	26.5	41.6	37.9	...	...	...	...
7039-T6	381	55.3	Bend	T-L	32.3	29.4	33.5	30.5	...	...	...	...
7075-T651	536	77.7	Bend	T-L	22.5	20.5	27.6	25.1	...	...	...	...
7075-T7351	403	58.5	Bend	T-L	35.9	32.7	32.1	29.2	...	...	...	...
7075-T7351	392	56.8	Bend	T-L	31.0	28.2	30.9	28.1	...	...	...	...

(a) 2124 is similar to 2024, but with higher purity base and special processing to improve fracture toughness. (b) $K_{Ic}(J)$.

Table 24 Results of fatigue-life tests on aluminum alloys

Data from Ref 2, 40-42

Alloy and condition	Stressing mode	Stress ratio, R	K_t	Fatigue strength at 10^6 cycles, at: 24 °C (75 °F) MPa	ksi	−196 °C (−320 °F) MPa	ksi	−253 °C (−423 °F) MPa	ksi
2014-T6 sheet	Axial	−1.0	1	115	17	170	25	315	46
		+0.01	1	215	31	325	47	435	63
2014-T6 sheet, GTA welded, 2319 filler	Axial	−1.0	1	83	12	105	15	125	18
2219-T62 sheet	Axial	−1.0	1	130	19	15	22	255	37
			3.5	52	7.5	45	6.5	62	9
2219-T87 sheet	Axial	−1.0	1	150	22	115-170	17-25	275	40
			3.5	52	7.5	48	7	55	8
2219-T87 sheet, GTA welded, 2319 filler	Axial	−1.0	1	69	10	83	12	150	22
5083-H113 plate	Flex	−1.0	1	140	20.5	190	27.5	...	...
5083-H113 plate, GMA welded, 5183 filler	Flex	−1.0	1	90	13	130	18.8	...	...
6061-T6 sheet (a)	Flex	−1.0	1	160	23	220	32	235	34
(b)	Flex	−1.0	1	165	24	230	33	230	33
7039-T6 sheet	Axial	−1.0	1	140	20	215	31	275	40
		+0.01	1	230	33	330	48	440	64
		−1.0	3.5	48	7	48	7	62	9
7075-T6 sheet	Axial	−1.0	1	96	14	145	21	250	36

(a) Surface finish, 150 μin. rms. (b) Surface finish, 20 μin. rms.

Table 25 Nominal compositions of copper and copper alloys

UNS number	Common name	Composition, % Cu	Zn	Sn	Ni	Others
C10200	Oxygen-free copper	99.95(Cu+Ag)	...	...	...	...
C12200	Phosphorus deoxidized, high residual phosphorus copper	99.90	...	...	...	0.02 P
C17200	Beryllium copper	98.1	...	...	...	1.9 Be
C22000	Commercial bronze, 90%	90.00	10.00	...	...	...
C26000	Cartridge brass, 70%	70.00	30.00	...	...	...
C51000	Phosphor bronze, 5% A	94.80	...	5.00	...	0.2 P
C70600	Copper nickel, 10%	88.60	...	...	10.00	1.4 Fe
C71500	Copper nickel, 30%	69.50	...	...	30.00	0.5 Fe

austenite to martensite.

Type 304 stainless steel usually is used in the annealed condition for tubing, pipes and valves employed in transfer of cryogens; for Dewar flasks and storage tanks; and for structural components that do not require high strength.

Types 310 and 310S are considered to be metallurgically stable for all conditions of cryogenic exposure. Therefore, these steels are used for structural components in which maximum stability

Table 26 Typical tensile properties of copper and copper alloys(a)

Data from Ref 2, 6, 43-47

Temperature		Tensile strength		Yield strength		Elongation,	Reduction in area,	Notch tensile strength(b)		Young's modulus	
°C	°F	MPa	ksi	MPa	ksi	%	%	MPa	ksi	GPa	10^6 psi
C10200, bar, O61 temper(c)											
24	75	220	32.2	75	10.9	54	86	...	...	...	...
−78	−108	270	39.0	80	11.6	53	84	...	...	...	...
−196	−320	360	52.2	88	12.8	60	84	...	...	...	...
−253	−423	420	60.7	90	13.1	69	83	...	...	...	...
C12200, bar, O61 temper(c)											
24	75	215	31.3	46	6.7	45	76	300	43.3	105	15.1
−78	−108	265	38.3	46	6.6	56	87	345	50.4	110	16.0
−196	−320	350	50.6	51	7.4	62	84	430	62.3	110	16.2
−253	−423	440	63.8	58	8.4	68	83	495	72.0	110	16.3
−269	−452	415	60.4	54	7.9	65	81	515	74.7	115	16.4
C22000, bar, O61 temper(c)											
24	75	265	38.5	66	9.6	56	84	345	49.9	105	15.1
−78	−108	290	41.8	70	10.2	57	80	385	55.6	115	16.4
−196	−320	380	55.2	91	13.2	86	78	475	69.2	120	17.7
−253	−423	505	73.2	110	15.6	95	73	525	76.3	125	18.0
−269	−451	470	68.2	105	15.0	91	73	545	78.9	125	18.1
C26000, bar, H03 temper(d)											
24	75	655	95.2	420	60.9	14	58	...	...	...	...
−78	−108	695	101	445	64.3	17	62	...	...	...	...
−196	−320	805	117	475	68.6	28	63	...	...	...	...
−253	−423	910	132	505	73.4	32	58	...	...	...	...
C70600, bar, O61 temper(c)											
24	75	340	49.6	150	21.4	38	79	450	65.0	120	17.7
−78	−108	375	54.7	170	24.7	42	77	505	73.1	...	...
−196	−320	495	72.0	170	24.8	50	77	600	87.2	135	19.5
−253	−423	570	82.5	210	30.2	50	73	670	97.1	...	...
−269	−452	555	80.6	170	24.9	53	73	690	100	140	20.5
C71500, bar, O61 temper(c)											
24	75	400	57.8	130	18.7	47	68	545	79.4	145	20.9
−78	−108	470	68.0	155	22.2	48	70	625	90.5	150	22.1
−196	−320	620	89.8	220	31.6	52	70	780	113	130	18.8
−253	−423	710	103	265	38.1	51	66	885	128	150	21.7
−269	−452	725	105	275	40.1	48	65	895	130	135	19.5
C51000, bar, H08 temper(e)											
24	75	535	77.4	495	72.0	18	78	940	136	110	15.6
−78	−108	590	85.6	545	78.7	30	78	1010	147	115	16.5
−196	−320	725	105	615	89.2	34	67	1150	167	115	16.7
−253	−423	905	131	725	105	39	62	1280	185	115	16.5
−269	−452	800	116	690	100	34	58	1280	185	115	16.4
C17200 sheet, TD02 temper(f)											
24	75	620	90	550	80	15	...	...	...	120	17.5
−78	−108	655	95	600	87	20	...	...	...	120	17.6
−196	−320	805	117	690	100	37	...	...	...	130	18.8
−253	−423	945	137	750	109	45	...	...	...	135	19.5
C17200 sheet, TH02 temper(g)											
24	75	1320	191	1140	166	2.8	...	...	...	130	19
−253	−423	1640	238	1230	178	3.5	...	...	...	145	21

(a) All alloys have a longitudinal base-metal orientation. (b) K_t = 5. (c) Annealed. (d) 3/4 hard. (e) Spring temper. (f) Solution treated, cold worked to 1/2 hard. (g) Aged 2 h at 350 °C (600 °F).

and a high degree of toughness are required at cryogenic temperatures.

Type 316 stainless steel is less stable than type 310, but tensile specimens of type 316 pulled to 0.2% offset (at the yield load) at −269 °C (−452 °F) showed no indication of martensite formation in the deformed regions (Ref 71). However, when tensile specimens of type 316 were pulled to fracture at −269 °C (−452 °F), the metallographic structures in the areas of the fractures transformed to approximately 50% martensite (Ref 72). Type 316 stainless steel is an important candidate material for structural components of superconducting and magnetic fusion machinery.

For higher-strength components of cryogenic structures, there are several stainless steels that contain significant amounts of manganese in place of some of the nickel, along with small additions of nitrogen and other elements that increase strength. Among these stainless steels are 21-6-9, Pyromet 538, Nitronic 40, Nitronic 60 and Kromarc 58.

Cast stainless steels of the compositions shown in Table 38 have been used for bubble chambers, for cylindrical magnet tubes for superconducting magnets, for valve bodies, and for other components that are cooled to −269 °C (−452 °F) in service. For such applications, castings are more appropriate than wrought products.

Tensile Properties. Typical tensile properties of annealed 300 series austenitic stainless steels at room temperature and at subzero temperatures are presented in Table 39, and tensile properties of cold worked 300 series stainless steels are given in Table 40. Cold working substantially increases yield and tensile strengths and reduces ductility, but ductility and notch toughness of the cold worked alloy often are sufficient for cryogenic applications. Tensile properties of other stainless steels are presented in Table 41. For the annealed alloys, the greatest effect of the nitrogen addition to produce an increase in yield strength at cryogenic temperatures. Kromarc 58 has been used for several structural applications in prototype superconducting generators. Other nitrogen-strengthened stainless steels have comparable properties. The data for cold worked AISI 202 and Kromarc sheet indicate how these alloys can be strengthened by cold working that results in reduced ductility. Solution treating and aging A-286 alloy develops good strength with good ductility and notch tough-

Fig. 13 Poisson's ratios for ferritic nickel alloy steels as determined ultrasonically (Ref 30)

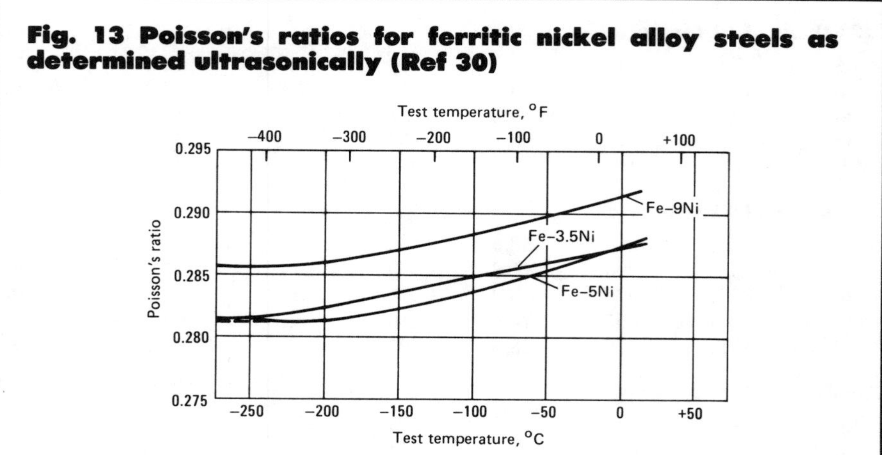

Table 27 Results of fatigue-life tests on copper alloys

Data from Ref 2, 49, 50

K_t	Fatigue strength, at 10^6 cycles(a), at: 24 °C (75 °F) MPa	ksi	−196 °C (−320 °F) MPa	ksi	−253 °C (−423 °F) MPa	ksi
C26000, H08 temper(b)						
1	235	34	495	72	705	102
3.1	180	26	270	39	340	49
C17200, TB00 temper(c)						
1	415	60	515	75	725	105
C17200, TH02 temper(d)						
1	385	56	560	81	580	84
3.1	215	31	325	47	310	45

(a) Flex stressing mode, R = −1.0 at 29 to 86 Hz. (b) Aged at 200 °C (400 °F). (c) Solution heat treated. (d) Solution treated and aged to 1/2 hard.

Table 28 Nominal compositions of high-nickel alloys

UNS number	Alloy designation	Nominal composition, % Ni	Cr	Fe	Mn	Si	C	Others
N05500	Monel K-500	Rem	...	1.0	0.6	0.15	0.15	29.5Cu, 2.8Al, 0.5Ti
	Hastelloy B	Rem	0.6	5.0	0.8	0.7	0.1	2.5Co, 28.0Mo, 0.2-0.6V
	Hastelloy C	Rem	15.5	5.5	0.5	0.5	0.07	1.5Co, 16 Mo, 4.0W
N06600	Inconel 600	Rem	15.8	7.2	0.2	0.2	0.04	0.10 Cu
N09706	Inconel 706	39 to 44	16	Rem	0.10	0.10	0.04	0.35 max Al, 3.0 (Nb+Ta), 1.7Ti
N07718	Inconel 718	Rem	18.6	18.5	...	...	0.04	0.4Al, 0.9Ti, 5.0Nb, 3.1Mo
N07750	Inconel X-750	Rem	15.0	6.8	0.7	...	0.04	0.8Al, 2.5Ti, 0.85Nb
	Invar 36	36	...	Rem	...	...	...	...

Table 29 Typical tensile properties of nickels and high-nickel alloys

Data from Ref 2, 6, 7, 10, 44, 51-55

Temperature °C	Temperature °F	Tensile strength MPa	Tensile strength ksi	Yield strength MPa	Yield strength ksi	Elongation, %	Reduction in area, %	Notch tensile strength(a) MPa	Notch tensile strength(a) ksi	Young's modulus GPa	Young's modulus 10^6 psi
Nickel 200 sheet, annealed, longitudinal orientation											
24	75	425	62	97	14	43	...	...	...	200	29
−253	−423	740	107	230	33	36	...	...	...	230	33
Nickel 270 bar, hot finished, longitudinal orientation											
24	75	365	53.1	175	25.1	...	...	...	...	205	30.0
−196	−320	495	71.5	220	31.8	...	...	...	...	225	32.3
−269	−452	730	106	225	32.7	...	...	...	...	225	32.5
K-500 sheet, longitudinal orientation(b)											
24	75	1030	150	710	103	22	...	940	136	...	...
−78	−108	1080	156	765	111	24	...	1000	145	...	...
−196	−320	1230	178	855	124	30	...	1120	163	...	...
−253	−423	1340	194	925	134	30	...	1190	173	...	...
K-500 sheet, transverse orientation(b)											
24	75	1020	148	715	104	22	...	...	...	...	...
−78	−108	1100	159	785	114	24	...	...	...	...	...
−196	−320	1260	183	910	132	30	...	...	...	...	...
−253	−423	1360	197	935	136	29	...	...	...	...	...
K-500 bar, longitudinal orientation(c)											
24	75	1080	157	705	102	28	54	...	...	...	...
−78	−108	1230	178	895	130	29	54	...	...	...	...
−196	−320	1300	188	86	125	32	54	...	...	...	...
−253	−423	1420	206	940	136	36	52	...	...	...	...
Hastelloy B sheet, cold rolled 40%, longitudinal orientation											
24	75	1320	191	1220	177	3	...	...	...	...	...
−73	−100	1530	222	1430	207	5	...	...	...	...	...
−196	−320	1570	228	1430	208	12	...	...	...	...	...
−253	−423	1950	283	1650	240	16	...	...	...	...	...
Hastelloy C sheet, cold rolled 20%, longitudinal orientation											
24	75	1140	165	1000	145	13	...	1250	182	...	...
−196	−320	1520	220	1280	186	32	...	1560	226	...	...
−253	−423	1740	252	1380	200	33	...	1690	245	...	...
Inconel 600 sheet, hard, cold rolled, longitudinal orientation											
24	75	910	132	885	128	4	...	...	...	...	...
−253	−423	1210	176	910	132	22	...	...	...	...	...
Inconel 600 bar, cold drawn, longitudinal orientation											
24	75	940	136	890	129	15	56	1230	179	170	25
−78	−108	985	143	910	132	20	58	...	...	...	...
−196	−320	1160	168	1030	150	26	62	...	...	...	...
−253	−423	1250	181	1100	160	30	56	...	...	...	...
−257	−430	1280	186	1210	176	20	56	1530	222	220	32
Inconel 706 forged billets(d)											
24	75	1260	183	1050	152	24	33	1880	272	...	...
−196	−320	1570	228	1200	174	29	33	2170	315	...	...
−269	−452	1680	243	1250	181	30	33	2250	326	...	...
Inconel 718 sheet, longitudinal orientation(e)											
24	75	1330	193	1090	158	18	...	1330	193	205	29.9
−78	−108	1490	216	1190	172	17	...	1470	213	220	31.7
−196	−320	1730	251	1310	190	21	...	1560	226	225	32.5
−253	−423	1740	252	1340	194	16	...	1500	217	225	32.6

(continued)

(a) K_t = 10 for K-500 sheet, Inconel 706 forged billets, Inconel 718 sheet and Inconel X-750 forged billets; K_t = 6.3 for Hastelloy C sheet, Inconel 718 forgings and Inconel X-750; K_t = 6.4 for Inconel 600 sheet. (b) Aged 16 h at 594 °C (1100 °F) plus controlled cooling. (c) Aged 21 h at 594 °C (1100 °F), 8 h at 538 °C (1000 °F), AC. (d) Aged 1 h at 980 °C (1800 °F), AC, 8 h at 730 °C (1350 °F), FC to 620 °C (1150 °F), hold 8 h, AC. (e) Aged 1 h at 955 °C (1750 °F), AC, 8 h at 720 °C (1325 °F), FC to 620 °C (1150 °F), hold 10 h, AC. (f) Aged 3/4 h at 980 °C (1800 °F), AC, 8 h at 720 °C (1325 °F), FC to 620 °C (1150 °F), hold 10 h, AC. (g) Annealed and aged 20 h at 700 °C (1300 °F), AC.

Table 29 (continued)

Temperature °C	Temperature °F	Tensile strength MPa	Tensile strength ksi	Yield strength MPa	Yield strength ksi	Elongation, %	Reduction in area, %	Notch tensile strength(a) MPa	Notch tensile strength(a) ksi	Young's modulus GPa	Young's modulus 10^6 psi
Inconel 718 sheet, transverse orientation(e)											
24	75	1320	192	1100	159	18	...	1300	188	200	28.8
−78	−108	1480	214	1210	176	12	...	1450	210	210	30.8
−196	−320	1700	246	1300	189	21	...	1500	217	230	33.5
−253	−423	1770	256	1370	198	16	...	1500	218	240	34.5
Inconel 718 bar, longitudinal orientation(f)											
24	75	1410	204	1170	170	15	18	...	...	...	...
−196	−320	1650	239	1340	197	21	20	...	...	...	...
−269	−452	1810	263	1410	204	21	20	...	...	...	...
Inconel 718 forgings, longitudinal orientation(f)											
24	75	1340	194	1150	167	24	35	2030	295	...	...
−78	−108	1350	196	1190	172	29	45	2170	314	...	...
−196	−320	1630	237	1300	188	26	34	2350	341	...	...
−253	−423	1680	244	1320	192	28	42	2390	347	...	...
−269	−452	1810	263	1410	204	21	20	...	...	...	...
Inconel 718 forgings, transverse orientation(f)											
24	75	1290	187	1150	167	18	28	1930	280	...	...
−253	−423	1740	253	1350	196	24	30	2300	333	...	...
Inconel 718 forgings, S-T orientation(f)											
24	75	1290	187	1140	166	17	23	1860	270	...	...
−253	−423	1630	237	1340	195	14	12	1970	286	...	...
Inconel X-750 sheet, longitudinal orientation(g)											
24	75	1220	177	815	118	24	...	1120	162	210	30.4
−78	−108	1320	192	875	127	28	...	1200	174	...	...
−196	−320	1500	217	905	131	32	...	1270	184	225	32.4
−253	−423	1590	230	940	136	32	...	1370	199	...	...
Inconel X-750 sheet, transverse orientation(g)											
24	75	1230	178	850	123	25	...	1160	168	...	...
−78	−108	1340	194	925	134	26	...	1210	175	...	...
−196	−320	1500	217	950	138	32	...	1270	184	...	...
−253	−423	1630	236	985	143	32	...	1390	201	...	...
Inconel X-750 bar, longitudinal orientation(g)											
24	75	1340	194	985	143	25	49	...	...	...	...
−196	−320	1570	228	1050	152	32	45	...	...	...	...
−253	−423	1700	246	1090	158	33	42	...	...	...	...
−257	−430	1720	249	1080	157	33	46	...	...	...	...
Inconel X-750 forged billet(d)											
24	75	985	143	665	96.2	18	18	1200	174	...	...
−196	−320	1090	158	770	112	16	14	1340	195	...	...
−269	−452	1020	148	735	107	14	13	1410	205	...	...
Invar 36 bar, cold drawn 12 to 15%, longitudinal orientation											
24	75	650	94	625	91	21	62	...	...	...	...
−78	−108	785	114	725	105	29	60	...	...	...	...
−196	−320	1080	156	915	133	27	61	...	...	...	...
−253	−423	1190	172	1120	162	23	58	...	...	...	...
−269	−452	1230	178	1110	161	20	52	...	...	...	...

(a) K_t = 10 for K-500 sheet, Inconel 706 forged billets, Inconel 718 sheet and Inconel X-750 forged billets; K_t = 6.3 for Hastelloy C sheet, Inconel 718 forgings and Inconel X-750; K_t = 6.4 for Inconel 600 sheet. (b) Aged 16 h at 594 °C (1100 °F) plus controlled cooling. (c) Aged 21 h at 594 °C (1100 °F), 8 h at 538 °C (1000 °F), AC. (d) Aged 1 h at 980 °C (1800 °F), AC, 8 h at 730 °C (1350 °F), FC to 620 °C (1150 °F), hold 8 h, AC. (e) Aged 1 h at 955 °C (1750 °F), AC, 8 h at 720 °C (1325 °F), FC to 620 °C (1150 °F), hold 10 h, AC. (f) Aged 3/4 h at 980 °C (1800 °F), AC, 8 h at 720 °C (1325 °F), FC to 620 °C (1150 °F), hold 10 h, AC. (g) Annealed and aged 20 h at 700 °C (1300 °F), AC.

Table 30 Typical tensile properties of high-nickel alloy GTAW weldments(a)

Data from Ref 2, 55

Temperature °C	Temperature °F	Tensile strength MPa	Tensile strength ksi	Yield strength MPa	Yield strength ksi	Elongation, %	Reduction in area, %	Notch tensile strength(b) MPa	Notch tensile strength(b) ksi
K-500 sheet with K Monel filler metal(c)									
24	75	1050	152	760	110	19	...	...	...
−78	−108	1110	161	825	120	21	...	...	...
−196	−320	1250	182	925	134	26	...	...	...
−253	−423	1370	198	1010	146	24	...	...	...
Inconel 706 forging with 718 filler metal(d)									
24	75	1110	161	1000	145	2	5	1630	236
−196	−320	1300	188	1170	170	4	4	1760	255
−269	−452	1370	199	1220	177	4	6	1880	273
Inconel 718 sheet, no filler metal(e)									
24	75	1320	191	1150	167	5	...	1200	174
−196	−320	1560	226	1290	187	4	...	1320	192
−253	−423	1730	251	1410	205	5	...	1440	209
Inconel 718 forging with 718 filler metal(d)									
24	75	1260	183	2000	159	2	6	1390	202
−196	−320	1430	208	1280	186	2	4	1470	213
−269	−452	1650	239	1280	185	28	33	2280	331
Inconel X-750 sheet with X-750 filler metal(f)									
24	75	1290	187	860	125	22	...	...	...
−78	−108	1340	195	915	133	24	...	...	...
−196	−320	1540	224	945	137	30	...	...	...
−253	−423	1660	241	1020	148	28	...	...	...
Inconel X-750 forged billet with F69 filler metal(d)									
24	75	1100	159	855	124	9	12	1570	228
−196	−320	1110	161	930	135	6	9	1710	248
−269	−452	1120	163	960	139	6	9	1630	236

(a) Base metal has longitudinal orientation. (b) For Inconel 706, Inconel 718 forging with 718 filler, and Inconel X-750 forged billet with F69 filler, K_t = 10; for Inconel 718 sheet, no filler, K_t = 6.3. (c) Annealed sheet, welded and aged at 594 °C (1100 °F). (d) Weldment heat treatment: 1 h at 980 °C (1800 °F), AC, 8 h at 730 °C (1325 °F), FC to 620 °C (1150 °F), hold 8 h, AC. (e) Weldment aged 8 h at 720 °C (1325 °F), FC to 620 °C (1150 °F), hold 10 h, AC. (f) Weldment aged 20 h at 700 °C (1300 °F), AC.

Fig. 14 Young's modulus for four austenitic stainless steels as determined ultrasonically (Ref 86)

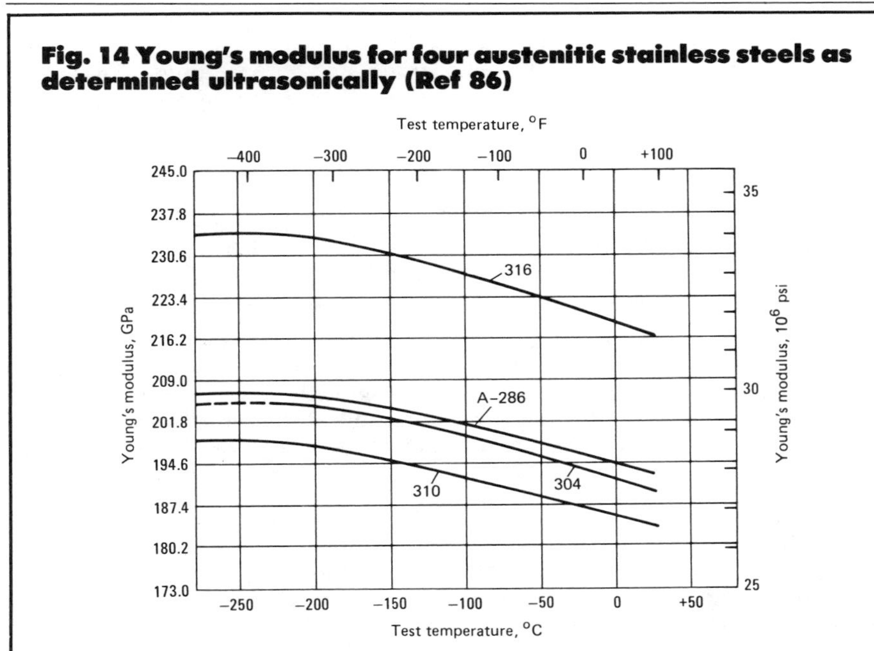

ness in the cryogenic range. Because of its low ductility, alloy 416 is not recommended for use below −196 °C (−320 °F) except in nonstressed applications.

Results of tensile tests on stainless steel weldments at subzero temperatures, given in Table 42, may be significant in selecting stainless steels for cryogenic applications. For annealed plate tested in the as-welded condition, Kromarc 58 has the most favorable properties of the alloys in Table 42. Tensile properties of A-286 weldments can be improved by age hardening; however, there is a significant advantage in being able to use weldments in the as-welded condition, as can be done with Kromarc weldments. Moreover, Kromarc is not as difficult to weld as A-286.

Tensile properties of cast stainless steels at subzero temperatures are presented in Table 43. These data indicate that alloys CF8 and CF8M have yield strengths at subzero temperatures

Table 31 Fracture toughness of high-nickel alloys and weldments

Data from 2, 52, 54, 55, 58, 60

Alloy and condition(a)	Form	Room temperature yield strength MPa	ksi	Specimen design	Orientation	Fracture toughness, K_{Ic} or K_{Ic} (J), at: 24 °C (75 °F) MPa$\sqrt{m}$	ksi$\sqrt{in.}$	−196 °C (−320 °F) MPa$\sqrt{m}$	ksi$\sqrt{in.}$	−269 °C (−452 °F) MPa$\sqrt{m}$	ksi$\sqrt{in.}$
Inconel 706 (VIM-VAR) STDA	Forging	1063	154	CT	C-R.....	134(b)	121(b)	...	...	158(b)	143(b)
Inconel 706 (VIM-VAR) GTA weld, ST/W/STDA	Forging-weldment	1063	154	CT	....	...	...	...	...	58.7(b)(c)	53.0(b)(c)
Inconel 718 STDA	Bar	1172	170	CT	T-S.....	96.3	87.8	103	94.0	112	102
Inconel 718 (VIM-VAR) STDA	Forging	1164	169	CT	C-R.....	61.5(b)	55.6(b)	...	...	75.5(b)	68.2(b)
Inconel 718 (VIM-VAR) GTA weld, ST/W/STDA	Forging-weldment	1164	169	CT	....	...	...	51.5(b)(c) 66.6(b)(d)	46.5(b)(c) 60.2(b)(d)	52.1(b)(c) 61.3(b)(d)	47.1(b)(c) 55.4(b)(d)
Inconel X-750 (VIM-VAR) STDA	Forging	824	120	CT	C-R.....	...	...	...	...	76.1(b)	69.2(b)(c)
Inconel X-750 (VIM) STDA	Forging	917	133	CT	C-R.....	...	...	...	...	145(b)	132(b)
Inconel X-750 (AAM-VAR) STDA	Forging	848	123	CT	C-R.....	...	...	...	...	237(b)	216(b)
Inconel X-750 (VIM-VAR) ST/W/STDA	Forging-weldment	824	120	CT	....	...	...	...	...	134(b)(c)(f) 176(b)(c)(g)	122(b)(c)(f) 160(b)(c)(g)

(a) VIM, vacuum induction melted; VAR, vacuum arc remelted; AAM, air arc melted; ST, solution treated; W, welded. STDA for Inconel 706 and Inconel X-750: 980 °C (1800 °F) 1 h, AC, 730 °C (1350 °F) 8 h, FC to 620 °C (1150 °F), hold 8 h, AC. STDA for Inconel 718: 980 °C (1800 °F) 1 h, AC, 720 °C (1325 °F) 8 h, FC to 620 °C (1150 °F), hold 8 h, AC. Filler metals: F-718 for Inconel 706 and Inconel 718; Inco F69 for Inconel X-750. (b) K_{Ic}. (c) Fusion zone. (d) Heat-affected zone. (e) This heat of Inconel X-750 had carbide precipitates at the grain boundaries, which caused abnormally low fracture toughness. (f) Gas tungsten arc weld. (g) Vacuum electron beam weld.

comparable to those of equivalent wrought alloys. Yield and tensile strengths of as-cast Kromarc 55, however, are considerably lower than those of the wrought alloy Kromarc 58.

Results of ultrasonic determinations of Young's modulus and Poisson's ratio for four stainless steels are shown in Fig. 14 and 15 and serve to supplement the tensile data.

Fracture Toughness. Fracture-toughness data for stainless steels are limited because steels of this type that are suitable for use at cryogenic temperatures have very high toughness. The fracture-toughness data that are available were obtained by the J-integral method and converted to $K_{Ic}(J)$ values. Such data for base metal and weldments are shown in Table 44. Fracture toughness of base metals are relatively high even at −269 °C (−452 °F); fracture toughness of fusion zones (FZ) of welds may be lower or higher than that of the base metal.

Fatigue-Crack-Growth Rates. Available data for determining fatigue-crack-growth rates at room temperature and at subzero temperatures for austenitic stainless steels and weldments are presented in Table 45. The fatigue-crack-growth rates of the base metals are generally higher at room temperature than at subzero temperatures, or about equal at room temperature and at subzero temperatures, except for 21-6-9 stainless steel. For 21-6-9, fatigue-crack-growth rates are

Table 32 Fatigue-crack-growth-rate data for compact specimens of high-nickel alloys

Data from Ref 52, 54, 58, 61, 62

Alloy and condition(a)	Orientation	Frequency, Hz	Stress ratio, R	Test temperature °C	Test temperature °F	C(b) da/dN: mm/cycle ΔK: MPa$\sqrt{m}$	C(b) da/dN: in./cycle ΔK: ksi$\sqrt{in.}$	n	Estimated range for ΔK MPa$\sqrt{m}$	Estimated range for ΔK ksi$\sqrt{in.}$
Inconel 706 forgings	C-R......	10	0.1	24	75	7.42×10^{-11}	4.2×10^{-12}	4.11	22 to 44	20 to 40
VIM-EFR STDA				−196	−320	5.89×10^{-10}	3.18×10^{-11}	3.35	22 to 88	20 to 80
				−269	−452	8.93×10^{-10}	4.72×10^{-11}	3.12	24 to 77	22 to 70
VIM-VAR STDA	C-R......	10	0.1	24	75	1.02×10^{-9}	5.47×10^{-11}	3.27	22 to 53	20 to 48
				−196	−320	2.87×10^{-10}	1.57×10^{-11}	3.5	25 to 77	23 to 70
				−269	−452	2.33×10^{-9}	1.19×10^{-10}	2.77	25 to 88	23 to 80
VIM-VAR, GTA weld	C-R......	10	0.1	24	75	1.02×10^{-10}	6.02×10^{-12}	4.34	22 to 35	20 to 32
+ STDA (FZ)				−269	−452	2.65×10^{-11}	1.57×10^{-12}	4.35	23 to 40	21 to 36
Inconel 718 forging	C-R......	10	0.1	24	75	1.52×10^{-10}	8.79×10^{-12}	4.10	22 to 48	20 to 44
VIM-VAR STDA				−196, −269	−320, −452	7.31×10^{-12}	4.42×10^{-13}	4.55	22 to 55	20 to 50
Inconel 718 forging	T-S......	20-28	0.1	24	75	8×10^{-11}	4.59×10^{-12}	4.0	20 to 70	18 to 64
STDA				−78, −196, −269	−108, −320, −452	4.8×10^{-11}	2.75×10^{-12}	4.0	25 to 90	23 to 82
Inconel X-750 forging	T-S......	20-28	0.1	24	75	2.4×10^{-9}	1.25×10^{-10}	3.0	26 to 80	24 to 73
STDA				−196, −269	−320, −452	6.6×10^{-11}	3.72×10^{-12}	3.8	30 to 92	27 to 84
VIM-VAR STDA	C-R......	10	0.1	24	75	6.34×10^{-15}	4.83×10^{-16}	7.0	23 to 44	21 to 40
				−269	−452	2.44×10^{-17}	2.04×10^{-18}	8.0	33 to 55	30 to 50
AAM-VAR STDA	C-R......	10	0.1	24, −196	75, −320	4.62×10^{-10}	2.52×10^{-11}	3.45	33 to 66	30 to 60
				−269	−452	1.52×10^{-10}	8.26×10^{-12}	3.45	38 to 82	35 to 75
VIM STDA	C-R......	10	0.1	24, −196, −269	75, −320, −452	2.63×10^{-11}	1.57×10^{-12}	4.4	27 to 55	25 to 50

(a) STDA for Inconel 706: 980 °C (1800 °F) 1 h, AC, 730 °C (1350 °F) 8 h, FC to 620 °C (1150 °F), hold at 620 °C (1150 °F) 8 h, AC. STDA for Inconel 718 and Inconel X-750: same as for Inconel 706. (b) Conversion factors for C in equation $da/dN = C\Delta K^n$. If da/dN is in in./cycle and ΔK is in ksi$\sqrt{in.}$, multiply C by $25.4/1.0989^n$ to obtain mm/cycle with ΔK in MPa$\sqrt{m}$. If da/dN is in mm/cycle and ΔK is in MPa $\sqrt{m}$, multiply C by $1.0989^n/25.4$ to obtain in./cycle with ΔK in ksi$\sqrt{in.}$

Table 33 Results of fatigue-life tests on unnotched specimens of high-nickel alloys

Data from Ref 2, 42

Alloy and condition(a)	Stressing mode	Stress ratio, R	Cyclic frequency, Hz	Fatigue strengths at 10^6 cycles 24 °C (75 °F) MPa	24 °C (75 °F) ksi	−196 °C (−320 °F) MPa	−196 °C (−320 °F) ksi	−253 °C (−423 °F) MPa	−253 °C (−423 °F) ksi
K Monel Sheet	Flex(b)	−1.0	...	380	55	395	57	475	69
	Flex(c)	−1.0	...	380	55	470	68	580	84
K Monel bar, STA	Axial	0	28	615	89	800	116	840	122
Inconel X-750 bar, STA	Axial	0	28	745	108	1010	147	1060	154
Inconel X-750 Sheet, STA	Flex(d)	−1.0	30-40	400	58	455	66	525	76
	Flex(e)	−1.0	30-40	495	72	580	84	705	102
Inconel 718 bar, STA	Axial	0	28	760	110	965	140	1075	156

(a) STA for K Monel: 815 °C (1500 °F), WQ, 590 °C (1100 °F) 16 h plus controlling cooling. STA for Inconel X-750: 980 °C (1800 °F), FC to 700 °C (1300 °F), hold at 700 °C (1300 °F) 20 h, AC. STA for Inconel 718: 1060 °C (1950 °F), AC, 760 °C (1400 °F) 10 h, FC to 650 °C (1200 °F), hold at 650 °C (1200 °F) 10 h, AC. (b) Surface finish, 90 μin. rms. (c) Surface finish, 16 μin. rms. (d) Surface finish, 64 μin. rms. (e) Surface finish, 11 μin. rms.

Table 34 Compositions of ferritic nickel steel plate for use at subzero temperatures

ASTM specification	Compositions of plates up to 50 mm (2 in.) thick, %(a) C	Mn	P	S	Si	Ni	Mo	Others
A203 A	0.17	0.70	0.035	0.040	0.15-0.30	2.10-2.50	...	...
A203 B	0.21	0.70	0.035	0.040	0.15-0.30	2.10-2.50	...	...
A203 C	0.17	0.70	0.035	0.040	0.15-0.30	3.25-3.75	...	...
A203 D	0.20	0.70	0.035	0.040	0.15-0.30	3.25-3.75	...	...
A645	0.13	0.30-0.60	0.025	0.025	0.20-0.35	4.75-5.25	0.20-0.35	0.02-0.12 Al, 0.020 N
A353	0.13	0.90	0.035	0.040	0.15-0.30	8.5-9.5	...	...
A553 I	0.13	0.90	0.035	0.040	0.15-0.30	8.5-9.5	...	...
A553 II	0.13	0.90	0.035	0.040	0.15-0.30	7.5-8.5	...	...

(a) Single values are maximum limits.

Table 35 Typical tensile properties of ferritic nickel steels

Data from Ref 2, 44, 64, 66-69

Temperature °C	°F	Tensile strength MPa	ksi	Yield strength MPa	ksi	Elongation, %	Reduction in area, %	Notch tensile strength(a) MPa	ksi	Young's modulus GPa	10^6 psi
A645 plate, longitudinal orientation(b)											
24	75	715	104	530	76.8	32	72	...	...	200	28.7
−168	−270	930	135	570	82.9	28	68	...	...	205	30.2
−196	−320	1130	164	765	111	30	62	...	...	210	30.7
A353 plate, longitudinal orientation(c)											
24	75	780	113	680	98.6	28	70	945	137	...	...
−151	−240	1030	149	850	123	17	61	...	...	...	...
−196	−320	1190	172	950	138	25	58	...	...	...	...
−253	−423	1430	208	1320	192	18	43	1310	190	...	...
−269	−452	1590	231	1430	208	21	59	...	...	...	...
A553-I plate, longitudinal orientation(d)											
24	75	770	112	695	101	27	69	...	...	...	...
−151	−240	995	144	885	128	18	42	...	...	...	...
−196	−320	1150	167	960	139	27	38	...	...	...	...

(a) $K_t = 6.4$. (b) Quenched, tempered, reversion annealed. (c) Double normalized and tempered: 900 °C (1650 °F) 1 h/in. of thickness, AC; 790 °C (1450 °F) 1 h/in. of thickness, AC; 570 °C (1050 °F) 1 h/in. of thickness, AC or WQ. (d) Quenched and tempered: 800 °C (1475 °F), WQ; 570 °C (1050 °F) 30 min/in. of thickness, AC or WQ.

Table 36 Fracture toughness of 5% and 9% Ni steel plate for compact tension specimens in T-L orientation

Data from Ref 65-70

Alloy and condition	Yield strength(a) MPa	ksi	Fracture toughness, K_{Ic} (J), at: −162 °C (−260 °F) MPa$\sqrt{m}$	ksi$\sqrt{in.}$	−196 °C (−320 °F) MPa$\sqrt{m}$	ksi$\sqrt{in.}$	−269 °C (−452 °F) MPa$\sqrt{m}$	ksi$\sqrt{in.}$
5Ni steel (A645) quenched, temperized, reversion annealed	534	77.5	196	178	87.1	79.3	58.4	53.2
9Ni steel (A553, Type I) quenched and tempered	689	99.9	...	...	184	167	...	...

(a) At room temperature.

higher at −269 °C (−452 °F) than at room temperature. A log-log plot of the da/dN data for type 304 stainless steel is shown in Fig. 16. For this steel, fatigue-crack-growth rates are nearly the same, at the same values of ΔK, for room-temperature and cryogenic-temperature tests. Fatigue-crack-growth rates in the fusion zones of welds tend to be higher than in the base metal.

Fatigue Strength. Results of flexural and axial fatigue tests at 10^6 cycles on austenitic stainless steels at room temperature and at subzero temperatures are presented in Table 46.

Table 37 Fatigue-crack-growth-rate data for ferritic steels for compact specimens in T-L orientation

Data from Ref 61, 69

Alloy and condition	Frequency, Hz	Stress ratio, R	Test temperature °C	Test temperature °F	C da/dN: mm/cycle ΔK: MPa√m	C da/dN: in./cycle ΔK: ksi√in.	n	Estimated range for ΔK MPa√m	Estimated range for ΔK ksi√in.
3.5Ni steel plate, ASTM A203E, quenched and tempered	20-28	0.1	24	75	1.3×10^{-8}	6.9×10^{-10}	3.2	18-60	16-54
5Ni steel plate, ASTM A645, austenitized, quenched, temperized, reversion annealed			−78	−108	1.3×10^{-8}	6.9×10^{-10}	3.2	30-70	27-63
			−101	−150	1.0×10^{-9}	5.3×10^{-11}	3.2	30-60	27-54
			−196	−320	1.6×10^{-14}	1.3×10^{-15}	7.6	20-30	18-27
	20-28	0.1	24	75	1.1×10^{-8}	5.6×10^{-10}	2.7	25–90	23-82
9Ni steel plate, ASTM A553, austenitized, quenched and tempered			−162	−260	1.1×10^{-8}	5.6×10^{-10}	2.7	25-60	23-54
			−196	−320	2.0×10^{-10}	1.2×10^{-11}	4.0	27-80	24-72
	20-28	0.1	24	75	2.0×10^{-8}	1.0×10^{-9}	2.7	16-70	14-63
			−162	−260	1.0×10^{-9}	5.4×10^{-11}	3.4	17-80	15-72
			−196	−320	4.8×10^{-11}	2.9×10^{-12}	4.4	17-64	15-58
			−269	−452	1.4×10^{-11}	9.1×10^{-13}	5.3	24-35	22-32

Fig. 15 Poisson's ratios for four austenitic stainless steels as determined ultrasonically (Ref 86)

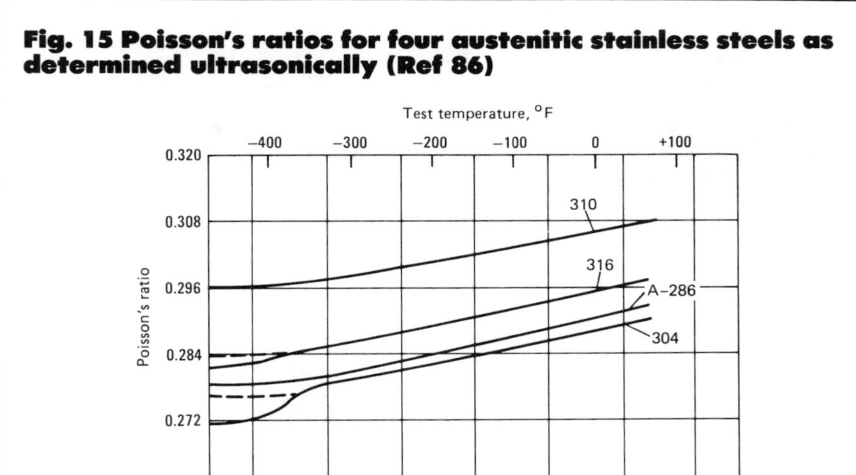

Fatigue strength increases as exposure temperature is decreased. Notched specimens have substantially lower fatigue strengths than corresponding unnotched specimens at all testing temperatures. Reducing surface roughness of unnotched specimens improves fatigue strength.

Titanium and Titanium Alloys

Unalloyed titanium and alpha titanium alloys have close-packed hexagonal (cph) crystal structures, which accounts for the fact that the properties of these metals do not follow the same trends at subzero temperatures as do the properties of metals with fcc or bcc structures.

Many of the available titanium alloys have been evaluated at subzero temperatures, but service experience at such temperatures has been gained only for Ti-5Al-2.5Sn and Ti-6Al-4V alloys. These alloys have very high strength-to-weight ratios at cryogenic temperatures and have been the preferred alloys for special applications at temperatures from −196 to −269 °C (−320 to −452 °F). Among these applications are spherical pressure vessels that are part of the propulsion and reaction-control systems for the Atlas and Centaur rockets, the Apollo and Saturn launch boosters, and the lunar module. These pressure vessels were fabricated by welding together hemispherical forgings that had been machined to the desired thickness. The Ti-5Al-2.5Sn alloy also has been used for fuel-pump impellers for pumping liquid hydrogen.

Commercially pure titanium may be used for tubing and other small-scale cryogenic applications that involve only low stresses in service.

The Ti-5Al-2.5Sn alloy usually is used in the mill-annealed condition and has a 100% alpha microstructure. The Ti-6Al-4V alloy may be used in the annealed condition or in the solution

Table 38 Nominal compositions or composition ranges for selected austenitic stainless steels

UNS number	Alloy designation	Nominal composition, % C	Mn	P max	S max	Si	Cr	Ni	Mo	Others
Wrought alloys										
S20200	AISI 202	0.15 max	7.5 to 10	...	...	1.0 max	17 to 19	4 to 6	...	0.25 N max
S30100	AISI 301	0.15 max	2.0 max	0.045	0.03	1.0 max	16 to 18	6 to 8	...	...
S30300	AISI 303	0.15 max	2.0 max	0.20	0.15 min	1.0 max	17 to 19	8 to 10	0.6 max	...
S30400	AISI 304	0.08 max	2.0 max	0.045	0.03	1.0 max	18 to 20	8 to 12	...	...
S30403	AISI 304L	0.03 max	2.0 max	0.045	0.03	1.0 max	18 to 20	8 to 12	...	...
S31000	AISI 310	0.25 max	2.0 max	0.045	0.03	1.5 max	24 to 26	19 to 22	...	...
S31008	AISI 310S	0.08 max	2.0 max	0.045	0.03	1.5 max	24 to 26	19 to 22	...	...
S31600	AISI 316	0.08 max	2.0 max	0.045	0.03	1.0 max	16 to 18	10 to 14	2.0 to 3.0	...
S32100	AISI 321	0.08 max	2.0 max	0.045	0.03	1.0 max	17 to 19	9 to 12	...	(5 × C) Ti min
S34700	AISI 347	0.08 max	2.0 max	0.045	0.03	1.0 max	17 to 19	9 to 13	...	(10 × C) (Nb + Ta)
S41600	AISI 416	0.15 max	1.25 max	0.06	0.15 min	1.0 max	12 to 14	...	...	...
	21-6-9	0.04 max	8 to 10	...	...	1.0 max	19 to 21.5	5.5 to 7.5	...	0.2 to 0.35N
	Pyromet 538	0.02	9.5	...	...	0.15	20	7.0	...	0.2N
	Nitronic 40	0.03	9.2	...	...	0.6	19.5	7.0	...	0.3N
	Nitronic 60	0.07	8.0	...	...	3.5	17	8.5	...	0.1N
K66286	A-286	0.05	1.4	...	...	0.4	15	26	1.25	0.2Al, 2.0Ti, 0.3V, 0.005B
	Kromarc 58	0.03	9.3	0.005	0.005	0.05	15.5	23	2.2	0.02Al, 0.16V, 0.17N, 0.008Zr, 0.016B
Cast alloys										
	CF-8	0.08 max	1.5 max	...	...	2.0 max	18 to 21	8 to 11	...	...
	CF-8M	0.08 max	1.5 max	...	...	2.0 max	18 to 21	9 to 12	2 to 3	...
	Kromarc 55	0.04	9.5	...	...	0.3	16	20	2.25	...

(a) Other specifications for 310 give lower limits on carbon content.

treated and aged condition, but for maximum toughness in cryogenic applications the annealed condition usually is preferred. The Ti-6Al-4V alloy is an alpha-beta alloy that has significantly higher yield and ultimate tensile strengths than the all-alpha alloy.

Interstitial impurities such as iron, oxygen, carbon, nitrogen and hydrogen tend to reduce the toughness of these alloys at both room and subzero temperatures. Therefore, for maximum toughness, extra-low-interstitial (ELI) grades are specified for critical applications. The composition limits for these alloys are given in Table 47. Note that the iron and oxygen contents of the ELI grades are substantially lower than those of the normal interstitial (NI) grades. Iron is a strong stabilizer of the beta phase, which has a bcc crystal structure (Ref 89). The NI grades are suitable for service to −195 °C (−320 °F); for temperatures below −195 °C, ELI grades generally are specified. For ELI grades, reduced creep strength at room temperature must be considered in design for pressure-vessel service. In Ti-5Al-2.5Sn, stress-rupture occurs at stresses below the yield strength.

Fig. 16 Fatigue-crack-growth-rate data for type 304 austenitic stainless steel (annealed) at room temperature and at subzero temperatures (Ref 61)

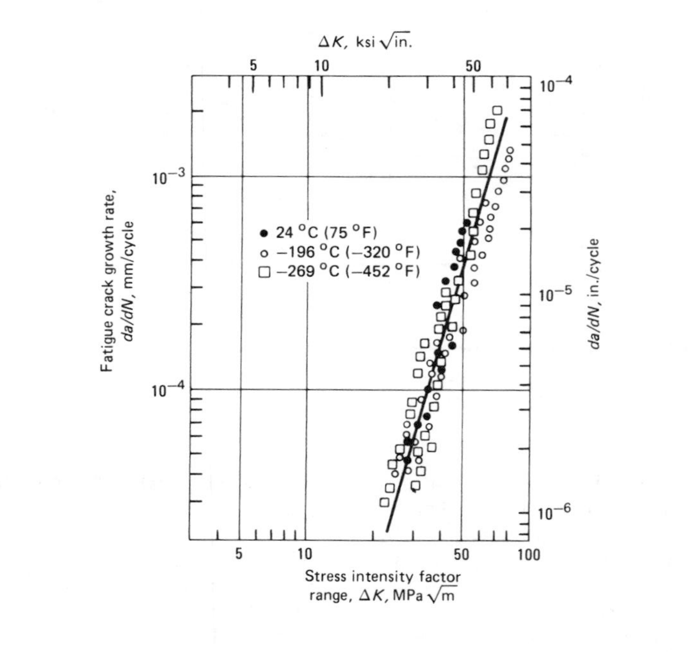

Table 39 Typical tensile properties of annealed type 300 austenitic stainless steels

Data from Ref 2, 10, 17, 24, 44, 73–78

Temperature		Tensile strength		Yield strength		Elongation, %	Reduction in area, %	Notch tensile strength(a)		Young's modulus	
°C	°F	MPa	ksi	MPa	ksi			MPa	ksi	GPa	10^6 psi
303 bar, longitudinal orientation											
24	75	730	106	425	61.4	67	70	...	...	...	...
−78	−108	1190	172	435	63.3	43	60	...	...	...	...
−196	−320	1660	240	465	67.3	36	54	...	...	...	...
−253	−423	2060	298	570	82.6	33	...	...	...	...	...
−269	−452	1830	266	...	...	30	37	...	...	...	...
304 sheet, longitudinal orientation											
24	75	660	95.5	295	42.5	75	...	715	104	...	...
−196	−320	1625	236	380	55.0	42	...	1450	210	...	...
−253	−423	1800	261	425	62.0	31	...	1160	168	...	...
−269	−452	1700	247	570	82.5	30	...	1230	178	...	...
304 plate, longitudinal orientation											
24	75	590	85.9	330	47.6	64	...	...	...	...	...
−253	−423	1720	250	410	59.4	...	...	...	...	...	...
304 bar, longitudinal orientation											
24	75	640	92.8	235	33.9	76	82	710	103	...	...
−78	−108	1150	167	300	43.2	50	76	...	...	...	...
−196	−320	1520	221	280	40.9	45	66	1060	153	...	...
−253	−423	1860	270	420	60.6	27	54	1120	162	...	...
−269	−452	1720	250	400	58.2	30	55	...	...	...	...
304L sheet, longitudinal orientation											
24	75	660	95.9	295	42.8	56	...	730	106	...	...
−78	−108	980	142	250	36.0	43	...	1030	150	...	...
−196	−320	1460	212	275	39.6	37	...	1420	206	...	...
−253	−423	1750	254	305	44.5	33	...	1290	187	...	...
−269	−452	1590	230	405	58.5	29	...	1460	212	...	...
304L sheet, transverse orientation											
−269	−452	1540	223	410	59.5	35	...	...	...	...	...
304L bar, longitudinal orientation											
24	75	660	95.5	405	58.9	78	81	...	...	190	27.6
−78	−108	1060	153	435	62.8	70	74	...	...	...	...
−196	−320	1510	219	460	66.6	43	66	...	...	205	29.7
−253	−423	1880	273	525	75.8	42	41	...	...	...	...
−269	−452	1660	241	545	79.4	34	56	...	...	200	29.2
310 sheet, longitudinal orientation											
24	75	570	83.0	240	35.0	50	...	645	93.9	...	...
−196	−320	1080	156	545	79.1	68	...	1070	155	...	...
−253	−423	1300	188	715	104	56	...	1250	182	...	...
−269	−452	1230	178	770	112	58	...	...	...	...	...
310 sheet, transverse orientation											
24	75	600	86.8	240	34.8	46	...	630	91.6	...	...
−269	−452	1280	186	800	116	58	...	...	...	...	...
310 bar, longitudinal orientation											
24	75	585	84.8	340	49.1	50	76	770	112	...	...
−78	−108	740	107	305	43.9	72	68	...	...	...	...
−196	−320	1090	158	520	75.5	68	50	...	...	205	29.9
−253	−423	1390	202	855	124	44	48	1305	189	...	...
−269	−452	1300	189	715	104	50	41	...	...	205	29.9

(continued)

(a) K_t = 5.2 for 304 and 304L sheet; K_t = 14 for 304 bar; K_t = 6.3 for 310 sheet; K_t = 6.4 for 310 bar; K_t = 10 for 310S forging; K_t = 3.5 for 321 sheet.

Table 39 (continued)

Temperature °C	Temperature °F	Tensile strength MPa	Tensile strength ksi	Yield strength MPa	Yield strength ksi	Elongation, %	Reduction in area, %	Notch tensile strength(a) MPa	Notch tensile strength(a) ksi	Young's modulus GPa	Young's modulus 10^6 psi
310S forging, transverse orientation											
24	75	585	84.8	260	37.9	54	71	800	116	...	...
−196	−320	1100	159	605	87.6	72	52	1350	196	...	...
−269	−452	1300	189	815	118	64	45	1600	232	...	...
316 sheet, longitudinal orientation											
24	75	595	86.4	275	39.8	60	...	...	...	...	...
−253	−423	1580	229	665	96.6	55	...	...	...	...	...
321 sheet, longitudinal orientation											
24	75	620	89.6	225	32.4	55	...	625	90.4	180	26.0
−196	−320	1380	200	315	45.6	46	...	1520	220	205	29.5
−253	−423	1650	239	375	54.5	36	...	1460	212	210	30.7
321 bar, longitudinal orientation											
24	75	675	97.6	430	62.2	55	79	...	...	...	...
−78	−108	1060	153	385	55.9	46	73	...	...	...	...
−196	−320	1540	223	450	65.4	38	60	...	...	...	...
−253	−423	1860	270	405	58.5	35	44	...	...	...	...
347 sheet, longitudinal orientation											
24	75	650	94	255	37	52	...	...	...	...	...
−196	−320	1365	198	420	61	47	...	...	...	...	...
−253	−423	1610	234	435	63	35	...	...	...	...	...
347 bar											
24	75	670	97.4	340	49.3	57	76	...	...	...	...
−78	−108	995	144	475	68.8	51	71	...	...	...	...
−196	−320	1470	214	430	62.2	43	60	...	...	...	...
−253	−423	1850	268	525	76.4	38	45	...	...	...	...

(a) K_t = 5.2 for 304 and 304L sheet; K_t = 14 for 304 bar; K_t = 6.3 for 310 sheet; K_t = 6.4 for 310 bar; K_t = 10 for 310S forging; K_t = 3.5 for 321 sheet.

Fig. 17 Young's modulus for Ti-5Al-2.5Sn and Ti-6Al-4V alloys as determined ultrasonically (Ref 95)

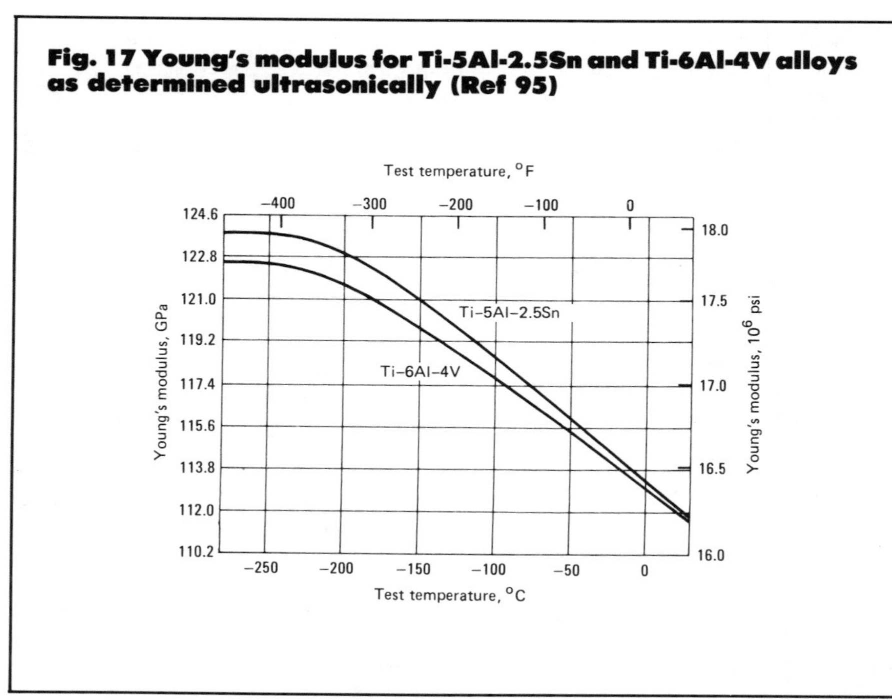

There are two precautions that should be emphasized in considering titanium and titanium alloys for service at cryogenic temperatures: titanium and titanium alloys must not be used for transfer or storage of liquid oxygen, and titanium must not be used where it will be exposed to air while below the temperature at which oxygen will condense on its surfaces. Any abrasion or impact of titanium that is in contact with liquid oxygen will cause ignition. Pressure vessels in contact with liquid oxygen in the Apollo launch vehicles were produced from Inconel 718 rather than from Ti-6Al-4V alloy to avoid this problem.

Tensile Properties. Typical tensile properties of titanium and of titanium alloys Ti-5Al-2.5Sn and Ti-6Al-4V at room temperature and at subzero temperatures are presented in Table 48. Marked increases in yield and tensile strengths are evident for commercial titanium and for titanium alloys as test

Table 40 Typical tensile properties of cold worked type 300 austenitic stainless steel sheet

Data from Ref 2, 44, 77

Temperature °C	Temperature °F	Tensile strength MPa	Tensile strength ksi	Yield strength MPa	Yield strength ksi	Elongation, %	Notch tensile strength(a) MPa	Notch tensile strength(a) ksi	Young's modulus GPa	Young's modulus 10^6 psi
301, hard, cold rolled (42 to 60% reduction), longitudinal orientation										
24	75	1310	190	1200	174	18	1390	201	...	...
−78	−108	1560	226	1130	164	23	1460	212	...	...
−196	−320	2020	293	1380	200	19	1660	241	...	...
−253	−423	2110	306	1610	233	14	1830	265	...	...
301, hard, cold rolled (42 to 60% reduction), transverse orientation										
24	75	1310	190	1060	153	10	1430	207	...	...
−78	−108	1560	226	1070	155	28	1430	208	...	...
−196	−320	2060	299	1310	190	28	1670	243	...	...
−253	−423	1900	275	1570	227	8	1360	197	...	...
301, extra hard, cold rolled (over 60% reduction), longitudinal orientation										
24	75	1500	217	1370	198	9	1600	232	175	25.6
−78	−108	1710	248	1400	203	22	1680	244	180	26.3
−196	−320	2220	322	1610	234	22	1940	282	180	26.2
−253	−423	2220	322	1810	262	13	1890	274	190	27.6
−269	−452	2140	310	1930	280	2	...	...	...	...
301, extra hard, cold rolled (over 60% reduction), transverse orientation										
24	75	1590	230	1280	186	8	1520	220	...	...
−78	−108	1770	257	1250	181	18	1590	230	...	...
−196	−320	2190	318	1560	226	18	1680	244	...	...
−253	−423	2180	316	1830	266	5	1340	194	...	...
304, hard, cold rolled, longitudinal orientation										
24	75	1320	191	1190	173	3	1460	212	180	25.9
−78	−108	1470	213	1300	188	10	1590	231	185	26.9
−196	−320	1900	276	1430	208	29	1910	277	200	29.1
−253	−423	2010	292	1560	226	2	2160	313	210	30.5
304, hard, cold rolled, transverse orientation										
24	75	1440	209	1180	171	5	1200	174	195	28.0
−78	−108	1600	232	1330	193	7	1400	203	200	28.9
−196	−320	1870	271	1480	214	23	1690	245	205	30.0
−253	−423	2160	313	1560	226	1	1900	276	215	31.1
304L, 70% cold reduced, longitudinal orientation										
24	75	1320	192	1080	156	3	...	...	...	...
−196	−320	1770	256	1530	222	14	...	...	...	...
−253	−423	1990	288	1770	256	2	...	...	...	...
304L, 70% cold reduced, transverse orientation										
24	75	1440	209	1220	177	4	...	...	...	...
−196	−320	1890	274	1630	236	12	...	...	...	...
−253	−423	2230	324	1940	282	1	...	...	...	...
310, 75% cold reduced, longitudinal orientation										
24	75	1180	171	1100	160	3	1360	197	175	25.4
−78	−108	1410	204	1290	187	4	1530	222	175	25.5
−196	−320	1720	249	1540	223	10	1900	276	180	26.4
−253	−423	2000	290	1790	259	10	2230	324	195	28.3
310, 75% cold reduced, transverse orientation										
24	75	1370	199	1110	161	4	1370	199	195	28.1
−78	−108	1540	224	1290	187	8	1640	238	190	27.6
−196	−320	1880	272	1520	221	10	2050	297	195	28.2
−253	−423	2140	311	1790	260	9	2190	318	200	29.1

(a) K_t = 6.3.

Table 41 Typical tensile properties of stainless steels other than type 300 series

Data from Ref 2, 6, 7, 11, 44, 73, 75, 79–81

Temperature		Tensile strength		Yield strength		Elongation,	Reduction in area,	Notch tensile strength(a)		Young's modulus	
°C	°F	MPa	ksi	MPa	ksi	%	%	MPa	ksi	GPa	10^6 psi
202 sheet, annealed, longitudinal orientation											
24	75	705	102	325	47.1	57	...	...	...	...	...
−73	−100	1080	156	485	70.2	41	...	...	...	...	...
−196	−320	1590	231	610	88.3	52	...	...	...	...	...
−268	−450	1420	206	765	111	25	...	...	...	...	...
202 sheet, cold reduced 50%, longitudinal orientation											
24	75	1080	156	965	140	21	...	...	...	...	...
−196	−320	1970	286	1070	155	28	...	...	...	...	...
−268	−450	1950	283	1240	180	20	...	...	...	...	...
21-6-9 plate, longitudinal orientation(b)											
24	75	705	102	385	55.9	54	80	...	...	...	...
−78	−108	895	130	590	85.4	60	75	...	...	...	...
−196	−320	1510	219	970	141	41	33	...	...	...	...
−253	−423	1660	241	1220	177	16	26	...	...	...	...
−269	−452	1700	247	1350	196	22	30	...	...	...	...
Pyromet 538 plate, longitudinal orientation(c)											
24	75	675	97.9	340	49.0	75	81	...	...	...	...
−196	−320	1370	199	800	116	76	73	...	...	...	...
−269	−452	1490	216	1010	147	52	59	...	...	...	...
Nitronic 40 plate, electroslag remelted; as rolled											
24	75	1010	146	840	122	35	72	...	...	...	...
−73	−100	1170	169	945	137	36	71	...	...	...	...
−196	−320	1830	266	1540	223	31	64	...	...	...	...
Nitronic 60 bar, annealed, longitudinal orientation											
24	75	750	109	400	58.1	66	79	1080	157	165	24.0
−73	−100	1020	148	535	77.9	70	81	1480	215	165	24.2
−196	−320	1500	218	695	101	60	66	1900	275	170	24.8
−253	−423	1410	204	860	125	24	27	1870	271	170	24.8
Kromarc 58 sheet, longitudinal orientation(d)											
24	75	695	101	285	41.5	62	...	...	...	...	...
−78	−108	825	120	395	57.6	59	...	...	...	...	...
−196	−320	1280	185	695	101	82	...	...	...	...	...
−253	−423	1450	210	880	128	56	...	...	...	...	...
Kromarc 58 sheet, longitudinal orientation(e)											
24	75	1280	186	1210	175	6	...	1370	198	...	...
−78	−108	1510	219	1430	207	7	...	1640	238	...	...
−196	−320	1880	272	1740	252	9	...	2000	290	...	...
−253	−423	2100	304	1990	288	1	...	2170	315	...	...

(continued)

(a) K_t = 7 for Nitronic 60 bar; K_t = 6.3 for Kromarc 58 sheet and A 286 sheet; K_t = 10 for Kromarc plate; K_t = 6.4 for A 286 bar. (b)Annealed 1 h at 1065 °C (1950 °F), WQ. (c) Annealed 1 h at 1095 °C (2000 °F), WQ. (d) Annealed 1 h at 1065 °C (1950 °F). (e)Annealed 1065 °C (1950 °F), cold rolled 80%. (f) Annealed 1 h at 980 °C (1800 °F), WQ. (g) Heat treatment: ½ h at 980 °C (1800 °F), WQ, aged 16 h at 595 °C (1100 °F), AC. (h) Heat treatment: ½ h at 980 °C (1800 °F), WQ or AQ, aged 16 h at 720 °C (1325 °F) AC. (j) Heat treatment: 1.5 h at 960 °C (1800 °F), AC, aged 16 h at 720 to 730 °C (1325 to 1350 °F), AC. (k) Heat treatment: 1 h at 980 °C (1800 °F), OQ, tempered 4 h at 370 °C (700 °F), AC.

Table 41 (continued)

Temperature °C	°F	Tensile strength MPa	ksi	Yield strength MPa	ksi	Elongation, %	Reduction in area, %	Notch tensile strength(a) MPa	ksi	Young's moduius GPa	10^6 psi
Kromarc 58 plate, longitudinal orientation(f)											
24	75	705	102	370	53.8	46	67	930	135	...	...
−196	−320	1300	188	785	114	50	58	1580	229	...	...
−269	−452	1320	192	1100	159	42	55	1880	272	...	...
A 286 sheet, longitudinal orientation(g)											
24	75	860	125	410	59.9	36	...	855	124	...	...
−196	−320	1230	179	615	89.4	44	...	1200	174	...	...
−253	−423	1460	212	745	108	34	...	1370	198	...	...
A 286 sheet, transverse orientation(g)											
24	75	840	122	405	58.9	40	...	855	124	...	...
−196	−320	1210	176	600	87.4	46	...	1160	168	...	...
−253	−423	1450	210	735	107	42	...	1330	193	...	...
A 286 sheet, longitudinal orientation(h)											
24	75	1040	151	690	99.8	21	...	1100	160	...	...
−78	−108	1050	153	695	101	26	...	...	...	...	...
−196	−320	1360	197	820	119	33	...	1380	200	...	...
−253	−423	1510	219	915	133	26	...	1480	214	...	...
A 286 sheet, transverse orientation(h)											
24	75	1050	152	725	105	23	...	1130	164	...	...
−78	−108	1120	163	785	114	27	...	...	...	...	...
−196	−320	1350	196	840	122	33	...	1370	198	...	...
−253	−423	1520	221	930	135	35	...	1450	210	...	...
A 286 bar, longitudinal orientation(j)											
24	75	1080	157	760	110	28	48	1250	181	180	26.4
−78	−108	1170	170	780	113	32	48	...	...	...	...
−196	−320	1410	204	860	125	40	48	...	...	195	28.4
−253	−423	1610	234	1030	149	41	46	...	...	...	...
−257	−430	1620	235	1030	150	34	46	1490	216	...	...
416 bar, longitudinal orientation(k)											
24	75	1400	203	1200	174	15	53	...	...	...	...
−78	−108	1500	218	1260	183	15	52	...	...	...	...
−196	−320	1800	261	1600	232	9	24	...	...	...	...
−253	−423	2020	293	2020	293	0.4	2	...	...	...	...

(a) K_t = 7 for Nitronic 60 bar; K_t = 6.3 for Kromarc 58 sheet and A 286 sheet; K_t = 10 for Kromarc 58 plate; K_t = 6.4 for A 286 bar. (b) Annealed 1 h at 1065 °C (1950 °F), WQ. (c) Annealed 1 h at 1095 °C (2000 °F), WQ. (d) Annealed 1 h at 1065 °C (1950 °F). (e) Annealed 1065 °C (1950 °F), cold rolled 80%. (f) Annealed 1 h at 980 °C (1800 °F), WQ. (g) Heat treatment: ½ h at 980 °C (1800 °F), WQ, aged 16 h at 595 °C (1100 °F), AC. (h) Heat treatment: ½ h at 980 °C (1800 °F), WQ or AQ, aged 16 h at 720 °C (1325 °F) AC. (j) Heat treatment: 1.5 h at 960 °C (1800 °F), AC, aged 16 h at 720 to 730 °C (1325 to 1350 °F), AC. (k) Heat treatment: 1 h at 980 °C (1800 °F), OQ, tempered 4 h at 370 °C (700 °F), AC.

Table 42 Typical tensile properties of stainless steel weldments

Data from Ref 2, 7, 15, 75, 79, 82

Alloy condition	Welding process	Filler	Form	Base metal orientation	Test temperature °C	Test temperature °F	Yield strength MPa	Yield strength ksi	Tensile strength MPa	Tensile strength ksi	Elongation, %	Reduction in area, %	Notch tensile strength (a) MPa	Notch tensile strength (a) ksi
Type 301, cold rolled 60%; tested as welded	GTA	None	Sheet	L	24	75	...	...	1034	150	7	...	...	...
					−78	−108	...	...	1489	216	13	...	...	...
					−196	−320	...	...	2006	291	16	...	...	...
					−253	−423	...	...	1675	243	6	...	...	...
Type 310, 3/4 hard; tested as welded	GTA	310	Sheet	L	24	75	380	55.1	530	76.8	4	...	...	...
					−78	−108	523	75.9	723	105	4	...	...	...
					−196	−320	752	109	1026	149	4	...	...	...
AISI, 310S, annealed	SMA	310S	Plate	...	24	75	334	48.5	582	84.4	40	76	841	122
					−196	−320	660	96.6	1066	155	46	67	1428	207
					−269	−452	829	120	1102	160	26	24	1672	242
21-6-9, annealed	SMA	Inconel 625	Plate	Weld(b)	−269	−452	878	127	1276	183	31	27	...	...
				HAZ(b)	−269	−452	1728	251	1873	272	21	33	...	...
	GTA	Inconel 625	Plate	Weld(b)	−269	−452	951	138	1222	177	18	20	...	...
				HAZ(b)	−269	−452	1740	252	1921	279	17	37	...	...
	GMA	Inconel 625	Plate	Weld(b)	−269	−452	833	121	1087	158	19	27	...	...
				HAZ(b)	−269	−452	1689	245	1866	271	15	27	...	...
Pyromet 528, annealed	GTA	Pyromet 538	Plate	...	24	75	414	60.0	725	105	51	74	1238	180
					−196	−320	1009	146	1456	211	48	61	2119	307
					−269	−452	1240	180	1646	239	31	24	1841	267
	GMA	In-182	Plate	...	24	75	413	59.9	729	106	53	75	1018	148
					−196	−320	800	116	1045	152	6	37	1416	205
					−269	−452	805	117	1086	158	6	40	1419	206
Kromarc 58, annealed plate; tested as welded	GTA	Kromarc 58	Plate	L	24	75	498	72.3	916	133	36	61	1153	167
					−78	−320	852	124	1321	192	46	41	1908	277
					−269	−452	1060	154	1438	209	33	40	2173	315
A-286 annealed sheet; welded and age hardened	GTA	A-286	Sheet	L	24	75	601	87.2	861	125	11	...	...	...
					−78	−108	610	88.9	931	135	13	...	...	...
					−196	−320	744	108	1145	166	16	...	...	...
					−253	−423	866	126	1286	186	15	...	...	...
Age hardened sheet; tested as welded	GTA	A-286	Sheet	L	24	75	386	56.0	685	99.3	9	...	...	...
					−78	−108	472	68.4	780	113	8	...	...	...
					−196	−320	601	87.2	948	138	9	...	...	...
					−253	−423	717	104	1069	155	8	...	...	...
Solution treated and aged plate; tested as welded	GTA	Inconel 92	Plate	...	24	75	305	44.3	605	87.7	39	63	...	...
					−196	−320	473	68.6	847	123	39	29	...	...
					−269	−452	583	84.6	914	133	27	27	...	...

(a) K_t = 10. (b) Weld parallel with specimen axis; weld specimens were all weld metal; HAZ specimens contained HAZ plus some weld metal and some base metal.

temperature is reduced from room temperature to −253 °C (−423 °F). In the cryogenic temperature range, these alloys have the highest strength-to-weight ratios of all fusion-weldable alloys that retain nearly the same strength in the weld metal as in the base metal. Yield and tensile strengths of an electron-beam weldment of Ti-5Al-2.5Sn(ELI) sheet are presented in Table 48.

The notch strengths given in Table 48 indicate that these two alloys retain sufficient notch toughness for use to −253 °C (−423 °F). However, the tensile data do not show any substantial improvement in ductility or notch toughness for the ELI grade of Ti-5Al-2.5Sn sheet over the normal interstitial grade except at very low temperatures. The recrystallization annealing treatment used for the Ti-6Al-4V(ELI) forging was developed as a means of improving fracture toughness in large forgings and thick plate.

Values of Young's modulus for titanium alloys increase substantially as test temperature is decreased, as shown in Table 48 and by the data obtained ultrasonically and plotted in Fig. 17. Values of Poisson's ratio for the two alloys in Fig. 17 are plotted in Fig. 18.

Fracture Toughness. Available data on plane-strain fracture toughness (K_{Ic}) at subzero temperatures for alloys Ti-5Al-2.5Sn and Ti-6Al-4V are presented in Table 49 along with corresponding data for weldments. These data indicate that there is a modest reduction in fracture toughness as test temperature is reduced from room temperature to subzero temperatures. However, the ELI grades have better toughness than the corresponding normal interstitial grades at subzero temperatures. The limited data for electron-beam weldments indicate that at −196 °C (−320 °F) there is a slight reduction in toughness in both fusion and heat-affected zones when compared to the base metal in Ti-6Al-4V(ELI) weldments.

Fatigue-Crack-Growth Rates. Data on fatigue-crack-growth rates for Ti-5Al-2.5Sn and Ti-6Al-4V alloys are plotted in Fig. 19. The corresponding data for determining the da/dN data from the equation $da/dN = C(\Delta K)^n$ are presented in Table 50. These data indicate that the exposure temperature has no effect on the fatigue-crack-growth rates for Ti-5Al-2.5Sn and Ti-6Al-4V(NI). However, over part of the ΔK range, the fatigue-crack-growth rates for Ti-6Al-4V(ELI) are higher at cryogenic temperatures than at room temperature at the same ΔK values.

Fatigue Strength. Values of fatigue strength at 10^6 cycles for titani-

Fig. 18 Poisson's ratios for Ti-5Al-2.5Sn and Ti-6Al-4V alloys as determined ultrasonically (Ref 95)

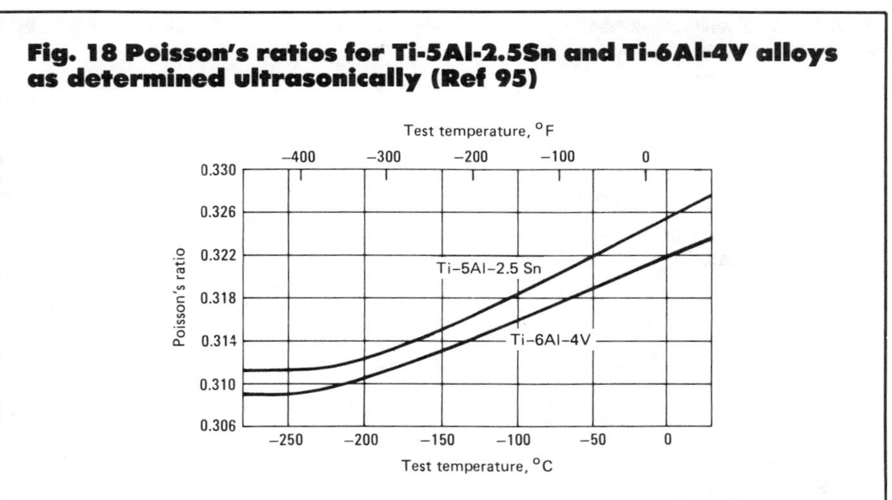

Table 43 Typical tensile properties of cast stainless steels

Data from Ref 83-85

Alloy	Temperature °C	Temperature °F	Tensile strength MPa	Tensile strength ksi	Yield strength MPa	Yield strength ksi	Elongation, %	Reduction in area, %
Type CF8 (8% ferrite), centrifugal castings(a)								
	24	75	...	...	240	35	70	...
	−78	−108	...	...	330	48	65	...
	−196	−320	...	...	380	55	55	...
	−269	−452	...	...	495	72	32	...
Type CF8M (24% ferrite)(a)								
	24	75	480	70	305	44	63	...
	−78	−108	670	97	460	67	60	...
	−196	−320	...	...	625	91	58	...
	−269	−452	...	...	750	109	53	...
Kromarc 55 (b)								
	24	75	475	69	230	33	50	...
	−196	−320	850	123	460	67	66	49
	−253	−423	880	128	560	81.2	42	30

(a) Solution treated 2 h at 1040 °C (1900 °F). (b) Cast, then solution treated 1 h at 1090 °C (2000 °F).

Table 44 Fracture toughness of austenitic stainless steels and weldments for compact tension specimens

Data from Ref 70, 75, 76, 87

Alloy and condition(a)	Form	Room temperature yield strength MPa	Room temperature yield strength ksi	Orientation	Fracture toughness, $K_{IC}(J)$, at: 24 °C MPa$\sqrt{m}$	(75 °F) ksi$\sqrt{in.}$	−196 °C MPa$\sqrt{m}$	(−320 °F) ksi$\sqrt{in.}$	−269 °C MPa$\sqrt{m}$	(−452 °F) ksi$\sqrt{in.}$
Type 310S, annealed	Plate	261	37.9	T-L	...	...	...	...	262	236
	Weldment(b)	...	...		...	...	...	...	118(c)	106(c)
Pyromet 538, STQ	Plate	338	49	T-L	...	...	275(b)	250(b)	182	165
	Weldment(d)	...	...		...	...	...	...	82.4(c)	74.4(c)
	Weldment(b)	...	...		...	...	...	...	176(c)	159(c)
Kromarc 58, STQ	Plate	371	53.8	T-L	...	...	...	...	216	195
	Weldment(d)	...	...		...	...	...	...	156(c)	141(c)
A-286, STA	Bar	608	88.2	T-S	125	114	123	112	118	107
	Plate	822	119	L-T	161	146	...	...	180	163
	Weldment(d)	...	...		...	...	...	...	249(c)	225(c)

(a) STQ = solution treated and quenched. STA for A-286: 900 °C (1650 °F) 5 h, OQ; age at 720 °C (1325 °F) 20 h, AC. Filler wires for 310S: E 310-16; for Kromarc 58: K-58; for Pyromet 538: 21-6-9; for A-286: Inconel 92.

Table 45 Fatigue-crack-growth-rate (*da/dN*) data for compact tension specimens of austenitic stainless steels

Data from Ref 61, 75

Alloy and condition(a)	Orientation	Frequency, Hz	Stress ratio, *R*	Test temperature or temperature range °C	Test temperature or temperature range °F	*C* *da/dN*:mm/cycle ΔK:MPa$\sqrt{m}$	*C* *da/dN*:in./cycle ΔK:ksi$\sqrt{in.}$	*n*	Estimated range for ΔK MPa$\sqrt{m}$	Estimated range for ΔK ksi$\sqrt{in.}$
Type 304 annealed plate	T-L	20 to 28	0.1	24 to −269	75 to −452	2.7×10^{-9}	1.4×10^{-10}	3.0	22 to 80	20 to 73
Type 304L annealed plate	T-L	20 to 28	0.1	24	75	2.0×10^{-10}	1.2×10^{-11}	4.0	22 to 54	20 to 49
				−196, −269	−320, −452	3.4×10^{-11}	2.0×10^{-12}	4.0	26 to 80	24 to 73
Type 310S annealed plate	T-L	20 to 28	0.1	24	75	3.5×10^{-11}	2.1×10^{-12}	4.4	24 to 35	22 to 32
				24	75	4.7×10^{-9}	2.4×10^{-10}	3.0	35 to 60	32 to 55
				−196, −269	−320, −452	1.1×10^{-10}	6.1×10^{-12}	3.7	25 to 80	23 to 73
	...	10	0.1	−196, −269	−320, −452	1.4×10^{-10}	7.9×10^{-12}	3.75	24 to 71	22 to 65
Type 310S, SMA weld with E310-16 filler	...	10	0.1	−196, −269	−320, −452	7.8×10^{-13}	5.0×10^{-14}	5.15	27 to 66	25 to 60
Type 316 annealed plate	T-L	20 to 28	0.1	24 to −269	75 to −452	2.1×10^{-10}	1.2×10^{-11}	3.8	19 to 16	17 to 14
21-6-9 annealed plate	T-L	20 to 28	0.1	24, −196	75, −320	1.9×10^{-10}	1.1×10^{-11}	3.7	25 to 80	23 to 73
				−269	−452	3.6×10^{-11}	2.2×10^{-12}	4.4	25 to 70	23 to 64
Pyromet 538, GTA weld in annealed plate using 21-6-9 filler	T-L	10	0.1	24	75	1.8×10^{-10}	9.9×10^{-12}	3.7	26 to 55	24 to 50
				−196, −269	−320, −452	7.6×10^{-14}	5.47×10^{-15}	6.36	24 to 44	22 to 40
Pyromet 538, SMA weld in annealed plate using Inconel 182 filler	T-L	10	0.1	24 to −269	75 to −452	2.5×10^{-12}	1.6×10^{-13}	5.13	25 to 55	23 to 50
Kromarc 58 annealed plate	...	10	0.1	24	75	2.3×10^{-10}	1.3×10^{-11}	3.9	31 to 44	28 to 40
				−196, −269	−320, −452	2.0×10^{-9}	1.04×10^{-10}	3.0	27 to 77	25 to 70
Kromarc 58, GTA weld in annealed plate	...	10	0.1	−196, −269	−320, −452	1.3×10^{-11}	7.6×10^{-13}	4.45	27 to 60	25 to 55
A-286 forging, STA	T-S	20 to 28	0.1	24	75	2.5×10^{-9}	1.3×10^{-10}	3.0	25 to 90	23 to 82
				−196, −269	−320, −452	2×10^{-12}	1.1×10^{-13}	4.0	32 to 90	29 to 82
A-286 plate, STA	...	10	0.1	24	75	1.3×10^{-8}	6.6×10^{-10}	2.7	35 to 55	32 to 50
				−196	−320	3.6×10^{-8}	1.76×10^{-9}	2.18	33 to 55	30 to 50
				−269	−452	2.7×10^{-8}	1.3×10^{-9}	2.18	33 to 55	30 to 50
A-286, GTA weld in ST plate using Inconel 92 filler	T-L	10	0.1	24	75	1.3×10^{-11}	7.66×10^{-13}	4.63	25 to 37	23 to 34
				−196, −269	−320, −452	1.8×10^{-12}	1.16×10^{-13}	5.0	30 to 55	27 to 50

(a) STA for A-286: 900 °C (1650 °F) 5 h, OQ; 720 °C (1325 °F) 20 h, AC. ST for A-286: 900 °C (1650 °F) 5 h, OQ.

um alloy base metal and weldments at room temperature and at subzero temperatures, based on results of axial and flexural fatigue tests, are presented in Table 51. For the unnotched specimens of parent metal, fatigue strength increased substantially when the test temperature was reduced from room temperature to −196 °C (−320 °F). When the test temperature was reduced to −253 °C (−423 °F), the fatigue strengths for some series of alloys were lower than at −196 °C (−320 °F). Fatigue strengths were much lower in the notched specimens than in the corresponding unnotched specimens. At −196 and −253 °C (−320 and −423 °F), the welded specimens had lower fatigue strengths than the base-metal specimens. Therefore, in designing welded structures of titanium alloys that will be subjected to fatigue loading at subzero temperatures, the weld areas usually should be thicker than the remaining areas. Hemispheres for spherical pressure vessels are machined so that the butting sections for the equatorial welds are thicker than the remaining sections, excluding inlet and discharge ports.

REFERENCES

1. *Cryogenic Materials Data Handbook,* Vol I and II, by F. R. Schwartzberg, S. H. Osgood, and R. G. Herzog: AFML-TDR-64-280, Martin Marietta Corporation, Denver, Aug 1978
2. *Handbook on Materials for Superconducting Machinery,* by K. R. Hanby, *et al.*: MCIC-HB-04, Metals and Ceramics Information Center, Battelle Columbus Laboratories, Columbus, OH, Jan 1977
3. New Data on Aluminum Alloys for Cryogenic Applications, by J. G. Kaufman and E. W. Johnson: in *Advances in Cryogenic Engineering,* edited by K. D. Timmerhaus, Vol 6, Plenum Press, 1961, p 637-649
4. Aluminum Alloys for Cryogenic Service, by J. E. Campbell: *Materials Research and Standards,* Vol 4, No. 10, Oct 1964, p 540-548
5. Sharp Notch Behavior of Some High-Strength Sheet Aluminum Alloys and Welded Joints at 75,

Table 46 Results of fatigue-life tests on austenitic stainless steels

Data from Ref 2, 41, 42, 49, 50

Alloy and condition	Stressing mode	Stress ratio, R	Cyclic frequency, Hz	K_t	Fatigue strengths at 10^6 cycles 24 °C MPa	(75 °F) ksi	−196 °C MPa	(−320 °F) ksi	−253 °C MPa	(−423 °F) ksi
Type 301 sheet, extra full hard	Flex	−1.0	29, 86	1.......	496	72	793	115	669	97
				3.1	172	25	303	44	...	...
Type 304L bar, annealed	Axial	−1.0	...	1.......	269	39	483	70	552(a)	80(a)
				3.1	193	28	207	30	228(a)	33(a)
Type 310 sheet, annealed	Flex(b)	−1.0	...	1.......	186	27	455	66	597	84
	Flex(c)	−1.0	...	1.......	213	31	490	71	662	96
Type 310 bar, annealed	Axial	−1.0	...	1.......	255	37	469	68	607(a)	88(a)
				3.1	186	27	234	34	352(a)	51(a)
Type 321 sheet, annealed	Axial	−1.0	...	1.......	221	32	303	44	372	54
				3.5	124	18	154	22.3	181	26.3
	Flex(b)	−1.0	30 to 40	1.......	172	25	303	44	358	52
Type 347 sheet, annealed	Flex(b)	−1.0	30 to 40	1.......	221	32	421	61	386	56
	Flex(c)	−1.0	30 to 40	1.......	241	35	469	68	510	74
A286 sheet, STA(f)	Flex(d)	−1.0	30 to 40	1.......	427	62	579	84	586	85
	Flex(e)	−1.0	30 to 40	1.......	496	72	703	102	779	113
A286 bar, STA(f)	Axial	−1.0	...	1.......	414	60	579	84	655	95

(a) Tested at −269 °C (−452 °F). (b) Surface finish 64 rms. (c) Surface finish 11 rms. (d) Surface finish 72 rms. (e) Surface finish 10 rms. (f) STA = 980 °C (1800 °F), WQ; 720 °C (1325 °F) 16 h, AC.

Fig. 19 Fatigue-crack-growth rates for Ti-5Al-2.5Sn and Ti-6Al-4V (Ref 61)

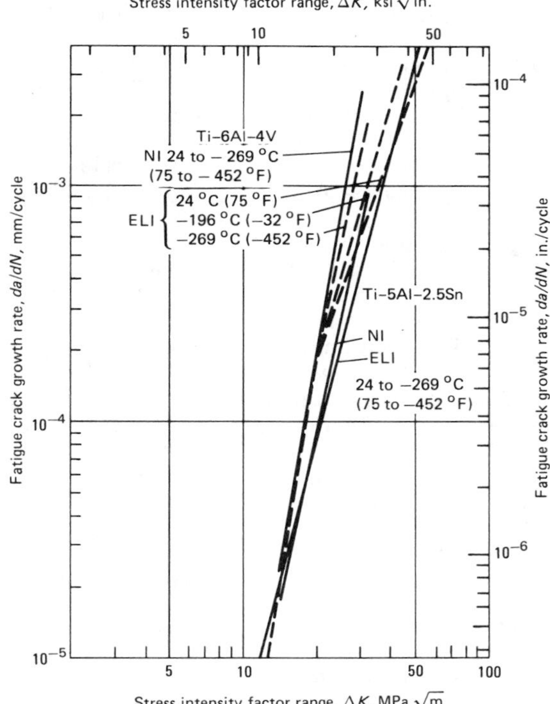

NI = normal interstitial content; ELI = extra-low interstitial content. See Table 50 for C and n values for fatigue-crack-growth rate equations.

−320 and −423 F, by M. P. Hanson, G. W. Stickley, and H. T. Richards: in *Low-Temperature Properties of High-Strength Aircraft and Missile Materials,* STP 287, American Society for Testing and Materials, Philadelphia, 1961, p 3-15

6. Materials for Use at Liquid Hydrogen Temperature, by J. H. Bolton, L. L. Godby, and B. L. Taft: in *Low-Temperature Properties of High-Strength Aircraft and Missile Materials,* STP 287, American Society for Testing and Materials, Philadelphia, 1961, p 108-120
7. Effects of Low Temperatures on the Mechanical Properties of Structural Metals, by H. L. Martin, *et al.*: NASA SP-5012 (01), Office of Technology Utilization, National Aeronautics and Space Administration, Washington, DC, 1968
8. Mechanical Properties of Several 2000 and 6000 Series Aluminum Alloys, by J. L. Christian and J. F. Watson: in *Advances in Cryogenic Engineering,* edited by K. D. Timmerhaus, Vol 10, Plenum Press, 1965, p 63-76
9. Tensile Properties and Notch Toughness of Aluminum Alloys at −452 °F in Liquid Helium, by J. G. Kaufman, K. O. Bogardus, and E. T. Wanderer: in *Advances*

Table 47 Compositions of titanium alloys

Alloy	Al	Sn	V	Composition, % Fe max	O max	C max	N max	H max	M max
Ti-75A	...	...	...	...	0.40	0.20	0.07	0.0125	...
Ti-5Al-2.5Sn	4.0-6.0	2.0-3.0	...	0.50	0.20	0.15	0.07	0.020	0.30
Ti-5Al-2.5Sn(ELI)(a)	4.7-5.6	2.0-3.0	...	0.20	0.12	0.08	0.05	0.0175	...
Ti-6Al-4V	5.5 to 6.75	...	3.5 to 4.5	...	...	...	...	...	...
Ti-6Al-4V(ELI)(a)	5.5 to 6.5	...	3.5 to 4.5	0.15	0.13	0.08	0.05	0.015	...

(a) Extra low interstitial.

in Cryogenic Engineering, edited by K. D. Timmerhaus, Vol 13, Plenum Press, 1968, p 294-308

10. Strain Cycling Fatigue Behavior of Ten Structural Metals Tested in Liquid Helium, Liquid Nitrogen, and Ambient Air, by A. J. Nachtigall: in *Properties of Materials for Liquified Natural Gas Tankage,* STP 579, American Society for Testing and Materials, Philadelphia, Sept 1975, p 378-396
11. Temperature Dependences of the Elastic Constants of Precipitation-Hardened Aluminum Alloys 2014 and 2219, by D. T. Read and H. M. Ledbetter: in *Journal of Engineering Materials and Technology,* Transactions of the ASME, Series H, Vol 99, No. 2, Apr 1977, p 181-184
12. The Properties of Aluminum Alloy 2219 Sheet, Plate and Welded Joints at Low Temperatures, by J. G. Kaufman, F. G. Nelson, and E. W. Johnson: in *Advances in Cryogenic Engineering,* edited by K. D. Timmerhaus, Vol 8, Plenum Press, 1963, p 661-670
13. "Determination of Design Allowable Properties, Fracture of 2219-T87 Aluminum Alloy", by W. L. Engstrom: NASA CR-115388, The Boeing Company, Seattle, Mar 1972
14. "Effects of Proof Loads and Combined Mode Loading on Fracture and Flaw Growth Characteristics of Aerospace Alloys", by R. C. Shah: NASA CR-134611, The Boeing Aerospace Company, Seattle, Mar 1974
15. "Proof Test Criteria for Thin-Walled 2219 Aluminum Pressure Vessels", by R. W. Finger: Vol I, NASA CR-135036, Vol II, NASA CR-135037, The Boeing Aerospace Company, Seattle, Aug 1976
16. Effects of Specimen Thickness on Fracture Toughness of an Aluminum Alloy, by D. T. Read and R. P. Reed: in *International Journal of Fracture,* Vol 13, No. 2, Apr 1977, p 201-213
17. Cryogenic Tensile Properties of Selected Aerospace Materials, by W. Weleff, H. S. McQueen, and W. F. Emmons: in *Advances in Cryogenic Engineering,* edited by K. D. Timmerhaus, Vol 10, Plenum Press, 1965, p 14-15
18. "Analysis and Test of Deep Flaws in Thin Sheets of Aluminum and Titanium, Vol I—Program Summary and Data Analysis, Vol II—Crack Opening Displacement and Stress-Strain Data", by R. W. Finger: NASA CR-135369 and NASA CR-135370, The Boeing Aerospace Company, Seattle, Apr 1978
19. Tensile Properties and Notch Toughness of Groove Welds in Wrought and Cast Aluminum Alloys at Cryogenic Temperatures, by F. G. Nelson, J. G. Kaufman, and F. T. Wanderer: in *Advances in Cryogenic Engineering,* edited by K. D. Timmerhaus, Vol 14, Plenum Press, 1969, p 71-82
20. Hard Tooling for the Fabrication of the External Tank of the Space Shuttle, by G. W. Oyler and F. R. Clover: *Welding Journal,* Vol 56, No. 12, Dec 1977, p 23-30
21. LNG Materials and Fluids: A Users Manual of Property Data in Graphic Format, First Edition, edited by D. Mann: National Bureau of Standards, Boulder, CO, 1977
22. Fracture Toughness of Cryogenic Alloys, by A. W. Pense, R. D. Stout, and B. R. Somers: in *Advances in Cryogenic Engineering,* edited by K. D. Timmerhaus, *et al.,* Vol 24, Plenum Press, 1978, p 548-559
23. Burst Tests of Pre-Flawed Welded Aluminum Alloy Pressure Vessels at −220 F, by R. L. Lake, F. W. DeMoney, and R. J. Eiber: in *Advances in Cryogenic Engineering,* edited by K. D. Timmerhaus, Vol 13, Plenum Press, 1968, p 278-293
24. Tensile Behavior of Parent-Metal and Welded 5000-Series Aluminum Alloy Plate at Room and Cryogenic Temperatures, by L. P. Rice, J. E. Campbell, and W. F. Simmons: in *Advances in Cryogenic Engineering,* edited by K. D. Timmerhaus, Vol 7, Plenum Press, 1962, p 478-489
25. The Tensile Property Evaluation of One 5000-Series Aluminum Alloy at the Temperature of Liquid Helium, by L. P. Rice, J. E. Campbell, and W. F. Simmons: in *Advances in Cryogenic Engineering,* edited by K. D. Timmerhaus, Vol 8, Plenum Press, 1963, p 671-677
26. Large-Scale Fracture Toughness Tests of Thick 5083-O Plate and 5183 Welded Panels at Room Temperature, −260, and −320 F, by J. G. Kaufman, F. G. Nelson, and R. H. Wygonik; in *Fatigue and Fracture Toughness—Cryogenic Behavior,* STP 556, American Society for Testing and Materials, Philadelphia, 1974, p 125-158
27. Fracture Mechanics Parameters for a 5083-O Aluminum Alloy at Low Temperatures, by R. L. Tobler and R. P. Reed: *Journal of Engineering Materials and Technology,* Transactions of the ASME, Series H, Vol 99, No. 4, Oct 1977, p 306-312
28. Crack Growth and Fracture of Thick 5083-O Plate Under Liquified Natural Gas Ship Spectrum Loading, by R. A. Kelsey, R. H. Wygonik, and Per Tenge: in *Properties of Materials for Liquified Natural Gas Tankage,* STP 579, American Society for Testing and Materials, Philadelphia, Sept 1975, p 44-79
29. Mechanical Properties of US/USSR Al-Mg Plate and Welds for LNG Applications, by R. A. Kelsey and F. G. Nelson: in *Advances in Cryogenic Engineering,* edited

Table 48 Typical tensile properties of titanium and two titanium alloys

Data from Ref 2, 7, 14, 44, 60, 89-93

Temperature		Tensile strength		Yield strength		Elongation,	Reduction in area,	Notch tensile strength(a)		Young's modulus	
°C	°F	MPa	ksi	MPa	ksi	%	%	MPa	ksi	GPa	10^6 psi
Ti-75A sheet, annealed, longitudinal orientation											
24	75	580	84.3	465	67.6	25	...	785	114	...	...
−78	−108	750	109	615	89.2	25	...	...	...	...	...
−196	−320	1050	152	940	136	18	...	1100	159	...	...
−253	−423	1280	186	1190	173	8	...	875	127	...	...
Ti-75A sheet, annealed, transverse orientation											
24	75	585	85.1	475	69.0	25	...	800	116	...	...
−78	−108	760	110	645	93.4	20	...	905	131	...	...
−196	−320	1060	153	965	140	14	...	1120	163	...	...
−253	−423	1340	194	1260	182	7	...	880	128	...	...
Ti-5Al-2.5Sn sheet, nominal interstitial annealed, longitudinal orientation											
24	75	850	123	795	115	16	...	1130	164	105	15.4
−78	−108	1080	156	1020	148	13	...	1310	190	115	16.6
−196	−320	1370	199	1300	188	14	...	1630	236	120	17.7
−253	−423	1700	246	1590	231	7	...	1430	208	130	18.5
Ti-5Al-2.5Sn sheet, nominal interstitial annealed, transverse orientation											
24	75	895	130	860	125	14	...	1170	170	...	...
−78	−108	1050	152	1020	148	12	...	1250	181	...	...
−196	−320	1430	208	1370	198	12	...	1630	236	...	...
−253	−423	1670	242	1610	234	6	...	1290	187	...	...
−268	−450	1590	231	...	...	1.5	...				
Ti-5Al-2.5Sn (ELI) sheet, annealed, longitudinal orientation											
24	75	800	116	740	107	16	...	1060	154	115	16.4
−78	−108	960	139	880	128	14	...	1190	173	125	18.0
−196	−320	1300	188	1210	175	16	...	1560	226	130	18.6
−253	−423	1570	228	1450	210	10	...	1670	242	130	19.2
Ti-5Al-2.5Sn (ELI) sheet, annealed, transverse orientation											
24	75	805	117	760	110	14	...	1100	159	110	16.0
−78	−108	950	138	895	130	12	...	1260	182	125	18.1
−196	−320	1300	188	1230	179	14	...	1570	228	130	18.9
−253	−423	1570	228	1480	214	8	...	1530	222	140	20.1
Ti-5Al-2.5Sn (ELI) sheet/weldment, annealed, EB weld											
24	75	815	118	785	114	...	...	...	...	...	...
−196	−320	1300	189	1210	176	...	...	...	...	...	...
−253	−423	1510	219	1380	200	...	...	...	...	...	...
Ti-5Al-25Sn (ELI) plate, annealed, longitudinal orientation											
24	75	765	111	705	102	33	43	...	...	...	...
−253	−423	1430	208	1390	202	17	32	...	...	...	...
Ti-5Al-25Sn (ELI) forgings, as forged, tangential orientation											
24	75	835	121	760	110	15	36	...	...	...	...
−78	−108	980	142	905	131	12	31	...	...	...	...
−196	−320	1260	182	1100	159	15	30	...	...	...	...
−253	−423	1420	206	1260	182	13	22	...	...	...	...
Ti-6Al-4V (ELI) sheet, annealed, longitudinal orientation											
24	75	960	139	890	129	12	...	1120	162	110	16.2
−78	−108	1160	168	1100	160	9	...	1220	177	115	16.6
−196	−320	1500	217	1420	206	10	...	1460	211	120	17.5
−253	−423	1770	256	1700	246	4	...	1500	217	130	18.6

(continued)

(a) K_t = 6.3 for all three sheet forms; K_t = 5 to 8 for Ti-6Al-4V (ELI) forgings. (b) Recrystallization annealing treatment: 930 °C (1700 °F) 4 h, FC to 760 °C (1400 °F) in 3 h, cooled to 480 °C (900 °F) in 3/4 h, AC.

Table 48 (continued)

Temperature °C	°F	Tensile strength MPa	ksi	Yield strength MPa	ksi	Elongation, %	Reduction in area, %	Notch tensile strength(a) MPa	ksi	Young's modulus GPa	10^6 psi
Ti-6Al-4V (ELI) sheet, annealed, transverse orientation											
24	75	960	139	895	130	12	...	1130	164	110	16.0
−78	−108	1170	169	1100	160	12	...	1260	183	115	16.5
−196	−320	1500	218	1460	212	11	...	1440	209	125	18.2
−253	−423	1750	254	1700	246	4	...	1550	225	130	19.2
Ti-6Al-4V (ELI) plate, annealed, longitudinal orientation											
24	75	890	129	840	122	15	37	...	...	...	...
−253	−423	1640	238	1600	232	...	8	...	...	...	...
Ti-6Al-4V (ELI) forgings, as forged, longitudinal orientation											
24	75	970	141	915	133	14	40	1330	193	...	...
−78	−108	1160	168	1120	163	13	31	1560	226	...	...
−196	−320	1570	227	1480	214	11	31	1900	276	...	...
−253	−423	1650	239	1570	227	11	24	1820	264	...	...
Ti-6Al-4V (ELI) forging, recrystallization annealed(b)											
24	75	890	129	825	120	14	41	...	...	110	16.1
−196	−320	1430	207	1370	198	10	16	...	...	120	17.5

(a) K_t = 6.3 for all three sheet forms; K_t = 5 to 8 for Ti-6Al-4V (ELI) forgings. (b) Recrystallization annealing treatment: 930 °C (1700 °F) 4 h, FC to 760 °C (1400 °F) in 3 h, cooled to 480 °C (900 °F) in 3/4 h, AC.

by K. D. Timmerhaus, *et al.,* Vol 24, Plenum Press, 1978, p 505-518

30. Low Temperature Elastic Properties of Aluminum 5083-O and Four Ferritic Nickel Steels, by W. F. Weston and H. M. Ledbetter: in *Properties of Materials for Liquified Natural Gas Tankage,* STP 579, American Society for Testing and Materials, Philadelphia, Sept 1975, p 397-420
31. Design Stresses for Aluminum Alloy 5083-O and 5183 Welds at Cryogenic Temperatures, by K. O. Bogardus and R. C. Malcolm: in *Properties of Materials for Liquified Natural Gas Tankage,* STP 579, American Society for Testing and Materials, Philadelphia, Sept 1975, p 190-204
32. Current Developments in Liquified Natural Gas Presented at International Conference: *Industrial Heating,* Nov 1977, p 37, 38
33. Tensile Properties and Notch Toughness of Some 7XXX Alloys at −452 F, by J. G. Kaufman and E. T. Wanderer: in *Advances in Cryogenic Engineering,* edited by K. D. Timmerhaus, Vol 16, Plenum Press, 1971, p 27-36
34. Performance of a New Cryogenic Aluminum Alloy, 7039, by F. W. DeMoney: in *Advances in Cryogenic Engineering,* edited by K. D. Timmerhaus, Vol 9, Plenum Press, 1964, p 112-123
35. Effect of Aging on the Tensile Properties of Alloy 7039 GTA Welds at Low Temperatures, by F. W. DeMoney: in *Advances in Cryogenic Engineering,* edited by K. D. Timmerhaus, Vol 12, Plenum Press, 1967, p 500-507
36. Properties of 7000 Series Aluminum Alloys at Cryogenic Temperatures, by J. L. Christian and F. W. Watson: in *Advances in Cryogenic Engineering,* edited by K. D. Timmerhaus, Vol 6, Plenum Press, 1961, p 604-621
37. Notch Toughness of Some Aluminum Alloy Castings at Cryogenic Temperatures, by J. W. Coursen, J. G. Kaufman, and W. E. Sicha: in *Advances in Cryogenic Engineering,* edited by K. D. Timmerhaus, Vol 12, Plenum Press, 1967, p 473-483
38. *Damage Tolerant Design Handbook,* Parts I and II, by J. E. Campbell, *et al.*: MCIC-HB-01, Metals and Ceramics Information Center, Battelle Columbus Laboratories, Columbus, OH, Jan 1975
39. Environment-Assisted Fatigue Crack Propagation in 3003-O Aluminum, by R. Roberts, K. Wnek, and J. C. Tafuri: in *Advances in Cryogenic Engineering,* edited by K. D. Timmerhaus, *et al.,* Vol 24, Plenum Press, 1978, p 187-196
40. Fatigue Behavior of Aluminum and Titanium Sheet Materials Down to −423 F, by F. R. Schwartzberg, T. F. Kiefer, and R. D. Keys: in *Advances in Cryogenic Engineering,* edited by K. D. Timmerhaus, Vol 10, Plenum Press, 1965, p 1-13
41. "Determination of Low-Temperature Fatigue Properties of Structural Metal Alloys", by T. F. Kiefer, R. D. Keys, and F. R. Schwartzberg: Final Report, The Martin Company, Denver, Oct 1965
42. "Fatigue Properties of Sheet, Bar, and Cast Metallic Materials for Cryogenic Applications", by E. H. Schmidt: Report R-7564, Rocketdyne Division, North American Rockwell Corporation, Canoga Park, CA, 30 Aug 1968
43. Low-Temperature Mechanical Properties of Welded and Brazed Copper, by R. P. Reed: in *Advances in Cryogenic Engineering,* edited by K. D. Timmerhaus, Vol 14, Plenum Press, 1969, p 83-87
44. *Tensile and Impact Properties of Selected Materials from 20° to 300 °K,* by K. A. Warren and R. P. Reed: National Bureau of Standards Monograph 63, 28 June 1963

Table 49 Fracture toughness of two titanium alloys and weldments

Data from Ref 2, 38, 41, 91, 92, 95, 97-99

Alloy and condition(a)	Form	Room temperature yield strength, MPa	ksi	Specimen design	Orientation	Fracture toughness, K_{Ic}: 24 °C(75 °F), MPa$\sqrt{m}$	24 °C(75 °F), ksi$\sqrt{in.}$	−196 °C(−320 °F), MPa$\sqrt{m}$	−196 °C(−320 °F), ksi$\sqrt{in.}$	−253 °C(−423 °F), MPa$\sqrt{m}$	−253 °C(−423 °F), ksi$\sqrt{in.}$	−269 °C(−452 °F), MPa$\sqrt{m}$	−269 °C(−452 °F), ksi$\sqrt{in.}$
Ti-5Al-2.5Sn(NI), annealed	Plate	876	127	CT	L-T	71.8	65.4	53.4	48.6	...	...	...	...
		876	127	Bend	L-T	...	...	...	...	51.4	46.8	...	...
		876	127	Bend	L-S	...	...	...	...	50.2	45.7	...	...
	Bar	871	126	CT	T-S	77.2	70.3	42.1	38.3	...	...	42.0	38.2
Ti-5Al-2.5Sn(ELI), annealed	Plate	703	102	CT	L-T	...	...	111	101	...	...	...	...
		703	102	Bend	L-T	...	...	...	...	89.6	81.5	...	...
Ti-5Al-2.5Sn(ELI), as forged	Forging	760	110	CT	R-L	...	...	...	...	79.4	72.3	...	...
					R-C	...	...	...	...	58.5	53.2	...	...
Ti-5Al-2.5Sn(ELI)	Forging(b)	779	113	CT	...	...	...	...	...	54.4 to 75.3	49.5 to 68.5	...	...
Ti-6Al-4V (NI), annealed	Bar	942	136	CT	T-L	47.4	43.2	38.8	35.3	...	...	38.5	35.1
Ti-6Al-4V (ELI), as forged	Forging	830	120	CT	T-L	...	...	61.0	55.5	...	...	54.1	49.2
Ti-6Al-4V (ELI), RA	Forging	830	120	CT	M-L(c)	...	...	62.8	57.2	...	...	...	...
					M-R(c)	...	...	62.0	56.4	...	...	...	...
Ti-6Al-4V (ELI), RA, electron beam welded, SR	Forging	830	120	CT	M-R(c)	...	...	61.1(d)	55.6(d)	...	...	...	...
	Weldment	...	...	...	M-L(c)	...	...	56.9(d)	51.8(d)	...	...	...	...
					M-R(c)	...	...	57.1(e)	52.0(e)	...	...	...	...
					M-R(c)	...	...	51.0(f)	46.4(f)	...	...	...	...

(a) SR = stress relieved: 540 °C (1000 °F) 50 h, AC. FC = furnace cool. AC = air cool. NI = normal interstitial content. ELI = extra low interstitial content. RA = recrystallization annealed: 930 °C (1700 °F) 4 h, FC to 810 °C (1400 °F) in 3 h, cooled to 480 °C (900 °F) in 3/4 h, AC. (b) Range for 18 tests. (c) M-L and M-R are specific orientations in a spherical forging. (d) Fusion zone. (e) Heat affected zone. (f) Heat affected zone boundary.

Table 50 Fatigue crack growth rate (*da/dN*) data for two titanium alloys(a)

Data from Ref 61

Alloy and condition (b)	Orientation	Test temperature, °C	Test temperature, °F	C: *da/dN*:mm/cycle, ΔK:MPa $\sqrt{m}$	C: *da/dN*:in./cycle, ΔK:ksi $\sqrt{in.}$	*n*	Estimated range for ΔK, MPa $\sqrt{m}$	Estimated range for ΔK, ksi $\sqrt{in.}$
Ti-5Al-2.5Sn (NI), annealed bar	T-S	24, −196, −269	75, −320, −452	5.1×10^{-11}	3.2×10^{-12}	4.8	14 to 30	13 to 27
Ti-5Al-2.5Sn (LI), annealed bar	T-L	24, −196, −269	75, −320, −452	4.9×10^{-10}	2.8×10^{-11}	4.0	10 to 60	9 to 54
Ti-6Al-4V (NI), annealed bar	T-L	24, −196, −269	75, −320, −452	3.1×10^{-12}	2.2×10^{-13}	6.0	14 to 30	13 to 27
Ti-6Al-4V (ELI), recrystallization annealed bar	T-L	24, −196, −269	75, −320, −452	1.9×10^{-13}	1.4×10^{-14}	7.0	10 to 20	9 to 18
		24, −196	75, −320	3.0×10^{-8}	1.6×10^{-9}	3.0	20 to 40	18 to 36

(a) Stress ratio: $R = 0.1$, at 20 to 28 Hz; compact specimens. (b) NI = normal interstitial, LI = low interstitial, ELI = extra low interstitial.

45. Tensile Properties of Copper, Nickel, and Some Copper-Nickel Alloys at Low Temperatures, by G. W. Geil and N. L. Carwile: NBS Circular 520, National Bureau of Standards, Washington, DC, 7 May 1952, p 67-96
46. Low Temperature Tensile Properties of Copper and Four Bronzes, by R. M. McClintock, D. A. Van-Gundy, and R. H. Kropschot: *ASTM Bulletin,* Vol 240, Sept 1959, p 47-50
47. Low-Temperature (295 ° to 4 °K) Mechanical Properties of Selected

Table 51 Results of fatigue-life tests on two titanium alloys

Data from Ref 41, 42, 49, 50

Alloy and condition	Stressing mode	Stress ratio, R	K_t	Fatigue strengths at 10^6 cycles: 24 °C(75 °F) MPa	ksi	−196 °C(−320 °F) MPa	ksi	−253 °C(−423 °F) MPa	ksi
Ti-5Al-2.5Sn (ELI) sheet, annealed	Axial	0.01	1	495	72	815	118	760	110
			3.5	220	32	205	30	160	23
Ti-5Al-2.5Sn (ELI) sheet(a)	Axial	0.01	1	485	70	565	82	425	62
Ti-5Al-2.5Sn (ELI) bar, annealed(b)	Axial	0	1	760	110	985	143	925	134
Ti-6Al-4V (ELI) sheet(c)	Axial	0.01	1	505	73	675	98	895	130
			3.5	285	41	295	43	275	40
Ti-6Al-4V (ELI) sheet(a)	Axial	0.01	1	600	87	595	86	560	81
Ti-6Al-4V sheet, annealed	Flex	−1.0	1	345	50	550	80	530	77
			3.1	170	25	185	27	255	37

(a) Gas tungsten arc welded, base metal filler. (b) Cyclic frequency, 28 Hz. (c) STA: 900 °C (1650 °F) 5 min, WQ; 540 °C (1000 °F) 4 h, AC.

Copper Alloys, by R. P. Reed and R. P. Mikesell: *Journal of Materials,* Vol 2, No. 2, June 1976, p 370-392

48. "Low Temperature Elastic Properties of Some Copper-Nickel Alloys", by H. M. Ledbetter and W. F. Weston: Materials Research for Superconducting Machinery-IV, Semi Annual Report, National Bureau of Standards, Boulder, CO, 10 Oct 1975
49. The Fatigue Behavior of Certain Alloys in the Temperature Range from Room Temperature to −423 F, by D. N. Gideon, *et al.*: in *Advances in Cryogenic Engineering,* edited by K. D. Timmerhaus, Vol 7, Plenum Press, 1962, p 503-508
50. "Investigation of Notch Fatigue Behavior of Certain Alloys in the Temperature Range of Room Temperature to −423 F", by D. N. Gideon, *et al.*: ASD-TR-62-351, Battelle Memorial Institute, Columbus, OH, Aug 1962
51. Mechanical Properties of Several Nickel-Base Alloys at Room and Cryogenic Temperatures, by J. L. Christian: in *Advances in Cryogenic Engineering,* edited by K. D. Timmerhaus, Vol 12, Plenum Press, 1967, p 520-531
52. Cryogenic Fracture Toughness and Fatigue Crack Growth Rate Properties of Inconel 706 Base Material and Gas Tungsten Arc Weldments, by W. A. Logsdon, J. M. Wells, and R. Kossowsky: paper presented at the Cryogenic Engineering Conference, Madison, WI, Aug 21-24, 1979
53. Structural Alloys for Cryogenic Service, by J. L. Christian, *et al.*: *Metal Progress,* Vol 83, No. 3, Mar 1963, p 101-104
54. Low Temperature Effects on the Fracture Behavior of a Nickel Base Superalloy, by R. L. Tobler: *Cryogenics,* Vol 16, No. 11, Nov 1976, p 669-674
55. Evaluation of Inconel X750 Weldments for Cryogenic Applications, by J. M. Wells: in *Advances in Cryogenic Engineering,* edited by K. D. Timmerhaus, *et al.,* Vol 22, Plenum Press, 1977, p 80-90
56. Low-Temperature Elastic Properties of a Nickel-Chromium-Iron-Molybdenum Alloy, by W. F. Weston and H. M. Ledbetter: *Materials Science and Engineering,* Vol 20, No. 3, Sept 1975, p 287-290
57. Dynamic Low-Temperature Elastic Properties of Two Austenitic Nickel-Chromium-Iron Alloys, by W. F. Weston, H. M. Ledbetter, and E. R. Naimon: *Materials Science and Engineering,* Vol 20, No. 1, July 1975, p 95-104
58. Cryogenic Fracture Mechanics Properties of Several Manufacturing Process/Heat Treatment Combinations of Inconel X750, by W. A. Logsdon: in *Advances in Cryogenic Engineering,* edited by K. D. Timmerhaus, *et al.,* Vol 22, Plenum Press, 1977, p 47-58
59. The Development of Titanium and Inconel Cryogenic Pressure Vessels, by R. J. Balthazar and H. E. Sutton: in *Advances in Cryogenic Engineering,* edited by K. D. Timmerhaus, Vol 11, Plenum Press, 1966, p 437-446
60. Fracture of Structural Alloys at Temperatures Approaching Absolute Zero, by R. L. Tobler: *Proceedings of the Fourth International Conference on Fracture,* University of Waterloo, Ontario, Canada, June 1977
61. Fatigue Crack Growth Resistance of Structural Alloys at Cryogenic Temperatures, by R. L. Tobler and R. P. Reed: in *Advances in Cryogenic Engineering,* edited by K. D. Timmerhaus, *et al.,* Vol 24, Plenum Press, 1978, p 82-90
62. The Influence of Processing and Heat Treatment on the Cryogenic Fracture Mechanics Properties of Inconel 718, by W. A. Logsdon, R. Kossowsky, and J. M. Wells: in *Advances in Cryogenic Engineering,* edited by K. D. Timmerhaus, *et al.,* Vol 24, Plenum Press, 1978, p 197-209
63. "Fracture Toughness and Related Characteristics of the Cryogenic Nickel Steels", by A. W. Pense and R. D. Stout: WRC Bulletin 205, Welding Research Council, New York, May 1975
64. An Evaluation of Three Steels for Cryogenic Service, by J. P. Bruner and D. A. Sarno: in *Advances in Cryogenic Engineering,* edited by K. D. Timmerhaus, *et al.,* Vol 24, Plenum Press, 1978, p 529-539
65. Fatigue and Fracture Toughness Properties of 9 Percent Nickel Steel at LNG Temperatures, by D. A. Sarno, D. E. McCabe, and T. G. Heberling: in *Journal of Engineering for Industry,* Trans-

actions of the ASME, Series B, Vol 95, No. 4, Nov 1973, p 1069-1075

66. "Strength and Fracture Toughness of Nickel Containing Steels", by A. G. Haynes, *et al.*: STP 579, American Society for Testing and Materials, Philadelphia, 1975, p 288-323
67. "Fracture Toughness and Related Characteristics of the Cryogenic Nickel Steels", by A. W. Pense and R. D. Stout: WRC Bulletin No. 205, Welding Research Council, May 1975
68. Fracture Toughness of 5% Nickel Steel Weldments, by D. A. Sarno, J. P. Bruner, and G. E. Kampschaefer: *Welding Journal,* Vol 39, No. 11, Nov 1974, p 486s-494s
69. Fracture Behavior of the Heat-Affected Zone in 5% Ni Steel Weldments, by H. I. McHenry and R. P. Reed: *Welding Journal,* Vol 56, No. 4, Apr 1977, p 104s-112s
70. Low Temperature Fracture Behavior of Iron Nickel Alloy Steels, by R. L. Tobler, *et al.*: in *Properties of Materials for Liquified Natural Gas Tankage,* STP 579, American Society for Testing and Materials, Philadelphia, Sept 1975, p 261-287
71. "Temperature Dependence of Yielding in Austenitic Stainless Steels", by R. L. Tobler, R. P. Reed, and D. S. Burkhalter, National Bureau of Standards, Boulder, CO
72. Austenitic Stainless Steels at Cryogenic Temperatures, l-Structural Stability and Magnetic Properties, by D. C. Larbalestier and H. W. King: *Cryogenics,* Vol 13, No. 3, Mar 1973, p 160-168
73. Mechanical Properties of Four Austenitic Stainless Steels at Temperatures Between 300 ° and 20 °K, by C. J. Guntner and R. P. Reed: in *Advances in Cryogenic Engineering,* edited by K. D. Timmerhaus, Vol 6, Plenum Press, 1961, p 565-576
74. Low Temperature Mechanical Properties of 300 Series Stainless Steels and Titanium, by T. S. DeSisto and L. C. Carr: in *Advances in Cryogenic Engineering,* edited by K. D. Timmerhaus, Vol 6, Plenum Press, 1961, p 577-586
75. Evaluations of Weldments in Austenitic Stainless Steels for Cryogenic Applications, by J. M. Wells, W. A. Logsdon, and R. Kossowsky: in *Advances in Cryogenic Engineering,* edited by K. D. Timmerhaus, *et al.,* Vol 24, Plenum Press, 1978, p 150-159
76. Fracture Mechanics Properties of Austenitic Stainless Steels for Advanced Applications, by W. A. Logsdon, J. M. Wells, and R. Kossowsky: in *Proceedings Second International Conference on Mechanical Behavior of Materials,* American Society for Metals, Metals Park, OH, 1976, p 1283-1289
77. Low Temperature Properties of Cold-Rolled AISI Types 301, 302, 304ELC, and 310 Stainless Steel Sheet, by J. F. Watson and J. L. Christian: in *Low Temperature Properties of High-Strength Aircraft and Missile Materials,* STP 287, American Society for Testing and Materials, Philadelphia, 1961, p 170-193
78. The Effect of Experimental Variables Including the Martensitic Transformation on the Low-Temperature Mechanical Properties of Austenitic Stainless Steels, by C. J. Guntner and R. P. Reed: in *Transactions of the ASM,* Vol 55, American Society for Metals, Metals Park, OH, Sept 1962, p 399-419
79. Mechanical Properties of Inconel 625 Welds in 21-6-9 Stainless Steel, by R. R. Vandervoort: in *Cryogenics,* Vol 18, No. 8, Aug 1979, p 448-452
80. "The Stress Corrosion Resistance and the Cryogenic Temperature Mechanical Properties of Annealed Nitronic 60 Bar Material", by J. W. Montano: NASA TM X-73359, George C. Marshall Space Flight Center, AL, Jan 1977
81. Properties of Some Precipitation-Hardening Stainless Steels at Very Low Temperatures, by J. E. Campbell and L. P. Rice: in *Low-Temperature Properties of High-Strength Aircraft and Missile Materials,* STP 287, American Society for Testing and Materials, Philadelphia, 1961, p 158-167
82. Mechanical Properties of High-Strength 301 Stainless Steel Sheet at 70, −320, and −423 F in Base Metal and Welded Joint Configuration, by J. F. Watson and J. L. Christian: in *Low-Temperature Properties of High-Strength Aircraft and Missile Materials,* STP 287, American Society for Testing and Materials, Philadelphia, 1961, p 136-149
83. Tensile and Impact Properties of Cast Stainless Steels at Cryogenic Temperatures, by E. R. Hall: in *Evaluation of Metallic Materials in Design for Low-Temperature Service,* STP 302, American Society for Testing and Materials, Philadelphia, 1962, p 85-93
84. Evaluation of Stainless Steel Casting Alloys for Cryogenic Service in the 80-Inch Liquid Hydrogen Bubble Chamber, by C. L. Goodzeit: in *Advances in Cryogenic Engineering,* edited by K. D. Timmerhaus, Vol 10, Plenum Press, 1965, p 26-36
85. "The Choice of Steel for the ISABELLE Magnet Tubes", by D. Dew-Hughes and K. S. Lee: presented at the Cryogenic Engineering Conference, Madison, WI, Aug 21-24, 1979
86. Low-Temperature Elastic Properties of Four Austenitic Stainless Steels, by H. M. Ledbetter, W. F. Weston, and E. R. Naimon: *Journal of Applied Physics,* Vol 6, No. 9, Sept 1975, p 3855-3860
87. The Fracture Toughness and Fatigue Crack Growth Rate of an Fe-Ni-Cr Superalloy at 298, 76 and 4K, by R. P. Reed, R. L. Tobler, and R. P. Mikesell: in *Advances in Cryogenic Engineering,* edited by K. D. Timmerhaus, *et al.,* Vol 22, Plenum Press, 1977, p 68-79
88. Titanium Alloys for Cryogenic Service, by R. G. Broadwell and R. A. Wood: *Materials Research and Standards,* Vol 4, No. 10, Oct 1964, p 549-554
89. Sharp-Edge-Notch Tensile Characteristics of Several High-Strength Titanium Sheet Alloys at Room and Cryogenic Temperatures, by G. B. Espey, M. H. Jones, and W. F. Bown, Jr.: in *Low-Temperature Properties of High-Strength Aircraft and Missile Materials,* STP 287, American Society for Testing and Materials, Philadelphia, 1961, p 74-95
90. Mechanical Properties of Titanium Alloys at Cryogenic Temperatures, by J. L. Christian and A. Hurlich: in *Advances in Cryogenic Engineering,* edited by K. D. Timmerhaus, Vol 13, Plenum Press, 1968, p 318-333
91. Low Temperature Fracture Behavior of a Ti-6Al-4V Alloy and

Its Electron Beam Welds, by R. L. Tobler: in *Toughness and Fracture Behavior of Titanium,* STP 651, American Society for Testing and Materials, Philadelphia, July 1978, p 267-294

92. Influence of Composition, Annealing Treatment and Texture on the Fracture Toughness of Ti-5Al-2.5Sn Plate at Cryogenic Temperatures, by R. H. Van Stone, *et al.*: in *Toughness and Fracture Behavior of Titanium,* STP 651, American Society for Testing and Materials, Philadelphia, July 1978, p 154-179

93. "Behavior of Ti-5Al-2.5Sn ELI Titanium Alloy Sheet Parent and Weld Metal in the Presence of Cracks at 20K", by T. L. Sullivan: NASA TN D-6544, Lewis Research Center, Cleveland, Nov 1971

94. Elastic Properties of Two Titanium Alloys at Low Temperatures, by E. R. Naimon, W. F. Weston, and H. M. Ledbetter: *Cryogenics,* Vol 14, No. 5, May 1974, p 246-249

95. Fracture Testing and Results for a Ti-6Al-4V Alloy at Liquid Helium Temperature, by C. W. Fowlkes and R. L. Tobler: *Engineering Fracture Mechanics,* Vol 8, No. 3, 1976, p 487-500

96. "Fracture Toughness of Ti-5Al-2.5Sn ELI Forgings", by W. G. Reuter: Memoranda and Material R&D Report, Aerojet-General Corporation, 26 July 1971

97. "Fracture Control of H-O Engine Components", by J. T. Ryder: NASA CR-135137, Lockheed-California Company, Burbank, CA, Feb 1977

98. "Fatigue and Fracture Toughness Testing at Cryogenic Temperatures", by R. L. Tobler, *et al.*: in *Materials Research in Support of Superconducting Machinery-II,* Second Semi-Annual Technical Report, edited by A. F. Clark, *et al.,* National Bureau of Standards, Boulder, CO, 1 Sept 1974

99. "Fatigue Crack Growth and J-Integral Fracture Parameters of Ti-6Al-4V at Ambient and Cryogenic Temperatures", by R. L. Tobler: in *Materials Research in Support of Superconducting Machinery-III,* Third Semi-Annual Technical Report, edited by R. P. Reed, *et al.,* National Bureau of Standards, Boulder, CO, 1 March 1975

Industrial Uses of Depleted Uranium

By Paul Loewenstein
Vice President and Technical Director
Nuclear Metals, Inc.

DEPLETED URANIUM (sometimes referred to as DU) is a by-product of the process by which the fissionable isotope U-235 is extracted from natural uranium, and thus can be considered a by-product of the nuclear industry. From an engineering standpoint, the most singular property of uranium is its great density—almost twice that of lead, and nearly as great as those of gold and tungsten. Because of this high density, thin layers of uranium are capable of absorbing as much penetrating radiation, such as gamma rays, as could be absorbed by much thicker layers of less dense metals such as lead and iron.

Uranium is much easier to fabricate than dense metals such as tungsten and rhenium, and is much less costly than heavy metals such as gold and platinum. These qualities make uranium a good candidate for objects that must be small, yet very heavy for their size. Depleted uranium is relatively abundant and available. Consequently, industrial non-nuclear usage has increased steadily during recent years.

Applications

There are three principal non-nuclear uses of depleted uranium: radiation shielding; counterweights in airplanes, helicopters and missiles; and armor-piercing projectiles for military ordnance (Ref 1 and 2). Unalloyed uranium is used mainly in shielding and counterweight applications. Heat treated uranium alloys are used in ballistic or armor-piercing ordnance applications. Besides these three chief uses, depleted uranium has been used in several specific applications where its combination of great density, good fabricability and relatively good mechanical properties give it an advantage over alternative materials.

Radiation Shielding. Shipping containers made of depleted uranium are used as spent fuel casks for transportation of highly radioactive spent fuel elements from nuclear reactors to disposal sites. Casks of this type are very heavy (up to several thousand kilograms) and have to withstand and dissipate the heat generated by the spent fuel elements. The uranium containers usually are clad with stainless steel to limit corrosion and contamination.

Containers used in transporting radioactive isotopes for medical and industrial applications are similar in purpose to those used for spent fuel elements, but are much smaller and lighter in weight.

In many devices for medical radiation therapy, depleted uranium is used as shielding against stray radiation from the radioactive isotope inside the device. Depleted uranium is used instead of alternative shielding materials such as lead, which is considerably more bulky, or tungsten, which is more costly and more difficult to fabricate into complex shapes. Uranium can function as both a shielding material and a structural material, greatly reducing the size and improving the mobility of these devices.

Uranium is used extensively in industrial radiographic equipment to house and shield isotope sources such as Ir-192, Co-60 and Cs-137. Two types of equipment are common. In one, the radioactive source is stationary within the uranium shield, and radiation is allowed to escape by sliding or rotating a uranium plug. In the other, generally known as a "football", the radioactive source moves out of the uranium shield to expose radiographic film. For storage and transportation, the source slides back into the center of the shield through an S-shape tube surrounded by depleted uranium. Motion of the radioactive source is controlled remotely from a safe distance by means of a flexible cable. Figure 1 shows the relation of shielding effect to thickness of depleted uranium for the three gamma-ray sources most widely used.

Fig. 1 Effectiveness of uranium as a shielding material for three common radioisotopes

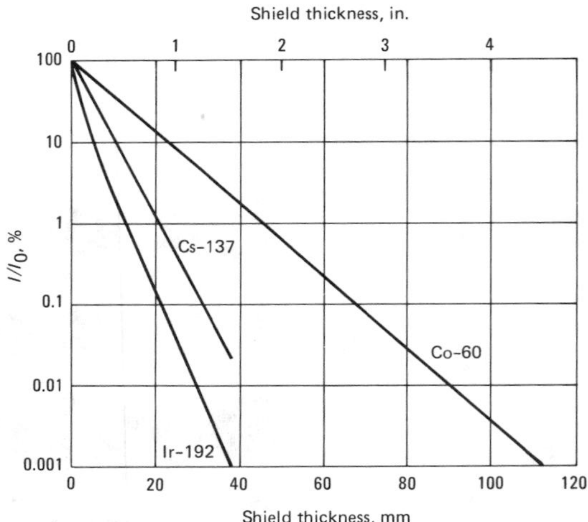

Relative intensity of an attenuated beam of gamma rays vs thickness of uranium shield; I is the intensity of radiation on the side of the shield away from the source, and I_0 is the intensity of radiation on the side of the shield toward the source.

Counterweights are used in aerodynamic control devices of airplanes, missiles and helicopters to maintain the center of gravity when such devices are moved. Counterweights frequently are complex in shape to fit control-surface contours. High density is important in order to keep the counterweight small compared to the control surface. Depleted uranium is well suited for this application, and uranium counterweights have been used in many civilian and military aircraft. For example, 1500 kg of uranium counterweights are used in each Boeing 747.

Armor-piercing Projectiles. Kinetic-energy projectiles constitute by far the largest single use of depleted uranium. In addition to high density, depleted uranium alloys offer high penetrator effectiveness against single and multiple targets, postpenetration pyrophoricity, ease of fabrication, abundant availability and low cost.

Two alloys, U-0.75 and U-2Mo, are used for penetrators. Penetrators are heat treated to obtain the best possible performance against specific targets. Each of the three military services has one weapons system employing depleted uranium penetrators. The U. S. Navy uses the Phalanx penetrator (U-2Mo, weighing about 0.07 kg) to defend against ship-to-ship missiles; the Air Force uses the GAU-8/A penetrator (U-0.75%Ti, weighing 0.272 kg), which is intended to be fired from A-10 aircraft against armored tanks, and the Army uses the XM774 penetrator (U-0.75Ti, weighing about 3.3 kg) in projectiles for the 105-mm battle-tank gun. Other ordnance applications are under development.

Oil-well Sinker Bars. Depleted uranium is used extensively in oil-well logging. Heavy uranium weights encapsulated in steel (sinker bars) are used to help lower logging instruments into oil wells against the dense, high-pressure fluids present in these wells. Again, the high density of uranium is important so that sinker bars are small but heavy.

Other Applications. Depleted uranium alloys have been used for special high-performance gyroscope rotors. A gyroscope rotor having a U-8Mo rim and a lightweight beryllium hub has been produced and tested with satisfactory results. Depleted uranium flywheels have been produced for large inertial energy-storage devices. Uranium also has been used for vibration damping, especially in boring bars and machine tools.

Properties of Depleted Uranium

Uranium is an allotropic metal having three phases in the solid state. Below 688 °C, the metal is in the alpha phase (orthorhombic), which exhibits increasing ductility from room temperature to the phase-transformation temperature. From 688 to 775 °C, the metal is in the beta phase (tetragonal), which is brittle and unworkable. From 775 °C to the melting point, it is in the gamma phase (body-centered cubic), which exhibits great ductility and very low strength.

Uranium is a highly anisotropic material. Properties can vary extensively, depending on fabrication history and orientation with respect to the direction of working. Impurities such as carbon, iron, silicon and aluminum have strong effects on mechanical properties. The properties given in this section are typical of production material for unalloyed depleted uranium of standard purity and for U-0.75Ti and U-2Mo, the two depleted uranium alloys most extensively produced for nonnuclear applications.

- **Unalloyed uranium (as cast)**
  Melting point: 1130 °C
  Density: 19 Mg/m^3
  Tensile strength: 450 MPa (65 ksi)
  Yield strength (0.2% offset): 207 MPa (30 ksi)
  Modulus of elasticity (tension): 172 GPa (25×10^6 psi)
  Elongation: 1 to 5%
  Reduction in area: 1 to 10%
  Hardness: 50 to 100 HRB

Hardness and strength of unalloyed uranium can be increased by warm or cold working.

- **Uranium alloys U-0.75Ti and U-2Mo**

	Melting point, °C	Density, Mg/m^3
U-0.75Ti	1200	18.6
U-2Mo	1150	18.5

Mechanical properties of these two alloys vary widely with heat treatment. The standard heat treatment for ordnance applications consists of heating into the gamma phase (about 850 °C), quenching in water or oil, and then aging at any of various temperatures in the range 350 to 450 °C. Strength increases and ductility decreases with increasing aging temperature until the material reaches a peak-aged condition at about 450 °C. Above this temperature, the material becomes overaged and loses strength but gains in ductility. Typical tensile data are given in Tables 1 and 2 for underaged, peak aged and overaged material.

Table 1 Tensile properties of U-0.75% Ti (Ref 3)

Yield strength		Tensile strength		Elongation, %
MPa	ksi	MPa	ksi	
Underaged				
700	101	1350	196	14
850	123	1450	210	13
1000	145	1525	221	7½
Peak aged				
1200	174	1650	239	2½
Overaged				
1000	145	1450	210	3
850	123	1300	188	4
700	101	1175	170	7

Table 2 Tensile properties of U-2% Mo (Ref 3)

Yield strength		Tensile strength		Elongation, %
MPa	ksi	MPa	ksi	
Underaged				
700	101	1150	167	4
850	123	1200	174	4
1000	145	1250	181	2½
1150	167	1350	196	1½
Peak aged				
1350	196	1600	232	1½
Overaged				
1150	167	1375	199	1½
1000	145	1400	203	3½
850	123	1225	178	8
700	101	1125	163	17
550	80	925	134	24

Table 3 Estimated availability of depleted uranium (a)

	'78	'79	'80	'81	'82	'83	'84	'85	'86	'87	'88
UF_6	184.8	202.1	213.0	228.5	250.8	273.2	297.8	324.5	351.1	377.8	405.3
UF_4	67.7	63.5	50.2	45.7	44.6	37.4	35.0	32.5	27.9	27.9	27.9
Total	252.5	265.6	263.2	274.2	295.4	310.6	332.8	357.0	379.0	405.7	433.2

(a) Metric tons (Mg) of metal. Data from U.S. Department of Energy (Feb 1979).

Production and Availability

Natural uranium (NU) contains about 0.7% of the fissionable isotope U-235, the remainder being comprised almost entirely of the isotope U-238. Power reactors of the type built in the United States require a U-235 content of 3%. Uranium is enriched from 0.7 to 3% U-235 by the gaseous diffusion process, in which the uranium is present as uranium hexafluoride (UF_6). Five to six kilograms of depleted uranium containing 0.2 to 0.3% U-235 are produced for each kilogram of uranium that is enriched to 3% U-235.

Depleted uranium (DU) is available mainly from government sources in the form of uranium hexafluoride (UF_6) or uranium tetrafluoride (UF_4); UF_4 also is known as "green salt." The amounts of depleted uranium estimated to be available for non-nuclear uses in 1978 through 1988 are shown in Table 3.

Green salt (UF_4) is obtained by chemically reducing UF_6 with hydrogen. Green salt is reduced to metal by an exothermic reaction with magnesium in a closed vessel. The product of this reaction is high-purity unalloyed uranium in the shape of a short cylinder, known as a "derby", weighing between 150 and 500 kg. Figure 2 shows schematically the steps involved in producing depleted uranium metal from the ore. At present there are five industrial producers of depleted uranium products for non-nuclear use in the United States and one in Canada.

Methods of Fabrication

The starting material for all depleted uranium products is derby metal. The usual methods of fabrication of DU products include casting, extrusion, rolling, and forging and swaging; these methods are discussed below. Almost all other conventional metalworking processes (including drawing, spinning, tube drawing, die forging and roll straightening) have been applied to DU, but few are of commercial significance.

Melting and Casting. Uranium can be melted by any of several different techniques. However, because uranium is very reactive when heated in air, melting must be done either under a protective inert atmosphere or in vacuum. Also, because uranium reacts with most ordinary crucible materials, it must be melted in a graphite crucible.

Uranium for industrial non-nuclear uses is melted in cold-wall vacuum induction furnaces (Ref 4). Crucibles and molds are made of high-density graphite. To prevent the uranium from contamination by carbon picked up

Fig. 2 Production of depleted uranium metal

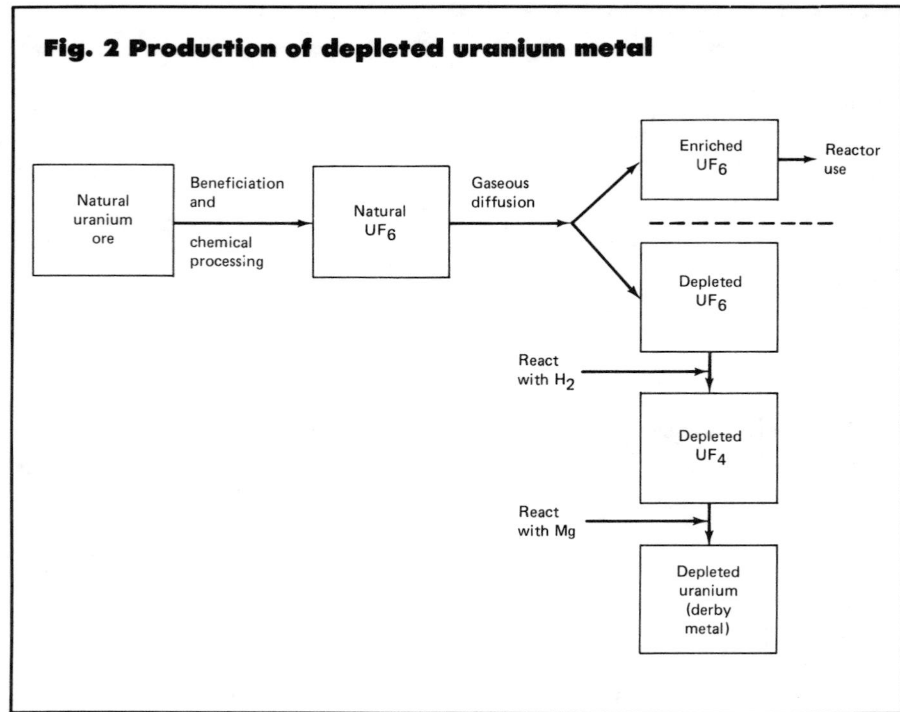

from the graphite crucible, crucibles are coated with zirconia, which is applied either by plasma spraying or by brushing on a slurry containing the zirconia. Zirconia coatings are effective in preventing extensive carbon contamination at temperatures up to 1400 °C.

Typically, uranium is melted and poured at temperatures between 1300 and 1400 °C. Because of the high density of uranium, dross and oxides float to the top of the melt. Uranium is bottom poured rather than tilt poured so that dross and oxides will not be trapped as inclusions in the castings. Alloy segregation can be a problem because uranium is so much more dense than the usual alloy ingredients. Castings weighing as much as 4500 kg can be produced commercially.

Cylindrical or rectangular graphite molds are used to produce billets or slabs for further processing by extrusion or rolling. Semipermanent split molds made of graphite are used for casting of shields, counterweights and other complex shapes. A zirconium tube in the shape of an "S" is placed in a graphite mold, and molten uranium is poured around this tube to produce a "football" isotope shield.

A recent development of commercial importance is use of investment casting for producing depleted uranium parts in large numbers (Ref 5, 6). Investment casting can be used to produce precision parts, minimizing costly machining for parts of complex shape. Parts weighing up to 45 kg are made by investment casting; each mold produces from a few to 1000 parts, depending on weight.

Extrusion. Rod, tubing and similar long shapes of depleted uranium and its alloys are produced by extrusion (Ref 7, 8). Shapes having cross-sectional dimensions of up to 125 mm are available.

Two basic extrusion techniques are used: bare extrusion and canned extrusion. For bare extrusion, billets are heated in a salt bath. The salt prevents oxidation and serves as a lubricant during extrusion. Bare extrusion is carried out either in the gamma phase (ordinarily at 750 to 875 °C) or the alpha phase (at 530 to 675 °C) for either unalloyed or alloyed uranium. For canned extrusions, billets are placed in a copper tube and enclosed by welding copper plates over the ends of the tube. The closed can may be evacuated before heating for extrusion. The copper can prevents oxidation of the uranium during heating and covers the extruded section completely, preventing contamination. Canned extrusion is carried out in the alpha phase (at 530 to 675 °C). Only the stronger alloys (such as U-2Mo) can be extruded in the gamma phase when canned in copper, because unalloyed uranium and most dilute uranium alloys are too soft to coextrude with the copper can.

The main problem with bare extrusion is that uranium picks up hydrogen from the salt bath or a contaminated press area. The disadvantages of copper-canned extrusion are the cost of copper and the cost of removing the copper can by chemical pickling.

Rolling of uranium and uranium alloys is a commercial production method for rod and plate (Ref 7). Rolling is carried out in the alpha phase (ordinarily at 500 to 650 °C). For large sections and heavy reductions, rolling speeds have to be controlled in order to avoid self-heating and transformation to the brittle beta phase.

Forging and swaging of uranium shapes are carried out at elevated temperatures high in the alpha phase to take advantage of the increased ductility of alpha uranium at these temperatures. Forging or swaging of rod can increase yield and decrease machining, because either process produces a near net shape close to finished size. Both processes are used extensively in production of U-0.75Ti penetrators for the U. S. Air Force's GAU-8 projectile (Ref 9).

Heat Treating

For most applications, unalloyed uranium does not require heat treatment. Heating into the beta phase followed by rapid quenching is sufficient to refine grain size of castings, and to partly remove preferred orientation and anisotropy of wrought uranium.

The U-0.75Ti and U-2Mo alloys are heat treated by solution treating at 800 to 850 °C, quenching in oil or water, and aging. Two problems interfere with heat treatment. First, poor thermal conductivity makes it difficult to fully dissolve the titanium in U-0.75Ti sections that are more than 30 mm in either thickness or diameter. Second, in sizes over 20 mm there is a tendency for centerline voids to form during rapid quenching from the gamma phase, chiefly because of the extensive shrinkage that takes place during transformations from gamma to beta and from beta to alpha. Both problems have been partly overcome for rods 35 mm in diameter by lowering the rods axially into agitated room-temperature water at a controlled rate of 80 mm/min. Slower rates of directional quenching produce material which will not respond to aging; faster rates lead to formation of centerline voids.

Depending on property requirements and size of end product, solution treating is carried out in vacuum furnaces with internal or external water or oil quenching systems or in atmosphere furnaces or salt baths with external oil or water quenching. Control of hydrogen in the uranium during solution treating is vital for applications that require high ductility.

Aging of solution treated and quenched U-0.75Ti and U-2Mo alloys starts at about 350 °C and peaks at about 450 °C. Overaging and softening occur at higher temperatures. Aging is carried out in vacuum furnaces, inert-atmosphere furnaces or lead-tin baths. Aging times are usually 2 to 16 h, depending on property requirements. Typical aging curves for the two alloys are shown in Fig. 3 and 4.

Machining

Depleted uranium and its alloys are considered difficult to machine; nevertheless, depleted uranium is machined on a large scale at very high rates of production. The problems associated with machining are due to a combination of characteristics, including toughness and stringiness, abrasiveness, galling, work hardening, pyrophoricity, low modulus, high density, reactivity with coolants, reactivity with tools and grinding wheels, and toxicity. Some of these characteristics vary with alloy composition and heat treatment.

Health and safety considerations override all other problems because of uranium's high toxicity and pyrophoricity (see sections on health hazards and pyrophoricity, below). Uranium is heavy, and therefore it does not become airborne unless finely divided. Almost all machining of uranium results in some sparking or burning. Fine chips or finely divided oxides can become airborne, and each machine should be enclosed and heavily ventilated.

It is also important to prevent any metal that falls on the shop floor from being tracked throughout the plant. In general, protective footgear and clothing are used in the machining area.

Machining of uranium requires equipment having extra rigidity and ruggedness. Machine tools for uranium should have 50 to 100% greater capacity than tools for machining similar parts from steel.

Fig. 3 Aging behavior of U-0.75Ti (Ref 3)

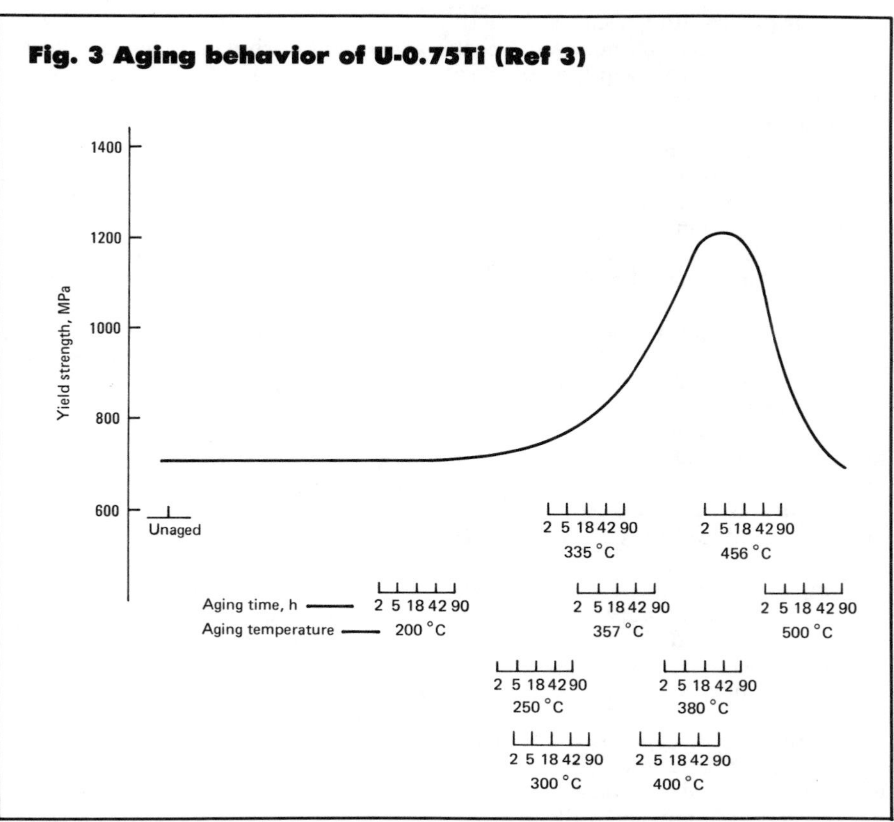

Fig. 4 Aging behavior of U-2Mo (Ref 3)

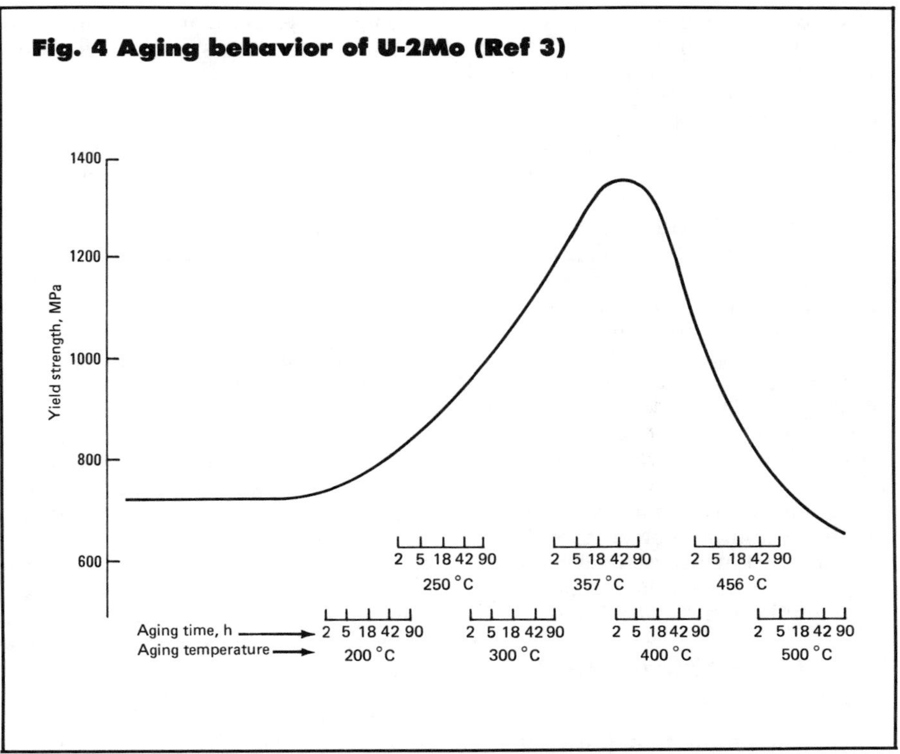

Turning. Uranium is turned on both conventional and numerically controlled heavy-duty equipment. In large-scale production, surface speeds up to 1.5 m/s are used for roughing, and speeds up to 3.0 m/s are used in finishing. A C-2 general-purpose grade of carbide performs satisfactorily for most turning operations; coated carbides normally used for machining steel perform well for most finishing operations.

Cutting tools having a positive rake angle are more free-cutting than negative-rake tools and put the least pressure on the workpiece. Accumulation of large quantities of chips should be avoided to diminish the potential for a pyrophoric reaction to occur.

Drilling and Tapping. Oil-hole drills, in combination with heavy flow of soluble-oil coolant, are favored for drilling uranium. Drills made of tool steels with high cobalt contents perform better than drills made of standard high speed tool steels. Drills must be kept sharp, and positive feeds must be used.

Uranium is difficult to tap. Body drills larger than those recommended for other materials are used to facilitate tapping. A tap generally can be used only once before it is reground. It is difficult to achieve thread depths greater than 50%.

Grinding. A series of tests was performed to determine optimum conditions for centerless plunge grinding of U-0.75Ti at hardness levels of 42 to 46 HRC (Ref 10). Best performance (best grinding ratio) was obtained with an A-80K-12 wheel (aluminum oxide with a vitrified binder), a 20% solution of soluble oil with chlorine and sulfur additions, an infeed rate of 0.27 mm/s, a grinding-wheel surface speed of 29.8 m/s and a regulating-wheel speed of 33.5 mm/s. Optimum operational settings will be slightly different for other uranium alloys and other types of grinding. Wheel wear, although greater than for most other metals, is considered acceptable for production grinding.

The most important consideration in grinding uranium is disposal of the fine grinding dust, which will react with the coolant and thus should not be allowed to accumulate in the machine.

Special Problems and Precautions

Depleted uranium requires special precautions during fabrication and sometimes during use. Ownership, production and use of depleted uranium are subject to state and federal regulations. These regulations are concerned mainly with three properties of the metal: radioactivity, toxicity and pyrophoricity. These problems can be handled routinely and have not constituted serious barriers to manufacture and use of depleted uranium products for commercial applications.

Health Hazards. Depleted uranium is only mildly radioactive (specific activity of 3.6×10^{-7} Ci/g vs 6.77×10^{-7} Ci/g for natural uranium) and is listed with natural uranium and thorium as a "low specific activity" (LSA) material in shipping regulations. Like lead and like metals with atomic numbers higher than lead, depleted uranium is a heavy-metal poison that can be lethal if a sufficient amount of dust or fumes is ingested.

The main hazard to health occurs in those fabrication steps where finely divided particles (dust or oxides) can become airborne. In operations such as melting and casting, machining, grinding, pickling, and heating without using a protective atmosphere or vacuum, it is essential to provide extensive ventilation and to monitor workers' breathing zones. Vents and fume hoods that protect workers are exhausted through carefully monitored filter systems. Workers must change footwear and clothing when leaving areas where finely divided uranium is present.

Users of depleted uranium objects generally do not have to be concerned with the health hazards presented by the metal. Solid pieces of depleted uranium are not sufficiently radioactive to be hazardous; neither do they present the kind of toxic hazard associated with finely divided dust or fumes.

Pyrophoricity. Large pieces of uranium will oxidize rapidly and will sustain slow combustion when heated in air to temperatures about 500 °C. The metal becomes truly pyrophoric only when finely divided. Because pyrophoric reactions take place at the surface of the metal, surface condition and the amount of exposed surface area are critical. Solid metal, particularly with a smoothly machined surface, reacts slowly; within several days the silvery as-machined surface turns to a tea color, and within a month turns black. Machine turnings, particularly fine turnings having literally hundreds of square metres of surface area per kilogram, may react sufficiently to generate enough heat to cause ignition if they are not kept cool under water. Grinding sludge, with still larger surface area, may react even under copious quantities of water.

Finely divided scrap is kept inert by storing it under water or mineral oil. Scrap prepared for shipment to disposal sites may be mixed with an inert insulating material such as sand, or may be mixed into concrete to ensure that no reaction occurs during transport.

Fires are extinguished by cooling the uranium and by restricting access of oxygen to the uranium by covering it with graphite powder or with a dry powdered chemical extinguisher. Water should never be used on uranium fires, because water reacts with the hot metal and generates hydrogen, which adds to the combustion.

Corrosion. The reactivity of uranium promotes corrosion, especially in severe environments. Figure 5 shows corrosion rates of unalloyed uranium and two uranium alloys in high-humidity, high-temperature air (Ref 11). Under such severe conditions, unalloyed depleted uranium corrodes rapidly, U-2Mo corrodes at about one-half the rate of unalloyed uranium and U-0.75Ti corrodes at a much slower rate.

Under normal storage conditions, uranium and uranium alloys have shelf lives of many years. Uranium and its alloys are shiny as machined, but in the presence of oxygen will acquire a dark oxide film in a few hours or a few days. This film serves as a protective coating.

Corrosion protection for most unalloyed uranium objects (such as radiation shields) is obtained by painting with epoxy paints or by plating. A typical plating system consists of a copper flash, followed by nickel and finally cadmium.

Because uranium alloys corrode more slowly than unalloyed uranium, they generally do not require painting or plating. For example, U-0.75Ti and U-2Mo kinetic penetrators employed by the three military services are used without protective coatings, yet have projected storage lives of more than ten years and have passed all military atmospheric exposure tests.

Considerable data are available on corrosion of uranium and its alloys in water, steam and liquid metals (Ref 12).

Scrap Disposal and Transportation. Depleted uranium scrap is buried

Fig. 5 Average corrosion rates of unalloyed uranium and uranium alloys in hot, humid air

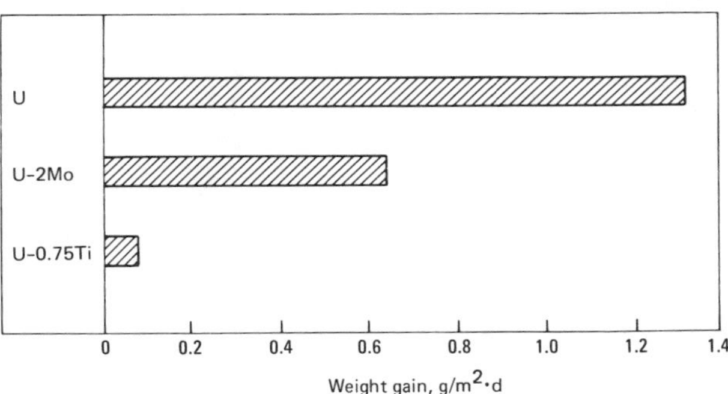

Specimens were tested at 74 °C and 75% relative humidity. Uranium alloys were tested in the solution treated and aged condition.

at designated licensed disposal sites in various parts of the United States. Methods of packaging scrap for transportation to disposal sites are determined mainly by government regulations. A typical method of packaging consists of placing the scrap in a 30-gallon drum and placing this drum inside a 55-gallon drum. Large pieces of scrap are simply placed within the inner drum. Chips are mixed with vermiculite or sand and placed in the inner drum. Very fine particulate material, especially grinding sludge, is mixed into concrete and placed in the drum in the form of concrete blocks.

Shipment of low specific activity (LSA) materials in interstate commerce is regulated by Title 49 of the Department of Transportation, which prescribes labeling and packaging. For depleted uranium the main consideration is that boxes and other containers be able to withstand a prescribed amount of mechanical shock and exposure to fire without releasing the uranium.

Licensing. Possession of more than 6.8 kg (15 lb) of depleted uranium in any form requires a license from the U. S. Nuclear Regulatory Commission. Title 10, Part 40, of Federal Regulations describes the requirements and the necessary steps for obtaining such a license. In addition, other local, state and federal regulations may apply to possession and use of uranium objects. Licenses are granted only to those who can satisfy requirements for technical competency, including the ability to control exposure of operating personnel, and to keep the concentration of particulate uranium in air and liquid effluents below statutory limits.

Certain users are excused from licensing requirements. If a DU product is used solely because of the high density of uranium, the user is excused from licensing requirements, and need only register with the U. S. Nuclear Regulatory Commission and dispose of the DU by returning it to a licensed recipient. (The user is *not* excused if he performs any metallurgical processing or mechanical working of the uranium.) If the DU is a counterweight or balance weight in an airplane, helicopter or missile, the user is totally excused from regulatory control. Users of DU-shielded shipping or storage containers or of DU-shielded equipment for cancer therapy or industrial radiography also are totally excused from regulatory control.

REFERENCES

1. "Trends in the Use of Depleted Uranium", Report NMAB-275, National Materials Advisory Board, National Academy of Sciences, Washington, June 1971
2. E. G. Blasch *et al,* The Use of Uranium as a Shielding Material, *Nuclear Engineering and Design,* Vol 13, 1970, p 146-182
3. K. H. Eckelmeyer, Aging Phenomena in Dilute Uranium Alloys, *Physical Metallurgy of Uranium Alloys,* Proceedings of the Third Army Materials Technology Conference, Vail CO, 1974, 2nd Ed., edited by J. J. Burke, D. A. Colling, A. E. Gorum and J. Greenspan, Metals and Ceramics Information Center, Columbus OH and Brooke Hill Publishing Co., Chestnut Hill MA, 1976, p 463-509
4. J. L. Cadden *et al,* Melting and Casting of Uranium Alloys, *Physical Metallurgy of Uranium Alloys,* Proceedings of the Third Army Materials Technology Conference, Vail CO, 1974, 2nd Ed., edited by J. J. Burke, D. A. Colling, A. E. Gorum and J. Greenspan, Metals and Ceramics Information Center, Columbus OH and Brooke Hill Publishing Co., Chestnut Hill MA, 1976, p 3-81
5. R. F. Huber, P. Loewenstein and N. E. Weare, "Production Process for Low Cost Uranium Alloy Penetrators", Nuclear Metals, Inc., Concord MA, Air Force Materials Laboratory Report AFML-TR-77-1, February 1977
6. E. J. Tenerini, "Feasibility of Investment Casting Preforms for Phalanx Penetrators", Navy Report N60921-77-C-0097, Nuclear Metals, Inc., Concord MA, November 1977
7. P. Loewenstein *et al,* "Fabrication of Core Materials", Chapter 11 in *Nuclear Reactor Fuel Elements, Metallurgy and Fabrication,* edited by A. R. Kaufmann, Interscience Publishers, New York, 1962, p 363-426
8. N. E. Weare, Extrusion of DU Penetrator Alloys Using the Canned Billet Technique, *Proceedings of the High Density Alloy Penetrator Materials Conference,* SP 77-3, Army Materials and Mechanics Research Center, Watertown MA, April 1977, p 195-202
9. R. A. Schell, p 111-116 in *Proceedings of the High Density Alloy Penetrator Materials Conference,* SP 77-3, Army Materials and Mechanics Research Center, Watertown MA, April 1977
10. E. J. Tenerini, "Grinding Optimization", Final Report, DU Alloy Penetrator Casting Manufacturing Technology Program, Nuclear Metals, Inc., December 1977

11. D. J. Sandstrom, A Review of the Early AP Penetrator Work at LASL which led to the Selection of U-3/4 Ti Alloy, *Proceedings of the High Density Alloy Penetrator Materials Conference,* SP 77-3, Army Materials and Mechanics Research Center, Watertown MA, April 1977, p 423
12. Session V, "Corrosion and Its Control", *Physical Metallurgy of Uranium Alloys,* Proceedings of the Third Army Materials Technology Conference, Vail CO, 1974, 2nd Ed., edited by J. J. Burke, D. A. Colling, A. E. Gorum and J. Greenspan, Metals and Ceramics Information Center, Columbus OH and Brooke Hill Publishing Co., Chestnut Hill MA, 1976, p 773-1002

Introduction to Zirconium and Its Alloys

By John H. Schemel
Sandvik Special Metals Corp.

ZIRCONIUM and most of its alloys exhibit strong anisotropy because of two characteristics of the metal—zirconium has a close-packed hexagonal crystal structure at room temperature and it undergoes allotropic transformation to a body-centered cubic structure at about 870 °C (1600 °F). The strong anisotropy profoundly influences the engineering properties of zirconium and its alloys and must be taken into account when selecting and processing a zirconium metal. The most common alloys are rather dilute alpha alloys whose characteristics are generally similar to those of unalloyed zirconium.

The Allotropic Transformation

In zirconium, the low-temperature alpha phase has a close-packed hexagonal crystal structure. This phase transforms to a body-centered cubic structure at about 870 °C (1600 °F). Small amounts of impurities, particularly oxygen, strongly affect the transformation temperature, as shown in Table 1.

The transformation on cooling generally results in a Widmanstätten structure of alpha zirconium; beta phase cannot be retained even by rapid quenching. The platelets of the Widmanstätten structure are finer, the more rapid the cooling rate. Other elements affect phase stability as follows:

Alpha-stabilizing elements raise the temperature of the allotropic alpha-to-beta transformation. These elements include Al, Sb, Sn, Be, Pb, Hf, N, O and Cd. Phase diagrams for many of the binary alloy systems formed between these various elements and zirconium exhibit a peritectic or a peritectoid reaction at the Zr-rich end.

Beta-stabilizing elements lower the alpha-to-beta transformation temperature. Typical beta stabilizers include Fe, Cr, Ni, Mo, Cu, Nb, Ta, V, Th, U, W, Ti, Mn, Co and Ag. For binary alloy systems between zirconium and these elements, there usually is a eutectoid, and often a eutectic reaction as well, at the Zr-rich end of the phase diagram.

Low-solubility intermetallic compound formers carbon, silicon and phosphorus have very low solubility in zirconium, even at temperatures in excess of 1000 °C (1830 °F). They readily form stable intermetallic compounds and are relatively insensitive to heat treatment.

Most alloying elements and impurities are soluble in beta zirconium but relatively insoluble in alpha zirconium, where they exist primarily as interme-

Table 1 Variation in allotropic transformation temperature with oxygen content of unalloyed zirconium

Temperature °C	Temperature °F	Phases present at oxygen content of: 1640 ppm	1370 ppm	970 ppm
954	1750	β	β	β
932	1710	α + β	β	β
927	1700	α + β	β	β
921	1690	α + β	β	β
915	1680	α	α + β	β
910	1670	α	α + β	β
904	1660	α	α + β	β
893	1640	α	α + β	α + β
888	1630	α	α	α + β
885	1625	α	α	α + β
865	1590	α	α	α
857	1575	α	α	α

Fig. 1 Typical pole figures for showing anisotropy in zirconium

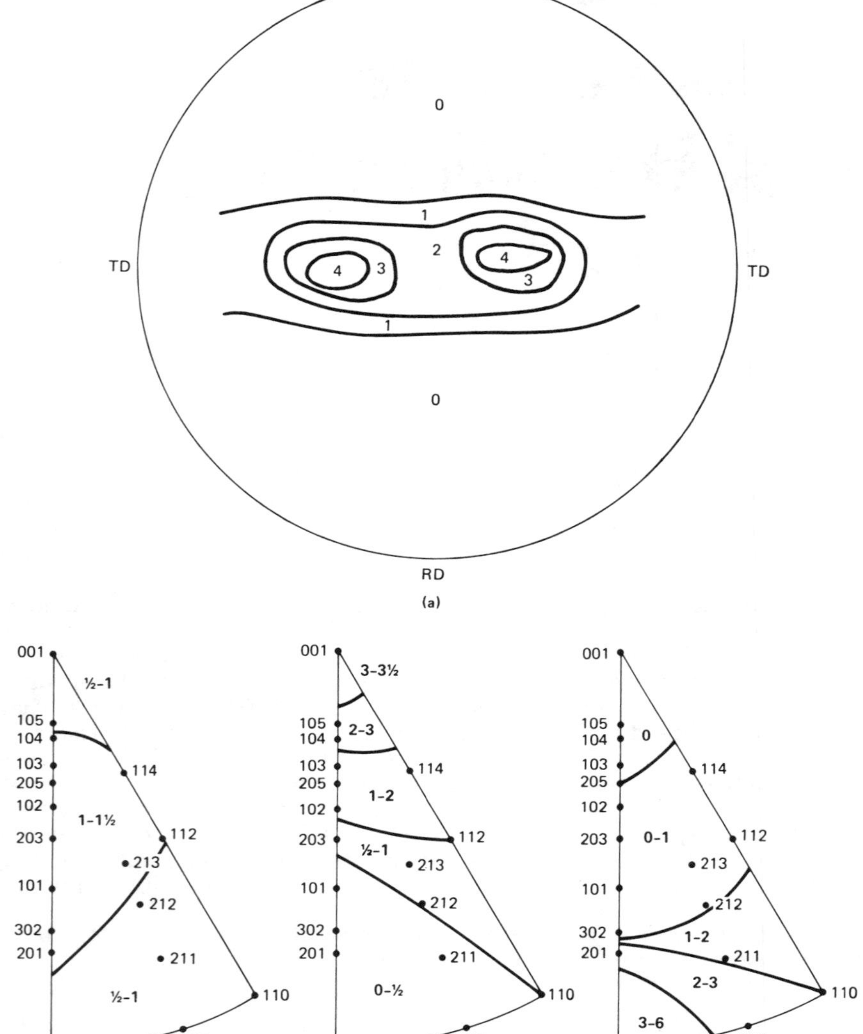

(a) Stereographic basal pole figure for hot rolled Zircaloy-2 plate. Numbers indicate relative density of basal poles in multiples of random occurrence. (b) Inverse pole figure for a typical Zircaloy-2 tubing sample.

tallic compounds. Size and distribution of these secondary phases are largely governed by reactions that take place during the last transformation from beta to alpha and by subsequent mechanical working at lower temperatures.

Heating at temperatures near the alpha-beta transus, or in the alpha-plus-beta region, causes migration of many impurities to grain boundaries, which impairs ductility and corrosion resistance, particularly in zirconium alloys.

Cold Work and Recrystallization

The degree to which unalloyed zirconium may be cold worked depends on both metal purity and method of reduction. Zirconium work hardens rapidly, reaching maximum hardness and strength after cold reduction of only about 20%. But during cold rolling, reductions of about 50% are common and reductions of 80% can be accomplished in some instances. Maximum reductions of 80% are obtained by starting with very soft metal and using machines that feature multiaxial loading (such as cold Pilger machines or Sendzimir rolling mills). Reductions during cold drawing are generally about 15 to 30%. Initially, deformation results in twinning, which reorients the lattice for slip—the primary mechanism for cold working.

Recrystallization is a function of amount of cold work, temperature and time, with time playing a relatively small role. In heavily cold worked material, recrystallization commences at about 510 °C (950 °F), and process annealing of such material usually is conducted at 620 to 790 °C (1150 to 1450 °F). Recrystallization will occur in times as short as 15 min, but much longer times normally are used to ensure that the entire furnace load reaches temperature. Grain growth is nearly nonexistent at the usual annealing temperatures; times of 100 h or more are required to effect grain growth of 2 to 3 ASTM sizes.

Large grains can be grown by annealing after cold reduction of about 5 to 8%. The most common source of large grains is reannealing after a straightening or forming operation that imparts only a small amount of cold work.

Anisotropy and Preferred Orientation

The relationships between preferred orientation of zirconium crystals, the working practice that caused it and the properties that result from it are complex and have been studied in great detail. In most engineering applications, it is important to understand that wrought forms of zirconium and its alloys have different tensile properties in the rolling direction (or longitudinal direction) than they have in the transverse direction. Yield strength is higher in the transverse direction; tensile strength is slightly higher in the rolling direction.

When crystallographic texture of zirconium is discussed, it is usually presented in the form of a stereographic basal pole figure of the type shown in Fig. 1(a) for hot rolled plate, or as an inverse pole figure of the type shown in Fig. 1(b) for tubing.

The Role of Oxygen

Originally, oxygen was considered a troublesome impurity in zirconium, and considerable effort was devoted to its elimination. But when oxygen levels were finally reduced below 1000 ppm, it was found that required strength levels in Zircaloy could no longer be met. The status of oxygen then changed to one of a controlled solid-solution alloying agent. Early methods for determining oxygen content were crude and relatively imprecise, so hardness (which is roughly related to oxygen content but much easier to measure) became the controlled attribute and is still widely used to express the purity or grade of both unalloyed zirconium and zirconium alloys.

The oxygen content of Kroll process sponge varies from about 500 to 2000 ppm, depending on the number of purification steps and the effectiveness of each step. A hardness of 125 HB indicates "soft" sponge with an oxygen content of about 800 ppm, whereas 165 HB is considered "hard" and indicates an oxygen content of about 1600 ppm. Crystal bar zirconium generally has hardness below 100 HB and contains less than 100 ppm oxygen.

Oxygen is a potent strengthener at room temperature but much of its effectiveness is lost at elevated temperature.

Zirconium Alloys

The most common alloys, Zircaloy-2 and Zircaloy-4, contain the strong alpha stabilizers tin and oxygen, plus the beta stabilizers iron, chromium and nickel. There is an extensive alpha-plus-beta field from about 790 to 1010 °C (1450 to 1850 °F). Most alloying elements form intermetallic compounds, and distribution of these compound phases is critical to corrosion resistance in steam and hot water. Generally these alloys are forged in the beta region, then solution treated at about 1065 °C (1950 °F) and water quenched. Subsequent hot working and heat treating is done in the alpha region (below 790 °C) to preserve the fine, uniform distribution of intermetallic compounds that results from solution treating and quenching.

Except for being somewhat stronger and less ductile than unalloyed grades, the Zircaloys are quite similar to unalloyed zirconium in all aspects of metallurgical behavior.

Zr-2.5Nb is the only other alloy of zirconium that has significant commercial importance. In zirconium, niobium is a mild beta stabilizer, and induces a eutectoid reaction when the niobium content exceeds about 1%. (The eutectoid point occurs at 20% Nb.) The mechanical and physical properties of Zr-2.5Nb are very similar to those of the Zircaloys; its corrosion resistance is slightly inferior to that of the Zircaloys.

Zirconium ores generally contain a substantial amount of zirconium's sister element, hafnium. Consequently, zirconium metal produced directly from sponge may contain several percent hafnium. Hafnium has chemical and metallurgical properties similar to those of zirconium, but its nuclear properties are markedly different: hafnium is a neutron absorber whereas zirconium is not. As a result, there are nuclear and non-nuclear grades of zirconium and zirconium alloys, the nuclear grades being essentially hafnium free and the non-nuclear grades containing as much as 4.5% hafnium. Properly speaking, Zircaloy-2, Zircaloy-4 and Zr-2.5Nb are alloy names that apply only to nuclear-grade materials. ASTM specifications for non-nuclear grades list R60704 as the alloy corresponding approximately to Zircaloy-2 and R60705 as the alloy corresponding approximately to Zr-2.5Nb.

Corrosion Resistance of Zirconium and Its Alloys

UNALLOYED ZIRCONIUM is generally more resistant to corrosion than stainless steels, but less resistant than tantalum, in many chemical mediums. Zirconium resists most inorganic and organic acids and is totally resistant to alkalis. It is attacked, however, by fluoride ions, wet chlorine, aqua regia, concentrated sulfuric acid, and ferric or cupric chlorides. Zirconium and its alloys derive their corrosion resistance from a regenerating, adherent oxide film that forms in most mediums.

The alloys of zirconium were developed specifically for use in nuclear reactors because they have good mechanical strength and ductility at reactor service temperatures; low neutron absorption; sufficient corrosion resistance to withstand pressurized water and steam; and adequate heat conductivity. As an added benefit, these properties are relatively stable even after extensive irradiation in a reactor core.

Stress-Corrosion Cracking. Zirconium and its alloys are generally resistant to stress-corrosion cracking in seawater and in most aqueous chemical mediums. Stress-corrosion cracking has been reported, however, in (*a*) concentrated methanol, (*b*) solutions containing heavy-metal chlorides, (*c*) organic solutions with small quantities of chlorides and (*d*) gaseous iodine or fused salts containing iodine. The related phenomenon of liquid metal embrittlement has been reported for zirconium in contact with molten cesium and with liquid or sodium cadmium.

Pitting. The regenerating oxide film on zirconium usually keeps pitting to a minimum. Still, severe pitting has been observed in hydrochloric acid solutions containing ferric or cupric ions.

Crevice Corrosion. Zirconium is highly resistant to accelerated corrosion under gaskets and fasteners or in corners and overlaps.

Oxidation. Zirconium forms a visible oxide film at about 200 °C (about 400 °F) and begins to exhibit a loose white scale with prolonged exposure over 425 °C (800 °F). Oxidation rates at several temperatures from 425 to 1200 °C (800 to 2190 °F) are given in Table 1. When heated in air, zirconium reacts primarily with oxygen rather than with nitrogen.

Corrosion in Chemical Solutions

Resistance to attack by both strong acids and strong alkalis is one of the most useful attributes of zirconium. In this regard, it surpasses even tantalum.

Table 2 presents corrosion-test data for unalloyed zirconium. This table should be used only as a guide, because impurities in the metal or in the corroding medium can have pronounced effects on the corrosion rate.

Table 1 Corrosion rates of zirconium in air, oxygen and nitrogen

Temperature		Weight gain, $g/m^2 \cdot h$			
°C	°F	Air	Dry air	Oxygen	Nitrogen
425	800	0.76	...	...	...
500	930	3.6	3.5	1.9	...
600	1110	9.5	6.3	3.8	...
700	1290	16.9	16.0	7.5	0.47
800	1470	...	...	...	0.72
900	1650	...	...	...	1.5
1000	1830	...	...	...	2.7
1100	2010	...	...	...	6.5
1200	2190	...	...	...	10.2

Inorganic Acids. Unalloyed zirconium has excellent resistance to sulfuric acid in concentrations up to 80% at room temperature, and up to 60% at the boiling point. The transition from low to high corrosion rate occurs abruptly over a span of acid concentration of only a few percent. Weld zones and heat-affected zones are attacked at lower acid concentrations than the recrystallized base metal. Figure 1 shows the region of resistant behavior for both weld metal and base metal in sulfuric acid, and also indicates the transition from resistant to nonresistant behavior. When it occurs, attack is rapid and intergranular and creates a highly pyrophoric surface layer that is easily ignited.

Corrosion rates of zirconium in boiling hydrochloric acid solutions are less than 25 μm/yr (1 mil/yr) at acid concentrations up to 20%, and less than 125 μm/yr (5 mils/yr) at concentrations up to 37%. Accelerated attack of weld heat-affected zones has been observed in boiling HC1 solutions at concentrations greater than 20% and in HC1 solutions at all concentrations at temperatures above the normal boiling point. Such attack can be minimized if the zirconium contains less than 500 ppm Fe + Cr. Severe pitting attack occurs in hydrochloric acid solutions containing ferric or cupric ions.

Zirconium is resistant to nitric acid solutions in all concentrations at all temperatures up to 200 °C (400 °F). It is generally resistant to red or white fuming nitric acid, but a strong pyrophoric reaction occurs with dry red fuming nitric acid (see Fig. 2).

The corrosion rate of zirconium in phosphoric acid increases with temperature, especially at higher acid concentrations (see Fig. 3).

Zirconium is attacked by hydrofluoric acid or fluoride ions even at concentrations as low as 0.001%.

Caustic Solutions. Zirconium is resistant to potassium hydroxide (KOH) solutions, and to fused KOH up to the temperature at which failure occurs by oxidation rather than by chemical attack. Zirconium crucibles are widely used for analytical laboratory work in which caustic fusion is followed by leaching with strong acids.

Organic Chemicals. Generally, the resistance of zirconium to organic compounds is good. Unalloyed zirconium is used extensively in equipment for synthesis of urea, for azo dye coupling, for production of alcohol and for

Table 2 Corrosion resistance of unalloyed zirconium

Chemical substance(a)	Concentration, wt %	Temperature	Typical corrosion rate μm/yr	mils/yr
Acetaldehyde	100	Boiling	<50	<2
Acetic acid				
Solution	<99.5	Boiling	<25	<1
Vapor	33	Boiling	<125	<5
Glacial acid	99.7	Boiling	<125	<5
Anhydride	99	Room to boiling	<50	<2
Aluminum chloride	5, 10	60 °C (140 °F)	<50	<2
	25	Room	<50	<2
	40	100 °C (212 °F)	<50	<2
Aluminum sulfate	60	100 °C (212 °F)	<50	<2
Ammonia (aqueous solution)	...	38 °C (100 °F)	<125	<5
Ammonium chloride	1, 40	100 °C (212 °F)	<125	<5
Ammonium hydroxide	28	27 °C (80 °F)	<125	<5
Ammonium sulfate	5, 10	100 °C (212 °F)	<125	<5
Aniline hydrochloride	5, 20	100 °C (212 °F)	<125	<5
Aqua regia(b)	...	77 °C (170 °F)	>1250	>50
Barium chloride	5, 20	100 °C (212 °F)	<125	<5
	25	Boiling	125-1250	5-50
Bromine water	...	Room	>1250	>50
Calcium chloride	5, 10, 20	100 °C (212 °F)	<125	<5
	75	Boiling	<125	<5
Calcium hypochlorite	2, 6, 20	100 °C (212 °F)	<125	<5
	Saturated	Room	125-1250	5-50
Carbolic acid	Saturated	100 °C (212 °F)	<125	<5
Carbon tetrachloride (liquid)	...	Boiling	<125	<5
Chlorine(c)	...	(d)	>1250	>50
Gas(e)	100	93 °C (200 °F)	>1250	>50
Gas, dry	100	Room	125	5
Chloracetic acid	30	82 °C (180 °F)	125-1250	5-50
Chromic acid	10 to 50	Boiling	<25	<1
Citric acid	10, 25, 50	100 °C (212 °F)	<25	<1
Cupric chloride	20, 40, 50	Boiling	>1250	>50
Cupric cyanide	Saturated	Room	>1250	>50
Dichloroacetic acid	100	(f)	125-1250	5-50
Ethylene dichloride	100	Boiling	<125	<5
Ferric chloride	5, 10, 20, 30	(g)	>1250	>50
	5, 10, 20, 30, 40, 50	Boiling	>1250	>50
Fluoboric acid	5-20	Elevated	>1250	>50
Fluorosilicic acid	10	Room	>1250	>50
Formic acid	10, 25, 50, 90	100 °C (212 °F)	<25	<1
Air-free	25	100 °C (212 °F)	<125	<5
Aerated	10-90	100 °C (212 °F)	<125	<5
Hydrobromic acid	40	Room	>1250	>50
Hydrochloric acid	5	Room	<25	<1
	10	35 °C (95 °F)	<125	<5
	20	35 °C (95 °F)	<125	<5
Aerated	5	35 °C (95 °F)	<125	<5
	10	35 °C (95 °F)	<125	<5
	20	35 °C (95 °F)	<125	<5
Hydrofluoric acid	48	Room	>1250	>50
Hydrogen peroxide	50	100 °C (212 °F)	<50	<2
Hydroxyacetic acid	...	40 °C (104 °F)	<125	<5
Lactic acid	10-100	149 °C (300 °F)	<25	<1
Magnesium chloride	5-40	100 °C (212 °F)	<125	<5
Manganous chloride	5-20	100 °C (212 °F)	<50	<2
Mercuric chloride	1, 5, 10, 55	100 °C (212 °F)	<25	<1
	Saturated	(h)	<25	<1
Nickel chloride	5-20	100 °C (212 °F)	<25	<1
Nitric acid	10, 20, 40, 69, 75	260 °C (500 °F)	<25	<1
	65, 75	Boiling	<25	<1
Oxalic acid	All concentrations through 100%	100 °C (212 °F)	<25	<1
Phenol	Saturated	Room	<125	<5

(continued)

Table 2 (continued)

Chemical substance(a)	Concentration, wt %	Temperature	Typical corrosion rate µm/yr	mil/yr
Phosphoric acid	5-30	Room	<125	<5
	35-50	Room	<125	<5
	85	100 °C (212 °F)	125-500	5-20
	5-35	60 °C (140 °F)	<125	<5
	5-50	100 °C (212 °F)	<125	<5
Potassium chloride	Saturated	Room	<125	<5
Potassium hydroxide	10	Boiling	<25	<1
	25	Boiling	<25	<1
	50	Boiling	<125	<5
Silver nitrate	50	Room	<125	<5
Sodium chloride	29	Boiling	<25	<1
	Saturated	(j)	<25	<1
Sodium hydroxide	10, 25	Boiling	<25	<1
	28	Room	<25	<1
	40	100 °C (212 °F)	<25	<1
Sodium hypochlorite	6	100 °C (212 °F)	<125	<5
Stannic chloride	5	100 °C (212 °F)	<25	<1
	24	Boiling	<25	<1
Sulfuric acid				
Aerated	1-60	100 °C (212 °F)	<125	<5
Air-free	15	Room	<100	<20
Sulfurous acid	6	Room	<25	<5
	Saturated	191 °C (375 °F)	25-250	5-50
Tannic acid	25	100 °C (212 °F)	<25	<1
Tartaric acid	10-50	100 °C (212 °F)	<25	<1
	10, 25, 50	60 °C (140 °F)	<25	<1
Tetrachloroethane	100	Boiling	<125	<5
Trichloroacetic	100	100 °C (212 °F)	>250	>50
	10-40	Room	<50	<5
Trichloroethylene	99	Boiling	<125	<5
Trisodium phosphate	5, 20	100 °C (212 °F)	<125	<5
Zinc chloride	10	Boiling	<125	<5
	20	100 °C (212 °F)	<125	<5

(a) Aqueous solution unless otherwise noted. (b) 3 parts HCl to 1 part HNO_3. (c) Water saturated. (d) At room temperature and at 75 °C (167 °F). (e) Containing more than 0.13% water vapor. (f) At 100 °C (212 °F) and at the boiling point. (g) At room temperature and at 100 °C (212 °F). (h) At room temperature and at 93 °C (200 °F). (j) At room temperature and at the boiling point.

processes employing lactic acid, and in nuclear reactors cooled with organic fluids (see section on nuclear applications, below). Application problems in organic chemicals generally are similar to the problems encountered in organic coolants for nuclear reactors; hydrogen embrittlement and intergranular attack resulting from trace impurities in the organic mediums are the major problems that can arise.

Nuclear Applications

Corrosion of unalloyed zirconium in water and in steam is reported to be very erratic (Ref 2, 3). This erratic behavior probably is caused by variations in the concentrations of impurities. Nitrogen and carbon are particularly detrimental to corrosion resistance; nitrogen in concentrations greater than about 40 ppm or carbon in concentrations greater than about 300 ppm markedly increases corrosion rate (Ref 2, 3). Corrosion data for a large number of heats of high-quality crystal bar zirconium are plotted in Fig. 4. An indication of the erratic nature of zirconium corrosion can be seen in the curves for 316 and 360 °C. The former must be plotted as a band rather than a single line because there is too much scatter in the data. The latter has three bands extending upward from it, each band representing a change in corrosion rate from the basic rate represented by the single line and each band

Fig. 1 Resistance of unalloyed zirconium to corrosion by sulfuric acid

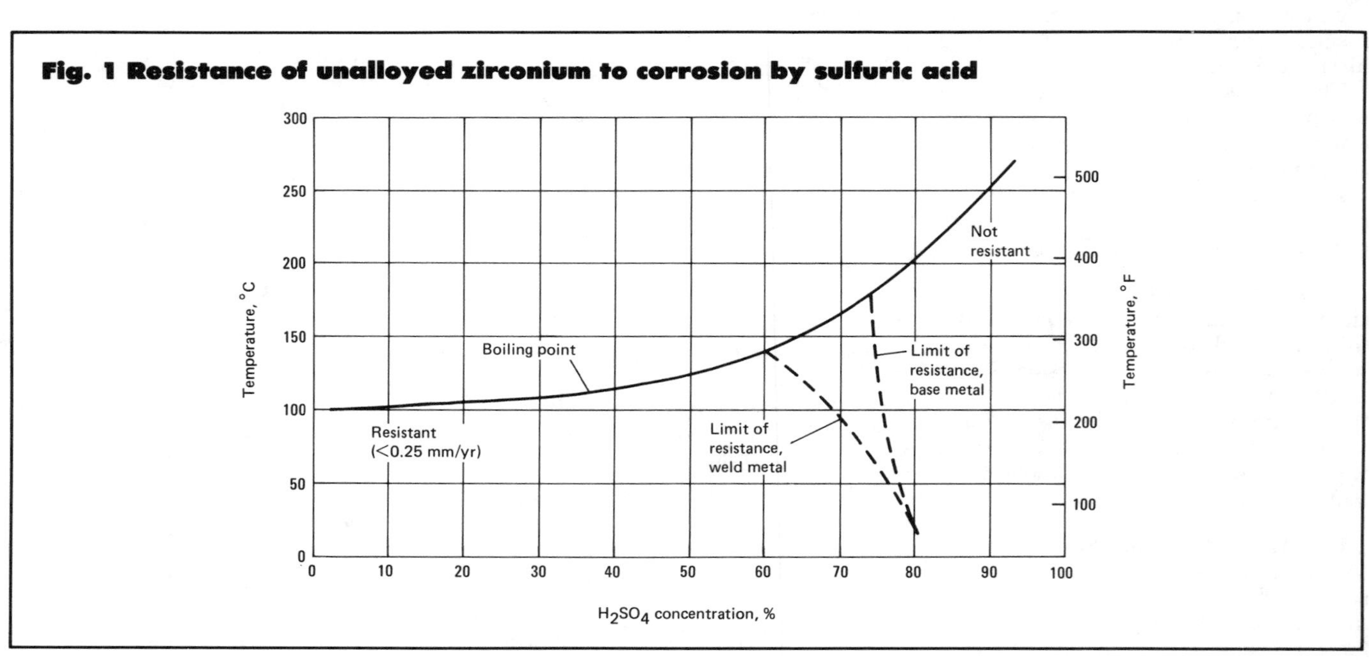

Fig. 2 Effect of nitrogen peroxide and water on the limit for pyrophoric behavior of zirconium in red fuming nitric acid

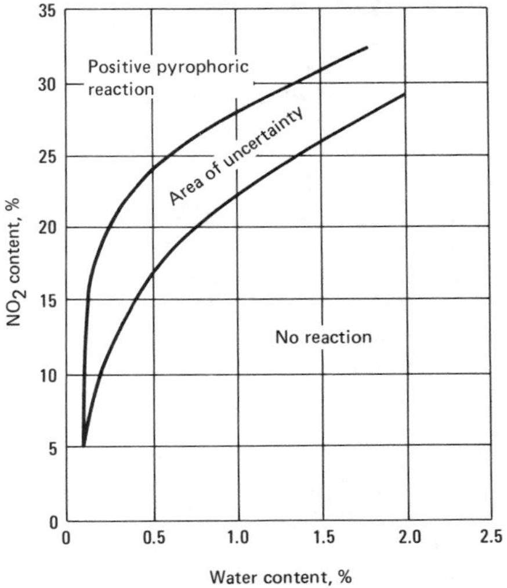

Fig. 3 Corrosion of zirconium in phosphoric acid solutions

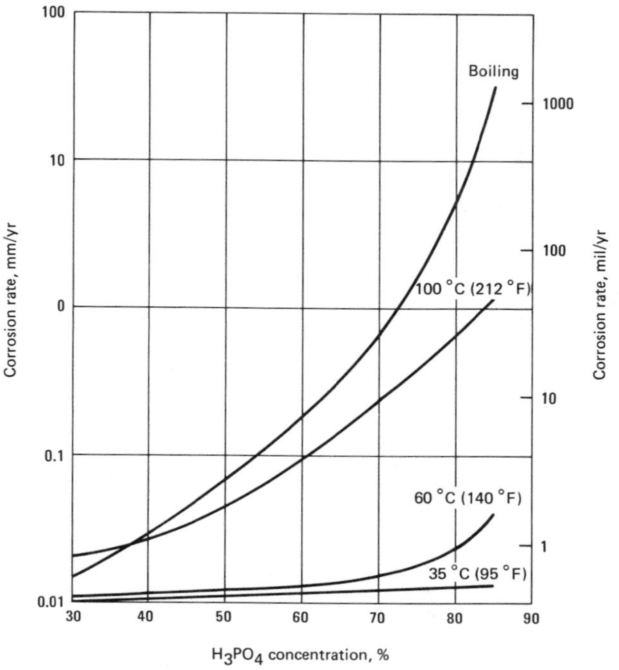

representing data from a different set of corrosion specimens.

The erratic corrosion behavior of "pure" zirconium, caused largely by sensitivity to variations in very low concentrations of impurities, provided the impetus for development of zirconium alloys with more reliable corrosion resistance. Zircaloy-2, Zircaloy-4 and Zr-2.5Nb are the most important zirconium alloys developed for use in water and steam.

Zircaloy-2. The corrosion resistance of Zircaloy-2 in high-temperature water is vastly superior to that of unalloyed zirconium. A tightly adherent oxide film forms on this alloy at a rate that at first is quasicubic but after an initial period undergoes transition to linear behavior. Unlike the oxide film on unalloyed zirconium, the oxide film on Zircaloy-2 remains dark and adherent throughout transition and in the post-transition regime. Corrosion data for Zircaloy-2 in pressurized water at three temperatures and in steam at 400 °C (750 °F) are plotted in Fig. 5.

In a series of tests conducted for 14 days, the aqueous corrosion behavior of Zircaloy-2 was found to be only slightly affected by metallurgical treatment; however, resistance to steam at 400 °C and above was markedly decreased by heating in, or cooling slowly through, the two-phase (alpha-plus-beta) field. Heating Zircaloy-2 for 8 h at or below 750 °C (1380 °F) and furnace cooling had no effect on the 14 day corrosion

Fig. 4 Corrosion of arc-melted crystal bar zirconium in pressurized water and in steam (Ref 3)

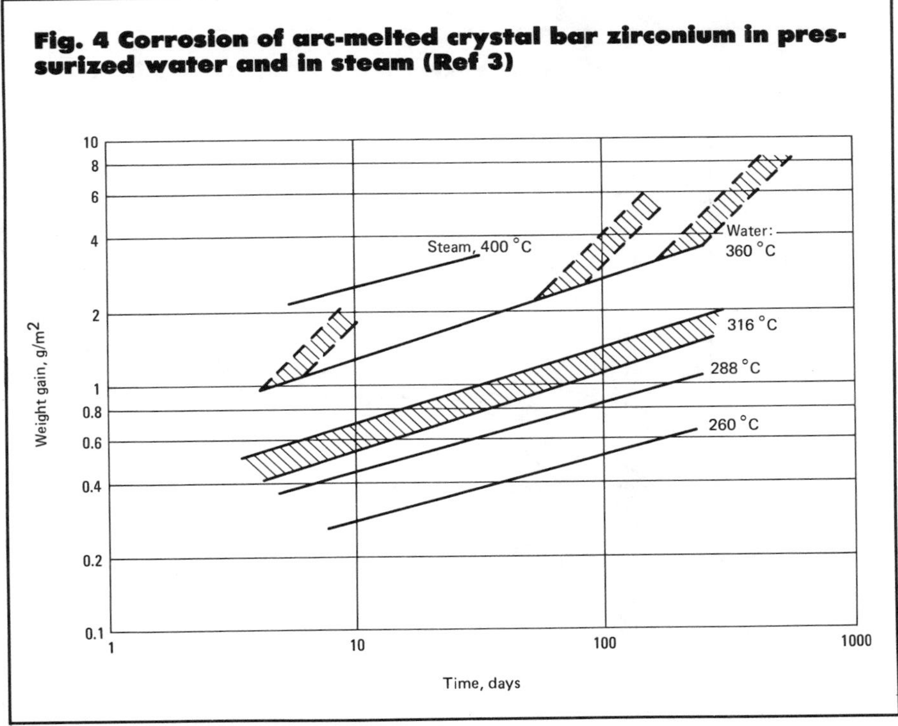

Fig. 5 Corrosion of Zircaloy-2 in pressurized water and in steam (Ref 3)

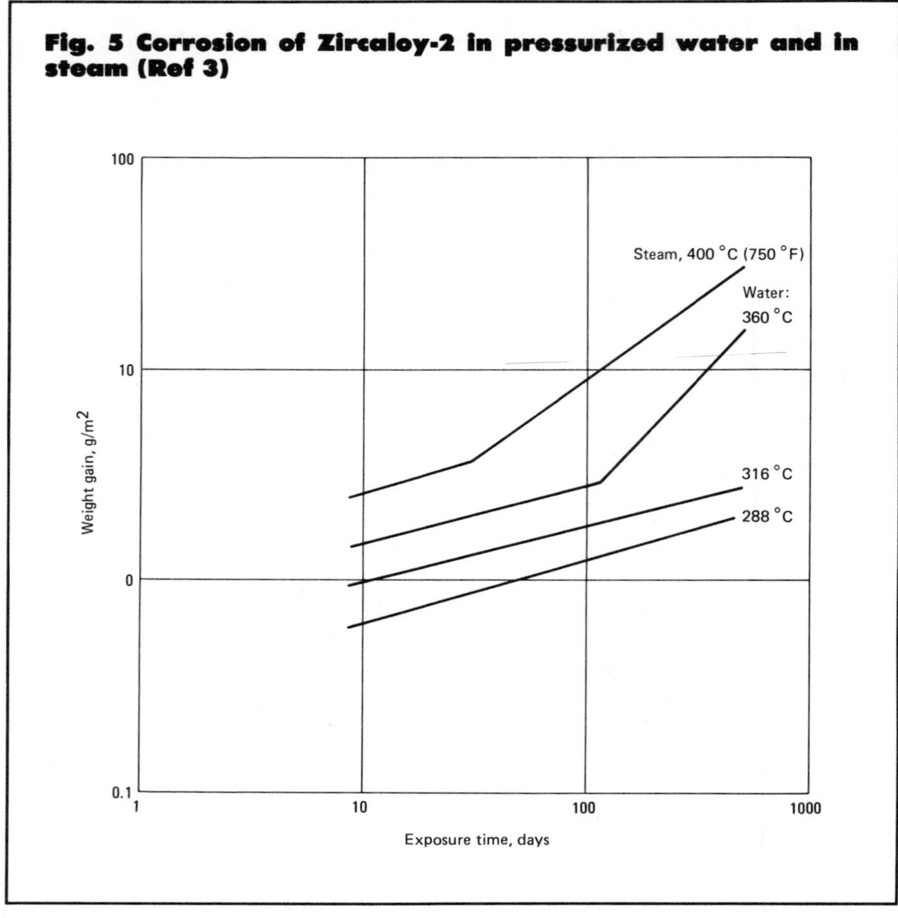

Fig. 6 Hydrogen pickup characteristics of Zircaloy-2 (Ref 2)

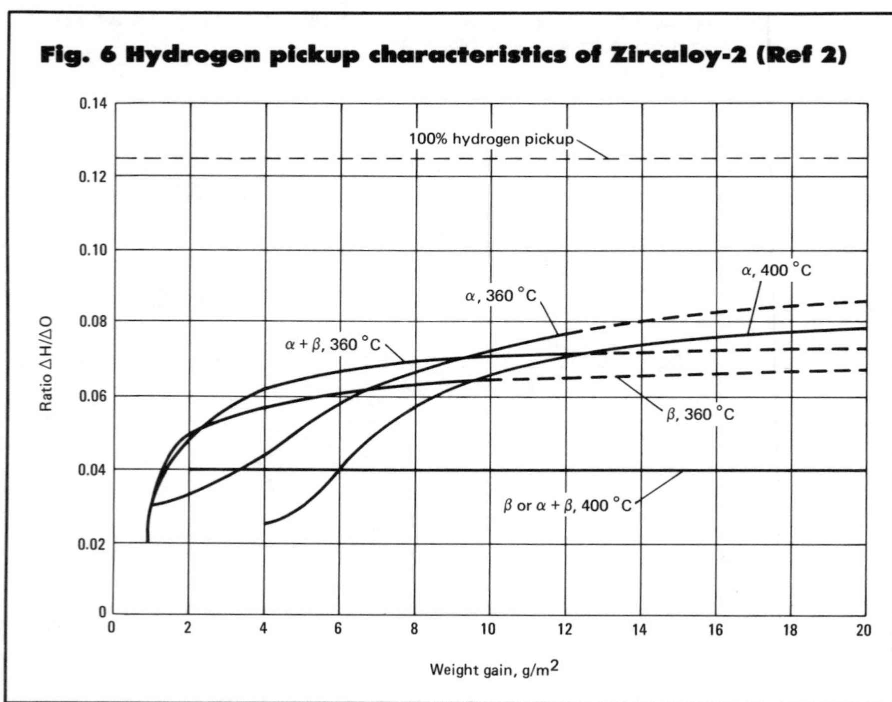

behavior. However, heating 8 h at 850 °C (1560 °F) produced a 20% increase in weight gain on subsequent exposure to steam at 400 °C; heating at 900 °C (1650 °F), a 40% increase; heating at 950 °C (1740 °F), an 80% increase; and heating at 1010 °C (1850 °F), an increase of more than 150%. Similarly, material heated above 1010 °C and cooled through the temperature range 1010 to 800 °C (1850 to 1470 °F) at a rate of less than 50 °C (90 °F) per minute exhibited a subsequent corrosion rate in steam at 400 °C (750 °F) greater than that for as-received or air quenched material. The lower the cooling rate (below 50 °C/min), the greater the corrosion rate. For instance, weight gain after 14 days for material cooled at 50 °C/min was only slightly greater than for metal not exposed to high temperature, whereas weight gain for material cooled at 20 °C (36 °F) per minute was more than twice that of unheated material. At temperatures above 400 °C, the corrosion behavior of Zircaloy-2 was erratic and extremely sensitive to metallurgical structure.

In aqueous environments, the corrosion reaction between water and zirconium releases hydrogen, a portion of which is absorbed into the metal. Hydrogen absorption is important because hydrogen embrittles zirconium and its alloys by forming precipitates of zirconium hydride. Embrittlement is manifested as a drop in tensile ductility, and the degree of embrittlement depends on the amount of hydrogen absorbed, the crystallographic texture of the zirconium and the orientation of hydride platelets in relation to the axis of applied tensile stress. As little as 30 ppm absorbed hydrogen can embrittle zirconium if the hydride platelets are perpendicular to the tensile axis.

The amount of hydrogen absorbed may be expressed as percent of theoretical, which is the actual amount picked up divided by the amount released by the corrosion reaction, or as the ratio $\Delta H/\Delta O$ (for 100% pickup, $\Delta H/\Delta O = 0.125$). The ratio $\Delta H/\Delta O$ can be plotted as a function of weight gain because weight gain is directly related to the amount of hydrogen released by the corrosion reaction. A typical plot for Zircaloy-2 is given in Fig. 6; it shows that hydrogen pickup levels off at about 50 to 70% except for β-annealed material, or material heated in the $\alpha+\beta$ field, subsequently used at 400 °C (750 °F). At this temperature, hydrogen

Fig. 7 Corrosion behavior of Zircaloy-4 in water and steam (Ref 2)

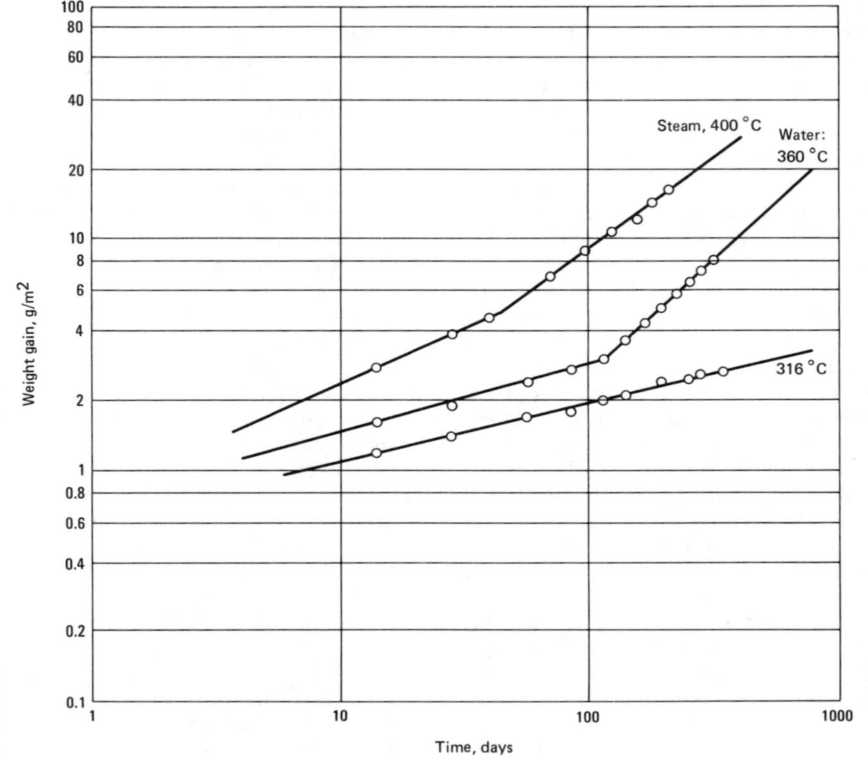

Fig. 8 Hydrogen pickup versus hydrogen overpressure for Zircaloy-2 and Zircaloy-4

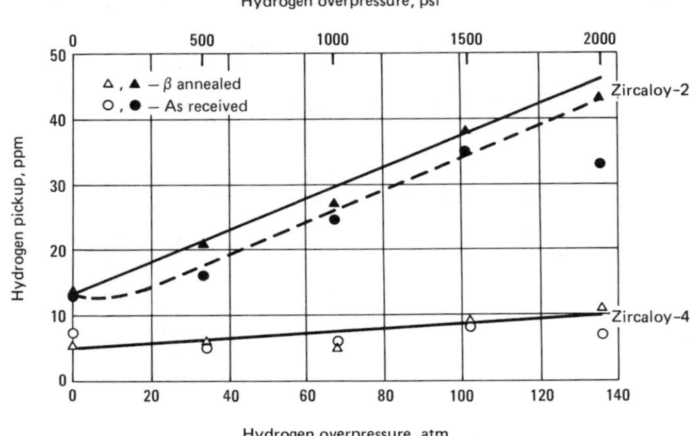

Specimens were immersed for 14 days in water pressurized by admitting compressed hydrogen into the vapor space above the waterline; temperature of the water was 343 °C (Ref 2).

pickup for β or α + β material is stable at about 30%.

Zircaloy-4 differs in composition from Zircaloy-2 only in having no nickel and a slightly greater iron content. Both variations are intended to reduce hydrogen pickup. Corrosion behavior of Zircaloy-4 is very similar to that of Zircaloy-2 (compare Fig. 7 and Fig. 5), but hydrogen pickup for Zircaloy-4 is significantly lower—particularly when the metal is exposed to water at 360 °C (680 °F). At this temperature, hydrogen pickup for Zircaloy-4 is about 25% of theoretical, or less than half that for Zircaloy-2. As an added benefit, hydrogen pickup for Zircaloy-4 is much less sensitive to hydrogen overpressure than that for Zircaloy-2 (see Fig. 8).

For both Zircaloys, hydrogen pickup is markedly decreased when dissolved oxygen is present in the corroding medium. For example, in an experiment in which Zircaloy-4 was placed in water for 200 days at 360 °C (680 °F), hydrogen pickup was 18% at an oxygen overpressure of essentially zero but only 4% at an overpressure of 6.9 MPa (68 atmospheres, or 1000 psi).

Experiments conducted in LiOH solutions at 360 °C indicate that the corrosion resistance of the Zircaloys is independent of hydroxyl-ion (OH^-) concentration up to about 0.002 *M* (pH 11.3), but is significantly reduced at concentrations of 0.005 *M* (pH 11.7) and higher (Ref 2, 4). Hydrogen pickup, when measured as percent of theoretical, is not affected by hydroxyl-ion concentration.

Zr-2.5Nb is considered to be somewhat less resistant to corrosion than the Zircaloys. Nevertheless, Zr-2.5Nb is an acceptable alloy for many applications; a prime example is the use of Zr-2.5Nb pressure tubes in the primary loops of CANDU reactors.

In one corrosion investigation, Zr-2.5Nb strip β annealed for 30 min at 900 °C (1650 °F) and then subjected to various heat treatments was tested in water at various temperatures. The heat treatments studied were:

1. α-β anneal 2 h at 650 °C (1200 °F) and furnace cool
2. Cold work 25 to 50%; α-β anneal 6 h at 700 °C (1290 °F) and air cool
3. Cold work 25 to 50%; α-β anneal 1/2 h at 800 °C (1470 °F) and air cool
4. β anneal 10 to 60 min at 880 to 960 °C (1620 to 1760 °F), quench in

water or oil, then temper 24 h at 500 °C (930 °F)
5 β anneal 10 min at 960 °C and rapid cool
6 β anneal 10 min at 960 °C and water quench

When the corrosion data for Zr-2.5Nb were analyzed and compared with corrosion data for untreated Zircaloy-2, the post-transition corrosion rates paralleled those for Zircaloy-2, except for material β annealed and water quenched (treatment 6 in list above). Corrosion rates for Zr-2.5Nb subjected to treatment 6 were about the same at all temperatures from 316 to 400 °C (600 to 750 °F) and consistently much higher than those for the other treatments. Corrosion rates for the other treatments were two to three times the rates for Zircaloy-2 at all temperatures (Ref 5). The plotted data converge with increasing temperature, which suggests that Zr-2.5Nb would be superior at temperatures above 400 °C. Such a reversal in relative corrosion rates has been observed (Ref 6) in steam at 482 °C (900 °F).

The corrosion resistance of β-annealed and water-quenched Zr-2.5Nb can be improved substantially by aging at 500 °C (930 °F), and can be improved even more by cold working followed by aging. For service in water, the corrosion resistance of Zr-2.5Nb cold worked 10 to 30% and then aged 24 h at 500 °C approaches that of Zircaloy-2 for short exposure times, and exceeds it for long exposure times even at temperatures as low as 316 °C (600 °F) (Ref 7). It is believed that the intermediate cold work increases the effectiveness of the aging treatment by increasing aging kinetics.

The hydriding resistance of Zr-2.5Nb is similar to that of Zircaloy-4 when expressed as percent of theoretical. But the consequences of hydrogen pickup are not quite as serious as for the Zircaloys because Zr-2.5Nb has a higher solubility limit for hydrogen and thus can tolerate higher levels of hydrogen than the Zircaloys before becoming embrittled. In the study discussed above, with one exception, hydrogen pickup ranged from 8 to 22% of theoretical for treatments 1 to 6, with exact values depending on heat treatment and corrosion temperature. The exception was as-quenched material (treatment 6) corroded at 400 °C (750 °F), for which hydrogen pickup was 32% of theoretical.

Corrosion in Organic Coolants. The organic coolants used in reactors are polyphenyls and commonly are circulated at temperatures of 350 to 400 °C (660 to 750 °F). Fuel cladding surfaces in contact with the coolant reach temperatures of 465 °C (870 °F). Corrosion rates of the reactor-grade zirconium alloys in organic coolants are similar to the rates in low-pressure steam. A long-term rate of 27 mg/m²·d has been reported for Zircaloy-4 and Zr-2.5Nb at 400 °C in a coolant containing 80 to 150 ppm water (Ref 8). The use of zirconium alloys in organic coolants is not governed by corrosion per se, but rather by hydriding due to hydrogen pickup from the coolant. Hydriding remains at a minimum when small amounts of moisture are present in the coolant, when the zirconium is preoxidized, and when low concentrations of chlorine and dissolved hydrogen are maintained in the coolant. The data in Table 3 illustrate the effects of these three techniques.

Alloying appears to have a significant effect on hydriding in organic coolants. In tests performed in a terphenyl coolant (HB-40), Zircaloy-4 was more resistant to hydriding than Zircaloy-2, Zr-2.5Nb was more resistant than Zircaloy-4, and the resistance of Ozhennite-0.5 was clearly superior to those of the other three alloys at all temperatures from about 350 to 500 °C (660 to 930 °F). In these tests, the water content of the terphenyl was 80 to 150 ppm, the dissolved hydrogen content 60 to 150 mL/kg and the chlorine content less than 0.5 ppm (Ref 10).

Table 3 Hydriding rates for Zr-base alloys in organic coolant (Ref 9)

Impurity concentration in coolant	Hydriding rate, mg/m²·h — Zircaloy-2 Pickled	Zircaloy-2 Preoxidized	Zr-2.5Nb Pickled	Zr-2.5Nb Preoxidized
Water(a)				
<40 ppm	5	0.32	10	0.12
40-60 ppm	4	0.11	...	0.01
60-1000 ppm	0.45	0.10	0.25	0.01
Dissolved hydrogen(b)				
~1 mL/kg	0.25	0.05	...	...
8 mL/kg	1.4	0.06	...	...
100 mL/kg	70	0.30	...	...
Chlorine(c)				
None added(d)	3.9	0.1	...	...
50 ppm C_2HCl_3 added(e)	94	21	...	...

(a) Hydrogen dissolved in organic coolant, ~100 mL/kg; temperature, 400 °C. (b) Water content of organic coolant, <40 ppm; temperature, 425 °C. (c) Temperature, 370 to 375 °C. (d) Chlorine concentration at 370 °C, 0.8 ppm at start of test and 0.5 ppm after 67 h. (e) Chlorine concentration at 370 °C, 8.2 ppm at start of test and 1.6 ppm after 67 h.

Liquid-Metal Corrosion. Although data are available on corrosion of zirconium in many liquid metals (Ref 11), the liquid metals of primary importance, because of their application in nuclear reactors, are sodium and sodium-potassium alloys—especially the sodium-potassium eutectic. Corrosion of zirconium in sodium or sodium-potassium alloys proceeds by the mechanism of differential solubility, in which zirconium dissolves in the liquid metal at high-temperature regions of the circuit and precipitates out at the lower-temperature regions. (The solubility of zirconium in liquid sodium is 3.6 ppm at 1380 °C or 2515 °F, but only 0.09 ppm at 720 °C or 1330 °F.)

Oxygen dissolved in liquid sodium in amounts greater than about 5 ppm stifles the dissolution reaction. At such oxygen concentrations, an adherent layer of ZrO_2 forms on the zirconium, thus preventing direct contact between the zirconium and the liquid metal and thereby inhibiting dissolution. The oxidation of zirconium in liquid sodium follows a parabolic rate law (Ref 12 and 13), although near-cubic behavior also has been reported (Ref 14). Between the limits of 20 and 200 ppm, oxidation behavior is not significantly affected by variations in oxygen concentration.

There appears to be no advantage of Zircaloy-2 over unalloyed zirconium in liquid sodium.

REFERENCES

1. J. H. Schemel, *Manual on Zirconium and Hafnium,* STP 639,

American Society for Testing and Materials, Philadelphia, 1977

2. S. Kass, The Development of the Zircaloys, *Corrosion of Zirconium Alloys,* STP 368, American Society for Testing and Materials, Philadelphia, 1964
3. D. E. Thomas, Corrosion in Water and Steam, *Metallurgy of Zirconium,* B. Lustman and F. Kerze, Jr. (Ed.), McGraw-Hill, New York, 1955
4. E. Hillner and J. N. Chirigos, "The Effect of Lithium Hydroxide and Related Solutions on the Corrosion Rate of Zircaloy in 680 °F Water", WAPD-TM-307, Westinghouse Atomic Power Division, Aug 1962
5. S. B. Dalgaard, Corrosion and Hydriding Behavior of Some Zr-2.5 wt % Nb Alloys in Water, Steam and Various Gases at High Temperature, *Proceedings of Conference on Corrosion of Reactor Materials (Vol 2),* IAEA, Vienna, 1962
6. H. H. Klepfer, Zirconium-Niobium Binary Alloys for Boiling Water Reactor Service: Part I—Corrosion Resistance, *Journal of Nuclear Materials,* Vol 9, 1963, p 65
7. J. E. LeSurf, The Corrosion Behavior of 2.5 Nb Zirconium Alloy, *Applications-Related Phenomena in Zirconium and Its Alloys,* STP 458, American Society for Testing and Materials, Philadelphia, 1969
8. J. Boulton and M. G. Wright, Ozhennite 0.5—Its Potential and Development, *Applications-Related Phenomena in Zirconium and Its Alloys,* STP 458, American Society for Testing and Materials, Philadelphia, 1969
9. W. M. Campbell *et al,* Development of Organic Liquid Coolants, *UNICPUAE Vol 8,* United Nations, New York, 1964, p 3
10. A. J. Mooradian *et al,* "Current Status of Canadian Organic Cooled Reactor Technology", AECL-2943, U. S. Atomic Energy Commission, Washington DC, Sept 1967
11. R. F. Koenig, Corrosion of Zirconium and Its Alloys in Liquid Metals, *Metallurgy of Zirconium,* B. Lustman and F. Kerze, Jr. (Ed.), McGraw-Hill, New York, 1955
12. R. L. Carter, R. L. Eichelberger and S. Siegel, Recent Developments in the Technology of Sodium-Graphite Materials, *UNICPUAE Vol 7,* United Nations, New York, 1958, p. 72
13. T. L. MacKay, Oxidation of Zirconium and Zirconium Alloys in Liquid Sodium, *Journal of the Electrochemical Society,* Vol 110, 1963, p 960
14. M. Davis and A. Draycott, Compatibility of Reactor Materials in Flowing Sodium, *UNICPUAE Vol 7,* United Nations, New York, 1958, p 94

Low Expansion Alloys

Edited by D. Wenschhof
Market Coordinator
Huntington Alloys, Inc.

LOW-EXPANSION ALLOYS of iron and nickel, and their various modifications, are used for absolute standards of length, such as rods and tapes for geodetic work. Other applications include:

- Compensating pendulums and balance wheels for clocks and watches
- Moving parts that require control of expansion, such as pistons for some internal-combustion engines
- Bimetal strip
- Glass-to-metal seals
- Thermostatic strip
- Vessels and piping for storage and transportation of liquefied natural gas
- Superconducting systems in power transmission
- Integrated-circuit lead frames
- Components for radios and other electronic devices.

Alloys of iron and nickel have many inconsistent properties, depending on the relative proportions of the two elements. The coefficients of linear expansion range from a small negative value (−0.5 μm/m·K) to a large positive value (20 μm/m·K).

Alloys that contain less than 36% nickel have much higher coefficients of expansion than alloys containing 36% nickel or more. Alloys that contain less than 36% nickel are of the so-called "irreversible" type and are excluded from the present discussion.

One alloy, containing 36% nickel with small quantities of manganese, silicon and carbon amounting to a total of less than 1%, has a coefficient of expansion so low that its length is almost invariable for ordinary changes in temperature. For this reason, the alloy was named "Invar".

After the discovery of Invar, an intensive study was made of the thermal and elastic properties of several similar alloys. Those with higher nickel contents were found to have higher coefficients of expansion. The alloy containing 39% nickel has a coefficient of expansion corresponding to that of low-expansion glasses. The 46Ni alloy has a coefficient equivalent to that of platinum (9.0 μm/m·K) and has been named "Platinite". "Dumet wire" is an alloy containing 42% nickel. Covered with copper to prevent gassing at the seal, it is used to replace platinum as the "seal-in" wire in incandescent lamps and vacuum tubes. The 56Ni alloy has a coeffi-

Fig. 1 Coefficient of linear expansion at 20 °C vs Ni content for Fe-Ni alloys containing 0.4% Mn and 0.1% C

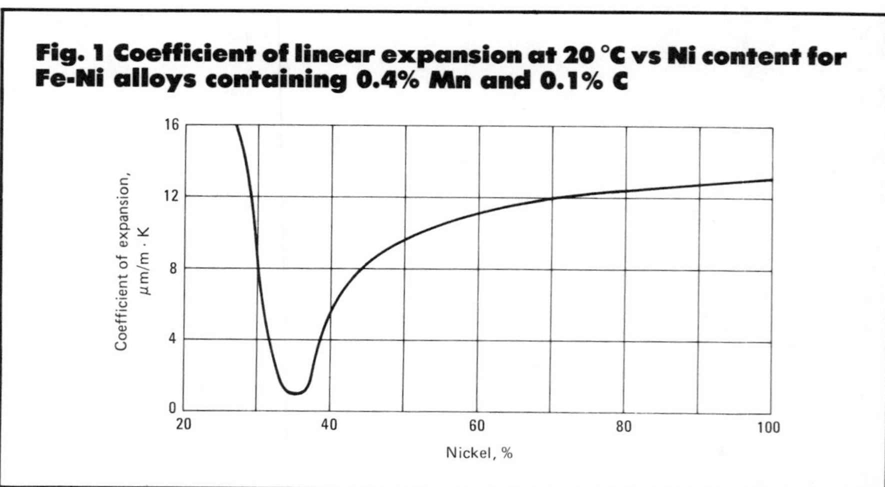

cient approaching that of ordinary steel (11 μm/m·K). Nilvar is identical with Invar (36% Ni).

Elinvar, containing 36% nickel and 12% chromium, has an invariable modulus of elasticity over a considerable range of temperature. It also has low thermal expansivity.

There is an advantage in replacing some of the nickel with cobalt in the 36Ni alloy. Substitution of 5% cobalt for 5% nickel provides an alloy with an expansion coefficient even lower than that of Invar. The 31Ni-5Co alloy is also less susceptible to variations in heat treatment.

Effect of Composition on Expansivity

The effect of variation in nickel content on linear expansivity is shown in Fig. 1. Minimum expansivity occurs at about 36% Ni. Small additions of other metals have considerable influences on the position of this minimum. The effects of additions of manganese, chromium, copper and carbon are shown in Fig. 2.

Minimum expansivity shifts toward higher nickel contents when manganese or chromium is added and toward lower nickel contents when copper or carbon is added. The value of the minimum expansivity for any of these ternary alloys is, in general, greater than that of a typical Invar.

Additions of silicon, tungsten and molybdenum produce effects similar to those caused by additions of manganese and chromium; the composition of minimum expansivity shifts toward higher contents of nickel.

Addition of carbon is said to produce instability in Invar. This instability is attributed to the changing solubility of carbon in the austenitic matrix during heat treatment.

Fig. 2 Effect of alloying elements on expansion characteristics of Fe-Ni alloys

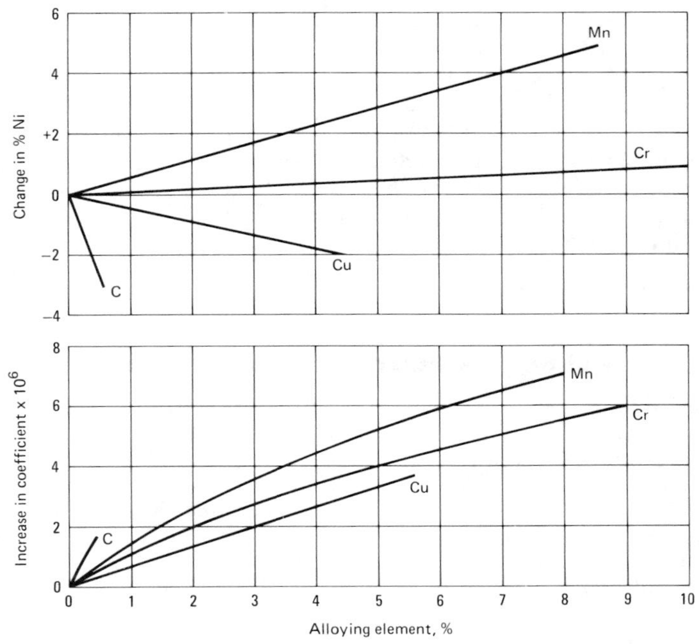

Top: Displacement of nickel content caused by additions of manganese, chromium, copper and carbon to alloy of minimum expansivity. Bottom: Change in value of minimum coefficient of expansion caused by additions of manganese, chromium, copper and carbon.

Invar

Invar and related alloys have low coefficients of expansion only over a rather narrow range of temperature (see Fig. 3). At low temperatures, in the region from *A* to *B* in Fig. 3, the coefficient of expansion is high. In the interval between *B* and *C*, the coefficient decreases, reaching a minimum in the region from *C* to *D*. With increasing temperature, the coefficient begins again to increase in the range from *D* to *E*, and thereafter (from *E* to *F*) the expansion curve follows a trend similar

Fig. 3 Change in length of a typical Invar over different ranges of temperature

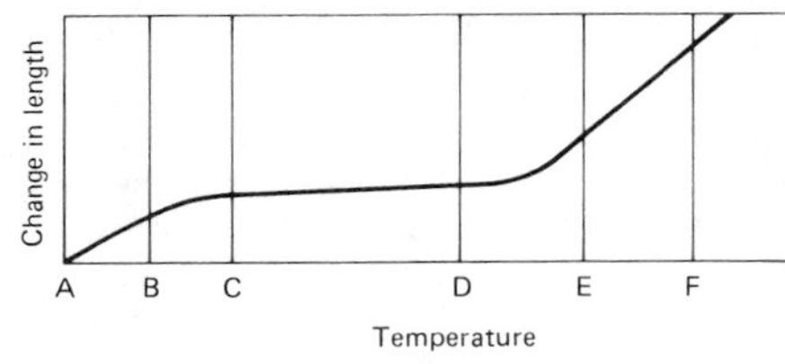

to that of the nickel or iron of which the alloy is composed. The minimum expansivity prevails only in the range from *C* to *D*.

Between *D* and *E*, the coefficient is changing rapidly to a higher value. The temperature limits for a well-annealed 64 alloy are 162 and 271 °C (324 and 520 °F). These temperatures correspond to the initial and final losses of magnetism in the material. The slope of the curve between *C* and *D* is then a measure of the coefficient of expansion over a limited range of temperature.

Table 1 gives coefficients of linear expansion of iron-nickel alloys between 0 and 38 °C (32 and 100 °F). The expansion behavior of several Fe-Ni alloys over wider ranges of temperature is represented by curves 1 to 5 in Fig. 4; for comparison, Fig. 4 also includes the curve of length ratio versus temperature for an ordinary steel.

Table 1 Thermal expansion of Fe-Ni alloys between 0 and 38 °C

Ni, %	Mean coefficient, μm/m·K
31.4	3.395 + 0.00885 *t*
34.6	1.373 + 0.00237 *t*
35.6	0.877 + 0.00127 *t*
37.3	3.457 − 0.00647 *t*
39.4	5.357 − 0.00448 *t*
43.6	7.992 − 0.00273 *t*
44.4	8.508 − 0.00251 *t*
48.7	9.901 − 0.00067 *t*
50.7	9.984 + 0.00243 *t*
53.2	10.045 + 0.00031 *t*

Heat treatment and cold work change the expansivity of Invar (or Nilvar) considerably. The effect of heat treatment is shown in Table 2. The

Table 2 Effect of heat treatment on coefficient of thermal expansion of Invar

Condition	Temperature °C	°F	Mean coefficient, μm/m·K
As forged	17 to 100	63 to 212	1.66
	17 to 250	63 to 480	3.11
Quenched from 830 °C (1560 °F)	18 to 100	65 to 212	0.64
	18 to 250	65 to 480	2.53
Quenched from 830 °C and tempered	16 to 100	60 to 212	1.02
	16 to 250	60 to 480	2.43
Cooled from 830 °C to room temperature in 19 h	16 to 100	60 to 212	2.01
	16 to 250	60 to 480	2.89

Table 3 Physical and mechanical properties of Invar

Solidus temperature	1425 °C (2600 °F)
Density	8.0 Mg/m^3 (0.29 lb/in.^{3})
Tensile strength	450 to 585 MPa (65 to 85 ksi)
Yield strength	275 to 415 MPa (40 to 60 ksi)
Elastic limit	140 to 205 MPa (20 to 30 ksi)
Elongation	30 to 45%
Reduction in area	55 to 70%
Scleroscope hardness	19
Brinell hardness	160
Modulus of elasticity	150 GPa (21.4 × 10^6 psi)
Thermoelastic coefficient	500 μm/m·K
Specific heat	515 J/kg (0.123 Btu/ft·°F) at 25 to 100 °C (78 to 212 °F)
Thermal conductivity	11 W/m·K (6.4 Btu/ft·h·°F) at 20 to 100 °C (68 to 212 °F)
Thermoelectric potential (against copper)	9.8 μV/K at −96 °C (−140 °F)

expansivity is greatest in well-annealed material and least in quenched material.

Cold drawing also decreases expansivity. The values for the coefficients in the following table are from experiments on two heats of Invar:

Material condition	Expansivity, μm/m·K(a)
Direct from hot mill	1.4
	1.4
Annealed and quenched	0.5
	0.8
Quenched and cold drawn(b)	0.14
	0.3

(a) Individual measurements for two heats of Invar. (b) 3.2 to 6.4 mm (0.125 to 0.250 in.) diam.

By cold working after quenching, it is possible to produce material with a zero, or even a negative, coefficient of expansion. A negative coefficient may be increased to zero by careful annealing at a low temperature. However, these artificial methods of securing an exceptionally low coefficient produce instability in the material. With lapse of time and variation in temperature, exceptionally low coefficients usually revert to normal values. For special applications (geodetic tapes, for example), it is essential to stabilize the material by cooling it slowly from 100 to 20 °C (212 to 68 °F) over a period of many months, followed by prolonged aging at room temperature. However, unless the material is to be used within the limits of normal atmospheric variation in temperature, such stabilization is of no value. Although these variations in heat treating practice are important in special applications, they are of little significance for ordinary uses.

Magnetic Properties. Invar and all similar iron-nickel alloys are ferromagnetic at room temperature. They become paramagnetic at higher temperatures. The points of inflection in the curves in Fig. 4 indicate the loss of magnetism. The loss of magnetism in a well-annealed sample of a true Invar begins at 162 °C (324 °F) and ends at 271 °C (520 °F). In a quenched sample, the loss begins at 205 °C (400 °F) and ends at 271 °C (520 °F).

Fig. 4 Thermal expansion of Fe-Ni alloys

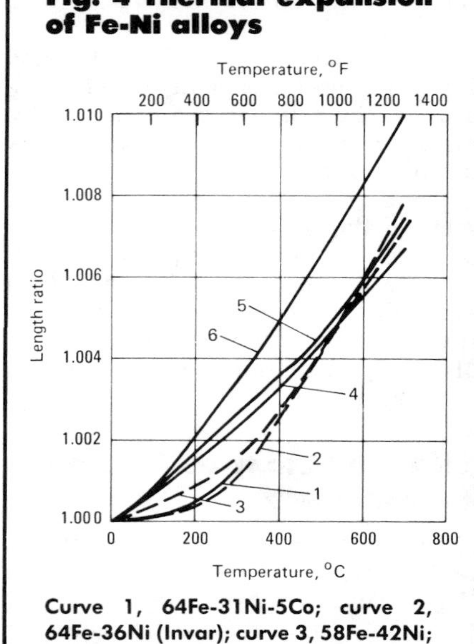

Curve 1, 64Fe-31Ni-5Co; curve 2, 64Fe-36Ni (Invar); curve 3, 58Fe-42Ni; curve 4, 53Fe-47Ni; curve 5, 48Fe-52Ni; curve 6, carbon steel (0.25% C).

Slow cooling through this range of temperature eliminates to a large extent the troublesome variability in properties of materials of this class.

Electrical Properties. The electrical resistivity of Invar is between 750 and 850 nΩ·m at ordinary temperatures. The temperature coefficient of electrical resistivity is about 1.2 mΩ/Ω·K over the range of low expansivity. The thermoelectric potential versus pure copper is about 10 μV/K.

Other Physical and Mechanical Properties. Table 3 presents data on miscellaneous properties of Invar in the hot rolled and forged conditions. The effects of temperature on mechanical properties of forged 66Fe-34Ni are illustrated in Fig. 5.

Processing. Considerable care must be used in hot working of iron-nickel alloys, because at hot working temperature they have a tendency to check and break up when carelessly handled. Invar and related alloys should be annealed in a reducing atmosphere. Because they are susceptible to intercrystalline attack during annealing, they should be processed in an atmosphere that contains a large percentage of a

Table 4 Expansion characteristics of Fe-Ni alloys

Composition, % Mn	Si	Ni	Inflection temperature °C	°F	Mean coefficient of expansion, from 20 °C to inflection temperature, μm/m·K
0.11	0.02	30.14	155	320	9.2
0.15	0.33	35.65	215	430	1.54
0.12	0.07	38.70	340	660	2.50
0.24	0.03	41.88	375	725	4.85
...	...	42.31	380	735	5.07
...	...	43.01	410	790	5.71
...	...	45.16	425	820	7.25
0.35	...	45.22	425	820	6.75
0.24	0.11	46.00	465	890	7.61
...	...	47.37	465	890	8.04
0.09	0.03	48.10	497	950	8.79
0.75	0.00	49.90	500	960	8.84
...	...	50.00	515	985	9.18
0.25	0.20	50.05	527	1010	9.46
0.01	0.18	51.70	545	1040	9.61
0.03	0.16	52.10	550	1050	10.28
0.35	0.04	52.25	550	1050	10.09
0.05	0.03	53.40	580	1105	10.63
0.12	0.07	55.20	590	1125	11.36
0.25	0.05	57.81	None		12.24
0.22	0.07	60.60	None		12.78
0.18	0.04	64.87	None		13.62
0.00	0.05	67.98	None		14.37

Fig. 5 Mechanical properties of a forged 34% Ni alloy

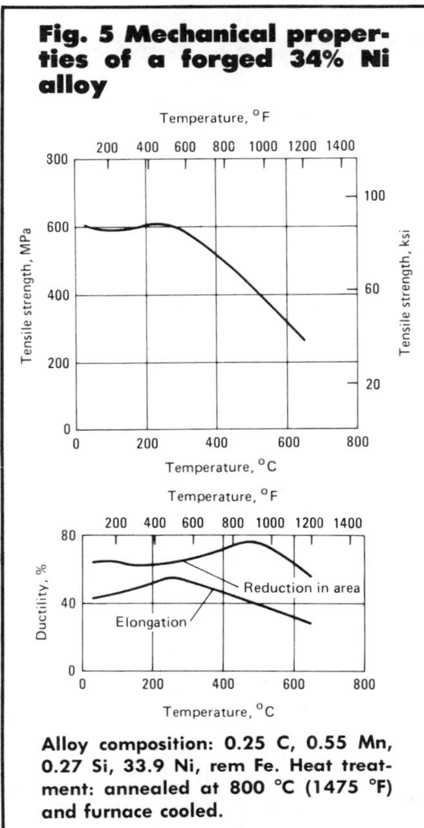

Alloy composition: 0.25 C, 0.55 Mn, 0.27 Si, 33.9 Ni, rem Fe. Heat treatment: annealed at 800 °C (1475 °F) and furnace cooled.

Table 6 Mechanical properties of low-expansion Fe-Ni alloys containing 2.4 Ti and 0.06 C

Condition	Tensile strength MPa	ksi	Yield strength MPa	ksi	Elongation(a), %	Hardness, HB
42Ni-55.5Fe-2.4Ti-0.06C (b)						
Solution treated	620	90	275	40	32	140
Solution treated and age hardened	1140	165	825	120	14	330
Solution treated, cold rolled 50% and age hardened	1345	195	1140	165	5	385
52Ni-45.5Fe-2.4Ti-0.06C (c)						
Solution treated	585	85	240	35	27	125
Solution treated and age hardened	825	120	655	95	17	305

(a) In 50 mm or 2 in. (b) Inflection temperature, 220 °C (430 °F); minimum coefficient of expansion, 3.2 μm/m·K. (c) Inflection temperature, 440 °C (824 °F); minimum coefficient of expansion, 9.5 μm/m·K.

Table 5 Minimum coefficient of expansion in low-expansion Fe-Ni alloys containing titanium

Ti	Optimum Ni	Minimum coefficient of expansion, μm/m·K
0%	36.5%	1.4
2	40.0	2.9
3	42.5	3.6

neutral gas (such as nitrogen) and a small percentage of a reducing gas. Cold rolling and drawing of iron-nickel alloys are quite similar to corresponding processing procedures for nickel.

Heat Treatment. Annealing is done at 750 to 850 °C (1380 to 1560 °F). When the alloy is quenched in water from these temperatures, expansivity is decreased, but instability is induced both in actual length and in coefficient of expansion. To overcome these deficiencies and to stabilize the material, it is necessary to anneal Invar at a low temperature such as 95 to 150 °C (200 to 300 °F) and then cool it to room temperature over a period of several months. Slow cooling through the magnetic transformation range also may be satisfactory.

Where exacting specifications call for definite and invariable properties, alloys with low coefficients of expansion are being made by powder metallurgy. Where free machining is desirable, addition of 0.25% selenium produces satisfactory results.

Iron-Nickel Alloys Other Than Invar

Iron-nickel alloys that have nickel contents higher than that of Invar retain to some extent the expansion characteristics of Invar. Because further additions of nickel raise the temperature at which the inherent magnetism of the alloy disappears, the inflection temperature in the expansion curve rises with increasing nickel content. Although this increase in range is an advantage in some circumstances, it is accompanied by an increase in coefficient of expansion.

Table 4 and Fig. 6 present informa-

Fig. 6 Effect of nickel content on expansion of Fe-Ni alloys

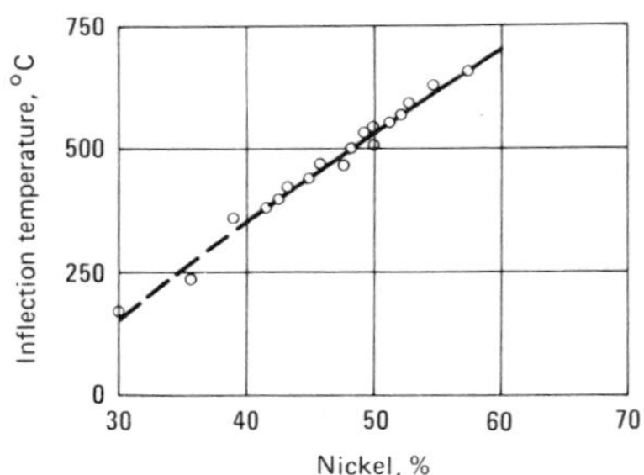

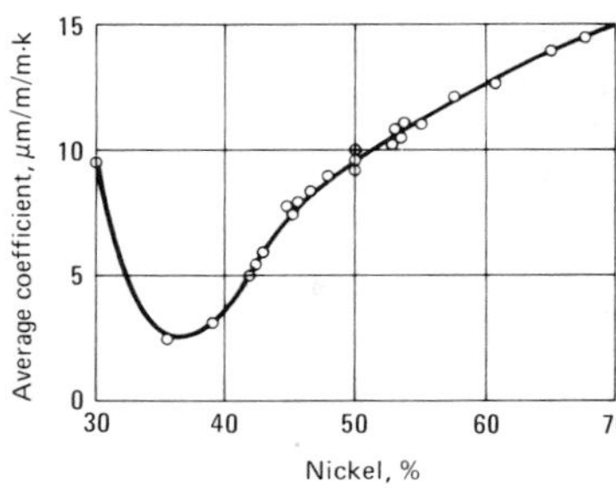

Top: Variation of inflection temperature. Bottom: Variation of average coefficient of expansion between room temperature and inflection temperature.

Table 7 Thermoelastic coefficients of constant modulus Fe-Ni-Cr-Ti alloys

Composition, % Ni	Cr	C	Ti	Thermoelastic coefficient, annealed condition, μm/m·K	Range of possible coefficients(a), μm/m·K
42	5.4	0.06	2.4	0	+18 to −23
42	6.0	0.06	2.4	+36	+54 to +13
42	6.3	0.06	2.4	−36	−18 to −60

(a) Any value in this range can be obtained by varying the heat treatment.

tion on the coefficients of expansion of iron-nickel alloys at temperatures up to the inflection temperature. They also give data on alloys containing up to 68% Ni.

Special Alloys

Certain ternary alloys that contain nickel and iron are interesting because of their special properties and characteristics.

Super Nilvar. Replacement of some of the nickel by cobalt in an alloy of the Invar composition lowers in linear coefficient of expansion. An alloy known as Super Nilvar, which contains 31% nickel and 4 to 6% cobalt, has a zero coefficient of expansion in the hot rolled condition. Annealing and quenching appear to raise the coefficient somewhat, but in general the coefficient of alloys containing both nickel and cobalt, regardless of condition, is lower than that of Invar (unless the Invar is in the cold worked condition).

Fe-Ni-Co Alloys. Cobalt has been added to Fe-Ni alloys in amounts as high as 40%. Such additions increase the coefficient of expansion at room temperature. However, because they also raise the inflection temperature, they produce an alloy with a moderately low coefficient of expansion over a wider range of temperature. These observations are expressed in the equations below.

If θ is inflection temperature in degrees Celsius, X is nickel content, Y is cobalt content, Z is manganese content and W is carbon content, the inflection temperature of any low-expansion Fe-Ni-Co alloy is given by:

$$\theta = 19.5\,(X + Y) - 22Z - 465$$

If we introduce the restriction that the nickel and iron contents shall be so related as to depress the Ar_3 point to about −100 °C (−150 °F), then:

$$Y = 0.0795\theta + 4.8Z + 19W - 18.1$$

and

$$X = 41.9 - 0.0282\theta - 3.7Z - 19W$$

for inflection temperatures between 200 and 600 °C (400 and 1100 °F).

The expansivities in μm/m·K obtainable with these compositions are given by:

$$\alpha_1 = 0.024\theta + 0.38Z - 1.2W - 6.65$$

and

$$\alpha_2 = 0.024\theta + 0.38Z - 1.2W - 5.6$$

Here, α_1 represents minimum expansivity, and α_2 mean expansivity, when θ is between 350 and 600 °C (660 and 1100 °F).

The alloys Kovar and Fernico contain approximately 54 Fe, 28 Ni and 18 Co. They are used for sealing wires in glass.

Fe-Co-Cr Alloys. An alloy containing 36.5 to 37 Fe, 53 to 54.5 Co and 9 to 10 Cr has an exceedingly low coefficient of expansion, at times negative over the range from 0 to 100 °C (32 to 212 °F). Fernichrome, a similar alloy containing 37 Fe, 30 Ni, 25 Co and 8 Cr, has been used for seal-in wires in electronic components sealed in special glasses.

Elinvar. Invar has the highest thermoelastic coefficient of all the low-expansion iron-nickel alloys. However, two Fe-Ni alloys containing 29 and 45% nickel have zero thermoelastic coefficient. This means that the modulus of elasticity does not change with variations in temperature. Because small variations in nickel content produce large variations in thermoelastic coefficient, commercial application of straight Fe-Ni alloys having zero ther-

Table 8 Mechanical properties of constant-modulus alloy 50Fe-42Ni-5.4Cr-2.4Ti

Condition	Tensile strength MPa	Tensile strength ksi	Yield strength MPa	Yield strength ksi	Elongation(a), %	Hardness, HB	Modulus of elasticity GPa	Modulus of elasticity 10^6 psi
Solution treated	620	90	240	35	40	145	165	24
Solution treated and aged 3 h at 730 °C (1345 °F)	1240	180	795	115	18	345	185	26.5
Solution treated and cold worked 50%	930	135	895	130	6	275	175	25.5
Solution treated, cold worked 50% and aged 1 h at 730 °C (1345 °F)	1380	200	1240	180	7	395	185	27

(a) In 50 mm or 2 in.

Table 9 Typical tensile properties of Incoloy 903

Temperature °C	Temperature °F	Tensile strength MPa	Tensile strength ksi	Yield strength MPa	Yield strength ksi	Elongation, %	Reduction in area, %
21	70	1310	190	1100	160	14	40
650	1200	1000	145	895	130	18	55

Note: Material heat treated 1 h at 845 °C (1550 °F) and water quenched; then 8 h at 720 °C (1325 °F), furnace cooled at 55 °C/h (100 °F/h) to 620 °C (1150 °F), held 8 h and air cooled.

Table 10 Dynamic modulus of elasticity of age-hardened Incoloy 903

Temperature °C	Temperature °F	Tensile modulus GPa	Tensile modulus 10^6 psi	Torsional modulus GPa	Torsional modulus 10^6 psi
−196	−320	148.9	21.59	...	...
−129	−200	147.7	21.42	...	...
−73	−100	147.1	21.34	...	...
−18	0	146.8	21.29	...	...
+38	+100	146.9	21.30	59.5	8.63
93	200	147.2	21.35	59.0	8.56
149	300	147.7	21.42	59.4	8.62
204	400	148.4	21.52	60.3	8.75
260	500	149.4	21.67	61.0	8.84
316	600	150.6	21.84	61.2	8.88
371	700	151.7	22.00	61.3	8.89
427	800	152.9	22.18	61.0	8.84
482	900	154.0	22.34	59.6	8.65
538	1000	152.4	22.10	58.0	8.41
593	1100	150.0	21.75	55.8	8.10
649	1200	147.8	21.43	52.7	7.64

moelastic coefficient is not practical. Addition of 12% chromium to an alloy containing 36% nickel produces Elinvar, an alloy with zero thermoelastic coefficient and no susceptibility to the small variations in nickel content expected in commercial melting.

Elinvar is used for such articles as hairsprings and balance wheels for clocks and watches, and tuning forks used in radio synchronization. It is particularly advantageous where an invariable modulus of elasticity is required. It has the further advantage of being comparatively rustproof.

The composition of Elinvar has been modified somewhat from its original specification. The limits now used are 33 to 35 Ni, 61 to 53 Fe, 4 to 5 Cr, 1 to 3 W, 0.5 to 2 Mn, 0.5 to 2 Si and 0.5 to 2 C.

Hardenable Alloys. Alloys that have low coefficients of expansion and alloys with constant modulus of elasticity can be made age hardenable by adding titanium.

In low-expansion alloys, nickel content must be increased when titanium is added. The higher nickel content is required because any titanium that has not combined with the carbon in the alloy will neutralize more than twice its own weight in nickel by forming an intermetallic compound during the hardening operation. As shown in Table 5, addition of titanium raises the lowest attainable rate of expansion and raises the nickel content at which the minimum expansion occurs. Titanium also lowers the inflection temperature. Mechanical properties of alloys containing 2.4% titanium and 0.06% carbon are given in Table 6.

In alloys of the constant-modulus type, addition of titanium allows thermoelastic coefficient to be varied by adjustment of heat treating schedules. The alloys in Table 7 are the three most widely used compositions. The recommended solution treatment for the alloys that contain 2.4% Ti is 950 to 1000 °C (1740 to 1830 °F) for 20 to 90 min, depending on section size. Recommended duration of aging varies from 48 h at 600 °C (1110 °F) to 3 h at 730 °C (1345 °F) for solution-treated material.

For material that has been solution treated and subsequently cold worked 50%, aging time varies from 4 h at 600 °C to 1 h at 730 °C. Table 8 gives mechanical properties of a constant-modulus alloy containing 42 Ni, 5.4 Cr and 2.4 Ti. Heat treatment and cold work markedly affect these properties.

Incoloy 903 is a high-strength, precipitation-hardenable alloy that has both a constant low coefficient of thermal expansion and a constant modulus of elasticity. It is hardened by addition of titanium and niobium and contains 38 Ni, 15 Co, 0.7 Al, 3 Nb, 1.4 Ti, rem Fe. Table 9 shows the high strength obtainable with precipitation-hardened material. As shown in Fig. 7, the alloy typically exhibits a coefficient of expansion of about 7.2 μm/m·K from room temperature to around 425 °C (800 °F). The modulus of elasticity remains almost constant from −196 to +650 °C (−320 to +1200 °F). Modulus values are listed in Table 10. The com-

Fig. 7 Typical coefficients of thermal expansion for Incoloy 903

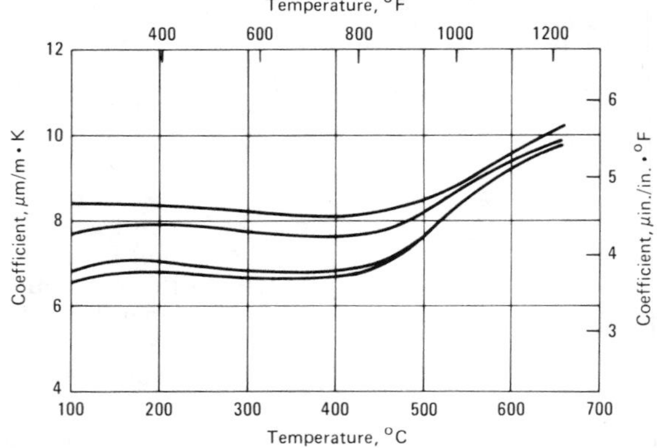

Data are for four individual heats of Incoloy 903, and are plotted for expansion from room temperature to the indicated temperature.

bination of exceptional strength and constant low coefficient of expansion makes Incoloy 903 useful for applications that require close operating tolerances over a range of temperatures. Some components for jet engines fall into this category.

Engineering Applications

Use of alloys with low coefficients of expansion has been confined mainly to such applications as geodetic tape, bimetal strip, glass-to-metal seals and electronic and radio components. Almost all variable condensers are made of Invar. Struts on jet engines are made of Invar to ensure rigidity with temperature changes. Contact tabs of 47% nickel alloy, used in electrical resistors made of enamel, match the expansion characteristics of the enamel.

Fusible Alloys

Edited by Joseph B. Long
Tin Research Institute, Inc.

The term "FUSIBLE ALLOYS" refers to any of the more than 100 white metal alloys that melt at relatively low temperatures. Most commercial fusible alloys contain bismuth, lead, tin, cadmium, indium and antimony, and special alloys of this class may also contain significant amounts of zinc, silver, thallium or gallium.

Many of the fusible alloys used in industrial applications are based on eutectic compositions (see Table 1). Use of such alloys is important in automatic safety devices such as fire sprinklers, boiler plugs and controls for furnaces. Under ambient temperature, such an alloy has sufficient strength to hold parts together, but at a specific elevated temperature the fusible-alloy link will melt, thus disconnecting the parts. In fire sprinklers, the links melt when dangerous temperatures are reached, releasing water from piping systems and extinguishing the fire. Boiler plugs and furnace controls react similarly because an increase in temperature beyond the safety limits of the furnace or boiler operation will melt the plug. When the fusible-alloy link melts, pressure or heat in the boiler can be dissipated, or fuel-supply feeding ceased, and operation can be reduced to a safe level.

In addition to eutectic alloys, each of which melts at a specific temperature, there are numerous noneutectic alloys, which melt over a range of temperatures. Some of the more important non-

Table 1 Compositions and melting temperatures of several eutectic fusible alloys

Alloy (a)	Melting temperature °C	°F	Bi	Pb	Composition,% Sn	Cd	Other
A	46.8	117	44.70	22.60	8.30	5.30	19.10 In
B	58	136	49.00	18.00	12.00	...	21.00 In
C	70	158	50.00	26.70	13.30	10.00	...
D	91.5	197	51.60	40.20	...	8.20	...
E	95	203	52.50	32.00	15.50	...	...
F	102.5	217	54.00	...	26.00	20.00	...
G	124	255	55.50	44.50	...	...	...
H	138.5	281	58.00	...	42.00	...	...
I	142	288	...	30.60	51.20	18.20	...
J	144	291	60.00	...	...	40.00	...
K	177	351	...	...	67.75	32.25	...
L	183	362	...	38.14	61.86	...	...
M	199	390	...	...	91.00	...	9.00 Zn
N	221.3	430	...	...	96.00	...	3.50 Ag
O	236	457	...	79.7	...	17.7	2.60 Sb
P	247	477	...	87.0	...	...	13.00 Sb

(a) Letter designations are intended only for identification of alloys in Table 3.

Table 2 Compositions, yield temperatures and melting-temperature ranges of some noneutectic fusible alloys

Alloy (a)	Yield temperature °C	°F	Melting-temperature range °C	°F	Bi	Composition,% Pb	Sn	Cd
Q	70.5	159	70 to 73	158 to 163	50.50	27.8	12.40	9.30
R	72.0	162	70 to 79	158 to 174	50.00	34.5	9.30	6.20
S	72.5	163	70 to 84	158 to 183	50.72	30.91	14.97	3.40
T	72.5	163	70 to 90	158 to 194	42.50	37.70	11.30	8.50
U	75	167	70 to 101	158 to 214	35.10	36.40	19.06	9.44
V	96	205	95 to 104	203 to 219	56.00	22.00	22.00	...
W	96	205	95 to 149	203 to 300	67.00	16.00	17.00	...
X	101	214	101 to 143	214 to 289	33.33	33.34	33.33	...
Y(b)	116	241	103 to 227	217 to 440	48.00	28.50	14.50	...
Z	150	302	138 to 170	281 to 338	40.00	...	60.00	...

(a) Letter designations are intended only for identification of alloys in Table 3. (b) Also contains 9.00% Sb.

Table 3 Fusible alloys suitable for various applications

Application	Suitable alloys (a)
Matrix metal for:	
Large bearings	X
Punch and die assemblies	Y
Small bearings	Y
Anchoring for:	
Rods and tubular members	Y
Bushings in drill jigs, plates, fixtures	C, Y, Z
Inserts and parts in ceramics, plastics, metal powders, wood	G, Z
Locator members in aircraft and automotive assembly, drilling, inspection and welding fixtures	C, H, Z
Hold-down bolts in floors	G, Z
Magnets in chucks, instruments, workholding devices	H, Z
Needles in lace and textile machinery	G, H, T
Patterns in foundry match plates	G, H, T
Precision parts for testing and inspection	A, B, C, G, Z
Reamers for axial and concentric alignment in turret tool holders	C, H
Shafts in permanent magnet rotors	H, Y
Chucking for:	
Lens buffing and grinding	A, C
Gem cutting	G, Z
Metal spinning re-entrant and bottleneck shapes	C, G
Fusible cores for:	
Electroforming	G, H, Z
Founding	C, G
Holding and forming fiberglass and plastic laminates	C, G
Compound wax patterns	B, C
Punch and die applications:	
Light sheet metal embossing dies	C, G, H, Y, Z
Form blocks for bending, forming and joggling of extruded shapes, sheet metal and tubing	H, Y
Soft metal dies for lost-wax patterns	H, Z
Pilots for die-casting and plastic trim dies	H
Repairing broken dies	Y
Stretch press form blocks	G, H
Stripper plates in stamping dies	Y
Guerin process dies	C, G, H, Y
Mold applications:	
Duplicating plaster or plastic patterns	G, H, Z
Casting plastics, phenolics, elastomers, epoxys, polyesters	G, H, Z
Forming sheet and tubular polyethylene plastics by vacuum or air pressure	H, Z
Plastic teeth (dental)	H
Prosthetic development work	A, H, Z
Encapsulating avionic, electrical and electronic components	H, Z
Pattern applications:	
Dental models-special compositions	A
Duplicate patterns for foundry match plates, ceramics, plastics and pottery	C, H, Z
Lining and sealing core boxes	H
Model airplane, railroad and ship parts	G, H, T
Patterns and core boxes spray-metallized, altered, repaired	T, Z
Tracer models for pantograph engraving and duplicating	H, Z
Miscellaneous applications:	
Filler for tube and mold bending	C, G
Low-melting solders	
Heat-transfer mediums in constant-temperature baths	
Seals for bright annealing and nitriding furnaces	
Seals for adjustment screws on torque wrenches, instruments	
Hold-down clamp pads	
Ammunition composites	
Fire-detection apparatus and alarm systems	
Safety plugs for tanks and cylinders for compressed gas	
Fusible elements in automatic sprinkler heads and fire-door release links	
Automatic shut-offs for gas and electric water-heating systems	
Selenium rectifiers such as the counterelectrode	
For addition of lead and bismuth to aluminum and other metals and alloys to obtain free-cutting materials	

(a) See Tables 1 and 2.

eutectic alloy compositions are listed in Table 2.

A fusible alloy with a long melting range is useful in staking rods and tubing in assemblies because the alloy is distributed around part surfaces while still molten and provides a firm anchorage after it solidifies.

Properties. Most fusible alloys are heavy, bright, silvery, nontarnishing metals that can be melted repeatedly with negligible loss of elemental constituents. They are ageable alloys and thus their mechanical properties often are dependent on the period of time that has elapsed since casting, as well as on conditions of casting and rate of solidification. The test conditions employed also affect mechanical-property values. For example, many fusible alloys may appear brittle when subjected to sudden shock but exhibit high ductility under slow rates of strain.

In certain alloys, normal thermal contraction due to cooling after solidification can be partly, completely, or more than compensated for by expansion due to aging. For example, bismuth alloys containing 33 to 66% lead exhibit net expansion after solidification and during subsequent aging. Some fusible alloys show no contraction (shrinkage) and expand rapidly while still warm; others show slight shrinkage during the first few minutes after solidification and then begin to expand; in still others, expansion does not commence until some time after the fusible alloy casting has cooled to room temperature.

All three characteristics—net expansion, net contraction, and little or no volume change—provide specific advantages, depending on the application. For instance, a wood pattern used for making molds must be of somewhat greater dimensions than those desired in a casting in order to compensate for shrinkage of the casting on solidification and during cooling to room temperature. Where metal patterns are cast from a master wood pattern, two such allowances will have to be made unless the alloy used for the metal patterns possesses zero shrinkage. Fusible alloys with eutectic compositions are often used for casting metal patterns from wood masters because they undergo definite growth that is sufficient to allow for "cleaning" of production castings without reducing dimensions below required values. The growth characteristics of fusible alloys are often used to advantage when a metal part such as a turbine blade is to be firmly anchored in a lathe chuck. After the part is machined, the fusible alloy is melted away.

Growth of fusible alloys that exhibit this characteristic varies from 0.01 to 0.7% in linear dimensions, depending primarily on alloy composition.

Generally, load-bearing capacity of fusible alloys is good, although some deformation will occur under prolonged stress. In addition, hardness and other mechanical properties of many fusible alloys change gradually with time, probably due to the same microstructural changes that are responsible for growth or shrinkage on aging.

Applications. Because of their unusual mechanical and physical properties, low-melting alloys are now used in a wide variety of applications. As previously indicated, however, attainment of certain specific properties is largely dependent on alloy composition. Table 3 presents a list of typical applications and the fusible alloys commonly used for each application or group of applications. Where more than one alloy is indicated, suitability depends on operating temperature or other controlling factors, such as conditions of service or the nature of the materials to be joined.

Materials for Sliding Bearings

By the ASM Committee on Sliding Bearings*

A SLIDING BEARING (plain bearing) is a machine element designed to transmit loads or reaction forces to a shaft that rotates relative to the bearing. Journal bearings are cylindrical (full cylinders or segments of cylinders) and are used when the load or reaction force is essentially radial—that is, perpendicular to the axis of the shaft. Thrust bearings are ring-shape bearings (full rings or segments of rings) and are used when the load or reaction force is parallel to the direction of the shaft axis. Both radial and axial loads can be accommodated by flange bearings, which are journal bearings constructed with one or two integral thrust-bearing surfaces. The sliding movement of the shaft surface or thrust-collar surface relative to the bearing surface is characteristic of all plain bearings. In many applications, plain bearings offer advantages over rolling-contact bearings—advantages such as lower cost, smaller space requirements, ability to operate with marginal lubricants, resistance to corrosion and ability to sustain high specific loads.

Historical Development. Plain-bearing principles have been employed in rotating mechanical devices since about 3000 B.C. However, the first use of a metal alloy for its special properties in bearing service was probably Isaac Babbitt's adaptation of a pewter composition in 1839. Design and manufacture of metallic plain bearings as separate machinery components that are mechanically assembled in precisely machined housings are recent developments. Until early in the twentieth century, machinery bearings were machined from massive tin alloy or bronze castings, or were made by pouring molten babbitt metal into cast or machined holes in iron or steel machine members. The final bearing bore was then produced by cutting away excess material, leaving the hole with a relatively thick lining of bearing metal, completely bonded to the housing.

Since the 1930's, bearing developments have proliferated with respect to materials and materials systems, mechanical design and manufacturing technology. Single-metal cast-in-place bearings have been largely supplanted by replaceable-insert bearings made from multilayer laminated metals, and as a result mechanical design of bearings has become an increasingly refined engineering specialty. Manufacturing technology has evolved into a remarkably sophisticated complex of metallurgical, chemical and mechanical processes. The scale of manufacturing operations ranges from production of small lots by job-shop methods to extremely high-volume mass production of automotive bearings.

In 1979, production of plain bearings of all types in the United States exceeded one billion pieces, with a total value in excess of $400 million.

Classifications

Functions and Applications. When considered in terms of function or location in a machine, bearings are

*See page XIV for committee list.

commonly called by the following names, which describe their use:

- Connecting-rod bearing
- Main bearing
- Piston-pin bushing
- Camshaft bearing or bushing
- Crankshaft thrust washer
- Leaf-spring bushing
- Front-axle bushing
- Idler-gear bushing
- Brake-pedal shaft bushing
- Rocker-arm bushing
- Transmission-gear bushing
- Reverse-pinion bushing
- Differential-trunnion bushing
- Intermediate gear thrust washer
- Countershaft thrust washer
- Electric-motor shaft bushing.

The terms "bearing" and "bushing" are used interchangeably, and do not have meanings that are significantly different in terms of function or location in a machine.

A major distinction can be drawn, however, between (*a*) connecting-rod and main bearings and (*b*) the remaining items in the list. In a reciprocating engine the connecting-rod and main bearings must support the entire firing load on the pistons, which is transmitted through the connecting rods to the crankshaft. These bearings therefore are subject to exceptionally severe cyclic loading, shaft-deflection effects, and variations in lubricant-film thickness. The conditions imposed on the other bearings listed differ mainly in type of load (steady or intermittent), magnitude of load, speed of operation, direction of operation (undirectional or oscillating) and operating temperature.

Structural Characteristics. Bearings are also frequently classified according to material construction as solid (single-metal), bimetal (two-layer) or trimetal (three-layer) bearings. These terms indicate the number of principal mechanically functional layers, not including cosmetic surface finishes, bonding layers, diffusion zones or other microscopically thin interface layers that may also be present. These various material constructions are illustrated in the section on microstructure of sleeve bearing materials, pages 231 to 240 in Volume 7 of the 8th Edition of this Handbook.

All three of these constructions are in widespread commercial use, in a wide variety of applications. Use of two or three separate layers provides a means of developing combinations of properties that cannot be obtained with single metals, and has proven to be one of the most important factors in the successful development of cost-effective bearing material improvements.

In current practice, connecting-rod and main bearings of reciprocating engines are usually of either bimetal or trimetal construction, but single-metal bearings are employed occasionally. Bearings for other machinery applications are most often of single-metal or bimetal construction, although trimetal construction is sometimes required.

Size, Configuration and Manufacturing Method. With respect to size, bearings commonly are classified as either thin-wall or heavy-wall bearings; in general, bearings greater than about 5 mm (0.2 in.) in wall thickness and not less than 150 mm (6 in.) in diameter are considered heavy-wall bearings. Configuration may be further described as half round, full round, flanged or washer. SAE standards classify bearings used in mass-produced machinery (virtually all thin-wall bearings) into three groups: sleeve-type half bearings, split-type bushings and thrust washers. Because of the high-volume, low-cost nature of this market, such bearings are almost always made by high-speed forming and machining processes from materials in flat strip form; the terms "strip-type bearing" and "sheet metal bearing" are used occasionally.

Heavy-wall bearings are manufactured in small lots (from one piece to a few thousand pieces) by more conventional machine-shop manufacturing methods. Starting materials include both flat slabs and tubular shapes of single-metal, bimetal and trimetal construction. The terms "slab bearings" and "shell-cast bearings" sometimes are employed to distinguish between bearings made from these two forms of starting material.

Operating Conditions

It is necessary to analyze both the mechanical design and the material requirements of a bearing in terms of the system in which it will be used. The important mechanical components of a system include the bearing housing, the lubricant film, the surface and subsurface of the bearing itself, and the surface of the journal or thrust collar. The interactions among these components in an operating machine can be exceedingly complex, and they are not subject to rigorous analysis on the basis of any established general theory.

However, a substantial body of empirical knowledge and a growing fund of theoretical understanding exist, permitting engineering decisions to be made on questions of bearing design and material selection with a minimum of trial-and-error testing.

The technical factors that are most important from the standpoint of material selection are discussed briefly in the following subsections.

Loads and Speeds. The approximate magnitude of the specific load to which a bearing will be subjected should be known, so that materials of insufficient strength can be eliminated from consideration. However, precise values of actual bearing unit loads in operating machinery are not easily obtained. The most common approximation is the mean unit load, P, which is calculated from the equation:

$$P = \frac{F}{L \times D} \qquad \text{(Eq 1)}$$

where F is maximum total force acting on the bearing, L is axial bearing length and D is bearing diameter.

Total force usually can be estimated with reasonable accuracy from known machine design parameters, and sometimes can be determined experimentally. Values of P for bearings in commercial machinery vary from nearly zero to 105 MPa (15 ksi). Where the loading on a bearing is steady or varies in a simple pattern, the P value is a satisfactory reference for selecting bearing materials of adequate strength.

For connecting-rod and main bearings in reciprocating machinery, the load cycle is quite complex and the P value becomes a less reliable indicator of material-strength requirements. It has recently become feasible, with the aid of a suitable computer program, to analyze the forces acting on connecting-rod and main bearings in terms of peak oil-film pressure (POF), which approaches a measure of the true unit compressive stress to which the bearing liner will be subjected. Values of POF for reciprocating engines may be three to nine times the P value, and POF values as high as 115 MPa (60 ksi) are not uncommon.

The magnitude of bearing unit loading (P or POF) can vary considerably at different engine speeds. When the loading is cyclic, as it is in connecting-rod and main bearings and also in piston pin bushings, rotating speed determines the number of load cycles that will be endured in a given time period, and consequently affects the rate at which damage may occur from fatigue or wear.

For these reasons, speed is an important factor in selection of bearing materials. It has been common design practice to define bearing operating conditions in terms of the *PV* (pressure × velocity) factor. Although this factor is inadequate as a basis for material selection except in very simple applications, it does recognize the importance of both loads and speeds in the selection of bearing materials.

A much more complete expression is given by the Sommerfeld number, which also takes into account bearing size and configuration, clearance between bearing and journal surfaces, and lubricant viscosity. Calculations based on this expression are extremely useful in mechanical design of bearings and lubricating systems, and have contributed substantially to better understanding of the material requirements in specific applications.

Lubrication and Friction. In order to minimize the friction associated with the sliding movement of plain bearings, a fluid lubricant is almost invariably used. Oils and greases are the most commonly used lubricants; but other fluids, including hydraulic fluids, water and even air, may be used in special applications.

Every effort should be made to design any bearing so that a continuous film of lubricant will be maintained between the sliding surfaces during operation.

Proper clearance between the bearing and the shaft, and an adequate supply of lubricant with suitable viscosity characteristics, are important factors in fluid-film lubrication. Normally, this type of lubrication is associated with relatively high shaft speeds and light to moderate loads, which favor formation of a full oil film. However, even at sliding velocities as low as a few feet per minute, a substantial load can be carried on a partial film of oil or grease as long as the proper bearing and shaft materials are being used. Oil grooves of various designs often are provided at the bearing surface to ensure uniform distribution of the oil in the areas where it is most needed.

Under ideal conditions of lubrication, a continuous oil film would be maintained between the sliding surfaces, and thus the material used for the bearing surface would be unimportant. Under actual conditions of bearing operation, however, some direct contact between shaft and bearing surface is unavoidable. As a result, the bearing properties of the material forming the bearing surface are actually of great practical importance.

The oil film may be thought of as a combination of the fluid lubricant and reaction products formed between the lubricant and the bearing surface, so that a soft, semifluid, greaselike film is the layer that prevents metal-to-metal contact. Although various bearing materials do show significant differences in the ability to maintain a continuous oil film at the working surface, all those in commercial use can be successfully lubricated with petroleum-base or synthetic hydrocarbon oils.

In the event of oil-film failure, direct contact between shaft surface and bearing surface can occur, resulting in frictional heating and temperature increases sufficient to cause breakdown of the lubricant film. Contact area between surface asperities also increases under these conditions, because the strength of the bearing material decreases at the higher temperature. Under such conditions, wear of the bearing surface, and even of the shaft, may become excessive. More severe consequences of oil-film failure include catastrophic failure due to seizure or partial melting of the bearing liner. The most frequent causes of such catastrophic failure include overloading, oil-supply failure, or the presence of foreign abrasive debris in the clearance area.

Much of the technology of bearing materials and their selection is related to the need for bearings to operate under conditions that fall short of full-film (hydrodynamic) lubrication. The states of boundary lubrication and thin-film lubrication both involve imperfect separation of the bearing and shaft surfaces, and either or both states will exist, at least momentarily, in virtually any operating machine.

Heat and Temperature. Even in the absence of metallic contact, generation of some frictional heat is unavoidable in machinery bearings. In any heat engine, additional heating occurs by conduction from the working fluid. Artificial cooling of bearings generally is accomplished by use of a heat exchanger installed in the lubricant system, by means of which excessively high oil and bearing temperatures can be avoided. In machinery other than heat engines, it is not usually necessary to employ artificial lubricant cooling.

In general, successful lubrication and bearing operation require that the inlet lubricant temperature be maintained below about 135 °C (275 °F) if the lubricant is a hydrocarbon oil. Under these conditions, bearing-back temperatures of 175 °C (350 °F) or less usually can be maintained. Temperatures as high as 260 °C (500 °F) can be tolerated by some synthetic lubricating fluids, with correspondingly higher bearing-back temperatures. Actual surface temperatures of bearing liners in operation are rarely known, but must be assumed to be higher than corresponding bearing-back temperatures.

The most important considerations here are (*a*) the reduced viscosity of most lubricating fluids at higher temperatures and the resulting tendency toward thinner and less-perfect lubricant films, and (*b*) the reduced mechanical strength of bearing-liner materials at elevated temperatures. Both of these factors are significant in specifying a bearing material for a given application.

In general, lowest material cost and highest bearing reliability result from low lubricant and bearing-back temperatures; cost-effective mechanical designs of bearings and of lubrication systems are those that avoid unnecessarily high temperatures.

Corrosion. Except in pumps that handle corrosive fluids, plain bearings usually are not required to operate in extremely hostile chemical environments. It is possible, however, for corrosive problems to develop in lubricating oils as a result of oxidation and/or by reaction with engine coolants and combustion products. Fatty acids, alcohols, aldehydes and ketones formed in this way can cause corrosion of bearing metals. Acidic sulfur compounds in lubricating oils, which may be initially present or which may result from oxidation or from contamination by combustion products, also can act as corrodents.

Properties of Bearing Materials

Surface and Bulk Properties. The nature of the conditions under which plain bearings must operate and the wide ranges over which these conditions can vary lead to concern for bearing-material properties of two kinds: (*a*) surface properties (those associated with the bearing surface and immediate subsurface layers) and (*b*) bulk properties.

Conventional engineering definitions of material properties are not sufficient to characterize all the essential attributes of bearing materials. Although there is no universally accepted system of nomenclature, measurement or testing for these properties, they can be defined and studied in terms of the following characteristics:

- *Compatibility:* the antiwelding and antiscoring characteristics of a bearing material when operated with a given mating material
- *Conformability:* the ability of a material to yield to and compensate for slight misalignment and to conform to variations in the shape of the shaft or of the bearing-housing bore
- *Embeddability:* the ability of a material to embed dirt or foreign particles and thus prevent them from scoring and wearing shaft and bearing surfaces
- *Load capacity:* the maximum unit pressure under which a material can operate without excessive friction, wear and fatigue damage
- *Fatigue strength:* the ability of a material to function under cyclic loading below its elastic limit without developing cracks or surface pits
- *Corrosion resistance:* the ability of a material to withstand chemical attack by uninhibited or contaminated lubricating oils
- *Hardness:* the ability to resist plastic flow under high unit compressive loads, conventionally measured by indentation hardness testing
- *Strength:* the ability to resist elastic and plastic deformation under load, conventionally measured by compression, shear and tensile testing.

Compatibility can be regarded as a purely surface characteristic. Conformability and embeddability involve the surface and immediate subsurface, but are strongly related to the bulk properties of strength and hardness. The other characteristics relate principally to bulk properties.

Measurement and Testing. Of all the characteristics in the list above, only hardness and strength can be measured satisfactorily by standard laboratory test methods. Many special dynamic test rigs and test methods have been developed in the plain-bearing industry to evaluate and measure the other characteristics and their interactions. Although much useful information has been developed through laboratory rig testing, it still is often necessary to test bearing materials and designs in full-size operating machines in order to clearly establish their overall suitability. Such testing is necessarily expensive and time-consuming. It should be undertaken only after careful study of the conditions under which the bearing will operate and of prior experience in similar applications.

An appreciation of the relative importance of the various bearing-material characteristics in the application at hand and an understanding of the trade-offs that are available usually will limit the viable choices to a manageable number. Full-scale bearing tests will then serve more for confirmation of the validity of design and material selections than for the exploration of design and material variables.

Compatibility is an antiseizure/antiscore characteristic that relates principally to the ease with which the bearing material surface will adhere or weld to a steel journal surface under the influence of pressure and heat and in the absence of a lubricant or some other interfering surface film. Some potential for scoring and seizure exists under all boundary and thin-film lubrication conditions. Factors of theoretical concern here are (*a*) the atomic diameter of the bearing material relative to that of iron and (*b*) the nature of the bonds (metallic or covalent) that will form between bearing-material atoms and iron atoms.

Compatibility with steel should be good if the bearing metal has an atomic diameter greater by at least 15% than that of iron (that is, not less than 0.285 nm) and is a b-subgroup metal. Only seven commercially significant metals meet these criteria: silver, cadmium, indium, tin, gold, lead and bismuth. Of these, tin and lead offer the most attractive combinations of cost, availability and engineering properties. Both are widely employed in bearing alloys because of their contributions to compatibility—either as the alloy base, such as in lead and tin babbitts, or as major microconstituents, such as in copper-lead and aluminum-tin alloys. Silver, cadmium, indium and bismuth are also utilized, but less extensively.

Useful compatibility with steel is also exhibited by many nonmetals, including synthetic resins, carbon, cemented carbides, intermetallic compounds and ceramics. For all of these materials, compatibility probably can be attributed to a substantial lack of unsatisfied surface-atom force fields.

Conformability and Embeddability Versus Load Capacity and Fatigue Strength. Both conformability and embeddability depend on yielding and plastic flow of the bearing material under load. Thus it is not surprising that soft, weak, low-modulus metals such as tin and lead exhibit the best conformability and embeddability. Conversely, high load capacity and high fatigue strength are exhibited by harder, stronger, higher-modulus metals such as copper and aluminum. Useful compromises between these sets of opposed properties are effected by alloying to produce polyphase structures with intermediate properties, and by employing layered constructions in which softer and weaker surface layers are reinforced by one or more harder, stronger backing layers.

Bearing-Material Structures. All commercially significant bearing metals except silver are used as polyphase alloys. Typical microstructures are shown beginning on page 232 in Volume 7 of the 8th Edition of this Handbook. Four general microstructural types can be recognized:

- *Soft Matrix with Discrete Hard Particles.* Lead babbitts and tin babbitts are of this type. The compatibility, conformability and embeddability of these alloys are somewhat lower than those of unalloyed lead or tin due to the presence of hard intermetallic and metalloid particles, which effectively increase bulk strength properties.
- *Interlocked Soft and Hard Continuous Phases.* Many copper-lead alloys are of this type. These structures consist of continuous, mutually supporting copper and lead sponges. A large volume of lead contributes to compatibility. Conformability, embeddability, hardness and strength are intermediate between those of lead and those of copper.
- *Strong Matrix with Discrete Soft-Phase Pockets.* Leaded bronzes, aluminum-tin alloys and aluminum-lead alloys are of this type. These structures consist of a continuous copper-base or aluminum-base metallic matrix of intermediate to high strength that contains discrete pools or pockets of lead and/or tin. Con-

formability, embeddability, strength and hardness are dictated by the strength of the matrix phase. Compatibility is enhanced if soft-phase material is exposed at the bearing-alloy surface.

- *Laminated Construction.* One of the most useful concepts in bearing-material design came with the recognition in 1941 that the effective load capacities and fatigue strengths of lead and tin alloys were sharply increased when these alloys were used as thin layers intimately bonded to strong bearing backs of bronze or steel. Use is made of this principle (Fig. 1) in both two-layer and three-layer constructions, in which the surface layer is composed of a lead or tin alloy, usually no more than 0.13 mm (0.005 in.) thick. Unimpaired compatibility is provided by such a layer, together with reasonably high levels of conformability and embeddability. Other useful compromises can be effected between surface and bulk properties by employing an intermediate copper alloy or aluminum alloy layer between the surface alloy layer and a steel back. In these three-layer constructions, use of surface-layer thicknesses as low as 0.013 mm (0.0005 in.) offers even more favorable compromises between surface and bulk properties than are possible with two-layer constructions.

Corrosion Resistance. Bearing failure due to corrosion alone is rare. Corrosion usually interacts with mechanical and thermal factors to produce failure by fatigue, wear or seizure under conditions the bearing normally would be able to tolerate. To a considerable extent, bearing corrosion can be avoided by use of oxidation inhibitors in commercial lubricating oils, and by periodic oil changes. There are, however, many situations in which neither of these practices is dependable and where bearing materials with inherently high corrosion resistance should be used.

Commercially pure lead is susceptible to corrosion by fatty acids. Lead-base and copper-lead bearing alloys can suffer severe corrosion damage in acidic lubricating oils. Tin additions in excess of about 5% provide effective protection against this kind of corrosion, and for this reason tin is used extensively in lead-base bearing alloys. Both copper and lead are attacked by acidic oils that contain sulfur. This is of particular concern with copper-lead bearing alloys. Effective protection can be obtained by employing layered construction, with a surface layer of a lead alloy containing tin, or a tin alloy. As long as the corrosion-resistant surface layer is intact, the underlying copper-lead alloy will not suffer damage by corrosion.

Tin and aluminum bearing alloys are substantially impervious to corrosion by the products of oil oxidation, and are used extensively in applications where the potential for lubricating-oil corrosion is known to be high. Although lubricating oil oxidation and contamination are the most common causes of bearing damage by corrosion, other sources of bearing corrosion, such as seawater, animal and vegetable oils, and corrosive gas, should be recognized. Selection and specification of a bearing material for a specific application should take into account the anticipated service conditions under which the bearings will have to operate, and the potentials for corrosion that these conditions may involve.

Bearing-Material Systems

Because of the widely varying conditions under which bearings must operate, commercial bearing materials have evolved as specialized engineering materials systems rather than as commodity products. They are used in relatively small tonnages and are produced by a relatively small number of manufacturers. Much proprietary technology is involved in alloy formulation and processing methods. Successful selection of a bearing material for a specific application often requires close technical cooperation between the user and the bearing producer.

Single-Metal Systems

Virtually all single-metal sliding bearings are made of either copper alloys or aluminum alloys (Table 1). Considerable ranges of compositions and properties are available in the older copper group. Brasses and bronzes have been widely used in bearing applications since the mid-1800's. Interest in the use of aluminum alloys was stimulated by World War II metal shortages and greatly accelerated by the commercial introduction of aluminum-tin bearing alloys in 1946. Since then, metal economics have encouraged the use of aluminum alloy bearings, but brasses and bronzes continue to be preferred by many designers of heavy and special-purpose machinery.

Fig. 1 Variation of bearing life with babbitt thickness for lead or tin babbitt bearings

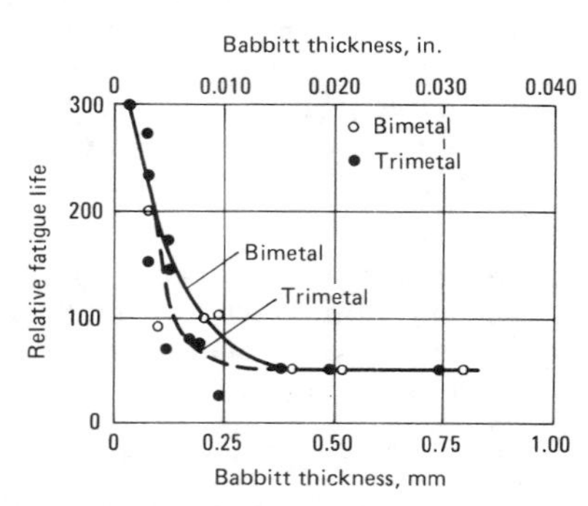

Bearing load: 14 MPa (2000 psi) for all tests.

Single-metal systems do not exhibit outstandingly good surface properties, and their tolerance of boundary and thin-film lubrication conditions is limited. As a result, the load-capacity rating for a single-metal bearing usually is low relative to the fatigue strength of the material from which it was made. Because of their metallurgical simplicity, these materials are well suited to small-lot manufacturing from cast tubes or bars, using conventional machine-shop processes.

Copper Alloys. Except for commercial bronze and low-lead tin bronze, copper alloys in single-metal systems are almost always used in cast form. This provides thick bearing walls (3.20 mm, or 0.125 in.) strong enough so that the bearing is retained in place when press fitted into the housing.

Commercial bronze and low-lead tin bronze (alloys C22000 and C47600) are used extensively in the form of wrought strip for thin-wall bushings, which are made in large volumes by high-speed press forming. The poor compatibility of these alloys can be improved by embedding a graphite-resin paste in rolled or pressed-in indentations, so that the running surface of the bushing consists of interspersed areas of graphite and bronze. Such bushings are widely used in automotive engine starting motors.

The lead in leaded tin bronzes is present in the form of free lead, dispersed throughout a copper-tin matrix so that the bearing surface consists of interspersed areas of lead and bronze.

Table 1 Single-metal bearing-material systems

Class	Material	Bearing-performance characteristics(a) Compatibility	Conformability	Embeddability	Fatigue strength	Corrosion resistance(b)	Load-capacity rating(c) MPa	ksi	Typical applications
1	Commercial bronze (10% Zn)	F	E	F	D	B	28	4	Bushings, washers
2	Tin bronze								
	High lead (16-25%)	D	D	D	D	E	21	3	Mill-machinery bearings, pump bearings, railroad-car bearings
	Medium lead (4-10%)	E	E	E	C	D	28	4	Wrist pin bushings, pump bushings, electric-motor bushings, track-roller bushings, farm-equipment gear bushings, mill-machinery bearings, machine-tool bearings
	Low lead (1-4%)	F	F	F	B	B	34	5	Wrist pin bushings, mill-machinery bearings, machine-tool bearings, earth-moving-machinery bearings, farm-equipment gear bushings
	Unleaded	F	F	F	A	B	34	5	Wrist pin bushings, mill-machinery bearings, machine-tool bearings, railroad-car wheel bearings
3	Aluminum alloy, low tin	D	D	D	D	A	28	4	Connecting-rod main bearings, bushings, mill-machinery bearings

(a) Bearing-performance characteristics rated on scale A through F, where A is highest (best) and F is lowest (poorest). (b) Corrosion resistance refers to corrosion by fatty acids of the kind that can form in petroleum-base oils. (c) Load-capacity rating approximates maximum safe unit loading for operation with steel journal under cyclic loading and excellent lubrication.

This improves compatibility, conformability and embeddability. In general, the best selection from this group of materials for a given application will be the highest-lead composition that can be used without risking excessive wear, plastic deformation or fatigue damage.

Low-lead and unleaded bronzes are also used in porous bushings produced by powder metallurgy methods. The sintered bushings are impregnated with oil, which provides a built-in supply of lubricant. Such bushings are widely used in applications involving light loads and requiring self-lubricating properties.

Aluminum Alloys. Virtually all solid aluminum bearings used in the United States are made from alloys containing from 5.5 to 7% tin, plus smaller amounts of copper, nickel, silicon and magnesium. Starting forms for bearing fabrication include cast and wrought tubes as well as rolled plate and strip, which can be press formed into half-round shapes. As in the case of solid bronze bearings, relatively thick bearing walls are employed.

The tin in these alloys is present in the form of free tin, dispersed throughout an aluminum matrix so that the bearing surface consists of interspersed areas of aluminum and tin. Surface properties are enhanced by the free tin in much the same way that those of bronze are improved by the presence of free lead.

The high thermal expansion of aluminum poses special problems in maintaining press-fit and running clearances. Various methods are employed for increasing yield strength, through heat treatment and cold work, to overcome plastic flow and permanent deformation under service temperatures and loads.

Bimetal Systems

All bimetal systems employ a strong bearing back to which a softer, weaker, relatively thin layer of a bearing alloy is metallurgically bonded. Low-carbon steel is by far the most widely used bearing-back material, although alloy steels, bronzes, brasses and (to a limited extent) aluminum alloys are also used. The bimetal bearing-material systems currently in significant commercial use in the United States are classified in Table 2.

The strengthening effect of a steel bearing back is illustrated clearly for classes 3 and 4 in Table 2, which may be compared with the aluminum and copper alloy single-metal systems in Table 1. When steel bearing backs are employed, load-capacity ratings for both copper and aluminum alloys are sharply increased above those of the corresponding single metals, without degrading any other properties. Similarly, in classes 1, 2, 5, 6 and 7, the strong bearing-back materials permit use of lead and tin alloys that have extremely good surface properties but that are so low in strength that they can be used only as single-metal bodies under very light loads.

Thin-layer construction has a strengthening effect on lead and tin alloys as illustrated in Table 2 (classes 1 and 2), where a 50% increase in load capacity is achieved by reducing babbitt layer thickness. Although similar behavior has been observed with aluminum and copper alloys, the thin liner effects are less dramatic. Liner thicknesses employed with these stronger alloys are established by metal economics and manufacturing process considerations, rather than by strength/thickness relationships.

Deterioration in surface properties with increasing liner-alloy fatigue strength is clearly seen by comparison of classes 1 and 2 with classes 3 and 4, and by comparisons within classes 3 and 4 (Table 2). In practice, only those

Table 2 Bimetal bearing-material systems

Class	Backing layer	Surface layer	Bearing-performance characteristics(a): Compat-ibility	Conform-ability	Embed-dability	Fatigue strength	Corrosion resistance(b)	Load-capacity rating(c): MPa	ksi	Typical applications
1	Steel	Tin babbitt:								
		0.25-0.50 mm (0.010-0.020 in.)	A	A	A	F	A	10	1.5	Connecting-rod and main bearings, camshaft bearings, electric-motor bushings, pump bushings, thrust washers
		0.102 mm (0.004 in.)	A	B	B	E	A	15	2.2	
2	Steel	Lead babbitt:								
		0.25-0.50 mm (0.010-0.020 in.)	A	A	A	F	B	10	1.5	Connecting-rod and main bearings, camshaft bearings, transmission bushings, pump bushings, thrust washers
		0.102 mm (0.004 in.)	A	B	B	E	B	15	2.2	
3	Steel	Aluminum alloy:								
		High-tin	B	C	C	D	A	31	4.5	Connecting-rod and main bearings, camshaft bearings, transmission bushings, pump bushings, thrust washers
		High-lead	B	C	C	D	A	41	6	Same as above
		Low-tin	D	D	D	C	A	41	6	Camshaft bearings, transmission bushings, thrust washers
		Tin-free	D	D	D	C	A	41	6	Same as above
4	Steel	Copper alloy:								
		Copper-lead	C	C	C	C	F	38	5.5	Connecting-rod and main bearings, camshaft bearings
		High-lead bronze	D	D	D	C	E	45	6.5	Camshaft bearings, turbine bearings, pump bushings, thrust washers
		Medium-lead bronze	E	E	E	B	D	55	8	Piston pin bushings, rocker-arm bushings, wear plates, steering-knuckle bushings, guide bushings, thrust washers
5	Medium-lead bronze	Tin babbitt, 0.25-0.50 mm (0.010-0.020 in.)	A	A	A	F	B	10	1.5	Connecting-rod and main bearings, thrust washers, railroad-car journal bearings, mill-machinery bearings
6	Medium-lead bronze	Lead babbitt, 0.25-0.50 mm (0.010-0.020 in.)	A	A	A	F	C	10	1.5	Connecting-rod and main bearings
7	Medium-lead bronze	Lead babbitt, 0.025 mm (0.001 in.)	A	C	B	C	C	48	7	Connecting-rod and main bearings
8	Aluminum alloy, low tin	Lead babbitt, 0.025 mm (0.001 in.)	A	C	B	D	C	41	6	Connecting-rod and main bearings

(a) Bearing-performance characteristics rated on scale A through F, where A is highest (best) and F is lowest (poorest). (b) Corrosion resistance refers to corrosion by fatty acids of the kind that can form in petroleum-base oils. (c) Load-capacity rating approximates maximum safe unit loading for operation with steel journal under cyclic loading and excellent lubrication.

Table 3 Trimetal bearing-material systems

Class	Backing layer	Intermediate layer	Surface layer	Bearing-performance characteristics(a) Compat-ibility	Conform-ability	Embed-dability	Fatigue strength	Corrosion resist-ance(b)	Load-capacity rating(c) MPa	ksi	Typical applications
1	Steel	Medium-lead bronze	Tin babbitt, 0.25-0.50 mm (0.010-0.020 in.)	A	A	A	F	B	10	1.5	Large connecting-rod and main bearings, bushings
2	Steel	High-lead bronze	Tin babbitt, 0.25-0.50 mm (0.010-0.020 in.)	A	A	A	F	B	10	1.5	Large connecting-rod and main bearings, bushings
3	Steel	Copper-lead	Lead babbitt, 0.075 mm (0.003 in.)	A	B	B	E	C	21	3	Connecting-rod and main bearings, camshaft bearings
4	Steel	Copper-lead	Lead babbitt, 0.025 mm (0.001 in.)	A	C	C	B	C	55	8	Connecting-rod and main bearings, bushings
5	Steel	High-lead bronze	Lead babbitt, 0.025 mm (0.001 in.)	A	C	C	B	C	69	10	Connecting-rod and main bearings, bushings, thrust washers
6	Steel	Medium-lead bronze	Lead babbitt, 0.025 mm (0.001 in.)	A	D	D	A	C	83	12	Connecting-rod and main bearings
7	Steel	Aluminum, low tin	Lead babbitt, 0.025 mm (0.001 in.)	A	C	C	B	B	48	7	Connecting-rod and main bearings
8	Steel	Aluminum, tin free, low alloy	Lead babbitt, 0.025 mm (0.001 in.)	A	C	C	B	B	55	8	Connecting-rod and main bearings
9	Steel	Aluminum, tin free, low alloy, precipitation hardened	Lead babbitt, 0.025 mm (0.001 in.)	A	C	C	B	B	69	10	Connecting-rod and main bearings
10	Steel	Aluminum, tin free, high alloy	Lead babbitt, 0.025 mm (0.001 in.)	A	C	C	B	B	69	10	Connecting-rod and main bearings
11	Steel	Silver	Lead babbitt, 0.025 mm (0.001 in.)	A	D	D	A	B	83	12	Connecting-rod and main bearings for aircraft reciprocating engines
12	Steel	Silver-lead	Lead babbitt, 0.025 mm (0.001 in.)	A	D	D	A	B	83	12	Connecting-rod and main bearings for aircraft reciprocating engines

(a) Bearing-performance characteristics rated on scale A through F, where A is highest (best) and F is lowest (poorest). (b) Corrosion resistance refers to corrosion by fatty acids of the kind that can form in petroleum-base oils. (c) Load-capacity rating approximates maximum safe unit loading for operation with steel journal under cyclic loading and excellent lubrication.

systems whose surface properties are rated "D" or better are successful under boundary and thin-film lubrication conditions. This restricts the use of bimetal materials in connecting-rod and main bearings to loads of 48 MPa (7 ksi) or less.

Bronze-back bearings (see Table 2, classes 5 and 6) do not exhibit combinations of performance characteristics substantially different from those of steel-back bearings. The practical advantages of bronze as a bearing-back material lie partly in the economics of small-lot manufacturing and partly in the relative ease with which worn bronze-back bearings can be salvaged by rebabbitting and remachining. From the standpoint of performance, the advantage of bronze over steel as a bearing-back material is the protection bronze affords against catastrophic bearing seizure in case of severe liner wear or fatigue. Similar protection is provided by the aluminum alloy bearing back in class 8.

Although the surface properties of bronze bearing-back materials are not impressive, they are superior to those of steel, and these "reserve" bearing properties can be of considerable practical importance in large, expensive machinery used in certain critical applications.

Trimetal Systems

All trimetal systems employ a steel bearing back, an intermediate layer of relatively high fatigue strength, and a tin alloy or lead alloy surface layer. The systems in current commercial use are listed by classes in Table 3. Most of these systems are derived from the steel-back bimetal systems in Table 2 (classes 3 and 4) by addition of a surface layer.

The strengthening effects of thin-layer construction are notable in those systems that incorporate electroplated lead alloy surface layers approximately 0.025 mm (0.001 in.) thick (Table 3, classes 4 to 12). Comparison of fatigue-strength and load-capacity ratings of these systems with those of corresponding bimetal systems in Table 2 shows that the thin lead alloy surface layer upgrades not only surface properties but also fatigue strength. The increase in fatigue strength can be attributed at least in part to the elimination of surface stress raisers, from which fatigue cracks can propagate.

Class 1 and class 2 trimetal systems comprise leaded bronze intermediate layers with relatively thick tin alloy surface layers, and represent an evolution from bronze-back babbitt construction wherein steel has replaced most of the bronze. This produces the expected economy and bearing-back yield strength, but retains the desirable "reserve" bearing properties exhibited by bronze-back construction.

Class 11 and class 12 trimetal systems, which have silver and silver-lead alloy intermediate layers, are too costly for most commercial applications. However, they provide an unequalled combination of high load capacity and corrosion resistance. They continue to have limited use in radial piston engines for aircraft.

Trimetal systems with electroplated lead-base surface layers and copper or aluminum alloy intermediate layers provide the best available combinations of cost, fatigue strength and surface properties. Such bearings have high tolerances for boundary and thin-film lubrication conditions, and thus can be used under higher loads than can any of the bimetal systems. Although more costly than the corresponding steel-back bimetal systems, they are used in high-volume automotive applications as well as in larger mobile and stationary engines. A highly developed body of mechanical, metallurgical and chemical manufacturing technology has been established in the plain-bearing industry, and this technology permits mass production of precision trimetal bearings without a severe cost penalty.

Casting Processes

Single-Metal Systems

Except for porous bronze oil-impregnated bushings, all the single-metal systems listed in Table 1 are commercially produced by casting, either with or without subsequent mechanical working. Plate, strip and sheet forms of commercial bronze, of low-lead and lead-free tin bronzes and of aluminum-tin alloys are initially cast as ingots, slabs or bars by static and continuous casting methods similar to those used for other brass and aluminum mill products. Subsequent rolling and annealing operations are also similar to those used for conventional mill products. Because of the extreme hot shortness of leaded tin bronzes and aluminum-tin alloys, these alloys must be rolled either cold or at only slightly elevated temperatures, with frequent intermediate annealing.

The recrystallized wrought structures of bronze and aluminum-tin bearing alloys are substantially different from the initial cast structures, with respect to the configurations of the copper and aluminum phases and of the free-lead and free-tin phases. The improvements in ductility and forming characteristics that result from these structural changes are of great importance in subsequent bearing manufacturing operations. Bearing performance properties are not strongly affected by these changes. Both the as-cast and wrought forms of these alloys are in commercial use and are equally acceptable in bearing applications.

Tubular and cylindrical bronze and aluminum-tin alloy shapes are produced by static, centrifugal and continuous casting methods, and subsequently are machined into bearings. The medium- and high-lead bronzes are used only in the as-cast condition because of their low ductility and extreme hot shortness, which preclude any substantial amount of plastic deformation of cast shapes. Cast aluminum-tin alloy tubes can withstand a limited amount of cold work, however, and in some instances cold compression of 4 to 5% is employed to increase yield strength and improve press-fit retention in the finished bearings.

Bimetal Systems

Specialized casting methods are widely employed for producing bimetal bearing materials in both tubular and flat strip forms. Except for aluminum alloy systems (Table 2, classes 3 and 8), all bimetal systems in commercial use can, at least in principle, be produced by casting methods, and systems that incorporate tin and lead babbitt liners greater than about 0.1 mm (0.004 in.) thick are universally produced by casting.

Babbitt Centrifugal Casting. Short tubular steel and bronze shapes (bearing shells) are commonly lined with tin or lead alloys by various forms of centrifugal casting. In these processes, a machined steel or bronze shell is first preheated and coated by immersion in molten tin or tin alloy. The prepared shell is then placed in a lathelike "spinner" and rotated at a controlled speed about its axis. Molten babbitt is admitted through one end and is uniformly distributed around the inside wall of the shell by centrifugal action. The molten layer then is cooled and solidified by spraying water against the outside of the rotating bearing shell. When properly controlled, these processes produce fine-grain liner layers of reasonably uniform thickness, completely bonded to the steel or bronze bearing-back material. Centrifugal casting methods are especially well-suited to large-diameter, thick-wall bearings, which are made in relatively small quantities, and to full-round seamless bearings, which cannot be fabricated from flat strip.

Bronze Centrifugal Casting. Leaded tin bronzes also can be applied to the inner walls of steel shells by centrifugal casting. Various methods of shell preparation are employed, including both molten-salt and controlled-atmosphere preheating. Complete absence of oxidation of the inner wall of the steel shell is always a fundamental requirement for complete metallurgical bonding. Centrifugal casting of bronzes is most successful with alloys containing more than about 3% tin and not more than about 15% lead. Outside this composition range, leaded tin bronze and copper-lead alloys are sensitive to lead segregation and consequent nonuniform "centrifuged" microstructures. Within these composition limits and under well-controlled process conditions, mechanically sound, well-bonded bronze layers with reasonably uniform microstructures can be produced.

Bronze Gravity Casting. All copper-lead alloys and leaded bronzes containing up to about 35% lead can be successfully cast in and bonded to steel shells by gravity casting methods, in which centrifugal forces are not a factor. In these processes, a core usually is used to form an annular space inside the shell, into which molten bronze or copper-lead alloy is poured. Several different processes of this kind are in commercial use, utilizing a variety of preheating methods, core materials, pouring methods and quenching procedures.

As in centrifugal casting, absence of oxides on the inner shell wall is necessary for complete bonding of the alloy layer to the steel back. Liner microstructures produced by gravity shell casting methods generally are more uniform than those obtained by centrifugal casting. For low-tin and/or high-lead compositions, gravity casting is preferred because of the absence of centrifuging effects on the solidifying alloy.

Babbitt Strip Casting. Steel-back tin alloy and lead alloy bearing strip materials are commonly produced by continuous casting in specially designed process lines in which separate cleaning, etching, hot tinning, liner-alloy casting and quenching operations are carried out continuously on a moving steel bearing-back strip. One or more in-line machining operations may also be incorporated so that the strip emerges with a closely controlled thickness, suitable for bearing fabrication.

Bronze Strip and Slab Casting. The oldest commercial processes for producing steel-back copper-lead and leaded bronze bearing strip also utilize continuous casting on a moving steel strip. Steel preheating, alloy casting and quenching operations are performed under a strongly reducing atmosphere to ensure freedom from oxidation. Some in-line machining also can be done, but the cast strip usually is machined in a separate line for close control of thickness. Additional cold rolling and annealing operations are also employed—particularly with the low and medium lead-tin bronze alloys, in which recrystallized structures are frequently preferred for their superior fabrication properties.

Strip casting of copper alloys comprises a difficult technology, requiring close process control, a high level of operator skill, and relatively expensive special-purpose equipment. It is used by only a few bearing manufacturers, but with considerable commercial success. It is employed not only for thin-gage coiled materials but also for heavy-gage slabs with steel thicknesses as high as 15 mm (0.60 in.).

Trimetal Systems

Trimetal materials with relatively thick surface layers (Table 3, classes 1, 2 and 3) are used mostly in large bearings. These bearings are produced in relatively low volumes from steel shells initially lined by casting with copper-lead alloys or bronze. After intermediate machining to remove excess liner alloy, such shells are commonly relined with tin or lead babbitt by centrifugal casting. Methods used are essentially the same as those for casting in bare steel or solid bronze shells.

Powder Metallurgy Processes

Single-Metal Systems. The only commercial use of powder metallurgy methods for making single-metal bearing materials is in fabrication of porous metal bushings, which are subsequently impregnated with oil. Low-lead and lead-free tin bronzes, and at least one iron-base alloy, are used. Methods employed are similar to those used for making structural powder metallurgy shapes—that is, pressing in a closed die followed by sintering under a reducing atmosphere. Postsinter coining and re-pressing operations sometimes are used for close control of final dimensions.

Bimetal and Trimetal Systems. No powder metallurgy processes are in commercial use with lead-base or tin-base bearing alloys, nor are there at present any commercially developed processes for lining bearing shells by means of powder metallurgy methods. In manufacture of steel-back copper-lead alloy and leaded bronze strip, however, powder metallurgy methods are employed more extensively than any other.

A wide variety of steel-back copper alloy materials, including counterparts of all of the cast copper-lead and leaded bronze bearing alloys (Table 2, class 4), can be produced by continuous sintering on a steel backing strip. In these processes, prealloyed powder particles are spread uniformly onto moving steel strip. As the strip passes through a furnace under a reducing atmosphere, the particles become sintered together, forming an open grid bonded to the steel strip. After cooling, this bimetal is rolled to densify the liner alloy and then resintered to develop complete interparticle and alloy/steel bonds. After resintering, the strip material may receive further rolling—to attain finish stock size, and sometimes to strain harden the alloy liner for increased strength.

Strip sintering technology makes possible the production of steel-core "sandwich" material, which is especially suitable for applications requiring two bearing surfaces (such as in some thrust washers). In this instance, powder spreading, sintering, cooling and rolling are repeated on the opposite side, after which the strip is finally resintered. Sintered strip for most automotive and truck bearing applications is processed in coils up to approximately 5 mm (0.2 in.) in over-all thickness. Thick-wall materials with steel layers up to about 16 mm (5/8 in.) thick also can be processed in flat slab lengths.

Both bimetal and trimetal bearing materials also can be made by impregnation or infiltration of a lower-melting lead alloy into a layer of sintered copper alloy powder. In impregnation, a bilayer strip made from prealloyed copper-lead alloy or leaded bronze powder is immersed in a bath of molten lead-tin alloy heated above the melting point of lead. During immersion, some of the lead at the surface of the strip is replaced by the lead-tin alloy. In infiltration, the copper alloy powder layer is free-sintered and not compacted after sintering. The open-grid sintered layer is then infiltrated with material having a lower melting temperature than that of the grid alloy. The infiltrant is usually molten lead or a lead alloy, but it may be a nonmetallic material such as Teflon.

It will be noted that by combining the technologies of sintering and casting it becomes possible to produce a class of unique metallic structures that cannot be obtained by casting alone. Several unique trimetal compositions are produced commercially, in which a copper-nickel, copper-tin or copper-nickel-tin grid is sintered and then infiltrated and overcast with lead babbitt, which is subsequently machined away to leave a surface layer about 0.025 to 0.08 mm (0.001 to 0.003 in.) thick. Also produced commercially are several self-lubricating trilayer structures in which the infiltrant is PTFE (Teflon), alone or in combination with lead, graphite or MoS_2 powder.

Powder Rolling. One very useful application of direct powder rolling that has been developed commercially in the plain-bearing industry is production of an aluminum-lead alloy strip for subsequent bonding to a steel back (see second item under class 3 in Table 2). In this method, prealloyed lead-aluminum powder and unalloyed aluminum powder are fed separately to a powder rolling mill and continuously compacted into a bilayer aluminum strip. After sintering, this strip is roll bonded to low-carbon steel, with the unalloyed aluminum bonding layer next to the steel. This steel-back strip is used as a bimetal material for bearing applications where the unit loading is beyond the capability of tin or lead babbitt bimetal material.

Roll Bonding Processes

Virtually all commercial manufacture of bimetal aluminum alloy bearing strip materials (see Table 2, class 3) is currently done by roll bonding the liner alloy to a steel backing strip. Both batch and continuous processes are employed, the latter being favored for economical high-volume processing of lighter-gage material.

In all roll bonding processes, whether batch or continuous, very clean aluminum and steel surfaces are forced together under intense pressure in a rolling mill, so that solid-phase bonding (cold welding) can occur between the two metals at many sites in the interface. Heat, which may be applied simultaneously with pressure in hot rolling and/or subsequently in postroll annealing, serves to develop complete diffusion bonding from the initial weld sites and to recrystallize the aluminum alloys so that the final bimetallic strip product exhibits useful liner-alloy ductility and complete bonding.

Tin-aluminum alloys usually are not bonded directly to steel because of undesirable interactions between the free tin constituent and the steel backing. A layer of electrolytic nickel plating on the steel surface is commonly used to alleviate these effects with both low-tin and high-tin alloy compositions.

Another method commonly used with tin-aluminum alloys employs a tin-free aluminum interlayer. This is accomplished by the use of alclad tin-aluminum alloy strip. The tin-free cladding layer serves as the bonding surface and is present as a distinct bond interlayer in the finished bimetal strip.

Direct roll bonding to steel is most commonly employed for tin-free aluminum alloys (see fourth item under class 3 in Table 2) and for lead-aluminum strip materials produced with an integral pure-aluminum bonding layer.

Electroplating Processes

Plated Overlays. Lead alloy surface layers (overlays) whose thickness must be limited to less than about 0.05 mm or 0.002 in. (Table 3, classes 4 to 12) are most commonly produced by electroplating the lead alloy on bimetallic bearings that have previously been finish machined. Specially designed plating racks are used to ensure uniform distribution of plating current over the bearing surface. With close control of current, critical dimensional tolerances often can be maintained so precisely that no machining of the electrodeposited alloy surface is required. One manufacturer has recently commercialized a process in which the lead alloy electroplating is applied continuously to precision-rolled bimetal strip. In this process, all forming and machining operations are done after electroplating.

Electroplated lead babbitts comprise both binary lead-tin and ternary lead-tin-copper compositions, all of which are commercially deposited from fluoborate electrolytes. To ensure against bond and plate defects, extreme care is exercised in preparing the basis metal.

In addition to cleaning and etching treatments, preplating basis-metal preparation usually includes deposition of one or more very thin metallic interlayers. A thin layer of nickel is most frequently used over copper-lead alloys and bronzes to prevent diffusion of tin from the plated surface layer into the copper basis metal. Copper is most often used over aluminum alloys to ensure complete adhesion of the plated lead alloy layer, and nickel sometimes is plated over the copper to prevent diffusion of tin from the lead alloy layer into the copper layer.

Plated Silver Intermediate Layers. Unalloyed silver and silver-lead alloy bearing liners are applied to steel shells by electrodeposition from cyanide plating baths. Final machining usually is done after plating, leaving a substantially thick layer (typically 0.25 to 0.38 mm, or 0.010 to 0.015 in.) of bonded silver liner material. Although as-plated thickness tolerances are not critical, special racking and masking techniques are employed to restrict plating to the surfaces where it is required and to eliminate local concentrations of high current density. If the structure of the plated layer is to be uniform, and the bond strength of the liner uniformly high, the steel basis metal must be prepared very carefully, and plating-bath compositions and cleanness must be properly controlled. Although the basic principles involved in silver plating of bearing liners are the same as for decorative silver plating, the unusually thick deposits involved (normally in excess of 0.50 mm, or 0.020 in.) and the extremely high quality requirements for bond and plated-metal soundness have led to development of several unique operating and control practices.

Tin Alloys

Tin-base bearing materials (babbitts) are alloys of tin, antimony and copper that contain limited amounts of zinc, aluminum, arsenic, bismuth, and iron. The compositions of tin-base bearing alloys, according to ASTM B23 and SAE specifications, are shown in Table 4.

The presence of zinc in these bearing metals generally is not favored. Arsenic increases resistance to deformation at all temperatures; zinc has a similar effect at 38 °C (100 °F), but causes little or no change at room temperature. Zinc has a marked effect on the microstructures of some of these alloys. Small quantities of aluminum (even less than 1%) will modify their microstructures. Bismuth is objectionable because, in combination with tin, it forms a eutectic that melts at 137 °C (279 °F). At temperatures above this eutectic, alloy strength is decreased appreciably.

In high-tin alloys, such as ASTM grades 1, 2 and 3, and SAE 11 and 12, lead content is limited to 0.50% or less because of the deleterious effect of higher percentages on the strength of these alloys at temperatures of 149 °C (300 °F) or above. Lead and tin form a eutectic that melts at 183 °C (361 °F). At higher temperatures, bearings become fragile as a result of formation of a liquid phase within them. The mechanical properties of ASTM grades 1 to 3 are shown in Table 5.

The mechanical-property values obtained from massive cast specimens are dependent on temperature. Hardness and compression tests are sensitive also to duration of the load because of the plastic nature of these materials. Bulk properties may be of some value in initial screening of materials, but they do not accurately predict behavior of the material in the form of a thin layer bonded to a strong backing, which is the manner in which the babbitts are normally used. The relationship that exists between bearing life and thickness of babbitt is shown in Fig. 2, which also shows the marked influence of operating temperature.

Fig. 2 Variation of bearing life with temperature for SAE 12 bimetal bearings

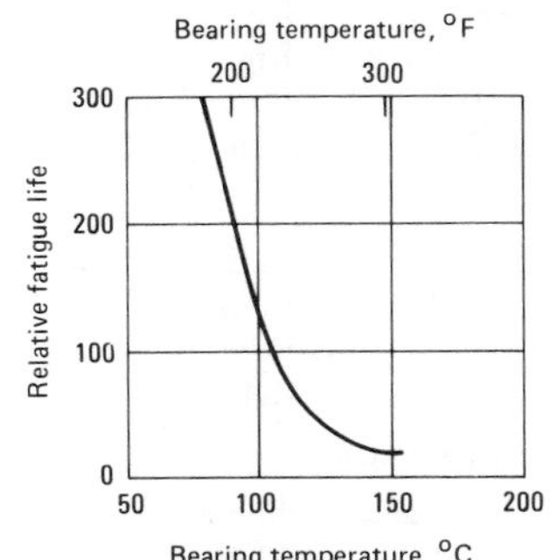

SAE 12 alloy lining, 0.05 to 0.13 mm (0.002 to 0.005 in.) thick, on steel backing. Bearing load: 14 MPa (2000 psi).

Compared with other bearing materials, tin alloys have low resistance to fatigue, but their strength is sufficient to warrant their use under low-load conditions. These alloys are commercially easy to bond and handle,

Table 4 Compositions of tin-base bearing alloys

Designation	Sn(a)	Sb	Pb (max)	Cu	Fe (max)	As (max)	Bi (max)	Zn (max)	Al (max)	Others (max total)
	Nominal composition, %									
ASTM B23 alloys										
Alloy 1	91.0	4.5	0.35	4.5	0.08	0.10	0.08	0.005	0.005	...
Alloy 2	89.0	7.5	0.35	3.5	0.08	0.10	0.08	0.005	0.005	...
Alloy 3	84.0	8.0	0.35	8.0	0.08	0.10	0.08	0.005	0.005	...
Alloy 11	87.5	6.8	0.50	5.8	0.08	0.10	0.08	0.005	0.005	...
SAE alloys										
SAE 11	86.0	6.0-7.5	0.50	5.0-6.5	0.08	0.10	0.08	0.005	0.005	0.20
SAE 12	88.0	7.0-8.0	0.50	3.0-4.0	0.08	0.10	0.08	0.005	0.005	0.20

(a) Desired in ASTM alloys; specified minimum in SAE alloys.

Table 5 Properties of selected ASTM B23 tin-base bearing alloys

Designation	Specific gravity	Compressive yield strength(a)(b) At 20 °C (68 °F) MPa	At 20 °C (68 °F) ksi	At 100 °C (212 °F) MPa	At 100 °C (212 °F) ksi	Compressive ultimate strength(a)(c) At 20 °C (68 °F) MPa	At 20 °C (68 °F) ksi	At 100 °C (212 °F) MPa	At 100 °C (212 °F) ksi	Hardness(d), HB At 20 °C	Hardness(d), HB At 100 °C	Solidus temperature °C	Solidus temperature °F	Liquidus temperature °C	Liquidus temperature °F	Pouring temperature °C	Pouring temperature °F
Alloy 1	7.34	30.3	4.40	18.3	2.65	88.6	12.85	47.9	6.95	17.0	8.0	223	433	371	700	440	825
Alloy 2	7.39	42.1	6.10	20.7	3.00	102.7	14.90	60.0	8.70	24.5	12.0	241	466	354	669	425	795
Alloy 3	7.46	45.5	6.60	21.7	3.15	121.3	17.60	68.3	9.90	27.0	14.5	240	464	422	792	490	915

(a) The compression-test specimens were cylinders 1½ in. long and ½ in. in diameter, machined from chill castings 2 in. long and ¾ in. in diameter. (b) Values for yield point were taken from stress-strain curves at a deformation of 0.125% reduction of gage length. (c) Values for ultimate strength were taken as the unit load necessary to produce a deformation of 25% of the length of the specimen. (d) Tests were made on the bottom face of parallel machined specimens cast at room temperature in a steel mold 2 in. in diameter by ⅝ in. deep. The Brinell hardness values listed are the averages of three impressions on each alloy, using a 10-mm ball and applying a 500-kg load for 30 s.

Table 6 Nominal compositions of lead-base bearing alloys

Designation	Pb	Sb	Sn	Cu (max)	Fe (max)	As	Bi (max)	Zn (max)	Al (max)	Cd (max)	Others
	Nominal composition, %										
ASTM B23 alloys											
Alloy 7 (a)	Rem	15.0	10.0	0.50	0.1	0.45	0.10	0.005	0.005	0.05	...
Alloy 8	Rem	15.0	5.0	0.50	0.1	0.45	0.10	0.005	0.005	0.05	...
Alloy 13 (b)	Rem	10.0	6.0	0.50	0.1	0.25(a)	0.10	0.005	0.005	0.05	...
Alloy 15 (c)	Rem	16.0	1.0	0.50	0.1	1.10	0.10	0.005	0.005	0.05	...
Other alloys											
SAE 16	Rem	3.5	4.5	0.10	...	0.05(a)	0.10	0.005	0.005	0.005	...
AAR M501 (d)	Rem	8.75	3.5	0.50	...	0.20(a)	...	...	...	...	...
SAE 19	Rem	...	10.0	...	...	...	...	...	...	...	...
SAE 190	Rem	...	7.0	3.0	...	...	...	...	...	...	...
Proprietary alloys											
A	95.65	...	3.35	0.08	...	...	...	...	...	...	0.67 Ca
B	83.30	12.54	0.84	0.10	...	3.05	...	...	...	...	...
C	Rem	10.0	3.0	0.20	...	...	...	...	...	...	2.0 Ag

(a) Also SAE 14. (b) Also SAE 13. (c) Also SAE 5. (d) Association of American Railroads, Specification M501; also ASTM B67.

Table 7 Properties of selected ASTM B23 lead-base bearing alloys

Designation	Specific gravity	Compressive yield strength(a)(b) At 20 °C (68 °F) MPa	ksi	At 100 °C (212 °F) MPa	ksi	Compressive ultimate strength(a)(c) At 20 °C (68 °F) MPa	ksi	At 100 °C (212 °F) MPa	ksi	Hardness(d), HB At 20 °C	At 100 °C	Solidus temperature °C	°F	Liquidus temperature °C	°F	Pouring temperature °C	°F
Alloy 7	9.73	24.5	3.55	11.0	1.60	107.9	15.65	42.4	6.15	22.5	10.5	240	464	268	514	338	640
Alloy 8	10.04	23.4	3.40	12.1	1.75	107.6	15.60	42.4	6.15	20.0	9.5	237	459	272	522	340	645
Alloy 15	10.05	...	...	...	...	...	...	...	...	21.0	13.0	248	479	281	538	350	662

(a) The compression-test specimens were cylinders 1.5 in. long, 0.5 in. in diameter, machined from chill castings 2 in. long, 0.75 in. in diameter. (b) Values were taken from stress-strain curves at a deformation of 0.125% reduction of gage length. (c) Values were taken as the unit load necessary to produce a deformation of 25% of the length of the specimen. (d) Tests were made on the bottom face of parallel-machined specimens that had been cast at room temperature in a steel mold, 2 in. in diameter by 0.625 in. deep. Values listed are the averages of three impressions on each alloy, using a 10-mm ball and applying a 500-kg load for 30 s.

and they have excellent antiseizure qualities. Furthermore, they are much more resistant to corrosion than lead-base bearing alloys.

These alloys vary in microstructure in accordance with their composition. Alloys that contain about 0.5 to 8% Cu and less than about 8% Sb are characterized by a solid-solution matrix in which are distributed needles of a copper-rich constituent and fine, rounded particles of precipitated SbSn. The proportion of the copper-rich constituent increases with copper content. SAE 12 (ASTM grade 2) has a structure of this type in which the needles often assume a characteristic hexagonal starlike pattern. Alloys that contain about 0.5 to 8% Cu and more than about 8% Sb exhibit primary cuboids of SbSn, as well as needles of the copper-rich constituent in the solid-solution matrix. In alloys with about 8% Sb and about 0.5 to 8% Cu, rapid cooling suppresses formation of the SbSn cuboids. This is particularly true of alloys containing lower percentages of copper.

Lead Alloys

There are two types of lead babbitts: (*a*) alloys of lead, tin, antimony and in many instances arsenic, and (*b*) alloys of lead, calcium, tin and one or more of the alkaline earth metals. Many alloys of the first group have been used for centuries as type metals, and were probably employed as bearing materials because of the properties they were known to possess. The advantages of arsenic additions in this type of bearing alloy have been generally recognized since 1938. Alloys of the second type were developed early in the present century.

Typical compositions of lead-base bearing alloys covered by ASTM specifications, and their corresponding SAE designations, are listed in Table 6 along with compositions of other proprietary alloys. Additional information on the mechanical properties of some of these alloys is given in Table 7.

In the absence of arsenic, the microstructures of these alloys comprise cuboid primary crystals of SbSn, or of antimony embedded in a ternary mixture of Pb-Sb-SbSn in which lead forms the matrix. The number of these cuboids per unit volume of alloy increases as antimony content increases. If antimony content is more than about 15%, the total amount of the hard constituents increases to such an extent that the alloys become too brittle to be useful as bearing materials.

Arsenic is added to lead babbitts to improve their mechanical properties, particularly at elevated temperatures. All lead babbitts are subject to softening or loss of strength during prolonged exposure to the temperatures (95 to 150 °C; 200 to 300 °F) at which they serve as bearings in internal-combustion engines. Addition of arsenic minimizes such softening. Under suitable casting conditions, the arsenical lead babbitts—for example, SAE 15 (ASTM grade 15)—develop remarkably fine and uniform structures. They also have better fatigue strength than arsenic-free alloys.

Arsenical babbitts give satisfactory service in many applications. Use of these alloys increased greatly during the second world war, particularly in the automobile industry and in diesel engines. The alloy most widely used is SAE 15 (ASTM grade 15), which contains 1% arsenic. Automobile bearings of this alloy usually are made from continuously cast bimetal (steel/babbitt) strip. When properly handled, this alloy can withstand the considerable strain that results from forming the bimetal strip into bearings.

Diesel-engine bearings often are cast as individual bearing shells by either centrifugal or gravity methods. In applications where higher hardness is required and where formability requirements are less severe (rolling-mill bearings, for instance), an alloy that contains 3% arsenic has been used successfully (alloy B in Table 6).

For many years, lead-base bearing alloys were considered to be only low-cost substitutes for tin alloys. However, the two groups of alloys do not differ greatly in antiseizure characteristics, and when lead-base alloys are used with steel backs and in thicknesses below 0.75 mm (0.03 in.), they have fatigue resistance that is equal to, if not better than, that of tin alloys. Bearings of any of these alloys remain serviceable longest when they are no more than 0.13 mm (0.005 in.) thick (see Fig. 1). The superiority of lead alloys over tin alloys becomes more marked as operating temperature increases. For this reason, automotive engineers generally favor lead-base alloys of compositions that approximate ASTM alloys 7 and 15 and SAE alloy 16. SAE alloy 16 is cast into and on a porous sintered matrix, usually copper-nickel, bonded to steel. The surface layer of babbitt is 0.025 to 0.13 mm (0.001 to 0.005 in.) thick.

The fatigue resistance of bearing materials depends to a great extent on the design of the bearing. The strength and rigidity of the supporting structure, the thickness of the backing metal (steel or bronze), the thickness of the bearing material and the character of the bond between the bearing material and the backing are all factors of consequence in bearings for use in high-speed reciprocating engines, such as the main and connecting-rod bearings of automobile and aircraft engines.

Resistance to fatigue is somewhat less important in bearings that operate under static load—for example, journal bearings in traction-motor supports for diesel locomotives and in railway freight cars. In such bearings, antiseizure characteristics, conformability, compressive strength, and resistance to abrasion and corrosion are of greater significance. The lining metal generally employed in such journal bearings is the low-arsenic AAR alloy (ASTM B67) cast onto a leaded bronze back.

Pouring temperature and rate of cooling markedly influence the microstructures and properties of lead alloys, particularly when they are used in the form of heavy liners for railway journals. High pouring temperatures and low cooling rates, such as result from use of overly hot mandrels, promote segregation and formation of a coarse structure. A coarse structure may cause brittleness, low compressive strength and low hardness. Therefore, low pouring temperatures (325 to 345 °C; 620 to 650 °F) usually are recommended. Because these alloys remain relatively fluid almost to the point of complete solidification (about 240 °C, or 465 °F, for most compositions), they are easy to manipulate and can be handled with no great loss of metal from drossing.

Use of lead babbitts containing calcium and alkaline earth metals is confined almost entirely to railway applications, although these babbitts also are employed to some extent in certain diesel-engine bearings. One of the more widely used alloys contains 1.0 to 1.5% tin, 0.50 to 0.75% calcium, and small amounts of various other elements. The strength of this alloy approximates that of a tin alloy containing 90% Sn, 8% Sb and 2% Cu. Hardness of this lead alloy is about 20 HB, and the solidus is 321 °C (610 °F). The liquidus is probably near 338 °C (640 °F). The pouring temperature, which varies from 500 to 520 °C (930 to 970 °F), is relatively high and accounts for the formation of a larger volume of dross than that encountered in melting of Pb-Sb-Sn alloys. Care must be taken to avoid contamination of the alloy with antimonial lead babbitts, and vice versa. Deformability and resistance to wear are of the same order as those of the other lead babbitts. Most alloys of this type are subject to corrosion by acidic oils.

Overlays

The improvement in fatigue life that can be achieved by decreasing babbitt-layer thickness has already been noted. Economically as well as mechanically, it is difficult to consistently achieve very thin uniform babbitt layers bonded to bimetal shells by casting techniques. Therefore, the process of electroplating a thin precision babbitt layer on a very accurately machined bimetal shell was perfected. A specially designed plating rack allows the thickness of the coplated babbitt layer to be regulated so accurately that machining usually is not required.

Coplated tin babbitts were found to be inferior in performance to lead babbitts. Plated babbitts are somewhat different in structure and composition from their cast counterparts. SAE alloy 190 is the most widely used overlay plate. The tin content of this alloy gives it better wear resistance than that of pure lead, and is necessary to protect the lead from corrosion; the copper content increases fatigue life.

When an SAE 190 overlay is plated directly onto a copper-lead bimetal surface, the tin has a tendency to migrate to the copper-lead interface, forming a brittle copper-tin intermetallic compound and/or diffusing into the lead phase. This decreases the corrosion resistance of the overlay and causes embrittlement along the bond line. To avoid this deterioration, a thin, continuous barrier layer, preferably nickel about 1.3 μm (0.05 mil) thick, is plated onto the copper-lead surface just prior to plating of the overlay. In addition to providing better surface behavior, overlays improve fatigue performance of the intermediate layer by preventing cracking in this layer. Plated overlays generally range in thickness from 0.013 to 0.05 mm (0.0005 to 0.002 in.), with fatigue life increasing markedly as overlay thickness decreases. In order to take full advantage of the improved fatigue life achieved with thin overlays, it is necessary to minimize assembly imperfections (such as misalignment) and to maintain close tolerances on machined shafts and bearing bores. Engine components must be thoroughly cleaned before assembly, and adequate air and lubricant filtration must be maintained if the overlay is to survive during the useful life of the bearing. Under adverse wear conditions, however, premature removal of the overlay will not necessarily impair operation of the bearing, because the exposed intermediate bearing alloy layer should continue to function satisfactorily.

A surface layer of SAE 190 0.025 mm (0.001 in.) thick is commonly used in trimetal bearings such as those listed in Table 3, classes 4 to 10. Such trimetal bearings constitute the highest load-capacity bearings so far developed. In classes 11 and 12 trimetal bearings, the surface layer usually is SAE 19.

Copper Alloys

Copper-base bearing alloys comprise a large family of materials with a considerable range of properties. They include commercial bronze, copper-lead alloys, and leaded and unleaded tin bronzes. They are used alone in single-metal bearings, as bearing backs with babbitt surface layers, as bimetal layers bonded to steel backs, and as intermediate layers in steel-backed trimetal bearings (see Tables 1, 2 and 3).

Pure copper is a relatively soft, weak metal. The principal alloying element used to harden and strengthen it is tin, with which it forms a solid solution. However, when tin content is higher than about 8%, a hard constituent (the alpha-delta eutectoid) develops in cast copper-tin alloys because of deviation from true equilibrium. This constituent is quite hard, and its presence causes a considerable improvement in wear resistance. Lead is present in all cast copper-base bearing alloys as a nearly pure, discrete phase, because its solid solubility in the matrix is practically nil. The lead phase, which is exposed on the running surface of a bearing, constitutes a site vulnerable to corrosive attack under certain operating conditions.

The antifriction behavior of copper-base bearing alloys improves as lead content increases, although at the same time strength is degraded because of increased interruption of the continuity of the copper alloy matrix by the soft, weak lead. Thus, through judicious control of tin content, lead content and microstructure, an entire family of bearing alloys has evolved to suit a wide variety of bearing applications.

Table 8 gives specification numbers and nominal compositions of copper-base bearing alloys, as well as the forms in which the alloys are used and general notations on typical applications. The information in this table should be used in conjunction with the appropriate portions of Tables 1 to 3 and with the brief descriptions that follow.

Commercial Bronze. Lead-free copper alloys are characterized by poor antifriction properties but fairly good load-carrying ability. Wrought commercial bronze strip (SAE 795) with 10% zinc can be readily press formed into cylindrical bushings and thrust washers. Strength can be increased by cold working this inexpensive material.

Unleaded Tin Bronzes. The unleaded copper-tin alloys are known as phosphor bronzes because they are deoxidized with phosphorus. They are used principally in cast form as shapes for specific applications, or as rods or tubes from which solid bearings are machined. They have excellent strength and wear resistance, both of which improve with increasing tin content, but poor surface properties. They are used for bridge turntables and trunnions in contact with high-strength steel, and in other slow-moving applications.

Low-Lead Tin Bronzes. The inherently poor machinability of tin bronzes can be improved by adding small amounts of lead. Such additions do not significantly improve surface properties such as lubricity, however, and applications for these alloys are essentially the same as those for unleaded tin bronzes.

Medium-Lead Tin Bronzes. The only wrought strip material in this group of alloys is SAE 791, which is press formed into solid bushings and thrust washers. C83600 is used in cast form as bearing backs in bimetal bearings. SAE 793 and 798 are chemically similar low-tin, medium-lead materials that are cast or sintered on steel backs and used as surface layers for medium-load bimetal bushings. SAE 792 and 797 are higher in tin and slightly higher in lead, are cast or sintered on steel backs, and are used for heavy-duty applications such as wrist pin bushings and heavy-duty thrust surfaces.

High-lead tin bronzes contain medium to high amounts of tin, and relatively high lead contents to markedly improve antifriction characteristics. SAE 794 and 799 are widely used as bushings for rotating loads, and have the same chemical matrix as 793 and 798 but with three times as much free lead. Both are generally cast or sintered on steel backs and are used for somewhat higher speeds and lower loads than alloys 793 and 798. The 3Sn-25Pb alloy cast on a steel back provides a much stronger bronze matrix than plain 75-25 copper-lead alloy, and is used with a plated overlay as the intermediate layer in heavy-duty trimetal bearing applications such as main and connecting-rod bearings in diesel truck engines. This construction provides the highest load-carrying ability available at the present time in copper alloy trimetal bearings.

Copper-lead alloys are used extensively in automotive, aircraft and general engineering applications. These alloys are usually cast or sintered to a steel backing strip from which parts are blanked and formed into full-round or half-round shapes depending on final application. Copper-lead alloys continuously cast on steel strip typically consist of copper dendrites perpendicular and securely anchored to the steel back, with an interdendritic lead phase. In contrast, sintered copper-lead alloys of similar composition are composed of more equiaxed copper grains with an intergranular lead phase.

The high-lead alloys (28 to 40% Pb) may be used bare on steel or cast iron journals as medium-duty automotive bearings. Tin content in these alloys is restricted to a low value to maintain a soft copper matrix, which along with the higher lead improves the antifriction/antiseizure properties. Bare bimetal copper-lead bearings are used less frequently today than they were some years ago because the lead phase, present as nearly pure unalloyed lead in all cast copper alloys, is susceptible to attack by corrosive products that can form in the crankcase lubricant during the longer oil-change periods now in use. Therefore, many of the copper-lead alloys with lead contents near 25% are used as bases for plated overlays in trimetal bearings for automotive and diesel engines.

Other alloys included in this group are the special sintered and infiltrated or impregnated materials SAE 482, 484, and 485, which were described in the section on powder metallurgy processing. The last two items in Table 8 consist of an open copper-nickel or copper-nickel-tin grid, which is sintered onto a steel back, then infiltrated with a Pb-Sn-Sb alloy (SAE 16) to make bimetal grid bearings. Alternatively, the lead-base alloy may be overcast so that it completely covers the grid. Ex-

Table 8 Designations and nominal compositions of copper-base bearing alloys

Designations				Nominal composition, %						
UNS number	SAE	Other	Former SAE	Cu	Sn	Pb	Zn	Other	Form	Use
Commercial bronze										
C22000	795	...	...	90	0.5	...	9.5	...	Wrought strip	Solid bronze bushings and washers
Unleaded tin bronzes										
C90300	C90300	...	620	88	8	0	4	...	Cast tubes	Solid bronze bearings
C90500	C90500	...	62	88	10	0	2	...	Cast tubes	Solid bronze bearings
C91100	...	...	...	84	16	0	0	...	Cast tubes	Solid bronze bearings
C91300	...	...	...	81	19	0	0	...	Cast tubes	Solid bronze bearings
Low-lead tin bronzes										
C92200	C92200	...	622	88.5	6	1.5	4	...	Cast tubes	Solid bronze bearings
C92300	C92300	...	621	87.0	8.5	0.5	4	...	Cast tubes	Solid bronze bearings
C92700	C92700	...	63	87.5	10	2	0.5	...	Cast tubes	Solid bronze bearings
Medium-lead tin bronzes										
C54400	791	...	...	88	4	4	4	...	Wrought strip	Solid bronze bushings and washers
		F32/62	...	87	4	4	3	2 Fe	Cast on steel back	Bimetal bushings and washers; trimetal intermediate layer
C83600	C83600	...	40	85	5	5	5	...	Cast tubes	Solid bronze bearings and bronze bearing backs
C93200	C93200	...	660	83	7	7	3	...	Cast tubes	Solid bronze bearings
C93600	793	...	...	85	4	8	3	...	Cast on steel back	Bimetal surface layer
... .	798	...	...	88	4	8	...	...	Sintered on steel back	Bimetal surface layer
C93700	C93700	...	64	80	10	10	...	...	Cast tubes	Solid bronze bearings and bronze bearing backs
... .	792	...	...	80	10	10	...	...	Cast on steel back	Bimetal surface layer and trimetal intermediate layer
... .	797	...	...	80	10	10	...	...	Sintered on steel back	Bimetal surface layer
High-lead tin bronze										
C93800	C93800	...	67	78	6	16	...	...	Cast tubes	Solid bronze bearings and bronze bearing backs
... .	...	AMS 4825	...	74	10	16	...	...	Cast on steel back	Bimetal surface layer
... .	794	...	...	71.5	3.5	23	2	...	Cast on steel back	Bimetal surface layer
... .	799	...	...	74	3	23	...	...	Sintered on steel back	Bimetal surface layer
... .	...	AMS 4824	...	75	1	24	...	...	Cast on steel back	Trimetal intermediate layer
... .	...	F780	...	74	2.5	23.5	...	...	Sintered on steel back	Trimetal intermediate layer
... .	...	F15/112	...	72	3	25	...	...	Cast on steel back	Bimetal surface layer and trimetal intermediate layer
... .	...	AMS 4840	...	70	5	25	...	...	Cast tubes	Solid bronze bearings

(continued)

(a) Composition of dense alloy after infiltration. (b) Composition of open grid before infiltration

cess babbitt can then be machined off, leaving a very thin layer covering the grid, to make a medium-duty trimetal bearing.

Aluminum Alloys

Successful commercial use of aluminum alloys in plain bearings dates back to about 1940, when low-tin aluminum alloy castings were introduced to replace solid bronze bearings for heavy machinery. Production of steel-backed strip materials by roll bonding became commercially successful about 1950, permitting the development of practical bimetal and trimetal bearing-material systems using aluminum alloys in place of babbitts and copper alloys.

The ready availability of aluminum and its relatively stable cost have provided an incentive for continuing development of its use in plain bearings.

Table 8 (continued)

UNS number	Designations SAE	Other	Former SAE	Cu	Nominal composition, % Sn	Pb	Zn	Other	Form	Use
Copper-lead alloys										
... .	49	...	...	75.5	0.5	24	...	...	Cast on steel back	Trimetal intermediate layer
... .	49	...	...	75.5	0.5	24	...	...	Sintered on steel back	Trimetal intermediate layer
... .	48	...	...	70	...	28	...	1.5 Ag	Cast on steel back	Bimetal surface layer and trimetal intermediate layer
... .	482	...	...	67	5	28	...	...	Sintered on steel impregnated with Pb-Sn	Bimetal surface layer
... .	480	...	...	65	...	35	...	...	Cast on steel back	Bimetal surface layer
... .	480	...	...	65	...	35	...	...	Sintered on steel back	Bimetal surface layer
... .	481	...	...	55	0.25	40	...	5 Ag	Cast on steel back	Bimetal surface layer
... .	484	...	...	55	3	42	...	(a)	Sintered on steel back infiltrated with Pb	Bimetal surface layer
... .	485	...	...	48	1	51	...	(a)	Sintered on steel back, infiltrated with Pb	Bimetal surface layer
				98	2	...	...	(b)	Sintered on steel back, infiltrated with Pb	Bimetal surface layer
... .	...	F510	...	41	2	48	...	7 Ni, 2 Sb(a)	Sintered on steel back, infiltrated with Pb-Sn-Sb alloy (SAE 16)	Bimetal surface layer and trimetal intermediate layer
				86	2	...	...	12 Ni(b)	Sintered on steel back, infiltrated with Pb-Sn-Sb alloy (SAE 16)	Bimetal surface layer and trimetal intermediate layer
... .	...	M100A	...	40	1	48	...	9 Ni, 2 Sb(a)	Sintered on steel back, infiltrated with Pb-Sn-Sb alloy (SAE 16)	Bimetal surface layer and trimetal intermediate layer
				85	...	...	...	15 Ni(b)	Sintered on steel back, infiltrated with Pb-Sn-Sb alloy (SAE 16)	Bimetal surface layer and trimetal intermediate layer

(a) Composition of dense alloy after infiltration. (b) Composition of open grid before infiltration.

Aluminum single-metal, bimetal and trimetal systems now can be used in the same load ranges as babbitts, copper-lead alloys and high-lead tin bronzes. Moreover, the outstanding corrosion resistance of aluminum has become an increasingly important consideration in recent years, and has led to widespread use of aluminum alloy materials in automotive engine bearings in preference to copper-lead alloys and leaded bronzes.

Designations and Compositions. Alloy designations and nominal compositions of the commercial aluminum-base bearing alloys used most extensively in the United States are listed in Table 9. In these alloys, additions of silicon, copper, nickel, magnesium and manganese function to strengthen the aluminum through solid-solution and precipitation mechanisms. Fatigue strength and the opposing properties of conformability and embeddability are largely controlled by these elements. Tin, cadmium and lead are instrumental in upgrading the inherently poor compatibility of aluminum. Silicon has a beneficial effect on compatibility in addition to a moderate strengthening effect. Although not well understood theoretically, this compatibility-improving mechanism is of considerable practical value and is utilized effectively in the high-lead and tin-free alloys.

Microstructural Characteristics. The cast low-tin alloys (alloys 4, 5, 6 and 7 in Table 9) all display similar microstructures consisting of equiaxed aluminum grains with $NiAl_3$, free silicon (if present) and free tin precipitated in the grain boundaries. Tin forms a nearly complete envelope around each aluminum grain. The copper and magnesium are mostly or completely in solid solution in the aluminum and are not visible under a microscope. Microstructures of the wrought low-tin and high-tin alloys (1, 8 and 9 in Table 9) exhibit the expected effects of rolling and annealing in that the as-cast aluminum grains are replaced by new recrystallized aluminum grains, with

Table 9 Designations and nominal composition of aluminum-base bearing alloys

	Designation			Nominal composition, %									
No.	SAE	AA	Other	Al	Si	Cu	Ni	Mg	Sn	Cd	Other	Form	Typical applications
High-tin aluminum alloy													
1	783	8081	...	79	...	1	...	...	20	...	...	Wrought strip, O temper, bonded to steel back	Bimetal surface layer
High-lead aluminum alloys													
2	...	...	F-66	85	4	1	...	...	1.5	...	8.5 Pb	Powder rolled and sintered strip, O temper, bonded to steel back	Bimetal surface layer
3	...	...	AL-6	88	4	0.5	...	0.5	1	...	6 Pb	Wrought strip, O temper, bonded to steel back	Bimetal surface layer
Low-tin aluminum alloys													
4	770	850.0	...	91.5	0.7	1	1	...	6.5	...	...	Cast tubes, T101 temper(a)	Solid aluminum bearings; aluminum bearing backs
5	...	A850.0	...	89.5	2.5	1	0.5	...	6.5	...	...	Cast tubes, T101 temper(a)	Solid aluminum bearings; aluminum bearing backs
6	...	B850.0	...	89.5	...	2	1.2	0.8	6.5	...	...	Cast tubes, T5 temper(b)	Solid aluminum bearings; aluminum bearing backs
7	...	...	MB-7	89	0.7	1	1.7	1	7	...	...	Cast tubes, T5 temper(b)	Solid aluminum bearings; aluminum bearing backs
8	780	828.0	...	90.5	1.5	1	0.5	...	6.5	...	...	Wrought strip and plate, H12(c) temper	Solid aluminum bearings; aluminum bearing backs
												Wrought strip, O temper, bonded to steel back	Bimetal surface layer; trimetal intermediate layer
9	...	...	A300	91	1	2	...	...	6	...	...	Wrought strip, O temper, bonded to steel back	Trimetal intermediate layer
Tin-free aluminum alloys													
10	781	4002	...	95	4	0.1	...	0.1		1	...	Wrought strip, O temper, bonded to steel back	Bimetal surface layer; trimetal intermediate layer
11	782	...	...	95	...	1	1	...		3	...	Wrought strip, O temper, bonded to steel back	Bimetal surface layer; trimetal intermediate layer
12	...	...	A250	...	...	1	1	...		3	1.5 Mn	Wrought strip, O temper, bonded to steel back	Trimetal intermediate layer
13	...	...	AS78	88	11	1	...	...		...	...	Wrought strip, O temper, bonded to steel back	Trimetal intermediate layer
14	...	4002	F-154	95	4	0.1	...	0.1		1	...	Wrought strip, T6 temper(d), bonded to steel back	Trimetal intermediate layer

(a) Artificially aged and cold pressed. (b) Artificially aged. (c) Strain hardened, approximately 25% cold reduction. (d) Solution treated and artificially aged.

the insoluble phases ($NiAl_3$ and silicon) uniformly redistributed throughout. The original continuous grain-boundary envelope of free tin assumes a completely new configuration, the tin now appearing as somewhat elongated, discontinuous "lakes". This characteristic structure, often termed "reticular", results in much greater ductility than that of as-cast alloys. This is of considerable practical importance, especially in the high-tin alloy. The wrought tin-free alloys (alloys 10 to 14 in Table 9) exhibit very simple microstructures, each consisting of a recrystallized aluminum matrix with a fine, uniform dispersion of free silicon or $NiAl_3$ particles. Cadmium (if present) is barely discernible under a microscope; copper and magnesium (if present) are in solid solution and cannot be discerned at all.

The lead-aluminum alloys (2 and 3 in Table 9) exhibit much the same microstructural characteristics as the tin-free alloys, but with the addition of thin stringers or ribbons of Pb-Sn constituent, elongated in the rolling direction. During recrystallization, this constituent does not coalesce into lakes as does free tin, and the ribbonlike configuration persists in finished bearings. The effectiveness of the modest lead concentrations in these alloys in imparting surface compatibility probably is related to the favorable orientation of the Pb-Sn ribbons relative to the bearing surface.

Study of the microstructural characteristics of aluminum-base bearing alloys is more valuable in predicting their fabrication behavior than their performance capabilities, and, with a few exceptions, microstructural quality standards are not routinely applied during manufacture of aluminum-base bearings and bearing materials.

Mechanical Properties and Alloy Tempers. Conventional mechanical properties, somewhat like microstructural features, are of more value in understanding fabrication behavior of aluminum-base bearing alloys than in predicting their bearing performance. With the exception of solid aluminum alloy bearings, in which there is no steel back and where press-fit

Table 10 Approximate mechanical properties of aluminum-base bearing alloys

Classification	Tensile strength MPa	ksi	Yield strength MPa	ksi	Hardness, HB(a)
High-tin aluminum strip	114	16.5	41	6	30
High-lead aluminum strip	117	17	62	9	32
Low-tin aluminum strip	117 to 138	17 to 20	48 to 124	7 to 18	32 to 40
Tin-free aluminum strip	124 to 207	18 to 30	62 to 138	9 to 20	38 to 48
Low-tin aluminum castings	159 to 234	23 to 34	117 to 172	17 to 25	54 to 74

(a) 500-kg load, 10-mm ball.

retention depends entirely on the strength of the aluminum alloy, mechanical properties of finished bearings are rarely specified—and then usually for control purposes only. Consideration of some of these properties (Table 10) does, however, contribute to an understanding of these alloys as a family of related engineering materials, and of their relationship to the better-known structural aluminum alloys. The wrought alloys as a group are low in hardness and strength compared with conventional aluminum structural alloys. With one exception (No. 14, Table 9), no use is made of heat treatment or cold working for increasing mechanical strength.

Cast aluminum-base bearing alloys are low in hardness and strength compared with conventional cast aluminum alloys, but are heat treated and cold worked to increase their yield-strength levels above as-cast values.

The majority of current commercial applications of aluminum-base bearing alloys involve steel-backed bimetal or trimetal bearings. To determine the most cost-effective aluminum material for any specific application, consideration should be given to the economic advantages of bimetal versus trimetal systems. The higher cost of the high-tin and high-lead alloys usually is offset by eliminating the cost of the lead alloy overlay plate. If the higher load capacity of a trimetal material is required, it then becomes important to select an aluminum liner alloy that provides adequate but not excessive strength, so that conformability and embeddability are not sacrificed unnecessarily. The tin-free alloy group (alloys 10 to 14 in Table 9) offers a wide range of strength properties, and the most economical choice usually is found in this group.

Silver Alloys

Use of silver in bearings is largely confined to unalloyed silver (AMS 4815) electroplated on steel shells, which then are machined to very close dimensional tolerances and finally precision plated to size with a thin overlay of soft metal. The overlay may be lead-tin-copper or lead-tin alloy. In some aircraft applications, the overlay consists of a plated layer of lead with a final layer of indium. Such bearings are then heat treated to diffuse the indium into the lead.

As a bearing material for heavy-duty applications, plated silver is invariably used with an overlay because silver itself possesses poor surface characteristics. Silver on steel with an overlay is regarded as the ultimate fatigue-resistant bearing material and is superior to all other bearing alloys in corrosion resistance. Load-carrying tests have indicated that silver is suitable for loads up to 83 MPa (12 ksi).

Silver-lead alloy liners, with 0.5 or 2.5% lead for improved surface characteristics, have been supplanted by pure silver intermediate layers in all but a few applications. Silver-lead alloy liners, when used, are also applied by electrodeposition, as discussed earlier in this article. The use of silver-lead is now limited to a locomotive application where bearings are finished with a very thin plated layer of pure lead. In another locomotive application, a slipper bearing is finished with a thin lead plate on pure silver.

Silver was widely used during and after World War II in aircraft applications, where its high cost could be justified. With the phasing out of piston engines, however, the use of silver in bearings has greatly declined. Current applications are specialized, chiefly in the aircraft and locomotive industries. In view of the rapidly rising cost of silver, any increase in demand for this material would stimulate the search for a comparable, less expensive substitute.

Other Metallic Bearing Materials

Gray Cast Irons. Cast irons are standard materials for certain applications involving friction and wear, such as brake drums, piston rings, cylinder liners and gears. Cast irons perform well in such applications, and thus should be given consideration as bearing materials. Gray iron bearings have proved successful in refrigeration compressors where bearing pressures seldom exceed 4500 kPa (650 psi) for main bearings and 5500 kPa (800 psi) for connecting-rod bearings. Normally, the journals in refrigeration compressors are made either of steel, carburized and hardened to 55 to 60 HRC, or of pearlitic malleable or ductile iron, hardened to 44 to 48 HRC and having a surface finish of 0.3 μm (12 μin.) rms or better. Because of occasional dilution of the oil with liquid refrigerant and heavy foaming of the oil, lubrication may become marginal for short periods of time. Fine-grain iron with uniformly distributed No. 6 (or finer) graphite flakes usually performs well during these periods. Often the bearings are phosphate coated to improve their seizure resistance. This type of coating also creates a spongelike surface that promotes retention of oil.

For good wear resistance, gray cast iron should be pearlitic with randomly distributed graphite flakes. Cast irons have been heat treated to martensitic structures for use as cylinder liners, but the benefits of such heat treatment have not been economically justifiable. Hardened cast iron has been used successfully for the ways on machine tools.

Cemented Carbides. Hard materials such as cemented tungsten carbide have been used successfully for various specialized bearing and seal applications. Cemented carbides are available in many grades, adaptable to a variety of applications. One manufacturer offers 17 standard materials in the tungsten carbide family, a cemented oxide material, a high-density sintered tungsten-copper-nickel alloy and a cemented chromium carbide material. With proper design and materials selection, performance of sleeve-type antifriction bearings, mechanical rubbing-face seals, and seals employing packings can be improved by making them from carbide.

With grinder spindles, it was found

that the life of diamond grinding wheels was increased eight times when cemented tungsten carbide bearings were used. The surface finish, out-of-roundness and taper of the workpiece also showed marked improvement. These improvements result from the high modulus of elasticity, high stability and high abrasion resistance of the cemented tungsten carbide bearings. As a result of these properties, sleeve-type cemented tungsten carbide bearings can be designed with greatly reduced running clearance, and thus can produce higher accuracy in the spindle. The reduced clearance, and the ability of carbide to take and maintain a finish of 0.025 μm (1 μin.) and lower, keep an appreciable amount of grit from filtering between the bearing and the shaft. The high hardness resists scoring by the grit particles that may work through. Cemented tungsten carbides are highly resistant to surface damage from lubrication failure, thus permitting the use of very-low-viscosity lubricants and mist-lubrication systems. Another application for cemented carbides is pump bearings that must resist certain corrosive conditions in the chemical industry.

Nonmetallic Bearing Materials

Today, nonmetallic bearing materials are widely used. They have many inherent advantages over metals, including better corrosion resistance, lighter weight, better resistance to mechanical shock, and the ability to function with very marginal lubrication or with no lubricant at all. The major disadvantages of most nonmetallics are their high coefficients of thermal expansion and low thermal conductivity characteristics. For many years, carbon-graphites, wood, rubber and laminated phenolics dominated the field of nonmetallic bearing materials. In the early 1940's, development of nylon and Teflon gave engineering designers two new nonmetallics with very unique characteristics, particularly the ability to operate dry.

A wide variety of plastic composites in now being used very successfully in bearing applications. Addition of fiber reinforcements and fillers such as solid lubricants and metal powders to the resin matrix can significantly improve the physical, thermal and tribological properties of these plastics. This is illustrated in Table 11, where a few of the more promising materials are listed. The PV values (stress × velocity) are for dry operation. Even with marginal lubricants, such as water, some of these compounds have truly outstanding load-carrying capacities. It should be noted, however, that certain plastics wear at a higher rate when a lubricant is present. When the effectiveness of a lubricant depends on formation of a transferred film, its use may inhibit material transfer, resulting in higher wear.

Table 11 Typical PV values for plastics sliding unlubricated on steel at a velocity of 100 ft/min

Type of plastic	Operating temperature limit(a) °C	°F	Typical PV(b) values
Teflon (unfilled)	230 to 260	450 to 500	1800
Nylon (unfilled)	120 to 150	250 to 300	2000 to 4000
Acetal (unfilled)	82	180	3000 to 4500
Polysulfone (unfilled)	150 to 175	300 to 350	4000 to 5000
Teflon + 30% bronze powder	230 to 260	450 to 500	28 000
Acetal + 30% glass fiber + 15% Teflon fiber	82	180	12 000
Polyester + 30% glass fiber + 15% Teflon fiber	135	275	30 000
Formulated polyphenylene sulfide (proprietary)	175	350	35 000 to 45 000
Formulated polyamide-imide (proprietary)	205	400	12 000 to 12 900
Polyimide + 15% graphite powder	260	500	>300 000

(a) Continuous use. (b) PV, stress in psi × sliding velocity in ft/min.

The following paragraphs present more detailed discussions of some nonmetallic materials typically used for bearings.

Nylon. The low melting point of nylon limits its use to temperatures below about 150 °C (300 °F). To obtain dimensional stability, nylon should be stress relieved at a temperature at least 28 °C (50 °F) above the maximum temperature expected in service. This is usually accomplished by heat treating the nylon in oil or some other suitable liquid. Graphite, molybdenum disulfide and other fillers are added to nylon to improve its bearing properties. The static coefficient of friction for nylon against nylon is more than twice the kinetic friction. The friction values for steel against nylon are lower than for nylon against nylon.

Teflon, a polytetrafluoroethylene resin, is a thermoplastic material. It is used as a bearing material mainly for two reasons: (*a*) chemical inertness; and (*b*) at low speeds, an extremely low coefficient of friction with sliding metals (about 0.05). Use of Teflon as a bearing material is limited, however, because of its low thermal conductivity, high thermal expansion, thermal instability and poor resistance to cold flow. In designing Teflon bearings, consideration must be given to the fact that Teflon has a transition point at 21 °C (70 °F), which results in a linear increase of 4 mm/m (0.004 in./in.). When Teflon is heated to about 340 °C (650 °F) or higher, it gives off toxic fumes, and therefore it must be cooled during fabrication operations such as machining. It can be used at service temperatures (260 °C; 500 °F) higher than those for nylon, and it is not hygroscopic.

Teflon is used in bearings in several ways: (*a*) as a film applied by water dispersion and then cured; (*b*) as an impregnant in a metal matrix; and (*c*) as a woven layer, supported by a woven layer of glass bonded to a metal surface. Teflon applied by water dispersion generally is used as a means of preventing fretting corrosion where there is intermittent oscillation. Woven Teflon is recommended for oscillating or low-speed use, although it has been used successfully at loads as high as 400 MPa (60 ksi). However, a rule of thumb for application of this material is to work to a maximum PV value of 30 000 (load in pounds per square inch multiplied by velocity in feet per minute). Laboratory tests on metal-filled Teflon (60% Teflon; 40% metal) proved that bearings with bronze as the metal filler were greatly superior in wear resistance to those with either lead or aluminum. At low speeds, the dynamic coefficient of friction of unmodified Teflon is lower than that of the filled types; this is not true at high speeds. However, because of the increased strength provided by the fillers, it may be desirable to accept a slightly higher coefficient of friction to gain strength.

Carbon-graphite is used extensively in bearing and brush applications. Its excellent performance as a brush material confirms its desirable bearing qualities. Its service temperature, usually limited to about 370 °C (700 °F), has recently been increased by processing techniques. Carbon-graphite has good wear resistance at temperatures too high for conventional lubricants. Carbon-graphite can also operate as a bearing in water, gasoline and other nondestructive solvents. It has a low coefficient of expansion, and its thermal conductivity is between those of copper and cast iron. Although it possesses reasonable strength, its edges are likely to chip or crack during machining or installation. This material is not usually considered for applications involving high impact loads. It is used in packing rings, seals, instrument bearings, and sleeve and thrust bearings.

Wood. Lignum vitae, one of the hardest woods known, has been used for centuries as a lining for various underwater bearings in ships, where metal corrodes severely. It is inexpensive and readily obtainable. Oil-impregnated wood is also used in some bearing applications.

A composition material that has a base of either paper, fabric or asbestos may be substituted for wood in applications such as ship-rudder bearings, liners for rolling mills, inking-roll bearings, bushings and pump sleeves.

Rubber often is used in bearings for devices that operate in water. Rubber can absorb shock and has fairly good resistance to abrasion and other types of wear. Rubber bearings usually are backed by a metal shell that provides additional strength. If the rubber bearing is properly designed, much of the solid contaminating material in the water can be washed out through longitudinal passages fabricated in the bearing surface.

Bearing-Material Selection

It must be emphasized that selection of a bearing-material system for a specific application and of a mechanical design for the bearing itself are closely interrelated processes. Neither process is entirely straightforward, neither can be approached independently, and both require a good understanding of other interacting components of the machine system.

Although we have considered the principles involved in bearing operation, we have not attempted to present a detailed discussion of mechanical design factors in this article. The reader should therefore not expect to make final decisions on materials for specific applications on the basis of this text alone.

Most manufacturers of plain bearings have experienced engineering staff personnel available to aid potential users with both mechanical design and material selection. Because of the wide material selection offered by most of these experienced specialized producers and their background of experience in practical applications, full advantage should be taken of the engineering services they can provide.

Special Topics in Materials Engineering

Concepts and Criteria in Materials Engineering

By Charles O. Smith
Professor of Engineering
University of Nebraska-Omaha
and
Bruce E. Boardman
Senior Engineer
Deere & Co. Technical Center

ENGINEERING DESIGN of metal components is a complex task requiring consideration of many interrelated factors, not all of which are necessarily compatible. Thus, compromises and tradeoffs among various design factors are routinely made. A designer must know the relative importance of these factors and how they interact before intelligent choices between compatible requirements can be made. For convenience in discussion, design factors have been somewhat arbitrarily grouped into three categories: functional requirements, analysis of total life cycle and other major factors; the factors are listed in Tables 1, 2 and 3, respectively. These categories intersect and overlap, which constitutes a major problem in engineering design.

Functional Requirements in Design

It should be obvious that any design must necessarily meet performance specifications; therefore, this item heads the list in Table 1. These specifications must reflect a full and complete analysis of the functions required of the product. An important distinction must be made between performance specifications, which enumerate the basic functional requirements of the product, and product specifications, which list requirements for configurations, tolerances, materials, manufacturing methods and the like. Performance specifications represent the basic parameters from which the design can be formulated; product specifications are codifications of designs, used for purchase or manufacture of the product. Excellence in design or product specification is not possible without complete and adequate performance specifications.

Performance specifications must reflect thorough consideration of the factors listed in Tables 2 and 3. Consequences and risks involved in possible product failures caused by predictable misuse or overload or by imperfections in workmanship or material must be considered in establishing performance specifications. Situations in which the consequences of product failure would be dire or in which only the very lowest risks of failure can be tolerated dictate the use of stringent performance specifications. When product failure does not involve a risk of personal injury

Table 1 Functional requirements in design

Performance specifications
- Definition of need
- Risks and consequences of underspecification
- Consequences of overspecification

Design configuration
- Probabilistic or deterministic approach
- Stress or load considerations
- Restrictions on size, weight or volume
- Service hazards, such as cyclic loading or aggressive environment
- Failure anticipation
- Reliability, maintainability, availability and repairability
- Quantity to be produced
- Value analysis
- Candidate materials and manufacturing processes

Redesign
- Design review
- Simplification and standardization
- Functional substitution

Table 2 Total life cycle in design

Material selection
Producibility
Durability
Feasibility of recycling
Energy requirements
- For production
- During use
- For reclamation

Environmental compatibility
- Effect of product on environment
- Effect of environment on product

Inspection and quality-assurance testing
Handling
Packaging
Shipping and storage
Scrap value

and is not likely to result in great financial loss to the user, economic considerations usually imply that performance specifications be no more stringent than necessary to meet functional requirements. Realistic performance specifications result in design and manufacture of products that perform their required functions with little risk of failure, and that at the same time can be produced at the lowest possible cost.

As an example of setting performance specifications to suit the application, resistance to corrosion can be specified at any of three levels: (*a*) avoiding contamination by corrosion

Table 3 Other major factors in design

State of the art
- Prior knowledge
- Possible patent infringement
- Competitive products

Conformance to standards
- Codes for specific products, such as pressure vessels
- Safety requirements
  - Products—Consumer Products Safety Commission
    - Warnings
    - Unintended uses
    - Labels
  - Manufacturing—Occupational Safety and Health Administration
- Environmental requirements—Environmental Protection Agency
- Industry standards
  - ANSI
  - ASTM
  - SAE
  - UL
  - ISO

Human factors
- Ease of operation
- Ease of maintenance

Aesthetics
Cost

products, (*b*) preventing leaks into or out of closed containers, and (*c*) maintaining structural integrity and other mechanical and physical properties in spite of corrosive attack. For food-processing equipment, the first of these considerations has paramount importance. For a bridge, the third factor is critical; furthermore, a bridge must retain its structural integrity for many years. In a petrochemical plant, all three considerations are important; chemical process equipment may be designed for continuous operation for two or three years, and any breakdown between scheduled maintenance periods would be extremely expensive. Furthermore, leakage of dangerous chemicals from process equipment is unacceptable. In this case, the cost of a breakdown and the damage caused by leakage can justify use of expensive materials if their performance reduces the probability of leaks or breakdown to very low levels. On an automobile body, low-carbon steel with a corrosion-resistant surface treatment and coating provides corrosion resistance consistent with the anticipated lifetime of the vehicle. Thus, for these various applications, the appropriate criteria for corrosion resistance depend on one or more of the following factors: the degree of contamination permitted by the application, the intended lifetime of the object, the corrosion characteristics of the environment, and the consequences and risks associated with corrosion failure.

Considering the specific example of an automotive exhaust system, the performance specifications for that system must provide for the following basic functions:

- Conducting engine exhaust gases away from the engine
- Preventing noxious fumes from entering the automobile
- Cooling the exhaust gases
- Reducing engine noise
- Reducing exposure of automobile body parts to exhaust gases
- Affecting engine performances as little as possible
- Helping control undesirable exhaust emissions
- Having a service life that is acceptably long
- Having a reasonable cost, both as original equipment and as a replacement part.

In its simplest form, the exhaust system consists of a series of tubes that collect the gases at the engine and convey them to the rear of the automobile. The size of the tubing is determined by the volume of exhaust gases to be carried away and the extent to which the exhaust system can be allowed to impede the flow of these gases from the engine. An additional device, the muffler, is necessary to satisfy the requirement for noise reduction. Under current practice, the system must contain a catalyst to convert noxious gases to less-harmful emissions. The basic lifetime requirement is that the system must resist the attack of hot, moist exhaust gases for some specified period, whether that period be defined in years, miles or number of cold starts. The practical requirement that the exhaust system be placed under the automobile imposes additional corrosion hazards: the system must resist attack not only by the exhaust gases, but also by the atmosphere, water, mud and road salt. The location also requires that the exhaust system be designed as complex shapes that will not interfere with running gear of the car, road clearance or the passenger compartment. The number of automobiles produced each year requires that material used in exhaust systems be readily available at minimum cost. The choice of welded low-carbon steel tubing for most components of the system adequately meets every requirement.

The design process includes determination of the configuration of the product and its various component parts, and selection of the materials from which and processes by which it is to be made. In its early stages, the process consists of evaluating various combinations of preliminary configuration, candidate material and potential method of manufacture, and comparing these with the previously established performance specifications. The relationship between configuration and material (processed in some specific way) is that every configuration places certain demands on the material, which has certain capabilities to meet these demands. A common specific relationship is one between the stress imposed by the configuration and the strength of the material. Of course, there are other relationships as well. Changes in processing can change the properties of a material, and certain combinations of configuration and material cannot be made by some manufacturing processes.

Quantitative relationships between configurational demands and material capabilities can be established by deterministic methods—the better-known approach—or probabilistic methods. In the former, nominal or average values of stress, dimensions and strength are used in design calculations; appropriate safety factors are used to compensate for expected variations in these parameters and for discontinuities in the material. In the probabilistic approach, described in greater detail in Ref 1, each design parameter is accorded a statistical distribution of values. From these distributions and from an allowable limit on probability of failure, minimum acceptable dimensions in critical areas (or minimum strength levels for critical components) can be calculated. Compared with deterministic methods, the probabilistic approach requires greater sophistication on the part of the designer and more elaborate calculations, but offers the potential for more compact parts that use less material. A significant handicap to use of probabilistic methods is the fact that statistical distributions of properties are not widely available and often must be determined before these methods can be applied to a specific design problem.

In either approach to design, the effects of notches and stress concentrations must be considered, because such features increase the vulnerability of all types of parts to failure. However, studies in fracture mechanics have shown that, under some circumstances, notches and discontinuities in the material may be benign, and therefore of little consequence in some design applications.

Cyclic loading, use at extreme temperatures and the presence of agents that cause either general corrosion or stress-corrosion cracking are special hazards that must be considered during the materials-selection process. Some common failure modes, and those mechanical properties most related to particular modes, are illustrated in Fig. 1. Cyclic loading, in particular, is a very common factor in the design of anything that has moving parts; it is also widely recognized that fatigue is responsible for a large portion of all service failures. In 1943, Almen (Ref 2) made an observation that is still valid about fatigue:

"Fully 90 percent of all fatigue failures occurring in service or during laboratory and road tests are traceable to design and production defects, and only the remaining 10 percent are primarily the responsibility of the metallurgist as defects in material, material specification, or heat treatment.

"Study of fatigue of materials is the joint duty of metallurgical, engineering, and production departments. There is no definite line between mechanical and metallurgical factors that contribute to fatigue. This overlapping of responsibility is not sufficiently understood.

"Hence, the engineers are constantly demanding new metallurgical miracles instead of correcting their own faults. Until metallurgists are less willing to look for metallurgical causes of fatigue and insist that equally competent examination for mechanical causes be made, we cannot hope to make full use of our engineering material."

Although Almen spoke specifically about fatigue, his comments can be applied to engineering design on a far broader basis. His comments must not be construed to excuse materials engineers from a proper share of the overall design responsibility, nor to allow them to slacken their efforts to find better materials for specific applications. Almen's comments do imply that all aspects of design must be considered, because even apparently insignificant factors can have far-reaching effects. (For example, there is at least one known instance in which fatigue failure of an aircraft in flight was traced to an inspection stamp that was imprinted on a component using too heavy a hammer blow.)

Even with parts for which loading in service is known accurately and stress analysis is straightforward, gross design deficiencies may arise from reliance on static load-carrying capacity based solely on tensile and yield strengths. The possibility of failure in other modes, such as fatigue, stress-corrosion cracking and brittle fracture caused by impact loading, must be considered in the design process.

Any discussion of potential failures in service should include careful consideration of the possible consequences of failure. Those failure modes that might endanger life or limb, or destroy other components of the apparatus, are to be avoided if at all possible. Sometimes, a piece of equipment is designed so that one component will fail in a relatively harmless fashion and thus avoid the potentially more serious consequences of the failure of another. For example, a piece of earth-moving equipment might be designed so that the engine will stall if the operator attempts to lift a load so heavy that it might upset the equipment or damage any of its structural components. A blowout plug in a pressure vessel is another example. Conformance to codes and standards, such as those listed in Table 3, may preclude serious consequences of service failure, but designers still must exercise careful judgment in studying (and designing against) the consequences of possible modes of failure.

The size and weight of a part can affect the choice of both material and manufacturing process. Small parts often can be economically machined from solid bar stock, even in fairly large quantities. The material cost of a small part may be far less than the cost of manufacturing it, perhaps making relatively expensive materials feasible. Large parts may be difficult or impossible to heat treat to high strength levels. There are also limits on size for parts that can be formed by various manufacturing processes. Die castings, investment castings and powder metallurgy parts are generally limited to a few

Fig. 1 Relationships between failure modes and material properties

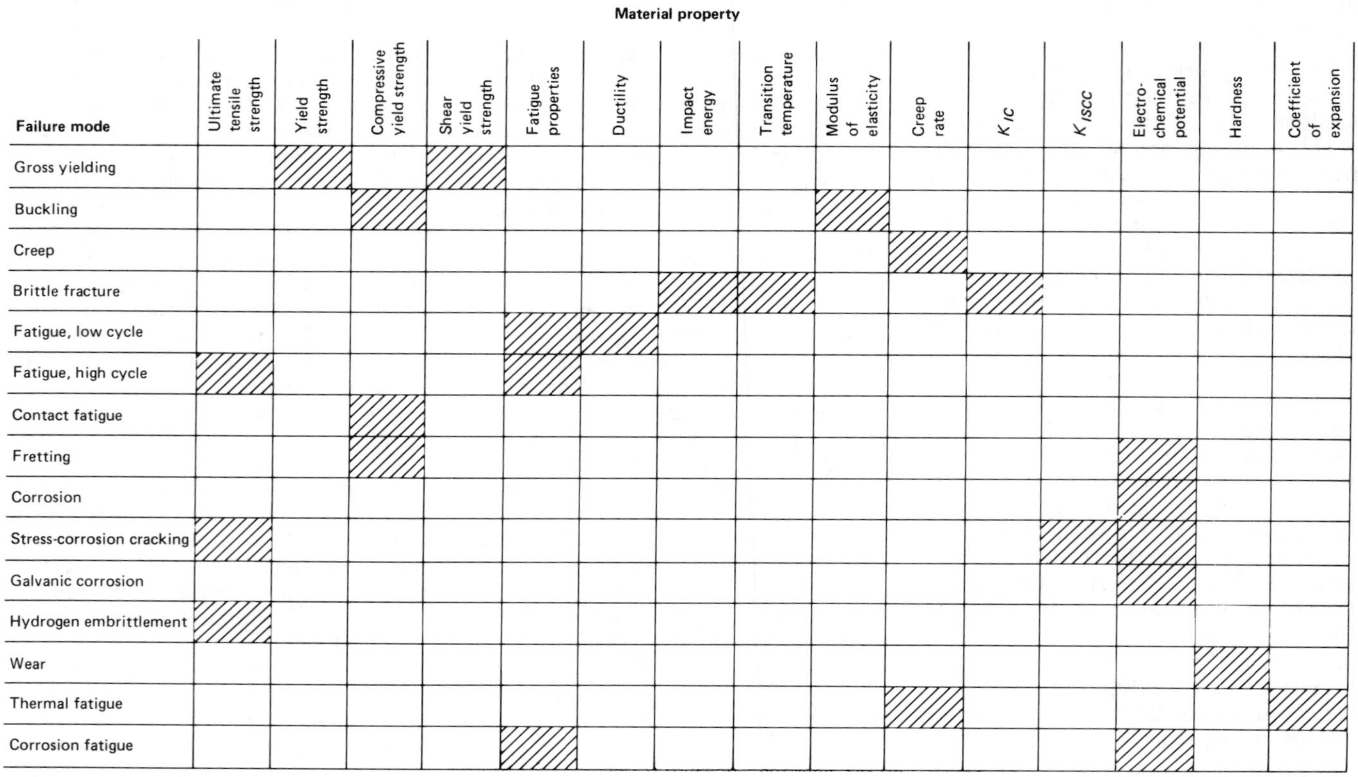

Material property

Failure mode	Ultimate tensile strength	Yield strength	Compressive yield strength	Shear yield strength	Fatigue properties	Ductility	Impact energy	Transition temperature	Modulus of elasticity	Creep rate	K_{IC}	K_{ISCC}	Electro-chemical potential	Hardness	Coefficient of expansion
Gross yielding		X		X											
Buckling			X						X						
Creep										X					
Brittle fracture							X	X			X				
Fatigue, low cycle					X	X									
Fatigue, high cycle	X				X										
Contact fatigue			X												
Fretting			X										X		
Corrosion													X		
Stress-corrosion cracking	X											X	X		
Galvanic corrosion													X		
Hydrogen embrittlement	X														
Wear														X	
Thermal fatigue										X					X
Corrosion fatigue					X								X		

Shaded block at intersection of material property and failure mode indicates that a particular material property is influential in controlling a particular failure mode.

kilograms. When weight is a critical factor, parts often are made from materials having high strength-to-weight ratios.

The quantity of parts to be made can affect all aspects of the engineering design process. Low-quantity production runs can seldom justify the investment in tooling required by production processes such as forging or die casting, and may limit the choice of materials to those already in the designer's factory or those stocked by service centers. High-quantity production runs may be affected by the capability of materials producers to supply the required quantity. Mass-produced parts sometimes can be designed and redesigned, requiring a large expenditure for engineering and evaluation, but providing enough savings, considering the large quantity involved, to make the effort worthwhile. Design for small-quantity parts may be limited to finding the first design and material that serve the required purpose.

Products that may be manufactured in several locations can present additional problems for designers because the cost and availability of materials can vary from place to place. If the product is to be made in different countries, the nearest equivalent grades of steel, for example, might be different enough to affect service performance. In some areas that have low prevailing labor costs, it may be desirable to design a labor-intensive product; in a high-cost labor market, the designer often attempts to design the product to fit the capabilities of automated manufacturing equipment.

It is relatively late in the design process before designers, materials engineers and manufacturing engineers, by working together, can establish those factors listed in Table 4 on a detailed basis. Only then can the field of candidate materials and manufacturing processes be narrowed to a manageable number of alternatives. The implications of each of these alternative materials and manufacturing processes can then be evaluated, and any required changes in configuration can be made.

One of the complicating factors in materials selection is that virtually all materials properties, including fabricability, are interrelated. Substituting one material for another, or changing some aspect of processing in order to effect a change in one particular property, generally affects other properties simultaneously. Similar interrelations that are more difficult to characterize exist among the various mechanical and physical properties and variables associated with manufacturing processes. For example, cold drawing a wire to increase its strength also increases its electrical resistivity. Steels

Table 4 General factors in materials selection

Functional requirements and constraints
Mechanical properties
Design configuration
Available and alternative materials
Fabricability
Corrosion and degradation resistance
Stability
Properties of unique interest
Cost

Table 5 Tests for value

Every material, every part, every operation should pass these tests—a negative response is passing:

- Can we do without it?
- Does it do more than is required?
- Does it cost more than it is worth?
- Is there something that does the job better?
- Can it be made by a less-costly method?
- Can a standard item be used?
- Considering the quantities used, could a less-costly tooling method be used?
- Does it cost more than the total of reasonable labor, overhead, material and profit?
- Can someone else provide it at less cost without affecting dependability?
- If it were your money, would you refuse to buy the item because it costs too much?

that have high carbon and alloy contents for high hardenability and strength generally are difficult to machine and weld. Additions of alloying elements such as lead to enhance machinability generally lower long-life fatigue strength and make welding and cold forming difficult. The list of these relationships is nearly limitless.

Value analysis using criteria such as those listed in Table 5 can provide both designers and managers with assurance that the final combination of configuration, material and manufacturing process is a good one.

Redesign often can improve performance and reduce cost, and may occur informally during the early stages of design. It may also result from formal design-review procedures. Failure-mode analysis is particularly useful in redesign to reduce the likelihood of further failures. These techniques often have resulted in greatly improved product performance.

Standardization and simplification of design can lead to substantial savings without loss of performance. Retaining only the most efficient sizes, types, grades and models of a product is an example; during World War II, the number of "standard" types of steel, brass and bronze valves in the United States was reduced from 4080 to 2500. Utilization of standard off-the-shelf components can also lead to significant savings. It might be less costly, for example, to use a 1/4-20 bolt 2 in. long, which is a widely stocked size, even though design requirements could be satisfied by a bolt 1⅞ in. long, which would probably be a special-order item. Further savings could be realized if the assembly were to be redesigned to allow use of 1½-in. or 1¾-in. bolts.

Functional substitution offers great opportunity for improvement and cost reduction through redesign. The goal of functional substitution is to find a new and different way to meet a design requirement. For example, a bolted assembly might be redesigned for assembly by welding, by pressing mating parts together or by adhesive bonding. The scope of a functional redesign program might be rather modest, as in the example just mentioned, or it might entail complete redesign of the product.

Total Life Cycle in Design

It is an accepted principle of engineering to design a product for minimum cost, consistent with fulfilling the functional requirements of the product. It is tempting to define the cost of a product strictly on the basis of component and labor costs (plus allowances for overhead, selling costs and profit). However, such accounting neglects several important items in the total cost of the product to the user, such as cost of energy to operate the product, cost of maintaining the product, depreciation and cost of disposal. In addition to the cost of a product to its producer and user, there is the cost to society at large; some of the components of this cost are consumption of raw materials and energy and the impact of the product on the environment. The relative importance of considering the cost to the user and to society during design depends in part on the nature of the product. An automobile, for instance, has considerable user and societal costs; for a bolt, these costs are smaller, harder to identify and easier to neglect.

To a materials engineer, the concept of total life cycle must include the total life of the product, as discussed above and in Table 2, but should also include the total life cycle of the components and materials in the product. Whenever possible, a product should be designed so that components that do not wear out can be reused, or so that the materials in the product can be recycled.

Energy Requirements. Reuse of components and materials obviously conserves raw materials, such as the ore from which a metal is made, but it also conserves the energy required to extract the metal from the ore. Basic metal-production operations are highly energy-intensive. It has been estimated that 95% of the energy required to manufacture aluminum beverage cans from ore can be saved if metal in the cans is recycled. It may be possible to conserve energy by choosing materials that do not require heat treatment but still have adequate strength. Heavily drafted cold-drawn-and-stress-relieved steel bars can have yield strengths of 690 MPa (100 ksi); for many purposes, such bars can be used in place of hardened-and-tempered bars. The potential for conservation of scarce resources, particularly those that might be subject to manipulation for economic or political gain, is apparent in these two examples.

As suggested above, the energy required to operate a mechanical device can represent a significant portion of its total life-cycle cost. For example, the cost of electricity used to operate an air conditioner for the duration of the warranty period may be greater than its purchase price.

Producibility. The question of producibility affects engineering design at several levels. The most basic question is whether the technology required to produce economically the desired quantity of the item exists. A product should be designed in accordance with established technological practice whenever possible; working at the limit of technology leads to higher scrap levels and difficulties in production, and advancing the limit of technology requires a developmental program before production can begin. A second question concerns the capability of the factory (or entire industry) to produce the desired number of parts. The third question concerns utilization of existing equipment and personnel. Some of the answers to these questions are managerial prerogatives, but it is usually desirable for designers to analyze the various possibilities and present a set of alternatives, together with possible consequences, advantages and disadvantages of each, to management.

Durability, or intended service lifetime, is one of the parameters on which a design must be based. In general, basic goals for service lifetime are established for, rather than by, designers. As in the case of producibility, it may be desirable for a designer to develop alternative combinations of material and design that could significantly affect the anticipated service life of the product.

Quality assurance should be an integral part of the manufacturing process of any product, especially because the possible consequences of permitting deficient products to reach the marketplace can be dire. This article does not attempt to describe a quality assurance program, or how to develop one (see Ref 3); rather, it emphasizes the importance of such a program and points out how design and quality assurance are interrelated.

A designer must know the types and severity of discontinuities that can be detected in a quality assurance program and have an appropriate level of confidence in the detection methods. By knowing what can be expected from the quality assurance program, the designer can then provide a margin of safety to compensate for the existence of discontinuities that cannot be detected. This is part of the basic philosophy of fracture mechanics. A similar approach is applicable to design of redundant systems in the product. In either case, the product should be designed with the capabilities and limitations of the quality assurance program in mind.

Handling requirements of various products are too frequently neglected in engineering design. Almost without exception, a product is made in one location and used in another. Thus, it is essential to provide means for transporting it from the manufacturer to the seller to the user; the means of transportation can affect the design of the product in several ways. The most obvious concern is to ensure that the product will not be degraded in handling, transit, or storage. Whenever feasible, a product should be designed against the possibility of damage caused by normal handling. A product that may be easily damaged in handling must be carefully (and expensively) protected by packaging or special shipping procedures. It is also important to design a product to minimize shipping and storage space requirements. For example, wastebaskets are usually designed to be nested during handling; many wastebaskets have lugs or other design features that limit nesting in order to permit easy separation. Particularly for consumer products, the design of packages may not be considered a part of product design, but it is still appropriate for a product designer to consider packaging requirements that might result from various design alternatives.

Other Major Factors in Design

Besides the functional factors listed in Table 1 and the total life-cycle considerations described in Table 2, several other major factors affect design and materials selection; some of them are listed in Table 3.

State of the Art. Most of the time, the state of the art in a particular field can be inferred from an engineering evaluation of products currently on the market. Existing products may lack capabilities that could be provided by a new product. The extent of improvement might range from purely cosmetic (which might be the case for a new product intended to capitalize on the market acceptance of an established product) to complete redesign resulting in an entirely new product unlike anything previously produced. The state of the art can be defined not only by existing products but also by industrial and societal standards, technical publications and patents. Patent infringement is often considered a matter for legal departments, but technical personnel are often able to analyze technological aspects of patents on devices and processes that might be applicable to new products.

Conformance to Standards. Many products are subject to either mandatory or voluntary standards. In general, the former are imposed through an act of law; fuel-economy requirements on automobiles and minimum pressure ratings for plumbing fittings are examples of legislated design standards. The requirement that small household appliances operate on 115-V, 60-Hz electric power is mandatory for products to be sold in the United States, not because of an act of law but because almost all electric utilities produce such electricity. Voluntary standards include those for standard nuts and bolts; grades of various product forms of aluminum, copper, steel and other metals; and methods of rating products.

Standards that pertain to the safety of either the builder or user of a product are particularly important. Several applicable safety standards for consumer products are acts of law. Other laws cover the safety of working conditions. Some standards, particularly those pertaining to consequences of the unintended use of products, have evolved from bodily injury litigation. All of these types of standards must be considered during the design process.

Conformance to standards is usually, but not necessarily, evidence of proper design. A designer must determine that a particular standard is appropriate to the product under consideration, and that the standard provides adequate assurance of satisfactory performance in service. Many standards, especially voluntary standards, have been derived from designs that have proven to be acceptable for most applications and that reflect the levels of quality and performance capability most manufacturers desire to produce.

Human factors in design describe the interactions between a device and anyone who uses or maintains it. Examples of human factors in design include designing the handle of a portable electric drill to fit comfortably into either hand of the user, arranging gages and controls on an automobile dashboard to be easily seen or reached, and designing an ejector to remove the beaters of an electric mixer. One branch of industrial engineering deals specifically with human factors; contact with persons involved in such research would be useful to designers, as would publications dealing with this subject, such as Ref 4.

Aesthetics is an important aspect of the design of almost any product, but one that should not be allowed to compromise functional requirements. Some consumer products are sold almost exclusively by aesthetic appeal; design of such products is usually assigned to artists. However, aesthetic appeal is important even for industrial products. A smooth or shiny finish, an artistically pleasing shape, or an appearance of ruggedness and durability may have little or no influence on the ability of the product to perform its function but may, nevertheless, have a considerable effect on the attitude of a potential buyer or on worker acceptance when the item is put into service.

Cost. Assuming that a product meets the basic functional requirements established for it, the most important single factor in its design and manufacture is cost. The cost of any product must be competitive with the cost of comparable products already on the market. Whether there is a directly comparable product or not, the cost must be low enough to convince a prospective purchaser that the benefits to be derived from the product exceed its cost. Cost-benefit studies almost always precede the purchase of major pieces of manufacturing equipment.

Cost generally enters the design process as one of several criteria against which the merit of a design is judged. Whenever a choice exists between different materials, designs or manufacturing processes, the least costly alternative will be chosen, provided that the basic functional requirements for the product can still be met. In some instances, the possibility of significant cost reduction will justify re-evaluation of basic functional requirements and performance specifications; it may be possible to modify the design goals slightly and thereby significantly reduce costs. However, it is important to remember that any modification of functional requirements and performance specifications may adversely affect utility and marketability.

Factors in Materials Selection

Two often-cited reasons for selecting a certain material for a particular application are (*a*) the material has always been used in that application and (*b*) the material has the right properties. Neither is evidence of original thinking or even careful analysis of the application. The collective experience gained from common usage of a material in a particular application is useful information, but not justification in itself for selecting a material. The time has passed when each application has its preferred material and a particular material its secure market. In the context in which it is frequently mentioned, the term "property" connotes something that a material inherently possesses. On the contrary, a property should be regarded as the response of the material to a given set of imposed conditions. It must also be recognized that this property should be that of the material in its final available and processed form. Tabulated properties data, such as those available from the sources listed in Table 6, are helpful, but such information must be used judiciously and must be relevant to a particular application.

Regardless of specific expertise, every engineer concerned with hardware of any description (and this includes essentially all engineers) must deal constantly with selecting an appropriate material (or combination of materials) for his design. Except in trivial applications, it certainly is not sufficient to indicate that the components should be "steel" or "aluminum" or "plastic". Rather, the engineer must focus his attention, knowledge and skill on the general factors in materials selection listed in Table 4. It is obvious that these are inseparable and interwoven with the factors listed in Tables 1, 2 and 3.

In principle, one could write a mathematical expression describing the merit of an engineering design as a function of all these variables, differentiate it with respect to each of the criteria for evaluation and solve the resulting differential equations to obtain the ideal solution. No one has any illusion that we are in a position to do this. The principle is valid, however, and should provide a basis for action by the designer. In some instances, a standard, readily available component may be much less costly than, and yet nearly as effective as, a component of optimized, nonstandard design.

Mechanical Properties. One question usually asked in selecting materials is whether strength is adequate to withstand the stresses imposed by service loading. Although the primary selection criterion is often strength, it also may be toughness, corrosion resistance, electrical conductivity, magnetic characteristics, thermal conductivity, specific gravity, strength-to-weight ratio or some other property. For example, in residential water service, where the water pressure is relatively low, weaker and more expensive copper tubing might be a better choice than stronger steel pipe. Steel pipe comes in sections and is joined by threaded connections with elbows at corners, whereas soft copper can be obtained in coils and can be easily bent around corners. Thus, the lower installation cost of copper would overcome its higher material cost. Also, because of the relatively low water pressure, the strength of copper would be adequate, and the greater strength of steel unnecessary. Furthermore, in the event of freezing, copper usually yields instead of bursting; in many regions of the country, it is more resistant to corrosion by the local water than is steel. In general, the usual criterion for selection is not just one property, such as strength, but some combination of properties, manufacturing characteristics and cost.

Ultimate tensile strength is a commonly measured and widely reported indication of the ability of a material to withstand loads. However, the direct application of tensile-strength data to design problems is extremely difficult. First of all, the definition of "ultimate tensile strength" is the maximum stress (based on initial cross-sectional area) that a specimen can withstand before failure; it occurs at the onset of plastic instability. Thus, any compo-

Table 6 Sources of reference data in the United States

American Society for Metals
American Iron and Steel Institute
American Society for Testing and Materials
Society of Automotive Engineers
Government-Industry Data Exchange Program (GIDEP), Corona, CA
Defense Documentation Center (DDC), Alexandria, VA
National Technical Information Service (NTIS), Springfield, VA
Machinability Data Center (MDC), Cincinnati, OH
Mechanical Properties Data Center (MPDC), Columbus, OH
Metals and Ceramics Information Center (MCIC), Columbus, OH
Nondestructive Testing Information Analysis Center (NTIAC), San Antonio, TX
Thermophysical and Electronic Properties Information Analysis Center (TEPIAC), West Lafayette, IN
Chemical Propulsion Information Agency (CPIA), Laurel, MD
Infrared Information and Analysis Center (IRIA), Ann Arbor, MI
Coordinating Agency for Supplier Evaluation (CASE), Sacramento, CA
National Bureau of Standards (NBS), Gaithersburg, MD
Smithsonian Science Information Exchange (SSIE), Washington, DC

nent loaded to its ultimate tensile strength is likely to fracture immediately. Secondly, if the design is based on a fraction of the ultimate strength, there is the question of what fraction provides adequate strength and safety, together with efficient use of the material. Finally, there seems to be only a rough correlation between tensile strength and material properties such as hardness and fatigue strength at a specified number of cycles, and no correlation whatsoever between tensile strength and properties such as resistance to crack propagation, impact resistance or proportional limit.

Yield strength indicates the lowest stress at which measurable permanent deformation occurs. This information is necessary to estimate the forces required for forming operations. Yield strength is also useful in considering the effects of a single-application overload; most structures must be designed so that a foreseeable overload will not exceed the yield strength.

Hardness is another widely measured material property and is useful for estimating the wear resistance of materials and estimating approximate strength of steels. Its most widespread application is for quality assurance in heat treating. However, only rough correlation can be made between hardness and other mechanical properties or between hardness and behavior of materials in service.

Ductility of a metal, usually measured as the percent reduction in area or elongation that occurs during a tensile test, is often considered an important factor in material selection. It is assumed that, if a metal has a certain minimum elongation in tensile testing, it will not fail in service through brittle fracture. It is also assumed that if a little ductility is good, a lot is better. Neither of these assumptions is accurate. Metals normally considered very ductile can fail in a seemingly brittle manner, such as under fatigue or stress-corrosion conditions, or when the service temperature is below the ductile-to-brittle transition temperature. How much ductility is actually usable under service conditions and how best to measure it are very controversial. Several estimates of usable ductility (such as the ability of materials to absorb the movement within a large structure that is necessary to equalize the load among all of the members of the structure) fall in the range of 1 to 2% elongation in tensile testing. Larger amounts of ductility only indicate the possibility of more extensive permanent deformation of the structure. In many structures, any permanent deformation destroys the usefulness of the structure; for example, the aerodynamic efficiency of an aircraft wing can be substantially reduced by only 1 to 1½% deformation. The amount of ductility required for processing may far exceed that actually usable in service. Steel sheet, for example, often must have considerable ductility, far more than a part might need in service, but nevertheless enough to allow the part to be formed to its required shape.

Design Configuration. The shape of an object is partly responsible for service demands placed on the material from which it is made. Cross-sectional dimensions, for example, may determine the stresses imposed on the material by service loads. These apparent stresses may be increased through the effects of notches and changes in section size. Design configuration determines whether or not a component will be subjected to particular hazards, such as wear or exposure to a corrosive environment. It may also determine the product form from which the item is made. A long, slender part is often made from strip or bar simply because the raw material has the same general shape. A part with a high degree of rotational symmetry is often turned on a lathe. A hollow sphere may be made by joining two hemispheres stamped from flat sheet.

Available and Alternative Materials. Regardless of the merits of a material for a given application, if that material is not readily available in the desired form and quantity, it is a poor choice for that application. The question of material availability is compounded by shortages caused by market fluctuations, inability or reluctance of producers to supply the desired material in the desired form or quantity, and changes in the relative costs of various materials. A further complication may exist if a particular part is made in several locations, especially if these locations are in different countries. Thus, it is desirable to select more than one material for an application and, if possible, to give an order of preference.

Sometimes, an alternative material will be necessary as a substitute for a preferred but unavailable material. In other instances, the alternative material will be selected because of superiority in the specific situation. For example, a tool manufacturer had been purchasing AISI 1078 and 1086 carbon steel bars, which have overlapping composition ranges. This manufacturer decided to use only 1078 and thus reduce his inventory. Even more important, because it is a more widely used grade, 1078 is produced more frequently than 1086, and thus is more available and easier to schedule.

An alternative material for a pressure vessel may have slightly greater strength than a material it can replace. Even a small increase in strength, however, can mean a significant reduction in wall thickness. It also may permit fabrication of slightly larger vessels. The pressure-vessel designer, however, must be sure that the increase in strength does not concurrently reduce fracture resistance below acceptable values.

The desired form can also cause availability problems. A material that can be obtained only in castings obviously cannot readily be used in applications requiring drawn tubing, extrusions, wire cloth or other fabricated shapes.

Fabricability describes the capability of a metal to be fabricated by various manufacturing processes, such as machining, casting, forging, welding and stamping. Whenever possible, product design should incorporate those materials that can be readily fabricated using the desired process(es) without special precautions. Materials that can be fabricated by several methods offer convenient alternatives in the event unforeseen conditions suddenly preclude use of the principal fabrication method.

Questions regarding production, such as equipment and manufacturing capacity available for fabricating components, are directly related to fabricability, as is the quantity required. If several thousand duplicate parts are necessary, the high cost of dies or other specialized equipment for quantity production may be economically justified. If few pieces are required, hand production from relatively expensive materials in stock may be less expensive than use of more elaborate methods and less expensive materials. Even so, an engineering estimate should be made of the relation between product cost and quantity for each proposed production method.

Design and fabricability are closely related. For example, a manufacturer of equipment for pleasure boats made lightweight anchors from 6061-T6 extrusions to replace steel anchors. The new aluminum alloy anchors had four major advantages over those made of steel: (1) more holding power for the same size, (2) half the weight, (3) freedom from rusting and staining and (4) interlocking construction rather than welded construction, which allowed shipping before assembly, thus saving transportation costs.

A material may not be commercially available in the desired mill form, but it may be possible, with relatively small-scale development, to produce it in the desired form. This entails considerable expense, but circumstances may justify the extra cost, as exemplified by the development of fabrication procedures for beryllium and zirconium components for nuclear reactors.

Fabricability and mechanical properties are interrelated, and they quite often act in opposite directions. As strength levels are increased, often by increasing carbon or alloy content, weldability and machinability frequently decrease. Although most materials can be welded, higher-strength materials generally require special techniques; the designer must be aware of the added cost built into the part in order to meet the welding requirements. In some instances, there may be no practical method of regaining the strength of the base metal lost through welding. As machinability is increased by additions of sulfur or lead, long-life fatigue resistance is decreased.

Corrosion and Degradation in Service. A material may be regarded as either resistant or susceptible to corrosion and degradation in service, depending on the nature of the application. Sometimes, it may be preferable to use a low-cost, frequently replaced component and assume that it will be corroded or degraded in a short time; in other situations, however, this approach may not be acceptable because of high potential danger to personnel or for other reasons.

There are other factors that should be considered when dealing with corrosive media, because neglect of these may lead to erroneous interpretations of corrosion tests and handbook data:

- A sample in a simple static test at a given temperature may corrode in a manner and at a rate significantly different than if the material were simultaneously transferring heat to the corrosive medium or if there were significant relative movement between the material and the corrosive medium.
- The test medium may become contaminated during the experiment, and its corrosive characteristics may change.
- Lack of correlation between laboratory tests and operating conditions may be caused by a limited ratio of solution volume to surface area of test material in the laboratory test.
- Pressure of gas or vapor above a corrosive liquid may appreciably affect the amounts of oxygen or other gases that may be dissolved in the solution.
- Alloys that owe their corrosion resistance to development of passive films are particularly susceptible to development of concentration cells.
- The ability of the material to withstand stress-corrosion cracking or corrosion fatigue in service is not always accurately predicted by standard laboratory tests for these qualities. If the material is, in fact, susceptible to either failure process in the service environment, sudden and unpredictable service failures may occur because cracks propagate so rapidly under the combined action of stress and corrosion.

Stability of material in service can be affected by temperature, fluctuations in temperature, length of time at temperature and, in some applications, exposure to radiation. Elevated temperature not only reduces strength and induces creep but also can produce changes in microstructure, such as tempering of martensitic steels and overaging of precipitation-hardening alloys. Obviously, duration of exposure is important in determining the extent to which these phenomena occur and, consequently, in determining appropriate stability requirements. A rocket motor, for instance, may be required to operate only briefly, whereas a steam turbine is expected to operate for many years. In many applications, it may be essential to avoid any and all failures that would require the equipment to be shut down for repairs. In others, especially those involving mechanical wear, replacement at regular intervals is anticipated, and the affected part is designed to be readily replaced.

Other aspects of stability are the consequences of failure. For instance, a leak in a teakettle may be only a nuisance, but a leak in a vessel containing a flammable or radioactive fluid is critical. It should be noted that many designs for long-term operation involve extrapolation or educated guessing because the best available data typically represent lifetimes much shorter than those anticipated in actual service. The need for conservatism in design for such service should be obvious.

Properties of Unique Interest. For most problems in materials selection, the choice represents a compromise among a large number of properties and fabrication characteristics of the material; each of these properties and fabrication characteristics has a significant effect on final choice. Often, however, one or two properties are far more important than all others. Some examples of applications where a single property of unique interest dominates the selection of a material are:

- *High density:* gyroscope rotors, and winding weights in self-winding watches
- *High stiffness-to-weight ratio:* aircraft frame components
- *Low melting point:* fusible links in sprinkler systems and fire doors
- *Expansion during freezing:* type metal
- *Special thermal-expansion characteristics:* metal-to-glass seals
- *Electrical conductivity (including superconductivity):* power-distribution systems
- *Wear resistance:* many applications, from plowshares and ore-crusher jaws to gear teeth and cam surfaces.

These are only a few examples of applications where a requirement for one or two properties of special interest severely limits the number of candidate materials.

A requirement for a particular combination of properties further limits the number of candidate materials. Beryllia (BeO) conducts heat as well as aluminum alloys but is an electrical insulator, the only known material with high thermal and low electrical conductivities. For applications demanding high toughness at low temperatures, the choice of materials is limited to metals with face-centered-cubic lattices and certain high-nickel steels. Turbine blades in the hot stages of gas turbine engines must resist deformation and corrosion at operating temperatures and must not fail in thermal fatigue.

Cost. In almost every situation, final selection of a material for a specific application depends on a compromise. In some applications, there are specialized requirements that restrict the choice to relatively few materials. Even then, there is a compromise among the contending factors. The compromise and final selection also involve economic considerations. Initial cost of a piece of equipment involves raw-material, fabrication, assembly and installation costs. The cost of the material is generally a relatively small part of the total life-cycle cost of the equipment. After the equipment has been placed in service, there may be additional costs, such as required rate of return on investment, depreciation, taxes, product-liability costs, replacements due to failure, shutdown expenses when equipment is undergoing repair or replacement, and production loss.

The tests for value listed in Table 5 are directly pertinent to economic considerations. If the answer to any of these questions (and perhaps other similar ones) is affirmative, the selection task is incomplete.

No industry is immune to savings through more effective application of materials. At least three major approaches may be taken to reduce cost through better use: (1) reconsider the material and mill form selected, (2) reconsider the shape of the part and its method of fabrication and (3) redesign to take full advantage of properties. Although these apply primarily to production-line parts rather than tailor-made parts, the philosophy is applicable to all products.

In many situations, definite savings can be realized by the simple expedient of changing from one material to another without substantially changing the form or processing procedure. There are other applications for which two or more materials can be considered as alternates, with the choice at any given time dictated by current market prices.

Great savings can often be realized by changing the fabrication procedure or form in which material is used. One prominent example is the automobile-engine crankshaft. For many years, crankshafts were machined from forged steel that had to meet rather stringent specifications of strength and toughness. However, high toughness is an unnecessary requirement because a bent crankshaft is just as useless as a broken one, and impact loading on a crankshaft is not severe. Realization of these facts made it possible to effect a change to cast crankshafts. Today, there are millions of cast crankshafts in successful operation, all incapable of appreciable plastic bending, but having adequate rigidity and wear resistance. The cost of a cast and machined crankshaft is significantly less than that of a forged and machined one.

Summary

Materials selection, which is a part of the over-all design process, places several requirements on the designer. First, the designer must make a systematic study of the intended purpose of the product under design, noting all explicit and implied constraints on the design. The principal result of this study is characterization of the functional requirements of the product, including the relative priorities of potentially conflicting requirements. Another result of this study is a listing of the criteria to be used in evaluating various possible designs; cost is almost always one of the most important. The second phase of the design process is to determine the required capabilities of each component of the product, then to compare each of these requirements with characteristics of the candidate materials. These characteristics, however, should not be considered merely as numbers in published tables of properties. They are better described as the response of that material *in the form of a specific component* to the imposed service conditions. (The effects of manufacturing processes on material properties are implicitly considered when the material is evaluated as a specific component.)

The third step in the design process is to consider various combinations of configuration, material and fabrication process, evaluating each against previously established criteria. Several methods of evaluation may be used, including, among others, design review, value analysis and simulated service testing of prototype and pilot-lot products. The final analysis in the material-selection process might include studies of field service reports and cost analysis of the production model.

REFERENCES

1. Probabilistic Design, by E. B. Haugen and P. H. Wirsching: *Machine Design,* Vol 47, No. 9, 17 Apr 1975, p 98-104; No. 10, 1 May 1975, p 80-85; No. 12, 15 May 1975, p 83-87; No. 13, 29 May 1975, p 54-58; No. 14, 12 June 1975, p 108-112. Also, *Probabilistic Approaches to Design,* by E. B. Haugen: Wiley, 1968
2. Probe Failures by Fatigue To Unmask Mechanical Causes, by J. O. Almen: *SAE Journal,* Vol 51, May 1943
3. *Quality Control Handbook,* 3rd Ed., edited by J. M. Juran: McGraw-Hill, 1974
4. *Human Factors in Engineering and Design,* 4th Ed., by E. J. McCormick: McGraw-Hill, 1976

Guidelines for Selection of Material

By B. P. Bardes
Materials Engineer
General Electric Co.
and
L. J. Korb
Supervisor, Metals & Ceramics Group
Rockwell International

MANY ENGINEERS view material selection as a simple, straightforward task: one merely chooses the metal or alloy that has traditionally been used for the application at hand (or for similar or related applications); if there is no precedent, and if no special property such as high electrical conductivity or superior corrosion resistance is required, one chooses low-carbon steel. Such a simplistic view is gradually being supplanted by a more sophisticated approach based on rigorous testing and analysis. Of course, the purpose of any such analysis is to provide a basis for predicting service performance, and thus evaluation of any material selected for a given application ultimately depends on whether or not the component performs as intended.

The first task in the process of selecting a material for a particular application should be to question the suitability of the material or class of materials whose use in that application is widely accepted. Such questioning is intended to (*a*) confirm that the material is indeed the "logical choice" and (*b*) ensure that recent innovations in materials (including the substitution of nonmetals for metals) are not automatically overlooked. Similar questioning of product form and production process are intimately related to this task, and often are performed in conjunction with it. Selection of material class, specific alloy, specific product form and production process should always be done according to a logical step-by-step process. This ensures that each important facet of material selection will always be considered. A checklist of some of the more important facets is presented below.

Performance requirements and service environments for a particular part generally are analyzed very early in the design process. Analysis of environments should include, in addition to obvious service-related degradation such as that due to corrosion or wear, potential exposure hazards such as damage to computer components during transportation and field assembly or installation, or ingestion of foreign objects into aircraft jet engines.

The materials engineer must know the type and magnitude of loading and the nature of the environment that the part will or may encounter in service. Loading may be cyclic or sustained. It may be imposed rapidly or slowly. It may be critical in tension, compression, torsion, shear or bending. It may be distributed uniformly or may be concentrated in a localized area. Magnitude of loading and configuration of the part jointly determine the stresses to which the part will be subjected. Finite-element analysis may be useful for calculating stress distributions. Wherever possible, calculated stresses should be compared with experimentally determined stresses. Service environments such as high temperatures and corrosive media present special problems in material selection.

Data accumulated from testing at room temperature in benign environments generally do not apply to conditions involving the same material at different temperatures or in other environments, but often the required data are not available and thus room-temperature test data must be used as a basis for predicting performance until appropriate data have been determined in the laboratory or until actual operating experience has been accumulated. Because the deformation and fracture mechanisms that influence behavior under operating conditions may not affect room-temperature behavior, such predictions must be made with great care.

Correlation of Performance Requirements and Service Environments with Measurable Properties. The purpose of analyzing performance requirements and service environments is to determine exactly what the material used in a particular product must be able to withstand. The next two steps in the material-selection process, which may be the most important, are to decide (*a*) which properties best characterize the ability of a material to withstand the conditions expected in service and (*b*) what values of these properties are required in the particular application.

The property that correlates best with service behavior may seem an unlikely choice; for example, behavior of materials in strain-controlled low-cycle fatigue tests often correlates well with thermal fatigue behavior of the same materials in service.

It is important to consider not only mechanical properties, but physical and chemical properties as well. In some applications, a physical property such as coefficient of thermal expansion, electrical conductivity or density (or strength-to-density ratio) may be more important than mechanical behavior. In other applications, a chemical property such as pyrophoricity or corrosion resistance, or the effect of the material on some other component of a larger system, may be the controlling factor in material selection.

It is then necessary to establish priorities among the various property requirements. It is usually convenient to evaluate candidate materials first on the basis of the most critical property, and then to consider the less critical properties. On occasion, property requirements are mutually exclusive, and some sort of compromise in performance requirements or product configuration must be made. Figure 1 of the preceding article may be useful in relating modes of possible failures to material properties.

Heat Treatment. One of the major considerations in material selection is whether the material, or the part made from it, will need to be heat treated, and if so, what type of heat treatment will be used. Heat treatments may be specified to harden a material such as steel throughout its cross section, to harden only the surface of a part, to relieve internal stresses, to soften the material, to enhance some physical property such as electrical resistance, or to facilitate a subsequent fabrication operation.

Two factors—cost of heat treatment and uncertainty about the availability of energy—have made it increasingly more desirable to choose materials that need not be heat treated. The decision to use a particular heat treatment will often be based on the required yield strength. Yield strength does not always correlate directly with functional and service requirements, but it does provide a convenient index to other properties of the material that, like yield strength, vary with heat treated condition.

Once it has been determined that heat treating a particular component is unavoidable, the response of candidate materials to their required heat treatments becomes an important factor in selection. A steel part that must be through hardened, for example, must contain sufficient carbon to achieve the required as-quenched hardness and sufficient quantities of alloying elements to ensure an acceptable percentage of martensite in critical sections, assuming that the part is heat treated according to the usual production heat treating procedure. The factors that apply to the selection of steel for its response to heat treatment are discussed on pages 455 to 497 in Volume 1 of the 9th Edition of this Handbook. Responses of nonferrous alloys are discussed in Volume 2 of the 9th Edition and elsewhere in this volume. Specific heat treating practices are described in the Metals Handbook volume on heat treating.

A requirement for selective surface hardening, such as by induction hardening or carburizing, can cause additional problems in steel selection. The steel selected must be capable of providing the desired properties in both the case and core of the finished part. These problems are discussed in the articles on surface hardening, pages 527 to 542 in Volume 1 of the 9th Edition.

Configuration of components can affect the selection of a material in two ways: it can determine the most appropriate mill form from which the part can be made, and it can indicate which fabrication process is best suited to producing the part. Parts that are much thinner in one dimension than in the other two are often made from sheet or plate, even though forming or welding operations might make the shape of the finished part something other than flat. Long and slender parts often can be made from rod or wire; here, machining, bending or heading operations might be used to achieve the desired shape. Tubing or pipe might be appropriate for hollow parts. Irregular parts can often be advantageously cast or forged. (In the context of material selection, a casting or forging can represent both the product form and the fabricating process.) The effects of manufacturing practices on the properties of the various mill commodities are discussed in articles on product forms throughout the first three volumes of the 9th Edition of this Handbook.

Parts with high degrees of rotational symmetry often can be turned on a lathe or an automatic screw machine; for such an application, bars of a material produced specifically for desirable machining characteristics should be seriously considered. Storage tanks, on the other hand, often are made by bending and welding plate or sheet; thus, the material specified for the tank should be capable of being formed and welded readily. Parts designed to be cold headed normally are not made from materials formulated for good machinability or hardenability because such materials are apt to crack during severe cold forming. Parts to be hardened by heat treatment must contain appropriate amounts of the alloying elements required to produce the needed response.

Fabrication methods are frequently specified by the designer of the part. As indicated above, choice of fabrication method can affect choice of material. The materials engineer must ensure that candidate materials capable of meeting other requirements for the part also can be fabricated using the indicated process.

For parts that require several fabrication operations, the materials engineer must consider the requirements placed on the material by each operation. There may be restrictions on the sequence of fabrication operations. For example, steel parts that have been hardened to 40 HRC or more cannot be extensively cold formed; thus in the preferred sequence of operations, cold forming is done before hardening. Certain combinations of fabrication operations place limitations on material selection, possibly requiring a material that is not ideally suited to any one operation. Consider the piston rod in a hydraulic cylinder. It must be induction hardened to prevent nicks, score marks and wear, yet both ends must be machined to provide means of attaching the rod to the piston and whatever the cylinder will operate. To keep the piston rod straight and free from surface oxidation during heat treatment and to permit construction of cylinder assemblies of any desired length, the rod is normally induction hardened and centerless ground in long lengths, then cut to size. The material used in this application, which is often 1045 steel, is lower in carbon content than might be considered ideal for induction hardening, but higher than desirable for easy machining.

Joining of the component parts of an assembly must be considered in choosing the material for each part. Weldability (whether of similar or dissimilar metals), selection of a brazing alloy or selection of a fastener system is critical. The possibility of corrosion couples between the component parts and fasteners used to join them must also be considered. Mechanical joints such as those made by bending or crimping generally require that at least one component be ductile enough to permit a controlled amount of deformation during assembly. Before specifying materials and manufacturing processes, it must be verified that the capability of performing the chosen process does exist, either in house or at a supplier's plant. If not, provision must be made for developing the necessary expertise or acquiring the necessary equipment.

Availability is an important aspect of selection. For it to be a viable candidate for selection, a material must be available in the required quantity within a reasonable time and in the desired mill product form and size, with the required tolerances and surface finish. For example, most steel producers require a minimum quantity (typically, several tons) for each item on an order. Thus, lesser quantities of steel products must be obtained from a steel service center or custom processor, such as a firm that rerolls, slits or coats coils of steel sheet or strip. For applications requiring small quantities of steel, therefore, a materials engineer should choose the grade and product form from those stocked by a local service center.

On the other hand, for applications involving very large quantities of a particular alloy in a particular product form it is necessary to determine if producers have the capability of producing the required quantity of the desired product. The time required for a mill to fill an order may be several weeks or months; a service center can often fill small orders for stock items within a few days. Standard product forms already in the user's plant are available immediately.

Material producers cannot produce every grade in every product form. In some cases, it is not technologically feasible; in others, market demand has not justified the investment in necessary facilities. For the same reasons, there are size limitations on the various product forms. Tolerances and surface finishes normally applicable to mill forms are discussed in the appropriate sections of trade-association publications such as "AISI Steel Products Manual", "Aluminum Standards and Data" and "Copper Development Association Standards Handbook".

Another aspect of the availability of materials is the vulnerability of the supply of a material (even one used as an alloying element) to economic or political pressures.

Properties in the fabricated component determine the suitability of a given alloy for that component. The first test of any engineering design is whether or not the product performed as intended. The same criterion applies to the selection of materials in that design. The performance of a particular material in test specimens is useful information in evaluating candidate materials or confirming that certain fabrication processes were properly performed, but useful only to the extent that the laboratory results can be correlated with performance of the component in actual service.

Checklist for Material Selection. The items listed below are some of the factors that must be considered in selecting a material for a particular application.

- Analysis of performance requirements and service environments
- Correlation of performance requirements and service environments with measurable mechanical, physical and/or chemical properties of candidate materials
- Response to heat treatment
- Configuration of components
- Fabrication methods, including sequence and complexity of fabrication operations, as well as manufacturing experience at the intended production facility
- Availability
- Properties in the fabricated component
- Cost of raw material
- Cost of finished goods.

Selection for Economy in Manufacture

By the ASM Committee on Selection for Economy in Manufacture*

FOR MAXIMUM ECONOMY IN MANUFACTURE, selection of the material for a production part must be done in conjunction with selection of the manufacturing method. Of course, it must be assumed that the material will be selected from among those candidate materials that meet all design and engineering requirements. Throughout this article, it is further assumed that the design characteristics of the components discussed meet all established functional performance standards, and that factors such as strength, durability and safety have been duly considered during design of the parts. When all of these preconditions have been met, selecting a material for economy in manufacture involves consideration of several, if not all, of the following factors:

Raw-material factors

- Chemical composition
- Form of mill product
- Size of mill product
- Material condition or temper
- Surface finish
- Quality characteristics

Processing factors

- Formability
- Machinability
- Weldability
- Response to heat treatment
- Coatability

General factors

- Quantity of material required
- Availability of grade and product form
- Plant standardization of grades and sizes
- Energy consumption
- Availability of required processing equipment.

Because these factors are interdependent and strongly influenced by variables in manufacture of the material and of the component made from it, none should be considered singly. Once all pertinent factors have been assembled, the different options should be evaluated for their effect on total manufacturing cost (including procurement, storage, handling and distribution) and final selection should be made on the basis of greatest economy in manufacture.

Standard specifications, such as ASTM, ASME or AMS, may establish unified requirements for any or all of the selection factors described below under the headings "Raw Material Factors" and "Processing Factors". For certain applications, such as pressure vessels, it is necessary to use a material that conforms to the applicable standard specification so that the finished product meets all requirements of the appropriate code.

Raw-Material Factors

Chemical Composition. Determination of chemical composition may entail more than straightforward selection of a particular alloy. It may be necessary to order to a restricted range of composition, such as a narrow range of carbon or manganese content to facilitate such processes as induction hardening or welding of steel parts.

Product Form. Most wrought metals are produced in the forms of bar, rod, wire, tube, plate, sheet, strip and a wide variety of structural and special shapes. However, before any of these forms is selected, consideration should be given to the use of forgings, castings or powder metallurgy products, which have dimensions closer to the finished dimensions of the part. The form select-

*See page XIV for committee list.

ed must be the one most compatible with the manufacturing processes required for fabrication, as well as with the design of the part.

Size. All mill forms are made to certain standard dimensional tolerances in terms of length, width, diameter and thickness. Unless otherwise specified in the purchase agreement, producers will furnish products to their own standard tolerances. For maximum economy and availability, standard sizes and tolerances should be used whenever possible.

Condition. Metals may be produced to particular conditions or tempers, such as cold rolled, cold drawn, hot rolled, annealed, or quenched and tempered. Selection of the appropriate condition may eliminate one or more fabricating processes, such as heat treating, cleaning, straightening or machining.

Surface Finish. Many materials can be produced with special surface finishes that may eliminate the need for one or more fabrication processes. In some instances, the higher cost of material with a special surface finish may be more than offset by the resulting simplification in processing.

Quality descriptors are used to indicate that a material has been manufactured in such a way as to make it particularly well suited to specific applications or fabrication processes. Steel wire, for example, may be specified as "valve spring quality". Steel sheet may be produced as "drawing quality special killed". Highly restrictive quality descriptors usually imply a significant increase in material cost, but the extra cost may be justified if scrap rate is reduced, if one or more fabricating operations are simplified or eliminated, or, more importantly, if the parts being made (such as valve springs or ball bearings) require optimum properties.

Processing Factors

Formability is the relative ease with which a material can be plastically deformed, such as by forging, cold heading, bending, drawing or extruding. It is determined by chemical composition, condition, and the millprocessing variables implied by the quality descriptor.

Machinability is the relative ease with which a material can be cut or removed by various machining processes such as turning, milling or drilling. This characteristic is determined primarily by chemical composition and microstructure.

Weldability is the relative ease with which a metal or combination of metals can be satisfactorily joined by welding, with or without filler metal. It is determined largely by chemical composition and by section shape and thickness.

Response to heat treatment is often the most important factor in selecting a material for a particular application. Response to heat treatment includes mechanical-property levels attainable through heat treatment, hardenability, and suitability for carburizing or nitriding. It is determined mainly by chemical composition and the size of the section to be heat treated.

Coatability is the relative ease with which coatings can be applied to a material; it is strongly affected by surface cleanness.

General Factors

Quantity is an important factor in production of any item. It affects both availability and the relative economy of various fabricating processes.

Availability is the relative ease with which a material can be obtained in the desired shape, size, tolerances, quality and quantity. It includes availability from mills, casting or forging shops, service centers, secondary processors (such as custom coaters and wiredrawing shops) and the manufacturer's own inventory.

Plant standardization of raw-material compositions, sizes and shapes sometimes can reduce over-all operating costs for a plant even though it increases the cost of manufacturing parts of one or more specific designs.

In most manufacturing operations, it is more efficient to standardize on a few grades and sizes of material from which to manufacture a wide variety of products or components than it is to choose a different grade and size for each manufactured item. Selection of grades and sizes on an individual basis can lead to serious problems, including limited availability of certain grades and sizes; excessive costs of procuring, storing, handling, distributing and accounting for the various items of raw material; and unnecessary expenditures for floor space and capital equipment.

In the other extreme, making all parts from the same alloy, mill form and size is illogical except in rare instances. The best policy is to stock the smallest number of alloys, mill forms and sizes that will satisfy plant-wide needs at the lowest over-all cost.

Energy Consumption. During product manufacture, energy consumption is important not only because energy is expensive, but also because the availability of energy in certain forms is subject to occasional or periodic shortages. The most obvious energy costs are those associated with fabricating operations; these costs must be directly (and visibly) borne by the manufacturer.

In some instances, it is now preferable to consider total energy consumption over the entire life cycle of the product, including energy consumed in reducing ore to metal, in manufacturing mill products, and in fabricating, operating and recycling the desired consumer product.

Interactive Effects

Interactive effects among the selection factors mentioned above are quite common. A change in any of the factors concerning raw material may affect processing. For example, a change in chemical composition may affect surface finish, machinability or response to heat treatment. Likewise, a change in processing often requires a change in raw material. The general factors are also interactive with raw-material and processing factors. For example, a change in quality requirements may affect availability. Thus, in changing any of these factors affecting selection of a material for a particular part, the possibility of interactive effects on other factors must also be considered.

Assuming that both processing and operating requirements have been satisfied, the predominant factor in selection is cost. To some degree, all of the selection factors entail cost. Yet, individual cost factors (both higher and lower) may be balanced in such a way as to result in an over-all reduction in the cost of the finished part.

This balance of factors may lower final product cost by: (*a*) decreasing the cost of the raw material, (*b*) increasing or enhancing adaptability to specific manufacturing operations, (*c*) simplifying or eliminating one or more manufacturing operations, (*d*) simplifying or eliminating one or more inspection operations and (*e*) reducing or eliminating scrap losses.

Composition

In planning for economy in manufacture, choice of chemical composition must be correlated with mechanical-property requirements and performance requirements. For many applications, the specific composition itself is unimportant, except for the mechanical properties or heat treatment characteristics that it represents. More than one composition may meet the specified requirements (see Example 1). Maximum economy often can be achieved by specifying only the required mechanical and physical properties, and allowing purchasing and manufacturing personnel the greatest possible freedom in making decisions regarding factors such as form and size of purchased stock, quantity-price relationships, and substitution of thermally or mechanically treated rough stock so that an in-plant heat treating step can be eliminated. This freedom may allow, for example, a carbon steel with high residuals to be substituted for a more expensive alloy steel without compromising hardenability requirements. On the other hand, lack of such freedom may prevent savings if, for example, it prohibits use of alloys to which sulfur or lead has been added to improve machinability.

Economies may be realized by specifying nonstandard compositions (see Examples 2 and 3). This generally requires purchase of relatively large quantities. If small quantities are required, it is economically unfeasible to use a nonstandard alloy; also, some standard grades are made only on order and cannot be purchased in quantities less than a full heat.

Consultation with various metal producers often can result in economies in specification of alloy composition. Many producers and users have developed alloys that contain greater quantities of low-cost alloying elements and smaller quantities of expensive alloying elements (see Example 4).

Because the pricing structure of alloys and commodities is constantly changing and because the relative costs of various fabricating operations are also subject to change, purchase specifications must be frequently reviewed to ensure that the grade being purchased remains the most economical. Example 5 illustrates how the price differential between two grades of steel changed over a period of 17 years; in this instance, however, the money saved by virtually eliminating a scrap rate of 40% far exceeded the cost increase for the substitute material. Example 6 illustrates how the cost of steel can be minimized by adjusting composition limits to take advantage of pricing policies of steel producers. Example 7 illustrates how substitution of an alloy steel for a plain carbon steel provided the most economical means of meeting a new service requirement for a large gear. Examples 8 and 9 illustrate the substitution of nonmetals for metals.

Form

For parts of simple shape, such as bolts or straight shafts, the most economical raw material and method of manufacture are readily apparent. For parts more complex in shape, applicability of two or more forms and fabrication methods adds complexity to the process of selection. A small gear, for example, may be completely machined from bar stock. On the other hand, depending largely on the total number of parts to be produced, it may be more economical to begin with a close-tolerance, forged gear blank. Gears for small mechanical devices may be most economically made by powder metallurgy.

Among such alternatives, final selection should be based on comparison of over-all costs. Cost studies should always be used to analyze the relative economic merits of all the various forms and methods of manufacture—machining from tube or bar, casting, hot forging, cold forging, powder metallurgy, extruding or welding.

Production quantity is the factor most likely to determine the form selected. For a small gear, the total cost of making 100 pieces might favor machining from bar. If the quantity were increased to 10 000 or more, use of forged gear blanks might constitute a considerable economic advantage. With a larger gear, if only 500 pieces were to be made, it might be more economical to leave the hub solid and drill the bore. If the required quantity were 5000 pieces, it would probably be advantageous to have the hub pierced in the forging operation. With even larger quantities, forging of minor contours might be justified. As quantity increases, it becomes progressively easier to amortize the initial cost of forging dies, and savings in the costs of metal and machining increase accordingly.

Selection of product form may include consideration of bar or wire that is drawn to a special shape. Use of special shapes can sometimes reduce the amount of machining necessary to make a part from standard shapes and thereby reduce total cost.

Examples 10 through 20 illustrate how selection of the proper product form can help minimize the cost of the finished part. Examples 10 and 11 describe two instances where significant savings were realized by changing to powder metallurgy parts. Examples 12 and 13 illustrate the fact that single-piece construction is cheaper in some instances whereas use of assemblies is more economical in others. Tubing is the most economical starting material for many parts (see Examples 14 and 15), but not always (see Examples 16 and 17). Likewise, castings may or may not be the most economical form for a part, as illustrated in Examples 18, 19 and 20.

Size

The design of the part determines the size of the raw material within a fairly narrow range. Nevertheless, there is some flexibility in selection, and savings depend on judicious choice of final size.

To illustrate the problem, consider a simple shaft with a constant diameter that must be held to close tolerances. Such a shaft may be made most economically from bar purchased in the turned-and-centerless-ground condition. This is particularly true if the shaft requires a surface finish that can be achieved only by a secondary operation such as grinding. Neither hot rolled nor cold drawn bars could satisfy this condition without additional machining.

However, for a more intricate shaft with several different diameters along its length, paying a premium for ground bar of special size would be uneconomical; either hot rolled or cold drawn bar would fit the need better. Because most of the surface will be machined off, there is no justification for a stringent surface requirement on the raw bar.

Ideally, the size of stock selected should be as close as possible to the size of the finished part. There are instances, however, where some additional removal of stock must be allowed to avoid an excessive scrap rate. When bar is slightly out-of-round or otherwise warped, additional stock serves as a safety factor against rejection. Even though extra stock removal is wasteful to some extent, it may be cheaper to purchase slightly larger stock than to pay for scrapped parts. In other instances, it may be more economical to select a bar that is oversize but lower priced because it carries a lower size extra.

The possibility of parts being rejected due to surface imperfections such as decarburization or seams is an additional reason for buying oversize stock, as illustrated in Example 21. The excess stock should be adequate to permit complete removal of all decarburized metal and seams. It should also eliminate the need for magnetic-particle inspection to detect surface imperfections.

Conventional diametric machining allowances for hot rolled round bars are 3.2 mm (1/8 in.) for bars 38.1 to 76.2 mm (1½ to 3 in.) in diameter and 6.4 mm (1/4 in.) for bars over 76.2 mm (3 in.) in diameter. These allowances result in the following percentage increases in cross-sectional area:

Diameter before allowance mm	in.	Diameter after allowance mm	in.	Increase in area, %
38.1	1.50	41.3	1.625	17.4
40.0	1.57	43.2	1.700	16.6
50.8	2.00	54.0	2.125	14.8
60.0	2.36	63.2	2.49	11.0
75.0	2.95	78.2	3.08	8.8
88.9	3.50	95.3	3.75	14.8
100.0	3.94	106.3	4.185	12.9
127.0	5.00	133.3	5.25	10.2

Thus, for some applications, it may be advantageous for the user to order bar of a smaller size and accept a scrap rate a few percentage points above normal in machining or some other operation, rather than to purchase 8 to 17% excess stock. In other applications, it might be preferable to use cold finished bars, as in Example 22. Cold finished bars are more expensive, but may not have to be machined to size and often require little or no stock removal to eliminate surface imperfections.

Condition

Many metals are available in a wide variety of conditions. In almost all instances, a higher cost is associated with any condition other than that of the basic, commercial-quality product of the particular form being considered. It is the responsibility of the manufacturer of a part to determine if the additional material cost can be justified by savings in manufacturing. It is assumed that all functional requirements can be met by all materials under consideration.

Conditions for bar include hot drawn, cold drawn, and cold drawn and stress relieved. Hot rolled steel sheet is available in yield strengths as high as 690 MPa (100 ksi), whereas cold rolled steel sheet can be obtained with tensile strengths as high as 1380 MPa (200 ksi). Bars and sheets can be ordered to special internal-soundness requirements if normal inclusion content or internal cleanness is not acceptable. Carbon and alloy steel plate is available either quenched and tempered or normalized. Many other examples could be cited.

Choosing a different condition generally implies a change in the properties of the material and, consequently, in its reaction to manufacturing processes. For example, a quenched-and-tempered product may be specified to eliminate production-line heat treating and cleaning. The potential savings must be weighed against the higher material premium and the higher costs of machining the hardened stock. If it is possible to use heat treated stock, additional savings may be realized if distortion or quench cracking during heat treatment has been causing high rejection levels.

Sometimes savings are more difficult to evaluate. Two alternative designs for an automotive underbody subassembly involved a choice between HSLA (high-strength, low-alloy) 550-MPa (80-ksi) hot rolled steel and low-carbon cold rolled steel for the frame rails. Weight differences between the two proposals were minimal, but the design that incorporated HSLA steel gave a greater safety factor in crash testing. The higher cost of the HSLA material was offset by the fact that fewer parts were needed in the subassembly, thereby reducing the amount of capital equipment for fabrication and assembly and simplifying quality-control procedures. In many cases where strength-level changes in material are contemplated, the cost-benefit analysis must be carried beyond the individual part to consider potential benefits to the entire structure.

Often consideration of condition can be fairly simple and easily overlooked. In many instances, developed blanks for stamped parts are laid out from a coil strictly on geometric considerations to minimize blanking scrap losses. Sometimes, a more judicious orientation of the blank relative to the rolling direction (such as rolling direction transverse to the critical bending axes) can allow utilization of a less expensive alloy, even though blanking scrap losses may be higher. In this instance, the manufacturer takes advantage of the fact that mechanical properties in sheet are generally better in the longitudinal direction.

Several steels, particularly medium-carbon, plain carbon and medium-carbon alloy bars, are commercially available in a pretreated condition intended to enhance mechanical properties. Steels such as 1050, 1141 and 1144 are specially processed by cold drawing with heavy drafts. The bars are then stress relieved at about 370 to 510 °C (700 to 950 °F), depending on steel composition and mill practice.

As a result of cold drawing, tensile-strength values up to about 1035 MPa (150 ksi) can be obtained with hardnesses up to about 300 HB. The price of cold drawn stress-relieved steel bars is only slightly above that of conventionally cold drawn material. Certain steels (including those mentioned above) are also obtainable in the hot drawn condition. With hardnesses up to about 350 HB, these materials have even higher mechanical properties. (Hot drawn bars are passed through dies at about 370 °C, or 700 °F.)

At higher cost, carbon and alloy steels are also available in the quenched-and-tempered condition. In this condition, a common hardness range is 250 to 300 HB; hardnesses above or below this range can be special-ordered.

Whether or not the purchase of steel in the hardened-and-tempered condition is economical must be determined on the basis of direct costs. The premiums paid for steels in this condition, along with the added cost of machining at higher hardness, must be weighed against the cost of heat treating and cleaning parts in the production line. This is particularly true of those steels for which hardness is known to markedly decrease machining speeds or tool life, or both.

Part design is most likely to influence the decision to use hardened-and-tempered steel. Where little machining is required, such pretreatment may have almost no effect on final cost. When machining operations are extensive, the added time, effort and tool wear involved may rule out the use of pretreated steels.

If a part is likely to distort excessively in heat treatment, the use of steels that are hardened and tempered before machining is attractive. Savings realized through scrap reduction may more than offset the extra costs of raw material and fabrication.

Example 23 illustrates pricing of steel sheet, which is dictated by the requirements of a particular operation. Examples 24 and 25 describe two specific applications where changes in condition reduced manufacturing costs.

Surface Finish and Texture

Surface finish and texture are important features of metal products—not only for appearance, but also because they may affect performance. For example, excessive asperities in the surfaces of two parts in rolling or rubbing contact can increase friction, overheating and wear. As a result, operational costs are increased, and the product may fail prematurely. Evaluation and selection of optimum finishes and textures requires careful analysis of (*a*) the functional demands of the application on surface condition and (*b*) relative costs of producing or procuring specific finishes and textures, and of maintaining them throughout subsequent manufacturing steps. (Relative costs change because of fluctuations in price and availability, and thus should be frequently reviewed.)

A finish or texture is usually characteristic of a specific manufacturing process. For example, turned, milled and ground surfaces have different characteristic appearances (when viewed microscopically). Surface roughness is defined as the degree of deviation from an ideally smooth surface; specialized instrumentation and measuring techniques are necessary for control and measurement of surface roughness. Generally, the cost of producing various finishes and textures increases substantially as surface roughness is decreased.

The surface finish of a mill product can be affected by many variables in manufacture, including melting practice, hot rolling practice, employment of special techniques to remove injurious surface blemishes, technique of scale removal, and heat treatment and cold finishing practice (if either is used). The special measures required to produce mill products with low surface roughness represent additional costs. Because of the extra cost, the choice of cold finished rather than hot finished products must be justified by the functional requirements of the fabricated product (such as the appearance requirements for automobile body components) or by reduction in the number and cost of fabricating operations. Specifying restricted surface roughness, such as by requiring ground and polished bars or a reflective surface finish on sheet, requires further justification for the additional cost involved. Example 26 illustrates a choice between class 1 and class 2 steel sheet. To take full advantage of special surface finishes on mill products, the largest possible portion of the surface with the special finish should be retained in the finished part.

The surface finish of sheet can affect its formability. Lubricants do not adhere well to very smooth sheet. The appearance of rough sheet is objectionable for many applications, particularly if the sheet is finished with glossy paint or porcelain. Imperfections in the surface of the sheet, such as fluting, rolled-in dirt or scale, and skin lamination, make forming operations more difficult if not impossible.

For some applications, the surface of a product is deliberately textured. Hot rolled floor plates, for example, have a raised pattern to make them less slippery when wet or oily. Embossed sheet, which is made in a wide variety of patterns, is produced by passing the sheet through a set of rolls, of which one is smooth and the other is engraved with the reverse of the desired pattern. Rigidized sheet contains three-dimensional patterns, which are produced by passing the sheet through a set of synchronized rolls that have mating patterns cut into their surfaces. Both types of textured sheet are stronger and more rigid than smooth sheet of the same thickness; use of textured sheet results in increases in strength and stiffness and a significant decrease in elongation, and can reduce the weight of parts.

Some mill products are textured by grit blasting or shot peening. These processes, which can be readily applied to fabricated parts, are used both to induce residual compressive stresses at the surface and for cosmetic purposes. Example 27 describes an application in which shot peening was used to induce residual compressive stresses and thus improve fatigue resistance.

Quality Descriptors

Quality descriptors are a convenient means of identifying mill products that have been produced so as to be especially well suited to certain applications or fabrication processes. The various quality descriptors imply that special manufacturing practices and more restrictive inspections have been utilized by the manufacturer to ensure the customer that the product will perform as intended. These tests and manufacturing practices are not necessarily spelled out in a discussion of quality, and thus it might appear that the purchaser is forced to buy on faith. However, the system has been proved workable through many years of usage. The buyer does not have to be familiar with mill processing, nor does the mill have to divulge proprietary information, but the buyer is able to obtain material that is suited for a particular application (or fabrication process).

There is a price extra associated with most quality descriptors, and the amount of extra cost increases as more restrictions are placed on the product. Thus, it is logical to purchase commercial quality materials unless the particular application requires the special attributes implied by some other quality descriptor. Examples 28 and 29 illustrate the importance of choosing stock of a quality appropriate for the manufacturing process being used, even if that choice results in higher cost for starting stock.

Specifications

A specification is a statement of the attributes that a particular item must possess. By incorporating a specification into a purchase agreement for that item, the buyer and seller agree on what constitutes grounds for acceptability of that item. Specifications may be widely used published documents, such as those of ASTM or AMS, or they may be documents prepared for the benefit of a single user, such as government specifications or proprietary company specifications. The advantages of using a standard specification are that most producers are familiar with its requirements and that, furthermore, those requirements were probably written to be consistent with the capabilities of most producers. By comparison, before accepting orders that refer to proprietary specifications, the producers must analyze the requirements and compare them with their own capabilities. In some circumstances, a producer may accept an order to a standard specification, but decline an order (or charge a higher price) for the same item made to a proprietary specification.

A large proportion of steel bar stock is produced and marketed to AISI-SAE designations. These designations contain ranges and limits of chemical composition (and hardenability requirements for H-steels), but they are not specifications, for they contain no specific requirements for dimensional tolerances or mechanical properties. Incorporated in these designations, however, are implied specifications in the form of standard tolerances and practices that are published in various sections of the AISI Steel Products Manual. This common practice eliminates voluminous and repetitive specifications.

Formability

The formability of a material is the relative ease with which it can be shaped by various manufacturing processes into a component part. Formability depends on the mechanical properties of the material at various temperatures, the character of the forming process and the shape of the part. Significant savings in production costs may be achieved by proper matching of the manufacturing process, such as stamping, roll forming or extrusion, with the material's ability to be stretched, drawn, bent or extruded. Simplification of component shape can lessen formability requirements, thus permitting use of less costly forming processes and materials. For example, a requirement for drawing quality (DQ) steel sheet may be changed to one for commercial quality (CQ) sheet, which is less costly, by minor design modification. Application of a forming-analysis technique, such as the use of a forming-limit diagram (FLD) or strain analysis, in either the blueprint or the prototype stage, may save manufacturing costs by avoiding forming problems that increase scrap rates (see Example 30).

Forming-analysis techniques are useful to die shop personnel because they provide a visual image of metal flow in the stamping operation, and to die and tooling engineers because they allow an exact method of evaluating the variables involved in any stamping operation. Judicious use of FLD and strain analysis and concurrent base-material evaluation makes it possible to quantify the effects of blank development, lubrication, tooling modifications, press modifications and material substitutions. Strain analysis can be an invaluable aid for die development, quality-assurance programs and trouble-shooting.

Significant cost savings can sometimes be achieved by substituting thinner-gage high-strength materials for lower-strength materials. Although high-strength materials generally cost more, gage reduction sometimes is sufficient to effect cost savings. Formability of high-strength materials such as HSLA steel or nitrogenized steel is generally lower than that of lower-strength materials such as low-carbon steels. Through minor design modifications and better manufacturing techniques, these less-formable high-strength materials can be substituted for lower-strength materials, and savings in both cost and weight can be achieved (see Example 31). In applications that involve corrosion, however, it may not be desirable to make a part from a high-strength material; a given amount of corrosion will cause a proportionately greater loss of load-carrying capacity in a thin part made of high-strength material than in a thicker part made of low-strength material.

Generally, stamping from sheet metal is one of the least costly high-volume forming processes. Parts made from tubing can sometimes be changed to stampings made from cold rolled sheet to reduce the final piece cost (see Example 17). Forming cost decreases as volume of production increases. Cold headed processes are often found to be less expensive than machining or hot forging processes (see Example 32), even though a premium quality mill product may be required for cold heading.

Lubrication has a strong influence on formability. With better lubrication compounds and practice, less formable material can be used in parts where costly, highly formable material would otherwise be required (see Examples 33 and 34).

Many factors other than material selection also affect metal forming. In many instances, the manufacturer can realize significant savings by manipulating all factors involved in manufacturing a part (lubrication, blank design, tooling, pressing speed and others), rather than assuming that the process is a constant and that material must be found to fit the process.

Machinability

The cost of producing many machined parts is strongly dependent on the machinability of the materials selected for those parts. The machinability requirements of a particular part are determined by many factors, including the amount of metal that must be removed, type of machining operation, required surface finish, difficult machining steps such as deep-hole drilling, quantity to be produced and relative costs of material and machining time. If extensive or difficult machining operations are required, selection of materials with high machinability ratings is justified, as illustrated in Examples 35 and 36. Ferrous and nonferrous alloys that contain small amounts of insoluble second phases such as lead, sulfides or selenides are widely preferred when the utmost in machinability is needed. Leaded steels, leaded brasses and free-machining stainless steels containing sulfur or selenium, for instance, are commonly specified for screw-machine products.

Generally, the harder the material, the more difficult it is to machine. However, machinability is directly affected more by microstructure than by hardness. The machinability of many classes of alloys can be improved if the microstructure is a two-phase structure consisting of either a brittle or an easily sheared second phase dispersed throughout a moderately ductile matrix. In turning 4140 steel for example, far greater tool life and productive efficiency can be expected if the steel has a coarse lamellar pearlite-ferrite structure resulting from annealing than if it has been hardened and tempered to approximately 300 HB. Similarly, a high-carbon alloy steel having approximately 0.60% carbon will be more easily machined after spheroidize annealing than after a standard annealing treatment that produces a pearlitic structure. In alloys whose microstructures can be readily altered by heat treatment, such as many steels, it may prove more economical to purchase the material in a readily machinable condition (such as spheroidized) and to heat treat the material to another condition more compatible with service requirements after it has been machined to final configuration. In some instances, particularly with steel components that require close tolerances or fine finishes, parts are rough machined in a highly machinable condition, heat treated to final hardened condition and then finish machined (or, more often, ground) to final dimensional and surface-finish requirements.

The extent to which a metal has been cold worked can affect its machining behavior. Cold working a very soft steel can improve its machining characteristics, but extensive cold working, particularly of medium-carbon or alloy steels, can harden the steel sufficiently to greatly accelerate tool wear. Cold working causes residual stresses in the steel, which in turn can cause warping and loss of dimensional stability if a machining or thermal operation changes the distribution of the residual stresses. Residual stresses can be reduced by stress-relief annealing, but such a treatment may alter the mechanical properties of the part.

Examples 15, 16, 22 and 37 illustrate how the form and size of the starting material can affect the cost of machining.

Weldability

Weldability is the relative ease with which a metal or combination of metals can be joined by welding, with or without a filler metal. It is a complicated property, determined principally by (*a*) chemical composition of the metal(s) to be welded, (*b*) section shape and thickness, (*c*) cleanness of the surfaces to be joined and (*d*) the mechanical properties of the metals.

Low-carbon steels are, in general, readily welded by all of the common welding techniques. For best weldability, the following ranges and limits of composition are preferred:

Carbon	up to 0.20%
Manganese	0.30 to 0.60%
Silicon	0.10 to 0.20%
Sulfur	0.04% max
Phosphorus	0.04% max
Residual elements	0.10% max total

A composition that meets these ranges and limits should not be considered as a firm requirement for acceptable weldability, but as a preferred composition for which there are virtually no restrictions on welding method, type of joint, type of electrode, welding current, or position or speed of welding. Steels with compositions in these ranges are readily available in a variety of forms.

Most steels of other compositions are also weldable. However, special techniques, equipment and precautions may be required, all of which will probably raise welding costs and rejection rates.

Certain other metals are readily welded, but on the whole it cannot be assumed that materials other than low-carbon steel can be welded satisfactorily without special techniques or precautions.

Response to Heat Treatment

Satisfactory response to heat treatment is the capability of a material, fashioned into a particular part, to be heat treated to a desired microstructure, hardness and strength level without undergoing cracking, distortion or excessive size change. The most important factors affecting response to heat treatment are the recrystallization, phase-decomposition and/or transformation characteristics of the material, the configuration and mass of the part, and control of the heat treating operation.

Before determining what response to heat treatment is needed for a particular application, it is important to determine that heat treatment is absolutely necessary. The economy that may be realized by eliminating heat treatment can be considerable, but is often neglected due to the automatic assumption that a part—particularly a steel part—will be heat treated.

Because heat treatment can create substantial thermal stresses in the part (even if the part is not quenched), the part should be designed so that the deleterious effects of these thermal stresses are minimized; thermal stresses are greatest during quenching. The basic objective is to design the part so that the temperature distribution during heating and cooling is as uniform as possible. Sharp corners, particularly internal corners, must be avoided, because they concentrate thermal stresses, thus accentuating their effect on the part. The largest possible radii should be used in corners. Uniform section thickness facilitates uniform temperature distribution. Designing a part so as to minimize thermal stresses during heat treatment usually permits choice of the least costly material that can provide the required properties.

Perhaps the most common heat treatment is the austenitize, quench and temper sequence, which is the dominant process for hardening carbon and alloy steels. The critical aspect of this sequence is that the hardened steel part must contain a minimum percentage of martensite in order to meet the specified hardness. In other words, the hardenability of the steel must be sufficient to provide that percentage of martensite when the part is quenched into a medium that is appropriate to the size, shape and distortion tolerance of the part.

To achieve particular combinations of mechanical properties, it may be necessary to specify other heat treatments, such as austempering, induction hardening, flame hardening, carburizing or nitriding. Each of these processes places certain restrictions on the choice of steel; the restrictions may include carbon content, hardenability or the need for certain alloying elements.

Seemingly minor changes in the heat treating operation can often produce significant reduction in manufacturing costs, particularly if the change reduces the scrap rate, permits a change to a less costly alloy, or facilitates a subsequent fabricating operation. See Examples 2, and 38 to 42, which illustrate how response to heat treatment can affect cost of manufacture.

Coatings

Coating of materials, either by the supplier or by the user during manufacture of parts, must be carefully considered when assessing the cost of producing a given part or assembly. Obvious costs are those of the coating material and of the process for applying it. Less obvious are effects of the coating processes on the properties of the material being coated, which can adversely affect the quality and usefulness of the end product and produce high scrap costs. A few of the more important coatings used in manufacturing processes are:

Organic	Inorganic	Metallic
Paints	Phosphates	Metal spraying
Lubricants	Chromates	Dip-coating
		Plating

Application of most of these coatings involves elevated temperatures ranging from 65 °C (150 °F) to as high as 455 °C (850 °F) for zinc coating or well over 650 °C (1200 °F) for aluminum coating. Metal spraying normally does not raise the substrate temperature much over 95 °C (200 °F).

Phosphate coatings and various lubricants are used to aid forming of sheet and cold heading of bar and rod. Low-carbon steel sheet can be adversely affected by the temperatures involved in phosphating and lubricant application (see Example 34). High scrap costs can result if these processes are not considered during initial specification of the sheet.

Paints are frequently used for corrosion protection and for enhancing the appearance of the finished part. Paint can be the least expensive way to achieve these desired features. However, the costs of applying paint must be carefully evaluated to avoid undue labor or process costs. Sometimes, use of an entirely different coating process may save money by eliminating costly hand operations.

Chromate dip coatings often are applied at the mill for corrosion protection of galvanized steel sheet. A chromate dip coated galvanized product is said to be "passivated" or "stabilized". In many manufacturing processes, the presence of a chromate coating causes no difficulties. However, chromate coating must not be used on galvanized steel parts that are to be resistance welded, or phosphated and painted, after forming. Once applied, a chromate coating on galvanized steel is almost impossible to remove without damaging the galvanized coating. Any chromate present will interfere with proper phosphating and thereby cause poor paint adhesion in the final product. Chromate also causes problems in resistance welding. Purchase agreements for galvanized steel sheet must be carefully written to avoid these problems.

A coating process such as metal spraying, which is most often done during the final stages of component production, normally does not adversely affect material selection unless the temperatures involved are capable of changing the mechanical properties of the part. The possibility of corrosion due to strong galvanic couples between the coating metal and the substrate metal or alloy also should be considered.

Higher-temperature dip coatings (galvanizing and aluminizing) have marked effects on the mechanical properties of the substrate material. The properties of commercially available coated sheet products are well defined, but the use of these processes on parts should be approached cautiously.

Mill-product form, size, condition and surface finish are not critical to coating processes, except where coating adhesion can be impaired by an improper condition of surface roughness (as in metal spraying), where irregularities can cause imperfections or objectionable appearance in the finish (as in plating), or where one surface coating can affect another coating (as in painting chromate-coated galvanized sheet).

Electroplating, electroless plating or mechanical plating of one metal on another is used to alter the appearance, wear resistance or corrosion resistance of both large and small parts. Before plating is specified, the designer must carefully consider the possibilities that changes in properties will be caused by plating and that the thickness of the plating may affect critical dimensions.

Electroplating of through-hardened or case hardened steel parts, or severely cold worked carbon steel parts that have not been stress relieved, can result in hydrogen embrittlement. However, preplating operations such as cathodic cleaning, acid pickling, and stripping of defective electroplate can cause as much or more hydrogen embrittlement as the electroplating process itself. To avoid delayed fracture of stressed parts due to such embrittlement, it is necessary to bake the plated parts at about 200 °C (400 °F), for at least 4 h. This must be done within a few hours of the plating operation to be truly effective. Baking is an additional operation that raises the cost of producing the parts, but it is an essential operation when there is a need to ensure freedom from hydrogen embrittlement.

Electroless plating is a chemical plating process rather than an electroplating process; it produces virtually no hydrogen embrittlement and is frequently used to plate case hardened screws and bolts that must not exhibit embrittlement in service. The chemical baths used for electroless plating require strict control of composition and temperature. Electroless plating generally is more expensive than electroplating. Not all metal plating systems are available as electroless processes; nickel and copper are most often plated by such processes. Imparting hardness and wear resistance to electroless nickel requires heat treatment at 370 to 425 °C (700 to 800 °F) under a protective atmosphere.

Mechanical plating is a proprietary process by which metal powders are mechanically welded to the surfaces of parts. The process requires activation of the surface during tumbling in a medium of glass beads and metal powder. Soft metals such as zinc, cadmium, tin and their alloys can be readily plated on steel by this process. Because electrical current is not required, hydrogen embrittlement does not occur.

Many organic paint systems are being converted from solvent-base to dry-powder or water-base systems to comply with restrictions on solvent emissions in environmental protection laws. Powdered paint, applied electrostatically, presents no pollution problems but is considerably more expensive than the older solvent-base systems. Use of electrophoretic paint is increasing rapidly—especially in the automotive industry. This type of paint system poses only minimal pollution problems and has the advantage of providing coverage that is uniform and complete, even in small openings and deep recesses. Some solvent-base paint systems use "exempt" solvents, which are subject to fewer restrictions. These solvents generally are more expensive than the hydrocarbon-base solvents used in earlier systems. In general, there appears to be no cost differential involved in the choice among electrostatic, electrophoretic and exempt-solvent paint systems.

Quantity and Availability

The criteria of quantity and availability are considered together because the quantity of material required for a particular product can limit the number of possible sources for that material and, indirectly, limit the number of product forms from which the designer can choose.

The number of parts to be made affects decisions regarding choice of fabrication process; but, because mill products are marketed by weight, it is the weight of material required that affects availability. For products that require large quantities of material, the critical factors in availability include (*a*) the ability of one or more mills to produce the required quantity, (*b*) production and shipping schedules and (*c*) whether or not the material can be purchased from multiple sources to ensure adequate supply and provide price competition. For products that require very small quantities of material, the most significant issue is where a product reasonably close to the preferred grade, shape and size can be obtained.

To establish what procurement options are available, it is necessary to determine the minimum order quantity for each purchased item and then to estimate the length of time such a quantity would last in normal production. If that time period is less than a few months, purchase from the mill should be considered a reasonable option; if not, the material should be selected from in-plant supplies or from items stocked in service centers.

The two principal reasons for ordering mill products directly from the producer are the possibilities of obtaining uncommon items (such as nonstandard sizes and shapes or seldom-produced grades) and of obtaining the lowest possible purchase price. An additional advantage is that large quantities can be purchased on a single order and delivered in partial shipments to fabrication facilities (either captive or vendor) at scattered locations. Disadvantages of ordering from the mill include the delay of several weeks or months between ordering and receiving the material, the requirement that relatively large quantities be ordered at one time, and costs of possessing the material from the time it is received to the time it is fabricated into a product. The costs of possession, which are all too easily neglected in estimating the relative costs of materials obtained from various sources, include the costs of storage facilities, wages and overhead for personnel who handle or guard the material, cost of money to buy the material (either as interest on borrowed money or interest lost by not investing capital), taxes on inventory, and losses from inventory due to material obsolescence, pilferage or atmospheric corrosion.

Certain types of material, such as prepainted sheet or coated wire, may be advantageously purchased from a converter or secondary processor. These firms specialize in buying mill products from manufacturers, applying coatings or performing some other work to the customer's specifications and reselling the items.

Anticipated production quantities can affect the cost of manufacture by limiting the choice of production methods, as illustrated in Examples 43 and 44. Some operations, such as cold heading, stamping, welding and automatic screw machining, require extensive tooling and setup time, the costs of which must be spread over a large number of fabricated parts for economical production. Other operations, such as machining, casting and hand welding, require less tooling and setup time, and may be economical alternatives if only a few parts are to be made. However, these latter operations are relatively labor-intensive, and for production of large quantities of parts may be more costly than one of the more heavily automated processes.

When the parts to be produced are complex in configuration and either small or moderate in size, it is generally worthwhile to consider making the basic part from a casting or forging. Certain very large parts—such as casings for large pumps, and landing gear for jet aircraft—can be more economically produced from castings or forgings as well. Powder metallurgy parts are also attractive alternatives in many instances, especially for small parts made of materials of low to moderate strength. Powder metallurgy also makes it possible to use materials that are impossible or prohibitively expensive to make by conventional alloying techniques. Frequently, a detailed analysis of procurement options will be needed to establish the savings that can be achieved by using one of these "near net shape" product forms. Such an analysis must include amortization of the costs of dies, patterns, fixtures and other tooling that would not be needed if the part were made by another process. Depending on configuration, required strength and number of pieces, aluminum or iron castings frequently are the most economical.

Special Requirements

The compositional and other related factors that determine suitability for carburizing, nitriding or similar case hardening processes may be classified as special requirements in selection. A carburized gear that is subject to high loads, for example, will require selection of a low-carbon medium-alloy steel with fairly high hardenability. Hardenability is particularly important in achieving a sufficiently high core hardness to support the compressive loads on the carburized case.

Plain carbon steels generally are not nitrided, because they build up an excessively brittle "white layer" that cracks and spalls readily. Obtaining a satisfactorily nitrided case requires selection of steels with particular ranges of alloy content.

There are many other situations that demand special requirements. Perhaps the most common of these are the surface conditions necessary for decorative electroplating or similar metal-finishing processes, and the compositions and material conditions necessary to ensure satisfactory welds in many materials.

Energy Considerations

The availability and cost of energy are problems that are assuming increasing importance in the determination of economy in manufacturing operations. Because of rapid and ongoing changes in this area, energy costs should be re-evaluated on a regular basis. Recent shortages of natural gas and oil illustrate the nature and severity of the problem. Also, substantial increases in the costs of all energy supplies have occurred for various and complex reasons. Thus, manufacturers of all classes of products are looking for new and better ways to use energy more efficiently. Most industries start with some basic raw materials; a finished product is then manufactured and marketed. What may be a finished product for one company becomes a raw material for another. Hence, energy consumption and attendant cost usually is cumulative during the total manufacturing process of an end product.

Refining and conversion of most metals into semifinished products (such as billets, castings, bar, sheet and strip) requires high energy consumption. These products are then further processed (machined, stamped, formed, welded or heat treated) into finished products by other industries. In each step of this complex process, energy is consumed. Because energy costs are usually considered part of indirect manufacturing cost (burden), it is difficult to directly relate energy savings to reduction of over-all product costs. Nevertheless, it is very desirable to determine the amount of energy consumed in each manufacturing step and to minimize total energy consumption. Whenever an operation in the manufacture of raw material or in subsequent processing can be eliminated or reduced, savings in energy and product cost can result. With the multitude of raw material, equipment and processes used throughout the industry, specific guidelines for determining actual energy cost savings are most difficult to generate.

This article contains several examples that illustrate how energy can be conserved by eliminating heat treatment (Examples 25, 36, 37, and 45 through 47), by reducing the amount of machining (Example 28) or by reducing the amount of scrap (Example 29).

Examples of Selection

The examples in this section illustrate the principles of selection for economy in manufacture discussed on the preceding pages. Some examples illustrate only a single principle; others illustrate more than one. In some instances, only the material or product form is changed to make the part more economical; in others, only the processing is changed; but sometimes both material and processing must be changed to improve economy while maintaining the required properties in the finished part.

Example 1. A side gear (Fig. 1) was formerly made from 4118 steel; it was carburized to an effective case depth of 1.25 mm (0.050 in.) and oil quenched. A change to 1524 steel did not affect either machining or heat treatment. Hardenability of both case and core was acceptable. The properties of the finished gear did not change, but the change in grade reduced the cost of the steel by about 10%.

Fig. 1 Side gear (Example 1)

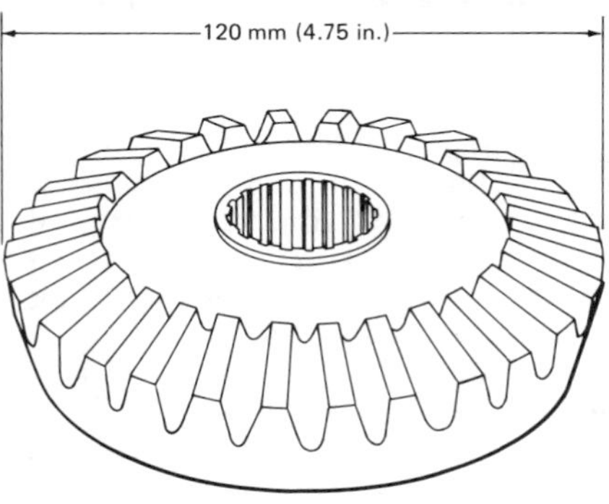

Example 2. A gear manufacturer used SAE 1053 steel for induction hardened gears. Because gears of some configurations cracked during water quenching, it was proposed that the material be changed to an oil-hardening standard alloy steel. Instead, it was decided to substitute a modified 1053 steel with a minimum chromium content of 0.18%. This nonstandard steel was about 7% cheaper than the proposed standard alloy steel.

Example 3. Fuel-injection camshafts were made from annealed 1050 steel forgings. Cams were similar in design to that shown in Fig. 2, but varied in size from 64 mm (2.50 in.) to as little as 32 mm (1.25 in.) across the cam contour. The number of cams integral with the shaft varied from one to eight. Lobe contours were induction hardened to a minimum hardness of 60 HRC. Scrap loss due to hardening cracks in the steep portions of lobes ran from as high as 5% for some heats to as low as 0.5% for others, with an over-all average of 1%. The scrap loss was not regarded as serious, but hardening cracks in the cam lobes resulted in fatigue failures early in service life. All shafts were therefore subjected to magnetic-particle inspection. Continued examination of cracked shafts proved that high manganese content (allowable range, 0.60 to 0.90%) and excessive residual amounts of chromium and molybdenum were responsible for cracking. A modified grade of 1050 steel with a maximum manganese content of 0.50% was purchased in 5-ton lots at no extra cost.

Fig. 2 Fuel-injection camshaft (Example 3)

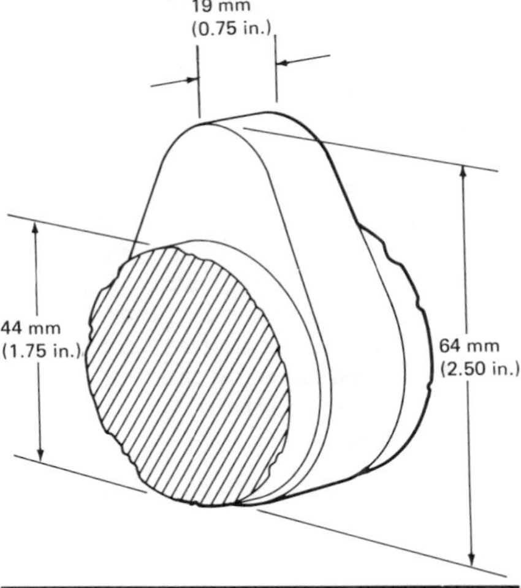

No harmful cracks had appeared 3 months after the material was changed to the modified 1050. Therefore, magnetic-particle inspection was discontinued, resulting in an average cost saving of $0.15 per shaft. Additional savings resulted from reductions in scrap loss and handling.

Example 4. A major corporation developed a series of steels that have the same case and core hardenabilities as standard grades but that are less costly. The substitute steels contain lower amounts of the more expensive alloying additions and greater amounts of the less costly additions. The average cost reduction brought about by substituting these steels for the corresponding standard grades is about 5%. Annual usage of these steels by this corporation is typically 20 Gg (20 000 tons), so that the amount saved is considerable. The substitute steels, which have been assigned EX numbers by SAE, are:

EX grade	Equivalent standard grade
EX-15H	8620H
EX-17H	8625H
EX-18H	8627H
EX-21H	8617H
EX-34H	8630H
EX-36H	8640H
EX-39H	8650H
EX-40H	8655H
EX-41H	8660H

Example 5. An unacceptably high scrap rate was encountered in making washer and screw assemblies from 1022 steel because of erratic hardenability of that grade. A change to 1524 steel with restricted manganese content reduced the scrap rate to near zero. This decision was reviewed periodically over a period of 17 years, and at the end of that time, the substitution was evaluated as summarized below:

	Original material	Substitute material
Grade	1022	1524
Scrap rate	40%	Near 0
Initial cost differential	...	+5%
Cost differential after 17 years	...	+13%

In spite of the 8% increase in cost differential, the near-zero scrap rate justified continued use of 1524.

Example 6. On occasion, the cost of steel can be reduced by specifying composition ranges so as to take advantage of steel company pricing policies. The composition extras for certain ranges of carbon and manganese are substantially higher than those for other ranges, and thus slight changes in specified composition may result in considerable reductions in price. One example, based on 1026 steel, is given in the table below. The extra for fine grain practice is required by most steel producers for steel containing more than 0.28% carbon.

	Grade	
	1026	1026 Mod
Chemical composition, %		
Carbon	0.22-0.28	0.22-0.33
Manganese	0.60-0.90	0.50-0.80
Phosphorus	0.035 max	0.040 max
Sulfur	0.045 max	0.050 max
Silicon	0.10 max	0.20 max
Price per 100 lb (1978 prices)		
Base price, carbon steel	$16.15	$16.15
Composition extra	1.00	0.20
Fine-grain extra	None	0.65
Size extra	0.90	0.90
Total	$18.05	$17.90

Example 7. A large spur gear with a pitch diameter of 465 mm (18.3 in.) and a face width of 53 mm (2.1 in.) was made from 1046 steel; the teeth were induction hardened by heating in a multiturn inductor and spray quenching. The moderate amount of distortion that occurred during quenching was acceptable for the original design. However, redesign of the equipment for which the gear was made placed higher loads on the gear and made this distortion unacceptable. To compensate, the specification for the gear was changed to include grinding of the teeth after hardening, which raised the relative cost of the gear from 100 to 139. Single-tooth induction hardening reduced the distortion to acceptable amounts and reduced the relative cost to 132. Changing the material to 8622 steel and changing the heat treatment to carburizing and hardening reduced the relative cost to 120; the performance of the carburized gear was acceptable. In this example, an alloy steel provided a lower-cost gear than a carbon steel, because a tooth grinding operation was eliminated.

Example 8. An aluminum headlamp panel and a steel headlight can were replaced by a single part made of fiberglass-reinforced polyester plastic. The change resulted in a one-time savings of $50 000 in tooling plus annual savings of approximately $50 000 due to lower part cost and reduced assembly labor. The additional benefits of more attractive styling and elimination of in-service corrosion also were accomplished by the change.

Example 9. An accelerator pedal was made of rubber-clad steel. When this part was made of injection-molded polypropylene, costs were reduced by about $80 000 per year. The plastic part was lighter in weight, which allowed the pedal-return spring to be made of lighter-gage wire, resulting in a pedal assembly that required less effort to operate.

Example 10. Grooved hubs (Fig. 3) had been machined from 12L14 cold finished bars. The parts were redesigned for manufacture by powder metallurgy methods, using FC-0200 powder. The groove was machined in a secondary operation. For parts 50 mm (2 in.) in outside diameter, a cost reduction of 34% was achieved; for similar parts 25 mm (1 in.) in outside diameter, cost was reduced only 6% because the reduction in material scrap was much less significant.

Example 11. Initial production lots of wrought steel pawls (Fig. 4) were plagued with quality problems and high scrap rates. Burrs from blanking, broaching, milling and grinding operations caused tooling location errors that resulted in a scrap rate of about 20%. Two deburring operations were necessary to minimize this scrap problem, and the cost of the part was significantly increased. Furthermore, it was difficult to maintain the desired tooth profile by broaching. A change to a powder metallurgy part eliminated several manufacturing steps, which reduced the cost of the part by more than 25%. The change to a powder metallurgy part halved the labor requirement for each part, which more than compensated for a 20% increase in cost of raw material. However, the most important benefit was the improvement in quality of the parts, which brought about additional savings by reducing inspection and assembly costs as well as warranty costs resulting from field failures.

Example 12. A fuel-pump operating lever had been made from a screw-machine part and two stampings (Fig. 5, left) that were welded together, machined and cyanide hardened. The part was redesigned as a powder metallurgy part (Fig. 5, right) made in one piece from FN-0208-S powder and heat treated to the required minimum tensile strength of 690 MPa (100 ksi). Substantial cost reduction was achieved through this change in material and manufacturing method.

Example 13. Lack of consistent quality in die cast aluminum bellcrank assemblies (Fig. 6, left) was the principal reason for considering alternate designs and materials for this part. Some lots of die castings had excessive dimensional variations, porosity in critical sections, and microstructural conditions that caused problems in ma-

Fig. 3 Grooved hub (Example 10)

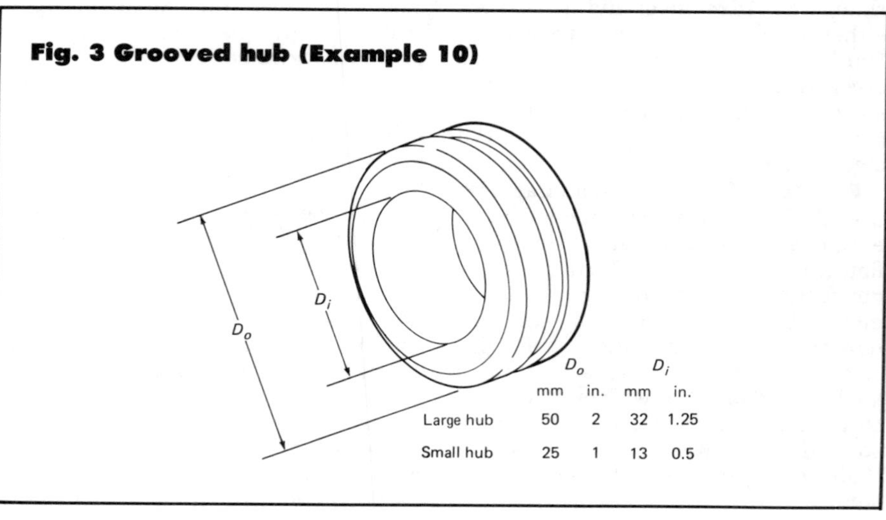

	D_o mm	D_o in.	D_i mm	D_i in.
Large hub	50	2	32	1.25
Small hub	25	1	13	0.5

Fig. 4 Pawl (Example 11)

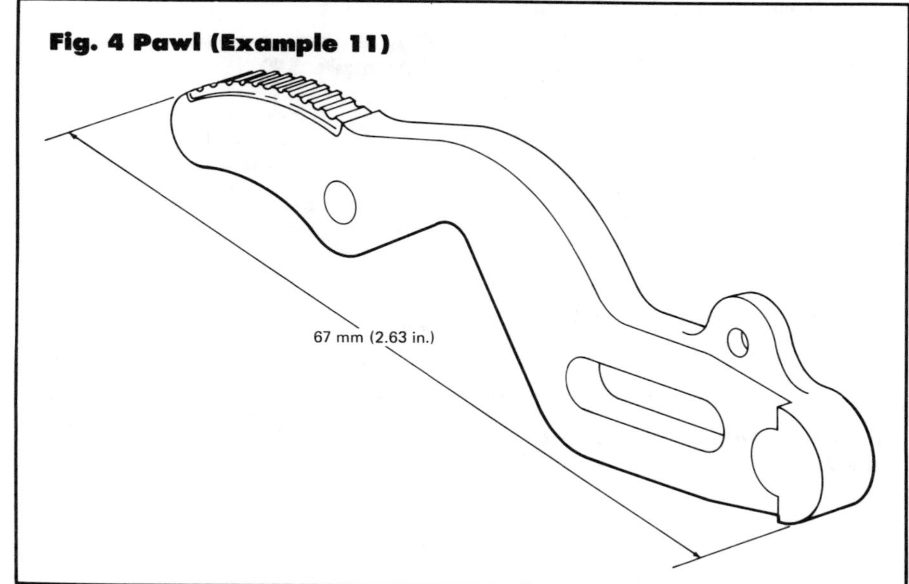

Fig. 5 Fuel-pump lever (Example 12)

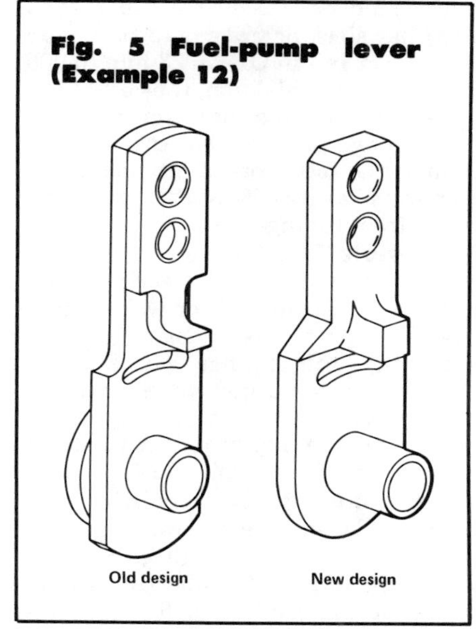

Fig. 6 Bellcrank assembly (Example 13)

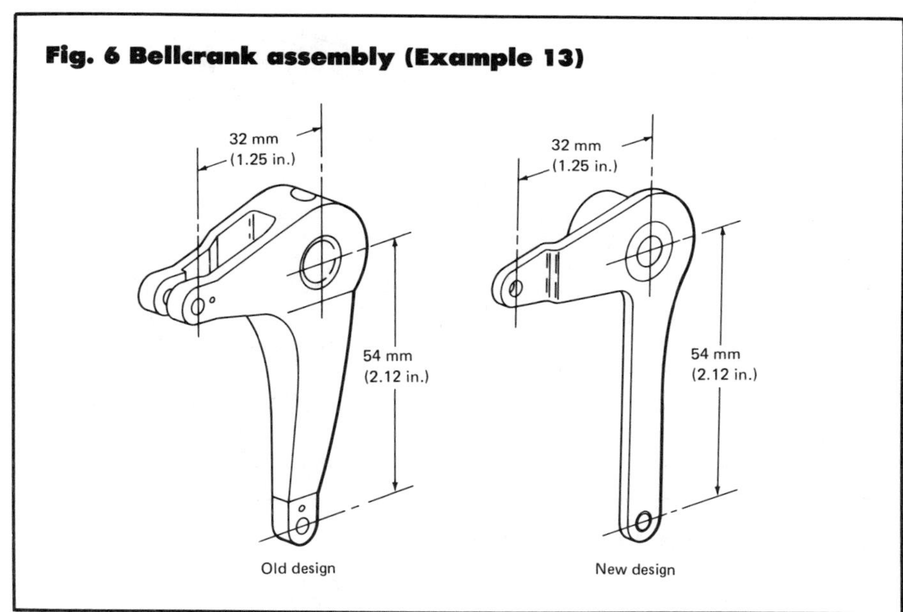

Fig. 7 Relative cost of 1018 steel round bar and 1015 steel mechanical tubing (Example 14)

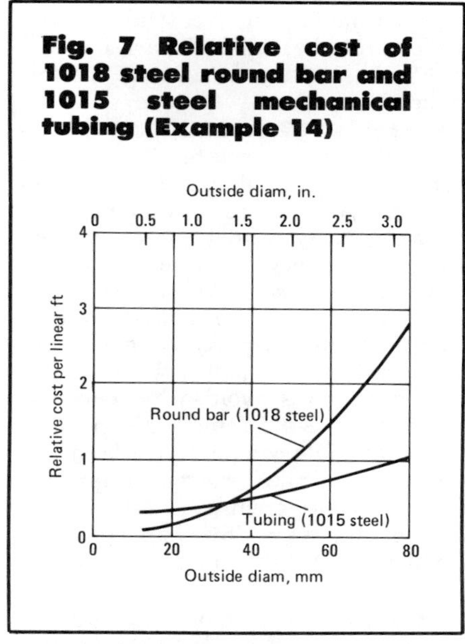

chining. Scrap rates for these lots were excessive. A fabricated assembly (Fig. 6, right) made from a stamped steel part and a separate hub (of self-locking design) had the same load-deflection characteristics as the die cast bellcrank. The cost of the fabricated assembly was about 32% less than that of the die cast part, and the scrap problem was virtually eliminated.

Example 14. Relative material cost as a function of diameter is shown graphically in Fig. 7 for 1018 steel round bar and 1015 steel mechanical tubing with a 3.2-mm (0.125-in.) wall. Both the bar and the tubing are assumed to be of random lengths and standard tolerances. Either round bar or mechanical tubing can be used as the starting stock for hollow parts such as rings, sleeves and fittings. For these applications, 1015 tubing is more economical when outside diameter exceeds 35 mm (1.375 in.). Less fabrication time is required for producing hollow parts from tubing, which results in additional savings.

Example 15. Spacers 49.0 mm (1.93 in.) in outside diameter, 40.6 mm (1.60 in.) in inside diameter and 3.2 mm (0.125 in.) thick had been machined from 1018 bar stock. A change to 1015 mechanical tubing raised the cost of the starting material by 45%, but reduced the costs of labor and burden by 36 and 38%, respectively. Because the costs of labor and burden represented a far larger portion of total cost than did the cost of material, the change reduced the cost of the part by 34%.

Example 16. Cylindrical receivers for sporting rifles, 214 mm (8.44 in.)

long, could have been machined from 1137 bar stock or from 1118 tubing 33 mm (1.312 in.) in OD by 23 mm (0.906 in.) in ID. For this item, the cost of bar stock was 16% less than that of seamless tubing, but the labor cost was 5% higher; the total cost of producing the part from bar was 2% less than producing it from tubing.

Example 17. Control-arm restrictors used in automotive steering gears had been machined from welded tubing. The part was redesigned as a stamping. This change was possible because the large quantities involved justified the investment in progressive die tooling; the change reduced the cost of the part by more than 85%.

Example 18. A connecting rod made of a low-alloy steel was originally designed as a forging (Fig. 8, left). Changing to a cast connecting rod of the same material (Fig. 8, right) resulted in a 20% reduction in weight, from 40 to 32 kg (89 to 70½ lb). The change from forging to casting reduced the cost of the finished part by 46% with no sacrifice in service performance.

Example 19. Brake-shoe components for large rock-hauling trucks were originally produced as weldments (Fig. 9, top), but the weldments were not sufficiently rigid to be considered fully satisfactory. In order to make individual components of the weldments from thicker stock, it would have been necessary to purchase more rugged forming equipment. This capital expenditure was avoided by redesigning the part as a casting. Pattern costs were negligible compared to the alternative cost of new forming equipment. The cast parts were fully satisfactory in service and were 36% less costly to produce than the original welded parts.

Example 20. A mounting bracket was made of SAE grade 35018 ferritic malleable cast iron, then machined and nitrided (Fig. 10, left). It was difficult to maintain the required tolerances with this procedure, and the scrap rate was excessive. The part was redesigned as a stamping (Fig. 10, right) made from plain carbon steel sheet; the part was then cyanide hardened. The cost of the finished part was reduced more than 15% by the change.

Example 21. The part illustrated in Fig. 11 was made from 52100 steel bar 19.0 mm (¾ in.) in diameter. After being machined, every part was subjected to magnetic-particle inspection to identify those parts in which surface imperfections in the original bar stock had not been machined away. Rejection rate was about 28%. Increasing the stock size to 19.4 mm (49/64 in.) increased the cost of raw material by 4% and the cost of machining by 5%, but decreased the rejection rate to zero. The manufacturer had sufficient confidence in the zero rejection rate to discontinue 100% magnetic-particle inspection and use a common statistical basis for quality assurance. The total cost of the part was reduced by 37% as a result of the increase in stock size.

Fig. 8 Connecting rod (Example 18)

Forging Casting

Fig. 9 Brake-shoe component (Example 19)

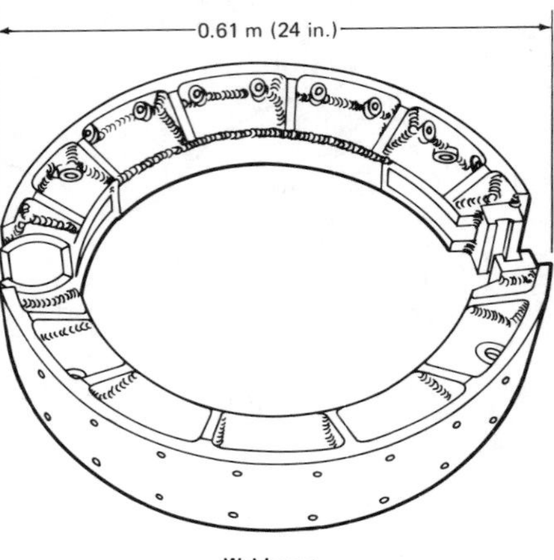

Weldment

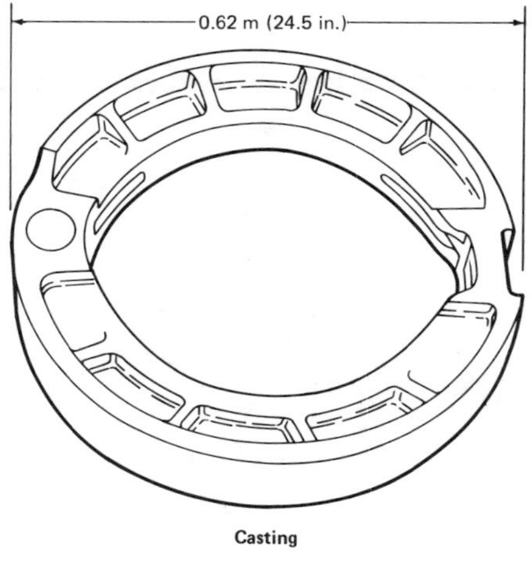

Casting

Fig. 10 Mounting bracket (Example 20)

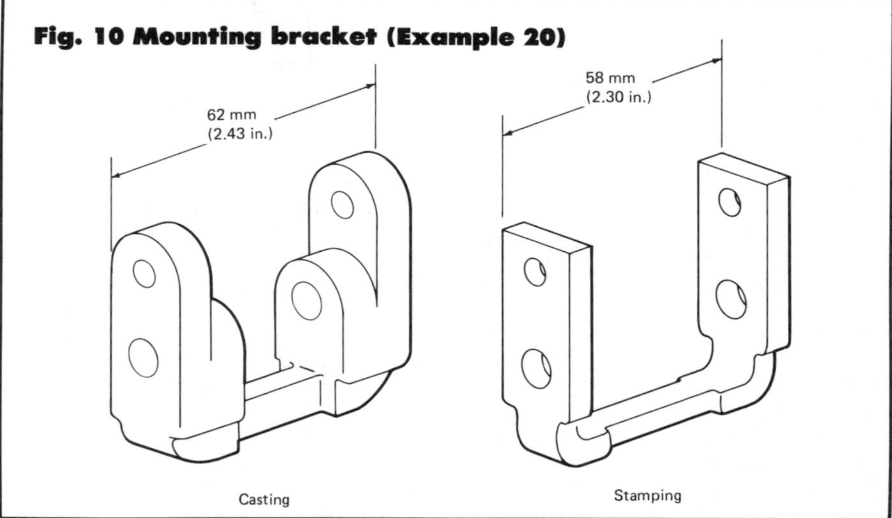

Casting Stamping

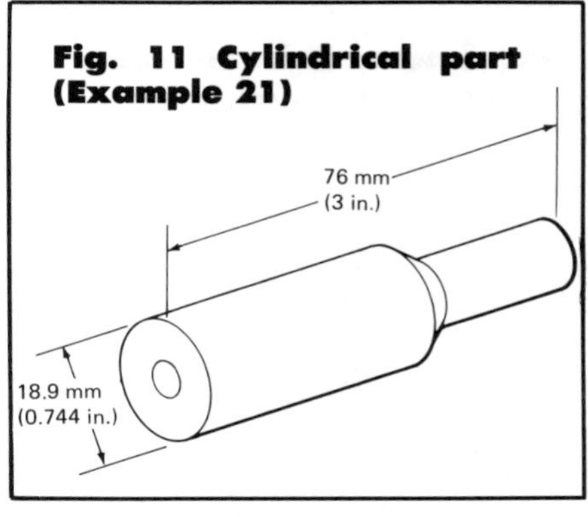

Fig. 11 Cylindrical part (Example 21)

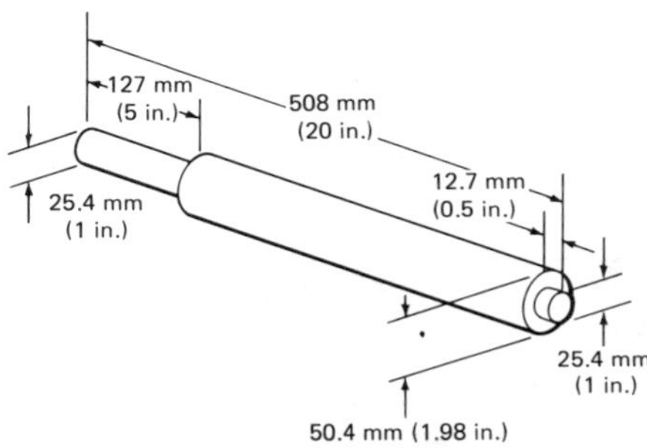

Fig. 12 Spindle (Example 22)

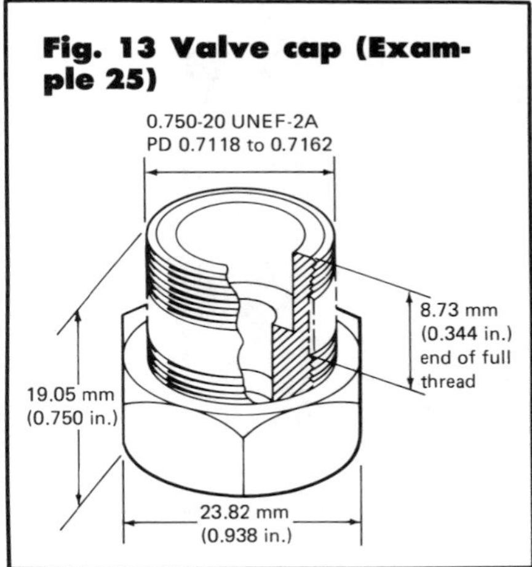

Fig. 13 Valve cap (Example 25)

Example 22. A spindle (Fig. 12) had been made from 1113 steel bar stock 50.8 mm (2 in.) in diameter—the standard size nearest the diameter of the largest section of the spindle (50.4 mm or 1.98 in.). It was determined that total cost could be reduced almost 13% by substituting cold finished 50.4-mm-diam bar stock, which could meet both surface-finish requirements and size tolerances over a large portion of the surface of the spindle without machining. The odd-size extra increased the cost of the steel by 0.7% but, because only the ends of the part now required machining, the cost of machining was reduced by almost 40%.

Example 23. High-strength low-alloy steels may be classified in the following groups:

- Fully killed, with controlled sulfide shape (relative cost, 100)
- Fully killed (relative cost, 97)
- Semikilled (relative cost, 89)
- Nitrogenized (relative cost, 84).

These steels are listed in order of decreasing formability and decreasing cost. The relative costs given are for representative steels with minimum yield strengths of 345 MPa (50 ksi). Provided the formability of a less costly grade is adequate, considerable savings can be realized by using that grade.

Example 24. Total cost of manufacturing high-quality, high-strength fasteners, 1/4–28 by 1 in. long was reduced by changing the material specification from cold heading quality to seam-free, decarburization-free cold heading quality 8740 steel. Before the change in material, each fastener had to be ground to eliminate surface imperfections traceable to the raw material. The change in material eliminated the need for grinding. Although the new steel was 25% more expensive, the total cost of the fastener was reduced 10% by the change in material.

Example 25. Threaded valve caps (Fig. 13) had been made from 1141 steel, hardened and tempered to 250 to 300 HB. To reduce over-all cost, the material was changed to a cold drawn and stress-relieved steel having a minimum yield strength of 700 MPa (100 ksi); both 1141 and 1144 were considered, but 1144 was selected for its superior machinability. Although the cost of the cold drawn and stress-relieved 1144 steel was about 8% higher than that of the 1141 originally used, elimination of heat treatment and improved machinability reduced the total cost of the part by 42%.

Example 26. Low-carbon cold rolled steel sheet is produced in two classes, depending on required surface appearance. Class 1 sheet is temper rolled after annealing; class 2 sheet normally is produced without temper rolling. Class 2 sheet may be temper rolled very lightly during oiling or rewinding. Minor surface imperfections such as slight surface scratches or light stretcher strains are permitted in class 2 sheet, but not in class 1 sheet. Temper rolling of class 1 sheet reduces its ductility slightly below that of class 2 sheet. The cost of class 1 sheet is about 3% higher than that of class 2 sheet.

An internal component for a household clothes dryer was finished with a two-coat stippled porcelain enamel. Although this part is classified as an appearance part, the enamel finish adequately covered the surface imperfections normally encountered in class 2 sheet. The choice of class 2 sheet for this application saved the 3% price differential between the two classes of sheet.

Example 27. A transmission flex plate had been made from SAE 950X HSLA steel and then nitrided. Although the mean fatigue life of these parts was acceptable, the frequency of early fatigue failures was too high. Several HSLA steels and several processing methods were evaluated in an attempt to reduce the early fatigue failures. A change to 950BK HSLA steel and shot peening provided the desired effect. The new steel was slightly more expensive than its predecessor, but shot peening was less costly than nitriding, and the thickness of the flex plate was reduced by about 8%; the total cost of production was significantly reduced.

Example 28. A transmission shaft was made from 4024 steel bar in an automatic screw machine. Over 40% of the bar stock was converted into chips in the machining process. For quantities of 100 000 shafts, cost was reduced 18% when the basic shaft was hot upset forged from a smaller bar. Figure 14 shows the starting stock for each process, along with a sketch of the finished part and of the intermediate hot forged blank. Additional savings might have been realized if a different intermediate blank were produced by cold extrusion; the further reduction in machining scrap might reduce cost by another 10%. These reductions in cost take into account the fact that a change in steel quality is required for either alternative manufacturing process: forging quality stock for the hot upset blank, and cold extrusion quality stock for the cold extruded blank. The reduction in the amount of steel converted into chips more than compensates for the extra cost of premium quality stock.

Example 29. Machine screws made from capped steels, such as 1010 and 1017, were being produced with a scrap

Fig. 14 Transmission shaft (Example 28)

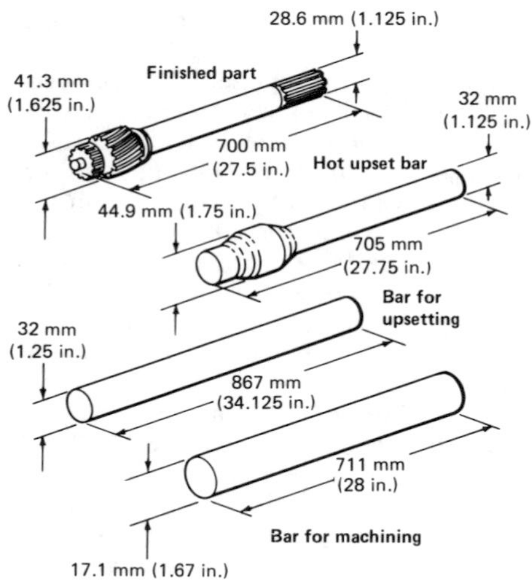

Fig. 15 Refrigerator liner with embossments (Example 30)

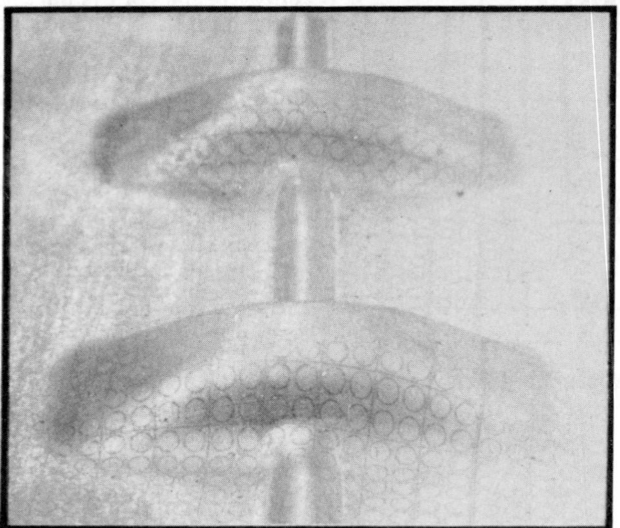

Original design

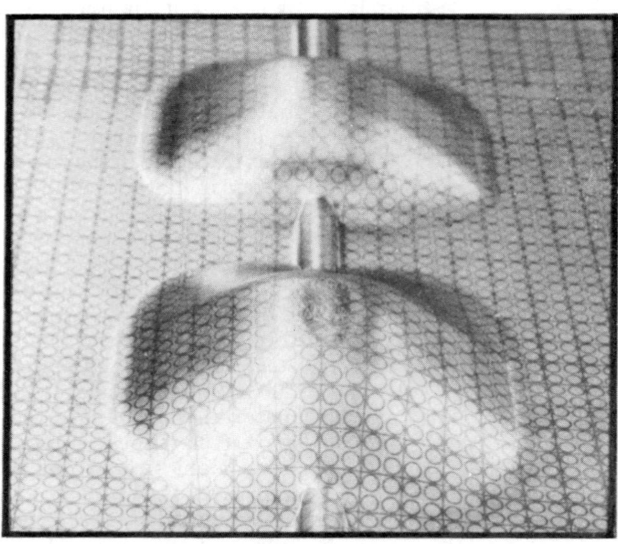

New design

rate of about 7%. A change to killed steels reduced the scrap rate to less than 1%. The 2.5% price extra for the killed steel was more than justified by the reduction in cost due to the reduction in scrap rate.

Example 30. To make the most economical selection of steel sheet for parts requiring extensive press forming or drawing, it is important to know the locations and extents of the most severe forming or drawing strains. A refrigerator liner made from 0.62-mm (0.025-in.) sheet had two severely formed embossments (see Fig. 15), which required the use of drawing quality steel. Even with the drawing quality steel, the scrap rate periodically rose to unacceptable levels. A strain analysis of the part showed that only the embossments, which accounted for only 0.2% of the total surface area, required the drawing quality steel. Furthermore, these two areas were the most likely locations for forming failures. The part was redesigned to reduce the severity of forming around the embossments. The redesign eliminated the periodic scrap problem and permitted use of commercial quality steel sheet, which reduced material cost by about 3%.

Example 31. An automotive looseline bumper assembly (Fig. 16) had been made from two pieces, a face bar 2.67 mm (0.105 in.) thick and a reinforcement 3.86 mm (0.152 in.) thick, both of which were made from low-carbon steel with a yield strength of 205 MPa (30 ksi). Substitution of HSLA steel with a yield strength of 310 MPa (45 ksi) and a thickness of 3.43 mm (0.135 in.) allowed elimination of the reinforcement. Although the HSLA steel cost about 20% more than the plain carbon steel, the weight reduction of 10 kg (23 lb) and the elimination of the reinforcement reduced the cost of the assembly by more than 20%.

Example 32. Quantity is the deciding factor in selecting a method for producing the special round-headed bolt shown in the sketch in Fig. 17. As indicated in the accompanying graph, these bolts can be cold headed more economically than they can be machined, except for very small quantities.

Example 33. An automotive hood was originally made from drawing quality steel sheet. Because of erratic behavior of the material during forming, upgrading to aluminum-killed steel was considered. But first, the effect of lubricant on forming behavior was studied. Two lubricants, the normal production lubricant and a new

Fig. 16 Automotive bumper assembly (Example 31)

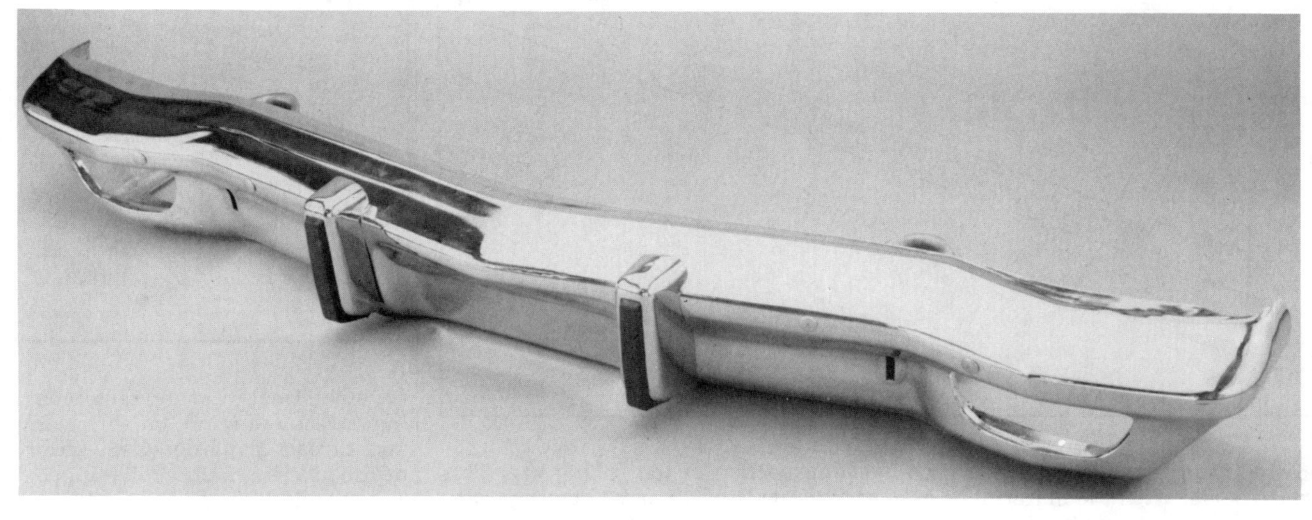

Fig. 17 Effect of quantity on relative costs of cold headed and machined bolts (Example 32)

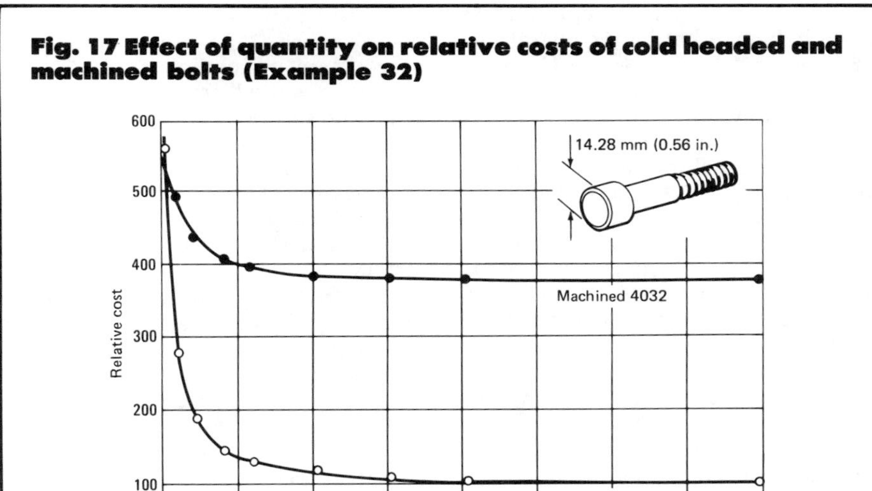

lubricant, were evaluated by means of circular grid strain analysis. Changing to the new lubricant resulted in such a dramatic decrease in peak strain on the part that the material, instead of being upgraded to aluminum-killed steel, was downgraded to commercial quality steel.

Example 34. A welded cylindrical shell of rimmed drawing quality steel sheet 0.84 mm (0.033 in.) thick, was formed into a square shell in an expander-die operation. Strains on certain regions of this part were severe enough to justify supplementary mechanical property requirements for the steel sheet. However, during pilot production, unacceptable and unexplainable scrap rates were encountered.

Testing of the material at various stages of manufacture indicated that the mechanical properties of the steel were being changed during processing, as follows:

	Yield strength MPa	ksi	Elongation, %
As received from mill	194	28.1	45.5
Immediately after lubricant coating	195	28.3	44
After recoiling and storage	212	30.7	41

Application and curing of the lubricant required short-term exposure of the sheet to temperatures of about 95 °C (200 °F). The coating process itself did not affect the properties of the steel. However, because the coated sheet was recoiled immediately, it retained the heat used to cure the lubricant, which accelerated the strain aging process. Use of a lubricant that can be cured at room temperature, and institution of minor tooling changes, reduced the scrap rate significantly and saved the energy that would have been required for elevated-temperature curing. A change in material to a more costly aluminum-killed grade (not susceptible to strain aging) was not necessary.

Example 35. A wheel spindle for a front-end loader had been made from 4140H steel. A change to 41L37 reduced the cost of machining by 45%, which far exceeded the small cost premium paid for the leaded steel.

Example 36. The functional requirements for a dual-pitch rack (Fig. 18) dictated that the part be straight

Fig. 18 Dual pitch rack (Example 36)

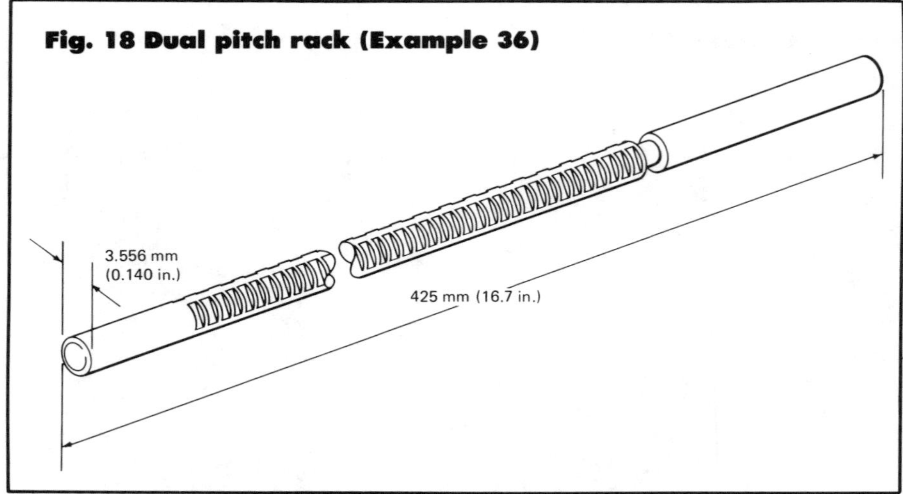

Fig. 19 Acme screw (Example 41)

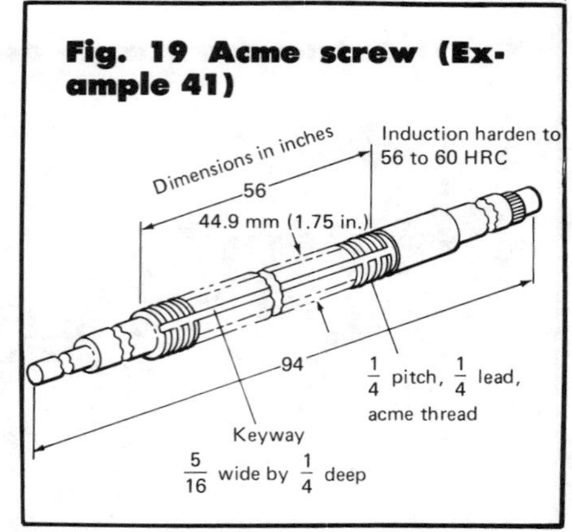

and free from burrs and have a good surface finish. With the material originally chosen—an air-hardening tool steel—the teeth could be ground without producing burrs, and the part did not warp either in heat treatment or in grinding. Because the high hardness of a heat treated steel was not needed except to allow burr-free grinding, a change was made to 12L14 free-machining steel. With this steel, the teeth could be milled rather than ground, and the part could still meet all functional requirements. The change reduced the cost of material by 80%, eliminated a heat treatment and reduced the total cost of the part by 52%.

Example 37. A gear blank 200 mm (7.875 in.) in diameter had been made from a forging. The roughness of the forged surface and the draft needed for forging caused occasional manufacturing problems. The product form was changed to bar, which was machined on a numerically controlled turning machine. About 10% more machining was necessary, but the bar cost 39% less than the forging, and a normalizing heat treatment was eliminated. Total cost of the blank was reduced by 30%. Hobbing of teeth and finish grinding were the same for both materials.

Example 38. Machine-tool boring bars had been made from carburized and hardened 1118 steel. Service life of these boring bars was unsatisfactory because of cracks at the ends, where cutting tools were clamped in place. Localized tempering of the ends of the bars (by immersion in a lead bath) only partly alleviated the problem. The material was changed to 1141 steel, which was induction hardened only on the shank portion of the boring bar. This part provided fully satisfactory performance in service, and it was substantially less costly to produce than the original part because carburizing and selective tempering were replaced by a single induction hardening operation.

Example 39. A motor shaft, made from 1050 steel, included a bearing journal that was induction hardened to 57 HRC minimum. Induction heating and water quenching caused cracks in approximately 10% of these shafts. Detection of all such quench cracks required 100% dye-penetrant inspection. Several possible solutions to the cracking problem were considered; the solution that was adopted—addition of polyalkylene glycol to the quench water—required no changes in other manufacturing operations or in material, and eliminated the need for 100% inspection. As a result, the cost of the part was reduced by more than 30%.

Example 40. A wear plate in a fifth-wheel mechanism had been made from 1045 steel, water quenched and tempered. Of the total cost of these parts, 25% was for scrap caused by quench cracks and 20% was for a straightening step. Material cost was only 13% of the total expense. Changing the steel to 50B50, which is 37% more costly than 1045, permitted a change to oil quenching, which eliminated both the quench cracking and the need for straightening. The total cost of the part was thereby reduced by 40%.

Example 41. This example illustrates how substituting of a less costly steel that can be induction hardened can decrease cost. An Acme screw (Fig. 19) was made from a hot rolled nitriding steel. Heat treating, rough machining, stress relieving, finish machining and thread grinding were performed before nitriding. After nitriding, extensive straightening was necessary because the portion incorporating the long keyway bowed an appreciable amount during nitriding.

Substitution of stress-relieved carbon-corrected 1151 steel effected an over-all cost reduction of 28.4%. This savings resulted from a substantial reduction in the number of operations, from substitution of induction hardening for nitriding, and from use of a less expensive steel. Field tests showed that the mechanical properties of the induction hardened parts were entirely satisfactory.

Example 42. Cold forged pins, 25 mm (1 in.) in diameter by 100 mm (4 in.) long, could be made from any of several medium-carbon alloy steels. The pins were carburized, hardened and tempered. The rather unusual practice of carburizing a medium-carbon steel was necessitated by the requirements that the surface be extremely wear-resistant and that the core be very strong. Three grades having adequate hardenability are 8740, 4140 and 5140; at the time in question (1978), 4140 cost about 8% less, and 5140 about 16% less, than 8740. However, because the relative costs of different alloying additions fluctuate, the relative costs of different grades of alloy steels are subject to change. Thus, 5140, which was most economical in 1978, will not necessarily remain the most economical; specifications for alloy steels must be reviewed frequently to determine if the relative costs of different grades have changed.

Example 43. The 8630 steel aircraft roller support illustrated in the

Fig. 20 Relative cost of aircraft roller supports machined from two forms of starting stock (Example 43)

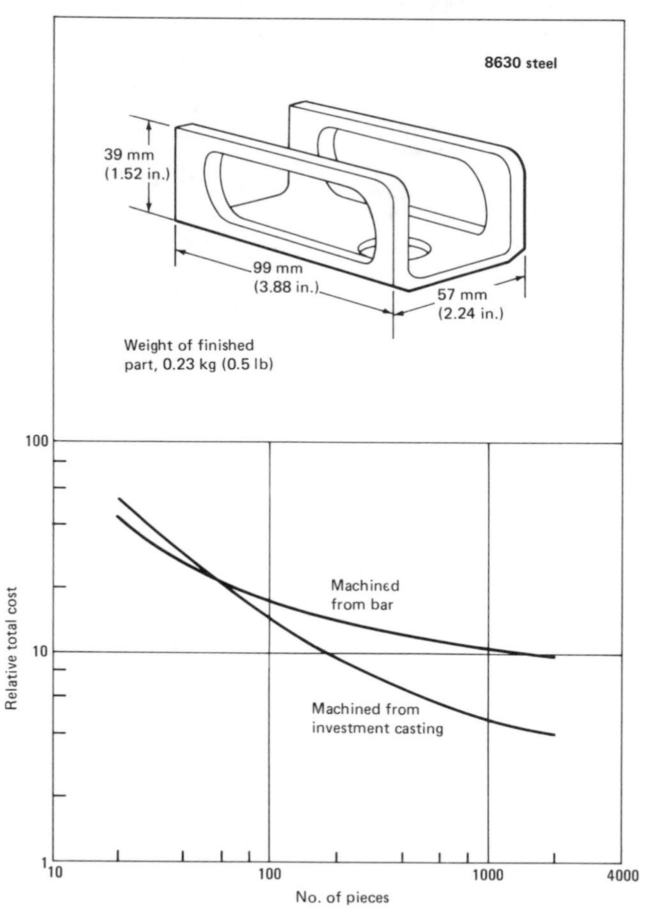

Fig. 21 Cluster gear (Example 44)

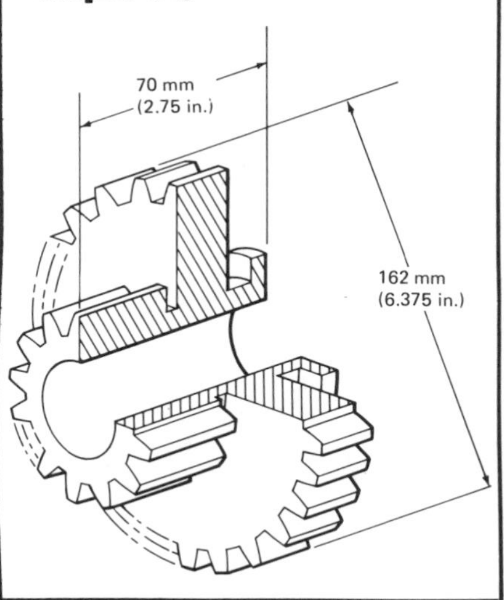

Fig. 22 Pump gear (Example 46)

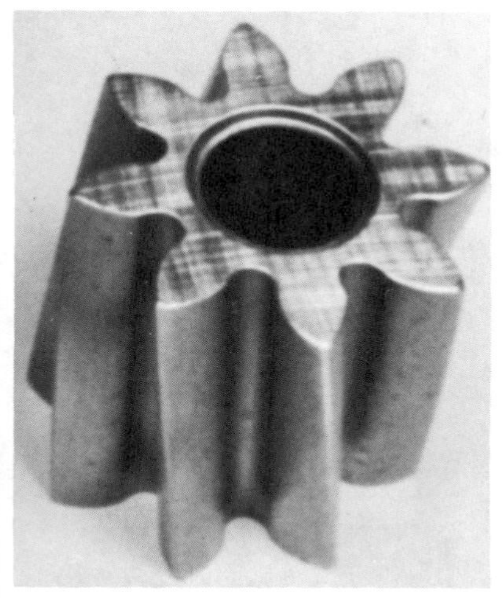

Fig. 23 Blower gear (Example 47)

sketch in Fig. 20 can be made by machining from bar stock or by finish machining an investment casting. As shown in the graph accompanying the sketch, the total cost of making the part from an investment casting is lower than the cost of making it from bar stock if the quantity produced is more than about 60 pieces. For smaller quantities, the cost of tooling for making the investment castings cannot be offset by savings in material and machining costs, so it is more economical to machine the part from solid bar stock. At a quantity of 4000 pieces, on the other hand, making the part from an investment casting provides a savings of about 60%.

Example 44. The 8650 steel cluster gear shown in Fig. 21 can be made from a forging or from bar stock. The costs of making various quantities of these gears from bar stock, relative to the costs of using forgings, are as follows:

	Relative cost(a) for quantities of:		
	50	125	250
Material cost	49	64	69
Machining cost	142	148	156
Total cost(b)	72	85	90

(a) Cost of a part machined from bar stock, expressed as a percentage of the cost of an equivalent part made from a forging. (b) Cost of forging includes allowance for die amortization.

These data indicate that, for quantities up to at least 250 parts, making the gears from bar stock costs less than making them from forgings, but that the cost advantage decreases as quantity is increased: it costs more to machine the bar stock and the penalty for machining increases as quantity is increased. To determine which method of manufacture is less expensive, it is necessary to know the total expected production quantity and the accounting procedures that will be used in amortizing the cost of the tooling for forging.

Example 45. A shaft 19 mm (3/4 in.) in diameter by 165 mm (6.5 in.) long had been made from 1141 steel, then hardened and tempered to 24 to 30 HRC. Substitution of cold drawn and stress-relieved 1144 steel eliminated both heat treatment and straightening. The energy consumed in making one shaft from each material is estimated as follows:

	Energy consumed, kJ	
	1141, heat treated	1144, stress relieved
Drawing and stress relieving (at mill)	5	348
Machining	29	27
Hardening and tempering	927	0
Total	961	375

Example 46. A pump gear 44 mm (1.75 in.) in diameter (Fig. 22) was made from hardened-and-tempered 41L40 steel. The total energy consumed was 7670 kJ. Substitution of hot drawn 1144 steel reduced energy consumption to 2350 kJ—a reduction of 69%. Substitution of the 1144 steel was acceptable because of the relatively low strength requirement of the gear.

Example 47. The energy consumed in making a blower gear 76 mm (3 in.) in diameter (Fig. 23) from 1144 steel, and then heat treating the gear, was 2720 kJ. Changing to cold drawn and stress-relieved 1144 reduced total energy consumption to 850 kJ—a reduction of nearly 70%.

Système Internationale d'Unités (SI)

SI Base Units

Quantity	Unit	Symbol	Quantity	Unit	Symbol	Quantity	Unit	Symbol
Length	metre	m	Amount of substance	mole	mol	Plane angle(a)	radian	rad
Mass	kilogram	kg				Solid angle(a)	steradian	sr
Time	second	s	Luminous intensity	candela	cd			
Electric current	ampere	A						
Thermodynamic temperature	kelvin	K						

(a) Supplementary unit

SI Derived Units(a)

Quantity	Unit	Symbol	Formula	Quantity	Unit	Symbol	Formula
Frequency (of a periodic phenomenon)	hertz	Hz	s^{-1}	Capacitance	farad	F	C/V
				Electric resistance	ohm	Ω	V/A
Force	newton	N	$kg \cdot m/s^2$	Conductance	siemens	S	A/V
Pressure, stress	pascal	Pa	N/m^2	Magnetic flux	weber	Wb	V·s
Energy, work, quantity of heat	joule	J	N·m	Magnetic flux density	tesla	T	Wb/m^2
Power, radiant flux	watt	W	J/s	Inductance	henry	H	Wb/A
Quantity of electricity, electric charge	coulomb	C	A·s	Luminous flux	lumen	lm	cd·sr
				Illuminance	lux	lx	lm/m^2
Electric potential, potential difference, electromotive force	volt	V	W/A	Activity (of radionuclides)	becquerel	Bq	s^{-1}
				Absorbed dose	gray	Gy	J/kg

(a) Derived units in this list include only those units for which special names and symbols have been approved by the General Conference on Weights and Measures (CGPM)

SI Prefixes(a)

Prefix	Multiplication factor	Symbol	Prefix	Multiplication factor	Symbol
exa	1 000 000 000 000 000 000 = 10^{18}	E	deci(b)	0.1 = 10^{-1}	d
peta	1 000 000 000 000 000 = 10^{15}	P	centi(c)	0.01 = 10^{-2}	c
			milli	0.001 = 10^{-3}	m
tera	1 000 000 000 000 = 10^{12}	T			
giga	1 000 000 000 = 10^{9}	G	micro	0.000 001 = 10^{-6}	μ
mega	1 000 000 = 10^{6}	M	nano	0.000 000 001 = 10^{-9}	n
			pico	0.000 000 000 001 = 10^{-12}	p
kilo	1 000 = 10^{3}	k			
hecto(b)	100 = 10^{2}	h	femto	0.000 000 000 000 001 = 10^{-15}	f
deka(b)	10 = 10^{1}	da	atto	0.000 000 000 000 000 001 = 10^{-18}	a

(a) Used to form multiples and decimal fractions of the base and derived SI units. (b) Normally avoided. (c) Use not recommended.

Abbreviations and Symbols

ac alternating current

AC air cooled

ACI Alloy Casting Institute

A_{cm} Solubility limit between austenite and austenite plus cementite

AFS American Foundrymen's Society

AISI American Iron and Steel Institute

AMS Aerospace Material Specification (of SAE)

ANSI American National Standards Institute

Ar_3 temperature at which transformation of austenite to ferrite begins on cooling

ASME American Society of Mechanical Engineers

ASTM American Society for Testing and Materials

atm atmosphere (pressure)

AWG American wire gage

AWS American Welding Society

B bar

B flux density

BHMA British Hard Metal Association

B_r remanence

B_s saturation induction

B&S Brown and Sharpe (gage)

Btu British thermal unit

CBN cubic boron nitride

CCPA Cemented Carbide Producers Association

CH cold work hardened

Ci Curie

cph close-packed hexagonal

CT compact tension

CVD chemical vapor deposition

da/dN crack growth rate

dB decibel

DBTT ductile-to-brittle transition temperature

dc direct current

diam diameter

DU depleted uranium

E emf (electromotive force)

EB electron beam

ELI extra-low interstitial

emf electromotive force

Eq equation

FC furnace cooled

fcc face centered cubic

Fig. figure

ft foot

g gram

GMAW gas metal-arc welding

GTAW gas tungsten-arc welding

h hour

HB Brinell hardness

H_c coercive force

HFRSc forged roll Scleroscope numbers (hardness)

HK Knoop hardness

hp horsepower

HR hot rolled

HRA Rockwell A hardness

HRB Rockwell B hardness

HRC Rockwell C hardness

HSc Scleroscope numbers (hardness)

HV Vickers hardness

I current, intensity of magnetization

IACS International Annealed Copper Standard (electrical conductivity)

in. inch

ipr inches per revolution

IPTS International Practical Temperature Scale

ISA Instrument Society of America

ISO International Standards Organization

ITS International Temperature Scale

J joules

kg kilogram

K_{Ic} plane-strain fracture toughness

K_{Iscc} threshold stress intensity for strain-corrosion cracking

ksi kips (1000 pounds) per square inch

K_t stress-concentration factor

lb pound

LCCA load-current-contact-aiding

LNG liquefied natural gas

max maximum

min minute; minimum

mod modified

M_s martensite start temperature

NBS National Bureau of Standards

No. number

N_v electron vacancy number

OD outside diameter

OQ oil quenched

P plate; penetration

pH negative logarithm of hydrogen activity

P/M powder metallurgy

ppm parts per million

psi pounds per square inch

R electrical resistivity, fatigue stress ratio, universal gas constant

RC resistance-capacitance

RD rolling direction

Ref reference

rem remainder

rev revolution

rf radio frequency

RH refrigeration hardened

rms root mean square

RT room temperature

SAE Society of Automotive Engineers

sfm surface feet per minute

SMAW shielded metal-arc welding

STA solution treated and aged

std standard

STDA solution treated and double aged

STOA solution treated and overaged

STQ solution treated and quenched

T temperature

TAPPI Technical Association of Pulp and Paper Industry

TCP topologically close-packed

TCR temperature coefficient of resistance

TD transverse direction

TH transformation hardened

TS tensile strength

TWI Taber wear index

UL Underwriters Laboratory

UNS Unified Numbering System

V voltage

VIM-VAR vacuum induction melted and vacuum arc remelted

vol volume

WQ water quenched

wt weight

YS yield strength

°C degree Celsius (centigrade)

°F degree Fahrenheit

= equals

> greater than

< less than

\+ plus; positive ion charge

− minus, negative ion charge

× diameters (magnification); times

÷ divided by

% percent

Δ difference

ζ thermoelectric power of the thermoelement

μ viscosity; magnetic permeability

Index

B

C

N

O

P

Q

R

S

T

U

V

W

Z